Mathematics for CBSE Class XI

Vikash Khatri
PhD (Structural/Civil Engineering),
MSc (Structural/Civil Engineering)

Universities Press

UNIVERSITIES PRESS MATHEMATICS FOR CBSE CLASS XI

UNIVERSITIES PRESS (INDIA) PRIVATE LIMITED

Registered Office
3-6-747/1/A & 3-6-754/1, Himayatnagar, Hyderabad 500 029, Telangana, India
info@universitiespress.com; www.universitiespress.com

Distributed by
Orient Blackswan Private Limited

Registered Office
3-6-752 Himayatnagar, Hyderabad 500 029 Telangana, India

Other Offices
Bengaluru / Chennai / Guwahati / Hyderabad / Kolkata
Mumbai / New Delhi / Noida / Patna / Visakhapatnam

ISBN 978-93-93330-09-3

501808

Typeset in Times Ten LT Std 10 points *by*
Cameo Corporate Services Limited, Coimbatore

Printed in India by
B.B. Press, Noida

Published by
Universities Press (India) Private Limited
3-6-747/1/A & 3-6-754/1, Himayatnagar, Hyderabad 500 029, Telangana, India

Preface

This first edition of *Universities Press Mathematics for CBSE Class XI* has been written in accordance with the latest CBSE syllabus and NCERT Exam Pattern. The whole curriculum has been demarcated into a 240-day segregation contained within 16 chapters, so that students can complete their exam preparation in an easy, step-by-step manner, requiring only about 1–2 hours of study per day.

The book is written in simple and easy-to-follow yet efficient language, so that all students can grasp the concepts behind the numerical sections. Solutions to all problems have been provided for conceptual understanding of the theoretical and numerical portions of each chapter. Detailed solutions have also been presented for all NCERT Unsolved Exercises and NCERT Exemplar Problems.

The questions included comprise the very short answer, short answer and long answer types, satisfying the format specified by the NCERT Exam Pattern. For every chapter, there is a revision section, to aid students in recapping what they have studied, as well as a chapter test to help them assess their knowledge. Ten full-syllabus sample question papers with solutions have also been provided.

This book will also be very helpful for students of various state boards and provides a strong foundation for students preparing for IIT-JEE and other engineering entrance examinations.

This book will help you, the student, realise your hidden potential, develop your mathematical abilities and significantly improve your academic performance.

Vikash Khatri

Acknowledgements

I extend my sincere gratitude to Mr. Karan Sadhwani (Banaras Hindu University, Varanasi) for checking and correcting the numerical sections with precision.

I have put in my most earnest efforts and invested ample time to present an error-free text. Nevertheless, I wholeheartedly welcome all feedback from readers. All suggestions, comments and constructive criticism, as well as errors, if any, can be sent to khatri.hbtik@gmail.com.

I appreciate the continuous and unwavering support and cooperation provided by the editorial team at Universities Press (India) Private Limited.

This book has been put together exclusively for aspirants preparing for the Class XI (CBSE and other state boards), IIT-JEE and other engineering entrance exams. I heartily hope that it proves useful and rewarding to students.

As air is necessary for life, Mathematics is essential for livelihood.

Vikash Khatri

Contents

CHAPTER 1

Sets

Day 1

Sets and their Representation

Set: A set is a well-defined collection of objects. Well-defined means, if a question is asked to different people, then the answer given by all the people must be the same.

For example, if a question is posed to students asking them to write the names of the three toppers of Class XI, then the answer given by different students will be the same. However, if the question is to write the names of three smart students of Class XI, then the answers given by different students may not be the same.

The three toppers are the same but the selection of three smart students may be different with respect to different students. Thus, the names of the three toppers of Class XI represent a set while the names of three smart students of Class XI do not represent a set.

Other examples of sets (well-defined collection of objects) are:

(i) The vowels in the English alphabet, that is, *a, e, i, o, u*.
(ii) Letters in the word 'CAREER', that is, *A, C, E, R*.
(iii) Even natural numbers less than 10, that is, 2, 4, 6, 8.
(iv) Prime factors of 30, that is, 2, 3, 5.
(v) All natural numbers which divide 30, that is, 1, 2, 3, 5, 6, 10, 15, 30.
(vi) Various types of triangles, that is, acute, obtuse, right triangle.
(vii) Solution of the quadratic equation: $x^2 - x - 2 = 0$, that is, $-2, 1$.
(viii) Names of all the students of your class.
(ix) Names of the rivers or states of India
(x) Names of the days in a week or months in a year.

Note:

(i) In a set, object, element and member are synonymous terms.
(ii) Sets are usually denoted by capital letters *A, B, C, …, X, Y, Z*, etc.
(iii) The elements of a set are represented by small letters *a, b, c, …, x, y, z*, etc.
(iv) If *a* is an element of a set *A*, we may read it as, '*a* belongs to *A*', represented by $a \in A$, where the Greek symbol $\in$ (epsilon) is used to denote the phrase 'belongs to'.
(v) If *b* is not an element of a set *A*, then we may read it as, '*b* does not belong to *A*', represented by $b \notin A$, where the Greek symbol $\notin$ (cut epsilon) is used to denote the phrase 'does not belong to'.

Some well-known symbols used to represent sets:

***N*:** the set of all natural numbers; $\{1, 2, 3, ...\}$

***W*:** the set of all whole numbers; $\{0, 1, 2, 3, ...\}$

***Z*:** the set of all integers; $\{..., -3, -2, -1, 0, 1, 2, 3, ...\}$

***Z+*:** the set of positive integers; $\{1, 2, 3, ...\}$

***Z–*:** the set of negative integers; $\{..., -3, -2, -1\}$

***Q*:** the set of all rational numbers; $\{-2, \frac{-3}{2}, -1, \frac{-1}{5}, 0, 3, \frac{7}{2}, \frac{9}{4} ...\}$

***Q+*:** the set of positive rational numbers; $\{2, \frac{3}{2}, 1, \frac{1}{5}, 0, 3, \frac{7}{2}, \frac{9}{4} ...\}$

***Q–*:** the set of negative rational numbers; $\{-2, \frac{-3}{2}, -1, \frac{-1}{5}, -3, -\frac{7}{2}, -\frac{9}{4} ...\}$

***R*:** the set of real numbers; $\{-2, \frac{-3}{2}, -1, \frac{-1}{5}, 0, \sqrt{2}, \sqrt{3}, \frac{7}{2}, \frac{9}{4}, \pi, ...\}$

***R+*:** the set of positive real numbers; $\{2, \frac{3}{2}, 1, \frac{1}{5}, 0, \sqrt{2}, \sqrt{3}, \frac{7}{2}, \frac{9}{4}, \pi, ...\}$

***R–*:** the set of negative real numbers; $\{-2, \frac{-3}{2}, -1, \frac{-1}{5}, -\sqrt{2}, -\sqrt{3}, -\frac{7}{2}, -\frac{9}{4}, -\pi, ...\}$

Two Methods of Representing a Set:

(i) Roster or tabular form
(ii) Set-builder form

Roster or Tabular Form: In the roster or tabular form, all the elements of a set are listed, separated by a comma and are enclosed within braces {}.

Example 1: Let set A be the collection of all the vowels in the English alphabet. In roster form, set A is given by $A = \{a, e, i, o, u\}$

Note: In the roster form, the order in which the elements are listed is irrelevant.
Therefore, the above set A may also be written as
$A = \{e, o, i, a, u\}$ or $A = \{u, a, e, i, o\}$

Example 2: Let set B be the collection of all the letters in the word 'career'. Then, in the roster form, set B is given by $B = \{a, c, e, r\}$.

Note: In the roster form, repetition of elements is irrelevant. Thus, in the roster form, elements are generally not repeated, that is, all the elements are taken as distinct. Hence, in the above set B, the repeated elements e and r must be written only once.
Therefore, the above set B may be written as $B = \{a, c, e, r\}$ or $B = \{c, a, r, e\}$ or $B = \{r, a, e, c\}$.

Example 3: Let set C be the collection of all the integers greater than –4 and less than or equal to 4. Then in the roster form, set C is given by $C = \{-3, -2, -1, 0, 1, 2, 3, 4\}$.

Example 4: Let set D be the collection of all the natural numbers that divide 30. Then in the roster form, set D is given by $D = \{1, 2, 3, 5, 6, 10, 15, 30\}$.

Example 5: Let set E be the names of all the months having 30 days. Then in the roster form, set E is given by $E = \{\text{Feb, April, June, Sept, Nov}\}$.

Example 6: Let set F be the collection of all the two-digit natural numbers, the sum of whose digits is 6. Then in the roster form, set F is given by $F = \{06, 15, 24, 33, 42, 51, 60\}$.

Example 7: Let set G be the collection of all the prime natural numbers less than 20. Then in the roster form, set G is given by $G = \{2, 3, 5, 7, 11, 13, 17, 19\}$.

Example 8: Let set H be the collection of all the letters in the word 'MATHEMATICA'. Then in the roster form, set H is given by $H = \{M, A, T, H, E, I, C\}$

Example 9: Let set I be the collection of all the numbers $3n$, where n is a natural number from 1 to 5. Then in the roster form, set I is given by $I = \{3, 6, 9, 12, 15\}$.

Example 10: Let set J be the collection of all the numbers 3^n, where n is a whole number from 0 to 4. Then in the roster form, set J is given by $J = \{1, 3, 9, 27, 81\}$.

Example 11: Let set K be the collection of all the integers whose magnitude is less than or equal to 3. Then in the roster form, set K is given by $K = \{-3, -2, -1, 0, 1, 2, 3\}$.

Example 12: Let set L be the collection of all the real numbers that satisfy the quadratic equation $x^2 - x - 12 = 0$. Then in the roster form, set L is given by $L = \{-3, 4\}$
$\left(\because x^2 - x - 12 = 0 \Rightarrow (x+3)(x-4) = 0 \Rightarrow x = -3, 4\right)$.

Set-Builder Form: Here, we write the property of all the elements in a set, that is, all the elements of a set possess a single common property that is not possessed by any other element outside the set.
Thus, in the set-builder form, we describe the element of the set by using a symbol x (or any other symbol such as the letters y, z, etc.), followed by a colon ':'. After the colon, we write the characteristic property possessed by all the elements of the set and then enclose the whole description within braces.
$X = \{x : x \text{ is the ....property of all the elements}\}$
The above description of the set X is read as 'X is the set of all x such that x is the ... property of set'. In this description, the braces stand for 'the set of all', and the colon stands for 'such that.

Example 13: The set-builder form of all the above sets is written as:

$A = \{x : x \text{ is a vowel in the English alphabet}\}$

$B = \{x : x \text{ is a letter in the word } \textit{career}\}$

$C = \{x : x \text{ is an integer and } -4 < x \leq 4\}$

$D = \{x : x \text{ is a natural number that divides } 30\}$

$E = \{x : x \text{ is a month of a year having 30 days}\}$

$F = \{x : x$ is a two-digit natural number such that the sum of its digits is $6\}$

$G = \{x : x \text{ is a prime natural number less than } 20\}$

$H = \{x : x \text{ is a letter in the word 'MATHEMATICA'}\}$

$I = \{x : x = 3n,\ n \in N,\ 1 \leq n \leq 5\}$

$J = \{x : x = 3^n,\ n \in W,\ 0 \leq n \leq 4\}$

$K = \{x : x \text{ is an integer and } |x| \leq 3\}$

$L = \{x : x$ is a real number that satisfies the quadratic equation $x^2 - x - 12 = 0\}$

Day 2

NCERT Exercise 1.1 (With Solutions)

Q1. Which of the following are sets? Justify your answer.

(i) The collection of all months of a year beginning with the letter J.
(ii) The collection of the ten most talented writers of India.
(iii) A team of the eleven best cricket batsmen of the world.
(iv) The collection of all the boys in your class.
(v) The collection of all natural numbers less than 100.
(vi) A collection of novels written by the writer Munshi Prem Chand.
(vii) The collection of all even integers.
(viii) The collection of questions in this chapter.
(ix) A collection of the most dangerous animals in the world.

Solution:

(i) Since the collection of all the months in a year beginning with the letter J is a well-defined collection (containing January, June, July), this collection is a set.
(ii) The collection of the ten most talented writers of India is not a well-defined collection because it depends on person-to-person criteria. This collection is not a set.
(iii) A team of the eleven best cricket batsmen of the world is not a well-defined collection because it depends on person-to-person criteria. This collection is not a set.
(iv) Since the collection of all the boys in your class is a well-defined collection (the class and boys in it are well defined), this collection is a set.
(v) Since the collection of all natural numbers less than 100 is a well-defined collection (1, 2, 3, …, 98, 99), this collection is a set.
(vi) Since the collection of novels written by the writer Munshi Prem Chand is a well-defined collection, this collection is a set.
(vii) Since the collection of all even integers is a well-defined collection, this collection is a set.
(viii) Since the collection of questions in this chapter is well-defined, this collection is a set.
(ix) Since the collection of the most dangerous animals in the world is not a well-defined collection (because its depends on person-to-person criteria), this collection is not a set.

Q2. Let $A = \{1,2,3,4,5,6\}$. Insert the appropriate symbol $\in$ or $\notin$ in the blank spaces.

(i) 5____A (ii) 8____A (iii) 0____A
(iv) 4____A (v) 2____A (vi) 10____A

Solution:

(i) $5 \in A$ (ii) $8 \notin A$
(iii) $0 \notin A$ (iv) $4 \in A$
(v) $2 \in A$ (vi) $10 \notin A$

Q3. Write the following sets in the roster form.

(i) $A = \{x : x \text{ is an integer and } -3 < x < 7\}$.
(ii) $B = \{x : x \text{ is a natural number less than } 6\}$
(iii) $C = \{x : x \text{ is a two-digit natural number such that the sum of its digits is } 8\}$
(iv) $D = \{x : x \text{ is a prime number which is a divisor of } 60\}$
(v) E = The set of all letters in the word TRIGONOMETRY.
(vi) F = The set of all letters in the word BETTER.

Solution:

(i) $A = \{x : x \text{ is an integer and } -3 < x < 7\}$:
The elements of the set A are $-2, -1, 0, 1, 2, 3, 4, 5, 6$
Therefore, in roster form, the set is
$A = \{-2, -1, 0, 1, 2, 3, 4, 5, 6\}$

(ii) $B = \{x : x \text{ is a natural number less than } 6\}$.
The elements of set B are 1, 2, 3, 4 and 5. Therefore, in roster form, the set is $B = \{1,2,3,4,5\}$.

(iii) $C = \{x : x \text{ is a two-digit natural number such that the sum of its digits is } 8\}$.
The elements of set C are 17, 26, 35, 44, 53, 62, 71 and 80. Therefore, in roster form, the set is $C = \{17, 26, 35, 44, 53, 62, 71, 80\}$.

(iv) $D = \{x : x \text{ is a prime number which is a divisor of } 60\}$.
The elements of set D are 2, 3 and 5 $(60 = 2 \times 2 \times 3 \times 5)$
Therefore, in roster form, the set is $D = \{2,3,5\}$.

(v) E = The set of all letters in the word TRIGONOMETRY.
There are 9 different letters $\{T, R, I, G, O, N, M, E, Y\}$ in the word TRIGONOMETRY, and letters T, R and O are repeated. Therefore, in roster form, the set is $E = \{T, R, I, G, O, N, M, E, Y\}$.

(vi) F = The set of all letters in the word BETTER.
There are 4 different letters $\{B, E, T, R\}$ in the word BETTER, out of which letters E and T are repeated. Therefore, in roster form, the set is $F = \{B, E, T, R\}$.

Q4. Write the following sets in the set-builder form.

(i) {3, 6, 9, 12} (ii) {2, 4, 8, 16, 32}
(iii) {5, 25, 125, 625} (iv) {2, 4, 6, ...}
(v) {1, 4, 9,...100}

Solution:

(i) Since 3,6,9,12 are multiples of 3, for natural numbers 1 to 4, in set-builder form, it is written as $\{x : x = 3n,\ n \in N,\ 1 \le n \le 4\}$.

(ii) Since 2, 4, 8, 16, 32 are powers of 2, for natural numbers 1 to 5, in set-builder form, it is written as $\{x : x = 2^n,\ n \in N,\ 1 \le n \le 5\}$.

(iii) Since 5, 25, 125, 625 are powers of 5, for natural numbers 1 to 4, in set-builder form, it is written as $\{x : x = 5^n,\ n \in N,\ 1 \le n \le 4\}$.

(iv) Since 2, 4, 6, ... form the set of all even natural numbers, in set-builder form, it is written as $\{x : x$ is an even natural number$\}$.

(v) Since 1, 4, 9, ..., 100 are squares of natural numbers from 1 to 10, in set-builder form, it is written as $\{x : x = n^2,\ n \in N,\ 1 \le n \le 10\}$.

Q5. List all the elements of the following sets.

(i) $A = \{x : x$ is an odd natural number$\}$

(ii) $B = \{x : x$ is an $-\frac{1}{2} < x < \frac{9}{2}$ integer$\}$

(iii) $C = \{x : x$ is an $x^2 \le 4$ integer$\}$

(iv) $D = \{x : x$ is a letter in the word 'LOYAL'$\}$

(v) $E = \{x : x$ is a month of a year not having 31 days$\}$

(vi) $F = \{x : x$ is a constant in the English alphabet which precedes $k\}$

Solution:

(i) $A = \{x : x$ is an odd natural number$\}$

$\Rightarrow A = \{1, 3, 5, ...\}$

(ii) $B = \{x : x$ is an integer; $-\frac{1}{2} < n < \frac{9}{2}\}$

$\Rightarrow B = \{0, 1, 2, 3, 4\}$

(iii) $C = \{x : x$ is an integer; $x^2 \le 4\}$

$\because (-1)^2 = (1)^2 = 1 \le 4;$

$(-2)^2 = (2)^2 = 4 \le 4;\quad 0^2 = 0 \le 4$

$\therefore C = \{-2, -1, 0, 1, 2\}$

(iv) $D = \{x : x$ is a letter in the word 'LOYAL'$\}$

$\Rightarrow D = \{L, O, Y, A\}$

(v) $E = \{x : x$ is a month of a year not having 31 days$\}$

$\Rightarrow E =$ {February, April, June, September, November}

(vi) $F = \{x : x$ is a consonant in the English alphabet which precedes $k\}$

$\Rightarrow F = \{b, c, d, f, g, h, j\}$

Q6. Match each of the sets on the left in the roster form with the same set on the right described in set-builder form.

(i) {1, 2, 3, 6}
(ii) {2, 3}
(iii) {M, A, T, H, E, I, C, S}
(iv) {1, 3, 5, 7, 9}

(A) $\{x : x$ is a prime number and a divisor of 6$\}$
(B) $\{x : x$ is an odd natural number less than 10$\}$
(C) $\{x : x$ is natural number and divisor of 6$\}$
(D) $\{x : x$ is a letter of the word MATHEMATICS]

Solution:

(i) – (C) All the elements of this set are natural as well as divisors of 6.

(ii) – (A) 2 and 3 are prime numbers and also divisors of 6.

(iii) – (D) All the elements of this set are letters of the word MATHEMATICS.

(iv) – (B) All the elements of this set are odd natural numbers less than 10.

Day 3

Types of Sets

Empty Set: A set that does not contain any element is called an empty set or a null set or a void set, and is denoted by the symbol ϕ or {}.

A set which contains at least a single element is called a non-empty set.

Examples of empty sets:

(i) $A = \{x : x < -2$ and $x > 2,\ x \in R\}$.

The set A is an empty set because there is no real number which is less than –2 and greater than 2.

(ii) $B = \{x : x^2 + 1 = 0, x$ is a real number$\}$.

The set B is an empty set because the equation $x^2 + 1 = 0$ is not satisfied by any real value of x.

(iii) $C = \{x : x^2 = 9, x \text{ is even}\}$.

Set C is an empty set because the equation $x^2 = 9$ is not satisfied by any even value of x.

(iv) $D = \{x : x$ is a letter in the word MATH and PHYSICS$\}$.

Set D is an empty set because there is no common letter that is in both MATH and PHYSICS.

(v) $E = \{x : x \text{ is an integer and } |x| \le -3\}$.

Set E is an empty set because the modulus function $|x| \ge 0$.

(vi) $F = \{x : x$ is a month of a year having 32 days$\}$.

Set F is an empty set because there is no month having 32 days.

(vii) $G = \{x : x$ is a two-digit natural number such that the sum of its digits is 19$\}$.

Set G is an empty set because the sum of the digits of a two-digit number will not be more than 18.

(viii) $H = \{x : x \text{ is even and odd}, x \in N\}$.

Set H is an empty set because a natural number is either an odd number or an even number.

(ix) $I = \{x : \sin x > 1\}$.

Set I is an empty set because the value of sin function cannot be greater than 1, that is, $-1 \le \sin x \le 1$.

(x) $J = \{x : 2^x = 3^{nx},\ n, x \in N\}$

Set J is an empty set because power of 2 always gives an even value while power of 3 is always gives an odd value.

Finite and Infinite Sets: An empty set or a set which contains a definite number of distinct elements is called a finite set. And if the number of distinct elements is infinite (not countable), then the set is called an infinite set.

For a set S, we denote the number of distinct elements by $n(S)$. Thus, if $n(S)$ is a natural number, then set S will be a non-empty finite set.

Note: All infinite sets cannot be described in the roster form. For example, the set of real numbers cannot be described in the roster form because the elements of this set do not follow any particular pattern.

The infinite sets which follow some definite pattern are written in the roster form by writing a few elements which clearly indicates the pattern of the set followed (or preceded) by three dots. For example, the set of natural numbers in roster form can be written as $N = \{1, 2, 3, \ldots\}$.

Examples of finite and infinite sets:

(i) $A = \{x : x$ is all the colours of the rainbow$\}$.

The set A is a finite set and equal to

A = {Violet, Indigo, Blue, Green, Yellow, Orange, Red}

(ii) $A = \{x : x$ is all days of a week$\}$.

The set A is a finite set and equal to

A = {Sunday, Monday, Tuesday, Wednesday, Thursday, Friday, Saturday}

(iii) $A = \{x : x$ is all the months of a year that start with letter $J\}$.

The set A is a finite set and equal to

A = {January, June, July}

(iv) A set of all points on a line.

A set of all triangles.

A set of all circles passing through a given point.

A set of all the lines parallel to a given line.

All of the above sets are infinite sets because the number of points on a line are infinite, the set of number of triangles is infinite, the set of all circles passing through a given point are infinite, and the set of all the lines parallel to a given line are infinite.

(v) $A = \{x : 1 < x < 2,\ x \in N\}$.

The set A is an empty set, but an empty set is a finite set. Therefore, set $A = \phi$ is a finite set.

(vi) $A = \{x : 1 \le x \le 2,\ x \in N\}$.

The set A is a finite set and equal to $A = \{1, 2\}$

(vii) $A = \{x : 1 \le x \le 2,\ x \in R\}$.

The set A is an infinite set because there are infinite real numbers between 1 and 2.

(viii) Consider the following sets:

$A = \{x : x \in N\} = \{1, 2, 3, \ldots\}$

$A = \{x : x \in W\} = \{0, 1, 2, 3, \ldots\}$

$A = \{x : x \in Z\} = \{\ldots, -3, -2, -1, 0, 1, 2, 3, \ldots\}$

$A = \{x : x \in R\} = (-\infty, \infty)$

The set of natural numbers, whole numbers and integers are infinite sets and follow some pattern, and are therefore written in the roster form. The set of real numbers is an infinite set but does not follow any particular pattern, and therefore it cannot be written in the roster form.

(ix) $A = \{x : x \in N \text{ and } x \text{ is prime}\}$.

$A = \{x : x \in N \text{ and } x \text{ is odd}\}$.

$A = \{x : x \in N \text{ and } x \text{ is even}\}$.

All of the above sets are infinite sets because prime numbers or odd numbers or even numbers are infinite.

(x) $A = \{x : x$ is all the human beings living in this world$\}$.

$A = \{x : x$ is all the leaves on a tree$\}$.

$A = \{x : x \text{ is all the trees in a forest}\}$.

$A = \{x : x \text{ is all the stars on a sky}\}$.

It is difficult to calculate the number of human beings living in this world or number of leaves on a tree or number of trees in a forest or number of stars in the sky living but they are all natural numbers. So, all the above sets are finite sets.

Equal Sets: Two sets A and B are said to be equal ($A = B$) if and only if they have exactly the same elements. Thus, two sets A and B are said to be equal if every element of A is also an element of B and if every element of B is also an element of A. Otherwise, the sets are said to be unequal ($A \neq B$).

As examples, consider the following sets:

$A = \{1, 2, 3\}$,

$B = \{2, 1, 3, 1, 2\}$,

$C = \{x : 1 \leq x \leq 3,\ x \in N\}$

$D = \{x : (x-1)(x-2)(x-3) = 0,\ x \in R\}$

All the above sets are equal, that is,

$A = B = C = D = \{1, 2, 3\}$

Now consider the sets

$E = \{x : 1 < x \leq 3,\ x \in N\} = \{2, 3\}$ and

$F = \{x : 1 \leq x \leq 3,\ x \in R\}$

The set $E \neq A$ because the element $1 \in A$ but $1 \notin E$

The set $F \neq A$ because set A contains only natural numbers 1, 2, 3 as elements while set F contains infinite real numbers between 1 and 3.

Note: Sets A and B are equal because the set does not change if one or more elements of the set are repeated. That is why we generally do not repeat any element in a set.

Day 4

NCERT Exercise 1.2 (With Solutions)

Q1. Which of the following are examples of a null set?

(i) Set of odd natural divisors divisible by 2
(ii) Set of even prime numbers
(iii) $\{x : x \text{ is a natural number}, x < 5 \text{ and } x > 7\}$
(iv) $\{y : y \text{ is a point common to any two parallel lines}\}$

Solution:

(i) Since all odd numbers are not divisible by 2, it is a null set.
(ii) Since 2 is an even prime number, the set of even prime numbers is not a null set.
(iii) Since a number cannot be simultaneously less than 5 and greater than 7, $\{x : x \text{ is a natural number}, x < 5 \text{ and } x > 7\}$ is a null set.
(iv) Since parallel lines never intersect, they have no common point. Thus $\{y : y \text{ is a point common to any two parallel lines}\}$ is a null set.

Q2. Which of the following sets are finite or infinite?

(i) The set of months of a year
(ii) {1, 2, 3..........}
(iii) {1, 2, 3...90, 100}
(iv) The set of positive integers greater than 100
(v) The set of prime numbers less than 99

Solution:

(i) The set of months of a year is a finite set because it has 12 elements.
(ii) {1, 2, 3, ...} is an infinite set because it has an infinite number of natural numbers.
(iii) {1, 2, 3, ... 99, 100} is a finite set because 1 to 100 is finite in number.
(iv) The set of positive integers greater than 100 is an infinite set because positive integers greater than 100 are infinite in number.
(v) Since prime numbers less than 99 are finite in number, the set of prime numbers less than 99 is a finite set.

Q3. State whether each of the following set is finite or infinite.

(i) The set of lines which are parallel to the x-axis.
(ii) The set of letters in the English alphabet.
(iii) The set of numbers which are multiples of 5.
(iv) The set of animals living on the earth.
(v) The set of circles passing through the origin (0, 0).

Solution:

(i) Since lines parallel to the x-axis are infinite in number, the set of lines that are parallel to the x-axis is an infinite set.
(ii) Since the English alphabet contains 26 elements, this is a finite set.
(iii) Since multiples of 5 are infinite in number, the set of numbers which are multiples of 5 is an infinite set.
(iv) Since the number of animals living on the earth is finite (very big number but countable), the set of animals living on the earth is a finite set.
(v) Since an infinite number of circles (of different radius) can pass through the origin, the set of circles passing through the origin (0, 0) is an infinite set.

Q4. In the following state whether $A = B$ or not.

(i) $A = \{a,b,c,d\}; B = \{d,c,b,a\}$

Solution: Since the order of elements is not significant for a set, $A = B$.

(ii) $A = \{4,8,12,16\}; B = \{8,4,16,18\}$

Solution: Since $12 \in A$ but $12 \notin B$, $A \neq B$.

(iii) $A = \{2,4,6,8,10\}$,

$B = \{x : x \text{ is a positive even integer and } x \leq 10\}$

Solution:

$\because B = \{2,4,6,8,10\}$

$\therefore A = B$

(iv) $A = \{x : x \text{ is a multiple of } 10\}$

$B = \{10,15,20,25,30\ldots\}$

Solution:

$\because A = \{10,20,30\ldots\}$ and thus $15 \in B$ but $15 \notin A$.

$\therefore A \neq B$

Q5. Are the following pair of sets equal? Give reasons.

(i) $A = \{2,2\}$;

$B = \{x : x \text{ is solution of } x^2 + 5x + 6 = 0\}$

(ii) $A = \{x : x \text{ is a letter in the word FOLLOW}\}$;

$B = \{y : y \text{ is a letter in the word WOLF}\}$

Solution:

(i) The equation $x^2 + 5x + 6 = 0$ can be solved as $(x+2)(x+3) = 0 \Rightarrow x = -2$ or -3

$\therefore A = \{2,3\}; B = \{-2,-3\}$, thus $A \neq B$

(ii) $A = \{x : x \text{ is a letter in the word FOLLOW}\}$

$= \{\text{F, O, L, W}\}$

$B = \{y : y \text{ is a letter in the WOLF}\} = \{\text{W, O, L, F}\}$

Since both sets A and B contain the same elements, $A = B$.

Q6. From the sets given below, select equal sets.

$A = \{2,4,8,12\}$

$B = \{1,2,3,4\}$

$C = \{4,8,12,14\}$

$D = \{3,1,4,2\}$

$E = \{-1,1\}$

$F = \{0,a\}$

$G = \{1,-1\}$

$H = \{0,1\}$

Solution:

$B = \{1,2,3,4\}$, $D = \{3,1,4,2\}$

Since both sets B and D contain the same elements $\{3,1,4,2\}$ and the order of elements is not significant, $B = D$.

$E = \{-1,1\}$, $G = \{1,-1\}$,

Since both sets E and G contain the same elements $\{1,-1\}$ and the order of elements is not significant, $E = G$.

$A = \{2,4,8,12\}$, $C = \{4,8,12,14\}$,

$2 \in A$ but $2 \notin C$; therefore $A \neq C$.

$F = \{0,a\}$, $H = \{0,1\}$

Since $1 \in H$ but $1 \notin F$, $F \neq H$.

F will be equal to H, when $a = 1$.

Day 5

Subsets, Power Set and Universal Set

Subsets: A set B is said to be a subset of a set A if every element of B is also an element of A. It is written as $B \subset A$ if whenever $a \in B$ then $a \in A$. Or $B \subset A$ if $a \in B \Rightarrow a \in A$. The statement is read as 'B is a subset of A if a is an element of B implies that a is also an element of A'.

If B is not a subset of A, then we write $B \not\subset A$.

<u>**Example 1:**</u> Let A be the set of all divisors of 30, $A = \{1,2,3,5,6,10,15,30\}$ and B the set of all prime divisors of 30, $B = \{2,3,5\}$, and C the set of all prime numbers less than 30, $C = \{2,3,5,7,11,13,17,19,23,29\}$. Then

B is a subset of A and written as $B \subset A$.

B is a subset of C and written as $B \subset C$.

C is not a subset of A and written as $C \not\subset A$.

A is not a subset of C and written as $A \not\subset C$.

Note:

- Every set A is a subset of itself, that is, $A \subset A$.
- Empty set ϕ is a subset of every set (since ϕ has no elements).
- For B to be a subset of A, all that is needed is that every element of B is in A, but it is possible that every element of A may or may not be in B.

(ii) $A=\{x : x \in R, -2 < x \le 5\} \Rightarrow A=(-2,5]$

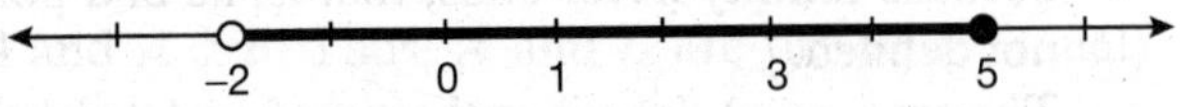

(iii) $A=\{x : x \in R, 1 \le x < 8\} \Rightarrow A=[1,8)$

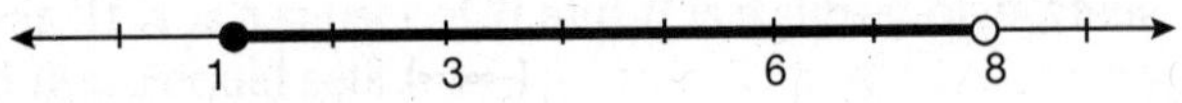

(iv) $A=\{x : x \in R, -1 < x < \infty\} \Rightarrow A=(-1,\infty)$

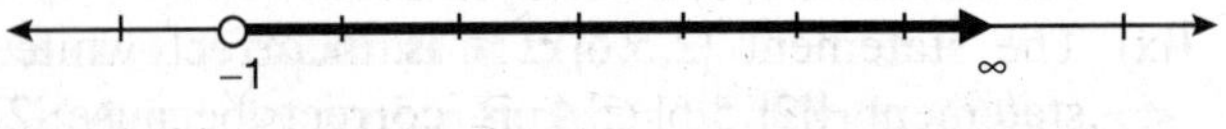

(v) $A=\{x : x \in R, -\infty < x \le 1\} \Rightarrow A=(-\infty,1]$

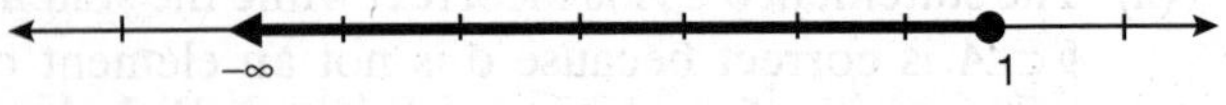

Day 6

NCERT Exercise 1.3 (With Solutions)

Q1. Make correct statements by filling in the symbols $\subset$ or $\not\subset$ in the blank spaces.

(i) $\{2,3,4\}....\{1,2,3,4,5\}$

(ii) $\{a,b,c\}.....\{b,c,d\}$

(iii) $\{x : x$ is a student of class XI of your school$\}$ _______ $\{x : x$ student of your school$\}$

(iv) $\{x : x$ is a circle in the plane$\}$ _______ $\{x : x$ is a circle in the same plane with radius 1 unit$\}$

(v) $\{x : x$ is a triangle in a plane$\}$ _______ $\{x : x$ is a rectangle in the plane$\}$

(vi) $\{x : x$ is an equilateral triangle in a plane$\}$ _______ $\{x : x$ is a triangle in the same plane$\}$

(vii) $\{x : x$ is an even natural number$\}$ _______ $\{x : x$ is an integer$\}$

Solution:

(i) $\{2,3,4\} \subset \{1,2,3,4,5\}$

(ii) $\{a,b,c\} \not\subset \{b,c,d\}$

(iii) $\{x : x$ is a student of class XI of your school$\} \subset \{x : x$ is student of your school$\}$

(iv) $\{x : x$ is a circle in the plane$\} \not\subset \{x : x$ is a circle in the same plane with radius 1 unit$\}$

(v) $\{x : x$ is a triangle in a plane$\} \not\subset \{x : x$ is a rectangle in the plane$\}$

(vi) $\{x : x$ is an equilateral triangle in a plane$\} \subset \{x : x$ in a triangle in the same plane$\}$

(vii) $\{x : x$ is an even natural number$\} \subset \{x : x$ is an integer$\}$

Q2. Examine whether the following statements are true or false.

(i) $\{a,b\} \not\subset \{b,c,a\}$

(ii) $\{a,e\} \subset \{x : x$ is a vowel in English alphabet$\}$

(iii) $\{1,2,3\} \subset \{1,3,5\}$

(iv) $\{a\} \subset \{a,b,c\}$

(v) $\{a\} \in \{a,b,c\}$

(vi) $\{x : x$ is an even natural number less than 6$\}$ $\subset \{x : x$ is a natural number which divides 36$\}$

Solution:

(i) False: $\{a,b\} \subset \{b,c,a\}$ since each element of $\{a, b\}$ is also an element of $\{b, c, a\}$.

(ii) True: Since a, e, are two vowels of the English alphabet.

(iii) False: $\because 2 \in \{1,2,3\}$ but $2 \notin \{1,3,5\}$

$\therefore \{1,2,3\} \not\subset \{1,3,5\}$

(iv) True: $\because a \in \{a,b,c\}$ $\quad \therefore \{a\} \subset \{a,b,c\}$

(v) False: {a} represents a set not an element. Therefore $a \in \{a,b,c\}$ while $\{a\} \notin \{a,b,c\}$.

(vi) True: $A=\{x : x$ is an even natural number less than 6$\} = \{2, 4\}$.

$B=\{x : x$ is a natural number which divides 36$\}$

$= \{1, 2, 3, 4, 6, 9, 12, 18, 36\}$.

$2, 4 \in A$ and $2, 4 \in B$. Therefore $A \subset B$.

Q3. Let $A=\{1,2,\{3,4\},5\}$. Which of the following statements are incorrect and why?

(i) $\{3,4\} \subset A$ (ii) $\{3,4\} \in A$

(iii) $\{\{3,4\}\} \subset A$ (iv) $1 \in A$

(v) $1 \subset A$ (vi) $\{1,2,5\} \subset A$

(vii) $\{1,2,5\} \in A$ (viii) $\{1,2,3\} \subset A$

(ix) $\Phi \in A$ (x) $\Phi \subset A$

(xi) $\{\Phi\} \subset A$

Solution:

(i) The statement $\{3,4\} \subset A$ is incorrect because $\{3,4\}$ is not a subset while it is an element $\{3,4\} \in A$.

(ii) The statement $\{3,4\} \in A$ is correct because $\{3,4\}$ is an element of A.

(iii) Statement $\{\{3,4\}\} \subset A$ is correct because $\{3,4\}$ is an element of A implies $\{\{3,4\}\}$ is a subset of A.

(iv) The statement $1 \in A$ is correct because 1 is an element of A.

(v) The statement $1 \subset A$ is incorrect because 1 is element of A, and in subset it is written as $\{1\} \subset A$

(vi) The statement $\{1,2,5\} \subset A$ is correct because each element of {1, 2, 5} is also an element of A.

(vii) The statement $\{1,2,5\} \in A$ is incorrect because $\{1,2,5\}$ represents a set not an element, while each element 1, 2, 5 belongs to A.

(viii) The statement $\{1,2,3\} \subset A$ is incorrect because $3 \notin A$.

(ix) The statement $\Phi \in A$ is incorrect because Φ is not an element of A.

(x) The statement $\Phi \subset A$ is correct because the empty set Φ is a subset of every set.

(xi) The statement $\Phi \subset A$ is incorrect because $\Phi \in \{\Phi\}$; however, $\Phi \in A$.

Note: The empty set is not an element of every set $(\Phi \subset A)$. It may be an element of some sets; if we can write $A = \{\phi, 1, 2\}$, then the set has the empty set as one of its elements. However, the set does not contain the empty set as an element.

Q4. Write down all the subsets of the following sets:

(i) {a} (ii) {a, b}

(iii) {1, 2, 3} (iv) Φ

Solution:

(i) The subsets of {a} are $\Phi, \{a\}$.

(ii) The subsets of {a, b} are $\Phi, \{a\}, \{b\}, \{a, b\}$.

(iii) The subsets of {1, 2, 3} are $\Phi, \{1\}, \{2\}, \{3\}, \{1,2\}, \{2,3\}, \{1,3\}, \{1,2,3\}$.

(iv) The only subset of Φ is Φ.

Q5. How many elements have $P(A)$, if $A = \Phi$?

Solution:

If A is a set with m elements, then

$n(A) = m \Rightarrow n[P(A)] = 2^m$.

Thus, for $A = \Phi$,

$n(A) = 0 \Rightarrow n[P(A)] = 2^\circ = 1$

Thus, $P(A)$ has only one element, Φ.

Q6. Write the following as intervals:

(i) $\{x : x \in R, -4 < x \le 6\}$

(ii) $\{x : x \in R, -12 < x < -10\}$

(iii) $\{x : x \in R, 0 \le x < 7\}$

(iv) $\{x : x \in R, 3 \le x \le 4\}$

Solution:

(i) $\{x : x \in R, -4 < x \le 6\} = (-4, 6]$

(ii) $\{x : x \in R, -12 < x < -10\} = (-12, -10)$

(iii) $\{x : x \in R, 0 \le x < 7\} = [0, 7)$

(iv) $\{x : x \in R, 3 \le x \le 4\} = [3, 4]$

Q7. Write the following intervals in set-builder form:

(i) (–3, 0) (ii) [6, 12]

(iii) (6, 12] (iv) [–23, 5)

Solution:

(i) $(-3, 0) = \{x : x \in R, -3 < x < 0\}$

(ii) $[6, 12] = \{x : x \in R, 6 \le x \le 12\}$

(iii) $(6, 12] = \{x : x \in R, 6 < x \le 12\}$

(iv) $[-23, 5) = \{x : x \in R, -23 \le x < 5\}$

Q8. What universal set(s) would you propose for each of the following?

(i) The set of right triangles

(ii) The set of isosceles triangles

Solution:

(i) For the set of right triangles, the universal set can be the set of all triangles.

(ii) For the set of isosceles triangles, the universal set can be the set of triangles or the set of polygons.

Q9. Given the sets $A = \{1,3,5\}$, $B = \{2,4,6\}$, $C = \{0,2,4,6,8\}$, which of the following may be considered as universals set (s) for all three sets A, B and C?

(i) {0, 1, 2, 3, 4, 5, 6}

(ii) Φ

(iii) {0, 1, 2, 3, 4, 5, 6, 7, 8, 9, 10}

(iv) {1, 2, 3, 4, 5, 6, 7, 8}

Solution:

(i) $\because A \subset \{0,1,2,3,4,5,6\}$, $B \subset \{0,1,2,3,4,5,6\}$

while, $C \not\subset \{0,1,2,3,4,5,6\}$

Therefore, the set {0, 1, 2, 3, 4, 5, 6} cannot be the universal set for the sets A, B, C.

(ii) $\because A \not\subset \Phi, B \not\subset \Phi, C \not\subset \Phi$, therefore, Φ cannot be the universal set for the sets A, B, C.

(iii) $A \subset \{0,1,2,3,4,5,6,7,8,9,10\}$

$B \subset \{0,1,2,3,4,5,6,7,8,9,10\}$

$C \subset \{0,1,2,3,4,5,6,7,8,9,10\}$

Therefore, the set $\{0,1,2,3,4,5,6,7,8,9,10\}$ is the universal set for the sets A, B and C.

(iv) $A \subset \{1,2,3,4,5,6,7,8\}$,

$B \subset \{1,2,3,4,5,6,7,8\}$ while $C \not\subset \{1,2,3,4,5,6,7,8\}$

Therefore, the set $\{1,2,3,4,5,6,7,8\}$ cannot be the universal set for the sets A, B and C.

Day 7

Venn Diagrams and Operations on Sets

Venn Diagrams: Relationships between sets can also be represented by means of diagrams which are known as Venn diagrams. The universal set is usually represented by a rectangle and its subsets by circles. In Venn diagrams, the elements of the sets are written in their respective circles.

Example 1: Let, $U = \{1, 2, 3, 4, 5, 6, 7, 8, 9, 10\}$ be the universal set. The different subsets are

$A = \{1, 3, 5, 7, 8\}$

$B = \{1, 2, 4, 6, 8\}$

$C = \{3, 5\}$

Since A, B and C are subsets, and also $C \subset A$, the above relation can be shown using a Venn diagram (Fig. 1.1).

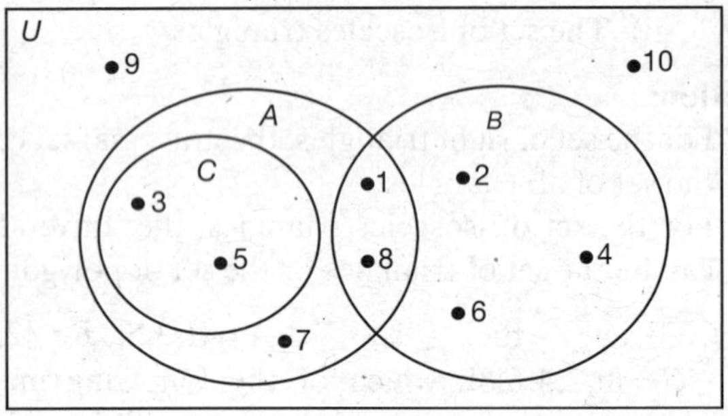

Fig. 1.1

- Elements 1, 3, 5, 7, 8 are in subset A, and therefore, must be written in diagram (circle) A.
- Elements 1, 2, 4, 6, 8 are in subset B, and therefore, must be written in diagram (circle) B.
- Elements 3, 5 are in subset C and also $C \subset A$, and therefore, diagram (circle) C must be within diagram (circle) A, and elements 3, 5 must be written in diagram (circle) C.
- Elements 9, 10 are not in any subset but they are in the universal set, and therefore, they must be written inside the rectangle but outside all the circles.
- Elements 1, 8 are in both subsets A and B, and therefore, subsets A and B must intersect each other and elements 1, 8 must be written in this intersection part (common to both A and B).

Operations on Sets: Different algebraic operations (addition, subtraction, multiplication and division) on two numbers give rise to another number. Similarly, different operations on two sets give rise to another set.

Usually we shade that portion which represents the operations on sets and their properties.

Union of Sets: Let A and B be any two sets. The union of sets A and B is the set which consists of all the elements of A and all the elements of B. The common element is taken only once, since repetition of elements in sets is irrelevant.

That is, the union of two sets A and B is the set which consists of all those elements which are either in A or in B (including those which are in both).

The union is denoted by symbol '$\cup$' and thus $A \cup B$ is read as 'A union B', and written as

$A \cup B = \{x : x \in A \text{ or } x \in B\}$.

The Venn diagrams showing the union of two and three sets are represented by the shaded portion as shown in Figs 1.2 to 1.4.

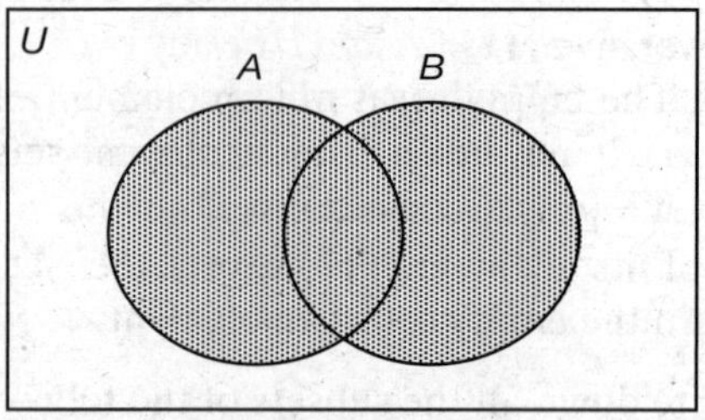

Fig. 1.2 $A \cup B = \{x : x \in A \text{ or } x \in B\}$

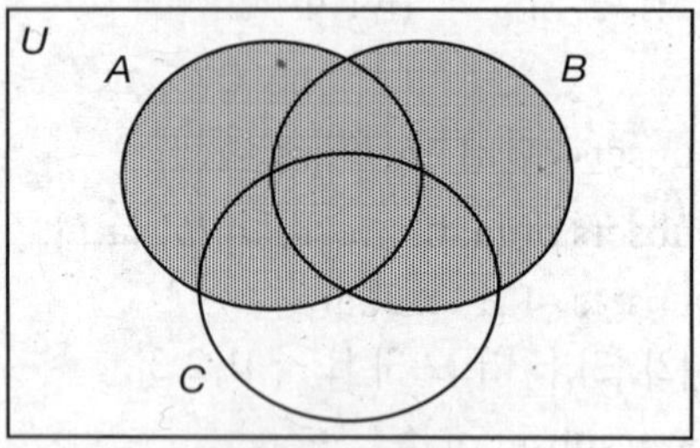

Fig. 1.3 $A \cup B$

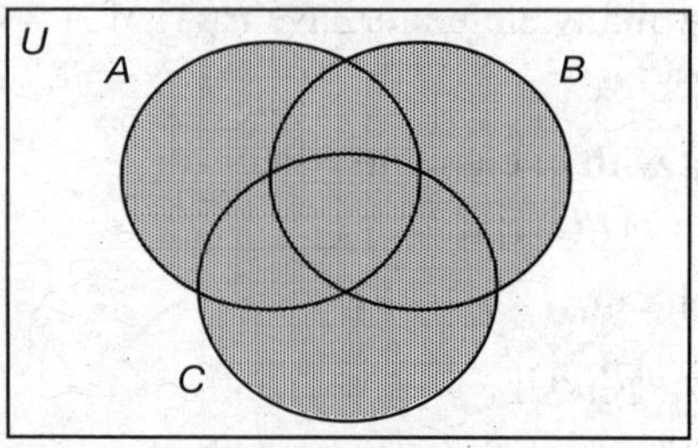

Fig. 1.4 $A \cup B \cup C = A \cup (B \cup C) = (A \cup B) \cup C$

Some properties of the operation of union:

(i) Idempotent law: $A \cup A = A$

(ii) Commutative law: $A \cup B = B \cup A$

(iii) Associative law: $(A \cup B) \cup C = A \cup (B \cup C)$

(iv) Law of identity element: $A \cup \phi = \phi \cup A = A$

(ϕ is the identity of U)

(v) Law of U: $A \cup U = U \cup A = U$

(vi) $A \times (B \cup C) = (A \times B) \cup (A \times C)$

Example 2: Let A be the set of all prime divisors of 30,

$\Rightarrow A = \{2, 3, 5\}$,

B is the set of all divisors of 15,

$\Rightarrow B = \{1, 3, 5, 15\}$

$C = \{x : x \text{ is all unit digit prime numbers}\}$

$\Rightarrow C = \{2, 3, 5, 7\}$

Then, $A \cup B = \{2, 3, 5\} \cup \{1, 3, 5, 15\} = \{1, 2, 3, 5, 15\}$

$B \cup C = \{1, 3, 5, 15\} \cup \{2, 3, 5, 7\} = \{1, 2, 3, 5, 7, 15\}$

$A \cup C = \{2, 3, 5\} \cup \{2, 3, 5, 7\} = \{2, 3, 5, 7\}$

$A \cup B \cup C = \{2, 3, 5\} \cup \{1, 3, 5, 15\} \cup \{2, 3, 5, 7\}$

$= \{1, 2, 3, 5, 15\} \cup \{2, 3, 5, 7\} = \{1, 2, 3, 5, 7, 15\}$

Intersection of Sets: Let A and B be any two sets. The intersection of A and B is the set which consists of all the common elements of A and B. The common element is taken only once, since repetition of elements in sets is irrelevant.

That is, the intersection of two sets A and B is the set which consists of all those elements which are in both A and B.

The intersection is denoted by symbol '$\cap$' and thus $A \cap B$ is read as 'A intersection B', and written as $A \cap B = \{x : x \in A \text{ and } x \in B\}$.

The Venn diagram showing the intersection of two and three sets are represented by the shaded portion as shown in the figures given below:

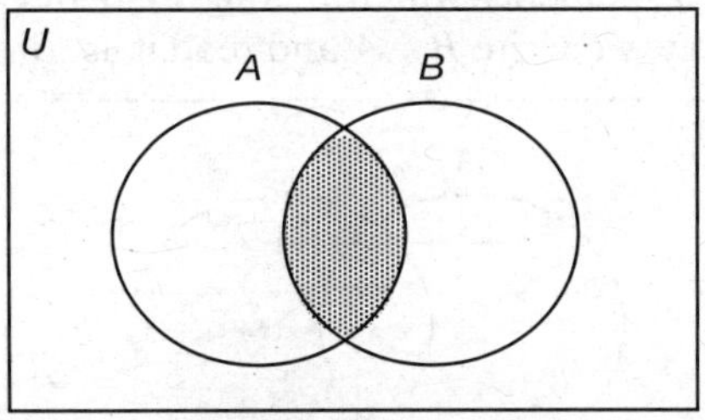

Fig. 1.5 $A \cap B = \{x : x \in A \text{ and } x \in B\}$

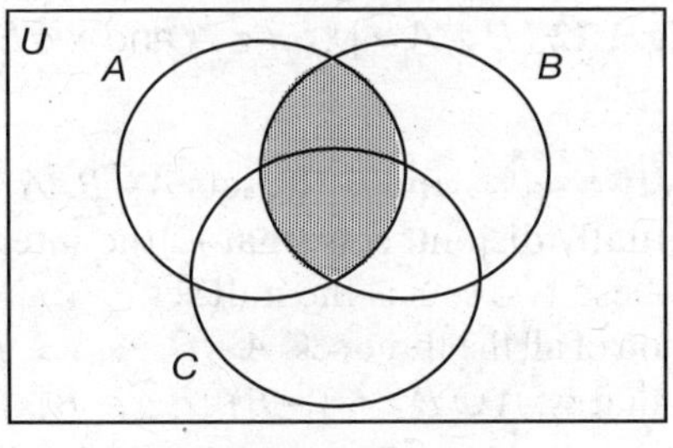

Fig. 1.6 $A \cap B$

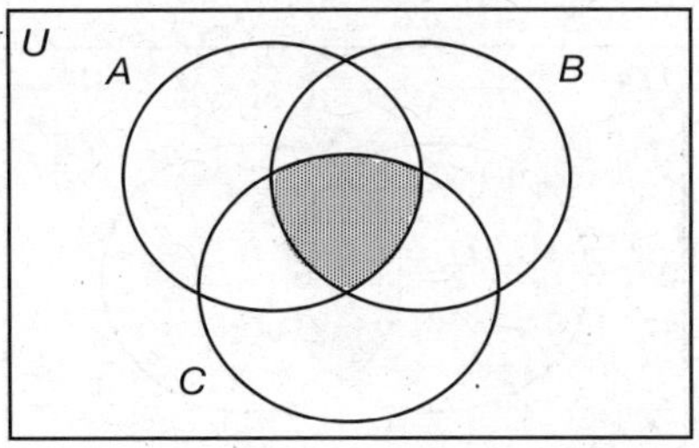

Fig. 1.7 $A \cap B \cap C = A \cap (B \cap C) = (A \cap B) \cap C$

Some properties of the operation of intersection:

(i) Idempotent law: $A \cap A = A$

(ii) Commutative law: $A \cap B = B \cap A$

(iii) Associative law: $(A \cap B) \cap C = A \cap (B \cap C)$

(iv) Law of nullity element: $A \cap \phi = \phi \cap A = \phi$

(ϕ is the nullity of A)

(v) Law of U: $A \cap U = U \cap A = A$

(vi) $A \times (B \cap C) = (A \times B) \cap (A \times C)$

Example 3: Let A be the set of all prime divisors of 30,

$\Rightarrow A = \{2, 3, 5\}$,

B is the set of all divisors of 15,

$\Rightarrow B = \{1, 3, 5, 15\}$

$C = \{x : x \text{ is all unit digit prime numbers}\}$,

$\Rightarrow C = \{2, 3, 5, 7\}$

Then, $A \cap B = \{2, 3, 5\} \cap \{1, 3, 5, 15\} = \{3, 5\}$

$B \cap C = \{1, 3, 5, 15\} \cap \{2, 3, 5, 7\} = \{3, 5\}$

$A \cap C = \{2, 3, 5\} \cap \{2, 3, 5, 7\} = \{2, 3, 5\}$

$A \cap B \cap C = \{2, 3, 5\} \cap \{1, 3, 5, 15\} \cap \{2, 3, 5, 7\}$

$= \{2, 3, 5\} \cap \{3, 5\} = \{3, 5\}$

Some More Properties

Distributive Law over Intersection ($\cap$):

$A \cap (B \cup C) = (A \cap B) \cup (A \cap C)$

A Venn diagram showing the distributive law over $\cap$ is represented by the shaded portion as shown in Fig. 1.8.

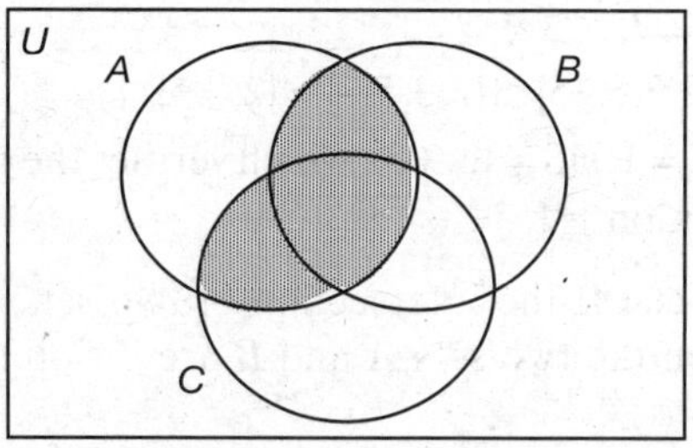

Fig. 1.8 $A \cap (B \cup C) = (A \cap B) \cup (A \cap C)$

Distributive Law over Union ($\cup$):

$A \cup (B \cap C) = (A \cup B) \cap (A \cup C)$

A Venn diagram showing the distributive law over $\cup$ is represented by the shaded portion as shown in Fig. 1.9.

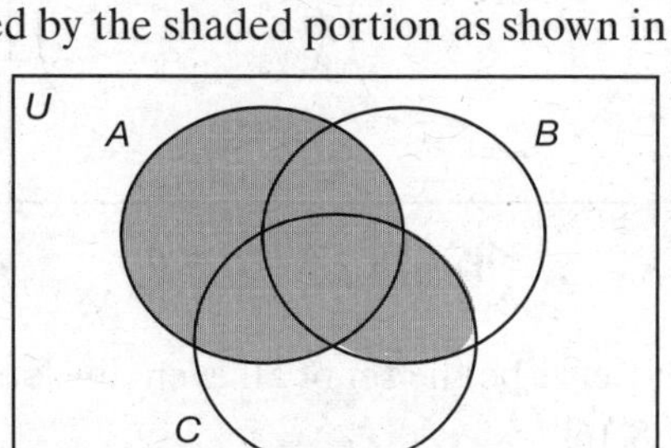

Fig. 1.9 $A \cup (B \cap C) = (A \cup B) \cap (A \cup C)$

Example 4: Let A be the set of all prime divisors of 30,

$\Rightarrow A = \{2, 3, 5\}$,

B is the set of all divisors of 15,

$\Rightarrow B = \{1, 3, 5, 15\}$

$C = \{x : x$ is all unit digit prime numbers$\}$,

$\Rightarrow C = \{2, 3, 5, 7\}$

Since

$A \cap B = \{2, 3, 5\} \cap \{1, 3, 5, 15\} = \{3, 5\}$

$A \cup B = \{2, 3, 5\} \cup \{1, 3, 5, 15\} = \{1, 2, 3, 5, 15\}$

$A \cap C = \{2, 3, 5\} \cap \{2, 3, 5, 7\} = \{2, 3, 5\}$

$A \cup C = \{2, 3, 5\} \cup \{2, 3, 5, 7\} = \{2, 3, 5, 7\}$

$B \cap C = \{1, 3, 5, 15\} \cap \{2, 3, 5, 7\} = \{3, 5\}$

$B \cup C = \{1, 3, 5, 15\} \cup \{2, 3, 5, 7\} = \{1, 2, 3, 5, 7, 15\}$

Distributive Law over Intersection ($\cap$):

$\text{LHS} = A \cap (B \cup C)$

$= \{2, 3, 5\} \cap \{1, 2, 3, 5, 7, 15\} = \{2, 3, 5\}$

$\text{RHS} = (A \cap B) \cup (A \cap C)$

$= \{3, 5\} \cup \{2, 3, 5\} = \{2, 3, 5\}$

Thus, LHS = RHS = $\{2, 3, 5\}$, which verifies the distributive law over intersection.

Distributive Law over Union ($\cup$):

$\text{LHS} = A \cup (B \cap C)$

$= \{2, 3, 5\} \cup \{3, 5\} = \{2, 3, 5\}$

$\text{RHS} = (A \cup B) \cap (A \cup C)$

$= \{1, 2, 3, 5, 15\} \cap \{2, 3, 5, 7\} = \{2, 3, 5\}$

Thus, LHS = RHS = $\{2, 3, 5\}$, which verifies the distributive law over union.

Disjoint Sets: If the intersection of two sets A and B is empty, then the two sets A and B are disjoint sets. That is, $A \cap B = \phi$

The Venn diagram of two disjoint sets is shown in Fig. 1.10.

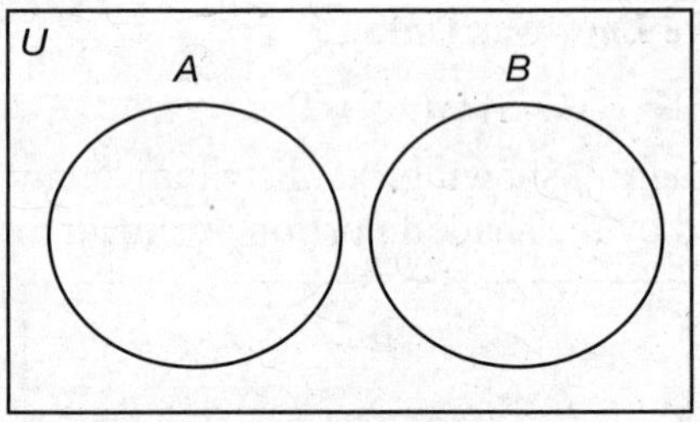

Fig. 1.10 $A \cap B = \phi$

Example 5: Let A be the set of all even divisors of 16,

$\Rightarrow A = \{2, 4, 8, 16\}$

B is the set of all divisors of 15,

$\Rightarrow B = \{1, 3, 5, 15\}$

C is the set of all prime divisors of 77,

$\Rightarrow C = \{7, 11\}$

Since $A \cap B = \{2, 4, 8, 16\} \cap \{1, 3, 5, 15\} = \phi$,

sets A and B are disjoint sets.

$B \cap C = \{1, 3, 5, 15\} \cap \{7, 11\} = \phi$

Thus, sets A and B are disjoint sets.

$A \cap C = \{2, 4, 8, 16\} \cap \{7, 11\} = \phi$

Thus, sets A and B are disjoint sets.

Difference of Sets: The difference of the sets A and B in this order is the set of elements which belong to A but not to B. Symbolically, we write $A - B$ and read it as 'A minus B'.

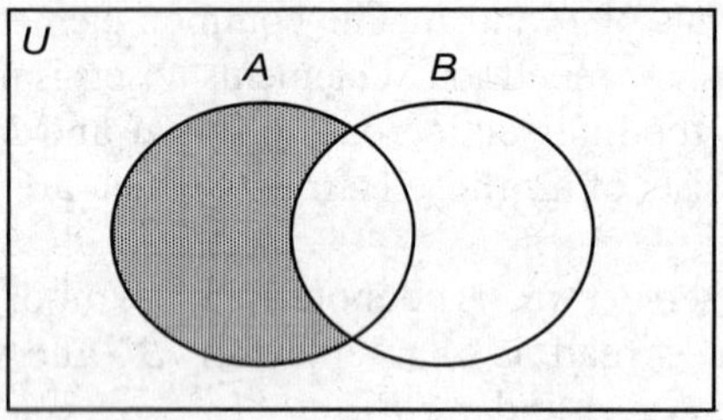

Fig. 1.11 $A - B = \{x : x \in A \text{ and } x \notin B\}$

Similarly, the difference of the sets B and A in this order is the set of elements which belong to B but not to A. Symbolically, we write $B - A$ and read it as 'B minus A'.

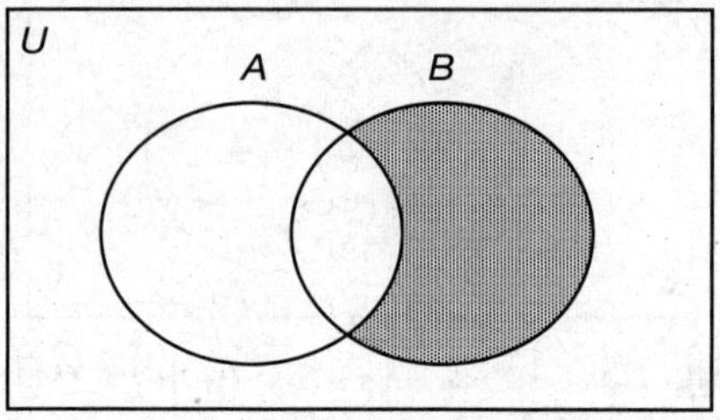

Fig. 1.12 $B - A = \{x : x \notin A \text{ and } x \in B\}$

Note:

- For any two sets A and B, the sets $A - B$, $A \cap B$, $B - A$ are mutually disjoint sets, that is, the intersection of any of these two sets is the null set.
- The union of all the three sets $A - B$, $A \cap B$, $B - A$ gives $A \cup B$, that is, $A \cup B = (A - B) \cup (A \cap B) \cup (B - A)$
- The Venn diagram of sets $A - B$, $A \cap B$, $B - A$ is shown in Fig. 1.13.

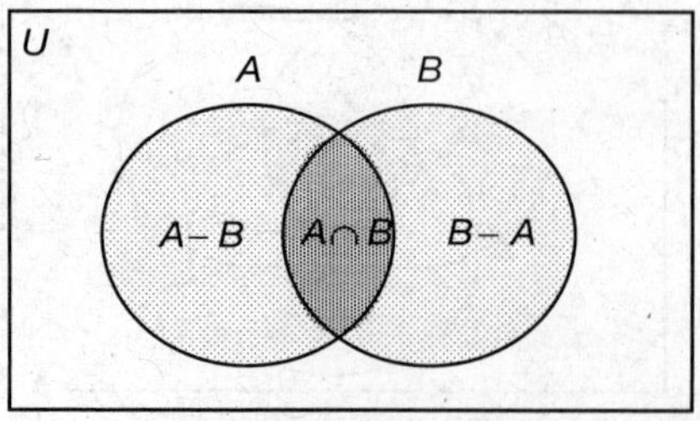

Fig. 1.13 $A \cup B = (A - B) \cup (A \cap B) \cup (B - A)$

- $A-(B\cup C)=(A-B)\cap(A-C)$
  $A-(B\cap C)=(A-B)\cup(A-C)$
  $(A\cap B)-C=(A-C)\cap(B-C)$
  $(A\cup B)-C=(A-C)\cup(B-C)$

Example 6: Let A be the set of all prime divisors of 30,
$\Rightarrow A=\{2,3,5\}$,
B is the set of all divisors of 15,
$\Rightarrow B=\{1,3,5,15\}$
$C=\{x: x$ is all unit digit prime numbers$\}$,
$\Rightarrow C=\{2,3,5,7\}$
Then, $A-B=\{2,3,5\}-\{1,3,5,15\}=\{2\}$
$B-A=\{1,3,5,15\}-\{2,3,5\}=\{1,15\}$
$B-C=\{1,3,5,15\}-\{2,3,5,7\}=\{1,5\}$
$C-B=\{2,3,5,7\}-\{1,3,5,15\}=\{2,7\}$
$A-C=\{2,3,5\}-\{2,3,5,7\}=\phi$
$C-A=\{2,3,5,7\}-\{2,3,5\}=\{7\}$
Also, we can see that the sets $A-B=\{2\}$, $A\cap B=\{3,5\}$ and $B-A=\{1,15\}$ are mutually disjoint sets, that is, the intersection of any of these two sets is the null set.
Also, we can see that the union of all the three sets $A-B$, $A\cap B$, $B-A$ is equal to $A\cup B$, that is,
$(A-B)\cup(A\cap B)\cup(B-A)$
$=\{2\}\cup\{3,5\}\cup\{1,15\}=\{1,2,3,5,15\}=A\cup B$
Also we can prove that,
For $A-(B\cup C)=(A-B)\cap(A-C)$
LHS: $A-(B\cup C)=\{2,3,5\}-\{1,2,3,5,7,15\}=\{\phi\}$
RHS: $(A-B)\cap(A-C)=\{2\}\cap\{\phi\}=\{\phi\}$
For $A-(B\cap C)=(A-B)\cup(A-C)$
LHS: $A-(B\cap C)=\{2,3,5\}-\{3,5\}=\{2\}$
RHS: $(A-B)\cup(A-C)=\{2\}\cup\{\phi\}=\{2\}$
Thus, LHS = RHS, verifying the above two results.

Day 8

NCERT Exercise 1.4 (With Solutions)

Q1. Find the union of each of the following pairs of sets:
(i) $X=\{1,3,5\}; Y=\{1,2,3\}$
(ii) $A=\{a,e,i,o,u\}; B=\{a,b,c\}$
(iii) $A=\{x: x$ is a natural number and multiple of 3$\}$
$B=\{x: x$ is a natural number less than 6$\}$
(iv) $A=\{x: x$ is a natural number and $1<x\le 6\}$
$B=\{x: x$ is a natural number and $6<x\le 10\}$
(v) $A=\{1,2,3\}; B=\Phi$

Solution:
(i) $X=\{1,3,5\}; Y=\{1,2,3\}$
$X\cup Y=\{1,2,3,5\}$
(ii) $A=\{a,e,i,o,u\}; B=\{a,b,c\}$
$A\cup B=\{a,b,c,i,o,u\}$
(iii) $A=\{x: x$ is a natural number and multiple of 3$\}=\{3, 6, 9,\ldots.\}$
$B=\{x: x$ is a natural number less than 6$\}$
$=\{1, 2, 3, 4, 5\}$
$\therefore A\cup B=\{1,2,4,5,3,6,9,12\ldots$multiple of 3$\}$
$\Rightarrow A\cup B=\{x: x=1,2,4,5$ or a multiple of 3$\}$
(iv) $A=\{x: x$ is a natural number and $1<x\le 6\}$
$=\{2,3,4,5,6\}$
$B=\{x: x$ is a natural number and $6<x<10\}$
$=\{7,8,9\}$
$\therefore A\cup B=\{2,3,4,5,6,7,8,9\}$
$\Rightarrow A\cup B=\{x: x\in N, 1\leqslant x<10\}$
(v) $A=\{1,2,3\}, B=\Phi$
$A\cup B=\{1,2,3\}$, $(\because S\cup\Phi=S)$

Q2. Let $A=\{a,b\}, B=\{a,b,c\}$. Is $A\subset B$? What is $A\cup B$?

Solution:
Since all elements of A are in B also, $A\subset B$ and $A\cup B=\{a,b,c\}$.

Q3. If A and B are two sets such that $A\subset B$, then what is $A\cup B$?

Solution:
If all elements of A are in B also, then $A\subset B$ and $A\cup B=B$.

Q4. If $A=\{1,2,3,4\}, B=\{3,4,5,6\}, C=\{5,6,7,8\}$ and $D=\{7,8,9,10\}$, find
(i) $A\cup B$ (ii) $A\cup C$
(iii) $B\cup C$ (iv) $B\cup D$
(v) $A\cup B\cup C$ (vi) $A\cup B\cup D$
(vii) $B\cup C\cup D$

Solution:
(i) $A\cup B=\{1,2,3,4,5,6\}$

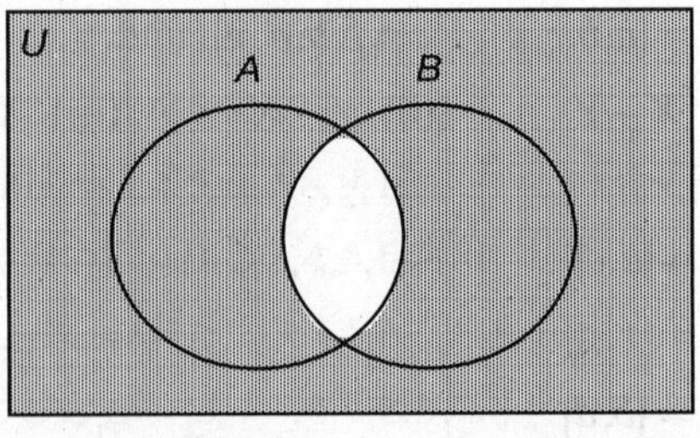

Fig. 1.16 $(A \cap B)' = \{x : x \in U \text{ and } x \notin (A \cap B)\}$

Some properties of complement of sets:

(a) Complement laws:
(i) $A \cup A' = U$ (ii) $A \cap A' = \phi$

(b) De Morgan's law:
(i) $(A \cup B)' = A' \cap B'$ (ii) $(A \cap B)' = A' \cup B'$

Note: $(A - B)' \neq A' - B'$

(c) Law of double complementation: $(A')' = A$

(d) Laws of empty set and universal set:
(i) $\phi' = U$ (ii) $U' = \phi$
(iii) $\phi' \cap A = U \cap A = A$ (iv) $\phi' \cup A = U \cup A = U$
(v) $U' \cap A = \phi \cap A = \phi$ (vi) $U' \cup A = \phi \cup A = A$

Example 1: U : Set of all students in your school
A : Set of all boys in your school
A' : Set of all girls in your school

Example 2: U : Set of all letters in the English alphabet
A : Set of all vowels in the English alphabet
A' : Set of all consonants in the English alphabet

Example 3: U : Set of all triangles in a plane
A : Set of all acute triangles and right triangles
A' : Set of all obtuse triangles

Example 4: U : Set of all real numbers
A : Set of all rational numbers
A' : Set of all irrational numbers

Example 5: U : Set of all integers
$A = \{x : x \text{ is an integer and } -4 < x \leq 4\}$
$A' = \{x : x \text{ is an integer and } x \leq -4 \text{ and } x > 4\}$

Example 6: U : Set of all natural numbers
$A = \{x : x = 3n,\ n \in N\}$
$A' = \{x : x = 3n - 1 \text{ and } x = 3n - 2,\ n \in N\}$

Example 7: U : Set of all integers
$A = \{x : x \text{ is an integer and } |x| \leq 3\}$
$A' = \{x : x \text{ is an integer and } |x| > 3\}$

Example 8: Let the universal set be $U = \{1,2,3,5,7,9,11,13,15\}$,
A is the set of all prime divisors of 30,
$\Rightarrow A = \{2,3,5\}$,
B is the set of all divisors of 15,
$\Rightarrow B = \{1,3,5,15\}$
$C = \{x : x \text{ is all unit digit prime numbers}\}$,
$\Rightarrow C = \{2,3,5,7\}$

Then

$A' = U - A$
$= \{1,2,3,5,7,9,11,13,15\} - \{2,3,5\}$
$= \{1,7,9,11,13,15\}$

$B' = U - B$
$= \{1,2,3,5,7,9,11,13,15\} - \{1,3,5,15\}$
$= \{2,7,9,11,13\}$

$C' = U - C$
$= \{1,2,3,5,7,9,11,13,15\} - \{2,3,5,7\}$
$= \{1,9,11,13,15\}$

$(A')' = (U - A)'$
$= (\{1,2,3,5,7,9,11,13,15\} - \{2,3,5\})'$
$= (\{1,7,9,11,13,15\})'$
$= \{1,2,3,5,7,9,11,13,15\} - \{1,7,9,11,13,15\}$
$= \{2,3,5\} = A$

$A \cup A' = \{2,3,5\} \cup \{1,7,9,11,13,15\}$
$= \{1,2,3,5,7,9,11,13,15\} = U$

$A \cap A' = \{2,3,5\} \cap \{1,7,9,11,13,15\} = \phi$

$(A \cup B)' = U - (A \cup B)$
$= \{1,2,3,5,7,9,11,13,15\} - \{1,2,3,5,15\}$
$= \{7,9,11,13\}$

$(A \cap C)' = U - (A \cap C)$
$= \{1,2,3,5,7,9,11,13,15\} - \{2,3,5\}$
$= \{1,7,9,11,13,15\}$

$(B - C)' = U - (B - C)$
$= \{1,2,3,5,7,9,11,13,15\} - \{1,15\}$
$= \{2,3,5,7,9,11,13\}$

De Morgan's Law: $(A \cup B)' = A' \cap B'$

LHS $= (A \cup B)'$
$= (\{2,3,5\} \cup \{1,3,5,15\})' = (\{1,2,3,5,15\})'$
$= \{1,2,3,5,7,9,11,13,15\} - \{1,2,3,5,15\}$
$= \{7,9,11,13\}$

RHS $= A' \cap B'$
$= (\{2,3,5\})' \cap (\{1,3,5,15\})'$
$= \{1,7,9,11,13,15\} \cap \{2,7,9,11,13\}$
$= \{7,9,11,13\}$

Thus, LHS = RHS = $\{7,9,11,13\}$, verifies De Morgan's Law: $(A \cup B)' = A' \cap B'$

De Morgan's Law: $(A \cap B)' = A' \cup B'$

LHS $= (A \cap B)'$

$= (\{2,3,5\} \cap \{1,3,5,15\})' = (\{3,5\})'$

$= \{1,2,3,5,7,9,11,13,15\} - \{3,5\}$

$= \{1,2,7,9,11,13,15\}$

RHS $= A' \cup B'$

$= (\{2,3,5\})' \cup (\{1,3,5,15\})'$

$= \{1,7,9,11,13,15\} \cup \{2,7,9,11,13\}$

$= \{1,2,7,9,11,13,15\}$

Thus, LHS = RHS = $\{1,2,7,9,11,13,15\}$, verifies De Morgan's Law: $(A \cap B)' = A' \cup B'$

Proof: $(A - B)' \neq A' - B'$

LHS $= (A - B)'$

$= (\{2,3,5\} - \{1,3,5,15\})' = (\{2\})'$

$= \{1,2,3,5,7,9,11,13,15\} - \{2\}$

$= \{1;3,5,7,9,11,13,15\}$

RHS $= A' - B'$

$= (\{2,3,5\})' - (\{1,3,5,15\})'$

$= \{1,7,9,11,13,15\} - \{2,7,9,11,13\}$

$= \{1,15\}$

Since LHS ≠ RHS, $(A - B)' \neq A' - B'$

Day 10

NCERT Exercise 1.5 (With Solutions)

Q1. Let $U = \{1,2,3;4,5,6,7,8,9\}$, $A = \{1,2,3,4\}$ $B = \{2,4,6,8\}$ and $C = \{3,4,5,6\}$. Find

(i) A' (ii) B'

(iii) $(A \cup C)'$ (iv) $(A \cup B)'$

(v) $(A')'$ (vi) $(B - C)'$

Solution:

(i) $A' = U - A = \{5,6,7,8,9\}$

(ii) $B' = U - B = \{1,3,5,7,9\}$

(iii) $A \cup C = \{1,2,3,4,5,6\}$ $\Rightarrow (A \cup C)' = \{7,8,9\}$

(iv) $A \cup B = \{1,2,3,4,6,8\}$ $\Rightarrow (A \cup B)' = \{5,7,9\}$

(v) $(A')' = A = \{1,2,3,4\}$

(vi) $B - C = \{2,8\}$ $\Rightarrow (B - C)' = \{1,3,4,5,6,7,9\}$

Q2. If $U = \{a,b,c,d,e,f,g,h\}$, find the complements of the following sets:

(i) $A = \{a,b,c\}$ (ii) $B = \{d,e,f,g\}$

(iii) $C = \{a,c,e,g\}$ (iv) $D = \{f,g,h,a\}$

Solution:

(i) $A' = U - A = \{d,e,f,g,h\}$

(ii) $B' = U - B = \{a,b,c,d\}$

(iii) $C' = U - C = \{b,d,f,h\}$

(iv) $D' = U - D = \{b,c,d,e\}$

Q3. Taking the set of natural numbers as the universal set, write down the complements of the following sets:

(i) $\{x : x$ is an even natural number$\}$

(ii) $\{x : x$ is an odd natural number$\}$

(iii) $\{x : x$ is a positive multiple of 3$\}$

(iv) $\{x : x$ is a prime number$\}$

(v) $\{x : x$ is a natural number divisible by 3 and 5$\}$

(vi) $\{x : x$ is a perfect square$\}$

(vii) $\{x : x$ is perfect cube$\}$

(viii) $\{x : x + 5 = 8\}$

(ix) $\{x : 2x + 5 = 9\}$

(x) $\{x : x \geq 7\}$

(xi) $\{x : x \in N, 2x + 1 > 10\}$

Solution:

$U = N$: set of natural numbers

(i) $A = \{x : x$ is an even natural number$\}$

$\Rightarrow A' = \{x : x$ is an odd natural number$\}$

(ii) $A = \{x : x$ is an odd natural number$\}$

$\Rightarrow A' = \{x : x$ is an even natural number$\}$

(iii) $A = \{x : x$ is a positive multiple of 3$\}$

$\Rightarrow A' = \{x : x \in N, x$ is not a multiple of 3$\}$

(iv) $A = \{x : x$ is a prime number$\}$

$\Rightarrow A' = \{x : x$ is not a prime number$\}$

(v) $A = \{x : x$ is a natural number divisible by 3 and 5$\}$

$\Rightarrow A' = \{x : x$ is a positive complement number and x is not divisible by 3 or 5$\}$

(vi) $A = \{x : x$ is a perfect square$\}$

$\Rightarrow A' = \{x : x$ is not a perfect square$\}$

(vii) $A = \{x : x$ is a perfect cube$\}$

$\Rightarrow A' = \{x : x$ is not a perfect cube$\}$

(viii) $A = \{x : x + 5 = 8\} \Rightarrow A = \{3\}$

$\Rightarrow A' = \{x : x \in N, x \neq 3\}$

(ix) $A = \{x : 2x + 5 = 9\} \Rightarrow A = \{2\}$

$\Rightarrow A' = \{x : x \in N, x \neq 2\}$

(x) $A = \{x : x \geq 7\}$

$\Rightarrow A' = \{x : x \in N, x < 7\}$

(xi) $A = \{x : x \in N, 2x + 1 > 10\}$

$\Rightarrow A' = \{x : x \in N, x \leq 9/2\}$

Q4. If $U = \{1,2,3,4,5,6,7,8,9\}, A = \{2,4,6,8\}$, $B = \{2,3,5,7\}$ verify that

(i) $(A \cup B)' = A' \cap B'$

(ii) $(A \cap B)' = A' \cup B'$

Solution:

$U = \{1,2,3,4,5,6,7,8,9\}, A = \{2,4,6,8\}$,

$B = \{2,3,5,7\}$

(i) $(A \cup B) = \{2,3,4,5,6,7,8\} \Rightarrow (A \cup B)' = \{1,9\}$

$A' \cap B' = \{1,3,5,7,9,\} \cap \{1,4,6,8,9\} = \{1,9\}$

$\therefore (A \cup B)' = A' \cap B'$

(ii) $(A \cap B) = \{2\} \Rightarrow (A \cap B)' = \{1,3,4,5,6,7,8,9\}$

$A' \cup B' = \{1,3,5,7,9\} \cup \{1,4,6,8,9\}$

$= \{1,3,4,5,6,7,8,9\}$

$\therefore (A \cap B)' = A' \cup B'$

Q5. Draw appropriate Venn diagrams for each of the following:

(i) $(A \cup B)'$ (ii) $A' \cap B'$

(iii) $(A \cap B)'$ (iv) $A' \cup B'$

Solution:

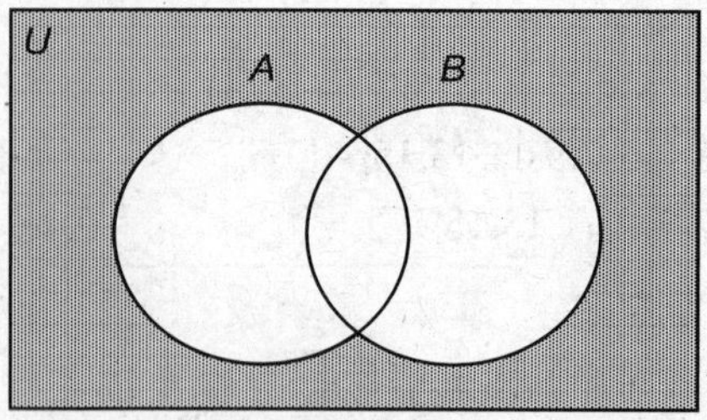

$(A \cup B)' = A' \cap B'$

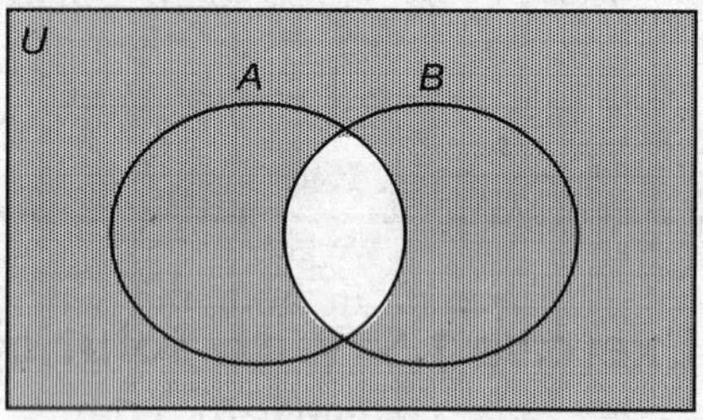

$(A \cap B)' = A' \cup B'$

Q6. Let U be the set of all triangles in a plane. If A is the set of all triangles with at least one angle different from 60°, what is A' ?

Solution:

A' is the set of all equilateral triangles.

Q7. Fill in the blanks to make each of the following a true statement:

(i) $A \cup A' =$ ______ (ii) $\Phi' \cap A =$ ______

(iii) $A \cap A' =$ ______ (iv) $U' \cup A =$ ______

Solution:

(i) $A \cup A' = U$

(ii) $\Phi' \cap A = U \cap A = A$

(iii) $A \cap A' = \Phi$

(iv) $U' \cap A = \Phi \cap A = \Phi$

Day 11

Practical Problems on Union and Intersection of Sets

Two Sets Problems: Let A and B be finite sets, and let $n(A)$ and $n(B)$ be the number of distinct elements in sets A and B, respectively. Then

$$n(A \cup B) = n(A) + n(B) - n(A \cap B)$$

And if A and B are disjoint sets, then $A \cap B = \phi$. Thus $n(A \cup B) = n(A) + n(B)$

Proof: Since the sets $A - B, A \cap B, B - A$ are disjoint and their union is equal to $A \cup B$,

$A \cup B = (A - B) \cup (A \cap B) \cup (B - A)$

$\Rightarrow n(A \cup B) = n(A - B) + n(A \cap B) + n(B - A)$

$\Rightarrow n(A \cup B) = n(A - B) + n(A \cap B)$
$\qquad + n(B - A) + n(A \cap B) - n(A \cap B)$

$\Rightarrow n(A \cup B) = n(A) + n(B) - n(A \cap B)$

$\left(\because n(A) = n(A - B) + n(A \cap B) \text{ and } n(B) = n(B - A) + n(A \cap B)\right)$

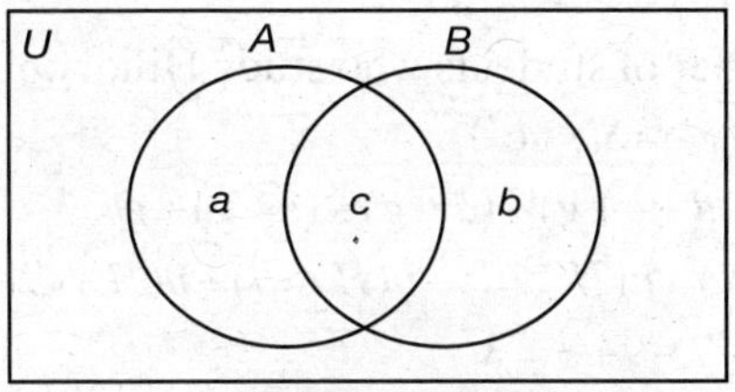

Fig. 1.17

Thus, as shown in Fig. 1.17, we get

Total number of elements in both the sets A and B: $n(A \cup B) = a + b + c$

Number of elements common to both A and B: $n(A \cap B) = c$

Number of elements in A: $n(A) = a + c$

Number of elements in B: $n(B) = b + c$

a = Number of elements only in A (means not in B)
$(a + c) - c = n(A) - n(A \cap B)$

b = Number of elements only in B (means not in A)
$(b + c) - c = n(B) - n(A \cap B)$

Three Sets Problems: Similarly for three finite sets A, B and C:

$n(A \cup B \cup C) = n(A) + n(B) + n(C) - n(A \cap B)$
$\qquad - n(B \cap C) - n(A \cap C) + n(A \cap B \cap C)$

Proof:

$n(A \cup B \cup C) = n[A \cup (B \cup C)]$

$= n(A) + n(B \cup C) - n[A \cap (B \cup C)]$

$= n(A) + n(B) + n(C) - n(B \cap C) - n[A \cap (B \cup C)]$

Since $A \cap (B \cup C) = (A \cap B) \cup (A \cap C)$

$\Rightarrow n[A \cap (B \cup C)] = n[(A \cap B) \cup (A \cap C)]$

$= n(A \cap B) + n(A \cap C) - n[(A \cap B) \cap (A \cap C)]$

$= n(A \cap B) + n(A \cap C) - n(A \cap B \cap C)$

Therefore

$n(A \cup B \cup C) = n(A) + n(B) + n(C) - n(A \cap B)$
$-n(B \cap C) - n(A \cap C) + n(A \cap B \cap C)$

Note: For solving problems in three sets, first we draw the Venn diagram and then we consider different parts of the diagram as a, b, c, ... as shown in Fig. 1.18. And finally, we find the solution using formulae or from the Venn diagram.

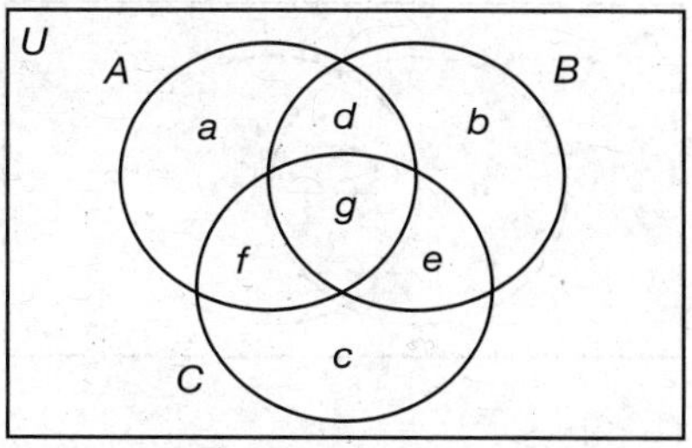

Fig. 1.18

Thus, as shown in Fig. 1.18, we get

Total number of elements in all the three sets A, B and C: $n(A \cup B \cup C) = a + b + c + d + e + f + g$

Number of elements in A: $n(A) = a + d + f + g$

Number of elements in B: $n(B) = b + d + e + g$

Number of elements in C: $n(C) = c + e + f + g$

Number of elements common to both A and B: $n(A \cap B) = d + g$

Number of elements common to both B and C: $n(B \cap C) = e + g$

Number of elements common to both A and C: $n(A \cap C) = f + g$

Number of elements common to all the three sets A, B and C: $n(A \cap B \cap C) = g$

a = Number of elements only in A (means neither in B and nor in C)
$= n(A) - n(A \cap B) - n(A \cap C) + n(A \cap B \cap C)$

b = Number of elements only in B (means neither in A and nor in C)
$= n(B) - n(A \cap B) - n(B \cap C) + n(A \cap B \cap C)$

c = Number of elements only in C (means neither in A and nor in B)
$= n(C) - n(A \cap C) - n(B \cap C) + n(A \cap B \cap C)$

d = Number of elements common to both A and B but not in $C = n(A \cap B) - n(A \cap B \cap C)$

e = Number of elements common to both B and C but not in $A = n(B \cap C) - n(A \cap B \cap C)$

f = Number of elements common to both A and C but not in $B = n(A \cap C) - n(A \cap B \cap C)$

Example 1: In a group of 50 students, the number of students studying Hindi and English were found to be as follows: Hindi = 25, English = 35, Hindi and English = 15. Find the number of students who study

(i) only Hindi
(ii) only English
(iii) at least one language
(iv) none of the languages

Solution: Let H be the set of students who study Hindi and E be the set of students who study English.

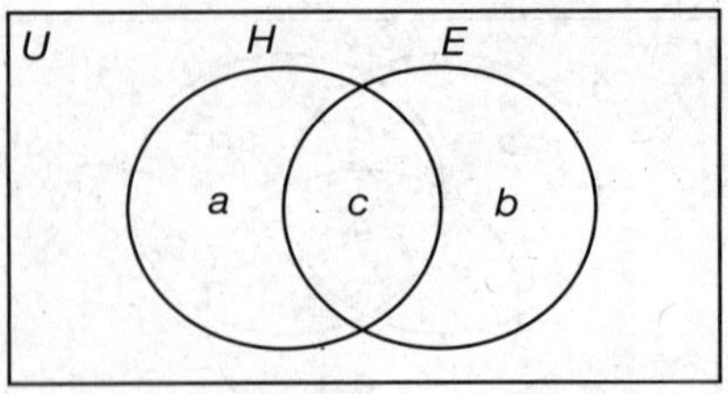

Then, $n(U) = 50$

$n(H) = a + c = 25$

$n(E) = b + c = 35$

$n(H \cap E) = c = 15$

(i) Number of students who study Hindi only

$a = n(H \cap E') = n(H) - n(H \cap E)$

$= (a + c) - (c) = 25 - 15 = 10$

(ii) Number of students who study English only

$b = n(E \cap H') = n(E) - n(H \cap E)$

$= (b + c) - (c) = 35 - 15 = 20$

(iii) Number of students who study at least one language

$= a + b + c$

$= 10 + 20 + 15 = 45$

(iv) Number of students who study none of the languages = Total students – Students who study at least one language = 50 – 45 = 5

Example 2: In a group of 50 students, the number of students studying Hindi, English, Sanskrit were found to be as follows: Hindi = 16, English = 18, Sanskrit = 14 Hindi and English = 7, English and Sanskrit = 5, Hindi and Sanskrit = 9, English, Hindi and Sanskrit = 3. Find the number of students who study

(i) only Hindi
(ii) only English
(iii) only Sanskrit
(iv) English and Sanskrit but not Hindi
(v) Hindi and Sanskrit but not English
(vi) Hindi and English but not Sanskrit
(vii) at least one of the three languages
(viii) none of the three languages

Solution: Let H be the set of students who study Hindi, E be the set of students who study English and S be the set of students who study Sanskrit.

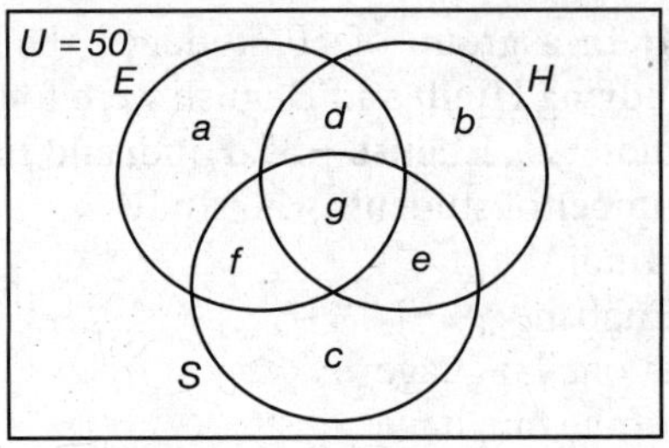

Then, $n(U) = 50$

$n(E) = a + d + f + g = 18$

$n(H) = b + d + e + g = 16$

$n(S) = c + e + f + g = 14$

$n(E \cap H) = d + g = 7$

$n(H \cap S) = e + g = 9$

$n(E \cap S) = f + g = 5$

$n(E \cap M \cap S) = g = 3$

(i) Number of students who study Hindi only

$b = n(H \cap S' \cap E')$

$= (b + d + e + g) - (d + g) - (e + g) + g$

$= n(H) - n(H \cap E) - n(H \cap S) + n(H \cap S \cap E)$

$= 16 - 7 - 9 + 3 = 3$

(ii) Number of students who study English only

$a = (a + d + f + g) - (d + g) - (f + g) + g$

$= n(E) - n(E \cap H) - n(E \cap S) + n(H \cap S \cap E)$

$= 18 - 7 - 5 + 3 = 9$

(iii) Number of students who study Sanskrit only

$c = (c + e + f + g) - (e + g) - (f + g) + g$

$= n(S) - n(S \cap H) - n(S \cap E) + n(H \cap S \cap E)$

$= 14 - 9 - 5 + 3 = 3$

(iv) Number of students who study English and Sanskrit but not Hindi

$f = n(E \cap S \cap H') = n(E \cap S) - n(E \cap S \cap H)$

$= (f + g) - g = 5 - 3 = 2$

(v) Number of students who study Hindi and Sanskrit but not English

$e = n(H \cap S \cap E') = n(H \cap S) - n(E \cap S \cap H)$

$= (e + g) - g = 9 - 3 = 6$

(vi) Number of students who study Hindi and English but not Sanskrit

$d = n(H \cap E \cap S') = (d + g) - g$

$= n(H \cap E) - n(E \cap S \cap H) = 7 - 3 = 4$

(vii) Number of students who study at least one of the three languages $= a + b + c + d + e + f + g$

$= 9 + 3 + 3 + 4 + 6 + 2 + 3 = 30$

(viii) Number of students who study none of the three languages = Total students – Students who study at least one of the three languages = 50 – 30 = 20

Day 12

NCERT Exercise 1.6 (With Solutions)

Q1. If X and Y are two sets such that $n(X)=17, n(Y)=23, n(XUY)=38$, find $n(X\cap Y)$.

Solution:

$n(X)=17, n(Y)=23, n(XUY)=38$

$n(X\cup Y)=n(X)+n(Y)-n(X\cap Y)$

$\Rightarrow 38=17+23-n(X\cap Y)$

$\Rightarrow n(X\cap Y)=40-38=2$

Q2. If X and Y are two sets such that $X\cup Y$ has 18 elements, X has 8 elements and Y has 15 elements, how many elements does $X\cap Y$ have?

Solution:

$n(X\cup Y)=18, n(X)=8, n(Y)=15$

$n(X\cup Y)=n(X)+n(Y)-n(X\cap Y)$

$\Rightarrow 18=8+15-n(X\cap Y)$

$\Rightarrow n(X\cap Y)=23-18=5$

Q3. In a group of 400 people, 250 can speak Hindi and 200 can speak English. How many people can speak both Hindi and English?

Solution:

Let H be the set of people who speak Hindi and E be the set of people who speak English. Then

$n(H\cup E)=400, n(H)=250, n(E)=200$

$n(H\cap E)=?$

$\because n(H\cup E)=e(E)-n(H\cap E)$

$\therefore 400=250+200-n(H\cap E)$

$\Rightarrow n(H\cap E)=450-400=50$

Thus, 50 people can speak both Hindi and English.

Q4. If S and T are two sets such that S has 21 elements, T has 32 elements, and $S\cap T$ has 11 elements, how many elements does $S\cup T$ have?

Solution:

$n(S)=21, n(T)=32, n(S\cap T)=11$

$\because n(S\cup T)=n(S)+n(T)-n(S\cap T)$

$\therefore n(S\cup T)=21+32-11=42$

Thus, the set $S\cup T$ has 42 elements.

Q5. If X and Y are two sets such that X has 40 elements, $X\cup Y$ has 60 elements and $X\cap Y$ has 10 elements, how many elements does Y have?

Solution:

$n(X)=40,\ n(X\cup Y)=60,\ n(X\cap Y)=10$

$\because n(X\cup Y)=n(X)+n(Y)-n(X\cap Y)$

$\therefore 60=40+n(Y)-10$

$\Rightarrow n(Y)=60-(40-10)=30$

Thus, the set Y has 30 elements.

Q6. In a group of 70 people, 37 like coffee, 52 like tea, and each person likes at least one of the two drinks. How many people like both coffee and tea?

Solution:

Let C denote the set of people who like coffee, and T denote the set of people who like tea. Then

$n(C\cup T)=70,\ n(C)=37,\ n(T)=52$

$\because n(C\cup T)=n(C)+n(T)-n(C\cap T)$

$\therefore 70=37+52-n(C\cap T)$

$\Rightarrow n(C\cap T)=89-70=19$

Thus, 19 people like both coffee and tea.

Q7. In a group of 65 people, 40 like cricket, 10 like both cricket and tennis. How many like only tennis and not cricket? How many like tennis?

Solution:

Let C denote the set of people who like cricket and T denote the set of people who like tennis. Then

$n(C\cup T)=65,\ n(C)=40,\ n(C\cap T)=10$

$\because n(C\cup T)=n(C)+n(T)-n(C\cap T)$

$\therefore 65=40+n(T)-10$

$\Rightarrow n(T)=65-30=35$

Thus, 35 people like tennis.

Now since $(T-C)\cup(T\cap C)=T$

And $(T-C)\cap(T\cap C)=\Phi$

$\therefore n(T)=n(T-C)+n(T\cap C)$

$\Rightarrow 35=n(T-C)+10 \Rightarrow n(T-C)=35-10=25$

Thus, 25 people like only tennis.

Q8. In a committee, 50 speak French, 20 speak Spanish and 10 speak both Spanish and French. How many speak at least one of these two languages?

Solution:

Let F be the set of people in the committee who speak French, and S be the set of people in the committee who speak Spanish. Then

$n(F)=50,\ n(S)=20,\ n(S\cap F)=10$

$\because n(S\cup F)=n(S)+n(F)-n(S\cap F)$

$\Rightarrow n(S\cup F)=20+50-10=60$

Thus, 60 people in the committee speak at least one of these two languages.

Day 13

NCERT Miscellaneous Exercise (With Solutions)

Q1. Decide, among the following sets, which sets are subsets of one another.
$A = \{x : x \in R \text{ and } x \text{ satisfy } x^2 - 8x + 12 = 0\}$,
$B = \{2, 4, 6\}$, $C = \{2, 4, 6, 8, ...\}$, $D = \{6\}$.

Solution:
$A = \{x : x \in R \text{ and } x \text{ satisfy } x^2 - 8x + 12 = 0\}$
$\because x^2 - 8x + 12 = 0$
$\Rightarrow (x-2)(x-6) = 0 \Rightarrow x = 2, 6$
$\therefore A = \{2, 6\}$, $B = \{2, 4, 6\}$,
$C = \{2, 4, 6, 8...\}$, $D = \{6\}$
Thus, $D \subset A \subset B \subset C$
$\Rightarrow D \subset A, D \subset B, D \subset C, A \subset B, A \subset C, B \subset C$

Q2. In each of the following, determine whether the statement is true or false. If it is true, prove it. If it is false, given an example.
(i) If $x \in A, A \in B$, then $x \in B$
(ii) If $A \subset B, B \in C$, then $A \in C$
(iii) If $A \subset B, B \subset C$, then $A \subset C$
(iv) If $A \not\subset B, B \not\subset C$, then $A \not\subset C$
(v) If $x \in A, A \not\subset B$, then $x \in B$
(vi) If $A \subset B, x \notin B$, then $x \notin A$

Solution:
(i) False:
Let $A = \{1, 2\}$, $B = \{\{1, 2\}, 3\}$
$2 \in A$, $\{1, 2\} \in B$, but $2 \notin B$
(ii) False:
Let $A = \{1\}$, $B = \{1, 2\}$, $C = \{\{1, 2\}, 3\}$
$A \subset B \Rightarrow B \in C$ but, $A \notin C$
(iii) True:
Let $A = \{1\}$, $B = \{1, 2\}$, $C = \{1, 2, 3\}$
$A \subset B$, $B \subset C \Rightarrow A \subset C$
(iv) False:
Let $A = \{3\}$, $B = \{1, 2\}, C = \{2, 3\}$
$A \not\subset B, B \not\subset C$ but $A \subset C$
(v) False:
Let $A = \{1, 2\}$, $B = \{2, 3\}$
$1 \in A$, $A \not\subset B$, but $1 \notin B$
(vi) True.
Let $A = \{1, 2\}$, $B = \{1, 2, 3\}$
$\because 4 \notin B \Rightarrow 4 \notin A$

Q3. Let A, B and C be sets such that $A \cup B = A \cup C, A \cap B = A \cap C$. Show that $B = C$.

Solution:
Let $x \in B$
$\Rightarrow x \in A \cup B$ $\quad (\because B \subset A \cup B)$
$\Rightarrow x \in A \cup C$ $\quad (\because A \cup B = A \cup C)$
$\Rightarrow x \in A$ or $x \in C$
Case I: Let $x \in A$
Also, $x \in B \Rightarrow x \in A \cap B$
$\Rightarrow x \in A \cap C$ $\quad (\because A \cap B = A \cap C)$
$\Rightarrow x \in A$ and $x \in C$
$\Rightarrow x \in C \Rightarrow B \subset C$
Similarly, for **Case II:** Assuming, $x \in A$ we can show that $C \subset B$. Thus, $B = C$.

Q4. Show that the following four conditions are equivalent.
(i) $A \subset B$ (ii) $A - B = \Phi$
(iii) $A \cup B = B$ (iv) $A \cap B = A$

Solution:
To prove
Equivalent condition $A \subset B \Leftrightarrow A - B = \Phi$
Let $A \subset B$
And suppose $A - B \neq \Phi$
This means that there exists $x \in A, x \neq B$, which is not possible as $A \subset B$.
$\therefore A - B = \Phi$
Thus, $A \subset B \Rightarrow A - B = \Phi$
Now, let $A - B = \Phi$
And suppose $A \subset B$
Let $x \in A$ then $x \in B$ because if $x \notin B$, then $A - B \neq \Phi$
Thus, $A - B = \Phi \Rightarrow A \subset B$
Hence, the equivalent condition $A \subset B \Leftrightarrow A - B = \Phi$ is proved.

To prove
Equivalent condition $A \subset B \Leftrightarrow A \cup B = B$
Let $A \subset B$
Then, $B \subset A \cup B$
Let $x \in A \cup B \Rightarrow x \in A$ or $x \in B$

Case I:
$x \in A \Rightarrow x \in B$ $(\because A \subset B)$
$\Rightarrow A \cup B \subset B$

Case II: $x \in B$

Then, $A \cup B = B$

Conversely, let $A \cup B = B$

Let $x \in A$

$\Rightarrow x \in A \cup B \quad (\because A \subset A \cup B)$

$\Rightarrow x \in B \qquad (\because A \cup B = B)$

$\Rightarrow A \subset B$

Thus, the equivalent condition $A \subset B \Leftrightarrow A \cup B = B$ is proved.

To prove

Equivalent condition $A \subset B \Leftrightarrow A \cap B = A$

Let $A \subset B$

Then, $A \cap B \subset A$

Let $x \in A$

As $A \subset B$, $\therefore x \in B$

$\Rightarrow x \in A \cap B \Rightarrow A \subset A \cap B$

$\Rightarrow A \cap B = A$

Conversely, suppose $A \cap B = A$

Let $x \in A \Rightarrow x \in A \cap B$

$\Rightarrow x \in A, x \in B \Rightarrow x \in B$

$\therefore A \subset B$

Thus, the equivalent condition $A \subset B \Leftrightarrow A \cap B = A$ is proved.

Q5. Show that if $A \subset B$, then $C - B \subset C - A$.

Solution:

Let $A \subset B$

And let $x \in C - B$

$\Rightarrow x \in C, x \notin B$

$\Rightarrow x \in C, x \notin A \quad (\because A \subset B)$

$\Rightarrow x \in C - A$

$\Rightarrow C - B \subset C - A$

Q6. Assume that $P(A) = P(B)$. Show that $A = B$.

Solution:

Let $P(A) = P(B)$

And let $x \in A$

$A \in P(A) = P(B)$

$\therefore x \in C$, for some $C \in P(B)$

And since, $C \subset B \Rightarrow x \in B \Rightarrow A \subset B$

Similarly, we can show that $B \subset A$

Thus, $A = B$

Q7. Is it true for any sets A and B, $P(A) \cup P(B) = P(A \cup B)$? Justify your answer.

Solution: False.

Let $A = \{1,2\}$, $B = \{2,3\}$

$\therefore A \cup B = \{1,2,3\}$

$P(A) = \{\Phi, \{1\}, \{2\}, \{1,2\}\}$

$P(B) = \{\Phi, \{2\}, \{3\}, \{2,3\}\}$

$P(A \cup B) = \{\Phi, \{1\}, \{2\}, \{3\}, \{1,2\}, \{1,3\}, \{2,3\}, \{1,2,3\}\}$

$P(A) \cup P(B) = \{\Phi, \{1\}, \{2\}, \{1,2\}, \{3\}, \{2,3\}\}$

$\therefore P(A) \cup P(B) \neq P(A \cup B)$

Q8. Show that for any sets A and B,
$A = (A \cap B) \cup (A - B)$,
$A \cup (B - A) \subset (A \cup B)$

Solution:

To prove $A = (A \cap B) \cup (A - B)$

Let $x \in A$. Then

Case I: $x \in A \cap B$

$\Rightarrow x \in (A \cap B)$

$\Rightarrow A \subset (A \cap B) \cup (A - B)$

Case II: $x \notin A \cap B \Rightarrow x \notin A$ or $x \notin B$

$\Rightarrow x \notin B \quad (\because x \in A)$

$\Rightarrow x \notin A - B$

$\Rightarrow A \subset (A \cap B) \cup (A - B)$ (i)

Now, $A \cap B \subset A, (A - B) \subset A$

$\therefore (A \cap B) \cup (A - B) \subset A$ (ii)

From (1) and (2), we proved that $A = (A \cap B) \cup (A - B)$

To prove $A \cup (B - A) \subset A \cup B$

Let $x \in A \cup (B - A)$

$\Rightarrow x \in A$ or $x \in (B - A)$

$\Rightarrow x \in A$ or$(x \in B, x \notin A)$

$\Rightarrow (x \in A$ or $x \in B), (x \in A$ or $x \notin A)$

$\Rightarrow x \in (A \cup B)$

$\therefore A \cup (B - A) \subset (A \cup B)$ (iii)

Now for $(A \cup B) \subset A \cup (B - A)$.

Let $y \in A \cup B \Rightarrow y \in A$ or $y \in B$

$\Rightarrow (y \in A$ or $y \in B), (y \in A$ or $y \notin A)$

$\Rightarrow y \in A$ or $(y \in B, y \notin A)$

$\Rightarrow y \in A \cup (B - A)$

$\Rightarrow A \cup B \subset A \cup (B - A)$ (iv)

From (3) and (4), we have proved that $A \cup B(B - A) = A \cup B$

Q9. Show that
(i) $A \cup (A \cap B) = A$ (ii) $A \cap (A \cup B) = A$

Solution:

(i) $A \subset A \Rightarrow A \cap B \subset A$

$\Rightarrow A \cup (A \cap B) \subset A$ (i)

Also, $A \subset A \cup (A \cap B)$ (ii)

From (1) and (2), $A \cup (A \cap B) = A$

(ii) $\because A \cap (A \cup B) = (A \cap A) \cup (A \cap B)$

$= A \cup (A \cap B) = A$

Q10. Show that $A \cap B = A \cap C$ need not imply that $B = C$.

Solution:

Let $A = \{0,1\},\ B = \{0,2\},\ C = \{0,3\}$

Then, $A \cap B = \{0\},\ A \cap C = \{0\}$

Thus, $A \cap B = A \cap C = \{0\}$

But, $B \neq C$

Q11. Let A and B be two sets. If $A \cap X = B \cap X = \Phi$, $A \cup X = B \cup X$ for some set X, show that $A = B$.

Solution:

A and B are two sets such that $A \cap X = B \cap X,\ A \cup X = B \cup X$ for some set X.

Now, $A = A \cap (A \cup X)$

$= A \cap (B \cup X)$ $(\because A \cup X = B \cup X)$

$= (A \cap B) \cup (A \cap X)$ By Distributive Law

$= (A \cap B) \cup \Phi$ $(\because A \cap X = \Phi)$

$= A \cap B$ (i)

Similarly, $B = B \cap (A \cup X)$

$= B \cap (A \cup X)$ $(\because B \cup X = A \cup X)$

$= (B \cap A) \cup (B \cap X)$ By Distributive Law

$= (B \cap A) \cup \Phi$ $(\because B \cap X = \Phi)$

$= B \cap A = A \cap B$ (ii)

Thus, from (1) and (2), we get $A = B$.

Q12. Find sets A, B and C such that $A \cap B,\ B \cap C,\ A \cap C$, are non-empty sets and $A \cap B \cap C = \Phi$.

Solution:

Let $A = \{1,2\},\ B = \{2,3\},\ C = \{1,3\}$

Now, $A \cap B = \{2\},\ B \cap C = \{3\},\ A \cap C = \{1\}$.

$\therefore A \cap B, B \cap C, A \cap C$ are non-empty.

But, $A \cap B \cap C = \Phi$.

Q13. In a survey of 600 students in a school, 150 students were found to be drinking tea and 225 drinking coffee, 100 were drinking both tea and coffee. Find how many students were drinking neither tea nor coffee.

Solution:

Let U be the set of all students who took part in the survey, T be the set of students drinking tea and C be the set of students drinking coffee.

Then, $n(U) = 600,\ n(T) = 150,\ n(C) = 225,\ n(T \cap C) = 100$

Now, number of students drinking neither tea nor coffee is equal to $n(T' \cap C') = n(T \cup C)'$

$= n(U) - n(T \cup C)$

$= n(U) - [n(T) + n(C) - n(T \cap C)]$

$= 600 - (150 + 225 - 100) = 600 - 275 = 325$

Q14. In a group of students 100, students know Hindi, 50 know English and 25 know both. Each of the students knows either Hindi or English. How many students are there in the group?

Solution:

Let E be the set of all students who know English, H be the set of all students who know Hindi and U be the set of all students in the group. Then,

$U = H \cup E \Rightarrow n(U) = n(H \cup E)$

$\Rightarrow n(H \cup E) = n(H) + n(E) - n(H \cap E)$

$\Rightarrow n(U) = 100 + 50 - 25 = 125$

Thus, there are 125 students in the group.

Q15. In a survey of 60 people, it was found that 25 people read newspaper H, 26 read newspaper T, 26 read newspaper 1, 9 read both H and I, 11 read both H and T, 8 read both T and I, 3 read all three newspapers. Find:

(i) the number of people who read at least one of the newspapers.

(ii) the number of people who read exactly one newspaper.

Solution:

Let H be the set of people who read newspaper H, T be the set of people who read newspaper T and I be the set of people who read newspaper I.

Then, $n(H) = 25,\ n(T) = 26,\ n(I) = 26,\ n(T \cap I) = 9$,

$n(H \cap T) = 11,\ n(T \cap I) = 8,\ n(H \cap T \cap I) = 3$

Let U be the set of people who took part in the survey, then,

(i) Number of people who read at least one of the newspapers is equal to $n(H \cup T \cup I)$

$= n(H) + n(T) + n(I) - n(H \cap T)$

$- n(T \cap I) - n(H \cap I) + n(H \cap T \cap I)$

$= 25 + 26 + 26 - 11 - 8 - 9 + 3 = 52$

Thus, 52 people read at least one of the newspapers.

(ii) Number of people who read exactly one newspaper is,

$= n(H \cup T \cup I) - n(H \cap T)$

$- n(T \cap I) - n(H \cap I) + 2n(H \cap T \cap I)$

$= 52 - 11 - 8 - 9 + 2 \times 3 = 52 - 28 + 6 = 30$

Thus, 30 people read exactly one newspaper.

Q16. In a survey it was found that 21 people liked product A, 26 liked product B and 29 liked product

C. If 14 people liked products A and B, 12 people liked products C and A, 14 people liked products B and C and 8 liked all the three products. Find how many liked product C only.

Solution: Let A, B and C be the people who like product A, product B and product C, respectively.

Then, $n(A)=21,\ n(B)=26,\ n(C)=29,$
$n(A\cap B)=14,\ n(C\cap A)=12,$
$n(B\cap C)=14,\ n(A\cap B\cap C)=8$
Number of people who liked only product C is
$=n(C)-n(A\cap C)-n(B\cap C)+n(A\cap B\cap C)$
$=29-12-14+8=37-26=11$

Day 14

NCERT Exemplar Problems: Solutions (Q1–Q20)

Q1. Write the following sets in the roster form.
(i) $A=\{x : x\in R, 2x+11=15\}$
(ii) $B=\{x \mid x^2=x, x\in R\}$
(iii) $C=\{x \mid x \text{ is a positive factor of a prime number } p\}$

Solution:
(i) $A=\{x : x\in R, 2x+11=15\}$
$\therefore 2x+11=15 \Rightarrow 2x=4 \Rightarrow x=2$
Thus, $A=\{2\}$
(ii) $B=\{x \mid x^2=x,\ x\in R\}$
$\therefore x^2=x \Rightarrow x^2-x=0 \Rightarrow x(x-1)=0$
$\Rightarrow x=0,1$
Thus, $B=\{0,1\}$
(iii) $C=\{x \mid x \text{ is a positive factor of prime number } p\}$
Positive factors of a prime number are 1 and the number itself. Therefore
$C=\{1,p\}$

Q2. Write the following sets in the roster form.
(i) $D=\{t \mid t^3=t, t\in R\}$
(ii) $E=\{w \mid \frac{w-2}{w+3}=3,\ w\in R\}$
(iii) $F=(x \mid x^4-5x^2+6=0, x\in R\}$

Solution:
(i) $D=\{t \mid t^3=t, t\in R\}$
$\because t^3=t \Rightarrow t^3-t=0$
$\Rightarrow t(t-1)(t+1)=0 \Rightarrow t=0,1-1$
$\therefore D=\{-1,0,1\}$
(ii) We have, $E=\{w \mid \frac{w-2}{w+3}=3,\ w\in R\}$
$\because \frac{w-2}{w+3}=3 \Rightarrow w-2=3w+9 \Rightarrow 2w=-11$
$\Rightarrow w=\frac{-11}{2},\quad \therefore E=\left\{\frac{-11}{2}\right\}$
(iii) We have, $F=(x \mid x^4-5x^2+6=0, x\in R\}$
$\because x^4-5x^2+6=0 \Rightarrow x^4-3x^2-2x^2+6=0$
$\Rightarrow x^2(x^2-3)-2(x^2-3)=0 \Rightarrow (x^2-3)(x^2-2)=0$
$\Rightarrow x=\pm\sqrt{2},\pm\sqrt{3},\quad \therefore F=\{-\sqrt{3},-\sqrt{2},\sqrt{2},\sqrt{3}\}$

Q3. $Y=\{x \mid x$ is a positive factor of the number $2^{p-1}(2^p-1)$, where 2^p-1 is a prime number$\}$. Write Y in the roster form.

Solution:
$Y=\{x \mid x$ is a positive factor of the number $2^{p-1}(2^p-1)$, where 2^p-1 is a prime number$\}$
Since 2^p-1 is a prime number, the factors of $2^{p-1}(2^p-1)$ are $1,2,2^2,2^3,\dots,2^{p-1},2^p-1$
Thus, $Y=\{1,2,2^2,2^3,\dots,2^{p-1},2^p-1\}$

Q4. State which of the following statements are true and which are false. Justify your answer.
(i) $35\in\{x \mid x$ has exactly four positive factors$\}$.
(ii) $128\in\{y \mid$ the sum of all the positive factors of y is $2y\}$.
(iii) $3\notin\{x \mid x^4-5x^3+2x^2-112x+6=0\}$.
(iv) $496\notin\{y \mid$ the sum of all the positive factors of y is $2y\}$.

Solution:
(i) The factors of 35 are 1, 5, 7 and 35.
Therefore, statement (i) is true.
(ii) The factors of 128 are 1, 2, 4, 8, 16, 32, 64 and 128.
$\therefore$ Sum of factors $=1+2+4+8+16+32+64+128$
$=255\neq 2\times 128$
Therefore, statement (ii) is false.
(iii) $3\notin\{x \mid x^4-5x^3+2x^2-112x+6=0\}$
At $x=3$,
$(3)^4-5(3)^3+2(3)^2-112(3)+6=-346\neq 0$
Therefore, statement (iii) is true.
(iv) The factors of 496 are 1, 2, 4, 8, 16, 31, 62, 124, 248, and 496.
$\therefore$ Sum of factors
$=1+2+4+8+16+31+62+124+248+496$
$=992=2(496)$

Therefore, $496 \notin \{y |$ the sum of all the positive factors of y is $2y\}$.
Thus, statement (iv) is false.

Q5. If $L = \{1,2,3,4\}, M = \{3,4,5,6\}, N = \{1,3,5\}$, verify that $L-(M \cup N) = (L-M) \cap (L-N)$.

Solution:

$L = \{1,2,3,4\}, M = \{3,4,5,6\}, N = \{1,3,5\}$

$\therefore M \cup N = \{1,3,4,5,6\} \Rightarrow L-(M \cup N) = \{2\}$

Now $L-M = \{1,2\},\ L-N = \{2,4\}$

$\therefore (L-M) \cap (L-N) = \{2\}$

Thus, $L-(M \cup N) = (L-M) \cap (L-N)$

Q6. If A and B are subsets of the universal set U, then show that

(i) $A \subset A \cup B$ (ii) $A \subset B \Leftrightarrow A \cup B = B$
(iii) $(A \cap B) \subset A$

Solution:

(i) Let $x \in A \Rightarrow x \in A$ or $x \in B \Rightarrow x \in A \cup B$

Thus, $A \subset A \cup B$

(ii) If $A \subset B$

Let $x \in A \cup B \Rightarrow x \in A$ or $x \in B$

$\Rightarrow x \in B\ (\because A \subset B)$

$\Rightarrow A \cup B \subset B$ (i)

Also $B \subset A \cup B$ (ii)

Therefore, from equations (i) and (ii), $A \cup B = B$

Now, if $A \cup B = B$

Let $y \in A \Rightarrow y \in A \cup B$

$\Rightarrow y \in B\ (\because A \cup B = B)$

$\Rightarrow A \subset B$

Thus, $A \subset B \Leftrightarrow A \cup B = B$

(iii) Let $x \in A \cap B \Rightarrow x \in A$ and $x \in B$

$\Rightarrow x \in A$

Thus, $A \cap B \subset A$

Q7. Given that $N = \{1,2,3,...,100\}$, write

(i) the subset of N whose elements are even numbers.

(ii) the subset of N whose elements are perfect square numbers.

Solution:

$N = \{1,2,3,....,100\}$

(i) Subset of N whose elements are even numbers $= \{2,4,6,8,...100\}$

(ii) Subset of N whose elements are perfect square numbers

$= \{1,4,9,16,25,36,49,64,81,100\}$

Q8. If $X = \{1,2,3\}$, and if n represents any member of X, write the following sets containing all numbers represented by

(i) $4n$ (ii) $n+6$ (iii) $n/2$ (iv) $n-1$

Solution:

Since $X = \{1,2,3\}$,

(i) $\{4n | n \in N\} = \{4,8,12\}$

(ii) $\{n+6 | n \in X\} = \{7,8,9\}$

(iii) $\{n/2 | n \in X\} = \{1/2, 1, 3/2\}$

(iv) $\{n-1 | n \in X\} = \{0,1,2\}$

Q9. If $Y = \{1,2,3,.....10\}, a$ represents any element of Y, write the following sets, containing all the elements satisfying the given conditions.

(i) $a \in Y$ but $a^2 \notin Y$ (ii) $a+1=6,\ a \in Y$

(iii) a is less than 6 and $a \in Y$

Solution:

Since $Y = \{1,2,3,.....10\}$

(i) $\{a : a \in Y, a^2 \notin Y\} = \{4,5,6,7,8,9,10\}$

(ii) $\{a : a+1 = 6, a \in Y\} = \{5\}$

(iii) a is less than 6 and $a \in Y\ \} = \{1, 2, 3, 4, 5\}$

Q10. A, B and C are subsets of universal set U. If $A = \{2,4,6,8,12,20\}$, $B = \{3,6,9,12,15\}$, $C = \{5,10,15,20\}$, U is the set of all whole numbers, draw a Venn diagram showing the relation of U, A, B and C.

Solution:

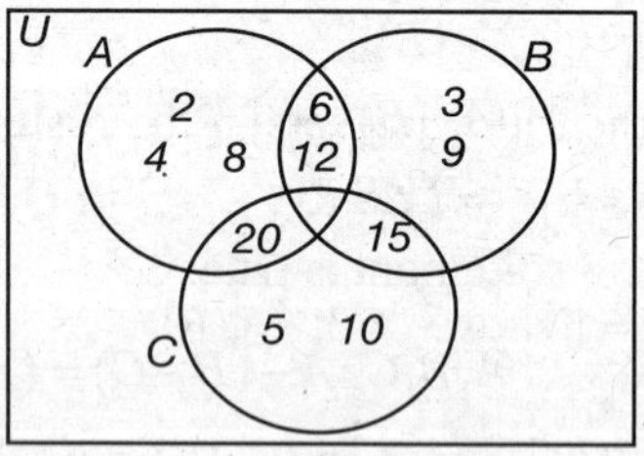

Q11. Let U be the set of all boys and girls in a school, G be the set of all girls in the school, B be the set of all boys in the school and S be the set of all students in the school who take swimming. Some, but not all, students in the school take swimming. Draw a Venn diagram showing one of the possible interrelationships among sets U, G, B are S.

Solution:

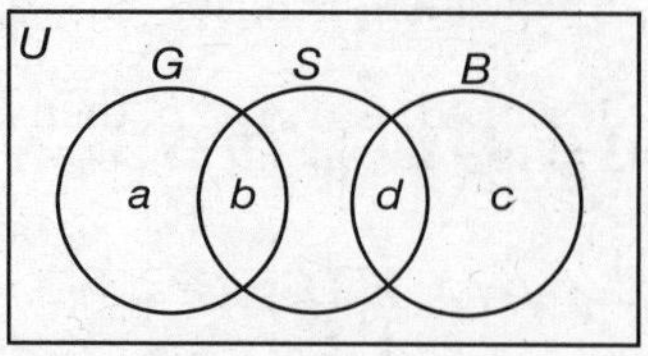

U: Set of all boys and girls in a school
G: Set of all girls in the school
B: Set of all boys in the school
S: Set of all students in the school who take swimming
a: Number of girls in the school who do not take swimming
b: Number of girls in the school who take swimming
c: Number of boys in the school who do not take swimming
d: Number of boys in the school who take swimming

Q12. For all sets A, B and C show that $(A-B)\cap(A-C)=A-(B\cup C)$.

Solution:
Let $x\in(A-B)\cap(A-C)$
$\Rightarrow x\in(A-B)$ and $x\in(A-C)$
$\Rightarrow (x\in A, x\notin B)$ and $(x\in A, x\notin C)$
$\Rightarrow x\in A, (x\notin B$ and $x\notin C) \Rightarrow x\in A, x\notin(B\cup C)$
$\Rightarrow x\in A-(B\cup C)$
$\Rightarrow (A-B)\cap(A-C)\subset A-(B\cup C)$ (i)
Now, let $y\in A-(B\cup C)\Rightarrow y\in A, y\notin(B\cup C)$
$\Rightarrow y\in A, (y\notin B$ and $y\notin C)$
$\Rightarrow (y\in A, y\notin B)$ and $(y\in A, y\notin C)$
$\Rightarrow y\in(A-B)$ and $y\in(A-C)$
$\Rightarrow y\in(A-B)\cap(A-C)$
$\Rightarrow A-(B\cup C)\subset(A-B)\cap(A-C)$ (ii)
Thus, from Eqs (i) and (ii),
$A-(B\cup C)=(A-B)\cap(A-C)$

Q13. For all sets A and B, $(A-B)\cup(A\cap B)=A$.

Solution:
$\text{LHS}=(A-B)\cup(A\cap B)$
$=[(A-B)\cup A]\cap[(A-B)\cup B]$
$=A\cap(A\cup B)=A=\text{RHS}$
Thus, the given statement is true.

Q14. For all sets A, B, C, $A-(B-C)=(A-B)-C$.

Solution:
Let $x\in A-(B-C)$
$\Rightarrow x\in A, x\notin(B-C)$
$\Rightarrow x\in A,\ x\notin B,\ x\in C$
And let $x\in(A-B)-C$
$\Rightarrow x\in(A-B),\ x\notin C$
$\Rightarrow x\in A,\ x\notin B,\ x\notin C$
Thus, the given statement is false.

Q15. For all sets A, B, C if $A\subset B$, then $A\cap C\subset B\cap C$.

Solution:
Let $x\in A\cap C\Rightarrow x\in A$ and $x\in C$
$\Rightarrow x\in B$ and $x\in C$ $(\because A\subset B)$
$\Rightarrow x\in(B\cap C)\Rightarrow(A\cap C)\subset(B\cap C)$
Thus, the given statement is true.

Q16. For all sets A, B, C if $A\subset B$, then $A\cup C\subset B\cup C$.

Solution:
Let $x\in A\cup C\Rightarrow x\in A$ or $x\in C$
$\Rightarrow x\in B$ or $x\in C$ $(\because A\subset B)$
$\Rightarrow x\in B\cup C\Rightarrow A\cup C\subset B\cup C$
Thus, the given statement is true.

Q17. For all sets A, B, C if $A\subset C, B\subset C$, then $A\cup B\subset C$.

Solution:
Let $x\in A\cup B\Rightarrow x\in A$ or $x\in B$
$\Rightarrow x\in C$ or $x\in C$ $(\because A\subset C, B\subset C)$
$\Rightarrow x\in C\Rightarrow A\cup B\subset C$
Thus, given statement is true.

Q18. For all sets A and B, $A\cup(B-A)=A\cup B$.

Solution:
$\text{LHS}=A\cup(B-A)=A\cup(B\cap A')$
$(\because A-B=A\cap B')$
$=(A\cup B)\cap(A\cup A')=(A\cup B)\cap U$
$(\because A\cup A'=U)$
$=A\cup B=\text{RHS}$ $(\because S\cap U=S)$

Q19. For all sets A, B $A-(A-B)=A\cap B$.

Solution:
$\text{LHS}=A-(A-B)=A-(A\cap B')$
$(\because A-B=A\cap B')$
$=A\cap(A\cap B')'=A\cap[A'\cup(B')']$
$(\because (A\cap B)'=A'\cup B')$
$=A\cap(A'\cup B)$ $(\because (A')'=A)$
$=(A\cap A')\cup(A\cap B)$
$=\phi\cup(A\cap B)=A\cap B=\text{RHS}$

Q20. For all sets A, B, $A-(A\cap B)=A-B$.

Solution: $\text{LHS}=A-(A\cap B)=A\cap(A\cap B)'$
$(\because A-B=A\cap B')$
$=A\cap(A'\cup B')$ $(\because (A\cap B)'=A'\cup B')$
$=(A\cap A')\cup(A\cap B')$
$=\phi\cup(A\cap B')$
$=A\cap B'=A-B=\text{RHS}$ $(\because \phi\cup A=A)$

Day 15

NCERT Exemplar Problems: Solutions (Q21–Q32)

Q21. For all sets A, B, $(A \cup B) - B = A - B$.

Solution:

LHS $= (A \cup B) - B = (A \cup B) \cap B'$ $(\because A - B = A \cap B')$

$= (A \cap B') \cup (B \cap B') = (A \cap B') \cup \phi$ $(\because B \cap B' = \phi)$

$= A \cap B'$ $(\because A \cup \phi = A)$

$= A - B = RHS$

Q22. Let $T = \left\{ x \middle| \frac{x+5}{x-7} - 5 = \frac{4x-40}{13-x} \right\}$. Is T an empty set? Justify your answer.

Solution:

$T = \left\{ x \middle| \frac{x+5}{x-7} - 5 = \frac{4x-40}{13-x} \right\}$

$\therefore \frac{x+5}{x-7} - 5 = \frac{4x-40}{13-x} \Rightarrow \frac{x+5-5(x-7)}{x-7} = \frac{4x-40}{13-x}$

$\Rightarrow \frac{-4x+40}{x-7} = \frac{4x-40}{13-x}$

$\Rightarrow -(4x-40)(13-x) = (4x-40)(x-7)$

$\Rightarrow (4x-40)(x-7) + (4x-40)(13-x) = 0$

$\Rightarrow (4x-40)(x-7+13-x) = 0$

$\Rightarrow 4(x-10)6 = 0 \Rightarrow 24(x-10) = 0$

$\Rightarrow x = 10, \quad \therefore T = \{10\}$

Thus, T is not an empty set.

Q23. If A, B, C be sets, show that $A \cap (B \cup C) = (A \cap B) \cup (A \cap C)$.

Solution:

Let $x \in A \cap (B \cup C)$

$\Rightarrow x \in A$ and $x \in (B \cup C)$

$\Rightarrow x \in A$ and $(x \in B$ or $x \in C)$

$\Rightarrow (x \in A$ and $x \in B)$ or $(x \in A$ and $x \in C)$

$\Rightarrow x \in A \cap B$ or $x \in A \cap C$

$\Rightarrow x \in (A \cap B) \cup (A \cap C)$

$\Rightarrow A \cap (B \cup C) \subset (A \cap B) \cup (A \cap C)$ (i)

Again, let $y \in (A \cap B) \cup (A \cap C)$

$\Rightarrow y \in (A \cap B)$ or $y \in (A \cap C)$

$\Rightarrow (y \in A$ and $y \in B)$ or $(y \in A$ and $y \in C)$

$\Rightarrow y \in A$ and $(y \in B$ or $y \in C)$

$\Rightarrow y \in A$ and $y \in B \cup C$

$\Rightarrow y \in A \cap (B \cup C)$

$\Rightarrow (A \cap B) \cup (A \cap C) \subset A \cap (B \cap C)$ (ii)

From equations (i) and (ii),

$A \cap (B \cup C) = (A \cap B) \cup (A \cap C)$

Q24. Out of 100 students, 15 passed in English, 12 passed in Mathematics, 8 in Science, 6 in English and Mathematics, 7 in Mathematics and Science, 4 in English and Science, 4 in all the three. Find how many passed in,

(i) English and Mathematics but not is Science.
(ii) Mathematics and Science but not in English.
(iii) Mathematics only.
(iv) More than one subject only.

Solution:

Let U be the total number of students, M be the set of students who passed in Mathematics, E be the set of students who passed in English and S be the set of students who passed in Science.

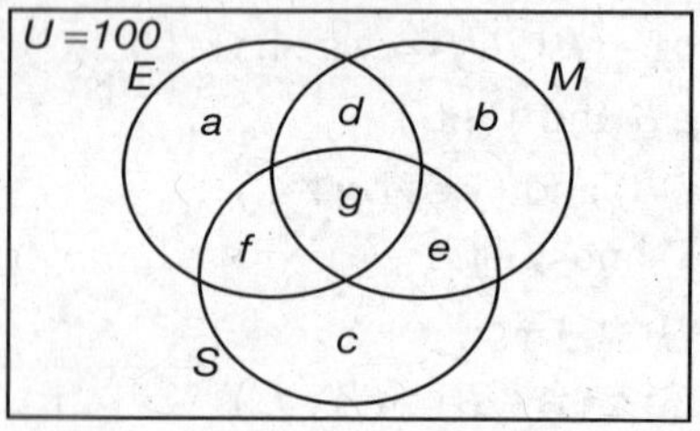

Then, $n(U) = 100$,

$n(E) = a + d + f + g = 15$

$n(M) = b + d + e + g = 12$

$n(S) = c + e + f + g = 8$

$n(E \cap M) = d + g = 6$

$n(M \cap S) = e + g = 7$

$n(E \cap S) = f + g = 4$

$n(E \cap M \cap S) = g = 4$

(i) Number of students who passed in English and Mathematics but not in Science is
$d = n(E \cap M \cap S') = (d + g) - g$
$= n(E \cap M) - n(E \cap M \cap S) = 6 - 4 = 2$

(ii) Number of students who passed in Mathematics and Science but not in English is
$e = n(M \cap S \cap E') = (e + g) - g$
$= n(M \cap S) - n(M \cap S \cap E) = 7 - 4 = 3$

(iii) Number of students who passed in Mathematics only is equal to $b = n(M \cap S' \cap E')$
$= (b + d + e + g) - (d + g) - (e + g) + g$
$= n(M) - n(M \cap E) - n(M \cap S) + n(M \cap S \cap E)$
$= 12 - 7 - 6 + 4 = 3$

(iv) Number of students who passed in more than one subject only is equal to $d + e + f + g$
$= (f + g) + (d + g) + (e + g) - 2g$

$n(E \cap M) + n(m \cap S) + n(E \cap S) - 2n(E \cap M \cap S)$
$= 6 + 7 + 4 - 2 \times 4 = 17 - 8 = 9$

Q25. In a class of 60 students, 25 students play cricket and 20 students play tennis and 100 students play both the games. Find the number of students who play neither.

Solution:
Let C be the set of students who play cricket and T be the set of students who play tennis. Then
$n(U) = 60,\ n(C) = 25,\ n(T) = 20,\ n(C \cap T) = 10$
$\therefore n(C \cup T) = n(C) + n(T) - n(C \cap T)$
$= 25 + 20 - 10 = 35$
$\therefore$ Number of students who play neither
$= n(U) - n(C \cup T) = 60 - 35 = 25$

Q26. In a survey of 200 students of a school, it was found that 120 study Mathematics, 90 study Physics and 70 study Chemistry, 40 study Mathematics and Physics, 30 study Physics and Chemistry, 50 study Chemistry and Mathematics and 20 none of these subjects. Find the number of students who study all the three subjects.

Solution:
Let M be the set of students who study Mathematics, P be the set of students who study Physics and C be the set of students who study Chemistry.
Then, $n(U) = 200,\ n(M) = 120,\ n(C) = 90,$
$n(M \cap P) = 40,\ n(P \cap C) = 30,\quad n(C \cap M) = 50,$
$n(M' \cap P' \cap C') = 20,$
$\Rightarrow n(U) - n(M \cup P \cup C) = 20$
$\Rightarrow n(M \cup P \cup C) = 200 - 20 = 180$
$\therefore n(M \cup P \cup C) = n(M) + n(P) + n(C)$
$-n(M \cap P) - n(P \cap C) - n(M \cap C) + n(M \cap P \cap C)$
$\Rightarrow 180 = 120 + 90 + 70 - 40 - 30 - 50 + n(M \cap P \cap C)$
$\Rightarrow 180 = 160 + n(M \cap P \cap C)$
$\Rightarrow n(M \cap P \cap C) = 180 - 160 = 20$
Thus, the number of students who study all the three subjects is 20.

Q27. In a town of 10000 families, 40% buy newspaper A, 20% buy newspaper B, 10% buy newspaper C, 5% buy A and B, 3% buy B and C and 4% buy A and C. If 2% of families buy all the three newspaper, then find
(i) number of families which buy newspaper A only
(ii) number of families which buy none of A, B, C.

Solution:
Let A be the set of families which buy newspaper A, B be set of families which buy newspaper B and C be the set of families which buy newspaper C.

Then, $n(U) = 10000$
$n(A) = 40\%,\ n(B) = 20\%,\ n(C) = 10\%$
$n(A \cap B) = 5\%,\ n(B \cap C) = 3\%$
$n(A \cap C) = 4\%,\ n(A \cap B \cap C) = 2\%$

(i) Number of families that buy newspaper A only
$= n(A) - n(A \cap B) - n(A \cap C) + n(A \cap B \cap C)$
$= (40 - 5 - 4 + 2)\% = 33\% = 10000 \times \frac{33}{100} = 3300$

(ii) Number of families that do not buy A, B or C newspaper $= n(U) - n(A \cup B \cup C)$
$= n(U) - [n(A) + n(B) + n(C) - n(A \cap B)$
$-n(B \cap C) - n(A \cap C) + n(A \cap B \cap C)]$
$= [100 - (40 + 20 + 10 - 5 - 3 - 4 + 2)]\%$
$= (100 - 60)\% = 40\% = 10000 \times \frac{40}{100} = 4000$

Q28. In a group of 50 students, the number of students studying French, English, Sanskrit were found to be: French = 17, English = 13, Sanskrit = 15, French and English = 9, English and Sanskrit = 4, French and Sanskrit = 5, English, French and Sanskrit = 3. Find the number of students who study
(i) only French (ii) only English
(iii) only Sanskrit
(iv) English and Sanskrit but not French
(v) French and Sanskrit but not English
(vi) French and English but not Sanskrit
(vii) at least one of the three languages
(viii) none of the three languages

Solution:
Let F be the set of students who study French, E be the set of students who study English and S be the set of students who study Sanskrit.

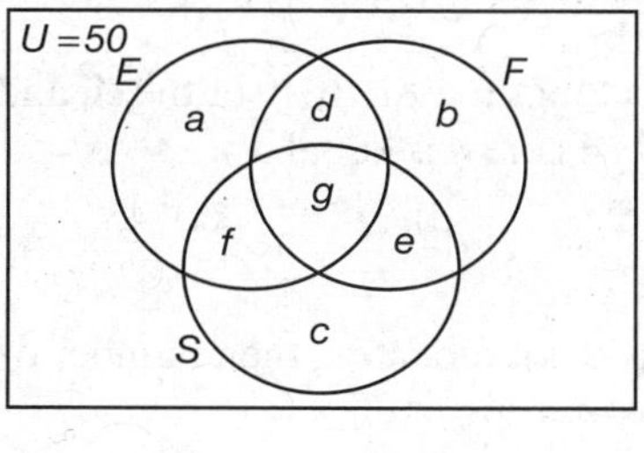

Then, $n(U) = 50$
$n(E) = a + d + f + g = 13$
$n(F) = b + d + e + g = 17$
$n(S) = c + e + f + g = 15$
$n(E \cap F) = d + g = 9$
$n(F \cap S) = e + g = 5$
$n(E \cap S) = f + g = 4$
$n(E \cap M \cap S) = g = 3$

(i) Number of students who study French only
$b = n(F \cap S' \cap E')$

$= (b+d+e+g)-(d+g)-(e+g)+g$
$= n(F)-n(F\cap E)-n(F\cap S)+n(F\cap S\cap E)$
$= 17-9-5+3=6$

(ii) Number of students who study English only
$a=(a+d+f+g)-(d+g)-(f+g)+g$
$= n(E)-n(E\cap F)-n(E\cap S)+n(F\cap S\cap E)$
$= 13-9-4+3=3$

(iii) Number of students who study Sanskrit only
$c=(c+e+f+g)-(e+g)-(f+g)+g$
$= n(S)-n(S\cap F)-n(S\cap E)+n(F\cap S\cap E)$
$= 15-5-4+3=9$

(iv) Number of students who study English and Sanskrit but not French
$f=n(E\cap S\cap F')=(f+g)-g$
$= n(E\cap S)-n(E\cap S\cap F)=4-3=1$

(v) Number of students who study French and Sanskrit but not English
$e=n(F\cap S\cap E')=(e+g)-g$
$= n(F\cap S)-n(E\cap S\cap F)=5-3=2$

(vi) Number of students who study French and English but not Sanskrit
$d=n(F\cap E\cap S')=(d+g)-g$
$= n(F\cap E)-n(E\cap S\cap F)=9-3=6$

(vii) Number of students who study atleast one of the three languages $= a+b+c+d+e+f+g$
$= 6+6+3+1+3+2+9=30$

(viii) Number of students who study none of the three languages = Total students – Students who study atleast one of the three languages = 50 – 30 = 20

Q29. Suppose $A_1, A_2, ..., A_{30}$ are thirty sets each having 5 elements and B_1, B_2, B_n are n sets each with 3 elements. Let $\bigcup_{i=1}^{30} A_i = \bigcup_{j=1}^{n} A_j = S$ and such elements of S belongs to exactly 10 of the A_i and exactly 9 of the B_j. Then, n is equal to

(a) 15 (b) 3 (c) 45 (d) 35

Solution: (c)
If elements are not repeated, then number of elements in $A_1\cup A_2\cup A_3, ..., \cup A_{30}$ is 30 × 5.

But each element is used 10 times, therefore $S=\frac{30\times 5}{10}=15$

If elements in $B_1, B_2, ...B_n$ are not repeated, then the total number of elements is $3n$ but each element is repeated 9 times, therefore $S=\frac{3n}{9}=15 \Rightarrow n=45$.

Q30. Two finite sets have m and n elements. The number of subsets of the first set is 112 more than that of the second set. The values of m and n are, respectively,

(a) 4, 7 (b) 7, 4 (c) 4, 4 (d) 7, 7

Solution: (a)
Since the number of subsets of set containing m elements is 112 more than the subsets of the set containing n elements,
$\because 2^m-2^n=112 \Rightarrow 2^n\cdot(2^{m-n}-1)=2^4\cdot 7$
$\Rightarrow 2^n=2^4$ and $2^{m-n}-1=7$
$\Rightarrow n=4$ and $2^{m-n}=8=2^3$
$\Rightarrow m-n=3 \Rightarrow m-4=3$
$\Rightarrow m=7$

Q31. The set $(A\cap B')'\cup(B\cap C)$ is equal to

(a) $A'\cup B\cup C$ (b) $A'\cup B$
(c) $A'\cup C'$ (d) $A'\cap B$

Solution: (b)
$(A\cap B')'\cup(B\cap C)$
$=[A'\cup(B')']\cup(B\cap C)$
$=(A'\cup B)\cup(B\cap C)=A'\cup B$
$\because (A\cap B')'=(A'\cup B)$ and $(A')'=A$

Q32. Let F_1 be the set of parallelograms, F_2 the set of rectangles, F_3 the set of rhombuses, F_4 the set of squares and F_5 the set of trapeziums in a plane. Then, F_1 may be equal to

(a) $F_2\cap F_3$ (b) $F_3\cap F_4$
(c) $F_2\cap F_5$ (d) $F_2\cap F_3\cup F_4\cup F_1$

Solution: (d)
Since every rectangle, rhombus, square in a plane is a parallelogram but every trapezium is not a parallelogram, F_1 is any of the F_1, F_2, F_3, F_4. Thus $F_1=F_2\cup F_3\cup F_4\cup F_1$

Day 16

NCERT Exemplar Problems: Solutions (Q33–Q57)

Q33. Let S = set of points inside the square, T = set of points inside the triangle and C = set of points inside the circle. If the triangle and circle intersect each other and are contained in a square, then

(a) $S\cap T\cap C=\phi$ (b) $S\cup T\cup C=C$
(c) $S\cup T\cup C=S$ (d) $S\cup T=S\cap C$

Solution: (c)

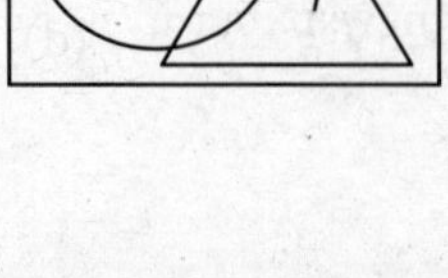

Using a Venn diagram, it can be represented as $S \cup T \cup C = S$

Q34. Let R be the set of points inside a rectangle of sides a and b $(a, b > 1)$ with two sides along the positive direction of X-axis and Y-axis. Then,

(a) $R = \{(x, y) : 0 \le x \le a, 0 \le y \le b\}$

(b) $R = \{(x, y) : 0 \le x < a, 0 \le y \le b\}$

(c) $R = \{(x, y) : 0 \le x \le a, 0 < y < b\}$

(d) $R = \{(x, y) : 0 < x < a, 0 < y < b\}$

Solution: (d)

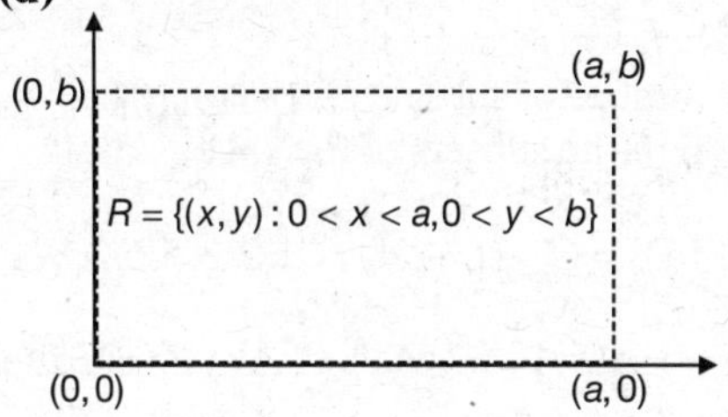

Since R is the set of points inside the rectangle of sides a and b, $R = \{(x, y) : 0 < x < a, 0 < y < b\}$.

Q35. In a town of 840 persons, 450 persons read Hindi, 300 read English and 200 read both. Then, the number of persons who read neither, is

(a) 210 (b) 290 (c) 180 (d) 260

Solution: (b)

Let H be the set of persons who read Hindi and E be the set of persons who read English.

Then, $n(U) = 840$

$n(H) = 450,\ n(E) = 300,\ n(H \cap E) = 200$

Number of persons who read neither is $= n(H' \cap E')$

$= n(U) - n(H \cup E)$

$= n(U) - [n(H) + n(E) - n(H \cap E)]$

$= 840 - (450 + 300 - 200) = 840 - 550 = 290$

Q36. $X = \{8^n - 7n - 1 | n \in N\}$ and

$Y = \{49n - 49 | n \in N\}$. Then,

(a) $X \subset Y$ (b) $Y \subset X$

(c) $X = Y$ (d) $X \cap Y = \phi$

Solution: (a)

$X = \{8^n - 7n - 1 | n \in N\} = \{0, 49, 490, ...\}$

$Y = \{49n - 49 | n \in N\} = \{0, 49, 147, ...\}$

Since every element of X is in Y but every element of Y is not in X, $X \subset Y$.

Q37. A survey shows that 63% of the people watch A news channel whereas 76% watch B news channel. If x% of the people watch both channel, then,

(a) $x = 35$ (b) $x = 63$

(c) $39 \le x \le 63$ (d) $x = 39$

Solution: (c)

Let A be the set of percentage of those people who watch A news channel and B be the set of percentage of those people who watch B news channel. Then,

$n(A) = 63,\ n(B) = 76,\ n(A \cap B) = x$

$\because n(A \cup B) \le 100$

$\therefore n(A) + n(B) - n(A \cap B) \le 100$

$\Rightarrow 63 + 76 - x \le 100 \Rightarrow 139 - x \le 100$

$\Rightarrow 139 - 100 \le x \Rightarrow 39 \le x$

$\because n(A \cap B) \le n(A)$

$\therefore x \le 63$

Thus, $39 \le x \le 63$

Q38. If sets A and B are defined as

$A = \{(x, y) | y = \frac{1}{x},\ x \ne 0,\ x \in R\}$

$B = \{(x, y) | y = -x,\ x \in R\}$, then

(a) $A \cap B = A$ (b) $A \cap B = B$

(c) $A \cap B = \phi$ (d) $A \cup B = A$

Solution: (c)

Since there is no $x \in R$ for which $-x = \frac{1}{x}$

$\Rightarrow A \cap B = \phi$

Q39. If A and B are two sets, then $A \cap (A \cup B)$ equals

(a) A (b) B (c) ϕ (d) $A \cap B$

Solution: (a)

Let $x \in A \cap (A \cup B) \Rightarrow x \in A$ and $x \in (A \cup B)$

$\Rightarrow x \in A$ and $(x \in A$ or $x \in B)$

$\Rightarrow (x \in A$ and $x \in A)$ or $(x \in A$ and $x \in B)$

$\Rightarrow x \in A$ or $x \in (A \cap B)$

$\Rightarrow x \in A \cup (A \cap B) \Rightarrow x \in A$

Q40. If $A = \{1, 3, 5, ..., 15, 17\}$, $B = \{2, 4, 6, ..., 16, 18\}$ and N the set of natural numbers is the universal set, then $(A' \cup (A \cup B) \cap B')$ is

(a) ϕ (b) N (c) A (d) B

Solution: (b)

$A' \cup ((A \cup B) \cap B')$

$= A' \cup [(A \cap B') \cup (B \cap B')]$

$= A' \cup [(A \cap B') \cup \phi]$

$= A' \cup (A \cap B')$

$= (A' \cup A) \cap (A' \cup B')$

$= N \cap (A' \cup B')$

$= A' \cup B' = (A \cap B)' = \phi' = N$

Q41. If $S = \{x | x$ is a positive multiple of 3 less than 100$\}$ and $P = \{x | x$ is a prime number less than 20$\}$.

Then, $n(S)+n(P)$ is equal to
(a) 34 (b) 31 (c) 33 (d) 41

Solution: (d)
$S=\{x|x$ is a positive multiple of 3 less than 100$\}$
$\Rightarrow S=\{3,6,9,...,96,99\}$
$\because t_n=a+(n-1)d \Rightarrow 99=3+(n-1)3 \Rightarrow n=33$
Thus, $n(S)=33$
$P=\{x|x$ is a prime number less than 20$\}$
$\Rightarrow P=\{2,3,5,7,11,13,17,19\} \Rightarrow n(P)=8$
Thus, $n(S)+n(P)=33+8=41$

Q42. If X and Y are two sets and X' denotes the complement of X, then $X\cap(X\cup Y)'$ is equal to
(a) X (b) Y (c) ϕ (d) $X\cap Y$

Solution: (c)
$X\cap(X\cup Y)'=X\cap(X'\cap Y')$
$=(X\cap X')\cap(X\cap Y')$
$=\phi\cap A=\phi$ (let $X\cap Y'=A$)

Q43. The set $\{x\in N:1\le x\le 2\}$ can be written as ______

Solution:
The set $\{x\in N:1\le x\le 2\}$ can be written as $x=\{1,2\}$.

Q44. When $A=\phi$, then number of elements in P(A) is ______

Solution:
$\because A=\phi \Rightarrow n(A)=0$
Therefore, the number of elements in P(A) is $n\{P(A)\}=2^{n(A)}=2^{\circ}=1$

Q45. If A and B are finite sets, such that $A\subset B$, then $n(A\cup B)$ is equal to ______

Solution:
If A and B are two finite sets such that $A\subset B$, then
$n(A\cup B)=n(B)$.

Q46. If A and B are any two sets, then $A-B$ is equal to ______

Solution:
If A and B are any two sets, then
$A-B=A\cap B' \Rightarrow \therefore A-B=A\cap B'$

Q47. Power set of the set $A=\{1,2\}$ is ______

Solution:
$\because A=\{1,2\}$
$\therefore$ Power set of the set $A=\{1,2\}$ is
$P(A)=\{\phi,\{1\},\{2\},\{1,2\}\}$

Q48. If the sets $A=\{1,3,5\}$, $B=\{2,4,6\}$, $C=\{0,2,4,6,8\}$. Then, the universal set of all the three sets A, B and C can be ______

Solution:
Universal set for A, B and C is given by $U=\{0,1,2,3,4,5,6,7,8\}$.

Q49. If $U=\{1,2,3,4,5,6,7,8,9,10\}, A=\{1,2,3,5\}$ and $B=\{2,4,6,7\}, C=\{2,3,4,8\}$. Then,
(i) $(B\cup C)'$ is______ (ii) $(C-A)'$ is______

Solution:
(i) $\because B\cup C=\{2,3,4,6,7,8\}$
$\therefore (B\cup C)'=U-(B\cup C)=\{1,5,9,10\}$
(ii) $\because C-A=\{4,8\}$
$\therefore (C-A)'=U-(C-A)=\{1,2,3,5,6,7,9,10\}$

Q50. For all sets A and B, $A-(A\cap B)$ is equal to ______

Solution:
$A-(A\cap B)=A\cap(A\cap B)'$
$(\because A-B=A\cap B')$
$=A\cap(A'\cup B')$ $(\because (A\cap B)'=A'\cup B')$
$=(A\cap A')\cup(A\cap B')$
$=\phi\cup(A\cap B')$
$=A\cap B'=A-B$ $(\because \phi\cup A=A)$

Q51. Match the following sets for all sets A, B and C.

Column I	Column II
(i) $((A'\cup B')-A)'$	(a) $A-B$
(ii) $[(B'\cup(B'-A)]'$	(b) A
(iii) $(A-B)-(B-C)$	(c) B
(iv) $(A-B)\cap(C-B)$	(d) $(A\times B)\cap(A\times C)$
(v) $A\times(B\cap C)$	(e) $(A\times B)\cup(A\times C)$
(vi) $A\times(B\cup C)$	(f) $(A\cap C)-B$

Solution:
(i) $((A'\cup B')-A)'=[(A'\cup B')\cap A']'$
$[\because A-B=A\cap B']$
$=[(A\cap B)'\cap A']'[\because (A\cap B)'=A'\cup B']$
$=[(A\cap B)']'=(A\cap B)\cup A[\because (A')'=A]$
$=A$
(ii) $[(B'\cup(B'-A)]'=[B'\cup(B'\cap A')]'$
$[\because A-B=A\cap B']$
$=[B'\cup(B\cup A)']'[\because A'\cap B'=(A\cup B)']$
$=(B')'\cap[(B\cup A)']'[\because (A\cup B)'=A'\cap B']$
$=B\cap(B\cup A)[\because (A\cup B)'=A'\cap B']$
$=B[\because (A')'=A]$
(iii) $(A-B)-(B-C)=(A\cap B')-(B\cap C')$

$[\because A-B=A\cap B']$
$=(A\cap B')\cap(B\cap C')'=(A\cap B')\cap[B'\cup(C')']$
$=(A\cap B')\cap(B'\cup C)[\because (A')'=A]$
$=[A\cap(B'\cup C)]\cap[B'\cap(B'\cup C)]$
$=[A\cap(B'\cup C)]\cap B'=(A\cap B')\cap[(B'\cup C)\cap B']$
$=(A\cap B')\cap B'=A\cap B'=A-B$

(iv) $(A-B)\cap(C-B)\Rightarrow(A\cap B')\cap(C\cap B')$
$[\because A-B=A\cap B']$
$\Rightarrow(A\cap C)\cap B'\Rightarrow(A\cap C)-B[\because A\cap B'=A-B]$

(v) $A\times B\cap C=(A\times B)\cap(A\times C)$

(vi) $A\times(B\cup C)=(A\times B)\cup(A\times C)$
Hence, the correct matches are
(i) $\leftrightarrow$ (b), (ii) $\leftrightarrow$ (c), (iii) $\leftrightarrow$ (a), (iv) $\leftrightarrow$ (f)
(v) $\leftrightarrow$ (d), (vi) $\leftrightarrow$ (e)

True/False

Q52. If A is any set, then $A\subset A$.

Solution: True.
Every set is the subset of itself.
Therefore, for any set $A\subset A$.

Q53. If $M=\{1,2,3,4,5,6,7,8,9\}$
$B=\{1,2,3,4,5,6,7,7,8,9\}$ then $B\not\subset M$.

Solution: False.
$M=\{1,2,3,4,5,6,7,8,9\}$
$B=\{1,2,3,4,5,6,7,7,8,9\}$
Every element of B is also in M
$\therefore B\subset M$ (repetition of elements does not affect the property of the set)

Q54. The sets {1, 2, 3, 4} and {3, 4, 5, 6} are equal.

Solution: False.
$2\in\{1,2,3,4\}$ but $2\notin\{3,4,5,6\}$
$\therefore\{1,2,3,4\}\neq\{3,4,5,6\}$

Q55. $Q\cup Z=Q$, where Q is the set of rational numbers and Z is the set of integers.

Solution: True.
Since every integer is also a rational number, $Q\cup Z=Q$

Q56. Let sets R and T be defined as
$R=\{x\in Z|x \text{ is divisible by } 2\}$
$T=\{x\in Z|x \text{ is divisible by } 6\}$. Then $T\subset R$

Solution: True.
$R=\{x\in Z|x \text{ is divisible by } 2\}$
$=\{....-6,-4,-2,0,2,4,6...\}$
$R=\{x\in Z|x \text{ is divisible by } 6\}$
$\{....,-12,-6,0,6,12....\}$
Since every element of T is also in $R, T\subset R$.

Q57. Given $A=\{0,1,2\}, B=\{x\in R|0\le x\le 2\}$. Then $A=B$.

Solution: False.
$A=\{0,1,2\}$
$B=\{x\in R|0\le x\le 2\}\Rightarrow x\in(0,2)$
A is finite with only three elements while in set B there are infinite real numbers between 0 to 2; therefore $A\neq B$.

Day 17

Very Short Answer (VSA): 1-Mark and Short Answer (SA): 2-Mark Questions

Q1. The set $\{x\in N:1\le x<3\}$ can be written as ______

Solution:
The set $\{x\in N:1\le x<3\}$ can be written as $x=\{1,2\}$.

Q2. When $A=\{0\}$ then number of elements in P(A) is______

Solution:
$\because A=\{0\}\Rightarrow n(A)=1$
Therefore, number of elements in P(A) is $n\{P(A)\}=2^{n(A)}=2^1=2$
That is, $P(A)=\{\phi,\{0\}\}$

Q3. Power set of the set $A=\{0,1\}$ is______

Solution:
$\because A=\{0,1\}$
$\therefore$ Power set of the set $A=\{0,1\}$ is
$P(A)=\{\phi,\{0\},\{1\},\{0,1\}\}$

Q4. If the sets $A=\{1,2\}$, $B=\{2,3\}$, $C=\{0,4\}$, then the universal set of all the three sets A, B and C can be ______

Solution:
Universal set for A, B and C is given by $U=\{0,1,2,3,4\}$

Q5. Let $A=\{x\in N|1\le x\le 2\}$, $B=\{x\in R|1\le x\le 2\}$. Then A______B.

Solution:
$A = \{1, 2\}$

$B = \{x \in R | 1 \le x \le 2\} \Rightarrow x \in [1, 2]$

Since A is finite with only two elements 1 and 2 while in set B there are infinite real numbers from 1 to 2, $A \ne B$.

Q6. Let sets R and T be defined as
$R = \{x \in Z | x \text{ is divisible by } 3\}$
$T = \{x \in Z | x \text{ is divisible by } 6\}$.
Then T______R

Solution:
$R = \{x \in Z | x \text{ is divisible by } 3\}$

$= \{..., -6, -3, 0, 3, 6, ...\}$

$R = \{x \in Z | x \text{ is divisible by } 6\}$

$\{...., -12, -6, 0, 6, 12....\}$

Since every element of T is also in R, $T \subset R$.

Q7. Suppose $A_1, A_2, ..., A_{30}$ are thirty sets each having 5 elements and B_1, B_2, B_n are n sets each with 3 elements. Let $\bigcup_{i=1}^{30} A_i = \bigcup_{j=1}^{n} A_j = S$ and such elements of S belongs to exactly 10 of the A_i and exactly 6 of the B_j. Then, n is equal to

(a) 15 (b) 30 (c) 45 (d) 35

Solution: (b)
If elements are not repeated, then the number of elements in $A_1 \cup A_2 \cup A_3, ..., \cup A_{30}$ is 30 × 5.

But each element is used 10 times. Therefore $S = \frac{30 \times 5}{10} = 15$

If elements in $B_1, B_2, ...B_n$ are not repeated, then total number of elements is $3n$. However, each element is repeated 9 times, and therefore $S = \frac{3n}{6} = 15 \Rightarrow n = 30$.

Q8. Two finites sets have m and n elements. The number of subsets of the first set is 120 more than of the second set. The values of m and n are, respectively

(a) 3, 7 (b) 4, 7 (c) 3, 4 (d) 4, 4

Solution: (a)
The number of subsets of set containing m elements is 120 more than the subsets of the set containing n elements,

$\because 2^m - 2^n = 120 \Rightarrow 2^n \cdot (2^{m-n} - 1) = 2^3 \cdot 15$

$\Rightarrow 2^n = 2^3$ and $2^{m-n} - 1 = 15$

$\Rightarrow n = 3$ and $2^{m-n} = 16 = 2^4$

$\Rightarrow m - n = 4 \Rightarrow m - 3 = 4$

$\Rightarrow m = 7$

Q9. If $X = \{2n | n \in N\}$, and $Y = \{3n | n \in N\}$, then

(a) $X \subset Y$ (b) $Y \subset X$
(c) $X = Y$ (d) $X \cap Y \ne \phi$

Solution: (a)
$X = \{2n | n \in N\} = \{2, 4, 6, 8, 10, 12, ...\}$

$Y = \{3n | n \in N\} = \{3, 6, 9, 12, ...\}$

Since every element of X is not in Y, also, every element of Y is not in X, $X \not\subset Y$ and $Y \not\subset X$

But, $X \cap Y = \{6, 12, 18, ...\} \ne \phi$.

Q10. A survey shows that 60% of the people watch A news channel whereas 80% watch B news channel. If x% of the people watch both channels, then

(a) $x = 40$ (b) $x = 60$
(c) $40 \le x \le 60$ (d) $60 \le x \le 80$

Solution: (c)
Let A be the set of percentage of those people who watch A news channel and B be the set of percentage of those people who watch B news channel. Then,

$n(A) = 60, \; n(B) = 80, \; n(A \cap B) = x$

$\because n(A \cup B) \le 100$

$\therefore n(A) + n(B) - n(A \cap B) \le 100$

$\Rightarrow 60 + 80 - x \le 100 \Rightarrow 140 - x \le 100$

$\Rightarrow 140 - 100 \le x \Rightarrow 40 \le x$

$\because n(A \cap B) \le n(A)$

$\therefore x \le 60$

Thus, $40 \le x \le 60$

Q11. If $S = \{x | x$ is a positive multiple of 5 less than or equal to 100$\}$ and $P = \{x | x$ is a prime number less than 20$\}$. Then, $n(S) - n(P)$ is equal to

(a) 34 (b) 31 (c) 33 (d) 41

Solution: (d)
$S = \{x | x$ is a positive multiple of 5 less than or equal to 100$\}$

$\Rightarrow S = \{5, 10, 15, ..., 95, 100\}$

$\because t_n = a + (n-1)d \Rightarrow 100 = 5 + (n-1)5 \Rightarrow n = 20$

Thus, $n(S) = 20$

$P = \{x | x$ is a prime number less than 20$\}$

$\Rightarrow P = \{2, 3, 5, 7, 11, 13, 17, 19\} \Rightarrow n(P) = 8$

Thus, $n(S) - n(P) = 20 - 8 = 12$

Q12. Let $T = \left\{ x \left| \frac{x+8}{x-4} = \frac{4-x}{8-x} \right. \right\}$, then the set T is

(a) $\{4\}$ (b) $\{8\}$ (c) $\{10\}$ (d) $\{12\}$

Solution: (c)

$T = \left\{ x \left| \frac{x+8}{x-4} = \frac{4-x}{8-x} \right. \right\}$

$\therefore \frac{x+8}{x-4}=\frac{4-x}{8-x} \Rightarrow (x^2-8^2)=(x-4)^2$

$\Rightarrow x^2-64=x^2-8x+16 \Rightarrow 8x=80$

$\Rightarrow x=10$

$\therefore T=\{10\}$

Q13. Write the following as intervals:

(i) $\{x: x \in R, -1 < x \leq 1\}$

(ii) $\{x: x \in R, -1 < x < 1\}$

Solution:

(i) $\{x: x \in R, -1 < x \leq 1\} = (-1,1]$

(ii) $\{x: x \in R, -1 < x < 1\} = (-1,1)$

Q14. Write the following intervals in set-builder form:

(i) $[-1,1)$ (ii) $[-1,1]$

Solution:

(i) $[-1,1) = \{x: x \in R, -1 \leq x < 1\}$

(ii) $[-1,1] = \{x: x \in R, -1 \leq x \leq 1\}$

Q15. Given that $N = \{1,2,3,...,100\}$, write

(i) the subset of N whose elements are multiples of 10,

(ii) the subset of N whose elements are divisible by 15.

Solution:

Since $N = \{1,2,3,.....,100\}$

(i) Subset of N whose elements are multiples of 10 is $=\{10,20,30,40,50,60,70,80,90,100\}$

(ii) Subset of N whose elements are divisible by 15 is $=\{15,30,45,60,75,90\}$

Q16. If $X = \{1,2,3\}$, if n represents any member of X, write the following sets containing all numbers represented by

(i) $3n$ (ii) $n+3$

Solution:

Since $X = \{1,2,3\}$,

(i) $\{3n | n \in N\} = \{3,6,9\}$

(ii) $\{n+3 | n \in X\} = \{4,5,6\}$

Q17. If $Y = \{1,2,3,.....10\}$, a represents any element of Y, write the following sets, containing all the elements satisfying the given conditions.

(i) $a \in Y$ and $a^2 \in Y$

(ii) $a^2 - 20a + 99 = 0$, $a \in Y$

(iii) a is less than 5 and greater than 7, and $a \in Y$

Solution:

Since $Y = \{1,2,3,.....10\}$

(i) $\{a: a \in Y, a^2 \in Y\} = \{1,2,3\}$

(ii) $a^2 - 20a + 99 = 0 \Rightarrow (a-9)(a-11) = 0$

$\Rightarrow a = 9,11 \because 9 \in Y$ but $11 \notin Y$

$\therefore \{a: a^2 - 20a + 99 = 0, a \in Y\} = \{9\}$

(iii) a is less than 5 and greater than or equal to 7, and $a \in Y = \{1, 2, 3, 4, 7, 8, 9, 10\}$

Q18. Decide, among the following sets, which sets are subsets of one another:

$A = \{x: x \in N$ and x satisfy $x^2 - 3x + 2 = 0\}$

$B = \{x: x \in R$ and x satisfy $x^2 - 1 = 0\}$

$C = \{x: x \in I$ and x satisfy $-2 < x < 2\}$

$D = \{x: x \in W$ and x satisfy $x \leq 2\}$

Solution:

$A = \{x: x \in N$ and x satisfy $x^2 - 3x + 2 = 0\}$

$\because x^2 - 3x + 2 = 0$

$\Rightarrow (x-1)(x-2) = 0 \Rightarrow x = 1,2$

$\therefore A = \{1,2\}$

$B = \{x: x \in R$ and x satisfy $x^2 - 1 = 0\}$

$\because x^2 - 1 = (x-1)(x+1) = 0 \Rightarrow x = -1,1$

$\therefore B = \{-1,1\}$

$C = \{x: x \in I$ and x satisfy $-2 < x < 2\}$

$\therefore C = \{-1,0,1\}$

$D = \{x: x \in W$ and x satisfy $x \leq 2\}$

$\therefore D = \{0,1,2\}$

Thus, $A \subset D$ and $B \subset C$

Q19. Write the following sets in the roster form.

(i) $A = \{x: x \in R, x^2 + x - 6 = 0\}$

(ii) $B = \{x | x^2 + x = 0, x \in R\}$

(iii) $C = \{x | x$ is a positive factor of 60$\}$

(iv) $D = \{x: x \in I,\ x^2 \leq 2\}$

Solution:

(i) $A = \{x: x \in R, x^2 + x - 6 = 0\}$

$\because x^2 + x - 6 = 0 \Rightarrow (x+3)(x-2) = 0 \Rightarrow x = -3,2$

Therefore, $A = \{-3,2\}$

(ii) $B = \{x | x^2 + x = 0,\ x \in R\}$

$\because x^2 + x = 0 \Rightarrow x(x-1) = 0 \Rightarrow x = 0,1$

Therefore, $B = \{0,1\}$

(iii) $C = \{x | x$ is a positive factor of 60$\}$

The factors of 60 are 1, 2, 3, 4, 15, 20, 30, 60

Therefore, $C = \{1,2,3,4,5,6,10,12,15,20,30,60\}$

(iv) $\because D = \{x: x \in I,\ x^2 \leq 2\}$

Therefore, $D = \{-2,-1,0,1,2\}$

Q20. A, B and C are subsets of universal set U. If $U = \{x : x \in W,\ x \le 10\}$
$A = \{1,4,7,8\},\ B = \{2,4,6,8\},\ C = \{1,3,6,9\}$,
Draw a Venn diagram showing the relation of U, A, B and C.

Solution:
$\because U = \{x : x \in W,\ x \le 10\} = \{0,1,2,3,4,5,6,7,8,9,10\}$
The Venn diagram showing the relation of U, A, B and C is given below:

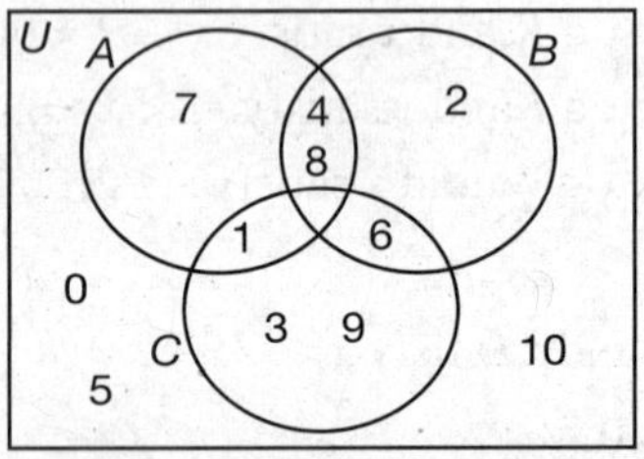

Q21. Let $A = \{\phi, 1, \{2\}\}$, which of the following statements are correct and why?
(i) $\phi \in A$ (ii) $\{\phi\} \subset A$
(iii) $\phi \subset A$ (iv) $1 \in A$
(v) $\{1\} \subset A$ (vi) $\{1\} \in A$
(vii) $2 \in A$ (viii) $\{2\} \subset A$
(ix) $\{2\} \in A$ (x) $\{\{2\}\} \subset A$

Solution:
(i) and (ii): The statements $\phi \in A$ and $\{\phi\} \subset A$ are correct because ϕ is an element of A implies that $\{\phi\} \subset A$.
(iii) The statement $\phi \subset A$ is correct because ϕ represents an empty set and every empty set is subset of a set.
(iv) and (v): The statements $1 \in A$ and $\{1\} \subset A$ are correct because 1 is an element of A implies that $\{1\} \subset A$.
(vi) The statement $\{1\} \in A$ is incorrect because $\{1\}$ is not an element of A.
(vii) and (viii): The statement $2 \in A$ and $\{2\} \subset A$ is incorrect because 2 is not an element of A.
(ix) and (x): The statements $\{2\} \in A$ and $\{\{2\}\} \subset A$ are correct because $\{2\}$ is an element of A implies that $\{\{2\}\} \subset A$.

Day 18

Long Answer-I (LA-I): 4-Mark and Long Answer-II (LA-II): 6-Mark Questions

Q1. $A = \{1,2,3\},\ B = \{2,3,4\}$ and $U = \{1,2,3,4,5\}$, then prove that
(i) $(A \cup B)' = A' \cap B'$ (ii) $(A \cap B)' = A' \cup B'$
(iii) $(A - B)' \ne A' - B'$

Solution:
Since $A = \{1,2,3\} \Rightarrow A' = U - A$
$\Rightarrow A' = \{1,2,3,4,5\} - \{1,2,3\} = \{4,5\}$
$B = \{2,3,4\} \Rightarrow B' = U - B$
$\Rightarrow B' = \{1,2,3,4,5\} - \{2,3,4\} = \{1,5\}$
(i) LHS $= (A \cup B)' = (\{1,2,3\} \cup \{2,3,4\})'$
$= (\{1,2,3,4\})' = \{1,2,3,4,5\} - \{1,2,3,4\} = \{5\}$
RHS $= A' \cap B' = \{4,5\} \cap \{1,5\} = \{5\}$
(ii) LHS $= (A \cap B)' = (\{1,2,3\} \cap \{2,3,4\})'$
$= (\{2,3\})' = \{1,2,3,4,5\} - \{2,3\} = \{1,4,5\}$
RHS $= A' \cup B' = \{4,5\} \cap \{1,5\} = \{1,4,5\}$
Thus, LHS = RHS, which verifies the above two results.
(iii) LHS $= (A - B)' = (\{1,2,3\} - \{2,3,4\})'$
$= (\{1\})' = \{1,2,3,4,5\} - \{1\} = \{2,3,4,5\}$
RHS $= A' - B' = \{4,5\} - \{1,5\} = \{4\}$
LHS $\ne$ RHS, therefore $(A - B)' \ne A' - B'$.

Q2. $A = \{1,2,3\}$, $B = \{2,3,4\}$ and $C = \{1,4,5\}$, then prove that
(i) $A - (B \cup C) = (A - B) \cap (A - C)$
(ii) $A - (B \cap C) = (A - B) \cup (A - C)$

Solution:
Since $A - B = \{1,2,3\} - \{2,3,4\} = \{1\}$
$A - C = \{1,2,3\} - \{1,4,5\} = \{2,3\}$
$B \cap C = \{2,3,4\} \cap \{1,4,5\} = \{4\}$
$B \cup C = \{2,3,4\} \cup \{1,4,5\} = \{1,2,3,4,5\}$
Thus,
(i) For $A - (B \cup C) = (A - B) \cap (A - C)$
LHS: $A - (B \cup C) = \{1,2,3\} - \{1,2,3,4,5\} = \{\phi\}$
RHS: $(A - B) \cap (A - C) = \{1\} \cap \{2,3\} = \{\phi\}$
(ii) For $A - (B \cap C) = (A - B) \cup (A - C)$
LHS: $A - (B \cap C) = \{1,2,3\} - \{4\} = \{1,2,3\}$
RHS: $(A - B) \cup (A - C) = \{1\} \cup \{2,3\} = \{1,2,3\}$
Thus, LHS = RHS, which verifies the above two results.

Q3. $A = \{1,2,3\}$, $B = \{2,3,4\}$ and $C = \{1,2,5\}$, then prove that
(i) $(A \cap B) - C = (A - C) \cap (B - C)$
(ii) $(A \cup B) - C = (A - C) \cup (B - C)$

Solution:
Since $A \cap B = \{1,2,3\} \cap \{2,3,4\} = \{2,3\}$

$A \cup B = \{1,2,3\} \cup \{2,3,4\} = \{1,2,3,4\}$
$A - C = \{1,2,3\} - \{1,2,5\} = \{3\}$
$B - C = \{2,3,4\} - \{1,2,5\} = \{3,4\}$
Thus,

(i) For $(A \cap B) - C = (A - C) \cap (B - C)$
LHS: $(A \cap B) - C = \{2,3\} - \{1,2,5\} = \{3\}$
RHS: $(A - C) \cap (B - C) = \{3\} \cap \{3,4\} = \{3\}$

(ii) For $(A \cup B) - C = (A - C) \cup (B - C)$
LHS: $(A \cup B) - C = \{1,2,3,4\} - \{1,2,5\} = \{3,4\}$
RHS: $(A - C) \cup (B - C) = \{3\} \cup \{3,4\} = \{3,4\}$
Thus, LHS = RHS, which verifies the above two results.

Q4. $A = \{1,2\}$, $B = \{2,3\}$ and $C = \{1,3\}$, then prove that
(i) $A \times (B \cap C) = (A \times B) \cap (A \times C)$
(ii) $A \times (B \cup C) = (A \times B) \cup (A \times C)$

Solution:
Since $B \cap C = \{2,3\} \cap \{1,3\} = \{3\}$
$B \cup C = \{2,3\} \cup \{1,3\} = \{1,2,3\}$
Thus,

(i) For $A \times (B \cap C) = (A \times B) \cap (A \times C)$
LHS: $A \times (B \cap C) = \{1,2\} \times \{3\} = \{(1,3),(2,3)\}$
RHS: $(A \times B) \cap (A \times C)$
$= \{(1,2),(1,3),(2,2),(2,3)\} \cap \{(1,1),(1,3),(2,1),(2,3)\}$
$= \{(1,3),(2,3)\}$

(ii) For $A \times (B \cup C) = (A \times B) \cup (A \times C)$
LHS: $A \times (B \cup C) = \{1,2\} \times \{1,2,3\}$
$= \{(1,1),(1,2),(1,3),(2,1),(2,2),(2,3)\}$
RHS: $(A \times B) \cup (A \times C)$
$= \{(1,2),(1,3),(2,2),(2,3)\} \cup \{(1,1),(1,3),(2,1),(2,3)\}$
$= \{(1,1),(1,2),(1,3),(2,1),(2,2),(2,3)\}$
Thus, LHS = RHS, verify above two results.

Q5. If X and Y are two sets such that $X \cap Y$ has 5 elements, X has 10 elements and Y has 15 elements; how many elements does $X \cup Y$ have?

Solution:
$\because n(X \cap Y) = 5,\ n(X) = 10,\ n(Y) = 15$
$\therefore n(X \cup Y) = n(X) + n(Y) - n(X \cap Y)$
$\Rightarrow n(X \cup Y) = 10 + 15 - 5 = 20$

Q6. If X and Y are two sets such that $n(X) = 10,\ n(Y) = 20,\ n(XUY) = 30$, find $n(X \cap Y)$.

Solution:
$\because n(X \cup Y) = n(X) + n(Y) - n(X \cap Y)$
$\therefore n(X \cap Y) = n(X) + n(Y) - n(X \cup Y)$
$\Rightarrow n(X \cap Y) = 10 + 20 - 30 = 0$
This implies that sets X and Y are disjoint sets.

Q7. In a group of 25 people, the number of people playing cricket and football were found to be as follows: cricket = 14, football = 11, cricket and football both = 8. Find the number of people who play
(i) only cricket
(ii) only football
(iii) at least one game
(iv) neither of the games

Solution:
Let C be the set of people who play cricket and F be the set of people who play football.

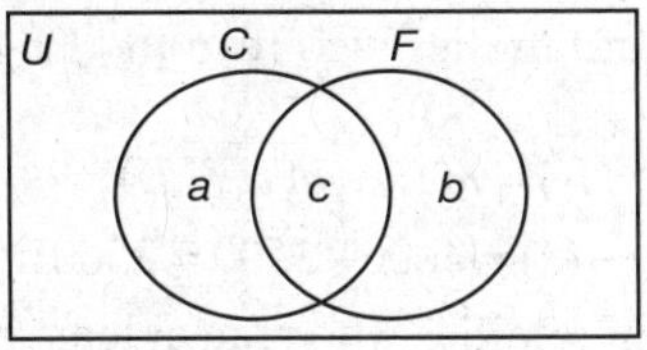

Then, $n(U) = 25$
$n(C) = a + c = 14$
$n(F) = b + c = 11$
$n(C \cap F) = c = 8$

(i) Number of people who play cricket only
$a = n(C \cap F') = n(C) - n(C \cap F)$
$= (a + c) - (c) = 14 - 8 = 6$

(ii) Number of people who play football only
$b = n(F \cap C') = n(F) - n(C \cap F)$
$= (b + c) - (c) = 11 - 8 = 3$

(iii) Number of people who play at least one game
$= a + b + c$
$= 6 + 3 + 8 = 17$

(iv) Number of people who play neither of the games = Total people – people who play at least one game
$= 25 - 17 = 8$.

Q8. In a town of 10,000 families, 60% read a Hindi newspaper, 40% read an English newspaper, 15% read both Hindi and English newspapers. Find the number of families who read
(i) only the Hindi newspaper
(ii) only the English newspaper
(iii) at least one newspaper
(iv) neither of the newspapers

Solution:
Let H be the set of families who read the Hindi newspaper and E be the set of families who read the English newspaper.

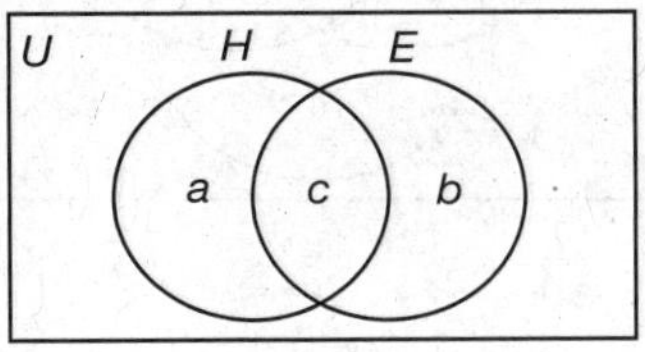

Then, $n(U) = 10,000$

$n(H) = a + c = 60\% = 60 \times \frac{10,000}{100} = 6000$

$n(E) = b + c = 40\% = 40 \times \frac{10,000}{100} = 4000$

$n(H \cap E) = c = 15\% = 15 \times \frac{10,000}{100} = 1500$

(i) Number of families who read the Hindi newspaper only
$a = n(H \cap E') = n(H) - n(H \cap E)$
$= (a + c) - (c) = 6000 - 1500 = 4500$

(ii) Number of families who read the English newsaper only
$b = n(E \cap H') = n(E) - n(H \cap E)$
$= (b + c) - (c) = 4000 - 1500 = 2500$

(iii) Number of families who read at least one newspaper
$= a + b + c$
$= 4500 + 2500 + 1500 = 8500$

(iv) Number of families who read none of the newspapers
= Total families – Families who read at least one newspaper $= 10000 - 8500 = 1500$

Q9. In a group of 2000 students of Indian Institute of Science, the number of students who like coffee, milk, tea were found to be as follows: coffee = 50%, milk = 75%, tea = 30%, coffee and milk = 35%, milk and tea = 25%, coffee and tea = 15%, coffee, milk and tea = 10%. Find the number of students who like

(i) only coffee
(ii) only milk
(iii) only tea
(iv) coffee and milk but not tea
(v) mlk and tea but not coffee
(vi) coffee and tea but not milk
(vii) at least one of the drinks
(viii) none of the three drinks

Solution:

Let C be the set of students who like coffee, M be the set of students who like milk and T be the set of students who like tea.

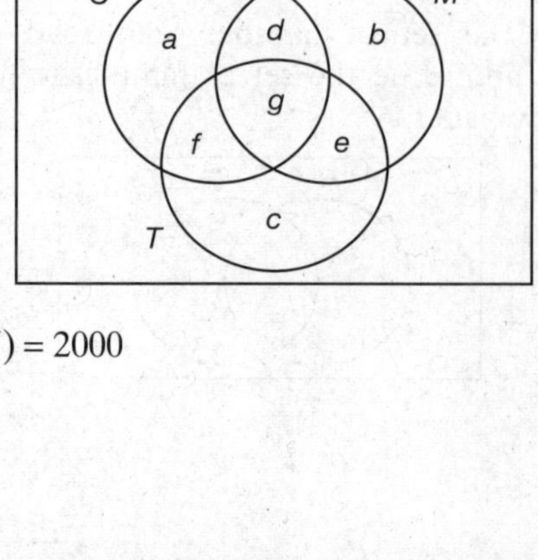

Then, $n(U) = 2000$

$n(C) = a + d + f + g = 50\% = 50 \times \frac{2000}{100} = 1000$

$n(M) = b + d + e + g = 75\% = 75 \times \frac{2000}{100} = 1500$

$n(T) = c + e + f + g = 30\% = 30 \times \frac{2000}{100} = 600$

$n(C \cap M) = d + g = 35\% = 35 \times \frac{2000}{100} = 700$

$n(M \cap T) = e + g = 25\% = 25 \times \frac{2000}{100} = 500$

$n(C \cap T) = f + g = 15\% = 15 \times \frac{2000}{100} = 300$

$n(C \cap M \cap T) = g = 10\% = 10 \times \frac{2000}{100} = 200$

(i) Number of students who like coffee only
$a = n(C \cap M' \cap T')$
$= (a + d + f + g) - (d + g) - (f + g) + g$
$= n(C) - n(C \cap M) - n(C \cap T) + n(C \cap M \cap T)$
$= 1000 - 700 - 300 + 200 = 200$

(ii) Number of students who like milk only
$b = n(M \cap C' \cap T')$
$= (b + d + e + g) - (d + g) - (e + g) + g$
$= n(M) - n(M \cap C) - n(M \cap T) + n(C \cap M \cap T)$
$= 1500 - 700 - 500 + 200 = 500$

(iii) Number of students who like tea only
$c = n(T \cap C' \cap M')$
$= (c + e + f + g) - (e + g) - (f + g) + g$
$= n(T) - n(T \cap C) - n(T \cap M) + n(C \cap M \cap T)$
$= 600 - 300 - 500 + 200 = 0$

(iv) Number of students who like coffee and milk but not tea
$d = n(C \cap M \cap T') = (d + g) - g$
$= n(C \cap M) - n(C \cap M \cap T) = 700 - 200 = 500$

(v) Number of students who like milk and tea but not coffee
$e = n(M \cap T \cap C') = (e + g) - g$
$= n(M \cap T) - n(C \cap M \cap T) = 500 - 200 = 300$

(vi) Number of students who like coffee and tea but not milk
$f = n(C \cap T \cap M') = (f + g) - g$
$= n(C \cap T) - n(C \cap M \cap T) = 300 - 200 = 100$

(vii) Number of students who like at least one of the three drinks $= a + b + c + d + e + f + g$
$= 200 + 500 + 0 + 500 + 300 + 100 + 200 = 1800$

(viii) Number of students who like none of the three drinks = Total students – Students who like at least one of the three drinks $= 2000 - 1800 = 200$

CHAPTER

2 Relations and Functions

Day 19

Cartesian Products of Sets

If P and Q are two non-empty sets, then the Cartesian product $P \times Q$ is the set of all ordered pairs (p,q) of elements from P and Q, such that $p \in P, q \in Q$; that is, $P \times Q = \{(p,q) : p \in P,\ q \in Q\}$.

Note:

(i) If either P or Q is an empty set, then $P \times Q$ will also be an empty set, that is, $P \times Q = \phi$

(ii) If either P or Q is a non-empty infinite set, then $P \times Q$ will also be an infinite set.

(iii) Two ordered pairs are equal if and only if the corresponding first elements are equal and the second elements are also equal.

(iv) If there are p elements in set P and q elements in set Q, then there will be pq elements in the Cartesian product $P \times Q$, that is, if $n(P) = p$, $n(Q) = q$, then $n(P \times Q) = pq$.

(v) If $A \times A \times A = \{(a,b,c) : a,b,c \in A\}$, then (a,b,c) is called an ordered triplet.

(vi) $A \times (B \cap C) = (A \times B) \cap (A \times C)$

And $A \times (B \cup C) = (A \times B) \cup (A \times C)$

Example 1: Let A be a set of 2 colours (Black and White) and B be a set of 3 toys (Elephant, Horse and Monkey). The sets A and B are written as A = {Black, White} and B ={Elephant, Horse, Monkey}

Thus, the Cartesian product $A \times B$ is the set of all pairs of coloured toys that can be made from these two sets, that is,

$A \times B$ = {(Black, Elephant), (White, Elephant),
(Black, Horse), (White, Horse),
(Black, Monkey), (White, Monkey)}

Similarly,

$B \times A$ = {(Elephant, Black),(Elephant, White),
(Horse, Black), (Horse, White),
(Monkey, Black), (Monkey, White)}

Example 2: Let us consider the two sets:
$P = \{D, M, U\}$, where D, M, U represent Delhi, Madhya Pradesh and Uttar Pradesh, respectively, and $Q = \{01, 02\}$, representing codes for the vehicle registration plates issued by D, M, U.

If the three states, Delhi, Madhya Pradesh and Uttar Pradesh, were making codes for the vehicle registration plates with the restriction that the code begins with an element from set P, then the total number of codes that can be formed from the given sets is given by the Cartesian product $P \times Q$, that is,

$$P \times Q = \{(D,01),(D,02),(M,01),(M,02),(U,01),(U,02)\}$$

Since there are 3 elements in set P and 2 elements in set Q, there will be a total of 6 possible codes. Also note that the order in which these elements are paired is crucial. That is, the code $(D, 01)$ will not be the same as the code $(01, D)$.

Example 3: Let us consider the sets $A = \{1,3,5\}$ and $B = \{2,4,6\}$. Then, the position of points in two dimensions using the sets X and Y is given by

$X \times Y = \{(x,y) : x \in A,\ y \in B\}$, that is,

$$X \times Y = \{(1,2),(1,4),(1,6),(3,2),(3,4),(3,6),(5,2),(5,4),(5,6),\}$$

Similarly, if we change the position of elements in the ordered pairs, then

$X \times Y = \{(x,y) : x \in B,\ y \in A\}$, that is,

$$X \times Y = \{(2,1),(2,3),(2,5),(4,1),(4,3),(4,5),(6,1),(6,3),(6,5),\}$$

Note that the order in which these elements are paired is crucial. That is, the point in the position $(2,3)$ will be distinct from the point in the position $(3,2)$.

Example 4: Let us consider the set $A = \{0,1\}$. Then, the position of points in two dimensions using the set $A = \{0,1\}$ is given by $X \times Y = \{(x,y) : x, y \in A\}$, that is, $X \times Y = \{(0,0),(0,1),(1,0),(1,1)\}$.

Similarly, the position of points in three dimensions using the set $A = \{0,1\}$ is given by $X \times Y \times Z = \{(x,y,z) : x,y,z \in A\}$, that is,

$$X \times Y \times Z = \{(0,0,0),(0,1,0),(1,0,0),(1,1,0),(0,0,1),(0,1,1),(1,0,1),(1,1,1)\}$$

Example 5: If $\left(\frac{x}{2}+1, \frac{1}{3}\right) = \left(\frac{2}{3}, y-\frac{4}{3}\right)$, then the ordered pairs $\left(\frac{x}{2}+1, \frac{1}{3}\right) = \left(\frac{2}{3}, y-\frac{4}{3}\right)$ if and only if the corresponding elements are equal, that is,

$\Rightarrow \frac{x}{2}+1=\frac{2}{3}$ and $\frac{1}{3}=y-\frac{4}{3}$

$\Rightarrow \frac{x}{2}=\frac{2}{3}-1=\frac{-1}{3}$ and $y=\frac{1}{3}+\frac{4}{3}$

$\Rightarrow x=\frac{-2}{3}$ and $y=\frac{5}{3}$

Example 6: Let $A=\{0,1\}, B=\{-1,0\}$ and $C=\{-1,1\}$. Prove that

(i) $A\times(B\cap C)=(A\times B)\cap(A\times C)$

(ii) $A\times(B\cup C)=(A\times B)\cup(A\times C)$

Proof:

(i) To verify $A\times(B\cap C)=(A\times B)\cap(A\times C)$

$B\cap C=\{-1,0\}\cap\{-1,1\}=\{-1\}$

$\therefore$ LHS $=A\times(B\cap C)=\{0,1\}\times\{-1\}=\{(0,-1),(1,-1)\}$

Now, $A\times B=\{(0,-1),(0,0),(1,-1),(1,0)\}$

$A\times C=\{(0,-1),(0,1),(1,-1),(1,1)\}$

$\therefore$ RHS $=(A\times B)\cap(A\times C)=\{(0,-1),(1,-1)\}$

LHS = RHS $=\{(0,-1),(1,-1)\}$

Therefore, $A\times(B\cap C)=(A\times B)\cap(A\times C)$

(ii) To verify $A\times(B\cup C)=(A\times B)\cup(A\times C)$

$B\cup C=\{-1,0\}\cup\{-1,1\}=\{-1,0,1\}$

$\therefore$ LHS $=A\times(B\cup C)=\{0,1\}\times\{-1,0,1\}$

$=\{(0,-1),(0,0),(0,1),(1,-1),(1,0),(1,1)\}$

Now, $A\times B=\{(0,-1),(0,0),(1,-1),(1,0)\}$

$A\times C=\{(0,-1),(0,1),(1,-1),(1,1)\}$

$\therefore$ RHS $=(A\times B)\cup(A\times C)$

$=\{(0,-1),(0,0),(0,1),(1,-1),(1,0),(1,1)\}$

LHS = RHS

$=\{(0,-1),(0,0),(0,1),(1,-1),(1,0),(1,1)\}$

Therefore, $A\times(B\cup C)=(A\times B)\cup(A\times C)$

Example 7: If the set A has 2 elements, the set B has 3 elements, the set C has 4 elements, the set D is an empty set, and the set E has infinite elements, then

Number of elements in $(A\times B)$ = (Number of elements in A) × (Number of elements in B) $=2\times3=6$

Number of elements in $(A\times B\times C)$ = (Number of elements in A) × (Number of elements in B) × (Number of elements in C) $=2\times3\times4=24$

Number of elements in $(A\times D)$ = (Number of elements in A) × (Number of elements in B) $=2\times0=0$

Thus, the set $(A\times D)$ will be an empty set.

Number of elements in $(A\times E)$ = (Number of elements in A) × (Number of elements in B) $=2\times\infty=\infty$

Thus, the set $(A\times E)$ has infinite elements.

Example 8: Let $A=\{0,1\}, B=\{-1,0\}$.

$\therefore A\times B=\{(0,-1),(0,0),(1,-1),(1,0)\}$

$\Rightarrow n(A\times B)=4$

Thus, the number of subsets of $A\times B$ are $n(A\times B)=2^4=16$. These are

$\{\Phi,\{(0,-1)\},\{(0,0)\},\{(1,-1)\},\{(1,0)\},$
$\{(0,-1),(0,0)\},\{(0,-1),(1,-1)\},\{(0,-1),(1,0)\},$
$\{(0,0),(1,-1)\},\{(0,0),(1,0)\},\{(1,-1),(1,0)\},$
$\{(0,0),(1,-1),(1,0)\},\{(0,-1),(1,-1),(1,0)\},$
$\{(0,-1),(0,0),(1,0)\},\{(0,-1),(0,0),(1,-1)\},$
$\{(0,-1),(0,0),(1,-1),(1,0)\}\}$

Example 9: If the Cartesian product of two sets is given by $A\times B=\{(1,2),(0,2),(1,1),(0,1),(1,-1),(0,-1)\}$, then find the sets A and B.

Solution:

The Cartesian product of two non-empty sets P and Q is given as

$P\times Q=\{(p,q):p\in P,q\in Q\}$

$\therefore A$ is the set of all first elements and B is the set of all second elements.

Thus, $A=\{1,0\},\ B=\{2,1,-1\}$

Example 10: The Cartesian product $A\times A$ has 9 elements among which are found (0, 0) and (1, –1). Find the set A and the remaining elements of $A\times A$.

Solution:

If $n(A)=p, n(B)=q$, then $n(A\times B)=pq$

$\therefore n(A\times A)=n(A)\times n(A)$

Thus, $n(A\times A)=9\Rightarrow n(A)\times n(A)=9\Rightarrow n(A)=3$

The ordered pairs (0, 0) and (1, –1) are two of the nine elements of $A\times A$.

$A\times A=\{(a,a):a\in A\}$. Therefore, –1, 0 and 1 are the elements of A.

$n(A)=3$; therefore $A=\{-1,0,1\}$.

Thus, the remaining elements of set $A\times A$ are

$(-1,-1),(-1,0),(-1,1),(0,-1),(0,1),(1,0),(1,1)$.

Day 20

NCERT Exercise 2.1 (With Solutions)

Q1. If $\left(\frac{x}{3}+1, y-\frac{2}{3}\right)=\left(\frac{5}{3},\frac{1}{3}\right)$, find the values of x and y.

Solution:

Since the ordered pairs $\left(\frac{x}{3}+1, y-\frac{2}{3}\right)=\left(\frac{5}{3},\frac{1}{3}\right)$ are equal, the corresponding elements will also be equal.

Thus, $\frac{x}{3}+1=\frac{5}{3}$ and $y-\frac{2}{3}=\frac{1}{3}$

$\Rightarrow \frac{x}{3}=\frac{5}{3}-1=\frac{2}{3}$ and $y=\frac{2}{3}+\frac{1}{3}=1$

$\Rightarrow x=2$ and $y=1$

Q2. If the set A has 3 elements and the set $B=\{3,4,5\}$ then find the number of elements in $(A\times B)$.

Solution:

Set A has 3 elements and the number of elements in set B is 3.

Therefore, the number of elements in $(A\times B)$

= (Number of elements in A) × (Number of elements in B) $=3\times3=9$

Thus, the number of elements in $(A\times B)$ is 9.

Q3. If $G=\{7,8\}, H=\{5,4,2\}$, find $G\times H$ and $H\times G$.

Solution:

The Cartesian product $P\times Q$ of two non-empty sets P and Q is defined as

$P\times Q=\{(p,q):p\in P, q\in Q\}$

$\therefore G\times H=\{(7,5),(7,4),(7,2),(8,5),(8,4),(8,2)\}$

$H\times G=\{(5,7),(5,8),(4,7),(4,8),(2,7),(2,8)\}$

Q4. State whether each of the following statement is true or false. If the statement is false, rewrite the given statement correctly.

(i) If $P=\{m,n\}, Q=\{n,m\}$, then $P\times Q=\{(m,n),(n,m)\}$.

(ii) If A and B are non-empty sets, then $A\times B$ is a non-empty set of ordered pairs (x, y) such that $x\in A, y\in B$.

(iii) If $A=\{1,2\}, B=\{3,4\}$, then $A\times(B\cap\Phi)=\Phi$.

Solution:

(i) False

If $P=\{m,n\}, Q=\{n,m\}$, then

$P\times Q=\{(m,n),(m,m),(n,n),(n,m)\}$

(ii) True

(iii) True

Q5. If $A=\{-1,1\}$, find $A\times A\times A$.

Solution:

For any non-empty set A, $A\times A\times A$ is defined as $A\times A\times A=\{(a,b,c);a,b,c\in A\}$.

Thus, for $A=\{-1,1\}$

$A\times A\times A=\{(-1,-1,-1),(-1,-1,1),$
$(-1,1,-1),(1,-1,-1,),(-1,1,1),$
$(1,-1,1),(1,1,-1),(1,1,1)\}$

Q6. If $A\times B=\{(a,x),(a,y),(b,x),(b,y)\}$, find A and B.

Solution:

$A\times B=\{(a,x),(a,y),(b,x),(b,y)\}$

The cartesian product of two non-empty sets P and Q is defined as $P\times Q=\{(p,q):p\in P, q\in Q\}$

$\therefore$ A is the set of all first elements and B is the set of all second elements.

Thus, $A=\{a,b\}$, $B=\{x,y\}$

Q7. Let $A=\{1,2\}, B=\{1,2,3,4\}, C=\{5,6\}$, and $D=\{5,6,7,8\}$. Verify that

(i) $A\times(B\cap C)=(A\times B)\cap(A\times C)$

(ii) $A\times C$ is a subset of $B\times D$

Solution:

(i) To verify $A\times(B\cap C)=(A\times B)\cap(A\times C)$

$B\cap C=\{1,2,3,4\}\cap\{5,6\}=\Phi$

$\therefore$ LHS $=A\times(B\cap C)=A\times\Phi=\Phi$

Now, $A\times B=\{(1,1),(1,2),(1,3),(1,4),$
$(2,1),(2,2),(2,3),(2,4)\}$

$A\times C=\{(1,5),(1,6),(2,5),(2,6)\}$

$\therefore$ RHS $=(A\times B)\cap(A\times C)=\Phi$

Thus, $A\times(B\cap C)=(A\times B)\cap(A\times C)$

(ii) To verify: $A\times C$ is a subset of $B\times D$

$A\times C=\{(1,5),(1,6),(2,5),(2,6)\}$

$B\times D=\{(1,5),(1,6),(1,7),(1,8),$
$(2,5),(2,6),(2,7),(2,8),$
$(3,5),(3,6),(3,7),(3,8),$
$(4,5),(4,6),(4,7)(4,8)\}$

All the elements of set $A\times C$ are the elements of set $B\times D$. Therefore, $A\times C$ is a subset of $B\times D$.

Q8. Let $A=\{1,2\}, B=\{3,4\}$. Write $A\times B$. How many subsets does $A\times B$ have? List them.

Solution:

$A=\{1,2\}, B=\{3,4\}$

$\therefore A\times B=\{(1,3),(1,4),(2,3),(2,4)\}$
$\Rightarrow n(A\times B)=4$
Thus, the number of subsets of $A\times B$ is $n[A\times B]=2^4=16$.
These are,
$\Phi,\{(1,3)\},\{(1,4)\},\{(2,3)\},\{(2,4)\},\{(1,3),(1,4)\},$
$\{(1,3),(2,3)\},\{(1,3),(2,4)\},\{(1,4),(2,3)\},$
$\{(1,4),(2,4)\},\{(2,3),(2,4)\},\{(1,3),(1,4),(2,3)\},$
$\{(1,3),(1,4),(2,4)\},\{(1,3),(2,3),(2,4)\}$
$\{(1,4),(2,3),(2,4)\},\{(1,3),(1,4),(2,3),(2,4)\}$

Q9. Let A and B be two sets such that $n(A)=3, n(B)=2$. If $(x, 1)$, $(y, 2)$, $(z, 1)$ are in $A \times B$, find A and B, where x, y, z are distinct elements.

Solution:
$n(A)=3, n(B)=2$, and $(x,1),(y,2),(z,1)$ are in $A \times B$.
Therefore,
A = Set of first elements of the ordered pair elements of $A \times B$.
B = Set of second elements of the ordered pair elements of $A \times B$.
$\therefore x, y, z$ are the elements of A; and 1 and 2 are the elements of B.
Since $n(A)=3, n(B)=2, A=\{x,y,z\}, B=\{1,2\}$.

Q10. The Cartesian product $A \times A$ has 9 elements among which are found (–1, 0) and (0, 1). Find the set A and the remaining elements of $A \times A$.

Solution:
If $n(A)=p, n(B)=q$, then $n(A\times B)=pq$
$\therefore n(A\times A)=n(A)\times n(A)$
$n(A\times A)=9$
$\therefore n(A)\times n(A)=9 \Rightarrow n(A)=3$
The ordered pairs (–1, 0) and (0, 1) are two of the nine elements of $A \times A$ and $A\times A=\{(a,a):a\in A\}$. Therefore, –1, 0 and 1 are elements of A.
Since $n(A)=3$, $A=\{-1,0,1\}$.
Thus, the remaining elements of set $A \times A$ are
$(-1,-1),(-1,1),(0,-1),(0,0),(1,-1),(1,0),(1,1)$.

Day 21

Relations: A relation R from a non-empty set A to a non-empty set B is a subset of the Cartesian product $A\times B$.

The subset derived is described by a relationship (R) between the first element (x) and the second element (y) of the ordered pairs (x, y) in $A\times B$, that is, $R=\{(x,y): x\in A$ and $y\in B$; x is related to y with some relation $R\}$. Also, denoted by $x\,R\,y$.

Note: A relation R from A to A is also stated as relation R on A.

Domain: The set of all first elements (x) of the ordered pairs (x, y) in a relation R from a set A to a set B is called the domain of the relation R.

Range: The set of all second elements (y) of the ordered pairs (x, y), obtained by the relation R from a set A to a set B is called the range of the relation R.
The second element is also known as the image of the first element.

Co-domain: The whole given set B is called the co-domain of the relation R.
Note: (i) An arrow diagram is a visual representation of a relation R from a set A to a set B.

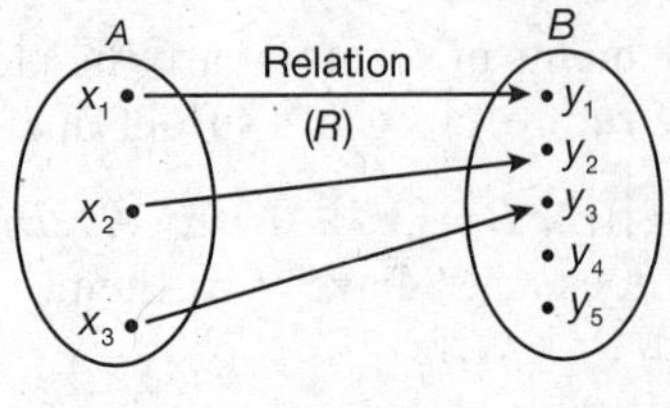

Fig. 2.1

(ii) The range of relation $R\subseteq$ co-domain (set B).
Domain: (x_1, x_2, x_3)
Range: (y_1, y_2, y_3)
Co-domain: $(y_1, y_2, y_3, y_4, y_5)$

(iii) A relation may be represented algebraically either by the roster method or by the set-builder method.
$R=\{(x,y): x\,R\,y,\ X\in A,\ y\in B\}$
or $R=\{(x_1,y_1),(x_2,y_2),(x_3,y_3)\}$

(iv) The total number of relations that can be defined from set A to set B is the number of possible subsets of $A\times B$. If $n(A)=p, n(B)=q$, then $n(A\times B)=pq$ and thus the total number of relations is 2^{pq}.

__Example 1:__ Relation R from A to A:

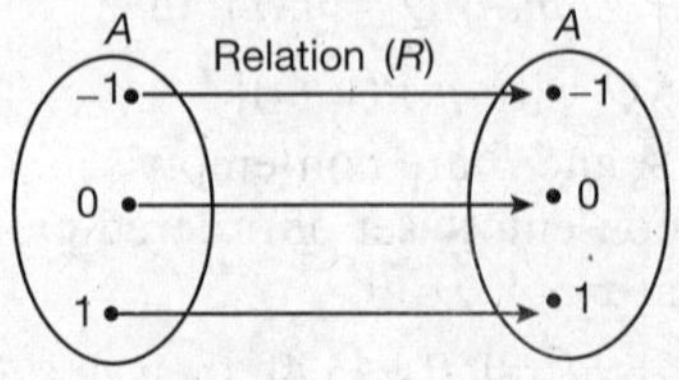

Set-builder form: $R=\{(x,y): y=x^3;\ x,y\in A\}$
Roster form: $R=\{(-1,-1),(0,0),(1,1)\}$
Domain of $R=\{-1,0,1\}$
Range of $R=\{-1,0,1\}$
Co-domain of $R=\{-1,0,1\}$

Example 2: Relation R from A to A:

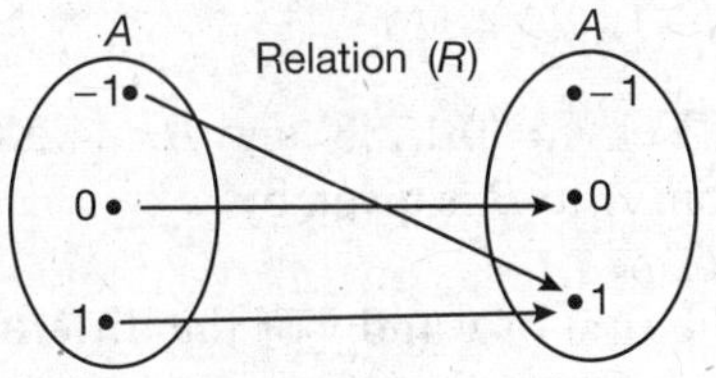

Set-builder form: $R = \{(x, y) : y = x^2;\ x, y \in A\}$

Roster form: $R = \{(-1,1),(0,0),(1,1)\}$

Domain of $R = \{-1,0,1\}$

Range of $R = \{0,1\}$

Co-domain of $R = \{-1,0,1\}$

Example 3: Relation R from A to B:

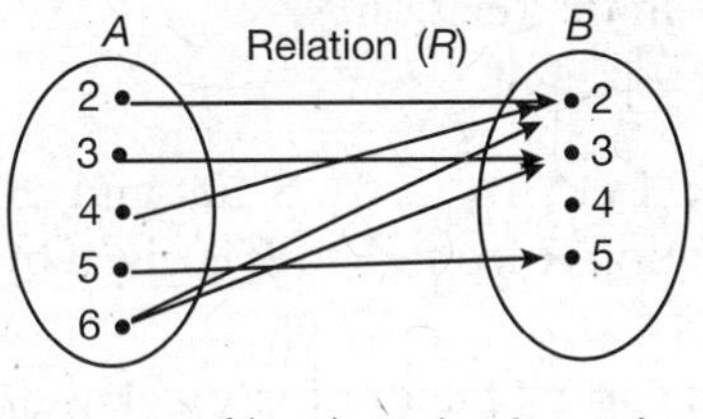

Set-builder form: $R = \{(x, y) : y$ is the prime factor of x, $x \in A\}$

Roster form: $R = \{(2,2),(3,3),(4,2),(5,5),(6,2),(6,3)\}$

Domain of $R = \{2,3,4,5,6\}$

Range of $R = \{2,3,5\}$

Co-domain of $R = \{2,3,4,5\}$

Example 4: $A = \{a,b,c\} \Rightarrow n(A) = 3$ and $B = \{x, y\} \Rightarrow n(B) = 2$

$\therefore A \times B = \{(a,x),(a,y),(b,x),(b,y),(c,x),(c,y)\}$

$\Rightarrow n(A \times B) = n(A) \times n(B) = 3 \times 2 = 6$

Thus, the number of subsets or relations of $A \times B$ is 2^6.

Example 5: Let $A = \{0,1,2,3,4,5,6\}$. Then, the relation R from A to A is given by

Set-builder form: $R = \{(x, y) : y = x + 1,\ x, y \in A\}$

Roster form: $R = \{(0,1),(1,2),(2,3),(3,4),(4,5),(5,6)\}$

The domain of R is the set of all first elements of the ordered pairs in the relation R.

$\therefore$ Domain of $R = \{0,1,2,3,4,5\}$

The whole set A is the co-domain of the relation R.

$\therefore$ Co-domain of $R = \{0,1,2,3,4,5,6\}$

The range of R is the set of all second elements of the ordered pairs in the relation R.

$\therefore$ Range of $R = \{1,2,3,4,5,6\}$

Example 6: Let $A = \{2,3,5\}$ and $B = \{2,3,4,5,6\}$. Then, relation R from A to B is given by,

Set-builder form: $R = \{(x, y) : y$ is exactly divisible by x; $x \in A, y \in B\}$.

Roster form: $R = \{(2,2),(2,4),(2,6),(3,3),(3,6),(5,5)\}$

The domain of R is the set of all first elements of the ordered pairs in the relation R.

$\therefore$ Domain of $R = \{2,3\}$

The whole set A is the co-domain of the relation R.

$\therefore$ Co-domain of $R = \{2,3,4,5,6\}$

The range of R is the set of all second elements of the ordered pairs in the relation R.

$\therefore$ Range of $R = \{2,3,4,5,6\}$

Example 7: Let relation R be defined on the set N of the natural numbers and be given by,

Set-builder form: $R = \{(x, y) : y = x - 1,\ x$ is a natural number less than 5; $x, y \in N\}$.

Since the natural numbers less than 5 are 1, 2, 3 and 4, but for $x = 1$ we get $y = x - 1 = 1 - 1 = 0 \notin N$, in the roster form:

$R = \{(2,1),(3,2),(4,3)\}$

The domain of R is the set of all first elements of the ordered pairs in the relation R.

$\therefore$ Domain of $R = \{2,3,4\}$

The whole set N is the co-domain of the relation R.

$\therefore$ Co-domain of $R = N = \{1,2,3,4,...\}$

The range of R is the set of all second elements of the ordered pairs in the relation R.

$\therefore$ Range of $R = \{1,2,3\}$

Example 8: Let $A = \{0,1,4,5\}$ and $B = \{1,2,3,4\}$. Then, the relation R from A to B is given by,

Set-builder form: $R = \{(x, y) :$ the difference between x and y is odd; $x \in A, y \in B\}$.

Roster form:

$R = \{(0,1),(0,3),(1,2),(1,4),(4,1),(4,3),(5,2),(5,4)\}$

The domain of R is the set of all first elements of the ordered pairs in the relation R.

$\therefore$ Domain of $R = \{0,1,4,5\}$

The whole set B is the co-domain of the relation R.

$\therefore$ Co-domain of $R = \{1,2,3,4\}$

The range of R is the set of all second elements of the ordered pairs in the relation R.

$\therefore$ Range of $R = \{1,2,3,4\}$

Example 9: Let $A = \{1,2,3,4,5,6\}$. Then, the relation R from A to A is given by,

Set-builder form: $R=\{(x,y): y+x\le 6,\ x,y\in A\}$

Roster form:

$R=\{(1,1),(1,2),(1,3),(1,4),(1,5),$
$(2,1),(2,2),(2,3),(2,4),(3,1),$
$(3,2),(3,3),(4,1),(4,2),(5,1)\}$

The domain of R is the set of all first elements of the ordered pairs in the relation R.

$\therefore$ Domain of $R=\{1,2,3,4,5\}$

The whole set A is the co-domain of the relation R.

$\therefore$ Co-domain of $R=\{1,2,3,4,5,6\}$

The range of R is the set of all second elements of the ordered pairs in the relation R.

$\therefore$ Range of $R=\{1,2,3,4,5\}$

Example 10: Let relation R defined on the set N of the natural numbers be given by

Set-builder form:

$R=\{(x,y): \frac{x}{y}\in N,\ x\le 6,\ y\in N\}$

The natural numbers less than or equal to 6 are 1, 2, 3, 4, 5 and 6, therefore, in the roster form:

$R=\{(1,1),(2,1),(2,2),(3,1),(3,3),$
$(4,1),(4,2),(4,4),(5,1),(5,5),$
$(6,1),(6,2),(6,3),(6,6)\}$

The domain of R is the set of all first elements of the ordered pairs in the relation R.

$\therefore$ Domain of $R=\{1,2,3,4,5,6\}$

The whole set N is the co-domain of the relation R.

$\therefore$ Co-domain of $R=N=\{1,2,3,4,...\}$

The range of R is the set of all second elements of the ordered pairs in the relation R.

$\therefore$ Range of $R=\{1,2,3,4,5,6\}$

Example 11: Let $A=\{0,1,2,3\}$ and $B=\{4,5,6\}$. Then, the relation R from A to B is given by,

Set-builder form:

$R=\{(x,y)$: the sum of x and y or the difference between x and y is multiple of 3; $x\in A, y\in B\}$.

Roster form: $R=\{(0,6),(1,4),(1,5),(2,4),(2,5),(3,6)\}$

The domain of R is the set of all first elements of the ordered pairs in the relation R.

$\therefore$ Domain of $R=\{0,1,2,3\}$

The whole set B is the co-domain of the relation R.

$\therefore$ Co-domain of $R=\{4,5,6\}$

The range of R is the set of all second elements of the ordered pairs in the relation R.

$\therefore$ Range of $R=\{4,5,6\}$

Example 12: Let $A=\{2,4,6,8,10\}$ and $B=\{1,2,3,4,5\}$. Then, the relation R from A to B is given by

Set-builder form:

$R=\{(x,y): y$ is prime factor of $x;\ x\in A, y\in B\}$.

Roster form:

$R=\{(2,2),(4,2),(6,2),(6,3),(8,2),(10,2),(10,5)\}$

The domain of R is the set of all first elements of the ordered pairs in the relation R.

$\therefore$ Domain of $R=\{2,4,6,8,10\}$

The whole set B is the co-domain of the relation R.

$\therefore$ Co-domain of $R=\{1,2,3,4,5\}$

The range of R is the set of all second elements of the ordered pairs in the relation R.

$\therefore$ Range of $R=\{2,3,5\}$

Day 22

NCERT Exercise 2.2 (With Solutions)

Q1. Let $A=\{1,2,3,....14\}$. Define a relation R from A to A by $R=\{(x,y): 3x-y=0$, where $x,y\in A\}$. Write down its domain, co-domain and range.

Solution:

The relation R from A to A is given by

$R=\{(x,y): 3x-y=0$, where $x,y\in A\}$

$\Rightarrow R=\{(x,y): 3x=y$, where $x,y\in A\}$

$\therefore R=\{(1,3),(2,6),(3,9),(4,12)\}$

The domain of R is the set of all first elements of the ordered pairs in the relation R.

$\therefore$ Domain of $R=\{1, 2, 3, 4\}$

The whole set A is the co-domain of the relation R.

$\therefore$ Co-domain of $R=\{1,2,3,......14\}$

The range of R is the set of all second elements of the ordered pairs in the relation R.

$\therefore$ Range of $R=\{3,6,9,12\}$

Q2. Define a relation R on the set N of the natural numbers by $R=\{(x,y): y=x+5$, x is a natural number less than $4; x,y\in N\}$. Define this relationship using the roster form. Write down the domain and the range.

Solution:

Given $R=\{(x,y): y=x+5, x$ is a natural number less than $4; x,y\in N\}$.

The natural numbers less than 4 are 1, 2, and 3.

$\therefore R=\{(1,6),(2,7),(3,8)\}$

The domain of R is the set of all first elements of the ordered pairs in the relation R.

$\therefore$ Domain of $R = \{1,2,3\}$

The range of R is the set of all second elements of the ordered pairs in the relation R.

$\therefore$ Range of $R = \{6,7,8\}$

Q3. $A = \{1,2,3,5\}, B = \{4,6,9\}$. Define a relation R from A to B by $R = \{(x, y)$: the difference between x and y is odd; $x \in A, y \in B\}$. Write R in the roster form.

Solution:

$A = \{1,2,3,5\}, B = \{4,6,9\}$

$R = \{(x, y)$: the difference between x and y is odd; $x \in A, y \in B\}$.

$\therefore R = \{(1,4),(1,6),(2,9),(3,4),(3,6),(5,4),(5,6)\}$

Q4. The given figure shows a relationship between the sets P and Q. Write this relation in the (i) set-builder, (ii) roster form. What is its domain and range?

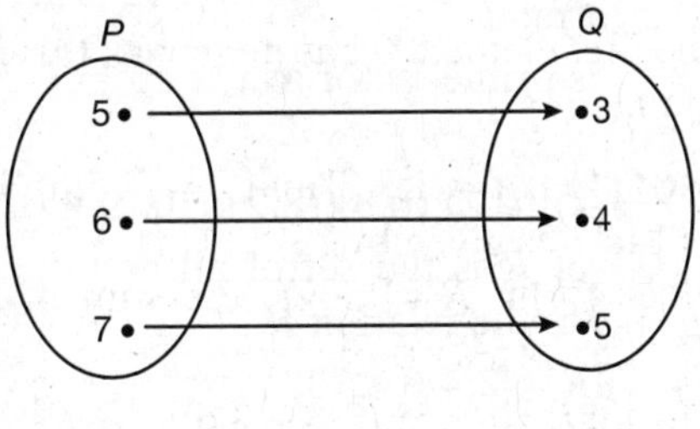

Solution:

According to the given figure,

$P = \{5,6,7\}, Q = \{3,4,5\}$

Therefore, $R = \{(x, y): y = x - 2; x \in P$ or $x = 5,6,7\}$

$\Rightarrow R = \{(5,3),(6,4),(7,5)\}$

Domain of $R = \{5,6,7\}$

Range of $R = \{3,4,5\}$

Q5. Let $A = \{1,2,3,4,6\}$. Let R be the relation on A defined by $R = \{(a,b): a, b \in A, b$ is exactly divisible by $a\}$. (i) Write R in the roster form, (ii) find the domain of R and (iii) find the range of R.

Solution:

$A = \{1,2,3,4,6\}$

$R = \{(a,b): a, b \in A, b$ is exactly divisible by $a\}$

Therefore,

$R = \{(1,1),(1,2),(1,3),(1,4),(1,6),(2,2),$
$(2,4),(2,6),(3,3),(3,6),(4,4),(6,6)\}$

(ii) Domain of $R = \{1,2,3,4,6\}$

(iii) Range of $R = \{1,2,3,4,6\}$

Q6. Determine the domain and range of the relation R defined by $R = \{(x, x+5): x \in \{0,1,2,3,4,5\}\}$.

Solution:

$\because R = \{(x, x+5): x \in \{0,1,2,3,4,5\}\}$

$\therefore R = \{(0,5),(1,6),(2,7),(3,8),(4,9),(5,10)\}$

Thus, domain of $R = \{0,1,2,3,4,5\}$

Range of $R = \{5,6,7,8,9,10\}$

Q7. Write the relation $R = \{(x, x^3): x$ is a prime number less than $10\}$ in the roster form.

Solution:

$R = \{(x, x^3): x$ is a prime number less than $10\}$

The prime numbers less than 10 are 2, 3, 5, 7.

$\therefore R = \{(2,8),(3,27),(5,125),(7,343)\}$

Q8. Let $A = \{x, y, z\}, B = \{1,2\}$. Find the number of relations from A to B.

Solution:

$A = \{x, y, z\}, B = \{1,2\}$

$\therefore A \times B = \{(x,1),(x,2),(y,1),(y,2),(z,1),(z,2)\}$

$n(A \times B) = 6$, the number of subsets of $A \times B$ is 2^6.

Therefore, the number of relations from A to B is 2^6.

Q9. Let R be the relation on Z defined by $R = \{(a,b): a, b \in Z, a - b$ is an integer$\}$. Find the domain and range of R.

Solution:

$R = \{(a,b): a, b \in Z, a - b$ is an integer$\}$

The difference between any two integers is always an integer,

$\therefore$ Domain of $R = Z$

Range of $R = Z$

Day 23

Functions

A special type of relation f from a non-empty set A to a non-empty set B is said to be a function if every element of set A has one and only one image in set B, that is, the domain of f is A and no two distinct ordered pairs in f have the same first element.

Thus, if f is a function from A to B and $(a,b) \in f$, then $f(a) = b$, where b is called the image (or range) of a under f and a is called the pre-image (or domain) of b under f.

- The function f from A to B is denoted by $f: A \rightarrow B$ or $b = f(a)$.

- f denotes the function (or special relation) between $a \in A$ and $b \in B$.
- A is the given domain of function f.
- a is the element of the domain of function f.
- B is given co-domain of function f.
- b is the range or image of function f, that is, all values of b obtained by the relation $b = f(a)$, for all $a \in A$ is defined as the range of function f.
- For a valid function, range must be less than or equal to co-domain, Range $\subseteq$ Co-domain.
- So, do not get confused between co-domain and range, that is, co-domain is given while range is obtained from the domain and function f.

Example 1: Consider the relation R from A to B given by $R = \{(a_1,b_1),(a_2,b_2),(a_3,b_3),(a_4,b_4),(a_5,b_3)\}$.

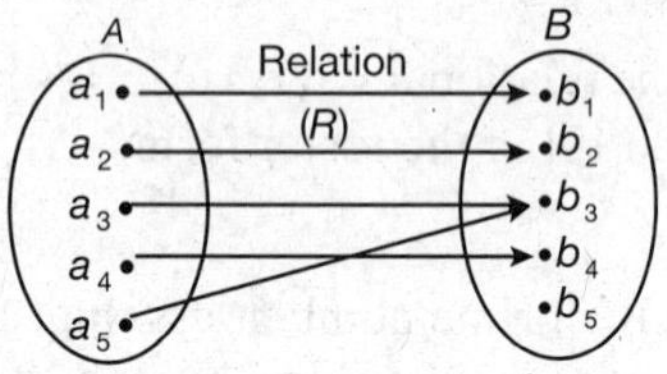

Every element (a_1,a_2,a_3,a_4,a_5) has one and only one image, therefore this relation is a function.

Thus,

Domain: (a_1,a_2,a_3,a_4,a_5)

Range: (b_1,b_2,b_3,b_4)

Co-domain: (b_1,b_2,b_3,b_4,b_5)

Note that the range of relation $R \subseteq$ co-domain.

Example 2: Consider the relation R from A to B given by $R = \{(-1,-1),(0,0),(1,1)\}$.

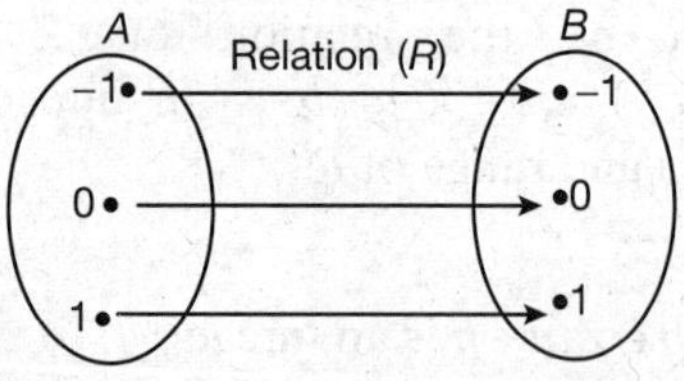

Since every element –1, 0 and 1 has one and only one image, this relation is a function.

Thus, domain = {–1, 0, 1} and range = {–1, 0, 1}

Example 3: Consider the relation R from A to B given by $R = \{(-1,1),(0,0),(1,1)\}$.

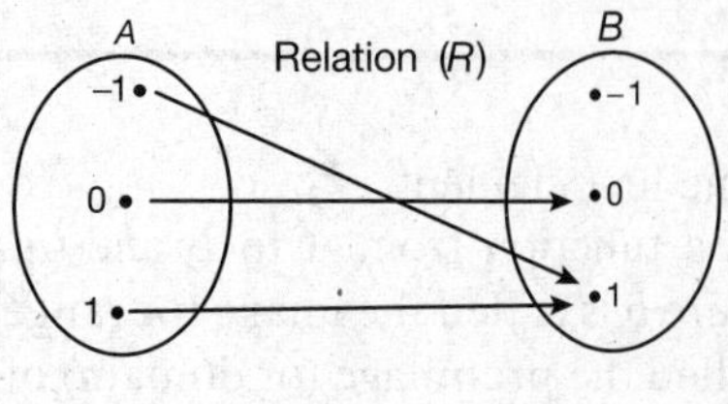

Since every element –1, 0 and 1 has one and only one image, this relation is a function.

Thus, domain = {–1, 0, 1} and range = {0, 1}.

Example 4: Consider the relation R from A to B given by $R = \{(0,0),(1,-1),(1,1)\}$.

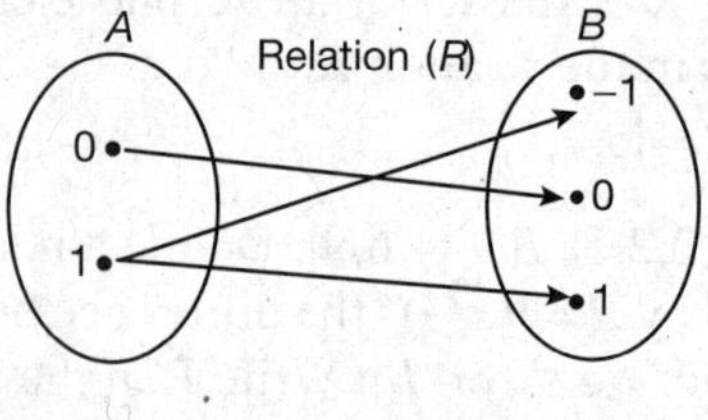

Since the first element 1 corresponds to two different images –1 and 1, this relation is not a function.

Example 5: Consider the relation R from A to B given by $R = \{(-1,-1),(1,1)\}$.

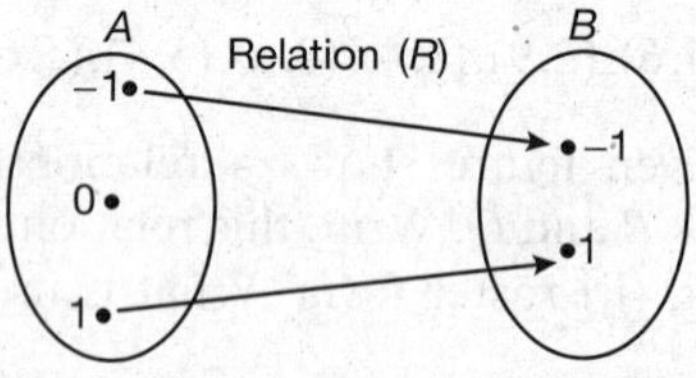

Since the first element 0 has no image, this relation is not a function.

Example 6: Let $A = \{1,2,3\}$ and $B = \{4,5,6\}$. Then, relation R from A to B is given by,

Set-builder form: $R = \{(x, y):$ the sum of x and y is a multiple of 3; $x \in A, y \in B\}$.

Roster form: $R = \{(1,5),(2,4),(3,6)\}$

Domain of $R = \{1,2,3\}$

Co-domain of $R = \{4,5,6\}$

Range of $R = \{4,5,6\}$

Since every element has one and only one image, this relation is a function.

Example 7: Let $A = \{2,3,4,5,7,8,9\}$ and $B = \{2,3,5,7\}$. Then, relation R from A to B is given by,

Set-builder form: $R = \{(x, y): x$ is exactly divisible by y; $x \in A, y \in B\}$.

Roster form:

$R = \{(2,2),(4,2),(8,2),(3,3),(9,3),(5,5),(7,7)\}$

Domain of $R = \{2,3,4,5,7,8,9\}$

Co-domain of $R = \{2,3,5,7\}$

Range of $R = \{2,3,5,7\}$

Since every element has one and only one image, this relation is a function.

Example 8: Let $A = \{0,1,2,3,4,5,6\}$. Then, relation R from A to A is given by,

Set-builder form: $R = \{(x, y): y = x+1,\ x, y \in A\}$

Roster form:

$R = \{(0,1),(1,2),(2,3),(3,4),(4,5),(5,6)\}$

Domain of $R = \{0,1,2,3,4,5\}$

Range of $R = \{1,2,3,4,5,6\}$

Since the first element 6 has no image, this relation is not a function.

Example 9: Let $A = \{0,1\}$. Then, relation R from A to A is given by,

Set-builder form: $R = \{(x, y) : y + x \le 1,\ x, y \in A\}$

Roster form: $R = \{(0,0),(0,1),(1,0)\}$

Domain of $R = \{0,1\}$

Co-domain of $R = \{0,1\}$

Range of $R = \{0,1\}$

Since the first element 0 corresponds to two different images 0 and 1, this relation is not a function.

Example 10: Let relation R defined on the set N of the natural numbers be given by,

Set-builder form: $R = \{(x, y) : y = x + 1,\ x$ is a natural number less than 5; $x, y \in N\}$.

Natural numbers less than 5 are 1, 2, 3 and 4,

$\therefore$ Roster form: $R = \{(1,2),(2,3),(3,4),(4,5)\}$

Domain of $R = \{1,2,3,4\}$

Co-domain of $R = N = \{1,2,3,4,...\}$

Range of $R = \{2,3,4,5\}$

Since every element has one and only one image, this relation is a function.

Example 11: Let relation R defined on the set N of the natural numbers be given by,

Set-builder form: $R = \{(x, y) : y = x - 1,\ x$ is a natural number less than 5; $x, y \in N\}$.

The natural numbers less than 5 are 1, 2, 3 and 4, but for $x = 1$ we get $y = x - 1 = 1 - 1 = 0 \notin N$,

$\therefore$ in the roster form: $R = \{(2,1),(3,2),(4,3)\}$

Domain of $R = \{2,3,4\}$

Co-domain of $R = N = \{1,2,3,4,...\}$

Range of $R = \{1,2,3\}$

Since every element has one and only one image, this relation is a function.

Since the first element 1 has no image, this relation is not a function.

Example 12: Let relation R defined on the set N of the natural numbers be given by,

Set-builder form:

$$R = \{(x, y) : \frac{x}{y} \in N,\ 1 < x \le 6,\ y \in N\}$$

The natural numbers greater than 1 and less than or equal to 6 are 2, 3, 4, 5 and 6, therefore in the roster form:

$$R = \{(2,2),(3,3),(4,2),(4,4),\\ (5,5),(6,2),(6,3),(6,6)\}$$

Domain of $R = \{2,3,4,5,6\}$

Co-domain of $R = N = \{1,2,3,4,...\}$

Range of $R = \{2,3,4,5,6\}$

Since the first element 4 corresponds to two different images 2 and 4, and also the first element 6 corresponds to three different images 2, 3 and 6, this relation is not a function.

Day 24

Real-Valued Functions

A function that has either R or one of its subsets as its range is called a real-valued function. And, if its domain is also either R or a subset of R, then it is called a real function.

A relation f from $X \to Y$ is said to be a real function if every $x \in X$ has one and only one image in $y \in Y$, that is, the domain of f is A and we should not get two distinct values of $y \in Y$ for the same value $x \in X$.

Thus, if f is a function from X to Y and $(x, y) \in f$, then $y = f(x)$, where y is called the image (or range) of x under f and x is called the pre-image (or domain) of y under f.

Representation of a Real Function

- The function f from X to Y is denoted by $f : X \to Y$ or $y = f(x)$.
- f denotes the real function between $x \in X$ and $y \in Y$.
- X is the given domain of function f.
- x is the element of the domain of function f.
- Y is the given co-domain of function f.
- y is the range of function f, that is, all values of y obtained by the relation $y = f(x)$, for all $x \in X$, is defined as the range of function f.
- For a valid function, the range must be less than or equal to the co-domain, Range $\subseteq$ Co-domain.
- So, do not get confused between co-domain and range, that is, co-domain is given while range is obtained from the domain and function f.

Note: R represents all real numbers.

R^+ means $[0, \infty)$ and R^- means $(-\infty, 0]$

Properties of functions: A function f may have the following possibilities:

(i) The distinct elements of X under the function f have distinct images in Y.
(ii) All the distinct elements of X under the function f have an image in Y. But it is not necessary that all distinct elements of Y under the function f have a pre-image in X. That is, Range $\subseteq$ Co-domain.
(iii) Two or more distinct elements of X under the function f must not have the same image in Y.

Graphical Representation of a Function: The property of a function is that for any value of x we get a single value of y. Graphically if we draw a line parallel to the y-axis and if it cuts the graph at two distinct points, then the graph is not a function.

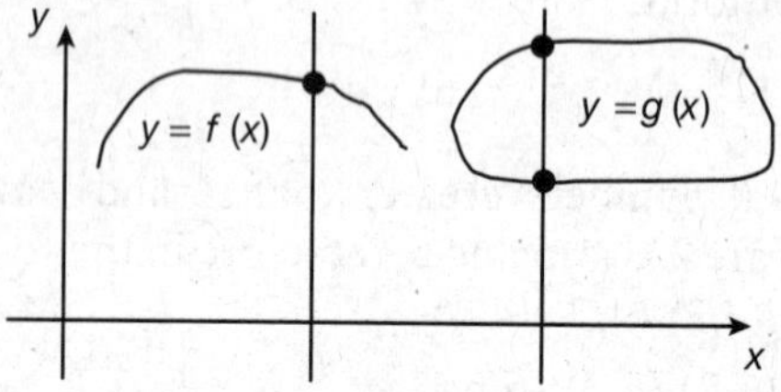

Fig. 2.2

Consider the graphs $y = f(x)$ and $y = g(x)$ as shown in Fig. 2.2.
A line parallel to the y-axis cuts the graph $y = f(x)$ only at one point while a line parallel to the y-axis cuts the graph $y = g(x)$ at two distinct points. Therefore, graph $y = f(x)$ represents a function while $y = g(x)$ does not represent a function.

__Example 1:__ Consider the equation of a circle $x^2 + y^2 = r^2$. This is an equation, not a function.

Now solving for $y = f(x)$, we get $y = \pm\sqrt{r^2 - x^2}$ which is again an equation not a function, because for any value of x we get two values of y.

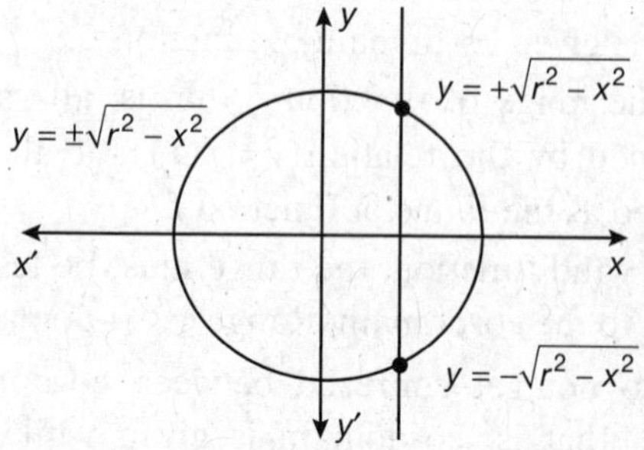

While the property of a function is that for any value of x we get a single value of y, graphically if we draw a line parallel to the y-axis and if it cuts the graph at two distinct points, then the graph is not a function.

Now if the function is given as $y = +\sqrt{r^2 - x^2}$ or (or $y = -\sqrt{r^2 - x^2}$) then this will be a function. If now we draw a line parallel to the y-axis, it will cut the graph only at a single point. $y = +\sqrt{r^2 - x^2}$ shows an upper semi-circle while $y = -\sqrt{r^2 - x^2}$ shows a lower semi-circle. Therefore, any line parallel to the y-axis will cut the upper (or lower) semi-circle only at a single point.

__Example 2:__ Consider the function $f : R \to R$ given by $y = x$.

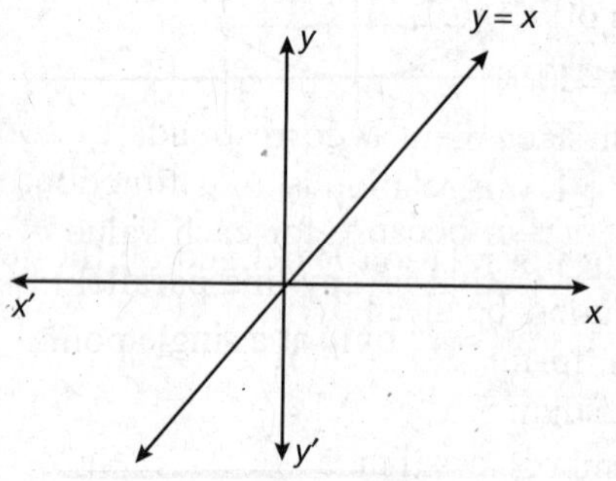

Every element has one and only one image. Therefore $f : R \to R$ given by $y = x$ is a function.
Thus, domain: R
Co-domain: R
Range: R (since for all x, $y \in R$)
Since range is equal to co-domain, this is a valid function.

__Example 3:__ $f : N \to R$ given by $f(x) = x$
Domain: N
Co-domain: R
Range: N (since for all x, $y \in N$)
Since range is less than co-domain, this is a valid function.

__Example 4:__ $f : R \to N$ given by $f(x) = x$
Domain: R
Co-domain: N
Range: R (since for all x, $y \in R$)
Since range is greater than co-domain, this is not a valid function.

__Example 5:__ Consider the function $f : R \to R - \{0\}$ given by $y = x$.
Or relation $R = \{(-1, -1), (1, 1)\}$

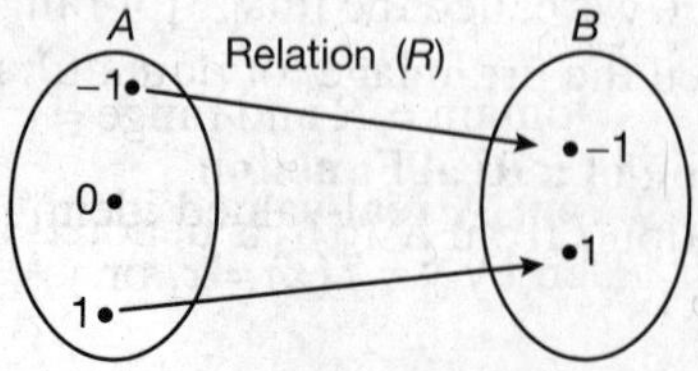

Since the first element 0 has no image, that is, element 0 is not in the co-domain, the range is greater than the co-domain. Thus, this relation is not a function.
Thus, for a function, the range must be less than or equal to the co-domain.

Example 6: Consider the function $f: R \to R$ given by $y = x^2$.

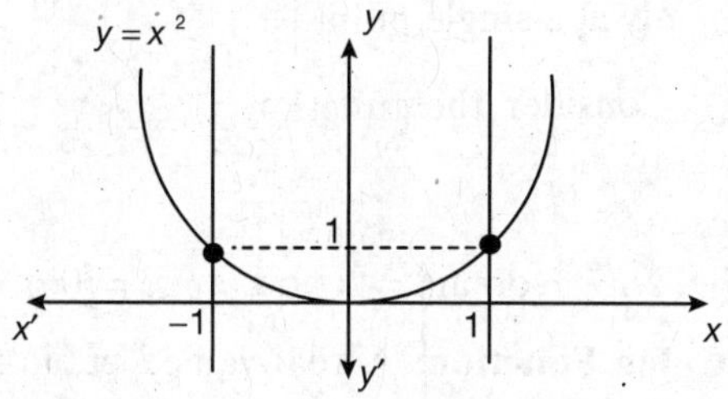

$y = x^2$ is a function because for each value of x, we get unique value of y, that is, any line parallel to the y-axis will cut the graph $y = x^2$ only at a single point.

Thus, domain = R and range = $[0, \infty)$

Example 7: Consider the function $f: R \to R$ given by $y^2 = x \Rightarrow y = \pm\sqrt{x}$.

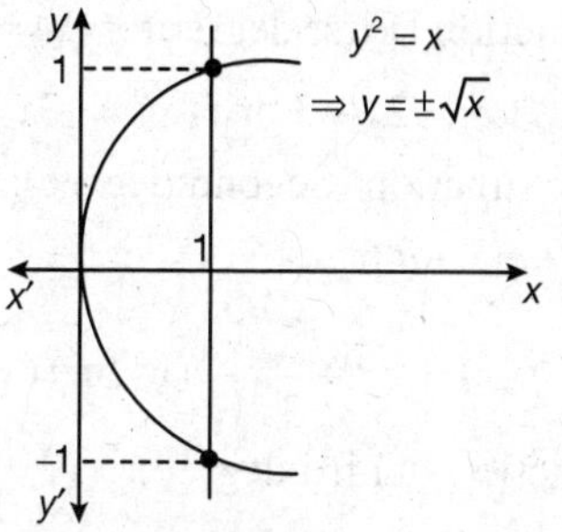

$y^2 = x \Rightarrow y = \pm\sqrt{x}$ is not a function because for each value of x, we get two distincts value of y, that is, any line parallel to the y-axis cuts the graph at two different points. For example, the line $x = 1$ cuts the graph at $y = \pm 1$.
Since the element 1 corresponds to two different images –1 and 1, this is not a function.

Day 25

Graph of Real-Valued Functions-I

(i) Constant Function: A real-valued constant function $f: R \to R$ is given by $y = f(x) = c$, where c is a constant and $x \in R$.

x	–3	–2	–1	0	1	2	3
y	c	c	c	c	c	c	c

The graph is a line parallel to the x-axis and intercepts the y-axis at $y = c$.

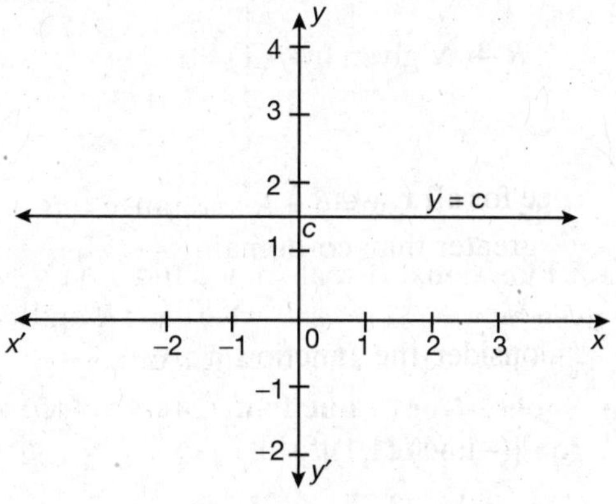

Fig. 2.3 Domain = R and range = $\{c\}$

(ii) Identity Function: A real-valued identity function $f: R \to R$ is given by $y = f(x) = x$, or $y = f(x) = -x$, where $x \in R$.

x	–3	–2	–1	0	1	2	3
$y = x$	–3	–2	–1	0	1	2	3
$y = -x$	3	2	1	0	–1	–2	–3

The graph is a straight line that passes through the origin and is equally inclined to the x and y axes.

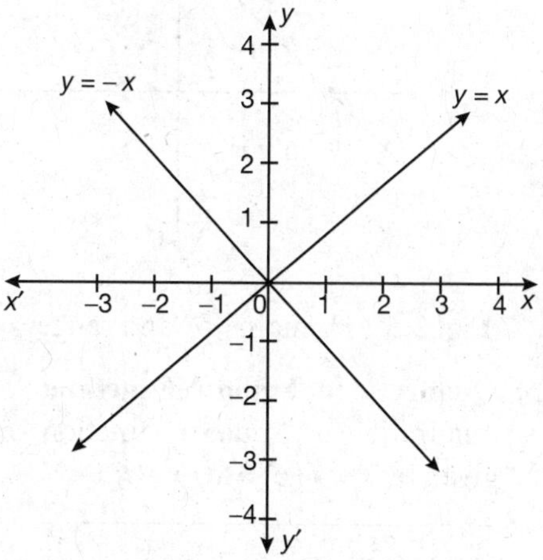

Fig. 2.4 Domain = R and range = R

(iii) Polynomial Function: A real-valued polynomial function $f: R \to R$ is given by $y = f(x) = a_0 + a_1x + a_2x^2 + \ldots + a_nx^n$, where $x \in R$ and $n \in Z^+$ (a non-negative integer) and constants $a_0, a_1, a_2, \ldots, a_n \in R$.

Example 1: For all $x \in R$:

Constant function (Zero degree): $f(x) = a_0$

Example: $f(x) = 2$ or $f(x) = -3.5$

Linear function (First degree): $f(x) = a_0 + a_1 x$

Example: $f(x) = 2x + 3$ or $f(x) = -3x$

Quadratic function (Second degree):

$f(x) = a_0 + a_1 x + a_2 x^2$

Example: $f(x) = \sqrt{2} + \frac{5}{3}x - 4x^2$ or $f(x) = 2x^2 - \frac{\sqrt{3}}{4}$

Cubic function (Third degree):

$f(x) = a_0 + a_1 x + a_2 x^2 + a_3 x^3$

Example: $f(x) = \sqrt{2} + 4x^2 - x^3$ and $f(x) = 2x^3 - \frac{\sqrt{3}}{4}x$

Note: The functions $f(x) = \sqrt{x}$ and $f(x) = x^{-2}$ are not polynomial functions because $1/2$ or $-2 \notin Z^+$.

(a) Linear Function: A real-valued identity function $f: R \to R$ is given by $y = mx + c$, where m and c are constants and $x \in R$.

The graph is a straight line with slope m and intersect the x and y axes at the points $(-c/m, 0)$ and $(0, c)$, respectively.

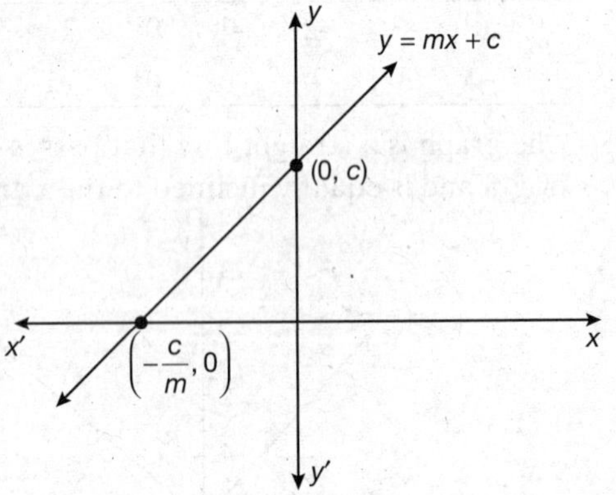

Fig. 2.5 Domain = R and range = R

(b) Quadratic (or Square) Function: A real-valued quadratic (or square) function $f: R \to R$ is given by $y = f(x) = x^2, x \in R$.

x	−3	−2	−1	0	1	2	3
y	9	4	1	0	1	4	9

The graph lies above the x-axis and is symmetrical about the y-axis with a minimum value of $y = 0$ at $x = 0$ and this graph is also known as a standard parabola.

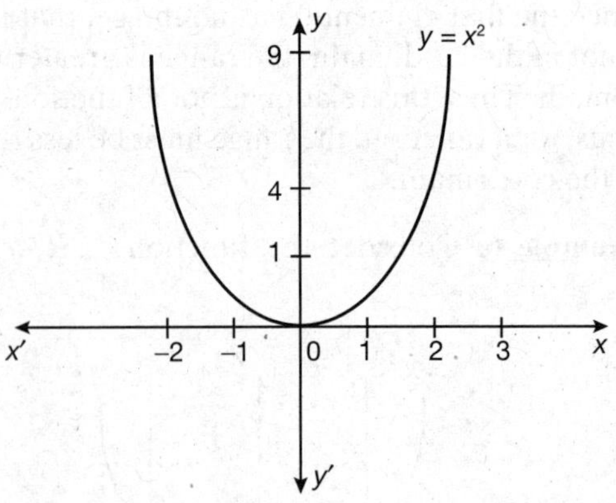

Fig. 2.6 Domain = R and range = $[0, \infty)$

(c) Cubic Function: A real-valued cubic function $f: R \to R$ is given by $y = f(x) = x^3, x \in R$.

x	−3	−2	−1	0	1	2	3
y	−27	−8	−1	0	1	8	27

The graph passes through the origin and lies in the 1st and 3rd quadrants symmetrically.

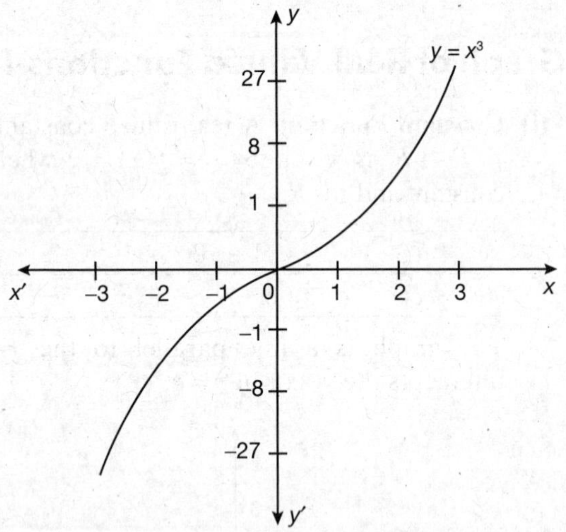

Fig. 2.7 Domain = R and range = R

(iv) Root Functions: A real-valued function $f: R \to R$ is given by $y = f(x) = ax^n$, where $x \in R$ and $n \notin Z$ (a non-integer real) and constant $a \in R$.

(a) Square-Root Function: A real-valued square-root function $f: [0, \infty) \to R$ is given by $y = f(x) = \sqrt{x}, x \in [0, \infty)$.

x	0	1	4	9	16	25
y	0	1	2	3	4	5

The graph lies only in the 1st quadrant.

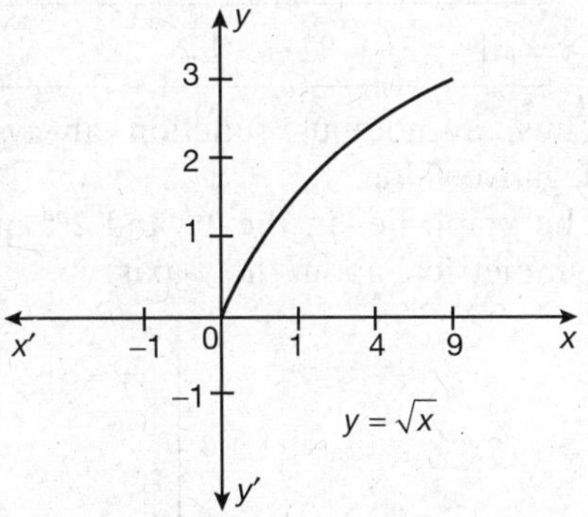

Fig. 2.8 Domain = $[0, \infty)$ and range = $[0, \infty)$

(b) Cube-Root Function: A real-valued cube-root function $f: R \to R$ is given by $y = f(x) = \sqrt[3]{x}$, $x \in R$.

x	–27	–8	–1	0	1	8	27
y	–3	–2	–1	0	1	2	3

The graph passes through the origin and lies in the 1st and 3rd quadrants symmetrically.

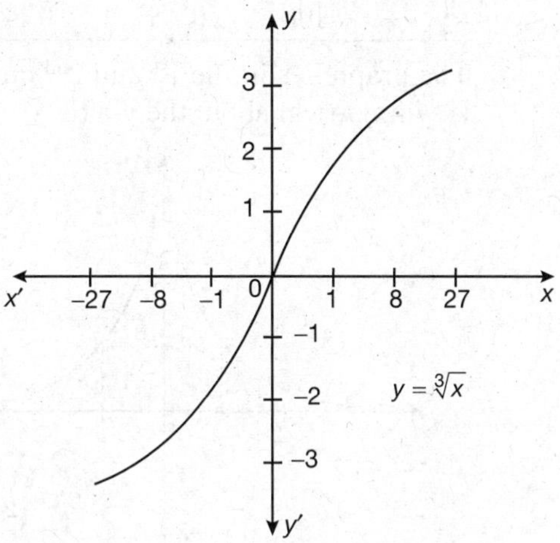

Fig. 2.9 Domain = R and range = R

Day 26

Graphs of Real-Valued Functions-II

(v) Rational Functions: Rational functions are functions of the type $\dfrac{f(x)}{g(x)}$, where $f(x), g(x)$ are polynomial functions of x defined in a domain, where $g(x) \neq 0$.

(a) Standard Reciprocal Function: A real-valued reciprocal function $f: R - \{0\} \to R$ is given by $y = f(x) = \dfrac{1}{x}$, $x \in R$ and $x \neq 0$.

x	0^+	1/100	1/10	1	10	100	$+\infty$
y	$+\infty$	100	10	1	0.1	0.01	0^+

x	0^-	–1/100	–1/10	–1	–10	–100	$-\infty$
y	$+\infty$	–100	–10	–1	–0.1	–0.01	0^-

The graph lies in the 1st and 3rd quadrants symmetrically.

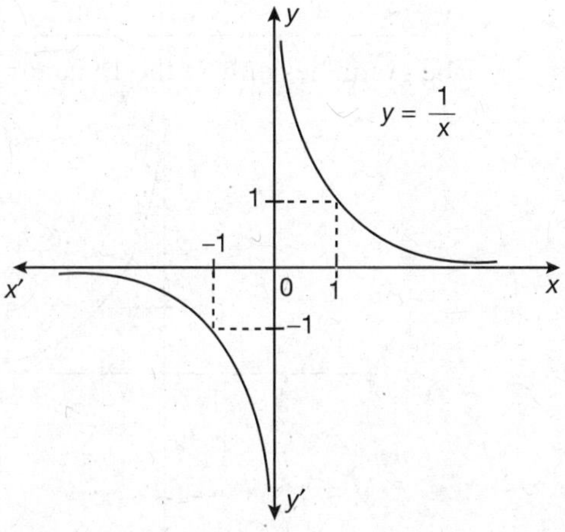

Fig. 2.10 Domain = $(-\infty, 0) \cup (0, \infty)$ and Range = $(-\infty, 0) \cup (0, \infty)$

(b) Reciprocal Squared Function: A real-valued reciprocal function $f: R - \{0\} \to R$ is given by $y = f(x) = \dfrac{1}{x^2}$, $x \in R$ and $x \neq 0$.

x	0^+	1/100	1/10	1	10	100	$+\infty$
y	$+\infty$	100	10	1	0.1	0.01	0^+

x	0^-	$-1/100$	$-1/10$	-1	-10	-100	$-\infty$
y	$+\infty$	100	10	1	0.1	0.01	0^+

The graph lies in the 1st and 2nd quadrants and is symmetrical about the y-axis.

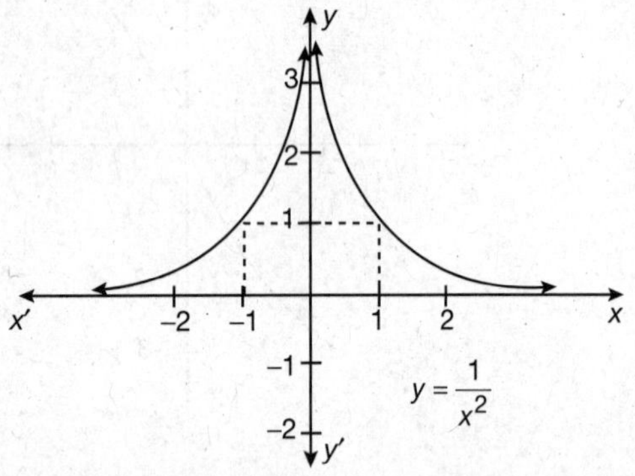

Fig. 2.11 Domain $= (-\infty, 0) \cup (0, \infty)$ and Range $= (0, \infty)$

(c) Reciprocal Square-Root Function: A real-valued square-root function $f: (0, \infty) \to R$ is given by $y = f(x) = \frac{1}{\sqrt{x}}$, $x \in (0, \infty)$.

x	0^+	1/9	1/4	1	4	9	∞
y	∞	3	2	1	1/2	1/3	0

The graph lies only in the 1st quadrant.

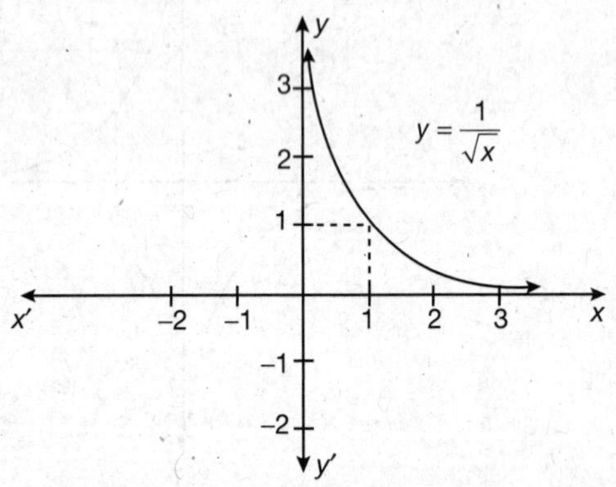

Fig. 2.12 Domain $= (0, \infty)$ and range $= (0, \infty)$

(vi) Modulus (or Absolute Value) Function: A real-valued modulus function $f: R \to R$ is given by $y = f(x) = |x|$, $x \in R$.

Modulus function is a piecewise or split function represented as,

$$y = f(x) = |x| = \begin{cases} x, & x \geq 0 \\ -x, & x < 0 \end{cases}$$

That is, it depends on the sign of the x-value. If the x-value is positive, $x \geq 0$, then the function is defined by $f(x) = x$. If the x-value is negative, $x < 0$, then the function is defined by $f(x) = -x$.

x	-3	-2	-1	0	1	2	3
$y = \lvert x \rvert$	3	2	1	0	1	2	3

Thus, a modulus function always gives a non-negative value.

The graph lies in the 1st and 2nd quadrants and is symmetrical about the y-axis.

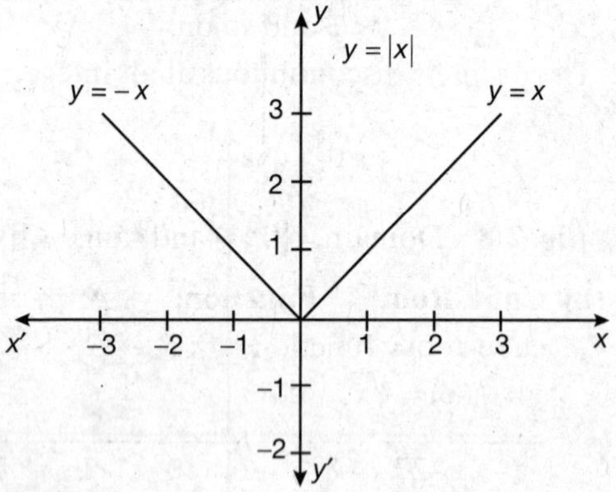

Fig. 2.13 Domain $= R$ and range $= [0, \infty)$

(vii) Signum Function: A real-valued signum function $f: R \to R$ is a piecewise or split function represented as,

$$y = f(x) = \begin{cases} \frac{|x|}{x} \text{ or } \frac{x}{|x|}, & x \neq 0 \\ 0, & x = 0 \end{cases}$$

$$\Rightarrow f(x) = \begin{cases} 1, & \text{if } x > 0 \\ 0, & \text{if } x = 0 \\ -1, & \text{if } x < 0 \end{cases}$$

The graph has horizontal rays $y = 1$ and $y = -1$ parallel to the x-axis in the 1st and 3rd quadrants, start from $y = 0^+$ and $y = 0^-$, respectively, and $y = 0$ at $x = 0$.

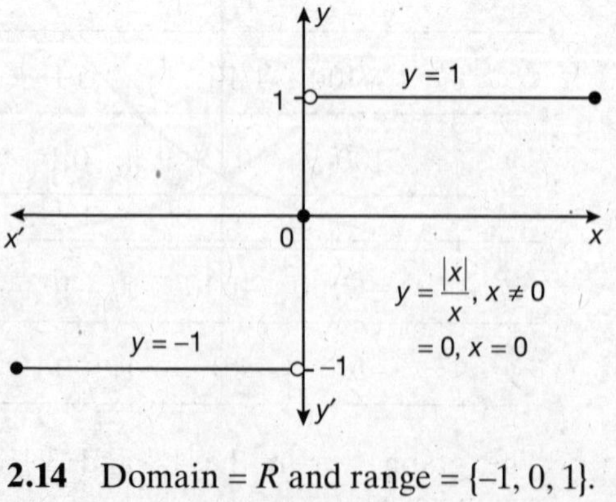

Fig. 2.14 Domain $= R$ and range $= \{-1, 0, 1\}$.

(viii) Greatest Integer Function: A real-valued greatest integer function (also known as floor function) $f: R \to R$ is given by $y = f(x) = [x]$, $x \in R$.

It is equal to the value of the greatest integer, less than or equal to x.
Thus,

$[x] = -2$ for $-2 \le x < -1$

$[x] = -1$ for $-1 \le x < 0$

$[x] = 0$ for $0 \le x < 1$

$[x] = 1$ for $1 \le x < 2$

$[x] = 2$ for $2 \le x < 3$ and so on.

The graph is discontinuous at all integers.

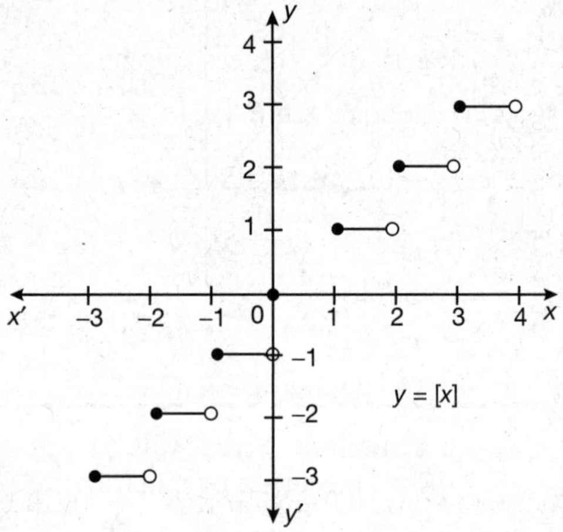

Fig. 2.15 Domain = R and range = Z

(ix) Exponential Function: A real-valued exponential function $f: R \to R$ is given by $y = a^x$, $x \in R$, $a > 0$, and $a \neq 1$.

$$\Rightarrow y = \begin{cases} a^x \text{ (rising curve)} & \text{if } a > 1 \\ a^x \text{ (falling curve)} & \text{if } 0 < a < 1 \end{cases}$$

The graph lies above the x-axis and it is a rising curve for $a > 1$ and a falling curve for $0 < a < 1$.

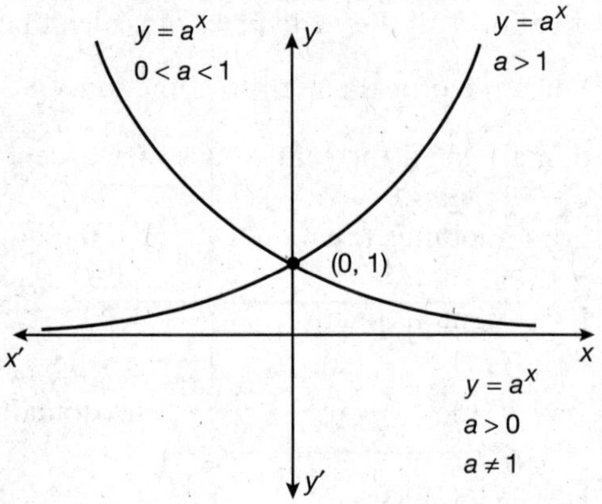

Fig. 2.16 Domain = R and range = $(0, \infty)$

<u>**Example 2:**</u> Graphs for $y = 2^x$ and $y = (0.5)^x$, $x \in R$ are shown below.

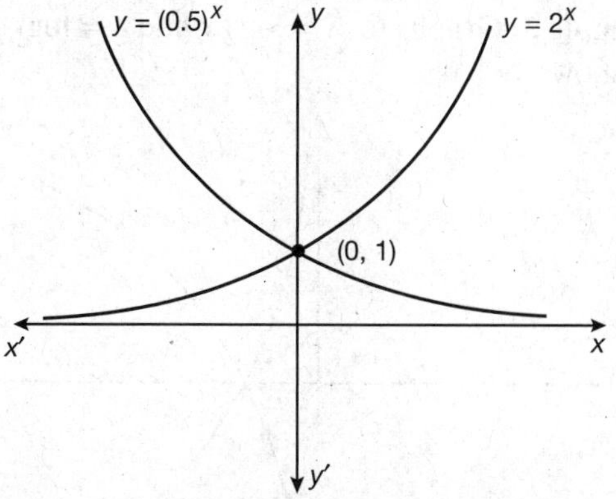

The graph is similar to the graphs for $y = a^x$, $a > 1$ and $y = a^x$, $0 < a < 1$, respectively.

Properties of exponential functions: Let x, y be real numbers, a, b be positive real numbers, and $a, b \neq 1$. Then

- $a^x = a^y$ if and only if $x = y$
- If $x > y$ then $\begin{cases} a^x > a^y & \text{if } a > 1 \\ a^x < a^y & \text{if } 0 < a < 1 \end{cases}$
- $a^x \cdot a^y = a^{x+y}$; $\dfrac{a^x}{a^y} = a^{x-y}$ and $(a^x)^y = a^{x \cdot y}$
- $(a \cdot b)^x = a^x \cdot b^x$ and $\left(\dfrac{a}{b}\right)^x = \dfrac{a^x}{b^x}$

(x) Logarithmic Function: A real-valued exponential function $f: R^+ \to R$ is given by $y = \log_a x$, $x \in R^+$, $a > 0$, and $a \neq 1$

$$\Rightarrow y = \begin{cases} \log_a x \text{ (rising curve)} & \text{if } a > 1 \\ \log_a x \text{ (falling curve)} & \text{if } 0 < a < 1 \end{cases}$$

The graph lies right of the y-axis and it is a rising curve for $a > 1$ and a falling curve for $0 < a < 1$.

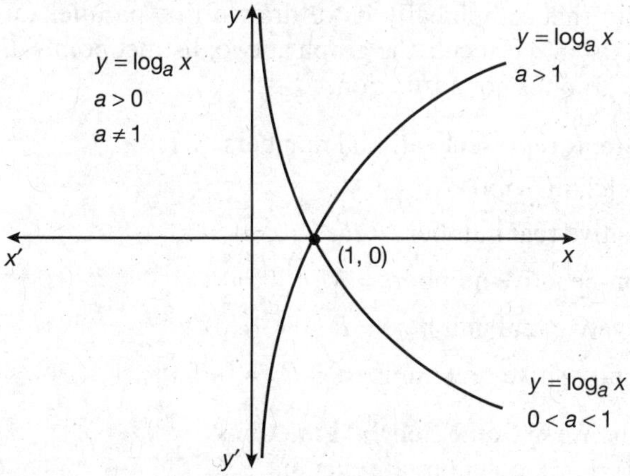

Fig. 2.17 Domain = $(0, \infty)$ and range = R

Example 3: Graphs for $y = \log_2 x$ and $y = \log_{1/2} x$, $x \in R^+$ are shown below.

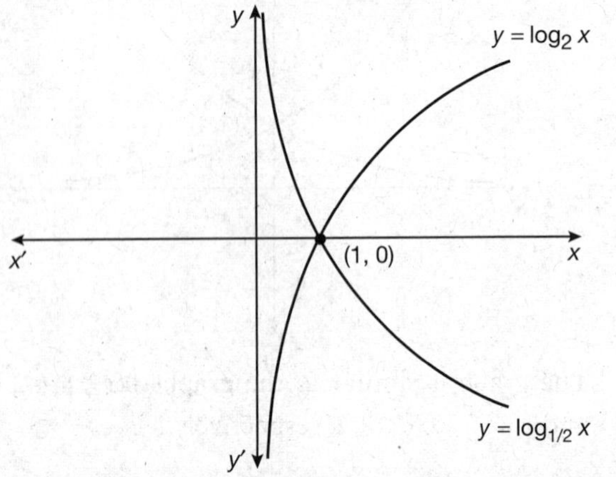

The graph is similar to graphs $y = \log_a x, a > 1$ and $y = \log_a x,\ 0 < a < 1$, respectively.

Properties of logarithmic functions: Let x, y be real numbers, a, b be positive real numbers, and $a, b \neq 1$. Then,

- $\log_a x = \log_a y$ if and only if $x = y$
- If $x > y$ then
  $$\begin{cases} \log_a x > \log_a y & \text{if } a > 1 \\ \log_a x < \log_a y & \text{if } 0 < a < 1 \end{cases}$$
- $\log_a (x \cdot y) = \log_a x + \log_a y$
  $$\log_a \left(\frac{x}{y}\right) = \log_a x - \log_a y$$
- $\log_x y = \dfrac{1}{\log_y x}$
- $\log_a x^n = n \log_a x,\ n \in R$
- $\log_{a^m} x^n = \dfrac{n}{m} \log_a x,\ n \in R$
- $a^{\log_a x} = \log_a a^x = x$
- $\log_a a = 1$ and $\log_a 1 = 0$

Day 27

Domain and Range of Real Functions-I

Consider a function f from X to Y and $(x, y) \in f$, such that $y = f(x)$. Then, all real values of x for which the function f exists and is defined is called the domain of function f. All real values of y obtained by the relation $y = f(x)$ for all real values of x (for which function f exists and is defined) is called the range of function f.
And all given real values of y together is called the co-domain of function f. For a valid function, the range must be less than or equal to the co-domain, Range $\subseteq$ Co-domain. So, do not get confused between co-domain and range. That is, co-domain is given while range is obtained from the domain and function f.
Note that, graphically if we draw a line parallel to the y-axis, and if it cuts the graph at two distinct points, then the graph is not a function.

Note: R represents all real numbers

$R - \{a\} = (-\infty, a) \cup (a, \infty)$

Positive real numbers $= R^+ = (0, \infty)$

Non-negative numbers $= R^+ = [0, \infty)$

Negative real numbers $= R^- = (-\infty, 0)$

Non-positive real numbers $= R^- = (-\infty, 0]$

Domain of Some Special Functions

(i) For a polynomial function $y = f(x) = a_0 + a_1 x + a_2 x^2 + \ldots + a_n x^n$, the domain is all the real values of x.

(ii) For a reciprocal function $y = \dfrac{1}{f(x)}$, the domain is all the real values of x such that $f(x) \neq 0$.

(iii) For a rational function $y = \dfrac{f(x)}{g(x)}$, the domain is all the real values of x such that $g(x) \neq 0$.

(iv) For a square-root function $y = \sqrt{f(x)}$, the domain is all the real values of x such that $f(x) \geq 0$.

(v) For a cube-root function $y = \sqrt[3]{f(x)}$, the domain is all the real values of x.

(vi) For a reciprocal square-root function $y = \dfrac{1}{\sqrt{f(x)}}$, the domain is all the real values of x such that $f(x) > 0$.

(vii) For a reciprocal cube-root function $y = \dfrac{1}{\sqrt[3]{f(x)}}$, the domain is all the real values of x except 0, that is, $R - \{0\} = (-\infty, 0) \cup (0, \infty)$.

(viii) For a modulus function $y = |f(x)|$, the domain is all the real values of x.

(ix) For a signum function
$$y = \begin{cases} \dfrac{|f(x)|}{f(x)} \text{ or } \dfrac{f(x)}{|f(x)|}, & x \neq 0 \\ 0, & x = 0 \end{cases}$$, the domain is all the real values of x.

(x) For the greatest integer function $y = [f(x)]$, the domain is all the real values of x.

(xi) For an exponential function $y = f(x)^{g(x)}$, the domain is all the real values of x such that $f(x) > 0$ and $f(x) \neq 1$.

(xii) For a logarithmic function $y = \log_{g(x)} f(x)$, the domain is all the real values of x such that $f(x) > 0,\ g(x) > 0$ and $g(x) \neq 1$.

How to Find the Range of a Function

Process-I: Substitute the different values of x in the expression of $y = f(x)$, and find all the possible values of y (minimum y-value to maximum y-value, critical points, boundary points).

Process-II: First find the inverse of function $y = f(x)$; let it be $x = f^{-1}(y) = g(y)$ and write it as $y = g(x)$. Then, the domain of function $y = g(x)$ will be the range of function $y = f(x)$.

Note: For simple and special functions, we normally use Process-I while for critical functions we use Process-II.

Example 1: Find the domain and range of the real function $f(x) = 2x^2 + 3$

Solution:

Domain:

Since $f(x)$ is defined for all $x \in R$, the domain of f is R.

Range:

Since for all $x \in R,\ x^2 \geq 0$

$\Rightarrow 2x^2 \geq 0 \Rightarrow 2x^2 + 3 \geq 0 + 3 \Rightarrow f(x) \geq 3$

$\therefore$ Range of $f = [3, \infty)$

Alternatively:

Let $y = 2x^2 + 3 \Rightarrow x^2 = \dfrac{y-3}{2} \Rightarrow x = \sqrt{\dfrac{y-3}{2}}$

The square-root function $y = \sqrt{f(x)}$ is defined for all real values of x such that $f(x) \geq 0$.

Therefore, function $x = \sqrt{\dfrac{y-3}{2}}$ gives a real value of x for all real values of y such that $f(y) \geq 0$.

$\dfrac{y-3}{2} \geq 0 \Rightarrow y - 3 \geq 0 \Rightarrow y \geq 3$

Thus, range of $f = (3, \infty)$

Example 2: Find the domain and range of the real function $f(x) = x^2 + 2x - 3$.

Solution:

Domain:

Since $f(x)$ is defined for all $x \in R$, the domain of f is R.

Range:

Rewrite $f(x) = x^2 + 2x - 3 = x^2 + 2x + 1 - 1 - 3$

$\Rightarrow f(x) = (x+1)^2 - 4$

Since for all $x \in R,\ (x+1)^2 \geq 0$

$\Rightarrow (x+1)^2 - 4 \geq 0 - 4 \Rightarrow f(x) \geq -4$

$\therefore$ Range of $f = (-4, \infty)$

Alternatively:

Let $y = x^2 + 2x - 3 \Rightarrow x^2 + 2x - 3 - y = 0$

Since the quadratic equation $ax^2 + bx + c = 0$ gives real roots, $x \in R$, if $b^2 - 4ac \geq 0$

$\therefore 2^2 - 4 \cdot 1 \cdot (-3 - y) \geq 0$

$\Rightarrow 4 + 12 + 4y \geq 0 \Rightarrow 4y \geq -16 \Rightarrow y \geq -4$

Thus, range of $f = [-4, \infty)$

Example 3: Find the domain and range of the real function $f(x) = \dfrac{1}{3-2x}$.

Solution:

Domain:

Reciprocal function $y = \dfrac{1}{f(x)}$ is defined for all real values of x such that $f(x) \neq 0$.

$\Rightarrow 3 - 2x \neq 0 \Rightarrow x \neq 3/2$

Thus, the domain of f is

$= R - \{3/2\} = (-\infty, 3/2) \cup (3/2, \infty)$

Range:

Let $y = \dfrac{1}{3-2x} \Rightarrow 3 - 2x = \dfrac{1}{y}$

$\Rightarrow 2x = 3 - \dfrac{1}{y} \Rightarrow 2x = \dfrac{3y-1}{y} \Rightarrow x = \dfrac{3y-1}{2y}$

Rational function $x = \dfrac{f(y)}{g(y)}$ is defined for all real values of y such that $g(y) \neq 0 \Rightarrow 2y \neq 0 \Rightarrow y \neq 0$.

Thus, the range of f is $= R - \{0\} = (-\infty, 0) \cup (0, \infty)$.

Example 4: Find the domain and range of the real function $f(x) = \dfrac{3x-2}{1+4x}$.

Solution:

Domain:

Rational function $y = \dfrac{f(x)}{g(x)}$ is defined for all real values of x such that $g(x) \neq 0$

$\Rightarrow 1 + 4x \neq 0 \Rightarrow x \neq -1/4$

Thus, the domain of f is

$= R - \{-1/4\} = (-\infty, -1/4) \cup (-1/4, \infty)$

Range:

Let $y = \dfrac{3x-2}{1+4x} \Rightarrow y + 4xy = 3x - 2$

$\Rightarrow y + 4xy = 3x - 2 \Rightarrow 3x - 4xy = y + 2$

$\Rightarrow x(3 - 4y) = y + 2 \Rightarrow x = \dfrac{y+2}{3-4y}$

Rational function $x = \dfrac{f(y)}{g(y)}$ is defined for all real values of y such that $g(y) \neq 0$

$\Rightarrow 3 - 4y \neq 0 \Rightarrow y \neq 3/4$

Thus, the range of *f* is
$= R - \{3/4\} = (-\infty, 3/4) \cup (3/4, \infty)$

Example 5: Find the domain and the range of the real function *f* defined by $f(x) = \sqrt{(x+1)}$.

Solution:
The given function $f(x) = \sqrt{(x+1)}$ is defined when $x + 1 \geq 0 \Rightarrow x \geq -1$.
Therefore, the domain of *f* is the set of all real numbers greater than or equal to –1, that is, the domain of $f = [-1, \infty)$.
Now, for $x \geq -1$
$\Rightarrow (x+1) \geq -1 + 1 \Rightarrow (x+1) \geq 0 \Rightarrow \sqrt{(x+1)} \geq 0$
Therefore, the range of *f* is the set of all real numbers greater than or equal to 0, that is, the range of $f = [0, \infty)$.

Example 6: Find the domain and the range of the real function *f* defined by $f(x) = \sqrt{1 - x^2}$.

Solution:
Since $f(x) = \sqrt{1 - x^2}$ is defined when $1 - x^2 \geq 0$ $\Rightarrow x^2 \leq 1 \Rightarrow -1 \leq x \leq 1$, the domain of $f(x)$ is $-1 \leq x \leq 1$ or [–1, 1].
For any value of *x* such that $-1 \leq x \leq 1$, the value of $f(x)$ will lie between 0 and 1.
$\therefore$ The range of $f(x)$ is $\{x : 0 \leq x \leq 1\}$ or [0,1].

Example 7: Find the domain and range of the function $f(x) = \dfrac{1}{\sqrt{x^2 - 1}}$.

Solution:
Since $f(x)$ is defined when $x^2 - 1 > 0$
$\Rightarrow (x-1)(x+1) > 0 \Rightarrow x < -1$ or $x > 1$
$\therefore$ Domain of *f* is $= (-\infty, -1) \cup (1, \infty)$
Now, in the domain of *x*,
$$x^2 - 1 > 0 \Rightarrow \sqrt{x^2 - 1} > 0 \Rightarrow 0 < \frac{1}{\sqrt{x^2 - 1}} < \infty$$
$\Rightarrow 0 < f(x) < \infty$, and therefore, the range of *f* is the set of all real numbers greater than or equal to 0, that is, the range of $f = (0, \infty) = R^+$.
Alternatively:
Let $f(x) = y = \dfrac{1}{\sqrt{x^2 - 1}}$
$$\Rightarrow \sqrt{x^2 - 1} = \frac{1}{y} \Rightarrow x^2 - 1 = \frac{1}{y^2}$$
$$\Rightarrow x^2 = \frac{1}{y^2} + 1$$
Since for all $x \in R$, $x^2 \geq 0$
$$\Rightarrow \frac{1}{y^2} + 1 \geq 0 \Rightarrow \frac{1}{y^2} \geq -1$$
Since $y^2 \geq 0 \therefore \Rightarrow \dfrac{1}{y^2} \geq 0$ and $y \neq 0$
Thus, the range of $f = (0, \infty) = R^+$.

Example 8: Find the domain and range of the real function $f(x) = \dfrac{1}{x^2 + x - 2}$.

Solution:
Factorise $f(x) = \dfrac{1}{x^2 + x - 2}$ as $f(x) = \dfrac{1}{(x-1)(x+2)}$.
Domain:
Reciprocal function $y = \dfrac{1}{f(x)}$ is defined for all real values of *x* such that $f(x) \neq 0$.
$\Rightarrow (x-1)(x+2) \neq 0 \Rightarrow x \neq 1, -2$
Thus, the domain of *f* is,
$= R - \{-2, 1\} = (-\infty, -2) \cup (-2, 1) \cup (1, \infty)$
Range:
$$\text{Let } y = \frac{1}{x^2 + x - 2} \Rightarrow x^2 + x - 2 - \frac{1}{y} = 0$$
$$\Rightarrow x^2 + x + \frac{1}{4} - \frac{1}{4} - 2 - \frac{1}{y} = 0$$
$$\Rightarrow \left(x + \frac{1}{2}\right)^2 - \frac{9}{4} - \frac{1}{y} = 0$$
$$\Rightarrow \left(x + \frac{1}{2}\right)^2 = \frac{9}{4} + \frac{1}{y}$$
Since for all $x \in R$, $\left(x + \dfrac{1}{2}\right)^2 \geq 0$
$$\Rightarrow \frac{9}{4} + \frac{1}{y} \geq 0 \Rightarrow \frac{9y + 4}{4y} \geq 0$$
$$\Rightarrow y \leq -\frac{4}{9} \text{ or } y \geq 0 \text{ and } y \neq 0$$
$$\Rightarrow y \leq -\frac{4}{9} \text{ or } y > 0$$
Therefore, the range of *f* is,
$= (-\infty, -4/9] \cup (0, \infty)$
Alternatively:
$$\text{Let } y = \frac{1}{x^2 + x - 2} \Rightarrow x^2 + x - 2 - \frac{1}{y} = 0$$
The quadratic equation $ax^2 + bx + c = 0$ gives real roots, $x \in R$, if $b^2 - 4ac \geq 0$
$\therefore 1^2 - 4 \cdot 1 \cdot (-2 - 1/y) \geq 0$
$\Rightarrow 1 + 8 + 4/y \geq 0 \Rightarrow 9 + 4/y \geq 0$
$$\Rightarrow \frac{9y + 4}{y} \geq 0 \Rightarrow y \leq -\frac{4}{9} \text{ or } y \geq 0 \text{ and } y \neq 0$$
$\Rightarrow y > 0$ or $y \leq -4/9$
Thus, the range of $f = (-\infty, -4/9] \cup (0, \infty)$

Day 28

Domain and Range of Real Functions-II

Example 9: Find the domain and range of the real function $f(x)=\dfrac{x^2+x-2}{x^2+x}$.

Solution:

Factorise $f(x)=\dfrac{x^2+x-2}{x^2+x}$ as $f(x)=\dfrac{(x-1)(x+2)}{x(x+1)}$.

Domain:

Rational function $y=\dfrac{f(x)}{g(x)}$ is defined for all real values of x such that $g(x)\neq 0$

$\Rightarrow x(x+1)\neq 0 \Rightarrow x\neq 0,-1$

Thus, the domain of f is

$=R-\{-1,0\}=(-\infty,-1)\cup(-1,0)\cup(0,\infty)$

Range:

Let $y=\dfrac{x^2+x-2}{x^2+x}\Rightarrow yx^2+yx=x^2+x-2$

$\Rightarrow (y-1)x^2+(y-1)x+2=0$

$\Rightarrow x^2+x+\dfrac{2}{y-1}=0$

$\Rightarrow x^2+x+\dfrac{1}{4}-\dfrac{1}{4}+\dfrac{2}{y-1}=0$

$\Rightarrow \left(x+\dfrac{1}{2}\right)^2+\dfrac{-y+1+8}{4(y-1)}=0$

$\Rightarrow \left(x+\dfrac{1}{2}\right)^2=\dfrac{y-9}{4(y-1)}$

Since for all $x\in R$, $\left(x+\dfrac{1}{2}\right)^2\geq 0$

$\Rightarrow \dfrac{y-9}{4(y-1)}\geq 0 \Rightarrow y\leq 1$ or $y\geq 9$ and $y\neq 1$

$\Rightarrow y<1$ or $y\geq 9$

Therefore, the range of f is,

$=(-\infty,1)\cup[9,\infty)$

Alternatively:

Let $y=\dfrac{x^2+x-2}{x^2+x}\Rightarrow yx^2+yx=x^2+x-2$

$\Rightarrow (y-1)x^2+(y-1)x+2=0$

$\Rightarrow x^2+x+\dfrac{2}{y-1}=0$

The quadratic equation $ax^2+bx+c=0$ gives real roots, $x\in R$, if $b^2-4ac\geq 0$

$\therefore 1^2-4\cdot 1\cdot\left(\dfrac{2}{y-1}\right)\geq 0$

$\Rightarrow \dfrac{y-1-8}{y-1}\geq 0$

$\Rightarrow \dfrac{y-9}{y-1}\geq 0$

$\Rightarrow y\leq 1$ or $y\geq 9$ and $y\neq 1$

$\Rightarrow y<1$ or $y\geq 9$

Thus, the range of $f=(-\infty,1)\cup[9,\infty)$

Example 10: Find the domain and range of the real function $f(x)=|x-1|-2$.

Solution:

Since $f(x)$ is defined for all $x\in R$, the domain of f is R.

$|x-1|$ is always positive real numbers,

$\therefore |x-1|\geq 0\Rightarrow |x-1|-2\geq 0-2\Rightarrow f(x)\geq -2$

Thus, the range of f is $[-2,\infty)$.

Example 11: Find the domain and range of the real function $f(x)=\dfrac{1}{\sqrt{|x-2|-1}}$.

Solution:

$$\because f(x)=\frac{1}{\sqrt{|x-2|-1}}=\begin{cases}\dfrac{1}{\sqrt{x-2-1}}=0, & x\geq 2\\[2mm] \dfrac{1}{\sqrt{-(x-2)-1}}, & x<2\end{cases}$$

$$\Rightarrow f(x)=\begin{cases}\dfrac{1}{\sqrt{x-3}}, & x\geq 2\\[2mm] \dfrac{1}{\sqrt{1-x}}, & x<2\end{cases}$$

The reciprocal function $y=\dfrac{1}{f(x)}$ is defined for all real values of x such that $f(x)\neq 0$.

$\therefore \sqrt{x-3}\neq 0$ or $\sqrt{1-x}\neq 0$

$\Rightarrow x\neq 3$ or $x\neq 1$

Thus, the domain of f is

$R-\{1,3\}=(-\infty,1)\cup(1,3)\cup(3,\infty)$

$|x-2|$ is always a positive real number

$\therefore |x-2|\geq 0\Rightarrow |x-2|-1\geq 0-1$

$\Rightarrow \sqrt{|x-2|-1}\geq 0\Rightarrow \dfrac{1}{\sqrt{|x-2|-1}}<\infty$

Thus, the range of f is $(0,\infty)$.

Example 12: Find the domain and range of the real function $f(x)=\sqrt{1-[x]}$ where [.] denotes the greatest integer function.

Solution:

Since the given function $f(x)=\sqrt{1-[x]}$ is defined when $1-[x]\ge 0 \Rightarrow [x]\le 1 \Rightarrow x<2$, the domain of f is $(-\infty,2)$.

Now, since for $1-[x]\ge 0 \Rightarrow \sqrt{1-[x]}\ge 0$, the range of f is the set of all real numbers greater than or equal to 0, that is, the range of $f=[0,\infty)$.

Example 13: Find the domain and range of the real function $f(x)=\dfrac{1}{[x]-2}$ where [.] denotes the greatest integer function.

Solution:

Domain:

Reciprocal function $y=\dfrac{1}{f(x)}$ is defined for all real values of x such that $f(x)\ne 0$

$\Rightarrow [x]-2\ne 0 \Rightarrow [x]\ne 2 \Rightarrow x\ne [2,3)$

Thus, the domain of f is,

$=R-[2,3)=(-\infty,2)\cup[3,\infty)$

Range:

Let $y=\dfrac{1}{[x]-2}\Rightarrow [x]-2=\dfrac{1}{y}\Rightarrow [x]=\dfrac{1}{y}+2$

Therefore, function $[x]=\dfrac{1}{y}+2$ gives the real value of x for all real values of y except at $y=0$.

Thus, the range of $f=R-\{0\}$ or $(-\infty,0)\cup(0,\infty)$.

Example 14: Find the domain and range of the real function $[f(x)]=\dfrac{1}{[x]^2+[x]-2}$, where [.] denotes the greatest integer function.

Solution:

Let $[f(x)]=[y]=\dfrac{1}{[x]^2+[x]-2}$

$\Rightarrow [y]=\dfrac{1}{([x]-1)([x]+2)}$

Domain:

Reciprocal function $y=\dfrac{1}{f(x)}$ is defined for all real values of x such that $f(x)\ne 0$.

$\Rightarrow ([x]-1)([x]+2)\ne 0$

$\Rightarrow [x]\ne 1,\ -2$

$\Rightarrow x\ne [1,2),\ [-2,-1)$

Thus, the domain of f is

$=R-[-2,-1)\cup[1,2)$

$=(-\infty,-2)\cup[-1,1)\cup[2,\infty)$

Range:

Let $[y]=\dfrac{1}{[x]^2+[x]-2}$

$\Rightarrow [x]^2+[x]-2-\dfrac{1}{[y]}=0$

$\Rightarrow [x]^2+[x]+\dfrac{1}{4}-\dfrac{1}{4}-2-\dfrac{1}{[y]}=0$

$\Rightarrow \left([x]+\dfrac{1}{2}\right)^2-\dfrac{9}{4}-\dfrac{1}{[y]}=0$

$\Rightarrow \left([x]+\dfrac{1}{2}\right)^2=\dfrac{9}{4}+\dfrac{1}{[y]}$

Since for all $x\in R$, $\left([x]+\dfrac{1}{2}\right)^2\ge 0$

$\Rightarrow \dfrac{9}{4}+\dfrac{1}{[y]}\ge 0 \Rightarrow \dfrac{9[y]+4}{4[y]}\ge 0$

$\Rightarrow [y]\le -\dfrac{4}{9}$ or $[y]\ge 0$ and $[y]\ne 0$

$\Rightarrow [y]>0$ or $[y]\le -4/9$

$\Rightarrow y\ge 1$ or $y<0$

Thus, the range of $f=(-\infty,0)\cup[1,\infty)$

Example 15: Find the domain and range of the real function $f(x)=\dfrac{1}{\sqrt{1-\sin x}}$.

Solution:

Domain:

Since $-1\le \sin x\le 1$, therefore $1-\sin x\ge 0$, $f(x)=\dfrac{1}{\sqrt{1-\sin x}}$ is defined for all values of x such that $1-\sin x\ne 0$

$\Rightarrow \sin x\ne 1$

$\Rightarrow x\ne 2n\pi+\dfrac{\pi}{2},\ \forall\ n\in Z$

Thus, the domain of $f=R-\{2n\pi+\dfrac{\pi}{2}: n\in Z\}$

Range:

$\because -1\le \sin x\le 1 \Rightarrow 1\ge -\sin x\ge -1$

$\Rightarrow 1+1\ge 1-\sin x\ge 1-1$

$\Rightarrow 2\ge 1-\sin x\ge 0$

$\Rightarrow \sqrt{2}\ge \sqrt{1-\sin x}\ge 0$

$\Rightarrow \dfrac{1}{\sqrt{1-\sin x}}\ge \dfrac{1}{\sqrt{2}}$ or $\dfrac{1}{0}\ge \dfrac{1}{\sqrt{1-\sin x}}$

$\Rightarrow f(x)\ge \dfrac{1}{\sqrt{2}}$ or $\infty\ge f(x)$

$\Rightarrow \infty\ge f(x)\ge \dfrac{1}{\sqrt{2}}$

Thus, the range of $f=\left[\dfrac{1}{\sqrt{2}},\infty\right)$

Alternatively:

Let $y = \dfrac{1}{\sqrt{1-\sin x}}$

$\Rightarrow \sqrt{1-\sin x} = \dfrac{1}{y}$

$\Rightarrow 1-\sin x = \dfrac{1}{y^2} \Rightarrow \sin x = 1 - \dfrac{1}{y^2}$

$\Rightarrow \sin x = \dfrac{y^2-1}{y^2}$

Now, since $-1 \le \sin x \le 1$

Therefore, $-1 \le \dfrac{y^2-1}{y^2} \le 1$

$\Rightarrow -1 \le \dfrac{y^2-1}{y^2} \le 1$

$\Rightarrow -y^2 \le y^2 - 1 \le y^2$

$\Rightarrow 1 \le 2y^2 \Rightarrow y^2 \ge \dfrac{1}{2}$

$\Rightarrow y \ge \dfrac{1}{\sqrt{2}}$ or $y \le -\dfrac{1}{\sqrt{2}}$

But, since $y = \dfrac{1}{\sqrt{1-\sin x}}$ will always be non-negative, $y \ge \dfrac{1}{\sqrt{2}}$.

Thus, the range of $f = \left[\dfrac{1}{\sqrt{2}}, \infty\right)$

Example 16: Find the domain and range of the real function $f(x) = \dfrac{1}{1+\cos 2x}$.

Solution:

Domain:

Since $-1 \le \cos 2x \le 1$, $f(x) = \dfrac{1}{1+\cos 2x}$ is defined for all values of x such that $1+\cos 2x \ne 0 \Rightarrow \cos 2x \ne -1$

$\Rightarrow 2x \ne 2n\pi \pm \pi, \ \forall\ n \in Z$

$\Rightarrow x \ne n\pi \pm \dfrac{\pi}{2}, \ \forall\ n \in Z$

Thus, the domain of $f = R - \{n\pi \pm \dfrac{\pi}{2} : n \in Z\}$

Range:

$\because -1 \le \cos 2x \le 1$

$\Rightarrow 1-1 \le 1+\cos 2x \le 1+1$

$\Rightarrow 0 \le 1+\cos 2x \le 2$

$\Rightarrow \dfrac{1}{1+\cos 2x} \le \dfrac{1}{0}$ or $\dfrac{1}{1+\cos 2x} \ge \dfrac{1}{2}$

$\Rightarrow f(x) \le \infty$ or $f(x) \ge \dfrac{1}{2}$

$\Rightarrow \infty \ge f(x) \ge \dfrac{1}{2}$

Thus, the range of $f = \left[\dfrac{1}{2}, \infty\right)$

Alternatively:

Let $y = \dfrac{1}{1+\cos 2x} \Rightarrow 1+\cos 2x = \dfrac{1}{y}$

$\Rightarrow \cos 2x = \dfrac{1}{y} - 1 = \dfrac{1-y}{y}$

Now, since $-1 \le \cos 2x \le 1$

Therefore, $-1 \le \dfrac{1-y}{y} \le 1$

$\Rightarrow 0 \le \dfrac{1-y}{y} + 1$ or $\dfrac{1-y}{y} - 1 \le 0$

$\Rightarrow 0 \le \dfrac{1-y+y}{y}$ or $\dfrac{1-y-y}{y} \le 0$

$\Rightarrow 0 \le \dfrac{1}{y}$ or $\dfrac{1-2y}{y} \le 0$

$\because -1 \le \cos 2x \le 1$, $y = \dfrac{1}{1+\cos 2x}$ will always be non-negative, thus $y \ge \dfrac{1}{2}$.

Thus, the range of $f = \left[\dfrac{1}{2}, \infty\right)$

Day 29

Algebra of Real Functions

Algebra of real functions means the addition of two real functions, or subtraction of a real function from another real function, or multiplication of a real function by a scalar constant, or multiplication of two real functions, or division of one real function by another.

(i) Addition of two real functions: Let $f : X \to R, g : X \to R$ be any two real functions, where $X \subset R$. Then, addition of functions f and g will be defined as $(f+g) : X \to R$ given by $(f+g)(x) = f(x) + g(x)$, for all $x \in X$.

(ii) Subtraction of a real function from another: Let $f : X \to R, g : X \to R$ be any two real functions, where $X \subset R$. Then, subtraction of a function g from a function f will be defined as $(f-g) : X \to R$ given by $(f-g)(x) = f(x) - g(x)$, for all $x \in X$.

(iii) Multiplication by a scalar: Let $f: X \rightarrow R$ be a real function, α be a scalar (real number). Then, multiplication of function f by a scalar α will be defined as $\alpha f: X \rightarrow R$ given by $(\alpha f)(x) = \alpha f(x)$, for all $x \in X$.

(iv) Multiplication of two real functions: Let $f: X \rightarrow R, g: X \rightarrow R$ be any two real functions, where $X \subset R$. Then, multiplication of functions f and g will be defined as $(f \cdot g): X \rightarrow R$ given by $(f \cdot g)(x) = f(x) \cdot g(x)$, for all $x \in X$.

(v) Quotient of two functions: Let $f: X \rightarrow R, g: X \rightarrow R$ be any two real functions, where $X \subset R$. Then, the quotient of functions f and g, that is, division of a function f by a function g will be defined as $\left(\frac{f}{g}\right): X \rightarrow R$ given by $\left(\frac{f}{g}\right)(x) = \frac{f(x)}{g(x)}$, for all $x \in X$, provided $g(x) \neq 0$.

Example 1: If a function f is defined by $f(x) = 2x - 3$, then find $f(-1)$.

Solution:

Since $f(x) = 2x - 3$,

$f(-1) = 2 \times (-1) - 3 = -2 - 3 = -5$

Example 2: If a function f is defined by $f(x) = x^2 + 2x - 1$, then find $f(3)$.

Solution:

Since $f(x) = x^2 + 2x - 1$,

$f(3) = 3^2 + 2 \times 3 - 1 = 9 + 6 - 1 = 14$

Example 3: If a function f is defined by $f(x) = \frac{x+2}{x^2 - 1}$, then find $f(0)$.

Solution:

Since $f(x) = \frac{x+2}{x^2-1}$, $f(0) = \frac{0+2}{0^2-1} = -2$

Example 4: If a function f is defined by $f(x) = \sqrt{\frac{1}{x^2+x}}$ then find $f(-2)$.

Solution:

Since $f(x) = \sqrt{\frac{1}{x^2+x}}$,

$f(-2) = \sqrt{\frac{1}{(-2)^2 + (-2)}} = \frac{1}{\sqrt{2}}$

Example 5: If $f(x) = x^2 - x + 1$, find $\frac{f(2) - f(1)}{(2-1)}$.

Solution:

$\because f(x) = x^2 - x + 1$

$\therefore \frac{f(2) - f(1)}{(2-1)} = \frac{(2^2 - 2 + 1) - (1^2 - 1 + 1)}{(2-1)} = \frac{3-1}{1} = 2$

Example 6: The function t which maps temperature in degree celsius from degree fahrenheit is defined by $t(F) = \frac{5}{9}(F - 32)$. Find (i) t (14) and (ii) value of F, when $t(F) = 25$.

Solution:

Since $t(F) = \frac{5}{9}(F - 32)$,

(i) $t(14) = \frac{5}{9}(14 - 32) = -10°\text{C}$

(ii) $t(F) = \frac{5}{9}(F - 32) = 25°\text{C}$

$$\Rightarrow F - 32 = \frac{25 \times 9}{5} \Rightarrow F - 32 = 45 \Rightarrow F = 77$$

Example 7: The relation f is defined by $f(x) = \begin{cases} x^2 + 1, & 0 \leq x \leq 2 \\ 3x - 1, & 2 \leq x \leq 5 \end{cases}$

The relation g is defined by $g(x) = \begin{cases} x^2 + 1, & 0 \leq x \leq 3 \\ 3x - 1, & 3 \leq x \leq 5 \end{cases}$

Show that f is a function and g is not a function.

Solution:

The relation f is defined as $f(x) = \begin{cases} x^2 + 1, & 0 \leq x \leq 2 \\ 3x - 1, & 2 \leq x \leq 5 \end{cases}$

Since, at $x = 2$,

$f(x) = x^2 + 1 \Rightarrow f(2) = 2^2 + 1 = 5$

or $f(x) = 3x - 1 \Rightarrow f(2) = 3 \times 2 - 1 = 5$

At $x = 2$, $f(x) = 5$, and also for $0 \leq x \leq 5$, the images of $f(x)$ are unique.

Therefore, f is a function.

The relation g is defined as $g(x) = \begin{cases} x^2 + 1, & 0 \leq x \leq 3 \\ 3x - 1, & 3 \leq x \leq 5 \end{cases}$

Since, at $x = 3$,

$g(x) = x^2 + 1 \Rightarrow g(3) = 3^2 + 1 = 10$

or $g(x) = 3x - 1 \Rightarrow g(3) = 3 \times 3 - 1 = 8$

For $x = 3$, we get two values of $g(x)$, that is, the element 3 of the domain of the relation g corresponds to two different images (10 and 8). Therefore, g is not a function.

Example 8: The relation f is defined by $f(x) = \begin{cases} |x|, & -2 \leq x \leq -1 \\ x, & -1 \leq x \leq 2 \end{cases}$

The relation g is defined by $g(x) = \begin{cases} |x|, & -2 \leq x \leq 0 \\ x, & 0 \leq x \leq 2 \end{cases}$

The relation h is defined by $h(x)=\begin{cases}|x|, & -2\le x\le 1\\ x, & 1\le x\le 2\end{cases}$

Show that f is not a function but g and h are functions.

Solution:

The relation f is defined as $f(x)=\begin{cases}|x|, & -2\le x\le -1\\ x, & -1\le x\le 2\end{cases}$

Since, at $x=-1$,
$f(x)=|x|\Rightarrow f(-1)=|-1|=1$
or $f(x)=x\Rightarrow f(-1)=-1$
Since, for $x=-1$ we get two values of $f(x)$, that is, the element -1 of the domain of the relation f corresponds to two different images (1 and -1), f is not a function.

The relation g is defined as $g(x)=\begin{cases}|x|, & -2\le x\le 0\\ x, & 0\le x\le 2\end{cases}$

Since, at $x=0$,
$g(x)=|x|\Rightarrow g(0)=|0|=0$
or $g(x)=x\Rightarrow g(0)=0$
Thus, at $x=0$, $g(x)=0$, and also for $-2\le x\le 2$, the images of $g(x)$ are unique.
Therefore, g is a function.

The relation h is defined as $h(x)=\begin{cases}|x|, & -2\le x\le 1\\ x, & 1\le x\le 2\end{cases}$

$h(x)=|x|\Rightarrow h(1)=|1|=1$
or $h(x)=x\Rightarrow h(1)=1$
Thus, at $x=1$, $h(x)=1$, and also for $-2\le x\le 2$, the images of $h(x)$ are unique.
Therefore, h is a function.

Example 9: Let $f,g:R\to R$ be defined respectively by $f(x)=x+1,\quad g(x)=x^2-1$. Find $f+g,\ g-f,\ f\cdot g,\ \frac{g}{f},\ \frac{f}{g}$.

Solution:

$\because f(x)=x+1,\quad g(x)=x^2-1.$

$(f+g)(x)=f(x)+g(x)=(x+1)+(x^2-1)$

$\Rightarrow (f+g)(x)=x^2+x$

$(g-f)(x)=g(x)-f(x)=(x^2-1)-(x+1)$

$\Rightarrow (f+g)(x)=x^2-x-2$

$(f\cdot g)(x)=f(x)\cdot g(x)=(x+1)\cdot(x^2-1)$

$\Rightarrow (f\cdot g)(x)=x^3+x^2-x-1$

$$\left(\frac{g}{f}\right)(x)=\frac{g(x)}{f(x)}=\frac{(x^2-1)}{(x+1)}=\frac{(x-1)(x+1)}{(x+1)}=x-1$$

$$\left(\frac{f}{g}\right)(x)=\frac{f(x)}{g(x)}=\frac{(x+1)}{(x^2-1)}=\frac{1}{x-1},\quad x\ne 1$$

Example 10: If $f(x)=x-\frac{1}{x}$, then find

(i) $f(-x)$ (ii) $f\left(\frac{1}{x}\right)$ (iii) $f\left(-\frac{1}{x}\right)$

Solution:

(i) $f(-x)=(-x)-\frac{1}{(-x)}=-\left(x-\frac{1}{x}\right)=-f(x)$

(ii) $f\left(\frac{1}{x}\right)=\frac{1}{x}-\frac{1}{1/x}=\frac{1}{x}-x=-\left(x-\frac{1}{x}\right)=-f(x)$

(iii) $f\left(-\frac{1}{x}\right)=-\frac{1}{x}-\frac{1}{-1/x}=-\frac{1}{x}+x=x-\frac{1}{x}=f(x)$

Example 11: Let f and g be real functions defined by $f(x)=x^2,\ g(x)=x$.

(i) For what real numbers x, $f(x)=g(x)$?

(ii) For what real number x, $f(x)<g(x)$?

Solution:

(i) $f(x)=g(x)\Rightarrow x^2=x\Rightarrow x^2-x=0$
$\Rightarrow x(x-1)=0\Rightarrow x=0,1$

(ii) $f(x)=g(x)\Rightarrow x^2<x\Rightarrow x^2-x<0$
$\Rightarrow x(1-x)<0\Rightarrow 0<x<1$

Example 12: Redefine the function
$f(x)=|x-1|+|1-x|+|x+1|,\ -2\le x\le 2$

Solution:

$\because f(x)=|x-1|+|1-x|+|x+1|,\ -2\le x\le 2$

$$\therefore f(x)=\begin{cases}-(x-1)+(1-x)-(x+1), & -2\le x<-1\\ -(x-1)+(1-x)+(x+1), & -1\le x<1\\ (x-1)-(1-x)+(x+1), & 1\le x\le 2\end{cases}$$

$$\Rightarrow f(x)=\begin{cases}-x+1+1-x-x-1, & -2\le x<-1\\ -x+1+1-x+x+1, & -1\le x<1\\ x-1-1+x+x+1, & 1\le x\le 2\end{cases}$$

$$\Rightarrow f(x)=\begin{cases}-3x+1, & -2\le x<-1\\ -x+3, & -1\le x<1\\ 3x-1, & 1\le x\le 2\end{cases}$$

Example 13: The relation f is defined by

$$f(x)=\begin{cases}ax-1, & -2\le x\le -1\\ x^2, & -1\le x\le 1\\ b-x, & 1\le x\le 2\end{cases}$$

If relation f is a function then find the value of a and b.

Solution:

The relation f is defined as $f(x)=\begin{cases}ax-1, & -2\le x\le -1\\ x^2, & -1\le x\le 1\\ b-x, & 1\le x\le 2\end{cases}$

The given relation f will be a function if it gives a unique value for all x in the given domain.

Thus, at $x = -1$,

$f(x) = ax - 1 \Rightarrow f(-1) = -a - 1$

or $f(x) = x^2 \Rightarrow f(-1) = (-1)^2 = 1$

For unique image at $x = -1$,

$-a - 1 = 1 \Rightarrow -a = 2 \Rightarrow a = -2$

Now, at $x = 1$,

$f(x) = x^2 \Rightarrow f(1) = (1)^2 = 1$

or $f(x) = b - x \Rightarrow f(1) = b - 1$

For unique image at $x = 1$, $b - 1 = 1 \Rightarrow b = 2$

Therefore, for $a = -2$ and $b = 2$ the relation f is a function.

Example 14: Let $f(x) = x + \frac{1}{x}$, $x \neq 0$ then justify the following statements.

(i) $f(x)\cdot f(y) \neq f(xy)$, for all $x, y \in R$

(ii) $f(x)\cdot f(y) > f(xy)$, if both $x, y > 0$ or $x, y < 0$

(iii) $f(x)\cdot f(y) < f(xy)$, if $x > 0$ and $y < 0$ or $x < 0$ and $y > 0$

Solution:

$\because f(x) = x + \frac{1}{x} \Rightarrow f(xy) = xy + \frac{1}{xy}$

$\therefore f(x)\cdot f(y) = \left(x + \frac{1}{x}\right)\cdot\left(y + \frac{1}{y}\right)$

$= xy + \frac{1}{xy} + \frac{x}{y} + \frac{y}{x} = xy + \frac{1}{xy} + \frac{x^2 + y^2}{xy}$

$\Rightarrow f(x)\cdot f(y) = f(xy) + \frac{x^2 + y^2}{xy}$

(i) $f(x)\cdot f(y) \neq f(xy)$, since $\frac{x^2 + y^2}{xy} \neq 0; x, y \in R$

(ii) When both $x, y > 0$ or $x, y < 0$, then

$\frac{x^2 + y^2}{xy} > 0 \Rightarrow f(x)\cdot f(y) > f(xy)$

(iii) When $x > 0$ and $y < 0$ or $x < 0$ and $y > 0$, then

$\frac{x^2 + y^2}{xy} < 0 \Rightarrow f(x)\cdot f(y) < f(xy)$

Day 30

NCERT Exercise 2.3 (With Solutions)

Q1. Which of the following relations are functions? Give reasons. If it is a function, determine its domain and range.

(i) $\{(2,1),(5,1),(8,1),(11,1),(14,1),(17,1)\}$

(ii) $\{(2,1),(4,2),(6,3),(8,4),(10,5),(12,6),(14,7)\}$

(iii) $\{(1,3),(1,5),(2,5)\}$

Solution:

(i) $\{(2,1),(5,1),(8,1),(11,1),(14,1),(17,1)\}$

Since 2, 5, 8, 11, 14, and 17 are the elements of the domain of the given relation having their unique images, this relation is a function.

Thus, domain = $\{2,5,8,11,14,17\}$ and range = $\{1\}$

(ii) $\{(2,1),(4,2),(6,3),(8,4),(10,5),(12,6),(14,7)\}$

Since every element 2, 4, 6, 8, 10, 12 and 14 has one and only one image, this relation is a function.

Thus, domain = {2, 4, 6, 8, 10, 12} and range = {1, 2, 3, 4, 5, 6, 7}

(iii) $\{(1,3),(1,5),(2,5)\}$

Since the same first element, that is, 1, corresponds to two different images, 3 and 5, this relation is not a function.

Q2. Find the domain and range of the following real function:

(i) $f(x) = -|x|$ (ii) $f(x) = \sqrt{9 - x^2}$

Solution:

(i) $f(x) = -|x|, x \in R$

$\because |x| = \begin{cases} x, & \text{if } x \geq 0 \\ -x, & \text{if } x < 0 \end{cases}$

$\therefore f(x) = -|x| = \begin{cases} -x, & \text{if } x \geq 0 \\ x, & \text{if } x < 0 \end{cases}$

Since $f(x)$ is defined for all $x \in R$, the domain of f is R.

Since $|x|$ is always a positive real number, $f(x) = -|x|$ will be a non-positive real number.

$\therefore$ The range of f is $(-\infty, 0]$.

(ii) $f(x) = \sqrt{9 - x^2}$

Since $f(x) = \sqrt{9 - x^2}$ is defined when $9 - x^2 \geq 0$ $\Rightarrow x^2 \leq 9 \Rightarrow -3 \leq x \leq 3$, the domain of $f(x)$ is $-3 \leq x \leq 3$ or $[-3, 3]$.

For any value of x such that $-3 \leq x \leq 3$, the value of $f(x)$ will lie between 0 and 3.

$\therefore$ The range of $f(x)$ is $\{x : 0 \leq x \leq 3\}$ or $[0, 3]$.

Q3. A function f is defined by $f(x) = 2x - 5$. Find
(i) $f(0)$ (ii) $f(7)$ (iii) $f(-3)$

Solution:
Since $f(x) = 2x - 5$,
(i) $f(0) = 2\times 0 - 5 = 0 - 5 = -5$
(ii) $f(7) = 2\times 7 - 5 = 14 - 5 = 9$
(iii) $f(-3) = 2\times(-3) - 5 = -6 - 5 = -11$

Q4. The function t which maps temperature in degree celsius into temperature in degree fahrenheit is defined by $t(C) = \frac{9C}{5} + 32$, find
(i) $t(0)$ (ii) $t(28)$ (iii) $t(-10)$
(iv) the value of C, when $t(C) = 212$

Solution:
Since $t(C) = \frac{9C}{5} + 32$,
(i) $t(0) = \frac{9\times 0}{5} + 32 = 0 + 32 = 32$
(ii) $t(28) = \frac{9\times 28}{5} + 32 = \frac{252+160}{5} = \frac{412}{5}$
(iii) $t(-10) = \frac{9\times(-10)}{5} + 32 = 9\times(-2) + 32 = 14$
(iv) Since $t(C) = 212$
$\therefore 212 = \frac{9C}{5} + 32 \Rightarrow \frac{9C}{5} = 180 \Rightarrow C = \frac{180\times 5}{9} = 100$
Thus, the value of t, when $t(C) = 212$, is 100.

Q5. Find the range of each of the following functions.
(i) $f(x) = 2 - 3x,\ x \in R,\ x > 0.$
(ii) $f(x) = x^2 + 2,\ x$, is a real number
(iii) $f(x) = x,\ x$ is real number

Solution:
(i) $f(x) = 2 - 3x,\ x \in R,\ x > 0.$
Since $x > 0 \Rightarrow 3x > 0 \Rightarrow -3x < 0 \Rightarrow 2 - 3x < 2 - 0$
$\Rightarrow 2 - 3x < 2 \Rightarrow f(x) < 2$
$\therefore$ Range of $f = (-\infty, 2)$
(ii) $f(x) = x^2 + 2,\ x$, is a real number.
Since x a real number, $x^2 \geq 0$
$\Rightarrow x^2 + 2 \geq 0 + 2 \Rightarrow x^2 + 2 \geq 2 \Rightarrow f(x) \geq 2$
$\therefore$ Range of $f = [2, \infty)$
(iii) $f(x) = x,\ x$ is a real number
Since x is a real number and $f(x) = x$, for all real numbers, the range of $f = R$.

Day 31

NCERT Miscellaneous Exercise (With Solutions)

Q1. The relation f is defined by $f(x) = \begin{cases} x^2, & 0 \leq x \leq 3 \\ 3x, & 3 \leq x \leq 10 \end{cases}$

The relation g is defined by $g(x) = \begin{cases} x^2, & 0 \leq x \leq 2 \\ 3x, & 2 \leq x \leq 10 \end{cases}$

Show that f is a function and g is not a function.

Solution:
The relation f is defined as $f(x) = \begin{cases} x^2, & 0 \leq x \leq 3 \\ 3x, & 3 \leq x \leq 10 \end{cases}$

Since, at $x = 3$,
$f(x) = x^2 \Rightarrow f(3) = 3^2 = 9$
or $f(x) = 3x \Rightarrow f(3) = 3\times 3 = 9$
Thus, at $x = 3$, $f(x) = 9$, and also for $0 \leq x \leq 10$, the images of $f(x)$ are unique.
Therefore, f is a function.

The relation g is defined as $g(x) = \begin{cases} x^2, & 0 \leq x \leq 2 \\ 3x, & 2 \leq x \leq 10 \end{cases}$

Since, at $x = 2$
$g(x) = x^2 \Rightarrow g(2) = 2^2 = 4$
or $g(x) = 2x \Rightarrow g(2) = 3\times 2 = 6$
Element 2 of the domain of the relation g corresponds to two different images, that is, 4 and 6.
Therefore, g is not a function.

Q2. If $f(x) = x^2$, find $\frac{f(1.1) - f(1)}{(1.1 - 1)}$.

Solution:
$\frac{f(1.1) - f(1)}{(1.1 - 1)} = \frac{(1.1)^2 - (1)^2}{(1.1 - 1)} \quad (\because f(x) = x^2)$
$= \frac{1.21 - 1}{0.1} = \frac{0.21}{0.1} = 2.1$

Q3. Find domain of function $f(x) = \frac{x^2 + 2x + 1}{x^2 - 8x + 12}$

Solution:
The given function $f(x) = \frac{x^2 + 2x + 1}{x^2 - 8x + 12}$ can be simplified as $f(x) = \frac{(x+1)^2}{(x-6)(x-2)}$

since function f is defined for all real numbers except when $x-6=0$ or $x-2=0$.

Thus, at $x = 2, 6$ function f is not defined.

Hence, the domain of f is $R-\{2,6\}$.

Q4. Find the domain and the range of the real function f defined by $f(x)=\sqrt{(x-1)}$.

Solution:

The given function $f(x)=\sqrt{(x-1)}$ is defined when $x-1\geq 0 \Rightarrow x\geq 1$.
Therefore, the domain of f is the set of all real numbers greater than or equal to 1, that is, the domain of $f=[1,\infty)$.
Now, for $x\geq 1 \Rightarrow (x-1)\geq 0 \Rightarrow \sqrt{(x-1)}\geq 0$
Therefore, the range of f is the set of all real numbers greater than or equal to 0, that is, the range of $f=(0,\infty)$.

Q5. Find the domain and the range of the real function f defined by $f(x)=|x-1|$.

Solution:

$|x-1|$ is defined for all real numbers.

$\therefore$ Domain of $f = R$

Also, since $x\in R$, $|x-1|$ will take all non-negative real numbers, the range of f is the set of all non-negative real numbers, range of $f=[0,\infty)$.

Q6. Let $f=\left\{\left(x,\dfrac{x^2}{1+x^2}\right): x\in R\right\}$ be a function from R into R. Determine the range of f.

Solution:

Let, for $x\in R$, $y=\dfrac{x^2}{1+x^2}$, where y is the range of f

Now, rearranging the equation in terms of x, we get

$$y=\frac{x^2}{1+x^2} \Rightarrow y+x^2y=x^2 \Rightarrow x^2(1-y)=y$$

$$\Rightarrow x^2=\frac{y}{1-y} \Rightarrow x=\sqrt{\frac{y}{1-y}}$$

Now, the function $g(y)=\sqrt{\dfrac{y}{1-y}}$ is defined when $\dfrac{y}{1-y}\geq 0$

and $1-y\neq 0$

$\Rightarrow \dfrac{y}{y-1}\leq 0$ and $y\neq 1$

$\Rightarrow 0\leq y\leq 1$ and $y\neq 1$

$\Rightarrow 0\leq y<1$ Thus, the range of $f=[0,1)$

Q7. Let $f,g: R\to R$ be defined, respectively, by $f(x)=x+1, g(x)=2x-3$. Find $f+g, f-g, \dfrac{f}{g}$.

Solution:

$\because f(x)=x+1,\ g(x)=2x-3$

$\Rightarrow (f+g)(x)=3x-2$

$\therefore (f+g)(x)=f(x)+g(x)=(x+1)+(2x-3)$

$(f-g)(x)=f(x)-g(x)=(x+1)-(2x-3)$

$\Rightarrow (f-g)(x)=x+1-2x+3=-x+4$

$\left(\dfrac{f}{g}\right)(x)=\dfrac{f(x)}{g(x)}, g(x)\neq 0, x\in R$

$\therefore \left(\dfrac{f}{g}\right)(x)=\dfrac{x+1}{2x-3}, 2x-3\neq 0 \text{ or } 2x\neq 3$

$\Rightarrow \left(\dfrac{f}{g}\right)(x)=\dfrac{x+1}{2x-3},\ x\neq 3/2$

Q8. Let $f=\{(1,1),(2,3),(0,-1),(-1,-3)\}$ be a function Z to Z defined by $f(x)=ax+b$, for some integers a, b. Determine a, b.

Solution:

$f=\{(1,1),(2,3),(0,-1),(-1,-3)\}$ and $f(x)=ax+b$,

$\therefore (1,1)\in f \Rightarrow f(1)=1 \Rightarrow a\times 1+b=1 \Rightarrow a+b=1$

$(0,-1)\in f \Rightarrow f(0)=-1 \Rightarrow a\times 0+b=-1 \Rightarrow b=-1$

Substituting the value of $b=-1$ in $a+b=1$, we get $a+(-1)=1 \Rightarrow a=2$.

Q9. Let R be a relation from N to N defined by $R=\{(a,b): a,b\in N, a=b^2\}$. Are the following true?

(i) $(a,a)\in R$, for all $a\in N$

(ii) $(a,b)\in R$, implies $(b,a)\in R$

(iii) $(a,b)\in R, (b,c)\in R$ implies $(a,c)\in R$.

Justify your answer in each case.

Solution:

$R=\{(a,b): a,b\in N, a=b^2\}$

(i) $2\in N$, but $2\neq 2^2$.
Therefore, the statement '$(a,a)\in R$, for all $a\in N$' is not true.

(ii) $(4,2)\in R$, as $4=2^2$, but $(2,4)\notin R$ as $2\neq 4^2$
Therefore, the statement '$(a,b)\in R$, implies $(b,a)\in R$' is not true.

(iii) $(4,2)\in R$, $(9,3)\in R$ but $(4,3)\notin R$ as $4\neq 3^2$
Therefore, the statement '$(a,b)\in R, (b,c)\in R$ implies $(a,c)\in R$' is not true.

Q10. Let $A=\{1,2,3,4\}, B=\{1,5,9,11,15,16\}$ and $f=\{(1,5),(2,9),(3,1),(4,5),(2,11)\}$. Are the following true?

(i) f is a relation from A to B

(ii) f is a function from A to B

Justify your answer in each case.

Solution:

$A = \{1,2,3,4\}, B = \{1,5,9,11,15,16\}$

$\therefore A \times B = \{(1,1),(1,5),(1,9),(1,11),(1,15),(1,16),$
$(2,1),(2,5),(2,9),(2,11),(2,15),(2,16),$
$(3,1),(3,5),(3,9),(3,11),(3,15),(3,16),$
$(4,1),(4,5),(4,9),(4,11),(4,15)(4,16)\}$

$f = \{(1,5),(2,9),(3,1),(4,5),(2,11)\}$

(i) A relation from a non-empty set A to a non-empty set B is a subset of the Cartesian product $A \times B$. Thus, f is a relation from A to B.

(ii) Since the same first element, that is 1, corresponds to different images 1, 5, ..., relation f is not a function.

Q11. Let f be the subset of $Z \times Z$ defined by $f = \{(ab, a+b) : a, b \in Z\}$. Is f a function from Z to Z: Justify your answer.

Solution:

The relation f is defined as $f = \{(ab, a+b) : a, b \in Z\}$.

A relation f from a set A to a set B is said to be a function if every element of set A has unique images in set B. Let, $(1,2),(-1,-2) \in Z$,

$\Rightarrow (1\times 2, 1+2), (-1\times -2, -1+(-2)) \in f$

$\Rightarrow (2,3),(2,-3) \in f$

Since element 2 corresponds to two different images 3 and –3, the relation f is not a function.

Q12. Let $A = \{9,10,11,12,13\}$ and let $f : A \to N$ be defined by $f(n) =$ the highest prime factor of n. Find the range of f.

Solution:

$A = \{9,10,11,12,13\}$ and $f : A \to N$ is defined as

$f(n) =$ The highest prime factor of n.

Now, since the prime factor of 9 = 3

Prime factors of 10 = 2, 5

Prime factor of 11 = 11

Prime factors of 12 = 2, 3

Prime factor of 13 = 13

$\therefore f(9) =$ The highest prime factor of 9 = 3

$f(10) =$ The highest prime factor of 10 = 5

$f(11) =$ The highest prime factor of 11 = 11

$f(12) =$ The highest prime factor of 12 = 3

$f(13) =$ The highest prime factor of 13 = 13

Thus, the range of f is the set of all $f(n)$, where $n \in A$, that is, range of $f = \{3,5,11,13\}$.

Day 32

NCERT Exemplar Problems: Solutions (Q1–Q11)

Q1. If $A = \{-1,2,3\}, B = \{1,3\}$, then determine

(i) $A \times B$ (ii) $B \times A$
(iii) $B \times B$ (iv) $A \times A$

Solution:

$A = \{-1,2,3\}, B = \{1,3)$

(i) $A \times B = \{(-1,1),(-1,3),(2,1),(2,3),(3,1),(3,3)\}$

(ii) $B \times A = \{(1,-1),(1,2),(1,3),(3,-1),(3,2),(3,3)\}$

(iii) $B \times B = \{(1,1),(1,3),(3,1),(3,3)\}$

(iv) $A \times A = \{(-1,-1),(-1,2),(1-,3),(2,-1),$
$(2,2),(2,3),(3,-1),(3,2),(3,3)\}$

Q2. If $P = \{x : x < 3, x \in N\}$, $Q = \{x : x \le 2, x \in W\}$, then find $(P \cup Q) \times (P \cap Q)$, where W is the set of whole numbers.

Solution:

Since $P = \{x : x < 3, \ x \in N\} = \{1,2\}$

$Q = \{x : x \le 2, x \in W\} = \{0,1,2\}$

$\therefore P \cup Q = \{0,1,2), \ P \cap Q = \{1,2\}$

Thus, $(P \cup Q) \times (P \cap Q)$

$= \{0,1,2) \times \{1,2\}$

$= \{(0,1),(0,2),(1,1),(1,2),(2,1),(2,2)\}$

Q3. If $A = \{x : x \in W, x < 2\}$, $B = \{x : x \in N, 1 < x < 5\}$ $C = \{3,5\}$, then find

(i) $A \times (B \cap C)$ (ii) $A \times (B \cup C)$

Solution:

$A = \{x : x \in W, x < 2\} = \{0,1\}$

$B = \{x : x \in N, 1 < x < 5\} = \{2,3,4\}$

$C = \{3,5\}$

(i) $\because B \cap C = \{3\}$

$\therefore A \times (B \cap C) = \{0,1\} \times \{3\} = \{(0,3),(1,3)\}$

(ii) $\because (B \cup C) = \{2,3,4,5\}$

$\therefore A \times (B \cup C) = \{0,1\} \times \{2,3,4,5\}$

$= \{(0,2),(0,3),(0,4),(0,5),(1,2),(1,3),(1,4),(1,5)\}$

Q4. In each of the following cases, find a and b.

(i) $(2a+b, a-b) = (8,3)$

(ii) $\left(\frac{a}{4}, a-2b\right) = (0, 6+b)$

Solution:

(i) $(2a+b, a-b) = (8,3) \Rightarrow 2a+b = 8,\ a-b = 3$

(Two ordered pairs are equal, if their corresponding elements are equal.)

Solving the above two equations, we get $a = \frac{11}{3},\ b = \frac{2}{3}$

(ii) $\left(\frac{a}{4}, a-2b\right) = (0, 6+b)$

$\Rightarrow \frac{a}{4} = 0 \Rightarrow a = 0$, and

$a - 2b = 6 + b \Rightarrow 0 - 2b = 6 + b \Rightarrow -3b = 6$

$\Rightarrow b = -2$

$\therefore a = 0, b = -2$

Q5. $A = \{1,2,3,4,5\}, S = \{(x,y) : x \in A, y \in A\}$, then find the ordered pairs which satisfy the conditions given below.

(i) $x+y=5$ (ii) $x+y<5$ (iii) $x+y>8$

Solution:

$A = \{1,2,3,4,5\},\ S = \{(x,y) : x \in A, y \in A\}$,

(i) The set of ordered pairs satisfying $x+y=5$ is {(1, 4), (2,3), (3,2), (4,1)}.

(ii) The set of ordered pairs satisfying $x+y<5$ is {(1,1), (1,2), (1,3), (2, 1), (2, 2), (3, 1)}.

(iii) The set of ordered pairs satisfying $x+y>8$ is {(4, 5), (5, 4), (5, 5)}.

Q6. If $R = \{(x,y) : x, y \in W,\ x^2 + y^2 = 25\}$, then find the domain and range of R.

Solution:

The set of ordered pairs satisfying

$R = \{(x,y) : x, y \in W,\ x^2 + y^2 = 25\}$, is

$= \{(0,5),(3,4),(4,3),(5,0)\}$

Domain of R = Set of first element of ordered pairs in R $= \{0,3,4,5\}$

Range of R = Set of second element of ordered pairs in $R = \{0,3,4,5\}$

Q7. If $R_1 = \{(x,y) | y = 2x+7$, where $x \in R, -5 \le x \le 5\}$ is a relation, then find the domain and range of R_1.

Solution:

$R_1 = \{(x,y) | y = 2x+7$, where $x \in R, -5 \le x \le 5\}$

Domain of R is $-5 \le x \le 5$

Therefore, $-5 \le x \le 5$

$\Rightarrow -10 \le 2x \le 10$

$\Rightarrow -10+7 \le 2x+7 \le 10+7$

$\Rightarrow -3 \le 2x+7 \le 17$

$\Rightarrow -3 \le y \le 17$

Thus, the range of $R_1 = \{-3 \le y \le 17, y \in R) = [-3, 17]$

Q8. If $R_2 = \{x, y) | x, y$ are integers and $x^2 + y^2 = 64\}$ is a relation, then find the value of R_2.

Solution:

$R_2 = \{x, y) | x, y$ are integers and $x^2 + y^2 = 64\}$

64 is the sum of squares of 0 and ±8.

$\therefore R_2 = \{(0,8),(0,-8),(8,0),(-8,0)\}$

Q9. If $R_3 = \{(x,|x|) | x$ is a real number$\}$ is a relation, then find domain and range of R_3.

Solution:

$R_3 = \{(x,|x|) | x$ is a real number$\}$

Since x is defined for all real x, the domain of $R_3 = R$.

Since $|x|$ always gives non-negative values for any real number, the range of $R_3 = R^+ \cup \{0\}$ or $(0, \infty)$.

Q10. Is the given relation a function? Give reasons for your answer.

(i) $h = \{(4,6),(3,9),(-11,6),(3,11)\}$

(ii) $f = \{(x,x) | x$ is a real number$\}$

(iii) $g = \left\{\left(x, \frac{1}{x}\right) | x\right.$ is a positive integer$\}$

(iv) $s = \{(x, x^2) | x$ is a positive integer$\}$

(v) $t = \{(x,3) | x$ is a positive number$\}$

Solution:

(i) $h = \{(4,6),(3,9),(-11,6),(3,11)\}$

Since 3 has two images 9 and 11, it is not a function.

(ii) $f = \{(x,x) | x$ is a real number

Since every element in the domain has a unique image, it is a function.

(iii) $g = \left\{\left(x, \frac{1}{x}\right) | x\right.$ is a positive integer$\}$

Since, for every positive integer x, $\frac{1}{x}$ is unique and distinct, it is a function.

(iv) $s = \{(x, x^2) | x$ is a positive integer$\}$

Since the square of any positive integer is unique, every element in the domain has a unique image. Thus, it is a function.

(v) $t = \{(x,3) | x$ is a real number$\}$

Since every element in the domain has the image 3 (unique), it is a function (constant function).

Q11. If f and g are real functions defined by $f(x) = x^2 + 7, g(x) = 3x + 5$ then, find each of the following.

(i) $f(3) + g(-5)$ (ii) $f\left(\frac{1}{2}\right) \times g(14)$

(iii) $f(-2) + g(-1)$ (iv) $f(t) - f(-2)$

(v) $\dfrac{f(t)-f(5)}{t-5}$, if $t \neq 5$

Solution:

Since $f(x)=x^2+7,\ g(x)=3x+5$,

(i) $f(3)=3^2+7=9+7=16$

$g(-5)=3(-5)+5=-15+5=-10$

$\therefore f(3)+g(-5)=16-10=6$

(ii) $f\left(\frac{1}{2}\right)=\left(\frac{1}{2}\right)^2+7=\frac{1}{4}+7=\frac{29}{4}$

$g(14)=3(14)+5=42+5=47$

$\therefore f\left(\frac{1}{2}\right)\times g(14)=\frac{29}{4}\times 47=\frac{1363}{4}$

(iii) $f(-2)=(-2)^2+7=4+7=11$

$g(-1)=3(-1)+5=-3+5=2$

$\therefore f(-2)+g(-1)=11+2=13$

(iv) $f(t)=t^2+7, f(-2)=(-2)^2+7=4+7=11$

$\therefore f(t)-f(-2)=t^2+7-11=t^2-4$

(v) $f(t)=t^2+7, f(5)=5^2+7=25+7=32$

$\therefore \dfrac{f(t)-f(5)}{t-5}=\dfrac{t^2+7-32}{t-5}$ if $t\neq 5$

$=\dfrac{t^2-25}{t-5}=\dfrac{(t-5)(t+5)}{(t-5)}=t+5 \quad (\because t\neq 5)$

Day 33

NCERT Exemplar Problems: Solutions (Q12–Q25)

Q12. Let f and g be real functions defined by $f(x)=2x+1,\ g(x)=4x-7$.

(i) For what real numbers x, $f(x)=g(x)$?

(ii) For what real number x, $f(x)<g(x)$?

Solution:

(i) $f(x)=g(x) \Rightarrow 2x+1=4x-7 \Rightarrow 2x=8 \Rightarrow x=4$

(ii) $f(x)<g(x) \Rightarrow 2x+1<4x-7$

$\Rightarrow 2x-2x+1<4x-7-2x \Rightarrow 1<2x-7$

$\Rightarrow 1+7<2x-7+7 \Rightarrow 8<2x$

$\Rightarrow 2x>8 \Rightarrow x>4$

Q13. If f and g are two real-valued functions defined as $f(x)=2x+1,\ g(x)=x^2+1$, then find

(i) $f+g$ (ii) $f-g$ (iii) fg (iv) $\dfrac{f}{g}$

Solution:

(i) $(f+g)(x)=f(x)+g(x)$

$=2x+1+x^2+1=x^2+2x+2$

(ii) $(f-g)=f(x)-g(x)=(2x+1)-(x^2+1)$

$=2x+1-x^2-1=2x-x^2=x(2-x)$

(iii) $(fg)(x)=f(x)\cdot g(x)=(2x+1)(x^2+1)$

$=2x^3+2x+x^2+1=2x^3+x^2+2x+1$

(iv) $\dfrac{f}{g}(x)=\dfrac{f(x)}{g(x)}=\dfrac{2x+1}{x^2+1}$

Q14. Express the following functions as sets of ordered pairs and determine their range. $f:x\rightarrow R$, $f(x)=x^3+1$, where $x=\{-1,0,3,9,7\}$

Solution:

$f:x\rightarrow R, f(x)=x^3+1$

$f(-1)=(-1)^3+1=-1+1=0$

$f(0)=(0)^3+1=0+1=1$

$f(3)=(3)^3+1=27+1=28$

$f(9)=(9)^3+1=729+1=730$

$f(7)=(7)^3+1=343+1=344$

$\therefore f=\{(-1,0),(0,1),(3,28),(9,730),(7,344)\}$

Thus, range of $f=\{0,1,28,730,344\}$

Q15. Find the values of x for which the functions $f(x)=3x^2-1, g(x)=3+x$ are equal.

Solution:

$f(x)=g(x)\Rightarrow 3x^2-1=3+x$

$\Rightarrow 3x^2-x-4=0 \Rightarrow (3x-4)(x+1)=0$

$\Rightarrow x=-1,\ 4/3$

Q16. Is $g=\{(1,1),(2,3),(3,5),(4,7)\}$ a function? Justify. If this is described by the relation $g(x)=\alpha x+\beta$, then what values should be assigned to α,β?

Solution:

$g=\{(1,1),(2,3),(3,5),(4,7)\}$

Since every element has a unique image under g, it is a function.

Now since, $g(x)=\alpha+\beta$

When $x=1$, then $g(1)=\alpha+\beta \Rightarrow 1=\alpha+\beta$ (i)

When $x=2$, then

$g(2)=\alpha(2)+\beta \Rightarrow 3=2\alpha+\beta$ (ii)

Solving above two equations, we get $\alpha=2,\beta=-1$.

Q17. Find the domain of each of the following functions given by

(i) $f(x)=\dfrac{1}{\sqrt{1-\cos x}}$ (ii) $f(x)=\dfrac{1}{\sqrt{x+|x|}}$

(iii) $f(x)=x|x|$ (iv) $f(x)=\dfrac{x^3-x+3}{x^2-1}$

(v) $f(x)=\dfrac{3x}{28-x}$

Solution:

(i) $f(x)=\dfrac{1}{\sqrt{1-\cos x}}$ is defined when

$1-\cos x>0 \Rightarrow \cos x<1$

$\because \cos x \geq -1$

$\therefore -1 \leq \cos x<1$

Thus, $f(x)$ is defined, for all x, except when $1-\cos x \neq 0 \Rightarrow \cos x \neq 1 \Rightarrow x \neq 2n\pi \ \forall n \in Z$

Hence, domain of $f=R-\{2n\pi : n \in Z\}$

(ii) $f(x)=\dfrac{1}{\sqrt{x+|x|}}$

$\because x+|x|=\begin{cases} =x-x=0, & x<0 \\ =x+x=2x, & x \geq 0 \end{cases}$

Since $f(x)$ is not defined when $x+|x|=0$,

Domain of $f=R^+$

(iii) $f(x)=x|x|$

$f(x)$ is defined for all $x \in R$.

$\therefore$ Domain of $f=R$

(iv) $f(x)=\dfrac{x^3-x+3}{x^2-1}$

Since $f(x)$ is not defined when $x^2-1=0 \Rightarrow (x-1)(x+1)=0 \Rightarrow x=-1,1$

$\therefore$ Domain of $f=R-\{-1,1\}$

(v) $f(x)=\dfrac{3x}{28-x}$

Since $f(x)$ is not defined when $28-x=0 \Rightarrow x=28$

$\therefore$ Domain of $f=R-\{28\}$

Q18. Find the range of the following functions

(i) $f(x)=\dfrac{3}{2-x^2}$ (ii) $f(x)=1-|x-2|$

(iii) $f(x)=|x-3|$ (iv) $f(x)=1+3\cos 2x$

Solution:

(i) $f(x)=\dfrac{3}{2-x^2}$

Let $y=f(x)$

$\Rightarrow y=\dfrac{3}{2-x^2} \Rightarrow 2-x^2=\dfrac{3}{y}$

$\Rightarrow x^2=2-\dfrac{3}{y} \Rightarrow x=\sqrt{\dfrac{2y-3}{y}}$

Since x is real, $\dfrac{2y-3}{y} \geq 0$ and $y \neq 0$

$\Rightarrow 2y-3 \geq 0$ and $y>0$

or $2y-3 \leq 0$ and $y<0$

$\Rightarrow y \geq \dfrac{3}{2}$ or $y<0$

Thus, range of $f=(-\infty,0) \cup \left[\dfrac{3}{2},\infty\right)$

(ii) $\because |x-2| \geq 0 \Rightarrow -|x-2| \leq 0$

$\Rightarrow 1-|x-2| \leq 1 \Rightarrow f(x) \leq 1$

$\therefore$ Range of $f=(-\infty,1]$

(iii) $\because |x-3| \geq 0 \Rightarrow f(x) \geq 0$

$\therefore$ Range of $f=[0,\infty)$

(iv) $-1 \leq \cos 2x \leq 1 \Rightarrow -3 \leq 3\cos 2x \leq 3$

$\Rightarrow 1-3 \leq 1+3\cos 2x \leq 1+3$

$\Rightarrow -2 \leq 1+3\cos 2x \leq 4$

$\Rightarrow -2 \leq f(x) \leq 4$

Thus, range of $f=[-2,4]$

Q19. Redefine the function

$f(x)=|x-2|+|2+x|, \ -3 \leq x \leq 3$

Solution:

$\because f(x)=|x-2|+|2+x|, -3 \leq x \leq 3$

$\therefore f(x)=\begin{cases} -(x-2)-(2+x), & -3 \leq x<-2 \\ -(x-2)+(2+x), & -2 \leq x<2 \\ (x-2)+(2+x), & 2 \leq x \leq 3 \end{cases}$

$\Rightarrow f(x)=\begin{cases} -2x, & -3 \leq x<-2 \\ 4, & -2 \leq x<2 \\ 2, & 2 \leq x \leq 3 \end{cases}$

Q20. If $f(x)=\dfrac{x-1}{x+1}$, then show that

(i) $f\left(\dfrac{1}{x}\right)=-f(x)$ (ii) $f\left(-\dfrac{1}{x}\right)=\dfrac{-1}{f(x)}$

Solution:

(i) $f\left(\dfrac{1}{x}\right)=\dfrac{\frac{1}{x}-1}{\frac{1}{x}+1}=\dfrac{1-x}{1+x}=-\dfrac{(x-1)}{x+1}=-f(x)$

(ii) $f\left(-\dfrac{1}{x}\right)=\dfrac{-\frac{1}{x}-1}{-\frac{1}{x}+1}=\dfrac{(-1-x)/x}{(-1+x)/x}$

$$= \frac{-(x+1)}{x-1} = \frac{\frac{-1}{x-1}}{x+1} = -\frac{1}{f(x)}$$

Q21. If $f(x) = \sqrt{x}$, $g(x) = x$ are two functions defined in the domain $R^+ \cup \{0\}$, then find the value of

(i) $(f+g)(x)$ (ii) $(f-g)(x)$

(iii) $(fg)(x)$ (iv) $\left(\frac{f}{g}\right)(x)$

Solution:

(i) $(f+g)(x) = f(x) + g(x) = \sqrt{x} + x$

(ii) $(f-g)(x) = f(x) - g(x) = \sqrt{x} - x$

(iii) $(fg)(x) = f(x) \cdot g(x) = \sqrt{x} \cdot x = x^{3/2}$

(iv) $\left(\frac{f}{g}\right)(x) = \frac{f(x)}{g(x)} = \frac{\sqrt{x}}{x} = \frac{1}{\sqrt{x}}$ defined in the domain R^+.

Q22. Find the domain and range of the function $f(x) = \frac{1}{\sqrt{x-5}}$.

Solution:

$f(x)$ is defined if $x - 5 > 0 \Rightarrow x > 5$

$\therefore$ Domain of $f = (5, \infty)$

Let $f(x) = y \Rightarrow y = \frac{1}{\sqrt{x-5}} \Rightarrow \sqrt{x-5} = \frac{1}{y}$

$\Rightarrow x - 5 = \frac{1}{y^2} \Rightarrow x = \frac{1}{y^2} + 5$

$\because x \in (5, \infty) \therefore y \in R^+$

Thus, range of $f = R^+$

Q23. If $f(x) = y = \frac{ax-b}{cx-a}$, then prove that $f(y) = x$

Solution:

$\because f(x) = y = \frac{ax-b}{cx-a}$

$$\therefore f(y) = \frac{ay-b}{cy-a} = \frac{a\left(\frac{ax-b}{cx-a}\right) - b}{c\left(\frac{ax-b}{cx-a}\right) - a}$$

$$= \frac{a(ax-b) - b(cx-a)}{c(ax-b) - a(cx-a)}$$

$$= \frac{a^2x - ab - bcx + ab}{acx - bc - acx + a^2}$$

$$= \frac{a^2x - bcx}{a^2 - bc} = \frac{x(a^2-bc)}{(a^2-bc)} = x$$

Thus, $f(y) = x$

Q24. Let $n(A) = m$, $n(B) = n$. Then, the total number of non-empty relations that can be defined from A to B is

(a) m^n (b) $n^m - 1$

(c) $mn - 1$ (d) $2^{mn} - 1$

Solution: (d)

$n(A) = m, n(B) = n$

$n(A \times B) = n(A) \cdot n(B) = mn$

Thus, the total number of relations from A to B is $2^{n(A \times B)} - 1 = 2^{mn} - 1$ (–1 for empty relation).

Q25. If $[x]^2 - 5[x] + 6 = 0$, where [.] denotes the greatest integer function, then

(a) $x \in [3, 4]$ (b) $x \in (2, 3]$

(c) $x \in [2, 3]$ (d) $x \in [2, 4)$

Solution: (d)

$[x]^2 - 5[x] + 6 = 0$

$\Rightarrow [x]^2 - 3[x] - 2[x] + 6 = 0$

$\Rightarrow [x]([x] - 3) - 2([x] - 3) = 0$

$([x] - 3)([x] - 2) = 0 \Rightarrow [x] = 2, 3$

$\Rightarrow x \in [2, 4)$

$\because$ For $x \in [2, 3) \Rightarrow [x] = 2$ and for $x \in [3, 4) \Rightarrow [x] = 3$

Day 34

NCERT Exemplar Problems: Solutions (Q26–Q42)

Q26. Range of $f(x) = \frac{1}{1 - 2\cos x}$ is

(a) $\left[\frac{1}{3}, 1\right]$ (b) $\left[-1, \frac{1}{3}\right]$

(c) $(-\infty, -1] \cup \left[\frac{1}{3}, \infty\right)$ (d) $\left[-\frac{1}{3}, 1\right]$

Solution: (b)

Let $f(x) = \frac{1}{1 - 2\cos x} = y$

$\Rightarrow 1 - 2\cos x = \frac{1}{y} \Rightarrow 2\cos x = 1 - \frac{1}{y}$

$\Rightarrow \cos x = \frac{1}{2} - \frac{1}{2y}$

$\because -1 \le \cos x \le 1$

$\therefore -1 \le \frac{1}{2} - \frac{1}{2y} \le 1$

$\Rightarrow -1 - \frac{1}{2} \le -\frac{1}{2y} \le 1 - \frac{1}{2}$

$\Rightarrow -\frac{3}{2} \le -\frac{1}{2y} \le \frac{1}{2}$

$\Rightarrow -3 \le -\frac{1}{y} \le 1 \Rightarrow 3 \ge \frac{1}{y} \ge -1$

$\Rightarrow \frac{1}{3} \ge y \ge -1$

Alternatively, $\because -1 \le \cos x \le 1$

$\therefore 1 - 2 \le 1 - 2\cos x \le 1 + 2$

$\Rightarrow -1 \le 1 - 2\cos x < 3$

$\Rightarrow -1 \le \frac{1}{1 - 2\cos x} \le \frac{1}{3}$

$\Rightarrow -1 \le f(x) \le \frac{1}{3}$

Thus, range of $f = [-1, 1/3]$

Q27. Let $f(x) = \sqrt{1 + x^2}$, then

(a) $f(xy) = f(x)\cdot f(y)$

(b) $f(xy) \ge f(x)\cdot f(y)$

(c) $f(xy) \le f(x)\cdot f(y)$

(d) None of these

Solution: (c)

$\because f(x) = \sqrt{1+x^2} \Rightarrow f(xy) = \sqrt{1 + x^2y^2}$

$\therefore f(x)\cdot f(y) = \sqrt{1+x^2}\cdot\sqrt{1+y^2}$

$= \sqrt{(1+x^2)(1+y^2)} = \sqrt{1 + x^2 + y^2 + x^2y^2}$

$\because \sqrt{1+x^2y^2} \le \sqrt{1+x^2+y^2+x^2y^2}$

$\therefore f(xy) \le f(x)\cdot f(y)$

Q28. Domain of $\sqrt{a^2 - x^2}$ $(a > 0)$ is

(a) $(-a, a)$ (b) $[-a, a]$

(c) $[0, a]$ (d) $(-a, 0)$

Solution: (b)

$f(x) = \sqrt{a^2 - x^2}$ is defined, if

$a^2 - x^2 \ge 0 \Rightarrow x^2 - a^2 \le 0$

$\Rightarrow (x - a)(x + a) \le 0 \Rightarrow -a \le x \le a$ $(\because a > 0)$

$\therefore$ Domain of $f = [-a, a]$

Q29. If $f(x) = ax + b$, where a and b are integers, $f(-1) = -5, f(3) = 3$, then a and b are equal to

(a) $a = -3, b = -1$ (b) $a = 2, b = -3$

(c) $a = 0, b = 2$ (d) $a = 2, b = 3$

Solution: (b)

$f(x) = ax + b \Rightarrow f(-1) = a(-1) + b$

$\Rightarrow -5 = -a + b$ (i)

$f(3) = a(3) + b \Rightarrow 3 = 3a + b$ (ii)

Solving equations (i) and (ii), we get $a = 2, b = -3$

Q30. The domain of the function f defined by $f(x) = \sqrt{4 - x} + \frac{1}{\sqrt{x^2 - 1}}$ is equal to

(a) $(-\infty, -1) \cup (1, 4]$ (b) $(-\infty, -1] \cup (1, 4]$

(c) $(-\infty, -1) \cup [1, 4]$ (d) $(-\infty, -1) \cup [1, 4)$

Solution: (a)

$f(x) = \sqrt{4 - x} + \frac{1}{\sqrt{x^2 - 1}}$, $f(x)$ is defined, if

$4 - x \ge 0$ and $x^2 - 1 > 0$

$\Rightarrow x - 4 \le 0$ and $(x + 1)(x - 1) > 0$

$\Rightarrow x \le 4$ and $x < -1$ or $x > 1$

$\therefore$ Domain of $(-\infty, -1) \cup (1, 4]$

Q31. The domain and range of the real function f defined by $f(x) = \frac{4 - x}{x - 4}$ is given by

(a) Domain = R, Range = $\{-1, 1\}$

(b) Domain = $R - \{1\}$, Range = R

(c) Domain = $R - \{4\}$, Range = $\{-1\}$

(d) Domain = $R - \{-4\}$, Range = $\{-1, 1\}$

Solution: (c)

$f(x) = \frac{4 - x}{x - 4}$ is defined, if $x - 4 \ne 0 \Rightarrow x \ne 4$

$\therefore$ Domain of $f = R - \{4\}$

Let $f(x) = \frac{4 - x}{x - 4} = y \Rightarrow xy - 4y = 4 - x$

$\Rightarrow xy + x = 4 + 4y \Rightarrow x(y + 1) = 4(1 + y)$

$\Rightarrow x = \frac{4(1 + y)}{y + 1}$

x will be real for all value of y, except when $y + 1 = 0 \Rightarrow y = -1$

Thus, range of $f = R - \{-1\}$

Q32. The domain and range of real function f defined by $f(x) = \sqrt{x - 1}$ is given by

(a) Domain = $(1, \infty)$, Range = $(0, \infty)$

(b) Domain = $[1, \infty)$, Range = $(0, \infty)$

(c) Domain = $(1, \infty)$, Range = $[0, \infty)$

(d) Domain = $[1, \infty)$, Range = $[0, \infty)$

Solution: (d)

$f(x) = \sqrt{x - 1}$ is defined, if $x - 1 \ge 0 \Rightarrow x \ge 1$

$\therefore$ Domain of $f = (1, \infty)$

Let $f(x) = \sqrt{x-1} = y$

$\Rightarrow x - 1 = y^2 \Rightarrow x = 1 + y^2$

x is defined for all real values of y

Also $y \geq 0$ because $x \geq 1$

$\therefore$ range of $f = (0, \infty)$

Q33. The domain of the function f is given by $f(x) = \dfrac{x^2+2x+1}{x^2-x-6}$

(a) $R - \{3, -2\}$ (b) $R - \{-3, 2\}$

(c) $R - [3, -2]$ (d) $R - (3, -2)$

Solution: (a)

$\because f(x) = \dfrac{x^2+2x+1}{x^2-x-6} = \dfrac{(x+1)^2}{(x-3)(x+2)}$

$\therefore f(x)$ is defined, if

$(x-3)(x+2) \neq 0 \Rightarrow x \neq -2, 3$

Thus, domain of $f = R - \{3, -2\}$

Q34. The domain and range of the function f is given by $f(x) = 2 - |x-5|$ is

(a) Domain $= R+$, Range $= (-\infty, 1]$

(b) Domain $= R$, Range $= [-\infty, 2]$

(c) Domain $= R$, Range $= (-\infty, 2]$

(d) Domain $= R+$, Range $= (-\infty, 2]$

Solution: (b)

$f(x) = 2 - |x-5|$ is defined for all $x \in R$

$\therefore$ Domain of $f = R$

Since $(x-5) \geq 0 \Rightarrow -|x-5| \leq 0$

$\Rightarrow 2 - |x-5| \leq 2 \Rightarrow f(x) \leq 2$

$\therefore$ Range of $f = [-\infty, 2]$

Q35. The domain for which the functions defined by $f(x) = 3x^2 - 1, g(x) = 3 + x$ are equal to

(a) $\left\{-1, \dfrac{4}{3}\right\}$ (b) $\left[1, \dfrac{4}{3}\right]$

(c) $\left[-1, -\dfrac{4}{3}\right]$ (d) $\left\{-1, -\dfrac{4}{3}\right\}$

Solution: (a)

The domain for which the functions defined by $f(x) = 3x^2 - 1, g(x) = 3 + x$ are equal means all values of x for which functions $f(x) = g(x)$. Thus

$3x^2 - 1 = 3 + x \Rightarrow 3x^2 - x - 4 = 0$

$\Rightarrow (3x-4)(x+1) = 0$

$\Rightarrow x = -1, \dfrac{4}{3}$

Thus, domain for which $f(x) = g(x)$ are equal to $\left\{-1, \dfrac{4}{3}\right\}$

Q36. Let f, g be two real functions given by

$f = \{(0,1), (2,0), (3,-4), (4,2), (5,1)\}$

$g = \{(1,0), (2,2), (3,-1), (4,4), (5,3)\}$

Then the domain of $f \cdot g$ is given by ______

Solution:

$f = \{(0,1), (2,0), (3,-4), (4,2), (5,1)\}$

$g = \{(1,0), (2,2), (3,-1), (4,4), (5,3)\}$

$\therefore$ Domain of $f = \{0, 2, 3, 4, 5\}$

And domain $g = \{1, 2, 3, 4, 5\}$

$\therefore$ Domain of $(f \cdot g) =$ Domain of $f \cap$ Domain of $g = \{2, 3, 4, 5\}$

Q37. Let $f = \{(2,4), (5,6), (8,-1), (10,-3)\}$ $g = \{(2,5), (7,1), (8,4), (10,13), (11,5)\}$ be two real functions. Then, match the following.

Column I	Column II
(i) $f - g$	(a $\left\{\left(2, \dfrac{4}{5}\right), \left(8, \dfrac{-1}{4}\right), \left(10, \dfrac{-3}{13}\right)\right\}$
(ii) $f + g$	(b) $\{(2,20), (8,-4), (10,-39)\}$
(iii) $f \cdot g$	(c) $\{(2,-1), (8,-5), (10,-16)\}$
(iv) $\dfrac{f}{g}$	(d) $\{(2,9), (8,3), (10,-10)\}$

Solution:

$f = \{(2,4), (5,6), (8,-1), (10,-3)\}$

$g = \{(2,5), (7,1), (8,4), (10,13), (11,5)\}$

Domain of $f = \{2, 5, 8, 10\}$

And domain $g = \{2, 7, 8, 10, 11\}$

The domain of $f - g,\ f + g,\ f \cdot g,\ \dfrac{f}{g}$ is domain of $f \cap$ domain of g. Thus,

$\therefore$ Domain of $f - g,\ f + g,\ f \cdot g,\ \dfrac{f}{g}$ is

$=$ Domain of $f \cap$ Domain of g

$= \{2, 5, 8, 10\} \cap \{2, 7, 8, 10, 11\} = \{2, 8, 10\}$

Thus,

(i) $(f-g)(2) = f(2) - g(2) = 4 - 5 = -1$

$(f-g)(8) = f(8) - g(8) = -1 - 4 = -5$

$(f-g)(10) = f(10) - g(10) = -3 - 13 = -16$

$\therefore f - g = \{(2,-1), (8,-5), (10,-16)\}$

(ii) $(f+g)(2) = f(2) + g(2) = 4 + 5 = 9$

$(f+g)(8) = f(8)+g(8) = -1+4 = 3$

$(f+g)(10) = f(10)+g(10) = -3+13 = 10$

$\therefore f+g = \{(2,9),(8,3),(10,10)\}$

(iii) $(f\cdot g)(2) = f(2)\cdot g(2) = 4\times 5 = 20$

$(f\cdot g)(8) = f(8)\cdot g(8) = -1\times 4 = -4$

$(f\cdot g)(10) = f(10)\cdot g(10) = -3\times 13 = -39$

$\therefore fg = \{(2,20),(8,-4),(10,-39)\}$

(iv) $\left(\frac{f}{g}\right)(2) = \frac{f(2)}{g(2)} = \frac{4}{5}$

$\left(\frac{f}{g}\right)(8) = \frac{f(8)}{g(8)} = \frac{-1}{4}$

$\left(\frac{f}{g}\right)(10) = \frac{f(10)}{g(10)} = \frac{-3}{13}$

$\therefore \frac{f}{g} = \left\{\left(2,\frac{4}{5}\right),\left(8,-\frac{1}{4}\right),\left(10,\frac{-3}{13}\right)\right\}$

Q38. The ordered pair (5, 2) belongs to the relation $R = \{(x,y): y = x-5, x, y \in Z\}$

Solution: False

$R = \{(x,y): y = x-5, x, y \in Z\}$

If $x = 5$, then $y = 5-5 = 0$

Thus, (5, 2) does not belong to R.

Q39. If $P = \{1,2\}$, then $P\times P\times P = \{(1,1,1),(2,2,2),(1,2,2),(2,1,1)\}$

Solution: False

$P = \{1,2\} \Rightarrow n(P) = 2$

$\therefore n(P\times P\times P) = n(P)\times n(P)\times n(P) = 2\times 2\times 2 = 8$ elements while the given $P\times P\times P$ has 4 elements.

Q40. If $A = \{1,2,3\}, B = \{3,4\}, C = \{4,5,6\}$, then $(A\times B)\cup(A\times C) = \{(1,3),(1,4),(1,5),(1,6),(2,3),(2,4),(2,5),(2,6),(3,3),(3,4),(3,5),(3,6)\}$

Solution: True

$A = \{1,2,3\}, B = \{3,4\}, C = \{4,5,6\}$

$A\times B = \{(1,3),(1,4),(2,3),(2,4),(3,3),(3,4)\}$

$A\times C = \{(1,4),(1,5),(1,6),(2,4),(2,5),(2,6),(3,4),(3,5),(3,6)\}$

$(A\times B)\cup(A\times C) = \{(1,3),(1,4),(1,5),(1,5),(1,6),(2,3),(2,4)(2,5),(2,6),(3,3),(3,4),(3,5),(3,6)\}$

Q41. If $(x-2, y+5) = \left(-2,\frac{1}{3}\right)$ are two equal ordered pairs, then $x = 4, y = \frac{-14}{3}$

Solution: False

$\because (x-2, y+5) = \left(-2,\frac{1}{3}\right)$

$\therefore x-2 = -2$ and $y+5 = \frac{1}{3}$

$\Rightarrow x = -2+2$ and $y = \frac{1}{3}-5$

$\Rightarrow x = 0,\ y = \frac{-14}{3}$

Q42. If $A\times B = \{(a,x),(a,y),(b,x),(b,y)\}$, then $A = \{a,b\}, B = \{x,y\}$.

Solution: True

$A\times B = \{(a,x),(a,y),(b,x),(b,y)\}$

A = set of first element of ordered pairs in $A\times B = \{a,b\}$

B = set of second element of ordered pairs in $A\times B = \{x,y\}$

Day 35

Very Short Answer (VSA): 1-Mark and Short Answer (SA): 2-Mark Questions

Q1. If $[x]^2+[x]-6=0$, where [.] denotes the greatest integer function, then

(a) $x\in[2,3)$ (b) $x\in[2,3)\cup[-3,-2)$

(c) $x\in[-3,-2)$ (d) $x\in(2,3)\cup(-3,-2)$

Solution: (b)

$[x]^2+[x]-6=0$

$\Rightarrow [x]^2+3[x]-2[x]+6=0$

$\Rightarrow [x]([x]+3)-2([x]+3)=0$

$\Rightarrow ([x]+3)([x]-2)=0$

$\Rightarrow [x]=2,-3$

$\because$ For $[x]=2 \Rightarrow x\in[2,3)$

and for $[x]=-3 \Rightarrow x\in[-3,-2)$

Q2. Range of $f(x) = \frac{2}{1-\sin x}$ is

(a) $[-1,1]$ (b) $[-1,2]$
(c) $[1,\infty)$ (d) $(-\infty,-1)$

Solution: (c)

Let $f(x)=\dfrac{2}{1-\sin x}=y$

$\Rightarrow 1-\sin x=\dfrac{2}{y}\Rightarrow \sin x=1-\dfrac{2}{y}$

$\because -1\le \sin x\le 1$

$\therefore -1\le 1-\dfrac{2}{y}\le 1$

$\Rightarrow -1-1\le -\dfrac{2}{y}\le 1-1$

$\Rightarrow -2\le -\dfrac{2}{y}\le 0\Rightarrow -1\le -\dfrac{1}{y}\le 0$

$\Rightarrow 1\ge \dfrac{1}{y}\ge 0\Rightarrow y\ge 1$ and $y<\infty$

$\Rightarrow 1\le y<\infty$

Thus, range of $f=[1,\infty)$

Q3. Domain of $\dfrac{1}{\sqrt{a^2-x^2}}$ $(a>0)$ is

(a) $(-a,a)$ (b) $[-a,a]$
(c) $[0,a]$ (d) $(-a,0)$

Solution: (a)

$f(x)=\dfrac{1}{\sqrt{a^2-x^2}}$ is defined, if

$a^2-x^2>0\Rightarrow x^2-a^2<0$

$\Rightarrow (x-a)(x+a)<0\Rightarrow -a<x<a$ $(\because a>0)$

$\therefore$ Domain of $f=(-a,a)$

Q4. If $f(x)=ax+b$, where a and b are integers, $f(-2)=1$, $f(-1)=2$, then a and b are equal to

(a) $a=-3, b=-1$ (b) $a=2, b=-3$
(c) $a=0, b=2$ (d) $a=2, b=3$

Solution: (b)

$\because f(x)=ax+b$

$\therefore f(-2)=a(-2)+b=1$

and $f(-1)=a(-1)+b=2$

$\Rightarrow -2a+b=1$ and $-a+b=2$

Solving the above two equations, we get $a=1$, $b=3$.

Q5. The domain of the function f defined by $f(x)=\sqrt{2-x}+\dfrac{1}{\sqrt{x^2-1}}$ is equal to

(a) $(-\infty,-1)\cup(1,2]$ (b) $(-\infty,-1]\cup(1,2]$
(c) $(-\infty,-1)\cup[1,2]$ (d) $(-\infty,-1)\cup[1,2)$

Solution: (a)

$f(x)=\sqrt{2-x}+\dfrac{1}{\sqrt{x^2-1}}$; $f(x)$ is defined, if

$2-x\ge 0$ and $x^2-1>0$

$\Rightarrow x-2\le 0$ and $(x+1)(x-1)>0$

$\Rightarrow x\le 2$ and $x<-1$ or $x>1$

$\therefore$ Domain of $(-\infty,-1)\cup(1,2]$

Q6. The domain and range of the real function f defined by $f(x)=\dfrac{1-x}{x-2}$ is given by

(a) Domain $=R$, Range $=\{-1,1\}$
(b) Domain $=R-\{1\}$, Range $=$ R
(c) Domain $=R-\{2\}$, Range $=\{-1\}$
(d) Domain $=R-\{-2\}$, Range $=\{-1,1\}$

Solution: (c)

$f(x)=\dfrac{1-x}{x-2}$ is defined, if $x-2\ne 0\Rightarrow x\ne 2$

$\therefore$ Domain of $f=R-\{2\}$

Let $f(x)=\dfrac{1-x}{x-2}=y$

$\Rightarrow xy-2y=1-x$

$\Rightarrow xy+x=1+2y$

$\Rightarrow x(y+1)=1+2y$

$\Rightarrow x=\dfrac{1+2y}{y+1}$

x will be real for all value of y, except when $y+1=0\Rightarrow y=-1$

Thus, range of $f=R-\{-1\}$

Q7. The domain and range of real function f define by $f(x)=\sqrt{1-x}$ is given by

(a) Domain $=(1,\infty)$, Range $=(0,\infty)$
(b) Domain $=[1,\infty)$, Range $=(0,\infty)$
(c) Domain $=(1,\infty)$, Range $=[0,\infty)$
(d) Domain $=[1,\infty)$, Range $=[0,\infty)$

Solution: (d)

$f(x)=\sqrt{1-x}$ is defined, if $1-x\ge 0\Rightarrow x\le 1$

$\therefore$ Domain of $f=(-\infty,1)$

Let $f(x)=\sqrt{1-x}=y$

$\Rightarrow 1-x=y^2\Rightarrow x=1-y^2$

Since x is defined for all real values of y, but $y=\sqrt{1-x}$ gives only non-negative values, the range of $f=[0,\infty)$.

Q8. Domain of the function f given by $f(x) = \dfrac{x}{x^2-1}$

(a) $f = R - \{-1,1\}$ (b) $R - (-1,1)$
(c) $R - [-1,1]$ (d) $R - \{-1,0,1\}$

Solution: (a)

$\because f(x) = \dfrac{x}{x^2-1} = \dfrac{x}{(x-1)(x+1)}$

$\therefore f(x)$ is defined, if $(x-1)(x+1) \neq 0 \Rightarrow x \neq -1,1$

Thus, domain of $f = R - \{-1,1\}$

Q9. The domain and range of the function f is given by $f(x) = 2 - |x-1|$ is

(a) Domain $= R+$, Range $= (-\infty, 1]$
(b) Domain $= R$, Range $= [-\infty, 2]$
(c) Domain $= R$, Range $= (-\infty, 2]$
(d) Domain $= R+$, Range $= (-\infty, 2]$

Solution: (b)

Since $f(x) = 2 - |x-1|$ is defined for all $x \in R$

$\therefore$ Domain of $f = R$

Since $|x-1| \geq 0 \Rightarrow -|x-1| \leq 0$

$\Rightarrow 2 - |x-1| \leq 2 \Rightarrow f(x) \leq 2$

$\therefore$ Range of $f = [-\infty, 2]$

Q10. The range of the function f is given by $f(x) = |1-x| + |1+x|$ is

(a) $(-\infty,-1) \cup (1,\infty)$ (b) $(2,\infty)$
(c) $(-\infty,-1) \cup (1,\infty)$ (d) $[2,\infty)$

Solution: (d)

$$\text{Since } f(x) = \begin{cases} (1-x)-(1+x), & x \leq -1 \\ (1-x)+(1+x), & -1 \leq x \leq 1 \\ -(1-x)+(1+x), & x \geq 1 \end{cases}$$

$$\Rightarrow f(x) = \begin{cases} -2x, & x \leq -1 \\ 2, & -1 \leq x \leq 1 \\ 2x, & x \geq 1 \end{cases}$$

$\therefore$ Range of $f = (2,\infty)$

Q11. The domain for which the functions defined by $f(x) = x^2 - 1$, $g(x) = x - 1$ are equal to

(a) $\{0,1\}$ (b) $\{-1,1\}$
(c) $(0,1)$ (d) $[0,1]$

Solution: (a)

The domain for which the functions defined by $f(x) = x^2-1$, $g(x) = x-1$ are equal means all values of x for which functions $f(x) = g(x)$. Thus

$x^2 - 1 = x - 1 \Rightarrow x^2 - x = 0$

$\Rightarrow x(x-1) = 0 \Rightarrow x = 0,1$

Thus, domain for which $f(x) = g(x)$ are equal is $\{0,1\}$.

Q12. If $P = \{1,2\}$, then find $P \times P \times P$

Solution: $P = \{1,2\} \Rightarrow n(P) = 2$

$\therefore n(P \times P \times P) = n(P) \times n(P) \times n(P) = 2 \times 2 \times 2 = 8$ elements given by $P \times P \times P = \{(1,1,1),(1,1,2),(1,2,1),(2,1,1),(2,2,1),(2,1,2),(1,2,2),(2,2,2)\}$

Q13. If $A \times B = \{(a,x),(a,y),(b,x),(b,y)\}$, then find A and B.

Solution:

$A \times B = \{(a,x),(a,y),(b,x),(b,y)\}$,

$A =$ set of first element of ordered pairs in $A \times B = \{a,b\}$

$B =$ set of second element of ordered pairs in $A \times B = \{x,y\}$

Q14. Does the ordered pair (3, 2) and (2, 3) belong to the relation $R = \{(x,y) : y = x-1, x,y \in Z\}$?

Solution:

$R = \{(x,y) : y = x-1,\ x,y \in Z\}$

If $x = 3$, then $y = 3-1 = 2$

Thus, (3, 2) belongs to R.

If $x = 2$, then $y = 2-1 = 1$

Thus, (2, 3) does not belong to R.

Q15. If $A = \{1,2,3\}, B = \{3,4\}, C = \{4,5,6\}$, then find $(A \times B) \cap (A \times C)$.

Solution:

$A = \{1,2,3\}, B = \{3,4\}, C = \{4,5,6\}$

$A \times B = \{(1,3),(1,4),(2,3),(2,4),(3,3),(3,4)\}$

$A \times C = \{(1,4),(1,5),(1,6),(2,4),(2,5),(2,6),(3,4),(3,5),(3,6)\}$

$\therefore (A \times B) \cap (A \times C) = \{(1,4),(2,4)\}$

Q16. Let $f = \{(-1,0),(0,-1),(1,c),(d,1),\}$ be a function Z to Z defined by $f(x) = ax+b$, for some integers a, b. Determine a, b and also find elements c and d.

Solution:

$\because f = \{(1,1),(0,-1),(-1,0)\}$ and $f(x) = ax+b$

$\therefore (-1,0) \in f \Rightarrow f(-1) = 0$

$\Rightarrow a \times (-1) + b = 0 \Rightarrow -a + b = 0$

$(0,-1) \in f \Rightarrow f(0) = -1$

$\Rightarrow a \times 0 + b = -1 \Rightarrow b = -1 \Rightarrow a = -1$

Thus, $f(x) = -x - 1$

Now, $c = f(1) = -1 - 1 = -2$

And, $1 = f(d) \Rightarrow 1 = -d - 1 \Rightarrow d = -1 - 1 = -2$

Q17. If $P=\{x: x\le 2,\ x\in N\}$, $Q=\{x: x<2,\ x\in W\}$, then find $(P\cup Q)\times(P\cap Q)$, where N is the set of natural numbers and W is the set of whole numbers.

Solution:

Since $P=\{x: x\le 2,\ x\in N\}=\{1,2\}$

$Q=\{x: x<2,\ x\in W\}=\{0,1\}$

$\therefore P\cup Q=\{0,1,2),\ P\cap Q=\{1\}$

Thus, $(P\cup Q)\times(P\cap Q)$

$=\{0,1,2)\times\{1\}=\{(0,1),(1,1),(2,1)\}$

Q18. Find a and b if $\left(\frac{a}{2},1+b\right)=(2b,a-1)$

Solution:

Two ordered pairs are equal if their corresponding elements are equal. Therefore

$\left(\frac{a}{2},1+b\right)=(2b,a-2)$

$\Rightarrow \frac{a}{2}=2b$ and $1+b=a-2$

$\Rightarrow a=4b$ and $a-b=3$

$\Rightarrow 4b-b=3 \Rightarrow b=1$

$\Rightarrow a=4$

Q19. If $R=\{(x,y): x+y\le 3,\ x\in N$ and $y\in W\}$, then find the domain and range of R.

Solution:

The set of ordered pairs satisfying

$R=\{(x,y): x+y\le 3,\ x\in N$ and $y\in W\}$ is

$=\{(1,0),(1,1),(1,2),(2,0),(2,1),(3,0)\}$

Domain of R = Set of first element of ordered pairs in R $=\{1,2,3\}$

Range of R = Set of second element of ordered pairs in $R=\{0,1,2\}$

Q20. If $R_1=\{(x,y)|y=2x-3$, where $x\in R, -1\le x\le 1\}$ is a relation, then find the domain and range of R_1.

Solution:

$R_1=\{(x,y)|y=2x-3$, where $x\in R, -1\le x\le 1\}$

Domain of R is $-1\le x\le 1$

Therefore, $-1\le x\le 1$

$\Rightarrow -2\le 2x\le 2$

$\Rightarrow -2-3\le 2x-3\le 2-3$

$\Rightarrow -5\le 2x-3\le -1$

$\Rightarrow -5\le y\le -1$

Thus, the range of $R_1=\{-5\le y\le -1,\ y\in R)=[-5,-1]$

Q21. If $R_3=\{(x,[x])|\ x$ is a real number$\}$ is a relation, then find domain and range of R_3.

Solution:

$R_3=\{(x,[x])|\ x$ is a real number$\}$

Since x is defined for all real x, the domain of $R_3=R$.

Since $[x]$ always gives integer values for any real number, the range of $R_3=Z$.

Q22. Express the following functions as set of ordered pairs and determine their range. $f: x\to R$, $f(x)=x^2-1$, where $x=\{-1,0,1\}$

Solution:

$f: x\to R, f(x)=x^2-1$

$f(-1)=(-1)^2-1=1-1=0$

$f(0)=(0)^2-1=0-1=-1$

$f(1)=(1)^2-1=1-1=0$

$\therefore f=\{(-1,0),(0,-1),(1,0)\}$

Thus, range of $f=\{0,-1\}$

Day 36

Long Answer-I (LA-I): 4-Mark and Long Answer-II (LA-II): 6-Mark Questions

Q1. Let $f,g: R\to R$ be defined, respectively, by $f(x)=x+1,\ g(x)=x-1$.

Find $f+g, f-g, fg, \frac{f}{g}$ and $\frac{g}{f}$.

Solution:

(i) $(f+g)(x)=f(x)+g(x)$

$\Rightarrow (f+g)(x)=(x+1)+(x-1)=2x$

(ii) $(f-g)(x)=f(x)-g(x)$

$\Rightarrow (f-g)(x)=(x+1)-(x-1)=2$

(iii) $(fg)(x)=f(x)\cdot g(x)=(x+1)(x-1)=x^2-1$

(iv) $\left(\frac{f}{g}\right)(x)=\frac{f(x)}{g(x)},\ g(x)\ne 0,\ x\in R$

$\Rightarrow \left(\frac{f}{g}\right)(x)=\frac{x+1}{x-1},\ x-1\ne 0$ or $x\ne 1$

(v) $\left(\frac{g}{f}\right)(x)=\frac{g(x)}{f(x)},\ f(x)\ne 0,\ x\in R$

$\Rightarrow \left(\frac{g}{f}\right)(x) = \frac{x-1}{x+1},\ x+1 \neq 0$ or $x \neq -1$

Q2. Let $A = \{1,2,3\}$, $B = \{1,5,6\}$ and $f = \{(1,5),(2,6),(3,1)\}$. Are the following true?

(i) f is a relation from A to B

(ii) f is a function from A to B

Justify your answer in each case.

Solution:

$A = \{1,2,3\}, B = \{1,5,6\}$

$\therefore A \times B = \{(1,1),(1,5),(1,6),$
$(2,1),(2,5),(2,6),$
$(3,1),(3,5),(3,6)\}$

$f = \{(1,5),(2,6),(3,1)\}$

(i) A relation from a non-empty set A to a non-empty set B is a subset of the cartesian product A × B. Thus, f is a relation from A to B.

(ii) Since the same first element 1, corresponds to different images 1, 5, 6, the relation f is not a function.

Q3. If $A = \{-1,0\}, B = \{0,1\}$, then determine

(i) $A \times B$ (ii) $B \times A$
(iii) $B \times B$ (iv) $A \times A$

Solution:

(i) $A \times B = \{(-1,0),(-1,1),(0,0),(0,1)\}$

(ii) $B \times A = \{(0,-1),(0,0),(1,-1),(1,0)\}$

(iii) $B \times B = \{(0,0),(1,0),(0,1),(1,1)\}$

(iv) $A \times A = \{(-1,-1),(-1,0),(0,0),(0,-1)\}$

Q4. Let f be the subset of $Z \times Z$ defined by $f = \{(a/b, a+b) : a,b \in Z\}$. Is f a function from Z to Z: Justify your answer.

Solution:

The relation f is defined as $f = \{(a/b, a+b) : a,b \in Z\}$.

A relation f from a set A to a set B is said to be a function if every element of set A has unique images in set B.

Let $(1,1),(-1,-1) \in Z$

$\Rightarrow (1/1, 1+1),(-1/-1, -1+(-1)) \in f$

$\Rightarrow (1,2),(1,-2) \in f$

Since element 1 corresponds to two different images 2 and –2, the relation f is not a function.

Q5. Let $A = \{6,7,8,9,10\}$ and let $f : A \to N$ be defined by $f(n) =$ the highest prime factor of n. Find the range of f.

Solution:

$A = \{6,7,8,9,10\}$ and $f : A \to N$ is defined as

$f(n) =$ The highest prime factor of n

Now, since the prime factors of 6 = 2, 3

Prime factors of 7 = 7

Prime factor of 8 = 2

Prime factor of 9 = 3

Prime factors of 10 = 2, 5

$\therefore f(6) =$ The highest prime factor of 6 = 3

$f(7) =$ The highest prime factor of 7 = 7

$f(8) =$ The highest prime factor of 8 = 2

$f(9) =$ The highest prime factor of 9 = 3

$f(10) =$ The highest prime factor of 10 = 5

Thus, the range of f is the set of all $f(n)$, where $n \in A$, that is, the range of $f = \{2,3,5,7\}$.

Q6. $A = \{0,1,2,3,4\}, S = \{(x,y) : x \in A, y \in A\}$, then find the order which satisfies the conditions given below.

(i) $x + y = 4$ (ii) $x \cdot y = 4$ (iii) $x + y \leq 4$

Solution:

$A = \{0,1,2,3,4\},\ S = \{(x,y) : x \in A, y \in A\}$

(i) The set of ordered pairs satisfying $x + y = 4$ is {(0, 4), (1, 3), (2, 2), (3, 1), (4, 0)}

(ii) The set of ordered pairs satisfying $x \cdot y = 4$ is {(1, 4), (2, 2), (4, 1)}

(iii) The set of ordered pairs satisfying $x + y \leq 4$ is {(0, 0), (0, 1), (0, 2), (0, 3), (0, 4), (1, 0), (1, 1), (1, 2), (1, 3), (2, 0), (2, 1), (2, 2), (3, 0) (3, 1), (4, 0)}

Q7. Let $f = \{(0,0),(1,2),(2,4)\}$ and $g = \{(1,3),(2,6)\}$ be two real functions. Then find

(i) $f - g$ (ii) $f + g$
(iii) $f \cdot g$ (iv) $\frac{f}{g}$

Solution:

Since $f = \{(0,0),(1,2),(2,4)\}$

Domain of $f = \{0,1,2\}$

and $g = \{(1,3),(2,6)\}$

Domain of $g = \{1,2\}$

The domain of $f - g,\ f + g,\ f \cdot g,\ \frac{f}{g}$ is the domain of $f \cap$ domain of g.

$\therefore$ Domain of $f - g,\ f + g,\ f \cdot g,\ \frac{f}{g}$ is

= Domain of $f \cap$ Domain of g

$= \{0,1,2\} \cap \{1,2\} = \{1,2\}$

Thus,

(i) $(f - g)(1) = f(1) - g(1) = 2 - 3 = -1$

$(f-g)(2) = f(2) - g(2) = 4 - 6 = -2$

$\therefore f - g = \{(1,-1),(2,-2)\}$

(ii) $(f+g)(1) = f(1) + g(1) = 2 + 3 = 5$

$(f+g)(2) = f(2) + g(2) = 4 + 6 = 10$

$\therefore f - g = \{(1,5),(2,10)\}$

(iii) $(f \cdot g)(1) = f(1) \cdot g(1) = 2 \cdot 3 = 6$

$(f \cdot g)(2) = f(2) \cdot g(2) = 4 \cdot 6 = 24$

$\therefore f \cdot g = \{(1,6),(2,24)\}$

(iv) $\left(\frac{f}{g}\right)(1) = \frac{f(1)}{g(1)} = \frac{2}{3}$

$\left(\frac{f}{g}\right)(2) = \frac{f(2)}{g(6)} = \frac{4}{6} = \frac{2}{3}$

$\therefore \frac{f}{g} = \left\{\left(1, \frac{2}{3}\right), \left(2, \frac{2}{3}\right)\right\}$

Q8. If $f(x) = \sqrt{x}$, $g(x) = x^2$ are two functions defined in the domain $R^+ \cup \{0\}$, then find

(i) $(f+g)(x)$ (ii) $(f-g)(x)$

(iii) $(fg)(x)$ (iv) $\left(\frac{f}{g}\right)(x)$ (v) $\left(\frac{g}{f}\right)(x)$

Solution:

(i) $(f+g)(x) = f(x) + g(x) = \sqrt{x} + x^2$

(ii) $(f-g)(x) = f(x) - g(x) = \sqrt{x} - x^2$

(iii) $(fg)(x) = f(x) \cdot g(x) = \sqrt{x} \cdot x^2 = x^{5/2}$

(iv) $\left(\frac{f}{g}\right)(x) = \frac{f(x)}{g(x)} = \frac{\sqrt{x}}{x^2} = \frac{1}{x\sqrt{x}}$ defined in the domain R^+

(v) $\left(\frac{g}{f}\right)(x) = \frac{g(x)}{f(x)} = \frac{x^2}{\sqrt{x}} = x\sqrt{x}$ defined in the domain $R^+ \cup \{0\}$

Q9. Let $f = \left\{\left(x, \frac{x^2+1}{x^2-1}\right) : x \in R\right\}$ be a function from R into R. Determine the range of f.

Solution:

Let, for $x \in R$, $y = \frac{x^2+1}{x^2-1}$, where y is the range of f

Now, rearranging the equation in terms of x, we get

$y = \frac{x^2+1}{x^2-1} \Rightarrow x^2 y - y = x^2 + 1$

$\Rightarrow x^2 y - x^2 = 1 + y$

$\Rightarrow x^2(y-1) = 1 + y$

$\Rightarrow x^2 = \frac{1+y}{y-1} \Rightarrow x = \sqrt{\frac{1+y}{y-1}}$

Now, the function $g(y) = \sqrt{\frac{1+y}{y-1}}$ is defined when $\frac{1+y}{y-1} \geq 0$ and $y - 1 \neq 0$

$\Rightarrow y \leq -1$ or $y \geq 1$ and $y \neq 1$

$\Rightarrow y \leq -1$ or $y > 1$

Thus, the range of $f = (-\infty, -1] \cup (1, \infty)$.

Q10. Find the domain and the range of the real function f defined by $f(x) = \frac{1}{|x^2 - 1|}$.

Solution:

Function $f(x) = \frac{1}{|x^2-1|}$ is defined for all real numbers except when $x^2 - 1 = 0 \Rightarrow x^2 = 1$

$\Rightarrow x = 1, -1$

$\therefore$ Domain of $f = R - \{-1, 1\}$

Also, since $|x^2 - 1|$ will take all non-negative real numbers, the range of f is the set of all positive real numbers, that is, range of $f = (0, \infty)$.

Q11. Find the domain and the range of the real function f defined by $f(x) = \sqrt{(x^2 - 1)}$.

Solution:

Since the given function $f(x) = \sqrt{(x^2-1)}$ is defined when $x^2 - 1 \geq 0 \Rightarrow x^2 \geq 1$

$\Rightarrow x \leq -1$ or $x \geq 1$

Therefore, the domain of f is the set of all real numbers less than or equal to –1, or greater than or equal to 1, that is, the domain of $f = (-\infty, -1) \cup (1, \infty)$.

Now, since in the domain of x, $(x^2 - 1) \geq 0 \Rightarrow \sqrt{(x^2-1)} \geq 0 \Rightarrow f(x) \geq 0$

Therefore, the range of f is the set of all real numbers greater than or equal to 0, that is, range of $f = [0, \infty)$.

Q12. Find domain of the following functions:

(i) $f(x) = \frac{1}{\sqrt{1 - \sin x}}$ (ii) $f(x) = \frac{x^2 - 2x + 1}{x^2 - x - 6}$

(iii) $f(x) = \frac{1}{x - |x|}$ (iv) $f(x) = \frac{1}{x - [x]}$

Solution:

(i) $f(x) = \frac{1}{\sqrt{1-\sin x}}$ is defined when $1 - \sin x > 0 \Rightarrow \sin x < 1$

Also, $\sin x \geq -1$ and since $-1 \leq \sin x < 1$

Thus, $f(x)$ is defined, for all x, except when

$1-\sin x \neq 0 \Rightarrow \sin x \neq 1 \Rightarrow x \neq (2n+1)\frac{\pi}{2}\ \forall n \in Z$

Hence, domain of $f = R-\{(2n+1)\frac{\pi}{2} : n \in Z\}$

(ii) The given function $f(x)=\dfrac{x^2-2x+1}{x^2-x-6}$ can be simplified as $f(x)=\dfrac{(x-1)^2}{(x-3)(x+2)}$

Function f is defined for all real numbers except when $x-3=0$ or $x+2=0$

Thus, at $x = 3, -2$ function f is not defined

Hence, the domain of f is $R-\{-2,3\}$.

(ii) $f(x)=\dfrac{1}{x-|x|}$

$$\Rightarrow x-|x|=\begin{cases} x-x=0, & x\geq 0 \\ x+x=2x, & x<0 \end{cases}$$

Since $f(x)$ is not defined when $x-|x|=0$,

Domain of $f = R^-$

(iii) $f(x)=\dfrac{1}{x-[x]}$

Let $x=[x]+f$, where f is the fraction part

$$\because x=\begin{cases} [x], & x=Z \\ [x]+f, & x\neq Z \end{cases}$$

$$\therefore x-[x]=\begin{cases} 0, & x=Z \\ f, & x\neq Z \end{cases}$$

Since $f(x)$ is not defined when $x-[x]=0$, it means that the function is defined at all real values of x except at all integers. Thus, the domain of $f = R-Z$.

Q13. Find the range of the following functions:

(i) $f(x)=\dfrac{1}{2-x^2}$ (ii) $f(x)=2-|x-1|$

(iii) $f(x)=1+2\cos 2x$

Solution:

(i) $f(x)=\dfrac{1}{2-x^2}$

Let $y = f(x)$

$\Rightarrow y=\dfrac{1}{2-x^2} \Rightarrow 2-x^2=\dfrac{1}{y}$

$\Rightarrow x^2=2-\dfrac{1}{y} \Rightarrow x=\sqrt{\dfrac{2y-1}{y}}$

Since x is real, $\dfrac{2y-1}{y}\geq 0$ and $y\neq 0$

$\Rightarrow y\geq \frac{1}{2}$ or $y\leq 0$ and $y\neq 0$

$\Rightarrow y\geq \frac{1}{2}$ or $y<0$

Thus, range of $f = (-\infty,0)\cup\left[\frac{1}{2},\infty\right)$

(ii) $\because |x-1|\geq 0 \Rightarrow -|x-1|\leq 0$

$\Rightarrow 2-|x-1|\leq 2 \Rightarrow f(x)\leq 2$

$\therefore$ Range of $f = (-\infty,1]$

(iii) $-1\leq \cos 2x\leq 1 \Rightarrow -2\leq 2\cos 2x\leq 2$

$\Rightarrow 1-2\leq 1+2\cos 2x\leq 1+2$

$\Rightarrow -1\leq 1+2\cos 2x\leq 3 \Rightarrow -1\leq f(x)\leq 3$

Thus, range of $f = [-1,3]$

CHAPTER

3 Trigonometric Functions

Day 37

Angles and their Measurement

Angle: An angle is the measurement of rotation of a given ray about its initial position. The point of rotation is called the vertex.

The position of the original ray is called the *initial side* and the position of the ray after rotation is called the *terminal side* of the angle.

The measure of rotation or the measure of an angle is the amount of rotation from the initial side to the terminal side, about the point of rotation.

The measure of an angle is said to be positive if the direction of rotation is anti-clockwise, and it said to be negative if the direction of rotation is clockwise (Fig. 3.1).

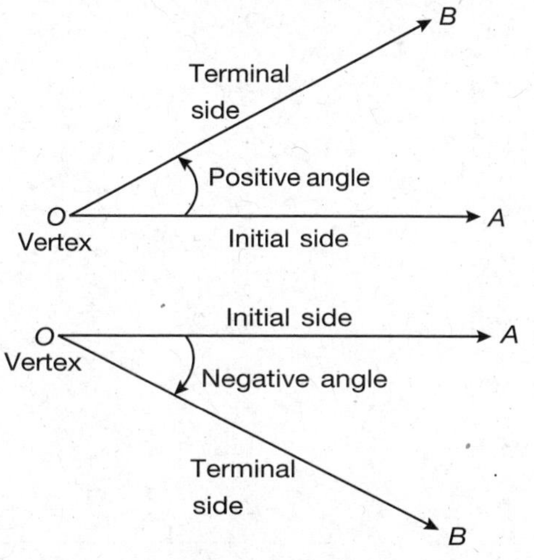

Fig. 3.1 Positive and negative angles

There are several units for measuring angles, among which the most commonly used units are degree measure and radian measure. In general, the measured angles are represented by the symbol $\angle\theta$ or $\angle AOB$.

Degree Measure: If a rotation from the initial side to the terminal side is $\left(\frac{1}{360}\right)^{\text{th}}$ of a complete revolution, the measurement of the angle is said to be one degree, written as $1°$.

Thus, one complete rotation or revolution is measured to be an angle of 360° [Fig. 3.2(i)].

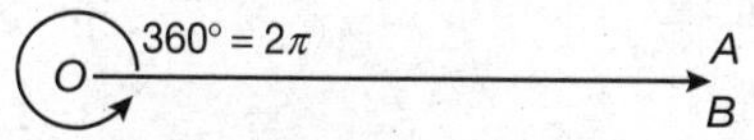

Fig. 3.2(i) Angle 360°

A half rotation or revolution is measured to be an angle of 180° [Fig. 3.2(ii)].

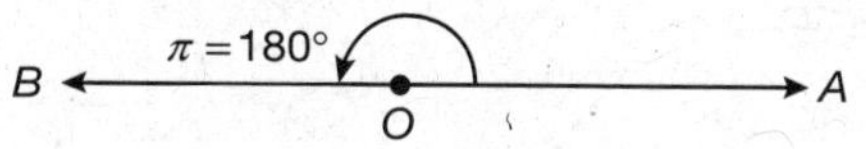

Fig. 3.2(ii) Angle 180°

A one-fourth rotation or revolution is measured to be an angle of 90° [Fig. 3.2(iii)].

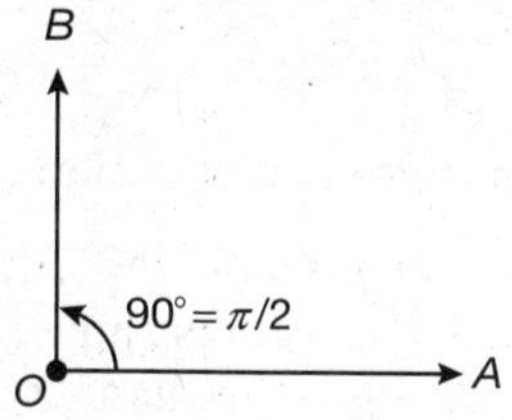

Fig. 3.2(iii) Angle 90°

A three-fourth rotation or revolution is measured to be an angle of 270° [Fig. 3.2(iv)].

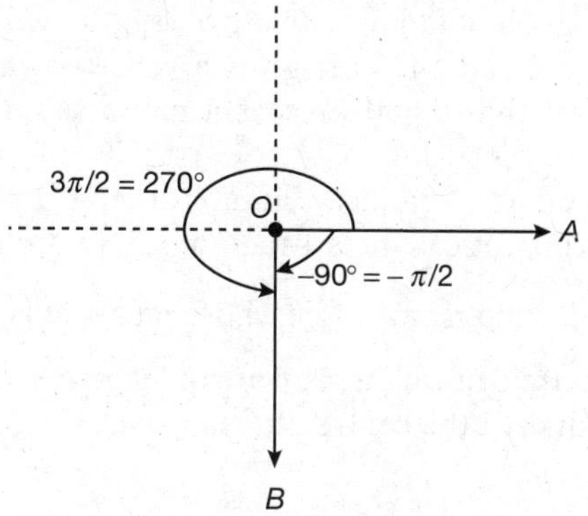

Fig. 3.2(iv) Angle 270°

Note: A degree is divided into 60 minutes and a minute is divided into 60 seconds. That is, one-sixtieth of a degree is called a minute, written as $1'$ and one-sixtieth of a minute is called a second, written as $1''$.

Thus, $1° = 60'$ and $1' = 60'' \Rightarrow 1° = 3600''$.

Radian Measure: A radian is the angle subtended by a circular arc at the centre of the circle.
Thus, the angle subtended at the centre of a unit circle (circle of radius 1 unit) by a circular arc of length 1 unit is said to be a measure of 1 radian.

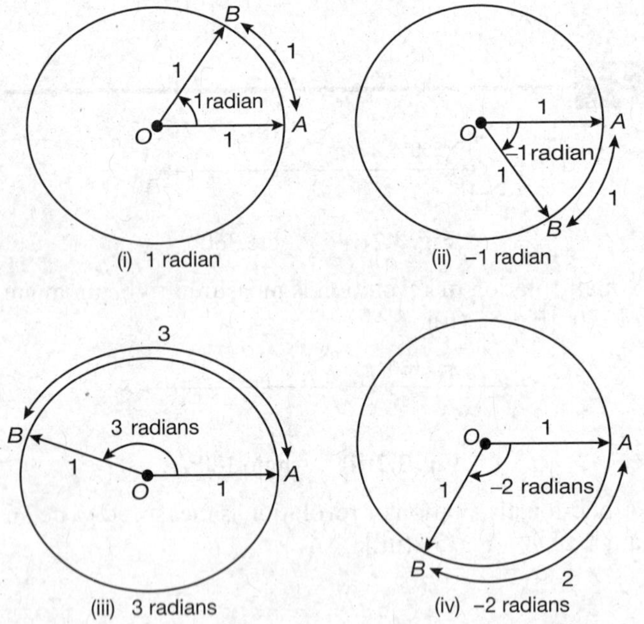

Fig. 3.3 Representation of radian measurement

In the Figs 3.3(i) to (iv), *OA* is the initial side and *OB* is the terminal side. The figures show the angles whose measures are 1 radian, –1 radian, 3 radians and –2 radians.

Note: Since the circumference of a circle of radius 1 unit is 2π, one complete revolution of the initial side subtends an angle of 2π radians.
Thus, in a circle of radius *r*, an arc of length *r* will subtend an angle of 1 radian. This gives a well-known relation which states that equal arcs of a circle subtend equal angles at the centre.
Now, in a circle of radius *r*, an arc of length *r* subtends an angle whose measure is 1 radian. Therefore, an arc of length *l* will subtend an angle whose measure is $\frac{l}{r}$ radian.
Thus, in a circle of radius *r*, an arc of length *l* subtends an angle θ radian at the centre, that is,

$$\theta = \frac{l}{r} \text{ or } l = r\theta$$

Note: Every real number will correspond to a radian measure and vice versa. Thus, radian measures and real numbers can be considered as one and the same.

Relation between Degree and Radian: Since a revolution around a circle measures the degree angle as 360° and a circle subtends an angle at the centre whose radian measure is 2π, it follows that 2π radians = 360° or π radians = 180°.
From the above, we obtain the relationship of radians in terms of degrees and degrees in terms of radians, as follows.

$$\text{Radian measure} = \frac{\pi}{180} \times \text{Degree measure}$$

$$\text{Degree measure} = \frac{180}{\pi} \times \text{Radian measure}$$

The relation between degrees and radians for some common angles are given in the table below

Degree	**Radian**
0	0
30	$\pi/6$
45	$\pi/4$
60	$\pi/3$
90	$\pi/2$
120	$2\pi/3$
135	$3\pi/4$
150	$5\pi/6$
180	π
210	$7\pi/6$
225	$5\pi/4$
270	$3\pi/2$
300	$5\pi/3$
330	$11\pi/6$
360	2π

Note: Since angles are measured either in degree or in radians, we adopt the convention that whenever we write the angle as $\theta°$, we mean the angle whose degree measure is θ, and whenever we write the angle as only θ (without the degree symbol), we mean the angle whose radian measure is θ. Generally, when an angle is expressed in radians, the word, 'radian' is frequently omitted.

Angles measured as $\pi = 180°$ or $\frac{\pi}{2} = 90°$ or $\frac{\pi}{3} = 60°$ or $\frac{\pi}{4} = 45°$ are written with the understanding that $\pi, \frac{\pi}{2}, \frac{\pi}{3}, \frac{\pi}{4}$ are radian measures while $180°, 90°, 60°, 45°$

are degree measures. For example, angle 1° represents an angle measure of one degree, that is, $1° = \frac{\pi}{180}$ radian = 0.01746 radians approximately.

Angle 1 represents an angle measure of one radian, that is, $1 = 1 \text{ radian} = \frac{180°}{\pi} = 57°16'$ approximately (taking the approximate value of π as 22/7).

Representation of Angles in Degree and Radian: Representation of some angles in degrees and radians is shown below (Fig. 3.4).

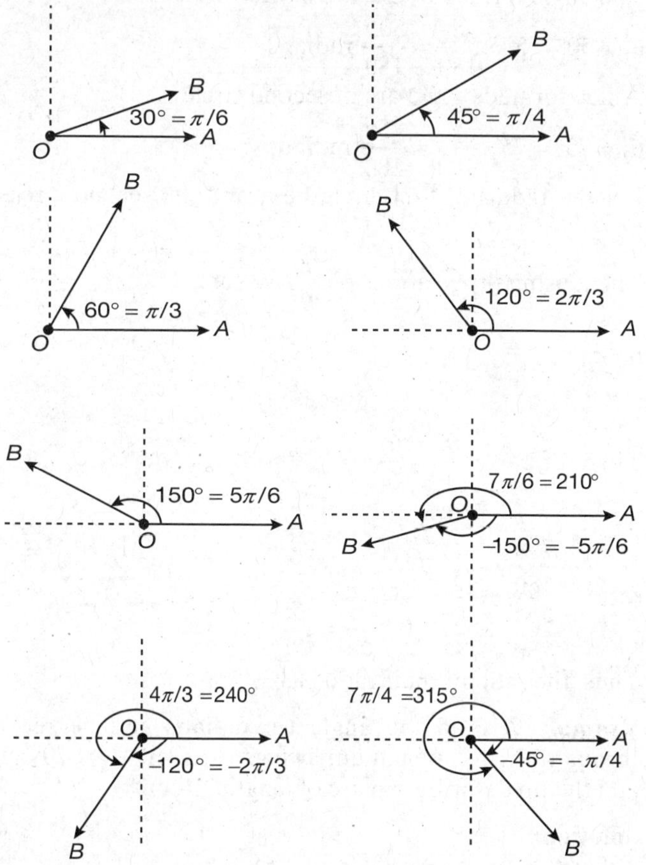

Fig. 3.4 Representation of some angles in degrees and radians

Example 1: Find the radian measure corresponding to the following degree measures:

(i) 500° (ii) –630°

(iii) –115°30′ (iv) 75°45′30″

Solution:

(i) 180° = π radians

Therefore, $500° = \frac{\pi}{180} \times 500 = \frac{25\pi}{9}$ radians

(ii) 180° = π radians

Therefore, $-630° = \frac{\pi}{180} \times (-630) = -\frac{7\pi}{2}$ radians

(iii) $-115°30' = -\left(115\frac{30}{60}\right)^{\circ}$

$= -\left(115\frac{1}{2}\right)^{\circ} = -\left(\frac{231}{2}\right)^{\circ}$

Therefore, $-\left(\frac{231}{2}\right)^{\circ} = \frac{\pi}{180} \times \left(-\frac{231}{2}\right) = -\frac{77\pi}{120}$ radians

(iv) $5°10'30'' = 5°(10\frac{30}{60})' = 5°(10\frac{1}{2})'$

$= 5°(\frac{21}{2})' = \left(5\frac{21}{2} \times \frac{1}{60}\right)^{\circ} = \left(5\frac{7}{40}\right)^{\circ} = \left(\frac{207}{40}\right)^{\circ}$

Therefore, $\left(\frac{207}{40}\right)^{\circ} = \frac{\pi}{180} \times \frac{207}{40} = \frac{23\pi}{800}$ radians

Example 2: Find the degree measures corresponding to the following radian measures (use $\pi = \frac{22}{7}$).

(i) $\frac{4\pi}{3}$ (ii) $-\frac{5\pi}{6}$ (iii) $\frac{11}{12}$ (iv) –3

Solution:

(i) π radians = 180°

Therefore, $\frac{4\pi}{3}$ radians $= \frac{180}{\pi} \times \frac{4\pi}{3}$ degree

= 240 degree = 240°

(ii) π radians = 180°

Therefore, $-\frac{5\pi}{6}$ radians $= \frac{180}{\pi} \times \left(-\frac{5\pi}{6}\right)$ degree

= –150 degree = –150°

(iii) π radians = 180°

Therefore, $\frac{11}{12}$ radians $= \frac{180}{\pi} \times \frac{11}{12}$ degree

$= \frac{180 \times 7}{22} \times \frac{11}{12} = \frac{105}{2}$ degree $= 52\frac{1}{2}$ degree

$= 52° + \frac{1}{2} \times 60$ minutes $[\because 1° = 60']$

= 52° + 30 minutes = 52°32′

(iv) π radians = 180°

Therefore, –3 radians $= -\frac{180}{\pi} \times 3$ degree

$= -\frac{180 \times 7}{22} \times 3$ degree $= -\frac{1890}{11}$ degree

$= -171\frac{9}{11}$ degree

$= -\left[171° + \frac{9}{11} \times 60 \text{ minutes}\right]$ $(\because 1° = 60')$

$= -\left[171° + \frac{540}{11} \text{ minutes}\right] = -\left[171° + 49\frac{1}{11} \text{ minutes}\right]$

$= -[171° + 49' + \frac{1}{11} \times 60''] \quad (\because 1' = 60'')$

$= -[171° + 49' + 5''] = -171°49'5''$

Example 3: A wheel makes 100 revolutions in one minute. Through how many degrees or radians does it turn in one second?

Solution:
The number of revolutions in one minute (60 seconds) = 100

Therefore, number of revolutions in 1 second $= \frac{100}{60} = \frac{5}{3}$

The angle formed in one revolution = 360° = 2π radians
Therefore, the angle formed in 5 revolutions

$= 360° \times \frac{5}{3} = 600° = 2\pi \times \frac{5}{3} = \frac{10}{3}\pi$ radians

Example 4: Find the degree and radian measure of the angle subtended at the centre of a circle of radius 80 cm by an arc of length 33 cm (use $\pi = \frac{22}{7}$).

Solution:
Given radius $r = 80\,\text{cm}$, length of arc $l = 33\,\text{cm}$

Since angle in radians, $\theta = \frac{l}{r}$

$\Rightarrow \theta = \frac{33}{80}$ radians

Now, π radians = 180°

Therefore, $\frac{33}{80}$ radians $= \frac{180}{\pi} \times \frac{33}{80}$ degree

$= \frac{180 \times 7}{22} \times \frac{33}{80}$ degree $= \frac{189}{8}$ degree

$= 23\frac{5}{8}$ degree

$= 23° + \frac{5}{8} \times 60$ minutes $[\because 1° = 60']$

$= 23° + \frac{75}{2}$ minutes

$= 23° + 37' + \frac{1}{2} \times 60'' = 23° + 37' + 30'' = 23°37'30''$

Thus, the angle formed by the arc at the centre is $= 23°37'30''$.

Example 5: In a circle of diameter 30 cm, the length of a chord is 15 cm. Find the length of the minor arc of the chord.

Solution:

Given radius $r = \frac{30}{2} = 15\,\text{cm}$

Length of chord $AB = 15\,\text{cm}$
Since the radius is equal to the length of the chord, the triangle formed by the chord with centre is an equilateral triangle. That is, the angle subtended = 60°

$= 60 \times \frac{\pi}{180} = \frac{\pi}{3}$ radians

Now, using the relations $l = \theta \times r$,

$\Rightarrow l = \frac{\pi}{3} \times 15\,\text{cm} = 5\pi\,\text{cm}$

Thus, the length of the minor arc of the chord is 5π cm.

Example 6: If in two circles, arcs of the same length subtend angles 50° and 75° at the centre, find the ratio of their radii.

Solution:
The angle formed by the arc of first circle is

$\theta_1 = 50° = 50 \times \frac{\pi}{180} = \frac{50\pi}{180}$ radians

Angle formed by the arc of second circle is

$\theta_2 = 75° = 75 \times \frac{\pi}{180} = \frac{75\pi}{180}$ radians

Let the radius of first circle be r_1 and the second circle be r_2.

Then, using the relations $r = \frac{l}{\theta}$, we get

$r_1 = \frac{l}{\theta_1},\ r_2 = \frac{l}{\theta_2}$

Therefore, $\frac{r_1}{r_2} = \frac{\frac{l}{\theta_1}}{\frac{l}{\theta_2}} = \frac{\theta_2}{\theta_1}$

$\Rightarrow \frac{r_1}{r_2} = \frac{\frac{75\pi}{180}}{\frac{50\pi}{180}} = \frac{75}{50} = \frac{3}{2} = 3:2$

Thus, the ratio of their radii is 3:2.

Example 7: Find the angle (in radians and degrees) through which a pendulum swings if its length is 70 cm and the tip describes an arc of length 20 cm.

Solution:

(i) Given the length of arc $l = 20\,\text{cm}$
Length of the pendulum = radius of circle, $r = 70\,\text{cm}$

Thus, using the relations $\theta = \frac{l}{r}$ we get

$\theta = \frac{20}{70} = \frac{2}{7}$ radians

Now, π radians = 180°

Therefore, $\frac{2}{7}$ radians $= \frac{180}{\pi} \times \frac{2}{7}$ degree

$= \frac{180 \times 7}{22} \times \frac{2}{7} = \frac{180}{11}$ degree

$= 16\frac{4}{11}$ degree

$= 16° + \frac{4}{11} \times 60$ minutes $[\because 1° = 60']$

$= 16° + \frac{240}{11}$ minutes

$= 16° + 21' + \frac{9}{11} \times 60''$

$= 16° + 21' + 49'' = 16°21'49''$

Thus, the angle formed by the pendulum is 2/7 radians and 16°21′49″.

Day 38

NCERT Exercise 3.1 (With Solutions)

Q1. Find the radian measure corresponding to the following degree measures:

(i) 25° (ii) –47°30′
(iii) 240° (iv) 520°

Solution:

(i) 180° = π radians

Therefore, $25° = \frac{\pi}{180} \times 25$ radian $= \frac{5\pi}{36}$ radians

(ii) 180° = π radians

Therefore, $-47°30' = -47\frac{1}{2}$ degree $= -\frac{95}{2}$ degree

$= -\frac{\pi}{180} \times \frac{95}{2}$ radians $= -\frac{19\pi}{72}$ radians

(iii) 180° = π radians

Therefore, $240° = \frac{\pi}{180} \times 240$ radians $= \frac{4\pi}{3}$ radians

(iv) 180° = π radians

Therefore, $520° = \frac{\pi}{180} \times 520$ radians $= \frac{26\pi}{9}$ radians

Q2. Find the degree measures corresponding to the following radian measures (use $\pi = \frac{22}{7}$):

(i) 11/16 (ii) –4
(iii) $\frac{5\pi}{3}$ (iv) $\frac{7\pi}{6}$

Solution:

(i) π radians = 180°

Therefore, 11/16 radians $= \frac{180}{\pi} \times \frac{11}{16}$ degree

$= \frac{180 \times 7}{22} \times \frac{11}{16}$ degree $= \frac{315}{6}$ degree

$= 39\frac{3}{8}$ degree

$= 39° + \frac{3}{8} \times 60$ minutes $[\because 1° = 60']$

$= 39° + 45/2$ minutes $= 39° + 22\frac{1}{2}$ minutes

$= 39° + 22' + \frac{1}{2} \times 60''$ $[\because 1' = 60'']$

$= 39° + 22' + 30'' = 39°22'30''$

Thus, 11/16 radians = 39°22′30″

(ii) π radians = 180°

Therefore, –4 radians $= -\frac{180}{\pi} \times 4$ degree

$= -\frac{180 \times 7}{22} \times 4$ degree $= -\frac{2250}{11}$ degree

$= -229\frac{1}{11}$ degree

$= -\left[229° + \frac{1}{11} \times 60 \text{minutes}\right]$ $[\because 1° = 60']$

$= -\left[229° + \frac{60}{11} \text{minutes}\right]$

$= -\left[229° + 5\frac{5}{11} \text{minutes}\right]$

$= -[229° + 5' + \frac{5}{11} \times 60'']$ $[\because 1' = 60'']$

$= -[229° + 5' + 27''] = -229°5'27''$

Thus, –4 radians = –229°5′27″

(iii) π radians = 180°

Therefore, $\frac{5\pi}{3}$ radians $= \frac{180}{\pi} \times \frac{5\pi}{3}$ degree

= 300 degree = 300°

Thus, $\frac{5\pi}{3}$ radians = 300°

(iv) π radians = 180°

Therefore, $\frac{7\pi}{6}$ radians $= \frac{180}{\pi} \times \frac{7\pi}{6}$ degree = 210 degree = 210°

Thus, $\frac{7\pi}{6}$ radians = 210°

Q3. A wheel makes 360 revolutions in one minute. Through how many radians does it turn in one second?

Solution:

The number of revolutions in one minute (60 seconds) = 360

Therefore, number of revolutions in 1 second $= \frac{360}{60} = 6$

The angle formed in one revolution = 360° = 2π radians.

Therefore, the angle formed in 6 revolutions = $6 \times 2\pi = 12\pi$ radians

Q4. Find the degree measure of the angle subtended at the centre of a circle of radius 100 cm by an arc of length 22 cm (use $\pi = \frac{22}{7}$).

Solution:

Given radius $r = 100$ cm, length of arc, $l = 22$ cm

Angle in radians, $\theta = \frac{l}{r}$

$\Rightarrow \theta = \frac{22}{100}$ radians $= \frac{11}{50}$ radians

π radians $= 180°$

Therefore, $= \frac{11}{50}$ radians $= \frac{180}{\pi} \times \frac{11}{50}$ degree

$= \frac{180 \times 7}{22} \times \frac{11}{50}$ degree $= \frac{63}{5}$ degree $= 12\frac{3}{5}$ degree

$= 12° + \frac{3}{5} \times 60$ minutes $[\because 1° = 60']$

$= 12° + 36$ minutes

$= 12° + 36' = 12°36'$

Thus, the angle formed by arc at the centre is $12°36'$.

Q5. In a circle of diameter 40 cm, the length of a chord is 20 cm. Find the length of minor arc of the chord.

Solution:

Given radius $r = \frac{40}{2} = 20$ cm

Length of chord $AB = 20$ cm

The radius is equal to the length of the chord. Therefore, the triangle formed by the chord with the centre is an equilateral triangle. That is, the angle subtended = 60 degree

$= 60 \times \frac{\pi}{180} = \frac{\pi}{3}$ radians

Now, using the relation $l = \theta \times r$

$\Rightarrow l = \frac{\pi}{3} \times 20$ cm $= \frac{20\pi}{3}$ cm

Thus, the length of minor arc of the chord is $\frac{20\pi}{3}$ cm.

Q6. If in two circles, arcs of the same length subtend angles 60° and 75° at the centre, find the ratio of their radii.

Solution:

The angle formed by the arc of the first circle is

$\theta_1 = 60° = 60 \times \frac{\pi}{180} = \frac{60\pi}{180}$ radians

Angle formed by the arc of the second circle is

$\theta_2 = 75° = 75 \times \frac{\pi}{180} = \frac{75\pi}{180}$ radians

Let the radius of first circle be r_1 and the second circle be r_2.

Then, using the relation $r = \frac{l}{\theta}$, we get

$r_1 = \frac{l}{\theta_1}$, $r_2 = \frac{l}{\theta_2}$

Therefore, $\frac{r_1}{r_2} = \frac{\frac{l}{\theta_1}}{\frac{l}{\theta_2}} = \frac{\theta_2}{\theta_1}$

$\Rightarrow \frac{r_1}{r_2} = \frac{\frac{75\pi}{180}}{\frac{60\pi}{180}} = \frac{75}{60} = \frac{5}{4} = 5:4$

Thus, the ratio of their radii is 5:4.

Q7. Find the angle (in radians) through which a pendulum swings if its length is 75 cm and the tip describes an arc of length

(i) 10 cm (ii) 15 cm (iii) 21 cm

Solution:

(i) Given the length of the arc $l = 10$ cm

Length of the pendulum = radius of circle, $r = 75$ cm

Thus, using the relation $\theta = \frac{l}{r}$ we get

$\theta = \frac{10}{75} = \frac{2}{15}$ radians

Thus, the angle formed by the pendulum is 2/15 radians.

(ii) Given the length of the arc $l = 15$ cm

Length of the pendulum = radians of circle $r = 75$ cm

Thus, using the relation $\theta = \frac{l}{r}$ we get

$\theta = \frac{15}{75} = \frac{1}{5}$ radians

Thus, the angle formed by the pendulum is 1/5 radians.

(iii) Given the length of the arc, $l = 21$ cm

Length of the pendulum = radius of circle $r = 75$ cm

Thus, using the relation $\theta = \frac{l}{r}$ we get

$\theta = \frac{21}{75} = \frac{7}{25}$ radians

Thus, the angle formed by the pendulum is 7/25 radians.

Day 39

Trigonometric Functions

In this chapter, we extend our previous knowledge of trigonometric ratios (trigonometric ratios of acute angles is the ratio of the sides of a right-angled triangle) to any angle in terms of radian measure and study them as trigonometric functions.

Consider a unit circle with its centre at the origin of the coordinate axes (Fig. 3.5).

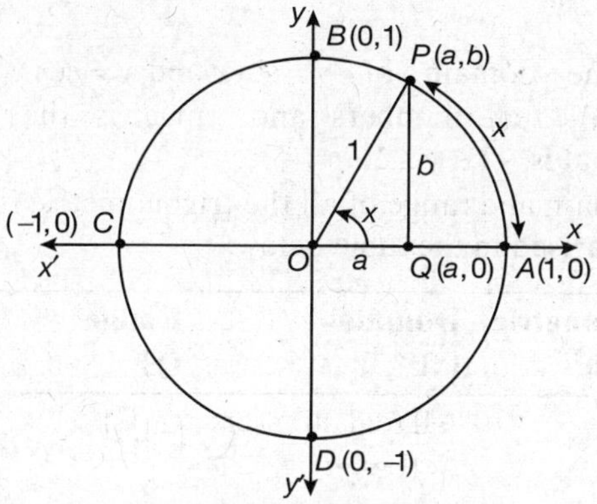

Fig. 3.5 Unit circle

Let $P(a,b)$ be any point on the circle and point $Q(a,0)$ just below point P on the x-axis, and angle $POQ = x$ radian, that is, length of arc $AP = x$.

Now, since ΔOQP is a right-angled triangle at Q and radius $OP = 1$,

$\sin x = \frac{b}{1} = b$ and $\cos x = \frac{a}{1} = a$

Thus, using the hypotenuse theorem: $OQ^2 + PQ^2 = OP^2$ or $a^2 + b^2 = 1$ or $\cos^2 x + \sin^2 x = 1$, for every point on the unit circle.

One complete revolution subtends an angle of 2π radians at the centre of the circle. Therefore, in the quadrilateral $ABCD$ with coordinates $(1,0),(0,1),(-1,0),(0,-1)$, respectively, we get

$\angle AOB = \frac{\pi}{2}$, $\angle AOC = \pi$, $\angle AOD = \frac{3\pi}{2}$

Since $\cos x = a$ where a gives the value of the x-coordinate on the unit circle, we get $\cos 0° = 1$, $\cos\frac{\pi}{2} = 0$, $\cos\pi = -1$, $\cos\frac{3\pi}{2} = 0$. Similarly, since $\sin x = b$, where b gives the value of the y-coordinate on the unit circle, we get $\sin 0° = 0$, $\sin\frac{\pi}{2} = 1$, $\sin\pi = 0$, $\sin\frac{3\pi}{2} = -1$.

Now, if we take one complete revolution or rotation of 2π radian from the point P, we again come back to the same point P.

Thus, as x increases (or decreases) by any integral multiple of 2π, the values of the sine and cosine functions do not change. Hence, we get

$\sin(2n\pi \pm x) = \sin x, \quad n \in Z$

$\cos(2n\pi \pm x) = \cos x, \quad n \in Z$

Also, $\sin x = 0$, if $x = 0, \pm\pi, \pm 2\pi, \pm 3\pi, \ldots$, that is, when x is an integral multiple of π and

$\cos x = 0$, if $x = \pm\frac{\pi}{2}, \pm\frac{3\pi}{2}, \pm\frac{5\pi}{2}, \ldots$, that is, $\cos x$ vanishes when x is an odd multiple of $\frac{\pi}{2}$.

Thus, $\sin x = 0$ implies $x = n\pi$, and $\cos x = 0$ implies $x = (2n+1)\frac{\pi}{2}$, where n is any integer.

Similarly, other trigonometric functions in terms of the sine and cosine functions can be defined as:

$\operatorname{cosec} x = \frac{1}{\sin x}, \quad x \neq n\pi$, where n is any integer

$\sec x = \frac{1}{\cos x}, \quad x \neq (2n+1)\frac{\pi}{2}$, where $n \in Z$

$\tan x = \frac{\sin x}{\cos x}, \quad x \neq (2n+1)\frac{\pi}{2}$, where $n \in Z$

$\cot x = \frac{\cos x}{\sin x}, \quad x \neq n\pi$, where $n \in Z$

Now, dividing equation $\sin^2 x + \cos^2 x = 1$ (where x is any real number) by $\cos^2 x$ and $\sin^2 x$ simultaneously, we get $1 + \tan^2 x = \sec^2 x$ and $1 + \cot^2 x = \operatorname{cosec}^2 x$.

Values of Some Trigonometric Functions: The values of some trigonometric functions or ratios are given in the table below.

Angle	**0 (0°)**	**$\pi/6$ (30°)**	**$\pi/4$ (45°)**	**$\pi/3$ (60°)**	**$\pi/2$ (90°)**
sin	0	½	$\frac{1}{\sqrt{2}}$	$\frac{\sqrt{3}}{2}$	1
cos	1	$\frac{\sqrt{3}}{2}$	$\frac{1}{\sqrt{2}}$	1/2	0
tan = sin/cos	0	$\frac{1}{\sqrt{3}}$	1	$\sqrt{3}$	Not defined
cosec = 1/sin	Not defined	2	$\sqrt{2}$	$\frac{2}{\sqrt{3}}$	1
sec = 1/cos	1	$\frac{2}{\sqrt{3}}$	$\sqrt{2}$	2	Not defined
cot = cos/sin	Not defined	$\sqrt{3}$	1	$\frac{1}{\sqrt{3}}$	0

Signs of Trigonometric Functions:

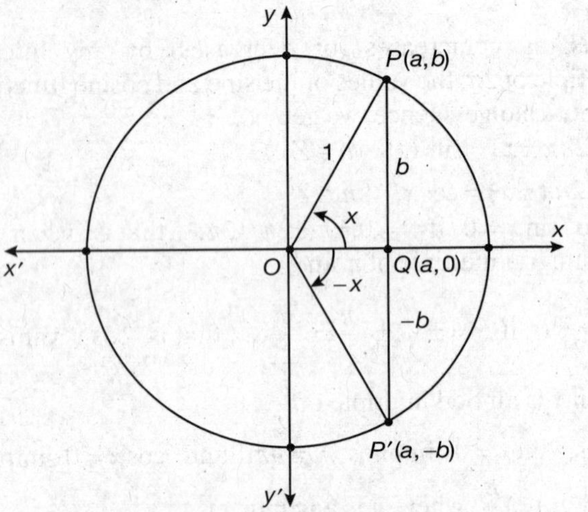

Fig. 3.6 Sign of a point on a unit circle

Let $P(a,b)$ and $P'(a,-b)$ be two points on the circle (Fig. 3.6). Then, angle $POQ = x$ radian and angle $P'OQ = -x$ radian (since anti-clockwise angles should be taken as positive and clockwise angles as negative).

In ΔPOQ, $\sin x = \frac{b}{1} = b$ and $\cos x = \frac{a}{1} = a$, while in $\Delta P'OQ$,

$\sin(-x) = \frac{-b}{1} = -b$ and $\cos(-x) = \frac{a}{1} = a$

Therefore, we get $\sin(-x) = -\sin x$ and $\cos(-x) = \cos x$.

Now, $\sin x = b$ and $\cos x = a$, and therefore, the sign of the trigonometric functions will depend on the sign of a and b, and the sign of a and b will depend on the coordinate axes. Therefore,

(i) In the first quadrant $(0 < x < \frac{\pi}{2})$, both a, b are positive, and thus, both $\cos x$ and $\sin x$ will be positive.

(ii) In the second quadrant $(\frac{\pi}{2} < x < \pi)$, a is negative and b is positive. Therefore, $\cos x$ will be negative while $\sin x$ will be positive.

(iii) In the third quadrant $(\pi < x < \frac{3\pi}{2})$, both a, b are negative, and thus, both $\cos x$ and $\sin x$ will be negative.

(iv) In the fourth quadrant $(\frac{3\pi}{2} < x < 2\pi)$, a is positive and b is negative. Therefore, $\cos x$ will be positive while $\sin x$ will be negative.

Similarly we can find the signs of the other trigonometric functions in different quadrants, as shown in the table.

	I	II	III	IV
sin x	+	+	−	−
cos x	+	−	−	+
tan x	+	−	+	−
cosec x	+	+	−	−
sec x	+	−	−	+
cot x	+	−	+	−

Domain and Range of Trigonometric Functions: For every point $P(a,b)$ on the unit circle, $-1 \le a \le 1$ and $-1 \le b \le 1$. Therefore, we get $-1 \le \cos x \le 1$ and $-1 \le \sin x \le 1$ for all x.

Thus, the domain of $y = \sin x$ and $y = \cos x$ is the set of all real numbers and range is the interval $[-1,1]$, that is, $-1 \le y \le 1$.

The domain and range of all the trigonometric functions is summarised in the table below.

Trigonometric function	Domain (x)	Range (y)
$y = \sin x$	All real numbers $R: (-\infty, \infty)$	$[-1,1]$
$y = \cos x$	All real numbers $R: (-\infty, \infty)$	$[-1,1]$
$y = \tan x$	$x \in R$ $x \ne (2n+1)\frac{\pi}{2}$ $n \in Z$	All real numbers $R: (-\infty, \infty)$
$y = \text{cosec}\, x$	$x \in R$ $x \ne n\pi$ $n \in Z$	$R-(-1,1)$, that is, $(-\infty,-1] \cup [1,\infty)$
$y = \sec x$	$x \in R$ $x \ne (2n+1)\frac{\pi}{2}$ $n \in Z$	$R-(-1,1)$, i.e., $(-\infty,-1] \cup [1,\infty)$
$y = \cot x$	$x \in R$ $x \ne n\pi$ $n \in Z$	All real numbers $R: (-\infty, \infty)$

$\text{cosec}\, x = \frac{1}{\sin x}$; therefore, the domain of $y = \text{cosec}\, x$ is the set $\{x : x \in R, x \ne n\pi, n \in Z\}$ and the range is the set $\{y : y \in R,\ y \le -1 \text{ or } y \ge 1\}$.

Since $\sec x = \frac{1}{\cos x}$, the domain of $y = \sec x$ is the set $\{x : x \in R, x \ne (2n+1)\frac{\pi}{2}, n \in Z\}$ and the range is the set $\{y : y \in R, y \le -1 \text{ or } y \ge 1\}$.

Since $\tan x = \dfrac{\sin x}{\cos x}$, the domain of $y = \tan x$ is the set $\{x : x \in R, x \neq (2n+1)\dfrac{\pi}{2}, n \in Z\}$ and the range is the set of all real numbers, that is, $-\infty \leq y \leq \infty$.

Since $\cot x = \dfrac{\cos x}{\sin x}$, the domain of $y = \cot x$ is the set $\{x : x \in R, x \neq n\pi, n \in Z\}$ and the range is the set of all real numbers, $-\infty \leq y \leq \infty$.

Graphs of Trigonometric Functions

Graph of sin *x*: In the first quadrant, as x increases from 0 to $\dfrac{\pi}{2}$, $\sin x$ increases from 0 to 1; in the second quadrant, as x increases from $\dfrac{\pi}{2}$ to π, $\sin x$ decreases from 1 to 0; in the third quadrant, as x increases from π to $\dfrac{3\pi}{2}$, $\sin x$ decreases from 0 to –1 and finally, in the fourth quadrant, as x increases from $\dfrac{3\pi}{2}$ to 2π, $\sin x$ increases from –1 to 0.

Also, we know that the values of $\sin x$ repeats after an interval of 2π.

Using this knowledge and the previous knowledge of the domain (all real numbers) and the range $[-1,1]$ of function $y = \sin x$, the behaviour and graph of function $y = \sin x$ is shown in Fig. 3.7.

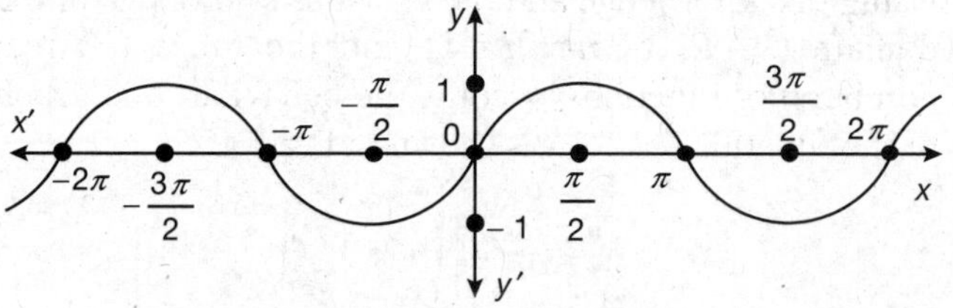

Fig. 3.7 sin *x*

Graph of cos *x*: In the first quadrant, as x increases from 0 to $\dfrac{\pi}{2}$, $\cos x$ decreases from 1 to 0; in the second quadrant, as x increases from $\dfrac{\pi}{2}$ to π, $\cos x$ decreases from 0 to –1; in the third quadrant, as x increases from π to $\dfrac{3\pi}{2}$, $\cos x$ increases from –1 to 0 and finally, in the fourth quadrant, as x increases from $\dfrac{3\pi}{2}$ to 2π, $\cos x$ increases from 0 to 1.

Also, we know that the values of $\cos x$ repeat after an interval of 2π.

Using this knowledge and the previous knowledge of the domain (all real numbers) and range $[-1,1]$ of function $y = \cos x$, the behaviour and graph of function $y = \cos x$ is shown in Fig. 3.8.

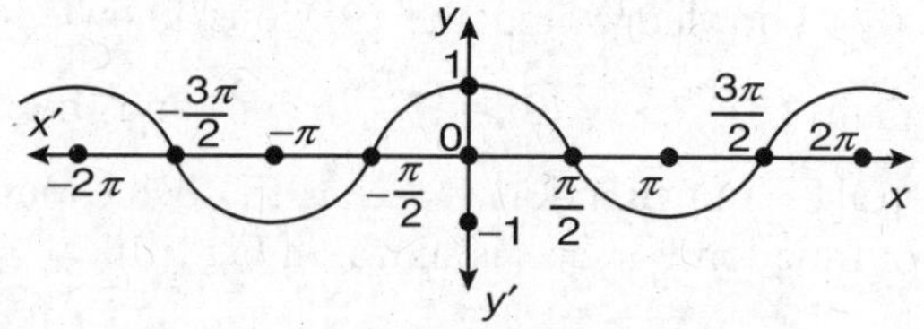

Fig. 3.8 cos *x*

Graph of cosec *x*: In the first quadrant, as x increases from 0 to $\dfrac{\pi}{2}$, $\operatorname{cosec} x$ decreases from ∞ (infinity) to 1; in the second quadrant, as x increases from $\dfrac{\pi}{2}$ to π, $\operatorname{cosec} x$ increases from 1 to ∞; in the third quadrant, as x increases from π to $\dfrac{3\pi}{2}$, $\operatorname{cosec} x$ increases from $-\infty$ to 1 and finally, in the fourth quadrant, as x increases from $\dfrac{3\pi}{2}$ to 2π, $\operatorname{cosec} x$ decreases from 1 to $-\infty$.

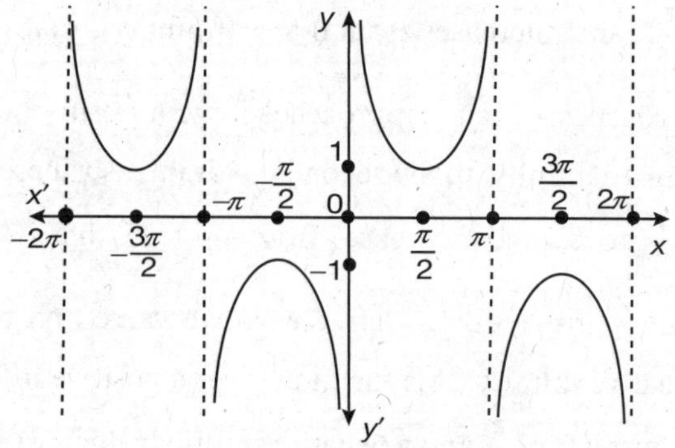

Fig. 3.9 cosec *x*

Also, $\operatorname{cosec} x$ is the reciprocal of $\sin x$, and therefore, the values of $\operatorname{cosec} x$ repeat after an interval of 2π.

Using this knowledge and the previous knowledge of the domain $(x \in R,\ x \neq n\pi,\ n \in Z)$ and range $(-\infty,-1] \cup [1,\infty)$ of function $y = \operatorname{cosec} x$, the behaviour and graph of function $y = \operatorname{cosec} x$ is shown in Fig. 3.9.

Graph of sec *x*: In the first quadrant, as x increases from 0 to $\dfrac{\pi}{2}$, $\sec x$ increases from 1 to ∞; in the second quadrant, as x increases from $\dfrac{\pi}{2}$ to π, $\sec x$ increases from $-\infty$ to 1; in the third quadrant, as x increases from π to $\dfrac{3\pi}{2}$, $\sec x$ decreases from 1 to $-\infty$ and finally, in the fourth quadrant, as x increases from $\dfrac{3\pi}{2}$ to 2π, $\sec x$ decreases from ∞ (infinity) to 1.

Also, since $\sec x$ is the reciprocal of $\cos x$, the values of $\sec x$ repeat after an interval of 2π.

Using this knowledge and the previous knowledge of the domain $(x \in R,\ x \neq (2n+1)\frac{\pi}{2},\ n \in Z)$ and the range $(-\infty,-1] \cup [1,\infty)$ of function $y = \sec x$, the behaviour and graph of function $y = \sec x$ is shown in Fig. 3.10.

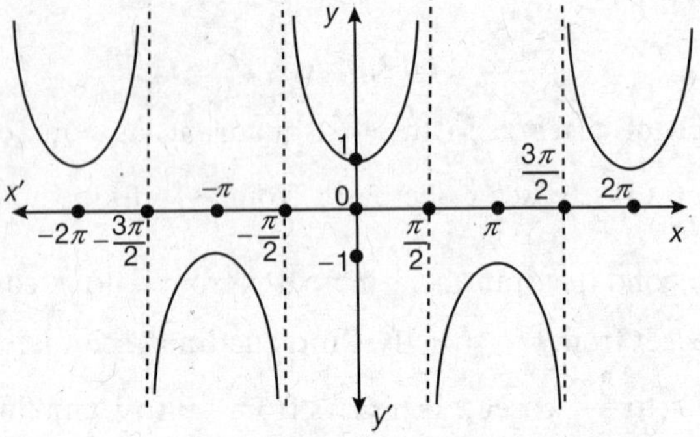

Fig. 3.10 sec x

Graph of tan x: In the first quadrant, as x increases from 0 to $\frac{\pi}{2}$, $\tan x$ increases from 0 to ∞ (infinity), that is, as x approaches $\frac{\pi}{2}$, $\tan x$ approaches large a positive value tending to infinity; in the second quadrant, as x increases from $\frac{\pi}{2}$ to π, $\tan x$ decreases from $-\infty$ to 0, that is, as x crosses $\frac{\pi}{2}$, $\tan x$ takes a large negative value and tends to negative infinity (this function gives a positive infinity value just left of $\frac{\pi}{2}$ and a negative infinity value just right of $\frac{\pi}{2}$).

Also, note that (we shall see in the next section) the period of function of $\tan x$ is π, that is, the values of $\tan x$ repeat after an interval of π.

Thus, the graph function $y = \tan x$ in the third quadrant repeats as in the first quadrant; and the graph in the fourth quadrant repeats as in the second quadrant.

Using this knowledge and the previous knowledge of the domain $(x \in R,\ x \neq (2n+1)\frac{\pi}{2},\ n \in Z)$ and the range (all real numbers) of function $y = \tan x$, the behaviour and graph of function $y = \tan x$ is shown in Fig. 3.11.

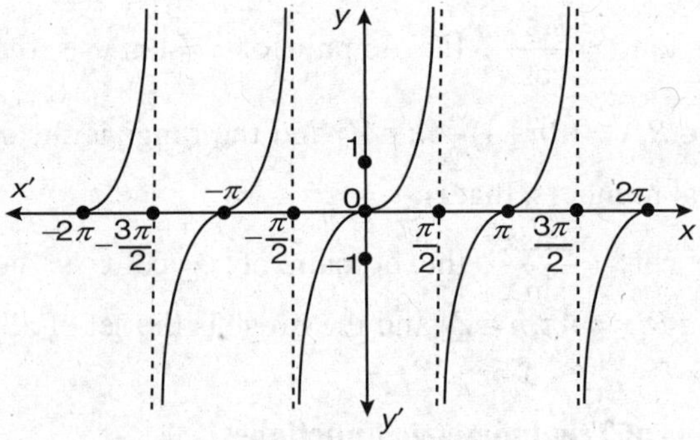

Fig. 3.11 tan x

Graph of cot x: In the first quadrant, as x increases from 0 to $\frac{\pi}{2}$, $\cot x$ decreases from ∞ (infinity) to 0; in the second quadrant, as x increases from $\frac{\pi}{2}$ to π, $\cot x$ decreases from 0 to $-\infty$ (this function gives a negative infinity value just left of $n\pi$ and a positive infinity value just right of $n\pi$, $n \in Z$).

Also, note that (we shall see in the next section) the period of the function $\cot x$ is π, that is, the values of $\cot x$ repeat after an interval of π.

Thus, the graph function $y = \cot x$ in the third quadrant repeats as in the first quadrant; and the graph in the fourth quadrant repeats as in the second quadrant.

Using this knowledge and the previous knowledge of the domain $(x \in R,\ x \neq n\pi,\ n \in Z)$ and the range (all real numbers) of function $y = \cot x$, the behaviour and graph of the function $y = \cot x$ is shown in Fig. 3.12.

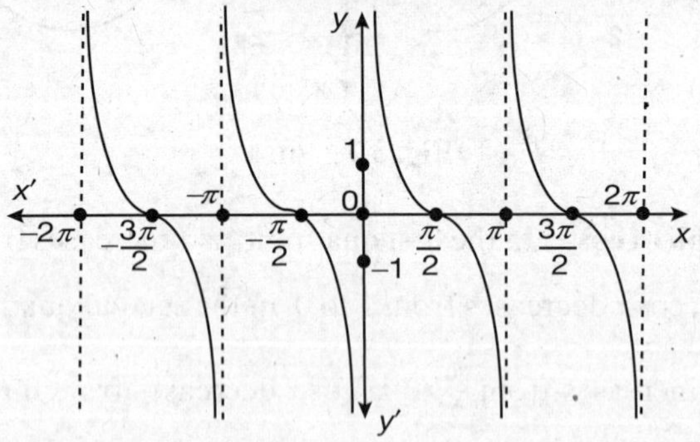

Fig. 3.12 cot x

Day 40

NCERT Exercise 3.2 (With Solutions)

Find the values of the other five trigonometric functions in Exercises 1 to 5.

Q1. $\cos x = -\frac{1}{2}, x$ lies in the third quadrant.

Solution:

When x lies in the third quadrant:

$\sin x = \sqrt{1-\cos^2 x}$

$= \sqrt{1-\left(-\frac{1}{2}\right)^2} = \sqrt{\frac{3}{4}} = \pm\frac{\sqrt{3}}{2}$

Therefore, $\sin x = -\frac{\sqrt{3}}{2}$

($\because \sin x$ is negative in the third quadrant)

$\operatorname{cosec} x = \frac{1}{\sin x} = -\frac{2}{\sqrt{3}}$

($\because \operatorname{cosec} x$ is negative in the third quadrant)

$\sec x = \frac{1}{\cos x} = -\frac{2}{1} = -2$

($\because \cos x$ is negative in the third quadrant)

$\tan x = \frac{\sin x}{\cos x} = \frac{-\sqrt{3}/2}{-1/2} = \sqrt{3}$

($\because \tan x$ is positive in the third quadrant)

$\cot x = \frac{1}{\tan x} = \frac{1}{\sqrt{3}}$

($\because \cot x$ is positive in the third quadrant)

Q2. $\sin x = \frac{3}{5}$, x lies in the second quadrant.

Solution:

When x lies in the second quadrant:

$\cos x = \sqrt{1-\sin^2 x}$

$= \sqrt{1-\left(\frac{3}{5}\right)^2} = \sqrt{\frac{16}{25}} = \pm\frac{4}{5}$

Therefore, $\cos x = -\frac{4}{5}$

($\because \cos x$ is negative in the second quadrant)

$\operatorname{cosec} x = \frac{1}{\sin x} = \frac{5}{3}$

($\because \operatorname{cosec} x$ is positive in the second quadrant)

$\sec x = \frac{1}{\cos x} = -\frac{5}{4}$

($\because \sec x$ is negative in the second quadrant)

$\tan x = \frac{\sin x}{\cos x} = \frac{3/5}{-4/5} = -\frac{3}{4}$

($\because \tan x$ is negative in the second quadrant)

$\cot x = \frac{1}{\tan x} = -\frac{4}{3}$

($\because \cot x$ is negative in the second quadrant)

Q3. $\cot x = \frac{3}{4}$, x lies in the third quadrant.

Solution:

When x lies in the third quadrant:

$\operatorname{cosec} x = \sqrt{1+\cot^2 x}$

$= \sqrt{1+\left(\frac{3}{4}\right)^2} = \sqrt{\frac{25}{16}} = \pm\frac{5}{4}$

$\therefore \operatorname{cosec} x = -\frac{5}{4}$

($\because \operatorname{cosec} x$ is negative in the third quadrant)

$\sin x = \frac{1}{\operatorname{cosec} x} = -\frac{4}{5}$

($\because \sin x$ is negative in the third quadrant)

$\cos x = \sqrt{1-\sin^2 x} = \sqrt{1-\left(-\frac{4}{5}\right)^2} = \sqrt{\frac{9}{25}} = \pm\frac{3}{5}$

Therefore, $\cos x = -\frac{3}{5}$

($\because \cos x$ is negative in the third quadrant)

$\sec x = \frac{1}{\cos x} = -\frac{5}{3}$

($\because \sec x$ is negative in the third quadrant)

$\tan x = \frac{1}{\cot x} = \frac{4}{3}$

($\because \tan x$ is positive in the third quadrant)

Q4. $\sec x = \frac{13}{5}$, x lies in the fourth quadrant

Solution:

When x lies in the fourth quadrant:

$\sec x = \frac{13}{5} \Rightarrow \cos x = \frac{5}{13}$

($\because \cos x$ is positive in the fourth quadrant)

$\sin x = \sqrt{1-\cos^2 x}$

$= \sqrt{1-\left(\frac{5}{13}\right)^2} = \sqrt{\frac{144}{169}} = \pm\frac{12}{13}$

Therefore, $\sin x = -\frac{12}{13}$

($\because \sin x$ is negative in the fourth quadrant)

$\operatorname{cosec} x = \frac{1}{\sin x} = -\frac{13}{12}$

($\because \operatorname{cosec} x$ is negative in the fourth quadrant)

$\tan x = \frac{\sin x}{\cos x} = \frac{-12/13}{5/13} = -\frac{12}{5}$

($\because \tan x$ is negative in the fourth quadrant)

$\cot x = \frac{1}{\tan x} = -\frac{5}{12}$

($\because \cot x$ is negative in the fourth quadrant)

Q5. $\tan x = -\frac{5}{12}$, x lies in the second quadrant.

Solution:

For x lies in the second quadrant:

$\sec x = \sqrt{1+\tan^2 x}$

$= \sqrt{1+\left(-\frac{5}{12}\right)^2} = \sqrt{\frac{169}{144}} = \pm\frac{13}{12}$

Therefore, $\sec x = -\frac{13}{12}$

($\because \sec x$ is negative in the second quadrant)

$\cos x = \frac{1}{\sec x} = -\frac{12}{13}$

($\because \cos x$ is negative in the second quadrant)

$\sin x = \sqrt{1-\cos^2 x}$

$= \sqrt{1-\left(-\frac{12}{13}\right)^2} = \sqrt{\frac{25}{169}} = \pm\frac{5}{13}$

Therefore, $\sin x = \frac{5}{13}$

($\because \sin x$ is positive in the second quadrant)

$\text{cosec}\, x = \frac{1}{\sin x} = \frac{13}{5}$

($\because \text{cosec}\, x$ is positive in the second quadrant)

$\cot x = \frac{1}{\tan x} = \frac{1}{-5/12} = -\frac{12}{5}$

($\because \cot x$ is negative in the second quadrant)

Find the values of the trigonometric functions in Exercises 6 to 10.

Q6. $\sin 765°$

Solution:

$\sin 765° = \sin(2\times 360° + 45°) = \sin 45° = \frac{1}{\sqrt{2}}$

[$\because$ sin is positive in the first quadrant]

Q7. $\text{cosec}(-1410°)$

Solution:

$\text{cosec}(-1410°) = -\text{cosec}(1410°)$

($\because \text{cosec}(-\theta) = -\text{cosec}\,\theta$)

$= -\text{cosec}(4\times 360° - 30°) = -\text{cosec}(-30°)$

[$\because$ cosec is negative in the fourth quadrant]

$= \text{cosec}\, 30° = 2$

Q8. $\tan\frac{19\pi}{3}$

Solution:

$\tan\frac{19\pi}{3} = \tan\left(6\pi + \frac{\pi}{3}\right) = \tan\frac{\pi}{3} = \sqrt{3}$

[$\because$ tan is positive in the first quadrant]

Q9. $\sin\left(-\frac{11\pi}{3}\right)$

Solution:

$\sin\left(-\frac{11\pi}{3}\right) = -\sin\left(\frac{11\pi}{3}\right)$

($\because \sin(-\theta) = -\sin\theta$)

$= -\sin\left(4\pi - \frac{\pi}{3}\right) - [-\sin\frac{\pi}{3}] = \sin\frac{\pi}{3} = \frac{\sqrt{3}}{2}$

[$\because$ sin is negative in the fourth quadrant]

Q10. $\cot\left(-\frac{15\pi}{4}\right)$

Solution:

$\cot\left(-\frac{15\pi}{4}\right) = -\cot\left(\frac{15\pi}{4}\right)$

($\because \cot(-\theta) = -\cot\theta$)

$= -\cot\left(4\pi - \frac{\pi}{4}\right) = -[-\cot\frac{\pi}{4}] = \cot\frac{\pi}{4} = 1$

[$\because$ cot is negative in the fourth quadrant]

Some more examples

Example 1: Prove that

$\sqrt{\frac{1-\cos x}{1+\cos x}} = \text{cosec}\, x - \cot x$

Solution:

$\text{LHS} = \sqrt{\frac{1-\cos x}{1+\cos x}} = \sqrt{\frac{(1-\cos x)(1-\cos x)}{(1+\cos x)(1-\cos x)}}$

$= \sqrt{\frac{(1-\cos x)^2}{(1-\cos^2 x)}} = \frac{1-\cos x}{\sqrt{1-\cos^2 x}} = \frac{1-\cos x}{\sin x}$

$= \frac{1}{\sin x} - \frac{\cos x}{\sin x} = \text{cosec}\, x - \cot x = \text{RHS}$

Example 2: Prove that

$\sin^6 x + \cos^6 x = 1 - 3\sin^2 x\cos^2 x$

Solution:

$\text{LHS} = \sin^6 x + \cos^6 x$

$= (\sin^2 x + \cos^2 x)^3 - 3\sin^2 x\cos^2 x(\sin^2 x + \cos^2 x)$

($\because \sin^2 x + \cos^2 x = 1$)

$= 1 - 3\sin^2 x\cos^2 x = \text{RHS}$

Example 3: Prove that

$(\sin x - \text{cosec}\, x)(\cos x - \sec x)(\tan x + \cot x) = 1$

Solution:

$\text{LHS} = (\sin x - \text{cosec}\, x)(\cos x - \sec x)(\tan x + \cot x)$

$= \left(\sin x - \frac{1}{\sin x}\right)\left(\cos x - \frac{1}{\cos x}\right)\left(\frac{\sin x}{\cos x} + \frac{\cos x}{\sin x}\right)$

$$=\left(\frac{\sin^2 x-1}{\sin x}\right)\left(\frac{\cos^2 x-1}{\cos x}\right)\left(\frac{\sin^2 x+\cos^2 x}{\sin x\cos x}\right)$$

$$=\left(\frac{-\cos^2 x}{\sin x}\right)\left(\frac{-\sin^2 x}{\cos x}\right)\left(\frac{1}{\sin x\cos x}\right)\ (\because \sin^2 x+\cos^2 x=1)$$

$$=\left(\frac{\cos^2 x\sin^2 x}{\sin x\cos x\sin x\cos x}\right)=1=\text{RHS}$$

Example 4: Prove that $\dfrac{\operatorname{cosec} x}{\tan x+\cot x}=\cos x$

Solution:

$$\text{LHS}=\frac{\operatorname{cosec} x}{\tan x+\cot x}=\frac{\frac{1}{\sin x}}{\frac{\sin x}{\cos x}+\frac{\cos x}{\sin x}}$$

$$=\frac{\sin x\cos x}{\sin x(\sin^2 x+\cos^2 x)}=\cos x=\text{RHS}$$

$(\because \sin^2 x+\cos^2 x=1)$

Example 5: Prove that

$\sec^4 x-\tan^4 x=\sec^2 x+\tan^2 x$

Solution:

$\text{LHS}=\sec^4 x-\tan^4 x=(\sec^2 x)^2-\tan^4 x$

$=(1+\tan^2 x)^2-\tan^4 x\quad (\because \sec^2 x-\tan^2 x=1)$

$=1+\tan^4 x+2\tan^2 x-\tan^4 x$

$=1+2\tan^2 x=1+\tan^2 x+\tan^2 x$

$=\sec^2 x+\tan^2 x=\text{RHS}$

Example 6: Prove that

$\sec^2 x\cdot\operatorname{cosec}^2 x-2=\tan^2 x+\cot^2 x$

Solution:

$\text{LHS}=\sec^2 x\cdot\operatorname{cosec}^2 x-2=\tan^2 x+\cot^2 x$

$=(1+\tan^2 x)(1+\cot^2 x)-2$

$(\because \sec^2 x-\tan^2 x=1 \text{ and } \operatorname{cosec}^2 x-\cot^2 x=1)$

$=1+\tan^2 x+\cot^2 x+\tan^2 x\cdot\cot^2 x-2$

$=\tan^2 x+\cot^2 x=\text{RHS}\quad (\because \tan^2 x\cdot\cot^2 x=1)$

Example 7: Prove that

$\dfrac{\tan x+\sec x-1}{\tan x-\sec x+1}=\dfrac{1+\sin x}{\cos x}$

Solution:

$$\text{LHS}=\frac{\tan x+\sec x-1}{\tan x-\sec x+1}=\frac{(\tan x)+(\sec x-1)}{(\tan x)-(\sec x-1)}$$

$$=\frac{(\tan x)+(\sec x-1)}{(\tan x)-(\sec x-1)}\times\frac{(\tan x)+(\sec x-1)}{(\tan x)+(\sec x-1)}$$

$$=\frac{[\tan x+\sec x-1]^2}{(\tan x)^2-(\sec x-1)^2}$$

$$=\frac{\tan^2 x+\sec^2 x+1-2\sec x-2\tan x+2\sec x\tan x}{\tan^2 x-\sec^2 x-1+2\sec x}$$

$$=\frac{2\sec^2 x-2\sec x-2\tan x+2\sec x\tan x}{-1-1+2\sec x}$$

$$=\frac{\sec^2 x-\sec x-\tan x+\sec x\tan x}{-1+\sec x}$$

$(\because \sec^2 x-\tan^2 x=1)$

$$=\frac{\sec x(\sec x-1)+\tan x(\sec x-1)}{\sec x-1}$$

$$=\frac{(\sec x+\tan x)(\sec x-1)}{\sec x-1}=\sec x+\tan x$$

$$=\frac{1}{\cos x}+\frac{\sin x}{\cos x}=\frac{1+\sin x}{\cos x}=\text{RHS}$$

Example 8: Prove that

$(\sin x+\operatorname{cosec} x)^2+(\cos x+\sec x)^2-\tan^2 x-\cot^2 x=7$

Solution:

$\text{LHS}=(\sin x+\operatorname{cosec} x)^2+(\cos x+\sec x)^2$

$-\tan^2 x-\cot^2 x$

$=\sin^2 x+\operatorname{cosec}^2 x+2\sin x\operatorname{cosec} x+\cos^2 x$

$+\sec^2 x+2\cos x\sec x-\tan^2 x-\cot^2 x$

$=(\sin^2 x+\cos^2 x)+(\operatorname{cosec}^2 x-\cot^2 x)$

$+(\sec^2 x-\tan^2 x)+2\cos x\sec x+2\sin x\operatorname{cosec} x$

$=1+1+1+2+2=7=\text{RHS}$

$(\because \sin^2 x+\cos^2 x=1, \operatorname{cosec}^2 x-\cot^2 x=1,$

$\sec^2 x-\tan^2 x=1, \cos x\sec x=\sin x\operatorname{cosec} x=1)$

Example 9: If $\cot x=\sqrt{2}$, then find the value of $\dfrac{\operatorname{cosec}^2 x+\sec^2 x}{\operatorname{cosec}^2 x-\sec^2 x}$.

Solution:

Since $\cot x=\sqrt{2}$, $\tan x=\dfrac{1}{\cot x}=\dfrac{1}{\sqrt{2}}$

Now, $\dfrac{\operatorname{cosec}^2 x+\sec^2 x}{\operatorname{cosec}^2 x-\sec^2 x}=\dfrac{1+\cot^2 x+1+\tan^2 x}{1+\cot^2 x-1-\tan^2 x}$

$(\because \operatorname{cosec}^2 x-\cot^2 x=1,\ \sec^2 x-\tan^2 x=1)$

$$=\frac{2+\cot^2 x+\tan^2 x}{\cot^2 x-\tan^2 x}=\frac{2+2+1/2}{2-1/2}=\frac{9/2}{3/2}=3$$

Example 10: Find the value of $\dfrac{\sin^2 30°+\cos^2 45°}{\sin 60°-\cos 45°}$.

Solution:

$$\frac{\sin^2 30°+\cos^2 45°}{\sin^2 60°-\sin^2 45°}=\frac{(1/2)^2+(1/\sqrt{2})^2}{(\sqrt{3}/2)^2-(1/\sqrt{2})^2}$$

$$=\frac{1/4+1/2}{3/4-1/2}=\frac{3/4}{1/4}=3$$

Example 11: Find the value of $\tan^2 30° + \tan^2 60° - \cot^2 45°$.

Solution:

$\tan^2 30° + \tan^2 60° - \cot^2 45°$

$= (1/\sqrt{3})^2 + (\sqrt{3})^2 - (1)^2 = 1/3 + 3 - 1 = 7/3$

Example 12: Find the value of $\dfrac{\sec^2 30° + \sec^2 45°}{\text{cosec}^2 60° - \text{cosec}^2 45°}$.

Solution:

$$\frac{\sec^2 30° + \sec^2 45°}{\text{cosec}^2 60° - \text{cosec}^2 45°}$$

$$= \frac{(2/\sqrt{3})^2 + (\sqrt{2})^2}{(2/\sqrt{3})^2 - (\sqrt{2})^2} = \frac{4/3 + 2}{4/3 - 2} = \frac{10/3}{-2/3} = -5$$

Example 13:

$\sin 300° + \sin 750° + \cos 1050° + \cos 1830°$

Solution:

$\sin 300° + \sin 750° + \cos 1050° + \cos 1830°$

$= \sin(360° - 60°) + \sin(720° + 30°)$

$+ \cos(1080° - 30°) + \cos(1800° + 30°)$

$= \sin(360° - 60°) + \sin(2 \times 360° + 30°)$

$+ \cos(3 \times 360° - 30°) + \cos(5 \times 360° + 30°)$

$= \sin 60° + \sin 30° + \cos 30° + \cos 30°$

($\because$ sin and cosine functions repeat after 360°)

$$= \frac{\sqrt{3}}{2} + \frac{1}{2} + \frac{\sqrt{3}}{2} + \frac{\sqrt{3}}{2} = \frac{3\sqrt{3}+1}{2}$$

Example 14: $\sin\dfrac{25\pi}{6} + \cos\dfrac{35\pi}{6} + \tan\dfrac{55\pi}{6}$

Solution:

$\sin\dfrac{25\pi}{6} + \cos\dfrac{35\pi}{6} + \tan\dfrac{55\pi}{6}$

$$= \sin\left(4\pi + \frac{\pi}{6}\right) + \cos\left(6\pi - \frac{\pi}{6}\right) + \tan\left(9\pi + \frac{\pi}{6}\right)$$

($\because$ sin and cosine functions repeat after 2π and tan function repeats after π)

$$= \sin\left(\frac{\pi}{6}\right) + \cos\left(\frac{\pi}{6}\right) + \tan\left(\frac{\pi}{6}\right)$$

$$= \frac{1}{2} + \frac{\sqrt{3}}{2} + \frac{1}{\sqrt{3}} = \frac{\sqrt{3}+3+2}{2\sqrt{3}} = \frac{5+\sqrt{3}}{2\sqrt{3}}$$

Example 15: $\sin(-30°) + \cos(-60°) + \tan(-45°)$

Solution:

$\sin(-30°) + \cos(-60°) + \tan(-45°)$

$= -\sin 30° + \cos 60° - \tan 45°$

$\left(\begin{array}{l} \because \sin(-x) = -\sin x \\ \cos(-x) = \cos x \\ \tan(-x) = -\tan x \end{array}\right)$

$$= -\frac{1}{2} + \frac{1}{2} - \frac{1}{\sqrt{2}} = -\frac{1}{\sqrt{2}}$$

Example 16:

$$\sec\left(-\frac{15\pi}{4}\right) + \text{cosec}\left(-\frac{25\pi}{4}\right) + \cot\left(-\frac{35\pi}{4}\right)$$

Solution:

$$\sec\left(-\frac{15\pi}{4}\right) + \text{cosec}\left(-\frac{25\pi}{4}\right) + \cot\left(-\frac{35\pi}{4}\right)$$

$$= \sec\left(\frac{15\pi}{4}\right) - \text{cosec}\left(\frac{25\pi}{4}\right) - \cot\left(\frac{35\pi}{4}\right)$$

$\left(\begin{array}{l} \because \sec(-x) = \sec x \\ \text{cosec}(-x) = \text{cosec}\, x \\ \cot(-x) = -\cot x \end{array}\right)$

$$= \sec\left(4\pi - \frac{\pi}{4}\right) - \text{cosec}\left(6\pi + \frac{\pi}{4}\right) - \cot\left(9\pi - \frac{\pi}{4}\right)$$

$$= \sec\left(\frac{\pi}{4}\right) - \text{cosec}\left(\frac{\pi}{4}\right) - \cot\left(\frac{\pi}{4}\right)$$

($\because$ sec and cosec functions repeat after 2π and cot function repeats after π)

$= \sqrt{2} - \sqrt{2} - 1 = -1$

Day 41

Trigonometric Functions of Sum and Difference of Two Angles-I

In the previous chapter, we derived two trigonometric identities:

(i) $\sin(-x) = -\sin x$

(ii) $\cos(-x) = -\cos x$

Now, we will derive and prove other trigonometric identities, that is, expressions for trigonometric functions of the sum and difference of two angles and related expressions.

(iii) $\cos(x + y) = \cos x \cos y - \sin x \sin y$

Proof:

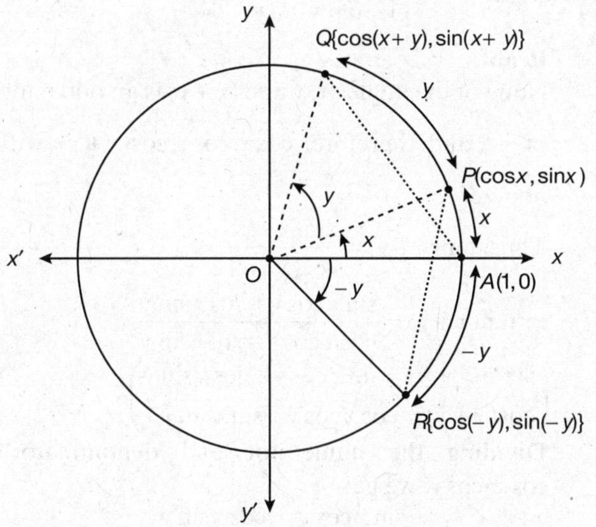

Fig. 3.13 Sum of two angles

Consider the centre of the unit circle to be at the origin (Fig. 3.13). Let x be the angle AOP, y be the angle POQ and $-y$ be the angle AOR, then $(x+y)$ is the angle AOQ.

Thus, the coordinates of different points are $A(1,0)$, $Q\{\cos(x+y), \sin(x+y)\}$, $P(\cos x, \sin x)$, $R\{\cos(-y), \sin(-y)\}$

Now, the triangles OAQ and OPR are congruent (two sides $OQ = OP = 1$, $OA = OR = 1$ and mid angles $\angle AOQ = \angle POR = x+y$), and therefore, the sides AQ and PR are also equal.

Thus, using the distance formula, we get

$(AQ)^2 = [\cos(x+y)-1]^2 + [\sin(x+y)-0]^2$

$= \cos^2(x+y) + 1 - 2\cos(x+y) + \sin^2(x+y)$

$= 2 - 2\cos(x+y) \quad (\because \sin^2\theta + \cos^2\theta = 1)$

Also, $(PR)^2 = [\cos x - \cos(-y)]^2 + [\sin x - \sin(-y)]^2$

$= (\cos x - \cos y)^2 + (\sin x + \sin y)^2$

$(\because \cos(-x) = \cos x \text{ and } \sin(-x) = -\sin x)$

$= \cos^2 x + \cos^2 y - 2\cos x\cos y + \sin^2 x$

$\quad + \sin^2 y + 2\sin x \sin y$

$= 2 - 2(\cos x\cos y - \sin x \sin y)$

$(\because \sin^2\theta + \cos^2\theta = 1)$

$AQ = PR$, and therefore, $(AQ)^2 = (PR)^2$

$\Rightarrow 2 - 2\cos(x+y) = 2 - 2(\cos x \cos y - \sin x \sin y)$

$\Rightarrow \cos(x+y) = \cos x\cos y - \sin x \sin y$

(iv) $\cos(x-y) = \cos x \cos y + \sin x \sin y$

Proof:

Replacing y with $-y$ in identity (iii), we get

$\cos(x+(-y)) = \cos x\cos(-y) - \sin x \sin(-y)$

$\Rightarrow \cos(x-y) = \cos x\cos y + \sin x \sin y$

$(\because \cos(-x) = \cos x \text{ and } \sin(-x) = -\sin x)$

(v) $\cos\left(\frac{\pi}{2} - x\right) = \sin x$

Proof:

Replacing x with $\frac{\pi}{2}$, y with x in identity (iv), we get

$\cos\left(\frac{\pi}{2} - x\right) = \cos\frac{\pi}{2}\cos x + \sin\frac{\pi}{2}\sin x = \sin x.$

$\left(\because \cos\frac{\pi}{2} = 0 \text{ and } \sin\frac{\pi}{2} = 1\right)$

(vi) $\sin\left(\frac{\pi}{2} - x\right) = \cos x$

Proof:

Replacing x with $\frac{\pi}{2} - x$, in the identity (v), we get

$\cos\left[\frac{\pi}{2} - \left(\frac{\pi}{2} - x\right)\right] = \sin\left(\frac{\pi}{2} - x\right)$

$\Rightarrow \cos x = \sin\left(\frac{\pi}{2} - x\right)$

(vii) $\sin(x+y) = \sin x\cos y + \cos x \sin y$

Proof:

$\sin(x+y) = \cos\left(\frac{\pi}{2} - (x+y)\right)$

[Using identity (v)]

$\therefore \sin(x+y) = \cos\left(\left(\frac{\pi}{2} - x\right) - y\right)$

$= \cos\left(\frac{\pi}{2} - x\right)\cos y + \sin\left(\frac{\pi}{2} - x\right)\sin y$

[Using identity (iv)]

$\Rightarrow \sin(x+y) = \sin x\cos y + \cos x \sin y$

(viii) $\sin(x-y) = \sin x\cos y - \cos x \sin y$

Proof:

Replacing y with $-y$ in identity (vii), we get

$\sin(x+(-y)) = \sin x\cos(-y) + \cos x\sin(-y)$

$\Rightarrow \sin(x-y) = \sin x \cos y - \cos x \sin y$

$(\because \cos(-y) = \cos y \text{ and } \sin(-y) = -\sin y)$

(ix) $\cos\left(\frac{\pi}{2} + x\right) = -\sin x$

Proof:

Replacing x with $-x$ in identity (v), we get

$\cos\left(\frac{\pi}{2} - (-x)\right) = \sin(-x) \Rightarrow \cos\left(\frac{\pi}{2} + x\right) = -\sin x$

$(\because \sin(-x) = -\sin x)$

(x) $\sin\left(\frac{\pi}{2}+x\right)=\cos x$

Proof:
Replacing x with $-x$ in identity (vi), we get

$\sin\left(\frac{\pi}{2}-(-x)\right)=\cos(-x)\Rightarrow \sin\left(\frac{\pi}{2}+x\right)=\cos x$

$(\because \cos(-x)=\cos x)$

(xi) $\cos(\pi-x)=-\cos x$

Proof:
Replacing x with π, y with x in identity (iv), we get

$\cos(\pi-x)=\cos\pi\cos x+\sin\pi\sin x=-\cos x$

$(\because \cos\pi=-1 \text{ and } \sin\pi=0)$

(xii) $\sin(\pi-x)=\sin x$

Proof:
Replacing x with π, y with x in identity (viii), we get

$\sin(\pi-x)=\sin\pi\cos x-\cos\pi\sin x=\sin x$

$(\because \cos\pi=-1 \text{ and } \sin\pi=0)$

(xiii) $\cos(\pi+x)=-\cos x$

Proof:
Replacing x with $-x$ in identity (xi), we get

$\cos(\pi+x)=-\cos(-x)=-\cos x$

$(\because \cos(-x)=\cos x)$

(xiv) $\sin(\pi+x)=-\sin x$

Proof:
Replacing x with $-x$ in identity (xii), we get

$\sin(\pi+x)=\sin(-x)=-\sin x$

$(\because \sin(-x)=-\sin x)$

(xv) $\cos(2\pi-x)=\cos x$

Proof:
Replacing x with 2π, y by x in identity (iv), we get

$\cos(2\pi-x)=\cos 2\pi\cos x+\sin 2\pi\sin x=\cos x$

$(\because \cos 2\pi=1 \text{ and } \sin 2\pi=0)$

(xvi) $\sin(2\pi-x)=-\sin x$

Proof:
Replacing x with 2π, y with x in identity (viii), we get

$\sin(2\pi-x)=\sin 2\pi\cos x-\cos 2\pi\sin x=-\sin x$

$(\because \cos 2\pi=1 \text{ and } \sin 2\pi=0)$

Note: Similarly results for other trigonometric functions $(\tan x, \cot x, \sec x, \operatorname{cosec} x)$ can be obtained from the results of $\sin x, \cos x$.
For example, dividing identity (i) by (ii), we get

$$\tan(-x)=\frac{\sin(-x)}{\cos(-x)}=\frac{-\sin x}{\cos x}=-\tan x$$

(xvii) $\tan(x+y)=\dfrac{\tan x+\tan y}{1-\tan x\tan y}$; $x, y, (x+y)\neq n\frac{\pi}{2}$

Proof:
None of the angles x, y and $(x+y)$ is an odd multiple of $\frac{\pi}{2}$, and therefore, $\cos x, \cos y, \cos(x+y)$ will be non-zero.

Thus, $\tan(x+y)=\dfrac{\sin(x+y)}{\cos(x+y)}$

$$\Rightarrow \tan(x+y)=\frac{\sin x\cos y+\cos x\sin y}{\cos x\cos y-\sin x\sin y}$$

$\left(\begin{array}{l}\because \sin(x+y)=\sin x\cos y+\cos x\sin y\\ \cos(x+y)=\cos x\cos y-\sin x\sin y\end{array}\right)$

Dividing the numerator and denominator by $\cos x\cdot\cos y$, we get

$$\tan(x+y)=\frac{\dfrac{\sin x\cos y}{\cos x\cos y}+\dfrac{\cos x\sin y}{\cos x\cos y}}{\dfrac{\cos x\cos y}{\cos x\cos y}-\dfrac{\sin x\sin y}{\cos x\cos y}}$$

$$\Rightarrow \tan(x+y)=\frac{\tan x+\tan y}{1-\tan x\tan y}$$

(xviii) $\tan(x-y)=\dfrac{\tan x-\tan y}{1+\tan x\tan y}$

Proof:
Replacing y with $-y$ in identity (xvii), we get

$\tan(x-y)=\tan[x+(-y)]$

$$\Rightarrow \tan(x-y)=\frac{\tan x+\tan(-y)}{1-\tan x\tan(-y)}$$

$$=\frac{\tan x-\tan y}{1+\tan x\tan y} \quad (\because \tan(-y)=-\tan y)$$

(xix) $\cot(x+y)=\dfrac{\cot x\cot y-1}{\cot y+\cot x}$

where $x, y, (x+y)\neq n\pi$

Proof:
None of the angles x, y and $(x+y)$ is an odd multiple of π, and therefore, $\sin x, \sin y, \sin(x+y)$ will be non-zero.

Thus, $\cot(x+y)=\dfrac{\cos(x+y)}{\sin(x+y)}$

$$=\frac{\cos x\cos y-\sin x\sin y}{\sin x\cos y+\cos x\sin y}$$

$\left(\begin{array}{l}\because \sin(x+y)=\sin x\cos y+\cos x\sin y\\ \cos(x+y)=\cos x\cos y-\sin x\sin y\end{array}\right)$

Dividing the numerator and denominator by $\sin x\cdot\sin y$, we get

$$\cot(x+y)=\frac{\dfrac{\cos x\cos y}{\sin x\sin y}-\dfrac{\sin x\sin y}{\sin x\sin y}}{\dfrac{\sin x\cos y}{\sin x\sin y}+\dfrac{\cos x\sin y}{\sin x\sin y}}$$

$$\Rightarrow \cot(x+y)=\frac{\cot x\cot y-1}{\cot y+\cot x}$$

(xx) $\cot(x-y)=\dfrac{\cot x\cot y+1}{\cot y-\cot x}$,

where $x, y, (x-y)\neq n\pi$

Proof:

Replacing y with $-y$ in identity (xix), we get

$$\cot(x+(-y))=\frac{\cot x\cot(-y)-1}{\cot(-y)+\cot x}$$

$$=\frac{-\cot x\cot y-1}{-\cot y+\cot x}\quad (\because \cot(-y)=-\cot y)$$

$$\Rightarrow \cot(x-y)=\frac{\cot x\cot y+1}{\cot y-\cot x}$$

Example 1: Find the value of

(i) $\sin 105°$ (ii) $\cos 15°$

(iii) $\tan 105°$ (iii) $\cot 75°$

Solution:

(i) $\sin 105°=\sin(45°+60°)$

$=\sin 45°\cos 60°+\cos 45°\sin 60°$

$(\because \sin(x+y)=\sin x\cos y+\cos x\sin y)$

$$=\frac{1}{\sqrt{2}}\cdot\frac{1}{2}+\frac{1}{\sqrt{2}}\cdot\frac{\sqrt{3}}{2}=\frac{1+\sqrt{3}}{2\sqrt{2}}=\frac{\sqrt{3}+1}{2\sqrt{2}}$$

(ii) $\cos 15°=\cos(45°-30°)$

$=\cos 45°\cdot\cos 30°+\sin 45°\cdot\sin 30°$

$(\because \cos(x-y)=\cos x\cos y+\sin x\sin y)$

$$=\frac{1}{\sqrt{2}}\cdot\frac{\sqrt{3}}{2}+\frac{1}{\sqrt{2}}\cdot\frac{1}{2}=\frac{\sqrt{3}+1}{2\sqrt{2}}$$

(iii) $\tan 105°=\tan(60°+45°)$

$$=\frac{\tan 60°+\tan 45°}{1-\tan 60°\cdot\tan 45°}$$

$$\left(\because \tan(x+y)=\frac{\tan x+\tan y}{1-\tan x\tan y}\right)$$

$$=\frac{\sqrt{3}+1}{1-\sqrt{3}\cdot 1}=\left(\frac{\sqrt{3}+1}{1-\sqrt{3}}\right)\left(\frac{1+\sqrt{3}}{1+\sqrt{3}}\right)$$

$$=\frac{(\sqrt{3}+1)^2}{1-3}=\frac{3+1+2\sqrt{3}}{-2}$$

$$=\frac{4+2\sqrt{3}}{-2}=-(2+\sqrt{3})$$

(iv) $\cot 75°=\cot(45°+30°)$

$$=\frac{\cot 45°\cot 30°-1}{\cot 45°+\cot 30°}$$

$$\left(\because \cot(x+y)=\frac{\cot x\cot y-1}{\cot y+\cot x}\right)$$

$$=\frac{1\cdot\sqrt{3}-1}{1+\sqrt{3}}=\left(\frac{\sqrt{3}-1}{1+\sqrt{3}}\right)\left(\frac{1-\sqrt{3}}{1-\sqrt{3}}\right)$$

$$=\frac{-(\sqrt{3}-1)^2}{1-3}=\frac{3+1-2\sqrt{3}}{2}$$

$$=\frac{4-2\sqrt{3}}{2}=2-\sqrt{3}$$

Example 2: If $\cos x=\dfrac{3}{5}$, $0<x<\dfrac{\pi}{2}$ and $\sin y=\dfrac{12}{13}$, $\dfrac{\pi}{2}<y<\pi$, then find the value of

(i) $\sin(x-y)$ (ii) $\cos(x+y)$ (iii) $\tan(x-y)$

Solution:

Since $0<x<\dfrac{\pi}{2}$, x lies in the 1st quadrant, we have

$\cos x=\dfrac{3}{5}$,

$$\Rightarrow \sin x=\sqrt{1-\cos^2 x}=\sqrt{1-(3/5)^2}=\sqrt{\frac{16}{25}}=\frac{4}{5}$$

and $\tan x=\dfrac{\sin x}{\cos x}=\dfrac{4}{3}$

(Since in the 1st quadrant, sin, cos and tan are all positive.)

$\dfrac{\pi}{2}<y<\pi$. Therefore, y lies in the 2nd quadrant. Thus, we have $\sin y=\dfrac{12}{13}$

$$\therefore \cos y=-\sqrt{1-\sin^2 y}=-\sqrt{1-\left(\frac{12}{13}\right)^2}=-\frac{5}{13}$$

and $\tan y=\dfrac{\sin y}{\cos y}=-\dfrac{12}{5}$

(Since in the 2nd quadrant, sin is positive while cos and tan are negative.)

(i) $\sin(x-y)=\sin x\cos y-\cos x\sin y$

$$=\frac{4}{5}\cdot\left(-\frac{5}{13}\right)-\frac{3}{5}\left(\frac{12}{13}\right)=-\frac{56}{65}$$

(ii) $\cos(x+y)=\cos x\cos y-\sin x\sin y$

$$=\frac{3}{5}\cdot\left(-\frac{5}{13}\right)-\frac{4}{5}\left(\frac{12}{13}\right)=-\frac{63}{65}$$

(iii) $$\tan(x-y)=\frac{\tan x-\tan y}{1+\tan x\tan y}=\frac{\dfrac{4}{3}-\left(-\dfrac{12}{5}\right)}{1+\dfrac{4}{3}\cdot\left(-\dfrac{12}{5}\right)}=-\frac{56}{33}$$

Example 3: Find the value of

(i) $\sin^2\frac{2\pi}{3}+\sec\frac{4\pi}{3}+2\cot\frac{5\pi}{4}$

(ii) $\cos^2\frac{5\pi}{6}+\text{cosec}\frac{7\pi}{6}-2\tan\frac{3\pi}{4}$

(iii) $\dfrac{\cos 6\pi-\sin\frac{7\pi}{2}}{\text{cosec}\frac{5\pi}{2}-\sec 5\pi}$

Solution:

(i) $\sin^2\frac{2\pi}{3}+\sec\frac{4\pi}{3}+2\cot\frac{5\pi}{4}$

$\because \sin\frac{2\pi}{3}=\sin\left(\pi-\frac{\pi}{3}\right)=\sin\frac{\pi}{3}=\frac{\sqrt{3}}{2}$

$\sec\frac{4\pi}{3}=\sec\left(\pi+\frac{\pi}{3}\right)=-\sec\frac{\pi}{3}=-2$

$\cot\frac{5\pi}{4}=\cot\left(\pi+\frac{\pi}{4}\right)=\cot\frac{\pi}{4}=1$

$\therefore \sin^2\frac{2\pi}{3}+\sec\frac{4\pi}{3}+2\cot\frac{5\pi}{4}$

$=\left(\frac{\sqrt{3}}{2}\right)^2+(-2)+2\times 1=\frac{3}{4}-2+2=\frac{3}{4}$

(ii) $\cos^2\frac{5\pi}{6}+\text{cosec}\frac{7\pi}{6}-2\tan\frac{3\pi}{4}$

$\because \cos\frac{5\pi}{6}=\cos\left(\pi-\frac{\pi}{6}\right)=-\cos\frac{\pi}{6}=-\frac{\sqrt{3}}{2}$

$\text{cosec}\frac{7\pi}{6}=\text{cosec}\left(\pi+\frac{\pi}{6}\right)=-\text{cosec}\frac{\pi}{6}=-2$

$\tan\frac{3\pi}{4}=\tan\left(\pi-\frac{\pi}{4}\right)=-\tan\frac{\pi}{4}=-1$

$\therefore \cos^2\frac{5\pi}{6}+\text{cosec}\frac{7\pi}{6}-2\tan\frac{3\pi}{4}$

$=\left(-\frac{\sqrt{3}}{2}\right)^2+(-2)-2\times(-1)=\frac{3}{4}-2+2=\frac{3}{4}$

(iii) $\dfrac{\cos 6\pi-\sin\frac{7\pi}{2}}{\text{cosec}\frac{5\pi}{2}-\sec 5\pi}$

$\because \cos 6\pi=\cos(0)=1$

$\sin\frac{7\pi}{2}=\sin\left(4\pi-\frac{\pi}{2}\right)=\sin\left(-\frac{\pi}{2}\right)=-1$

$\sec 5\pi=\sec(4\pi+\pi)=\sec\pi=-1$

$\text{cosec}\frac{5\pi}{2}=\text{cosec}\left(2\pi+\frac{\pi}{2}\right)=\text{cosec}\frac{\pi}{2}=1$

$\therefore \dfrac{\cos 6\pi-\sin\frac{7\pi}{2}}{\text{cosec}\frac{5\pi}{2}-\sec 5\pi}=\dfrac{1-(-1)}{1-(-1)}=\dfrac{2}{2}=1$

Example 4: Find the value of

(i) $\sin 70°\cos 10°-\cos 70°\sin 10°$

(ii) $\cos 100°\cos 20°-\sin 100°\sin 20°$

(iii) $\dfrac{\cot 40°\tan 80°+1}{\tan 80°-\cot 40°}$

(iv) $2\tan 50°+\tan 20°-\tan 70°$

Solution:

(i) $\sin 70°\cos 10°-\cos 70°\sin 10°$

$=\sin(70°-10°)=\sin 60°=\frac{\sqrt{3}}{2}$

$[\because \sin x\cos y-\cos x\sin y=\sin(x-y)]$

(ii) $\cos 100°\cos 20°-\sin 100°\sin 20°=\sin(100°+20°)$

$[\because \cos x\cos y-\sin x\sin y=\cos(x+y)]$

$=\cos 120°=\cos(90°+30°)=-\sin 30°=-\frac{1}{2}$

(iii) $\dfrac{\cot 40°\tan 80°+1}{\tan 80°-\cot 40°}=\dfrac{\cot 40°\tan(90°-10°)+1}{\tan(90°-10°)-\cot 40°}$

$=\dfrac{\cot 40°\cot 10°+1}{\cot 10°-\cot 40°}\left(\because \dfrac{\cot x\cot y+1}{\cot y-\cot x}=\cot(x-y)\right)$

$=\cot(40°-10°)=\cot 30°=\sqrt{3}$

(iv) $2\tan 50°+\tan 20°-\tan 70°$

$=\tan 50°+\tan 20°+\tan 50°-\tan 70°$

$=\tan(50°+20°)(1-\tan 50°\tan 20°)+\tan 50°-\tan 70°$

$\left[\because \tan(x+y)=\dfrac{\tan x+\tan y}{1-\tan x\tan y}\right]$

$=\tan 70°(1-\tan 50°\tan 20°)+\tan 50°-\tan 70°$

$=\tan 70°-\tan 70°\tan 50°\tan 20°+\tan 50°-\tan 70°$

$=\tan 50°(1-\tan 70°\tan 20°)$

$=\tan 50°[1-\tan(90°-20°)\tan 20°]$

$=\tan 50°[1-\cot 20°\tan 20°]=\tan 50°[1-1]=0$

Example 5: Prove that $\dfrac{\sin(x+y)}{\cos(x-y)}=\dfrac{\tan x+\tan y}{1+\tan x\tan y}$.

Solution:

$\text{LHS}=\dfrac{\sin(x+y)}{\cos(x-y)}=\dfrac{\sin x\cdot\cos y+\cos x\cdot\sin y}{\cos x\cdot\cos y+\sin x\cdot\sin y}$

Dividing the numerator and determinator by

$\cos x\cos y$, we get LHS $=\dfrac{\tan x+\tan y}{1+\tan x\tan y}=$ RHS

Example 6: Prove that

$\dfrac{\sin(x-y)}{\cos x\cos y}+\dfrac{\sin(y-z)}{\cos y\cos z}+\dfrac{\sin(z-x)}{\cos y\cos x}=0.$

Solution:

Since the first term of the LHS is

$$\frac{\sin(x-y)}{\cos x\cos y}=\frac{\sin x\cos y-\cos x\sin y}{\cos x\cos y}$$

$$=\frac{\sin x\cos y}{\cos x\cos y}-\frac{\cos x\sin y}{\cos x\cos y}=\tan x-\tan y$$

Similarly, we get the second and third terms as $\tan y-\tan z$ and $\tan z-\tan x$.

Thus, LHS $=(\tan x-\tan y)$
$+(\tan y-\tan z)+(\tan z-\tan x)=0$

Example 7: Find the value of $\tan x\tan y+\tan y\tan z+\tan z\tan x$ if $\frac{\cos(x-y)}{\cos x\cos y}+\frac{\cos(y-z)}{\cos y\cos z}+\frac{\cos(z-x)}{\cos y\cos x}=0.$

Solution:
Since the first term of the LHS is

$$\frac{\cos(x-y)}{\cos x\cos y}=\frac{\cos x\cos y-\sin x\sin y}{\cos x\cos y}$$

$$=\frac{\cos x\cos y}{\cos x\cos y}-\frac{\sin x\sin y}{\cos x\cos y}=1-\tan x\tan y$$

Similarly, we get the second and third terms as $1-\tan y\tan z$ and $1-\tan z\tan x$.

Thus, $\frac{\cos(x-y)}{\cos x\cos y}+\frac{\cos(y-z)}{\cos y\cos z}+\frac{\cos(z-x)}{\cos y\cos x}=0$

$\Rightarrow(1-\tan x\tan y)+(1-\tan y\tan z)+(1-\tan z\tan x)=0$

$\Rightarrow\tan x\tan y+\tan y\tan z+\tan z\tan x=3$

Example 8: If $\cos(x-y)+\cos(y-z)+\cos(z-x)=-\frac{3}{2}$ then prove that $\cos x+\cos y+\cos z=\sin x+\sin y+\sin z=0.$

Solution:

$2[\cos(x-y)+\cos(y-z)+\cos(z-x)]+3=0$

$\Rightarrow 2[\cos x\cdot\cos y+\sin x\cdot\sin y+\cos y\cdot\cos z$
$+\sin y\cdot\sin z+\cos z\cdot\cos x+\sin z\cdot\sin x]+3=0$

$\Rightarrow 2(\cos x\cdot\cos y+\cos y\cdot\cos z+\cos z\cdot\cos x)$
$+2(\sin x\cdot\sin y+\sin y\cdot\sin z+\sin z\cdot\sin x)+(\sin^2 x+\cos^2 x)$
$+(\sin^2 y+\cos^2 y)+(\sin^2 z+\cos^2 z)=0$

$\Rightarrow(\cos^2 x+\cos^2 y+\cos^2 z+2\cos x\cdot\cos y$
$+2\cos y\cdot\cos z+2\cos z\cdot\cos x)$
$+(\sin^2 x+\sin^2 y+\sin^2 z+2\sin x\cdot\sin y$
$+2\sin y\cdot\sin z+2\sin z\cdot\sin x)=0$

$\Rightarrow(\cos x+\cos y+\cos z)^2+(\sin x+\sin y+\sin z)^2=0$

$\Rightarrow\cos x+\cos y+\cos z=0$

and $\sin x+\sin y+\sin z=0$

Example 9: Prove that $\frac{\tan\left(\frac{\pi}{4}+x\right)}{\tan\left(\frac{\pi}{4}-x\right)}=\left(\frac{1+\tan x}{1-\tan x}\right)^2.$

Solution:

$$\text{LHS}=\frac{\tan\left(\frac{\pi}{4}+x\right)}{\tan\left(\frac{\pi}{4}-x\right)}=\frac{\left(\dfrac{\tan\frac{\pi}{4}+\tan x}{1-\tan\frac{\pi}{4}\tan x}\right)}{\left(\dfrac{\tan\frac{\pi}{4}-\tan x}{1+\tan\frac{\pi}{4}\tan x}\right)}$$

$$\left[\because\tan(A+B)=\frac{\tan A+\tan B}{1-\tan A\tan B},\ \tan(A-B)=\frac{\tan A-\tan B}{1+\tan A\tan B}\right]$$

$$=\frac{\left(\dfrac{1+\tan x}{1-\tan x}\right)}{\left(\dfrac{1-\tan x}{1+\tan x}\right)}=\left(\frac{1+\tan x}{1-\tan x}\right)^2=\text{RHS}$$

Example 10: Find the value of $\frac{\cos(\pi+x)\sin\left(\frac{3\pi}{2}+x\right)\tan^2(\pi-x)}{\sin(\pi-x)\cos\left(\frac{\pi}{2}+x\right)}.$

Solution:

$$\text{LHS}=\frac{\cos(\pi+x)\sin\left(\frac{3\pi}{2}+x\right)\tan^2(\pi-x)}{\sin(\pi-x)\cos\left(\frac{\pi}{2}+x\right)}$$

$$=\frac{-\cos(x)(-\cos x)(-\tan x)^2}{\sin(x)(-\sin x)}=\cot^2 x\tan^2 x=1$$

Example 11: Find the value of $\cos\left(\frac{3\pi}{2}+x\right)\cos(2\pi+x)\times\left[\cot\left(\frac{3\pi}{2}-x\right)+\cot(2\pi+x)\right].$

Solution:

$$\text{LHS}=\cos\left(\frac{3\pi}{2}+x\right)\cos(2\pi+x)\times\left[\cot\left(\frac{3\pi}{2}-x\right)+\cot(2\pi+x)\right]$$

$=\sin x\cos x[\tan x+\cot x]$

$=\sin x\cos x\left(\frac{\sin x}{\cos x}+\frac{\cos x}{\sin x}\right)$

$=\sin x\cos x\left(\frac{\sin^2 x+\cos^2 x}{\sin x\cos x}\right)$

$=\sin^2 x+\cos^2 x=1$

Example 12: Find the value of $\cot x\cot 2x+\cot 3x(\cot 2x-\cot x)=1.$

Solution:

LHS $= \cot x \cot 2x + \cot 3x(\cot(\pi - 2x) + \cot(2\pi - x))$

$= \cot x \cot 2x + \cot 3x(-\cot 2x - \cot x)$

$= \cot x \cot 2x - \cot 3x \cot 2x - \cot 3x \cot x$

$= \cot x \cot 2x - \cot 3x(\cot 2x + \cot x)$

$= \cot x \cot 2x - \cot(2x + x)(\cot 2x + \cot x)$

$= \cot x \cot 2x - \left[\dfrac{\cot 2x \cot x - 1}{\cot x + \cot 2x}\right](\cot 2x + \cot x)$

$\left[\because \cot(A+B) = \dfrac{\cot A \cot B - 1}{\cot A + \cot B}\right]$

$= \cot x \cot 2x - (\cot 2x \cot x - 1) = 1 =$ RHS

Example 13: Prove that

$\cos(45° - x)\cos(45° - y) - \sin(45° - x)\sin(45° - y)$

$= \sin(x + y)$

Solution:

LHS $= \cos(45° - x)\cos(45° - y)$

$- \sin(45° - x)\sin(45° - y)$

$= \cos\{(45° - x) + (45° - y)\}$

$[\because \cos x \cos y - \sin x \sin y = \cos(x + y)]$

$= \cos\{90° - (x + y)\}$

$= \sin(x + y) =$ RHS

Day 42

Trigonometric Functions of Sum and Difference of Two Angles-II

(xxi) $\sin 2x = 2\sin x \cos x = \dfrac{2\tan x}{1 + \tan^2 x}$

$\sin(x + y) = \sin x \cos y + \cos x \sin y$

Therefore, replacing y with x, we get

$\sin 2x = 2\sin x \cos x$

Rewrite $\sin 2x = \dfrac{2\sin x \cos x}{\cos^2 x + \sin^2 x}$

Dividing the numerator and denominator by $\cos^2 x$, we get

$\sin 2x = \dfrac{2\tan x}{1 + \tan^2 x}$

(xxii) $\cos 2x = \cos^2 x - \sin^2 x = 2\cos^2 x - 1$

$= 1 - 2\sin^2 x = \dfrac{1 - \tan^2 x}{1 + \tan^2 x}$

Since $\cos(x + y) = \cos x \cos y - \sin x \sin y$

Replacing y with x, we get

$\cos 2x = \cos^2 x - \sin^2 x$

Rewriting,

$\cos 2x = \cos^2 x - (1 - \cos^2 x) = 2\cos^2 x - 1$

Also, $\cos 2x = (1 - \sin^2 x) - \sin^2 x = 1 - 2\sin^2 x$

Again, $\cos 2x = \cos^2 x - \sin^2 x = \dfrac{\cos^2 x - \sin^2 x}{\cos^2 x + \sin^2 x}$

Dividing the numerator and denominator by $\cos^2 x$, we get

$\cos 2x = \dfrac{1 - \tan^2 x}{1 + \tan^2 x}$

(xxiii) $\tan 2x = \dfrac{2\tan x}{1 - \tan^2 x}$

$\tan(x + y) = \dfrac{\tan x + \tan y}{1 - \tan x \tan y}$

Therefore, replacing y with x, we get

$\tan 2x = \dfrac{2\tan x}{1 - \tan^2 x}$

Or, dividing identities (xx) by (xxi), we get

$\tan 2x = \dfrac{\dfrac{2\tan x}{1 + \tan^2 x}}{\dfrac{1 - \tan^2 x}{1 + \tan^2 x}} = \dfrac{2\tan x}{1 - \tan^2 x}$

(xxiv) $\sin 3x = 3\sin x - 4\sin^3 x$

Rewriting, $\sin 3x = \sin(2x + x)$

$= \sin 2x \cos x + \cos 2x \sin x$

$(\because \sin(x + y) = \sin x \cos y + \cos x \sin y)$

$= 2\sin x \cos x \cos x + 1(1 - 2\sin^2 x)\sin x$

$\left(\begin{array}{l}\because \sin 2x = 2\sin x \cos x \\ \cos 2x = 1 - 2\sin^2 x\end{array}\right)$

$= 2\sin x \cos^2 x + 1(1 - 2\sin^2 x)\sin x$

$= 2\sin x(1 - \sin^2 x) + \sin x - 2\sin^3 x$

$= 2\sin x - 2\sin^3 x + \sin x - 2\sin^3 x$

$= 3\sin x - 4\sin^3 x$

(xxv) $\cos 3x = 4\cos^3 x - 3\cos x$

Rewriting, $\cos 3x = \cos(2x + x)$

$= \cos 2x \cos x - \sin 2x \sin x$

$(\because \cos(x + y) = \cos x \cos y - \sin x \sin y)$

$= (2\cos^2 x - 1)\cos x - 2\sin x \cos x \sin x$

$\left(\begin{array}{l}\because \sin 2x = 2\sin x \cos x \\ \cos 2x = 2\cos^2 x - 1\end{array}\right)$

$= (2\cos^2 x - 1)\cos x - 2\cos x(1 - \cos^2 x)$

$= 2\cos^3 x - \cos x - 2\cos x + 2\cos^3 x$

$= 4\cos^3 x - 3\cos x$

(xxvi) $\tan 3x = \dfrac{3\tan x - \tan^3 x}{1-3\tan^2 x}$

Rewriting, $\tan 3x = \tan(2x+x)$

$$= \frac{\tan 2x + \tan x}{1-\tan 2x \tan x} \quad \left(\because \tan(x+y) = \frac{\tan x + \tan y}{1-\tan x \tan y}\right)$$

$$= \frac{\dfrac{2\tan x}{1-\tan^2 x} + \tan x}{1-\dfrac{2\tan x \cdot \tan x}{1-\tan^2 x}} \quad \left(\because \tan 2x = \frac{2\tan x}{1-\tan^2 x}\right)$$

$$= \frac{2\tan x + \tan x - \tan^3 x}{1-\tan^2 x - 2\tan^2 x} = \frac{3\tan x - \tan^3 x}{1-3\tan^2 x}$$

Example 1: Find the value of:

(a) $\sin 18°$ and (b) $\cos 36°$

Solution:

(a) $\sin 18°$:

$x = 18° \Rightarrow 5x = 90° \Rightarrow 2x + 3x = 90°$

$\Rightarrow 2x = 90° - 3x$

$\Rightarrow \sin 2x = \sin(90° - 3x)$

$\Rightarrow \sin 2x = \cos 3x$

$\Rightarrow 2\sin x \cos x = 4\cos^3 x - 3\cos x$

$\Rightarrow 2\sin x = 4\cos^2 x - 3$

$\Rightarrow 2\sin x = 4(1-\sin^2 x) - 3$

$\Rightarrow 4\sin^2 x + 2\sin x - 1 = 0$

$\Rightarrow \sin x = \dfrac{-2 \pm \sqrt{4+16}}{8} = \dfrac{-2 \pm 2\sqrt{5}}{8} = \dfrac{-1 \pm \sqrt{5}}{4}$

18° lies in the 1st quadrant, therefore

$\sin 18° = \sin x = \dfrac{-1+\sqrt{5}}{4}$

(b) $\cos 36°$:

$\cos 36° = 1 - 2\sin^2 18°$

$[\because \cos 2x = 1 - 2\sin^2 x]$

$\Rightarrow \cos 36° = 1 - 2\times\left(\dfrac{\sqrt{5}-1}{4}\right)^2 = 1 - 2\times\left(\dfrac{5+1-2\sqrt{5}}{16}\right)$

$\Rightarrow \cos 36° = 1 - \dfrac{3-\sqrt{5}}{4} = \dfrac{4-3+\sqrt{5}}{4} = \dfrac{\sqrt{5}+1}{4}$

Example 2: If $\sin x = \dfrac{3}{5}$, $0° < x < 90°$, find the values of $\sin 2x, \cos 2x, \tan 2x, \sin 3x, \cos 3x, \tan 3x$.

Solution:

In the 1st quadrant, sin, cos and tan are all positive, and therefore, $\cos x = \sqrt{1-\sin^2 x} = \sqrt{1-\dfrac{9}{25}} = \sqrt{\dfrac{16}{25}} = \dfrac{4}{5}$ and

$\tan x = \dfrac{\sin x}{\cos x} = \dfrac{3}{4}$.

Now, $\sin 2x = 2\sin x \cos x = 2 \times \dfrac{3}{5} \times \dfrac{4}{5} = \dfrac{24}{25}$

$\cos 2x = \cos^2 x - \sin^2 x = \dfrac{16}{25} - \dfrac{9}{25} = \dfrac{7}{25}$

$\tan 2x = \dfrac{2\tan x}{1-\tan^2 x} = \dfrac{2\times 3/4}{1-(3/4)^2} = \dfrac{3/2}{1-\dfrac{9}{16}} = \dfrac{3/2}{7/16} = \dfrac{24}{7}$

$\sin 3x = 3\sin x - 4\sin^3 x$

$= 3\times\dfrac{3}{5} - 4\times\dfrac{27}{125} = \dfrac{9}{5} - \dfrac{108}{125} = \dfrac{117}{125}$

$\cos 3x = 4\cos^3 x - 3\cos x$

$= 4\times\dfrac{64}{125} - 3\times\dfrac{4}{5} = \dfrac{256}{125} - \dfrac{12}{5} = -\dfrac{44}{125}$

$\tan 3x = \dfrac{3\tan x - \tan^3 x}{1-3\tan^2 x}$

$= \dfrac{2\times 3/4 - (3/4)^3}{1-3\times(3/4)^2} = \dfrac{6/8 - 27/64}{1-\dfrac{27}{16}} = \dfrac{21/64}{-11/16} = -\dfrac{21}{44}$

Example 3: Prove that

$\sin 6x = 6\sin x \cos x - 32\sin^3 x \cos^3 x$.

Solution:

LHS $= \sin 6x = \sin 2(3x)$

$= 2\sin 3x \cos 3x$ $[\because \sin 2A = 2\sin A \cos A]$

$= 2(3\sin x - 4\sin^3 x)(4\cos^3 x - 3\cos x)$

$[\because \sin 3A = 3\sin A - 4\sin^3 A$

and $\cos 3A = 4\cos^3 A - 3\cos A]$

$= 2(12\sin x \cos^3 x + 12\sin^3 x \cos x$

$\quad - 9\sin x \cos x - 16\sin^3 x \cos^3 x)$

$= 24\sin x \cos x(\cos^2 x + \sin^2 x)$

$\quad - 18\sin x \cos x - 32\sin^3 x \cos^3 x$

$= 24\sin x \cos x - 18\sin x \cos x - 32\sin^3 x \cos^3 x$

$= 6\sin x \cos x - 32\sin^3 x \cos^3 x =$ RHS

Example 4: Prove that $\tan 4x = \dfrac{4\tan x(1-\tan^2 x)}{1-6\tan^2 x + \tan^4 x}$.

Solution:

LHS $= \tan 4x = \tan 2(2x) = \dfrac{2\tan 2x}{1-\tan^2(2x)}$

$\left[\because \tan 2A = \dfrac{2\tan A}{1-\tan^2 A}\right]$

$$= \frac{2\left(\dfrac{2\tan x}{1-\tan^2 x}\right)}{1-\left(\dfrac{2\tan x}{1-\tan^2 x}\right)^2} = \frac{\left(\dfrac{4\tan x}{1-\tan^2 x}\right)}{\left[1-\dfrac{4\tan^2 x}{(1-\tan^2 x)^2}\right]}$$

$$= \frac{\left(\frac{4\tan x}{1-\tan^2 x}\right)}{\left[\frac{(1-\tan^2 x)^2 - 4\tan^2 x}{(1-\tan^2 x)^2}\right]}$$

$$= \frac{4\tan x(1-\tan^2 x)}{(1-\tan^2 x)^2 - 4\tan^2 x}$$

$$= \frac{4\tan x(1-\tan^2 x)}{1+\tan^4 x - 2\tan^2 x - 4\tan^2 x}$$

$$= \frac{4\tan x(1-\tan^2 x)}{1-6\tan^2 x + \tan^4 x}$$

$$= \frac{4\tan x(1-\tan^2 x)}{1-6\tan^2 x + \tan^4 x} = \text{RHS}$$

Example 5: Prove that $\tan\left(\frac{\pi}{4}+x\right)+\tan\left(\frac{\pi}{4}-x\right) = 2\sec 2x.$

Solution:

$$\tan\left(\frac{\pi}{4}+x\right)+\tan\left(\frac{\pi}{4}-x\right)$$

$$= \frac{\tan\frac{\pi}{4}+\tan x}{1-\tan\frac{\pi}{4}\tan x} + \frac{\tan\frac{\pi}{4}-\tan x}{1+\tan\frac{\pi}{4}\tan x}$$

$$= \frac{1+\tan x}{1-\tan x} + \frac{1-\tan x}{1+\tan x}$$

$$= \frac{(1+\tan x)^2 + (1-\tan x)^2}{(1-\tan x)(1+\tan x)}$$

$$= \frac{(1+\tan^2 x + 2\tan x) + (1+\tan^2 x - 2\tan x)}{1-\tan^2 x}$$

$$= \frac{2(1+\tan^2 x)}{1-\tan^2 x} = \frac{2}{\cos 2x} = 2\sec 2x$$

Example 6: Prove that $\frac{1}{\sin 10°} - \frac{\sqrt{3}}{\cos 10°} = 4.$

Solution:

$$\text{LHS} = \frac{1}{\sin 10°} - \frac{\sqrt{3}}{\cos 10°} = \frac{\cos 10° - \sqrt{3}\sin 10°}{\sin 10°\cos 10°}$$

$$= \frac{2\left(\frac{1}{2}\cos 10° - \frac{\sqrt{3}}{2}\sin 10°\right)}{\sin 10°\cos 10°}$$

$$= \frac{2(\sin 30°\cos 10° - \cos 30°\sin 10°)}{\sin 10°\cos 10°}$$

$$= \frac{2\sin(30°-10°)}{\sin 10°\cos 10°} = \frac{4\sin 20°}{2\sin 10°\cos 10°}$$

$$= \frac{4\sin 20°}{\sin 20°} = 4$$

Example 7: Prove that
$\tan x + \tan 2x + 4\tan 4x + 8\cot 8x = \cot x.$

Solution:

$\tan x + \tan 2x + 4\tan 4x + 8\cot 8x = \cot x$

$\Rightarrow \cot x - \tan x - 2\tan 2x - 4\tan 4x - 8\cot 8x = 0$

Now, $\cot x - \tan x = \frac{1}{\tan x} - \tan x = \frac{1-\tan^2 x}{\tan x}$

$= \frac{2(1-\tan^2 x)}{2\tan x} = \frac{2}{\tan 2x} = 2\cot 2x$

Substituting, we get

$\text{LHS} = \cot x - \tan x - 2\tan 2x - 4\tan 4x - 8\cot 8x$

$= 2\cot 2x - 2\tan 2x - 4\tan 4x - 8\cot 8x$

$= 2(\cot 2x - \tan 2x) - 4\tan 4x - 8\cot 8x$

$= 2\left(\frac{1}{\tan 2x} - \tan 2x\right) - 4\tan 4x - 8\cot 8x$

$= 2\frac{(1-\tan^2 2x)}{\tan 2x} - 4\tan 4x - 8\cot 8x$

$= 4\left(\frac{1-\tan^2 2x}{2\tan 2x}\right) - 4\tan 4x - 8\cot 8x$

$= \frac{4}{\tan 4x} - 4\tan 4x - 8\cot 8x$

$= 4\left(\frac{1}{\tan 4x} - \tan 4x\right) - 8\cot 8x$

$= 4\frac{(1-\tan^2 4x)}{\tan 4x} - 8\cot 8x = 8\frac{(1-\tan^2 4x)}{2\tan 4x} - 8\cot 8x$

$= \frac{8}{\tan 8x} - 8\cot 8x = 8\cot 8x - 8\cot 8x = 0 = \text{RHS}$

Example 8: Prove that
$\cos^3 x\cos 3x + \sin^3 x\sin 3x = \cos^3 2x.$

Solution:

$\because \sin 3x = 3\sin x - 4\sin^3 x \Rightarrow \sin^3 x = \frac{1}{4}(3\sin x - \sin 3x)$

$\cos 3x = 4\cos^3 x - 3\cos x \Rightarrow \cos^3 x = \frac{1}{4}(3\cos x + \cos 3x)$

$\therefore \text{LHS} = \cos^3 x\cos 3x + \sin^3 x\sin 3x = \cos^3 2x$

$\frac{1}{4}(3\cos x + \cos 3x)\cos 3x + \frac{1}{4}(3\sin x - \sin 3x)\sin 3x$

$= \frac{3}{4}(\cos 3x\cos x + \sin 3x\sin x) + \frac{1}{4}(\cos^2 3x - \sin^2 3x)$

$= \frac{3}{4}\cos(3x - x) + \frac{1}{4}\cos 2\cdot(3x) = \frac{3}{4}\cos 2x + \frac{1}{4}\cos 6x$

$= \frac{3}{4}\cos 2x + \frac{1}{4}\cos 3\cdot(2x)$

$= \frac{3}{4}\cos 2x + \frac{1}{4}(4\cos^3 2x - 3\cos 2x)$

$= \frac{3}{4}\cos 2x + \cos^3 2x - \frac{3}{4}\cos 2x = \cos^3 2x = \text{RHS}$

Example 9: If $2\tan x = 3\tan y$, then prove that $\tan(x+y) = \dfrac{5\sin 2y}{\cos 2y - 1}$.

Solution:

$\because 2\tan x = 3\tan y \Rightarrow \tan x = \frac{3}{2}\tan y$

$$\therefore \tan(x+y) = \frac{\tan x + \tan y}{1 - \tan x \cdot \tan y}$$

$$= \frac{\frac{3}{2}\tan y + \tan y}{1 - \frac{3}{2}\tan y \tan y} = \frac{5\tan y}{2 - 3\tan^2 y}$$

$$= \frac{5\frac{\sin y}{\cos y}}{2 - 3\frac{\sin^2 y}{\cos^2 y}} = \frac{5\sin y\cos y}{2\cos^2 y - 3\sin^2 y}$$

$$= \frac{\frac{5}{2}\cdot 2\sin y\cos y}{1 + \cos 2y - \frac{3}{2}(1 - \cos 2y)}$$

$$= \frac{\frac{5}{2}\sin 2y}{-\frac{1}{2} + \frac{5}{2}\cos 2y} = \frac{5\sin 2y}{\cos 2y - 1}$$

Day 43

Trigonometric Functions of Sum and Difference of Two Angles–III

(xxvii) Product formula

(a) $2\cos x\cos y = \cos(x+y) + \cos(x-y)$

(b) $-2\sin x\sin y = \cos(x+y) - \cos(x-y)$

(c) $2\sin x\cos y = \sin(x+y) + \sin(x-y)$

(d) $2\cos x\sin y = \sin(x+y) - \sin(x-y)$

Since

$\cos(x+y) = \cos x\cos y - \sin x\sin y \quad (1)$

$\cos(x-y) = \cos x\cos y + \sin x\sin y \quad (2)$

Adding and subtracting equations (1) and (2), we get

$\cos(x+y) + \cos(x-y) = 2\cos x\cos y \quad (3)$

$\cos(x+y) - \cos(x-y) = -2\sin x\sin y \quad (4)$

Further, since

$\sin(x+y) = \sin x\cos y + \cos x\sin y \quad (5)$

$\sin(x-y) = \sin x\cos y - \cos x\sin y \quad (6)$

Adding and subtracting equations (5) and (6), we get

$\sin(x+y) + \sin(x-y) = 2\sin x\cos y \quad (7)$

$\sin(x+y) - \sin(x-y) = 2\cos x\sin y \quad (8)$

(xxviii) Addition formula

(a) $\cos x + \cos y = 2\cos\frac{x+y}{2}\cos\frac{x-y}{2}$

(b) $\cos x - \cos y = -2\sin\frac{x+y}{2}\sin\frac{x-y}{2}$

(c) $\sin x + \sin y = 2\sin\frac{x+y}{2}\cos\frac{x-y}{2}$

(d) $\sin x - \sin y = 2\cos\frac{x+y}{2}\sin\frac{x-y}{2}$

In identity (xxvii), substituting $x + y = \theta$ and $x - y = \phi$

$\Rightarrow x = \left(\frac{\theta+\phi}{2}\right)$ and $y = \left(\frac{\theta-\phi}{2}\right)$, we get

$$\cos\theta + \cos\phi = 2\cos\left(\frac{\theta+\phi}{2}\right)\cos\left(\frac{\theta-\phi}{2}\right)$$

$$\cos\theta - \cos\phi = -2\sin\left(\frac{\theta+\phi}{2}\right)\sin\left(\frac{\theta-\phi}{2}\right)$$

$$\sin\theta + \sin\phi = 2\sin\left(\frac{\theta+\phi}{2}\right)\cos\left(\frac{\theta-\phi}{2}\right)$$

$$\sin\theta - \sin\phi = 2\cos\left(\frac{\theta+\phi}{2}\right)\sin\left(\frac{\theta-\phi}{2}\right)$$

Since θ, ϕ can take any real values, we can replace θ with x and ϕ with y.

Thus, we get

$$\cos x + \cos y = 2\cos\frac{x+y}{2}\cos\frac{x-y}{2}$$

$$\cos x - \cos y = -2\sin\frac{x+y}{2}\sin\frac{x-y}{2},$$

$$\sin x + \sin y = 2\sin\frac{x+y}{2}\cos\frac{x-y}{2}$$

$$\sin x - \sin y = 2\cos\frac{x+y}{2}\sin\frac{x-y}{2}$$

(xxix) (a) $\sin(x+y)\sin(x-y) = \sin^2 x - \sin^2 y$

(b) $\cos(x+y)\cos(x-y) = \cos^2 x - \sin^2 y$

In identity (xxvii), substituting $x + y = 2\theta$ and $x - y = 2\phi$

$\Rightarrow x = \theta + \phi$ and $y = \theta - \phi$, we get

(a) $-2\sin x\sin y = \cos(x+y) - \cos(x-y)$

$\Rightarrow -2\sin(\theta+\phi)\sin(\theta-\phi) = \cos 2\theta - \cos 2\phi$

$= (1 - 2\sin^2\theta) - (1 - 2\sin^2\phi)$

$= -2(\sin^2\theta - \sin^2\phi)$

$\Rightarrow \sin(\theta+\phi)\sin(\theta-\phi) = \sin^2\theta - \sin^2\phi$

(b) $2\cos x\cos y=\cos(x+y)+\cos(x-y)$

$\Rightarrow 2\cos(\theta+\phi)\cos(\theta-\phi)=\cos 2\theta+\cos 2\phi$

$=(2\cos^2\theta-1)+(1-2\sin^2\phi)$

$=2(\cos^2\theta-\sin^2\phi)$

$\Rightarrow \cos(\theta+\phi)\cos(\theta-\phi)=\cos^2\theta-\sin^2\phi$

Since θ,ϕ can take any real values, we can replace θ with x and ϕ with y.

Thus, we get

$\cos(x+y)\cos(x-y)=\cos^2 x-\sin^2 y$

Example 1: Express each of the following as the sum or difference of the cosines and sines:

(i) $2\sin x\cdot\cos 3x$ (ii) $\cos 4x\cdot\sin 2x$

(iii) $3\cos 5x\cdot\cos 7x$ (iv) $4\sin 6x\cdot\sin 2x$

Solution:

(i) $2\sin x\cdot\cos 3x$

$=\sin(x+3x)+\sin(x-3x)$

$[\because 2\sin x\cos y=\sin(x+y)+\sin(x-y)]$

$=\sin 4x+\sin(-2x)=\sin 4x-\sin 2x$

(ii) $\cos 4x\cdot\sin 2x$

$=\frac{1}{2}\{\sin(4x+2x)-\sin(4x-2x)\}$

$[\because 2\cos x\sin y=\sin(x+y)-\sin(x-y)]$

$=\frac{1}{2}\{\sin 6x-\sin 2x\}$

(iii) $3\cos 5x\cdot\cos 7x$

$=\frac{3}{2}\{\cos(3x+5x)+\cos(3x-5x)\}$

$[\because 2\cos x\cdot\cos y=\cos(x+y)+\cos(x-y)]$

$=\frac{3}{2}\{\cos 8x+\cos(-2x)\}=\frac{3}{2}\{\cos 8x+\cos 2x\}$

(iv) $4\sin 6x\cdot\sin 2x$

$=\frac{4}{(-2)}\{\cos(6x+2x)-\cos(6x-2x)\}$

$[\because -2\sin x\sin y=\cos(x+y)-\cos(x-y)]$

$=-2(\cos 8x-\cos 4x)=2(\cos 4x-\cos 8x)$

Example 2: Express each of the following as the product of the sines and cosines:

(i) $\sin 5x+\sin 3x$ (ii) $\sin 2x-\sin 4x$

(iii) $\cos x+\cos 7x$ (iv) $\cos 6x-\cos 8x$

Solution:

(i) $\sin 4x+\sin 3x$

$=2\sin\left(\frac{5x+3x}{2}\right)\cos\left(\frac{5x-3x}{2}\right)$

$\left[\because \sin x+\sin y=2\sin\frac{x+y}{2}\cos\frac{x-y}{2}\right]$

$=2\sin 4x\cos x$

(ii) $\sin 2x-\sin 4x$

$=2\sin\left(\frac{2x-4x}{2}\right)\cdot\cos\left(\frac{2x+4x}{2}\right)$

$\left[\because \sin x-\sin y=2\sin\frac{(x-y)}{2}\cos\frac{(x+y)}{2}\right]$

$=2\sin(-x)\cdot\cos 3x=-2\sin x\cos 3x$

(iii) $\cos x+\cos 7x$

$=2\cos\left(\frac{x+7x}{2}\right)\cdot\cos\left(\frac{x-7x}{2}\right)$

$\left[\because \cos x+\cos y=2\cos\frac{x+y}{2}\cdot\cos\frac{x-y}{2}\right]$

$=2\cos 4x\cdot\cos(-3x)=2\cos 4x\cos 3x$

(iv) $\cos 6x-\cos 8x$

$=-2\sin\left(\frac{6x+8x}{2}\right)\cdot\sin\left(\frac{6x-8x}{2}\right)$

$\left[\because \cos x-\cos y=-2\sin\frac{x+y}{2}\cdot\sin\frac{x-y}{2}\right]$

$=-2\sin 7x\cdot\sin(-x)=2\sin x\sin 7x$

Example 3: Prove that

$\sin 20^\circ\sin 40^\circ\sin 60^\circ\sin 80^\circ=\frac{3}{16}.$

Solution:

LHS $=\sin 20^\circ\sin 40^\circ\sin 60^\circ\sin 80^\circ$

$=\sin 20^\circ\sin 40^\circ\cdot\frac{\sqrt{3}}{2}\cdot\sin 80^\circ$

$=\frac{\sqrt{3}}{4}(2\sin 80^\circ\sin 40^\circ)\sin 20^\circ$

$=\frac{\sqrt{3}}{4}(\cos 40^\circ-\cos 120^\circ)\sin 20^\circ$

$[\because -2\sin x\sin y=\cos(x+y)-\cos(x-y)]$

$=\frac{\sqrt{3}}{4}\left[\cos 40^\circ\sin 20^\circ-\left(-\frac{1}{2}\right)\cdot\sin 20^\circ\right]$

$=\frac{\sqrt{3}}{8}[2\cos 40^\circ\sin 20^\circ+\sin 20^\circ]$

$=\frac{\sqrt{3}}{8}[\sin(40^\circ+20^\circ)-\sin(40^\circ-20^\circ)+\sin 20^\circ]$

$[\because 2\cos x\sin y=\sin(x+y)-\sin(x-y)]$

$=\frac{\sqrt{3}}{8}[\sin 60^\circ-\sin 20^\circ+\sin 20^\circ]$

$=\frac{\sqrt{3}}{8}\sin 60^\circ=\frac{\sqrt{3}}{8}\cdot\frac{\sqrt{3}}{2}=\frac{3}{16}$

Example 4: Prove that

$\cos 15^\circ\cos 30^\circ\cos 45^\circ\cos 60^\circ\cos 75^\circ=\frac{3}{16}.$

Solution:

LHS $= \cos 15° \cos 30° \cos 45° \cos 60° \cos 75°$

$= \cos 15° \cdot \frac{\sqrt{3}}{2} \cdot \frac{1}{\sqrt{2}} \cdot \frac{1}{2} \cdot \cos 75°$

$= \frac{\sqrt{3}}{4\sqrt{2}} \cos 75° \cos 15°$

$[\because 2\cos x \cdot \cos y = \cos(x+y) + \cos(x-y)]$

$= \frac{\sqrt{3}}{8\sqrt{2}}[\cos(75°+15°) + \cos(75°-15°)]$

$= \frac{\sqrt{3}}{8\sqrt{2}}[\cos 90° + \cos 60°]$

$= \frac{\sqrt{3}}{8\sqrt{2}}\left[0 + \frac{1}{2}\right] = \frac{\sqrt{3}}{16\sqrt{2}} = \frac{\sqrt{6}}{32}$

Example 5: Prove that

$\sin 55° + \sin 65° + \cos 110° + \cos 150° = 0.$

Solution:

LHS $= \sin 55° + \sin 65° + \cos 110° + \cos 150°$

$= (\sin 55° + \sin 65°) + (\cos 150° + \cos 110°)$

$= 2\sin\frac{65°+15°}{2}\cos\frac{65°-15°}{2}$

$+2\cos\frac{150°+110°}{2}\cos\frac{150°-110°}{2}$

$\left[\because \sin x + \sin y = 2\sin\frac{x+y}{2}\cos\frac{x-y}{2}\right]$

$\left[\because \cos x + \cos y = 2\cos\frac{x+y}{2}\cdot\cos\frac{x-y}{2}\right]$

$= 2\sin 40° \cos 20° + 2\cos 130° \cos 20°$

$= 2\cos 20°(\sin 40° - \sin 40°) = 0 =$ RHS

Example 6: Prove that

$\sin 55° + \cos 65° - \sin 25° - \cos 35° = 0$

Solution:

LHS $= \sin 55° + \cos 65° - \sin 25° - \cos 35°$

$= (\sin 55° - \sin 25°) + (\cos 65° - \cos 35°)$

$= 2\sin\frac{55°-25°}{2}\cos\frac{55°+25°}{2}$

$-2\sin\frac{65°+35°}{2}\sin\frac{65°-35°}{2}$

$\left[\because \sin x - \sin y = 2\sin\frac{(x-y)}{2}\cos\frac{(x+y)}{2}\right]$

$\left[\because \cos x - \cos y = -2\sin\frac{x+y}{2}\cdot\sin\frac{x-y}{2}\right]$

$= 2\sin 15° \cos 40° - 2\sin 50° \sin 15°$

$= 2\sin 15°(\sin 50° - \sin 50°) = 0 =$ RHS

Example 7: Prove that $\tan 20° \tan 40° \tan 80° = \sqrt{3}$.

Solution:

LHS $= \tan 20° \tan 40° \tan 80°$

$= \frac{\sin 20° \sin 40° \sin 80°}{\cos 20° \cos 40° \cos 80°}$

$= \frac{(2\sin 20° \cdot \sin 40°)\sin 80°}{(2\cos 20° \cdot \cos 40°)\cos 80°}$

$= \frac{(\cos 20° - \cos 60°)\sin 80°}{(\cos 60° + \cos 20°)\cos 80°}$

$= \frac{\sin 80° \cdot \cos 20° - \frac{1}{2}\sin 80°}{\frac{1}{2}\cos 80° + \cos 80° \cdot \cos 20°}$

$= \frac{2\sin 80° \cdot \cos 20° - \sin 80°}{\cos 80° + 2\cos 80° \cdot \cos 20°}$

$= \frac{\sin 100° + \sin 60° - \sin 80°}{\cos 80° + \cos 100° + \cos 60°}$

$= \frac{\sin(180°-80°) + \sin 60° - \sin 80°}{\cos 80° + \cos(180°-80°) + \cos 60°}$

$= \frac{\sin 80° + \sin 60° - \sin 80°}{\cos 80° - \cos 80° + \cos 60°} = \frac{\sin 60°}{\cos 60°}$

$= \tan 60° = \sqrt{3} =$ RHS

Example 8: Prove that

$\tan 6° \cdot \tan 42° \cdot \tan 66° \cdot \tan 78° = 1$

Solution:

$= \frac{\sin 6°}{\cos 6°} \cdot \frac{\sin 42°}{\cos 42°} \cdot \frac{\sin 66°}{\cos 66°} \cdot \frac{\sin 78°}{\cos 78°}$

$= \frac{2\sin 66° \sin 6°}{2\cos 66° \cos 6°} \cdot \frac{2\sin 78° \sin 42°}{2\cos 78° \cos 42°}$

$= \frac{\cos 60° - \cos 72°}{\cos 60° + \cos 72°} \cdot \frac{\cos 36° - \cos 120°}{\cos 36° + \cos 120°}$

$= \frac{\frac{1}{2} - \sin 18°}{\frac{1}{2} + \sin 18°} \cdot \frac{\cos 36° + \frac{1}{2}}{\cos 36° - \frac{1}{2}} = \frac{1 - 2\sin 18°}{1 + 2\sin 18°} \cdot \frac{2\cos 36° + 1}{2\cos 36° - 1}$

$= \frac{1 - 2\left(\frac{\sqrt{5}-1}{4}\right)}{1 + 2\left(\frac{\sqrt{5}-1}{4}\right)} \cdot \frac{2\left(\frac{\sqrt{5}+1}{4}\right) + 1}{2\left(\frac{\sqrt{5}+1}{4}\right) - 1}$

$= \frac{2 - \sqrt{5} + 1}{2 + \sqrt{5} - 1} \times \frac{2\sqrt{5} + 2 + 4}{2\sqrt{5} + 2 - 4} = \frac{3 - \sqrt{5}}{1 + \sqrt{5}} \times \frac{3 + \sqrt{5}}{\sqrt{5} - 1} = \frac{9 - 5}{5 - 1} = 1$

Example 9: Show that

$\sin x \cdot \sin(y-z) + \sin y \cdot \sin(z-x) + \sin z \cdot \sin(x-y) = 0.$

Solution:

$\sin x \cdot \sin(y-z) + \sin y \cdot \sin(z-x) + \sin z \cdot \sin(x-y)$

$$= \frac{1}{2}[2\sin x \cdot \sin(y-z) + 2\sin y \cdot \sin(z-x)$$
$$+ 2\sin z \cdot \sin(x-y)]$$
$$= \frac{1}{2}[\cos(x-y+z) - \cos(x+y-z) + \cos(y-z+x)$$
$$-\cos(y+z-x) + \cos(z-x+y) - \cos(z+x-y)] = 0$$

Example 10: Prove that

(i) $\frac{(1-\cos 4x)+\sin 4x}{(1+\cos 4x)+\sin 4x} = \tan 2x$

(ii) $\frac{\sin 5x + \sin 3x}{\cos 5x + \cos 3x} = \tan 4x$

(iii) $\frac{\sin 3x + \sin 5x + \sin 7x + \sin 9x}{\cos 3x + \cos 5x + \cos 7x + \cos 9x} = \tan 6x$

Solution:

(i) $\text{LHS} = \frac{(1-\cos 4x)+\sin 4x}{(1+\cos 4x)+\sin 4x}$

$$= \frac{2\sin^2 2x + 2\sin 2x \cos 2x}{2\cos^2 2x + 2\sin 2x \cos 2x}$$

$$= \frac{2\sin 2x(\sin 2x + \cos 2x)}{2\cos 2x(\cos 2x + \sin 2x)} = \tan 2x$$

(ii) $\text{LHS} = \frac{\sin 5x + \sin 3x}{\cos 5x + \cos 3x}$

$$= \frac{2\sin\left(\frac{5x+3x}{2}\right)\cdot\cos\left(\frac{5x-3x}{2}\right)}{2\cos\left(\frac{5x+3x}{2}\right)\cdot\cos\left(\frac{5x-3x}{2}\right)}$$

$$= \frac{2\sin 4x \cdot \cos x}{2\cos 4x \cdot \cos x} = \tan 4x = \text{RHS}$$

(iii) $\text{LHS} = \frac{\sin 3x + \sin 5x + \sin 7x + \sin 9x}{\cos 3x + \cos 5x + \cos 7x + \cos 9x} = \tan 6x$

$$= \frac{(\sin 7x + \sin 5x) + (\sin 9x + \sin 3x)}{(\cos 7x + \cos 5x) + (\cos 9x + \cos 3x)}$$

$$= \frac{\left[2\sin\left(\frac{7x+5x}{2}\right)\cos\left(\frac{7x-5x}{2}\right) + 2\sin\left(\frac{9x+3x}{2}\right)\cos\left(\frac{9x-3x}{2}\right)\right]}{\left[2\cos\left(\frac{7x+5x}{2}\right)\cdot\cos\left(\frac{7x-5x}{2}\right) + 2\cos\left(\frac{9x+3x}{2}\right)\cos\left(\frac{9x+3x}{2}\right)\right]}$$

$$= \frac{2\sin 6x \cdot \cos x + 2\sin 6x \cdot \cos 3x}{2\cos 6x \cdot \cos x + 2\cos 6x \cdot \cos 3x}$$

$$= \frac{\sin 6x(\cos x + \cos 3x)}{\cos 6x(\cos x + \cos 3x)} = \tan 6x = \text{RHS}$$

Example 11: Prove that

(i) $(\sin x + \sin y)^2 + (\cos x + \cos y)^2 = 4\cos^2\left(\frac{x-y}{2}\right)$

(ii) $\sin x + \sin y + \sin z - \sin(x+y+z)$
$= 4\sin\frac{x+y}{2}\sin\frac{y+z}{2}\sin\frac{z+x}{2}$

Solution:

(i) $\text{LHS} = (\sin x + \sin y)^2 + (\cos x + \cos y)^2$

$$= \left[2\sin\frac{x+y}{2}\cdot\cos\frac{x-y}{2}\right]^2 + \left[2\cos\frac{x+y}{2}\cdot\cos\frac{x-y}{2}\right]^2$$

$$= 4\sin^2\left(\frac{x+y}{2}\right)\cos^2\left(\frac{x-y}{2}\right)$$
$$+ 4\cos^2\left(\frac{x+y}{2}\right)\cos^2\left(\frac{x-y}{2}\right)$$

$$= 4\cos^2\left(\frac{x-y}{2}\right)\left[\sin^2\left(\frac{x+y}{2}\right) + \cos^2\left(\frac{x+y}{2}\right)\right]$$

$$= 4\cos^2.\left(\frac{x-y}{2}\right) = \text{RHS}$$

(ii) $\text{LHS} = \sin x + \sin y + \sin z - \sin(x+y+z)$
$= (\sin x + \sin y) - [\sin(x+y+z) - \sin z]$

$$= 2\sin\left(\frac{x+y}{2}\right)\cos\left(\frac{x-y}{2}\right)$$
$$-2\sin\left(\frac{x+y+z-z}{2}\right)\cos\left(\frac{x+y+z+z}{2}\right)$$

$$= 2\sin\left(\frac{x+y}{2}\right)\cos\left(\frac{x-y}{2}\right)$$
$$-2\sin\left(\frac{x+y}{2}\right)\cos\left(\frac{x+y+2z}{2}\right)$$

$$= 2\sin\left(\frac{x+y}{2}\right)\left[\cos\frac{(x-y)}{2} - \cos\frac{(x+y+2z)}{2}\right]$$

$$= 2\sin\left(\frac{x+y}{2}\right)\left[-2\sin\left(\frac{x-y+x+y+2z}{4}\right)\sin\left(\frac{x-y-x-y-2z}{4}\right)\right]$$

$$= -4\sin\left(\frac{x+y}{2}\right)\sin\left(\frac{x+z}{2}\right)\sin\left(\frac{-z-y}{2}\right)$$

$$= 4\sin\left(\frac{x+y}{2}\right)\sin\left(\frac{y+z}{2}\right)\sin\left(\frac{z+x}{2}\right) = \text{RHS}$$

Example 12: Prove that

$\tan(60° - x)\cdot\tan x\cdot\tan(60° + x) = \tan 3x$

Solution:

$$\text{LHS} = \frac{\sin(60° - x)}{\cos(60° - x)}\cdot\frac{\sin x}{\cos x}\cdot\frac{\sin(60° + x)}{\cos(60° + x)}$$

$$= \frac{\sin x(\sin^2 60° - \sin^2 x)}{\cos x(\cos^2 60° - \sin^2 x)}$$

$$\begin{bmatrix} \because \sin(x+y)\sin(x-y) = \sin^2 x - \sin^2 y \\ \cos(x+y)\cos(x-y) = \cos^2 x - \sin^2 y \end{bmatrix}$$

$$= \frac{\sin x\left(\frac{3}{4} - \sin^2 x\right)}{\cos x\left(\frac{1}{4} - \sin^2 x\right)} = \frac{\sin x(3 - 4\sin^2 x)}{\cos x(1 - 4\sin^2 x)}$$

$$= \frac{3\sin x - 4\sin^3 x}{\cos x[1 - 4(1-\cos^2 x]} = \frac{3\sin x - 4\sin^3 x}{\cos x(4\cos^2 x - 3)}$$

$$= \frac{3\sin x - 4\sin^3 x}{4\cos^3 x - 3\cos x} = \frac{\sin 3x}{\cos 3x} = \tan 3x$$

Example 13: Prove that

$\cos^2 x + \cos^2 y - 2\cos x \cos y \cdot \cos(x+y) = \sin^2(x+y)$

Solution:

$\text{LHS} = \cos^2 x + \cos^2 y - 2\cos x \cos y \cdot \cos(x+y)$

$= \cos^2 x + (1 - \sin^2 y) - 2\cos x \cos y \cos(x+y)$

$= 1 + (\cos^2 x - \sin^2 y) - 2\cos x \cos y \cos(x+y)$

$= 1 + \cos(x+y)\cdot\cos(x-y) - 2\cos x \cos y \cos(x+y)$

$[\because \cos^2 x - \sin^2 y = \cos(x+y)\cos(x-y)]$

$= 1 + \cos(x+y)[\cos(x-y) - 2\cos x \cos y]$

$= 1 + \cos(x+y)[\cos x \cdot \cos y + \sin x \cdot \sin y - 2\cos x \cdot \cos y]$

$= 1 - \cos(x+y)[\cos x \cdot \cos y - \sin x \cdot \sin y]$

$= 1 - \cos(x+y)\cos(x+y)$

$= 1 - \cos^2(x+y) = \sin^2(x+y)$

Example 14: Prove that

$\cos^2 x + \sin^2 y - 2\cos x \sin y \cdot \sin(x+y) = \cos^2(x+y)$

Solution:

$\text{LHS} = \cos^2 x + \sin^2 y - 2\cos x \sin y \sin(x+y)$

$= 1 - \sin^2 x + \sin^2 y - 2\cos x \sin y \sin(x+y)$

$= 1 - (\sin^2 x - \sin^2 y) - 2\cos x \sin y \sin(x+y)$

$= 1 - \sin(x+y)\cdot\sin(x-y) - 2\cos x \sin y \sin(x+y)$

$[\because \sin^2 x - \sin^2 y = \sin(x+y)\sin(x-y)]$

$= 1 - \sin(x+y)[\sin(x-y) + 2\cos x \sin y]$

$= 1 - \sin(x+y)[\sin x \cdot \cos y - \cos x \cdot \sin y + 2\cos x \cdot \sin y]$

$= 1 - \sin(x+y)[\sin x \cdot \cos y + \cos x \cdot \sin y]$

$= 1 - \sin(x+y)\sin(x+y)$

$= 1 - \sin^2(x+y) = \cos^2(x+y)$

Day 44

NCERT Exercise 3.3 (With Solutions)

Q1. Prove that $\sin^2\frac{\pi}{6} + \cos^2\frac{\pi}{3} - \tan^2\frac{\pi}{4} = -\frac{1}{2}$.

Solution:

$\text{LHS} = \sin^2\frac{\pi}{6} + \cos^2\frac{\pi}{3} - \tan^2\frac{\pi}{4}$

$= \left(\frac{1}{2}\right)^2 + \left(\frac{1}{2}\right)^2 - (1)^2$

$= \frac{1}{4} + \frac{1}{4} - 1 = -\frac{1}{2} = \text{RHS}$

Q2. Prove that $2\sin^2\frac{\pi}{6} + \text{cosec}^2\frac{7\pi}{6}\cos^2\frac{\pi}{3} = \frac{3}{2}$.

Solution:

$\text{LHS} = 2\sin^2\frac{\pi}{6} + \text{cosec}^2\frac{7\pi}{6}\cos^2\frac{\pi}{3}$

$= 2\left(\frac{1}{2}\right)^2 + \text{cosec}^2\left(\pi + \frac{\pi}{6}\right)\left(\frac{1}{2}\right)^2$

$= 2\times\frac{1}{4} + \left(-\text{cosec}\frac{\pi}{6}\right)^2\left(\frac{1}{4}\right)$

$= \frac{1}{2} + (-2)^2\left(\frac{1}{4}\right)$

$= \frac{1}{2} + \frac{4}{4} = \frac{3}{2} = \text{RHS}$

Q3. Prove that $\cot^2\frac{\pi}{6} + \text{cosec}\frac{5\pi}{6} + 3\tan^2\frac{\pi}{6} = 6$.

Solution:

$\text{LHS} = \cot^2\frac{\pi}{6} + \text{cosec}\frac{5\pi}{6} + 3\tan^2\frac{\pi}{6}$

$= (\sqrt{3})^2 + \text{cosec}\left(\pi - \frac{\pi}{6}\right) + 3\left(\frac{1}{\sqrt{3}}\right)$

$= 3 + \text{cosec}\frac{\pi}{6} + 3\times\frac{1}{3}$

$= 3 + 2 + 1 = 6 = \text{RHS}$

Q4. Prove that $2\sin^2\frac{3\pi}{4}+2\cos^2\frac{\pi}{4}+2\sec^2\frac{\pi}{3}=10.$

Solution:

$$\text{LHS}=2\sin^2\frac{3\pi}{4}+2\cos^2\frac{\pi}{4}+2\sec^2\frac{\pi}{3}$$

$$=2\left\{\sin\left(\pi-\frac{\pi}{4}\right)\right\}^2+2\left(\frac{1}{\sqrt{2}}\right)^2+2(2)^2$$

$$=2\left\{\sin\frac{\pi}{4}\right\}^2+2\times\frac{1}{2}+8$$

$$=2\left(\frac{1}{\sqrt{2}}\right)^2+1+8$$

$$=1+1+8=10=\text{RHS}$$

Q5. Find the value of: (i) sin 75° (ii) tan 15°.

Solution:

(i) $\sin 75°=\sin(45°+30°)$

$=\sin 45°\cos 30°+\cos 45°\sin 30°$

$[\because \sin(x+y)=\sin x\cos y+\cos x\sin y]$

$$\left(\frac{1}{\sqrt{2}}\right)\left(\frac{\sqrt{3}}{2}\right)+\left(\frac{1}{\sqrt{2}}\right)\left(\frac{1}{2}\right)$$

$$=\frac{\sqrt{3}}{2\sqrt{2}}+\frac{1}{2\sqrt{2}}=\frac{\sqrt{3}+1}{2\sqrt{2}}$$

(ii) $\tan 15°=\tan(45°-30°)$

$$=\frac{\tan 45°-\tan 30°}{1+\tan 45°\tan 30°}$$

$$\left[\because \tan(x-y)=\frac{\tan x-\tan y}{1+\tan x\tan y}\right]$$

$$=\frac{1-\frac{1}{\sqrt{3}}}{1+1\left(\frac{1}{\sqrt{3}}\right)}=\frac{\frac{\sqrt{3}-1}{\sqrt{3}}}{\frac{\sqrt{3}+1}{\sqrt{3}}}$$

$$=\frac{\sqrt{3}-1}{\sqrt{3}+1}=\frac{(\sqrt{3}-1)^2}{(\sqrt{3}+1)(\sqrt{3}-1)}=\frac{3+1-2\sqrt{3}}{(\sqrt{3})^2-(1)^2}$$

$$=\frac{4-2\sqrt{3}}{3-1}=2-\sqrt{3}$$

Q6. Prove that:

$$\cos\left(\frac{\pi}{4}-x\right)\cos\left(\frac{\pi}{4}-y\right)-\sin\left(\frac{\pi}{4}-x\right)\sin\left(\frac{\pi}{4}-y\right)$$

$$=\sin(x+y)$$

Solution:

$$\cos\left(\frac{\pi}{4}-x\right)\cos\left(\frac{\pi}{4}-y\right)-\sin\left(\frac{\pi}{4}-x\right)\sin\left(\frac{\pi}{4}-y\right)$$

$$=\frac{1}{2}\left[2\cos\left(\frac{\pi}{4}-x\right)\cos\left(\frac{\pi}{4}-y\right)\right]+\frac{1}{2}\left[-2\sin\left(\frac{\pi}{4}-x\right)\sin\left(\frac{\pi}{4}-y\right)\right]$$

$$=\frac{1}{2}\begin{bmatrix}\cos\left\{\left(\frac{\pi}{4}-x\right)+\left(\frac{\pi}{4}-y\right)\right\}+\\ \cos\left\{\left(\frac{\pi}{4}-x\right)-\left(\frac{\pi}{4}-y\right)\right\}\end{bmatrix}$$

$$+\frac{1}{2}\begin{bmatrix}\cos\left\{\left(\frac{\pi}{4}-x\right)+\left(\frac{\pi}{4}-y\right)\right\}\\ -\cos\left\{\left(\frac{\pi}{4}-x\right)-\left(\frac{\pi}{4}-y\right)\right\}\end{bmatrix}$$

$$\begin{bmatrix}\because 2\cos A\cos B=\cos(A+B)+\cos(A-B)\\ -2\sin A\sin B=\cos(A+B)-\cos(A-B)\end{bmatrix}$$

$$=\frac{1}{2}\left[\cos\left(\frac{\pi}{2}-(x+y)\right)+\cos(y-x)\right]$$

$$+\frac{1}{2}\left[\cos\left(\frac{\pi}{2}-(x+y)\right)-\cos(y-x)\right]$$

$$=2\times\frac{1}{2}\cos\left[\frac{\pi}{2}-(x+y)\right]=\sin(x+y)=\text{RHS}$$

Q7. Prove that $\dfrac{\tan\left(\frac{\pi}{4}+x\right)}{\tan\left(\frac{\pi}{4}-x\right)}=\left(\dfrac{1+\tan x}{1-\tan x}\right)^2.$

Solution:

$$\text{LHS}=\frac{\tan\left(\frac{\pi}{4}+x\right)}{\tan\left(\frac{\pi}{4}+x\right)}=\frac{\left(\dfrac{\tan\frac{\pi}{4}+\tan x}{1-\tan\frac{\pi}{4}\tan x}\right)}{\left(\dfrac{\tan\frac{\pi}{4}-\tan x}{1+\tan\frac{\pi}{4}\tan x}\right)}$$

$$\begin{bmatrix}\because \tan(A+B)=\dfrac{\tan A+\tan B}{1-\tan A\tan B}\\ \tan(A-B)=\dfrac{\tan A-\tan B}{1+\tan A\tan B}\end{bmatrix}$$

$$=\frac{\left(\dfrac{1+\tan x}{1-\tan x}\right)}{\left(\dfrac{1-\tan x}{1+\tan x}\right)}=\left(\frac{1+\tan x}{1-\tan x}\right)^2=\text{RHS}$$

Q8. Prove that $\dfrac{\cos(\pi+x)\cos(-x)}{\sin(\pi-x)\cos\left(\frac{\pi}{2}+x\right)}=\cot^2 x.$

Solution:

$$\text{LHS}=\frac{\cos(\pi+x)\cos(-x)}{\sin(\pi-x)\cos\left(\frac{\pi}{2}+x\right)}$$

$= \frac{[-\cos x][\cos x]}{(\sin x)(-\sin x)} = \frac{-\cos^2 x}{-\sin^2 x}$

$= \cot^2 x = \text{RHS}$

Q9. $\cos\left(\frac{3\pi}{2}+x\right)\cos(2\pi+x)\left[\cot\left(\frac{3\pi}{2}-x\right)+\cot(2\pi+x)\right]=1$

Solution:

$\text{LHS} = \cos\left(\frac{3\pi}{2}+x\right)\cos(2\pi+x)\left[\cot\left(\frac{3\pi}{2}-x\right)+\cot(2\pi+x)\right]$

$= \sin x\cos x[\tan x+\cot x]$

$= \sin x\cos\left(\frac{\sin x}{\cos x}+\frac{\cos x}{\sin x}\right)$

$= \sin^2 x+\cos^2 x = 1 = \text{RHS}$

Q10. Prove that $\sin(n+1)x\sin(n+2)x+\cos(n+1)x\cos(n+2)x=\cos x.$

Solution:

$\text{LHS} = \sin(n+1)x\sin(n+2)x+\cos(n+1)x\cos(n+2)x$

$= \frac{1}{2}[2\sin(n+1)x\times\sin(n+2)x$
$+2\cos(n+1)x\times\cos(n+2)x]$

$= \frac{1}{2}[\cos\{(n+1)x-(n+2)x\}+\cos\{(n+1)x+(n+2)x\}$
$+\cos\{(n+1)x+(n+2)x\}+\cos\{(n+1)x-(n+2)x\}]$

$\left[\begin{array}{l}\because -2\sin A\sin B = \cos(A+B)-\cos(A-B)\\ 2\cos A\cos B = \cos(A+B)+\cos(A-B)\end{array}\right]$

$= \frac{1}{2}\times 2\cos\{(n+1)x-(n+2)x\}$

$= \cos(-x) = \cos x = \text{RHS}$

Q11. Prove that $\cos\left(\frac{3\pi}{4}+x\right)-\cos\left(\frac{3\pi}{4}-x\right)=-\sqrt{2}\sin x.$

Solution:

$\text{LHS} = \cos\left(\frac{3\pi}{4}+x\right)-\cos\left(\frac{3\pi}{4}-x\right)$

$= -2\sin\left\{\frac{\left(\frac{3\pi}{4}+x\right)+\left(\frac{3\pi}{4}-x\right)}{2}\right\}\sin\left\{\frac{\left(\frac{3\pi}{4}+x\right)-\left(\frac{3\pi}{4}-x\right)}{2}\right\}$

$\left[\because \cos A-\cos B = -2\sin\left(\frac{A+B}{2}\right)\cdot\sin\left(\frac{A-B}{2}\right)\right]$

$= -2\sin(3\pi/4)\sin x = -2\sin\left(\pi-\frac{\pi}{4}\right)\sin x$

$= -2\sin\frac{\pi}{4}\sin x = -2\frac{1}{\sqrt{2}}\times\sin x = -\sqrt{2}\sin x = \text{RHS}$

Q12. Prove that $\sin^2 6x-\sin^2 4x = \sin 2x\sin 10x.$

Solution:

$\text{LHS} = \sin^2 6x-\sin^2 4x$

$= (\sin 6x+\sin 4x)(\sin 6x-\sin 4x)$

$= \left[2\sin\left(\frac{6x+4x}{2}\right)\cos\left(\frac{6x-4x}{2}\right)\right]\left[2\cos\left(\frac{6x+4x}{2}\right)\cdot\sin\left(\frac{6x-4x}{2}\right)\right]$

$\left[\begin{array}{l}\because \sin A+\sin B = 2\sin\left(\frac{A+B}{2}\right)\cos\left(\frac{A-B}{2}\right)\\ \sin A-\sin B = 2\cos\left(\frac{A+B}{2}\right)\sin\left(\frac{A-B}{2}\right)\end{array}\right]$

$= (2\sin 5x\cos x)(2\cos 5x\sin x)$

$= (2\sin x\cos x)(2\sin 5x\cos 5x)$

$= \sin 2x\sin 10x = \text{RHS}$

Q13. Prove that $\cos^2 2x-\cos^2 6x = \sin 4x\sin 8x.$

Solution:

$\text{LHS} = \cos^2 2x-\cos^2 6x$

$= (\cos 2x+\cos 6x)(\cos 2x-\cos 6x)$

$= \left[2\cos\left(\frac{2x+6x}{2}\right)\cos\left(\frac{2x-6x}{2}\right)\right]\left[-2\sin\left(\frac{2x+6x}{2}\right)\sin\frac{(2x-6x)}{2}\right]$

$\left[\begin{array}{l}\because \cos A+\cos B = 2\cos\left(\frac{A+B}{2}\right)\cos\left(\frac{A-B}{2}\right)\\ \cos A-\cos B = -2\sin\left(\frac{A+B}{2}\right)\sin\left(\frac{A-B}{2}\right)\end{array}\right]$

$= [2\cos 4x\cos(-2x)][-2\sin 4x\sin(-2x)]$

$= [2\cos 4x\cos 2x][2\sin 4x\sin 2x]$

$(\because \cos(-\theta) = \cos\theta \text{ and } \sin(-\theta) = -\sin\theta)$

$= (2\sin 2x\cos 2x)(2\sin 4x\cos 4x)$

$= \sin 4x\sin 8x = \text{RHS}$

Q14. Prove that $\sin 2x+2\sin 4x+\sin 6x = 4\cos^2 x\sin 4x.$

Solution:

$\text{LHS} = \sin 2x+2\sin 4x+\sin 6x$

$= [\sin 2x+\sin 6x]+2\sin 4x$

$$= \left[2\sin\left(\frac{2x+6x}{2}\right)\left(\frac{2x-6x}{2}\right)\right] + 2\sin 4x$$

$$\left[\because \sin A + \sin B = 2\sin\left(\frac{A+B}{2}\right)\cos\left(\frac{A-B}{2}\right)\right]$$

$$= 2\sin 4x\cos(-2x) + 2\sin 4x$$
$$= 2\sin 4x\cos 2x + 2\sin 4x$$
$$= 2\sin 4x(\cos 2x + 1)$$
$$= 2\sin 4x(2\cos^2 x)$$
$$= 4\cos^2 x\sin 4x = \text{RHS}$$

Q15. Prove that

$\cot 4x(\sin 5x + \sin 3x) = \cot x(\sin 5x - \sin 3x)$.

Solution:

$$\text{LHS} = \cot 4x(\sin 5x + \sin 3x)$$
$$= \frac{\cot 4x}{\sin 4x}\left[2\sin\left(\frac{5x+3x}{2}\right)\cos\left(\frac{5x-3x}{2}\right)\right]$$

$$\left[\because \sin A + \sin B = 2\sin\left(\frac{A+B}{2}\right)\cos\left(\frac{A-B}{2}\right)\right]$$

$$= \left(\frac{\cos 4x}{\sin 4x}\right)[2\sin 4x\cos x]$$
$$= 2\cos 4x\cos x$$
$$\text{RHS} = \cot x(\sin 5x - \sin 3x)$$
$$= \frac{\cos x}{\sin x}\left[2\cos\left(\frac{5x+3x}{2}\right)\sin\left(\frac{5x-3x}{2}\right)\right]$$

$$\left[\because \sin A - \sin B = 2\cos\left(\frac{A+B}{2}\right)\sin\left(\frac{A-B}{2}\right)\right]$$

$$= \frac{\cos x}{\sin x}[2\cos 4x\sin x] = 2\cos 4x\cdot\cos x$$
$$\therefore \text{LHS} = \text{RHS}$$

Q16. Prove that $\dfrac{\cos 9x - \cos 5x}{\sin 17x - \sin 3x} = -\dfrac{\sin 2x}{\cos 10x}$.

Solution:

$$\left[\begin{array}{l}\because \sin A - \sin B = 2\cos\left(\frac{A+B}{2}\right)\sin\left(\frac{A-B}{2}\right)\\ \cos A - \cos B = -2\sin\left(\frac{A+B}{2}\right)\sin\left(\frac{A-B}{2}\right)\end{array}\right]$$

$$\text{LHS} = \frac{\cos 9x - \cos 5x}{\sin 17x - \sin 3x}$$
$$= \frac{-2\sin\left(\frac{9x+5x}{2}\right)\cdot\sin\left(\frac{9x-5x}{2}\right)}{\left(\frac{17x+3x}{2}\right)\cdot\sin\left(\frac{17x-3x}{2}\right)}$$
$$= \frac{-2\sin 7x\cdot\sin 2x}{2\cos 10x\cdot\sin 7x}$$
$$= -\frac{\sin 2x}{\cos 10x} = \text{RHS}$$

Q17. Prove that: $\dfrac{\sin 5x + \sin 3x}{\cos 5x + \cos 3x} = \tan 4x$.

Solution:

$$\text{LHS} = \frac{\sin 5x + \sin 3x}{\cos 5x + \cos 3x}$$
$$= \frac{2\sin\left(\frac{5x+3x}{2}\right)\cdot\cos\left(\frac{5x-3x}{2}\right)}{2\cos\left(\frac{5x+2x}{2}\right)\cdot\cos\left(\frac{5x-3x}{2}\right)}$$

$$\left[\begin{array}{l}\because \sin A + \sin B = 2\sin\left(\frac{A+B}{2}\right)\cos\left(\frac{A-B}{2}\right)\\ \cos A + \cos B = -2\cos\left(\frac{A+B}{2}\right)\cos\left(\frac{A-B}{2}\right)\end{array}\right]$$

$$= \frac{2\sin 4x\cdot\cos x}{2\cos 4x\cdot\cos x} = \frac{\sin 4x}{\cos 4x}$$
$$= \tan 4x = \text{RHS}$$

Q18. Prove that $\dfrac{\sin x + \sin y}{\cos x + \cos y} = \tan\dfrac{x-y}{2}$.

Solution:

$$\text{LHS} = \frac{\sin x + \sin y}{\cos x + \cos y}$$
$$= \frac{2\cos\left(\frac{x+y}{2}\right)\cdot\sin\left(\frac{x-y}{2}\right)}{2\cos\left(\frac{x+y}{2}\right)\cdot\cos\left(\frac{x-y}{2}\right)}$$

$$\left[\begin{array}{l}\because \sin A - \sin B = 2\cos\left(\frac{A+B}{2}\right)\sin\left(\frac{A-B}{2}\right)\\ \cos A + \cos B = -2\cos\left(\frac{A+B}{2}\right)\cos\left(\frac{A-B}{2}\right)\end{array}\right]$$

$$= \frac{\sin\left(\frac{x-y}{2}\right)}{\cos\left(\frac{x-y}{2}\right)} = \tan\left(\frac{x-y}{2}\right) = \text{RHS}$$

Q19. Prove that $\dfrac{\sin x + \sin 3x}{\cos x + \cos 3x} = \tan 2x$.

Solution:

$$\text{LHS} = \frac{\sin x + \sin 3x}{\cos x + \cos 3x}$$
$$= \frac{2\sin\left(\frac{x+3x}{2}\right)\cos\left(\frac{x-3x}{2}\right)}{2\cos\left(\frac{x+3x}{2}\right)\cos\left(\frac{x-3x}{2}\right)}$$

$$\left[\begin{array}{l}\because \sin A + \sin B = 2\sin\left(\frac{A+B}{2}\right)\cos\left(\frac{A-B}{2}\right)\\ \cos A + \cos B = -2\cos\left(\frac{A+B}{2}\right)\cos\left(\frac{A-B}{2}\right)\end{array}\right]$$

$$= \frac{\sin 2x}{\cos 2x} = \tan 2x = \text{RHS}$$

Q20. Prove that $\frac{\sin x - \sin 3x}{\sin^2 x - \cos^2 x} = 2\sin x$.

Solution:

$$\therefore \text{LHS} = \frac{\sin x - \sin 3x}{\sin^2 x - \cos^2 x}$$

$$= \frac{2\cos\left(\frac{x+3x}{2}\right)\sin\left(\frac{x-3x}{2}\right)}{-\cos 2x}$$

$$\left[\begin{array}{l} \because \cos^2 A - \sin^2 A = \cos 2A \\ \sin A - \sin B = 2\cos\left(\frac{A+B}{2}\right)\sin\left(\frac{A-B}{2}\right) \end{array}\right]$$

$$= \frac{2\cos 2x \sin(-x)}{-\cos 2x}$$

$$= -2\times(-\sin x) = 2\sin x = \text{RHS}$$

Q21. Prove that $\frac{\cos 4x + \cos 3x + \cos 2x}{\sin 4x + \sin 3x + \sin 2x} = \cot 3x$.

Solution:

$$\text{LHS} = \frac{\cos 4x + \cos 3x + \cos 2x}{\sin 4x + \sin 3x + \sin 2x}$$

$$= \frac{(\cos 4x + \cos 2x) + \cos 3x}{(\sin 4x + \sin 2x) + \sin 3x}$$

$$= \frac{2\cos\left(\frac{4x+2x}{2}\right)\cos\left(\frac{4x-2x}{2}\right) + \cos 3x}{2\sin\left(\frac{4x+2x}{2}\right)\cos\left(\frac{4x-2x}{2}\right) + \sin 3x}$$

$$\left[\begin{array}{l} \because \cos A + \cos B = 2\cos\left(\frac{A+B}{2}\right)\cos\left(\frac{A-B}{2}\right) \\ \sin A + \sin B = 2\sin\left(\frac{A+B}{2}\right)\cos\left(\frac{A-B}{2}\right) \end{array}\right]$$

$$= \frac{2\cos 3x \cos x + \cos 3x}{2\sin 3x \cos x + \sin 3x}$$

$$= \frac{\cos 3x(2\cos x + 1)}{\sin 3x(2\cos x + 1)}$$

$$= \cot 3x = \text{RHS}$$

Q22. Prove that

$$\cot x \cot 2x - \cot 2x \cot 3x - \cot 3x \cot x = 1.$$

Solution:

$$\text{LHS} = \cot x \cot 2x - \cot 2x \cot 3x - \cot 3x \cot x$$

$$= \cot x \cot 2x - \cot 3x(\cot 2x + \cot x)$$

$$= \cot x \cot 2x - \cot(2x + x)(\cot 2x + \cot x)$$

$$= \cot x \cot 2x - \left[\frac{\cot 2x \cot x - 1}{\cot x + \cot 2x}\right](\cot 2x + \cot x)$$

$$\left[\because \cot(A+B) = \frac{\cot A \cot B - 1}{\cot A + \cot B}\right]$$

$$= \cot x \cot 2x - (\cot 2x \cot x - 1) = 1 = \text{RHS}$$

Q23. Prove that $\tan 4x = \frac{4\tan x(1-\tan^2 x)}{1 - 6\tan^2 x + \tan^4 x}$.

Solution:

$$\text{LHS} = \tan 4x = \tan 2(2x) = \frac{2\tan 2x}{1 - \tan^2(2x)}$$

$$\left[\because \tan 2A = \frac{2\tan A}{1 - \tan^2 A}\right]$$

$$= \frac{2\left(\frac{2\tan x}{1-\tan^2 x}\right)}{1 - \left(\frac{2\tan x}{1-\tan^2 x}\right)}$$

$$= \frac{\left(\frac{4\tan x}{1-\tan^2 x}\right)}{\left[1 - \frac{4\tan^2 x}{(1-\tan^2 x)^2}\right]}$$

$$= \frac{\left(\frac{4\tan x}{1-\tan^2 x}\right)}{\left[\frac{(1-\tan^2 x)^2 - 4\tan^2 x}{(1-\tan^2 x)^2}\right]}$$

$$= \frac{4\tan x(1-\tan^2 x)}{(1-\tan^2 x)^2 - 4\tan^2 x}$$

$$= \frac{4\tan x(1-\tan^2 x)}{1 + \tan^4 x - 2\tan^2 x - 4\tan^2 x}$$

$$= \frac{4\tan x(1-\tan^2 x)}{1 - 6\tan^2 x + \tan^4 x} = \text{RHS}$$

Q24. Prove that $\cos 4x = 1 - 8\sin^2 x \cos^2 x$.

Solution:

$$\text{LHS} = \cos 4x = \cos 2(2x)$$

$$= 1 - 2\sin^2 2x \qquad [\because \cos 2A = 1 - 2\sin^2 A]$$

$$= 1 - 2(2\sin x \cos x)^2 \qquad [\because \sin 2A = 2\sin A \cos A]$$

$$= 1 - 8\sin^2 x = \cos^2 x = \text{RHS}$$

Q25. Prove that:
$\cos 6x = 32\cos^6 x - 48\cos^4 x + 18\cos^2 x - 1$.

Solution:

$$\text{LHS} = \cos 6x = \cos 3(2x)$$

$= 4\cos^3 2x - 3\cos 2x$

$[\because \cos 3A = 4\cos^3 A - 3\cos A]$

$= 4[(2\cos^2 x - 1)^3 - 3(2\cos^2 x - 1)$

$[\because \cos 2x = 2\cos^2 x - 1]$

$= 4[(2\cos^2 x)^3 - (1)^3 - 3(2\cos^2 x)^2 + 3(2\cos^2 x)]$

$-6\cos^2 x + 3$

$= 4[8\cos^6 x - 1 - 12\cos^4 x + 6\cos^2 x] - 6\cos^2 x + 3$

$= 32\cos^6 x - 4 - 48\cos^4 x + 24\cos^2 x - 6\cos^2 x + 3$

$= 32\cos^6 x - 48\cos^4 x + 18\cos^2 x - 1 = \text{RHS}$

Day 45

Trigonometric Equations

Equations involving trigonometric functions of a variable angle are called trigonometric equations.

Since the values of $\sin x$ and $\cos x$ repeat after an interval of 2π and the values of $\tan x$ repeat after an interval of π, the solutions of a trigonometric equation for which $0 \le x < 2\pi$ are called principal solutions.

The expression involving integer $n \in Z$, which gives all solutions of a trigonometric equation, is called the general solution, where Z denotes the set of integers.

The general solutions of some trigonometric equations are given in the table below.

Trigonometric function	General solution
$\sin x = \sin y$	$x = n\pi + (-1)^n y,\ n \in Z$
$\cos x = \cos y$	$x = 2n\pi \pm y,\ n \in Z$
$\tan x = \tan y$	$x = n\pi + y,\ n \in Z$ $x, y \ne (2n+1)\frac{\pi}{2}$
$\text{cosec}\, x = \text{cosec}\, y$	$x = n\pi + (-1)^n y,\ n \in Z$ $x, y \ne n\pi$
$\sec x = \sec y$	$x = 2n\pi \pm y,\ n \in Z$ $x, y \ne (2n+1)\frac{\pi}{2}$
$\cot x = \cot y$	$x = n\pi + y,\ n \in Z$ $x, y \ne n\pi$
$\sin^2 x = \sin^2 y$ or $\cos^2 x = \cos^2 y$ or $\sec^2 x = \sec^2 y$ or $\text{cosec}^2 x = \text{cosec}^2 y$	$x = n\pi \pm y,\ n \in Z$
$\tan^2 x = \tan^2 y$	$x = n\pi \pm y,\ n \in Z$ $x, y \ne (2n+1)\frac{\pi}{2}$
$\cot^2 x = \cot^2 y$	$x = n\pi \pm y,\ n \in Z$ $x, y \ne n\pi$
$\sin x = \sin y$ and $\cos x = \cos y$	$x = 2n\pi + y,\ n \in Z$

The general and principal solutions of some trigonometric functions are given in the table below.

Trigonometric function	Principal solution (x)	General solution
$\sin x = 0$	$x = 0$ or π	$x = n\pi$
$\sin x = 1$	$x = \frac{\pi}{2}$	$x = n\pi + (-1)^n \frac{\pi}{2}$
$\sin x = -1$	$x = \frac{3\pi}{2}$	$x = n\pi + (-1)^n \frac{3\pi}{2}$
$\sin x = -\frac{1}{2}$	$x = \frac{7\pi}{6}$ or $\frac{11\pi}{6}$	$x = n\pi + (-1)^n \frac{7\pi}{6}$
$\sin x = \frac{\sqrt{3}}{2}$	$x = \frac{\pi}{3}$ or $\frac{2\pi}{3}$	$x = n\pi + (-1)^n \frac{\pi}{3}$
$\cos x = 0$	$x = \frac{\pi}{2}$ or $\frac{3\pi}{2}$	$x = 2n\pi \pm \frac{\pi}{2}$
$\cos x = 1$	$x = 0$	$x = 2n\pi$
$\cos x = -1$	$x = \pi$	$x = 2n\pi \pm \pi$
$\cos x = \frac{1}{\sqrt{2}}$	$x = \frac{\pi}{4}$ or $\frac{7\pi}{4}$	$x = 2n\pi \pm \frac{\pi}{4}$

Trigonometric function	Principal solution (x)	General solution
$\cos x=-\dfrac{1}{\sqrt{2}}$	$x=\dfrac{3\pi}{4}$ or $\dfrac{5\pi}{4}$	$x=2n\pi\pm\dfrac{3\pi}{4}$
$\tan x=1$	$x=\dfrac{\pi}{4}$ or $\dfrac{5\pi}{4}$	$x=n\pi+\dfrac{\pi}{4}$
$\tan x=-1$	$x=\dfrac{3\pi}{4}$	$x=n\pi+\dfrac{3\pi}{4}$

Example 1: Find the principal and general solutions of the following equations.

(i) $\sin x=-\dfrac{1}{\sqrt{2}}$ (ii) $\cos x=\dfrac{\sqrt{3}}{2}$

(iii) $\tan x=-\sqrt{3}$ (iv) $\operatorname{cosec}(2x)=\dfrac{2}{\sqrt{3}}$

(v) $\sec\left(\dfrac{x}{3}\right)=-\sqrt{2}$ (vi) $\cot\left(\dfrac{3x}{2}\right)=0$

Solution:

(i) $\sin x=-\dfrac{1}{\sqrt{2}}$

$\because \sin\dfrac{\pi}{4}=\dfrac{1}{\sqrt{2}}$

$\therefore \sin\left(\pi+\dfrac{\pi}{4}\right)=-\sin\dfrac{\pi}{4}=-\dfrac{1}{\sqrt{2}}=\sin\dfrac{5\pi}{4}$

$\sin\left(2\pi-\dfrac{\pi}{4}\right)=-\sin\dfrac{\pi}{4}=-\dfrac{1}{\sqrt{2}}=\sin\dfrac{7\pi}{4}$

Thus, $\sin\dfrac{5\pi}{4}=\sin\dfrac{7\pi}{4}=-\dfrac{1}{\sqrt{2}}$

Therefore, the principal solutions are $x=\dfrac{5\pi}{4},\dfrac{7\pi}{4}$.

Now, $\sin x=\sin\dfrac{5\pi}{4}\Rightarrow x=n\pi+(-1)^n\dfrac{5\pi}{4}$

Therefore, the general solution is

$x=n\pi+(-1)^n\dfrac{5\pi}{4}$, where, $n\in Z$.

(ii) $\cos x=\dfrac{\sqrt{3}}{2}$

$\because \cos\dfrac{\pi}{6}=\dfrac{\sqrt{3}}{2}$

$\therefore \cos\left(2\pi-\dfrac{\pi}{6}\right)=\cos\dfrac{11\pi}{6}=\dfrac{\sqrt{3}}{2}$

Thus, $\cos\dfrac{\pi}{6}=\cos\dfrac{11\pi}{6}=\dfrac{\sqrt{3}}{2}$

Therefore, the principal solutions are $x=\dfrac{\pi}{6},\dfrac{11\pi}{6}$.

Now, $\cos x=\cos\dfrac{\pi}{6}\Rightarrow x=2n\pi\pm\dfrac{\pi}{6}$

Therefore, the general solution is $x=2n\pi\pm\dfrac{\pi}{6}$, where, $n\in Z$.

(iii) $\tan x=-\sqrt{3}$

$\because \tan\dfrac{\pi}{3}=\sqrt{3}$

$\therefore \tan\left(\pi-\dfrac{\pi}{3}\right)=-\tan\dfrac{\pi}{3}=-\sqrt{3}=\tan\dfrac{2\pi}{3}$

$\tan\left(2\pi-\dfrac{\pi}{3}\right)=-\tan\dfrac{\pi}{3}=-\sqrt{3}=\tan\dfrac{5\pi}{3}$

Thus, $\tan\dfrac{2\pi}{3}=\tan\dfrac{5\pi}{3}=-\sqrt{3}$

Therefore, the principal solutions are $x=\dfrac{2\pi}{3},\dfrac{5\pi}{3}$.

Now, $\tan x=\tan\dfrac{2\pi}{3}\Rightarrow x=n\pi+\dfrac{2\pi}{3}$

Therefore, the general solution is $x=n\pi+\dfrac{2\pi}{3}$, where, $n\in Z$.

(iv) $\operatorname{cosec}(2x)=\dfrac{2}{\sqrt{3}}$

$\because \operatorname{cosec}\dfrac{\pi}{3}=\dfrac{2}{\sqrt{3}}$

$\therefore \operatorname{cosec}\left(\pi-\dfrac{\pi}{3}\right)=\operatorname{cosec}\dfrac{2\pi}{3}=\dfrac{2}{\sqrt{3}}$

Therefore, the principal solutions are $x=\dfrac{\pi}{3},\dfrac{2\pi}{3}$.

Now, $\operatorname{cosec}(2x)=\operatorname{cosec}\dfrac{\pi}{3}$

$\Rightarrow \sin(2x)=\sin\dfrac{\pi}{3}$ $\left[\because \operatorname{cosec}x=\dfrac{1}{\sin x}\right]$

$\Rightarrow 2x=n\pi+(-1)^n\dfrac{\pi}{3}\Rightarrow x=\dfrac{n\pi}{2}+(-1)^n\dfrac{\pi}{6}$

Therefore, the general solution is $x=\dfrac{n\pi}{2}+(-1)^n\dfrac{\pi}{6}$, where, $n\in Z$.

(v) $\sec\left(\dfrac{x}{3}\right)=-\sqrt{2}$

$\because \sec\dfrac{\pi}{4}=\sqrt{2}$

$\therefore \sec\left(\pi-\dfrac{\pi}{4}\right)=-\sec\dfrac{\pi}{4}=-\sqrt{2}=\sec\dfrac{3\pi}{4}$

$\sec\left(\pi+\dfrac{\pi}{4}\right)=-\sec\dfrac{\pi}{4}=-\sqrt{2}=\sec\dfrac{5\pi}{4}$

Thus, $\sec\dfrac{3\pi}{4}=\sec\dfrac{5\pi}{4}=-\sqrt{2}$

Therefore, the principal solutions are $x=\dfrac{3\pi}{4},\dfrac{5\pi}{4}$.

Now, $\sec\left(\frac{x}{3}\right) = \sec\frac{3\pi}{4}$

$\Rightarrow \cos\left(\frac{x}{3}\right) = \cos\frac{3\pi}{4}$ $\left[\because \sec x = \frac{1}{\cos x}\right]$

$\Rightarrow \frac{x}{3} = n\pi + \frac{3\pi}{4} \Rightarrow x = 3n\pi + \frac{9\pi}{4}$

Therefore, the general solution is $x = 3n\pi + \frac{9\pi}{4}$, where, $n \in Z$.

(vi) $\cot\left(\frac{3x}{2}\right) = 0$

$\because \cot\frac{\pi}{2} = 0$

$\therefore \cot\left(\pi + \frac{\pi}{2}\right) = \cot\frac{3\pi}{2}$

Therefore, the principal solutions are $x = \frac{\pi}{2}, \frac{3\pi}{2}$.

Now, $\cot\left(\frac{3x}{2}\right) = 0 = \cot\frac{\pi}{2}$

$\Rightarrow \tan\left(\frac{3x}{2}\right) = \tan\frac{\pi}{2}$ $\left[\because \cot x = \frac{1}{\tan x}\right]$

$\Rightarrow \frac{3x}{2} = n\pi + \frac{\pi}{2} \Rightarrow x = \frac{2n\pi}{3} + \frac{\pi}{3}$

Therefore, the general solution is $x = \frac{2n\pi}{3} + \frac{\pi}{3}$, where, $n \in Z$.

Example 2: Find the general solution of the following equations.

(i) $\sin 3x = \sin x$

$\because \sin 3x = \sin x \Rightarrow \sin 3x - \sin x = 0$

$\Rightarrow 2\cos\left(\frac{3x+x}{2}\right)\sin\left(\frac{3x-x}{2}\right) = 0$

$\Rightarrow \cos(2x)\sin x = 0$

$\left[\because \sin A - \sin B = 2\cos\left(\frac{A+B}{2}\right)\sin\left(\frac{A-B}{2}\right)\right]$

$\Rightarrow \sin x = 0$ or $\cos(2x) = 0$

Thus, $\sin x = 0 = \sin 0° \Rightarrow x = n\pi$, where $n \in Z$.

And $\cos(2x) = 0 = \cos\left(\frac{\pi}{2}\right)$

$\Rightarrow 2x = 2n\pi \pm \frac{\pi}{2} \Rightarrow x = n\pi \pm \frac{\pi}{4}$ where $n \in Z$.

Therefore, the general solution of the equation $\sin 3x = \sin x$ is $x = n\pi$ or $x = n\pi \pm \frac{\pi}{4}$ where $n \in Z$.

(ii) $\cos 4x = \cos 2x$

$\because \cos 4x = \cos 2x \Rightarrow \cos 4x - \cos 2x = 0$

$\Rightarrow -2\sin\left(\frac{4x+2x}{2}\right)\sin\left(\frac{4x-2x}{2}\right) = 0$

$\Rightarrow \sin(3x)\sin x = 0 \Rightarrow \sin(3x) = 0$ or $\sin x = 0$

$\left[\because \cos A - \cos B = -2\sin\left(\frac{A+B}{2}\right)\sin\left(\frac{A-B}{2}\right)\right]$

Thus, $\sin x = 0 = \sin 0° \Rightarrow x = n\pi$, where $n \in Z$.

And $\sin 3x = 0 = \sin 0° \Rightarrow 3x = n\pi \Rightarrow x = \frac{n\pi}{3}$

Therefore, the general solution of the equation $\cos 4x = \cos 2x$ is $x = n\pi$ or $x = \frac{n\pi}{3}$ where $n \in Z$

(iii) $\sin 2x + \cos 2x = 1$

$\because \sin 2x + \cos 2x = 1$

$\Rightarrow 2\sin x\cos x - 2\sin^2 x = 0$

$\Rightarrow 2\sin x(\cos x - \sin x) = 0$

$\Rightarrow \sin x = 0$ or $\cos x = \sin x$

Thus, $\sin x = 0 = \sin 0° \Rightarrow x = n\pi$, where $n \in Z$.

And $\cos x = \sin x \Rightarrow \tan x = 1 = \tan\left(\frac{\pi}{4}\right)$

$\Rightarrow x = n\pi + \left(\frac{\pi}{4}\right)$, where $n \in Z$.

Therefore, the general solution of the equation $\sin 2x + \cos 2x = 1$ is $x = n\pi$ or $x = n\pi + \frac{\pi}{4}$ where $n \in Z$.

(iv) $\sin 3x = \cos 2x$

$\because \cos 2x = \sin 3x \Rightarrow \cos 2x = \cos\left(\frac{\pi}{2} - 3x\right)$

$\Rightarrow 2x = 2n\pi \pm \left(\frac{\pi}{2} - 3x\right)$

Taking positive sign, we get $2x = 2n\pi + \frac{\pi}{2} - 3x$

$\Rightarrow 5x = 2n\pi + \frac{\pi}{2} \Rightarrow x = (4n+1)\frac{\pi}{10}$

Taking negative sign, we get $2x = 2n\pi - \frac{\pi}{2} + 3x$

$\Rightarrow -x = (4n-1)\frac{\pi}{2} \Rightarrow x = -(4n-1)\frac{\pi}{2}$

Therefore, the general solution of the equation $\sin 3x = \cos 2x$ is $x = (4n+1)\frac{\pi}{10}$ or $x = -(4n-1)\frac{\pi}{2}$ where $n \in Z$.

Alternatively:

We can also solve it by converting cos into sin as explained below.

$\because \sin 3x = \cos 2x \Rightarrow \sin 3x = \sin\left(\frac{\pi}{2} - 2x\right)$

$\Rightarrow 3x = n\pi + (-1)^n\left(\frac{\pi}{2} - 2x\right)$

$\Rightarrow 3x + (-1)^n 2x = n\pi + (-1)^n\frac{\pi}{2}$

For even integers, $n = 2m$ we get

$$3x + (-1)^{2m} 2x = 2m\pi + (-1)^{2m} \frac{\pi}{2}$$

$$3x + 2x = 2m\pi + \frac{\pi}{2} \quad (\because (-1)^n = (-1)^{2m} = 1)$$

$$\Rightarrow x = (4m+1)\frac{\pi}{10}$$

For odd integers, $n = 2m - 1$ we get

$$3x + (-1)^{2m+1} 2x = (2m+1)\pi + (-1)^{2m+1} \frac{\pi}{2}$$

$$3x - 2x = (2m+1)\pi - \frac{\pi}{2} \quad (\because (-1)^n = (-1)^{2m+1} = -1)$$

$$\Rightarrow x = (4m+1)\frac{\pi}{2}$$

Therefore, the general solution of the equation $\sin 3x = \cos 2x$ is $x = (4m+1)\frac{\pi}{10}$ or $x = (4m+1)\frac{\pi}{2}$ where $m \in Z$

(v) $\sin x = \tan x$

$$\because \sin x = \tan x \Rightarrow \sin x - \frac{\sin x}{\cos x} = 0$$

$$\Rightarrow \sin x(\cos x - 1) = 0$$

$$\Rightarrow \sin x = 0 = \sin 0° \text{ or } \cos x = 1 = \cos 0°$$

$$\Rightarrow x = n\pi \text{ or } x = 2n\pi$$

Therefore, the general solution of the equation $\sin x = \tan x$ is $x = n\pi$ or $x = 2n\pi$, where $n \in Z$.

(vi) $\tan 2x = -\cot\left(x + \frac{\pi}{6}\right)$

$$\because \tan 2x = -\cot\left(x + \frac{\pi}{3}\right) = \tan\left(\frac{\pi}{2} + x + \frac{\pi}{3}\right)$$

$$\Rightarrow \tan 2x = \tan\left(\frac{5\pi}{6} + x\right)\left[\because \tan\left(\frac{\pi}{2} + \theta\right) = -\cot\theta\right]$$

$$\Rightarrow 2x = n\pi + \left(\frac{5\pi}{6} + x\right) \Rightarrow x = n\pi + \frac{5\pi}{6}$$

Therefore, the general solution of the equation $\tan 2x = -\cot\left(x + \frac{\pi}{6}\right)$ is $x = n\pi + \frac{5\pi}{6}$, where $n \in Z$.

(vii) $\tan x - \frac{\cot x}{\sqrt{3}} = \frac{1}{\sqrt{3}} - 1$

$$\because \tan x - \frac{\cot x}{\sqrt{3}} = \frac{1}{\sqrt{3}} - 1$$

$$\Rightarrow \tan x - \frac{1}{\sqrt{3}\tan x} = \frac{1 - \sqrt{3}}{\sqrt{3}}$$

$$\Rightarrow \sqrt{3}\tan^2 x - 1 = \tan x - \sqrt{3}\tan x$$

$$\Rightarrow \sqrt{3}\tan^2 x + \sqrt{3}\tan x - \tan x - 1 = 0$$

$$\Rightarrow \sqrt{3}\tan x(\tan x + 1) - 1(\tan x + 1) = 0$$

$$\Rightarrow (\tan x + 1)(\sqrt{3}\tan x - 1) = 0$$

$$\Rightarrow \tan x = -1 \text{ or } \tan x = \frac{1}{\sqrt{3}}$$

$$\Rightarrow x = n\pi + \frac{3\pi}{4} \text{ or } x = n\pi + \frac{\pi}{6}$$

Therefore, the general solution of the equation $\tan 2x = -\cot\left(x + \frac{\pi}{6}\right)$ is $x = n\pi + \frac{3\pi}{4}$ or $x = n\pi + \frac{\pi}{6}$, where $n \in Z$.

Example 3: Find the general solution of the equation $\tan(x - 45°) = \tan(x + 45°)$.

Solution:

$$\because \tan(x - 45°) = \tan(x + 45°)$$

$$\Rightarrow \frac{\tan x - \tan 45°}{1 + \tan x \tan 45°} = \frac{\tan x + \tan 45°}{1 - \tan x \tan 45°}$$

$$\Rightarrow \frac{\tan x - 1}{1 + \tan x} = \frac{\tan x + 1}{1 - \tan x}$$

$$\Rightarrow -(1 - \tan x)^2 = (1 + \tan x)^2$$

$$\Rightarrow -1 - \tan^2 x + 2\tan x = 1 + \tan^2 x + 2\tan x$$

$$\Rightarrow 2(1 + \tan^2 x) = 0 \Rightarrow \tan^2 x = -1$$

Since $\tan^2 x \neq -1$, the equation $\tan(x - 45°) = \tan(x + 45°)$ does not have any solution.

Day 46

NCERT Exercise 3.4 (With Solutions)

Q1. Find the principal and general solutions of the equation $\tan x = \sqrt{3}$.

Solution:

$$\tan x = \sqrt{3} \Rightarrow \tan\frac{\pi}{3} = \sqrt{3}$$

$$\text{Also, } \tan\left(\frac{4\pi}{3}\right) = \tan\left(\pi + \frac{\pi}{3}\right) = \tan\frac{\pi}{3} = \sqrt{3}$$

Therefore, the principal solutions are $x = \frac{\pi}{3}, \frac{4\pi}{3}$.

Now, $\tan x = \tan\frac{\pi}{3} \Rightarrow x = n\pi + \frac{\pi}{3}$, where $n \in Z$

Therefore, the general solution is $x = n\pi + \frac{\pi}{3}$, where $n \in Z$.

Q2. Find the principal and general solutions of the equation $\sec x = 2$.

Solution:

$\sec x = 2 \Rightarrow \sec\frac{\pi}{3} = 2$

Also, $\sec\frac{5\pi}{3} = \sec\left(2\pi - \frac{\pi}{3}\right) = \sec\frac{\pi}{3} = 2$

Therefore, the principal solutions are $x = \frac{\pi}{3}, \frac{5\pi}{3}$

Now, $\sec x = \sec\frac{\pi}{3}$

$\Rightarrow \cos x = \cos\frac{\pi}{3}$ $\left[\because \sec x = \frac{1}{\cos x}\right]$

$\Rightarrow x = 2n\pi \pm \frac{\pi}{3}$, where $n \in Z$

Therefore, the general solution is $x = 2n\pi \pm \frac{\pi}{3}$, where $n \in Z$.

Q3. Find the principal and general solutions of the equation $\cot x = -\sqrt{3}$.

Solution:

$\because \cot\frac{\pi}{6} = \sqrt{3}$

$\therefore \cot\left(\pi - \frac{\pi}{6}\right) = -\cot\frac{\pi}{6} = -\sqrt{3} = \cot\frac{5\pi}{6}$

or $\cot\left(2\pi - \frac{\pi}{6}\right) = -\cot\frac{\pi}{6} = -\sqrt{3} = \cot\frac{11\pi}{6}$

Therefore, the principal solutions are $x = \frac{5\pi}{6}, \frac{11\pi}{6}$

Now, $\cot x = \cot\frac{5\pi}{6}$

$\Rightarrow \tan x = \tan\frac{5\pi}{6}$ $\left[\because \cot x = \frac{1}{\tan x}\right]$

$\Rightarrow x = n\pi + \frac{5\pi}{6}$, where $n \in Z$

Therefore, the general solution is $x = n\pi + \frac{5\pi}{6}$, where $n \in Z$.

Q4. Find the general solution of $\text{cosec}\, x = -2$.

Solution:

$\because \text{cosec}\frac{\pi}{6} = 2$

$\therefore \text{cosec}\left(\pi + \frac{\pi}{6}\right) = -\text{cosec}\frac{\pi}{6} = -2 = \text{cosec}\frac{7\pi}{6}$

$\text{cosec}\left(2\pi - \frac{\pi}{6}\right) = -\text{cosec}\frac{\pi}{6} = -2 = \text{cosec}\frac{11\pi}{6}$

$\text{cosec}\frac{7\pi}{6} = -2, \text{cosec}\frac{11\pi}{6} = -2$

Therefore, the principal solutions are $x = \frac{7\pi}{6}, \frac{11\pi}{6}$.

Now, $\text{cosec}\, x = \text{cosec}\frac{7\pi}{6}$

$\Rightarrow \sin x = \sin\frac{7\pi}{6}$ $\left[\because \text{cosec}\, x = \frac{1}{\sin x}\right]$

$\Rightarrow x = n\pi + (-1)^n \frac{7\pi}{6}$, where $n \in Z$

Therefore, the general solution is $x = n\pi + (-1)^n \frac{7\pi}{6}$, where $n \in Z$.

Q5. Find the general solution of the equation $\cos 4x = \cos 2x$.

Solution:

$\cos 4x = \cos 2x \Rightarrow \cos 4x - \cos 2x = 0$

$\Rightarrow -2\sin\left(\frac{4x + 2x}{2}\right)\sin\left(\frac{4x - 2x}{2}\right) = 0$

$\left[\because \cos A - \cos B = -2\sin\left(\frac{A+B}{2}\right)\sin\left(\frac{A-B}{2}\right)\right]$

$\Rightarrow \sin 3x \sin x = 0$

$\Rightarrow \sin 3x = 0$ or $\sin x = 0$

$\therefore 3x = n\pi$ or $x = n\pi$,

$\Rightarrow x = \frac{n\pi}{3}$ or $x = n\pi$, where $n \in Z$

Q6. Find the general solution of the equation $\cos 3x + \cos x - \cos 2x = 0$.

Solution:

$\cos 3x + \cos x - \cos 2x = 0$

$\Rightarrow 2\cos\left(\frac{3x + x}{2}\right)\cos\left(\frac{3x - x}{2}\right) - \cos 2x = 0$

$\left[\cos A + \cos B = 2\cos\left(\frac{A+B}{2}\right)\cos\left(\frac{A-B}{2}\right)\right]$

$\Rightarrow 2\cos 2x \cos x - \cos 2x = 0$

$\Rightarrow \cos 2x(2\cos x - 1) = 0$

$\Rightarrow \cos 2x = 0$ or $2\cos x - 1 = 0$

$\Rightarrow \cos 2x = 0$ or $\cos x = \frac{1}{2}$

$\Rightarrow 2x = (2n+1)\frac{\pi}{2}$ or $\cos x = \cos\left(2n\pi \pm \frac{\pi}{3}\right)$

$\therefore x = (2n+1)\frac{\pi}{4}$ or $x = 2n\pi \pm \frac{\pi}{3}$, where $n \in Z$

Q7. Find the general of the equation $\sin 2x + \cos x = 0$.

Solution:

$\sin 2x + \cos x = 0 \Rightarrow 2\sin x \cos x + \cos x = 0$

$\Rightarrow \cos x(2\sin x + 1) = 0$

$\Rightarrow \cos x = 0$ or $2\sin x + 1 = 0$

$\cos x = 0 \Rightarrow \cos x = (2n+1)\frac{\pi}{2}$, where $n \in Z$

$2\sin x + 1 = 0 \Rightarrow \sin x = \frac{-1}{2} = -\sin\frac{\pi}{6}$

$= \sin\left(\pi + \frac{\pi}{6}\right) = \sin\left(\pi + \frac{\pi}{6}\right) = \sin\frac{7\pi}{6}$

$\Rightarrow x = n\pi + (-1)^n \frac{7\pi}{6}$, where $n \in Z$

Therefore, the general solution is

$(2n+1)\frac{\pi}{2}$ or $n\pi + (-1)^n \frac{7\pi}{6}, n \in Z$

Q8. Find the general solution of the equation $\sec^2 2x = 1 - \tan 2x$.

Solution:

$\sec^2 2x = 1 - \tan 2x$

$\Rightarrow 1 + \tan^2 2x = 1 - \tan 2x$

$\Rightarrow \tan^2 2x + \tan 2x = 0$

$\Rightarrow \tan 2x(\tan 2x + 1) = 0$

$\Rightarrow \tan 2x = 0$ or $\tan 2x + 1 = 0$

Now, $\tan 2x = 0 \Rightarrow \tan 2x = \tan(n\pi)$

$\Rightarrow 2x = n\pi \Rightarrow x = \frac{n\pi}{2}$, where $n \in Z$

And, $\tan 2x + 1 = 0$

$\Rightarrow \tan 2x = -1 = -\tan\frac{\pi}{4} = \tan\left(\pi - \frac{\pi}{4}\right) = \tan\frac{3\pi}{4}$

$\Rightarrow 2x = n\pi + \frac{3\pi}{4} \Rightarrow x = \frac{n\pi}{2} + \frac{3\pi}{8}$, where $n \in Z$

Therefore, the general solution is

$\frac{n\pi}{2}$ or $\frac{n\pi}{2} + \frac{3\pi}{8}$, $n \in Z$

Q9. Find the general solution of the equation $\sin x + \sin 3x + \sin 5x = 0$.

Solution:

$(\sin x + \sin 5x) + \sin 3x = 0$

$\Rightarrow \left[2\sin\left(\frac{x+5x}{2}\right)\cos\left(\frac{x-5x}{2}\right)\right] + \sin 3x = 0$

$\left[\sin A + \sin B = 2\sin\left(\frac{A+B}{2}\right)\cos\left(\frac{A-B}{2}\right)\right]$

$\Rightarrow 2\sin 3x\cos(-2x) + \sin 3x = 0$

$\Rightarrow 2\sin 3x\cos 2x + \sin 3x = 0$

$\Rightarrow \sin 3x(2\cos 2x + 1) = 0$

$\Rightarrow \sin 3x = 0$ or $2\cos 2x + 1 = 0$

Now, $\sin 3x = 0 \Rightarrow 3x = n\pi \Rightarrow x = \frac{n\pi}{3}$, where $n \in Z$.

And, $2\cos 2x + 1 = 0$

$\Rightarrow \cos 2x = \frac{-1}{2} = -\cos\frac{\pi}{3} = \cos\left(\pi - \frac{\pi}{3}\right) = \cos\frac{2\pi}{3}$

$\Rightarrow 2x = 2n\pi \pm \frac{2\pi}{3} \Rightarrow x = n\pi \pm \frac{\pi}{3}$, where $n \in Z$

Therefore, the general solution is

$\frac{n\pi}{3}$ or $n\pi \pm \frac{\pi}{3}, n \in Z$

Some More Examples

Example 1: Find the general solution of the equation $\cos 4x + \cos 2x - \cos 3x = 0$.

Solution:

$\cos 4x + \cos 2x - \cos 3x = 0$

$\Rightarrow 2\cos\left(\frac{4x+2x}{2}\right)\cos\left(\frac{4x-2x}{2}\right) - \cos 3x = 0$

$\left[\because \cos A + \cos B = 2\cos\left(\frac{A+B}{2}\right)\cos\left(\frac{A-B}{2}\right)\right]$

$\Rightarrow 2\cos 3x\cos 2x - \cos 3x = 0$

$\Rightarrow \cos 3x(2\cos 2x - 1) = 0$

$\Rightarrow \cos 3x = 0$ or $2\cos 2x - 1 = 0$

$\Rightarrow \cos 3x = \cos\left(\frac{\pi}{2}\right)$ or $\cos 2x = \frac{1}{2} = \cos\left(\frac{\pi}{3}\right)$

$\Rightarrow \cos 3x = \left(2n\pi \pm \frac{\pi}{2}\right) \Rightarrow 3x = 2n\pi \pm \frac{\pi}{2}$

or $\cos 2x = \cos\left(2n\pi \pm \frac{\pi}{3}\right) \Rightarrow 2x = 2n\pi \pm \frac{\pi}{3}$

Therefore, the general solution of the equation $\cos 4x + \cos 2x - \cos 3x = 0$ is $x = \frac{(4n \pm 1)\pi}{6}$

or $x = \frac{(6n \pm 1)\pi}{6}$, where $n \in Z$.

Example 2: Find the general solution of the equation $\sin x + \sin 3x + \sin 2x = 0$.

Solution:

$(\sin x + \sin 3x) + \sin 2x = 0$

$\Rightarrow \left[2\sin\left(\frac{x+3x}{2}\right)\cos\left(\frac{x-3x}{2}\right)\right] + \sin 2x = 0$

$\left[\sin A + \sin B = 2\sin\left(\frac{A+B}{2}\right)\cos\left(\frac{A-B}{2}\right)\right]$

$\Rightarrow 2\sin 2x\cos(-2x) + \sin 2x = 0$

$\Rightarrow 2\sin 2x\cos 2x + \sin 2x = 0$

$\Rightarrow \sin 2x(2\cos 2x + 1) = 0$

$\Rightarrow \sin 2x = 0$ or $2\cos 2x + 1 = 0$

Now, $\sin 2x = 0 \Rightarrow 2x = n\pi \Rightarrow x = \frac{n\pi}{2}$

And, $2\cos 2x + 1 = 0 \Rightarrow \cos 2x = \frac{-1}{2} = -\cos\frac{\pi}{3}$

$\Rightarrow \cos 2x = \cos\left(\pi - \frac{\pi}{3}\right) = \cos\frac{2\pi}{3}$

$\Rightarrow 2x = 2n\pi \pm \frac{2\pi}{3} \Rightarrow x = n\pi \pm \frac{\pi}{3}$

Therefore, the general solution of the equation $\sin x + \sin 3x + \sin 2x = 0$ is $\frac{n\pi}{2}$ or $n\pi \pm \frac{\pi}{3}$, where $n \in Z$.

Example 3: Find the general solution of the equation $4\cos x \cos 2x \cos 3x - 1 = 0$.

Solution:

$4\cos x \cos 2x \cos 3x - 1 = 0$

$\Rightarrow (2\cos 3x \cos x) 2\cos 2x - 1 = 0$

$\Rightarrow (\cos 4x + \cos 2x) 2\cos 2x - 1 = 0$

$\Rightarrow 2\cos 4x \cos 2x + 2\cos^2 2x - 1 = 0$

$\Rightarrow 2\cos 4x \cos 2x + \cos 4x = 0$

$\Rightarrow \cos 4x(2\cos 2x + 1) = 0$

Now, $\cos 4x = 0 = \cos\left(\frac{\pi}{2}\right)$

$\Rightarrow 4x = 2n\pi \pm \frac{\pi}{2} \Rightarrow x = \frac{(4n \pm 1)\pi}{8}$

And, $2\cos 2x + 1 = 0 \Rightarrow \cos 2x = \frac{-1}{2} = -\cos\frac{\pi}{3}$

$\Rightarrow \cos 2x = \cos\left(\pi - \frac{\pi}{3}\right) = \cos\frac{2\pi}{3}$

$\Rightarrow 2x = 2n\pi \pm \frac{2\pi}{3} \Rightarrow x = n\pi \pm \frac{\pi}{3}$

Therefore, the general solution of the equation $4\cos x \cos 2x \cos 3x - 1 = 0$ is

$x = \frac{(4n \pm 1)\pi}{8}$ or $x = n\pi \pm \frac{\pi}{3}$ where $n \in Z$.

Example 4: Find the general solution of the equation $4\sin x \sin 2x \sin 5x = \sin 6x$.

Solution:

$4\sin x \sin 2x \sin 5x = \sin 6x$

$\Rightarrow 4\sin 2x \sin x \sin 5x = \sin 6x$

$\Rightarrow 4\sin 2x \sin(3x - 2x) \cdot \sin(3x + 2x) = \sin 6x$

$\Rightarrow 4[\sin 2x(\sin^2 3x - \sin^2 2x)] = 3\sin 2x - 4\sin^3 2x$

$\Rightarrow 4\sin 2x \sin^2 3x - 4\sin^3 2x = 3\sin 2x - 4\sin^3 2x$

$\Rightarrow 4\sin 2x \sin^2 3x = 3\sin 2x$

$\Rightarrow \sin 2x(4\sin^2 3x - 3) = 0$

$\Rightarrow \sin 2x = 0$ or $4\sin^2 3x - 3 = 0$

$\Rightarrow \sin 2x = 0$ or $\sin^2 3x = \frac{3}{4}$

Now, $\sin 2x = 0 \Rightarrow 2x = n\pi \Rightarrow x = \frac{n\pi}{2}$

And, $\sin^2 3x = \frac{3}{4} \Rightarrow \sin 3x = \pm\frac{\sqrt{3}}{2}$

$\Rightarrow \sin 3x = \sin\left(\pm\frac{\pi}{3}\right) \Rightarrow 3x = n\pi + (-1)^n\left(\pm\frac{\pi}{3}\right)$

$\Rightarrow 3x = n\pi \pm \frac{\pi}{3}, n \in I \Rightarrow x = \frac{n\pi}{3} \pm \frac{\pi}{9}$

Therefore, the general solution of the equation $4\sin x \sin 2x \sin 5x = \sin 6x$ is

$x = \frac{n\pi}{2}$ or $x = \frac{n\pi}{3} \pm \frac{\pi}{9}$, where $n \in Z$.

Example 5: Find the general solution of the equation $4\sin x \cos x + 2\sin x - 2\cos x - 1 = 0$.

Solution:

$4\sin x \cos x + 2\sin x - 2\cos x - 1 = 0$

$\Rightarrow 2\sin x(2\cos x + 1) - 1(2\cos x + 1) = 0$

$\Rightarrow (2\cos x + 1)(2\sin x - 1) = 0$

$\Rightarrow 2\cos x + 1 = 0$ or $2\sin x - 1 = 0$

Now, $2\cos x + 1 = 0 \Rightarrow \cos x = -\frac{1}{2} = \cos\frac{2\pi}{3}$

$\Rightarrow x = 2n\pi \pm \frac{2\pi}{3}$

And, $2\sin x - 1 = 0 \Rightarrow \sin x = \frac{1}{2} = \sin\left(\frac{\pi}{6}\right)$

$\Rightarrow x = n\pi + (-1)^n \frac{\pi}{6}$

Therefore, the general solution of the equation $4\sin x \cos x + 2\sin x - 2\cos x - 1 = 0$ is

$x = 2n\pi \pm \frac{2\pi}{3}$ or $x = n\pi + (-1)^n \frac{\pi}{6}$, where $n \in Z$.

Example 6: Find the general solution of the equation $\sec^2 2x + \tan 2x = 1$.

Solution:

$\sec^2 2x + \tan 2x = 1$

$\Rightarrow 1 + \tan^2 2x + \tan 2x - 1 = 0$

$\Rightarrow \tan^2 2x + \tan 2x = 0$

$\Rightarrow \tan 2x(\tan 2x + 1) = 0$

$\Rightarrow \tan 2x = 0$ or $\tan 2x + 1 = 0$

Now, $\tan 2x = 0 \Rightarrow \tan 2x = \tan(n\pi)$

$\Rightarrow 2x = n\pi \Rightarrow x = \frac{n\pi}{2}$

And, $\tan 2x + 1 = 0$

$\Rightarrow \tan 2x = -1 = -\tan\frac{\pi}{4} = \tan\left(\pi - \frac{\pi}{4}\right) = \tan\frac{3\pi}{4}$

$\Rightarrow 2x = n\pi + \frac{3\pi}{4} \Rightarrow x = \frac{n\pi}{2} + \frac{3\pi}{8}$, where $n \in Z$.

Therefore, the general solution of the equation $\sec^2 2x + \tan 2x = 1$ is

$x = \frac{n\pi}{2}$ or $x = \frac{n\pi}{2} + \frac{3\pi}{8}$ where $n \in Z$.

Example 7: Find the general solution of the equation $\sec^2 x = 2 + \tan x - \cot x$.

Solution:

$\sec^2 x = 2 + \tan x - \cot x$

$\Rightarrow \sec^2 x - \tan x + \cot x - 2 = 0$

$\Rightarrow \sec^2 x - 1 - \tan x + \cot x - 1 = 0$

$\Rightarrow \tan^2 x - \tan x + \cot x - 1 = 0$

$\Rightarrow \tan^2 x - \tan x + \cot x - \tan x \cot x = 0$

$\Rightarrow \tan x(\tan x - 1) - \cot x(\tan x - 1) = 0$

$\Rightarrow (\tan x - \cot x)(\tan x - 1) = 0$

$\Rightarrow \tan x - \cot x = 0$ or $\tan x - 1 = 0$

$\Rightarrow \tan^2 x = 0$ or $\tan x = 1$

Now, $\tan x = 0 = \tan(n\pi) \Rightarrow x = n\pi$

And, $\tan x = 1 = \tan\frac{\pi}{4} \Rightarrow x = n\pi + \frac{\pi}{4}$

Therefore, the general solution of the equation $\sec^2 x = 2 + \tan x - \cot x$ is

$x = n\pi$ or $x = n\pi + \frac{\pi}{4}$, where $n \in Z$.

Example 8: Find the general solution of the equation $\sin x - \tan x = 1 - \sin x \tan x$.

Solution:

$\because \sin x - \tan x = 1 - \sin x \tan x$

$\Rightarrow \sin x \tan x - \tan x + \sin x - 1 = 0$

$\Rightarrow \tan x(\sin x - 1) + 1(\sin x - 1) = 0$

$\Rightarrow (\sin x - 1)(\tan x + 1) = 0$

$\Rightarrow \sin x - 1 = 0$ or $\tan x + 1 = 0$

$\Rightarrow \sin x = 1$ or $\tan x = -1$

$\Rightarrow \sin x = 1 = \sin\frac{\pi}{2} \Rightarrow x = n\pi + (-1)^n \frac{\pi}{2}$

or $\tan x = -1 = \tan\left(-\frac{\pi}{4}\right) \Rightarrow x = n\pi + \left(-\frac{\pi}{4}\right)$

Therefore, the general solution of the equation $\sin x - \tan x = 1 - \sin x \tan x$ is

$x = n\pi + (-1)^n \frac{\pi}{2}$ or $x = n\pi + \left(-\frac{\pi}{4}\right)$, where $n \in Z$.

Example 9: Find the general solution of the equation $\tan x - \cos x = \cos x - \sec x$.

Solution:

$\because \tan x - \cos x = \cos x - \sec x$

$\Rightarrow \frac{\sin x}{\cos x} - \cos x = \cos x - \frac{1}{\cos x}$

$\Rightarrow 1 + \sin x = 2\cos^2 x = 2(1 - \sin^2 x)$

$\Rightarrow 2\sin^2 x + \sin x - 1 = 0$

$\Rightarrow 2\sin^2 x + 2\sin x - 1\sin x - 1 = 0$

$\Rightarrow (\sin x + 1)(2\sin x - 1) = 0$

$\Rightarrow \sin x = -1$ or $\sin x = \frac{1}{2}$

$\Rightarrow \sin x = \sin\frac{3\pi}{2}$ or $\sin x = \sin\frac{\pi}{6}$

$\Rightarrow x = n\pi + (-1)^n \frac{3\pi}{2}$ or $x = n\pi + (-1)^n \frac{\pi}{6}$

But, since $x = n\pi + (-1)^n \frac{3\pi}{2} = [2n + (-1)^n 3]\frac{\pi}{2}$ is an odd multiple of $\frac{\pi}{2}$, where $\tan x + \sec x$ is not defined, $x = n\pi + (-1)^n \frac{3\pi}{2}$ is not accepted.

Therefore, the general solution of the equation $\tan x - \cos x = \cos x - \sec x$ is $x = n\pi + (-1)^n \frac{\pi}{6}$, where $n \in Z$.

Example 10: If $\tan x + \tan y - \tan x \tan y + 1 = 0$, then find all values of $x + y$.

Solution:

$\because \tan x + \tan y - \tan x \tan y + 1 = 0$

$\Rightarrow \tan x + \tan y = -1 + \tan x \tan y$

$\Rightarrow \frac{\tan x + \tan y}{1 - \tan x \tan y} = -1$

$\Rightarrow \tan(x + y) = -1 = -\tan\left(\frac{\pi}{4}\right) = \tan\left(\pi - \frac{\pi}{4}\right)$

$\Rightarrow x + y = n\pi + \frac{3\pi}{4}$, where $n \in Z$

Example 11: Find the general solution of the equation $\tan 2x = \frac{2\tan x}{1 - \tan x}$.

Solution:

$\because \tan 2x = \frac{2\tan x}{1 - \tan x}$

$\Rightarrow \tan 2x = \frac{2\tan x(1 + \tan x)}{(1 - \tan x)(1 + \tan x)}$

$\Rightarrow \tan 2x = \frac{2\tan x(1 + \tan x)}{(1 - \tan^2 x)}$

$\Rightarrow \tan 2x(1 - \tan^2 x) = 2\tan x(1 + \tan x)$

$\Rightarrow \tan 2x(1 - \tan x)(1 + \tan x) - 2\tan x(1 + \tan x) = 0$

$\Rightarrow (1 + \tan x)\{\tan 2x(1 - \tan x) - 2\tan x\} = 0$

$\Rightarrow (1 + \tan x)\{\tan 2x - \tan x \tan 2x - 2\tan x\} = 0$

$\Rightarrow (1+\tan x)\left\{\left(1-\frac{2\tan x}{\tan 2x}\right)-\tan x\right\}=0$

$\Rightarrow (1+\tan x)(\tan^2 x-\tan x)=0$

$\Rightarrow \tan x(\tan x-1)(1+\tan x)=0$

$\Rightarrow \tan x(\tan^2 x-1)=0$

$\Rightarrow \tan x=0$ or $\tan^2 x=1$

$\Rightarrow \tan x=0$ or $\tan x=\pm 1$

Now, $\tan x=0 \Rightarrow x=n\pi$

or $\tan x=\pm 1=\tan\left(\pm\frac{\pi}{4}\right)\Rightarrow x=n\pi\pm\frac{\pi}{4}$

Therefore, the general solution of the equation $\tan 2x=\frac{2\tan x}{1-\tan x}$ is $x=n\pi$ or $x=n\pi-\frac{\pi}{4}$ or $x=n\pi+\frac{\pi}{4}$, where $n\in Z$.

Example 12: Find the general solution of the equation $\tan x+\tan\left(x+\frac{\pi}{3}\right)+\tan\left(x+\frac{2\pi}{3}\right)+3=0.$

Solution:

$\because \tan x+\tan\left(x+\frac{\pi}{3}\right)+\tan\left(x+\frac{2\pi}{3}\right)+3=0$

$\Rightarrow \tan x+\frac{\tan x+\tan\frac{\pi}{3}}{1-\tan x\tan\frac{\pi}{3}}+\frac{\tan x+\tan\frac{2\pi}{3}}{1-\tan x\tan\frac{2\pi}{3}}=-3$

$\Rightarrow \tan x+\frac{\tan x+\sqrt{3}}{1-\sqrt{3}\tan x}+\frac{\tan x-\sqrt{3}}{1+\sqrt{3}\tan x}=-3$

$\Rightarrow \tan x+\frac{8\tan x}{1-3\tan^2 x}=-3$

$\Rightarrow \frac{\tan x-3\tan^3 x+8\tan x}{1-3\tan^2 x}=-3$

$\Rightarrow \frac{3(3\tan x-\tan^3 x)}{1-3\tan^2 x}=-3$

$\Rightarrow 3\tan 3x=-3\Rightarrow \tan 3x=-1=\tan\left(-\frac{\pi}{4}\right)$

$\Rightarrow 3x=n\pi+\left(-\frac{\pi}{4}\right)\Rightarrow x=\frac{n\pi}{3}-\frac{\pi}{12}$

Therefore, the general solution of the equation $\tan x+\tan\left(x+\frac{\pi}{3}\right)+\tan\left(x+\frac{2\pi}{3}\right)+3=0$ is $x=\frac{n\pi}{3}-\frac{\pi}{12}$, where $n\in Z$.

Day 47

Sine and Cosine Formulae

Let ABC be a triangle with interior angles A, B, C (which lie between 0° and 180°), respectively. The angle A is the interior angle between the sides AB and AC. Similarly, the angle B is the interior angle between the sides BA and BC, and the angle C is the interior angle between the sides CA and CB.

The sides AB, BC and CA are opposite to the vertices C, A and B and are denoted by c, a, b, respectively (Fig. 3.14).

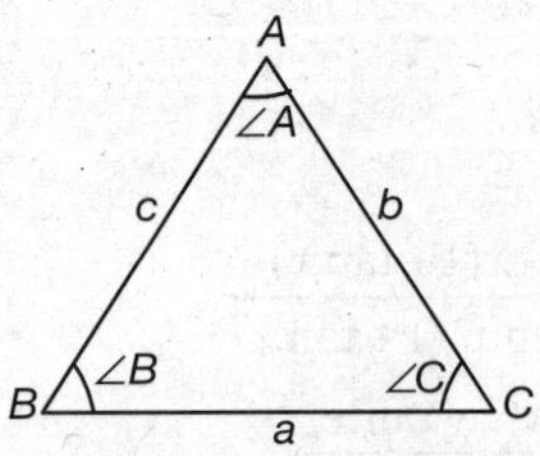

Fig. 3.14 Triangle with sides a, b and c

Note: A, B, C are the interior angles of the triangle ABC. Therefore, it must be noted that the sum of the interior angles must be equal to 180°, that is, $A+B+C=\pi$.

Theorem 1: Sine Formula: In any triangle, the sides are proportional to the sines of the opposite angles. In a triangle ABC,

$$\frac{\sin A}{a}=\frac{\sin B}{b}=\frac{\sin C}{c}$$

Proof:

Let ABC be either of the triangles shown in Figs 3.15 (i) or (ii).

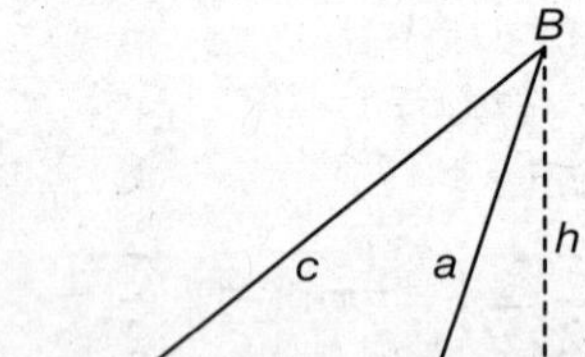

Fig. 3.15(i) Sine and cosine formulae

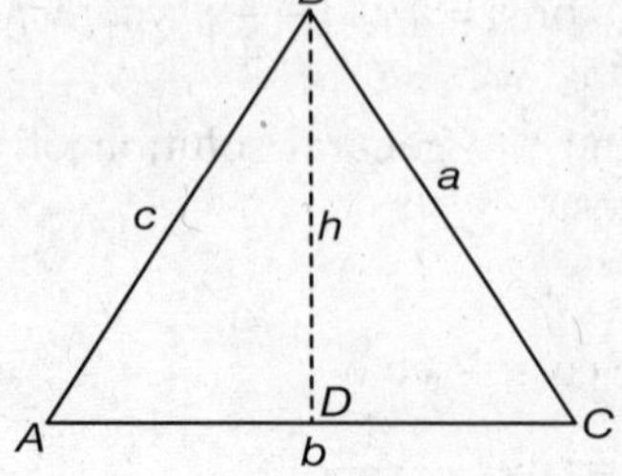

Fig. 3.15(ii) Sine and cosine formulae

The altitude h is drawn from the vertex B to meet the side AC at point D [in Fig. 3.15 (i), AC is produced to meet the altitude at D].

From the right-angled triangle ABD in Fig. 3.15(i), we get $\sin A = \frac{h}{c} \Rightarrow h = c\sin A$ (1)

From the right-angled triangle CBD in Fig 3.15(ii), we get $\sin C = \frac{h}{a} \Rightarrow h = a\sin C$ (2)

or in Fig. 3.15 (ii),

$\sin(180° - C) = \frac{h}{a} \Rightarrow h = a\sin C$ (3)

$(\because \sin(180° - C) = \sin C)$

Therefore, from equations (1) and (2) or (3), we get

$h = c\sin A = a\sin C \Rightarrow \frac{\sin A}{a} = \frac{\sin C}{c}$ (4)

Similarly, we can prove that

$\frac{\sin A}{a} = \frac{\sin B}{b}$ (5)

Thus, from equations (4) and (5), we get

$$\frac{\sin A}{a} = \frac{\sin B}{b} = \frac{\sin C}{c}$$

Theorem 2: Cosine Formulae: Let A, B, C be the vertices of a triangle and a, b, c be the lengths of the sides opposite the angles A, B and C, respectively. Then

$$a^2 = b^2 + c^2 - 2bc\cos A$$

$$b^2 = c^2 + a^2 - 2ca\cos B$$

$$c^2 = a^2 + b^2 - 2ab\cos C$$

Proof:

Again referring to triangle ABC in Fig. 3.15 (i), we get

$BC^2 = BD^2 + DC^2 = BD^2 + (AD - AC)^2$

$= BD^2 + AD^2 + AC^2 - 2AD \cdot AC$

$= AB^2 + AC^2 - 2(AB\cos A) \cdot AC$

$(\because \text{In triangle } ABD,\ BD^2 + AD^2 = AB^2)$

Or $\Rightarrow a^2 = c^2 + b^2 - 2bc\cos A$

Similarly, we can prove that

$b^2 = c^2 + a^2 - 2ca\cos B$

$c^2 = a^2 + b^2 - 2ab\cos C$

Similarly, we can prove all of the above equations using Fig. 3.15 (ii), where C is an acute angle.

Note: In an angle format, the cosine formulae can be written as

$$\cos A = \frac{b^2 + c^2 - a^2}{2bc}$$

$$\cos B = \frac{c^2 + a^2 - b^2}{2ac}$$

$$\cos C = \frac{a^2 + b^2 - c^2}{2ab}$$

Theorem 3: Projection Formulae: Let A, B, C be the vertices of a triangle and a, b, c be the lengths of the sides opposite the angles A, B and C, respectively. Then

$$a = b\cos C + c\cos B$$

$$b = c\cos A + a\cos C$$

$$c = a\cos B + b\cos A$$

Proof:

From the sine formulae, we have

$\frac{a}{\sin A} = \frac{b}{\sin B} = \frac{c}{\sin C} = k$ (say)

Therefore, $b = \frac{a\sin B}{\sin A}$ and $c = \frac{a\sin C}{\sin A}$

Thus, first projection formula, $b\cos C + c\cos B$

$\frac{a\sin B}{\sin A}\cos C + \frac{a\sin C}{\sin A}\cos B$

$= \frac{a}{\sin A}(\sin B\cos C + \cos B\sin C)$

$= \frac{a\sin(B+C)}{\sin A} = \frac{a\sin(\pi - A)}{\sin A}$ $(\because A + B + C = \pi)$

$= \frac{a\sin A}{\sin A} = a$

Similarly, we can prove other results of the projection formulae.

Theorem 4: Napier's Analogies: Let A, B, C be the vertices of a triangle and a, b, c be the lengths of the sides opposite the angles A, B, and C, respectively. Then

$$\tan\frac{B-C}{2} = \frac{b-c}{b+c}\cot\frac{A}{2}$$

$$\tan\frac{C-A}{2} = \frac{c-a}{c+a}\cot\frac{B}{2}$$

$$\tan\frac{A-B}{2} = \frac{a-b}{a+b}\cot\frac{C}{2}$$

Proof:

From the sine formulae, we have

$\frac{a}{\sin A} = \frac{b}{\sin B} = \frac{c}{\sin C} = k$ (say)

Therefore, $\frac{b-c}{b+c} = \frac{k(\sin B - \sin C)}{k(\sin B + \sin C)}$

$\Rightarrow \frac{b-c}{b+c} = \frac{2\cos\frac{B+C}{2}\sin\frac{B-C}{2}}{2\sin\frac{B+C}{2}\cos\frac{B-C}{2}}$

$\Rightarrow \frac{b-c}{b+c} = \cot\frac{(B+C)}{2}\tan\frac{(B-C)}{2}$

$\Rightarrow \frac{b-c}{b+c} = \cot\left(\frac{\pi}{2} - \frac{A}{2}\right)\tan\left(\frac{B-C}{2}\right)$

$\left(\because A + B + C = \pi \Rightarrow \frac{B+C}{2} = \frac{\pi - A}{2}\right)$

$$\Rightarrow \frac{b-c}{b+c} = \tan\left(\frac{A}{2}\right)\tan\left(\frac{B-C}{2}\right)$$

$$\Rightarrow \frac{b-c}{b+c} = \frac{\tan\frac{B-C}{2}}{\cot A/2}$$

$$\Rightarrow \tan\frac{B-C}{2} = \frac{b-c}{b+c}\cot\left(\frac{A}{2}\right)$$

Similarly, we can prove the other results of Napier's analogies.

Example 1: In any triangle ABC, prove that $a(\sin B - \sin C) + b(\sin C - \sin A) + c(\sin A - \sin B) = 0$.

Solution:

From the sine formulae,

$$\frac{\sin A}{a} = \frac{\sin B}{b} = \frac{\sin C}{c} = k(\text{say})$$

$\Rightarrow \sin A = ak$, $\sin B = bk$ and $\sin C = ck$

Therefore, substituting the values of $\sin A$, $\sin B$, $\sin C$ in Equation (1), we get

$$\text{LHS} = a(\sin B - \sin C)$$
$$+ b(\sin C - \sin A) + c(\sin A - \sin B)$$
$$= a(bk - ck) + b(ck - ak) + c(ak - bk)$$
$$= k(ab - ac + bc - ba + ca - cb) = 0 = \text{RHS}$$

Example 2: In any triangle ABC, prove that $(b+c)\cos A + (c+a)\cos B + (a+b)\cos C = a+b+c$.

Solution:

$$\text{LHS} = (b+c)\cos A + (c+a)\cos B + (a+b)\cos C$$
$$= (b\cos A + c\cos A) + (c\cos B + a\cos B)$$
$$+ (b\cos C + a\cos C)$$
$$= (b\cos C + c\cos A) + (a\cos C + c\cos A)$$
$$+ (a\cos B + b\cos A)$$
$$= a+b+c = \text{RHS (using projection formulae)}$$

Example 3: In any triangle ABC, prove that $(a+b)^2\sin^2\frac{C}{2} + (a-b)^2\cos^2\frac{C}{2} = c^2$.

Solution:

$$\text{LHS} = (a+b)^2\sin^2\frac{C}{2} + (a-b)^2\cos^2\frac{C}{2}$$
$$= (a+b)^2\left(1-\cos^2\frac{C}{2}\right) + (a-b)^2\cos^2\frac{C}{2}$$
$$(a+b)^2 - 4ab\cos^2\frac{C}{2}$$
$$= (a+b)^2 - 2ab(\cos C + 1) \quad (\because 2\cos^2 x = \cos 2x + 1)$$
$$= a^2 + b^2 + 2ab - 2ab\cos C - 2ab$$
$$= a^2 + b^2 - 2ab\cos C = c^2 = \text{RHS}$$

(using cosine formula)

Example 4: In any triangle ABC, prove that

$$\frac{\cos A}{b\cos C + c\cos B} + \frac{\cos B}{c\cos A + a\cos C} + \frac{\cos C}{a\cos B + b\cos A} = \frac{a^2+b^2+c^2}{2abc}.$$

Solution:

$$\frac{\cos A}{b\cos C + c\cos B} + \frac{\cos B}{c\cos A + a\cos C} + \frac{\cos C}{a\cos B + b\cos A}$$
$$= \frac{\cos A}{a} + \frac{\cos B}{b} + \frac{\cos C}{c} \text{ (using projection formula)}$$
$$\frac{b^2+c^2-a^2}{2abc} + \frac{a^2+c^2-b^2}{2abc} + \frac{a^2+b^2-c^2}{2abc}$$
$$= \frac{a^2+b^2+c^2}{2abc} \text{ (using cosine formula)}$$

Example 5: In any triangle ABC, prove that

$$2\left(a\cos^2\frac{B}{2} + b\cos^2\frac{A}{2}\right) = 2\left(b\cos^2\frac{C}{2} + c\cos^2\frac{B}{2}\right)$$
$$= 2\left(c\cos^2\frac{A}{2} + a\cos^2\frac{C}{2}\right) = a+b+c.$$

Solution:

$$\text{LHS} = 2\left(a\cos^2\frac{B}{2} + b\cos^2\frac{A}{2}\right)$$
$$= a(\cos B + 1) + b(\cos A + 1) \quad (\because 2\cos^2 x = \cos 2x + 1)$$
$$= a\cos B + a + b\cos A + b = a + b + (a\cos B + b\cos A)$$
$$= a+b+c \text{ (using projection formula)}$$

Similarly, we can prove other results.

Example 6: In any triangle ABC, prove that $a\sin(B-C) + b\sin(C-A) + C\sin(A-B) = 0$.

Solution:

$$\because a\sin(B-C) = a(\sin B\cos C - \cos B\sin C) \quad (1)$$

Now, from sine formulae

$$\frac{\sin A}{a} = \frac{\sin B}{b} = \frac{\sin C}{c} = k(\text{say})$$

$\Rightarrow \sin B = bk$ and $\sin C = ck$

And from cosine formulae

$$\cos B = \frac{c^2+a^2-b^2}{2ac} \text{ and } \cos C = \frac{a^2+b^2-c^2}{2ab}$$

Therefore, substituting the values of $\sin B$, $\sin C$, $\cos B$, $\cos C$ in Equation (1), we get

$$a\sin(B-C)$$
$$= a\left[bk\left(\frac{a^2+b^2-c^2}{2ab}\right) - ck\left(\frac{c^2+a^2-b^2}{2ac}\right)\right]$$
$$= \frac{k}{2}(a^2+b^2-c^2-c^2-a^2+b^2) = k(b^2-c^2)$$

Similarly, $b\sin(C-A)=k(c^2-a^2)$ and $c\sin(A-B)=k(a^2-b^2)$, thus,
LHS $= a\sin(B-C)+b\sin(C-A)+C\sin(A-B)$
$= k(b^2-c^2+c^2-a^2+a^2-b^2)=0=$ RHS

Example 7: In any triangle ABC, prove that
(i) $\dfrac{\cos B}{\cos C}=\dfrac{c-b\cos A}{b-c\cos A}$ (ii) $\dfrac{\sin B}{\sin C}=\dfrac{c-a\cos B}{b-a\cos C}$

Solution:

(i) Using the projection formulae,
$c=b\cos A+a\cos B$ and $b=a\cos C+c\cos A$
$\Rightarrow c-b\cos A=a\cos B$ and $b-c\cos A=a\cos C$
Dividing both equations, we get
$$\text{RHS}=\frac{c-b\cos A}{b-c\cos A}=\frac{a\cos B}{a\cos C}=\frac{\cos B}{\cos C}$$

(ii) Using the projection formulae,
$c=b\cos A+a\cos B$ and $b=a\cos C+c\cos A$
$\Rightarrow c-a\cos B=b\cos A$ and $b-a\cos C=c\cos A$
Dividing both the equations, we get
$$\text{RHS}=\frac{c-a\cos B}{b-a\cos C}=\frac{b\cos A}{c\cos A}=\frac{b}{c}$$
Using the sine formulae
$$\because \frac{a}{\sin A}=\frac{b}{\sin B}=\frac{c}{\sin C}=k\text{ (say)}$$
$\Rightarrow b=k\sin B$ and $c=k\sin C$
$$\text{Therefore, RHS}=\frac{b}{c}=\frac{k\sin B}{k\sin C}=\frac{\sin B}{\sin C}=\text{LHS}$$

Example 8: In any triangle ABC, prove that
$a\cos A+b\cos B+c\cos C=2a\sin B\sin C$
$=2b\sin A\sin C=2c\sin A\sin B$

Solution:

From the sine formulae
$$\frac{a}{\sin A}=\frac{b}{\sin B}=\frac{c}{\sin C}=k\text{ (say)}$$
$\Rightarrow a=k\sin A,\ b=k\sin B$ and $c=k\sin C$

Thus, LHS $=a\cos A+b\cos B+c\cos C$
$=k\sin A\cos A+k\sin B\cos B+k\sin C\cos C$
$=\dfrac{k}{2}(2\sin A\cos A+2\sin B\cos B+2\sin C\cos C)$
$=\dfrac{k}{2}(\sin 2A+\sin 2B+\sin 2C)$
$=\dfrac{k}{2}(4\sin A\sin B\sin C)=2k\sin A\sin B\sin C$
$=2a\sin B\sin C\quad (\because k\sin A=a)$
$=2b\sin A\sin C\quad (\because k\sin B=b)$
$=2c\sin A\sin B\quad (\because k\sin C=c)$

Example 9: In any triangle ABC, prove that
$$\frac{a}{b-c}=\frac{\tan\frac{B}{2}+\tan\frac{C}{2}}{\tan\frac{B}{2}-\tan\frac{C}{2}}.$$

Solution:
$$\frac{a}{b-c}=\frac{k\sin A}{k\sin B-k\sin C}=\frac{\sin A}{\sin B-\sin C}$$
$$\left(\begin{array}{l}\because \dfrac{a}{\sin A}=\dfrac{b}{\sin B}=\dfrac{c}{\sin C}=k\text{ (say)}\\ \Rightarrow a=k\sin A,\ b=k\sin B,\ c=k\sin C\end{array}\right)$$
$$=\frac{\sin(B+C)}{\sin B-\sin C}=\frac{2\sin\frac{B+C}{2}\cos\frac{B+C}{2}}{2\cos\frac{B+C}{2}\sin\frac{B-C}{2}}$$
$$\left(\begin{array}{l}\because A+B+C=\pi\Rightarrow A=\pi-(B+C)\\ \therefore \sin A=\sin\{\pi-(B+C)\}=\sin(B+C)\end{array}\right)$$
$$\frac{\sin\frac{B+C}{2}}{\sin\frac{B-C}{2}}=\frac{\sin\frac{B}{2}\cos\frac{C}{2}+\cos\frac{B}{2}\sin\frac{C}{2}}{\sin\frac{B}{2}\cos\frac{C}{2}-\cos\frac{B}{2}\sin\frac{C}{2}}$$
$$=\frac{\tan\frac{B}{2}+\tan\frac{C}{2}}{\tan\frac{B}{2}-\tan\frac{C}{2}}$$

Example 10: In any triangle ABC, prove that
$$\frac{a-b}{a+b}=\frac{\tan\left(\frac{A-B}{2}\right)}{\tan\left(\frac{A+B}{2}\right)}.$$

Solution:

From Napier's analogies,
$$\tan\frac{A-B}{2}=\frac{a-b}{a+b}\cot\frac{C}{2}$$
Also, A, B, C are the interior angles of the triangle ABC, and therefore, the sum of the interior angles must be equal to 180°, that is, $A+B+C=\pi$.
$$\Rightarrow A+B=\pi-C\Rightarrow\frac{A+B}{2}=\frac{\pi-C}{2}=\frac{\pi}{2}-\frac{C}{2}$$
Thus,
$$\tan\left(\frac{A-B}{2}\right)=\frac{a-b}{a+b}\cot\frac{C}{2}=\frac{a-b}{a+b}\tan\left(\frac{\pi}{2}-\frac{C}{2}\right)$$
$$\Rightarrow\tan\left(\frac{A-B}{2}\right)=\frac{a-b}{a+b}\tan\left(\frac{A+B}{2}\right)$$
$$\Rightarrow\frac{\tan\left(\frac{A-B}{2}\right)}{\tan\left(\frac{A+B}{2}\right)}=\frac{a-b}{a+b}$$

Example 11: In any triangle ABC, prove that

$$\frac{b^2-c^2}{\cos B+\cos C}+\frac{c^2-a^2}{\cos C+\cos A}+\frac{a^2-b^2}{\cos A+\cos B}=0.$$

Solution:

From the sine formulae,

$$\frac{a}{\sin A}=\frac{b}{\sin B}=\frac{c}{\sin C}=k\,(\text{say})$$

$$\Rightarrow a=k\sin A,\ b=k\sin B \text{ and } c=k\sin C$$

Thus, $\dfrac{b^2-c^2}{\cos B+\cos C}+\dfrac{c^2-a^2}{\cos C+\cos A}+\dfrac{a^2-b^2}{\cos A+\cos B}$

$$=\frac{k^2\sin^2 B-k^2\sin^2 C}{\cos B+\cos C}+\frac{k^2\sin^2 C-k^2\sin^2 A}{\cos C+\cos A}+\frac{k^2\sin^2 A-k^2\sin^2 B}{\cos A+\cos B}$$

$$=k^2\left(\frac{\sin^2 B-\sin^2 C}{\cos B+\cos C}+\frac{\sin^2 C-\sin^2 A}{\cos C+\cos A}+\frac{\sin^2 A-\sin^2 B}{\cos A+\cos B}\right)$$

$$=k^2\left(\frac{\cos^2 C-\cos^2 B}{\cos B+\cos C}+\frac{\cos^2 A-\cos^2 C}{\cos C+\cos A}+\frac{\cos^2 B-\cos^2 A}{\cos A+\cos B}\right)$$

$$=k^2(\cos C-\cos B+\cos A-\cos C+\cos B-\cos A)=0$$

Example 12: In any triangle ABC, prove that

$$\frac{a^2\sin(B-C)}{\sin B+\sin C}+\frac{b^2\sin(C-A)}{\sin C+\sin A}+\frac{c^2\sin(A-B)}{\sin A+\sin B}=0.$$

Solution:

From the sine formulae,

$$\frac{a}{\sin A}=\frac{b}{\sin B}=\frac{c}{\sin C}=k\,(\text{say})$$

$$\Rightarrow a=k\sin A,\ b=k\sin B \text{ and } c=k\sin C$$

Thus, $\dfrac{a^2\sin(B-C)}{\sin B+\sin C}+\dfrac{b^2\sin(C-A)}{\sin C+\sin A}+\dfrac{c^2\sin(A-B)}{\sin A+\sin B}$

$$=\frac{k^2\sin^2 A\sin(B-C)}{\sin B+\sin C}+\frac{k^2\sin^2 B\sin(C-A)}{\sin B+\sin C}+\frac{k^2\sin^2 C\sin(A-B)}{\sin B+\sin C}$$

$$=\frac{k^2\sin A\sin(B+C)\sin(B-C)}{\sin B+\sin C}+\frac{k^2\sin B\sin(C+A)\sin(C-A)}{\sin C+\sin A}+\frac{k^2\sin C\sin(A+B)\sin(A-B)}{\sin A+\sin B}$$

$$=\frac{k^2\sin A(\sin^2 B-\sin^2 C)}{\sin B+\sin C}+\frac{k^2\sin B(\sin^2 C-\sin^2 A)}{\sin C+\sin A}+\frac{k^2\sin C(\sin^2 A-\sin^2 B)}{\sin A+\sin B}$$

$$\left(\because \sin(x+y)\sin(x-y)=\sin^2 x-\sin^2 y\right)$$

$$=k^2\sin A(\sin B-\sin C)+k^2\sin B(\sin C-\sin A)+k^2\sin C(\sin A-\sin B)$$

$$=k^2(\sin A\sin B-\sin A\sin C+\sin B\sin C-\sin B\sin A+\sin C\sin A-\sin C\sin B)=0$$

Example 13: In any triangle ABC, prove that

$$(b-c)\cot\frac{A}{2}+(c-a)\cot\frac{B}{2}+(a-b)\cot\frac{C}{2}.$$

Solution:

Using Napier's analogies,

$$\tan\frac{B-C}{2}=\frac{b-c}{b+c}\cot\frac{A}{2}$$

$$\Rightarrow (b-c)\cot\frac{A}{2}=(b+c)\tan\frac{B-C}{2}$$

$$\tan\frac{C-A}{2}=\frac{c-a}{c+a}\cot\frac{B}{2}$$

$$\Rightarrow (c-a)\cot\frac{B}{2}=(c+a)\tan\frac{C-A}{2}$$

$$\tan\frac{A-B}{2}=\frac{a-b}{a+b}\cot\frac{C}{2}$$

$$\Rightarrow (a-b)\cot\frac{C}{2}=(a+b)\tan\frac{A-B}{2}$$

Thus, $(b-c)\cot\dfrac{A}{2}+(c-a)\cot\dfrac{B}{2}+(a-b)\cot\dfrac{C}{2}$

$$=(b+c)\tan\frac{B-C}{2}+(c+a)\tan\frac{C-A}{2}+(a+b)\tan\frac{A-B}{2}$$

$$=(k\sin B+k\sin C)\tan\frac{B-C}{2}+(k\sin C+k\sin A)\tan\frac{C-A}{2}+(k\sin A+k\sin B)\tan\frac{A-B}{2}$$

$$\left(\begin{array}{l}\because \dfrac{a}{\sin A}=\dfrac{b}{\sin B}=\dfrac{c}{\sin C}=k\,(\text{say})\\ \Rightarrow a=k\sin A,\ b=k\sin B,\ c=k\sin C\end{array}\right)$$

$$=k\begin{pmatrix}2\sin\dfrac{B+C}{2}\cos\dfrac{B-C}{2}\tan\dfrac{B-C}{2}\\ +2\sin\dfrac{C+A}{2}\cos\dfrac{C-A}{2}\tan\dfrac{C-A}{2}\\ 2+\sin\dfrac{A+B}{2}\cos\dfrac{A-B}{2}\tan\dfrac{A-B}{2}\end{pmatrix}$$

$$= 2k\begin{pmatrix} \sin\frac{B+C}{2}\sin\frac{B-C}{2} \\ +\sin\frac{C+A}{2}\sin\frac{C-A}{2} \\ +\sin\frac{A+B}{2}\sin\frac{A-B}{2} \end{pmatrix}$$

$$= k\begin{pmatrix} \sin^2\frac{B}{2}-\sin^2\frac{C}{2}+\sin^2\frac{B}{2}-\sin^2\frac{C}{2} \\ +\sin^2\frac{B}{2}-\sin^2\frac{C}{2} \end{pmatrix} = 0$$

$(\because \sin(x+y)\sin(x-y) = \sin^2 x - \sin^2 y)$

Day 48

Sine and Cosine Formulae—Applications

Example 1: In any triangle ABC, if $a = 3,\ b = 4,\ c = 5$, find: $\cos A, \cos B, \cos C, \sin A, \sin B, \sin C$.

Solution:
Using the cosine formula,

$$\cos A = \frac{b^2+c^2-a^2}{2bc} = \frac{4^2+5^2-3^2}{2\times4\times5} = \frac{4}{5}$$

$$\cos B = \frac{c^2+a^2-b^2}{2ca} = \frac{5^2+3^2-4^2}{2\times5\times3} = \frac{3}{5}$$

$$\cos C = \frac{a^2+b^2-c^2}{2ab} = \frac{3^2+4^2-5^2}{2\times3\times4} = \frac{0}{24} = 0$$

$$\sin A = \sqrt{1-\cos^2 A} = \sqrt{1-\left(\frac{4}{5}\right)^2} = \sqrt{1-\frac{16}{25}} = \sqrt{\frac{9}{25}} = \frac{3}{5}$$

$$\sin B = \sqrt{1-\cos^2 B} = \sqrt{1-\left(\frac{3}{5}\right)^2} = \sqrt{\frac{16}{25}} = \frac{4}{5}$$

$$\sin C = \sqrt{1-\cos^2 C} = \sqrt{1-(0)^2} = \sqrt{1-0} = \sqrt{1} = 1$$

Example 2: In any triangle ABC, if $a = 3,\ c = 3,\ \angle A = 30°$ find: b, $\angle B$ and $\angle C$.

Solution:
Using the sine formula,

$$\frac{a}{\sin A} = \frac{b}{\sin B} = \frac{c}{\sin C} \Rightarrow \frac{3}{1/2} = \frac{b}{\sin B} = \frac{3}{\sin C}$$

$$\Rightarrow \sin C = \frac{3}{6} = \frac{1}{2} \Rightarrow \angle C = 30°$$

$$\therefore \angle B = 180° - 30° - 30° = 120°$$

$$\Rightarrow b = 6\sin 120° = 3\sqrt{3}$$

Example 3: In any triangle ABC, if $b = 6,\ \angle B = 30°,\ \angle C = 45°$ find: a, c and $\angle A$.

Solution:

$$\because \angle A = 180° - 30° - 45° = 105°$$

Using the sine formula,

$$\frac{a}{\sin A} = \frac{b}{\sin B} = \frac{c}{\sin C}$$

$$\Rightarrow \frac{a}{\sin 105°} = \frac{6}{\sin 30°} = \frac{c}{\sin 45°}$$

$$\Rightarrow a = \frac{6\sin 105°}{\sin 30°} = \frac{6(\sqrt{3}-1)/2\sqrt{2}}{1/2}$$

$$\Rightarrow a = 3\sqrt{2}(\sqrt{3}-1)$$

And, $c = \dfrac{6\sin 45°}{\sin 30°} = \dfrac{6/\sqrt{2}}{1/2} \Rightarrow c = 6\sqrt{2}$

Example 4: A tree stands vertically on a hill side which makes an angle of 15° with the horizontal. From a point on the ground 30 m down the hill from the base of the tree, the angle of elevation of the top of the tree is 45°. Find the height of the tree.

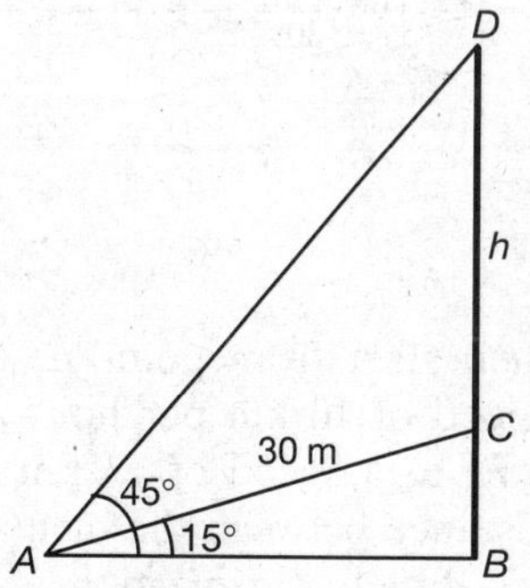

Solution:

Let BC be the height of the hill and CD be the height of the tree. Then, according to the question,

$$\because \tan 15° = \frac{BC}{AB} \text{ and } \cos 15° = \frac{AB}{AC} = \frac{AB}{30}$$

$$\therefore \tan 45° = \frac{h+BC}{AB} = \frac{h}{AB} + \frac{BC}{AB}$$

$$\Rightarrow 1 = \frac{h}{30\cos 15°} + \tan 15°$$

$$\Rightarrow 1 = \frac{h}{30\cos(45°-30°)} + \tan(45°-30°)$$

$$\Rightarrow 1=\frac{h}{30\left(\frac{1}{\sqrt{2}}\frac{\sqrt{3}}{2}+\frac{1}{\sqrt{2}}\frac{1}{2}\right)}+\left(\frac{1-\frac{1}{\sqrt{3}}}{1+1\cdot\frac{1}{\sqrt{3}}}\right)$$

$$\left(\begin{array}{l}\because \cos(x-y)=\cos x\cos y+\sin x\sin y\\ \text{and}\ \ \tan(x-y)=\dfrac{\tan x-\tan y}{1+\tan x\tan y}\end{array}\right)$$

$$\Rightarrow 1=\frac{2\sqrt{2}h}{30(\sqrt{3}+1)}+\left(\frac{\sqrt{3}-1}{\sqrt{3}+1}\right)$$

$$\Rightarrow \frac{2\sqrt{2}h}{30(\sqrt{3}+1)}=1-\left(\frac{\sqrt{3}-1}{\sqrt{3}+1}\right)=\frac{2}{\sqrt{3}+1}$$

$$\Rightarrow \frac{2\sqrt{2}h}{30}=2\Rightarrow h=15\sqrt{2}\text{ m is the height of tree}$$

Example 5: Two men leave a point at the same time. One goes 10 km per hour in the direction E60°N and other travels 15 km per hour in the direction S60°E. Find the distance between them at the end of 2 hours.

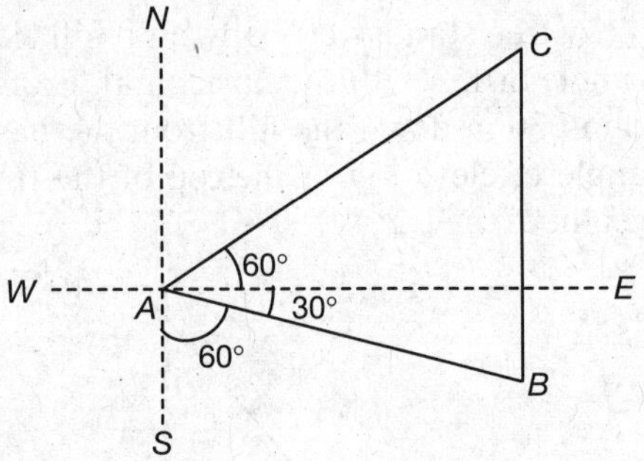

Solution:
Let both the men start from point A, one along AC (E60°N) at a speed of 10 km per hour and other one along AB (S60°E) at a speed of 15 km per hour, and let BC be the distance between the men at the end of 2 hours. Then, according to the question,

Distance $AC = 10\times 2 = 20\text{km}$

Distance $AB = 15\times 2 = 30\text{km}$

Angle BAC = 90°, and therefore, by the cosine formula,

$$\cos 90^\circ=\frac{(AB)^2+(AC)^2-(BC)^2}{2(AB)(AC)}$$

$$\Rightarrow 0=\frac{(30)^2+(20)^2-(BC)^2}{2(30)(20)}$$

$$\Rightarrow \frac{1300-(BC)^2}{1200}=0\Rightarrow BC=\sqrt{1300}=10\sqrt{13}$$

$\Rightarrow BC = 36\text{km}$ is the distance between the men at the end of 2 hours.

Example 6: Two buildings, A and B, are on the same side of a straight road. From a point C opposite the road, the distance of the buildings A and B is 12 m and 17 m, respectively. If the angle C is 60°, find the distance between the buildings (use $\sqrt{2}=1.414$).

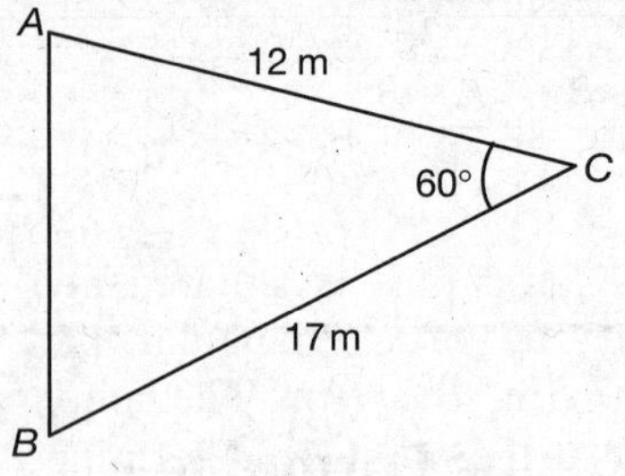

Solution:
According to the question, angle ACB = 60°, and therefore, by the cosine formula,

$$\cos 60^\circ=\frac{(BC)^2+(AC)^2-(AB)^2}{2(BC)(AC)}$$

$$\Rightarrow \frac{1}{2}=\frac{(17)^2+(12)^2-(AB)^2}{2(17)(12)}$$

$$\Rightarrow 408=433-(AB)^2\Rightarrow (AB)^2=25$$

$\Rightarrow AB = 5\text{m}$ is the distance between the trees.

Example 7: The angle of elevation of the top point P of the vertical tower PQ of height h at a point A is 45° and from a point B, the angle of elevation is 60°, where B is a point at a distance 10 m from the point A measured along the line AB which makes an angle of 30° with AQ. Find height of the tower.

Solution:
Since $\angle PAQ = 45°, \angle BAQ = 30°, \angle PBH = 45°$

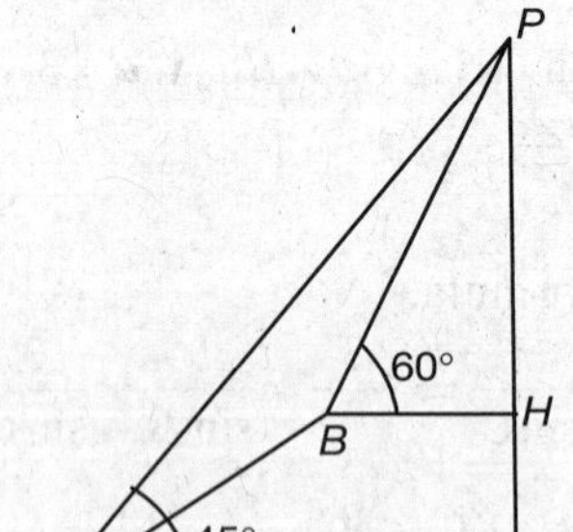

$\angle PAB = 15°$, $\angle BPH = 30°$

$\angle APB = 15°$ and $\angle ABP = 150°$

Thus, triangle APQ will be an isocleles right triangle, and hence, $AQ = PQ = h$ and

$AP^2 = AQ^2 + PQ^2 = h^2 + h^2 = 2h^2 \Rightarrow AP = \sqrt{2}h$

Also, triangle APB will be an isocleles triangle, and hence, $PB = AB = 10$ m

Applying the sine formulae in ΔABP, we get

$$\frac{AB}{\sin 15^\circ} = \frac{AP}{\sin 150^\circ} \Rightarrow \frac{10}{\sin 15^\circ} = \frac{\sqrt{2}h}{\sin 150^\circ}$$

$$\Rightarrow h = \frac{10\sin(180^\circ - 30^\circ)}{\sqrt{2}\sin 15^\circ} = \frac{10(1/2)}{\sqrt{2}(\sqrt{3}-1)/2\sqrt{2}}$$

$$\Rightarrow h = \frac{10}{\sqrt{3}-1} = 5(\sqrt{3}+1)$$

Example 8: A lamp post is situated at the midpoint Q of the side AC of a triangular plot ABC with $AB = 4\,\text{m}$, $AC = 5\,\text{m}$, $BC = 6\,\text{m}$. The lamp post subtends an angle of 30° at the point B. Determine the height of the lamp post.

Solution:

$AB = c = 4\,\text{m}$, $AC = b = 5\,\text{m}$, $BC = a = 6\,\text{m}$

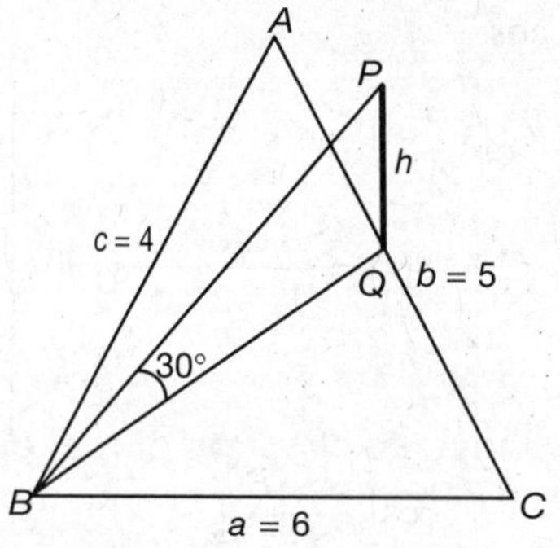

Q is the midpoint of the side AC at which the lamp post PQ of height h is located at Q and the lamp post subtends an angle of 30° at B.

Therefore, by the cosine formulae in ΔABC, we get

$$\cos A = \frac{b^2+c^2-a^2}{2bc} = \frac{25+16-36}{2\times5\times4} = \frac{1}{8}$$

Similarly using the cosine formulae in ΔABQ, we get

$$\cos A = \frac{AB^2+AQ^2-BQ^2}{2\times AB\times AQ} = \frac{4^2+(2.5)^2-BQ^2}{2\times4\times2.5}$$

$$\Rightarrow \frac{1}{8} = \frac{16+6.25-BQ^2}{20}$$

$$\Rightarrow BQ^2 = 22.25 - 2.5 = 19.75$$

$$\Rightarrow BQ = 4.44\text{ m}$$

($AQ = 2.5$ since Q is the midpoint of AC)

Now, from ΔPQB, right angled at Q, we get

$$\tan 30^\circ = \frac{PQ}{BQ} = \frac{h}{4.44}$$

$$\Rightarrow \frac{1}{\sqrt{3}} = \frac{h}{4.44} \Rightarrow h = 2.56\text{ m}$$

Example 9: To measure the height of a hill, a man measures the angle of elevation to the top of the hill to be 30° and then he moves 20 m forward towards the hill and measures the angle of elevation to be 45°. Find the height of the hill.

Solution:

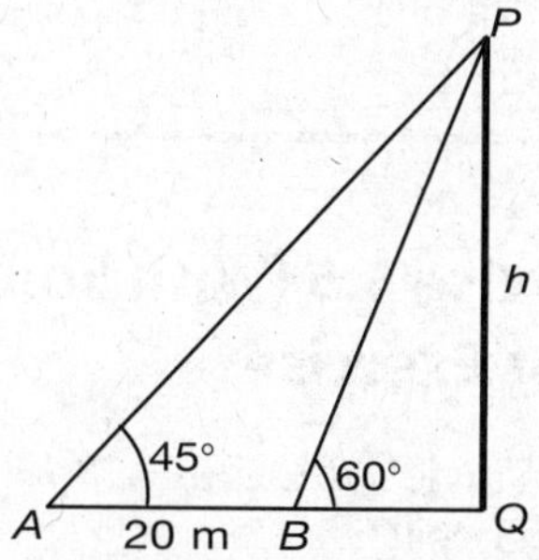

Let the height of the hill be $PQ = h$, angle of elevation from the 1st point A be $\angle PAQ = 45^\circ$ and angle of elevation from the 2nd point B ($AB = 20$ m) be $\angle PBQ = 60^\circ$.

Let $BQ = x$; then $\tan 45^\circ = \dfrac{h}{20+x}$ and $\tan 60^\circ = \dfrac{h}{x}$

$$\Rightarrow 1 = \frac{h}{20+x} \text{ and } \sqrt{3} = \frac{h}{x}$$

$$\Rightarrow h = 20 + x = 20 + \frac{h}{\sqrt{3}}$$

$$\Rightarrow h = \frac{20\sqrt{3}}{(\sqrt{3}-1)} = 10\sqrt{3}(\sqrt{3}+1)$$

Thus, the height of hill is $10\sqrt{3}(\sqrt{3}+1)$ m.

Example 10: To measure the height of a tower, two men standing 100 m apart (opposite to the tower) measure the angle of elevation to the top of the hill to be 30° and 60°. Find the height of the tower.

Solution:

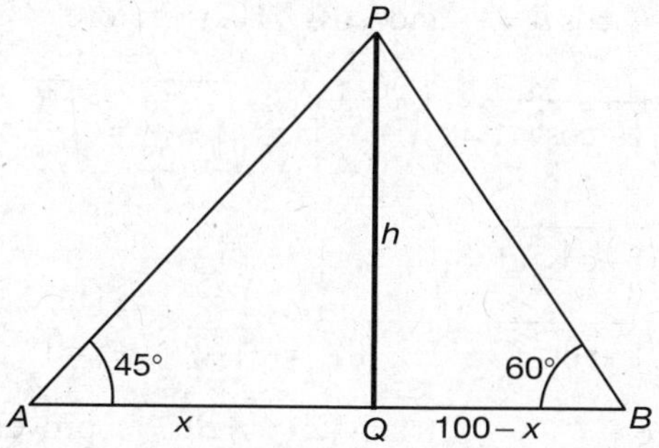

Let the height of the tower be $PQ = h$, angle of elevation from the first man A be $\angle PAQ = 45^\circ$ and angle of elevation from the second man B ($AB = 100$ m) be $\angle PBQ = 60^\circ$.

Let $AQ = x \Rightarrow BQ = 100 - x$; then

$$\tan 45^\circ = \frac{h}{x} \text{ and } \tan 60^\circ = \frac{h}{100-x}$$

$\Rightarrow 1 = \frac{h}{x}$ and $\sqrt{3} = \frac{h}{100-x}$

$\Rightarrow h = x = \sqrt{3}(100-x) \Rightarrow$

$\Rightarrow h = \frac{100\sqrt{3}}{(\sqrt{3}+1)} = 50\sqrt{3}(\sqrt{3}-1)$

Thus, the height of tower is $10\sqrt{3}(\sqrt{3}+1)$ m.

Day 49

NCERT Exercise 3.5 (With Solutions) Supplement Exercise

Q1. In any triangle ABC, if $a = 18, b = 24, c = 30$, find $\cos A, \cos B, \cos C$.

Solution:
Using the cosine formula:

$$\cos A = \frac{b^2+c^2-a^2}{2bc} = \frac{24^2+30^2-18^2}{2\times 24\times 30}$$

$$= \frac{576+900-324}{1440} = \frac{1152}{1440} = \frac{4}{5}$$

$$\cos B = \frac{c^2+a^2-b^2}{2ca} = \frac{30^2+18^2-24^2}{2\times 30\times 18}$$

$$= \frac{900+324-576}{1080} = \frac{648}{1080} = \frac{3}{5}$$

$$\cos C = \frac{a^2+b^2-c^2}{2ab} = \frac{18^2+24^2-30^2}{2\times 18\times 24}$$

$$= \frac{324+576-900}{864} = \frac{0}{864} = 0$$

Q2. In any triangle ABC, if $a = 18, b = 24, c = 30$, find $\sin A, \sin B, \sin C$.

Solution: Using the cosine formula in Q1, we have $\cos A = \frac{4}{5}$, $\cos B = \frac{3}{5}$ and $\cos C = 0$, therefore

$$\sin A = \sqrt{1-\cos^2 A} = \sqrt{1-\left(\frac{4}{5}\right)^2} = \sqrt{1-\frac{16}{25}} = \sqrt{\frac{9}{25}} = \frac{3}{5}$$

$$\sin B = \sqrt{1-\cos^2 B} = \sqrt{1-\left(\frac{3}{5}\right)^2} = \sqrt{\frac{16}{25}} = \frac{4}{5}$$

$$\sin C = \sqrt{1-\cos^2 C} = \sqrt{1-(0)^2} = \sqrt{1-0} = \sqrt{1} = 1$$

Q3. For any triangle ABC, prove that

$$\frac{a+b}{c} = \frac{\cos\left(\frac{A-B}{2}\right)}{\sin\frac{C}{2}}.$$

Solution:

From the sine formula, $\frac{a}{\sin A} = \frac{b}{\sin B} = \frac{c}{\sin C} = k$

$$\therefore \text{LHS} = \frac{a+b}{c} = \frac{k\sin A + k\sin B}{k\sin C}$$

$$= \frac{k(\sin A+\sin B)}{k\sin C} = \frac{2\sin\frac{A+B}{2}\cos\frac{A-B}{2}}{\sin C}$$

$$\left[\because \sin A+\sin B = 2\sin\frac{A+B}{2}\cos\frac{A-B}{2}\right]$$

$$= \frac{2\sin\left(90-\frac{C}{2}\right)\cos\frac{A-B}{2}}{2\sin\frac{C}{2}\cos\frac{C}{2}}$$

$$\left[\begin{array}{l}\because \sin\theta = 2\sin\frac{\theta}{2}\cos\frac{\theta}{2}\\ \text{Also, } A+B+C = 180° \Rightarrow \frac{A+B+C}{2} = 90°\\ \Rightarrow \frac{A+B}{2} = 90° - \frac{C}{2}\end{array}\right]$$

$$= \frac{\cos\frac{C}{2}\cos\frac{A-B}{2}}{\sin\frac{C}{2}\cos\frac{C}{2}} = \frac{\cos\left(\frac{A-B}{2}\right)}{\sin\frac{C}{2}} = \text{RHS}$$

Q4. For any triangle ABC, prove that

$$\frac{a-b}{c} = \frac{\sin\left(\frac{A-B}{2}\right)}{\cos\frac{C}{2}}.$$

Solution:

From the sine formula, $\frac{a}{\sin A} = \frac{b}{\sin B} = \frac{c}{\sin C} = k$

$$\therefore \text{LHS} = \frac{a-b}{c} = \frac{k\sin A - k\sin B}{k\sin B}$$

$$= \frac{k(\sin A-\sin B)}{k\sin C}$$

$$= \frac{2\cos\frac{A+B}{2}\sin\frac{A-B}{2}}{\sin C}$$

$$\left[\because \sin A-\sin B = 2\cos\frac{A+B}{2}\sin\frac{A-B}{2}\right]$$

$$= \frac{2\cos\left(90-\frac{C}{2}\right)\sin\frac{A-B}{2}}{2\sin\frac{C}{2}\cos\frac{C}{2}}$$

$$\left[\begin{array}{l}\because \sin\theta = 2\sin\frac{\theta}{2}\cos\frac{\theta}{2} \\ \text{Also, } A+B+C = 180^\circ \Rightarrow \frac{A+B+C}{2} = 90^\circ \\ \Rightarrow \frac{A+B}{2} = 90^\circ - \frac{C}{2}\end{array}\right]$$

$$= \frac{\sin\frac{C}{2}\sin\frac{A-B}{2}}{\sin\frac{C}{2}\cos\frac{C}{2}} = \frac{\sin\left(\frac{A-B}{2}\right)}{\cos\frac{C}{2}} = \text{RHS}$$

Q5. For any triangle ABC, prove that $\sin\frac{B-C}{2} = \frac{b-c}{a}\cos\frac{A}{2}$.

Solution:

$$\text{RHS} = \frac{b-c}{a}\cos\frac{A}{2} = \frac{k\sin B - k\sin C}{k\sin A}\cos\frac{A}{2}$$

$$\left[\text{Assuming } \frac{a}{\sin A} = \frac{b}{\sin B} = \frac{c}{\sin C} = k\right]$$

$$= \frac{k(\sin B - \sin C)}{k\sin A}\cos\frac{A}{2}$$

$$= \frac{2\cos\frac{B+C}{2}\sin\frac{B-C}{2}}{\sin A}\cos\frac{A}{2}$$

$$\left[\because \sin A - \sin B = 2\cos\frac{A+B}{2}\sin\frac{A-B}{2}\right]$$

$$= \frac{2\cos\left(90 - \frac{A}{2}\right)\sin\frac{B-C}{2}}{2\sin\frac{A}{2}\cos\frac{A}{2}}\cos\frac{A}{2}$$

$$\left[\begin{array}{l}\because \frac{A+B}{2} = 90^\circ - \frac{C}{2} \\ \sin\theta = 2\sin\frac{\theta}{2}\cos\frac{\theta}{2}\end{array}\right]$$

$$= \frac{\sin\frac{A}{2}\sin\frac{B-C}{2}}{\sin\frac{A}{2}} = \sin\frac{B-C}{2} = \text{LHS}$$

Q6. For any triangle ABC, prove that $a(b\cos C - c\cos B) = b^2 - c^2$.

Solution:

$$\text{LHS} = a(b\cos C - c\cos B)$$

$$= a\left[b\left(\frac{a^2+b^2-c^2}{2ab}\right) - c\left(\frac{a^2+c^2-b^2}{2ac}\right)\right]$$

$$\left[\because \cos A = \frac{b^2+c^2-a^2}{2bc}\right]$$

$$= a\left[\frac{a^2+b^2-c^2}{2a} - \frac{a^2+c^2-b^2}{2a}\right]$$

$$= a\left[\frac{a^2+b^2-c^2-a^2-c^2+b^2}{2a}\right]$$

$$= a\left[\frac{2(b^2-c^2)}{2a}\right] = b^2 - c^2 = \text{RHS}$$

Q7. For any triangle ABC, prove that $a(\cos C - \cos B) = 2(b-c)\cos^2\frac{A}{2}$.

Solution:

$$\text{LHS} = a(\cos C - \cos B)$$

$$= a\left[\left(\frac{a^2+b^2-c^2}{2ab}\right) - \left(\frac{a^2+c^2-b^2}{2ac}\right)\right]$$

$$\left[\because \cos A = \frac{b^2+c^2-a^2}{2bc}\right]$$

$$= a\left[\frac{ca^2+cb^2-c^3-ba^2-bc^2+b^3}{2abc}\right]$$

$$= \frac{b^3-c^3+b^2c-bc^2+a^2c-a^2b}{2bc}$$

$$= \frac{(b-c)(b^2+bc+c^2)+bc(b-c)-a^2(b-c)}{2bc}$$

$$= (b-c)\left[\frac{b^2+bc+c^2+bc-a^2}{2bc}\right]$$

$$= (b-c)\left[\frac{2bc}{2bc} + \frac{b^2+c^2-a^2}{2bc}\right]$$

$$= (b-c)(1+\cos A) \quad \left[\because \cos A = \frac{b^2+c^2-a^2}{2bc}\right]$$

$$= 2(b-c)\cos^2\frac{A}{2} = \text{RHS}$$

$$\left[\because 1+\cos A = 2\cos^2\frac{A}{2}\right]$$

Q8. For any triangle ABC, prove that $\frac{\sin(B-C)}{\sin(B+C)} = \frac{b^2-c^2}{a^2}$.

Solution:

$$\text{LHS} = \frac{\sin(B-C)}{\sin(B+C)} = \frac{\sin B\cos C - \cos B\sin C}{\sin B\cos C + \cos B\sin C}$$

$$= \frac{kb\left(\frac{a^2+b^2-c^2}{2ab}\right) - \left(\frac{a^2+c^2-b^2}{2ac}\right)kc}{kb\left(\frac{a^2+b^2-c^2}{2ab}\right) + \left(\frac{a^2+c^2-b^2}{2ac}\right)kc}$$

$$\left[\because \frac{a}{\sin A}=\frac{b}{\sin B}=\frac{c}{\sin C}=k \quad \cos A=\frac{b^2+c^2-a^2}{2bc}\right]$$

$$=\frac{\left(\frac{a^2+b^2-c^2}{2a}-\frac{a^2+c^2-b^2}{2a}\right)}{\left(\frac{a^2+b^2-c^2}{2a}+\frac{a^2+c^2-b^2}{2a}\right)}$$

$$=\frac{a^2+b^2-c^2-a^2-c^2+b^2}{a^2+b^2-c^2+a^2+c^2-b^2}$$

$$=\frac{2(b^2-c^2)}{2(a^2)}=\frac{b^2-c^2}{a^2}=\text{RHS}$$

Q9. For any triangle ABC, prove that $(b+c)\cos\frac{B+C}{2}=a\cos\frac{B-C}{2}$.

Solution:

$$\text{LHS}=(b+c)\cos\frac{B+C}{2}$$

$$=(k\sin B+k\sin C)\cos\frac{B+C}{2}$$

$$\left[\text{Assuming } \frac{a}{\sin A}=\frac{b}{\sin B}=\frac{c}{\sin C}=k\right]$$

$$=k(\sin B+\sin C)\cos\frac{B+C}{2}$$

$$=k\left(2\sin\frac{B+C}{2}\cos\frac{B-C}{2}\right)\cos\frac{B+C}{2}$$

$$\left[\because \sin A+\sin B=2\sin\frac{A+B}{2}\cos\frac{A-B}{2}\right]$$

$$=2k\sin\left(90-\frac{A}{2}\right)\cos\frac{B-C}{2}\cos\left(90-\frac{A}{2}\right)$$

$$\left[\because \frac{B+C}{2}=90^\circ-\frac{A}{2}\right]$$

$$=2k\cos\frac{A}{2}\cos\frac{B-C}{2}\sin\frac{A}{2}$$

$$=k\sin A\cos\frac{B-C}{2}$$

$$\left[\because 2\sin\frac{\theta}{2}\cos\frac{\theta}{2}=\sin\theta\right]$$

$$=a\cos\frac{B-C}{2}=\text{RHS}$$

Q10. For any triangle ABC, prove that $a\cos A+b\cos B+c\cos C=2a\sin B\sin C$.

Solution:

$$\text{LHS}=a\cos A+b\cos B+c\cos C$$

$$=k\sin A\cos A+k\sin B\cos B+k\sin C\cos C$$

$$\left[\text{Assuming } \frac{a}{\sin A}=\frac{b}{\sin B}=\frac{c}{\sin C}=k\right]$$

$$=\frac{k}{2}(2\sin A\cos A+2\sin B\cos B+2\sin C\cos C)$$

$$=\frac{k}{2}(\sin 2A+\sin 2B+2\sin C\cos C)$$

$$[\because 2\sin\theta\cos\theta=\sin 2\theta]$$

$$=\frac{k}{2}[2\sin(A+B)\cos(A-B)+2\sin C\cos C]$$

$$\left[\because \sin A+\sin B=2\sin\frac{A+B}{2}\cos\frac{A-B}{2}\right]$$

$$=\frac{k}{2}[2\sin(180-C)\cos(A-B)+2\sin C\cos C]$$

$$[\because A+B=180^\circ-C]$$

$$=\frac{k}{2}[2\sin C\cos(A-B)+2\sin C\cos C]$$

$$=k\sin C[\cos(A-B)+\cos C]$$

$$=k\sin C[\cos(A-B)+\cos\{180-(A+B)\}]$$

$$[\because A+B+C=180^\circ]$$

$$=k\sin C[\cos(A-B)-\cos(A+B)]$$

$$=k\sin C[2\sin A\sin B]$$

$$[\because \cos(A-B)-\cos(A+B)=2\sin A\sin B]$$

$$=2a\sin B\sin C=\text{RHS} \quad [\because k\sin A=a]$$

Q11. For any triangle ABC, prove that $\frac{\cos A}{a}+\frac{\cos B}{b}+\frac{\cos C}{c}=\frac{a^2+b^2+c^2}{2abc}$.

Solution:

$$\text{LHS}=\frac{\cos A}{a}+\frac{\cos B}{b}+\frac{\cos C}{c}$$

$$=\frac{1}{a}\left(\frac{b^2+c^2-a^2}{2bc}\right)+\frac{1}{b}\left(\frac{a^2+c^2-b^2}{2ac}\right)+\frac{1}{c}\left(\frac{a^2+b^2-c^2}{2ab}\right)$$

$$\left[\because \cos A=\frac{b^2+c^2-a^2}{2bc}\right]$$

$$=\frac{b^2+c^2-a^2+a^2+c^2-b^2+a^2+b^2-c^2}{2abc}$$

$$=\frac{a^2+b^2+c^2}{2abc}=\text{RHS}$$

Q12. For any triangle ABC, prove that $(b^2-c^2)\cot A+(c^2-a^2)\cot B+(a^2-b^2)\cot C=0$.

Solution:

$$\text{LHS}=(b^2-c^2)\cot A+(c^2-a^2)\cot B+(a^2-b^2)\cot C$$

$$=(b^2-c^2)\frac{\cos A}{\sin A}+(c^2-a^2)\frac{\cos B}{\sin B}+(a^2-b^2)\frac{\cos C}{\sin C}$$

$$= (b^2 - c^2)\left[\frac{1}{ka}\left(\frac{b^2 + c^2 - a^2}{2bc}\right)\right]$$

$$+(c^2 - a^2)\left[\frac{1}{kb}\left(\frac{a^2 + c^2 - b^2}{2ac}\right)\right]$$

$$+(a^2 - b^2)\left[\frac{1}{kc}\left(\frac{a^2 + b^2 - c^2}{2ab}\right)\right]$$

$$\left[\begin{array}{l}\because \dfrac{\sin A}{a} = \dfrac{\sin B}{b} = \dfrac{\sin C}{c} = k, \\ \text{Also, } \cos A = \dfrac{b^2 + c^2 - a^2}{2bc}\end{array}\right]$$

$$= \frac{1}{kabc}\left[\begin{array}{l}(b^2 - c^2)(b^2 + c^2 - a^2) \\ +(c^2 - a^2)(a^2 + c^2 - b^2) \\ +(a^2 - b^2)(a^2 + b^2 - c^2)\end{array}\right]$$

$$= \frac{1}{kabc}\left[\begin{array}{l}(b^4 + b^2c^2 - a^2b^2 - b^2c^2 - c^4 + c^2a^2) \\ +(c^2a^2 + c^4 - c^2b^2 - a^4 - c^2a^2 + a^2b^2) \\ +(a^4 + a^2b^2 - c^2a^2 - a^2b^2 - b^4 + b^2c^2)\end{array}\right]$$

$$= \frac{1}{kabc}(0) = 0 = \text{RHS}$$

Q13. For any triangle ABC, prove that

$$\frac{b^2 - c^2}{a^2}\sin 2A + \frac{c^2 - a^2}{b^2}\sin 2B + \frac{a^2 - b^2}{c^2}\sin 2C = 0.$$

Solution:

$$\text{LHS} = \frac{b^2 - c^2}{a^2}\sin 2A + \frac{c^2 - a^2}{b^2}\sin 2B + \frac{a^2 - b^2}{c^2}\sin 2C$$

$$= \frac{b^2 - c^2}{a^2} 2\sin A\cos A + \frac{c^2 - a^2}{b^2} 2\sin B\cos B$$

$$+ \frac{a^2 - b^2}{c^2} 2\sin C\cos C$$

$[\because \sin 2A = 2\sin A\cos A]$

$$= \left(\frac{b^2 - c^2}{a^2}\right)\left[ka\left(\frac{b^2 - c^2 - a^2}{2bc}\right)\right]$$

$$+ \left(\frac{c^2 - a^2}{b^2}\right)\left[kb\left(\frac{a^2 + c^2 - b^2}{2ac}\right)\right]$$

$$+ \left(\frac{a^2 - b^2}{c^2}\right)\left[kc\left(\frac{a^2 + b^2 - c^2}{2ab}\right)\right]$$

$$\left[\begin{array}{l}\because \dfrac{\sin A}{a} = \dfrac{\sin B}{b} = \dfrac{\sin C}{c} = k, \\ \text{Also, } \cos A = \dfrac{b^2 + c^2 - a^2}{2bc}\end{array}\right]$$

$$= \frac{k}{abc}\left[\begin{array}{l}(b^2 - c^2)(b^2 + c^2 - a^2) \\ +(c^2 - a^2)(a^2 + c^2 - b^2) \\ +(a^2 - b^2)(a^2 + b^2 - c^2)\end{array}\right]$$

$$= \frac{k}{abc}\left[\begin{array}{l}(b^4 + b^2c^2 - a^2b^2 - b^2c^2 - c^4 + c^2a^2) + \\ (c^2a^2 + c^4 - b^2c^2 - a^4 - c^2a^2 + a^2b^2) + \\ (a^4 + a^2b^2 - c^2a^2 - a^2b^2 - b^4 + b^2c^2)\end{array}\right]$$

$$= \frac{k}{abc}(0) = 0 = \text{RHS}$$

Q14. A tree stands vertically on a hill side which makes an angle of 15° with the horizontal. From a point on the ground 35 m down the hill from the base of the tree, the angle of elevation of the top of the tree is 60°. Find the height of the tree.

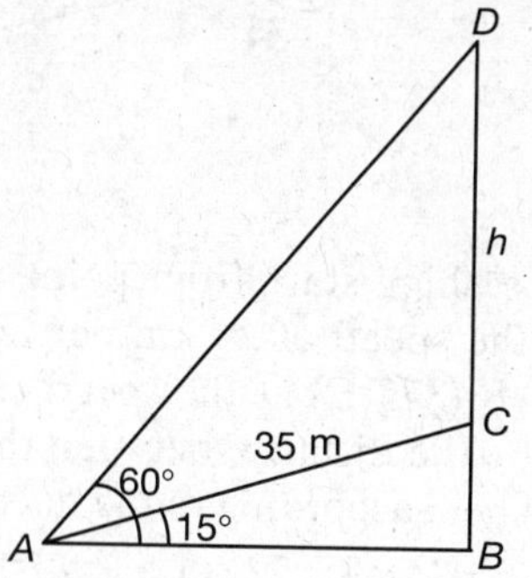

Solution:

Let BC be the height of the hill and CD be the height of the tree. Then, according to question

$$\because \tan 15° = \frac{BC}{AB} \text{ and } \cos 15° = \frac{AB}{AC} = \frac{AB}{35}$$

$$\therefore \tan 60° = \frac{h + BC}{AB} = \frac{h}{AB} + \frac{BC}{AB}$$

$$\Rightarrow \sqrt{3} = \frac{h}{35\cos 15°} + \tan 15°$$

$$\Rightarrow \sqrt{3} = \frac{h}{35\cos(45° - 30°)} + \tan(45° - 30°)$$

$$\Rightarrow \sqrt{3} = \frac{h}{35\left(\frac{1}{\sqrt{2}}\frac{\sqrt{3}}{2} + \frac{1}{\sqrt{2}}\frac{1}{2}\right)} + \left(\frac{1 - \frac{1}{\sqrt{3}}}{1 + 1\cdot\frac{1}{\sqrt{3}}}\right)$$

$$\left(\begin{array}{l}\because \cos(x - y) = \cos x\cos y + \sin x\sin y \\ \text{and } \tan(x - y) = \dfrac{\tan x - \tan y}{1 + \tan x\tan y}\end{array}\right)$$

$$\Rightarrow \sqrt{3} = \frac{2\sqrt{2}h}{35(\sqrt{3} + 1)} + \left(\frac{\sqrt{3} - 1}{\sqrt{3} + 1}\right)$$

$$\Rightarrow \frac{2\sqrt{2}h}{35(\sqrt{3} + 1)} = \sqrt{3} - \left(\frac{\sqrt{3} - 1}{\sqrt{3} + 1}\right)$$

$$\Rightarrow \frac{2\sqrt{2}h}{35(\sqrt{3} + 1)} = \frac{3 + \sqrt{3} - \sqrt{3} + 1}{\sqrt{3} + 1} = \frac{4}{\sqrt{3} + 1}$$

$\Rightarrow h = 35\sqrt{2}$ m is the height of tree

Q15. Two ships leave a port at the same time. One goes 24 km per hour in the direction N45°E and the other travels 32 km per hour in the direction S75°E. Find the distance between the ships at the end of 3 hours.

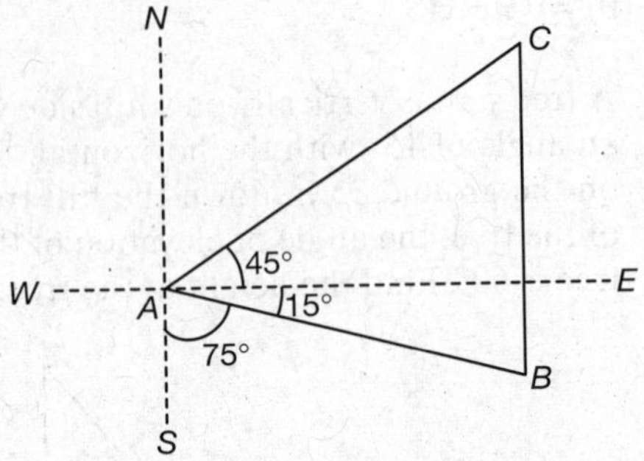

Solution:
Let both the ships start from point A, one along AC (N45°E) at the speed of 24 km per hour and the other one along AB (S75°E) at the speed of 32 km per hour, and let BC be the distance between the ships at the end of 3 hours. Then, according to the question,

Distance $AC = 24 \times 3 = 72\,\text{km}$

Distance $AB = 32 \times 3 = 96\,\text{km}$

Since angle $BAC = 60°$, by cosine formula,

$$\cos 60° = \frac{(AB)^2 + (AC)^2 - (BC)^2}{2(AB)(AC)}$$

$$\Rightarrow \frac{1}{2} = \frac{(96)^2 + (72)^2 - (BC)^2}{2(96)(72)}$$

$\Rightarrow BC = 86.53\,\text{km}$ is the distance between the ships at the end of 3 hours.

Q16. Two trees, A and B, are on the same side of a river. From a point C in the river, the distances of the trees A and B are 250 m and 300 m, respectively. If the angle C is 45°, find the distance between the trees (use $\sqrt{2} = 1.414$).

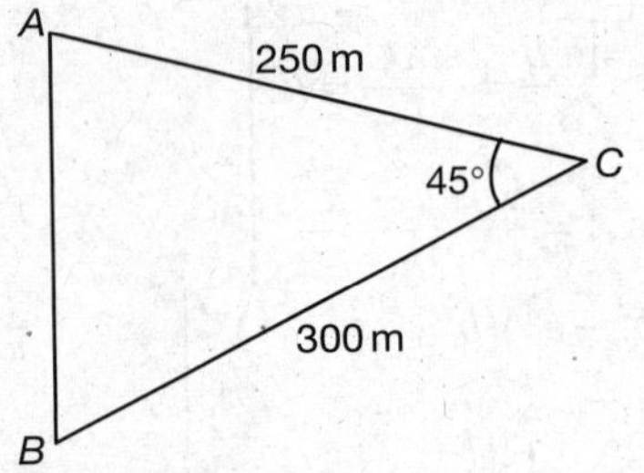

Solution:
According to the question, angle ACB = 60°, and therefore, by cosine formula,

$$\cos 45° = \frac{(BC)^2 + (AC)^2 - (AB)^2}{2(BC)(AC)}$$

$$\Rightarrow \frac{1}{\sqrt{2}} = \frac{(300)^2 + (250)^2 - (AB)^2}{2(300)(250)}$$

$\Rightarrow AB = 215.5$ m is the distance between the trees.

Day 50

NCERT Miscellaneous Exercise (With Solutions)

Q1. Prove that

$$2\cos\frac{\pi}{13}\cos\frac{9\pi}{13} + \cos\frac{3\pi}{13} + \cos\frac{5\pi}{13} = 0.$$

Solution:

$$\text{LHS} = 2\cos\frac{\pi}{13}\cos\frac{9\pi}{13} + \cos\frac{3\pi}{13} + \cos\frac{5\pi}{13}$$

$$= 2\cos\frac{\pi}{13}\cos\frac{9\pi}{13} + 2\cos\left(\frac{\frac{3\pi}{13}+\frac{5\pi}{13}}{2}\right)\cos\left(\frac{\frac{3\pi}{13}-\frac{5\pi}{13}}{2}\right)$$

$$\left[\cos x + \cos y = 2\cos\left(\frac{x+y}{2}\right)\cos\left(\frac{x-y}{2}\right)\right]$$

$$= 2\cos\frac{\pi}{13}\cos\frac{9\pi}{13} + 2\cos\frac{4\pi}{13}\cos\left(\frac{-\pi}{13}\right)$$

$$= 2\cos\frac{\pi}{13}\cos\frac{9\pi}{13} + 2\cos\frac{4\pi}{13}\cos\frac{\pi}{13}$$

$$= 2\cos\frac{\pi}{13}\left[\cos\frac{9\pi}{13} + \cos\frac{4\pi}{13}\right]$$

$$= 2\cos\frac{\pi}{13}\left[2\cos\frac{\left(\frac{9\pi}{13}+\frac{4\pi}{13}\right)}{2}\cos\left(\frac{\frac{9\pi}{13}-\frac{4\pi}{13}}{2}\right)\right]$$

$$= 2\cos\frac{\pi}{13}\left[2\cos\frac{\pi}{2}\cos\frac{5\pi}{26}\right]$$

$$= 2\cos\frac{\pi}{13} \times 2 \times 0 \times \cos\frac{5\pi}{26} = 0 = \text{RHS}$$

Q2. Prove that
$(\sin 3x + \sin x)\sin x + (\cos 3x - \cos x)\cos x = 0.$

Solution:

$$\text{LHS} = (\sin 3x + \sin x)\sin x + (\cos 3x - \cos x)\cos x$$

$$= \sin 3x \sin x + \sin^2 x + \cos 3x \cos x - \cos^2 x$$

$$= \cos 3x \cos x + \sin 3x \sin x - (\cos^2 x - \sin^2 x)$$

$= \cos(3x - x) - \cos 2x$

$[\cos(A - B) = \cos A \cos B + \sin A \sin B]$

$= \cos 2x - \cos 2x = 0 = \text{RHS}$

Q3. Prove that

$$(\cos x + \cos y)^2 + (\sin x - \sin y)^2 = 4\cos^2 \frac{x+y}{2}.$$

Solution:

$\text{LHS} = (\cos x + \cos y)^2 + (\sin x - \sin y)^2$

$= \cos^2 x + \cos^2 y + 2\cos x \cos y + \sin^2 x$

$+ \sin^2 y - 2\sin x \sin y$

$= (\cos^2 x + \sin^2 x) + (\cos^2 y + \sin^2 y)$

$+2(\cos x \cos y - \sin x \sin y)$

$= 1 + 1 + 2\cos(x + y)$

$[\cos(A + B) = (\cos A \cos B - \sin A \sin B)]$

$= 2[1 + \cos(x + y)]$

$= 2\left[2\cos^2\left(\frac{x+y}{2}\right)\right] \quad [\because 1 + \cos 2A = 2\cos^2 A]$

$= 4\cos^2 \frac{x+y}{2} = \text{RHS}$

Q4. Prove that

$$(\cos x - \cos y)^2 + (\sin x - \sin y)^2 = 4\sin^2 \frac{x-y}{2}.$$

Solution:

$\text{LHS} = (\cos x - \cos y)^2 + (\sin x - \sin y)^2$

$= \cos^2 x + \cos^2 y - 2\cos x \cos y$

$+ \sin^2 x + \sin^2 y - 2\sin x \sin y$

$= (\cos^2 x + \sin^2 x) + (\cos^2 y + \sin^2 y)$

$-2[\cos x \cos y + \sin x \sin y]$

$= 1 + 1 - 2[\cos(x - y)]$

$[\cos(A - B) = \cos A \cos B + \sin A \sin B]$

$= 2[1 - \cos(x - y)]$

$= 2\left[2\sin^2\left(\frac{x-y}{2}\right)\right] \quad [1 - \cos 2A = 2\sin^2 A]$

$= 4\sin^2\left(\frac{x-y}{2}\right) = \text{RHS}$

Q5. Prove that

$\sin x + \sin 3x + \sin 5x + \sin 7x$

$= 4\cos x \cos 2x \sin 4x$

Solution:

$\text{LHS} = \sin x + \sin 3x + \sin 5x + \sin 7x$

$= (\sin x + \sin 5x) + (\sin 3x + \sin 7x)$

$= 2\sin\left(\frac{x+5x}{2}\right) \cdot \cos\left(\frac{x-5x}{2}\right) +$

$2\sin\left(\frac{3x+7x}{2}\right)\cos\left(\frac{3x-7x}{2}\right)$

$\left[\because \sin A + \sin B = 2\sin\left(\frac{A+B}{2}\right) \cdot \cos\left(\frac{A-B}{2}\right)\right]$

$= 2\sin 3x \cos(-2x) + 2\sin 5x \cos(-2x)$

$= 2\sin 3x \cos 2x + 2\sin 5x \cos 2x$

$= 2\cos 2x[\sin 3x + \sin 5x]$

$= 2\cos 2x\left[2\sin\left(\frac{3x+5x}{2}\right) \cdot \cos\left(\frac{3x-5x}{2}\right)\right]$

$= 2\cos 2x[2\sin 4x \cdot \cos(-x)]$

$= 4\cos x \cos 2x \sin 4x = \text{RHS}$

Q6. Prove that

$$\frac{(\sin 7x + \sin 5x) + (\sin 9x + \sin 3x)}{(\cos 7x + \cos 5x) + (\cos 9x + \cos 3x)} = \tan 6x.$$

Solution:

$\text{LHS} = \frac{(\sin 7x + \sin 5x) + (\sin 9x + \sin 3x)}{(\cos 7x + \cos 5x) + (\cos 9x + \cos 3x)}$

$= \frac{[2\sin 6x \cdot \cos x] + [2\sin 6x \cdot \cos 3x]}{[2\cos 6x \cdot \cos x] + [2\cos 6x \cdot \cos 3x]}$

$\left[\begin{array}{l} \because \sin A + \sin B = 2\sin\left(\frac{A+B}{2}\right) \cdot \cos\left(\frac{A-B}{2}\right), \\ \cos A + \cos B = 2\cos\left(\frac{A+B}{2}\right) \cdot \cos\left(\frac{A-B}{2}\right) \end{array}\right]$

$= \frac{2\sin 6x[\cos x + \cos 3x]}{2\cos 6x[\cos x + \cos 3x]} = \tan 6x = \text{RHS}$

Q7. Prove that

$$\sin 3x + \sin 2x - \sin x = 4\sin x \cos\frac{x}{2} \cos\frac{3x}{2}.$$

Solution:

$\text{LHS} = \sin 3x + \sin 2x - \sin x$

$= \sin 3x + (\sin 2x - \sin x)$

$= \sin 3x + \left[2\cos\left(\frac{2x+x}{2}\right)\sin\left(\frac{2x-x}{2}\right)\right]$

$\left[\sin A - \sin B = 2\cos\left(\frac{A+B}{2}\right)\sin\left(\frac{A-B}{2}\right)\right]$

$= \sin 3x + \left[2\cos\left(\frac{3x}{2}\right)\sin\left(\frac{x}{2}\right)\right]$

$= 2\sin\left(\frac{3x}{2}\right)\cos\left(\frac{3x}{2}\right) + 2\cos\frac{3x}{2}\sin\frac{x}{2}$

$[\because \sin 2A = 2\sin A \cdot \cos A]$

$= 2\cos\left(\frac{3x}{2}\right)\left[\sin\left(\frac{3x}{2}\right) + \sin\left(\frac{x}{2}\right)\right]$

$= 2\cos\left(\frac{3x}{2}\right)\left[2\sin\left\{\frac{\left[\left(\frac{3x}{2}\right) + \left(\frac{x}{2}\right)\right]}{2}\right\}\cos\left\{\frac{\left[\left(\frac{3x}{2}\right) - \left(\frac{x}{2}\right)\right]}{2}\right\}\right]$

$\left[\because \sin A+\sin B=2\sin\left(\frac{A+B}{2}\right)\cos\left(\frac{A-B}{2}\right)\right]$

$=2\cos\left(\frac{3x}{2}\right)2\sin x\cos\left(\frac{x}{2}\right)$

$=4\sin x\cos\left(\frac{x}{2}\right)\cos\left(\frac{3x}{2}\right)=$ RHS

Q8. Find $\sin\frac{x}{2}$, $\cos\frac{x}{2}$, $\tan\frac{x}{2}$, if $\tan x=-\frac{4}{3}, x$ in quadrant II.

Solution:

Since x is in quadrant II,

$\frac{\pi}{2}<x<\pi \Rightarrow \frac{\pi}{4}<\frac{x}{2}<\frac{\pi}{2}$

Thus, $\sin\frac{x}{2}$, $\cos\frac{x}{2}$, $\tan\frac{x}{2}$, all lie in the first quadrant. And, in the first quadrant, all sin, cos, tan are positive.

Now, since $\tan x=-\frac{4}{3}$.

$\Rightarrow \sec x=\sqrt{1+\tan^2 x}=\sqrt{1+\left(\frac{-4}{3}\right)^2}=\sqrt{1+\frac{16}{9}}=\pm\frac{5}{3}$

$\Rightarrow \cos x=-\frac{3}{5}$ ($\because$ cos is negative in quadrant II)

Now, $\because \cos x=2\cos^2\frac{x}{2}-1 \Rightarrow \frac{-3}{5}=2\cos^2\frac{x}{2}-1$

$\Rightarrow 2\cos^2\frac{x}{2}=1-\frac{3}{5} \Rightarrow 2\cos^2\frac{x}{2}=\frac{2}{5}$

$\Rightarrow \cos^2\frac{x}{2}=\frac{1}{5} \Rightarrow \cos\frac{x}{2}=\frac{1}{\sqrt{5}}$

($\because$ cos is positive in quadrant I)

Now, $\sin\frac{x}{2}=\sqrt{1-\cos^2\frac{x}{2}}=\sqrt{1-\left(\frac{1}{\sqrt{5}}\right)^2}=\sqrt{\frac{4}{5}}=\frac{2}{\sqrt{5}}$

($\because$ sin is positive in quadrant I)

And, $\tan\frac{x}{2}=\frac{\sin\frac{x}{2}}{\cos\frac{x}{2}}=\frac{(2/\sqrt{5})}{(1/\sqrt{5})}=2$

Thus, the respective values of $\sin\frac{x}{2},\cos\frac{x}{2},\tan\frac{x}{2}$ are

$\frac{2}{\sqrt{5}},\frac{1}{\sqrt{5}},2 \Rightarrow \frac{2\sqrt{5}}{5},\frac{\sqrt{5}}{5},2.$

Q9. Find, $\sin\frac{x}{2},\cos\frac{x}{2},\tan\frac{x}{2}$ for $\cos x=-\frac{1}{3}$, x in quadrant III.

Solution:

Since x is in quadrant III,

$\pi<x<\frac{3\pi}{2} \Rightarrow \frac{\pi}{2}<\frac{x}{2}<\frac{3\pi}{4}$

Therefore, $\sin\frac{x}{2},\cos\frac{x}{2},\tan\frac{x}{2}$ are in quadrant II, and thus, $\cos\frac{x}{2}$ and $\tan\frac{x}{2}$ are negative and $\sin\frac{x}{2}$ is positive.

$\because \sin^2\frac{x}{2}=\frac{1-\cos x}{2}$

$\therefore \sin^2\frac{x}{2}=\frac{1-\left(-\frac{1}{3}\right)}{2}=\frac{\left(1+\frac{1}{3}\right)}{2}=\frac{\frac{4}{3}}{2}=2/3$

$\Rightarrow \sin\frac{x}{2}=\frac{\sqrt{2}}{\sqrt{3}}=\frac{\sqrt{2}}{\sqrt{3}}\times\frac{\sqrt{3}}{\sqrt{3}}=\frac{\sqrt{6}}{3}$

($\because$ sin is positive in quadrant II)

Now, since $\cos^2\frac{x}{2}=\frac{1+\cos x}{2}$

$\Rightarrow \cos^2\frac{x}{2}=\frac{1+\left(-\frac{1}{3}\right)}{2}=\frac{\left(\frac{3-1}{3}\right)}{2}=\frac{\frac{2}{3}}{2}=\frac{1}{3}$

$\Rightarrow \cos\frac{x}{2}=-\frac{1}{\sqrt{3}}=-\frac{1}{\sqrt{3}}\times\frac{\sqrt{3}}{\sqrt{3}}=\frac{-\sqrt{3}}{3}$

($\because$ cos is negative in quadrant II)

$\tan\frac{x}{2}=\frac{\sin\frac{x}{2}}{\cos\frac{x}{2}}=\left(\frac{\frac{\sqrt{2}}{\sqrt{3}}}{\frac{-1}{\sqrt{3}}}\right)=-\sqrt{2}$

($\because$ tan is negative in quadrant II)

Thus, the respective values of $\sin\frac{x}{2},\cos\frac{x}{2},\tan\frac{x}{2}$ are $\frac{\sqrt{6}}{3},\frac{-\sqrt{3}}{3},-\sqrt{2}$.

Q10. Find $\sin\frac{x}{2},\cos\frac{x}{2},\tan\frac{x}{2}$ for $\sin x=\frac{1}{4}$, x in quadrant II.

Solution:

Since x is in quadrant II,

$\Rightarrow \frac{\pi}{2}<x<\pi \Rightarrow \frac{\pi}{4}<\frac{x}{2}<\frac{\pi}{2}$

Therefore, $\sin\frac{x}{2},\cos\frac{x}{2},\tan\frac{x}{2}$ are all positive.

Now for $\sin x=\frac{1}{4}$.

$\cos x=\sqrt{1-\sin^2 x}=\sqrt{1-\left(\frac{1}{4}\right)^2}=\pm\sqrt{\frac{15}{16}}=-\frac{\sqrt{15}}{4}$

($\because$ cos is negative in quadrant II)

Now, $\sin\frac{x}{2}=\sqrt{\frac{1-\cos x}{2}}=\sqrt{\frac{1-\left(-\frac{\sqrt{15}}{4}\right)}{2}}=\sqrt{\frac{4+\sqrt{15}}{8}}$

$\Rightarrow \sin\frac{x}{2}=\sqrt{\frac{4+\sqrt{15}}{8}}=\sqrt{\frac{8+2\sqrt{15}}{16}}=\frac{\sqrt{8+2\sqrt{15}}}{4}$

($\because$ sin is positive in quadrant I)

Similarly, $\cos\frac{x}{2}=\sqrt{\frac{1+\cos x}{2}}=\sqrt{\frac{1+\left(-\frac{\sqrt{15}}{4}\right)}{2}}=\sqrt{\frac{4-\sqrt{15}}{8}}$

$=\sqrt{\frac{4-\sqrt{15}}{8}\times\frac{2}{2}}=\sqrt{\frac{8-2\sqrt{15}}{16}}=\frac{\sqrt{8-2\sqrt{15}}}{4}$

($\because$ cos is positive in quadrant I)

And, $\tan\frac{x}{2}=\frac{\sin\frac{x}{2}}{\cos\frac{x}{2}}=\frac{\left(\frac{\sqrt{8+2\sqrt{15}}}{4}\right)}{\left(\frac{\sqrt{8-2\sqrt{15}}}{4}\right)}=\frac{\sqrt{8+2\sqrt{15}}}{\sqrt{8-2\sqrt{15}}}$

$=\sqrt{\frac{8+2\sqrt{15}}{8-2\sqrt{15}}\times\frac{8+2\sqrt{15}}{8+2\sqrt{15}}}=\sqrt{\frac{(8+2\sqrt{15})^2}{64-60}}$

$=\frac{8+2\sqrt{15}}{2}=4+\sqrt{15}$

Thus, the respective values of $\sin\frac{x}{2},\cos\frac{x}{2},\tan\frac{x}{2}$ are $\frac{\sqrt{8+2\sqrt{15}}}{4},\frac{\sqrt{8-2\sqrt{15}}}{4},4+\sqrt{15}$

Day 51

NCERT Exemplar Problems: Solutions (Q1–Q23)

Q1. Prove that $\frac{\tan A+\sec A-1}{\tan A-\sec A+1}=\frac{1+\sin A}{\cos A}$.

Solution:

$\text{LHS}=\frac{\tan A+\sec A-1}{\tan A-\sec A+1}$

$=\frac{\tan A+\sec A-(\sec^2 A-\tan^2 A)}{(\tan A-\sec A+1)}$

$[\because \sec^2 A-\tan^2 A=1]$

$=\frac{(\tan A+\sec A)-(\sec A+\tan A)(\sec A-\tan A)}{(1-\sec A+\tan A)}$

$=\frac{(\sec A+\tan A)(1-\sec A+\tan A)}{1-\sec A+\tan A}$

$=\sec A+\tan A=\frac{1}{\cos A}+\frac{\sin A}{\cos A}$

$=\frac{1+\sin A}{\cos A}=\text{RHS}$

Q2. If $\frac{2\sin\alpha}{1+\cos\alpha+\sin\alpha}=y$, then prove that $\frac{1-\cos\alpha+\sin\alpha}{1+\sin\alpha}$ is also equal to y.

Solution:

$\frac{1-\cos\alpha+\sin\alpha}{1+\sin\alpha}$

$=\frac{(1-\cos\alpha+\sin\alpha)}{(1+\sin\alpha)}\cdot\frac{(1+\cos\alpha+\sin\alpha)}{(1+\cos\alpha+\sin\alpha)}$

$=\frac{\{(1+\sin\alpha)-\cos\alpha\}}{(1+\sin\alpha)}\cdot\frac{\{(1+\sin\alpha)+\cos\alpha\}}{(1+\cos\alpha+\sin\alpha)}$

$=\frac{(1+\sin\alpha)^2-\cos^2\alpha}{(1+\sin\alpha)(1+\sin\alpha+\cos\alpha)}$

$=\frac{1+\sin^2\alpha+2\sin\alpha-1+\sin^2\alpha}{(1+\sin\alpha)(1+\sin\alpha+\cos\alpha)}$

$=\frac{2\sin^2\alpha+2\sin\alpha}{(1+\sin\alpha)(1+\sin\alpha+\cos\alpha)}$

$=\frac{2\sin\alpha(1+\sin\alpha)}{(1+\sin\alpha)(1+\sin\alpha+\cos\alpha)}$

$=\frac{2\sin\alpha}{1+\sin\alpha+\cos\alpha}=y$

Q3. If $m\sin\theta=n\sin(\theta+2\alpha)$, then prove that $\tan(\theta+\alpha)\cot\alpha=\frac{m+n}{m-n}$.

Solution:

$m\sin\theta=n\sin(\theta+2\alpha)$

$\therefore \frac{m}{n}=\frac{\sin(\theta+2\alpha)}{\sin\theta}$

Using componendo and dividendo, we get

$\frac{m+n}{m-n}=\frac{\sin(\theta+2\alpha)+\sin\theta}{\sin(\theta+2\alpha)-\sin\theta}$

$\Rightarrow \frac{m+n}{m-n}=\frac{2\sin\left(\frac{\theta+2\alpha+\theta}{2}\right)\cdot\cos\left(\frac{\theta+2\alpha-\theta}{2}\right)}{\left(\frac{\theta+2\alpha+\theta}{2}\right)\cdot\sin\left(\frac{\theta+2\alpha-\theta}{2}\right)}$

$\left[\begin{array}{l}\because \sin x+\sin y=2\sin\frac{x+y}{2}\cdot\cos\frac{x-y}{2}\\ \sin x-\sin y=2\cos\frac{x+y}{2}\sin\cdot\frac{x-y}{2}\end{array}\right]$

$\Rightarrow \dfrac{m+n}{m-n} = \dfrac{\sin(\theta+\alpha)\cdot\cos\alpha}{\cos(\theta+\alpha)\cdot\sin\alpha}$

$\Rightarrow \dfrac{m+n}{m-n} = \tan(\theta+\alpha)\cdot\cot\alpha$

Q4. If $\cos(\alpha+\beta)=\dfrac{4}{5}$, $\sin(\alpha-\beta)=\dfrac{5}{13}$, α lies between $0, \dfrac{\pi}{4}$, then find that value of $\tan 2\alpha$.

Solution:

$\because \cos(\alpha+\beta)=\dfrac{4}{5}$

$\therefore \sin(\alpha+\beta)=\sqrt{1-\dfrac{16}{25}}=\sqrt{\dfrac{9}{25}}=\dfrac{3}{5}$

$\because \sin(\alpha-\beta)=\dfrac{5}{13}$

$\therefore \cos(\alpha-\beta)=\sqrt{1-\dfrac{25}{169}}=\sqrt{\dfrac{144}{169}}=\dfrac{12}{13}$

$\therefore \cos(\alpha-\beta)=\dfrac{12}{13}$

Now, $\tan(\alpha+\beta)=\dfrac{\sin(\alpha+\beta)}{\cos(\alpha+\beta)}=\dfrac{3/5}{4/5}=3/4$

[since α lies between $0, \dfrac{\pi}{4}$]

And, $\tan(\alpha-\beta)=\dfrac{\sin(\alpha-\beta)}{\cos(\alpha-\beta)}=\dfrac{5/13}{12/13}=\dfrac{5}{12}$

Thus, $\tan 2\alpha = \tan(\alpha+\beta+\alpha-\beta)$

$=\dfrac{\tan(\alpha+\beta)+\tan(\alpha-\beta)}{1-\tan(\alpha+\beta)\cdot\tan(\alpha-\beta)}$

$\left[\because \tan(x+y)=\dfrac{\tan x+\tan y}{1-\tan x\cdot\tan y}\right]$

$=\dfrac{3/4+5/12}{1-\dfrac{3}{4}\cdot\dfrac{5}{12}}=\dfrac{\dfrac{9+5}{12}}{\dfrac{16-5}{16}}=\dfrac{14\times16}{12\times11}=\dfrac{56}{33}$

Q5. If $\tan x=\dfrac{b}{a}$, then find the value of $\sqrt{\dfrac{a+b}{a-b}}+\sqrt{\dfrac{a-b}{a+b}}$.

Solution:

$\sqrt{\dfrac{a+b}{a-b}}+\sqrt{\dfrac{a-b}{a+b}}=\dfrac{\sqrt{(a+b)^2}+\sqrt{(a-b)^2}}{\sqrt{(a-b)(a+b)}}$

$=\dfrac{(a+b)+(a-b)}{\sqrt{a^2-b^2}}=\dfrac{2a}{\sqrt{a^2+b^2}}$

$=\dfrac{2}{\sqrt{1-\left(\dfrac{b}{a}\right)^2}}=\dfrac{2}{\sqrt{1-\tan^2 x}}\quad\left[\because \dfrac{b}{a}=\tan x\right]$

$=\dfrac{2\cos x}{\sqrt{1-\dfrac{\sin^2 x}{\cos^2 x}}}=\dfrac{2\cos x}{\sqrt{\cos^2 x-\sin^2 x}}$

$=\dfrac{2\cos x}{\sqrt{\cos 2x}}\quad[\because \cos 2x=\cos^2 x-\sin^2 x]$

Q6. Prove that

$$\cos\theta\cos\frac{\theta}{2}-\cos 3\theta\cos\frac{9\theta}{2}=\sin\frac{7\theta}{2}\sin 4\theta.$$

Solution:

$\text{LHS}=\cos\theta\cos\dfrac{\theta}{2}-\cos 3\theta\cos\dfrac{9\theta}{2}$

$=\dfrac{1}{2}\left[2\cos\theta\cdot\cos\dfrac{\theta}{2}-2\cos 3\theta\cdot\cos\dfrac{9\theta}{2}\right]$

$=\dfrac{1}{2}\left[\begin{array}{l}\cos\left(\theta+\dfrac{\theta}{2}\right)+\cos\left(\theta-\dfrac{\theta}{2}\right)-\\ \cos\left(3\theta+\dfrac{9\theta}{2}\right)+\cos\left(3\theta-\dfrac{9\theta}{2}\right)\end{array}\right]$

$[\because 2\cos A\cos B=\cos(A+B)+\cos(A-B)]$

$=\dfrac{1}{2}\left(\cos\dfrac{3\theta}{2}+\cos\dfrac{\theta}{2}-\cos\dfrac{15\theta}{2}-\cos\dfrac{3\theta}{2}\right)$

$=\dfrac{1}{2}\left[\cos\dfrac{\theta}{2}-\cos\dfrac{15\theta}{2}\right]$

$=-\dfrac{1}{2}\left[2\sin\left(\dfrac{(\theta+15\theta)/2}{2}\right)\cdot\sin\left(\dfrac{(\theta-15\theta)/2}{2}\right)\right]$

$\left[\because \cos x-\cos y=-2\sin\dfrac{x+y}{2}\cdot\sin\dfrac{x-y}{2}\right]$

$=\sin 4\theta\cdot\sin\dfrac{7\theta}{2}=\text{RHS}$

Q7. If $a\cos\theta+b\sin\theta=m$, $a\sin\theta-b\cos\theta=n$, then show that $a^2+b^2=m^2+n^2$.

Solution:

Since $a\cos\theta+b\sin\theta=m$ (i)

$a\sin\theta-b\cos\theta=n$ (ii)

Squaring and adding both equations, we get

$m^2+n^2=(a\cos\theta+b\sin\theta)^2+(a\sin\theta-b\cos\theta)^2$

$=a^2\cos^2\theta+b^2\sin^2\theta+2ab\sin\theta\cdot\cos\theta$

$+a^2\sin^2\theta+b^2\cos^2\theta-2ab\sin\theta\cdot\cos\theta$

$\Rightarrow m^2+n^2=a^2(\cos^2\theta+\sin^2\theta)+b^2(\sin^2\theta+\cos^2\theta)$

$\Rightarrow m^2+n^2=a^2+b^2\quad(\because \cos^2\theta+\sin^2\theta=1)$

Q8. Find the value of $\tan 22°30'$.

Solution:

$\tan\dfrac{\theta}{2}=\dfrac{\sin\dfrac{\theta}{2}}{\cos\dfrac{\theta}{2}}=\dfrac{2\sin\dfrac{\theta}{2}\cdot\cos\dfrac{\theta}{2}}{2\cos^2\dfrac{\theta}{2}}=\dfrac{\sin\theta}{1+\cos\theta}$

Let $\theta = 45°$

$$\therefore \tan 22°30' = \frac{\sin 45°}{1+\cos 45°} = \frac{1/\sqrt{2}}{1+\frac{1}{\sqrt{2}}} = \frac{1}{\sqrt{2}+1}$$

Q9. Prove that $\sin 4A = 4\sin A\cos^3 A - 4\cos A\sin^3 A$.

Solution:

LHS $= \sin 4A = 2\sin 2A \cdot \cos 2A$

$= 2(2\sin A \cdot \cos A)(\cos^2 A - \sin^2 A)$

$= 4\sin A \cdot \cos^3 A - 4\cos A \sin^3 A$

$$\left[\begin{array}{l}\because \sin 2A = 2\sin A \cdot \cos A \\ \cos 2A = \cos^2 A - \sin^2 A\end{array}\right]$$

Q10. If $\tan\theta + \sin\theta = m$, $\tan\theta - \sin\theta = n$, then prove that $m^2 - n^2 = 4\sin\theta\tan\theta$.

Solution:

Since $\tan\theta + \sin\theta = m$ (i)

$\tan\theta - \sin\theta = n$ (ii)

Therefore,

$m+n = \tan\theta + \sin\theta + \tan\theta - \sin\theta = 2\tan\theta$

And, $m-n = \tan\theta + \sin\theta - \tan\theta + \sin\theta = 2\sin\theta$

Multiplying both equations, we get

$(m+n)(m-n) = 4\sin\theta \cdot \tan\theta$

$\Rightarrow m^2 - n^2 = 4\sin\theta \cdot \tan\theta$

Q11. If $\tan(A+B) = p$, $\tan(A-B) = q$, then show that $\tan 2A = \dfrac{p+q}{1-pq}$.

Solution:

Since $\tan(A+B) = p$ (i)

$\tan(A-B) = q$ (ii)

$\therefore \tan 2A = \tan(A+B+A-B)$

$$= \frac{\tan(A+B)+\tan(A-B)}{1-\tan(A+B)\tan(A-B)} = \frac{p+q}{1-pq}$$

$$\left[\because \tan(x+y) = \frac{\tan x + \tan y}{1-\tan x\tan y}\right]$$

Q12. If $\cos\alpha + \cos\beta = 0 = \sin\alpha + \sin\beta$, then prove that $\cos 2\alpha + \cos 2\beta = -2\cos(\alpha+\beta)$.

Solution:

$\because \cos\alpha + \cos\beta = 0 = \sin\alpha + \sin\beta$

$\Rightarrow (\cos\alpha + \cos\beta)^2 = (\sin\alpha + \sin\beta)^2$

$\Rightarrow \cos^2\alpha + \cos^2\beta + 2\cos\alpha\cos\beta =$

$\sin^2\alpha + \sin^2\beta + 2\sin\alpha\sin\beta$

$\Rightarrow \cos^2\alpha - \sin^2\alpha + \cos^2\beta - \sin^2\beta$

$= 2(\sin\alpha\sin\beta - \cos\alpha\cos\beta)$

$\Rightarrow \cos 2\alpha + \cos 2\beta = -2\cos(\alpha+\beta)$

Q13. If $\dfrac{\sin(x+y)}{\sin(x-y)} = \dfrac{a+b}{a-b}$, then show that $\dfrac{\tan x}{\tan y} = \dfrac{a}{b}$.

Solution:

$$\because \frac{\sin(x+y)}{\sin(x-y)} = \frac{a+b}{a-b}$$

Using componendo and dividendo,

$$\Rightarrow \frac{\sin(x+y)+\sin(x-y)}{\sin(x-y)-\sin(x-y)} = \frac{a+b+a-b}{a+b-a+b}$$

$$\Rightarrow \frac{2\sin\left(\frac{x+y+x-y}{2}\right)\cdot\cos\left(\frac{x+y-x+y}{2}\right)}{2\cos\left(\frac{x+y+x-y}{2}\right)\cdot\sin\left(\frac{x+y-x+y}{2}\right)} = \frac{2a}{2b}$$

$$\left[\begin{array}{l}\because \sin x + \sin y = 2\sin\frac{x+y}{2}\cdot\cos\frac{x-y}{2}, \\ \sin - \sin y = 2\cos\frac{x+y}{2}\cdot\sin\frac{x-y}{2}\end{array}\right]$$

$$\Rightarrow \frac{\sin x\cdot\cos y}{\cos x\cdot\sin y} = \frac{a}{b} \Rightarrow \frac{\tan x}{\tan y} = \frac{a}{b}$$

Q14. If $\tan\theta = \dfrac{\sin\alpha - \cos\alpha}{\sin\alpha + \cos\alpha}$, then show that $\sin\alpha + \cos\alpha = \sqrt{2}\cos\theta$.

Solution:

$$\text{Since } \tan\theta = \frac{\sin\alpha - \cos\alpha}{\sin\alpha + \cos\alpha}$$

$$\Rightarrow \tan\theta = \frac{\cos\alpha(\tan\alpha - 1)}{\cos\alpha(\tan\alpha + 1)}$$

$$\Rightarrow \tan\theta = \frac{\tan\alpha - \tan\frac{\pi}{4}}{1+\tan\frac{\pi}{4}\cdot\tan\alpha}, \quad \left[\because \tan\frac{\pi}{4} = 1\right]$$

$$\Rightarrow \tan\theta = \tan\left(\alpha - \frac{\pi}{4}\right)$$

$$\left[\because \tan(x-y) = \frac{\tan x - \tan y}{1+\tan x\tan y}\right]$$

$$\Rightarrow \theta = \alpha - \frac{\pi}{4} \Rightarrow \alpha = \theta + \frac{\pi}{4}$$

Now, $\sin\alpha + \cos\alpha$

$$= \sin\left(\theta + \frac{\pi}{4}\right) + \cos\left(\theta + \frac{\pi}{4}\right)$$

$$= \sin\theta\cdot\cos\frac{\pi}{4} + \cos\theta\cdot\sin\frac{\pi}{4} + \cos\theta\cdot\cos\frac{\pi}{4} - \sin\theta\cdot\sin\frac{\pi}{4}$$

$$= \frac{1}{\sqrt{2}}\sin\theta + \frac{1}{\sqrt{2}}\cos\theta + \frac{1}{\sqrt{2}}\cos\theta - \frac{1}{\sqrt{2}}\sin\theta$$

$$= \frac{2}{\sqrt{2}}\cdot\cos\theta = \sqrt{2}\cos\theta$$

Q15. If $\sin\theta+\cos\theta=1$, then find the general value of θ.

Solution:

$\sin\theta+\cos\theta=1$,

Squaring both sides, we get

$\sin^2\theta+\cos^2\theta+2\sin\theta\cdot\cos\theta=1$

$\Rightarrow 1+2\sin\theta\cdot\cos\theta=1 \Rightarrow \sin 2\theta=0$

$[\because \sin 2x=2\sin x\cos x]$

$\Rightarrow \sin 2\theta=0 \Rightarrow 2\theta=n\pi+(-1)^n\cdot 0$

$\therefore \theta=\dfrac{n\pi}{2}$

Q16. Find the most general value of θ satisfying the equation $\tan\theta=-1,\ \cos\theta=\dfrac{1}{\sqrt{2}}$.

Solution:

Since $\tan\theta=-1 \Rightarrow \tan\theta=-\tan\dfrac{\pi}{4}$

$\Rightarrow \tan\theta=\tan\left(2\pi-\dfrac{\pi}{4}\right) \Rightarrow \tan\theta=\tan\dfrac{7\pi}{4}$

$\Rightarrow \theta=\dfrac{7\pi}{4}$

Now, since $\cos\theta=\dfrac{1}{\sqrt{2}} \Rightarrow \cos\theta=\cos\dfrac{\pi}{4}$

$\Rightarrow \cos\theta=\cos\left(2\pi-\dfrac{\pi}{4}\right) \Rightarrow \cos\theta=\cos\dfrac{7\pi}{4}$

$\therefore \theta=\dfrac{7\pi}{4}$

Thus, the general value of $\theta=2n\pi+\dfrac{7\pi}{4}$.

Q17. If $\cot\theta+\tan\theta=2\operatorname{cosec}\theta$, then find the general value of θ.

Solution:

$\cot\theta+\tan\theta=2\operatorname{cosec}\theta$

$\Rightarrow \dfrac{\cos\theta}{\sin\theta}+\dfrac{\sin\theta}{\cos\theta}=\dfrac{2}{\sin\theta}$

$\Rightarrow \dfrac{\cos^2+\sin^2\theta}{\sin\theta\cdot\cos\theta}=\dfrac{2}{\sin\theta}$

$\Rightarrow \dfrac{1}{\cos\theta}=2 \Rightarrow \cos\theta=\dfrac{1}{2}$

$\Rightarrow \cos\theta=\cos\dfrac{\pi}{3} \Rightarrow \theta=2n\pi\pm\dfrac{\pi}{3}$

Q18. If $2\sin^2\theta=3\cos\theta$, where $0\le\theta\le 2\pi$, then find the value of θ.

Solution:

$2\sin^2\theta=3\cos\theta \Rightarrow 2-2\cos^2\theta=3\cos\theta$

$\Rightarrow 2\cos^2\theta+3\cos\theta-2=0$

$\Rightarrow (\cos\theta+2)(2\cos\theta-1)=0$

$\Rightarrow 2\cos\theta=1 \quad (\because \cos\theta\ne -2)$

$\Rightarrow \cos\theta=\dfrac{1}{2} \Rightarrow \cos\theta=\cos\dfrac{\pi}{3} \Rightarrow \theta=\dfrac{\pi}{3}$

Also, $\cos\theta=\cos\left(2\pi-\dfrac{\pi}{3}\right)$

$\Rightarrow \cos\theta=\cos\dfrac{5\pi}{6} \Rightarrow \theta=\dfrac{5\pi}{6}$

Thus, the values of θ are $\dfrac{\pi}{3},\dfrac{5\pi}{6}$.

Q19. If $\sec x\cos 5x+1=0$, where $0\le x\le\dfrac{\pi}{2}$, then find the value of x.

Solution:

$\sec x\cos 5x+1=0 \Rightarrow \dfrac{\cos 5x+\cos x}{\cos x}=0$

$\Rightarrow \cos 5x+\cos x=0$, but $\cos x\ne 0$

Now, $\cos 5x+\cos x=0$

$\Rightarrow 2\cos\left(\dfrac{5x+x}{2}\right)\cdot\cos\left(\dfrac{5x-x}{2}\right)=0$

$\left[\because \cos x+\cos y=2\cos\dfrac{x+y}{2}\cdot\cos\dfrac{x-y}{2}\right]$

$\Rightarrow 2\cos 3x\cdot\cos 2x=0$

$\Rightarrow \cos 3x=0$ or $\cos 2x=0$

$\Rightarrow \cos 3x=\cos\dfrac{\pi}{2}$ or $\cos 2x=\cos\dfrac{\pi}{2}$

$\Rightarrow 3x=\dfrac{\pi}{2}$ or $2x=\dfrac{\pi}{2}$

$\Rightarrow x=\dfrac{\pi}{6}$ or $x=\dfrac{\pi}{4}$

Also,

$\Rightarrow \cos 3x=\cos\dfrac{3\pi}{2}$ or $\cos 2x=\cos\dfrac{3\pi}{2}$

$\Rightarrow 3x=\dfrac{3\pi}{2}$ or $2x=\dfrac{3\pi}{2}$

$\Rightarrow x=\dfrac{\pi}{2}$ or $x=\dfrac{3\pi}{4}$

But since, $\cos x\ne 0,\ x\ne\dfrac{\pi}{2}$

Hence, the solutions are $\dfrac{\pi}{4},\dfrac{\pi}{6}$, for $0\le x\le\dfrac{\pi}{2}$.

Q20. If $\sin(\theta+\alpha)=a$, $\sin(\theta+\beta)=b$, then prove that $\cos 2(\alpha-\beta)-4ab\cos(\alpha-\beta)=1-2a^2-2b^2$.

Solution:

$\sin(\theta+\alpha)=a$ and $\sin(\theta+\beta)=b$

$\therefore \cos(\theta+\alpha)=\sqrt{1-a^2}$ and $\cos(\theta+\beta)=\sqrt{1-b^2}$

Now, $\cos(\alpha-\beta)=\cos\{(\theta+\alpha)-(\theta+\beta)\}$

$= \cos(\theta+\beta)\cos(\theta+\alpha)+\sin(\theta+\alpha)\sin(\theta+\beta)$

$= \sqrt{1-a^2}\sqrt{1-b^2}+a\cdot b = ab+\sqrt{(1-a^2)(1-b^2)}$

$= ab+\sqrt{1-a^2-b^2+a^2b^2}$

Thus, $\cos 2(\alpha-\beta)-4ab\cos(\alpha-\beta)$

$= 2\cos^2(\alpha-\beta)-1-4ab\cos(\alpha-\beta)$

$= 2\cos(\alpha-\beta)(\cos(\alpha-\beta)-2ab)-1$

$= 2(ab+\sqrt{1-a^2-b^2+a^2b^2})$

$(ab+\sqrt{1-a^2-b^2+a^2b^2}-2ab)-1$

$= 2[(\sqrt{1-a^2-b^2+a^2b^2}+ab)$

$(\sqrt{1-a^2-b^2+a^2b^2}-ab)]-1$

$= 2[1-a^2-b^2+a^2b^2-a^2b^2]-1$

$= 2-2a^2-2b^2-1 = 1-2a^2-2b^2$

Q21. If $\cos(\theta+\phi) = m\cos(\theta-\phi)$, then prove that

$$\tan\theta = \frac{1-m}{1+m}\cot\phi.$$

Solution:

Since $\cos(\theta+\phi) = m\cos(\theta-\phi)$,

$$\Rightarrow \frac{\cos(\theta+\phi)}{\cos(\theta-\phi)} = \frac{m}{1}$$

Using the componendo and dividendo rule,

$$\frac{\cos(\theta-\phi)-\cos(\theta+\phi)}{\cos(\theta-\phi)+\cos(\theta+\phi)} = \frac{1-m}{1+m}$$

$$\Rightarrow \frac{-2\sin\left(\frac{\theta-\phi+\theta+\phi}{2}\right)\cdot\sin\left(\frac{\theta-\phi-\theta-\phi}{2}\right)}{2\cos\left(\frac{\theta-\phi+\theta+\phi}{2}\right)\cdot\cos\left(\frac{\theta-\phi-\theta-\phi}{2}\right)} = \frac{1-m}{1+m}$$

$$\Rightarrow \frac{\sin\theta\cdot\sin\phi}{\cos\theta\cdot\cos\phi} = \frac{1-m}{1+m}$$

$[\because \sin(-\theta) = -\sin\theta,\ \cos(-\theta) = \cos\theta]$

$$\Rightarrow \tan\theta\cdot\tan\phi = \frac{1-m}{1+m}$$

$$\Rightarrow \tan\theta = \left(\frac{1-m}{1+m}\right)\cot\phi$$

Q22. Find the value of the expression

$$3\left[\sin^4\left(\frac{3\pi}{2}-\alpha\right)+\sin^4(3\pi+\alpha)\right]$$
$$-2\left[\sin^6\left(\frac{\pi}{2}+\alpha\right)+\sin^6(5\pi-\alpha)\right]$$

Solution:

$$3\left[\sin^4\left(\frac{3\pi}{2}-\alpha\right)+\sin^4(3\pi+\alpha)\right]$$
$$-2\left[\sin^6\left(\frac{\pi}{2}+\alpha\right)+\sin^6(5\pi-\alpha)\right]$$

$= 3[\cos^4\alpha+\sin^4\alpha]-2[\cos^6\alpha+\sin^6\alpha]$

$= 3[(\cos^2\alpha+\sin^2\alpha)^2-2\cos^2\alpha\sin^2\alpha]-2$

$[(\cos^2\alpha+\sin^2\alpha)^3-3\cos^2\alpha\sin^2\alpha(\cos^2\alpha+\sin^2\alpha)]$

$= 3[1-2\cos^2\alpha\sin^2\alpha]-2[1-3\cos^2\alpha\sin^2\alpha]$

$= 3-6\cos^2\alpha\sin^2\alpha-2+6\cos^2\alpha\sin^2\alpha]$

$= 3-2 = 1$

Q23. If $a\cos 2\theta+b\sin 2\theta = c$ has α, β as its roots, then prove that $\tan\alpha+\tan\beta = \dfrac{2b}{a+c}$.

Solution:

$a\cos 2\theta+b\sin 2\theta = c$

$$\Rightarrow a\left(\frac{1-\tan^2\theta}{1+\tan^2\theta}\right)+b\left(\frac{2\tan\theta}{1+\tan^2\theta}\right) = c$$

$$\left[\because \sin 2\theta = \frac{2\tan\theta}{1+\tan^2\theta},\ \cos 2\theta = \frac{1-\tan^2\theta}{1+\tan^2\theta}\right]$$

$\Rightarrow a(1-\tan^2\theta)+2b\tan\theta = c(1+\tan^2\theta)$

$\Rightarrow a-a-\tan^2\theta+2b\tan\theta = c+c\tan^2\theta$

$\Rightarrow (a+c)\tan^2\theta-2b\tan\theta+c-a = 0$

This equation has $\tan\alpha, \tan\beta$ as its roots, and therefore, the sum of the roots,

$$\tan\alpha+\tan\beta = \frac{-(-2b)}{a+c} = \frac{2b}{a+c}$$

Day 52

NCERT Exemplar Problems: Solutions (Q24–Q52)

Q24. If $x = \sec\phi-\tan\phi$, $y = \operatorname{cosec}\phi+\cot\phi$, then show that $xy+x-y+1 = 0$.

Solution:

Since $x = \sec\phi-\tan\phi$ (i)

$y = \operatorname{cosec}\phi+\cot\phi$ (ii)

Therefore, $xy = (\sec\phi-\tan\phi)(\operatorname{cosec}\phi+\cot\phi)$

$\Rightarrow xy = \sec\phi\cdot\operatorname{cosec}\phi-\operatorname{cosec}\phi\cdot\tan\phi$

$+\sec\phi\cdot\cot\phi-\tan\phi\cdot\cot\phi$

$$\Rightarrow xy = \sec\phi\cdot\operatorname{cosec}\phi-\frac{1}{\cos\phi}+\frac{1}{\sin\phi}-1$$

$\Rightarrow 1+xy = \sec\phi\operatorname{cosec}\phi-\sec\phi+\operatorname{cosec}\phi$ (iii)

Also,

$x-y=\sec\phi-\tan\phi-\operatorname{cosec}\phi-\cot\phi$

$\Rightarrow x-y=\sec\phi-\operatorname{cosec}\phi-\dfrac{\sin\phi}{\cos\phi}-\dfrac{\cos\phi}{\sin\phi}$

$\Rightarrow x-y=\sec\phi-\operatorname{cosec}\phi-\left(\dfrac{\sin^2\phi+\cos^2\phi}{\sin\phi\cdot\cos\phi}\right)$

$\Rightarrow x-y=\sec\phi-\operatorname{cosec}\phi-\dfrac{1}{\sin\phi\cdot\cos\phi}$

$\Rightarrow x-y=\sec\phi-\operatorname{cosec}\phi-\operatorname{cosec}\phi\cdot\cos\phi$

$\Rightarrow x-y=-(\sec\phi\cdot\operatorname{cosec}\phi-\sec\phi+\operatorname{cosec}\phi)$

$\Rightarrow x-y=-(xy-1)$

$\Rightarrow xy+x-y+1=0$

Q25. If θ lies in the first quadrant and $\cos\theta=\dfrac{8}{17}$, then find the value of $\cos(30°+\theta)+\cos(45°-\theta)+\cos(120°-\theta)$.

Solution:

$\because \cos\theta=\dfrac{8}{17}\Rightarrow \sin\theta=\sqrt{1-\dfrac{64}{289}}=\pm\dfrac{15}{17}$

$\Rightarrow \sin\theta=\dfrac{15}{17}$ [since θ lies in the first quadrant]

Now, $\cos(30°+\theta)+\cos(45°-\theta)+\cos(120°-\theta)$

$=\cos(30°+\theta)+\cos(45°-\theta)+\cos(90°+30°-\theta)$

$=\cos(30°+\theta)+\cos(45°-\theta)-\sin(30°-\theta)$

$=\cos 30°\cos\theta-\sin 30°\sin\theta+\cos 45°\cos\theta$

$+\sin 45°\sin\theta-\sin 30°\cos\theta+\cos 30°\sin\theta$

$=\dfrac{\sqrt{3}}{2}\cos\theta-\dfrac{1}{2}\sin\theta+\dfrac{1}{\sqrt{2}}\cos\theta+\dfrac{1}{\sqrt{2}}$

$\sin\theta-\dfrac{1}{2}\cos\theta\dfrac{\sqrt{3}}{2}\sin\theta$

$=\left(\dfrac{\sqrt{3}}{2}+\dfrac{1}{\sqrt{2}}-\dfrac{1}{2}\right)\cos\theta+\left(\dfrac{1}{\sqrt{2}}-\dfrac{1}{2}+\dfrac{\sqrt{3}}{2}\right)\sin\theta$

$=\left(\dfrac{\sqrt{6}+2-\sqrt{2}}{2\sqrt{2}}\right)\cos\theta+\left(\dfrac{2-\sqrt{2}+\sqrt{6}}{2\sqrt{2}}\right)\sin\theta$

$=\left(\dfrac{\sqrt{6}+2-\sqrt{2}}{2\sqrt{2}}\right)\dfrac{8}{17}+\left(\dfrac{2-\sqrt{2}+\sqrt{6}}{2\sqrt{2}}\right)\dfrac{15}{17}$

$=\dfrac{1}{17(2\sqrt{2})}(8\sqrt{6}+16-8\sqrt{2}+30-15\sqrt{2}+15\sqrt{6})$

$=\dfrac{1}{17(2\sqrt{2})}(23\sqrt{6}-23\sqrt{2}+46)$

$=\dfrac{23\sqrt{6}}{17(2\sqrt{2})}-\dfrac{23\sqrt{2}}{17(2\sqrt{2})}+\dfrac{46}{17(2\sqrt{2})}$

$=\dfrac{23\sqrt{3}}{17(2)}-\dfrac{23}{17(2)}+\dfrac{23}{17\sqrt{2}}$

$=\dfrac{23}{17}\left(\dfrac{\sqrt{3}-1}{2}+\dfrac{1}{\sqrt{2}}\right)$

Q26. Find the value of $\cos^4\dfrac{\pi}{8}+\cos^4\dfrac{3\pi}{8}+\cos^4\dfrac{5\pi}{8}+\cos^4\dfrac{7\pi}{8}$.

Solution:

$\cos^4\dfrac{\pi}{8}+\cos^4\dfrac{3\pi}{8}+\cos^4\dfrac{5\pi}{8}+\cos^4\dfrac{7\pi}{8}$

$=\cos^4\dfrac{\pi}{8}+\cos^4\dfrac{3\pi}{8}+\cos^4\left(\pi-\dfrac{3\pi}{8}\right)+\cos^4\left(\pi-\dfrac{\pi}{8}\right)$

$=\cos^4\dfrac{\pi}{8}+\cos^4\dfrac{3\pi}{8}+\cos^4\dfrac{3\pi}{8}+\cos^4\dfrac{\pi}{8}$

$=2\left[\cos^4\dfrac{\pi}{8}+\cos^4\dfrac{3\pi}{8}\right]$

$=2\left[\cos^4\dfrac{\pi}{8}+\cos^4\left(\dfrac{\pi}{2}-\dfrac{\pi}{8}\right)\right]$

$=2\left[\cos^4\dfrac{\pi}{8}+\sin^4\dfrac{\pi}{8}\right]$

$=2\left[\left(\cos^2\dfrac{\pi}{8}+\cos^2\dfrac{\pi}{8}\right)^2-2\cos^2\dfrac{\pi}{8}\cdot\sin^2\dfrac{\pi}{8}\right]$

$=2\left[1-2\cos^2\dfrac{\pi}{8}\cdot\sin^2\dfrac{\pi}{8}\right]$

$=2-\left(2\sin\dfrac{\pi}{8}\cdot\cos\dfrac{\pi}{8}\right)^2=2-\left(\sin\dfrac{2\pi}{8}\right)^2$

$=2-\left(\dfrac{1}{\sqrt{2}}\right)^2=2-\dfrac{1}{2}=\dfrac{3}{2}$

Q27. Find the general solution of the equation $5\cos^2+7\sin^2\theta-6=0$.

Solution:

$5\cos^2+7\sin^2\theta-6=0$

$\Rightarrow 5\cos^2+7(1-\cos^2\theta)-6=0$

$\Rightarrow 5\cos^2\theta+7-7\cos^2\theta-6=0$

$\Rightarrow -2\cos^2\theta+1=0\Rightarrow\cos^2\theta=1/2$

$\Rightarrow \cos^2\theta=\cos^2\dfrac{\pi}{4}\Rightarrow\theta=n\pi\pm\dfrac{\pi}{4}$

$\left[\because \cos^2\theta=\cos^2\alpha\quad\therefore\theta=n\pi\pm\alpha\right]$

Q28. Find the general solution of the equation $\sin x-3\sin 2x+\sin 3x=\cos x-3\cos 2x+\cos 3x$.

Solution:

$\sin x-3\sin 2x+\sin 3x=\cos x-3\cos 2x+\cos 3x$

$\Rightarrow 2\sin\left(\frac{x+3x}{2}\right)\cdot\cos\left(\frac{3x-x}{2}\right)-3\sin 2x$

$= 2\cos\left(\frac{3x+x}{2}\right)\cdot\cos\left(\frac{3x-x}{2}\right)-3\cos 2x$

$\Rightarrow 2\sin 2x\cos x - 3\sin 2x = 2\cos 2x\cdot\cos x - 3\cos 2x$

$\Rightarrow \sin 2x(2\cos x - 3) = \cos 2x(2\cos x - 3)$

$\Rightarrow \frac{\sin 2x}{\cos 2x} = 1 \Rightarrow \tan 2x = 1 \Rightarrow \tan 2x = \tan\frac{\pi}{4}$

$\Rightarrow 2x = n\pi + \frac{\pi}{4} \Rightarrow x = \frac{n\pi}{2}+\frac{\pi}{8}$

$[\because \tan\theta = \tan\alpha \quad \therefore \theta = n\pi + \alpha]$

Q29. Find the general solution of the equation $(\sqrt{3}-1)\cos\theta + (\sqrt{3}+1)\sin\theta = 2$.

Solution:

$(\sqrt{3}-1)\cos\theta + (\sqrt{3}+1)\sin\theta = 2$

Put $\sqrt{3}-1 = r\sin\alpha,\ \ \sqrt{3}+1 = r\cos\alpha$

$\therefore r^2 = (\sqrt{3}-1)^2 + (\sqrt{3}+1)^2 = 3+1-2\sqrt{3}+3+1+2\sqrt{3}$

$\Rightarrow r^2 = 8 \Rightarrow r = 2\sqrt{2}$

Now, $\tan\alpha = \frac{\sqrt{3}-1}{\sqrt{3}+1} = \frac{\tan\frac{\pi}{3}-\tan\frac{\pi}{4}}{1+\tan\frac{\pi}{3}\cdot\frac{\pi}{4}}$

$\Rightarrow \tan\alpha = \tan\left(\frac{\pi}{3}-\frac{\pi}{4}\right)$

$\Rightarrow \tan\alpha = \tan\frac{\pi}{12} \Rightarrow \alpha = \frac{\pi}{12}$

Also, $r\sin\alpha\cos\theta + r\cos\alpha\sin\theta = 2$

$r[\sin(\theta+\alpha)] = 2$

$\Rightarrow \sin(\theta+\alpha) = \frac{2}{2\sqrt{2}}$

$\Rightarrow \sin(\theta+\alpha) = \frac{1}{\sqrt{2}}$

$\Rightarrow \sin(\theta+\alpha) = n\pi + (-1)^n\frac{\pi}{4}$

$\theta = n\pi + (-1)^n\cdot\frac{\pi}{4}-\frac{\pi}{12}$

Q30. If $\sin\theta + \text{cosec}\,\theta = 2$, then $\sin^2\theta + \text{cosec}^2\theta$ is equal to

(a) 1 (b) 4
(c) 2 (d) None of these

Solution: (c)

$\sin\theta + \text{cosec}\,\theta = 2$

$\Rightarrow \sin^2\theta + \text{cosec}^2\theta + 2\sin\theta\cdot\text{cosec}\,\theta = 4$

$\Rightarrow \sin^2\theta + \text{cosec}^2\theta + 2\cdot 1 = 4$

$\Rightarrow \sin^2\theta + \text{cosec}^2\theta = 4-2$

$\Rightarrow \sin^2\theta + \text{cosec}^2\theta = 2$

Q31. If $f(x) = \cos^2 x + \sec^2 x$, then

(a) $f(x) < 1$ (b) $f(x) = 1$
(c) $2 < f(x) < 1$ (d) $f(x) \geq 2$

Solution: (d)

$\because AM \geq GM$

$\Rightarrow \frac{\cos^2 x + \sec^2 x}{2} \geq \sqrt{\cos^2 x\cdot\sec^2 x}$

$\Rightarrow \cos^2 x + \sec^2 x \geq 2 \quad [\because \cos x\cdot\sec x = 1]$

$\Rightarrow f(x) \geq 2$

Q32. If $\tan\theta = \frac{1}{2}, \tan\phi = \frac{1}{3}$, then the value of $\theta+\phi$ is

(a) $\frac{\pi}{6}$ (b) π (c) 0 (d) $\frac{\pi}{4}$

Solution: (d)

$\tan(\theta+\phi) = \frac{\tan\theta+\tan\phi}{1-\tan\theta\cdot\tan\phi}$

$\Rightarrow \tan(\theta+\phi) = \frac{\frac{1}{2}+\frac{1}{3}}{1-\frac{1}{2}\cdot\frac{1}{3}} = \frac{\frac{3+2}{6}}{\frac{6-1}{6}} = \frac{5}{5} = 1$

$\Rightarrow \tan(\theta+\phi) = \tan\frac{\pi}{4} \Rightarrow \theta+\phi = \frac{\pi}{4}$

Q33. Which of the following is not correct?

(a) $\sin\theta = -\frac{1}{5}$ (b) $\cos\theta = 1$
(c) $\sec\theta = \frac{1}{2}$ (d) $\tan\theta = 2$

Solution: (c)

Since range of $\sec\theta$ is, $\sec\theta$ cannot be equal to 1/2.

Q34. The value of $\tan 1°\tan 2°\tan 3°$ ______ $\tan 89°$ is

(a) 0 (b) 1
(c) 1/2 (d) Not defined

Solution: (b)

$\tan 1°\tan 2°\tan 3°...\tan 89°$

$= \tan 1°\tan 2°\tan 3°...\tan 44°\tan 45°$

$\tan(90°-44°)\tan(90°-43°)...\tan(90°-1°)$

$= \tan 1°\cdot\cot 1°\tan 2°\cdot\cot 2°...\tan 44°\cdot\cot 44°\tan 45°$

$= 1\cdot 1...1\cdot 1 = 1$

Q35. The value of $\frac{1-\tan^2 15°}{1+\tan^2 15°}$ is

(a) 1 (b) $\sqrt{3}$ (c) $\frac{\sqrt{3}}{2}$ (d) 2

Solution: (c)

Let $\theta = 15°$

$$\cos 2\theta = \frac{1-\tan^2\theta}{1+\tan^2\theta}$$

$$\therefore \cos 30° = \frac{1-\tan^2 15°}{1+\tan^2 15°}$$

$$\Rightarrow \frac{1-\tan^2 15°}{1+\tan^2 15°} = \cos 30° = \frac{\sqrt{3}}{2}$$

Q36. The value of $\cos 1° \cos 2° \cos 3°$ ______ $\cos 179°$ is

(a) $\frac{1}{\sqrt{2}}$ (b) 0 (c) 1 (d) –1

Solution: (b)

$\cos 1° \cos 2° \cos 3° .... \cos 179°$
$= \cos 1° \cos 2° ..... \cos 90° .... \cos 179°$
$= 0 \quad [\because \cos 90° = 0]$

Q37. If $\tan\theta = 3$, θ lies in the third quadrant, then the value of $\sin\theta$ is

(a) $\frac{1}{\sqrt{10}}$ (b) $-\frac{1}{\sqrt{10}}$ (c) $\frac{-3}{\sqrt{10}}$ (d) $\frac{3}{\sqrt{10}}$

Solution: (c)

$\because \sec\theta = \sqrt{1+\tan^2\theta} = \sqrt{1+9} = \pm\sqrt{10}$

$\Rightarrow \sec\theta = -\sqrt{10} \Rightarrow \cos\theta = -\frac{1}{\sqrt{10}}$

(since sec is negative in quadrant III)

$\therefore \sin\theta = \sqrt{1-\cos^2\theta} = \sqrt{1-\frac{1}{10}} = \pm\frac{3}{\sqrt{10}}$

$\Rightarrow \sin\theta = -\frac{3}{\sqrt{10}}$

(since sin is negative in quadrant III)

Q38. The value of $\tan 75° - \cot 75°$ is

(a) $2\sqrt{3}$ (b) $2+\sqrt{3}$ (c) $2-\sqrt{3}$ (d) 1

Solution: (a)

$$\tan 75° - \cot 75° = \frac{\sin 75°}{\cos 75°} - \frac{\cos 75°}{\sin 75°}$$

$$= \frac{\sin^2 75° - \cos^2 75°}{\sin 75° \cdot \cos 75°} = \frac{-2\cos 150°}{\sin 150°}$$

$$= \frac{-2\cos(90°+60°)}{\sin(90°+60°)} = \frac{2\sin 60°}{\cos 60°}$$

$= 2\tan 60° = 2\sqrt{3}$

Q39. Which of the following is correct?

(a) $\sin 1° > \sin 1$ (b) $\sin 1° < \sin 1$

(c) $\sin 1° = \sin 1$ (d) $\sin 1° = \frac{\pi}{180°}\sin 1$

Solution: (b)

If θ is increasing, then $\sin\theta$ is also increasing.

$\therefore \sin 1° < \sin 1 \quad [\because 1 \text{ rad} = 57°30']$

Q40. If $\tan\alpha = \frac{m}{m+1}$, $\tan\beta = \frac{1}{2m+1}$, then $\alpha+\beta$ is equal to

(a) $\frac{\pi}{2}$ (b) $\frac{\pi}{3}$ (c) $\frac{\pi}{6}$ (d) $\frac{\pi}{4}$

Solution: (d)

$$\tan(\alpha+\beta) = \frac{\tan\alpha + \tan\beta}{1-\tan\alpha\cdot\tan\beta}$$

$$\Rightarrow \tan(\alpha+\beta) = \frac{\frac{m}{m+1}+\frac{1}{2m+1}}{1-\left(\frac{m}{m+1}\right)\left(\frac{1}{2m+1}\right)}$$

$$= \frac{m(2m+1)+m+1}{(m+1)(2m+1)-m}$$

$$= \frac{2m^2+m+m+1}{2m^2+2m+m+1-m} = \frac{2m^2+2m+1}{2m^2+2m+1}$$

$$\Rightarrow \tan(\alpha+\beta) = 1 \Rightarrow \alpha+\beta = \frac{\pi}{4}$$

Q41. The minimum value of $3\cos x + 4\sin x + 8$ is

(a) 5 (b) 9 (c) 7 (d) 3

Solution: (d)

Let $y = 3\cos x + 4\sin x + 8 \Rightarrow y - 8 = 3\cos x + 4\sin x$

Since maximum and minimum value of $a\cos x + b\sin x$ is $\pm\sqrt{a^2+b^2}$

$\therefore$ Minimum value of $y - 8 = -\sqrt{9+16}$

$\Rightarrow y - 8 = -5 \Rightarrow y_{min} = -5+8 = 3$

Q42. The value of $\tan 3A - \tan 2A - \tan A$ is

(a) $\tan 3A \tan 2A \tan A$
(b) $-\tan 3A \tan 2A \tan A$
(c) $\tan A \tan 2A - \tan 2A \tan 3A - \tan 3A \tan A$
(d) None of the above

Solution: (a)

Let $3A = A + 2A \Rightarrow \tan 3A = \tan(A+2A)$

$$\Rightarrow \tan 3A = \frac{\tan A + \tan 2A}{1-\tan A\cdot\tan 2A}$$

$\Rightarrow \tan A + \tan 2A = \tan 3A - \tan 3A\cdot\tan 2A\cdot\tan A$

$\Rightarrow \tan 3A - \tan 2A - \tan A = \tan 3A\cdot\tan 2A\cdot\tan A$

Q43. The value of $\sin(45°+\theta) - \cos(45°-\theta)$ is

(a) $2\cos\theta$ (b) $2\sin\theta$ (c) 1 (d) 0

Solution: (d)

$\sin(45°+\theta)-\cos(45°-\theta)$
$= \sin 45° \cdot \cos\theta + \cos 45° \cdot \sin\theta$
$-\cos 45° \cdot \cos\theta - \sin 45° \cdot \sin\theta$
$= \frac{1}{\sqrt{2}} \cdot \cos\theta + \frac{1}{\sqrt{2}} \cdot \sin\theta - \frac{1}{\sqrt{2}} \cdot \cos\theta - \frac{1}{\sqrt{2}} \sin\theta = 0$

Q44. The value of $\cot\left(\frac{\pi}{4}+\theta\right)\cot\left(\frac{\pi}{4}-\theta\right)$ is
(a) –1 (b) 0
(c) 1 (d) not defined

Solution: (c)
$\cot\left(\frac{\pi}{4}+\theta\right)\cot\left(\frac{\pi}{4}-\theta\right)$
$= \left(\frac{\cot\frac{\pi}{4}\cot\theta - 1}{\cot\frac{\pi}{4} + \cot\theta}\right) \cdot \left(\frac{\cot\frac{\pi}{4}\cot\theta + 1}{\cot\theta - \cot\frac{\pi}{4}}\right)$
$= \left(\frac{\cot\theta - 1}{\cot\theta + 1}\right) \cdot \left(\frac{\cot\theta + 1}{\cot\theta - 1}\right) = 1$

Q45. $\cos 2\theta \cos 2\phi + \sin^2(\theta-\phi) - \sin^2(\theta+\phi)$ is equal to
(a) $\sin 2(\theta+\phi)$ (b) $\cos 2(\theta+\phi)$
(c) $\sin 2(\theta-\phi)$ (d) $\cos 2(\theta-\phi)$

Solution: (b)
$\cos 2\theta \cos 2\phi + \sin^2(\theta-\phi) - \sin^2(\theta+\phi)$
$= \cos 2\theta \cdot \cos 2\phi + \sin(\theta-\phi+\theta+\phi) \cdot \sin(\theta-\phi-\theta-\phi)$
$\left[\because \sin^2(x-y) - \sin^2(x+y) = \sin(x+y)\sin(x-y)\right]$
$= \cos 2\theta \cdot \cos 2\phi + \sin 2\theta \cdot \sin(-2\phi)$
$= \cos 2\theta \cdot \cos 2\phi - \sin 2\theta \cdot \sin 2\phi$
$= \cos(2\theta + 2\phi) = \cos 2(\theta+\phi)$

Q46. The value of
$\cos 12° + \cos 84° + \cos 150° + \cos 132°$ is
(a) 1/2 (b) 1 (c) –1/2 (d) 1/8

Solution: (c)
$\cos 12° + \cos 150° + \cos 84° + \cos 132°$
$= 2\cos\left(\frac{12°+150°}{2}\right) \cdot \cos\left(\frac{12°-150°}{2}\right) +$
$2\cos\left(\frac{84°+132°}{2}\right) \cdot \cos\left(\frac{84°-132°}{2}\right)$
$= 2\cos 84° \cos 72° + 2\cos 108° \cdot \cos 24°$
$= 2\cos 84° \cos(90°-18°) + 2\cos(90°+18°) \cdot \cos 24°$
$= 2\cos 84° \sin 18° - 2\sin 18° \cdot \cos 24°$
$= 2\sin 18°(\cos 84° - \cos 24°)$
$= 2\sin 18° \cdot 2\sin\left(\frac{84°+24°}{2}\right) \cdot \sin\left(\frac{84°-24°}{2}\right)$
$= -4\sin 18° \cdot \sin 54° \sin 30°$
$= -4\left(\frac{\sqrt{5}-1}{4}\right) \cdot \cos 36° \cdot \frac{1}{2}$
$= -(\sqrt{5}-1)\left(\frac{\sqrt{5}+1}{4}\right) \cdot \frac{1}{2} = -\left(\frac{5-1}{8}\right) = \frac{-4}{8} = \frac{-1}{2}$

Q47. If $\tan A = \frac{1}{2}, \tan B = \frac{1}{3}$, then $\tan(2A+B)$ is equal to
(a) 1 (b) 2 (c) 3 (d) 4

Solution: (c)
Since, $\tan 2A = \frac{2\tan A}{1-\tan^2 A} = \frac{2 \cdot \frac{1}{2}}{1-\frac{1}{4}} = \frac{4}{3}$
Therefore, $\tan(2A+B) = \frac{\tan 2A + \tan B}{1 - \tan 2A \cdot \tan B}$
$= \frac{\frac{4}{3}+\frac{1}{3}}{1-\frac{4}{3}\cdot\frac{1}{3}} = \frac{\frac{5}{3}}{\frac{5}{9}} = 3$

Q48. The value of $\sin\frac{\pi}{10}\sin\frac{13\pi}{10}$ is
(a) 1/2 (b) –1/2 (c) –1/4 (d) 1

Solution: (c)
$\sin\frac{\pi}{10}\sin\frac{13\pi}{10} = \sin\frac{\pi}{10}\sin\left(\pi + \frac{3\pi}{10}\right)$
$= -\sin\frac{\pi}{10}\sin\frac{3\pi}{10} = -\sin 18° \cdot \sin 54°$
$= -\sin 18° \cdot \cos 36° = -\left(\frac{\sqrt{5}-1}{4}\right)\left(\frac{\sqrt{5}+1}{4}\right)$
$= -\left(\frac{5-1}{16}\right) = -\frac{1}{4}$

Q49. The value of $\sin 50° - \sin 70° + \sin 10°$
(a) 1 (b) 0 (c) 1/2 (d) 2

Solution: (b)
$\sin 50° - \sin 70° + \sin 10°$
$= 2\cos\left(\frac{50°+70°}{2}\right) \cdot \sin\left(\frac{50°-70°}{2}\right) + \sin 10°$
$= -2\cos 60° \sin 10° + \sin 10°$
$= -2 \cdot \frac{1}{2} \sin 10° + \sin 10° = 0$

Q50. If $\sin\theta - \cos\theta = 1$, then the value of $\sin 2\theta$
(a) 1 (b) 1/2 (c) 0 (d) –1

Solution: (c)
$\because \sin\theta - \cos\theta = 1$

Squaring both sides, we get

$\sin^2\theta+\cos^2\theta-2\sin\theta\cdot\cos\theta=1 \Rightarrow 1-\sin 2\theta=1$

$\therefore \sin 2\theta=0$

Q51. If $\alpha+\beta=\frac{\pi}{4}$, then the value of $(1+\tan\alpha)(1+\tan\beta)$ is

(a) 1 (b) 2
(c) –2 (d) not defined

Solution: (b)

Since $\tan(\alpha+\beta)=\dfrac{\tan\alpha+\tan\beta}{1-\tan\alpha\cdot\tan\beta}$

$\Rightarrow 1=\dfrac{\tan\alpha+\tan\beta}{1-\tan\alpha\cdot\tan\beta}\quad (\because \alpha+\beta=\frac{\pi}{4})$

$\Rightarrow \tan\alpha+\tan\beta=1-\tan\alpha\tan\beta$

$\Rightarrow \tan\alpha+\tan\beta+\tan\alpha\tan\beta=1$

$\therefore (1+\tan\alpha)(1+\tan\beta)$

$=1+\tan\alpha+\tan\beta+\tan\alpha\tan\beta$

$=1+1=2$

Q52. If $\sin\theta=\frac{-4}{5}$, θ lies in the third quadrant, then the value of $\cos\frac{\theta}{2}$ is

(a) 1/5 (b) $-\frac{1}{\sqrt{10}}$ (c) $-\frac{1}{\sqrt{5}}$ (d) $\frac{1}{\sqrt{10}}$

Solution: (c)

$\cos\theta=\sqrt{1-\frac{16}{25}}=\sqrt{\frac{25-16}{25}}=\pm\frac{3}{5}\Rightarrow \cos\theta=\frac{-3}{5}$ [since θ lies in the third quadrant]

$\Rightarrow 2\cos^2\frac{\theta}{2}-1=\frac{-3}{5}\Rightarrow 2\cos^2\frac{\theta}{2}=1-\frac{3}{5}$

$\Rightarrow 2\cos^2\frac{\theta}{2}=\frac{2}{5}\Rightarrow \cos\frac{\theta}{2}=\pm\frac{1}{\sqrt{5}}$

$\Rightarrow \cos\frac{\theta}{2}=-\frac{1}{\sqrt{5}}$

[since θ lies in the third quadrant]

Day 53

NCERT Exemplar Problems: Solutions (Q53–Q76)

Q53. The number of solutions of equation $\tan x+\sec x=2\cos x$, x lying in the interval $[0,2\pi]$ is

(a) 0 (b) 1 (c) 2 (d) 3

Solution: (c)

$\tan x+\sec x=2\cos x$

$\Rightarrow \dfrac{\sin x}{\cos x}+\dfrac{1}{\cos x}=2\cos x\Rightarrow 1+\sin x=2\cos^2 x$

$\Rightarrow 1+\sin x=2(1-\sin^2 x)\Rightarrow 1+\sin x=2-2\sin^2 x$

$\Rightarrow 2\sin^2 x+\sin x-1=0$

$\Rightarrow (2\sin x-1)(\sin x+1)=0$

$\Rightarrow \sin x=-1$ or $\sin x=\frac{1}{2}$

$\therefore x=\frac{3\pi}{2}, x=\frac{\pi}{6}$

Thus, the number of possible solutions are 2.

Q54. The value of $\sin\frac{\pi}{18}+\sin\frac{\pi}{9}+\sin\frac{2\pi}{9}+\sin\frac{5\pi}{18}$ is

(a) $\sin\frac{7\pi}{18}+\sin\frac{4\pi}{9}$ (b) 1

(c) $\cos\frac{\pi}{6}+\cos\frac{3\pi}{7}$ (d) $\cos\frac{\pi}{9}+\cos\frac{\pi}{9}$

Solution: (a)

$\sin\frac{\pi}{18}+\sin\frac{\pi}{9}+\sin\frac{2\pi}{9}+\sin\frac{5\pi}{18}$

$=\sin 10^\circ+\sin 20^\circ+\sin 40^\circ+\sin 50^\circ$

$=\sin 50^\circ+\sin 10^\circ+\sin 40^\circ+\sin 20^\circ$

$=\sin 130^\circ+\sin 10^\circ+\sin 140^\circ+\sin 20^\circ$

$=2\sin 70^\circ\cos 60^\circ+2\sin 80^\circ\cdot\cos 60^\circ$

$\left[\because \sin x+\sin y=2\sin\frac{x+y}{2}\cdot\cos\frac{x-y}{2}\right]$

$=2\cdot\frac{1}{2}\sin 70^\circ+2\cdot\frac{1}{2}\sin 80^\circ\quad\left[\because \cos 60^\circ=\frac{1}{2}\right]$

$=\sin 70^\circ+\sin 80^\circ=\sin\frac{7\pi}{18}+\sin\frac{4\pi}{9}$

Q55. If A lies in the second quadrant and $3\tan A+4=0$, then the value of $2\cot A-5\cos A+\sin A$ is

(a) $\frac{-53}{10}$ (b) $\frac{23}{10}$ (c) $\frac{37}{10}$ (d) $\frac{7}{10}$

Solution: (b)

$3\tan A+4=0\Rightarrow \tan A=\frac{-4}{3}\Rightarrow \cot A=\frac{-3}{4}$

$\Rightarrow \sec A=\sqrt{1+\frac{16}{9}}=\sqrt{\frac{25}{9}}=\pm\frac{5}{3}$

$\Rightarrow \sec A=\frac{-5}{3}\Rightarrow \cos A=\frac{-3}{5}$

(since A lies in the second quadrant)

$\sin A=\sqrt{1-\frac{9}{25}}=\frac{\sqrt{25-9}}{25}=\pm\frac{4}{5}\Rightarrow \sin A=\frac{4}{5}$

(since A lies in the second quadrant)

$\therefore 2\cot A - 5\cos A + \sin A$

$$= 2\left(\frac{-3}{4}\right) - 5\left(\frac{-3}{5}\right) + \frac{4}{5} = \frac{-6}{4} + 3 + \frac{4}{5} = \frac{23}{10}$$

Q56. The value of $\cos^2 48° - \sin^2 12°$ is

(a) $\frac{\sqrt{5}+1}{8}$ (b) $\frac{\sqrt{5}-1}{8}$

(c) $\frac{\sqrt{5}+1}{5}$ (d) $\frac{\sqrt{5}+1}{2\sqrt{2}}$

Solution: (a)

$\cos^2 48° - \sin^2 12°$

$= \cos(48° + 12°) - \cos(48° - 12°)$

$$= \cos 60° \cdot \cos 36° = \frac{1}{2} \cdot \frac{\sqrt{5}+1}{4} = \frac{\sqrt{5}+1}{8}$$

Q57. If $\tan\alpha = \frac{1}{7}, \tan\beta = \frac{1}{3}$, then $\cos 2\alpha$ is equal to

(a) $\sin 2\beta$ (b) $\sin 4\beta$

(c) $\sin 3\beta$ (d) $\cos 2\beta$

Solution: (b)

$$\cos 2\alpha = \frac{1-\tan^2\alpha}{1+\tan^2\alpha} = \frac{1-\frac{1}{49}}{1+\frac{1}{49}} = \frac{48}{50} = \frac{24}{25}$$

$$\tan 2\beta = \frac{2\tan\beta}{1-\tan^2\beta} = \frac{2\times\frac{1}{3}}{1-\frac{1}{9}} = \frac{2/3}{8/9} = \frac{3}{4}$$

$$\sin 4\beta = \frac{2\tan 2\beta}{1+\tan^2 2\beta} = \frac{2\times\frac{3}{4}}{1+\frac{9}{16}} = \frac{\frac{6}{4}}{\frac{25}{16}} = \frac{24}{25}$$

$\therefore \cos 2\alpha = \sin 4\beta$

Q58. If $\tan\theta = \frac{a}{b}$, then $b\cos 2\theta + a\sin 2\theta$ is equal to

(a) a (b) b

(c) a/b (d) none of these

Solution: (b)

$$b\cos 2\theta + a\sin 2\theta = b\left(\frac{1-\tan^2\theta}{1+\tan^2\theta}\right) + a\left(\frac{2\tan\theta}{1+\tan^2\theta}\right)$$

$$= b\left(\frac{1-\frac{a^2}{b^2}}{1+\frac{a^2}{b^2}}\right) + a\left(\frac{\frac{2a}{b}}{1+\frac{a^2}{b^2}}\right) = b\left(\frac{b^2-a^2}{b^2+a^2}\right) + \frac{2a^2b}{a^2+b^2}$$

$$= \frac{b}{a^2+b^2}[b^2 - a^2 + 2a^2] = \frac{(a^2+b^2)b}{(a^2+b^2)} = b$$

Q59. If for real values of $x, \cos\theta = x + \frac{1}{x}$, then

(a) θ is an acute angle

(b) θ is a right angle

(c) θ is an obtuse angle

(d) No value of θ is possible

Solution: (d)

$$\cos\theta = x + \frac{1}{x} \Rightarrow \cos\theta = \frac{x^2+1}{x}$$

$\Rightarrow x^2 - x\cos\theta + 1 = 0$

For a real value of x,

$(-\cos\theta)^2 - 4\times 1 \geq 0 \Rightarrow \cos^2\theta \geq 4$

But, $\cos\theta \geq 2$ is not possible

Q60. The value of $\frac{\sin 50°}{\sin 130°}$ is______

Solution:

$$\frac{\sin 50°}{\sin 130°} = \frac{\sin(180° - 130°)}{\sin 130°} = \frac{\sin 130°}{\sin 130°} = 1$$

Q61. If $k = \sin\left(\frac{\pi}{18}\right)\sin\left(\frac{5\pi}{18}\right)\sin\left(\frac{7\pi}{18}\right)$, then the numerical value of k is______

Solution:

$$k = \sin\left(\frac{\pi}{18}\right)\sin\left(\frac{5\pi}{18}\right)\sin\left(\frac{7\pi}{18}\right)$$

$= \sin 10° \sin 50° \sin 70° = \sin 10° \cos 40° \cdot \cos 20°$

$$= \frac{1}{2}\sin 10°[2\cos 40° \cdot \cos 20°]$$

$$= \frac{1}{2}\sin 10°[\cos 60° + \cos 20°]$$

$[\because 2\cos x \cdot \cos y = \cos(x+y) + \cos(x-y)]$

$$= \frac{1}{2}\sin 10° \cdot \frac{1}{2} + \frac{1}{2}\sin 10° \cos 20°$$

$$= \frac{1}{4}\sin 10° + \frac{1}{4}[\sin 30° - \sin 10°] = \frac{1}{8}$$

Q62. If $\tan A = \frac{1-\cos B}{\sin B}$, then $\tan 2A =$ ______

Solution:

$$\tan A = \frac{1-\cos B}{\sin B} = \frac{1-1+2\sin^2\frac{B}{2}}{2\sin\frac{B}{2}\cdot\cos\frac{B}{2}} = \tan\frac{B}{2}$$

$$\therefore \tan 2A = \frac{2\tan A}{1-\tan^2 A}$$

$$= \frac{2\cdot\tan(B/2)}{1-\tan^2(B/2)} \Rightarrow \tan 2A = \tan B$$

Q63. If $\sin x + \cos x = a$, then
(i) $\sin^6 x + \cos^6 x =$ ______
(ii) $|\sin x - \cos x| =$ ______

Solution:

$\sin x + \cos x = a$

Squaring both sides, we get

$(\sin x + \cos x)^2 = a^2$

$\Rightarrow \sin^2 x + \cos^2 x + 2\sin x \cos x = a^2$

$\Rightarrow \sin x \cdot \cos x = \frac{1}{2}(a^2 - 1)$

(i) $\sin^6 x + \cos^6 x = (\sin^2 x)^3 + (\cos^2 x)^3$

$= (\sin^2 x + \cos^2 x)(\sin^4 x - \sin^2 x \cos^2 x + \cos^4 x)$

$= \sin^4 x + \cos^4 x - \frac{1}{4}(a^2 - 1)^2$

$= (\sin^2 x + \cos^2 x)^2 - 2\sin^2 x \cos^2 x - \frac{1}{4}(a^2 - 1)^2$

$= 1 - 2 \cdot \frac{1}{4}(a^2 - 1)^2 - \frac{1}{4}(a^2 - 1)^2 = \frac{1}{4}[4 - 3(a^2 - 1)^2]$

(ii) $|\sin x - \cos x| = \sqrt{(\sin x - \cos x)^2}$

$= \sqrt{\sin^2 x + \cos^2 x - 2\sin x \cos x}$

$= \sqrt{1 - 2\frac{1}{2}(a^2 - 1)} = \sqrt{1 - a^2 + 1} = \sqrt{2 - a^2}$

Q64. In right-angled ΔABC with $\angle C = 90°$ the equation whose roots are $\tan A$, $\tan B$ is ______

Solution:

In right-angled ΔABC, $\angle C = 90°$

$\therefore \tan(A + B) = \frac{\tan A + \tan B}{1 - \tan A \tan B} = \tan(90°)$

$\Rightarrow \frac{1}{0} = \frac{\tan A + \tan B}{1 - \tan A \tan B} \Rightarrow \tan A \tan B = 1$

Also, $\tan A + \tan B = \frac{\sin A}{\cos A} + \frac{\sin B}{\cos B}$

$= \frac{\sin A}{\cos A} + \frac{\sin(90° - A)}{\cos(90° - A)} [\because \angle C = 90°, \angle B = 90° - A]$

$= \frac{\sin A}{\cos A} + \frac{\cos A}{\sin A} = \frac{\sin^2 A + \cos^2 A}{\sin A \cdot \cos A}$

$= \frac{1}{\sin A \cdot \cos A} = \frac{2}{2 \cdot \sin A \cdot \cos A} = \frac{2}{\sin 2A}$

Thus, the required equation is $x^2 - \left(\frac{2}{\sin A}\right)x + 1 = 0$

Q65. $3(\sin x - \cos x)^4 + 6(\sin x + \cos x)^2 + 4(\sin^6 x + \cos^6 x) =$ ______

Solution:

$3(\sin x - \cos x)^4 + 6(\sin x + \cos x)^2 + 4(\sin^6 x + \cos^6 x)$

$= 3[\sin^2 x + \cos^2 x - 2\sin x \cos x]^2$

$+6[\sin^2 x + \cos^2 x + 2 \cdot \sin x \cos x]$

$+4[(\sin^2 x)^3 + (\cos^2 x)^3]$

$= 3(1 - \sin 2x)^2 + 6(1 + \sin 2x)$

$+4(\sin^2 + \cos^2 x)(\sin^4 x - \sin^2 x \cos^2 x + \cos^4 x)$

$= 3(1 + \sin^2 2x - 2\sin 2x) + 6 + 6\sin 2x$

$+4[(\sin^2 x + \cos^2 x)^2 - 3\sin^2 x \cos^2 x]$

$= 3 + 3\sin^2 2x - 6\sin 2x + 6 + 6\sin 2x + 4 - 3\sin^2 2x$

$= 13$

Q66. Given $x > 0$, the value of $f(x) = -3\cos\sqrt{3 + x + x^2}$ lies in the interval ______

Solution:

$f(x) = -3\cos\sqrt{3 + x + x^2}$

Let $\sqrt{3 + x + x^2} = \theta$,

And since, $-1 \le \cos\theta \le 1 \Rightarrow -3 \le 3\cos\theta \le 3$

$\Rightarrow -3 \le -3\cos\theta \le 3$

Thus, the value of $f(x)$ lies in $[-3, 3]$.

Q67. The maximum distance of a point on the graph of the function $y = \sqrt{3}\sin x + \cos x$ from the x-axis is______

Solution:

$y = \sqrt{3}\sin x + \cos x$

$y = 2\left[\frac{\sqrt{3}}{2}\sin x + \frac{1}{2}\cos x\right] = 2[\sin x \cdot \cos\frac{\pi}{6} + \cos x \sin\frac{\pi}{6}]$

$= 2\sin(x + \pi/6)$

Thus, the maximum distance of $y = \sqrt{3}\sin x + \cos x$ from the x-axis is 2 units.

In each of the questions 68 to 75, state whether the statement is true or false. Also give justification.

Q68. If $\tan A = \frac{1 - \cos B}{\sin B}$ then $\tan 2A = \tan B$

Solution: True

$$\tan A = \frac{1 - \cos B}{\sin B} = \frac{1 - 1 + 2\sin^2\frac{B}{2}}{2\sin\frac{B}{2} \cdot \cos\frac{B}{2}} = \tan\frac{B}{2}$$

$$\text{Now, } \tan 2A = \frac{2\tan A}{1 - \tan^2 A} = \frac{2 \cdot \tan\frac{B}{2}}{1 - \tan^2 B/2} = \tan B$$

Q69. The equality $\sin A+\sin 2A+\sin 3A=3$ holds for some real value of A.

Solution: False

Since the maximum value of $\sin\theta=1$, $\sin A+\sin 2A+\sin 3A=3$ is possible if and only if $\sin A=\sin 2A=\sin 3A=1$, which is not possible for any real value of A.

Q70. $\sin 10°$ is greater than $\cos 10°$.

Solution: False

$\sin 10°=\sin(90°-80°)=\cos 80°$

$\because \cos 80°<\cos 10° \therefore \sin 10°<\cos 10°$

Q71. $\cos\frac{2\pi}{15}\cos\frac{4\pi}{15}\cos\frac{8\pi}{15}\cos\frac{16\pi}{15}=\frac{1}{16}$

Solution: True

$$\text{LHS}=\cos\frac{2\pi}{15}\cos\frac{4\pi}{15}\cos\frac{8\pi}{15}\cos\frac{16\pi}{15}$$

$$=\cos 24°\cos 48°\cos 96°\cos 192°$$

$$=\frac{1}{16\sin 24°}\begin{bmatrix}(2\sin 24°\cos 24°)(2\cos 48°)\\(2\cos 96°)(2\cos 192°)\end{bmatrix}$$

$$=\frac{1}{16\sin 24°}[2\sin 48°\cos 48°(2\cos 96°)(2\cos 192°)]$$

$$=\frac{1}{16\sin 24°}[(2\sin 96°\cos 96°)(2\cos 192°)]$$

$$=\frac{1}{16\sin 24°}[2\sin 192°\cos 192°)$$

$$=\frac{1}{16\sin 24°}\sin 384°$$

$$=\frac{\sin(360°+24°)}{16\sin 24°}=\frac{1}{16}=\text{RHS}$$

Q72. One value of θ which satisfies the equation $\sin^4\theta-2\sin^2\theta-1$ lies between 0 and 2π.

Solution: False

$$\sin^4\theta-2\sin^2\theta-1=0$$

$$\Rightarrow \sin^2\theta=\frac{2\pm\sqrt{4+4}}{2}$$

$$\Rightarrow \sin^2\theta=\frac{2\pm 2\sqrt{2}}{2}=1\pm\sqrt{2}$$

$$\Rightarrow \sin^2\theta=(1+\sqrt{2}) \text{ or } (1-\sqrt{2})$$

$$\because -1\le\sin\theta\le 1 \therefore 0\le\sin^4\theta\le 1$$

Thus, $\Rightarrow \sin^2\theta=1+\sqrt{2}$ or $(1-\sqrt{2})$ is not possible.

Q73. If $\operatorname{cosec} x=1+\cot x$, then $x=2n\pi, 2n\pi+\frac{\pi}{2}$

Solution: True

$$\operatorname{cosec} x=1+\cot x$$

$$\Rightarrow \frac{1}{\sin x}=1+\frac{\cos x}{\sin x}\Rightarrow\frac{1}{\sin x}=\frac{\sin x+\cos x}{\sin x}$$

$$\Rightarrow \sin x+\cos x=1$$

$$\Rightarrow \frac{1}{\sqrt{2}}\cdot\sin x+\frac{1}{\sqrt{2}}\cdot\cos x=\frac{1}{\sqrt{2}}$$

$$\Rightarrow \sin\frac{\pi}{4}\sin x+\cos x\cos\frac{\pi}{4}=\frac{1}{\sqrt{2}}$$

$$\Rightarrow \cos\left(x-\frac{\pi}{4}\right)=\cos\frac{\pi}{4}$$

$$\therefore x-\frac{\pi}{4}=2n\pi\pm\frac{\pi}{4}$$

For positive sign, $x=2n\pi+\frac{\pi}{4}+\frac{\pi}{4}=2n\pi+\frac{\pi}{2}$

For negative sign, $x=2n\pi-\frac{\pi}{4}+\frac{\pi}{4}=2n\pi$

Q74. If $\tan\theta+\tan 2\theta+\sqrt{3}\tan\theta\tan 2\theta=\sqrt{3}$, then $\theta=\frac{n\pi}{3}+\frac{\pi}{9}$.

Solution: True

$$\tan\theta+\tan 2\theta+\sqrt{3}\tan\theta\tan 2\theta=\sqrt{3}$$

$$\Rightarrow \tan\theta+\tan 2\theta=\sqrt{3}-\sqrt{3}\tan\theta\tan 2\theta$$

$$\Rightarrow \tan\theta+\tan 2\theta=\sqrt{3}(1-\tan\theta\tan 2\theta)$$

$$\Rightarrow \frac{\tan\theta+\tan 2\theta}{1-\tan\theta\tan 2\theta}=\sqrt{3}$$

$$\Rightarrow \tan(\theta+2\theta)=\tan\frac{\pi}{3}\Rightarrow\tan 3\theta=\tan\frac{\pi}{3}$$

$$\therefore 3\theta=n\pi+\frac{\pi}{3}\Rightarrow\theta=\frac{n\pi}{3}+\frac{\pi}{9}$$

Q75. If $\tan(\pi\cos\theta)=\cot(\pi\sin\theta)$, then $\cos\left(\theta-\frac{\pi}{4}\right)=\pm\frac{1}{2\sqrt{2}}$.

Solution: True

$$\tan(\pi\cos\theta)=\cot(\pi\sin\theta)$$

$$\Rightarrow \tan(\pi\cos\theta)=\tan\left[\frac{\pi}{2}-(\pi\sin\theta)\right]$$

$$\Rightarrow \pi\cos=\frac{\pi}{2}-\pi\sin\theta\Rightarrow\pi(\sin\theta+\cos\theta)=\frac{\pi}{2}$$

$$\Rightarrow \sin\theta+\cos\theta=\frac{1}{2}\Rightarrow\frac{1}{\sqrt{2}}\cdot\sin\theta+\frac{1}{\sqrt{2}}\cdot\cos\theta=\frac{1}{2\sqrt{2}}$$

$$\Rightarrow \sin\theta\cdot\sin\frac{\pi}{4}+\cos\theta\cdot\cos\frac{\pi}{4}=\pm\frac{1}{2\sqrt{2}}$$

$$\therefore \cos\left(\theta-\frac{\pi}{4}\right)=\pm\frac{1}{2\sqrt{2}}$$

Q76. In the following, match each item given under column I to its correct answer given in column II.

Column I

(i) $\sin(x+y)\sin(x-y)$

(ii) $\cos(x+y)\cos(x-y)$

(iii) $\cot\left(\frac{\pi}{4}+\theta\right)$

(iv) $\tan\left(\frac{\pi}{4}+\theta\right)$

Column II

(a) $\cos^2 x-\sin^2 y$

(b) $1-\tan\theta/1+\tan\theta$

(c) $1+\tan\theta/1-\tan\theta$

(d) $\sin^2 x-\sin^2 y$

Solution:

(i) $\sin(x+y)\sin(x-y)=\sin^2 x-\sin^2 y$

(ii) $\cos(x+y)\cos(x-y)=\cos^2 x-\sin^2 y$

(iii) $\cot\left(\frac{\pi}{4}+\theta\right)=\dfrac{\cot\frac{\pi}{4}\cot\theta-1}{\cot\frac{\pi}{4}+\cot\theta}$

$=\dfrac{-1+\cot\theta}{1+\cot\theta}=\dfrac{1-\tan\theta}{1+\tan\theta}$

(iv) $\tan\left(\frac{\pi}{4}+\theta\right)=\dfrac{\tan\frac{\pi}{4}+\tan\theta}{1-\tan\frac{\pi}{4}\tan\theta}=\dfrac{1+\tan\theta}{1-\tan\theta}$

Day 54

Very Short Answer (VSA): 1-Mark and Short Answer (SA): 2-Mark Questions

Q1. If $\sin\theta+\operatorname{cosec}\theta=3$, then $\sin^2\theta+\operatorname{cosec}^2\theta$ is equal to

(a) 1 (b) 4
(c) 2 (d) None of these

Solution: (a)

$\sin\theta+\operatorname{cosec}\theta=3$

$\Rightarrow \sin^2\theta+\operatorname{cosec}^2\theta+2\sin\theta\cdot\operatorname{cosec}\theta=3$

$\Rightarrow \sin^2\theta+\operatorname{cosec}^2\theta+2\cdot 1=3$

$\Rightarrow \sin^2\theta+\operatorname{cosec}^2\theta=3-2$

$\Rightarrow \sin^2\theta+\operatorname{cosec}^2\theta=1$

Q2. If $\tan\theta=\frac{1}{2}, \tan\phi=\frac{1}{3}$, then the value of $\tan(\theta-\phi)$ is

(a) $\frac{1}{7}$ (b) 1 (c) 0 (d) $\frac{6}{7}$

Solution: (d)

$\tan(\theta-\phi)=\dfrac{\tan\theta-\tan\phi}{1+\tan\theta\cdot\tan\phi}$

$\Rightarrow \tan(\theta-\phi)=\dfrac{\frac{1}{2}-\frac{1}{3}}{1+\frac{1}{2}\cdot\frac{1}{3}}=\dfrac{\frac{3-2}{6}}{\frac{6+1}{6}}=\frac{1}{7}$

Q3. Which of the following is correct?

(a) $\sin\theta=\frac{1}{2}$ (b) $\tan\theta=\frac{1}{2}$

(c) $\operatorname{cosec}\theta=\frac{1}{2}$ (d) $\tan\theta=2$

Solution: (c)

The range of $\operatorname{cosec}\theta$ is $R-(-1,1)$

Therefore, $\operatorname{cosec}\theta$ cannot be equal to 1/2.

Q4. The value of $\sin 1°+\sin 2°+\sin 3°+...+\sin 358°+\sin 359°$ is

(a) $\frac{1}{\sqrt{2}}$ (b) 0 (c) 1 (d) –1

Solution: (b)

$\sin 1°+\sin 2°+\sin 3°+...+\sin 357°+\sin 358°+\sin 359°$

$=\sin 1°+\sin 2°+...+\sin 179°+\sin 180°$

$+\sin(180°+1°)+\sin(180°+2°)+...+\sin(180°+179°)$

$=\sin 1°+\sin 2°+...+\sin 179°+\sin 180°$

$-\sin 1°-\sin 2°-...-\sin 179°=0 \quad (\because \sin 180°=0)$

Q5. The value of $\dfrac{1-\tan^2 75°}{1+\tan^2 75°}$ is

(a) 1 (b) $\sqrt{3}$ (c) $\frac{\sqrt{3}}{2}$ (d) 2

Solution: (c)

Let $\theta=75°$

$\cos 2\theta=\dfrac{1-\tan^2\theta}{1+\tan^2\theta}$

$\therefore \cos 150°=\dfrac{1-\tan^2 75°}{1+\tan^2 75°}$

$\Rightarrow \dfrac{1-\tan^2 75°}{1+\tan^2 75°}=\cos(90°+60°)=\sin 60°=\frac{\sqrt{3}}{2}$

Q6. If $\tan\theta = -2$, θ lies in the second quadrant, then the value of $\sin\theta$ is

(a) $\frac{1}{\sqrt{10}}$ (b) $-\frac{1}{\sqrt{10}}$ (c) $\frac{-2}{\sqrt{5}}$ (d) $\frac{2}{\sqrt{5}}$

Solution: (d)

$\because \sec\theta = \sqrt{1+\tan^2\theta} = \sqrt{1+(-2)^2} = \pm\sqrt{5}$

$\Rightarrow \sec\theta = -\sqrt{5} \Rightarrow \cos\theta = -\frac{1}{\sqrt{5}}$

(since sec is negative in the second quadrant)

$\therefore \sin\theta = \sqrt{1-\cos^2\theta} = \sqrt{1-\frac{1}{5}} = \pm\frac{2}{\sqrt{5}}$

$\Rightarrow \sin\theta = +\frac{2}{\sqrt{5}}$

(since sin is positive in the second quadrant)

Q7. The value of $\tan 15° + \cot 15°$ is

(a) 4 (b) 3 (c) 2 (d) 1

Solution: (a)

$\tan 15° + \cot 15° = \frac{\sin 15°}{\cos 15°} + \frac{\cos 15°}{\sin 15°}$

$= \frac{\sin^2 15° + \cos^2 15°}{\sin 15° \cdot \cos 15°} = \frac{2}{2\sin 15° \cdot \cos 15°}$

$= \frac{2}{\sin 30°} = 4$

Q8. If $\tan\alpha = \frac{m+1}{m}$, $\tan\beta = \frac{m-1}{m}$, $\tan(\alpha+\beta) = 1$, then m is equal to

(a) $\pm\sqrt{2}$ (b) ± 1 (c) $\pm\frac{1}{\sqrt{2}}$ (d) 0

Solution: (d)

$\tan(\alpha+\beta) = \frac{\tan\alpha + \tan\beta}{1-\tan\alpha\cdot\tan\beta}$

$\Rightarrow 1 = \frac{\frac{m+1}{m} + \frac{m-1}{m}}{1-\left(\frac{m+1}{m}\right)\left(\frac{m-1}{m}\right)}$

$\Rightarrow 1-\left(\frac{m+1}{m}\right)\left(\frac{m-1}{m}\right) = \frac{m+1}{m} + \frac{m-1}{m}$

$\Rightarrow 1-\left(\frac{m^2-1}{m^2}\right) = \frac{2m}{m}$

$\Rightarrow m^2 - m^2 + 1 = 2m^2 \Rightarrow m = \pm 1/\sqrt{2}$

Q9. If $\tan\theta = 2$, θ lies in first quadrant, then $\cos 2\theta + \sin 2\theta - \tan 2\theta$ is equal to

(a) $\frac{13}{15}$ (b) $\frac{23}{15}$ (c) $\frac{3}{5}b$ (d) $\frac{7}{5}$

Solution: (b)

$\cos 2\theta + \sin 2\theta - \tan 2\theta$

$= \left(\frac{1-\tan^2\theta}{1+\tan^2\theta}\right) + \left(\frac{2\tan\theta}{1+\tan^2\theta}\right) - \left(\frac{2\tan\theta}{1-\tan^2\theta}\right)$

$= \left(\frac{1-4}{1+4}\right) + \left(\frac{4}{1+4}\right) - \left(\frac{4}{1-4}\right)$

$= \left(\frac{-3}{5}\right) + \left(\frac{4}{5}\right) - \left(\frac{4}{-3}\right) = \frac{1}{5} + \frac{4}{3} = \frac{23}{15}$

Q10. The value of $\tan 3A - \tan 2A - \tan A$ is

(a) $\tan 3A \tan 2A \tan A$

(b) $-\tan 3A \tan 2A \tan A$

(c) $\tan A \tan 2A - \tan 2A \tan 3A - \tan 3A \tan A$

(d) None of the above

Solution: (a)

Let $3A = A + 2A \Rightarrow \tan 3A = \tan(A+2A)$

$\Rightarrow \tan 3A = \frac{\tan A + \tan 2A}{1-\tan A\cdot\tan 2A}$

$\Rightarrow \tan A + \tan 2A = \tan 3A - \tan 3A\cdot\tan 2A\cdot\tan A$

$\Rightarrow \tan 3A - \tan 2A - \tan A = \tan 3A\cdot\tan 2A\cdot\tan A$

Q11. The value of $\sin(45°+\theta) - \cos(45°-\theta)$ is

(a) $2\cos\theta$ (b) $2\sin\theta$ (c) 1 (d) 0

Solution: (d)

$\sin(45°+\theta) - \cos(45°-\theta)$

$= \sin 45°\cdot\cos\theta + \cos 45°\cdot\sin\theta$

$-\cos 45°\cdot\cos\theta - \sin 45°\cdot\sin\theta$

$= \frac{1}{\sqrt{2}}\cdot\cos\theta + \frac{1}{\sqrt{2}}\cdot\sin\theta - \frac{1}{\sqrt{2}}\cdot\cos\theta - \frac{1}{\sqrt{2}}\sin\theta = 0$

Q12. The value of $\cot\left(\frac{\pi}{4}+\theta\right)\cot\left(\frac{\pi}{4}-\theta\right)$ is

(a) –1 (b) 0

(c) 1 (d) Not defined

Solution: (c)

$\cot\left(\frac{\pi}{4}+\theta\right)\cot\left(\frac{\pi}{4}-\theta\right)$

$= \left(\frac{\cot\frac{\pi}{4}\cot\theta - 1}{\cot\frac{\pi}{4} + \cot\theta}\right)\cdot\left(\frac{\cot\frac{\pi}{4}\cot\theta + 1}{\cot\theta - \cot\frac{\pi}{4}}\right)$

$= \left(\frac{\cot\theta - 1}{\cot\theta + 1}\right)\cdot\left(\frac{\cot\theta + 1}{\cot\theta - 1}\right) = 1$

Q13. The value of $\sin 15° + \cos 15°$ is
(a) 1/2 (b) 1 (c) –1/2 (d) 1/8

Solution: (c)
$\sin 15° + \cos 15° = \cos 75° + \cos 15°$
$= 2\cos\left(\frac{75°+15°}{2}\right)\cdot\cos\left(\frac{75°-15°}{2}\right)$
$= 2\cos 45° \cos 30° = 2\times\frac{1}{\sqrt{2}}\times\frac{\sqrt{3}}{2} = \sqrt{\frac{3}{2}}$

Q14. If $\tan A = \frac{1}{2}, \tan B = \frac{1}{3}$, then $\tan(A+2B)$ is equal to
(a) 1 (b) 2 (c) 3 (d) 4

Solution: (b)
$\tan 2B = \frac{2\tan B}{1-\tan^2 B} = \frac{2\cdot\frac{1}{3}}{1-\frac{1}{9}} = \frac{3}{4}$

Therefore, $\tan(A+2B) = \frac{\tan A + \tan 2B}{1-\tan A\cdot\tan 2B}$
$= \frac{\frac{1}{2}+\frac{3}{4}}{1-\frac{1}{2}\cdot\frac{3}{4}} = \frac{\frac{5}{4}}{\frac{5}{8}} = 2$

Q15. The value of $\sin\frac{\pi}{12}\sin\frac{17\pi}{12}$ is
(a) 1/2 (b) –1/2 (c) –1/4 (d) 1

Solution: (c)
$\sin\frac{\pi}{12}\sin\frac{17\pi}{12} = \sin\frac{\pi}{12}\sin\left(\pi+\frac{5\pi}{12}\right)$
$= -\sin\frac{\pi}{12}\sin\frac{5\pi}{12} = -\sin 15°\cdot\sin 75°$
$= -\sin 15°\cdot\cos 15° = -\frac{1}{2}\sin 30° = -\frac{1}{4}$

Q16. The value of $\sin 15° + \sin 45° - \sin 75°$
(a) 1 (b) 0 (c) 1/2 (d) 2

Solution: (b)
$\sin 15° + \sin 45° - \sin 75°$
$= \sin 45° - \sin 75° + \sin 15°$
$= 2\cos\left(\frac{75°+45°}{2}\right)\cdot\sin\left(\frac{45°-75°}{2}\right) + \sin 15°$
$= -2\cos 60° \sin 15° + \sin 15°$
$= -2\cdot\frac{1}{2}\sin 15° + \sin 15° = 0$

Q17. If for real values of $x, 2\sin\theta = x + \frac{1}{x}$, then
(a) $\theta = (2n-1)\frac{\pi}{2},\ n\in Z$
(b) $\theta = n\pi,\ n\in Z$
(c) Possible for all real values of θ
(d) No value of θ is possible

Solution: (a)
$2\sin\theta = x + \frac{1}{x} \Rightarrow 2\sin\theta = \frac{x^2+1}{x}$
$\Rightarrow x^2 - 2x\sin\theta + 1 = 0$
For real values of x,
$(-2\sin\theta)^2 - 4\times 1 \ge 0 \Rightarrow \sin^2\theta \ge 1$
But, the maximum value of $\sin\theta = 1$, and therefore
$\sin^2\theta \ge 1 \Rightarrow \sin\theta = \pm 1 \Rightarrow \theta = (2n-1)\frac{\pi}{2}$.

Q18. If $\sin\theta = \frac{-4}{5}, \theta$ lies in the fourth quadrant, then the value of $\sin\frac{\theta}{2}$ is
(a) $\frac{1}{\sqrt{5}}$ (b) $-\frac{1}{\sqrt{10}}$ (c) $-\frac{1}{\sqrt{5}}$ (d) $\frac{1}{\sqrt{10}}$

Solution: (a)
$\cos\theta = \sqrt{1-\frac{16}{25}} = \sqrt{\frac{25-16}{25}} = \pm\frac{3}{5} \Rightarrow \cos\theta = \frac{3}{5}$
(since θ lies in the fourth quadrant)
$\therefore \cos\theta = 1 - 2\sin^2\frac{\theta}{2} \Rightarrow \frac{3}{5} \Rightarrow 1 - 2\sin^2\frac{\theta}{2}$
$\Rightarrow 2\sin^2\frac{\theta}{2} = \frac{2}{5} \Rightarrow \sin\frac{\theta}{2} = \pm\frac{1}{\sqrt{5}}$
$\Rightarrow \sin\frac{\theta}{2} = \frac{1}{\sqrt{5}}$ (since θ lies in the fourth quadrant, $\theta/2$ lies in the second quadrant)

Q19. The number of solutions of equation $\tan x + \sec x = 2\cos x$, x lying in the interval $[0, 2\pi]$ is
(a) 0 (b) 1 (c) 2 (d) 3

Solution: (c)
$\sin x - \text{cosec}\, x = \cot^2 x$
$\Rightarrow \sin x - \text{cosec}\, x = \text{cosec}^2 x - 1$
$\Rightarrow \sin x + 1 = \frac{1}{\sin x} + \frac{1}{\sin^2 x}$
$\Rightarrow \sin x + 1 = \frac{\sin^2 x + 1}{\sin^2 x}$
$\Rightarrow \sin^3 x + \sin^2 x = \sin x + 1$
$\Rightarrow \sin^3 x + \sin^2 x - \sin x - 1 = 0$
$\Rightarrow (\sin x + 1)(\sin^2 x - 1) = 0$
$\Rightarrow \sin x = -1$ or $\sin^2 x = 1$
$\Rightarrow \sin x = \pm 1 \Rightarrow x = \frac{\pi}{2}, \frac{3\pi}{2}$

Thus, the number of possible solutions are 2.

Q20. If $\tan\theta = \frac{1}{3}$, θ lies in the first quadrant, then $\cos 4\theta$ is equal to

(a) $\frac{24}{25}$ (b) $\frac{3}{4}$ (c) $\frac{9}{16}$ (d) $\frac{9}{25}$

Solution: (a)

$$\tan 2\theta = \frac{2\tan\theta}{1-\tan^2\theta} = \frac{2\times\frac{1}{3}}{1-\frac{1}{9}} = \frac{2/3}{8/9} = \frac{3}{4}$$

$$\sin 4\theta = \frac{2\tan 2\theta}{1+\tan^2 2\theta} = \frac{2\times\frac{3}{4}}{1+\frac{9}{16}} = \frac{\frac{6}{4}}{\frac{25}{16}} = \frac{24}{25}$$

Q21. The value of $\frac{\sin 130°}{\sin 220°}$ is______

Solution:

$$\frac{\sin 130°}{\sin 220°} = \frac{\sin(180°-50°)}{\sin(180°+50°)} = \frac{\sin 50°}{-\sin 50°} = -1$$

Q22. If $2\cos\left(\frac{4\pi}{9}\right)\cos\left(\frac{2\pi}{9}\right)\cos\left(\frac{\pi}{9}\right)$, then the numerical value of k is______

Solution:

$k = 2\cos\left(\frac{4\pi}{9}\right)\cos\left(\frac{2\pi}{9}\right)\cos\left(\frac{\pi}{9}\right)$

$= 2\cos 80°\cos 40°\cdot\cos 20°$

$= 2\sin 10°\cos 40°\cdot\cos 20°$

$= \sin 10°[2\cos 40°\cdot\cos 20°]$

$= \sin 10°[\cos 60° + \cos 20°]$

$[\because 2\cos x\cdot\cos y = \cos(x+y)+\cos(x-y)]$

$= \sin 10°\cdot\frac{1}{2} + \sin 10°\cos 20°$

$= \frac{1}{2}\sin 10° + \frac{1}{2}\cdot 2\sin 10°\cos 20°$

$= \frac{1}{2}\sin 10° + \frac{1}{2}[\sin 30° - \sin 10°] = \frac{1}{4}$

Q23. If $\sin x + \cos x = a$, then $\sin^3 x + \cos^3 x =$ ______

Solution:

$\because \sin x + \cos x = a$,

Squaring both sides, we get

$(\sin x + \cos x)^2 = a^2$

$\Rightarrow \sin^2 x + \cos^2 x + 2\sin x\cos x = a^2$

$\Rightarrow \sin x\cdot\cos x = \frac{1}{2}(a^2-1)$

Therefore, $\sin^3 x + \cos^3 x$

$= (\sin x + \cos x)^3 - 3\sin x\cos x(\sin x + \cos x)$

$= a^3 - \frac{3}{2}(a^2-1)a = a^3 - \frac{3}{2}a^3 + \frac{3}{2}a$

$= -\frac{1}{2}a^3 + \frac{3}{2}a = \frac{a(3-a^2)}{2}$

Q24. In a right-angled ΔABC with $\angle C = 90°$, the equation whose roots are $\sin A$, $\cos B$ is ______

Solution:

In the right-angled ΔABC

$\angle C = 90° \Rightarrow A + B = 90°$

$\therefore \cos B = \cos(90° - A) = \sin A$

Thus, the equation must have two common roots, that is, $\sin A$.

Hence, the required equation is

$x^2 - 2\sin A\, x + \sin^2 A = 0 \Rightarrow (x - \sin A)^2 = 0$

Q25. If $(\sin x - \cos x)^4 + 2(\sin x + \cos x)^2 = 4$, then $x =$ ______ where $x \in [0, 2\pi]$

Solution:

$(\sin x - \cos x)^4 + 2(\sin x + \cos x)^2 = 4$

$\Rightarrow [\sin^2 x + \cos^2 x - 2\sin x\cos x]^2$

$+2[\sin^2 x + \cos^2 x + 2\cdot\sin x\cos x] = 4$

$\Rightarrow (1 - \sin 2x)^2 + 2(1 + \sin 2x) = 4$

$\Rightarrow 1 + \sin^2 2x - 2\sin 2x + 2 + 2\sin 2x = 4$

$\Rightarrow \sin^2 2x = 1 \Rightarrow \sin 2x = \pm 1$

$\Rightarrow 2x = \frac{\pi}{2}, \frac{3\pi}{2} \Rightarrow x = \frac{\pi}{4}, \frac{3\pi}{4}$

Q26. The maximum distance of a point on the graph of the function $y = \sin x + \cos x$ from x-axis is______

Solution:

$y = \sin x + \cos x = \sqrt{2}\left[\frac{1}{\sqrt{2}}\sin x + \frac{1}{\sqrt{2}}\cos x\right]$

$= \sqrt{2}[\sin x\cdot\cos\frac{\pi}{4} + \cos x\sin\frac{\pi}{4}]$

$= \sqrt{2}\sin(x + \pi/4)$

Thus, the maximum distance of $y = \sin x + \cos x$ from the x-axis is $\sqrt{2}$ units.

Q27. If $\tan\theta + \tan 2\theta + \tan\theta\tan 2\theta = 1$, then $\theta =$ ______

Solution:

$\tan\theta + \tan 2\theta + \tan\theta\tan 2\theta = 1$

$\Rightarrow \tan\theta + \tan 2\theta = 1 - \tan\theta\tan 2\theta$

$\Rightarrow \dfrac{\tan\theta + \tan 2\theta}{1 - \tan\theta\tan 2\theta} = 1$

$\Rightarrow \tan(\theta + 2\theta) = \tan\dfrac{\pi}{4} \Rightarrow \tan 3\theta = \tan\dfrac{\pi}{4}$

$\therefore 3\theta = n\pi + \dfrac{\pi}{4} \Rightarrow \theta = \dfrac{n\pi}{4} + \dfrac{\pi}{16}$

Q28. The minimum value of $3\cos x + 4\sin x + 8$ is ______

Solution:

Let $y = 3\cos x + 4\sin x + 8 \Rightarrow y - 8 = 3\cos x + 4\sin x$

The maximum and minimum value of $a\cos x + b\sin x$ is $\pm\sqrt{a^2 + b^2}$

$\therefore$ Minimum value of $y - 8 = -\sqrt{9 + 16}$

$\Rightarrow y - 8 = -5 \Rightarrow y_{\min} = -5 + 8 = 3$

Day 55

Long Answer-I (LA-I): 4-Mark and Long Answer-II (LA-II): 6-Mark Questions

Q1. Find the radian measure corresponding to the following degree measures:

(i) 36° (ii) −145°30′

(iii) 240° (iv) 520°

Solution:

(i) $180° = \pi$ radians

Therefore, $36° = \dfrac{\pi}{180} \times 36$ radian $= \dfrac{\pi}{5}$ radians

(ii) $180° = \pi$ radians

Therefore, $-145°30' = -145\dfrac{1}{2}$ degree $= -\dfrac{291}{2}$ degree

$= -\dfrac{\pi}{180} \times \dfrac{291}{2}$ radians $= -\dfrac{97\pi}{120}$ radians

Q2. Find the degree measures corresponding to the following radian measures (use $\pi = \dfrac{22}{7}$):

(i) 11/8 (ii) $\dfrac{5\pi}{4}$

Solution:

(i) π radians $= 180°$

Therefore, 11/8 radians $= \dfrac{180}{\pi} \times \dfrac{11}{8}$ degree

$= \dfrac{180 \times 7}{22} \times \dfrac{11}{8}$ degree $= \dfrac{315}{4} =$ degree

$= 78\dfrac{3}{4}$ degree

$= 78° + \dfrac{3}{4} \times 60$ minutes $[\because 1° = 60']$

$= 78° + 45$ minutes $= 78°45'$

Thus, 11/8 radians $= 78°45'$

(ii) π radians $= 180°$

Therefore, $\dfrac{5\pi}{4}$ radians $= \dfrac{180}{\pi} \times \dfrac{5\pi}{4}$ degree

$= 225$ degree $= 225°$

Thus, $\dfrac{5\pi}{4}$ radians $= 225°$

Q3. Find the angle in radians and degrees through which a pendulum swings if its length is 70 cm and the tip describes an arc of length 11 cm.

Solution:

Given the length of arc $l = 11$ cm

Length of the pendulum = radius of circle, $r = 70$ cm

Thus, using the relation $\theta = \dfrac{l}{r}$ we get

$\theta = \dfrac{11}{70}$ radians

π radians $= 180°$

Therefore, 11/70 radians $= \dfrac{180}{\pi} \times \dfrac{11}{70}$ degree

$= \dfrac{180 \times 7}{22} \times \dfrac{11}{70}$ degree $= 9$ degree

Thus, the angle formed by the pendulum is 11/70 radians or 9°.

Q4. Find the values of the other five trigonometric functions if $\cos x = -\dfrac{1}{2}$, x lies in the second quadrant.

Solution:

For x lies in the second quadrant:

$\sin x = \sqrt{1 - \cos^2 x}$

$= \sqrt{1 - \left(-\dfrac{1}{2}\right)^2} = \sqrt{\dfrac{3}{4}} = \pm\dfrac{\sqrt{3}}{2}$

Therefore, $\sin x = \dfrac{\sqrt{3}}{2}$

($\because \sin x$ is positive in the second quadrant)

$$\text{cosec}\, x = \frac{1}{\sin x} = \frac{2}{\sqrt{3}}$$

($\because \text{cosec}\, x$ is positive in the second quadrant)

$$\sec x = \frac{1}{\cos x} = \frac{2}{1} = 2$$

($\because \cos x$ is positive in the second quadrant)

$$\tan x = \frac{\sin x}{\cos x} = \frac{\sqrt{3}/2}{-1/2} = -\sqrt{3}$$

($\because \tan x$ is negative in the second quadrant)

$$\cot x = \frac{1}{\tan x} = -\frac{1}{\sqrt{3}}$$

($\because \cot x$ is negative in the second quadrant)

Q5. Find the value of the trigonometric function cosec(−1035°).

Solution:

$\text{cosec}(-1035°) = -\text{cosec}(1035°)$

($\because \text{cosec}(-\theta) = -\text{cosec}\,\theta$)

$= -\text{cosec}(3\times 360° - 45°) = -\text{cosec}(-45°)$

[$\because$ cosec is negative in the fourth quadrant]

$= \text{cosec}\, 45° = \sqrt{2}$

Q6. Prove that $\sin^8 x + \cos^8 x = 1 - 4\sin x\cos x - 6\sin^2 x\cos^2 x.$

Solution:

$\text{LHS} = \sin^8 x + \cos^8 x = (\sin^2 x + \cos^2 x)^4$

$-4\sin^3 x\cos x - 6\sin^2 x\cos^2 x - 4\sin x\cos^3 x$

$= (\sin^2 x + \cos^2 x)^4 - 4\sin x\cos x(\sin^2 x + \cos^2 x)$

$-6\sin^2 x\cos^2 x \quad (\because \sin^2 x + \cos^2 x = 1)$

$= 1 - 4\sin x\cos x - 6\sin^2 x\cos^2 x = \text{RHS}$

Q7. Prove that $(\sin x + \text{cosec}\, x)^2 + (\cos x + \sec x)^2 - (\tan x + \cot x)^2 = 5$

Solution:

$\text{LHS} = (\sin x + \text{cosec}\, x)^2 + (\cos x + \sec x)^2$

$\quad - (\tan x + \cot x)^2$

$= \sin^2 x + \text{cosec}^2 x + 2\sin x\,\text{cosec}\, x$

$\quad + \cos^2 x + \sec^2 x + 2\cos x\sec x$

$\quad - \tan^2 x - \cot^2 x - 2\tan x\cot x$

$= (\sin^2 x + \cos^2 x) + (\text{cosec}^2 x - \cot^2 x)$

$\quad + (\sec^2 x - \tan^2 x) + 2\sin x\,\text{cosec}\, x$

$\quad + 2\cos x\sec x - 2\tan x\cot x$

$= 1 + 1 + 1 + 2 + 2 - 2 = 5 = \text{RHS}$

($\because \sin^2 x + \cos^2 x = 1,\ \text{cosec}^2 x - \cot^2 x = 1$

$\sec^2 x - \tan^2 x = 1, \sin x\,\text{cosec}\, x = 1$

$\cos x\sec x = 1,\ \tan x\cot x = 1$)

Q8. Find the value of $\dfrac{\sin^2 30° + \cos^2 45°}{\text{cosec}^2 60° - \sec^2 45°}$

Solution:

$$\frac{\sin^2 30° + \cos^2 45°}{\text{cosec}^2 60° - \sec^2 45°} = \frac{(1/2)^2 + (1/\sqrt{2})^2}{(2/\sqrt{3})^2 - (\sqrt{2})^2}$$

$$= \frac{1/4 + 1/2}{4/3 - 2} = \frac{3/4}{-2/3} = -\frac{9}{8}$$

Q9. Find the value of $\sin 390° + \tan 855° - \cos 1330°$.

Solution:

$\sin 390° + \tan 855° - \cos 1330°$

$= \sin(360° + 30°) + \tan(720° + 135°)$

$\quad - \cos(1080° + 250°)$

$= \sin(360° + 30°) + \tan(2\times 360° + 135°)$

$\quad - \cos(3\times 360° + 250°)$

$= \sin 30° + \tan(90° + 45°) - \cos(180° + 60°)$

$= \sin 30° - \cot 45° + \cos 60° = \frac{1}{2} - 1 + \frac{1}{2} = 0$

Q10. $\sin\dfrac{13\pi}{6} + \cos\dfrac{11\pi}{3} + \tan\dfrac{15\pi}{4}$

Solution:

$$\sin\frac{13\pi}{6} + \cos\frac{11\pi}{3} + \tan\frac{15\pi}{4}$$

$$= \sin\left(2\pi + \frac{\pi}{6}\right) + \cos\left(4\pi - \frac{\pi}{3}\right) + \tan\left(3\pi + \frac{3\pi}{4}\right)$$

$$= \sin\left(2\pi + \frac{\pi}{6}\right) + \cos\left(4\pi - \frac{\pi}{3}\right) + \tan\left(3\pi + \frac{\pi}{2} + \frac{\pi}{4}\right)$$

($\because$ sin and cosine functions repeat after 2π and tan function repeats after π)

$$= \sin\left(\frac{\pi}{6}\right) + \cos\left(\frac{\pi}{3}\right) - \cot\left(\frac{\pi}{4}\right)$$

$$= \frac{1}{2} + \frac{1}{2} - 1 = 0$$

Q11. Find the value of $\dfrac{\sin(\pi + x)\cos(-x)}{\cos(\pi - x)\sin\left(\dfrac{\pi}{2} + x\right)}$.

Solution:

$$\text{LHS} = \frac{\sin(\pi + x)\cos(-x)}{\cos(\pi - x)\sin\left(\frac{\pi}{2}+x\right)} = \frac{[-\sin x][\cos x]}{(-\sin x)(\cos x)} = 1$$

Q12. Prove that

$$\cos\left(\frac{\pi}{4}+x\right) - \cos\left(\frac{\pi}{4}-x\right) = -\sqrt{2}\sin x.$$

Solution:

$$\text{LHS} = \cos\left(\frac{\pi}{4}+x\right) - \cos\left(\frac{\pi}{4}-x\right)$$

$$= -2\sin\left\{\frac{\left(\frac{\pi}{4}+x\right)+\left(\frac{\pi}{4}-x\right)}{2}\right\}\sin\left\{\frac{\left(\frac{\pi}{4}+x\right)-\left(\frac{\pi}{4}-x\right)}{2}\right\}$$

$$\left[\because \cos A - \cos B = -2\sin\left(\frac{A+B}{2}\right)\cdot\sin\left(\frac{A-B}{2}\right)\right]$$

$$= -2\sin(\pi/4)\sin x = -2\times\frac{1}{\sqrt{2}}\sin x = -\sqrt{2}\sin x$$

Q13. Prove that $\sin^2 3x - \sin^2 x = \sin 2x \sin 4x$.

Solution:

$$\text{LHS} = \sin^2 3x - \sin^2 x$$

$$= (\sin 3x + \sin x)(\sin 3x - \sin x)$$

$$= \left[2\sin\left(\frac{3x+x}{2}\right)\cos\left(\frac{3x-x}{2}\right)\right]$$

$$\times\left[2\cos\left(\frac{3x+x}{2}\right)\sin\left(\frac{3x-x}{2}\right)\right]$$

$$\left[\begin{array}{l}\because \sin A + \sin B = 2\sin\left(\frac{A+B}{2}\right)\cos\left(\frac{A-B}{2}\right)\\ \sin A - \sin B = 2\cos\left(\frac{A+B}{2}\right)\sin\left(\frac{A-B}{2}\right)\end{array}\right]$$

$$= (2\sin 2x\cos x)(2\cos 2x \sin x)$$

$$= (2\sin x\cos x)(2\sin 2x\cos 2x)$$

$$= \sin 2x\sin 4x = \text{RHS}$$

Q14. Find the general solution of the equation $\cos 3x + \cos x - \sin 2x = 0$.

Solution:

$$\cos 3x + \cos x - \cos 2x = 0$$

$$\Rightarrow 2\cos\left(\frac{3x+x}{2}\right)\cos\left(\frac{3x-x}{2}\right) - \sin 2x = 0$$

$$\left[\cos A + \cos B = 2\cos\left(\frac{A+B}{2}\right)\cos\left(\frac{A-B}{2}\right)\right]$$

$$\Rightarrow 2\cos 2x\cos x - 2\sin x\cos x = 0$$

$$\Rightarrow 2\cos x(\cos 2x - \sin x) = 0$$

$$\Rightarrow \cos x = 0 \Rightarrow x = (2n+1)\frac{\pi}{2}$$

$$\text{or } \cos 2x = \sin x = \cos\left(\frac{\pi}{2}-x\right)$$

$$\Rightarrow 2x = 2n\pi \pm\left(\frac{\pi}{2}-x\right) \Rightarrow 2x \pm x = 2n\pi \pm \frac{\pi}{2}$$

$$\Rightarrow 3x = 2n\pi + \frac{\pi}{2} \text{ or } x = 2n\pi - \frac{\pi}{2}$$

Q15. Find the general solution of the equation $\sec^2 2x = 1 - \tan 2x$.

Solution:

$$\sec^2 2x = 1 + \tan 2x$$

$$\Rightarrow 1 + \tan^2 2x = 1 + \tan 2x$$

$$\Rightarrow \tan^2 2x - \tan 2x = 0$$

$$\Rightarrow \tan 2x(\tan 2x - 1) = 0$$

$$\Rightarrow \tan 2x = 0 \text{ or } \tan 2x - 1 = 0$$

Now, $\tan 2x = 0 \Rightarrow \tan 2x = \tan(n\pi)$

$$\Rightarrow 2x = n\pi \Rightarrow x = \frac{n\pi}{2}, \text{ where } n \in Z$$

And, $\tan 2x - 1 = 0$

$\Rightarrow \tan 2x = 1 = \tan\frac{\pi}{4} \Rightarrow 2x = n\pi + \frac{\pi}{4} \Rightarrow x = \frac{n\pi}{2} + \frac{\pi}{8}$, where $n \in Z$

Therefore, the general solution is

$$\frac{n\pi}{2} \text{ or } \frac{n\pi}{2} + \frac{\pi}{8},\ n \in Z$$

Q16. Find the general solution of the equation $\sin x + \sin 5x = 0$.

Solution:

$$\sin x + \sin 5x = 0$$

$$\Rightarrow 2\sin\left(\frac{x+5x}{2}\right)\cos\left(\frac{x-5x}{2}\right) = 0$$

$$\left[\sin A + \sin B = 2\sin\left(\frac{A+B}{2}\right)\cos\left(\frac{A-B}{2}\right)\right]$$

$$\Rightarrow 2\sin 3x\cos(-2x) = 0$$

$$\Rightarrow 2\sin 3x\cos 2x = 0$$

$$\Rightarrow \sin 3x = 0 \text{ or } \cos 2x = 0$$

Now, $\sin 3x = 0 \Rightarrow 3x = n\pi \Rightarrow x = \frac{n\pi}{3}$, where $n \in Z$

And, $\cos 2x = 0$

$\Rightarrow \cos 2x = \cos\frac{\pi}{2}$

$\Rightarrow 2x = 2n\pi \pm \frac{\pi}{2} \Rightarrow x = n\pi \pm \frac{\pi}{4}$, where $n \in Z$

Therefore, the general solution is

$\frac{n\pi}{3}$ or $n\pi \pm \frac{\pi}{2}, n \in Z$

Q17. In any triangle ABC, if $a = 2,\ b = 3,\ c = 4$, find $\cos A, \cos B, \cos C$.

Solution:

Using the cosine formula,

$$\cos A = \frac{b^2 + c^2 - a^2}{2bc} = \frac{3^2 + 4^2 - 2^2}{2 \times 3 \times 4} = \frac{21}{24} = \frac{7}{8}$$

$$\cos B = \frac{c^2 + a^2 - b^2}{2ca} = \frac{4^2 + 2^2 - 3^2}{2 \times 4 \times 2} = \frac{11}{16}$$

$$\cos C = \frac{a^2 + b^2 - c^2}{2ab} = \frac{2^2 + 3^2 - 4^2}{2 \times 2 \times 3} = \frac{-3}{12} = \frac{-1}{4}$$

Q18. For any triangle ABC, prove that:

$$\cos\left(\frac{A-B}{2}\right) = \frac{a+b}{c}\sin\frac{C}{2}.$$

Solution:

From the sine formula, $\frac{a}{\sin A} = \frac{b}{\sin B} = \frac{c}{\sin C} = k$

$$\therefore \frac{a+b}{c} = \frac{k\sin A + k\sin B}{k\sin C} = \frac{\sin A + \sin B}{\sin C}$$

$$= \frac{2\sin\frac{A+B}{2}\cos\frac{A-B}{2}}{\sin C}$$

$$\left[\because \sin A + \sin B = 2\sin\frac{A+B}{2}\cos\frac{A-B}{2}\right]$$

$$\Rightarrow \cos\frac{A-B}{2} = \frac{\frac{a+b}{c}\sin C}{2\sin\frac{A+B}{2}}$$

$$\Rightarrow \cos\frac{A-B}{2} = \frac{\frac{a+b}{c}2\sin\frac{C}{2}\cos\frac{C}{2}}{2\sin\left(90 - \frac{C}{2}\right)}$$

$$\left[\begin{array}{l} \because \sin\theta = 2\sin\frac{\theta}{2}\cos\frac{\theta}{2} \\ \text{Also, } A+B+C = 180° \Rightarrow \frac{A+B+C}{2} = 90° \\ \Rightarrow \frac{A+B}{2} = 90° - \frac{C}{2} \end{array}\right]$$

$$\Rightarrow \cos\frac{A-B}{2} = \frac{\frac{a+b}{c}2\sin\frac{C}{2}\cos\frac{C}{2}}{2\cos\left(\frac{C}{2}\right)}$$

$$\Rightarrow \cos\frac{A-B}{2} = \frac{a+b}{c}\sin\frac{C}{2} = \text{RHS}$$

Q19. Find $\sin\frac{x}{2}$, $\cos\frac{x}{2}$, $\tan\frac{x}{2}$, if $\tan x = \frac{3}{4}$, x in quadrant III.

Solution:

Since x is in quadrant III,

$$\pi < x < \frac{3\pi}{2} \Rightarrow \frac{\pi}{2} < \frac{x}{2} < \frac{3\pi}{4}$$

Thus, $\sin\frac{x}{2}$, $\cos\frac{x}{2}$, $\tan\frac{x}{2}$, all lie in the second quadrant. And in the second quadrant, sin is positive while cos and tan are negative.

Now, since $\tan x = \frac{3}{4}$

$$\Rightarrow \sec x = \sqrt{1 + \tan^2 x} = \sqrt{1 + \left(\frac{3}{4}\right)^2} = \sqrt{1 + \frac{9}{16}} = \pm\frac{5}{4}$$

$\Rightarrow \cos x = -\frac{4}{5}$ ($\because$ cos is negative in quadrant III)

$$\because \cos x = 2\cos^2\frac{x}{2} - 1 \Rightarrow \frac{-4}{5} = 2\cos^2\frac{x}{2} - 1$$

$$\Rightarrow 2\cos^2\frac{x}{2} = 1 - \frac{4}{5} \Rightarrow 2\cos^2\frac{x}{2} = \frac{1}{5}$$

$$\Rightarrow \cos^2\frac{x}{2} = \frac{1}{10} \Rightarrow \cos\frac{x}{2} = -\frac{1}{\sqrt{10}}$$

($\because$ cos is negative in quadrant II)

$$\sin\frac{x}{2} = \sqrt{1 - \cos^2\frac{x}{2}} = \sqrt{1 - \left(\frac{1}{\sqrt{10}}\right)^2} = \sqrt{\frac{9}{10}} = \frac{3}{\sqrt{10}}$$

($\because$ sin is positive in quadrant II)

$$\text{And, } \tan\frac{x}{2} = \frac{\sin\frac{x}{2}}{\cos\frac{x}{2}} = \frac{(3/\sqrt{10})}{(-1/\sqrt{10})} = -3$$

($\because$ tan is negative in quadrant II)

Thus, the respective values of $\sin\frac{x}{2}, \cos\frac{x}{2}, \tan\frac{x}{2}$ are

$\frac{3}{\sqrt{10}}, \frac{-1}{\sqrt{10}}, -3.$

Q20. Prove that

$$\frac{(\sin 4x + \sin 2x)\cdot(\sin 3x - \sin x)}{(\cos 4x + \cos 2x)\cdot(\cos 3x - \cos x)} = \tan x \tan 3x.$$

Solution:

$$\text{LHS} = \frac{(\sin 4x + \sin 2x)\cdot(\sin 3x - \sin x)}{(\cos 4x + \cos 2x)\cdot(\cos 3x - \cos x)}$$

$$= \frac{[2\sin 3x \cdot \cos x]\cdot[2\cos 2x \cdot \sin x]}{[2\cos 3x \cdot \cos x]\cdot[-2\cos 2x \cdot \cos x]}$$

$$\left[\begin{array}{l} \because \sin A + \sin B = 2\sin\left(\frac{A+B}{2}\right)\cdot\cos\left(\frac{A-B}{2}\right) \\ \sin A - \sin B = 2\cos\left(\frac{A+B}{2}\right)\sin\left(\frac{A-B}{2}\right) \\ \cos A + \cos B = 2\cos\left(\frac{A+B}{2}\right)\cdot\cos\left(\frac{A-B}{2}\right) \\ \cos A - \cos B = -2\sin\left(\frac{A+B}{2}\right)\cdot\sin\left(\frac{A-B}{2}\right) \end{array}\right]$$

$$= -\frac{\sin 3x \cdot \cos x \cdot \cos 2x \cdot \sin x}{\cos 3x \cdot \cos x \cdot \cos 2x \cdot \cos x}$$

$$= -\frac{\sin 3x \cdot \sin x}{\cos 3x \cdot \cos x} = \tan x \tan 3x = \text{RHS}$$

Revision and Unit Test 1

Day 56

Revision (Chapter I)

Sets

A set is a well-defined collection of objects. Well-defined indicates that if a question is asked to different people, then the answer given by all the persons must be the same.

Note:

(i) In a set, objects, elements and members are synonymous terms.

(ii) Sets are usually denoted by capital letters *A*, *B*, *C*, ..., *X*, *Y*, *Z*.

(iii) The elements of a set are represented by small letters *a*, *b*, *c*, ..., *x*, *y*, *z*.

(iv) If *a* is an element of a set *A*, then we may read '*a* belongs to *A*', represented by $a \in A$, where the Greek symbol $\in$ (epsilon) is used to denote the phrase 'belongs to'.

(v) If *b* is not an element of a set *A*, then we may read '*b* does not belong to *A*', represented by $b \notin A$, where the Greek symbol $\notin$ (cut epsilon) is used to denote the phrase 'does not belong to'.

Some Well-known Symbols Used to Represent Sets

N: the set of all natural numbers; $\{1, 2, 3, ...\}$

W: the set of all whole numbers; $\{0, 1, 2, 3, ...\}$

Z: the set of all integers; $\{..., -3, -2, -1, 0, 1, 2, 3, ...\}$

Z+: the set of positive integers; $\{1, 2, 3, ...\}$

Z–: the set of negative integers; $\{..., -3, -2, -1\}$

Q: the set of all rational numbers; $\{-2, \frac{-3}{2}, -1, \frac{-1}{5}, 0, 3, \frac{7}{2}, \frac{9}{4} ...\}$

Q+: the set of positive rational numbers;
$\{2, \frac{3}{2}, 1, \frac{1}{5}, 0, 3, \frac{7}{2}, \frac{9}{4} ...\}$

Q–: the set of negative rational numbers;
$\{-2, \frac{-3}{2}, -1, \frac{-1}{5}, -3, -\frac{7}{2}, -\frac{9}{4} ...\}$

R: the set of real numbers;
$\{-2, \frac{-3}{2}, -1, \frac{-1}{5}, 0, \sqrt{2}, \sqrt{3}, \frac{7}{2}, \frac{9}{4}, \pi, ...\}$

R+: the set of positive real numbers;
$\{2, \frac{3}{2}, 1, \frac{1}{5}, 0, \sqrt{2}, \sqrt{3}, \frac{7}{2}, \frac{9}{4}, \pi, ...\}$

R–: the set of negative real numbers;
$\{-2, \frac{-3}{2}, -1, \frac{-1}{5}, -\sqrt{2}, -\sqrt{3}, -\frac{7}{2}, -\frac{9}{4}, -\pi, ...\}$

Two Methods of Representing a Set:

(i) Roster or tabular form

(ii) Set-builder form

Roster or Tabular Form: In the roster or tabular form, all the elements of a set are listed, separated by a comma and are enclosed within braces {}.

Note: In the roster form, the order in which the elements are listed is irrelevant.

Note: In the roster form, repetition of elements is irrelevant; thus, elements are generally not repeated. That is, all the elements are taken as distinct.

Set-builder Form: In the set-builder form, we write the property of all the elements in a set, that is, all the elements of a set possess a single common property which is not possessed by any other element outside the set.

Thus, in the set-builder form, we describe the element of the set by using a symbol *x* (or any other symbol like the letters *y*, *z*) which is followed by a colon ':'. After the colon, we write the characteristic property possessed by all the elements of the set and then enclose the whole description within braces.

$X = \{x : x \text{ is the ....property of all the elements}\}$

The above description of the set *X* is read as '*X* is the set of all *x* such that *x* is the ... property of the set'. In this description, the braces stand for 'the set of all', and the colon stands for 'such that'.

Empty Set: A set that does not contain any element is called an empty set or a null set or a void set, denoted by the symbols ϕ or {}.

A set that contains at least a single element is called a non-empty set.

Finite and Infinite Sets: An empty set or a set which contains a definite number of distinct elements is called a finite set. If the number of distinct elements is infinite (or not countable) then the set is called an infinite set.
For a set *S*, we denote the number of distinct elements as *n*(S). Thus, if *n*(S) is a natural number, set *S* will be a non-empty finite set.

Note: All infinite sets cannot be described in the roster form. For example, the set of real numbers cannot be described in the roster form because the elements of this set do not follow any particular pattern.

The infinite sets which follow some definite pattern are written in the roster form by writing a few elements which clearly indicate the pattern of the set followed (or preceded) by three dots.

Equal Sets: Two sets A and B are said to be equal ($A = B$) if and only if they have exactly the same elements. Thus, two sets A and B are said to be equal if every element of A is also an element of B and if every element of B is also an element of A.
Otherwise, the sets are said to be unequal ($A \neq B$).

Subsets: A set B is said to be subset of a set A if every element of B is also an element of A.
Symbolically it can be represented as $B \subset A$ if whenever $a \in B$, then $a \in A$.
Or $B \subset A$ if $a \in B \Rightarrow a \in A$. The statement is read as 'B is a subset of A if a is an element of B implies that a is also an element of A'.
If B is not a subset of A, then we write $B \not\subset A$.

Note:

- Every set A is a subset of itself, $A \subset A$.
- Empty set ϕ is a subset of every set (since ϕ has no element).
- For B to be a subset of A, every element of B must be in A, but it is possible that every element of A may or may not be in B.

Equal Sets in Subsets: If every element of B is in A and every element of A is also in B, then we say that $B \subset A$ and $A \subset B$; thus, A and B are the same (equal) sets.
Symbolically it can be written as,
$A \subset B$ and $B \subset A \Rightarrow A = B$
Read as 'If A is a subset of B and B is a subset of A, then A and B are equal sets'.
or $A = B \Rightarrow A \subset B$ and $B \subset A$
Read as 'If A and B are equal sets, then A is a subset of B and B is also a subset of A'.
In general, it is written as
$A \subset B$ and $B \subset A \Leftrightarrow A = B$ where symbol $\Leftrightarrow$ represents if and only if (or vice versa) condition (in short, written as 'iff').

Note: $N \subset Z \subset Q \subset R,\ T \subset R,\ N \not\subset T$.

Proper Subset and Superset: Let A and B be two sets. If $A \subset B$ and $A \neq B$, then A is called a proper subset of B, and B is called a superset of A.

Singleton Set: If a set A has only one element, then it is called a singleton set.

Power Set: The collection of all the subsets (including empty set) of a set A is called the power set of A, denoted by $P(A)$. Thus, $P(A)$ is the power set that contains all the subsets of A as elements of $P(A)$.

Also, if A is a set with a number of elements, $n(A) = n$, then the number of elements in the power set $P(A)$ is $n[P(A)] = 2^n$.

Universal Set: A universal set is the set which contains all the sets under consideration. The universal set is usually denoted by U, and all its subsets by the letters A, B, C...

For example, while studying the system of numbers, sometimes we consider the set of natural numbers as the universal set, the set of integers as the universal set or the set of real numbers as the universal set.

Note:

- The difference between two symbols $\in$ 'belongs to' and $\subset$ 'subset' is that $\in$ represents an element which is in a set while $\subset$ represents a subset of a set.
- An element of a set can never be a subset of itself.

Intervals as Subsets of R

Let $a, b \in R,\ a < b$. Then

Open Interval: The set of all real numbers $\{x : x \in R,\ a < x < b\}$ is called an open interval and it is denoted by (a, b). All the points between a and b belong to the open interval (a, b) except points a and b, that is, points a and b do not belong to this interval. On a number line, it is represented as

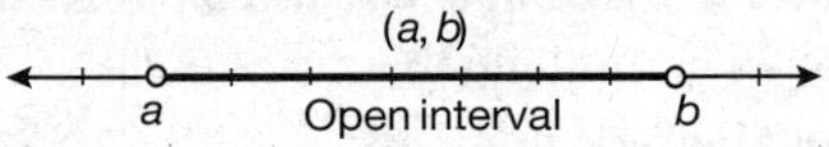

Closed Interval: The set of all real numbers $\{x : x \in R,\ a \le x \le b\}$ is called a closed interval and it is denoted by $[a, b]$. All the points between a and b belong to the closed interval $[a, b]$ including points a and b, that is, points a and b also belong to this interval. On a number line, it is represented as

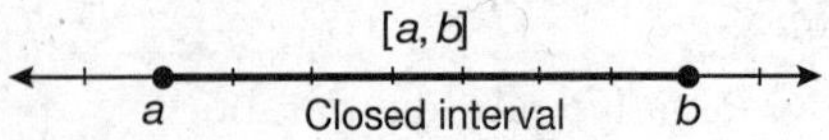

Half Open and Half Closed Interval: The interval which is open at one end and closed at the other end is called a half open and half closed interval. That is,
$[a, b) = \{x : x \in R,\ a \le x < b\}$ is a half open and half closed interval from a to b, including point a but excluding point b.
$(a, b] = \{x : x \in R,\ a < x \le b\}$ is a half open and half closed

interval from a to b, including point b but excluding point a. On a number line, it is represented as

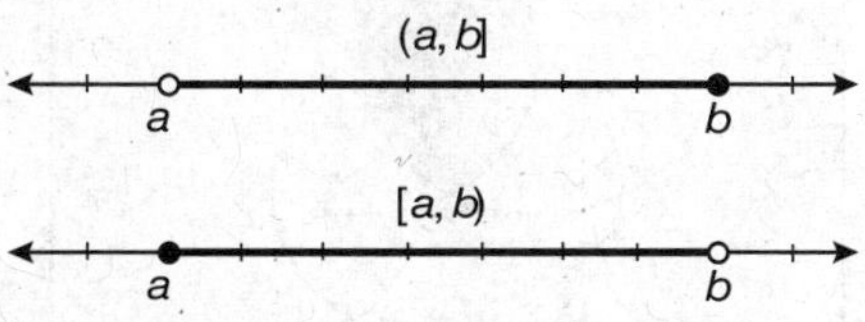

Half open and half closed intervals

Note:

- The number $|b-a|$ is called the length of any of the intervals $(a,b),[a,b],[a,b)$ or $(a,b]$.
- $x \in R$; therefore intervals (a,b), $[a,b]$, $[a,b)$ or $(a,b]$ contain infinitely many points.
- Infinity ∞ is always represented with an open interval, because infinity never ends (its end point is not defined).

The set $(-\infty,\infty)$ describes the set of real numbers in relation to a line extending from $-\infty$ to ∞, represented by,

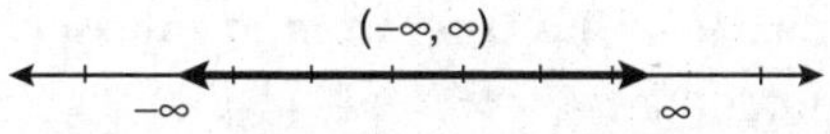

The set (a,∞) describes the set of real numbers in relation to a line extending from a to ∞, excluding point a, represented by,

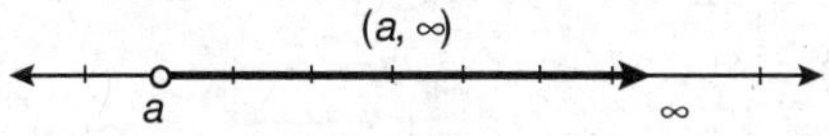

The set $(-\infty,b)$ describes the set of real numbers in relation to a line extending from $-\infty$ to b, excluding point b, represented by,

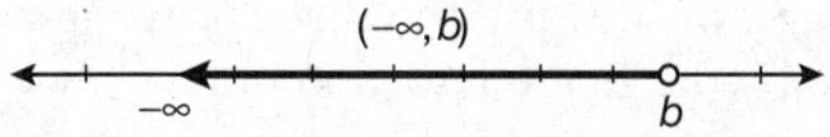

The set $[a,\infty)$ describes the set of real numbers in relation to a line extending from a to ∞, including point a, represented by,

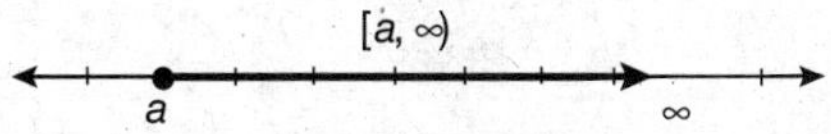

The set $(-\infty,b]$ describes the set of real numbers in relation to a line extending from $-\infty$ to b, including point b, represented by,

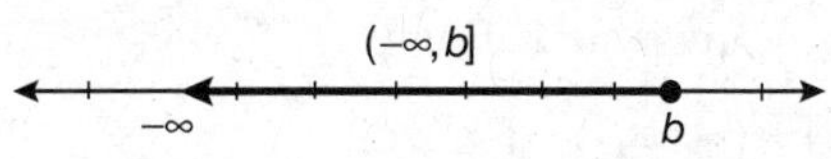

The set $[0,\infty)$ defines the set of non-negative real numbers, represented by,

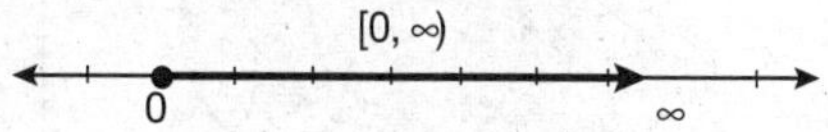

The set $(-\infty,0)$ defines the set of negative real numbers, represented by,

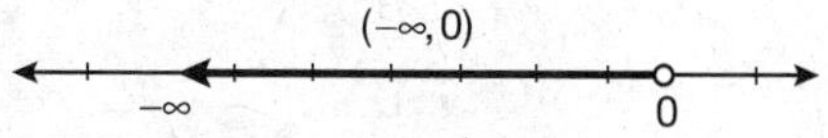

Venn Diagrams: Relationships between sets can also be represented by means of diagrams which are known as Venn diagrams. The universal set is usually represented by a rectangle and its subsets by circles.

In Venn diagrams, the elements of the sets are written in their respective circles.

Operations on Sets: Just as different algebraic operations (addition, subtraction, multiplication and division) on two numbers give rise to another number, different operations on two sets also give rise to another set. Usually, we shade that portion which represents the operations on sets and their properties.

Union of Sets: Let A and B be any two sets. The union of sets A and B is the set which consists of all the elements of A and all the elements of B. The common element is taken only once, since repetition of elements in sets is irrelevant.

The union of two sets A and B is the set which consists of all those elements which are either in A or in B (including those which are in both). Union is denoted by the symbol '$\cup$' and thus $A \cup B$ is read as 'A union B', and is written as

$A \cup B = \{x : x \in A \text{ or } x \in B\}$.

Venn diagrams showing the union of two and three sets are represented by the shaded portions in the figures below.

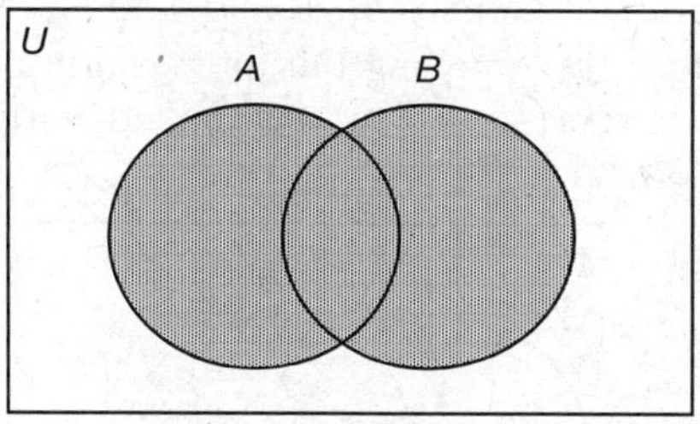

$A \cup B = \{x : x \in A \text{ or } x \in B\}$

Fig. 1.2

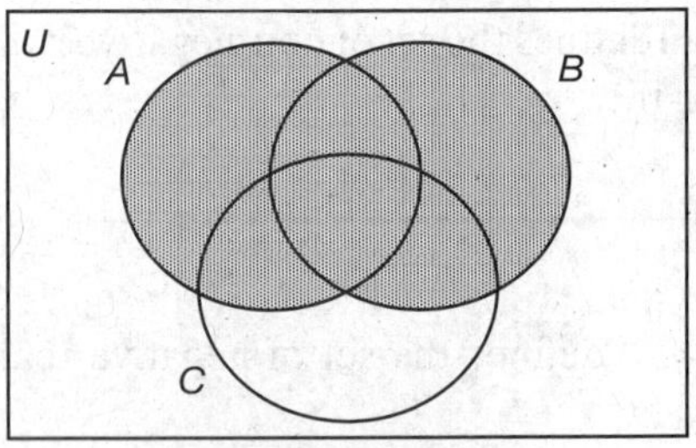

$A \cup B$

Fig. 1.3

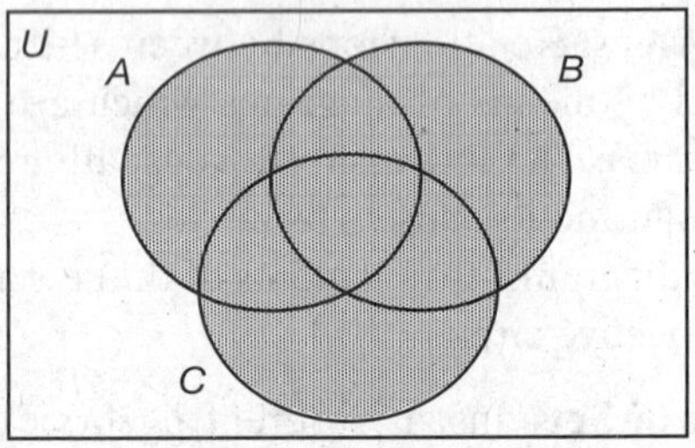

$A \cup B \cup C = A \cup (B \cup C) = (A \cup B) \cup C$

Fig. 1.4

Some Properties of the Operation of Union

(i) Idempotent law: $A \cup A = A$

(ii) Commutative law: $A \cup B = B \cup A$

(iii) Associative law: $(A \cup B) \cup C = A \cup (B \cup C)$

(iv) Law of identity element: $A \cup \phi = \phi \cup A = A$

(ϕ is the identity of U)

(v) Law of U: $A \cup U = U \cup A = U$

(vi) $A \times (B \cup C) = (A \times B) \cup (A \times C)$

Intersection of Sets: Let A and B be any two sets. The intersection of A and B is the set which consists of all the common elements of A and B. The common element is taken only once, since repetition of elements in sets is irrelevant.

The intersection of two sets A and B is the set which consists of all those elements which are in both A and B. Intersection is denoted by the symbol '$\cap$' and thus $A \cap B$ is read as 'A intersection B', and written as $A \cap B = \{x : x \in A \text{ and } x \in B\}$

Venn diagrams showing the intersection of two and three sets is represented by the shaded portion in the figures below.

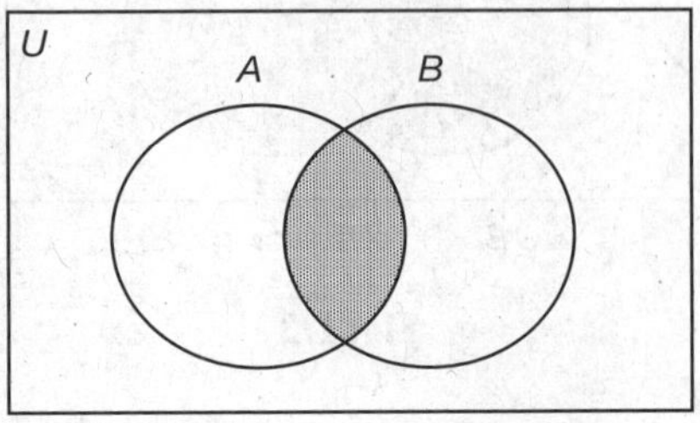

$A \cap B = \{x : x \in A \text{ and } x \in B\}$

Fig. 1.5

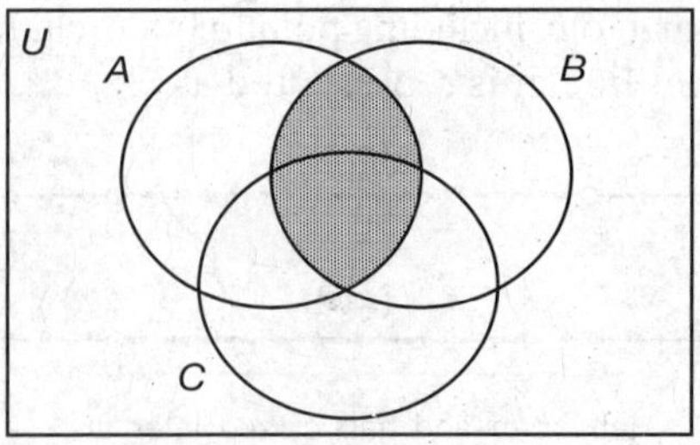

$A \cap B$

Fig. 1.6

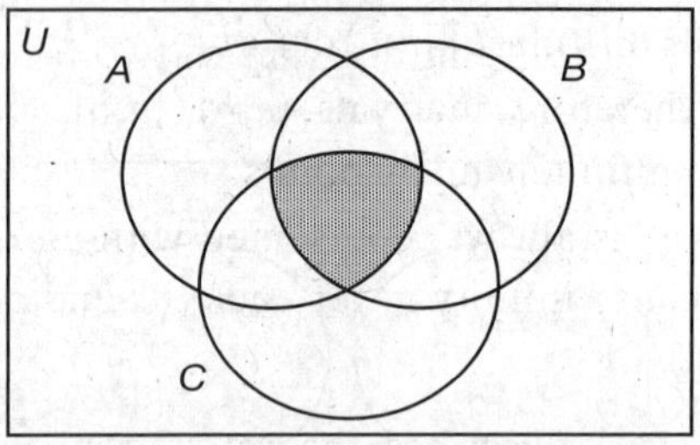

$A \cap B \cap C = A \cap (B \cap C) = (A \cap B) \cap C$

Fig. 1.7

Some Properties of the Operation of Intersection

(i) Idempotent law: $A \cap A = A$

(ii) Commutative law: $A \cap B = B \cap A$

(iii) Associative law: $(A \cap B) \cap C = A \cap (B \cap C)$

(iv) Law of nullity element: $A \cap \phi = \phi \cap A = \phi$

(ϕ is the nullity of A)

(v) Law of U: $A \cap U = U \cap A = A$

(vi) $A \times (B \cap C) = (A \times B) \cap (A \times C)$

Distributive Law over Intersection ($\cap$)

$A \cap (B \cup C) = (A \cap B) \cup (A \cap C)$

A Venn diagram showing the distributive law over $\cap$ is represented by the shaded portion in Fig. 1.8.

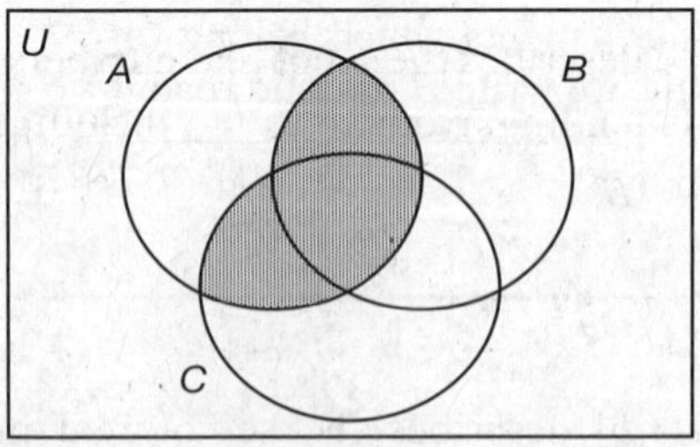

$A \cap (B \cup C) = (A \cap B) \cup (A \cap C)$

Fig. 1.8

Distributive Law over Union ($\cup$)

$A \cup (B \cap C) = (A \cup B) \cap (A \cup C)$

A Venn diagram showing distributive law over $\cup$ is represented by the shaded portion in Fig. 1.9.

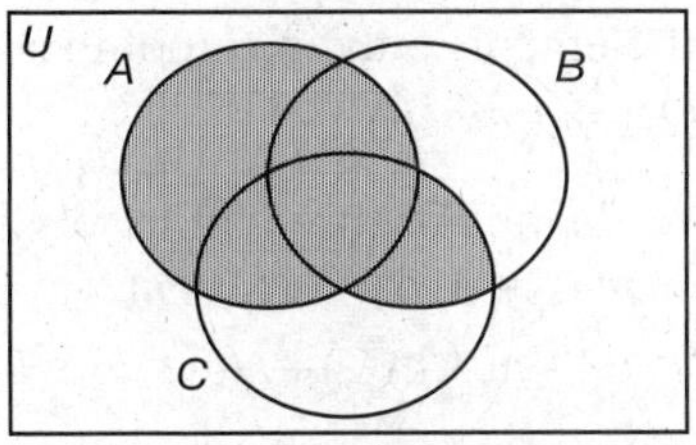

$A \cup (B \cap C) = (A \cup B) \cap (A \cup C)$

Fig. 1.9

Disjoint Sets: If the intersection of two sets A and B is empty, then the two sets A and B are disjoint sets. That is, $A \cap B = \phi$. The Venn diagram of two disjoint sets is shown in Fig. 1.10.

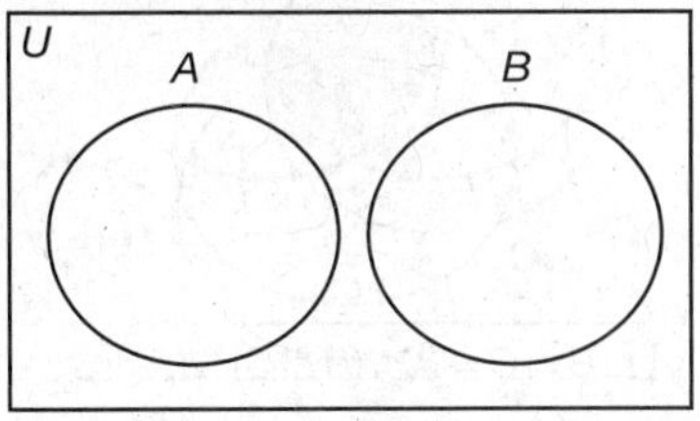

$A \cap B = \phi$

Fig. 1.10

Difference of Sets: The difference of the sets A and B in this order is the set of elements which belong to A but not to B. Symbolically, we write $A - B$ and read it as 'A minus B'.

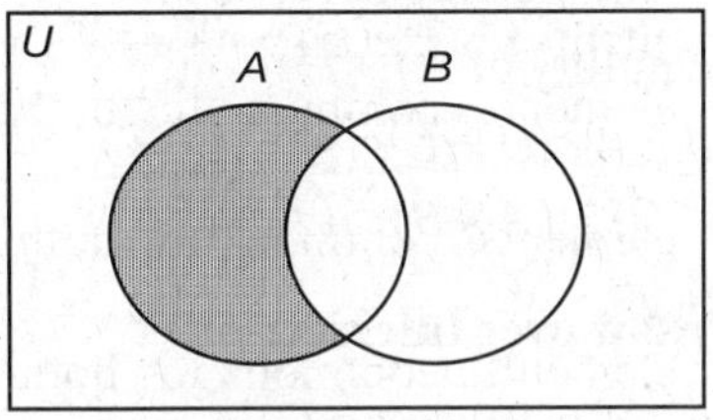

$A - B = \{x : x \in A \text{ and } x \notin B\}$

Fig. 1.11

Similarly, the difference of the sets B and A in this order is the set of elements which belong to B but not to A. Symbolically, we write $B - A$ and read it as 'B minus A'.

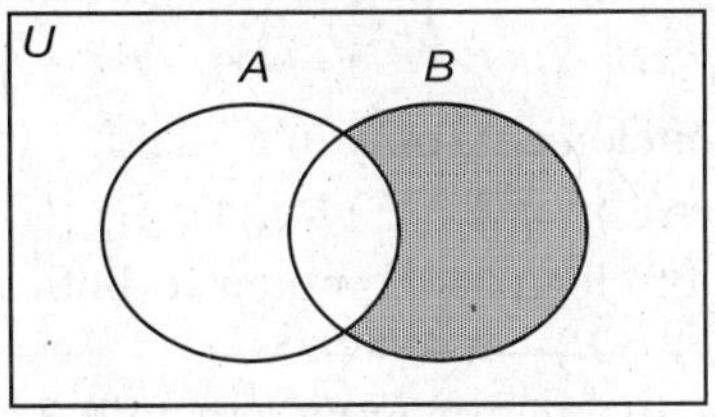

$B - A = \{x : x \notin A \text{ and } x \in B\}$

Fig. 1.12

Note:

- For any two sets A and B, the sets $A - B$, $A \cap B$, $B - A$ are mutually disjoint sets, that is, the intersection of any of these two sets is the null set.
- The union of all the three sets $A - B$, $A \cap B$, $B - A$ gives $A \cup B$, that is, $A \cup B = (A - B) \cup (A \cap B) \cup (B - A)$.
- The Venn diagram of the sets $A - B$, $A \cap B$, $B - A$ is shown in Fig. 1.13.

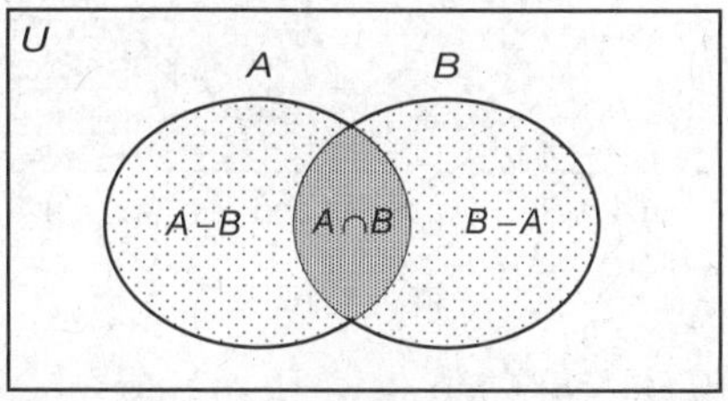

$A \cup B = (A - B) \cup (A \cap B) \cup (B - A)$

Fig. 1.13

- $A - (B \cup C) = (A - B) \cap (A - C)$
  $A - (B \cap C) = (A - B) \cup (A - C)$
  $(A \cap B) - C = (A - C) \cap (B - C)$
  $(A \cup B) - C = (A - C) \cup (B - C)$

Complement of a Set: Let U be the universal set and A be a subset of U. Then, the complement of set A is the set of all elements of U which do not belong to A.
Symbolically, we write A' or A^C, read as complement of A with respect to U. Given by
A' or $A^C = \{x : x \in U \text{ and } x \notin A\} = U - A$
Thus, the complement of a set A is the difference between a universal set U and the set A.

Note: If A is a subset of the universal set U, then its complement A' or A^C is also a subset of U.
The Venn diagram showing the complement of set A with respect to U is represented by the shaded portion in Figs 1.14 to 1.16.

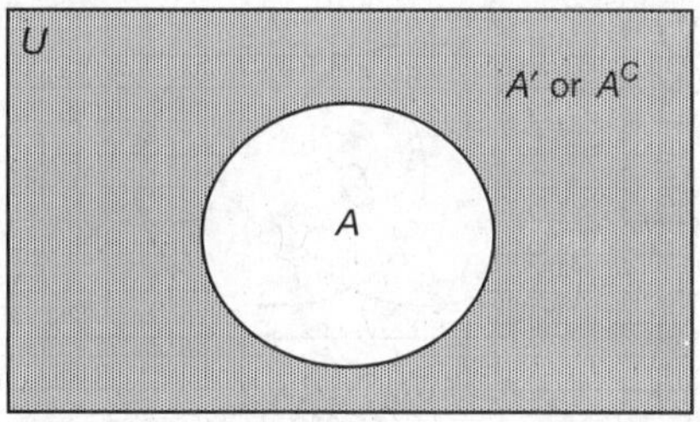

A' or $A^C = \{x : x \in U \text{ and } x \notin A\} = U - A$

Fig. 1.14

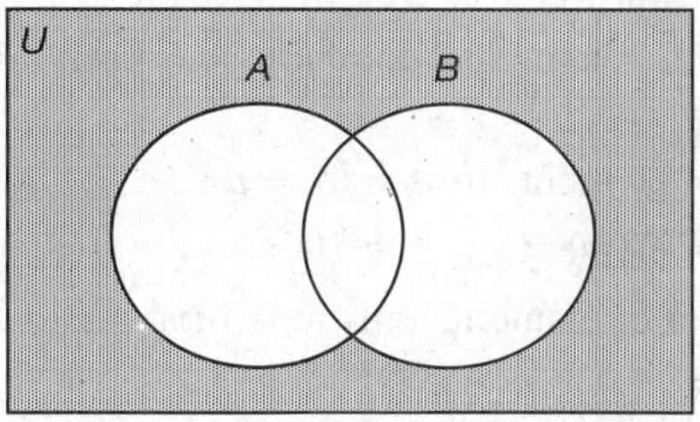

$(A \cup B)' = \{x : x \in U \text{ and } x \notin (A \cup B)\}$

Fig. 1.15

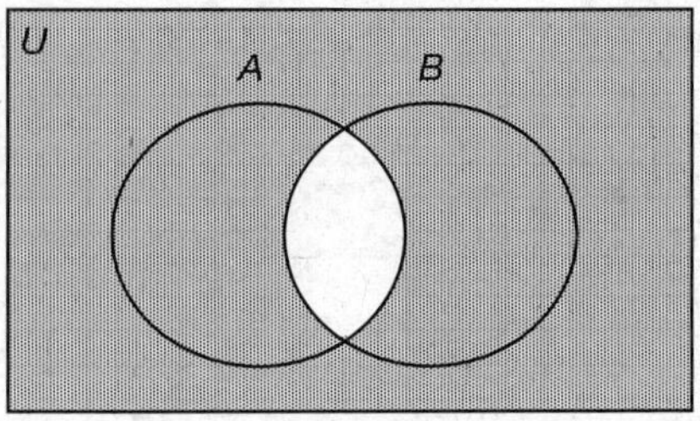

$(A \cap B)' = \{x : x \in U \text{ and } x \notin (A \cap B)\}$

Fig. 1.16

Some Properties of Complement of Sets

(a) Complement laws:

(i) $A \cup A' = U$ (ii) $A \cap A' = \phi$

(b) De Morgan's law:

(i) $(A \cup B)' = A' \cap B'$ (ii) $(A \cap B)' = A' \cup B'$

Note: $(A - B)' \neq A' - B'$

(c) Law of double complementation: $(A')' = A$

(d) Laws of empty set and universal set:

(i) $\phi' = U$ (ii) $U' = \phi$

(iii) $\phi' \cap A = U \cap A = A$ (iv) $\phi' \cup A = U \cup A = U$

(v) $U' \cap A = \phi \cap A = \phi$

(vi) $U' \cup A = \phi \cup A = A$

Two Sets Problems: Let A and B be finite sets, and let $n(A)$ and $n(B)$ be the number of distinct elements in sets A and B, respectively.

Then,

$n(A \cup B) = n(A) + n(B) - n(A \cap B)$

And if A and B are disjoint sets, then $A \cap B = \phi$; thus $n(A \cup B) = n(A) + n(B)$

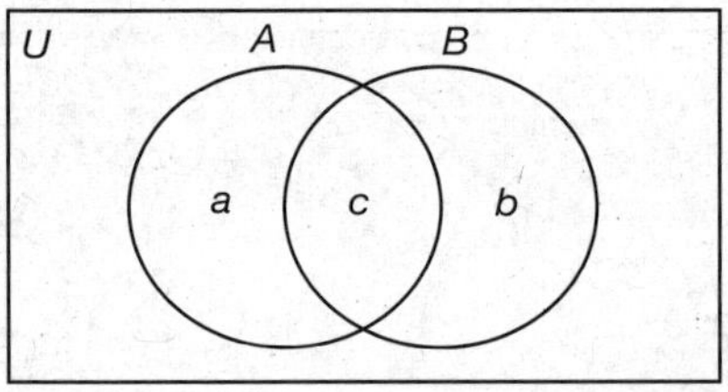

Fig. 1.17

Thus, as shown in Fig. 1.17, we get

Total number of elements in both the sets A and B: $n(A \cup B) = a + b + c$

Number of elements common to both A and B: $n(A \cap B) = c$

Number of elements in A: $n(A) = a + c$

Number of elements in B: $n(B) = b + c$

a = Number of elements only in A (not in B)

$(a + c) - c = n(A) - n(A \cap B)$

b = Number of elements only in B (not in A)

$(b + c) - c = n(B) - n(A \cap B)$

Three Sets Problems: For three finite sets A, B and C:

$$n(A \cup B \cup C) = n(A) + n(B) + n(C) - n(A \cap B) - n(B \cap C) - n(A \cap C) + n(A \cap B \cap C)$$

Note: For solving problems in three sets, first we draw the Venn diagram and then we consider different parts of the diagram as a, b, c, ..., as shown in Fig. 1.18. Finally, we find the solution using formulae or by using the Venn diagram.

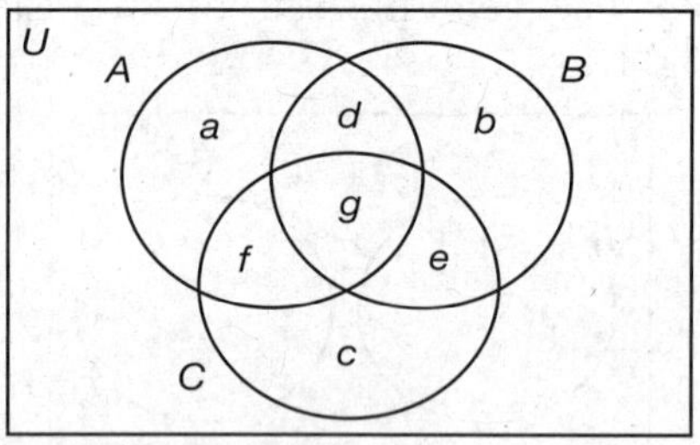

Fig. 1.18

Thus, as shown in Fig. 1.18, we get

Total number of elements in all the three sets A, B and C: $n(A \cup B \cup C) = a + b + c + d + e + f + g$

Number of elements in A: $n(A) = a + d + f + g$

Number of elements in B: $n(B) = b + d + e + g$

Number of elements in C: $n(C) = c + e + f + g$

Number of elements common to both A and B: $n(A \cap B) = d + g$

Number of elements common to both B and C: $n(B \cap C) = e + g$

Number of elements common to both A and C: $n(A \cap C) = f + g$

Number of elements common to all the three sets A, B and C: $n(A \cap B \cap C) = g$

a = Number of elements only in A (neither in B nor in C) $= n(A) - n(A \cap B) - n(A \cap C) + n(A \cap B \cap C)$

b = Number of elements only in B (neither in A nor in C) $= n(B) - n(A \cap B) - n(B \cap C) + n(A \cap B \cap C)$

c = Number of elements only in C (neither in A nor in B) $= n(C) - n(A \cap C) - n(B \cap C) + n(A \cap B \cap C)$

d = Number of elements common to both A and B but not in $C = n(A \cap B) - n(A \cap B \cap C)$

e = Number of elements common to both B and C but not in $A = n(B \cap C) - n(A \cap B \cap C)$

f = Number of elements common to both A and C but not in $B = n(A \cap C) - n(A \cap B \cap C)$

Day 57

Revision (Chapter II)

Relations and Functions

Cartesian Products of Sets: Consider two non-empty sets P and Q. The Cartesian product $P\times Q$ is the set of all ordered pairs (p,q) of elements from P and Q, such that $p\in P, q\in Q$; that is, $P\times Q=\{(p,q): p\in P,\ q\in Q\}$.

Note:

(i) If either P or Q is the empty set, then $P\times Q$ will also be an empty set, that is, $P\times Q=\phi$

(ii) If either P or Q is a non-empty infinite set, then $P\times Q$ will also be an infinite set.

(iii) Two ordered pairs are equal if and only if the corresponding first elements are equal and the second elements are also equal.

(iv) If there are p elements in set P and q elements in set Q, then there will be pq elements in Cartesian product $P\times Q$, that is, if $n(P)=p$, $n(Q)=q$, then $n(P\times Q)=pq$.

(v) If $A\times A\times A=\{(a,b,c): a,b,c\in A\}$, then (a,b,c) is called an ordered triplet.

(vi) $A\times(B\cap C)=(A\times B)\cap(A\times C)$
And $A\times(B\cup C)=(A\times B)\cup(A\times C)$

Relations: A relation R from a non-empty set A to a non-empty set B is a subset of the Cartesian product $A\times B$.
The subset is derived by describing a relationship (R) between the first element (x) and the second element (y) of the ordered pairs (x,y) in $A\times B$, that is, $R=\{(x,y): x\in A$ and $y\in B$; x relates to y with some relation $R\}$. Also denoted by $x\,R\,y$.

Note: A relation R from A to A is also stated as relation R on A.

Domain: The set of all first elements (x) of the ordered pairs (x,y) in a relation R from a set A to a set B is called the domain of the relation R.

Range: The set of all second elements (y) of the ordered pairs (x,y), obtained by the relation R from a set A to a set B is called the range of the relation R.
The second element is also known as the image of the first element.

Co-domain: The whole given set B is called the co-domain of the relation R.

Note:

(i) An arrow diagram is a visual representation of a relation R from a set A to a set B.

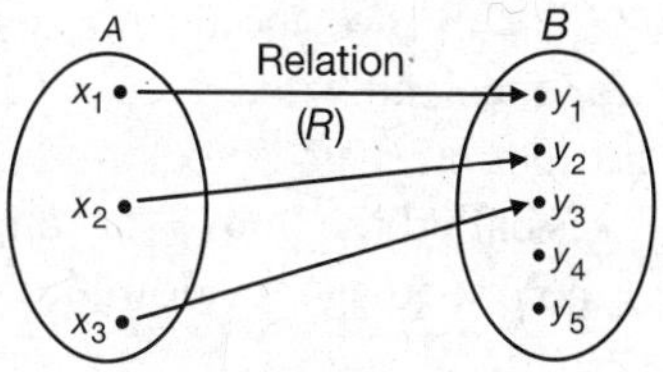

Fig. 2.1

(ii) The range of relation $R\subseteq$ co-domain (set B).
Domain: (x_1,x_2,x_3)
Range: (y_1,y_2,y_3)
Co-domain: (y_1,y_2,y_3,y_4,y_5)

(iii) A relation may be represented algebraically either by the roster method or by the set-builder method.
$R=\{(x,y): x\,R\,y,\ X\in A,\ y\in B\}$
or $R=\{(x_1,y_1),(x_2,y_2),(x_3,y_3)\}$

(iv) The total number of relations that can be defined from set A to set B is the number of possible subsets of $A\times B$. If $n(A)=p, n(B)=q$, then $n(A\times B)=pq$ and thus the total number of relations is 2^{pq}.

Functions: A special type of relation f from a non-empty set A to a non-empty set B is said to be a function if every element of set A has one and only one image in set B. That is, the domain of f is A and no two distinct ordered pairs in f have the same first element.
Thus, if f is a function from A to B and $(a,b)\in f$, then $f(a)=b$, where b is called the image (or range) of a under f and a is called the pre-image (or domain) of b under f.

- The function f from A to B is denoted by $f: A\to B$ or $b=f(a)$.
- f denotes the function (or special relation) between $a\in A$ and $b\in B$.
- A is the given domain of function f.
- a is the element of the domain of function f.
- B is the given co-domain of function f.
- b is the range or image of function f, that is, all values of b obtained by the relation $b=f(a)$, for all, is defined as the range of function f.
- For a valid function, the range must be less than or equal to the co-domain, Range $\subseteq$ Co-domain.
- So, do not get confused with co-domain and range. Co-domain is given while range is obtained from the domain and function f.

Real-valued Function: A function which has either R or one of its subsets as its range is called a real valued function. If its domain is also either R or a subset of R, then it is called a real function.

A relation f from $X \to Y$ is said to be a real function if every $x \in X$ has one and only one image in $y \in Y$, that is, the domain of f is A and we should not get two distinct values of $y \in Y$ for the same value $x \in X$.

Thus, if f is a function from X to Y and $(x, y) \in f$, then $y = f(x)$, where y is called the image (or range) of x under f and x is called the pre-image (or domain) of y under f.

Representation of a Real Function

- The function f from X to Y is denoted by $f : X \to Y$ or $y = f(x)$.
- f denotes the real function between $x \in X$ and $y \in Y$.
- X is the given domain of function f.
- x is the element of the domain of function f.
- Y is the given co-domain of function f.
- y is the range of function f, that is, all values of y obtained by the relation $y = f(x)$ for all $x \in X$ is defined as the range of function f.
- For a valid function, the range must be less than or equal to the co-domain, Range $\subseteq$ Co-domain.
- So, do not get confused with co-domain and range. Co-domain is given while range is obtained from the domain and function f.

Note: R represents all real numbers.
R^+ means $[0, \infty)$ and R^- means $(-\infty, 0]$.

Properties of Functions: A function f may have following possibilities:

(i) The distinct elements of X under the function f have distinct images in Y.
(ii) All the distinct elements of X under the function f have an image in Y. But, it is not necessary that all distinct elements of Y under the function f have a pre-image in X. That is, Range $\subseteq$ Co-domain.
(iii) Two or more distinct elements of X under the function f must not have the same image in Y.

If we graphically draw a line parallel to the y-axis and it cuts the graph at two distinct points, then the graph is not a function.

Graphical Representation of a Function: Since the property of a function is that for any value of x we get a single value of y, graphically if we draw a line parallel to the y-axis, if it cuts the graph at two distinct points, then the graph is not a function.

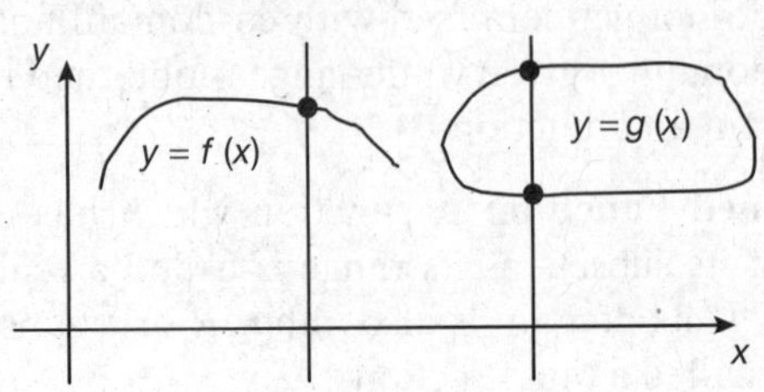

Fig. 2.2

Consider the graphs $y = f(x)$ and $y = g(x)$ as shown in Fig. 2.2.

Since a line parallel to the y-axis cuts the graph $y = f(x)$ only at one point, while a line parallel to y-axis cuts the graph $y = g(x)$ at two distinct points, the graph $y = f(x)$ represents a function while $y = g(x)$ does not represent a function.

(i) Constant Function: A real-valued constant function $f : R \to R$ is given by $y = f(x) = c$, where c is a constant and $x \in R$.
The graph is a line parallel to the x-axis and intercepts the y-axis at $y = c$.
Domain = R and range = $\{c\}$.

(ii) Identity Function: A real-valued identity function $f : R \to R$ is given by $y = f(x) = x$, or $y = f(x) = -x$, where $x \in R$.
The graph is a straight line that passes through the origin and is equally inclined to the x and y axes.
Domain = R and range = R.

(iii) Polynomial Function: A real-valued polynomial function $f : R \to R$ is given by $y = f(x) = a_0 + a_1x + a_2x^2 + ... + a_nx^n$, where $x \in R$ and $n \in Z^+$ (a non-negative integer) and constants $a_0, a_1, a_2, ..., a_n \in R$.
Constant function (Zero degree): $f(x) = a_0$
Linear function (First degree): $f(x) = a_0 + a_1x$
Quadratic function (Second degree):
$f(x) = a_0 + a_1x + a_2x^2$
Cubic function (Third degree):
$f(x) = a_0 + a_1x + a_2x^2 + a_3x^3$

(a) Linear Function: A real-valued identity function $f : R \to R$ is given by $y = mx + c$, where m and c are constants and $x \in R$.
The graph is a straight line with a slope m and it intersects the x and y axes at the points $(-c/m, 0)$ and $(0, c)$, respectively.
Domain = R and range = R.

(b) Quadratic (or Square) Function: A real-valued quadratic (or square) function $f : R \to R$ is given by $y = f(x) = x^2, x \in R$.
The graph lies above the x-axis and is symmetrical about the y-axis with a minimum value of $y = 0$ at $x = 0$, and this graph is also known as a standard parabola.
Domain = R and range = $[0, \infty)$.

(c) Cubic Function: A real-valued cubic function $f : R \to R$ is given by $y = f(x) = x^3, x \in R$.
The graph passes through the origin and lies in the 1st and 3rd quadrants symmetrically.
Domain = R and range = R.

(iv) Root Functions: A real-valued function $f: R \to R$ is given by $y = f(x) = ax^n$, where $x \in R$ and $n \notin Z$ (a non-integer, real) and constant $a \in R$.

(a) Square-Root Function: A real-valued square-root function $f:[0,\infty) \to R$ is given by $y = f(x) = \sqrt{x}$, $x \in [0,\infty)$.
The graph lies only in the 1st quadrant.
Domain = $[0,\infty)$ and range = $[0,\infty)$.

(b) Cube-Root Function: A real-valued cube-root function $f: R \to R$ is given by
$y = f(x) = \sqrt[3]{x}$, $x \in R$.
The graph passes through the origin and lies in the 1st and 3rd quadrants symmetrically.
Domain = R and range = R.

(v) Rational Functions: Rational functions are functions of the type $\dfrac{f(x)}{g(x)}$, where $f(x), g(x)$ are polynomial functions of x defined in a domain, where $g(x) \neq 0$.

(a) Standard Reciprocal Function: A real-valued reciprocal function $f: R-\{0\} \to R$ is given by $y = f(x) = \dfrac{1}{x}$, $x \in R$ and $x \neq 0$.
The graph lies in the 1st and 3rd quadrants symmetrically.
Domain = $(-\infty,0)\cup(0,\infty)$ and
Range = $(-\infty,0)\cup(0,\infty)$

(b) Reciprocal Squared Function: A real-valued reciprocal function $f: R-\{0\} \to R$ is given by $y = f(x) = \dfrac{1}{x^2}$, $x \in R$ and $x \neq 0$.
The graph lies in the 1st and 2nd quadrants and is symmetrical about the y-axis.
Domain = $(-\infty,0)\cup(0,\infty)$ and
Range = $(0,\infty)$

(c) Reciprocal Square-Root Function: A real-valued square-root function $f:(0,\infty) \to R$ is given by $y = f(x) = \dfrac{1}{\sqrt{x}}$, $x \in (0,\infty)$.
The graph lies only in the 1st quadrant.
Domain = $(0,\infty)$ and range = $(0,\infty)$.

(vi) Modulus (or Absolute Value) Function: A real-valued modulus function $f: R \to R$ is given by $y = f(x) = |x|$, $x \in R$
Modulus function is a piecewise or split function represented as,

$$y = f(x) = |x| = \begin{cases} x, & x \geq 0 \\ -x, & x < 0 \end{cases}$$

That is, it depends on the sign of the x-value. If the x-value is positive, $x \geq 0$, then the function is defined by $f(x) = x$. And if the x-value is negative, $x < 0$, then the function is defined by $f(x) = -x$.
Thus, a modulus function always gives a non-negative value.
The graph lies in the 1st and 2nd quadrants and is symmetrical about the y-axis.
Domain = R and range = $[0,\infty)$.

(vii) Signum Function: A real-valued signum function $f: R \to R$ is a piecewise or split function represented as,

$$y = f(x) = \begin{cases} \dfrac{|x|}{x} \text{ or } \dfrac{x}{|x|}, & x \neq 0 \\ 0, & x = 0 \end{cases}$$

$$\Rightarrow f(x) = \begin{cases} 1, & \text{if } x > 0 \\ 0, & \text{if } x = 0 \\ -1, & \text{if } x < 0 \end{cases}$$

The graph is a horizontal ray with $y = 1$ and $y = -1$, parallel to the x-axis in the 1st and 3rd quadrants, starts from $y = 0^+$ and $y = 0^-$, respectively, and $y = 0$ at $x = 0$. Domain = R and range = $\{-1, 0, 1\}$.

(viii) Greatest Integer Function: A real-valued greatest integer function (also known as the floor function) $f: R \to R$ is given by $y = f(x) = [x]$, $x \in R$.
It is equal to the value of the greatest integer, less than or equal to x.
The graph is discontinuous at all integers.
Domain = R and range = Z.

(ix) Exponential Function: A real-valued exponential function $f: R \to R$ is given by $y = a^x$, $x \in R$, $a > 0$, and $a \neq 1$

$$\Rightarrow y = \begin{cases} a^x \text{ (rising curve)} & \text{if } a > 1 \\ a^x \text{ (falling curve)} & \text{if } 0 < a < 1 \end{cases}$$

The graph lies above the x-axis and it is a rising curve for $a > 1$ and a falling curve for $0 < a < 1$.
Domain = R and range = $(0,\infty)$.

Properties of Exponential Functions: Let x, y be real numbers, and a, b be positive real numbers, and $a, b \neq 1$. Then,

- $a^x = a^y$ if and only if $x = y$
- If $x > y$ then $\begin{cases} a^x > a^y & \text{if } a > 1 \\ a^x < a^y & \text{if } 0 < a < 1 \end{cases}$
- $a^x \cdot a^y = a^{x+y}$; $\dfrac{a^x}{a^y} = a^{x-y}$ and $(a^x)^y = a^{x.y}$

- $(a \cdot b)^x = a^x \cdot b^x$ and $\left(\frac{a}{b}\right)^x = \frac{a^x}{b^x}$

(x) Logarithmic Function: A real-valued exponential function $f: R^+ \to R$ is given by $y = \log_a x,\ x \in R^+,\ a > 0$, and $a \neq 1$

$$\Rightarrow y = \begin{cases} \log_a x \text{ (rising curve)} & \text{if } a > 1 \\ \log_a x \text{ (falling curve)} & \text{if } 0 < a < 1 \end{cases}$$

The graph lies to the right of the y-axis and it is a rising curve for $a > 1$ and a falling curve for $0 < a < 1$.

Domain $= (0, \infty)$ and range $= R$.

Properties of Logarithmic Functions: Let x, y be real numbers, and a, b be positive real numbers, and $a, b \neq 1$. Then,

- $\log_a x = \log_a y$ if and only if $x = y$
- If $x > y$ then
  $\begin{cases} \log_a x > \log_a y & \text{if } a > 1 \\ \log_a x < \log_a y & \text{if } 0 < a < 1 \end{cases}$
- $\log_a (x \cdot y) = \log_a x + \log_a y$
- $\log_a \left(\frac{x}{y}\right) = \log_a x - \log_a y$
- $\log_x y = \frac{1}{\log_y x}$
- $\log_a x^n = n \log_a x,\ n \in R$
- $\log_{a^m} x^n = \frac{n}{m} \log_a x,\ n \in R$
- $a^{\log_a x} = \log_a a^x = x$
- $\log_a a = 1$ and $\log_a 1 = 0$

Domain and Range of Real Functions: Consider a function f from X to Y and $(x, y) \in f$, such that $y = f(x)$; then all real values of x for which function f exists and is defined is called the domain of function f.

All real values of y obtained by the relation $y = f(x)$ for all real values of x (for which function f exists and is defined) is called the range of function f.

All given real values of y is called the co-domain of function f. For a valid function, the range must be less than or equal to the co-domain, Range $\subseteq$ Co-domain.

So, do not get confused with co-domain and range. Co-domain is given while range is obtained from the domain and function f.

Note that, graphically if we draw a line parallel to the y-axis, and if it cuts the graph at two distinct points, then the graph is not a function.

Note: R represents all real numbers

$R - \{a\} = (-\infty, a) \cup (a, \infty)$

Positive real numbers $= R^+ = (0, \infty)$

Non-negative numbers $= R^+ = [0, \infty)$

Negative real numbers $= R^- = (-\infty, 0)$

Non-positive real numbers $= R^- = (-\infty, 0]$

Domain of Some Special Functions

(i) For a polynomial function $y = f(x) = a_0 + a_1 x + a_2 x^2 + \ldots + a_n x^n$, the domain is all real values of x.

(ii) For a reciprocal function $y = \frac{1}{f(x)}$, the domain is all real values of x such that $f(x) \neq 0$.

(iii) For a rational function $y = \frac{f(x)}{g(x)}$, the domain is all real values of x such that $g(x) \neq 0$.

(iv) For a square-root function $y = \sqrt{f(x)}$, the domain is all real values of x such that $f(x) \geq 0$.

(v) For a cube-root function $y = \sqrt[3]{f(x)}$, the domain is all real values of x.

(vi) For a reciprocal square-root function $y = \frac{1}{\sqrt{f(x)}}$, the domain is all real values of x such that $f(x) > 0$.

(vii) For a reciprocal cube-root function $y = \frac{1}{\sqrt[3]{f(x)}}$, the domain is all real values of x except 0, that is, $R - \{0\} = (-\infty, 0) \cup (0, \infty)$.

(viii) For a modulus function $y = |f(x)|$, the domain is all real values of x.

(ix) For a signum function
$$y = \begin{cases} \frac{|f(x)|}{f(x)} \text{ or } \frac{f(x)}{|f(x)|}, & x \neq 0 \\ 0, & x = 0 \end{cases}$$, the domain is all real values of x.

(x) For a greatest integer function $y = [f(x)]$, the domain is all real values of x.

(xi) For an exponential function $y = f(x)^{g(x)}$, the domain is all real values of x such that $f(x) > 0$ and $f(x) \neq 1$.

(xii) For a logarithmic function $y = \log_{g(x)} f(x)$, the domain is all real values of x such that $f(x) > 0$, $g(x) > 0$ and $g(x) \neq 1$.

How to Find the Range of a Function

Process-I: Substitute the different values of x in the expression of $y = f(x)$, and find all possible values of y (minimum y-value to maximum y-value, critical points, boundary points).

Process-II: First find the inverse of the function $y = f(x)$; let it be $x = f^{-1}(y) = g(y)$ and write it as $y = g(x)$. Then, the domain of function $y = g(x)$ will be the range of function $y = f(x)$.

Note: For simple and special functions, we generally use Process-I while for critical functions we use Process-Ii.

Algebra of Real Functions: Algebra of real functions means the addition of two real functions, or subtraction a real function from another real function, or multiplication of a real function by a scalar constant, or multiplication of two real functions, or division of one real function by another.

(i) Addition of Two Real Functions: Let $f: X \to R, g: X \to R$ be any two real functions, where $X \subset R$. Then, the addition of functions f and g will be defined as $(f+g): X \to R$ given by $(f+g)(x) = f(x) + g(x)$, for all $x \in X$.

(ii) Subtraction of a Real Function from Another: Let $f: X \to R, g: X \to R$ be any two real functions, where $X \subset R$. Then, the subtraction of function g from a function f will be defined as $(f-g): X \to R$ given by
$(f-g)(x) = f(x) - g(x)$, for all $x \in X$.

(iii) Multiplication by a Scalar: Let $f: X \to R$ be a real function and α be a scalar (real number). Then, the multiplication of function f by a scalar α will be defined as $\alpha f: X \to R$ given by $(\alpha f)(x) = \alpha f(x)$, for all $x \in X$.

(iv) Multiplication of Two Real Functions: Let $f: X \to R, g: X \to R$ be any two real functions, where $X \subset R$. Then, the multiplication of functions f and g will be defined as $(f \cdot g): X \to R$ given by $(f \cdot g)(x) = f(x) \cdot g(x)$, for all $x \in X$.

(v) Quotient of two functions
Let $f: X \to R, g: X \to R$ be any two real functions, where $X \subset R$. Then, quotient of functions f and g that is, division of a function f by a function g will be defined as $\left(\frac{f}{g}\right): X \to R$ given by $\left(\frac{f}{g}\right)(x) = \frac{f(x)}{g(x)}$, for all $x \in X$, provided $g(x) \neq 0$.

Day 58

Revision (Chapter III)

Trigonometric Functions

Angle: An angle is a measurement of rotation of a given ray about its initial position. The point of rotation is called the vertex.

The position of the original ray is called the initial side and the position of the ray after rotation is called the terminal side of the angle.

The measurement of rotation or the measure of an angle is the amount of rotation from the initial side to the terminal side, about the point of rotation.

The measurement of the angle is said to be positive if the direction of rotation is anti-clockwise, and it said to be negative if the direction of rotation is clockwise.
In general, the measured angles are represented by the symbol $\angle\theta$ or $\angle AOB$.

Degree Measure: If a rotation from the initial side to the terminal side is $\left(\frac{1}{360}\right)^{\text{th}}$ of a complete revolution, the measure of the angle is said to be one degree; it is written as 1°. Thus, the measure of one complete rotation or revolution is 360°.

Note: A degree is divided into 60 minutes and a minute is divided into 60 seconds. That is, $1° = 60'$ and $1' = 60'' \Rightarrow 1° = 3600''$.

Radian Measure: Radian measure is the angle subtended by a circular arc at the centre of the circle.
Thus, the angle subtended at the centre of a unit circle (circle of radius 1 unit) by a circular arc of length 1 unit is said to be the measure of 1 radian.

Note: The circumference of a circle of radius 1 unit is 2π; therefore one complete revolution of the initial side subtends an angle of 2π radians.

Thus, in a circle of radius r, an arc of length r will subtend an angle of 1 radian. This leads to a well-known relation that equal arcs of a circle subtend equal angles at the centre.

Now, in a circle of radius r, an arc of length r subtends an angle whose measure is 1 radian; therefore an arc of length l will subtend an angle whose measure is $\frac{l}{r}$ radian.

Thus, in a circle of radius r, an arc of length l subtends an angle θ radian at the centre, that is, $\theta = \frac{l}{r}$ or $l = r\theta$.

Note: Every real number will correspond to a radian measure and vice versa. Thus, radian measures and real numbers can be considered as one and the same.

Relation between Degree and Radian: Since one revolution around a circle measures the degree angle as 360° and the circle subtends at the centre an angle whose radian measure is 2π, it follows that 2π radian = 360° or π radian = 180°.

From the above relation, we get a relation of the radian measure in terms of the degree measure and the degree measure in terms of the radian measure:

$$\text{Radian measure} = \frac{\pi}{180} \times \text{Degree measure}$$

$$\text{Degree measure} = \frac{180}{\pi} \times \text{Radian measure}$$

The relation between degree and radian measure of some common angles are given in the table below.

Degree	**Radian**
0°	0
30°	$\pi/6$
45°	$\pi/4$
60°	$\pi/3$
90°	$\pi/2$
120°	$2\pi/3$
135°	$3\pi/4$
150°	$5\pi/6$
180°	π
210°	$7\pi/6$
225°	$5\pi/4$
270°	$3\pi/2$
300°	$5\pi/3$
330°	$11\pi/6$
360°	2π

Note:

- Since angles are measured either in degrees or in radians, we adopt the convention that whenever we write angle $\theta°$, we mean the angle whose degree measure is θ, and whenever we write angle only θ (without degree symbol), we mean the angle whose radian measure is θ. Normally when an angle is expressed in radians, the word, 'radian' is frequently omitted.
- Angle measures

$\pi = 180°$ or $\frac{\pi}{2} = 90°$ or $\frac{\pi}{3} = 60°$ or $\frac{\pi}{4} = 45°$ are written with the understanding that $\pi, \frac{\pi}{2}, \frac{\pi}{3}, \frac{\pi}{4}$ are radian measures while $180°, 90°, 60°, 45°$ are degree measures.

Values of Some Trigonometric Functions: The values of some trigonometric functions or ratios is given in the table below.

Angle	0 (0°)	$\pi/6$ (30°)	$\pi/4$ (45°)	$\pi/3$ (60°)	$\pi/2$ (90°)
sin	0	1/2	$\frac{1}{\sqrt{2}}$	$\frac{\sqrt{3}}{2}$	1
cos	1	$\frac{\sqrt{3}}{2}$	$\frac{1}{\sqrt{2}}$	1/2	0
tan = sin/cos	0	$\frac{1}{\sqrt{3}}$	1	$\sqrt{3}$	Not defined
cosec = 1/sin	Not defined	2	$\sqrt{2}$	$\frac{2}{\sqrt{3}}$	1
sec = 1/cos	1	$\frac{2}{\sqrt{3}}$	$\sqrt{2}$	2	Not defined
cot = cos/sin	Not defined	$\sqrt{3}$	1	$\frac{1}{\sqrt{3}}$	0

The domain and range of all the trigonometric functions are summarised in the table below.

Trigonometric function	**Domain** (x)	**Range** (y)
$y = \sin x$	All real numbers $R: (-\infty, \infty)$	$[-1, 1]$
$y = \cos x$	All real numbers $R: (-\infty, \infty)$	$[-1, 1]$
$y = \tan x$	$x \in R$ $x \neq (2n+1)\frac{\pi}{2}$ $n \in Z$	All real numbers $R: (-\infty, \infty)$
$y = \text{cosec}\, x$	$x \in R$ $x \neq n\pi$ $n \in Z$	$R - (-1, 1)$, that is, $(-\infty, -1] \cup [1, \infty)$

Trigonometric function	Domain (x)	Range (y)
$y = \sec x$	$x \in R$, $x \neq (2n+1)\dfrac{\pi}{2}$, $n \in Z$	$R-(-1,1)$, that is, $(-\infty,-1] \cup [1,\infty)$
$y = \cot x$	$x \in R$, $x \neq n\pi$, $n \in Z$	All real numbers $R: (-\infty,\infty)$

Signs of Trigonometric Functions

Quadrant	I	II	III	IV
$\sin x$	+	+	–	–
$\cos x$	+	–	–	+
$\tan x$	+	–	+	–
$\text{cosec}\, x$	+	+	–	–
$\sec x$	+	–	–	+
$\cot x$	+	–	+	–

Trigonometric Functions

$\sin(2n\pi \pm x) = \sin x, \quad n \in Z,$

$\cos(2n\pi \pm x) = \cos x, \quad n \in Z,$

$\tan(n\pi \pm x) = \tan x, \quad n \in Z,$

$\text{cosec}\, x = \dfrac{1}{\sin x}, \quad x \neq n\pi$, where n is any integer

$\sec x = \dfrac{1}{\cos x}, \quad x \neq (2n+1)\dfrac{\pi}{2}$, where $n \in Z$

$\tan x = \dfrac{\sin x}{\cos x}, \quad x \neq (2n+1)\dfrac{\pi}{2}$, where $n \in Z$

$\cot x = \dfrac{\cos x}{\sin x}, \quad x \neq n\pi$, where $n \in Z$

Graph of Trigonometric Functions

Graph of sin x

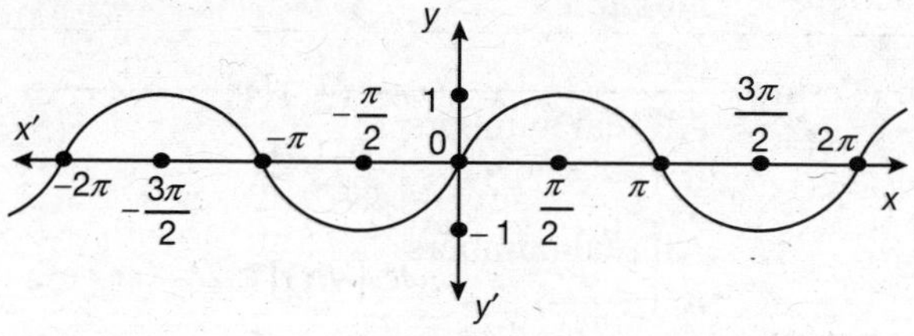

Fig. 3.7

Graph of cos x

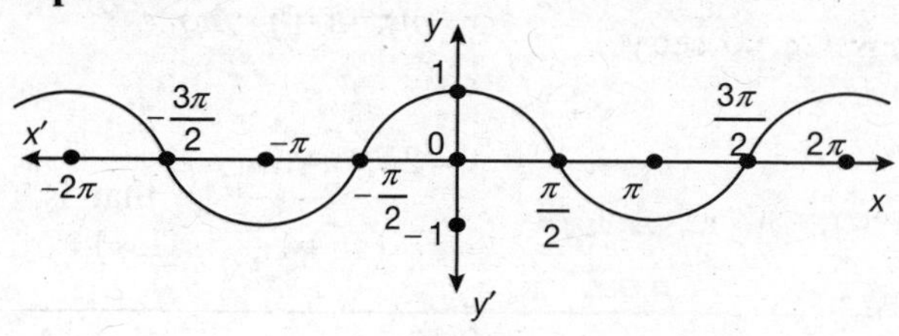

Fig. 3.8

Graph of cosec x

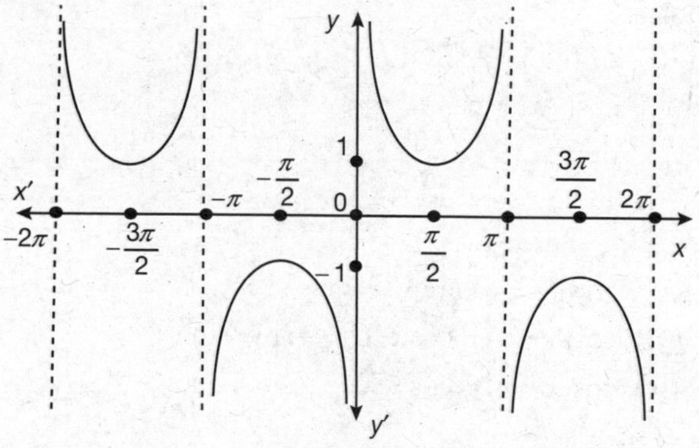

Fig. 3.9

Graph of sec x

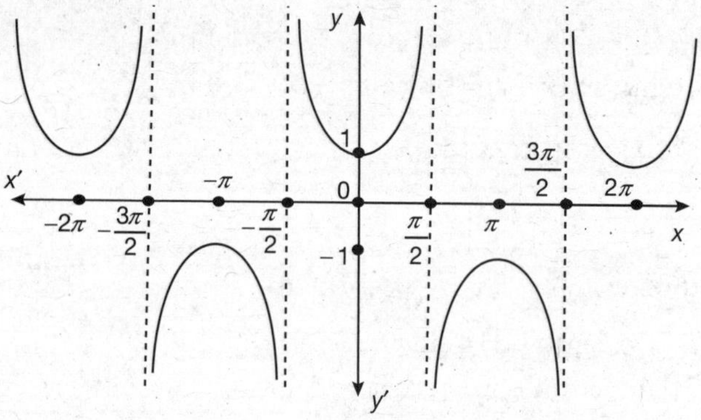

Fig. 3.10

Graph of tan x

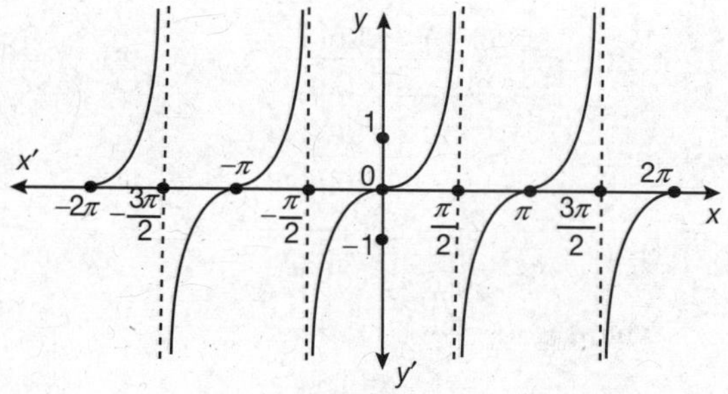

Fig. 3.11

Graph of cot x

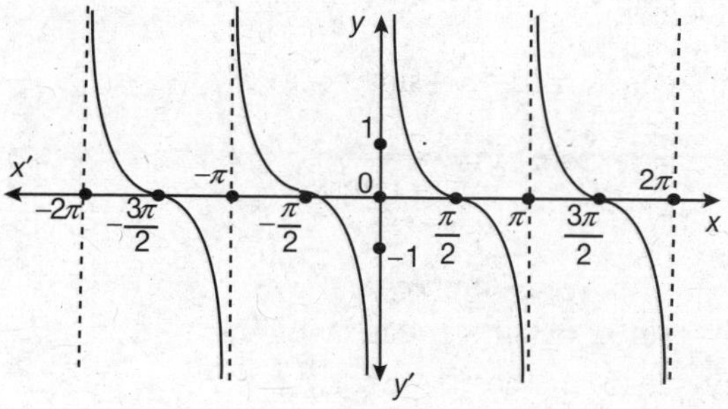

Fig. 3.12

Trigonometric Functions of Sum and Difference of Two Angles:

(i) $\sin(-x) = -\sin x$

$\cos(-x) = \cos x$

$\tan(-x) = -\tan x$

(ii) $\sin^2 x + \cos^2 x = 1$

$\sec^2 x - \tan^2 x = 1$

$\text{cosec}^2 x - \cot^2 x = 1$

(iii) $\cos(x+y) = \cos x \cos y - \sin x \sin y$

(iv) $\cos(x-y) = \cos x \cos y + \sin x \sin y$

(v) $\cos\left(\frac{\pi}{2} - x\right) = \sin x$

(vi) $\sin\left(\frac{\pi}{2} - x\right) = \cos x$

(vii) $\sin(x+y) = \sin x \cos y + \cos x \sin y$

(viii) $\sin(x-y) = \sin x \cos y - \cos x \sin y$

(ix) $\cos\left(\frac{\pi}{2} + x\right) = -\sin x$

(x) $\sin\left(\frac{\pi}{2} + x\right) = \cos x$

(xi) $\cos(\pi - x) = -\cos x$

(xii) $\sin(\pi - x) = \sin x$

(xiii) $\cos(\pi + x) = -\cos x$

(xiv) $\sin(\pi + x) = -\sin x$

(xv) $\cos(2\pi - x) = \cos x$

(xvi) $\sin(2\pi - x) = -\sin x$

(xvii) $\tan(x+y) = \dfrac{\tan x + \tan y}{1 - \tan x \tan y}$; $x, y, (x+y) \neq n\dfrac{\pi}{2}$

(xviii) $\tan(x-y) = \dfrac{\tan x - \tan y}{1 + \tan x \tan y}$

(xix) $\cot(x+y) = \dfrac{\cot x \cot y - 1}{\cot y + \cot x}$

where $x, y, (x+y) \neq n\pi$

(xx) $\cot(x-y) = \dfrac{\cot x \cot y + 1}{\cot y - \cot x}$

where $x, y, (x-y) \neq n\pi$

(xxi) $\sin 2x = 2\sin x \cos x = \dfrac{2\tan x}{1 + \tan^2 x}$

(xxii) $\cos 2x = \cos^2 x - \sin^2 x = 2\cos^2 x - 1$

$= 1 - 2\sin^2 x = \dfrac{1 - \tan^2 x}{1 + \tan^2 x}$

(xxiii) $\tan 2x = \dfrac{2\tan x}{1 - \tan^2 x}$

(xxiv) $\sin 3x = 3\sin x - 4\sin^3 x$

(xxv) $\cos 3x = 4\cos^3 x - 3\cos x$

(xxvi) $\tan 3x = \dfrac{3\tan x - \tan^3 x}{1 - 3\tan^2 x}$

(xxvii) Product formulae

(a) $2\cos x \cos y = \cos(x+y) + \cos(x-y)$

(b) $-2\sin x \sin y = \cos(x+y) - \cos(x-y)$

(c) $2\sin x \cos y = \sin(x+y) + \sin(x-y)$

(d) $2\cos x \sin y = \sin(x+y) - \sin(x-y)$.

(xxviii) Addition formulae

(a) $\cos x + \cos y = 2\cos\dfrac{x+y}{2}\cos\dfrac{x-y}{2}$

(b) $\cos x - \cos y = -2\sin\dfrac{x+y}{2}\sin\dfrac{x-y}{2}$

(c) $\sin x + \sin y = 2\sin\dfrac{x+y}{2}\cos\dfrac{x-y}{2}$

(d) $\sin x - \sin y = 2\cos\dfrac{x+y}{2}\sin\dfrac{x-y}{2}$

(xxix)

(a) $\sin(x+y)\sin(x-y) = \sin^2 x - \sin^2 y$

(b) $\cos(x+y)\cos(x-y) = \cos^2 x - \sin^2 y$

Trigonometric Equations: Equations involving trigonometric functions of a variable angle are called trigonometric equations.

Since the values of sin x and cos x repeat after an interval of 2π and the values of $\tan x$ repeat after an interval of π, the solutions of a trigonometric equation for which $0 \le x < 2\pi$ are called principal solutions.

The expression involving integer $n \in Z$ which gives all solutions of a trigonometric equation is called the general solution, where Z denotes the set of integers.

The general solutions of some trigonometric functions are given below.

Trigonometric function	General solution
$\sin x = \sin y$	$x = n\pi + (-1)^n y,\ n \in Z$
$\cos x = \cos y$	$x = 2n\pi \pm y,\ n \in Z$
$\tan x = \tan y$	$x = n\pi + y,\ n \in Z$ $x, y \neq (2n+1)\dfrac{\pi}{2}$
$\text{cosec}\, x = \text{cosec}\, y$	$x = n\pi + (-1)^n y,\ n \in Z$ $x, y \neq n\pi$
$\sec x = \sec y$	$x = 2n\pi \pm y,\ n \in Z$ $x, y \neq (2n+1)\dfrac{\pi}{2}$

Trigonometric function	General solution
$\cot x = \cot y$	$x = n\pi + y,\ n \in Z$ $x, y \neq n\pi$
$\sin^2 x = \sin^2 y$ or $\cos^2 x = \cos^2 y$ or $\sec^2 x = \sec^2 y$ or $\text{cosec}^2 x = \text{cosec}^2 y$	$x = n\pi \pm y,\ n \in Z$
$\tan^2 x = \tan^2 y$	$x = n\pi \pm y,\ n \in Z$ $x, y \neq (2n+1)\frac{\pi}{2}$
$\cot^2 x = \cot^2 y$	$x = n\pi \pm y,\ n \in Z$ $x, y \neq n\pi$
$\sin x = \sin y$ and $\cos x = \cos y$	$x = 2n\pi + y,\ n \in Z$

The general and principal solutions of some trigonometric functions are given in the table below.

Trigonometric function	Principal solution (x)	General solution
$\sin x = 0$	$x = 0$ or π	$x = n\pi$
$\sin x = 1$	$x = \frac{\pi}{2}$	$x = n\pi + (-1)^n \frac{\pi}{2}$
$\sin x = -1$	$x = \frac{3\pi}{2}$	$x = n\pi + (-1)^n \frac{3\pi}{2}$
$\sin x = -\frac{1}{2}$	$x = \frac{7\pi}{6}$ or $\frac{11\pi}{6}$	$x = n\pi + (-1)^n \frac{7\pi}{6}$
$\sin x = \frac{\sqrt{3}}{2}$	$x = \frac{\pi}{3}$ or $\frac{2\pi}{3}$	$x = n\pi + (-1)^n \frac{\pi}{3}$
$\cos x = 0$	$x = \frac{\pi}{2}$ or $\frac{3\pi}{2}$	$x = 2n\pi \pm \frac{\pi}{2}$
$\cos x = 1$	$x = 0$	$x = 2n\pi$
$\cos x = -1$	$x = \pi$	$x = 2n\pi \pm \pi$

Trigonometric function	Principal solution (x)	General solution
$\cos x = \frac{1}{\sqrt{2}}$	$x = \frac{\pi}{4}$ or $\frac{7\pi}{4}$	$x = 2n\pi \pm \frac{\pi}{4}$
$\cos x = -\frac{1}{\sqrt{2}}$	$x = \frac{3\pi}{4}$ or $\frac{5\pi}{4}$	$x = 2n\pi \pm \frac{3\pi}{4}$
$\tan x = 1$	$x = \frac{\pi}{4}$ or $\frac{5\pi}{4}$	$x = n\pi + \frac{\pi}{4}$
$\tan x = -1$	$x = \frac{3\pi}{4}$	$x = n\pi + \frac{3\pi}{4}$

Sine Formulae: In any triangle, the sides are proportional to the sines of the opposite angles, that is, in a triangle ABC,

$$\frac{\sin A}{a} = \frac{\sin B}{b} = \frac{\sin C}{c}$$

Cosine Formulae

$$a^2 = b^2 + c^2 - 2bc\cos A$$

$$b^2 = c^2 + a^2 - 2ca\cos B$$

$$c^2 = a^2 + b^2 - 2ab\cos C$$

Or

$$\cos A = \frac{b^2 + c^2 - a^2}{2bc}$$

$$\cos B = \frac{c^2 + a^2 - b^2}{2ac}$$

$$\cos C = \frac{a^2 + b^2 - c^2}{2ab}$$

Projection Formulae

$$a = b\cos C + c\cos B$$

$$b = c\cos A + a\cos C$$

$$c = a\cos B + b\cos A$$

Napier's Analogies

$$\tan\frac{B-C}{2} = \frac{b-c}{b+c}\cot\frac{A}{2}$$

$$\tan\frac{C-A}{2} = \frac{c-a}{c+a}\cot\frac{B}{2}$$

$$\tan\frac{A-B}{2} = \frac{a-b}{a+b}\cot\frac{C}{2}$$

Day 59

Unit Test I (Chapters I–III)

Time: 3 Hrs Maximum **Marks:** 80

General Instructions

(i) All the questions are compulsory.
(ii) The question paper consists of 36 questions divided into 4 sections A, B, C and D.
(iii) Section A comprises 20 questions of 1 mark each. Section B comprises 6 questions of 2 marks each. Section C comprises 6 questions of 4 marks each. Section D comprises 4 questions of 6 marks each.
(iv) There is no overall choice. However, an internal choice has been provided in four questions of 1 mark each, two questions of 2 marks each, two questions of 4 marks each, and two questions of 6 marks each. You have to attempt only one of the alternatives in all such questions.
(v) Use of calculators is not permitted.

SECTION A

Q1–Q10 are multiple choice type questions. Select the correct option. (1 mark each)

Q1. If $\sin\theta = \cos\theta$, then the value of $\sin 2\theta$
(a) 1 (b) ±1 (c) 0 (d) –1

Q2. The equality $\sin A + \sin 2A = 2$ holds for some real value of A.
(a) $A = (2n-1)\frac{\pi}{2},\ n \in Z$
(b) $A = n\pi,\ n \in Z$
(c) Possible for all real values of A
(d) No value of A is possible

Q3. Two finites sets have m and n elements. The number of subsets of the first set is 112 more than of the second set. The values of m and n are, respectively,
(a) 4, 7 (b) 7, 4 (c) 2, 5 (d) 7, 7

Q4. In a town of 100 persons, 45 persons read Hindi, 30 read English and 20 read both. Then, the number of persons who read neither is
(a) 25 (b) 45 (c) 35 (d) 55

Q5. $X = \{2^n - 1 | n \in N\}$, and
$Y = \{2n - 1 | n \in N\}$. Then
(a) $X \subset Y$ (b) $Y \subset X$
(c) $X = Y$ (d) $X \cap Y = \phi$

Q6. A survey shows that 60% of people watch A news channel whereas 70% watch B news channel. If x% of people watch both channels, then
(a) $x = 30$ (b) $x = 60$
(c) $30 \le x \le 60$ (d) $60 \le x \le 70$

Q7. If sets A and B are defined as
$A = \{(x, y) | y = \frac{1}{x},\ x \ne 0,\ x \in I\}$,
$B = \{(x, y) | y = x,\ x \in I\}$, then
(a) $A \cap B = 1$ (b) $A \cap B = -1$
(c) $A \cap B = \phi$ (d) $A \cap B = \{-1, 1\}$

Q8. $S = \{x | x$ is a positive multiple of 5 less than and equal to 100$\}$ and $P = \{x | x$ is a prime number less than 25$\}$. Then, $n(S) + n(P)$ is equal to
(a) 329 (b) 31 (c) 33 (d) 41

Q9. If $[x]^2 - 4 = 0$, where [.] denotes the greatest integer function, then
(a) $x \in [-2, -1) \cup [2, 3)$ (b) $x \in [-2, -1) \cup [3, 4)$
(c) $x \in [-3, -2) \cup [2, 3)$ (d) $x \in [-3, -2) \cup [3, 4)$

Q10. If $f(x) = ax^2 + bx - 1$, where a and b are integers, and $f(-1) = f(1) = 0$, then a and b are equal to
(a) $a = 1,\ b = 0$ (b) $a = -1,\ b = 0$
(c) $a = 0, b = 2$ (d) $a = 0,\ b = 1$

Q11–Q15 are fill in the blanks. (1 mark each)

Q11. Power set of the set $A = \{\phi, 0\}$ is______

Q12. $U = \{1, 2, 3, 4, 5, 6, 7, 8, 9\}$, $A = \{1, 3, 5, 7, 9\}$ and $B = \{2, 4, 6, 8\}$, $C = \{1, 2, 8, 9\}$. Then,
(i) $(B \cup C)'$ is______
(ii) $(A - C)'$ is______

Q13. If $\sin\theta + \cos\theta = 0$, then the general value of θ is______

OR

Q13. The domain of the function given by $f(x) = \frac{1}{\sqrt{x - |x|}}$ is______

Q14. The range of the function given by $f(x) = 1 - 2\cos x$ is ______

OR

Q14. If $\cot\theta + \tan\theta = \text{cosec}\,\theta$, then the general value of θ is ______

Q15. If $(xy - 1, x + y) = (-1, 2)$ are two equal ordered pairs, then the value of the ordered pair (x, y) is ______

OR

Q15. The domain and the range of the real function f defined by $f(x)=\sqrt{(1-x)}$ is ______

(Q16–Q20) Answer the following questions.
(1 mark each)

Q16. Write the following sets in the roster form:
(i) $A=\{x : x \text{ is an integer and } -2 < x \leq 2\}$.
(ii) $B=\{x : x \text{ is a prime number which is a divisor of } 20\}$

Q17. Write the following sets in the set-builder form:
(i) {-3, 0, 3, 6} (ii) {3, 5, 9, 17, 33}

Q18. Express the following functions as sets of ordered pairs and determine their range. $f : x \to R$, $f(x)=x^2-1$, where $x=\{-1,0,1\}$

Q19. Let $A=\{1,2\}, B=\{2,3\}, C=\{1,3\}$;
verify that $A\times(B\cap C)=(A\times B)\cap(A\times C)$

OR

Q19. Redefine the function
$f(x)=|x^2-1|+|x|,\ -2\leq x\leq 2$

Q20. Find the value of tan 15°.

OR

Q20. Find the value of
$4\sin 15^\circ\cos^3 15^\circ-4\cos 15^\circ\sin^3 15^\circ$

SECTION B

(Q21–Q26) Solve the following questions.
(2 marks each)

Q21. Let $f, g : R\to R$ be defined by
$f(x)=x-1,\quad g(x)=x^2-1, x\neq 1$. Find $g-f$ and $\dfrac{g}{f}$.

Q22. Let $A=\{1,2,3,....9\}$. Define a relation R from A to A by $R=\{(x,y): 2x-y=0, \text{ where } x,y\in A\}$. Write down its domain, co-domain and range. And show that it is a function.

Q23. If θ lies in the fourth quadrant and $\cos\theta=\dfrac{4}{5}$, then find the value of
$\cos(45^\circ-\theta)+\cos(135^\circ-\theta)+\cos(225^\circ-\theta)$.

Q24. Find the general solution of the equation
$4\cos^2+8\sin^2\theta-7=0$.

Q25. Let $A=\{x : x=3n,\ n\in W\}$. Insert the appropriate symbol $\in$ or $\notin$ in the blank spaces:
(i) 0 ____ A (ii) 2 ____ A (iii) 3 ____ A
(iv) 5 ____ A (v) 6 ____ A

OR

Q25. Let the sets $U=\{-2,-1,0,1,2\}$
$A=\{x : x \text{ is a natural number and } 0<x\leq 2\}$
$B=\{x : x \text{ is an integer and } x^2\leq 1\}$
Find $A\cap B,\ A\cup B,\ A',\ B-A$.

Q26. If $2\sin^2\theta=\cos\theta+2$, where $0\leq\theta\leq 2\pi$, then find the value of θ.

OR

Q26. If $\tan(\pi\cos\theta)=\cot\left(\dfrac{\pi}{2}\sin\theta\right)$, then $\theta=$ ____ where $\theta\in[0,\pi/2]$.

SECTION C

(Q27–Q32) Solve the following questions.
(4 marks each)

Q27. If $\cos(\alpha+\beta)=\dfrac{3}{5}$, $\sin(\alpha-\beta)=\dfrac{4}{5}$, α lie between $0, \dfrac{\pi}{4}$, then find that value of $\tan 2\alpha$.

Q28. Prove that
$\left(\cos\dfrac{3\theta}{2}-\cos\dfrac{5\theta}{2}\right)\cos\dfrac{\theta}{2}=\sin 2\theta\sin\theta$.

Q29. If $\dfrac{\sin(x+y)}{\sin(x-y)}=3$, then find the value of $\dfrac{\tan x}{\tan y}$.

Q30. Find the general solution of the equation $\operatorname{cosec} x=\cot x-1$.

Q31. If $\cos 5x+\cos x=0$ where $0\leq x\leq\dfrac{\pi}{2}$, then find the value of x.

OR

Q31. Find the domain and range of the real function $f(x)=\dfrac{1}{1+\sin x}$.

Q32. Find the domain and range of the real function $f(x)=\dfrac{x-2}{1+x}$.

OR

Q32. Find the general of the equation
$\sin x-3\sin 2x+\sin 3x=\cos x-3\cos 2x+\cos 3x$.

SECTION D

(Q33–Q36) Solve the following questions. **(6 marks each)**

Q33. Let $A = \{1, \{2, 3\}\}$. Which of the following statements are incorrect and why?

(i) $1 \in A$ (ii) $1 \subset A$

(iii) $\{2, 3\} \subset A$ (iv) $\{2, 3\} \in A$

(v) $\{\{2, 3\}\} \subset A$ (vi) $\{1, 2\} \in A$

(vii) $\{1, 2\} \subset A$

Q34. In a survey of 50 people, it was found that 20 people read the Hindi newspaper, 17 read the English newspaper, 15 read the local language newspaper, 9 read both Hindi and English, 10 read both English and the local language, 8 read both Hindi and the local language, 5 read all three newspapers. Find:

(i) the number of people who read at least one of the newspapers.

(ii) the number of people who read exactly one newspaper.

Q35. Find the value of

$\sin^4 \dfrac{3\pi}{8} + \sin^4 \dfrac{7\pi}{8} + \cos^4 \dfrac{5\pi}{8} + \cos^4 \dfrac{7\pi}{8}$.

OR

Q35. If $\dfrac{1 - \cos\alpha + \sin\alpha}{1 + \sin\alpha} = x$, then find the value of $\dfrac{\sin\alpha}{1 + \cos\alpha + \sin\alpha}$.

Q36. Find the value of $\cos\dfrac{2\pi}{7} \cos\dfrac{4\pi}{7} \cos\dfrac{8\pi}{7}$.

OR

Q36. For any triangle ABC, prove that:

$\dfrac{\sin(B - C)}{\sin A} = \dfrac{b^2 - c^2}{a^2}$.

Day 60

Unit Test 1 (Solutions)

SECTION A

Q1–Q10 are multiple choice type questions. Select the correct option. **(1 mark each)**

Q1. **Solution: (b)**

$\because \sin\theta = \cos\theta$. Squaring both sides, we get

$\sin^2\theta = \cos^2\theta \Rightarrow \cos^2\theta - \sin^2\theta = 0$

$\Rightarrow \cos 2\theta = 0 \Rightarrow \sin 2\theta = \pm 1$

Q2. **Solution: (d)**

The maximum value of $\sin\theta = 1$; therefore $\sin A + \sin 2A = 2$ is possible if and only if $\sin A = \sin 2A = 1$, which is not possible for any real value of A.

Q3. **Solution: (c)**

Since the number of subsets of the set containing m elements is 112 more than the subsets of the set containing n elements,

$\because 2^m - 2^n = 28 \Rightarrow 2^n \cdot (2^{m-n} - 1) = 2^2 \cdot 7$

$\Rightarrow 2^n = 2^2$ and $2^{m-n} - 1 = 7$

$\Rightarrow n = 2$ and $2^{m-n} = 8 = 2^3$

$\Rightarrow m - n = 3 \Rightarrow m - 2 = 3 \Rightarrow m = 5$

Q4. **Solution: (b)**

Let H be the set of persons who read Hindi and E be the set of persons who read English; then,

$n(U) = 100, n(H) = 45, n(E) = 30, n(H \cap E) = 20$

Number of persons who read neither is $= n(H' \cap E')$

$= n(U) - n(H \cup E) = n(U) - [n(H) + n(E) - n(H \cap E)]$

$= 100 - (45 + 30 - 20) = 120 - 75 = 45$

Q5. **Solution: (a)**

$X = \{2^n - 1 | n \in N\} = \{1, 3, 7, 15, ...\}$

$Y = \{2n - 1 | n \in N\} = \{1, 3, 5, 7, 9, 11, 13, 15, ...\}$

Every element of X is in Y but every element of Y is not in X; therefore $X \subset Y$.

Q6. **Solution: (c)**

Let A be the set of percentage of people who watch the A news channel and B be the set of percentage of people who watch the B news channel. Then

$n(A) = 60, \ n(B) = 70, \ n(A \cap B) = x$

$\because n(A \cup B) \le 100$

$\therefore n(A) + n(B) - n(A \cap B) \le 100$

$\Rightarrow 60 + 70 - x \le 100 \Rightarrow 130 - x \le 100$

$\Rightarrow 130 - 100 \le x \Rightarrow 30 \le x$

Also, $n(A \cap B) \le n(A) \Rightarrow x \le 60$

Thus, $30 \le x \le 60$

Q7. Solution: (d)

Since for $x \in I$, $x = \frac{1}{x} \Rightarrow x^2 = 1 \Rightarrow x = \pm 1$, that is,
$A = \{...-1/2, -1, 1, 1/2, ...\}$, $B = \{...-2, -1, 1, 2, ...\}$
$\Rightarrow A \cap B = \{-1, 1\}$

Q8. Solution: (a)

$S = \{x | x$ is a positive multiple of 5 less than and equal to 100$\} \Rightarrow S = \{5, 10, 15, ..., 95, 100\}$
$\because t_n = a + (n-1)d \Rightarrow 100 = 5 + (n-1)5 \Rightarrow n = 20$
Thus, $n(S) = 33$
$P = \{x | x$ is a prime number less than 25$\}$
$\Rightarrow P = \{2, 3, 5, 7, 11, 13, 17, 19, 23\} \Rightarrow n(P) = 9$
Thus, $n(S) + n(P) = 20 + 9 = 29$

Q9. Solution: (a)

$[x]^2 - 4 = 0 \Rightarrow ([x]-2)([x]+2) = 0$
$\Rightarrow [x] = -2, 2$
For $x \in [-2, -1) \Rightarrow [x] = -2$
and For $x \in [2, 3) \Rightarrow [x] = 2$

Q10. Solution: (a)

$f(x) = ax^2 + bx - 1$
$\Rightarrow f(-1) = a(-1)^2 + b(-1) - 1 = a - b - 1 = 0$
$f(1) = a(1)^2 + b(1) - 1 = a + b - 1 = 0$
$\Rightarrow a = 1$ and $b = 0$

Q11–Q15 are fill in the blanks. (1 mark each)

Q11. Solution:

Power set of the set A is $P(A) = \{\phi, \{\phi\}, \{0\}, \{\phi, 0\}\}$

Q12. Solution:

(i) $\because B \cup C = \{1, 2, 4, 6, 8, 9\}$
$\therefore (B \cup C)' = U - (B \cup C) = \{3, 5, 7\}$

(ii) $\because A - C = \{3, 5, 7\}$
$\therefore (A - C)' = U - (A - C) = \{1, 2, 4, 6, 8, 9\}$

Q13. Solution:

$\sin\theta + \cos\theta = 0$. Squaring both sides, we get
$\sin^2\theta + \cos^2\theta + 2\sin\theta \cdot \cos\theta = 0$
$\Rightarrow 1 + 2\sin\theta \cdot \cos\theta = 0 \Rightarrow \sin 2\theta = -1$
$[\because \sin 2x = 2\sin x \cos x]$
$\Rightarrow 2\theta = n\pi + (-1)^n \cdot \frac{3\pi}{2} \Rightarrow \theta = \frac{n\pi}{2} + (-1)^n \cdot \frac{3\pi}{4}$

OR

Q13. Solution:

$$f(x) = \frac{1}{\sqrt{x - |x|}}$$

$$\because x - |x| = \begin{cases} = x - (-x) = 2x, & x < 0 \\ = x - x = 0, & x \geq 0 \end{cases}$$

Since $f(x)$ is not defined when $x - |x| = 0$,
Domain of $f = R^-$

Q14. Solution:

$\because -1 \leq \cos x \leq 1 \Rightarrow -2 \leq 2\cos x \leq 2$
$\Rightarrow 2 \geq -2\cos x \geq -2$
$\Rightarrow 1 + 2 \geq 1 - 2\cos x \geq 1 - 2$
$\Rightarrow 3 \geq f(x) \geq -1$
Thus, range of $f = [-1, 3]$

OR

Q14. Solution:

$$\cot\theta + \tan\theta = \operatorname{cosec}\theta \Rightarrow \frac{\cos\theta}{\sin\theta} + \frac{\sin\theta}{\cos\theta} = \frac{1}{\sin\theta}$$

$$\Rightarrow \frac{\cos^2 + \sin^2\theta}{\sin\theta \cdot \cos\theta} = \frac{1}{\sin\theta} \Rightarrow \frac{1}{\cos\theta} = 1 \Rightarrow \cos\theta = 1$$

$\Rightarrow \cos\theta = \cos 0° \Rightarrow \theta = n\pi$

Q15. Solution:

$\because (xy - 1, x + y) = (-1, 2)$
$\therefore xy - 1 = -1$ and $x + y = 2$
$\Rightarrow x(2 - x) = 0 \Rightarrow x = 0, 2 \Rightarrow y = 2, 0$
$\Rightarrow (x, y) = (0, 2), (2, 0)$

OR

Q15. Solution:

The given function $f(x) = \sqrt{(1-x)}$ is defined when $1 - x \geq 0 \Rightarrow x \leq 1$.
Therefore, the domain of f is the set of all real numbers greater than or equal to –1, that is, the domain of $f = (-\infty, 1]$.
Now, for $x \leq 1$, $\sqrt{(1-x)} \geq 0$
Therefore, the range of f is the set of all real numbers greater than or equal to 0, that is, the range of $f = [0, \infty)$.

(Q16–Q20) Answer the following questions. (1 mark each)

Q16. Solution:

(i) $A = \{x : x$ is an integer and $-2 < x \leq 2\}$
The elements of the set A are $-1, 0, 1, 2$; therefore, in the roster form, set $A = \{-1, 0, 1, 2\}$.

(ii) $B = \{x : x$ is a prime number which is a divisor of 20$\}$
The elements of set B are 2, 3 and 5 $(36 = 2 \times 2 \times 3 \times 3)$; therefore, in the roster form, set $B = \{2, 3\}$.

Q17. Solution:

(i) –3, 0, 3, 6, are multiples of 3 for integers –1 to 2; therefore, in the set-builder form, it is written as $\{x : x = 3n,\ n \in I,\ -1 \le n \le 2\}$.

(ii) 3, 5, 9, 17, 33 are powers of 2 for natural numbers 1 to 5 plus 1; therefore, in the set-builder form, it is written as $\{x : x = 2^n + 1,\ n \in N,\ 1 \le n \le 5\}$.

Q18. Solution:

$f : x \to R, f(x) = x^2 - 1,$

$f(-1) = (-1)^2 - 1 = 1 - 1 = 0$

$f(0) = (0)^2 - 1 = 0 - 1 = -1$

$f(1) = (1)^2 - 1 = 1 - 1 = 0$

$\therefore f = \{(-1,0),(0,-1),(1,0)\}$

Thus, the range of $f = \{-1, 0\}$.

Q19. Solution:

To verify $A \times (B \cap C) = (A \times B) \cap (A \times C)$

$B \cap C = \{2,3\} \cap \{1,3\} = \{3\}$

$\therefore$ LHS $= A \times (B \cap C) = \{1,2\} \times \{3\} = \{(1,3),(2,3)\}$

Now, $A \times B = \{(1,2),(1,3),(2,2),(2,3)\}$ and

$A \times C = \{(1,1),(1,3),(2,2),(2,3)\}$

$\therefore$ RHS $= A \times (B \cap C) = (A \times B) \cap (A \times C)$

$= \{(1,2),(1,3),(2,2),(2,3)\} \cap \{(1,1),(1,3),(2,2),(2,3)\}$

$= \{(1,3),(2,3)\}$

Thus, $A \times (B \cap C) = (A \times B) \cap (A \times C)$.

OR

Q19. Solution:

$\because f(x) = |x^2 - 1| + |x|,\ -2 \le x \le 2$

$$\therefore f(x) = \begin{cases} (x^2-1)-(x), & -2 \le x < -1 \\ -(x^2-1)-(x), & -1 \le x < 0 \\ -(x^2-1)+(x), & 0 \le x \le 1 \\ (x^2-1)+(x), & 1 \le x < 2 \end{cases}$$

$$\Rightarrow f(x) = \begin{cases} x^2-x-1, & -2 \le x < -1 \\ -x^2-x+1, & -1 \le x < 0 \\ -x^2+x+1, & 0 \le x \le 1 \\ x^2+x-1, & 1 \le x < 2 \end{cases}$$

Q20. Solution:

$$\tan\frac{\theta}{2} = \frac{\sin\frac{\theta}{2}}{\cos\frac{\theta}{2}} = \frac{2\sin\frac{\theta}{2}\cdot\cos\frac{\theta}{2}}{2\cos^2\frac{\theta}{2}} = \frac{\sin\theta}{1+\cos\theta}$$

Let $\theta = 15°$

$$\therefore \tan 15° = \frac{\sin 30°}{1+\cos 30°} = \frac{1/2}{1+\frac{\sqrt{3}}{2}} = \frac{1}{2+\sqrt{3}} = 2 - \sqrt{3}$$

OR

Q20. Solution:

$\because 4\sin 15°\cos^3 15° - 4\cos 15°\sin^3 15°$

$= 2(2\sin 15°\cdot\cos 15°)(\cos^2 15° - \sin^2 15°)$

$= 2\sin 30°\cdot\cos 30° = \sin 60° = \sqrt{3}/2$

$$\begin{bmatrix} \because \sin 2A = 2\sin A\cdot\cos A \\ \cos 2A = \cos^2 A - \sin^2 A \end{bmatrix}$$

SECTION B

(Q21–Q26) Solve the following questions. **(2 marks each)**

Q21. Solution:

$\because f(x) = x - 1,\ \ g(x) = x^2 - 1$; therefore,

$(g - f)(x) = g(x) - f(x) = (x^2 - 1) - (x - 1)$

$\Rightarrow (f + g)(x) = x^2 - x$

$$\left(\frac{g}{f}\right)(x) = \frac{g(x)}{f(x)} = \frac{(x^2-1)}{(x-1)} = \frac{(x-1)(x+1)}{(x-1)} = x+1$$

$$\left(\frac{f}{g}\right)(x) = \frac{f(x)}{g(x)} = \frac{(x+1)}{(x^2-1)} = \frac{1}{x-1}$$

Q22. Solution:

The relation R from A to A is given by

$R = \{(x, y) : 2x - y = 0$, where $x, y \in A\}$

$\Rightarrow R = \{(x, y) : 2x = y$, where $x, y \in A\}$

$\therefore R = \{(1,2),(2,4),(3,6),(4,8)\}$

The domain of R is the set of all first elements of the ordered pairs in the relation R.

$\therefore$ Domain of $R = \{1, 2, 3, 4\}$

The whole set A is the co-domain of the relation R.

$\therefore$ Co-domain of $R = \{1,2,3,\ldots,9\}$

The range of R is the set of all second elements of the ordered pairs in the relation R.

$\therefore$ Range of $R = \{2,4,6,8\}$

Since every element has one and only one image, this relation is a function.

Q23. Solution:

$$\because \cos\theta = \frac{4}{5} \Rightarrow \sin\theta = \sqrt{1 - \frac{16}{25}} = \pm\frac{3}{5}$$

$\Rightarrow \sin\theta = -\frac{3}{5}$ [since θ lies in the fourth quadrant]

Now, $\cos(45° - \theta) + \cos(135° - \theta) + \cos(225° - \theta)$

$= \cos(45° - \theta) + \cos(90° + 45° - \theta) + \cos(180° + 45° - \theta)$

$= \cos(45° - \theta) - \sin(45° - \theta) - \cos(45° - \theta)$

$= -\sin(45° - \theta) = -\sin 45° \cos\theta - \cos 45° \sin\theta$

$= -\frac{1}{\sqrt{2}} \times \frac{4}{5} - \frac{1}{\sqrt{2}} \times \left(-\frac{3}{5}\right) = -\frac{1}{5\sqrt{2}}$

Q24. Solution:

$4\cos^2 + 8\sin^2\theta - 7 = 0$

$\Rightarrow 4\cos^2 + 8(1 - \cos^2\theta) - 7 = 0$

$\Rightarrow 4\cos^2\theta + 8 - 8\cos^2\theta - 7 = 0$

$\Rightarrow -4\cos^2\theta + 1 = 0 \Rightarrow \cos^2\theta = 1/4$

$\Rightarrow \cos^2\theta = \cos^2\frac{\pi}{3} \Rightarrow \theta = n\pi \pm \frac{\pi}{3}$

$\left[\because \cos^2\theta = \cos^2\alpha \quad \therefore \theta = n\pi \pm \alpha\right]$

Q25. Solution:

$A = \{x : x = 3n,\ n \in W\}$

$\Rightarrow A = \{0, 3, 6, 9, ...\}$, therefore

(i) $0 \in A$ (ii) $2 \notin A$

(iii) $3 \in A$ (iv) $5 \notin A$

(v) $6 \in A$

OR

Q25. Solution:

$A = \{1, 2\}$, $B = \{-1, 0, 1\}$ and

$U = \{-2, -1, 0, 1, 2\}$

$\therefore A \cap B = \{1\}$ $A \cup B = \{-1, 0, 1, 2\}$

$B - A = \{-1, 0\}$ $A' = U - A = \{-2, 2\}$

Q26. Solution:

$2\sin^2\theta = \cos\theta + 2 \Rightarrow 2\sin^2\theta - 2 = \cos\theta$

$\Rightarrow -2\cos^2\theta = \cos\theta \Rightarrow \cos\theta(1 + 2\cos\theta) = 0$

Thus, $\cos\theta = 0 \Rightarrow \theta = \frac{\pi}{2}, \frac{3\pi}{2}$

And, $1 + 2\cos\theta = 0 \Rightarrow \cos\theta = \frac{-1}{2}$

$\Rightarrow \theta = \frac{\pi}{2} + \frac{\pi}{6},\ \pi + \frac{\pi}{3} = \frac{2\pi}{3}, \frac{4\pi}{3}$

Thus, the values of θ are $\frac{\pi}{2}, \frac{3\pi}{2}, \frac{2\pi}{3}, \frac{4\pi}{3}$.

OR

Q26. Solution:

$\tan\left(\frac{\pi}{2}\cos\theta\right) = \cot\left(\frac{\pi}{2}\sin\theta\right)$

$\Rightarrow \tan\left(\frac{\pi}{2}\cos\theta\right) = \tan\left[\frac{\pi}{2} - \left(\frac{\pi}{2}\sin\theta\right)\right]$

$\Rightarrow \frac{\pi}{2}\cos\theta = \frac{\pi}{2} - \frac{\pi}{2}\sin\theta \Rightarrow \cos\theta = 1 - \sin\theta$

$\Rightarrow \sin\theta + \cos\theta = 1$

$\Rightarrow \frac{1}{\sqrt{2}} \cdot \sin\theta + \frac{1}{\sqrt{2}} \cdot \cos\theta = \frac{1}{\sqrt{2}}$

$\Rightarrow \sin\theta \cdot \cos\frac{\pi}{4} + \cos\theta \cdot \sin\frac{\pi}{4} = \pm\frac{1}{\sqrt{2}}$

$\Rightarrow \sin\left(\theta + \frac{\pi}{4}\right) = \pm\frac{1}{\sqrt{2}} = \sin\left(\pm\frac{\pi}{4}\right)$

$\Rightarrow \theta = 0, \frac{\pi}{2}$

SECTION C

(Q27–Q32) Solve the following questions.

(4 marks each)

Q27. Solution:

$\because \cos(\alpha + \beta) = \frac{3}{5}$

$\therefore \sin(\alpha + \beta) = \sqrt{1 - \frac{9}{25}} = \sqrt{\frac{16}{25}} = \frac{4}{5}$

$\because \sin(\alpha - \beta) = \frac{4}{5}$

$\therefore \cos(\alpha - \beta) = \sqrt{1 - \frac{16}{25}} = \sqrt{\frac{9}{25}} = \frac{3}{5}$

Now, $\tan(\alpha + \beta) = \frac{\sin(\alpha + \beta)}{\cos(\alpha + \beta)} = \frac{4/5}{3/5} = 4/3$

[since α lies between $0, \frac{\pi}{4}$]

And, $\tan(\alpha - \beta) = \frac{\sin(\alpha - \beta)}{\cos(\alpha - \beta)} = \frac{4/5}{3/5} = \frac{4}{3}$

Thus, $\tan 2\alpha = \tan(\alpha + \beta + \alpha - \beta)$

$= \frac{\tan(\alpha + \beta) + \tan(\alpha - \beta)}{1 - \tan(\alpha + \beta) \cdot \tan(\alpha - \beta)}$

$\left[\because \tan(x + y) = \frac{\tan x + \tan y}{1 - \tan x \cdot \tan y}\right]$

$= \frac{4/3 + 4/3}{1 - \frac{4}{3} \cdot \frac{4}{3}} = \frac{8/3}{-7/9} = -\frac{24}{7}$

Q28. Solution:

$$\text{LHS} = \left(\cos\frac{3\theta}{2} - \cos\frac{5\theta}{2}\right)\cos\frac{\theta}{2}$$

$$= \frac{1}{2}\left[2\cos\frac{3\theta}{2}\cos\frac{\theta}{2} - 2\cos\frac{5\theta}{2}\cos\frac{\theta}{2}\right]$$

$$= \frac{1}{2}\left[\begin{array}{l}\cos\left(\frac{3\theta}{2}+\frac{\theta}{2}\right)+\cos\left(\frac{3\theta}{2}-\frac{\theta}{2}\right)- \\ \cos\left(\frac{5\theta}{2}+\frac{\theta}{2}\right)-\cos\left(\frac{5\theta}{2}-\frac{\theta}{2}\right)\end{array}\right]$$

$[\because 2\cos A\cos B = \cos(A+B)+\cos(A-B)]$

$$= \frac{1}{2}[\cos 2\theta + \cos\theta - \cos 3\theta - \cos 2\theta]$$

$$= \frac{1}{2}[\cos\theta - \cos 3\theta]$$

$$= -\frac{1}{2}\left[2\sin\left(\frac{\theta+3\theta}{2}\right)\cdot\sin\left(\frac{\theta-3\theta}{2}\right)\right]$$

$$\left[\because \cos x - \cos y = -2\sin\frac{x+y}{2}\cdot\sin\frac{x-y}{2}\right]$$

$$= \sin 2\theta\cdot\sin\theta = \text{RHS}$$

Q29. Solution:

$$\because \frac{\sin(x+y)}{\sin(x-y)} = 3 = \frac{3}{1}$$

Using componendo and dividendo,

$$\Rightarrow \frac{\sin(x+y)+\sin(x-y)}{\sin(x-y)-\sin(x-y)} = \frac{3+1}{3-1} = 2$$

$$\Rightarrow \frac{2\sin\left(\frac{x+y+x-y}{2}\right)\cdot\cos\left(\frac{x+y-x+y}{2}\right)}{2\cos\left(\frac{x+y+x-y}{2}\right)\cdot\sin\left(\frac{x+y-x+y}{2}\right)} = 2$$

$$\left[\begin{array}{l}\because \sin x + \sin y = 2\sin\frac{x+y}{2}\cdot\cos\frac{x-y}{2}, \\ \sin - \sin y = 2\cos\frac{x+y}{2}\cdot\sin\frac{x-y}{2}\end{array}\right]$$

$$\Rightarrow \frac{\sin x\cdot\cos y}{\cos x\cdot\sin y} = 2 \Rightarrow \frac{\tan x}{\tan y} = 2$$

Q30. Solution:

$\operatorname{cosec} x = \cot x - 1$

$$\Rightarrow \frac{1}{\sin x} = \frac{\cos x}{\sin x} - 1 \Rightarrow \frac{1}{\sin x} = \frac{\cos x - \sin x}{\sin x}$$

$$\Rightarrow \cos x - \sin x = 1$$

$$\Rightarrow \frac{1}{\sqrt{2}}\cdot\cos x - \frac{1}{\sqrt{2}}\cdot\sin x = \frac{1}{\sqrt{2}}$$

$$\Rightarrow \cos x\cos\frac{\pi}{4} - \sin x\sin\frac{\pi}{4} = \frac{1}{\sqrt{2}}$$

$$\Rightarrow \cos\left(x+\frac{\pi}{4}\right) = \cos\frac{\pi}{4}$$

$$\therefore x + \frac{\pi}{4} = 2n\pi \pm \frac{\pi}{4}$$

For positive sign, $x = 2n\pi + \frac{\pi}{4} - \frac{\pi}{4} = 2n\pi$

For negative sign, $x = 2n\pi - \frac{\pi}{4} - \frac{\pi}{4} = 2n\pi - \frac{\pi}{2}$

Q31. Solution:

Now, $\cos 5x + \cos x = 0$

$$\Rightarrow 2\cos\left(\frac{5x+x}{2}\right)\cdot\cos\left(\frac{5x-x}{2}\right) = 0$$

$$\left[\because \cos x + \cos y = 2\cos\frac{x+y}{2}\cdot\cos\frac{x-y}{2}\right]$$

$$\Rightarrow 2\cos 3x\cdot\cos 2x = 0 \Rightarrow \cos 3x = 0 \text{ or } \cos 2x = 0$$

$$\Rightarrow \cos 3x = \cos\frac{\pi}{2} \text{ or } \cos 2x = \cos\frac{\pi}{2}$$

$$\Rightarrow 3x = \frac{\pi}{2} \text{ or } 2x = \frac{\pi}{2}$$

$$\Rightarrow x = \frac{\pi}{6} \text{ or } x = \frac{\pi}{4}$$

Also,

$$\Rightarrow \cos 3x = \cos\frac{3\pi}{2} \text{ or } \cos 2x = \cos\frac{3\pi}{2}$$

$$\Rightarrow 3x = \frac{3\pi}{2} \text{ or } 2x = \frac{3\pi}{2}$$

$$\Rightarrow x = \frac{\pi}{2} \text{ or } x = \frac{3\pi}{4}$$

Hence, the solutions are $\frac{\pi}{2}, \frac{\pi}{4}, \frac{\pi}{6}$, for $0 \le x \le \frac{\pi}{2}$.

OR

Q31. Solution:

Domain:

$-1 \le \sin x \le 1$; therefore, $f(x) = \frac{1}{1+\sin x}$ is defined for all values of x such that $1+\sin x \ne 0 \Rightarrow \sin x \ne -1$

$$\Rightarrow x \ne (2n+1)\pi + \frac{\pi}{2},\ \forall\, n \in Z$$

Thus, the domain of $f = R - \{(2n+1)\pi + \frac{\pi}{2} : n \in Z\}$

Range:

$$\because -1 \le \sin x \le 1 \Rightarrow 1-1 \le 1+\sin x \le 1+1$$

$$\Rightarrow 0 \le 1+\sin x \le 2$$

$$\Rightarrow \frac{1}{1+\sin x} \le \frac{1}{0} \text{ or } \frac{1}{1+\sin x} \ge \frac{1}{2}$$

$$\Rightarrow f(x) \le \infty \text{ or } f(x) \ge \frac{1}{2}$$

$$\Rightarrow \infty \ge f(x) \ge \frac{1}{2}$$

Thus, the range of $f = \left[\frac{1}{2}, \infty\right)$

Q32. Solution:

Domain:

Rational function $y = \dfrac{f(x)}{g(x)}$ is defined for all real values of x such that $g(x) \neq 0$

$\Rightarrow 1 + x \neq 0 \Rightarrow x \neq -1$

Thus, the domain of f is

$= R - \{-1\} = (-\infty, -1) \cup (-1, \infty)$

Range:

Let $y = \dfrac{x-2}{1+x} \Rightarrow y + xy = x - 2$

$\Rightarrow x - xy = y + 2 \Rightarrow x(1-y) = y + 2 \Rightarrow x = \dfrac{y+2}{1-y}$

Rational function $x = \dfrac{f(y)}{g(y)}$ is defined for all real values of y such that $g(y) \neq 0$

$\Rightarrow 1 - y \neq 0 \Rightarrow y \neq 1$

Thus, the range of f is

$= R - \{1\} = (-\infty, 1) \cup (1, \infty)$

OR

Q32. Solution:

$\sin x - 3\sin 2x + \sin 3x = 3\cos 2x - \cos x - \cos 3x$

$\Rightarrow 2\sin\left(\dfrac{x+3x}{2}\right)\cdot\cos\left(\dfrac{3x-x}{2}\right) - 3\sin 2x$

$= -2\cos\left(\dfrac{3x+x}{2}\right)\cdot\cos\left(\dfrac{3x-x}{2}\right) + 3\cos 2x$

$\Rightarrow 2\sin 2x\cos x - 3\sin 2x = -2\cos 2x\cdot\cos x + 3\cos 2x$

$\Rightarrow \sin 2x(2\cos x - 3) = -\cos 2x(2\cos x - 3)$

$\Rightarrow \dfrac{\sin 2x}{\cos 2x} = -1 \Rightarrow \tan 2x = -1 \Rightarrow \tan 2x = \tan\left(-\dfrac{\pi}{4}\right)$

$\Rightarrow 2x = n\pi - \dfrac{\pi}{4} \Rightarrow x = \dfrac{n\pi}{2} - \dfrac{\pi}{8}$

$[\because \tan\theta = \tan\alpha \quad \therefore \theta = n\pi + \alpha]$

SECTION D

(Q33–Q36) Solve the following questions.

(6 marks each)

Q33. Solution:

(i) The statement $1 \in A$ is correct because 1 is an element of A.

(ii) The statement $1 \subset A$ is incorrect because 1 is element of A, and in subset it is written as $\{1\} \subset A$.

(iii) The statement $\{2,3\} \subset A$ is incorrect because $\{2,3\}$ is not a subset while it is an element $\{2,3\} \in A$.

(iv) The statement $\{2,3\} \in A$ is correct because $\{2,3\}$ is an element of A.

(v) Statement $\{\{2,3\}\} \subset A$ is correct because $\{2,3\}$ is an element of A implies $\{\{2,3\}\}$ is a subset of A.

(vi) The statement $\{1,2\} \in A$ is incorrect because 2 not an element of A.

(vii) The statement $\{1,2\} \subset A$ is incorrect because 2 not an element of A.

Q34. Solution:

Let H be the set of people who read the Hindi newspaper, E be the set of people who read the English newspaper and L be the set of people who read the local language newspaper.

Then, $n(H) = 20,\ n(E) = 17,\ n(L) = 15$

$n(H \cap E) = 9, \quad n(E \cap L) = 10$

$n(H \cap L) = 8, \quad n(H \cap E \cap L) = 5$

Let U be the set of people who took part in the survey, then,

(i) Number of people who read at least one of the newspapers is equal to $n(H \cup E \cup L)$

$= n(H) + n(E) + n(L) - n(H \cap E)$

$-n(E \cap L) - n(H \cap L) + n(H \cap E \cap L)$

$= 20 + 17 + 15 - 9 - 10 - 8 + 5 = 30$

Thus, 30 people read at least one of the newspapers.

(ii) Number of people who read exactly one newspaper is

$= n(H \cup T \cup I) - n(H \cap T)$

$-n(T \cap I) - n(H \cap I) + 2n(H \cap T \cap I)$

$= 30 - 9 - 10 - 8 + 2 \times 5 = 13$

Thus, 13 people read exactly one newspaper.

Q35. Solution:

$\sin^4\dfrac{3\pi}{8} + \sin^4\dfrac{7\pi}{8} + \cos^4\dfrac{5\pi}{8} + \cos^4\dfrac{7\pi}{8}$

$= \sin^4\left(\dfrac{\pi}{2} - \dfrac{\pi}{8}\right) + \sin^4\left(\dfrac{\pi}{2} + \dfrac{3\pi}{8}\right)$

$+ \cos^4\left(\pi - \dfrac{3\pi}{8}\right) + \cos^4\left(\pi - \dfrac{\pi}{8}\right)$

$= \cos^4\dfrac{\pi}{8} + \cos^4\dfrac{3\pi}{8} + \cos^4\dfrac{3\pi}{8} + \cos^4\dfrac{\pi}{8}$

$= 2\left[\cos^4\dfrac{\pi}{8} + \cos^4\dfrac{3\pi}{8}\right] = 2\left[\cos^4\dfrac{\pi}{8} + \cos^4\left(\dfrac{\pi}{2} - \dfrac{\pi}{8}\right)\right]$

$= 2\left[\cos^4\dfrac{\pi}{8} + \sin^4\dfrac{\pi}{8}\right]$

$$= 2\left[\left(\cos^2\frac{\pi}{8}+\cos^2\frac{\pi}{8}\right)^2 - 2\cos^2\frac{\pi}{8}\cdot\sin^2\frac{\pi}{8}\right]$$

$$= 2\left[1-2\cos^2\frac{\pi}{8}\cdot\sin^2\frac{\pi}{8}\right] = 2-\left(2\sin\frac{\pi}{8}\cdot\cos\frac{\pi}{8}\right)^2$$

$$= 2-\left(\sin\frac{2\pi}{8}\right)^2 = 2-\left(\frac{1}{\sqrt{2}}\right)^2 = 2-\frac{1}{2}=\frac{3}{2}$$

OR

Q35. Solution:

$$\because \frac{1-\cos\alpha+\sin\alpha}{1+\sin\alpha} = x$$

$$= \frac{(1-\cos\alpha+\sin\alpha)}{(1+\sin\alpha)}\cdot\frac{(1+\cos\alpha+\sin\alpha)}{(1+\cos\alpha+\sin\alpha)}$$

$$= \frac{\{(1+\sin\alpha)-\cos\alpha\}}{(1+\sin\alpha)}\cdot\frac{\{(1+\sin\alpha)+\cos\alpha\}}{(1+\cos\alpha+\sin\alpha)}$$

$$= \frac{(1+\sin\alpha)^2-\cos^2\alpha}{(1+\sin\alpha)(1+\sin\alpha+\cos\alpha)}$$

$$= \frac{1+\sin^2\alpha+2\sin\alpha-1+\sin^2\alpha}{(1+\sin\alpha)(1+\sin\alpha+\cos\alpha)}$$

$$= \frac{2\sin^2\alpha+2\sin\alpha}{(1+\sin\alpha)(1+\sin\alpha+\cos\alpha)}$$

$$= \frac{2\sin\alpha(1+\sin\alpha)}{(1+\sin\alpha)(1+\sin\alpha+\cos\alpha)}$$

$$= \frac{2\sin\alpha}{1+\sin\alpha+\cos\alpha} = x$$

$$\therefore \frac{\sin\alpha}{1+\sin\alpha+\cos\alpha} = \frac{x}{2}$$

Q36. Solution:

$$\cos\frac{2\pi}{7}\cos\frac{4\pi}{7}\cos\frac{8\pi}{7}$$

$$= \frac{2\sin\frac{2\pi}{7}\cos\frac{2\pi}{7}}{2\sin\frac{2\pi}{7}}\cos\frac{4\pi}{7}\cos\frac{8\pi}{7}$$

$$= \frac{\sin\frac{4\pi}{7}}{2\sin\frac{2\pi}{7}}\cos\frac{4\pi}{7}\cos\frac{8\pi}{7}$$

$$= \frac{2\sin\frac{4\pi}{7}\cos\frac{4\pi}{7}}{4\sin\frac{2\pi}{7}}\cos\frac{8\pi}{7} = \frac{\sin\frac{8\pi}{7}}{4\sin\frac{2\pi}{7}}\cos\frac{8\pi}{7}$$

$$= \frac{2\sin\frac{8\pi}{7}\cos\frac{8\pi}{7}}{8\sin\frac{2\pi}{7}} = \frac{\sin\frac{16\pi}{7}}{8\sin\frac{2\pi}{7}}$$

$$= \frac{\sin\left(2\pi+\frac{2\pi}{7}\right)}{8\sin\frac{2\pi}{7}} = \frac{\sin\frac{2\pi}{7}}{8\sin\frac{2\pi}{7}} = \frac{1}{8}$$

OR

Q36. Solution:

$$\text{LHS} = \frac{\sin(B-C)}{\sin A} = \frac{\sin(B-C)}{\sin(B+C)}$$

$$\left[\begin{array}{l}\because A+B+C=180^\circ \Rightarrow B+C=180^\circ-A\\ \Rightarrow \sin(B+C)=\sin(180^\circ-A)=\sin A\end{array}\right]$$

$$= \frac{\sin B\cos C-\cos B\sin C}{\sin B\cos C+\cos B\sin C}$$

$$= \frac{kb\left(\frac{a^2+b^2-c^2}{2ab}\right)-\left(\frac{a^2+c^2-b^2}{2ac}\right)kc}{kb\left(\frac{a^2+b^2-c^2}{2ab}\right)+\left(\frac{a^2+c^2-b^2}{2ac}\right)kc}$$

$$\left[\begin{array}{l}\because \frac{a}{\sin A}=\frac{b}{\sin B}=\frac{c}{\sin C}=k\\ \cos A=\frac{b^2+c^2-a^2}{2bc}\end{array}\right]$$

$$= \frac{\left(\frac{a^2+b^2-c^2}{2a}-\frac{a^2+c^2-b^2}{2a}\right)}{\left(\frac{a^2+b^2-c^2}{2a}+\frac{a^2+c^2-b^2}{2a}\right)}$$

$$= \frac{a^2+b^2-c^2-a^2-c^2+b^2}{a^2+b^2-c^2+a^2+c^2-b^2}$$

$$= \frac{2(b^2-c^2)}{2(a^2)} = \frac{b^2-c^2}{a^2} = \text{RHS}$$

CHAPTER

4

Principle of Mathematical Induction

Day 61

Mathematical Induction-I

Let us check whether a given statement $P(n)$ for a natural number n is true or false.

Process:

1. Check the statement $P(n)$ for $n=1$. If $P(1)$ is true, proceed to the next step.
2. Assume that the statement is true for $n=k$ (where k is some positive integer), that is, $P(k)$ is true.
3. Now using statement $P(k)$ to be true, check the statement for $n=k+1$, that is, if $P(k+1)$ is true, then the statement $P(n)$ is true for all natural numbers n.

Note:

(i) If it is given that the statement is true for all $n \geq 4$, then step 1 will start from $n=4$ and we check that the statement is true for $n=4$, that is, $P(4)$.

(ii) The process of mathematical induction does not state that the given statement is true for $n=k$, but it only states that if the statement is true for $n=k$, then it is also true for $n=k+1$.

Example 1: Prove the following by using the principle of mathematical induction for all $n \in N$:

$$P(n): 2+4+6+...+2n = \frac{2n(n+1)}{2}$$

Solution:

Since the given statement is

$$P(n): 2+4+6+...+2n = \frac{2n(n+1)}{2}$$

For $n=1$,

$$2 = \frac{2\cdot 1(1+1)}{2} = 2, \text{ thus } P(1) \text{ is true.}$$

Let $P(k)$ be true for some positive integer k. Then

$$2+4+6+...+2k = \frac{2k(k+1)}{2} \text{ is true}$$

Now we shall prove that $P(k+1)$ is true

$$2+4+6+...+2k+2(k+1)$$

$$= \frac{2k(k+1)}{2} + 2(k+1), \quad [\because P(k) \text{ is true}]$$

$$= \frac{2k(k+1)+4(k+1)}{2}$$

$$= \frac{2(k+1)(k+2)}{2} = \frac{2(k+1)(k+1+1)}{2}$$

Since $P(k+1)$ is true whenever $P(k)$ is true, by the principle of mathematical induction, the statement $P(n)$ is true for all $n \in N$.

Example 2: Prove the following by using the principle of mathematical induction for all $n \in N$:

$$1^2+2^2+3^2+...+n^2 = \frac{n(n+1)(2n+1)}{6}$$

Solution:

Since the given statement is

$$P(n): 1^2+2^2+3^2+...+n^2 = \frac{n(n+1)(2n+1)}{6}$$

For $n=1$,

$$P(1): 1^2 = 1 = \frac{1(1+1)(2\cdot 1+1)}{6} = \frac{2\cdot 3}{6} = 1, \text{ is true.}$$

Let $P(k)$ be true for some positive integer k. Then

$$1^2+2^2+3^2+...+k^2 = \frac{k(k+1)(2k+1)}{6} \text{ is true.}$$

Now we shall prove that $P(k+1)$ is true

$$1^2+2^2+3^2+...+k^2+(k+1)^2$$

$$= (1^2+2^2+3^2+...+k^2)+(k+1)^2$$

$$= \frac{k(k+1)(2k+1)}{6}+(k+1)^2 \quad [\because P(k) \text{ is true}]$$

$$= \frac{k(k+1)(2k+1)+6(k+1)^2}{6}$$

$$= \frac{(k+1)\{k(2k+1)+6(k+1)\}}{6}$$

$$= \frac{(k+1)\{2k^2+7k+6)\}}{6}$$

$$= \frac{(k+1)\{(k+2)(2k+3)\}}{6}$$

$$= \frac{(k+1)(k+1+1)(2(k+1)+1)}{6}$$

Since $P(k+1)$ is true whenever $P(k)$ is true, by the principle of mathematical induction, the statement $P(n)$ is true for all $n \in N$.

Example 3: Prove the following by using the principle of mathematical induction for all $n \in N$:

$$2^2+5^2+8^2+...+(3n-1)^2 = \frac{n(6n^2+3n-1)}{2}$$

Solution:

Since the given statement is

$$P(n): 2^2+5^2+8^2+...+(3n-1)^2 = \frac{n(6n^2+3n-1)}{2}$$

For $n = 1$,

$$P(1): 2^2 = 4 = \frac{1\cdot(6\cdot1+3\cdot1-1)}{2} = \frac{8}{2} = 4, \text{ is true.}$$

Let $P(k)$ be true for some positive integer k. Then

$$2^2+5^2+8^2+...+(3k-1)^2 = \frac{k(6k^2+3k-1)}{2}$$

Now we shall prove that $P(k+1)$ is true

$$2^2+5^2+8^2+...+(3k-1)^2+(3(k+1)-1)^2$$

$$= \{2^2+5^2+8^2+...+(3k-1)^2\}+(3k+2)^2$$

$$= \frac{k(6k^2+3k-1)}{2}+(3k+2)^2 \quad [\because P(k) \text{ is true}]$$

$$= \frac{k(6k^2+3k-1)}{2}+(3k+2)^2$$

$$= \frac{k(6k^2+3k-1)+2(3k+2)^2}{2}$$

$$= \frac{6k^3+3k^2-k+18k^2+24k+8}{2}$$

$$= \frac{6k^3+21k^2+23k+8}{2}$$

$$= \frac{(k+1)(6k^2+15k+8)}{2}$$

$$= \frac{(k+1)\{(6(k+1)^2+3(k+1)-1)}{2}$$

Since $P(k+1)$ is true whenever $P(k)$ is true, by the principle of mathematical induction, the statement $P(n)$ is true for all $n \in N$.

Example 4: Prove the following by using the principle of mathematical induction for all $n \in N$:

$$\frac{1}{1\cdot3}+\frac{1}{3\cdot5}+\frac{1}{5\cdot7}+...+\frac{1}{(2n-1)(2n+1)} = \frac{n}{2n+1}$$

Solution:

Since the given statement is

$$P(n): \frac{1}{1\cdot3}+\frac{1}{3\cdot5}+\frac{1}{5\cdot7}+...+\frac{1}{(2n-1)(2n+1)} = \frac{n}{2n+1}$$

For $n = 1$,

$$P(1): \frac{1}{1\cdot3} = \frac{1}{3} = \frac{1}{2\cdot1+1} = \frac{1}{3}, \text{ is true.}$$

Let $P(k)$ be true for some positive integer k. Then

$$\frac{1}{1\cdot3}+\frac{1}{3\cdot5}+\frac{1}{5\cdot7}+...+\frac{1}{(2k-1)(2k+1)} = \frac{k}{2k+1}$$

Now we shall prove that $P(k+1)$ is true

$$\frac{1}{1\cdot3}+\frac{1}{3\cdot5}+\frac{1}{5\cdot7}+...+\frac{1}{(2k-1)(2k+1)}$$

$$+\frac{1}{\{2(k+1)-1\}\{2(k+1)+1\}}$$

$$\left(\frac{1}{1\cdot3}+\frac{1}{3\cdot5}+...+\frac{1}{(2k-1)(2k+1)}\right)+\frac{1}{(2k+1)(2k+3)}$$

$$= \frac{k}{2k+1}+\frac{1}{(2k+1)(2k+3)} \quad [\because P(k) \text{ is true}]$$

$$= \frac{k(2k+3)+1}{(2k+1)(2k+3)}$$

$$= \frac{2k^2+3k+1}{(2k+1)(2k+3)}$$

$$= \frac{(k+1)(2k+1)}{(2k+1)(2k+3)}$$

$$= \frac{(k+1)}{(2k+3)} = \frac{(k+1)}{\{2(k+1)+1\}}$$

Since $P(k+1)$ is true whenever $P(k)$ is true, by the principle of mathematical induction, the statement $P(n)$ is true for all $n \in N$.

Example 5: Prove the following by using the principle of mathematical induction for $n \geq 2$:

$$\left(1-\frac{1}{4}\right)\left(1-\frac{1}{9}\right)\left(1-\frac{1}{16}\right)...\left(1-\frac{1}{n^2}\right) = \frac{(n+1)}{2n}$$

Solution:

Since the given statement is

$$\left(1-\frac{1}{4}\right)\left(1-\frac{1}{9}\right)\left(1-\frac{1}{16}\right)...\left(1-\frac{1}{n^2}\right) = \frac{(n+1)}{2n}, \quad n \geq 2$$

For $n = 2$,

$$P(2): \left(1-\frac{1}{4}\right) = \frac{3}{4} = \frac{(2+1)}{2\cdot2} = \frac{3}{4}, \text{ is true.}$$

Let $P(k)$ be true for some positive integer k. Then

$$\left(1-\frac{1}{4}\right)\left(1-\frac{1}{9}\right)\left(1-\frac{1}{16}\right)...\left(1-\frac{1}{k^2}\right) = \frac{(k+1)}{2k}, \quad k \geq 2$$

Now we shall prove that $P(k+1)$ is true

$$\left(1-\frac{1}{4}\right)\left(1-\frac{1}{9}\right)\left(1-\frac{1}{16}\right)...\left(1-\frac{1}{k^2}\right)\left(1-\frac{1}{(k+1)^2}\right)$$

$$\left(\frac{k+1}{2k}\right)\left(1-\frac{1}{(k+1)^2}\right) \quad [\because P(k) \text{ is true}]$$

$$= \left(\frac{k+1}{2k}\right)\left(\frac{(k+1)^2-1}{(k+1)^2}\right)$$

$$= \frac{k^2+2k+1-1}{2k(k+1)} = \frac{k(k+2)}{2k(k+1)}$$

$$= \frac{(k+2)}{2(k+1)} = \frac{(k+1+1)}{2(k+1)}$$

Since $P(k+1)$ is true whenever $P(k)$ is true, by the principle of mathematical induction, statement $P(n)$ is true for all $n \in N$.

Example 6: Prove the following by using the principle of mathematical induction for all $n \in N$:

$$\frac{3}{1\cdot 2\cdot 2^1}+\frac{4}{2\cdot 3\cdot 2^2}+\frac{5}{3\cdot 4\cdot 2^3}+\ldots\frac{n+2}{n(n+1)\cdot 2^n}$$
$$=1-\frac{1}{(n+1)\cdot 2^n}$$

Solution:
Since the given statement is

$$P(n)=\frac{3}{1\cdot 2\cdot 2^1}+\frac{4}{2\cdot 3\cdot 2^2}+\ldots\frac{n+2}{n(n+1)\cdot 2^n}=1-\frac{1}{(n+1)\cdot 2^n}$$

For $n = 1$,

$$P(1)=\frac{3}{1\cdot 2\cdot 2^1}=\frac{3}{4}=1-\frac{1}{(1+1)\cdot 2^1}=1-\frac{1}{4}=\frac{3}{4} \text{ is true.}$$

Let $P(k)$ be true for some positive integer k. Then

$$\frac{3}{1\cdot 2\cdot 2^1}+\frac{4}{2\cdot 3\cdot 2^2}+\frac{5}{3\cdot 4\cdot 2^3}+\ldots\frac{k+2}{k(k+1)\cdot 2^k}=1-\frac{1}{(k+1)\cdot 2^k}$$

Now we shall prove that $P(k+1)$ is true

$$\frac{3}{1\cdot 2\cdot 2^1}+\frac{4}{2\cdot 3\cdot 2^2}+\ldots+\frac{k+2}{k(k+1)\cdot 2^k}+\frac{k+3}{(k+1)(k+2)\cdot 2^{k+1}}$$
$$=1-\frac{1}{(k+1)\cdot 2^k}+\frac{k+3}{(k+1)(k+2)\cdot 2^{k+1}} \quad [\because P(k) \text{ is true}]$$
$$=1-\frac{1}{(k+1)\cdot 2^k}\left(1-\frac{k+3}{(k+2)\cdot 2}\right)$$
$$=1-\frac{1}{(k+1)\cdot 2^k}\left(\frac{2k+4-k-3}{(k+2)\cdot 2}\right)$$
$$=1-\frac{1}{(k+1)\cdot 2^k}\left(\frac{k+1}{(k+2)\cdot 2}\right)$$
$$=1-\frac{1}{(k+2)\cdot 2^{k+1}}=1-\frac{1}{(k+1+1)\cdot 2^{k+1}}$$

Since $P(k+1)$ is true whenever $P(k)$ is true, by the principle of mathematical induction, the statement $P(n)$ is true for all $n \in N$.

Example 7: A sequence $a_1, a_2, a_3 ....$ is defined by letting $a_1 = 2,\ a_k = 3a_{k-1}$, for all natural numbers $k \geq 2$. Show that $a_n = 2\cdot 3^{n-1}$ for all natural numbers.

Solution:
Since the given statement is $a_1 = 2,\ a_k = 3a_{k-1}$, for all natural numbers $k \geq 2$.
For $n = 2$,
$\text{LHS} = a_2 = 3a_{2-1} = 3a_1 = 3\times 2 = 6,\ (\because a_1 = 2)$
$\text{RHS} = a_2 = 2\cdot 3^{2-1} = 2\cdot 3^1 = 6$
Since LHS = RHS, $P(2)$ is true.
Let $P(k)$ be true for some positive integer k. Then
$a_k = 2\cdot 3^{k-1}$
Now we shall prove that $P(k+1)$ is true
$a_{k+1} = 2\cdot 3^{k+1-1} = 2\cdot 3^{(k+1)-1}$
Since $P(k+1)$ is true whenever $P(k)$ is true, by the principle of mathematical induction, the statement $P(n)$ is true for all $n \in N$.

Example 8: A sequence $b_0, b_1, b_2 ....$ is defined by letting $b_0 = 1,\ b_k = 2 + b_{k-1}$, for all natural numbers k. Show that $b_n = 1 + 2n$, for all natural numbers n using mathematical induction.

Solution:
Since the given statement is, $b_n = 1 + 2n$ for all natural numbers given that $b_0 = 1,\ b_k = 2 + b_{k-1}$
For $n = 1$,
$\text{LHS} = b_1 = 2 + b_0 = 2 + 1 = 3$
$\text{RHS} = b_1 = 1 + 2\cdot 1 = 3$
Since LHS = RHS, $P(2)$ is true.
Let $P(k)$ be true for some positive integer k. Then
$b_k = 1 + 2k$
Now we shall prove that $P(k+1)$ is true
$P(k+1) : b_{k+1} = 1 + 2(k+1) = 2k + 3$
$\text{LHS} = b_{k+1} = 2k + 3$
$\text{RHS} = b_{k+1} = 2 + b_{k+1-1} = 2 + b_k = 2 + 1 + 2k = 2k + 3$
Since LHS = RHS, $P(k+1)$ is true whenever $P(k)$ is true, by the principle of mathematical induction, the statement $P(n)$ is true for all $n \in N$.

Day 62

Mathematical Induction-II

Example 1: $17n^3 + 103n$ is divisible by 6, for all natural numbers n.

Solution:
The statement is $P(n) : 17n^3 + 103n$ is divisible by 6, for all natural numbers n,

For $n = 1$,
$P(1) = 17\cdot(1)^3 + 103\cdot 1 = 120$ is divisible by 6, and thus, $P(1)$ is true.
Let $P(k)$ be true for some positive integer k. Then $17k^3 + 103k$ is divisible by 6. Consider
$17k^3 + 103k = 6m, \quad \text{where } m \in N$
Now we shall prove that $P(k+1)$ is true

$17(k+1)^3+103(k+1)$
$=17k^3+17+51k^2+51k+103k+103$
$=(17k^3+103k)+51k^2+51k+120$
$=6m+51k^2+51k+120$ [$\because P(k)$ is true]
$=6m+3(17k^2+17k+40)$
Now, when k is odd then $17k^2+17k+40$ will be even, since $17k^2+17k$ is even (odd + odd = even). And when k is even then $17k^2+17k+40$ will be even. Thus $17k^2+17k+40$ will be even for all k.
Therefore, let $17k^2+17k+40=2k'$, $k'\in N$
Hence, $6m+3(17k^2+17k+40)=6m+3\cdot2\cdot k'$
$=6(m+k')$ is divisible by 3
Since $P(k+1)$ is true whenever $P(k)$ is true, by the principle of mathematical induction, the statement $P(n)$ is true for all $n\in N$.

Example 2: $7^{2n}-1$ is divisible by 48, for all natural numbers n.

Solution:
The statement is $P(n): 7^{2n}-1$ is divisible by 48, for all natural numbers n.
For n = 1, $P(1)=7^{2\cdot1}-1=49-1=48$ is divisible by 48, and thus, $P(1)$ is true.
Let $P(k)$ be true for some positive integer k. Then $7^{2k}-1$ is divisible by 48. Consider
$7^{2k}-1=48m$, where $m\in N$
Now we shall prove that $P(k+1)$ is true
$7^{2(k+1)}-1=7^{2k}\cdot7^2-1$
$=(48m+1)\cdot7^2-1$ [$\because P(k)$ is true]
$=48m\cdot49+49-1=48m\cdot49+48$
$=48(49m+1)=48k'$ is divisible by 48
Since $P(k+1)$ is true whenever $P(k)$ is true, by the principle of mathematical induction, the statement $P(n)$ is true for all $n\in N$.

Example 3: $5^{2n+1}+2^{2n+1}$ is divisible by 7, for all natural numbers n.

Solution:
The statement is $P(n): 5^{2n+1}+2^{2n+1}$ is divisible by 7, for all natural numbers n.
For n = 1,
$P(1)=5^{2n+1}+2^{2n+1}=5^{2\cdot1+1}+2^{2\cdot1+1}=125+8=133$ is divisible by 7, and thus, $P(1)$ is true.
Let $P(k)$ be true for some positive integer k. Then $5^{2k+1}+2^{2k+1}$ is divisible by 7. Consider
$5^{2k+1}+2^{2k+1}=7m$, where $m\in N$
Now we shall prove that $P(k+1)$ is true
$5^{2(k+1)+1}+2^{2(k+1)+1}=5^{2k+1}5^2+2^{2k}2^3$
$=(7m-2^{2k+1})5^2+2^{2k+1}2^2$ [$\because P(k)$ is true]
$=(7m-2^{2k+1})\cdot25+2^{2k+1}\cdot4$
$=7m\cdot25-25\cdot2^{2k+1}+4\cdot2^{2k+1}$
$=7\cdot25m-21\cdot2^{2k+1}$
$=7(25m-3\cdot2^{2k+1})=7k'$ is divisible by 7
Since $P(k+1)$ is true whenever $P(k)$ is true, by the principle of mathematical induction, the statement $P(n)$ is true for all $n\in N$.

Example 4: Prove the following by using the principle of mathematical induction for all $n\in N$:
11^n-4^n is a multiple of 7.

Solution:
The given statement is
$P(n): 11^n-4^n$ is a multiple of 7,
For n = 1,
$11^1-4^1=7$ is a multiple of 7, $P(1)$ is true.
Let $P(k)$ be true for some positive integer k. Then 11^k-4^k is a multiple of 7, thus consider
$11^k-4^k=7m$, where $m\in N$
Now we shall prove that $P(k+1)$ is true
$11^{k+1}-4^{k+1}$
$=11^k\cdot11-4\cdot4^k$
$=11(7m+4^k)-4\cdot4^k$ [$\because P(k)$ is true]
$=11\cdot7m+4^k(11-4)$
$=7\cdot11m+7\cdot4^k=7(11m+4^k)$
$=7\times r$, where $r=(11m+4^k)$ is a natural number
Thus, $11^{k+1}-4^{k+1}$ is a multiple of 7.
Since $P(k+1)$ is true whenever $P(k)$ is true, by the principle of mathematical induction, the statement $P(n)$ is true for all $n\in N$.

Example 5: $\frac{1}{2^2}+\frac{1}{3^2}+\frac{1}{4^2}+\ldots+\frac{1}{n^2}<1$ for natural numbers $n\geq2$.

Solution:
The statement is $P(n): \frac{1}{2^2}+\frac{1}{3^2}+\frac{1}{4^2}+\ldots+\frac{1}{n^2}<1$ for natural numbers $n\geq2$.
For n = 2,
$P(2): \frac{1}{2^2}=\frac{1}{4}<1$, and thus, $P(2)$ is true.
Let $P(k)$ be true for some positive integer k. Then
$\frac{1}{2^2}+\frac{1}{3^2}+\ldots+\frac{1}{k^2}<1$ is true.
Now we shall prove that $P(k+1)$ is true, that is,
$\frac{1}{2^2}+\frac{1}{3^2}+\ldots+\frac{1}{k^2}+\frac{1}{(k+1)^2}<1$

$\because \frac{1}{2^2}+\frac{1}{3^2}+...+\frac{1}{k^2}<1$ [$\because P(k)$ is true]

$\Rightarrow \frac{1}{2^2}+\frac{1}{3^2}+...+\frac{1}{k^2}+\frac{1}{(k+1)^2}<1+\frac{1}{(k+1)^2}$

$\Rightarrow \frac{1}{2^2}+\frac{1}{3^2}+...+\frac{1}{k^2}+\frac{1}{(k+1)^2}<1$ $(\because \frac{1}{(k+1)^2}>0)$

Since $P(k+1)$ is true whenever $P(k)$ is true, by the principle of mathematical induction, the statement $P(n)$ is true for natural numbers $n \geq 2$.

Example 6: $\frac{1}{\sqrt{1}}+\frac{1}{\sqrt{2}}+....+\frac{1}{\sqrt{n}}<2\sqrt{n}$ for all natural numbers $n \in N$.

Solution:
The statement is

$P(n): \frac{1}{\sqrt{1}}+\frac{1}{\sqrt{2}}+....+\frac{1}{\sqrt{n}}<2\sqrt{n}$ for all natural numbers $n \in N$,

For $n = 1$, $P(1): \frac{1}{\sqrt{1}}<2\sqrt{1}$ is true.

Let $P(k)$ is true for some positive integer k. Then

$\frac{1}{\sqrt{1}}+\frac{1}{\sqrt{2}}+....+\frac{1}{\sqrt{k}}<2\sqrt{k}$

Now we shall prove that $P(k+1)$ is true, that is,

$\frac{1}{\sqrt{1}}+\frac{1}{\sqrt{2}}+....+\frac{1}{\sqrt{k}}+\frac{1}{\sqrt{k+1}}<2\sqrt{k+1}$

$\because \frac{1}{\sqrt{1}}+\frac{1}{\sqrt{2}}+....+\frac{1}{\sqrt{k}}<2\sqrt{k}$ [$\because P(k)$ is true]

$\Rightarrow \frac{1}{\sqrt{1}}+\frac{1}{\sqrt{2}}+....+\frac{1}{\sqrt{k}}+\frac{1}{\sqrt{k+1}}<2\sqrt{k}+\frac{1}{\sqrt{k+1}}$

Now we have to prove that

$\Rightarrow 2\sqrt{k}+\frac{1}{\sqrt{k+1}}<2\sqrt{k+1}$

$\Rightarrow \frac{2\sqrt{k}\sqrt{k+1}+1}{\sqrt{k+1}}<2\sqrt{k+1}$

$\Rightarrow 2\sqrt{k}\sqrt{k+1}+1<2\sqrt{k+1}\sqrt{k+1}$

$\Rightarrow 2\sqrt{k}\sqrt{k+1}<2\sqrt{k+1}\sqrt{k+1}$

$\Rightarrow \sqrt{k}<\sqrt{k+1}$ is true

Since $P(k+1)$ is true whenever $P(k)$ is true, by the principle of mathematical induction, the statement $P(n)$ is true for all $n \in N$.

Example 7: $\left(1+\frac{1}{1^2}\right)\left(1+\frac{1}{2^2}\right)\left(1+\frac{1}{3^2}\right)...\left(1+\frac{1}{n^2}\right)\leq(n+1)$ for natural numbers $n \in N$.

Solution:
The statement is

$P(n): \left(1+\frac{1}{1^2}\right)\left(1+\frac{1}{2^2}\right)\left(1+\frac{1}{3^2}\right)...\left(1+\frac{1}{n^2}\right)\leq(n+1)$ for natural numbers $n \in N$.

For $n = 1$,

$P(1): 1+\frac{1}{1^2}=2=1+1=2$, and thus, $P(1)$ is true.

Let $P(k)$ be true for some positive integer k. Then

$\left(1+\frac{1}{1^2}\right)\left(1+\frac{1}{2^2}\right)\left(1+\frac{1}{3^2}\right)...\left(1+\frac{1}{k^2}\right)\leq(k+1)$ is true.

Now we shall prove that $P(k+1)$ is true, that is,

$\left(1+\frac{1}{1^2}\right)\left(1+\frac{1}{2^2}\right)...\left(1+\frac{1}{k^2}\right)\left(1+\frac{1}{(k+1)^2}\right)\leq(k+1+1)$

$\because \left(1+\frac{1}{1^2}\right)\left(1+\frac{1}{2^2}\right)...\left(1+\frac{1}{k^2}\right)\leq(k+1)$

[$\because P(k)$ is true]

$\Rightarrow \left(1+\frac{1}{1^2}\right)\left(1+\frac{1}{2^2}\right)...\left(1+\frac{1}{k^2}\right)\left(1+\frac{1}{(k+1)^2}\right)$

$\leq(k+1)\left(1+\frac{1}{(k+1)^2}\right)$

$\Rightarrow \left(1+\frac{1}{1^2}\right)\left(1+\frac{1}{2^2}\right)...\left(1+\frac{1}{k^2}\right)\left(1+\frac{1}{(k+1)^2}\right)$

$\leq(k+1)+\frac{(k+1)}{(k+1)^2}$

$\Rightarrow \left(1+\frac{1}{1^2}\right)\left(1+\frac{1}{2^2}\right)...\left(1+\frac{1}{k^2}\right)\left(1+\frac{1}{(k+1)^2}\right)$

$\leq(k+1)+\frac{1}{(k+1)}\leq(k+1)+1$

$\left(\because \frac{1}{(k+1)}\leq 1\right)$

Since $P(k+1)$ is true whenever $P(k)$ is true, by the principle of mathematical induction, the statement $P(n)$ is true for natural numbers $n \in N$.

Example 8: Prove that for all $n \in N$

$\sin\alpha+\sin(\alpha+\beta)+\sin(\alpha+2\beta)+...+\sin[\alpha+(n-1)\beta]$

$=\frac{\sin\left[\alpha+\left(\frac{n-1}{2}\right)\beta\right]\sin\left(\frac{n\beta}{2}\right)}{\sin(\beta/2)}$

Solution:
Let $P(n): \sin\alpha+\sin(\alpha+\beta)+\sin(\alpha+2\beta)+...$

$+\sin[\alpha+(n-1)\beta]=\frac{\sin\left[\alpha+\left(\frac{n-1}{2}\right)\beta\right]\sin\left(\frac{n\beta}{2}\right)}{\sin(\beta/2)}$

For $n = 1$, $P(1): \frac{\sin\left[\alpha+\left(\frac{1-1}{2}\right)\beta\right]\sin\left(\frac{1\cdot\beta}{2}\right)}{\sin(\beta/2)}=\sin\alpha$

Thus, $P(1)$ is true.

Let $P(k)$ be true for some positive integer k. Then
$P(k): \sin\alpha + \sin(\alpha+\beta) + \sin(\alpha+2\beta) + \ldots$

$$+\sin[\alpha+(k-1)\beta] = \frac{\sin\left[\alpha+\left(\frac{k-1}{2}\right)\beta\right]\sin\left(\frac{k\beta}{2}\right)}{\sin(\beta/2)}$$

Now we shall prove that $P(k+1)$ is true
$P(k+1): \sin\alpha + \sin(\alpha+\beta) + \sin(\alpha+2\beta) + \ldots$
$+\sin[\alpha+(k-1)\beta] + \sin[\alpha+k\beta]$

$$= \frac{\sin\left[\alpha+\left(\frac{k}{2}\right)\beta\right]\sin\left(\frac{(k+1)\beta}{2}\right)}{\sin(\beta/2)}$$

$\text{LHS} = P(k+1): \sin\alpha + \sin(\alpha+\beta) + \sin(\alpha+2\beta) + \ldots$
$+\sin[\alpha+(k-1)\beta] + \sin[\alpha+k\beta]$

$$= \frac{\sin\left[\alpha+\left(\frac{k-1}{2}\right)\beta\right]\sin\left(\frac{k\beta}{2}\right)}{\sin(\beta/2)} + \sin[\alpha+k\beta]$$

$$= \frac{\sin\left[\alpha+\left(\frac{k-1}{2}\right)\beta\right]\sin\frac{k\beta}{2} + \sin(\alpha+k\beta)\sin\frac{\beta}{2}}{\sin(\beta/2)}$$

$$= \frac{\cos\left(\alpha+\frac{k\beta}{2}-\frac{\beta}{2}-\frac{k\beta}{2}\right)-\cos\left(\alpha+\frac{k\beta}{2}-\frac{\beta}{2}+\frac{k\beta}{2}\right) + \cos\left(\alpha+k\beta-\frac{\beta}{2}\right)-\cos\left(\alpha+k\beta+\frac{\beta}{2}\right)}{2\sin(\beta/2)}$$

$$= \frac{\cos\left(\alpha-\frac{\beta}{2}\right)-\cos\left(\alpha+k\beta-\frac{\beta}{2}\right) + \cos\left(\alpha+k\beta-\frac{\beta}{2}\right)-\cos\left(\alpha+k\beta+\frac{\beta}{2}\right)}{2\sin(\beta/2)}$$

$$= \frac{\cos\left(\alpha-\frac{\beta}{2}\right)-\cos\left(\alpha+k\beta+\frac{\beta}{2}\right)}{2\sin(\beta/2)}$$

$$= \frac{2\sin\left(\frac{\alpha-\frac{\beta}{2}+\alpha+k\beta+\frac{\beta}{2}}{2}\right)\sin\left(\frac{\alpha+k\beta+\frac{\beta}{2}-\alpha+\frac{\beta}{2}}{2}\right)}{2\sin(\beta/2)}$$

$$= \frac{\sin\left(\alpha+\frac{k\beta}{2}\right)\sin\left(\frac{k\beta+\beta}{2}\right)}{\sin(\beta/2)}$$

$$= \frac{\sin\left[\alpha+\left(\frac{k}{2}\right)\beta\right]\sin\left(\frac{(k+1)\beta}{2}\right)}{\sin(\beta/2)} = \text{RHS}$$

So, $P(k+1)$ is true. Hence, $P(n)$ is true.
Since LHS = RHS, $P(k+1)$ is true whenever $P(k)$ is true, by the principle of mathematical induction, the statement $P(n)$ is true for all $n \in N$.

Day 63

NCERT Exercise 4.1 (With Solutions Q1–Q8)

Q1. Prove the following by using the principle of mathematical induction for all $n \in N$:

$$1+3+3^2+\ldots.3^{n-1} = \frac{(3^n-1)}{2}$$

Solution:
Since the given statement is

$$P(n): 1+3+3^2+\ldots.+3^{n-1} = \frac{(3^n-1)}{2}$$

For $n = 1$, $P(1) = \frac{(3^1-1)}{2} = \frac{3-1}{2} = \frac{2}{2} = 1$ is true

Let $P(k)$ be true for some positive integer k. Then

$$1+3+3^2+\ldots..+3^{k-1} = \frac{(3^k-1)}{2} \text{ is true}$$

Now we shall prove that $P(k+1)$ is true

$$1+3+3^2+\ldots..+3^{(k+1)-1}$$
$$= 1+3+3^2+\ldots..+3^{k-1}+3^k$$
$$= \frac{(3^k-1)}{2} + 3^k \quad (\because P(k) \text{ is true})$$
$$= \frac{(3^k-1)+2\cdot 3^k}{2} = \frac{3\cdot 3^k-1}{2} = \frac{3^{k+1}-1}{2}$$

Since $P(k+1)$ is true whenever $P(k)$ is true, by the principle of mathematical induction, the statement $P(n)$ is true for all $n \in N$.

Q2. Prove the following by using the principle of mathematical induction for all $n \in N$:

$$1^3+2^3+3^3+\ldots\ldots+n^3 = \left(\frac{n(n+1)}{2}\right)^2$$

Solution:
Since the given statement is

$$P(n): 1^3+2^3+3^3+\ldots+n^3 = \left(\frac{n(n+1)}{2}\right)^2$$

For $n = 1$,

$$P(1): 1^3 = 1 = \left(\frac{1(1+1)}{2}\right)^2 = \left(\frac{1\cdot 2}{2}\right)^2 = 1^2 = 1, \text{ is true.}$$

Let $P(k)$ be true for some positive integer k. Then

$1^3+2^3+3^3+......+k^3=\left(\frac{k(k+1)}{2}\right)^2$ is true

Now we shall prove that $P(k+1)$ is true

$$1^3+2^3+3^3+....+k^3+(k+1)^3$$
$$=(1^3+2^3+3^3+.....+k^3)+(k+1)^3$$
$$=\left(\frac{k(k+1)}{2}\right)^2+(k+1)^3 \quad (\because P(k) \text{ is true})$$
$$=\frac{k^2(k+1)^2+4(k+1)^3}{4}$$
$$=\frac{(k+1)^2\{k^2+4(k+1)\}}{4}$$
$$=\frac{(k+1)^2(k+2)^2}{4}=\left(\frac{(k+1)(k+1+1)}{2}\right)^2$$

Since $P(k+1)$ is true whenever $P(k)$ is true, by the principle of mathematical induction, the statement $P(n)$ is true for all $n \in N$.

Q3. Prove the following by using the principle of mathematical induction for all $n \in N$:

$$1+\frac{1}{(1+2)}+\frac{1}{(1+2+3)}+....+\frac{1}{(1+2+3+....n)}=\frac{2n}{(n+1)}$$

Solution:

Since the given statement is

$$P(n): 1+\frac{1}{(1+2)}+\frac{1}{(1+2+3)}+....+\frac{1}{1+2+3+....n}=\frac{2n}{n+1}$$

For $n=1$, $P(1): 1=\frac{2\cdot 1}{1+1}=\frac{2}{2}=1$, is true.

Let $P(k)$ be true for some positive integer k. Then

$$1+\frac{1}{(1+2)}+\frac{1}{(1+2+3)}+....+\frac{1}{1+2+3+....k}=\frac{2k}{k+1}$$

Now we shall prove that $P(k+1)$ is true

$$1+\frac{1}{1+2}+\frac{1}{1+2+3}+....+\frac{1}{1+2+3+.....+k}+\frac{1}{1+2+3+....+k+(k+1)}$$
$$=\left(1+\frac{1}{1+2}+\frac{1}{1+2+3}+...+\frac{1}{1+2+3+....k}\right)+\frac{1}{1+2+3+.....+k+(k+1)}$$
$$=\frac{2k}{k+1}+\frac{1}{1+2+3+....+k+(k+1)} \quad (\because P(k) \text{ is true})$$
$$=\frac{2k}{k+1}+\frac{1}{\left(\frac{(k+1)(k+2)}{2}\right)}$$
$$=\frac{2k}{(k+1)}+\frac{2}{(k+1)(k+2)}$$
$$=\frac{2}{(k+1)}\left(k+\frac{1}{k+2}\right)=\frac{2}{(k+1)}\left(\frac{k^2+2k+1}{k+2}\right)$$
$$=\frac{2}{(k+1)}\left(\frac{(k+1)^2}{k+2}\right)=\frac{2(k+1)}{(k+1+1)}$$

Since $P(k+1)$ is true whenever $P(k)$ is true, by the principle of mathematical induction, the statement $P(n)$ is true for all $n \in N$.

Q4. Prove the following by using the principle of mathematical induction for all $n \in N$:

$$1\cdot 2\cdot 3+2\cdot 3\cdot 4+.....+n(n+1)(n+2)=\frac{n(n+1)(n+2)(n+3)}{4}$$

Solution:

Since the given statement is

$$P(n): 1\cdot 2\cdot 3+2\cdot 3\cdot 4+.....+n(n+1)(n+2)=\frac{n(n+1)(n+2)(n+3)}{4}$$

For $n=1$,

$$P(1): 1\cdot 2\cdot 3=\frac{1(1+1)(1+2)(1+3)}{4}=\frac{1\cdot 2\cdot 3\cdot 4}{4}=6 \text{ is true.}$$

Let $P(k)$ be true for some positive integer k. Then

$$1\cdot 2\cdot 3+2\cdot 3\cdot 4+.....+k(k+1)(k+2)=\frac{k(k+1)(k+2)(k+3)}{4} \quad \text{(i)}$$

Now we shall prove that $P(k+1)$ is true

$$1\cdot 2\cdot 3+2\cdot 3\cdot 4+....+k(k+1)(k+2)+(k+1)(k+2)(k+3)$$
$$=\{1\cdot 2\cdot 3+2\cdot 3\cdot 4+...+k(k+1)(k+2)\}+(k+1)(k+2)(k+3)$$
$$=\frac{k(k+1)(k+2)(k+3)}{4}+(k+1)(k+2)(k+3)$$

$[\because P(k) \text{ is true}]$

$$=(k+1)(k+2)(k+3)\left(\frac{k}{4}+1\right)$$
$$=\frac{(k+1)(k+2)(k+3)(k+4)}{4}$$
$$=\frac{(k+1)(k+1+1)(k+1+2)(k+1+3)}{4}$$

Since $P(k+1)$ is true whenever $P(k)$ is true, by the principle of mathematical induction, statement $P(n)$ is true for all $n \in N$.

Q5. Prove the following by using the principle of mathematical induction for all $n \in N$:

$$1\cdot 3+2\cdot 3^2+3\cdot 3^3+...+n\cdot 3^n=\frac{(2n-1)3^{n+1}+3}{4}$$

Solution:
Since the given statement is

$$P(n): 1\cdot 3 + 2\cdot 3^2 + 3\cdot 3^3 + \ldots + n\cdot 3^n = \frac{(2n-1)3^{n+1}+3}{4}$$

For $n = 1$,

$$P(1): 1\cdot 3 = 3 = \frac{(2\cdot 1-1)3^{1+1}+3}{4} = \frac{3^2+3}{4} = 3, \text{ is true.}$$

Let $P(k)$ be true for some positive integer k. Then

$$1\cdot 3 + 2\cdot 3^2 + 3\cdot 3^3 + \ldots + k\cdot 3^n = \frac{(2k-1)3^{k+1}+3}{4}$$

Now we shall prove that $P(k+1)$ is true

$$1\cdot 3 + 2\cdot 3^2 + 3\cdot 3^3 + \ldots\ldots + k\cdot 3^k + (k+1)\cdot 3^{k+1}$$

$$= (1\cdot 3 + 2\cdot 3^2 + 3\cdot 3^3 + \ldots. + k\cdot 3^k) + (k+1)\cdot 3^{k+1}$$

$$= \frac{(2k-1)3^{k+3}+3}{4} + (k+1)3^{k+1} (\because P(k) \text{ is true})$$

$$= \frac{(2k-1)3^{k+1}+3+4(k+1)3^{k+1}}{4}$$

$$= \frac{3^{k+1}(6k+3)+3}{4} = \frac{3^{k+1}\cdot 3(2k+1)+3}{4}$$

$$= \frac{3^{(k+1)+1}(2k+1)+3}{4} = \frac{\{2(k+1)-1\}3^{(k+1)+1}+3}{4}$$

Since $P(k+1)$ is true whenever $P(k)$ is true, by the principle of mathematical induction, the statement $P(n)$ is true for all $n \in N$.

Q6. Prove the following by using the principle of mathematical induction for all $n \in N$:

$$1\cdot 2 + 2\cdot 3 + 3\cdot 4 + \ldots + n\cdot (n+1) = \left[\frac{n(n+1)(n+2)}{3}\right]$$

Solution:
Since the given statement is

$$P(n): 1\cdot 2 + 2\cdot 3 + 3\cdot 4 + \ldots + n\cdot (n+1) = \left[\frac{n(n+1)(n+2)}{3}\right]$$

For $n = 1$, $P(1): 1\cdot 2 = \frac{1(1+1)(1+2)}{3} = \frac{1\cdot 2\cdot 3}{3} = 2$, is true.

Let $P(k)$ be true for some positive integer k. Then

$$1\cdot 2 + 2\cdot 3 + 3\cdot 4 + \ldots + k\cdot (k+1) = \frac{k(k+1)(k+2)}{3}$$

Now we shall prove that $P(k+1)$ is true

$$1\cdot 2 + 2\cdot 3 + 3\cdot 4 + \ldots\ldots + k\cdot (k+1) + (k+1)\cdot (k+2)$$

$$= [1\cdot 2 + 2\cdot 3 + 3\cdot 4 + \ldots. + k\cdot (k+1)] + (k+1)\cdot (k+2)$$

$$= \frac{k(k+1)(k+2)}{3} + (k+1)(k+2) \quad [\because P(k) \text{ is true}]$$

$$= (k+1)(k+2)\left(\frac{k}{3}+1\right)$$

$$= \frac{(k+1)(k+2)(k+3)}{3}$$

$$= \frac{(k+1)(k+1+1)(k+1+2)}{3}$$

Since $P(k+1)$ is true whenever $P(k)$ is true, by the principle of mathematical induction, the statement $P(n)$ is true for all $n \in N$.

Q7. Prove the following by using the principle of mathematical induction for all $n \in N$:

$$1\cdot 3 + 3\cdot 5 + 5\cdot 7 + \ldots. + (2n-1)(2n+1) = \frac{n(4n^2+6n-1)}{3}$$

Solution:
Since the given statement is

$$P(n): 1\cdot 3 + 3\cdot 5 + 5\cdot 7 + \ldots. + (2n-1)(2n+1)$$

$$= \frac{n(4n^2+6n-1)}{3}$$

For $n = 1$,

$$P(1): 1\cdot 3 = 3 = \frac{1(4\cdot 1^2+6\cdot 1-1)}{3} = \frac{4+6-1}{3} = \frac{9}{3} = 3, \text{ is true.}$$

Let $P(k)$ be true for some positive integer k. Then

$$1\cdot 3 + 3\cdot 5 + 5\cdot 7 + \ldots. + (2k-1)(2k+1)$$

$$= \frac{k(4k^2+6k-1)}{3}$$

Now we shall prove that $P(k+1)$ is true

$$(1\cdot 3 + 3\cdot 5 + 5\cdot 7 + \ldots. + (2k-1)(2k+1)$$

$$+ \{2(k+1)-1\}\{2(k+1)+1\}$$

$$= \frac{k(4k^2+6k-1)}{3} + (2k+2-1)(2k+2+1)$$

$[\because P(k)$ is true$]$

$$= \frac{k(4k^2+6k-1)}{3}(2k+1)(2k+3)$$

$$= \frac{k(4k^2+6k-1)+3(4k^2+8k+3)}{3}$$

$$= \frac{4k^3+18k^2+23k+9}{3}$$

$$= \frac{4k^3+14k^2+9k+4k^2+14k+9}{3}$$

$$= \frac{k(4k^2+14k+9)+1(4k^2+14k+9)}{3}$$

$$= \frac{(k+1)(4k^2+14k+9)}{3}$$

$$= \frac{(k+1)\{4k^2+8k+4+6k+6-1\}}{3}$$

$$= \frac{(k+1)\{4(k^2+2k+1)+6(k+1)-1\}}{3}$$

$$= \frac{(k+1)\{4(k+1)^2+6(k+1)-1\}}{3}$$

Since $P(k+1)$ is true whenever $P(k)$ is true, by the principle of mathematical induction, the statement $P(n)$ is true for all $n \in N$.

Q8. Prove the following by using the principle of mathematical induction for all $n \in N$:
$1\cdot 2+2\cdot 2^2+3\cdot 2^2+....+n\cdot 2^n=(n-1)2^{n+1}+2$

Solution:
Since the given statement is
$P(n): 1\cdot 2+2\cdot 2^2+3\cdot 2^2+....+n\cdot 2^n=(n-1)2^{n+1}+2$
For $n = 1$,
$P(1): 1\cdot 2=2=(1-1)2^{1+1}+2=0+2=2$, is true.
Let $P(k)$ be true for some positive integer k. Then
$1\cdot 2+2\cdot 2^2+3\cdot 2^2+....+k\cdot 2^k=(k-1)2^{k+1}+2$
Now we shall prove that $P(k+1)$ is true
$\{1\cdot 2+2\cdot 2^2+3\cdot 2^3+...+k\cdot 2^k\}+(k+1)\cdot 2^{k+1}$
$=(k-1)2^{k+1}+2+(k+1)2^{k+1}$ [$\because P(k)$ is true]
$=2^{k+1}\{(k-1)+(k+1)\}+2$
$=2^{k+1}\cdot 2k+2=k\cdot 2^{(k+1)+1}+2$
$=\{(k+1)-1\}2^{(k+1)+1}+2$
Since $P(k+1)$ is true whenever $P(k)$ is true, by the principle of mathematical induction, the statement $P(n)$ is true for all $n \in N$.

Day 64

NCERT Exercise 4.1 (With Solutions Q9–Q15)

Q9. Prove the following by using the principle of mathematical induction for all $n \in N$:
$$\frac{1}{2}+\frac{1}{4}+\frac{1}{8}+....+\frac{1}{2^n}=1-\frac{1}{2^n}$$

Solution:
Since the given statement is
$$P(n): \frac{1}{2}+\frac{1}{4}+\frac{1}{8}+....+\frac{1}{2^n}=1-\frac{1}{2^n}$$
For $n = 1$,
$P(1): \frac{1}{2}=1-\frac{1}{2^1}=\frac{1}{2}$, is true.
Let $P(k)$ be true for some positive integer k. Then
$$\frac{1}{2}+\frac{1}{4}+\frac{1}{8}+....+\frac{1}{2^k}=1-\frac{1}{2^k}$$
Now we shall prove that $P(k+1)$ is true
$$\left(\frac{1}{2}+\frac{1}{4}+\frac{1}{8}+....+\frac{1}{2^k}\right)+\frac{1}{2^{k+1}}$$
$$=\left(1-\frac{1}{2^k}\right)+\frac{1}{2^{k+1}} \quad [\because P(k) \text{ is true}]$$
$$=1-\frac{1}{2^k}+\frac{1}{2\cdot 2^k}$$
$$=1-\frac{1}{2^k}\left(1-\frac{1}{2}\right)$$
$$=1-\frac{1}{2^k}\left(\frac{1}{2}\right)$$
$$=1-\frac{1}{2^{k+1}}$$
Since $P(k+1)$ is true whenever $P(k)$ is true, by the principle of mathematical induction, the statement $P(n)$ is true for all $n \in N$.

Q10. Prove the following by using the principle of mathematical induction for all $n \in N$:
$$\frac{1}{2\cdot 5}+\frac{1}{5\cdot 8}+\frac{1}{8\cdot 11}+...+\frac{1}{(3n-1)(3n+2)}=\frac{n}{(6n+4)}$$

Solution:
Since the given statement is
$$P(n): \frac{1}{2\cdot 5}+\frac{1}{5\cdot 8}+...+\frac{1}{(3n-1)(3n+2)}=\frac{n}{(6n+4)}$$
For $n = 1$,
$P(1)=\frac{1}{2\cdot 5}=\frac{1}{10}=\frac{1}{6\cdot 1+4}=\frac{1}{10}$, is true.
Let $P(k)$ be true for some positive integer k. Then
$$\frac{1}{2\cdot 5}+\frac{1}{5\cdot 8}+\frac{1}{8\cdot 11}+...+\frac{1}{(3k-1)(3k+2)}=\frac{k}{(6k+4)}$$
Now we shall prove that $P(k+1)$ is true
$$\frac{1}{2\cdot 5}+\frac{1}{5\cdot 8}+\frac{1}{8\cdot 11}+...+\frac{1}{(3k-1)(3k+2)}$$
$$+\frac{1}{\{3(k+1)-1\}\{3(k+1)+2\}}$$
$$=\frac{k}{6k+4}+\frac{1}{(3k+2)(3k+5)} \quad [\because P(k) \text{ is true}]$$
$$=\frac{1}{(3k+2)}\left(\frac{k}{2}+\frac{1}{3k+5}\right)$$
$$=\frac{1}{(3k+2)}\left(\frac{k(3k+5)+2}{2(3k+5)}\right)$$
$$=\frac{1}{(3k+2)}\left(\frac{3k^2+5k+2}{2(3k+5)}\right)$$
$$=\frac{1}{(3k+2)}\left(\frac{(3k+2)(k+1)}{2(3k+5)}\right)$$
$$=\frac{(k+1)}{6k+10}=\frac{k+1}{6(k+1)+4}$$

Since $P(k+1)$ is true whenever $P(k)$ is true, by the principle of mathematical induction, the statement $P(n)$ is true for all $n \in N$.

Q11. Prove the following by using the principle of mathematical induction for all $n \in N$:

$$\frac{1}{1\cdot2\cdot3}+\frac{1}{2\cdot3\cdot4}+\frac{1}{3\cdot4\cdot5}+...+\frac{1}{n(n+1)(n+2)} = \frac{n(n+3)}{4(n+1)(n+2)}$$

Solution:

Since the given statement is

$$P(n): \frac{1}{1\cdot2\cdot3}+\frac{1}{2\cdot3\cdot4}+\frac{1}{3\cdot4\cdot5}+...+\frac{1}{n(n+1)(n+2)} = \frac{n(n+3)}{4(n+1)(n+2)}$$

For $n = 1$,

$$P(1): \frac{1}{1\cdot2\cdot3} = \frac{1\cdot(1+3)}{4(1+1)(1+2)} = \frac{1\cdot4}{4\cdot2\cdot3} = \frac{1}{1\cdot2\cdot3}, \text{ is true.}$$

Let $P(k)$ be true for some positive integer k. Then

$$\frac{1}{1\cdot2\cdot3}+\frac{1}{2\cdot3\cdot4}+\frac{1}{3\cdot4\cdot5}+...+\frac{1}{k(k+1)(k+2)} = \frac{k(k+3)}{4(k+1)(k+2)}$$

Now we shall prove that $P(k+1)$ is true,

$$\frac{1}{1\cdot2\cdot3}+\frac{1}{2\cdot3\cdot4}+\frac{1}{3\cdot4\cdot5}+....+\frac{1}{k(k+1)(k+2)}+\frac{1}{(k+1)(k+2)(k+3)}$$

$$= \frac{k(k+3)}{4(k+1)(k+2)}+\frac{1}{(k+1)(k+2)(k+3)} \quad (\because P(k) \text{ is true})$$

$$= \frac{1}{(k+1)(k+2)}\left\{\frac{k(k+3)}{4}+\frac{1}{k+3}\right\}$$

$$= \frac{1}{(k+1)(k+2)}\left\{\frac{k^3+6k^2+9k+4}{4(k+3)}\right\}$$

$$= \frac{1}{(k+1)(k+2)}\left\{\frac{k^3+2k^2+k+4k^2+8k+4}{4(k+3)}\right\}$$

$$= \frac{1}{(k+1)(k+2)}\left\{\frac{(k+4)(k^2+2k+1)}{4(k+3)}\right\}$$

$$= \frac{(k+1)^2(k+4)}{4(k+1)(k+2)(k+3)}$$

$$= \frac{(k+1)\{(k+1)+3\}}{4\{(k+1)+1\}\{(k+1)+2\}}$$

Since $P(k+1)$ is true whenever $P(k)$ is true, by the principle of mathematical induction, the statement $P(n)$ is true for all $n \in N$.

Q12. Prove the following by using the principle of mathematical induction for all $n \in N$:

$$a+ar+ar^2+...+ar^{n-1} = \frac{a(r^n-1)}{r-1}$$

Solution:

Since the given statement is

$$P(n): a+ar+ar^2+...+ar^{n-1} = \frac{a(r^n-1)}{r-1}$$

For $n = 1$,

$$P(1): a = \frac{a(r^1-1)}{(r-1)} = a, \text{ is true.}$$

Let $P(k)$ be true for some positive integer k. Then

$$a+ar+ar^2+...+ar^{k-1} = \frac{a(r^k-1)}{r-1}$$

Now we shall prove that $P(k+1)$ is true,

$$a+ar+ar^2+.....+ar^{k-1}+ar^{(k+1)-1}$$

$$= \frac{a(r^k-1)}{r-1}+ar^k \; [\because P(k) \text{ is true}]$$

$$= \frac{a(r^k-1)+ar^k(r-1)}{r-1}$$

$$= \frac{ar^k-a+ar^{k+1}-ar^k}{r-1}$$

$$= \frac{ar^{k+1}-a}{r-1}$$

$$= \frac{a(r^{k+1}-1)}{r-1}$$

Since $P(k+1)$ is true whenever $P(k)$ is true, by the principle of mathematical induction, the statement $P(n)$ is true for all $n \in N$.

Q13. Prove the following by using the principle of mathematical induction for all $n \in N$:

$$\left(1+\frac{3}{1}\right)\left(1+\frac{5}{4}\right)\left(1+\frac{7}{9}\right).....\left(1+\frac{(2n+1)}{n^2}\right) = (n+1)^2$$

Solution:

Since the given statement is

$$P(1): \left(1+\frac{3}{1}\right)\left(1+\frac{5}{4}\right)\left(1+\frac{7}{9}\right).....\left(1+\frac{(2n+1)}{n^2}\right) = (n+1)^2$$

For $n = 1$,

$$P(1): \left(1+\frac{3}{1}\right) = 4 = (1+1)^2 = 2^2 = 4, \text{ is true.}$$

Let $P(k)$ be true for some positive integer k. Then

$$\left(1+\frac{3}{1}\right)\left(1+\frac{5}{4}\right)\left(1+\frac{7}{9}\right)...\left(1+\frac{(2k+1)}{k^2}\right) = (k+1)^2$$

Now we shall prove that $P(k+1)$ is true

$$\left(1+\frac{3}{1}\right)\left(1+\frac{5}{4}\right)...\left(1+\frac{(2k+1)}{k^2}\right)\left(1+\frac{\{2(k+1)+1}{(k+1)^2}\right)$$

$= (k+1)^2\left(1+\dfrac{2(k+1)+1}{(k+1)^2}\right)$ [$\because P(k)$ is true]

$= (k+1)^2\left(\dfrac{(k+1)^2+2(k+1)+1}{(k+1)^2}\right)$

$= (k+1)^2+2(k+1)+1$

$= ((k+1)+1)^2$

Since $P(k+1)$ is true whenever $P(k)$ is true, by the principle of mathematical induction, the statement $P(n)$ is true for all $n \in N$.

Q14. Prove the following by using the principle of mathematical induction for all $n \in N$:

$$\left(1+\frac{1}{1}\right)\left(1+\frac{1}{2}\right)\left(1+\frac{1}{3}\right)....\left(1+\frac{1}{n}\right)=(n+1)$$

Solution:

Since the given statement is

$$P(n):\left(1+\frac{1}{1}\right)\left(1+\frac{1}{2}\right)\left(1+\frac{1}{3}\right)....\left(1+\frac{1}{n}\right)=(n+1)$$

For $n = 1$,

$P(1):\left(1+\dfrac{1}{1}\right)=2=(1+1)$, is true.

Let $P(k)$ be true for some positive integer k. Then

$$P(k):\left(1+\frac{1}{1}\right)\left(1+\frac{1}{2}\right)\left(1+\frac{1}{3}\right)....\left(1+\frac{1}{k}\right)=(k+1) \qquad \text{(i)}$$

Now we shall prove that $P(k+1)$ is true

$$\left(1+\frac{1}{1}\right)\left(1+\frac{1}{2}\right)\left(1+\frac{1}{3}\right)...\left(1+\frac{1}{k}\right)\left(1+\frac{1}{k+1}\right)$$

$= (k+1)\left(1+\dfrac{1}{k+1}\right)$ [$\because P(k)$ is true]

$= (k+1)\left(\dfrac{(k+1)+1}{(k+1)}\right)=(k+1)+1$

Since $P(k+1)$ is true whenever $P(k)$ is true, by the principle of mathematical induction, the statement $P(n)$ is true for all $n \in N$.

Q15. Prove the following by using the principle of mathematical induction for all $n \in N$:

$$1^2+3^2+5^2+....+(2n-1)^2=\frac{n(2n-1)(2n+1)}{3}$$

Solution:

Since the given statement is

$$P(n)=1^2+3^2+5^2+....+(2n-1)^2=\frac{n(2n-1)(2n+1)}{3}$$

For $n = 1$,

$P(1)=1^2=1=\dfrac{1(2\cdot1-1)(2\cdot1+1)}{3}=\dfrac{1\cdot1\cdot3}{3}=1$, is true.

Let $P(k)$ be true for some positive integer k. Then

$$1^2+3^2+5^2+....+(2k-1)^2=\frac{k(2k-1)(2k+1)}{3} \qquad (1)$$

Now we shall prove that $P(k+1)$ is true

$\{1^2+3^2+5^2+....+(2k-1)^2\}+\{2(k+1)-1\}^2$

$=\dfrac{k(2k-1)(2k+1)}{3}+(2k+1)^2$ ($\because P(k)$ is true)

$=\dfrac{(2k+1)\{k(2k-1)+3(2k+1)\}}{3}$

$=\dfrac{(2k+1)\{2k^2+5k+3\}}{3}$

$=\dfrac{(2k+1)\{2k(k+1)+3(k+1)\}}{3}$

$=\dfrac{(2k+1)(k+1)(2k+3)}{3}$

$=\dfrac{(k+1)\{2(k+1)-1\}\{2(k+1)+1\}}{3}$

Since $P(k+1)$ is true whenever $P(k)$ is true, by the principle of mathematical induction, statement $P(n)$ is true for all $n \in N$.

Day 65

NCERT Exercise 4.1 (With Solutions Q16–Q24)

Q16. Prove the following by using the principle of mathematical induction for all $n \in N$:

$$\frac{1}{1\cdot4}+\frac{1}{4\cdot7}+\frac{1}{7\cdot10}+...+\frac{1}{(3n-2)(3n+1)}=\frac{n}{(3n+1)}$$

Solution:

Since the given statement is

$$P(n):\frac{1}{1\cdot4}+\frac{1}{4\cdot7}+\frac{1}{7\cdot10}+...+\frac{1}{(3n-2)(3n+1)}=\frac{n}{(3n+1)}$$

For $n = 1$, $P(1)=\dfrac{1}{1\cdot4}=\dfrac{1}{3\cdot1+1}=\dfrac{1}{4}=\dfrac{1}{4\cdot1}$, is true.

Let $P(k)$ be true for some positive integer k. Then

$$P(k)=\frac{1}{1\cdot4}+\frac{1}{4\cdot7}+\frac{1}{7\cdot10}+...+\frac{1}{(3k-2)(3k+1)}$$

$$=\frac{n}{(3k+1)}$$

Now we shall prove that $P(k+1)$ is true

$$\frac{1}{1\cdot4}+\frac{1}{4\cdot7}+\frac{1}{7\cdot10}+...+\frac{1}{(3k-2)(3k+1)}$$

$$+\frac{1}{\{3(k+1)-2\}\{3(k+1)+1\}}$$

$= \frac{k}{3k+1} + \frac{1}{(3k+1)(3k+4)}$ [$\because P(k)$ is true]

$= \frac{1}{(3k+1)}\left\{k + \frac{1}{(3k+4)}\right\}$

$= \frac{1}{(3k+1)}\left\{\frac{3k^2+4k+1}{(3k+4)}\right\}$

$= \frac{1}{(3k+1)}\left\{\frac{3k^2+3k+k+1}{(3k+4)}\right\}$

$= \frac{(3k+1)(k+1)}{(3k+1)(3k+4)}$

$= \frac{(k+1)}{3(k+1)+1}$

Since $P(k+1)$ is true whenever $P(k)$ is true, by the principle of mathematical induction, the statement $P(n)$ is true for all $n \in N$.

Q17. Prove the following by using the principle of mathematical induction for all $n \in N$:

$$\frac{1}{3\cdot5} + \frac{1}{5\cdot7} + \frac{1}{7\cdot9} + ... + \frac{1}{(2n+1)(2n+3)} = \frac{n}{3(2n+3)}$$

Solution:

Since the given statement is

$P(n): \frac{1}{3\cdot5} + \frac{1}{5\cdot7} + \frac{1}{7\cdot9} + ... + \frac{1}{(2n+1)(2n+3)} = \frac{n}{3(2n+3)}$

For $n = 1$,

$P(1): \frac{1}{3\cdot5} = \frac{1}{3(2\cdot1+3)} = \frac{1}{3\cdot5}$, is true.

Let $P(k)$ be true for some positive integer k. Then

$P(k): \frac{1}{3\cdot5} + \frac{1}{5\cdot7} + \frac{1}{7\cdot9} + ... + \frac{1}{(2k+1)(2k+3)} = \frac{k}{3(2k+3)}$

Now we shall prove that $P(k+1)$ is true

$\frac{1}{3\cdot5} + \frac{1}{5\cdot7} + \frac{1}{7\cdot9} + ... + \frac{1}{(2k+1)(2k+3)}$

$+ \frac{1}{\{2(k+1)+1\}\{2(k+1)+3\}}$

$= \frac{k}{3(2k+3)} + \frac{1}{(2k+3)(2k+5)}$ [$\because P(k)$ is true]

$= \frac{1}{(2k+3)}\left[\frac{k}{3} + \frac{1}{(2k+5)}\right]$

$= \frac{1}{(2k+3)}\left[\frac{2k^2+5k+3}{3(2k+5)}\right]$

$= \frac{1}{(2k+3)}\left[\frac{2k(k+1)+3(k+1)}{3(2k+5)}\right]$

$= \frac{(k+1)(2k+3)}{3(2k+3)(2k+5)} = \frac{(k+1)}{3\{2(k+1)+3\}}$

Since $P(k+1)$ is true whenever $P(k)$ is true, by the principle of mathematical induction, the statement $P(n)$ is true for all $n \in N$.

Q18. Prove the following by using principle of mathematical induction for all $n \in N$:

$$1+2+3+....+n < \frac{1}{8}(2n+1)^2$$

Solution:

Since the given statement is

$P(n): 1+2+3+....+n < \frac{1}{8}(2n+1)^2$

For $n = 1$,

$1 < \frac{1}{8}(2\cdot1+1)^2 = \frac{9}{8}$ is true.

Let $P(k)$ be true for some positive integer k. Then

$1+2+3+....+k < \frac{1}{8}(2k+1)^2$

Now we shall prove that $P(k+1)$ is true

$(1+2+...+k)+(k+1) < \frac{1}{8}(2k+1)^2 + (k+1)$

($\because P(k)$ is true)

$\Rightarrow (1+2+...+k)+(k+1) < \frac{1}{8}\{(2k+1)^2 + 8(k+1)\}$

$\Rightarrow (1+2+...+k)+(k+1) < \frac{1}{8}\{4k^2+12k+9\}$

$\Rightarrow (1+2+...+k)+(k+1) < \frac{1}{8}(2k+3)^2$

$\Rightarrow (1+2+...+k)+(k+1) < \frac{1}{8}\{2(k+1)+1\}^2$

Since $P(k+1)$ is true whenever $P(k)$ is true, by the principle of mathematical induction, the statement $P(n)$ is true for all $n \in N$.

Q19. Prove the following by using the principle of mathematical induction for all $n \in N$:

$n(n+1)(n+5)$ is multiple of 3.

Solution:

Since the given statement is

$P(n): n(n+1)(n+5)$, which is a multiple of 3.

For $n = 1$,

$1(1+1)(1+5) = 12$, is multiple of 3, $P(1)$ is true.

Let $P(k)$ is true for some positive integer k. Then

$k(k+1)(k+5)$ is multiple of 3, and thus, consider

$k(k+1)(k+5) = 3m$, where $m \in N$

Now we shall prove that $P(k+1)$ is true

$(k+1)\{(k+1)+1\}\{(k+1)+5\}$

$= (k+1)\{(k+1)+1\}\{(k+5)+1\}$

$= (k+1)\{(k+1)(k+5)+(2k+7)\}$

$= k(k+1)(k+5)+(k+1)(k+5)+(k+1)(2k+7)$

$= 3m+(k+1)\{2(k+5)+(k+2)\}$ [$\because P(k)$ is true]
$= 3m+(k+1)(3k+12)$
$= 3\{m+(k+1)(k+4)\}$
$= 3\times r$, where $r \in N$
Thus $(k+1)\{(k+1)+1\}\{(k+1)+5\}$ is a multiple of 3.
Since $P(k+1)$ is true whenever $P(k)$ is true, by the principle of mathematical induction, the statement $P(n)$ is true for all $n \in N$.

Q20. Prove the following by using principle of mathematical induction for all $n \in N$:
$10^{2n-1}+1$ is divisible by 11.

Solution:
Since the given statement is
$P(n): 10^{2n-1}+1$ is divisible by 11.
For $n = 1$,
$P(1) = 10^{2\cdot1-1}+1 = 11$, is divisible by 11, $P(1)$ is true.
Let $P(k)$ be true for some positive integer k. Then
$10^{2k-1}+1$ is divisible by 11, and thus, consider
$10^{2k-1}+1 = 11m$, where $m \in N$
Now we shall prove that $P(k+1)$ is true
$10^{2(k+1)-1}+1$
$= 10^{2k+1}+1$
$= 10^2 \cdot 10^{2k-1}+1$
$= 10^2 \cdot (11m-1)+1$ [$\because P(k)$ is true]
$= 10^2 \cdot 11m - 100 + 1$
$= 100 \times 11m - 99$
$= 11(100m-9)$
$= 11r$, where $r \in N$
Thus, $10^{2(k+1)-1}+1$ is divisible by 11.
Since $P(k+1)$ is true whenever $P(k)$ is true, by the principle of mathematical induction, the statement $P(n)$ is true for all $n \in N$.

Q21. Prove the following by using the principle of mathematical induction for all $n \in N$:
$x^{2n} - y^{2n}$ is divisible by $x+y$.

Solution:
Since the given statement is
$P(n): x^{2n} - y^{2n}$ is divisible by $x+y$.
For $n = 1$
$x^2 - y^2 = (x+y)(x-y)$ is divisible by $(x+y)$.
Let $P(k)$ be true for some positive integer k. Then
$x^{2k} - y^{2k}$ is divisible by $x+y$, and thus, consider
$x^{2k} - y^{2k} = m(x+y)$, where $m \in N$
Now we shall prove that $P(k+1)$ is true
$x^{2(k+1)} - y^{2(k+1)}$
$= x^{2k}\cdot x^2 - y^{2k}\cdot y^2$
$= x^2\{y^{2k}+m(x+y)\} - y^{2k}\cdot y^2$ [$\because P(k)$ is true]
$= m(x+y)x^2 + y^{2k}\cdot x^2 - y^{2k}\cdot y^2$
$= m(x+y)x^2 + y^{2k}(x^2-y^2)$
$= m(x+y)x^2 + y^{2k}(x+y)(x-y)$
$= (x+y)\{mx^2 + y^{2k}(x-y)\}$, is a factor of $(x+y)$.
Thus, $x^{2n} - y^{2n}$ is divisible by $x+y$.
Since $P(k+1)$ is true whenever $P(k)$ is true, by the principle of mathematical induction, the statement $P(n)$ is true for all $n \in N$.

Q22. Prove the following by using the principle of mathematical induction for all $n \in N$:
$3^{2n+2} - 8n - 9$ is divisible by 8.

Solution:
Since the given statement is
$P(n): 3^{2n+2} - 8n - 9$ is divisible by 8.
For $n = 1$,
$3^{2\times1+2} - 8\times1 - 9 = 64$, is divisible by 8, $P(1)$ is true
Let $P(k)$ be true for some positive integer k. Then
$3^{2k+2} - 8k - 9$ is divisible by 8, and thus, consider
$3^{2k+2} - 8k - 9 = 8m$, where $m \in N$
Now we shall prove that $P(k+1)$ is true
$3^{2(k+1)+2} - 8(k+1) - 9$
$= 3^{2k+2}\cdot 3^2 - 8k - 8 - 9$
$= 3^2(8m+8k+9) - 8k - 17$ [$\because P(k)$ is true]
$= 9\cdot 8\cdot(m+k) + 81 - 8k - 17$
$= 9\cdot 8\cdot(m+k) + 64 - 8k$
$= 8(9m+9k+8-k)$
$= 8(9m+8k+8)$
$= 8r$, where $r \in N$
Thus, $3^{2(k+1)+2} - 8(k+1) - 9$ is divisible by 8.
Since $P(k+1)$ is true whenever $P(k)$ is true, by the principle of mathematical induction, the statement $P(n)$ is true for all $n \in N$.

Q23. Prove the following by using the principle of mathematical induction for all $n \in N$:
$41^n - 14^n$ is a multiple of 27.

Solution:
Since the given statement is
$P(n): 41^n - 14^n$ is multiple of 27.
For $n = 1$,
$41^1 - 14^1 = 27$, is a multiple of 27, $P(1)$ is true.
Let $P(k)$ be true for some positive integer k. Then

$41^k - 14^k$ is a multiple of 27, and thus, consider

$41^k - 14^k = 27m$, where $m \in N$

Now we shall prove that $P(k+1)$ is true

$41^{k+1} - 14^{k+1}$

$= 41^k \cdot 41 - 14^k$

$= 41(27m + 14^k) - 14^k$ [$\because P(k)$ is true]

$= 41 \cdot 27m + 14^k(41 - 14)$

$= 41 \cdot 27m + 27 \cdot 14^k$

$= 27(41m - 14^k)$

$= 27 \times r$, where $r = (41m - 14^k)$ is a natural number

Thus, $41^{k+1} - 14^{k+1}$ is a multiple of 27.

Since $P(k+1)$ is true whenever $P(k)$ is true, by the principle of mathematical induction, statement $P(n)$ is true for all $n \in N$.

Q24. Prove the following by using the principle of mathematical induction for all $n \in N$:

$(2n+7) < (n+3)^2$

Solution:

Since the given statement is

$P(n): (2n+7) < (n+3)^2$

For $n = 1$,

$2 \cdot 1 + 7 = 9 < (1+3)^2 = 16$, is true.

Let $P(k)$ be true for some positive integer k. Then

$(2k+7) < (k+3)^2$

Now we shall prove that $P(k+1)$ is true

$\{2(k+1)+7\} = (2k+7)+2$

$\Rightarrow \{2(k+1)+7\} < (k+3)^2 + 2$, [$\because P(k)$ is true]

$\Rightarrow 2(k+1)+7 < k^2+6k+11$

$\Rightarrow 2(k+1)+7 < k^2+6k+16$

$(\because k^2+6k+11 < k^2+8k+16)$

$\Rightarrow 2(k+1)+7 < (k+4)^2$

$\Rightarrow 2(k+1)+7 < \{(k+1)+3\}^2$

Since $P(k+1)$ is true whenever $P(k)$ is true, by the principle of mathematical induction, the statement $P(n)$ is true for all $n \in N$.

Day 66

NCERT Exemplar Problems: Solutions (Q1–Q16)

Q1. Give an example of a statement $P(n)$ which is true for all $n \geq 4$ but $P(1)$, $P(2)$ and $P(3)$ are not true. Justify your answer.

Solution:

Let the statement $P(n)$: $3n < n^2$

For $n = 1$, $3 \times 1 < 1^2 \Rightarrow 3 < 1$ (False)

For $n = 2$, $3 \times 2 < 2^2 \Rightarrow 6 < 4$ (False)

For $n = 3$, $3 \times 3 < 3^2 \Rightarrow 9 < 9$ (False)

For $n = 4$, $3 \times 4 < 4^2 \Rightarrow 12 < 16$ (True)

Q2. Give an example of a statement $P(n)$ which is true for all n. Justify your answer.

Solution:

Consider the statement

$P(n): 1+2+3+...+n = \dfrac{n(n+1)}{2}$

For $n = 1$,

$1 = \dfrac{1(1+1)}{2} = 1$, thus $P(1)$ is true

Let $P(k)$ be true for some positive integer k. Then

$1+2+3+...+k = \dfrac{k(k+1)}{2}$ is true

Now we shall prove that $P(k+1)$ is true,

$1+2+3+...+k+(k+1)$

$= \dfrac{k(k+1)}{2} + (k+1)$, [$\because P(k)$ is true]

$= \dfrac{k(k+1)+2(k+1)}{2}$

$= \dfrac{(k+1)(k+2)}{2} = \dfrac{(k+1)(k+1+1)}{2}$

Since $P(k+1)$ is true whenever $P(k)$ is true, by the principle of mathematical induction, the statement $P(n)$ is true for all $n \in N$.

Q3. $4^n - 1$ is divisible by 3, for each natural number n.

Solution:

Since the given statement is $P(n): 4^n - 1$ is divisible by 3 for each natural number n,

For $n = 1$,

$P(1): 4^1 - 1 = 3$ is divisible by 3, thus $P(1)$ is true.

Let $P(k)$ is true for some positive integer k. Then

$4^k - 1$ is divisible by 3, and thus, consider

$4^k - 1 = 3m$, where $m \in N$

Now we shall prove that $P(k+1)$ is true

$4^{k+1} - 1 = 4 \cdot 4^k - 1$

$4(3m+1) - 1$ [$\because P(k)$ is true]

$= 3 \cdot 4m + 3 = 3(4m+1)$ is divisible by 3

Since $P(k+1)$ is true whenever $P(k)$ is true, by the principle of mathematical induction, the statement $P(n)$ is true for all $n \in N$.

Q4. $2^{3n} - 1$ is divisible by 7, for all natural numbers n.

Solution:

Since the given statement is $P(n): 2^{3n} - 1$ is divisible by 7

For $n = 1$,

$P(1): 2^{3\times1} - 1 = 2^3 - 1 = 8 - 1 = 7$ is divisible by 7, thus $P(1)$ is true.

Let $P(k)$ be true for some positive integer k. Then

$P(k): 2^{3k} - 1$ is divisible by 7. Thus, consider

$2^{3k} - 1 = 7m$, where $m \in N$

Now we shall prove that $P(k+1)$ is true

$2^{3(k+1)} - 1 = 2^3 \cdot 2^{3k} - 1$

$= 8(7m+1) - 1$ [$\because P(k)$ is true]

$= 7(8m+1)$ is divisible by 7

Since $P(k+1)$ is true whenever $P(k)$ is true, by the principle of mathematical induction, the statement $P(n)$ is true for all $n \in N$.

Q5. $n^3 - 7n + 3$ is divisible by 3, for all natural numbers n.

Solution:

Since the statement is $P(n): n^3 - 7n + 3$ is divisible by 3, for all natural number n.

For $n = 1$,

$P(1) = (1)^3 - 7(1) + 3 = -3$ is divisible by 3, thus P(1) is true.

Let $P(k)$ be true for some positive integer k. Then $k^3 - 7k + 3$ is divisible by 3, and thus, consider

$k^3 - 7k + 3 = 3m$, where $m \in N$

Now we shall prove that $P(k+1)$ is true

$(k+1)^3 - 7(k+1) + 3$

$= k^3 + 1 + 3k^2 + 3k - 7k - 7 + 3$

$= 3m + 7k - 3 + 3k^2 - 4k - 3$ [$\because P(k)$ is true]

$= 3m + 3k^2 + 3k - 6$

$= 3(m + k^2 + k - 2)$ is divisible by 3

Since $P(k+1)$ is true whenever $P(k)$ is true, by the principle of mathematical induction, the statement $P(n)$ is true for all $n \in N$.

Q6. $3^{2n} - 1$ is divisible by 8, for all natural numbers n.

Solution:

The given statement is $3^{2n} - 1$ is divisible by 8, for all natural numbers.

For $n = 1$,

$3^{2\cdot1} - 1 = 9 - 1 = 8$ is divisible by 8, thus P(1) is true.

Let $P(k)$ be true for some positive integer k. Then $3^{2k} - 1$ is divisible by 8, and thus, consider

$3^{2k} - 1 = 8m$, where $m \in N$

Now we shall prove that $P(k+1)$ is true

$3^{2(k+1)} - 1 = 3^2 \cdot 3^{2k} - 1$

$= 9(8m+1) - 1$ [$\because P(k)$ is true]

$= 8 \cdot 9m + 9 - 1$

$= 8(9m+1)$ is divisible by 8

Since $P(k+1)$ is true whenever $P(k)$ is true, by the principle of mathematical induction, the statement $P(n)$ is true for all $n \in N$.

Q7. For any natural numbers $n, 7^n - 2^n$ is divisible by 5.

Solution:

Since the statement is $P(n): 7^n - 2^n$ is divisible by 5, for any natural number n.

For $n = 1$, $7^1 - 2^1 = 5$, is divisible by 5, thus $P(1)$ is true.

Let $P(k)$ be true for some positive integer k. Then $7^k - 2^k$ is divisible by 5, and thus, consider

$7^k - 2^k = 5m$ where $m \in N$

Now we shall prove that $P(k+1)$ is true

$7^{k+1} - 2^{k+1} = 7 \cdot 7^k - 2 \cdot 2^k$

$= 7(5m + 2^k) - 2 \cdot 2^k$ [$\because P(k)$ is true]

$= 5 \cdot 7m + 7 \cdot 2^k - 2 \cdot 2^k$

$= 5 \cdot 7m + 5 \cdot 2^k$

$= 5(7m + 2^k)$ is divisible by 5.

Since $P(k+1)$ is true whenever $P(k)$ is true, by the principle of mathematical induction, the statement $P(n)$ is true for all $n \in N$.

Q8. For any natural numbers $n, x^n - y^n$ is divisible by $x - y$ where x and y are any integers with $x \neq y$.

Solution:

Since the given statement is $x^n - y^n$ is divisible by $x - y$ where x and y are any integers with $x \neq y$.

For $n = 1$,

$x^1 - y^1 = x - y$ is divisible by $x - y$, thus $P(1)$ is true.

Let $P(k)$ be true for some positive integer k. Then $x^k - y^k$ is divisible by $x - y$, and thus, consider

$x^k - y^k = m(x - y)$, where $m \in N$

Now we shall prove that $P(k+1)$ is true

$P(k+1): x^{k+1} - y^{k+1}$

$= x \cdot x^k - y \cdot y^k$

$= x\{m(x-y) + y^k\} - y \cdot y^k$ [$\because P(k)$ is true]

$= mx(x-y) + x \cdot y^k - y \cdot y^k$

$= mx(x-y) + y^k(x-y)$

$= (x-y)(mx+y^k)$ is divisible by $(x-y)$

Since $P(k+1)$ is true whenever $P(k)$ is true, by the principle of mathematical induction, the statement $P(n)$ is true for all $n \in N$.

Q9. $n^3 - n$ is divisible by 6, for each natural number $n \geq 2$.

Solution:

Since the statement is $P(n): n^3 - n$ is divisible by 6, for each natural number $n \geq 2$.

For $n = 2$, $P(2): (2)^3 - 2 \Rightarrow 8 - 2 = 6$, is divisible by 6, thus $P(1)$ is true.

Let $P(k)$ be true for some positive integer k. Then

$k^3 - k$ is divisible by 6, and thus, consider

$k^3 - k = 6m,$ where $m \in N$

Now we shall prove that $P(k+1)$ is true

$(k+1)^3 - (k+1)$

$= k^3 + 1 + 3k^2 + 3k - k - 1$

$= k^3 + 3k^2 + 2k$

$= 6m + k + 3k^2 + 2k$ [$\because P(k)$ is true]

$= 6m + 3k^2 + 3k = 6m + 3(k^2 + k)$

$k^2 + k$ will always be an even number since the square of an even (or odd) number is even (or odd) number and the sum of two even (or odd) numbers is always an even number. Therefore, $k^2 + k$ is divisible by 2, and thus, consider $k^2 + k = 2q$, where $q \in N$.

Thus, $6m + 3(k^2 + k) = 6m + 3 \cdot 2 \cdot q = 6(m+q)$ is divisible by 6.

Since $P(k+1)$ is true whenever $P(k)$ is true, by the principle of mathematical induction, the statement $P(n)$ is true for all $n \in N$.

Q10. $n(n^2 + 5)$ is divisible by 6, for each natural number n.

Solution:

Since the statement is $n(n^2 + 5)$ is divisible by 6, for each natural number n.

For $n = 1$, $1(1^2 + 5) = 6$, is divisible by 6, thus $P(1)$ is true.

Let $P(k)$ be true for some positive integer k. Then

$k(k^2 + 5)$ is divisible by 6, and thus, consider

$k(k^2 + 5) = 6m \Rightarrow k^3 + 5k = 6m,$ where $m \in N$

Now we shall prove that $P(k+1)$ is true

$(k+1)\{(k+1)^2 + 5\}$

$= (k+1)(k^2 + 2k + 6)$

$= k^3 + 3k^2 + 8k + 6$

$= (6m - 5k) + 3k^2 + 8k + 6$ [$\because P(k)$ is true]

$= (6m - 5k) + 3k^2 + 8k + 6$

$= 6m + 3k^2 + 3k + 6$

$= 6m + 3(k^2 + k + 2)$

$k^2 + k + 2$ will always be an even number since the square of an even (or odd) number is even (or odd) and the sum of two even (or odd) numbers is always an even number. Therefore, $k^2 + k + 2$ is divisible by 2, and thus, consider $k^2 + k + 2 = 2q$, where $q \in N$.

Thus, $6m + 3(k^2 + k + 2) = 6m + 3 \cdot 2 \cdot q = 6(m+q)$ is divisible by 6.

Since $P(k+1)$ is true whenever $P(k)$ is true, by the principle of mathematical induction, the statement $P(n)$ is true for all $n \in N$.

Q11. $n^2 < 2^n$, for all natural numbers $n \geq 5$.

Solution:

Since the statement is $P(n): n^2 < 2^n$ for all natural numbers $n \geq 5$.

For $n = 5$,

$P(5): 5^2 < 2^5 \Rightarrow 25 < 32$, thus $P(5)$ is true.

Let $P(k)$ be true for some positive integer k. Then

$P(k) = k^2 < 2^k$ is true.

Now we shall prove that $P(k+1)$ is true

$(k+1)^2 = k^2 + 2k + 1$

$\Rightarrow (k+1)^2 < 2^k + 2k + 1$ [$\because P(k)$ is true]

$\Rightarrow (k+1)^2 < 2^k + 2^k,$ ($\because 2k + 1 < 2^k$, for $k \geq 3$)

$\Rightarrow (k+1)^2 < 2 \cdot 2^k$

$\Rightarrow (k+1)^2 < 2^{k+1}$

Since $P(k+1)$ is true whenever $P(k)$ is true, by the principle of mathematical induction, the statement $P(n)$ is true for all $n \geq 5$.

Q12. $2n < (n+2)!$ for all natural numbers n.

Solution:

The statement is $P(n): 2n < (n+2)!$ for all natural numbers n.

For $n = 1$, $P(1): 2(1) < (1+2)! \Rightarrow 2 < 3! \Rightarrow 2 < 6$, is true.

Let $P(k)$ be true for some positive integer k. Then

$2k < (k+2)!$

Now we shall prove that $P(k+1)$ is true

$2(k+1) = 2k + 2$

$\Rightarrow 2(k+1) < (k+2)! + 2$ [$\because P(k)$ is trues]

$\Rightarrow 2(k+1) < (k+3)!$

[$\because (k+2)! + 2 < (k+3)!$]

Since $P(k+1)$ is true whenever $P(k)$ is true, by the principle of mathematical induction, the statement $P(n)$

is true for all $n \in N$.

Q13. $\sqrt{n} < \frac{1}{\sqrt{1}} + \frac{1}{\sqrt{2}} + ... + \frac{1}{\sqrt{n}}$ for all natural numbers $n \geq 2$.

Solution:
Since the statement is

$P(n): \sqrt{n} < \frac{1}{\sqrt{1}} + \frac{1}{\sqrt{2}} + ... + \frac{1}{\sqrt{n}}$ for all natural numbers $n \geq 2$.

For $n = 2$, $P(2): \sqrt{2} < \frac{1}{\sqrt{1}} + \frac{1}{\sqrt{2}}$ is true.

Let $P(k)$ be true for some positive integer k. Then

$\sqrt{k} < \frac{1}{\sqrt{1}} + \frac{1}{\sqrt{2}} + ... + \frac{1}{\sqrt{k}}$

Now we shall prove that $P(k+1)$ is true, that is,

$\sqrt{k+1} < \frac{1}{\sqrt{1}} + \frac{1}{\sqrt{2}} + ... + \frac{1}{\sqrt{k}} + \frac{1}{\sqrt{k+1}}$

$\because \sqrt{k} < \frac{1}{\sqrt{1}} + \frac{1}{\sqrt{2}} + ... + \frac{1}{\sqrt{k}}$ [$\because P(k)$ is true]

$\Rightarrow \sqrt{k} + \frac{1}{\sqrt{k+1}} < \frac{1}{\sqrt{1}} + \frac{1}{\sqrt{2}} + ... + \frac{1}{\sqrt{k}} + \frac{1}{\sqrt{k+1}}$

Now we have to prove that

$\Rightarrow \sqrt{k+1} < \sqrt{k} + \frac{1}{\sqrt{k+1}}$

$\Rightarrow \sqrt{k+1} < \frac{\sqrt{k}\sqrt{k+1}+1}{\sqrt{k+1}}$

$\Rightarrow k+1 < \sqrt{k}\sqrt{k+1}+1$

$\Rightarrow k < \sqrt{k(k+1)} \Rightarrow \sqrt{k} < \sqrt{k+1}$ is true

Since $P(k+1)$ is true whenever $P(k)$ is true, by the principle of mathematical induction, the statement $P(n)$ is true for all $n \in N$.

Q14. $2+4+6+...+2n = n^2 + n$, for all natural numbers n.

Solution:
Since the given statement is

$P(n): 2+4+6+...+2n = n^2 + n$ for all natural numbers n.

For $n = 1$, $P(1): 2 = 1^2 + 1 = 2$, is true.

Let $P(k)$ be true for some positive integer k. Then

$2+4+6+...+2k = k^2 + k$

Now we shall prove that $P(k+1)$ is true

$P(k+1): 2+4+8+...+2k+2(k+1)$

$= k^2 + k + 2(k+1)$ [$\because P(k)$ is true]

$= k^2 + k + 2k + 2$

$= k^2 + 2k + 1 + k + 1$

$= (k+1)^2 + k + 1$

Since $P(k+1)$ is true whenever $P(k)$ is true, by the principle of mathematical induction, the statement $P(n)$ is true for all $n \in N$.

Q15. $1+2+2^2+....+2^n = 2^{n+1}-1$ for all whole numbers n.

Solution:
Since the given statement is

$P(n): 1+2+2^2+....+2^n = 2^{n+1}-1$ for all whole numbers n.

For $n = 0$, $P(0): 1 = 2^{0+1} - 1 = 1$, is true.

Let $P(k)$ be true for some positive integer k. Then

$1+2+2^2+....+2^k = 2^{k+1}-1$ is true.

Now we shall prove that $P(k+1)$ is true

$P(k+1): 1+2+2^2+....+2^k+2^{k+1}$

$= 2^{k+1} - 1 + 2^{k+1}$ [$\because P(k)$ is true]

$= 2 \cdot 2^{k+1} - 1 = 2^{k+2} - 1$

$= 2^{(k+1)+1} - 1$

Since $P(k+1)$ is true whenever $P(k)$ is true, by the principle of mathematical induction, the statement $P(n)$ is true for all $n \in N$.

Q16. $1+5+9+....+(4n-3) = n(2n-1)$, for all natural numbers n.

Solution:
Since the given statement is

$P(n): 1+5+9+....+(4n-3) = n(2n-1)$, for all natural numbers n.

For $n = 1$, $P(1): 1 = 1(2 \times 1 - 1) = 1$, is true.

Let $P(k)$ be true for some positive integer k. Then

$1+5+9+...+(4k-3) = k(2k-1)$

Now we shall prove that $P(k+1)$ is true

$P(k+1): 1+5+9+...+(4k-3)+4(k+1)-3$

$= k(2k-1)+4(k+1)-3$ [$\because P(k)$ is true]

$= k(2k-1)+4(k+1)-3$

$= 2k^2+3k+1$

$= 2k^2+2k+k+1$

$= 2k(k+1)+1(k+1)$

$= (k+1)(2k+1)$

$= (k+1)\{2(k+1)-1\}$

Since $P(k+1)$ is true whenever $P(k)$ is true, by the principle of mathematical induction, the statement $P(n)$ is true for all $n \in N$.

Day 67

NCERT Exemplar Problems: Solutions (Q17–Q30)

Q17. A sequence $a_1, a_2, a_3 ....$ is defined by letting $a_1 = 3,\ a_k = 7a_{k-1}$, for all natural numbers $k \geq 2$. Show that $a_n = 3\cdot 7^{n-1}$ for all natural numbers.

Solution:

Since the given statement is $a_1 = 3,\ a_k = 7a_{k-1}$, for all natural numbers $k \geq 2$.

For $n = 2$,

$\text{LHS} = a_2 = 7a_{2-1} = 7a_1 = 7\times 3 = 21,\ (\because a_1 = 3)$

$\text{RHS} = a_2 = 3\cdot 7^{2-1} = 3\cdot 7^1 = 21$

Since LHS = RHS, $P(2)$ is true.

Let $P(k)$ be true for some positive integer k. Then

$a_k = 3\cdot 7^{k-1}$

Now we shall prove that $P(k+1)$ is true

$a_{k+1} = 3\cdot 7^{k+1-1} = 3\cdot 7^k = 3\cdot 7^{(k+1)-1}$

Since $P(k+1)$ is true whenever $P(k)$ is true, by the principle of mathematical induction, the statement $P(n)$ is true for all $n \in N$.

Q18. A sequence $b_0, b_1, b_2 ....$ is defined by letting $b_0 = 5,\ b_k = 4 + b_{k-1}$, for all natural numbers k. Show that $b_n = 5 + 4n$, for all natural number n using mathematical induction.

Solution:

Since the given statement is, $b_n = 5 + 4n$ for all natural numbers given that $b_0 = 5,\ b_k = 4 + b_{k-1}$

For $n = 1$, $\text{LHS} = b_1 = 4 + b_0 = 4 + 5 = 9$

$\text{RHS} = b_1 = 5 + 4\cdot 1 = 5 + 4 = 9$

Since, LHS = RHS, $P(2)$ is true.

Let $P(k)$ be true for some positive integer k. Then

$b_k = 5 + 4k$

Now we shall prove that $P(k+1)$ is true

$P(k+1): b_{k+1} = 5 + 4(k+1)$

$\text{LHS} = b_{k+1} = 5 + 4(k+1)$

$\text{RHS} = b_{k+1} = 4 + b_{k+1-1} = 4 + b_k = 4 + 5 + 4k = 5 + 4(k+1)$

Since LHS = RHS, $P(k+1)$ is true whenever $P(k)$ is true, by the principle of mathematical induction, the statement $P(n)$ is true for all $n \in N$.

Q19. A sequence $d_1, d_2, d_3 ....$ is defined by letting $d_1 = 2,\ d_k = \dfrac{d_{k-1}}{k}$, for all natural numbers $k \geq 2$.

Show that $d_n = \dfrac{2}{n!}$, for all $n \in N$.

Solution:

Since the statement is $P(n): d_n = \dfrac{2}{n!}, \forall \in N$,

For $n = 2$, LHS $d_2 = \dfrac{2}{2!} = \dfrac{2}{2\times 1} = 1$

RHS $d_2 = \dfrac{d_{2-1}}{2} = \dfrac{d_1}{2} = \dfrac{2}{2} = 1$

Since LHS = RHS, $P(2)$ is true.

Let $P(k)$ be true for some positive integer k. Then

$d_k = \dfrac{2}{k!}$

Now we shall prove that $P(k+1)$ is true

$\text{LHS} = d_{k+1} = \dfrac{2}{(k+1)!}$

$\text{RHS} = d_{k+1} = \dfrac{d_{k+1-1}}{k+1} = \dfrac{d_k}{k+1} = \dfrac{2}{k!k} = \dfrac{2}{(k+1)!}$

Since LHS = RHS, $P(k+1)$ is true whenever $P(k)$ is true, by the principle of mathematical induction, the statement $P(n)$ is true for all $n \in N$.

Q20. Prove that for all $n \in N$

$$\cos\alpha + \cos(\alpha+\beta) + \cos(\alpha+2\beta) + ... + \cos[\alpha+(n-1)\beta] = \frac{\cos\left[\alpha + \left(\frac{n-1}{2}\right)\beta\right]\sin\left(\frac{n\beta}{2}\right)}{\sin\frac{\beta}{2}}$$

Solution:

$P(n): \cos\alpha + \cos(\alpha+\beta) + \cos(\alpha+2\beta) + ...$
$+\cos[\alpha+(n-1)\beta]$

$$= \frac{\cos\left[\alpha + \left(\frac{n-1}{2}\right)\beta\right]\sin\left(\frac{n\beta}{2}\right)}{\sin\frac{\beta}{2}}$$

For $n = 1$,

$$P(1): \frac{\cos\left[\alpha + \left(\frac{n-1}{2}\right)\beta\right]\sin\frac{\beta}{2}}{\sin\frac{\beta}{2}} = \frac{\cos(\alpha)\sin\frac{\beta}{2}}{\sin\frac{\beta}{2}} = \cos\alpha$$

Thus, $P(1)$ is true.

Let $P(k)$ be true for some positive integer k. Then

$P(k): \cos\alpha + \cos(\alpha+\beta) + \cos(\alpha+2\beta) + ...$
$+\cos[\alpha+(k-1)\beta]$

$$= \frac{\cos\left[\alpha + \left(\frac{k-1}{2}\right)\beta\right]\sin\left(\frac{k\beta}{2}\right)}{\sin\frac{\beta}{2}}$$

Now we shall prove that $P(k+1)$ is true

$P(k+1): \cos\alpha + \cos(\alpha+\beta) + \cos(\alpha+2\beta) + \ldots$
$+ \cos[\alpha+(k-1)\beta] + \cos[\alpha+(k+1-1)\beta]$

$$= \frac{\cos\left[\alpha+\left(\frac{k}{2}\right)\beta\right]\sin\left(\frac{(k+1)\beta}{2}\right)}{\sin\frac{\beta}{2}}$$

$\text{LHS} = \cos\alpha + \cos(\alpha+\beta) + \cos(\alpha+2\beta) + \ldots.$
$+ \cos[\alpha+(k-1)\beta] + \cos(\alpha+k\beta)$

$$= \frac{\cos\left[\alpha+\left(\frac{k-1}{2}\right)\beta\right]\sin\frac{k\beta}{2}}{\sin\frac{\beta}{2}} + \cos(\alpha+k\beta)$$

$$= \frac{\cos\left[\alpha+\left(\frac{k-1}{2}\right)\beta\right]\sin\frac{k\beta}{2} + \cos(\alpha+k\beta)\sin\frac{\beta}{2}}{\sin\frac{\beta}{2}}$$

$$= \frac{\sin\left(\alpha+\frac{k\beta}{2}-\frac{\beta}{2}+\frac{k\beta}{2}\right) - \sin\left(\alpha+\frac{k\beta}{2}-\frac{\beta}{2}-\frac{k\beta}{2}\right) + \sin\left(\alpha+k\beta+\frac{\beta}{2}\right) - \sin\left(\alpha+k\beta-\frac{\beta}{2}\right)}{2\sin\frac{\beta}{2}}$$

$$= \frac{\sin\left(\alpha+k\beta-\frac{\beta}{2}\right) - \sin\left(\alpha-\frac{\beta}{2}\right) + \sin\left(\alpha+k\beta+\frac{\beta}{2}\right) - \sin\left(\alpha+k\beta-\frac{\beta}{2}\right)}{2\sin\frac{\beta}{2}}$$

$$= \frac{\sin\left(\alpha+k\beta+\frac{\beta}{2}\right) - \sin\left(\alpha-\frac{\beta}{2}\right)}{2\sin\frac{\beta}{2}}$$

$$= \frac{2\cos\left(\frac{2\alpha+k\beta}{2}\right)\sin\left(\frac{k\beta+\beta}{2}\right)}{2\sin\frac{\beta}{2}}$$

$$= \frac{\cos\left(\frac{2\alpha+k\beta}{2}\right)\sin\left(\frac{k\beta+\beta}{2}\right)}{\sin\frac{\beta}{2}}$$

$$= \frac{\cos\left(\alpha+\frac{k\beta}{2}\right)\sin\left((k+1)\frac{\beta}{2}\right)}{\sin\frac{\beta}{2}} = \text{RHS}$$

So, $P(k+1)$ is true. Hence, $P(n)$ is true.

Since LHS = RHS, $P(k+1)$ is true whenever $P(k)$ is true, by the principle of mathematical induction, the statement $P(n)$ is true for all $n \in N$.

Q21. Prove that

$$\cos\theta\cos 2\theta\cos 2^2\theta \ldots \cos 2^{n-1}\theta = \frac{\sin 2^n\theta}{2^n \sin\theta}, \quad \forall n \in N.$$

Solution:

Since the given statement is

$$\cos\theta\cos 2\theta\cos 2^2\theta \ldots \cos 2^{n-1}\theta = \frac{\sin 2^n\theta}{2^n \sin\theta}, \quad \forall n \in N.$$

For $n = 1$,

$$P(1): \cos\theta = \frac{\sin 2^1\theta}{2^1\sin\theta} = \frac{\sin 2\theta}{2\sin\theta} = \frac{2\sin\theta\cos\theta}{2\sin\theta} = \cos\theta \text{ is true.}$$

Let $P(k)$ is true for some positive integer k, then

$$P(k): \cos\theta\cos 2\theta\cos 2^2\theta \ldots \cos 2^{k-1}\theta = \frac{\sin 2^k\theta}{2^k\sin\theta} \text{ is true.}$$

Now we shall prove that $P(k+1)$ is true,

$P(k+1): \cos\theta\cdot\cos 2\theta\cdot\cos 2^2\theta \ldots \cos 2^{k-1}\theta\cdot\cos 2^k\theta$

$$= \frac{\sin 2^k\theta}{2^k\sin\theta}\cdot\cos 2^k\theta = \frac{2\sin 2^k\theta\cdot\cos 2^k\theta}{2\cdot 2^k\sin\theta}$$

$$= \frac{\sin 2\cdot 2^k\theta}{2^{k+1}\sin\theta} = \frac{\sin 2^{(k+1)}\theta}{2^{k+1}\sin\theta}$$

Since $P(k+1)$ is true whenever $P(k)$ is true, by the principle of mathematical induction, the statement $P(n)$ is true for all $n \in N$.

Q22. Prove that, for all $n \in N$,

$$\sin\theta + \sin 2\theta + \sin 3\theta + \ldots + \sin n\theta = \frac{\sin\frac{n\theta}{2}\sin\frac{(n+1)}{2}\theta}{\sin\frac{\theta}{2}},$$

Solution:

Since the given statement is

$P(n): \sin\theta + \sin 2\theta + \sin 3\theta + \ldots + \sin n\theta$

$$= \frac{\sin\frac{n\theta}{2}\sin\frac{(n+1)}{2}\theta}{\sin\frac{\theta}{2}} \quad \text{for all } n \in N$$

$$\text{For } n = 1,\ P(1): \frac{\sin\frac{\theta}{2}\cdot\sin\frac{(1+1)}{2}\theta}{\sin\frac{\theta}{2}} = \frac{\sin\frac{\theta}{2}\cdot\sin\theta}{\sin\frac{\theta}{2}} = \sin\theta$$

thus, $P(1)$ is true.

Let $P(k)$ be true for some positive integer k. Then

$$P(k): \sin\theta + \sin 2\theta + \sin 3\theta + \ldots + \sin k\theta = \frac{\sin\frac{k\theta}{2}\sin\frac{(k+1)}{2}\theta}{\sin(\theta/2)}$$

Now we shall prove that $P(k+1)$ is true

$P(k+1): \sin\theta + \sin 2\theta + \ldots + \sin k\theta + \sin(k+1)\theta$

$$= \frac{\sin\frac{(k+1)\theta}{2}\sin\frac{(k+1+1)}{2}\theta}{\sin(\theta/2)}$$

LHS $= \sin\theta + \sin 2\theta + \ldots + \sin k\theta + \sin(k+1)\theta$

$$= \frac{\sin\frac{k\theta}{2}\sin\left(\frac{k+1}{2}\right)\theta}{\sin(\theta/2)} + \sin(k+1)\theta$$

$$= \frac{\sin\frac{k\theta}{2}\sin\left(\frac{k+1}{2}\right)\theta + \sin(k+1)\theta\cdot\sin\frac{\theta}{2}}{\sin(\theta/2)}$$

$$= \frac{\cos\left(\frac{k\theta}{2} - \frac{k+1}{2}\theta\right) - \cos\left(\frac{k\theta}{2} + \frac{k+1}{2}\theta\right) + \cos\left((k+1)\theta - \frac{\theta}{2}\right) - \cos\left((k+1)\theta + \frac{\theta}{2}\right)}{2\sin(\theta/2)}$$

$$= \frac{\cos\left(\frac{\theta}{2}\right) - \cos\left(k\theta + \frac{\theta}{2}\right) + \cos\left(k\theta + \frac{\theta}{2}\right) - \cos\left(k\theta + \frac{3\theta}{2}\right)}{2\sin(\theta/2)}$$

$$= \frac{\cos\left(\frac{\theta}{2}\right) - \cos\left(k\theta + \frac{3\theta}{2}\right)}{2\sin(\theta/2)}$$

$$= \frac{2\sin\left(\dfrac{\frac{\theta}{2} + k\theta + \frac{3\theta}{2}}{2}\right)\cdot\sin\left(\dfrac{\frac{\theta}{2} + k\theta - \frac{3\theta}{2}}{2}\right)}{\sin(\theta/2)}$$

$$= \frac{\sin\left(\frac{k\theta + 2\theta}{2}\right)\cdot\sin\left(\frac{k\theta + \theta}{2}\right)}{\sin(\theta/2)} = \frac{\sin\left((k+1)\frac{\theta}{2}\right)\cdot\sin\left((k+1+1)\frac{\theta}{2}\right)}{\sin(\theta/2)}$$

So, $P(k+1)$ is true. Hence, $P(n)$ is true.

Since LHS = RHS, $P(k+1)$ is true whenever $P(k)$ is true, by the principle of mathematical induction, the statement $P(n)$ is true for all $n \in N$.

Q23. Show that $P(n): \frac{n^5}{5} + \frac{n^3}{3} + \frac{7n}{15}$ is a natural number, for all $n \in N$.

Solution:

Since the given statement is $P(n): \frac{n^5}{5} + \frac{n^3}{3} + \frac{7n}{15}$ is a natural number, for all $n \in N$.

For $n = 1$, $P(1): \frac{1^5}{5} + \frac{1^3}{3} + \frac{7\cdot 1}{15} = \frac{3+5+7}{15} = \frac{15}{15} = 1$, is a natural number, thus $P(1)$ is true.

Let $P(k)$ be true for some positive integer k. Then

$P(k): \frac{k^5}{5} + \frac{k^3}{3} + \frac{7k}{15}$ is natural number, and thus, consider

$$\frac{k^5}{5} + \frac{k^3}{3} + \frac{7k}{15} = m, \quad m \in N$$

Now we shall prove that $P(k+1)$ is true

$$\frac{(k+1)^5}{5} + \frac{(k+1)^3}{3} + \frac{7(k+1)}{15}$$

$$= \frac{k^5 + 5k^4 + 10k^3 + 10k^2 + 5k + 1}{5} + \frac{k^3 + 1 + 3k(k+1)}{3} + \frac{7k + k}{15}$$

$$= \frac{k^5 + 5k^4 + 10k^3 + 10k^2 + 5k + 1}{5} + \frac{k^3 + 1 + 3k^2}{3} + \frac{7k + k}{15}$$

$$= \frac{k^5}{5} + \frac{k^3}{3} + \frac{7k}{15} + \frac{5k^4 + 10k^3 + 10k^2 + 5k + 1}{5} + \frac{3k^2 + 3k + 1}{3} + \frac{7k + 7}{15}$$

$$= m + k^4 + 2k^3 + 2k^2 + k + k^2 + k + \frac{1}{5} + \frac{1}{3} + \frac{7}{15}$$

$= m + k^4 + 2k^3 + 3k^2 + 2k + 1$, is a natural number

Since $P(k+1)$ is true whenever $P(k)$ is true, by the principle of mathematical induction, the statement $P(n)$ is true for all $n \in N$.

Q24. Prove that $\frac{1}{n+1} + \frac{1}{n+2} + \ldots + \frac{1}{2n} > \frac{13}{24}$, for all natural numbers $n > 1$.

Solution:

Since the given statement is

$P(n): \frac{1}{n+1} + \frac{1}{n+2} + \ldots + \frac{1}{2n} > \frac{13}{24}$, for all natural numbers $n > 1$.

For $n = 2$, $P(2): \frac{1}{2+1} + \frac{1}{2+2} = \frac{1}{3} + \frac{1}{4} = \frac{7}{12} = \frac{14}{24} > \frac{13}{24}$, thus $P(2)$ is true.

Let $P(k)$ be true for some positive integer $k > 1$. Then

$$P(k): \frac{1}{k+1} + \frac{1}{k+2} + \ldots + \frac{1}{2k} > \frac{13}{24}$$

$$\frac{1}{k+1} + \frac{1}{k+2} + \ldots + \frac{1}{2k} - \frac{13}{24} > 0$$

Now we shall prove that $P(k+1)$ is true

$$P(k+1) = \frac{1}{k+2} + \frac{1}{k+3} + \ldots \frac{1}{2k} + \frac{1}{2(k+1)}$$

$$\Rightarrow P(k+1) = \frac{1}{k+1} + \frac{1}{k+2} + \frac{1}{k+3} + \ldots \frac{1}{2k} - \frac{13}{24} + \frac{1}{2(k+1)} - \frac{1}{k+1} + \frac{13}{24}$$

$$\Rightarrow P(k+1) > \frac{1}{2(k+1)} - \frac{1}{k+1} + \frac{13}{24}$$

$\Rightarrow P(k+1) > \frac{13}{24} - \frac{1}{2(k+1)}$

$\Rightarrow P(k+1) > \frac{13}{24} - \frac{1}{2(2+1)}, \quad (\because k > 1)$

$\Rightarrow P(k+1) > \frac{13}{24} - \frac{1}{6} \Rightarrow P(k+1) > \frac{9}{24} > 0$

Since $P(k+1)$ is true whenever $P(k)$ is true, by the principle of mathematical induction, the statement $P(n)$ is true for all $n \in N$.

Q25. Prove that number of subsets of a set containing n distinct elements is 2^n, for all $n \in N$.

Solution:
Since the given statement is $P(n)$: Number of subsets of a set containing n distinct elements is 2^n, for all $n \in N$.
For $n = 1$, Number of subsets of a set that contain 1 element is $2^1 = 2$, $\{1\}, \{\phi\}$ which is true.
Let $P(k)$ be true for some positive integer k. Then
$P(k)$: Number of subsets of a set containing k distinct elements is 2^k, which is true.
Now we shall prove that $P(k+1)$ is true
$P(k+1)$: Number of subsets of a set containing $(k+1)$ distinct elements is 2^{k+1}.
Since the addition of one element in the set, the number of subsets becomes double.
$\therefore$ Number of subsets of a set containing $(k+1)$ distinct elements $= 2 \times 2^k = 2^{k+1}$.
Since $P(k+1)$ is true whenever $P(k)$ is true, by the principle of mathematical induction, the statement $P(n)$ is true for all $n \in N$.

Q26. If $10^n + 3 \cdot 4^{n+2} + k$ is divisible by 9, for all $n \in N$, then the least positive integral value of k is
(a) 5 (b) 3 (c) 7 (d) 1

Solution: (a)
Since the statement is $P(n): 10^n + 3 \cdot 4^{n+2} + k$ is divisible by 9, for all $n \in N$.
Therefore, for $n = 1$, the given statement is also true, and thus, $10^1 + 3 \cdot 4^{1+2} + k$ is divisible by 9.
$\Rightarrow 10^1 + 3 \times 4^{1+2} + k = 9m, \quad m \in N$
$\Rightarrow 10 + 3 \times 64 + k = 9m \Rightarrow 202 + k = 9m$
For $m = 23$, 207 is the least value larger than 202, which is divisible by 9. Therefore
$202 + k = 207 \Rightarrow k = 5$ is the least value of k for which $10^n + 3 \cdot 4^{n+2} + k$ is divisible by 9.

Q27. For all $n \in N$, $3 \cdot 5^{2n+1} + 2^{3n+1}$ is divisible by
(a) 19 (b) 17 (c) 23 (d) 25

Solution: (b, c)
For $n = 1, 3 \cdot 5^{2n+1} + 2^{3n+1} = 3 \cdot 5^{2 \cdot 1+1} + 2^{3 \cdot 1+1}$
$= 3 \cdot 5^3 + 2^4 = 3 \times 125 + 16 = 391 = 17 \times 23$
Thus, $3 \cdot 5^{2n+1} + 2^{3n+1}$ is divisible by both 17 and 23.

Q28. If $x^n - 1$ is divisible by $(x - k)$, then the least positive integral value of k is
(a) 1 (b) 2 (c) 3 (d) 4

Solution: (a)
For $n = 1$, $x^1 - 1$ is divisible by $(x - k)$.
If $x - 1$ is divisible by $(x - k)$, the least possible integral value of k is 1.

Q29. If $P(n): 2n < n!, \quad n \in N$, then $P(n)$ is true for all $n \geq$ ______

Solution:
$P(n): 2n < n!$
$\Rightarrow n(n-1)(n-2)...3 \cdot 2 \cdot 1 > 2n \Rightarrow (n-1)! > 2$
$\because 3! > 2 \therefore n - 1 > 2 \Rightarrow n > 3 \Rightarrow n \geq 4$

Q30. Let $P(n)$ be a statement and let $P(k) \Rightarrow P(k+1)$, for some natural number k, then $P(n)$ is true for all $n \in N$.

Solution: False
Because the given statement must be true for $P(1)$.

Day 68

Long Answer-I (LA-I): 4-Mark and Long Answer-II (LA-II): 6-Mark Questions

Q1. Using the principle of mathematical induction, prove that the sum of the arithmetic series with the first term a and common difference d is given by $S_n = a + (n-1)d$ for all natural numbers n.

Solution:
Since the sum of arithmetic series with first term a and common difference d is given by $S_n = a + (n-1)d$
For $n = 1$,
LHS $= S_1 = a$
RHS $= S_1 = a + (1-1)d = a$
Since LHS = RHS, $P(1)$ is true.

Let $P(k)$ be true for some positive integer k. Then

$P(k): S_k = a+(k-1)d$

Now we shall prove that $P(k+1)$ is true, that is,

$P(k+1): S_{k+1} = a+(k+1-1)d = a+kd$

$\text{LHS} = S_{k+1} = a_1 + a_2 + ... + a_k + a_{k+1}$

$= a+(k-1)d+d$

$= a+(k-1+1)d$

$= a+kd = \text{RHS}$

Since LHS = RHS, $P(k+1)$ is true whenever $P(k)$ is true, by the principle of mathematical induction, the statement $P(n)$ is true for all $n \in N$.

Q2. Using the principle of mathematical induction prove that the sum of the geometrical series with first term a and common ratio r is given by

$$S_n = \frac{a(r^n-1)}{r-1} \text{ for all natural numbers } n.$$

Solution:

Since the sum of the geometrical series with the first term a and common ratio r is given by $S_n = \dfrac{a(r^n-1)}{r-1}$

For $n = 1$,

$\text{LHS} = S_1 = a$

$$\text{RHS} = S_1 = \frac{a(r^1-1)}{r-1} = a$$

Since LHS = RHS, $P(1)$ is true.

Let $P(k)$ be true for some positive integer k. Then

$$P(k): S_k = \frac{a(r^k-1)}{r-1}$$

Now we shall prove that $P(k+1)$ is true, that is,

$$P(k+1): S_{k+1} = \frac{a(r^{k+1}-1)}{r-1}$$

$\text{LHS} = S_{k+1} = a_1 + a_2 + ... + a_k + a_{k+1}$

$$= \frac{a(r^k-1)}{r-1} + ar^k = \frac{a(r^k-1)+ar^k(r-1)}{r-1}$$

$$= \frac{ar^k - a + ar^{k+1} - ar^k}{r-1}$$

$$= \frac{ar^{k+1}-a}{r-1} = \frac{a(r^{k+1}-1)}{r-1} = \text{RHS}$$

Since LHS = RHS, $P(k+1)$ is true whenever $P(k)$ is true, by the principle of mathematical induction, the statement $P(n)$ is true for all $n \in N$.

Q3. Prove the following by using the principle of mathematical induction for all $n \in N$

$$\frac{1}{\sqrt{1}+\sqrt{2}} + \frac{1}{\sqrt{2}+\sqrt{3}} + ... + \frac{1}{\sqrt{n}+\sqrt{n+1}} = \frac{n}{\sqrt{1}+\sqrt{n+1}}$$

Solution:

Since the given statement is $P(n)$:

$$\frac{1}{\sqrt{1}+\sqrt{2}} + \frac{1}{\sqrt{2}+\sqrt{3}} + ... + \frac{1}{\sqrt{n}+\sqrt{n+1}} = \frac{n}{\sqrt{1}+\sqrt{n+1}}$$

For $n = 1$,

$$P(1) = \frac{1}{\sqrt{1}+\sqrt{1+1}} = \frac{1}{\sqrt{1}+\sqrt{2}} \text{ is true.}$$

Let $P(k)$ be true for some positive integer k. Then

$$P(k) = \frac{1}{\sqrt{1}+\sqrt{2}} + \frac{1}{\sqrt{2}+\sqrt{3}} + ... + \frac{1}{\sqrt{k}+\sqrt{k+1}} = \frac{k}{\sqrt{1}+\sqrt{k+1}}$$

Now we shall prove that $P(k+1)$ is true

$$P(k+1) = \frac{1}{\sqrt{1}+\sqrt{2}} + \frac{1}{\sqrt{2}+\sqrt{3}} + ... + \frac{1}{\sqrt{k}+\sqrt{k+1}} + \frac{1}{\sqrt{k+1}+\sqrt{k+2}} = \frac{k+1}{\sqrt{1}+\sqrt{k+2}}$$

$\text{LHS} = P(k+1)$

$$= \frac{1}{\sqrt{1}+\sqrt{2}} + ... + \frac{1}{\sqrt{k}+\sqrt{k+1}} + \frac{1}{\sqrt{k+1}+\sqrt{k+2}}$$

$$= \frac{k}{\sqrt{1}+\sqrt{k+1}} + \frac{1}{\sqrt{k+1}+\sqrt{k+2}} \quad [\because P(k) \text{ is true}]$$

$$= \left(\frac{k}{\sqrt{1}+\sqrt{k+1}}\right)\left(\frac{\sqrt{1}-\sqrt{k+1}}{\sqrt{1}-\sqrt{k+1}}\right) + \left(\frac{1}{\sqrt{k+1}+\sqrt{k+2}}\right)\left(\frac{\sqrt{k+1}-\sqrt{k+2}}{\sqrt{k+1}-\sqrt{k+2}}\right)$$

$$= \frac{k(\sqrt{1}-\sqrt{k+1})}{1-(k+1)} + \frac{\sqrt{k+1}-\sqrt{k+2})}{(k+1)-(k+2)}$$

$$= \frac{k-k\sqrt{k+1}}{-k} + \frac{\sqrt{k+1}-\sqrt{k+2})}{-1}$$

$$= \sqrt{k+1} - 1 + \sqrt{k+2} - \sqrt{k+1}$$

$$= \sqrt{k+2} - 1 = \left(\sqrt{k+2}-1\right)\left(\frac{\sqrt{k+2}+1}{\sqrt{k+2}+1}\right)$$

$$= \frac{k+2-1}{\sqrt{k+2}+1} = \frac{k+1}{\sqrt{1}+\sqrt{k+2}} = \text{RHS}$$

Since $P(k+1)$ is true whenever $P(k)$ is true, by the principle of mathematical induction, the statement $P(n)$ is true for all $n \in N$.

Q4. $n^3 + 2n$ is divisible by 3, for all natural numbers n.

Solution:

Since the statement is $P(n): n^3 + 2n$ is divisible by 3, for all natural number n.

For $n = 1$,

$P(1) = (1)^3 + 2 \cdot 1 = 3$ is divisible by 3, thus $P(1)$ is true.

Let $P(k)$ be true for some positive integer k. Then k^3+2k is divisible by 3, and thus, consider

$k^3+2k=3m, \quad$ where $m \in N$

Now we shall prove that $P(k+1)$ is true

$(k+1)^3+2(k+1)$

$=k^3+1+3k^2+3k+2k+2$

$=3m-2k+1+3k^2+3k+2k+2 \quad [\because P(k) \text{ is true}]$

$=3m+3k^2+3k+3$

$=3(m+k^2+k+1)$ is divisible by 3

Since $P(k+1)$ is true whenever $P(k)$ is true, by the principle of mathematical induction, the statement $P(n)$ is true for all $n \in N$.

Q5. $7^{2n}-48n-1$ is divisible by 2304, for all natural numbers n.

Solution:

Since the statement is $P(n): 7^{2n}-48n-1$ is divisible by 2304, for all natural number n.

For $n=1$,

$P(1)=7^{2\cdot 1}-48-1=49-49=0$ is divisible by 2304, thus $P(1)$ is true.

Let $P(k)$ be true for some positive integer k. Then $7^{2k}-48k-1$ is divisible by 2304, and thus, consider

$7^{2k}-48k-1=2304m, \quad$ where $m \in N$

Now we shall prove that $P(k+1)$ is true

$7^{2(k+1)}-48(k+1)-1=7^{2k}\cdot 7^2-48k-48-1$

$=(2304m+48k+1)\cdot 7^2-48k-49 \quad (\because P(k) \text{ is true})$

$=2304m\cdot 49+49\cdot 48k+49-48k-49$

$=2304(49m)+2304k$

$=2304(49m+k)$ is divisible by 2304

Since $P(k+1)$ is true whenever $P(k)$ is true, by the principle of mathematical induction, the statement $P(n)$ is true for all $n \in N$.

Q6. $\frac{1}{\sqrt{1}}+\frac{1}{\sqrt{2}}+....+\frac{1}{\sqrt{n}}>2(\sqrt{n+1}-1)$ for all natural numbers $n \in N$.

Solution:

Since the statement is

$P(n): \frac{1}{\sqrt{1}}+\frac{1}{\sqrt{2}}+....+\frac{1}{\sqrt{n}}>2(\sqrt{n+1}-1)$ for all natural numbers $n \in N$

For $n=1$, $P(1): 1>2(\sqrt{2}-1)$ is true.

Let $P(k)$ be true for some positive integer k. Then

$\frac{1}{\sqrt{1}}+\frac{1}{\sqrt{2}}+....+\frac{1}{\sqrt{k}}>2(\sqrt{k+1}-1)$

Now we shall prove that $P(k+1)$ is true

$\frac{1}{\sqrt{1}}+\frac{1}{\sqrt{2}}+....+\frac{1}{\sqrt{k}}+\frac{1}{\sqrt{k+1}}>2(\sqrt{k+2}-1)$

$\because \frac{1}{\sqrt{1}}+\frac{1}{\sqrt{2}}+....+\frac{1}{\sqrt{k}}>2(\sqrt{k+1}-1)$

$(\because P(k) \text{ is true})$

$\Rightarrow \frac{1}{\sqrt{1}}+\frac{1}{\sqrt{2}}+....+\frac{1}{\sqrt{k}}+\frac{1}{\sqrt{k+1}}>2(\sqrt{k+1}-1)+\frac{1}{\sqrt{k+1}}$

Now we have to prove that

$\Rightarrow 2(\sqrt{k+1}-1)+\frac{1}{\sqrt{k+1}}>2(\sqrt{k+2}-1)$

$\Rightarrow \frac{2(\sqrt{k+1}-1)\sqrt{k+1}+1}{\sqrt{k+1}}>2(\sqrt{k+2}-1)$

$\Rightarrow \frac{2(k+1)-2\sqrt{k+1}+1}{\sqrt{k+1}}>2(\sqrt{k+2}-1)$

$\Rightarrow 2k-2\sqrt{k+1}+3>2(\sqrt{k+2}-1)\sqrt{k+1}$

$\Rightarrow 2k-2\sqrt{k+1}+3>2\sqrt{k+2}\sqrt{k+1}-2\sqrt{k+1}$

$\Rightarrow 2k+3>2\sqrt{k+2}\sqrt{k+1}>2\sqrt{k+1}\sqrt{k+1}$

$\Rightarrow 2k+3>2(k+1) \Rightarrow 3>2$ is true

Since $P(k+1)$ is true whenever $P(k)$ is true, by the principle of mathematical induction, the statement $P(n)$ is true for all $n \in N$.

Q7. $\frac{1}{2}\cdot\frac{3}{4}\cdot\frac{5}{6}\cdots\frac{2n-1}{2n}\le\frac{1}{\sqrt{3n+1}}$ for natural numbers $n \in N$.

Solution:

Since the statement is

$P(n): \frac{1}{2}\cdot\frac{3}{4}\cdot\frac{5}{6}\cdots\frac{2n-1}{2n}\le\frac{1}{\sqrt{3n+1}}$ for natural numbers $n \in N$.

For $n=1$, $P(1): \frac{1}{2}=\frac{1}{\sqrt{3\cdot 1+1}}=\frac{1}{2}$, thus $P(1)$ is true.

Let $P(k)$ be true for some positive integer k. Then

$\frac{1}{2}\cdot\frac{3}{4}\cdot\frac{5}{6}\cdots\frac{2k-1}{2k}\le\frac{1}{\sqrt{3k+1}}$ is true.

Now we shall prove that $P(k+1)$ is true

$\frac{1}{2}\cdot\frac{3}{4}\cdot\frac{5}{6}\cdots\frac{2k-1}{2k}\cdot\frac{2k+1}{2k+2}\le\frac{1}{\sqrt{3(k+1)+1}}$

$\because \frac{1}{2}\cdot\frac{3}{4}\cdot\frac{5}{6}\cdots\frac{2k-1}{2k}\le\frac{1}{\sqrt{3k+1}}$

$[\because P(k) \text{ is true}]$

$\therefore \frac{1}{2}\cdot\frac{3}{4}\cdot\frac{5}{6}\cdots\frac{2k-1}{2k}\cdot\frac{2k+1}{2k+2}\le\frac{1}{\sqrt{3k+1}}\cdot\frac{2k+1}{2k+2}$

Now we have to prove that

$\Rightarrow \frac{1}{\sqrt{3k+1}}\cdot\frac{2k+1}{2k+2}\le\frac{1}{\sqrt{3(k+1)+1}}$

$\Rightarrow (2k+1)\sqrt{3(k+1)+1} \le 2(k+1)\sqrt{3k+1}$

$\Rightarrow (2k+1)\sqrt{3k+4} \le 2(k+1)\sqrt{3k+1}$

$\Rightarrow (2k+1)^2(3k+4) \le 4(k+1)^2(3k+1)$

$\Rightarrow (4k^2+4k+1)(3k+4) \le (4k^2+8k+4)(3k+1)$

$\Rightarrow 12k^3+12k^2+3k+16k^2+16k+4$

$\le 12k^3+24k^2+12k+4k^2+8k+4$

$\Rightarrow 19k \le 20k$ is true.

Since $P(k+1)$ is true whenever $P(k)$ is true, by the principle of mathematical induction, the statement $P(n)$ is true for natural numbers $n \in N$.

Q8. $\left(1-\frac{1}{\sqrt{2}}\right)\left(1-\frac{1}{\sqrt{3}}\right)\ldots\left(1-\frac{1}{\sqrt{n}}\right) < \frac{2}{n^2}$ for natural numbers $n \ge 2$.

Solution:

Since the statement is

$P(n): \left(1-\frac{1}{\sqrt{2}}\right)\left(1-\frac{1}{\sqrt{3}}\right)\ldots\left(1-\frac{1}{\sqrt{n}}\right) < \frac{2}{n^2}$ for natural numbers $n \ge 2$.

For $n = 1$,

$P(1): 1-\frac{1}{\sqrt{2}} < \frac{2}{4} \Rightarrow 1-\frac{1}{\sqrt{2}} < \frac{1}{2}$, thus $P(1)$ is true.

Let $P(k)$ be true for some positive integer k. Then

$\left(1-\frac{1}{\sqrt{2}}\right)\left(1-\frac{1}{\sqrt{3}}\right)\ldots\left(1-\frac{1}{\sqrt{k}}\right) < \frac{2}{k^2}$ is true.

Now we shall prove that $P(k+1)$ is true

$$\left(1-\frac{1}{\sqrt{2}}\right)\left(1-\frac{1}{\sqrt{3}}\right)\ldots\left(1-\frac{1}{\sqrt{k}}\right)\left(1-\frac{1}{\sqrt{k+1}}\right) < \frac{2}{(k+1)^2}$$

$$\because \left(1-\frac{1}{\sqrt{2}}\right)\left(1-\frac{1}{\sqrt{3}}\right)\ldots\left(1-\frac{1}{\sqrt{k}}\right) < \frac{2}{k^2}$$

[$\because P(k)$ is true]

$$\therefore \left(1-\frac{1}{\sqrt{2}}\right)\ldots\left(1-\frac{1}{\sqrt{k}}\right)\left(1-\frac{1}{\sqrt{k+1}}\right) < \frac{2}{k^2}\left(1-\frac{1}{\sqrt{k+1}}\right)$$

Now we have to prove that

$\Rightarrow \frac{2}{k^2}\left(1-\frac{1}{\sqrt{k+1}}\right) < \frac{2}{(k+1)^2}$

$\Rightarrow 1-\frac{1}{\sqrt{k+1}} < \frac{k^2}{(k+1)^2}$

$\Rightarrow 1-\frac{k^2}{(k+1)^2} < \frac{1}{\sqrt{k+1}}$

$\Rightarrow \frac{k^2+1+2k-k^2}{(k+1)^2} < \frac{1}{\sqrt{k+1}}$

$\Rightarrow 1+2k < (k+1)^{3/2}$

$\Rightarrow (1+2k)^2 < (k+1)^3$

$\Rightarrow 1+4k^2+4k < k^3+1+3k^2+3k$

$\Rightarrow k^2+k < k^3$ is true ($\because k \ge 2$)

Since $P(k+1)$ is true whenever $P(k)$ is true, by the principle of mathematical induction, the statement $P(n)$ is true for natural numbers $n \in N$.

CHAPTER

5

Complex Numbers and Quadratic Equations

Day 69

Algebra of Complex Numbers

A complex number is introduced to solve a quadratic equation $ax^2+bx+c=0$, when the discriminant $D=b^2-4ac<0$.

The quadratic equation $x^2-1=0$ has real solutions ($\because D=0-4(-1)=4>0$), but the quadratic equation $x^2+1=0$ has no real solution ($\because D=0-4\cdot 1=-4<0$) because $x^2+1=0$ gives $x^2=-1$ and the square of every real number is non-negative.

Thus, for solving these ($x^2+1=0$) quadratic equations where the discriminant $D=b^2-4ac<0$, we extend the real number system to the complex number system.

Complex Number: A number of the form $z=a+ib$, where a and b are real numbers, is defined to be a complex number. The complex number $z=a+ib$ cannot be represented on a real number line because it is an imaginary number.

Thus, here we introduce a new symbol 'i' known as iota (an imaginary number) such that $i^2=-1 \Rightarrow i=\sqrt{-1}$.

Therefore, we can say that i is a solution of the equation $x^2+1=0$.

Note:

1. For the complex number $z=a+ib$, a is called the real part, denoted by Re z, and b is called the imaginary part denoted by Im z of the complex number z. As an example, if $z=2+3i$, then $\text{Re}\,z=2$, $\text{Im}\,z=3$.
2. For the complex number $z=a+ib$, if $\text{Re}\,z=a=0$, then the complex number is known as a purely imaginary number, and if $\text{Im}\,z=b=0$, then the complex number is known as a purely real number.
   As an example, $z=0+3i$ is a purely imaginary number, while $z=2+0i$ is a purely real number.
3. In general, we write the imaginary part of z before i if it was given in a number and we write it after i if it was a variable. As an example, we write $z=2+3i$, that is, 3 before i since 3 is a number and we write $z=a+ib$, that is, b after i since b is a variable.
4. A real number can be represented in a complex number system but a complex number cannot be represented in a real number system. As an example, real number 2 in a complex number system is represented as $z=2+0i$, where $\text{Re}\,z=2$, $\text{Im}\,z=0$.
5. Two complex numbers $z_1=a+ib$ and $z_2=c+id$ are equal if and only if $a=c$ and $b=d$. As an example, if $z_1=2+3i$ and $z_2=x+iy$, then $z_1=z_2 \Rightarrow 2+3i=x+iy \Rightarrow x=2$ and $y=3$.

Addition of Complex Numbers: Let $z_1=a+ib, z_2=c+id$ be any two complex numbers. Then their sum is defined as,

$$z_1+z_2=(\text{Re}\,z_1+\text{Re}\,z_2)+i(\text{Im}\,z_1+\text{Im}\,z_2)$$
$$=(a+c)+i(b+d)$$

As an example,

$$(2+3i)+(1-2i)=(2+1)+(3-2)i=3-i$$
$$(3+2i)+(1-2i)=(3+1)+(2-2)i=4+0i$$
$$(2+3i)+(-2+i)=(2-2)+(3+1)i=0+4i$$

Properties of addition of complex numbers:

(i) Closure law: The sum z_1+z_2 of two complex numbers z_1, z_2 is a complex number. Note that if we get z_1+z_2 as a real number, even then we write it in the complex number system.
As an example,
$(3+2i)+(1-2i)=(3+1)+(2-2)i=4+0i$

(ii) Commutative law: For any two complex numbers z_1, z_2 : $z_1+z_2=z_2+z_1$.

(iii) Associative law: For any three complex numbers z_1, z_2, z_3 : $z_1+(z_2+z_3)=(z_1+z_2)+z_3$.

(iv) Additive identity: For any complex number z, there exists a zero complex number $0+i0$ (denoted by 0) such that $z+0=z$ is known as the additive identity.

(v) Additive inverse: For any complex number $z=a+ib$, there exists a negative of the complex number $-z(=-a-ib)$ such that $z+(-z)=0$, and this is known as the additive inverse.

Difference of Two Complex Numbers: For any two complex numbers $z_1=a+ib$, $z_2=c+id$, the difference,

$$z_1-z_2=(\text{Re}\,z_1-\text{Re}\,z_2)+i(\text{Im}\,z_1-\text{Im}\,z_2)$$
$$=(a-c)+i(b-d)$$

As an example,
$(2+3i)-(1-2i)=(2-1)+(3+2)i=1+5i$

Multiplication of Two Complex Numbers: Let $z_1=a+ib$, $z_2=c+id$ be any two complex numbers. Then their product $z_1\cdot z_2$ is defined as,

$$z_1z_2=(a+ib)\cdot(c+id)$$
$$=ac+iad+ibc+i^2bd$$
$$=ac+iad+ibc-bd \quad (\because i^2=-1)$$
$$=(ac-bd)+i(ad+bc)$$

As an example, $(3+2i)\cdot(1-2i)$
$=(3\cdot1-2\cdot(-2))+(3\cdot(-2)+2\cdot1)i=7-4i$

Properties of multiplication of complex numbers:

(i) Closure law: The product $z_1\cdot z_2$ of two complex numbers z_1, z_2 is a complex number. Note that if we get $z_1\cdot z_2$ as a real number, even then we write it in the complex number system.
As an example, $(1+2i)\cdot(1-2i)$
$=(1\cdot1-2\cdot(-2))+(2\cdot1+1\cdot(-2))i=5+0i$

(ii) Commutative law: For any two complex numbers $z_1,z_2:\ z_1\cdot z_2=z_2\cdot z_1$.

(iii) Associative law: For any three complex numbers $z_1,z_2,z_3:\ z_1\cdot(z_2\cdot z_3)=(z_1\cdot z_2)\cdot z_3$.

(iv) Distributive law: For any three complex numbers z_1,z_2,z_3
(a) $z_1\cdot(z_2+z_3)=z_1\cdot z_2+z_1\cdot z_3$
(b) $(z_1+z_2)\cdot z_3=z_1\cdot z_3+z_2\cdot z_3$

(v) Multiplicative identity: For any complex number z there exists an identity complex number $1+i0$ (denoted by 1) such that $z\cdot1=z$ is known as the multiplicative identity.

(vi) Multiplicative inverse: For any non-zero complex number $z=a+ib$, there exists an inverse of complex number z (denoted by $\frac{1}{z}$ or z^{-1}), such that $z\cdot\frac{1}{z}=1$, and is known as the multiplicative inverse. It is given by $z^{-1}=\frac{a}{a^2+b^2}+i\frac{-b}{a^2+b^2}$. As an example, the multiplicative inverse of the complex number $z=3+4i$ is

$$z^{-1}=\frac{3}{3^2+4^2}+i\frac{-4}{3^2+4^2}=\frac{3}{25}+i\frac{-4}{25}=\frac{1}{25}(3-4i)$$

Division of Two Complex Numbers: For any two complex numbers z_1,z_2, where $z_2\neq0$, the division z_1/z_2 is defined by $\frac{z_1}{z_2}=z_1\frac{1}{z_2}$.

As an example, let $z_1=1+3i$, $z_2=2-i$

$$\text{Then } \frac{z_1}{z_2}=\left((1+3i)\times\frac{1}{2-i}\right)=(1+3i)(2-i)^{-1}$$
$$=(1+3i)\cdot\left(\frac{2}{2^2+(-1)^2}+i\frac{-(-1)}{2^2+(-1)^2}\right)$$
$$=(1+3i)\left(\frac{2+i}{5}\right)=\frac{1}{5}[2-3+i(1+6)]=\frac{1}{5}(-1+7i)$$

Power of *i*

$$\because i=\sqrt{-1}\Rightarrow i^2=-1$$
$$\Rightarrow i^3=i^2\cdot i=(-1)\cdot i=-i$$
$$\Rightarrow i^4=(i^2)^2=(-1)^2=1$$
$$\Rightarrow i^5=i^4\cdot i=1\cdot i=i$$
$$\Rightarrow i^6=i^5\cdot i=i\cdot i=-1$$
$$\Rightarrow i^7=i^6\cdot i=-1\cdot i=-i$$
$$\Rightarrow i^8=(i^2)^4=(-1)^4=1$$

And so on ... ...
Similarly, we have

$$i^{-1}=\frac{1}{i}=\frac{1}{i}\times\frac{i}{i}=\frac{i}{-1}=-i$$
$$\Rightarrow i^{-2}=\frac{1}{i^2}=\frac{1}{-1}=-1$$
$$\Rightarrow i^{-3}=\frac{1}{i^3}=\frac{1}{i^2}\cdot\frac{1}{i}=-1\cdot(-i)=i$$
$$\Rightarrow i^{-4}=\frac{1}{i^4}=\frac{1}{(i^2)^2}=\frac{1}{(-1)^2}=\frac{1}{1}=1$$

And so on ... ...
Thus, in general, for any integer,
$i^{4n}=1,\ i^{4n+1}=i,\ i^{4n+2}=-1,\ i^{4n+3}=-i$, for all integers $n\in Z$.

Note: Summation of four consecutive powers of i is equal to zero, that is,
$i^{4n}+i^{4n+1}+i^{4n+2}+i^{4n+3}=1+i-1-i=0$, for all integers $n\in Z$.

Example 1:

(1) i^{50}

Solution:
$i^{50}=i^{4\times12+2}=(i^4)^{12}\cdot i^2=1\cdot(-1)=-1 \quad (\because i^4=1,\ i^2=-1)$

(2) i^{-101}

Solution:
$$i^{-101}=\frac{1}{i^{101}}=\frac{1}{i^{4\times25+1}}=\frac{1}{(i^4)^{25}i}=\frac{1}{i}=\frac{i}{i^2}=\frac{i}{-1}=-i$$
$(\because i^4=1,\ i^2=-1)$

(3) $(1+i)^4$

Solution:
$$(1+i)^4=[(1+i)^2]^2=[1^2+i^2+2i]^2$$
$$=[1-1+2i]^2=4i^2=-4 \quad (\because i^2=-1)$$

(4) $(1+i)^n\left(1+\frac{1}{i}\right)^n$

Solution:

$(1+i)^n\left(1+\frac{1}{i}\right)^n = (1+i)^n\left(1+\frac{1}{i}\times\frac{i}{i}\right)^n$

$= (1+i)^n\left(1+\frac{i}{i^2}\right)^n = (1+i)^n(1-i)^n$

$= \left(1-i^2\right)^n = (1-(-1))^n = 2^n \quad (\because i^2 = -1)$

(5) $\sum_{n=0}^{101}(i^{n+1})$, where $n \in N$.

Solution: The summation of four consecutive powers of i is equal to zero, and there are 102 terms in the summation. Therefore, the summation of 100 (4×25) terms will be equal to zero. Thus, we will be left with two terms (either from the beginning or from the end).

Thus, $\sum_{n=0}^{101}(i^{n+1}) = i^{0+1} + i^{1+1} + \sum_{n=2}^{101}(i^{n+1})$

$= i^1 + i^2 + 0 \quad (\because i^n + i^{n+1} + i^{n+2} + i^{n+3} = 0)$

$= i - 1 \quad (\because i^2 = -1)$

Square Roots of a Negative Real Number: $\sqrt{a}\times\sqrt{b} = \sqrt{ab}$ for all real numbers a and b, when both $a \geq 0, b \geq 0$ or $a < 0, b \geq 0$ or $a \geq 0, b < 0$.

Now, for the condition when $a < 0, b < 0$

Let assume a and b are positive real numbers. Then

$\sqrt{-a} = \sqrt{(-1)\times a} = \sqrt{-1}\sqrt{a} = i\sqrt{a} \quad (\because i^2 = -1)$

Similarly,

$\sqrt{-b} = \sqrt{(-1)\times b} = \sqrt{-1}\sqrt{b} = i\sqrt{b}$

Thus, $\sqrt{-a}\times\sqrt{-b} = i\sqrt{a}\times i\sqrt{b} = i^2\sqrt{ab} = -\sqrt{ab}$

$(\because \sqrt{a}\times\sqrt{b} = \sqrt{ab}$ for $a \geq 0, b \geq 0$ and $i^2 = -1)$

Thus, note that $\sqrt{a}\times\sqrt{b} \neq \sqrt{ab}$ if both a and b are negative real numbers.

Note: Since, $i^2 = -1$ and also $(-i)^2 = i^2 = -1$, we get square roots of -1 as $i, -i$. But in general for the symbol $\sqrt{-1}$, we mean i.

As an example: $i, -i$ are the solutions of the equation $x^2 + 1 = 0$.

Also, $(\sqrt{2}i)^2 = (\sqrt{2})^2 i^2 = 2(-1) = -2$

And $(-\sqrt{2}i)^2 = (-\sqrt{2})^2 i^2 = -2$

Therefore, we get the square roots of -2 as $\sqrt{2}i, -\sqrt{2}i$. But in general, the symbol $-\sqrt{2}$ means $\sqrt{2}i$, that is, $\sqrt{-2} = \sqrt{2}i$, not $-\sqrt{2}i$.

Identities in complex numbers:

(i) $(z_1+z_2)^2 = z_1^2 + z_2^2 + 2z_1z_2$

(ii) $(z_1-z_2)^2 = z_1^2 - 2z_1z_2 + z_2^2$

(iii) $(z_1+z_2)^3 = z_1^3 + 3z_1^2z_2 + 3z_1z_2^2 + z_2^3$

(iv) $(z_1-z_2)^3 = z_1^3 - 3z_1^2z_2 + 3z_1z_2^2 - z_2^3$

(v) $z_1^2 - z_2^2 = (z_1+z_2)(z_1-z_2)$

Note: Many other identities which are true for all real numbers are also true for all complex numbers.

Example 2: Express the following in the form of $x+iy$.

(1) $2(1+i)+i(1-i)$

Solution:

$2(1+i)+i(1-i) = 2+2i+i-i^2 = 2+3i-(-1) = 3+3i$

(2) $\left(2i-\frac{1}{i}\right)^2$

Solution:

$\left(2i-\frac{1}{i}\right)^2 = \left(2i-\frac{1}{i}\times\frac{i}{i}\right)^2 = (2i+i)^2 = (3i)^2 = -9 = -9+0i$

(3) $\left(\frac{i}{1-i}\right)^2$

Solution:

$\left(\frac{i}{1-i}\right)^2 = \left(\frac{i}{1-i}\times\frac{1+i}{1+i}\right)^2 = \left(\frac{i+i^2}{1-i^2}\right)^2 = \left(\frac{i-1}{2}\right)^2$

$= \frac{i^2+1-2i}{4} = \frac{-i}{2} = 0 - \frac{1}{2}i$

Example 3: If $a = \sin\theta + i\cos\theta$, then find the value of $\frac{1-a}{1+a}$.

Solution:

$\frac{1-a}{1+a} = \frac{1-\sin\theta - i\cos\theta}{1+\sin\theta + i\cos\theta} \quad (\because a = \sin\theta + i\cos\theta)$

$= \frac{(1-\sin\theta)-i\cos\theta}{(1+\sin\theta)+i\cos\theta}\times\frac{(1+\sin\theta)-i\cos\theta}{(1+\sin\theta)-i\cos\theta}$

$= \frac{1-\sin^2\theta-\cos^2\theta-i(2\cos\theta)}{(1+\sin\theta)^2+\cos^2\theta}$

$= \frac{1-(\sin^2\theta+\cos^2\theta)-i(2\cos\theta)}{1+2\sin\theta+(\sin^2\theta+\cos^2\theta)}$

$= \frac{1-1-i(2\cos\theta)}{1-2\sin\theta+1} = \frac{-i\cos\theta}{1-\sin\theta}$

Example 4: Express $\frac{(1+i\sqrt{2})}{(\sqrt{2}+i)}+\frac{(1-i\sqrt{2})}{(\sqrt{2}-i)}$ in the form of $x+iy$.

Solution:

$$\frac{(1+i\sqrt{2})}{(\sqrt{2}+i)}+\frac{(1-i\sqrt{2})}{(\sqrt{2}-i)}=\frac{(1+i\sqrt{2})(\sqrt{2}-i)+(\sqrt{2}+i)(1-i\sqrt{2})}{(\sqrt{2}+i)(\sqrt{2}-i)}$$

$$=\frac{\sqrt{2}-i+2i-i^2\sqrt{2}+\sqrt{2}+i-2i-i^2\sqrt{2}}{(\sqrt{2})^2-i^2}$$

$$=\frac{2\sqrt{2}+2\sqrt{2}}{2-(-1)}=\frac{4\sqrt{2}}{3} \quad (\because i^2=-1)$$

Example 5: Evaluate: $\left[i^{11}+\left(\frac{1}{i}\right)^{22}\right]^3$

Solution:

$$\left[i^{11}+\left(\frac{1}{i}\right)^{22}\right]^3=\left[i^{4\times2+3}+\frac{1}{i^{4\times5+2}}\right]^3$$

$$=\left[(i^4)^2\cdot i^3+\frac{1}{(i^4)^5\cdot i^2}\right]^3=\left[i^3+\frac{1}{i^2}\right]^3$$

$$=[-i-1]^3 \quad (\because i^2=-1,\ i^3=-i,\ i^4=1)$$

$$=(-1)^3(1+i)^3$$

$$=-(1^3+i^3+3\cdot1^2\cdot i+3\cdot1\cdot i^2)$$

$$=-(1-i+3i-3)=-(-2+2i)=2-2i$$

Example 6: If $\left(\frac{1-i}{1+i}\right)^n+\left(\frac{1+i}{1-i}\right)^n=2$, then find the least positive integral value of n.

Solution:

$$\left(\frac{1-i}{1+i}\right)^n+\left(\frac{1+i}{1-i}\right)^n=2$$

$$\Rightarrow\left(\frac{1-i}{1+i}\times\frac{1-i}{1-i}\right)^n+\left(\frac{1+i}{1-i}\times\frac{1+i}{1+i}\right)^n=2$$

$$\Rightarrow\left(\frac{1^2+i^2-2i}{1^2-i^2}\right)^n+\left(\frac{1^2+i^2+2i}{1^2-i^2}\right)^n=2$$

$$\Rightarrow\left(\frac{1-1-2i}{1-(-1)}\right)^n+\left(\frac{1-1+2i}{1-(-1)}\right)^n=2 \quad (\because i^2=-1)$$

$$\Rightarrow\left(\frac{-2i}{2}\right)^n+\left(\frac{2i}{2}\right)^n=2$$

$$\Rightarrow(-i)^n+(i)^n=2$$

$$\Rightarrow i^n\left((-1)^n+1^n\right)=2\Rightarrow n=4$$

Thus, the least positive integral value of n is 4.

Example 7: Reduce $\left(\frac{1}{1-2i}-\frac{i}{1+i}\right)\left(\frac{1+3i}{1-i}\right)$ to the standard form.

Solution:

$$\left(\frac{1}{1-2i}-\frac{i}{1+i}\right)\left(\frac{1+3i}{1-i}\right)=\left[\frac{(1+i)-i(1-2i)}{(1-2i)(1+i)}\right]\left[\frac{1+3i}{1-i}\right]$$

$$=\left[\frac{1+i-i+2i^2}{1+i-2i-2i^2}\right]\left[\frac{1+3i}{1-i}\right]$$

$$=\left[\frac{-1}{3-i}\right]\left[\frac{1+3i}{1-i}\right] \quad (\because i^2=-1)$$

$$=\left[\frac{-1-3i}{3-3i-i+i^2}\right]$$

$$=\frac{-1-3i}{2-4i}$$

$$=\frac{-1-3i}{2-4i}\times\frac{2+4i}{2+4i}=\frac{-2-4i-6i-12i^2}{\left[(2)^2-(4i)^2\right]}$$

$$=\frac{10-10i}{(4+16)}=\frac{10(1-i)}{20}$$

$$=\frac{1}{2}(1-i)$$

Day 70

Conjugate and Modulus of a Complex Number

Conjugate of a Complex Number: If $z=a+ib$ is a complex number, then its conjugate is given by $\bar{z}=a-ib$.

As an example,

$z=2+i\Rightarrow\bar{z}=\overline{2+i}=2-i$

$z=1-2i\Rightarrow\bar{z}=\overline{1-2i}=1+2i$

Properties of conjugate:

(i) $\bar{\bar{z}}=z$

(ii) $\overline{z_1z_2}=\bar{z}_1\bar{z}_2$

(iii) $\overline{z_1\pm z_2}=\bar{z}_1\pm\bar{z}_2$

(iv) $\overline{\left(\frac{z_1}{z_2}\right)}=\frac{\bar{z}_1}{\bar{z}_2}$ provided $z_2\neq0$.

(v) $\text{Re}(z)=\frac{1}{2}\left(z+\bar{z}\right)$ and $\text{Im}(z)=\frac{1}{2i}\left(z-\bar{z}\right)$

Modulus of a Complex Number: If $z = a + ib$ is a complex number, then its modulus is $|z| = \sqrt{a^2 + b^2}$ (non-negative real number).

As an example, $|2+i| = \sqrt{2^2 + 1^2} = \sqrt{5}$

$|1-2i| = \sqrt{1^2 + (-2)^2} = \sqrt{5}$

Properties of modulus:

(i) $|z| = |\bar{z}|$

(ii) $|z\bar{z}| = |z|^2$

(iii) $|z_1 z_2| = |z_1||z_2|$

(iv) $\left|\dfrac{z_1}{z_2}\right| = \dfrac{|z_1|}{|z_2|}$ provided $|z_2| \neq 0$

Multiplicative Inverse of a Complex Number: If $z = a + ib$ is a complex number, then its inverse is $z^{-1} = \dfrac{\bar{z}}{|z|^2} = \dfrac{a - ib}{a^2 + b^2}$

Note: For two complex numbers,

$$\frac{z_1}{z_2} = z_1 z_2^{-1} = z_1 \frac{\overline{z_2}}{|z_2|^2} = \frac{z_1 \overline{z_2}}{|z_2|^2}$$

Example 1: Find the multiplicative inverse of the following complex numbers (i) $2 - 3i$ (ii) $-\sqrt{3} + 2i$.

Solution:

(i) $z = 2 - 3i \Rightarrow \bar{z} = 2 + 3i$

$|z|^2 = 2^2 + (-3)^2 = 4 + 9 = 13$

Therefore, the multiplicative inverse of $2 - 3i$ is

$$z^{-1} = \frac{\bar{z}}{|z|^2} = \frac{2+3i}{13} = \frac{2}{13} + \frac{3}{13}i$$

(ii) $z = -\sqrt{3} + 2i \Rightarrow \bar{z} = -\sqrt{3} - 2i$

$|z|^2 = (-\sqrt{3})^2 + 2^2 = 3 + 4 = 7$

Therefore, the multiplicative inverse of $-\sqrt{3} + 2i$ is

$$z^{-1} = \frac{\bar{z}}{|z|^2} = \frac{-\sqrt{3} - 2i}{7} = \frac{-\sqrt{3}}{7} - \frac{2}{7}i$$

Example 2: For any two complex numbers z_1, z_2, prove that $\text{Im}(z_1 z_2) = \text{Re}(z_1)\text{Im}(z_2) + \text{Re}(z_2)\text{Im}(z_1)$.

Solution:

Let $z_1 = x_1 + iy_1, \ z_2 = x_2 + iy_2$

$\text{LHS} = \text{Im}(z_1 z_2) = \text{Im}[(x_1 + iy_1)(x_2 + iy_2)]$

$= \text{Im}\left[x_1 x_2 + i x_1 y_2 + i y_1 x_2 + i^2 y_1 y_2\right]$

$= \text{Im}[x_1 x_2 + i x_1 y_2 + i y_1 x_2 - y_1 y_2] \quad (\because i^2 = -1)$

$= \text{Im}[(x_1 x_2 - y_1 y_2) + i(x_1 y_2 + y_1 x_2)]$

$= x_1 y_2 + x_2 y_1$

$= \text{Re}(z_1)\text{Im}(z_2) + \text{Re}(z_2)\text{Im}(z_1) = \text{RHS}$

Example 3: Let $z_1 = 1 - i, \ z_2 = -1 + i$.

(i) Show that $\dfrac{z_1 z_2}{\overline{z_1} z_2}$ is purely real

(ii) Find $\left|\dfrac{z_1 \overline{z_2}}{\overline{z_1} z_2}\right|$

(iii) Find $\text{Im}\left(\dfrac{1}{\overline{z_1}} + \dfrac{1}{z_2}\right)$

Solution:

$z_1 = 1 - i \Rightarrow \overline{z_1} = 1 + i$

$z_2 = -1 + i \Rightarrow \overline{z_2} = -1 - i$

(i) $\dfrac{z_1 z_2}{\overline{z_1} z_2} = \dfrac{(1-i)(-1+i)}{(1+i)(-1-i)}$

$= \dfrac{-1 + i + i - i^2}{-1 - i - i - i^2}$

$= \dfrac{-1 + 2i + 1}{-1 - 2i + 1} = \dfrac{2i}{-2i} = -1 \quad (\because i^2 = -1)$

(ii) $\left|\dfrac{z_1 \overline{z_2}}{\overline{z_1} z_2}\right| = \left|\dfrac{(1-i)(-1-i)}{(1+i)(-1+i)}\right|$

$= \left|\dfrac{-1 - i + i + i^2}{-1 + i - i + i^2}\right|$

$= \left|\dfrac{-1-1}{-1-1}\right| = 1 \quad (\because i^2 = -1)$

(iii) $\text{Im}\left(\dfrac{1}{\overline{z_1}} + \dfrac{1}{z_2}\right) = \text{Im}\left(\dfrac{1}{1+i} + \dfrac{1}{-1+i}\right)$

$= \text{Im}\left(\dfrac{-1 + i + 1 + i}{(1+i)(-1+i)}\right)$

$= \text{Im}\left(\dfrac{2i}{i^2 - 1}\right) \quad (\because i^2 = -1)$

$= \text{Im}\left(\dfrac{2i}{-2}\right) = \text{Im}(0 + (-1)i) = -1$

Example 4: If $z = x + iy$, then show that $z\bar{z} + (z + \bar{z}) + 1 = 0$ represents a circle.

Solution:

Let $z = x + iy$

$z\bar{z} + (z + \bar{z}) + 1 = 0$

$\Rightarrow (x + iy)(x - iy) + \{(x + iy) + (x - iy)\} + 1 = 0$

$\Rightarrow x^2 + y^2 + 2x + 1 = 0$ represents the equation of a circle.

Example 5: If the imaginary part of $\dfrac{\bar{z} + 1}{\bar{z} - 1}$ is 2, then show that the locus of the point representing z in the complex plane is a circle.

Solution:

Let $z = x + iy$

$\text{Im}\left(\dfrac{\bar{z}+1}{\bar{z}-1}\right)=2$

$\Rightarrow \text{Im}\left(\dfrac{x-iy+1}{x+iy-1}\right)=2$

$\Rightarrow \text{Im}\left(\dfrac{[(x+1)-iy][(x-1)+iy]}{[(x-1)-iy][(x-1)+iy]}\right)=2$

$\Rightarrow \text{Im}\left(\dfrac{(x+1)(x-1)+y^2+i[y(x+1)-y(x-1)]}{(x-1)^2+y^2}\right)=2$

$\Rightarrow \text{Im}\left(\dfrac{(x^2-1)+y^2}{(x-1)^2+y^2}+\dfrac{2y}{(x-1)^2+y^2}i\right)=2$

$\Rightarrow \dfrac{2y}{(x-1)^2+y^2}=2$

$\Rightarrow y=x^2+1-2x+y^2$

$\Rightarrow x^2+y^2-2x-y+1=0$ represents the equation of a circle.

Example 6: Solve the equation $|z|=z+1+i$.

Solution:

Let $z=x+iy$,

$|z|=z+1+i$

$\Rightarrow |x+iy|=x+iy+1+i$

$\Rightarrow \sqrt{x^2+y^2}=x+1+i(y+1)$

$\Rightarrow (x+1)-\sqrt{x^2+y^2}+i(y+1)=0+0i$

On comparing the real and imaginary parts, we get

$y+1=0 \Rightarrow y=-1$

Also, $(x+1)-\sqrt{x^2+y^2}=0$

$\Rightarrow x^2+1+2x=x^2+y^2$

$\Rightarrow x=\dfrac{y^2-1}{2}=\dfrac{(-1)^2-1}{2}=0$

Thus, $z=x+iy=0-i$

Example 7: Show that $\left|\dfrac{z-2}{z+2}\right|=1$ represents the imaginary axis.

Solution:

Let $z=x+iy$

$\left|\dfrac{z-2}{z+2}\right|=1 \Rightarrow \left|\dfrac{x+iy-2}{x+iy+2}\right|=1$

$\Rightarrow |x-2+iy|=|x+2+iy|$

$\Rightarrow \sqrt{(x-2)^2+y^2}=\sqrt{(x+2)^2+y^2}$

$\Rightarrow x^2-4x+4+y^2=(x^2+4x+4+y^2)$

$\Rightarrow 8x=0 \Rightarrow x=0$ represents the imaginary axis.

Example 8: Solve the system of equations $\text{Im}(z^2)=0,\ |z|=1$.

Solution:

Let $z=x+iy$

$\text{Im}(z^2)=0$

$\Rightarrow \text{Im}(x^2+2ixy-y^2)=0$

$\Rightarrow \text{Im}\{(x^2-y^2)+2ixy\}=0$

$\Rightarrow xy=0$

Either $x=0$ or $y=0$

Also, $|z|=\sqrt{x^2+y^2}=1$

$\Rightarrow x^2+y^2=1$

Thus, when $x=0 \Rightarrow y=\pm 1$

And, when $y=0 \Rightarrow x=\pm 1$

Therefore, $z=x+iy=\pm 1+0i,\ 0\pm i$

Example 9: Find the complex number satisfying the equation $z+|(z+1)|-i=0$.

Solution:

Let $z=x+iy$

$z+|(z+1)|-i=0$

$\Rightarrow x+iy+|x+iy+1|-i=0$

$\Rightarrow x+i(y-1)+\left(\sqrt{(x+1)^2+y^2}\right)=0$

$\Rightarrow x+\left(\sqrt{(x+1)^2+y^2}\right)+iy(1+y)=0+i0$

Equating the real and imaginary parts, we get

$x+\left(\sqrt{(x+1)^2+y^2}\right)=0$ and $y(1+y)=0$

Thus,

$y(1+y)=0 \Rightarrow y=-1$

$(\because z$ is a complex number $\therefore y\neq 0)$

Therefore,

$x+\left(\sqrt{(x+1)^2+y^2}\right)=0$

$\Rightarrow x+\left(\sqrt{(x+1)^2+(-1)^2}\right)=0 \quad (\because y=-1)$

$\Rightarrow x+\left(\sqrt{(x+1)^2+1}\right)=0$

$\Rightarrow x^2=x^2+2x+1+1$

$\Rightarrow 2x+2=0 \Rightarrow x=-1$

$\therefore z=x+iy=-1-i$

Example 10: Find the real value of α for which the expression $\dfrac{1-i\sin\alpha}{1+2i\sin\alpha}$ is purely imaginary.

Solution:

Since $\dfrac{1-i\sin\alpha}{1+2i\sin\alpha}$ is purely imaginary,

$$\text{Re}\left(\frac{1-i\sin\alpha}{1+2i\sin\alpha}\right)=0$$

$$\Rightarrow \text{Re}\left(\frac{(1-i\sin\alpha)(1-2i\sin\alpha)}{(1+2i\sin\alpha)(1-2i\sin\alpha)}\right)=0$$

$$\Rightarrow \text{Re}\left(\frac{1-i\sin\alpha-2i\sin\alpha+2i^2\sin^2\alpha}{1-4i^2\sin^2\alpha}\right)=0$$

$$\Rightarrow \text{Re}\left(\frac{1-2\sin^2\alpha}{1+4\sin^2\alpha}-\frac{3i\sin\alpha}{1+4\sin^2\alpha}i\right)=0$$

$$\Rightarrow \frac{1-2\sin^2\alpha}{1+4\sin^2\alpha}=0\Rightarrow \sin^2\alpha=\frac{1}{2}$$

$$\Rightarrow \sin\alpha=\pm\frac{1}{\sqrt{2}}\Rightarrow \alpha=n\pi\pm(-1)^n\frac{\pi}{4}$$

Example 11: Show that the complex number z which satisfies the condition $|z-1|+|z-i|=1$ lies on a straight line equally inclined to both the real and imaginary axes.

Solution:

Let $z=x+iy$

$|z-1|+|z-i|=1$

$\Rightarrow |x-1+iy)|=|x+i(y-1)|$

$\Rightarrow (x-1)^2+y^2=x^2+(y-1)^2$

$\Rightarrow x^2+1-2x+y^2=x^2+y^2+1-2y$

$\Rightarrow y=x$ represents a straight line equally inclined to both the real and imaginary axes.

Example 12: Solve the equation $\dfrac{\bar{z}}{|z|^2}=z-z^{-1}$.

Solution:

Let $z=x+iy$,

$$\frac{\bar{z}}{|z|^2}=z-z^{-1}$$

$$\Rightarrow \frac{\bar{z}}{|z|^2}=z-\frac{\bar{z}}{|z|^2}$$

$$\Rightarrow 2\frac{\bar{z}}{|z|^2}=z\Rightarrow 2\bar{z}=z|z|^2$$

$\Rightarrow 2(x-iy)=(x+iy)|x+iy|^2$

$\Rightarrow 2x+i2y=x(x^2+y^2)+iy(x^2+y^2)$

$\Rightarrow x(x^2+y^2-2)+iy(x^2+y^2-2)=0+0i$

On comparing the real and imaginary parts, we get

$\Rightarrow x(x^2+y^2-2)=0$

And $y(x^2+y^2-2)=0$

$\Rightarrow x=0\Rightarrow y=\pm\sqrt{2}$

or $y=0\Rightarrow x=\pm\sqrt{2}$

or $x^2+y^2=2$, for all values of x and y

Thus, $z=x+iy=0-i$

Day 71

NCERT Exercise 5.1 (With Solutions)

Q1. Express the given complex number in the form $a+ib:(5i)\left(-\dfrac{3}{5}i\right)$.

Solution:

$(5i)\left(-\dfrac{3}{5}i\right)=-5\times\dfrac{3}{5}\times i\times i=-3i^2=3 \quad (\because i^2=-1)$

$\Rightarrow a+ib=3+0i$

Q2. Express the given complex number in the form $a+ib:i^9+i^{19}$.

Solution:

$i^9+i^{19}=i^{4\times2+1}+i^{4\times4+3}$

$=(i^4)^2\cdot i+(i^4)^4\cdot i^3=i+i^3 \quad (\because i^4=1)$

$=i-i=0 \quad (\because i^3=-i)$

$\Rightarrow a+ib=0+0i$

Q3. Express the given complex number in the form $a+ib:i^{-39}$.

Solution:

$$i^{-39}=\frac{1}{i^{39}}=\frac{1}{i^{4\times9+3}}=\frac{1}{(i^4)^9 i^3}$$

$$=\frac{1}{-i} \quad (\because i^4=1,\ i^3=-i)$$

$$=\frac{i}{-i^2}=i \quad (\because i^2=-1)$$

$\Rightarrow a+ib=0+i$

Q4. Express the given complex number in the form $a+ib:3(7+i7)+i(7+i7)$.

Solution:

$3(7+i7)+i(7+i7)=21+21i+7i+7i^2$

$=21+28i+7\times(-1) \quad (\because i^2=-1)$

$=14+28i$

Q5. Express the given complex number in the form $a+ib:(1-i)-(-1+i6)$.

Solution:

$(1-i)-(-1+i6)=1-i+1-6i=2-7i$

Q6. Express the given complex number in the form $a+ib:\left(\frac{1}{5}+i\frac{2}{5}\right)-\left(4+i\frac{5}{2}\right)$.

Solution:

$$\left(\frac{1}{5}+i\frac{2}{5}\right)-\left(4+i\frac{5}{2}\right)=\frac{1}{5}+\frac{2}{5}i-4-\frac{5}{2}i=\left(\frac{1}{5}-4\right)+\left(\frac{2}{5}-\frac{5}{2}\right)i$$

$$=\frac{-19}{5}+\left(\frac{-21}{10}\right)i=\frac{-19}{5}-\frac{21}{10}i$$

Q7. Express the given complex number in the form $a+ib:\left[\left(\frac{1}{3}+i\frac{7}{3}\right)+\left(4+i\frac{1}{3}\right)\right]-\left(-\frac{4}{3}+i\right)$.

Solution:

$$\left[\left(\frac{1}{3}+i\frac{7}{3}\right)+\left(4+i\frac{1}{3}\right)\right]-\left(-\frac{4}{3}+i\right)=\frac{1}{3}+\frac{7}{3}i+4+\frac{1}{3}i+\frac{4}{3}-i$$

$$=\left(\frac{1}{3}+4+\frac{4}{3}\right)+i\left(\frac{7}{3}+\frac{1}{3}-1\right)=\frac{17}{3}+i\frac{5}{3}$$

Q8. Express the given complex number in the form $a+ib:(1-i)^4$.

Solution:

$(1-i)^4=[(1-i)^2]^2$

$=[1^2+i^2-2i]^2$

$=[1-1-2i]^2 \quad (\because i^2=-1)$

$=4i^2=-4$

$\Rightarrow a+ib=-4+0i$

Q9. Express the given complex number in the form $a+ib:\left(\frac{1}{3}+3i\right)^3$.

Solution:

$$\left(\frac{1}{3}+3i\right)^3=\left(\frac{1}{3}\right)^3+(3i)^3+3\left(\frac{1}{3}\right)^2(3i)+3\left(\frac{1}{3}\right)(3i)^2$$

$$=\frac{1}{27}+27i^3+i+9i^2$$

$$=\frac{1}{27}+27\times(-i)+i+9\times(-1). \quad (\because i^2=-1,\ i^3=-i)$$

$$=\frac{1}{27}-27i+i-9=\left(\frac{1}{27}-9\right)+(-27+1)i$$

$$=\frac{-242}{27}-26i$$

Q10. Express the given complex number in the form $a+ib:\left(-2-\frac{1}{3}i\right)^3$.

Solution:

$$\left(-2-\frac{1}{3}i\right)^3=(-1)^3\left(2+\frac{1}{3}i\right)^3$$

$$=-\left(2^3+\left(\frac{i}{3}\right)^3+3(2)^2\left(\frac{i}{3}\right)+3(2)\left(\frac{i}{3}\right)^2\right)$$

$$=-\left(8+\frac{i^3}{27}+4i+\frac{2}{3}i^2\right)$$

$$=-\left(8-\frac{i}{27}+4i-\frac{2}{3}\right) \quad (\because i^2=-1,\ i^3=-i)$$

$$=-8+\frac{i}{27}-4i+\frac{2}{3}=-\frac{22}{3}-\frac{107i}{27}$$

Q11. Find the multiplicative inverse of the complex number $4-3i$.

Solution:

$z=4-3i\Rightarrow \bar{z}=4+3i$

$|z|^2=4^2+(-3)^2=16+9=25$

Therefore, the multiplicative inverse of $4-3i$ is

$$z^{-1}=\frac{\bar{z}}{|z|^2}=\frac{4+3i}{25}=\frac{4}{25}+\frac{3}{25}i$$

Q12. Find the multiplicative inverse of the complex number $\sqrt{5}+3i$.

Solution:

$z=\sqrt{5}+3i\Rightarrow \bar{z}=\sqrt{5}-3i$

$|z|^2=(\sqrt{5})^2+3^2=5+9=14$

Therefore, the multiplicative inverse of $\sqrt{5}+3i$

$$z^{-1}=\frac{\bar{z}}{|z|^2}=\frac{\sqrt{5}+3i}{14}=\frac{\sqrt{5}}{14}+\frac{3i}{14}$$

Q13. Find the multiplicative inverse of the complex number $-i$.

Solution:

$z=-i\Rightarrow \bar{z}=i$

$|z|^2=1^2=1$

Therefore, the multiplicative inverse of $-i$ is

$$z^{-1}=\frac{\bar{z}}{|z|^2}=\frac{i}{1}=i$$

Q14. Express the following expression in the form of $a+ib$.

$$\frac{(3+i\sqrt{5})(3-i\sqrt{5})}{(\sqrt{3}+\sqrt{2}i)-(\sqrt{3}-i\sqrt{2})}$$

Solution:

$$\frac{(3+i\sqrt{5})(3-i\sqrt{5})}{(\sqrt{3}+\sqrt{2}i)-(\sqrt{3}-i\sqrt{2})}=\frac{(3)^2-(i\sqrt{5})^2}{\sqrt{3}+\sqrt{2}i-\sqrt{3}+\sqrt{2}i}$$

$$=\frac{9-5i^2}{2\sqrt{2}i}=\frac{9-5(-1)}{2\sqrt{2}i}\quad (\because i^2=-1)$$

$$=\frac{14}{2\sqrt{2}i}=\frac{7}{\sqrt{2}i}=\frac{7}{\sqrt{2}i}\times\frac{i}{i}$$

$$=\frac{7i}{\sqrt{2}i^2}=\frac{7i}{\sqrt{2}(-1)}=\frac{-7i}{\sqrt{2}}=0-\frac{7\sqrt{2}}{2}i$$

Day 72

Argand Plane and Polar Representation of a Complex Number

Argand Plane (or Complex Plane): In a real plane, an ordered pair of real numbers (*x*, *y*) represents a unique point in the XY-plane with reference to a set of mutually perpendicular lines known as the *x*-axis and the *y*-axis. Similarly, in a complex plane (known as Argand plane), the complex number $x+iy$ which corresponds to the ordered pair (*x*, *y*) represents a unique point in the XY-plane (or say Argand plane) with reference to a set of mutually perpendicular lines known as the *x*-axis (known as the real axis) and the *y*-axis (known as the imaginary axis).
The points on the *x*-axis correspond to complex numbers of the form $x+0i$ and the points on the *y*-axis correspond to complex numbers of the form $0+iy$.

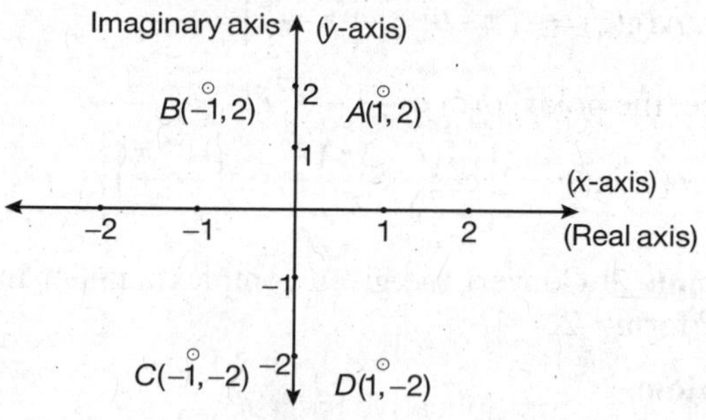

Fig. 5.1

As an example, complex numbers such as $1+2i$, $-1+2i$, $-1-2i$, $1-2i$, which correspond to the ordered pairs, respectively, are represented geometrically by the points *A*, *B*, *C* and *D* in the Argand plane.
Note: In the Argand plane, the modulus of the complex number $|z|=|x+iy|=\sqrt{x^2+y^2}$ is the distance between the point *P*(*x*, *y*) and the origin *O*(0, 0).

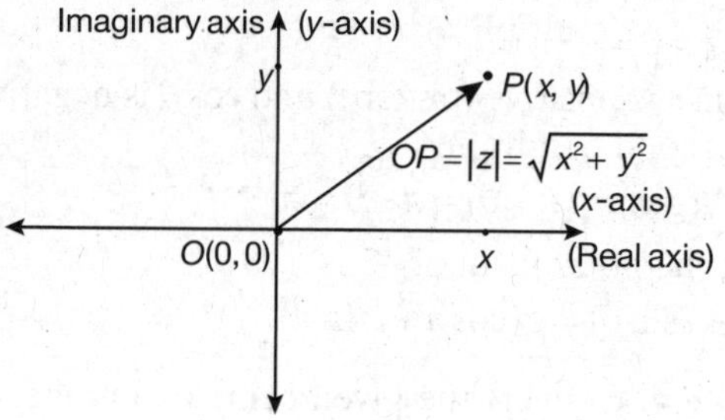

Fig. 5.2

The complex number $z=x+iy$ and its conjugate $\bar{z}=x-iy$ are represented geometrically by the points $P(x,y)$, $Q(x,-y)$ in the Argand plane. The point $Q(x,-y)$ is the mirror of the point $P(x,y)$ on the real axis.

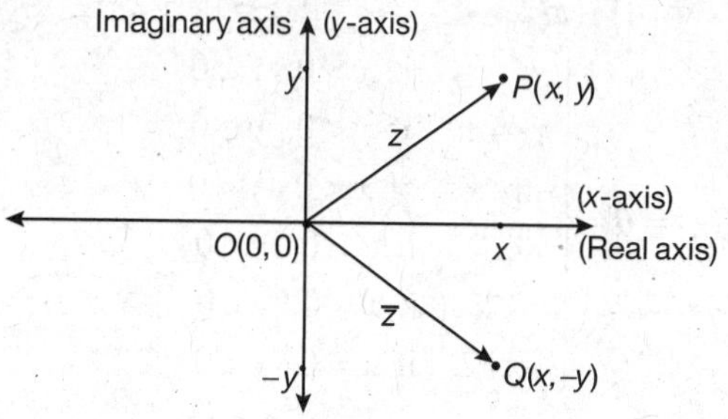

Fig. 5.3

Polar Representation of a Complex Number: Let the point *P* represent the non-zero complex number $z=x+iy$. Let the length of the directed line segment OP be *r* and the angle which OP makes with the positive direction of the *x*-axis be θ.

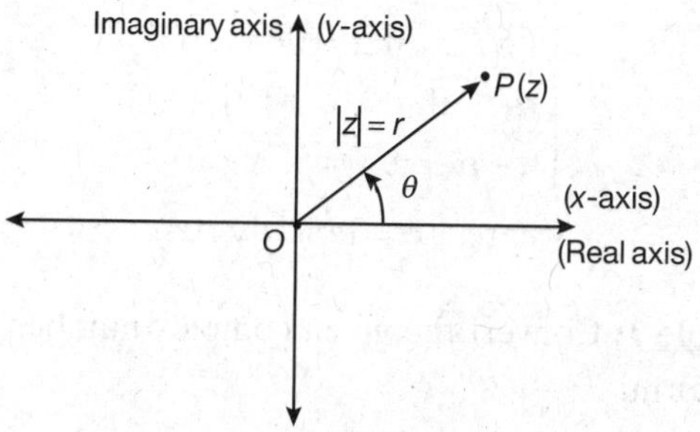

Fig. 5.4

Then the point $P(x,y)$ is uniquely determined by the ordered pair of real numbers (r,θ), known as the polar coordinates of the point *P* (considering the origin as the pole and the positive direction of the *x*-axis as the initial line).
Thus, in polar coordinates, we may write $x=r\cos\theta$, $y=r\sin\theta$ and therefore $z=r(\cos\theta+i\sin\theta)$, known as the polar form of the complex number $z=x+iy$.

And, the modulus of *z* is $|z|=\sqrt{x^2+y^2}=r$

Argument (or amplitude) of *z* is denoted by $\arg z=\theta$.
Since $\sin\theta$ and $\cos\theta$ are functions of the period 2π, any complex number $z\neq 0$ takes only one value of θ in the

interval $0 \le \theta \le 2\pi$, that is, $-\pi < \theta \le \pi$ or any other interval of length 2π.

Normally we take the value of θ between $-\pi < \theta \le \pi$, known as the principal argument of z and it is denoted by arg z, unless specified otherwise.

Thus, if $z = x + iy = r(\cos\theta + i\sin\theta)$. Then,

$$\arg(z) = \theta = \begin{cases} \tan^{-1}(y/x), & \text{if } x > 0,\ y \ge 0 \\ \pi - \tan^{-1}(y/|x|), & \text{if } x < 0,\ y \ge 0 \\ -\pi + \tan^{-1}(|y|/|x|), & \text{if } x < 0,\ y < 0 \\ -\tan^{-1}(|y|/x), & \text{if } x > 0,\ y < 0 \\ \pi/2, & \text{if } x = 0,\ y > 0 \\ -\pi/2, & \text{if } x = 0,\ y < 0 \end{cases}$$

$$\Rightarrow \arg(z) = \theta = \begin{cases} \alpha, & \text{if } x > 0,\ y \ge 0 \\ \pi - \alpha, & \text{if } x < 0,\ y \ge 0 \\ -\pi + \alpha, & \text{if } x < 0,\ y < 0 \\ -\alpha, & \text{if } x > 0,\ y < 0 \\ \pi/2, & \text{if } x = 0,\ y > 0 \\ -\pi/2, & \text{if } x = 0,\ y < 0 \end{cases}$$

where, $\alpha = \tan^{-1}(|y|/|x|)$

$$\Rightarrow \arg(z) = \theta = \begin{cases} \alpha, & \text{if } z \text{ lies in quadrant I} \\ \pi - \alpha, & \text{if } z \text{ lies in quadrant II} \\ -\pi + \alpha, & \text{if } z \text{ lies in quadrant III} \\ -\alpha, & \text{if } z \text{ lies in quadrant IV} \\ \pi/2, & \text{if } x = 0,\ y > 0 \\ -\pi/2, & \text{if } x = 0,\ y < 0 \\ \alpha, & \text{if } y = 0,\ x > 0 \\ \pi - \alpha, & \text{if } y = 0,\ x < 0 \end{cases}$$

where, $\alpha = \tan^{-1}(|y|/|x|)$

Example 1: Convert the given complex number into the polar form: $\sqrt{3} - i$.

Solution:

Since $r\cos\theta = \sqrt{3}$, $\quad r\sin\theta = -1$

On squaring and adding, we get

$r^2\cos^2\theta + r^2\sin^2\theta = (\sqrt{3})^2 + (-1)^2$

$\Rightarrow r^2(\cos^2\theta + \sin^2\theta) = 3 + 1 \quad (\because \sin^2\theta + \cos^2\theta = 1)$

$\Rightarrow r^2 = 4 \Rightarrow r = 2 \Rightarrow |z| = 2 \qquad (\because r > 0)$

$\because 2\cos\theta = \sqrt{3},\ \ 2\sin\theta = -1$

$\therefore \cos\theta = \dfrac{\sqrt{3}}{2}, \quad \sin\theta = \dfrac{-1}{2}$

$\Rightarrow |\tan\theta| = \dfrac{1}{\sqrt{3}} \Rightarrow \theta = \dfrac{\pi}{6}$

Since sin θ is positive and cos θ is negative, θ lies in the fourth quadrant.

Thus, $\text{Arg}(z) = -\theta = -\dfrac{\pi}{6}$

Hence, the polar form of $\sqrt{3} - i$ is

$r(\cos\theta + i\sin\theta) = 2\left[\cos\left(-\dfrac{\pi}{6}\right) + i\sin\left(-\dfrac{\pi}{6}\right)\right]$

Example 2: Convert the given complex number into the polar form $-1 - \sqrt{3}i$.

Solution:

Since $r\cos\theta = -1$, $\quad r\sin\theta = -\sqrt{3}$

On squaring and adding, we obtain

$r^2\cos^2\theta + r^2\sin^2\theta = (-1)^2 + (-\sqrt{3})^2$

$\Rightarrow r^2(\cos^2\theta + \sin^2\theta) = 1 + 3 \quad (\because \sin^2\theta + \cos^2\theta = 1)$

$\Rightarrow r^2 = 4 \Rightarrow r = 2 \qquad (\because r > 0)$

$\Rightarrow |z| = 2$

$\because 2\cos\theta = -1,\ \ 2\sin\theta = -\sqrt{3}$

$\therefore \cos\theta = \dfrac{-1}{2}, \quad \sin\theta = \dfrac{-\sqrt{3}}{2}$

$\Rightarrow |\tan\theta| = \sqrt{3} \Rightarrow \theta = \dfrac{\pi}{3}$

Since sin θ and cos θ are both negatives, θ lies in the third quadrant.

Thus, $\text{Arg}(z) = -(\pi - \theta) = -\left(\pi - \dfrac{\pi}{3}\right) = \dfrac{-2\pi}{3}$

Hence, the polar form of $-1 - \sqrt{3}i$ is,

$r(\cos\theta + i\sin\theta) = 2\left[\cos\left(-\dfrac{2\pi}{3}\right) + i\sin\left(-\dfrac{2\pi}{3}\right)\right]$

Example 3: Convert the given complex number into the polar form: –2.

Solution:

Since complex number $z = -2 = -2 + 0i$

$r\cos\theta = -2, \quad r\sin\theta = 0$

On squaring and adding, we get

$r^2\cos^2\theta + r^2\sin^2\theta = (-2)^2 + (0)^2$

$\Rightarrow r^2(\cos^2\theta + \sin^2\theta) = 4 \quad (\because \sin^2\theta + \cos^2\theta = 1)$

$\Rightarrow r^2 = 4 \Rightarrow r = 2 \Rightarrow |z| = 2 \qquad (\because r > 0)$

$\because 2\cos\theta = -2, \quad 2\sin\theta = 0$

$\therefore \cos\theta = -1, \quad \sin\theta = 0$

$\Rightarrow |\tan\theta| = 0 \Rightarrow \theta = 0$

Since sin θ is positive (or zero) and cos θ is negative, θ lies in the second quadrant.

Thus, $\text{Arg}(z) = (\pi - \theta) = \pi - 0 = \pi$

Hence, the polar form of –2 is

$r(\cos\theta + i\sin\theta) = 2(\cos\pi + i\sin\pi)$

Example 4: Convert the given complex number into the polar form: $-i$.

Solution:

The complex number $z = -i = 0 - i$

Therefore, $r\cos\theta = 0, \quad r\sin\theta = -1$

On squaring and adding, we get

$r^2\cos^2\theta + r^2\sin^2\theta = (0)^2 + (-1)^2$

$\Rightarrow r^2(\cos^2\theta + \sin^2\theta) = 1 \quad (\because \sin^2\theta + \cos^2\theta = 1)$

$\Rightarrow r^2 = 1 \Rightarrow r = 1 \Rightarrow |z| = 1 \qquad (\because r > 0)$

$\because \cos\theta = 0, \quad \sin\theta = -1$

$\therefore |\tan\theta| = \infty \Rightarrow \theta = \frac{\pi}{2}$

Since $\sin\theta$ is negative and $\cos\theta$ is positive (or zero), θ lies in the third quadrant.

Thus, $\text{Arg}(z) = \theta = -\frac{\pi}{2}$

Hence, the polar form of $-i$ is

$r(\cos\theta + i\sin\theta) = \cos\left(-\frac{\pi}{2}\right) + i\sin\left(-\frac{\pi}{2}\right)$

Example 5: Convert the given complex number into the polar form: $\frac{7-i}{(2-i)^2}$.

Solution:

$\frac{7-i}{(2-i)^2} = \frac{7-i}{4+i^2-4i}$

$= \frac{7-i}{3-4i} \quad (\because i^2 = -1)$

$= \frac{7-i}{3-4i} \times \frac{3+4i}{3+4i} = \frac{21+28i-3i-4i^2}{3^2+4^2}$

$= \frac{-3+25i+28}{3^2+4^2} = \frac{25+25i}{25}$

$= 1+i$

Now let $r\cos\theta = 1, \quad r\sin\theta = 1$

On squaring and adding, we get

$r^2(\cos^2\theta + \sin^2\theta) = (1)^2 + 1^2$

$\Rightarrow r^2 = 2 \quad (\because \sin^2\theta + \cos^2\theta = 1)$

$\Rightarrow r = \sqrt{2} \quad (\because r > 0)$

$r\cos\theta = 1, \quad r\sin\theta = 1$

$\Rightarrow |\tan\theta| = 1 \Rightarrow \theta = \frac{\pi}{4}$

Since $\sin\theta$ and $\cos\theta$ are both positive, θ lies in the first quadrant.

Thus, $\text{Arg}(z) = \theta = \frac{\pi}{4}$

Hence, the polar form of $\frac{7-i}{(2-i)^2} = 1+i$ is

$= r(\cos\theta + i\sin\theta) = \sqrt{2}\left(\cos\frac{\pi}{4} + i\sin\frac{\pi}{4}\right)$

Example 6: Write the complex number $z = \frac{\cos\frac{2\pi}{3} + i\sin\frac{2\pi}{3}}{1+i}$ in the polar form.

Solution:

$$z = \frac{\cos\frac{2\pi}{3} + i\sin\frac{2\pi}{3}}{1+i} = \frac{\cos\frac{2\pi}{3} + i\sin\frac{2\pi}{3}}{\sqrt{2}\left[\frac{1}{\sqrt{2}} + i\frac{1}{\sqrt{2}}\right]}$$

$$= \frac{\cos\frac{2\pi}{3} + i\sin\frac{2\pi}{3}}{\sqrt{2}[\cos(\pi/4) + i\sin(\pi/4)]}$$

$$= \frac{1}{\sqrt{2}}\left[\cos\left(\frac{2\pi}{3} - \frac{\pi}{4}\right) + i\sin\left(\frac{\pi}{3} - \frac{\pi}{4}\right)\right]$$

$$= \frac{1}{\sqrt{2}}\left[\cos\left(\frac{5\pi}{12}\right) + i\sin\left(\frac{5\pi}{12}\right)\right]$$

Example 7: If $\arg(z-i) = \arg(z+1)$, then find the value of $y-x$, where $z = x+iy$.

Solution:

$\arg(z-i) = \arg(z+1)$

$\Rightarrow \arg(x+iy-i) = \arg(x+iy+1)$

$\Rightarrow \arg[x+i(y-1)] = \arg[(x+1)+iy]$

$\Rightarrow \tan^{-1}\frac{y-1}{x} = \tan^{-1}\frac{y}{x+1}$

$\Rightarrow \frac{y-1}{x} = \frac{y}{x+1} \Rightarrow xy + y - x - 1 = xy$

$\Rightarrow y - x = 1$

Example 8: If z_1, z_2 are two complex numbers such that $|z_1| = |z_2|$ and $\arg(z_2) - \arg(z_1) = \frac{\pi}{2}$, then show that $\bar{z}_2 = i \cdot z_1$.

Solution:

Let $z_1 = r_1(\cos\theta_1 + i\sin\theta_1)$, $z_2 = r_2(\cos\theta_2 + i\sin\theta_2)$ be two complex numbers.

Since $|z_1| = |z_2|$ and $\arg(z_2) - \arg(z_1) = \frac{\pi}{2}$,

$r_1 = r_2$ and $\theta_2 - \theta_1 = \frac{\pi}{2}$

Now,

$\text{RHS} = \bar{z}_2 = r_2(\cos\theta_2 - i\sin\theta_2)$

$= r_1[\cos(\frac{\pi}{2} + \theta_1) + i\sin(\frac{\pi}{2} + \theta_1)]$

$(\because r_1 = r_2, \ \theta_2 = \frac{\pi}{2} + \theta_1)$

$= r_1(-\sin\theta_1 + i\cos\theta_1)$

$= i \cdot r_1(i\sin\theta_1 + \cos\theta_1)$

$= i \cdot r_1(\cos\theta_1 + i\sin\theta_1) = i \cdot z_1 = \text{LHS}$

Example 9: If z_1, z_2 and z_3, z_4 are two pairs of conjugate complex numbers, then find $\arg(z_1 z_3)+\arg(z_2 z_4)$.

Solution:

Let $z_1 = r(\cos\theta + i\sin\theta)$. Then,

$z_2 = \overline{z_1} = r(\cos\theta - i\sin\theta) = r[\cos(-\theta)+i\sin(-\theta)]$

And let $z_3 = R(\cos\alpha + i\sin\alpha)$. Then,

$z_4 = \overline{z_3} = R(\cos\alpha - i\sin\alpha) = R[\cos(-\alpha)+i\sin(-\alpha)]$

$\therefore \arg(z_1 z_3)+\arg(z_2 z_4)$

$= [\arg(z_1)+\arg(z_3)]+[\arg(z_2)+\arg(z_4)]$

$= [\theta+\alpha]+[(-\theta)+(-\alpha)] = 0$

Example 10: If z and w are two complex numbers such that $|zw| = 1$, $\arg(w)-\arg(z) = \frac{\pi}{2}$, then show that $\bar{z}\,w = i$.

Solution:

Let $z = r_1(\cos\theta_1 + i\sin\theta_1)$, $w = r_2(\cos\theta_2 + i\sin\theta_2)$

Since $|zw| = 1 \Rightarrow |z||w| = r_1 r_2 = 1$

Also, $\arg(z)-\arg(w) = \frac{\pi}{2} \Rightarrow \theta_2 - \theta_1 = \frac{\pi}{2}$

Now, LHS $= \bar{z}w$

$= r_1(\cos\theta_1 - i\sin\theta_1)\cdot r_2(\cos\theta_2 + i\sin\theta_2)$

$= r_1 r_2[\cos(\theta_2-\theta_1)+i\sin(\theta_2-\theta_1)]$

$= r_1 r_2[\cos(\pi/2)+i\sin(\pi/2)]$

$= 1(0+i) = i =$ RHS

Example 11: If $|z| = 2$, $\arg(z) = \frac{4\pi}{3}$, then find the value z.

Solution:

Let $z = x+iy = r(\cos\theta + i\sin\theta)$. Then

$|z| = r = 2$ and $\arg(z) = \theta = \frac{4\pi}{3}$

$\therefore z = 2\left(\cos\frac{4\pi}{3} + i\sin\frac{4\pi}{3}\right)$

$= 2[\cos(\pi+\pi/3)+i\sin(\pi+\pi/3)]$

$= 2\left[-\cos\frac{\pi}{3} - i\sin\frac{\pi}{3}\right] = 2\left[\frac{-1}{2} - i\frac{\sqrt{3}}{2}\right] = -1 - i\sqrt{3}$

Example 12: Find the principal argument of $(1-i\sqrt{3})^2$.

Solution:

$z = (1-i\sqrt{3})^2$

$\Rightarrow z = 1-3-2i\sqrt{3} \Rightarrow z = -2-i2\sqrt{3}$

$\Rightarrow \arg(z) = \tan\theta = \left|\frac{\text{Im}(z)}{\text{Re}(z)}\right|$

$\Rightarrow \tan\theta = \left|\frac{-2\sqrt{3}}{-2}\right| = |\sqrt{3}| = \sqrt{3}$

$\Rightarrow \tan\theta = \tan\frac{\pi}{3} \Rightarrow \theta = \frac{\pi}{3}$

$\because \text{Re}(z) < 0,\ \text{Im}(z) < 0$

Therefore, θ lies in the third quadrant.

Thus, $\text{Arg}(z) = (-\pi+\theta) = -\pi + \frac{\pi}{3} = -\frac{2\pi}{3}$

Day 73

NCERT Exercise 5.2 (With Solutions)

Q1. Find the modulus and the argument of the complex number $z = -1 - i\sqrt{3}$.

Solution:

Since $r\cos\theta = -1$, $r\sin\theta = -\sqrt{3}$

On squaring and adding, we get

$(r\cos\theta)^2 + (r\sin\theta)^2 = (-1)^2 + (-\sqrt{3})^2$

$\Rightarrow r^2(\cos^2\theta + \sin^2\theta) = 1+3$

$\Rightarrow r^2 = 4 \quad (\because \sin^2\theta + \cos^2\theta = 1)$

$\Rightarrow r = \sqrt{4} = 2 \Rightarrow |z| = 2 \quad (\because r > 0)$

$\because 2\cos\theta = -1,\ 2\sin\theta = -\sqrt{3}$

$\therefore \cos\theta = \frac{-1}{2},\quad \sin\theta = \frac{-\sqrt{3}}{2}$

$\Rightarrow |\tan\theta| = \sqrt{3} \Rightarrow \theta = \frac{\pi}{3}$

Since both $\sin\theta, \cos\theta$ are negative, θ lies in the third quadrant.

Thus, $\text{Arg}(z) = -(\pi-\theta) = -\left(\pi - \frac{\pi}{3}\right) = \frac{-2\pi}{3}$

Hence, the modulus and argument of the complex number $-1-\sqrt{3}i$ are 2 and $\frac{-2\pi}{3}$, respectively.

Q2. Find the modulus and the argument of the complex number $z = -\sqrt{3}+i$.

Solution:

Since $r\cos\theta = -\sqrt{3}$, $r\sin\theta = 1$

On squaring and adding, we obtain

$r^2\cos^2\theta + r^2\sin^2\theta = \left(-\sqrt{3}\right)^2 + 1^2$

$\Rightarrow r^2 = 3+1 = 4 \quad (\because \sin^2\theta + \cos^2\theta = 1)$

$\Rightarrow r = \sqrt{4} = 2 \quad (\because r > 0)$

$\Rightarrow |z| = 2$

$\because 2\cos\theta = -\sqrt{3}, \quad 2\sin\theta = 1$

$\therefore \cos\theta = \frac{-\sqrt{3}}{2}, \quad \sin\theta = \frac{1}{2}$

$\Rightarrow |\tan\theta| = \frac{1}{\sqrt{3}} \Rightarrow \theta = \frac{\pi}{6}$

Since $\sin\theta$ is positive and $\cos\theta$ is negative, θ lies in the second quadrant.

Thus, $\text{Arg}(z) = (\pi - \theta) = \pi - \frac{\pi}{6} = \frac{5\pi}{6}$

Hence, the modulus and argument of the complex number $-\sqrt{3}+i$ are 2 and $\frac{5\pi}{6}$, respectively.

Q3. Convert the given complex number into the polar form $1-i$.

Solution:

Since $r\cos\theta = 1, \quad r\sin\theta = -1$

On squaring and adding, we obtain

$r^2\cos^2\theta + r^2\sin^2\theta = 1^2 + (-1)^2$

$\Rightarrow r^2(\cos^2\theta + \sin^2\theta) = 1 + 1 \quad (\because \sin^2\theta + \cos^2\theta = 1)$

$\Rightarrow r^2 = 2 \Rightarrow r = \sqrt{2} \qquad (\because r > 0)$

$\Rightarrow |z| = \sqrt{2}$

$\because \sqrt{2}\cos\theta = 1, \quad \sqrt{2}\sin\theta = -1$

$\therefore \cos\theta = \frac{1}{\sqrt{2}}, \quad \sin\theta = -\frac{1}{\sqrt{2}}$

$|\tan\theta| = 1 \Rightarrow \theta = \frac{\pi}{4}$

Since $\sin\theta$ is negative and $\cos\theta$ is positive, θ lies in the fourth quadrant.

Thus, $\text{Arg}(z) = -(\theta) = -\frac{\pi}{4}$

Hence, the polar form of $1-i$ is

$r(\cos\theta + i\sin\theta) = \sqrt{2}\left[\cos\left(-\frac{\pi}{4}\right) + i\sin\left(-\frac{\pi}{4}\right)\right]$

Q4. Convert the given complex number into the polar form: $-1+i$.

Solution:

Since $r\cos\theta = -1, \quad r\sin\theta = 1$

On squaring and adding, we obtain

$r^2\cos^2\theta + r^2\sin^2\theta = (-1)^2 + 1^2$

$\Rightarrow r^2(\cos^2\theta + \sin^2\theta) = 1 + 1 \quad (\because \sin^2\theta + \cos^2\theta = 1)$

$\Rightarrow r^2 = 2 \Rightarrow r = \sqrt{2} \qquad (\because r > 0)$

$\Rightarrow |z| = \sqrt{2}$

$\because \sqrt{2}\cos\theta = -1, \quad \sqrt{2}\sin\theta = 1$

$\therefore \cos\theta = \frac{-1}{\sqrt{2}}, \quad \sin\theta = \frac{1}{\sqrt{2}}$

$|\tan\theta| = 1 \Rightarrow \theta = \frac{\pi}{4}$

Since $\sin\theta$ is positive and $\cos\theta$ is negative, θ lies in the second quadrant.

Thus, $\text{Arg}(z) = (\pi - \theta) = \pi - \frac{\pi}{4} = \frac{3\pi}{4}$

Hence, the polar form of $-1+i$ is

$r(\cos\theta + i\sin\theta) = \sqrt{2}\left(\cos\frac{3\pi}{4} + i\sin\frac{3\pi}{4}\right)$

Q5. Convert the given complex number into the polar form $-1-i$.

Solution:

Since $r\cos\theta = -1, \quad r\sin\theta = -1$

On squaring and adding, we obtain

$r^2\cos^2\theta + r^2\sin^2\theta = (-1)^2 + (-1)^2$

$\Rightarrow r^2(\cos^2\theta + \sin^2\theta) = 1 + 1 \quad (\because \sin^2\theta + \cos^2\theta = 1)$

$\Rightarrow r^2 = 2 \Rightarrow r = \sqrt{2} \qquad (\because r > 0)$

$\Rightarrow |z| = \sqrt{2}$

$\because \sqrt{2}\cos\theta = -1, \quad \sqrt{2}\sin\theta = -1$

$\therefore \cos\theta = \frac{-1}{\sqrt{2}}, \quad \sin\theta = \frac{-1}{\sqrt{2}}$

$|\tan\theta| = 1 \Rightarrow \theta = \frac{\pi}{4}$

Since $\sin\theta$ and $\cos\theta$ are both negative, θ lies in the third quadrant.

Thus, $\text{Arg}(z) = -(\pi - \theta) = -\left(\pi - \frac{\pi}{4}\right) = \frac{-3\pi}{4}$

Hence, the polar form of $-1-i$ is

$r(\cos\theta + i\sin\theta) = \sqrt{2}\left[\cos\left(-\frac{3\pi}{4}\right) + i\sin\left(-\frac{3\pi}{4}\right)\right]$

Q6. Convert the given complex number into the polar form: -3.

Solution:

Since complex number $z = -3 = -3 + 0i$

Therefore, $r\cos\theta = -3, \quad r\sin\theta = 0$

On squaring and adding, we get

$r^2\cos^2\theta + r^2\sin^2\theta = (-3)^2 + (0)^2$

$\Rightarrow r^2(\cos^2\theta + \sin^2\theta) = 9 \quad (\because \sin^2\theta + \cos^2\theta = 1)$

$\Rightarrow r^2 = 9 \Rightarrow r = 3 \qquad (\because r > 0)$

$\Rightarrow |z| = 3$

$\because 3\cos\theta = -3, \quad 3\sin\theta = 0$

$\therefore \cos\theta = -1, \quad \sin\theta = 0$

$\Rightarrow |\tan\theta| = 0 \Rightarrow \theta = 0$

Since $\sin\theta$ is positive (or zero) and $\cos\theta$ is negative, θ lies in the second quadrant.

Thus, $\text{Arg}(z) = (\pi - \theta) = \pi - 0 = \pi$

Hence, the polar form of -3 is

$r(\cos\theta + i\sin\theta) = 3(\cos\pi + i\sin\pi)$

Q7. Convert the given complex number into the polar form: $\sqrt{3}+i$.

Solution:

Since $r\cos\theta=\sqrt{3}, \quad r\sin\theta=1$

On squaring and adding, we get

$r^2\cos^2\theta+r^2\sin^2\theta=(\sqrt{3})^2+(1)^2$

$\Rightarrow r^2(\cos^2\theta+\sin^2\theta)=3+1 \quad (\because \sin^2\theta+\cos^2\theta=1)$

$\Rightarrow r^2=4\Rightarrow r=2 \quad (\because r>0)$

$\Rightarrow |z|=2$

$\because 2\cos\theta=\sqrt{3}, \quad 2\sin\theta=1$

$\therefore \cos\theta=\dfrac{\sqrt{3}}{2}, \quad \sin\theta=\dfrac{1}{2}$

$|\tan\theta|=\dfrac{1}{\sqrt{3}}\Rightarrow\theta=\dfrac{\pi}{6}$

Since $\sin\theta$ and $\cos\theta$ are both positive, θ lies in the first quadrant.

Thus, $\text{Arg}(z)=\theta=\dfrac{\pi}{6}$

Hence, the polar form of $\sqrt{3}+i$ is

$r(\cos\theta+i\sin\theta)=2\left(\cos\dfrac{\pi}{6}+i\sin\dfrac{\pi}{6}\right)$

Q8. Convert the given complex number into the polar form: i.

Solution:

Since complex number $z=i=0+i$

$r\cos\theta=0, \quad r\sin\theta=1$

On squaring and adding, we get

$r^2\cos^2\theta+r^2\sin^2\theta=(0)^2+(1)^2$

$\Rightarrow r^2(\cos^2\theta+\sin^2\theta)=1 \quad (\because \sin^2\theta+\cos^2\theta=1)$

$\Rightarrow r^2=1\Rightarrow r=1 \quad (\because r>0)$

$\Rightarrow |z|=1$

$\because \cos\theta=0, \quad \sin\theta=1$

$\therefore |\tan\theta|=\infty\Rightarrow\theta=\dfrac{\pi}{2}$

Since $\sin\theta$ is positive and $\cos\theta$ is positive (or zero), θ lies in the first quadrant.

Thus, $\text{Arg}(z)=\theta=\dfrac{\pi}{2}$

Hence, the polar form of i is

$r(\cos\theta+i\sin\theta)=\left(\cos\dfrac{\pi}{2}+i\sin\dfrac{\pi}{2}\right)$

Day 74

Quadratic Equations with Complex Roots

Let us consider a quadratic equation $ax^2+bx+c=0$ with real coefficient a, b, c and $a\neq 0$, where the discriminant $D=b^2-4ac<0$.

Since in these cases, the quadratic equation has no real solution ($\because D<0$), the solution of these types of equations is carried out in the complex number system.

Thus, the solution to the above equation in the set of complex numbers is obtained as

$$x=\frac{-b\pm\sqrt{b^2-4ac}}{2a}=\frac{-b\pm\sqrt{(4ac-b^2)\times(-1)}}{2a}$$

$$\Rightarrow x=\frac{-b\pm\sqrt{(4ac-b^2)}\,i}{2a}$$

Note: In a polynomial equation, the complex roots always occur in a conjugate pair, that is, if $x=a+ib$ is one of the roots of a polynomial equation, then $x=a-ib$ will also be a root of the equation.

Example 1: Solve the equation $x^4+3x^2+2=0$.

Solution:

Substituting $x^2=y$, we get

$y^2+3y+2=0\Rightarrow(y+1)(y+2)=0$

Solving, we get

$y=-1$ and $y=-2$

$\Rightarrow x^2=-1$ and $x^2=-2$

$\Rightarrow x=\sqrt{-1}$ and $x=\sqrt{-2}$

$\Rightarrow x=\pm i$ and $x=\pm i\sqrt{2}$

Example 2: Solve the equation $x^4+x^3+2x^2+x+1=0$.

Solution:

By factorising the equation $x^4+x^3+2x^2+x+1=0$, we get $(x^2+x+1)(x^2+1)=0$

Solving, we get

$x^2+1=0$ and $x^2+x+1=0$

Thus, $x^2+1=0\Rightarrow x^2=-1\Rightarrow x=\sqrt{-1}=\pm i$

Now, for equation $x^2+x+1=0$

The discriminant

$D=b^2-4ac=1^2-4\times1\times4=-3<0$

Therefore, the solution of the given equation will be imaginary and given by

$$x=\frac{-b\pm\sqrt{D}}{2a}=\frac{-1\pm\sqrt{-3}}{2\times1}=\frac{-1\pm\sqrt{3}\,i}{2}, \ (\because \sqrt{-1}=i)$$

Hence, the solution of the equation is $x^4 + x^3 + 2x^2 + x + 1 = 0$ is $x = \pm i, \frac{1}{2}(-1 \pm \sqrt{3}i)$.

NCERT Exercise 5.3 (With Solutions)

Q1. Solve the equation $x^2 + 3 = 0$.

Solution:
The discriminant
$D = b^2 - 4ac = 0^2 - 4\times1\times3 = -12 < 0$
Therefore, the solution of the given equation is

$$x = \frac{-b \pm \sqrt{D}}{2a} = \frac{\pm\sqrt{-12}}{2\times1} = \frac{\pm\sqrt{12}\,i}{2} \quad (\because \sqrt{-1} = i)$$

$$\Rightarrow x = \frac{\pm2\sqrt{3}\,i}{2} = \pm\sqrt{3}\,i$$

Q2. Solve the equation $2x^2 + x + 1 = 0$.

Solution:
The discriminant
$D = b^2 - 4ac = 1^2 - 4\times2\times1 = -7 < 0$
Therefore, solution of the given equation is

$$x = \frac{-b \pm \sqrt{D}}{2a} = \frac{-1\pm\sqrt{-7}}{2\times2} = \frac{-1\pm\sqrt{7}\,i}{4} \quad (\because \sqrt{-1} = i)$$

Q3. Solve the equation $x^2 + 3x + 9 = 0$.

Solution:
The discriminant
$D = b^2 - 4ac = 3^2 - 4\times1\times9 = -27 < 0$
Therefore, the solution of the given equation is

$$x = \frac{-b \pm \sqrt{D}}{2a} = \frac{-3\pm\sqrt{-27}}{2\times1} = \frac{-3\pm3\sqrt{3}i}{2}, (\because \sqrt{-1} = i)$$

Q4. Solve the equation $-x^2 + x - 2 = 0$.

Solution:
The discriminant
$D = b^2 - 4ac = 1^2 - 4\times(-1)\times(-2) = -7 < 0$
Therefore, the solution of the given equation is

$$x = \frac{-b \pm \sqrt{D}}{2a} = \frac{-1\pm\sqrt{-7}}{2\times(-1)} = \frac{1\pm\sqrt{7}i}{2}, \quad (\because \sqrt{-1} = i)$$

Q5. Solve the equation $x^2 + 3x + 5 = 0$.

Solution:
The discriminant
$D = b^2 - 4ac = 3^2 - 4\times1\times5 = -11 < 0$
Therefore, the solution of the given equation is

$$x = \frac{-b \pm \sqrt{D}}{2a} = \frac{-3\pm\sqrt{-11}}{2\times1} = \frac{-3\pm\sqrt{11}i}{2}, (\because \sqrt{-1} = i)$$

Q6. Solve the equation $x^2 - x + 2 = 0$.

Solution:
The discriminant
$D = b^2 - 4ac = (-1)^2 - 4\times1\times2 = 1 - 8 = -7$
Therefore, the solution of the given equation is

$$x = \frac{-b \pm \sqrt{D}}{2a} = \frac{-(-1)\pm\sqrt{-7}}{2\times1} = \frac{1\pm\sqrt{7}i}{2}, \ (\because \sqrt{-1} = i)$$

Q7. Solve the equation $\sqrt{2}x^2 + x + \sqrt{2} = 0$.

Solution:
The discriminant
$D = b^2 - 4ac = 1^2 - 4\times\sqrt{2}\times\sqrt{2} = -7 < 0$
Therefore, the solution of the given equation is

$$x = \frac{-b \pm \sqrt{D}}{2a} = \frac{-(-1)\pm\sqrt{-7}}{2\times\sqrt{2}} = \frac{1\pm\sqrt{7}i}{2\sqrt{2}}, \ (\because \sqrt{-1} = i)$$

Q8. Solve the equation $\sqrt{3}x^2 - \sqrt{2}x + 3\sqrt{3} = 0$.

Solution:
The discriminant
$D = b^2 - 4ac = (-\sqrt{2})^2 - 4(\sqrt{3})(3\sqrt{3}) = -34 < 0$
Therefore, the solution of the given equation is

$$x = \frac{-b \pm \sqrt{D}}{2a} = \frac{-(-\sqrt{2})\pm\sqrt{-34}}{2\times\sqrt{3}}$$

$$\Rightarrow x = \frac{\sqrt{2}\pm\sqrt{34}i}{2\sqrt{3}}, \ (\because \sqrt{-1} = i)$$

Q9. Solve the equation $x^2 + x + \frac{1}{\sqrt{2}} = 0$.

Solution:
The discriminant
$D = b^2 - 4ac = (\sqrt{2})^2 - 4\times(\sqrt{2})\times1 = 2 - 4\sqrt{2} < 0$
Therefore, the solution of the given equation is

$$x = \frac{-b \pm \sqrt{D}}{2a} = \frac{-\sqrt{2}\pm\sqrt{2-4\sqrt{2}}}{2\times\sqrt{2}} = \frac{-1\pm\sqrt{1-2\sqrt{2}}}{\sqrt{2}}$$

$$= \frac{-1\pm\left(\sqrt{2\sqrt{2}-1}\right)i}{\sqrt{2}}, \quad (\because \sqrt{-1} = i)$$

Q10. Solve the equation $x^2 + \frac{x}{\sqrt{2}} + 1 = 0$.

Solution:
The discriminant
$D = b^2 - 4ac = 1^2 - 4\times\sqrt{2}\times\sqrt{2} = -7 < 0$
Therefore, the solution of the given equation is

$$x = \frac{-b \pm \sqrt{D}}{2a} = \frac{-1\pm\sqrt{-7}}{2\sqrt{2}} = \frac{-1\pm\sqrt{7}i}{2\sqrt{2}}, \ (\because \sqrt{-1} = i)$$

Day 75

Square-Root of a Complex Number

Let us find the square-root of complex number $z = a + ib$, where a and b are real numbers and $b \neq 0$.
Now, let the square-root of the complex number $z = a + ib$ be $x + iy$, then $x + iy = \sqrt{a+ib}$.
Squaring both sides, we get
$(x+iy)^2 = a + ib \Rightarrow (x^2 - y^2) + i(2xy) = a + ib$
Comparing the real and imaginary parts, we get
$\Rightarrow x^2 - y^2 = a$ and $2xy = b$
We know that $(x^2 + y^2)^2 = (x^2 - y^2)^2 + (2xy)^2$
$\Rightarrow (x^2 + y^2)^2 = a^2 + b^2$
$\Rightarrow x^2 + y^2 = \sqrt{a^2 + b^2}$
Thus, adding equations $x^2 + y^2 = \sqrt{a^2+b^2}$ and $x^2 - y^2 = a$, we get

$$2x^2 = \sqrt{a^2+b^2} + a \Rightarrow x = \pm\sqrt{\frac{\sqrt{a^2+b^2}+a}{2}}$$

And by subtracting the above equations, we get

$$2y^2 = \sqrt{a^2+b^2} - a \Rightarrow y = \pm\sqrt{\frac{\sqrt{a^2+b^2}-a}{2}}$$

Hence, square-root of $a + ib = \sqrt{a+ib} = x + iy$

$$= \pm\sqrt{\frac{\sqrt{a^2+b^2}+a}{2}} \pm i\sqrt{\frac{\sqrt{a^2+b^2}-a}{2}}$$

$$= \pm\sqrt{\frac{|z|+a}{2}} \pm i\sqrt{\frac{|z|-a}{2}} \qquad (\because |z| = \sqrt{a^2+b^2})$$

$$= \pm\sqrt{\frac{|z|+\text{Re}(z)}{2}} \pm i\sqrt{\frac{|z|-\text{Re}(z)}{2}}$$

Note: If b is negative, then the product xy will also be negative, and therefore, in this case, we take the value of x as positive and y as negative, or vice versa.
And if b is positive, then the product xy will also be positive, and therefore, in this case, we take the value of both x and y as positive or negative.

NCERT Exercise 5.4 (With Solutions) Supplement Exercise

Q1. Find the square-root of $-15 - 8i$.

Solution:
Let $x + iy = \sqrt{a+ib}$, where $a = -15$ and $b = -8$.
Then, $\text{Re}(z) = a = -15$ and
$|z| = \sqrt{a^2+b^2} = \sqrt{(-15)^2 + (-8)^2} = 17$
The product xy (or b) is negative, and therefore, we take the value of x as positive and y as negative, or vice versa.
Thus, the square-root of $-15 - 8i$ is given by $\sqrt{-15-8i}$

$$= \pm\sqrt{\frac{|z|+\text{Re}(z)}{2}} \mp i\sqrt{\frac{|z|-\text{Re}(z)}{2}}$$

$$= \pm\sqrt{\frac{17+(-15)}{2}} \mp i\sqrt{\frac{17-(-15)}{2}}$$

$$= \pm\sqrt{\frac{2}{2}} \mp i\sqrt{\frac{34}{2}}$$

$$= \pm 1 \mp 4i = (1 - 4i, -1 + 4i)$$

Q2. Find the square-root of $-8 - 6i$.

Solution:
Let $x + iy = \sqrt{a+ib}$, where $a = -8$ and $b = -6$.
Then, $\text{Re}(z) = a = -8$ and
$|z| = \sqrt{a^2+b^2} = \sqrt{(-8)^2 + (-6)^2} = 10$
The product xy (or b) is negative, and therefore, we take the value of x as positive and y as negative, or vice versa.
Thus, the square-root of $-8 - 6i$ is given by $\sqrt{-8-6i}$

$$= \pm\sqrt{\frac{|z|+\text{Re}(z)}{2}} \mp i\sqrt{\frac{|z|-\text{Re}(z)}{2}}$$

$$= \pm\sqrt{\frac{10+(-8)}{2}} \mp i\sqrt{\frac{10-(-8)}{2}}$$

$$= \pm\sqrt{\frac{2}{2}} \mp i\sqrt{\frac{18}{2}}$$

$$= \pm 1 \mp 3i = (1 - 3i, -1 + 3i)$$

Q3. Find the square-root of $1 - i$.

Solution:
Let $x + iy = \sqrt{a+ib}$, where $a = 1$ and $b = -1$.
Then, $\text{Re}(z) = a = 1$ and
$|z| = \sqrt{a^2+b^2} = \sqrt{(1)^2 + (-1)^2} = \sqrt{2}$
Since the product xy (or b) is negative, we take the value of x as positive and y as negative, or vice versa.
Thus, the square-root of $1 - i$ is given by $\sqrt{1-i}$

$$= \pm\sqrt{\frac{|z|+\text{Re}(z)}{2}} \mp i\sqrt{\frac{|z|-\text{Re}(z)}{2}}$$

$$= \pm\sqrt{\frac{\sqrt{2}+(1)}{2}} \mp i\sqrt{\frac{\sqrt{2}-(1)}{2}}$$

$$= \pm\sqrt{\frac{\sqrt{2}+1}{2}} \mp i\sqrt{\frac{\sqrt{2}-(1)}{2}}$$

$$= \left(\sqrt{\frac{\sqrt{2}+1}{2}} - i\sqrt{\frac{\sqrt{2}-(1)}{2}},\; -\sqrt{\frac{\sqrt{2}+1}{2}} + i\sqrt{\frac{\sqrt{2}-(1)}{2}}\right)$$

Q4. Find the square-root of $-i$.

Solution:

Let $x+iy=\sqrt{a+ib}$, where $a=0$ and $b=-1$.

Then, $\text{Re}(z)=a=0$ and

$|z|=\sqrt{a^2+b^2}=\sqrt{(0)^2+(-1)^2}=1$

Since the product xy (or b) is negative, we take the value of x as positive and y as negative, or vice versa.

Thus, the square-root of $-i$ is given by $\sqrt{0-i}$

$$=\pm\sqrt{\frac{|z|+\text{Re}(z)}{2}}\mp i\sqrt{\frac{|z|-\text{Re}(z)}{2}}$$

$$=\pm\sqrt{\frac{1+(0)}{2}}\mp i\sqrt{\frac{1-(0)}{2}}$$

$$=\pm\frac{1}{\sqrt{2}}\mp i\frac{1}{\sqrt{2}}$$

$$=\begin{pmatrix}\frac{1}{\sqrt{2}}-i\frac{1}{\sqrt{2}}, \\ -\frac{1}{\sqrt{2}}+i\frac{1}{\sqrt{2}}\end{pmatrix}$$

Q5. Find the square-root of i.

Solution:

Let $x+iy=\sqrt{a+ib}$, where $a=0$ and $b=1$.

Then, $\text{Re}(z)=a=0$ and

$|z|=\sqrt{a^2+b^2}=\sqrt{(0)^2+(1)^2}=1$

Since the product xy (or b) is positive, we take the value of both x and y as positive or negative.

Thus, the square-root of i is given by $\sqrt{0+i}$

$$=\pm\sqrt{\frac{|z|+\text{Re}(z)}{2}}\pm i\sqrt{\frac{|z|-\text{Re}(z)}{2}}$$

$$=\pm\sqrt{\frac{1+(0)}{2}}\pm i\sqrt{\frac{1-(0)}{2}}$$

$$=\pm\frac{1}{\sqrt{2}}\pm i\frac{1}{\sqrt{2}}$$

$$=\left(\frac{1}{\sqrt{2}}+i\frac{1}{\sqrt{2}},\ -\frac{1}{\sqrt{2}}-i\frac{1}{\sqrt{2}}\right)$$

Q6. Find the square-root of $1+i$.

Solution:

Let $x+iy=\sqrt{a+ib}$, where $a=1$ and $b=1$.

Then, $\text{Re}(z)=a=1$ and

$|z|=\sqrt{a^2+b^2}=\sqrt{(1)^2+(1)^2}=\sqrt{2}$

Since the product xy (or b) is positive, we take the value of both x and y as positive or negative.

Thus, the square-root of $1+i$ is given by $\sqrt{1+i}$

$$=\pm\sqrt{\frac{|z|+\text{Re}(z)}{2}}\pm i\sqrt{\frac{|z|-\text{Re}(z)}{2}}$$

$$=\pm\sqrt{\frac{\sqrt{2}+(1)}{2}}\pm i\sqrt{\frac{\sqrt{2}-(1)}{2}}$$

$$=\pm\sqrt{\frac{\sqrt{2}+1}{2}}\pm i\sqrt{\frac{\sqrt{2}-(1)}{2}}$$

$$=\begin{pmatrix}\sqrt{\frac{\sqrt{2}+1}{2}}+i\sqrt{\frac{\sqrt{2}-(1)}{2}} \\ -\sqrt{\frac{\sqrt{2}+1}{2}}-i\sqrt{\frac{\sqrt{2}-(1)}{2}}\end{pmatrix}$$

Day 76

NCERT Miscellaneous Exercise (With Solutions)

Q1. Evaluate: $\left[i^{18}+\left(\frac{1}{i}\right)^{25}\right]^3$

Solution:

$$\left[i^{18}+\left(\frac{1}{i}\right)^{25}\right]^3=\left[i^{4\times4+2}+\frac{1}{i^{4\times6+1}}\right]^3$$

$$=\left[(i^4)^4\cdot i^2+\frac{1}{(i^4)^6\cdot i}\right]^3$$

$$=\left[i^2+\frac{1}{i}\right]^3 \quad (\because i^4=1)$$

$$=\left[-1+\frac{1}{i}\times\frac{i}{i}\right]^3 \quad (\because i^2=-1)$$

$$=\left[-1+\frac{i}{i^2}\right]^3=[-1-i]^3 \quad (\because i^2=-1)$$

$$=(-1)^3(1+i)^3=-(1^3+i^3+3\cdot1^2\cdot i+3\cdot1\cdot i^2)$$

$$=-(1-i+3i-3) \quad (\because i^3=-i)$$

$$=-(-2+2i)=2-2i$$

Q2. For any two complex numbers z_1, z_2, prove that $\text{Re}(z_1z_2)=\text{Re}\,z_1\,\text{Re}\,z_2-\text{Im}\,z_1\,\text{Im}\,z_2$.

Solution:

Let $z_1=x_1+iy_1,\ z_2=x_2+iy_2$

LHS $= \text{Re}(z_1z_2) = \text{Re}[(x_1+iy_1)(x_2+iy_2)]$

$= \text{Re}[x_1x_2 + ix_1y_2 + iy_1x_2 + i^2y_1y_2]$

$= \text{Re}[x_1x_2 + ix_1y_2 + iy_1x_2 - y_1y_2] \quad (\because i^2 = -1)$

$= \text{Re}[(x_1x_2 - y_1y_2) + i(x_1y_2 + y_1x_2)]$

$= x_1x_2 - y_1y_2$

$= \text{Re}(z_1)\text{Re}(z_2) - \text{Im}(z_1)\text{Im}(z_2) =$ RHS

Q3. Reduce $\left(\frac{1}{1-4i} - \frac{2}{1+i}\right)\left(\frac{3-4i}{5+i}\right)$ to the standard form.

Solution:

$$\left(\frac{1}{1-4i} - \frac{2}{1+i}\right)\left(\frac{3-4i}{5+i}\right)$$

$$= \left[\frac{(1+i)-2(1-4i)}{(1-4i)(1+i)}\right]\left[\frac{3-4i}{5+i}\right]$$

$$= \left[\frac{1+i-2+8i}{1+i-4i-4i^2}\right]\left[\frac{3-4i}{5+i}\right]$$

$$= \left[\frac{-1+9i}{5-3i}\right]\left[\frac{3-4i}{5+i}\right] \quad (\because i^2 = -1)$$

$$= \left[\frac{-3+4i+27i-36i^2}{25+5i-15i-3i^2}\right] \quad (\because i^2 = -1)$$

$$= \frac{33+31i}{28-10i} = \frac{33+31i}{2(14-5i)}$$

$$= \frac{(33+31i)}{2(14-5i)} \times \frac{(14+5i)}{(14+5i)}$$

$$= \frac{462+165i+434i+155i^2}{2[(14)^2-(5i)^2]}$$

$$= \frac{307+599i}{2(196-25i^2)} \quad (\because i^2 = -1)$$

$$= \frac{307+599i}{2(221)} \quad (\because i^2 = -1)$$

$$= \frac{307+599i}{442}$$

$$= \frac{307}{442} + \frac{599}{442}i$$

Q4. If $x - iy = \sqrt{\frac{a-ib}{c-id}}$, prove that $(x^2+y^2)^2 = \frac{a^2+b^2}{c^2+d^2}$.

Solution:

$$x - iy = \sqrt{\frac{a-ib}{c-id}}$$

On squaring both sides, we get

$$(x-iy)^2 = \frac{a-ib}{c-id}$$

$$\Rightarrow x^2 - y^2 - 2ixy = \frac{a-ib}{c-id} \quad (\because i^2 = -1)$$

On multiplying the numerator and denominator by $c+id$,

$$\Rightarrow x^2 - y^2 - 2ixy = \frac{a-ib}{c-id} \times \frac{c+ib}{c+id}$$

$$\Rightarrow x^2 - y^2 - 2ixy = \frac{(ac+bd)+i(ad-bc)}{c^2+d^2}$$

$$\Rightarrow x^2 - y^2 - 2ixy = \frac{(ac+bd)}{c^2+d^2} + \frac{i(ad-bc)}{c^2+d^2}$$

On comparing the real and imaginary parts, we get

$$x^2 - y^2 = \frac{ac+bd}{c^2+d^2} \text{ and}$$

$$-2xy = \frac{ad-bc}{c^2+d^2}$$

LHS $= (x^2+y^2)^2 = (x^2-y^2)^2 + 4x^2y^2$

$= (x^2-y^2)^2 + (-2xy)^2$

$$= \left(\frac{ac+bd}{c^2+d^2}\right)^2 + \left(\frac{ad-bc}{c^2+d^2}\right)^2$$

$$= \frac{a^2c^2+b^2d^2+2acbd+a^2d^2+b^2c^2-2adbc}{(c^2+d^2)^2}$$

$$= \frac{a^2c^2+b^2d^2+a^2d^2+b^2c^2}{(c^2+d^2)^2}$$

$$= \frac{a^2(c^2+d^2)+b^2(c^2+d^2)}{(c^2+d^2)^2}$$

$$= \frac{(c^2+d^2)(a^2+b^2)}{(c^2+d^2)} = \frac{a^2+b^2}{c^2+d^2} = \text{RHS}$$

Q5. Convert the following into the polar form:

(i) $\frac{1+7i}{(2-i)^2}$ (ii) $\frac{1+3i}{1-2i}$

Solution:

(i) $$\frac{1+7i}{(2-i)^2} = \frac{1+7i}{4+i^2-4i}$$

$$= \frac{1+7i}{3-4i} \quad (\because i^2 = -1)$$

$$= \frac{1+7i}{3-4i} \times \frac{3+4i}{3+4i} = \frac{3+4i+21i+28i^2}{3^2+4^2}$$

$$= \frac{3+25i-28}{3^2+4^2} = \frac{-25+25i}{25}$$

$= -1+i$

Now let $r\cos\theta = -1$, $r\sin\theta = 1$

On squaring and adding, we get

$r^2(\cos^2\theta + \sin^2\theta) = (-1)^2 + 1^2$

$\Rightarrow r^2 = 2 \quad (\because \sin^2\theta + \cos^2\theta = 1)$

$\Rightarrow r = \sqrt{2} \quad (\because r > 0)$

$r\cos\theta = -1$, $r\sin\theta = 1$

$$\Rightarrow |\tan\theta| = 1 \Rightarrow \theta = \frac{\pi}{4}$$

Since $\sin\theta$ is positive and $\cos\theta$ is negative, θ lies in the second quadrant.

Thus, $\text{Arg}(z) = (\pi - \theta) = \pi - \frac{\pi}{4} = \frac{3\pi}{4}$

Hence, the polar form of $-1+i$ is

$= r(\cos\theta + i\sin\theta) = \sqrt{2}\left(\cos\frac{3\pi}{4} + i\sin\frac{3\pi}{4}\right)$

(ii) $\frac{1+3i}{1-2i}$

$= \frac{1+3i}{1-2i} \times \frac{1+2i}{1+2i}$

$= \frac{1+2i+3i-6}{1+4} \quad (\because i^2 = -1)$

$= \frac{-5+5i}{5} = -1+i$

Hence, the polar form of $-1+i$ is

$= r(\cos\theta + i\sin\theta) = \sqrt{2}\left(\cos\frac{3\pi}{4} + i\sin\frac{3\pi}{4}\right)$

Q6. Solve the equation $3x^2 - 4x + \frac{20}{3} = 0$.

Solution:

The given quadratic equation is $3x^2 - 4x + \frac{20}{3} = 0$

$\Rightarrow 9x^2 - 12x + 20 = 0$

The discriminant of the given equation is

$D = b^2 - 4ac = (-12)^2 - 4\times 9 \times 20 = -576 < 0$

Therefore, solution of the given equation is

$x = \frac{-b \pm \sqrt{D}}{2a} = \frac{-(-12) \pm \sqrt{-576}}{2\times 9}$

$= \frac{12 \pm \sqrt{576}\,i}{18} \quad (\because \sqrt{-1} = i)$

$= \frac{12 \pm 24i}{18} = \frac{2 \pm 4i}{3} = \frac{2}{3} \pm \frac{4}{3}i$

Q7. Solve the equation $x^2 - 2x + \frac{3}{2} = 0$.

Solution:

The given quadratic equation is $x^2 - 2x + \frac{3}{2} = 0$

$\Rightarrow 2x^2 - 4x + 3 = 0$

The discriminant of the given equation is

$D = b^2 - 4ac = (-4)^2 - 4\times 2\times 3 = -8 < 0$

Therefore, the solution of the given equation is

$x = \frac{-b \pm \sqrt{D}}{2a} = \frac{-(-4) \pm \sqrt{-8}}{2\times 2}$

$= \frac{4 \pm 2\sqrt{2}i}{4} \quad (\because \sqrt{-1} = i)$

$= \frac{2 \pm \sqrt{2}i}{2} = 1 \pm \frac{\sqrt{2}}{2}i$

Q8. Solve the equation $27x^2 - 10x + 1 = 0$.

Solution:

The given quadratic equation is $27x^2 - 10x + 1 = 0$

The discriminant of the given equation is

$D = b^2 - 4ac = (-10)^2 - 4\times 27\times 1 = -8 < 0$

Therefore, the solution of the given equation is

$x = \frac{-b \pm \sqrt{D}}{2a} = \frac{-(-10) \pm \sqrt{-8}}{2\times 27}$

$= \frac{10 \pm 2\sqrt{2}i}{54} \quad (\because \sqrt{-1} = i)$

$= \frac{5 \pm \sqrt{2}i}{27} = \frac{5}{27} \pm \frac{\sqrt{2}}{27}i$

Q9. Solve the equation $21x^2 - 28x + 10 = 0$.

Solution:

The given quadratic equation is $21x^2 - 28x + 10 = 0$

The discriminant of the given equation is

$D = b^2 - 4ac = (-28)^2 - 4\times 21\times 10 = -56 < 0$

Therefore, the solution of the given equation is

$x = \frac{-b \pm \sqrt{D}}{2a} = \frac{-(-28) \pm \sqrt{-56}}{2\times 21}$

$= \frac{28 \pm 56i}{42} \quad (\because \sqrt{-1} = i)$

$= \frac{28 \pm 2\sqrt{14}i}{42} = \frac{28}{42} \pm \frac{2\sqrt{14}}{42}i$

$= \frac{2}{3} \pm \frac{\sqrt{14}}{21}i$

Q10. If $z_1 = 2 - i$, $z_2 = 1 + i$, find $\left|\frac{z_1 + z_2 + 1}{z_1 - z_2 + i}\right|$

Solution:

$\left|\frac{z_1 + z_2 + 1}{z_1 - z_2 + i}\right| = \left|\frac{(2-i)+(1+i)+1}{(2-i)-(1+i)+1}\right|$

$= \left|\frac{4}{2-2i}\right| = \left|\frac{2}{1-i}\right|$

$= \left|\frac{2}{1-i} \times \frac{1+i}{1+i}\right| = \left|\frac{2(1+i)}{1^2 - i^2}\right|$

$= \left|\frac{2(1+i)}{1+1}\right| \quad (\because i^2 = -1)$

$= \left|\frac{2(1+i)}{2}\right| = |1+i|$

$= \sqrt{1^2 + 1^2} = \sqrt{2}$

Q11. If $a + ib = \frac{(x+i)^2}{2x^2+1}$, prove that $a^2 + b^2 = \frac{(x^2+1)^2}{(2x+1)^2}$.

Solution:

$\because a + ib = \frac{(x+i)^2}{2x^2+1}$

$= \dfrac{x^2 + i^2 + 2xi}{2x^2 + 1} \quad (\because i^2 = -1)$

$= \dfrac{x^2 - 1 + i2x}{2x^2 + 1}$

$= \dfrac{x^2 - 1}{2x^2 + 1} + i\left(\dfrac{2x}{2x^2 + 1}\right)$

$\therefore a - ib = \dfrac{x^2 - 1}{2x^2 + 1} - i\left(\dfrac{2x}{2x^2 + 1}\right)$

Multiplying both equations, we get

$(a + ib)(a - ib)$

$= \left[\dfrac{x^2 - 1}{2x^2 + 1} + i\left(\dfrac{2x}{2x^2 + 1}\right)\right]\left[\dfrac{x^2 - 1}{2x^2 + 1} - i\left(\dfrac{2x}{2x^2 + 1}\right)\right]$

$\Rightarrow a^2 + b^2 = \left(\dfrac{x^2 - 1}{2x^2 + 1}\right)^2 + \left(\dfrac{2x}{2x^2 + 1}\right)^2 \quad (\because i^2 = -1)$

$\Rightarrow a^2 + b^2 = \dfrac{x^4 + 1 - 2x^2 + 4x^2}{(2x^2 + 1)^2}$

$\Rightarrow a^2 + b^2 = \dfrac{x^4 + 1 + 2x^2}{(2x^2 + 1)^2}$

$\Rightarrow a^2 + b^2 = \dfrac{(x^2 + 1)^2}{(2x^2 + 1)^2}$

Q12. Let $z_1 = 2 - i,\ z_2 = -2 + i$. Find

(i) $\text{Re}\left(\dfrac{z_1 z_2}{\bar{z}_1}\right)$ (ii) $\text{Im}\left(\dfrac{1}{z_1 \bar{z}_1}\right)$

Solution:

(i) $z_1 = 2 - i,\ z_2 = -2 + i.$

$\because z_1 z_2 = (2 - i)(-2 + i) = -4 + 4i - i^2$

$= -3 + 4i \quad (\because i^2 = -1)$

and $z_1 = 2 - i \Rightarrow \bar{z}_1 = 2 + i$

$\therefore \dfrac{z_1 z_2}{\bar{z}_1} = \dfrac{-3 + 4i}{2 + i} = \dfrac{(-3 + 4i)(2 - i)}{(2 + i)(2 - i)}$

$= \dfrac{-6 + 3i + 8i - 4i^2}{2^2 + 1^2} = \dfrac{-2 + 11i}{5}$

$\Rightarrow \dfrac{z_1 z_2}{\bar{z}_1} = \dfrac{-2}{5} + \dfrac{11}{5}i$

$\Rightarrow \text{Re}\left(\dfrac{z_1 z_2}{\bar{z}_1}\right) = \dfrac{-2}{5}$

(ii) $\dfrac{1}{z_1 \bar{z}_1} = \dfrac{1}{(2 - i)(2 + i)}$

$= \dfrac{1}{(2)^2 + (1)^2} = \dfrac{1}{5} = \dfrac{1}{5} + 0i$

$\Rightarrow \text{Im}\left(\dfrac{1}{z_1 \bar{z}_1}\right) = 0$

Q13. Find the modulus and argument of the complex number $\dfrac{1 + 3i}{1 - 3i}$.

Solution:

$z = \dfrac{1 + 3i}{1 - 3i} = \dfrac{1 + 3i}{1 - 3i} \times \dfrac{1 + 3i}{1 + 3i}$

$= \dfrac{1 + 3i + 2i + 6i^2}{1^2 + 3^2} = \dfrac{-5 + 5i}{10}$

$= \dfrac{-1 + i}{2} = \dfrac{-1}{2} + \dfrac{1}{2}i$

Let $r\cos\theta = \dfrac{-1}{2}, \quad r\sin\theta = \dfrac{1}{2}$

On squaring and adding, we get

$r^2(\cos^2\theta + \sin^2\theta) = \left(\dfrac{-1}{2}\right)^2 + \left(\dfrac{1}{2}\right)^2$

$\Rightarrow r^2 = \dfrac{1}{2} \quad (\because \sin^2\theta + \cos^2\theta = 1)$

$\Rightarrow r = \dfrac{1}{\sqrt{2}} \quad (\because r > 0)$

Since $|\tan\theta| = 1 \Rightarrow \theta = \dfrac{\pi}{4}$

Since $\sin\theta$ is positive and $\cos\theta$ is negative, θ lies in the second quadrant.

Thus, $\text{Arg}(z) = (\pi - \theta) = \pi - \dfrac{\pi}{4} = \dfrac{3\pi}{4}$

Therefore, the modulus and argument of the given complex number are $\dfrac{1}{\sqrt{2}}, \dfrac{3\pi}{4}$, respectively.

Q14. Find the real numbers x and y if $(x - iy)(3 + 5i)$ is the conjugate of $-6 - 24i$.

Solution:

Let $z = (x - iy)(3 + 5i)$

$z = 3x + 5xi - 3yi - 5yi^2 = 3x + 5xi - 3yi + 5y$

$= (3x - 5y) + i(5x - 3y)$

$\Rightarrow \bar{z} = (3x + 5y) - i(5x - 3y)$

The conjugate of $(x - iy)(3 + 5i)$ is $-6 - 24i$, and therefore

$(3x + 5y) - i(5x - 3y) = -6 - 24i$

Equating the real and imaginary parts, we get

$3x + 5y = -6$ and $5x - 3y = 24$

Solving the above two equations, we get

$x = 3$ and $y = -3$

Q15. Find the modulus of $\dfrac{1 + i}{1 - i} - \dfrac{1 - i}{1 + i}$.

Solution:

$\left|\dfrac{1 + i}{1 - i} - \dfrac{1 - i}{1 + i}\right| = \left|\dfrac{(1 + i)^2 - (1 - i)^2}{(1 - i)(1 + i)}\right|$

$$= \left|\frac{1+i^2+2i-1-i^2+2i}{1^2+1^2}\right| = |2i|$$

$$\Rightarrow \left|\frac{1+i}{1-i} - \frac{1-i}{1+i}\right| = |2i| = \sqrt{2^2} = 2$$

Q16. If $(x+iy)^3 = u+iv$, then show that: $\frac{u}{x}+\frac{v}{y} = 4(x^2-y^2)$.

Solution:

$(x+iy)^3 = u+iv$

$\Rightarrow x^3+(iy)^3+3x^2(iy)+3x(iy)^2 = u+iv$

$\Rightarrow x^3+i^3y^3+3x^2yi+3xy^2i^2 = u+iv$

$\Rightarrow x^3-iy^3+3x^2yi-3xy^2 = u+iv$

$\Rightarrow (x^3-3xy^2)+i(3x^2y-y^3) = u+iv$

On equating the real and imaginary parts, we get

$u = x^3-3xy^2$ and $v = 3x^2y-y^3$

$\Rightarrow \frac{u}{x} = x^2-3y^2$ and $\frac{v}{y} = 3x^2-y^2$

$\therefore \frac{u}{x}+\frac{v}{y} = x^2-3y^2+3x^2-y^2 = 4(x^2-y^2)$

Q17. If α, β are different complex numbers with $|\beta| = 1$, then find $\left|\frac{\beta-\alpha}{1-\bar{\alpha}\beta}\right|$.

Solution:

Let $\left|\frac{\beta-\alpha}{1-\bar{\alpha}\beta}\right| = k$. Then

$$\left(\frac{\beta-\alpha}{1-\bar{\alpha}\beta}\right)\overline{\left(\frac{\beta-\alpha}{1-\bar{\alpha}\beta}\right)} = k^2 \quad (\because |z|^2 = z\bar{z})$$

$$\Rightarrow k^2 = \frac{(\beta-\alpha)(\bar{\beta}-\bar{\alpha})}{(1-\bar{\alpha}\beta)(1-\alpha\bar{\beta})} \quad \left(\because \overline{\left(\frac{z_1}{z_2}\right)} = \frac{\overline{z_1}}{\overline{z_2}},\ \overline{(\bar{z})} = z\right)$$

$$\Rightarrow k^2 = \frac{|\beta|^2+|\alpha|^2-\alpha\bar{\beta}-\beta\bar{\alpha}}{1-\beta\bar{\alpha}-\alpha\bar{\beta}+|\beta|^2|\alpha|^2}$$

$$\Rightarrow k^2 = \frac{1+|\alpha|^2-\alpha\bar{\beta}-\beta\bar{\alpha}}{1-\beta\bar{\alpha}-\alpha\bar{\beta}+|\alpha|^2} \quad (\text{Given } |\beta| = 1)$$

$$\Rightarrow k^2 = \frac{1+|\alpha|^2-\alpha\bar{\beta}-\beta\bar{\alpha}}{1+|\alpha|^2-\beta\bar{\alpha}-\alpha\bar{\beta}} = 1$$

$\Rightarrow k = 1$ ($\because$ Modulus $k > 0$)

Thus, $\left|\frac{\beta-\alpha}{1-\bar{\alpha}\beta}\right| = k = 1$

Aleternative Method

Let $\alpha = a+ib$, $\beta = x+iy$

Given $|\beta| = 1 \Rightarrow \sqrt{x^2+y^2} = 1 \Rightarrow x^2+y^2 = 1$

Thus,

$$\left|\frac{\beta-\alpha}{1-\bar{\alpha}\beta}\right| = \left|\frac{(x+iy)-(a+ib)}{1-(a-ib)(x+iy)}\right|$$

$$= \left|\frac{(x-a)+i(y-b)}{1-(ax+aiy-ibx+by)}\right| = \left|\frac{(x-a)+i(y-b)}{(1-ax-by)+i(bx-ay)}\right|$$

$$= \frac{|(x-a)+i(y-b)|}{|(1-ax-by)+i(bx-ay)|} \quad \left(\because \left|\frac{z_1}{z_2}\right| = \frac{|z_1|}{|z_2|}\right)$$

$$= \frac{\sqrt{(x-a)^2+(y-b)^2}}{\sqrt{(1-ax-by)^2+(bx-ay)^2}}$$

$$= \frac{\sqrt{(x^2+y^2)+a^2+b^2-2ax-2by}}{\sqrt{1+a^2(x^2+y^2)+b^2(y^2+x^2)-2ax-2by}}$$

$$= \frac{\sqrt{1+a^2+b^2-2ax-2by}}{\sqrt{1+a^2+b^2-2ax-2by}} = 1 \quad (x^2+y^2 = 1)$$

Thus, $\left|\frac{\beta-\alpha}{1-\bar{\alpha}\beta}\right| = 1$

Q18. Find the number of non-zero integral solutions of the equation $|1-i|^x = 2^x$.

Solution:

$|1-i|^x = 2^x \Rightarrow |1+(-i)|^x = 2^x$

$\Rightarrow \left(\sqrt{1^2+(-1)^2}\right)^x = 2^x$

$\Rightarrow \left(\sqrt{2}\right)^x = 2^x \Rightarrow 2^{\frac{x}{2}} = 2^x$

$\Rightarrow \frac{x}{2} = x \Rightarrow \frac{x}{2} = 0 \Rightarrow x = 0$

Thus, 0 is the only integral solution of the given equation. Therefore, the number of non-zero integral solutions of the given equation is 0.

Q19. If $(a+ib)(c+id)(e+if)(g+ih) = A+iB$, then show that $(a^2+b^2)(c^2+d^2)(e^2+f^2)(g^2+h^2) = A^2+B^2$.

Solution:

$(a+ib)(c+id)(e+if)(g+ih) = A+iB$

$\Rightarrow |(a+ib)(c+id)(e+if)(g+ih)| = |A+iB|$

$\Rightarrow |(a+ib)|\times|(c+id)|\times|(e+if)|\times|(g+ih)| = |A+iB|$

$(\because |z_1z_2| = |z_1||z_2|)$

$\Rightarrow \sqrt{a^2+b^2}\times\sqrt{c^2+d^2}\times\sqrt{e^2+f^2}\times\sqrt{g^2+h^2}$

$= \sqrt{A^2+B^2}$

On squaring both sides, we get

$(a^2+b^2)(c^2+d^2)(e^2+f^2)(g^2+h^2) = A^2+B^2$.

Q20. If $\left(\frac{1+i}{1-i}\right)^m = 1$, then find the least positive integral value of m.

Solution:

$$\left(\frac{1+i}{1-i}\right)^m = 1$$

$$\Rightarrow \left(\frac{1+i}{1-i}\times\frac{1+i}{1+i}\right)^m = 1$$

$$\Rightarrow \left(\frac{1^2+i^2+2i}{1^2-i^2}\right)^m = 1$$

$$\Rightarrow \left(\frac{1-1+2i}{1-(-1)}\right)^m = 1 \quad (\because i^2=-1)$$

$$\Rightarrow \left(\frac{2i}{2}\right)^m = 1 \Rightarrow i^m = 1$$

$$\Rightarrow m = 4$$

Thus, the least positive integral value of m is 4.

Day 77

NCERT Exemplar Problems: Solutions (Q1–Q20)

Q1. For a positive integer n, find the value of

$$(1-i)^n\left(1-\frac{1}{i}\right)^n.$$

Solution:

$$(1-i)^n\left(1-\frac{1}{i}\right)^n = (1-i)^n\left(\frac{i-1}{i}\right)^n$$

$$=\left(\frac{(1-i)(i-1)}{i}\right)^n = \left(\frac{i-1-i^2+i}{i}\right)^n$$

$$=\left(\frac{2i}{i}\right)^n = 2^n \qquad (\because i^2=-1)$$

Q2. Evaluate $\sum_{n=1}^{13}(i^n+i^{n+1})$, where $n\in N$.

Solution: Since the summation of four consecutive powers of i is equal to zero, and there are 13 terms in the summation, the summation of 12 (4×3) terms will be equal to zero. Therefore, we will have only one term (either from the beginning or from the end).

Thus,

$$\sum_{n=1}^{13}(i^n+i^{n+1}) = i^1+i^{1+1}+\sum_{n=2}^{13}(i^n+i^{n+1})$$

$$= i^1+i^2+0 \quad (\because i^n+i^{n+1}+i^{n+2}+i^{n+3}=0)$$

$$= i-1 \quad (\because i^2=-1)$$

Alternative Method

$$\sum_{n=1}^{13}(i^n+i^{n+1})$$

$$=\sum_{n=1}^{4}(i^n+i^{n+1})+\sum_{n=5}^{8}(i^n+i^{n+1})+\sum_{n=9}^{12}(i^n+i^{n+1})+i^{13}+i^{13+1}$$

$$= i^{13}+i^{14} \quad (\because i^n+i^{n+1}+i^{n+2}+i^{n+3}=0)$$

$$= i^{4\cdot3+1}+i^{4\cdot2+2} = (i^4)^3\cdot i+(i^4)^3\cdot i^2$$

$$= i+i^2 = i-1 \quad (\because i^4=1,\ i^2=-1)$$

Alternative Method

$$\sum_{n=1}^{13}(i^n+i^{n+1})$$

$$=(i+i^2+i^3+i^4+i^5+i^6+i^7+i^8+i^9+i^{10}+i^{11}+i^{12}+i^{13})$$

$$+(i^2+i^3+i^4+i^5+i^6+i^7+i^8+i^9+i^{10}+i^{11}+i^{12}+i^{13}+i^{14})$$

$$= i+i^{14}+2(i^2+i^3+i^4+i^5+i^6+i^7+i^8+i^9+i^{10}+i^{11}+i^{12}+i^{13})$$

$$= i+i^{14}+2(-1-i+1+i-1-i+1+i-1-i+1+i)$$

$$= i+(i^4)^3\cdot i^2+2\cdot 0 = i-1 \quad (i^2=-1,\ i^3=-i,\ i^4=1)$$

Q3. If $\left(\frac{1+i}{1-i}\right)^3-\left(\frac{1-i}{1+i}\right)^3 = x+iy$, then find (x, y).

Solution:

Given that $\left(\frac{1+i}{1-i}\right)^3-\left(\frac{1-i}{1+i}\right)^3 = x+iy \qquad$ (i)

$$\left(\frac{1+i}{1-i}\right)^3-\left(\frac{1-i}{1+i}\right)^3 = \left(\frac{1+i}{1-i}\times\frac{1+i}{1+i}\right)^3-\left(\frac{1-i}{1+i}\times\frac{1-i}{1-i}\right)^3$$

$$=\left(\frac{1+i^2+2i}{1-i^2}\right)^3-\left(\frac{1+i^2-2i}{1-i^2}\right)^3$$

$$=\left(\frac{1+(-1)+2i}{1-(-1)}\right)^3-\left(\frac{1+(-1)-2i}{1-(-1)}\right)^3 \quad (\because i^2=-1)$$

$$=\left(\frac{2i}{2}\right)^3-\left(\frac{-2i}{2}\right)^3$$

$$= i^3-(-i)^3 = i^3+i^3 = 2i^3 = -2i \quad (\because i^3=-i)$$

$$= 0-2i = x+iy,$$

$$\Rightarrow x=0,\quad y=-2$$

Thus, $(x, y) = (0, -2)$

Q4. If $\frac{(1+i)^2}{2-i} = x+iy$, then find the value of $x+y$.

Solution:

$$\frac{(1+i)^2}{2-i} = \frac{1+i^2+2i}{2-i} = \frac{2i}{2-i} \quad (\because i^2=-1)$$

$$\frac{2i}{2-i}\times\frac{2+i}{2+i} = \frac{4i-2}{4+1} = \frac{-2}{5}+\frac{4}{5}i = x+iy$$

$\Rightarrow x = \frac{-2}{5}, \quad y = \frac{4}{5}$

$\therefore x + y = \frac{-2}{5} + \frac{4}{5} = \frac{2}{5}$

Q5. If $\left(\frac{1-i}{1+i}\right)^{100} = a + ib$, then find (a, b).

Solution:

$\left(\frac{1-i}{1+i}\right)^{100} \Rightarrow \left[\frac{(1-i)}{(1+i)} \times \frac{(1-i)}{(1-i)}\right]^{100}$

$= \left(\frac{1+i^2-2i}{1-i^2}\right)^{100} = \left(\frac{-2i}{2}\right)^{100} \quad (\because i^2 = -1)$

$= (-i)^{100} = (-1)^{100}(i^4)^{25} = 1 \cdot 1 \quad (\because i^4 = 1)$

$= 1 + 0i = a + ib \Rightarrow a = 1, \; b = 0$

Thus, $(a, b) = (1, 0)$

Q6. If $a = \cos\theta + i\sin\theta$, then find the value of $\frac{1+a}{1-a}$.

Solution:

$\frac{1+a}{1-a} = \frac{1+\cos\theta + i\sin\theta}{1-\cos\theta - i\sin\theta} \quad (\because a = \cos\theta + i\sin\theta)$

$= \frac{(1+\cos\theta) + i\sin\theta}{(1-\cos\theta) - i\sin\theta} \times \frac{(1-\cos\theta) + i\sin\theta}{(1-\cos\theta) + i\sin\theta}$

$= \frac{(1-\cos^2\theta) - \sin^2\theta + i(2\sin\theta)}{(1-\cos\theta)^2 + \sin^2\theta}$

$= \frac{1-(\cos^2\theta + \sin^2\theta) + i(2\sin\theta)}{1 - 2\cos\theta + (\cos^2\theta + \sin^2\theta)}$

$= \frac{1-1+i(2\sin\theta)}{1-2\cos\theta+1} = \frac{i\sin\theta}{1-\cos\theta}$

$= \frac{2i\sin(\theta/2)\cos(\theta/2)}{2\sin^2(\theta/2)}$

$= \frac{i\cos(\theta/2)}{\sin(\theta/2)} = i\cot(\theta/2)$

Q7. If $(1+i)z = (1-i)\bar{z}$, then show that $z = -i\bar{z}$.

Solution:

$(1+i)z = (1-i)\bar{z}$

$\Rightarrow \frac{z}{\bar{z}} = \frac{(1-i)}{(1+i)} = \frac{(1-i)(1-i)}{(1+i)(1-i)} = \frac{1+i^2-2i}{1-i^2}$

$\Rightarrow \frac{z}{\bar{z}} = \frac{1-1-2i}{2} \quad (\because i^2 = -1)$

$\Rightarrow \frac{z}{\bar{z}} = -i$

$\Rightarrow z = -i\bar{z}$

Q8. If $z = x + iy$, then show that $z\bar{z} + 2(z + \bar{z}) + b = 0$, where $b \in R$, represents a circle.

Solution:

Let $z = x + iy$

$z\bar{z} + 2(z + \bar{z}) + b = 0$

$\Rightarrow (x+iy)(x-iy) + 2\{(x+iy)+(x-iy)\} + b = 0$

$\Rightarrow x^2 + y^2 + 4x + b = 0$ represents the equation of a circle.

Q9. If the real part of $\frac{\bar{z}+2}{\bar{z}-1}$ is 4, then show that the locus of the point representing z in the complex plane is a circle.

Solution:

Let $z = x + iy$

$\text{Re}\left(\frac{\bar{z}+2}{\bar{z}-1}\right) = 4$

$\Rightarrow \text{Re}\left(\frac{x-iy+2}{x+iy-1}\right) = 4$

$\Rightarrow \text{Re}\left(\frac{[(x+2)-iy][(x-1)+iy]}{[(x-1)-iy][(x-1)+iy]}\right) = 4$

$\Rightarrow \text{Re}\left(\frac{(x+2)(x-1)+y^2+i[y(x+2)-y(x-1)]}{(x-1)^2+y^2}\right) = 4$

$\Rightarrow \frac{(x+2)(x-1)+y^2}{(x-1)^2+y^2} = 4$

$\Rightarrow x^2 - x + 2x - 2 + y^2 = 4(x^2 - 2x + 1 + y^2)$

$\Rightarrow 3x^2 + 3y^2 - 9x + 6 = 0$

$\Rightarrow x^2 + y^2 - 3x + 2 = 0$ represents the equation of a circle.

Q10. Show that the complex number z satisfying the condition $\arg\left(\frac{z-1}{z+1}\right) = \frac{\pi}{4}$ lies on a circle.

Solution:

Let $z = x + iy$

$\arg\left(\frac{z-1}{z+1}\right) = \frac{\pi}{4} \Rightarrow \arg\left(\frac{x-1+iy}{x+1+iy}\right) = \frac{\pi}{4}$

$\Rightarrow \arg\left(\frac{x-1+iy}{x+1+iy} \times \frac{x+1-iy}{x+1-iy}\right) = \frac{\pi}{4}$

$\Rightarrow \arg\left(\frac{x^2-1+y^2+2iy}{(x+1)^2+y^2}\right) = \frac{\pi}{4}$

$\Rightarrow \frac{2y}{x^2-1+y^2} = \tan\frac{\pi}{4} \Rightarrow \frac{2y}{x^2-1+y^2} = 1$

$\Rightarrow 2y = x^2 - 1 + y^2$

$\Rightarrow x^2 + y^2 - 2y - 1 = 0$ represents the equation of a circle.

Q11. Solve the equation $|z| = z + 1 + 2i$.

Solution:

Let $z = x + iy$

$|z| = z+1+2i \Rightarrow |x+iy| = x+iy+1+2i$

$\Rightarrow \sqrt{x^2+y^2} = x+1+i(y+2)$

$\Rightarrow (x+1)-\sqrt{x^2+y^2}+i(y+2) = 0+0i$

On comparing the real and imaginary parts, we get $y+2=0 \Rightarrow y=-2$

Also, $(x+1)-\sqrt{x^2+y^2}=0$

$x^2+1+2x = x^2+y^2$

$\Rightarrow x = \frac{y^2-1}{2} = \frac{(-2)^2-1}{2} = \frac{3}{2}$

Thus, $z = x+iy = 3/2-2i$

Q12. If $|z+1| = z+2(1+i)$, then find the value of z.

Solution:
Let $z = x+iy$

$|x+iy+1| = x+iy+2(1+i)$

$\Rightarrow |x+1+iy| = (x+2)+i(y+2)$

$\Rightarrow \sqrt{(x+1)^2+y^2} = (x+2)+i(y+2)$

$\Rightarrow (x+2)-\sqrt{(x+1)^2+y^2}+i(y+2)=0$

On comparing the real and imaginary parts, we get
$y+2=0 \Rightarrow y=-2$

Also, $(x+2)-\sqrt{(x+1)^2+y^2}=0$

$\Rightarrow (x+2)^2 = (x+1)^2+y^2$

$\Rightarrow x^2+4+4x = x^2+1+2x+y^2$

$\Rightarrow x = \frac{y^2-3}{2} = \frac{(-2)^2-3}{2} = \frac{1}{2}$

Thus, $z = x+iy = \frac{1}{2}-2i$

Q13. If $\arg(z-1) = \arg(z+3i)$, then find $x-1:y$, where $z = x+iy$.

Solution:
$\arg(z-1) = \arg(z+3i)$

$\Rightarrow \arg(x+iy-1) = \arg(x+iy+3i)$

$\Rightarrow \arg(x-1+iy) = \arg[x+i(y+3)]$

$\Rightarrow \tan^{-1}\frac{y}{x-1} = \tan^{-1}\frac{y+3}{x}$

$\Rightarrow \frac{y}{x-1} = \frac{y+3}{x} \Rightarrow xy = xy+3x-y-3$

$\Rightarrow 3(x-1) = y \Rightarrow \frac{x-1}{y} = \frac{1}{3}$

Thus, $(x-1):y = 1:3$

Q14. Show that $\left|\frac{z-2}{z-3}\right| = 2$ represents a circle. Find its centre and radius.

Solution: Let $z = x+iy$

$\left|\frac{z-2}{z-3}\right| = 2 \Rightarrow \left|\frac{x+iy-2}{x+iy-3}\right| = 2$

$\Rightarrow |x-2+iy| = 2|x-3+iy|$

$\Rightarrow \sqrt{(x-2)^2+y^2} = 2\sqrt{(x-3)^2+y^2}$

$\Rightarrow x^2-4x+4+y^2 = 4(x^2-6x+9+y^2)$

$\Rightarrow 3x^2+3y^2-20x+32=0$

$\Rightarrow x^2+y^2-\frac{20}{3}x+\frac{32}{3}=0$

On comparing the above equation with the general equation of a circle $x^2+y^2+2gx+2fy+c=0$, we get $g=\frac{-10}{3}$, $f=0$ and $c=\frac{32}{3}$.

$\therefore$ Centre $= (-g,-f) = (10/3, 0)$

Radius $r = \sqrt{g^2+f^2-c} = \sqrt{(10/3)^2+0-32/3} = 2/3$

Q15. If $\frac{z-1}{z+1}$ is a purely imaginary number $(z \neq -1)$, then find the value of $|z|$.

Solution:
Since $\frac{z-1}{z+1}$ is a purely imaginary number,

$\text{Re}\left(\frac{z-1}{z+1}\right) = 0$

Let $z = x+iy$

$\text{Re}\left(\frac{z-1}{z+1}\right) = 0$

$\Rightarrow \text{Re}\left(\frac{x+iy-1}{x+iy+1}\right) = 0 \Rightarrow \text{Re}\left(\frac{x-1+iy}{x+1+iy}\right) = 0$

$\Rightarrow \text{Re}\left(\frac{(x-1+iy)(x+1-iy)}{(x+1+iy)(x+1-iy)}\right) = 0$

$\Rightarrow \text{Re}\left(\frac{(x^2-1)+y^2+2iy}{(x+1)^2+y^2}\right) = 0$

$\Rightarrow (x^2-1)+y^2 = 0 \Rightarrow x^2+y^2 = 1$

$\Rightarrow |z| = 1$

Q16. If z_1, z_2 are two complex numbers such that $|z_1| = |z_2|$ and $\arg(z_1)+\arg(z_2) = \pi$, then show that $z_1 = -\bar{z}_2$.

Solution:
Let $z_1 = r_1(\cos\theta_1 + i\sin\theta_1)$, $z_2 = r_2(\cos\theta_2 + i\sin\theta_2)$ be two complex numbers.

$|z_1| = |z_2|$ and $\arg(z_1)+\arg(z_2) = \pi$

Therefore, $r_1 = r_2$ and $\theta_1+\theta_2 = \pi$

Now,

$\text{RHS} = -\bar{z}_2 = -r_2(\cos\theta_2 - i\sin\theta_2)$

$= -r_1[\cos(\pi-\theta_1)+i\sin(\pi-\theta_1)] \quad (\because r_1 = r_2,\ \theta_2 = \pi-\theta_1)$

$= -r_1(-\cos\theta_1 - i\sin\theta_1)$
$= r_1(\cos\theta_1 + i\sin\theta_1) = z_1 = \text{LHS}$

Q17. If $|z_1| = 1$ $(z_1 \neq -1)$, $z_2 = \dfrac{z_1 - 1}{z_1 + 1}$, then show that the real part of z_2 is zero.

Solution:
Let $z_1 = x + iy$
$\Rightarrow |z_1| = \sqrt{x^2 + y^2} = 1 \quad (\because |z_1| = 1)$
$\Rightarrow x^2 + y^2 = 1$
Now, $z_2 = \dfrac{z_1 - 1}{z_1 + 1} = \dfrac{x + iy - 1}{x + iy + 1}$
$= \dfrac{x - 1 + iy}{x + 1 + iy} = \dfrac{(x - 1 + iy)(x + 1 - iy)}{(x + 1 + iy)(x + 1 - iy)}$
$= \dfrac{x^2 - 1 + y^2 + 2iy}{(x+1)^2 + y^2}$
$= \dfrac{1 - 1 + 2iy}{(x+1)^2 + y^2} \quad (\because x^2 + y^2 = 1)$
$= 0 + i\left(\dfrac{2y}{(x+1)^2 + y^2}\right)$
Thus, the real part of z_2 is zero.

Q18. If z_1, z_2 and z_3, z_4 are two pairs of conjugate complex numbers, then find $\arg\left(\dfrac{z_1}{z_4}\right) + \arg\left(\dfrac{z_2}{z_3}\right)$.

Solution:
Let $z_1 = r(\cos\theta + i\sin\theta)$
$z_2 = \bar{z}_1 = r(\cos\theta - i\sin\theta) = r[\cos(-\theta) + i\sin(-\theta)]$
And let $z_3 = R(\cos\alpha + i\sin\alpha)$
$z_4 = \bar{z}_3 = R(\cos\alpha - i\sin\alpha) = R[\cos(-\alpha) + i\sin(-\alpha)]$
$\therefore \arg\left(\dfrac{z_1}{z_4}\right) + \arg\left(\dfrac{z_2}{z_3}\right) = \arg(z_1) - \arg(z_4) + \arg(z_2) - \arg(z_3)$
$= \theta - (-\alpha) + (-\theta) - \alpha = 0$

Q19. If $|z_1| = |z_2| = \ldots\ldots |z_n| = 1$, then show that
$$|z_1 + z_2 + z_3 + \ldots\ldots + z_n| = \left|\frac{1}{z_1} + \frac{1}{z_2} + \frac{1}{z_3} + \ldots + \frac{1}{z_n}\right|.$$

Solution:
$\text{LHS} = |z_1 + z_2 + z_3 + \ldots + z_n|$
$= |\overline{z_1 + z_2 + z_3 + \ldots + z_n}| \quad (\because |z| = |\bar{z}|)$
$= |\bar{z}_1 + \bar{z}_2 + \bar{z}_3 + \ldots \bar{z}_n|$
$= \left|\dfrac{z_1\bar{z}_1}{z_1} + \dfrac{z_2\bar{z}_2}{z_2} + \dfrac{z_3\bar{z}_3}{z_3} + \ldots + \dfrac{z_n\bar{z}_n}{z_n}\right|$
$= \left|\dfrac{|z_1|}{z_1} + \dfrac{|z_2|}{z_2} + \dfrac{|z_3|}{z_3} + \ldots + \dfrac{|z_n|}{z_n}\right|$
$= \left|\dfrac{1}{z_1} + \dfrac{1}{z_2} + \dfrac{1}{z_3} + \ldots + \dfrac{1}{z_n}\right| = \text{RHS}$
$(\because |z_1| = |z_2| = \ldots |z_n| = 1 \Rightarrow z_1\bar{z}_1 = z_2\bar{z}_2 = \ldots = z_n\bar{z}_n = 1)$

Q20. If the complex numbers z_1, z_2 are such that $\arg(z_1) - \arg(z_2) = 0$, then show that $|z_1 - z_2| = |z_1| - |z_2|$.

Solution:
Let $z_1 = r_1(\cos\theta + i\sin\theta), z_2 = r_2(\cos\theta + i\sin\theta)$
$\because \arg(z_1) - \arg(z_2) = 0 \Rightarrow \theta_1 = \theta_2 = \theta$
$\text{LHS} = |z_1 - z_2| = |r_1(\cos\theta + i\sin\theta) - r_2(\cos\theta + i\sin\theta)|$
$= |(r_1 - r_2)\cos\theta + i(r_1 - r_2)\sin\theta| = (r_1 - r_2)|\cos\theta + i\sin\theta|$
$= (r_1 - r_2)\sqrt{\cos^2\theta + \sin^2\theta}$
$= (r_1 - r_2) = |z_1| - |z_2| \quad (\because r = |z|)$

Day 78

NCERT Exemplar Problems: Solutions (Q21–Q27)

Q21. Solve the system of equations $\text{Re}(z^2) = 0$, $|z| = 2$.

Solution:
Let $z = x + iy$
$\text{Re}(z^2) = 0$
$\Rightarrow \text{Re}(x^2 + 2ixy - y^2) = 0$
$\Rightarrow \text{Re}\{(x^2 - y^2) + 2ixy\} = 0$
$\Rightarrow x^2 - y^2 = 0 \Rightarrow x^2 = y^2$
Also, $|z| = \sqrt{x^2 + y^2} = 2$
$\Rightarrow x^2 + y^2 = 4$
Solving the above two equations, we get
$x^2 + y^2 = 4 \Rightarrow 2x^2 = 4 \Rightarrow x^2 = 2$
$\Rightarrow x = \pm\sqrt{2} \Rightarrow y = \pm\sqrt{2} \quad (\because x^2 = y^2)$
$\Rightarrow z = x + iy = \pm\sqrt{2} \pm i\sqrt{2}$

Q22. Find the complex number satisfying the equation $z + \sqrt{2}|(z + 1)| + i = 0$.

Solution:
Let $z = x + iy$

$z+\sqrt{2}|(z+1)|+i=0$

$\Rightarrow x+iy+\sqrt{2}|x+iy+1|+i=0$

$\Rightarrow x+i(1+y)+\sqrt{2}\left(\sqrt{(x+1)^2+y^2}\right)=0$

$\Rightarrow x+\sqrt{2}\left(\sqrt{(x+1)^2+y^2}\right)+i(1+y)=0+i0$

Equating the real and imaginary parts, we get

$x+\sqrt{2}\left(\sqrt{(x+1)^2+y^2}\right)=0$ and $(1+y)=0$

Thus,

$1+y=0 \Rightarrow y=-1$

Also,

$x+\sqrt{2}\left(\sqrt{(x+1)^2+y^2}\right)=0$

$\Rightarrow x+\sqrt{2}\left(\sqrt{(x+1)^2+(-1)^2}\right)=0 \quad (\because y=-1)$

$\Rightarrow x+\sqrt{2}\left(\sqrt{(x+1)^2+1}\right)=0$

$\Rightarrow x^2=2(x^2+2x+1+1) \Rightarrow x^2+4x+4=0$

$\Rightarrow (x+2)^2=0 \Rightarrow x=-2$

$\therefore z=x+iy=-2-i$

Q23. Write the complex number $z=\dfrac{1-i}{\cos\frac{\pi}{3}+i\sin\frac{\pi}{3}}$ in the polar form.

Solution:

$$z=\frac{1-i}{\cos\frac{\pi}{3}+i\sin\frac{\pi}{3}}=\frac{\sqrt{2}\left[\frac{1}{\sqrt{2}}-i\frac{1}{\sqrt{2}}\right]}{\cos\frac{\pi}{3}+i\sin\frac{\pi}{3}}$$

$$=\frac{\sqrt{2}[\cos(-\pi/4)+i\sin(-\pi/4)]}{\cos\pi/3+i\sin\pi/3}$$

$$=\sqrt{2}\left[\cos\left(\frac{-\pi}{4}-\frac{\pi}{3}\right)+i\sin\left(\frac{-\pi}{4}-\frac{\pi}{3}\right)\right]$$

$$=\sqrt{2}\left[\cos\left(\frac{-7\pi}{12}\right)+i\sin\left(\frac{-7\pi}{12}\right)\right]$$

$$=\sqrt{2}\left[\cos\left(\frac{7\pi}{12}\right)-i\sin\left(\frac{7\pi}{12}\right)\right]$$

$$=\sqrt{2}\left[-\cos\left(\pi-\frac{5\pi}{12}\right)-i\sin\left(\pi-\frac{5\pi}{12}\right)\right]$$

$$=-\sqrt{2}\left[\cos\frac{5\pi}{12}+i\sin\frac{5\pi}{12}\right]$$

Q24. If z and w are two complex numbers such that $|zw|=1$, $\arg(z)-\arg(w)=\dfrac{\pi}{2}$ then show that $\bar{z}\,w=-i$

Solution:

Let $z=r_1(\cos\theta_1+i\sin\theta_1)$, $w=r_2(\cos\theta_2+i\sin\theta_2)$

Since, $|zw|=1 \Rightarrow |z||w|=r_1r_2=1$

Also, $\arg(z)-\arg(w)=\dfrac{\pi}{2} \Rightarrow \theta_1-\theta_2=\dfrac{\pi}{2}$

Now,

LHS $=\bar{z}w$

$=r_1(\cos\theta_1-i\sin\theta_1)\cdot r_2(\cos\theta_2+i\sin\theta_2)$

$=r_1r_2[\cos(\theta_2-\theta_1)+i\sin(\theta_2-\theta_1)]$

$=r_1r_2[\cos(-\pi/2)+i\sin(-\pi/2)]$

$=1(0-i)$

$=-i=$ RHS

Q25. Fill in the blanks for the following.

(i) For any two complex numbers z_1, z_2 and any real numbers a, b, $|az_1-bz_2|^2+|bz_1+az_2|^2=$ ______

Solution:

$|az_1-bz_2|^2+|bz_1+az_2|^2$

$=|az_1|^2+|bz_2|^2-2\,\mathrm{Re}(az_1\cdot b\bar{z}_2)+|bz_1|^2$

$+|az_2|^2+2\,\mathrm{Re}(az_1\cdot b\bar{z}_2)$

$=(a^2+b^2)|z_1|^2+(a^2+b^2)|z_2|^2$

$=(a^2+b^2)\left(|z_1|^2+|z_2|^2\right)$

(ii) The value of $\sqrt{-25}\times\sqrt{-9}$ is______

Solution:

$\sqrt{-25}\times\sqrt{-9}=i\sqrt{25}\times i\sqrt{9}=i^2(5\times 3)=-15$

(iii) The number $\dfrac{(1-i)^3}{1-i^3}$ is equal to______

Solution:

$$\frac{(1-i)^3}{1-i^3}=\frac{1-i^3+3i^2-3i}{1-i^3}$$

$$=\frac{1+i-3-3i}{1+i}=\frac{-2-2i}{1+i}=\frac{-2(1+i)}{1+i}=-2$$

(iv) The sum of the series $i+i^2+i^3+\ldots$ up to 1000 terms is______

Solution:

Since $1000=4\times 250$ and $i^n+i^{n+1}+i^{n+2}+i^{n+3}=0$

$i+i^2+i^3+\ldots$ up to 1000 terms

$\therefore i+i^2+i^3+i^4+\ldots i^{1000}=0$

(v) Multiplicative inverse of $1+i$ is ______

Solution:

Multiplicative inverse of $1+i$ is

$$=\frac{1}{1+i}=\frac{1-i}{1-i^2}=\frac{1}{2}(1-i)$$

(vi) If z_1, z_2 are complex numbers such that z_1+z_2 is real number, then $z_1=$ ______

Solution:

Let $z_1=x_1+iy_1, z_2=x_2+iy_2$

$z_1 + z_2 = [(x_1 + x_2) + i(y_1 + y_2)]$

$z_1 + z_2$ is real, and therefore, $y_1 + y_2 = 0 \Rightarrow y_2 = -y_1$

Thus,

$z_2 = x_2 + iy_2 = x_2 - iy_1$

or $z_2 = \bar{z}_1$ if $x_1 = x_2$

(vii) $\arg(z) + \arg \bar{z}$ where, $(\bar{z} \neq 0)$ is______

Solution:

$\arg(z) + \arg(\bar{z}) = \theta + (-\theta) = 0$

(viii) If $|z+4| \leq 3$, then the greatest and least values of $|z+1|$ are ______ and ______

Solution:

For the greatest value of $|z+1|$

$|z+1| = |z+4-3| \leq |z+4| + |-3|$

$\Rightarrow |z+1| \leq 3+3 \quad (\because |z+4| \leq 3)$

$\Rightarrow |z+1| \leq 6$

Thus, the greatest value of $|z+1|$ is 6.

The least value of any modulus will be equal to zero. Thus, let $|z+1| = 0 \Rightarrow z = -1$

And for $z = -1$, $|z+4| = |-1+4| = 3$,

satisfy the given condition.

(ix) If $\left|\frac{z-2}{z+2}\right| = \tan\frac{\pi}{6}$, then the locus of z is______

Solution:

Let $z = x+iy$. Then $\left|\frac{z-2}{z+2}\right| = \tan\frac{\pi}{6}$

$\Rightarrow \left|\frac{x+iy-2}{x+iy+2}\right| = \frac{1}{\sqrt{2}} \Rightarrow \frac{|x-2+iy|}{|x+2+iy|} = \frac{1}{\sqrt{2}}$

$\Rightarrow \sqrt{2}|x-2+iy| = |x+2+iy|$

$\Rightarrow 2\left((x-2)^2 + y^2\right) = (x+2)^2 + y^2$

$\Rightarrow 2[x^2 + 4 - 4x + y^2] = [x^2 + 4x + 4 + y^2]$

$\Rightarrow x^2 + y^2 - 12x + 4 = 0$ represents a circle.

(x) If $|z| = 4$, $\arg(z) = \frac{5\pi}{6}$, then $z =$ ______

Solution:

Let $z = x+iy = r(\cos\theta + i\sin\theta)$

$|z| = r = 4$ and $\arg(z) = \theta = \frac{5\pi}{6} \Rightarrow \tan\theta = \frac{5\pi}{6}$

$\therefore z = 4\left(\cos\frac{5\pi}{6} + i\sin\frac{5\pi}{6}\right)$

$= 4[\cos(\pi - \pi/6) + i\sin(\pi - \pi/6)]$

$= 4\left[-\cos\frac{\pi}{6} + i\sin\frac{\pi}{6}\right] = 4\left[\frac{-\sqrt{3}}{2} + i\frac{1}{2}\right] = -2\sqrt{3} + 2i$

Q26. State true or false for the following.

(i) The order relation is defined on the set of complex numbers.

Solution: False: The order relation is not defined on the set of complex numbers because comparison of complex numbers is not possible.

(ii) Multiplication of a non-zero complex number by $-i$ rotates the point about the origin through a right angle in the anti-clockwise direction.

Solution: False: Let $z = x+iy$ then $-iz = -ix + y = y - ix$, which is false.

(iii) For any complex number z, the minimum value of $|z| + |z-1|$ is 1.

Solution: True

$|z| + |z-1| \geq |z + z - 1|$

$\Rightarrow |z| + |z-1| \geq |2z-1|$

Since the minimum value of $|z| + |z-1|$ is at $z = 0$, the minimum value of $|z| + |z-1| = 1$.

(iv) The locus represented by $|z-1| = |z-i|$ is a line perpendicular to the join of the points (1, 0) and (0, 1).

Solution: True

Let $z = x+iy$

$|z-1| = |z-i|$

$\Rightarrow |x+iy-1| = |x+iy-i|$

$\Rightarrow |x-1+iy| = |x+i(y-1)|$

$\Rightarrow (x-1)^2 + y^2 = x^2 + (y-1)^2$

$\Rightarrow x^2 - 2x + 1 + y^2 = x^2 + 1 + y^2 - 2y$

$\Rightarrow y = x$

And equation of a line through the points, (1, 0) and (0, 1) is

$y - 0 = \frac{1-0}{0-1}(x-1) \Rightarrow y = -x+1$

Thus, the two lines are perpendicular to each other.

(v) If z is a complex number such that $z \neq 0$, $\text{Re}(z) = 0$, then $\text{Im}(z^2) = 0$.

Solution: False

Let $z = x+iy$

$\because \text{Re}(z) = 0 \Rightarrow x = 0$

$\because z \neq 0 \Rightarrow y \neq 0$

$\text{Im}(z^2) = i^2y^2 = -y^2$ which is a real number.

(vi) The inequality $|z-4|<|z-2|$ represents the region given by $x>3$.

Solution: True

$|z-4|<|z-2|$

$\Rightarrow |x-4+iy|<|x-2+iy|$

$\Rightarrow \sqrt{(x-4)^2+y^2}<\sqrt{(x-2)^2+y^2}$

$\Rightarrow (x-4)^2+y^2<(x-2)^2+y^2$

$\Rightarrow x^2-8x+16+y^2<x^2-4x+4+y^2$

$\Rightarrow 12<4x \Rightarrow x>3$

(vii) Let z_1, z_2 be two complex numbers such that $|z_1+z_2|=|z_1|+|z_2|$. Then $\arg(z_1-z_2)=0$.

Solution: True

Let $z_1=r_1(\cos\theta+i\sin\theta), z_2=r_2(\cos\theta+i\sin\theta)$

and $\arg(z_1)-\arg(z_2)=0 \Rightarrow \theta_1=\theta_2=\theta$

LHS $=|z_1+z_2|$

$=|r_1(\cos\theta+i\sin\theta)+r_2(\cos\theta+i\sin\theta)|$

$=|(r_1+r_2)\cos\theta+i(r_1+r_2)\sin\theta|$

$=(r_1+r_2)|\cos\theta+i\sin\theta|=(r_1+r_2)\sqrt{\cos^2\theta+\sin^2\theta}$

$=(r_1+r_2)=|z_1|+|z_2| \quad (\because r=|z|)$

(viii) 2 is not a complex number.

Solution: True. Since the imaginary part is equal to zero, 2 is not a complex number, and thus, 2 is a real number.

Q27. Match the statements of column A and column B.

Column A

(i) The polar form of $i+\sqrt{3}$ is

(ii) The argument of $-1+\sqrt{-3}$ is

(iii) If $|z+2|=|z-2|$, then locus of z is

(iv) If $|z+2i|=|z-2i|$, then locus of z is

(v) Region represented by $|z+4i|\geq 3$

(vi) Region represented by $|z+4|\leq 3$ is

(vii) Conjugate $\dfrac{1+2i}{1-i}$ lies in

(viii) Reciprocal of $1-i$ lies in

Column B

(a) Perpendicular bisector segment joining (–2, 0) and (2, 0)

(b) On or outside the circle having centre at (0, –4) and radius 3.

(c) $\dfrac{2\pi}{3}$

(d) Perpendicular bisector of segment joining (0, –2) and (0, 2).

(e) $2\left(\cos\dfrac{\pi}{6}+i\sin\dfrac{\pi}{6}\right)$

(f) On or inside the circle having centre (–4, 0) and radius 3 units.

(g) First quadrant

(h) Third quadrant

Solution:

(i) – (e)

$$z=i+\sqrt{3}=2\left(\frac{\sqrt{3}}{2}+i\frac{1}{2}\right)=2\left(\cos\frac{\pi}{6}+i\sin\frac{\pi}{6}\right)$$

(ii) – (c)

$$z=-1+\sqrt{-3}=-1+i\sqrt{3}=2\left(\frac{-1}{2}+i\frac{\sqrt{3}}{2}\right)$$

$$=2\left[\cos\left(\pi-\frac{\pi}{3}\right)+i\sin\left(\pi-\frac{\pi}{3}\right)\right]=2\left[\cos\frac{2\pi}{3}+i\sin\frac{2\pi}{3}\right]$$

(iii) – (a)

$|z+2|=|z-2|$

$\Rightarrow |x+2+iy|=|x-2+iy|$

$\Rightarrow (x+2)^2+y^2=(x-2)^2+y^2$

$\Rightarrow x^2+4x+4=x^2-4x+4$

$\Rightarrow 8x=0 \Rightarrow x=0$

It is a straight line which is a perpendicular bisector of the segment joining the points (–2, 0) and (2, 0).

(iv) – (a)

$|z+2i|=|z-2i|$

$\Rightarrow |x+i(y+2)|=|x+i(y-2)|$

$\Rightarrow x^2+(y+2)^2=x^2+(y-2)^2$

$\Rightarrow 4y=0 \Rightarrow y=0$

It is a straight line, which is a perpendicular bisector of the segment joining (0, –2), and (0, 2).

(v) – (b)

$|z+4i|\geq 3=|x+iy+4i|\geq 3$

$\Rightarrow |x+i(y+4)|\geq 3$

$\Rightarrow \sqrt{x^2+(y+4)^2}\geq 3$

$\Rightarrow x^2+(y+4)^2\geq 9$

$\Rightarrow x^2+y^2+8y+7\geq 0$

This represents a circle, on or outside having centre (0, –4) and radius 3.

(vi) – (f)

$|z+4|\leq 3 \Rightarrow |x+iy+4|\leq 3$

$\Rightarrow |x+4+iy|\leq 3 \Rightarrow \sqrt{(x+4)^2+y^2}\leq 3$

$\Rightarrow (x+4)^2+y^2\leq 9$

$\Rightarrow x^2+8x+y^2+7\leq 0$

This represents a circle, on or inside the circle having centre (–4, 0) and radius 3.

(vii) — (h)

$$z=\frac{1+2i}{1-i}=\frac{(1+2i)(1+i)}{(1-i)(1+i)}=\frac{1+2i+i+2i^2}{1-i^2}$$

$$=\frac{1-2+3i}{1+1}=\frac{-1}{2}+\frac{3}{2}i$$

$$\Rightarrow \bar{z}=\frac{-1}{2}-\frac{3}{2}i$$

Thus, $\left(\frac{-1}{2},\frac{-3}{2}\right)$ lies in the third quadrant.

(viii) — (g)

$$z=1-i\Rightarrow\frac{1}{z}=\frac{1}{1-i}=\frac{1+i}{(1-i)(1+i)}=\frac{1+i}{1-i^2}=\frac{1}{2}(1+i)$$

Thus, $\left(\frac{1}{2},\frac{1}{2}\right)$ lies in the first quadrant.

Day 79

NCERT Exemplar Problems: Solutions (Q28–Q50)

Q28. What is the conjugate of $\frac{2-i}{(1-2i)^2}$?

Solution:

$$z=\frac{2-i}{(1-2i)^2}=\frac{2-i}{1+4i^2-4i}$$

$$=\frac{2-i}{1-4-4i}=\frac{2-i}{-3-4i}=\frac{i-2}{3+4i}$$

$$=\frac{(i-2)(3-4i)}{(3+4i)(3-4i)}$$

$$=\frac{3i-6-4i^2+8i}{9+16}=\frac{-2+11i}{25}$$

$$\Rightarrow \bar{z}=\frac{1}{25}(-2-11i)=\frac{-2}{25}-\frac{11}{25}i$$

Q29. If $|z_1|=|z_2|$, is it necessary that $z_1=z_2$?

Solution:

Let $z_1=x_1+iy_1,\ z_2=x_2+iy_2$

$|z_1|=|z_2|\Rightarrow|x_1+iy_1|=|x_2+iy_2|$

$\Rightarrow x_1^2+y_1^2=x_2^2+y_2^2$ means the sum of squares of the real and imaginary parts is equal, and that does not mean that $z_1=z_2$. While $z_1=z_2$ if and only if the real parts and imaginary parts are equal.

Q30. If $\frac{(a^2+1)^2}{2a-i}=x+iy$, then what is the value of x^2+y^2?

Solution:

$$\frac{(a^2+1)^2}{(2a-i)}=x+iy$$

$$\Rightarrow\frac{(a^2+1)^2(2a+i)}{(2a-i)(2a+i)}=x+iy$$

$$\Rightarrow\frac{(a^2+1)^2(2a+i)}{4a^2+1}=x+iy$$

$$\Rightarrow x=\frac{2a(a^2+1)^2}{4a^2+1},\ y=\frac{(a^2+1)^2}{4a^2+1}$$

$$\therefore x^2+y^2=4a^2\left[\frac{(a^2+1)^2}{4a^2+1}\right]^2+\left[\frac{(a^2+1)^2}{4a^2+1}\right]^2$$

$$=\frac{(4a^2+1)(a^2+1)^4}{(4a^2+1)^2}=\frac{(a^2+1)^2}{(4a^2+1)}$$

Q31. Find the value of z, if $|z|=4,\ \arg(z)=\frac{5\pi}{6}$.

Solution:

Let $z=r(\cos\theta+i\sin\theta)$

And, $|z|=r=4,\quad \arg(z)=\theta=\frac{5\pi}{6}$

$$\therefore z=4\left[\cos\frac{5\pi}{6}+i\sin\frac{5\pi}{6}\right]$$

$$=4\left[\cos\left(\pi-\frac{\pi}{6}\right)+i\sin\left(\pi-\frac{\pi}{6}\right)\right]$$

$$=4\left[-\cos\frac{\pi}{6}+i\sin\frac{\pi}{6}\right]$$

$$=4\left[-\frac{\sqrt{3}}{2}+i\frac{1}{2}\right]=-2\sqrt{3}+2i$$

Q32. Find the value of $\left|(1+i)\frac{(2+i)}{(3+i)}\right|$.

Solution:

$$\left|(1+i)\frac{(2+i)}{(3+i)}\right|=\left|\frac{(2+i+2i+i^2)}{(3+i)}\right|=\left|\frac{1+3i}{3+i}\right|$$

$$=\left|\frac{(1+3i)(3-i)}{(3+i)(3-i)}\right|$$

$$=\left|\frac{3+9i-i-3i^2}{9-i^2}\right|=\left|\frac{6+8i}{10}\right|$$

$$=\sqrt{\left(\frac{6}{10}\right)^2+\left(\frac{8}{10}\right)^2}=\sqrt{\frac{36+64}{100}}=\sqrt{\frac{100}{100}}=1$$

Q33. Find the principal argument of $(1+i\sqrt{3})^2$.

Solution:
$z=(1+i\sqrt{3})^2$

$\Rightarrow z=1-3+2i\sqrt{3} \Rightarrow z=-2+i2\sqrt{3}$

$\Rightarrow \arg(z)=\tan\theta=\left|\frac{\text{Im}(z)}{\text{Re}(z)}\right|$

$\Rightarrow \tan\theta=\left|\frac{2\sqrt{3}}{-2}\right|=\left|-\sqrt{3}\right|=\sqrt{3}$

$\Rightarrow \tan\theta=\tan\frac{\pi}{3} \Rightarrow \theta=\frac{\pi}{3}$

$\because \text{Re}(z)<0,\ \text{Im}(z)>0$

$\Rightarrow \arg(z)=\pi-\frac{\pi}{3}=\frac{2\pi}{3}$

Q34. Where does z lie, if $\left|\frac{z-5i}{z+5i}\right|=1$?

Solution: Let $z=x+iy$

$\left|\frac{z-5i}{z+5i}\right|=1$

$\Rightarrow \left|\frac{x+iy-5i}{x+iy+5i}\right|=1$

$\Rightarrow \frac{|x+i(y-5)|}{|x+i(y+5)|}=1$

$\Rightarrow \sqrt{x^2+(y-5)^2}=\sqrt{x^2+(y+5)^2}$

$\Rightarrow x^2+y^2+25-10y=x^2+y^2+25+10y$

$\Rightarrow 20y=0 \Rightarrow y=0$

Thus, z lies on the real axis.

Q35. $\sin x+i\cos 2x,\ \cos x-i\sin 2x$ are conjugate to each other for

(a) $x=n\pi$ (b) $x=\left(n+\frac{1}{2}\right)\frac{\pi}{2}$

(c) $x=0$ (d) No value of x

Solution: (d)
Let

$\sin x+i\cos 2x=\overline{\cos x-i\sin 2x}$

$\Rightarrow \sin x+i\cos 2x=\cos x+i\sin 2x$

$\therefore \sin x=\cos x,\quad \cos 2x=\sin 2x$

$\Rightarrow \tan x=1,\quad \tan 2x=1$

$\Rightarrow \tan x=\tan\frac{\pi}{4}, \tan 2x=\tan\frac{\pi}{4}$

$\Rightarrow x=2x=\frac{\pi}{4}$, which is not possible for any value of x.

Q36. The real value of α for which the expression $\frac{1-i\sin\alpha}{1+2i\sin\alpha}$ is purely real is

(a) $(n+1)\frac{\pi}{2}$ (b) $(2n+1)\frac{\pi}{2}$

(c) $n\pi$ (d) None of these

Solution: (c)
$\frac{1-i\sin\alpha}{1+2i\sin\alpha}$ is purely real. Therefore

$\text{Im}\left(\frac{1-i\sin\alpha}{1+2i\sin\alpha}\right)=0$

$\Rightarrow \text{Im}\left(\frac{(1-i\sin\alpha)(1-2i\sin\alpha)}{(1+2i\sin\alpha)(1-2i\sin\alpha)}\right)=0$

$\Rightarrow \text{Im}\left(\frac{1-i\sin\alpha-2i\sin\alpha+2i^2\sin^2\alpha}{1-4i^2\sin^2\alpha}\right)=0$

$\Rightarrow \text{Im}\left(\frac{1-2\sin^2\alpha-3i\sin\alpha}{1+4\sin^2\alpha}\right)=0$

$\Rightarrow \frac{3\sin\alpha}{1+4\sin^2\alpha}=0 \Rightarrow \sin\alpha=0 \Rightarrow \alpha=n\pi$

Q37. If $z=x+iy$ lies in the third quadrant, then $\frac{\bar{z}}{z}$ also lies in the third quadrant, if

(a) $x>y>0$ (b) $x<y<0$

(c) $y<x<0$ (d) $y>x>0$

Solution: (c)
Since $z=x+iy$ lies in the third quadrant

$x<0,\ y<0$

Now, $\frac{\bar{z}}{z}=\frac{x-iy}{x+iy}=\frac{(x-iy)(x-iy)}{(x+iy)(x-iy)}=\frac{x^2-y^2-2ixy}{x^2+y^2}$

Thus, $\frac{\bar{z}}{z}=\frac{x^2-y^2}{x^2+y^2}-\frac{2ixy}{x^2+y^2}$ lies in the third quadrant if

$\therefore \frac{x^2-y^2}{x^2+y^2}<0$ and $\frac{2xy}{x^2+y^2}>0$

$x^2-y^2<0$ and $2xy>0$

$\Rightarrow x^2<y^2$ and $xy>0$

$\Rightarrow y<x<0\quad (\because x<0,\ y<0)$

Q38. The value of $(z+3)(\bar{z}+3)$ is equivalent to

(a) $|z+3|^2$ (b) $|z-3|$

(c) z^2+3 (d) None of these

Solution: (a)
Let $z=x+iy$. Then $(z+3)(\bar{z}+3)$

$=(x+iy+3)(x+3-iy)$

$=(x+3)^2-(iy)^2=(x+3)^2+y^2$

$=|x+3+iy|^2=|z+3|^2$

Q39. If $\left(\frac{1+i}{1-i}\right)^x=1$, then

(a) $x = 2n+1$ (b) $x = 4n$
(c) $x = 2n$ (d) $x = 4n+1$

Solution: (b)

$$\left(\frac{1+i}{1-i}\right)^x = 1 \Rightarrow \left(\frac{(1+i)(1+i)}{(1-i)(1-i)}\right)^x = 1$$

$$\Rightarrow \left(\frac{1+2i+i^2}{1-i^2}\right)^x = 1 \Rightarrow \left(\frac{2i}{2}\right)^x = 1$$

$$\Rightarrow i^x = 1 \Rightarrow i^x = (i^{4n}) \Rightarrow x = 4n \quad (\because i^{4n} = 1, n \in N)$$

Q40. A real value of x satisfies the equation $\left(\frac{3-4ix}{3+4ix}\right) = \alpha - i\beta \ (\alpha, \beta \in R)$, if $\alpha^2 + \beta^2$ is equal to

(a) 1 (b) –1 (c) 2 (d) –2

Solution: (a)

$$\alpha - i\beta = \frac{3-4ix}{3+4ix}$$

$$= \frac{(3-4ix)(3-4ix)}{(3+4ix)(3-4ix)} = \frac{9+16i^2x^2-24ix}{9-16i^2x^2}$$

$$\Rightarrow \alpha - i\beta = \frac{9-16x^2}{9+16x^2} - \frac{24x}{9+16x^2}i$$

$$\Rightarrow \alpha + i\beta = \frac{9-16x^2}{9+16x^2} + \frac{24x}{9+16x^2}i$$

Thus,

$$\alpha^2 + \beta^2 = (\alpha - i\beta)(\alpha + i\beta)$$

$$= \left(\frac{9-16x^2}{9+16x^2}\right)^2 + \left(\frac{24x}{9+16x^2}\right)^2$$

$$= \frac{81+256x^4-288x^2+576x^2}{(9+16x^2)^2}$$

$$= \frac{81+256x^4+288x^2}{(9+16x^2)^2} = \frac{(9+16x^2)^2}{(9+16x^2)^2} = 1$$

Q41. Which of the following is correct for any two complex numbers z_1, z_2?

(a) $|z_1 z_2| = |z_1||z_2|$
(b) $\arg(z_1 z_2) = \arg(z_1) \cdot \arg(z_2)$
(c) $|z_1 + z_2| = |z_1| + |z_2|$
(d) $|z_1 + z_2| = |z_1| - |z_2|$

Solution: (a)

Let $z_1 = r_1(\cos\theta_1 + i\sin\theta_1) \Rightarrow |z_1| = r_1$

$z_2 = r_2(\cos\theta_2 + i\sin\theta_2) \Rightarrow |z_2| = r_2$

$z_1 z_2 = z_1 = r_1 r_2(\cos\theta_1 + i\sin\theta_1)(\cos\theta_2 + i\sin\theta_2)$

$\Rightarrow z_1 z_2 = r_1 r_2[\cos(\theta_1+\theta_2) + i\sin(\theta_1+\theta_2)]$

$\therefore |z_1 z_2| = r_1 r_2$ and $\arg(z_1 z_2) = \arg(z_1) + \arg(z_2)$

Q42. The point represented by the complex number $(2-i)$ is rotated about the origin through an angle $\frac{\pi}{2}$ in the clockwise direction, the new position of the point is

(a) $1+2i$ (b) $-1-2i$ (c) $2+i$ (d) $-1+2i$

Solution: (b)

If a complex number is rotated about the origin through an angle θ in the clockwise direction, then the new position of z is $z' = ze^{i(-\theta)}$.

Thus, the new position of $(2-i)$ when rotated about the origin through an angle $\frac{\pi}{2}$ in the clockwise direction is

$$z' = ze^{i\pi/2} = (2-i)e^{i\pi/2}$$

$$= (2-i)\left[\cos\left(\frac{-\pi}{2}\right) + i\sin\left(\frac{-\pi}{2}\right)\right]$$

$$= (2-i)(0-i) = -1-2i$$

Q43. If $x, y \in R$, then $x+iy$ is a non-real complex number, if

(a) $x = 0$ (b) $y = 0$ (c) $x \neq 0$ (d) $y \neq 0$

Solution: (d)

$x+iy$ is non-real complex number if and only if $y \neq 0$.

Q44. If $a+ib = c+id$, then

(a) $a^2 + c^2 = 0$ (b) $b^2 + c^2 = 0$
(c) $b^2 + d^2 = 0$ (d) $a^2 + b^2 = c^2 + d^2$

Solution: (d)

Two complex numbers $z_1 = a+ib$ and $z_2 = c+id$ are equal if and only if $a = c$ and $b = d$.

Thus, we get $\Rightarrow a^2 = c^2$ and $b^2 = d^2$

$\Rightarrow a^2 + b^2 = c^2 + d^2$

Q45. The complex number z which satisfies the condition $\left|\frac{i+z}{i-z}\right| = 1$ lies on

(a) circle $x^2 + y^2 = 1$ (b) the x-axis
(c) the y-axis (d) the line $x+y=1$

Solution: (b)

Let $z = x+iy$

$$\left|\frac{i+z}{i-z}\right| = 1 \Rightarrow |i+z| = |i-z|$$

$$\Rightarrow |x+i(y+1)| = |-x-i(y-1)|$$

$$\Rightarrow x^2 + (y+1)^2 = x^2 + (y-1)^2$$

$$\Rightarrow 4y = 0 \Rightarrow y = 0$$

Thus, z lies on the x-axis.

Q46. If z is a complex number, then

(a) $|z^2| > |z|$ (b) $|z^2| = |z|^2$
(c) $|z^2| < |z|^2$ (d) $|z^2| \geq |z|^2$

Solution: (b)

Let $z = x+iy$

$|z| = |x+iy| = \sqrt{x^2+y^2}$

$\Rightarrow |z|^2 = x^2+y^2$

$z^2 = (x+iy)^2 = x^2+i^2y^2+2ixy = x^2-y^2+i2xy$

$\Rightarrow |z^2| = \sqrt{(x^2-y^2)^2+(2xy)^2}$

$\Rightarrow |z^2| = \sqrt{x^4+y^4-2x^2y^2+4x^2y^2}$

$\Rightarrow |z^2| = \sqrt{x^4+y^4+2x^2y^2} = \sqrt{(x^2+y^2)^2}$

$\Rightarrow |z^2| = x^2+y^2$

Thus, $|z^2| = |z|^2$

Q47. $|z_1+z_2| = |z_1|+|z_2|$ is possible, if

(a) $z_2 = \bar{z}_1$ (b) $z_2 = \dfrac{1}{z_1}$

(c) $\arg(z_1) = \arg(z_2)$ (d) $|z_1| = |z_2|$

Solution: (c)

$|z_1+z_2| = |z_1|+|z_2|$

$\Rightarrow |r_1\cos\theta_1 + i\sin\theta_1) + r_2(\cos\theta_2 + i\sin\theta_2)|$

$= |r_1(\cos\theta_1 + i\sin\theta_1)| + |r_2(\cos\theta_2 + i\sin\theta_2)|$

$\Rightarrow \sqrt{r_1^2+r_2^2+2r_1r_2[\cos(\theta_1-\theta_2)]} = r_1+r_2$

$\Rightarrow r_1^2+r_2^2+2r_1r_2\cos(\theta_1-\theta_2) = r_1^2+r_2^2+2r_1r_2$

$\Rightarrow 2r_1r_2[1-\cos(\theta_1-\theta_2)] = 0$

$\Rightarrow \cos(\theta_1-\theta_2) = 1 \Rightarrow \cos(\theta_1-\theta_2) = \cos 0°$

$\Rightarrow \theta_1-\theta_2 = 0° \Rightarrow \theta_1 = \theta_2$

$\Rightarrow \arg(z_1) = \arg(z_2)$

Q48. The real value of θ for which the expression $\dfrac{1+i\cos\theta}{1-2i\cos\theta}$ is real number is

(a) $n\pi+\dfrac{\pi}{4}$ (b) $n\pi+(-1)^n\dfrac{\pi}{4}$

(c) $2n\pi\pm\dfrac{\pi}{2}$ (d) None of these

Solution: (c)

$\dfrac{1+i\cos\theta}{1-2i\cos\theta}$ is a real number,

$\text{Im}\left(\dfrac{1+i\cos\theta}{1-2i\cos\theta}\right) = 0$

$\Rightarrow \text{Im}\left(\dfrac{(1+i\cos\theta)(1+2i\cos\theta)}{(1-2i\cos\theta)(1+2i\cos\theta)}\right) = 0$

$\Rightarrow \text{Im}\left(\dfrac{1-2\cos^2\theta+3i\cos\theta}{1+4\cos^2\theta}\right) = 0$

$\Rightarrow \dfrac{3\cos\theta}{1+4\cos^2\theta} = 0 \Rightarrow 3\cos\theta = 0$

$\Rightarrow \cos\theta = \cos\dfrac{\pi}{2} \Rightarrow \theta = 2n\pi\pm\dfrac{\pi}{2}$

Q49. The value of $\arg(x)$ when $x<0$ is

(a) 0 (b) $\dfrac{\pi}{2}$ (c) π (d) None of these

Solution: (c)

$\because x = x+0i$ and $x<0$

Therefore, x lies on the negative side of real axis, and thus, the principal $\arg(x) = \pi$

Q50. If $f(z) = \dfrac{7-z}{1-z^2}$, where $z = 1+2i$, then $|f(z)|$ is equal to

(a) $\dfrac{|z|}{2}$ (b) $|z|$ (c) $2|z|$ (d) None of these

Solution: (a)

$f(z) = \dfrac{7-z}{1-z^2} = \dfrac{7-(1+2i)}{1-(1+2i)^2}$

$= \dfrac{6-2i}{1-1-4i^2-4i} = \dfrac{3-i}{2-2i}$

$= \dfrac{(3-i)(2+2i)}{(2-2i)(2+2i)} = \dfrac{6-2i+6i-2i^2}{4-4i^2}$

$\Rightarrow f(z) = \dfrac{8+4i}{8} = 1+\dfrac{1}{2}i$

$\Rightarrow |f(z)| = \sqrt{1+\dfrac{1}{4}} = \dfrac{\sqrt{5}}{2} = \dfrac{|z|}{2}$ $(\because |z| = \sqrt{1^2+2^2} = \sqrt{5})$

Day 80

Very Short Answer (VSA): 1-Mark and Short Answer (SA): 2-Mark Questions

Q1. The value of $\sqrt{-4}\times\sqrt{-9}$ is______

Solution:

$\sqrt{-4}\times\sqrt{-9} = i\sqrt{4}\times i\sqrt{9} = i^2(2\times 3) = -6$

Q2. The sum of the series $i+i^2+i^3+...$ up to 2000 terms is______

Solution:

$2000 = 4\times 500$ and $i^n+i^{n+1}+i^{n+2}+i^{n+3} = 0$

$\therefore i+i^2+i^3+i^4+...i^{2000} = 0$

Q3. If $|z+2|\le 3$, then the greatest value of $|z+1|$ is______

Solution:

For the greatest value of $|z+1|$

$|z+1| = |z+2-1| \le |z+2| + |-1|$

$\Rightarrow |z+1| \le 3+1 \Rightarrow |z+1| \le 4 \quad (\because |z+4| \le 3)$

Thus, the greatest value of $|z+1|$ is 4.

Q4. If $\left|\dfrac{z-2}{z+2}\right| = \tan\dfrac{\pi}{4}$, then the locus of z is______

Solution:

Let $z = x+iy$. Then $\left|\dfrac{z-2}{z+2}\right| = \tan\dfrac{\pi}{4}$

$\Rightarrow \left|\dfrac{x+iy-2}{x+iy+2}\right| = 1 \Rightarrow \left|\dfrac{x-2+iy}{x+2+iy}\right| = 1$

$\Rightarrow |x-2+iy| = |x+2+iy|$

$\Rightarrow (x-2)^2 + y^2 = (x+2)^2 + y^2$

$\Rightarrow x^2 + 4 - 4x + y^2 = x^2 + 4x + 4 + y^2$

$\Rightarrow 8x = 0 \Rightarrow x = 0$ represents the imaginary axis.

Q5. If $|z| = 2$, $\arg(z) = \dfrac{3\pi}{4}$, then $z =$ ______

Solution:

Let $z = x+iy = r(\cos\theta + i\sin\theta)$

$|z| = r = 2$ and $\arg(z) = \theta = \dfrac{3\pi}{4} \Rightarrow \tan\theta = \dfrac{3\pi}{4}$

$\therefore z = 2\left(\cos\dfrac{3\pi}{4} + i\sin\dfrac{3\pi}{4}\right)$

$= 2[\cos(\pi - \pi/4) + i\sin(\pi - \pi/4)]$

$= 2\left[-\cos\dfrac{\pi}{4} + i\sin\dfrac{\pi}{4}\right] = 2\left[\dfrac{-1}{\sqrt{2}} + i\dfrac{1}{\sqrt{2}}\right] = -\sqrt{2} + i\sqrt{2}$

Q6. The multiplicative inverse of the complex number $3+4i$ is ______

Solution:

$z = 3+4i \Rightarrow \bar{z} = 3-4i$

$|z|^2 = 3^2 + (-4)^2 = 9+16 = 25$

Therefore, the multiplicative inverse of $4-3i$ is

$z^{-1} = \dfrac{\bar{z}}{|z|^2} = \dfrac{3-4i}{25} = \dfrac{3}{25} - \dfrac{4}{25}i$

Q7. The modulus of $\dfrac{1+i}{1-i} + \dfrac{1-i}{1+i}$ is ______

Solution:

$\left|\dfrac{1+i}{1-i} + \dfrac{1-i}{1+i}\right| = \left|\dfrac{(1+i)^2 + (1-i)^2}{(1-i)(1+i)}\right|$

$= \left|\dfrac{1+i^2+2i+1+i^2-2i}{1^2+1^2}\right| = 0$

Q8. The value of $(1-i)^4\left(1-\dfrac{1}{i}\right)^4$ is ______

Solution:

$(1-i)^4\left(1-\dfrac{1}{i}\right)^4 = \left(\dfrac{(1-i)(i-1)}{i}\right)^4$

$= \left(\dfrac{i-1-i^2+i}{i}\right)^4 = \left(\dfrac{2i}{i}\right)^4 = 2^4 = 16 \qquad (\because i^2 = -1)$

Q9. The value of $\sum_{n=1}^{14}(i^n + i^{n+1})$, is ______

Solution:

The summation of four consecutive powers of i is equal to zero, and there are 13 terms in the summation, and therefore, the summation of 12 (4×3) terms will be equal to zero. Therefore, we will be left with only one term (either from the beginning or from the end).

Thus, $\sum_{n=1}^{14}(i^n + i^{n+1}) = i^1 + i^{1+1} + i^2 + i^{2+1} + \sum_{n=3}^{14}(i^n + i^{n+1})$

$= i^1 + i^2 + i^2 + i^3 + 0 \quad (\because i^n + i^{n+1} + i^{n+2} + i^{n+3} = 0)$

$= i - 1 - 1 - i = -2 \quad (\because i^2 = -1)$

Q10. The number $\dfrac{(1+i)^3}{1+i^3}$ is equal to______

Solution:

$\dfrac{(1+i)^3}{1+i^3} = \dfrac{1+i^3+3i^2+3i}{1+i^3}$

$= \dfrac{1-i-3+3i}{1-i} = \dfrac{-2+2i}{1-i} = \dfrac{-2(1-i)}{1-i} = -2$

Q11. If $\left(\dfrac{1+i}{1-i}\right)^{100} = a+ib$, then (a, b) is equal to

(a) (1,0) (b) (1,1) (c) (0,1) (d) (−1,−1)

Solution: (a)

$\left(\dfrac{1+i}{1-i}\right)^{100} \Rightarrow \left[\dfrac{(1+i)}{(1-i)} \times \dfrac{(1+i)}{(1+i)}\right]^{100}$

$= \left(\dfrac{1+i^2+2i}{1-i^2}\right)^{100} = \left(\dfrac{2i}{2}\right)^{100} \qquad (\because i^2 = -1)$

$= (i)^{100} = (i^4)^{25} = 1 \quad (\because i^4 = 1)$

$= 1 + 0i = a+ib \Rightarrow a = 1,\ b = 0$

Q12. $\sin x + i\cos 2x$, $\cos x + i\sin 2x$ are conjugate to each other for

(a) $x = n\pi$ (b) $x = \left(n+\dfrac{1}{2}\right)\dfrac{\pi}{2}$

(c) $x = 0$ (d) No value of x

Solution: (d)

$\because \sin x + i\cos 2x = \overline{\cos x + i\sin 2x}$

$\Rightarrow \sin x + i\cos 2x = \cos x - i\sin 2x$

$\therefore \sin x = \cos x, \quad \cos 2x = -\sin 2x$

$\Rightarrow \tan x = 1, \quad \tan 2x = -1$

$\Rightarrow \tan x = \tan\frac{\pi}{4}, \tan 2x = \tan(-\frac{\pi}{4})$

$\Rightarrow x = \frac{\pi}{4}$ and $2x = -\frac{\pi}{4}$, is not possible for any value of x.

Q13. The value of $(z+2)(\bar{z}+2)$ is equivalent to

(a) $|z+2|^2$ (b) $|z-2|^2$ (c) z^2+2 (d) z^2-4

Solution: (a)

Let $z = x+iy$ then $(z+2)(\bar{z}+2)$

$= (x+iy+2)(x+2-iy)$

$= (x+2)^2 - (iy)^2 = (x+2)^2 + y^2$

$= |x+2+iy|^2 = |z+2|^2$

Q14. If $\left(\frac{1+i}{1-i}\right)^x = i$, then

(a) $x = 2n+1$ (b) $x = 4n$

(c) $x = 2n$ (d) $x = 4n+1$

Solution: (d)

$\left(\frac{1+i}{1-i}\right)^x = i \Rightarrow \left(\frac{(1+i)(1+i)}{(1-i)(1-i)}\right)^x = i$

$\Rightarrow \left(\frac{1+2i+i^2}{1-i^2}\right)^x = i \Rightarrow \left(\frac{2i}{2}\right)^x = i$

$\Rightarrow i^x = i \Rightarrow i^x = (i^{4n+1})$

$\Rightarrow x = 4n+1 \quad (\because i^{4n} = 1, n \in N)$

Q15. Which of the following is correct for any two complex numbers z_1, z_2?

(a) $|z_1 z_2| = |z_1||z_2|$

(b) $\arg(z_1 z_2) = \arg(z_1) + \arg(z_2)$

(c) $|z_1 + z_2| = |z_1| + |z_2|$

(d) $\arg(z_1 z_2) = \arg(z_1) - \arg(z_2)$

Solution: (a, b)

Let $z_1 = r_1(\cos\theta_1 + i\sin\theta_1) \Rightarrow |z_1| = r_1$

$z_2 = r_2(\cos\theta_2 + i\sin\theta_2) \Rightarrow |z_2| = r_2$

$z_1 z_2 = z_1 = r_1 r_2(\cos\theta_1 + i\sin\theta_1)(\cos\theta_2 + i\sin\theta_2)$

$\Rightarrow z_1 z_2 = r_1 r_2[\cos(\theta_1+\theta_2) + i\sin(\theta_1+\theta_2)]$

$\therefore |z_1 z_2| = r_1 r_2$ and $\arg(z_1 z_2) = \arg(z_1) + \arg(z_2)$

Q16. The point represented by the complex number $(1-i)$ is rotated about origin through an angle $\frac{\pi}{2}$ in the anti-clockwise direction, the new position of points is

(a) $1-i$ (b) $-1-i$

(c) $1+i$ (d) $-1+i$

Solution: (c)

If a complex number is rotated about the origin through an angle θ in the anti-clockwise direction, then the new position of z is $z' = ze^{i\theta}$.

Thus, the new position of $(1-i)$ when rotated about the origin through an angle $\frac{\pi}{2}$ in the anti-clockwise direction is $z' = ze^{i\pi/2} = (1-i)e^{i\pi/2}$.

$= (1-i)\left[\cos\left(\frac{\pi}{2}\right) + i\sin\left(\frac{\pi}{2}\right)\right]$

$= (1-i)(0+i) = i - i^2 = 1+i$

Q17. $|z_1 - z_2| = |z_1| - |z_2|$ is possible, if

(a) $z_2 = \bar{z}_1$ (b) $z_2 = \frac{1}{z_1}$

(c) $\arg(z_1) = \arg(z_2)$ (d) $|z_1| = |z_2|$

Solution: (c)

$|z_1 - z_2| = |z_1| - |z_2|$

$\Rightarrow |r_1\cos\theta_1 + i\sin\theta_1) - r_2(\cos\theta_2 + i\sin\theta_2)|$

$= |r_1(\cos\theta_1 + i\sin\theta_1)| - |r_2(\cos\theta_2 + i\sin\theta_2)|$

$\Rightarrow \sqrt{r_1^2 + r_2^2 - 2r_1r_2[\cos(\theta_1-\theta_2)]} = r_1 - r_2$

$\Rightarrow r_1^2 + r_2^2 - 2r_1r_2\cos(\theta_1-\theta_2) = r_1^2 + r_2^2 - 2r_1r_2$

$\Rightarrow 2r_1r_2[1-\cos(\theta_1-\theta_2)] = 0$

$\Rightarrow \cos(\theta_1-\theta_2) = 1 \Rightarrow \cos(\theta_1-\theta_2) = \cos 0°$

$\Rightarrow \theta_1 - \theta_2 = 0° \Rightarrow \theta_1 = \theta_2 \Rightarrow \arg(z_1) = \arg(z_2)$

Q18. If $f(z) = \frac{1-z}{1-z^2}$, where $z = 1+i$, then $|f(z)|$ is equal to

(a) $\frac{|z|}{4}$ (b) $|z|$

(c) $2|z|$ (d) None of these

Solution: (a)

$f(z) = \frac{1-z}{1-z^2} = \frac{1-(1+2i)}{1-(1+2i)^2}$

$= \frac{-2i}{1-1-4i^2-4i} = \frac{-2i}{4-4i}$

$= \frac{2}{4i+4} = \frac{1}{2+2i}$

$\Rightarrow f(z) = \frac{1}{2+2i} \times \frac{2+2i}{2-2i} = \frac{2+2i}{8} = \frac{1+i}{4}$

$\Rightarrow |f(z)| = \frac{\sqrt{2}}{4} = \frac{|z|}{4} \quad (\because |z| = \sqrt{1^2+1^2} = \sqrt{2})$

Q19. Express the given complex number in the form $a+ib: i^{11}+i^{14}$.

Solution:

$i^{11}+i^{14}=i^{4\times2+3}+i^{4\times3+2}=(i^4)^2\cdot i^3+(i^4)^3\cdot i^2$

$=i^3+i^2=-i-1=0 \qquad (\because i^4=1,\ i^3=-i,\ i^2=-1)$

$\Rightarrow a+ib=-1-i$

Q20. Find the region represented by $|z+i|\geq 2$.

Solution:

$|z+i|\geq 2=|x+iy+i|\geq 2$

$\Rightarrow |x+i(y+1)|\geq 2 \Rightarrow \sqrt{x^2+(y+1)^2}\geq 2$

$\Rightarrow x^2+(y+1)^2\geq 2 \Rightarrow x^2+y^2+2y-1\geq 0$

This represents a circle, on or outside having centre (0, –1) and radius $\sqrt{2}$.

Q21. Express the given complex number in the form $a+ib: i^{-33}$.

Solution:

$i^{-33}=\frac{1}{i^{33}}=\frac{1}{i^{4\times8+1}}=\frac{1}{(i^4)^8 i}$

$=\frac{1}{i}=\frac{i}{i^2}=-i \qquad (\because i^4=1,\ i^2=-1)$

$\Rightarrow a+ib=0-i$

Q22. Express the given complex number in the form $a+ib: (1-i)^{10}$.

Solution:

$(1-i)^{10}=[(1-i)^2]^5=[1^2+i^2-2i]^5$

$=[1-1-2i]^5=-32i^5=-32i^4\cdot i=-32i \quad (\because i^2=-1,\ i^4=1)$

$\Rightarrow a+ib=0-32i$

Q23. Solve the equation $x^2+4=0$.

Solution:

The discriminant

$D=b^2-4ac=0^2-4\times1\times4=-16<0$

Therefore, the solution of the given equation is

$x=\frac{-b\pm\sqrt{D}}{2a}=\frac{\pm\sqrt{-16}}{2\times1}=\frac{\pm4i}{2}=\pm2i \quad (\because \sqrt{-1}=i)$

Q24. Reduce $\frac{1+2i}{(2-i)^2}$ to the standard form.

Solution:

$\frac{1+2i}{(2-i)^2}=\frac{1+2i}{4+i^2-4i}=\frac{1+2i}{3-4i} \qquad (\because i^2=-1)$

$=\frac{1+2i}{3-4i}\times\frac{3+4i}{3+4i}=\frac{3+6i+4i+8i^2}{3^2+4^2}$

$=\frac{3+10i-8}{25}=\frac{-5+10i}{25}=\frac{-1}{5}+\frac{2}{5}i$

Q25. Find the real numbers x and y if $(x-iy)(3+5i)$ is the conjugate of $8-2i$.

Solution:

Let $z=(x-iy)(3+5i)$

$z=3x+5xi-3yi-5yi^2=3x+5xi-3yi+5y$

$=(3x-5y)+i(5x-3y)$

$\Rightarrow \bar{z}=(3x+5y)-i(5x-3y)$

The conjugate of $(x-iy)(3+5i)$ is $8-2i$. Therefore $(3x+5y)-i(5x-3y)=8-2i$

Equating the real and imaginary parts, we get

$3x+5y=8$ and $5x-3y=2$

Solving the above two equations, we get

$x=1$ and $y=1$

Q26. If $(1-i)z=(1+i)\bar{z}$, then show that $z=i\bar{z}$.

Solution:

$(1-i)z=(1+i)\bar{z}$

$\Rightarrow \frac{z}{\bar{z}}=\frac{(1+i)}{(1-i)}=\frac{(1+i)(1+i)}{(1-i)(1+i)}=\frac{1+i^2+2i}{1-i^2}$

$\Rightarrow \frac{z}{\bar{z}}=\frac{1-1+2i}{2} \qquad (\because i^2=-1)$

$\Rightarrow \frac{z}{\bar{z}}=i \Rightarrow z=i\bar{z}$

Q27. If $z=x+iy$, then show that $z\bar{z}+(z+\bar{z})+1=0$, where $b\in R$, represents a circle.

Solution:

Let $z=x+iy$

$z\bar{z}+(z+\bar{z})+1=0$

$\Rightarrow (x+iy)(x-iy)+\{(x+iy)+(x-iy)\}+1=0$

$\Rightarrow x^2+y^2+2x+1=0$ represents the equation of a circle.

Q28. If $\arg(z-1)=\arg(z+i)$, then find the value of $x-y$, where $z=x+iy$.

Solution:

$\arg(z-1)=\arg(z+i)$

$\Rightarrow \arg(x+iy-1)=\arg(x+iy+i)$

$\Rightarrow \arg(x-1+iy)=\arg[x+i(y+1)]$

$\Rightarrow \tan^{-1}\frac{y}{x-1}=\tan^{-1}\frac{y+1}{x}$

$\Rightarrow \frac{y}{x-1}=\frac{y+1}{x} \Rightarrow xy=xy+x-y-1$

$\Rightarrow x-y=1$

Q29. Show that the inequality $|z+1|<|z-2|$ represents the region given by $x<1/2$.

Solution:

$|z+1|<|z-2| \Rightarrow |x+1+iy|<|x-2+iy|$

$\Rightarrow \sqrt{(x+1)^2+y^2} < \sqrt{(x-2)^2+y^2}$
$\Rightarrow (x+1)^2+y^2 < (x-2)^2+y^2$
$\Rightarrow x^2+2x+1+y^2 < x^2-4x+4+y^2$
$\Rightarrow 6x < 3 \Rightarrow x < 1/2$

Q30. What is the conjugate of $\dfrac{4-3i}{(1-2i)^2}$?

Solution:

$z = \dfrac{4-3i}{(1-2i)^2} = \dfrac{4-3i}{1+4i^2-4i}$

$= \dfrac{4-3i}{1-4-4i} = \dfrac{4-3i}{-3-4i} = \dfrac{3i-4}{3+4i}$

$= \dfrac{(3i-4)(3-4i)}{(3+4i)(3-4i)}$

$= \dfrac{9i-12-12i^2+16i}{9+16} = \dfrac{0+25i}{25}$

$\Rightarrow z = 0+i \Rightarrow \bar{z} = 0-i$

Q31. Find the value of $\left|\dfrac{(1+i)(2+i)}{(3-i)}\right|$.

Solution:

$\left|\dfrac{(1+i)(2+i)}{(3-i)}\right|$

$= \left|\dfrac{(2+i+2i+i^2)}{(3-i)}\right| = \left|\dfrac{1+3i}{3-i}\right|$

$= \left|\dfrac{(1+3i)(3+i)}{(3+i)(3+i)}\right|$

$= \left|\dfrac{3+9i+i+3i^2}{9-i^2}\right| = \left|\dfrac{0+10i}{10}\right| = |i| = 1$

Day 81

Long Answer-I (LA-I): 4-Mark and Long Answer-II (LA-II): 6-Mark Questions

1. Find the modulus and the argument of the complex number $z = 1-i$.

Solution:

Since $r\cos\theta = 1,\quad r\sin\theta = -1$
On squaring and adding, we obtain
$r^2\cos^2\theta + r^2\sin^2\theta = 1^2 + (-1)^2$
$\Rightarrow r^2(\cos^2\theta + \sin^2\theta) = 1+1 \quad (\because \sin^2\theta + \cos^2\theta = 1)$
$\Rightarrow r^2 = 2 \Rightarrow r = \sqrt{2} \quad (\because r > 0)$
Modulus $|z| = \sqrt{2}$
$\because \sqrt{2}\cos\theta = 1,\quad \sqrt{2}\sin\theta = -1$
$\therefore \cos\theta = \dfrac{1}{\sqrt{2}},\quad \sin\theta = -\dfrac{1}{\sqrt{2}}$

$|\tan\theta| = 1 \Rightarrow \theta = \dfrac{\pi}{4}$

Since $\sin\theta$ is negative and $\cos\theta$ is positive, θ lies in the fourth quadrant.

Thus, $\text{Arg}(z) = -(\theta) = -\dfrac{\pi}{4}$

Hence, the modulus and argument of the complex number $1-i$ are $\sqrt{2}$ and $-\dfrac{\pi}{4}$, respectively.

Q2. Find square-root of $3-4i$.

Solution:

Let $x+iy = \sqrt{a+ib}$, where $a = 3$ and $b = -4$.

Then, $\text{Re}(z) = a = 3$ and

$|z| = \sqrt{a^2+b^2} = \sqrt{(3)^2+(-4)^2} = 5$

Since the product xy (or b) is negative, we take the value of x as positive and y as negative, or vice versa.

Thus, the square-root of $3-4i$ is given by $\sqrt{3-4i}$

$= \pm\sqrt{\dfrac{|z|+\text{Re}(z)}{2}} \mp i\sqrt{\dfrac{|z|-\text{Re}(z)}{2}}$

$= \pm\sqrt{\dfrac{5+3}{2}} \mp i\sqrt{\dfrac{5-3}{2}}$

$= \pm\sqrt{4} \mp i\sqrt{1} = \pm 2 \mp i$

$= (2-i, -2+i)$

Q3. If the imaginary part of $\dfrac{\bar{z}+1}{\bar{z}-1}$ is 2, then show that the locus of the point representing z in the complex plane is a circle.

Solution:

Let $z = x+iy$

$\text{Im}\left(\frac{\bar{z}+1}{\bar{z}-1}\right)=2$

$\Rightarrow \text{Im}\left(\frac{x-iy+1}{x+iy-1}\right)=2$

$\Rightarrow \text{Im}\left(\frac{[(x+1)-iy][(x-1)+iy]}{[(x-1)-iy][(x-1)+iy]}\right)=2$

$\Rightarrow \text{Im}\left(\frac{(x+1)(x-1)+y^2+i[y(x+1)-y(x-1)]}{(x-1)^2+y^2}\right)=2$

$\Rightarrow \text{Im}\left(\frac{x^2-1+y^2+i(2y)}{(x-1)^2+y^2}\right)=2$

$\Rightarrow \frac{2y}{(x-1)^2+y^2}=2$

$\Rightarrow y=x^2+1-2x+y^2$

$\Rightarrow x^2+y^2-2x-y+1=0$

$\Rightarrow x^2+y^2-2x-y+1=0$ represents the equation of a circle.

Q4. Evaluate: $\left[i^{25}-\left(\frac{1}{i}\right)^{25}\right]^2$

Solution:

$\left[i^{25}-\left(\frac{1}{i}\right)^{25}\right]^2=\left[i^{4\times4+2}-\frac{1}{i^{4\times6+1}}\right]^2$

$=\left[(i^4)^6\cdot i-\frac{1}{(i^4)^6\cdot i}\right]^2$

$=\left[i-\frac{1}{i}\right]^2 \quad (\because i^4=1)$

$=\left[i-\frac{1}{i}\times\frac{i}{i}\right]^3 \quad (\because i^2=-1)$

$=[i+i]^2=[2i]^2=-4 \quad (\because i^2=-1)$

Q5. Reduce $\left(\frac{1}{1-4i}-\frac{2}{1+i}\right)$ to the standard form.

Solution:

$\left(\frac{1}{1-4i}-\frac{2}{1+i}\right)$

$=\left[\frac{(1+i)-2(1-4i)}{(1-4i)(1+i)}\right]$

$=\left[\frac{1+i-2+8i}{1+i-4i-4i^2}\right]$

$=\left[\frac{-1+9i}{5-3i}\right] \quad (\because i^2=-1)$

$=\frac{-1+9i}{5-3i}\times\frac{5+3i}{5+3i}$

$=\frac{-5+27i^2-3i+45i}{\left[(5)^2-(3i)^2\right]}$

$=\frac{-32+42i}{34} \quad (\because i^2=-1)$

$=\frac{-16}{17}+\frac{21}{17}i$

Q6. If $z_1=1-i,\ z_2=1+i$, find $\left|\frac{z_1+z_2+1}{z_1-z_2+i}\right|$.

Solution:

$\left|\frac{z_1+z_2+1}{z_1-z_2+i}\right|=\left|\frac{(1-i)+(1+i)+1}{(1-i)-(1+i)+1}\right|$

$=\left|\frac{3}{1-2i}\right|$

$=\left|\frac{3}{1-2i}\times\frac{1+2i}{1+2i}\right|=\left|\frac{3(1+2i)}{1^2-(2i)^2}\right|$

$=\left|\frac{3(1+2i)}{1+4}\right| \quad (\because i^2=-1)$

$=\frac{3}{5}|1+2i|=\frac{3}{5}\sqrt{1^2+2^2}=\frac{3}{\sqrt{5}}$

Q7. Let $z_1=2-i,\ z_2=-2+i$. Find $\text{Im}\left(\frac{z_1z_2}{\bar{z}_1}\right)$.

Solution:

$z_1=2-i,\ z_2=-2+i$

$\because z_1z_2=(2-i)(-2+i)=-4+4i-i^2$

$=-3+4i \quad (\because i^2=-1)$

and $z_1=2-i\Rightarrow \bar{z}_1=2+i$

$\therefore \frac{z_1z_2}{\bar{z}_1}=\frac{-3+4i}{2+i}=\frac{(-3+4i)(2-i)}{(2+i)(2-i)}$

$=\frac{-6+3i+8i-4i^2}{2^2+1^2}=\frac{-2+11i}{5}$

$\Rightarrow \frac{z_1z_2}{\bar{z}_1}=\frac{-2}{5}+\frac{11}{5}i$

$\Rightarrow \text{Im}\left(\frac{z_1z_2}{\bar{z}_1}\right)=\frac{11}{5}$

Q8. Find the modulus and argument of the complex number $\frac{1+i}{1-i}$.

Solution:

$$z=\frac{1+i}{1-i}=\frac{1+i}{1-i}\times\frac{1+i}{1+i}$$

$$=\frac{1+i+i+i^2}{1^2+1^2}=\frac{0+2i}{2}=0+i$$

Let $r\cos\theta=0,\quad r\sin\theta=1$

On squaring and adding, we get

$$r^2(\cos^2\theta+\sin^2\theta)=(0)^2+(1)^2$$

$$\Rightarrow r^2=1 \qquad (\because \sin^2\theta+\cos^2\theta=1)$$

$$\Rightarrow r=1 \qquad (\because r>0)$$

Since $|\tan\theta|=\infty\Rightarrow\theta=\frac{\pi}{2}$

Since $\sin\theta$ is positive and $\cos\theta$ is zero, θ lies in the first quadrant.

Thus, $\text{Arg}(z)=(\theta)=\frac{\pi}{2}$

Therefore, the modulus and argument of the given complex number are $1,\frac{\pi}{2}$, respectively.

Q9. If $\left(\frac{1+i}{1-i}\right)^3+\left(\frac{1-i}{1+i}\right)^3=x+iy$, then find (x, y).

Solution:

Given that $\left(\frac{1+i}{1-i}\right)^3+\left(\frac{1-i}{1+i}\right)^3=x+iy$ (i)

$$\left(\frac{1+i}{1-i}\right)^3+\left(\frac{1-i}{1+i}\right)^3=\left(\frac{1+i}{1-i}\times\frac{1+i}{1+i}\right)^3+\left(\frac{1-i}{1+i}\times\frac{1-i}{1-i}\right)^3$$

$$=\left(\frac{1+i^2+2i}{1-i^2}\right)^3+\left(\frac{1+i^2-2i}{1-i^2}\right)^3$$

$$=\left(\frac{1+(-1)+2i}{1-(-1)}\right)^3+\left(\frac{1+(-1)-2i}{1-(-1)}\right)^3 \qquad (\because i^2=-1)$$

$$=\left(\frac{2i}{2}\right)^3+\left(\frac{-2i}{2}\right)^3$$

$$=i^3+(-i)^3=i^3-i^3=0 \qquad (\because i^3=-i)$$

$$=0+0i=x+iy\Rightarrow x=0,\quad y=0$$

Thus, $(x,y)=(0,0)$

Q10. Solve the equation $|z|=\bar{z}+1+i$.

Solution:

Let $z=x+iy$. Then $\bar{z}=x-iy$

Thus,

$$|z|=\bar{z}+1+2i\Rightarrow|x+iy|=x-iy+1+i$$

$$\Rightarrow\sqrt{x^2+y^2}=x+1-i(y-1)$$

$$\Rightarrow\sqrt{x^2+y^2}-(x+1)+i(y-1)=0+0i$$

On comparing the real and imaginary parts, we get

$$y-1=0\Rightarrow y=1$$

Also, $\sqrt{x^2+y^2}-(x+1)=0$

$$\Rightarrow x^2+1+2x=x^2+y^2$$

$$\Rightarrow x=\frac{y^2-1}{2}=\frac{(1)^2-1}{2}=0$$

Thus, $z=x+iy=0+i$

Q11. If z_1, z_2 are two complex numbers such that $|z_1|=|z_2|$ and $\arg(z_2)-\arg(z_1)=\frac{\pi}{2}$, then show that $z_2=iz_1$.

Solution:

Let $z_1=r_1(\cos\theta_1+i\sin\theta_1)$, $z_2=r_2(\cos\theta_2+i\sin\theta_2)$ be two complex numbers.

$|z_1|=|z_2|$ and $\arg(z_2)-\arg(z_1)=\frac{\pi}{2}$

Therefore, $r_1=r_2$ and $\theta_2-\theta_1=\frac{\pi}{2}$

Now,

$$\text{RHS}=z_2=r_2(\cos\theta_2+i\sin\theta_2)$$

$$=r_1\left[\cos\left(\frac{\pi}{2}+\theta_1\right)+i\sin\left(\frac{\pi}{2}+\theta_1\right)\right] \qquad \left(\because r_1=r_2,\ \theta_2=\frac{\pi}{2}+\theta_1\right)$$

$$=r_1(-\sin\theta_1+i\cos\theta_1)$$

$$=r_1(i^2\sin\theta_1+i\cos\theta_1) \qquad (\because i^2=-1)$$

$$=ir_1(i\sin\theta_1+\cos\theta_1)$$

$$=ir_1(\cos\theta_1+i\sin\theta_1)=iz_1=\text{LHS}$$

Q12. Write the complex number $z=\dfrac{1-i}{\cos\frac{\pi}{4}+i\sin\frac{\pi}{4}}$ in the polar form.

Solution:

$$z=\frac{1-i}{\cos\frac{\pi}{4}+i\sin\frac{\pi}{4}}=\frac{\sqrt{2}\left[\frac{1}{\sqrt{2}}-i\frac{1}{\sqrt{2}}\right]}{\cos\frac{\pi}{4}+i\sin\frac{\pi}{4}}$$

$$=\frac{\sqrt{2}[\cos(-\pi/4)+i\sin(-\pi/4)]}{\cos\pi/4+i\sin\pi/4}$$

$$=\sqrt{2}\left[\cos\left(\frac{-\pi}{4}-\frac{\pi}{4}\right)+i\sin\left(\frac{-\pi}{4}-\frac{\pi}{4}\right)\right]$$

$$=\sqrt{2}\left[\cos\left(\frac{-2\pi}{4}\right)+i\sin\left(\frac{-2\pi}{4}\right)\right]$$

$$=\sqrt{2}\left[\cos\left(\frac{\pi}{2}\right)-i\sin\left(\frac{\pi}{2}\right)\right]$$

$$=\sqrt{2}\left[-\cos\left(\pi-\frac{\pi}{2}\right)-i\sin\left(\pi-\frac{\pi}{2}\right)\right]$$

$$=-\sqrt{2}\left[\cos\frac{\pi}{2}+i\sin\frac{\pi}{2}\right]$$

Q13. If z and w are two complex numbers such that $|zw| = 1$, $\arg(z) - \arg(w) = \pi$, then show that $\bar{z}w = -1$.

Solution:

Let $z = r_1(\cos\theta_1 + i\sin\theta_1)$, $w = r_2(\cos\theta_2 + i\sin\theta_2)$

Since $|zw| = 1 \Rightarrow |z||w| = r_1 r_2 = 1$

Also, $\arg(z) - \arg(w) = \pi \Rightarrow \theta_1 - \theta_2 = \pi$

Now,

LHS $= \bar{z}w$

$= r_1(\cos\theta_1 - i\sin\theta_1)\cdot r_2(\cos\theta_2 + i\sin\theta_2)$

$= r_1 r_2[\cos(\theta_2 - \theta_1) + i\sin(\theta_2 - \theta_1)]$

$= r_1 r_2[\cos(-\pi) + i\sin(-\pi)]$

$= 1(-1+0) = -1 =$ RHS

Q14. Find the complex number satisfying the equation $|(z+1)| + z + i = 0$.

Solution:

Let $z = x + iy$. Then $|(z+1)| + z + i = 0$

$\Rightarrow |x + iy + 1| + x + iy + i = 0$

$\Rightarrow \sqrt{(x+1)^2 + y^2} + x + i(1+y) = 0$

$\Rightarrow \sqrt{(x+1)^2 + y^2} + x + i(1+y) = 0 + i0$

Equating the real and imaginary parts, we get

$\sqrt{(x+1)^2 + y^2} + x = 0$ and $1 + y = 0$

Thus, $1 + y = 0 \Rightarrow y = -1$

Also,

$\sqrt{(x+1)^2 + y^2} + x = 0$

$\Rightarrow \sqrt{(x+1)^2 + (-1)^2} + x = 0 \quad (\because y = -1)$

$\Rightarrow \sqrt{(x+1)^2 + 1} = -x$

$\Rightarrow x^2 + 2x + 2 = x^2$

$\Rightarrow x = -1$

$\therefore z = x + iy = -1 - i$

Q15. If $|z_1| = |z_2| = 1$, then show that

$$|z_1 + z_2| = \left|\frac{1}{z_1} + \frac{1}{z_2}\right|.$$

Solution:

LHS $= |z_1 + z_2|$

$= |\overline{z_1 + z_2}| = |\bar{z}_1 + \bar{z}_2| \quad (\because |z| = |\bar{z}|)$

$= \left|\frac{z_1\bar{z}_1}{z_1} + \frac{z_2\bar{z}_2}{z_2}\right| = \left|\frac{|z_1|}{z_1} + \frac{|z_2|}{z_2}\right|$

$= \left|\frac{1}{z_1} + \frac{1}{z_2}\right| =$ RHS

$(\because |z_1| = |z_2| = 1 \Rightarrow z_1\bar{z}_1 = z_2\bar{z}_2 = 1)$

Q16. If the complex numbers z_1, z_2 such that $\arg(z_1) - \arg(z_2) = 0$, then show that $|z_1 + z_2| = |z_1| + |z_2|$.

Solution:

Let $z_1 = r_1(\cos\theta + i\sin\theta), z_2 = r_2(\cos\theta + i\sin\theta)$

$\because \arg(z_1) - \arg(z_2) = 0 \Rightarrow \theta_1 = \theta_2 = \theta$

LHS $= |z_1 + z_2|$

$= |r_1(\cos\theta + i\sin\theta) + r_2(\cos\theta + i\sin\theta)|$

$= |(r_1 + r_2)\cos\theta + i(r_1 + r_2)\sin\theta|$

$= (r_1 + r_2)|\cos\theta + i\sin\theta|$

$= (r_1 + r_2)\sqrt{\cos^2\theta + \sin^2\theta}$

$= (r_1 + r_2) = |z_1| + |z_2| \quad (\because r = |z|)$

Q17. Solve the system of equations $\text{Re}(z^2) = 0$, $|z| = 2\sqrt{2}$.

Solution:

Let $z = x + iy$

$\text{Re}(z^2) = 0$

$\Rightarrow \text{Re}(x^2 + 2ixy - y^2) = 0$

$\Rightarrow \text{Re}\{(x^2 - y^2) + 2ixy\} = 0$

$\Rightarrow x^2 - y^2 = 0 \Rightarrow x^2 = y^2$

Also, $|z| = \sqrt{x^2 + y^2} = 2\sqrt{2}$

$\Rightarrow x^2 + y^2 = 8$

Solving the above two equations, we get

$x^2 + y^2 = 8 \Rightarrow 2x^2 = 8 \Rightarrow x^2 = 4$

$\Rightarrow x = \pm 2 \Rightarrow y = \pm 2 \quad (\because x^2 = y^2)$

$\Rightarrow z = x + iy = \pm 2 \pm 2i$

Q18. If $|z_1| = 1$ $(z_1 \neq -1)$, $z_2 = \frac{z_1 + 1}{z_1 - 1}$, then show that the real part of z_2 is zero.

Solution: Let $z_1 = x + iy$

$\Rightarrow |z_1| = \sqrt{x^2 + y^2} = 1 \quad (\because |z_1| = 1)$

$\Rightarrow x^2 + y^2 = 1$

Now, $z_2 = \frac{z_1 + 1}{z_1 - 1} = \frac{x + iy + 1}{x + iy - 1}$

$= \frac{x + 1 + iy}{x - 1 + iy} = \frac{(x + 1 + iy)(x - 1 - iy)}{(x - 1 + iy)(x - 1 - iy)}$

$= \frac{x^2 - 1 + y^2 + 2iy}{(x-1)^2 + y^2}$

$= \frac{1 - 1 + 2iy}{(x+1)^2 + y^2} \quad (\because x^2 + y^2 = 1)$

$= 0 + i\left(\frac{2y}{(x+1)^2 + y^2}\right)$

Thus, the real part of z_2 is zero.

Q19. Convert the given complex number into the polar form: $-i$.

Solution:

The complex number $z = i = 0 - i$

Therefore, $r\cos\theta = 0, \quad r\sin\theta = -1$

On squaring and adding, we get

$r^2\cos^2\theta + r^2\sin^2\theta = (0)^2 + (-1)^2$

$\Rightarrow r^2(\cos^2\theta + \sin^2\theta) = 1 \quad (\because \sin^2\theta + \cos^2\theta = 1)$

$\Rightarrow r^2 = 1 \Rightarrow r = 1 \Rightarrow |z| = 1 \qquad (\because r > 0)$

$\because \cos\theta = 0, \quad \sin\theta = -1$

$\therefore |\tan\theta| = \infty \Rightarrow \theta = \dfrac{\pi}{2}$

Since $\sin\theta$ is negative and $\cos\theta$ is positive (or zero), θ lies in the fourth quadrant.

Thus, $\text{Arg}(z) = -\theta = -\dfrac{\pi}{2}$

Hence, the polar form of $-i$ is

$r(\cos\theta + i\sin\theta) = \cos\left(-\dfrac{\pi}{2}\right) + i\sin\left(-\dfrac{\pi}{2}\right)$

Q20. Show that the locus represented by $|z+1| = |z+i|$ is a line perpendicular to the line obtained by joining the points (1, 0) and (0, 1).

Solution:

Let $z = x + iy$. Then

$|z+1| = |z+i|$

$\Rightarrow |x + iy + 1| = |x + iy + i|$

$\Rightarrow |x + 1 + iy| = |x + i(y+1)|$

$\Rightarrow (x+1)^2 + y^2 = x^2 + (y+1)^2$

$\Rightarrow x^2 + 2x + 1 + y^2 = x^2 + 1 + y^2 + 2y$

$\Rightarrow y = x$

And the equation of a line through the points, (1, 0) and (0, 1) is

$y - 0 = \dfrac{1-0}{0-1}(x-1) \Rightarrow y = -x + 1$

Thus, the two lines are perpendicular to each other.

Q21. If $\dfrac{z+1}{z-1}$ is a purely imaginary number $(z \neq -1)$, then find the value of $|z|$.

Solution:

Since $\dfrac{z+1}{z-1}$ is a purely imaginary number, $\text{Re}\left(\dfrac{z+1}{z-1}\right) = 0$

Let $z = x + iy$. Then

$\text{Re}\left(\dfrac{z+1}{z-1}\right) = 0$

$\Rightarrow \text{Re}\left(\dfrac{x+iy+1}{x+iy-1}\right) = 0 \Rightarrow \text{Re}\left(\dfrac{x+1+iy}{x-1+iy}\right) = 0$

$\Rightarrow \text{Re}\left(\dfrac{(x+1+iy)(x-1-iy)}{(x-1+iy)(x-1-iy)}\right) = 0$

$\Rightarrow \text{Re}\left(\dfrac{(x^2-1) + y^2 - 2iy}{(x-1)^2 + y^2}\right) = 0$

$\Rightarrow (x^2 - 1) + y^2 = 0 \Rightarrow x^2 + y^2 = 1$

$\Rightarrow |z| = 1$

Q22. Find the real value of α for which the expression $\dfrac{1 + i\sin\alpha}{1 + 2i\sin\alpha}$ is purely real.

Solution:

$\dfrac{1 + i\sin\alpha}{1 + 2i\sin\alpha}$ is purely real. Therefore

$\text{Im}\left(\dfrac{1 + i\sin\alpha}{1 + 2i\sin\alpha}\right) = 0$

$\Rightarrow \text{Im}\left(\dfrac{(1 + i\sin\alpha)(1 - 2i\sin\alpha)}{(1 + 2i\sin\alpha)(1 - 2i\sin\alpha)}\right) = 0$

$\Rightarrow \text{Im}\left(\dfrac{1 + i\sin\alpha - 2i\sin\alpha - 2i^2\sin^2\alpha}{1 - 4i^2\sin^2\alpha}\right) = 0$

$\Rightarrow \text{Im}\left(\dfrac{1 + 2\sin^2\alpha - i\sin\alpha}{1 + 4\sin^2\alpha}\right) = 0$

$\Rightarrow \dfrac{-\sin\alpha}{1 + 4\sin^2\alpha} = 0 \Rightarrow \sin\alpha = 0 \Rightarrow \alpha = n\pi$

Q23. Find the principal argument of $(1 - i\sqrt{3})^2$.

Solution:

$z = (1 - i\sqrt{3})^2$

$\Rightarrow z = 1 - 3 - 2i\sqrt{3} \Rightarrow z = -2 - i2\sqrt{3}$

$\Rightarrow \arg(z) = \tan\theta = \left|\dfrac{\text{Im}(z)}{\text{Re}(z)}\right|$

$\Rightarrow \tan\theta = \left|\dfrac{-2\sqrt{3}}{-2}\right| = |\sqrt{3}| = \sqrt{3}$

$\Rightarrow \tan\theta = \tan\dfrac{\pi}{3} \Rightarrow \theta = \dfrac{\pi}{3}$

$\because \text{Re}(z) < 0, \ \text{Im}(z) < 0$

$\Rightarrow \arg(z) = -\left(\pi - \dfrac{\pi}{3}\right) = -\dfrac{2\pi}{3}$

CHAPTER 6

Linear Inequalities

Day 82

Linear Inequalities in One Variable

Inequality: Two real numbers or two algebraic expressions related by the symbol '<', '>', '≤', or '≥' represent an inequality.

Examples:

Numerical inequalities: $2<3$, $-2>-3$

Literal inequalities: $x<2$, $y>2$, $x\geq -2$, $y\leq -3$

Double inequalities:

$1<2<3$ (read as 2 is greater than 1 and less than 3)

$a\leq x\leq b$ (read as x is greater than or equal to a and less than or equal to b)

$1<x\leq 2$ (read as x is greater than 1 and less than or equal to 2)

$a\leq x<2$ (read as x is greater than or equal to a and less than 2)

Strict inequalities in one variable:

$ax+b<0$

$ax+b>0$

Slack inequalities in one variable:

$ax+b\leq 0$

$ax+b\geq 0$

Strict inequalities in two variables:

$ax+by<c$

$ax+by>c$

Slack inequalities in two variables:

$ax+by\leq c$

$ax+by\geq c$

Quadratic inequalities in one variable:

$ax^2+bx+c\leq 0$

$ax^2+bx+c>0$

Algebraic solution of linear inequalities in one variable and their graphical representation: Any solution of an inequality in one variable is a value of the variable which makes the linear inequality a true statement.

Rules of Inequality

Rule 1: Equal numbers may be added (or subtracted) on both sides of an inequality without affecting the sign of the inequality.

Rule 2: Both sides of an inequality can be multiplied (or divided) by the same positive non-zero number without affecting the sign of the inequality. But when both sides are multiplied or divided by a negative number, then the sign of the inequality is reversed.

Example 1: The solution set of the inequality $x<5$ is

$x=\{1,2,3,4\}$ when x is a natural number

$x=\{0,1,2,3,4\}$ when x is a whole number

$x=\{...,-3,-2,-1,0,1,2,3,4\}$ when x is an integer

$x=(-\infty,5)$ when x is a real number, is graphically represented as

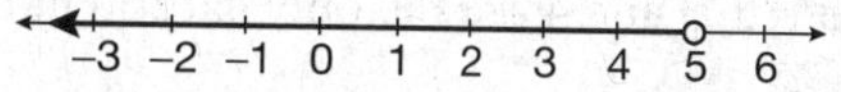

Example 2: The solution set of the inequality $x\leq 5$ is

$x=\{1,2,3,4,5\}$ when x is a natural number

$x=\{0,1,2,3,4,5\}$ when x is a whole number

$x=\{...,-3,-2,-1,0,1,2,3,4,5\}$ when x is an integer

$x=(-\infty,5]$ when x is a real number, is graphically represented as

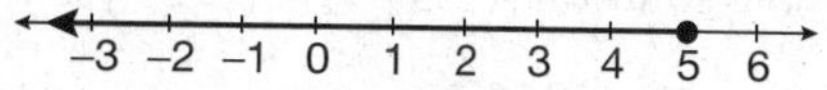

Example 3: The solution set of the inequality $x>-3$ is

$x=\{-2,-1,0,1,2,3,4,...\}$ when x is an integer

$x=(-3,\infty)$ when x is a real number, is graphically represented as

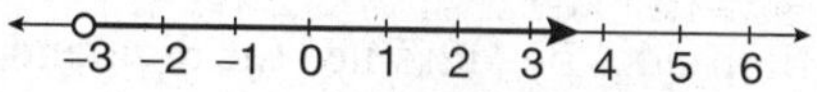

Example 4: The solution set of the inequality $x\geq -3$ is

$x=\{-3,-2,-1,0,1,2,3,4,...\}$ when x is an integer

$x=[-3,\infty)$ when x is a real number, is graphically represented as

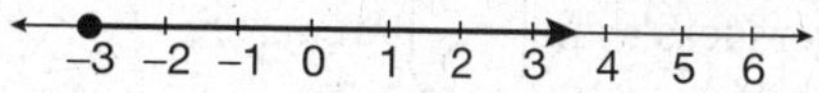

Example 5: The solution set of the inequality $-3\leq x<5$ is

$x=\{-3,-2,-1,0,1,2,3,4\}$ when x is an integer

$x = [-3, 5)$ when x is a real number, is graphically represented as

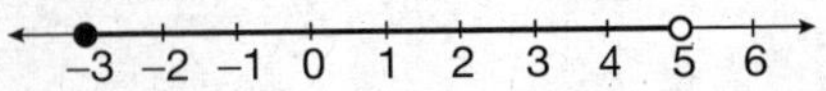

Example 6: The solution set of the inequality $-3 < x \leq 5$ is
$x = \{-2, -1, 0, 1, 2, 3, 4, 5\}$ when x is an integer
$x = (-3, 5]$ when x is a real number, is graphically represented as

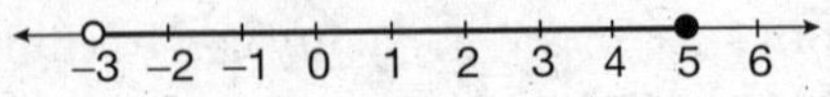

NCERT Exercise 6.1 (With Solutions Q1–Q11)

Q1. Solve $24x < 100$, when
(i) x is natural number (ii) x is an integer

Solution:

$24x < 100 \Rightarrow \frac{24x}{24} < \frac{100}{24} \Rightarrow x < \frac{25}{6}$
(Dividing both sides by 24)

(i) Since 1, 2, 3 and 4 are the only natural numbers less than $\frac{25}{6}$, when x is natural number, the solutions of the given inequality are $x = 1, 2, 3, 4$.
Thus, the solution set is $\{1, 2, 3, 4\}$

(ii) The integers less than $\frac{25}{6}$ are $\ldots, -3, -2, -1, 0, 1, 2, 3, 4$.
Therefore, when x is an integer, the solutions of the given inequality are $\ldots, -3, -2, -1, 0, 1, 2, 3, 4$.
Thus, the solution set is $\{\ldots, -3, -2, -1, 0, 1, 2, 3, 4\}$

Q2. Solve $-12x > 30$, when
(i) x is a natural number
(ii) x is an integer

Solution:

$-12x > 30 \Rightarrow \frac{-12x}{-12} < \frac{30}{-12} \Rightarrow x < -\frac{5}{2}$
(Dividing both sides by –12; since the dividend number is negative, the inequality sign will reverse)

(i) There is no natural number less than $\left(-\frac{5}{2}\right)$.
Therefore, when x is a natural number, there is no solution for the given inequality.

(ii) Integers less than $\left(-\frac{5}{2}\right)$ are $\ldots, -5, -4, -3$.
Therefore, when x is an integer, the solutions of the given inequality are $\ldots, -5, -4, -3$.
Thus, the solution set is $\{\ldots, -5, -4, -3\}$.

Q3. Solve $5x - 3 < 7$, when
(i) x is an integer (ii) x is a real number

Solution:

$5x - 3 < 7 \Rightarrow 5x - 3 + 3 < 7 + 3 \Rightarrow 5x < 10$
(Adding 3 to both the sides)
$\Rightarrow \frac{5x}{5} < \frac{10}{5} \Rightarrow x < 2$ (Dividing both sides by 5)

(i) Integers less than 2 are $\ldots, -3, -2, -1, 0, 1$
Thus, when x is an integer, the solutions of the given inequality are $\ldots, -3, -2, -1, 0, 1$.
Thus, the solution set is $\{\ldots, -3, -2, -1, 0, 1\}$.

(ii) When x is a real number, then the solutions of the given inequality are all real numbers less than 2, that is, $x < 2$.
Thus, the solution set of the given inequality is $x \in (-\infty, 2)$.

Q4. Solve $3x + 8 > 2$, when
(i) x is an integer (ii) x is a real number

Solution:

$3x + 8 > 2 \Rightarrow 3x + 8 - 8 > 2 - 8 \Rightarrow 3x > -6$
(Subtracting 8 from both sides)
$\Rightarrow \frac{3x}{3} > \frac{-6}{3} \Rightarrow x > -2$
(Dividing both sides by 3)

(i) Integers greater than –2 are –1, 0, 1, 2,...
Therefore, when x is an integer, the solutions of the given inequality are –1, 0, 1, 2, ...
Thus, the solution set is $\{-1, 0, 1, 2, \ldots\}$.

(ii) When x is a real number, then the solutions of the given inequality are all the real numbers greater than –2, that is, $x > -2$.
Thus, the solution set is $x \in (-2, \infty)$.

Q5. Solve the given inequality for real $x : 4x + 3 < 5x + 7$.

Solution:

$4x + 3 < 5x + 7 \Rightarrow 4x + 3 - 3 < 5x + 7 - 3$
$\Rightarrow 4x < 5x + 4$
(Subtracting 3 from both sides)
$\Rightarrow 4x - 5x < 5x + 4 - 5x \Rightarrow -x < 4$
(Subtracting $5x$ from both sides)
$\Rightarrow x > -4$
(Multiplying both sides by –1; since the multiplication number is negative, the inequality sign will reverse)
Therefore, all real numbers x which are greater than –4 are the solutions of the given inequality.
Thus, the solution set of the given inequality is $x \in (-4, \infty)$.

Q6. Solve the given inequality for real $x: 3x-7>5x-1$.

Solution:

$3x-7>5x-1 \Rightarrow 3x-7+7>5x-1+7$

$\Rightarrow 3x>5x+6$

(Adding 7 to both sides)

$\Rightarrow 3x-5x>5x+6-5x \Rightarrow -2x>6$

(Subtracting $5x$ from both sides)

$\Rightarrow \frac{-2x}{-2}<\frac{6}{-2} \Rightarrow x<-3$

(Dividing both sides by –2; since the dividend number is negative, the inequality sign will reverse)

Therefore, all real numbers x which are less than –3 are solutions of the given inequality.

Thus, the solution set of the given equality is $x \in (-\infty,-3)$.

Q7. Solve the given inequality for real x: $3(x-1) \le 2(x-3)$.

Solution:

$3(x-1) \le 2(x-3) \Rightarrow 3x-3 \le 2x-6$

$\Rightarrow 3x-3+3 \le 2x-6+3 \Rightarrow 3x \le 2x-3$

(Adding 3 to both sides)

$\Rightarrow 3x-2x \le 2x-3-2x \Rightarrow x \le -3$

(Subtracting $2x$ from both sides)

Therefore, all real numbers x which are less than or equal to –3 are the solutions of the given inequality.

Thus, the solution set of the given inequality is $x \in (-\infty,-3]$.

Q8. Solve the given inequality for real x: $3(2-x) \ge 2(1-x)$.

Solution:

$3(2-x) \ge 2(1-x) \Rightarrow 6-3x \ge 2-2x$

$\Rightarrow 6-3x-6 \ge 2-2x-6 \Rightarrow -3x \ge -2x-4$

(Subtracting 6 from both sides)

$\Rightarrow -3x+2x \ge -2x-4+2x \Rightarrow -x \ge -4$

(Adding $2x$ to both sides)

$\Rightarrow x \le 4$

(Multiplying both sides by –1; since the multiplication number is negative, the inequality sign will reverse)

Therefore, all real numbers x which are less than or equal to 4 are the solutions of the given inequality.

Thus, the solution set of the given inequality is $x \in (-\infty,4]$.

Q9. Solve the given inequality for real x:

$x+\frac{x}{2}+\frac{x}{3}<11$

Solution:

$x+\frac{x}{2}+\frac{x}{3}<11 \Rightarrow x\left(1+\frac{1}{2}+\frac{1}{3}\right)<11$

$\Rightarrow x\left(\frac{6+3+2}{6}\right)<11 \Rightarrow \frac{11x}{6}<11$

$\Rightarrow \frac{11x}{6}\times\frac{6}{11}<11\times\frac{6}{11} \Rightarrow x<6$

(Multiplying both sides by $\frac{6}{11}$)

Therefore, all real numbers x which are less than 6 are solutions of the given inequality.

Thus, the solution set of the given inequality is $x \in (-\infty,6)$.

Q10. Solve the given inequality for real x:

$\frac{x}{3}>\frac{x}{2}+1.$

Solution:

$\frac{x}{3}>\frac{x}{2}+1 \Rightarrow \frac{x}{3}-\frac{x}{2}>1$

$\Rightarrow \frac{2x-3x}{6}>1 \Rightarrow \frac{-x}{6}>1$

(Subtracting $\frac{x}{2}$ from both sides)

$\Rightarrow \frac{-x}{6}\times 6>1\times 6 \Rightarrow -x>6$

(Multiplying both sides by 6)

$\Rightarrow x<-6$

(Multiplying both sides by –1; since the multiplication number is negative, the inequality sign will reverse)

Therefore, all real numbers x which are less than –6 are solutions of the given inequality.

Thus, the solution set of the given inequality is $x \in (-\infty,-6)$.

Q11. Solve the given inequality for real x:

$\frac{3(x-2)}{5} \le \frac{5(2-x)}{3}$.

Solution:

$\frac{3(x-2)}{5} \le \frac{5(2-x)}{3} \Rightarrow 9(x-2) \le 25(2-x)$

$\Rightarrow 9x-18 \le 50-25x \Rightarrow 9x+25x \le 50+18$

$\Rightarrow 34x \le 68 \Rightarrow x \le 2$

Therefore, all real numbers x which are less than or equal to 2 are solutions of the given inequality.

Thus, the solution set of the given inequality is $x \in (-\infty,2]$.

Day 83

NCERT Exercise 6.1 (With Solutions Q12–Q26)

Q12. Solve the given inequality for real x:
$$\frac{1}{2}\left(\frac{3x}{5}+4\right)\geq\frac{1}{3}(x-6)$$

Solution:
$$\frac{1}{2}\left(\frac{3x}{5}+4\right)\geq\frac{1}{3}(x-6)\Rightarrow 3\left(\frac{3x}{5}+4\right)\geq 2(x-6)$$
$$\Rightarrow\frac{9x}{5}+12\geq 2x-12\Rightarrow\frac{9x}{5}-2x\geq -12-12$$
$$\Rightarrow\frac{10x-9x}{5}\geq -24\Rightarrow -\frac{x}{5}\geq -24$$
$$\Rightarrow -x\geq -120\Rightarrow x\leq 120$$
Therefore, all real numbers x which are less than or equal to 120 are solutions of the given inequality.
Thus, the solution set of the given inequality is $x\in(-\infty,120]$.

Q13. Solve the given inequality for real x:
$$2(2x+3)-10<6(x-2)$$

Solution:
$$2(2x+3)-10<6(x-2)$$
$$\Rightarrow 4x+6-10<6x-12$$
$$\Rightarrow 4x-4<6x-12$$
$$\Rightarrow 4x-6x<-12+4$$
$$\Rightarrow -2x<-8\Rightarrow -x<-4\Rightarrow x>4$$
Therefore, all real numbers x which are greater than or equal to 4 are the solutions of the given inequality.
Thus, the solution set of the given inequality is $x\in[4,\infty)$.

Q14. Solve the given inequality for real x:
$$37-(3x+5)\geq 9x-8(x-3).$$

Solution:
$$37-(3x+5)\geq 9x-8(x-3)$$
$$\Rightarrow 37-3x-5\geq 9x-8x+24$$
$$\Rightarrow 32-3x\geq x+24$$
$$\Rightarrow -3x-x\geq 24-32$$
$$\Rightarrow -4x\geq -8\Rightarrow -x\geq -2\Rightarrow x\leq 2$$
Therefore, all real numbers x which are less than or equal to 2 are solutions of the given inequality.
Thus, the solution set of the given inequality is $x\in(-\infty,2]$.

Q15. Solve the given inequality for real x:
$$\frac{x}{4}<\frac{(5x-2)}{3}-\frac{(7x-3)}{5}$$

Solution:
$$\frac{x}{4}<\frac{(5x-2)}{3}-\frac{(7x-3)}{5}$$
$$\Rightarrow\frac{x}{4}<\frac{5(5x-2)-3(7x-3)}{15}$$
$$\Rightarrow\frac{x}{4}<\frac{25x-10-21x+9}{15}$$
$$\Rightarrow\frac{x}{4}<\frac{4x-1}{15}\Rightarrow 15x<4(4x-1)$$
$$\Rightarrow 15x<16x-4\Rightarrow 15x-16x<-4$$
$$\Rightarrow -x<-4\Rightarrow x>4$$
Therefore, all real numbers x which are greater than 4 are solutions of the given inequality.
Thus, the solution set of the given inequality is $x\in(4,\infty)$.

Q16. Solve the given inequality for real x:
$$\frac{(2x-1)}{3}\geq\frac{(3x-2)}{4}-\frac{(2-x)}{5}$$

Solution:
$$\frac{(2x-1)}{3}\geq\frac{(3x-2)}{4}-\frac{(2-x)}{5}$$
$$\Rightarrow\frac{(2x-1)}{3}\geq\frac{5(3x-2)-4(2-x)}{20}$$
$$\Rightarrow\frac{(2x-1)}{3}\geq\frac{15x-10-8+4x}{20}$$
$$\Rightarrow\frac{(2x-1)}{3}\geq\frac{19x-18}{20}$$
$$\Rightarrow 20(2x-1)\geq 3(19x-18)$$
$$\Rightarrow 40x-20)\geq 57x-54$$
$$\Rightarrow 40x-57x\geq -54+20$$
$$\Rightarrow -17x\geq -34\Rightarrow -x\geq -2\Rightarrow x\leq 2$$
Therefore, all real numbers x which are less than or equal to 2 are solutions of the given inequality.
Thus, the solution set of the given inequality is $x\in(-\infty,2]$.

Q17. Solve the given inequality and show the graph of the solution on the number line: $3x-2<2x+1$.

Solution:
$$3x-2<2x+1\Rightarrow 3x-2x<1+2\Rightarrow x<3$$
Thus, the solution set of the given inequality is $x\in(-\infty,3)$.
The graphical (number line) representation of the solutions of the given inequality is

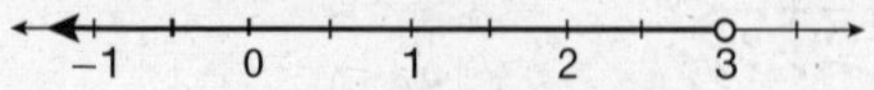

Q18. Solve the given inequality and show the graph of the solution on the number line: $5x-3\geq 3x-5$.

Solution:

$5x-3\geq 3x-5 \Rightarrow 5x-3x\geq -5+3$

$\Rightarrow 2x\geq -2 \Rightarrow x\geq -1$

Thus, the solution set of the given inequality is $x\in[-1,\infty)$.

The graphical (number line) representation of the solutions of the given inequality is

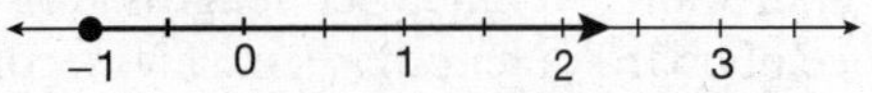

Q19. Solve the given inequality and show the graph of the solution on number line: $3(1-x)<2(x+4)$.

Solution:

$3(1-x)<2(x+4) \Rightarrow 3-3x<2x+8$

$\Rightarrow -3x-2x<8-3$

$\Rightarrow -5x<5 \Rightarrow -x<1$

$\Rightarrow x>-1$

Thus, the solution set of the given inequality is $x\in(-1,\infty)$.

The graphical (number line) representation of the solutions of the given inequality is

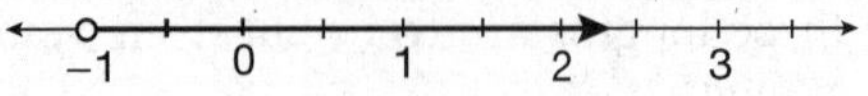

Q20. Solve the given inequality and show the graph of the solution on number line: $\frac{x}{2}\geq\frac{(5x-2)}{3}-\frac{(7x-3)}{5}$.

Solution:

$\frac{x}{2}\geq\frac{(5x-2)}{3}-\frac{(7x-3)}{5}$

$\Rightarrow \frac{x}{2}\geq\frac{5(5x-2)-3(7x-3)}{15}$

$\Rightarrow \frac{x}{2}\geq\frac{25x-10-21x+9}{15}$

$\Rightarrow \frac{x}{2}\geq\frac{4x-1}{15}$

$\Rightarrow 15x\geq 2(4x-1) \Rightarrow 15x\geq 8x-2$

$\Rightarrow 15x-8x\geq -2 \Rightarrow 7x\geq -2 \Rightarrow x\geq -\frac{2}{7}$

Thus, the solution set of the given inequality is $x\in[-\frac{2}{7},\infty)$.

The graphical (number line) representation of the solutions of the given inequality is

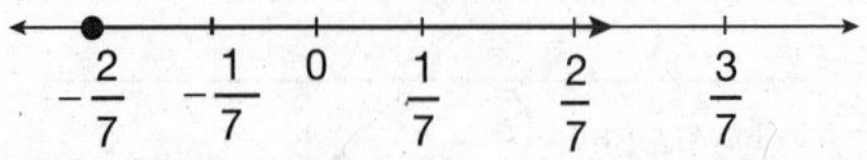

Q21. Ravi obtained 70 and 75 marks in the first two unit tests. Find the minimum marks he should get in the third test to have an average of at least 60 marks.

Solution:

Let x be the marks obtained by Ravi in the third unit test, so that the student should have an average of at least 60 marks. Then

$\frac{70+75+x}{3}\geq 60 \Rightarrow 145+x\geq 180$

$\Rightarrow x\geq 180-145 \Rightarrow x\geq 35$

Thus, Ravi should obtain a minimum of 35 marks to have an average of at least 60 marks.

Q22. To receive Grade 'A' in a course one must obtain an average of 90 marks or more in five examinations (each of 100 marks). If Sunita's marks in the first four examinations are 87, 92, 94 and 95, find the minimum marks that Sunita must obtain in the fifth examination to get grade 'A' in the course.

Solution:

Let x be the marks obtained by Sunita in the fifth examination.

For receiving grade 'A' in the course, she must obtain an average of 90 marks or more in five examinations. Therefore

$\frac{87+92+94+95+x}{5}\geq 90$

$\Rightarrow \frac{368+x}{5}\geq 90 \Rightarrow 368+x\geq 450$

$\Rightarrow x\geq 450-368 \Rightarrow x\geq 82$

Thus, Sunita must obtain a minimum of 82 marks in the fifth examination to get grade 'A' in the course.

Q23. Find all pairs of consecutive odd positive integers both of which are smaller than 10 such that their sum is more than 11.

Solution:

Let x be the smaller of the two consecutive odd positive integers. Then, the other larger integer must be $x+2$.

Both the integers are smaller than 10. Therefore

$x+2<10 \Rightarrow x<10-2 \Rightarrow x<8$

Also, the sum of the two integers is more than 11. Therefore

$x+(x+2)>11$

$\Rightarrow 2x+2>11 \Rightarrow 2x>11-2$

$\Rightarrow 2x>9 \Rightarrow x>\frac{9}{2}$

Therefore, from the inequalities, $x<8$ and $x>\frac{9}{2}$, we get

$x=5,6,7$

But since x is an odd number, x can take only values 5 and 7.

Thus, all pairs of consecutive odd positive integers are (5, 7) and (7, 9).

Q24. Find all pairs of consecutive even positive integers, both of which are larger than 5 such that their sum is less than 23.

Solution:
Let x be the smaller of the two consecutive even positive integers.
Then, the other larger integer is $x+2$.
Both the integers are larger than 5. Therefore
$x>5$
Also, the sum of the two integers is less than 23. Therefore
$x+(x+2)<23$
$\Rightarrow 2x+2<23 \Rightarrow 2x<23-2$
$\Rightarrow 2x<21 \Rightarrow x<\frac{21}{2}$
Therefore, from the inequalities, $x>5$ and $x<\frac{21}{2}$, we get
$x=6,7,8,9,10$
But x is an even number, and therefore, x can take only the values 6, 8, and 10.
Thus, the required possible pairs are (6, 8), (8, 10) and (10, 12).

Q25. The longest side of a triangle is 3 times the shortest side and the third side is 2 cm shorter than the longest side. If the perimeter of the triangle is at least 61 cm, find the minimum length of the shortest side.

Solution:
Let the length of the shortest side of the triangle be x cm.
Then, length of the longest side = $3x$ cm
And length of the third side = $(3x-2)$ cm
The perimeter of the triangle is at least 61 cm. Therefore
$x+3x+(3x-2)\geq 61$
$\Rightarrow 7x-2\geq 61$
$\Rightarrow 7x\geq 63 \Rightarrow x\geq 9$
Thus, the minimum length of the shortest side is 9 cm.

Q26. A man wants to cut three lengths from a single piece of board of length 91 cm. The second length is to be 3 cm longer than the shortest and the third length is to be twice as long as the shortest. What are the possible lengths of the shortest board if the third piece is to be at least 5 cm longer than the second?
[Hint: If x is the length of the shortest board, then $x,(x+3),2x$ are the lengths of the second and third piece, respectively. Thus, $x+(x+3)+2x\leq 91$, and $2x\geq(x+3)+5$]

Solution:
Let x be the length of the shortest board. Then $x,(x+3),2x$ are the lengths of the second and third piece, respectively. Since the three lengths are to be cut from a single piece of board of length,
$x+(x+3)+2x\leq 91$
$\Rightarrow 4x\leq 88 \Rightarrow x\leq 22$
Now, the third piece is at least 5 cm longer than the second piece. Therefore
$2x\geq(x+3)+5$
$\Rightarrow 2x\geq x+8 \Rightarrow x\geq 8$
Therefore, from the inequalities, $x\leq 22$ and $x\geq 8$, we get $8\leq x\leq 22$.
Thus, the possible length of the shortest board is greater than or equal to 8 cm and less than or equal to 22 cm.

Day 84

Linear Inequalities in Two Variables

Graphical Solution of Linear Inequalities in Two Variables: Since a line divides the Cartesian plane into two parts, each part is known as a half plane.
Thus, a vertical line will divide the plane into two, known as the left and right half planes.

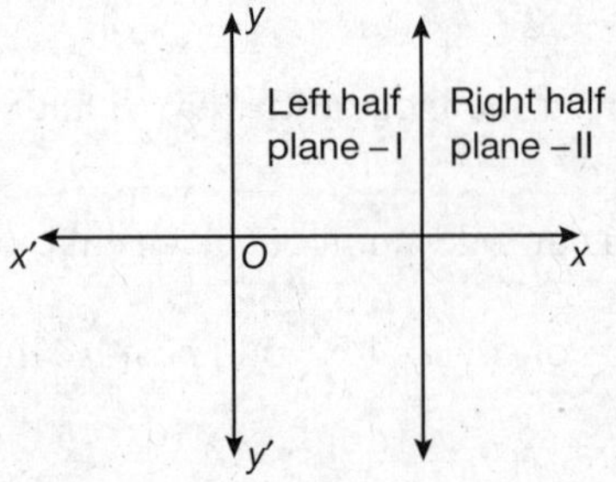

Fig. 6.1

And a non-vertical line will divide the plane into upper and lower half planes.

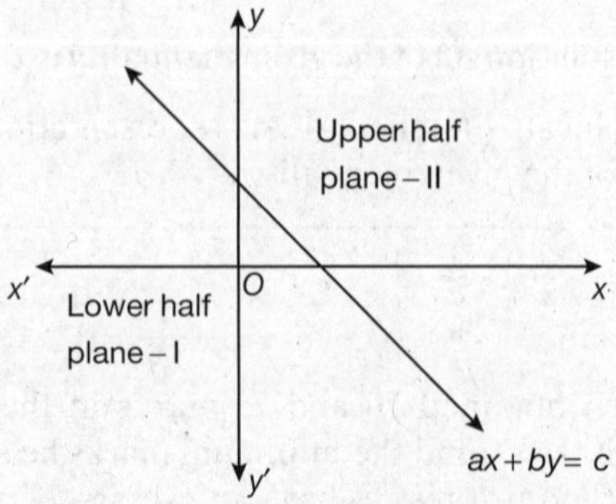

Fig. 6.2

A point in the Cartesian plane will either lie on a line or will lie in either of the half planes I or II.

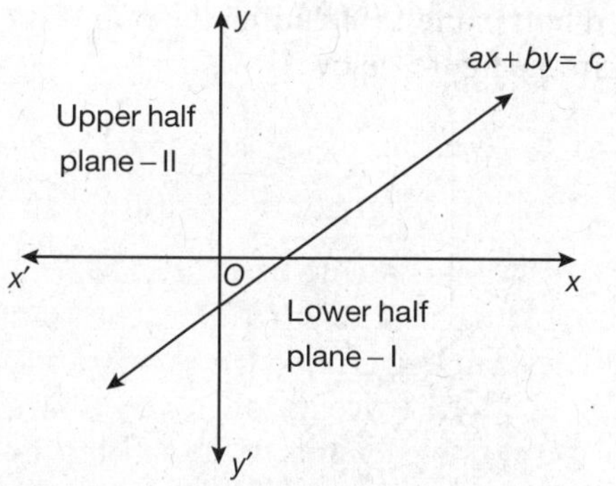

Fig. 6.3

Consider the line $ax+by=c,\ a\neq 0, b\neq 0$, for which three possibilities may arise:

(i) $ax+by=c$ means all points (x, y) satisfying the equation of line lie on the line it represents, and vice versa.

(ii) $ax+by>c$

(iii) $ax+by<c$

Note:

1. The region containing all the solutions of an inequality is called the solution region. The solution of an inequality can be represented by a shaded region.
2. In order to identity the half plane represented by an inequality, it is sufficient to take any point (not on the line) and check whether it satisfies the inequality or not. For convenience, the point (0, 0) is preferred, that is, if the origin $O(0,0)$ satisfies the inequality $ax+by>c$ or $ax+by<c$, then the portion divided by the line $ax+by=c$ or $ax+by=c$ containing the origin represents the solution of the inequality $ax+by>c$ or $ax+by<c$. And if the origin does not satisfy the inequality, then the portion of the line opposite to the origin represents the solution of the inequality.
3. In an inequality of the form $ax+by>c$ or $ax+by<c$, the points on the line $ax+by=c$ are not included in the solution region. Thus, a broken or dotted line is drawn in the solution region.
4. In an inequality of the type $ax+by\geq c$ or $ax+by\leq c$, the points on the line $ax+by=c$ are also included in the solution region. Thus, a dark line is drawn in the solution region.

<u>**Example 1:**</u>

(i) $x+y<1$

Since the origin $O(0,0)$ satisfies the inequality $x+y\leq 1$, the portion divided by the line $x+y=1$ containing the origin represents the solution of the inequality $x+y<1$, which is shown by the shaded region.

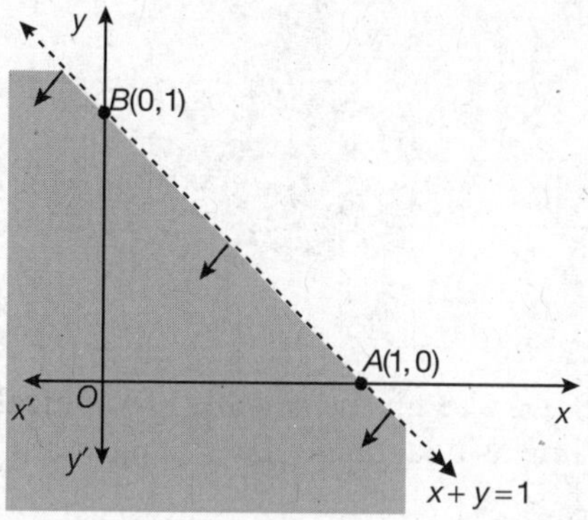

(ii) $x+y\geq 1$

Since the origin $O(0,0)$ does not satisfy the inequality $x+y\geq 1$, the portion divided by the line $x+y=1$ opposite to the origin represents the solution of the inequality $x+y\geq 1$, which is shown by the shaded region.

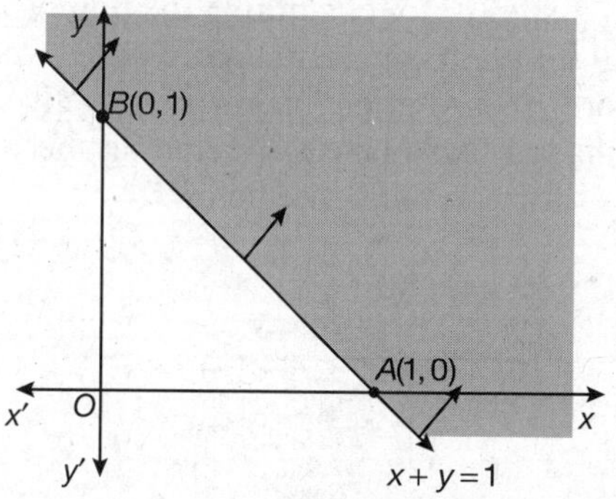

NCERT Exercise 6.2 (With Solutions)

Q1. Solve the given inequality graphically in a two-dimensional plane: $x+y<5$.

Solution:

The line $x+y=5$ divides the xy-plane in two half planes, I and II.

Let us select point (0, 0) (not on the line), which lies in one of the half planes.

Substituting this point in the given inequality, we get $0+0<5\Rightarrow 0<5$, which is true.

Therefore, the solution region of the given inequality is the half plane I [containing the point (0, 0)].

Also, since any point on the line does not satisfy the given strict inequality, the solution region of the given inequality is the shaded half plane I (excluding the points on the line), as shown in the figure below.

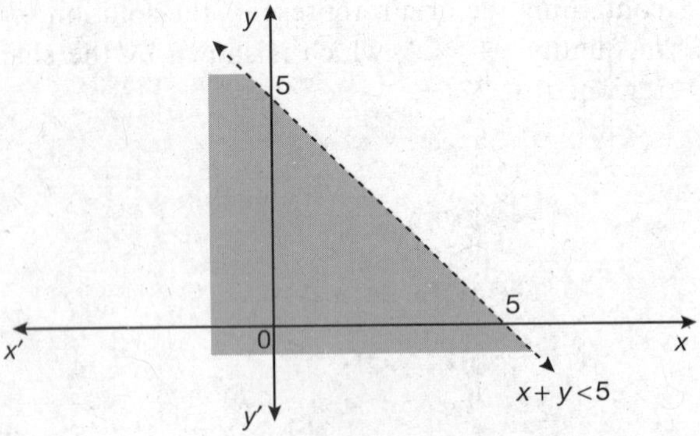

Q2. Solve the given inequality graphically in a two-dimensional plane: $2x+y \geq 6$.

Solution:
The line $2x+y=6$ divides the xy-plane in two half planes, I and II.
Let us select point (0, 0) (not on the line), which lies in one of the half planes.
Substituting this point in the given inequality, we get $2(0)+0 \geq 6 \Rightarrow 0 \geq 6$, which is false.
Therefore, the solution region of the given inequality is the half plane II [not containing the point (0, 0)].
Also, all the points on the line satisfy the given inequality.
Therefore, the solution region of the given inequality is the shaded half plane II (including the points on the line), as shown in the figure below.

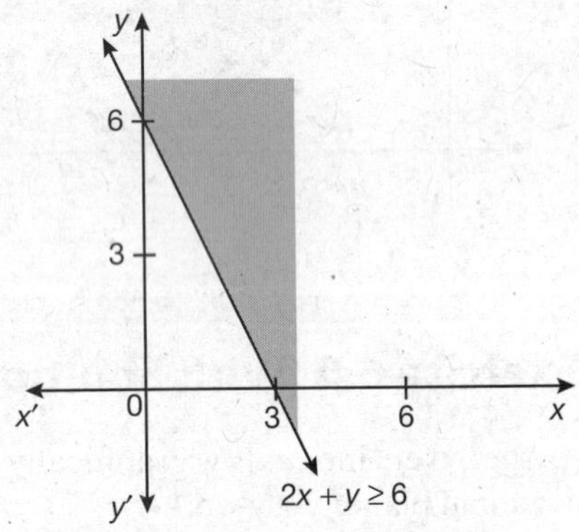

Q3. Solve the given inequality graphically in a two-dimensional plane: $3x+4y \leq 12$.

Solution:
The line $3x+4y=12$ divides the xy-plane in two half planes, I and II.
Let us select point (0, 0) (not on the line), which lies in one of the half planes.
Substituting this point in the given inequality, we get $3(0)+4(0) \leq 12 \Rightarrow 0 \leq 12$ which is true.
Therefore, the solution region of the given inequality is the half plane I [containing the point (0, 0)].
Also, since all the points on the line satisfy the given inequality, the solution region of the given inequality is the shaded half plane I (including the points on the line), as shown in the figure below.

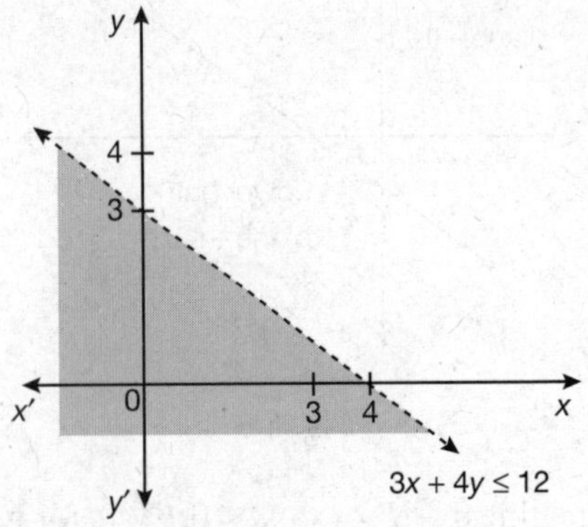

Q4. Solve the given inequality graphically in a two-dimensional plane: $y+8 \geq 2x$.

Solution:
The line $y+8=2x$ divides the xy-plane in two half planes, I and II.
Let us select point (0, 0) (not on the line), which lies in one of the half planes.
Substituting this point in the given inequality, we get $0+8 \geq 2(0) \Rightarrow 8 \geq 0$, which is true.
Therefore, the solution region of the given inequality is the half plane I [containing the point (0, 0)].
Also, all the points on the line satisfy the given inequality.
Therefore, the solution region of the given inequality is the shaded half plane I (including the points on the line), as shown in the figure below.

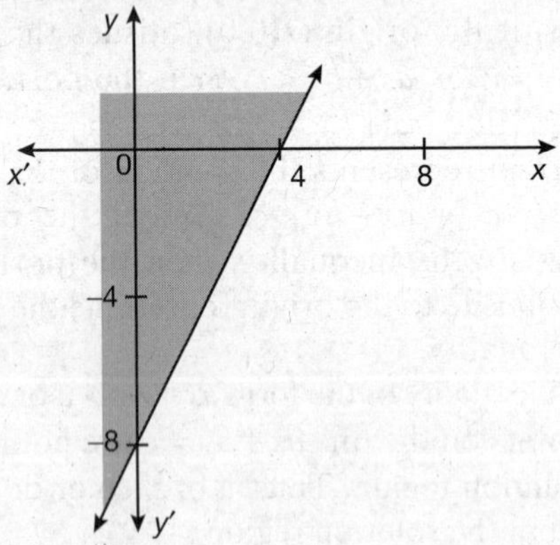

Q5. Solve the given inequality graphically in a two-dimensional plane: $x-y \leq 2$.

Solution:
The line $x-y=2$ divides the xy-plane in two half planes, I and II.
Let us select point (0, 0) (not on the line), which lies in one of the half planes.
Substituting this point in the given inequality, we get $0-0 \leq 2 \Rightarrow 0 \leq 2$ which is true.

Therefore, the solution region of the given inequality is the half plane I [containing the point (0, 0)].
Also, all the points on the line satisfy the given inequality. Therefore, the solution region of the given inequality is the shaded half plane I (including the points on the line), as shown in the figure below.

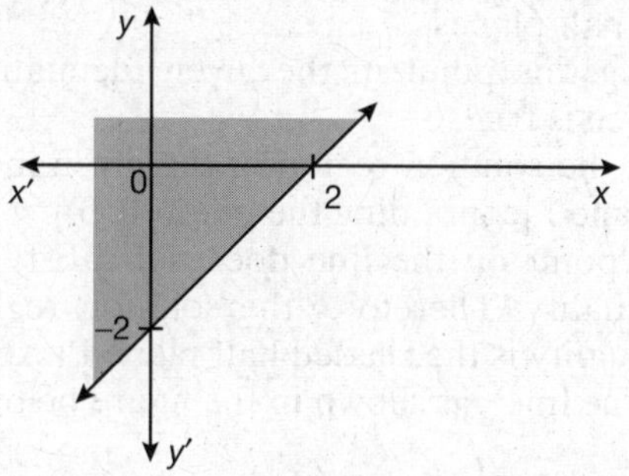

Q6. Solve the given inequality graphically in a two-dimensional plane: $2x - 3y > 6$.

Solution:
The line $2x - 3y = 6$ divides the xy-plane in two half planes, I and II.
Let us select point (0, 0) (not on the line), which lies in one of the half planes.
Substituting this point in the given inequality, we get $2(0) - 3(0) > 6 \Rightarrow 0 > 6$, which is false.
Therefore, the solution region of the given inequality is the half plane II [not containing the point (0, 0)].
Also, any point on the line does not satisfy the given strict inequality. Therefore, the solution region of the given inequality is the shaded half plane II (excluding the points on the line), as shown in the figure below.

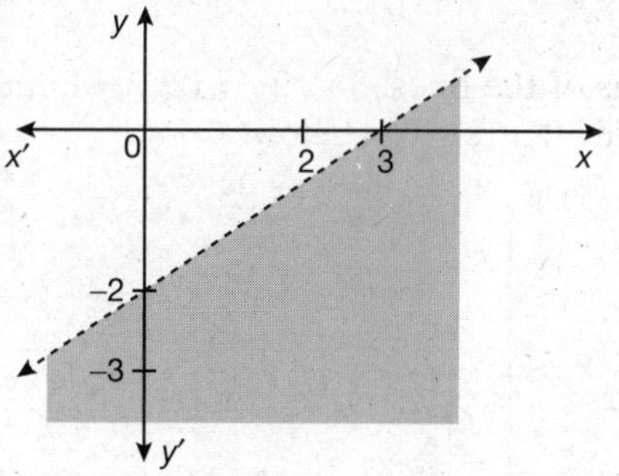

Q7. Solve the given inequality in a two-dimensional plane: $-3x + 2y \geq -6$.

Solution:
The line $-3x + 2y = -6$ divides the xy-plane in two half planes, I and II.
Let us select point (0, 0) (not on the line), which lies in one of the half planes.
Substituting this point in the given inequality, we get $-3(0) + 2(0) \geq -6 \Rightarrow 0 \geq -6$, which is true.
Therefore, the solution region of the given inequality is the half plane I [containing the point (0, 0)].
Also, all the points on the line satisfy the given inequality. Therefore, the solution region of the given inequality is the shaded half plane I (including the points on the line), as shown in the figure below.

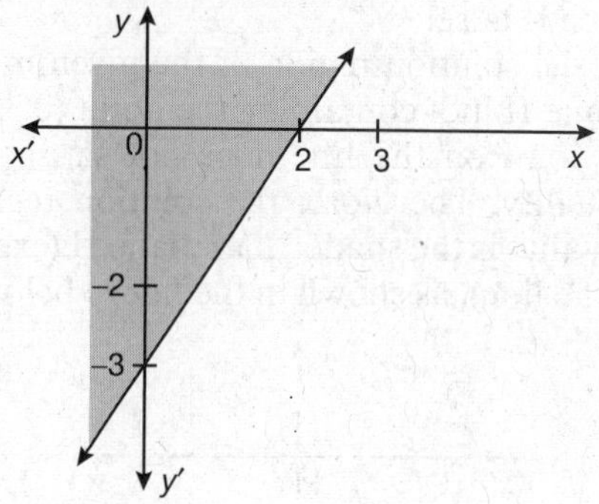

Q8. Solve the given inequality graphically in a two-dimensional plane: $3y - 5x < 30$.

Solution:
The line $3y - 5x = 30$ divides the xy-plane in two half planes, I and II.
Let us select point (0, 0) (not on the line), which lies in one of the half planes.
Substituting this point in the given inequality, we get $3(0) - 5(0) < 30 \Rightarrow 0 < 30$, which is false.
Therefore, the solution region of the given inequality is the half plane II [not containing the point (0, 0)].
Also, any point on the line does not satisfy the given strict inequality. Therefore, the solution region of the given inequality is the shaded half plane II (excluding the points on the line), as shown in the figure below.

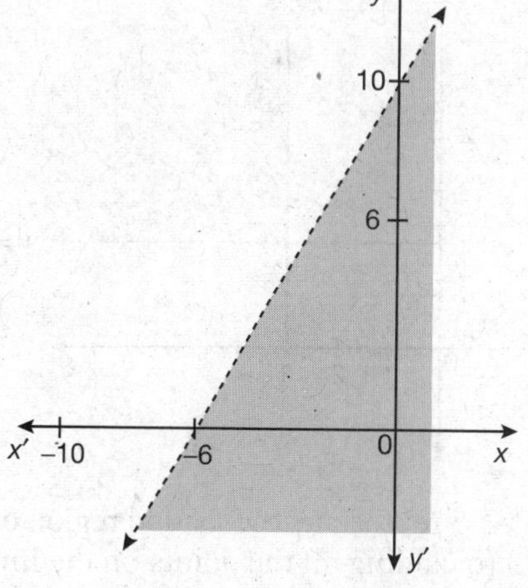

Q9. Solve the given inequality graphically in a two-dimensional plane $y < -2$.

Solution:
The line $y = -2$ divides the xy-plane in two half planes, I and II.
Let us select point (0, 0) (not on the line), which lies in one of the half planes.
Substituting this point in the given inequality, we get $0 < -2$, which is false.
Therefore, the solution region of the given inequality is the half plane II [not containing the point (0, 0)].
Also, any point on the line does not satisfy the given strict inequality. Therefore, the solution region of the given inequality is the shaded half plane II (excluding the points on the line), as shown in the figure below.

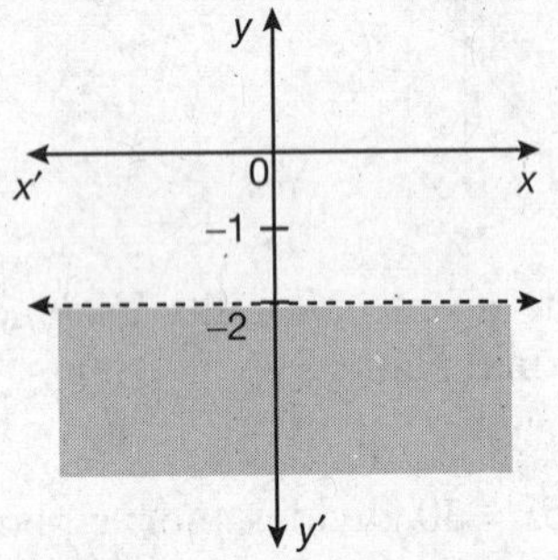

Q10. Solve the given inequality graphically in a two-dimensional plane: $x > -3$.

Solution:
The line $x = -3$ divides the xy-plane in two half planes, I and II.
Let us select point (0, 0) (not on the line), which lies in one of the half planes.
Substituting this point in the given inequality, we get $0 > -3$, which is true.
Therefore, the solution region of the given inequality is the half plane I [containing the point (0, 0)].
Also, any point on the line does not satisfy the given strict inequality. Therefore, the solution region of the given inequality is the shaded half plane I (excluding the points on the line), as shown in the figure below.

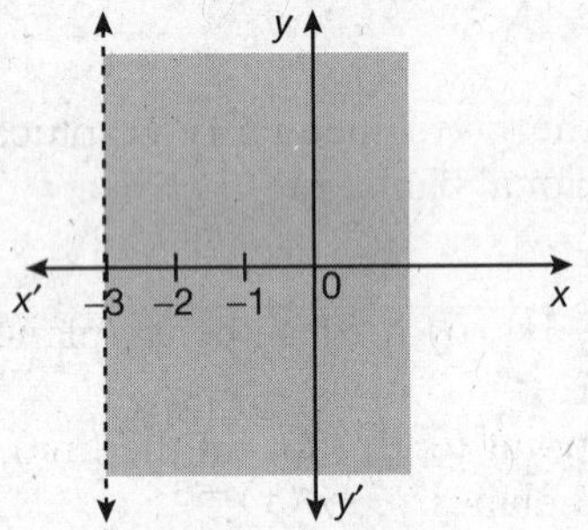

Day 85

NCERT Exercise 6.3 (With Solutions)

Q1. Solve the following system of inequalities graphically: $x \geq 3, y \geq 2$.

Solution:
The graphs of the lines, $x = 3$ and $y = 2$, are shown in the figure given below.

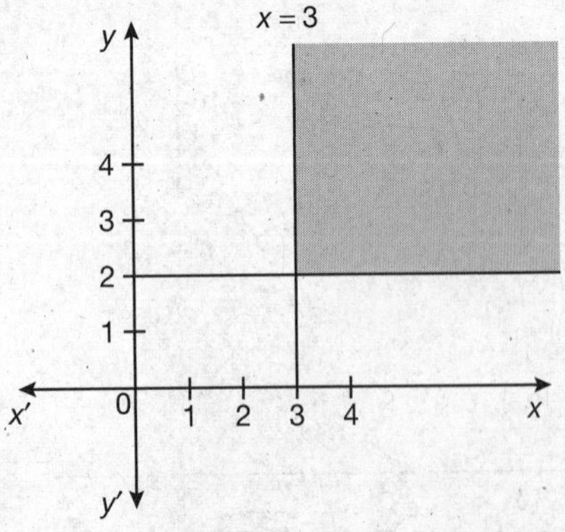

Inequality $x \geq 3$ represents the shaded region on the right of line $x = 3$ (including all the points on the line).
Inequality $y \geq 2$ represents the shaded region above the line $y = 2$ (including all the points on the line).
Thus, the solution of the given system of linear inequalities is represented by the common double shaded region including the points on the respective lines.

Q2. Solve the following system of inequalities graphically: $3x + 2y \leq 12, x > 1, y \geq 2$.

Solution:
The graphs of the lines, $3x + 2y = 12$, $x = 1$ and $y = 2$, are shown in the figure given below.

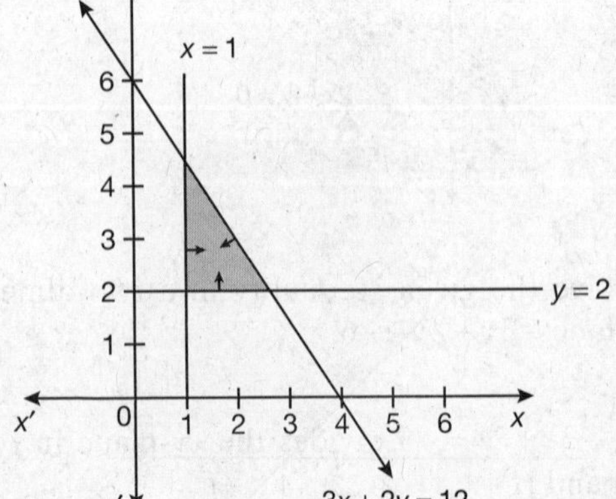

Inequality $3x + 2y \leq 12$ represents the shaded region below the line $3x + 2y = 12$ (including all the points on the line).
Inequality $x > 1$ represents the shaded region on the right of line $x = 1$ (excluding all the points on the line).

Inequality $y \geq 2$ represents the shaded region above the line $y = 2$ (including all the points on the line).

Thus, the solution of the given system of linear inequalities is represented by the common shaded region excluding the points on the line $x = 1$.

Q3. Solve the following system of inequalities graphically: $2x + y \geq 6, 3x + 4y \leq 12$.

Solution:

The graphs of the lines, $2x + y = 6$ and $3x + 4y = 12$ are shown in the figure given below.

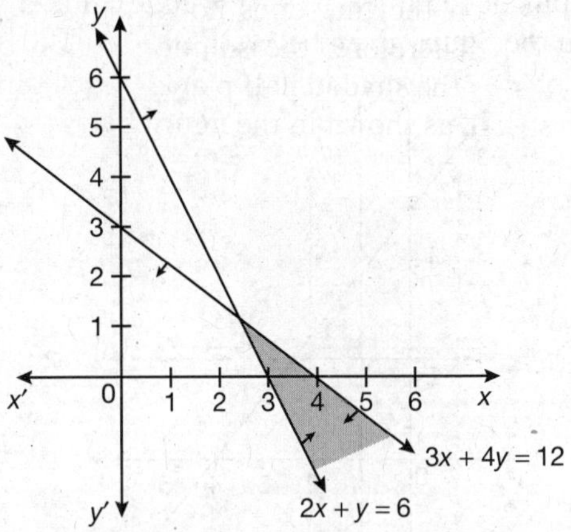

Inequality $2x + y \geq 6$ represents the shaded region above the line $2x + y = 6$ (including all the points on the line).

Inequality $3x + 2y \leq 12$ represents the shaded region below the line $3x + 2y = 12$ (including all the points on the line).

Thus, the solution of the given system of linear inequalities is represented by the common shaded region including the points on the respective lines.

Q4. Solve the following system of inequalities graphically: $x + y \geq 4, 2x - y > 0$.

Solution:

The graphs of the lines, $x + y = 4$ and $2x - y = 0$ are shown in the figure given below.

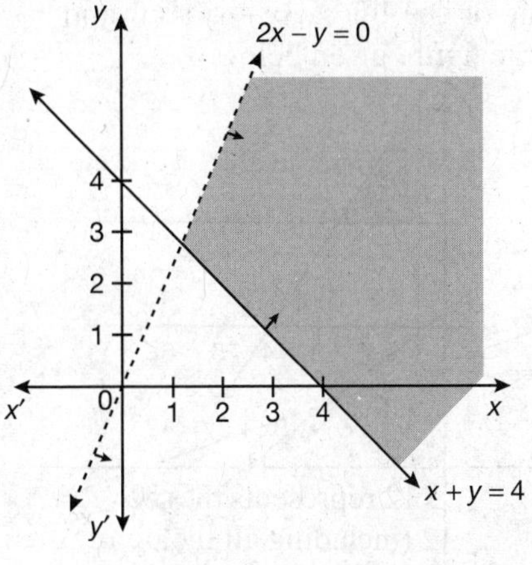

Inequality $x + y \geq 4$ represents the shaded region above the line $x + y = 4$ (including all the points on the line).

Inequality $2x - y > 0$ represents the shaded region below the line $2x - y = 0$ (excluding all the points on the line). This can be checked by taking any point like (1,0), other than (0,0) because the line passes through (0,0).

Thus, the solution of the given system of linear inequalities is represented by the common shaded region excluding the points on the line $2x - y = 0$.

Q5. Solve the following system of inequalities graphically: $2x - y > 1, x - 2y < -1$.

Solution:

The graphs of the lines, $2x - y = 1$ and $x - 2y = -1$ are shown in the figure given below.

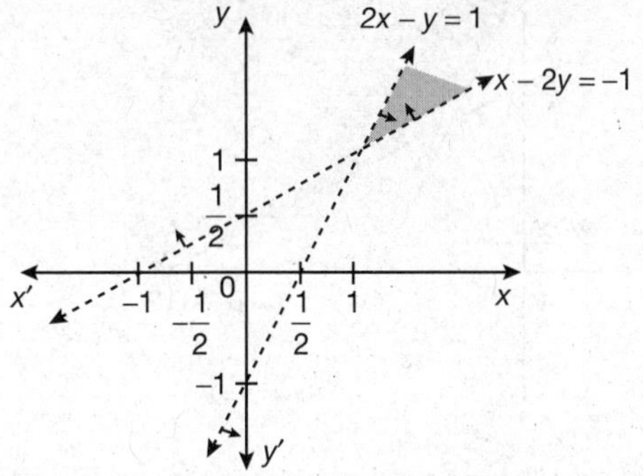

Inequality $2x - y > 1$ represents the shaded region above the line $2x - y = 1$ (excluding all the points on the line).

Inequality $x - 2y < -1$ represents the shaded region below the line $x - 2y = -1$ (excluding all the points on the line).

Thus, the solution of the given system of linear inequalities is represented by the common shaded region excluding the points on the respective lines.

Q6. Solve the following system of inequalities graphically: $x + y \leq 6, x + y \geq 4$.

Solution:

The graphs of the lines, $x + y = 6$ and $x + y = 4$ are shown in the figure given below.

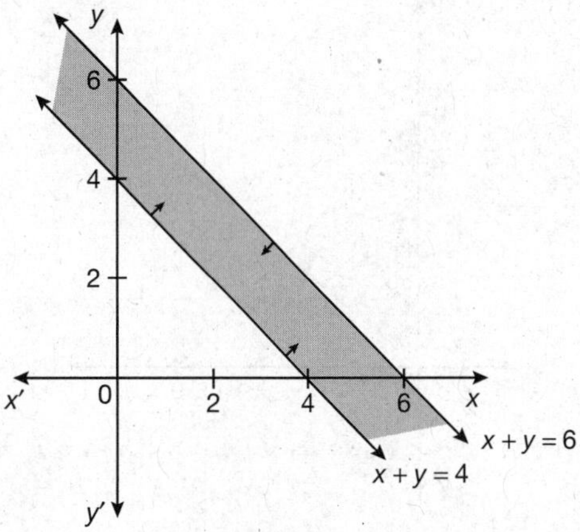

Inequality $x + y \leq 6$ represents the shaded region above the line $x + y = 6$ (including all the points on the line).

Inequality $x + y \geq 4$ represents the shaded region above the line $x + y = 4$ (including all the points on the line).

Thus, the solution of the given system of linear inequalities is represented by the common shaded region including the points on the respective lines.

Q7. Solve the following system of inequalities graphically: $2x + y \geq 8, x + 2y \geq 10$.

Solution:
The graphs of the lines, $2x + y = 8$ and $x + 2y = 10$ are shown in the figure given below.

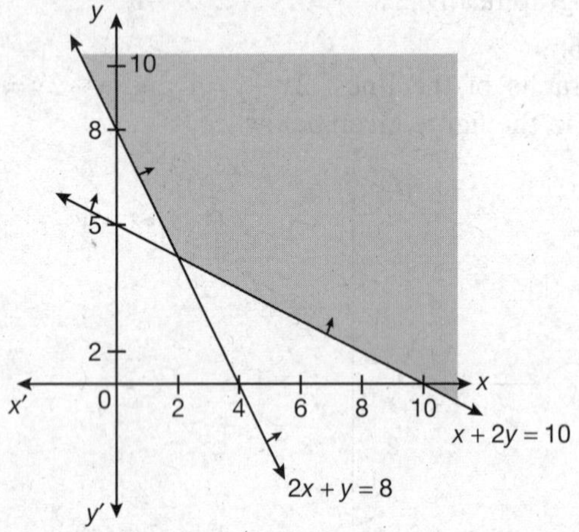

Inequality $2x + y \geq 8$ represents the shaded region above the line $2x + y = 8$ (including all the points on the line).
Inequality $x + 2y \geq 10$ represents the shaded region above the line $x + 2y = 10$ (including all the points on the line).
Thus, the solution of the given system of linear inequalities is represented by the common shaded region including the points on the respective lines.

Q8. Solve the following system of inequalities graphically: $x + y \leq 9, y > x, x \geq 0$.

Solution:
The graphs of the lines, $x + y = 9$ and $y = x$ are shown in the figure given below.

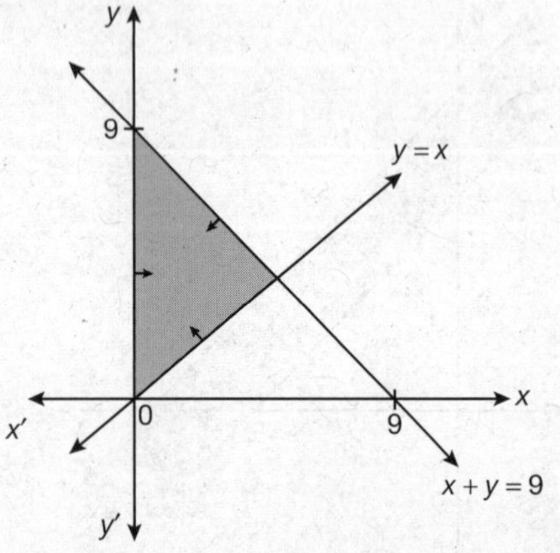

Inequality $x + y \leq 9$ represents the shaded region below the line $x + y = 9$ (including all the points on the line).
Inequality $y > x$ represents the shaded region above the line $y = x$ (excluding all the points on the line).
Inequality $x \geq 0$ represents the region on the right side of the line, $x = 0$ or the y-axis (including all the points on the y-axis).
Thus, the solution of the given system of linear inequalities is represented by the common shaded region including the points on the lines, $x + y = 9, x = 0$, and excluding the points on line $y = x$.

Q9. Solve the following system of inequalities graphically: $5x + 4y \leq 20, x \geq 1, y \geq 2$.

Solution:
The graphs of the lines, $5x + 4y = 20$, $x = 1$ and $y = 2$ are shown in the figure given below.

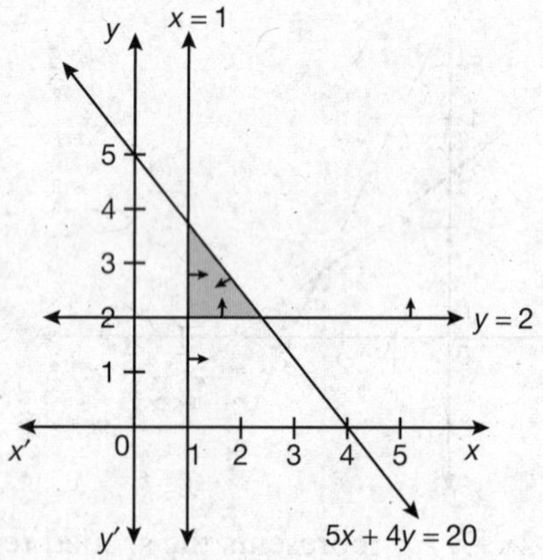

Inequality $5x + 4y \leq 20$ represents the shaded region below the line $5x + 4y = 20$ (including all the points on the line).
Inequality $x \geq 1$ represents the region on the right side of the line, $x = 1$ (including all the points on the line).
Inequality $y \geq 2$ represents the shaded region above the line $y = 2$ (including all the points on the line).
Thus, the solution of the given system of linear inequalities is represented by the common shaded region including the points on the respective lines.

Q10. Solve the following system of inequalities graphically: $3x + 4y \leq 60, x + y \leq 30, x \geq 0, y \geq 0$.

Solution:
The graphs of the lines, $3x + 4y \leq 60$ and $x + y \leq 30$ are shown in the figure given below.

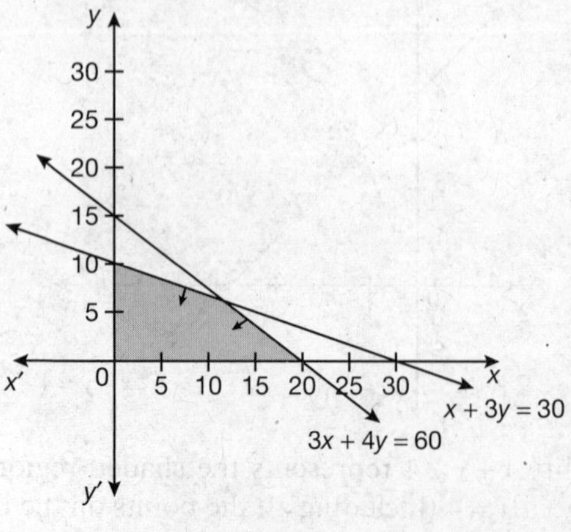

Inequality $3x+4y \le 60$ represents the shaded region below the line $3x+4y=60$ (including all the points on the line).
Inequality $x+y \le 30$ represents the shaded region below the line $x+y=30$ (including all the points on the line).
Inequality $x \ge 0$, $y \ge 0$ represents every point in the common shaded region in the first quadrant including the points on the axes.
Thus, the solution of the given system of linear inequalities is represented by the common shaded region including the points on the respective lines in the first quadrant.

Q11. Solve the following system of inequalities graphically: $2x+y \ge 4, x+y \le 3, 2x-3y \le 6$.

Solution:
The graphs of the lines, $2x+y=4$, $x+y=3$ and $2x-3y=6$ are shown in the figure given below.

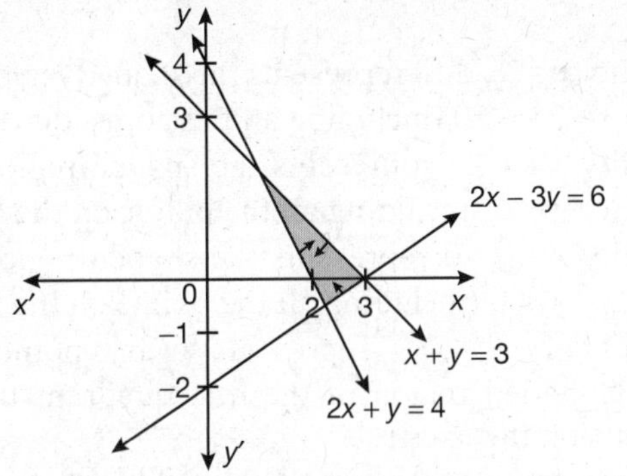

Inequality $2x+y \ge 4$ represents the shaded region above the line $2x+y=4$ (including all the points on the line).
Inequality $x+y \le 3$ represents the shaded region below the line $x+y=3$ (including all the points on the line).
Inequality $2x-3y \le 6$ represents the shaded region above the line $2x-3y=6$ (including all the points on the line).
Thus, the solution of the given system of linear inequalities is represented by the common shaded region including the points on the respective lines.

Q12. Solve the following system of inequalities graphically: $x-2y \le 3$, $3x+4y \ge 12$, $x \ge 0$, $y \ge 1$.

Solution:
The graphs of the lines, $x-2y=3$, $3x+4y=12$, $x=0$ and $y=1$ are shown in the figure given below.

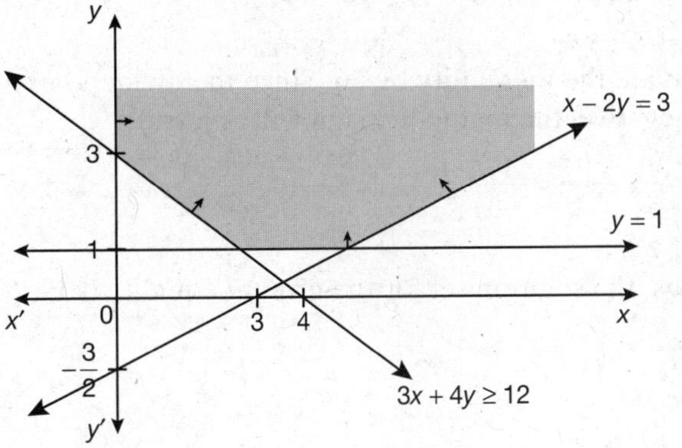

Inequality $x-2y \le 3$ represents the shaded region below the line $x-2y=3$ (including all the points on the line).
Inequality $3x+4y \ge 12$ represents the shaded region below the line $3x+4y=12$ (including all the points on the line).
Inequality $y \ge 1$ represents the shaded region above the line $y=1$ (including all the points on the line).
Inequality $x \ge 0$ represents the region on the right side of the line, $x=0$ or the y-axis (including all the points on the y-axis).
Thus, the solution of the given system of linear inequalities is represented by the common shaded region including the points on the respective lines.

Q13. Solve the following system of inequalities graphically: $4x+3y \le 60$, $y \ge 2x$, $x \ge 3$, $x, y \ge 0$.

Solution:
The graphs of the lines $4x+3y \le 60$, $y \ge 2x$ and $x \ge 3$ are shown in the figure given below.

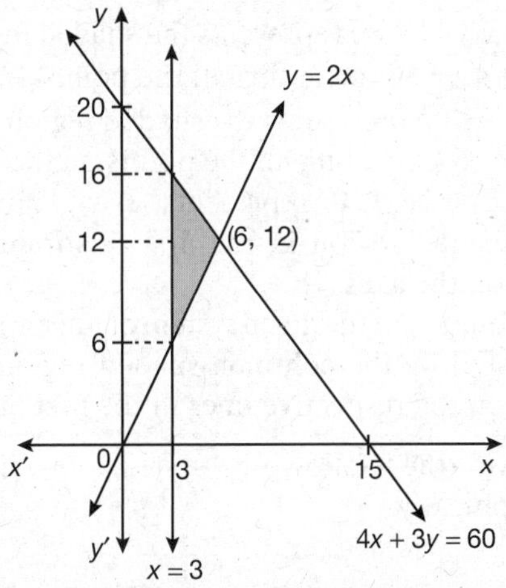

Inequality $4x+3y \le 60$ represents the shaded region below the line $4x+3y=60$ (including all the points on the line).
Inequality $y \ge 2x$ represents the shaded region above the line $y=2x$ (including all the points on the line).
Inequality $x \ge 3$ represents the shaded region right side of the line $x=3$ (including all the points on the line).
Inequality $x \ge 0$, $y \ge 0$ represents every point in the common shaded region in the first quadrant including the points on the axes.
Thus, the solution of the given system of linear inequalities is represented by the common shaded region including the points on the respective lines in the first quadrant.

Q14. Solve the following system of inequalities graphically:
$3x+2y \le 150, x+4y \le 80, x \le 15, y \ge 0, x \ge 0$.

Solution:
The graphs of the lines $3x+2y \le 150$, $x+4y \le 80$ and $x \le 15$ are shown in the figure given below.

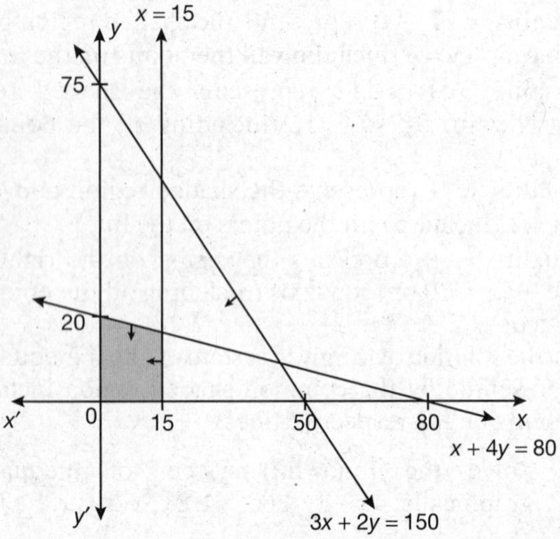

Inequality $3x+2y \le 150$ represents the shaded region below the line $3x+2y=150$ (including all the points on the line).
Inequality $x+4y \le 80$ represents the shaded region below the line $x+4y=80$ (including all the points on the line).
Inequality $x \le 15$ represents the shaded region left side of the line $x=15$ (including all the points on the line).
Inequality $x \ge 0$, $y \ge 0$ represents every point in the common shaded region in the first quadrant including the points on the axes.
Thus, the solution of the given system of linear inequalities is represented by the common shaded region including the points on the respective lines in the first quadrant.

Q15. Solve the following system of inequalities graphically:

$x+2y \le 10, x+y \ge 1, x-y \le 0, x \ge 0, y \ge 0.$

Solution:
The graphs of the lines $x+2y \le 10$, $x+y \ge 1$ and $x-y \le 0$ are shown in the figure given below.

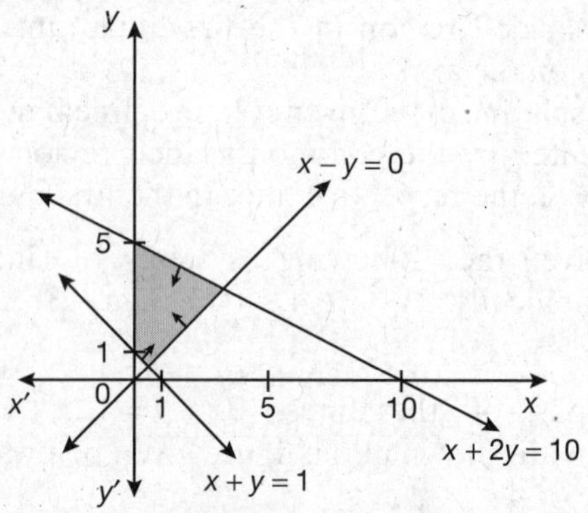

Inequality $x+2y \le 10$ represents the shaded region below the line $x+2y=10$ (including all the points on the line).
Inequality $x+y \ge 1$ represents the shaded region above the line $x+y=1$ (including all the points on the line).
Inequality $x-y \le 0$ represents the shaded region above the line $x-y=0$ (including all the points on the line).
Inequality $x \ge 0$, $y \ge 0$ represents every point in the common shaded region in the first quadrant including the points on the axes.
Thus, the solution of the given system of linear inequalities is represented by the common shaded region including the points on the respective lines in the first quadrant.

Day 86

NCERT Miscellaneous Exercise (With Solutions)

Q1. Solve the inequality $2 \le 3x-4 \le 5$.

Solution:
$2 \le 3x-4 \le 5 \Rightarrow 2+4 \le 3x-4+4 \le 5+4$
(Adding 4 in the inequality)
$\Rightarrow 6 \le 3x \le 9 \Rightarrow 2 \le x \le 3$
(Dividing the inequality by 3)
Therefore, all real numbers x which are greater than or equal to 2 but less than or equal to 3 are the solutions of the given inequality.
Thus, the solution set of the given inequality is $x \in [2,3]$.

Q2. Solve the inequality $6 \le -3(2x-4) < 12$.

Solution:
$6 \le -3(2x-4) < 12$
$\Rightarrow 6 \le -6x+12 < 12$
(Subtracting 12 from the inequality)
$\Rightarrow 6-12 \le -6x+12-12 < 12-12$
$\Rightarrow -6 \le -6x < 0$
(Divide the inequality by –6. Since the dividend number is negative, the inequality sign will reverse)
$\Rightarrow \frac{-6}{-6} \ge \frac{-6x}{-6} > \frac{0}{-6}$
$\Rightarrow 1 \ge x > 0$
Thus, the solution set for the given inequality is $x \in (0,1]$.

Q3. Solve the inequality $-3 \le 4 - \frac{7x}{2} \le 18$.

Solution:

$-3 \le 4 - \frac{7x}{2} \le 18 \Rightarrow -3 - 4 \le 4 - \frac{7x}{2} - 4 \le 18 - 4$

(Subtracting 4 from the inequality)

$\Rightarrow -7 \le -\frac{7x}{2} \le 14 \Rightarrow -7 \times \frac{2}{7} \le -\frac{7x}{2} \times \frac{2}{7} \le 14 \times \frac{2}{7}$

(Multiplying the inequality by $\frac{2}{7}$)

$\Rightarrow -2 \le -x \le 4 \Rightarrow 2 \ge x \ge -4$

(Multiply the inequality by –1. Since the multiplication number is negative, the inequality sign will reverse.)

Thus, the solution set for the given inequality is $x \in [-4, 2]$.

Q4. Solve the inequality $-15 < \frac{3(x-2)}{5} \le 0$.

Solution:

$-15 < \frac{3(x-2)}{5} \le 0 \Rightarrow -15 < \frac{3x}{5} - \frac{6}{5} \le 0$

$\Rightarrow -15 + \frac{6}{5} < \frac{3x}{5} - \frac{6}{5} + \frac{6}{5} \le 0 + \frac{6}{5}$

(Adding $\frac{6}{5}$ in the inequality)

$\Rightarrow -\frac{69}{5} < \frac{3x}{5} \le \frac{6}{5}$

(Multiplying the inequality by $\frac{5}{3}$)

$\Rightarrow -\frac{69}{5} \times \frac{5}{3} < \frac{3x}{5} \times \frac{5}{3} \le \frac{6}{5} \times \frac{5}{3}$

$\Rightarrow -23 < x \le 2$

Thus, the solution set for the given inequality is $x \in (23, 2]$.

Q5. Solve the inequality $-12 < 4 - \frac{3x}{-5} \le 2$.

Solution:

$-12 < 4 - \frac{3x}{-5} \le 2 \Rightarrow -12 < 4 + \frac{3x}{5} \le 2$

$\Rightarrow -12 - 4 < 4 + \frac{3x}{5} - 4 \le 2 - 4$

(Subtracting 4 from the inequality)

$\Rightarrow -16 < \frac{3x}{5} \le -2 \Rightarrow -16 \times \frac{5}{3} < \frac{3x}{5} \times \frac{5}{3} \le -2 \times \frac{5}{3}$

(Multiplying the inequality by $\frac{5}{3}$)

$\Rightarrow \frac{-80}{3} < x \le \frac{-10}{3}$

Thus, the solution set for the given inequality is

$x \in \left(\frac{-80}{3}, \frac{-10}{3}\right)$

Q6. Solve the inequality $7 \le \frac{(3x+11)}{2} \le 11$.

Solution:

$7 \le \frac{(3x+11)}{2} \le 11 \Rightarrow 7 \le \frac{3x}{2} + \frac{11}{2} \le 11$

$\Rightarrow 7 - \frac{11}{2} \le \frac{3x}{2} + \frac{11}{2} - \frac{11}{2} \le 11 - \frac{11}{2}$

(Subtracting $\frac{11}{2}$ from the inequality)

$\Rightarrow \frac{3}{2} \le \frac{3x}{2} \le \frac{11}{2} \Rightarrow \frac{3}{2} \times \frac{2}{3} \le \frac{3x}{2} \times \frac{2}{3} \le \frac{11}{2} \times \frac{2}{3}$

(Multiplying the inequality by $\frac{2}{3}$)

$\Rightarrow 1 \le x \le \frac{11}{3}$

Thus, the solution set for the given inequality is

$x \in \left[1, \frac{11}{3}\right]$

Q7. Solve the inequalities and represent the solution graphically on the number line:

$5x + 1 > -24, \quad 5x - 1 < 24$

Solution:

Inequality $5x - 1 < 24$

$\Rightarrow 5x < 24 + 1 \Rightarrow 5x < 25 \Rightarrow x < 5$

Inequality $5x + 1 > -24$

$\Rightarrow 5x > -24 - 1 \Rightarrow 5x > -25 \Rightarrow x > -5 \Rightarrow x < 5$

Thus, the solution set of the given inequality is $x \in (-5, 5)$. The graphical (number line) representation of the solutions of the given inequality is,

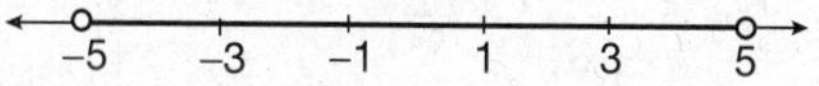

Q8. Solve the inequalities and represent the solution graphically on number line:

$2(x-1) < x + 5, \quad 3(x+2) > 2 - x$

Solution:

Inequality $2(x-1) < x + 5$

$\Rightarrow 2x - 2 < x + 5 \Rightarrow 2x - x < 5 + 2 \Rightarrow x < 7$

Inequality $3(x+2) > 2 - x \Rightarrow 3x + 6 > 2 - x$

$\Rightarrow 3x + x > 2 - 6 \Rightarrow 4x > -4 \Rightarrow x > -1$

Thus, the solution set of the given inequality is $x \in (-1, 7)$. The graphical (number line) representation of the solutions of the given inequality is,

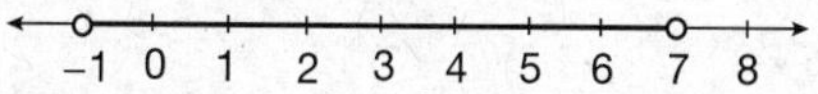

Q9. Solve the following inequalities and represent the solution graphically on the number line:

$3x-7>2(x-6),\ 6-x>11-2x$.

Solution:

Inequality $3x-7>2(x-6)$

$\Rightarrow 3x-7>2x-12 \Rightarrow 3x-2x>-12+7 \Rightarrow x>-5$

Inequality $6-x>11-2x$

$\Rightarrow -x+2x>11-6 \Rightarrow x>5$

Thus, the solution set of the given inequality is $x \in (5,\infty)$. The graphical (number line) representation of the solutions of the given inequality is,

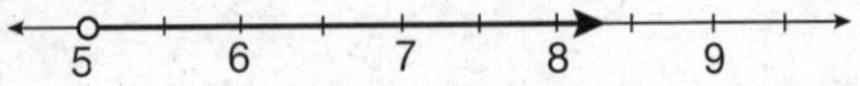

Q10. Solve the inequalities and represent the solution graphically on the number line:
$5(2x-7)-3(2x+3)\le 0,\ 2x+19\le 6x+47$

Solution:

Inequality $5(2x-7)-3(2x+3)\le 0$

$\Rightarrow 10x-35-6x-9\le 0 \Rightarrow 4x-44\le 0 \Rightarrow x\le 11$

Inequality $2x+19\le 6x+47$

$\Rightarrow 19-47\le 6x-2x \Rightarrow -28\le 4x \Rightarrow -7\le x$

Thus, the solution set of the given inequality is $x\in(-7,11)$. The graphical (number line) representation of the solutions of the given inequality is

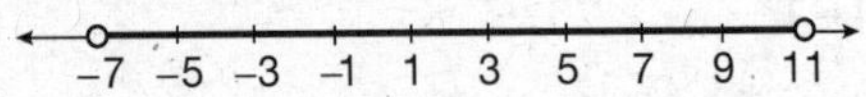

Q11. A solution is to be kept between 68°F and 77°F. What is the range in temperature in celsius (C) if the celsius/fahrenheit (F) conversion formula is given by $F=\frac{9}{8}C+32$?

Solution:

The solution is to be kept between 68°F and 77°F,

$68<F<77$

Substituting $F=\frac{9}{8}C+32$, we get

$68<\frac{9}{8}C+32<77$

$\Rightarrow 68-32<\frac{9}{5}C<77-32$

$\Rightarrow 36<\frac{9}{5}C<45$

$\Rightarrow 36\times\frac{5}{9}<C<45\times\frac{5}{9}$

$\Rightarrow 20<C<25$

Thus, the required range of temperature in degree celsius is between 20°C and 25°C.

Q12. A solution of 8% boric acid is to be diluted by adding a 2% boric acid solution to it. The resulting mixture is to be more than 4% but less than 6% boric acid. If we have 640 litres of the 8% solution, how many litres of the 2% solution will have to be added?

Solution:

Let x litres of 2% boric acid solution be required to be added.

Then, total mixture = $(x+640)$ litres

Since this resulting mixture is to be more than 4% but less than 6% boric acid,

$4\%\text{ of }(x+640)<2\%\text{ of }x+8\%\text{ of }640$

$\Rightarrow \frac{4}{100}(x+640)<\frac{2}{100}x+\frac{8}{100}(640)$

$\Rightarrow 4x+2560<2x+5120$

$\Rightarrow 2x<5120-2560$

$\Rightarrow 2x<2560$

$\Rightarrow x<1280$

Also,

$2\%\text{ of }x+8\%\text{ of }640<6\%\text{ of }(x+640)$

$\frac{2}{100}x+\frac{8}{100}(640)<\frac{6}{100}(x+640)$

$\Rightarrow 2x+5120<6x+3840$

$\Rightarrow 5120-3840<6x-2x$

$\Rightarrow 1280<4x \Rightarrow 320<x$

Thus, $320<x<1280$

Hence, the number of litres of 2% of boric solution that is to be added will have to be more than 320 litres but less than 1280 litres.

Q13. How many litres of water will have to be added to 1125 litres of the 45% solution of acid so that the resulting mixture will contain more than 25% but less than 30% acid content?

Solution:

Let x litres of water is added, then the total mixture = $(x+1125)$ litres

The amount of acid contained in the resulting mixture is 45% of 1125 litres. And this resulting mixture will contain more than 25% but less than 30% acid content. Therefore

$30\%\text{ of }(1125+x)>45\%\text{ of }1125$

$\Rightarrow \frac{30}{100}(1125+x)>\frac{45}{100}\times 1125$

$\Rightarrow 30(1125+x)>45\times 1125$

$\Rightarrow 30\times 1125+30x>45\times 1125$

$\Rightarrow 30x > 45\times 1125 - 30\times 1125$

$\Rightarrow 30x > (45-30)\times 1125$

$\Rightarrow x > \dfrac{15\times 1125}{30} = 562.5$

Also, 25% of $(1125+x) < 45\%$ of 1125

$\Rightarrow \dfrac{25}{100}(1125+x) < \dfrac{45}{100}\times 1125$

$\Rightarrow 25(1125+x) > 45\times 1125$

$\Rightarrow 25\times 1125 + 25x > 45\times 1125$

$\Rightarrow 25x > 45\times 1125 - 25\times 1125$

$\Rightarrow 25x > (45-25)\times 1125$

$\Rightarrow x > \dfrac{20\times 1125}{25} = 900$

$\therefore 562.5 < x < 900$

Thus, the required number of litres of water that is to be added will have to be more than 562.5 but less than 900.

Q14. The IQ of a person is given by the formula $IQ = \dfrac{MA}{CA}\times 100$, where MA is the mental age and CA is the chronological age. If $80 \le IQ \le 140$ for a group of 12-year-old children, find the range of their mental age.

Solution:

For a group of 12-year-old children,

$80 \le IQ \le 140$

And for a group of 12-year-old children,

CA = 12 years

$\therefore IQ = \dfrac{MA}{CA}\times 100 = \dfrac{MA}{12}\times 100 = \dfrac{25}{3}MA$

Thus,

$80 \le IQ \le 140$

$\Rightarrow 80 \le \dfrac{25}{3}MA \le 140$

$\Rightarrow 80\times\dfrac{3}{25} \le MA \le 140\times\dfrac{3}{25}$

$\Rightarrow 9.6 \le MA \le 16.8$

Thus, the range of mental age of the group of 12-year-old children is $9.6 \le MA \le 16.8$.

Day 87

NCERT Exemplar Problems: Solutions (Q1–Q18)

Q1. $\dfrac{4}{x-1} \le 3 \le \dfrac{6}{x+1}, \quad (x>0)$

Solution:

Consider the first two inequalities,

$\dfrac{4}{x-1} \le 3 \Rightarrow 4 \le 3(x+1)$

$\Rightarrow 4 \le 3x+3 \Rightarrow 4-3 \le 3x$

$\Rightarrow 1 \le 3x \Rightarrow \dfrac{1}{3} \le x$

Now, consider the last two inequalities,

$3 \le \dfrac{6}{x+1} \Rightarrow 3(x+1) \le 6$

$\Rightarrow 3x+3 \le 6 \Rightarrow 3x \le 6-3$

$\Rightarrow 3x \le 3 \Rightarrow x \le 1$

Thus, the solution set of the given inequality is $x \in [\frac{1}{3}, 1]$.

Q2. $\dfrac{|x-2|-1}{|x-2|-2} \le 0$

Solution:

Consider case I: $x \ge 2$

Then, $|x-2| = x-2$

$\therefore \dfrac{|x-2|-1}{|x-2|-2} \le 0$

$\Rightarrow \dfrac{x-2-1}{x-2-2} \le 0 \Rightarrow \dfrac{x-3}{x-4} \le 0$

$\Rightarrow x-3 \le 0$ and $x-4 > 0$

or $x-3 \ge 0$ and $x-4 < 0$

$\Rightarrow x \le 3$ and $x > 4$

or $x \ge 3$ and $x < 4$

$\Rightarrow x \le 3$ and $x > 4$ does not give any solution

And $x \ge 3$ and $x < 4 \Rightarrow x \in [3,4)$ is the solution

Consider case II: $x < 2$

Then, $|x-2| = -(x-2) = 2-x$

$\therefore \dfrac{|x-2|-1}{|x-2|-2} \le 0$

$\Rightarrow \dfrac{2-x-1}{2-x-2} \le 0 \Rightarrow \dfrac{1-x}{-x} \le 0$

$\Rightarrow 1-x \le 0$ and $-x > 0$

or $1-x > 0$ and $-x < 0$

$\Rightarrow 1 \le x$ and $x < 0$

or $1 > x$ and $x > 0$

$\Rightarrow 1 \le x$ and $x < 0$ does not give any solution

And $1 > x$ and $x > 0 \Rightarrow x \in (0,1)$ is the solution

Thus, the solution set of the given inequality is $x \in (0,1) \cup [3,4)$.

Q3. $\dfrac{1}{|x|-3} \le \dfrac{1}{2}$

Solution:

$\Rightarrow \dfrac{1}{|x|-3} - \dfrac{1}{2} \le 0 \Rightarrow \dfrac{2-(|x|-3)}{2(|x|-3)} \le 0$

$\Rightarrow \dfrac{-|x|+5}{2(|x|-3)} \le 0 \Rightarrow \dfrac{5-|x|}{|x|-3} \le 0$

Consider case I: $x \ge 0$

Then, $\dfrac{5-|x|}{|x|-3} \le 0 \Rightarrow \dfrac{5-x}{x-3} \le 0$

$\Rightarrow 5-x \le 0$ and $x-3>0$

or $5-x \ge 0$ and $x-3<0$

$\Rightarrow 5 \le x$ and $x>3$

or $5 \ge x$ and $x<3$

$\Rightarrow x \in [5,\infty) \cup [0,3)$, since $x \ge 0$

Now consider case II: $x<0$

Then, $\dfrac{5-|x|}{|x|-3} \le 0$

$\Rightarrow \dfrac{5-(-x)}{(-x)-3} \le 0 \Rightarrow \dfrac{5+x}{-x-3} \le 0$

$\Rightarrow 5+x \le 0$ and $-x-3>0$

or $5+x \ge 0$ and $-x-3<0$

$\Rightarrow x \le -5$ and $-3>x$

or $x \ge -5$ and $-3<x$

$\Rightarrow x \in (-\infty,-5] \cup (-3,0)$, since $x<0$

Thus, the solution set of the given inequality is

$x \in (-\infty,-5] \cup (-3,0) \cup [0,3) \cup [5,\infty)$

$\Rightarrow x \in (-\infty,-5] \cup (-3,3) \cup [5,\infty)$

Q4. $|x-1| \le 5,\ |x| \ge 2$

Solution:

Consider case I: $x \ge 1$

Then, $|x-1| = x-1$ and $|x| = x$

$\therefore |x-1| \le 5$ and $|x| \ge 2$

$\Rightarrow x-1 \le 5$ and $x \ge 2$

$\Rightarrow x \le 6$ and $x \ge 2$

$\Rightarrow 2 \le x \le 6$ is the solution, since $x \ge 1$.

Consider case II: $0 \le x < 1$

Then, $|x-1| = -(x-1) = 1-x$ and $|x| = x$

$\therefore |x-1| \le 5$ and $|x| \ge 2$

$\Rightarrow 1-x \le 5$ and $x \ge 2$

$\Rightarrow -4 \le x$ and $x \ge 2$

Does not give any solution, since $0 \le x < 1$

Consider case III: $x<0$

Then, $|x-1| = -(x-1) = 1-x$ and $|x| = -x$

$\therefore |x-1| \le 5$ and $|x| \ge 2$

$\Rightarrow 1-x \le 5$ and $-x \ge 2$

$\Rightarrow -4 \le x$ and $-2 \ge x$

$\Rightarrow -4 \le x \le -2$ is the solution,

Thus, the solution set of the given inequality is

$x \in [-4,-2] \cup [2,6]$.

Q5. $-5 \le \dfrac{2-3x}{4} \le 9$

Solution:

Consider the first two inequalities,

$-5 \le \dfrac{2-3x}{4} \Rightarrow -20 \le 2-3x$

$\Rightarrow 3x \le 2+20 \Rightarrow 3x \le 22$

$\Rightarrow x \le \dfrac{22}{3}$

Now, consider last two inequalities,

$\Rightarrow \dfrac{2-3x}{4} \le 9 \Rightarrow 2-3x \le 36$

$\Rightarrow -3x \le 36-2 \Rightarrow -3x \le 34$

$\Rightarrow 3x \ge -34 \Rightarrow x \ge -\dfrac{34}{3}$

Thus, the solution set of the given inequality is

$-\dfrac{34}{2} \le x \le \dfrac{22}{3} \Rightarrow x \in \left[\dfrac{-34}{3}, \dfrac{22}{3}\right]$

Q6. $4x+3 \ge 2x+17,\ 3x-5<-2$

Solution:

Inequality $4x+3 \ge 2x+17$

$\Rightarrow 4x-2x \ge 17-3 \Rightarrow 2x \ge 14$

$\Rightarrow x \ge \dfrac{14}{2} \Rightarrow x \ge 7$

Inequality $3x-5<-2$

$\Rightarrow 3x < -2+5 \Rightarrow 3x<3 \Rightarrow x<1$

Thus, the solution set of the given inequality is

$x<1 \cup x \ge 7 \Rightarrow x \in (-\infty,1) \cup [7,\infty)$

Q7. A company manufactures cassettes. Its cost and revenue functions are $C(x) = 26000+30x$, $R(x) = 43x$, respectively, where x is the number of cassettes produced and sold in a week. How many cassettes must be sold by the company to realise some profit?

Solution:

Cost function, $C(x) = 26000+30x$

And revenue function, $R(x) = 43x$

For profit, $R(x) > C(x)$

$\Rightarrow 43x > 26000 + 30x$

$\Rightarrow 43x - 30x > 26000$

$\Rightarrow 13x > 26000 \Rightarrow x > \frac{26000}{13} \Rightarrow x > 2000$

Thus, more than 2000 cassettes must be produced to get some profit.

Q8. The water acidity in a pool is considered normal when the average pH reading of three daily measurements is between 8.2 and 8.5. If the first two pH readings are 8.48 and 8.35, then find the range of pH value for the third reading that will result in the acidity level being normal.

Solution:

First pH value = 8.48

Second pH value = 8.35

The average pH value lies between 8.2 and 8.5.

$\therefore 8.2 < \frac{8.48 + 8.35 + x}{3} < 8.5$

$\Rightarrow 8.2 < \frac{16.83 + x}{3} < 8.5$

$\Rightarrow 24.6 < 16.83 + x < 25.5$

$\Rightarrow 24.6 - 16.83 < x < 25.5 - 16.83$

$\Rightarrow 7.77 < x < 8.67$

Thus, the third pH value lies between 7.77 and 8.67.

Q9. A solution of 9% acid is to be diluted by adding 3% acid solution to it. The resulting mixture is to be more than 5% but less than 7% acid. If there is 460 L of the 9% solution, how many litres of the 3% solution will have to be added?

Solution:

Let x L of 3% solution be added to 460 L of 9% solution of acid.

Then, total quantity of mixture = $(460 + x)$L

Total acid content in the $(460 + x)$L of mixture

$= \left(460 \times \frac{9}{100} + x \times \frac{3}{100}\right)$

The acid content in the resulting mixture must be more than 5% but less than 7% acid. Therefore

$5\% \text{ of } (460 + x) < 460 \times \frac{9}{100} + \frac{3x}{100} < 7\% \text{ of } (460 + x)$

$\Rightarrow \frac{5}{100}(460 + x) < 460 \times \frac{9}{100} + \frac{3}{100}x < \frac{7}{100}(460 + x)$

$\Rightarrow 5 \times (460 + x) < 460 \times 9 + 3x < 7 \times (460 + x)$

$\Rightarrow 2300 + 5x < 4140 + 3x < 3220 + 7x$

Taking first two inequalities,

$2300 + 5x < 4140 + 3x$

$\Rightarrow 5x - 3x < 4140 - 2300$

$\Rightarrow 2x < 1840 \Rightarrow x < 920$

And taking last two inequalities,

$4140 + 3x < 3220 + 7x$

$\Rightarrow 7x - 3x > 4140 - 3220$

$\Rightarrow 4x > 920 \Rightarrow x > 230$

Thus, the number of litres of the 3% solution of acid must be more than 230 L and less than 920 L.

Q10. A solution is to be kept between 40°C and 45°C. What is the range of temperature in degree fahrenheit, if the conversion formula is $F = \frac{9}{5}C + 32$?

Solution:

$F = \frac{9}{5}C + 32 \Rightarrow 5F = 9C + 32 \times 5$

$\Rightarrow 9C = 5F - 32 \times 5 \Rightarrow C = \frac{5F - 160}{9}$

The temperature in degree celsius lies between 40°C and 45°C. Therefore, $40 < \frac{5F - 160}{9} < 45$

$\Rightarrow 40 \times 9 < 5F - 160 < 45 \times 9$

$\Rightarrow 360 + 160 < 5F < 405 + 160$

$\Rightarrow 520 < 5F < 565 \Rightarrow 104 < F < 113$

Thus, the range of temperature in degree fahrenheit is 104 F to 113 F.

Q11. The longest side of a triangle is twice the shortest side and third side is 2 cm longer than the shortest side. If the perimeter of the triangle is more than 166 cm, then find the minimum length of the shortest side.

Solution:

Let the length of shortest side be x cm. Then

Longest side = = $2x$ cm

And third side =2+ Shortest side

$= (2 + x)$cm

Perimeter of triangle $= x + 2x + (x + 2) = 4x + 2$

According to the question,

Perimeter >166 cm

$\Rightarrow 4x + 2 > 166 \Rightarrow 4x > 166 - 2$

$\Rightarrow 4x > 164 \Rightarrow x > \frac{164}{4} \Rightarrow x > 41$ cm

Thus, the minimum length of the shortest side is 41 cm.

Q12. In drilling the world's deepest hole, it was found that the temperature T in degree celsius, x km below the earth's surface was given by $T = 30 + 25(x - 3)$, $3 \le x \le 15$. At what depth will the temperature be between 155°C and 205°C?

Solution:

$T = 30 + 25(x - 3)$, $3 \le x \le 15$,

Therefore, according to the question,

$155 < T < 205$

$\Rightarrow 155 < 30 + 25(x-3) < 205$

$\Rightarrow 125 < 25(x-3) < 175$

$\Rightarrow 5 < x-3 < 7 \Rightarrow 8 < x < 10$

Thus, at a depth of 8 to 10 km, the temperature lies between 155 and 205°C.

Q13. Solve the following system of inequalities

$$\frac{2x+1}{7x-1} > 5, \quad \frac{x+7}{x-8} > 2.$$

Solution:

Inequality, $\frac{2x+1}{7x-1} > 5 \Rightarrow \frac{2x+1}{7x-1} - 5 > 0$

$\Rightarrow \frac{(2x+1)-5(7x-1)}{7x-1} > 0$

$\Rightarrow \frac{2x+1-35x+5}{7x-1} > 0$

$\Rightarrow \frac{-33x+6}{7x+1} > 0 \Rightarrow \frac{33x-6}{7x-1} < 0$

$\Rightarrow \frac{11x-2}{7x-1} < 0 \Rightarrow x \in \left(\frac{1}{7}, \frac{2}{11}\right)$

Inequality, $\frac{x+7}{x-8} > 2 \Rightarrow \frac{x+7}{x-8} - 2 > 0$

$\Rightarrow \frac{x+7-2(x-8)}{x-8} > 0$

$\Rightarrow \frac{x+7-2x+16}{x-8} > 0$

$\Rightarrow \frac{-x+23}{x-8} > 0 \Rightarrow \frac{x-23}{x-8} < 0$

$\Rightarrow x \in (8, 23)$

The intersection of the solutions, $x \in \left(\frac{1}{7}, \frac{2}{11}\right)$ and $x \in (8, 23)$ is the null set. Therefore, the given system of equations has no solution.

Q14. Find the linear inequalities for which shaded region in the given figure is the solution set.

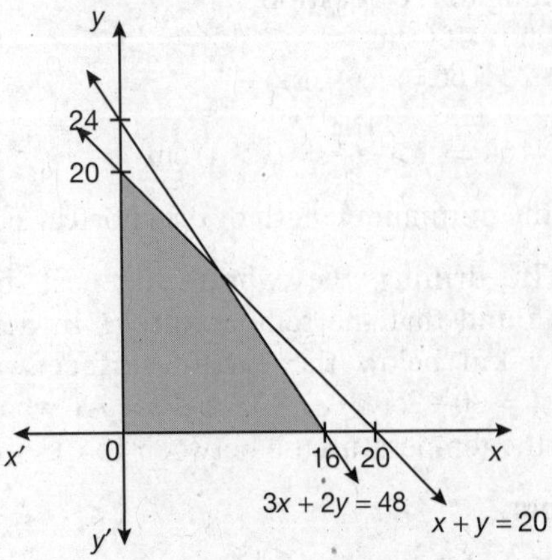

Solution: Consider the line $3x + 2y = 48$

As the shaded region and the origin are on the same side of the line $3x + 2y = 48$, it means that (0,0) satisfies the linear constraints $3x + 2y \leq 48$. Therefore, we get one inequality as $3x + 2y \leq 48$.

Now, consider the line $x + y = 20$,

As the shaded region and the origin are on the same side of the line $x + y = 20$, it means that (0,0) satisfies the constraints $x + y \leq 20$.

Therefore, we get the second inequality as $x + y \leq 20$.

The shaded region is above X-axis and is on the right side of Y-axis, so we must have $x \geq 0, y \geq 0$.

Thus, the linear inequalities corresponding to the given solution set are

$3x + 2y \leq 48, \; x + y \leq 20, \; x \geq 20, \; y \geq 0.$

Q15. Find the linear inequalities for which the shaded region in the given figure is the solution set.

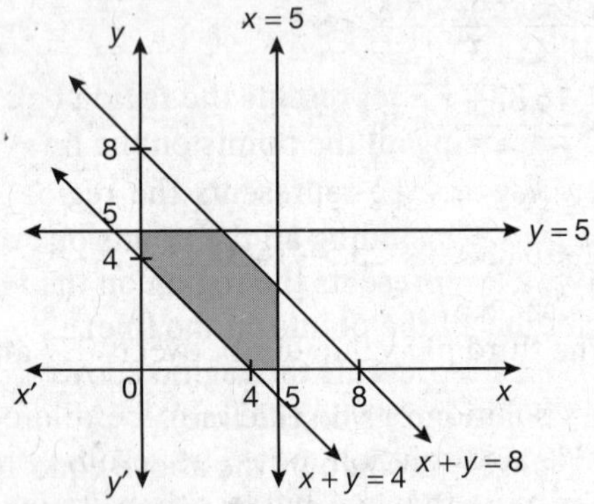

Solution:

Consider the line $x + y = 4$.

As the shaded region and the origin lie on the opposite side of this line, it means (0, 0) does not satisfy the linear constraint $x + y \geq 4$.

Therefore, we get one linear inequality as $x + y \geq 4$.

Now, consider the line $x + y = 8$. Since the shaded region and origin lie on the same side of this line, it means that (0, 0) satisfies the linear constraints $x + y \leq 8$. Therefore, we get the second linear inequality as $x + y \leq 8$.

Now, consider the line $x = 5$. Since the shaded region and origin are on the left of this line, it means that (0, 0) satisfies the constraints $x \leq 5$.

Therefore, we get the third linear inequality as $x \leq 5$.

Now, consider the line $y = 5$. Since the shaded region and origin are below this line, it means that (0, 0) satisfies the constraints $y \leq 5$.

Therefore, we get the fourth linear inequality as $y \leq 5$.

Also the shaded region is above the X-axis and on the right of the Y-axis, that is, the shaded region is in the first quadrant. So, we must have $x \geq 0, y \geq 0$.

Thus, the linear inequalities corresponding to the given solution set are $x + y \geq 4; \; x + y \leq 8; x \leq 5; \; y \leq 5; x \geq 0, \; y \geq 0$.

Q16. Show that the following system of linear inequalities has no solution $x+2y \le 3, 3x+4y \ge 12, x \ge 0, y \ge 1$.

Solution:

The graphs of the lines, $x+2y=3$, $3x+4y=12$, $x=0$ and $y=1$, are shown in the figure given below.

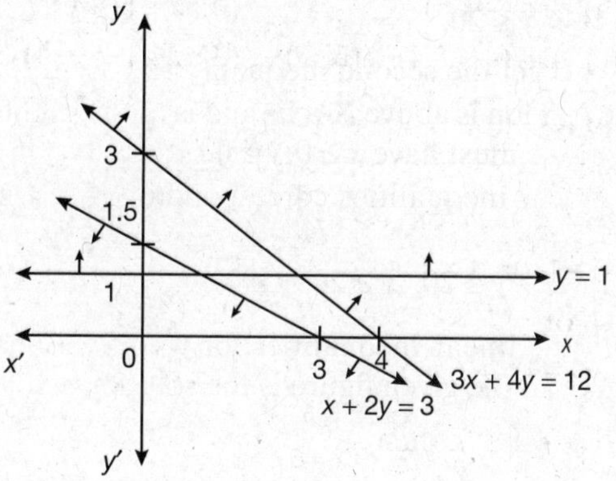

Inequality $x+2y \le 3$ represents the region below the line $x+2y=3$ (including all the points on the line).

Inequality $3x+2y \ge 12$ represents the region above the line $3x+2y=12$ (including all the points on the line).

Inequality $x \ge 0$ represents the region on the right of line $x=0$ (including all the points on the line).

Inequality $y \ge 1$ represents the region above the line $y=1$ (including all the points on the line).

Since the regions obtained by the above equations do not have a common region in all, the solution set is a null set. Thus, the given system of inequalities has no solution.

Q17. Solve the following system of linear inequalities $3x+2y \ge 24$, $3x+y \le 15$, $x \ge 4$.

Solution:

The graphs of the lines, $3x+2y=24$, $3x+y=15$, and $x=4$, are shown in the figure given below.

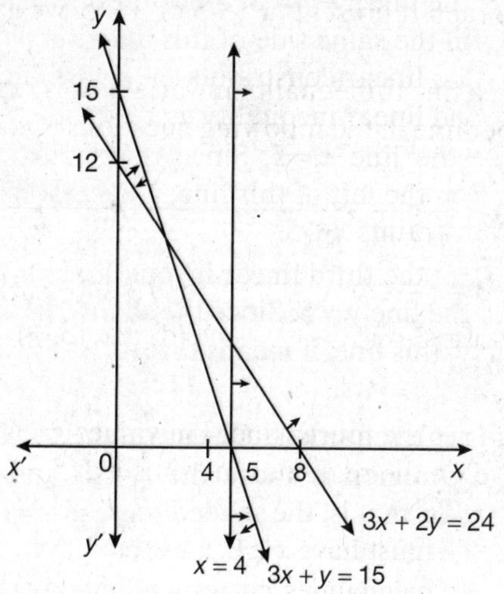

Inequality $3x+2y \ge 24$ represents the region above the line $3x+2y=24$ (including all the points on the line).

Inequality $3x+y \le 15$ represents the region below the line $3x+y=15$ (including all the points on the line).

Inequality $x \ge 4$ represents the region on the right of line $x=4$ (including all the points on the line).

Since the region obtained by the above equations do not have a common region in all, the solution set is a null set. Thus, the given system of inequalities has no solution.

Q18. Show that the solution set of the following system of linear inequalities is an unbounded region $2x+y \ge 8, x+2y \ge 10, x \ge 0, y \ge 0$.

Solution:

The graphs of the lines, $2x+y=8$, $x+2y=10$, $x=0$ and $y=0$, are shown in the figure given below.

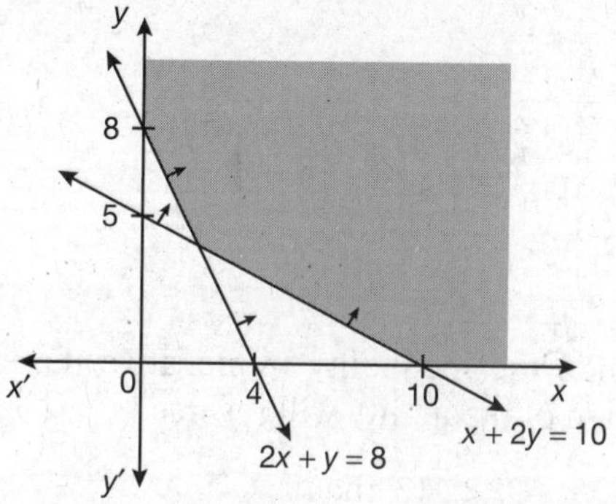

Inequality $2x+y \ge 8$ represents the shaded region above the line $2x+y=8$ (including all the points on the line).

Inequality $x+2y \ge 10$ represents the shaded region above the line $x+2y=10$ (including all the points on the line).

Inequalities $x \ge 0$, $y \ge 0$ represent the region in the first quadrant.

Thus,. the graph of the common shaded portion is unbounded.

Day 88

NCERT Exemplar Problems: Solutions (Q19–Q32)

Q19. If $x < 5$, then

(a) $-x < -5$ (b) $-x \le -5$
(c) $-x > -5$ (d) $-x \ge -5$

Solution: (c)

$x < 5 \Rightarrow -x > -5$

[If we multiply the inequality by negative numbers (–1), then the sign of inequality will reverse.]

Q20. If x, y and b are real numbers and $x < y$, $b < 0$, then

(a) $\frac{x}{b} < \frac{y}{b}$ (b) $\frac{x}{b} \le \frac{y}{b}$
(c) $\frac{x}{b} > \frac{y}{b}$ (d) $\frac{x}{b} \ge \frac{y}{b}$

Solution: (c)

$x < y \Rightarrow \frac{x}{b} > \frac{y}{b}$

[If we divide the inequality by negative numbers ($-b$), then the sign of inequality will reverse.]

Q21. If $-3x + 17 < -13$, then

(a) $x \in (10, \infty)$ (b) $x \in [10, \infty)$
(c) $x \in (-\infty, 10]$ (d) $x \in [-10, 10)$

Solution: (a)

$-3x + 17 < -13 \Rightarrow 3x - 17 > 13$

(Multiplying both sides by –1, since multiplication number is negative, inequality sign will reverse.)

$\Rightarrow 3x > 13 + 17 \Rightarrow 3x > 30 \Rightarrow x > 10$

Q22. If x is a real number and $|x| < 3$, then

(a) $x \ge 3$ (b) $-3 < x < 3$
(c) $x \le -3$ (d) $-3 \le x \le 3$

Solution: (b)

$|x| < 3 \Rightarrow -3 < x < 3$

$(\because |x| < a \Rightarrow -a < x < a)$

Q23. Let x and b be real numbers. If $b > 0$, $|x| > b$, then

(a) $x \in (-b, \infty)$ (b) $x \in [-\infty, b)$
(c) $x \in (-b, b)$ (d) $x \in (-\infty, -b) \cup (b, \infty)$

Solution: (d)

$|x| > b \Rightarrow x < -b$ or $x > b$ $(\because b > 0)$

$\Rightarrow x \in (-\infty, -b) \cup (b, \infty)$

Q24. If $|x - 1| > 5$, then

(a) $x \in (-4, 6)$ (b) $x \in [-4, 6]$
(c) $x \in (-\infty, -4) \cup (6, \infty)$ (d) $x \in [-\infty, -4) \cup [6, \infty)$

Solution: (c)

$|x - 1| > 5 \Rightarrow (x - 1) < -5$ or $(x - 1) > 5$

$(\because |x| > a \Rightarrow x < -a$ or $x > a)$

$\Rightarrow x < -5 + 1$ or $x > 5 + 1$

$\Rightarrow x < -4$ or $x > 6$

$\Rightarrow x \in (-\infty, -4) \cup (6, \infty)$

Q25. If $|x + 2| \le 9$, then

(a) $x \in (-7, 11)$
(b) $x \in [-11, 7]$
(c) $x \in (-\infty, -7) \cup (11, \infty)$
(d) $x \in (-\infty, -7) \cup [11, \infty)$

Solution: (b)

$|x + 2| \le 9 \Rightarrow -9 \le x + 2 \le 9$

$(\because |x| \le a \Rightarrow -a \le x \le a)$

$\Rightarrow -9 - 2 \le x \le 9 - 2$

$\Rightarrow -11 \le x \le 7 \Rightarrow x \in [-11, 7]$

Q26. The inequality representing the following graph is

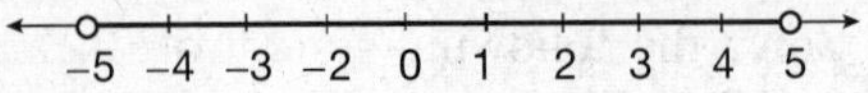

(a) $|x| < 5$ (b) $|x| \le 5$
(c) $|x| > 5$ (d) $|x| \ge 5$

Solution: (a)

Since the graph represents $x > -5$ or $x < 5$, we get $|x| < 5$.

Solution of a linear inequality in variable x is represented on a number line in the following questions (Q27 to Q30).

Q27.

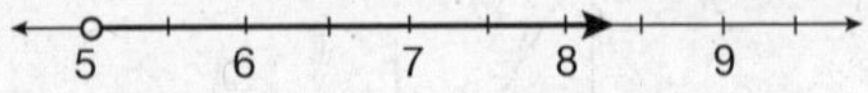

(a) $x \in (-\infty, 5)$ (b) $x \in (-\infty, 5]$
(c) $x \in [5, \infty)$ (d) $x \in (5, \infty)$

Solution: (d)

The given graph represents all the values greater than 5 except $x = 5$ on the real line, and thus $x \in (5, \infty)$.

Q28.

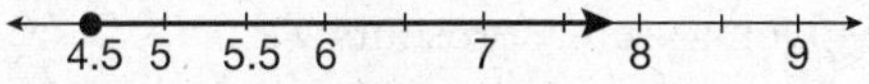

(a) $x \in \left(\frac{9}{2}, \infty\right)$ (b) $x \in \left[\frac{9}{2}, \infty\right)$

(c) $x \in -\left[\infty, \frac{9}{2}\right)$ (d) $x \in \left(-\infty, \frac{9}{2}\right]$

Solution: (b)

The given graph represents all the values greater than $\frac{9}{2}$ including $\frac{9}{2}$ as the real line $x \in \left[\frac{9}{2}, \infty\right)$.

Q29.

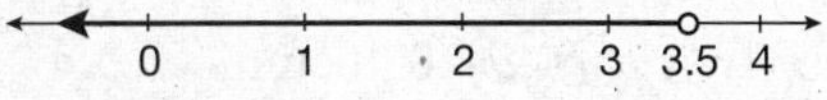

(a) $x \in \left(-\infty, \frac{7}{2}\right)$ (b) $x \in \left(-\infty, \frac{7}{2}\right]$

(c) $x \in \left[\frac{7}{2}, -\infty\right)$ (d) $x \in \left(\frac{7}{2}, \infty\right)$

Solution: (a) The given graph represents all the values less than 7/2 on the real line.

$\Rightarrow x \in \left(-\infty, \frac{7}{2}\right)$

Q30.

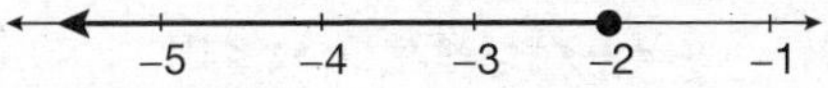

(a) $x \in (-\infty, -2)$ (b) $x \in (-\infty, -2]$

(c) $x \in (-2, \infty]$ (d) $x \in [-2, \infty)$

Solution: (b) The given graph represents all values less than –2 including –2.

$\Rightarrow x \in (-\infty, -2]$

Q31. State whether the following statements are true or false.

(i) If $x < y, b < 0$, then $\frac{x}{b} < \frac{y}{b}$.

(ii) If $xy > 0$, then $x > 0, y < 0$

(iii) If $xy < 0$, then $x < 0, y < 0$

(iv) If $xy < 0$, then $x < 0, y < 0$

(v) If $x < -5, x < -2$, then $x \in (-\infty, -5)$

(vi) If $x < -5, x > 2$, then $x \in (-5, 2)$

(vii) If $x > -2, x < 9$, then $x \in (-2, 9)$

(viii) If $|x| > 5$, then $x \in (-\infty, -5) \cup [5, \infty)$

(ix) If $|x| \geq 5$, then $x \in [-4, 4]$

(x) Graph of $x < 3$ is

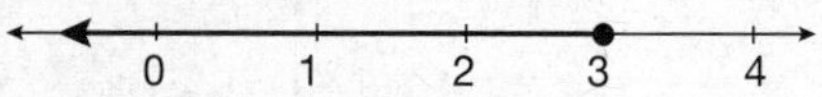

(xi) Graph of $x \geq 0$ is

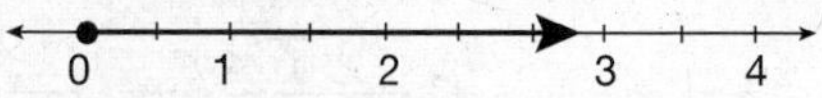

(xii) Graph of $y \leq 0$ is

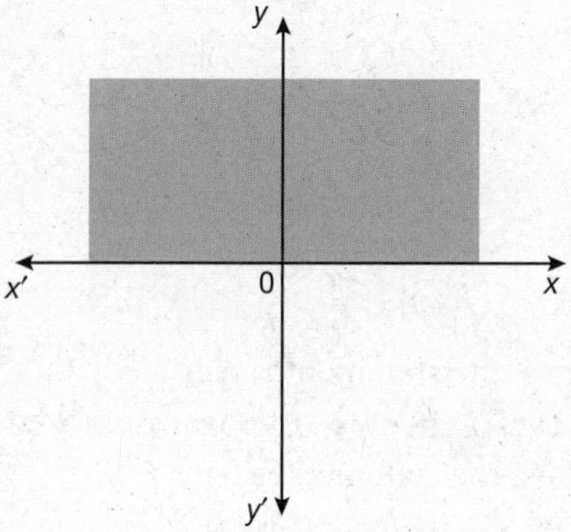

(xiii) Solution set of $x \geq 0, y \leq 0$ is

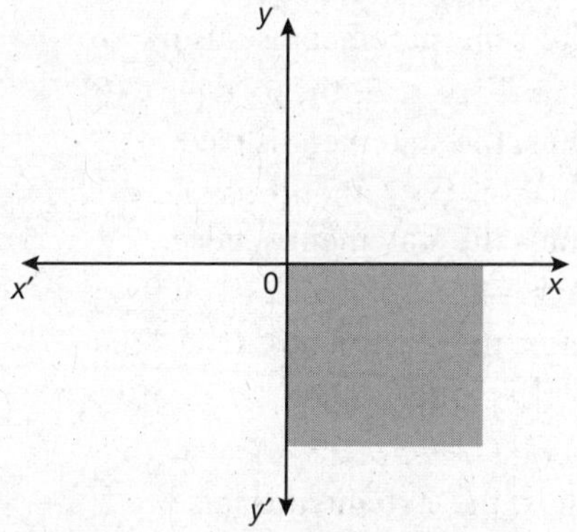

(xiv) Solution set of $x \geq 0, y \leq 1$ is

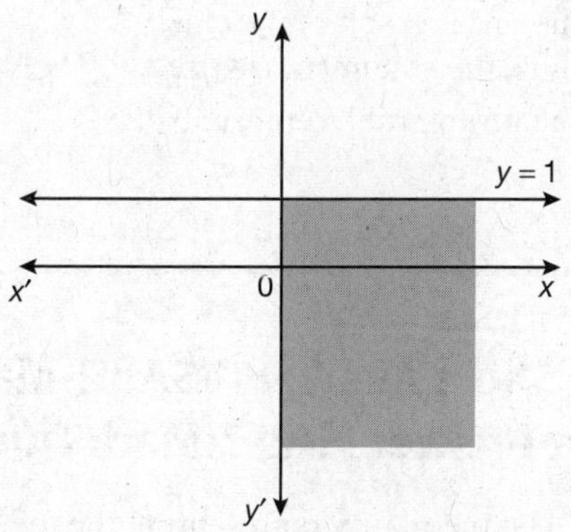

(xv) Solution set of $x + y \geq 0$ is

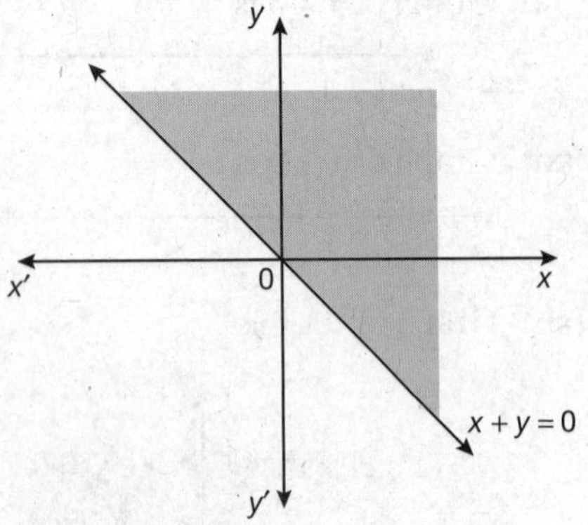

Solution:

(i) If $x < y, b < 0 \Rightarrow \frac{x}{b} > \frac{y}{b}$
Thus, the statement is true.

(ii) If $xy > 0$, then $x > 0, y > 0$ or $x < 0, y < 0$.
Thus, the statement is true.

(iii) If $xy > 0$, then $x < 0, y < 0$.
Thus, the statement is true.

(iv) If $xy < 0 \Rightarrow x < 0, y > 0$ or $x > 0, y < 0$.
Thus, the statement is false.

(v) If $x < -5, y < -2$, then $x \in (-\infty, -5)$
Thus, the statement is true.

(vi) If $x < -5, x > 2$, then x has no value.
Thus, the statement is false.

(vii) If $x > -2, x < 9$, then $x \in (-2, 9)$.
Thus, the statement is true.

(viii) If $|x| > 5$, then either $x < -5$, or $x > 5$.
$\Rightarrow x \in (-\infty, -5) \cup (5, \infty)$
Thus, the statement is false.

(ix) If $|x| > 4$, then $-4 \le x \le 4 \Rightarrow x \in [-4, 4]$
Thus, the statement is true.

(x) The given graph represents $x \le 3$.
Thus, the statement is false.

(xi) The given graph represents $x \ge 0$.
Thus, the statement is true.

(xii) The given graph represents $y \ge 0$.
Thus, the statement is false.

(xiii) The statement is true.

(xiv) The statement is true.

(xv) The statement is true.

Q32. Fill in the blanks

(i) If $-4x \ge 12$, then x.... -3.

(ii) If $\frac{-3}{4}x \le -3$, then x....4.

(iii) If $\frac{2}{x+2} > 0$, then x.... -2.

(iv) If $x > -5$, then $4x$....... -20.

(v) If $x > y, z < 0$, then $-xy$..... $-yz$.

(vi) if $p > 0, q < 0$, then $p - q$......p.

(vii) If $|x+2| > 5$, then x... -7 or x...3.

(viii) If $-2x + 1 \ge 9$, then x....4.

Solution:

(i) $-4x \ge 12 \Rightarrow x \le -3$

(ii) $\frac{-3}{4}x \le -3 \Rightarrow x \ge (-3) \times \frac{4}{3} \Rightarrow x \ge 4$

(iii) $\frac{2}{x+2} > 0 \Rightarrow x > -2$

(iv) $x > -5 \Rightarrow 4x > -20$

(v) if $x > y, z < 0$, then
$xy < yz \Rightarrow -xz > -yz \quad (\because z < 0)$

(vi) If $p > 0, q < 0$, then $p - q > p$

(vii) If $|x+2| > 5$, then
$x + 2 < -5$ or $x + 2 > 5$
$\Rightarrow x < -5 - 2$ or $x > 5 - 2$
$\Rightarrow x < -7$ or $x > 3$
$\Rightarrow x \in (-\infty, -7) \cup (3, \infty)$

(viii) If $-2x + 1 \ge 9$, then
$-2x \ge 9 - 1 \Rightarrow -2x \ge 8$
$\Rightarrow 2x \le -8 \Rightarrow x \le -4$

Day 89

Very Short Answer (VSA): 1-Mark and Short Answer (SA): 2-Mark Questions

Q1. The inequality representing the following graph is

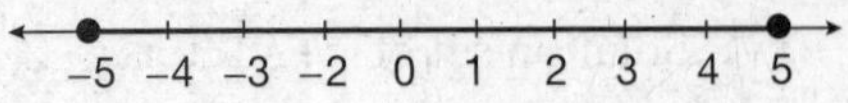

(a) $|x| < 5$ (b) $|x| \le 5$
(c) $|x| > 5$ (d) $|x| \ge 5$

Solution: (b)
Since the graph represents $x > -5$ or $x < 5$. we get $|x| < 5$.

Q2. If $|x+2| \le -2$, then

(a) $x \in (-2, 2)$ (b) $x \in [-2, 2]$
(c) $x \in (-\infty, -4) \cup (0, \infty)$ (d) $x \in \phi$

Solution: (d)
Since the modulus function is always greater and equal to zero, the solution set will be empty.

Q3. If $|x-1|>1$, then

(a) $x \in (0,2)$ (b) $x \in [0,2]$

(c) $x \in (-\infty,0) \cup (2,\infty)$ (d) $x \in [-\infty,0) \cup [2,\infty)$

Solution: (c)

$|x-1|>1 \Rightarrow (x-1)<-1$ or $(x-1)>1$

$(\because |x|>a \Rightarrow x<-a$ or $x>a)$

$\Rightarrow x<-1+1$ or $x>1+1$

$\Rightarrow x<0$ or $x>2 \Rightarrow x \in (-\infty,0)\cup(2,\infty)$

Q4. If x is a real number and $|x|<1$, then

(a) $x<1$ (b) $-1<x<1$

(c) $x<-1$ (d) $-3 \le x \le 3$

Solution: (b)

$|x|<1 \Rightarrow -1<x<1$ $(\because |x|<a \Rightarrow -a<x<a)$

Q5. Solve the given inequality for real x:

$$\frac{3(x-1)}{5} \le \frac{5(1-x)}{3}$$

Solution:

$\frac{3(x-1)}{5} > \frac{5(1-x)}{3} \Rightarrow 9(x-1) > 25(1-x)$

$\Rightarrow 9x-9 > 25-25x \Rightarrow 9x+25x > 25+9$

$\Rightarrow 34x > 34 \Rightarrow x>1$

Therefore, all real numbers x which are greater than 1 are the solutions of the given inequality.

Thus, the solution set of the given inequality is $x \in (1,\infty)$.

Q6. Solve the given inequality for real x:.

$3(1-x) \ge 2(2-x)$

Solution:

$3(1-x) \ge 2(2-x) \Rightarrow 3-3x \ge 4-2x$

$\Rightarrow 3-3x-3 \ge 2-2x-3 \Rightarrow -3x \ge -2x-1$

(Subtracting 3 from both sides)

$\Rightarrow -3x+2x \ge -2x-1+2x \Rightarrow -x \ge -1$

(Adding $2x$ to both sides)

$\Rightarrow x \le 1$ [Multiplying both sides by -1 (since the multiplication number is negative, the inequality sign will reverse)]

Therefore, all real numbers x which are less than or equal to 1 are solutions of the given inequality.

Thus, the solution set of the given inequality is $x \in (-\infty,1]$.

Q7. Solve the given inequality for real x:

$$\frac{x}{15} \le \frac{(5x-2)}{3} - \frac{(7x-3)}{5}$$

Solution:

$\frac{x}{15} \le \frac{(5x-2)}{3} - \frac{(7x-3)}{5}$

$\Rightarrow \frac{x}{15} \le \frac{5(5x-2)-3(7x-3)}{15}$

$\Rightarrow \frac{x}{15} \le \frac{25x-10-21x+9}{15}$

$\Rightarrow \frac{x}{15} \le \frac{4x-1}{15} \Rightarrow x \le 4x-1$

$\Rightarrow 3x \ge 1 \Rightarrow x \ge \frac{1}{3}$

Therefore, all real numbers x which are greater than or equal to $\frac{1}{3}$ are solutions of the given inequality.

Thus, the solution set of the given inequality is $x \in \left[\frac{1}{3},\infty\right)$.

Q8. Solve the given inequality for real x:

$9-(3x+5) \ge 9x-8(x-3)$

Solution:

$9-(3x+5) \ge 9x-8(x-3)$

$\Rightarrow 9-3x-5 \ge 9x-8x+24$

$\Rightarrow 4-3x \ge x+24$

$\Rightarrow -3x-x \ge 24-4$

$\Rightarrow -4x \ge 20 \Rightarrow -x \ge 5 \Rightarrow x \le -5$ +

Therefore, all real numbers x which are less than or equal to -5 are the solutions of the given inequality.

Thus, the solution set of the given inequality is $x \in (-\infty,-5)$.

Q9. Solve the given inequality for real x:

$$\frac{(2x+1)}{3} \ge \frac{(3x-2)}{4} - \frac{(2-x)}{5} + \frac{1}{10}$$

Solution:

$\frac{(2x+1)}{3} \ge \frac{(3x-2)}{4} - \frac{(2-x)}{5} + \frac{1}{10}$

$\Rightarrow \frac{(2x+1)}{3} \ge \frac{5(3x-2)-4(2-x)+2}{20}$

$\Rightarrow \frac{(2x+1)}{3} \ge \frac{15x-10-8+4x+2}{20}$

$\Rightarrow \frac{(2x+1)}{3} \ge \frac{19x-16}{20}$

$\Rightarrow 20(2x+1) \ge 3(19x-16)$

$\Rightarrow 40x+20 \ge 57x-48$

$\Rightarrow 40x-57x \ge -48-20$

$\Rightarrow -17x \ge -68 \Rightarrow -x \ge -4 \Rightarrow x \le 4$

Therefore, all real numbers x which are less than or equal to 4 are solutions of the given inequality.

Thus, the solution set of the given inequality is $x \in (-\infty,4]$.

Q10. Solve the inequality $-3 \le 4 - \frac{7x}{2} \le 11$.

Solution:

$-3 \le 4-\frac{7x}{2} \le 11 \Rightarrow -3-4 \le 4-\frac{7x}{2}-4 \le 11-4$

(Subtracting 4 from the inequality)

$\Rightarrow -7 \le -\frac{7x}{2} \le 7 \Rightarrow -7 \times \frac{2}{7} \le -\frac{7x}{2} \times \frac{2}{7} \le 7 \times \frac{2}{7}$

(Multiplying the inequality by $\frac{2}{7}$)

$\Rightarrow -2 \le -x \le 2 \Rightarrow 2 \ge x \ge -2$

(Multiplying the inequality by –1; as the multiplication number is negative, the inequality sign will reverse)

The solution set for the given inequality is $x \in [-2, 2]$.

Q11. Solve the inequality $0 \le -3(2x-4) < 12$

Solution:

$0 \le -3(2x-4) < 12$

$\Rightarrow 0 \le -6x + 12 < 12$

(Subtracting 12 from the inequality)

$\Rightarrow 0 - 12 \le -6x + 12 - 12 < 12 - 12$

$\Rightarrow -12 \le -6x < 0$

(Dividing the inequality by –6; as the dividend number is negative, the inequality sign will reverse)

$\Rightarrow \frac{-12}{-6} \ge \frac{-6x}{-6} > \frac{0}{-6}$

$\Rightarrow 2 \ge x > 0$

Thus, the solution set for the given inequality is $x \in (0, 2]$.

Q12. $|x-1| \le 2, \ |x| > 2$

Solution:

Consider case I: $x \ge 1$

Then, $|x-1| = x - 1$ and $|x| = x$

$\therefore |x-1| \le 2$ and $|x| > 2$

$\Rightarrow x - 1 \le 2$ and $x > 2$

$\Rightarrow x \le 3$ and $x > 2$

$\Rightarrow 2 < x \le 3$ is the solution, since $x \ge 1$.

Consider case II: $0 \le x < 1$

Then, $|x-1| = -(x-1) = 1 - x$ and $|x| = x$

$\therefore |x-1| \le 2$ and $|x| > 2$

$\Rightarrow 1 - x \le 2$ and $x > 2$

$\Rightarrow -1 \le x$ and $x > 2$

Does not give any solution, since intersection $-1 \le x$ and $x > 2$ is empty.

Consider case III: $x < 0$

Then, $|x-1| = -(x-1) = 1 - x$ and $|x| = -x$

$\therefore |x-1| \le 2$ and $|x| > 2$

$\Rightarrow 1 - x \le 2$ and $-x > 2$

$\Rightarrow -1 \le x$ and $-2 > x$

Does not give any solution, since intersection $-1 \le x$ and $-2 > x$ is empty.

Thus, the solution set of the given inequality is

$x \in (2, 3]$

Q13. Solve the following system of inequalities

$\frac{3}{x-2} > -1, \ \frac{2x-1}{x+1} \le 1.$

Solution:

Inequality $\frac{3}{x-2} > -1 \Rightarrow \frac{3}{x-2} + 1 > 0$

$\Rightarrow \frac{3 + 1(x-2)}{x-2} > 0$

$\Rightarrow \frac{x+1}{x-2} > 0$

$\Rightarrow x < -1$ or $x > 2$

$\Rightarrow x \in (-\infty, -1) \cup (2, \infty)$

Inequality, $\frac{2x-1}{x+1} \le 1 \Rightarrow \frac{2x-1}{x+1} - 1 \le 0$

$\Rightarrow \frac{2x - 1 - 1(x+1)}{x+1} \le 0$

$\Rightarrow \frac{x-2}{x+1} \le 0$

$\Rightarrow -1 \le x \le 2$

$\Rightarrow x \in [-1, 2]$

Since the intersection of the solutions $x \in (-\infty, -1) \cup (2, \infty)$ and $x \in [-1, 2]$ is a null set, the given system of equations has no solution.

Q14. Ravi obtained 60, 70 and 80 marks in the first three unit tests. Find the minimum marks he should get in the fourth test to have an average of at least 65 marks and at most 75 marks.

Solution:

Let x be the marks obtained by Ravi in the fourth unit test, so that he should have an average of at least 65 marks and at most 75 marks. Then

$65 \le \frac{60 + 70 + 80 + x}{4} \le 75$

$\Rightarrow 260 \le 210 + x \le 300$

$\Rightarrow 50 \le x \le 90$

Thus, Ravi should obtain a minimum of 50 marks and a maximum of 90 marks to have an average of at least 65 marks and at most 75 marks.

Q15. The longest side of a triangle is 3 times the shortest side and the third side is 2 cm shorter than the longest side. If the perimeter of the triangle is at most 54 cm, find maximum length of the shortest side.

Solution:

Let the length of the shortest side of the triangle be x cm

Then, length of the longest side = $3x$ cm

And length of the third side = $(3x - 2)$ cm

The perimeter of the triangle is at most 54 cm. Therefore

$x + 3x + (3x - 2) \leq 54$

$\Rightarrow 7x - 2 \leq 54$

$\Rightarrow 7x \leq 56 \Rightarrow x \leq 8$

Thus, the maximum length of the shortest side is 8 cm.

Q16. A solution is to be kept between 80°C and 120°C. What is the range in temperature in fahrenheit (F) if the celsius (C)/fahrenheit (F) conversion formula is given by $C = \frac{8F - 256}{9}$.

Solution:

The solution is to be kept between 80°C and 120°C; therefore, $80 < C < 120$.

Substituting $C = \frac{8F - 256}{9}$ we get

$80 < \frac{8F - 256}{9} < 120$

$\Rightarrow 720 + 256 < 8F < 1080 + 256$

$\Rightarrow 976 < 8F < 1336$

$\Rightarrow 122 < F < 167$

Thus, the required range of temperature in fahrenheit (F) is between 122 F and 167 F.

Q17. A solution of 8% boric acid is to be diluted by adding a 2% boric acid solution to it. The resulting mixture is to be more than 4% but less than 6% boric acid. If we have 1000 litrers of the 8% solution, how many litres of the 2% solution will have to be added?

Solution:

Let x litres of 2% boric acid solution be required to be added.

Then, total mixture $= (x + 1000)$ litres

This resulting mixture is to be more than 3% but less than 5% boric acid.

Therefore,

$3\% \text{ of } (x + 1000) < 2\% \text{ of } x + 8\% \text{ of } 1000$

$\Rightarrow \frac{3}{100}(x + 1000) < \frac{2}{100}x + \frac{8}{100}(1000)$

$\Rightarrow \frac{1}{100}x < 80 - 30$

$\Rightarrow x < 5000$

Also,

$2\% \text{ of } x + 8\% \text{ of } 1000 < 5\% \text{ of } (x + 1000)$

$\frac{2}{100}x + \frac{8}{100}(1000) < \frac{5}{100}(x + 1000)$

$\Rightarrow 80 - 50 < \frac{3}{100}x \Rightarrow 1000 < x$

Thus, $1000 < x < 5000$

Hence, the number of litres of 2% of boric solution that is to be added will have to be more than 1000 litres but less than 5000 litres.

Q18. The IQ of a person is given by the formula $IQ = \frac{MA}{CA} \times 100$, where MA is mental age and CA is chronological age. If $75 \leq IQ \leq 100$ for a group of 12-year-old children, find the range of their mental age.

Solution:

For a group of 12-year-old children,

$75 \leq IQ \leq 100$

And for a group of 12-year-old children,

CA = 12 years

$\therefore IQ = \frac{MA}{CA} \times 100 = \frac{MA}{12} \times 100 = \frac{25}{3} MA$

Thus,

$75 \leq IQ \leq 100$

$\Rightarrow 75 \leq \frac{25}{3} MA \leq 100$

$\Rightarrow 9 \leq MA \leq 12$

Thus, the range of mental age of the group of 12-year-old children is $9 \leq MA \leq 12$.

Q19. The longest side of a rectangle is 10 cm longer than the shortest side. If the perimeter of the rectangle is more than 100 cm and less than 140 cm, then find the minimum and maximum length of the shortest side.

Solution:

Let the length of shortest side be x cm. Then

Longest side $= (x + 10)$ cm

Perimeter of rectangle $= 2 \times (x + (x + 10)) = 4x + 20$

According to the question,

$100 < \text{Perimeter} < 140$

$\Rightarrow 100 < 4x + 20 < 140$

$\Rightarrow 80 < 4x < 120$

$\Rightarrow 20 < x < 30$

Thus, the minimum length of shortest side should be 20 cm and maximum side be 30 cm.

Q20. In drilling the world's deepest hole, it was found that the temperature T in degree celsius, x km below the earth's surface was given by $T = 30 + 25(x - 3)$, $3 \leq x \leq 15$. At what depth will the temperature be between 130°C and 180°C?

Solution:

$T = 30 + 25(x - 3)$, $3 \leq x \leq 15$,

Therefore, according to the question,

$130 < T < 180$
$\Rightarrow 130 < 30 + 25(x-3) < 180$
$\Rightarrow 100 < 25(x-3) < 150$
$\Rightarrow 4 < x-3 < 6$
$\Rightarrow 7 < x < 9$
Thus, at a depth of 7 km to 9 km, the temperature lies between 130 and 180°C.

Q21. Find all pairs of consecutive even positive integers both of which are smaller than 12 such that their sum is greater than or equal to 14.

Solution:
Let x be the smaller of the two consecutive even positive integers. Then, the other larger integer must be $x+2$.
Both the integers are smaller than 12. Therefore
$x+2 < 12 \Rightarrow x < 12-2 \Rightarrow x < 10$
Also, the sum of the two integers is more than 15. Therefore
$x+(x+2) \geq 14$
$\Rightarrow 2x+2 \geq 14$
$\Rightarrow 2x \geq 12$
$\Rightarrow x \geq 6$
Therefore, from inequalities, $x < 10$ and $x \geq 6$ we get $x = 6, 7, 8, 9$
But since x is an even number, x can take only the values 6 and 8.
Thus, the pairs of consecutive even positive integers are (4, 6) and (6, 8).

Day 90

Long Answer-I (LA-I): 4-Mark and Long Answer-II (LA-II): 6-Mark Questions

Q1. Solve the given inequality graphically in a two-dimensional plane: $2x - y \leq 8$.

Solution:
The line $2x - y = 8$ divides the xy-plane in two half planes, I and II.
Let us select a point (0, 0) (not on the line), which lies in one of the half planes.
Substituting this point in the given inequality, we get $2(0) - 0 \leq 8 \Rightarrow 0 \leq 8$, which is true.
Therefore, the solution region of the given inequality is the half plane I [containing the point (0, 0)].
Also, all the points on the line satisfy the given inequality. Therefore, the solution region of the given inequality is the shaded half plane I (including the points on the line), as shown in the figure below.

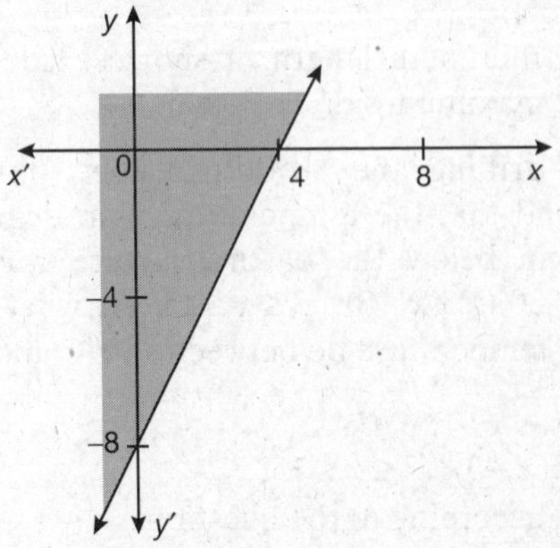

Q2. Solve the given inequality graphically in the two-dimensional plane: $3y - 2x < -6$.

Solution:
The line $3y - 2x = -6$ divides the xy-plane in two half planes, I and II.
Let us select a point (0, 0) (not on the line), which lies in one of the half planes.
Substituting this point in the given inequality, we get $3(0) - 2(0) < -6 \Rightarrow 0 < -6$, which is false.
Therefore, the solution region of the given inequality is the half plane II [not containing the point (0, 0)].
Also, any point on the line does not satisfy the given strict inequality. Therefore, the solution region of the given inequality is the shaded half plane II (excluding the points on the line), as shown in the figure below.

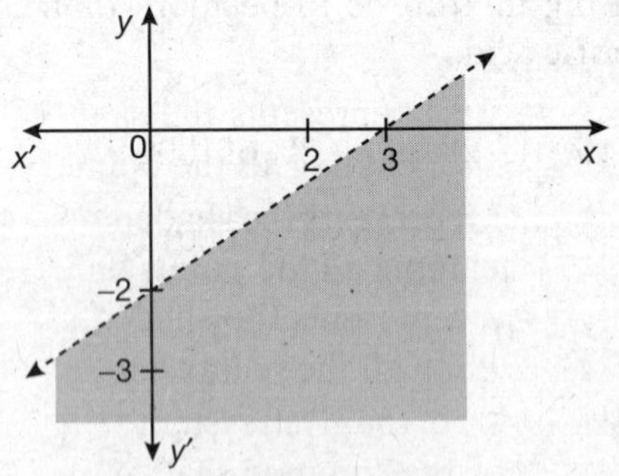

Q3. Solve the following system of inequalities graphically: $x \geq 2, y \geq 1$.

Solution:
The graphs of the lines, $x = 2$ and $y = 1$, are shown in the figure given below.

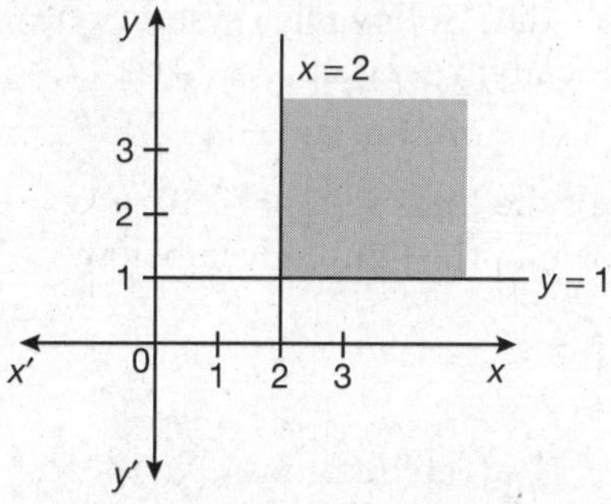

Inequality $x \geq 2$ represents the shaded region on the right of line $x = 2$ (including all the points on the line).
Inequality $y \geq 1$ represents the shaded region above the line $y = 1$ (including all the points on the line).
Thus, the solution of the given system of linear inequalities is represented by the common double-shaded region including the points on the respective lines.

Q4. Solve the following system of inequalities graphically: $x + y \leq 4, x \geq 1, y \geq 1$.

Solution:
The graphs of the lines, $x + y = 4$, $x = 1$ and $y = 2$, are shown in the figure given below.

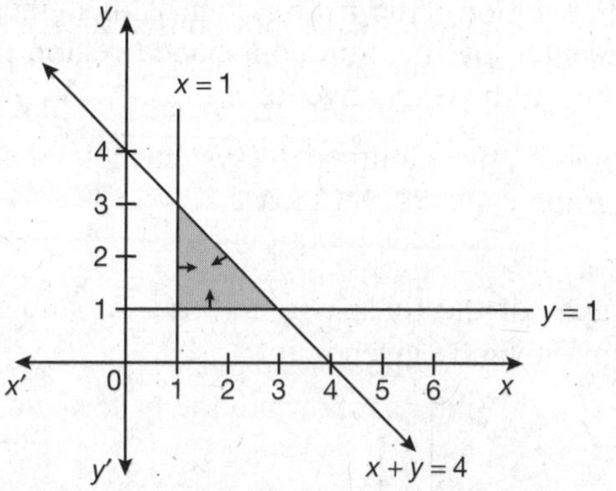

Inequality $x + y \leq 4$ represents the shaded region below the line $x + y = 4$ (including all the points on the line).
Inequality $x \geq 1$ represents the shaded region on the right of line $x = 1$ (including all the points on the line).
Inequality $y \geq 2$ represents the shaded region above the line $y = 2$ (including all the points on the line).
Thus, the solution of the given system of linear inequalities is represented by the common shaded region.

Q5. Solve the following system of inequalities graphically: $x + y \geq 4, x - y > 0$.

Solution:
The graphs of the lines, $x + y = 4$ and $x - y = 0$ are shown in the figure given below.

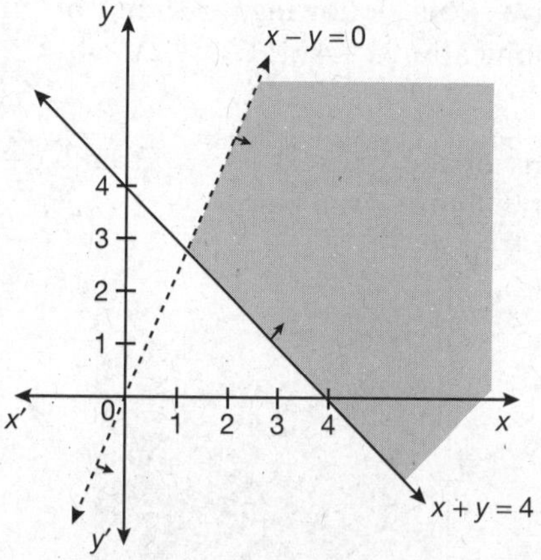

Inequality $x + y \geq 4$ represents the shaded region above the line $x + y = 4$ (including all the points on the line).
Inequality $x - y > 0$ represents the shaded region below the line $x - y = 0$ (excluding all the points on the line). This can be checked by taking any point like $(1, 0)$, other than $(0, 0)$ because the line passes through $(0, 0)$.
Thus, the solution of the given system of linear inequalities is represented by the common shaded region excluding the points on the line $x - y = 0$.

Q6. Solve the following system of inequalities graphically: $x + y \leq 3, x + y \geq 2$.

Solution:
The graphs of the lines, $x + y = 6$ and $x + y = 4$ are shown in the figure given below.

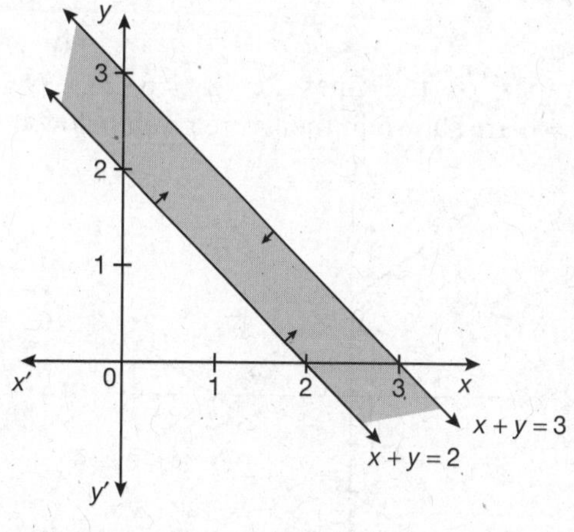

Inequality $x + y \leq 3$ represents the shaded region above the line $x + y = 3$ (including all the points on the line).
Inequality $x + y \geq 2$ represents the shaded region above the line $x + y = 2$ (including all the points on the line).
Thus, the solution of the given system of linear inequalities is represented by the common shaded region including the points on the respective lines.

Q7. Solve the following system of inequalities graphically: $3x+4y \geq 12, x+2y \geq 4$.

Solution:
The graphs of the lines $3x+4y=12$ and $x+2y=4$ are shown in the figure given below.

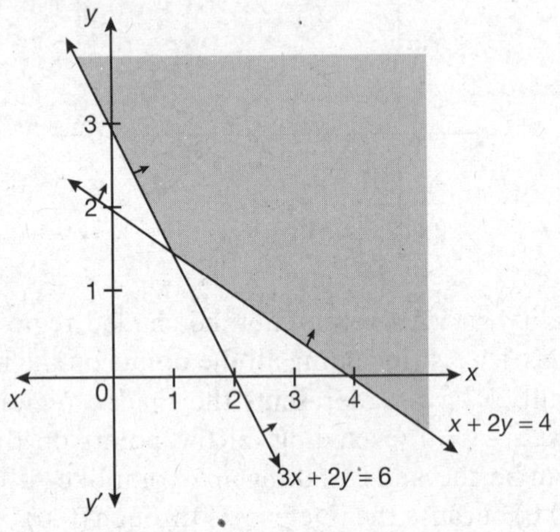

Inequality $3x+4y \geq 12$ represents the shaded region above the line $3x+4y=12$ (including all the points on the line). Inequality $x+2y \geq 4$ represents the shaded region above the line $x+2y=4$ (including all the points on the line). Thus, the solution of the given system of linear inequalities is represented by the common shaded region including the points on the respective lines.

Q8. Solve the following system of inequalities graphically: $3x+2y \geq 6, 2x+3y \leq 6, 2x-3y \leq 6$.

Solution:
The graphs of the lines $3x+2y=6$, $2x+3y=6$ and $2x-3y=6$ are shown in the figure given below.

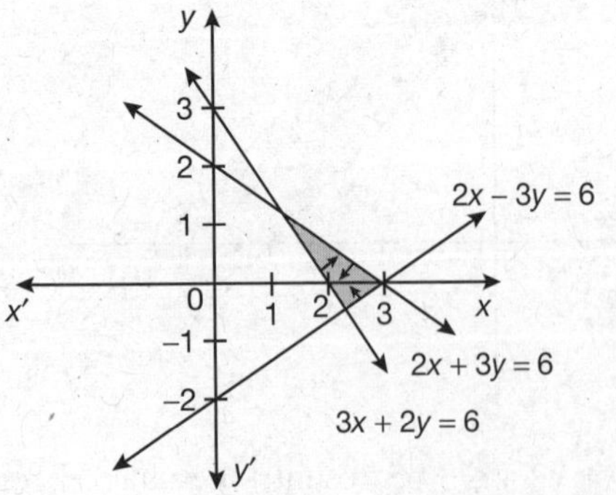

Inequality $3x+2 \geq 6$ represents the shaded region above the line $3x+2y=6$ (including all the points on the line). Inequality $2x+3y \leq 6$ represents the shaded region below the line $2x+3y=6$ (including all the points on the line). Inequality $2x-3y \leq 6$ represents the shaded region above the line $2x-3y=6$ (including all the points on the line). Thus, the solution of the given system of linear inequalities is represented by the common shaded region including the points on the respective lines.

Q9. Solve the following system of inequalities graphically: $x-2y \leq 3, 3x+4y \geq 12, y \geq 1$.

Solution:
The graphs of the lines $x-2y=3$, $3x+4y=12$, $x=0$ and $y=1$ are shown in the figure given below.

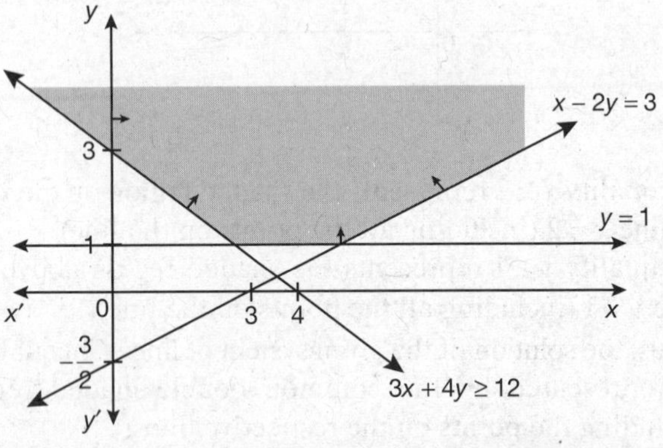

Inequality $x-2y \leq 3$ represents the shaded region below the line $x-2y=3$ (including all the points on the line). Inequality $3x+4y \geq 12$ represents the shaded region below the line $3x+4y=12$ (including all the points on the line). Inequality $y \geq 1$ represents the shaded region above the line $y=1$ (including all the points on the line). Thus, the solution of the given system of linear inequalities is represented by the common shaded region including the points on the respective lines.

Q10. Solve the following system of inequalities graphically: $x+y \leq 18$, $y \geq 2x$, $x \geq 3$, $x, y \geq 0$.

Solution:
The graphs of the lines $x+y \leq 18$, $y \geq 2x$ and $x \geq 3$ are shown in the figure given below.

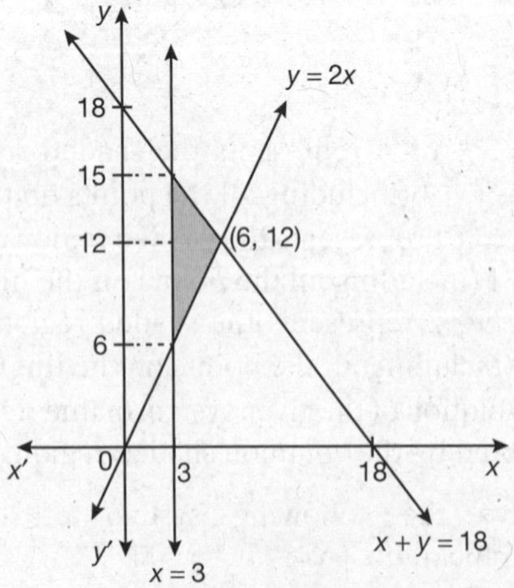

Inequality $x+y \leq 18$ represents the shaded region below the line $x+y=18$ (including all the points on the line). Inequality $y \geq 2x$ represents the shaded region above the line $y=2x$ (including all the points on the line). Inequality $x \geq 3$ represents the shaded region on the right side of the line $x=3$ (including all the points on the line).

Inequality $x \geq 0, y \geq 0$ represents every point in the common shaded region in the first quadrant including the points on the axes.

Thus, the solution of the given system of linear inequalities is represented by the common shaded region including the points on the respective lines in the first quadrant.

Q11. Solve the following system of inequalities graphically:
$x+2y \leq 4, x+y \geq 1, x-y \leq 0, x \geq 0, y \geq 0.$

Solution:
The graphs of the lines $x+2y \leq 10$, $x+y \geq 1$ and $x-y \leq 0$ are shown in the figure given below.

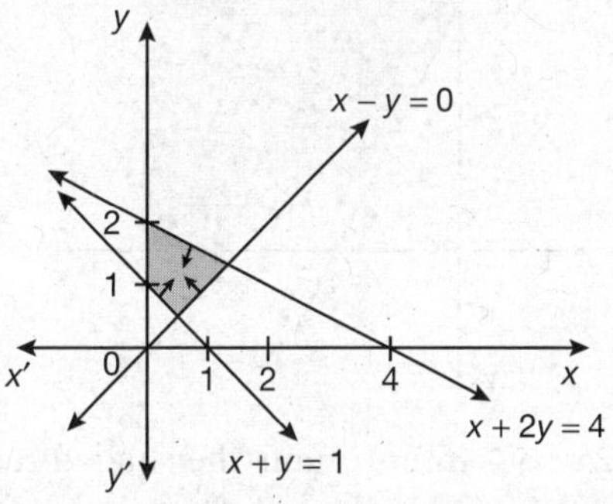

Inequality $x+2y \leq 4$ represents the shaded region below the line $x+2y=4$ (including all the points on the line).
Inequality $x+y \geq 1$ represents the shaded region above the line $x+y=1$ (including all the points on the line).
Inequality $x-y \leq 0$ represents the shaded region above the line $x-y=0$ (including all the points on the line).
Inequality $x \geq 0, y \geq 0$ represents every point in the common shaded region in the first quadrant including the points on the axes.
Thus, the solution of the given system of linear inequalities is represented by the common shaded region including the points on the respective lines in the first quadrant.

Q12. Find the linear inequalities for which shaded region in the given figure is the solution set.

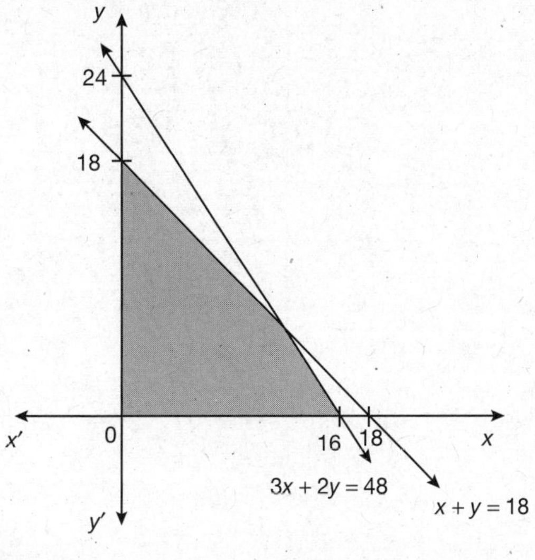

Solution:
Consider the line $3x+2y=48$.
The shaded region and the origin are on the same side of the line $3x+2y=48$, that is, $(0,0)$ satisfies the linear constraints $3x+2y \leq 48$. Therefore, we get one inequality as $3x+2y \leq 48$.
Now, consider the line $x+y=18$.
The shaded region and the origin are on the same side of the line $x+y=18$, that is, $(0,0)$ satisfies the constraints $x+y \leq 18$.
Therefore, we get the second inequality as $x+y \leq 18$.
And since the shaded region is above the x-axis and is on the right side of the y-axis, we must have $x \geq 0, y \geq 0$.
Thus, the linear inequalities corresponding to the given solution set are,
$3x+2y \leq 48,\ x+y \leq 18,\ x \geq 20,\ y \geq 0.$

Q13. Find the linear inequalities for which the shaded region in the given figure is the solution set.

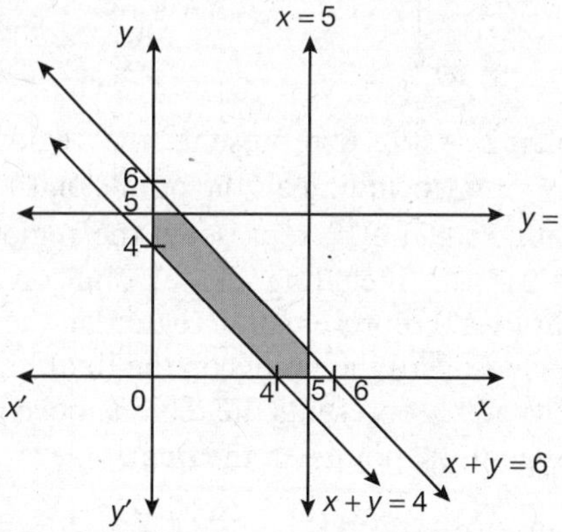

Solution:
Consider the line $x+y=4$.
The shaded region and the origin lie on the opposite sides of this line, that is, (0, 0) does not satisfy the linear constraint $x+y \geq 4$.
Therefore, we get one linear inequality as $x+y \geq 4$.
Now, consider the line $x+y=6$, since the shaded region and origin lie on the same side of this line, (0, 0) satisfies the linear constraints $x+y \leq 6$. Therefore, we get the second linear inequality as $x+y \leq 6$.
Now, consider the line $x=5$, since the shaded region and origin are on the left of this line, (0, 0) satisfies the constraints $x \leq 5$.
Therefore, we get the third linear inequality as $x \leq 5$.
Now, consider the line $y=5$, since the shaded region and origin are below this line, (0, 0) satisfies the constraints $y \leq 5$.
Therefore, we get the fourth linear inequality as $y \leq 5$.
Also the shaded region is above the x-axis and on the right of the y-axis, that is, the shaded region is in the first quadrant. So, we must have $x \geq 0, y \geq 0$.

Thus, the linear inequalities corresponding to the given solution set are $x+y \geq 4$; $x+y \leq 6; x \leq 5$; $y \leq 5; x \geq 0$, $y \geq 0$.

Q14. Show that the following system of linear inequalities has no solution $2x+3y \leq 6$, $3x+4y \geq 12$, $x \geq 0, y \geq 1$.

Solution:
The graphs of the lines, $x+2y=3$, $3x+4y=12$, $x=0$ and $y=1$, are shown in the figure given below.

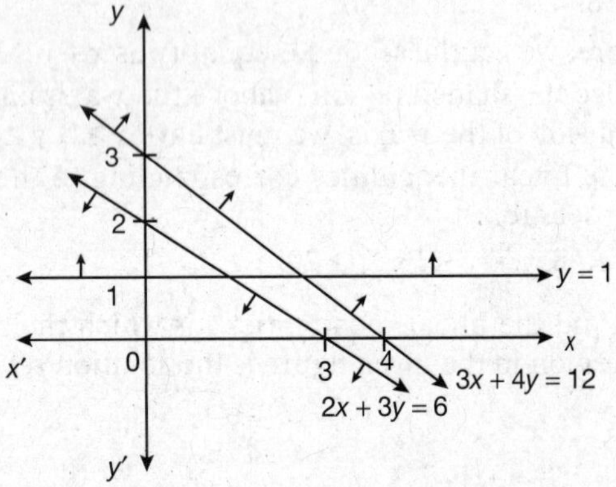

Inequality $2x+3y \leq 6$ represents the region below the line $2x+y=6$ (including all the points on the line).
Inequality $3x+2y \geq 12$ represents the region above the line $3x+2y=12$ (including all the points on the line).
Inequality $x \geq 0$ represents the region on the right of line $x=0$ (including all the points on the line).
Inequality $y \geq 1$ represents the region above the line $y=1$ (including all the points on the line).

Since the regions obtained by the above equations do not have a common region in all, the solution set is a null set. Thus, the given system of inequalities has no solution.

Q15. Show that the solution set of the following system of linear inequalities, is an unbounded region $2x+y \geq 8, x+2y \geq 8, x \geq 0, y \geq 0$.

Solution:
The graphs of the lines, $2x+y=8$, $x+2y=8$, $x=0$ and $y=0$, are shown in the figure given below.

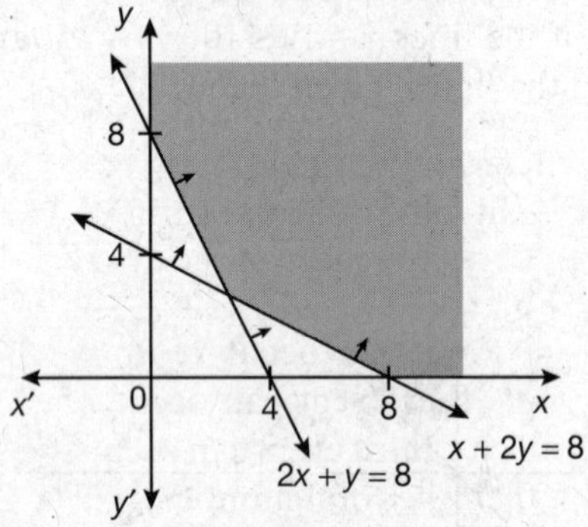

Inequality $2x+y \geq 8$ represents the shaded region above the line $2x+y=8$ (including all the points on the line).
Inequality $x+2y \geq 8$ represents the shaded region above the line $x+2y=8$ (including all the points on the line).
Inequalities $x \geq 0$, $y \geq 0$ represents the region in the first quadrant.
Thus, the graph of the common shaded portion is unbounded.

CHAPTER

7 Permutations and Combinations

Day 91

Fundamental Principle of Counting

The fundamental principle of counting (or multiplication principle) states that,

'If an event can occur in m different ways, following which another event can occur in n different ways, then the total number of occurrences of the events in the given order is $m \times n$'.

In general, 'If an event can occur in n_1 different ways, following which another event can occur in n_2 different ways, following which a third event can occur in n_3 different ways, and so on, then the total number of occurrences of the events in the given order is $n_1 \times n_2 \times n_3 \times \ldots$'.

As an example, let us choose three books, one out of 4 math books, one out of 6 physics books and one out of 3 chemistry books (in this order). Then, the required number of ways of choosing three books, 1 math, 1 physics and 1 chemistry, is $4 \times 6 \times 3 = 72$.

Note: The event of choosing three books could occur in various possible orders. But, we have to choose any one of the possible orders and count the number of different ways of the occurrence of the events in this chosen order.

Example 1: Three coins are tossed together and the outcomes are recorded. How many possible outcomes are there?

Solution:

When a coin is tossed once, the number of outcomes is 2 (head and tail), that is, in a throw each coin will show two different faces. Thus, by the fundamental principle of counting, the number of possible outcomes when three coins are tossed is $2 \times 2 \times 2 = 8$.

Example 2: Given 4 flags of different colours, how many different signals can be generated by arranging the flags on a vertical staff,

(a) if each signal requires the use of 2 flags, one below the other?

(b) if each signal requires the use of at least 2 flags in order (one below the other)?

Solution:

(a) Since each signal requires the use of 2 flags (one below the other), the top place can be filled in 4 different ways by any one of the 4 flags, following which, the bottom place can be filled in 3 different ways by any one of the remaining 3 different flags.
Thus, by the fundamental principle of counting, the number of different signals that can be generated is $4 \times 3 = 12$.

(b) Since each signal requires the use of at least 2 flags (one below the other), we have to count the total number of possible signals consisting of 2 flags, 3 flags and 4 flags.
Two-flag signals: As in part (a), the number of different signals that can be generated by 2 flags is $4 \times 3 = 12$.
Three-flag signals: Similarly, the number of different signals that can be generated by 3 flags is $4 \times 3 \times 2 = 24$.
Four-flag signals: Similarly, the number of different signals that can be generated by 4 flags is $4 \times 3 \times 2 \times 1 = 24$.
Thus, the total number of different signals that can be generated with at least 2 flags is $12 + 24 + 24 = 60$.

Example 3: How many 4-letter words (with or without meaning) can be formed out of the letters of the word MATH

(a) if repetition of letters is not allowed?

(b) if repetition of letters is allowed?

Solution:

(a) Formation of 4-letter words with 4 given letters (letters of the word MATH) means filling of 4 vacant places.
Since repetition of the letters is not allowed, the first place can be filled in 4 different ways by any of the letters of the word MATH, and then the second, third and fourth places can be filled in any of the remaining letters in 3, 2 and 1 different ways, respectively.
Thus, by the fundamental principle of counting, the number of 4-letter words (with or without meaning) that can be formed out of the letters of the word MATH, if no letter is repeated, is $4 \times 3 \times 2 \times 1 = 24$.

(b) Formation of 4-letter words with 4 given letters (letters of the word MATH) means filling of 4 vacant places.
Since repetition of the letters is allowed, the first place can be filled in 4 different ways by any of the letters of the word MATH, and similarly the second, third and fourth places can also be filled in 4 different ways.
Thus, by the fundamental principle of counting, the number of 4-letter words (with or without meaning) that can be formed out of the letters of the

word MATH, if repetition of letters is allowed, is $4\times4\times4\times4=256$.

Example 4: How many 4-letter codes can be formed using the vowels of the English alphabet,
(a) if repetition of the vowels is not allowed?
(b) if repetition of the vowels is allowed?

Solution:

(a) Formation of 4-letter codes with 5 given letters (vowels of the English alphabet a, e, i, o, u) means filling of 4 vacant places using 5 given objects.
Since repetition of the letters is not allowed, the first place can be filled in 5 different ways by any of the vowels (a, e, i, o, u) of the English alphabet and then the second, third and fourth places can be filled with any of the remaining vowels in 4, 3 and 2 different ways, respectively.
Thus, by the fundamental principle of counting, the number of 4-letter codes that can be formed out of the vowels (a, e, i, o, u), if no vowels are repeated, is $5\times4\times3\times2=120$.

(b) Formation of 4-letter codes with 5 given letters (vowels of the English alphabet a, e, i, o, u) means filling of 4 vacant places using 5 given objects.
Since repetition of the letters is allowed, the first place can be filled in 5 different ways by any of the vowels (a, e, i, o, u) of the English alphabet, similarly the second, third and fourth places can also be filled in 5 different ways.
Thus, by the fundamental principle of counting, the number of 4-letter codes that can be formed out of the vowels (a, e, i, o, u), if repetition of vowels is allowed, is $5\times5\times5\times5=625$.

Example 5: How many 4-digit numbers can be formed from the digits 1, 2, 3, 4 and 5 assuming that repetition of the digits is not allowed
(i) if there is no condition?
(ii) if the numbers are even?
(iii) if the numbers start with digit 3 and end with digit 1?

Solution:

(i) Formation of 4-digit number with 5 given digits (1, 2, 3, 4, 5) means filling of 4 vacant places using 5 given objects.
Since repetition of the digits is not allowed, the units place can be filled in 5 different ways by any of the digits (1, 2, 3, 4, 5) and then the tens, hundreds and thousands places can be filled in any of the remaining digits in 4, 3 and 2 different ways, respectively.
Thus, by the fundamental principle of counting, the number of 4-digit numbers that can be formed out of the digits (1, 2, 3, 4, 5), if no digit is repeated, is $5\times4\times3\times2=120$.

(ii) Since the number has to be even, the units place can be filled in two different ways by any of the digits (2 or 4).
Since repetition of the digits is not allowed, the tens, hundreds and thousands places can be filled in by any of the remaining digits in 4, 3 and 2 different ways, respectively.
Thus, by the fundamental principle of counting, the number of 4-digit even numbers that can be formed out of the digits (1, 2, 3, 4, 5), if no digit is repeated, is $2\times4\times3\times2=48$.

(iii) Since the number has to start with the digit 3 and end with the digit 1, the units place will be filled by the digit 1 and the thousands place will be filled by the digit 3. This is a fixed case and can be done only in a single way.
Since repetition of the digits is not allowed, the tens and hundreds places can be filled in by any of the remaining digits in 3 and 2 (since two digits 3 and 1 already occupy first and last place) different ways, respectively.
Thus, by the fundamental principle of counting, the number of 4-digit numbers starting with the digit 3 and ending with the digit 1 that can be formed out of the digits (1, 2, 3, 4, 5), if no digit is repeated, is $1\times1\times3\times2=6$.

Example 6: How many 4-digit numbers can be formed from the digits 1, 2, 3, 4 and 5 assuming that repetition of the digits is allowed
(i) if there is no condition?
(ii) if numbers are even?
(iii) if numbers start with the digit 3 and end with the digit 1?

Solution:

(i) Formation of 4-digit numbers with the 5 given digits (1, 2, 3, 4, 5) means filling of 4 vacant places using the 5 given objects.
Since repetition of the digits is allowed, the units place can be filled in 5 different ways by any of the digits (1, 2, 3, 4, 5) and similarly the tens, hundreds and thousands places can also be filled in 5 different ways by any of the digits (1, 2, 3, 4, 5).
Thus, by the fundamental principle of counting, the number of 4-digit numbers that can be formed out of the digits (1, 2, 3, 4, 5), if repetition of digits is allowed, is $5\times5\times5\times5=625$.

(ii) Since the number has to be even, the units place can be filled in two different ways by any of the digits (2 or 4).
Since repetition of the digits is allowed, the tens, hundreds and thousands places can be filled in 5 different ways by any of the digits (1, 2, 3, 4, 5).
Thus, by the fundamental principle of counting, the number of 4-digit even numbers that can be formed out of the digits (1, 2, 3, 4, 5), if repetition of digits is allowed, is $2\times5\times5\times5=250$.

(iii) Since the number has to start with the digit 3 and end with the digit 1, the units place will be filled by

the digit 1 and the thousands place will be filled by the digit 3. This is a fixed case and can be done only in a single way.
Since repetition of the digits is allowed, the tens and hundreds places can be filled in 5 different ways by any of the digits (1, 2, 3, 4, 5).

Thus, by the fundamental principle of counting, the number of 4-digit numbers starting with the digit 3 and ending with the digit 1 that can be formed out of the digits (1, 2, 3, 4, 5), if repetition of digits is allowed, is $1\times1\times5\times5=25$.

Example 7: How many 4-digit telephone numbers can be formed using the digits 0 to 9 assuming that repetition of the digits is not allowed

(i) if there is no condition?
(ii) if numbers start with 09?
(iii) if numbers do not end with 0?

Solution:

(i) Formation of 4-digit telephone numbers with 10 given digits (0–9) means filling of 4 vacant places using the 10 given objects.
Since repetition of digits is not allowed, the units place can be filled in 10 different ways using any of the digits from 0–9, and then the tens, hundreds and thousands places can be filled using any of the remaining digits in 9, 8 and 7 different ways, respectively.
Thus, by the fundamental principle of counting, the number of 4-digit telephone numbers that can be formed out of the digits (0–9), if no digit is repeated, is $10\times9\times8\times7=5040$.

(ii) Since the telephone number has to start with 09, the thousands and hundreds place will be filled by the digits 0 and 9, respectively. This is a fixed case and can be done only in a single way.
Since repetition of digits is not allowed, the units and tens places can be filled using any of the remaining digits in 8 and 7 (since two digits 09 already occupy two places) different ways, respectively.
Thus, by the fundamental principle of counting, the number of 4-digit telephone numbers starting with 09 that can be formed out of the digits (0–9), if no digit is repeated, is $1\times1\times8\times7=56$.

(iii) Since the telephone number does not end with 0, the units place can be filled in 9 different ways using any of the digits (1–9, except 0).
Since repetition of digits is not allowed, the tens, hundreds and thousands places can be filled using any of the remaining digits in 9, 8 and 7 different ways, respectively.
Thus, by the fundamental principle of counting, the number of 4-digit telephone numbers that can be formed out of the digits (0–9), if no digit is repeated, is $9\times9\times8\times7=4536$.

Example 8: How many 4-digit telephone numbers can be formed using the digits 0 to 9 assuming that repetition of the digits is allowed

(i) if there is no condition?
(ii) if numbers start with 09?
(iii) if numbers does not end with 0?

Solution:

(i) Formation of 4-digit telephone numbers with 10 given digits (0–9) means filling of 4 vacant places using the 10 given objects.
Since repetition of digits is allowed, the units place can be filled in 10 different ways by any of the digits from 0–9, and similarly the tens, hundreds and thousands places can also be filled in 10 different ways by any of the digits from 0–9.
Thus, by the fundamental principle of counting, the number of 4-digit telephone numbers that can be formed out of the digits (0–9), if repetition of digits is allowed, is $10\times10\times10\times10=10000$.

(ii) Since the telephone number has to start with 09, the thousands and hundreds place will be filled by the digits 0 and 9, respectively. This is a fixed case and can be done only in a single way.
Since repetition of digits is allowed, the units and tens places can be filled in 10 different ways.
Thus, by the fundamental principle of counting, the number of 4-digits telephone numbers starting with 09 that can be formed out of the digits (0–9), if repetition of digits is allowed, is $1\times1\times10\times10=100$.

(iii) Since the telephone numbers do not end with 0, the units place can be filled in 9 different ways by any of the digits (1–9, except 0).
Since repetition of digits is allowed, the tens, hundreds and thousands places can be filled any digits 0–9 in 10 different ways.
Thus, by the fundamental principle of counting, the number of 4-digit telephone numbers that can be formed out of the digits (0–9), if repetition of digits is allowed, is $9\times10\times10\times10=9000$.

NCERT Exercise 7.1 (With Solutions)

Q1. How many 3-digit numbers can be formed from the digits 1, 2, 3, 4 and 5 assuming that
(i) repetition of the digits is allowed?
(ii) repetition of the digits is not allowed?

Solution:

(i) Since repetition of digits is allowed, the units, tens and hundreds digits' places can be filled by any of the given five digits.
Thus, by the multiplication principle, the number of ways in which 3-digit numbers can be formed from the given is $5\times5\times5=125$.

(ii) Since repetition of digits is not allowed, if the units place is filled first, then it can be filled by any of the given five digits.

Then, the tens place can be filled with any of the remaining four digits and the hundreds place can be filled with any of the remaining three digits.

Thus, by the multiplication principle, the number of ways in which a 3-digit number can be formed without repeating is $5 \times 4 \times 3 = 60$.

Q2. How many 3-digit even numbers can be formed from the digits 1, 2, 3, 4, 5, 6 if the digits can be repeated?

Solution:
Since the number is even, the units place can be filled by 2 or 4 or 6 only (the units place can be filled in 3 ways). Then, the tens and hundreds places can be filled by any of the 6 digits in 6 different ways, as the digits can be repeated.

Therefore, by the multiplication principle, the required number of 3-digit even numbers is $3 \times 6 \times 6 = 108$.

Q3. How many 4-letter codes can be formed using the first 10 letters of the English alphabet, if no letter can be repeated?

Solution:
The first place can be filled in 10 different ways by any of the first 10 letters of the English alphabet. Then the second, third and fourth code places can be filled in any of the remaining letters in 9, 8, 7 and 6 different ways, respectively, since repetition is not allowed.

Therefore, by the multiplication principle, the number of 4-letter codes that can be formed using the first 10 letters of the English alphabet, if no letter is repeated, is $10 \times 9 \times 8 \times 7 = 5040$.

Q4. How many 5-digit telephone numbers can be constructed using the digits 0 to 9 if each number starts with 67 and no digit appears more than once?

Solution:
Since the 5-digit telephone numbers always start with 67, the first two places will be filled by digits 6, 7, _, _, _ and the remaining 3 places can be filled in 8, 7 and 6 different ways, respectively, since repetition is not allowed.

Thus, by the multiplication principle, the required number of ways in which 5-digit telephone numbers can be constructed is $8 \times 7 \times 6 = 336$.

Q5. A coin is tossed 3 times and the outcomes are recorded. How many possible outcomes are there?

Solution:
When a coin is tossed once, the number of outcomes is 2 (head and tail), that is, in each throw, the number of ways of showing a different face is 2.

Thus, by the multiplication principle, the required number of possible outcomes in 3 tossed is $2 \times 2 \times 2 = 8$.

Q6. Given 5 flags of different colours, how many different signals can be generated if each signal requires the use of 2 flags, one below the other?

Solution:
Since each signal requires the use of 2 flags, the top place can be filled in 5 different ways by any one of the 5 flags, following which, the bottom place can be filled in 4 different ways by any one of the remaining 4 different flags.

Thus, by the multiplication principle, the number of different signals that can be generated is $5 \times 4 = 20$.

Day 92

Permutations

Factorial notation ($n!$): The notation $n!$ (read as factorial n or n factorial) represents the product of the first n natural numbers, that is, $n!$ is the product $1 \times 2 \times 3 \times \ldots \times (n-1) \times n$.

Thus, $n! = 1 \times 2 \times 3 \times 4 \ldots\ldots \times (n-1) \times n$

Usually, we write it as $n! = n \times (n-1) \times \ldots \times 3 \times 2 \times 1$

$0! = 1$ (special property)

$1! = 1$

$2! = 2 \times 1 = 2$

$3! = 3 \times 2 \times 1 = 6$

$4! = 4 \times 3 \times 2 \times 1 = 24$

And so on.

Note: For a natural number n, we can write

$n! = n(n-1)!$ (provided $n \geq 1$)

$= n(n-1)(n-2)!$ (provided $n \geq 2$)

$= n(n-1)(n-2)(n-3)!$ (provided $n \geq 3$), and so on

As an example, we can write

$5! = 5 \times 4! = 5 \times 4 \times 3! = 5 \times 4 \times 3 \times 2!$

$= 5 \times 4 \times 3 \times 2 \times 1!$

Permutations: A permutation means the arrangement of the given objects in a definite order taken some or all at a time. It is denoted by ${}^nP_r = \frac{n!}{(n-r)!}$, $0 \leq r \leq n$, which represents the permutations (or arrangements) in an order of r objects from a given n objects. Given by

$${}^nP_r = \frac{n!}{(n-r)!}$$

$$= \frac{n(n-1)\ldots(n-r+1)(n-r)(n-r-1)\ldots 3 \times 2 \times 1}{(n-r)(n-r-1)..3 \times 2 \times 1}$$

$$\Rightarrow {}^nP_r = n(n-1)(n-2)\ldots\ldots(n-r+1)$$

Note:

(i) When $r = n$, $^nP_n = \frac{n!}{(n-n)!} = \frac{n!}{0!} = n!$

It means the arrangement of all the n objects in a row, that is, filling of n vacant places out of n given objects. This can be done as follows: for the first place we have n choices out of n different objects. Similarly, for the second place we have $(n - 1)$ choices out of the remaining $(n-1)$ different objects, and so on. Thus, by the fundamental principle of counting, the total number of arrangements will be $n\times(n-1)\times\ldots\times3\times2\times1 = n!$

(ii) When $r = 0$, $^nP_0 = \frac{n!}{(n-0)!} = \frac{n!}{n!} = 1$

It means no arrangement is done of any object, that is, leaving all the objects as they are without any arrangement. This can be done only in one way.

Example 1: Evaluate

(i) 7! (ii) 5!−4! (iii) 4!+3! (iv) $\frac{6!}{3!\times2!}$

Solution:

(i) $7! = 7\times6\times5\times4\times3\times2\times1 = 5040$

(ii) $5! - 4! = 5\cdot4! - 4! = 4!(5-1) = 4\cdot3\cdot2\cdot1\cdot4 = 96$

(iii) $4! + 3! = 4\cdot3! + 3! = 3!(4+1) = 3\cdot2\cdot1\cdot5 = 30$

(iv) $\frac{6!}{3!\times2!} = \frac{6\times5\times4\times3!}{3!\times2\times1} = \frac{6\times5\times4}{2} = 60$

Example 2: If $3!\cdot4!\cdot n = 7!$, find n.

Solution:

$3!\cdot4!\cdot n = 7!$

$\Rightarrow n\cdot3!\cdot4! = 7\times6\times5\times4!$

$\Rightarrow n\times3\times2\times1 = 7\times6\times5$

$\Rightarrow n = 35$

Example 3: If $n(3!+4!) = 7!$, find n.

Solution:

$n(3!+4!) = 7!$

$\Rightarrow n\cdot3!\cdot(1+4) = 7\times6\times5\times4\times3!$

$\Rightarrow n\cdot5 = 7\times6\times5\times4 \Rightarrow n = 168$

Example 4: If $\frac{1}{3!}+\frac{1}{4!}=\frac{n}{7!}$, find n.

Solution:

$\frac{1}{3!}+\frac{1}{4!}=\frac{n}{7!}$

$\Rightarrow \frac{1}{3!}+\frac{1}{4\times3!}=\frac{n}{7\times6\times5\times4\times3!}$

$\Rightarrow \frac{1}{3!}\left(1+\frac{1}{4}\right)=\frac{1}{3!}\left(\frac{n}{7\times6\times5\times4}\right)$

$\Rightarrow 1+\frac{1}{4}=\frac{n}{7\times6\times5\times4}$

$\Rightarrow \frac{5}{4}=\frac{n}{7\times6\times5\times4}$

$\Rightarrow n = 7\times6\times5\times5 \Rightarrow n = 1050$

Example 5: Evaluate $\frac{n!}{(n-r)!}$, when

(i) $n = 6$, $r = 2$ and (ii) $n = 6$, $r = 3$

Solution:

(i) When $n = 6$, $r = 2$:

$\frac{n!}{(n-r)!} = \frac{6!}{(6-2)!} = \frac{6!}{4!} = \frac{6\times5\times4!}{4!} = 30$

(ii) When $n = 6$, $r = 3$:

$\frac{n!}{(n-r)!} = \frac{6!}{(6-3)!} = \frac{6!}{3!} = \frac{6\times5\times4\times3!}{3!}$

$= 6\times5\times4 = 120$

Example 6: Find n, if $^{n+1}P_4 : {^nP_3} = 4:1$

Solution:

$^{n-1}P_4 : {^nP_3} = 4:1$

$\Rightarrow \frac{^{n+1}P_4}{^nP_3} = 4$

$\Rightarrow \frac{\left[\frac{(n+1)!}{(n+1-4)!}\right]}{\left[\frac{n!}{(n-3)!}\right]} = 4$

$\Rightarrow \frac{(n+1)!}{(n-3)!}\times\frac{(n-3)!}{n!} = 4$

$\Rightarrow \frac{(n+1)!}{n!} = 4$

$\Rightarrow \frac{(n+1)\cdot n!}{n!} = 4$

$\Rightarrow n+1 = 4 \Rightarrow n = 3$

Example 7: Find r if $^4P_r = 2\,{^3P_{r+1}}$

Solution:

$^4P_r = 2\,{^3P_{r+1}}$

$\Rightarrow \frac{4!}{(4-r)!} = 2\times\frac{3!}{(3-r-1)!}$

$\Rightarrow \frac{4\times3!}{(4-r)!} = \frac{2\times3!}{(2-r)!}$

$\Rightarrow \frac{2}{(4-r)!} = \frac{1}{(2-r)!}$

$\Rightarrow \frac{2}{(4-r)(3-r)(2-r)!} = \frac{1}{(2-r)!}$

$\Rightarrow \frac{2}{(4-r)(3-r)} = 1$

$\Rightarrow (4-r)(3-r) = 2$

$\Rightarrow 12-4r-3r+r^2=2$

$\Rightarrow r^2-7r+10=0$

$\Rightarrow (r-2)(r-5)=0$

$\Rightarrow r=2$ or $r=5$

For ${}^nP_r=\dfrac{n!}{(n-r)!}, 0\le r\le n$

Therefore, for ${}^4P_r,\ 0\le r\le 4$

Thus, $r=2,\ (\because r\ne 5)$

Example 8: Find n and r, if ${}^nP_{r-1}=20$ and ${}^{n+1}P_r=120$

Solution:

${}^nP_{r-1}=20$

$$\Rightarrow \frac{n!}{(n-r+1)!}=20 \qquad \text{(i)}$$

Also,

${}^{n+1}P_r=120$

$$\Rightarrow \frac{(n+1)!}{(n-r+1)!}=120 \qquad \text{(ii)}$$

From equations (i) and (ii),

$$\Rightarrow \frac{(n+1)!}{n!}=\frac{120}{20}$$

$\Rightarrow n+1=6 \Rightarrow n=5$

Substituting in equation (i), we get

$$\frac{n!}{(n-r+1)!}=20$$

$$\Rightarrow \frac{5!}{(5-r+1)!}=20$$

$$\Rightarrow (6-r)!=\frac{5\cdot4\cdot3\cdot2\cdot1}{20}=6$$

$\Rightarrow (6-r)!=3\cdot2\cdot1=3!$

$\Rightarrow 6-r=3$

$\Rightarrow r=3$

Example 9: Find n and r, if ${}^nP_{r-1}:{}^nP_{r+1}=1:2$ and ${}^{n+1}P_r:{}^{n-1}P_r=10:1$

Solution:

${}^nP_{r-1}:{}^nP_{r+1}=1:2$

$$\Rightarrow \frac{{}^nP_{r-1}}{{}^nP_{r+1}}=\frac{1}{2}$$

$$\Rightarrow \frac{\left[\dfrac{n!}{(n-r+1)!}\right]}{\left[\dfrac{n!}{(n-r-1)!}\right]}=\frac{1}{2}$$

$$\Rightarrow \frac{n!}{(n-r+1)!}\times\frac{(n-r-1)!}{n!}=\frac{1}{2}$$

$$\Rightarrow \frac{(n-r+1)!}{(n-r+1)(n-r)(n-r+1)!}=\frac{1}{2}$$

$$\Rightarrow \frac{1}{(n-r+1)(n-r)}=\frac{1}{2}$$

$$\Rightarrow (n-r+1)(n-r)=2 \qquad \text{(i)}$$

Also, ${}^{n+1}P_r:{}^{n-1}P_r=10:1$

$$\Rightarrow \frac{\left[\dfrac{(n+1)!}{(n+1-r)!}\right]}{\left[\dfrac{(n-1)!}{(n-1-r)!}\right]}=10$$

$$\Rightarrow \frac{(n+1)!}{(n+1-r)!}\times\frac{(n-r-1)!}{(n-1)!}=10$$

$$\Rightarrow \frac{(n+1)\times n\times(n-r-1)!}{(n-r+1)(n-r)(n-r-1)!}=10$$

$$\Rightarrow \frac{(n+1)\times n}{(n-r+1)(n-r)}=10 \qquad \text{(ii)}$$

From equations (i) and (ii),

$$\Rightarrow \frac{(n+1)\times n}{(n-r+1)(n-r)}=10$$

$$\Rightarrow \frac{(n+1)\times n}{2}=10$$

$\Rightarrow (n+1)\times n=20=5\times4$

$\Rightarrow n=4$

Substituting in equation (i), we get

$(n-r+1)(n-r)=2$

$\Rightarrow (4-r+1)(4-r)=2$

$\Rightarrow (5-r)(4-r)=2=2\times1$

$\Rightarrow 4-r=1$

$\Rightarrow r=3$

NCERT Exercise 7.2 (With Solutions)

Q1. Evaluate

(i) $8!$ (ii) $4!-3!$

Solution:

(i) $8!=1\times2\times3\times4\times5\times6\times7\times8=40320$

(ii) $4!-3!=4\cdot3!-3!=3!\cdot(4-1)=3\cdot2\cdot1\cdot3=18$

Q2. Is $3!+4!=7!$?

Solution:

$3!=1\times2\times3=6$ and $4!=1\times2\times3\times4=24$

$\therefore 3!+4!=6+24=30$

While, $7!=1\times2\times3\times4\times5\times6\times7=5040$

$\therefore 3!+4!\ne 7!$

Q3. Compute $\dfrac{8!}{6!\times 2!}$

Solution:

$$\frac{8!}{6!\times 2!}=\frac{8\times 7\times 6!}{6!\times 2\times 1}=\frac{8\times 7}{2}=28$$

Q4. If $\frac{1}{6!}+\frac{1}{7!}=\frac{x}{8!}$, find x.

Solution:

$$\frac{1}{6!}+\frac{1}{7!}=\frac{x}{8!}\Rightarrow\frac{1}{6!}+\frac{1}{7\times 6!}=\frac{x}{8\times 7\times 6!}$$

$$\Rightarrow\frac{1}{6!}\left(1+\frac{1}{7}\right)=\frac{x}{8\times 7\times 6!}$$

$$\Rightarrow 1+\frac{1}{7}=\frac{x}{8\times 7}\Rightarrow\frac{8}{7}=\frac{x}{8\times 7}$$

$$\Rightarrow x=\frac{8\times 8\times 7}{7}\Rightarrow x=64$$

Q5. Evaluate $\frac{n!}{(n-r)!}$, when

(i) $n=6,\ r=2$ (ii) $n=9,\ r=5$

Solution:

(i) When $n=6, r=2$:

$$\frac{n!}{(n-r)!}=\frac{6!}{(6-2)!}=\frac{6!}{4!}=\frac{6\times 5\times 4!}{4!}=30$$

(ii) When $n=9, r=5$:

$$\frac{n!}{(n-r)!},=\frac{9!}{(9-5)!}=\frac{9!}{4!}=\frac{9\times 8\times 7\times 6\times 5\times 4!}{4!}$$

$$=9\times 8\times 7\times 6\times 5=15160$$

Day 93

Theorems of Permutations

When all the objects are distinct

Theorem 1: The number of permutations of n different objects taken r at a time, where $0<r\le n$ and the objects do not repeat, is given by

$^nP_r=n(n-1)(n-2)...(n-r+1)$

Theorem 2: The number of permutations of n different objects taken r at a time, when repetition is allowed, is n^r.

When all the objects are not distinct

Theorem 3: The number of permutations of n objects, where p objects are of the same kind and the rest are all different is $\frac{n!}{p!}$.

Theorem 4: The number of permutations of n objects, where p_1 objects are of one kind, p_2 are of the second kind,..., p_k are of the kth kind and the rest, if any, are of different kinds is $\frac{n!}{p_1!\,p_2!\,...\,p_k!}$.

Example 1: How many 4-letter words, with or without meaning, can be formed from the letters of the words FORMULA, assuming that no letter is repeated,

(i) with no condition? (ii) for words starting with A?
(iii) where the first letter is a vowel?
(iv) for words starting with a consonant and ending with U?

Solution:

There are 7 different letters in the word FORMULA.

(i) Number of 4-letter words that can be formed from the letter of the word FORMULA without repetition of letters is $^7P_4=\frac{7!}{(7-4)!}=\frac{7!}{3!}=\frac{7\times 6\times 5\times 4\times 3!}{3!}=840$

(ii) Since the word has to start with A, the first place is occupied by A. This is a fixed case and can be done only in a single way.
The first place is already occupied with a letter (A) and the letters cannot be repeated, therefore the remaining three places (of a 4-letter word) are to be occupied by the remaining 6 letters in $^6P_3=\frac{6!}{(6-3)!}=\frac{6!}{3!}=\frac{6\times 5\times 4\times 3!}{3!}=120$ ways
Thus, by the multiplication principle, the number of words that can be formed is $1\times 120=120$.

(iii) Since there are 3 different vowels in the given word FORMULA and repetition of the letters is not allowed, the first place can be occupied in 3 different ways by any of the vowels (A, O or U).
The first place is already occupied with a letter (A, O or U) and the letters cannot be repeated, therefore the remaining three places (of a 4-letter word) are to be occupied by the remaining 6 letters in $^6P_3=\frac{6!}{(6-3)!}=\frac{6!}{3!}=\frac{6\times 5\times 4\times 3!}{3!}=120$ ways
Thus, by the multiplication principle, the number of words that can be formed is $3\times 120=360$.

(iv) Since there are 4 different consonants in the given word FORMULA and repetition of the letters is not allowed, the first place is occupied in 4 different ways by any of the consonant (F, L, M or R).
Since the word has to end with U, the last place is occupied by U. This is a fixed case and can be done only in a single way.
Now, since the first place is already occupied with a letter (F, L, M or R) and the last place with letter U, and the letters cannot be repeated, the remaining two places (of a 4-letter word) are to be occupied by the remaining 5 letters in

$$^{5}P_2 = \frac{5!}{(5-2)!} = \frac{5!}{3!} = \frac{5\times4\times3!}{3!} = 20 \text{ ways}$$

Thus, by the multiplication principle, the number of words that can be formed is $4\times1\times20 = 80$.

Example 2: How many words, with or without meaning, can be made from all the letters of the words FORMULA, assuming that no letter is repeated,

(i) with no condition?
(ii) when the words start with A and end with a consonant?
(iii) when there are always two letters between A and R?
(iv) when R occurs after A?
(v) when R occurs just after A?
(vi) when all vowels always occur together?
(vii) when all vowels do not occur together?

Solution:

There are 7 different letters in the word FORMULA.

(i) The number of 7-letter words that can be formed from the letter of the word FORMULA without repetition of letters is

$$^{7}P_7 = \frac{7!}{(7-7)!} = \frac{7!}{0!} = \frac{7\times6\times5\times4\times3\times2\times1}{1} = 5040$$

(ii) Since the word starts with A, the first place is occupied by A. This is a fixed case and can be done only in a single way.

Since there are 4 different consonants in the given word FORMULA and repetition of the letters is not allowed, the last place can be occupied in 4 different ways by any of the consonants (F, L, M or R).

Now, since the first place is occupied with letter A and the last place is occupied with a letter (F, L, M or R), and the letters cannot be repeated, the remaining five places are occupied by the remaining 5 letters in $^{5}P_5 = \frac{5!}{(5-5)!} = \frac{5\times4\times3\times2\times1}{1} = 120$ ways.

Thus, by the multiplication principle, the number of words that can be formed is $1\times4\times120 = 480$.

(iii) Since the letters have to be arranged in such a way that there are always 2 letters between A and R, the places of A and R are fixed in such a way that there are 2 letters between them.

The remaining 5 letters can be arranged in 5! ways. Also, the letters A and R can be placed such that there are 2 letters between them in $2\times4 = 8$ ways (A and R take the 1st and 4th positions, or 2nd and 5th positions, or 3rd and 6th positions, or 4th and 7th positions, and vice versa).

Therefore, by the multiplication principle, the number of arrangements is $= 5!\times8 = 960$.

(iv) The total number of 7-letter words that can be formed from the letter of the word FORMULA, without repeating of letters is $^{7}P_7 = 5040$ [as calculated in (i)]. Now, in these 5040 cases, half the time, R will come before A and in the other half, R will come after A.

Thus, the number of arrangements when R occurs after A is $\frac{5040}{2} = 2520$.

(v) Since R has to occur just after A, we take 'RA' as a single letter. Therefore, this single letter 'RA' together with the 5 remaining letters (F, O, M, U, L) of the word FORMULA forms 6 letters. Thus, the number of arrangements without repetition of letters when R occurs just after A is

$$^{6}P_6 = \frac{6!}{(6-6)!} = \frac{6!}{0!} = \frac{6\times5\times4\times3\times2\times1}{1} = 720$$

(vi) Since there are 3 different vowels in the given word FORMULA and repetition of the letters is not allowed, we take 'OUA' as a single letter. Therefore, this single letter 'OUA' together with the 4 remaining letters (F, R, M, L) of the word FORMULA forms 5 letters. Thus, the arrangement of these 5 letters is

$$^{5}P_5 = \frac{5!}{(5-5)!} = \frac{5!}{0!} = \frac{5\times4\times3\times2\times1}{1} = 120$$

Also, the single assumed letter 'OUA' contains 3 letters which can also be arranged in

$$^{3}P_3 = \frac{3!}{(3-3)!} = \frac{3!}{0!} = \frac{3\times2\times1}{1} = 6 \text{ ways}$$

Thus, by the multiplication principle, the number of words that can be formed is $120\times6 = 720$.

(vii) The total number of 7-letter words that can be formed from the letter of the word FORMULA without repetition of letters is $^{7}P_7 = 5040$ as calculated in (i). The total number of 7-letter words formed when all vowels always occur together is 720 [as in (vi)].

Thus, the total number of 7-letter words formed when all vowels do not occur together is $5040 - 720 = 4320$.

Example 3: How many words, with or without meaning can be made from all the letters of the words MINIMIZES, assuming that no letter is repeated,

(i) with no condition?
(ii) when words start with I and end with a consonant?
(iii) when there are always two letters between I and M?
(iv) when Z occurs after S?
(v) when all Ms always occur just after I?
(vi) when all vowels always occur together?
(vii) when all vowels do not occur together?

Solution:

There are 9 letters in the word MINIMIZES, 4 vowels $(1E, 3I)$ and 5 consonants $(2M, 1N, 1S, 1Z)$.

(i) Number of 9-letter words that can be formed from the letter of word MINIMIZES, without repeating letters is $= \frac{9!}{3!2!} = 30240$ (we divided by 3! and 2!, because I appears 3 times and M appears 2 times)

(ii) Since the word starts with I, the first place is occupied by I. This is a fixed case and can be done only in a single way.

There are 5 different consonants in the given word MINIMIZES and repetition of the letters is not allowed. Therefore, for the last place, we have consonants ($2M$, $1N$, $1S$, $1Z$).
For that, we take two cases,
Case-1: The last place is occupied by repeated letter M (which is a fixed case and can be done only in a single way).

Now, since the first place is occupied with letter I and last place is occupied with letter M, the remaining 7 places are occupied by the remaining 7 letters (2I, 1E, 1M, $1N$, $1S$, $1Z$) in $= \frac{7!}{2!} = 2520$ ways.
(We divided by 2!, because I appears 2 times, the remaining 1 I is fixed at the first place; for M we do not divide by 2! because 1 M is fixed at the last place and now we are left with only 1 M.)
Case-2: The last place is occupied in 3 different ways by any of the consonants (N, S or Z).

The first place is occupied with letter I and the last place is occupied with a letter (N, S, or Z), the letters cannot be repeated. Therefore, the remaining 7 places are occupied by the remaining 7 letters (2 I, 1 E, 2 M, 2(N, S, Z) in $= \frac{7!}{2!2!} = 1260$ ways. (We divided by 2!.2!, because I and M appear 2 times).
Thus, by the multiplication principle, the number of words that can be formed is

$$\underbrace{1\times1\times2520}_{\text{case-1}}+\underbrace{1\times3\times1260}_{\text{case-2}} = 6300$$

(iii) The letters have to be arranged in such a way that there are always 2 letters between I and M.
Therefore, the places of 1 I and 1 M are fixed in such a way that there are 2 letters between them.
The remaining 7 letters (2 I, 1 E, 1 M, 1 N, 1 S, 1 Z) can be arranged in $= \frac{7!}{2!} = 2520$ ways.

Also, the letters I and M can be placed such that there are 2 letters between them in $2\times6 = 12$ ways, (A and R take the 1st and 4th positions, or 2nd and 5th positions, or 3rd and 6th positions, or 4th and 7th positions, 5th and 8th positions, or 6th and 9th positions, and vice versa).
Therefore, by the multiplication principle, the number of arrangements is $= 2520\times12 = 30240$.

(iv) The total number of 9-letter words that can be formed from the letter of the word MINIMIZES without repetition of letters is $= \frac{9!}{3!2!} = 30240$ [as calculated in (i)].
Now, in these 30240 cases, half the time, Z will come before S and in the other half, Z will come after S. Thus, the number of arrangements when Z occurs after S is $\frac{30240}{2} = 15120$.

(v) M has to occur just after I. Therefore, we take 'MI' as a single letter. Thus, two pairs of letter 'MI' together with the 5 remaining letters (1 I, 1 E, 1 N, 1 S, 1 Z) of the word MINIMIZES, forms 7 letters. Thus, the number of arrangements without repetition of letters when M occurs just after I is $= \frac{7!}{2!} = 2520$ (we divided by 2! because there are two pairs of the letters 'MI').

(vi) Since there are 4 vowels (1 E, 3 I) in the given word MINIMIZES, and repetition of the letters is not allowed, we take 'EIII' as a single letter. Therefore, this single letter 'EIII' together with the 5 remaining letters (2 M, 1 N, 1 S, 1 Z) of the word MINIMIZES, forms 6 letters. Thus, the arrangement of these 6 letters can be done in $= \frac{6!}{2!} = 360$ ways. (We have divided by 2! because M appears 2 times.)
Also, the single assumed letter 'EIII' contains 4 letters which can also be arranged in $= \frac{4!}{3!} = 4$ ways.
(We have divided by 3! because I appears 3 times.)
Thus, by the multiplication principle, the number of words that can be formed is $360\times4 = 1440$.

(vii) The total number of 9-letter words that can be formed from the letters of the word MINIMIZES, without repeatition of letters is 30240 [as calculated in (i)]. The total number of 9-letter words that can be formed when all the vowels always occur together is 1440 [as calculated in (vi)].

Thus, the total number of 9-letter words that can be formed when all the vowels do not occur together is $30240 - 1440 = 28800$.

Day 94

NCERT Exercise 7.3 (With Solutions)

Q1. How many 3-digit numbers can be formed by using the digits 1 to 9 if no digit is repeated?

Solution:
Since no digits are repeated, the number of 3-digit numbers formed using the digits 1 to 9, is

$$= {}^9P_3 = \frac{9!}{(9-3)!} = \frac{9!}{6!} = \frac{9\times8\times7\times6!}{6!} = 9\times8\times7 = 507$$

Q2. How many 4-digit numbers are there with no digit repeated?

Solution:
Since the number consists of 4 digits, the thousands place can be filled with any of the digits from 1 to 9 as the digit

0 cannot be included. Therefore, the number of ways in which the thousands place can be filled is 9.
And the hundreds, tens and units places can be filled by any of the digits from 0 to 9.

But, the digits cannot be repeated in the 4-digit numbers, and the thousands place is already occupied by a digit. Therefore, the hundreds, tens and units places are to be filled by the remaining 9 digits.
Thus, number of such 3-digit numbers

$$= {}^9P_3 = \frac{9!}{(9-3)!} = \frac{9!}{6!} = \frac{9\times8\times7\times6!}{6!} = 504$$

Hence, by the multiplication principle, the required number of 4-digit numbers is $9\times504 = 4536$.

Q3. How many 3-digit even numbers can be made using the digits 1, 2, 3, 4, 6, 7 if no digit is repeated?

Solution:
Since the digits cannot be repeated in the 3-digit numbers and the units place is already occupied by a digit (2, 4 or 6), the hundreds and tens places are to be filled by the remaining 5 digits.
Therefore, the number of ways in which the hundreds and tens place can be filled with the remaining 5 digits is

$$= {}^5P_2 = \frac{5!}{(5-2)!} = \frac{5!}{3!} = \frac{5\times4\times3!}{3!} = 20$$

Thus, by the multiplication principle, the required number of 3-digit numbers is $3\times20 = 60$.

Q4. Find the number of 4-digit numbers that can be formed using the digits 1, 2, 3, 4, 5 if no digit is repeated. How many of these will be even?

Solution:
The 4-digit numbers formed using the digits, 1, 2, 3, 4, and 5 is $= {}^5P_4 = \frac{5!}{(5-4)!} = \frac{5!}{1!} = 1\times2\times3\times4\times5 = 120.$
Now, 4-digit even numbers formed by using the digits, 1, 2, 3, 4, 5:
The number of ways in which units place is filled with digits is 2 (2 or 4).

Since the digits are not repeated and the units place is already occupied with a digit (2 or 4), the remaining places are to be filled by the remaining 4 digits. Therefore, the number of ways in which the remaining places can be filled is

$$= {}^4P_3 = \frac{4!}{(4-3)!} = \frac{4!}{1!} = 4\times3\times2\times1 = 24$$

Thus, by the multiplication principle, the number of even numbers is $24\times2 = 48$.

Q5. From a committee of 8 persons, in how many ways can a chairman and a vice chairman be chosen, assuming one person cannot hold more than one position?

Solution: The number of ways of choosing a chairman and a vice chairman out of 8 persons is

$$= {}^8P_2 = \frac{8!}{(8-2)!} = \frac{8!}{6!} = \frac{8\times7\times6!}{6!} = 8\times7 = 56$$

Q6. Find n if ${}^{n-1}P_3 : {}^nP_4 = 1:9$

Solution:

$${}^{n-1}P_3 : {}^nP_4 = 1:9 \Rightarrow \frac{{}^{n-1}P_3}{{}^nP_4} = \frac{1}{9}$$

$$\Rightarrow \frac{\left[\frac{(n-1)!}{(n-1-3)!}\right]}{\left[\frac{n!}{(n-4)!}\right]} = \frac{1}{9}$$

$$\Rightarrow \frac{(n-1)!}{(n-4)!}\times\frac{(n-4)!}{n!} = \frac{1}{9}$$

$$\Rightarrow \frac{(n-1)!}{n\times(n-1)!} = \frac{1}{9}$$

$$\Rightarrow \frac{1}{n} = \frac{1}{9} \Rightarrow n = 9$$

Q7. Find r if
(i) ${}^5P_r = 2\,{}^6P_{r-1}$ (ii) ${}^5P_r = {}^6P_{r-1}$

Solution:
(i) ${}^5P_r = 2\,{}^6P_{r-1}$

$$\Rightarrow \frac{5!}{(5-r)!} = 2\times\frac{6!}{(6-r+1)!}$$

$$\Rightarrow \frac{5!}{(5-r)!} = \frac{2\times6!}{(7-r)!}$$

$$\Rightarrow \frac{5!}{(5-r)!} = \frac{2\times6\times5!}{(7-r)(6-r)(5-r)!}$$

$$\Rightarrow 1 = \frac{2\times6}{(7-r)(6-r)}$$

$$\Rightarrow (7-r)(6-r) = 12$$

$$\Rightarrow 42-6r-7r+r^2 = 12$$

$$\Rightarrow r^2-13r+30 = 0$$

$$\Rightarrow (r-3)(r-10) = 0$$

$$\Rightarrow r = 3 \text{ or } r = 10$$

For ${}^5P_r = \frac{n!}{(n-r)!}, 0\le r\le n$

Therefore, $0\le r\le 5$
Thus, $r = 3$, $(r\ne10)$

(ii) ${}^5P_r = {}^6P_{r-1}$

$$\Rightarrow \frac{5!}{(5-r)!} = \frac{6!}{(6-r+1)!}$$

$$\Rightarrow \frac{5!}{(5-r)!} = \frac{6}{(7-r)(6-r)(5-r)!}$$

$$\Rightarrow 1 = \frac{6}{(7-r)(6-r)}$$

$\Rightarrow (7-r)(6-r)=6$

$\Rightarrow 42-7r-6r+r^2-6=0$

$\Rightarrow r^2-13r+36=0$

$\Rightarrow (r-4)(r-9)=0$

$\Rightarrow r=4$ or $r=9$

For ${}^5P_r=\dfrac{n!}{(n-r)!}, 0\le r\le n$

Therefore, for $0\le r\le 5$

Thus, $r=4$, $(r\ne 9)$

Q8. How many words, with or without meaning can be formed using all the letters of the word EQUATION, using each letter exactly once?

Solution:

Since there are no repeating letters in the word EQUATION, the number of words that can be formed using all the letters of the word EQUATION, using each letter exactly once, is ${}^8P_8=8!=40320$

Q9. How many words, with or without meaning can be made from the letters of the word MONDAY, assuming that no letter is repeated, if

(i) 4 letters are used at a time?

(ii) all the letters are used at a time?

(iii) all the letters are used but the first letter is a vowel?

Solution:

There are 6 different letters in the word MONDAY.

(i) Number of 4-letter words that can be formed from the letter of the word MONDAY, without repeating of letters is ${}^6P_4=\dfrac{6!}{(6-4)!}=\dfrac{6!}{2!}=\dfrac{6\times5\times4\times3\times2!}{2!}=360$

(ii) Number of words that can be formed by using all the letters of the word MONDAY at a time is ${}^6P_6=6!=6\times5\times4\times3\times2\times1=720$

(iii) Since, in the given word, there are 2 different vowels which have to occupy the first place, it can be done in 2 ways (A or O).
Since the letters cannot be repeated and the first place is already occupied with a letter (A or O), the remaining five places are to be filled by the remaining 5 letters. This can be done in 5! ways.
Thus, by the multiplication principle, the number of words that can be formed is $5!\times2=120\times2=240$.

Q10. In how many of the distinct permutations of the letters in MISSISSIPPI do the four Is not come together?

Solution:

In the given word MISSISSIPPI, I appears 4 times, S appears 4 times, P appears 2 times, and M appears just once.

Therefore, the number of distinct permutations of the letters in the given word is $=\dfrac{11!}{4!4!2!}=34650$

Since there are 4 Is is in the given word, when they occur together, they are treated as a single letter (IIII) for the time being. This single letter together with the remaining 7 letters will account as 8 letters.

Thus, the number of arrangements where all Is occur together is $\dfrac{8!}{4!2!}=840$ ways.

Hence, the number of distinct permutations of the letters in MISSISSIPPI in which the four Is do not occur together $=34650-840=33810$.

Q11. In how many ways can the letters of the word PERMUTATIONS be arranged if

(i) the words start with P and end with S?

(ii) the vowels are all together?

(iii) there are always 4 letters between P and S?

Solution:

In the word PERMUTATIONS, there are 2 Ts and all the other letters appear only once.

(i) If P and S are fixed at the extreme ends (P at the end and S at the right end), then the remaining 10 letters can be arranged in $=\dfrac{10!}{2!}=1814400$ ways.

(ii) There are 5 vowels in the given word, each appearing only once. Since they have to always occur together, they are treated as a single letter. Thus, this single letter together with the remaining 7 letters will account as 8 letters.
Thus, 8 letters in which there are 2 Ts can be arranged in $\dfrac{8!}{2!}$ ways.
Corresponding to each of these arrangements, the 5 different vowels can be arranged in 5! ways.
Therefore, by the multiplication principle, the number of arrangements is $=\dfrac{8!}{2!}\times5!=2419200$.

(iii) The letters have to be arranged in such a way that there are always 4 letters between P and S.
Therefore, the places of P and S are fixed in such a way that there are 4 letters between them.
The remaining 10 letters in which there are 2 Ts can be arranged in $\dfrac{10!}{2!}$ ways.
Also, the letters P and S can be placed such that there are 4 letters between them in $2\times7=14$ ways, (P and S take the 1st and 5th position, or 2nd and 6th position, and so on ...7th and 12th position, and vice versa).
Therefore, by the multiplication principle, the number of arrangements is $=\dfrac{10!}{2!}\times14=25401600$.

Day 95

Combinations

Permutation is the arrangement of objects while *combination* means selection of objects.

Thus, combination (or selection) of r different objects from given n different objects is denoted by $^nC_r = \frac{n!}{r!(n-r)!}$.

Now, these r different objects can be rearranged in $r!$ ways. If we multiply ^nC_r with $r!$, we get $^nC_r \times r! = \frac{n!}{(n-r)!} = {}^nP_r$, that is, permutation of r different objects from given n different objects.

Hence, we conclude that permutation consists of two steps: first combination (or selection) and then arrangement. So, selection ^nC_r and arrangement $r!$, together represent permutation ^nP_r.

Theorem: The number of combinations of n objects, if some or all objects are taken at a time, is $2^n - 1$. (2^n means, for each of the n objects we have two choices, that is, the object is chosen or the object is not chosen and -1 is used for the case when all the objects are not chosen).

This can also be explained as, out of n objects, only 1 object can be chosen in $^nC_1$ ways, similarly 2 objects can be chosen in $^nC_2$ ways, and so on. n objects can be chosen in ^nC_n ways.

Thus, the total number of combinations of n objects, if some or all objects are taken at a time is

$^nC_1 + {}^nC_2 + {}^nC_3 + \ldots {}^nC_n = 2^n - 1$

$(\because {}^nC_0 + {}^nC_1 + {}^nC_2 + {}^nC_3 + \ldots {}^nC_n = 2^n)$

Note:

(i) For $r = 0$, $^nC_0 = \frac{n!}{0!n!} = 1$ represents selecting nothing at all (or leaving behind all the objects) which can be done only in one way.

(ii) For $r = n$, $^nC_n = \frac{n!}{n!0!} = 1$ represents selection of all the objects, which can be done only in one way.

(iii) $^nC_{n-r} = \frac{n!}{(n-r)!(n-(n-r))!} = \frac{n!}{(n-r)!r!} = {}^nC_r$ represents that selecting r objects out of n objects is the same as rejecting $(n-r)$ objects, and vice versa.

(iv) $^nC_a = {}^nC_b$ if and only if $a = b$ or $b = n - a \Rightarrow n = a + b$.

(v) $^nC_{r-1} + {}^nC_r = {}^{n+1}C_r$

Proof:

$$^nC_{r-1} + {}^nC_r = \frac{n!}{(r-1)!(n-r+1)!} + \frac{n!}{r!(n-r)!}$$

$$= \frac{n!}{(r-1)!(n-r+1)!(n-r)!} + \frac{n!}{r\times(r-1)!(n-r)!}$$

$$= \frac{n!}{(r-1)!(n-r)!}\left[\frac{1}{n-r+1} + \frac{1}{r}\right]$$

$$= \frac{n!}{(r-1)!(n-r)!} \times \frac{n-r+1+r}{r(n-r+1)}$$

$$= \frac{(n+1)n!}{r(r-1)!(n-r+1)(n-r)!}$$

$$= \frac{(n+1)!}{r!(n+1-r)!} = {}^{n+1}C_r$$

Example 1: Find n in the following cases:

(i) If $^nC_7 = {}^nC_2$ (ii) $^{2n}C_3 : {}^nC_2 = 44:3$

(iii) $^nP_r = 336$ and $^nC_r = 56$

(iv) $^{15}C_6 + {}^{15}C_7 + {}^{15}C_8 + {}^{15}C_9 = {}^nC_r$

Solution:

(i) $^nC_a = {}^nC_b \Rightarrow a = b$ or $n = a + b$

Therefore, $^nC_7 = {}^nC_2 \Rightarrow n = 7 + 2 = 9$

(ii) $^{2n}C_3 : {}^nC_2 = 44:3 \Rightarrow \frac{^{2n}C_3}{^nC_2} = \frac{3}{2}$

$$\Rightarrow \frac{(2n)!}{3!(2n-3)!} \times \frac{2!(n-2)!}{n!} = \frac{3}{2}$$

$$\Rightarrow \frac{(2n)(2n-1)(2n-2)(2n-3)!}{6\cdot(2n-3)!} \times \frac{2\cdot(n-2)!}{n(n-1)(n-2)!} = \frac{3}{2}$$

$$\Rightarrow \frac{4(n)(2n-1)(n-1)(2n-3)!}{6\cdot(2n-3)!} \times \frac{2\cdot(n-2)!}{n(n-1)(n-2)!} = \frac{3}{2}$$

$$\Rightarrow \frac{8}{6}(2n-1) = \frac{44}{3}$$

$$\Rightarrow 2n - 1 = 11 \Rightarrow n = 6$$

(iii) $^nP_r = 336$ and $^nC_r = 56$

$\because {}^nP_r = {}^nC_r \cdot r!$

$\therefore 336 = 56 \times r! \Rightarrow r! = \frac{336}{56} = 6$

$\Rightarrow r! = 3\times2\times1 = 3! \Rightarrow r = 3$

Now, $\because {}^nP_r = 336$

$\therefore \frac{n!}{(n-r)!} = 336 \Rightarrow \frac{n!}{(n-3)!} = 336$

$\Rightarrow \frac{n(n-1)(n-2)(n-3)!}{(n-3)!} = 336$

$\Rightarrow n(n-1)(n-2) = 336 = 8\times7\times6$

$\Rightarrow n = 8$

(iv) $^{15}C_6 + {}^{15}C_7 + {}^{15}C_8 + {}^{15}C_9 = 2{}^nC_9$

LHS $= {}^{15}C_6 + {}^{15}C_7 + {}^{15}C_8 + {}^{15}C_9$

$= {}^{15}C_{15-6} + {}^{15}C_{15-7} + {}^{15}C_8 + {}^{15}C_9$ $[\because {}^nC_r = {}^nC_{n-r}]$

$= {}^{15}C_9 + {}^{15}C_8 + {}^{15}C_8 + {}^{15}C_9$

$= 2({}^{15}C_8 + {}^{15}C_9) = 2{}^{16}C_9$ $[\because {}^nC_{r-1} + {}^nC_r = {}^{n+1}C_r]$

Example 2: How many 4-letter words, with or without meaning, can be made from the letters of the words FORMULA, assuming that no letter is repeated,

(i) consisting of 2 vowels and 2 consonants?
(ii) consisting of at least one vowel?

Solution:
There are 3 vowels (A, O or U) and 4 consonants (F, L, M or R) in the word FORMULA.

(i) 2 consonants can be selected from 4 consonants in ${}^4C_2$ ways.
   2 vowels can be selected from 3 vowels in ${}^3C_2$ ways.
   Thus, by the multiplication principle, the number of words that can be formed is ${}^4C_2 \times {}^3C_2 \times 4! = 432$ (multiplication with 4! is done for permutation of 4 selected letters).
(ii) Selection of at least one vowel can be solved in two different ways:
   (a) Selection of at least one vowel = Selection with no condition – Selection with no vowel (only consonant)
   $= ({}^7C_4 - {}^4C_4) \times 4! = (35-1) \times 4! = 816$
   (multiplication with 4! is done for permutation of 4 selected letters).
   (b) Selection of at least one vowel = Selection of 1 vowel and 3 consonants + 2 vowels and 2 consonants + 3 vowels and 1 consonant
   $({}^3C_1 \times {}^4C_3 + {}^3C_2 \times {}^4C_2 + {}^3C_3 \times {}^4C_1) \times 4!$
   $= (3\times 4 + 3\times 6 + 1\times 4)\times 4! = 816$ ways (multiplication with 4! is done for permutation of 4 selected letters).

Example 3: How many words, with or without meaning, can be made from all the letters of the words FORMULA, assuming that no letter is repeated,

(i) when the vowels occupy odd places?
(ii) when no two vowels come together?

Solution:
There are 3 vowels (A, O or U) and 4 consonants (F, L, M or R) in the word FORMULA.

(i) Out of 4 odd places (1, 3, 5, 7), 3 vowels (A, O or U) can be arranged in ${}^4C_3 \times 3!$ ways.
   In the remaining 4 places (1 odd place and 3 even places) 4 consonants (F, L, M or R) can be arranged in 4! ways.
   Thus, by the multiplication principle, the number of words that can be formed is ${}^4C_3 \times 3! \times 4! = 576$.
(ii) Since, no two vowels can be together, first arrange the 4 consonants (F, L, M or R) in such a way that we get one gap between each of them as shown, _ C _ C _ C _ C _. In these 4 places, 4 consonants (F, L, M or R) can be arranged in 4! ways.
   Now, out of 5 blank places, 3 vowels (A, O or U) can be arranged in ${}^5C_3 \times 3!$ ways.
   Thus, by the multiplication principle, the number of words that can be formed is $4! \times {}^5C_3 \times 3! = 1440$.
   (This method is known as the gap method.)

Example 4: How many 4-letter words consisting of 2 vowels and 2 consonants, with or without meaning, can be made from the letters of the words MINIMIZES, assuming that no letter is repeated?

Solution:
There are 4 vowels (3 I, 1 E) and 5 consonants (2 M, 1 N, 1 S, 1 Z) in the word MINIMIZES.

(i) Since I is repeated, the selection of two vowels can be done in two different ways—either both letters are the same (selection of 2 I), which can be done in one way only or both letters are different (selection of 1 I, 1 E), which can be done in ${}^2C_2$ ways.
   Similarly, since M is repeated, the selection of two consonants can be done in two different ways—either both letters are the same (selection of 2 M), which can be done in one way only or both letters are different (selection of any two different letters from 4 consonants (1 M, 1 N, 1 S, 1 Z), which can be done in ${}^4C_2$ ways.
   Thus, by the multiplication principle, the number of words that can be formed is
   (2 I, 2 M or 2 I, 2 different consonants or 2 different vowels, 2 M or 2 different vowels, 2 different consonants)
   $$1\times 1\times\frac{4!}{2!2!} + 1\times {}^4C_2 \times\frac{4!}{2!} + {}^2C_2 \times 1\times\frac{4!}{2!} + {}^2C_2 \times {}^4C_2 \times 4!$$
   $6+72+12+144 = 234$
   [multiplication with 4! is done for permutation of 4 selected letters and division by 2! is done for repeated letters (2 I or 2 M)].

Example 5: How many words, with or without meaning, can be made from all the letters of the words MINIMIZES, assuming that no letter is repeated,

(i) when the vowels occupy odd places?
(ii) when no two vowels come together?

Solution:
There are 4 vowels (3 I, 1 E) and 5 consonants (2 M, 1 N, 1 S, 1 Z) in the word MINIMIZES.

(i) Out of 5 odd places (1, 3, 5, 7, 9), 4 vowels (3 I, 1 E) can be arranged in ${}^5C_4 \times \frac{4!}{3!}$ ways
   [multiplication with 4! is done for permutation of 4 selected letters and division by 3! is done for repeated letters (3 I)].

In the remaining 5 places (1 odd place and 4 even places) 5 consonants (2 M, 1 N, 1 S, 1 Z) can be arranged in $\frac{5!}{2!}$ ways

[multiplication with 5! is done for permutation of 5 selected letters and division by 2! is done for repeated letters (2 M)].

Thus, by the multiplication principle, the number of words that can be formed is ${}^5C_4 \times \frac{4!}{3!} \times \frac{5!}{2!} = 1200.$

(ii) Since no two vowels can be together, first arrange 5 consonants (2 M, 1 N, 1 S, 1 Z) in such a way that we get one gap between each of them as shown _ C _ C _ C _ C _ C _. In these 5 places, 5 consonants (2 M, 1 N, 1 S, 1 Z) can be arranged in $\frac{5!}{2!}$ ways.

Now, out of 6 blank places, 4 vowels (3 I, 1 E) can be arranged in ${}^6C_4 \times \frac{4!}{3!}$ ways.

Thus, by the multiplication principle, the number of words that can be formed is $\frac{5!}{2!} \times {}^6C_4 \times \frac{4!}{3!} = 3600.$

(This method is known as the gap method.)

Example 6: Find the number of ways of selecting 6 balls from 4 red, 5 green and 6 blue balls (all balls are different),

(i) no condition,

(ii) if selection consists of 1 red, 2 green and 3 blue balls,

(iii) if each selection consists of 2 balls of each colour,

(iv) if each selection consists of at least 1 red ball.

Solution:

(i) 6 balls can be selected from 15 balls (without any condition) in ${}^{15}C_6$ ways.

(ii) 1 red ball can be selected from 4 red balls in ${}^4C_1$ ways.
2 green balls can be selected from 5 green balls in ${}^5C_2$ ways.

3 blue balls can be selected from 6 blue balls in ${}^6C_3$ ways.

Thus, by the multiplication principle, the number of ways of selecting 6 balls is $= {}^4C_1 \times {}^5C_2 \times {}^6C_3$

$= \frac{4!}{1!3!} \times \frac{5!}{2!3!} \times \frac{6!}{3!3!} = 4 \times 10 \times 20 = 800$

(iii) 2 red balls can be selected from 4 red balls in ${}^4C_2$ ways.

2 green balls can be selected from 5 green balls in ${}^5C_2$ ways.

2 blue balls can be selected from 6 blue balls in ${}^6C_2$ ways.

Thus, by the multiplication principle, the number of ways of selecting 6 balls is $= {}^4C_2 \times {}^5C_2 \times {}^6C_2$

$= \frac{4!}{2!2!} \times \frac{5!}{2!3!} \times \frac{6!}{2!4!} = 6 \times 10 \times 15 = 900$

(iv) Selection of at least 1 red ball consist of the following cases:

1 red ball and 5 balls from green or blue + 2 red balls and 4 balls from green or blue + 3 red balls and 3 balls from green or blue + 4 red balls and 2 balls from green or blue.

Thus, by the multiplication principle, the number of ways of selecting 6 balls is

$= {}^4C_1 \times {}^{11}C_5 + {}^4C_2 \times {}^{11}C_4 + {}^4C_3 \times {}^{11}C_3 + {}^4C_4 \times {}^{11}C_2$

$= \frac{4!}{1!3!} \times \frac{11!}{5!6!} + \frac{4!}{2!2!} \times \frac{11!}{4!7!} + \frac{4!}{3!1!} \times \frac{11!}{3!8!} + \frac{4!}{4!0!} \times \frac{11!}{2!9!}$

$= 4 \times 462 + 6 \times 330 + 4 \times 165 + 1 \times 55 = 4543$

Example 7: There are 2 red balls, 3 green balls and 4 blue balls (all balls are different) in a bag. How many different collections (or combinations) can be made such that the

(i) collection is non-empty?

(ii) collection consists of at least 1 blue ball?

(iii) collection consists of at least 1 ball of each colour?

Solution:

(i) The number of ways of selecting a non-empty collection is $= 2^9 - 1 = 511$ ways. (2^9 means, for each of the 9 balls we have two choices—the ball is chosen or ball is not chosen and –1 is used for one empty selection in which there is no ball is selected)

(ii) The number of ways of selecting at least 1 blue ball is $= (2^4 - 1) \times 2^5 = 480$ ways. (2^4 means, for each of the 4 blue balls we have two choices—blue ball is chosen or not chosen. Similarly 2^5 means, for each of the 5 (red or green) balls we have two choices— (red or green) ball is chosen or not chosen and –1 is used for one empty selection in which no blue ball is selected).

(iii) The number of ways of selecting at least 1 ball of each colour is $= (2^2 - 1) \times (2^3 - 1) \times (2^4 - 1) = 315$ ways. ($2^2, 2^3, 2^4$ is done for each of the 2 red balls, 3 green balls and 4 blue balls we have two choices— ball is chosen or not chosen, and –1 is used in each bracket for empty selection in which there is no red ball, or no green ball or no blue ball selected).

Example 8: There are 4 women and 5 men, out of whom a committee of 2 women and 3 men is to be formed. Find the number of ways in which a committee can be formed

(i) with no condition,

(ii) if a particular woman is included and a particular man is excluded,

(iii) if a particular woman is included then a particular man is excluded, and vice versa.

Solution:

(i) 2 women can be selected from 4 women in ${}^4C_2$ ways.
3 men can be selected from 5 men in ${}^5C_3$ ways.
Thus, by the multiplication principle, the required number of ways of selecting 2 women and 3 men with no condition is

$= {}^4C_2 \times {}^5C_3 = \frac{4!}{2!2!} \times \frac{5!}{3!2!} = 6 \times 10 = 60$

(ii) Let a woman W_1 be included in the committee. Then the remaining 1 woman can be selected from the remaining 3 women (since woman W_1 is already selected) in ${}^3C_1$ ways.

Let a man M_1 be excluded from the selection (or committee). Then 3 men can be selected from the remaining 4 men (since man M_1 is excluded from the selection) in ${}^4C_3$ ways.

Thus, by the multiplication principle, the number of ways of selecting 2 women and 3 men when a particular woman is included and a particular man is excluded is $= {}^3C_1 \times {}^4C_3 = \frac{3!}{1!2!} \times \frac{4!}{3!2!} = 3 \times 2 = 6$

(iii) Let us assume that if the woman W_1 is included in the committee, then the man M_1 is excluded from the committee and if the man M_1 is included in the committee, then the woman W_1 is excluded from the committee.

This can be solved in two different ways:

(a) W_1 is included but M_1 is not + M_1 is included but W_1 is not + neither W_1 nor M_1 is included in the committee. Thus, by the multiplication principle, the number of ways is $= {}^3C_1 \times {}^4C_3 + {}^3C_2 \times {}^4C_2 + {}^3C_2 \times {}^4C_3$

$= \frac{3!}{1!2!} \times \frac{4!}{3!1!} + \frac{3!}{2!1!} \times \frac{4!}{2!2!} + \frac{3!}{2!1!} \times \frac{4!}{3!1!}$

$= 3 \times 4 + 3 \times 6 + 3 \times 4 = 12 + 18 + 12 = 42$

(b) Total selection with no condition – both W_1 and M_1 are included in the committee

Thus, by the multiplication principle, the required number of ways is $= {}^4C_2 \times {}^5C_3 - {}^3C_1 \times {}^4C_2$

$= \frac{4!}{2!2!} \times \frac{5!}{3!2!} - \frac{3!}{1!2!} \times \frac{4!}{2!2!} = 6 \times 10 - 3 \times 6 = 42$

Day 96

NCERT Exercise 7.4 (With Solutions)

Q1. If ${}^nC_8 = {}^nC_2$, find ${}^nC_2$.

Solution:

${}^nC_a = {}^nC_b \Rightarrow a = b$ or $n = a + b$

Therefore, ${}^nC_8 = {}^nC_2 \Rightarrow n = 8 + 2 = 10$

$\therefore {}^nC_2 = {}^{10}C_2 = \frac{10!}{2!(10-2)!} = \frac{10!}{8!} = \frac{10 \times 9 \times 8!}{8!} = 45$

Q2. Determine n if

(i) ${}^{2n}C_3 = {}^nC_3 = 12:1$

(ii) ${}^{2n}C_3 : {}^nC_3 = 11:1$

Solution:

(i) $\frac{{}^{2n}C_3}{{}^nC_3} = \frac{12}{1}$

$\Rightarrow \frac{(2n)!}{3!(2n-3)!} \times \frac{3!(n-3)!}{n!} = \frac{12}{1}$

$\Rightarrow \frac{(2n)(2n-1)(2n-2)(2n-3)!}{(2n-3)!}$

$\times \frac{(n-3)!}{n(n-1)(n-2)(n-3)!} = 12$

$\Rightarrow \frac{2(2n-1)(2n-2)}{(n-1)(n-2)} = 12$

$\Rightarrow \frac{4(2n-1)(n-1)}{(n-1)(n-2)} = 12$

$\Rightarrow \frac{(2n-1)}{(n-2)} = 3 \Rightarrow 2n - 1 = 3(n-2)$

$\Rightarrow 2n - 1 = 3n - 6 \Rightarrow n = 5$

(ii) $\frac{{}^{2n}C_3}{{}^nC_3} = \frac{11}{1}$

$\Rightarrow \frac{(2n)!}{3!(2n-3)!} \times \frac{3!(n-3)!}{n!} = 11$

$\Rightarrow \frac{(2n)(2n-1)(2n-2)(2n-3)!}{(2n-3)!}$

$\times \frac{(n-3)!}{n(n-1)(n-2)(n-3)!} = 11$

$\Rightarrow \frac{2(2n-1)(2n-2)}{(n-1)(n-2)} = 11$

$\Rightarrow \frac{4(2n-1)(n-1)}{(n-1)(n-2)} = 11 \Rightarrow \frac{4(2n-1)}{n-2} = 11$

$\Rightarrow 4(2n-1) = 11(n-2)$

$\Rightarrow 8n - 4 = 11n - 22 \Rightarrow 3n = 18 \Rightarrow n = 6$

Q3. How many chords can be drawn through 21 points on a circle?

Solution:

For drawing one chord on a circle, only 2 points are required. Therefore, the number of chords that can be drawn through the given 21 points on a circle is

$= {}^{21}C_3 = \frac{21!}{2!(21-2)!} = \frac{21!}{2!19!} = \frac{21 \times 20}{2} = 210$

Q4. In how many ways can a team of 3 boys and 3 girls be selected from 5 boys and 4 girls?

Solution:

3 boys can be selected from 5 boys in ${}^5C_3$ ways.

3 girls can be selected from 4 girls in ${}^4C_3$ ways.

Therefore, by the multiplication principle, the number of ways in which a team of 3 boys and 3 girls can be selected is = ${}^5C_3 \times {}^4C_3$

$$= \frac{5!}{3!2!} \times \frac{4!}{3!2!} = \frac{5\times4\times3!}{3!\times2} \times \frac{4\times3!}{3!} = 10\times4 = 40$$

Q5. Find the number of ways of selecting 9 balls from 6 red, 5 white balls and 5 blue balls if each selection consists of 3 balls of each colour.

Solution:
There are a total of 6 red balls, 5 white balls and 5 blue balls, and 9 balls have to be selected in such a way that each selection consists of 3 balls of each colour.

Therefore, 3 balls can be selected from 6 red balls in ${}^6C_3$ ways.
3 balls can be selected from 5 white balls in ${}^5C_3$ ways.
3 balls can be selected from 5 blue balls in ${}^5C_3$ ways.
Thus, by the multiplication principle, the number of ways of selecting 9 balls is = ${}^6C_3 \times {}^5C_3 \times {}^5C_3$

$$= \frac{6!}{3!3!} \times \frac{5!}{3!2!} \times \frac{5!}{3!2!}$$

$$= \frac{6\times5\times4\times3!}{3!\times3\times2} \times \frac{5\times4\times3!}{3!\times2\times1} \times \frac{5\times4\times3!}{3!\times2\times1}$$

$$= 20\times10\times10 = 2000$$

Q6. Determine the number of 5-card combinations out of a deck of 52 cards if there is exactly one ace in each combination.

Solution:
In a deck of 52 cards, there are 4 aces. Combinations of 5 cards have to be made in which there is exactly one ace.

Then, one ace can be selected in ${}^4C_1$ ways and the remaining 4 cards can be selected out of the 48 cards in ${}^{48}C_4$ ways.

Thus, by the multiplication principle, the number of card combinations is = ${}^{48}C_4 \times {}^4C_1$

$$= \frac{48!}{4!\,44!} \times \frac{4!}{1!3!} = \frac{48\times47\times46\times45}{3\times2\times1} = 778032$$

Q7. In how many ways can one select a cricket team of eleven from 17 players in which only 5 players can bowl if each cricket team of 11 must include exactly 4 bowlers?

Solution:
Out of 17 players, 5 players are bowlers, therefore out of 5 bowlers a cricket team of 11 players with exactly 4 bowlers can be selected in ${}^5C_4$ ways and the remaining 7 players can be selected out of the 12 players in ${}^{17}C_2$ ways. Thus, by the multiplication principle, the number of ways of selecting the cricket team is = ${}^5C_4 \times {}^{12}C_7$

$$= \frac{5!}{4!1!} \times \frac{12!}{7!5!} = \frac{12\times11\times10\times9\times8}{4\times3\times2\times1} = 3960$$

Q8. A bag contains 5 black and 6 red balls. Determine the number of ways in which 2 black and 3 red balls can be selected.

Solution:
There are 5 black and 6 red balls in the bag, therefore 2 black balls can be selected out of 5 black balls in ${}^5C_2$ ways and 3 red balls can be selected out of 6 red balls in ${}^6C_3$ ways.

Thus, by the multiplication principle, the number of ways of selecting 2 black and 3 red balls is

$$= {}^5C_2 \times {}^6C_3 = \frac{5!}{2!3!} \times \frac{6!}{3!3!} = \frac{5\times4}{2} \times \frac{6\times5\times4}{3\times2\times1} = 200$$

Q9. In how many ways can a student choose a programme of 5 courses if 9 courses are available and 2 specific courses are compulsory for every student?

Solution:
There are 9 courses available out of which 2 specific courses are compulsory for every student.
Therefore, every student has to choose 3 courses out of the remaining 7 courses. This can be chosen in ${}^7C_3$ ways.
Thus, required number of ways of choosing the program is = ${}^7C_3 = \frac{7!}{3!4!} = \frac{7\times6\times5\times4!}{3\times2\times1\times4!} = 35$

Day 97

NCERT Miscellaneous Exercise (With Solutions)

Q1. How many words, with or without meaning, each containing 2 vowels and 3 consonants, can be formed from the letters of the word DAUGHTER?

Solution:
In the word DAUGHTER, there are 3 vowels (A, E and U) and 5 consonants (D, G, H, R, T). Therefore Number of ways of selecting 2 vowels out of 3 vowels is = ${}^3C_2 = 3$

Number of ways of selecting 3 consonants out of 5 consonants is = ${}^5C_3 = 10$

Thus, number of combinations of 2 vowels and 3 consonants = $3\times10 = 30$

Each of these 30 combinations of 2 vowels and 3 consonants can be arranged among themselves in 5! ways.

Hence, the total number of different words formed is $= 30 \times 5! = 3600$

Q2. How many words, with or without meaning, can be formed using all the letters of the word EQUATION at a time so that vowels and consonants occur together?

Solution:
In the word EQUATION, there are 5 vowels (A, E, I, O, U) and 3 consonants (N, Q, T).

Since, all the vowels and consonants have to occur together, both (AEIOU) and (QTN) can be assumed as a single letter each. Then, the permutations of these 2 letters taken all at a time is $^2P_2 = 2!$

Also, each corresponding letter can be permutated as 5! permutations of the five vowels taken all at a time and 3! permutations of the consonants taken all at a time.

Thus, by multiplication principle, total number of words $= 2! \times 5! \times 3! = 1440$

Q3. A committee of 7 has to be formed from 9 boys and 4 girls. In how many ways can this be done when the committee consists of:
(i) exactly 3 girls?
(ii) at least 3 girls?
(iii) at most 3 girls?

Solution:
A committee of 7 has to be formed from 9 boys and 4 girls.

(i) Since exactly 3 girls have to be there in every committee, each committee must consist of $(7-3) = 4$ boys only.

Thus, total number of ways is $= {}^4C_3 \times {}^9C_4$

$$= \frac{4!}{3!1!} \times \frac{9!}{4!5!} = \frac{9\times8\times7\times6\times5!}{3\times2\times1\times5!} = 504$$

(ii) Since at least 3 girls have to be there in every committee, the committee can consist of
(a) 3 girls and 4 boys or
(b) 4 girls and 3 boys

3 girls and 4 boys can be selected in ${}^4C_3 \times {}^9C_4$ ways.
4 girls and 3 boys can be selected in ${}^4C_4 \times {}^9C_3$ ways.
Therefore, the total number of ways is
$= {}^4C_3 \times {}^9C_4 + {}^4C_4 \times {}^9C_3 = 504 + 84 = 588$

(iii) Since at most 3 girls have to be there in every committee, the committee can consist of
(a) 3 girls and 4 boys
(b) 2 girls and 5 boys
(c) 1 girl and 6 boys
(d) No girl and 7 boys

3 girls and 4 boys can be selected in ${}^4C_3 \times {}^9C_4$ ways.
2 girls and 5 boys can be selected in ${}^4C_2 \times {}^9C_5$ ways.
1 girl and 6 boys can be selected in ${}^4C_1 \times {}^9C_6$ ways.
No girl and 7 boys can be selected in ${}^4C_0 \times {}^9C_7$ ways.
Therefore, the total number of ways is

$= {}^4C_3 \times {}^9C_4 + {}^4C_2 \times {}^9C_5 + {}^4C_1 \times {}^9C_6 + {}^4C_0 \times {}^9C_7$

$$= \frac{3!}{3!1!} \times \frac{9!}{4!5!} + \frac{4!}{2!2!} \times \frac{9!}{5!4!} + \frac{4!}{1!3!} \times \frac{9!}{6!3!} + \frac{4!}{0!4!} \times \frac{9!}{7!2!}$$

$= 504 + 756 + 336 + 36 = 1632$

Q4. If the different permutations of all the letters of the word EXAMINATION are listed in a dictionary, how many words are there in this list before the first word starting with E?

Solution:
In the given word EXAMINATION, there are 11 letters out of which, A, I and N appear 2 times and all the other letters appear only once.

The words that will be listed before the words starting with E in a dictionary will be the words that start with A only.

Therefore, to get the number of words starting with A, the letter A is fixed at the first position, and the remaining 10 letters taken all at a time are arranged.
Since there are 2 Is and 2 Ns in the remaining 10 letters,

the number of words starting with A is $= \frac{10!}{2!2!} = 907200$

Q5. How many 6-digit numbers can be formed from the digits 0, 1, 3, 7 and 9 which are divisible by 10 and no digit is repeated?

Solution: A number is divisible by 10 if its unit digit is 0, therefore, 0 is fixed at the units place.

Thus, the remaining 5 digits (1, 3, 5, 7 and 9) will be placed in the remaining 5 places in 5! ways.

Hence, the total number of 6-digit numbers which are divisible by 10 is $= 5! = 120$

Q6. The English alphabet has 5 vowels and 21 consonants. How many words with two different vowels and 2 different consonants can be formed from the alphabet?

Solution:
Two different vowels and 2 different consonants are to be selected from the English alphabet.

Since there are 5 vowels in the English alphabet, the number of ways of selecting 2 different vowels from the 5 vowels is $= {}^5C_2 = \frac{5!}{2!3!} = 10$

Since there are 21 consonants in the English alphabet, the number of ways of selecting 2 different consonants from the 21 consonants is

$= {}^{21}C_2 = \frac{21!}{2!19!} = 210$

Thus, the total number of ways of selecting 2 different vowels and 2 different consonants is $= 10 \times 210 = 2100$

Each of these 2100 combinations has 4 letters, which can be arranged among themselves in 4! ways. Therefore, the total number of words formed is
$2100 \times 4! = 50400$

Q7. In an examination, a question paper consists of 12 questions divided into two parts—Part I and II, containing 5 and 7 questions, respectively. A student is required to attempt 8 questions in all, selecting at least 3 from each part. In how many ways can a student select the questions?

Solution:
The questions paper consists of 12 questions divided into two parts—Part I and II, containing 5 and 7 questions respectively. A student has to attempt 8 questions, selecting at least 3 from each part.
This can be done in the following ways:
(a) 3 questions from Part I and 5 questions from Part II.
(b) 4 questions from Part I and 4 questions from Part II.
(c) 5 questions from Part I and 3 questions from Part II
Thus, 3 questions from Part I and 5 questions from Part II can be selected in ${}^5C_3 \times {}^7C_5$ ways.
4 questions from Part I and 4 questions from Part II can be selected in ${}^5C_4 \times {}^7C_4$ ways.
5 questions from Part I and 3 questions from Part II can be selected in ${}^5C_5 \times {}^7C_3$ ways.
Hence, the total number of ways of selecting questions is

$$= {}^5C_3 \times {}^7C_3 + {}^5C_4 \times {}^7C_4 + {}^5C_5 \times {}^7C_3$$

$$= \frac{5!}{2!3!} \times \frac{7!}{2!5!} + \frac{5!}{4!1!} \times \frac{7!}{4!3!} + \frac{5!}{5!0!} \times \frac{7!}{3!4!}$$

$$= 210 + 175 + 35 = 420$$

Q8. Determine the number of 5-card combinations out of a deck of 52 cards if each selection of 5 cards has exactly one king.

Solution:
From a deck of 52 cards, 5-card combinations have to be made in such a way that in each selection of 5 cards, there is exactly one king.

In a deck of 52 cards, there are 4 kings. Therefore, 1 king can be selected out of 4 kings in ${}^4C_1$ ways.

4 cards out of the remaining 48 cards can be selected in ${}^{48}C_4$ ways.

Thus, the total number of 5 card combinations is ${}^4C_1 \times {}^{48}C_4$

Q9. It is required to seat 5 men and 4 women in a row so that the women occupy the even places. How many such arrangements are possible?

Solution:
Five men and 4 women are to seated in a row such that the women occupy the even places.

There are 5 odd places (1, 3, 4, 7, 9), therefore 5 men can be seated in 5! ways.

And, there are 4 even places (2, 4, 6, 8). Therefore, 4 women can be seated in 4! ways.
Thus, the total number of arrangements is

$$= 4! \times 5! = 24 \times 120 = 2880$$

Q10. From a class of 25 students, 10 are to be chosen for an excursion party. There are 3 students who decide that either all of them will join or none of them will join. In how many ways can the excursion party be chosen?

Solution:
There are 3 students who decide that either all of them will join or none of them will join. Therefore, two cases arise:
Case I: All the three students join:
In this case, the remaining 7 students can be chosen from the remaining 22 students in ${}^{22}C_7$ ways (since 3 students already joined the party).
Case II: None of the three students join:
In this case 10 students can be chosen from the remaining 22 students in ${}^{22}C_{10}$ ways (since 3 students will not join the party).

Thus, total number of ways of choosing the excursion party is ${}^{22}C_7 + {}^{22}C_{10}$

Q11. In how many ways can be letters of the word ASSASSINATION be arranged so that all the Ss are together?

Solution:
In the given word ASSASSINATION, the letter A appears 3 times, S appears 4 times, I appears 2 times, N appears 2 times, and all the other letters appear only once.

All the words have to be arranged in such a way that all the Ss are together, therefore all 4 Ss (SSSS) is treated as a single letter.

Thus, this single letter together with the remaining 9 letters will account as 10 letters.
Therefore, these 10 letters in which there are 3 As, 2 Is and 2 Ns can be arranged in $\frac{10!}{3!2!2!} = 151200$ ways.

Day 98

NCERT Exemplar Problems: Solutions (Q1–Q18)

Q1. Eight chairs are numbered 1 to 8. Two women and 3 men wish to occupy one chair each. First the women choose the chairs from amongst the chairs 1 to 4 and then the men select from the remaining chairs. Find the total number of possible arrangements.

Solution:
Women choose the chairs first from among 1 to 4 chairs and there are two women. Therefore, the number of arrangements = ${}^4P_2$ ways.

Then, the men have to choose chairs from the remaining 6 chairs.
There are 3 men, therefore the number of arrangements $= {}^6P_3$ ways.
Thus, total number of possible arrangements

$$= {}^4P_2 \times {}^6P_3 = \frac{4!}{(4-2)!} \times \frac{6!}{(6-3)!}$$

$$= \frac{4!}{2!} \times \frac{6!}{3!} = \frac{4 \times 3 \times 2!}{2!} \times \frac{6 \times 5 \times 4 \times 3!}{3!}$$

$$= 4 \times 3 \times 6 \times 5 \times 4 = 1440$$

Q2. If the letters of the word 'RACHIT' are arranged in all possible ways as listed in dictionary. Then, what is the rank of the word 'RACHIT'?

Solution:
The letters of the word 'RACHIT' in alphabetical order are A, C, H, I, T.
Words beginning with A = 5!
Words beginning with C = 5!
Words beginning with H = 5!
Words beginning with I = 5!
Word beginning with R is RACHIT = 1
∴ Rank of the word 'RACHIT' in the dictionary
$= 4 \times 5! + 1 = 4 \times 120 + 1 = 481$

Q3. A candidate is required to answer 7 questions out of 12 questions, which are divided into two groups, each containing 6 questions. He is not permitted to attempt more than 5 questions from either group. Find the number of different ways of choosing questions.

Solution:
A candidate cannot attempt more than 5 questions from either group.
Therefore, total number of questions attempted from each group is given as

Group I	5	4	3	2
Group II	2	3	4	5

∴ Total number of possible ways
$= {}^6C_5 \times {}^6C_2 + {}^6C_4 \times {}^6C_3 + {}^6C_3 \times {}^6C_4 + {}^6C_4 \times {}^6C_5$
$= 2[{}^6C_5 \times {}^6C_2 + {}^6C_4 \times {}^6C_3] = 2[6 \times 15 + 15 \times 20]$
$= 2[90 + 300] = 2 \times 390 = 780$

Q4. Out of 18 points in a plane, no three are in the same line except five points which are collinear. Find the number of lines that can be formed joining the points.

Solution:
Total number of points =18
∴ Number of straight lines formed by joining the 18 points taking 2 at a time $= {}^{18}C_2$
Number of straight lines formed by joining 5 points taking 2 at a time $= {}^5C_2$
Out of 18 points, 5 points are collinear, and these 5 collinear points, when joined pair-wise will give only one line.
Therefore, the total number of straight lines formed
$= {}^{18}C_2 - {}^5C_2 + 1 \Rightarrow 153 - 10 + 1 = 44$

Q5. We wish to select 6 persons from 8 persons, but if the person A is chosen, then B must be chosen. In how many ways can the selections be made?

Solution:
Total number of persons = 8
Number of persons to be selected = 6
If A is chosen then, B must be chosen.
Therefore, the following cases will arise:
Case I: When A and B are both chosen
Number of ways $= {}^{8-2}C_{6-2} = {}^6C_4$
Case II: When A is not chosen, but B may or may not be chosen
Number of ways $= {}^{8-1}C_6 = {}^7C_6$
Thus, the total number of ways
$= {}^6C_4 + {}^7C_6 = 15 + 7 = 22$

Q6. How many committees of five persons with a chairperson can be selected from 12 persons?

Solution:
Total number of persons = 12
Number of persons to be selected = 5
Out of 12 persons, a chairperson is selected $= {}^{12}C_1 = 12$ ways
The remaining 4 persons are selected out of 11 persons as $= {}^{11}C_4 = 330$
Thus, the total number of ways to form a committee of 5 persons $= 12 \times 330 = 3960$

Q7. How many automobile licence plates can be made, if each plate contains two different letters followed by three different digits?

Solution:
There are 26 letters in the English alphabet and 10 digits (0 to 9).
Each plate contains two different letters followed by three different digits.
∴ Arrangement of 26 letters, taken 2 at a time

$$= {}^{26}P_2 = \frac{26!}{24!} = 26 \times 25 = 650$$

Arrangements of three-digit numbers that can be formed out of the 10 digits $= {}^{10}P_3 = 10 \times 9 \times 8 = 720$ ways.
∴ Total number of license plates $= 650 \times 720 = 468000$

Q8. A bag contains 5 black and 6 red balls. Determine the number of ways in which 2 black and 3 red balls can be selected from the lot.

Solution:
The bag contains 5 black and 6 red balls.
Therefore, 2 black balls are selected from 5 black balls in ${}^5C_2$ ways.

3 red balls are selected from 6 red balls in ${}^6C_3$ ways.
∴ Total number of ways in which 2 black and 3 red balls are selected is $= {}^5C_2 \times {}^6C_3 = 10 \times 20 = 200$ ways.

Q9. Find the number of permutations of n distinct things taken r together, in which 3 particular things must occur together.

Solution:
Total number of things $= n$
Therefore, combination of n things taken r at a time in which 3 things always occur $= {}^{n-3}C_{r-3}$
If three things are taken together, then it is considered as 1 group, and arrangement of these three things = 3!
Thus, total number of arrangements of $= (r-3+1) = (r-2)$ things is $= (r-2)!$
Thus, the total number of arrangements
$= {}^{n-3}C_{r-3} \times 3! \times (r-2)!$

Q10. Find the number of different words that can be formed from the letters of the word 'TRIANGLE', so that no vowels are together.

Solution:
The word 'TRIANGLE' contains 5 consonants and 3 vowels.
Now, 5 consonants can be arranged in =5! =120 ways

x	C	x	C	x	C	x	C	X	C	x

Since 3 vowels are not together, they can be arranged in between consonants (at x positions) at 6 places in ${}^6P_3$

$$= \frac{6!}{(6-3)!} = \frac{6!}{3!} = \frac{6 \times 5 \times 4 \times 3!}{3!} = 120 \text{ ways.}$$

Thus, the total number of arrangements
$= 120 \times 120 = 14400$

Q11. Find the number of positive integers greater than 6000 and less than 7000 which are divisible by 5, provided no digit is repeated.

Solution:
The number is divisible by 5. Therefore, at the units place, the digit will be 0 or 5. Therefore, the units digit can be filled in 2 ways.

Now, since the 4-digit number is greater than 6000 and less than 7000, the thousands digit can be filled only in 1 way (digit 6).

Since repetition is not allowed, the tens place can be filled in 8 ways, similarly the hundreds place can be filled in 7 ways.
Thus, the total number of integers formed is
$= 1 \times 8 \times 7 \times 2 = 14 \times 8 = 112$

Q12. There are 10 persons named $P_1, P_2, P_3, \ldots\ldots . P_{10}$. Out of 10 persons, 5 persons are to be arranged in a line such that in each arrangement P_1 must occur whereas P_4, P_5 do not occur. Find the number of such possible arrangements.

Solution:
Out of 10 persons $P_1, P_2, P_3, \ldots\ldots . P_{10}$, 5 persons are to be arranged in which P_1 must occur whereas P_4, P_5 never occur. Therefore, select only 4 persons out of 7 persons $(P_2, P_3, P_6, \ldots, P_{10})$
Thus, total number of selections is

$$= {}^7C_4 = \frac{7!}{4!(7-4)!} = \frac{7!}{4!3!} = \frac{5040}{24 \times 6} = 35$$

And hence, the number of arrangements of 5 persons is $= 35 \times 5! = 35 \times 120 = 4200$

Q13. There are 10 lamps in a hall. Each one of them can be switched on independently. Find the number of ways in which the hall can be illuminated.

Solution:
Each lamp has two choices (on or off). Therefore, the total number of ways in which the hall can be illuminated is $= 2^{10} - 1 = 1024 - 1 = 1023$
(–1 if all lamps switched off).

Q14. A box contains two white, three black and four red balls. In how many ways can three balls be drawn from the box, if at least one black ball is to be included in the draw?

Solution:
There are 2 white, 3 black and 4 red balls.
Selection to draw 3 balls out of these 9 balls in which at least one black ball is included can be done in the following ways.

Black balls	1	2	3
Other than black	2	1	0

Therefore, the total number of selections is
$= {}^3C_1 \times {}^6C_2 + {}^3C_2 \times {}^6C_1 + {}^3C_3 \times {}^6C_0$
$= 3 \times 15 + 3 \times 6 + 1 = 45 + 18 + 1 = 64$

Q15. If ${}^nC_{r-1} = 36$, ${}^nC_r = 84$, ${}^nC_{r+1} = 126$, then find the value of ${}^rC_2$.

Solution:
From the first two equations, we get

$$\frac{{}^nC_{r-1}}{{}^nC_r} = \frac{36}{84} \Rightarrow \frac{n!}{(r-1)!\{n-(r-1)\}!} \cdot \frac{r!(n-r)!}{n!} = \frac{3}{7}$$

$$\Rightarrow \frac{r!(n-r)!}{(r-1)!(n-r+1)!} = \frac{3}{7}$$

$$\Rightarrow \frac{r(r-1)!(n-r)!}{(r-1)!(n-r+1)(n-r)!} = \frac{3}{7}$$

$$\Rightarrow \frac{r}{(n-r+1)} = \frac{3}{7}$$

$\Rightarrow 7r = 3n - 3r + 3 \Rightarrow 10r - 3n = 3$

And from the last two equations, we get

$$\frac{^nC_r}{^nC_{r+1}} = \frac{84}{126} \Rightarrow \frac{n!}{r!(n-r)!} \cdot \frac{(r+1)!(n-r-1)!}{n!} = \frac{2}{3}$$

$$\Rightarrow \frac{(r+1)!(n-r-1)!}{r!(n-r)!} = \frac{2}{3}$$

$$\Rightarrow \frac{(r+1)r!(n-r-1)!}{r!(n-r)(n-r-1)!} = \frac{2}{3} \Rightarrow \frac{r+1}{n-r} = \frac{2}{3}$$

$\Rightarrow 3r + 3 = 2n - 2r \Rightarrow 2n - 5r = 3$

Solving the above two equations, we get

$r = 3$ and $n = 9$

Therefore, $^rC_2 = {^3C_2} = \frac{3!}{2!1!} = 3$

Q16. Find the number of integers greater than 7000 that can be formed with the digits 3, 5, 7, 8, 9 where no digits are repeated.

Solution:

The numbers greater than 7000 will be 4 digits and 5 digits numbers, with the digits 3, 5, 7, 8, 9 because repetition is not allowed.

All the five-digit numbers are greater than 7000. Therefore, the number of ways of forming 5-digit numbers is $5! = 5 \times 4 \times 3 \times 2 \times 1 = 120$

Four-digit numbers greater than 7000 can be formed when the thousands place can be filled in 3 ways (7, 8, 9), the hundreds place can be filled in 4 ways and the tens place can be filled in 3 ways. The units place can be filled in 2 ways.

Thus, the total number of 4-digit numbers greater than 7000 is $= 3 \times 4 \times 3 \times 2 = 72$

Thus, total number of integers greater than 7000 $= 120 + 72 = 192$

Q17. If 20 lines are drawn in a plane such that no two of them are parallel and no three are concurrent, in how many points will they intersect each other?

Solution:

No two lines are parallel means all lines intersect. No three lines are concurrent means three lines intersect at a point.

For one point of intersection, we require two lines. Therefore, the number of points of intersection is

$$= {^{20}C_2} = \frac{20!}{2!8!} = \frac{20 \times 19 \times 18!}{2 \times 1 \times 18!} = \frac{20 \times 19}{2} = 190$$

Q18. In a certain city, all telephone numbers have six digits, the first two digits always being 41 or 42 or 46 or 62 or 64. How many telephone numbers have all six digits distinct?

Solution:

The first two digits will always be 41 or 42 or 46 or 62 or 64, that is, 5 ways.

If the first two digits are 41, the remaining 4 digits can be arranged in

$$= {^8P_4} = \frac{8!}{8-4!} = \frac{8!}{4!} = \frac{8 \times 7 \times 6 \times 5 \times 4!}{4!} = 1680$$

Thus, the total number of telephone numbers having all six digits distinct $= 5 \times 1680 = 8400$

Day 99

NCERT Exemplar Problems: Solutions (Q19–Q40)

Q19. In an examination, a student has to answer 4 questions out of 5 questions; questions 1 and 2 are however compulsory. Determine the number of ways in which the student can make the choice.

Solution:

Two questions are compulsory out of 5. Therefore, selection of the other 2 out of the remaining 3 questions is $= {^{5-2}C_{4-2}} = {^3C_2} = 3$

Q20. If a convex polygon has 44 diagonals, then find the number of its sides.

Solution:

Let the convex polygon have n sides.

$\therefore$ Number of diagonals $= {^nC_2} - n$

According to the question,

$$^nC_2 - n = 44 \Rightarrow \frac{n!}{2!(n-2)!} - n = 44$$

$$\Rightarrow \frac{n(n-1)}{2} - n = 44$$

$\Rightarrow n(n-1) - 2n = 44 \times 2 \Rightarrow n(n-3) = 88$

$\Rightarrow n^2 - 3n - 88 = 0 \Rightarrow (n-11)(n+8) = 0$

$\Rightarrow n = 11, -8$

$\therefore n = 11$ $[\because n \neq -8]$

Q21. Eighteen mice were placed in two experimental groups and one control group, with all the groups being equally large. In how many ways can the mice be placed into three groups?

Solution:

Eighteen mice were placed equally in two experimental groups and one control group. Therefore

$$= \frac{\text{Total arrangement}}{\text{Equally likely arrangement}} = \frac{18!}{6!6!6!}$$

Q22. A bag contains six white marbles and five red marbles. Find the number of ways in which four marbles be drawn from the bag, if
(i) they can be of any colour.
(ii) two must be white and two red.
(iii) they must all be of the same colour.

Solution:
Total number of marbles = 6 white + 5 red = 11 marbles
(i) If they can be of any colour means, we have to select 4 marbles out of 11.
$\therefore$ Total number of ways = ${}^{11}C_4$
(ii) If two must be white, then selection of two white marbles will be ${}^6C_2$ and two must be red, and selection of two red marbles will be ${}^5C_2$.
$\therefore$ Total number of ways = ${}^6C_2 \times {}^5C_2$
(iii) If they all must be of same colour, then selection of 4 white marbles out of 6 = ${}^6C_4$ and selection of 4 red marbles out of 5 = ${}^5C_4$
$\therefore$ Total number of ways = ${}^6C_4 + {}^5C_4$

Q23. In how many ways can a football team of 11 players be selected from 16 players? How many of them will
(i) include 2 particular players?
(ii) exclude 2 particular players?

Solution:
Total number of players =16
(i) If 2 particular players are included
$= {}^{16-2}C_{11-2} = {}^{14}C_9$
(Selection of n objects taken r at a time in which p objects are always included is ${}^{n-p}C_{r-p}$)
(ii) If 2 particular players are excluded
$= {}^{16-2}C_{11} = {}^{14}C_{11}$
(Selection of n objects taken r at a time in which p objects are never included is ${}^{n-p}C_r$)

Q24. A sports team of 11 students is to be constituted, choosing at least 5 from Class XI and at least 5 from Class XII. If there are 20 students in each of these classes, in how many ways can the team be constituted?

Solution:
Total students in each class = 20
Choosing at least 5 from Class XI and at least 5 from Class XII can be done in the following ways

Class XI	5	6
Class XII	6	5

Therefore, total number of ways of selecting a team of 11 players is
$= {}^{20}C_5 \times {}^{20}C_6 + {}^{20}C_6 \times {}^{20}C_5 = 2 \times {}^{20}C_5 \times {}^{20}C_6$

Q25. A group consists of 4 girls and 7 boys. In how many ways can a team of 5 members be selected, if the team has
(i) no girls?
(ii) at least one boy and one girl?
(iii) at least three girls?

Solution:
Number of girls = 4 and number of boys = 7
Selection of a team of 5 members provided that
(i) the team has no girls
Total selection = ${}^7C_5 = \dfrac{7!}{5!2!} = \dfrac{7\times6}{2} = 21$
(ii) the team has at least one boy and one girl
Total selection
$= {}^7C_1 \times {}^4C_4 + {}^7C_2 \times {}^4C_3 + {}^7C_3 \times {}^4C_2 + {}^7C_4 \times {}^4C_1$
$= 7\times1+21\times4+35\times6+35\times4 = 441$
(iii) the team has at least three girls
Total selection
$= {}^4C_3 \times {}^7C_2 + {}^4C_4 \times {}^7C_1 = 4\times21+7 = 91$

Q26. A committee of 6 is to be chosen from 10 men and 7 women, so as to contain at least 3 men and 2 women. In how many different ways can this be done, if two particular women refuse to serve on the same committee?

Solution:
Total number of men = 10
And total number of women = 7
Selection of committee containing at least 3 men and 2 women can be done in the following ways:
(3 men and 3 women or 4 men and 2 women)
Total number of ways = ${}^{10}C_3 \times {}^7C_3 + {}^{10}C_4 \times {}^7C_2$
$= (120\times35+210\times21 = 4200+4410 = 8610$
If two particular women to be always there, then total number of ways = ${}^{10}C_3 \times {}^5C_1 + {}^{10}C_4 \times {}^5C_0$
${}^{10}C_4 \times {}^5C_0 + {}^{10}C_3 \times {}^5C_1 = 210+120\times5 = 810$
Total number of committee when two particular women are never together
$=$ Total$-$ Together $= 8610-810 = 7800$

Q27. If ${}^nC_{12} = {}^nC_8$, then n is equal to
(a) 20 (b) 12 (c) 6 (d) 30

Solution: (a)
${}^nC_{12} = {}^nC_8 \Rightarrow {}^nC_{n-12} = {}^nC_8 [\because {}^nC_r = {}^nC_{n-r}]$
$\Rightarrow n-12 = 8 \Rightarrow n = 12+8 = 20$

Q28. The number of possible outcomes when a coin is tossed 6 times is
(a) 36 (b) 64 (c) 12 (d) 32

Solution: (b)
Number of outcomes when tossing a coin 1 times = 2 (head and tail)

∴ Total possible outcomes when a coin tossed 6 times $= 2^6 = 64$

Q29. The number of different four-digit numbers that can be formed with the digits 2, 3, 4, 7 and using each digit only once is
(a) 120 (b) 96 (c) 24 (d) 100

Solution: (c)
Number of different four-digit numbers that can be formed with the digits 2, 3, 4, 7 and using each digit only once is $= {}^4P_4 = 4! = 24$

Q30. The sum of the digits in unit place of all the numbers formed with the help of 3, 4, 5, 6 taken all at a time is
(a) 432 (b) 108 (c) 36 (d) 18

Solution: (b)
If we fixed digit 3 at the units place, then total possible number formed with remaining 3 digits is 3!

Similarly, if we fixed 4, 5 and 6 at the units place, in each case the total possible numbers is 3!.

Thus, the sum of the digits in the units place of all these numbers $= (3+4+5+6)\times 3! = 18\times 3! = 108$

Q31. The total number of words formed by 2 vowels and 3 consonants taken from 4 vowels and 5 consonants is
(a) 60 (b) 120 (c) 7200 (d) 720

Solution: (c)
Number of selections of 2 vowels and 3 consonants from 4 vowels and 5 consonants is $= {}^4C_2 \times {}^5C_3$

$$= \frac{4!}{2!2!}\times\frac{5!}{3!2!} = \frac{4\times 3!}{2!2!}\times\frac{5\times 4\times 3\times 2!}{3!\times 2!} = 60$$

Therefore, the total number of words formed from these 2 vowels and 3 consonants is
$= 60\times 5! = 60\times 120 = 7200$

Q32. If a five-digit number divisible by 3 is to be formed using the numbers 0, 1, 2, 3, 4, 5 without repetitions, then the total number of ways this can be done is
(a) 216 (b) 600 (c) 240 (d) 325

Solution: (a)
A number is divisible by 3, when sum of the digits in the number must be divisible by 3.

Thus, 5 digits with their sum divisible by 3 can be chosen in the following as (1, 2, 3, 4, 5) or (0, 1, 2, 4, 5).

Now, consider the digits 1, 2, 3, 4, 5. Then the five-digit numbers that can be formed using theses digits is 5!

And, when we consider the digits 0, 1, 2, 4, 5 then, the five-digit numbers that can be formed using these digits is $= 4\cdot 4!$ (since 0 cannot be at the first place)
Thus, the total number of ways
$= 5! + 4\cdot 4! = 120 + 96 = 216$

Q33. Everybody in a room shakes hands with everybody else. If the total number of handshakes is 66, then the total number of persons in the room is
(a) 11 (b) 12 (c) 13 (d) 14

Solution: (b)
Let the total number of persons in the room be n.
Two persons are required for 1 handshake, and therefore, the number of hands shaken by n persons is ${}^nC_2$
Thus, according to the question

$${}^nC_2 = 66 \Rightarrow \frac{n!}{2!(n-2)!} = 66$$

$$= \frac{n(n-1)}{2} = 66 \Rightarrow n(n-1) = 132$$

$$\Rightarrow n^2 - n - 132 = 0 \Rightarrow (n-12)(n+11) = 0$$

$$\Rightarrow n = 12, -11$$

$$\therefore n = 12$$

Q34. The number of triangles that are formed by choosing the vertices form a set of 12 points, seven of which lie on the same line is
(a) 105 (b) 15 (c) 175 (d) 185

Solution: (d)
Total number of triangles formed from 12 points taking 3 at a time $= {}^{12}C_3$
Out of 12 points, 7 are collinear. So, these 7 points form no triangle.
Thus, the total number of triangles
$= {}^{12}C_3 - {}^7C_3 = 220 - 35 = 185$

Q35. The number of parallelograms that can be formed from a set of four parallel lines intersecting another set of three parallel lines is
(a) 6 (b) 18 (c) 12 (d) 9

Solution: (b)
To form a parallelogram we require a pair of lines from a set of 4 lines and another pair of lines from another set of 3 lines.
Thus, the total number of parallelograms
$= {}^4C_2 \times {}^3C_2 = 6\times 3 = 18$

Q36. The number of ways in which a team of eleven players can be selected from 22 players always including 2 of them and excluding 4 of them is
(a) ${}^{16}C_{11}$ (b) ${}^{16}C_5$ (c) ${}^{16}C_9$ (d) ${}^{20}C_9$

Solution: (c)
Selection of 11 players when 2 of them are always included and 4 are never included is $= {}^{22-2-4}C_{11-2} = {}^{16}C_9$

Q37. The number of 5-digit telephone numbers having at least one of their digits repeated is
(a) 90000 (b) 10000 (c) 30240 (d) 69760

Solution: (d)
If all the digits repeat, then 5-digit telephone numbers can be formed in 10^5 ways and if no digit repeated, then 5-digit telephone numbers can be formed in ${}^{10}P_5$ ways.

Therefore, number of 5-digit telephone numbers having at least one of their digits repeated is

$= 10^5 - {}^{10}P_5 = 10000 - \frac{10!}{5!} = 100000 - 10\times9\times8\times7\times6$

$= 100000 - 30240 = 69760$

Q38. The number of ways in which we can choose a committee from four men and six women, so that the committee includes at least two men and exactly twice as many women as men is
(a) 94 (b) 126
(c) 128 (d) None of these

Solution: (a)
Choosing a committee (four men and six women) which includes at least two men and exactly twice as many women as men can be done as 2 men and 4 women or 3 men and 6 women
Thus, total number of ways the committee is formed
$= {}^4C_2 \times {}^6C_4 + {}^4C_3 \times {}^6C_6 = 6\times15 + 4\times1 = 94$

Q39. The total number of 9-digit numbers which have all different digits is
(a) 10! (b) 9!
(c) $9\times9!$ (d) $10\times10!$

Solution: (c)
For 9-digit numbers with the digits 0, 1, 2, 3, 4, 5, 6, 7, 8, 9, zero cannot be placed at the first place.

Therefore, for the first place we have 9 choices (1, 2, 3, 4, 5, 6, 7, 8, 9), and the remaining 8 places can be filled in $= 9\times8! = 9!$ ways.
Thus, total number of ways $= 9\times9!$

Q40. The number of words which can be formed out of the letters of the word 'ARTICLE' so that vowels occupy the even place is
(a) 1440 (b) 144
(c) 7! (d) ${}^4C_4 \times {}^3C_3$

Solution: (b)
There are 3 even places (2, 4, 6) and three vowels (A, E, I). Therefore, they are arranged in $3! = 6$ ways.

There are 4 remaining odd places (1, 3, 5, 7) and four consonants (C, L, R, T). Therefore, they are arranged in $4! = 24$ ways.
Thus, total number of words $= 6\times24 = 144$

Day 100

NCERT Exemplar Problems: Solutions (Q41–Q64)

Q41. Given five different green dyes, four different blue dyes and three different red dyes, the number of combinations of dyes which can be chosen taking at least one green and one blue dye is
(a) 3600 (b) 3720 (c) 3800 (d) 3600

Solution: (b)
Choosing at least one blue and one green dye = all combination of selection minus those combinations in which there is no green or blue dyes.
Possible number of choosing green dyes $= 2^5$
Possible number of choosing blue dyes $= 2^4$
Possible number of choosing red dyes $= 2^3$
Therefore, total possibilities of selecting at least one blue and one green dye is
$= (2^5 - 1)(2^4 - 1)\times 2^3 = 3720$

Q42. If ${}^nP_r = 840$, ${}^nC_r = 35$, then r is equal to______

Solution:
$\because {}^nP_r = {}^nC_r \cdot r!$

$\therefore 840 = 35\times r! \Rightarrow r! = \frac{840}{35} = 24$

$\Rightarrow r! = 4\times3\times2\times1 = 4! \Rightarrow r = 4$

Q43. ${}^{15}C_8 + {}^{15}C_9 - {}^{15}C_6 - {}^{15}C_7$ is equal to______

Solution:
${}^{15}C_8 + {}^{15}C_9 - {}^{15}C_6 - {}^{15}C_7$

$= {}^{15}C_{15-8} + {}^{15}C_{15-9} - {}^{15}C_6 - {}^{15}C_7 \quad [\because {}^nC_r = {}^nC_{n-r}]$

$= {}^{15}C_7 + {}^{15}C_6 - {}^{15}C_6 - {}^{15}C_7 = 0$

Q44. The number of permutations of n different objects, taken r at a time, when repetitions are allowed, is______

Solution:
Number of permutations of n different objects taken r at a time when repetition is allowed $= n^r$

Q45. The number of different words that can be formed from the letters of the word 'INTERMEDIATE' such that two vowels never come together is ______

Solution:
In the word 'INTERMEDIATE' there are 6 consonants and 6 vowels.
Two vowels never come together. Therefore, the arrangement of 6 consonants out of which 2 are alike can be placed in 6!/2! ways

x C x C x C x C x C x C x

Now, six vowels out of which 3 Es are alike and 2 Is are alike can be arranged in places between 7 (marked as x) in ${}^7P_6 \times \frac{1}{3!} \times \frac{1}{2!}$ ways.

Thus, total number of words formed is

$= \frac{6!}{2!} \times {}^7P_6 \times \frac{1!}{3!} \times \frac{1!}{2!} = 151200$

Q46. Three balls are drawn from a bag containing 5 red, 4 white and 3 black balls. Find the number of ways in which this can be done, if at least 2 are red.

Solution:

Total number of ways $= {}^5C_2 \times {}^7C_1 + {}^5C_3$

$= 10 \times 7 + 10 = 70 + 10 = 80$

Q47. The number of six-digit numbers all digits of which are odd, is______

Solution:

Among the digits 0, 1, 2, 3, 4, 5, 6, 7, 8, 9 only digits 1, 3, 5, 7, 9 are odd. Therefore, the total number of six-digit numbers is $= 5^6$

Q48. In a football championship, 153 matches were played. Every two teams played one match with each other. The number of teams, participating in the championship is______

Solution:

Let the number of teams participating in the championship be n.

Every two teams played one match with each other. Therefore, the total matches played $= {}^nC_2 = 153$

$\Rightarrow \frac{n(n-1)}{2} = 153 \Rightarrow n^2 - n = 306$

$\Rightarrow n^2 - n - 306 = 0 \Rightarrow (n-18)(n+17) = 0$

$\Rightarrow n = 18, -17, \therefore n = 18$

Q49. The total number of ways in which six '+' and four '–' signs can be arranged in a line such that no two signs '–' occur together, is______

Solution:

No two '–' signs occur together. Therefore, first arrange '+' signs.

– + – + – + – + – + – + –

Then, '+' signs can be arranged in 1 way because all are identical.

And 4 '–' signs can be arranged at 7 places in ${}^7C_4$ ways.

Thus, total number of arrangements is

$= {}^7C_7 \times 1 = \frac{7!}{4!3!} = \frac{7\times6\times5\times4!}{3!\times4!} = \frac{7\times6\times5}{3\times2\times1} = 35$

Q50. A box contains 2 white balls, 3 black balls and 4 red balls. The number of ways three balls can be drawn from the box, if at least one black ball is to be included in the draw is ______

Solution:

There are 2 white, 3 black and 4 red balls. It is given that at least one black ball is to be included in the draw. Therefore, the total number of possible draws is

$= {}^3C_1 \times {}^6C_2 + {}^3C_2 \times {}^6C_1 + {}^3C_3$

$= 3\times15 + 3\times6 + 1 = 45 + 18 + 1 = 64$

Answer if the following questions are true or false

Q51. There are 12 points in a plane of which 5 points are collinear, then the number of lines obtained by joining these points in pairs is ${}^{12}C_2 - {}^5C_2$.

Solution: False

Total number of lines $= {}^{12}C_2 - {}^5C_2 + 1$

Q52. Three letters can be posted in five letter boxes in 3^5 ways.

Solution: False

Total number of ways $= 5^3 = 125$

Q53. In the permutations of n things, r taken together, the number of permutations in which m particular things occur together is ${}^{n-m}P_{r-m} \times {}^rP_m$.

Solution: False

Arrangement of n things, taken r at a time in which m things occur together, (considering these m things as 1 group) can be done as,

Number of ways of selecting object excluding those m objects $= (n-m)$

Number of arrangements of $(n-m+1)$ objects is $(n-m+1)!$ and m objects which were considered as 1 group, can be arranged in $m!$ ways.

Therefore, total number of arrangements $= (n-m+1)!\ m!$

Q54. In a steamer there are stalls for 12 animals and there are horse, cows and calves (not less than 12 each) ready to be shipped. They can be loaded in 3^{12} ways.

Solution: True

There are three types of animals and stalls available for 12 animals, therefore the total number of ways of loading $= 3^{12}$.

Q55. If some or all of n objects are taken at a time, then the number of combinations $2^n - 1$.

Solution: True.

If some or all objects are taken at a time, then the number of selections is ${}^nC_1 + {}^nC_2 + {}^nC_3 + ... + {}^nC_n$

$({}^nC_0 + {}^nC_1 + {}^nC_2 + {}^nC_3 + ... + {}^nC_n) - {}^nC_0 = 2^n - 1$

Q56. There will be only 24 selections containing at least one red ball out of a bag containing 4 red and 5 black balls. It is being given that the balls of the same colour are identical.

Solution: True

Total number of selections

$= [(4+1)(5+1)-1]-5 = (5\times6-1)-5 = 24$

Q57. Eighteen guests are to be seated, half on each side of a long table. Four particular guests desire to sit on one particular side and three others on other side of the table. The number of ways in which the seating arrangement can be made is $\frac{11!}{5!6!}(9!)(9!)$.

Solution: True

After seating 4 on one side and 3 on the other side, we have to select out of 11, 5 on one side and 6 on the other side.

Thus, remaining selection of the first side is $= {}^{(18-4-3)}C_5 = {}^{11}C_5$ and the remaining selection of the other side is ${}^{(11-5)}C_6 = {}^6C_6$

Total arrangements $= {}^{11}C_5 \times 9! \times {}^6C_6 \times 9!$

$= \frac{11!}{5!6!}\times 9! \times 1 \times 9! = \frac{11!}{5!6!}\times 9! \times 9!$

Q58. A candidate is required to answer 7 questions, out of 12 questions which are divided into two groups, each containing 6 questions. He is not permitted to attempt more than 5 questions from either group. He can choose the seven questions in 650 ways.

Solution: False

Attempting of questions can be done in the following ways

Group (A)	**2**	**3**	**4**	**5**
Group (B)	5	4	3	2

Number of ways of attempting 7 questions is

$= {}^6C_2 \times {}^6C_5 + {}^6C_3 \times {}^6C_4 + {}^6C_4 \times {}^6C_3 + {}^6C_5 \times {}^6C_2$

$= 2({}^6C_2 \times {}^6C_5 + {}^6C_3 \times {}^6C_4) = 2(15\times6 + 20\times15) = 780$

Q59. To fill 12 vacancies, there are 25 candidates of which 5 are from scheduled castes. If 13 of the vacancies are reserved for scheduled caste candidates while the rest are open to all, the number of ways in which the selection can be made is ${}^5C_3 \times {}^{20}C_9$.

Solution: False

Selection of 3 scheduled castes candidate out of 5 is ${}^5C_3$.

Selection of 9 other candidates out of 22 is ${}^{22}C_9$.

Thus, total number of selections $= {}^5C_3 \times {}^{22}C_9$

Match the following

Q60. There are 3 books on mathematics, 4 on physics and 5 on English. How many different selections can be made such that each selection consists...

Column I	Column II
(i) One book of each subject	(a) 3968
(ii) At least one book of each subject	(b) 60
(iii) At least book of English	(c) 3255

Solution:

There are 3 books of mathematics, 4 of physics and 5 on English.

(i) Selection of one book of each subject $= {}^3C_1 \times {}^4C_1 \times {}^5C_1 = 3\times4\times5 = 60$

(ii) Selection of at least one book of each subject $= (2^3-1)\times(2^4-1)\times(2^5-1) = 7\times15\times31 = 3255$

(iii) Selection of at least one book of English is based on following manner

English book	1	2	3	4	**5**
Others	11	10	9	8	7

Selection of at least one book of English is $= (2^5-1)\times 2^7 = 128\times31 = 3968$

Q61. Five boys and five girls form a line. Find the number of ways of making the seating arrangement under the following condition.

Column I	Column II
(i) Boys and girls alternate	(a) $5!\times6!$
(ii) No two girls sit together	(b) $10!-5!6!$
(iii) All the girls sit together	(c) $(5!)^2+(5!)^2$
(iv) All the girls are never together	(d) 2! 5! 5!

Solution:

(i) Boys and girls alternate $= (5!)^2+(5!)^2$

(ii) No two girls sit together =5! 6!

(iii) All the girls sit together =2! 5! 5!

(iv) All the girls are never together $= 10!-5!6!$

Q62. There are 10 professors and 20 lecturers, out of whom a committee of 2 professors and 3 lecturers is to be formed. Find

Column I

(i) In how many ways can a committee be formed?

(ii) In how many ways can a particular professor be included?

(iii) In how many ways can a particular lecturer be included?

(iv) In how many ways can a particular lecturer be excluded?

Column II

(a) ${}^{10}C_2 \times {}^{19}C_3$

(b) ${}^{10}C_2 \times {}^{19}C_2$

(c) ${}^9C_1 \times {}^{20}C_3$

(d) ${}^{10}C_2 \times {}^{20}C_3$

Solution:

(i) Selection of 2 professors out of 10 and 3 lecturers out of 20 = ${}^{10}C_2 \times {}^{20}C_3$

(ii) When a particular professor is included $= {}^{10-1}C_1 \times {}^{20}C_3 = {}^{9}C_1 \times {}^{20}C_3$

(iii) When a particular lecturer is included $= {}^{10}C_2 \times {}^{19}C_2$

(iv) When a particular is lecturer excluded $= {}^{10}C_2 \times {}^{19}C_3$

Q63. Using the digits 1, 2, 3, 4, 5, 6, 7, a number of 4 different digits is formed. Find

Column I

(i) How many numbers are formed?

(ii) How many numbers are exactly divisible by 2?

(iii) How many numbers are exactly divisible 25?

(iv) how many of these are exactly divisible by 4?

Column II

(a) 840

(b) 200

(c) 360

(d) 40

Solution:

(i) Total numbers of 4 digits formed with digits 1, 2, 3, 4, 5, 6, 7 is ${}^{7}C_4 \cdot 4! = \frac{7!}{4!3!} \cdot 4! = 7 \times 6 \times 5 \times 4 = 840$

(ii) A number is divisible by 2 when its units place is occupied by digits (2, 4, 6).

Total number is

$$3 \cdot {}^{6}C_3 \cdot 3! = 3 \cdot \frac{6!}{3!3!} \cdot 3! = \frac{3 \times 6 \times 5 \times 4}{3 \times 2 \times 1} = 360$$

(iii) A number is divisible by 25 when its last two digits will be (25 or 75).

Total number is $2 \cdot {}^{5}C_2 \cdot 2! = 2 \cdot \frac{5!}{2!3!} \cdot 2! = 2 \times 5 \times 4 = 40$

1234567

(iv) A number is divisible by 4 if its last two digits are divisible by 4, (12, 16, 24, 32, 36, 52, 56, 64, 72, 76)

Total number is

$$10 \cdot {}^{5}C_2 \cdot 2! = 10 \cdot \frac{5!}{2!3!} \cdot 2! = 10 \times 5 \times 4 = 200$$

Q64. How many words (with or without dictionary meaning) can be made from the letters of the word MONDAY, assuming that no letter is repeated, if

Column I

(i) 4 letters are used at a time

(ii) All letters are used at a time

(iii) All letters are used but the first is a vowel

Column II

(a) 720

(b) 240

(c) 360

Solution:

(i) 4 letters are used at a time

$$= {}^{6}P_4 = \frac{6!}{2!} = 6 \times 5 \times 4 \times 3 = 360$$

(ii) All letters used at a time

$= 6! = 6 \times 5 \times 4 \times 3 \times 2 \times 1 = 720$

(iii) All letters used but first is vowel (A or O)

$= 2 \times 5! = 2 \times 5 \times 4 \times 3 \times 2 \times 1 = 240$

Day 101

Very Short Answer (VSA 1): Mark and Short Answer (SA): 2-Mark Questions

Q1. Ten chairs are numbered 1 to 10. Three women and 4 men wish to occupy one chair each. First the women choose the chairs from amongst the chairs 1 to 5 and then men select from the remaining chairs. Find the total number of possible arrangements.

Solution:

Women choose the chairs first from among 1 to 5 chairs, and there are 3 women. Therefore, the number of arrangements $= {}^{5}P_3$ ways.

Now, arrangement of 4 men out of remaining 7 chairs is ${}^{7}P_4$ ways.

Thus, total number of possible arrangements is

$$= {}^{5}P_3 \times {}^{7}P_4 = \frac{5!}{(5-3)!} \times \frac{7!}{(7-4)!} = \frac{5!}{2!} \times \frac{7!}{3!} = 50400$$

Q2. We wish to select 3 persons from 6 persons, but if the person A is chosen, then B must not be chosen. In how many ways can selections be made?

Solution:

Total number of persons = 6

Number of persons to be selected = 4

If A is chosen then, B must not be chosen.

Therefore, following cases will arise:

Case I: When A is chosen, but B must not be chosen

Number of ways $= {}^{6-2}C_2 = {}^{4}C_2 = 6$

Case II: When A is not chosen, but B may or may not be chosen

Number of ways $= {}^{6-1}C_3 = {}^{5}C_3 = 10$

Thus, total number of ways is $= 6 + 10 = 16$

Q3. How many committees of six persons with a chairperson and a chairleader can be selected from 10 persons?

Solution:
Total number of persons = 10
Number of persons to be selected = 6
Out of 10 persons a chairperson is selected = $^{10}C_1 = 10$ ways
Out of remaining 9 persons, a chairleader is selected = $^9C_1 = 9$ ways
And remaining 4 persons are selected out of 8 persons as = $^8C_4 = 70$
Thus, total number of ways to form a committee of 6 persons = $10 \times 9 \times 70 = 6300$

Q4. How many secret codes can be made, if each code contains two different capital letters followed by two small letters?

Solution:
There are 26 letters in the English alphabet.
Therefore, arrangements of two different capital letters out of 26 letters is = $^{26}C_2 \times 2! = 26 \times 25 = 650$
And, arrangement of two small letters out of 26 letters is = both letters are same + both are different letters = $^{26}C_1 \times 1! + {}^{26}C_2 \times 2! = 26 + 26 \times 25 = 676$
Thus, the total number of secret codes containing two different capital letters followed by two small letters is = $650 \times 676 = 439400$

Q5. A bag contains 4 black and 6 red balls, determine the number of ways in which 2 black and 3 red balls or 3 black and 2 red balls can be selected from the lot.

Solution:
The bag contains 4 black and 6 red balls.
2 black balls are selected from 4 black balls in $^4C_2$ ways.
3 red balls are selected from 6 red balls in $^6C_3$ ways.
3 black balls are selected from 4 black balls in $^4C_3$ ways.
2 red balls are selected from 6 red balls in $^6C_2$ ways.
∴ Total number of ways in which the balls are selected is = $^4C_2 \times {}^6C_3 + {}^6C_2 \times {}^4C_3 = 6 \times 20 + 20 \times 6 = 240$

Q6. Find the number of permutations of n distinct things taken r together, in which m particular things must not occur.

Solution:
Total number of things = n
Therefore, permutation of n things taken r at a time in which m things must not occur = $^{n-m}C_r \times r!$

Q7. Find the number of permutations of n distinct things taken r together, in which m particular things must occur together.

Solution:
Total number of things = n
Therefore, combination of n things taken r at a time in which m things always occur = $^{n-m}C_{r-m}$
If m things are taken together, then it is considered as 1 group, and arrangements of these m things = $m!$
Thus, the total number of arrangements of = $(r-m+1)$ things is = $(r-m+1)!$
Thus, the total number of arrangements is = $^{n-m}C_{r-m} \times m! \times (r-m+1)!$

Q8. Find the number of positive integers greater than 2000 and less than 4000 which are divisible by 5, provided no digit is repeated.

Solution:
Since the number is divisible by 5, at the units place the digit will be 0 or 5. Therefore, units digit can be filled in 2 ways.

Now, since the 4-digit number is greater than 2000 and less than 4000, the thousands digit can be filled in 2 ways (digit 2 or 3).

Since repetition is not allowed, the tens place can be filled in 8 ways, and similarly, the hundreds place can be filled in 7 ways.

Thus, the total number of integers formed is
= $2 \times 2 \times 8 \times 7 = 224$

Q9. There are 10 persons named $P_1, P_2, P_3, ..., P_{10}$. Out of 10 persons, 5 persons are to be arranged in a line such that in each arrangement P_1 must occur whereas P_{10} does not occur. Find the number of such possible arrangements.

Solution:
Out of 10 persons $P_1, P_2, P_3, ... P_{10}$, 5 persons are to be arranged in which P_1 must occur whereas P_{10} never occurs. Therefore, we have to select only 4 persons out of the remaining 8 persons $(P_2, P_3, P_6, ..., P_9)$
Thus, the total number of selections is
= $^8C_4$
And thus, the number of arrangements of 5 persons is
$$= {}^8C_4 \times 5! = \frac{8!}{4!(8-4)!} \times 5! = 8 \times 7 \times 6 \times 5 \times 5 = 8400$$

Q10. A box contains two white, three black and four red balls. In how many ways can four balls be drawn from the box, if at least one ball of each colour must be included in the draw?

Solution:
Selection to draw 4 balls out of these 9 balls in which at least one ball of each colour is included, can be done in the following ways
2W1B1R + 1W2B1R + 1W1B2R
Therefore, the total number of selections is
= $^2C_2 \times {}^3C_1 \times {}^4C_1 + {}^2C_1 \times {}^3C_2 \times {}^4C_1 + {}^2C_1 \times {}^3C_1 \times {}^4C_2$
= $1 \times 3 \times 4 + 2 \times 3 \times 4 + 2 \times 3 \times 6 = 12 + 24 + 36 = 72$

Q11. Find the number of integers greater than 5000 that can be formed with the digits 3, 4, 5, 6, 7, where no digits are repeated.

Solution:

The numbers greater than 5000 will have 4 digits or 5 digits, with the digits 3, 4, 5, 6, 7 because repetition is not allowed.

Since all the five digits numbers are greater than 5000, the number of ways of forming 5-digit numbers is $5! = 5\times4\times3\times2\times1 = 120$

Now, four-digit numbers greater than 5000 can be formed when the thousands place is filled in 3 ways (5, 6, 7), and the hundreds place can be filled in 4 ways, the tens place can be filled in 3 ways and the units place can be filled in 2 ways.

Thus, the total number of 4-digits number greater than 5000 is $= 3\times4\times3\times2 = 72$

Thus, total number of integers greater than 5000 $= 120 + 72 = 192$

Q12. In a certain city, all telephone numbers have six digits, the first two digits always being 81 or 91. How many telephone numbers have all six digits distinct?

Solution:

The first two digits are always 81 or 91, that is, 2 ways.

And if first the two digits are 81, the remaining 4 digits can be arranged in

$$= {}^8P_4 = \frac{8!}{8-4!} = \frac{8!}{4!} = \frac{8\times7\times6\times5\times4!}{4!} = 1680$$

Thus, total number of telephone numbers having all six digits distinct $= 2\times1680 = 3360$

Q13. A bag contains four white marbles, five black marbles and six red marbles. Find the number of ways in which four marbles can be drawn from the bag, if they must all be of the same colour.

Solution:

If they must all be of the same colour, then selection of 4 marbles out of 4 white marbles is ${}^4C_4$; similarly selection of 4 marbles out of 5 black marbles is ${}^5C_4$ and selection of 4 marbles out of 6 red marbles is ${}^6C_4$

$\therefore$ Total number of ways $= {}^4C_4 + {}^5C_4 + {}^6C_4 = 21$

Q14. In how many ways can a cricket team of 11 players be selected from 15 players? How many of them will

(i) include 3 particular players?
(ii) exclude 2 particular players?

Solution:

(i) If 3 particular players are included

$= {}^{15-3}C_{11-3} = {}^{12}C_8$

(Selection of n objects taken r at a time in which p objects are always included is ${}^{n-p}C_{r-p}$)

(ii) If 2 particular players are excluded

$= {}^{15-2}C_{11} = {}^{13}C_{11}$

(Selection of n objects taken r at a time in which p objects are never included is ${}^{n-p}C_r$)

Q15. A cricket team of 11 students is to be constituted, choosing at least 4 from Class XI and at least 6 from Class XII. If there are 10 students in each of these classes, in how many ways can the team be constituted?

Solution:

Total students in each class = 10

Choosing at least 4 from class XI and at least 6 from class XII can be done in the following ways

Class XI	4	5
Class XII	7	6

Therefore, the total number of ways of selecting a team of 11 players is $= {}^{10}C_4 \times {}^{10}C_7 + {}^{10}C_5 \times {}^{10}C_6$

$= 210\times120 + 252\times210 = 78120$

Q16. A group consists of 4 girls and 6 boys. In how many ways can a team of 5 members be selected, if the team has at least one girl and at least two boys.

Solution:

Number of girls = 4 and number of boys = 6

Selection of a team of 5 members with at least one girl and at least two boys is

$= {}^4C_1 \times {}^6C_4 + {}^4C_2 \times {}^6C_3 + {}^4C_3 \times {}^6C_2$

$= 4\times15 + 6\times20 + 4\times15 = 240$

Q17. A committee of 8 is to be chosen from 10 men and 6 women, so as to contain at least 4 men and 3 women. In how many different ways can this be done, if one particular woman and man refuse to serve on the same committee?

Solution:

Total number of men = 10

And total number of women = 6

Selection of committee containing at least 4 men and 3 women can be done in the following ways:

(4 men and 4 women or 5 men and 3 women)

Total number of ways $= {}^{10}C_4 \times {}^6C_4 + {}^{10}C_5 \times {}^6C_3$

$= (210\times15 + 252\times20 = 8190$

If one particular woman and man must always be included in the committee, then total number of ways

$= {}^9C_3 \times {}^5C_3 + {}^9C_4 \times {}^5C_2 = 840 + 1260 = 2100$

Thus, total number of committees when one particular woman and man are never together

$=$ Total $-$ Together $= 8190 - 2100 = 6090$

Q18. The sum of the digits in the units place of all the numbers formed with the help of 1, 2, 3, 4 taken all at a time is

(a) 40 (b) 60 (c) 80 (d) 90

Solution: (b)
If we fixed digit 1 at the units place, then the total possible number formed with remaining 3 digits is 3!

Similarly, if we fixed 2, 3 and 4 at the units place, in each case the total possible numbers are 3!.
Thus, the sum of the digits in units place of all these numbers $= 1\times 3! + 2\times 3! + 3\times 3! + 4\times 3! = 60$

Q19. The sum of all the numbers formed with the help of digits 1, 2, 3, 4 taken all at a time is
(a) 40 (b) 60 (c) 80 (d) 90

Solution: (b)
The total number of four digits numbers formed by using all the digits at a time is $4! = 24$
And in these 24 numbers each digit occurs $24/4 = 6$ times at each (units, tens, hundrends and thousands) place.
Thus, the sum of the digits at each place is
$= 4\times 6 + 3\times 6 + 2\times 6 + 1\times 6 = 60$
Hence, the sum of all the numbers formed is
$= 60\times 1000 + 60\times 100 + 60\times 10 + 60\times 1 = 66660$

Q20. Twelve guests are to be seated, half on each side of a long table. Three particular guests desire to sit on one particular side and four others on other side of the table. Find the number of ways in which the seating arrangement can be made.

Solution:
After seating 3 on one side and 4 on the other side, we have to select out of 5, 3 on one side and 2 on the other side.
Thus, total arrangement is
$= {}^5C_3 \times 6! + {}^2C_2 \times 6! = \frac{5!}{3!2!}\times 6! + \frac{2!}{2!0!}\times 6! = 7920$

Day 102

Long Answer-I (LA-I): 4-Mark and Long Answer-II (LA-II): 6-Mark Questions

Q1. How many ways are there of choosing 4 cards from a pack of 52 playing cards, if
(i) 4 cards are of the same suit?
(ii) 4 cards are of the same colour?
(iii) 4 cards belong to four different suits?
(iv) 2 are red cards and 2 are black cards
(v) 2 belong to diamonds and 2 belong to clubs?
(vi) they contain at least 2 kings?
(vii) they contain at most 1 king and 1 black joker?

Solution:
In a pack of 52 playing cards:
There are 4 suits (club, diamond, heart and spade) containing 13 cards each.
There are 2 colours, black (club and spade) and red (diamond and heart) containing 26 cards each.
There are 4 cards of each king, queen and joker.

(i) Number of ways of choosing 4 cards of the same suit is $= {}^4C_1 \times {}^{13}C_4$ (${}^4C_1$ is for choosing 1 suit out of four suits and ${}^{13}C_4$ is for choosing 4 cards from the chosen suit).
(ii) Number of ways of choosing 4 cards of the same colour is $= {}^2C_1 \times {}^{26}C_4$ (${}^2C_1$ is for choosing 1 colour out of two colours and ${}^{26}C_4$ is for choosing 4 cards from the chosen colour)
(iii) Number of ways of choosing 4 cards from four different suits is $= {}^{13}C_1 \times {}^{13}C_1 \times {}^{13}C_1 \times {}^{13}C_1$ (${}^{13}C_1$ is for choosing 1 card from a suit and it is four times because of choosing 1 card from each of the 4 suits).
(iv) Number of ways of choosing 2 red cards and 2 black cards is $= {}^{26}C_2 \times {}^{26}C_2$ (${}^{26}C_2$ is for choosing 2 cards from a colour and it is two times because of choosing 2 cards from each of the 2 colours).
(v) Number of ways of choosing 2 diamond cards and 2 club cards is $= {}^{13}C_2 \times {}^{13}C_2$ (${}^{13}C_2$ is for choosing 2 cards from diamonds and clubs each).
(vi) Number of ways of choosing 4 cards with at least 2 kings is $= {}^4C_2 \times {}^{48}C_2 + {}^4C_3 \times {}^{48}C_1 + {}^4C_4 \times {}^{48}C_0$ (2 kings and 2 others or 3 kings and 1 other or 4 kings).
(vii) Number of ways of choosing 4 cards with at most 1 king and 1 black joker is
$= {}^4C_0 \times {}^2C_1 \times {}^{46}C_3 + {}^4C_1 \times {}^2C_1 \times {}^{46}C_2$ (no king, 1 black joker and 3 other or 1 king, 1 black joker and 2 other).

Q2. There are 4 women and 5 men, out of whom a committee of 3 persons is to be formed. Find the number of ways in which the committee can be formed
(i) if it consists of at least 1 woman
(ii) if a particular woman is included and a particular man is excluded
(iii) if a particular woman is included then a particular man is excluded, and vice versa.

Solution:
(i) Selection of 3 persons with at least 1 woman consists of the following cases:
1 woman and 2 Men + 2 women and 1 Man + 3 women and no man
Thus, by the multiplication principle, required number of ways is $= {}^4C_1 \times {}^5C_2 + {}^4C_2 \times {}^5C_1 + {}^4C_3 \times {}^5C_0$
$= \frac{4!}{1!3!}\times\frac{5!}{2!3!} + \frac{4!}{2!2!}\times\frac{5!}{1!4!} + \frac{4!}{3!1!}\times\frac{5!}{0!5!}$
$= 4\times 10 + 6\times 5 + 4\times 1 = 40 + 30 + 4 = 74$

(ii) Let a woman W_1 be included in the committee and a man M_1 be excluded from the selection (or committee). Then the remaining 2 persons (since woman W_1 is already selected) can be selected from the remaining 7 persons (since man M_1 is excluded from the selection) in ${}^7C_2 = 21$ ways.

(iii) If woman W_1 is included in the committee then man M_1 is excluded from the committee and if man M_1 is included in the committee then woman W_1 is excluded from the committee.
This can be solved in two different ways:

(a) W_1 is included but M_1 is not + M_1 is included but W_1 is not + neither W_1 nor M_1 is included in the committee
Thus, by the multiplication principle, the required number of ways is
$= {}^7C_2 + {}^7C_2 + {}^7C_3 = 21 + 21 + 35 = 77$

(b) Total selection with no condition – both W_1 and M_1 is included in the committee.
Thus, by the multiplication principle, the required number of ways is $= {}^9C_3 - {}^7C_1 = 84 - 7 = 77$.

Q3. A candidate is required to answer 6 questions out of 10 questions, which are divided into two groups, each containing 5 questions. Find the number of different ways of doing questions if at least 2 questions are compulsory from either group.

Solution:
At least 2 questions are compulsory from each group. Therefore, the total number of questions attempted from each group is given as

Group I	2	3	4
Group II	4	3	2

Thus, total number of possible ways
$= {}^5C_2 \times {}^5C_4 + {}^5C_3 \times {}^5C_3 + {}^5C_4 \times {}^5C_2$
$= 10 \times 5 + 10 \times 10 + 5 \times 10 = 200$

Q4. If the letters of the word 'BOARD' are arranged in all possible ways as listed in dictionary, then, what is the rank of the word 'BOARD'?

Solution:
The letters of the word 'BOARD' in alphabetical order are A, B, D, O, R.
Words starting with A = 4!
Word starting with BA = 3!
Word starting with BD = 3!
Next word will be BOADR = 1
Next word will be BOARD = 1
$\therefore$ Rank of the word 'BOARD ' in the dictionary is
$= 4! + 2 \times 3! + 2 = 24 + 12 + 2 = 38$

Q5. If the letters of the word 'FORMULA' are arranged in all possible ways as listed in dictionary, then, what is the rank of the word 'FORMULA'?

Solution:
The letters of the word 'FORMULA' in alphabetical order are A, F, L, M, O, R, U.
Words starting with A = 6!
Words starting with FA = 5!
Words starting with FL = 5!
Words starting with FM = 5!
Words starting with FOA = 4!
Words starting with FOL = 4!
Words starting with FOM = 4!
Words starting with FORA = 3!
Words starting with FORL = 3!
Words starting with FORMA = 2!
Words starting with FORML = 2!
Next word will be FORMUAL = 1
Next word will be FORMULA = 1
$\therefore$ Rank of the word 'FORMULA' in dictionary is
$= 6! + 3 \times 5! + 3 \times 4! + 2 \times 3! + 2 \times 2! + 2 = 1170$

Q6. If the letters of the word 'MANNER' are arranged in all possible ways as listed in dictionary, then, what is the rank of the word 'MANNER'?

Solution:
The letters of the word 'MANNER' in alphabetical order are 1 A, 1 E, 1 M, 2 N, 1 R.
Words starting with A $= \frac{5!}{2!} = 60$
Words starting with E $= \frac{5!}{2!} = 60$
Words starting with MAE $= \frac{3!}{2!} = 3$
Words starting with MANE $= 2! = 2$
Next word will be MANNER = 1
$\therefore$ Rank of the word 'MANNER' in the dictionary is
$= 60 + 60 + 3 + 2 + 1 = 126$

Q7. If the letters of the word 'GENERATE' are arranged in all possible ways as listed in dictionary, then, what is the rank of the word 'GENERATE'?

Solution:
The letters of the word 'GENERATE' in alphabetical order are 1 A, 3 E, 1 G, 1 N, 1 R, 1 T.
Words starting with A $= \frac{7!}{3!} = 840$
Words starting with E $= \frac{7!}{2!} = 1260$
Words starting with GA $= \frac{6!}{3!} = 120$
Words starting with GEE $= 5! = 120$
Words starting with GENA $= \frac{4!}{2!} = 12$
Words starting with GENEA $= 3! = 6$
Words starting with GENEE $= 3! = 6$

Next word will be GENERAET = 1
Next word will be GENERATE = 1
∴ Rank of the word 'GENERATE' in the dictionary is
$= 840 + 1260 + 120 + 120 + 12 + 6 + 6 + 1 + 1 = 2366$

Q8. If the letters of the word 'MINIMIZES' are arranged in all possible ways as listed in dictionary, then, what is the rank of the word 'MINIMIZES'?

Solution:
The letters of the word 'MINIMIZES' in alphabetical order are 1 E, 3 I, 2 M, 1 N, 1 S, 1 Z.

Words starting with E $= \frac{8!}{3!2!} = 3360$

Words starting with I $= \frac{8!}{2!2!} = 5040$

Words starting with ME $= \frac{7!}{3!} = 840$

Words starting with MIE $= \frac{6!}{2!} = 360$

Words starting with MII $= 6! = 720$

Words starting with MIM $= \frac{6!}{2!} = 360$

Words starting with MINE $= \frac{5!}{2!} = 60$

Words starting with MINIE $= 4! = 24$
Words starting with MINII $= 4! = 24$
Words starting with MINIME $= 3! = 6$
Words starting with MINIMIE $= 2! = 2$
Words starting with MINIMIS $= 2! = 2$
Next word will be MINIMIZES = 1
∴ Rank of the word 'MINIMIZES' in the dictionary is
$= 3360 + 5040 + 840 + 360 + 720 + 360$
$+ 60 + 24 + 24 + 6 + 2 + 2 + 1 = 10799$

Q9. Out of n points in a plane, m are in the same line and are collinear. Find the number of lines that can be formed by joining two points.

Solution:
The number of lines that can be formed by joining two points from given n points is $= {}^nC_2$

Since m points are collinear, and these m collinear points, when joined pair wise will give only one line, the total number of straight lines formed
$= {}^nC_2 - {}^mC_2 + 1$

Q10. If n lines are drawn in a plane such that no two of them are parallel and no three are concurrent, in how many points will they intersect each other?

Solution:
No two lines are parallel, therefore, all lines are intersecting pair-wise No three lines are concurrent means, no three lines will intersect at a point.

Thus, for one point of intersection, we require two lines, therefore the number of points of intersection are $= {}^nC_2$

Q11. Out of n points in a plane, m are in the same line and are collinear. Find the number of triangles that can be formed by joining the points.

Solution:
The number of triangles that can be formed by joining three points from given n points is $= {}^nC_3$

m points are collinear, and out of these m collinear points, no three points will give any triangle.

Therefore, total number of triangles formed
$= {}^nC_3 - {}^mC_3$

Q12. Find the number of parallelograms that can be formed from a set of n parallel lines intersecting another set of m parallel lines.

Solution:
To form a parallelogram, we require a pair of lines from a set of n parallel lines and another pair of lines from another set of m parallel lines.
Thus, the total number of parallelograms formed is $= {}^nC_2 \times {}^mC_2$

Q13. How many chords can be drawn through n points on a circle?

Solution:
For drawing one chord on a circle, only 2 points are required.

Therefore, number of chords that can be drawn through the given n points on a circle is $= {}^nC_2$

Q14. How many diagonals can be drawn through the vertices of a n sided polygon?

Solution:
For drawing one diagonal of a polygon, only 2 points are required.

Therefore, number of diagonals that can be drawn through n vertices of a polygon is $= {}^nC_2 - n$ (n is subtracted because two adjacent vertices do not give diagonal of a polygon while it gives side of a polygon).

Q15. How many triangles can be drawn through vertices of a n sided polygon?

Solution:
For drawing one triangle of a polygon, 3 points are required.

Therefore, number of diagonals that can be drawn through n vertices of a polygon is $= {}^nC_3$

CHAPTER

8 Binomial Theorem

Day 103

Pascal's Triangle

Let us consider some expansions

$(a+b)^0 = 1, \quad a+b \neq 0$

$(a+b)^1 = a+b$

$(a+b)^2 = a^2+2ab+b^2$

$(a+b)^3 = a^3+3a^2b+3ab^2+b^3$

$(a+b)^4 = a^4+4a^3b+6a^2b^2+4ab^3+b^4$

$(a+b)^5 = a^5+5a^4b+10a^3b^2+10a^2b^3+5ab^4+b^4$

In all these expansions, we may note some similarities:

(i) The total number of terms in the expansion is one more than the index. As an example, for index 2, in the expansion of $(a+b)^2$, the number of terms is 3.

(ii) Powers of the first quantity a keep on decreasing by 1 whereas the powers of the second quantity b increase by 1, in successive terms starting from the index value.

(iii) In each term of the expansion, the sum of the indices of a and b is the same and is equal to the index of $a+b$. As an example, for index 2, in the expansion of $(a+b)^2$, the sum of the power of a and b is 2 in each term.

Now, if we arrange the coefficient of each term in a triangular shape, we get a triangle with 1 at the top vertex and running down the two slanting sides. This array of numbers is known as Pascal's triangle.

Index	Coefficients
0	1
1	1 1
2	1 2 1
3	1 3 3 1
4	1 4 6 4 1
5	1 5 10 10 5 1

In this Pascal's triangle, it is seen that the addition of 1s in the row for index 1 gives rise to 2 in the row for index 2. The addition of 1, 2 and 2, 1 in the row for index 2, gives rise to 3 and 3 in the row for index 3 and so on. Also 1 is present at the beginning and at the end of each row. This can be continued till any index.

Similarly, the expansion for the higher powers of a binomial expansion is also possible by using Pascal's triangle.

For a higher index, expansion using Pascal's triangle can be done by writing all the rows of the Pascal's triangle, but it is a slightly lengthy process.

Binomial Theorem for Positive Integral Indices (*n*): Thus, to simplify it, we make use of the concept of combinations for writing all the numbers in the Pascal's triangle. This is known as the binomial theorem.

Now, if we observe the terms of Pascal's triangle, they can be written as

$${}^nC_r = \frac{n!}{r!(n-r)!}, \quad 0 \leq r \leq n$$

where n is a non-negative integer. Also, ${}^nC_0 = 1 = {}^nC_n$.

For example, $(a+b)^1 = a+b$ can also be written as $(a+b)^1 = {}^1C_0a + {}^1C_1b$

Similarly, $(a+b)^2 = a^2+2ab+b^2$ can also be written as $(a+b)^2 = {}^2C_0a^2 + {}^2C_1ab + {}^2C_2b^2$

$(a+b)^3 = a^3+3a^2b+3ab^2+b^3$ can also be written as $(a+b)^3 = {}^3C_0a^3 + {}^3C_1a^2b + {}^3C_2ab^2 + {}^3C_3b^3$

$(a+b)^4 = a^4+4a^3b+6a^2b^2+4ab^3+b^4$

$= {}^4C_0a^4 + {}^4C_1a^3b + {}^4C_2a^2b^2 + {}^4C_3ab^3 + {}^4C_4b^4$

In general, the binomial theorem is

$$(a+b)^n = {}^nC_0a^n + {}^nC_1a^{n-1}b + {}^nC_2a^{n-2}b^2 + \ldots + {}^nC_{n-2}a^2b^{n-2} + {}^nC_{n-1}ab^{n-1} + {}^nC_nb^n$$

Note:

(i) The coefficients ${}^nC_r$ occurring in the binomial theorem are known as binomial coefficients.

(ii) $(a+b)^n = \sum_{k=0}^{n} {}^nC_k a^{n-k} b^k$ represents

${}^nC_0a^nb^0 + {}^nC_1a^{n-1}b^1 + \ldots + {}^nC_ra^{n-r}b^r + \ldots + {}^nC_na^{n-n}b^n$

(iii) $\because {}^nC_r = {}^nC_{n-r}$

$\therefore (a+b)^n = \sum_{k=0}^{n} {}^nC_k a^{n-k} b^k = \sum_{k=0}^{n} {}^nC_k a^k b^{n-k}$

$= {}^nC_0a^0b^n + {}^nC_1a^1b^{n-1} + \ldots + {}^nC_ra^rb^{n-r} + \ldots + {}^nC_na^nb^{n-n}$

(iv) The number of terms in the expansion of $(a+b)^n$ is $(n+1)$, that is, one more than the index.

(v) In the successive terms of the expansion, the index of a keeps on decreasing by unity, starting from n to 0 in the last term. And, the index of b keeps on

Solution:

Using the binomial theorem, the expression $\left(\frac{2}{x}-\frac{x}{2}\right)^5$ can be expanded as,

$$= {}^5C_0\left(\frac{2}{x}\right)^5 - {}^5C_1\left(\frac{2}{x}\right)^4\left(\frac{x}{2}\right) + {}^5C_2\left(\frac{2}{x}\right)^3\left(\frac{x}{2}\right)^2$$

$$-{}^5C_3\left(\frac{2}{x}\right)^2\left(\frac{x}{2}\right)^3 + {}^5C_4\left(\frac{2}{x}\right)\left(\frac{x}{2}\right)^4 - {}^5C_5\left(\frac{x}{2}\right)^5$$

$$= \frac{32}{x^5} - 5\left(\frac{16}{x^4}\right)\left(\frac{x}{2}\right) + 10\left(\frac{8}{x^3}\right)\left(\frac{x^2}{4}\right)$$

$$-10\left(\frac{4}{x^2}\right)\left(\frac{x^3}{8}\right) + 5\left(\frac{2}{x}\right)\left(\frac{x^2}{16}\right) - \frac{x^5}{32}$$

$$= \frac{32}{x^5} - \frac{40}{x^3} + \frac{20}{x} - 5x + \frac{5}{8}x^3 - \frac{x^5}{32}$$

Q3. Expand the expression $(2x-3)^6$.

Solution:

Using the binomial theorem, the expression $(2x-3)^6$ can be expanded as,

$$= {}^6C_0(2x)^6 - {}^5C_1(2x)^5(3) + {}^6C_2(2x)^4(3)^2$$

$$-{}^6C_3(2x)^3(3)^3 + {}^6C_4(2x)^2(3)^4 - {}^6C_5(2x)(3)^5 + {}^6C_6(3)^6$$

$$= 64x^6 - 6(32x^5)(3) + 15(16x^4)(9) - 20(8x^3)(27)$$

$$+15(4x^2)(81) - 6(2x)(243) + 729$$

$$= 64x^6 - 576x^5 + 2160x^4 - 4320x^3$$

$$+4860x^2 - 2916x + 729$$

Q4. Expand the expression $\left(\frac{x}{3}+\frac{1}{x}\right)^5$.

Solution:

Using the binomial theorem, the expression $\left(\frac{x}{3}+\frac{1}{x}\right)^5$ can be expanded as,

$$= {}^5C_0\left(\frac{x}{3}\right)^5 + {}^5C_1\left(\frac{x}{3}\right)^4\left(\frac{1}{x}\right) + {}^5C_2\left(\frac{x}{3}\right)^3\left(\frac{1}{x}\right)^2$$

$$+{}^5C_3\left(\frac{x}{3}\right)^2\left(\frac{1}{x}\right)^3 + {}^5C_4\left(\frac{x}{3}\right)\left(\frac{1}{x}\right)^4 + {}^5C_5\left(\frac{1}{x}\right)^3$$

$$= \frac{x^5}{243} + 5\left(\frac{x^4}{81}\right)\left(\frac{1}{x}\right) + 10\left(\frac{x^3}{27}\right)\left(\frac{1}{x^2}\right)$$

$$+10\left(\frac{x^2}{9}\right)\left(\frac{1}{x^3}\right) + 5\left(\frac{x}{3}\right)\left(\frac{1}{x^4}\right) + \frac{1}{x^5}$$

$$= \frac{x^5}{243} + \frac{5x^3}{81} + \frac{10x}{27} + \frac{10}{9x} + \frac{5}{3x^3} + \frac{1}{x^5}$$

Q5. Expand $\left(x+\frac{1}{x}\right)^6$.

Solution:

Using the binomial theorem, the expression $\left(x+\frac{1}{x}\right)^6$ can be expanded as,

$$= {}^6C_0(x)^6 + {}^6C_1(x)^5\left(\frac{1}{x}\right) + {}^6C_2(x)^4\left(\frac{1}{x}\right)^2 + {}^6C_3(x)^3\left(\frac{1}{x}\right)^3$$

$$+{}^6C_4(x)^2\left(\frac{1}{x}\right)^4 + {}^6C_5(x)\left(\frac{1}{x}\right)^5 + {}^6C_6\left(\frac{1}{x}\right)^6$$

$$= x^6 + 6(x)^5\left(\frac{1}{x}\right) + 15(x)^4\left(\frac{1}{x^2}\right) + 20(x)^3\left(\frac{1}{x^3}\right)$$

$$+15(x)^2\left(\frac{1}{x^4}\right) + 6(x)\left(\frac{1}{x^5}\right) + \frac{1}{x^6}$$

$$= x^6 + 6x^4 + 15x^2 + 20 + \frac{15}{x^2} + \frac{6}{x^4} + \frac{1}{x^6}$$

Q6. Using the binomial theorem, evaluate $(96)^3$.

Solution:

Since 96 can be written as $96 = 100 - 4$

$\therefore (96)^3 = (100-4)^3$

$$= {}^3C_0(100)^3 - {}^3C_1(100)^2(4) + {}^3C_2(100)(4)^2 - {}^3C_3(4)^3$$

$$= (100)^3 - 3(100)^2(4) + 3(100)(4)^2 - (4)^3$$

$$= 1000000 - 120000 + 4800 - 64$$

$$= 884736$$

Q7. Using the binomial theorem, evaluate $(102)^5$.

Solution:

Since 102 can be written as $102 = 100 + 2$

$\therefore (102)^5 = (100+2)^5$

$$= {}^5C_0(100)^5 + {}^5C_1(100)^4(2) + {}^5C_2(100)^3(2)^2$$

$$+{}^5C_3(100)^2(2)^3 + {}^5C_4(100)(2)^4 + {}^5C_5(2)^5$$

$$= (100)^5 + 5(100)^4(2) + 10(100)^3(2)^2$$

$$+10(100)^2(2)^3 + 5(100)(2)^4 + (2)^5$$

$$= 10000000000 + 1000000000 + 40000000$$

$$+800000 + 8000 + 32 = 11040808032$$

Q8. Using the binomial theorem, evaluate $(101)^4$.

Solution:

Since 101 can be written as $101 = 100 + 1$

$\therefore (101)^4 = (100+1)^4$

$$= {}^4C_0(100)^4 + {}^4C_1(100)^3(1) + {}^4C_2(100)^2(1)^2$$

$$+{}^4C_3(100)(1)^3 + {}^4C_4(1)^4$$

$$= (100)^4 + 4(100)^3 + 6(100)^2 + 4(100) + (1)^4$$

$$= 100000000 + 4000000 + 60000 + 400 + 1$$

$$= 104060401$$

Q9. Using the binomial theorem, evaluate $(99)^5$.

Solution:

Since 99 can be written as $99 = 100 - 1$

$\therefore (99)^5 = (100-1)^5$

$= {}^5C_0(100)^5 - {}^5C_1(100)^4(1) + {}^5C_2(100)^3(1)^2$

$- {}^5C_3(100)^2(1)^3 + {}^5C_4(100)(1)^4 - {}^5C_5(1)^5$

$= (100)^5 - 5(100)^4 + 10(100)^3 - 10(100)^2 + 5(100) - 1$

$= 10000000000 - 500000000 + 10000000$

$-100000 + 500 - 1 = 9509900499$

Q10. Using the binomial theorem, indicate which number is larger, $(1.1)^{10000}$ or 10000.

Solution:

1.1 can be written as $1.1 = 1 + 0.1$

Therefore, by the binomial theorem,

$(1.1)^{10000} = (1+0.1)^{10000}$

$= {}^{10000}C_0 + {}^{10000}C_1(1.1) +$ other positive terms

$= 1 + 10000 \times 1.1 +$ other positive terms

$= 1 + 11000 +$ other positive terms

$= 10000 + 1001$ other positive terms

$= 10000 +$ some positive value > 10000

Thus, $(1.1)^{10000} > 10000$

Q11. Find $(a+b)^4 - (a-b)^4$. Hence, evaluate $(\sqrt{3}+\sqrt{2})^4 - (\sqrt{3}-\sqrt{2})^4$.

Solution:

Using the binomial theorem, the expressions of $(a+b)^4$ and $(a-b)^4$ can be expanded as,

$(a+b)^4 = {}^4C_0a^4 + {}^4C_1a^3b + {}^4C_2a^2b^2 + {}^4C_3ab^3 + {}^4C_4b^4$

$(a-b)^4 = {}^4C_0a^4 - {}^4C_1a^3b + {}^4C_2a^2b^2 - {}^4C_3ab^3 + {}^4C_4b^4$

$\therefore (a+b)^4 - (a-b)^4 = 2\left({}^4C_1a^3b + {}^4C_3ab^3\right)$

$= 2(4a^3b + 4ab^3) = 8ab(a^2+b^2)$

Thus, substituting $a = \sqrt{3}, b = \sqrt{2}$, we get

$(\sqrt{3}+\sqrt{2})^4 - (\sqrt{3}-\sqrt{2})^4 = 8(\sqrt{3})(\sqrt{2})\{(\sqrt{3})^2 + (\sqrt{2})^2\}$

$= 8(\sqrt{6})(3+2) = 40\sqrt{6}$

Q12. Find $(x+1)^6 + (x-1)^6$. Hence, or otherwise evaluate $(\sqrt{2}+1)^6 + (\sqrt{2}-1)^6$.

Solution:

Using the binomial theorem, the expressions of $(x+1)^6$ and $(x-1)^6$ can be expanded as,

$(x+1)^6 = {}^6C_0x^6 + {}^6C_1x^5 + {}^6C_2x^4 + {}^6C_3x^3$

$+ {}^6C_4x^2 + {}^6C_5x + {}^6C_6$

$(x-1)^6 = {}^6C_0x^6 - {}^6C_1x^5 + {}^6C_2x^4 - {}^6C_3x^3$

$+ {}^6C_4x^2 - {}^6C_5x + {}^6C_6$

$\therefore (x+1)^6 + (x-1)^6$

$= 2[{}^6C_0x^6 + {}^6C_2x^4 + {}^6C_4x^2 + {}^6C_6]$

$= 2[x^6 + 15x^4 + 15x^2 + 1]$

Substituting $x = \sqrt{2}$, we get

$(\sqrt{2}+1)^6 + (\sqrt{2}-1)^6 = 2[(\sqrt{2})^6 + 15(\sqrt{2})^4 + 15(\sqrt{2})^2 + 1]$

$= 2(8 + 15\times 4 + 15\times 2 + 1) = 2(99) = 198$

Q13. Show that $9^{n+1} - 8n - 9$ is divisible by 64, whenever n is a positive integer.

Solution:

Using the binomial theorem,

$(1+x)^n = {}^nC_0 + {}^nC_1x + {}^nC_2x^2 + \ldots + {}^nC_nx^n$

Substituting $a = 8$, we get

$(9)^n = (1+8)^n$

$= {}^nC_0 + {}^nC_1(8) + {}^nC_2(8)^2 + {}^nC_3(8)^3 + \ldots + {}^nC_n(8)^n$

$= 1 + 8n + 8^2[{}^nC_2 + {}^nC_2 \cdot 8 + \ldots + {}^nC_n(8)^{n-2}]$

$= 1 + 8n + 64k$

where $k = {}^nC_2 + {}^nC_2 \cdot 8 + \ldots + {}^nC_n(8)^{n-2}$

Thus, $9^{n+1} - 8n - 9$

$= 9 \cdot 9^n - 8n - 9$

$= 9 \cdot (1 + 8n + 64k) - 8n - 9$

$= 9 + 72n + 9 \cdot 64k - 8n - 9$

$= 64n + 9 \cdot 64k = 64(n + 9k) = 64k'$

Thus, $9^{n+1} - 8n - 9$ is divisible by 64 for all positive integers n.

Q14. Prove that $\sum_{r=0}^{n} 3^r\,{}^nC_r = 4^n$.

Solution:

Using the binomial theorem,

$(a+b)^n = \sum_{r=0}^{n} {}^nC_r\,a^r\,b^{n-r}$

Substituting, $a = 3$, $b = 1$, we get

$(1+3)^n = \sum_{r=0}^{n} {}^nC_r\,3^r\,1^{n-r} \Rightarrow \sum_{r=0}^{n} 3^r\,{}^nC_r = 4^n$

Day 105

General and Middle Terms in a Binomial Expansion

General Term in a Binomial Expansion

$(a+b)^n = {}^nC_0a^n + {}^nC_1a^{n-1}b + {}^nC_2a^{n-2}b^2 + ...$

$+ {}^nC_{n-2}a^2b^{n-2} + {}^nC_{n-1}ab^{n-1} + {}^nC_nb^n$

1st term is ${}^nC_0a^n$

2nd term is ${}^nC_1a^{n-1}b$

3rd term is ${}^nC_2a^{n-2}b^2$, and so on

$(r+1)^{\text{th}}$ term (or general term) is $T_{r+1} = {}^nC_ra^{n-r}b^r$

Or we may also write as $T_{r+1} = {}^nC_ra^rb^{n-r}$

Similarly, General Term in a Binomial Expansion

$(a-b)^n = {}^nC_0a^n + {}^nC_1a^{n-1}b + {}^nC_2a^{n-2}b^2 + ...$

$+ {}^nC_{n-2}a^2b^{n-2} + {}^nC_{n-1}ab^{n-1} + {}^nC_nb^n$

is the $(r+1)^{\text{th}}$ term (or general term) given by

$T_{r+1} = (-1)^r\, {}^nC_ra^{n-r}b^r$

Middle Term/s in a Binomial Expansion

(i) When *n* is even: The number of terms in the expansion will be $n+1$, that is, odd.
Therefore, the middle term is

$$\left(\frac{n+1+1}{2}\right)^{\text{th}} \Rightarrow \left(\frac{n}{2}+1\right)^{\text{th}} \text{ term}$$

$$T_{\text{middle/even}} = {}^nC_{(n/2+1)}a^{n-(n/2+1)}b^{(n/2+1)}$$

$$= {}^nC_{(n/2+1)}a^{n/2-1}b^{n/2+1}$$

(ii) When *n* is odd: The number of terms in the expansion will be $n+1$, that is, even.
Therefore, there will be two middle terms in the expansion given by $\left(\frac{n+1}{2}\right)^{\text{th}}$ term and $\left(\frac{n+1}{2}+1\right)^{\text{th}}$ term.

Note: In the expansion of $\left(x+\frac{1}{x}\right)^{2n}$, where $x \neq 0$, the middle term is $\left(\frac{2n+1+1}{2}\right)^{\text{th}}$, that is, $(n+1)^{\text{th}}$ term, (since $2n$ is even).

Thus, the middle term is given by ${}^{2n}C_nx^n\left(\frac{1}{x}\right)^n = {}^{2n}C_n$ (constant); it is also termed as independent of x or the constant term.

Example 1: Write the general term in the expansion of $\left(\frac{x^2}{y}-\frac{y^2}{x}\right)^6$.

Solution:

The $(r+1)^{\text{th}}$ term in the binomial expansion of $(a+b)^n$ is $T_{r+1} = {}^nC_ra^{n-r}b^r$

Substituting $n = 6$, $a = \left(\frac{x^2}{y}\right)$ and $b = \left(-\frac{y^2}{x}\right)^6$, we get the general term $T_{r+1} = {}^6C_r\left(\frac{x^2}{y}\right)^{6-r}\left(-\frac{y^2}{x}\right)^r$

$= (-1)^r\, {}^6C_r\left(x^{12-2r}\cdot y^{r-6}\right)\left(y^{2r}\cdot x^{-r}\right)$

$= (-1)^r\, {}^6C_rx^{12-3r}y^{3r-6}$

Example 2: Find the 5th term in the expansion of $\left(3x^2-\frac{1}{3\sqrt{x}}\right)^{10}, x \neq 0$.

Solution:

The $(r+1)^{\text{th}}$ term in the binomial expansion of $(a+b)^n$ is $T_{r+1} = {}^nC_ra^{n-r}b^r$.

Substituting $n = 10$, $a = 3x^2$ and $b = -\frac{1}{3\sqrt{x}}$, we get the general term $T_{r+1} = {}^{10}C_r(3x^2)^{10-r}\left(-\frac{1}{3\sqrt{x}}\right)^r$

Thus, the 5th term in the expansion of $\left(3x^2-\frac{1}{3\sqrt{x}}\right)^{10}$ is obtained by putting $r = 4$.

$\therefore T_5 = T_{4+1} = {}^{10}C_4(3x^2)^{10-4}\left(-\frac{1}{3\sqrt{x}}\right)^4$

$= (-1)^1\frac{10!}{6!4!}(3)^6(x)^{12}\left(\frac{1}{3}\right)^4\left(\frac{1}{\sqrt{x}}\right)^4$

$= \frac{10\cdot 9\cdot 8\cdot 7}{4\cdot 3\cdot 2}\cdot 3^2x^{12-2} = 1890x^{10}$

Example 3: Find the coefficient of x^{10} in $\left(2x^2+\frac{3}{x}\right)^8$.

Solution:

The $(r+1)^{\text{th}}$ term in the binomial expansion of $(a+b)^n$ is $T_{r+1} = {}^nC_ra^{n-r}b^r$.

Substituting $n = 8$, $a = 2x^2$ and $b = 3/x$, we get

$T_{r+1} = {}^8C_r\left(2x^2\right)^{8-r}\left(\frac{3}{x}\right)^r = {}^8C_r(2)^{8-r}(3)^r(x)^{16-3r}$

Comparing the indices of x in x^{10} and in T_{r+1}, we get $10 = 16-3r \Rightarrow r = 2$

Therefore, $T_{2+1} = {}^8C_2(2)^{8-2}(3)^2(x)^{16-6} \Rightarrow T_3 = 16128x^{10}$

Thus, the coefficient of x^{10} is 16128.

Example 4: Find the coefficient of x^8y^3 in $\left(2x^2 - \frac{\sqrt{y}}{2}\right)^{10}$.

Solution:

The $(r+1)^{th}$ term in the binomial expansion of $(a+b)^n$ is $T_{r+1} = {}^nC_r a^{n-r} b^r$.

Substituting $n = 10$, $a = 2x^2$ and $b = \frac{\sqrt{y}}{2}$, we get

$$T_{r+1} = {}^{10}C_r\left(2x^2\right)^{10-r}\left(-\frac{\sqrt{y}}{2}\right)^r$$

$$= (-1)^r\, {}^{10}C_r\,(2)^{10-2r}\,(x)^{20-2r}\,(y)^{r/2}$$

Comparing the indices of x and y in x^8y^3 and in T_{r+1}, we get $r = 6$

Therefore,

$$T_{6+1} = (-1)^6\, {}^{10}C_6\,(2)^{10-12}\,(x)^{20-12}\,(y)^{6/2} = \frac{{}^{10}C_6}{4}x^8y^3$$

Thus, the coefficient of x^8y^3 is

$$\frac{{}^{10}C_6}{4} = \frac{10!}{6!4!\cdot 4} = \frac{10\cdot 9\cdot 8\cdot 7}{4\cdot 3\cdot 2\cdot 4} = \frac{105}{2}$$

Example 5: Find the term independent of x in $\left(2x^2 + \frac{3}{x^3}\right)^{10}$ and also find its coefficient.

Solution:

The $(r+1)^{th}$ term in the binomial expansion of $(a+b)^n$ is $T_{r+1} = {}^nC_r a^{n-r} b^r$.

Substituting $n = 10$, $a = 2x^2$ and $b = \frac{3}{x^3}$, we get

$$T_{r+1} = {}^{10}C_r\left(2x^2\right)^{10-r}\left(\frac{3}{x^3}\right)^r = {}^{10}C_r\,(2)^{10-r}\,(3)^r\,(x)^{20-5r}$$

For terms independent of x, the indices must be 0, that is, x^0. Thus, comparing with T_{r+1}, we get $0 = 20 - 5r \Rightarrow r = 4$, Thus, the 5th term will be independent of x.

Also,

$$T_{4+1} = {}^{10}C_4\,(2)^{10-4}\,(3)^4\,(x)^{20-20} \Rightarrow T_5 = {}^{10}C_4\,(2)^6\,(3)^4\,(x)^0$$

The coefficient of the term independent of x is ${}^{10}C_4\,(2)^6\,(3)^4$.

Example 6: Find the middle term/s of the expression $\left(x^2 + \frac{1}{x}\right)^6$.

Solution:

Since $n = 6$ is even, the number of terms in the expansion will be $n + 1 = 7$, that is, odd.

Therefore, the middle term is $\left(\frac{n}{2}+1\right)^{th} \Rightarrow \left(\frac{6}{2}+1\right)^{th} = 4^{th}$ term given by

$$T_4 = T_{3+1} = {}^6C_3(x^2)^{6-3}\left(\frac{1}{x}\right)^3 = {}^6C_3x^3$$

Example 7: Find the middle term/s of the expression $\left(x^2 - \frac{1}{x^2}\right)^5$.

Solution:

Since $n = 5$ is odd, the number of terms in the expansion will be $n + 1 = 6$, that is, even.

Therefore, the middle terms are

$$\left(\frac{n+1}{2}\right)^{th} \Rightarrow \left(\frac{5+1}{2}\right)^{th} = 3^{rd} \text{ term and}$$

$$\left(\frac{n+1}{2}+1\right)^{th} \Rightarrow \left(\frac{5+1}{2}+1\right)^{th} \Rightarrow 4^{th} \text{ term}$$

Thus, the middle terms are

$$T_3 = T_{2+1} = (-1)^2\, {}^4C_2(x^2)^{5-2}\left(\frac{1}{x^2}\right)^2 \Rightarrow {}^4C_2x^2$$

$$T_4 = T_{3+1} = (-1)^3\, {}^5C_3(x^2)^{5-3}\left(\frac{1}{x^2}\right)^3 \Rightarrow -{}^5C_3\frac{1}{x^2}$$

Example 8: Find a positive value of m for which the coefficient of x^3 in the expansion $(1+x)^m$ is 10.

Solution:

The $(r+1)^{th}$ term in the binomial expansion of $(a+b)^n$ is $T_{r+1} = {}^nC_r a^{n-r} b^r$.

Let us assume that x^3 occurs in the $(r+1)^{th}$ term of the expansion $(1+x)^m$. Then, we get

$$T_{r+1} = {}^mC_r(1)^{m-r}(x)^r = {}^mC_3(x)^3$$

Comparing the indices of x in x^3 and in (T_{r+1}), we get $r = 3$.

Therefore, the coefficient of x^3 is ${}^mC_3$

The coefficient of x^3 in the expansion $(1+x)^m$ is 10, therefore ${}^mC_3 = 10$.

$$\Rightarrow \frac{m!}{3!(m-3)!} = 10 \Rightarrow \frac{m(m-1)(m-2)}{6} = 10$$

$$\Rightarrow m(m-1)(m-2) = 60 = 5\cdot 4\cdot 3 \Rightarrow m = 5$$

Thus, the positive value of $m = 5$, the coefficient of x^3 in the expansion $(1+x)^m$ is 10.

Example 9: Find a, b and n in the expansion of $(a+b)^n$ if the first three terms of the expansion are 32, 240 and 720, respectively.

Solution:

The $(r+1)^{\text{th}}$ term in the binomial expansion of $(a+b)^n$ is $T_{r+1} = {}^nC_r a^{n-r} b^r$.

The first three terms of the expansion are 32, 240 and 720, respectively. Therefore

$$T_1 = {}^nC_0 a^{n-0} b^0 = a^n = 32 \quad \text{(i)}$$

$$T_2 = {}^nC_1 a^{n-1} b^1 = na^{n-1}b = 240 \quad \text{(ii)}$$

$$T_3 = {}^nC_2 a^{n-2} b^2 = \frac{n(n-1)}{2} a^{n-2} b^2 = 720 \quad \text{(iii)}$$

Dividing (ii) by (i), we get

$$\frac{na^{n-1}b}{a^n} = \frac{240}{32} \Rightarrow \frac{nb}{a} = \frac{15}{2} \quad \text{(iv)}$$

Dividing (iii) by (ii), we get

$$\frac{n(n-1)a^{n-2}b^2}{2na^{n-1}b} = \frac{720}{240}$$

$$\Rightarrow \frac{(n-1)b}{a} = 6 \Rightarrow \frac{nb}{a} - \frac{b}{a} = 6 \quad \text{(v)}$$

From equations (iv) and (v), we get

$$\Rightarrow \frac{15}{2} - \frac{b}{a} = 6 \Rightarrow \frac{b}{a} = \frac{15}{2} - 6 = \frac{3}{2}$$

Thus, $n \cdot \frac{b}{a} = \frac{15}{2} \Rightarrow n \cdot \frac{3}{2} = \frac{15}{2} \Rightarrow n = 5$

Now from equation (i), we get

$a^n = 32 \Rightarrow a^5 = 2^5 \Rightarrow a = 2$

And from equation (iv), we get

$$\frac{nb}{a} = \frac{15}{2} \Rightarrow \frac{5 \times b}{2} = \frac{15}{2} \Rightarrow b = 3$$

Example 10: The coefficient of the $(r-1)^{\text{th}}, r^{\text{th}}, (r+1)^{\text{th}}$ terms in the expansion of $(x+1)^n$ are in the ratio 6:15:20. Find n and r.

Solution:

The $(r+1)^{\text{th}}$ term in the binomial expansion of $(a+b)^n$ is $T_{r+1} = {}^nC_r a^{n-r} b^r$.

Therefore, the $(r-1)^{\text{th}}$ term in the expansion of $(x+1)^n$ is $T_{r-1} = {}^nC_{r-2}(x)^{n-(r-2)}(1)^{(r-2)} = {}^nC_{r-2}x^{n-r+2}$

And, the rth term in the expansion of $(x+1)^n$ is $T_r = {}^nC_{r-1}(x)^{n-(r-1)}(1)^{(r-1)} = {}^nC_{r-1}x^{n-r+1}$

And, $(r+1)^{\text{th}}$ term in the expansion of $(x+1)^n$ is $T_{r+1} = {}^nC_r(x)^{n-r}(1)^r = {}^nC_r x^{n-r}$

Therefore, the coefficient of the $(r-1)^{\text{th}}, r^{\text{th}}, (r+1)^{\text{th}}$ terms in the expansion of $(x+1)^n$ are ${}^nC_{r-2}, {}^nC_{r-1}, {}^nC_r$, respectively.

The coefficients of the $(r-1)^{\text{th}}, r^{\text{th}}, (r+1)^{\text{th}}$ terms are in the ratio 1:3:5; therefore

$$\frac{{}^nC_{r-2}}{{}^nC_{r-1}} = \frac{2}{5} \quad \text{and} \quad \frac{{}^nC_{r-1}}{{}^nC_r} = \frac{3}{4}$$

Now, solving equation $\frac{{}^nC_{r-2}}{{}^nC_{r-1}} = \frac{2}{5}$

$$\Rightarrow \frac{n!}{(r-2)!(n-r+2)!} \times \frac{(r-1)!(n-r+1)!}{n!} = \frac{2}{5}$$

$$\Rightarrow = \frac{(r-1)(r-2)!(n-r+1)!}{(r-2)!(n-r+2)(n-r+1)!} = \frac{2}{5}$$

$$\Rightarrow \frac{r-1}{n-r+2} = \frac{2}{5} \Rightarrow 5r-5 = 2n-2r+4 \Rightarrow 2n-7r+9=0$$

Similarly, solving equation $\frac{{}^nC_{r-1}}{{}^nC_r} = \frac{3}{4}$

$$\Rightarrow \frac{n!}{(r-1)!(n-r+1)!} \times \frac{r!(n-r)!}{n!} = \frac{3}{4}$$

$$\Rightarrow \frac{r(r-1)!(n-r)!}{(r-1)!(n-r+1)(n-r)!} = \frac{3}{4}$$

$$\Rightarrow \frac{r}{n-r+1} = \frac{3}{4} \Rightarrow 4r = 3n-3r+3 \Rightarrow 3n-7r+3=0$$

Solving the above two equations, we get $n = 6$, $r = 3$.

Day 106

NCERT Exercise 8.2 (With Solutions)

Q1. Find the coefficient of x^5 in $(x+3)^8$.

Solution:

The $(r+1)^{\text{th}}$ term in the binomial expansion of $(a+b)^n$ is $T_{r+1} = {}^nC_r a^{n-r} b^r$.

Substituting $n = 8$, $a = x$ and $b = 3$, we get $T_{r+1} = {}^8C_r x^{8-r} 3^r$.

Comparing the indices of x in x^5 and in T_{r+1}, we get $r = 3$,

Therefore, $T_{3+1} = {}^8C_3 x^{8-3} 3^3 \Rightarrow T_4 = 27\,{}^8C_3 x^5$

Thus, the coefficient of x^5 is

$$27\,{}^8C_3 = 27 \cdot \frac{8!}{3!5!} = 1512$$

Q2. Find the coefficient of a^5b^7 in $(a-2b)^{12}$.

Solution:

The $(r+1)^{\text{th}}$ term in the binomial expansion of $(a+b)^n$ is $T_{r+1} = {}^nC_r a^{n-r} b^r$.

Substituting $n = 12$, $a = a$ and $b = -2b$, we get

$$T_{r+1} = {}^{12}C_r(a)^{12-r}(-2b)^r = {}^{12}C_r(-2)^r(a)^{12-r}(b)^r$$

Comparing the indices of a and b in a^5b^7 and in T_{r+1}, we get $r = 7$.
Therefore, $T_{7+1} = {}^{12}C_7(a)^{12-7}(-2b)^7 = (-2)^7\,{}^{12}C_7(a)^5(b)^7$
Thus, the coefficient of a^5b^7 is

$$(-2)^7\,{}^{12}C_7 = -2^7\frac{12!}{7!5!} = -2^7\frac{12\cdot11\cdot10\cdot9\cdot8\cdot7!}{5\cdot4\cdot3\cdot2\cdot7!} = -101376$$

Q3. Write the general term in the expansion of $(x^2 - y)^6$.

Solution:
The $(r+1)^{\text{th}}$ term in the binomial expansion of $(a+b)^n$ is $T_{r+1} = {}^nC_r a^{n-r}b^r$.
Substituting $n = 6$, $a = x^2$ and $b = -y$, we get the general term

$$T_{r+1} = {}^6C_r(x^2)^{6-r}(-y)^r = (-1)^r\,{}^6C_r\cdot x^{12-2r}\cdot y^r$$

Q4. Write the general term in the expansion of $(x^2 - xy)^{12}, x \neq 0$.

Solution:
The $(r+1)^{\text{th}}$ term in the binomial expansion of $(a+b)^n$ is $T_{r+1} = {}^nC_r a^{n-r}b^r$.
Substituting $n = 12$, $a = x^2$ and $b = -xy$, we get the general term $T_{r+1} = {}^{12}C_r(x^2)^{12-r}(-xy)^r$

$$= (-1)^r\,{}^{12}C_r x^{24-2r}\cdot y^r\cdot x^r = (-1)^r\,{}^{12}C_r\cdot x^{24-r}\cdot y^r$$

Q5. Find the 4th term in the expansion of $(x-2y)^{12}$.

Solution:
The $(r+1)^{\text{th}}$ term in the binomial expansion of $(a+b)^n$ is $T_{r+1} = {}^nC_r a^{n-r}b^r$.
Substituting $n = 12$, $a = x$ and $b = -2y$, we get the general term $T_{r+1} = {}^{12}C_r(x)^{12-r}(-2y)^r$.
Thus, the 4th term in the expansion of $(x-2y)^{12}$ is obtained by putting $r = 3$.

$$T_4 = T_{3+1} = {}^{12}C_3(x)^{12-3}(-2y)^3 = (-1)^3\cdot\frac{12!}{3!9!}\cdot x^9\cdot(2)^3$$

$$= -\frac{12\cdot11\cdot10}{3\cdot2}\cdot(2)^3x^9y^3 = -1760x^9y^3$$

Q6. Find the 13th term in the expansion of $\left(9x - \frac{1}{3\sqrt{x}}\right)^{18}, x \neq 0$.

Solution:
The $(r+1)^{\text{th}}$ term in the binomial expansion of $(a+b)^n$ is $T_{r+1} = {}^nC_r a^{n-r}b^r$.

Substituting $n = 18$, $a = 9x$ and $b = -\frac{1}{3\sqrt{x}}$, we get the general term $T_{r+1} = {}^{18}C_4(9x)^{18-4}\left(-\frac{1}{3\sqrt{x}}\right)^r$

Thus, the 13th term in the expansion of $\left(9x - \frac{1}{3\sqrt{x}}\right)^{18}$ is obtained by putting $r = 12$.

$$T_{13} = T_{12+1} = {}^{18}C_{12}(9x)^{18-12}\left(-\frac{1}{3\sqrt{x}}\right)^{12}$$

$$= (-1)^{12}\frac{18!}{12!6!}(9)^6(x)^6\left(\frac{1}{3}\right)^{12}\left(\frac{1}{\sqrt{x}}\right)^{12}$$

$$= \frac{18\cdot17\cdot16\cdot15\cdot14\cdot13\cdot12!}{12!\cdot6\cdot5\cdot4\cdot3\cdot2}\cdot x^6\cdot\left(\frac{1}{x^6}\right)\cdot3^{12}\left(\frac{1}{3^{12}}\right) = 18564$$

Q7. Find the middle term in the expansion of $\left(3 - \frac{x^3}{6}\right)^7$.

Solution:
In the expansion of $(a+b)^n$ if n is odd, then there are two middle terms, namely, the

$\left(\frac{n+1}{2}\right)^{\text{th}}$ term and $\left(\frac{n+1}{2}+1\right)^{\text{th}}$ term.

Therefore, the middle terms in the expansion $\left(3 - \frac{x^3}{6}\right)^7$ are $\left(\frac{7+1}{2}\right)^{\text{th}} = 4^{\text{th}}$ term and $\left(\frac{7+1}{2}+1\right)^{\text{th}} = 5^{\text{th}}$ term.

Thus, the middle terms in the expansion of $\left(3 - \frac{x^3}{6}\right)^7$ are

$$T_4 = T_{3+4} = {}^7C_3(3)^{7-3}\left(-\frac{x^3}{6}\right)^3 = (-1)^3\frac{7!}{3!4!}\cdot3^4\cdot\frac{x^9}{6^3}$$

$$= -\frac{7\cdot6\cdot5\cdot4!}{3\cdot2\cdot4!}\cdot3^4\cdot\frac{1}{2^3\cdot3^3}\cdot x^9 = -\frac{105}{8}x^9$$

$$T_5 = T_{4+1} = {}^7C_4(3)^{7-4}\left(-\frac{x^3}{6}\right)^4 = (-1)^4\frac{7!}{3!4!}\cdot3^3\cdot\frac{x^{12}}{6^4}$$

$$= \frac{7\cdot6\cdot5\cdot4!}{4!\cdot3\cdot2}\cdot\frac{3^3}{2^4\cdot3^4}\cdot x^{12} = \frac{35}{48}x^{12}$$

Q8. Find the middle terms in the expansion of $\left(\frac{x}{3}+9y\right)^{10}$.

Solution:
In the expansion $(a+b)^n$ if n even, then the middle term is $\left(\frac{n}{2}+1\right)^{\text{th}}$ term.

Therefore, the middle term in the expansion of $\left(\frac{x}{3}+9y\right)^{10}$ is $\left(\frac{10}{2}+1\right)^{\text{th}} = 6^{\text{th}}$ term

Thus, the middle term in the expansion of $\left(\frac{x}{3}+9y\right)^{10}$ is

$T_6 = T_{5+1} = {}^{10}C_5\left(\frac{x}{3}\right)^{10-5}(9y)^5$

$= \frac{10!}{5!5!}\cdot\frac{x^5}{3^5}\cdot 9^5\cdot y^5 = \frac{10\cdot 9\cdot 8\cdot 7\cdot 6\cdot 5!}{5\cdot 4\cdot 3\cdot 2\cdot 5!}\cdot\frac{1}{3^5}\cdot 3^{10}\cdot x^5y^5$

$= 61236x^5y^5$

Q9. In the expansion of $1+a)^{m+n}$ prove that the coefficients of a^m, a^n are equal.

Solution:

The $(r+1)^{\text{th}}$ term in the binomial expansion of $(a+b)^n$ is $T_{r+1} = {}^nC_r a^{n-r}b^r$.

Let us assume that a^m occurs in the $(r+1)^{\text{th}}$ term of the expansion $(1+a)^{m+n}$. Then we get

$T_{r+1} = {}^{m+n}C_r(1)^{m+n-r}(a)^r = {}^{m+n}C_r a^r$

Comparing the indices of a in a^m and in (T_{r+1}), we get $r = m$.

Therefore, the coefficient of a^m is

$${}^{m+n}C_m = \frac{(m+n)!}{m!(m+n-m)!} = \frac{(m+n)!}{m!n!} \quad \text{(i)}$$

Similarly, assuming that a^n occurs in the $(k+1)^{\text{th}}$ term of the expansion $(1+a)^{m+n}$, we get

$T_{k+1} = {}^{m+n}C_k(1)^{m+n-k}(a)^k = {}^{m+n}C_k(a)^k$

Comparing the indices of a in a^n and in T_{k+1}, we get $k = n$

Therefore, the coefficient of a^n is

$${}^{m+n}C_n = \frac{(m+n)!}{n!(m+n-n)!} = \frac{(m+n)!}{n!m!} \quad \text{(ii)}$$

Comparing equations (i) and (ii), the coefficients of a^m and a^n in the expansion of $(1+a)^{m+n}$ are equal to $\frac{(m+n)!}{m!n!}$.

Q10. The coefficients of the $(r-1)^{\text{th}}, r^{\text{th}}, (r+1)^{\text{th}}$ terms in the expansion of $(x+1)^n$ are in the ratio 1:3:5. Find n and r.

Solution:

The $(r+1)^{\text{th}}$ term in the binomial expansion of $(a+b)^n$ is $T_{r+1} = {}^nC_r a^{n-r}b^r$.

Therefore, the $(r-1)^{\text{th}}$ term in the expansion of $(x+1)^n$ is

$T_{r-1} = {}^nC_{r-2}(x)^{n-(r-2)}(1)^{(r-2)} = {}^nC_{r-2}x^{n-r+2}$

And the rth term in the expansion of $(x+1)^n$ is

$T_r = {}^nC_{r-1}(x)^{n-(r-1)}(1)^{(r-1)} = {}^nC_{r-1}x^{n-r+1}$

And, the $(r+1)^{\text{th}}$ term in the expansion of $(x+1)^n$ is

$T_{r+1} = {}^nC_r(x)^{n-r}(1)^r = {}^nC_r x^{n-r}$

Therefore, the coefficients of the $(r-1)^{\text{th}}, r^{\text{th}}, (r+1)^{\text{th}}$ terms in the expansion of $(x+1)^n$ are ${}^nC_{r-2}, {}^nC_{r-1}, {}^nC_r$, respectively.

The coefficients of the $(r-1)^{\text{th}}, r^{\text{th}}, (r+1)^{\text{th}}$ terms are in the ratio 1:3:5; therefore

$\frac{{}^nC_{r-2}}{{}^nC_{r-1}} = \frac{1}{3}$ and $\frac{{}^nC_{r-1}}{{}^nC_r} = \frac{3}{5}$

Now, solving equation $\frac{{}^nC_{r-2}}{{}^nC_{r-1}} = \frac{1}{3}$

$\Rightarrow \frac{n!}{(r-2)!(n-r+2)!}\times\frac{(r-1)!(n-r+1)!}{n!} = \frac{1}{3}$

$\Rightarrow = \frac{(r-1)(r-2)!(n-r+1)!}{(r-2)!(n-r+2)(n-r+1)!} = \frac{1}{3}$

$\Rightarrow \frac{r-1}{n-r+2} = \frac{1}{3} \Rightarrow 3r-3 = n-r+2 \Rightarrow n-4r+5 = 0$

Similarly, solving equation $\frac{{}^nC_{r-1}}{{}^nC_r} = \frac{3}{5}$

$\Rightarrow \frac{n!}{(r-1)!(n-r+1)!}\times\frac{r!(n-r)!}{n!} = \frac{3}{5}$

$\Rightarrow \frac{r(r-1)!(n-r)!}{(r-1)!(n-r+1)(n-r)!} = \frac{3}{5}$

$\Rightarrow \frac{r}{n-r+1} = \frac{3}{5} \Rightarrow 5r = 3n-3r+3 \Rightarrow 3n-8r+3 = 0$

Solving the above two equations we get $n = 7$, $r = 3$.

Q11. Prove that the coefficient of x^n in the expansion of $(1+x)^{2n}$ is twice the coefficient of x^n in the expansion of $(1+x)^{2n-1}$.

Solution:

The $(r+1)^{\text{th}}$ term in the binomial expansion of $(a+b)^n$ is $T_{r+1} = {}^nC_r a^{n-r}b^r$.

Let us assume that x^n occurs in the $(r+1)^{\text{th}}$ term of the expansion of $(1+x)^{2n}$. Then, we get

$T_{r+1} = {}^{2n}C_r(1)^{2n-r}(x)^r = {}^{2n}C_r(x)^r$

Comparing the indices of x in x^n and in T_{r+1}, we get $r = n$.

Therefore, the coefficient of x^n in the expansion of $(1+x)^{2n}$ is ${}^{2n}C_r = \frac{(2n)!}{n!(2n-n)!} = \frac{(2n)!}{n!n!} = \frac{(2n)!}{(n!)^2}$

Similarly assume that x^n occurs in the $(k+1)^{\text{th}}$ term of the expansion $(1+x)^{2n-1}$; then we get

$T_{k+1} = {}^{2n-1}C_k(1)^{2n-1-k}(x)^k = {}^{2n-1}C_k(x)^k$

Comparing the indices of x in x^n and in T_{k+1}, we get $k = n$.

Therefore, the coefficient of x^n in the expansion

of $(1+x)^{2n-1}$ is ${}^{2n-1}C_n = \frac{(2n-1)!}{n!(2n-1-n)!} = \frac{(2n-1)!}{n!(n-1)!}$

$\frac{(2n-1)!}{n!(n-1)!} \times \frac{2n}{2n} = \frac{1}{2} \cdot \frac{2n \cdot (2n-1)!}{n! \cdot n \cdot (n-1)!} = \frac{1}{2} \cdot \frac{(2n)!}{n!n!}$

Thus, we get $\frac{1}{2}({}^{2n}C_n) = {}^{2n-1}C_n \Rightarrow {}^{2n}C_n = 2({}^{2n-1}C_n)$

Therefore, the coefficient of x^n in the expansion of $(1+x)^{2n}$ is twice the coefficient of x^n in the expansion of $(1+x)^{2n-1}$.

Q12. Find a positive value of m for which the coefficient of x^2 in the expansion $(1+x)^m$ is 6.

Solution:

The $(r+1)^{th}$ term in the binomial expansion of $(a+b)^n$ is $T_{r+1} = {}^nC_r a^{n-r} b^r$.

Let us assume that x^2 occurs in the $(r+1)^{th}$ term of the expansion $(1+x)^m$. Then, we get

$T_{r+1} = {}^mC_r (1)^{m-r} (x)^r = {}^mC_r (x)^r$

Comparing the indices of x in x^2 and in (T_{r+1}), we get $r = 2$.

Therefore, the coefficient of x^2 is ${}^mC_2$.

The coefficient of x^2 in the expansion $(1+x)^m$ is 6, therefore ${}^mC_2 = 6$.

$\Rightarrow \frac{m!}{2!(m-2)!} = 6 \Rightarrow \frac{m(m-1)(m-2)!}{2(m-2)!} = 6$

$\Rightarrow m(m-1) = 12 \Rightarrow m^2 - m - 12 = 0$

$\Rightarrow (m-4)(m+3) = 0 \Rightarrow m = 4,\ m = -3$

Thus, the positive value of $m = 4$, the coefficient of x^2 in the expansion $(1+x)^m$ is 6.

Day 107

Binomial Theorem for Any Index

Binomial Theorem for Negative or Fraction Index

Case I: If $\left|\frac{b}{a}\right| < 1$, then

$$(a+b)^m = \left\{a \cdot \left(1+\frac{b}{a}\right)\right\}^m = a^m \left(1+\frac{b}{a}\right)^m$$

$$= a^m \left[1 + m\left(\frac{b}{a}\right) + \frac{m(m-1)}{1 \cdot 2}\left(\frac{b}{a}\right)^2 + \frac{m(m-1)(m-2)}{1 \cdot 2 \cdot 3}\left(\frac{b}{a}\right)^3 + \ldots \right]$$

$$= a^m + ma^{m-1}b + \frac{m(m-1)}{1 \cdot 2} a^{m-2} b^2 + \frac{m(m-1)(m-2)}{1 \cdot 2 \cdot 3} a^{m-3} b^3 + \ldots$$

Case II: If $\left|\frac{a}{b}\right| < 1$, then

$$(a+b)^m = \left\{b \cdot \left(1+\frac{a}{b}\right)\right\}^m = b^m \left(1+\frac{a}{b}\right)^m$$

$$= b^m \left[1 + m\left(\frac{a}{b}\right) + \frac{m(m-1)}{1 \cdot 2}\left(\frac{a}{b}\right)^2 + \frac{m(m-1)(m-2)}{1 \cdot 2 \cdot 3}\left(\frac{a}{b}\right)^3 + \ldots \right]$$

$$= b^m + mb^{m-1}a + \frac{m(m-1)}{1 \cdot 2} b^{m-2} a^2 + \frac{m(m-1)(m-2)}{1 \cdot 2 \cdot 3} b^{m-3} a^3 + \ldots$$

The general term in the expansion of $(a+b)^m$ for negative or fraction index is

$$T_r = \frac{m(m-1)(m-2)\ldots(m-r+1)}{1 \cdot 2 \cdot 3 \ldots r} a^{m-r} b^r;$$

$r = 0, 1, 2\ldots$

Similarly, expansion $(1+x)^m$

$$= 1 + mx + \frac{m(m-1)}{2!} x^2 + \frac{m(m-1)(m-2)}{3!} x^3 + \ldots$$

Note:

(i) Binomial theorem for negative or fraction index is valid only when $\left|\frac{b}{a}\right| < 1$ or $|x| < 1$.

(ii) For negative or fraction index, ${}^nC_r$ does not make any sense.

(iii) Binomial theorem for negative or fraction gives infinite terms. Therefore, we may write only certain terms because higher power of $|x|$ tends to zero.

Special cases for negative or fraction index $|x| < 1$:

(i) $\frac{1}{(1+x)} = (1+x)^{-1}$

$= 1 - x + x^2 - x^3 + \ldots + (-1)^r x^r + \ldots$

(ii) $\frac{1}{(1-x)} = (1-x)^{-1}$

$= 1 + x + x^2 + x^3 + \ldots + x^r + \ldots$

(iii) $\frac{1}{(1+x)^2} = (1+x)^{-2}$

$= 1 - 2x + 3x^2 - 4x^3 + \ldots + (-1)^r (r+1) x^r + \ldots$

(iv) $\frac{1}{(1-x)^2}=(1-x)^{-2}$

$=1+2x+3x^2+4x^3+...+(r+1)x^r+...$

(v) $\frac{1}{(1+x)^3}=(1+x)^{-3}$

$=1-3x+6x^2-10x^3+...+(-1)^r\frac{(r+1)(r+2)}{2}x^r+...$

(vi) $\frac{1}{(1-x)^3}=(1-x)^{-3}$

$=1+3x+6x^2+10x^3+...+\frac{(r+1)(r+2)}{2}x^r+...$

(vii) $\sqrt{(1+x)}=(1+x)^{1/2}$

$=1+\frac{1}{2}x+\frac{(1/2)(-1/2)}{2!}x^2+\frac{(1/2)(-1/2)(-3/2)}{3!}x^3+...$

$=1+\frac{1}{2}x-\frac{1}{8}x^2+\frac{1}{16}x^3+...$

(viii) $\sqrt{(1-x)}=(1-x)^{1/2}=(1+(-x))^{1/2}=1+\frac{1}{2}(-x)+$

$\frac{(1/2)(-1/2)}{2!}(-x)^2+\frac{(1/2)(-1/2)(-3/2)}{3!}(-x)^3+...$

$=1-\frac{1}{2}x-\frac{1}{8}x^2-\frac{1}{16}x^3-...$

(ix) $\frac{1}{\sqrt{(1+x)}}=(1+x)^{-1/2}=1-\frac{1}{2}x+$

$\frac{(-1/2)(-3/2)}{2!}x^2+\frac{(-1/2)(-3/2)(-5/2)}{3!}x^3+...$

$=1-\frac{1}{2}x+\frac{3}{8}x^2-\frac{5}{16}x^3+...$

(x) $\frac{1}{\sqrt{(1-x)}}=(1-x)^{-1/2}=(1+(-x))^{-1/2}=1-\frac{1}{2}(-x)+$

$\frac{(-1/2)(-3/2)}{2!}(-x)^2+\frac{(-1/2)(-3/2)(-5/2)}{3!}(-x)^3+...$

$=1+\frac{1}{2}x+\frac{3}{8}x^2+\frac{5}{16}x^3+...$

Example 1: Expand $(1-2x)^{-3/4}$ up to the term of x^3.

Solution:

$(1-2x)^{-3/4}$

$=1+\left(-\frac{3}{4}\right)(-2x)+\frac{(-3/4)(-3/4-1)}{2!}(-2x)^2$

$+\frac{(-3/4)(-3/4-1)(-3/4-2)}{3!}(-2x)^3+...$

$=1+\frac{3}{2}x+\frac{21}{8}x^2+\frac{231}{48}x^3+...$

Example 2: Expand $(1+x+x^2+x^3+...)^2$ up to the term of x^3 and write the general term if $|x|<1$.

Solution:

$\because \frac{1}{(1-x)}=(1-x)^{-1}=1+x+x^2+x^3+...+x^r+...$

$\therefore (1+x+x^2+x^3+...)^2=\left\{(1-x)^{-1}\right\}^2=(1-x)^{-2}$

$=1+2x+3x^2+4x^3+...+(r+1)x^r+...$

Thus, the general term of $(1+x+x^2+x^3+...)^2$ is $(r+1)x^r$.

Example 3: Expand $\frac{1}{\sqrt{(1-x)(1+x)^3}}$ up to the term of x^3.

Solution:

$\frac{1}{\sqrt{(1-x)}(1+x)^3}=\frac{1}{\sqrt{(1-x)}}\times\frac{1}{(1+x)^3}$

$=\left(1+\frac{1}{2}x+\frac{3}{8}x^2+\frac{5}{16}x^3\right)\times\left(1-3x+6x^2-10x^3\right)$

$=1+\frac{1}{2}x-3x+\frac{3}{8}x^2+6x^2+\frac{1}{2}x\times(-3x)+$

$+\frac{5}{16}x^3-10x^3+\frac{1}{2}x\times 6x^2+\frac{3}{8}x^2\times(-3x)$

$=1-\frac{5}{2}x+\frac{39}{8}x^2-\frac{125}{16}x^3$

Example 4: Expand $5(2-3x-2x^2)^{-1}$ up to the term of x^3 and write the general term if $|x|<1/2$.

Solution:

$\because 5(2-3x-2x^2)^{-1}=\frac{5}{(1-2x)(2+x)}$

$=\frac{4+2x+1-2x}{(1-2x)(2+x)}=\frac{2}{1-2x}+\frac{1}{2+x}$

$=\frac{2}{1-2x}+\frac{1/2}{1+x/2}$

$=2(1-2x)^{-1}-\frac{1}{2}\{1+(x/2)\}^{-1}$

$\because (1+x)^{-1}=1-x+x^2-x^3+...+(-1)^r x^r+...$

and $(1-x)^{-1}=1+x+x^2+x^3+...+x^r+...$

$\therefore 2(1-2x)^{-1}-\frac{1}{2}\{1+(x/2)\}^{-1}$

$=2\{1+(2x)+(2x)^2+(2x)^3+...+(2x)^r+...\}$

$-\frac{1}{2}\{1-\left(\frac{x}{2}\right)+\left(\frac{x}{2}\right)^2-\left(\frac{x}{2}\right)^3+...+(-1)^r\left(\frac{x}{2}\right)^r+...\}$

$=2+2^2x+2^3x^2+2^4x^3+...+2^{r+1}x^r+...$

$-\frac{1}{2}+\frac{x}{2^2}-\frac{x^2}{2^3}+\frac{x^3}{2^4}+...+(-1)^{r+1}\frac{x^r}{2^{r+1}}+...$

$=\left(2-\frac{1}{2}\right)+\left(2^2+\frac{1}{2^2}\right)x+\left(2^3-\frac{1}{2^3}\right)x^2$

$+\left(2^4+\frac{1}{2^4}\right)x^3+...+\left(2^{r+1}+\frac{(-1)^{r+1}}{2^{r+1}}\right)x^r+...$

Thus, the general term of $5(2-3x-2x^2)^{-1}$ is $\left(2^{r+1}+\frac{(-1)^{r+1}}{2^{r+1}}\right)x^r$.

Example 5: Expand $\frac{1}{1+2x}$ up to the term of x^4 and state the value for which the expression is valid. Also use the result to find the value of $\frac{1}{1.02}$ to 6 decimal places.

Solution:

$$\because \frac{1}{(1+x)}=(1+x)^{-1}=1-x+x^2-x^3+...+(-1)^r x^r+...$$

$$\therefore \frac{1}{(1+2x)}=(1+2x)^{-1}$$

$$=1-(2x)+(2x)^2-(2x)^3+(2x)^4$$

$$=1-2x+4x^2-8x^3+16x^4$$

The expression is valid for $|2x|<1 \Rightarrow |x|<\frac{1}{2}$.

Substituting $x=0.01$, we get

$$\frac{1}{1.02}=\frac{1}{(1+2\times 0.01)}$$

$$=1-2(0.01)+4(0.01)^2-8(0.01)^3+16(0.01)^4$$

$$=1-0.02+0.0004-0.000008+0.00000016$$

$=0.980392$ to 6 decimal places

Example 6: Expand $\frac{1}{(2-x)^2}$ up to the term of x^4 and state the value for which the expression is valid. Also use the result to find the value of $\frac{1}{(1.98)^2}$ to 6 decimal places.

Solution:

$$\because \frac{1}{(1-x)^2}=(1-x)^{-2}=1+2x+3x^2+4x^3+...+(r+1)x^r+...$$

$$\therefore \frac{1}{(2-x)^2}=\frac{1}{4}\times\frac{1}{(1-x/2)^2}$$

$$=\frac{1}{4}\times\left\{1+2(x/2)+3(x/2)^2+4(x/2)^3+5(x/2)^4\right\}$$

$$=\frac{1}{4}\left\{1+x+\frac{3}{4}x^2+\frac{1}{2}x^3+\frac{5}{16}x^4\right\}$$

The expression is valid for $|x/2|<1 \Rightarrow |x|<2$

Substituting $x=0.02$, we get

$$\frac{1}{(1.98)^2}=\frac{1}{(2-0.02)^2}$$

$$=\frac{1}{4}\times\left\{1+2(x/2)+3(x/2)^2+4(x/2)^3+5(x/2)^4\right\}$$

$$=\frac{1}{4}\left\{1+x+\frac{3}{4}x^2+\frac{1}{2}x^3+\frac{5}{16}x^4\right\}$$

$$=\frac{1}{4}\left\{1+0.02+\frac{3}{4}(0.02)^2+\frac{1}{2}(0.02)^3+\frac{5}{16}(0.02)^4\right\}$$

$$=\frac{1}{4}\{1+0.02+0.0003+0.000004+0.00000005\}$$

$=\frac{1}{4}\{1.020304\}=0.255076$ to 6 decimal places

Example 7: Expand $\sqrt{(1+x)}$ up to the term of x^3 and state the value for which the expression is valid. Also use the result to find the value of $\sqrt{30}$ to 3 decimal places.

Solution:

$$\sqrt{(1+x)}=(1+x)^{1/2}=1+\frac{1}{2}x-\frac{1}{8}x^2+\frac{1}{16}x^3$$

The expression is valid for $|x|<1$

$$\because \sqrt{30}=\sqrt{25\times 1.2}=\sqrt{25}\times\sqrt{1.2}=5\sqrt{1.2}$$

Substituting $x=0.2$, we get

$$\sqrt{30}=5\sqrt{1.2}=5\sqrt{(1+0.2)}$$

$$=5\left\{1+\frac{1}{2}(0.2)-\frac{1}{8}(0.2)^2+\frac{1}{16}(0.2)^3\right\}$$

$$=5\left\{1+0.1-\frac{0.01}{2}+\frac{0.001}{2}\right\}$$

$=5\times\{1.0955\}=5.4772=5.477$ to 3 decimal places

Example 8: Expand $\frac{1}{\sqrt{(1+x)}}$ up to the term of x^3 and state the value for which the expression is valid. Also use the result to find the value of $\frac{1}{\sqrt{80}}$ to 4 decimal places.

Solution:

$$\frac{1}{\sqrt{(1-x)}}=(1-x)^{-1/2}=1+\frac{1}{2}x+\frac{3}{8}x^2+\frac{5}{16}x^3+...$$

The expression is valid for $|x|<1$

$$\because \frac{1}{\sqrt{80}}=\frac{1}{\sqrt{100\times 8}}=\frac{1}{\sqrt{100}}\times\frac{1}{\sqrt{0.8}}=\frac{1}{10}\frac{1}{\sqrt{1-0.2}}$$

Substituting $x=0.2$, we get

$$\because \frac{1}{\sqrt{80}}=\frac{1}{10}\frac{1}{\sqrt{1-0.2}}$$

$$=\frac{1}{10}\left\{1+\frac{1}{2}(0.2)+\frac{3}{8}(0.2)^2+\frac{5}{16}(0.2)^3\right\}$$

$$=\frac{1}{10}\left\{1+0.1+\frac{0.03}{2}+\frac{0.005}{2}\right\}$$

$$=\frac{1}{10}\{1+0.1+0.015+0.0025\}$$

$=\frac{1}{10}\times 1.1175=0.11175=0.1118$ to 4 decimal places

Day 108

NCERT Miscellaneous Exercise (With Solutions)

Q1. Find a, b and n in the expansion of $(a+b)^n$ if the first three terms of the expansion are 729, 7290 and 30375, respectively.

Solution:

The $(r+1)^{\text{th}}$ term in the binomial expansion of $(a+b)^n$ is $T_{r+1} = {}^nC_r a^{n-r} b^r$.

The first three terms of the expansion are 729, 7290 and 3075, respectively. Therefore

$$T_1 = {}^nC_0 a^{n-0} b^0 = a^n = 729 \quad \text{(i)}$$

$$T_2 = {}^nC_1 a^{n-1} b^1 = na^{n-1}b = 7290 \quad \text{(ii)}$$

$$T_3 = {}^nC_2 a^{n-2} b^2 = \frac{n(n-1)}{2} a^{n-2} b^2 = 30375 \quad \text{(iii)}$$

Dividing (ii) by (i), we get

$$\frac{na^{n-1}b}{a^n} = \frac{7290}{729} \Rightarrow \frac{nb}{a} = 10 \quad \text{(iv)}$$

Dividing (iii) by (ii), we get

$$\frac{n(n-1)a^{n-2}b^2}{2na^{n-1}b} = \frac{30375}{7290}$$

$$\Rightarrow \frac{(n-1)b}{a} = \frac{25}{3} \Rightarrow \frac{nb}{a} - \frac{b}{a} = \frac{25}{3} \quad \text{(v)}$$

From equations (iv) and (v), we get

$$\Rightarrow 10 - \frac{b}{a} = \frac{25}{3} \Rightarrow \frac{b}{a} = 10 - \frac{25}{3} = \frac{5}{3}$$

Thus, $n \cdot \frac{b}{a} = 10 \Rightarrow n \cdot \frac{5}{3} = 10 \Rightarrow n = 6$

Now from equation (i), we get

$a^n = 729 \Rightarrow a^6 = 729 \Rightarrow a = \sqrt[6]{729} = 3$

And from equation (iv), we get

$$\frac{nb}{a} = 10 \Rightarrow \frac{6 \times b}{3} = 10 \Rightarrow b = 5$$

Q2. Find a if the coefficient of x^2, x^3 in the expansion of $(3+ax)^9$ are equal.

Solution:

The $(r+1)^{\text{th}}$ term in the binomial expansion of $(a+b)^n$ is $T_{r+1} = {}^nC_r a^{n-r} b^r$.

Let us assume that x^2 occurs in the $(r+1)^{\text{th}}$ term in the expansion of $(3+ax)^9$. Then we get

$T_{r+1} = {}^9C_r (3)^{9-r} (ax)^r = {}^9C_r (3)^{9-r} a^r x^r$

Comparing the indices of x in x^2 and in (T_{r+1}), we get $r = 2$.

Thus, the coefficient of x^2 is

$${}^9C_2 (3)^{9-2} a^2 = \frac{9!}{2!7!}(3)^7 a^2 = 36(3)^7 a^2$$

Similarly, assume that x^3 occurs in the $(k+1)^{\text{th}}$ term in the expansion of $(3+ax)^9$. Then we get

$T_{k+1} = {}^9C_k (3)^{9-k} (ax)^k = {}^9C_k (3)^{9-k} a^k x^k$

Comparing the indices of x in x^3 and in (T_{r+1}), we get $k = 3$.

Thus, the coefficient of x^3 is

$${}^9C_3 (3)^{9-3} a^3 = \frac{9!}{3!6!}(3)^6 a^3 = 84(3)^6 a^3$$

The coefficients of x^2 and x^3 are equal. Therefore

$$84(3)^6 a^3 = 36(3)^7 a^2 \Rightarrow 84a = 104 \Rightarrow a = \frac{9}{7}.$$

Q3. Find the coefficient of x^5 in the product $(1+2x)^6(1-x)^7$ using the binomial theorem.

Solution:

Using the binomial theorem, the expansion of $(1+2x)^6$ is

$$(1+2x)^6 = {}^6C_0 + {}^6C_1(2x) + {}^6C_2(2x)^2 + {}^6C_3(2x)^3 + {}^6C_4(2x)^4 + {}^6C_5(2x)^5 + {}^6C_6(2x)^6$$

$$= 1 + 12x + 60x^2 + 160x^3 + 240x^4 + 192x^5 + 64x^6$$

Using the binomial theorem, the expansion of $(1-x)^7$ is

$$(1-x) = {}^7C_0 - {}^7C_1(x) + {}^7C_2(x)^2 - {}^7C_3(x)^3 + {}^7C_4(x)^4 - {}^7C_5(x)^5 + {}^7C_6(x)^6 - {}^7C_7(x)^7$$

$$= 1 - 7x + 21x^2 - 35x^3 + 35x^4 - 21x^5 + 7x^6 - x^7$$

$$\therefore (1+2x)^6(1-x)^7$$

$$= (1 + 12x + 60x^2 + 160x^3 + 240x^4 + 192x^5 + 64x^6)$$

$$(1 - 7x + 21x^2 - 35x^3 + 35x^4 - 21x^5 + 7x^6 - x^7)$$

We need the terms which involve x^5; therefore complete multiplication of the two brackets is not required.

Thus, the terms containing x^5 are

$$1(-21x^5) + (12x)(35x^4) + (60x^2)(-35x^3)$$

$$+(160x^3)(21x^2) + (240x^4)(-7x) + (192x^5)(1)$$

$$= 171x^5$$

Thus, the coefficient of x^5 in the given product is 171.

Q4. If a and b are distinct integers, prove that $a-b$ is a factor of $a^n - b^n$ whenever n is a positive integer. [**Hint**: write $a^n = (a-b+b)^n$ and expand]

Solution:

Since we can write $a = a - b + b$

$$\therefore a^n = (a-b+b)^n = [(a-b)+b]^n$$

$= {}^nC_0(a-b)^n + {}^nC_1(a-b)^{n-1}b + ... + {}^nC_nb^n$

$= (a-b)^n + {}^nC_1(a-b)^{n-1}b + ... + {}^nC_{n-1}(a-b)b^{n-1} + b^n$

$= (a-b)^n + {}^nC_1(a-b)^{n-1}b + ... + {}^nC_{n-1}(a-b)b^{n-1}$

$$\Rightarrow a^n - b^n = (a-b)\begin{Bmatrix}(a-b)^{n-1} + {}^nC_1(a-b)^{n-2}b \\ + ... + {}^nC_{n-1}b^{n-1}\end{Bmatrix}$$

$\Rightarrow a^n - b^n = (a-b)k$

where, $k = (a-b)^{n-1} + {}^nC_1(a-b)^{n-2}b + ... + {}^nC_{n-1}b^{n-1}$

$a^n - b^n = k(a-b)$, where k is some natural number. Therefore, $(a-b)$ is a factor of $\left(a^n - b^n\right)$ for all positive integers n.

Q5. Evaluate $(\sqrt{3}+\sqrt{2})^6 - (\sqrt{3}-\sqrt{2})^6$.

Solution:

Using the binomial theorem,

$(a+b)^6 = {}^6C_0a^6 + {}^6C_1a^5b + {}^6C_2a^4b^2 + {}^6C_3a^3b^3 + {}^6C_4a^2b^4 + {}^6C_5a^1b^5 + {}^6C_6b^6$

$= a^6 + 6a^5b + 15a^4b^2 + 20a^3b^3 + 15a^2b^4 + 6ab^5 + b^6$

And, $(a-b)^6 = {}^6C_0a^6 - {}^6C_1a^5b + {}^6C_2a^4b^2 - {}^6C_3a^3b^3 + {}^6C_4a^2b^4 - {}^6C_5a^1b^5 + {}^6C_6b^6$

$= a^6 - 6a^5b + 15a^4b^2 - 20a^3b^3 + 15a^2b^4 - 6ab^5 + b^6$

$\therefore (a+b)^6 - (a-b)^6 = 2[6a^5b + 20a^3b^3 + 6ab^5]$

Putting $a = \sqrt{3}, b = \sqrt{2}$ we get

$(\sqrt{3}+\sqrt{2})^6 - (\sqrt{3}-\sqrt{2})^6$

$= 2[6(\sqrt{3})^5(\sqrt{2}) + 20(\sqrt{3})^3(\sqrt{2})^3 + 6(\sqrt{3})(\sqrt{2})^5]$

$= 2[54\sqrt{6} + 120\sqrt{6} + 24\sqrt{6}] = 396\sqrt{6}$

Q6. Find the value of $(a^2 + \sqrt{a^2-1})^4 + (a^2 - \sqrt{a^2-1})^4$.

Solution:

Using the binomial theorem,

$(x+y)^4 = {}^4C_0x^4 + {}^4C_1x^3y + {}^4C_2x^2y^2 + {}^4C_3xy^3 + {}^4C_4y^4$

$= x^4 + 4x^3y + 6x^2y^2 + 4xy^3 + y^4$

$(x-y)^4 = {}^4C_0x^4 - {}^4C_1x^3y + {}^4C_2x^2y^2 - {}^4C_3xy^3 + {}^4C_4y^4$

$= x^4 - 4x^3y + 6x^2y^2 - 4xy^3 + y^4$

And,

$\therefore (x+y)^4 + (x-y)^4 = 2(x^4 + 6x^2y^2 + y^4)$

Putting $x = a^2, y = \sqrt{a^2-1}$, we get

$(a^2 + \sqrt{a^2-1})^4 + (a^2 - \sqrt{a^2-1})^4$

$= 2[(a^2)^4 + 6(a^2)^2(\sqrt{a^2-1})^2 + (\sqrt{a^2-1})^4]$

$= 2[a^8 + 6a^4(a^2-1) + (a^2-1)^2]$

$= 2[a^8 + 6a^6 - 6a^4 + a^4 - 2a^2 + 1]$

$= 2a^8 + 12a^6 - 10a^4 - 4a^2 + 2$

Q7. Find an approximation of $(0.99)^5$ using the first three terms of its expansion.

Solution:

We can write $0.99 = 1 - 0.01$

$\therefore (0.99)^5 = (1-0.01)^5$

$= {}^5C_0(1)^5 - {}^5C_1(1)^4(0.01) + {}^5C_2(1)^3(0.01)^2 + ...$

(Approximately neglecting smaller value terms)

$= 1 - 5(0.01) + 10(0.01)^2 = 1 - 0.05 + 0.001 = 0.951$

Thus, the value of $(0.99)^5$ is approximately 0.951.

Q8. Find n, if the ratio of the fifth term from the beginning to the fifth term from the end in the expansion of $\left(\sqrt[4]{2} + \frac{1}{\sqrt[4]{3}}\right)^n$ is $\sqrt{6}:1$.

Solution:

In the expansion,

$(a+b)^n = {}^nC_0a^n + {}^nC_1a^{n-1}b + ... + {}^nC_{n-1}ab^{n-1} + {}^nC_nb^n$

Fifth term from the beginning $= {}^nC_4a^{n-4}b^4$

Fifth term from the end $= {}^nC_{n-4}a^4b^{n-4}$

Therefore, in the expansion of $\left(\sqrt[4]{2} + \frac{1}{\sqrt[4]{3}}\right)^n$

Fifth term from the beginning is ${}^nC_4(\sqrt[4]{2})^{n-4}\left(\frac{1}{\sqrt[4]{3}}\right)^4$

Fifth term from the end is ${}^nC_{n-4}(\sqrt[4]{2})^4\left(\frac{1}{\sqrt[4]{3}}\right)^{n-4}$

The ratio of the fifth term from the beginning to the fifth term the end is $\sqrt{6}:1$

Therefore,

$$\frac{{}^nC_4(\sqrt[4]{2})^{n-4}\left(\frac{1}{\sqrt[4]{3}}\right)^4}{{}^nC_{n-4}(\sqrt[4]{2})^4\left(\frac{1}{\sqrt[4]{3}}\right)^{n-4}} = \frac{\sqrt{6}}{1}$$

$$\Rightarrow {}^nC_4\frac{(\sqrt[4]{2})^n}{(\sqrt[4]{2})^4}\cdot\frac{1}{3} = \sqrt{6}\,{}^nC_{n-4}2\cdot\frac{(\sqrt[4]{3})^4}{(\sqrt[4]{3})^n}$$

$$\Rightarrow {}^nC_4\frac{(\sqrt[4]{2})^n}{2}\cdot\frac{1}{3} = \sqrt{6}\,{}^nC_{n-4}2\cdot\frac{3}{(\sqrt[4]{3})^n}$$

$$\Rightarrow (\sqrt[4]{2})^n = 36\sqrt{6}\cdot\frac{1}{(\sqrt[4]{3})^n} \quad (\because {}^nC_4 = {}^nC_{n-4})$$

$$\Rightarrow (\sqrt[4]{2})^n(\sqrt[4]{3})^n = 36\sqrt{6} \Rightarrow (\sqrt[4]{6})^n = 6^2\sqrt{6}$$

$$\Rightarrow 6^{\frac{n}{4}} = 6^{\frac{5}{2}} \Rightarrow \frac{n}{4} = \frac{5}{2} \Rightarrow n = 4\times\frac{5}{2} = 10$$

Q9. Expand using the binomial theorem $\left(1+\frac{x}{2}-\frac{2}{x}\right)^4, x \neq 0.$

Solution:

The given expression $\left(1+\frac{x}{2}-\frac{2}{x}\right)^4$ can be written as $\left[\left(1+\frac{x}{2}\right)-\frac{2}{x}\right]^4$. Therefore, using the binomial theorem,

$$\left(1+\frac{x}{2}-\frac{2}{x}\right)^4 = {}^4C_0\left(1+\frac{x}{2}\right)^4 - {}^4C_1\left(1+\frac{x}{2}\right)^3\left(\frac{2}{x}\right)$$

$$+{}^4C_2\left(1+\frac{x}{2}\right)^2\left(\frac{2}{x}\right)^2 - {}^4C_3\left(1+\frac{x}{2}\right)\left(\frac{2}{x}\right)^3 + {}^4C_4\left(\frac{2}{x}\right)^4$$

$$=\left(1+\frac{x}{2}\right)^4 - 4\left(1+\frac{x}{2}\right)^3\left(\frac{2}{x}\right)+6\left(1+x+\frac{x^2}{4}\right)\left(\frac{4}{x^2}\right)$$

$$-4\left(1+\frac{x}{2}\right)\left(\frac{8}{x^3}\right)+\frac{16}{x^4}$$

$$=\left(1+\frac{x}{2}\right)^4 - \frac{8}{x}\left(1+\frac{x}{2}\right)^3+\frac{24}{x^2}+\frac{24}{x}+6-\frac{32}{x^3}-\frac{16}{x^2}+\frac{16}{x^4}$$

$$=\left(1+\frac{x}{2}\right)^4 - \frac{8}{x}\left(1+\frac{x}{2}\right)^3+\frac{8}{x^2}+\frac{24}{x}+6-\frac{32}{x^3}+\frac{16}{x^4}$$

Again using binomial theorem, we expand

$$\left(1+\frac{x}{2}\right)^4 = {}^4C_0(1)^4 + {}^4C_1(1)^3\left(\frac{x}{2}\right) + {}^4C_2(1)^2\left(\frac{x}{2}\right)^2$$

$$+ {}^4C_3(1)^1\left(\frac{x}{2}\right)^3 + {}^4C_4\left(\frac{x}{2}\right)^4$$

$$=1+2x+\frac{3x^2}{2}+\frac{x^3}{2}+\frac{x^4}{16}$$

And, $\left(1+\frac{x}{2}\right)^3$

$$= {}^3C_0(1)^3 + {}^3C_1(1)^2\left(\frac{x}{2}\right) + {}^3C_2(1)\left(\frac{x}{2}\right)^2 + {}^3C_3\left(\frac{x}{2}\right)^3$$

$$=1+\frac{3x}{2}+\frac{3x^2}{4}+\frac{x^3}{8}$$

Thus, $\left(\left(1+\frac{x}{2}\right)-\frac{2}{x}\right)^4$

$$=1+2x+\frac{3x^2}{2}+\frac{x^3}{2}+\frac{x^1}{16}$$

$$-\frac{8}{x}\left(1+\frac{3x}{2}+\frac{3x^2}{4.}+\frac{x^3}{8}\right)+\frac{8}{x^2}+\frac{24}{x}+6-\frac{32}{x^3}+\frac{16}{x^4}$$

$$=1+2x+\frac{x}{2}x^2+\frac{x^3}{2}+\frac{x^4}{16}-\frac{8}{x}$$

$$-12-6x-x^2+\frac{8}{x^2}+\frac{24}{x}+6-\frac{32}{x^3}+\frac{16}{x^4}$$

$$=\frac{16}{x}+\frac{8}{x^2}-\frac{32}{x^3}+\frac{16}{x^4}-4x+\frac{x^2}{2}+\frac{x^3}{2}+\frac{x^4}{16}-5$$

Q10. Find the expansion of $(3x^2-2ax+3a^2)^3$ using the binomial theorem.

Solution:

The given expression can be written as $\left[(3x^2-2ax)+3a^2\right]^3$.
Therefore, using the binomial theorem,

$$(3x^2-2ax+3a^2)^3$$

$$= {}^3C_0(3x^2-2ax)^3 + {}^3C_1(3x^2-2ax)^2(3a)^2$$

$$+{}^3C_2(3x^2-2ax)(3a^2)^2 + {}^3C_3(3a^2)^2$$

$$=(3x^2-2ax)^3+3(9x^4-12ax^3-12ax^3+4a^2x^2)(3a^2)$$

$$+3(3x^2-2ax)(9a^4)+27a^6$$

$$=(3x^2-2ax)^3+81a^2a^4-108a^3x^3$$

$$+36a^4x^2+81a^4x^2-54a^5x+27a^6$$

$$=(3x^2-2ax)^3+81a^2x^4-108a^3x^3$$

$$+117a^4x^2-54a^4x+27a^6$$

Again using the binomial theorem, we get

$$(3x^2-2ax)^3 = {}^3C_0(3x^2)^3 - {}^3C_1(3x^2)^2(2ax)$$

$$+{}^3C_2(3x^2)(2ax)^2 - {}^3C_3(2ax)^3$$

$$=27x^6-54ax^5+36a^2x^4-8a^3x^3$$

Thus,

$$(3x^2-2ax+3a^2)^3$$

$$=27x^6-54ax^5+36a^2x^4-8a^3x^3+81a^2x^4$$

$$-108a^3x^3+117a^4x^2-54a^5x+27a^6$$

$$=27x^6-54ax^5+117a^2x^4-116a^3x^3$$

$$+117a^4x^2-54a^5x+27a^6$$

Day 109

NCERT Exemplar Problems: Solutions (Q1–Q16)

Q1. Find the term independent of x, where $x \neq 0$, in the expansion of $\left(\frac{3x^2}{2}-\frac{1}{3x}\right)^{15}$.

Solution:

The general term of the given expansion $\left(\frac{3x^2}{2}-\frac{1}{3x}\right)^{15}$ is

$$T_{r+1} = {}^{15}C_r\left(\frac{3x^2}{2}\right)^{15-r}\left(-\frac{1}{3x}\right)^r$$

$$= {}^{15}C_r 3^{15-r}x^{30-2r}2^{r-15}(-1)^r\cdot 3^{-r}\cdot x^{-r}$$

$$= {}^{15}C_r(-1)^r 3^{15-2r}2^{r-15}x^{30-3r}$$

For the term independent of x, $30-3r=0 \Rightarrow r=10$

Therefore, $T_{r+1}=T_{10+1}=11^{th}$ term is independent of x, thus $T_{11}=T_{10+1}= {}^{15}C_{10}(-1)^{10}3^{15-20}2^{10-15}$

$$= {}^{15}C_{10}3^{-5}2^{-5} = {}^{15}C_{10}\left(\frac{1}{6}\right)^5$$

Q2. If the term free from x in the expansion of $\left(\sqrt{x}-\frac{k}{x^2}\right)^{10}$ is 405, then find the value of k.

Solution:

The general term of the given expansion $\left(\sqrt{x}-\frac{k}{x^2}\right)^{10}$ is

$$T_{r+1} = {}^{10}C_r(\sqrt{x})^{10-r}\left(\frac{-k}{x^2}\right)^r.$$

$$= {}^{10}C_r(x)^{\frac{1}{2}(10-r)}(-k)^r\cdot x^{-2r} = {}^{10}C_r(-k)^r x^{\frac{10-5r}{2}}$$

For the term free from x, $\frac{10-5r}{2}=0 \Rightarrow r=2$

Thus, $T_{2+1}=T_3$ is free from x.

The term free from x is equal to 405. Therefore

$$T_{2+1} = {}^{10}C_2(-k)^2 = 405 \Rightarrow \frac{10!}{2!\times 8!}(-k)^2 = 405$$

$$\Rightarrow 45k^2 = 405 \Rightarrow k^2 = \frac{405}{45} = 9 \Rightarrow k = \pm 3$$

Q3. Find the coefficient of x in the expansion of $(1-3x+7x^2)(1-x)^{16}$.

Solution:

Since $(1-3x+7x^2)(1-x)^{16}$

$$= (1-3x+7x^2)\left({}^{16}C_0 1^{16} - {}^{16}C_1 1^{15}x^1 \text{ +terms of higher power of } x\right)$$

$$= (1-3x+7x^2)\left(1-16x \text{ +terms of higher power of } x\right)$$

$$= 1-16x-3x + \text{terms of higher power of } x$$

$$= 1-19x + \text{terms of higher power of } x$$

Therefore, the coefficient of x is –19.

Q4. Find the term independent of x in the expansion of $\left(3x-\frac{2}{x^2}\right)^{15}$.

Solution:

The general term of the expansion $\left(3x-\frac{2}{x^2}\right)^{15}$ is

$$T_{r+1} = {}^{15}C_r(3x)^{15-r}\left(\frac{-2}{x^2}\right)^r$$

$$= {}^{15}C_r(3x)^{15-r}(-2)^r x^{-2r} = {}^{15}C_r 3^{15-r}(-2)^r x^{15-3r}$$

For the term independent of x, $15-3r=0 \Rightarrow r=5$

Therefore, the term $T_{5+1}=T_6$ is independent of x.

Thus, $T_6 = T_{5+1} = {}^{15}C_5 3^{15-5}(-2)^5 = -{}^{15}C_5\, 3^{10}\, 2^5$

Q5. Find the middle term (terms) in the expansion of

(i) $\left(\frac{x}{a}-\frac{a}{x}\right)^{10}$ (ii) $\left(3x-\frac{x^3}{6}\right)^9$

Solution:

(i) The power of the binomial expansion $\left(\frac{x}{a}-\frac{a}{x}\right)^{10}$ is $n = 10$ is even.

Therefore, it has one middle term, that is, the $\left(\frac{10}{2}+1\right)$ term = 6th term.

Thus, $T_6 = T_{5+1} = {}^{10}C_5\left(\frac{x}{a}\right)^{10-5}\left(\frac{-a}{x}\right)^5$

$$= -{}^{10}C_5\left(\frac{x}{a}\right)^5\left(\frac{a}{x}\right)^5 = -{}^{10}C_5 = -252$$

(ii) The power of the binomial expansion $\left(3x-\frac{x^3}{6}\right)^9$ is $n=9$ is odd.

Therefore, it has two middle terms, that is,

$\left(\frac{9+1}{2}\right)$ and $\left(\frac{9+1}{2}+1\right)$ = 5th term and 6th term.

Thus, $T_5 = T_{(4+1)} = {}^9C_4(3x)^{9-4}\left(-\frac{x^3}{6}\right)^4$

$= \frac{9\times8\times7\times6\times5!}{4\times3\times2\times1\times5!}3^5x^5x^{12}6^{-4}$

$= \frac{7\times6\times3\times3^1}{2^4}x^{17} = \frac{189}{8}x^{17}$

And $T_6 = T_{5+1} = {}^9C_5(3x)^{9-5}\left(-\frac{x^3}{6}\right)^5$

$= -\frac{9\times8\times7\times6\times5!}{5!\times4\times3\times2\times1}\cdot3^4\cdot x^4\cdot x^{15}\cdot6^{-5}$

$= \frac{-21\times6}{3\times2^5}x^{19} = \frac{-21}{16}x^{19}$

Q6. Find the coefficient of x^{15} in the expansion of $(x-x^2)^{10}$.

Solution:

The general term of the given expansion is $T_{r+1} = {}^{10}C_r x^{10-r}(-x^2)^r = (-1)^r\cdot{}^{10}C_r\cdot x^{10-r}\cdot x^{2r}$

$= (-1)^r\,{}^{10}C_r x^{10+r}$

For the coefficient of x^{15}, $10+r = 15 \Rightarrow r = 5$

Thus, $T_{5+1} = (-1)^5\,{}^{10}C_5x^{15}$

Hence, coefficient of x^{15} is

$= (-1)^5\,{}^{10}C_5 = -1\frac{10\times9\times8\times7\times6\times5!}{5\times4\times3\times2\times1\times5!} = -252$

Q7. Find the coefficient of $\frac{1}{x^{17}}$ in the expansion of $\left(x^4-\frac{1}{x^3}\right)^{15}$.

Solution:

The general term of the given expansion $\left(x^4-\frac{1}{x^3}\right)^{15}$ is

$T_{r+1} = {}^{15}C_r(x^4)^{15-r}\left(-\frac{1}{x^3}\right)^r$

$= {}^{15}C_r x^{60-4r}(-1)^r x^{-3r} = (-1)^r\,{}^{15}C_r x^{60-7r}$

For the coefficient of $\frac{1}{x^{17}}$ or x^{-17}, $60-7r = -17 \Rightarrow r = 11$

Thus, $T_{11+1} = (-1)^{11}\,{}^{15}C_{11}x^{60-77}$

Hence, the coefficient of $\frac{1}{x^{17}}$ or x^{-17} is

$= (-1)^{11}\,{}^{15}C_{11} = \frac{-15\times14\times13\times12\times11!}{11!\times4\times3\times2\times1} = -1365$

Q8. Find the sixth term of the expansion $(y^{1/2}+x^{1/3})^n$, if the binomial coefficient of the third term from the end is 45.

Solution:

The sixth term of the given expansion $(y^{1/2}+x^{1/3})^n$ is $T_6 = T_{5+1} = {}^nC_5(y^{1/2})^{n-5}(x^{1/3})^5$.

The binomial coefficient of third term from the end = binomial coefficient of third term from the beginning $= {}^nC_2$.

And, it is given that binomial coefficient of the third term from the end is 45.

$\therefore {}^nC_2 = 45 \Rightarrow \frac{n(n-1)(n-2)!}{2!(n-2)!} = 45$

$\Rightarrow \frac{n(n-1)(n-2)!}{2!(n-2)!} = 45 \Rightarrow n(n-1) = 90$

$\Rightarrow n^2-n-90 = 0 \Rightarrow (n-10)(n+9) = 0$

$\Rightarrow (n+9) = 0$ or $(n-10) = 0$

$\Rightarrow n = 10\quad [\because n \neq -9]$

Thus, $T_6 = {}^{10}C_5y^{5/2}x^{5/3} = 252y^{5/2}x^{5/3}$

Q9. Find the value of r, if the coefficient of $(2r+4)^{th}$, $(r-2)^{th}$ terms in the expansion of $(1+x)^{18}$ are equal.

Solution:

The $(2r+4)^{th}$ term of the expansion $(1+x)^{18}$ is

$T_{2r+3+1} = {}^{18}C_{2r+3}(1)^{18-2r-3}(x)^{2r+3} = {}^{18}C_{2r+3}x^{2r+3}$

And the $(r-2)^{th}$ term of the expansion $(1+x)^{18}$ is

$T_{r-3+1} = {}^{18}C_{r-3}(1)^{18-r+3}(x)^{r-3}$

The coefficients of $(2r+4)^{th}$, $(r-2)^{th}$ terms in the expansion of $(1+x)^{18}$ are equal. Therefore

${}^{18}C_{2r+3} = {}^{18}C_{r-3} \Rightarrow {}^{18}C_{2r+3} = {}^{18}C_{18-(r-3)}$

$\Rightarrow {}^{18}C_{2r+3} = {}^{18}C_{21-r}\quad [\because {}^nC_r = {}^nC_{n-r}]$

$\Rightarrow 2r+3 = 21-r \Rightarrow 3r = 18 \Rightarrow r = 6$

Q10. If the coefficient of second, third and fourth terms in the expansion of $(1+x)^{2n}$ are in AP, then show that $2n^2-9n+7 = 0$.

Solution:

In the given expansion $(1+x)^{2n}$

Coefficient of the 2nd term = ${}^{2n}C_1$

Coefficient of the 3rd term = ${}^{2n}C_2$

Coefficient of the 4th term = ${}^{2n}C_3$

The coefficients of the second, third and fourth terms are in AP, therefore ${}^{2n}C_1$, ${}^{2n}C_2$, ${}^{2n}C_3$ are in AP.

Thus, $2\cdot {}^{2n}C_2 = {}^{2n}C_1 + {}^{2n}C_3$

$$\Rightarrow 2\left(\frac{2n!}{2!\times(2n-2)!}\right) = \frac{2n!}{1!(2n-1)!} + \frac{2n!}{3!(2n-3)!}$$

$$\Rightarrow \frac{1}{(2n-2)(2n-3)!} = \frac{1}{(2n-1)(2n-2)(2n-3)!} + \frac{1}{6(2n-3)!}$$

$$\Rightarrow \frac{1}{(2n-2)} = \frac{1}{(2n-1)(2n-2)} + \frac{1}{6}$$

$$\Rightarrow \frac{1}{(n-1)} = \frac{1}{(2n-1)(n-1)} + \frac{1}{3}$$

$$\Rightarrow 3(2n-1) = 3 + (2n-1)(n-1)$$

$$\Rightarrow 6n-3 = 3+2n^2-2n-n+1$$

$$\Rightarrow 2n^2-9n+7=0$$

Q11. Find the coefficient of x^4 in the expansion of $(1+x+x^2+x^3)^{11}$.

Solution:

The given expansion can be written as

$= (1+x+x^2+x^3)^{11} = [(1+x)+x^2(1+x)]^{11}$

$= [(1+x)(1+x^2)]^{11} = (1+x)^{11}\cdot(1+x^2)^{11}$

Using the binomial theorem,

$= ({}^{11}C_0 + {}^{11}C_1x + {}^{11}C_2x^2 + {}^{11}C_3x^3 + {}^{11}C_4x^4 + ...)$

$\times({}^{11}C_0 + {}^{11}C_1x^2 + {}^{11}C_2x^4 + ...)$

$= (1+11x+55x^2+165x^3+330x^4+...)$

$\times(1+11x^2+55x^4+...)$

Thus, the coefficient of x^4 is

$= 55+55\cdot 11+330 = 990$

Q12. If p is a real number and the middle term in the expansion of $\left(\frac{p}{2}+2\right)^8$ is 1120, then find the value of p.

Solution:

The power of the given expansion $\left(\frac{p}{2}+2\right)^8$ is $n=8$ (even); therefore, it has only one middle term, namely, the $\left(\frac{8}{2}+1\right)$th $= 5^{\text{th}}$ term

Thus, $T_5 = T_{4+1} = {}^8C_4\left(\frac{P}{2}\right)^{8-4}\cdot 2^4 = 1120$

$$\Rightarrow {}^8C_4p^4\cdot\frac{1}{2^4}2^4 = 1120$$

$$\Rightarrow \frac{8\times7\times6\times5\times4!}{4!\times4\times3\times2\times1}p^4 = 1120$$

$$\Rightarrow p^4 = \frac{1120}{70} = 16 \Rightarrow p^4 = 2^4$$

$$\Rightarrow p^2 = 4 \Rightarrow p = \pm 2$$

Q13. Show that the middle term in the expansion of $\left(x-\frac{1}{x}\right)^{2n}$ is $\frac{1\times3\times5\times........\times(2n-1)}{n!}(-2)^n$.

Solution:

The binomial expansion $\left(x-\frac{1}{x}\right)^{2n}$ has even power. Therefore, it has only one middle term, namely, the $\left(\frac{2n}{2}+1\right)^{\text{th}}$ term $= (n+1)^{\text{th}}$ term

Thus, $T_{n+1} = {}^{2n}C_n(x)^{2n-n}\left(-\frac{1}{x}\right)^n = {}^{2n}C_nx^n(-1)^nx^{-n}$

$$= {}^{2n}C_n(-1)^n = \frac{(2n)!}{n!n!}(-1)^n$$

$$= \frac{1\cdot2\cdot3\cdot4\cdot5....(2n-1)(2n)}{n!n!}(-1)^n$$

$$= \frac{1\cdot3\cdot5....(2n-1)\cdot2\cdot4\cdot6.....(2n)}{n!n!}(-1)^n$$

$$= \frac{1\cdot3\cdot5..(2n-1)\cdot2^n(1\cdot2\cdot3...n)(-1)^n}{n!n!}$$

$$= \frac{1\cdot3\cdot5..(2n-1)\cdot2^nn!(-1)^n}{n!n!}$$

$$= \frac{1\cdot3\cdot5..(2n-1)}{n!}2^n(-1)^n = \frac{1\cdot3\cdot5..(2n-1)}{n!}(-2)^n$$

Q14. Find n in the binomial $\left(\sqrt[3]{2}+\frac{1}{\sqrt[3]{3}}\right)^n$, if the ratio of the 7th term from the beginning to the 7th from the end is 1/6.

Solution:

The given binomial expansion is $\left(\sqrt[3]{2}+\frac{1}{\sqrt[3]{3}}\right)^n$; therefore, the 7th term from the beginning is

$$T_{7,B} = T_{6+1} = {}^nC_6(\sqrt[3]{2})^{n-6}\left(\frac{1}{\sqrt[3]{3}}\right)^6$$

Similarly, the 7th term from the end is

$$T_{7,E} = {}^nC_{n-6}(\sqrt[3]{2})^6\left(\frac{1}{\sqrt[3]{3}}\right)^{n-6}$$

Thus, $\dfrac{{}^nC_6(\sqrt[3]{2})^{n-6}\left(\dfrac{1}{\sqrt[3]{3}}\right)^6}{{}^nC_{n-6}(\sqrt[3]{2})^6\left(\dfrac{1}{\sqrt[3]{3}}\right)^{n-6}} = \dfrac{1}{6}$

$\Rightarrow \dfrac{(\sqrt[3]{2})^{n-6}\left(\dfrac{1}{\sqrt[3]{3}}\right)^6}{(\sqrt[3]{2})^6\left(\dfrac{1}{\sqrt[3]{3}}\right)^{n-6}} = \dfrac{1}{6} \quad (\because {}^nC_6 = {}^nC_{n-6})$

$\Rightarrow \dfrac{2^{\frac{n-6}{3}}\cdot 3^{-6/3}}{2^{6/3}\cdot 3^{-\left(\frac{n-6}{3}\right)}} = \dfrac{1}{6}$

$\Rightarrow \left(2^{\frac{n-6}{3}-\frac{6}{3}}\right)\left(3^{\frac{-6}{3}+\frac{(n-6)}{3}}\right) = 6^{-1}$

$\Rightarrow \left(2^{\frac{n-12}{3}}\right)\cdot\left(3^{\frac{n-12}{3}}\right) = 6^{-1} \Rightarrow (6)^{\frac{n-12}{3}} = 6^{-1}$

$\Rightarrow \dfrac{n-12}{3} = -1 \Rightarrow n-12 = -3 \Rightarrow n = 9$

Q15. In the expansion of $(x+a)^n$, if the sum of the odd terms is denoted by O and the sum of even terms by E, then, prove that
(i) $O^2 - E^2 = (x^2 - a^2)^n$
(ii) $4OE = (x+a)^{2n} - (x-a)^{2n}$

Solution:

(i)
$(x+a)^n = {}^nC_0x^n + {}^nC_1x^{n-1}a^1 + \ldots + {}^nC_{n-1}x^1a^{n-1} + {}^nC_na^n$

Since the sum of odd terms

$O = {}^nC_0x^n + {}^nC_2x^{n-2}a^2 + \ldots$

and the sum of even terms

$E = {}^nC_1x^{n-1}a + {}^nC_3x^{n-3}a^3 + \ldots$

$\therefore (x+a)^n = O + E$

Now, replacing $a \to -1$, we get

$(x-a)^n = {}^nC_0x^n - {}^nC_1x^{n-1}a^1 + \ldots + {}^nC_{n-1}x^1a^{n-1} + {}^nC_na^n$

$\therefore (x-a)^n = O - E$

Thus, multiplying the above two equations, we get

$(O+E)(O-E) = (x+a)^n(x-a)^n$

$\Rightarrow O^2 - E^2 = (x^2-a^2)^n$

(ii) Now, $4OE = (O+E)^2 - (O-E)^2$

$= [(x+a)^n]^2 - [(x-a)^n]^2 = (x+a)^{2n} - (x-a)^{2n}$

Q16. If x^p occurs in the expansion of $\left(x^2+\dfrac{1}{x}\right)^{2n}$, then prove that its coefficient is $\dfrac{(2n)!}{\left(\dfrac{4n-p}{3}\right)!\left(\dfrac{2n+p}{3}\right)!}$.

Solution:

The general term of the given expansion $\left(x^2+\dfrac{1}{x}\right)^{2n}$ is

$T_{r+1} = {}^{2n}C_r(x^2)^{2n-r}\left(\dfrac{1}{x}\right)^r = {}^{2n}C_rx^{4n-2r}x^{-r} = {}^{2n}C_rx^{4n-3r}$

For the coefficient of x^p, $4n-3r = p \Rightarrow r = \dfrac{4n-p}{3}$

$\therefore$ Coefficient of x^p is $= {}^{2n}C_r = \dfrac{(2n)!}{r!(2n-r)!}$

$= \dfrac{2n!}{\left(\dfrac{4n-p}{3}\right)!\left(2n-\dfrac{4n-p}{3}\right)!}$

$= \dfrac{2n!}{\left(\dfrac{4n-p}{3}\right)!\left(\dfrac{6n-4n+p}{3}\right)!}$

$= \dfrac{(2n)!}{\left(\dfrac{4n-p}{3}\right)!\left(\dfrac{2n+p}{3}\right)!}$

Day 110

NCERT Exemplar Problems: Solutions (Q17–Q40)

Q17. Find the term independent of x in the expansion of $(1+x+2x^3)\left(\dfrac{3}{2}x^2-\dfrac{1}{3x}\right)^9$.

Solution:

The general term of the expansion $\left(\dfrac{3}{2}x^2-\dfrac{1}{3x}\right)^9$ is

$T_{r+1} = {}^9C_r\left(\dfrac{3}{2}x^2\right)^{9-r}\left(-\dfrac{1}{3x}\right)^r$

$= {}^9C_r\left(\dfrac{3}{2}\right)^{9-r}x^{18-2r}\left(-\dfrac{1}{3}\right)^r x^{-r}$

$= {}^9C_r\left(\dfrac{3}{2}\right)^{9-r}\left(-\dfrac{1}{3}\right)^r x^{18-3r}$

Therefore, the general term in the expansion $(1+x+2x^3)\left(\dfrac{3}{2}x^2-\dfrac{1}{3x}\right)^9$ is

$= {}^9C_r\left(\frac{3}{2}\right)^{9-r}\left(-\frac{1}{3}\right)^r x^{18-3r} + {}^9C_r\left(\frac{3}{2}\right)^{9-r}\left(-\frac{1}{3}\right)^r x^{19-3r}$

$+2\cdot{}^9C_r\left(\frac{3}{2}\right)^{9-r}\left(-\frac{1}{3}\right)^r x^{21-3r}$

For the term independent of x, $18-3r=0$, $19-3r=0$, $21-3r=0$, we get $r=6, r=19/3, r=7$

Now, the possible values of r are 6 and 7, and $r=19/3$ is not independent of x.

Therefore, the term independent of x is

$${}^9C_6\left(\frac{3}{2}\right)^{9-6}\left(-\frac{1}{3}\right)^6 + 2\cdot{}^9C_7\left(\frac{3}{2}\right)^{9-7}\left(-\frac{1}{3}\right)^7$$

$$= \frac{9\times8\times7\times6!}{6!\times3\times2}\cdot\frac{3^3}{2^3}\cdot\frac{1}{3^6} - 2\cdot\frac{9\times8\times7!}{7!\times2\times1}\cdot\frac{3^2}{2^2}\cdot\frac{1}{3^7} = \frac{17}{54}$$

Q18. The total number of terms in the expansion of $(x+a)^{100}+(x-a)^{100}$ after simplification is

(a) 50 (b) 202
(c) 51 (d) None of these

Solution: (c)

Fifty terms of $(x+a)^{100}$ cancel out with 50 terms of $(x-a)^{100}$ and 51 terms of $(x+a)^{100}$ get added to the 51 terms of $(x-a)^{100}$, like

$(x+a)^{100}+(x-a)^{100}$

$= {}^{100}C_0x^{100} + {}^{100}C_1x^{99}a + \ldots\ldots + {}^{100}C_{100}a^{100} +$

${}^{100}C_0x^{100} - {}^{100}C_1x^{99}a + \ldots\ldots. + {}^{100}C_{100}a^{100}$

$= 2\underbrace{[{}^{100}C_0x^{100} + {}^{100}C_2x^{98}a^2 + \ldots. + {}^{100}C_{100}a^{100}]}_{51\text{terms}}$

Thus, in the expansion of $(x+a)^n+(x-a)^n$ the total number of terms will be $\frac{n}{2}+1$ when n is even.

Q19. If the integer $r>1, y>2$ and coefficients of the $(3r)^{\text{th}}$ and $(r+2)^{\text{th}}$ terms in the binomial expansion of $(1+x)^{2n}$ are equal, then

(a) $n=2r$ (b) $n=3r$
(c) $n=2r+1$ (d) None of these

Solution: (a)

The coefficients of $(3r)^{\text{th}}$ and $(r+2)^{\text{th}}$ terms are equal in the expansion of $(1+x)^{2n}$; therefore,

$T_{3r}=T_{r+2} \Rightarrow T_{3r-1+1}=T_{r+1+1}$

$\Rightarrow {}^{2n}C_{3r-1}x^{3r-1} = {}^{2n}C_{r+1}x^{r+1} \Rightarrow {}^{2n}C_{3r-1} = {}^{2n}C_{r+1}$

$\Rightarrow 3r-1+r+1=2n$ ($\because {}^nC_x={}^nC_y \Rightarrow x+y=n$)

$\Rightarrow 4r=2n \Rightarrow n=2r$

Q20. The two successive terms in the expansion of $(1+x)^{24}$ whose coefficients are in the ratio 1:4 are

(a) 3rd and 4th (b) 4th and 5th
(c) 5th and 6th (d) 6th and 7th

Solution: (c)

Let two successive terms in the expansion of $(1+x)^{24}$ be $(r+1)^{\text{th}}$ and $(r+2)^{\text{th}}$ terms. Then

$\therefore T_{r+1}={}^{24}C_rx^r,\ T_{r+2}={}^{24}C_{r+1}x^{r+1}$

Since it is given that ratio of two successive terms in the expansion of $(1+x)^{24}$ is 1:4,

$$\therefore \frac{{}^{24}C_r}{{}^{24}C_{r+1}}=\frac{1}{4}\Rightarrow \frac{\frac{(24)!}{r!(24-r)!}}{\frac{(24)!}{(r+1)!(24-r-1)!}}=\frac{1}{4}$$

$$\Rightarrow \frac{(r+1)!(23-r)!}{r!(24-r)!}=\frac{1}{4}\Rightarrow \frac{(r+1)r!(23-r)!}{r!(24-r)23-r)!}=\frac{1}{4}$$

$$\Rightarrow \frac{r+1}{24-r}=\frac{1}{4}\Rightarrow 4r+4=24-r\Rightarrow 5r=20\Rightarrow r=4$$

$\therefore T_{4+1}=T_5,\ T_{4+2}=T_6$

Thus, the ratio of the 5th and 6th terms is 1:4.

Q21. The coefficients of x^n in the expansion of $(1+x)^{2n},(1+x)^{2n-1}$ are in the ratio

(a) 1:2 (b) 1:3 (c) 3:1 (d) 2:1

Solution: (d)

The coefficient of x^n in the expansion of $(1+x)^{2n}={}^{2n}C_n$.

And the coefficient of x^n in the expansion of $(1+x)^{2n-1}={}^{2n-1}C_n$.

$$\because \frac{{}^{2n}C_n}{{}^{2n-1}C_n}=\frac{\frac{(2n)!}{n!n!}}{\frac{(2n-1)!}{n!(n-1)!}}=\frac{(2n)!n!(n-1)!}{n!n!(2n-1)!}$$

$$=\frac{2n(2n-1)!(n-1)!}{n(n-1)!(2n-1)!}=\frac{2n}{n}=\frac{2}{1}=2:1$$

Q22. If the coefficients of the 2nd, 3rd and 4th terms in the expansion of $(1+x)^n$ are in AP, then the value of n is,

(a) 2 (b) 7 (c) 11 (d) 14

Solution: (b)

The expansion of $(1+x)^n$ is

${}^nC_0+{}^nC_1x+{}^nC_2x^2+{}^nC_3x^3+\ldots.+{}^nC_nx^n$

$\therefore$ Coefficient of 2nd term $={}^nC_1$

Coefficient of 3rd term $={}^nC_2$

Coefficient of 4th term $={}^nC_3$

${}^nC_1, {}^nC_2, {}^nC_3$ are in AP, $\therefore 2{}^nC_2={}^nC_1+{}^nC_3$

$$\Rightarrow 2\left(\frac{(n)!}{(n-2)!2!}\right)=\frac{(n)!}{(n-1)!}+\frac{(n)!}{3!(n-3)!}$$

$$\Rightarrow \frac{n(n-1)(n-2)!}{(n-2)!}=\frac{n(n-1)!}{(n-1)!}+\frac{n(n-1)(n-2)(n-3)!}{6(n-3)!}$$

$$\Rightarrow n(n-1)=n+\frac{n(n-1)(n-2)}{6}$$

$$\Rightarrow 6(n-1)=6+(n-1)(n-2)$$

$$\Rightarrow 6n-6=6+n^2-3n+2 \Rightarrow n^2-9n+14=0$$

$$\Rightarrow (n-7)(n-2)=0 \Rightarrow n=2 \text{ or } n=7$$

Since $n = 2$ is not possible, therefore $n = 7$.

Q23. If A and B are coefficients of x^n in the expansion of $(1+x)^{2n}$ and $(1+x)^{2n-1}$, respectively, then A/B is equal to

(a) 1 (b) 2 (c) 1/2 (d) $1/n$

Solution: (b)

The coefficient of x^n in the expansion of $(1+x)^{2n}$ is $A = {}^{2n}C_n$.

The coefficient of x^n in the expansion of $(1+x)^{2n-1}$ is $B = {}^{2n-1}C_n$.

$$\text{Thus, } \frac{A}{B}=\frac{{}^{2n}C_n}{{}^{2n-1}C_n}=\frac{\frac{(2n)!}{n!n!}}{\frac{(2n-1)!}{n!(n-1)!}}=\frac{(2n)!n!(n-1)!}{n!n!(2n-1)!}$$

$$=\frac{2n(2n-1)!(n-1)!}{n(n-1)!(2n-1)!}=\frac{2n}{n}=\frac{2}{1}=2:1$$

Q24. If the middle term of $\left(\frac{1}{x}+x\sin x\right)^{10}$ is equal to $7\frac{7}{8}$, then the value of x is

(a) $2n\pi+\frac{\pi}{6}$ (b) $n\pi+\frac{\pi}{6}$

(c) $n\pi+(-1)^n\frac{\pi}{6}$ (d) $n\pi+(-1)^n\frac{\pi}{3}$

Solution: (c)

In the expansion of $\left(\frac{1}{x}+x\sin x\right)^{10}$, $n=10$ (even); therefore, it has only one middle term, namely, the 6th term.

$$\therefore T_6 = T_{5+1} = {}^{10}C_5\left(\frac{1}{x}\right)^{10-5}(x\sin x)^5$$

$$\Rightarrow {}^{10}C_5 x^{-5}x^5\sin^5 x = 7\frac{7}{8}$$

$$\Rightarrow \frac{10\cdot9\cdot8\cdot7\cdot6\cdot5!}{5\cdot4\cdot3\cdot2\cdot1\cdot5!}\sin^5 x=\frac{63}{8}$$

$$\Rightarrow \sin^5 x=\frac{1}{32}=\left(\frac{1}{2}\right)^5$$

$$\Rightarrow \sin x=\frac{1}{2}\Rightarrow x=n\pi+(-1)^n\frac{\pi}{6}$$

Q25. The largest coefficient in the expansion of $(1+x)^{30}$ is ______

Solution:

The middle term/s will have the largest coefficient/s in the expansion. Therefore, in the expansion of $(1+x)^{30}$, the largest coefficient is $= {}^{30}C_{30/2} = {}^{30}C_{15}$.

Q26. The number of terms in the expansion of $(x+y+z)^n$ ______

Solution:

$$(x+y+z)^n=[x+(y+z)]^n={}^nC_0x^n+{}^nC_1x^{n-1}(y+z)$$
$$+{}^nC_2x^{n-2}(y+z)^2+\dots+{}^nC_n(y+z)^n$$

The number of terms in $(x+y)^n$ is $n+1$; therefore, the number of terms in $(x+y+z)^n$ is

$$=1+2+3+\dots+n+(n+1)=\frac{(n+1)(n+2)}{2}.$$

Q27. In the expansion of $\left(x^2-\frac{1}{x^2}\right)^{16}$, the value of the constant term is______

Solution:

Let r^{th} term be the constant term. Then

$$T_{r+1}={}^{16}C_r(x^2)^{16-r}\left(-\frac{1}{x^2}\right)^r={}^{16}C_rx^{32-2r}(-1)^rx^{-2r}$$

$$=(-1)^r\,{}^{16}C_r\,x^{32-4r}$$

For the constant term, $32-4r=0\Rightarrow r=8$

$$\therefore T_{8+1}={}^{16}C_8$$

Q28. If the seventh term from the beginning and the end in the expansion of $\left(\sqrt[3]{2}+\frac{1}{\sqrt[3]{3}}\right)^n$ are equal, then n is equal to______

Solution:

Since the seventh term from the beginning and from the end in the expansion of $\left(\sqrt[3]{2}+\frac{1}{\sqrt[3]{3}}\right)^n$ are equal

$${}^nC_6(\sqrt[3]{2})^{n-6}\left(\frac{1}{\sqrt[3]{3}}\right)^6={}^nC_{n-6}(\sqrt[3]{2})^6\left(\frac{1}{\sqrt[3]{3}}\right)^{n-6}$$

$$\Rightarrow (2)^{\frac{n-6}{3}}(3)^{-6/3}=2^{6/3}(3)^{-\frac{(n-6)}{3}}\quad(\because {}^nC_6={}^nC_{n-6})$$

$$\Rightarrow (2)^{\frac{n-12}{3}}=\left(\frac{1}{3}\right)^{\frac{n-12}{3}}$$

This is possible only when, $\frac{n-12}{3}=0\Rightarrow n=12$

Q29. The coefficient of $a^{-6}b^4$ in the expansion of $\left(\frac{1}{a}-\frac{2b}{3}\right)^{10}$ is______

Solution:

Let the rth term have a coefficient $a^{-6}b^4$; then $T_{r+1} = {}^{10}C_r\left(\frac{1}{a}\right)^{10-r}\left(-\frac{2b}{3}\right)^r$. Thus, for the coefficient of $a^{-6}b^4$, $10-r=6 \Rightarrow r=4$.

Therefore, coefficient of $a^{-6}b^4 = {}^{10}C_4(-2/3)^4 = \frac{1120}{27}$

Q30. The middle term in the expansion of $(a^3+ba)^{28}$ is______

Solution:

The middle term of the expansion $(a^3+ba)^{28}$ is

$=\left(\frac{28}{2}+1\right)$th term $=15^{\text{th}}$ term

$\therefore T_{15} = T_{14+1} = {}^{28}C_{14}(a^3)^{28-14}(ba)^{14}$

$= {}^{28}C_{14}a^{42}b^{14}a^{14} = {}^{28}C_{14}a^{56}b^{14}$

Q31. The ratio of the coefficient of x^p, x^q in the expansion of $(1+x)^{p+q}$ is______

Solution:

In the expansion of $(1+x)^{p+q}$,

Coefficient of $x^p = {}^{p+q}C_p$

Coefficient of $x^q = {}^{p+q}C_q$

$\therefore \frac{{}^{p+q}C_p}{{}^{p+q}C_q} = \frac{{}^{p+q}C_p}{{}^{p+q}C_p} = 1:1$

Q32. The position of the term independent of x in the expansion of $\left(\sqrt{\frac{x}{3}}+\frac{3}{2x^2}\right)^{10}$ is______

Solution:

Let the constant term in the expansion of $\left(\sqrt{\frac{x}{3}}+\frac{3}{2x^2}\right)^{10}$ be,

$$T_{r+1} = {}^{10}C_r\left(\sqrt{\frac{x}{3}}\right)^{10-r}\left(\frac{3}{2x^2}\right)^r$$

$$= {}^{10}C_r \cdot x^{\frac{10-r}{2}} \cdot 3^{\frac{-10+r}{2}} \cdot 3^r \cdot 2^{-r} \cdot x^{-2r}$$

$$= {}^{10}C_r 3^{\frac{-10+3r}{2}} 2^{-r} x^{\frac{10-5r}{2}}$$

For constant term, $10-5r=0 \Rightarrow r=2$

Thus, the third term is independent of x.

Q33. If 25^{15} is divided by 13, then the remainder is______

Solution:

$25^{15} = (26-1)^{15}$

$= {}^{15}C_0 26^{15} - {}^{15}C_1 26^{14} + ... - {}^{15}C_{15}$

$= {}^{15}C_0 26^{15} - {}^{15}C_1 26^{14} + ... -1-13+13$

$= {}^{15}C_0 26^{15} - {}^{15}C_1 26^{14} + ... -13+12$

Therefore, when 25^{15} is divided by 13, the remainder will be 12.

Q34. The sum of the series $\sum_{r=0}^{10} {}^{20}C_r$ is $2^{19} + \frac{{}^{20}C_{10}}{2}$.

Solution: True

$\sum_{r=0}^{10} {}^{20}C_r = {}^{20}C_0 + {}^{20}C_1 + {}^{20}C_2 + ... + {}^{20}C_{10}$

$= \frac{1}{2}\left({}^{20}C_0 + {}^{20}C_1 + ... + {}^{20}C_{10} + {}^{20}C_{10} + {}^{20}C_{11} + ... {}^{20}C_{20}\right)$

$(\because {}^nC_r = {}^nC_{n-r})$

$= \frac{2^{20}}{2} + \frac{{}^{20}C_{10}}{2} = 2^{19} + \frac{{}^{20}C_{10}}{2}$

$(\because {}^nC_0 + {}^nC_1 + ... + {}^nC_n = 2^n)$

Q35. The expression $7^9 + 9^7$ is divisible by 64.

Solution: True

$7^9 + 9^7 = (8-1)^9 + (1+8)^7 = (1+8)^7 - (1-8)^9$

$= ({}^7C_0 + {}^7C_1 8 + {}^7C_2 8^2 + {}^7C_3 8^3 ... + {}^7C_7 8^7)$

$-({}^9C_0 - {}^9C_1 8 + {}^9C_2 8^2 - {}^9C_3 8^3 ... - {}^9C_9 8^9)$

$= {}^7C_0 + {}^7C_1 8 + {}^7C_2 8^2 - {}^9C_0 + {}^9C_1 8 - {}^9C_2 8^2 + 64k$

where, $k = ({}^7C_3 8^3 ... + {}^7C_7 8^7) - (-{}^9C_3 8^3 ... - {}^9C_9 8^9)$

$\Rightarrow 7^9 + 9^7 = 1 + 7\cdot 8 + 21\cdot 64 - 1 + 9\cdot 8 - 36\cdot 64 + 64k$

$= 16\cdot 8 - 15\cdot 64 + 64k = 64(2-15+k) = 64k'$

which is divisible by 64.

Q36. The number of terms in the expansion of $[(2x+y^3)^4]^7$ is 8.

Solution: False

The given expansion is $[(2x+y^3)^4]^7 = (2x+y^3)^{28}$, therefore, the number of terms is 29.

Q37. The sum of the coefficients of the two middle terms in the expansion of $(1+x)^{2n-1}$ is equal to ${}^{2n-1}C_n$.

Solution: False

The expansion of $(1+x)^{2n-1}$ has two middle terms; therefore $\left(\frac{2n-1+1}{2}\right)$ and $\left(\frac{2n-1+1}{2}+1\right)$, that is, the nth term and $(n+1)^{th}$ term, are the middle terms.

Thus, the coefficient of the nth term = ${}^{2n-1}C_{n-1}$

Coefficient of the $(n+1)^{th}$ term = ${}^{2n-1}C_n$

Hence, sum of coefficients = ${}^{2n-1}C_{n-1} + {}^{2n-1}C_n$

$= {}^{2n-1+1}C_n = {}^{2n}C_n \quad (\because {}^nC_r + {}^nC_{r-1} = {}^{n+1}C_r)$

Q38. The last two digits of the numbers 3^{400} are 01.

Solution: True

$3^{400} = 9^{200} = (10-1)^{200}$

$= {}^{200}C_0 10^{200} - {}^{200}C_1 10^{199} + ... - {}^{200}C_{199} 10^1 + {}^{200}C_{200} 1^{200}$

$= 10^{200} - 200\times 10^{199} + ... - 10\times 200 + 1$

$= 10^{200} - 200\times 10^{199} + ... - 2000 + 1$

$= 10^{200} - 200\times 10^{199} + ... - 1999$

$= 10^{200} - 200\times 10^{199} + ... + 01$

Thus, the last two digits are 01.

Q39. If the expansion of $\left(x-\frac{1}{x^2}\right)^{2n}$ contains a term independent of x, then n is a multiple of 2.

Solution: False

Let the independent term of x in the binomial expansion $\left(x-\frac{1}{x^2}\right)^{2n}$ be $T_{r+1} = {}^{20}C_r(x)^{2n-r}\left(-\frac{1}{x^2}\right)^r$

$= {}^{2n}C_r x^{2n-r}(-1)^r x^{-2r} = {}^{2n}C_r x^{2n-3r}(-1)^r$

For independent of x, $2n-3r=0 \Rightarrow r=\frac{2n}{3}$ which is not an integer, n is a multiple of 2; therefore the term independent of x is not possible.

Q40. The number of terms in the expansion of $(a+b)^n$, where $n\in N$, is one less than the power n.

Solution: False

The number of terms in the expansion of $(a+b)^n$, where $n\in N$, is one more than the power n.

Day 111

Very Short Answer (VSA): 1-Mark and Short Answer (SA): 2-Mark Questions

Q1. The largest coefficient in the expansion of $(1+x)^{20}$ is______

Solution:

The middle term/s will have the largest coefficient/s in the expansion; therefore in the expansion of $(1+x)^{20}$ largest coefficient is = ${}^{20}C_{20/2} = {}^{20}C_{10}$.

Q2. The number of terms in the expansion of $(x+y+z)^9$ ______

Solution:

The number of terms in the expansion of $(x+y+z)^n$ is $=\frac{(n+1)(n+2)}{2}$; therefore the number of terms in the expansion of $(x+y+z)^9$ is $=\frac{(9+1)(9+2)}{2} = 55$.

Q3. In the expansion of $\left(x^3-\frac{1}{x}\right)^{16}$, the value of the constant term is______.

Solution:

Let the rth term be the constant term. Then

$T_{r+1} = {}^{16}C_r(x^3)^{16-r}\left(-\frac{1}{x}\right)^r = (-1)^r\,{}^{16}C_r x^{48-3r-r}$

$= (-1)^r\,{}^{16}C_r x^{48-4r}$

For the constant term, $48-4r=0 \Rightarrow r=12$; therefore, the value of constant term is

$T_{12+1} = (-1)^{12}\,{}^{16}C_{12} = {}^{16}C_{12}$

Q4. If the seventh term from the beginning and the end in the expansion of $\left(\sqrt[3]{2}+\sqrt[3]{3}\right)^n$ are equal, then n is equal to______

Solution:

The seventh term from the beginning and from the end in the expansion of $\left(\sqrt[3]{2}+\sqrt[3]{3}\right)^n$ are equal; therefore,

${}^nC_6(\sqrt[3]{2})^{n-6}\left(\sqrt[3]{3}\right)^6 = {}^nC_{n-6}(\sqrt[3]{2})^6\left(\sqrt[3]{3}\right)^{n-6}$

$\Rightarrow (2)^{\frac{n-6}{2}}(3)^{6/2} = 2^{6/2}(3)^{\frac{n-6}{2}} \quad (\because {}^nC_6 = {}^nC_{n-6})$

$\Rightarrow (2)^{\frac{n-6}{2}-3} = (3)^{\frac{n-6}{2}-3} \Rightarrow (2)^{\frac{n-12}{2}} = (3)^{\frac{n-12}{2}}$

This is possible only when $\frac{n-12}{2}=0 \Rightarrow n=12$.

Q5. The coefficient of $a^{-5}b^5$ in the expansion of $\left(\frac{2}{a}-\frac{b}{2}\right)^{10}$ is______

Solution:

Let r^{th} term have a coefficient of $a^{-5}b^5$. Then $T_{r+1} = {}^{10}C_r\left(\frac{2}{a}\right)^{10-r}\left(-\frac{b}{2}\right)^r$, thus for the coefficient of $a^{-5}b^5$, $10-r=5 \Rightarrow r=5$.

Therefore, coefficient of $a^{-5}b^5 = {}^{10}C_5(-1)^5 = -{}^{10}C_5$

Q6. The middle term in the expansion of $(a^2b+b^2a)^{18}$ is______

Solution:

The middle term of the expansion of $(a^2b+b^2a)^{18}$ is

$=\left(\frac{18}{2}+1\right)^{\text{th}}$ term $=10^{\text{th}}$ term

$\therefore T_{10} = T_{9+1} = {}^{18}C_9(a^2b)^{18-9}(b^2a)^9 = {}^{18}C_9a^{27}b^{27}$

Q7. The ratio of the coefficient of x^m, x^n in the expansion of $(1+x)^{m+n}$ is______

Solution:

In the expansion of $(1+x)^{m+n}$,

Coefficient of $x^m = {}^{m+n}C_m$

Coefficient of $x^n = {}^{m+n}C_n$

$$\therefore \frac{{}^{m+n}C_m}{{}^{m+n}C_n} = \frac{{}^{m+n}C_m}{{}^{m+n}C_m} = 1:1 \quad \left(\because {}^nC_r = {}^nC_{n-r}\right)$$

Q8. The position of the term independent of x in the expansion of $\left(\frac{\sqrt{x}}{3}+\frac{3}{x}\right)^{15}$ is______

Solution:

Let the constant term in expansion of $\left(\frac{\sqrt{x}}{3}+\frac{3}{x}\right)^{15}$ be

$$T_{r+1} = {}^{15}C_r\left(\frac{\sqrt{x}}{3}\right)^{15-r}\left(\frac{3}{x}\right)^r$$

$$= {}^{15}C_r \cdot x^{\frac{15-r}{2}} \cdot 3^{r-15} \cdot 3^r \cdot x^{-r} = {}^{13}C_r 3^{2r-15} x^{\frac{15-3r}{2}}$$

For constant term, $15-3r=0 \Rightarrow r=5$

Thus, $T_{5+1} = T_6$, sixth term is independent of x.

Q9. The number of terms in the expansion of $[\{(x+y)^2\}^3]^4$ is ______

Solution:

The given expansion is

$[\{(x+y)^2\}^3]^4 = (x+y)^{2.3.4} = (x+y)^{24}$; therefore, the number of terms is 25.

Q10. If 23^{23} is divided by 11, then the remainder is______

Solution:

$23^{23} = (1+22)^{23}$

$= {}^{23}C_0 + {}^{23}C_1 22^1 + {}^{23}C_2 22^2 \ldots {}^{23}C_{23} 22^{23}$

$= 1 + 11\cdot 2\cdot({}^{23}C_1 + {}^{23}C_2 22^1 \ldots {}^{23}C_{23} 22^{22})$

$= 1 + 11k$ {where, $k = 2\cdot({}^{23}C_1 + {}^{23}C_2 22^1 \ldots {}^{23}C_{23} 22^{22})$}

Therefore, when 23^{23} is divided by 11, the remainder will be 1.

Q11. The sum of the series $\sum_{r=0}^{10} {}^{21}C_r$ is ______

Solution:

$$\sum_{r=0}^{10} {}^{21}C_r = {}^{21}C_0 + {}^{21}C_1 + {}^{21}C_2 + \ldots + {}^{21}C_{10}$$

$$= \frac{1}{2}\begin{pmatrix} {}^{21}C_0 + {}^{21}C_1 + {}^{21}C_2 + \ldots + {}^{21}C_{10} \\ + {}^{21}C_0 + {}^{21}C_1 + {}^{21}C_2 + \ldots + {}^{21}C_{10} \end{pmatrix}$$

$$= \frac{1}{2}\begin{pmatrix} {}^{21}C_0 + {}^{21}C_1 + {}^{21}C_2 + \ldots + {}^{21}C_{10} \\ + {}^{21}C_{21} + {}^{21}C_{20} + {}^{21}C_{19} + \ldots + {}^{21}C_{11} \end{pmatrix}$$

$(\because {}^nC_r = {}^nC_{n-r})$

$$= \frac{1}{2}\left(2^{21}\right) = 2^{20} \quad \left(\because {}^nC_0 + {}^nC_1 + \ldots + {}^nC_n = 2^n\right)$$

Q12. The sum of the coefficient of the two middle terms in the expansion of $(1+x)^{2n+1}$ is equal to ...

Solution:

The expansion of $(1+x)^{2n+1}$ has two middle terms; therefore $\left(\frac{2n+1+1}{2}\right)$ and $\left(\frac{2n+1+1}{2}+1\right)$, that is, the $(n+1)^{\text{th}}$ term and $(n+2)^{\text{th}}$ term, are the middle terms.

Thus, the coefficient of the $(n+1)^{\text{th}}$ term $= {}^{2n+1}C_{n+1}$

Coefficient of the $(n+2)^{\text{th}}$ term $= {}^{2n+1}C_{n+2}$

Hence, sum of coefficients $= {}^{2n+1}C_{n+1} + {}^{2n+1}C_{n+2}$

$= {}^{2n+2}C_{n+2} = {}^{2n+2}C_n$

$(\because {}^nC_r + {}^nC_{r-1} = {}^{n+1}C_r$ and ${}^nC_r = {}^nC_{n-r})$

Q13. The last two digits of the number 7^{70} are ______

Solution:

$7^{70} = 49^{35} = (50-1)^{35}$

$= {}^{35}C_0 50^{35} - {}^{35}C_1 50^{34} + \ldots - {}^{35}C_{33} 50^2 + {}^{35}C_{34} 50^1 - {}^{35}C_{35}$

$$= \underbrace{({}^{35}C_0 50^{35} - {}^{35}C_1 50^{34} + \ldots - {}^{35}C_{33} 50^2)}_{\text{last two digits be 00}} + 35\cdot 50 - 1$$

= number with last two digits $00 + 1750 - 1$
= number with last two digits $00 + 1749$
Thus, the last two digits are 49.

Q14. If the expansion of $\left(\sqrt{x}-\frac{1}{x^2}\right)^{2n}$ contains a term independent of x, then n is a multiple of ______

Solution:
Let the independent term of x in the binomial expansion $\left(\sqrt{x}-\frac{1}{x^2}\right)^{2n}$ be

$$T_{r+1} = {}^{2n}C_r(\sqrt{x})^{2n-r}\left(-\frac{1}{x^2}\right)^r$$

$$= {}^{2n}C_r x^{\frac{2n-r}{2}}(-1)^r x^{-2r} = (-1)^r\, {}^{2n}C_r x^{\frac{2n-5r}{2}}$$

For independent of x, $2n-5r=0 \Rightarrow r=\frac{2n}{5}$ which will be an integer, only when n is a multiple of 5.

Q15. The coefficient of x^{25} in the expansion of $(x^3-x^2)^{10}$ is ______

Solution:
The general term of the given expansion is $T_{r+1} = {}^{10}C_r(x^3)^{10-r}(-x^2)^r = (-1)^r \cdot {}^{10}C_r \cdot x^{30-3r+2r}$

$= (-1)^r\, {}^{10}C_r x^{30-r}$

For the coefficient of x^{25}, $30-r=25 \Rightarrow r=5$
Thus, $T_{5+1} = (-1)^5\, {}^{10}C_5 x^{25}$
Hence, the coefficient of x^{25} is

$$= (-1)^5\, {}^{10}C_5 = -1\frac{10\times9\times8\times7\times6\times5!}{5\times4\times3\times2\times1\times5!} = -252$$

Q16. The middle term (terms) in the expansion of $\left(\frac{x}{a}-\frac{a}{x}\right)^8$ is/are ______

Solution:

The power of the binomial expansion $\left(\frac{x}{a}-\frac{a}{x}\right)^8$ is $n=8$ is even.

Therefore, it has one middle term, that is, the $\left(\frac{8}{2}+1\right)$ term = 5th term.

Thus, $T_5 = T_{4+1} = {}^8C_4\left(\frac{x}{a}\right)^{8-4}\left(\frac{-a}{x}\right)^4 = {}^8C_4 = 70$

Q17. The total number of terms in the expansion of $(x+a)^{99}+(x-a)^{99}$ after simplification is
(a) 50 (b) 198
(c) 51 (d) None of these

Solution: (a)
Fifty terms of $(x+a)^{99}$ cancel out with 50 terms of $(x-a)^{99}$ and 50 terms of $(x+a)^{99}$ get added to the 50 terms of $(x-a)^{99}$, like

$(x+a)^{99}+(x-a)^{99}$

$= {}^{99}C_0x^{99} + {}^{99}C_1x^{98}a + {}^{99}C_2x^{97}a^2 + \ldots + {}^{99}C_{98}a^{98}x + {}^{99}C_{99}a^{99}$

$+ {}^{99}C_0x^{99} - {}^{99}C_1x^{98}a + {}^{99}C_2x^{97}a^2 - \ldots + {}^{99}C_{98}a^{98}x - {}^{99}C_{99}a^{99}$

$= \underbrace{2[{}^{99}C_0x^{99} + {}^{99}C_2x^{97}a^2 \ldots + {}^{99}C_{98}a^{98}x]}_{50\text{ terms}}$

Thus, in the expansion of $(x+a)^n+(x-a)^n$, the total number of terms will be $\frac{n+1}{2}$ when n is odd.

Q18. If the integer $r>1, y>2$ and the coefficients of the $(3r-2)^{th}$ and $(r+2)^{th}$ terms in the binomial expansion of $(1+x)^{2n}$ are equal, then
(a) $n=2r-1$ (b) $n=3r$
(c) $n=2r+1$ (d) None of these

Solution: (a)
The coefficients of the $(3r-2)^{th}$ and $(r+2)^{th}$ terms are equal in the expansion of $(1+x)^{2n}$; therefore $T_{3r-2} = T_{r+2} \Rightarrow T_{3r-3+1} = T_{r+1+1}$

$\Rightarrow {}^{2n}C_{3r-3}x^{3r-3} = {}^{2n}C_{r+1}x^{r+1} \Rightarrow {}^{2n}C_{3r-3} = {}^{2n}C_{r+1}$

$\Rightarrow 3r-3+r+1 = 2n$ $(\because {}^nC_x = {}^nC_y \Rightarrow x+y=n)$

$\Rightarrow 4r = 2n+2 \Rightarrow n = 2r-1$

Q19. The two successive terms in the expansion of $(1+x)^{15}$ whose coefficients are in ratio 1:3 are
(a) 3rd and 4th (b) 4th and 5th
(c) 5th and 6th (d) 6th and 7th

Solution: (b)
Let two successive terms in the expansion of $(1+x)^{15}$ be the $(r+1)^{th}$ and $(r+2)^{th}$ terms.

$\therefore T_{r+1} = {}^{15}C_rx^r,\ T_{r+2} = {}^{15}C_{r+1}x^{r+1}$

It is given that the ratio of two successive terms in the expansion of $(1+x)^{24}$ is 1:3

$$\therefore \frac{{}^{15}C_r}{{}^{15}C_{r+1}} = \frac{1}{3} \Rightarrow \frac{\dfrac{(15)!}{r!(15-r)!}}{\dfrac{(15)!}{(r+1)!(15-r-1)!}} = \frac{1}{4}$$

$$\Rightarrow \frac{r+1}{15-r} = \frac{1}{3} \Rightarrow 3r+3 = 15-r \Rightarrow 4r = 12 \Rightarrow r = 3$$

$\therefore T_{3+1} = T_4,\ T_{3+2} = T_5$

Thus, the ratio of the 4th and 5th terms is 1:3.

Q20. The coefficient of x^n in the expansion of $(1+x)^{2n+1}, (1+x)^{2n-1}$ are in the ratio

(a) $\dfrac{2(2n+1)}{(n+1)}$ (b) $\dfrac{(2n+1)}{(n+1)}$

(c) $\dfrac{2(2n+1)}{(2n-1)}$ (d) $\dfrac{(2n+1)}{(2n-1)}$

Solution: (a)

The coefficient of x^n in the expansion of $(1+x)^{2n+1} = {}^{2n+1}C_n$.

And the coefficient of x^n in the expansion of $(1+x)^{2n-1} = {}^{2n-1}C_n$.

$$\therefore \frac{{}^{2n+1}C_n}{{}^{2n-1}C_n} = \frac{\dfrac{(2n+1)!}{n!(n+1)!}}{\dfrac{(2n-1)!}{n!(n-1)!}} = \frac{(2n+1)(2n)}{n(n+1)} = \frac{2(2n+1)}{(n+1)}$$

Q21. If the sum of the coefficients of the 1st and 4th terms is equal to sum of coefficient of 2nd and 3rd terms in the expansion of $(1+x)^n$, then the value of n is

(a) 2 (b) 6 (c) 11 (d) 14

Solution: (b)

The expansion of $(1+x)^n$ is

${}^nC_0 + {}^nC_1x + {}^nC_2x^2 + {}^nC_3x^3 + ... + {}^nC_nx^n$

$\therefore$ Coefficient of 1st term = ${}^nC_0$

Coefficient of 2nd term = ${}^nC_1$

Coefficient of 3rd term = ${}^nC_2$

Coefficient of 4th term = ${}^nC_3$

The sum of the coefficients of the 1st and 4th terms is equal to the sum of coefficients of the 2nd and 3rd terms. Therefore

${}^nC_0 + {}^nC_3 = {}^nC_1 + {}^nC_2$

$\Rightarrow 1 + \dfrac{(n)!}{3!(n-3)!} = \dfrac{(n)!}{(n-1)!} + \dfrac{(n)!}{(n-2)!2!}$

$\Rightarrow 1 + \dfrac{n(n-1)(n-2)}{6} = n + \dfrac{n(n-1)}{2}$

$\Rightarrow 6 + n(n-1)(n-2) = 6n + 3n(n-1)$

$\Rightarrow n(n-1)(n-2) - 3n(n-1) - 6(n-1) = 0$

$\Rightarrow (n-1)(n^2 - 2n - 3n - 6) = 0$

$\Rightarrow (n-1)(n^2 - 5n - 6) = 0$

$\Rightarrow (n-1)(n+1)(n-6) = 0$

$\Rightarrow n = 1$ or $n = -1$ or $n = 6$

Since $n = 1$ or $n = -1$ is not possible, therefore, $n = 6$.

Q22. If the middle term of $\left(\dfrac{1}{x} - x\cos x\right)^8$ is equal to $4\dfrac{3}{8}$, then the value of x is

(a) $2n\pi \pm \dfrac{\pi}{6}$ (b) $n\pi + \dfrac{\pi}{6}$

(c) $n\pi + (-1)^n \dfrac{\pi}{6}$ (d) $2n\pi \pm \dfrac{\pi}{3}$

Solution: (d)

In the expansion $\left(\dfrac{1}{x} - x\cos x\right)^8$, $n = 8$ (even); therefore it has only one middle term, namely, the 5th term.

$\therefore T_5 = T_{4+1} = {}^8C_4\left(\dfrac{1}{x}\right)^{8-4}(-x\cos x)^4$

$\Rightarrow (-1)^4\,{}^8C_4x^{-4}x^4\cos^4 x = 4\dfrac{3}{8}$

$\Rightarrow \dfrac{8\cdot7\cdot6\cdot5}{4\cdot3\cdot2\cdot1}\cos^4 x = \dfrac{35}{8}$

$\Rightarrow \dfrac{7\cdot2\cdot5}{1}\cos^4 x = \dfrac{35}{8}$

$\Rightarrow \cos^4 x = \dfrac{1}{16} = \left(\dfrac{1}{2}\right)^4$

$\Rightarrow \cos x = \dfrac{1}{2} \Rightarrow x = 2n\pi \pm \dfrac{\pi}{3}$

Q23. The approximation of $(1.01)^5$ using the first three terms of its expansion is

(a) 1.051 (b) 1.271 (c) 1.512 (d) 1.412

Solution: (a)

We can write $1.01 = 1 + 0.01$

$\therefore (1.01)^5 = (1+0.01)^5$

$= {}^5C_0(1)^5 + {}^5C_1(1)^4(0.01) + {}^5C_2(1)^3(0.01)^2 + ...$

(Approximately, neglecting smaller value terms)

$= 1 + 5(0.01) + 10(0.01)^2 = 1 + 0.05 + 0.001 = 1.051$

Thus, the value of $(1.01)^5$ is approximately 1.051.

Q24. Show that the expression $4^{16} - 6^{14}$ is divisible by

(a) 13 (b) 17 (c) 19 (d) 25

Solution: (d)

$4^{16} + 6^{14} = (5-1)^{16} - (1+5)^{14}$

$= ({}^{16}C_0 5^{16} - {}^{16}C_1 5^{15} + ... {}^{16}C_{14} 5^2 - {}^{16}C_{15} 5 + {}^{16}C_{16})$

$-({}^{14}C_0 + {}^{14}C_1 5 + {}^{14}C_2 5^2 + {}^{14}C_{13} 5^{13} + {}^{14}C_{14} 5^{14})$

$= ({}^{16}C_0 - {}^{16}C_1 5 + {}^{16}C_2 5^2 - ... - {}^{16}C_{15} 5^{15} + {}^{16}C_{16} 5^{16})$

$-({}^{14}C_0 + {}^{14}C_1 5 + {}^{14}C_2 5^2 + ... + {}^{14}C_{13} 5^{13} + {}^{14}C_{14} 5^{14})$

$= ({}^{16}C_0 - {}^{16}C_1 5 - {}^{14}C_0 - {}^{14}C_1 5) + 25k$, where,

$k = ({}^{16}C_2 - ... + {}^{16}C_{16} 5^{14} - {}^{14}C_2 - ... - {}^{14}C_{14} 5^{12})$

$\Rightarrow 4^{16} - 6^{14} = ({}^{16}C_0 - {}^{16}C_1 5 - {}^{14}C_0 - {}^{14}C_1 5) + 25k$

$= 1 - 16\cdot5 - 1 - 14\cdot5) + 25k$

$= -30\cdot5 + 25k = 25(-6+k) = 25k'$

which is divisible by 25.

Q25. The coefficient of x^{10} in the product of $(x^2-x)^4(1+x)^6$ using the binomial theorem is
(a) 2 (b) 6 (c) 11 (d) 14

Solution: (a)
Since the product $(x^2-x)^4(1+x)^6$
$= x^4(x-1)^4(1+x)^6$
$= x^4(-1)^4(1-x)^4(1+x)^6$
$= x^4(1-x^2)^4(1+x)^2$
$= x^4(1-x^2)^4(1+2x+x^2)$
$= (x^4+2x^5+x^6)(1-x^2)^4$
Using the binomial theorem, the expansions of $(1-x^2)^4$ is
$= {}^4C_0 - {}^4C_1(x^2) + {}^4C_2(x^2)^2 - {}^4C_3(x^2)^3 - {}^4C_4(x^2)^4$
$= 1-4x^2+6x^4-4x^6-x^8$
$\therefore (x^2-x)^4(1+x)^6$
$= (x^4+2x^5+x^6)(1-x^2)^4$
$= (x^4+2x^5+x^6)(1-4x^2+6x^4-4x^6-x^8)$
We need the terms which involve x^{10}, therefore complete multiplication of the two brackets is not required.
Thus, the terms containing x^{10} are
$= (x^4)(-4x^6)+(x^6)(6x^4) = 2x^{10}$
Thus, the coefficient of x^{10} in the given product is 2.

Day 112

Long Answer-I (LA-I): 4-Mark and Long Answer-II (LA-II): 6-Mark Questions

Q1. Find a if the coefficients of x^2, x^3 in the expansion of $(a+2x)^9$ are equal.

Solution:
The $(r+1)^{\text{th}}$ term in the binomial expansion of $(a+b)^n$ is $T_{r+1} = {}^nC_r a^{n-r} b^r$.
Let us assume that x^2 occurs in the $(r+1)^{\text{th}}$ term in the expansion of $(a+2x)^9$. Then we get,
$T_{r+1} = {}^9C_r(a)^{9-r}(2x)^r = {}^9C_r(a)^{9-r}2^r x^r$
Comparing the indices of x in x^2 and in (T_{r+1}), we get $r = 2$.
Thus, the coefficient of x^2 is ${}^9C_2(a)^{9-2}2^2 = 4\cdot{}^9C_2a^7$
Similarly, assume that x^3 occurs in the $(k+1)^{\text{th}}$ term in the expansion of $(a+2x)^9$, then we get
$T_{k+1} = {}^9C_k(a)^{9-k}(2x)^k = {}^9C_k(a)^{9-k}2^k x^k$
Comparing the indices of x in x^3 and in (T_{r+1}), we get $k = 3$.
Thus, the coefficient of x^3 is
${}^9C_3(a)^{9-3}2^3 = 8\cdot{}^9C_3a^6$
Since the coefficients of x^2 and x^3 are equal
$4\cdot{}^9C_2a^7 = 8\cdot{}^9C_3a^6$
$$\Rightarrow a = \frac{8\cdot 9\cdot 8\cdot 7}{3\cdot 2}\times\frac{2}{4\cdot 9\cdot 8} \Rightarrow a = \frac{14}{3}$$

Q2. Find the coefficient of x^5 in the product $(1-x)^4(1+2x)^6$ using the binomial theorem.

Solution:
Using the binomial theorem, the expansion of $(1+2x)^6$ is
$(1+2x)^6 = {}^6C_0 + {}^6C_1(2x) + {}^6C_2(2x)^2 + {}^6C_3(2x)^3 + {}^6C_4(2x)^4 + {}^6C_5(2x)^5 + {}^6C_6(2x)^6$
$= 1+12x+60x^2+160x^3+240x^4+192x^5+64x^6$
Using the binomial theorem, the expansion of $(1-x)^4$ is
$(1-x)^4 = {}^4C_0 - {}^4C_1(x) + {}^4C_2(x)^2 - {}^4C_3(x)^3 + {}^4C_4(x)^4$
$= 1-4x+6x^2-4x^3+x^4$
$\therefore (1-x)^4(1+2x)^6 = (1-4x+6x^2-4x^3+x^4)\times(1+12x+60x^2+160x^3+240x^4+192x^5+64x^6)$
We need the terms which involve x^5; therefore complete multiplication of the two brackets is not required.
Thus, the terms containing x^5 are
$= (1-4x+6x^2-4x^3+x^4)\times(1+12x+60x^2+160x^3+240x^4+192x^5+64x^6)$
$= (1)(192x^5)+(-4x)(240x^4)+(6x^2)(160x^3)+(-4x^3)(60x^2)+(x^4)(12x)$
$= (192-960+960-240+12) = -36x^5$
Thus, the coefficient of x^5 in the given product is –36.

Q3. Evaluate $(\sqrt{3}+\sqrt{2})^5+(\sqrt{3}-\sqrt{2})^5$.

Solution:
Using the binomial theorem,
$(a+b)^5 = {}^5C_0a^5 + {}^5C_1a^4b + {}^5C_2a^3b^2 + {}^5C_3a^2b^3 + {}^5C_4ab^4 + {}^5C_5b^5$
$= a^5+5a^4b+10a^3b^2+10a^2b^3+5ab^4+b^5$

And,

$(a-b)^5 = {}^5C_0a^5 - {}^5C_1a^4b + {}^5C_2a^3b^2 - {}^5C_3a^2b^3 + {}^5C_4ab^4 - {}^5C_5b^5$

$= a^5 - 5a^4b + 10a^3b^2 - 10a^2b^3 + 5ab^4 - b^5$

$\therefore (a+b)^5 + (a-b)^5$

$= 2[a^5 + 10a^3b^2 + 5ab^4]$

Putting $a = \sqrt{3}, b = \sqrt{2}$, we get

$(\sqrt{3}+\sqrt{2})^5 + (\sqrt{3}-\sqrt{2})^5$

$= 2[5(\sqrt{3})^5 + 10(\sqrt{3})^3(\sqrt{2})^2 + 5(\sqrt{3})(\sqrt{2})^4]$

$= 2[45\sqrt{3} + 60\sqrt{3} + 20\sqrt{3}] = 250\sqrt{3}$

Q4. Find the value of $(\sqrt{a^2+1}+\sqrt{a^2-1})^4 + (\sqrt{a^2+1}-\sqrt{a^2-1})^4$.

Solution:

Using the binomial theorem,

$(x+y)^4 = {}^4C_0x^4 + {}^4C_1x^3y + {}^4C_2x^2y^2 + {}^4C_3xy^3 + {}^4C_4y^4$

$= x^4 + 4x^3y + 6x^2y^2 + 4xy^3 + y^4$

$(x-y)^4 = {}^4C_0x^4 - {}^4C_1x^3y + {}^4C_2x^2y^2 - {}^4C_3xy^3 + {}^4C_4y^4$

$= x^4 - 4x^3y + 6x^2y^2 - 4xy^3 + y^4$

And,

$\therefore (x+y)^4 + (x-y)^4 = 2(x^4 + 6x^2y^2 + y^4)$

Putting $x = \sqrt{a^2+1}, y = \sqrt{a^2-1}$, we get

$(\sqrt{a^2+1}+\sqrt{a^2-1})^4 + (\sqrt{a^2+1}-\sqrt{a^2-1})^4$

$= 2[(\sqrt{a^2+1})^4 + 6(\sqrt{a^2+1})^2(\sqrt{a^2-1})^2 + (\sqrt{a^2-1})^4]$

$= 2[(a^2+1)^2 + 6(a^2+1)(a^2-1) + (a^2-1)^2]$

$= 2[a^4 + 2a^2 + 1 + 6a^4 - 6 + a^4 - 2a^2 + 1]$

$= 16a^4 - 8$

Q5. Find n, if the ratio of the fourth term from the beginning to the fourth term from the end in the expansion of $\left(\sqrt[3]{2}+\sqrt[2]{3}\right)^n$ is 4 : 27.

Solution:

In the expansion,

$(a+b)^n = {}^nC_0a^n + {}^nC_1a^{n-1}b + ... + {}^nC_{n-1}ab^{n-1} + {}^nC_nb^n$

Fourth term from the beginning $= {}^nC_4a^{n-4}b^4$

Fourth term from the end $= {}^nC_{n-4}a^4b^{n-4}$

Therefore, in the expansion of $\left(\sqrt[3]{2}+\sqrt[2]{3}\right)^n$

Fourth term from the beginning is ${}^nC_3(\sqrt[3]{2})^{n-3}\left(\sqrt[2]{3}\right)^3$

Fourth term from the end is ${}^nC_{n-3}(\sqrt[3]{2})^3\left(\sqrt[2]{3}\right)^{n-3}$

The ratio of the fourth term from the beginning to the fourth term the end is 4 : 27.

Therefore,

$$\frac{{}^nC_3(\sqrt[3]{2})^{n-3}\left(\sqrt[2]{3}\right)^3}{{}^nC_{n-3}(\sqrt[3]{2})^3\left(\sqrt[2]{3}\right)^{n-3}} = \frac{4}{27}$$

$$\Rightarrow \frac{(\sqrt[3]{2})^{n-3}\left(\sqrt[2]{3}\right)^3}{(\sqrt[3]{2})^3\left(\sqrt[2]{3}\right)^{n-3}} = \frac{4}{27} \quad (\because {}^nC_3 = {}^nC_{n-3})$$

$$\Rightarrow \frac{(2)^{n/3-2}}{(3)^{n/2-3}} = \frac{4}{27} = \frac{2^2}{3^3} \Rightarrow n = 12$$

Q6. Expand using the binomial theorem $\left(2+x-\frac{1}{x}\right)^3, x \neq 0$.

Solution:

The given expression $\left(2+x-\frac{1}{x}\right)^3$ can be written as $\left[2+\left(x-\frac{1}{x}\right)\right]^3$. Therefore, the using binomial theorem,

$\left[2+\left(x-\frac{1}{x}\right)\right]^3$.

$= {}^3C_02^3 + {}^3C_12^2\left(x-\frac{1}{x}\right) + {}^3C_22\left(x-\frac{1}{x}\right)^2 + {}^3C_3\left(x-\frac{1}{x}\right)^3$

$= 8 + 12\left(x-\frac{1}{x}\right) + 6\left(x-\frac{1}{x}\right)^2 + \left(x-\frac{1}{x}\right)^3$

Again, using the binomial theorem, $\left(x-\frac{1}{x}\right)^3$

$= {}^3C_0x^3 - {}^3C_1x^2\left(\frac{1}{x}\right) + {}^3C_2x\left(\frac{1}{x}\right)^2 - {}^3C_3\left(\frac{1}{x}\right)^3$

$= x^3 - 3x + \frac{3}{x} - \frac{1}{x^3}$

Thus, $\left(2+x-\frac{1}{x}\right)^3$

$= 8 + 12x - \frac{12}{x} + 6x^2 + \frac{1}{x^2} - 12 + x^3 - 3x + \frac{3}{x} - \frac{1}{x^3}$

$= x^3 + 6x^2 + 9x - 4 - \frac{9}{x} + \frac{1}{x^2} - \frac{1}{x^3}$

Q7. Find the term independent of x, where $x \neq 0$, in the expansion of $\left(2x^2 - \frac{1}{2x}\right)^{12}$.

Solution:

The general term of the given expansion $\left(2x^2-\frac{1}{2x}\right)^{12}$ is

$T_{r+1}={}^{12}C_r\left(2x^2\right)^{12-r}\left(-\frac{1}{2x}\right)^r$

$={}^{12}C_r 2^{12-r}x^{24-2r}(-1)^r 2^{-r}\cdot x^{-r}$

$=(-1)^r\,{}^{12}C_r 2^{12-2r}x^{24-3r}$

For the term independent of x, $24-3r=0\Rightarrow r=8$

Therefore, $T_{r+1}=T_{8+1}=9^{\text{th}}$ term is independent of x, and thus $T_9=T_{8+1}=(-1)^8\,{}^{12}C_8 2^{12-2\cdot 8}$

$=\frac{{}^{12}C_8}{2^4}=\frac{12\cdot 11\cdot 10\cdot 9}{2^4\cdot 4\cdot 3\cdot 2}=\frac{495}{16}$

Q8. If the term free from x in the expansion of $\left(\sqrt{x}-\frac{k}{x}\right)^9$ is 84, then find the value of k.

Solution:

The general term of the given expansion $\left(\sqrt{x}-\frac{k}{x}\right)^9$ is

$T_{r+1}={}^9C_r(\sqrt{x})^{9-r}\left(\frac{-k}{x}\right)^r={}^9C_r(-k)^r x^{\frac{9-3r}{2}}$

For the term free from x, $\frac{9-3r}{2}=0\Rightarrow r=3$

Thus, $T_{3+1}=T_4$ term is free from x.

The term free from x is equal to 405; therefore

$T_{3+1}={}^9C_3(-k)^3=84\Rightarrow\frac{9\cdot 8\cdot 7}{3\cdot 2}(-k)^3=84$

$\Rightarrow -84k^3=84\Rightarrow k^3=-1\Rightarrow k=-1$

Q9. Find the coefficient of x^2 in the expansion of $(1-2x+3x^2)(1-x)^{11}$.

Solution:

Since $(1-2x+3x^2)(1-x)^{11}$

$=(1-2x+3x^2)\begin{pmatrix}{}^{11}C_0-{}^{11}C_1x^1-{}^{11}C_2x^2\\ \text{+terms of higher power of } x\end{pmatrix}$

$=(1-2x+3x^2)\begin{pmatrix}1-11x-55x^2\\ \text{+terms of higher power of } x\end{pmatrix}$

$=-55x^2+22x^2+3x^2+\text{other terms of } x$

$=-30x^2+\text{other terms of } x$

Therefore, coefficient of x is –30.

Q10. Find the middle term (terms) in the expansion of $\left(3x-\frac{x^3}{3}\right)^5$.

Solution:

The power of the binomial expansion $\left(3x-\frac{x^3}{6}\right)^5$ is $n=5$ is odd.

Therefore, it has two middle terms, namely,

$\left(\frac{5+1}{2}\right)$ and $\left(\frac{5+1}{2}+1\right)=3^{\text{rd}}$ term and 4th term.

Thus, $T_3=T_{(2+1)}={}^5C_2(3x)^{5-2}\left(-\frac{x^3}{3}\right)^2$

$=10\cdot 3^3x^3x^6 3^{-2}=30x^9$

And $T_4=T_{(3+1)}={}^5C_3(3x)^{5-3}\left(-\frac{x^3}{3}\right)^3$

$=10\cdot 3^2x^2x^9 3^{-3}(-1)^3=-\frac{10}{3}x^{11}$

Q11. Find the coefficient of $\frac{1}{x^{15}}$ in the expansion of $\left(x^2-\frac{1}{x^3}\right)^{10}$.

Solution:

The general term of the given expansion $\left(x^2-\frac{1}{x^3}\right)^{10}$ is

$T_{r+1}={}^{10}C_r(x^2)^{10-r}\left(-\frac{1}{x^3}\right)^r$

$={}^{10}C_rx^{20-2r}(-1)^r x^{-3r}=(-1)^r\,{}^{10}C_rx^{20-5r}$

For the coefficient of $\frac{1}{x^{15}}$ or x^{-15}, $20-5r=-15\Rightarrow r=7$

Thus, $T_{7+1}=(-1)^7\,{}^{10}C_7x^{20-35}=-{}^{10}C_7x^{-15}$

Hence, coefficient of $\frac{1}{x^{15}}$ or x^{-15} is

$=-{}^{10}C_7=\frac{-10\times 9\times 8}{3\times 2\times 1}=-120$

Q12. Find the seventh term of the expansion of $(y^{1/2}+x^{1/3})^n$, if the binomial coefficient of the fourth term from the end is 220.

Solution:

The seventh term of the given expansion is $(y^{1/2}+x^{1/3})^n$ is $T_7=T_{6+1}={}^nC_6(y^{1/2})^{n-6}(x^{1/3})^6$.

The binomial coefficient of the fourth term from the end = binomial coefficient of fourth term from the beginning $={}^nC_3$

And it is given that binomial coefficient of the fourth term from the end is 220.

$\therefore {}^nC_3 = 220 \Rightarrow \dfrac{n(n-1)(n-2)}{3\cdot 2} = 220$

$\Rightarrow n(n-1)(n-2) = 1320 = 12\times 11\times 10$

$\Rightarrow n = 12$

Thus, $T_7 = {}^{12}C_6(y^{1/2})^{12-6}(x^{1/3})^6 = {}^{12}C_6 y^3 x^2$

Q13. Find the value of r, if the coefficient of the $(2r+5)^{\text{th}}$, $(2r-3)^{\text{th}}$ terms in the expansion of $(1+x)^{10}$ are equal.

Solution:

The $(2r+5)^{\text{th}}$ term of the expansion $(1+x)^{10}$ is

$T_{2r+4+1} = {}^{10}C_{2r+4}(x)^{2r+4} = {}^{10}C_{2r+4}x^{2r+4}$

And the $(2r-3)^{\text{th}}$ term of the expansion $(1+x)^{10}$ is

$T_{2r-4+1} = {}^{10}C_{2r-4}(x)^{2r-4} = {}^{10}C_{2r-4}x^{2r-4}$

The coefficients of the $(2r+3)^{\text{th}}$, $(2r-3)^{\text{th}}$ terms in the expansion of $(1+x)^{10}$ are equal. Therefore

${}^{10}C_{2r+4} = {}^{10}C_{2r-4}$

$\Rightarrow {}^{10}C_{2r+4} = {}^{10}C_{10-2r+4} \quad [\because {}^nC_r = {}^nC_{n-r}]$

$\Rightarrow {}^{10}C_{2r+4} = {}^{10}C_{16-2r}$

$\Rightarrow 2r+4 = 16-2r \Rightarrow 4r = 12 \Rightarrow r = 3$

Q14. If the coefficients of the first, second and third terms in the expansion of $(1+x)^{2n}$ are in AP, then show that $2n^2-5n+1=0$.

Solution:

In the given expansion $(1+x)^{2n}$

Coefficient of 1st term = ${}^{2n}C_0$

Coefficient of 2nd term = ${}^{2n}C_1$

Coefficient of 3rd term = ${}^{2n}C_2$

The coefficients of the first, second and third terms are in AP; therefore ${}^{2n}C_0, {}^{2n}C_1, {}^{2n}C_2$ are in AP.

Thus, $2\cdot {}^{2n}C_1 = {}^{2n}C_0 + {}^{2n}C_2$

$\Rightarrow 2(2n) = 1 + (2n)(2n-1)/2$

$\Rightarrow 4n = 1 + 2n^2 - n$

$\Rightarrow 2n^2 - 5n + 1 = 0$

Q15. Find the coefficient of x^4 in the expansion of $(1+x+x^2+x^3)^6$.

Solution:

The given expansion can be written as

$= (1+x+x^2+x^3)^6 = [(1+x)+x^2(1+x)]^6$

$= [(1+x)(1+x^2)]^6 = (1+x)^6\cdot(1+x^2)^6$

Using the binomial theorem

$= ({}^6C_0 + {}^6C_1x + {}^6C_2x^2 + {}^6C_3x^3 + {}^6C_4x^4 + \ldots)$

$\times({}^6C_0 + {}^6C_1x^2 + {}^6C_2x^4 + \ldots.)$

$= {}^6C_0\,{}^6C_2x^4 + {}^6C_2\,{}^6C_1x^4 + {}^6C_4\,{}^6C_0x^4$

Thus, the coefficient of x^4 is

$= {}^6C_0\,{}^6C_2 + {}^6C_2\,{}^6C_1 + {}^6C_4\,{}^6C_0 = 15 + 15\cdot 6 + 15 = 120$

Q16. Find n in the binomial $\left(\sqrt[3]{2} + \dfrac{1}{\sqrt[3]{3}}\right)^n$, if the ratio of the 7th term from the beginning to the 7th from the end is 1.

Solution:

The given binomial expansion is $\left(\sqrt[3]{2} + \dfrac{1}{\sqrt[3]{3}}\right)^n$. Therefore, the 7th term from beginning is

$T_{7,B} = T_{6+1} = {}^nC_6(\sqrt[3]{2})^{n-6}\left(\dfrac{1}{\sqrt[3]{3}}\right)^6$

Similarly, the 7th term from the end is

$T_{7,E} = {}^nC_{n-6}(\sqrt[3]{2})^6\left(\dfrac{1}{\sqrt[3]{3}}\right)^{n-6}$

Thus, $\dfrac{{}^nC_6(\sqrt[3]{2})^{n-6}\left(\dfrac{1}{\sqrt[3]{3}}\right)^6}{{}^nC_{n-6}(\sqrt[3]{2})^6\left(\dfrac{1}{\sqrt[3]{3}}\right)^{n-6}} = 1$

$\Rightarrow \dfrac{(\sqrt[3]{2})^{n-6}\left(\dfrac{1}{\sqrt[3]{3}}\right)^6}{(\sqrt[3]{2})^6\left(\dfrac{1}{\sqrt[3]{3}}\right)^{n-6}} = 1 \quad (\because {}^nC_6 = {}^nC_{n-6})$

$\Rightarrow \left(2^{\frac{n-6}{3}-\frac{6}{3}}\right)\left(3^{\frac{-6}{2}+\frac{(n-6)}{2}}\right) = 1$

$\Rightarrow \left(2^{\frac{n-12}{3}}\right)\cdot\left(3^{\frac{-12+n}{2}}\right) = 1$

$\Rightarrow \left(2^{\frac{n-12}{3}}\right)\cdot\left(3^{\frac{n-12}{2}}\right) = 1 = 2^0 3^0 \Rightarrow n = 12$

Q17. Find the term independent of x in the expansion of $(1+x+x^2+x^3)\left(\sqrt{x} - \dfrac{1}{\sqrt{x}}\right)^9$.

Solution:

The general term of the expansion $\left(\sqrt{x} - \dfrac{1}{\sqrt{x}}\right)^9$ is

$$T_{r+1} = {}^9C_r\left(\sqrt{x}\right)^{9-r}\left(-\frac{1}{\sqrt{x}}\right)^r$$

$$= {}^9C_r x^{\frac{9-r}{2}}(-1)^r x^{-r/2} = (-1)^r\, {}^9C_r x^{9-2r}$$

Therefore, the general term in the expansion

$(1+x+x^2+x^3)\left(\sqrt{x}-\frac{1}{\sqrt{x}}\right)^9$ is

$$= (-1)^r\, {}^9C_r x^{9-2r} + (-1)^r\, {}^9C_r x^{10-2r}$$
$$+ (-1)^r\, {}^9C_r x^{11-2r} + (-1)^r\, {}^9C_r x^{12-2r}$$

For the term independent of x, $9-2r=0$, $10-2r=0$, $11-2r=0$, $12-2r=0$, we get $r=9/2$, $r=5$, $r=11/2$, $r=6$

Now, the possible values of r are 5 and 6, and $r=9/2$, $11/2$ is not independent of x.

Therefore, the term independent of x is

$$= (-1)^5\, {}^9C_5 + (-1)^6\, {}^9C_6 = -\frac{9\times8\times7\times6}{4\times3\times2} + \frac{9\times8\times7}{3\times2\times1} = -42$$

CHAPTER

9 Sequences and Series

Day 113

Sequence

- Arrangement of numbers in a definite pattern (or a specified rule) is known as a sequence.
- The numbers occurring in a sequence are called its terms; they are denoted by $a_1, a_2, a_3, ...a_n, ...$, and so on.
- The subscripts denote the position of the term in a sequence. For example, a_1 is the 1st term, a_2 is the 2nd term, and so on; a_n is the nth (also called the general) term of the sequence.
- A sequence containing a finite number of terms is called a finite sequence.
- A sequence is called infinite if it is not a finite sequence or contains infinite terms.

As an example,

(i) Consider the sequence of even natural numbers 2, 4, 6, ... In this sequence,

$a_1 = 2 = 2\times 1$

$a_2 = 4 = 2\times 2$

$a_3 = 6 = 2\times 3$

... and so on

$a_n = 2n = 2\times n$ is the general term of the sequence.

(ii) Consider the sequence of odd natural numbers 1, 3, 5, ... In this sequence,

$a_1 = 1 = 2\times 1 - 1$

$a_2 = 3 = 2\times 2 - 1$

$a_3 = 5 = 2\times 3 - 1$

... and so on

$a_n = 2n - 1$ is the general term of the sequence.

(iii) Consider the sequence 1, 1, 2, 3, 5, 8, ..., which has no visible pattern, but the sequence is generated by the recurrence relation given by

$a_1 = a_2 = 1, \quad a_3 = a_1 + a_2$

$a_n = a_{n-2} + a_{n-1}$ $(n > 2)$, is the general term of the sequence called the Fibonacci sequence.

(iv) Consider the sequence of primes 2, 3, 5, 7, ... There is no formula for the prime sequence, and thus, it can only be described by a verbal description.

Note: In a sequence, we should not expect that its terms will necessarily be given by a specific formula, but a rule for generating the terms $a_1, a_2, a_3, ..., a_n, ...$ will be in succession.

Series

- For a given sequence $a_1, a_2, a_3, ...a_n$, the expression $a_1 + a_2 + a_3 + ... + a_n + ...$ is called the series associated with the given sequence.
- The series is finite or infinite according to whether the given sequence is finite or infinite.
- In general, we represent a series with the symbol sigma, using the Greek letter $\sum$ as a means of indicating the summation involved. Thus, the series

$$\sum_{k=1}^{n} a_k = a_1 + a_2 + a_3 + ... + a_n$$

Example 1: Write the first four terms of the following sequences whose n^{th} term is given as (i) $a_n = (2n+1)(n+2)$

(ii) $a_n = \dfrac{2-n}{n+1}$

(iii) $a_n = n^2 2^n$

(iv) $a_n = (-1)^{n+1} 2^{n-1}$

(v) $a_n = (-1)^n (n^2 - 3^2)$

Solution:

(i) $a_n = (2n+1)(n+2)$. Therefore, substituting $n = 1,2,3,4$, we get

$a_1 = (2\cdot 1 - 1)(1+2) = 3$

$a_2 = (2\cdot 2 - 1)(2+2) = 12$

$a_3 = (2\cdot 3 - 1)(3+2) = 25$

$a_4 = (2\cdot 4 - 1)(4+2) = 42$

(ii) $a_n = \dfrac{2-n}{n+1}$. Therefore, substituting $n = 1,2,3,4$, we get

$a_1 = \dfrac{2-1}{1+1} = \dfrac{1}{2}, \qquad a_2 = \dfrac{2-2}{2+1} = 0$

$a_3 = \dfrac{2-3}{3+1} = -\dfrac{1}{4}, \qquad a_4 = \dfrac{2-4}{4+1} = -\dfrac{2}{5}$

(iii) $a_n = n^2 2^n$. Therefore, substituting $n = 1,2,3,4$, we get

$a_1 = 1^2 2^1 = 2, \quad a_2 = 2^2 2^2 = 8$

$a_3 = 3^2 2^3 = 72, \quad a_4 = 4^2 2^4 = 256$

(iv) $a_n = (-1)^{n+1} 2^{n-1}$. Therefore, substituting $n = 1,2,3,4$, we get

$a_1 = (-1)^{1+1} 2^{1-1} = 2^0 = 1$

$a_2 = (-1)^{2+1} 2^{2-1} = -2^1 = -2$

$a_3 = (-1)^{3+1} 2^{3-1} = 2^2 = 4$

$a_4 = (-1)^{4+1} 2^{4-1} = -2^3 = -8$

(v) $a_n = (-1)^n (n^2 - 3^2)$. Therefore, substituting $n=1,2,3,4$, we get

$a_1 = (-1)^1 (1^2 - 3^2) = 8$

$a_2 = (-1)^2 (2^2 - 3^2) = -5$

$a_3 = (-1)^3 (3^2 - 3^2) = 0$

$a_4 = (-1)^4 (4^2 - 3^2) = 7$

Example 2: Find the 5th term and 10th term in the following sequence whose n^{th} term is

(i) $a_n = \dfrac{n(n-1)}{2}$

(ii) $a_n = (-1)^n \dfrac{2^n}{n^2}$

(iii) $a_n = \dfrac{n^2 - 50}{n+5}$

Solution:

(i) $a_n = \dfrac{n(n-1)}{2}$

Substituting $n = 5$, we get

$a_5 = \dfrac{5(5-1)}{2} = 10$

And substituting $n = 10$, we get

$a_{10} = \dfrac{10(10-1)}{2} = 45$

(ii) $a_n = (-1)^n \dfrac{2^n}{n^2}$

Substituting $n = 5$, we get

$a_5 = (-1)^5 \dfrac{2^5}{5^2} = -\dfrac{32}{25}$

And substituting $n = 10$, we get

$a_{10} = (-1)^{10} \dfrac{2^{10}}{10^2} = \dfrac{32 \cdot 32}{5 \cdot 5 \cdot 2 \cdot 2} = \dfrac{256}{25}$

(iii) $a_n = \dfrac{n^2 - 50}{n+5}$

Substituting $n = 5$, we get

$a_5 = \dfrac{5^2 - 50}{5+5} = \dfrac{-25}{10} = -\dfrac{5}{2}$

And substituting $n = 10$, we get

$a_{10} = \dfrac{10^2 - 50}{5+10} = \dfrac{50}{15} = \dfrac{10}{3}$

Example 3: Write the first four terms of the following sequences and obtain the corresponding series.

(i) $a_1 = 2,\ a_n = 2a_{n-1} - 1$ for all $n > 1$

(ii) $a_1 = -2,\ a_n = a_{n-1} + n$ for all $n \geq 2$

(iii) $a_1 = 2,\quad a_2 = 4,\ a_n = (-1)^n \dfrac{a_{n-1}}{a_{n-2}}$ for all $n > 2$

Solution:

(i) $a_1 = 2,\ a_n = 2a_{n-1} - 1$ for all $n > 1$

$a_2 = 2a_1 - 1 = 2(2) - 1 = 3$

$a_3 = 2a_2 - 1 = 2(3) - 1 = 5$

$a_4 = 2a_3 - 1 = 2(5) - 1 = 9$

Thus, the first four terms of the sequence are 2, 3, 5 and 9, and the corresponding series is

$2 + 3 + 5 + 9 + \ldots\ldots\ldots$

(ii) $a_1 = -2,\ a_n = a_{n-1} + n$ for all $n \geq 2$

$a_2 = a_1 + 1 = -2 + 1 = -1$

$a_3 = a_2 + 2 = -1 + 2 = 1$

$a_4 = a_3 + 3 = 1 + 3 = 4$

Thus, the first four terms of the sequence are –2, –1, 1 and 4, and the corresponding series is

$-2 - 1 + 1 + 4\ldots\ldots\ldots$

(iii) $a_1 = 2,\quad a_2 = 4,\ a_n = (-1)^n \dfrac{a_{n-1}}{a_{n-2}}$ for all $n > 2$.

$a_3 = (-1)^3 \dfrac{a_{3-1}}{a_{3-2}} = -\dfrac{a_2}{a_1} = -\dfrac{4}{2} = -2$

$a_4 = (-1)^4 \dfrac{a_{4-1}}{a_{4-2}} = \dfrac{a_3}{a_2} = \dfrac{-2}{2} = -1$

$a_5 = (-1)^5 \dfrac{a_{5-1}}{a_{5-2}} = -\dfrac{a_4}{a_3} = -\dfrac{(-1)}{(-2)} = -\dfrac{1}{2}$

Thus, the first four terms of the sequence are 2, 4, –2, –1 and –1/2, and the corresponding series is

$2 + 4 - 2 - 1 - 1/2\ldots\ldots\ldots$

Day 114

NCERT Exercise 9.1 (With Solutions)

Q1. Write the first five terms of the sequence whose n^{th} term is $a_n = n(n+2)$.

Solution:

$a_n = n(n+2)$. Therefore, substituting $n = 1, 2, 3, 4, 5$, we get

$a_1 = 1(1+2) = 3$

$a_2 = 2(2+2) = 8$

$a_3 = 3(3+2) = 15$

$a_4 = 4(4+2) = 24$

$a_5 = 5(5+2) = 35$

Q2. Write the first five terms of the sequence whose n^{th} term is $a_n = \frac{n}{n+1}$.

Solution:

$a_n = \frac{n}{n+1}$. Therefore, substituting $n = 1, 2, 3, 4, 5$, we get

$a_1 = \frac{1}{1+1} = \frac{1}{2}$, $\quad a_2 = \frac{2}{2+1} = \frac{2}{3}$

$a_3 = \frac{3}{3+1} = \frac{3}{4}$, $\quad a_4 = \frac{4}{4+1} = \frac{4}{5}$

$a_5 = \frac{5}{5+1} = \frac{5}{6}$

Q3. Write the first five terms of the sequence whose n^{th} term is $a_n = 2^n$.

Solution:

$a_n = 2^n$. Therefore, substituting $n = 1, 2, 3, 4, 5$, we get

$a_1 = 2^1 = 2$, $\quad a_2 = 2^2 = 4$

$a_3 = 2^3 = 8$, $\quad a_4 = 2^4 = 16$

$a_5 = 2^5 = 32$

Q4. Write the first five terms of the sequence whose n^{th} term is $a_n = \frac{2n-3}{6}$.

Solution:

$a_n = \frac{2n-3}{6}$. Therefore, substituting $n = 1, 2, 3, 4, 5$, we get

$a_1 = \frac{2\times1-3}{6} = \frac{-1}{6}$

$a_2 = \frac{2\times2-3}{6} = \frac{1}{6}$

$a_3 = \frac{2\times3-3}{6} = \frac{1}{2}$

$a_4 = \frac{2\times4-3}{6} = \frac{5}{6}$

$a_5 = \frac{2\times5-3}{6} = \frac{7}{6}$

Q5. Write the first five terms of the second whose n^{th} term is $a_n = (-1)^{n-1}5^{n+1}$.

Solution:

$a_n = (-1)^{n-1}5^{n+1}$. Therefore, substituting $n = 1, 2, 3, 4, 5$, we get

$a_1 = (-1)^{1-1}5^{1+1} = 5^2 = 25$

$a_2 = (-1)^{2-1}5^{2+1} = -5^3 = -125$

$a_3 = (-1)^{3-1}5^{3+1} = 5^4 = 625$

$a_4 = (-1)^{4-1}5^{4+1} = 5^5 = -3125$

$a_5 = (-1)^{5-1}5^{5+1} = 5^6 = 15625$

Q6. Write the first five terms of the sequence whose n^{th} term is $a_n = n\frac{n^2+5}{4}$.

Solution:

$a_n = n\frac{n^2+5}{4}$. Therefore, substituting $n = 1, 2, 3, 4, 5$, we get

$a_1 = 1\cdot\frac{1^2+5}{4} = \frac{6}{4} = \frac{3}{2}$

$a_2 = 2\cdot\frac{2^2+5}{4} = 2\cdot\frac{9}{4} = \frac{9}{2}$

$a_3 = 3\cdot\frac{3^2+5}{4} = 3\cdot\frac{14}{4} = \frac{21}{2}$

$a_4 = 4\cdot\frac{4^2+5}{4} = 4\cdot\frac{21}{4} = 21,$

$a_5 = 5\cdot\frac{5^2+5}{4} = 5\cdot\frac{30}{4} = \frac{75}{2}$

Q7. Find the 17th term and 24th term in the following sequence whose n^{th} term is $a_n = 4n-3$.

Solution:

Substituting $n = 17$, we get

$a_{17} = 4(17) - 3 = 68 - 3 = 65$

And substituting $n = 24$, we get

$a_{24} = 4(24) - 3 = 96 - 3 = 93$

Q8. Find the 7th term in the following sequence whose n^{th} term is $a_n = \frac{n^2}{2n}$.

Solution:

Substituting $n = 7$, we get $a_7 = \frac{7^2}{2^7} = \frac{49}{128}$.

Q9. Find the 9th term in the following sequence whose nth term is $a_n = (-1)^{n-1}n^3$.

Solution:

Substituting $n = 9$, we get

$a_9 = (-1)^{9-1}(9)^3 = (9)^3 = 729$

Q10. Find the 20th term in the following sequence whose n^{th} term is $a_n = \frac{n(n-2)}{n+3}$.

Solution:

Substituting $n = 20$, we get

$$a_{20} = \frac{20(20-2)}{20+3} = \frac{20(18)}{23} = \frac{360}{23}$$

Q11. Write the first five terms of the following sequence and obtain the corresponding series;

$a_1 = 3,\ a_n = 3a_{n-1} + 2$ for all $n > 1$.

Solution:

$a_1 = 3,\ a_n = 3a_{n-1} + 2$ for all $n > 1$.

$a_2 = 3a_1 + 2 = 3(3) + 2 = 11$

$a_3 = 3a_2 + 2 = 3(11) + 2 = 35$

$a_4 = 3a_3 + 2 = 3(35) + 2 = 107$

$a_5 = 3a_4 + 2 = 3(107) + 2 = 323$

Thus, the first five terms of the sequence are 3, 11, 35, 107 and 323, and the corresponding series is

$3 + 11 + 35 + 107 + 323 + ...$

Q12. Write the first five terms of the following sequence and obtain the corresponding series:

$a_1 = -1,\ a_n = \frac{a_{n-1}}{n},\ n \geq 2.$

Solution:

$a_1 = -1,\ a_n = \frac{a_{n-1}}{n},\ n \geq 2$

$a_2 = \frac{a_1}{2} = \frac{-1}{2}$

$a_3 = \frac{a_2}{3} = \frac{-1}{6}$

$a_4 = \frac{a_3}{4} = \frac{-1}{24}$

$a_5 = \frac{a_4}{4} = \frac{-1}{120}$

Thus, the first five terms of the sequence are

$-1, \frac{-1}{2}, \frac{-1}{6}, \frac{-1}{24}, \frac{-1}{120}$, and the corresponding series is

$$(-1) + \left(\frac{-1}{2}\right) + \left(\frac{-1}{6}\right) + \left(\frac{-1}{24}\right) + \left(\frac{-1}{120}\right) + ...$$

Q13. Write the first five terms of the following sequence and obtain the corresponding series:

$a_1 = a_2 = 2,\ a_n = a_{n-1} - 1,\ n > 2.$

Solution:

$a_1 = a_2 = 2,\ a_n = a_{n-1} - 1,\ n > 2$

$a_3 = a_2 - 1 = 2 - 1 = 1$

$a_4 = a_3 - 1 = 1 - 1 = 0$

$a_5 = a_4 - 1 = 0 - 1 = -1$

Thus, the first five terms of the sequence are 2, 2, 1, 0, –1, and the corresponding series is $2 + 2 + 1 + 0 + (-1) + ...$

Q14. The Fibonacci sequence is defined by $1 = a_1 = a_2,\ a_n = a_{n-1} + a_{n-2},\ n > 2$. Find $\frac{a_{n-1}}{a_n}$ for $n = 1, 2, 3, 4, 5$.

Solution:

$1 = a_1 = a_2,\ a_n = a_{n-1} + a_{n-2},\ n > 2$

$a_3 = a_2 + a_1 = 1 + 1 = 2$

$a_4 = a_3 + a_2 = 2 + 1 = 3$

$a_5 = a_4 + a_3 = 3 + 2 = 5$

$a_6 = a_5 + a_4 = 5 + 3 = 8$

For $n = 1, \frac{a_n + 1}{a_n} = \frac{a_2}{a_1} = \frac{1}{1} = 1$

For $n = 2, \frac{a_n + 1}{a_n} = \frac{a_3}{a_2} = \frac{2}{1} = 2$

For $n = 3, \frac{a_n + 1}{a_n} = \frac{a_4}{a_3} = \frac{3}{2}$

For $n = 4, \frac{a_n + 1}{a_n} = \frac{a_5}{a_4} = \frac{5}{3}$

For $n = 5, \frac{a_n + 1}{a_n} = \frac{a_6}{a_5} = \frac{8}{5}$

Day 115

Arithmetic Progression

In an arithmetic progression, the difference of any two consecutive numbers is a constant.

A sequence $a_1, a_2, a_3, \ldots, a_n, \ldots$ is called an arithmetic sequence or an arithmetic progression if $a_{n+1} = a_n + d,\ n \in N$, where a_1 is called the first term and the constant term d is called the common difference of the AP.

Consider an AP with its first term a and common difference d, that is, the AP will be

$a, a+d, a+2d, \ldots$ and so on.

Thus, 1st term $= a_1 = a$

2nd term $= a_2 = a+d$

3rd term $= a_3 = a+2d$

4th term $= a_4 = a+3d$

In general, the nth term of the AP is $a_n = a+(n-1)d$.

For a finite sequence, we say the nth term of the AP as the last term of AP is denoted by $l = a+(n-1)d$.

Sum of *n* terms of an AP

S_n = the sum to n terms of AP

= the sum of $a, a+d, a+2d, \ldots, a+(n-1)d$

$$\Rightarrow S_n = \frac{n}{2}[2a+(n-1)d] = \frac{n}{2}[a+l]$$

Properties of an AP

(i) If a constant is added (or subtracted) to each term of an AP, then the resulting sequence is also an AP.

(ii) If each term of an AP is multiplied (or divided) by a non-zero constant, then the resulting sequence is also an AP.

Arithmetic Mean (AM): A number A is called the arithmetic mean (AM) of two given numbers a and b, if a, A, b is in AP, That is, $A-a = b-A \Rightarrow A = \frac{a+b}{2}$

AM between two numbers a and b is interpreted as their average $\frac{a+b}{2}$.

Similarly, we can insert two AM (A_1 and A_2) between a and b, such that a, A_1, A_2, b is in AP.

$$A_1 - a = A_2 - A_1 = b - A_2$$

$$\Rightarrow A_1 = \frac{2a+b}{3} \text{ and } A_2 = \frac{a+2b}{3}$$

In general, for any two given numbers a and b, we can insert as many numbers as we like between them such that the resulting sequence is an AP.

Let $A_1, A_2, A_3, \ldots, A_n$ be n AM numbers between a and b such that $a, A_1, A_2, A_3, \ldots, A_n, b$ forms an AP.

Now, the above sequence has $(n+2)$ terms, and therefore, b will be the $(n+2)^{th}$ term, that is,

$$b = a+[(n+2)-1]d = a+(n+1)d.$$

Or, the common difference of the above sequence

$$d = \frac{b-a}{n+1}$$

Thus, n numbers between a and b are as follows:

$$A_1 = a+d = a+\frac{b-a}{n+1}$$

$$A_2 = a+2d = a+\frac{2(b-a)}{n+1}$$

$$A_3 = a+3d = a+\frac{3(b-a)}{n+1} \text{ and so on}$$

$$A_n = a+nd = a+\frac{n(b-a)}{n+1}$$

Example 1: In an AP, the first term is 1 and the sum of the first six terms is equal to the sum of the next two terms. Show that the 8th term is –13.

Solution:

First term, $a = 1$

Let d be the common difference of the AP.

Therefore, AP is $1, 1+d, 1+2d, 1+3d \ldots$

Sum of the first six terms is

$$S_6 = \frac{6}{2}[2\cdot 1+(6-1)d] = 6+15d$$

Sum of the next four terms is

$$S_{8-6} = \frac{8}{2}[2\cdot 1+(8-1)d]-[6+15d]$$

$$= 8+28d-6-15d = 2+13d$$

Since the sum of the first six terms is equal to the sum of next two terms,

$$S_6 = S_{8-6} \Rightarrow 6+15d = 2+13d \Rightarrow d = -2$$

Thus, the 8th term of the AP is

$$a_8 = a+(n-1)d = 1+(8-1)(-2) = -13$$

Example 2: Find the sum of all natural numbers from 100 to 1000, which are a multiple of 2 but not a multiple of 3.

Solution:

The natural numbers from 100 to 1000, which are multiples of 2, are 100, 102, 104, … 1000.

This sequence forms an AP, with the first term, $a = 100$, last term, $l = 1000$ and common difference, $d = 2$.

Let n be the total number of terms. Then

$$a+(n-1)d = l \Rightarrow 100+(n-1)2 = 1000$$

$$\Rightarrow 2(n-1) = 900 \Rightarrow n-1 = 450 \Rightarrow n = 451$$

Thus, the sum of all natural numbers from 100 to 1000 which are multiples of 2 is

$S_{n,2} = \frac{n}{2}[2a+(n-1)d]$

$= \frac{451}{2}[2(100)+(451-1)(2)]$

$= \frac{451}{2}[200+900] = 248050$

Now, we need the sum of all natural numbers from 100 to 1000 which are multiples of 2 but not multiples of 3. Therefore, we subtract the sum of all the natural numbers from 100 to 1000 which are multiples of 6.

Thus, the natural numbers from 100 to 1000, which are multiples of 6, are 102, 108, … 996.

This sequence forms an AP, with first term, $a = 102$, last term, $l = 996$ and common difference, $d = 6$.

Let n be the total number of terms. Then

$a+(n-1)d = l \Rightarrow 102+(n-1)6 = 996$

$\Rightarrow 6(n-1) = 894 \Rightarrow n-1 = 149 \Rightarrow n = 150$

Thus, the sum of all natural numbers from 100 to 1000 which are multiples of 6 is

$S_{n,6} = \frac{n}{2}[2a+(n-1)d]$

$= \frac{150}{2}[2(102)+(150-1)(6)]$

$= \frac{150}{2}[204+894] = 82350$

Thus, the sum of all natural numbers from 100 to 1000 which are multiples of 2 but not multiples of 3 is

$S_2 - S_6 = 248050 - 82350 = 165700$

Example 3: How many terms of the AP $-20, -\frac{35}{2}, -15, \ldots$ are needed to give the sum equal to zero?

Solution:

Common difference of the given AP is $d = -\frac{35}{2} - (-20) = \frac{40-35}{2} = \frac{5}{2}$

Let the sum of n terms of the given AP be equal to zero. Then

$S_n = \frac{n}{2}[2a+(n-1)d] = 0$

$\Rightarrow \frac{n}{2}\left[2\times(-20)+(n-1)\left(\frac{5}{2}\right)\right] = 0$

$\Rightarrow (n-1)\left(\frac{5}{2}\right) = 40 \Rightarrow n-1 = 16 \Rightarrow n = 17$

Example 4: In an AP, if the pth term is q and the qth term is p, then find the sum of first pq terms; $p \neq q$.

Solution:

The nth term of an AP is $a_n = a+(n-1)d$, and therefore

p^{th} term $= a_p = a+(p-1)d = q$

q^{th} term $= a_q = a+(q-1)d = p$

Solving the above two equations, we get

$(p-1)d-(q-1)d = q-p$

$\Rightarrow (p-1-q+1)d = q-p$

$\Rightarrow (p-q)d = q-p$

$\Rightarrow d = -1$

And putting the value of d, we get

$a+(p-1)(-1) = q \Rightarrow a = p+q-1$

Thus, the sum of the first pq terms of the AP is

$S_{pq} = \frac{pq}{2}[2a+(pq-1)d]$

$S_{pq} = \frac{pq}{2}[2(p+q-1)+(pq-1)(-1)]$

$= \frac{pq}{2}[2p+2q-2-pq+1]$

$= \frac{pq}{2}[2p+2q-pq-1]$

Example 5: If the sum of a certain number of terms of the AP –19, –16, –13,… is 190, find the last term.

Solution:

Since $a = -19$, $d = -16-(-19) = 3$, and the sum of n terms of the given AP is 190,

$S_n = \frac{n}{2}[2a+(n-1)d] = 190$

$\Rightarrow \frac{n}{2}[2\times(-19)+(n-1)(3)] = 190$

$\Rightarrow \frac{n}{2}[-38+3n-3] = 190$

$\Rightarrow -41n+3n^2 = 380$

$\Rightarrow 3n^2-41n-380 = 0$

$\Rightarrow (3n+19)(n-20) = 0$

$\Rightarrow n = 20$ or $n = -\frac{19}{3}$

n cannot be equal to $\frac{-19}{3}$ and therefore, $n = 20$

Thus, the last term of the AP is

$a_l = a_{20} = -19+(20-1)(3) = 38$

Example 6: Find the sum to n terms of the AP whose k^{th} term is $2k-1$.

Solution:

The kth term of the AP is $2k-1$, and therefore

$a_k = a+(k-1)d$

$\Rightarrow a+(k-1)d = 2k-1$

$\Rightarrow a-d+kd = 2k-1$

Comparing the constant coefficient and coefficient of k, we get

$d = 2$ and $a - d = -1 \Rightarrow a = 2 - 1 = 1$

Thus, sum to n terms of the AP is

$$S_n = \frac{n}{2}[2a + (n-1)d]$$

$$= \frac{n}{2}[2(1) + (n-1)(2)] = \frac{n}{2}(2n) = n^2$$

Example 7: If the sum of n terms of an AP is $(2n^2 - 3n)$, find the common difference.

Solution:

The sum of n terms of an AP is $S_n = \frac{n}{2}[2a + (n-1)d]$, and therefore

$$\frac{n}{2}[2a + (n-1)d] = 2n^2 - 3n$$

$$\Rightarrow \frac{n}{2}[2a - d + nd] = 2n^2 - 3n$$

$$\Rightarrow \frac{d}{2}n^2 + \frac{2a-d}{2}n = 2n^2 - 3n$$

Comparing the coefficient of n and n^2, we get

$\frac{d}{2} = 2 \Rightarrow d = 4$ and

$\frac{2a-d}{2} = -3 \Rightarrow 2a - 4 = -6 \Rightarrow a = -1$

Thus, the common difference of the AP is 4.

Example 8: The sum of n terms of two arithmetic progressions are in the ratio $4n+3 : 8n+5$. Find the ratio of their 15th terms.

Solution:

Let a_1, a_2, d_1, d_2 be the first terms and all common difference of the first and second arithmetic progression, respectively.

Then according to the given condition,

$$\frac{\text{Sum of } n \text{ terms of first AP}}{\text{Sum of } n \text{ terms of second AP}} = \frac{4n+3}{8n+5}$$

$$\Rightarrow \frac{\frac{n}{2}[2a_1 + (n-1)d_1]}{\frac{n}{2}[2a_2 + (n-1)d_2]} = \frac{4n+3}{8n+5}$$

$$\Rightarrow \frac{2a_1 + (n-1)d_1}{2a_2 + (n-1)d_2} = \frac{4n+3}{8n+5} \quad \text{(i)}$$

Substituting $n = 28$ in equation (i), we get

$$\frac{2a_1 + 28d_1}{2a_2 + 28d_2} = \frac{4 \cdot 28 + 3}{8 \cdot 28 + 5}$$

$$\Rightarrow \frac{a_1 + 14d_1}{a_2 + 14d_2} = \frac{115}{325} = \frac{23}{65} \quad \text{(ii)}$$

Now, the ratio of the 15th terms of both the APs is

$$\frac{15^{\text{th}} \text{ term of first AP}}{15^{\text{th}} \text{ term of second AP}}$$

$$= \frac{a_1 + 14d_1}{a_2 + 14d_2} = \frac{23}{65} \text{ from eq. (ii)}$$

Example 9: Insert four numbers between 10 and 35 such that the resulting sequence is an AP.

Solution:

Let A_1, A_2, A_3, A_4 be the four numbers between 10 and 35 such that 10, A_1, A_2, A_3, A_4, 35 is an AP.

Thus, $a = 10$, $l = 35$, $n = 6$

Therefore,

$l = a + (n-1)d \Rightarrow 35 = 10 + (6-1)d \Rightarrow d = 5$

$\therefore A_1 = a + d = 10 + 5 = 15$

$A_2 = a + 2d = 10 + 2 \times 5 = 20$

$A_3 = a + 3d = 10 + 3 \times 5 = 25$

$A_4 = a + 4d = 10 + 4 \times 5 = 30$

Hence, the required four numbers between 10 and 35 are 15, 20, 25, 30 which makes an AP.

Example 10: If $\frac{a^{n+1} + b^{n+1}}{a^n + b^n}$ is the AM between a and b, then find the value of n.

Solution:

The AM of a and b is $= \frac{a+b}{2}$

Therefore, $\frac{a+b}{2} = \frac{a^{n+1} + b^{n+1}}{a^n + b^n}$

$$\Rightarrow (a+b)(a^n + b^n) = 2(a^{n+1} + b^{n+!})$$

$$\Rightarrow a^{n+1} + ab^n + ba^n + b^{n+1} = 2a^{n+1} + 2b^{n+1}$$

$$\Rightarrow ab^n + a^n b = a^{n+1} + b^{n+1}$$

$$\Rightarrow ab^n - b^{n+1} = a^{n+1} - a^n b$$

$$\Rightarrow b^n(a-b) = a^n(a-b)$$

$$\Rightarrow b^n = a^n \Rightarrow \left(\frac{b}{a}\right)^n = 1 = \left(\frac{b}{a}\right)^0$$

$$\Rightarrow n = 0$$

Example 11: Between 11 and 55, m numbers have been inserted in such a way that the resulting sequence is an AP and the ratio of $(m+1)^{\text{th}}$ and $(m-1)^{\text{th}}$ numbers is 55:47. Find the value of m.

Solution:

Let $A_1, A_2 \ldots\ldots A_m$ be m numbers such that 11, $A_1, A_2 \ldots\ldots A_m$, 55 is an AP.

Thus, $a = 11$, $l = 55$, $n = m + 2$

$\therefore l = a + (n-1)d \Rightarrow 55 = 11 + (m+2-1)(d)$

$$\Rightarrow 44 = (m+1)d \Rightarrow d = \frac{44}{m+1}$$

Let, $A_1 = a + d$

$A_2 = a + 2d$

$A_3 = a + 3d, \ldots$

$A_{m-1} = a + (m-1)d$

$A_{m+1} = a + (m+1)d$

The ratio of $(m+1)^{\text{th}}$ and $(m-1)^{\text{th}}$ numbers is 55:47. Therefore

$$\frac{a+(m+1)d}{a+(m-1)d} = \frac{55}{47}$$

$$\Rightarrow \frac{11+44}{11+(m-1)\left(\frac{44}{m+1}\right)} = \frac{55}{47}$$

$$\Rightarrow \frac{55(m+1)}{11(m+1)+44(m-1)} = \frac{55}{47}$$

$$\Rightarrow \frac{(m+1)}{55m-33} = \frac{1}{47}$$

$$\Rightarrow 47m + 47 = 55m - 33$$

$$\Rightarrow 8m = 80 \Rightarrow m = 10$$

Thus, the value of m is 10.

Example 12: A man starts repaying a loan with a first instalment of Rs 100. If he increases the instalment by Rs 5 every month, and thus, pays the total loan in 5 years, then what is the total amount of loan he pays?

Solution:

The first instalment of the loan is Rs 100. Therefore, the second instalment of the loan is Rs 105 and so on.

Thus the amount that the man repays every month forms an AP, given by 100, 105, 110,..., where the first term $a = 100$, common difference, $d = 5$.

Also, he pays the total loan amount in 5 years (or 60 months). Therefore, the total loan he pays is

$$S_{60} = \frac{n}{2}\{2a + (n-1)d\} = \frac{60}{2}\{2 \times 100 + (60-1)5\}$$

$= \text{Rs } 14850$

Day 116

NCERT Exercise 9.2 (With Solutions)

Q1. Find the sum of odd integers from 1 to 2001.

Solution:

The odd integers from 1 to 2001 are 1, 3, 5, ..., 1999, 2001. This sequence forms an AP, with the first term, $a = 1$, last term, $l = 2001$ and common difference, $d = 2$.

Let n be the total number of terms. Then

$a + (n-1)d = l \Rightarrow 1 + (n-1)(2) = 2001$

$\Rightarrow 2n - 2 = 2000 \Rightarrow n = 1001$

Thus, the sum of odd integers from 1 to 2001 is

$$S_n = \frac{n}{2}[2a + (n-1)d]$$

$$= \frac{1001}{2}[2 \times 1 + (1001-1) \times 2]$$

$$= \frac{1001}{2} \times 2002 = 1001 \times 1001 = 1002001$$

Q2. Find the sum of all natural numbers between 100 and 1000, which are multiples of 5.

Solution:

The natural numbers between 100 and 1000, which are multiples of 5, are 105, 110, ... 995.

This sequence forms an AP, with the first term, $a =$ 105, last term, $l = 995$ and common difference, $d = 5$.

Let n be the total number of terms. Then

$a + (n-1)d = l \Rightarrow 105 + (n-1)5 = 995$

$\Rightarrow (n-1)5 = 995 - 105 = 890$

$\Rightarrow n - 1 = 178 \Rightarrow n = 179$

Thus, the sum of all natural numbers between 100 and 1000 which are multiples of 5 is

$$S_n = \frac{n}{2}[2a + (n-1)d]$$

$$= \frac{179}{2}[2(105) + (179-1)(5)]$$

$$= \frac{179}{2}[2(105) + (178)(5)]$$

$$= (179)(550) = 98450$$

Q3. In an AP, the first term is 2 and the sum of the first five terms is one-fourth of the next five terms. Show that the 20th term is –112.

Solution:

First term, $a = 2$

Let d be the common difference of the AP.

Therefore, the AP is $2, 2+d, 2+2d, 2+3d \ldots$

Sum of the first five terms is

$$S_5 = \frac{5}{2}[2 \cdot 2 + (5-1)d] = 10 + 10d$$

Sum of the next five terms is

$$S_{10-5} = \frac{10}{2}[2 \cdot 2 + (10-1)d] - \frac{5}{2}[2 \cdot 2 + (5-1)d]$$

$= 10 + 35d$

The sum of the first five terms is one-fourth the sum of the next five terms. Therefore $S_5 = \frac{1}{4} S_{10-5}$

$\Rightarrow 10+10d=\frac{1}{4}(10+35d)$

$\Rightarrow 40+40d=10+35d$

$\Rightarrow 5d=-30 \Rightarrow d=-6$

Thus, the 20th term of the AP is

$a_{20}=a+(n-1)d=2+(20-1)(-6)=-112$

Q4. How many terms of the AP $-6,-\frac{11}{2},-5,....$ are needed to give the sum –25?

Solution:

Common difference of the given AP is

$d=-\frac{11}{2}-(-6)=\frac{-11+12}{2}=\frac{1}{2}$

Let the sum of n terms of the given AP be –25. Then

$S_n=\frac{n}{2}[2a+(n-1)d]$

$\Rightarrow -25=\frac{n}{2}\left[2\times(-6)+(n-1)\left(\frac{1}{2}\right)\right]$

$\Rightarrow -50=n\left[-12+\frac{n}{2}-\frac{1}{2}\right]$

$\Rightarrow -100=n(-25+n) \Rightarrow n^2-25n+100=0$

$\Rightarrow (n-5)(n-20)=0$

$\Rightarrow n=5$ or 20

Q5. In an AP, if the pth term is $1/q$ and the qth term is $1/p$, prove that the sum of first pq terms is $1/2(pq+1)$, where $p\neq q$.

Solution:

The nth term of an AP is $a_n=a+(n-1)d$, and therefore

p^{th} term $=a_p=a+(p-1)d=\frac{1}{q}$

q^{th} term $=a_q=a+(q-1)d=\frac{1}{p}$

Solving the above two equations, we get

$(p-1)d-(q-1)d=\frac{1}{q}-\frac{1}{p}$

$\Rightarrow (p-1-q+1)d=\frac{p-q}{pq}$

$\Rightarrow (p-q)d=\frac{p-q}{pq}$

$\Rightarrow d=\frac{1}{pq}$

And putting the value of d, we get

$a+(p-1)\frac{1}{pq}=\frac{1}{q} \Rightarrow a=\frac{1}{q}-\frac{1}{q}+\frac{1}{pq}=\frac{1}{pq}$

Thus, the sum of first pq terms of the AP is

$S_{pq}=\frac{pq}{2}[2a+(pq-1)d]$

$=\frac{pq}{2}\left[\frac{2}{pq}+(pq-1)\frac{1}{pq}\right]$

$=1+\frac{1}{2}(pq-1)=\frac{1}{2}(pq+1)$

Q6. If the sum of a certain number of terms of the AP 25, 22, 19, … is 116, find the last term.

Solution:

$a=25,\ d=22-25=-3$, and the sum of n terms of the given AP is 116. Then,

$S_n=\frac{n}{2}[2a+(n-1)d]$

$\Rightarrow 116=\frac{n}{2}[2\times 25+(n-1)(-3)]$

$\Rightarrow 116=\frac{n}{2}[53-3n] \Rightarrow 232=53n-3n^2$

$\Rightarrow 3n^2-53n+232=0$

$\Rightarrow (n-8)(3n-29)=0$

$\Rightarrow n=8 \text{ or } n=\frac{29}{3}$

n cannot be equal to $\frac{29}{3}$ and therefore, $n=8$

Thus, the last term of the AP is

$a_l=a+(n-1)d=25+(8-1)(-3)=4$

Q7. Find the sum to n terms of the AP whose k^{th} term is $5k+1$.

Solution:

The kth term of the AP is $5k+1$. Therefore

$a_k=a+(k-1)d$

$\Rightarrow a+(k-1)d=5k+1$

$\Rightarrow a-d+kd$

Comparing the constant coefficient and coefficient of k, we get

$d=5$ and $a-d=1 \Rightarrow a=1+5=6$

Thus, the sum to n terms of the AP is

$S_n=\frac{n}{2}[2a+(n-1)d]$

$=\frac{n}{2}[2(6)+(n-1)(5)]=\frac{n}{2}(5n+7)$

Q8. If the sum of n terms of an AP is $(pn+qn^2)$, where p and q are constant, find the common difference.

Solution:

The sum of n terms of an AP is

$S_n=\frac{n}{2}[2a+(n-1)d]$, and therefore

$$\frac{n}{2}[2a+(n-1)d]=pn+qn^2$$

$$\Rightarrow \frac{n}{2}[2a-d+nd]=pn+qn^2$$

$$\Rightarrow \frac{2a-d}{2}n+\frac{d}{2}n^2=pn+qn^2$$

Comparing the coefficient of n and n^2, we get

$$\Rightarrow p=\frac{2a-d}{2} \text{ and } q=\frac{d}{2} \Rightarrow d=2q$$

Thus, the common difference of the AP is $2q$.

Q9. The sum of n terms of two arithmetic progressions are in the ratio $5n+4:9n+6$. Find the ratio of their 18th terms.

Solution:

Let a_1, a_2, d_1, d_2 be the first terms and all common difference of the first and second arithmetic progression respectively.

Then according to the given condition,

$$\frac{\text{Sum of } n \text{ terms of first AP}}{\text{Sum of } n \text{ terms of second AP}}=\frac{5n+4}{9n+6}$$

$$\Rightarrow \frac{\frac{n}{2}[2a_1+(n-1)d_1]}{\frac{n}{2}[2a_2+(n-1)d_2]}=\frac{5n+4}{9n+6}$$

$$\Rightarrow \frac{2a_1+(n-1)d_1}{2a_2+(n-1)d_2}=\frac{5n+4}{9n+6} \quad \text{(i)}$$

Substituting $n=35$ in equation (i), we get

$$\frac{2a_1+34d_1}{2a_2+34d_2}=\frac{5(35)+4}{9(35)+9}$$

$$\Rightarrow \frac{a_1+17d_1}{a_2+17d_2}=\frac{179}{321} \quad \text{(ii)}$$

Now, the ratio of the 18th terms of both the APs is

$$\frac{18^{th} \text{ term of first AP}}{18^{th} \text{ term of second AP}}$$

$$=\frac{a_1+17d_1}{a_2+17d_2}=\frac{179}{321} \text{ from equation (ii)}$$

Q10. If the sum of the first p terms of an AP is equal to the sum of the first q terms, then find the sum of the first $(p+q)$ terms.

Solution:

Let a and d be the first terms and the common difference of the AP, respectively. Then

$$S_p=\frac{p}{2}[2a+(n-1)d]$$

$$S_q=\frac{q}{2}[2a+(q-1)d]$$

The sum of the first p terms of an AP is equal to the sum of the first q terms of an AP, and therefore

$$\frac{p}{2}[2a+(p-1)d]=\frac{q}{2}[2a+(q-1)d]$$

$$\Rightarrow p[2a+(p-1)d]=q[2a+(q-1)d]$$

$$\Rightarrow 2ap+pd(p-1)=2aq+qd(q-1)$$

$$\Rightarrow 2a(p-q)+d[p(p-1)-q(q-1)]=0$$

$$\Rightarrow 2a(p-q)+d[p^2-p+q^2+q]=0$$

$$\Rightarrow 2a(p-q)+d[(p-q)(p+q-1)]=0$$

$$\Rightarrow (p-q)\{2a+d(p+q-1)\}=0$$

$$\Rightarrow d=\frac{-2a}{p+q-1}, \quad (\because p\neq q)$$

Thus, the sum of the first $(p+q)$ terms of the AP is

$$S_{p+q}=\frac{p+q}{2}[2a+(p+q-1)\cdot d]$$

$$=\frac{p+q}{2}\left[2a+(p+q-1)\left(\frac{-2a}{p+q-1}\right)\right]$$

$$=\frac{p+q}{2}[2a-2a]=0$$

Q11. The sums of the first p, q and r terms of an AP are a, b, c, respectively. Prove that $\frac{a}{p}(q-r)+\frac{b}{q}(r-p)+\frac{c}{r}(p-q)=0$.

Solution:

Let a_1 and d be the first term and the common difference of the AP, respectively. Then

$$S_p=\frac{p}{2}[2a_1+(p-1)d]=a$$

$$\Rightarrow 2a_1+(p-1)d=\frac{2a}{p} \quad \text{(i)}$$

$$S_q=\frac{q}{2}[2a_1+(q-1)d]=b$$

$$\Rightarrow 2a_1+(q-1)d=\frac{2b}{q} \quad \text{(ii)}$$

$$S_r=\frac{r}{2}[2a_1+(r-1)d]=c$$

$$\Rightarrow 2a_1+(r-1)d=\frac{2c}{r} \quad \text{(iii)}$$

Subtracting equation (ii) from equation (i), we get

$$(p-1)d-(q-1)d=\frac{2a}{p}-\frac{2b}{q}$$

$$\Rightarrow d(p-1-q+1)=\frac{2aq-2bq}{pq}$$

$$\Rightarrow d(p-q)=\frac{2aq-2np}{pq}$$

$$\Rightarrow d=\frac{2(aq-bp)}{pq(p-q)} \quad \text{(iv)}$$

Subtracting equation (iii) from equation (ii), we get

$$(q-1)d-(r-1)d=\frac{2b}{q}-\frac{2c}{r}$$

$$\Rightarrow d(q-1-r+1)=\frac{2b}{q}-\frac{2c}{r}$$

$$\Rightarrow d(q-r)=\frac{2br+2qc}{qr}$$

$$\Rightarrow d=\frac{2(br-qc)}{qr(q-r)} \quad \text{(v)}$$

Equating equations (iv) and (v) for the value of d, we get

$$\frac{aq-bp}{pq(p-q)}=\frac{br-qc}{qr(q-r)}$$

$$\Rightarrow qr(q-r)(aq-bq)=pq(p-q)(br-qc)$$

$$\Rightarrow r(aq-bp)(q-r)=p(br-qc)(p-q)$$

$$\Rightarrow (aqr-bqr)(q-r)=(bpr-pqc)(p-q)$$

Dividing both sides by pqr, we get

$$\left(\frac{a}{p}-\frac{b}{q}\right)(q-r)=\left(\frac{b}{q}-\frac{c}{r}\right)(p-q)$$

$$\Rightarrow \frac{a}{p}(q-r)-\frac{b}{q}(q-r+p-q)+\frac{c}{r}(p-q)=0$$

$$\Rightarrow \frac{a}{p}(q-r)+\frac{b}{q}(r-p)+\frac{c}{r}(p-q)=0$$

Q12. The ratio of the sums of m and n terms of an AP is $m^2:n^2$. Show that the ratio of the mth and n^{th} term is $(2m-1):(2n-1)$.

Solution:

Let a and d be the first term and the common difference of the AP, respectively. Then

$$\frac{\text{Sum of } m \text{ terms of an AP}}{\text{Sum of } n \text{ terms of an AP}}=\frac{m^2}{n^2}$$

$$\Rightarrow \frac{\frac{m}{2}[2a+(m-1)d]}{\frac{n}{2}[2a+(n-1)d]}=\frac{m^2}{n^2}$$

$$\Rightarrow \frac{2a+(m-1)d}{2a+(n-1)d}=\frac{m}{n} \quad \text{(i)}$$

Substituting $m=2m-1,\ n=2n-1$ in Eq. (i), we get

$$\frac{2a+(2m-2)d}{2a+(2n-2)d}=\frac{2m-1}{2n-1}$$

$$\Rightarrow \frac{a+(m-1)d}{a+(n-1)}=\frac{2m-1}{2n-1}$$

$$\Rightarrow \frac{m^{\text{th}} \text{ term of AP}}{n^{\text{th}} \text{ term of AP}}$$

$$=\frac{a+(m-1)d}{a+(n-1)d}=\frac{2m-1}{2n-1}$$

Q13. If the sum of n terms of an AP is $2n^2+5n$ and its mth term is 164, find the value of m.

Solution:

Let a and d be the first term and the common difference of the AP, respectively. Then

Sum of n terms of an AP is

$$S_n=\frac{n}{2}[2a+(n-1)d]$$

$$\Rightarrow \frac{n}{2}[2a-d+nd]=3n^2+5n$$

$$\Rightarrow \frac{2a-d}{2}n+\frac{d}{2}n^2=5n+3n^2$$

Comparing the coefficient of n and n^2, we get

$$\Rightarrow \frac{2a-d}{2}=5 \text{ and } \frac{d}{2}=3 \Rightarrow d=6$$

Also, $a-\frac{d}{2}=5 \Rightarrow a-3=5 \Rightarrow a=8$

The mth term is equal to 164, and therefore $a_m=a+(m-1)d=164$

$$\Rightarrow 8+(m-1)6=164$$

$$\Rightarrow (m-1)6=156$$

$$\Rightarrow m-1=26 \Rightarrow m=27$$

Q14. Insert five numbers between 8 and 26 such the resulting sequence is an AP.

Solution:

Let A_1,A_2,A_3,A_4,A_5 be the five numbers between 8 and 26 such that 8, A_1,A_2,A_3,A_4,A_5 26 is an AP.

Thus, $a=8,\ l=26,\ n=7$

Therefore, $l=a+(n-1)d \Rightarrow 26=8+(7-1)d$

$\Rightarrow 6d=26-8=18 \Rightarrow d=3$

$\therefore A_1=a+d=8+3=11$

$A_2=a+2d=8+2\times3=8+6=14$

$A_3=a+3d=8+3\times3=8+9=17$

$A_4=a+4d=8+4\times3=8+12=20$

$A_5=a+5d=8+5\times3=8+15=23$

Hence, the required five numbers between 8 and 26 are 11, 14, 17, 20, 23, which makes an AP.

Q15. If $\frac{a^n+b^n}{a^{n-1}+b^{n-1}}$ is the AM between a and b, then find the value of n.

Solution:

The AM of a and b is $=\frac{a+b}{2}$

Therefore, $\frac{a+b}{2}=\frac{a^n+b^n}{a^{n-1}+b^{n-1}}$

$\Rightarrow (a+b)(a^{n-1}+b^{n-1}) = 2(a^n+b^n)$

$\Rightarrow a^n + ab^{n-1} + ba^{n-1} + b^n = 2a^n + 2b^n$

$\Rightarrow ab^{n-1} + a^{n-1}b = a^n + b^n$

$\Rightarrow ab^{n-1} - b^n = a^n - a^{n-1}b$

$\Rightarrow b^{n-1}(a-b) = a^{n-1}(a-b)$

$\Rightarrow b^{n-1} = a^{n-1} \Rightarrow \left(\frac{b}{a}\right)^{n-1} = 1 = \left(\frac{b}{a}\right)^0$

$\Rightarrow n-1 = 0 \Rightarrow n = 1$

Q16. Between 1 and 31, m numbers have been inserted in such a way that the resulting sequence is an AP and the ratio of the 7th and $(m-1)$th numbers is 5:9. Find the value of m.

Solution:

Let $A_1, A_2 ....A_m$ be m numbers such that $1, A_1, A_2 ....A_m$, 31 is an AP.

Thus, $a = 1,\ l = 31,\ n = m+2$

$\therefore l = a + (n-1)d \Rightarrow 31 = 1 + (m+2-1)(d)$

$\Rightarrow 30 = (m+1)d \Rightarrow d = \frac{30}{m+1}$

We get $A_1 = a + d$

$A_2 = a + 2d$

$A_3 = a + 3d, ...$

$A_7 = a + 7d$

$A_{m-1} = a + (m-1)d$

The ratio of the 7th and $(m-1)$th numbers is 5:9, and therefore

$$\frac{a+7d}{a+(m-1)d} = \frac{5}{9}$$

$$\Rightarrow \frac{1+7\left(\frac{30}{(m+1)}\right)}{1+(m-1)\left(\frac{30}{m+1}\right)} = \frac{5}{9}$$

$$\Rightarrow \frac{m+1+210}{m+1+30m-30} = \frac{5}{9}$$

$$\Rightarrow \frac{m+211}{31m-29} = \frac{5}{9}$$

$\Rightarrow 9m + 1899 = 155m - 145$

$\Rightarrow 146m = 2044 \Rightarrow m = 14$

Thus, the value of m is 14.

Q17. A man starts repaying a loan with a first instalment of Rs. 100. If he increases the instalment by Rs 5 every month, what amount he will pay in the 30th instalment?

Solution:

The first installment of the loan is Rs 100. Therefore, the second installment of the loan is Rs 105 and so on.

Thus the amount that the man repays every month forms an AP, given by 100, 105, 110,..., where the first term $a = 100$, common difference, $d = 5$.

Hence, the amount to be paid in the 30th instalment is $A_{30} = a + (30-1)d = 100 + (29)(5) = \text{Rs } 245$

Q18. The difference between any two consecutive interior angles of a polygon is 5°. If the smallest angle is 120°, find the number of the sides of the polygon.

Solution:

The angles of the polygon will form an AP with a common difference $d = 5°$ and first term $a = 120°$.

Also, the sum of all the angles of a polygon with n sides is $180° (n-2)$.

$\therefore S_n = 180°(n-2)$

$\Rightarrow \frac{n}{2}[2a + (n-1)d] = 180°(n-2)$

$\Rightarrow \frac{n}{2}[2 \times 120° + (n-1)5°] = 180°(n-2)$

$\Rightarrow 240n + 5n^2 - 5n = 360 - 720$

$\Rightarrow 5n^2 - 125n + 720 = 0$

$\Rightarrow n^2 - 25n + 144 = 0$

$\Rightarrow (n-9)(n-16) = 0$

$\Rightarrow n = 9$ or $n = 16$

Day 117

Geometric Progression and Geometric Mean

Geometric Progression (GP): In a geometric progression, the ratio of any two consecutive numbers is a constant.

A sequence $a_1, a_2, a_3, ..., a_n, ...$ is called a geometric sequence or geometric progression if $\frac{a_{n+1}}{a_n} = r,\ n \in N$, where a_1 is called the first term and the constant term r is called the common ratio of the GP.

Consider a GP with its first term a and common ratio r. The GP will be,

$a, ar, ar^2, ar^3, ...ar^{n-1}$ ...; and so on

Thus, 1st term $= a_1 = a$

2nd term $= a_2 = ar$

3rd term $= a_3 = ar^2$

4th term $= a_4 = ar^3$

In general, the nth term of the GP is $a_n = ar^{n-1}$.

Sum of *n* terms of a GP

S_n = the sum to n terms of GP

$\Rightarrow S_n = a + ar + ar^2 + ... + ar^{n-1}$ (i)

Case 1: When $r = 1$,

$S_n = a + a + a + ... + a$ (n terms) $= na$

Case 2: When $r \neq 1$,

Multiplying equation (i) by r, we get

$rS_n = ar + ar^2 + ar^3 + ... + ar^n$ (ii)

Subtracting equation (ii) from equation (i), we get

$(1-r)S_n = a - ar^n = a(1-r^n)$

$$\Rightarrow S_n = \frac{a(1-r^n)}{1-r} \text{ or } S_n = \frac{a(r^n-1)}{r-1}$$

Properties of a GP

(i) If each term of a GP is multiplied (or divided) by a non-zero constant, then the resulting sequence is also a GP.

(ii) But, if a constant is added (or subtracted) to each term of a GP, then the resulting sequence may not be a GP.

Geometric Mean (GM): A number G is called the geometric mean (GM) of the two given numbers a and b, if a, G, b are in GP, that is, $\frac{G}{a} = \frac{b}{G} \Rightarrow G = \sqrt{ab}$.

Similarly, we can insert two GM (G_1 and G_2) between a and b, such that a, G_1, G_2, b are in GP.

$$\frac{G_1}{a} = \frac{G_2}{G_1} = \frac{b}{G_2}$$

$$\Rightarrow G_1 = a\left(\frac{b}{a}\right)^{1/3} \text{ and } G_2 = a\left(\frac{b}{a}\right)^{2/3}$$

In general, for any two given numbers a and b, we can insert as many numbers as we like between them such that the resulting sequence is a GP.

Let $G_1, G_2, G_3, ..., G_n$ be n GM numbers between a and b such that $a, G_1, G_2, G_3, .., G_n, b$ forms a GP. Now, since the above sequence has $(n+2)$ terms, b will be the $(n+2)^{th}$ term, that is,

$$b = ar^{n+1} \Rightarrow r = \left(\frac{b}{a}\right)^{\frac{1}{n+1}}$$

Or, common ratio of above sequence $r = \left(\frac{b}{a}\right)^{\frac{1}{n+1}}$

Thus, n numbers between a and b are as follows:

$$G_1 = ar = a\left(\frac{b}{a}\right)^{\frac{1}{n+1}}$$

$$G_2 = ar^2 = a\left(\frac{b}{a}\right)^{\frac{2}{n+1}}$$

$$G_3 = ar^3 = a\left(\frac{b}{a}\right)^{\frac{3}{n+1}}, \text{ and so on}$$

$$G_n = ar^n = a\left(\frac{b}{a}\right)^{\frac{n}{n+1}}$$

Relationship Between AM and GM: Let A and G be the AM and GM of two given positive real numbers a and b, respectively. Then

$$A = \frac{a+b}{2} \text{ and } G = \sqrt{ab}$$

Rearranging, we get

$$A - G = \frac{a+b}{2} - \sqrt{ab} = \frac{a+b-2\sqrt{ab}}{2}$$

$$\Rightarrow A - G = \frac{\left(\sqrt{a}-\sqrt{b}\right)^2}{2} \geq 0$$

$$\Rightarrow A - G \geq 0 \Rightarrow A \geq G$$

Thus, we conclude that the arithmetic mean is always greater than or equal to the geometric mean.

NCERT Exercise 9.3 (With Solutions Q1–Q11)

Q1. Find the 20th and n^{th} terms of the GP $\frac{5}{2}, \frac{5}{4}, \frac{5}{8}, ...$

Solution:

The given GP is $\frac{5}{2}, \frac{5}{4}, \frac{5}{8}, ...$

First term, $a = \frac{5}{2}$

Common ratio, $r = \frac{5/4}{5/2} = \frac{1}{2}$

$$a_{20} = ar^{20-1} = \frac{5}{2}\left(\frac{1}{2}\right)^{19} = \frac{5}{(2)(2)^{19}} = \frac{5}{(2)^{20}}$$

$$a_n = ar^{n-1} = \frac{5}{2}\left(\frac{1}{2}\right)^{n-1} = \frac{5}{(2)(2)^{n-1}} = \frac{5}{(2)^n}$$

Q2. Find the 12th term of a GP whose 8th term is 192 and the common ratio is 2.

Solution:
Common ratio $r = 2$
Let a be the first term of the GP. Then
$a_8 = ar^{8-1} \Rightarrow ar^7 = 192 \Rightarrow a(2)^7 = 192$

$$\Rightarrow a = \frac{192}{2^7} = \frac{2^6 \times 3}{2^7} = \frac{3}{2}$$

$$\therefore a_{12} = ar^{12-1} = \left(\frac{3}{2}\right)(2)^{11} = (3)(2)^{10} = 3072$$

Q3. The 5th, 8th and 11th terms of a GP are p, q, s, respectively. Show that $q^2 = ps$.

Solution:
Let a be the first term and r be the common ratio of the GP. Then

$a_5 = ar^{5-1} = ar^4 = p$ (i)

$a_8 = ar^{8-1} = ar^7 = q$ (ii)

$a_{11} = ar^{11-1} = ar^{10} = s$ (iii)

Dividing equation (ii) by equation (i), we get

$$\frac{ar^7}{ar^4} = \frac{p}{q} \Rightarrow r^3 = \frac{p}{q} \quad \text{(iv)}$$

Dividing equation (3) by equation (2), we get

$$\frac{ar^{10}}{ar^7} = \frac{s}{q} \Rightarrow r^3 = \frac{s}{q} \quad \text{(v)}$$

Equating equations (iv) and (v) for the values of r^3, we get

$$\frac{q}{p} = \frac{s}{q} \Rightarrow q^2 = ps$$

Q4. The 4th term of a GP is the square of its second term, and the first term is –3. Determine its 7th term.

Solution:
First term of a GP is $a = -3$
Let r be the common ratio of the GP. Then
$a_n = ar^{n-1}$
The 4th term of the GP is the square of its 2nd term.
Therefore $a_4 = (a_2)^2 \Rightarrow ar^{4-1} = (ar^{2-1})^2$
$\Rightarrow ar^3 = a^2r^2 \Rightarrow r = a = -3$
Thus, the 7th term of the GP is
$a_7 = ar^{7-1} = ar^6 = (-3)(-3)^6 = -(3)^7 = -2187$

Q5. Which term of the following sequences:
(a) $2, 2\sqrt{2}, 4, \ldots$ is 128?
(b) $\sqrt{3}, 3, 3\sqrt{3}, \ldots$ is 729?
(c) $\frac{1}{3}, \frac{1}{9}, \frac{1}{27}, \ldots$ is $\frac{1}{19683}$?

Solution:
(a) The given sequence is $2, 2\sqrt{2}, 4, \ldots$ is 128

where $a = 2$, $r = (2\sqrt{2})/2 = \sqrt{2}$
Let the nth term of the given sequence be 128. Then
$a_n = ar^{n-1} \Rightarrow (2)(\sqrt{2})^{n-1} = 128$

$$\Rightarrow (2)(2)^{\frac{n-1}{2}} = (2)^7 \Rightarrow (2)^{\frac{n+2}{2}} = (2)^7$$

$$\Rightarrow \frac{n+1}{2} = 7 \Rightarrow n+1 = 14 \Rightarrow n = 13$$

Thus, the 13th term of the given sequence is 128.

(b) The given sequence is $\sqrt{3}, 3, 3\sqrt{3}, \ldots$, where
$a = \sqrt{3}$, $r = \frac{3}{\sqrt{3}} = \sqrt{3}$.
Let the nth term of the given sequence be 729. Then
$a_n = ar^{n-1} \Rightarrow ar^{n-1} = 729$

$$\Rightarrow (\sqrt{3})(\sqrt{3})^{n-1} = 729 \Rightarrow (3)^{\frac{1}{2}}(3)^{\frac{n-1}{2}} = (3)^6$$

$$\Rightarrow (3)^{\frac{n}{2}} = (3)^6 \Rightarrow \frac{n}{2} = 6 \Rightarrow n = 12$$

Thus, the 12th term of the given sequence is 729.

(c) The given sequence is $\frac{1}{3}, \frac{1}{9}, \frac{1}{27}, \ldots$, where
$a = \frac{1}{3}$, $r = \frac{1/9}{1/3} = \frac{1}{3}$
Let the nth term of the given sequence be $\frac{1}{19683}$. Then

$$a_n = ar^{n-1} = \frac{1}{19683}$$

$$\Rightarrow \left(\frac{1}{3}\right)\left(\frac{1}{3}\right)^{n-1} = \frac{1}{3^9}$$

$$\Rightarrow \left(\frac{1}{3}\right)^n = \left(\frac{1}{3}\right)^9 \Rightarrow n = 9$$

Thus, the 9th term of the given sequence is $\frac{1}{19683}$.

Q6. For what values of x, are the numbers $\frac{-2}{7}, x, \frac{-7}{2}$ in GP?

Solution:
If the given numbers $\frac{-2}{7}, x, \frac{-7}{2}$ are in GP, then

$$x^2 = \left(\frac{-2}{7}\right) \times \left(\frac{-7}{2}\right) \Rightarrow x^2 = 1 \Rightarrow x = \pm 1.$$

Thus, for $x = \pm 1$, the given numbers will be in GP.

Q7. Find the sum to 20 terms in the geometric progression 0.15, 0.015, 0.0015……

Solution:
Since $a = 0.15$, $r = \frac{0.015}{0.15} = 0.1$

$\therefore S_{20} = \dfrac{0.15[1-(0.1)^{20}]}{1-0.1}$

$= \dfrac{0.15}{0.9}[1-(0.1)^{20}] = \dfrac{1}{6}[1-(0.1)^{20}]$

Q8. Find the sum to n terms in the geometric progression $\sqrt{7}, \sqrt{21}, 3\sqrt{7}, \ldots\ldots$

Solution:

The given GP is $\sqrt{7}, \sqrt{21}, 3\sqrt{7}, \ldots\ldots$

Therefore, $a = \sqrt{7},\ r = \dfrac{\sqrt{21}}{\sqrt{7}} = \sqrt{3}$

Thus, the sum to n terms is $S_n = \dfrac{a(r^n - 1)}{r-1}$

$\Rightarrow S_n = \dfrac{\sqrt{7}[(\sqrt{3})^n - 1]}{\sqrt{3}-1}$

$\Rightarrow S_n = \dfrac{\sqrt{7}[(\sqrt{3})^n - 1]}{\sqrt{3}-1} \times \dfrac{\sqrt{3}+1}{\sqrt{3}+1}$

$\Rightarrow S_n = \dfrac{\sqrt{7}(\sqrt{3}+1)[(\sqrt{3})^n - 1]}{3-1}$

$\Rightarrow S_n = \dfrac{\sqrt{7}(\sqrt{3}+1)[(\sqrt{3})^n - 1]}{2}$

Q9. Find the sum to n terms in the geometric progression $1, -a, a^2, -a^3 \ldots\ldots$ (if $a \neq -1$).

Solution:

The given GP is $1, -a, a^2, -a^3 \ldots\ldots$

Therefore, first term $= a_1 = 1$

Common ratio $= r = -a$

Thus, sum to n terms is $S_n = \dfrac{a_1(1-r^n)}{1-r}$

$\Rightarrow S_n = \dfrac{1\times[1-(-a)^n]}{1-(-a)} = \dfrac{[1-(-a)^n]}{1+a}$

Q10. Find the sum to n terms in the geometric progression $x^3, x^5, x^7, \ldots\ldots$ (if $x \neq -1$).

Solution:

The given GP is $x^3, x^5, x^7, \ldots\ldots$

Therefore, $a = x^3,\ r = \dfrac{x^5}{x^3} = x^2$

Thus, sum to n terms is

$S_n = \dfrac{a(1-r^n)}{1-r} = \dfrac{x^3[1-(x^2)^n]}{1-x^2} = \dfrac{x^3(1-x^{2n})}{1-x^2}$

Q11. Evaluate $\sum_{k=1}^{11}(2+3^k)$.

Solution:

$\sum_{k=1}^{11}(2+3^k) = \sum_{k=1}^{11}(2) + \sum_{k=1}^{11}3^k = 2\times 11 + \sum_{k=1}^{11}3^k$

$= 22 + \sum_{k=1}^{11}3^k = 22 + (3^1 + 3^2 + 3^3 + \ldots. + 3^{11}) = 22 + S_{11}$

The terms of this sequence $3, 3^2, 3^3, \ldots.$ form a GP.

Therefore $S_{11} = \dfrac{3[(3)^{11}-1]}{3-1} = \dfrac{3}{2}(3^{11}-1)$.

Thus, $\sum_{k=1}^{11}(2+3^k) = 22 + S_{11} = 22 + \dfrac{3}{2}(3^{11}-1)$

Day 118

NCERT Exercise 9.3 (With Solutions Q12–Q32)

Q12. The sum of the first three terms of a GP is $\dfrac{39}{10}$ and their product is 1. Find the common ratio and the terms.

Solution:

Let $\dfrac{a}{r}, a, ar$ be the first three terms of the GP. Then

$\dfrac{a}{r} + a + ar = \dfrac{39}{10} \Rightarrow a\left(\dfrac{1}{r}+1+r\right) = \dfrac{39}{10}$

and $\left(\dfrac{a}{r}\right)(a)(ar) = 1 \Rightarrow a^3 = 1 \Rightarrow a = 1$

Thus, we get

$\dfrac{1}{r} + 1 + r = \dfrac{39}{10} \Rightarrow 1 + r + r^2 = \dfrac{39}{10}r$

$\Rightarrow 10r^2 - 29r + 10 = 0$

$\Rightarrow (5r-2)(2r-5) = 0 \Rightarrow r = \dfrac{2}{5}$ or $r = \dfrac{5}{2}$

Hence, the first three terms of the GP are $\dfrac{5}{2}, 1, \dfrac{2}{5}$ or $\dfrac{2}{5}, 1, \dfrac{5}{2}$.

Q13. How many terms of the GP $3, 3^2, 3^3 \ldots$ are needed to give the sum 120?

Solution:

The given GP is $3, 3^2, 3^3 \ldots$ Then, $a = 3$, $r = 3$.

Let the sum of n terms of this GP be equal to 120. Then

$$S_n = \frac{a(1-r^n)}{1-r} \Rightarrow 120 = \frac{3(3^n-1)}{3-1}$$

$\Rightarrow 80 = 3^n - 1 \Rightarrow 3^n = 81$

$\Rightarrow 3^n = 3^4 \Rightarrow n = 4$

Thus, four terms of the given GP are required to obtain the sum of 120.

Q14. The sum of the first three terms of a GP is 16 and the sum of the next three terms is 128. Determine the first term, the common ratio and the sum to n terms of the GP.

Solution:

Let the GP be $a, ar, ar^2, ar^3, ...$

The sum of the first three terms of the GP is 16 and the sum of the next three terms is 128. Therefore

$a + ar + ar^2 = 16$, and $ar^3 + ar^4 + ar^5 = 128$

$\Rightarrow a(1+r+r^2) = 16$ and $ar^3(1+r+r^2) = 128$

Solving the above two equations, we get

$$\frac{ar^3(1+r+r^2)}{a(1+r+r^2)} = \frac{128}{16} \Rightarrow r^3 = 8 \Rightarrow r = 2$$

Substituting $r = 2$, we get

$$a(1+2+4) = 16 \Rightarrow a = \frac{16}{7}$$

Thus, the sum to n terms of the GP is

$$S_n = \frac{a(r^n-1)}{r-1} = \frac{16/7(2^n-1)}{2-1} = \frac{16}{7}(2^n-1)$$

Q15. Given a GP with $a = 729$ and 7th term 64, determine the summation S_7.

Solution:

$a = 729$, $a_7 = 64$, and let r be the common ratio of the GP.

Then $a_n = ar^{n-1}$

$\Rightarrow a_7 = ar^{7-1} \Rightarrow 64 = 729r^6$

$$\Rightarrow r^6 = \frac{64}{729} \Rightarrow r^6 = \left(\frac{2}{3}\right)^6 \Rightarrow r = \frac{2}{3}$$

Therefore, summation $S_n = \dfrac{a(1-r^n)}{1-r}$

$$\Rightarrow S_7 = \frac{729\left[1-\left(\frac{2}{3}\right)^7\right]}{1-\frac{2}{3}} = 3\times 729\left[1-\left(\frac{2}{3}\right)^7\right]$$

$$\Rightarrow S_7 = (3)^7\left[\frac{(3)^7-(2)^7}{(3)^7}\right] = (3)^7-(2)^7 = 2059$$

Q16. Find a GP for which sum of the first two terms is -4 and the fifth term is 4 times the third term.

Solution:

Let a be the first term and r be the common ratio of the GP. Then $S_n = \dfrac{a(1-r^n)}{1-r}$

The fifth term of the GP is 4 times the third term. Therefore $a_5 = 4 \times a_3$

$\Rightarrow ar^4 = 4ar^2 \Rightarrow r^2 = 4 \Rightarrow r = \pm 2$

The sum of the first two terms of the GP is -4. Therefore

$$S_2 = 4 = \frac{a(1-r^2)}{1-r}).$$

For $r = 2$

$$\Rightarrow -4 = \frac{a[1-(2)^2]}{1-2} \Rightarrow -4 = \frac{a(1-4)}{-1} \Rightarrow a = \frac{-4}{3}$$

For $r = -2$

$$-4 = \frac{a[1-(2)^2]}{1-(-2)} \Rightarrow -4 = \frac{a(1-4)}{1+2} \Rightarrow a = 4$$

Thus, the required GP is

$$\frac{-4}{3}, \frac{-8}{3}, \frac{-16}{3} .... \text{or } 4, -8, 16, -32...$$

Q17. If the 4th, 10th and 16th terms of a GP are x, y, z, respectively, prove that x, y, z are in GP.

Solution:

Let a be the first term and r be the common ratio of the GP. Then

$a_4 = ar^3 = x$ (i)

$a_{10} = ar^9 = y$ (ii)

$a_{16} = ar^{15} = z$ (iii)

Dividing equation (ii) by equation (i), we get

$$\frac{y}{x} = \frac{ar^9}{ar^3} \Rightarrow \frac{y}{x} = r^6$$

Dividing equation (iii) by equation (ii), we get

$$\frac{z}{y} = \frac{ar^{15}}{ar^9} \Rightarrow \frac{z}{y} = r^6$$

$\therefore \dfrac{y}{x} = \dfrac{z}{y}$, and thus, x, y, z are in GP.

Q18. Find the sum to n terms of the sequence, 8, 88, 888, 8888.......

Solution:

The given sequence 8, 88, 888, 888, ..., can be written as $8 + 88 + 888 + ...$ to n terms.

$$= \frac{8}{9}[9 + 99 + 999 + ... \text{ to } n \text{ terms}]$$

$$= \frac{8}{9}[(10-1) + (10^2-1) + (10^3-1) + ... \text{to } n \text{ terms}]$$

$$= \frac{8}{9}[(10 + 10^2 + ...n \text{ terms}) - (1+1+1+...n \text{ terms})]$$

$$= \frac{8}{9}\left[\frac{10(10^n-1)}{10-1} - n\right] = \frac{8}{9}\left[\frac{10(10^n-1)}{9} - n\right]$$

$$= \frac{80}{81}(10^n-1) - \frac{8}{9}n$$

Q19. Find the sum of the products of the corresponding terms of the sequences 2, 4, 8, 16, 32 and 128, 32, 8, 2, 1/2.

Solution:
Sum of the products of the corresponding terms

$S_P = 2\times128+4\times32+8\times8+16\times2+32\times\frac{1}{2}$

$= 64\left[4+2+1+\frac{1}{2}+\frac{1}{2^2}\right] = 64S_5$

Since $4,2,1,\frac{1}{2},\frac{1}{2^2}$ is a GP with the first term, a = 4, common ratio $r = \frac{1}{2}$ and number of terms $n = 5$.

Thus, $S_n = \frac{a(1-r^n)}{1-r}$

$\Rightarrow S_5 = \frac{4\left[1-\left(\frac{1}{2}\right)^5\right]}{1-\frac{1}{2}} = \frac{4\left[1-\frac{1}{32}\right]}{1/2} = \frac{31}{4}$

Hence, $S_P = 64S_5 = 64\left(\frac{31}{4}\right) = 496$

Q20. Show that the products of the corresponding terms of the sequence $a, ar, ar^2, ..., ar^{n-1}$, and $A, Ar, Ar^2, ..., AR^{n-1}$ form a GP and find the common ratio.

Solution:
The products of the corresponding terms are $aA, arAr, ar^2Ar^2, ..., ar^{n-1}AR^{n-1}$, and are in GP. Therefore,

$\because \frac{\text{Second term}}{\text{First term}} = \frac{arAR}{aA} = rR$

Also $\frac{\text{Third term}}{\text{Second term}} = \frac{ar^2AR^2}{arAR} = rR$

Thus, the above sequence forms a GP with a common ratio rR.

Q21. Find four numbers forming a geometric progression in which the third term is greater than the first term by 9, and the second term is greater than the 4th by 18.

Solution:
Let a be the first term and r be the common ratio of the GP. Then $a_n = ar^{n-1}$

Thus,

$a_3 = a_1 + 9 \Rightarrow ar^2 = a + 9 \Rightarrow a(r^2-1) = 9$

$a_2 = a_4 + 18 \Rightarrow ar = ar^3 + 18 \Rightarrow ar(r^2-1) = -18$

Solving above two equations, we get

$\frac{ar(r^2-1)}{a(r^2-1)} = \frac{-18}{9} \Rightarrow r = -2$

Substituting the value of r, we get

$a(r^2-1) = 9 \Rightarrow a(4-1) = 9 \Rightarrow a = 3$

Thus, the first four numbers of the GP are

$3, 3(-2), 3(-2)^2, 3(-2)^3 \Rightarrow 3, -6, 12, -24$

Q22. If the $p^{\text{th}}, q^{\text{th}} r^{\text{th}}$ terms of a GP are a, b, c, respectively, prove that $a^{q-r}\times b^{r-p}\times c^{p-q} = 1$.

Solution:
Let A be the first term and R be the common ratio of the GP. Then

$AR^{p-1} = a, \quad Ar^{q-1} = b, \quad AR^{r-1} = c$

Therefore, $a^{q-r}\times b^{r-p}\times c^{p-q}$

$= \left(A\times R^{(p-1)}\right)^{(q-r)} \times \left(A\times R^{(q-1)}\right)^{(r-p)} \times \left(A\times R^{(r-1)}\right)^{(p-q)}$

$= A^{q-r}\times R^{(p-1)(q-r)} \times A^{r-p}\times R^{(q-1)(r-p)} \times A^{p-q}\times R^{(r-1)(p-q)}$

$= A^{q-r+r-p+p-q}\times R^{(pr-pr-q+r)+(rq-r+p-pq)+(pr-p-qr+q)}$

$= A^0 \times R^0 = 1$

Q23. If the first and the nth term of a GP are a and b, respectively, and if p is the product of n terms, prove that $P^2 = (ab)^n$.

Solution:
The first term of the GP is a and the nth term is b.
Therefore, the GP is $a, ar, ar^2 ... ar^{n-1}$, and $b = ar^{n-1}$ where r is the common ratio.
Thus, p = product of n terms

$\Rightarrow p = (a)(ar)(ar^2).....(ar^{n-1})$

$= (a\times a\times .....a)(r\times r^2\times ......r^{n-1})$

$= a^n\times r^{1+2+....(n-1)} = a^n r^{\frac{n(n-1)}{2}}$

$\Rightarrow p^2 = a^{2n}r^{n(n-1)} = [a\cdot ar^{(n-1)}]^n = a^n\cdot[ar^{(n-1)}]^n$

$\Rightarrow p^2 = a^n b^n = (ab)^n$

Q24. Show that the ratio of the sum of the first n terms of a GP to the sum of terms from $(n+1)^{\text{th}}$ to $(2n)^{\text{th}}$ term is $1/r^n$.

Solution:
Let a be the first term and r be the common ratio of the GP. Then

Sum of first n terms $= \frac{a(1-r^n)}{(1-r)}$

There are n terms from $(n+1)^{\text{th}}$ to $(2n)^{\text{th}}$ term. Therefore, the sum of terms from the $(n+1)^{\text{th}}$ to $(2n)^{\text{th}}$ term is

$S_n = \frac{a_{n+1}(1-r^n)}{1-r} = \frac{ar^n(1-r^n)}{1-r}.$

Thus, ratio $= \frac{a(1-r^n)}{(1-r)} \times \frac{(1-r)}{ar^n(1-r^n)} = \frac{1}{r^n}$

Thus, the ratio of the sum of first n terms of a GP to the sum of terms from the $(n+1)^{\text{th}}$ to the $(2n)^{\text{th}}$ term is $1/r^n$.

Q25. If a, b, c and d are in GP show that $(a^2+b^2+c^2)(b^2+c^2+d^2) = (ab+bc+cd)^2$.

Solution:

a, b, c and d are in GP. Therefore

$bc = ad, \ b^2 = ac, \ c^2 = bd$

Now, RHS $= (ab+bc+cd)^2 = (ab+ad+cd)^2$

$= [ab+d(a+c)]^2$

$= a^2b^2 + 2abd(a+c) + d^2(a+c)^2$

$= a^2b^2 + 2a^2bd + 2acbd + d^2(a^2+2ac+c^2)$

$= a^2b^2 + 2a^2c^2 + 2b^2c^2 + d^2a^2 + 2d^2b^2 + d^2c^2$

$= a^2b^2 + a^2c^2 + a^2c^2 + b^2c^2 + b^2c^2$

$+ d^2a^2 + d^2b^2 + d^2b^2 + d^2c^2$

$= a^2b^2 + a^2c^2 + a^2d^2 + b^2 \times b^2 + b^2c^2$

$+ b^2d^2 + c^2b^2 + c^2 \times c^2 + c^2d^2$

$= a^2(b^2+c^2+d^2) + b^2(b^2+c^2+d^2) + c^2(b^2+c^2+d^2)$

$= (a^2+b^2+c^2)(b^2+c^2+d^2) =$ LHS

Q26. Insert two numbers between 3 and 81 so that the resulting sequence is a GP.

Solution:

Let G_1 and G_2 be two numbers between 3 and 81 such that the series 3, G_1, G_2, 81, forms a GP.

Let a be the first term and r be the common ratio of the GP, then $a = 3$ and $81 = (3)(r)^3 \Rightarrow r^3 = 27 \Rightarrow r = 3$.

Thus, for $r = 3$,

$G_1 = ar = (3)(3) = 9$

$G_2 = ar^2 = (3)(3) = 27$

Thus, the two numbers are 9 and 27.

Q27. Find the value of n so that $\frac{a^{n+1}+b^{n+1}}{a^n+b^n}$ may be the geometric mean between a and b.

Solution:

$\frac{a^{n+1}+b^{n+1}}{a^n+b^n}$ is the geometric mean between a and b.

Therefore $\frac{a^{n+1}+b^{n+1}}{a^n+b^n} = \sqrt{ab}$.

Squaring both sides, we get

$\left(a^{n+1}+b^{n+1}\right)^2 = ab\left(a^n+b^n\right)^2$

$\Rightarrow a^{2n+2} + 2a^{n+1}b^{n+1} + b^{2n+2} = (ab)(a^{2n} + 2a^nb^b + b^{2n})$

$\Rightarrow a^{2n+2} + 2a^{n+1}b^{n+1} + b^{2n+1} = a^{2n+1}b + 2a^{n+1}b^{n+1} + ab^{2n+1}$

$\Rightarrow a^{2n+2} + b^{2n+2} = a^{2n+1}b + ab^{2n+1}$

$\Rightarrow a^{2n+2} - a^{2n+1}b = ab^{2n+1} - b^{2n+2}$

$\Rightarrow a^{2n+1}(a-b) = b^{2n+1}(a-b)$

$\Rightarrow \left(\frac{a}{b}\right)^{2n+1} = 1 = \left(\frac{a}{b}\right)^0$

$\Rightarrow 2n+1 = 0 \Rightarrow n = \frac{-1}{2}$

Q28. The sum of two numbers is 6 times their geometric mean, show that numbers are in the ratio $(3+2\sqrt{2}):(3-2\sqrt{2})$.

Solution:

Let the two numbers be a and b. Then GM $= \sqrt{ab}$

$\Rightarrow a+b = 6\sqrt{ab} \Rightarrow (a+b)^2 = 36(ab)$

$\Rightarrow a^2+b^2-34ab = 0$

Dividing by ab, we get

$\frac{a}{b} + \frac{b}{a} - 34 = 0$

Let us assume that the ratio $r = \frac{a}{b}$; then

$r + \frac{1}{r} - 34 = 0 \Rightarrow r^2 - 34r + 1 = 0$

$\therefore r = \frac{34 \pm \sqrt{34^2-4}}{2} \Rightarrow r = 17 \pm 12\sqrt{2}$

$\Rightarrow r = 17 + 12\sqrt{2}$ (considering positive value)

$\Rightarrow r = (3+2\sqrt{2})^2 = \frac{(3+2\sqrt{2})^2}{3^2-(2\sqrt{2})^2}$

$\Rightarrow \frac{a}{b} = \frac{3+2\sqrt{2}}{3-2\sqrt{2}}$

Thus, the required ratio is $(3+2\sqrt{2}):(3-2\sqrt{2})$.

Q29. If A and G be AM and GM, respectively between two positive numbers, prove that the numbers are $A \pm \sqrt{(A+G)(A-G)}$.

Solution:

A and G are the AM and GM between two positive numbers.

Let these two positive number be a and b. Then

AM $= A = \frac{a+b}{2}$ and GM $= G = \sqrt{ab}$

$\Rightarrow a+b = 2A$ and $ab = G^2$

Solving the above two equations, we get

$a + \frac{G^2}{a} = 2A \Rightarrow a^2 - (2A)a + G^2 = 0$

$\therefore a = \frac{2A \pm \sqrt{4A^2-4G^2}}{2}$

$\Rightarrow a = A \pm \sqrt{(A+G)(A-G)}$

Thus, if $a = A+\sqrt{(A+G)(A-G)}$. Then $b = A-\sqrt{(A+G)(A-G)}$, and vice versa.

Thus, the two numbers are $A \pm \sqrt{(A+G)(A-G)}$.

Q30. The number of bacteria in a certain culture doubles every hour. If there were 30 bacteria present in the culture originally, how many bacteria will be present at the end of the 2nd hour, 4th hour and n^{th} hour?

Solution:

The number of bacteria doubles every hour. Therefore, the number of bacteria after every hour will form a GP with $a = 30$ and $r = 2$.

Therefore, the number of bacteria present at the end of the 2nd hour is $a_3 = ar^2 = (30)(2)^2 = 120$

And the number of bacteria at the end of 4th hour is $a_5 = ar^4 = (30)(2)^4 = 480$

Thus, number of bacteria at the end of nth hour will be $a_{n+1} = ar^n = 30(2)^n$.

Q31. What will Rs 500 amount to in 10 years after it is deposited in a bank which pays an annual interest rate of 10% compounded annually?

Solution:

The amount deposited in the bank is Rs 500, and the bank pays an annual interest rate of 10% compounded annually. Therefore, the amount after n years will be $P_n = A(1+r)^n$.

Thus, at the end of ten years, amount

$$P_{10} = A(1+r)^{10} = 500\left(1+\frac{1}{10}\right)^{10} = 500(1.1)^{10}$$

Q32. If AM and GM of the roots of a quadratic equation are 8 and 5, respectively, obtain the quadratic equation.

Solution:

Let the root of the quadrant equation be α and β. Then

$$\text{AM} = \frac{\alpha+\beta}{2} = 8 \Rightarrow \alpha+\beta = 16$$

$$\text{GM} = \sqrt{\alpha\beta} = 5 \Rightarrow \alpha\beta = 25$$

Thus, the quadratic equation will be given by

$x^2 - x$ (Sum of roots) + (Product of roots) = 0

$$x^2 - x(\alpha+\beta) + (\alpha\beta) = 0$$

$$\Rightarrow x^2 - 16x + 25 = 0$$

Day 119

Special Series

(i) $\sum_{k=1}^{n} k = 1+2+3+4+...+n = \frac{n(n+1)}{2}$

(ii) $1^2+2^2+3^2+.....+n^2 = \left\{\frac{n(n+1)}{2}\right\}^2$

(iii) $1^3+2^3+3^3+.....+n^3 = \left[\frac{n(n+1)}{2}\right]^2$

(i) Sum of the first n natural numbers:

$$\sum_{k=1}^{n} k = 1+2+3+4+...+n = \frac{n(n+1)}{2}$$

Proof:

Series $1+2+3+4+...+n$ is an AP with the first term 1 and the number of terms as n; the sum of this AP is

$$S_n = \frac{n}{2}\{2a+(n-1)d\}$$

$$= \frac{n}{2}\{2\times1+(n-1)\times1\}$$

$$\Rightarrow S_n = 1+2+3+4+...+n = \frac{n(n+1)}{2}$$

(ii) Sum of squares of the first n natural numbers:

$$1^2+2^2+3^2+.....+n^2 = \left\{\frac{n(n+1)}{2}\right\}^2$$

Proof:

Let $S_n = 1^2+2^2+3^2+.....+n^2 = \sum_{k=1}^{n} k^2$

Using the identity $k^3-(k-1)^3 = 3k^2-3k+1$

Putting $k = 1, 2, ...n$, successively, we get

$$1^3-0^3 = 3(1)^2-3(1)+1$$

$$2^3-1^3 = 3(2)^2-3(2)+1$$

$$3^3-2^3 = 3(3)^2-3(3)+1$$

$$n^3-(n-1)^3 = 3(n)^2-3(n)+1$$

Adding both sides, we get

$$n^3-0^3 = 3(1^2+2^2+3^2+..+n^2) - 3(1+2+3+...+n)+n$$

$$\Rightarrow n^3 = 3\sum_{k=1}^{n} k^2 - 3\sum_{k=1}^{n} k + n$$

$$\Rightarrow \sum_{k=1}^{n} k^2 = \frac{1}{3}\left[n^3 + 3\sum_{k=1}^{n} k - n\right]$$

$$\Rightarrow \sum_{k=1}^{n} k^2 = \frac{1}{3}\left[n^3 + \frac{3n(n+1)}{2} - n\right]$$

$$\Rightarrow \sum_{k=1}^{n} k^2 = \frac{1}{3}\left[\frac{2n^3 + 3n^2 + n}{2}\right]$$

$$\Rightarrow \sum_{k=1}^{n} k^2 = \frac{1}{6}(2n^3 + 3n^2 + n)$$

$$\Rightarrow S_n = \sum_{k=1}^{n} k^2 = \frac{n(n+1)(2n+1)}{6}$$

(iii) Sum of cubes of the first n natural numbers:

$$1^3 + 2^3 + 3^3 + \ldots.. + n^3 = \left[\frac{n(n+1)}{2}\right]^2$$

Proof:

Let $S_n = 1^3 + 2^3 + 3^3 + \ldots.. + n^3 = \sum_{k=1}^{n} k^3$

Using the identity,

$(k+1)^4 - k^4 = 4k^3 + 6k^2 + 4k + 1$

Putting $k = 1, 2, 3, \ldots., n$, we get

$2^4 - 1^4 = 4(1)^3 + 6(1)^2 + 4(1) + 1$

$3^4 - 2^4 = 4(2) + 6(2)^2 + 4(2) + 1$

$4^4 - 3^4 = 4(3) + 6(3)^2 + 4(3) + 1$

____

____

$n^4 - (n-1)^4 = 4(n-1)^3 + 6(n-1)^2 + 4(n-1) + 1$

$(n+1)^4 - n^4 = 4n^3 + 6n^2 + 4n + 1$

Adding both sides, we get

$(n+1)^4 - 1^4 = 4(1^3 + 2^3 + .. + n^3) +$
$6(1^2 + 2^2 + 3^2 + \ldots + n^2) + 4(1 + 2 + 3 + \ldots + n) + n$

$$\Rightarrow (n+1)^4 - 1^4 = 4\sum_{k=1}^{n} k^3 + 6\sum_{k=1}^{n} k^2 + 4\sum_{k=1}^{n} k + n$$

$$\Rightarrow \{(n+1)^2 - 1^2\}\{(n+1)^2 + 1^2\}$$

$$= 4\sum_{k=1}^{n} k^3 + 6\frac{n(n+1)(2n+1)}{6} + 4\frac{n(n+1)}{2} + n$$

$$\Rightarrow 4\sum_{k=1}^{n} k^3 = n(n+2)\{(n+1)^2 + 1^2\}$$
$$-n(n+1)(2n+1) - 2n(n+1) - n$$

Simplifying and solving, we get $4\sum_{k=1}^{n} k^3 = n^2(n+1)^2$

$$\Rightarrow S_n = \sum_{k=1}^{n} k^3 = \left[\frac{n(n+1)}{2}\right]^2$$

Example 1: Find the sum to n terms of the series whose n^{th} term is given by $n(n+1)(n-2)$.

Solution:

The nth term of a series is

$a_n = n(n+1)(n-2) = n^3 - n^2 - 2n$

Therefore, $S_n = \sum_{k=1}^{n} a_k$

$$= \sum_{k=1}^{n} k^3 - \sum_{k=1}^{n} k^2 - 2\sum_{k=1}^{n} k$$

$$= \frac{n^2(n+1)^2}{4} - \frac{n(n+1)(2n+1)}{6} - 2\frac{n(n+1)}{2}$$

$$= \frac{n(n+1)}{2}\left[\frac{n(n+1)}{2} - \frac{(2n+1)}{3} - 2\right]$$

$$= \frac{n(n+1)}{2}\left[\frac{3n^2 + 3n - 4n - 2 - 12}{6}\right]$$

$$= \frac{n(n+1)}{2}\left[\frac{3n^2 - n - 14}{6}\right]$$

$$= \frac{n(n+1)(3n^2 + 6n - 7n - 14)}{12}$$

$$= \frac{n(n+1)(n+2)(3n-7)}{12}$$

Example 2: Find the sum to n terms of the series $1\times1^2 + 3\times2^2 + 5\times3^2 + 7\times4^2 + \ldots.$

Solution:

n^{th} term of the series $1\times1^2 + 3\times2^2 + 5\times3^2 + 7\times4^2 + \ldots.$ is

$a_n = (2n-1)n^2 = 2n^3 - n^2$

Therefore, $S_n = \sum_{k=1}^{n} a_k$

$$= \sum_{k=1}^{n}(2k^3 - k^2) = 2\sum_{k=1}^{n} k^3 - \sum_{k=1}^{n} k^2$$

$$= 2\times\frac{n^2(n+1)^2}{4} - \frac{n(n+1)(2n+1)}{6}$$

$$= \frac{n(n+1)}{2}\left\{n(n+1) - \frac{(2n+1)}{3}\right\}$$

$$= \frac{n(n+1)}{2}\left\{\frac{3n^2 + 3n - 2n - 1}{3}\right\}$$

$$= \frac{n(n+1)(3n^2 + n - 1)}{6}$$

NCERT Exercise 9.4 (With Solutions)

Q1. Find the sum to n terms of the series $1\times2+2\times3+3\times4+4\times5+\ldots\ldots$

Solution:

The nth term of the series $1\times2+2\times3+3\times4+\ldots$ is $a_n=n(n+1)$

Therefore, $S_n=\sum_{k=1}^{n}a_k$

$$\sum_{k=1}^{n}k(k+1)=\sum_{k=1}^{n}k^2+\sum_{k=1}^{n}k$$

$$=\frac{n(n+1)(2n+1)}{6}+\frac{n(n+1)}{2}$$

$$=\frac{n(n+1)}{2}\left(\frac{2n+1}{3}+1\right)$$

$$=\frac{n(n+1)}{2}\left(\frac{2n+4}{3}\right)$$

$$=\frac{n(n+1)(n+2)}{3}$$

Q2. Find the sum to n terms of the series $1\times2\times3+2\times3\times4+3\times4\times5+\ldots\ldots$

Solution:

The nth term of the series $1\times2\times3+2\times3\times4+3\times4\times5+\ldots\ldots$ is

$a_n=n(n+1)(n+2)=n^3+3n^2+2n$

Therefore, $S_n=\sum_{k=1}^{n}a_k$

$$=\sum_{k=1}^{n}k^3+3\sum_{k=1}^{n}k^2+2\sum_{k=1}^{n}k$$

$$=\left[\frac{n(n+1)}{2}\right]^2+\frac{3n(n+1)(2n+1)}{6}+\frac{2n(n+1)}{2}$$

$$=\frac{n(n+1)}{2}\left[\frac{n(n+1)}{2}+(2n+1)+2\right]$$

$$=\frac{n(n+1)}{2}\left[\frac{n^2+n+4n+6}{2}\right]$$

$$=\frac{n(n+1)}{4}(n^2+5n+6)$$

$$=\frac{n(n+1)}{4}(n+2)(n+3)=\frac{n(n+1)(n+2)(n+3)}{4}$$

Q3. Find the sum to n terms of the series $3\times1^2+5\times2^2+7\times3^2+\ldots\ldots\ldots$

Solution:

The nth term of the series $3\times1^2+5\times2^2+7\times3^2+\ldots$ is $a_n=(2n+1)n^2=2n^3+n^2$.

Therefore, $S_n=\sum_{k=1}^{n}a_k$

$$=\sum_{k=1}^{n}(2k^3+k^2)=2\sum_{k=1}^{n}k^3+\sum_{k=1}^{n}k^2$$

$$=2\left[\frac{n(n+1)}{2}\right]^2+\frac{n(n+1)(2n+1)}{6}$$

$$=\frac{n^2(n+1)^2}{2}+\frac{n(n+1)(2n+1)}{6}$$

$$=\frac{n(n+1)}{2}\left[n(n+1)+\frac{2n+1}{3}\right]$$

$$=\frac{n(n+1)}{2}\left[\frac{3n^2+3n+2n+1}{3}\right]$$

$$=\frac{n(n+1)}{2}\left[\frac{3n^2+5n+1}{3}\right]$$

$$=\frac{n(n+1)(3n^2+5n+1)}{6}$$

Q4. Find the sum to n terms of the series $\frac{1}{1\times2}+\frac{1}{2\times3}+\frac{1}{3\times4}+\ldots$

Solution:

The nth term of the series $\frac{1}{1\times2}+\frac{1}{2\times3}+\frac{1}{3\times4}+\ldots$ is

$$a_n=\frac{1}{n(n+1)}=\frac{1}{n}-\frac{1}{n+1}$$

Therefore, $a_1=\frac{1}{1}-\frac{1}{2}$, $a_2=\frac{1}{2}-\frac{1}{3}$.

$a_3=\frac{1}{3}-\frac{1}{4},\ldots$ $a_n=\frac{1}{n}-\frac{1}{n+1}$

Thus, $S_n=\sum_{k=1}^{n}a_k=a_1+a_2+\ldots+a_n$

$$=\left(\frac{1}{1}-\frac{1}{2}\right)+\left(\frac{1}{2}-\frac{1}{3}\right)+\left(\frac{1}{3}-\frac{1}{4}\right)+\ldots+\left(\frac{1}{n}-\frac{1}{n+1}\right)$$

$$\Rightarrow S_n=1-\frac{1}{n+1}=\frac{n+1-1}{n+1}=\frac{n}{n+1}$$

Q5. Find the sum to n terms of the series $5^2+6^2+7^2+\ldots+20^2$.

Solution:

The nth term of the series $5^2+6^2+7^2+\ldots+20^2$ is $a_n=(n+4)^2=n^2+8n+16$.

Therefore, $S_n = \sum_{k=1}^{n} a_k$

$= \sum_{k=1}^{n} (k^2 + 8k + 16)$

$= \sum_{k=1}^{n} k^2 + 8\sum_{k=1}^{n} k + \sum_{k=1}^{n} 16$

$= \frac{n(n+1)(2n+1)}{6} + \frac{8n(n+1)}{2} + 16n$

Now, the 16th term is $(16+4)^2 = 20^2$

$\therefore S_{16} = \frac{16(16+1)(2\times16+1)}{6} + \frac{8\times16\times(16+1)}{2} + 16\times16$

$= \frac{(16)(17)(33)}{6} + \frac{(8)(16)(17)}{2} + 256 = 2840$

Thus, $5^2 + 6^2 + 7^2 + ... + 20^2 = 2840$

Q6. Find the sum to n terms of the series $3\times8 + 6\times11 + 9\times14 + ...$

Solution:

The nth term of the series $3\times8 + 6\times11 + 9\times14 + ...$ is $a_n = (3n)(3n+5) = 9n^2 + 15n$

Therefore, $S_n = \sum_{k=1}^{n} a_k$

$= \sum_{k=1}^{n} (9k^2 + 15k) = 9\sum_{k=1}^{n} k^2 + 15\sum_{k=1}^{n} k$

$= 9\times\frac{n(n+1)(2n+1)}{6} + 15\times\frac{n(n+1)}{2}$

$= \frac{3n(n+1)(2n+1)}{2} + \frac{15n(n+1)}{2}$

$= \frac{3n(n+1)}{2}(2n+1+5)$

$= \frac{3n(n+1)}{2}(2n+6) = 3n(n+1)(n+3)$

Q7. Find the sum to n terms of the series $1^2 + (1^2+2^2) + (1^2+2^2+3^2) + ...$

Solution:

The nth term of the series $1^2 + (1^2+2^2) + (1^2+2^2+3^2) + ...$ is

$a_n = (1^2 + 2^2 + 3^2 + ... + n^2) = \frac{n(n+1)(2n+1)}{6}$

$= \frac{n^3 + 3n^2 + n}{6} = \frac{1}{3}n^3 + \frac{1}{2}n^2 + \frac{1}{6}n$

Therefore, $S_n = \sum_{k=1}^{n} a_k$

$= \sum_{k=1}^{n}\left(\frac{1}{3}k^3 + \frac{1}{2}k^2 + \frac{1}{6}k\right)$

$= \frac{1}{3}\sum_{k=1}^{n} k^3 + \frac{1}{2}\sum_{k=1}^{n} k^2 + \frac{1}{6}\sum_{k=1}^{n} k$

$= \frac{1}{3}\frac{n^2(n+1)^2}{(2)^2} + \frac{1}{2}\times\frac{n(n+1)(2n+1)}{6} + \frac{1}{6}\times\frac{n(n+1)}{2}$

$= \frac{n(n+1)}{6}\left[\frac{n(n+1)}{2} + \frac{(2n+1)}{2} + \frac{1}{2}\right]$

$= \frac{n(n+1)}{6}\left[\frac{n^2+3n+2}{2}\right]$

$= \frac{n(n+1)}{6}\left[\frac{(n+1)(n+2)}{2}\right]$

$= \frac{n(n+1)^2(n+2)}{12}$

Q8. Find the sum to n terms of the series whose n^{th} term is given by $n(n+1)(n+4)$.

Solution:

The nth term of the series is

$a_n = n(n+1)(n+4) = n^3 + 5n^2 + 4n$

Therefore, $S_n = \sum_{k=1}^{n} a_k$

$= \sum_{k=1}^{n} k^3 + 5\sum_{k=1}^{n} k^2 + 4\sum_{k=1}^{n} k$

$= \frac{n^2(n+1)^2}{4} + \frac{5n(n+1)(2n+1)}{6} + \frac{4n(n+1)}{2}$

$= \frac{n(n+1)}{2}\left[\frac{n(n+1)}{2} + \frac{5(2n+1)}{3} + 4\right]$

$= \frac{n(n+1)}{2}\left[\frac{3n^2 + 3n + 20n + 10 + 24}{6}\right]$

$= \frac{n(n+1)}{2}\left[\frac{3n^2 + 23n + 34}{6}\right]$

$= \frac{n(n+1)(3n^2 + 23n + 34)}{12}$

Q9. Find the sum to n terms of the series whose n^{th} term is given by $n^2 + 2^n$.

Solution:

The nth term of a series is $a_n = n^2 + 2^n$

Therefore, $S_n = \sum_{k=1}^{n} a_k$

$= \sum_{k=1}^{n}(k^2+2^k) = \sum_{k=1}^{n}k^2 + \sum_{k=1}^{n}2^k$

$\sum_{k=1}^{n}2^k = 2^1+2^2+3^3+....$ forms a GP with both the first term and common ratio equal to 2. Therefore

$\sum_{k=1}^{n}2^k = \frac{(2)[(2)^n-1]}{2-1} = 2(2^n-1)$

Thus, $S_n = \sum_{k=1}^{n}a_k = \frac{n(n+1)(2n+1)}{6} + 2(2^n-1)$

Q10. Find the sum to n terms of the series whose n^{th} term is given by $(2n-1)^2$.

Solution:

The nth term of the series is $a_n = (2n-1)^2 = 4n^2-4n+1$

Therefore, $S_n = \sum_{k=1}^{n}a_k$

$S_n = \sum_{k=1}^{n}a_k = \sum_{k=1}^{n}(4k^2-4k+1)$

$= 4\sum_{k=1}^{n}k^2 - 4\sum_{k=1}^{n}k + \sum_{k=1}^{n}1$

$= \frac{4n(n+1)(2n+1)}{3} - \frac{4n(n+1)}{2} + n$

$= \frac{2n(n+1)(2n+1)}{3} - 2n(n+1) + n$

$= n\left[\frac{2(2n^2+3n+1)-6(n+1)+3}{3}\right]$

$= n\left[\frac{4n^2+6n+2-6n-6+3}{3}\right]$

$= n\left[\frac{4n^2-1}{3}\right]$

$= \frac{n(2n+1)(2n-1)}{3}$

Day 120

Infinite Geometric Progression (GP)

Geometric progression $a, ar, ar^2, ar^3, ...$ with the first term a and common ratio r ($|r|<1$), is known as an infinite geometric progression.

$|r|<1$, and therefore, $r^n \to 0$ when n is sufficiently large, $n \to \infty$ and thus, the summation of infinite geometrical series is

$S_n = a + ar + ar^2 + ... + ar^{n-1}$

$\Rightarrow S_n = \frac{a(1-r^n)}{1-r}$

$\Rightarrow S_\infty = \frac{a}{1-r}$

Since $r^n \to 0$ when $n \to \infty$

NCERT Exercise 9.5 (With Solutions) Supplement Exercise

Find the sum to infinity in each of the following geometric progressions

Q1. $1, \frac{1}{3}, \frac{1}{9}.....$

Solution: Since $a = 1,\ r = \frac{1}{3} < 1$

Summation, $S = 1 + \frac{1}{3} + \frac{1}{9} + ...$

$\Rightarrow S_\infty = \frac{a}{1-r} = \frac{1}{1-\frac{1}{3}} = \frac{3}{2} = 1.5$

Q2. 6, 1.2, 0.24........

Solution:

Since $a = 6,\ r = 0.2 < 1$

Summation, $S = 6 + 1.2 + 0.24 + ...$

$\Rightarrow S_\infty = \frac{a}{1-r} = \frac{6}{1-0.2} = \frac{3}{0.4} = 7.5$

Q3. $5, \frac{20}{7}, \frac{80}{49},..........$

Solution:

Since $a = 5,\ r = \frac{4}{7} < 1$

Summation, $S = 5 + \frac{20}{7} + \frac{80}{49} + ...$

$\Rightarrow S_\infty = \frac{a}{1-r} = \frac{5}{1-\frac{4}{7}} = \frac{5\times7}{7-4} = \frac{35}{3}$

Q4. $\frac{-3}{4}, \frac{3}{16}, \frac{-3}{64},........$

Solution:

Since $a = \frac{-3}{4},\ r = \frac{-1}{4} \Rightarrow |r| = \frac{1}{4} < 1$

Therefore summation, $S = \frac{-3}{4} + \frac{3}{16} + \frac{-3}{64} + \ldots$

$$\Rightarrow S_{\infty} = \frac{a}{1-r} = \frac{-3/4}{1-\left(-\frac{1}{4}\right)} = \frac{-3}{5}$$

Q5. Prove that: $3^{\frac{1}{2}} \times 3^{\frac{1}{4}} \times 3^{\frac{1}{8}} \times \ldots = 3$

Solution:

$$\text{LHS} = 3^{\frac{1}{2}} \times 3^{\frac{1}{4}} \times 3^{\frac{1}{8}} \ldots = 3^{\frac{1}{2}+\frac{1}{4}+\frac{1}{8}+\ldots}$$

The power of 3 forms of a GP with $a = \frac{1}{2}, r = \frac{1}{2}$, and

therefore $\text{LHS} = 3^{\frac{1}{2}+\frac{1}{4}+\frac{1}{8}+\ldots} = 3^{\left(\frac{1/2}{1-1/2}\right)} = 3 = \text{RHS}$

(Since $S_{\infty} = \frac{1}{2} + \frac{1}{4} + \frac{1}{8} + \ldots = \frac{a}{1-r} = \frac{1/2}{1-1/2} = 1/2$)

Q6. Let $x = 1 + a + a^2 + \ldots\ldots, y = 1 + b + b^2 + \ldots.$ where $|a| < 1, |b| < 1$. Prove that

$$1 + ab + a^2b^2 + \ldots\ldots = \frac{xy}{x+y-1}.$$

Solution:

$$x = 1 + a + a^2 + \ldots = \frac{1}{1-a} \quad (\because S_{\infty} = \frac{a}{1-r} = \frac{a}{1-a})$$

Similarly,

$$y = 1 + b + b^2 + \ldots = \frac{1}{1-b}$$

Thus, $\text{LHS} = 1 + ab + a^2b^2 + \ldots = \frac{1}{1-ab}$

$$\text{And, RHS} = \frac{xy}{x+y-1} = \frac{\left(\frac{1}{1-a}\right)\left(\frac{1}{1-b}\right)}{\left(\frac{1}{1-a}\right)+\left(\frac{1}{1-b}\right)-1}$$

$$= \frac{1}{1-b+1-a-(1-a)(1-b)}$$

$$= \frac{1}{1-b+1-a-1+a+b-ab}$$

$$= \frac{1}{1-ab} = \text{LHS}$$

Exponential Series: Exponential series with a variable x can be expressed as $e^x = 1 + \frac{x}{1!} + \frac{x^2}{2!} + \frac{x^3}{3!} + \ldots \frac{x^n}{n!} + \ldots$

Substituting $x = 1$, we get value of e as

$$e = 1 + \frac{1}{1!} + \frac{1}{2!} + \frac{1}{3!} + \frac{1}{4!} + \ldots$$

Example 1: Find the value of e^2, to one decimal place.

Solution:

Since $e^x = 1 + \frac{x}{1!} + \frac{x^2}{2!} + \frac{x^3}{3!} + \frac{x^4}{4!} + \ldots$

Substituting $x = 2$, we get

$$e^2 = 1 + \frac{2}{1!} + \frac{2^2}{2!} + \frac{2^3}{3!} + \frac{2^4}{4!} + \ldots$$

$$= 1 + 2 + 2 + \frac{4}{3} + \frac{2}{3} + \frac{4}{15} + \ldots$$

$= 7.355 = 7.4$ up to one decimal place.

Example 2: Find the coefficient of x^2 in the expansion of e^{3+2x+x^2} as a series in powers of x.

Solution:

Since $e^x = 1 + \frac{x}{1!} + \frac{x^2}{2!} + \frac{x^3}{3!} + \frac{x^4}{4!} + \ldots$

Therefore, $e^{3+2x+x^2} = e^3 e^{2x} e^{x^2}$

$$= e^3\left(1 + \frac{2x}{1!} + \frac{(2x)^2}{2!} + \ldots\right)\left(1 + \frac{x^2}{1!} + \frac{(x^2)^2}{2!} + \ldots\right)$$

$$= e^3\left(1 + 2x + 2x^2 + \ldots\right)\left(1 + x^2 + \ldots\right)$$

$$= e^3\left(1 + 2x + 3x^2 + \ldots\right)$$

Coefficient of x^2 in the expansion of e^{3+2x+x^2} is $3e^3$.

Example 3: Expand $e^{(2x+x^2)}$ up to the power of x^3.

Solution:

Since $e^x = 1 + \frac{x}{1!} + \frac{x^2}{2!} + \frac{x^3}{3!} + \frac{x^4}{4!} + \ldots$

Therefore, $e^{(2x+x^2)}$

$$= 1 + \frac{2x+x^2}{1!} + \frac{(2x+x^2)^2}{2!} + \frac{(2x+x^2)^3}{3!} \ldots$$

$$= 1 + 2x + x^2 + \frac{4x^2+4x^3+x^4}{2} + \frac{8x^3+12x^4+6x^5+x^6}{6} \ldots$$

$$= 1 + 2x + x^2 + 2x^2 + 2x^3 + \frac{4}{3}x^3 \ldots$$

$$= 1 + 2x + 3x^2 + \frac{10}{3}x^3 \ldots$$

Logarithmic Series:

Logarithmic series with a variable x ($|x| \le 1$) can be expressed as $\log_e(1+x) = x - \frac{x^2}{2} + \frac{x^3}{3} - \frac{x^4}{4} + \ldots$

Substituting $x = 1$, we get

$$\log_e 2 = 1 - \frac{1}{2} + \frac{1}{3} - \frac{1}{4} + \ldots$$

Example 1: Find the value of $\log_e 3 - \log_e 2$, rounded off to one decimal place.

Solution:

$\log_e(1+x) = x - \frac{x^2}{2} + \frac{x^3}{3} - \frac{x^4}{4} + ...$

$\therefore \log_e 3 - \log_e 2 = \log_e 3/2 = \log_e(1+1/2)$

$= \frac{1}{2} - \frac{(1/2)^2}{2} + \frac{(1/2)^3}{3} - ... = \frac{1}{2} - \frac{1}{8} + \frac{1}{24} - ...$

$= 0.5 - 0.125 + 0.04 = 0.4$ up to one decimal place.

Example 2: Find the coefficient of x^2 in the expansion of $\log\frac{(1+3x+2x^2)}{(1-x)}$ as a series in powers of x.

Solution:

$\log_e(1+x) = x - \frac{x^2}{2} + \frac{x^3}{3} - \frac{x^4}{4} + ...$

$\therefore \log\frac{(1+3x+2x^2)}{(1-x)} = \log\{(1+x)(1+2x)/(1-x)\}$

$= \log(1+x) + \log(1+2x) - \log(1-x)$

$= \left(x - \frac{x^2}{2} + ...\right) + \left(2x - \frac{(2x)^2}{2} + ...\right) - \left((-x) - \frac{(-x)^2}{2} + ...\right)$

$= x - \frac{x^2}{2} + 2x - \frac{4x^2}{2} + x + \frac{x^2}{2} = 4x - 2x^2$

Thus, the coefficient of x^2 is –2.

Example 3: Expand $\log(1+2x+3x^2)$ up to the power of x^3.

Solution:

$\log_e(1+x) = x - \frac{x^2}{2} + \frac{x^3}{3} - \frac{x^4}{4} + ...$

Therefore, $\log(1+2x+3x^2) = \log\{1+(2x+3x^2)\}$

$= (2x+3x^2) - \frac{(2x+3x^2)^2}{2} + \frac{(2x+3x^2)^3}{3} + ...$

$= 2x + 3x^2 - \frac{4x^2+12x^3+9x^4}{2} + \frac{8x^3+...}{3} + ...$

$= 2x + x^2 - \frac{10}{3}x^3 + ...$

Day 121

NCERT Miscellaneous Exercise (With Solutions Q1–Q18)

Q1. Show that the sum of the $(m+n)^{\text{th}}, (m-n)^{\text{th}}$ terms of an AP is equal to twice the mth term.

Solution:
Let a and d be the first term and the common difference of the AP, respectively.
Then, the kth term of the AP is given by

$a_k = a + (k-1)d$

$\therefore a_{m+n} = a + (m+n-1)d$

$a_{m-n} = a + (m-n-1)d$

$a_m = a + (m-1)d$

$\therefore a_{m+n} + a_{m-n} = a + (m+n-1)d + a + (m-n-1)d$

$= 2a + (m+n-1+m-n-1)d$

$= 2a + (2m-2)d = 2[a + (m-1)d] = 2a_m$

Thus, the sum of the $(m+n)^{\text{th}}, (m-n)^{\text{th}}$ terms of the AP is equal to twice the mth term.

Q2. If the sum of three numbers in an AP is 24 and their product is 440, find the numbers.

Solution:
Let the three numbers in the AP be $a-d$, a, $a+d$. Then
$(a-d)+(a)+(a+d) = 24 \Rightarrow 3a = 24 \Rightarrow a = 8$

Also,
$(a-d)a(a+d) = 440 \Rightarrow (8-d)(8)(8+d) = 440$

$\Rightarrow (8-d)(8+d) = 55 \Rightarrow 64 - d^2 = 55$

$\Rightarrow d^2 = 9 \Rightarrow d = \pm 3$

Therefore, for $d = 3$, the numbers are 5, 8, and 11 and for $d = -3$, the numbers are 11, 8 and 5.
Thus, the three numbers are 5, 8, 11.

Q3. Let the sum of the n, $2n$, $3n$ terms of an AP be S_1, S_2, S_3, respectively. Show that $S_3 = 3(S_2 - S_1)$.

Solution:
Let a and d be the first term and the common difference of the AP, respectively. Then,

$S_1 = \frac{n}{2}[2a + (n-1)d]$

$S_2 = \frac{2n}{2}[2a + (2n-1)d] = n[2a + (2n-1)d]$

$S_3 = \frac{3n}{2}[2a + (3n-1)d]$

Now, RHS $= 3(S_2 - S_1)$

$= 3n[2a + (2n-1)d] - \frac{3n}{2}[2a + (n-1)d]$

$= 3n\left\{\frac{4a + 4nd - 2d - 2a - nd + d}{2}\right\}$

$= 3n\left[\dfrac{2a+3nd-d}{2}\right]$

$= \dfrac{3n}{2}[2a+(3n-1)d] = S_3 = \text{LHS}$

Q4. Find the sum of all numbers between 200 and 400 that are divisible by 7.

Solution:

The numbers between 200 and 400 that are divisible by 7 are 203, 210, 217, …, 399

Thus, First term, $a = 203$, Last term, $l = 399$

Common difference, $d = 7$

Let the number of terms of the AP be n. Then

$l = a+(n-1)d \Rightarrow 399 = 203+(n-1)7$

$\Rightarrow 7(n-1) = 196 \Rightarrow n = 29$

Thus, the sum of all numbers between 200 and 400 which are divisible by 7 is

$S_n = \dfrac{n}{2}(a+l) \Rightarrow S_{29} = \dfrac{29}{2}(203+399) = 8729$

Q5. Find the sum of integers from 1 to 100 that are divisible by 2 or 5.

Solution:

The integers from 1 to 100, that are divisible by 2, are 2, 4, 6……100.

This forms an AP with both the first term and common difference equal to 2.

Let n be the number of terms. Then

$l = a+(n-1)d \Rightarrow 100 = 2+(n-1)2 \Rightarrow n = 50$

Thus, the sum of integers from 1 to 100 that are divisible by 2 is,

$S_2 = 2+4+6+....+100 = \dfrac{50}{2}[2(2)+(50-1)(2)] = 2550$

The integers from 1 to 100 that are divisible by 5 are 5, 10…..100.

This forms an AP with both the first term and common difference equal to 5.

Let n be the number of terms. Then

$l = a+(n-1)d \Rightarrow 100 = 5+(n-1)5 \Rightarrow 5n = 100 \Rightarrow n = 20$

Thus, the sum of integers from 1 to 100 that are divisible by 5 is

$S_5 = 5+10+...+100 = \dfrac{20}{2}[2(5)+(20-1)5] = 1050$

The integers, which are divisible by both 2 and 5, are 10, 20,……100.

This also forms an AP with both the first term and common difference equal to 10.

Let n be the number of terms. Then

$l = a+(n-1)d \Rightarrow 100 = 10+(n-1)(10)$

$\Rightarrow 100 = 10n \Rightarrow n = 10$

Thus, the sum of integers from 1 to 100 that are divisible by 10 is

$S_{10} = 10+20+...+100 = \dfrac{10}{2}[2(10)+(10-1)(10)] = 550$

Since the numbers that are divisible by 2 and 5 are also divided by 10, the total numbers divisible by 2 or 5 is

$S = S_2 + S_5 - S_{10} = 2550+1050-550 = 3050$

Q6. Find the sum of all two-digit numbers which when divided by 4 yields 1 as remainder.

Solution:

The two-digit numbers, which when divided by 4, yield 1 as remainder, are 13, 17, …, 97.

This forms an AP series with first term 13 and common difference 4.

Let n be the number of terms in this AP. Then

$a_n = a+(n-1)d \Rightarrow 97 = 13+(n-1)(4)$

$\Rightarrow 4(n-1) = 84 \Rightarrow n-1 = 21 \Rightarrow n = 22$

Thus, the sum of all two digit numbers which when divided by 4, yields 1 as remainder is

$S_{22} = \dfrac{22}{2}[22(13)+(22-1)(4)] = 1210$

Q7. If f is a function satisfying $f(x+y) = f(x)\cdot f(y)$ for all $x, y \in N$, such that $f(1) = 3$ and $\sum_{1}^{n} f(x) = 120$, find the value of n.

Solution:

Since $f(x+y) = f(x)\cdot f(y)$ and $f(1) = 3$

Putting $x = y = 1$, we get

$f(1+1) = f(2) = f(1)f(1) = 3\times 3 = 9$

Similarly, $f(3) = f(1+2) = f(1)f(2) = 3\times 9 = 27$

$f(4) = f(1+3) = f(1)f(3) = 3\times 27 = 81$

Thus, $f(1), f(2), f(3),...,$ gives a GP series 3, 9, 27,…., with both the first term and common ratio equal to 3.

Thus, $\sum_{x=1}^{n} f(x) = 120 \Rightarrow S_n = \dfrac{a(r^n-1)}{r-1} = 120$

$\Rightarrow \dfrac{3(3^n-1)}{3-1} = 120 \Rightarrow 3^n - 1 = 80$

$\Rightarrow 3^n = 81 = 3^4 \Rightarrow n = 4$

Q8. The sum of some terms of a GP is 315 whose first term and the common ratio are 25 and 2, respectively. Find last term and the number of terms.

Solution:

Let the first term a be 5, common ratio r be 2, and the sum of n terms of the GP be 315.

Therefore, $S_n = \dfrac{a(r^n - 1)}{r-1}$

$\Rightarrow 315 = \dfrac{5(2^n - 1)}{2-1} \Rightarrow 2^n - 1 = 63$

$\Rightarrow 2^n = 64 = (2)^6 \Rightarrow n = 6$

Thus, the last term of the GP is the 6th term given by

$a_6 = ar^{6-1} = (5)(2)^5 = (5)(32) = 160.$

Q9. The first term of a GP is 1. The sum of the third term and fifth term is 90. Find the common ratio of the GP.

Solution:

Let a and r be the first term and the common ratio of the GP, respectively. Then

$a_1 = a = 1, \quad a_3 = ar^2 = r^2, \quad a_5 = ar^4 = r^4$

$\therefore r^2 + r^4 = 90 \Rightarrow r^4 + r^2 - 90 = 0$

$\Rightarrow r^2 = \dfrac{-1+\sqrt{1+360}}{2} = \dfrac{-1 \pm \sqrt{361}}{2} = \dfrac{-1 \pm 19}{2}$

$\Rightarrow r^2 = -10 \text{ or } 9 \Rightarrow r = \pm 3$

Thus, the common ratio of the GP is ±3.

Q10. The sum of three numbers in a GP is 56. If we subtract 1, 7, 21 from these numbers in that order, we obtain an arithmetic progression. Find the numbers.

Solution:

Let the three numbers in GP be a, ar, ar^2. Then

$a + ar + ar^2 = 56 \Rightarrow a(1 + r + r^2) = 56$

And $a-1, ar-7, ar^2-21$ forms an AP. Therefore

$2(ar-7) = (a-1) + (ar^2 - 21)$

$\Rightarrow 2ar - 14 = ar^2 + a - 22$

$\Rightarrow ar^2 - 2ar + a = 8$

$\Rightarrow a(r^2 - 2r + 1) = 8$

Solving the above two equations, we get

$\Rightarrow \dfrac{56}{1+r+r^2} = \dfrac{8}{r^2 - 2r + 1}$

$\Rightarrow 7r^2 - 14r + 7 = 1 + r + r^2$

$\Rightarrow 6r^2 - 15r + 6 = 0$

$\Rightarrow 3(2r-1)(r-2) = 0$

$\Rightarrow r = 1/2 \text{ and } r = 2$

When $r = 1/2 \Rightarrow a = 32$, the three numbers in GP are 32, 16, 8.

When $r = 2 \Rightarrow a = 8$, the three numbers in GP are 8, 16, 32.

Q11. A GP consists of an even number of terms. If the sum of all the terms is 5 times the sum of terms occupying odd places, then find its common ratio.

Solution:

Let the GP be $T_1, T_2, T_3 .... T_{2n}$.

Number of terms = $2n$

The sum of all the terms is 5 times the sum of terms occupying odd places. Therefore

$T_1 + T_2 + T_3 + ... + T_{2n} = 5[T_1 + T_3 + ... + T_{2n-1}]$

$\Rightarrow T_2 + T_4 + ... + T_{2n} = 4[T_1 + T_3 + ... + T_{2n-1}]$

Let the GP be $a, ar, ar^2, ar^3 ....$ Then

$T_2 + T_4 + ... + T_{2n} = ar + ar^3 + ... + ar^{2n-1} = \dfrac{ar(r^n - 1)}{r^2 - 1}$

$T_1 + T_3 + ... + T_{2n-1} = a + ar^2 + ... + ar^{2n-2} = \dfrac{a(r^n - 1)}{r^2 - 1}$

$\therefore \Rightarrow \dfrac{ar(r^n - 1)}{r^2 - 1} = \dfrac{4a(r^n - 1)}{r^2 - 1}$

$\Rightarrow ar = 4a \Rightarrow r = 4$

Thus, the common ratio of the GP is 4.

Q12. The sum of the first four terms of an AP is 56. The sum of the last four terms is 112. If its first term is 11, then find the number of terms.

Solution:

Let the AP be $a, a+d, a+2d, ..., a+(n-1)d$, where the first term $a = 11$. Then

Sum of first four terms = 56

$\Rightarrow a + (a+d) + (a+2d) + (a+3d) = 56$

$\Rightarrow 4a + 6d = 56$

$\Rightarrow a + (a+d) + (a+2d) + (a+3d) = 56$

$\Rightarrow 4a + 6d = 56$

$\Rightarrow 4 \times 11 + 6d = 56 \Rightarrow d = 2$

Sum of last four terms = 112

$\Rightarrow [a + (n-4)d] + [a + (n-3)d]$

$+ [a + (n-2)d] + [a + n - 1)d] = 12$

$\Rightarrow 4a + (4n - 10)d = 112$

$\Rightarrow 4 \times 11 + (4n - 10) \times 2 = 112$

$\Rightarrow 4n - 10 = 34 \Rightarrow 4n = 44 \Rightarrow n = 11$

Thus, the number of terms of the AP is 11.

Q13. If $\dfrac{a+bx}{a-bx} = \dfrac{b+cx}{b-cx} = \dfrac{c+dx}{c-dx} (x \neq 0)$, then show that a, b, c and d are in GP.

Solution:

Since $\dfrac{a+bx}{a-bx} = \dfrac{b+cx}{b-cx}$

$\Rightarrow (a+bx)(b-cx) = (b+cx)(a-bx)$

$\Rightarrow ab - acx + b^2x - bcx^2 = ab - b^2x + acx - bcx^2$

$\Rightarrow 2b^2x = 2acx \Rightarrow b^2 = ac \Rightarrow \dfrac{b}{a} = \dfrac{c}{b}$

Also, $\dfrac{b+cx}{b-cx} = \dfrac{c+dx}{c-dx}$

$\Rightarrow (b+cx)(c-dx)=(b-cx)(c+dx)$

$\Rightarrow bc-bdx+c^2x-cdx^2=bc+bdx-c^2x-cdx^2$

$\Rightarrow 2c^2x=2bdx \Rightarrow c^2=bd \Rightarrow \frac{c}{b}=\frac{d}{c}$

Thus, from the above two equations, we get $\frac{b}{a}=\frac{c}{b}=\frac{d}{c}$

Thus, a, b, c and d are in GP.

Q14. Let S be the sum, P the product and R the sum of reciprocals of n terms in a GP. Prove that $P^2R^n=S^n$.

Solution:

Let the GP be $a, ar, ar^2, ar^3 ......ar^{n-1}$. Then

$S=a+ar+ar^2+...+ar^{n-1}=\frac{a(r^n-1)}{r-1}$

$P=a\times ar\times ar^2\times...\times ar^{n-1}=a^n\times r^{1+2+....n-1}=a^n r^{\frac{n(n-1)}{2}}$

$R=\frac{1}{a}+\frac{1}{ar}+.....+\frac{1}{ar^{n-1}}$

$=\frac{r^{n-1}+r^{n-2}+....r+1}{ar^{n-1}}$

$=\frac{1(r^n-1)}{(r-1)}\times\frac{1}{ar^{n-1}}=\frac{r^n-1}{ar^{n-1}(r-1)}$

Now, LHS $=P^2R^n=\left(a^n r^{\frac{n(n-1)}{2}}\right)^2\left(\frac{r^n-1}{ar^{n-1}(r-1)}\right)^n$

$=a^{2n}r^{n(n-1)}\frac{(r^n-1)^n}{a^n r^{n(n-1)}(r-1)^n}$

$=\frac{a^n(r^n-1)^n}{(r-1)^n}=\left[\frac{a(r^n-1)}{(r-1)}\right]^n=S^n$

Thus, $P^2R^n=S^n$

Q15. The $p^{\text{th}}, q^{\text{th}}, r^{\text{th}}$ terms of an AP are a, b, c, respectively. Show that $(q-r)a+(r-p)b+(p-q)c=0$.

Solution:

Let a_1 and d be the first term and the common difference of the AP, respectively. Then the nth term of an AP is given by, $a_n=a_1+(n-1)d$

$\therefore a_p=a_1+(p-1)d=a$ (i)

$a_q=a_1+(q-1)d=b$ (ii)

$a_r=a_1+(r-1)d=c$ (iii)

Subtracting equation (ii) from equation (i), we get

$(p-1-q+1)d=a-b \Rightarrow (p-q)d=a-b$

$\Rightarrow d=\frac{a-b}{p-q}$

Subtracting equation (iii) from equation (ii), we get

$(q-1-r+1)d=b-c \Rightarrow (q-r)d=b-c$

$\Rightarrow d=\frac{b-c}{q-r}$

Equating both the values of d, we get

$\frac{a-b}{p-q}=\frac{b-c}{q-r}\Rightarrow (a-b)(q-r)=(b-c)(p-q)$

$\Rightarrow aq-bq-ar+br=bp-bq-cp+cq$

$\Rightarrow aq-ar+br=bp-cp+cq$

$\Rightarrow (aq-ar)+(br-bp)+(cp-cq)=0$

$\Rightarrow a(q-r)+b(r-p)+c(p-q)=0$

Q16. If $a\left(\frac{1}{b}+\frac{1}{c}\right), b\left(\frac{1}{c}+\frac{1}{a}\right), c\left(\frac{1}{a}+\frac{1}{b}\right)$ are in AP, prove that a, b, c are in AP.

Solution:

$a\left(\frac{1}{b}+\frac{1}{c}\right), b\left(\frac{1}{c}+\frac{1}{a}\right), c\left(\frac{1}{a}+\frac{1}{b}\right)$ are in AP,

$\therefore b\left(\frac{1}{c}+\frac{1}{a}\right)-a\left(\frac{1}{b}+\frac{1}{c}\right)=c\left(\frac{1}{a}+\frac{1}{b}\right)-b\left(\frac{1}{c}+\frac{1}{a}\right)$

$\Rightarrow \frac{b(a+c)}{ac}-\frac{a(b+c)}{bc}=\frac{c(a+b)}{ab}-\frac{b(a+c)}{ac}$

$\Rightarrow \frac{b^2a+b^2c-a^2b-a^2c}{abc}=\frac{c^2a+c^2b-b^2a-b^2c}{abc}$

$\Rightarrow b^2a-a^2b+b^2c-a^2c=c^2a-b^2a+c^2b-b^2c$

$\Rightarrow ab(b-a)+c(b^2-a^2)=a(c^2-b^2)+bc(c-b)$

$\Rightarrow ab(b-a)+c(b-a)(b+a)$

$=a(c-b)(c+b)+bc(c-b)$

$\Rightarrow (b-a)(ab+cb+ca)=(c-b)(ac+ab+bc)$

$\Rightarrow b-a=c-b \Rightarrow 2b=a+c$

Thus, a, b, c are in AP.

Q17. If a, b, c, d are in GP prove that $(a^n+b^n), (b^n+c^n), (c^n+d^n)$ are in GP.

Solution:

a, b, c, d are in GP. Therefore

$b^2=ac,\ c^2=bd,\ ad=bc$

To prove that $(a^n+b^n), (b^n+c^n), (c^n+d^n)$ are in GP, we have to prove that $(b^n+c^n)^2=(a^n+b^n)(c^n+d^n)$

Now, LHS $=(b^n+c^n)^2=b^{2n}+2b^nc^n+c^{2n}$

$=(b^2)^n+2b^nc^n+(c^2)^n$

$=(ac)^n+2b^nc^n+(bd)^n$

$=a^nc^n+b^nc^n+a^nd^n+b^nd^n$

$=a^nc^n+b^nc^n+a^nd^n+b^nd^n$

$=c^n(a^n+b^n)+d^n(a^n+b^n)$

$=(a^n+b^n)(c^n+d^n)=$ RHS

$\therefore (b^n+c^n)^2=(a^n+b^n)(c^n+d^n)$

Thus, $(a^n+b^n),(b^n+c^n),(c^n+d^n)$ are in GP.

Q18. If a and b are the roots of $x^2-3x+p=0$, c,d are roots of $x^2-12x+q=0$, where a, b, c, d forms a GP, prove that $(q+p):(q-p)=17:15$.

Solution:

a and b are the roots of $x^2-3x+p=0$. Therefore $a+b=3,\ ab=p$.

Also, c and d are the roots of $x^2-12x+q=0$. Therefore $c+d=12,\ cd=q$

Now, since a, b, c, d are in GP, then assume that $b=ar$, $c=ar^2$, $d=ar^3$, and thus

$a+b=3 \Rightarrow a+ar=3 \Rightarrow a(1+r)=3$

$c+d=12 \Rightarrow ar^2+ar^3=12 \Rightarrow ar^2(1+r)=12$

On dividing, we get $\dfrac{ar^2(1+r)}{a(1+r)}=\dfrac{12}{3} \Rightarrow r^2=4$

Now, $\dfrac{q+p}{q-p}=\dfrac{cd+ab}{cd-ab}=\dfrac{ar^2\cdot ar^3+a\cdot ar}{ar^2\cdot ar^3-a\cdot ar}$

$=\dfrac{r^4+1}{r^4-1}=\dfrac{4^2+1}{4^2-1}=\dfrac{17}{15}$

Day 122

NCERT Miscellaneous Exercise (With Solutions Q19–Q32)

Q19. The ratio of the AM and GM of two positive numbers a and b, is $m:n$. Show that

$$a:b=\left(m+\sqrt{m^2-n^2}\right):\left(m-\sqrt{m^2-n^2}\right).$$

Solution:

The two numbers are a and b. Therefore

$\text{AM}=\dfrac{a+b}{2}$ and $\text{GM}=\sqrt{ab}$

The ratio of the AM and GM of two positive numbers a and b, is $m:n$. Therefore

$\dfrac{a+b}{2\sqrt{ab}}=\dfrac{m}{n} \Rightarrow \dfrac{(a+b)^2}{4(ab)}=\dfrac{m^2}{n^2}$

$\Rightarrow (a+b)^2=\dfrac{4abm^2}{n^2} \Rightarrow (a+b)=\dfrac{2\sqrt{abm}}{n}$ (i)

Using identity $(a-b)^2=(a+b)^2-4ab$, we get

$(a-b)^2=\dfrac{4abm^2}{n^2}-4ab=\dfrac{4ab(m^2-n^2)}{n^2}$

$\Rightarrow (a-b)=\dfrac{2\sqrt{ab}\sqrt{m^2-n^2}}{n}$ (ii)

Adding equations (i) and (ii), we get

$2a=\dfrac{2\sqrt{ab}}{n}(m+\sqrt{m^2-n^2})$

$\Rightarrow a=\dfrac{\sqrt{ab}}{n}(m+\sqrt{m^2-n^2})$

Substituting the value of a in equation (i), we get

$b=\dfrac{2\sqrt{ab}}{n}m-\dfrac{\sqrt{ab}}{n}(m+\sqrt{m^2-n^2})$

$=\dfrac{\sqrt{ab}}{n}m-\dfrac{\sqrt{ab}}{n}\sqrt{m^2-n^2}=\dfrac{\sqrt{ab}}{n}(m-\sqrt{m^2-n^2})$

$\therefore \dfrac{a}{b}=\dfrac{\dfrac{\sqrt{ab}}{n}(m+\sqrt{m^2-n^2})}{\dfrac{\sqrt{ab}}{n}(m-\sqrt{m^2-n^2})}=\dfrac{(m+\sqrt{m^2-n^2})}{(m-\sqrt{m^2-n^2})}$

Q20. If a, b, c are in AP, b, c, d are in GP and $\frac{1}{c},\frac{1}{d},\frac{1}{e}$ are in AP, prove that a, c, e are in GP.

Solution:

a, b, c are in AP. Therefore

$2b=a+c$ (i)

Since, b, c, d are in GP,

$c^2=bd$ (ii)

Also, $\frac{1}{c},\frac{1}{d},\frac{1}{c}$ are in AP. Therefore

$\dfrac{2}{d}=\dfrac{1}{c}+\dfrac{1}{e}$ (iii)

Now, $\dfrac{2}{d}=\dfrac{1}{c}+\dfrac{1}{e} \Rightarrow \dfrac{2}{(c^2/b)}=\dfrac{1}{c}+\dfrac{1}{e}$

$\Rightarrow \dfrac{2b}{c^2}=\dfrac{1}{c}+\dfrac{1}{e} \Rightarrow \dfrac{2(a+c)}{2c^2}=\dfrac{1}{c}+\dfrac{1}{e}$

$\Rightarrow \dfrac{a+c}{c^2}=\dfrac{e+c}{ce} \Rightarrow \dfrac{a+c}{c}=\dfrac{e+c}{e}$

$\Rightarrow (a+c)e=(e+c)c \Rightarrow ae+ce=ec+c^2$

$\Rightarrow c^2=ae$, Thus, a, c, e are in GP.

Q21. Find the sum of the following series up to n terms:

(i) $5+55+555+\ldots\ldots$ (ii) $.6+.66+.666+\ldots\ldots$

Solution:

(i) $5+55+555+\ldots\ldots$

Let $S_n=5+55+555+\ldots\ldots$ to n terms

$$=\frac{5}{9}[9+99+999+\ldots\ldots \text{to } n \text{ terms}]$$

$$=\frac{5}{9}[(10-1)+(10^2-1)+(10^3-1)+\ldots \text{to } n \text{ terms}]$$

$$=\frac{5}{9}[(10+10^2+10^3+\ldots n \text{ terms})-(1+1+\ldots n \text{ terms})]$$

$$=\frac{5}{9}\left[\frac{10(10^n-1)}{10-1}-n\right]=\frac{5}{9}\left[\frac{10(10^n-1)}{9}-n\right]$$

$$=\frac{50}{81}(10^n-1)-\frac{5n}{9}$$

(ii) $.6+.66+.666+\ldots\ldots$

Let $S_n=0.6+0.66+0.666+\ldots\ldots$ to n terms

$$=6[0.1+0.11+0.111+\ldots \text{to } n \text{ terms}]$$

$$=\frac{6}{9}[0.9+0.99+0.999+\ldots \text{to } n \text{ terms}]$$

$$=\frac{6}{9}\left[\left(1-\frac{1}{10}\right)+\left(1-\frac{1}{10^2}\right)+\left(1-\frac{1}{10^3}\right)+\ldots \text{to } n \text{ terms}\right]$$

$$=\frac{2}{3}\left[(1+1+\ldots n \text{ terms})-\frac{1}{10}\left(1+\frac{1}{10}+\frac{1}{10^2}+\ldots n \text{ terms}\right)\right]$$

$$=\frac{2}{3}\left[n-\frac{1}{10}\left(\frac{1-\left(\frac{1}{10}\right)^n}{1-\frac{1}{10}}\right)\right]$$

$$=\frac{2}{3}n-\frac{2}{30}\times\frac{10}{9}(1-10^{-n})=\frac{2}{3}n-\frac{2}{27}(1-10^{-n})$$

Q22. Find the 20th term of the series $2\times4+4\times6+6\times8+\ldots.+n$ terms.

Solution:

The nth term of the given series $2\times4+4\times6+6\times8+\ldots.+n$ terms is

$a_n=2n\times(2n+2)=4n^2+4n$

Therefore, the 20th term of the series is

$a_{20}=4(20)^2+4(20)=4(400)+80=1680$

Q23. Find the sum of the first n terms of the series: $3+7+13+21+31+\ldots..$

Solution:

The given series can be written as

$S=3+7+13+21+31+\ldots.+a_{n-1}+a_n$

$S=\quad 3+7+13+21+\ldots.+a_{n-2}+a_{n-1}+a_n$

Subtracting both the equations, we get

$$S-S=3+[(7-3)+(13-7)+(21-13)+\ldots+(a_{n-1}-a_{n-2})+(a_n-a_{n-1})]-a_n$$

$$\Rightarrow 0=3+[4+6+8+\ldots(n-1)\text{ terms}]-a_n$$

$$\Rightarrow a_n=3+[4+6+8+\ldots.(n-1)\text{ terms}]$$

$$\Rightarrow a_n=3+\left(\frac{n-1}{2}\right)[2\times4+(n-1-1)2]$$

$$\Rightarrow a_n=3+\frac{(n-1)}{2}(2n+4)=n^2+n+1$$

Thus, the summation of first n terms of the series is

$S=3+7+13+21+31+\ldots..$

$$\Rightarrow S=\sum_{k=1}^{n}a_k=\sum_{k=1}^{n}k^2+\sum_{k=1}^{n}k+\sum_{k=1}^{n}1$$

$$=\frac{n(n+1)(2n+1)}{6}+\frac{n(n+1)}{2}+n$$

$$=n\left[\frac{2n^2+3n+1+3n+3+6}{6}\right]$$

$$=n\left[\frac{2n^2+6n+10}{6}\right]$$

$$=\frac{n}{3}(n^2+3n+5)$$

Q24. If S_1, S_2, S_3 are the sums of the first n natural numbers, their squares and their cubes, respectively, show that $9S_2^2=S_3(1+8S_1)$.

Solution:

S_1, S_2, S_3 are sums of the first n natural numbers, their squares and their cubes, respectively. Then

$$S_1=\frac{n(n+1)}{2},\ S_2=\frac{n(n+1)(2n+1)}{6},\ S_3=\frac{n^2(n+1)^2}{4}$$

Now, LHS $=S_3(1+8S_1)$

$$=\frac{n^2(n+1)^2}{4}\left[1+\frac{8n(n+1)}{2}\right]$$

$$=\frac{n^2(n+1)^2}{4}[1+4n^2+4n]=\frac{n^2(n+1)^2}{4}(2n+1)^2$$

$$=\left(\frac{n(n+1)(2n+1)}{2}\right)^2=\left(\frac{n(n+1)(2n+1)}{2}\times\frac{3}{3}\right)^2$$

$$=\left(3\frac{n(n+1)(2n+1)}{6}\right)^2=(3S_2)^2=9S_2^2=\text{RHS}$$

Q25. Find the sum of the following series up to n terms

$$\frac{1^3}{1}+\frac{1^3+2^3}{1+3}+\frac{1^3+2^3+3^3}{1+2+5}+\ldots.$$

Solution:

The nth of the given series $\frac{1^3}{1}+\frac{1^3+2^3}{1+3}+\frac{1^3+2^3+3^3}{1+2+5}+\ldots.$ is

$$a_n = \frac{1^3+2^3+3^3+...+n^3}{1+3+5+...+(2n-1)} = \frac{\left[\frac{n(n+1)}{2}\right]^2}{\frac{n}{2}[2\times 1+(n-1)2]}$$

$$= \frac{n^2(n+1)^2}{4n^2} = \frac{(n+1)^2}{4} = \frac{1}{4}n^2+\frac{1}{2}n+\frac{1}{4}$$

$$\therefore S_n = \sum_{k=1}^{n} a_k = \sum_{k=1}^{n}\left(\frac{1}{4}k^2+\frac{1}{2}k+\frac{1}{4}\right)$$

$$= \frac{1}{4}\frac{n(n+1)(2n+1)}{6}+\frac{1}{2}\frac{n(n+1)}{2}+\frac{n}{4}.$$

$$= \frac{n(n+1)(2n+1)+6n(n+1)+6n}{24}$$

$$= \frac{n(2n^2+9n+13)}{24}$$

Q26. Show that

$$\frac{1\times 2^2+2\times 3^2+...+n\times(n+1)^2}{1^2\times 2+2^2\times 3+...+n^2\times(n-1)} = \frac{3n+5}{3n+1}.$$

Solution:

The nth term of the numerator is

$= n(n+1)^2 = n^3+2n^2+n$

And the nth term of the denominator is

$= n^2(n+1) = n^3+n^2$

$$\therefore \frac{1\times 2^2+2\times 3^2+...+n\times(n+1)^2}{1^2\times 2+2^2\times 3+...+n^2\times(n+1)}$$

$$= \frac{\sum_{k=1}^{n} a_k}{\sum_{k=1}^{n} a_k} = \frac{\sum_{k=1}^{n}(k^3+2k^2+k)}{\sum_{k=1}^{n}(k^3+k^2)}$$

Now, $\sum_{k=1}^{n}(k^3+2k^2+k)$

$$= \frac{n^2(n+1)^2}{4}+\frac{2n(n+1)(2n+1)}{6}+\frac{n(n+1)}{2}$$

$$= \frac{n(n+1)}{2}\left[\frac{n(n+1)}{2}+\frac{2}{3}(2n+1)+1\right]$$

$$= \frac{n(n+1)}{2}\left[\frac{3n^2+3n+8n+4+6}{6}\right]$$

$$= \frac{n(n+1)}{12}[3n^2+11n+10] = \frac{n(n+1)}{12}[(n+2)(3n+5)]$$

$$= \frac{n(n+1)(n+2)(3n+5)}{12}$$

And, $\sum_{k=1}^{n}(k^3+k^2) = \frac{n^2(n+1)^2}{4}+\frac{n(n+1)(2n+1)}{6}$

$$= \frac{n(n+1)}{2}\left[\frac{n(n-1)}{2}+\frac{2n+1}{3}\right]$$

$$= \frac{n(n+1)}{2}\left[\frac{3n^2+3n+4n+2}{6}\right]$$

$$= \frac{n(n+1)}{12}[3n^2+7n+2] = \frac{n(n+1)}{12}[(n+2)(3n+1)]$$

$$= \frac{n(n+1)(n+2)(3n+1)}{12}$$

Thus, $\frac{1\times 2^2+2\times 3^2+...+n\times(n+1)^2}{1^2\times 2+2^2\times 3+...+n^2\times(n+1)}$

$$= \frac{\sum_{k=1}^{n}(k^3+2k^2+k)}{\sum_{k=1}^{n}(k^3+k^2)}$$

$$= \frac{n(n+1)(n+2)(3n+5)/12}{n(n+1)(n+2)(3n+1)/12} = \frac{3n+5}{3n+1}$$

Q27. A farmer buys a used tractor for Rs 12000. He pays Rs 6000 cash and agrees to pay the balance in annual instalments of Rs 500 plus 12% interest on the unpaid amount. How much will be the tractor cost him?

Solution:

The farmer pays Rs 6000 in cash.

Therefore, unpaid amount = Rs 12000 – Rs 6000 = Rs 6000

The interest paid annually is 12% of 6000, 12% of 5500, 12% of 5000, …, 12% of 500

Therefore, the total interest to be paid is

= 12% of 6000 + 12% of 5500 + 12% of 5000 + … + 12% of 500

= 12% of (6000 + 5500 + 500 + … +500)

=12% of (500 + 1000 + 1500 + … + 6000)

500, 1000, 1500, …, 6000 form an AP series with the first term equal to 500 and the common difference equal to 5000. Let the number of terms be n. Then

$6000 = 500+(n-1)(500) \Rightarrow n = 12$

Therefore,

$\therefore 500+1000+...+6000$

$$= \frac{12}{2}[2(500)+(12-1)(500)$$

$= 6[1000+5500] = 39000$

Thus, total interest to be paid is

= 12% of 39000 = 4680

Hence, cost of tractor is (12000 + 4680) = Rs 16680

Q28. Shamshad Ali buys a scooter for 22000. He pays 4000 cash and agrees to pay the balance in annual instalments of 1000 plus 10% interest on the unpaid amount. How much will the scooter cost him?

Solution:

Shamshad Ali buys a scooter for 22000 and pays and 4000 in cash. Therefore

Unpaid amount = 22000 – 4000 = 18000

The interest paid annually is 10% of 18000, 10% of 17000, 10% of 16000, …, 10% of 1000

Thus, total interest to be paid is

= 10% of 18000 + 10% of 17000 + 10% of 16000 + … + 10% of 1000

= 10% of (18000 + 17000 + 16000 + … + 1000)

= 10% of (1000 + 2000 + 3000 + … + 18000)

1000, 2000, 3000, …, 18000 form an AP with the first term equal to 1000 and the common difference equal to 1000. Let the number of terms be n. Then

$$18000 = 1000 + (n-1)(1000) \Rightarrow n = 18$$

Therefore,

$$\therefore 1000 + 2000 + ... + 18000$$

$$= \frac{18}{2}[2(1000) + (18-1)(1000)]$$

$$= 9[2000 + 17000] = 171000$$

Thus, total interest to be paid

= 10% of Rs 171000 = Rs 17180

Hence, cost of scooter is 22000 + 17100 = Rs 39100

Q29. A person writes a letter to four of his friends. He asks each one of them to copy the letter and mail it to four different persons with instructions that they move the chain similarly. Assuming that the chain is not broken and that it costs 50 paise to mail one letter, find the amount spent on postage when the 8th set of letters is mailed.

Solution:

The number of letters mailed forms a GP as $4, 4^2, \ldots, 4^8$, with the first term = 4, common ratio = 4 and number of terms = 8.

Therefore, the sum of n terms of the GP is $S_n = \dfrac{a(r^n - 1)}{r - 1}$

$$\therefore S_8 = \frac{4(4^8 - 1)}{4 - 1} = \frac{4(65536 - 1)}{3} = \frac{4(65535)}{3} = 87380$$

Since it costs 50 paise to mail one letter, the total cost of mailing 87380 letters is

$$= 87380 \times \frac{50}{100} = 43690$$

Q30. A man deposited 10000 in a bank at the rate of 5% simple interest annually. Find the amount in the 15th year since he deposited the amount and also calculate the total amount after 20 years.

Solution:

The man deposited 10000 in a bank at the rate of 5% simple interest annually, and therefore, interest is

$$= \frac{5}{100} \times 10000 = 500$$

Total amount with interest in fifteen years

$$= 10000 + \underbrace{500 + 500 + .. + 500}_{14 \text{ times}}$$

$$= 10000 + 14 \times 500 = 17000$$

Total amount with interest in twenty years

$$= 10000 + \underbrace{500 + 500 + .. + 500}_{20 \text{ times}}$$

$$= 10000 + 20 \times 500 = 20000$$

Q31. A manufacturer reckons that the value of a machine which costs him 15625 will depreciate each year by 20%, find the estimated value at the end of 5 years.

Solution:

Initial cost of machine = 15625

Since the machine depreciates by 20% every year, its value after every year is 80% of the original cost. That is, 4/5 of the original cost.

Thus, the value at the end of 5 years is

$$= 15625 \times \underbrace{\frac{4}{5} \times \frac{4}{5} \times .. \times \frac{4}{5}}_{5 \text{times}} = 15625 \times \left(\frac{4}{5}\right)^5 = 5120$$

Thus, the value of the machine at the end of 5 years is 5120.

Q32. A hundred and fifty workers were engaged to finish a job in a certain number of days. 4 workers dropped out on the second day, 4 more workers dropped out on the third day and so on. It took 8 more days to finish the work. Find the number of days in which the work was completed.

Solution:

Let x be the number of days in which 150 workers finish the work.

Let $(x + 8)$ be the number of days required to finish the work, if workers dropped out.

According to the question,

$$150x = 150 + 146 + 142 + ...(x + 8) \text{ terms}$$

The series $150 + 146 + 142 + ...(x + 8)$ terms is an AP with a first term of 150, a common difference of –4 and the number of terms as $(x+8)$. Therefore

$$\Rightarrow 150x = \frac{(x+8)}{2}[2(150) + (x + 8 - 1)(-4)]$$

$$\Rightarrow 150x = (x + 8)(136 - 2x)$$

$$\Rightarrow 75x = 68x - x^2 + 544 - 8x$$

$$\Rightarrow x^2 + 15x - 544 = 0$$

$$\Rightarrow x^2 + 32x - 17x - 544 = 0$$

$$\Rightarrow (x - 17)(x + 32) = 0$$

$$\Rightarrow x = 17 \text{ or } x = -32$$

x cannot be negative, and therefore, $x = 17$.

Thus, the number of days in which the work was completed is 17.

Thus, required number of days = $(17 + 8) = 25$

Day 123

NCERT Exemplar Problems: Solutions (Q1–Q17)

Q1. The first term of an AP is a and the sum of the first p terms is zero. Show that the sum of its next q terms is $\frac{-a(p+q)q}{p-1}$.

Solution:
Let the common difference of an AP be d. Then according to the question,

$$S_p = 0 \Rightarrow \frac{p}{2}[2a+(p-1)d] = 0$$

$$\Rightarrow 2a+(p-1)d = 0 \Rightarrow d = \frac{-2a}{p-1}$$

Now, sum of the next q terms $= S_{p+q} - S_p = S_{p+q} - 0$

$$= \frac{p+q}{2}[2a+(p+q-1)d]$$

$$= \frac{p+q}{2}[2a+(p-1)d+qd]$$

$$= \frac{p+q}{2}\left[2a+(-2a)+\frac{q(-2a)}{p-1}\right]$$

$$= \frac{p+q}{2}\left[-\frac{2aq}{p-1}\right]$$

$$= \frac{p+q}{2}\left[\frac{-2aq}{p-1}\right] = \frac{-a(p+q)q}{(p-1)}$$

Q2. A man saved 66000 in 20 years. In each succeeding year after the first year, he saved 200 more than what he saved in the previous year. How much did he save in the first year?

Solution:
Let us assume he saved Rs a in the first year.
In each succeeding year an increment of 200 has been made, which forms an AP with first term $= a$, common difference $d = 200$ and $n = 20$ years. Therefore

$$S_{20} = \frac{20}{2}[2a+(20-1)d]$$

$$\Rightarrow 66000 = 10(2a+19d)$$

$$\Rightarrow 66000 = 20a+190d$$

$$\Rightarrow 66000 = 20a+190\times 200$$

$$\Rightarrow 20a = 66000-38000$$

$$\Rightarrow 20a = 28000 \Rightarrow a = \frac{28000}{20} = 1400$$

Thus, he saved Rs 1400 in the first year.

Q3. A man accepts a position with an initial salary of Rs 5200 per month. It is understood that he will receive an automatic increase of Rs 320 in the next month and each month thereafter.
(i) Find his salary for the tenth month.
(ii) What is his total earning during the first year?

Solution:
The man gets a fixed increment of 320 each month. Therefore, this forms an AP whose first term is $a = 5200$ and common difference $d = 320$.
(i) Salary for tenth month (for $n = 10$)

$$a_{10} = a+(n-1)d = 5200+(10-1)\times 320$$
$$= 5200+9\times 320 = 8080$$

(ii) Total earning during the first year ($n = 12$)

$$S_{12} = \frac{12}{2}[2\times 5200+(12-1)320]$$
$$= 6[10400+11\times 320] = 83520$$

Q4. If the pth and q^{th} terms of a GP are q and p, respectively, then show that its $(p+q)^{th}$ term is $\left(\frac{q^p}{p^q}\right)^{\frac{1}{p-q}}$.

Solution:
Let the first term and common ratio of GP be a and r, respectively. Then

p^{th} term $= q \Rightarrow a\cdot r^{p-1} = q$

And q^{th} term $= p \Rightarrow ar^{q-1} = p$

Solving the above two equations, we get

$$\frac{ar^{p-1}}{ar^{q-1}} = \frac{q}{p} \Rightarrow r^{p-1-q+1} = \frac{q}{p}$$

$$\Rightarrow r^{p-q} = \frac{q}{p} \Rightarrow r = \left(\frac{q}{p}\right)^{\frac{1}{p-q}}$$

Substituting the value of r, we get

$$a\left(\frac{q}{p}\right)^{\frac{p-1}{p-q}} = q \Rightarrow a = q\left(\frac{q}{p}\right)^{\frac{1-p}{p-q}}$$

Therefore, $(p+q)^{th}$ term is $T_{p+q} = a\cdot r^{p+q-1}$

$$= q\cdot\left(\frac{q}{p}\right)^{\frac{1-p}{p-q}}\left[\left(\frac{q}{p}\right)^{\frac{1}{p-q}}\right]^{p+q-1} = q\cdot\left(\frac{q}{p}\right)^{\frac{1-p}{p-q}}\left(\frac{q}{p}\right)^{\frac{p+q-1}{p-q}}$$

$$= q\cdot\left(\frac{q}{p}\right)^{\frac{1-p}{p-q}+\frac{p+q-1}{p-q}} = q\cdot\left(\frac{q}{p}\right)^{\frac{q}{p-q}} = q\cdot\left(\frac{q^q}{p^q}\right)^{\frac{1}{p-q}}$$

$$= \left(q^{p-q}\frac{q^q}{p^q}\right)^{\frac{1}{p-q}} = \left(\frac{q^{q+p-q}}{p^q}\right)^{\frac{1}{p-q}} = \left(\frac{q^p}{p^q}\right)^{\frac{1}{p-q}}$$

Q5. A carpenter was hired to build 192 window frames. The first day he made five frames and each day thereafter, he made two more frames than he made the day before. How many days did it take him to finish the job?
$(p+q)^{\text{th}}$

Solution:
The carpenter made five frames in the first day and each day thereafter he made two frames more than the previous day. Therefore it makes an AP series with first term $a = 5$ and common difference $d = 2$.

Let us assume the carpenter finished the job in n days. Then

$$S_n = 192 = \frac{n}{2}[2a+(n-1)d]$$

$$\Rightarrow 192 = \frac{n}{2}[2\times 5+(n-2)2]$$

$$\Rightarrow 192 = \frac{n}{2}[8+2n] \Rightarrow 192 = 4n+n^2$$

$$\Rightarrow n^2+4n-192=0 \Rightarrow (n-12)(n+16)=0$$

$$\Rightarrow n = 12 \quad [\because n \neq -16]$$

Q6. The sum of interior angles of a triangle is 180°. Show that the sum of the interior angles of polygons with 3, 4, 5, 6, ... sides form an arithmetic progression. Find the sum of the interior angles for a 21 sided polygon.

Solution:
The sum of interior angles of a polygon of side n is $= (2n-4)\times 90° = (n-2)\times 180°$

Thus, the sum of the interior angles of a polygon with sides 3 is $= (3-2)\times 180° = 180°$.

Sum of interior angles of polygon with side 4 is $= (4-2)\times 180° = 360°$

Similarly, the sum of interior angles of polygons with sides 5, 6, 7, ... are $540°, 720°, 900°, ...$

Thus, the series $180°, 360°, 540°, 720°, 900°, ...$ forms an AP with $a = 180°$, $d = 180°$.

Therefore, the sum of interior angles of a 21 sided polygon $=$ 19th term of the above series.

$$\therefore a_{19} = a+(19-1)d = 180+18\times 180 = 3420°$$

Q7. A side of an equilateral triangle is 20 cm long. A second equilateral triangle is inscribed in it by joining the mid-points of the sides of the first triangle. The process is continued as shown in the accompanying diagram. Find the perimeter of the sixth inscribed equilateral triangle.

Solution:
Side of equilateral $\Delta ABC = 20$ cm. By joining the mid-points of this triangle, we get another equilateral triangle of side equal to half of the length of side of ΔABC.

Continuing in this way, we get a set of equilateral triangles with sides equal to half of the side of the previous triangle.

Thus, perimeter of first triangle $= 20\times 3 = 60$ cm

Perimeter of second triangle $= 10\times 3 = 30$ cm

Perimeter of third triangle $= 5\times 3 = 15$ cm

This forms a GP series 60, 30, 15,... with $a = 60$ and $r = \frac{1}{2}$.

Thus, the perimeter of sixth inscribed triangle is the sixth term of the series.

$$\Rightarrow a_6 = ar^{6-1} = 60\times\left(\frac{1}{2}\right)^5 = \frac{60}{32} = \frac{15}{8}\text{ cm}$$

Q8. In a potato race, 20 potatoes are placed in a line at intervals of 4 m with the first potato, 24 m from the starting point. A contestant is required to bring the potatoes back to the starting place one at a time. How far would he run in bringing back all the potatoes?

Solution:
The contestant is required to bring the potatoes back to the starting place one at a time. Therefore, the distance travelled to bring first potato is

$= 24+24 = 2\times 24 = 48$ m

Distance travelled to bring the second potato is

$= 2(24+4) = 2\times 28 = 56$ m

Distance travelled to bring the third potato is

$= 2(24+4+4) = 2\times 32 = 64$ m

Thus, this makes an AP series 48, 56, 64, ... with $a = 48$, $d = 56-48 = 8$, $n = 20$

Hence, the total distance that he runs in bringing back all potatoes is the sum of 20 terms of the above series, that is,

$$S_{20} = \frac{20}{2}[2\times 48+19\times 8] = 10[96+152] = 2480 \text{ m}$$

Q9. In a cricket tournament, 16 school teams participated. A sum of Rs 8000 is to be awarded among the teams as prize money. If the last placed team is awarded Rs 275 in prize money and the award increases by the same amount for successive finishing places, how much amount will the first place team receive?

Solution:
Let the first place team receive Rs a.

The award money increases by the same amount for successive finishing places; therefore this makes an AP series with $l = 275$, $n = 16$, $S_{16} = 8000$. Let the constant increase amount be d. Then

$$l = a+(n-1)(-d)$$

$$\Rightarrow 275 = a+(16-1)(-d) \Rightarrow 275 = a-15d$$

(Take the common difference as negative because the series is decreasing.)

$\therefore S_{16} = \frac{16}{2}[2a+(n-1)\cdot(-d)]$

$\Rightarrow 8000 = 8[2a+(16-1)(-d)]$

$\Rightarrow 1000 = 2a-15d$

Solving the above two equations, we get

$(2a-15d)-(a-15d) = 100-275$

$\Rightarrow 2a-15d-a+15d = 275 \Rightarrow a = 725$

Thus, the first place team receives Rs 725.

Q10. If $a_1, a_2, a_3, \ldots, a_n$ are in AP, where $a_i > 0$ for all I, show that

$$\frac{1}{\sqrt{a_1}+\sqrt{a_2}}+\frac{1}{\sqrt{a_2}+\sqrt{a_3}}+\ldots..+\frac{1}{\sqrt{a_{n-1}}+\sqrt{a_n}}=\frac{n-1}{\sqrt{a_1}+\sqrt{a_n}}$$

Solution:

$a_1, a_2, a_3, \ldots, a_n$ are in AP. Therefore, the common difference $d = a_2 - a_1 = a_3 - a_2 = \ldots. = a_n - a_{n-1}$.

Now, $a_2 - a_1 = d$, can be written as

$(\sqrt{a_2})^2 - (\sqrt{a_1})^2 = d$

$\Rightarrow (\sqrt{a_2}-\sqrt{a_1})(\sqrt{a_2}+\sqrt{a_1}) = d$

$\Rightarrow \frac{1}{\sqrt{a_1}+\sqrt{a_2}} = \frac{\sqrt{a_2}-\sqrt{a_1}}{d}$

Similarly, $\frac{1}{\sqrt{a_2}+\sqrt{a_3}} = \frac{\sqrt{a_3}-\sqrt{a_2}}{d}$

... ...

$\frac{1}{\sqrt{a_{n-1}}+\sqrt{a_n}} = \frac{\sqrt{a_n}-\sqrt{a_{n-1}}}{d}$

Adding all these terms, we get

$\frac{1}{\sqrt{a_1}+\sqrt{a_2}}+\frac{1}{\sqrt{a_2}+\sqrt{a_3}}+\ldots..+\frac{1}{\sqrt{a_{n-1}}+\sqrt{a_n}}$

$= \frac{1}{d}[\sqrt{a_2}-\sqrt{a_1}+\sqrt{a_3}-\sqrt{a_2}+\ldots+\sqrt{a_n}-\sqrt{a_{n-1}}$

$= \frac{1}{d}\left(\sqrt{a_n}-\sqrt{a_1}\right)$ (i)

Also, $a_n = a_1 + (n-1)d$

$\Rightarrow a_n - a_1 = (n-1)d$

$\Rightarrow (\sqrt{a_n})^2 - (\sqrt{a_1})^2 = (n-1)d$

$\Rightarrow (\sqrt{a_n}-\sqrt{a_1})(\sqrt{a_n}+\sqrt{a_1}) = (n-1)d$

$\Rightarrow \frac{1}{d}\left(\sqrt{a_n}-\sqrt{a_1}\right) = \frac{(n-1)}{\sqrt{a_n}+\sqrt{a_1}}$ (ii)

From the above two equations, we get

$\frac{1}{\sqrt{a_1}+\sqrt{a_2}}+\frac{1}{\sqrt{a_2}+\sqrt{a_3}}+\ldots+\frac{1}{\sqrt{a_{n-1}}+\sqrt{a_n}}$

$= \frac{1}{d}\left(\sqrt{a_n}-\sqrt{a_1}\right) = \frac{n-1}{\sqrt{a_n}+\sqrt{a_1}}$

Q11. Find the sum of the series

$(3^3-2^3)+(5^3-4^3)+(7^3-6^3)+\ldots$

(i) n terms (ii) 10 terms

Solution:

$(3^3-2^3)+(5^3-4^3)+(7^3-6^3)+\ldots$

$= (3^3+5^3+7^3+\ldots)-(2^3+4^3+6^3+\ldots)$

Let T_n be the nth term of the series. Then

$T_n = (2n+1)^3-(2n)^3$

$= (2n+1-2n)[(2n+1)^2+(2n+1)2n+(2n)^2]$

$[\because a^3-b^3 = (a-b)(a^2+ab+b^2)]$

$= 4n^2+1+4n+4n^2+2n+4n^2$

$\Rightarrow T_n = 12n^2+6n+1$

(i) Sum of the n terms of the series is

$S_n = \sum T_n = \sum(12n^2+6n+1)$

$= 12\sum n^2+6\sum n+\sum n$

$= 12\cdot\frac{n(n+1)(2n+1)}{6}+\frac{6n(n+1)}{2}+n$

$= 2n(n+1)(2n+1)+3n(n+1)+n$

$= n(2(n+1)(2n+1)+3(n+1)+1)$

$= n\left(4n^2+6n+2+3n+3+1\right)$

$= 4n^3+9n^2+6n$

(ii) Sum of 10 terms of the series is

$S_n = 4n^3+9n^2+6n$

$\Rightarrow S_{10} = 4\times(10)^3+9\times(10)^2+6\times10$

$= 4\times1000+9\times100+60 = 4960$

Q12. Find the nth term of an AP if the sum of its first n terms is $2n+3n^2$.

Solution:

The sum of n terms of the AP is $S_n = 2n+3n^2$

Therefore, the sum of (n – 1) terms of the AP is $S_{n-1} = 2(n-1)+3(n-1)^2 = 3n^2-4n+1$.

Thus, the nth term of the AP is $T_n = S_n - S_{n-1}$ $= (2n+3n^2)-(3n^2-4n+1) = 6n-1$

Q13. If A is the arithmetic mean and G_1, G_2 be two geometric means between any two numbers, then prove that $2A = \frac{G_1^2}{G_2}+\frac{G_2^2}{G_1}$.

Solution:

Let the two numbers be a and b. Then $A = \frac{a+b}{2} \Rightarrow 2A = a+b$

If G_1, G_2 are the geometric means between a and b, then $a, G_1, G_2\ b$ are in GP.

Let r be the common ratio of this GP. Then

$b = ar^{4-1} \Rightarrow b = ar^3 \quad (\because a_n = ar^{n-1})$

$$\Rightarrow \frac{b}{a} = r^3 \Rightarrow r = \left(\frac{b}{a}\right)^{1/3}$$

Thus, $G_1 = ar = a\left(\frac{b}{a}\right)^{1/3}$ and $G_2 = ar^2 = a\left(\frac{b}{a}\right)^{2/3}$

Now, RHS $= \frac{G_1^2}{G_2} + \frac{G_2^2}{G_1}$

$$= \frac{\left[a\left(\frac{b}{a}\right)^{1/3}\right]^2}{a\left(\frac{b}{a}\right)^{2/3}} + \frac{\left[a\left(\frac{b}{a}\right)^{2/3}\right]^2}{a\left(\frac{b}{a}\right)^{1/3}} = \frac{a^2\left(\frac{b}{a}\right)^{2/3}}{a\left(\frac{b}{a}\right)^{2/3}} + \frac{a^2\left(\frac{b}{a}\right)^{4/3}}{a\left(\frac{b}{a}\right)^{1/3}}$$

$$= a + a\left(\frac{b}{a}\right) = a + b = 2A = \text{LHS}$$

Q14. If $\theta_1, \theta_2, \theta_3, ..., \theta_n$ are in AP with a common difference is d, show that

$$\sec\theta_1 + \sec\theta_2 + \sec\theta_3 + ... + \sec\theta_{n-1}\sec\theta_n = \frac{\tan\theta_n - \tan\theta_1}{\sin d}.$$

Solution:

$\theta_1, \theta_2, \theta_3, ..., \theta_n$ are in AP. Therefore, the common difference $d = \theta_2 - \theta_1 = \theta_3 - \theta_2 = ... = \theta_n - \theta_{n-1}$.

Now, equation $\sec\theta_1 + \sec\theta_2 + ... + \sec\theta_{n-1}\sec\theta_n = \frac{\tan\theta_n - \tan\theta_1}{\sin d}$ can be written as

$\sin d[\sec\theta_1\sec\theta_2 + \sec\theta_2\sec\theta_3 + ... + \sec\theta_{n-1}\sec\theta_n]$
$= \tan\theta_n - \tan\theta_1$

Thus, taking only the first term of the LHS, we get

$$\sin d\sec\theta_1\sec\theta_2 = \frac{\sin d}{\cos\theta_1\cos\theta_2} = \frac{\sin(\theta_2 - \theta_1)}{\cos\theta_1\cos\theta_2}$$

$$= \frac{\sin\theta_2\cos\theta_1 - \cos\theta_2\sin\theta_1}{\cos\theta_1\cos\theta_2}$$

$$= \frac{\sin\theta_2\cos\theta_1}{\cos\theta_1\cos\theta_2} - \frac{\cos\theta_2\sin\theta_1}{\cos\theta_1\cos\theta_2}$$

$\Rightarrow \sin d\sec\theta_1\sec\theta_2 = \tan\theta_2 - \tan\theta_1$

Similarly,

$\sin d\sec\theta_2\sec\theta_3 = \tan\theta_3 - \tan\theta_2$

... ...

$\sin d\sec\theta_{n-1}\sec\theta_n = \tan\theta_n - \tan\theta_{n-1}$

Hence, LHS

$= \sin d[\sec\theta_1\sec\theta_2 + \sec\theta_2\sec\theta_3 + ... + \sec\theta_{n-1}\sec\theta_n]$

$= \tan\theta_2 - \tan\theta_1 + \tan\theta_3 - \tan\theta_2 + ... + \tan\theta_n - \tan\theta_{n-1}$

$= \tan\theta_n - \tan\theta_1 = \text{RHS}$

Q15. If the sum of p terms of an AP is q and the sum of q terms is p, then show that the sum of $p+q$ terms is $-(p+q)$. Also, find the sum of first $p-q$ terms (where, $p > q$).

Solution:

Let the first term and common difference of the AP be a and d, respectively.

Then, $S_p = q \Rightarrow \frac{p}{2}[2a + (p-1)d] = q$

$$\Rightarrow 2a + (p-1)d = \frac{2q}{p} \qquad \text{(i)}$$

And $S_q = p \Rightarrow \frac{q}{2}[2a + (q-1)d] = p$

$$\Rightarrow 2a + (q-1)d = \frac{2p}{q} \qquad \text{(ii)}$$

Subtracting equation (ii) from equation (i), we get

$$2a + (p-1)d - 2a - (q-1)d = \frac{2q}{p} - \frac{2p}{q}$$

$$\Rightarrow [(p-1) - (q-1)]d = \frac{2q^2 - 2p^2}{pq}$$

$$\Rightarrow [p - 1 - q + 1]d = \frac{2(q^2 - p^2)}{pq}$$

$$\Rightarrow (p-q)d = \frac{2(q^2 - p^2)}{pq}$$

$$\Rightarrow d = \frac{-2(p+q)}{pq} \qquad \text{(iii)}$$

Substituting the value of d in equation (i), we get

$$2a + (p-1)\left(\frac{-2(p+q)}{pq}\right) = \frac{2q}{p}$$

$$\Rightarrow 2a = \frac{2q}{p} + \frac{2(p+q)(p-1)}{pq}$$

$$\Rightarrow a = \left[\frac{q}{p} + \frac{(p+q)(p-1)}{pq}\right] \qquad \text{(iv)}$$

Now, $S_{p+q} = \frac{p+q}{2}[2a + (p+q-1)d]$

$$= \frac{p+q}{2}\left[\frac{2q}{p} + \frac{2(p+q)(p-1)}{pq} - \frac{(p+q-1)2(p+1)}{pq}\right]$$

$$= (p+q)\left[\frac{q}{p} + \frac{(p+q)(p-1) - (p+q-1)(p+q)}{pq}\right]$$

$$= (p+q)\left[\frac{q}{p} + \frac{(p+q)(p-1-p-q+1)}{pq}\right]$$

$$= p+q\left[\frac{q}{p} - \frac{p+q}{p}\right] = (p+q)\left[\frac{q-p-q}{p}\right]$$

$$\Rightarrow S_{p+q} = -(p+q)$$

Now, $S_{p-q} = \frac{p-q}{2}[2a+(p-q-1)d]$

$= \frac{p-q}{2}\left[\frac{2q}{2} + \frac{2(p+q)(p-1)}{pq} - \frac{(p-q-1)2(p+q)}{pq}\right]$

$= (p-q)\left[\frac{q}{p} + \frac{p+q(p-1-p+q+1)}{pq}\right]$

$= (p-q)\left[\frac{q}{p} + \frac{(p+q)q}{pq}\right]$

$= (p-q)\left[\frac{q}{p} + \frac{p+q}{p}\right] = (p-q)\frac{(p+2q)}{p}$

Q16. If the pth, q^{th}, r^{th} terms of an AP and GP are both a, b and c, respectively, then show that $a^{b-c} \cdot b^{c-a} \cdot c^{a-b} = 1$.

Solution:
Let A, d be the first term and common difference of the AP and G, R be the first term and common ratio of the GP, respectively.
According to the given condition,

$A+(p-1)d = a$ (i)

$A+(q-1)d = b$ (ii)

$A+(r-1)d = c$ (iii)

$a = GR^{p-1}$ (iv)

$b = GR^{q-1}$ (v)

$c = GR^{r-1}$ (vi)

Subtracting equation (ii) from equation (i), we get
$d(p-1-q+1) = a-b \Rightarrow a-b = d(q-r)$
And subtracting equation (iii) from equation (ii), we get
$d(q-1-r+1) = b-c \Rightarrow c-a = d(r-p)$
Subtracting equation (i) from equation (iii), we get
$d(r-1-p+1) = c-a \Rightarrow c-a = d(r-p)$
Now, LHS $= a^{b-c} \cdot b^{c-a} \cdot c^{a-b}$

LHS $= (GR^{p-1})^{d(q-r)}(GR^{q-1})^{d(r-p)}(GR^{r-1})^{d(p-q)}$

$= G^{d(q-r)+d(r-p)+d(p-q)} R^{(p-1)d(q-r)+(q-1)d(r-p)+(r-1)d(p-q)}$

$= G^{d(q-r+r-p+p-q)} R^{d(pq-pr-q+r+qr-pq-r+p+rp-rq-p+q)}$

$= x^{\circ}R^{\circ} = 1 = \text{RHS}$

Q17. If the sum of n terms of an AP is given by $S_n = 3n+2n^2$, then the common difference of the AP is
(a) 3 (b) 2 (c) 6 (d) 4

Solution: (d)

$\because S_n = 3n+2n^2$

$\therefore S_1 = 3\times1+2(1)^2 = 5 = T_1$

$S_2 = 3\times2+2(2)^2 = 14$

$T_2 = S_2 - S_1 = [3\times2+2\times(2)^2]-[3\times1+2\times(1)^2]$

$= 14-5 = 9$

Thus, common difference $(d) = T_2 - T_1 = 9-5 = 4$

Day 124

NCERT Exemplar Problems: Solutions (Q18–Q36)

Q18. If the third term of a GP is 4, then the product of its first 5 terms is
(a) 4^3 (b) 4^4
(c) 4^5 (d) None of these

Solution: (c)
Let a and r be the first term and the common ratio, respectively. Then $T_3 = 4 \Rightarrow ar^2 = 4$.
Thus, the product of the first 5 terms is
$= a\cdot ar\cdot ar^2\cdot ar^3\cdot ar^4 = a^5r^{10} = (ar^2)^5 = (4)^5$

Q19. If 9 times the 9th term of an AP is equal to 13 times the 13th term, then the 22nd term of the AP is
(a) 0 (b) 22 (c) 198 (d) 220

Solution: (a)
Let a and d be the first term and common difference, respectively. Then $9\cdot T_9 = 13\cdot T_{13}$.

$\Rightarrow 9(a+8d) = 13(a+12d)$

$\Rightarrow 9a+72d = 13a+156d$

$\Rightarrow 4a = -84d \Rightarrow a = -21d \Rightarrow a+21d = 0$

Thus, the 22nd term is $T_{22} = a+(22-1)d = a+21d = 0$.

Q20. If x, $2y$ and $3z$ are in AP where the distinct numbers x, y, z are in GP, then the common ratio of the GP is
(a) 3 (b) 1/3 (c) 2 (d) 1/2

Solution: (b)
Let x, y, z be in GP with a common ratio r. Then $y = xr$ and $z = xr^2$.
x, $2y$, $3z$ are in AP. Therefore

$2y = \frac{x+3z}{2} \Rightarrow 4y = x+3z$

$4(xr) = x+3(xr^2) \Rightarrow 4r = 1+3r^2$

$\Rightarrow 3r^2-4r+1 = 0 \Rightarrow (3r-1)(r-1) = 0$

$\Rightarrow r = 1/3$ or $r = 1$

Q21. If in an AP, $S_n = qn^2$, $S_m = qm^2$, where S_r denotes the sum of r terms of the AP, then S_q equals to

(a) q (b) q^2 (c) q^3 (d) q^4

Solution: (c)

$\because S_n = \frac{n}{2}[2\times a+(n-1)d] = qn^2 \Rightarrow a+\frac{(n-1)d}{2} = qn$

$S_m = \frac{m}{2}[2\times a+(m-1)d] = qm^2 \Rightarrow a+\frac{(m-1)d}{2} = qm$

Solving the above two equations, we get

$\frac{(n-m)d}{2} = q(n-m) \Rightarrow d = 2q$

And, $a+\frac{(n-1)2q}{2} = qn \Rightarrow a = q$

Thus, $S_q = \frac{q}{2}[2\times a+(q-1)d]$

$= \frac{q}{2}[2\times q+(q-1)2q]$

$= \frac{q}{2}[2q+2q^2-2q] = \frac{q}{2}\times 2q^2 = q^3$

Q22. Let S_n denote the sum of the first n terms of an AP, if $S_{2n} = 3S_n$, then $S_{3n} : S_n$ is equal to

(a) 4 (b) 6 (c) 8 (d) 10

Solution: (b)

Let A, d be the first term and the common difference of AP, respectively. Then

$S_n = \frac{n}{2}[2a+(n-1)d]$

$\Rightarrow S_{2n} = \frac{2n}{2}[2a+(2n-1)d] = n[2a+(2n-1)d]$

$\Rightarrow S_{3n} = \frac{3n}{2}[2a+(3n-1)d]$

Since $S_{2n} = 3S_n$,

$\Rightarrow n[2a+(2n-1)d] = 3\frac{n}{2}[2a+(n-1)d]$

$\Rightarrow 4a+(4n-2)d = 6a+(3n-3)d$

$\Rightarrow -2a+(4n-2-3n+3)d = 0$

$\Rightarrow -2a+(n+1)d = 0 \Rightarrow d = \frac{2a}{n+1}$

Now, $\frac{S_{3n}}{S_n} = \frac{\frac{3n}{2}[2a+(3n-1)d]}{\frac{n}{2}[2a+(n-1)d]}$

$\Rightarrow \frac{S_{3n}}{S_n} = \frac{6a+(9n-3)\frac{2a}{n+1}}{2a+(n-1)\frac{2a}{n+1}}$

$\Rightarrow \frac{S_{3n}}{S_n} = \frac{6an+6a+18an-6a}{2an+2a+2an-2a} = \frac{24an}{4an} = 6$

Q23. The minimum value of $4^x + 4^{1-x}, x \in R$ is

(a) 2 (b) 4 (c) 1 (d) 0

Solution: (b)

Since AM $\geq$ GM

$\Rightarrow \frac{4^x + 4^{1-x}}{2} \geq \sqrt{4^x.4^{1-x}}$

$\Rightarrow 4^x + 4^{1-x} \geq 2\sqrt{4} \Rightarrow 4^x + 4^{1-x} \geq 4$

Q24. Let S_n denote the sum of the cubes of the first n natural numbers and s_n denote the sum of the first n natural numbers, then $\sum_{r=1}^{n}\frac{S_r}{s_r}$ equals to

(a) $\frac{n(n+1)(n+2)}{6}$ (b) $\frac{n(n+1)}{2}$

(c) $\frac{n^2+3n+2}{2}$ (d) None of these

Solution: (a)

$\sum_{r=1}^{n}\frac{S_r}{s_r} = \frac{S_1}{s_1}+\frac{S_2}{s_2}+\frac{S_3}{s_3}+....+\frac{S_n}{s_n}$

Let T_n be the nth term of the above series. Then

$T_n = \frac{S_n}{s_n} = \frac{\left[\frac{n(n+1)}{2}\right]^2}{\frac{n(n+1)}{2}} = \frac{n(n+1)}{2} = \frac{1}{2}[n^2+n]$

Thus, summation $\sum_{r=1}^{n}\frac{S_r}{s_r} = \sum T_n = \frac{1}{2}[\sum n^2 + \sum n]$

$= \frac{1}{2}\left[\frac{n(n+1)(2n+1)}{6}+\frac{n(n+1)}{2}\right]$

$= \frac{1}{2}\cdot\frac{n(n+1)}{2}\left[\frac{(2n+1)}{3}+1\right]$

$= \frac{1}{4}n(n+1)\left[\frac{2n+1+3}{3}\right]$

$= \frac{1}{6}n(n+1)(n+2)$

Q25. If t_n denotes the nth term of the series $2+3+6+11+18+...$, then t_{50} is

(a) 49^2-1 (b) 49^2

(c) 50^2+1 (d) 49^2+2

Solution: (d)

Let S_n be sum of the series $2+3+6+11+18+...+t_{50}$. Then

$S_n = 2+3+6+11+18+...+t_{50}...$

$\Rightarrow S_n = 2+3+6+11+18+...+t_{49}+t_{50}...$

Subtracting, we get

$0 = 2+1+3+5+7+...-t_{50}$

$\Rightarrow t_{50} = 2+1+3+5+7+....$ up to 49 terms

$\therefore t_{50} = 2+[1+3+5+7+.....$ up to 49 terms

$= 2+\frac{49}{2}[2\times1+48\times2] = 2+\frac{49}{2}[98] = 2+(49)^2$

Q26. The lengths of three unequal edges of a rectangular solid block are in GP. If the volume of the block is 216 cm^3 and the total surface area is 252 cm^2, then the length of the longest edge is

(a) 12 cm (b) 6 cm (c) 18 cm (d) 3 cm

Solution: (a)

Let the length, breadth and height of a rectangular solid block be $\frac{a}{r}$, a, ar, respectively. Then

Volume of solid block $= \frac{a}{r}\times a\times ar = 216$ cm^3

$\Rightarrow a^3 = 216 \Rightarrow a^3 = 6^3 \Rightarrow a = 6$

Surface area $= 2\left(\frac{a^2}{r}+a^2+a^2r\right) = 252$

$\Rightarrow 2a^2\left(\frac{1}{r}+r+1\right) = 252$

$\Rightarrow 2\times36\left(\frac{1+r+r^2}{r}\right) = 252$

$\Rightarrow 1+r+r^2 = \frac{21}{6}r \Rightarrow 6r^2-15r+6=0$

$\Rightarrow (2r-1)(r-2)=0 \Rightarrow r = \frac{1}{2}, 2$

For $r = \frac{1}{2}$,

Length $= \frac{a}{r} = \frac{6\times2}{1} = 12$

Breadth $= a = 6$

Height $= ar = 6\times\frac{1}{2} = 3$

For $r = 2$,

Length $= \frac{a}{r} = \frac{6}{2} = 3$

Breadth $= a = 6$

Height $= ar = 6\times2 = 12$

Q27. If a, b and c are in GP, then the value of $\frac{a-b}{b-c}$ is equal to______

Solution:

Let a, b, c be in GP with a common ratio r. Then $b = ar$, $c = ar^2$.

$\therefore \frac{a-b}{b-c} = \frac{a-ar}{ar-ar^2} = \frac{a(1-r)}{ar(1-r)} = \frac{1}{r} = \frac{a}{b} = \frac{b}{c}$

Q28. The sum of terms equidistant from the beginning and end in an AP is equal to______

Solution:

Let the AP be $a, a+d, a+2d, ..., a+(n-1)d$

$\therefore a_1 + a_n = a+a+(n-1)d = 2a+(n-1)d$

$a_2 + a_{n-1} = (a+d)+[a+(n-1)d] = 2a+(n-1)d$

$a_3 + a_{n-2} = (a+2d)+[a+(n-3)d] = 2a+(n-1)d$

Thus, following this pattern, we see that the sum of terms equidistant from the beginning and end in an AP is equal to [first term + last term].

Q29. The third term of a GP is 4, the product of the first five terms is______

Solution:

Let a and r be the first term and common ratio, respectively. Then $T_3 = 4 \Rightarrow ar^2 = 4$.

Thus, the product of first 5 terms is

$= ar\cdot ar\cdot ar^2\cdot ar^3\cdot ar^4 = a^5r^{10} = (ar^2)^5 = (4)^5$

Q30. Two sequences cannot be in both AP and GP together.

Solution: True

Consider an AP $a, a+d, a+2d, ...$

Now, $\frac{a_2}{a_1} = \frac{a+d}{a} \neq \frac{a+2d}{a+d}$. Thus, two sequences cannot be in both AP and GP together.

Q31. Every progression is a sequence but the converse, that is, every sequence is also a progression need not necessarily be true.

Solution: True

Consider the sequence of prime numbers 2, 3, 5, 7, 11,... which is not a progression because it does not follow a specific pattern.

Q32. Any term of an AP (except the first) is equal to half the sum of terms which are equidistant from it.

Solution: True

Consider an AP $a, a+d, a+2d, ...$

Then, $\frac{a_2+a_4}{2} = \frac{a+d+a+3d}{2} = \frac{2a+4d}{2} = a+2d = a_3$

Similarly, $\frac{a_3+a_5}{2} = \frac{a+2d+a+4d}{2} = \frac{2a+6d}{2} = a+3d = a_4$

Hence, the statement is true.

Q33. The sum or difference of two GPs is again a GP.

Solution: False

Let two GP be $a, ar_1, ar_1^2, ar_1^3 ...;\ b, br_2, br_2^2, br_2^3 ...$

Now, sum of the two GPs

$a+b,(ar_1+br_2),(ar_1^2+br_2^2)...$

Now, $\frac{T_2}{T_1}=\frac{ar_1+br_2}{a+b},\frac{T_3}{T_2}=\frac{ar_1^2+br_2^2}{ar_1+br_2}$

$\Rightarrow \frac{T_2}{T_1}\neq\frac{T_3}{T_2}$

Similarly, the difference of the two GPs is

$a-b,ar_1-br_2,ar_1^2-br_2^2,.....$

Now, $\frac{T_2}{T_1}=\frac{ar_1-br_2}{a-b},\frac{T_3}{T_2}=\frac{ar_1^2-br_2^2}{ar_1-br_2}$

$\Rightarrow \frac{T_2}{T_1}\neq\frac{T_3}{T_2}$

Thus, the sum or difference of two GP is not a GP.

Q34. If the sum of n terms of a sequence is a quadratic expression, then it always represents an AP.

Solution: False

Let $S_n=an^2+bn+c$

$S_1=a+b+c$

$a_1=a+b+c$

$S_2=4a+2b+c$

$a_2=S_2-S_1=4a+4b+c-(a+b+c)=3a+b$

$S_3=9a+3b+c\Rightarrow a_3=S_3-S_2=5a+b$

$a_2-a_1=(3a+b)-(a+b+c)=2a-c$

$a_3-a_2=(5a+b)-(3a+b)=2a$

$a_2-a_1\neq a_3-a_2$ therefore, statement is false.

Q35. Match the following.

Column I	Column II
(i) $4,1,\frac{1}{4},\frac{1}{16}$	(a) AP
(ii) 2, 3, 5, 7	(b) Sequence
(iii) 13, 8, 3, -2, -7	(c) GP

Solution:

(i) $4,1,\frac{1}{4},\frac{1}{16}\Rightarrow\frac{T_2}{T_1}=\frac{1}{4}\Rightarrow\frac{T_3}{T_2}=\frac{1}{4}\Rightarrow\frac{T_4}{T_3}=\frac{1/16}{1/4}=\frac{1}{4}$

Thus, it is a GP.

(ii) 2, 3, 5, 7

$\because T_2-T_1=3-2=1$

$T_3-T_2=5-3=2$

$\because T_2-T_1\neq T_3-T_2$

Thus, it is not an AP.

Again, $\frac{T_2}{T_1}=3/2\Rightarrow\frac{T_3}{T_2}=5/3$

$\because \frac{T_2}{T_1}\neq\frac{T_3}{T_2}$ Thus, it is not a GP.

(iii) 13, 8, 3, −2, −7

$T_2-T_1=8-13=-5$

$T_3-T_2=3-8=-5$

$\because T_2-T_1=T_3-T_2$ Thus, it is an AP.

Q36. Match the following.

Column I

(i) $1^2+2^2+3^2+.....+n^2$

(ii) $1^3+2^3+3^3+.....+n^3$

(iii) $2+4+6+.......+2n$

(iv) $1+2+3+....+n$

Column II

(a) $\left[\frac{n(n+1)}{2}\right]^2$ (b) $n(n+1)$

(c) $\frac{n(n+1)(n+2)}{6}$ (d) $\frac{n(n+1)}{2}$

Solution:

(i) $1^2+2^2+3^2+.....+n^2$

Consider the identity, $(k+1)^3-k^3=3k^2+3k+1$

Substituting $k=1,2,3,...(n-1),n$ successively, we get

$2^3-1^3=3\cdot1^2+3\cdot1+1$

$3^3-2^3=3\cdot2^2+3\cdot2+1$

$4^3-3^3=3\cdot3^2+3\cdot3+1$

$n^3-(n-1)^3=3\cdot(n-1)^2+3\cdot(n-1)+1$

$(n+1)^3-n^3=3\cdot n^2+3\cdot n+1$

Adding column-wise, we get

$(n+1)^3-1=3\left(\sum_{r+1}^{n}r^2\right)+3\left(\sum_{r+1}^{n}r\right)+n$

$n^3+3n^2+3n=3\left(\sum_{r+1}^{n}r^2\right)+3\frac{n(n+1)}{2}+n$

$\Rightarrow 3\left(\sum_{r=1}^{n}r^2\right)=n^3+3n^2+3n-\frac{3n(n+1)}{2}-n$

$\Rightarrow 3\left(\sum_{r=1}^{n}r^2\right)=\frac{2n^3+6n^2+6n-3n^2-3n-2n}{2}$

$\Rightarrow 3\left(\sum_{r=1}^{n}r^2\right)=\frac{2n^3+3n^2+n}{2}=\frac{n(n+1)(2n+1)}{2}$

$\Rightarrow \sum_{r=1}^{n}r^2=\frac{n(n+1)(2n+1)}{6}$

(ii) $1^3+2^3+3^3+.....+n^3$

Consider the identity

$(k+1)^4-k^4=4k^3+6k^2+4k+1$

Substituting $k=1,2,3,...(n-1),n$ successively, we get

$$2^4 - 1^4 = 4\cdot 1^3 + 6\cdot 1^2 + 4\cdot 1 + 1$$
$$3^4 - 2^4 = 4\cdot 2^3 + 6\cdot 2^2 + 4\cdot 2 + 1$$
$$4^4 - 3^4 = 4\cdot 3^3 + 6\cdot 3^2 + 4\cdot 3 + 1$$
$$n^4 - (n-1)^4 = 4(n-1)^3 + 6(n-1)^2 + 4(n-1) + 1$$
$$(n+1)^4 - n^4 = 4\cdot n^3 + 6\cdot n^2 + 4\cdot n + 1$$

Adding column-wise, we get

$$(n+1)^4 - 1^4 = 4\cdot(1^3+2^3+...+n^3) + 6(1^2+2^2+...+n^2) + 4(1+2+...+n) + (1+1+...+1)n \text{ terms}$$

$$\Rightarrow 4\left(\sum_{r=1}^{n} r^3\right) + 6\left(\sum_{r=1}^{n} r^2\right) + 4\left(\sum_{r=1}^{n} r\right) + n$$
$$= n^4 + 4n^3 + 6n^2 + 4n$$

$$\Rightarrow n^4 + 4n^3 + 6n^2 + 4n = 4\left(\sum_{r=1}^{n} r^3\right) + 6\left[\frac{n(n+1)(2n+1)}{6}\right] + 4\left[\frac{n(n+1)}{2}\right] + n$$

$$\Rightarrow \sum_{r=1}^{n} r^3 = \frac{n^2(n+1)^2}{4} = \left[\frac{n(n+1)}{2}\right]^2$$

(iii) $2+4+6+...+2n = 2[1+2+3+...+n]$

$$= 2\times\frac{n(n+1)}{2} = n(n+1)$$

(iv) $S_n = 1+2+3+...+n = \dfrac{n(n+1)}{2}$

Day 125

Very Short Answer (VSA): 1-Mark and Short Answer (SA): 2-Mark Questions

Q1. Coefficient of x^3 in the expansion of $e^{(2x)}$ is ______

Solution:

$$e^x = 1 + \frac{x}{1!} + \frac{x^2}{2!} + \frac{x^3}{3!} + \frac{x^4}{4!} + ...$$

Therefore, $e^{(2x)}$

$$= 1 + \frac{2x}{1!} + \frac{(2x)^2}{2!} + \frac{(2x)^3}{3!} ...$$
$$= 1 + 2x + 2x^2 + \frac{4}{3}x^3 ...$$

Thus, the coefficient of x^3 in the expansion of $e^{(2x)}$ is $\frac{4}{3}$.

Q2. Coefficient of x^3 in the expansion of $\log(1+2x^2)$ is ______

Solution:

$$\log_e(1+x) = x - \frac{x^2}{2} + \frac{x^3}{3} - \frac{x^4}{4} + ...$$

Therefore, $\log(1+2x^2)$

$$= (2x^2) - \frac{(2x^2)^2}{2} + \frac{(2x^2)^3}{3} + ...$$
$$= 2x^2 - 2x^4 + \frac{8}{3}x^6 + ...$$

Thus, the coefficient of x^3 in the expansion of $\log(1+2x^2)$ is 0.

Q3. Value of the expansion $2^{\frac{1}{3}} \times 2^{\frac{1}{9}} \times 2^{\frac{1}{27}} \times ...$ is ______

Solution:

$$\text{LHS} = 2^{\frac{1}{3}} \times 2^{\frac{1}{9}} \times 2^{\frac{1}{27}} \times ... = 2^{\frac{1}{3}+\frac{1}{9}+\frac{1}{27}+...}$$

The power of 3 forms of a GP with $a = \frac{1}{3}, r = \frac{1}{3}$. Therefore

$$\text{LHS} = 2^{\frac{1}{3}+\frac{1}{9}+\frac{1}{27}+...} = 2^{\left(\frac{1/3}{1-1/3}\right)} = 2^{\left(\frac{1}{2}\right)} = \sqrt{2}$$

(Since $S_\infty = \frac{1}{3} + \frac{1}{9} + \frac{1}{27} + ... = \frac{a}{1-r} = \frac{1/3}{1-1/3} = 1/2$)

Q4. Summation of series $\frac{-3}{2}, 1, \frac{-2}{3}, ...$ up to infinite terms is ______

Solution:

$$a = \frac{-3}{2},\ r = \frac{-2}{3} \Rightarrow |r| = \frac{2}{3} < 1$$

Therefore summation, $S = \frac{-3}{2} + 1 - \frac{2}{3} + \frac{4}{9} - ...$

$$\Rightarrow S_\infty = \frac{a}{1-r} = \frac{-3/2}{1-(-2/3)} = \frac{-3/2}{5/3} = \frac{-9}{10}$$

Q5. The sum to 9 terms in the geometric progression 0.9, 0.09, 0.009,... is ______

Solution:

Since $a = 0.9$, $r = \frac{0.09}{0.9} = 0.1$

$\therefore S_9 = \frac{0.9[1-(0.1)^9]}{1-0.1}$

$= \frac{0.9}{0.9}[1-(0.1)^9] = 1-(0.1)^9$

Q6. If a, b and c are in GP with a common ratio r, then the value of $\frac{b-a}{c-b}$ (in terms of r) is equal to______

Solution:
Let a, b, c be in GP, with a common ratio r. Then $b = ar$, $c = ar^2$

$\therefore \frac{b-a}{c-b} = \frac{ar-1}{ar^2-ar} = \frac{a(r-1)}{ar(r-1)} = \frac{1}{r}$

Q7. A man started its saving starting from Rs. 100. If he increases the instalment by Rs 5 every month, then amount he will save in 5 years is ______

Solution:
In the the first month he saves Rs 100.
Therefore, in the second first month he will save Rs 105, and so on.
Thus the amount that the man saves every month forms an AP, given by 100, 105, 110,..., where the first term is a = 100 and the common difference is, d = 5.
Hence, the amount he will save in 5 years (or 60 months) is,

$S_{60} = \frac{60}{2}[2\times 100+(60-1)5] = 30[200+295] = 14850$

Q8. The number terms of a GP $2, 2^2, 2^3$... needed to give the sum 126 is ______

Solution:
The given GP is $2, 2^2, 2^3$.... Then, $a = 2$, $r = 2$.
Let sum of n terms of this GP be equal to 126. Then

$S_n = \frac{a(1-r^n)}{1-r} \Rightarrow 126 = \frac{2(2^n-1)}{2-1}$

$\Rightarrow 63 = 2^n - 1$

$\Rightarrow 2^n = 64 = 2^6 \Rightarrow n = 6$

Thus, four terms of the given GP are required to obtain the sum as 120.

Q9. If the sum of n terms of an AP is given by $S_n = 3n - 2n^2$, then the common difference of the AP is
(a) 3 (b) –3 (c) 4 (d) –4

Solution: (d)

$\because S_n = 3n - 2n^2$

$\therefore S_1 = 3\times 1 - 2(1)^2 = 1 = T_1$

$S_2 = 3\times 2 - 2(2)^2 = -2$

$T_2 = S_2 - S_1 = -2-1 = -3$

Thus, common difference $(d) = T_2 - T_1 = -3-1 = -4$

Q10. For a positive real number a, the minimum value of $a^x + a^{1-x}, x \in R$, is
(a) $\sqrt{a}$ (b) $2\sqrt{a}$ (c) $2a$ (d) 0

Solution: (b)

$\text{AM} \geq \text{GM} \Rightarrow \frac{a^x + a^{1-x}}{2} \geq \sqrt{a^x \cdot a^{1-x}}$

$\Rightarrow a^x + a^{1-x} \geq 2\sqrt{a}$

Q11. If the third term of a GP is 2, then the product of its first 5 terms is
(a) 8 (b) 16 (c) 32 (d) 64

Solution: (c)
Let a and r be the first term and the common ratio, respectively. Then $T_3 = 2 \Rightarrow ar^2 = 2$.
Thus, the product of first 5 terms is

$= a\cdot ar\cdot ar^2\cdot ar^3\cdot ar^4 = a^5 r^{10} = (ar^2)^5 = (2)^5 = 32$

Q12. If 5 times the 5th term of an AP is equal to 10 times the 10th term, then the 15th term of the AP is
(a) 0 (b) 22 (c) 198 (d) 220

Solution: (a)
Let a and d be the first term and the common difference, respectively. Then $5\cdot T_5 = 10\cdot T_{10}$.

$\Rightarrow 5(a+4d) = 10(a+9d) \Rightarrow 5a+20d = 10a+90d$

$\Rightarrow 5a = -70d \Rightarrow a = -14d \Rightarrow a+14d = 0$

Thus, the 15th term is $T_{15} = a+(15-1)d = a+14d = 0$

Q13. If t_n denotes the nth term of the series $2+4+8+14+22+...$, then t_{21} is
(a) 422 (b) 22^2 (c) 504 (d) 20^2

Solution: (a)
Let S_n be sum of the series $2+4+8+14+...+t_{21}$. Then

$S_n = 2+4+8+14+22+...+t_{21}...$

$\Rightarrow S_n = 2+4+8+14+...+t_{20}+t_{21}...$

Subtracting, we get $0 = 2+2+4+6+8+...-t_{21}$

$\Rightarrow t_{21} = 2+2+4+6+8+....$ up to 20 terms

$\therefore t_{21} = 2+[2+4+6+8+.....$ up to 20 terms]

$= 2+\frac{20}{2}[2\times 2+19\times 2] = 2+10\times 42 = 422$

Q14. Let S_n denote the sum of the squares of the first n natural numbers and s_n denote the sum of the first n natural numbers. Then $\sum_{r=1}^{n} \frac{S_r}{s_r}$ equals to

(a) $n(n+1)(n+2)$ (b) $n(n+1)$
(c) $(n^2+3n+2)/2$ (d) $n(n^2+n+2)/6$

Solution: (d)

$$\sum_{r=1}^{n} \frac{S_r}{s_r} = \frac{S_1}{s_1} + \frac{S_2}{s_2} + \frac{S_3}{s_3} + ... + \frac{S_n}{s_n}$$

Let T_n be the nth term of the above series. Then

$$T_n = \frac{S_n}{s_n} = \frac{n(n+1)(2n+1)/6}{n(n+1)/2} = \frac{(2n+1)}{3} = \frac{2}{3}n + \frac{1}{3}$$

Thus, summation $\sum_{r=1}^{n} \frac{S_r}{s_r} = \sum T_n = \frac{2}{3}\sum n + \sum \frac{1}{3}$

$$= \frac{2}{3}\left[\frac{n(n+1)}{2}\right]^2 + \frac{1}{3}n = \frac{1}{3}n\left[\frac{n(n+1)}{2} + 1\right]$$

$$= \frac{1}{3}n\left[\frac{n^2+n+2}{2}\right] = \frac{n(n^2+n+2)}{6}$$

Q15. If x, $3/2y$ and $2z$ are in AP where the distinct number x, y, z are in GP, then the common ratio of the GP is

(a) 3 (b) 1/3 (c) 2 (d) 1/2

Solution: (d)

Let x, y, z be in GP with a common ratio r. Then $y = xr$ and $z = xr^2$.

x, $3/2y$, $2z$ are in AP. Therefore

$$\frac{3}{2}y = \frac{x+2z}{2} \Rightarrow 3y = x + 2z$$

$$3(xr) = x + 2(xr^2) \Rightarrow 3r = 1 + 2r^2$$

$$\Rightarrow 2r^2 - 3r + 1 = 0 \Rightarrow (2r-1)(r-1) = 0$$

$$\Rightarrow r = 1/2 \text{ or } r = 1$$

Q16. Write the first four terms of the sequence whose n^{th} term is $a_n = n(n-2) + (-1)^{n+1}2^{n-1}$.

Solution:

$a_n = n(n-2) + (-1)^{n+1}2^{n-1}$

Therefore, substituting $n = 1, 2, 3, 4$ we get,

$$a_1 = 1(1-2) + (-1)^{1+1}2^{1-1} = -1 + 1 = 0$$

$$a_2 = 2(2-2) + (-1)^{2+1}2^{2-1} = 0 - 2 = -2$$

$$a_3 = 3(3-2) + (-1)^{3+1}2^{3-1} = 3 + 4 = 7$$

$$a_4 = 4(4-2) + (-1)^{4+1}2^{4-1} = 8 - 8 = 0$$

Q17. Insert two numbers between 8 and 27 so that the resulting sequence is a GP.

Solution:

Let G_1 and G_2 be two numbers between 3 and 81 such that the series 8, G_1, G_2, 27 forms a GP.

Let a be the first term and r be the common ratio of the GP. Then $a = 8$ and

$$27 = (8)(r)^3 \Rightarrow r^3 = \frac{27}{8} \Rightarrow r = \frac{3}{2}$$

Thus, $G_1 = ar = 8 \times \frac{3}{2} = 12$

$$G_2 = ar^2 = 8 \times \left(\frac{3}{2}\right)^2 = 18$$

Thus, the two numbers are 12 and 18.

Q18. Find the sum of all natural numbers from 40 to 400, which are multiples of 4.

Solution:

The natural numbers from 40 to 4000, which are multiples of 4, are 40, 44, 48,..., 4000.

This sequence forms an AP, with a first term $a = 40$, last term $l = 400$ and a common difference, $d = 4$.

Let, n be the total number of terms. Then

$$a + (n-1)d = l \Rightarrow 40 + (n-1)4 = 400$$

$$\Rightarrow (n-1)4 = 360 \Rightarrow n - 1 = 90 \Rightarrow n = 91$$

Thus, the sum of all natural numbers from 40 to 4000 which are multiples of 4 is

$$S_n = \frac{n}{2}[2a + (n-1)d] = \frac{91}{2}[2(40) + (91-1)(4)]$$

$$\Rightarrow S_n = \frac{91}{2}[80 + 360] = 20020$$

Q19. Write the first six terms of the following sequence and obtain the corresponding series

$$a_1 = 1,\ a_2 = 2,\ a_n = \frac{a_{n-1}}{a_{n-2}},\ n > 2.$$

Solution:

$$a_1 = 1,\ a_2 = 2,\ a_n = \frac{a_{n-1}}{a_{n-2}},\ n > 2$$

$$a_3 = \frac{a_{3-1}}{a_{3-2}} = \frac{a_2}{a_1} = \frac{2}{1} = 2$$

$$a_4 = \frac{a_{4-1}}{a_{4-2}} = \frac{a_3}{a_2} = \frac{2}{2} = 1$$

$$a_5 = \frac{a_{5-1}}{a_{5-2}} = \frac{a_4}{a_3} = \frac{1}{2}$$

$$a_6 = \frac{a_{6-1}}{a_{6-2}} = \frac{a_5}{a_4} = \frac{1/2}{1} = \frac{1}{2}$$

Thus, the first six terms of the sequence are 1, 2, 2, 1, 1/2 and 1/2, and the corresponding series is

$1+2+2+1+1/2+1/2+...$

Q20. The sum of some terms of a GP is 155 whose first term and the common ratio are 5 and 2, respectively. Find the last term and the number of terms.

Solution:

Let the first term a be 5, the common ratio r be 2, and the sum of n terms of the GP be 155.

Therefore, $S_n = \dfrac{a(r^n - 1)}{r-1}$

$\Rightarrow 155 = \dfrac{5(2^n - 1)}{2-1} \Rightarrow 2^n - 1 = 31$

$\Rightarrow 2^n = 32 = (2)^5 \Rightarrow n = 5$

Thus, the last term of the GP is the 5th term given by

$a_5 = ar^{5-1} = (5)(2)^4 = (5)(16) = 80$

Q21. If the sum of three numbers in an AP is 21 and their product is 231, find the numbers.

Solution:

Let the three numbers in the AP be $a-d$, a, $a+d$. Then

$(a-d)+(a)+(a+d) = 21 \Rightarrow 3a = 21 \Rightarrow a = 7$

Also,

$(a-d)a(a+d) = 231 \Rightarrow (7-d)(7)(7+d) = 231$

$\Rightarrow 47 - d^2 = 33$

$\Rightarrow d^2 = 16 \Rightarrow d = \pm 4$

Therefore, for $d = 4$, the numbers are 3, 7, and 11 and for $d = -4$ the numbers are 11, 7 and 3.

Thus, the three numbers are 3, 7, 11.

Day 126

Long Answer-I (LA-I): 4-Mark and Long Answer-II (LA-II): 6-Mark Questions

Q1. In an AP, if the first term is 5 and the sum of the first five terms is one-third of the next five terms, then find 10th term.

Solution:

First term $a = 5$

Let d be the common difference of the AP.

Therefore, AP is $5, 5+d, 5+2d, 5+3d...$

Sum of the first five terms is

$S_5 = \frac{5}{2}[2\cdot 5 + (5-1)d] = 25 + 10d$

Sum of the next five terms is

$S_{10-5} = \frac{10}{2}[2.5 + (10-1)d] - [25 + 10d]$

$[50 + 45d] - [25 + 10d] = 25 + 35d$

The sum of the first five terms is one-third of the next five terms; therefore $S_5 = \frac{1}{3} S_{10-5}$

$\Rightarrow 25 + 10d = \frac{1}{3}(25 + 35d)$

$\Rightarrow 75 + 30d = 25 + 35d \Rightarrow 5d = 50 \Rightarrow d = 10$

Thus, the 10th term of the AP is

$a_{10} = a + (n-1)d = 5 + (10-1)(10) = 95$

Q2. How many terms of the AP $-3, -\frac{3}{2}, 0, \frac{3}{2}....$ are needed to give the sum 132?

Solution:

Common difference of the given AP is

$d = -\frac{3}{2} - (-3) = \frac{-3+6}{2} = \frac{3}{2}$

Let the sum of n terms of the given AP be 132. Then

$S_n = \frac{n}{2}[2a + (n-1)d]$

$\Rightarrow 132 = \frac{n}{2}\left[2\times(-3) + (n-1)\left(\frac{3}{2}\right)\right]$

$\Rightarrow 264 = n\left[-6 + \frac{3n}{2} - \frac{3}{2}\right]$

$\Rightarrow 528 = n(-15 + 3n) \Rightarrow 3n^2 - 15n - 528 = 0$

$\Rightarrow n^2 - 5n - 176 = 0$

$\Rightarrow (n+11)(n-16) = 0 \Rightarrow n = -11$ or 16

$\therefore n = 16\ (\because n \neq -11)$

Q3. The sum of n terms of two arithmetic progressions are in the ratio $3n+2 : 2n+3$. Find the ratio of their 11th terms.

Solution:

Let a_1, a_2, d_1, d_2 be the first terms and the common difference of the first and second arithmetic progression, respectively.

Then, according to the given condition,

$$\frac{\text{Sum of } n \text{ terms of first AP}}{\text{Sum of } n \text{ terms of second AP}} = \frac{3n+2}{2n+3}$$

$$\Rightarrow \frac{\frac{n}{2}[2a_1+(n-1)d_1]}{\frac{n}{2}[2a_2+(n-1)d_2]} = \frac{3n+2}{2n+3}$$

$$\Rightarrow \frac{2a_1+(n-1)d_1}{2a_2+(n-1)d_2} = \frac{3n+2}{2n+3} \quad \text{(i)}$$

Substituting $n = 20$ in equation (i), we get

$$\frac{2a_1+20d_1}{2a_2+20d_2} = \frac{3(20)+2}{2(20)+3}$$

$$\Rightarrow \frac{a_1+10d_1}{a_2+10d_2} = \frac{62}{43} \quad \text{(ii)}$$

Now, the ratio of the 11th term of both the APs is

$$\frac{11^{\text{th}}\text{ term of first AP}}{11^{\text{th}}\text{ term of second AP}}$$

$$= \frac{a_1+10d_1}{a_2+10d_2} = \frac{62}{43} \text{ from equation (ii)}$$

Q4. Let S_n denote the sum of the first n terms of an AP. If $S_{3n} = 2S_n$, then find $S_{2n} : S_n$.

Solution:

Let A, d be the first term and the common difference of the AP, respectively. Then

$$S_n = \frac{n}{2}[2a+(n-1)d]$$

$$\Rightarrow S_{2n} = \frac{2n}{2}[2a+(2n-1)d] = n[2a+(2n-1)d]$$

$$\Rightarrow S_{3n} = \frac{3n}{2}[2a+(3n-1)d]$$

Since $S_{3n} = 2S_n$

$$\Rightarrow \frac{3n}{2}[2a+(3n-1)d] = 2\frac{n}{2}[2a+(n-1)d]$$

$$\Rightarrow 6a+3(3n-1)d = 4a+2(n-1)d$$

$$\Rightarrow 2a+(7n-5)d = 0 \Rightarrow d = \frac{-2a}{7n-5}$$

Now, $$\frac{S_{2n}}{S_n} = \frac{\frac{2n}{2}[2a+(3n-1)d]}{\frac{n}{2}[2a+(n-1)d]}$$

$$\Rightarrow \frac{S_{2n}}{S_n} = \frac{4a+(6n-2)\left(\frac{-2a}{7n-5}\right)}{2a+(n-1)\left(\frac{-2a}{7n-5}\right)}$$

$$\Rightarrow \frac{S_{2n}}{S_n} = \frac{28n-20-12n+4}{14n-10-2n+2}$$

$$= \frac{16n-16}{12n-8} = \frac{4n-4}{3n-2}$$

Q5. If the sum of n terms of an AP is $2n^2+3n$ and its mth term is 105, find the value of m.

Solution:

Let a and d be the first term and the common difference of the AP, respectively. Then

Sum of n terms of the AP is

$$S_n = \frac{n}{2}[2a+(n-1)d]$$

$$\Rightarrow \frac{n}{2}[2a-d+nd] = 2n^2+3n$$

$$\Rightarrow \frac{2a-d}{2}n+\frac{d}{2}n^2 = 3n+2n^2$$

Comparing the coefficient of n and n^2, we get

$$\Rightarrow \frac{2a-d}{2} = 3 \text{ and } \frac{d}{2} = 2 \Rightarrow d = 4$$

Also, $a-\frac{d}{2} = 3 \Rightarrow a-2 = 3 \Rightarrow a = 5$

The mth term is equal to 105. Therefore

$$a_m = a+(m-1)d = 105$$

$$\Rightarrow 5+(m-1)4 = 105 \Rightarrow (m-1)4 = 100$$

$$\Rightarrow m-1 = 25 \Rightarrow m = 26$$

Q6. The sum of the first three terms of a GP is $\frac{19}{6}$ and their product is 1. Find the common ratio and the terms.

Solution:

Let $\frac{a}{r}, a, ar$ be the first three terms of the GP. Then

$$\frac{a}{r}+a+ar = \frac{19}{6} \Rightarrow a\left(\frac{1}{r}+1+r\right) = \frac{19}{6}$$

and $\left(\frac{a}{r}\right)(a)(ar) = 1 \Rightarrow a^3 = 1 \Rightarrow a = 1$

Thus, we get

$$\frac{1}{r}+1+r = \frac{19}{6} \Rightarrow 1+r+r^2 = \frac{19}{6}r$$

$$\Rightarrow 6r^2+6r+6 = 19r \Rightarrow 6r^2-13r+6 = 0$$

$$\Rightarrow (3r-2)(2r-3) = 0 \Rightarrow r = \frac{2}{3} \text{ or } r = \frac{3}{2}$$

Hence, the first three terms of GP are $\frac{3}{2}, 1, \frac{2}{3}$ or $\frac{2}{3}, 1, \frac{3}{2}$.

Q7. Given a GP with $a = 64$ and the 7th term 729, determine summation S_7.

Solution:

$a = 64$, $a_7 = 729$, and let r be the common ratio of the GP.

Then $a_n = ar^{n-1}$.

$\Rightarrow a_7 = ar^{7-1} \Rightarrow 729 = 64r^6$

$\Rightarrow r^6 = \frac{729}{64} \Rightarrow r^6 = \left(\frac{3}{2}\right)^6 \Rightarrow r = \frac{3}{2}$

Therefore, summation $S_n = \frac{a(r^n - 1)}{r-1}$

$\Rightarrow S_7 = \frac{64\left[\left(\frac{3}{2}\right)^7 - 1\right]}{\frac{3}{2}-1} = 2\times 64\left[\left(\frac{3}{2}\right)^7 - 1\right]$

$\Rightarrow S_7 = (2)^7\left[\frac{(3)^7-(2)^7}{(2)^7}\right] = (3)^7-(2)^7 = 2059$

Q8. Find a GP for which the sum of the first two terms is –3 and the fifth term is 4 times the third term.

Solution:

Let a be the first term and r be the common ratio of the GP. Then $S_n = \frac{a(1-r^n)}{1-r}$.

The fifth term of the GP is 4 times the third term,

$\therefore a_5 = 4\times a_3 \Rightarrow ar^4 = 4ar^2 \Rightarrow r^2 = 4 \Rightarrow r = \pm 2$

The sum of the first two terms of the GP is – 3, and therefore $S_2 = -3 = \frac{a(1-r^2)}{1-r}$.

For, $r = 2$

$\Rightarrow -3 = \frac{a[1-(2)^2]}{1-2} \Rightarrow -3 = \frac{a(1-4)}{-1} \Rightarrow a = \frac{-3}{3} = -1$

For, $r = -2$

$-3 = \frac{a[1-(2)^2]}{1-(-2)} \Rightarrow -3 = \frac{a(1-4)}{1+2} \Rightarrow a = 3$

Thus, the required GP is

$-1, -2, -4, -8, -16...$ or $3, -6, 12, -24, 48, ...$

Q9. The lengths of three unequal edges of a rectangular solid block are in GP. If the volume of the block is 27 cm³ and the total surface area is 78 cm², then find the dimensions of the rectangular solid block.

Solution:

Let the length, breadth and height of the rectangular solid block be $\frac{a}{r}$, a, ar, respectively. Then

Volume of solid block $= \frac{a}{r}\times a\times ar = 27$ cm³

$\Rightarrow a^3 = 27 \Rightarrow a^3 = 3^3 \Rightarrow a = 3$

Surface area $= 2\left(\frac{a^2}{r} + a^2 + a^2r\right) = 78$

$\Rightarrow 2a^2\left(\frac{1}{r} + r + 1\right) = 78 \Rightarrow 2\times 3^2\left(\frac{1+r+r^2}{r}\right) = 78$

$\Rightarrow 1 + r + r^2 = \frac{13}{3}r \Rightarrow 3r^2 - 10r + 3 = 0$

$\Rightarrow (3r-1)(r-3) = 0 \Rightarrow r = \frac{1}{3}, 3$

For $r = \frac{1}{3}$,

Length $= \frac{a}{r} = \frac{3}{1/3} = 9$

Breadth $= a = 3$

Height $= ar = 3\times\frac{1}{3} = 1$

For $r = 3$,

Length $= \frac{a}{r} = \frac{3}{3} = 1$

Breadth $= a = 3$

Height $= ar = 3\times 3 = 9$

Q10. Find the sum to n terms of the series $\frac{1}{1\times 3} + \frac{1}{3\times 5} + \frac{1}{5\times 7} + ...$

Solution:

The nth term of the series $\frac{1}{1\times 3} + \frac{1}{3\times 5} + \frac{1}{5\times 7} + ...$ is

$a_n = \frac{1}{(2n-1)(2n+1)} = \frac{1}{2}\left(\frac{1}{2n-1} - \frac{1}{2n+1}\right)$

Therefore, $a_1 = \frac{1}{2}\left(\frac{1}{1} - \frac{1}{3}\right)$, $a_2 = \frac{1}{2}\left(\frac{1}{3} - \frac{1}{5}\right)$

$a_3 = \frac{1}{2}\left(\frac{1}{5} - \frac{1}{7}\right)$,.... $a_n = \frac{1}{2}\left(\frac{1}{2n-1} - \frac{1}{2n+1}\right)$

Thus, $S_n = \sum_{k=1}^{n} a_k = a_1 + a_2 + ... + a_n$

$= \frac{1}{2}\left\{\left(\frac{1}{1} - \frac{1}{3}\right) + \left(\frac{1}{3} - \frac{1}{5}\right) + ... + \left(\frac{1}{2n-1} - \frac{1}{2n+1}\right)\right\}$

$\Rightarrow S_n = \frac{1}{2}\left(1 - \frac{1}{2n+1}\right) = \frac{1}{2}\left(\frac{2n+1-1}{2n+1}\right) = \frac{n}{2n+1}$

Q11. Find the sum to n terms of the series $2\times 1^2 + 4\times 3^2 + 6\times 5^2 + ...$

Solution:

The nth term of the series $2\times 1^2 + 4\times 3^2 + 6\times 5^2 + ...$ is

$a_n = 2n(2n+1)^2 = 8n^3 + 8n^2 + 2n$

Therefore, $S_n = \sum_{k=1}^{n} a_k$

$$= \sum_{k=1}^{n}(8k^3+8k^2+2k) = 8\sum_{k=1}^{n}k^3 + 8\sum_{k=1}^{n}k^2 + 2\sum_{k=1}^{n}k$$

$$= 8\left[\frac{n(n+1)}{2}\right]^2 + 8\left[\frac{n(n+1)(2n+1)}{6}\right] + 2\left[\frac{n(n+1)}{2}\right]$$

$$= \frac{n(n+1)}{2}\left[4n(n+1) + 8\left(\frac{2n+1}{3}\right) + 2\right]$$

$$= \frac{n(n+1)}{2}\left[\frac{12n^2+12n+16n+8+6}{3}\right]$$

$$= \frac{n(n+1)}{2}\left[\frac{12n^2+28n+14}{3}\right]$$

$$= \frac{n(n+1)(6n^2+14n+7)}{3}$$

Q12: Expand $e^{(x^2+x)}$ up to the power of x^3.

Solution:

$$e^x = 1 + \frac{x}{1!} + \frac{x^2}{2!} + \frac{x^3}{3!} + \frac{x^4}{4!} + \ldots$$

Therefore, $e^{(x^2+x)}$

$$= 1 + \frac{x^2+x}{1!} + \frac{(x^2+x)^2}{2!} + \frac{(x^2+x)^3}{3!} \ldots$$

$$= 1 + x^2 + x + \frac{x^4+2x^3+x^2}{2} + \frac{x^6+3x^5+3x^4+x^3}{6} \ldots$$

$$= 1 + x^2 + 2x + x^3 + \frac{1}{2}x^2 + \frac{1}{6}x^3 \ldots$$

$$= 1 + 2x + \frac{3}{2}x^2 + \frac{7}{6}x^3 \ldots$$

Q13: Expand $\log(1+x+x^2)$ up to the power of x^3.

Solution:

$$\log_e(1+x) = x - \frac{x^2}{2} + \frac{x^3}{3} - \frac{x^4}{4} + \ldots$$

Therefore, $\log(1+x+x^2) = \log\{1+(x+x^2)\}$

$$= (x+x^2) - \frac{(x+x^2)^2}{2} + \frac{(x+x^2)^3}{3} + \ldots$$

$$= x + x^2 - \frac{x^2+2x^3+x^4}{2} + \frac{x^3+\ldots}{3} + \ldots$$

$$= x + x^2 - \frac{1}{2}x^2 + x^3 + \frac{1}{3}x^3 + \ldots$$

$$= x + \frac{1}{2}x^2 + \frac{4}{3}x^3 + \ldots$$

Q14. Find the sum of the series $0.3 + 0.33 + 0.333 + \ldots\ldots\ldots$ up to n terms.

Solution:

Let $S_n = 0.3 + 0.33 + 0.333 + \ldots\ldots\ldots$ to n terms.

$$= 3[0.1 + 0.11 + 0.111 + \ldots \text{to } n \text{ terms}]$$

$$= \frac{3}{9}[0.9 + 0.99 + 0.999 + \ldots \text{to } n \text{ terms}]$$

$$= \frac{1}{3}\left[\left(1-\frac{1}{10}\right) + \left(1-\frac{1}{10^2}\right) + \left(1-\frac{1}{10^3}\right) + \ldots \text{to } n \text{ terms}\right]$$

$$= \frac{1}{3}\left[(1+1+\ldots n \text{ terms}) - \frac{1}{10}\left(1 + \frac{1}{10} + \frac{1}{10^2} + \ldots n \text{ terms}\right)\right]$$

$$= \frac{1}{3}\left[n - \frac{1}{10}\left(\frac{1-\left(\frac{1}{10}\right)^n}{1-\frac{1}{10}}\right)\right]$$

$$= \frac{1}{3}n - \frac{1}{30} \times \frac{10}{9}(1-10^{-n}) = \frac{1}{3}n - \frac{1}{27}(1-10^{-n})$$

Q15. Find the sum of the following series up to n terms

$$\frac{1^2}{1} + \frac{1^2+2^2}{1+2} + \frac{1^2+2^2+3^2}{1+2+3} + \ldots$$

Solution:

The nth term of the given series

$\frac{1^2}{1} + \frac{1^2+2^2}{1+2} + \frac{1^2+2^2+3^2}{1+2+3} + \ldots$ is

$$a_n = \frac{1^2+2^2+3^2+\ldots+n^2}{1+2+3+\ldots+n} = \frac{\frac{n(n+1)(2n+1)}{6}}{\frac{n(n+1)}{2}} = \frac{2n+1}{3}$$

$$\therefore S_n = \sum_{k=1}^{n} a_k = \sum_{k=1}^{n}\left(\frac{2}{3}k + \frac{1}{3}\right)$$

$$= \frac{2}{3}\frac{n(n+1)}{2} + \frac{n}{3} = \frac{n(n+1)+n}{3} = \frac{n(n+2)}{3}$$

Q16. Find the sum of the series $(3^2-2^2)+(5^2-4^2)+(7^2-6^2)+\ldots$ up to n terms.

Solution:

Let T_n be the nth term of the series. Then

$$T_n = (2n+1)^2 - (2n)^2$$

$$= (2n+1-2n)[(2n+1)+(2n)] = 4n+1$$

$$[\because a^2-b^2 = (a-b)(a+b)]$$

Thus, the sum of the n terms of the series is

$$S_n = \sum T_n = \sum(4n+1)$$

$$= 4\sum n + \sum 1 = 4 \cdot \frac{n(n+1)}{2} + n$$

$$= 2n(n+1) + n = n(2n+3)$$

Q17. A farmer buys a used tractor for Rs 6000, in an annual instalment of Rs 500 plus 10% interest on the unpaid amount. How much will the tractor cost be for him?

Solution:
The farmer pays Rs 500 per month; therefore, the interest paid annually is 10% of 6000, 10% of 5500, 10% of 5000, ..., 10% of 500.
Therefore, the total interest to be paid is
= 10% of (6000 + 5500 + 500 + ... +500)
= 10% of (500 + 1000 + 1500 + ... + 6000)
500, 1000, 1500, ..., 6000 forms an AP series with the first term equal to 500 and the common difference equal to 5000. Let the number of terms be n. Then

$6000 = 500 + (n-1)(500) \Rightarrow n = 12$

Therefore,

$\therefore 500 + 1000 + ... + 6000$

$= \frac{12}{2}[2(500) + (12-1)(500) = 39000$

Thus, total interest to be paid is
= 10% of 39000 = 3900
Hence, the cost of tractor is (6000 + 3900) = Rs 9900.

Revision and Unit Test 2

Day 127

Revision (Chapters IV–VII)

Principle of Mathematical Induction: Let us assume we have to check if a given statement $P(n)$ for a natural number n is True or False.

Process:

(i) Check the statement $P(n)$ for $n=1$; if $P(1)$ is true proceed to the next step.

(ii) Assume that the statement is true for $n=k$ (where k is some positive integer), that is, $P(k)$ is true.

(iii) Now using statement $P(k)$ as true, check the statement for $n=k+1$, that is, if $P(k+1)$ is true then statement $P(n)$ is true for all natural numbers n.

Note: (i) If it is given that the statement is true for all $n \geq 4$, then step 1 will start from $n=4$ and we check that the statement is true for $n=4$, that is, $P(4)$.

(ii) The process of mathematical induction does not state that the given statement is true for $n=k$, but it only states that if the statement is true for $n=k$, then it is also true for $n=k+1$.

Complex Number System: A complex number is introduced to solve a quadratic equation $ax^2+bx+c=0$, when the discriminant $D=b^2-4ac<0$.

Quadratic equation $x^2-1=0$ has real solutions ($\because D=0-4(-1)=4>0$), but the quadratic equation $x^2+1=0$ has no real solution ($\because D=0-4\cdot 1=-4<0$) because $x^2+1=0$ gives $x^2=-1$ and the square of every real number is non-negative.

Thus, for solving this ($x^2+1=0$) type of quadratic equation where the discriminant $D=b^2-4ac<0$, we extend the real number system to the complex number system.

Complex Number: A number of the form $z=a+ib$, where a and b are real numbers, is defined as a complex number. These complex numbers $z=a+ib$ cannot be represented on a real number line because these are imaginary numbers.

Thus, we introduce a new symbol 'i' known as iota (an imaginary number) such that $i^2=-1 \Rightarrow i=\sqrt{-1}$. Therefore, we can say that i is a solution of the equation $x^2+1=0$.

Note:

1. For the complex number $z=a+ib$, a is called the real part, denoted by Re z, and b is called the imaginary part, denoted by Im z of the complex number z.
2. For the complex number $z=a+ib$, if $\text{Re}\, z=a=0$, then the complex number is known as a purely imaginary number, and if $\text{Im}\, z=b=0$, then the complex number is known as a purely real number.
3. In general, we write the imaginary part of z before i if it is given in a number and we write it after i if it is in a variable.
4. A real number can be represented in a complex number system but a complex number cannot be represented in a real number system.
5. Two complex numbers $z_1=a+ib$ and $z_2=c+id$ are equal if and only if $a=c$ and $b=d$.

Addition of Complex Numbers

Let $z_1=a+ib, z_2=c+id$ be any two complex numbers. Then their sum is defined as

$$z_1+z_2=(\text{Re}\, z_1+\text{Re}\, z_2)+i(\text{Im}\, z_1+\text{Im}\, z_2)$$
$$=(a+c)+i(b+d)$$

Properties of Addition of Complex Numbers

(i) Closure law: The sum z_1+z_2 of two complex numbers z_1, z_2 is a complex number. Note that if we get z_1+z_2 as a real number, even then we write it in the complex number system.

(ii) Commutative law: For any two complex numbers $z_1, z_2 : z_1+z_2=z_2+z_1$.

(iii) Associative law: For any three complex numbers $z_1, z_2, z_3 : z_1+(z_2+z_3)=(z_1+z_2)+z_3$.

(iv) Additive identity: For any complex number z, there exists a zero complex number $0+i0$ (denoted by 0) such that $z+0=z$ is known as the additive identity.

(v) Additive inverse: For any complex number $z=a+ib$, there exists a negative of a complex number $-z(=-a-ib)$ such that $z+(-z)=0$, is known as the additive inverse.

Difference of Two Complex Numbers

For any two complex numbers,

$z_1=a+ib,\ z_2=c+id$, the difference

$$z_1-z_2=(\text{Re}\, z_1-\text{Re}\, z_2)+i(\text{Im}\, z_1-\text{Im}\, z_2)$$
$$=(a-c)+i(b-d)$$

Multiplication of Two Complex Numbers

Let $z_1 = a+ib$, $z_2 = c+id$ be any two complex numbers. Then their product $z_1 \cdot z_2$ is defined as

$z_1 z_2 = (a+ib).(c+id) = (ac-bd)+i(ad+bc)$

Properties of Multiplication of Complex Numbers

(i) Closure law: The product $z_1 \cdot z_2$ of two complex numbers z_1, z_2 is a complex number. Note that if we get $z_1 \cdot z_2$ as a real number, even then we write it in the complex number system.

(ii) Commutative law: For any two complex numbers $z_1, z_2 : z_1 \cdot z_2 = z_2 \cdot z_1$.

(iii) Associative law: For any three complex numbers $z_1, z_2, z_3 : z_1 \cdot (z_2 \cdot z_3) = (z_1 \cdot z_2) \cdot z_3$.

(iv) Distributive law: For any three complex numbers z_1, z_2, z_3.

(a) $z_1 \cdot (z_2 + z_3) = z_1 \cdot z_2 + z_1 \cdot z_3$

(b) $(z_1 + z_2) \cdot z_3 = z_1 \cdot z_3 + z_2 \cdot z_3$

(v) Multiplicative identity: For any complex number z, there exists an identity complex number $1+i0$ (denoted by 1) such that $z \cdot 1 = z$ is known as the multiplicative identity.

(vi) Multiplicative inverse: For any non-zero complex number $z = a+ib$, there exists an inverse of the complex number z (denoted by $\frac{1}{z}$ or z^{-1}), such that $z \cdot \frac{1}{z} = 1$, is known as the multiplicative inverse. And it is given by $z^{-1} = \frac{a}{a^2+b^2} + i\frac{-b}{a^2+b^2}$

Division of Two Complex Numbers

For any two complex numbers z_1, z_2, where $z_2 \neq 0$, the division z_1 / z_2 is defined by $\frac{z_1}{z_2} = z_1 \frac{1}{z_2}$.

Power of i

$\because i = \sqrt{-1} \Rightarrow i^2 = -1$

$\Rightarrow i^3 = i^2 \cdot i = (-1) \cdot i = -i$

$\Rightarrow i^4 = (i^2)^2 = (-1)^2 = 1$

And so on ... ...

Similarly, we have

$$i^{-1} = \frac{1}{i} = \frac{1}{i} \times \frac{i}{i} = \frac{i}{-1} = -i$$

$$\Rightarrow i^{-2} = \frac{1}{i^2} = \frac{1}{-1} = -1$$

$$\Rightarrow i^{-3} = \frac{1}{i^3} = \frac{1}{i^2} \cdot \frac{1}{i} = -1 \cdot (-i) = i$$

$$\Rightarrow i^{-4} = \frac{1}{i^4} = \frac{1}{(i^2)^2} = \frac{1}{(-1)^2} = \frac{1}{1} = 1$$

And so on ... ...

Thus, in general, for any integer

$i^{4n} = 1$, $i^{4n+1} = i$, $i^{4n+2} = -1$, $i^{4n+3} = -i$, for all integers $n \in Z$.

Note: The summation of four consecutive powers of i is equal to zero, that is,

$i^{4n} + i^{4n+1} + i^{4n+2} + i^{4n+3} = 1 + i - 1 - i = 0$, for all integers $n \in Z$.

Square Roots of a Negative Real Number

$\sqrt{a} \times \sqrt{b} = \sqrt{ab}$ for all real numbers a and b, when both $a \geq 0, b \geq 0$ or $a < 0, b \geq 0$ or $a \geq 0, b < 0$.

Now, for the condition when $a < 0, b < 0$

Let assume a and b are positive real numbers. Then

$\sqrt{-a} = \sqrt{(-1) \times a} = \sqrt{-1}\sqrt{a} = i\sqrt{a}$, $(\because i^2 = -1)$

Similarly,

$\sqrt{-b} = \sqrt{(-1) \times b} = \sqrt{-1}\sqrt{b} = i\sqrt{b}$

Thus, $\sqrt{-a} \times \sqrt{-b} = i\sqrt{a} \times i\sqrt{b} = i^2\sqrt{ab} = -\sqrt{ab}$

$(\because \sqrt{a} \times \sqrt{b} = \sqrt{ab}$ for $a \geq 0, b \geq 0$ and $i^2 = -1)$

Thus, note that $\sqrt{a} \times \sqrt{b} \neq \sqrt{ab}$ if both a and b are negative real numbers.

Note: $i^2 = -1$ and also $(-i)^2 = i^2 = -1$; therefore, we get square roots of -1 as $i, -i$. However, in general for the symbol $\sqrt{-1}$, we mean i.

Identities in Complex Numbers

(i) $(z_1 + z_2)^2 = z_1^2 + z_2^2 + 2z_1z_2$

(ii) $(z_1 - z_2)^2 = z_1^2 - 2z_1z_2 + z_2^2$

(iii) $(z_1 + z_2)^3 = z_1^3 + 3z_1^2z_2 + 3z_1z_2^2 + z_2^3$

(iv) $(z_1 - z_2)^3 = z_1^3 - 3z_1^2z_2 + 3z_1z_2^2 - z_2^3$

(v) $z_1^2 - z_2^2 = (z_1 + z_2)(z_1 - z_2)$

Note: Many other identities which are true for all real numbers are also true for all complex numbers.

Conjugate of a Complex Number: If $z = a+ib$ is a complex number, then its conjugate is given by $\bar{z} = a - ib$.

Properties of Conjugate

(i) $\bar{\bar{z}} = z$

(ii) $\overline{z_1 z_2} = \bar{z}_1 \bar{z}_2$

(iii) $\overline{z_1 \pm z_2} = \bar{z}_1 \pm \bar{z}_2$

(iv) $\overline{\left(\frac{z_1}{z_2}\right)} = \frac{\bar{z}_1}{\bar{z}_2}$ provided $z_2 \neq 0$.

(v) $\text{Re}(z) = \frac{1}{2}(z + \bar{z})$ and $\text{Im}(z) = \frac{1}{2i}(z - \bar{z})$

Modulus of a Complex Number: If $z = a + ib$ is a complex number, then its modulus is $|z| = \sqrt{a^2 + b^2}$ (non-negative real number).

Properties of Modulus

(i) $|z| = |\bar{z}|$ (ii) $|z\bar{z}| = |z|^2$

(iii) $|z_1 z_2| = |z_1||z_2|$

(iv) $\left|\dfrac{z_1}{z_2}\right| = \dfrac{|z_1|}{|z_2|}$ provided $|z_2| \neq 0$

Multiplicative Inverse of a Complex Number: If $z = a + ib$ is a complex number, then its inverse is $z^{-1} = \dfrac{\bar{z}}{|z|^2} = \dfrac{a - ib}{a^2 + b^2}$.

Note: For two complex numbers,

$$\frac{z_1}{z_2} = z_1 z_2^{-1} = z_1 \frac{\bar{z}_2}{|z_2|^2} = \frac{z_1 \bar{z}_2}{|z_2|^2}$$

Argand Plane (or Complex Plane)

In a real plane, an ordered pair of real numbers (*x*, *y*) represents a unique point in the XY-plane with reference to a set of mutually perpendicular lines, known as the *x*-axis and *y*-axis.

Similarly, in a complex plane (known as Argand plane), the complex number $x + iy$, which corresponds to the ordered pair (*x*, *y*), represents a unique point in the XY-plane (Argand plane) with reference to a set of mutually perpendicular lines known as the *x*-axis (known as the real axis) and *y*-axis (known as the imaginary axis). Points on the *x*-axis correspond to the complex numbers of the form $x + 0i$ and points on the *y*-axis correspond to the complex numbers of the form $0 + iy$.

Note: In the Argand plane, the modulus of the complex number $|z| = |x + iy| = \sqrt{x^2 + y^2}$ is the distance between the point *P*(*x*, *y*) and the origin *O*(0, 0).

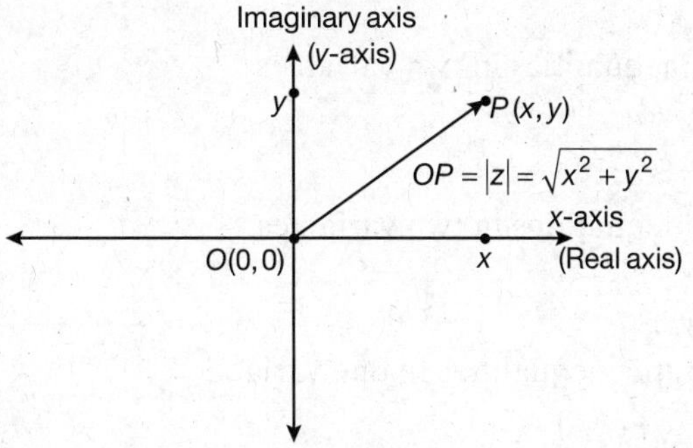

Fig. 5.2

The complex number $z = x + iy$ and its conjugate $z = x - iy$ are represented geometrically by the points $P(x, y)$, $Q(x, -y)$ in the Argand plane. The point $Q(x, -y)$ is the mirror of the point $P(x, y)$ on the real axis.

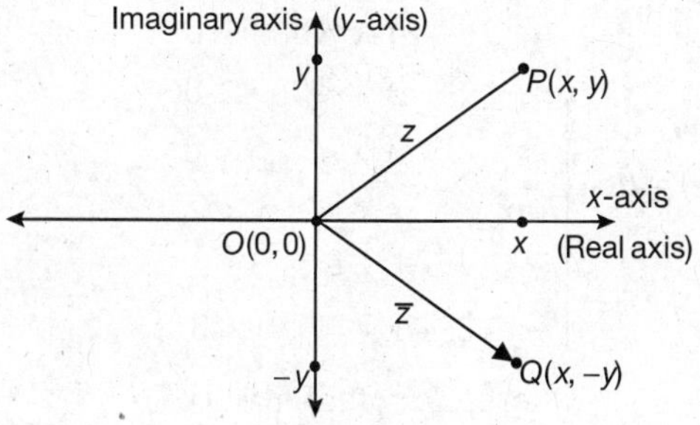

Fig. 5.3

Polar Representation of a Complex Number

Let the point *P* represent the non-zero complex number $z = x + iy$. Let the length of directed line segment *OP* be *r* and the angle which *OP* makes with the positive direction of the *x*-axis be θ.

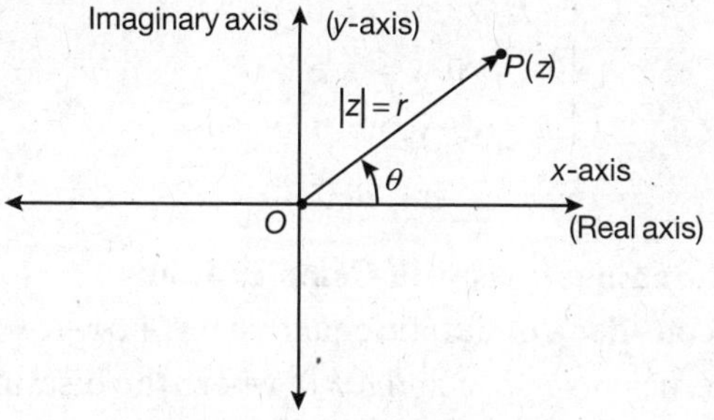

Fig. 5.4

Then the point $P(x, y)$ is uniquely determined by the ordered pair of real numbers (r, θ), known as the polar coordinates of the point P (considering the origin as the pole and the positive direction of the *x*-axis as the initial line).

Thus, in polar coordinates we may write $x = r\cos\theta$, $y = r\sin\theta$ and therefore $z = r(\cos\theta + i\sin\theta)$, known as the polar form of the complex number $z = x + iy$.

And, the modulus of *z* is $|z| = \sqrt{x^2 + y^2} = r$

Argument (or amplitude) of *z* is denoted by $\arg z = \theta$

Since $\sin\theta$ and $\cos\theta$ are functions of period 2π, any complex number $z \neq 0$ takes only one value of θ in the interval $0 \leq \theta \leq 2\pi$, that is, $-\pi < \theta \leq \pi$ or any other interval of length 2π.

Normally we take the value of θ between $-\pi < \theta \leq \pi$, known as the principal argument of *z* and denoted by arg *z*, unless specified otherwise.

Thus, if $z = x + iy = r(\cos\theta + i\sin\theta)$, then

$$\arg(z) = \theta = \begin{cases} \tan^{-1}(y/x), & \text{if } x > 0,\ y \geq 0 \\ \pi - \tan^{-1}(y/|x|), & \text{if } x < 0,\ y \geq 0 \\ -\pi + \tan^{-1}(|y|/|x|), & \text{if } x < 0,\ y < 0 \\ -\tan^{-1}(|y|/x), & \text{if } x > 0,\ y < 0 \\ \pi/2, & \text{if } x = 0,\ y > 0 \\ -\pi/2, & \text{if } x = 0,\ y < 0 \end{cases}$$

$$\Rightarrow \arg(z)=\theta=\begin{cases}\alpha, & \text{if } x>0,\ y\ge 0\\ \pi-\alpha, & \text{if } x<0,\ y\ge 0\\ -\pi+\alpha, & \text{if } x<0,\ y<0\\ -\alpha, & \text{if } x>0,\ y<0\\ \pi/2, & \text{if } x=0,\ y>0\\ -\pi/2, & \text{if } x=0,\ y<0\end{cases}$$

where, $\alpha=\tan^{-1}(|y|/|x|)$

$$\Rightarrow \arg(z)=\theta=\begin{cases}\alpha, & \text{if } z \text{ lies in I}^{\text{st}} \text{ quadrant}\\ \pi-\alpha, & \text{if } z \text{ lies in II}^{\text{nd}} \text{ quadrant}\\ -\pi+\alpha, & \text{if } z \text{ lies in III}^{\text{rd}} \text{ quadrant}\\ -\alpha, & \text{if } z \text{ lies in IV}^{\text{th}} \text{ quadrant}\\ \pi/2, & \text{if } x=0,\ y>0\\ -\pi/2, & \text{if } x=0,\ y<0\\ \alpha, & \text{if } y=0,\ x>0\\ \pi-\alpha, & \text{if } y=0,\ x<0\end{cases}$$

where, $\alpha=\tan^{-1}(|y|/|x|)$

Quadratic Equations with Complex Roots

Let us consider a quadratic equation $ax^2+bx+c=0$ with real coefficients a, b, c and $a\neq 0$, where the discriminant $D=b^2-4ac<0$.

In these cases, the quadratic equation has no real solution ($\because D<0$); therefore the solution of these types of equations must be obtained using the complex number system.

Thus, the solution to the above equation in the set of complex numbers is obtained as

$$x=\frac{-b\pm\sqrt{b^2-4ac}}{2a}=\frac{-b\pm\sqrt{(4ac-b^2)\times(-1)}}{2a}$$

$$\Rightarrow x=\frac{-b\pm\sqrt{(4ac-b^2)}\,i}{2a}$$

Note: In a polynomial equation, the complex roots always occur in a conjugate pair, that is, if $x=a+ib$ is one of the roots of a polynomial equation, then $x=a-ib$ will also be the root of the equation.

Square-root of a Complex Number: Let us assume we have to find the square-root of the complex number $z=a+ib$, where a and b are real numbers and $b\neq 0$.

Now, let the square-root of complex number $z=a+ib$ be $x+iy$; then $x+iy=\sqrt{a+ib}$.

Squaring both sides, we get,

$(x+iy)^2=a+ib$

$\Rightarrow (x^2-y^2)+i(2xy)=a+ib$

Comparing the real and imaginary parts, we get

$\Rightarrow x^2-y^2=a$ and $2xy=b$

Solving, we get

$$x=\pm\sqrt{\frac{\sqrt{a^2+b^2}+a}{2}} \text{ and } y=\pm\sqrt{\frac{\sqrt{a^2+b^2}-a}{2}}$$

Hence, square-root of $a+ib=\sqrt{a+ib}=x+iy$

$$=\pm\sqrt{\frac{\sqrt{a^2+b^2}+a}{2}}\pm i\sqrt{\frac{\sqrt{a^2+b^2}-a}{2}}$$

$$=\pm\sqrt{\frac{|z|+\text{Re}(z)}{2}}\pm i\sqrt{\frac{|z|-\text{Re}(z)}{2}}$$

Note: If b is negative, then the product xy will also be negative; therefore in this case, we take the value of x as positive and y as negative, or vice-versa.

If b is positive, then the product xy will also be positive; therefore in this case, we take value of both x and y as positive or negative.

Inequality: Two real numbers or two algebraic expressions related by the symbol '<', '>', '≤' or '≥' represent an inequality.

Examples:

Numerical inequalities: $2<3$, $-2>-3$

Literal inequalities: $x<2$, $y>2$, $x\ge -2$, $y\le -3$

Double inequalities:

$1<2<3$ (read as 2 is greater than 1 and less than 3)

$a\le x\le b$ (read as x is greater than or equal to a and less than or equal to b)

$1<x\le 2$ (read as x is greater than 1 and less than or equal to 2)

$a\le x<2$ (read as x is greater than or equal to a and less than 2)

Strict inequalities in one variable

$ax+b<0$

$ax+b>0$

Slack inequalities in one variable

$ax+b\le 0$

$ax+b\ge 0$

Strict inequalities in two variables

$ax+by<c$

$ax+by>c$

Slack inequalities in two variables

$ax+by\le c$

$ax+by\ge c$

Quadratic inequalities in one variable

$ax^2+bx+c\le 0$

$ax^2+bx+c>0$

Algebraic Solution of Linear Inequalities in one Variable and their Graphical Representation

Any solution of an inequality in one variable is a value of the variable which makes the linear inequality a true statement.

Rules of Inequality

Rule 1: Equal numbers may be added (or subtracted) on both sides of an inequality without affecting the sign of the inequality.

Rule 2: Both sides of an inequality can be multiplied (or divided) by the same positive non-zero number without affecting the sign of the inequality. But when both sides are multiplied or divided by a negative number, then the sign of the inequality is reversed.

Graphical Solution of Linear Inequalities in Two Variables

Since a line divides the Cartesian plane into two parts, each part is known as a half plane.

Thus, a vertical line will divide the plane into the left and right half planes.

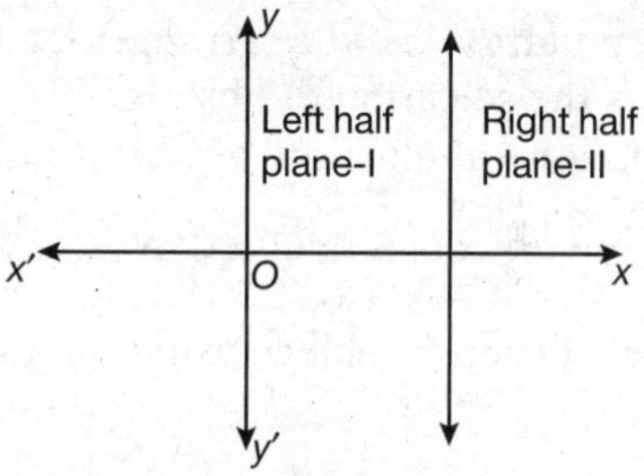

Fig. 6.1

A non-vertical line will divide the plane into the upper and lower half planes.

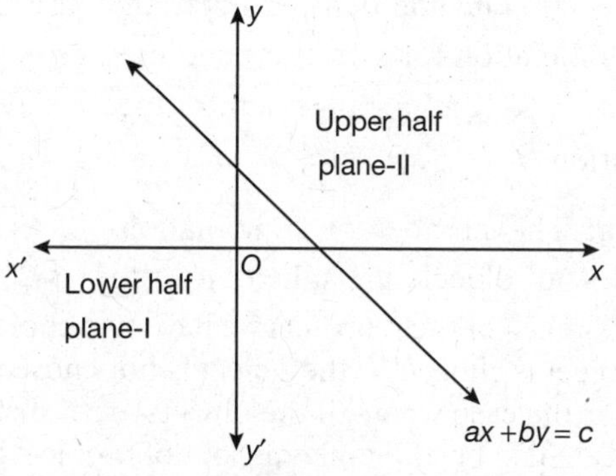

Fig. 6.2

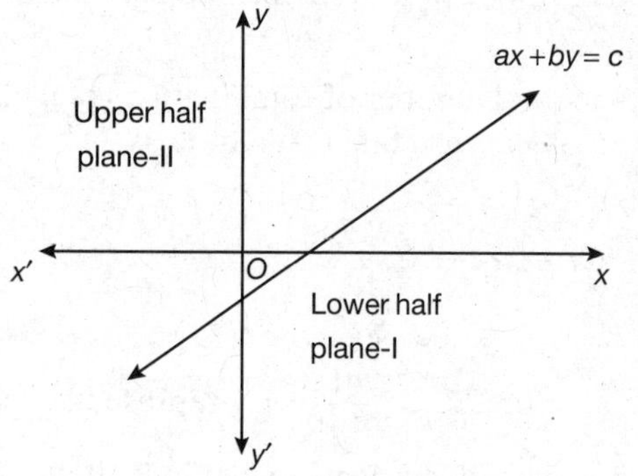

Fig. 6.3

A point in the Cartesian plane will either lie on a line or in either of the half planes I or II.

Consider the line $ax + by = c,\ a \neq 0, b \neq 0$, for which three possibilities may arise:

(i) $ax + by = c$ means all points (x, y) satisfying the equation of the line lie on the line it represents, and vice-versa.

(ii) $ax + by > c$ (iii) $ax + by < c$.

Note

(i) The region containing all the solutions of an inequality is called the solution region. The solution of an inequality can be represented by a shaded region.

(ii) In order to identity the half plane represented by an inequality, it is sufficient to take any point (not on the line) and check whether it satisfies the inequality or not. For convenience, the point (0, 0) is preferred, that is, if origin $O(0,0)$ satisfies the inequality $ax + by > c$ or $ax + by < c$, then the portion divided by the line $ax + by = c$ or $ax + by = c$ containing the origin represents the solution of the inequality $ax + by > c$ or $ax + by < c$. And if the origin does not satisfy the inequality, then the portion of the line opposite the origin represents the solution of the inequality.

(iii) In an inequality of the form $ax + by > c$ or $ax + by < c$, the points on the line $ax + by = c$ are not included in the solution region. Thus, a broken or dotted line is drawn in the solution region.

(iv) In an inequality of the type $ax + by \geq c$ or $ax + by \leq c$, the points on the line $ax + by = c$ are also included in the solution region. Thus, a dark line is drawn in the solution region.

Fundamental Principle of Counting: The fundamental principle of counting (or multiplication principle) states that, 'if an event can occur in *m* different ways, following which another event can occur in *n* different ways, then the total number of occurrences of the events in the given order is *m* x *n*'.

In general, 'if an event can occur in n_1 different ways, following which another event can occur in n_2 different ways, following which a third event can occur in n_3 different ways, and so on, then the total number of occurrences of the events in the given order is $n_1 \times n_2 \times n_3 \times ...$'.

Note: The event of choosing three books could occur in various possible orders. But, we have to choose any one of the possible orders and count the number of different ways of the occurrence of the events in this chosen order.

Factorial Notation ($n!$): The notation $n!$ (read as factorial *n* or *n* factorial) represents the product of the first *n* natural numbers, that is, $n!$ is the product $1 \times 2 \times 3 \times ... \times (n-1) \times n$.

Thus, $n! = 1\times2\times3\times4.....\times(n-1)\times n$

Usually, we write it as $n! = n\times(n-1)\times...\times3\times2\times1$

$0! = 1$ (special property)

$1! = 1$

$2! = 2\times1 = 2$

$3! = 3\times2\times1 = 6$

$4! = 4\times3\times2\times1 = 24$

And so on.

Note that for a natural number n, we can write it as

$n! = n(n-1)!$ (provided $n \geq 1$)

$= n(n-1)(n-2)!$ (provided $n \geq 2$)

$= n(n-1)(n-2)(n-3)!$ (provided $n \geq 3$)

And so on.

Permutations

A permutation means an arrangement of the given objects in a definite order taken some or all at a time. It is denoted by ${}^nP_r = \dfrac{n!}{(n-r)!}$, $0 \leq r \leq n$, which represents the permutations (or arrangements) in the order of r objects from the given n objects. Given by

$${}^nP_r = \frac{n!}{(n-r)!}$$

$$= \frac{n(n-1)...(n-r+1)(n-r)(n-r-1)...3\times2\times1}{(n-r)(n-r-1)..3\times2\times1}$$

$$\Rightarrow {}^nP_r = n(n-1)(n-2).....(n-r+1)$$

Note:

(i) When $r = n$, ${}^nP_n = \dfrac{n!}{(n-n)!} = \dfrac{n!}{0!} = n!$

It means the arrangement of all the n objects in a row, that is, filling of n vacant places out of n given objects. For the first place, we have n choices out of n different objects. Similarly, for the second place, we have $(n - 1)$ choices out of the remaining $(n - 1)$ different objects, and so on. Thus, by the fundamental principle of counting, the total number of arrangements will be $n\times(n-1)\times...\times3\times2\times1 = n!$

(ii) When $r = 0$, ${}^nP_0 = \dfrac{n!}{(n-0)!} = \dfrac{n!}{n!} = 1$

It means no arrangement of any object, that is, leaving behind all the objects as they are without any arrangement, can be done only in one way.

Permutations when all the objects are distinct

Theorem 1: The number of permutations of n different objects taken r at a time, where $0 < r \leq n$ and the objects do not repeat, is given by

${}^nP_r = n(n-1)(n-2)...(n-r+1)$

Theorem 2: The number of permutations of n different objects taken r at a time, when repetition is allowed, is n^r.

Permutations when all the objects are not distinct objects

Theorem 3: The number of permutations of n objects, where p objects are of the same kind and the rest are all different is $\dfrac{n!}{p!}$.

Theorem 4: The number of permutations of n objects, where p_1 objects are of one kind, p_2 are of the second kind, ..., p_k are of kth kind and the rest, if any, are of different kinds is $\dfrac{n!}{p_1!\,p_2!\,...\,p_k!}$.

Combinations

Just like permutation is the arrangement of objects, combination is the selection of objects.

Thus, combination (or selection) of r different objects from given n different objects is denoted by ${}^nC_r = \dfrac{n!}{r!(n-r)!}$.

Now, these r different objects can be rearranged in $r!$ ways.

If we multiply ${}^nC_r$ with $r!$, we get ${}^nC_r \times r! = \dfrac{n!}{(n-r)!} = {}^nP_r$, that is, permutation of r different objects from the given n different objects.

Hence, we conclude that permutation consists of two steps, combination (or selection) and then arrangement. So, selection ${}^nC_r$ and arrangement $r!$, together represent permutation ${}^nP_r$.

Theorem: The number of combinations of n objects, if some or all objects are taken at a time, is $2^n - 1$. (2^n means, for each of the n objects we have two choices, that is, the object is chosen or the object is not chosen and -1 is used for the case when all the objects are not chosen).

This can also be explained as, out of n objects, only 1 object can be chosen in ${}^nC_1$ ways; similarly 2 objects can be chosen in ${}^nC_2$ ways, and so on. n objects can be chosen in ${}^nC_n$ ways.

Thus, the total number of combinations of n objects, if some or all objects are taken at a time, is

${}^nC_1 + {}^nC_2 + {}^nC_3 + ... {}^nC_n = 2^n - 1$

$(\because {}^nC_0 + {}^nC_1 + {}^nC_2 + {}^nC_3 + ... {}^nC_n = 2^n)$

Note:

(i) For $r = 0$, ${}^nC_0 = \dfrac{n!}{0!n!} = 1$ represents selecting nothing at all (or leaving behind all the objects) which is done only in one way.

(ii) For $r = n$, ${}^nC_n = \dfrac{n!}{n!0!} = 1$ represents the selection of all the objects, which is done only in one way.

(iii) ${}^nC_{n-r} = \dfrac{n!}{(n-r)!(n-(n-r))!} = \dfrac{n!}{(n-r)!r!} = {}^nC_r$ represents that selection of r objects out of n objects, which is the same as rejecting $(n-r)$ objects, and vice versa.

(iv) ${}^nC_a = {}^nC_b$ if and only if $a = b$ or $b = n-a \Rightarrow n = a+b$

(v) ${}^nC_{r-1} + {}^nC_r = {}^{n+1}C_r$

Day 128

Revision (Chapters VIII–IX)

Pascal's Triangle: Consider some of the expansions,

$(a+b)^0 = 1, \quad a+b \neq 0$

$(a+b)^1 = a+b$

$(a+b)^2 = a^2 + 2ab + b^2$

$(a+b)^3 = a^3 + 3a^2b + 3ab^2 + b^3$

$(a+b)^4 = a^4 + 4a^3b + 6a^2b^2 + 4ab^3 + b^4$

$(a+b)^5 = a^5 + 5a^4b + 10a^3b^2 + 10a^2b^3 + 5ab^4 + b^4$

In all these expansions, we may notice some similarities:

(i) The total number of terms in the expansion is one more than the index. For example, for index 2 in the expansion of $(a+b)^2$, the number of terms is 3.

(ii) Powers of the first quantity 'a' keep on decreasing by 1 whereas the powers of the second quantity 'b' keep on increasing by 1, in successive terms, starting from the index value.

(iii) In each term of the expansion, the sum of the indices of a and b is the same and is equal to the index of $a+b$. As an example, for index 2, in the expansion of $(a+b)^2$, the sum of the power of a and b is 2 in each term.

Now, if we arrange the coefficient of each term in a triangular shape, we get a triangle with 1 at the top vertex and running down the two slanting sides. This array of numbers is known as Pascal's triangle.

Index	Coefficients
0	1
1	1 1
2	1 2 1
3	1 3 3 1
4	1 4 6 4 1
5	1 5 10 10 5 1

In this Pascal's triangle we can see that the addition of 1s in the row for index 1 gives rise to 2 in the row for index 2. The addition of 1, 2 and 2, 1 in the row for index 2 gives rise to 3 and 3 in the row for index 3, and so on. Also 1 is present at the beginning and at the end of each row. This can be continued till any index.

Similarly, expansion for the higher powers of a binomial expansion is also possible using Pascal's triangle.

For a higher index, expansion using Pascal's triangle can be performed by writing all the rows of the Pascal's triangle, but it is a slightly lengthy process.

Binomial Theorem for Positive Integral Indices (n): Thus, to simplify this process, we make use of the concept of combinations to write all the numbers in the Pascal's triangle, known as the binomial theorem.

Now, we see that the terms of the Pascal's triangle can be written by

${}^nC_r = \dfrac{n!}{r!(n-r)!}$, $0 \le r \le n$, where n is a non-negative integer. Also, ${}^nC_0 = 1 = {}^nC_n$.

As an example, $(a+b)^1 = a+b$ can also be written as $(a+b)^1 = {}^1C_0a + {}^1C_1b$.

Similarly, $(a+b)^2 = a^2 + 2ab + b^2$ can also be written as $(a+b)^2 = {}^2C_0a^2 + {}^2C_1ab + {}^2C_2b^2$.

$(a+b)^3 = a^3 + 3a^2b + 3ab^2 + b^3$ can also be written $(a+b)^3 = {}^3C_0a^3 + {}^3C_1a^2b + {}^3C_2ab^2 + {}^3C_3b^3$

$(a+b)^4 = a^4 + 4a^3b + 6a^2b^2 + 4ab^3 + b^4$

$= {}^4C_0a^4 + {}^4C_1a^3b + {}^4C_2a^2b^2 + {}^4C_3ab^3 + {}^4C_4b^4$

In general, the binomial theorem is

$(a+b)^n = {}^nC_0a^n + {}^nC_1a^{n-1}b + {}^nC_2a^{n-2}b^2 + \ldots$

$\quad + {}^nC_{n-2}a^2b^{n-2} + {}^nC_{n-1}ab^{n-1} + {}^nC_nb^n$

Note:

(i) The coefficients ${}^nC_r$ occurring in the binomial theorem are known as binomial coefficients.

(ii) $(a+b)^n = \sum_{k=0}^{n} {}^nC_ka^{n-k}b^k$ represents ${}^nC_0a^nb^0 + {}^nC_1a^{n-1}b^1 + \ldots + {}^nC_ra^{n-r}b^r + \ldots + {}^nC_na^{n-n}b^n$

(iii) $\because {}^nC_r = {}^nC_{n-r}$

$\therefore (a+b)^n = \sum_{k=0}^{n} {}^nC_ka^{n-k}b^k = \sum_{k=0}^{n} {}^nC_ka^kb^{n-k}$

$= {}^nC_0a^0b^n + {}^nC_1a^1b^{n-1} + \ldots + {}^nC_ra^rb^{n-r} + \ldots + {}^nC_na^nb^{n-n}$

(iv) The number of terms in the expansion of $(a+b)^n$ is $(n+1)$, that is, one more than the index.

(v) In successive terms of the expansion, the index of a keeps on decreasing by unity, from n to 0 in the last term. Similarly, the index of b keeps on increasing by unity, from 0 to n in the last term, or vice versa.

(vi) In the expansion of $(a+b)^n$, the sum of the indices of a and b is equal to n, in each term of the expansion.

Some Special Cases

$$(a+b)^n = {}^nC_0a^n + {}^nC_1a^{n-1}b + {}^nC_2a^{n-2}b^2 + \ldots + {}^nC_{n-2}a^2b^{n-2} + {}^nC_{n-1}ab^{n-1} + {}^nC_nb^n$$

(i) Taking $a = x,\ b = -y$, we get

$$(x-y)^n = {}^nC_0x^n - {}^nC_1x^{n-1}y + {}^nC_2x^{n-2}y^2 + \ldots + (-1)^n\,{}^nC_ny^n$$

(ii) Taking $a = 1, b = x$, we get

$$(1+x)^n = {}^nC_0 + {}^nC_1x + {}^nC_2x^2 + {}^nC_3x^3 + \ldots + {}^nC_nx^n$$

In particular, for $x = 1$ we get

$$2^n = {}^nC_0 + {}^nC_1 + {}^nC_2 + \ldots + {}^nC_n.$$

(iii) Taking $a = 1, b = -x$, we get

$$(1-x)^n = {}^nC_0 - {}^nC_1x + {}^nC_2x^2 - \ldots + (-1)^n\,{}^nC_nx^n$$

In particular, for $x = 1$, we get

$$0 = {}^nC_0 - {}^nC_1 + {}^nC_2 - \ldots + (-1)^n\,{}^nC_n$$

(iv) Taking $a = x,\ b = \frac{1}{x}$, we get

$$\left(x+\frac{1}{x}\right)^n = {}^nC_0x^n + {}^nC_1x^{n-2} + {}^nC_2x^{n-4} + {}^nC_3x^{n-6} + \ldots + {}^nC_{n-2}\left(\frac{1}{x}\right)^{n-4} + {}^nC_{n-1}\left(\frac{1}{x}\right)^{n-2} + {}^nC_n\left(\frac{1}{x}\right)^n$$

Similarly,

$$\left(x-\frac{1}{x}\right)^n = {}^nC_0x^n - {}^nC_1x^{n-2} + {}^nC_2x^{n-4} - {}^nC_3x^{n-6} + \ldots + {}^nC_{n-1}(-1)^{n-1}\left(\frac{1}{x}\right)^{n-2} + (-1)^n\,{}^nC_n\left(\frac{1}{x}\right)^n$$

General Term in a Binomial Expansion

$$(a+b)^n = {}^nC_0a^n + {}^nC_1a^{n-1}b + {}^nC_2a^{n-2}b^2 + \ldots + {}^nC_{n-2}a^2b^{n-2} + {}^nC_{n-1}ab^{n-1} + {}^nC_nb^n$$

1st term is ${}^nC_0a^n$; 2nd term is ${}^nC_1a^{n-1}b$, 3rd term is ${}^nC_2a^{n-2}b^2$, and so on. The $(r+1)^{\text{th}}$ term (or general term) is $T_{r+1} = {}^nC_ra^{n-r}b^r$.

We may also write it as $T_{r+1} = {}^nC_ra^rb^{n-r}$.

Similarly, General Term in a Binomial Expansion

$$(a-b)^n = {}^nC_0a^n + {}^nC_1a^{n-1}b + {}^nC_2a^{n-2}b^2 + \ldots + {}^nC_{n-2}a^2b^{n-2} + {}^nC_{n-1}ab^{n-1} + {}^nC_nb^n$$

is $(r+1)^{\text{th}}$ term (or general term) given by $T_{r+1} = (-1)^r\,{}^nC_ra^{n-r}b^r$

Middle Term/s in a Binomial Expansion

(i) When n is even: The number of terms in the expansion will be $n+1$, that is, odd.

Therefore, the middle term is

$$\left(\frac{n+1+1}{2}\right)^{\text{th}} \Rightarrow \left(\frac{n}{2}+1\right)^{\text{th}} \text{ term}$$

$$T_{middle/even} = {}^nC_{(n/2+1)}a^{n-(n/2+1)}b^{(n/2+1)} = {}^nC_{(n/2+1)}a^{n/2-1}b^{n/2+1}$$

(ii) When n is odd: The number of terms in the expansion will be $n+1$, that is, even.

Thus, there will be two middle terms in the expansion given by $\left(\frac{n+1}{2}\right)^{\text{th}}$ term and $\left(\frac{n+1}{2}+1\right)^{\text{th}}$ term.

Note: In the expansion of $\left(x+\frac{1}{x}\right)^{2n}$, where $x \neq 0$, the middle term is $\left(\frac{2n+1+1}{2}\right)^{\text{th}}$, that is, $(n+1)^{\text{th}}$ term, (since $2n$ is even).

Thus, the middle term is given by ${}^{2n}C_nx^n\left(\frac{1}{x}\right)^n = {}^{2n}C_n$ (constant). This term is also independent of x (constant term).

Binomial Theorem for Negative or Fractional Index

Case-I: If $\left|\frac{b}{a}\right| < 1$, then

$$(a+b)^m = a^m + ma^{m-1}b + \frac{m(m-1)}{1\cdot 2}a^{m-2}b^2 + \frac{m(m-1)(m-2)}{1\cdot 2\cdot 3}a^{m-3}b^3 + \ldots$$

Case-II: If $\left|\frac{a}{b}\right| < 1$, then

$$(a+b)^m = b^m + mb^{m-1}a + \frac{m(m-1)}{1\cdot 2}b^{m-2}a^2 + \frac{m(m-1)(m-2)}{1\cdot 2\cdot 3}b^{m-3}a^3 + \ldots$$

General Term in the expansion $(a+b)^m$ for Negative or Fractional Index is

$$T_r = \frac{m(m-1)(m-2)\ldots(m-r+1)}{1\cdot 2\cdot 3\ldots r}a^{m-r}b^r;\ r = 0, 1, 2\ldots$$

Similarly, expansion $(1+x)^m$

$$= 1 + mx + \frac{m(m-1)}{2!}x^2 + \frac{m(m-1)(m-2)}{3!}x^3 + \ldots$$

Note:

(i) Binomial theorem for a negative or fractional index is valid only when $\left|\frac{b}{a}\right| < 1$ or $|x| < 1$.

(ii) For a negative or fraction index, ${}^nC_r$ does not make any sense.
(iii) Binomial theorem for a negative or fractional index gives infinite terms; therefore we may write only certain terms because the higher powers of $|x|$ tend to zero.

Special Cases for Negative or Fractional Index $|x|<1$

(i) $(1+x)^{-1} = 1-x+x^2-x^3+...+(-1)^r x^r+...$

(ii) $(1-x)^{-1} = 1+x+x^2+x^3+...+x^r+...$

(iii) $(1+x)^{-2} = 1-2x+3x^2-...+(-1)^r(r+1)x^r+...$

(iv) $(1-x)^{-2} = 1+2x+3x^2+...+(r+1)x^r+...$

(v) $\dfrac{1}{(1+x)^3} = (1+x)^{-3}$

$= 1-3x+6x^2-10x^3+...+(-1)^r\dfrac{(r+1)(r+2)}{2}x^r+...$

(vi) $\dfrac{1}{(1-x)^3} = (1-x)^{-3}$

$= 1+3x+6x^2+10x^3+...+\dfrac{(r+1)(r+2)}{2}x^r+...$

(vii) $\sqrt{(1+x)} = (1+x)^{1/2} = 1+\dfrac{1}{2}x-\dfrac{1}{8}x^2+\dfrac{1}{16}x^3+...$

(viii) $\sqrt{(1-x)} = (1-x)^{1/2} = 1-\dfrac{1}{2}x-\dfrac{1}{8}x^2-\dfrac{1}{16}x^3-...$

(ix) $\dfrac{1}{\sqrt{(1+x)}} = (1+x)^{-1/2} = 1-\dfrac{1}{2}x+\dfrac{3}{8}x^2-\dfrac{5}{16}x^3+...$

(x) $\dfrac{1}{\sqrt{(1-x)}} = (1-x)^{-1/2} = 1+\dfrac{1}{2}x+\dfrac{3}{8}x^2+\dfrac{5}{16}x^3+...$

Sequence

- The arrangement of numbers in a definite pattern (or a specified rule) is known as a sequence.
- The various numbers occurring in a sequence are called its terms, denoted by $a_1, a_2, a_3, ...a_n, ...$.
- The subscripts denote the position of the term in a sequence. As an example, a_1 is the 1st term, a_2 is the 2nd term, and so on, a_n is the nth term (also called the general term) of the sequence.
- A sequence containing a finite number of terms is called a finite sequence.
- A sequence is said to be infinite if it is not a finite sequence if it contains infinite terms.

Note: In a sequence, generally we should not expect that its terms will necessarily be given by a specific formula, but a rule for generating the terms $a_1, a_2, a_3, ..., a_n, ...$ will be in succession.

Series

- For a given sequence $a_1, a_2, a_3, ...a_n$, the expression $a_1+a_2+a_3+.....+a_n+...$ is called the series associated with the given sequence.
- The series is finite or infinite depending on whether the given sequence is finite or infinite.
- In general, we represent a series with the symbol sigma, using the Greek letter $\sum$ (sigma), as a means of indicating the summation involved. Thus, the series $\sum_{k=1}^{n} a_k = a_1+a_2+a_3+.....+a_n$

Arithmetic Progression (AP)

In an arithmetic progression, the difference of any two consecutive numbers is a constant.

A sequence $a_1, a_2, a_3, ..., a_n, ...$ is called an arithmetic sequence or arithmetic progression if $a_{n+1} = a_n + d,\ n \in N$, where a_1 is called the first term and the constant term d is called the common difference of the AP.

Consider an AP with its first term being a and the common difference d. The AP will be,

$a, a+d, a+2d, ...$ and so on

Thus, 1st term $= a_1 = a$

2nd term $= a_2 = a+d$

3rd term $= a_3 = a+2d$

4th term $= a_4 = a+3d$

In general, the nth term of the AP is $a_n = a+(n-1)d$.

For a finite sequence, we say n^{th} term of the AP is the last term of the AP denoted by $l = a+(n-1)d$.

Sum of *n* terms of an AP

S_n = the sum to n terms of the AP

= the sum of $a, a+d, a+2d, ..., a+(n-1)d$

$\Rightarrow S_n = \dfrac{n}{2}[2a+(n-1)d] = \dfrac{n}{2}[a+l]$

Properties of an AP

(i) If a constant is added (or subtracted) to each term of an AP, then the resulting sequence is also an AP.
(ii) If each term of an AP is multiplied (or divided) by a non-zero constant, then the resulting sequence is also an AP.

Arithmetic Mean (AM)

A number A is called the arithmetic mean (AM) of two given numbers a and b, if a, A, b are in AP,

$A-a = b-A \Rightarrow A = \dfrac{a+b}{2}$

The AM between two numbers a and b is interpreted as their average $\dfrac{a+b}{2}$.

Similarly, we can insert two AM (A_1 and A_2) between a and b, such that a, A_1, A_2, b is in AP.

$A_1 - a = A_2 - A_1 = b - A_2$

$\Rightarrow A_1 = \dfrac{2a+b}{3}$ and $A_2 = \dfrac{a+2b}{3}$

In general, for any two given numbers a and b, we can insert as many numbers as we like between them such that the resulting sequence is an AP.

Let $A_1, A_2, A_3, \ldots, A_n$ be n AM numbers between a and b such that $a, A_1, A_2, A_3, \ldots, A_n, b$ forms an AP.

Now, since the above sequence has $(n+2)$ terms, b will be the $(n+2)^{\text{th}}$ term, that is,

$b = a + [(n+2)-1]d = a + (n+1)d.$

Or common difference of the above sequence $d = \dfrac{b-a}{n+1}$

Thus, n numbers between a and b are as follows:

$$A_1 = a + d = a + \frac{b-a}{n+1}$$

$$A_2 = a + 2d = a + \frac{2(b-a)}{n+1}$$

$$A_3 = a + 3d = a + \frac{3(b-a)}{n+1} \text{ and so on}$$

$$A_n = a + nd = a + \frac{n(b-a)}{n+1}$$

Geometric Progression (GP)

In a geometric progression, the ratio of any two consecutive numbers is a constant.

A sequence $a_1, a_2, a_3, \ldots, a_n, \ldots$ is called a geometric sequence or a geometric progression if $\dfrac{a_{n+1}}{a_n} = r,\ n \in N$, where a_1 is called the first term and the constant term r is called the common ratio of the GP.

Consider a GP with its first term a and common ratio r. The GP will be,

$a, ar, ar^2, ar^3, \ldots ar^{n-1} \ldots$; and so on

Thus, 1st term $= a_1 = a$

2nd term $= a_2 = ar$

3rd term $= a_3 = ar^2$

4th term $= a_4 = ar^3$

In general, the nth term of the GP is $a_n = ar^{n-1}$

Sum of *n* terms of a GP

S_n = the sum to n terms of the GP

$\Rightarrow S_n = a + ar + ar^2 + \ldots + ar^{n-1}$ (i)

Case 1: When $r = 1$,

$S_n = a + a + a + \ldots + a$ (n terms) $= na$

Case 2: When $r \neq 1$,

Multiplying equation (i) by r, we get

$rS_n = ar + ar^2 + ar^3 + \ldots + ar^n$ (ii)

Subtracting equation (ii) from equation (i), we get

$(1-r)S_n = a - ar^n = a(1-r^n)$

$$\Rightarrow S_n = \frac{a(1-r^n)}{1-r} \text{ or } S_n = \frac{a(r^n-1)}{r-1}$$

Properties of a GP

(i) If each term of a GP is multiplied (or divided) by a non-zero constant, then the resulting sequence is also a GP.

(ii) But, if a constant is added (or subtracted) to each term of a GP, then the resulting sequence may not be a GP.

Geometric Mean (GM)

A number G is called the geometric mean (GM) of the two given numbers a and b, if a, G, b is in GP. That is,

$$\frac{G}{a} = \frac{b}{G} \Rightarrow G = \sqrt{ab}$$

Similarly, we can insert two GM (G_1 and G_2) between a and b, such that a, G_1, G_2, b is in GP.

$$\frac{G_1}{a} = \frac{G_2}{G_1} = \frac{b}{G_2} \Rightarrow G_1 = a\left(\frac{b}{a}\right)^{1/3} \text{ and } G_2 = a\left(\frac{b}{a}\right)^{2/3}$$

In general, for any two given numbers a and b, we can insert as many numbers as we like between them such that the resulting sequence is a GP.

Let $G_1, G_2, G_3, \ldots, G_n$ be n GM numbers between a and b such that $a, G_1, G_2, G_3, \ldots, G_n, b$ forms a GP. Now, since the above sequence has $(n+2)$ terms, b will be the $(n+2)^{\text{th}}$ term, that is,

$$b = ar^{n+1} \Rightarrow r = \left(\frac{b}{a}\right)^{\frac{1}{n+1}}$$

The common ratio of the above sequence $r = \left(\dfrac{b}{a}\right)^{\frac{1}{n+1}}$.

Thus, n numbers between a and b are as follows:

$$G_1 = ar = a\left(\frac{b}{a}\right)^{\frac{1}{n+1}}$$

$$G_2 = ar^2 = a\left(\frac{b}{a}\right)^{\frac{2}{n+1}}$$

$$G_3 = ar^3 = a\left(\frac{b}{a}\right)^{\frac{3}{n+1}}, \text{ and so on}$$

$$G_n = ar^n = a\left(\frac{b}{a}\right)^{\frac{n}{n+1}}$$

Relationship Between AM and GM

Let A and G be the AM and GM of two given positive real numbers a and b, respectively. Then

$$A = \frac{a+b}{2} \text{ and } G = \sqrt{ab}$$

Rearranging, we get

$$A - G = \frac{a+b}{2} - \sqrt{ab} = \frac{a+b-2\sqrt{ab}}{2}$$

$$\Rightarrow A - G = \frac{\left(\sqrt{a}-\sqrt{b}\right)^2}{2} \geq 0$$

$$\Rightarrow A - G \geq 0 \Rightarrow A \geq G$$

Thus, we conclude that the arithmetic mean is always greater than or equal to the geometric mean.

Special Series

(i) $\sum_{k=1}^{n} k = 1+2+3+4+...+n = \frac{n(n+1)}{2}$

(ii) $1^2+2^2+3^2+.....+n^2 = \left\{\frac{n(n+1)}{2}\right\}^2$

(iii) $1^3+2^3+3^3+.....+n^3 = \left[\frac{n(n+1)}{2}\right]^2$

Infinite Geometric Progression (GP)

The geometric progression $a, ar, ar^2, ar^3, ...$ with the first term as a and a common ratio r ($|r|<1$), is known as an infinite geometric progression.

$|r|<1$; therefore $r^n \to 0$ when n is sufficiently large, $n \to \infty$ and thus, the summation of the infinite geometrical series is,

$S_n = a+ar+ar^2+...+ar^{n-1}$

$\Rightarrow S_n = \frac{a(1-r^n)}{1-r}$

$\Rightarrow S_\infty = \frac{a}{1-r}$

Since, $r^n \to 0$ when $n \to \infty$

Exponential Series

Exponential series with variable x can be expressed as

$e^x = 1+\frac{x}{1!}+\frac{x^2}{2!}+\frac{x^3}{3!}+...\frac{x^n}{n!}+...$

Substituting $x = 1$, we get the value of e as

$e = 1+\frac{1}{1!}+\frac{1}{2!}+\frac{1}{3!}+\frac{1}{4!}+...$

Logarithmic Series

Logarithmic series with variable x ($|x| \le 1$) can be expressed as $\log_e(1+x) = x-\frac{x^2}{2}+\frac{x^3}{3}-\frac{x^4}{4}+...$

Substituting $x = 1$, we get

$\log_e 2 = 1-\frac{1}{2}+\frac{1}{3}-\frac{1}{4}+...$

Day 129

Unit Test 2 (Chapters IV–IX)

Time: 3 Hrs Maximum **Marks:** 80

General Instructions

(i) All the questions are compulsory.

(ii) The question paper consists of 36 questions divided into 4 sections A, B, C, and D.

(iii) Section A comprises 20 questions of 1 mark each. Section B comprises 6 questions of 2 marks each. Section C comprises 6 questions of 4 marks each. Section D comprises 4 questions of 6 marks each.

(iv) There is no overall choice. However, an internal choice has been provided in four questions of 1 mark each, two questions of 2 marks each, two questions of 4 marks each, and two questions of 6 marks each. You have to attempt only one of the alternatives in all such questions.

(v) Use of calculators is not permitted.

SECTION A

Q1–Q10 are multiple choice type questions. Select the correct option. (1 mark each)

Q1. If the middle term of $\left(\frac{1}{x}-x\cos x\right)^6$ is equal to $4\frac{3}{8}$, then the value of x is

(a) $2n\pi \pm \pi$ (b) $2n\pi \pm \frac{\pi}{2}$

(c) $2n\pi$ (d) $2n\pi \pm \frac{\pi}{3}$

Q2. If t_n denotes the nth term of the series $2+4+8+14+22+...$, then t_{11} is

(a) 422 (b) 22^2 (c) 504 (d) 20^2

Q3. If the sum of n terms of an AP is given by $S_n = 2n-n^2$, then the common difference of the AP is

(a) 3 (b) –3 (c) 2 (d) –2

Q4. The approximation of $(1.1)^5$ up to two decimal places is

(a) 1.61 (b) 1.71 (c) 1.51 (d) 1.41

Q5. If $\left(\frac{1+i}{1-i}\right)^{75} = a+ib$, then (a, b) is equal to

(a) (1,0) (b) (0,–1)

(c) (0,1) (d) (–1,–1)

Q6. If $|x-1|>2$, then

(a) $x \in (0,2)$ (b) $x \in [0,2]$

(c) $x \in (-\infty,-1)\cup(3,\infty)$ (d) $x \in [-\infty,0)\cup[2,\infty)$

Q7. Summation of the series $\frac{-3}{2}, \frac{1}{2}, \frac{-1}{6}, \ldots\ldots$ up to infinite terms is ______

(a) $-9/8$ (b) $9/8$

(c) $-5/4$ (d) $5/4$

Q8. The coefficient of $\frac{1}{x^{15}}$ in the expansion of $\left(x^3 - \frac{1}{x^2}\right)^{10}$ is

(a) 10 (b) -10

(c) 20 (d) -20

Q9. If $|z| = \sqrt{2}$, $\arg(z) = \frac{3\pi}{4}$, then $z =$ ______

(a) $-1-i$ (b) $1-i$

(c) $-1+i$ (d) $1+i$

Q10. Ten chairs are numbered 1 to 10. Four women and 4 men wish to occupy one chair each. First the women choose the chairs from amongst the chairs 1 to 5 and then the men select from the remaining chairs. The total number of possible arrangements is

(a) 50 (b) 60 (c) 70 (d) 75

Q11–Q15 are fill in the blanks. (1 mark each)

Q11. If $|z+2| \geq 3$, then least value of $|z+1|$ is______

Q12. If $\left|\frac{z-1}{z+1}\right| = \tan\frac{\pi}{4}$, then the locus of z is______

Q13. In the expansion of $\left(x^2 - \frac{2}{x}\right)^9$, the value of the constant term is______

Q14. The coefficient of $a^{-6}b^4$ in the expansion of $\left(\frac{2}{a} - \frac{b}{2}\right)^{10}$ is______

OR

Q14. The sum of the series $\sum_{r=0}^{11} {}^{21}C_r$ is______

Q15. A candidate is required to answer 5 questions out of 10 questions, which are divided into two groups, each containing 5 questions. Find the number of different ways of selecting the questions if at least 1 question is compulsory from either group.

OR

Q15. Out of 10 points in a plane, 4 points are in the same line and are collinear. Find the number of lines that can be formed by joining two points.

(Q16–Q20) Answer the following questions. (1 mark each)

Q16. Find the real numbers x and y if $(x-iy)(3+5i)$ is the conjugate of $11-7i$.

Q17. Find the principal argument of $(1-i\sqrt{3})^2$.

Q18. Solve the inequality $-3 \leq 7 - \frac{5x}{2} \leq 2$.

Q19. The longest side of a rectangle is 10 cm longer than the shortest side. If the perimeter of the rectangle is more than 80 cm and less than 120 cm, then find the minimum and maximum length of the shortest side.

OR

Q19. The letters of the word 'POINT' are arranged in all possible ways as listed in the dictionary. Then, what is the rank of the word 'POINT'?

Q20. Find the number of positive integers greater than 1000 and less than 4000 which are divisible by 5, provided no digit is to be repeated.

OR

Q20. The sum of some terms of a GP is 155 whose first term and the common ratio are 5 and 2, respectively. Find the ratio of the last term and the number of terms.

SECTION B

(Q21–Q26) Solve the following questions. (2 marks each)

Q21. Ravi obtained 55, 65 and 80 marks in the first three unit tests. Find the minimum marks he should get in the fourth test to have an average of at least 60 marks and at most 75 marks.

Q22. Solve the given inequality for real x: $\frac{(2x+1)}{2} \geq \frac{(3x-2)}{4} - \frac{(2-x)}{5} + \frac{1}{2}$.

Q23. Evaluate $(\sqrt{2}+1)^5 + (\sqrt{2}-1)^5$.

Q24. Find the middle term (terms) in the expansion of $\left(2x - \frac{x^3}{2}\right)^5$.

Q25. If a bag contains 5 black and 6 red balls, determine the number of ways in which 2 black and 3 red balls or 3 black and 2 red balls can be selected from the lot.

OR

Q25. Ten guests are to be seated, half on each side of a long table. Three particular guests desire to sit on one particular side and four others on the other side of the table. Find the number of ways in which the seating arrangement can be made.

Q26. Find the sum to n terms of the series $1\times1+2^2\times3+3^2\times5+\ldots\ldots\ldots$

OR

Q26. Find the sum of the series $0.5+0.55+0.555+\ldots\ldots\ldots$ up to n terms.

SECTION C

(Q27–Q32) Solve the following questions. **(4 marks each)**

Q27. Find the complex number satisfying the equation $|(z+1)|+z-i=0$.

Q28. If $\left(\frac{1+i}{1-i}\right)^2+\left(\frac{1-i}{1+i}\right)^2=x+iy$, then find (x, y).

Q29. Solve the following system of inequalities $\frac{3}{x-2}>1,\ \frac{2x-1}{x+1}\le -1.$

Q30. Find the number of integers less than 5000 that can be formed with the digits 3, 4, 5, 6, 7 where no digits are repeated.

Q31. Expand using binomial theorem $\left(2+x-\frac{2}{x}\right)^3, x\neq 0.$

OR

Q31. Find n in the binomial $\left(\sqrt[3]{2}+\frac{1}{\sqrt[4]{3}}\right)^n$, if the ratio of 7th term from the beginning to the 7th from the end is 1.

Q32. The sum of n terms of two arithmetic progressions are in the ratio $3n-1:2n+1$. Find the ratio of their 8th terms.

OR

Q32. Find a GP for which the sum of the first two terms is –8 and the fifth term is 9 times the third term.

SECTION D

(Q33–Q36) Solve the following questions. **(6 marks each)**

Q33. Prove the following by using the principle of mathematical induction for all $n\in N$

$$\frac{1}{1\times3}+\frac{1}{3\times5}+\frac{1}{5\times7}+\ldots\frac{1}{(2n-1)(2n+1)}=\frac{n}{2n+1}.$$

Q34. If the imaginary part of $\frac{\bar{z}+1}{\bar{z}-1}$ is 1, then show that the locus of the point representing z in the complex plane is a circle.

Q35. Solve the following system of inequalities graphically: $3x+2y\ge 6, 2x+3y\le 6, 2x-3y\le 6.$

OR

Q35. What are the number of ways of choosing 4 cards from a pack of 52 playing cards, such that

(i) 4 cards are of the same suit
(ii) 4 cards are of the same colour
(iii) 4 cards belong to four different suits
(iv) 2 are red cards and 2 are black cards
(v) 2 cards belong to the diamonds suit and 2 belong to the clubs suit
(vi) contain at least 2 kings

Q36. Find the coefficient of x^4 in the product $(1-x)^4(1+2x)^6$ using the binomial theorem.

OR

Q36. In an AP, if the first term is 5 and the sum of the first five terms is one-third of the next five terms, find sum of the first 10 terms.

Day 130

Unit Test 2 (Solutions)

SECTION A

Q1–Q10 are multiple choice type questions. Select the correct option. **(1 mark each)**

Q1. Solution: (a)

In the expansion $\left(\frac{1}{x}-x\cos x\right)^8$, $n=6$ (even); therefore it has only one middle term, that is, the 4th term.

$$\therefore T_4=T_{3+1}={}^6C_3\left(\frac{1}{x}\right)^{6-3}(-x\cos x)^3$$

$$\Rightarrow (-1)^3\,{}^6C_3x^{-3}x^3\cos^3 x=4\frac{3}{8}$$

$$\Rightarrow -\frac{6\cdot5\cdot4}{3\cdot2\cdot1}\cos^3 x=20$$

$$\Rightarrow \cos^3 x=-1$$

$$\Rightarrow \cos x=-1\Rightarrow x=2n\pi\pm\pi$$

Q2. Solution: (a)

Let S_n be sum of the series $2+4+8+14+\ldots+t_{11}$. Then

$S_n = 2+4+8+14+22+...+t_{11}...$

$\Rightarrow S_n = 2+4+8+14+...+t_{10}+t_{11}...$

Subtracting, we get, $0 = 2+2+4+6+8+...-t_{11}$

$\Rightarrow t_{11} = 2+2+4+6+8+....$ up to 10 terms

$\therefore t_{11} = 2+[2+4+6+8+.....$ up to 10 terms]

$= 2+\frac{10}{2}[2\times2+9\times2] = 112$

Q3. Solution: (d)

$\because S_n = 2n-n^2$

$\therefore S_1 = 2\times1-(1)^2 = 1 = T_1$

$S_2 = 2\times2-(2)^2 = 0$

$T_2 = S_2 - S_1 = 0-1 = -1$

Thus, common difference $(d) = T_2 - T_1 = -1-1 = -2$

Q4. Solution: (a)

We can write $1.1 = 1+0.1$

$\therefore (1.1)^5 = (1+0.1)^5$

$= {}^5C_0(1)^5 + {}^5C_1(1)^4(0.1) + {}^5C_2(1)^3(0.1)^2 + {}^5C_3(1)^2(0.1)^3 + ...$

$= 1+0.5+0.1+0.01+... = 1.61$

(Approximately neglecting smaller value terms)

$= 1+5(0.01)+10(0.01)^2 = 1+0.05+0.001 = 1.051$

Thus, the value of $(1.01)^5$ is approximately 1.051.

Q5. Solution: (b)

$\left(\frac{1+i}{1-i}\right)^{75} \Rightarrow \left[\frac{(1+i)}{(1-i)}\times\frac{(1+i)}{(1+i)}\right]^{75}$

$= \left(\frac{1+i^2+2i}{1-i^2}\right)^{75} = \left(\frac{2i}{2}\right)^{75} \quad (\because i^2 = -1)$

$= (i)^{75} = (i^4)^{18} + i^3 = 1 \quad (\because i^4 = 1,\ i^3 = -i)$

$= 0 - i = a+ib \Rightarrow a = 0,\ b = -1$

Q6. Solution: (c)

$|x-1| > 2 \Rightarrow (x-1) < -2$ or $(x-1) > 2$

$(\because |x| > a \Rightarrow x < -a$ or $x > a)$

$\Rightarrow x < -2+1$ or $x > 2+1$

$\Rightarrow x < -1$ or $x > 3 \Rightarrow x \in (-\infty,-1)\cup(3,\infty)$

Q7. Solution: (a)

$a = \frac{-3}{2},\ r = \frac{-1}{3} \Rightarrow |r| = \frac{1}{3} < 1,$

Therefore summation, $S = \frac{-3}{2}+\frac{1}{2}-\frac{1}{6}+\frac{1}{18}-...$

$\Rightarrow S_\infty = \frac{a}{1-r} = \frac{-3/2}{1-(-1/3)} = \frac{-3/2}{4/3} = \frac{-9}{8}$

Q8. Solution: (b)

The general term of the given expansion $\left(x^3 - \frac{1}{x^2}\right)^{10}$ is $T_{r+1} = {}^{10}C_r(x^3)^{10-r}\left(-\frac{1}{x^2}\right)^r$

$= {}^{10}C_r x^{30-3r}(-1)^r x^{-2r} = (-1)^r\, {}^{10}C_r x^{30-5r}$

For the coefficient of $\frac{1}{x^{15}}$ or x^{-15},

$30-5r = -15 \Rightarrow r = 9$

Thus, $T_{9+1} = (-1)^9\, {}^{10}C_9 x^{30-45} = -{}^{10}C_9 x^{-15}$

Hence, coefficient of $\frac{1}{x^{15}}$ or x^{-15} is

$= -{}^{10}C_9 = -10$

Q9. Solution: (c)

Let $z = x+iy = r(\cos\theta + i\sin\theta)$. Then

$|z| = r = \sqrt{2}$ and $\arg(z) = \theta = \frac{3\pi}{4} \Rightarrow \tan\theta = \frac{3\pi}{4}$

$\therefore z = \sqrt{2}\left(\cos\frac{3\pi}{4} + i\sin\frac{3\pi}{4}\right)$

$= \sqrt{2}[\cos(\pi-\pi/4) + i\sin(\pi-\pi/4)]$

$= \sqrt{2}\left[-\cos\frac{\pi}{4} + i\sin\frac{\pi}{4}\right] = \sqrt{2}\left[\frac{-1}{\sqrt{2}} + i\frac{1}{\sqrt{2}}\right] = -1+i$

Q10. Solution: (d)

Since the women choose the chairs first from among 1 to 5 chairs, and there are 4 women, the number of arrangements = ${}^5P_4$ ways.

Now, arrangement of 4 men out of remaining 6 chairs is ${}^6P_4$ ways.

Thus, total number of possible arrangements is

$= {}^5P_4 \times {}^6P_4 = \frac{5!}{(5-4)!}\times\frac{6!}{(6-4)!} = 5\times\frac{6\times5}{2} = 75$

Q11–Q15 are fill in the blanks. (1 mark each)

Q11. Solution:

For the least value of $|z+1|$

$|z+1| = |z+2-1| \geq |z+2| - |1|$

$\Rightarrow |z+1| \geq 3-1 \Rightarrow |z+1| \geq 2 \quad (\because |z+2| \geq 3)$

Thus, the least value of $|z+1|$ is 2.

Q12. Solution:

Let $z = x+iy$. Then $\left|\frac{z-1}{z+1}\right| = \tan\frac{\pi}{4}$

$\Rightarrow \left|\frac{x+iy-1}{x+iy+1}\right| = 1 \Rightarrow \left|\frac{x-1+iy}{x+1+iy}\right| = 1$

$\Rightarrow |x-1+iy| = |x+1+iy|$

$\Rightarrow (x-1)^2 + y^2 = (x+1)^2 + y^2$

$\Rightarrow x^2 + 1 - 2x + y^2 = x^2 + 2x + 1 + y^2$

$\Rightarrow 4x = 0 \Rightarrow x = 0$ represent the imaginary axis.

Q13. Solution:

Let the rth term be the constant term. Then

$T_{r+1} = {}^9C_r(x^2)^{9-r}\left(-\frac{2}{x}\right)^r = (-1)^r 2^r\, {}^9C_r x^{18-2r-r}$

$= (-1)^r 2^r\, {}^9C_r\, x^{18-3r}$

For the constant term, $18-3r=0 \Rightarrow r=6$; therefore the value of constant term is

$T_{6+1} = (-1)^6 2^6\, {}^9C_6 = 32\times 84 = 2688$

Q14. Solution:

Let the rth term have a coefficient of $a^{-6}b^4$. Then

$T_{r+1} = {}^{10}C_r\left(\frac{2}{a}\right)^{10-r}\left(-\frac{b}{2}\right)^r$; thus for the coefficient of $a^{-6}b^4$, $10-r=6 \Rightarrow r=4$.

Therefore, coefficient of

$a^{-6}b^4 = {}^{10}C_4(-1)^4\frac{2^6}{2^4} = 4\,{}^{10}C_4.$

OR

Q14. Solution:

$\sum_{r=0}^{11} {}^{21}C_r = {}^{21}C_0 + {}^{21}C_1 + {}^{21}C_2 + \ldots + {}^{21}C_{10} + {}^{21}C_{11}$

$= \frac{1}{2}\begin{pmatrix} {}^{21}C_0 + {}^{21}C_1 + {}^{21}C_2 + \ldots + {}^{21}C_{10} \\ + {}^{21}C_0 + {}^{21}C_1 + {}^{21}C_2 + \ldots + {}^{21}C_{10} \end{pmatrix} + {}^{21}C_{11}$

$= \frac{1}{2}\begin{pmatrix} {}^{21}C_0 + {}^{21}C_1 + {}^{21}C_2 + \ldots + {}^{21}C_{10} \\ + {}^{21}C_{21} + {}^{21}C_{20} + {}^{21}C_{19} + \ldots + {}^{21}C_{11} \end{pmatrix} + {}^{21}C_{11}$

$(\because {}^nC_r = {}^nC_{n-r})$

$= \frac{1}{2}(2^{21}) + {}^{21}C_{11} = 2^{20} + {}^{21}C_{11}$

$\left(\because {}^nC_0 + {}^nC_1 + \ldots + {}^nC_n = 2^n\right)$

Q15. Solution:

At least 1 question is compulsory from either group; therefore the total number of questions attempted from each group is given as

Group I	1	2	3	4
Group II	4	3	2	1

Thus, total number of possible ways

$= {}^5C_1 \times {}^5C_4 + {}^5C_2 \times {}^5C_3 + {}^5C_3 \times {}^5C_2 + {}^5C_4 \times {}^5C_1$

$= 5\times 5 + 10\times 10 + 10\times 10 + 5\times 5 = 250$

OR

Q15. Solution:

The number of lines that can be formed by joining two points from given n points is $= {}^nC_2$

m points are collinear, and these m collinear points when joined pair-wise will give only one line.

Therefore, the total number of straight lines formed

$= {}^nC_2 - {}^mC_2 + 1 = {}^{10}C_2 - {}^4C_2 + 1 = 45 - 6 + 1 = 40$

(Q16–Q20) Answer the following questions.

(1 mark each)

Q16. Solution:

Let $z = (x-iy)(3+5i)$

$z = 3x + 5xi - 3yi - 5yi^2 = 3x + 5xi - 3yi + 5y$

$= (3x-5y) + i(5x-3y)$

$\Rightarrow \bar{z} = (3x+5y) - i(5x-3y)$

The conjugate of $(x-iy)(3+5i)$ is $4-2i$; therefore $(3x+5y) - i(5x-3y) = 11 - 7i.$

Equating the real and imaginary parts, we get

$3x+5y = 11$ and $5x-3y = 7$

Solving above two equations, we get

$x = 2$ and $y = 1$

Q17. Solution:

$z = (1-i\sqrt{3})^2$

$\Rightarrow z = 1 - 3 - 2i\sqrt{3} \Rightarrow z = -2 - i2\sqrt{3}$

$\Rightarrow \arg(z) = \tan\theta = \left|\frac{\text{Im}(z)}{\text{Re}(z)}\right|$

$\Rightarrow \tan\theta = \left|\frac{-2\sqrt{3}}{-2}\right| = |\sqrt{3}| = \sqrt{3}$

$\Rightarrow \tan\theta = \tan\frac{\pi}{3} \Rightarrow \theta = \frac{\pi}{3}$

$\because \text{Re}(z) < 0,\ \text{Im}(z) < 0$

$\Rightarrow \arg(z) = -\left(\pi - \frac{\pi}{3}\right) = -\frac{2\pi}{3}$

Q18. Solution:

$-3 \le 7 - \frac{5x}{2} \le 2 \Rightarrow -3-7 \le 7 - \frac{5x}{2} - 7 \le 2 - 7$

(Subtracting 7 from the inequality)

$\Rightarrow -10 \le -\frac{5x}{2} \le -5 \Rightarrow -10\times\frac{2}{5} \le -\frac{5x}{2}\times\frac{2}{5} \le -5\times\frac{2}{5}$

(Multiplying the inequality by $\frac{2}{5}$)

$\Rightarrow -4 \le -x \le -2 \Rightarrow 4 \ge x \ge 2$

(Multiplying the inequality by –1. Since the multiplication number is negative, the inequality sign will reverse.)

Thus, the solution set for the given inequality is $x \in [2, 4]$.

Q19. Solution:

Let the length of shortest side be x cm. Then

Longest side $= (x+10)$ cm

Perimeter of rectangle $= 2\times(x + (x+10)) = 4x + 20$

According to the question,

$80 <$ Perimeter < 120

$\Rightarrow 80 < 4x + 20 < 120$

$\Rightarrow 60 < 4x < 100$

$\Rightarrow 15 < x < 25$

Thus, the minimum length of shortest side is 15 cm and maximum side is 25 cm.

OR

Q19. Solution:

The letters of the word 'POINT' in alphabetical order are I, N, O, P, T.

Words starting with I = 4!

Words starting with N = 4!

Words starting with O = 4!

Word starting with PI = 3!

Word starting with PN = 3!

Word starting with POINT = 1

∴ Rank of the word 'POINT' in dictionary is

$= 3\times 4! + 2\times 3! + 1 = 72 + 12 + 1 = 85$

Q20. Solution:

The number is divisible by 5; therefore, at the units place, the number will be 0 or 5.

Therefore, the units digit can be filled in 2 ways.

Now, the 4-digit number is greater than 2000 and less than 4000; therefore the thousands place can be filled in 3 ways (digit 1 or 2 or 3).

Since repetition is not allowed, the tens place can be filled in 8 ways; similarly the hundreds place can be filled in 7 ways.

Thus, total number of integers formed is

$= 2\times 3\times 8\times 7 = 336$

OR

Q20. Solution:

Let the first term be *a* 5 and the common ratio *r* be 2, and the sum of *n* terms of the GP be 155.

Therefore, $S_n = \dfrac{a(r^n - 1)}{r-1}$

$\Rightarrow 155 = \dfrac{5(2^n - 1)}{2-1} \Rightarrow 2^n - 1 = 31$

$\Rightarrow 2^n = 32 = (2)^5 \Rightarrow n = 5$

The last term of the GP is the 5th term given by

$a_5 = ar^{5-1} = (5)(2)^4 = (5)(16) = 80$

Thus, the ratio of the last term and the number of terms is

$\dfrac{80}{5} = 16$

SECTION B

(Q21–Q26) Solve the following questions. (2 marks each)

Q21. Solution:

Let *x* be the marks obtained by Ravi in the fourth unit test so that he should have an average of at least 60 marks and at most 70 marks. Then

$$65 \le \frac{55+65+80+x}{4} \le 70$$

$\Rightarrow 260 \le 200 + x \le 280 \Rightarrow 60 \le x \le 80$

Thus, the student should obtain a minimum of 60 marks and a maximum of 80 marks to have an average of at least 60 marks and at most 70 marks.

Q22. Solution:

$$\frac{(2x+1)}{2} \ge \frac{(3x-2)}{4} - \frac{(2-x)}{5} + \frac{1}{2}$$

$$\Rightarrow \frac{(2x+1)}{2} \ge \frac{5(3x-2)-4(2-x)+10}{20}$$

$$\Rightarrow \frac{(2x+1)}{2} \ge \frac{15x-10-8+4x+10}{20}$$

$$\Rightarrow \frac{(2x+1)}{2} \ge \frac{19x-8}{20}$$

$\Rightarrow 10(2x+1) \ge (19x-8)$

$\Rightarrow 20x + 10 \ge 19x - 8 \Rightarrow x \ge -18$

Therefore, all real numbers *x* which are greater than or equal to –18 are the solutions of the given inequality.

Thus, the solution set of the given inequality is $x \in [-18, \infty)$.

Q23. Solution:

Using the binomial theorem,

$$(a+b)^5 = {}^5C_0a^5 + {}^5C_1a^4b + {}^5C_2a^3b^2 + {}^5C_3a^2b^3 + {}^5C_4ab^4 + {}^5C_5b^5$$

$= a^5 + 5a^4b + 10a^3b^2 + 10a^2b^3 + 5ab^4 + b^5$

And,

$$(a-b)^5 = {}^5C_0a^5 - {}^5C_1a^4b + {}^5C_2a^3b^2 - {}^5C_3a^2b^3 + {}^5C_4ab^4 - {}^5C_5b^5$$

$= a^5 - 5a^4b + 10a^3b^2 - 10a^2b^3 + 5ab^4 - b^5$

$\therefore (a+b)^5 + (a-b)^5$

$= 2[a^5 + 10a^3b^2 + 5ab^4]$

Putting $a = \sqrt{2}, b = 1$ we get

$(\sqrt{2}+1)^5 + (\sqrt{2}-1)^5$

$= 2[5(\sqrt{2})^5 + 10(\sqrt{2})^3(1)^2 + 5(\sqrt{2})(1)^4]$

$= 2[20\sqrt{2} + 20\sqrt{2} + 5\sqrt{2}] = 90\sqrt{2}$

Q24. Solution:

The power of the binomial expansion $\left(3x - \dfrac{x^3}{6}\right)^5$ is $n = 5$ is odd.

Therefore, it has two middle terms, that is,

$\left(\dfrac{5+1}{2}\right)$ and $\left(\dfrac{5+1}{2} + 1\right)$ = 3rd term and 4th term.

Thus, $T_3 = T_{(2+1)} = {}^5C_2(2x)^{5-2}\left(-\dfrac{x^3}{2}\right)^2$

$= 10\cdot 2^3 x^3 x^6 2^{-2} = 20x^9$

And $T_4 = T_{(3+1)} = {}^5C_3(2x)^{5-3}\left(-\frac{x^3}{2}\right)^3$

$= 10\cdot 2^2 x^2 x^9 2^{-3}(-1)^3 = -5x^{11}$

Q25. Solution:

The bag contains 5 black and 6 red balls. Two black balls are selected from 5 black balls in ${}^5C_2$ ways.

Three red balls are selected from 6 red balls in ${}^6C_3$ ways.

Three black balls are selected from 4 blacks balls in ${}^5C_3$ ways.

Two red balls are selected from 6 red balls in ${}^6C_2$ ways.

$\therefore$ Total number of ways in balls are selected is

$= {}^5C_2 \times {}^6C_3 + {}^6C_2 \times {}^5C_3 = 10\times 20 + 15\times 10 = 350$

OR

Q25. Solution:

After seating 3 on one side and 4 on the other side, we have to select out of 3, 2 on one side and 1 on the other side.

Thus, the total number of arrangements are

$= {}^3C_2 \times 5! + {}^1C_1 \times 5! = (3+1)\times 5! = 480$

Q26. Solution:

The nth term of the series $1\times 1 + 2^2\times 3 + 3^2 \times 5 + \ldots$ is $a_n = n^2(2n-1) = 2n^3 - n^2$

Therefore, $S_n = \sum_{k=1}^{n} a_k$

$= \sum_{k=1}^{n}(k^3 - k^2) = 2\sum_{k=1}^{n} k^3 - \sum_{k=1}^{n} k^2$

$= 2\left[\frac{n(n+1)}{2}\right]^2 - \left[\frac{n(n+1)(2n+1)}{6}\right]$

$= \frac{n(n+1)}{2}\left[n(n+1) - \left(\frac{2n+1}{3}\right)\right]$

$= \frac{n(n+1)}{2}\left[\frac{3n^2 + 3n - 2n - 1}{3}\right]$

$= \frac{n(n+1)}{2}\left[\frac{3n^2 + n - 1}{3}\right]$

$= \frac{n(n+1)(3n^2+n-1)}{6}$

OR

Q26. Solution:

Let $S_n = 0.5 + 0.55 + 0.555 + \ldots\ldots$ to n terms

$= 5[0.1 + 0.11 + 0.111 + \ldots \text{to } n \text{ terms}]$

$= \frac{5}{9}[0.9 + 0.99 + 0.999 + \ldots \text{to } n \text{ terms}]$

$= \frac{5}{9}\left[\left(1-\frac{1}{10}\right) + \left(1 - \frac{1}{10^2}\right) + \left(1 - \frac{1}{10^3}\right) + \ldots \text{to } n \text{ terms}\right]$

$= \frac{5}{9}\left[(1+1+\ldots n \text{ terms}) - \frac{1}{10}\left(1 + \frac{1}{10} + \frac{1}{10^2} + \ldots n \text{ terms}\right)\right]$

$= \frac{5}{9}\left[n - \frac{1}{10}\left(\frac{1-\left(\frac{1}{10}\right)^n}{1-\frac{1}{10}}\right)\right] = \frac{5}{9}n - \frac{5}{81}(1-10^{-n})$

SECTION C

(Q27–Q32) Solve the following questions.

(4 marks each)

Q27. Solution:

Let $z = x + iy$. Then $|(z+1)| + z - i = 0$

$\Rightarrow |x + iy + 1| + x + iy - i = 0$

$\Rightarrow \sqrt{(x+1)^2 + y^2} + x + i(y-1) = 0$

$\Rightarrow \sqrt{(x+1)^2 + y^2} + x + i(y-1) = 0 + i0$

Equating the real and imaginary parts, we get

$\sqrt{(x+1)^2 + y^2} + x = 0$ and $y - 1 = 0$

Thus, $y - 1 = 0 \Rightarrow y = 1$

Also, $\sqrt{(x+1)^2 + y^2} + x = 0$

$\Rightarrow \sqrt{(x+1)^2 + (1)^2} + x = 0 \quad (\because y = 1)$

$\Rightarrow \sqrt{(x+1)^2 + 1} = -x$

$\Rightarrow x^2 + 2x + 2 = x^2 \Rightarrow x = -1$

$\therefore z = x + iy = -1 + i$

Q28. Solution:

Given that $\left(\frac{1+i}{1-i}\right)^2 + \left(\frac{1-i}{1+i}\right)^2 = x + iy \qquad \text{(i)}$

$\left(\frac{1+i}{1-i}\right)^2 + \left(\frac{1-i}{1+i}\right)^2 = \left(\frac{1+i}{1-i}\times\frac{1+i}{1+i}\right)^2 + \left(\frac{1-i}{1+i}\times\frac{1-i}{1-i}\right)^2$

$= \left(\frac{1+i^2+2i}{1-i^2}\right)^2 + \left(\frac{1+i^2-2i}{1-i^2}\right)^2$

$= \left(\frac{1+(-1)+2i}{1-(-1)}\right)^2 + \left(\frac{1+(-1)-2i}{1-(-1)}\right)^2 \quad (\because i^2 = -1)$

$= \left(\frac{2i}{2}\right)^2 + \left(\frac{-2i}{2}\right)^2$

$= i^2 + (-i)^2 = i^2 + i^2 = -2 \quad (\because i^2 = -1)$

$= -2 + 0i = x + iy \Rightarrow x = -2,\ y = 0$

Thus, $(x, y) = (0,0)$

Q29. Solution:

Inequality $\frac{3}{x-2} > 1 \Rightarrow \frac{3}{x-2} - 1 > 0$

$\Rightarrow \frac{3-1(x-2)}{x-2} > 0 \Rightarrow \frac{5-x}{x-2} > 0$

$\Rightarrow \frac{x-5}{x-2} < 0 \Rightarrow 2 < x < 5 \Rightarrow x \in (2,5)$

Inequality $\frac{2x-1}{x+1} \le -1 \Rightarrow \frac{2x-1}{x+1} + 1 \le 0$

$\Rightarrow \frac{2x-1+1(x+1)}{x+1} \le 0$

$\Rightarrow \frac{3x}{x+1} \le 0 \Rightarrow -1 < x \le 0 \Rightarrow x \in (-1,0]$

The intersection of the solutions, $x \in (2,5)$ and $x \in (-1,0]$ is the null set. Therefore, the given system of equations has no solution.

Q30. Solution:

Numbers less than 5000 will be of 1 digit, 2 digits, 3 digits and 4 digits, with the digits 3, 4, 5, 6, 7 because repetition is not allowed.

All the 1-digit, 2-digit and 3-digit numbers are less than 5000; therefore number of ways of forming

1-digit numbers is 5

2-digit numbers is $5 \times 4 = 20$

3-digits numbers is $5 \times 4 \times 3 = 60$

Now, four-digit numbers less than 5000 can be formed when the thousands place is filled in 2 ways (3, 4), and the hundreds place in 4 ways, the tens place in 3 ways, the units place in 2 ways.

Thus, the total number of 4-digit numbers less than 5000 is $= 2 \times 4 \times 3 \times 2 = 48$

Thus, the total number of integers less than $5000 = 5 + 20 + 60 + 48 = 133$

Q31. Solution:

The given expression $\left(2 + x - \frac{2}{x}\right)^3$ can be written as $\left[2 + \left(x - \frac{2}{x}\right)\right]^3$; therefore using the binomial theorem, $\left[2 + \left(x - \frac{2}{x}\right)\right]^3$

$$= {}^3C_0 2^3 + {}^3C_1 2^2\left(x - \frac{2}{x}\right) + {}^3C_2 2\left(x - \frac{2}{x}\right)^2 + {}^3C_3\left(x - \frac{2}{x}\right)^3$$

$$= 8 + 12\left(x - \frac{2}{x}\right) + 6\left(x - \frac{2}{x}\right)^2 + \left(x - \frac{2}{x}\right)^3$$

Again using the binomial theorem, $\left(x - \frac{2}{x}\right)^3$

$$= {}^3C_0 x^3 - {}^3C_1 x^2\left(\frac{2}{x}\right) + {}^3C_2 x\left(\frac{2}{x}\right)^2 - {}^3C_3\left(\frac{2}{x}\right)^3$$

$$= x^3 - 6x + \frac{12}{x} - \frac{8}{x^3}$$

Thus, $\left(2 + x - \frac{2}{x}\right)^3$

$$= 8 + 12x - \frac{24}{x} + 6x^2 + \frac{4}{x^2} - 24 + x^3 - 6x + \frac{12}{x} - \frac{8}{x^3}$$

$$= x^3 + 6x^2 + 6x - 16 - \frac{12}{x} + \frac{4}{x^2} - \frac{8}{x^3}$$

OR

Q31. Solution:

The given binomial expansion is $\left(\sqrt[3]{2} + \frac{1}{\sqrt[4]{3}}\right)^n$; therefore the 7th term from the beginning is

$$T_{7,B} = T_{6+1} = {}^nC_6(\sqrt[3]{2})^{n-6}\left(\frac{1}{\sqrt[4]{3}}\right)^6$$

Similarly, the 7th term from the end is

$$T_{7,E} = {}^nC_{n-6}(\sqrt[3]{2})^6\left(\frac{1}{\sqrt[4]{3}}\right)^{n-6}$$

Thus, $\frac{{}^nC_6(\sqrt[3]{2})^{n-6}\left(\frac{1}{\sqrt[4]{3}}\right)^6}{{}^nC_{n-6}(\sqrt[3]{2})^6\left(\frac{1}{\sqrt[4]{3}}\right)^{n-6}} = 1$

$$\Rightarrow \frac{(\sqrt[3]{2})^{n-6}\left(\frac{1}{\sqrt[4]{3}}\right)^6}{(\sqrt[3]{2})^6\left(\frac{1}{\sqrt[4]{3}}\right)^{n-6}} = 1 \quad (\because {}^nC_6 = {}^nC_{n-6})$$

$$\Rightarrow \left(2^{\frac{n-6}{3} - \frac{6}{3}}\right)\left(3^{\frac{-6}{4} + \frac{(n-6)}{4}}\right) = 1$$

$$\Rightarrow \left(2^{\frac{n-12}{3}}\right)\cdot\left(3^{\frac{n-12}{4}}\right) = 1 = 2^0 3^0 \Rightarrow n = 12$$

Q32. Solution:

Let a_1, a_2, d_1, d_2 be the first terms and the common difference of the first and second arithmetic progression, respectively.

Then according to the given condition,

$$\frac{\text{Sum of } n \text{ terms of first AP}}{\text{Sum of } n \text{ terms of second AP}} = \frac{3n-1}{2n+1}$$

$$\Rightarrow \frac{\frac{n}{2}[2a_1 + (n-1)d_1]}{\frac{n}{2}[2a_2 + (n-1)d_2]} = \frac{3n-1}{2n+1}$$

$$\Rightarrow \frac{2a_1 + (n-1)d_1}{2a_2 + (n-1)d_2} = \frac{3n-1}{2n+1} \quad \text{(i)}$$

Substituting $n = 15$ in equation (i), we get

$$\frac{2a_1 + 14d_1}{2a_2 + 14d_2} = \frac{3(15) - 1}{2(15) + 1}$$

$$\Rightarrow \frac{a_1 + 7d_1}{a_2 + 7d_2} = \frac{44}{31} \quad \text{(ii)}$$

Now, the ratio of 8th terms of both the APs is

$$\frac{8^{\text{th}} \text{ term of first AP}}{8^{\text{th}} \text{ term of second AP}}$$

$$= \frac{a_1 + 7d_1}{a_2 + 7d_2} = \frac{44}{31} \text{ from equation (ii)}$$

OR

Q32. Solution:

Let a be the first term and r be the common ratio of the GP; then $S_n = \frac{a(1 - r^n)}{1 - r}$.

The 5th term of the GP is 9 times the 3rd term,

$\therefore a_5 = 9 \times a_3 \Rightarrow ar^4 = 9ar^2 \Rightarrow r^2 = 9 \Rightarrow r = \pm 3$

The sum of the first two terms of the GP is – 8;

therefore $S_2 = -8 = \frac{a(1 - r^2)}{1 - r}$

For, $r = 3$

$$\Rightarrow -8 = \frac{a[1 - (3)^2]}{1 - 3} \Rightarrow -8 = \frac{a(1 - 9)}{-2} \Rightarrow a = -2$$

For, $r = -3$

$$-8 = \frac{a[1 - (-3)^2]}{1 - (-3)} \Rightarrow -8 = \frac{a(1 - 9)}{1 + 3} \Rightarrow a = 4$$

Thus, the required GP is

$-1, -2, -4, -8, -16...$ or $3, -6, 12, -24, 48, ...$

SECTION D

(Q33–Q36) Solve the following questions.

(6 marks each)

Q33. Solution:

The given statement is $P(n)$:

$$\frac{1}{1 \times 3} + \frac{1}{3 \times 5} + \frac{1}{5 \times 7} + ... \frac{1}{(2n - 1)(2n + 1)} = \frac{n}{2n + 1}$$

For $n = 1$, $P(1)$: LHS $= \frac{1}{1 \times 3} = \frac{1}{3}$

RHS $= \frac{1}{2 \cdot 1 + 1} = \frac{1}{3}$

Thus, $P(1)$ is true.

Let $P(k)$ be true for some positive integer k. Then

$$\frac{1}{1 \times 3} + \frac{1}{3 \times 5} + \frac{1}{5 \times 7} + ... \frac{1}{(2k - 1)(2k + 1)} = \frac{k}{2k + 1}$$

Now we shall prove that $P(k + 1)$ is true,

$$P(k + 1): \frac{1}{1 \times 3} + \frac{1}{3 \times 5} + ... \frac{1}{(2k - 1)(2k + 1)} + \frac{1}{(2k + 1)(2k + 3)}$$

$$= \frac{k + 1}{2(k + 1) + 1} = \frac{k + 1}{2k + 3}$$

LHS $= P(k + 1)$

$$= \frac{1}{1 \times 3} + \frac{1}{3 \times 5} + ... \frac{1}{(2k - 1)(2k + 1)} + \frac{1}{(2k + 1)(2k + 3)}$$

$$= \frac{k}{2k + 1} + \frac{1}{(2k + 1)(2k + 3)} \quad (\because P(k) \text{ is true})$$

$$= \frac{k(2k + 3) + 1}{(2k + 1)(2k + 3)} = \frac{2k^2 + 3k + 1}{(2k + 1)(2k + 3)}$$

$$= \frac{(2k + 1)(k + 1)}{(2k + 1)(2k + 3)} = \frac{k + 1}{2k + 3} = \text{RHS}$$

$P(k + 1)$ is true whenever $P(k)$ is true.

Therefore, by the principle of mathematical induction, statement $P(n)$ is true for all $n \in N$.

Q34. Solution:

Let $z = x + iy$

$$\text{Im}\left(\frac{z + 1}{\bar{z} - 1}\right) = 1$$

$$\Rightarrow \text{Im}\left(\frac{x - iy + 1}{x + iy - 1}\right) = 1$$

$$\Rightarrow \text{Im}\left(\frac{[(x + 1) - iy][(x - 1) + iy]}{[(x - 1) - iy][(x - 1) + iy]}\right) = 1$$

$$\Rightarrow \text{Im}\left(\frac{(x + 1)(x - 1) + y^2 + i[y(x + 1) - y(x - 1)]}{(x - 1)^2 + y^2}\right) = 1$$

$$\Rightarrow \text{Im}\left(\frac{x^2 - 1 + y^2 + i(2y)}{(x - 1)^2 + y^2}\right) = 1$$

$$\Rightarrow \frac{2y}{(x - 1)^2 + y^2} = 1$$

$$\Rightarrow 2y = x^2 + 1 - 2x + y^2$$

$$\Rightarrow x^2 + y^2 - 2x - 2y + 1 = 0$$

$\Rightarrow x^2 + y^2 - 2x - 2y + 1 = 0$ represents the equation of a circle.

Q35. Solution:

The graphs of the lines, $3x + 2y = 6$, $2x + 3y = 6$ and $2x - 3y = 6$ are shown in the figure given below.

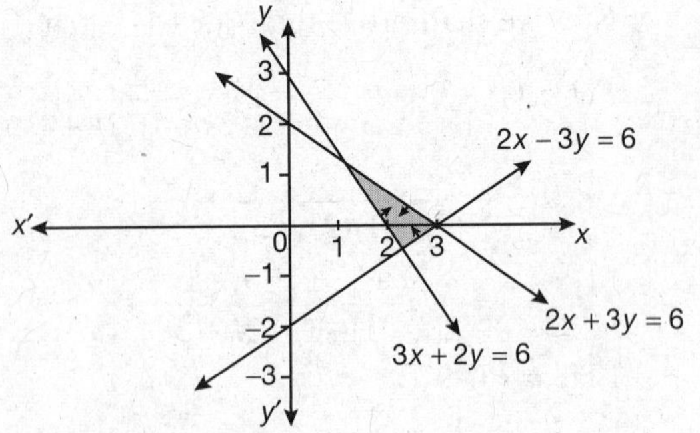

Inequality $3x+2\geq 6$ represents the shaded region above the line $3x+2y=6$ (including all the points on the line).

Inequality $2x+3y\leq 6$ represents the shaded region below the line $2x+3y=6$ (including all the points on the line).

Inequality $2x-3y\leq 6$ represents the shaded region above the line $2x-3y=6$ (including all the points on the line).

Thus, the solution of the given system of linear inequalities is represented by the common shaded region including the points on the respective lines.

OR

Q35. Solution:

In a pack of 52 playing cards:

There are 4 suits (clubs, diamonds, hearts and spades) which contain 13 cards each.

There are 2 colours, black (clubs and spades) and red (diamonds and hearts) which contain 26 cards each.

There are 4 cards of each—king, queen and jack.

(i) Number of ways of choosing 4 cards of the same suit is $= {}^4C_1\times{}^{13}C_4$ (${}^4C_1$ is for choosing 1 suit out of four suits and ${}^{13}C_4$ is for choosing 4 cards from the chosen suit).

(ii) Number of ways of choosing 4 cards of the same colour is $= {}^2C_1\times{}^{26}C_4$ (${}^2C_1$ is for choosing 1 colour out of two colours and ${}^{26}C_4$ is for choosing 4 cards from the chosen colour).

(iii) Number of ways of choosing 4 cards from four different suits is $= {}^{13}C_1\times{}^{13}C_1\times{}^{13}C_1\times{}^{13}C_1$ (${}^{13}C_1$ is for choosing 1 card from a suit and it is four times because of choosing 1 card from each of the 4 suits).

(iv) Number of ways of choosing 2 red cards and 2 black cards is $= {}^{26}C_2\times{}^{26}C_2$ (${}^{26}C_2$ is for choosing 2 cards from a colour and it is two times because of choosing 2 cards from each of the 2 colours).

(v) Number of ways of choosing 2 diamond cards and 2 clubs cards is $= {}^{13}C_2\times{}^{13}C_2$ (${}^{13}C_2$ is for choosing 2 cards from diamonds and clubs each).

(vi) Number of ways of choosing 4 cards with at least 2 kings is $= {}^4C_2\times{}^{48}C_2+{}^4C_3\times{}^{48}C_1+{}^4C_4\times{}^{48}C_0$ (2 kings and 2 others or 3 kings and 1 other or 4 kings).

Q36. Solution:

Using the binomial theorem, expansion of $(1+2x)^6$ is

$$(1+2x)^6 = {}^6C_0+{}^6C_1(2x)+{}^6C_2(2x)^2+{}^6C_3(2x)^3+{}^6C_4(2x)^4+\ldots$$

$$=1+12x+60x^2+160x^3+240x^4+\ldots$$

Using the binomial theorem, the expansion of $(1-x)^4$ is

$$(1-x)^4 = {}^4C_0-{}^4C_1(x)+{}^4C_2(x)^2-{}^4C_3(x)^3+{}^4C_4(x)^4$$

$$=1-4x+6x^2-4x^3+x^4$$

$$\therefore (1-x)^4(1+2x)^6=(1-4x+6x^2-4x^3+x^4)\times(1+12x+60x^2+160x^3+240x^4)$$

We need the terms which involve x^4; therefore complete multiplication of the two brackets is not required.

Thus, the terms containing x^4 are

$$=(1-4x+6x^2-4x^3+x^4)\times(1+12x+60x^2+160x^3+240x^4)$$

$$=(1)(240x^4)+(-4x)(160x^3)+(6x^2)(60x^2)+(-4x^3)(12x)+(x^4)(1)$$

$$=(240-640+360-48+1)=-87x^4$$

Thus, the coefficient of x^4 in the given product is -87.

OR

Q36. Solution:

First term, $a=5$

Let d be the common difference of the AP.

Therefore, the AP is $5, 5+d, 5+2d, 5+3d\ldots$

Sum of the first five terms is

$$S_5=\frac{5}{2}[2\cdot 5+(5-1)d]=25+10d$$

Sum of next five terms is

$$S_{10-5}=\frac{10}{2}[2\cdot 5+(10-1)d]-[25+10d]$$

$$[50+45d]-[25+10d]=25+35d$$

The sum of the first five terms is one-third of the next five terms; therefore $S_5=\frac{1}{3}S_{10-5}$

$$\Rightarrow 25+10d=\frac{1}{3}(25+35d)$$

$$\Rightarrow 75+30d=25+35d\Rightarrow 5d=50\Rightarrow d=10$$

Thus, the sum of the first 10 terms of the AP is

$$S_{10}=\frac{10}{2}[2\times 5+(10-1)\times 10]=5\times 100=500$$

CHAPTER

10 Basics of Two-Dimensional Coordinate Geometry

Day 131

Cartesian System

The Cartesian system, also known as the coordinate system, specifies each point uniquely by numerical coordinates.

The two-dimensional coordinate system, also known as the rectangular coordinate system, represents a point in a plane while the three-dimensional coordinates represent a point in a space.

In this chapter, we focus only on the rectangular (or two-dimensional) coordinate system. We will cover the three-dimensional coordinate system in later chapters.

Rectangular Coordinate System

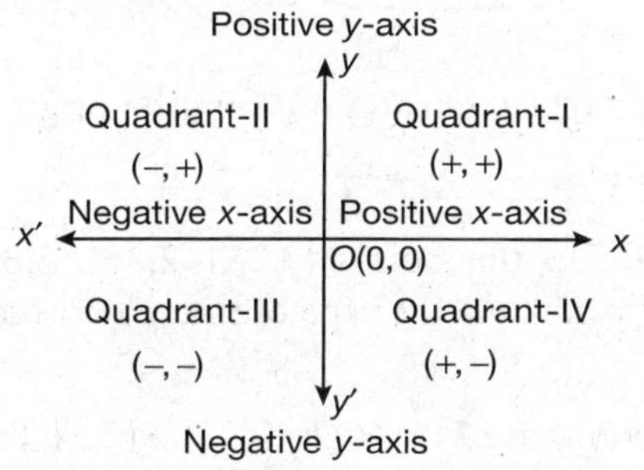

Fig. 10.1 Coordinate system

The rectangular coordinate system (Fig. 10.1) is shown by two perpendicular lines, known as coordinate axes. One line is horizontal and is represented as the x-axis and the other is vertical and is represented as the y-axis. The origin is the intersection of the x- and y-axes, given by $O(0,0)$.

In stating the coordinates of a point in the coordinate plane, the x-coordinate comes first, and then the y-coordinate, inside a small bracket (x, y).

The position of a point in a rectangular coordinate system is given by pair of numbers written as (x, y), with a specific sign, gives the distance of point from the x and y-axes, respectively. The values of the x and y-coordinates of a point are also known as the abscissa and ordinate, respectively.

The horizontal x-axis is represented by $X'X$, where OX' gives the negative direction of the x-axis and OX gives the positive direction of the x-axis.

The vertical y-axis is represented by $Y'Y$, where OY' gives the negative direction of the y-axis and OY gives the positive direction of the y-axis.

The x-coordinate of a point is its perpendicular distance from the y-axis measured along the x-axis (positive along the positive direction of the x-axis and negative along the negative direction of the x-axis).

The y-coordinate of a point is its perpendicular distance from the x-axis measured along the y-axis (positive along the positive direction of the y-axis and negative along the negative direction of the y-axis).

These two coordinate axes divide the plane into four parts. These four parts are called quadrants (one fourth part), numbered I, II, III and IV anticlockwise from OX.

Quadrant Plane-I: Quadrant plane-I is the quadrant plane XOY, where both the coordinate axes are positive, and thus for a point (x, y) in this plane, the sign of both coordinates x and y are positive, that is $(+,+)$.

Quadrant Plane-II: Quadrant plane-II is the quadrant plane YOX', where the x-axis is negative, while the y-axis is positive. Thus, for a point (x, y) in this plane, the sign of the x coordinate is negative and that of y is positive, that is $(-,+)$.

Quadrant Plane-III: Quadrant plane-III is the quadrant plane $X'OY'$, where both coordinate axes are negative. Thus, for a point (x, y) in this plane, the signs of both coordinates x and y are negative, that is $(-,-)$.

Quadrant Plane-IV: Quadrant plane-IV is the quadrant plane $Y'OX$, where the x-axis is positive, while the y-axis is negative. Thus, for a point (x, y) in this plane, the sign of the x coordinate is positive and that of y is negative, that is $(+,-)$.

Every point on the x-axis has no distance (zero distance) from the x-axis; therefore, the y-coordinate of every point lying on the x-axis is always zero. Thus, the coordinates of any point on the x-axis are of the form $(x, 0)$, where x is the distance of the point from the y-axis.

Every point on the y-axis has no distance (zero distance) from the y-axis; therefore, the x-coordinate of every point lying on the y-axis is always zero. Thus, the coordinates of any point on the y-axis are of the form (0, y), where y is the distance of the point from the x-axis.

Coordinates of the origin are (0, 0), that is, it has zero distance from both the axes, so its abscissa and ordinate are both zero.

Note: If $x \neq y$ then $(x, y) \neq (y, x)$, and $(x, y) = (y, x)$ if and only if $x = y$.

Example 1:

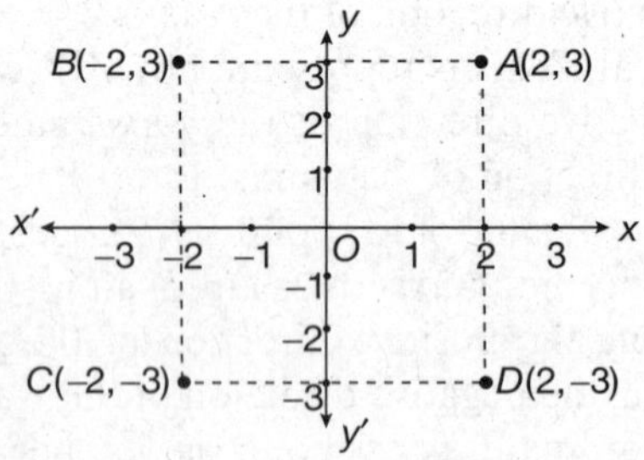

(i) Consider a point A, whose perpendicular distance from the y-axis measured along the positive direction of the x-axis is 2 units, and the perpendicular distance from the x-axis measured along the positive direction of the y-axis is 3 units.
This point is represented by $A(2,3)$ in 1st the quadrant.

(ii) Consider a point B, whose perpendicular distance from the y-axis measured along the negative direction of the x-axis is 2 units, and the perpendicular distance from the x-axis measured along the positive direction of the y-axis is 3 units.
This point is represented by $B(-2,3)$ in the 2nd quadrant.

(iii) Consider a point C, whose abscissa and ordinate are −2 and −3, respectively.
This point is represented by $C(-2,-3)$ in the 3rd quadrant.

(iv) Consider a point D, whose x-coordinate and y-coordinate are 2 and −3, respectively.
This point is represented by $D(2,-3)$ in the 4th quadrant.

Example 2:

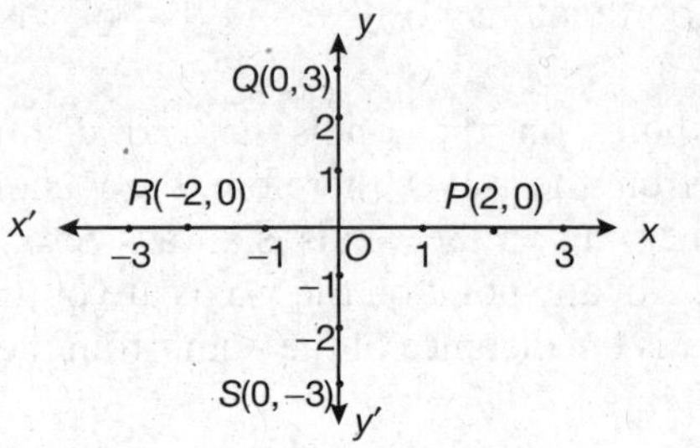

(v) Consider a point P, whose perpendicular distance from the y-axis measured along the positive direction of the x-axis is 2 units and is at a distance zero from the x-axis.
This point is represented by $P(2,0)$ at a distance of 2 units from the origin on the positive x-axis.

(vi) Consider a point Q, whose perpendicular distance from the x-axis measured along the positive direction of the y-axis is 3 units and is at a distance zero from the y-axis.
This point is represented by $Q(0,3)$ at a distance of 3 units from the origin on the positive y-axis.

(vii) Consider a point R, whose abscissa and ordinate are −2 and 0, respectively.
This point is represented by $R(-2,0)$ at a distance of 2 units from the origin on the negative x-axis.

(viii) Consider a point S, whose x-coordinate and y-coordinate are 0 and −3, respectively.
This point is represented by $S(0,-3)$ at a distance of 3 units from the origin on the negative y-axis.

Distance Formula

The distance between any two points $P(x_1, y_1)$ and $Q(x_2, y_2)$ in a plane is given by

$PQ = \sqrt{(x_2 - x_1)^2 + (y_2 - y_1)^2}$

The distance of a point $P(x, y)$ from the origin $O(0,0)$ is

$PO = \sqrt{x^2 + y^2}$.

Example 3: Do the points $(3,2), (-2,-3), (2,3)$ form a triangle? If yes, name the type of triangle formed.

Solution:

The given points are $P(3,2), Q(-2,-3), R(2,3)$; then by the distance formula,

$PQ = \sqrt{(3+2)^2 + (2+3)^2} = \sqrt{5^2 + 5^2} = \sqrt{50} = 7.07$

$QR = \sqrt{(-2-3)^2 + (-3-3)^2}$

$= \sqrt{(-4)^2 + (-6)^2} = \sqrt{52} = 7.21$

$PR = \sqrt{(3-2)^2 + (2-3)^2} = \sqrt{1^2 + (-1)^2} = \sqrt{2} = 1.41$

Since the sum of any two of these distances is greater than the third distance, the points P, Q and R form a triangle.

Also, since $PQ^2 + PR^2 = 50 + 2 = 52 = QR^2$, by the Pythagoras theorem, PQR is a right triangle with $\angle P = 90°$.

Example 4: Show that the points $(1,7), (4,2), (-1,-1), (-4,4)$ are the vertices of a square.

Solution:

The given points are $A(1,7), B(4,2), C(-1,-1), D(-4,4)$; then by the distance formula,

$AB = \sqrt{(1-4)^2 + (7-2)^2} = \sqrt{9+25} = \sqrt{34}$

$BC = \sqrt{(4+1)^2 + (2+1)^2} = \sqrt{25+9} = \sqrt{34}$

$CD = \sqrt{(-1+4)^2 + (-1-4)^2} = \sqrt{9+25} = \sqrt{34}$

$DA = \sqrt{(1+4)^2 + (7-2)^2} = \sqrt{25+9} = \sqrt{34}$

$AC = \sqrt{(1+1)^2 + (7+1)^2} = \sqrt{4+64} = \sqrt{68}$

$BD = \sqrt{(4+4)^2 + (2-4)^2} = \sqrt{64+4} = \sqrt{68}$

Note:

Square: All sides are equal and both diagonals are also equal.

Parallelogram: All sides are equal but both diagonals are not equal.

Rectangle: Opposite sides are equal and diagonals are also equal.

Now, since $AB = BC = CD = DA, AC = BD$, all the four sides of the quadrilateral ABCD are equal and its diagonals AC and BD are also equal. Therefore, $ABCD$ is a square.

Example 5: Show that the points A (3, 1), B (6, 4) and C (8, 6) are collinear (form a straight line).

Solution:

Using the distance formula, we get,

$AB = \sqrt{(6-3)^2 + (4-1)^2} = \sqrt{9+9} = \sqrt{18} = 3\sqrt{2}$

$BC = \sqrt{(8-6)^2 + (6-4)^2} = \sqrt{4+4} = \sqrt{8} = 2\sqrt{2}$

$AC = \sqrt{(8-3)^2 + (6-1)^2} = \sqrt{25+25} = \sqrt{50} = 5\sqrt{2}$

Since $AB + BC = 3\sqrt{2} + 2\sqrt{2} = 5\sqrt{2} = AC$, we can say that the points A, B and C are on a straight line and are thus collinear.

Example 6: Find a relation between x and y such that the point (x, y) is equidistant from the points $(-1,2)$ and $(3,-5)$.

Solution:

Let $P(x, y)$ be equidistant from the points $A(-1,2)$ and $B(3,-5)$; then $AP = BP \Rightarrow AP^2 = BP^2$.

By the distance formula, we get,

$(x+1)^2 + (y-2)^2 = (x-3)^2 + (y+5)^2$

$\Rightarrow x^2 + 1 + 2x + y^2 + 4 - 4y$

$= x^2 + 9 - 6x + y^2 + 25 + 10y$

$\Rightarrow 2x - 4y + 5 = -6x + 10y + 34$

$\Rightarrow 8x - 14y - 29 = 0$ is the required relation.

Example 7: Find a point on the y-axis which is equidistant from the points A (6, 5) and B (–4, 3).

Solution:

Let a point on the y-axis be P (0, y), which is equidistant from the points A (6, 5) and B (–4, 3); then $AP = BP \Rightarrow AP^2 = BP^2$.

By the distance formula,

$(6-0)^2 + (5-y)^2 = (-4-0)^2 + (3-y)^2$

$\Rightarrow 36 + 25 + y^2 - 10y = 16 + 9 + y^2 - 6y$

$\Rightarrow 4y = 36 \Rightarrow y = 9$

Thus, the point (0, 9) is equidistant from the points A (6, 5) and B (–4, 3).

Check:

$AP = \sqrt{(6-0)^2 + (5-9)^2} = \sqrt{36+16} = \sqrt{52}$

$BP = \sqrt{(-4-0)^2 + (3-9)^2} = \sqrt{16+36} = \sqrt{52}$

Section Formula

The coordinates of the point P (x, y) which divides the line segment joining the points $A(x_1, y_1)$ and $B(x_2, y_2)$ internally, in the ratio $m:n$ are $\left(\frac{mx_2 + nx_1}{m+n}, \frac{my_2 + ny_1}{m+n}\right)$.

Note: If the ratio in which P divides AB is 1:k then the coordinates of the point P will be $\left(\frac{x_2 + kx_1}{1+k}, \frac{y_2 + ky_1}{1+k}\right)$.

Note: If P is the mid-point of the line segment AB, then P divides AB in the ratio 1:1, and thus, the coordinates of the point P will be $\left(\frac{x_2 + x_1}{1+1}, \frac{y_2 + y_1}{1+1}\right) = \left(\frac{x_2 + x_1}{2}, \frac{y_2 + y_1}{2}\right)$.

Example 8: Find the coordinates of the point which divides the line segment joining the points (4, –3), and (8, 5) in the ratio 3:1 internally.

Solution:

Let P (x, y) be the point which divides the line segment joining the points (4, –3), and (8, 5) in the ratio 3:1 internally; then by section formula, we get $x = \frac{3(8)+1(4)}{3+1} = 7, y = \frac{3(5)+1(-3)}{3+1} = 3$

Therefore, P (7, 3) is the required point.

Example 9: In what ratio does the point (–4, 6) divide the line segment joining the points $A(-6,10), B(3,-8)$?

Solution:

Let (–4, 6) divide AB internally in the ratio $m:n$; then by the section formula, we get

$(-4,6) = \left(\frac{3m-6n}{m+n}, \frac{-8m+10n}{m+n}\right)$

$\Rightarrow -4 = \frac{3m-6n}{m+n}, \quad 6 = \frac{-8m+10n}{m+n}$

$\Rightarrow -4m - 4n = 3m - 6n \Rightarrow 7m = 2n$

$\Rightarrow m/n = 2/7 \Rightarrow m:n = 2:7$

Therefore, the point (–4, 6) divides the line segment joining the points $A(-6,10), B(3,-8)$ in the ratio 2:7.

Example 10: Find the coordinates of the points of trisection (points dividing in three equal parts) of the line segment joining the points $A(2,-2)$ and $B(-7,4)$.

Solution:
Let P and Q be the points of trisection of AB, that is $AP = PQ = QB$. Therefore, P divides AB internally in the ratio 1:2. Thus, the coordinates of P, using the section formula are,

$$\left(\frac{1(-7)+2(2)}{1+2}, \frac{1(4)+2(-2)}{1+2}\right) = (-1,0)$$

Similarly, Q divides AB internally in the ratio 2:1, and thus, the coordinates of Q, using the section formula, are

$$\left(\frac{2(-7)+1(2)}{2+1}, \frac{2(4)+1(-2)}{2+1}\right) = (-4,2).$$

Therefore, the coordinates of the points of trisection of the line segment joining A and B are (–1, 0) and (–4, 2).

Example 11: Find the ratio in which the y-axis divides the line segment joining the points (6, –6), and (–1, –4). Also find the point of intersection.

Solution:
Let the ratio in which y-axis divides the line segment joining the points (6, –6), and (–1, –4) be k:1; then by the section formula, the coordinates of the point which divides AB in the ratio k:1 are $\left(\frac{-k+5}{k+1}, \frac{-4k-6}{k+1}\right)$.

Since this point lies on the y-axis, its abscissa is 0,

$$\frac{-k+5}{k+1} = 0 \Rightarrow k = 5$$

Hence, the ratio is 5:1. Putting the value of k = 5, we get the point of intersection as $\left(0, \frac{-13}{3}\right)$.

Example 12: If the points $A(6,1), B(8,2), C(9,4), D(p,3)$ are the vertices of a parallelogram, taken in order, find the value of p.

Solution:
Since the diagonals of a parallelogram bisect each other, the coordinates of the mid-point of diagonal AC = coordinates of the mid-point of diagonal BD.

$$\Rightarrow \left(\frac{6+9}{2}, \frac{1+4}{2}\right) = \left(\frac{8+p}{2}, \frac{2+3}{2}\right) \Rightarrow \left(\frac{15}{2}, \frac{5}{2}\right) = \left(\frac{8+p}{2}, \frac{5}{2}\right)$$

$$\Rightarrow \frac{15}{2} = \frac{8+p}{2} \Rightarrow p = 7$$

Area of a Triangle

The area of the triangle whose vertices are $A(x_1, y_1), B(x_2, y_2), C(x_3, y_3)$ is given by

$$= \frac{1}{2}\left|x_1(y_2 - y_3) + x_2(y_3 - y_1) + x_3(y_1 - y_2)\right|$$

Note: If we change the order in which vertices are taken, it will only alter the sign but the numerical value will remain the same. Therefore, we take the absolute numerical value since the area is a positive quantity.

Example 13: Find the area of a triangle whose vertices are $(1,-1), (-4,6), (-3,5)$.

Solution:
The area of the triangle formed by the vertices $A(1,-1), B(-4,6), C(-3,5)$ is given by

$$\frac{1}{2}[1(6+5) + (-4)(-5+1) + (-3)(-1-6)]$$

$$= \frac{1}{2}(11+16+21) = 24$$

Thus, the area of the triangle is 24 square units.

Example 14: Find the area of a triangle formed by the points $A(5,2), B(4,7), C(7,-4)$.

Solution:
The area of the triangle formed by the vertices $A(5,2), B(4,7), C(7,-4)$ is given by

$$\frac{1}{2}[5(7+4) + 4(-4-2) + 7(2-7)]$$

$$= \frac{1}{2}(55 - 24 - 35) = \frac{-4}{2} = -2$$

Since the area cannot be negative, we take absolute numerical value. Thus, the area of the triangle is 2 square units.

Example 15: Find the area of the triangle formed by the points $P(-1.5,3), Q(6,-2), R(-3,4)$.

Solution:
The area of the triangle formed by the given points is equal to $\frac{1}{2}[-1.5(-2-4) + 6(4-3) + (-3)(3+2)]$.

$$= \frac{1}{2}(9+6-15) = 0$$

Since the area of the triangle is 0, the vertices are collinear, that is, they lie on a line.

Example 16: Find the value of k if the points $A(2,3), B(4,k), C(6,-3)$ are collinear.

Solution: Since the given points are collinear, the area of the triangle formed by them must be 0.

$\frac{1}{2}[2(k+3)+4(-3-3)+6(3-k)]=0$

$\Rightarrow \frac{1}{2}(-4k)=0 \Rightarrow k=0$

Thus, for $k = 0$, the points $A(2,3), B(4,0), C(6,-3)$ are collinear.

Check: Area of ΔABC

$=\frac{1}{2}[2(0+3)+4(-3-3)+6(3-0)]=0$

Example 17: If $A(-5,7), B(-4,-5), C(-1,-6), D(4,5)$ are the vertices of a quadrilateral, find the area of the quadrilateral $ABCD$.

Solution: By joining B and D, we get two triangles ABD and BCD.

Now, area of ΔABD

$=\frac{1}{2}[-5(-5-5)+(-4)(5-7)+4(7+5)]$

$=\frac{1}{2}(50+8+48)=\frac{106}{2}=53$ square units

And area of ΔBCD

$=\frac{1}{2}[-4(-6-5)-1(5+5)+4(-5+6)]$

$=\frac{1}{2}(44-10+4)=19$ square units

Thus, the area of quadrilateral $ABCD$ = area of ΔABD + area of ΔBCD = 53 + 19 = 72 square units.

Note: To find the area of a polygon, we divide it into triangular regions, which have no common area, and add the areas of these regions.

Day 132

Slope of a Line

In two dimensions, the angle (θ) made by the line l in the positive direction of the x-axis and measured anti-clockwise is called the inclination of the line. And the value $\tan\theta$ is called the gradient of the line or the slope of the line.

The slope of a line (Fig. 10.2) is denoted by m; thus, $m=\tan\theta,\ \theta \neq 90°$, that is, the slope of a line whose inclination is 90° is not defined.

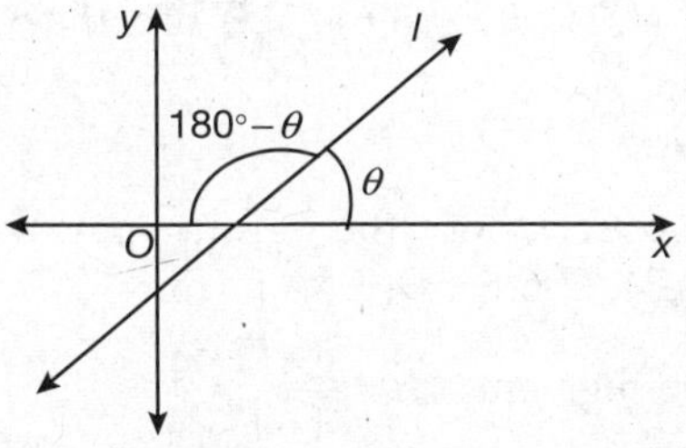

Fig. 10.2 Slope of a line

Note:

(i) $0° \le \theta \le 180°$

(ii) If the line is parallel to the x-axis or coinciding with the x-axis, then the slope of the line $\theta = 0°$.

(iii) The inclination of a vertical line (parallel to or coinciding with the y-axis) is 90°.

(iv) A line is said to be increasing if it goes up from left to right, that is, its slope is positive, $m > 0$.

(v) A line is said to be decreasing if it goes down from left to right, that is, its slope is negative, $m < 0$.

(vi) Slope of the x-axis is zero and slope of the y-axis is not defined.

Example 1: Find the slope of the line which makes an angle of 30° with the negative direction of the x-axis, measured clockwise.

Solution:

The line makes an angle of 30° with the negative direction of the x-axis measured clockwise. Therefore, the angle made by the line with the positive direction of the x-axis measured anticlockwise is $180° - 30° = 150°$. Thus, the slope of the given line is

$m=\tan 150° = \tan(180°-30°) = -\tan 30° = -1/\sqrt{3}$

Slope of a line when coordinates of any two points on the line are given

Let $P(x_1, y_1), Q(x_2, y_2)$ be the two points on a non-vertical line l whose inclination is θ (Fig. 10.3).

Note: $x_1 \neq x_2$, otherwise the line will become perpendicular to the x-axis and its slope will not be defined). The inclination of the line l may be acute or obtuse.

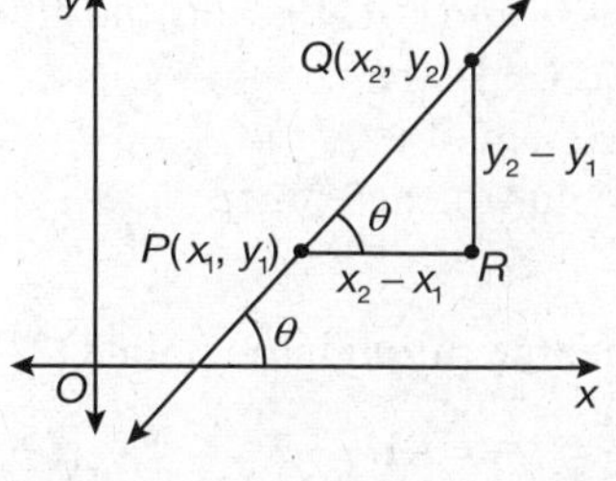

Fig. 10.3 Slope of a line passing through two points

Slope of the line l is given by $m = \tan\theta = \dfrac{QR}{PR}$

But in ΔPRQ, $\dfrac{QR}{PR} = \dfrac{y_2 - y_1}{x_2 - x_1}$

Therefore, the slope of the line l is $m = \dfrac{y_2 - y_1}{x_2 - x_1}$.

Example 2: Find the slope of a line which passes through the origin, and the mid-point of the line segment joining the points $P(0,-2)$, $B(2,0)$.

Solution:
The coordinates of the mid-point of the segment joining the points $P(0,-2)$, $B(2,0)$ are $\left(\dfrac{0+2}{2}, \dfrac{-2+0}{2}\right) = (1,-1)$.

Therefore, the slope of the line passing through the points (0, 0) and (1, –1) is $\dfrac{-1-0}{1-0} = \dfrac{-1}{1} = -1$

[since the slope of the line passing through the points $(x_1, y_1), (x_2, y_2)$ is given by $m = \dfrac{y_2 - y_1}{x_2 - x_1}, x_2 \neq x_1$].

Conditions for parallelism and perpendicularity of lines in terms of their slopes

In a coordinate plane, let us suppose that non-vertical lines l_1, l_2 have slopes m_1, m_2, respectively, and inclinations α, β, respectively.

Parallel lines: Two lines are parallel, if their inclinations are equal, and thus their slopes are equal. That is $\tan\alpha = \tan\beta \Rightarrow m_1 = m_2$.

Perpendicular lines: Two lines are perpendicular, if difference in inclination of the two lines is equal to 90°. That is, $\beta = \alpha + 90°$.

$\Rightarrow \tan\beta = \tan(\alpha + 90°) \Rightarrow \tan\beta = -\cot\alpha$

$\Rightarrow \tan\beta = -\dfrac{1}{\tan\alpha} \Rightarrow m_2 = -\dfrac{1}{m_1} \Rightarrow m_1 m_2 = -1$

Thus, two non-vertical lines are perpendicular to each other if and only if the product of their slopes is equal to –1.

Example 3: Find the value of x such that the line through the points (2, –3) and (–1, 0), and the line through the points (4, 0) and (x, 1) are
(i) parallel to each other,
(ii) perpendicular to each other.

Solution:
The slope of the line through the points (2, –3) and (–1, 0) is $m_1 = \dfrac{0-(-3)}{-1-2} = \dfrac{3}{-3} = -1$.

And, the slope of the line through the points (4, 0) and (x, 1) is $m_2 = \dfrac{1-0}{x-4} = \dfrac{1}{x-4}$.

Case (i) The two lines are parallel to each other,

$\therefore m_1 = m_2 \Rightarrow -1 = \dfrac{1}{x-4} \Rightarrow x - 4 = -1 \Rightarrow x = 3$

Case (ii) The two lines are perpendicular to each other,

$\therefore m_1 m_2 = -1 \Rightarrow (-1) \times \left(\dfrac{1}{x-4}\right) = -1 \Rightarrow x - 4 = 1 \Rightarrow x = 5$

Example 4: Without using the Pythagoras theorem, show that the points $(a,0), (0,b), (-1,1)$ are the vertices of a right-angled triangle.

Solution:
Let the vertices of the given triangle be $A(-2,0)$, $B(0,2)$, $C(1,1)$. The slope of a non-vertical line passing through the points $(x_1, y_1), (x_2, y_2)$ is given by $m = \dfrac{y_2 - y_1}{x_2 - x_1}, x_2 \neq x_1$.

$\therefore$ Slope of line segment AB $(m_1) = \dfrac{2-0}{0-(-2)} = 1$

Slope of segment BC $(m_2) = \dfrac{1-2}{1-0} = -1$

Slope of segment CA $(m_3) = \dfrac{-2-1}{0-1} = 3$

Since the product of the slopes of line segments AB and BC is $m_1 m_3 = -1$, the line segments AB and BC are perpendicular to each other, that is, the given triangle is right-angled triangle at vertex B (0, 2).

Example 5: Without using the distance formula, show that points $(1,2), (4,3), (6,6), (3,5)$ are the vertices of a parallelogram.

Solution:
Let the vertices of the given parallelogram be $A(1,2)$, $B(4,3)$, $C(6,6)$, $D(3,5)$. Then,

Slope of line segment $AB = \dfrac{3-2}{4-1} = \dfrac{1}{3}$

Slope of line segment $CD = \dfrac{5-6}{3-6} = \dfrac{1}{3}$

Slope of line segment $BC = \dfrac{6-3}{6-4} = \dfrac{3}{2}$

Slope of line segment $AD = \dfrac{5-2}{3-1} = \dfrac{3}{2}$

Now, the slope of line segment AB = slope of line segment CD; therefore the line segments AB and CD are parallel to each other.

Similarly, the slope of line segment BC = slope of line segment AD; therefore the line segments BC and AD are parallel to each other.

Since both pairs of opposite sides of quadrilateral $ABCD$ are parallel, ABCD is a parallelogram. Thus, the points $(1,2),(4,3),(6,6),(3,5)$ are the vertices of a parallelogram.

Angle between two lines

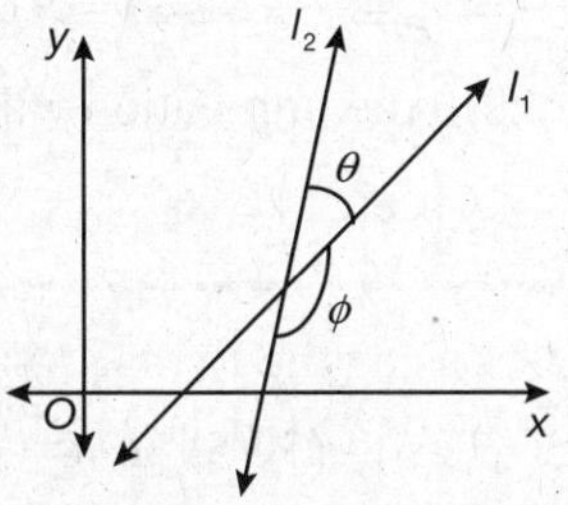

Figure 10.4 Angle between two lines

Let l_1, l_2 be two non-vertical lines with slopes m_1, m_2 respectively (Fig. 10.4); then the angle between the two lines is given by $\left|\frac{m_2 - m_1}{1+m_1m_2}\right|$, $(1+m_1m_2 \neq 0)$. Thus, the acute angle between the lines is $\tan\theta = \frac{m_2-m_1}{1+m_1m_2}$.

And, the obtuse angle between the lines is $\tan\phi = -\left(\frac{m_2-m_1}{1+m_1m_2}\right)$.

Note:$(\theta+\phi = 180° \Rightarrow \tan\phi = \tan(180° - \theta) = -\tan\theta)$

In general, if $\frac{m_2-m_1}{1+m_1m_2}$ is positive, then $\tan\theta$ will be positive and $\tan\phi$ will be negative, which means θ will be acute and ϕ will be obtuse.

And, if $\frac{m_2-m_1}{1+m_1m_2}$ is negative, then $\tan\theta$ will be negative and $\tan\phi$ will be positive, which means that θ will be obtuse and ϕ will be acute.

Example 6: Find the angle between the x-axis and the line joining the points $(2,-1),(1,-2)$.

Solution:

The slope of the line joining the points $(2,-1),(1,-2)$ is $m = \frac{-2-(-1)}{1-2} = 1$; therefore, the inclination (θ) is given by $\tan\theta = m = 1 \Rightarrow \theta = 45°$. Thus, the angle between the x-axis and the joining the points $(3,-1),(4,-2)$ is $45°$.

Example 7: Find the angle between the lines $2y+x-1=0$ and $y-3x+1=0$.

Solution:

The given lines can be written as $y = -\frac{1}{2}x+\frac{1}{2}$ and $y = 3x-1$. Therefore, the slopes of the lines are $m_1 = \frac{-1}{2}$ and $m_2 = 3$.

Let θ be the angle between the lines; then $\tan\theta = \frac{3-(-1/2)}{1+3\times(-1/2)} = \frac{7/2}{-1/2} = -7 \Rightarrow \theta = \tan^{-1}(-7)$ is the angle between the lines $2y+x-1=0$ and $y-3x+1=0$.

Example 8: The tangent of the angle between two lines is 1/2 and the slope of one line is 1/3. Find the slope of the other line.

Solution:

Let the slope of the other line be m and θ be the angle between the lines; then

$$\tan\theta = \left|\frac{m_1-m_2}{1+m_1m_2}\right| \Rightarrow \frac{1}{3} = \left|\frac{1/2-m}{1+(1/2)\cdot m}\right|$$

$$\Rightarrow \frac{1}{3} = \left|\frac{1-2m}{2+m}\right| \Rightarrow \frac{1}{3} = \pm\left(\frac{1-2m}{2+m}\right)$$

Case I:

$$\frac{1}{3} = +\left(\frac{1-2m}{2+m}\right) \Rightarrow 2+m = 3-6m \Rightarrow m = 1/7$$

Case II:

$$\frac{1}{3} = -\left(\frac{1-2m}{2+m}\right) \Rightarrow 2+m = -3+6m \Rightarrow m = 1$$

Thus, the slope of the other line can be either 1/7 or 1.

Collinearity of three points (using slope): Three points are collinear in a plane if the slope of line segment AB = slope of line segment BC (since two lines having the same slope passing through a common point are coinciding lines). Thus, the three given points A, B and C are collinear.

Example 9: Find the value of x for which the points $(x,-1),(2,1),(4,3)$ are collinear.

Solution:

If the points $A(x,-1), B(2,1), C(4,3)$ are collinear, then the slope of line segment AB = slope of line segment BC.

$$\Rightarrow \frac{1-(-1)}{2-x} = \frac{3-1}{4-2} \Rightarrow \frac{2}{2-x} = \frac{2}{2}$$

$$\Rightarrow 2 = 2-2x \Rightarrow 2x = 0 \Rightarrow x = 0$$

Example 10: Consider the given Population vs Year graph.

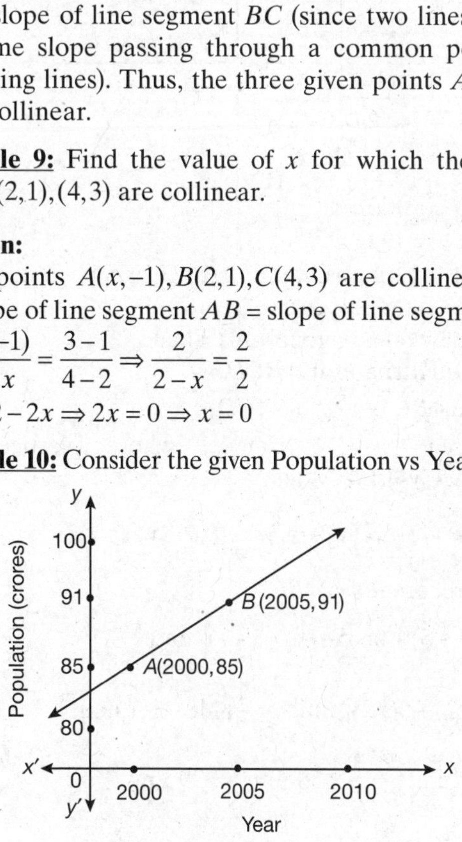

Year vs Population in crores

Find the slope of the line AB and using it, find what the population will be in the year 2010.

Solution:

The line AB passes through points A (2000, 85) and B (2005, 91); therefore, its slope is $\frac{91-85}{2005-2000}=\frac{6}{5}$.

Let y be the population in the year 2010. Then according to the given graph, line AB must pass through point C (2010, y). Thus,

Slope of line segment AB = slope of line segment BC

$\Rightarrow \frac{6}{5}=\frac{y-91}{2010-2005} \Rightarrow \frac{6}{5}=\frac{y-91}{5} \Rightarrow y=91+6=97$

Thus, in the year 2010 the population will be 97 crores.

Day 133

NCERT Exercise 10.1 (With Solutions)

Q1. Draw a quadrilateral in the Cartesian plane, whose variance are $(-4,5),(0,7),(5,-5),(-4,-2)$. Also find its area.

Solution:

The given quadrilateral with vertices $A(-4,5), B(0,7), C(5,-5), D(-4,-2)$ in the Cartesian plane is shown below.

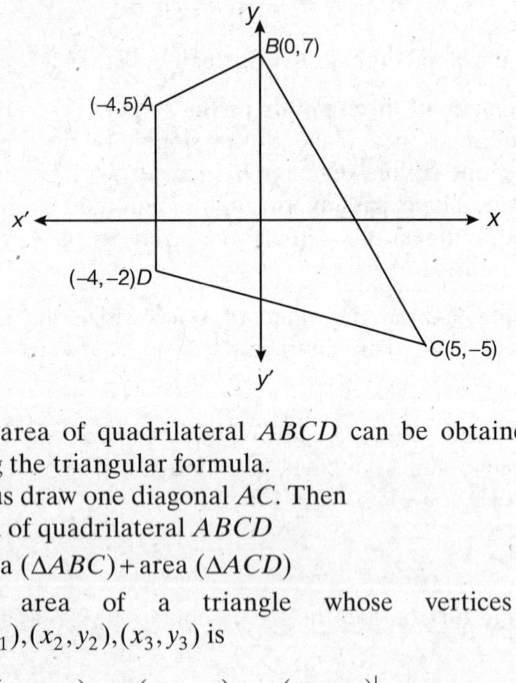

The area of quadrilateral $ABCD$ can be obtained by using the triangular formula.

Let us draw one diagonal AC. Then

Area of quadrilateral $ABCD$

= area (ΔABC) + area (ΔACD)

The area of a triangle whose vertices are $(x_1,y_1),(x_2,y_2),(x_3,y_3)$ is

$\frac{1}{2}|x_1(y_2-y_1)+x_2(y_3-y_1)+x_3(y_1-y_2)|$

Therefore, area of ΔABC

$=\frac{1}{2}|-4(7+5)+0(-5-)+5(5-7)|$ unit^2

$=\frac{1}{2}|-4(12)+5(-2)|$ unit^2 $=\frac{1}{2}|-48-10|$ unit^2

$=\frac{1}{2}|-58|$ unit^2 $=\frac{1}{2}\times 58$ unit^2 $=29$ unit^2

Area of ΔACD

$=\frac{1}{2}|-4(-5+2)+5(-2-5)+(5+5)|$ unit^2

$=\frac{1}{2}|-4(-3)+5(-7)-4(10)|$ unit^2

$=\frac{1}{2}|12-35-40|$ unit^2 $=\frac{1}{2}|-63|$ unit^2 $=\frac{63}{2}$ unit^2

Thus area $(ABCD)=\left(29+\frac{63}{2}\right)$ unit^2 $=\frac{121}{2}$ unit^2

Q2. The base of an equilateral triangle with side $2a$ lies along the y-axis such that the mid-point of the base is at the origin. Find the vertices of the triangle.

Solution:

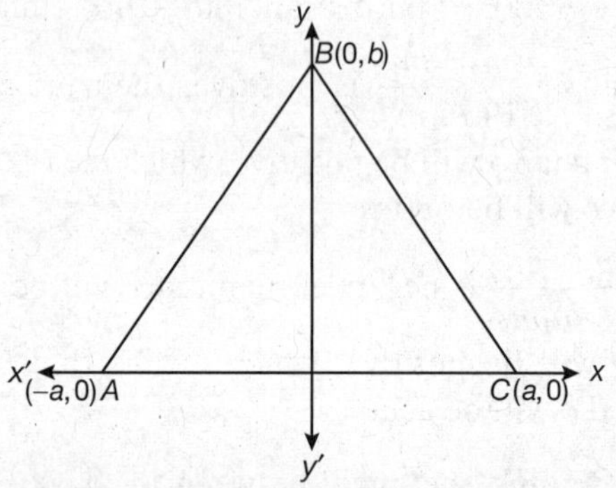

Let ABC be the given equilateral triangle with side $=2a$; then $= AB = BC = CA = 2a$.

The base AC lies along the y-axis such that the mid-point of AC is at the origin; then the coordinates of vertices A and B be $(-a,0)$ and $(a,0)$, respectively. And vertex B must lie on the x-axis; let its coordinate be $(0,b)$.

Now, since the angle BAC is 60°, $\tan 60° = \frac{b}{a} \Rightarrow b = a\sqrt{3}$.

Thus, coordinates of point $B=(0,\pm\sqrt{3}a)$

Thus, the vertices of the given equilateral triangle are $(-a,0),(a,0),(\sqrt{3}a,0)$ or $(-a,0),(a,0),(-\sqrt{3}a,0)$.

Q3. Find the distance between $P(x_1,y_1), Q(x_2,y_2)$ when:

(i) PQ is parallel to the y-axis
(ii) PQ is parallel to the x-axis

Solution:

The given points are $P(x_1, y_1), Q(x_2, y_2)$

(i) When PQ is parallel to the y-axis, then $x_1 = x_2$.
Distance between P and Q
$= \sqrt{(x_2 - x_1)^2 + (y_2 - y_1)^2} = \sqrt{(y_2 - y_1)^2} = |y_2 - y_1|$

(ii) When PQ is parallel to the x-axis, then $y_1 = y_2$.
Distance between P and Q
$= \sqrt{(x_2 - x_1)^2 + (y_2 - y_1)^2} = \sqrt{(x_2 - x_1)^2} = |x_2 - x_1|$

Q4. Find a point on the x-axis which is equidistant from the points (7, 6) and (3, 4).

Solution:

Let (a, 0) be the point on the x-axis which is equidistant from the points (7, 6) and (3, 4). Then

$\sqrt{(7-a)^2 + (6-0)^2} = \sqrt{(3-a)^2 + (4-0)^2}$

$\Rightarrow \sqrt{49 + a^2 - 14a + 36} = \sqrt{9 + a^2 - 6a + 16}$

$\Rightarrow \sqrt{a^2 - 14a + 85} = \sqrt{a^2 - 6a + 25}$

Squaring both sides, we get

$a^2 - 14a + 85 = a^2 - 6a + 25$

$\Rightarrow -8a = -60 \Rightarrow a = \dfrac{60}{8} = \dfrac{12}{2}$

Thus, the point on the x-axis which is equidistant from the points (7, 6) and (3, 4) is $\left(\dfrac{15}{2}, 0\right)$.

Q5. Find the slope of a line, which passes through the origin and the mid-point of the line segment joining the points $P(0, -4), B(8, 0)$.

Solution:

The coordinates of the mid-point of the segment joining the points $P(0, -4), B(8, 0)$ are $\left(\dfrac{0+8}{2}, \dfrac{-4+0}{2}\right) = (4, -2)$.

The slope of the line passing through the points (0, 0) and (4, –2) is $\dfrac{-2-0}{4-0} = \dfrac{-2}{4} = -\dfrac{1}{2}$ [since the slope of the line passing through the points $(x_1, y_1), (x_2, y_2)$ is given by $m = \dfrac{y_2 - y_1}{x_2 - x_1}, x_2 \neq x_1$)].

Q6. Without using the Pythagoras theorem, show that the points $(4, 4), (3, 5), (-1, -1)$ are the vertices of a right-angled triangle.

Solution:

Let the vertices of the given triangle be $A(4, 4), B(3, 5), C(-1, -1)$.

The slope of a non-vertical passing through the points $(x_1, y_1), (x_2, y_2)$ is given by $m = \dfrac{y_2 - y_1}{x_2 - x_1}, x_2 \neq x_1$.

$\therefore$ Slope of line segment $AB\ (m_1) = \dfrac{5-4}{3-4} = -1$

Slope of segment $BC\ (m_2) = \dfrac{-1-5}{-1-3} = \dfrac{-6}{-4} = \dfrac{3}{2}$

Slope of segment $CA\ (m_3) = \dfrac{4+1}{4+1} = \dfrac{5}{5} = 1$

The product of the slopes of the line segments AB and CA is $m_1 m_3 = -1$.

Therefore, the line segments AB and CA are perpendicular to each other, that is, the given triangle is a right-angled triangle at vertex A (4, 4).

Q7. Find the slope of the line, which makes an angle of 30° with the positive direction of the y-axis measured anticlockwise.

Solution:

The line makes an angle of 30° with the positive direction of the y-axis measured anticlockwise; therefore, the angle made by the line with the positive direction of the x-axis measured anticlockwise is $90° + 30° = 120°$.

Thus, the slope of the given line is

$m = \tan 120° = \tan(180° - 60°) = -\tan 60° = -\sqrt{3}$

Q8. Find the value of x for which the points $(x, -1), (2, 1), (4, 5)$ are collinear.

Solution:

If the points $A(x, -1), B(2, 1), C(4, 5)$ are collinear, then slope of line segment AB = slope of line segment BC.

$\Rightarrow \dfrac{1-(-1)}{2-x} = \dfrac{5-1}{4-2} \Rightarrow \dfrac{2}{2-x} = \dfrac{4}{2}$

$\Rightarrow 2 = 4 - 2x \Rightarrow 2x = 2 \Rightarrow x = 1$

Q9. Without using the distance formula, show that the points $(-2, -1), (4, 0), (3, 3), (-3, 2)$ are the vertices of a parallelogram.

Solution:

Let the vertices of the given parallelogram be $A(-2, -1)$, $B(4, 0)$, $C(3, 3)$, $D(-3, 2)$

Slope of line segment $AB = \dfrac{0+1}{4+2} = \dfrac{1}{6}$

Slope of line segment $CD = \dfrac{2-3}{-3-3} = \dfrac{-1}{-6} = \dfrac{1}{6}$

Slope of line segment $BC = \dfrac{3-0}{3-4} = \dfrac{3}{-1} = -3$

Slope of line segment $AD = \dfrac{2+1}{-3+2} = \dfrac{3}{-1} = -3$

Now, since the slope of line segment AB = slope of line segment CD, the line segments AB and CD are parallel to each other.

Similarly, since the slope of line segment BC = slope of line segment AD, the line segments BC and AD are parallel to each other.

Since both pairs of opposite sides of the quadrilateral $ABCD$ are parallel, $ABCD$ is a parallelogram. Thus, the points $(-2,-1),(4,0),(3,3),(-3,2)$ are the vertices of a parallelogram.

Q10. Find the angle between the x-axis and the line joining the points $(3,-1),(4,-2)$.

Solution:
Since the slope of the line joining the points $(3,-1),(4,-2)$ is $m=\dfrac{-2-(-1)}{4-3}=-2+1=-1$, the inclination (θ) is given by,

$\tan\theta = m = -1 \Rightarrow \theta = (90°+45°) = 135°$

Thus, the angle between the x-axis and the joining the points $(3,-1),(4,-2)$ is 135°.

Q11. The slope of a line is double that of another line. If the tangent of the angle between them is 1/3, find the slopes of the lines.

Solution:
Let $m, 2m$ be the slopes of the two given lines l_1, l_2, and let θ be the angle between them; then

$$\tan\theta = \left|\frac{m_1-m_2}{1+m_1m_2}\right| = \left|\frac{m-2m}{1+m\cdot 2m}\right| = \left|\frac{-m}{1+2m^2}\right|.$$

Since the tangent of the angle between the two lines is 1/3,

$$\therefore \frac{1}{3} = \left|\frac{-m}{1+2m^2}\right| \Rightarrow \frac{1}{3} = \pm\frac{m}{1+2m^2}$$

Case I:

$$\frac{1}{3} = \frac{m}{1+2m^2} \Rightarrow 2m^2-3m+1=0$$

$\Rightarrow (m-1)(2m-1)=0 \Rightarrow m=1, m=1/2$

If $m=1$, then the slopes of the lines are 1 and 2.

If $m=1/2$, then the slopes of the lines are $\frac{1}{2}, 1$.

Case II:

$$\Rightarrow \frac{1}{3} = \frac{-m}{1+2m^2} \Rightarrow 2m^2+3m+1=0$$

$\Rightarrow (m+1)(2m+1)=0 \Rightarrow m=-1, m=-1/2$

If $m=-1$, then the slopes of the lines are -1 and -2.

If $m=-1/2$, then the slopes of the lines are $-\frac{1}{2}, -1$.

Thus, the slopes of the two lines are 1 and 2 or 1/2 and 1 or -1 and -2 or $-\frac{1}{2}$ and -1.

Q12. A line passes through $(x_1, y_1), (h, k)$. If the slope of the line is m, then show that $k-y_1 = m(h-x_1)$.

Solution:
The slope of the lines passing through $(x_1,y_1),(h,k)$ is $\dfrac{k-y_1}{h-x_1}$

Since the slope of the given line is m,

$$\frac{k-y_1}{h-x_1} = m \Rightarrow k-y_1 = m(h-x_1)$$

Q13. If three points $(h,0),(a,b)(0,k)$ lie on a line, then show that $\dfrac{a}{h}+\dfrac{b}{k}=1$.

Solution:
Since the points $A(h,0), B(a,b), C(0,k)$ lie on a line, the slope of line segment AB = slope of line segment BC; therefore

$$\frac{b-0}{a-h} = \frac{k-b}{0-a} \Rightarrow \frac{b}{a-h} = \frac{k-b}{-a}$$

$\Rightarrow -ab = (k-b)(a-h)$

$\Rightarrow -ab = ka - kh - ab + bh \Rightarrow ka + bh = kh$

Dividing both sides by kh, we get

$$\frac{ka}{kh}+\frac{hh}{kh}=\frac{kh}{kh} \Rightarrow \frac{a}{h}+\frac{b}{k}=1$$

Q14. Consider the given population and year graph.

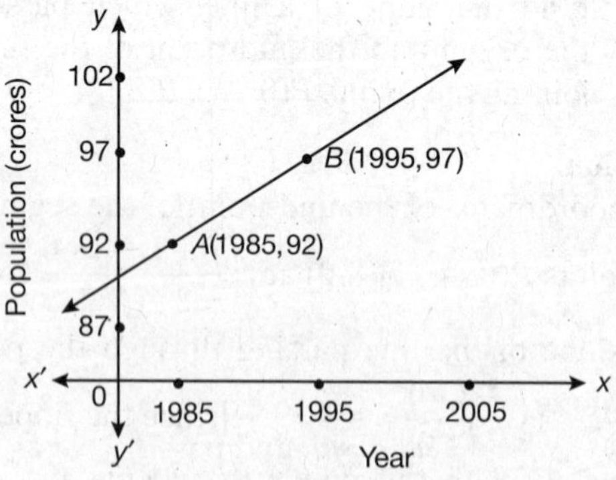

Year vs Population in crores

Find the slope of the line AB and using it, find what the populations will be in the year 2010.

Solution:
Since line AB passes through points A (1985, 92) and B (1995, 97), its slope is $\dfrac{97-92}{1995-1985}=\dfrac{5}{10}=\dfrac{1}{2}$.

Let y be the population in the year 2010. Then, according to the given graph, line AB must pass through point C (2010, y). Thus,

Slope of line segment AB = slope of line segment BC

$$\Rightarrow \frac{1}{2} = \frac{y-97}{2010-1995} \Rightarrow \frac{1}{2} = \frac{y-97}{15} \Rightarrow \frac{15}{2} = y-97$$

$$\Rightarrow y-97 = 7.5 \Rightarrow y = 7.5+97 = 104.5$$

Thus, in the year 2010, the population will be 104.5 crores.

Day 134

Various Forms of the Equation of a Line

Horizontal and Vertical Lines: If a horizontal line L is at a distance a from the x-axis (Fig. 10.5), then the ordinate of every point lying on the line is either a or $-a$. Therefore, equation of the line L is either $y = a$ or $y = -a$.

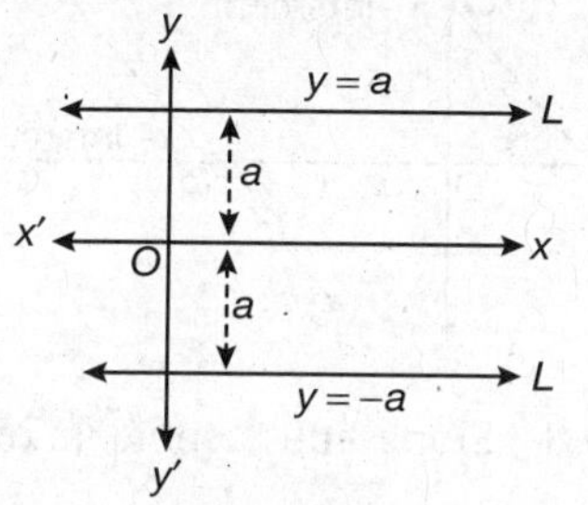

Fig. 10.5 Horizontal lines

Similarly, if a vertical line L is at a distance b from the y-axis (Fig. 10.6), then the abscissa of every point lying on the line is either b or $-b$. Therefore, the equation of the line L is either $x = b$ or $x = -b$.

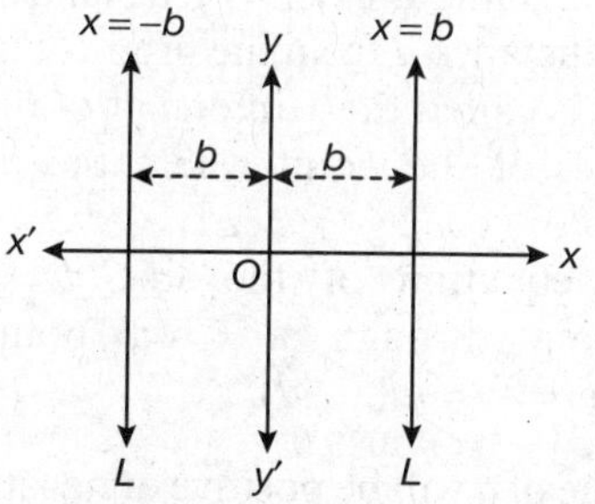

Fig. 10.6 Vertical lines

Point–Slope Form: Let there exist a non-vertical line L (Fig. 10.7) whose slope is m and passing through a fixed point $P_0(x_0, y_0)$.

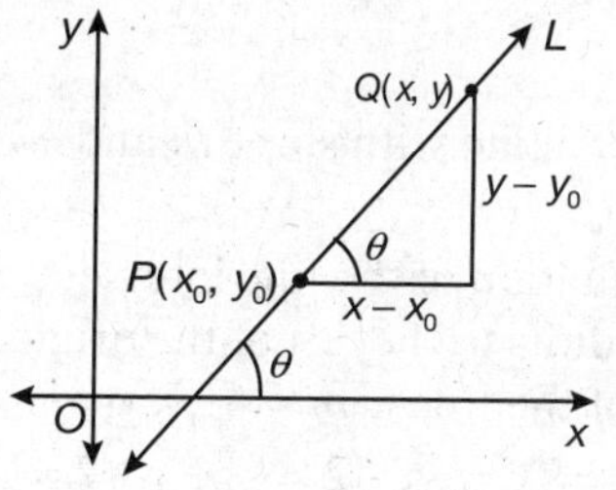

Fig. 10.7 Point–slope form of a line

Now, take an arbitrary point $Q(x, y)$ on the line. Then, by the definition of the slope of a line passing through two points, we get

$$\tan\theta = \frac{y-y_0}{x-x_0} = m \Rightarrow y - y_0 = m(x - x_0)$$

Thus, the equation of a line whose slope is m and passing through a fixed point is $y - y_0 = m(x - x_0)$.

Example 1: Find the equation of the line which passes through the point (2, –3) and

(a) slope = 1/3.

(b) inclined with the x-axis at an angle of 60°.

Solution:

The equation of the line that passes through point (x_0, y_0) whose slope is m is $(y - y_0) = m(x - x_0)$

(a) Equation of the line passing through point (2, –3) whose slope is 1/3 is $(y+3) = \frac{1}{3}(x-2)$

$$\Rightarrow 3(y+3) = x-2 \Rightarrow x - 3y - 11 = 0$$

(b) The slope of the line that inclines with the x-axis at an angle of 60° is $m = \tan 60° = \sqrt{3}$.

Therefore, the equation of the line that passes through (2, –3) and inclines with the x-axis at an angle of 60°, that is, whose slope is $\sqrt{3}$, is given by

$$(y+3) = \sqrt{3}(x-2) \Rightarrow y+3 = \sqrt{3}x - 2\sqrt{3}$$

$$\Rightarrow \sqrt{3}x - y - 3 - 2\sqrt{3} = 0$$

Example 2: Find the equation of the line passing through (–1, 1) and

(a) perpendicular to the line through the points $(2,4), (-2,6)$.

(b) parallel to the line $y = 2x - 3$.

Solution:

(a) The slope of the line joining the points $(2,4), (-2,6)$ is

$$m = \frac{6-4}{-2-2} = \frac{2}{-4} = -\frac{1}{2}.$$

And we know that two non-vertical lines are perpendicular to each other if and only if their slopes are the negative reciprocal of each other.

Therefore, the slope of the line perpendicular to the line through the points $(2,4), (-2,6)$ is,

$$= -\frac{1}{m} = -\frac{1}{(-1/2)} = 2$$

Thus, the equation of the line passing through the point (–1, 1), whose slope is 2, is $(y-1)=2(x+1)$

$\Rightarrow 2x - y + 3 = 0$

(b) The slope of the given line $y = 2x - 3$ is $m = 2$, and we know that two lines are parallel to each other if and only if their slopes are equal.

Therefore, slope of the line parallel to the given line $y = 2x - 3$ is $m = 2$.

Thus, the equation of the line passing through point (–1, 1), whose slope is 2, is $(y-1)=2(x+1)$.

$\Rightarrow 2x - y + 3 = 0$

Two-Point Form: Let the line L pass through two given points $P_1(x_1, y_1), P_2(x_2, y_2)$ and let an arbitrary point $Q(x, y)$ be on the line L.

All the three points lie on a line L. Therefore, by the definition of collinearity, the slope of line segment P_1P = slope of line segment P_1P_2.

$$\Rightarrow \frac{y-y_1}{x-x_1} = \frac{y_2-y_1}{x_2-x_1} \Rightarrow y - y_1 = \frac{y_2-y_1}{x_2-x_1}(x-x_1)$$

Thus, the equation of the line passing through the points $(x_1, y_1), (x_2, y_2)$ is $y - y_1 = \frac{y_2-y_1}{x_2-x_1}(x-x_1)$.

Example 3: Find the equation of the line which passes through the points $(-1,1),(2,-2)$.

Solution:
The equation of the line passing through points $(x_1, y_1), (x_2, y_2)$ is $y - y_1 = \frac{y_2-y_1}{x_2-x_1}(x-x_1)$.

Therefore, the equation of the line passing through the points $(-1,1),(2,-2)$ is $(y-1) = \frac{-2-1}{2+1}(x+1)$.

$$\Rightarrow (y-1) = \frac{-3}{3}(x+1) \Rightarrow y - 1 = -(x+1) \Rightarrow x + y = 0$$

Example 4: By using the concept of equation of a line, prove that the three points $(1,0),(0,-2),(2,2)$ are collinear.

Solution:
If the points $(1,0),(0,-2),(2,2)$ are collinear, then the line passing through points $(1,0),(0,-2)$ must also be passing through the point $(2,2)$.

Now, the equation of the line passing through points $(1,0),(0,-2)$ is $(y-0) = \frac{(-2-0)}{(0-1)}(x-1)$.

$y = 2(x-1) \Rightarrow 2x - y - 2 = 0$

Since the third point $x = 2, y = 2$ also satisfies the equation of line passing through points $(1,0),(0,-2)$, the three points $(1,0),(0,-2),(2,2)$ lie on a line, and hence, they are collinear.

Slope–Intercept Form

***y*-Intercept:** Let a non-vertical line L exist with slope m and cutting the y-axis at a distance c from the origin (Fig. 10.8). Then the distance c is called the y-intercept of the line L, and the coordinates of the point where the line meet the y-axis are $(0, c)$.

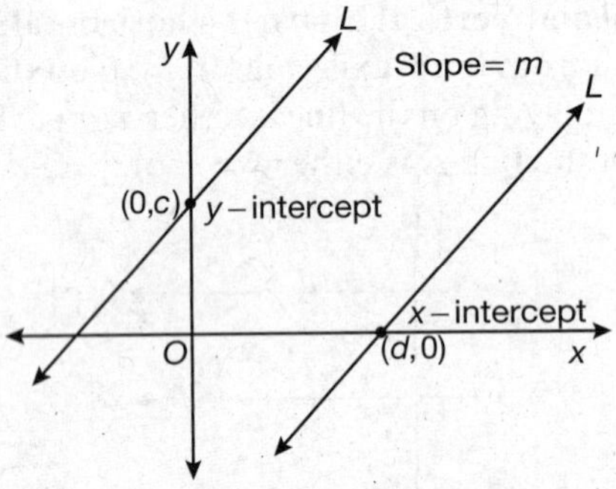

Fig. 10.8 Slope–intercept form of a line

Thus, the equation of the line L having slope m and passing through a fixed point $(0, c)$ is $y - c = m(x-0) \Rightarrow y = mx + c$.

Note: The value of c will be positive or negative according to whether the intercept is made on the positive or negative side of the y-axis, respectively.

***x*-Intercept:** Let a line L exist with a slope m and cutting the x-axis at a distance d from the origin (Fig. 10.8). Then the distance d is called the x-intercept of the line L, and the coordinates of the point where the line meet the x-axis are $(d, 0)$.

Thus, the equation of the line L having slope m and passing through a fixed point $(d, 0)$ is $y - 0 = m(x-d) \Rightarrow y = m(x-d)$.

Note: The value of d will be positive or negative according to whether the intercept is made on the positive or negative side of the x-axis, respectively.

Example 5: Find the equation of the line which intersects the x-axis at a distance of 2 units to the left of the origin with a slope of –1.

Solution:
The equation of a line with slope m and x-intercept is a is given by $y = m(x-a)$.

Thus, the equation of the line intersecting the x-axis at a distance of 2 units to the left of the origin, that is, $a = -2$ and the slope of the line is $m = -1$, is given by

$y = -1[x-(-2)] \Rightarrow y = -x - 2 \Rightarrow x + y + 2 = 0$

Example 6: Find the equation of the line which intersects the y-axis at a distance of 2 units below the origin and makes an angle of 135° with the positive direction of the x-axis.

Solution:

The equation of a line with a slope m and a y-intercept equal to b, is given $y-b=mx$.

Thus, the equation of the line intersecting the y-axis at a distance of 2 units below the origin, that is $b=-2$ and the slope of the line is $m=\tan 135°=-1$, is given by $y-(-2)=(-1)x \Rightarrow x+y+2=0$.

Intercept Form: Let a line L make x-intercept equal to a and y-intercept equal to b on the axes (Fig. 10.9). Therefore, the line L meets the x-axis at the point $(a, 0)$ and the y-axis at the point $(0, b)$.

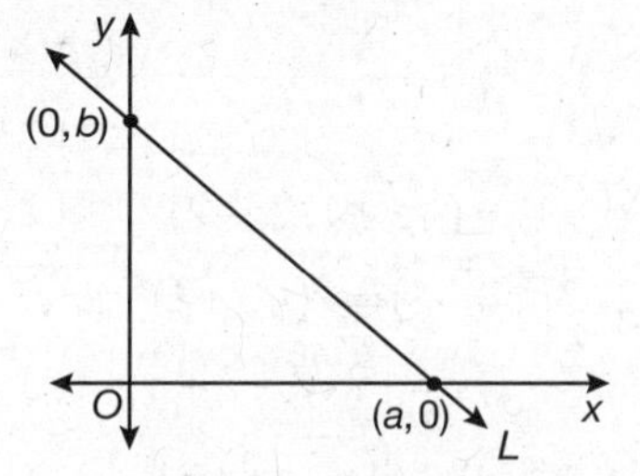

Fig. 10.9 Intercept form of a line

Thus, by the two-point form of the equation of the line, we get $y-0=\dfrac{b-0}{0-a}(x-a)$.

$ay=-bx+ab \Rightarrow bx+ay=ab \Rightarrow \dfrac{x}{a}+\dfrac{y}{b}=1$

Thus, the equation of the line making intercepts a and b on the x-and y-axes, respectively, is $\dfrac{x}{a}+\dfrac{y}{b}=1$.

Example 7: Find the equation of a line whose y-intercept is two times the x-intercept and which passes through the point (3, 4).

Solution:

The equation of a line in the intercept form is $\dfrac{x}{a}+\dfrac{y}{b}=1$, where a and b are the intercepts on the x and y axes, respectively.

The y-intercept of the line is two times the x-intercept; therefore $b=2a$.

Thus, the equation of the line in the intercept form is $\dfrac{x}{a}+\dfrac{y}{b}=1 \Rightarrow \dfrac{x}{a}+\dfrac{y}{2a}=1 \Rightarrow 2x+y=2a$.

The given line passes through point (3, 4); therefore it satisfies the equation of the line, and hence, $2\times 3+4=2a \Rightarrow a=5$.

Thus, $2x+y=10$ is the required equation of the line.

Normal Form: For a non-vertical line, the following data is given (Fig. 10.10).

(i) Length of the perpendicular (normal) from the origin to the line L is $OA=p$.

(ii) Angle which the normal OA makes with the positive direction of the x-axis is ω.

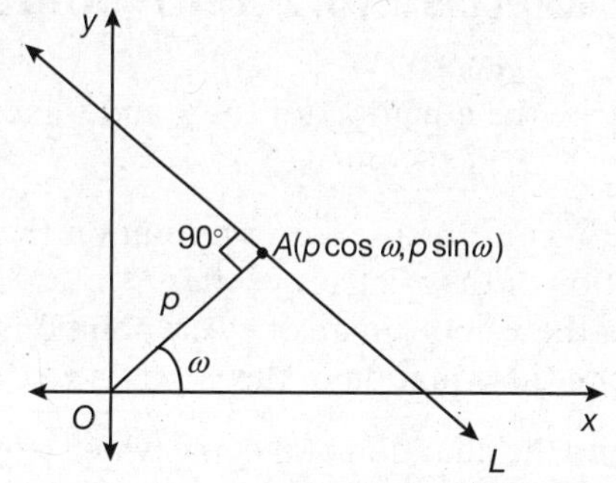

Fig. 10.10 Normal form of a line

Line segment OA is of length p and inclined to an angle of ω with the positive x-axis; therefore, the coordinates of point A are $(p\cos\omega, p\sin\omega)$.

And line L is perpendicular to OA; therefore, the slope of the line L is

$$=-\frac{1}{\text{slope of } OA}=-\frac{1}{\tan\omega}=-\frac{\cos\omega}{\sin\omega}$$

Thus, the equation of the line L having a slope of $-\dfrac{\cos\omega}{\sin\omega}$ and passing through the point $A(p\cos\omega, p\sin\omega)$ is

$$y-p\sin\omega=-\frac{\cos\omega}{\sin\omega}(x-p\cos\omega)$$

$$\Rightarrow y\sin\omega-p\sin^2\omega=-x\cos\omega+p\cos^2\omega$$

$$\Rightarrow x\cos\omega+y\sin\omega=p(\sin^2\omega+\cos^2\omega)$$

$$\Rightarrow x\cos\omega+y\sin\omega=p$$

Thus, the equation of the line having a normal distance p from the origin and angle ω which the normal makes with the positive direction of x-axis is given by $x\cos\omega+y\sin\omega=p$.

Example 8: Find the equation of the line which is at a perpendicular distance of 4 units from the origin and the angle made by the perpendicular with the positive x-axis is 45°.

Solution:
If p is the length of the normal from the origin to a line and α is the angle made by the normal with the positive direction of the x-axis, then the equation of the line is given by $x\cos\alpha + y\sin\alpha = p$.
Thus, the required equation of the given line is

$$x\cos 45° + y\sin 45° = 4 \quad (\because p = 4 \text{ and } \alpha = 45°)$$

$$\Rightarrow x\frac{1}{\sqrt{2}} + y\cdot\frac{1}{\sqrt{2}} = 4 \Rightarrow x + y = 4\sqrt{2}$$

Day 135

NCERT Exercise 10.2 (With Solutions)

Q1. Write the equation for the x and y axes.

Solution:
Since the y-coordinate of every point on the x-axis is 0, the equation of the x-axis is $y = 0$.
Similarly, the x-coordinate of every point on the y-axis is 0, therefore the equation of the y-axis is $x = 0$.

Q2. Find the equation of the line which passes through the point (–4, 3) with a slope of ½.

Solution:
The equation of the line that passes through point (x_0, y_0) whose slope is m is $(y - y_0) = m(x - x_0)$. Therefore, the equation of the line passing through the point (–4, 3) whose slope is 1/2 is

$$(y-3) = \frac{1}{2}(x+4) \Rightarrow 2(y-3) = x+4$$

$$\Rightarrow x - 2y + 10 = 0$$

Q3. Find the equation of the line which passes through (0, 0) with a slope m.

Solution:
We know that the equation of the line passing through (x_0, y_0) whose slope is m is $(y - y_0) = m(x - x_0)$. Thus, the equation of the line passing through the point (0, 0) whose slope is m is
$(y - 0) = m(x - 0)$
That is, $y = mx$

Q4. Find the equation of the line which passes through $(2, 2\sqrt{3})$ and is inclined with the x-axis at an angle of 75°.

Solution:
Slope of the line that inclines with the x-axis at an angle of 75° is $m = \tan 75°$

$$\Rightarrow m = \tan(45° + 30°) = \frac{\tan 45° + \tan 30°}{1 - \tan 45°\cdot\tan 30°}$$

$$= \frac{1 + \frac{1}{\sqrt{3}}}{1 - 1\cdot\frac{1}{\sqrt{3}}} = \frac{\frac{\sqrt{3}+1}{\sqrt{3}}}{\frac{\sqrt{3}-1}{\sqrt{3}}} = \frac{\sqrt{3}+1}{\sqrt{3}-1}$$

The equation of the line passing through point (x_0, y_0) whose slope is m is $(y - y_0) = m(x - x_0)$. Therefore, the equation of line passes through $(2, 2\sqrt{3})$ and inclines with the x-axis at an angle of 75°, that is, whose slope is $\frac{\sqrt{3}+1}{\sqrt{3}-1}$ is given by

$$\Rightarrow (y - 2\sqrt{3}) = \frac{\sqrt{3}+1}{\sqrt{3}-1}(x-2)$$

$$\Rightarrow (y - 2\sqrt{3})(\sqrt{3}-1) = (\sqrt{3}+1)(x-2)$$

$$\Rightarrow y(\sqrt{3}-1) - 2\sqrt{3}(\sqrt{3}-1) = x(\sqrt{3}+1) - 2(\sqrt{3}+1)$$

$$\Rightarrow (\sqrt{3}+1)x - (\sqrt{3}-1)y = -2\sqrt{3} - 2 + 6 - 2\sqrt{3}$$

$$\Rightarrow (\sqrt{3}+1)x - (\sqrt{3}-1)y + 4(\sqrt{3}-1) = 0$$

Q5. Find the equation of the line which intersects the x-axis at a distance of 3 units to the left of origin with a slope of –2.

Solution:
The equation of a line with slope m and x-intercept is a is given by $y = m(x - a)$.

Thus, the equation of the line intersecting the x-axis at a distance of 3 units to the left of the origin, that is, $a = -3$ and the slope of the line is $m = -2$ is given by
$y = -2[x - (-3)] \Rightarrow y = -2x - 6 \Rightarrow 2x + y + 6 = 0$

Q6. Find the equation of the line which intersects the y-axis at a distance of 2 units above the origin and makes an angle of 30° with the positive direction of the x-axis.

Solution:
The equation of the line with slope m and with y-intercept equal to b is given by $y - b = mx$. Thus, the equation of the line intersecting the y-axis at a distance of 2 units above the origin, that is, $b = 2$ and the slope of the line is $m = \tan 30° = \frac{1}{\sqrt{3}}$, is given by

$$y = \frac{1}{\sqrt{3}}x + 2 \Rightarrow y = \frac{x + 2\sqrt{3}}{\sqrt{3}}$$

$\sqrt{3}y = x + 2\sqrt{3} \Rightarrow$ that is, $x - \sqrt{3}y + 2\sqrt{3} = 0$

Q7. Find the equation of the line that passes through the points $(-1,1),(2,-4)$.

Solution:
The equation of the line passing through points $(x_1,y_1),(x_2,y_2)$ is $y-y_1=\dfrac{y_2-y_1}{x_2-x_1}(x-x_1)$.
Therefore, the equation of the line passing through the points $(-1,1),(2,-4)$ is $(y-1)=\dfrac{-4-1}{2+1}(x+1)$.

$\Rightarrow (y-1)=\dfrac{-5}{3}(x+1)\Rightarrow 3(y-1)=-5(x+1)$

$3y-3=-5x-5\Rightarrow 5x+3y+2=0$

Q8. Find the equation of the line which is at a perpendicular distance of 5 units from the origin and the angle made by the perpendicular with the positive x-axis is 30°.

Solution:
If p is the length of the normal from the origin to a line and α is the angle made by the normal with the positive direction of the x-axis, then the equation of the line is given by $x\cos\alpha+y\sin\alpha=p$.

Thus, the required equation of the given line is $x\cos 30°+y\sin 30°=5$ $(\because p=5 \text{ and } \alpha=30°)$

$\Rightarrow x\dfrac{\sqrt{3}}{2}+y\cdot\dfrac{1}{2}=5\Rightarrow \sqrt{3}x+y=10$

Q9. The vertices of the triangle ΔPQR are $P(2,1),Q(-2,3),R(4,5)$. Find the equation of the median through the vertex R.

Solution:
It is given that the vertices of the triangle ΔPQR are $P(2,1),Q(-2,3),R(4,5)$.
Let RL be the median through vertex R, where L is the mid-point of PQ. Then, the coordinates of point L are given by $\left(\dfrac{2-2}{2},\dfrac{1+3}{2}\right)=(0,2)$.
The equation of the line passing through points $(x_1,y_1),(x_2,y_2)$ is $y-y_1=\dfrac{y_2-y_1}{x_2-x_1}(x-x_1)$.
Therefore, the equation of median RL, where $R(4,5)$ and $L(0,2)$ is $y-5=\dfrac{2-5}{0-4}(x-4)$.

$\Rightarrow y-5=\dfrac{-3}{-4}(x-4)\Rightarrow 4(y-5)=3(x-4)$

$\Rightarrow 4y-20=3x-12\Rightarrow 3x-4y+8=0$

Thus, the required equation of the median through the vertex R is $3x-4y+8=0$.

Q10. Find the equation of the line passing through (–3, 5) and perpendicular to the line through the points $(2,5),(-3,6)$.

Solution:
The slope of the line joining the points $(2,5),(-3,6)$ is $m=\dfrac{6-5}{-3-2}=\dfrac{1}{-5}$. And, we know that two non-vertical lines are perpendicular to each other if and only if their slopes are negative reciprocals of each other.

Therefore, the slope of the line perpendicular to the line through the points $(2,5),(-3,6)$ is

$=-\dfrac{1}{m}=-\dfrac{1}{(-1/5)}=5$

Thus, the equation of the line passing through the point (–3, 5), whose slope is 5, is $(y-5)=5(x+3)$

$\Rightarrow y-5=5x+15\Rightarrow 5x-y+20=0$

Q11. A line perpendicular to the line segment joining the points (1, 0) and (2, 3) divides it in the ratio $1:n$. Find the equation of the line.

Solution:
Using the section formula, the coordinates of the point that divides the line segment joining the points (1, 0) and (2, 3) in the ratio $1:n$ is given by

$$\left(\frac{1\times n+2\times 1}{1+n},\frac{0\times n+3\times 1}{1+n}\right)=\left(\frac{n+2}{n+1},\frac{3}{n+1}\right)$$

Since the slope of the line joining the points $(1,0)$ and $(2,3)$ is $m=\dfrac{3-0}{2-1}=3$, the slope of the line that is perpendicular to the line joining the points (1, 0) and (2, 3) $=-\dfrac{1}{m}=-\dfrac{1}{3}$ (since two non-vertical lines are perpendicular to each other if and only if their slopes are negative reciprocals of each other).

Thus, the equation of the line passing through $\left(\dfrac{n+2}{n+1},\dfrac{3}{n+1}\right)$ and whose slope is $-\dfrac{1}{3}$ is,

$$\left(y-\frac{3}{n+1}\right)=\frac{-1}{3}\left(x-\frac{n+2}{n+1}\right)$$

$\Rightarrow 3[(n+1)y-3]=-[x(n+1)-(n+2)]$

$\Rightarrow 3(n+1)y-9=-(n+1)x+n+2$

$\Rightarrow (1+n)x+3(1+n)y=n+11$

Q12. Find the equation of a line that cuts off equal intercepts on the coordinate axes and passes through the point (2, 3).

Solution:

The equation of a line in the intercept form is $\frac{x}{a}+\frac{y}{b}=1$, where, a and b are the intercepts on the x and y axes, respectively.

The line cuts off equal intercepts on both the axes; therefore $a = b$, and thus, the equation of the line can be written as $\frac{x}{a}+\frac{y}{b}=1 \Rightarrow x+y=a$.

Since the given line passes through point (2, 3), it satisfies the equation of the line; hence, $2+3=a \Rightarrow a=5$. Thus $x+y=5$ is the required equation of the line.

Q13. Find the equation of the line passing through the point (2, 2) and cutting off intercepts on the axes whose sum is 9.

Solution:

The equation of a line in the intercept form is $\frac{x}{a}+\frac{y}{b}=1$, where a and b are the intercepts on the x and y axes, respectively.

It is given that $a+b=9 \Rightarrow b=9-a$; therefore the equation of the line can be written as $\frac{x}{a}+\frac{y}{9-a}=1$.

The given the line passes through the point (2, 2); therefore, it satisfies the equation of the line, and hence,

$$\frac{2}{a}+\frac{2}{9-a}=1 \Rightarrow 2\left(\frac{1}{a}+\frac{1}{9-a}\right)=1 \Rightarrow 2\left(\frac{9-a+a}{a(9-a)}\right)=1$$

$$\Rightarrow \frac{18}{9a-a^2}=1 \Rightarrow a^2-9a+18=0$$

$$\Rightarrow (a-3)(a-6)=0 \Rightarrow a=3,\ a=6$$

Thus, if $a=6 \Rightarrow b=9-6=3$, then the equation of the line is $\frac{x}{6}+\frac{y}{3}=1 \Rightarrow x+2y-6=0$.

And if $a=3 \Rightarrow b=9-3=6$ then the equation of the line is $\frac{x}{3}+\frac{y}{6}=1 \Rightarrow 2x+y-6=0$.

Q14. Find the equation of the line through the point (0, 2) making an angle of $\frac{2\pi}{3}$ with the positive x-axis. Also, find the equation of the line parallel to it and crossing the y-axis at a distance of 2 units below the origin.

Solution:

The slope of the line making an angle of $\frac{2\pi}{3}$ with the positive x-axis is $m=\tan\left(\frac{2\pi}{3}\right)=-\sqrt{3}$.

Therefore, the equation of the line passing through the point (0, 2) and having a slope of $-\sqrt{3}$ is $(y-2)=-\sqrt{3}(x-0) \Rightarrow \sqrt{3}x+y-2=0$.

Now, the slope of line parallel to the line $\sqrt{3}x+y-2=0$ is $-\sqrt{3}$.

And it is given that the line crosses the y-axis 2 units below the origin, that is, it passes through point (0, –2). Therefore, the equation of the line passing through the point (0, –2) and having a slope of $-\sqrt{3}$ is $y-(-2)=-\sqrt{3}(x-0) \Rightarrow \sqrt{3}x+y+2=0$.

Q15. The perpendicular from the origin to a line meets it at the point (–2, 9). Find the equation of the line.

Solution:

Let the slope of the line joining the origin (0, 0) and point (–2, 9) be $m=\frac{9-0}{-2-0}=-\frac{9}{2}$.

Then, the slope of the line perpendicular to the line joining the origin and point (–2, 9) is $=-\frac{1}{m}=-\frac{1}{-9/2}=\frac{2}{9}$.

Thus, the equation of the line passing through (–2, 9) and having a slope $\frac{2}{9}$ is $(y-9)=\frac{2}{9}(x+2)$.

$9y-81=2x+4 \Rightarrow 2x-9y+85=0$

Q16. The length L (in centimetre) of a copper rod is a linear function of its celsius temperature C. In an experiment, if $L = 124.942$ when $C = 20$, and $L = 125.134$ when $C = 110$, express L in terms of C.

Solution:

It is given that when $C = 20$, the value of L is 124.942, whereas when $C = 110$, the value of L is 125.134.

Since the points (C, L), assuming C along the x-axis and L along the y-axis satisfy the linear relation between C and L, the equation of the line passing through points the given points (20, 124.942) and (110, 125.134) in the XY plane is given by $(L-124.942)=\frac{125.134-124.942}{110-20}(C-20)$

$$L-124.942=\frac{0.192}{90}(C-20)$$

$\Rightarrow L=0.00213(C-20)+124.942$, is the required linear relation between C and L.

Q17. The owner of a milk store finds that he can sell 980 litres of milk each week at Rs 14/litre and 1220 litres of milk each week at Rs 16/litre. Assuming a linear relationship between selling price and demand, how many litres could he sell weekly at Rs 17/litre?

Solution:
The relationship between selling price and demand is linear; therefore, assume that the selling price per litre is along the x-axis and the demand is along the y-axis. Thus, the two given points, that is, (14, 980) and (16, 1220) in the XY plane will satisfy the linear relationship between selling price and demand.

Thus, the linear relationship between the selling price per litre and demand is the equation of the line passing through points (14, 980) and (16, 1220), given by

$$y-980=\frac{1220-980}{16-14}(x-14) \Rightarrow y-980=\frac{240}{2}(x-14)$$

$$\Rightarrow y-980=120(x-14) \Rightarrow y=120(x-14)+980$$

Now, when $x=$ Rs 17/litre, then

$$y=120(17-14)+980=1340$$

Thus, the owner of the milk store could sell 1340 litres of milk weekly at Rs 17/litre.

Q18. If $P(a,b)$ is the mid-point of a line segment between axes, show that equation of the line is $\frac{x}{a}+\frac{y}{b}=2$.

Solution: Let AB be the line segment with coordinates $A(0,y)$ on the y-axis and $B(x,0)$ on the x-axis and let $P(a,b)$ be its mid-point.

$P(a,b)$ is the mid-point of AB. Therefore

$$P(a,b)=P\left(\frac{0+x}{2},\frac{y+0}{2}\right) \Rightarrow (a,b)=\left(\frac{x}{2},\frac{y}{2}\right)$$

$$\Rightarrow a=\frac{x}{2},\ b=\frac{y}{2} \Rightarrow x=2a,\ y=2b$$

Thus, the respective coordinates of A and B are $(0,2b),(2a,0)$.

Hence, the equation of the line passing through the points $A(0,2b), B(2a,0)$ is $(y-2b)=\frac{(0-2b)}{(2a-0)}(x-0)$.

$$\Rightarrow y-2b=\frac{-2b}{2a}(x) \Rightarrow a(y-2b)=-bx$$

$$\Rightarrow bx+ay=2ab$$

Dividing both sides by ab, we get

$$\frac{bx}{ab}+\frac{ay}{ab}=\frac{2ab}{ab} \Rightarrow \frac{x}{a}+\frac{y}{b}=2$$

Thus, the equation of the line is $\frac{x}{a}+\frac{y}{b}=2$.

Q19. Point $R(h,k)$ divides a line segment between the axes in the ratio 1:2. Find the equation of the line.

Solution:
Let AB be the line segment between the axes such that the point $R(h,k)$ divides AB in the ratio 1:2.

Let AB be the line segment with coordinates $A(x,0)$ on the x-axis and $B(0,y)$ on the y-axis. Since the point $R(h,k)$ divides AB in the ratio 1:2, then according to the section formula,

$$(h,k)=\left(\frac{1\times 0+2\times x}{1+2},\frac{1\times y+2\times 0}{1+2}\right) \Rightarrow (h,k)=\left(\frac{2x}{3},\frac{y}{3}\right)$$

$$\Rightarrow h=\frac{2x}{3},\ k=\frac{y}{3} \Rightarrow x=\frac{3h}{2},\ y=3k$$

Therefore, the respective coordinates of A and B are $\left(\frac{3h}{2},0\right),(0,3k)$.

Thus, the equation of line AB passing through the points $\left(\frac{3h}{2},0\right),(0,3k)$ is $(y-0)=\frac{3k-0}{0-\frac{3h}{2}}\left(x-\frac{3h}{2}\right)$.

$$y=-\frac{2k}{h}\left(x-\frac{3h}{2}\right) \Rightarrow 2kx+hy=3hk$$

Hence, the required equation of the line is $2kx+hy=3hk$.

Q20. By using the concept of the equation of a line, prove that the three points $(3,0),(-2,-2),(8,2)$ are collinear.

Solution:
If the points $(3,0),(-2,-2),(8,2)$ are collinear, then the line passing through the points $(3,0),(-2,-2)$ must also be passing through the point $(8,2)$.

Now, the equation of the line passing through the points $(3,0),(-2,-2)$ is $(y-0)=\frac{(-2-0)}{(-2-3)}(x-3)$.

$$y=\frac{-2}{-5}(x-3) \Rightarrow 2x-5y=6$$

Since the third point $x=8, y=2$ also satisfies the equation of the line passing through the points (3, 0) and (–2, –2), the three points $(3,0),(-2,-2),(8,2)$ lie on a line, and hence, they are collinear.

Day 136

General Equation of a Line

Any equation of the form $Ax+By+C=0$, where A, B and C are real constants such that A and B are not zero simultaneously, is called the general equation of a line.

Various forms of the general equation of a line are given below.

(a) Slope–intercept form

$$Ax+By+C=0 \Rightarrow y=-\frac{A}{B}x-\frac{C}{B}, \text{ where } B\neq 0$$

Comparing the above equation with the slope–intercept form $y = mx + c$, we get $m = -\frac{A}{B}$, $c = -\frac{C}{B}$.

Thus, equation $Ax + By + C = 0$ is the slope–intercept form of the equation of a line whose slope is $-\frac{A}{B}$ and the y-intercept is $-\frac{C}{B}$.

Note: If $B = 0$, then $x = -\frac{C}{A}$, which is the vertical line whose slope is undefined and the x-intercept is $-\frac{C}{A}$.

Example 1: Reduce the equation $2x + 3y - 6 = 0$ into the slope–intercept form and find its slope and the y-intercept.

Solution:

The given equation $2x + 3y - 6 = 0$ in the slope–intercept form can be written as $y = \frac{1}{3}(-2x + 6) \Rightarrow y = -\frac{2}{3}x + 2$.

Comparing the above equation with $y = mx + c$, we get $m = -\frac{2}{3}$, $c = 2$.

Therefore, in the slope–intercept form, the slope and the y-intercept are $-\frac{2}{3}$, 2, respectively.

(b) Intercept form

$$Ax + By + C = 0 \Rightarrow \frac{x}{-\frac{C}{A}} + \frac{y}{-\frac{C}{B}} = 1, \quad \text{where} \quad C \neq 0$$

Comparing the above equation with the intercept form $\frac{x}{a} + \frac{y}{b} = 1$, we get $a = -\frac{C}{A}$, $b = -\frac{C}{B}$.

Thus, equation $Ax + By + C = 0$ is the intercept form of the equation of a line whose x-intercept is $-\frac{C}{A}$ and y-intercept is $-\frac{C}{B}$.

Note: If $C = 0$, then $Ax + By + C = 0 \Rightarrow Ax + By = 0$, which is a line passing through the origin; therefore there are zero intercepts on the axes.

Example 2: Reduce the equation $2x - 3y - 9 = 0$ to the intercept form and find its intercepts on the axes.

Solution:

The given equation $2x - 3y - 9 = 0$ in the intercept form can be written as $2x - 3y = 9 \Rightarrow \frac{2x}{9} - \frac{3y}{9} = 1$

$$\Rightarrow \frac{x}{(9/2)} + \frac{y}{(-3)} = 1$$

Comparing the above equation with $\frac{x}{a} + \frac{y}{b} = 1$, we get $a = 9/2$, $b = -3$.

Therefore, in the intercept form, the x- and y-intercepts are $9/2$ and -3, respectively.

(c) Normal form

Let $x\cos\omega + y\sin\omega = p$ be the normal form of the line represented by the equation

$Ax + By + C = 0 \Rightarrow Ax + By = -C$.

Since both equations are the same,

$$\frac{A}{\cos\omega} = \frac{B}{\sin\omega} = -\frac{C}{P}$$

$$\Rightarrow \cos\omega = -\frac{Ap}{C}, \quad \sin\omega = -\frac{Bp}{C}$$

$$\text{Also, } \sin^2\omega + \cos^2\omega = \left(-\frac{Ap}{C}\right)^2 + \left(-\frac{Bp}{C}\right)^2 = 1$$

$$\Rightarrow p^2 = \frac{C^2}{A^2 + B^2} \Rightarrow p = \pm\frac{C}{\sqrt{A^2 + B^2}}$$

Thus, we get

$$\cos\omega = \pm\frac{A}{\sqrt{A^2 + B^2}}, \quad \sin\omega = \pm\frac{B}{\sqrt{A^2 + B^2}}$$

Thus, the normal form of the general equation $Ax + By + C = 0$ is $x\cos\omega + y\sin\omega = p$, where

$$\cos\omega = \pm\frac{A}{\sqrt{A^2 + B^2}}, \quad \sin\omega = \pm\frac{B}{\sqrt{A^2 + B^2}},$$

$$p = \pm\frac{C}{\sqrt{A^2 + B^2}}.$$

Note: Keep the signs in such a way that p is positive.

Example 3: Reduce the equation $\sqrt{3}x - y + 2 = 0$ to the normal form. Find the perpendicular distance from the origin and angle between the perpendicular and the positive x-axis.

Solution:

The given equation $\sqrt{3}x - y + 2 = 0$ can be written as $-\sqrt{3}x + y = 2$.

Dividing both sides by $\sqrt{(-\sqrt{3})^2 + (-1)^2} = \sqrt{4} = 2$, we get $-\frac{\sqrt{3}x}{2} + \frac{1}{2}y = \frac{2}{2} \Rightarrow \left(-\frac{\sqrt{3}}{2}\right)x + \left(\frac{1}{2}\right)y = 1$

$\Rightarrow x\cos 150° + y\sin 150° = 1$ which is the normal form of the given equation.

Comparing above equation with the normal form of the equation of line by $x\cos\alpha + y\sin\alpha = p$, we get $\alpha = 150°$, $p = 1$

Thus, the perpendicular distance of the line from the

origin is 1, while the angle between the perpendicular and the positive x-axis is 150°.

Distance of a Point from a Line: Let a line $L: Ax + By + C = 0$ (Fig. 10.11), whose distance from the point $P(x_0, y_0)$ is d (length of the perpendicular drawn from the point to the line).

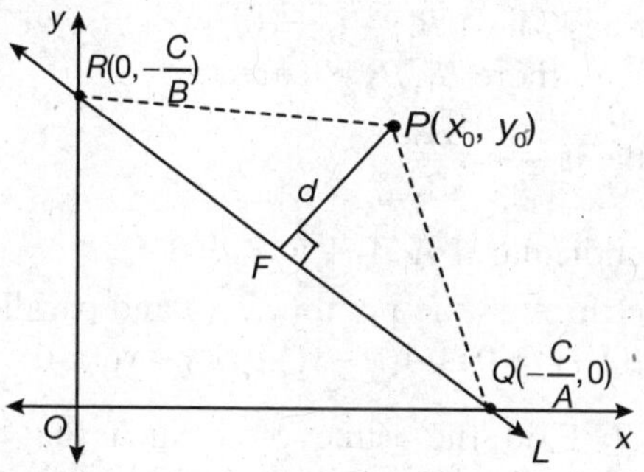

Fig. 10.11 Distance of a point from a line

Draw a perpendicular PF from the point P to the line L, where F is the foot of the perpendicular. Let the line meet the x and y axes at the points

$$Q\left(-\frac{C}{A}, 0\right), R\left(0, -\frac{C}{B}\right)$$

Now, the area of the triangle PQR is given by area

$$(\Delta PQR) = \frac{1}{2} PF \cdot QR$$

$$\Rightarrow d = PF = \frac{2\text{area}(\Delta PQR)}{QR} \quad \text{(i)}$$

Also, the area of the triangle PQR is given by (ΔPQR)

$$= \frac{1}{2}\left| x_0\left(0 + \frac{C}{B}\right) + \left(-\frac{C}{A}\right)\left(-\frac{C}{B} - y_0\right) + 0(y_0 - 0)\right|$$

$$= \frac{1}{2}\left| x_0 \frac{C}{B} + y_0 \frac{C}{A} + \frac{C^2}{AB}\right|$$

$$2 \text{ area} \Rightarrow 2\text{area}(\Delta PQR) = \left|\frac{C}{AB}\right| \cdot |Ax_0 + By_0 + C| \text{ and}$$

$$QR = \sqrt{\left(0 + \frac{C}{A}\right)^2 + \left(\frac{B}{C} - 0\right)^2} = \left|\frac{C}{AB}\right| \sqrt{A^2 + B^2}$$

Substituting the values of area $(\Delta PQR), QR$ in equation (i), we get $d = PF = \dfrac{|Ax_0 + By_0 + C|}{\sqrt{A^2 + B^2}}$.

Thus, the perpendicular distance (d) of a line $Ax + By + C = 0$ from a point (x_0, y_0) is given by

$$d = \frac{|Ax_0 + By_0 + C|}{\sqrt{A^2 + B^2}}$$

Example 4: Find the distance of the point $(-1, 1)$ from the line $2x + 3y - 1 = 0$.

Solution:

Comparing the given equation $2x + 3y - 1 = 0$ of the line with the general equation of the line $Ax + By + C = 0$, we get $A = 2$, $B = 3$, $C = -1$.

The perpendicular distance (d) of a line $Ax + By + C = 0$ from a point (x_1, y_1) is given by

$$d = \frac{|Ax_1 + By_1 + C|}{\sqrt{A^2 + B^2}}$$

The given point is $(x_1, y_1) = (-1, 1)$

Therefore, the distance of point $(-1, 1)$ from the given line $2x + 3y - 1 = 0$ is

$$= \frac{|2(-1) + (2)(1) - 1|}{\sqrt{(2)^2 + (3)^2}} = \frac{|-1|}{\sqrt{13}} = \frac{1}{\sqrt{13}} \text{ units}$$

Example 5: Find the points on the y-axis, whose distances from the line $\dfrac{x}{3} + \dfrac{y}{4} = 1$ are 3 units.

Solution:

The given equation of the line is $\dfrac{x}{3} + \dfrac{y}{4} = 1$

$\Rightarrow 4x + 3y - 12 = 0$

Comparing above equation with the general equation of the line $Ax + By + C = 0$, we get $A = 4$, $B = 3$, $C = -12$.

Let (0, b) be the point on the x-axis whose distance from the given line is 3 units.

The perpendicular distance (p) of a line $Ax + By + C = 0$ from a point (x_1, y_1) is given by

$$\frac{|Ax_1 + By_1 + C|}{\sqrt{A^2 + B^2}}; \text{ therefore } 3 = \frac{|4 \times 0 + 3 \times b - 12|}{\sqrt{4^2 + 3^2}}$$

$$\Rightarrow 3 = \frac{|3b - 12|}{5} \Rightarrow |3b - 12| = 15 \Rightarrow \pm(3b - 12) = 20$$

$$\Rightarrow (3b - 12) = 15 \text{ or } -(3b - 12) = 15$$

$$\Rightarrow 3b = 15 + 12 \text{ or } -3b = 15 - 12$$

$$\Rightarrow b = 9 \text{ or } b = -1$$

Thus, the required points on the y-axis are $(0, -1)$ or $(0, 9)$.

Distance between Two Parallel Lines: The slopes of two parallel lines are equal. Therefore, two parallel lines in the slope–intercept form can be taken as (Fig. 10.12),

$L_1: y = mx + c_1$

$L_2: y = mx + c_2$

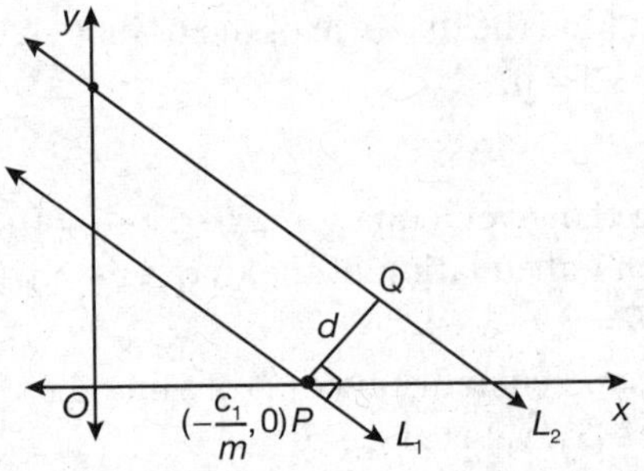

Fig. 10.12 Distance between two parallel lines

The line L_1 intersects the *x*-axis at the point $P\left(-\frac{c_1}{m}, 0\right)$; therefore the distance between the two lines is equal to the length of the perpendicular from point *P* to line L_2, that is,

$$d = PQ = \frac{\left|(-m)\left(-\frac{c_1}{m}\right) + (-c_2)\right|}{\sqrt{1+m^2}}$$

$\Rightarrow d = \frac{|c_1 - c_2|}{\sqrt{1+m^2}}$ is the distance *d* between two parallel lines $y = mx + c_1$ and $y = mx + c_2$.

Note: If the parallel lines are given in the general form, that is, $Ax + By + C_1 = 0, Ax + By + C_2 = 0$, then the distance between these parallel is given by $d = \frac{|C_1 + C_2|}{\sqrt{A^2 + B^2}}$.

Example 6: Find the distance between the parallel lines $3x + 4y - 2 = 0$, $3x + 4y + 6 = 0$.

Solution:

The distance (*d*) between the parallel lines $Ax + By + C_1 = 0$, $Ax + By + C_2 = 0$ is given by

$$d = \frac{|C_1 - C_2|}{\sqrt{A^2 + B^2}}$$

Therefore, the distance between the given parallel lines $3x + 4y - 2 = 0$, $3x + 4y + 6 = 0$ is (where $A = 3, B = 4, C_1 = -2, C_2 = 6$)

$$d = \frac{|C_1 - C_2|}{\sqrt{A^2 + B^2}} = \frac{|-2-6|}{\sqrt{(3)^2 + (4)^2}} = \frac{|-8|}{5} = \frac{8}{5} \text{ units}$$

Note:

(I) Two lines $A_1x + B_1y + C_1 = 0, A_2x + B_2y + C_2 = 0$, in general form, where $B_1, B_2 \neq 0$ are

(i) Parallel if $\frac{A_1}{B_1} = \frac{A_2}{B_2}$

(ii) Perpendicular if $A_1A_2 + B_1B_2 = 0$

(II) A line through the point (x_1, y_1) and parallel to the line $Ax + By + C = 0$ is $A(x - x_1) + B(y - y_1) = 0$.

Example 7: Find the value of *a*, such that the line equations of the lines $2x + 3y - 1 = 0$, $ax - 3y + 1 = 0$ (i) are parallel or (ii) are perpendicular to each other.

Solution:

Two lines $A_1x + B_1y + C_1 = 0, A_2x + B_2y + C_2 = 0$ are

(i) Parallel if $\frac{A_1}{B_1} = \frac{A_2}{B_2} \Rightarrow \frac{2}{3} = \frac{a}{-3} \Rightarrow a = -2$

(ii) Perpendicular if $A_1A_2 + B_1B_2 = 0$

$$\Rightarrow 2 \times a + 3 \times (-3) = 0 \Rightarrow a = \frac{9}{2}$$

Example 8: Find equation of the line parallel to the line $2x - 3y + 1 = 0$ and passing through the point $(-2, 2)$.

Solution:

The equation of the line through the point (x_1, y_1) and parallel to the line $Ax + By + C = 0$ is $A(x - x_1) + B(y - y_1) = 0$.

Comparing the given equation with $Ax + By + C = 0$, we get $A = 2$, $B = -3$.

Thus, the equation of the line through the point $(-2, 2)$ and parallel to the line $2x - 3y + 1 = 0$ is

$$2(x + 2) - 3(y - 2) = 0 \Rightarrow 2x - 3y + 10 = 0.$$

Day 137

NCERT Exercise 10.3 (With Solutions)

Q1. Reduce the following equations into the slope–intercept form and find their slopes and the *y* intercepts.

(i) $x + 7y = 0$ (ii) $6x + 3y - 5 = 0$

(iii) $y = 0$

Solution:

(i) The given equation $x + 7y = 0$, in the slope–intercept form can be written as $y = -\frac{1}{7}x + 0$.

Comparing the above equation with $y = mx + c$, we get $m = -\frac{1}{7}, c = 0$.

Therefore, in the slope–intercept form, the slope and the y-intercept are $-\frac{1}{7}$ and 0, respectively.

(ii) The given equation $6x+3y-5=0$ in the slope–intercept form can be written as $y=\frac{1}{3}(-6x+3) \Rightarrow y=-2x+\frac{5}{3}$.

Comparing the above equation with $y=mx+c$, we get $m=-2, c=\frac{5}{3}$.

Therefore, in the slope–intercept form, the slope and the y-intercept are $-2, \frac{5}{3}$, respectively.

(iii) The given equation $y=0$ in the slope–intercept form can be written as $y=0\cdot x+0$.

Comparing the above equation with $y=mx+c$, we get $m=0, c=0$.

Therefore, in the slope–intercept form, the slope and the y-intercept are 0, 0, respectively.

Q2. Reduce the following equations into the intercept form and find their intercepts on the axes.

(i) $3x+2y-12=0$ (ii) $4x-3y=6$

(iii) $3y+2=0$

Solution:

(i) The given equation $3x+2y-12=0$ in the intercept form can be written as $3x+2y=12 \Rightarrow \frac{3x}{12}+\frac{2y}{12}=1$

$\Rightarrow \frac{x}{4}+\frac{y}{6}=1$

Comparing the above equation with $\frac{x}{a}+\frac{y}{b}=1$, we get $a=4, b=6$.

Therefore, in the intercept form, the x- and y-intercepts are 4 and 6, respectively.

(ii) The given equation $4x-3y=6$, in the intercept form can be written as $\frac{4x}{6}-\frac{3y}{6}=1 \Rightarrow \frac{2x}{3}-\frac{y}{2}=1$

$\Rightarrow \frac{x}{(3/2)}+\frac{y}{(-2)}=1$

Comparing the above equation with $\frac{x}{a}+\frac{y}{b}=1$, we get $a=\frac{3}{2}, b=-2$.

Therefore, in the intercept form, the x- and y-intercepts are 3/2 and –2, respectively.

(iii) The given equation $3y+2=0$ in the intercept form can be written as $3y=-2 \Rightarrow \frac{y}{\left(-\frac{2}{3}\right)}=1$.

Comparing the above equation with $\frac{x}{a}+\frac{y}{b}=1$, we get $b=-\frac{2}{3}$.

Therefore, in the intercept form, the y-intercept is $-\frac{2}{3}$ and it has no intercept on the x-axis.

Q3. Reduce the following equations into the normal form. Find their perpendicular distance from the origin and angle between perpendicular and the positive x-axis.

(i) $x-\sqrt{3}y+8=0$ (ii) $y-2=0$

(iii) $x-y=4$

Solution:

(i) The given equation $x-\sqrt{3}y+8=0$ can be written as $x-\sqrt{3}y=-8 \Rightarrow -x+\sqrt{3}y=8$.

Dividing both sides by $\sqrt{(-1)^2+(\sqrt{3})^2}=\sqrt{4}=2$, we get $-\frac{x}{2}+\frac{\sqrt{3}}{2}y=\frac{8}{2} \Rightarrow \left(-\frac{1}{2}\right)x+\left(\frac{\sqrt{3}}{2}\right)y=4$

$\Rightarrow x\cos 120° + y\sin 120° = 4$ which is the normal form of the given equation.

Comparing the above equation with the normal form of the equation of line by $x\cos\alpha + y\sin\alpha = p$, we get $\alpha = 120°$, $p=4$.

Thus, the perpendicular distance of the line from the origin is 4, while the angle between the perpendicular and the positive x-axis is 120°.

(ii) The given equation $y-2=0$ can be written as $0\cdot x+1\cdot y=2$.

Dividing both sides by $\sqrt{0^2+1^2}=1$, we get $0\cdot x+1\cdot y=2 \Rightarrow x\cos 90° + y\sin 90° = 2$ which is the normal form of the given equation.

Comparing the above equation with the normal form of the equation of line by $x\cos\alpha + y\sin\alpha = p$, we get $\alpha = 90°$, $p=2$.

Thus, the perpendicular distance of the line from the origin is 2, while the angle between the perpendicular and the positive x-axis is 90°.

(iii) The given equation $x-y=4$ can be written as $1\cdot x+(-1)y=4$.

Dividing both sides by $\sqrt{1^2+(-1)^2}=\sqrt{2}$,

we get $\frac{1}{\sqrt{2}}x+\left(-\frac{1}{\sqrt{2}}\right)y=\frac{4}{\sqrt{2}}$

$\Rightarrow x\cos\left(2\pi-\frac{\pi}{4}\right)+y\sin\left(2\pi-\frac{\pi}{4}\right)=2\sqrt{2}$

$\Rightarrow x\cos 315° + y\sin 315° = 2\sqrt{2}$ which is the normal form of the given equation.

Comparing the above equation with the normal form of the equation of line by $x\cos\alpha + y\sin\alpha = p$, we get $\alpha = 315°$, $p = 2\sqrt{2}$.

Thus, the perpendicular distance of the line from the origin is $2\sqrt{2}$, while the angle between the perpendicular and the positive x-axis is 315°.

Q4. Find the distance of the point (–1,1) from the line $12(x+6) = 5(y-2)$.

Solution:

The given equation of the line is $12(x+6) = 5(y-2)$

$\Rightarrow 12x + 72 = 5y - 10 \Rightarrow 12x - 5y + 82 = 0$

Comparing the above equation with the general equation of the line $Ax + By + C = 0$, we get $A = 12$, $B = -5$, $C = 82$.

Since the perpendicular distance (d) of a line $Ax + By + C = 0$, from a point (x_1, y_1) is given by

$$d = \frac{|Ax_1 + By_1 + C|}{\sqrt{A^2 + B^2}}$$

The given point is $(x_1, y_1) = (-1, 1)$

Therefore, the distance of point (–1, 1) from the given line $12x - 5y + 82 = 0$ is

$$= \frac{|12(-1) + (-5)(1) + 82|}{\sqrt{(12)^2 + (-5)^2}} = \frac{|65|}{13} = 5 \text{ units}$$

Q5. Find the points on the x-axis, whose distances from the line $\frac{x}{3} + \frac{y}{4} = 1$ are 4 units.

Solution:

The given equation of the line is $\frac{x}{3} + \frac{y}{4} = 1$

$\Rightarrow 4x + 3y - 12 = 0$

Comparing the above equation with the general equation of the line $Ax + By + C = 0$, we get $A = 4$, $B = 3$, $C = -12$.

Let (a, 0) be the point on the x-axis whose distance from the given line is 4 units.

The perpendicular distance (p) of a line $Ax + By + C = 0$ from a point (x_1, y_1) is given by

$$\frac{|Ax_1 + By_1 + C|}{\sqrt{A^2 + B^2}}; \text{ therefore } 4 = \frac{|4a + 3 \times 0 - 12|}{\sqrt{4^2 + 3^2}}$$

$$\Rightarrow 4 = \frac{|4a - 12|}{5} \Rightarrow |4a - 12| = 20 \Rightarrow \pm(4a - 12) = 20$$

$\Rightarrow (4a - 12) = 20$ or $-(4a - 12) = 20$

$\Rightarrow 4a = 20 + 12$ or $-4a = 20 - 12$

$\Rightarrow a = 8$ or -2

Thus, the required points on the x-axis are $(-2,0), (8,0)$.

Q6. Find the distance between the parallel lines

(i) $15x + 8y - 34 = 0, 15x + 8y + 31 = 0$

(ii) $(x + y) + p = 0,\ l(x + y) - r = 0$

Solution:

The distance (d) between the parallel lines $Ax + By + C_1 = 0$, $Ax + By + C_2 = 0$ is given by

$$d = \frac{|C_1 - C_2|}{\sqrt{A^2 + B^2}}$$

(i) Therefore, the distance between the given parallel lines $15x + 8y - 34 = 0$, $15x + 8y + 31 = 0$ is (where $A = 15, B = 8, C_1 = -34, C_2 = 31$)

$$d = \frac{|C_1 - C_2|}{\sqrt{A^2 + B^2}} = \frac{|-34 - 31|}{\sqrt{(15)^2 + (8)^2}} = \frac{|-65|}{17} = \frac{65}{11} \text{ units}$$

(ii) Therefore, the distance between the given parallel lines $(x + y) + p = 0$, $l(x + y) - r = 0$ $\Rightarrow lx + ly + p = 0,\ lx + ly - r = 0$ is

$$d = \frac{|C_1 - C_2|}{\sqrt{A^2 + B^2}} = \frac{|P + r|}{\sqrt{l^2 + l^2}} = \frac{1}{\sqrt{2}}\left|\frac{p + r}{l}\right| \text{ units}$$

Q7. Find the equation of the line parallel to the line $3x - 4y + 2 = 0$ and passing through the point (–2, 3).

Solution:

The equation of the given line is $3x - 4y + 2 = 0$

$$\Rightarrow y = \frac{3}{4}x + \frac{1}{2}$$

Comparing with the slope–intercept form $y = mx + c$, we get the slope of the given line = ¾.

Thus, the slope of the other parallel lines is $= m = \frac{3}{4}$.

Now, the equation of the line that has a slope of 3/4 and passes through the point (–2, 3) is

$$(y - 3) = \frac{3}{4}\{x - (-2)\} \Rightarrow 4y - 12 = 3x + 6$$

$\Rightarrow 3x - 4y + 18 = 0$

Q8. Find the equation of the line perpendicular to the line $x - 7y + 5 = 0$, given that the x intercept is equal to 3.

Solution:

The given equation of line is $x - 7y + 5 = 0 \Rightarrow y = \frac{1}{7}x + \frac{5}{7}$

Comparing with the equation $y = mx + c$, we get the slope of the given line =1/7.

Thus, the slope of the line perpendicular to the line having a slope of 1/7 is $m = -\frac{1}{(1/7)} = -7$.

Therefore, the equation of the line with a slope of –7 and x-intercept equal to 3 is given by $y = m(x-d)$.

$\Rightarrow y = -7(x-3) \Rightarrow 7x + y - 21 = 0$

Q9. Find the angles between the lines $\sqrt{3}x + y = 1,\ x + \sqrt{3}y = 1$.

Solution:

The given equations of lines can be written as $\sqrt{3}x + y = 1 \Rightarrow y = -\sqrt{3}x + 1$

$x + \sqrt{3}y = 1 \Rightarrow y = -\frac{1}{\sqrt{3}}x + \frac{1}{\sqrt{3}}$

Thus, we get the slope first line, $m_1 = -\sqrt{3}$ and the slope of other line, $m_2 = -\frac{1}{\sqrt{3}}$.

Therefore, the acute angle (θ) between the two lines is given by

$$\tan\theta = \frac{|m_1 - m_2|}{|1 + m_1 m_2|} \Rightarrow \tan\theta = \left|\frac{-\sqrt{3} + \frac{1}{\sqrt{3}}}{1 + (-\sqrt{3})\left(-\frac{1}{\sqrt{3}}\right)}\right|$$

$$\tan\theta = \left|\frac{\frac{-3+1}{\sqrt{3}}}{1+1}\right| = \left|\frac{-2}{2\times\sqrt{3}}\right| \Rightarrow \tan\theta = 1/\sqrt{3} \Rightarrow \theta = 30^\circ$$

Thus, the angle between the given lines is either 30° or $180^\circ - 30^\circ = 150^\circ$.

Q10. The line through the points $(h,3), (4,1)$ intersects the line $7x - 9y - 19 = 0$ at right angles. Find the value of h.

Solution:

The slope of the line passing through points $(h,3),(4,1)$ is $m_1 = \frac{1-3}{4-h} = \frac{-2}{4-h}$.

And the slope of the line $7x - 9y - 19 = 0 \Rightarrow y = \frac{7}{9}x - \frac{19}{9}$ is $m_2 = 7/9$

Therefore, the two lines are perpendicular if $m_1 \times m_2 = -1 \Rightarrow \left(\frac{-2}{4-h}\right)\times\left(\frac{7}{9}\right) = -1 \Rightarrow \frac{-14}{36-9h} = -1$

$\Rightarrow 14 = 36 - 9h \Rightarrow 9h = 36 - 14 \Rightarrow h = 22/9$

Q11. Prove that the line through the point (x_1, y_1) and parallel to the line $Ax + By + C = 0$ is $A(x - x_1) + B(y - y_1) = 0$.

Solution:

The slope of the line $Ax + By + C = 0 \Rightarrow y = \left(\frac{-A}{B}\right)x + \left(\frac{-C}{B}\right)$ is $m = -A/B$.

Therefore, the slope of the other parallel lines is also $= m = -A/B$

Thus, the equation of the line passing through point (x_1, y_1) and having a slope $m = -A/B$ is

$y - y_1 = m(x - x_1) \Rightarrow y - y_1 = -\frac{A}{B}(x - x_1)$

$B(y - y_1) = -A(x - x_1) \Rightarrow A(x - x_1) + B(y - y_1) = 0$

Hence, the line through point (x_1, y_1) and parallel to line $Ax + By + C = 0$ is $A(x - x_1) + B(y - y_1) = 0$

Q12. Two lines passing through the point (2, 3) intersect each other at an angle of 60°. If the slope of one line is 2, find equation of the other line.

Solution:

Let the slope of the first line be $m_1 = 2$ and the slope of the other line be m_2.

Then, the angle between the two lines is given by

$$\frac{|m_1 - m_2|}{|1 + m_1 m_2|} = \tan 60^\circ = \sqrt{3} \Rightarrow \pm\left(\frac{2 - m_2}{1 + 2m_2}\right) = \sqrt{3}$$

$$\Rightarrow \frac{2 - m_2}{1 + 2m_2} = \sqrt{3} \text{ or } -\left(\frac{2 - m_2}{1 + 2m_2}\right) = \sqrt{3}$$

$$\Rightarrow 2 - m_2 = \sqrt{3}(1 + 2m_2) \text{ or } -(2 - m_2) = \sqrt{3}(1 + 2m_2)$$

$$\Rightarrow (2\sqrt{3}+1)m_2 = 2 - \sqrt{3} \text{ or } (2\sqrt{3}-1)m_2 = -2 - \sqrt{3}$$

$$\Rightarrow m_2 = \frac{2-\sqrt{3}}{(2\sqrt{3}+1)} \text{ or } m_2 = \frac{-(2+\sqrt{3})}{(2\sqrt{3}-1)}$$

Case 1: $m_2 = \frac{(2-\sqrt{3})}{(2\sqrt{3}-1)}$

The equation of the line passing through point (2, 3) and having a slope of $\frac{(2-\sqrt{3})}{(2\sqrt{3}+1)}$ is given by

$(y-3) = \frac{(2-\sqrt{3})}{2\sqrt{3}+1}(x-2)$

$(2\sqrt{3}+1)y - 3(2\sqrt{3}+1) = (2-\sqrt{3})x - 2(2-\sqrt{3})$

$(\sqrt{3}-2)x + (2\sqrt{3}+1)y = -1 + 8\sqrt{3}$

Case 2: $m_2 = \frac{-(2+\sqrt{3})}{(2\sqrt{3}-1)}$

The equation of the line passing through point (2, 3) and having a slope of $\frac{-(2+\sqrt{3})}{(2\sqrt{3}-1)}$ is

$(y-3) = \frac{-(2+\sqrt{3})}{(2\sqrt{3}-1)}(x-2)$

$(2\sqrt{3}-1)y - 3(2\sqrt{3}-1) = -(2+\sqrt{3})x + 2(2+\sqrt{3})$

$(2+\sqrt{3})x + (2\sqrt{3}-1)y = 1 + 8\sqrt{3}$

Thus, the required equations of the other line are $(\sqrt{3}-2)x + (2\sqrt{3}+1)y = -1 + 8\sqrt{3}$ or

$(2+\sqrt{3})x+(2\sqrt{3}-1)y=1+8\sqrt{3}$

Q13. Find the equation of the right bisector of the line segment joining the points (3, 4) and (–1, 2).

Solution:
The slope of the line segment passing through the points (3, 4) and (–1, 2) is $= \dfrac{2-4}{-1-3} = \dfrac{-2}{-4} = 1/2$.
Therefore, slope of the line perpendicular to this line is $= -\dfrac{1}{(1/2)} = -2$.
The right bisector of a line segment bisects the line segment at 90° and passes through the mid-point.
Now, the mid-point of the line is $= \left(\dfrac{3-1}{2}, \dfrac{4+2}{2}\right) = (1,3)$
Therefore, the equation of the line passing through (1, 3) and having a slope of –2 is $(y-3) = -2(x-1)$
$\Rightarrow 2x+y=5$

Q14. Find the coordinates of the foot of perpendicular from the point (–1, 3) to the line $3x-4y-16=0$.

Solution:
Let (*a*, *b*) be the coordinates of the foot of the perpendicular from the point (–1, 3) to the line $3x-4y-16=0$. Then
Slope of the line joining (–1, 3) and (*a*, *b*), is $m_1 = \dfrac{b-3}{a+1}$
Slope of the line $3x-4y-16=0 \Rightarrow y=\dfrac{3}{4}x-4$ is $m_2 = 3/4$
Since the above two lines are perpendicular to each other, $m_1m_2=-1$
$\Rightarrow \left(\dfrac{b-3}{a+1}\right)\times\left(\dfrac{3}{4}\right) = -1 \Rightarrow \dfrac{3b-9}{4a+4} = -1$
$\Rightarrow 3b-9=-4a-4 \Rightarrow 4a+3b=5$
Also, the point (*a*, *b*) lies on line $3x-4y=16$; therefore, $3a-4b=16$.
Solving the above two equations, we get $a=68/25$, $b=-49/25$ which are the required coordinates of the foot of the perpendicular to the line $3x-4y-16=0$.

Q15. The perpendicular from the origin to the line $y=mx+c$ meets it at the point (–1, 2). Find the values of *m* and *c*.

Solution:
The perpendicular from the origin to the line $y=mx+c$ meets it at the point (–1, 2). Therefore, the slope of the line joining $(0,0),(-1,2) = \dfrac{2}{-1} = -2$.
Since it is perpendicular to the given equation of line $y=mx+c$,
$m\times -2 = -1 \Rightarrow m = 1/2$
Since the point (–1, 2) lies on the given line, it will satisfy the equation $y=mx+c$. Therefore
$\therefore 2 = m(-1)+c \Rightarrow 2 = \dfrac{1}{2}(-1)+c \Rightarrow c = 2+\dfrac{1}{2} = \dfrac{5}{2}$
Thus, the values of *m* and *c* are $\dfrac{1}{2}$ and $\dfrac{5}{2}$, respectively.

Q16. If *p* and *q* are the lengths of the perpendicular from the origin to the lines $x\cos\theta - y\sin\theta = k\cos 2\theta$, $x\sec\theta + y\,\text{cosec}\,\theta = k$, respectively, prove that $p^2+4q^2=k^2$.

Solution:
The perpendicular distance (*d*) of a line $Ax+By+C=0$ from a point (x_1, y_1) is given by $d = \left|\dfrac{Ax_1+By_1+C}{\sqrt{A^2+B^2}}\right|$
Comparing equation $x\cos\theta - y\sin\theta = k\cos 2\theta$ with equation $Ax+By+C=0$, we get
$A=\cos\theta, B=-\sin\theta, C=-k\cos 2\theta$
Therefore, the perpendicular distance (*p*) of the origin (0, 0) from line $x\cos\theta - y\sin\theta = k\cos 2\theta$ is given by
$$p = \left|\frac{A(0)+B(0)+C}{\sqrt{A^2+B^2}}\right| = \frac{|C|}{\sqrt{A^2+B^2}}$$
$$\Rightarrow p = \frac{|-k\cos 2\theta|}{\sqrt{\cos^2\theta+(-\sin^2\theta)}} = |-k\cos 2\theta| \quad \text{(i)}$$
Similarly, comparing the equation $x\sec\theta + y\,\text{cosec}\,\theta = k$ with $Ax+By+C=0$, we get
$A=\sec\theta$, $B=\text{cosec}\,\theta$, $C=k$
Therefore, the perpendicular distance (*q*) of the origin (0, 0) from the line $x\sec\theta + y\,\text{cosec}\,\theta = k$ is given by
$$q = \left|\frac{A(0)+B(0)+C}{\sqrt{A^2+B^2}}\right| = \frac{|C|}{\sqrt{A^2+B^2}}$$
$$\Rightarrow q = \frac{|-k|}{\sqrt{\sec^2\theta+\text{cosec}^2\theta}} \quad \text{(ii)}$$
Thus, adding equations (i) and (ii), we get
$$p^2+4q^2 = (|-k\cos 2\theta|)^2 + 4\left(\frac{|-k|}{\sqrt{\sec^2\theta+\text{cosec}^2\theta}}\right)^2$$
$$= k^2\cos^2 2\theta + \frac{4k^2}{(\sec^2\theta+\text{cosec}^2\theta)}$$
$$= k^2\cos^2 2\theta + \frac{4k^2}{\left(\dfrac{1}{\cos^2\theta}+\dfrac{1}{\sin^2\theta}\right)}$$
$$= k^2\cos^2 2\theta + \frac{4k^2}{\left(\dfrac{\sin^2\theta+\cos^2\theta}{\sin^2\theta\cos^2\theta}\right)}$$

$$= k^2\cos^2 2\theta + \frac{4k^2}{\left(\frac{1}{\sin^2\theta\cos^2\theta}\right)}$$

$= k^2\cos^2 2\theta + 4k^2\sin^2\theta\cos^2\theta$

$= k^2\cos^2 2\theta + k^2(2\sin\theta\cos\theta)^2$

$= k^2\cos^2 2\theta + k^2\sin^2 2\theta$

$\Rightarrow p^2 + 4q^2 = k^2(\cos^2 2\theta + \sin^2 2\theta)$

$\Rightarrow p^2 + 4q^2 = k^2$

Q17. In the triangle ABC with vertices $A(2,3), B(4,-1), C(1,2)$, find the equation and length of altitude from the vertex A.

Solution:

Let AD be the altitude of triangle ABC from vertex A to BC, that is, $AD \perp BC$.

The slope of line BC is $= \frac{2+1}{1-4} = -1$, therefore, the slope of altitude AD will be equal to 1 ($\because AD \perp BC$).

Thus, the equation of the line passing through point (2, 3) and having a slope of 1 is $(y-3) = 1(x-2)$.

$\Rightarrow y - x = 1$

Therefore, the equation of the altitude from vertex A is $y - x = 1$.

And, the equation of BC is $(y+1) = \frac{2+1}{1-4}(x-4)$.

$\Rightarrow (y+1) = -1(x-4) \Rightarrow x + y - 3 = 0$

The perpendicular distance (d) of a line $Ax + By + C = 0$ from a point (x_1, y_1) is given by $d = \left|\frac{Ax_1 + By_1 + C}{\sqrt{A^2 + B^2}}\right|$

Comparing equation $x + y - 3 = 0$ with the equation of line $Ax + By + C = 0$, we get $A = 1$, $B = 1$, $C = -3$.

Therefore, length of AD = Length of the perpendicular from A (2, 3) to BC

$$= \frac{|1\times 2 + 1\times 3 - 3|}{\sqrt{1^2 + 1^2}} = \frac{|2|}{\sqrt{2}} = \sqrt{2} \text{ units}$$

Thus, the equation and the length of the altitude from vertex A are $y - x = 1$, $\sqrt{2}$ units, respectively.

Q18. If p is the length of the perpendicular from the origin to the line whose intercepts on the axes are a and b, then show that: $\frac{1}{p^2} = \frac{1}{a^2} + \frac{1}{b^2}$.

Solution:

The equation of a line whose intercepts on the axes are a and b is $\frac{x}{a} + \frac{y}{b} = 1 \Rightarrow bx + ay - ab = 0$, and the perpendicular distance (d) of a line $Ax + By + C = 0$ from a point (x_1, y_1) is given by $d = \left|\frac{Ax_1 + By_1 + C}{\sqrt{A^2 + B^2}}\right|$.

Comparing equation $bx + ay - ab = 0$ to the general equation of line $Ax + By + C = 0$ we get $A = 1$, $B = a$, $C = -ab$.

Therefore, perpendicular distance (p) from the origin (0,0) to the line $bx + ay - ab = 0$, we get

$$p = \frac{|A(0) + B(0) - ab|}{\sqrt{b^2 + a^2}} \Rightarrow p = \frac{|-ab|}{\sqrt{a^2 + b^2}}$$

Squaring both sides, we get

$$p^2 = \frac{(-ab)^2}{a^2 + b^2} \Rightarrow p^2(a^2 + b^2) = a^2b^2$$

$$\Rightarrow \frac{a^2 + b^2}{a^2b^2} = \frac{1}{p^2} \Rightarrow \frac{1}{p^2} = \frac{1}{a^2} + \frac{1}{b^2}$$

Day 138

Equation of Family of Lines Passing Through the Point of Intersection of Two Lines

Let the general equation of two intersecting lines l_1, l_2 be given by

$l_1 : A_1x + B_1y + C_1 = 0$

$l_2 : A_2x + B_2y + C_2 = 0$

Then, with the above two equations, we can form an equation $l_1 + k\,l_2 = 0$

$\Rightarrow A_1x + B_1y + C_1 + k(A_2x + B_2y + C_2) = 0$ represents a family of lines, where k is an arbitrary constant called parameter. For a particular value of k, we get a family of equations and the value of k may be obtained from the other given condition.

<u>**Example 1:**</u> Find the equation of the line through the intersection of lines $x - 2y + 3 = 0$, $3x + y - 2 = 0$ and parallel to the y-axis.

Solution:

The equation of any line through the point of intersection of the given lines is of the form

$x - 2y + 3 + k(3x + y - 2) = 0$

$\Rightarrow (1 + 3k)x + (-2 + k)y + 3 - 2k = 0$

This line is parallel to the y-axis; therefore the coefficient of y should be zero, $-2+k=0 \Rightarrow k=2$

Substituting the value of $k=2$, we get

$(1+3\times 2)x+(-2+2)y+3-2\times 2=0 \Rightarrow 7x-1=0$ is the required equation of the line.

Example 2: Find the equation of the line through the intersection of the lines $x-2y+3=0$, $3x+y-2=0$ and passes through the origin.

Solution:

The equation of any line through the point of intersection of the given lines is of the form

$x-2y+3+k(3x+y-2)=0$

$\Rightarrow (1+3k)x+(-2+k)y+3-2k=0$

This line passes through the origin; therefore (0,0) must satisfy the equation, that is, $3-2k=0 \Rightarrow k=3/2$

Substituting the value of $k=3/2$, we get

$(1+3\times\frac{3}{2})x+(-2+\frac{3}{2})y+3-2\times\frac{3}{2}=0 \Rightarrow 11x-y=0$ is the required equation of the line.

Example 3: Find the equation of the line through the intersection of lines $x-2y+3=0$, $3x+y-2=0$ and whose inclination to the positive direction of the x-axis is 45°.

Solution:

The equation of any line through the point of intersection of the given lines is of the form

$x-2y+3+k(3x+y-2)=0$

$\Rightarrow (1+3k)x+(-2+k)y+3-2k=0$

$\Rightarrow y=\frac{(1+3k)}{(2-k)}x+\frac{3-2k}{(2-k)}=0$

The inclination of this line to the positive direction of the x-axis is 45°; therefore the slope

$\tan 45° = \pm 1 = \frac{(1+3k)}{(2-k)}$

$\Rightarrow 2-k=1+3k$ or $-2+k=1+3k$

$\Rightarrow k=1/4$ or $k=-3/2$

Substituting the value of $k=1/4$, we get

$(1+3\times\frac{1}{4})x+(-2+\frac{1}{4})y+3-2\times\frac{1}{4}=0$

$\Rightarrow 7x-7y+10=0$ is the required equation of the line.

Also, substituting the value of $k=-3/2$, we get

$(1+3\times\frac{-3}{2})x+(-2+\frac{-3}{2})y+3-2\times\frac{-3}{2}=0$

$\Rightarrow -7x-7y+12=0$ is the required equation of the line.

Example 4: Find the equation of the line through the intersection of lines $x-2y+3=0$, $3x+y-2=0$ and (i) parallel to line $y=x+1$

(ii) perpendicular to line $y=x+1$

Solution:

The equation of any line through the point of intersection of the given lines is of the form

$x-2y+3+k(3x+y-2)=0$

$\Rightarrow (1+3k)x+(-2+k)y+3-2k=0$

$\Rightarrow y=\frac{(1+3k)}{(2-k)}x+\frac{3-2k}{(2-k)}=0$

(i) This line is parallel to line $y=x+1$; therefore, the slopes of both lines must be equal,

$1=\frac{(1+3k)}{(2-k)} \Rightarrow 2-k=1+3k \Rightarrow k=1/4$

Substituting the value of $k=1/4$, we get

$(1+3\times\frac{1}{4})x+(-2+\frac{1}{4})y+3-2\times\frac{1}{4}=0$

$\Rightarrow 7x-7y+10=0$ is the required equation of the line.

(ii) This line is perpendicular to the line $y=x+1$; therefore, the product of the slopes must be equal to −1, that is,

$-1=\frac{(1+3k)}{(2-k)}\times 1 \Rightarrow -2+k=1+3k \Rightarrow k=-3/2$

Substituting the value of $k=-3/2$, we get

$(1+3\times\frac{-3}{2})x+(-2+\frac{-3}{2})y+3-2\times\frac{-3}{2}=0$

$\Rightarrow -7x-7y+12=0$ is the required equation of the line.

NCERT Exercise 10.4 (With Solutions) Supplement Exercise

Q1. Find the equation of the line through the intersection of the lines $3x+4y=7$, $x-y+2=0$ and whose slope is 5.

Solution:

The equation of any line through the point of intersection of the given lines is of the form

$3x+4y-7+k(x-y+2)=0$

$\Rightarrow (3+k)x+(4-k)y-7+2k=0$

$\Rightarrow (k-4)y=(3+k)x-7+2k=0$

$\Rightarrow y=\frac{(3+k)}{(k-4)}x+\frac{-7+2k}{(k-4)}=0$

The slope of the line is 5; therefore slope

$\frac{(3+k)}{(k-4)}=5 \Rightarrow 3+k=5k-20 \Rightarrow k=23/4$

Substituting the value of $k=23/4$, we get

$(3+\frac{23}{4})x+(4-\frac{23}{4})y-7+2\times\frac{23}{4}=0$

$\Rightarrow 35x-7y+18=0$ is the required equation of the line.

Q2. Find the equation of the line through the intersection of lines $x+2y-3=0$, $4x-y+7=0$ and which is parallel to $5x+4y-20=0$.

Solution:

The equation of any line through the point of intersection of the given lines is of the form

$x+2y-3+k(4x-y+7)=0$

$\Rightarrow (1+4k)x+(2-k)y-3+7k=0$

$\Rightarrow (k-2)y=(1+4k)x-3+7k=0$

$\Rightarrow y=\frac{(1+4k)}{(k-2)}x+\frac{-3+k}{(k-2)}=0$

This line is parallel to the lines $5x+4y-20=0$ $y=\frac{-5}{4}x+5=0$; therefore the slopes of the both the lines must be equal, that is, $\frac{-5}{4}=\frac{(1+4k)}{(k-2)}\Rightarrow -5k+10=4+16k$

$\Rightarrow k=6/21$

Substituting the value of $k=6/21$, we get

$(1+4\times\frac{6}{21})x+(2-\frac{6}{21})y-3+7\times\frac{6}{21}=0$

$\Rightarrow 15x+12y-7=0$ is the required equation of the line.

Q3. Find the equation of the line through the intersection of lines $2x+3y-4=0$, $x-5y=7$ that has its x-intercept equal to -4.

Solution:

The equation of any line through the point of intersection of the given lines is of the form

$2x+3y-4+k(x-5y-7)=0$

$\Rightarrow (2+k)x+(3-5k)y-4-7k=0$

This line has its x-intercept equal to -4; therefore,

$\Rightarrow (2+k)\times(-4)+(3-5k)\times 0-4-7k=0$

$\Rightarrow -8-4k-4-7k=0\Rightarrow k=-12/11$

Substituting the value of $k=-12/11$, we get

$(2+\frac{-12}{11})x+(3-5\times\frac{-12}{11})y-4-7\times\frac{-12}{11}=0$

$\Rightarrow 10x+93y+40=0$ is the required equation of the line.

Q4. Find the equation of the line through the intersection of the lines $5x-3y=1$, $2x+3y-23=0$ and perpendicular to the line $5x-3y-1=0$.

Solution:

The equation of any line through the point of intersection of the given lines is of the form

$5x-3y-1+k(2x+3y-23)=0$

$\Rightarrow (5+2k)x+(-3+3k)y-1-23k=0$

$\Rightarrow (3-3k)y=(5+2k)x-1-23k=0$

$\Rightarrow y=\frac{(5+2k)}{(3-3k)}x+\frac{-1-23k}{(3-3k)}=0$

This line is perpendicular to line $5x-3y-1=0$ $\Rightarrow y=\frac{5}{3}x-\frac{1}{3}=0$; therefore, the product of slopes must be equal to -1, that is, $-1=\frac{(5+2k)}{(3-3k)}\times\frac{5}{3}$

$\Rightarrow -9+9k=25+10k\Rightarrow k=-34$

Substituting the value of $k=-34$, we get

$(5+2\times(-34))x+(-3+3\times(-34))y-1-23\times(-34)=0$

$\Rightarrow -63x-105y+781=0$ $\Rightarrow 63x+105y-781=0$ is the required equation of the line.

Day 139

Shifting of Origin

Translation of Axes: Translation of axes is a special case of transfer of axes, in which the new system of coordinate axes is obtained by transferring the coordinate axes parallel to original axes (Fig. 10.13).

Thus, in the new system of coordinate axes, each point (including the origin) is shifted to a new point.

OX and *OY* are the coordinate axes of the given coordinate system, where the intersection point *O* gives the origin with coordinate (0,0).

Now, let the given coordinate system be transferred to a new coordinate system by lateral transfer of the *OX* axis by h units and *OY* axis by k units. Thus, we get new axes $O'X'$ and $O'Y'$, and the intersection point O' will be the origin in the new coordinate system with coordinates (h,k).

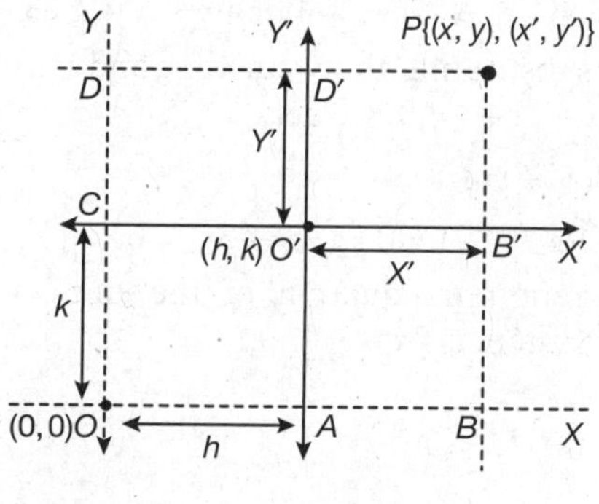

Fig. 10.13 Translation of axes

Therefore, any point $P(x, y)$ in the given coordinate system will be changed to $P(x', y')$ in the new coordinate system.

The relation between the given and the new coordinate systems will be obtained as

$\because OB = OA + AB \Rightarrow x = h + x' \Rightarrow x' = x - h$

Thus, the abscissa in the new coordinate system = abscissa in the given coordinate system – lateral transfer of the OX axis (h units).

Similarly, $OD = OC + CD \Rightarrow y = k + y' \Rightarrow y' = y - k$

Thus, the ordinate in the new coordinate system = ordinate in the given coordinate system – lateral transfer of the OY axis (k units).

Hence, any point $P(x, y)$ in the given coordinate system will be given by $P(x', y') = P(x - h, y - k)$ in the new coordinate system.

Example 1: Find the new coordinates of the point $(3, -4)$ if the origin is shifted to the point $(-2, 1)$ by a translation of axes.

Solution:
The coordinates of the new origin are $h = -2$, $k = 1$ and the original coordinates are $x = 3$, $y = -4$.

Any point $P(x, y)$ in the given coordinate system will be changed to $P(x', y')$ in the new coordinate system by the relation $x' = x - h$ and $y' = y - k$

Therefore, the point $P(3, -4)$ in the given coordinate system will be transferred to $x' = x - h = 3 - (-2) = 5$ and $y' = y - k = -4 - 1 = -5$ in the new coordinate system.

Thus, the coordinates of the point $(3, -4)$ in the new coordinate system are $(5, -5)$.

Example 2: Find what the equation of line $2x - 3y + 1 = 0$ becomes when the origin is shifted to the point $(-1, 1)$ by a translation of axes.

Solution:
The coordinates of the new origin are $h = -1$, $k = 1$ and any point $P(x, y)$ in the given coordinate system will be changed to $P(x', y')$ in the new coordinate system by the relation $x = x' + h \Rightarrow x = x' - 1$ and $y = y' + k \Rightarrow y = y' + 1$.

Therefore, substituting the value of x and y in the given equation $2x - 3y + 1 = 0$, we get

$$2(x' + 1) - 3(y' + 1) + 1 = 0$$

$$\Rightarrow 2x' + 2 - 3y' - 3 + 1 = 0 \Rightarrow 2x' - 3y' = 0$$

Thus, the required equation of the line in the new coordinate system is $2x - 3y = 0$.

NCERT Exercise 10.5 (With Solutions) Supplement Exercise

Q1. Find the new coordinates of the points in each of the following cases if the origin is shifted to the point $(-3, -2)$ by a translation of axes.

Solution:
The coordinates of the new origin are $h = -3$, $k = -2$ and any point $P(x, y)$ in the given coordinate system will be changed to $P(x', y')$ in the new coordinate system by the relation $x' = x - h$ and $y' = y - k$.

Therefore, the coordinates of the given points in new coordinate system are:

(i) Given coordinates $(1, 1) \Rightarrow (1 + 3, 1 + 2) = (4, 3)$ are the coordinates in the new coordinate system.

(ii) Given coordinates $(0, 1) \Rightarrow (0 + 3, 1 + 2) = (3, 3)$ are the coordinates in the new coordinate system.

(iii) Given coordinates $(5, 0) \Rightarrow (5 + 3, 0 + 2) = (8, 2)$ are the coordinates in the new coordinate system.

(iv) Given coordinates $(-1, -2) \Rightarrow (-1 + 3, -2 + 2) = (2, 0)$ are the coordinates in the new coordinate system.

(v) Given coordinates $(3, -5) \Rightarrow (3 + 3, -5 + 2) = (6, -3)$ are the coordinates in the new coordinate system.

Q2. Find what the following equations become when the origin is shifted to the point $(1, 1)$ by a translation of axes.

Solution:
The coordinates of the new origin are $h = 1$, $k = 1$ and any point $P(x, y)$ in the given coordinate system will be changed to $P(x', y')$ in the new coordinate system by the relation $x = x' + h \Rightarrow x = x' + 1$ and $y = y' + k \Rightarrow y = y' + 1$.

(i) Therefore, substituting the value of x and y in the given equation $x^2 + xy - 3y^2 - y + 2 = 0$, we get

$$(x' + 1)^2 + (x' + 1)(y' + 1) - 3(y' + 1)^2 - (y' + 1) + 2 = 0$$

$$\Rightarrow x'^2 + 1 + 2x' + x'y' + x' + y' + 1$$

$$- 3y'^2 - 3 - 6y' - y' - 1 + 2 = 0$$

$$\Rightarrow x'^2 - 3y'^2 + 1 + x'y' + 3x' - 6y' = 0$$

Thus, the required equation in the new coordinate system is $x^2 - 3y^2 + 1 + xy + 3x - 6y = 0$.

(ii) Therefore, substituting the value of x and y in the given equation $xy - y^2 - x + y = 0$, we get

$$(x' + 1)(y' + 1) - (y' + 1)^2 - (x' + 1) + (y' + 1) = 0$$

$$\Rightarrow x'y' + x' + y' + 1 - y'^2 - 1 - 2y' - x' - 1 + y' + 1 = 0$$

$\Rightarrow x'y' - y'^2 = 0$

Thus, the required equation in the new coordinate system is $xy - y^2 = 0$.

(iii) Therefore, substituting the value of x and y in the given equation $xy - x - y + 1 = 0$, we get $(x'+1)(y'+1) - (x'+1) - (y'+1) + 1 = 0$

$\Rightarrow x'y' + x' + y' + 1 - x' - 1 - y' - 1 + 1 = 0 \Rightarrow x'y' = 0$

Thus, the required equation in the new coordinate system is $xy = 0$.

Day 140

NCERT Miscellaneous Exercise (With Solutions Q1–Q13)

Q1. Find the values of k for which the line $(k-3)x - (4-k^2)y + k^2 - 7k + 6 = 0$ is

(a) parallel to the x-axis,
(b) parallel to the y-axis,
(c) passing through the origin.

Solution:

(a) The given equation of line $(k-3)x - (4-k^2)y + k^2 - 7k + 6 = 0$ will be parallel to the x-axis, if

Slope of the given line = Slope of the x-axis, that is, $m = 0$ for the line $y = mx + c$

Now, the given equation of line can be written as

$(4-k^2)y = (k-3)x + k^2 - 7k + 6 = 0$

$$y = \frac{(k-3)}{(4-k^2)}x + \frac{k^2 - 7k + 6}{(4-k^2)}$$

Comparing above equation with line $y = mx + c$, we get slope of the given line $m = \dfrac{(k-3)}{(4-k^2)} = 0$

(since slope of the x-axis = 0)

$$\therefore \frac{(k-3)}{(4-k^2)} = 0 \Rightarrow k - 3 = 0 \Rightarrow k = 3$$

Thus, for $k = 3$, the given line will be parallel to the x-axis.

(b) If the given line is parallel to the y-axis, it is vertical, and hence, its slope will be undefined (or mathematically we can say $m = \infty$ in the line $y = mx + c$).

Thus, the slope of the given line is $\dfrac{(k-3)}{(4-k^2)} = \infty$, that is, undefined if $4 - k^2 = 0 \Rightarrow k^2 = 4 \Rightarrow k = \pm 2$

Thus, for $k = \pm 2$, the line is parallel to the y-axis

(c) If the given line is passing through the origin, then point (0, 0) satisfies the given equation of the line. Then $(k-3)(0) - (4-k^2)(0) + k^2 - 7k + 6 = 0$

$k^2 - 7k + 6 = 0 \Rightarrow k^2 - 6k - k + 6 = 0$

$(k-6)(k-1) = 0 \Rightarrow k = 1, 6$

Thus, for $k = 1, 6$, the given line passes through the origin.

Q2. Find the values of θ, p if the equation $x\cos\theta + y\sin\theta = p$ is the normal form of the line $\sqrt{3}x + y + 2 = 0$.

Solution:

The given equation $\sqrt{3}x + y + 2 = 0$ of the line can be written as $-\sqrt{3}x - y = 2$.

Dividing both sides by $\sqrt{(-\sqrt{3})^2 + (-1)^2} = 2$, we get

$$-\frac{\sqrt{3}}{2}x - \frac{1}{2}y = \frac{2}{2} \Rightarrow \left(-\frac{\sqrt{3}}{2}\right)x + \left(-\frac{1}{2}\right)y = 1$$

Comparing above equation with the normal form $x\cos\theta + y\sin\theta = p$, we get

$$\cos\theta = -\frac{\sqrt{3}}{2}, \ \sin\theta = -\frac{1}{2}, \ p = 1$$

The values of $\cos\theta, \sin\theta$ are negative; therefore

$$\theta = \pi + \frac{\pi}{6} = \frac{7\pi}{6}$$

Thus, the values of θ and p are $\dfrac{7\pi}{6}$ and 1, respectively.

Q3. Find the equations of the lines which cut-off intercepts on the axes whose sum and product are 1 and –6, respectively.

Solution:

Let the intercepts cut by the given lines on the axes be a and b; then $a + b = 1$ and $ab = -1$.

Solving the above two equations, we get $a = 3,\ b = -2$ or $a = -2,\ b = 3$

The equation of the line whose intercept on the axes are a and b is $\dfrac{x}{a} + \dfrac{y}{b} = 1 \Rightarrow bx + ay = ab$; then substituting the values of a and b, we get

Case 1: $a = 3, b = -2$

In this case, the equation of the line is

$-2x + 3y = -6 \Rightarrow 2x - 3y - 6 = 0$

Case 2: $a = -2, b = 3$

In this case, the equations of the line is

$3x - 2y = -6 \Rightarrow 3x - 2y + 6 = 0$

Thus, the required equations of the lines are $2x-3y-6=0,\ 3x-2y+6=0$.

Q4. What are the points on the y-axis whose distance from the line $\frac{x}{3}+\frac{y}{4}=1$ is 4 units?

Solution:

Let $(0, b)$ be the point on the y-axis whose distance from the given line $\frac{x}{3}+\frac{y}{4}=1 \Rightarrow 4x+3y-12=0$ is 4 units.

Comparing equation $4x+3y-12=0$ to the general equation of the line $Ax+By+C=0$, we get $A=4, B=3, C=-12$.

The perpendicular distance (d) of a line $Ax+By+C=0$, from a point (x_1, y_1) is given by $d=\frac{|Ax_1+By_1+C|}{\sqrt{A^2+B^2}}$

And, it is given that the perpendicular distance of point $(0, b)$ from line $\frac{x}{3}+\frac{y}{4}=1$ is 4 units; therefore

$$\frac{|4(0)+3(b)-12|}{\sqrt{4^2+3^2}}=4 \Rightarrow \frac{|3b-12|}{5}=4$$

$$\Rightarrow |3b-12|=20 \Rightarrow \pm(3b-12)=20$$

$$\Rightarrow (3b-12)=20 \text{ or } -(3b-12)=20$$

$$\Rightarrow 3b=32 \text{ or } 3b=-8 \Rightarrow b=\frac{32}{3} \text{ or } b=-\frac{8}{3}$$

Thus, the required points on the y-axis are $\left(0,\frac{32}{3}\right),\left(0,-\frac{8}{3}\right)$

Q5. Find the perpendicular distance from the origin to the line joining the points $(\cos\theta, \sin\theta), (\cos\phi, \sin\phi)$

Solution:

Let the equation of the line joining the points $(\cos\theta, \sin\theta), (\cos\phi, \sin\phi)$ be given by

$$y-\sin\theta=\frac{\sin\phi-\sin\theta}{\cos\phi-\cos\theta}(x-\cos\theta)$$

$$\Rightarrow y(\cos\phi-\cos\theta)-\sin\theta(\cos\phi-\cos\theta)$$

$$=x(\sin\phi-\sin\theta)-\cos\theta(\sin\phi-\sin\theta)$$

$$\Rightarrow x(\sin\theta-\sin\phi)+y(\cos\phi-\cos\theta)$$

$$+\cos\theta\sin\phi-\cos\theta\sin\theta-\sin\theta\cos\phi+\sin\theta\cos\theta=0$$

$$x(\sin\theta-\sin\phi)+y(\cos\phi-\cos\theta)+\sin(\phi-\theta)=0$$

Comparing with the general equation of the line $Ax+By+C=0$, we get

$A=\sin\theta-\sin\phi,\ B=\cos\phi-\cos\theta,\ C=\sin(\phi-\theta)$

The perpendicular distance (d) of a line $Ax+By+C=0$ from a point (x_1, y_1) is given by

$$d=\frac{|Ax_1+By_1+C|}{\sqrt{A^2+B^2}}$$

Therefore, the perpendicular distance (d) of the given line from the origin $(0,0)$ is

$$d=\frac{|(\sin\theta-\sin\phi)(0)+(\cos\phi-\cos\theta)(0)+\sin(\phi-\theta)|}{\sqrt{(\sin\theta-\sin\phi)^2+(\cos\phi-\cos\theta)^2}}$$

$$=\frac{|\sin(\phi-\theta)|}{\sqrt{\sin^2\theta+\sin^2\phi-2\sin\theta\sin\phi+\cos^2\phi+\cos^2\theta-2\cos\phi\cos\theta}}$$

$$=\frac{|\sin(\phi-\theta)|}{\sqrt{2-2(\sin\theta\sin\phi+\cos\phi\cos\theta)}}$$

$$=\frac{|\sin(\phi-\theta)|}{\sqrt{2-2(\cos\phi-\theta)}}=\frac{|\sin(\phi-\theta)|}{\sqrt{2(1-\cos(\phi-\theta))}}$$

$$=\frac{|\sin(\phi-\theta)|}{\sqrt{2\left(2\sin^2\left(\frac{\phi-\theta}{2}\right)\right)}}=\frac{|\sin(\phi-\theta)|}{\left|2\sin\left(\frac{\phi-\theta}{2}\right)\right|}$$

$$=\left|\frac{2\sin\left(\frac{\phi-\theta}{2}\right)\cos\left(\frac{\phi-\theta}{2}\right)}{2\sin\left(\frac{\phi-\theta}{2}\right)}\right|=\left|\cos\left(\frac{\phi-\theta}{2}\right)\right|$$

Q6. Find the equation of the line parallel to the y-axis and drawn through the point of intersection of the lines $x-7y+5=0,\ 3x+y=0$.

Solution:

The equation of any line parallel to the y-axis is in the form of $x=a$, where a is a constant.

The intersection point of the two given lines $x-7y+5=0$ and $3x+y=0$ is $x=-\frac{5}{22}, y=\frac{15}{22}$.

And line $x=a$ passes through the point of intersection $\left(-\frac{5}{22},\frac{15}{22}\right)$; therefore it will satisfy the equation of line, and thus, we get $x=a \Rightarrow -\frac{5}{22}=a$.

Thus, the required equation of the line parallel to the y-axis is $x=-\frac{5}{22}$.

Q7. Find the equation of a line drawn perpendicular to the line $\frac{x}{4}+\frac{y}{6}=1$ through the point where it meets the y-axis.

Solution:

The given equation of the line is $\frac{x}{4}+\frac{y}{6}=1$.

$3x+2y-12=0 \Rightarrow y=\frac{-3}{2}x+6$ is the equation in the form $y=mx+c$.

$\therefore$ Slope of the line $= -\dfrac{3}{2}$

And slope of the line perpendicular to the given line

$= -\dfrac{1}{(-3/2)} = 2/3$

Let the given line intercept the y-axis at (0, y); then substituting $x = 0$ in the given equation of line, we get $\dfrac{y}{6} = 1 \Rightarrow y = 6.$

Thus, the given line intersects the y-axis at (0, 6).

Hence, the equation of the line that has a slope of 2/3 and passes through the point (0, 6) is $(y-6) = \dfrac{2}{3}(x-0).$

$3y - 18 = 2x \Rightarrow 2x - 3y + 18 = 0$

Q8. Find the area of the triangle made by the lines $y - x = 0,\ x + y = 0,\ x - k = 0.$

Solution:

Since the equations of the given lines are

$y - x = 0$ (i)

$x + y = 0$ (ii)

$x - k = 0$ (iii)

The point of intersection of the lines $y - x = 0$ and $x + y = 0$ is $x = 0, y = -k.$

The point of intersection of the lines $x + y = 0$ and $x - k = 0$ is $x = k, y = -k.$

And, the point of intersection of the lines $y - x = 0$ and $x - k = 0$ is $x = k, y = k.$

Therefore, the vertices of the triangle formed by the three given lines are $(0,0), (k,-k), (k,k).$

The area of the triangle whose vertices are $(x_1, y_1), (x_2, y_2), (x_3, y_3)$ is

$\dfrac{1}{2}\left|x_1(y_2 - y_1) + x_2(y_3 - y_1) + x_3(y_1 - y_2)\right|$

Therefore, the area of the triangle formed by the three given lines is

$= \dfrac{1}{2}\left|0(-k-k) + k(k-0) + k(0+k)\right|$

$= \dfrac{1}{2}\left|k^2 + k^2\right| = \dfrac{1}{2}\left|2k^2\right| = k^2$ square units

Q9. Find the value of p so that the three lines $3x + y - 2 = 0,\ px + 2y - 3 = 0,\ 2x - y - 3 = 0$ may intersect at one point.

Solution:

The point of intersection of the two given lines $3x + y - 2 = 0$ and $2x - y - 3 = 0$ is $x = 1, y = -1.$

Since all the three lines intersect at one point, $x = 1, y = -1$ must satisfy the third equation of the line $px + 2y - 3 = 0 \Rightarrow p \times 1 + 2 \times (-1) - 3 = 0$

$\Rightarrow p - 2 - 3 = 0 \Rightarrow p = 5$

Thus, for $p = 5$, all the three lines will intersect at a point.

Q10. If three lines whose equations are $y = m_1x + c_1, y = m_2x + c_2, y = m_3x + c_3$ are concurrent, then show that $m_1(c_2 - c_3) + m_2(c_3 - c_1) + m_3(c_1 - c_2) = 0.$

Solution:

Subtracting equation $y = m_1x + c_1$ from $y = m_2x + c_2$, we get $0 = (m_2 - m_1)x + (c_2 - c_1)$

$\Rightarrow (m_1 - m_2)x = c_2 - c_1 \Rightarrow x = \dfrac{c_2 - c_1}{m_1 - m_2}$

Substituting the value of x in equation $y = m_1x + c_1$, we get $y = m_1\left(\dfrac{c_2 - c_1}{m_1 - m_2}\right) + c_1 \Rightarrow y = \dfrac{m_1c_2 - m_1c_1}{m_1 - m_2} + c_1$

$y = \dfrac{m_1c_2 - m_1c_1 + m_1c_1 - m_2c_1}{m_1 - m_2} \Rightarrow y = \dfrac{m_1c_2 - m_2c_1}{m_1 - m_2}$

$\therefore \left(\dfrac{c_2 - c_1}{m_1 - m_2}, \dfrac{m_1c_2 - m_2c_1}{m_1 - m_2}\right)$ is the point of intersection of the first two given lines.

Now, if the given three lines are concurrent, then the point of intersection of the first two lines must satisfy the third point. Thus, putting $\left(\dfrac{c_2 - c_1}{m_1 - m_2}, \dfrac{m_1c_2 - m_2c_1}{m_1 - m_2}\right)$ in equation $y = m_3x + c_3$ we get

$\dfrac{m_1c_2 - m_2c_1}{m_1 - m_2} = m_3\left(\dfrac{c_2 - c_1}{m_1 - m_2}\right) + c_3$

$\Rightarrow \dfrac{m_1c_2 - m_2c_1}{m_1 - m_2} = \dfrac{m_3c_2 - m_3c_1 + c_3m_1 - c_3m_2}{m_1 - m_2}$

$\Rightarrow \dfrac{m_1c_2 - m_2c_1}{m_1 - m_2} = \dfrac{m_3c_2 - m_3c_1 + c_3m_1 - c_3m_2}{m_1 - m_2}$

$\Rightarrow m_1c_2 - m_2c_1 - m_3c_2 + m_3c_1 - c_3m_1 + c_3m_2 = 0$

$\Rightarrow m_1(c_2 - c_3) + m_2(c_3 - c_1) + m_3(c_1 - c_2) = 0$

Q11. Find the equation of the lines through the point (3, 2) which make an angle of 45° with the line $x - 2y = 3.$

Solution:

Let the slope of the required line be m_1, and writing the given line as $y = \dfrac{1}{2}x - \dfrac{3}{2}$, in the form of $y = mx + c$, we get slope of the given line $= m_2 = 1/2.$

It is given that the angle between the required line and the given line $x - 2y = 3$ is 45°.

The acute angle θ between lines l_1, l_2 with slope m_1, m_2 is given by $\tan\theta = \left|\dfrac{m_2 - m_1}{1 + m_1m_2}\right|$

Therefore, $\tan 45° = \frac{|m_2 - m_1|}{1+m_1m_2} \Rightarrow 1 = \left|\frac{\frac{1}{2}-m_1}{1+\frac{m_1}{2}}\right|$

$\Rightarrow 1 = \left|\frac{1-2m_1}{2+m_1}\right| \Rightarrow 1 = \pm\left(\frac{1-2m_1}{2+m_1}\right)$

$\Rightarrow 1 = \frac{1-2m_1}{2+m_1}$ or $1 = -\left(\frac{1-2m_1}{2+m_1}\right)$

$\Rightarrow 2+m_1 = 1-2m_1$ or $2+m_1 = -1+2m_1$

$\Rightarrow m_1 = -1/3$ or $m_1 = 3$

Case 1: $m_1 = -1/3$

The equation of the line passing through (3, 2) and having a slope of $-\frac{1}{3}$ is:

$y-2 = -\frac{1}{3}(x-3) \Rightarrow 3y-6 = -x+3 \Rightarrow x+3y = 9$ Case 2: $m_1 = 3$

The equation of the line passing through (3, 2) and having a slope of 3 is:

$y-2 = 3(x-3) \Rightarrow y-2 = 3x-9 \Rightarrow 3x-y = 7$

Thus, the equations of the lines are $3x-y=7$ and $x+3y=9$.

Q12. Find the equation of the line passing through the point of intersection of the lines $4x+7y-3=0$, $2x-3y+1=0$ that has an equal intercept on the axes.

Solution:

Solving equations $4x+7y-3=0, 2x-3y+1=0$, we get the point of intersection of the two given lines as equal to $x = \frac{1}{13}, y = \frac{5}{13}$.

Now, let the equation of the line having equal intercepts on the axes be $\frac{x}{a}+\frac{y}{a} = 1 \Rightarrow x+y = a$

And since line passes through point $\left(\frac{1}{13}, \frac{5}{13}\right)$, it will satisfy the equation of line, and hence $\frac{1}{13}+\frac{5}{13} = a \Rightarrow a = 6/13$

Thus, the required equation of the line is $x+y = \frac{6}{13} \Rightarrow 13x+13y = 6$.

Q13. Show that the equations of the line passing through the origin and making an angle θ with the line $y = mx+c$ is $\frac{y}{x} = \frac{m \pm \tan\theta}{1 \mp m\tan\theta}$.

Solution:

Let the equation of the passing through the origin be $y = m_1 x$.

Let this line makes an angle of θ with the line $y = m_2 x + c$. Then, angle θ is given by $\tan\theta = \left|\frac{m_1 - m_2}{1+m_1m_2}\right|$

$\Rightarrow \tan\theta = \left|\frac{\frac{y}{x}-m_2}{1+\frac{y}{x}m_2}\right| = \pm\left(\frac{\frac{y}{x}-m_2}{1+\frac{y}{x}m_2}\right)$

$\Rightarrow \tan\theta = \frac{\frac{y}{x}-m_2}{1+\frac{y}{x}m_2}$ or $\tan\theta = -\left(\frac{\frac{y}{x}-m_2}{1+\frac{y}{x}m_2}\right)$

Case 1: $\tan\theta = \frac{\frac{y}{x}-m_2}{1+\frac{y}{x}m_2}$

$\Rightarrow \tan\theta + \frac{y}{x}m_2 \tan\theta = \frac{y}{x} - m_2$

$\Rightarrow m_2 + \tan\theta = \frac{y}{x}(1 - m_2\tan\theta)$

$\Rightarrow \frac{y}{x} = \frac{m_2+\tan\theta}{1-m_2\tan\theta}$

Case 2: $\tan\theta = -\left(\frac{\frac{y}{x}-m_2}{1+\frac{y}{x}m_2}\right)$

$\Rightarrow \tan\theta + \frac{y}{x}m_2\tan\theta = -\frac{y}{x}+m_2$

$\Rightarrow \frac{y}{x}(1+m_2\tan\theta) = m_2 - \tan\theta$

$\Rightarrow \frac{y}{x} = \frac{m_2-\tan\theta}{1+m_2\tan\theta}$

Therefore, the required equations of lines are given by $\frac{y}{x} = \frac{m_2+\tan\theta}{1-m_2\tan\theta}$ and $\frac{y}{x} = \frac{m_2-\tan\theta}{1+m_2\tan\theta}$

Day 141

NCERT Miscellaneous Exercise (With Solutions Q14–Q24)

Q14. Show that the equation of the line joining the points $(-1,1),(5,7)$ is divided by the line $x+y=4$ in the ratio 1:2.

Solution:

The equation of the line joining the points $(-1,1),(5,7)$ is given by $y-1=\frac{7-1}{5+1}(x+1)$

$\Rightarrow x-y+2=0$

And, the equation of the given line is $x+y-4=0$.

Therefore, the point of intersection of above two lines is given by $x=1, y=3$.

Let point (1, 3) divide the line segment joining (–1, 1) and (5, 7) in the ratio 1: k; then by the section formula,

$$(1,3)=\left(\frac{k(-1)+1(5)}{1+k},\frac{k(1)+1(7)}{1+k}\right)$$

$$\Rightarrow (1,3)=\left(\frac{-k+5}{1+k},\frac{k+5}{1+k}\right)\Rightarrow \frac{-k+5}{1+k}=1, \frac{k+5}{1+k}=3$$

$$\therefore \frac{-k+5}{1+k}=1\Rightarrow -k+5=1+k\Rightarrow 2k=4\Rightarrow k=2$$

Thus, the line joining the points $(-1,1),(5,7)$ is divided by line $x+y=4$ in the ratio 1:2.

Q15. Find the distance of the line $4x+7y+5=0$ from the point (1, 2) along the line $2x-y=0$.

Solution:

Let point A (1, 2) be on the given line $2x-y=0$. Solving the two given lines $4x+7y+5=0$ and $2x-y=0$, we get the point of intersection, $x=\frac{-5}{18}, y=\frac{-5}{9}\Rightarrow B\left(\frac{-5}{18},\frac{-5}{9}\right)$.

Thus, by using the distance formula, the distance between points A and B is given by

$$AB=\sqrt{\left(1+\frac{5}{18}\right)^2+\left(2+\frac{5}{9}\right)^2}\text{ units}$$

$$=\sqrt{\left(\frac{23}{18}\right)^2+\left(\frac{23}{9}\right)^2}=\sqrt{\left(\frac{23}{9}\right)^2\left(\frac{1}{4}+1\right)}$$

$$=\frac{23}{9}\sqrt{\frac{5}{4}}=\frac{23\sqrt{5}}{18}\text{ units}$$

Q16. Find the direction in which a straight line must be drawn through the point (–1, 2) so that its point of intersection with the line $x+y=4$ may be at a distance of 3 units from this point.

Solution:

Let $y=mx+c$ be the line through point (–1, 2); then $2=m(-1)+c\Rightarrow 2=-m+c\Rightarrow c=m+2$

$\therefore y=mx+m+2$ (i)

And the given line is $x+y=4$ (ii)

Now, solving equations (i) and (ii), we get,

$$x=\frac{2-m}{m+1}, y=\frac{5m+2}{m+1}$$

$\therefore \left(\frac{2-m}{m+1},\frac{5m+2}{m+1}\right)$ is the point of intersection of the two lines.

Since this point is at a distance of 3 units from point (–1, 2), by the distance formula,

$$\sqrt{\left(\frac{5m+2}{m+1}-2\right)^2+\left(\frac{2-m}{m+1}+1\right)^2}=3$$

$$\Rightarrow \left(\frac{2-m+m+1}{m+1}\right)^2+\left(\frac{5m+2-2m-2}{m+1}\right)^2=3^2$$

$$\Rightarrow \frac{9}{(m+1)^2}+\frac{9m^2}{(m+1)^2}=9\Rightarrow \frac{1+m^2}{(m+1)^2}=1$$

$$\Rightarrow 1+m^2=m^2+1+2m\Rightarrow 2m=0\Rightarrow m=0$$

Thus, the slope of the required line must be zero. That is, the line $y=mx+m+2=2$ is parallel to the x-axis.

Q17. The hypotenuse of a right-angled triangle has its end at the points (1, 3) and (–4, 1). Find the equation of the legs (perpendicular sides).

Solution:

Let AC be the hypotenuse of a right-angled triangle (right angle at C), and the sides AC and BC be the legs of the triangle.

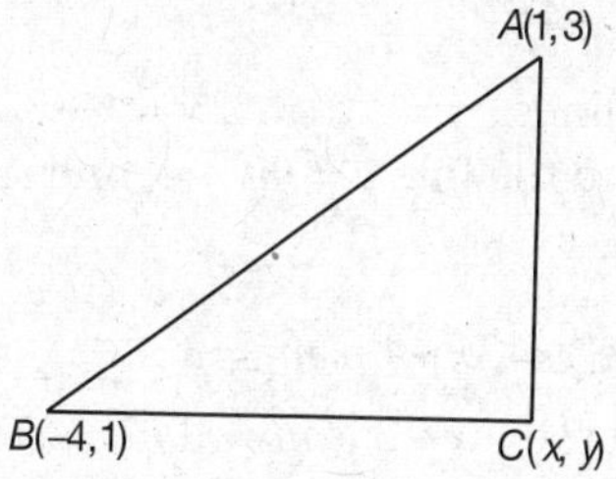

Let the slope of side AC be m; then the slope of the side BC will be $-1/m$ (since the sides are perpendicular).

Thus, the equation of the leg AC is $y-3=m(x-1) \Rightarrow y=mx-m+3$

And, the equation of the leg BC is

$y-1=\frac{-1}{m}(x+4) \Rightarrow my=-x-3$

Note: For different values of m we get different equations of the legs. There are infinite solutions that exist such that point C will lie on a circle with AB as a diameter.

In a special case, when the legs are parallel to the coordinate axes, we get $m=0$ or $m=\infty$.

Thus, for $m=0$, the equation of the legs is $y=3$ or $x=-3$ and $x=-3$.

For $m=\infty$, the equation of the legs is $x=1$ and $y=1$.

Q18. Find the image of the point (3, 8) with respect to the line $x+3y=7$ assuming the line to be a plane mirror.

Solution:

Let point B (a, b) be the image of point A (3, 8) with respect to the line $x+3y=7$.

The slope of the line $AB=\frac{b-8}{a-3}$ and the slope of the given line $y=-\frac{1}{3}x+\frac{7}{3}$ is $-\frac{1}{3}$.

Therefore, $\left(\frac{b-8}{a-3}\right)\times\left(-\frac{1}{3}\right)=-1$ (the given line must be perpendicular to the bisector line AB).

$\Rightarrow \frac{b-8}{3a-9}=1 \Rightarrow b-8=3a-9 \Rightarrow 3a-b=1$ (i)

Also, since the mid-point $\left(\frac{a+3}{2},\frac{b+8}{2}\right)$ of the line segment AB will lie on the given line $x+3y=7$; therefore

$\left(\frac{a+3}{2}\right)+3\left(\frac{b+8}{2}\right)=7.$

$\Rightarrow a+3+3b+24=14 \Rightarrow a+3b=-13$ (ii)

Solving equations (i) and (ii), we get

$a=-1, b=-4$

Thus, the image of the given point with respect to the given line is (–1, –4).

Q19. If the lines $y=3x+1$, $2y=x+3$ are equally inclined to the line $y=mx+4$, find the value of m.

Solution:

Slope of the line $y=3x+1$ is $m_1=3$

Slope of the line $2y=x+3$ is $m_2=1/2$

Slope of the line $y=mx+4$ is $m_3=m$

Since the third line is equally inclined to the other two lines, the angle between the 1st and 3rd lines must be equal to the angle between the 2nd and 3rd lines. Thus,

$$\therefore \left|\frac{m_1-m_2}{1+m_1m_2}\right|=\left|\frac{m_2-m_3}{1+m_2m_3}\right| \Rightarrow \left|\frac{3-m}{1+3m}\right|=\left|\frac{\frac{1}{2}-m}{1+\frac{1}{2}m}\right|$$

$$\Rightarrow \left|\frac{3-m}{1+3m}\right|=\left|\frac{1-2m}{m+2}\right| \Rightarrow \frac{3-m}{1+3m}=\pm\left(\frac{1-2m}{m+2}\right)$$

$$\Rightarrow \frac{3-m}{1+3m}=\frac{1-2m}{m+2} \text{ or } \frac{3-m}{1+3m}=-\left(\frac{1-2m}{m+2}\right)$$

Case 1: $\frac{3-m}{1+3m}=\frac{1-2m}{m+2}$

$\Rightarrow (3-m)(m+2)=(1-2m)(1+3m)$

$\Rightarrow -m^2+m+6=1+m-6m^2 \Rightarrow 5m^2+5=0$

$\Rightarrow (m^2+1)=0 \Rightarrow m=\sqrt{-1}$, which is not real.

Hence, this case is not possible.

Case 2: $\frac{3-m}{1+3m}=-\left(\frac{1-2m}{m+2}\right)$

$\Rightarrow (3-m)(m+2)=-(1-2m)(1+3m)$

$\Rightarrow -m^2+m+6=-(1+m-6m^2)$

$\Rightarrow 7m^2-2m-7=0$

$$\Rightarrow m=\frac{2\pm\sqrt{4-4(7)(-7)}}{2(7)}=\frac{2\pm2\sqrt{1+49}}{14}$$

$$\Rightarrow m=\frac{1\pm5\sqrt{2}}{7}$$

Thus, the required value of m is $\frac{1\pm5\sqrt{2}}{7}$.

Q20. If the sum of the perpendicular distances of a variable point P (x, y) from the lines $x+y-5=0$ and $3x-2y+7=0$ is always equal to 10, show that P must move on a line.

Solution:

We have to prove that P must move on a line. That is, we have to show that P (x, y) gives the equation of the line in the variables x and y.

Let the perpendicular distance of P (x, y) from the line $x+y-5=0$ be $d_1=\frac{|x+y-5|}{\sqrt{(1)^2+(1)^2}}$ and the perpendicular distance of P (x, y) from the line $3x-2y+7=0$ be $d_2=\frac{|3x-2y+7|}{\sqrt{(3)^2+(-2)^2}}$

Now, it is given that $d_1+d_2=10$

$$\therefore \frac{|x+y-5|}{\sqrt{2}}+\frac{|3x-2y+7|}{\sqrt{13}}=10$$

$$\Rightarrow \sqrt{13}|x+y-5|+\sqrt{2}|3x-2y+7|-10\sqrt{26}=0$$

$$\Rightarrow \sqrt{13}|X|+\sqrt{2}|Y|-10\sqrt{26}=0$$

(Assuming $x+y-5=X$ and $3x-2y+7=Y$)

$\Rightarrow \pm\sqrt{13}X \pm \sqrt{2}Y - 10\sqrt{26} = 0$ is the equation of the line, for any signs of X, Y.

Q21. Find the equation of the line which is equidistant from the parallel lines $9x+6y-7=0$, $3x+2y+6=0$.

Solution:

Let $P(x, y)$ be the arbitrary point that is equidistant from the lines $9x+6y-7=0$ and $3x+2y+6=0$.

Then the perpendicular distance of $P(x, y)$ from the line $9x+6y-7=0$ is $d_1 = \frac{|9x+6y-7|}{\sqrt{(9)^2+(6)^2}} = \frac{|9x+6y-7|}{3\sqrt{13}}$.

And, the perpendicular distance of $P(x, y)$ from the line $3x+2y+6=0$ is $d_2 = \frac{|3x+2y+6|}{\sqrt{(3)^2+(2)^2}} = \frac{|3x+2y+6|}{\sqrt{13}}$.

Since $P(x, y)$ is equidistant from both the lines,

$d_1 = d_2 \Rightarrow \frac{|9x+6y-7|}{3\sqrt{13}} = \frac{|3x+2y+6|}{\sqrt{13}}$

$\Rightarrow |9x+6y-7| = 3|3x+2y+6|$

$\Rightarrow (9x+6y-7) = \pm 3(3x+2y+6)$

$\Rightarrow 9x+6y-7 = 3(3x+2y+6)$

or $9x+6y-7 = -3(3x+2y+6)$

Case 1: $9x+6y-7 = 3(3x+2y+6)$

$\Rightarrow 9x+6y-7 = 9x+6y+18$

$\Rightarrow -7 = 18$ which is not possible.

Case 2: $9x+6y-7 = -3(3x+2y+6)$

$\Rightarrow 9x+6y-7 = -9x-6y-18$

$\Rightarrow 18x+12y+11=0$ is the required equation of the line.

Q22. A ray of light passing through the point (1, 2) reflects on the x-axis at point A and the reflected ray passes through the point (5, 3). Find the coordinates of A.

Solution:

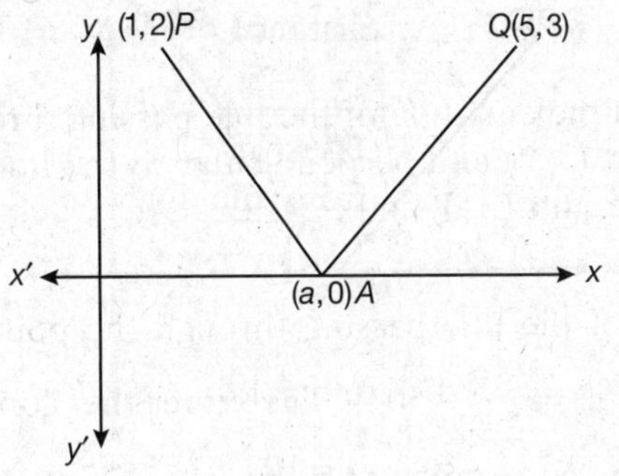

Let the coordinates of point A be $(a, 0)$. Let PA be the incident ray passing through $P(1, 2)$ and AQ be the reflected ray passing through $Q(5, 3)$.

The line is reflected at the point $A(a, 0)$ on the x-axis; therefore, the incident angle $\angle PAX$ must be equal to the reflected angle $\angle XAQ$. Thus,

$\angle PAX' = \angle QAX = \theta$

$\therefore \angle PAX = 180° - \angle QAX = 180° - \theta$

Now, the slope of the line

$AQ = \frac{3-0}{5-a} \Rightarrow \tan\theta = \frac{3}{5-a}$

Slope of the line

$PA = \frac{2-0}{1-a} \Rightarrow \tan(180° - \theta) = -\tan\theta = \frac{2}{1-a}$

Equating the above two equations, we get

$\frac{3}{5-a} = -\frac{2}{1-a} \Rightarrow 3-3a = -10+2a \Rightarrow a = \frac{13}{5}$

Thus, the coordinates of point A are $\left(\frac{13}{5}, 0\right)$.

Q23. Prove that the product of the lengths of the perpendicular drawn from the points $\left(\sqrt{a^2-b^2}, 0\right), \left(-\sqrt{a^2-b^2}, 0\right)$ to the line $\frac{x}{a}\cos\theta + \frac{y}{b}\sin\theta = 1$ is b^2.

Solution:

The equation of the given line is $\frac{x}{a}\cos\theta + \frac{y}{b}\sin\theta = 1$

$\Rightarrow bx\cos\theta + ay\sin\theta - ab = 0$

Length of the perpendicular from point $\left(\sqrt{a^2-b^2}, 0\right)$ to the line $bx\cos\theta + ay\sin\theta - ab = 0$ is

$$d_1 = \frac{\left|b\cos\theta\left(\sqrt{a^2-b^2}, 0\right) + a\sin\theta(0) - ab\right|}{\sqrt{b^2\cos^2\theta + a^2\sin^2\theta}}$$

$$\Rightarrow d_1 = \frac{\left|b\cos\theta\sqrt{a^2-b^2} - ab\right|}{\sqrt{b^2\cos^2\theta + a^2\sin^2\theta}}$$

Length of the perpendicular from point $\left(-\sqrt{a^2-b^2}, 0\right)$ to the line $bx\cos\theta + ay\sin\theta - ab = 0$ is

$$d_2 = \frac{\left|b\cos\theta\left(-\sqrt{a^2-b^2}, 0\right) + a\sin\theta(0) - ab\right|}{\sqrt{b^2\cos^2\theta + a^2\sin^2\theta}}$$

$$\Rightarrow d_2 = \frac{\left|b\cos\theta\sqrt{a^2-b^2} + ab\right|}{\sqrt{b^2\cos^2\theta + a^2\sin^2\theta}}$$

Thus, the product of the lengths of the perpendicular drawn from the points $\left(\sqrt{a^2-b^2}, 0\right), \left(-\sqrt{a^2-b^2}, 0\right)$ to the line $\frac{x}{a}\cos\theta + \frac{y}{b}\sin\theta = 1$ is

$$d_1d_2 = \frac{\left|b\cos\theta\sqrt{a^2-b^2}-ab\right|\left|\left(b\cos\theta\sqrt{a^2-b^2}+ab\right)\right|}{\left(\sqrt{b^2\cos^2\theta+a^2\sin^2\theta}\right)^2}$$

$$= \frac{\left|\left(b\cos\theta\sqrt{a^2-b^2}-ab\right)\left(b\cos\theta\sqrt{a^2-b^2}+ab\right)\right|}{b^2\cos^2\theta+a^2\sin^2\theta)}$$

$$= \frac{\left|\left(b\cos\theta\sqrt{a^2-b^2}\right)^2-(ab)^2\right|}{(b^2\cos^2\theta+a^2\sin^2\theta)}$$

$$= \frac{\left|b^2\cos^2\theta(a^2-b^2)-a^2b^2\right|}{(b^2\cos^2\theta+a^2\sin^2\theta)}$$

$$\frac{\left|a^2b^2\cos^2\theta-b^4\cos^2\theta-a^2b^2\right|}{b^2\cos^2\theta+a^2\sin^2\theta}$$

$$= \frac{b^2\left|a^2\cos^2\theta-b^2\cos^2\theta-a^2\right|}{b^2\cos^2\theta+a^2\sin^2\theta}$$

$$= \frac{b^2\left|a^2\cos^2\theta-b^2\cos^2\theta-a^2\sin^2\theta-a^2\cos^2\theta\right|}{b^2\cos^2\theta+a^2\sin^2\theta}$$

$[\sin^2\theta+\cos^2\theta=1]$

$$= \frac{b^2\left|-(b^2\cos^2\theta+a^2\sin^2\theta)\right|}{b^2\cos^2\theta+a^2\sin^2\theta}$$

$$= \frac{b^2(b^2\cos^2\theta+a^2\sin^2\theta)}{(b^2\cos^2\theta+a^2\sin^2\theta)} = b^2$$

Q24. A person standing at the junction (crossing) of two straight paths represented by the equations $2x-3y+4=0, 3x+4y-5=0$ wants to reach the path whose equation is $6x-7y+8=0$ in the least time. Find the equation of the path that he should follow.

Solution:

The equations of the given lines are

$2x-3y+4=0$ (i)

$3x+4y-5=0$ (ii)

$6x-7y+8=0$ (iii)

The person is standing at the junction of the paths represented by the lines (i) and (ii).

Thus, solving equations (i) and (ii), we get

$$x=-\frac{1}{17}, y=\frac{22}{17}$$

Thus, the person is standing at the point $\left(-\frac{1}{17},\frac{22}{17}\right)$.

The person can reach path (iii) in the least time if he walks along the perpendicular line to (iii) from point $\left(-\frac{1}{17},\frac{22}{17}\right)$

Slope of the line (iii) = 6/7

∴ Slope of the line perpendicular to line

$$\text{(iii)} = -\frac{1}{(6/7)} = -\frac{7}{6}$$

Thus, the equation of the line passing through $\left(-\frac{1}{17},\frac{22}{17}\right)$ and having a slope of $-\frac{7}{6}$ is given by

$$\left(y-\frac{1}{17}\right)=-\frac{7}{6}\left(x+\frac{1}{17}\right)\Rightarrow 6(17y-22)=-7(17x+1)$$

$102y-132=-119x-7 \Rightarrow 119x+102y=125$

Hence, the path that the person should follow is $119x+102y=125$.

Day 142

NCERT Exemplar Problems: Solutions (Q1–Q25)

Q1. Find the equation of the straight line which passes through the point (1, –2) and cuts off equal intercepts from the axes.

Solution:

Let the equal intercept along the *X* and *Y*-axes be *a*; then the equation of the line is $\frac{x}{a}+\frac{y}{a}=1 \Rightarrow x+y=a$

The point (1, –2) lies on the line; therefore

$1+(-2)=a \Rightarrow a=-1$

Thus, the required equation of the line is

$x+y=-1 \Rightarrow x+y+1=0$

Q2. Find the equation of the line passing through the point (5, 2) and perpendicular to the line joining the points (2, 3) and (3, –1).

Solution:

The slope of the line passing through the points (2, 3) and (3, –1) is $=\frac{-1-3}{3-2}=-4$. Therefore, the slope of the required line perpendicular to the line joining the points (2, 3) and (3, –1) is 1/4.

Thus, the equation of the line passing the point (5, 2) and having slope 1/4 is $y-2=\frac{1}{4}(x-5)$.

$\Rightarrow 4y-8=x-5 \Rightarrow x-4y+3=0$

Q3. Find the angle between the lines $y=(2-\sqrt{3})(x+5)$ and $y=(2+\sqrt{3})(x-7)$.

Solution:
Slope of the line $y=(2-\sqrt{3})(x+5)$ is $m_1=2-\sqrt{3}$

Slope of the line $y=(2+\sqrt{3})(x-7)$ is $m_2=2+\sqrt{3}$

Let θ be the angle between the two lines, then

$$\tan\theta=\left|\frac{m_1-m_2}{1+m_1m_2}\right|$$

$$\Rightarrow \tan\theta=\left|\frac{(2-\sqrt{3})-(2+\sqrt{3})}{1+(2-\sqrt{3})(2+\sqrt{3})}\right| \Rightarrow \tan\theta\left|\frac{-2\sqrt{3}}{1+4-3}\right|$$

$\Rightarrow \tan\theta=\sqrt{3}=\tan\pi/3 \Rightarrow \theta=\pi/3=60°$

And for obtuse angle $\theta=\pi-\pi/3=2\pi/3=120°$

Thus, the angle between the two lines is 60° or 120°.

Q4. Find the equation of the line which passes through the point (3, 4) and cuts off intercepts from the coordinates axes such that their sum is 14.

Solution:
Let the intercept along the X and Y axes be a and b, respectively. Then $a+b=14 \Rightarrow b=14-a$

Thus, the equation of the line is $\frac{x}{a}+\frac{y}{b}=1$

$$\Rightarrow \frac{3}{a}+\frac{y}{14-a}=1$$

Since the point (3, 4) lies on the line,

$$\frac{3}{a}+\frac{4}{14-a}=1 \Rightarrow \frac{42-3a+4a}{a(14-a)}=1$$

$\Rightarrow 42+a=14a-a^2 \Rightarrow a^2-13a+42=0$

$\Rightarrow (a-6)(a-7)=0 \Rightarrow a=6$ or $a=7$

Case 1: When $a=6$, then $b=14-6=8$

Thus, the equation of the line is $\frac{x}{6}+\frac{y}{8}=1$

Case 2: When $a=7$, then $b=14-7=7$

Thus, the equation of the line is

$\frac{x}{7}+\frac{y}{7}=1 \Rightarrow x+y=7$

Q5. Find the points on the line $x+y=4$ which lie at a unit distance from the line $4x+3y=10$.

Solution:
Let the required point (h, k) lie on the line $x+y=4$. Then $h+k=4 \Rightarrow k=4-h$

Thus, the distance of the point (h, k) from the line $4x+3y=10$ is $\left|\frac{4h+3k-10}{\sqrt{16+9}}\right|=1$

$\Rightarrow 4h+3k-10=\pm 5$

Case 1: $4h+3k-10=5$

$\Rightarrow 4h+3k=15 \Rightarrow 4h+3(4-h)=15$

$\Rightarrow h=15-12=3 \Rightarrow k=4-h=1$

Thus, the required point is (3, 1).

Case 2: $4h+3k-10=-5$

$\Rightarrow 4h+3k=5 \Rightarrow 4h+3(4-h)=5$

$\Rightarrow h=5-12=-7 \Rightarrow k=4-h=11$

Thus, the required point is (–7, 11)

Hence, the points on the line $x+y=4$ which lie at a unit distance from the line $4x+3y=10$ are (3, 1) and (–7, 11).

Q6. Show that the tangent of an angle between the lines $\frac{x}{a}+\frac{y}{b}=1$, $\frac{x}{a}-\frac{y}{b}=1$ is $\frac{2ab}{a^2-b^2}$.

Solution:

The slope of the line $\frac{x}{a}+\frac{y}{b}=1 \Rightarrow y=\frac{-b}{a}x+b$ is

$m_1=-b/a$

And, the slope of the line $\frac{x}{a}-\frac{y}{b}=1 \Rightarrow y=\frac{b}{a}x-b$ is

$m_2=b/a$

Let θ be the angle between the given lines then

$$\tan\theta=\left|\frac{m_1-m_2}{1+m_1m_2}\right| \Rightarrow \tan\theta=\left|\frac{-\frac{b}{a}-\frac{b}{a}}{1+\left(\frac{-b}{a}\right)\left(\frac{b}{a}\right)}\right|$$

$$\Rightarrow \tan\theta=\left|\frac{\frac{-2b}{a}}{\frac{a^2-b^2}{a^2}}\right| \Rightarrow \tan\theta=\frac{2ab}{a^2-b^2}$$

Q7. Find the equation of the line passing through (1, 2) and making an angle of 30° with the Y-axis.

Solution:
The line makes an angle of 30° with the Y-axis; therefore it will make an angle of 60° with the X-axis.

Thus, the slope of the line is $m=\tan 60°=\sqrt{3}$.

Hence, the equation of the line passing through (1, 2) and having slope $\sqrt{3}$ is $y-2=\sqrt{3}(x-1)$.

$\Rightarrow y-2=\sqrt{3}x-\sqrt{3} \Rightarrow y-\sqrt{3}x-2+\sqrt{3}=0$

Q8. Find the equation of the line passing through the point of intersection of $2x+y=5, x+3y+8=0$ and parallel to the line $3x+4y=7$.

Solution:
Equating equations of lines $2x+y=5$ and $x+3y+8=0$, we get the point of intersection as $x=\frac{23}{5}$ and $y=\frac{-21}{5}$.
The required line is parallel to the line $3x+4y=7$; therefore, slope of the line is $m=-\frac{3}{4}$.

Thus, the equation of the line passing through the point $\left(\frac{23}{5},\frac{-21}{5}\right)$ having slope $\frac{-3}{4}$ is

$y+\frac{21}{5}=\frac{-3}{4}\left(x-\frac{23}{5}\right) \Rightarrow 3x+4y+5=0.$

Q9. For what values of a and b are the intercepts cut off on the coordinate axes by the line $ax+by+8=0$ equal in length but opposite in sign to those cut off by the line $2x-3y+6=0$ on the axes?

Solution:
The given equation of the line is $ax+by+8=0$.

$\Rightarrow \frac{x}{-8/a}+\frac{y}{-8/b}=1$

Therefore, the intercepts are $\frac{-8}{a},\frac{-8}{b}$.
And for the other given equation of the line $2x-3y+6=0$

$\Rightarrow \frac{x}{-3}+\frac{y}{2}=1$, the intercepts are –3 and 2.
The intercepts of both the lines are equal in length but opposite in sign. Therefore

$\frac{-8}{a}=3, \frac{-8}{b}=-2 \Rightarrow a=-\frac{8}{3},\ b=4$

Q10. If the intercept of a line between the coordinates axes is divided by the point (–5, 4) in the ratio 1:2, then find the equation of the line.

Solution:
Let the point of interception of the line be (h, k); then the intercept coordinates on the x and y axes will be $(h, 0)$ and $(0, k)$, respectively. Then

$-5=\frac{1\times 0+2\times h}{1+2}$ and $4=\frac{1\times k+2\times 0}{1+2}$

$\Rightarrow h=-\frac{15}{2}$ and $k=12$

Therefore, the intercept coordinates are $\left(-\frac{15}{2},0\right)$ and $(0,12)$.

Thus, the equation of the passing through the points $\left(-\frac{15}{2},0\right)$ and $(0,12)$ is

$y-0=\frac{12-0}{0+15/2}\left(x+\frac{15}{2}\right) \Rightarrow y=\frac{12.2}{15}\left(x+\frac{15}{2}\right)$

$\Rightarrow 5y=8x+60 \Rightarrow 8x-5y+60=0$

Q11. Find the equation of a straight line on which length of the perpendicular from the origin is four units and the line makes an angle of 120° with the positive direction of the X-axis.

Solution:
The line makes an angle of 120° with the positive direction of the X-axis; therefore the normal from origin to the line makes an angle $\alpha=120°-90°=30°$.
Thus, the required equation of line in normal form is,

$x\cos 30°+y\sin 30°=4 \Rightarrow x\cdot\frac{\sqrt{3}}{2}+y\cdot\frac{1}{2}=4$

$\Rightarrow \sqrt{3}x+y=8$

Q12. Find the equation of one of the sides of an isosceles right-angled triangle whose hypotenuse is given by $3x+4y=4$ and the opposite vertex of the hypotenuse is (2, 2).

Solution:
Let the slope of one of the sides of an isosceles right-angled triangle be m.
The slope of the hypotenuse is $\frac{-3}{4}$ and hypotenuse makes an angle of 45° with one of the sides; therefore

$\tan 45°=\left|\frac{m+\frac{3}{4}}{1-\frac{3m}{4}}\right| \Rightarrow 1=\pm\left(\frac{m+3/4}{1-\frac{3m}{4}}\right)$

Case 1: $1-\frac{3m}{4}=+(m+3/4)$

$\Rightarrow m+\frac{3m}{4}=1-\frac{3}{4} \Rightarrow \frac{7m}{4}=\frac{1}{4} \Rightarrow m=\frac{1}{7}$

Case 2: $1-\frac{3m}{4}=-(m+3/4)$

$\Rightarrow 1-\frac{3m}{4}=-m-\frac{3}{4} \Rightarrow \frac{m}{4}=\frac{-7}{4} \Rightarrow m=-7$

Thus, the equation of one of the sides is an equation of a line passing through (2, 2) and having slope equal to 1/7, that is $y-2=\frac{1}{7}(x-2)$.

$\Rightarrow 7y-14=x-2 \Rightarrow x-7y+12=0$

Similarly, the equation of the other side is an equation of a line passing through (2, 2) and having a slope equal to –7, that is, $y-2=-7(x-2)$.

$\Rightarrow y-2=-7x+14 \Rightarrow 7x+y-16=0$

Q10. If the equation of the base of an equilateral triangle is $x+y=2$ and the vertex is (2, –1), then find the length of the side of the triangle.

Solution:

Let ΔABC be an equilateral triangle having the equation of the base BC as $x+y=2$ and the vertex A as (2, –1).

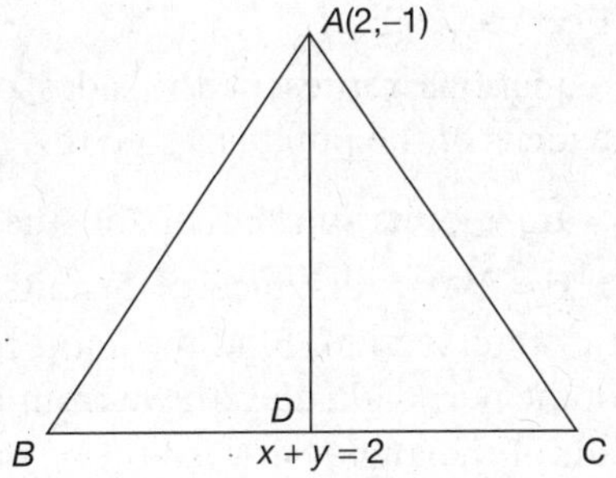

Now, the length of perpendicular from A (2, –1) to the line $x+y=2$ is given by $AD = \left|\frac{2+(-1)-2}{\sqrt{1^2+1^2}}\right| = \frac{1}{\sqrt{2}}$.

In ΔABD, $\sin 60° = \frac{AD}{AB}$, therefore, the length of the side of the equilateral triangle is

$$AB = \frac{AD}{\sin 60°} = \frac{1/\sqrt{2}}{\sqrt{3}/2} = \frac{\sqrt{2}}{\sqrt{3}} = \sqrt{\frac{2}{3}}$$

Q14. A variable line passes through a fixed point P. The algebraic sum of the perpendicular drawn from the points (2, 0), (0, 2) and (1, 1) on the line is zero. Find the coordinates of the point P.

Solution:

Let the slope of the line be m and the coordinates of fixed point P be (h,k); then the equation of the line is $y-k=m(x-h)$.

Now, the perpendicular distance of the point (2,0) from the line $y-k=m(x-h)$ is

$$d_1 = \frac{0-k-m(2-h)}{\sqrt{1+m^2}}$$

Similarly, the perpendicular distance of point (0,2) from the line $y-k=m(x-h)$ is $d_2 = \frac{2-k-m(0-h)}{\sqrt{1+m^2}}$

And, the perpendicular distance of point (1,1) from line $y-k=m(x-h)$ is $d_3 = \frac{1-k-m(1-h)}{\sqrt{1+m^2}}$

The algebraic sum of the perpendicular drawn from the points (2, 0), (0, 2) and (1, 1) on the line is zero; therefore $d_1+d_2+d_3=0$

$$\frac{-k-2m+mh+2-k+mh+1-k-m+mh}{\sqrt{1+m^2}}=0$$

$\Rightarrow -3k-3m+3mh+3=0 \Rightarrow -k-m+mh+1=0$

$\Rightarrow 1-k=m(1-h)$; this represents the family of straight lines, and thus, comparing this with the line $y-k=m(x-h)$, we get the coordinates of the fixed point P equal to (1, 1), [note that this point also lies on the line $y-1=m(x-1)$].

Q15. In what direction should a line be drawn through the point (1, 2), so that its point of intersection with the line $x+y=4$ is at a distance of $\frac{\sqrt{6}}{3}$ from the given point?

Solution:

Let the slope of the line be m and the line pass through the point A (1, 2); therefore, the equation of the line is $y-2=m(x-1) \Rightarrow mx-y+2-m=0$.

Solving this equation with the line $x+y-4=0$, we get the point of intersection

$$B(x,y) = \left(\frac{m+2}{m+1}, \frac{3m+2}{m+1}\right)$$

Now, $AB = \sqrt{\left(\frac{m+2}{m+1}-1\right)^2 + \left(\frac{3m+2}{m+1}-2\right)^2} = \frac{\sqrt{6}}{3}$

$$\Rightarrow \left(\frac{m+2-m-1}{m+1}\right)^2 + \left(\frac{3m+2-2m-2}{m+1}\right)^2 = \frac{6}{9}$$

$$\Rightarrow \left(\frac{1}{m+1}\right)^2 + \left(\frac{m}{m+1}\right)^2 = \frac{6}{9} \Rightarrow \frac{1+m^2}{(1+m)^2} = \frac{6}{9}$$

$$\Rightarrow 9+9m^2 = 6+6m^2+12m \Rightarrow m^2-4m+1=0$$

$$\therefore m = \frac{4\pm\sqrt{16-4}}{2} = 2\pm\sqrt{3} = 2+\sqrt{3} \text{ or } 2-\sqrt{3}$$

Thus, $\theta = 75°$ or $15°$

Q16. A straight line moves so that the sum of the reciprocal of its intercepts made on the axes is constant. Show that the line passes through a fixed point.

Solution:

The intercept form of a line is $\frac{x}{a}+\frac{y}{b}=1$.

Given that $\frac{1}{a}+\frac{1}{b}=$ constant $=\frac{1}{k} \Rightarrow \frac{k}{a}+\frac{k}{b}=1$, comparing this with line $\frac{x}{a}+\frac{y}{b}=1$, we see that a fixed point (k, k) lies on $\frac{x}{a}+\frac{y}{b}=1$.

Hence, the line passes through the fixed point (k, k).

Q17. Find the equation of the line which passes through the point (–4, 3) and the portion of the line intercepted between the axes is divided internally in the ratio 5:3 by this point.

Solution:

Let the intercepts on the X and Y-axes be $(x,0)$ and $(0, y)$, respectively. Then

$$-4=\frac{5\times0+3x}{5+3}\Rightarrow -4=\frac{3x}{8}\Rightarrow x=\frac{-32}{3}$$

$$\text{and } 3=\frac{5\cdot y+3\cdot 0}{5+3}\Rightarrow 3=\frac{5y}{8}\Rightarrow y=\frac{24}{5}$$

Thus, the intercept on the X and Y-axes are

$a=\frac{-32}{3},\ b=\frac{24}{5}.$

$\therefore$ Equation of required line is $\frac{x}{-32/3}+\frac{y}{24/5}=1$

$$\Rightarrow \frac{-3x}{32}+\frac{5y}{24}=1\Rightarrow 9x-20y+96=0$$

Q18. Find the equations of the lines through the point of intersection of the lines $x-y+1=0, 2x-3y+5=0$ and whose distance from the point (3, 2) is 7/5.

Solution:

Solving equations $x-y+1=0$ and $2x-3y+5=0$, we get the point of intersection equal to (2, 3).

Let the slope of the required line be m; then the equation of the line is $y-3=m(x-2)$.

$\Rightarrow mx-y+3-2m=0$

The distance of the line from the point (3, 2) is 7/5; therefore,

$$\therefore \frac{7}{5}=\left|\frac{3m-2+3-2m}{\sqrt{1+m^2}}\right|\Rightarrow \frac{49}{25}=\frac{(m+1)^2}{1+m^2}$$

$\Rightarrow 49+49m^2=25(m^2+2m+1)$

$\Rightarrow 24m^2-50m+24=0\Rightarrow 12m^2-25m+12=0$

$$\therefore m=\frac{25\pm\sqrt{625-4\cdot12\cdot12}}{24}=\frac{25\pm7}{24}=\frac{4}{3}\text{ or }\frac{3}{4}$$

Thus, for $m=\frac{4}{3}$ equation of line is $y-3=\frac{4}{3}(x-2)$

$\Rightarrow 3y-9=4x-8=0\Rightarrow 4x-3y+1=0$

And for $m=\frac{3}{4}$ equation of line is $y-2=\frac{3}{4}(x-2)$

$\Rightarrow 4y-12=3x-6\Rightarrow 3x-4y+6=0$

Q19. If the sum of the distance of a moving point in a plane from the axes is 1, then find the locus of the point.

Solution:

Let the coordinates of the moving point P be (x, y).

Given that the sum of distances of point $P(x, y)$ in a plane form of the axes is 1; therefore $|x|+|y|=1$

$\Rightarrow \pm x\pm y=1\Rightarrow x+y=1$ or $-x-y=1$

or $-x+y=1$ or $x-y=1$

These four equations represent the sides of a square; therefore the locus of the point is a square.

Q20. P_1, P_2 are points on either of the two lines $y-\sqrt{3}|x|=2$ at a distance of 5 units from their point of intersection. Find the coordinates of the foot of the perpendicular drawn from P_1, P_2 on the bisector of the angle between the given lines.

Solution:

The given equations of the lines are $y-\sqrt{3}x=2$ for $x\geq0$ and $y+\sqrt{3}x=2$ for $x\leq0$; therefore solving them we get the point of intersection A (0, 2).

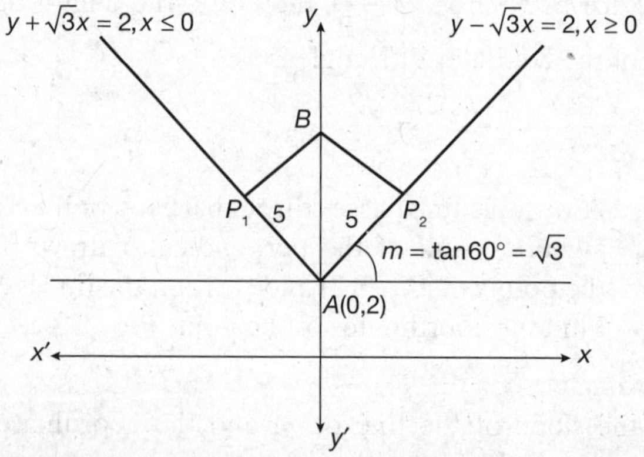

Since both the lines are symmetrical about the y-axis, the coordinates of the foot of perpendicular drawn from P_1, P_2 on the bisector of the angle between the given lines must lie on the y-axis at a common point; let it be $B(0, y)$.

Now in ΔABP_2, $\cos 30°=\frac{AP_2}{AB}$

$$\Rightarrow AB=AP_2\cos 30°=5\cdot\frac{\sqrt{3}}{2}$$

$$\therefore OB=OA+AB=2+5\frac{\sqrt{3}}{2}$$

Thus, the coordinates of the foot of the perpendicular are

$$\left(0, 2+\frac{5\sqrt{3}}{2}\right).$$

Q21. If p is the length of the perpendicular from the origin on the line $\frac{x}{a}+\frac{y}{b}=1, a^2, p^2, b^2$ are in AP, then show that $a^4+b^4=0$.

Solution:
The length of the perpendicular from the origin on the line $\frac{x}{a}+\frac{y}{b}=1 \Rightarrow bx+ay-ab=0$ is p; therefore

$$p=\left|\frac{-ab}{\sqrt{a^2+b^2}}\right| \Rightarrow p^2=\frac{a^2b^2}{a^2+b^2}$$

Given that, a^2, p^2, b^2 are in AP.

Q22. A line intercept of –3 on the Y-axis and the tangent at angle to the X-axis is 3/5, its equation is

(a) $5y-3x+15=0$ (b) $3y-5x+15=0$
(c) $5y-3x-15=0$ (d) none of the above

Solution: (a)
Let the equation of the line be $y=mx+c$
Since it has an intercept of –3 on the Y-axis,
$-3=0+c \Rightarrow c=-3$
And since the tangent at angle to the X-axis is $m=3/5$, the equation of the line is $y=\frac{3}{5}x-3$

$\Rightarrow 5y=3x-15 \Rightarrow 5y-3x+15=0$

Q23. The slope of the line which cuts off intercepts of equal lengths on the axes is

(a) –1 (b) 0 (c) 2 (d) $\sqrt{3}$

Solution: (a)
Let the intercept form of the equation of the line be $\frac{x}{a}+\frac{y}{b}=1 \Rightarrow \frac{x}{a}+\frac{y}{a}=1$ (since cuts off intercepts of equal lengths) $\Rightarrow x+y=a \Rightarrow y=-x+a$

$\therefore$ The required slope of the line is $m=-1$

Q24. The equation of the straight line passing through the point (3, 2) and perpendicular to the line $y=x$ is

(a) $x-y=5$ (b) $x+y=5$
(c) $x+y=1$ (d) $x-y=1$

Solution: (b)
The line passes through the point (3, 2) and is perpendicular to the line $y=x$; therefore the slope of the line is $m=\frac{-1}{1}=-1$ (since the line is perpendicular to the line $y=x$).

Thus, the equation of the line which passes through (3, 2) is $y-2=-1(x-3) \Rightarrow x+y=5$.

Q25. The equation of the line passing through the point (1, 2) and perpendicular to the line $x+y+1=0$ is

(a) $y-x+1=0$ (b) $y-x-1=0$
(c) $y-x+2=0$ (d) $y-x-2=0$

Solution: (b)
The slope of the given line $x+y+1=0 \Rightarrow y=-x-1$ is –1.
Therefore, the slope of the perpendicular line to the line $x+y+1=0$ is $m=\frac{-1}{-1}=1$.

Thus, the equation of the line passing through the point (1, 2) and perpendicular to the line $x+y+1=0$ is $y-2=1(x-1)$

$\Rightarrow y-2=x-1 \Rightarrow y-x-1=0$

$\therefore 2p^2=a^2+b^2 \Rightarrow \frac{2a^2b^2}{a^2+b^2}=a^2+b^2$

$\Rightarrow 2a^2b^2=(a^2+b^2)^2$

$\Rightarrow 2a^2+b^2=a^4+b^4+2a^2b^2 \Rightarrow a^4+b^4=0$

Day 143

NCERT Exemplar Problems: Solutions (Q26–Q59)

Q26. The tangent of the angle between the lines whose intercepts on the axes are $a, -b$ and $b, -a$, respectively, is

(a) $\frac{a^2-b^2}{ab}$ (b) $\frac{b^2-a^2}{ab}$
(c) $\frac{b^2-a^2}{2ab}$ (d) none of these

Solution: (c)
Equation of the line whose intercepts on the axes are $a, -b$, is $\frac{x}{a}-\frac{y}{b}=1 \Rightarrow \frac{y}{b}=\frac{x}{a}-1 \Rightarrow y=\frac{bx}{a}-b$

Thus, the slope of this line is $m_1=\frac{b}{a}$.

Equation of the line whose intercepts on the axes are $b, -a$ is $\frac{x}{b}-\frac{y}{a}=1 \Rightarrow \frac{y}{a}=\frac{x}{b}-1 \Rightarrow y=\frac{a}{b}x-a$

Thus, the slope of this line is $m_2=\frac{a}{b}$.

$$\therefore \tan\theta=\frac{\frac{b}{a}-\frac{a}{b}}{1+\frac{a}{b}\cdot\frac{b}{a}}=\frac{\frac{b^2-a^2}{ab}}{2}=\frac{b^2-a^2}{2ab}$$

Q27. If the line $\frac{x}{a}+\frac{y}{b}=1$ passes through the points (2, –3) and (4, –5), then (a, b) is

(a) (1, 1) (b) (–1, 1)
(c) (1, –1) (d) (–1, –1)

Solution: (d)

Since the points (2, –3) and (4, –5) lie on the line $\frac{x}{a}+\frac{y}{b}=1$, we get $\frac{2}{a}+\frac{3}{b}=1$ and $\frac{4}{a}+\frac{5}{b}=1$.

Solving the above two equations, we get $(a, b)=(-1,-1)$.

Q28. The distance of the point of intersection of the lines $2x-3y+5=0, 3x+4y=0$ from the line $5x-2y=0$ is

(a) $\frac{130}{17\sqrt{29}}$ (b) $\frac{13}{7\sqrt{29}}$
(c) $\frac{130}{7}$ (d) none of these

Solution: (a)

Solving the given two equations of lines $2x-3y+5=0$ and $3x+4y=0$, we get the point of intersection $\left(\frac{-20}{17}, \frac{15}{17}\right)$.

$\therefore$ Distance of point $\left(\frac{-20}{17}, \frac{15}{17}\right)$ from the line

$5x-2y=0$ is $d=\frac{\left|-5\times\frac{20}{11}-2\left(\frac{15}{17}\right)\right|}{\sqrt{25+4}}=\frac{130}{17\sqrt{29}}$

Q29. The equations of the lines which pass through the point (3, –2) and are inclined at 60° to the line $\sqrt{3}x+y=1$ is

(a) $y+2=0, \sqrt{3}x-y-2-3\sqrt{3}=0$
(b) $x-2=0, \sqrt{3}x-y+2+3\sqrt{3}=0$
(c) $x-2=0, \sqrt{3}x-y-2-3\sqrt{3}=0$
(d) None of the above

Solution: (a)

The slope of the line $\sqrt{3}x+y=1 \Rightarrow y=-\sqrt{3}x+1$ is $m_1=-\sqrt{3}$.

Let the slope of the required line be m_2. Then

$$\therefore \tan\theta=\left|\frac{-\sqrt{3}-m_2}{1-\sqrt{3}m_2}\right| \quad \left(\because \tan\theta=\left|\frac{m_1-m_2}{1+m_1m_2}\right|\right)$$

$$\Rightarrow \tan 60^\circ=\pm\left(\frac{-\sqrt{3}-m_2}{1-\sqrt{3}m_2}\right)$$

Case 1: Taking the positive sign,

$$\sqrt{3}=\left(\frac{-\sqrt{3}-m_2}{1-\sqrt{3}m_2}\right)$$

$$\Rightarrow \sqrt{3}-3m_2=-\sqrt{3}-m_2 \Rightarrow m_2=\sqrt{3}$$

Thus, the equation of the line passing through (3, –2) and having a slope of $\sqrt{3}$ is

$$y+2=\sqrt{3}(x-3) \Rightarrow \sqrt{3}x-y-2-3\sqrt{3}=0$$

Case 2: Taking the negative sign,

$$\sqrt{3}=-\left(\frac{-\sqrt{3}-m_2}{1-\sqrt{3}m_2}\right)$$

$$\Rightarrow \sqrt{3}-3m_2=\sqrt{3}+m_2 \Rightarrow m_2=0$$

Thus, the equation of the line passing through (3, –2) and having a slope of 0 is

$$y+2=0(x-3) \Rightarrow y+2=0$$

Hence, the required equations of lines are $\sqrt{3}x-y-2-3\sqrt{3}=0, y+2=0$.

Q30. The equations of the lines passing through the point (1, 0) and at a distance of $\frac{\sqrt{3}}{2}$ from the origin are

(a) $\sqrt{3}x+y-\sqrt{3}=0, \sqrt{3}x-y-\sqrt{3}=0$
(b) $\sqrt{3}x+y+\sqrt{3}=0, \sqrt{3}x-y+\sqrt{3}=0$
(c) $x+\sqrt{3}y-\sqrt{3}=0, x-\sqrt{3}y-\sqrt{3}=0$
(d) none of these

Solution: (a)

Let the slope of the line be m; then the equation of the line passing through (1, 0) is

$$y-0=m(x-1) \Rightarrow y-mx+m=0.$$

The distance of the line $y-mx+m=0$ from the origin is $\frac{\sqrt{3}}{2}$, therefore $\frac{\sqrt{3}}{2}=\left|\frac{0-0+m}{\sqrt{1+m^2}}\right|$

$\Rightarrow \frac{\sqrt{3}}{2} = \frac{m}{\sqrt{1+m^2}} \Rightarrow \frac{3}{4} = \frac{m^2}{1+m^2}$

$\Rightarrow 3+3m^2 = 4m^2 \Rightarrow m^2 = 3 \Rightarrow m = \pm\sqrt{3}$

Thus, for $m = +\sqrt{3}$ the equation of the line is

$y = \sqrt{3}(x-1) \Rightarrow \sqrt{3}x - y - \sqrt{3} = 0$

And for $m = -\sqrt{3}$ the equation of the line is

$y = -\sqrt{3}(x-1) \Rightarrow \sqrt{3}x + y - \sqrt{3} = 0$

Q31. The distance between the lines $y = mx + c_1, y = mx + c_2$ is

(a) $\frac{c_1 - c_2}{\sqrt{m^2+1}}$ (b) $\frac{|c_1 - c_2|}{\sqrt{1+m^2}}$

(c) $\frac{c_2 - c_1}{\sqrt{1+m^2}}$ (d) 0

Solution: (b)

The distance between the lines

$y = mx + c_1, y = mx + c_2$ is given by $d = \frac{|c_1 - c_2|}{\sqrt{1+m^2}}$

Q32. The coordinates of the foot of the perpendicular from the point (2, 3) on the line $y = 3x + 4$ is given by

(a) $\left(\frac{37}{10}, \frac{-1}{10}\right)$ (b) $\left(-\frac{1}{10}, \frac{37}{10}\right)$

(c) $\left(\frac{10}{37}, -10\right)$ (d) $\left(\frac{2}{3}, -\frac{1}{3}\right)$

Solution: (b)

The slope of the line $y = 3x + 4$ is $m = 3$; therefore the slope of the perpendicular line is $-\frac{1}{3}$.

Thus, the equation of the perpendicular line passing through the point (2, 3) is $y - 3 = -\frac{1}{3}(x-2)$

$\Rightarrow x + 3y - 11 = 0$

Now, solving the above two equations $y = 3x + 4$ and $y - 3 = -\frac{1}{3}(x-2) \Rightarrow x + 3y - 11 = 0$, we get the point of intersection $x = -\frac{1}{10}$ and $y = \frac{37}{10}$, and thus, the foot of perpendicular is $\left(-\frac{1}{10}, \frac{37}{10}\right)$.

Q33. If the coordinates of the middle point of the portion of a line intercepted between the coordinates axes is (3, 2), then the equation of the line will be

(a) $2x + 3y = 12$ (b) $3x + 2y = 12$

(c) $4x - 3y = 6$ (d) $5x - 2y = 10$

Solution: (a)

Let a and b be the intercepts on the X and Y axes, respectively. Since the coordinates of the middle point of intercept is (3, 2)

$3 = \frac{1\cdot 0 + 1\cdot a}{1+1} \Rightarrow 3 = \frac{a}{2} \Rightarrow a = 6$

and $2 = \frac{1\cdot b + 1\cdot 0}{1+1} \Rightarrow 2 = \frac{b}{2} \Rightarrow b = 4$

Thus, the equation of the line is $\frac{x}{6} + \frac{y}{4} = 1 \Rightarrow 2x + 3y = 12$

Q34. The equation of the line passing through (1, 2) and parallel to the line $y = 3x - 1$ is

(a) $y + 2 = x + 1$ (b) $y + 2 = 3(x+1)$

(c) $y - 2 = 3(x-1)$ (d) $y - 2 = x - 1$

Solution: (c)

The line passes through (1, 2) and is parallel to the line $y = 3x - 1$; therefore the slope of the required line 3. Thus, the equation of the line is $y - 2 = 3(x-1)$.

Q35. Equations of the diagonals of the square formed by the lines $x = 0, y = 0, x = 1, y = 1$ are

(a) $y = x, y + x = 1$ (b) $y = x, y + x = 2$

(c) $2y = x, y + x = 1/3$ (d) $y = 2x, y + 2x = 1$

Solution: (a)

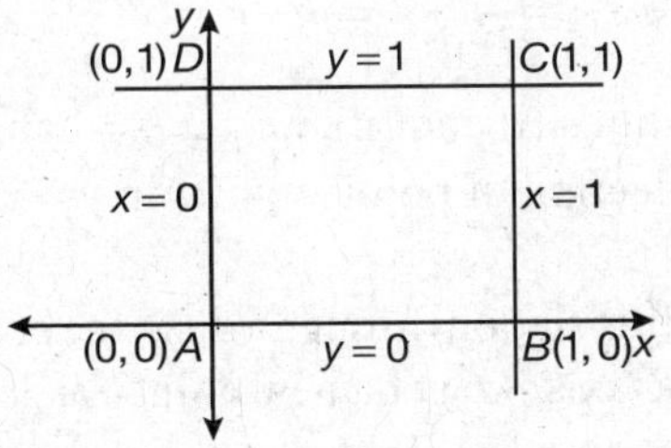

Equation of the diagonal AC is

$y - 0 = \frac{1-0}{1-0}(x-0) \Rightarrow y = x$

Equation of the line BD is $y - 0 = \frac{1-0}{0-1}(x-1)$

$\Rightarrow y = -x + 1 \Rightarrow x + y - 1 = 0$

Q36. For specifying a straight line, how many geometrical parameters should be known?

(a) 1 (b) 2 (c) 4 (d) 3

Solution: (b)

For specifying a straight line, two geometrical parameters should be known. Examples:

In equation $y = mx + c$, the unknown parameters are m and c.

In equation $\frac{x}{a} + \frac{y}{b} = 1$, the unknown parameters are a and b.

In equation $x\cos w + y\sin w = p$, the unknown parameters are w and p.

Q37. The point (4, 1) undergoes the following two successive transformations.

(i) Reflection about the line $y = x$.

(ii) Translation through a distance 2 units along the positive X-axis.

Then, the final coordinates of the point are

(a) (4, 3) (b) (3, 4)

(c) (1, 4) (d) $\left(\frac{7}{2}, \frac{7}{2}\right)$

Solution: (b)

Let the reflection of point A (4, 1) in $y = x$ be B (h, k); then the mid-point of $AB\left(\frac{4+h}{2}, \frac{1+k}{2}\right)$ must lie on $y = x$.

Therefore, $\frac{4+h}{2} = \frac{1+k}{2} \Rightarrow h - k = -3$ (i)

Also, since the slope of the line $y = x$ is 1, the slope of AB perpendicular to the line $y = x$ is –1; thus $\frac{k-1}{h-4} = -1 \Rightarrow k - 1 = -h + 4 \Rightarrow h + k = 5$ (ii)

Solving equations (i) and (ii), we get $h = 1$ and $k = 4$.

Thus, the reflection of point A (4, 1) in $y = x$ is the point B (1, 4).

Hence, after translation through a distance of 2 units along the positive X-axis, we get a new point $(1+2, 4) \Rightarrow (3, 4)$.

Q38. A point equidistant from the lines $4x + 3y + 10 = 0, 5x - 12y + 26 = 0, 7x + 24y - 50 = 0$ is

(a) (1, –1) (b) (1, 1) (c) (0, 0) (d) (0, 1)

Solution: (c)

Let the point (h, k) be equidistant from all the three given lines.

Distance of point (h, k) from the line $4x + 3y + 10 = 0$ is $= \frac{|4h + 3k + 10|}{\sqrt{16+9}}$

Distance of point (h, k) from the line $5x - 12y + 26 = 0$ is $= \frac{|5h - 12k + 26|}{\sqrt{25+144}}$

Distance of point (h, k) from the line $7x + 24y - 50 = 0$ is $= \frac{|7h + 24k - 50|}{\sqrt{7^2 + 24^2}}$

Since the point (h, k) is equidistant from all the three given lines,

$$\frac{|4h + 3k + 10|}{\sqrt{16+9}} = \frac{|5h - 12k + 26|}{\sqrt{25+144}} = \frac{|7h + 24k - 50|}{\sqrt{49+576}}$$

$$\Rightarrow \frac{|4h + 3k + 10|}{5} = \frac{|5h - 12k + 26|}{13} = \frac{|7h + 24k - 50|}{25}$$

Since $h = 0, k = 0$, satisfy the above equality, the required point is (0, 0) and the distance is $\frac{10}{5} = \frac{26}{13} = \frac{50}{25} = 2$ units

Q39. A line passes through (2, 2) and is perpendicular to the line $3x + 4y = 3$. Its y-intercept is

(a) 1/3 (b) 2/3 (c) 1 (d) 4/3

Solution: (d)

Since the slope of the given line $y = -3x + 3$ is –3, the slope of the required line $= \frac{1}{3}$.

Thus, the equation of the required line passing through (2, 2) is $y - 2 = \frac{1}{3}(x - 2) \Rightarrow x - 3y + 4 = 0$

Hence, substituting $x = 0$ we get the y-intercept as $0 - 3y + 4 = 0 \Rightarrow y = 4/3$.

Q40. The ratio in which the line $3x + 4y + 2 = 0$ divides the distance between the lines $3x + 4y + 5 = 0, 3x + 4y - 5 = 0$ is

(a) 1:2 (b) 3:7 (c) 2:3 (d) 2:5

Solution: (b)

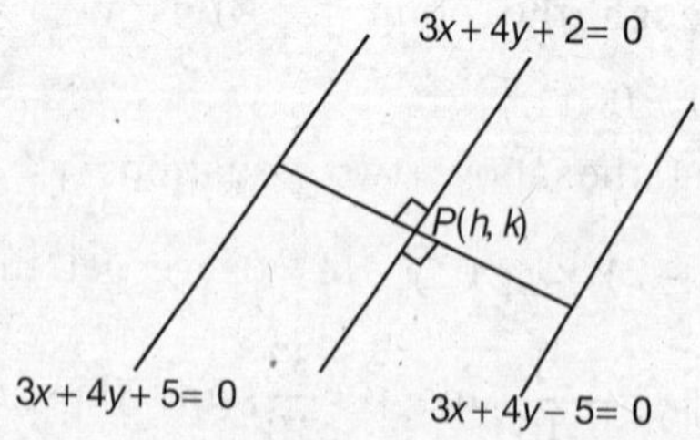

Let point $P(h,k)$ lie on the line $3x + 4y + 2 = 0$. Then, $3h + 4k + 2 = 0 \Rightarrow 3h + 4k = -2$

Now, the perpendicular distance of the point $P(h,k)$ from the line $3x + 4y + 5 = 0$ is

$$\frac{|3h+4k+5|}{\sqrt{9+16}}=\frac{|-2+5|}{\sqrt{9+16}}=\frac{3}{5}$$

Similarly, the perpendicular distance of the point $P(h,k)$ from the line $3x+4y-5=0$ is

$$\frac{|3h+4k-5|}{\sqrt{9+16}}=\frac{|-2-5|}{\sqrt{9+16}}=\frac{7}{5}$$

Thus, the required ratio is $\frac{3}{5}:\frac{7}{5}=3:7$.

Q41. One vertex of the equilateral triangle with the centroid at the origin and one side as $x+y-2=0$ is

(a) $(-1,-1)$ (b) $(2,2)$ (c) $(-2,-2)$ (d) $(2,-2)$

Solution: (c)

Let ABC be the triangle with the centroid $O(0,0)$ and the coordinates of the vertex $A(h,k)$ and the coordinates of perpendicular from A to the line BC $x+y-2=0$ be $G(a,b)$.

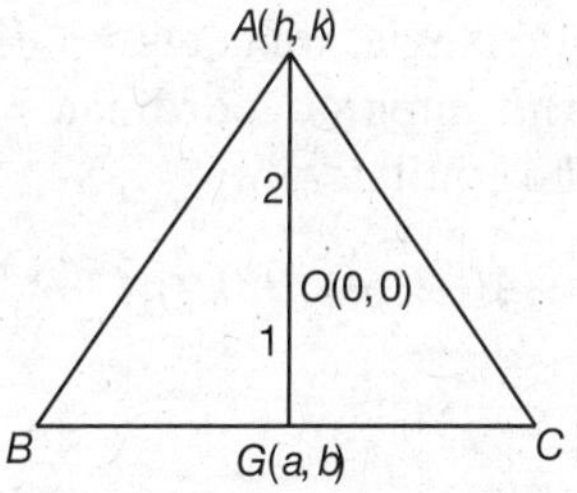

Since the centroid divides the perpendicular in the ratio of 2:1,

$0=\frac{2a+h}{3}$ and $0=\frac{2b+k}{3}$

$\Rightarrow 2a=-h$ and $2b=-k$ (i)

Also, the point $G(a,b)$ lies on the line $x+y-2=0$

$\therefore a+b-2=0$ (ii)

Since the slope of the line BC is $m_{BC}=-1$, the slope of the line AG is

$m_{AG}=\frac{k-0}{h-0}=\frac{k}{h}=1\Rightarrow k=h$ (iii) $(\because AG\perp BC)$

Thus, solving equations (i) and (iii), we get $a=b$

And, substituting in Eq. (ii), we get $2a-2=0\Rightarrow a=1\Rightarrow b=1\Rightarrow h=k=-2$

Hence, the vertex A is $(-2,-2)$.

Q42. If a, b, c are in AP, then the straight lines $ax+by+c=0$ will always pass through ... ...

Solution:

Since a, b, c are in AP, $b=\frac{a+c}{2}$

$\Rightarrow a-2b+c=0$

Comparing this with equation $ax+by+c=0$, we get $x=1, y=2$

Thus, point (1, –2) lies on the line $ax+by+c=0$.

Q43. The line which cuts off equal intercepts from the axes and pass through the point (1, –2) is ...

Solution:

Let the equation of the line which cuts off equal intercepts from the axes be $\frac{x}{a}+\frac{y}{a}=1\Rightarrow x+y=a$

Since this line passes through (1, –2),

$1+(-2)=a\Rightarrow a=-1$

Thus, the required equation of the line is $\Rightarrow x+y=-1\Rightarrow x+y+1=0$

Q44. Equation of the line through the point (3, 2) and making an angle of 45° with the line $x-2y=3$ is ... ...

Solution:

Let the slope of the required line be m. The line makes an angle of 45° with the line $x-2y=3\Rightarrow\frac{1}{2}x-\frac{3}{2}$.

$$\text{Therefore, } \tan 45^\circ=\left|\frac{m-\frac{1}{2}}{1+\frac{1}{2}m}\right|\Rightarrow 1=\pm\left(\frac{m-\frac{1}{2}}{1+\frac{m}{2}}\right)$$

Case 1: Taking the positive sign, we get $1=+\left(\frac{m-\frac{1}{2}}{1+\frac{m}{2}}\right)$

$$\Rightarrow 1+\frac{m}{2}=m-\frac{1}{2}\Rightarrow\frac{m}{2}=\frac{3}{2}\Rightarrow m=3$$

Thus, the equation of the line passing through the point (3, 2) is $x-2=3(x-3)\Rightarrow 3x-y-7=0$.

Case 2: Taking the negative sign, we get $1=-\left(\frac{m-\frac{1}{2}}{1+\frac{m}{2}}\right)$

$$\Rightarrow 1+\frac{m}{2}=-m+\frac{1}{2}\Rightarrow\frac{3m}{2}=\frac{-1}{2}\Rightarrow m=\frac{-1}{3}$$

Thus, the equation of the line passing through the point (3, 2) is $y-2=-\frac{1}{3}(x-3)\Rightarrow x+3y-9=0$.

Q45. The points (3, 4) and (2, –6) are situated on the ... ... of the line $3x-4y-8=0$.

Solution:

For point (3, 4), $3\times3-4\times4-8 \Rightarrow -15<0$

And for point (2, −6), $3\times2-4\times(-6)-8=22>0$

Since the values for points (3, 4) and (2, −6) are of opposite signs, the points lie on the opposite side to the line $3x-4y-8=0$.

Q46. A point moves so that the square of its distance from the point (3, −2) is numerically equal to its distance from the line $5x-12y=3$. The equation of its locus is ……….

Solution:

Let the coordinates of the point be (*h*, *k*); then the square of distance between the points (3,−2) and (h,k) is $d_1^2=(-2-k)^2+(3-h)^2$.

Now, the distance of the point (*h*, *k*) from the line $5x-12y=3$, is $d_2=\left|\dfrac{5h-12k-3}{\sqrt{25+144}}\right|=\left|\dfrac{5h-12k-3}{13}\right|$

Since $d_1^2=d_2$,

Therefore, $(3-h)^2+(2+k)^2=\dfrac{5h-12k-3}{13}$

$\Rightarrow 9-6h+h^2+4+4k+k^2=\dfrac{5h-12k-3}{13}$

$\Rightarrow h^2+k^2-6h+4k+13=\dfrac{5h-12k-3}{13}$

$\Rightarrow 13h^2+13k^2-78h+52k+169=5h-12k-3$

$\Rightarrow 13h^2+13k^2-83h+64k+172=0$

Thus, the locus of the point is

$13x^2+13y^2-83x+64y+172=0$

Q47. Locus of the midpoints of the portion of the line $x\sin\theta+y\cos\theta=p$ intercepted between the axes is ……….

Solution:

Let $p(h,k)$ be the midpoint of the line $x\sin\theta+y\cos\theta=p$ intercepted between the axes.

And writing equation $\dfrac{x}{p/\sin\theta}+\dfrac{y}{p/\cos\theta}=1$ in the normal form $\dfrac{x}{a}+\dfrac{y}{b}=1$, we get intersects on the axes $a=\dfrac{p}{\sin\theta}$ and $b=\dfrac{p}{\cos\theta}$.

Now, since the midpoint of the intercept is $\left(\dfrac{a}{2},\dfrac{b}{2}\right)$, we get midpoint

$h=\dfrac{a}{2}=\dfrac{p}{2\sin\theta}$ and $k=\dfrac{b}{2}=\dfrac{p}{2\cos\theta}$

$\Rightarrow \sin\theta=\dfrac{p}{2h}$ and $\cos\theta=\dfrac{p}{2k}$

Since, $\sin^2\theta+\cos^2\theta=1$

$\dfrac{p^2}{4h^2}+\dfrac{p^2}{4k^2}=1\Rightarrow\dfrac{1}{h^2}+\dfrac{1}{k^2}=\dfrac{4}{p^2}$, thus the locus of the midpoint is $\dfrac{1}{x^2}+\dfrac{1}{y^2}=\dfrac{4}{p^2}$

$\Rightarrow 4x^2y^2=p^2(x^2+y^2)$

True/False

Q48. If the vertices of a triangle have integral coordinates, then the triangle cannot be equilateral.

Solution: True

In an equilateral triangle, we get $60°=\sqrt{3}=$ slope of the line; therefore with integral coordinates as vertices, the triangle cannot be equilateral.

Q49. The points $A(-2,-1), B(0,5), C(-1,2)$ are collinear.

Solution: False

Slope of $AB=\dfrac{5-1}{0+2}=2$

Slope of $BC=\dfrac{2-5}{-1-0}=3$

The slopes of line segments *AB* and *BC* are different; therefore, the points *A*, *B* and C are not collinear.

Q50. Equation of the line passing through the point $(a\cos^3\theta, a\sin^3\theta)$ and the perpendicular to the line $x\sec\theta+y\operatorname{cosec}\theta=a$ is $x\cos\theta-y\sin\theta=a\sin 2\theta$.

Solution: False

The slope of the line $x\sec\theta+y\operatorname{cosec}\theta=a$

$\Rightarrow y=-x\dfrac{\sec\theta}{\operatorname{cosec}\theta}+a\Rightarrow y=-x\tan\theta+a$ is $-\tan\theta$

Therefore, the slope of the required perpendicular line is

$=-\dfrac{1}{(-\tan\theta)}=\cot\theta$

Thus, the equation of the line passing through the point $(a\cos^3\theta, a\sin^3\theta)$ and having a slope of $\cot\theta$ is

$y-a\sin^3\theta=\cot\theta(x-a\cos^3\theta)$

$\Rightarrow y\sin\theta-a\sin^4\theta=x\cos\theta-a\cos^4\theta$

$\Rightarrow x\cos\theta-y\sin\theta=a\cos^4\theta-a\sin^4\theta$

$\Rightarrow x\cos\theta - y\sin\theta$

$= a[(\cos^2\theta + \sin^2\theta)(\cos^2\theta - \sin^2\theta)]$

$\Rightarrow x\cos\theta - y\sin\theta = a\cos 2\theta$

Q51. The straight line $5x+4y=0$ passes through the point of intersection of the straight lines $x+2y-10=0, 2x+y+5=0$.

Solution: True

Solving equations $x+2y-10=0$ and $2x+y+5=0$, we get $x=-\frac{20}{3}$ and $y=\frac{25}{3}$; therefore, the point of intersection is $\left(-\frac{20}{3}, \frac{25}{3}\right)$.

If the line $5x+4y=0$ passes through the point $\left(-\frac{20}{3}, \frac{25}{3}\right)$, then this point should lie on this line.

Now, $5\left(\frac{-20}{3}\right)+\frac{4(25)}{3}=\frac{-100}{3}+\frac{100}{3}=0$

Therefore, this point lies on the line $5x+4y=0$.

Q52. The vertex of an equilateral triangle is (2, 3) and the equation of the opposite side is $x+y=2$. Then, the other two sides are $y-3=(2\pm\sqrt{3})(x-2)$.

Solution: True

Let ABC be an equilateral triangle with vertex A (2, 3), and the equation of opposite side BC be $x+y=2 \Rightarrow y=-x+2$; thus the slope of side BC is –1.

Let the slope of the two lines be m; then the angle between these lines and BC is 60°.

$\therefore \tan 60° = \left|\frac{m+1}{1-m}\right| \Rightarrow \sqrt{3} = \pm\left(\frac{m+1}{1-m}\right)$

Case 1: Taking the positive sign,

$\sqrt{3} = +\left(\frac{m+1}{1-m}\right) \Rightarrow \sqrt{3}-\sqrt{3}m = m+1$

$\Rightarrow m+\sqrt{3}m = \sqrt{3}-1 \Rightarrow m = \frac{\sqrt{3}-1}{\sqrt{3}+1}$

$\Rightarrow m = \frac{(\sqrt{3}-1)(\sqrt{3}-1)}{(\sqrt{3}+1)(\sqrt{3}-1)} = \frac{3+1-2\sqrt{3}}{3-1} = 2-\sqrt{3}$

Case 2: Taking the negative sign,

$\sqrt{3} = -\left(\frac{m+1}{1-m}\right) \Rightarrow \sqrt{3}-\sqrt{3}m = -m-1$

$\Rightarrow \sqrt{3}m - m = \sqrt{3}+1 \Rightarrow m = \frac{\sqrt{3}+1}{\sqrt{3}-1}$

$\Rightarrow m = \frac{(\sqrt{3}+1)(\sqrt{3}+1)}{(\sqrt{3}-1)(\sqrt{3}+1)} = \frac{3+1+2\sqrt{3}}{3-1} = 2+\sqrt{3}$

Thus, the equation of the other two sides is $y-3=(2\pm\sqrt{3})(x-2)$.

Q53. The equation of the line joining the point (3, 5) to the point of intersection of the lines $4x+y-1=0$, $7x-3y-35=0$ is equidistant from the points (0, 0) and (8, 34).

Solution: True

Solving the equations $4x+y-1=0$ and $7x-3y-35=0$, we get $x=2$ and $y=-7$; thus the point of intersection is (2, –7).

Now, the equation of the line passing through the points (3,5), (2,–7) is $y-5=\frac{-7-5}{2-3}(x-3)$.

$\Rightarrow y-5=12(x-3) \Rightarrow 12x-y-31=0$

The distance of the point (0, 0) from the line $12x-y-31=0$ is $d_1 = \frac{|0-0-31|}{\sqrt{144+1}} = \frac{31}{\sqrt{145}}$.

The distance of the point (8, 34) from the line $12x-y-31=0$ is $d_2 = \frac{|96-34-31|}{\sqrt{145}} = \frac{31}{\sqrt{145}}$.

Since $d_1 = d_2 = \frac{31}{\sqrt{145}}$, the statement is true.

Q54. The line $\frac{x}{a}+\frac{y}{b}=1$ moves in such a way that $\frac{1}{a^2}+\frac{1}{b^2}=\frac{1}{c^2}$, where c is a constant. The locus of the foot of the perpendicular from the origin on the given line is $x^2+y^2=c^2$.

Solution: True

The equation of the line passing through the origin and perpendicular to the line $\frac{x}{a}+\frac{y}{b}=1$ is $\frac{x}{b}-\frac{y}{a}=0$. Now, eliminating the variables a and b from the equations $\frac{x}{a}+\frac{y}{b}=1$ and $\frac{x}{b}-\frac{y}{a}=0$, we get the loci. Thus, squaring and adding above two equations, we get,

$\frac{x^2}{a^2}+\frac{y^2}{b^2}+\frac{2xy}{ab}+\frac{x^2}{b^2}+\frac{y^2}{a^2}-\frac{2xy}{ab}=1$

$\Rightarrow x^2\left(\frac{1}{a^2}+\frac{1}{b^2}\right)+y^2\left(\frac{1}{a^2}+\frac{1}{b^2}\right)=1 \Rightarrow (x^2+y^2)\left(\frac{1}{a^2}+\frac{1}{b^2}\right)=1$

$\Rightarrow (x^2+y^2)\frac{1}{c^2}=1 \Rightarrow x^2+y^2=c^2$

$\left[\because \frac{1}{a^2}+\frac{1}{b^2}=\frac{1}{c^2}\right]$

Q55. The lines $ax+2y+1=0, bx+3y+1=0$ and $cx+4y+1=0$ are concurrent, if a, b and c are in GP.

Solution: False

Solving equations $ax+2y+1=0$ and $ax+2y+1=0$, we get $x=\dfrac{1}{2b-3a}$ and $y=\dfrac{(a-b)}{2b-3a}$, and thus, the point of intersection is $\left(\dfrac{1}{2b-3a}, \dfrac{a-b}{2b-3a}\right)$.

This point lies on $cx+4y+1=0$; therefore

$$\frac{c}{2b-3a}+\frac{4(a-b)}{2b-3a}+1=0$$

$\Rightarrow c+4a-4b+2b-3a=0 \Rightarrow -2b+a+c=0 \Rightarrow 2b=a+c$, hence a, b and c are in AP not in GP.

Q56. The line joining the points $(3,-4),(-2,6)$ is perpendicular to the line joining the points $(-3,6),(9,-18)$.

Solution: False

The slope of line joining the points $A(3,-4), B(-2,6)$ is

$$AB=\frac{6+4}{-2-3}=-2.$$

And, the slope of line joining the points $P(-3,6), Q(9,-18)$ is $PQ=\dfrac{-18-6}{9+3}=-2.$

Thus, the line AB is parallel to line PQ, not perpendicular.

Q57. Match the following

Column I

(i) The coordinates of the points P and Q on the line $x+5y=13$ which are at a distance of 2 units from the line $12x-5y+26=0$ are

(ii) The coordinates of the point on the line $x+y=4$, which are at a unit distance from the line $4x+3y-10=0$ are

(iii) The coordinates of the point on the line joining $A(-2,5), B(3,1)$ such that point P divides the line joining the points A and B in the ratio 1:2.

Column II

(a) $(3,1),(-7,11)$

(b) $\left(-\dfrac{1}{3},\dfrac{11}{3}\right),\left(\dfrac{4}{3},\dfrac{7}{3}\right)$

(c) $\left(1,\dfrac{12}{5}\right),\left(-3,\dfrac{16}{5}\right)$

Solution:

(i)–(c) Let the coordinates of the point $P(x_1, y_1)$ on the line $x+5y=13$ be $P(13-5y_1, y_1)$.

$\therefore$ The distance of P from the line $12x-5y+26=0$ is

$$=2=\left|\frac{12(13-5y_1)-5y_1+26}{\sqrt{144+25}}\right|$$

$$\Rightarrow 2=\left|\frac{156-60y_1-5y_1+26}{13}\right|$$

$$\Rightarrow 2=|14-5y_1| \Rightarrow 2=\pm(14-5y_1)$$

Case 1: Taking the positive sign,

$$2=+(14-5y_1)\Rightarrow y_1=\frac{12}{5}$$

$$\Rightarrow x_1=13-5y_1=13-12=1$$

So, the coordinates of the point P are $P\left(1,\dfrac{12}{5}\right)$.

Case 2: Taking the negative sign,

$$2=-(14-5y_1)\Rightarrow y_1=\frac{16}{5}$$

$$\Rightarrow x_1=13-5y_1=13-16=-3$$

So, the coordinates of the point Q are $\left(-3,\dfrac{16}{5}\right)$.

(ii)–(a) Let the coordinates of the point on the line $x+y=4$ be $(4-y_1 \cdot y_1)$.

The distance of the point $(4-y_1 \cdot y_1)$ from the line $4x+3y-10=0$ is

$$1=\left|\frac{16-4y_1+3y_1-10}{5}\right| \Rightarrow 1=\left|\frac{6-y_1}{5}\right|$$

$$\Rightarrow 1=\pm\left(\frac{6-y_1}{5}\right)$$

Case 1: Taking the positive sign,

$$\Rightarrow 1=+\left(\frac{6-y_1}{5}\right)\Rightarrow 5=6-y_1 \Rightarrow y_1=1$$

$$x_1=4-y_1=4-1=3$$

Thus, the point is (3, 1).

Case 2: Taking the negative sign,

$$\Rightarrow 1=-\left(\frac{6-y_1}{5}\right)\Rightarrow 5=-6+y_1 \Rightarrow y_1=11$$

$$x_1=4-y_1=4-11=-7$$

Thus, the point is (–7, 11).

(iii)–(b) The given point is $A(-2,5), B(3,1)$.

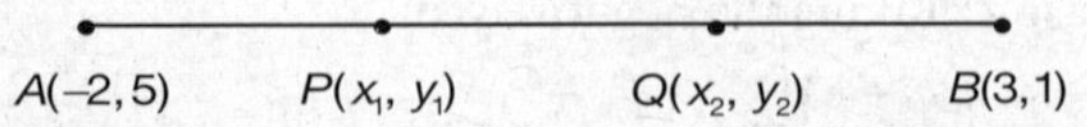

The point P divides the line joining the points A and B in ratio 1:2.

$$\therefore x_1=\frac{1\times3+2\times(-2)}{1+2}=\frac{-1}{3},\ y_1=\frac{1\times1+2\times5}{1+2}=\frac{11}{3}$$

Thus, the coordinates of point P are $\left(\dfrac{-1}{3},\dfrac{11}{3}\right)$.

Now, the point Q divides the line joining the points A and B in the ratio 2:1.

$$\therefore x_2 = \frac{2\times 3+1\times(-2)}{2+1} = \frac{4}{3},\ y_2 = \frac{2\times 1+1\times 5}{2+1} = \frac{7}{3}$$

Thus, the coordinates of point Q are $\left(\frac{4}{3}, \frac{7}{3}\right)$.

Q58. The value of the λ, if the lines $(2+3y+4)+\lambda(6x-y+12)=0$ are

Column I

(i) parallel to the Y-axis is

(ii) perpendicular to $7x+y-4=0$ is

(iii) passes through (1, 2) is

(iv) parallel to the X-axis is

Column II

(a) $\lambda = -\frac{3}{4}$ (b) $\lambda = -\frac{1}{3}$

(c) $\lambda = -\frac{17}{41}$ (d) $\lambda = 3$

Solution:

Since the given equation of the line is $(2+3y+4)+\lambda(6x-y+12)=0$

$\Rightarrow y(3-\lambda) = -x(2+6\lambda)-4-12\lambda$

$\Rightarrow y = -x\frac{(2+6\lambda)}{(3-\lambda)} - \frac{(4+12\lambda)}{(3-\lambda)}$

Therefore, the slope of the equation is $m = \frac{-(2+6\lambda)}{3-\lambda}$.

(i)—(d)

If the line $(2+3y+4)+\lambda(6x-y+12)=0$ is parallel to the Y-axis, that is, perpendicular to the X-axis, then the slope of the line, $m = \tan 90° = \infty$.

Thus, $\frac{-(2+6\lambda)}{3-\lambda} = \frac{1}{0} \Rightarrow \lambda = 3$

(ii)—(c) If the line $(2+3y+4)+\lambda(6x-y+12)=0$ is perpendicular to the line $7x+y-4=0 \Rightarrow y=-7x+4$

Then $\frac{-(2+6\lambda)}{(3-\lambda)}(-7) = -1 \Rightarrow 14+42\lambda = -3+\lambda$

$\Rightarrow 41\lambda = -17 \Rightarrow \lambda = -17/41$

(iii)—(a) If the line $(2+3y+4)+\lambda(6x-y+12)=0$ passes through the point (1, 2), then $(2+6+4)+\lambda(6-2+16)=0$

$\Rightarrow 12+16\lambda = 0 \Rightarrow \lambda = -3/4$

(iv)—(d) If the line $(2+3y+4)+\lambda(6x-y+12)=0$ is parallel to the X-axis, then slope $m = 0$.

$\Rightarrow \frac{-(2+6\lambda)}{3-\lambda} = 0 \Rightarrow -(2+6\lambda) = 0 \Rightarrow \lambda = -\frac{1}{3}$

Q59. The equation of the line through the intersection of the lines $2x-3y=0, 4x-5y=2$ and

Column I

(i) through the point (2, 1) is

(ii) perpendicular to the line $x+2y+1=0$

(iii) parallel to the line $3x-4y+5=0$ is

(iv) equally inclined to the axes is

Column II

(a) $2x-y=4$ (b) $x+y-5=0$

(c) $x-y-1=0$ (d) $3x-4y-1=0$

Solution:

Solving equations $2x-3y=0$ and $4x-5y=2$, we get $x=3$ and $y=2$, thus the point of intersection is (3, 2).

(i)—(c) The equation of the line passes through the point (3, 2) and (2, 1) is $y-2 = \frac{1-2}{2-3}(x-3)$

$\Rightarrow y-2 = (x-3) \Rightarrow x-y-1=0$

(ii)—(a) Since the required line is perpendicular to the line $x+2y+1=0 \Rightarrow y = -\frac{1}{2}x - \frac{1}{2}$, the slope of the required line is 2. Thus, the equation of the line passing through the point (3, 2) with a slope = 2 is $y-2 = 2(x-3) \Rightarrow 2x-y-4=0$

(iii)—(d) Since the required line is parallel to the line $3x-4y+5=0 \Rightarrow y = \frac{3}{4}x + \frac{5}{4}$, the slope of the line is $\frac{3}{4}$. Thus, the equation of the line passing through the point (3, 2) with a slope $= \frac{3}{4}$ is

$y-2 = \frac{3}{4}(x-3) \Rightarrow 4y-8 = 3x-9 \Rightarrow 3x-4y-1=0$

(iv)—(b,c) Since the line is equally inclined to the axes, the slope of the line is $m = \pm\tan 45° = \pm 1$.

Case 1: Taking the positive sign, $m=+1$

Equation of the line is $y-2 = 1(x-3)$

$\Rightarrow x-y-1=0$

Case 2: Taking the negative sign, $m=-1$

Equation of the line is $y-2 = -1(x-3)$

$\Rightarrow y-2 = -x+3 \Rightarrow x+y-5=0$

Day 144

Very Short Answer (VSA): 1-Mark and Short Answer (SA): 2-Mark Questions

Q1. Do the points $(-3,0),(3,0),(0,4)$ form a triangle? If yes, name the type of triangle formed.

Solution:
The given points are $P(-3,0)$, $Q(3,0)$, $R(0,4)$; then by the distance formula

$PQ=\sqrt{(3+3)^2+(0-0)^2}=\sqrt{9^2}=9$

$QR=\sqrt{(0-3)^2+(4-0)^2}=\sqrt{9+16}=\sqrt{25}=5$

$PR=\sqrt{(0+3)^2+(4-0)^2}=\sqrt{9+16}=\sqrt{25}=5$ Since two sides $PR=QR$, the triangle PQR is an isosceles triangle.

Q2. Show that the points A (3, 1), B (6, 4) and C (2, 0) are collinear (form a straight line).

Solution:
Using the distance formula, we get

$AB=\sqrt{(6-3)^2+(4-1)^2}=\sqrt{9+9}=\sqrt{18}=3\sqrt{2}$

$BC=\sqrt{(2-6)^2+(0-4)^2}=\sqrt{16+16}=4\sqrt{2}$

$AC=\sqrt{(2-3)^2+(0-1)^2}=\sqrt{1+1}=\sqrt{2}$

Now, since, $AB+AC=3\sqrt{2}+\sqrt{2}=4\sqrt{2}=BC$, we can say that the points A, B and C are on a straight line and are thus collinear.

Q3. Find a point on the y-axis which is equidistant from the points A (6, 5) and B (4, –3).

Solution:
Let a point on the y-axis be P (0, y), which is equidistant from the points A (6, 5) and B (4, –3); then

$AP=BP \Rightarrow AP^2=BP^2$

By the distance formula,

$(6-0)^2+(5-y)^2=(4-0)^2+(-3-y)^2$

$\Rightarrow 36+25+y^2-10y=16+9+y^2+6y$

$\Rightarrow 16y=36 \Rightarrow y=9/4$

Thus, the point (0, 9/4) is equidistant from the points A (6, 5) and B (4, –3).

Q4. Find the area of a triangle whose vertices are $(1,-1),(-4,2),(-3,5)$.

Solution:
The area of the triangle formed by the vertices $A(1,-1),B(-4,2),C(-3,5)$, is given by

$\frac{1}{2}[1(2+5)+(-4)(-5+1)+(-3)(-1-2)]$

$=\frac{1}{2}(7+16+9)=16$

Thus, the area of the triangle is 16 square units.

Q5. Find the coordinates of the points of trisection (points dividing the line in three equal parts) of the line segment joining the points $A(2,-2)$ and $B(-4,4)$.

Solution:
Let P and Q be the points of trisection of AB, that is, $AP=PQ=QB$

Therefore, P divides AB internally in the ratio 1:2. Thus, the coordinates of P, using the section formula, are

$\left(\frac{1(-4)+2(2)}{1+2},\frac{1(4)+2(-2)}{1+2}\right)=(0,0)$

Similarly, Q divides AB internally in the ratio 2:1. Thus, the coordinates of Q using the section formula, are

$\left(\frac{2(-4)+1(2)}{2+1},\frac{2(4)+1(-2)}{2+1}\right)=(-2,2)$

Therefore, the coordinates of the points of trisection of the line segment joining A and B are (0, 0) and (–2, 2).

Q6. Find the value of k if the points $A(2,3),B(4,0),C(k,-3)$ are collinear.

Solution:
Since the given points are collinear, the area of the triangle formed by them must be 0.

$\frac{1}{2}[2(0+3)+4(-3-3)+k(3-0)]=0$

$\Rightarrow \frac{1}{2}(6-24+3k)=0 \Rightarrow k=6$

Thus, for k = 6, the points $A(2,3),B(4,0),C(6,-3)$ are collinear.

Q7. If the points $A(6,1),B(8,2),C(9,4),D(7,q)$ are the vertices of a parallelogram, taken in order, find the value of q.

Solution:
Since the diagonals of a parallelogram bisect each other, the coordinates of the mid-point of diagonal AC = coordinates of the mid-point of diagonal BD.

$$\Rightarrow \left(\frac{6+9}{2}, \frac{1+4}{2}\right) = \left(\frac{8+7}{2}, \frac{2+q}{2}\right)$$

$$\Rightarrow \left(\frac{15}{2}, \frac{5}{2}\right) = \left(\frac{15}{2}, \frac{2+q}{2}\right)$$

$$\Rightarrow \frac{5}{2} = \frac{2+q}{2} \Rightarrow q = 3$$

Q8. Find the slope of the line, which makes an angle of 45° with the positive direction of the y-axis measured clockwise.

Solution:
The line makes an angle of 45° with the positive direction of the y-axis measured clockwise; therefore, the angle made by the line with the positive direction of the x-axis measured anticlockwise is $90° - 45° = 45°$.
Thus, the slope of the given line is
$m = \tan 45° = 1$

Q9. Find the value of x for which the points $(1,-1),(2,1),(4,x)$ are collinear.

Solution:
If the points $A(1,-1), B(2,1), C(4,x)$ are collinear; then the slope of line segment AB = slope of line segment BC.

$$\Rightarrow \frac{1-(-1)}{2-1} = \frac{x-1}{4-2} \Rightarrow 2 = \frac{x-1}{2} \Rightarrow x = 5$$

Q10. Find the angle between the x-axis and the line joining the points $(2,3),(4,1)$.

Solution:
The slope of the line joining the points $(2,3),(4,1)$ is $m = \frac{1-3}{4-2} = -1$; then the inclination (θ) given by $\tan\theta = m = -1 \Rightarrow \theta = (90° + 45°) = 135°$

Thus, the angle between the x-axis and the line joining the points $(2,3),(4,1)$ is 135°.

Q11. Find the equation of the line which intersects the x-axis at a distance of 2 units to the left of origin with a slope of –1.

Solution:
The equation of a line with slope m and x-intercept of a is given by $y = m(x-a)$.

Thus, the equation of the line intersecting the x-axis at a distance of 2 units to the left of the origin, that is, $a = -2$ and the slope of the line is $m = -1$ is given by
$y = -1[x-(-2)] \Rightarrow y = -x-3 \Rightarrow x+y+3 = 0$

Q12. Find the equation of the line which passes through the point (–4, 3) with a slope of ½.

Solution:
The equation of the line passes that through point (x_0, y_0) whose slope is m, is $(y-y_0) = m(x-x_0)$.

Therefore, the equation of the line passing through point (–4, 3) whose slope is 1/2 is

$$(y-3) = -\frac{1}{2}(x+4) \Rightarrow 2(y-3) = -x-4$$

$$\Rightarrow x+2y-2 = 0$$

Q13. Find the equation of the line which intersects the y-axis at a distance of 2 units above the origin and makes an angle of 30° with the positive direction of the x-axis.

Solution:
The equation of the line with a slope of m and making a y-intercept equal to b, is given $y - b = mx$.
Thus, the equation of the line intersecting the y-axis at a distance of 2 units above the origin, that is $b = 2$, and having a slope of $m = \tan 30° = \frac{1}{\sqrt{3}}$ is given by

$$y = \frac{1}{\sqrt{3}}x + 2 \Rightarrow y = \frac{x+2\sqrt{3}}{\sqrt{3}}$$

$$\sqrt{3}y = x + 2\sqrt{3} \Rightarrow \text{i.e.,} \ x - \sqrt{3}y + 2\sqrt{3} = 0$$

Q14. Find the equation of the line which is at a perpendicular distance of $\sqrt{2}$ units from the origin and the angle made by the perpendicular with the positive x-axis is 45°.

Solution:
If p is the length of the normal from the origin to a line and α is the angle made by the normal with the positive direction of the x-axis, then the equation of the line is given by $x\cos\alpha + y\sin\alpha = p$.

Thus, the required equation of the given line is

$$x\cos 45° + y\sin 45° = \sqrt{2} \quad \left(\because p = \sqrt{2} \text{ and } \alpha = 45°\right)$$

$$\Rightarrow x\cdot\frac{1}{\sqrt{2}} + y\cdot\frac{1}{\sqrt{2}} = \sqrt{2} \Rightarrow x + y = 2$$

Q15. Find the equation of the line passing through (1, 2) and perpendicular to the line through the points $(2,5),(-3,6)$.

Solution:

The slope of the line joining the points $(2,5),(-3,6)$ is $m=\frac{6-5}{-3-2}=\frac{1}{-5}$.

And, we know that two non-vertical lines are perpendicular to each other if and only if their slopes are negative reciprocals of each other.

Therefore, the slope of the line perpendicular to the line through the points $(2,5),(-3,6)$ is

$$=-\frac{1}{m}=-\frac{1}{(-1/5)}=5$$

Thus, the equation of the line passing through the point (1, 2), whose slope is 5, is $(y-2)=5(x-1)$.

$$\Rightarrow y-2=5x-5 \Rightarrow 5x-y-3=0$$

Q16. Find the equation of the line passing through the point (1, 1) and cutting off intercepts on the axes whose sum is 4.

Solution:

The equation of a line in the intercept form is $\frac{x}{a}+\frac{y}{b}=1$, where; a and b are the intercepts on the x and y axes, respectively.

It is given that $a+b=4 \Rightarrow b=4-a$, therefore the equation of the line can be written as $\frac{x}{a}+\frac{y}{4-a}=1$.

Given the line passes through point (1, 1), it therefore satisfies the equation of the line, hence

$$\frac{1}{a}+\frac{1}{4-a}=1 \Rightarrow 1\left(\frac{1}{a}+\frac{1}{4-a}\right)=1 \Rightarrow 1\left(\frac{4-a+a}{a(4-a)}\right)=1$$

$$\Rightarrow \frac{4}{4a-a^2}=1 \Rightarrow a^2-4a+4=0$$

$$\Rightarrow (a-2)^2 \Rightarrow a=2$$

Thus, the equation of the line is $\frac{x}{2}+\frac{y}{4-2}=1 \Rightarrow x+y-2=0$

Q17. The perpendicular from the origin to a line meets it at the point (–2, 4). Find the equation of the line.

Solution:

Let the slope of the line joining the origin (0, 0) and point (–2, 4) be $m=\frac{4-0}{-2-0}=-2$.

Then, the slope of the line perpendicular to the line joining the origin and the point (–2, 4) is $=-\frac{1}{m}=-\left(\frac{1}{-2}\right)=\frac{1}{2}$

Thus, the equation of the line passing through (–2, 4) and having a slope $\frac{1}{2}$ is $(y-4)=\frac{1}{2}(x+2)$.

$$2y-8=x+2 \Rightarrow x-2y+10=0$$

Q18. The length L (in cm) of a copper rod is a linear function of its celsius temperature C. In an experiment, if L = 102.64 when $C=20$, and $L=105.28$ when $C=120$, express L in terms of C.

Solution:

It is given that when C = 20, the value of L is 102.64, whereas when C = 120, the value of L is 105.28.

The points (C, L), assuming C is along the x-axis and L along the y-axis, satisfy the linear relation between C and L; therefore, the equation of the line passing through points the given points (20, 102.64) and (120, 105.28) in the XY plane is given by

$$(L-102.64)=\frac{105.28-102.64}{120-20}(C-20)$$

$$L-102.64=\frac{2.64}{100}(C-20)$$

$\Rightarrow L=0.0264(C-20)+102.64$, is the required linear relation between C and L.

Q19. By using the concept of the equation of a line, prove that the three points $(3,0),(-2,-2),(13,4)$ are collinear.

Solution:

If the points $(3,0),(-2,-2),(13,4)$ are collinear, then the line passing through points $(3,0),(-2,-2)$ must also be passing through the point $(13,4)$.

Now, the equation of the line passing through points $(3,0),(-2,-2)$ is $(y-0)=\frac{(-2-0)}{(-2-3)}(x-3)$

$$y=\frac{-2}{-5}(x-3) \Rightarrow 2x-5y=6$$

Since the third point $x=13, y=4$ also satisfies the equation of the line passing through points (3, 0) and (–2, –2), the three points $(3,0),(-2,-2),(13,4)$ lie on a line, and hence they are collinear.

Q20. Reduce the equation $5x+2y-10=0$ into the intercept form and find its intercepts on the axes.

Solution:

The given equation $5x+2y-10=0$, in the intercept form can be written as $5x+2y=10 \Rightarrow \frac{5x}{10}+\frac{2y}{10}=1 \Rightarrow \frac{x}{2}+\frac{y}{5}=1$

Comparing the above equation with $\frac{x}{a}+\frac{y}{b}=1$, we get $a=2, b=5$.

Therefore, in the intercept form, the x and y-intercepts are 2 and 5, respectively.

Q21. Find the distance of the point (–1,2) from the line $12(x+6)=5(y-2)$.

Solution:

The given equation of the line is $12(x+6)=5(y-2)$

$\Rightarrow 12x+72=5y-10 \Rightarrow 12x-5y+82=0$

Comparing the above equation with the general equation of the line $Ax+By+C=0$, we get $A=12$, $B=-5$, $C=82$.

The perpendicular distance (d) of a line $Ax+By+C=0$, from a point (x_1, y_1) is given by

$$d=\frac{|Ax_1+By_1+C|}{\sqrt{A^2+B^2}}$$

The given point is $(x_1, y_1)=(-1,2)$

Therefore, the distance of the point (–1, 2) from the given line $12x-5y+82=0$ is

$$=\frac{|12(-3)+(-5)(4)+82|}{\sqrt{(12)^2+(-5)^2}}=\frac{|-36-20+82|}{13}=2 \text{ units}$$

Q22. Find the equation of the line parallel to the line $3x-4y+2=0$ and passing through the point (–1, 4).

Solution:

The equation of the given line is $3x-4y+2=0$

$\Rightarrow y=\frac{3}{4}x+\frac{1}{2}$

Comparing with the slope intercept form $y=mx+c$, we get the slope of the given line = 3/4

Thus, the slope of the other parallel lines is $= m = \frac{3}{4}$

Now, the equation of the line that has a slope of 3/4 and passes through the point (–1, 4) is

$(y-4)=\frac{3}{4}\{x-(-1)\} \Rightarrow 4y-16=3x+3$

$\Rightarrow 3x-4y+19=0$

Q23. Find the angles between the lines $x+y=2,\ y=1$.

Solution:

The given equations of lines can be written as $x+y=2 \Rightarrow y=-x+2$ and $y=0\cdot x+1$

Thus, we get the slope of the first line as $m_1=-1$ and the slope of the other line as $m_2=0$.

Therefore, the acute angle (θ) between the two lines is given by

$$\tan\theta=\frac{|m_1-m_2|}{|1+m_1m_2|} \Rightarrow \tan\theta=\left|\frac{-1-0}{1+(-1)\times 0}\right|=1$$

$\Rightarrow \tan\theta=1 \Rightarrow \theta=45°$

Thus, the angle between the given lines is either $45°$ or $180°-45°=135°$.

Q24. Find the equation of the right bisector of the line segment joining the points (1, 2) and (3, 4).

Solution:

The slope of the line segment passing through the points (1, 2) and (3, 4) is $\frac{4-2}{3-1}=1$.

Therefore, the slope of the line perpendicular to this line is $-\frac{1}{(1)}=-1$.

The right bisector of a line segment bisects the line segment at 90° and passes through the mid-point. Now, the mid-point of the line is $=\left(\frac{3+1}{2}, \frac{4+2}{2}\right)=(2,3)$

Therefore, the equation of the line passing through (2, 3) and having a slope of –1 is $(y-3)=-1(x-2)$

$\Rightarrow x+y-5=0$

Q25. The line through the points $(h,3),(2,1)$ intersects the line $4x-5y-6=0$ at right angles. Find the value of h.

Solution:

The slope of the line passing through points $(h,3),(2,1)$ is $m_1=\frac{1-3}{2-h}=\frac{-2}{2-h}$.

And, the slope of the line $4x-5y-6=0 \Rightarrow y=\frac{4}{5}x-\frac{6}{5}$ is $m_2=4/5$.

Therefore, the two lines are perpendicular if $m_1 \times m_2 = -1 \Rightarrow \left(\frac{-2}{2-h}\right) \times \left(\frac{4}{5}\right) = -1 \Rightarrow \frac{-8}{10-5h} = -1$

$\Rightarrow -8 = -10 + 5h \Rightarrow 5h = 2 \Rightarrow h = 2/5$

Q26. The perpendicular from the origin to the line $y = mx + c$ meets it at the point (1, 2). Find the values of m and c.

Solution:

The perpendicular from the origin to the line $y = mx + c$ meets it at the point (1, 2); therefore, the slope of the line joining $(0,0),(1,2) = \frac{2-0}{1-0} = 2.$

It is perpendicular to the given equation of the line $y = mx + c$; therefore, $m \times 2 = -1 \Rightarrow m = -1/2.$

Point (1, 2) lies on the given line; therefore it will satisfy the equation $y = mx + c$, and thus,

$\therefore 2 = m \times 1 + c \Rightarrow 2 = -\frac{1}{2} \times 1 + c \Rightarrow c = 2 + \frac{1}{2} = \frac{5}{2}$

Thus, the values of m and c are $-\frac{1}{2}$ and $\frac{5}{2}$, respectively.

Q27. Find the equation of the line through the intersection of the lines $x - 2y + 3 = 0$, $2x + y - 3 = 0$ and parallel to the y-axis.

Solution:

The equation of any line through the point of intersection of the given lines is of the form

$x - 2y + 3 + k(2x + y - 3) = 0$

$\Rightarrow (1 + 2k)x + (-2 + k)y + 3 - 3k = 0$

Since this line is parallel to the y-axis, the coefficient of y should be zero, that is, $-2 + k = 0 \Rightarrow k = 2.$

Substituting the value of $k = 2$, we get

$(1 + 2 \times 2)x + (-2 + 2)y + 3 - 3 \times 2 = 0 \Rightarrow 5x - 3 = 0$

which is the required equation of the line.

Q28. Find the equation of the line through the intersection of lines $x - 2y + 3 = 0$, $2x + y - 3 = 0$ and parallel to the line $y = x + 1$.

Solution:

The equation of any line through the point of intersection of the given lines is of the form

$x - 2y + 3 + k(2x + y - 3) = 0$

$\Rightarrow (1 + 2k)x + (-2 + k)y + 3 - 3k = 0$

$\Rightarrow y = \frac{(1+2k)}{(2-k)}x + \frac{3-3k}{(2-k)} = 0$

Since this line is parallel to line $y = x + 1$, the slope of both lines must be equal, that is,

$1 = \frac{(1+2k)}{(2-k)} \Rightarrow 2 - k = 1 + 2k \Rightarrow k = 1/3$

Substituting the value of $k = 1/3$, we get

$(1 + 2 \times \frac{1}{3})x + (-2 + \frac{1}{3})y + 3 - 3 \times \frac{1}{3} = 0$

$\Rightarrow 5x - 5y + 6 = 0$ is required equation of the line.

Q29. Find the equation of the line through the intersection of the lines $3x + 4y = 7$, $x - y + 2 = 0$ and whose slope is 2.

Solution:

The equation of any line through the point of intersection of the given lines is of the form

$3x + 4y - 7 + k(x - y + 2) = 0$

$\Rightarrow (3 + k)x + (4 - k)y - 7 + 2k = 0$

$\Rightarrow (k - 4)y = (3 + k)x - 7 + 2k = 0$

$\Rightarrow y = \frac{(3+k)}{(k-4)}x + \frac{-7+2k}{(k-4)} = 0$

The slope of the line is 2, therefore, the slope $\frac{(3+k)}{(k-4)} = 2 \Rightarrow 3 + k = 2k - 8 \Rightarrow k = 11$

Substituting the value of $k = 11$, we get

$(3 + 11)x + (4 - 11)y - 7 + 2 \times 11 = 0$

$\Rightarrow 14x - 7y + 15 = 0$ is required equation of the line.

Q30. Find the new coordinates of the point (3,–2) if the origin is shifted to the point (2,–1) by a translation of axes.

Solution:

The coordinates of the new origin are $h = 2$, $k = -1$ and the original coordinates are $x = 3$, $y = -2$.

Any point $P(x, y)$ in the given coordinate system will be changed to $P(x', y')$ in the new coordinate system, by the relation $x' = x - h$ and $y' = y - k$.

Therefore, the point $P(3, -2)$ in the given coordinate system will be transferred to $x' = x - h = 3 - 2 = 1$ and $y' = y - k = -2 - (-1) = -1$ in the new coordinate system.

Thus, the coordinates of the point (3,–2) in the new coordinate system are (1,–1).

Day 145

Long Answer-I (LA-I): 4-Mark and Long Answer-II (LA-II): 6-Mark Questions

Q1. The points $(1,7),(4,2),(-1,-1),(-4,4)$ represent a

(a) square (b) rectangle
(c) parallelogram (d) rhombus

Solution:
The given points are
$A(1,7), B(4,2), C(-1,-1), D(-4,4)$; then by the distance formula,

$$AB = \sqrt{(1-4)^2 + (7-2)^2} = \sqrt{9+25} = \sqrt{34}$$
$$BC = \sqrt{(4+1)^2 + (2+1)^2} = \sqrt{25+9} = \sqrt{34}$$
$$CD = \sqrt{(-1+4)^2 + (-1-4)^2} = \sqrt{9+25} = \sqrt{34}$$
$$DA = \sqrt{(1+4)^2 + (7-2)^2} = \sqrt{25+9} = \sqrt{34}$$
$$AC = \sqrt{(1+1)^2 + (7+1)^2} = \sqrt{4+64} = \sqrt{68}$$
$$BD = \sqrt{(4+4)^2 + (2-4)^2} = \sqrt{64+4} = \sqrt{68}$$

Now, since $AB = BC = CD = DA, AC = BD,$ all the four sides of the quadrilateral $ABCD$ are equal and its diagonal AC and BD are also equal. Therefore, $ABCD$ is a square.

Note: Square: All sides are equal and both diagonals are also equal.
Parallelogram: All sides are equal but both diagonals are not equal.
Rectangle: Opposite sides are equal and diagonals are also equal.

Q2. In what ratio does the point $(-4, y)$ divide the line segment joining the points $A(-6,10), B(3,-8)$? Also find the value of y.

Solution:
Let $(-4, y)$ divide AB internally in the ratio $m:n$; then by the section formula, we get

$$(-4, y) = \left(\frac{3m-6n}{m+n}, \frac{-8m+10n}{m+n}\right)$$

$$\Rightarrow -4 = \frac{3m-6n}{m+n}, \quad y = \frac{-8m+10n}{m+n}$$

From the 1st equation, $-4m - 4n = 3m - 6n$

$\Rightarrow 7m = 2n \Rightarrow m/n = 2/7 \Rightarrow m:n = 2:7$

Now, from the 2nd equation, $y = \dfrac{-8m+10n}{m+n}$

$$\Rightarrow y = \frac{-8(m/n)+10}{(m/n)+1} = \frac{-8(2/7)+10}{(2/7)+1}$$

$$\Rightarrow y = \frac{-16+70}{2+7} = 6$$

Therefore, the point (–4, 6) divides the line segment joining the points $A(-6,10), B(3,-8)$ in the ratio 2:7.

Q3. If $A(-5,0), B(-4,-5), C(-1,-6), D(4,5)$ are the vertices of a quadrilateral find the area of the quadrilateral $ABCD$.

Solution:
By joining B and D, we get two triangles ABD and BCD.
Now, the area of ΔABD

$$= \frac{1}{2}[-5(-5-5) + (-4)(5-0) + 4(0+5)]$$

$$= \frac{1}{2}(50 - 20 + 20) = 25 \text{ square units}$$

And area of ΔBCD

$$= \frac{1}{2}[-4(-6-5) - 1(5+5) + 4(-5+6)]$$

$$= \frac{1}{2}(44 - 10 + 4) = 19 \text{ square units}$$

Thus, the area of quadrilateral $ABCD$ = area of ΔABD + area of ΔBCD = 25 + 19 = 44 square units.

Q4. Without using the Pythagoras theorem, find the value of a, such that the points $A(4,1), B(3,a), C(-1,-1)$ are the vertices of a right-angled triangle at angle B.

Solution:
Let the vertices of the given triangle be $A(4,1), B(3,a), C(-1,-1)$.

The slope of a non-vertical line passing through the points $(x_1, y_1), (x_2, y_2)$ is given by

$$m = \frac{y_2 - y_1}{x_2 - x_1}, x_2 \neq x_1$$

$\therefore$ Slope of line segment $AB\ (m_1) = \dfrac{a-1}{3-4} = 1 - a$

Slope of segment $BC\ (m_2) = \dfrac{-1-a}{-1-3} = \dfrac{1+a}{4}$

Slope of segment $CA\ (m_3) = \dfrac{4+1}{1+1} = \dfrac{5}{2}$

The points $A(4,1), B(3,a), C(-1,-1)$ are the vertices of a right-angled triangle at angle B; therefore the product of the slopes of the line segments AB and BC must be equal to –1.

Thus, $m_1 m_2 = -1 \Rightarrow \left(\frac{1+a}{4}\right) \times (1-a) = -1$

$\Rightarrow 1 - a^2 = -4 \Rightarrow a^2 = 5 \Rightarrow a = \pm\sqrt{5}$

Therefore, line segments AB and BC are perpendicular to each other, that is, the given triangle is a right-angled triangle at vertex $B\ (3, \pm\sqrt{5})$.

Q5. Find the value of x such that the points $(-2,-1), (x,y), (3,3), (-3,2)$ are the vertices of a parallelogram.

Solution:

Let the vertices of the given parallelogram be $A(-2,-1)$, $B(x,y)$, $C(3,3)$, $D(-3,2)$. Then

Slope of line segment $AB = \frac{y+1}{x+2} = \frac{1}{6}$

Slope of line segment $CD = \frac{2-3}{-3-3} = \frac{-1}{-6} = \frac{1}{6}$

Slope of line segment $BC = \frac{3-y}{3-x}$

Slope of line segment $AD = \frac{2+1}{-3+2} = \frac{3}{-1} = -3$

$ABCD$ represents a parallelogram; therefore the slope of the line segment AB = slope of the line segment CD

$\Rightarrow \frac{y+1}{x+2} = \frac{1}{6} \Rightarrow 6y - x + 4 = 0$

Also, the slope of the line segment BC = slope of the line segment $AD \Rightarrow \frac{3-y}{3-x} = -3 \Rightarrow y + 3x - 12 = 0$

Solving the above two equations, we get $x = 4$ and $y = 0$.

Thus, the points $(-2,-1), (4,0), (3,3), (-3,2)$ are the vertices of a parallelogram.

Q6. The owner of a milk store finds that, he can sell 980 litres of milk each week at Rs 14/litre and 1220 litres of milk each week at Rs 16/litre. Assuming a linear relationship between the selling price and the demand, how many litres could he sell weekly at Rs 18/litre?

Solution:

The relationship between the selling price and the demand is linear. Assume that the selling price per litre is along the x-axis and the demand is along the y-axis. Thus, the two given points that is (14, 980) and (16, 1220) in the XY plane will satisfy the linear relationship between the selling price and demand.

Thus, the linear relationship between the selling price per litre and demand is the equation of the line passing through points (14, 980) and (16, 1220) given by

$$y - 980 = \frac{1220-980}{16-14}(x-14) \Rightarrow y - 980 = \frac{240}{2}(x-14)$$

$$\Rightarrow y - 980 = 120(x-14) \Rightarrow y = 120(x-14) + 980$$

When x = Rs 18/litre, $y = 120(18-14) + 980 = 1460$

Thus, the owner of the milk store could sell 1460 litres of milk weekly at Rs 18/litre.

Q7. Consider the given population and year graph.

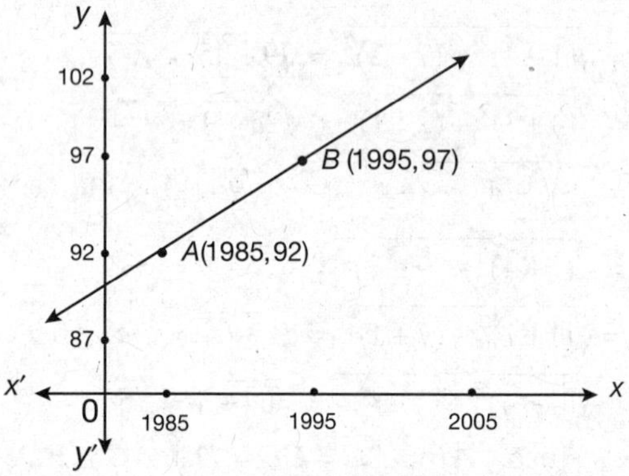

Find the slope of the line AB and using it, find what will be the population in the year 2005.

Solution:

The line AB passes through points A (1985, 92) and B (1995, 97); therefore its slope is $\frac{97-92}{1995-1985} = \frac{5}{10} = \frac{1}{2}$

Let y be the population in the year 2005. Then, according to the given graph, the line AB must pass through point C (2005, y). Thus,

Slope of line segment AB = slope of line segment BC

$\Rightarrow \frac{1}{2} = \frac{y-97}{2005-1995} \Rightarrow \frac{1}{2} = \frac{y-97}{10}$

$\Rightarrow 5 = y - 97 \Rightarrow y = 102$

Thus, in the year 2005 the population will be 102 crores.

Q8. Find the equation of the line which passes through $(2, \sqrt{3})$ and is inclined with the x-axis at an angle of 75°.

Solution:

Slope of the line that inclines with the x-axis at an angle of 75° is $m = \tan 75°$

$$\Rightarrow m = \tan(45° + 30°) = \frac{\tan 45° + \tan 30°}{1 - \tan 45° \cdot \tan 30°}$$

$$= \frac{1 + \frac{1}{\sqrt{3}}}{1 - 1 \cdot \frac{1}{\sqrt{3}}} = \frac{\frac{\sqrt{3}+1}{\sqrt{3}}}{\frac{\sqrt{3}-1}{\sqrt{3}}} = \frac{\sqrt{3}+1}{\sqrt{3}-1}$$

The equation of the line passing through point (x_0, y_0) whose slope is m, is $(y - y_0) = m(x - x_0)$. Therefore, the equation of the line passes through $(2, \sqrt{3})$ and inclines with the x-axis at an angle of 75°, that is, whose slope is $\frac{\sqrt{3}+1}{\sqrt{3}-1}$ is given by

$$\Rightarrow (y - \sqrt{3}) = \frac{\sqrt{3}+1}{\sqrt{3}-1}(x-2)$$

$$\Rightarrow (y - \sqrt{3})(\sqrt{3}-1) = (\sqrt{3}+1)(x-2)$$

$$\Rightarrow y(\sqrt{3}-1) - \sqrt{3}(\sqrt{3}-1) = x(\sqrt{3}+1) - 2(\sqrt{3}+1)$$

$$\Rightarrow (\sqrt{3}+1)x - (\sqrt{3}-1)y = -2\sqrt{3} - 2 + 3 - 2\sqrt{3}$$

$$\Rightarrow (\sqrt{3}+1)x - (\sqrt{3}-1)y = 1 - 4\sqrt{3}$$

Q9. The vertices of triangle ΔPQR are $P(2,1), Q(-2,3), R(3,1)$. Find equation of the median through the vertex R.

Solution:

It is given that the vertices of triangle ΔPQR are $P(2,1), Q(-2,3), R(3,1)$.

Let RL be the median through vertex R, where L is the mid-point of PQ.

Then, the coordinates of point L are given by $\left(\frac{2-2}{2}, \frac{1+3}{2}\right) = (0,2)$

The equation of the line passing through points $(x_1, y_1), (x_2, y_2)$ is $y - y_1 = \frac{y_2 - y_1}{x_2 - x_1}(x - x_1)$.

Therefore, the equation of median RL, where $R(3,1)$ and $L(0,2)$ is $y - 1 = \frac{2-1}{0-3}(x-3)$.

$$\Rightarrow y - 1 = \frac{-1}{3}(x-3) \Rightarrow 3(y-1) = -x + 3$$

$$\Rightarrow x + 3y - 6 = 0$$

Thus, the required equation of the median through the vertex R is $x + 3y - 6 = 0$.

Q10. A line perpendicular to the line segment joining the points (1, 0) and (2, 3) divides it in the ratio 1 : 2. Find the equation of the line.

Solution:

Using the section formula, the coordinates of the point that divides the line segment joining the points (1, 0) and (2, 3) in the ratio 1 : 2 is given by

$$\left(\frac{1\times2+2\times1}{1+2}, \frac{0\times2+3\times1}{1+2}\right) = \left(\frac{4}{3}, 1\right)$$

The slope of the line joining the points (1, 0) and (2, 3) is $m = \frac{3-0}{2-1} = 3$. Therefore, the slope of the line that is perpendicular to the line joining the points (1, 0) and (2, 3) $= -\frac{1}{m} = -\frac{1}{3}$ (since two non-vertical lines are perpendicular to each other if and only if their slopes are negative reciprocals of each other).

Thus, the equation of the line passing through $\left(\frac{4}{3}, 1\right)$ and whose slope is $-\frac{1}{3}$ is,

$$(y-1) = \frac{-1}{3}\left(x - \frac{4}{3}\right) \Rightarrow 9(y-1) = -(3x-4)$$

$$\Rightarrow 9y - 9 = -3x + 4 \Rightarrow 3x + 9y - 13 = 0$$

Q11. Find equation of the line through the point (0, 2) making an angle $\frac{2\pi}{3}$ with the positive x-axis. Also, find the equation of the line parallel to it and crossing the y-axis at a distance of 3 units above the origin.

Solution:

The slope of the line making angle $\frac{2\pi}{3}$ with the positive x-axis is $m = \tan\left(\frac{2\pi}{3}\right) = -\sqrt{3}$.

Therefore, the equation of the line passing through point (0, 2) and having a slope of $-\sqrt{3}$ is $(y-2) = -\sqrt{3}(x-0) \Rightarrow \sqrt{3}x + y - 2 = 0$.

Now, the slope of the line parallel to the line $\sqrt{3}x + y - 2 = 0$ is $-\sqrt{3}$.

It is given that the line crosses the y-axis 3 units above the origin, that is, it passes through point (0, 3).

Therefore, the equation of the line passing through point (0, 3) and having a slope $-\sqrt{3}$ is $y - 3 = -\sqrt{3}(x-0) \Rightarrow \sqrt{3}x + y - 3 = 0$

Q12. Point $R(h,k)$ divides a line segment between the axes in the ratio 2:3. Find the equation of the line.

Solution:

Let AB be the line segment between the axes such that point $R(h,k)$ divides AB in the ratio 2:3.

Let AB be the line segment with coordinates $A(x,0)$ on the x-axis and $B(0,y)$ on the y-axis.

Point $R(h,k)$ divides AB in the ratio 2:3, then according to the section formula,

$$(h,k)=\left(\frac{2\times 0+3\times x}{2+3},\frac{2\times y+3\times 0}{2+3}\right)\Rightarrow (h,k)=\left(\frac{3x}{5},\frac{2y}{5}\right)$$

$$\Rightarrow h=\frac{3x}{5},\ k=\frac{2y}{5}\Rightarrow x=\frac{5h}{3},\ y=\frac{5k}{2}$$

Therefore, the respective coordinates of A and B are $\left(\frac{5h}{3},0\right),\left(0,\frac{5k}{2}\right)$

Thus, the equation of the line AB passing through the points

$$\left(\frac{5h}{3},0\right),\left(0,\frac{5k}{2}\right)\text{ is }(y-0)=\frac{\frac{5k}{2}-0}{0-\frac{5h}{3}}\left(x-\frac{5h}{3}\right)$$

$$y=-\frac{3k}{2h}\left(x-\frac{5h}{3}\right)\Rightarrow 6hy=-9kx+15hk$$

Hence, the required equation of the line is $6hy=-9kx+15hk\Rightarrow \frac{3x}{5h}+\frac{2y}{5k}=1$

Q13. Reduce the equation $x-\sqrt{3}y-8=0$ into the normal form. Find its perpendicular distance from the origin and angle between the perpendicular and the positive x-axis.

Solution:

The given equation $x-\sqrt{3}y-8=0$ can be written as $x-\sqrt{3}y-8=0\Rightarrow x-\sqrt{3}y=8$.

Dividing both sides by $\sqrt{(-1)^2+(\sqrt{3})^2}=\sqrt{4}=2$, we get $\frac{x}{2}-\frac{\sqrt{3}}{2}y=\frac{8}{2}\Rightarrow\left(\frac{1}{2}\right)x+\left(-\frac{\sqrt{3}}{2}\right)y=4$

$\Rightarrow x\cos 300°+y\sin 300°=4$ which is the normal form of the given equation.

Comparing the above equation with the normal form of the equation of the line by $x\cos\alpha+y\sin\alpha=p$, we get $\alpha=300°$, $p=4$.

Thus, the perpendicular distance of the line from the origin is 4, while the angle between the perpendicular and the positive x-axis is 330°.

Q14. Find the points on the x-axis, whose distances from the line $\frac{x}{3}+\frac{y}{4}=1$ are 2 units.

Solution:

The given equation of the line is $\frac{x}{3}+\frac{y}{4}=1$

$\Rightarrow 4x+3y-12=0$

Comparing the above equation with the general equation of the line $Ax+By+C=0$, we get $A=4$, $B=3$, $C=-12$.

Let $(a, 0)$ be the point on the x-axis whose distance from the given line is 2 units.

The perpendicular distance (p) of a line $Ax+By+C=0$ from a point (x_1,y_1) is given by

$$\frac{|Ax_1+By_1+C|}{\sqrt{A^2+B^2}},\text{ therefore }2=\frac{|4a+3\times 0-12|}{\sqrt{4^2+3^2}}$$

$$\Rightarrow 2=\frac{|4a-12|}{5}\Rightarrow |4a-12|=10\Rightarrow \pm(4a-12)=20$$

$$\Rightarrow (4a-12)=10\text{ or }-(4a-12)=10$$

$$\Rightarrow 4a=10+12\text{ or }-4a=10-12\Rightarrow a=\frac{11}{2}\text{ or }\frac{1}{2}$$

Thus, the required points on the x-axis are $\left(\frac{11}{2},0\right),\left(\frac{1}{2},0\right)$.

Q15. Two lines passing through the origin intersect each other at an angle of 60°. If the slope of one line is –1, find the equation of the other line.

Solution:

Let the slope of the first line be $m_1=-1$ and the slope of the other line be m_2.

Then, the angle between the two lines is given by

$$\frac{|m_1-m_2|}{|1+m_1m_2|}=\tan 60°=\sqrt{3}\Rightarrow\pm\left(\frac{-1-m_2}{1+(-1)m_2}\right)=\sqrt{3}$$

$$\Rightarrow\frac{-1-m_2}{1-m_2}=\sqrt{3}\text{ or }\frac{1+m_2}{1-m_2}=\sqrt{3}$$

$$\Rightarrow -1-m_2=\sqrt{3}(1-m_2)\text{ or }1+m_2=\sqrt{3}(1-m_2)$$

$$\Rightarrow(\sqrt{3}-1)m_2=\sqrt{3}+1\text{ or }(\sqrt{3}+1)m_2=\sqrt{3}-1$$

$$\Rightarrow m_2=\frac{\sqrt{3}+1}{\sqrt{3}-1}\text{ or }m_2=\frac{\sqrt{3}-1}{\sqrt{3}+1}$$

Case 1: $m_2=\frac{\sqrt{3}+1}{\sqrt{3}-1}$

The equation of the line passing through the origin (0, 0) and having a slope of $\frac{\sqrt{3}+1}{\sqrt{3}-1}$ is given by $y=\frac{\sqrt{3}+1}{\sqrt{3}-1}x$

Case 2: $m_2=\frac{\sqrt{3}-1}{\sqrt{3}+1}$

The equation of the line passing through the origin (0, 0) and having a slope of $\frac{\sqrt{3}-1}{\sqrt{3}+1}$ is given by $y=\frac{\sqrt{3}-1}{\sqrt{3}+1}x$

Thus, the required equations of the other line are $(\sqrt{3}+1)x-(\sqrt{3}-1)y=0$ or

$(\sqrt{3}-1)x-(\sqrt{3}+1)y=0$

Q16. Find the coordinates of the foot of perpendicular from the point (1, 2) to the line $3x-4y-5=0$.

Solution:

Let (a,b) be the coordinates of the foot of the perpendicular from the point (1, 2) to the line $3x-4y-5=0$. Then

Slope of the line joining (1, 2) and (*a*, *b*), is $m_1=\frac{b-2}{a-1}$

Slope of the line $3x-4y-5=0 \Rightarrow y=\frac{3}{4}x-\frac{5}{4}$ is $m_2=3/4$

The above two lines are perpendicular to each other; therefore $m_1m_2=-1$.

$\Rightarrow \left(\frac{b-2}{a-1}\right)\times\left(\frac{3}{4}\right)=-1 \Rightarrow \frac{3b-6}{4a-4}=-1$

$\Rightarrow 3b-6=-4a+4 \Rightarrow 4a+3b=10$

Also, the point (*a*, *b*) lies on the line $3x-4y-5=0$, therefore $3a-4b=5$.

Solving the above two equations, we get $a=55/25=11/5$, $b=10/25=2/5$ which are the required coordinates of the foot of the perpendicular to the line $3x-4y-5=0$.

Q17. In the triangle *ABC* with vertices $A(5,3), B(4,-1), C(1,2)$ find the equation and length of altitude from the vertex *A*.

Solution:

Let *AD* be the altitude of the triangle *ABC* from the vertex *A* to *BC*, that is $AD \perp BC$.

The slope of the line *BC* is $=\frac{2+1}{1-4}=-1$; therefore, the slope of altitude *AD* will be equal to 1 ($\because AD \perp BC$).

Thus, the equation of the line passing through point (5, 3) and having a slope of 1 is $(y-3)=1(x-5)$

$\Rightarrow x-y=2$

Therefore, the equation of the altitude from vertex *A* is $x-y=2$.

And the equation of *BC* is $(y+1)=\frac{2+1}{1-4}(x-4)$

$\Rightarrow (y+1)=-1(x-4) \Rightarrow x+y-3=0$

Since the perpendicular distance (*d*) of a line $Ax+By+C=0$ from a point (x_1,y_1) is given by $d=\left|\frac{Ax_1+By_1+C}{\sqrt{A^2+B^2}}\right|$

Comparing equation $x+y-3=0$ with the equation of the line $Ax+By+C=0$, we get $A=1$, $B=1$, $C=-3$.

Therefore, length of AD = length of the perpendicular from *A* (5, 3) to *BC*.

$=\frac{|1\times5+1\times3-3|}{\sqrt{1^2+1^2}}=\frac{|5|}{\sqrt{2}}=\frac{5}{\sqrt{2}}$ units

Thus, the equation and the length of the altitude from vertex *A* are $x-y=2$, $\frac{5}{\sqrt{2}}$ units, respectively.

Q18. Find the equation of the line through the intersection of the lines $2x+3y-4=0$, $x-5y=7$ that has its *x*-intercept equal to –2.

Solution:

The equation of any line through the point of intersection of the given lines is of the form

$2x+3y-4+k(x-5y-7)=0$

$\Rightarrow (2+k)x+(3-5k)y-4-7k=0$

This line has its *x*-intercept equal to –2; therefore

$\Rightarrow (2+k)\times(-2)+(3-5k)\times0-4-7k=0$

$\Rightarrow -4-2k-4-7k=0 \Rightarrow k=-8/9$

Substituting the value of $k=-8/9$, we get

$\left(2+\frac{-8}{9}\right)x+\left(3-5\times\frac{-8}{9}\right)y-4-7\times\frac{-8}{9}=0$

$\Rightarrow 10x+67y+20=0$ is the required equation of the line.

Q19. Find what the equation $2x-3y+xy-1=0$ becomes when the origin is shifted to the point $(-1,1)$ by a translation of axes.

Solution:

The coordinates of the new origin are $h=-1$, $k=1$ and any point $P(x,y)$ in given coordinate system will be changed to $P(x',y')$ in new coordinate system by the relation $x = x' + h \Rightarrow x = x' - 1$ and $y = y' + k \Rightarrow y = y' + 1$. Therefore, substituting the value of x and y in the given equation $2x - 3y + xy - 1 = 0$, we get $2(x'+1) - 3(y'+1) + (x'+1)(y'+1) - 1 = 0$.

$\Rightarrow 2x' + 2 - 3y' - 3 + x'y' + x' + y' + 1 - 1 = 0$

$\Rightarrow 3x' - 2y' + x'y' - 1 = 0$

Thus, the required equation of the line in the new coordinate system is $3x - 2y + xy - 1 = 0$.

CHAPTER

11 Conic Sections

Day 146

Circle

A circle is the set of all points in a plane that are equidistant from a fixed point in the plane.

- The fixed point is called the centre of the circle and the distance from the centre to a point on the circle is called the radius of the circle.
- Let *P* (*x*, *y*) be any point on the circle (Fig. 11.1). Then the equation of the circle with centre *C* (*h*, *k*) and radius *r* is given by $|CP| = r$.

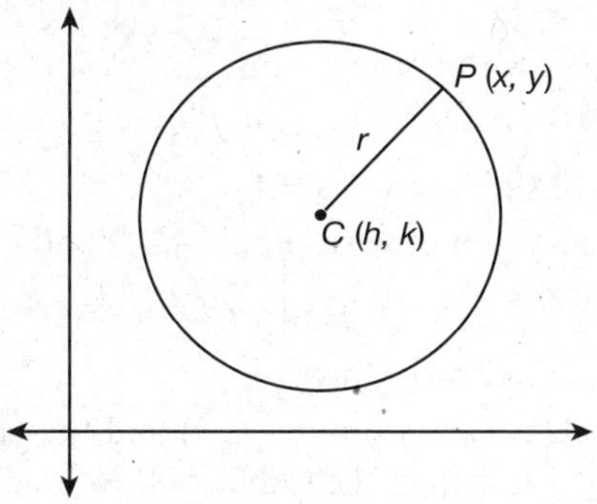

Fig. 11.1 Circle of radius *r* with centre at *C* (*h*, *k*)

- Using the distance formula, we get
$\sqrt{(x-h)^2+(y-k)^2} = r$
$\Rightarrow (x-h)^2+(y-k)^2 = r^2$ as the required equation of the circle with the centre at (*h*, *k*) and radius *r*.
- The equation of a circle (Fig. 11.1) with the centre at the origin *O* (0, 0) and radius *r* is given by $|OP| = r$.

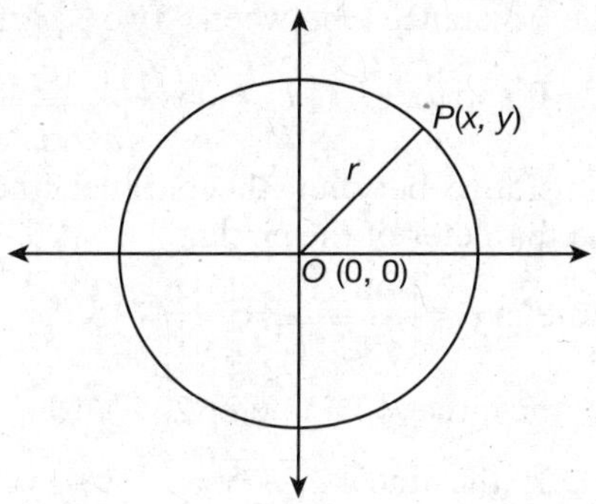

Fig. 11.2 Circle of radius *r* with centre at *O* (0, 0)

- Using the distance formula, we get
$\Rightarrow \sqrt{x^2+y^2} = r$
$\Rightarrow x^2+y^2 = r^2$ as the required equation of the circle with centre at origin (0, 0) and radius *r*.

Example 1: Find the equation of the circle with centre (–1, 2) and radius 3.

Solution:
The equation of a circle with centre (*h*, *k*) and radius *r* is
$(x-h)^2+(y-k)^2 = r^2$
Therefore, the equation of the circle with centre (–1, 2) and radius 3 is $(x+1)^2+(y-2)^2 = (3)^2$
$\Rightarrow x^2+2x+1+y^2-4y+4 = 9$
$\Rightarrow x^2+y^2+2x-4y-4 = 0$

Example 2: Find the centre and radius of the circle $(x+4)^2+(y-3)^2 = 36$.

Solution:
The given equation $(x+4)^2+(y-3)^2 = 36$ of the circle can be written as $\{x-(-4)\}^2+(y-3)^2 = 6^2$. Comparing this with the equation of the circle $(x-h)^2+(y-k)^2 = r^2$, we get $h = -4, k = 3, r = 6$.
Therefore, the centre of the given circle is (–4, 3) and the radius is 6.

Example 3: Find the centre and radius of the circle $x^2+y^2-2x-4y-4 = 0$.

Solution:
The given equation $x^2+y^2-2x-4y-4 = 0$ of the circle can be written as $(x^2-2x)+(y^2-4y) = 4$
$\Rightarrow \{x^2-2(x)(1)+1^2\}+\{y^2-2(y)(2)+2^2\}-1-4 = 4$
$\Rightarrow (x-1)^2+(y-2)^2 = 9 = (3)^2$
Comparing this with the equation of the circle $(x-h)^2+(y-k)^2 = r^2$, we get $h = 1, k = 2, r = 3$.
Therefore, the centre of the given circle is (1, 2) and the radius is 3.

Example 4: Find the equation of a circle with centre (2, 2) and passing through the point (4, 3).

Solution:

The centre of the circle is given as $(h,k) = (2,2)$.

The circle passes through the point (4, 5). Therefore, the radius (r) of the circle is the distance between the points (2, 2) and (4, 5). Then

$r = \sqrt{(2-4)^2 + (2-3)^2}$

$\Rightarrow r = \sqrt{(-2)^2 + (-1)^2} = \sqrt{5}$

Thus, the equation of the circle is

$(x-h)^2 + (y-k)^2 = r^2$

$\Rightarrow (x-2)^2 + (y-2)^2 = (\sqrt{5})^2$

$\Rightarrow x^2 - 4x + 4 + y^2 - 4y + 4 = 5$

$\Rightarrow x^2 + y^2 - 4x - 4y + 3 = 0$

Example 5: Find the equation of the circle passing through the point (4, 1) and (6, 5) and whose centre is on the line $2x + y = 7$.

Solution:

Let the equation of the required circle be $(x-h)^2 + (y-k)^2 = r^2$.

The circle passes through the points (4, 1) and (6, 5); therefore

$(4-h)^2 + (1-k)^2 = r^2$ (i)

$(6-h)^2 + (5-k)^2 = r^2$ (ii)

The centre (h, k) of the circle lies on the line $2x + y = 7$; therefore $2h + k = 7$ (iii)

Solving equations (i) and (ii), we get

$(4-h)^2 + (1-k)^2 = (6-h)^2 + (5-k)^2$

$\Rightarrow 16 - 8h + h^2 + 1 - 2k + k^2$

$= 36 - 12h + h^2 + 25 - 10k + k^2$

$\Rightarrow 16 - 8h + 1 - 2k = 36 - 12h + 25 - 10k$

$\Rightarrow h + 2k = 11$ (iv)

Now solving equations (iii) and (iv), we get $h = 1, k = 5$.

Substituting the values of h and k in Equation (i), we get

$(4-1)^2 + (1-5)^2 = r^2 \Rightarrow r^2 = 25$

Thus, the equation of the required circle is

$(x-1)^2 + (y-5)^2 = 25$

$\Rightarrow x^2 - 2x + 1 + y^2 - 10y + 25 = 25$

$\Rightarrow x^2 + y^2 - 2x - 10y + 1 = 0$

Example 6: Find the equation of the circle with radius 5 whose centre lies on the x-axis and passes through the point (2, 4).

Solution:

Let the equation of the required circle be $(x-h)^2 + (y-k)^2 = r^2$.

The radius of the circle is 5 and its centre lies on the x-axis; therefore $k = 0$, $r = 5$.

Thus, the equation of the circle is $(x-h)^2 + y^2 = 25$.

Now, the circle passes through point (2, 4); therefore $(2-h)^2 + 4^2 = 25$.

$\Rightarrow (2-h)^2 = 25 - 16 \Rightarrow 2 - h = \pm\sqrt{9} = \pm 3$

Case 1: For $2 - h = 3 \Rightarrow h = -1$

Thus, the equation of the circle is $(x+1)^2 + y^2 = 25$

$\Rightarrow x^2 + y^2 + 2x - 24 = 0$

Case 2: For $2 - h = -3 \Rightarrow h = 5$

Thus, the equation of the circle is $(x-5)^2 + y^2 = 25$

$\Rightarrow x^2 + y^2 - 10x = 0$

Example 7: Does the point $(-2,3)$ lie inside, outside or on the circle $x^2 + y^2 = 16$?

Solution:

The equation of the given circle is $x^2 + y^2 = 16 \Rightarrow (x-0)^2 + (y-0)^2 = 4^2$

Comparing with the equation the circle $(x-h)^2 + (y-k)^2 = r^2$, we get $h = 0, k = 0, r = 4$.

$\therefore$ Centre = (0,0) and radius = 4

Now, the distance between point $(-2,3)$ and centre (0, 0) is $d = \sqrt{(-2-0)^2 + (3-0)^2} = \sqrt{4+9} = \sqrt{13}$

$\Rightarrow d = \sqrt{13} = 3.6 < 4$ (radius)

The distance between point $(-2,3)$ and the centre (0, 0) of the circle is less than the radius of the circle. Therefore the point $(-2,3)$ lies inside the circle.

Example 8: If the lines $3x + 4y + 4 = 0$, $6x - 8y - 6 = 0$ are tangents to a circle, then find the radius of the circle.

Solution:

The given lines $3x + 4y + 4 = 0$ and $6x - 8y - 6 = 0 \Rightarrow 3x - 4y - 3 = 0$ are parallel and are tangents to the circle; therefore they must touch the circle diametrically.

(*Note:* The distance between two parallel lines $ax + by + c_1 = 0$ and $ax + by + c_2 = 0$ is $\frac{c_1 - c_2}{\sqrt{a^2 + b^2}}$)

Thus, the distance between the parallel lines must be equal to the diameter of the circle.

Hence, diameter $d = \left|\frac{4+3}{\sqrt{9+16}}\right| = \left|\frac{7}{5}\right| = 7/5$

$\therefore$ Radius of the circle = Diameter/2 $= 7/10$

Example 9: If the lines $2x - 3y = 5$, $3x - 4y = 7$ are the diameters of a circle of area 22 square units, then obtain the equation of the circle.

Solution:

Solving the given lines $2x-3y-5=0$ and $3x-4y-7=0$, we get $\frac{x}{21-20}=\frac{y}{-15+14}=\frac{1}{-8+9}$

$\Rightarrow \frac{x}{1}=\frac{y}{-1}=\frac{1}{+1} \Rightarrow x=1, y=-1$

Since the given lines are diameters of the circle, the intersection point of these lines will give the centre of the circle. Thus, the coordinates of the centre of the circle are (1, –1).

Let the radius of the circle be r; then

$\pi r^2 = 22 \Rightarrow \frac{22}{7}\times r^2 = 22$

$\Rightarrow r^2 = \frac{22\times 7}{22} \Rightarrow r^2 = \frac{2\times 7}{2} \Rightarrow r^2 = 7$

Thus, the equation of the circle is $(x-1)^2+(y+1)^2=7$

$\Rightarrow x^2-2x+1+y^2+2y+1=7$

$\Rightarrow x^2+y^2-2x+2y-5=0$

Example 10: Find the equation of the circle passing through (0, 0) and making intercepts a and $-a$ on the coordinate axes.

Solution:

Let the equation of the required circle be $(x-h)^2+(y-k)^2=r^2$.

The centre of the circle passes through (0, 0); therefore $(0-h)^2+(0-k)^2=r^2 \Rightarrow h^2+k^2=r^2$.

Thus, the equation of the circle is $(x-h)^2+(y-k)^2=h^2+k^2$.

Now, since the circle makes intercepts a and $-a$ on the coordinate axes, it means the circle passes through points $(a, 0)$ and $(0, -a)$. Therefore

$(a-h)^2+(0-k)^2=h^2+k^2$ (i)

$(0-h)^2+(-a-k)^2=h^2+k^2$ (ii)

From Equation (i), we get

$a^2-2ah+h^2+k^2=h^2+k^2$

$\Rightarrow a^2-2ah=0 \Rightarrow a(a-2h)=0$

$\Rightarrow a=0$ or $(a-2h)=0$

$\Rightarrow a-2h=0 \Rightarrow h=a/2\ (\because a\neq 0)$

And, from Equation (ii), we get

$h^2+a^2+2ak+k^2=h^2+k^2$

$\Rightarrow a^2+2ak=0 \Rightarrow a(a+2k)=0$

$\Rightarrow a=0$ or $(a+2k)=0$

$\Rightarrow a+2h=0 \Rightarrow k=-a/2\ (\because a\neq 0)$

Hence, the required equation of the circle is

$$\left(x-\frac{a}{2}\right)^2+\left(y+\frac{a}{2}\right)^2=\left(\frac{a}{2}\right)^2+\left(\frac{a}{2}\right)^2$$

$$\Rightarrow \left(\frac{2x-a}{2}\right)^2+\left(\frac{2y+a}{2}\right)^2=\frac{a^2+a^2}{4}$$

$\Rightarrow 4x^2-4ax+a^2+4y^2+4ay+a^2=2a^2$

$\Rightarrow 4x^2+4y^2-4ax+4ay=0$

$\Rightarrow x^2+y^2-ax+ay=0$

Day 147

NCERT Exercise 11.1 (With Solutions)

Q1. Find the equation of the circle with centre (0, 2) and radius 2.

Solution:

The equation of a circle with centre (h, k) and radius r is $(x-h)^2+(y-k)^2=r^2$.

Therefore, the equation of the circle with centre (0, 2) and radius 2 is $(x-0)^2+(y-2)^2=2^2$

$\Rightarrow x^2+y^2+4-4y=4 \Rightarrow x^2+y^2-4y=0$

Q2. Find the equation of the circle with centre (–2, 3) and radius 4.

Solution:

The equation of a circle with centre (h, k) and radius r is $(x-h)^2+(y-k)^2=r^2$.

Therefore, the equation of the circle with centre (–2, 3) and radius 4 is $(x+2)^2+(y-3)^2=(4)^2$

$\Rightarrow x^2+4x+4+y^2-6y+9=16$

$\Rightarrow x^2+y^2+4x-6y-3=0$

Q3. Find the equation of the circle with centre $\left(\frac{1}{2},\frac{1}{4}\right)$ and radius 1/12.

Solution:

The equation of a circle with centre (h, k) and radius r is $(x-h)^2+(y-k)^2=r^2$.

Therefore, the equation of the circle with centre $\left(\frac{1}{2},\frac{1}{4}\right)$ and radius $\frac{1}{12}$ is $\left(x-\frac{1}{2}\right)^2+\left(y-\frac{1}{4}\right)^2=\left(\frac{1}{12}\right)^2$

$\Rightarrow x^2-x+\frac{1}{4}+y^2-\frac{y}{2}+\frac{1}{16}=\frac{1}{144}$

$\Rightarrow x^2-x+\frac{1}{4}+y^2-\frac{y}{2}+\frac{1}{16}-\frac{1}{144}=0$

$\Rightarrow 144x^2-144x+144y^2-72y+44=0$

$\Rightarrow 36x^2 + 36y^2 - 36x - 18y + 11 = 0$

Q4. Find the equation of the circle with centre (1, 1) and radius $\sqrt{2}$.

Solution:

The equation of a circle with centre (*h*, *k*) and radius *r* is $(x-h)^2 + (y-k)^2 = r^2$.

Therefore, the equation of the circle with centre (1,1) and radius $\sqrt{2}$ is $(x-1)^2 + (y-1)^2 = (\sqrt{2})^2$

$\Rightarrow x^2 - 2x + 1 + y^2 - 2y + 1 = 2$

$\Rightarrow x^2 + y^2 - 2x - 2y = 0$

Q5. Find the equation of the circle with centre $(-a, -b)$ and radius $\sqrt{a^2 - b^2}$.

Solution:

The equation of a circle with centre (*h*, *k*) and radius *r* is $(x-h)^2 + (y-k)^2 = r^2$.

Therefore, the equation of the circle with centre $(-a, -b)$ and radius $\sqrt{a^2 - b^2}$ is

$$(x+a)^2 + (y+b)^2 = \left(\sqrt{a^2 - b^2}\right)^2$$

$\Rightarrow x^2 + 2ax + a^2 + y^2 + 2by + b^2 = a^2 - b^2$

$\Rightarrow x^2 + y^2 + 2ax + 2by + 2b^2 = 0$

Q6. Find the centre and radius of the circle $(x+5)^2 + (y-3)^2 = 36$.

Solution:

The given equation $(x+5)^2 + (y-3)^2 = 36$ of the circle can be written as $\{x-(-5)\}^2 + (y-3)^2 = 6^2$. Comparing this with the equation of the circle $(x-h)^2 + (y-k)^2 = r^2$, we get $h = -5, k = 3, r = 6$.

Therefore, the centre of the given circle is (–5, 3) and radius is 6.

Q7. Find the centre and radius of the circle $x^2 + y^2 - 4x - 8y - 45 = 0$.

Solution:

The given equation $x^2 + y^2 - 4x - 8y - 45 = 0$ of the circle can be written as $(x^2 - 4x) + (y^2 - 8y) = 45$

$\Rightarrow \{x^2 - 2(x)(2) + 2^2\} + \{y^2 - 2(y) + 4^2\} - 4 - 16 = 45$

$\Rightarrow (x-2)^2 + (y-4)^2 = 65 = (\sqrt{65})^2$

Comparing this with the equation of circle $(x-h)^2 + (y-k)^2 = r^2$, we get

$h = 2, k = 4, r = \sqrt{65}$

Therefore, the centre of the given circle is (2, 4) and radius is $\sqrt{65}$.

Q8. Find the centre and radius of the circle $x^2 + y^2 - 8x + 10y - 12 = 0$.

Solution:

The given equation $x^2 + y^2 - 8x + 10y - 12 = 0$ of the circle can be written as $(x^2 - 8x) + (y^2 + 10y) = 12$

$\Rightarrow \{x^2 - 2(x)(4) + 4^2\} + \{y^2 + 2(y)(5) + 5^2\} - 16 - 25 = 12$

$\Rightarrow (x-4)^2 + (y+5)^2 = 53$

$\Rightarrow (x-4)^2 + \{y-(-5)\}^2 = (\sqrt{53})^2$

Comparing this with the equation of circle $(x-h)^2 + (y-k)^2 = r^2$, we get

$h = 4, k = -5, r = \sqrt{53}$

Therefore, the centre of the given circle is (4, –5) and radius is $\sqrt{53}$.

Q9. Find the centre and radius of the circle $2x^2 + 2y^2 - x = 0$.

Solution:

The given equation $2x^2 + 2y^2 - x = 0$ of the circle can be written as $2x^2 + 2y^2 - x = 0$.

$2x^2 + 2y^2 - x = 0$

$$\Rightarrow (2x^2 - x) + 2y^2 - x = 0 \Rightarrow 2\left[\left(x^2 - \frac{x}{2}\right) + y^2\right] = 0$$

$$\Rightarrow \left\{x^2 - 2x\left(\frac{1}{4}\right) + \left(\frac{1}{4}\right)^2\right\} + y^2 - \left(\frac{1}{4}\right)^2 = 0$$

$$\Rightarrow \left(x - \frac{1}{4}\right)^2 + (y-0)^2 = \left(\frac{1}{4}\right)^2$$

Comparing this with the equation of the circle $(x-h)^2 + (y-k)^2 = r^2$, we get

$h = 1/4, k = 0, r = 1/4$.

Therefore, the centre of the given circle is (1/4, 0) and radius is 1/4.

Q10. Find the equation of the circle passing through the points (4, 1) and (6, 5) and whose centre is on the line $4x + y = 16$.

Solution:

Let the equation of the required circle be $(x-h)^2 + (y-k)^2 = r^2$.

The circle passes through points (4, 1) and (6, 5); therefore

$(4-h)^2 + (1-k)^2 = r^2$ (i)

$(6-h)^2 + (5-k)^2 = r^2$ (ii)

The centre (*h*, *k*) of the circle lies on line $4x + y = 16$; therefore $4h + k = 16$ (iii)

Solving equations (i) and (ii), we get

$(4-h)^2+(1-k)^2=(6-h)^2+(5-k)^2$

$\Rightarrow 16-8h+h^2+1-2k+k^2$

$=36-12h+h^2+25-10k+k^2$

$\Rightarrow 16-8h+1-2k=36-12h+25-10k$

$\Rightarrow h+2k=11$ (iv)

Now solving equations (iii) and (iv), we get $h=3, k=4$.

Substituting the values of h and k in Equation (i), we get

$(4-3)^2+(1-4)^2=r^2 \Rightarrow r^2=10$

Thus, the equation of the required circle is

$(x-3)^2+(y-4)^2=10$

$\Rightarrow x^2-6x+9+y^2-8y+16=0$

$\Rightarrow x^2+y^2-6x-8y+15=0$

Q11. Find the equation of circle passing through the points (2, 3) and (–1, 1) and whose centre is on the line $x-3y-11=0$.

Solution:

Let the equation of the required circle be $(x-h)^2+(y-k)^2=r^2$

The circle passes through points (2, 3) and (–1, 1); therefore

$(2-h)^2+(3-k)^2=r^2$ (i)

$(-1-h)^2+(1-k)^2=r^2$ (ii)

The centre (h, k) of the circle lies on the line $x-3y-11=0$; therefore $h-3k-11=0$ (iii)

Solving equations (i) and (ii), we get

$(2-h)^2+(3-k)^2=(-1-h)^2+(1-k)^2$

$\Rightarrow 4-4h+h^2+9-6k+k^2=1+2h+h^2+1-2k+k^2$

$\Rightarrow 6h+4k=11$ (iv)

Now, solving equations (iii) and (iv), we get $h=\frac{7}{2}, k=\frac{-5}{2}$

Substituting the values of h and k in Equation (i), we get

$$\left(2-\frac{7}{2}\right)^2+\left(3+\frac{5}{2}\right)^2=r^2$$

$$\Rightarrow\left(\frac{-3}{2}\right)^2+\left(\frac{11}{2}\right)^2=r^2 \Rightarrow \frac{9}{4}+\frac{121}{4}=r^2 \Rightarrow \frac{130}{4}=r^2$$

Thus, the equation of the required circle is

$$\left(x-\frac{7}{2}\right)^2+\left(y+\frac{5}{2}\right)^2=\frac{130}{4}$$

$$\Rightarrow\left(\frac{2x-7}{2}\right)^2+\left(\frac{2y+5}{2}\right)^2=\frac{130}{4}$$

$\Rightarrow 4x^2-28x+49+4y^2+20y+25=130$

$\Rightarrow 4(x^2+y^2-7x+5y-14)=0$

$\Rightarrow x^2+y^2-7x+5y-14=0$

Q12. Find the equation of the circle with radius 5 whose centre lies on the x-axis and passes through the point (2, 3).

Solution:

Let the equation of the required circle be $(x-h)^2+(y-k)^2=r^2$.

The radius of the circle is 5 and its centre lies on the x-axis; therefore $k=0, r=5$.

Thus, the equation of the circle is $(x-h)^2+y^2=25$.

Now, since the circle passes through point (2, 3)

$(2-h)^2+3^2=25$

$\Rightarrow (2-h)^2=25-9 \Rightarrow 2-h=\pm\sqrt{16}=\pm 4$

Case 1: For $2-h=4 \Rightarrow h=-2$

Thus, the equation of the circle is $(x+2)^2+y^2=25$

$\Rightarrow x^2+4x+4+y^2=25$

$\Rightarrow x^2+y^2+4x-21=0$

Case 2: For $2-h=-4 \Rightarrow h=6$

Thus, the equation of the circle is $(x-6)^2+y^2=25$

$\Rightarrow x^2-12x+36+y^2=25$

$\Rightarrow x^2+y^2-12x+11=0$

Q13. Find the equation of the circle passing through (0, 0) and making intercepts a and b on the coordinate axes.

Solution:

Let the equation of the required circle be $(x-h)^2+(y-k)^2=r^2$

The centre of the circle passes through (0, 0); therefore $(0-h)^2+(0-k)^2=r^2 \Rightarrow h^2+k^2=r^2$

Thus, the equation of the circle is $(x-h)^2+(y-k)^2=h^2+k^2$

Now, since the circle makes intercepts a and b on the coordinate axes, it means the circle passes through points $(a, 0)$ and $(0, b)$. Therefore

$(a-h)^2+(0-k)^2=h^2+k^2$ (i)

$(0-h)^2+(b-k)^2=h^2+k^2$ (ii)

From Equation (i), we get

$a^2-2ah+h^2+k^2=h^2+k^2$

$\Rightarrow a^2-2ah=0 \Rightarrow a(a-2h)=0$

$\Rightarrow a=0$ or $(a-2h)=0$

$\Rightarrow a-2h=0 \Rightarrow h=a/2$ $(\because a \neq 0)$

And from Equation (ii), we get

$h^2+b^2-2bk+k^2=h^2+k^2$

$\Rightarrow b^2-2bk=0 \Rightarrow b(b-2k)=0$

$\Rightarrow b=0$ or $(b-2k)=0$

$\Rightarrow b-2h=0 \Rightarrow k=b/2$ $(\because b \neq 0)$

Hence, the required equation of the circle is

$$\left(x-\frac{a}{2}\right)^2+\left(y-\frac{b}{2}\right)^2=\left(\frac{a}{2}\right)^2+\left(\frac{b}{2}\right)^2$$

$$\Rightarrow \left(\frac{2x-a}{2}\right)^2+\left(\frac{2y-b}{2}\right)^2=\frac{a^2+b^2}{4}$$

$$\Rightarrow 4x^2-4ax+a^2+4y^2-4by+b^2=a^2+b^2$$

$$\Rightarrow 4x^2+4y^2-4ax-4by=0 \Rightarrow x^2+y^2-ax-by=0$$

Q14. Find the equation of a circle with centre (2, 2) and passes through the point (4, 5).

Solution:

The centre of the circle is given as $(h,k)=(2,2)$.

The circle passes through point (4, 5); therefore the radius (r) of the circle is the distance between the points (2, 2) and (4, 5).

$$\Rightarrow r=\sqrt{(2-4)^2+(2-5)^2}=\sqrt{(-2)^2+(-3)^2}=\sqrt{13}$$

Thus, the equation of the circle is

$$(x-h)^2+(y-k)^2=r^2$$

$$\Rightarrow (x-2)^2+(y-2)^2=(\sqrt{13})^2$$

$$\Rightarrow x^2-4x+4+y^2-4y+4=13$$

$$\Rightarrow x^2+y^2-4x-4y-5=0$$

Q15. Does the point (–2.5, 3.5) lie inside, outside or on the circle $x^2+y^2=25$?

Solution:

The equation of the given circle is $x^2+y^2=25 \Rightarrow (x-0)^2+(y-0)^2=5^2$.

Comparing with equation of the circle $(x-h)^2+(y-k)^2=r^2$, we get $h=0, k=0, r=5$.

$\therefore$ Centre = (0,0) and radius = 5

Now, the distance between point (–2.5, 3.5) and the centre (0, 0) is

$$d=\sqrt{(-2.5-0)^2+(3.5-0)^2}=\sqrt{6.25+12.25}$$

$$\Rightarrow d=\sqrt{18.5}=4.3<5 \text{ (radius)}$$

The distance between point (–2.5, 3.5) and centre (0, 0) of the circle is less than the radius of the circle. Therefore, the point (–2.5, 3.5) lies inside the circle.

Day 148

Parabola

A parabola is the set of all points in a plane that are equidistant from a fixed line and a fixed point in the plane. As shown in Fig. 11.3,

$$|P_1D_1|=|P_1F|, |P_2D_2|=|P_2F|, |P_3D_3|=|P_3F|$$

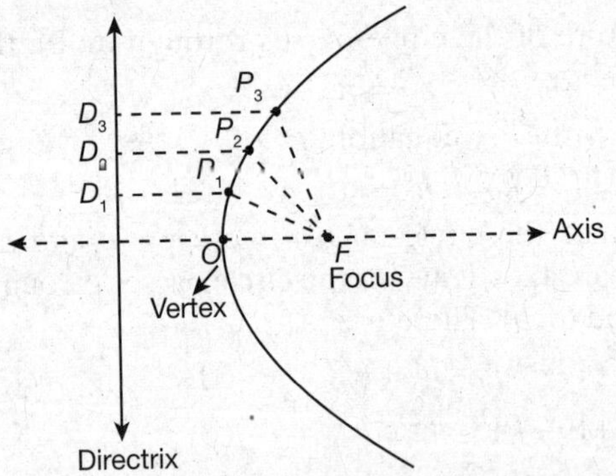

Fig. 11.3 Parabola

- The fixed line is called the directrix of the parabola.
- The fixed point is called the focus (F) of the parabola.
- A line through the focus and perpendicular to the directrix is called the axis of the parabola.
- The point of intersection of the parabola with the axis is called the vertex (O) of the parabola.

Standard Equations of a Parabola

- The standard equation of a parabola is simplest if the vertex is at the origin and the axis of symmetry is along the x-axis or y-axis.
- The four equations $y^2=4ax$ in Fig 11.4 (a), $y^2=-4ax$ in Fig 11.4 (b), $x^2=4ay$ in Fig 11.4 (c) and $x^2=-4ay$ in Fig 11.4 (d) are known as standard equations of parabolas.
- The standard equations of the parabola have the focus on one of the coordinate axes, vertex at the origin and the directrix parallel to the other coordinate axis.

Note: The study of the equations of parabolas with the focus at any point and any line as directrix is beyond the syllabus.

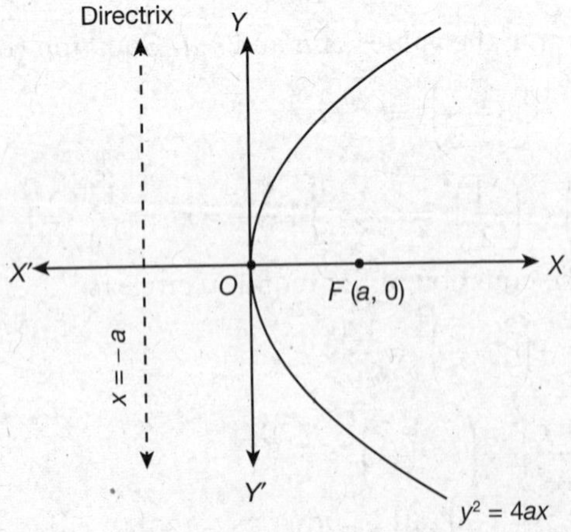

Fig. 11.4 (a) Parabola $y^2=4ax$, focus at $(a, 0)$ and directrix $x=-a$

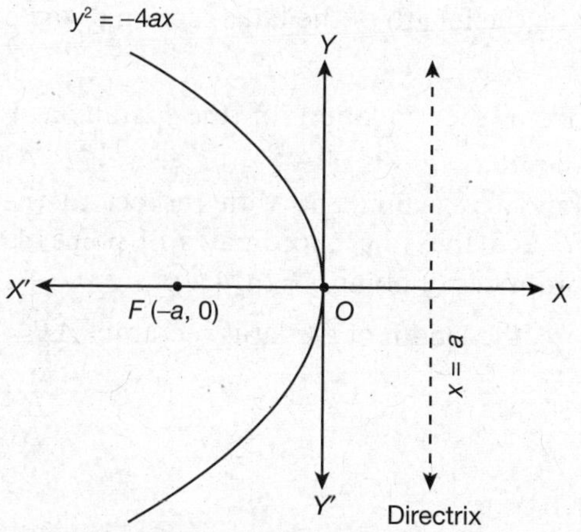

Fig. 11.4 (b) Parabola $y^2 = -4ax$, focus at $(-a, 0)$ and directrix $x = a$

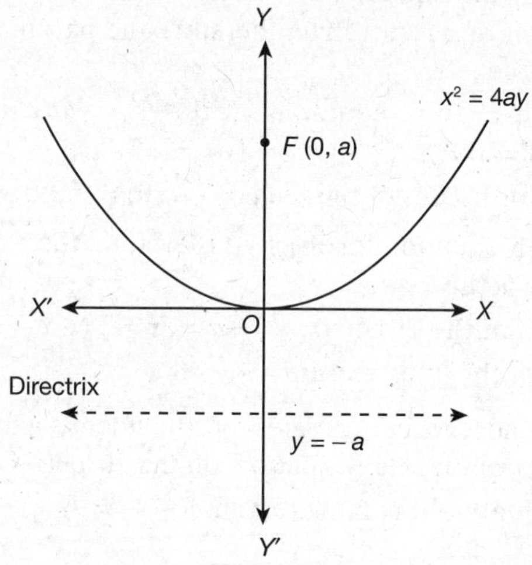

Fig. 11.4 (c) Parabola $x^2 = 4ay$, focus at $(0, a)$ and directrix $y = -a$

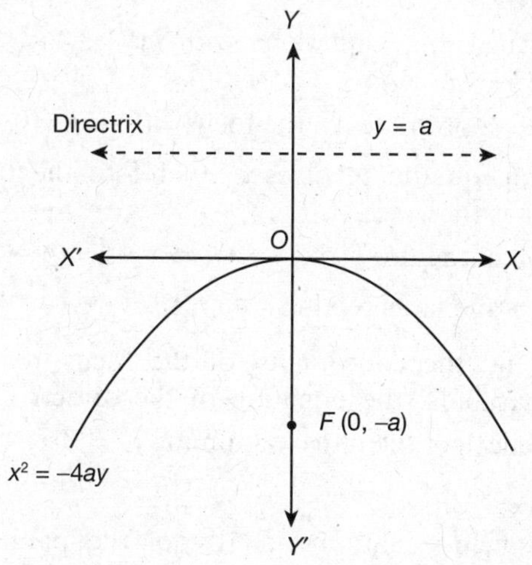

Fig. 11.4 (d) Parabola $x^2 = -4ay$, focus at $(0, -a)$ and directrix $y = a$

Derivation of Equation of Parabola

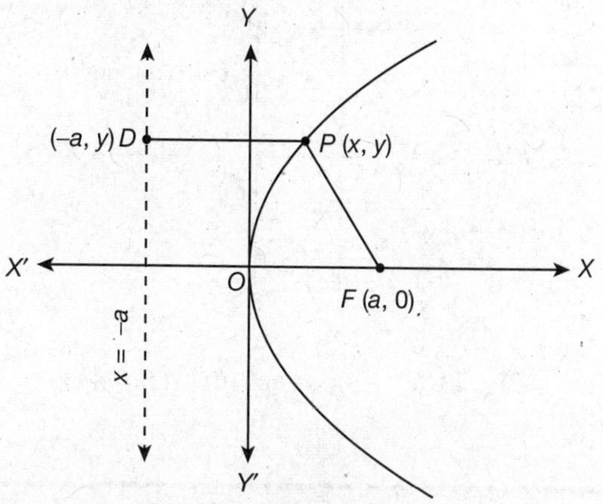

Fig. 11.5 Standard parabola $y^2 = 4ax$

Let a standard parabola with focus at $(a, 0)$ and directrix $x = -a$ be as shown in Fig. 11.5. Then, from the definition of parabola, the distance of point $P(x, y)$ from a point $D(-a, y)$ on the directrix is equal to the distance of point $P(x, y)$ from the focus $F(a, 0)$, that is, $|PD| = |PF|$.

Thus, by the distance formula, we get

$$\sqrt{(x+a)^2+(y-y)^2} = \sqrt{(x-a)^2+(y-0)^2}$$

$$\Rightarrow (x+a)^2 = (x-a)^2 + y^2$$

$$\Rightarrow x^2 + 2ax + a^2 = x^2 - 2ax + a^2 + y^2$$

$$\Rightarrow y^2 = 4ax \quad (a > 0).$$

Similarly, we can derive the other standard equations of parabolas.

Note: From the standard equations of parabolas, we may observe some points.

- A parabola is symmetric with respect to the axis of the parabola.
- If the equation has a y^2 term, then the axis of symmetry is along the x-axis and if the equation has an x^2 term, then the axis of symmetry is along the y-axis.
- When the axis of symmetry is along the x-axis, the parabola opens to the right if the coefficient of x is positive and to the left if the coefficient of x is negative.
- When the axis of symmetry is along the y-axis, the parabola opens upwards if the coefficient of y is positive and downwards if the coefficient of y is negative.

Latus Rectum: The latus rectum of a parabola is a line segment perpendicular to the axis of the parabola, passing through the focus and whose end points lie on the parabola (Fig. 11.6).

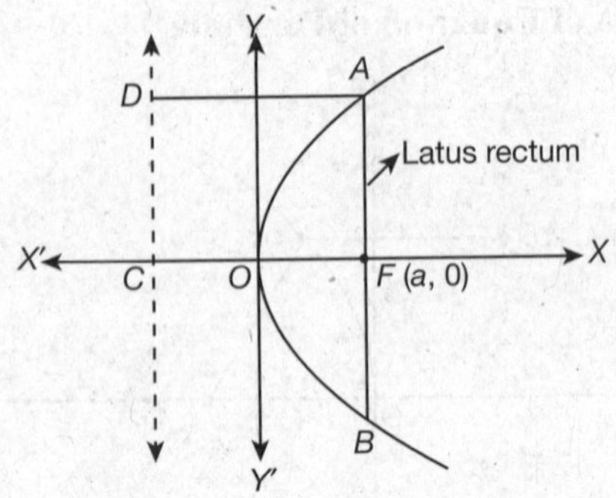

Fig. 11.6 Latus rectum ($AB = 4a$)

Let AB be the length of the latus rectum of the parabola $y^2 = 4ax$.

Then, from the equation of the parabola $y^2 = 4ax$, when we put $x = a$, we get $y^2 = 4a^2 \Rightarrow y = 2a$. And since the parabola is symmetric with respect to the x-axis, $AF = FB$ and thus the coordinates of point $A = (a, 2a)$ and coordinates of point $B = (a, -2a)$.

Therefore, the length of the latus rectum $= AB = 4a$.

Day 149

NCERT Exercise 11.2 (With Solutions)

Q1. Find the coordinates of the focus, axis of the parabola, the equation of the directrix and the length of the latus rectum for $y^2 = 12x$.

Solution:

In the given equation $y^2 = 12x$ of a parabola, the coefficient of x is positive; therefore the parabola opens towards the right.

Comparing this equation with $y^2 = 4ax$, we get $4a = 12 \Rightarrow a = 3$.

Thus, coordinates of the focus $= (a, 0) = (3, 0)$. The given equation involves y^2; therefore, the axis of the parabola is the x-axis.

Equation of the directrix, $x = -a \Rightarrow x = -3 \Rightarrow x + 3 = 0$

Length of the latus rectum $= 4a = 4 \times 3 = 12$

Q2. Find the coordinates of the focus, axis of the parabola, the equation of the directrix and the length of the latus rectum for $x^2 = 6y$.

Solution:

In the given equation $x^2 = 6y$ of the parabola, the coefficient of y is positive; therefore the parabola opens upwards.

Comparing this equation with $x^2 = 4ay$, we get $4a = 6 \Rightarrow a = 3/2$

Thus, coordinates of the focus $= (0, a) = \left(0, \dfrac{3}{2}\right)$

The given equation involves x^2; therefore, the axis of the parabola is the y-axis.

Equation of the directrix, $y = -3/2 \Rightarrow 2y + 3 = 0$

Length of the latus rectum $= 4a = 6$

Q3. Find the coordinates of the focus, axis of the parabola, the equation of the directrix and the length of the latus rectum for $y^2 = -8x$.

Solution:

In the given equation $y^2 = -8x$ of the parabola, the coefficient of x is negative; therefore the parabola opens towards the left.

Comparing this equation with $y^2 = -4ax$, we get $-4a = -8 \Rightarrow a = 2$.

Thus, coordinates of the focus $= (-a, 0) = (-2, 0)$

The given equation involves y^2; therefore, the axis of the parabola is the x-axis.

Equation of the directrix, $x = -a \Rightarrow x = 2 \Rightarrow x - 2 = 0$

Length of the latus rectum $= 4a = 8$

Q4. Find the coordinates of the focus, axis of the parabola, the equation of the directrix and the length of the latus rectum for $x^2 = -16y$.

Solution:

In the given equation $x^2 = -16y$ of the parabola, the coefficient of y is negative; therefore the parabola opens downwards.

Comparing this equation with $x^2 = -4ay$, we get $-4a = -16 \Rightarrow a = 4$.

Thus, the coordinates of the focus $= (0, -a) = (0, -4)$

The given equation involves x^2; therefore, the axis of the parabola is the y-axis.

Equation of the directrix, $y = -a \Rightarrow y = 4 \Rightarrow y - 4 = 0$

Length of the latus rectum $= 4a = 16$

Q5. Find the coordinates of the focus, axis of the parabola, the equation of the directrix and the length of the latus rectum for $y^2 = 10x$.

Solution:

In the given equation $y^2 = 10x$ of the parabola, the coefficient of x is positive; therefore the parabola opens towards the right.

Comparing this equation with $y^2 = 4ax$, we get $4a = 10 \Rightarrow a = 5/2$.

Thus, coordinates of the focus $= (a,0) = \left(\frac{5}{2},0\right)$

The given equation involves y^2; therefore, the axis of the parabola is the x-axis.

Equation of the directrix, $x = -5/2 \Rightarrow 2x+5=0$

Length of the latus rectum $= 4a = 10$

Q6. Find the coordinates of the focus, axis of the parabola, the equation of the directrix and the length of the latus rectum for $x^2 = -9y$.

Solution:

In the given equation $x^2 = -9y$ of the parabola, the coefficient of y is negative; therefore the parabola opens downwards.

Comparing this equation with $x^2 = -4ay$, we get $-4a = -9 \Rightarrow a = 9/4$.

Thus, the coordinates of the focus $= (0,-a) = \left(0,-\frac{9}{4}\right)$

The given equation involves x^2 therefore, the axis of the parabola is the y-axis.

Equation of the directrix,

$y = -a \Rightarrow y = 9/4 \Rightarrow 4y-9=0$

Length of the latus rectum $= 4a = 9$

Q7. Find the equation of the parabola that satisfies the following conditions: focus (6, 0); directrix $x = -6$.

Solution:

The focus (6, 0) lies on the x-axis; therefore the axis of the parabola is the x-axis.

Thus, the equation of the parabola is of the form $y^2 = 4ax$ or $y^2 = -4ax$.

Now, since the directrix is $x = -6$ and focus is (6, 0), the parabola is of the form $y^2 = 4ax$ with $a = 6$. Thus, the equation of the parabola is $y^2 = 24x$.

Q8. Find the equation of the parabola that satisfies the following conditions: focus (0, –3); directrix $y = 3$.

Solution:

Since the focus (0, –3) lies on the y-axis, the axis of the parabola is the y-axis.

Thus, the equation of the parabola is of the form $x^2 = 4ay$ or $x^2 = -4ay$.

Now, since the directrix is $y = 3$ and focus is (0, –3), the parabola is of the form $x^2 = -4ay$ with $a = 3$. Thus, the equation of the parabola is $x^2 = -12y$.

Q9. Find the equation of the parabola that satisfies the following conditions: vertex (0, 0); focus (3, 0).

Solution:

The focus (3, 0) lies on the x-axis; therefore the axis of the parabola is the x-axis.

Thus, the equation of the parabola is of the form $y^2 = 4ax$ or $y^2 = -4ax$.

Now, since the focus is (3, 0), the parabola is of the form $y^2 = 4ax$ with $a = 3$. Thus, the equation of the parabola is $y^2 = 12x$.

Q10. Find the equation of the parabola that satisfies the following conditions: vertex (0, 0) focus (–2, 0).

Solution:

The focus (–2, 0) lies on the x-axis; therefore the axis of the parabola is the x-axis.

Thus, the equation of the parabola is of the form $y^2 = 4ax$ or $y^2 = -4ax$.

Now, since the focus is (–2, 0), the parabola is of the form $y^2 = -4ax$ with $a = 2$. Thus, the equation of the parabola is $y^2 = -8x$.

Q11. Find the equation of the parabola that satisfies the following conditions: vertex (0, 0) passing through (2, 3) and axis is along the x-axis.

Solution:

The vertex of the parabola is (0, 0) and the axis of the parabola is the x-axis; therefore the equation of the parabola is of the form $y^2 = 4ax$ or $y^2 = -4ax$.

Now, since the parabola passes through point (2, 3) which lies in the first quadrant, the equation of the parabola is of the form $y^2 = 4ax$. Now, since the point (2, 3) must satisfy the equation $y^2 = 4ax$, $3^2 = 4a(2) \Rightarrow a = 9/8$

Thus, the equation of the parabola is

$y^2 = 4(9/8)x \Rightarrow y^2 = \frac{9}{2}x \Rightarrow 2y^2 = 9x$

Q12. Find the equation of the parabola that satisfies the following conditions: vertex (0, 0), passing through (5, 2) and symmetric with respect to the y-axis.

Solution:

The vertex of the parabola is (0, 0) and the axis of the parabola is the y-axis; therefore the equation of the parabola is of the form $x^2 = 4ay$ or $x^2 = -4ay$.

Now, since the parabola passes through point (5, 2) which lies in the first quadrant, the equation of the parabola is of the form $x^2 = 4ay$. Now, since point (5, 2) must satisfy the equation $x^2 = 4ay$, $(5)^2 = 4 \times a \times 2 \Rightarrow a = 25/8$

Thus, the equation of the parabola is

$x^2 = 4(25/8)y \Rightarrow 2x^2 = 25y$

Day 150

Ellipse

An ellipse is the set of all points (P) in a plane (Fig. 11.7), the sum of whose distances from two fixed points (F_1 and F_2) in the plane is a constant (which is always greater than the distance between the two fixed points).

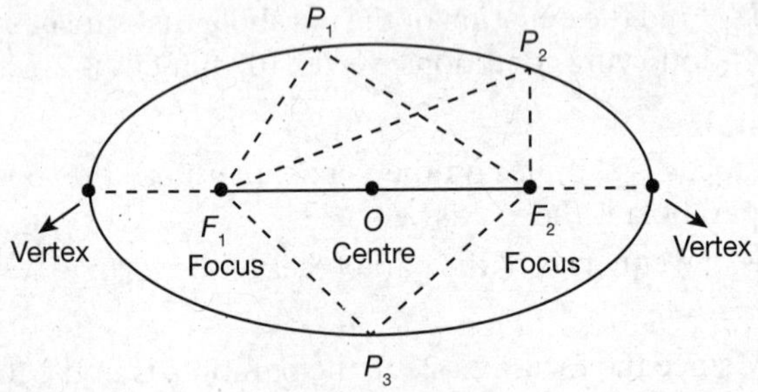

Fig. 11.7 Ellipse

That is, $|P_1F_1|+|P_1F_2|=|P_2F_1|+|P_2F_2|=|P_3F_1|+|P_3F_2|=k$

Note:

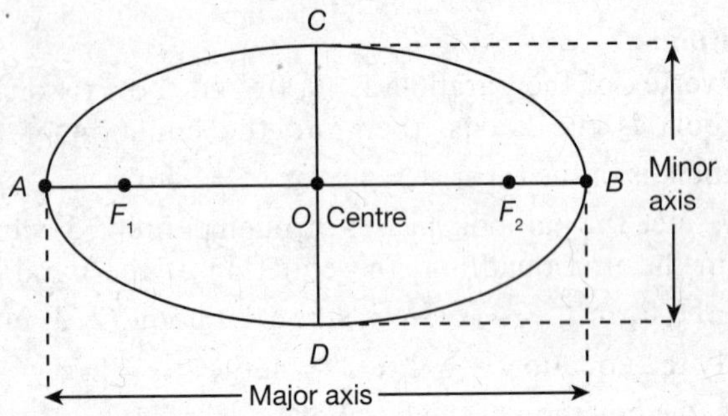

Fig. 11.8 Ellipse—foci, axes and vertices

- The two fixed points (F_1 and F_2) are called the foci (plural of 'focus') of the ellipse (Fig. 11.8).
- The mid-point of the line segment joining the foci is called the centre (O) of the ellipse.
- The line segment through the foci of the ellipse is called the major axis and the line segment through the centre and perpendicular to the major axis is called the minor axis.
- The end points of the major axis are called the vertices (A and B) of the ellipse.

Relationship between semi-major axis (*a*), semi-minor axis (*b*) and the distance of the focus from the centre of the ellipse (*c*)

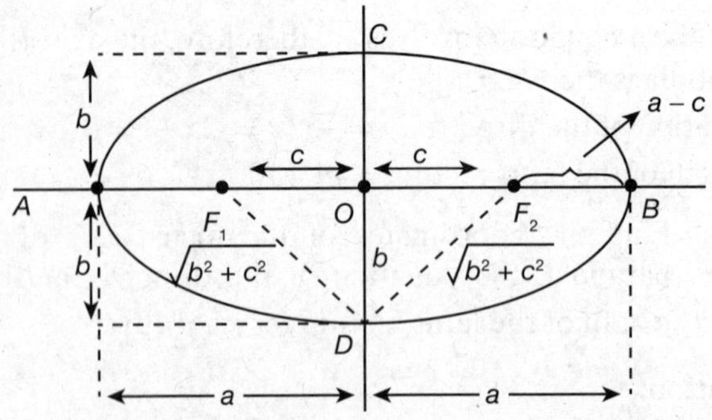

Fig. 11.9 Ellipse—major and minor axes

Let the length of the major axis be $2a$, the length of the minor axis be $2b$ and the distance between the foci (F_1 and F_2) be $2c$ (Fig. 11.9).

Then, the length of the semi-major axis is a and the semi-minor axis is b.

Let a point B be at the one end of the major axis; then the sum of the distances from the point B to the foci is

$BF_1+BF_2=OF_1+BO+BF_2$

$=c+a+a-c=2a$ (aince $F_1P=F_1O+OP$)

Now, take a point D at one end of the minor axis; then the sum of the distances from the point D to the foci is

$DF_1+DF_2=\sqrt{b^2+c^2}+\sqrt{b^2+c^2}=2\sqrt{b^2+c^2}$.

Since both points B and D lie on the ellipse, by the definition of ellipse,

$BF_1+BF_2=DF_1+DF_2$

$2a=2\sqrt{b^2+c^2}\Rightarrow a=\sqrt{b^2+c^2}$

$\Rightarrow a^2=b^2+c^2\Rightarrow c=\sqrt{a^2-b^2}$

Eccentricity (*e*): The eccentricity of an ellipse is the ratio of the distances from the centre of the ellipse to one focus and to one of the vertices of the ellipse. That is, $e=\dfrac{c}{a}$.

Thus, we can say that the focus of an ellipse is at a distance of $c=ae$ from the centre.

And the length of the minor axis is $b=\sqrt{a^2-c^2}=\sqrt{a^2-(ae)^2}=a\sqrt{1-e^2}$

Special Cases of Ellipse

From the equation $b=a\sqrt{1-e^2}$, if we keep a fixed and vary the eccentricity (e) from 0 to 1, the resulting ellipses will vary from circle to line segment.

Case 1: When $e=0\Rightarrow b=a\sqrt{1-0^2}=a$, both foci merge together with the centre of the ellipse, and thus, the ellipse becomes a circle with radius $r=a=b$ (Fig. 11.10).

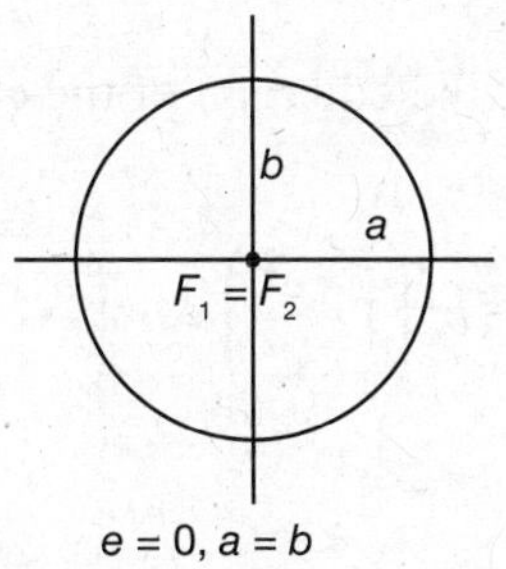

Fig. 11.10 Ellipse as a circle

Case 2: When $e = 1 \Rightarrow b = a\sqrt{1-1^2} = 0$, the minor axis becomes zero and the major axis is equal to the distance between the foci = $2a$. Therefore, the ellipse reduces to the line segment F_1F_2 joining the two foci (Fig. 11.11).

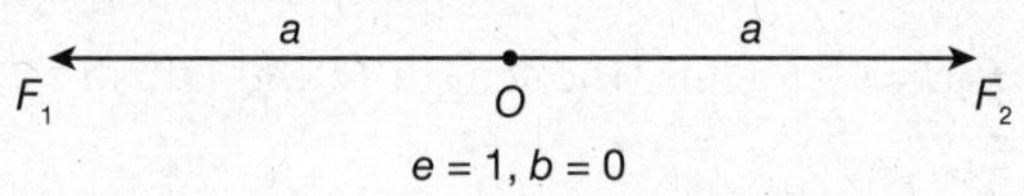

Fig. 11.11 Ellipse as a line segment

Standard Equations of an Ellipse

The equation of an ellipse is simplest if the centre of the ellipse is at the origin and the foci are on the x-axis (Fig. 11.12) or on the y-axis (Fig. 11.13).

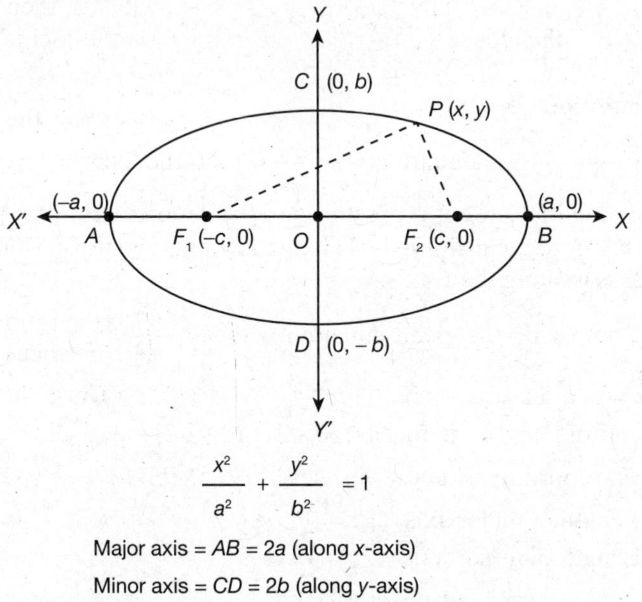

Fig. 11.12 Ellipse—foci on the x-axis

For deriving the equation of the ellipse with foci on the x-axis, let $F_1(-c,0)$ and $F_2(c,0)$ be the foci and $O(0,0)$ be the mid-point of the line segment F_1F_2.

Let $P(x, y)$ be any point on the ellipse. Then, by the definition of an ellipse, the sum of the distances from P to the two foci are equal to $2a$; therefore $PF_1 + PF_2 = 2a$

$\Rightarrow \sqrt{(x+c)^2 + y^2} + \sqrt{(x-c)^2 + y^2} = 2a$

(Using the distance formula)

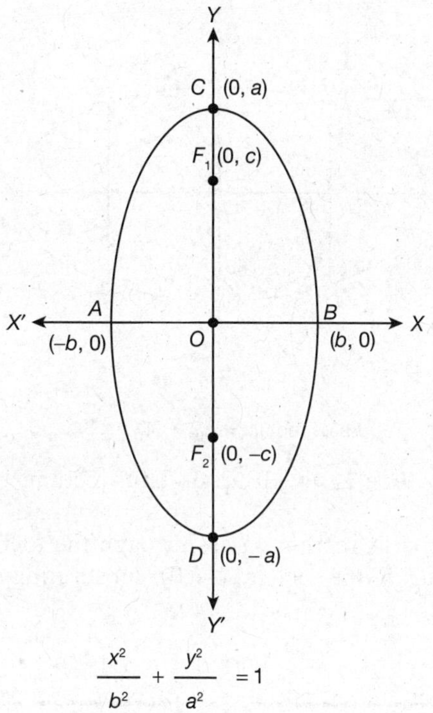

Fig. 11.13 Ellipse—foci on the y-axis

Squaring both sides and simplifying, we get

$\frac{x^2}{a^2} + \frac{y^2}{a^2 - c^2} = 1 \Rightarrow \frac{x^2}{a^2} + \frac{y^2}{b^2} = 1 \; (\because c^2 = a^2 - b^2)$ as the required equation of an ellipse with the foci on the x-axis.

Similarly, we can the derive equation of the ellipse $\frac{x^2}{b^2} + \frac{y^2}{a^2} = 1$ with the foci on the y-axis.

These two equations are known as standard equations of the ellipse.

Note:

- An ellipse is symmetric with respect to both the coordinate axes, since if point (x, y) is on the ellipse, then $(-x, y), (x, -y), (-x, -y)$ are also points on the ellipse.
- The foci always lie on the major axis.
- The major axis can be determined by finding the intercepts on the axes of symmetry.
- In standard equations of the ellipse, the major axis is along the x-axis if the coefficient of x^2 has the larger denominator and it is along the y-axis if the coefficient of y^2 has the larger denominator.

Latus rectum: The latus rectum of an ellipse is a line segment (AB or CD) perpendicular to the major axis through any of the foci and whose end points lie on the ellipse (Fig. 11.14).

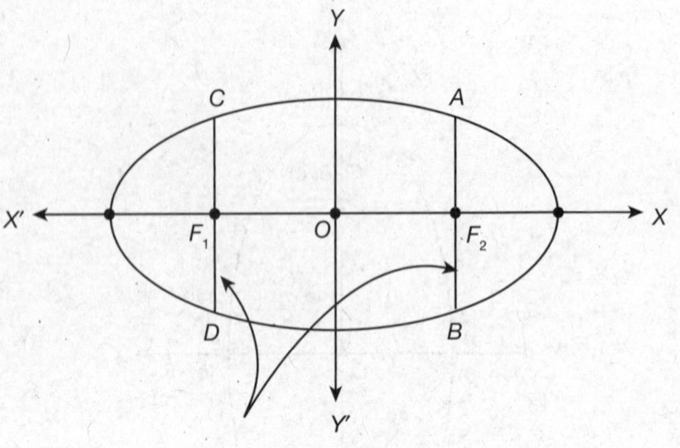

Fig. 11.14 Ellipse—latus rectum

Since the latus rectum passes through the foci, and the coordinates of the foci are $(\pm c, 0)$, substituting the value of $x = \pm c = \pm ae$ in the equation of the ellipse $\frac{x^2}{a^2}+\frac{y^2}{b^2}=1$, we get

$$\frac{c^2}{a^2}+\frac{y^2}{b^2}=1 \Rightarrow y^2 = b^2\left(1-\frac{c^2}{a^2}\right) = b^2\left(\frac{a^2-c^2}{a^2}\right)$$

$$\Rightarrow y^2 = \frac{b^4}{a^2} \qquad (\because b^2 = a^2 - c^2)$$

$\Rightarrow y = \pm\frac{b^2}{a}$, which is the intersection point of the latus rectum and the ellipse. That is, coordinates of A and B are $\left(c, \frac{b^2}{a}\right)$ and $\left(c, -\frac{b^2}{a}\right)$, respectively.

Since the ellipse is symmetric with respect to the y-axis, $AF_2 = F_2B = \frac{b^2}{a}$. Thus, the length of the latus rectum is $\frac{2b^2}{a}$.

Day 151

NCERT Exercise 11.3 (With Solutions)

Q1. Find the coordinates of the foci, the vertices, the length of the major axis, the minor axis, the eccentricity and the latus rectum of the ellipse $\frac{x^2}{36}+\frac{y^2}{16}=1$.

Solution:

In the given equation $\frac{x^2}{36}+\frac{y^2}{16}=1$ of the ellipse, the denominator of x^2 is greater than the denominator of y^2; therefore the major axis is along the x-axis and the minor axis is along the y-axis.

Comparing the given equation with $\frac{x^2}{a^2}+\frac{y^2}{b^2}=1$, we get $a = 6$, $b = 4$ and $c = \sqrt{a^2-b^2} = \sqrt{36-16} = 2\sqrt{5}$. Thus, the coordinates of the foci are $(2\sqrt{5}, 0), (-2\sqrt{5}, 0)$.

The coordinates of the vertices are (6, 0), (–6, 0).

Length of major axis $= 2a = 12$

Length of minor axis $= 2b = 8$

Eccentricity $e = \frac{c}{a} = \frac{2\sqrt{5}}{6} = \frac{\sqrt{5}}{3}$

Length of latus rectum $= \frac{2b^2}{a} = \frac{2\times 16}{6} = 16/3$

Q2. Find the coordinates of the foci, the vertices the length of the major axis, the minor axis, the eccentricity and the length of the latus rectum of the ellipse $\frac{x^2}{4}+\frac{y^2}{25}=1$.

Solution:

In the given equation $\frac{x^2}{4}+\frac{y^2}{25}=1$ of the ellipse, the denominator of y^2 is greater than the denominator of x^2; therefore the major axis is along the y-axis and the minor axis is along the x-axis.

Comparing the given equation with $\frac{x^2}{b^2}+\frac{y^2}{a^2}=1$, we get $a = 5$, $b = 2$ and $c = \sqrt{a^2-b^2} = \sqrt{25-4} = \sqrt{21}$. Thus, the coordinates of the foci are $(0, \sqrt{21}), (0, -\sqrt{21})$.

The coordinates of the vertices are $(0, 5), (0, -5)$

Length of major axis $= 2a = 10$

Length of minor axis $= 2b = 4$

Eccentricity $e = \frac{c}{a} = \frac{\sqrt{21}}{5}$

Length of latus rectum $= \frac{2b^2}{a} = \frac{2\times 4}{5} = 8/5$

Q3. Find the coordinates of the foci, the vertices the length of the major axis, the minor axis, the eccentricity and the length of the latus rectum of the ellipse $\frac{x^2}{16}+\frac{y^2}{9}=1$.

Solution:

In the given equation $\frac{x^2}{16}+\frac{y^2}{9}=1$ of the ellipse, the denominator of x^2 is greater than the denominator of y^2; therefore the major axis is along the x-axis and the minor axis is along the y-axis.

Comparing the given equation with $\frac{x^2}{a^2}+\frac{y^2}{b^2}=1$, we get $a=4$, $b=3$ and $c=\sqrt{a^2-b^2}=\sqrt{16-9}=\sqrt{7}$. Thus, the coordinates of the foci are $(\sqrt{7},0),(-\sqrt{7},0)$

The coordinates of the vertices are $(4,0)$, $(-4,0)$

Length of major axis $=2a=8$

Length of minor axis $=2b=6$

Eccentricity $e=\frac{c}{a}=\frac{\sqrt{7}}{4}$

Length of latus rectum $=\frac{2b^2}{a}=\frac{2\times 9}{4}=\frac{9}{2}$

Q4. Find the coordinates of the foci, the vertices the length of the major axis, the minor axis, the eccentricity and the length of the latus rectum of the ellipse

$$\frac{x^2}{25}+\frac{y^2}{100}=1.$$

Solution:

In the given equation $\frac{x^2}{25}+\frac{y^2}{100}=1$ of the ellipse, the denominator of y^2 is greater than the denominator of x^2; therefore the major axis is along the y-axis and the minor axis is along the x-axis.

Comparing the given equation with $\frac{x^2}{b^2}+\frac{y^2}{a^2}=1$, we get $a=10$, $b=5$ and $c=\sqrt{a^2-b^2}=\sqrt{100-25}=5\sqrt{3}$. Thus, the coordinates of the foci are $(0,5\sqrt{3}),(0,-5\sqrt{3})$.

The coordinates of the vertices are $(0,10)$, $(0,-10)$

Length of major axis $=2a=20$

Length of minor axis $=2b=10$

Eccentricity $e=\frac{c}{a}=\frac{5\sqrt{3}}{10}=\frac{\sqrt{3}}{2}$

Length of latus rectum $=\frac{2b^2}{a}=\frac{2\times 25}{10}=5$

Q5. Find the coordinates of the foci, the vertices, the length of the major axis, the minor axis, the eccentricity and the length of the latus rectum of the ellipse $\frac{x^2}{49}+\frac{y^2}{36}=1$.

Solution:

In the given equation $\frac{x^2}{49}+\frac{y^2}{36}=1$ of the ellipse, the denominator of x^2 is greater than the denominator of y^2; therefore the major axis is along the x-axis and the minor axis is along the y-axis.

Comparing the given equation with $\frac{x^2}{a^2}+\frac{y^2}{b^2}=1$, we get $a=7$, $b=6$ and $c=\sqrt{a^2-b^2}=\sqrt{49-36}=\sqrt{13}$. Thus, the coordinates of the foci are $(\sqrt{13},0),(-\sqrt{13},0)$.

The coordinates of the vertices are $(7,0)$, $(-7,0)$

Length of major axis $=2a=14$

Length of minor axis $=2b=12$

Eccentricity, $e=\frac{c}{a}=\frac{\sqrt{13}}{7}$

Length of latus rectum $=\frac{2b^2}{a}=\frac{2\times 36}{7}=72/5$

Q6. Find the coordinates of the foci, the vertices, the length of the major axis, the minor axis, the eccentricity and the length of the latus rectum of the ellipse $\frac{x^2}{100}+\frac{y^2}{400}=1$.

Solution:

In the given equation $\frac{x^2}{100}+\frac{y^2}{400}=1$ of the ellipse, the denominator of y^2 is greater than the denominator of x^2; therefore the major axis is along the y-axis and the minor axis is along the x-axis.

Comparing the given equation with $\frac{x^2}{b^2}+\frac{y^2}{a^2}=1$, we get $a=20$, $b=10$ and $c=\sqrt{a^2-b^2}=\sqrt{400-100}=10\sqrt{3}$. Thus, the coordinates of the foci are $(0,10\sqrt{3}),(0,-10\sqrt{3})$.

The coordinates of the vertices are $(0,20)$, $(0,-20)$

Length of major axis $=2a=40$

Length of minor axis $=2b=20$

Eccentricity, $e=\frac{c}{a}=\frac{10\sqrt{3}}{20}=\frac{\sqrt{3}}{2}$

Length of latus rectum $=\frac{2b^2}{a}=\frac{2\times 100}{20}=10$

Q7. Find the coordinates of the foci, the vertices, the length of the major axis, the minor axis, the eccentricity and the length of the latus rectum of the ellipse $36x^2+4y^2=144$.

Solution:

In the given equation $36x^2 + 4y^2 = 144$ $\Rightarrow \frac{36x^2}{144} + \frac{4y^2}{144} = 1 \Rightarrow \frac{x^2}{4} + \frac{y^2}{36} = 1$ of the ellipse, the denominator of y^2 is greater than the denominator of x^2; therefore the major axis is along the y-axis and the minor axis is along the x-axis.

Comparing the given equation with $\frac{x^2}{b^2} + \frac{y^2}{a^2} = 1$, we get $a = 6, b = 2$ and $c = \sqrt{a^2 - b^2} = \sqrt{36-4} = 4\sqrt{2}$. Thus, the coordinates of the foci are $(0, 4\sqrt{2}), (0, -4\sqrt{2})$.

The coordinates of the vertices are $(0,6)$, $(0,-6)$

Length of major axis $= 2a = 12$

Length of minor axis $= 2b = 4$

Eccentricity $e = \frac{c}{a} = \frac{4\sqrt{2}}{6} = \frac{2\sqrt{2}}{3}$

Length of latus rectum $= \frac{2b^2}{a} = \frac{2 \times 4}{6} = \frac{4}{3}$

Q8. Find the coordinates of the foci, the vertices, the length of the major axis, the minor axis, the eccentricity and the length of the latus rectum of the ellipse $16x^2 + y^2 = 16$.

Solution:

In the given equation $16x^2 + y^2 = 16$ $\Rightarrow \frac{16x^2}{16} + \frac{y^2}{16} = 1 \Rightarrow \frac{x^2}{1} + \frac{y^2}{16} = 1$ of the ellipse, the denominator of y^2 is greater than the denominator of x^2; therefore the major axis is along the y-axis and the minor axis is along the x-axis.

Comparing the given equation with $\frac{x^2}{b^2} + \frac{y^2}{a^2} = 1$, we get $a = 4, b = 1$ and $c = \sqrt{a^2 - b^2} = \sqrt{16-1} = \sqrt{15}$. Thus, the coordinates of the foci are $(0, \sqrt{15}), (0, -\sqrt{15})$.

The coordinates of the vertices are $(0,4)$, $(0,-4)$

Length of major axis $= 2a = 8$

Length of minor axis $= 2b = 2$

Eccentricity $e = \frac{c}{a} = \frac{\sqrt{15}}{4}$

Length of latus rectum $= \frac{2b^2}{a} = \frac{2 \times 1}{4} = \frac{1}{2}$

Q9. Find the coordinates of the foci, the vertices, the length of the major axis, the minor axis, the eccentricity and the length of the latus recturm of the ellipse $4x^2 + 9y^2 = 36$.

Solution:

In the given equation $4x^2 + 9y^2 = 36$ $\Rightarrow \frac{4x^2}{36} + \frac{9y^2}{36} = 1 \Rightarrow \frac{x^2}{9} + \frac{y^2}{4} = 1$ of the ellipse, the denominator of x^2 is greater than the denominator of y^2; therefore the major axis is along the x-axis and minor axis is along the y-axis.

Comparing the given equation with $\frac{x^2}{a^2} + \frac{y^2}{b^2} = 1$, we get $a = 7, b = 6$ and $c = \sqrt{a^2 - b^2} = \sqrt{9-4} = \sqrt{5}$. Thus, the coordinates of the foci are $(\sqrt{5}, 0), (-\sqrt{5}, 0)$.

The coordinates of the vertices are $(3,0)$, $(-3,0)$

Length of major axis $= 2a = 6$

Length of minor axis $= 2b = 4$

Eccentricity $e = \frac{c}{a} = \frac{\sqrt{5}}{3}$

Length of latus rectum $= \frac{2b^2}{a} = \frac{2 \times 4}{3} = \frac{8}{3}$

Q10. Find the equation of the ellipse that satisfies the given conditions: vertices $(\pm 5, 0)$, foci $(\pm 4, 0)$.

Solution:

The vertices $(\pm 5, 0)$ are on the x-axis; therefore the equation of the ellipse will be of the form $\frac{x^2}{a^2} + \frac{y^2}{b^2} = 1$ and the semi-major axis $a = 5$.

The foci are $(\pm 4, 0)$, therefore $c = 4$

Now, from equation $a^2 = b^2 + c^2$

$\Rightarrow 5^2 = b^2 + 4^2 \Rightarrow b^2 = 25 - 16 \Rightarrow b = \sqrt{9} = 3$

Thus, the equation of the ellipse is $\frac{x^2}{5^2} + \frac{y^2}{3^2} = 1$ or $\frac{x^2}{25} + \frac{y^2}{9} = 1$.

Q11. Find the equation of the ellipse that satisfies the given conditions: vertices $(0, \pm 13)$, foci $(0, \pm 5)$.

Solution:

The vertices $(0, \pm 13)$ are on the y-axis; therefore the equation of the ellipse will be of the form $\frac{x^2}{b^2} + \frac{y^2}{a^2} = 1$ and the semi-major axis $a = 13$.

The foci are $(0, \pm 5)$, therefore $c = 5$.

Now, from equation $a^2 = b^2 + c^2$

$\Rightarrow 13^2 = b^2 + 5^2 \Rightarrow b^2 = 169 - 25 \Rightarrow b = \sqrt{144} = 12$

Thus, the equation of the ellipse is $\frac{x^2}{12^2} + \frac{y^2}{13^2} = 1$ or

$\frac{x^2}{144}+\frac{y^2}{169}=1.$

Q12. Find the equation of the ellipse that satisfies the given condition: vertices $(\pm 6, 0)$, foci $(\pm 4, 0)$.

Solution:
The vertices $(\pm 6, 0)$ are on the x-axis; therefore the equation of the ellipse will be of the form $\frac{x^2}{a^2}+\frac{y^2}{b^2}=1$ and the semi-major axis $a = 6$.
The foci are $(\pm 4, 0)$, therefore $c = 4$
Now, from equation $a^2 = b^2 + c^2$
$\Rightarrow 6^2 = b^2 + 4^2 \Rightarrow b^2 = 36 - 16 \Rightarrow b = \sqrt{20}$
Thus, the equation of the ellipse is $\frac{x^2}{6^2}+\frac{y^2}{(\sqrt{20})^2}=1$ or $\frac{x^2}{36}+\frac{y^2}{20}=1.$

Q13. Find the equation of the ellipse that satisfies the given conditions: ends of major axis are $(\pm 3, 0)$ and ends of minor axis are $(0, \pm 2)$

Solution:
The ends of the major axis are $(\pm 3, 0)$; therefore the major axis is along the x-axis and $a = 3$.
The ends of the minor axis are $(0, \pm 2)$; therefore the minor axis is along the y-axis and $b = 2$.
Thus, the equation of the ellipse will be of the form

$$\frac{x^2}{a^2}+\frac{y^2}{b^2}=1 \Rightarrow \frac{x^2}{3^2}+\frac{y^2}{2^2}=1 \Rightarrow \frac{x^2}{9}+\frac{y^2}{4}=1$$

Q14. Find the equation of the ellipse that satisfies the given conditions: ends of major axis $(0, \pm\sqrt{5})$ and ends of minor axis $(\pm 1, 0)$.

Solution:
The ends of the major axis are $(0, \pm\sqrt{5})$; therefore the major axis is along the y-axis and $a = \sqrt{5}$.
The ends of the minor axis are $(\pm 1, 0)$; therefore the minor axis is along the y-axis and $b = 1$.
Thus, the equation of the ellipse will be of the form

$$\frac{x^2}{b^2}+\frac{y^2}{a^2}=1 \Rightarrow \frac{x^2}{1^2}+\frac{y^2}{(\sqrt{5})^2}=1 \Rightarrow \frac{x^2}{1}+\frac{y^2}{5}=1.$$

Q15. Find the equation of the ellipse that satisfies the given conditions: length of the major axis 26, foci $(\pm 5, 0)$.

Solution:
Length of the major axis, $2a = 26 \Rightarrow a = 13$
Since the foci $(\pm 5, 0)$ are on the x-axis, the major axis is along the x-axis and $c = 5$.
Now, from the equation $a^2 = b^2 + c^2$
$\Rightarrow 13^2 = b^2 + 5^2 \Rightarrow b^2 = 169 - 25 \Rightarrow b = \sqrt{144} = 12$
Thus, the equation of the ellipse will be of the form

$$\frac{x^2}{a^2}+\frac{y^2}{b^2}=1 \Rightarrow \frac{x^2}{13^2}+\frac{y^2}{12^2}=1 \Rightarrow \frac{x^2}{169}+\frac{y^2}{144}=1$$

Q16. Find the equation of the ellipse that satisfies the given conditions: length of the minor axis 16, foci $(0, \pm 6)$.

Solution:
Length of the minor axis, $2b = 16 \Rightarrow b = 8$
Since the foci $(0, \pm 6)$ are on the y-axis, the major axis is along the y-axis and $c = 6$.
Now, from the equation $a^2 = b^2 + c^2$
$\Rightarrow a^2 = 8^2 + 6^2 \Rightarrow a^2 = 100 \Rightarrow a = \sqrt{100} = 10$
Thus, the equation of the ellipse will be of the form

$$\frac{x^2}{b^2}+\frac{y^2}{a^2}=1 \Rightarrow \frac{x^2}{8^2}+\frac{y^2}{10^2}=1 \Rightarrow \frac{x^2}{64}+\frac{y^2}{100}=1$$

Q17. Find the equation of the ellipse that satisfies the given conditions: foci $(\pm 3, 0), a = 4$.

Solution:
The foci $(\pm 3, 0)$ are on the x-axis; therefore the major axis is along the x-axis and $c = 3$
Now, from the equation $a^2 = b^2 + c^2$
$\therefore 4^2 = b^2 + 3^2 \Rightarrow b^2 = 16 - 9 = 7 \Rightarrow b = \sqrt{7}$
Therefore, the equation of the ellipse will be of the form

$$\frac{x^2}{a^2}+\frac{y^2}{b^2}=1 \Rightarrow \frac{x^2}{4^2}+\frac{y^2}{(\sqrt{7})^2}=1 \Rightarrow \frac{x^2}{16}+\frac{y^2}{7}=1$$

Q18. Find the equation of the ellipse that satisfies the given conditions: $b = 3, c = 4$, centre at the origin; foci on the x-axis.

Solution:
It is given that $b = 3, c = 4$; therefore from the equation $a^2 = b^2 + c^2 \Rightarrow a^2 = 3^2 + 4^2 = 25 \Rightarrow a = 5$
Since the foci are on the x-axis, the major axis is along the x-axis.
Thus, the equation of the ellipse will be of the form

$$\frac{x^2}{a^2}+\frac{y^2}{b^2}=1 \Rightarrow \frac{x^2}{5^2}+\frac{y^2}{3^2}=1 \Rightarrow \frac{x^2}{25}+\frac{y^2}{9}=1$$

Q19. Find the equation of the ellipse that satisfies the given conditions: centre at (0, 0), major axis on the y-axis and passes through the points (3, 2) and (1, 6).

Solution:
The centre is at (0, 0) and the major axis is on the y-axis; therefore the equation of the ellipse will be of the form $\frac{x^2}{b^2}+\frac{y^2}{a^2}=1$, where a is the semi-major axis.

Since the ellipse passes through points (3, 2) and (1, 6), these points must satisfy the equation of ellipse $\frac{x^2}{b^2}+\frac{y^2}{a^2}=1$.

Thus, $\frac{9}{b^2}+\frac{4}{a^2}=1$ and $\frac{1}{b^2}+\frac{36}{a^2}=1$

Solving the above equations, we get $b^2 = 10$, $a^2 = 40$.

Thus, the equation of the ellipse is $\frac{x^2}{10}+\frac{y^2}{40}=1$

$\Rightarrow 4x^2 + y^2 = 40$

Q20. Find the equation of the ellipse that satisfies the given conditions: major axis is on the x-axis and passes through the points (4, 3) and (6, 2).

Solution:

The centre is at (0, 0) and the major axis is on the x-axis, therefore the equation of the ellipse will be of the form $\frac{x^2}{a^2}+\frac{y^2}{b^2}=1$, where a is the semi-major axis.

Since the ellipse passes through points (4, 3) and (6, 2), these points must satisfy the equation of ellipse $\frac{x^2}{a^2}+\frac{y^2}{b^2}=1$.

Thus, $\frac{16}{a^2}+\frac{9}{b^2}=1$ and $\frac{36}{a^2}+\frac{4}{b^2}=1$

Solving the above equations, we get $a^2 = 52, b^2 = 13$.

Thus, the equation of the ellipse is $\frac{x^2}{52}+\frac{y^2}{13}=1$

$\Rightarrow x^2 + 4y^2 = 52$

Day 152

Hyperbola

A hyperbola is the set of all points in a plane (Fig. 11.15), the magnitude of difference of whose distances from two fixed points (F_1 and F_2) in the plane is a constant (which is always less than the distance between the two fixed points).

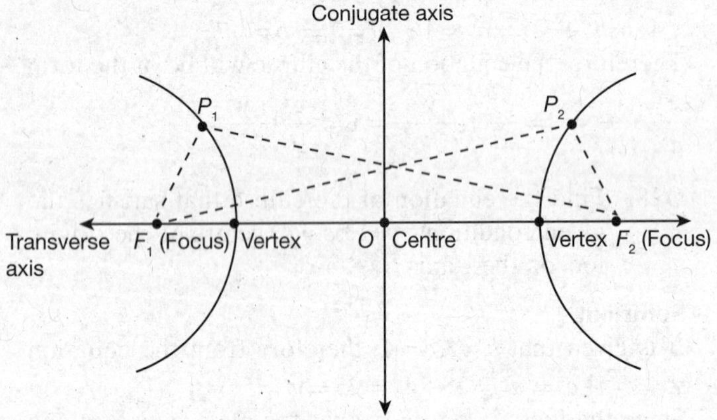

Fig. 11.15 Hyperbola

That is, $|P_1F_2 - P_1F_1| = |P_2F_2 - P_2F_1| = k$

Note:

- The two fixed points (F_1 and F_2) are called the foci (plural of 'focus') of the hyperbola.
- The mid-point of the line segment joining the foci is called the centre (O) of the hyperbola.
- The line through the foci of the hyperbola is called the transverse axis and the line through the centre and perpendicular to the transverse axis is called the conjugate axis.
- The points at which the hyperbola intersects the transverse axis are called the vertices of the hyperbola.

Relationship between length of transverse axis (2*a*), length of conjugate axis (2*b*) and the distance between the two foci of the hyperbola (2*c*)

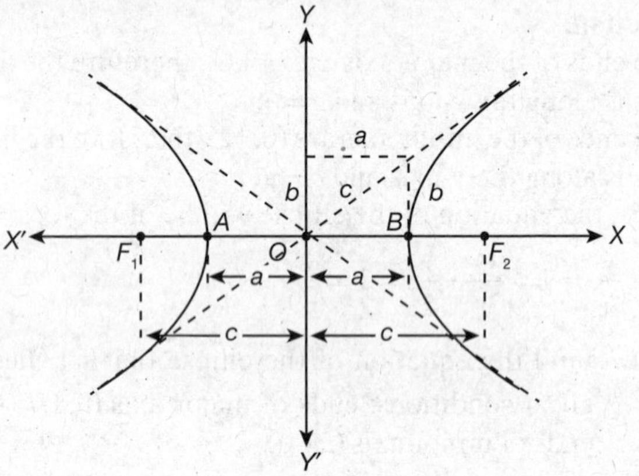

Fig. 11.16 Hyperbola: transverse and conjugate axes

Let the length of the transverse axis be $2a$, the length of the conjugate axis be $2b$ and the distance between the foci (F_1 and F_2) be $2c$ (Fig. 11.16).

Then, the length of the semi-transverse axis is a and the semi-conjugate axis is b.

Let a point B be at one end of the transverse axis; then the difference of the distances of the point B to the foci is

$BF_1 - BF_2 = BO + OF_1 - BF_2$

$= a + c - (c - a) = 2a$ (since $F_2B = F_2O - OB$)

Thus, the constant difference $|PF_2 - PF_1| = 2a$

And the quantity b is $b = \sqrt{c^2 - a^2}$, where $2b$ is the length of the conjugate axis.

Eccentricity (*e*): The eccentricity of a hyperbola is the ratio of the distances from the centre of the hyperbola to one focus and to one of the vertices of the hyperbola, that is, eccentricity, $e = \frac{c}{a}$.

Thus, we can say that the focus of a hyperbola is at a distance of $c = ae$ from the centre.

And, $b = \sqrt{c^2 - a^2} = \sqrt{(ae)^2 - a^2} = a\sqrt{e^2 - 1}$

Standard Equations of a Hyperbola

The equation of a hyperbola is simplest if the centre of the hyperbola is at the origin and the foci are on the x-axis (Fig. 11.17) or on the y-axis (Fig. 11.18).

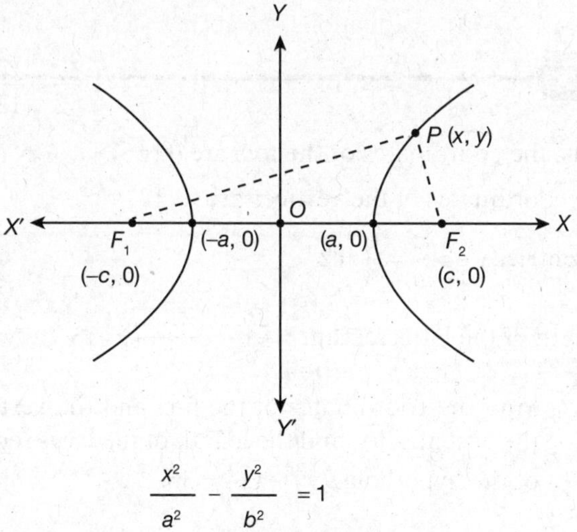

Fig. 11.17 Hyperbola: foci on x-axis

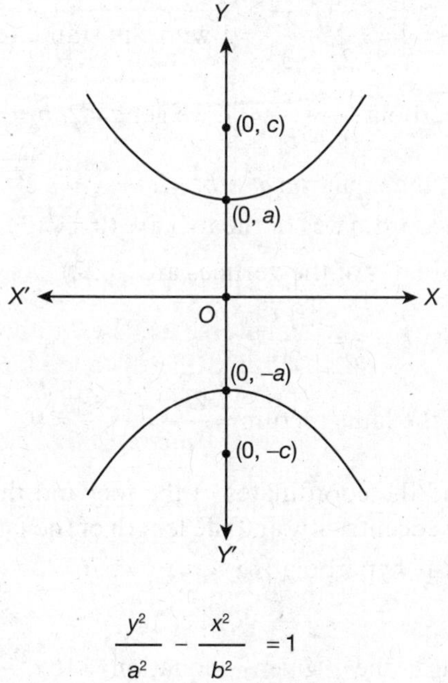

Fig. 11.18 Hyperbola: foci on y-axis

For derivation of the equation of the hyperbola with the foci on the x-axis, let $F_1(-c, 0)$ and $F_2(c, 0)$ be the foci and $O(0, 0)$ be the mid-point of the line segment F_1F_2 (Fig. 11.17).

Let $P(x, y)$ be any point on the hyperbola. Then by definition of a hyperbola, the magnitude of difference of the distances from P to the two foci is equal to $2a$; therefore $PF_1 - PF_2 = 2a$.

$\Rightarrow \sqrt{(x+c)^2 + y^2} - \sqrt{(x-c)^2 + y^2} = 2a$

(Using the distance formula)

Squaring both sides and simplifying, we get

$\frac{x^2}{a^2} - \frac{y^2}{c^2 - a^2} = 1 \Rightarrow \frac{x^2}{a^2} - \frac{y^2}{b^2} = 1 (\because c^2 - a^2 = b^2)$ as the required equation of the hyperbola with foci on the x-axis.

Similarly, we can derive the equation of the hyperbola $\frac{y^2}{a^2} - \frac{x^2}{b^2} = 1$ with foci on the y-axis.

These two equations are known as standard equations of the hyperbola.

Note:

- A hyperbola in which $a = b$ is called an equilateral hyperbola.
- A hyperbola is symmetric with respect to both the coordinates axes, since if point (x, y) is on the hyperbola then $(-x, y), (x, -y), (-x, -y)$ are also points on the hyperbola.
- The foci are always on the transverse axis.
- In standard equations of the hyperbola, the transverse axis is along the x-axis if the denominator of the coefficient of x^2 is positive and it is along the y-axis if the denominator of coefficient of y^2 is positive.
- From the equation of the hyperbola, $\frac{x^2}{a^2} = 1 + \frac{y^2}{b^2} \geq 1$

  $\Rightarrow \left|\frac{x}{a}\right| \geq 1 \Rightarrow x \leq -a$ or $x \geq a$.
- Therefore, no portion of the curve lies between the lines $x = -a$, $x = a$ (no intercept on the conjugate axis).

Latus rectum: The latus rectum of a hyperbola is a line segment (AB) perpendicular to the transverse axis through any of the foci and whose end points lie on the hyperbola (Fig. 11.19).

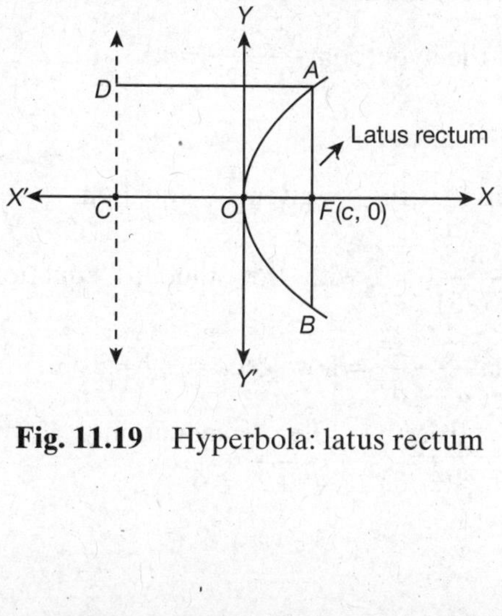

Fig. 11.19 Hyperbola: latus rectum

Since the latus rectum passes through the foci, and the coordinates of the foci are $(\pm c, 0)$, substituting the value of $x = \pm c = \pm ae$ in the equation of the hyperbola $\frac{x^2}{a^2} - \frac{y^2}{b^2} = 1$, we get,

$$\frac{c^2}{a^2} - \frac{y^2}{b^2} = 1 \Rightarrow y^2 = b^2\left(\frac{c^2}{a^2} - 1\right) = b^2\left(\frac{c^2 - a^2}{a^2}\right)$$

$$\Rightarrow y^2 = \frac{b^4}{a^2} \qquad (\because b^2 = c^2 - a^2)$$

$\Rightarrow y = \pm\frac{b^2}{a}$, which is intersection point of the latus rectum and the hyperbola. That is, the coordinates of A and B are $\left(c, \frac{b^2}{a}\right)$ and $\left(c, -\frac{b^2}{a}\right)$, respectively.

Since the hyperbola is symmetric with respect to the y-axis, $AF = FB = \frac{b^2}{a}$, and thus, the length of the latus rectum is $\frac{2b^2}{a}$.

Day 153

NCERT Exercise 11.4 (With Solutions)

Q1. Find the coordinates of the foci and the vertices, the eccentricity, and the length of the latus rectum of the hyperbola $\frac{x^2}{16} - \frac{y^2}{9} = 1$.

Solution:

Comparing the given equation $\frac{x^2}{16} - \frac{y^2}{9} = 1 \Rightarrow \frac{x^2}{4^2} - \frac{y^2}{3^2} = 1$ with the standard equation of hyperbola $\frac{x^2}{a^2} - \frac{y^2}{b^2} = 1$, we get $a = 4$, $b = 3$.

Now, from the equation $a^2 + b^2 = c^2$

$\Rightarrow c = \sqrt{a^2 + b^2} = \sqrt{16 + 9} = 5$

Thus, the coordinates of the foci are $(\pm 5, 0)$

The coordinates of the vertices are $(\pm 4, 0)$

Eccentricity $e = \frac{c}{a} = \frac{5}{4}$

Length of the latus rectum $= \frac{2b^2}{a} = \frac{2 \times 9}{4} = \frac{9}{2}$

Q2. Find the coordinates of the foci and the vertices, the eccentricity, and the length of the latus rectum of the hyperbola $\frac{y^2}{9} - \frac{x^2}{27} = 1$.

Solution:

Comparing the given equation $\frac{y^2}{9} - \frac{x^2}{27} = 1$

$\Rightarrow \frac{y^2}{3^2} - \frac{x^2}{(3\sqrt{3})^2} = 1$ with the standard equation of the hyperbola $\frac{y^2}{a^2} - \frac{x^2}{b^2} = 1$, we get $a = 3$, $b = 3\sqrt{3}$.

Now, from the equation $a^2 + b^2 = c^2$

$\Rightarrow c = \sqrt{3^2 + (3\sqrt{3})^2} = \sqrt{9 + 27} = 6$

Thus, the coordinates of the foci are $(0, \pm 6)$

The coordinates of the vertices are $(0, \pm 3)$

Eccentricity $e = \frac{c}{a} = \frac{6}{3} = 2$

Length of the latus rectum $= \frac{2b^2}{a} = \frac{2 \times 27}{3} = 18$

Q3. Find the coordinates of the foci and the vertices, the eccentricity, and the length of the latus rectum of the hyperbola $9y^2 - 4x^2 = 36$.

Solution:

Comparing the given equation $9y^2 - 4x^2 = 36$ $\Rightarrow \frac{y^2}{4} - \frac{x^2}{9} = 1 \Rightarrow \frac{y^2}{2^2} - \frac{x^2}{3^2} = 1$ with the standard equation of the hyperbola $\frac{y^2}{a^2} - \frac{x^2}{b^2} = 1$, we get $a = 2$, $b = 3$.

Now, from the equation $a^2 + b^2 = c^2 \Rightarrow c = \sqrt{2^2 + 3^2} = \sqrt{13}$

Thus, the coordinates of the foci are $(0, \pm\sqrt{13})$

The coordinates of the vertices are $(0, \pm 2)$

Eccentricity $e = \frac{c}{a} = \frac{\sqrt{13}}{2}$

Length of the latus rectum $= \frac{2b^2}{a} = \frac{2 \times 9}{2} = 9$

Q4. Find the coordinates of the foci and the vertices, the eccentricity, and the length of the latus rectum of the hyperbola $16x^2 - 9y^2 = 576$.

Solution:

Comparing the given equation $16x^2 - 9y^2 = 576$ $\Rightarrow \frac{x^2}{36} - \frac{y^2}{64} = 1 \Rightarrow \frac{x^2}{6^2} - \frac{y^2}{8^2} = 1$ with the standard equation of the hyperbola $\frac{x^2}{a^2} - \frac{y^2}{b^2} = 1$, we get $a = 6$, $b = 8$.

Now, from the equation $a^2+b^2=c^2$
$\Rightarrow c=\sqrt{a^2+b^2}=\sqrt{36+84}=10$
Thus,
The coordinates of the foci are $(\pm 10, 0)$

The coordinates of the vertices are $(\pm 6, 0)$

Eccentricity $e=\dfrac{c}{a}=\dfrac{10}{6}=\dfrac{5}{3}$

Length of the latus rectum $=\dfrac{2b^2}{a}=\dfrac{2\times 64}{6}=\dfrac{64}{3}$

Q5. Find the coordinates of the foci and the vertices, the eccentricity and the length of the latus rectum of the hyperbola $5y^2-9x^2=36$.

Solution:
Comparing the given equation $5y^2-9x^2=36$ $\Rightarrow \dfrac{y^2}{(36/5)}-\dfrac{x^2}{4}=1 \Rightarrow \dfrac{y^2}{(6/\sqrt{5})^2}-\dfrac{x^2}{2^2}$ with the standard equation of the hyperbola $\dfrac{y^2}{a^2}-\dfrac{x^2}{b^2}=1$, we get $a=\dfrac{6}{\sqrt{5}}, b=2$.

Now, from the equation $a^2+b^2=c^2$

$$\Rightarrow c=\sqrt{\left(\frac{6}{\sqrt{5}}\right)^2+2^2}=\sqrt{\frac{36}{5}+4}=\sqrt{\frac{56}{5}}=\frac{2\sqrt{14}}{\sqrt{5}}$$

Thus, the coordinates of the foci are $\left(0, \pm\dfrac{2\sqrt{14}}{\sqrt{5}}\right)$

The coordinates of the vertices are $\left(0, \pm\dfrac{6}{\sqrt{5}}\right)$

$$\text{Eccentricity } e=\frac{c}{a}=\frac{\left(\frac{2\sqrt{14}}{\sqrt{5}}\right)}{\left(\frac{6}{\sqrt{5}}\right)}=\frac{\sqrt{14}}{3}$$

$$\text{Length of the latus rectum}=\frac{2b^2}{a}=\frac{2\times 4}{\left(\frac{6}{\sqrt{5}}\right)}=\frac{4\sqrt{5}}{3}$$

Q6. Find the coordinates of the foci and the vertices, the eccentricity, and the length of the latus rectum of the hyperbola $49y^2-16x^2=784$.

Solution:
Comparing the given equation $49y^2-16x^2=784$ $\Rightarrow \dfrac{y^2}{16}-\dfrac{x^2}{49}=1 \Rightarrow \dfrac{y^2}{4^2}-\dfrac{x^2}{7^2}=1$ with the standard equation of the hyperbola $\dfrac{y^2}{a^2}-\dfrac{x^2}{b^2}=1$, we get $a=4$, $b=7$.

Now, from the equation $a^2+b^2=c^2$
$c=\sqrt{16+49}=\sqrt{65}$
Thus, the coordinates of the foci are $(0, \pm\sqrt{65})$

The coordinates of the vertices are $(0, \pm 4)$

Eccentricity $e=\dfrac{c}{a}=\dfrac{\sqrt{65}}{4}$

Length of the latus rectum $=\dfrac{2b^2}{a}=\dfrac{2\times 49}{4}=\dfrac{49}{2}$

Q7. Find the equation of the hyperbola satisfying the given conditions: vertices: $(\pm 2, 0)$, foci: $(\pm 3, 0)$.

Solution:
The vertices $(\pm 2, 0)$ are on the x-axis; therefore the equation of the hyperbola is of the form $\dfrac{x^2}{a^2}-\dfrac{y^2}{b^2}=1$. Since the vertices are $(\pm 2, 0) \Rightarrow a=2$ and the foci are $(\pm 3, 0) \Rightarrow c=3$, from the equation $a^2+b^2=c^2 \Rightarrow b^2=3^2-2^2 \Rightarrow b^2=9-4=5$

Thus, the equation of the hyperbola is $\dfrac{x^2}{4}-\dfrac{y^2}{5}=1$.

Q8. Find the equation of the hyperbola satisfying the given conditions: vertices: $(0, \pm 5)$, foci: $(0, \pm 8)$.

Solution:
The vertices are on the y-axis; therefore the equation of the hyperbola is of the form $\dfrac{y^2}{a^2}-\dfrac{x^2}{b^2}=1$.

Now, since the vertices are $(0, \pm 5) \Rightarrow a=5$ and the foci are $(0, \pm 8) \Rightarrow c=8$, from the equation $a^2+b^2=c^2 \Rightarrow b^2=8^2-5^2 \Rightarrow b^2=64-25=39$

Thus, the equation of the hyperbola is $\dfrac{y^2}{25}-\dfrac{x^2}{39}=1$.

Q9. Find the equation of the hyperbola satisfying the given conditions: vertices: $(0, \pm 3)$, foci: $(0, \pm 5)$.

Solution:
The vertices are on the y-axis; therefore the equation of the hyperbola is of the form $\dfrac{y^2}{a^2}-\dfrac{x^2}{b^2}=1$.

Since the vertices are $(0, \pm 3) \Rightarrow a=3$ and the foci are $(0, \pm 5) \Rightarrow c=5$, from the equation $a^2+b^2=c^2 \Rightarrow b^2=5^2-3^2 \Rightarrow b^2=25-9=16$

Thus, the equation of the hyperbola is $\dfrac{y^2}{9}-\dfrac{x^2}{16}=1$.

Q10. Find the equation of the hyperbola satisfying the given conditions: foci $(\pm5,0)$ and the transverse axis is of length 8.

Solution:
The foci $(\pm5,0)$ are on the x-axis; therefore the equation of the hyperbola is of the form $\frac{x^2}{a^2}-\frac{y^2}{b^2}=1$ and $c=5$.
Since the length of the transverse axis is 8, $2a=8 \Rightarrow a=4 \Rightarrow a^2=16$.
From the equation $a^2+b^2=c^2$
$\Rightarrow 4^2+b^2=5^2 \Rightarrow b^2=25-16=9$

Thus, the equation of the hyperbola $\frac{x^2}{16}-\frac{y^2}{9}=1$.

Q11. Find the equation of the hyperbola satisfying the given conditions: foci $(0,\pm13)$ and the conjugate axis is of length 24.

Solution:
The foci $(0,\pm13)$ are on the y-axis; therefore the equation of the hyperbola is of the form $\frac{y^2}{a^2}-\frac{x^2}{b^2}=1$ and $c=13$.
Since the conjugate axis is of length 24, $2b=24 \Rightarrow b=12 \Rightarrow b^2=144$
From the equation $a^2+b^2=c^2$
$\Rightarrow a^2+12^2=13^2 \Rightarrow a^2=169-144=25$

Thus, the equation of the hyperbola is $\frac{y^2}{25}-\frac{x^2}{144}=1$.

Q12. Find the equation of the hyperbola satisfying the given conditions: foci $(\pm3\sqrt{5},0)$ and the latus rectum is of length 8.

Solution:
The foci $(0,\pm13)$ are on the y-axis; theretore the equation of the hyperbola is of the form $\frac{y^2}{a^2}-\frac{x^2}{b^2}=1$ and $c=13$.

Since the length of latus rectum is $8 \Rightarrow \frac{2b^2}{a}=8 \Rightarrow b^2=4a$

From the equation $a^2+b^2=c^2$
$\Rightarrow a^2+4a=45 \Rightarrow a^2+4a-45=0$
$\Rightarrow (a+9)(a-5)=0 \Rightarrow a=-9,5$
Since a is non-negative, $a=5$
$\Rightarrow b^2=4a=4\times5=20$

Thus, the equation of the hyperbola is $\frac{x^2}{25}-\frac{y^2}{20}=1$.

Q13. Find the equation of the hyperbola satisfying the given conditions: foci $(\pm4,0)$ and the latus rectum is of length 12.

Solution:
The foci $(\pm4,0)$ are on the x-axis; therefore the equation of the hyperbola is of the form $\frac{x^2}{a^2}-\frac{y^2}{b^2}=1$ and $c=4$.

Since the length of the latus rectum is 8, $\frac{2b^2}{a}=12 \Rightarrow b^2=6a$

From the equation $a^2+b^2=c^2$
$\Rightarrow a^2+6a=12 \Rightarrow a^2+6a-16=0$
$\Rightarrow (a+8)(a-2)=0 \Rightarrow a=-8,2$
Since a is non-negative, $a=2$.
$\Rightarrow b^2=6a=6\times2=12$

Thus, the equation of the hyperbola is $\frac{x^2}{4}-\frac{y^2}{12}=1$.

Q14. Find the equation of the hyperbola satisfying the given conditions: vertices $(\pm7,0)$ and eccentricity $e=\frac{4}{3}$.

Solution:
The vertices $(\pm7,0)$ are on the x-axis; therefore the equation of the hyperbola is of the form $\frac{x^2}{a^2}-\frac{y^2}{b^2}=1$ and $a=7$.

Since the eccentricity $e=\frac{4}{3}$

$\Rightarrow \frac{c}{a}=\frac{4}{3} \Rightarrow \Rightarrow \frac{c}{7}=\frac{4}{3} \Rightarrow c=\frac{28}{3}$
From the equation $a^2+b^2=c^2$
$\Rightarrow 7^2+b^2=(28/3)^2 \Rightarrow b^2=\frac{784}{9}-49$

$\Rightarrow b^2=\frac{784-441}{9}=\frac{343}{9}$

Thus, the equation of the hyperbola is $\frac{x^2}{49}-\frac{9y^2}{343}=1$.

Q15. Find the equation of the hyperbola satisfying the given conditions: foci $(0,\pm\sqrt{10})$ and passing through (2, 3).

Solution:
The foci $(0,\pm\sqrt{10})$ are on the y-axis; therefore the equation of the hyperbola is of the form $\frac{y^2}{a^2}-\frac{x^2}{b^2}=1$ and $c=\sqrt{10}$.
From the equation $a^2+b^2=c^2$.
$\Rightarrow a^2+b^2=10 \Rightarrow b^2=10-a^2$
Since the hyperbola passes through the point (2, 3),
$\frac{9}{a^2}-\frac{4}{b^2}=1 \Rightarrow \frac{9}{a^2}-\frac{4}{(10-a^2)}=1$

$\Rightarrow 9(10-a^2)-4a^2 = a^2(10-a^2)$
$\Rightarrow 90-9a^2-4a^2 = 10a^2-a^4$
$\Rightarrow a^4-23a^2+90=0$
$\Rightarrow (a^2-5)(a^2-18)=0 \Rightarrow a^2 = 5,18$

In the hyperbola $a < c \Rightarrow a^2 < c^2 \Rightarrow a^2 < 10$; therefore $a^2 = 5 \Rightarrow b^2 = 10-a^2 = 10-5 = 5$

Thus, the equation of the hyperbola is $\frac{y^2}{5}-\frac{x^2}{5}=1.$

Day 154

NCERT Miscellaneous Exercise (With Solutions)

Q1. If a parabolic reflection is 20 cm in diameter and 5 cm deep, find the focus.

Solution:
Let the origin of the coordinate plane be taken as the vertex of the parabolic reflection in such a way that the parabola opens rightwards, that is, along the positive x-axis.

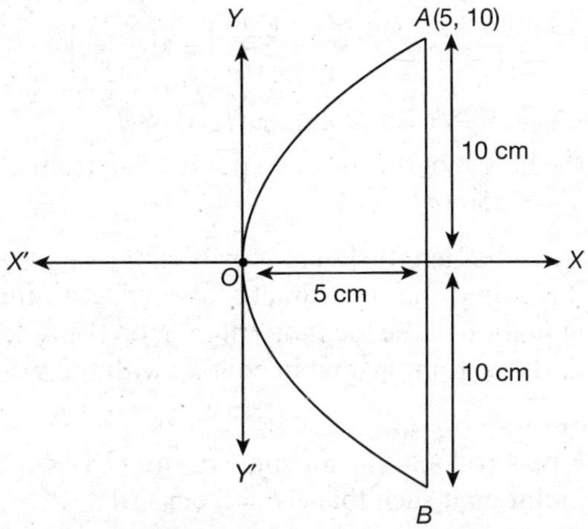

Therefore, the equation of the parabola is of the form $y^2 = 4ax$ (as it opens rightwards).
The parabola passes through the point $A(10,5)$; therefore $10^2 = 4a(5)$

$\Rightarrow 100 = 20a \Rightarrow a = \frac{100}{20} = 5$

Thus, the focus of the parabola is $(a,0) = (5,0)$, which is the mid-point of the diameter AB.

Q2. An arch is in the form of a parabola with its axis vertical. The arch is 10 m high and 5 m wide at the base. How wide is it 2 m from the vertex of the parabola?

Solution:
Let the origin of the coordinate plane be taken as the vertex of the parabolic arch in such a way that the parabola opens downwards, that is, along the negative y-axis.

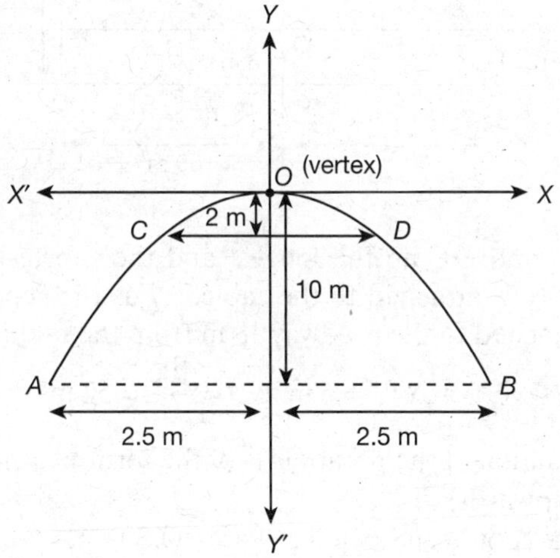

Thus, the equation of the parabola is of the form $x^2 = -4ay$ (as it opens downwards).

Since the parabola passes through the point $\left(\frac{5}{2},-10\right)$

$\Rightarrow \left(\frac{5}{2}\right)^2 = -4a(-10) \Rightarrow a = \frac{25}{4\times4\times10} = \frac{5}{32}$

Therefore, the equation of the parabolic arch is $x^2 = -\frac{5}{8}y$

Now, when $y = -2\text{ m} \Rightarrow x^2 = -\frac{5}{8}\times(-2) = \frac{5}{4}$

$\Rightarrow x = \sqrt{\frac{5}{4}} = \frac{\sqrt{5}}{2}\text{m}$

$\therefore AB = 2\times\frac{\sqrt{5}}{2}\text{m} = 2.23\text{m (approx)}$

Hence, when the arch is 2 m from the vertex of the parabola, its width is approximately 2.23 m.

Q3. The cable of a uniformly loaded suspension bridge hangs in the form of a parabola. The roadway which is horizontal and 100 m long is supported by vertical wires attached to the cable, the longest wire being 30 m and the shortest being 6 m. Find the length of the supporting wire attached to the roadway 18 m from the middle.

Solution:
Let the origin of the coordinate plane be taken as the vertex of the parabolic cable in such a way that the lowest

point of the cable is at the vertex and the parabola opens upwards, that is, along the positive y-axis.

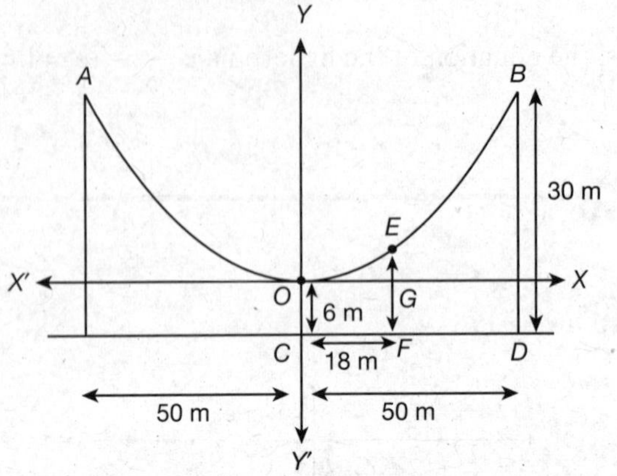

Let BD and OC be the longest and the shortest wires, respectively, attached to the cable. EF is the supporting wire attached to the roadway 18 m from the middle.

Since $BD = 30$ m, $OC = 6$ m, $CD = \frac{100}{2} = 50$ m

The equation of the parabola is of the form $y^2 = 4ax$ (as it opens upwards).

Since the coordinates of point A are $(50, 30-6) = A(50, 24)$, the cable passes through the point A (50, 24); therefore $(50)^2 = 4a(24)$

$$\Rightarrow a = \frac{50\times 50}{4\times 24} = \frac{625}{24}$$

Thus, the equation of the parabola is

$$x^2 = 4\times\frac{625}{24}\times y \Rightarrow 6x^2 = 625y$$

Now since the x-coordinate of point E is 18, at $x = 18, 6(18)^2 = 625y$

$$\Rightarrow y - \frac{6\times 18\times 18}{625} \Rightarrow y = 3.11 \Rightarrow EG = 3.11 \text{ m}$$

$\therefore EF = EG + GF = 3.11 \text{ m} + 6 \text{ m} = 9.11 \text{ m}$

Thus, the length of the supporting wire attached to the roadway 18 m from the middle is approximately 9.11 m.

Q4. An arch is in the form of a semi-ellipse. It is 8 m wide and 2 m high at the centre. Find the height of the arch at a point 1.5 m from one end.

Solution:

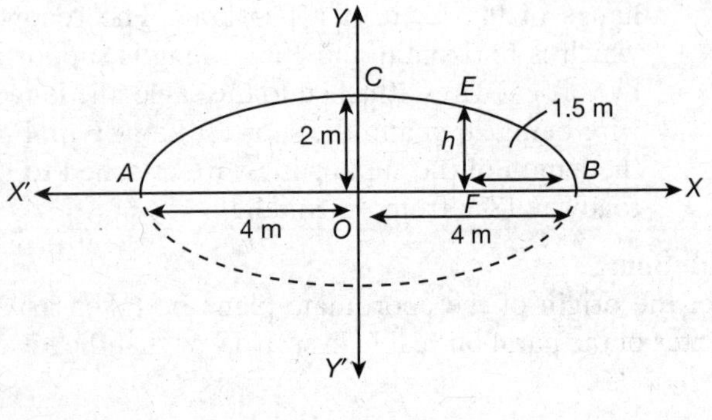

Let the origin of the coordinate plane be taken as the centre of the ellipse and the major axis be taken along the x-axis.

Since the width and height of the arch from the centre are 2 m and 8 m, respectively, the length of the major axis is 8 m and the length of the semi-minor axis is 2 m.

The equation of the semi-ellipse will be of the form $\frac{x^2}{a^2} + \frac{y^2}{b^2} = 1, y \geq 0$, where a is the semi-major axis.

Thus $2a = 8 \Rightarrow a = 4$ and $2b = 4 \Rightarrow b = 2$

Therefore, the equation of the semi-ellipse is

$$\frac{x^2}{4^2} + \frac{y^2}{2^2} = 1, \ y \geq 0$$

Let F be a point on the major axis such that $BF = 1.5$ m; then $OA = (4 - 1.5)$ m $= 2.5$ m

Thus, the x-coordinate of point E is 2.5, and on substituting the value of x with 2.5 in the equation of the ellipse, we get $\frac{(2.5)^2}{16} + \frac{y^2}{4} = 1.$

$$\Rightarrow \frac{6.25}{16} + \frac{y^2}{4} = 1 \Rightarrow y^2 = 4\left(1 - \frac{6.25}{16}\right) = 4\left(\frac{9.75}{15}\right)$$

$\Rightarrow y^2 = 2.4375 \Rightarrow y = 1.56$ (approx)

Thus, the height of the arch at a point 1.5 m from one end is $EF = y = 1.56$ (*approx*).

Q5. A rod of length 12 cm moves with its ends always touching the coordinate axes. Determine the equation of the locus of a point P on the rod, which is 3 cm from the end in contact with the x-axis.

Solution:

Let AB be a rod making an angle θ with OA and $P(x, y)$ be the point on it such that $AP = 3$ cm.

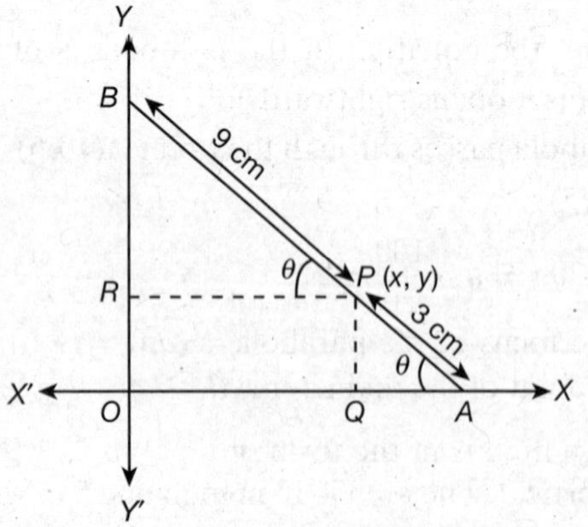

Then,

$PB = AB - AP = (12 - 3)$ cm $= 9$ cm $\quad [\because AB = 12 \text{ cm}]$

From P, draw $PQ \perp OA, PR \perp OB$. Therefore, from triangles PAQ and BOA

$3\cos\theta = 12\cos\theta - x \Rightarrow \cos\theta = x/9$

$3\sin\theta = y \Rightarrow \sin\theta = y/3$

Since $\sin^2\theta + \cos^2\theta = 1$,

$\left(\frac{y}{3}\right)^2 + \left(\frac{x}{9}\right)^2 = 1 \Rightarrow \frac{x^2}{81} + \frac{y^2}{9} = 1$ is the equation of the locus of point P on the rod.

Q6. Find the area of the triangle formed by joining the vertex of the parabola $x^2 = 12y$ to the ends of its latus rectum.

Solution:

Comparing the equation of the parabola $x^2 = 12y$ with $x^2 = 4ay$, we get $4a = 12 \Rightarrow a = 3$.

$\therefore$ The coordinates of foci are $F(0,a) = F(0,3)$

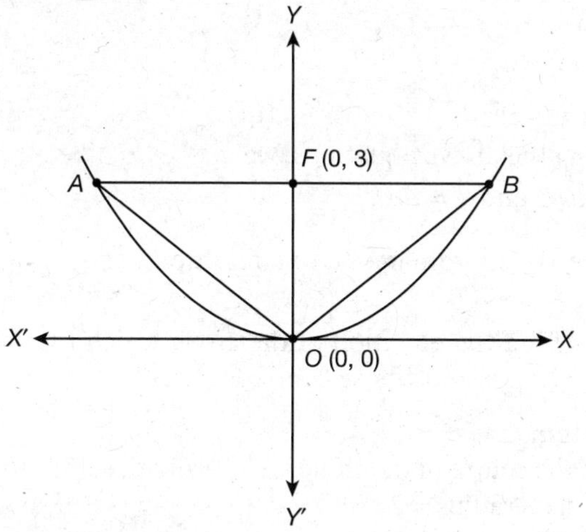

Let AB be the latus rectum of the given parabola; then at $y = 3, x^2 = 12(3) \Rightarrow x^2 = 36 \Rightarrow x = \pm 6$

$\therefore$ The coordinates of A are (–6, 3) and the coordinates of B are (6, 3). Thus, the vertices of ΔOAB are $O(0,0), A(-6,3), B(6,3)$.

Hence, area of ΔOAB

$= \frac{1}{2}|0(3-3)+(-6)(3-0)+6(0-3)|$ unit^2

$= \frac{1}{2}|(-6)(3)+6(-3)| = \frac{1}{2}|-18-18|$

$= \frac{1}{2}|-36| = \frac{1}{2}\times 36 = 18\,\text{unit}^2$

Q7. A man running a race course notes that the sum of the distance of the two flag posts from him is always 10 m and the distance between the flag posts is 8 m. Find the equation of the posts traced by the man.

Solution:

Let F_1 and F_2 be the positions of the two flag posts and P (x, y) be the position of the man. According to question, $PF_1 + PF_2 = 10$.

We know that if a point moves in a plane in such a way that the sum of its distance from two fixed points is constant, then the path is an ellipse and this constant value is equal to the length of the major axis of the ellipse.

Therefore, the path described by the man is an ellipse where the length of the major axis is 10 m, while points F_1 and F_2 are the foci.

Taking the origin of the coordinate plane as the centre of the ellipse, while taking the major axis along the x-axis, we get the equation of the ellipse as $\frac{x^2}{a^2} + \frac{y^2}{b^2} = 1$, where a is the semi-major axis.

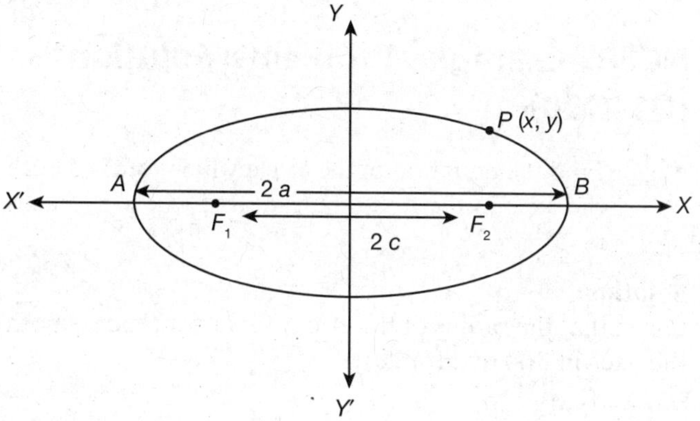

Since $2a = 10 \Rightarrow a = 5$

Distance between the foci $(2c) = 8 \Rightarrow c = 4$

Using the relation $c = \sqrt{a^2 - b^2} \Rightarrow 4 = \sqrt{25 - b^2}$

$\Rightarrow 16 = 25 - b^2 \Rightarrow b^2 = 25 - 16 = 9 \Rightarrow b = 3$

Thus, the equation of the path traced by the man is

$\frac{x^2}{25} + \frac{y^2}{9} = 1$

Q8. An equilateral triangle is inscribed in the parabola $y^2 = 4ax$, where one vertex is at the vertex of the parabola. Find the length of the side of the triangle.

Solution:

Let AOB be the equilateral triangle inscribed in the parabola $y^2 = 4ax$.

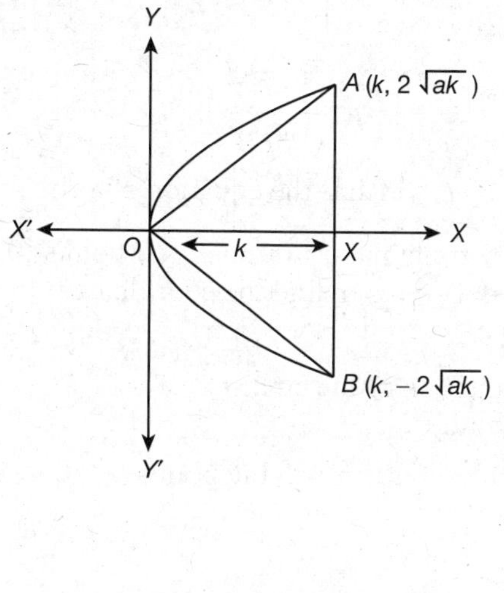

Let AB intersect the x-axis at point X, and $OX = k$. Therefore, from the equation of the given parabola,
$y^2 = 4ax \Rightarrow y^2 = 4ak \Rightarrow y = \pm 2\sqrt{ak}$

$\therefore$ The respective coordinates of points A and B are $(k, 2\sqrt{ak}), (k, -2\sqrt{ak})$.

Therefore, $AB = AX + XB = 2\sqrt{ak} + 2\sqrt{ak} = 4\sqrt{ak}$

Since OAB is an equilateral triangle, $OA^2 = AB^2$.

$\therefore k^2 + (2\sqrt{ak})^2 = (4\sqrt{ak})^2$

$\Rightarrow k^2 + 4ak = 16ak \Rightarrow k^2 = 12ak \Rightarrow k = 12a$

$\therefore AB = 4\sqrt{ak} = 4\sqrt{a \times 12a} = 4\sqrt{12a^2} = 8\sqrt{3}a$

Thus, the side of the equilateral triangle inscribed in the parabola $y^2 = 4ax$ is $8\sqrt{3}a$.

Day 155

NCERT Exemplar Problems: Solutions (Q1–Q20)

Q1. Find the equation of the circle which touches both the axes in the first quadrant and whose radius is a.

Solution:

Given that the radius of the circle is a and it touches both the axes in first quadrant,

$(h, k) = (a, a)$

So, the equation of required circle is

$(x-a)^2 + (y-a)^2 = a^2$

$\Rightarrow x^2 - 2ax + a^2 + y^2 + 2ay + a^2 = a^2$

$\Rightarrow x^2 + y^2 - 2ax - 2ay + a^2 = 0$

Q2. Show that the point (x, y) given by $x = \dfrac{2at}{1+t^2}, y = \dfrac{a(1-t^2)}{1+t^2}$ lies on a circle.

Solution:

Since $x = \dfrac{2at}{1+t^2}, y = \dfrac{a(1-t^2)}{1+t^2}$

Therefore, squaring and adding, we get

$$x^2 + y^2 = \frac{4a^2t^2}{(1+t^2)^2} + \frac{a^2(1-t^2)^2}{(1+t^2)^2}$$

$$= a^2\left(\frac{4t^2 + 1 + t^4 - 2t^2}{(1+t^2)^2}\right)$$

$$= a^2\left(\frac{1 + t^4 + 2t^2}{(1+t^2)^2}\right) = a^2\frac{(1+t^2)^2}{(1+t^2)^2}$$

$\Rightarrow x^2 + y^2 = a^2$ which is the equation of a circle.

Q3. If a circle passes through the points (0, 0), $(a, 0)$ and $(0, b)$, then find the coordinates of its centre.

Solution:

Let the equation of the circle be

$x^2 + y^2 + 2gx + 2fy + c = 0$

This circle passes through the points $(0,0)$, $(a,0)$, $(0,b)$; therefore, these points must satisfy the equation of the circle. Thus,

$c = 0$ (i)

$a^2 + 2ag = 0$ (ii)

$b^2 + 2bf = 0$ (iii)

Solving the above equations, we get

$g = -a/2$ and $f = -b/2$

Hence, the coordinates of the circle are $= (-g, -f) = \left(\dfrac{a}{2}, \dfrac{b}{2}\right)$

Q4. Find the equation of the circle which touches the X-axis and whose centre is (1, 2).

Solution:

Since the centre of the circle is $(1, 2)$ and the circle touches X-axis, its radius = 2

Thus, the equation of circle is $(x-1)^2 + (y-2)^2 = 2^2$

$\Rightarrow x^2 - 2x + 1 + y^2 - 4y + 4 = 4$

$\Rightarrow x^2 + y^2 - 2x - 4y + 1 = 0$

Q5. If the lines $3x + 4y + 4 = 0, 6x - 8y - 7 = 0$ are tangents to a circle, then find the radius of the circle.

Solution:

The given lines $3x + 4y + 4 = 0$ and $6x - 8y - 7 = 0 \Rightarrow 3x - 4y - 7/2 = 0$ are parallel and tangents to the circle; therefore they must touch the circle diametrically.

(*Note*: The distance between two parallel lines $ax + by + c_1 = 0$ and $ax + by + c_2 = 0$ is $\dfrac{c_1 - c_2}{\sqrt{a^2 + b^2}}$.)

Thus, the distance between the parallel lines must be equal to the diameter of the circle.

Hence, diameter $d = \left|\dfrac{4 + 7/2}{\sqrt{9+16}}\right| = \left|\dfrac{\frac{8+7}{2}}{5}\right| = 3/2$

$\therefore$ Radius of the circle = Diameter/2 = $3/4$

Q6. Find the equation of the circle which touches both the axes and the line $3x-4y+8=0$ and lies in the third quadrant.

Solution:

Let a be the radius of the circle. The circle touches both the axes and lies in the third quadrant; therefore the coordinates of the circle will be $(-a,-a)$.

Now, radius of the circle = perpendicular distance of centre $(-a,-a)$ from the tangent $3x-4y+8=0$. Thus,

$$a=\left|\frac{-3a+4a+8}{\sqrt{9+16}}\right|$$

$$\Rightarrow a=\left|\frac{a+8}{5}\right| \Rightarrow a=\pm\left(\frac{a+8}{5}\right)$$

Taking the positive sign, we get

$$a=\frac{a+8}{5} \Rightarrow 5a=a+8 \Rightarrow 4a=8 \Rightarrow a=2$$

Taking the negative sign, we get

$$a=\frac{-a-8}{5} \Rightarrow 5a=-a-8 \Rightarrow 6a=-8 \Rightarrow a=-4/3$$

But $a \neq -4/3$, $\therefore a=2$

Hence, the equation of the circle is

$(x+2)^2+(y+2)^2=2^2 \quad [\because a=2]$

$\Rightarrow x^2+4x+4+y^2+4y+4=4$

$\Rightarrow x^2+y^2+4x+4y+4=0$

Q7. If one end of the diameter of the circle $x^2+y^2-4x-6y+11=0$ is (3, 4), then find the coordinates of the other end of the diameter.

Solution:

Comparing the given equation $x^2+y^2-4x-6y+11=0$ of the circle with $x^2+y^2+2gx+2fy+c=0$, we get

$2g=-4,\ 2f=-6$

Thus, the centre of the circle is $(-g,-f)=(2,3)$.

Let the other end of the diameter be (x_1,y_1). The mid-point of two diametrical points lies on the centre of the circle; therefore

$$2=\frac{3+x_1}{2} \Rightarrow x_1=1 \text{ and } 3=\frac{4+y_1}{2} \Rightarrow y_1=2$$

Hence, the coordinates of the other end of the diameter will be (1, 2).

Q8. Find the equation of the circle having (1, –2) as its centre and passing through the point of intersection of the lines $3x+y=14, 2x+5y=18$.

Solution:

The circle passes through the point of intersection of the lines $3x+y=14, 2x+5y=18$; therefore solving the above equations of lines, we get

$2x+5y=18 \Rightarrow 2x+5(14-3x)=18$

$\Rightarrow -13x=-52 \Rightarrow x=4$

Putting the value of $x=4$, we get $12+y=14 \Rightarrow y=2$

Now, since the centre of the circle is (1, –2) and point (4, 2) lies on the circle,

Radius of the circle $=\sqrt{(4-1)^2+(2+2)^2}=5$

Thus, the equation of the circle is $(x-1)^2+(y+2)^2=5^2$

$\Rightarrow x^2-2x+1+y^2+4y+4=25$

$\Rightarrow x^2+y^2-2x+4y-20=0$

Q9. If the line $y=\sqrt{3}x+k$ touches the circle $x^2+y^2=16$, then find the value of k.

Solution:

For the given equation of a circle, $x^2+y^2=16$,

Radius = –4 and centre = (0, 0)

Now, since the line $y=\sqrt{3}x+k$ touches the circle, the perpendicular from (0, 0) to line $y=\sqrt{3}x+k$ = radius of the circle $\Rightarrow \left|\frac{0-0+k}{\sqrt{3+1}}\right|=4 \Rightarrow \pm\frac{k}{2}=4 \Rightarrow k=\pm 8$

(*Note*: Distance from the point $(m,\ n)$ to the line $Ax+By+k=0$ is $d=\left|\frac{Am+Bn+C}{A^2+B^2}\right|$)

Q10. Find the equation of a circle concentric with the circle $x^2+y^2-6x+12y+15=0$ and having double its area.

Solution:

For the given equation of the circle

$x^2+y^2-6x+12y+15=0$, we get

$2g=-6 \Rightarrow g=-3,\ \ 2f=12 \Rightarrow f=6,\ \ c=15$

$\therefore$ Centre $(-g,-f)=(3,-6)$

Radius of the given circle

$=\sqrt{g^2+f^2-c}=\sqrt{9+36-15}=\sqrt{30}$

Thus, the centre of the required circle will be (3, –6). (since the circles are concentric)

And let the radius of the required circle = R

$\therefore 2\times$ Area of the given circle = Area of the required circle

$\Rightarrow 2[\pi(\sqrt{30})^2]=\pi R^2 \Rightarrow 60=R^2 \Rightarrow R=\sqrt{60}$

$\therefore \sqrt{g^2+f^2-C}=\sqrt{60} \Rightarrow 9+36-C=60$

$\Rightarrow C=-15$

Hence, the required equation of the circle is

$x^2+y^2-6x+12y+15=0$

Q11. If the latus rectum of an ellipse is equal to half of the minor axis, then find its eccentricity.

Solution:

Let the equation of the ellipse be $\frac{x^2}{a^2}+\frac{y^2}{b^2}=1$

$\therefore$ Length of the major axis = 2a
Length of the minor axis = 2b

And length of the latus rectum = $\frac{2b^2}{a}$

The latus rectum of the ellipse is equal to half of the minor axis; therefore

$\frac{2b^2}{a}=\frac{2b}{2} \Rightarrow a=2b \Rightarrow b=a/2$

Also, $b^2=a^2(1-e^2) \Rightarrow \left(\frac{a}{2}\right)^2=a^2(1-e^2)$

$\Rightarrow \frac{a^2}{4}=a^2(1-e^2) \Rightarrow 1-e^2=\frac{1}{4}$

$\Rightarrow e^2=1-\frac{1}{4} \Rightarrow e=\sqrt{\frac{3}{4}}=\sqrt{\frac{3}{2}}$

Q12. For the ellipse with equation be $9x^2+25y^2=225$, find the eccentricity and foci.

Solution:
Given the equation of the ellipse, $9x^2+25y^2=225$,

$\Rightarrow \frac{x^2}{25}+\frac{y^2}{9}=1 \Rightarrow a=5, b=3$

Also, $b^2=a^2(1-e^2) \Rightarrow 9=25(1-e^2)$

$\Rightarrow \frac{9}{25}=1-e^2 \Rightarrow e^2=1-9/25$

$\Rightarrow e=\sqrt{1-9/25}=\sqrt{\frac{25-9}{25}}=\sqrt{\frac{16}{25}}=4/5$

Thus, foci $=(\pm ae,0)=(\pm 5\times 4/5,0)=(\pm 4,0)$

Q13. If the eccentricity of an ellipse is 5/8 and the distance between its foci is 10, then find the latus rectum of the ellipse.

Solution:

Let the equation of the ellipse be $\frac{x^2}{a^2}+\frac{y^2}{b^2}=1$,

Since eccentricity $e=\frac{5}{8}$

Therefore, the foci of the ellipse is $(\pm ae, 0)$.

Thus, distance between foci $= ae-(-ae)=2ae$

$\Rightarrow 2ae=10 \Rightarrow a\times\frac{5}{8}=5 \Rightarrow a=8$

$\because b^2=a^2(1-e^2)=64\left(1-\frac{25}{64}\right) \Rightarrow b^2=39$

$\therefore$ Length of the latus rectum of the ellipse

$=\frac{2b^2}{a}=2\left(\frac{39}{8}\right)=\frac{39}{4}$

Q14. Find the equation of the ellipse whose eccentricity is 2/3, latus rectum is 5 and centre is (0, 0).

Solution:
Since $e=2/3$ and latus rectum =5

Therefore, $\frac{2b^2}{a}=5 \Rightarrow b^2=\frac{5a}{2}$

And, $b^2=a^2(1-e^2) \Rightarrow \frac{5a}{2}=a^2\left(1-\frac{4}{9}\right)$

$\Rightarrow \frac{5}{2}=\frac{5a}{9} \Rightarrow a=\frac{9}{2} \Rightarrow a^2=\frac{81}{4}$

$\Rightarrow b^2=\frac{5\times 9}{2\times 2}=\frac{45}{4}$

Thus, the required equation of the ellipse is $\frac{4x^2}{81}+\frac{4y^2}{45}=1$.

Q15. Find the distance between the directrices of ellipse $\frac{x^2}{36}+\frac{y^2}{20}=1$.

Solution:

Comparing the give equation $\frac{x^2}{36}+\frac{y^2}{20}=1$ of the ellipse with $\frac{x^2}{a^2}+\frac{y^2}{b^2}=1$, we get $b^2=20$ and $a^2=36$

Also, $b^2=a^2(1-e^2) \Rightarrow 20=36(1-e^2)$

$\Rightarrow \frac{20}{36}=1-e^2 \Rightarrow e=\sqrt{1-\frac{20}{36}}=\sqrt{\frac{16}{36}}=\frac{2}{3}$

Now, directrices $=\left(+\frac{a}{e},-\frac{a}{e}\right)=\left(+\frac{6}{2/3},-\frac{6}{2/3}\right)$

$=(+9,-9)$

Thus, the distance between the directrices $=|9-(-9)|=18$

Q16. Find the coordinates of a point on the parabola $y^2=8x$, whose focal distance is 4.

Solution:
Comparing the given equation $y^2=8x$ of the parabola with $y^2=4ax$, we get $8=4a \Rightarrow a=2$

Therefore, focal distance $=|x+a|=4 \Rightarrow |x+2|=4$

$\Rightarrow x+2=\pm 4 \Rightarrow x=2,-6$

Since $x\neq -6$,

For $x=2, y^2=8\times 2 \Rightarrow y^2=16 \Rightarrow y=\pm 4$

Thus, the points are $(2,4),(2,-4)$.

Q17. Find the length of the line segment joining the vertex of the parabola $y^2 = 4ax$, and a point on the parabola, where the line segment makes an angle θ to the X-axis.

Solution:
Let the coordinates of any point P on the parabola be $y^2 = 4ax$ be $(at^2, 2at)$.

Therefore, $\tan\theta = \dfrac{2at}{at^2} = \dfrac{2}{t} \Rightarrow t = 2\cot\theta$

Thus, the length of the line segment

$= \sqrt{(at^2-0)^2+(2at-0)^2}$

$= \sqrt{a^2t^4+4a^2t^2} = at\sqrt{t^2+4} = 2a\cot\theta\sqrt{4\cot^2\theta+4}$

$= 4a\cot\theta\sqrt{1+\cot^2\theta} = 4a\cot\theta\cdot\operatorname{cosec}\theta$

$= \dfrac{4a\cos\theta}{\sin\theta}\cdot\dfrac{1}{\sin\theta} = \dfrac{4a\cos\theta}{\sin^2\theta}$

Q18. If the points (0, 4) and (0, 2) are, respectively, the vertex and focus of a parabola, find the equation of the parabola.

Solution:
The coordinates of the vertex of the parabola are V (0, 4) and the focus of the parabola is F (0, 2).

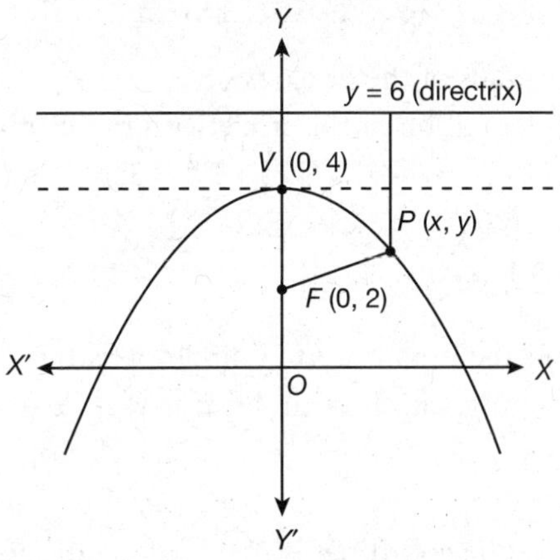

Therefore, the directrix of the parabola will be $y = 6$.

Let $P(x, y)$ be the point on parabola. Then the distance of point P from the directrix must be equal to the distance of point P from focus F, and thus

$\Rightarrow \left|\dfrac{0+y-6}{\sqrt{0+1}}\right| = \sqrt{(x-0)^2+(y-2)^2}$

$\Rightarrow |y-6| = \sqrt{x^2+y^2-4y+4}$

$\Rightarrow x^2+y^2-4y+4 = y^2+36-12y$

$\Rightarrow x^2+8y = 32$

Q19. If the line $y = mx+1$ is the tangent to the parabola $y^2 = 4x$, then find the value of m.

Solution:
The line $y = mx+1$ is a tangent to the parabola $y^2 = 4x$; therefore solving equations $y = mx+1$ (i) and $y^2 = 4x$ (ii), we get $(mx+1)^2 = 4x$ (iii)

$\Rightarrow m^2x^2+2mx+1 = 4x$

$\Rightarrow m^2x^2+x(2m-4)+1 = 0$

The line is a tangent to the parabola; therefore, the line and the parabola must meet at a single point. Thus, the discriminant is equal to zero.

Hence, $(2m-4)^2-4m^2\times 1 = 0$

$\Rightarrow 4m^2+16-16m-4m^2 = 0$

$\Rightarrow 16m = 16 \Rightarrow m = 1$

Q20. If the distance between the foci of a hyperbola is 16 and its eccentricity is $\sqrt{2}$, then obtain the equation of the hyperbola.

Solution:
$e = \sqrt{2}$ and distance between the foci $2ae = 16 \Rightarrow ae = 8$

Therefore, $a\sqrt{2} = 8 \Rightarrow a = 4\sqrt{2}$

Also, $b^2 = a^2(e^2-1)$

$\Rightarrow b^2 = (4\sqrt{2})^2[(\sqrt{2})^2-1] = 16\times 2(2-1) = 32(2-1)$

Thus, the equation of the hyperbola is

$\dfrac{x^2}{32}+\dfrac{y^2}{32} = 1 \Rightarrow x^2-y^2 = 32$

Day 156

NCERT Exemplar Problems: Solutions (Q21–Q40)

Q21. Find the eccentricity of the hyperbola $9y^2-4x^2 = 36$.

Solution:
The equation of the hyperbola is $9y^2-4x^2 = 36$

$\Rightarrow \dfrac{9y^2}{36}-\dfrac{4x^2}{36} = \dfrac{36}{36} \Rightarrow \dfrac{y^2}{4}-\dfrac{x^2}{9} = 1$

$\Rightarrow -\dfrac{x^2}{9}+\dfrac{y^2}{4} = 1$; comparing this equation with $-\dfrac{x^2}{a}+\dfrac{y^2}{b^2} = 1$, we get $a = 3, b = 2$.

$\therefore e = \sqrt{1+\left(\dfrac{a}{b}\right)^2} = \sqrt{1+\dfrac{9}{4}} = \dfrac{\sqrt{13}}{2}$

Q22. Find the equation of the hyperbola with eccentricity 3/2 and foci at $(\pm 2, 0)$.

Solution:

The eccentricity $e = 3/2$ and foci $(\pm ae, 0) = (\pm 2, 0)$.

Therefore, $ae = 2 \Rightarrow a \times \frac{3}{2} = 2 \Rightarrow a = 4/3$

Also, $b^2 = a^2(e^2 - 1)$

$$\Rightarrow b^2 = \frac{16}{9}\left(\frac{9}{4} - 1\right) = \frac{16}{4}\left(\frac{5}{4}\right) = \pm\frac{20}{9}$$

Thus, the equation of the hyperbola is $\frac{x^2}{16/9} + \frac{y^2}{20/9} = 1$

$$\Rightarrow \frac{x^2}{4} - \frac{y^2}{5} = 4/9$$

Q23. If the lines $2x - 3y = 5, 3x - 4y = 7$ are the diameters of a circle of area 154 square units, then obtain the equation of the circle.

Solution:

Solving the given lines $2x - 3y - 5 = 0$ and $3x - 4y - 7 = 0$,

we get $\frac{x}{21 - 20} = \frac{y}{-15 + 14} = \frac{1}{-8 + 9}$

$$\Rightarrow \frac{x}{1} = \frac{y}{-1} = \frac{1}{+1} \Rightarrow x = 1, y = -1$$

Since the given lines are the diameters of the circle, the intersection point of these lines will give the centre of the circle. Thus, the coordinates of the centre of the circle are (1, –1).

Let the radius of the circle be r; then

$$\pi r^2 = 154 \Rightarrow \frac{22}{7} \times r^2 = 154$$

$$\Rightarrow r^2 = \frac{154 \times 7}{22} \Rightarrow r^2 = \frac{14 \times 7}{2} \Rightarrow r^2 = 49$$

Thus, the equation of the circle is $(x - 1)^2 + (y + 1)^2 = 49$

$$\Rightarrow x^2 - 2x + 1 + y^2 + 2y + 1 = 49$$

$$\Rightarrow x^2 + y^2 - 2x + 2y = 47$$

Q24. Find the equation of the circle which passes through the points (2, 3) and (4, 5) and its centre lies on the straight line $y - 4x + 3 = 0$.

Solution:

Let the general equation of the circle be

$x^2 + y^2 + 2gx + 2fy + c = 0$ (i)

This circle passes through the points (2, 3) (4, 5); therefore

$4 + 9 + 4g + 6f + c = 0 \Rightarrow 4g + 6f + c = -13$ (ii)

and

$16 + 25 + 8g + 10f + c = 0 \Rightarrow 8g + 10f + c = -41$ (iii)

The centre of the circle $(-g, -f)$ lies on the straight line $y - 4x + 3 = 0$; therefore

$-f + 4g + 3 = 0$ (iv)

Solving the above three equations, we get

$g = -2,\ f = -5$ and $c = 25$

Thus, the equation of the circle is

$x^2 + y^2 - 4x - 10y + 25 = 0$

Q25. Find the equation of a circle whose centre is (3, –1) and which cuts off a chord of length 6 units on the line $2x - 5y + 18 = 0$.

Solution:

The given centre of the circle is (3, –1); therefore, the distance of the centre from the given line (or chord AB)

is $OP = \left|\frac{6 + 5 + 18}{\sqrt{4 + 25}}\right| = \frac{29}{\sqrt{29}} = \sqrt{29}$

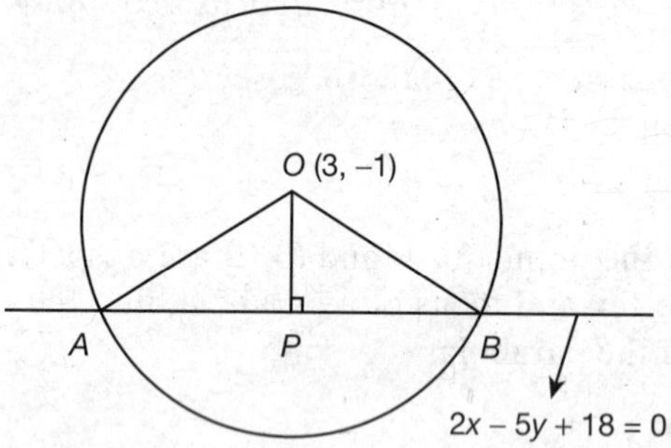

Now, in $\Delta OPB, OB^2 = OP^2 + PB^2$

$\Rightarrow OB^2 = 29 + 9 = 38 [\because AB = 6 \Rightarrow PB = 3]$

Thus, the radius of the circle is $\sqrt{38}$.

Hence, the equation of the circle with radius $r = \sqrt{38}$ and centre (3, –1) is $\Rightarrow (x - 3)^2 + (y + 1)^2 = 38$

$$\Rightarrow x^2 - 6x + 9 + y^2 + 2y + 1 = 38$$

$$\Rightarrow x^2 + y^2 - 6x + 2y - 28 = 0$$

Q26. Find the equation of a circle of radius 5 which is touching another circle $x^2 + y^2 - 2x - 4y - 20 = 0$ at (5, 5).

Solution:

Let (h, k) be the coordinates of the centre of the required circle. The centre of the given circle is (1, 2).

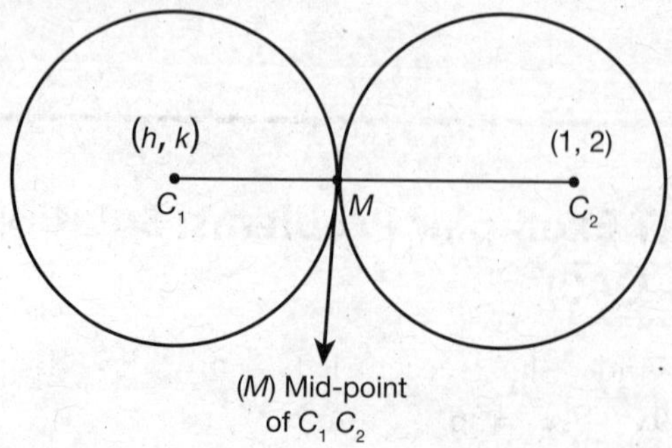

M is the mid-point of C_1C_2; therefore

$\therefore 5 = \frac{1+h}{2} \Rightarrow h = 9$

$5 = \frac{2+k}{2} \Rightarrow k = 8$

Hence, the equation of the required circle is

$(x-9)^2 + (y-8)^2 = 25$

$\Rightarrow x^2 - 18x + 81 + y^2 - 16y + 64 = 25$

$\Rightarrow x^2 + y^2 - 18x - 16y + 120 = 0$

Q27. Find the equation of a circle passing through the point (7, 3) having a radius of 3 units and whose centre lies on the line $y = x - 1$.

Solution:

Let (h, k) be the centre of the circle; then the equation of circle will be $(x-h)^2 + (y-k)^2 = r^2$

$\Rightarrow (x-h)^2 + (y-k)^2 = 9$ (i)

Since the centre (h, k) lies on the line $y = x - 1$

$\therefore k = h - 1$ (ii)

And since the circle passes through the point (7, 3),

$\therefore (7-h)^2 + (3-k)^3 = 9$

$\Rightarrow 49 - 14h + h^2 + 9 - 6k + k^2 = 9$

$\Rightarrow h^2 + k^2 - 14h - 6k + 49 = 0$ (iii)

Putting $k = h - 1$ in Equation (iii), we get

$h^2 + (h-1)^2 - 14h - 6(h-1) + 49 = 0$

$\Rightarrow h^2 + h^2 - 2h - 1 - 14h - 6h + 6 + 49 = 0$

$\Rightarrow 2h^2 - 22h + 56 = 0 \Rightarrow h^2 - 11h + 28 = 0$

$\Rightarrow (h-7)(h-4) = 0 \Rightarrow h = 4, 7$

When $h = 7$, then $k = 7 - 1 = 6$

$\therefore$ Centre = (7, 6)

When $h = 4$, then $k = 7 - 4 = 3$

$\therefore$ Centre = (4,3)

Thus, the equation of the circle when the centre (7, 6) is

$(x-7)^2 + (y-6)^2 = 9$

$\Rightarrow x^2 - 14x + 49 + y^2 - 12y + 36 = 9$

$\Rightarrow x^2 + y^2 - 14x - 12y + 76 = 0$

And, the equation of the circle when the centre (4, 3) is

$(x-4)^2 + (y-3)^2 = 9$

$\Rightarrow x^2 - 8x + 16x + y^2 - 6y + 9 = 9$

$\Rightarrow x^2 + y^2 - 8x - 6y + 16 = 0$

Q28. Find the equation of each of the following parabolas:

(i) directrix = 0, focus at (6, 0)

(ii) vertex at (0, 4), focus at (0, 2)

(iii) focus at (–1, –2), directrix $x - 2y + 3 = 0$

Solution:

(i) For directrix = 0 and focus = (6, 0)

The equation of the parabola is $(x-6)^2 + y^2 = x^2$

$\Rightarrow x^2 + 36 - 12x + y^2 = x^2$

$\Rightarrow y^2 - 12x + 36 = 0$

(ii) For vertex = (0, 4) and focus = (0, 2), the equation of the parabola is

$\sqrt{(x-0)^2 + (y-2)^2} = |y-6|$

$\Rightarrow x^2 + y^2 - 4y + 4 = y^2 - 12y + 36$

$\Rightarrow x^2 + 8y - 32 = 0$

(iii) For focus at (–1, –2) and directrix

$x - 2y + 3 = 0$

The equation of the parabola is

$\sqrt{(x+1)^2 + (y+2)^2} = \left|\frac{x-2y+3}{\sqrt{1+4}}\right|$

$\Rightarrow x^2 + 2x + 1 + y^2 + 4y + 4$

$= \frac{1}{5}[x^2 + 4y^2 + 9 - 4xy - 12y + 6x]$

$\Rightarrow 4x^2 + 4xy + y^2 + 4x + 32y + 16 = 0$

Q29. Find the equation of the set of all points, the sum of whose distances from the points (3, 0), (9, 0) is 12.

Solution:

Let the coordinates of the point be (x, y); then

$\sqrt{(x-3)^2 + y^2} + \sqrt{(x-9)^2 + y^2} = 12$

$\Rightarrow \sqrt{(x-3)^2 + y^2} = 12 - \sqrt{(x-9)^2 + y^2}$

On squaring both sides, we get

$x^2 - 6x + 9 + y^2 = 144 + (x^2 - 18x + 81 + y^2)$
$- 24\sqrt{(x-9)^2 + y^2}$

$\Rightarrow 12x - 216 = -24\sqrt{(x-9)^2 + y^2}$

$\Rightarrow x - 18 = -2\sqrt{(x-9)^2 + y^2}$

$\Rightarrow x^2 - 36x + 324 = 4(x^2 - 18x + 81 + y^2)$

$\Rightarrow 3x^2 + 4y^2 - 36x = 0$

Q30. Find the equation of the set of all points whose distance from (0, 4) are 2/3 of their distance from the line $y = 9$.

Solution:

Let the point be $P(x, y)$

$\therefore$ Distance from $(0,4) = \sqrt{x^2 + (y-4)^2}$

So, the distance from the line $y = 9$ is $\left|\frac{y-9}{\sqrt{1}}\right|$

$\therefore \sqrt{x^2+(y-4)^2} = \frac{2}{3}\left|\frac{y-9}{1}\right|$

$\Rightarrow x^2+y^2-8y+10 = \frac{4}{9}(y^2-18y+81)$

$\Rightarrow 9x^2+9y^2-72y+144 = 4y^2-72y+324$

$\Rightarrow 9x^2+5y^2 = 180$

Q31. Show that the set of all points such that the difference of their distance from (4, 0) and (–4, 0) is always equal to 2, represent a hyperbola.

Solution:

Let the coordinates of the point be $P(x, y)$; then

Distance of P from (4,0) is $\sqrt{(x-4)^2+y^2}$ (i)

Distance of P from (–4,0) is $\sqrt{(x+4)^2+y^2}$ (ii)

Now, $\sqrt{(x+4)^2+y^2} - \sqrt{(x-4)^2+y^2} = 2$

$\Rightarrow \sqrt{(x+4)^2+y^2} = 2+\sqrt{(x-4)^2+y^2}$

On squaring both sides, we get

$x^2+8x+16+y^2$

$= 4+x^2-8x+16+y^2+4\sqrt{(x-4)^2+y^2}$

$\Rightarrow 16x-4 = 4\sqrt{(x-4)^2+y^2}$

$\Rightarrow 4(x-1) = 4\sqrt{(x-4)^2+y^2}$

$\Rightarrow 16x^2-8x+1 = x^2+16-8x+y^2$

$\Rightarrow 15x^2-y^2 = 15$ which is the equation of a parabola.

Q32. Find the equation of the hyperbola with

(i) vertices $(\pm 5, 0)$, foci $(\pm 7, 0)$

(ii) vertices $(0, \pm 7)$, $e = 7/3$

(iii) foci $(0, \pm\sqrt{10})$, passing through (2, 3).

Solution:

(i) For vertices $(\pm 5, 0)$, foci $(\pm 7, 0)$ and $a = \pm 5$

$\therefore (\pm ae, 0) = (\pm 7, 0)$

$\Rightarrow ae = 7 \Rightarrow 5e = 7 \Rightarrow e = 7/5$

$\because b^2 = a^2(e^2-1) \Rightarrow b^2 = 25\left(\frac{49}{25}-1\right)$

$\Rightarrow b^2 = 25\left(\frac{49-25}{25}\right) \Rightarrow b^2 = 25$

Thus, the equation of the parabola is $\frac{x^2}{25}-\frac{y^2}{24} = 1$

(ii) For vertices $= (0, \pm 7), e = 4/3 \therefore b = 7$

$\because e^2 = 1+\frac{a^2}{b^2} \Rightarrow \frac{16}{9}-1 = \frac{a^2}{b^2} \Rightarrow \frac{7}{9} = \frac{a^2}{49} \Rightarrow a^2 = \frac{343}{9}$

Thus, the equation of the hyperbola is $-\frac{x^2 \times 9}{343}+\frac{y^2}{49} = 1$

$\Rightarrow -\frac{9x^2}{7}+y^2 = 49 \Rightarrow 9x^2-7y^2+343 = 0$

(iii) For foci $= (0, \pm\sqrt{10}) \Rightarrow be = \sqrt{10}$

$\Rightarrow a^2+b^2 = 10 \Rightarrow a^2 = 10-b^2$

$\therefore$ Equation of the hyperbola is $-\frac{x^2}{a^2}+\frac{y^2}{b^2} = 1$

This hyperbola passes through the point (2, 3).

$\therefore -\frac{4}{a^2}+\frac{9}{b^2} = 1 \Rightarrow \frac{-4}{10-b^2}+\frac{9}{b^2} = 1$

$\Rightarrow \frac{-4b^2+90-9b^2}{b^2(10-b^2)} = 1$

$\Rightarrow -13b^2+90 = 10b^2+b^4$

$\Rightarrow b^4-23b^2+90 = 0$

$\Rightarrow (b^2-18)(b^2-5) = 0$

For $b^2 = 18 \Rightarrow b = \pm 3\sqrt{2}$

And for $b^2 = 5 \Rightarrow b = \pm\sqrt{5}$

$\because b^2 = 18 \Rightarrow a^2 = -8$ which is not possible

And for $a^2 = 5 \Rightarrow b^2 = 5$

Thus, the equation of the hyperbola is $-\frac{x^2}{5}+\frac{y^2}{5} = 1$

$\Rightarrow y^2-x^2 = 5$

Q33. The line $x+3y = 0$ is a diameter of the circle $x^2+y^2+6x+2y = 0$.

Solution: False

Equation of the circle is $x^2+y^2+6x+2y = 0$.

$\therefore$ Centre (–3, –1)

The diameter passes through the centre; therefore the centre (–3, –1) must satisfy the equation of the diameter.

Now, for the given line $x+3y = 0 \Rightarrow -3+3\times(-1) \neq 0$

Therefore, the line is not a diameter of the circle.

Q34. The shortest distance from the point (2, –7) to the circle $x^2+y^2-14x-10y-151 = 0$ is equal to 5.

Solution: False

For the given circle $x^2+y^2-14x-10y-151 = 0$,

Centre = (7, 5) and

Radius $= \sqrt{49+25+151} = \sqrt{225} = 15$

Thus, the distance between the point (2, –7) and the centre of the circle is given by $= \sqrt{(2-7)^2+(-7-5)^2} = \sqrt{25+144} = 13$

$\therefore$ Shortest distance, $d = |13-15| = 2$

Q35. If the line $lx + my = 1$ is a tangent to the circle $x^2 + y^2 = a^2$ then the point (l, m) lies on a circle.

Solution: True

For the given circle $x^2 + y^2 = a^2$

Radius of circle = a and centre = (0, 0)

The distance from point (l, m) and the centre is

$\sqrt{(l-0)^2 + (m-0)^2} = a \Rightarrow l^2 + m^2 = a^2$

Thus, l m lies on the circle.

Q36. The point (1, 2) lies inside the circle $x^2 + y^2 - 2x + 6y + 1 = 0$.

Solution: False

For the given circle $S: x^2 + y^2 - 2x + 6y + 1 = 0$, the point (1, 2) gives $S: 1 + 4 - 2 + 12 + 1 = 16$

$\Rightarrow S > 0$; thus (1, 2) lies outside the circle.

Q37. The line $lx + my + n = 0$ will touch the parabola $y^2 = 4ax$, if $ln = am^2$.

Solution: True

Substituting the value of x from the equation of line $lx + my + n = 0 \Rightarrow x = -\left(\dfrac{my+n}{l}\right)$ in the equation of parabola $y^2 = 4ax$, we get $y^2 = -\dfrac{4a(my+n)}{l}$

$\Rightarrow ly^2 = -4amy - 4an \Rightarrow ly^2 + 4amy + 4an = 0$

For tangent, $D = 0 \Rightarrow 16a^2m^2 - 4l \times 4an = 0$

$\Rightarrow 16a^2m^2 = 16lan \Rightarrow am^2 = ln$

Q38. If P is a point on the ellipse $\dfrac{x^2}{16} + \dfrac{y^2}{25} = 1$ whose foci are S and S', then $PS + PS' = 8$.

Solution: False

Since the given equation of the ellipse $\dfrac{x^2}{16} + \dfrac{y^2}{25} = 1$

is in the form of $\dfrac{x^2}{a^2} + \dfrac{y^2}{b^2} = 1$ where $b > a$

And, $e = \sqrt{1 - \dfrac{a^2}{b^2}} = \sqrt{\dfrac{25-16}{25}} = 3/5$

Therefore, foci $S = (0, be), S'(0, -be)$

$\Rightarrow S = \left(0, \dfrac{3\times 5}{5}\right), S' = \left(0, -\dfrac{3\times 5}{5}\right)$

$\Rightarrow S = (0, 3), S' = (0, -3)$

Thus, $PS + PS' = 6$

Q39. The line $2x + 3y = 12$ touches the ellipse $\dfrac{x^2}{9} + \dfrac{y^2}{4} = 2$ at the point (3, 2).

Solution: True

The equation of the tangent to the ellipse $\dfrac{x^2}{9} + \dfrac{y^2}{4} = 2$ at (x_1, y_1) is $\dfrac{xx_1}{9} + \dfrac{yy_1}{4} = 2$.

$\therefore$ Tangent at (3, 2) is $\dfrac{3x_1}{9} + \dfrac{2y_1}{4} = 2 \Rightarrow \dfrac{x_1}{3} + \dfrac{y_1}{2} = 2$

$\Rightarrow 2x_1 + 3y_1 = 12$, which is the given equation of line $2x + 3y = 12$.

Thus, the statement is true.

Q40. The locus of the point of intersection of lines $\sqrt{3}x - y - 4\sqrt{3}k = 0, \sqrt{3}kx + ky - 4\sqrt{3} = 0$ for a different value of k is a hyperbola whose eccentricity is 2.

Solution: True

Since the given equations of the line are

$\sqrt{3}x - y - 4\sqrt{3}k = 0$ (i)

$\sqrt{3}kx + ky - 4\sqrt{3} = 0$ (ii)

From Equation (i), $4\sqrt{3}k = \sqrt{3}x - y \Rightarrow k = \dfrac{\sqrt{3}x - y}{4\sqrt{3}}$ put the value of k in Equation (ii), we get

$\sqrt{3}x\left(\dfrac{\sqrt{3}x-y}{4\sqrt{3}}\right) + \left(\dfrac{\sqrt{3}x-y}{4\sqrt{3}}\right)y - 4\sqrt{3} = 0$

$\Rightarrow \dfrac{1}{4}(\sqrt{3}x^2 - xy) + \dfrac{1}{4}\left(xy - \dfrac{y^2}{\sqrt{3}}\right) - 4\sqrt{3} = 0$

$\Rightarrow \dfrac{\sqrt{3}}{4}x^2 - \dfrac{y^2}{4\sqrt{3}} - 4\sqrt{3} = 0 \Rightarrow 3x^2 - y^2 - 48 = 0$

$\Rightarrow 3x^2 - y^2 = 48$, which is the equation of a hyperbola.

Day 157

NCERT Exemplar Problems: Solutions (Q41–Q59)

Q41. The equation of the circle having its centre at (3, –4) and touching the line $5x+12y-12=0$ is..........

Solution:

Radius of the circle = Perpendicular distance from the centre (3, –4) to the line $= r = \left|\dfrac{15-48-12}{\sqrt{25+144}}\right| = 45/13$

Thus, the required equation of the circle is

$$(x-3)^2+(y+4)^2=\left(\frac{45}{13}\right)^2$$

Q42. The equation of the circle circumscribing the triangle whose sides are the lines $y=x+2$, $3y=4x$, $2y=3x$ is.........

Solution:

The given equations of the line are

$y=x+2$ (i)

$3y=4x$ (ii)

$2y=3x$ (iii)

From equations (i) and (ii),

$\dfrac{4x}{3}=x+2 \Rightarrow 4x=3x+6 \Rightarrow x=6$

Substituting $x=6$ in Equation (i), we get $y=8$

$\therefore$ Point $A=(6,8)$

From equations (i) and (iii),

$\dfrac{3x}{2}=x+2 \Rightarrow 3x=2x+4 \Rightarrow x=4$

Substituting $x=4$ in Equation (i), we get $y=6$

$\therefore$ Point $B=(4,6)$

From equations (ii) and (iii) $x=0,\ y=0$

Therefore, $C=(0,0)$

Let the equation of the circle be

$x^2+y^2+2gx+2fy+c=0$

The points $A(6,8), B(4,6), C(0,0)$ lie on this circle $x^2+y^2+2gx+2fy+c=0$

Therefore, $36+64+12g+16f+c=0$

$\Rightarrow 12g+16f+c=-100$ (iv)

$16+36+8g+12f+c=0$

$\Rightarrow 8g+12f+c=-52$ (v)

$\Rightarrow c=0$ (vi)

Now solving equations (iv), (v) and (vi), we get $3g+4f+25=0$ and $2g+3f+13=0$

$\Rightarrow \dfrac{g}{+52-75}=\dfrac{f}{50-30}=\dfrac{1}{9-8} \Rightarrow \dfrac{g}{-23}=\dfrac{f}{11}=\dfrac{1}{1}$

$\Rightarrow g=-23, f=11$

Thus, the equation of the circle is $x^2+y^2-46x+22y+0=0$

$\Rightarrow x^2+y^2-46x+22y=0$

Q43. An ellipse is described by using an endless string which is passed over two pins. If the axes are 6 cm and 4 cm, the length of the string and distance between the pins are............

Solution:

Let the equation of the ellipse be $\dfrac{x^2}{a^2}+\dfrac{y^2}{b^2}=1.$

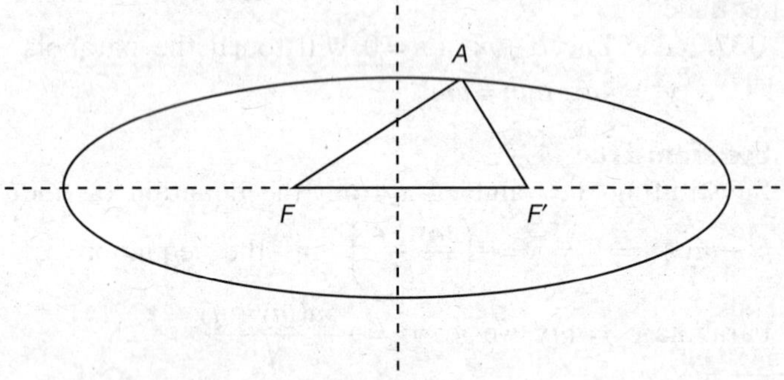

$\therefore 2a=6,\ 2b=4 \Rightarrow a=3, b=2$

Also, $c^2=a^2-b^2=(3)^2-(2)^2=9-4=5 \Rightarrow c=\sqrt{5}$

Length of the string is $L=AF'+F'F+AF$

$=a+c+2c+ac \Rightarrow 2a+2c=6+2\sqrt{5}$

$\therefore$ Distance between the points $cc'=2\sqrt{5}$

Q44. The equation of the ellipse having foci (0, 1), (0, –1) and minor axis of length 1 is...............

Solution:

The foci of the ellipse are $(0,\pm be) \Rightarrow be=1$

And length of minor axis, $2a=1 \Rightarrow a=\dfrac{1}{2}$

Also, $e^2=1-\dfrac{a^2}{b^2} \Rightarrow (be)^2=b^2-a^2$

$\Rightarrow 1=b^2-\dfrac{1}{4} \Rightarrow 1+\dfrac{1}{4}=b^2 \Rightarrow \dfrac{5}{4}=b^2$

Thus, the equation of the ellipse is

$$\frac{x^2}{1/4}+\frac{y^2}{5/4}=1 \Rightarrow \frac{4x^2}{1}+\frac{4y^2}{5}=1$$

Q45. The equation of the parabola having focus at (–1, –2) and directrix is $x-2y+3=0$, is.............

Solution:
The focus is at $F(-1,-2)$ and directrix is $x-2y+3=0$. Let any point on the parabola be (x, y); then $PF=\left|\frac{x-2y+3}{\sqrt{1+4}}\right|$

$$\Rightarrow (x+1)^2+(y+2)^2=\frac{(x-2y+3)^2}{5}$$
$$\Rightarrow 5[x^2+2x+1+y^2+4y+4]$$
$$=x^2+4y^2+9-4xy-12y+6x$$
$$\Rightarrow 4x^2+y^2+4x+32y+16=0$$

Q46. The equation of the hyperbola with vertices at $(0,\pm 6)$ and eccentricity 5/3 is........ and its foci are.........

Solution:
Let the equation of the hyperbola be $-\frac{x^2}{a^2}+\frac{y^2}{b^2}=1$.

Then vertices $=(0,\pm b)=(0,\pm 6)\Rightarrow b=6$

And since $e=\sqrt{1+\frac{a^2}{b^2}}=5/3\Rightarrow 1+\frac{a^2}{36}=\frac{25}{9}$

$\Rightarrow \frac{a^2}{4}=16\Rightarrow a^2=48$

Thus, the equation of the hyperbola is $\frac{-x^2}{48}+\frac{y^2}{36}=1\Rightarrow \frac{y^2}{36}-\frac{x^2}{48}=1$

Hence, foci $=(0,\pm be)=\left(0,\pm\frac{5}{3}\times 6\right)=(0,\pm 10)$

Q47. The area of the circle centre at (1, 2) and passing through the point (4, 6) is
(a) 5π (b) 10π
(c) 25π (d) None of these

Solution: (c)
Radius of the circle = distance between points (1, 2) and (4, 6) $r=\sqrt{9+16}=5$
$\therefore$ Required area $=\pi r^2=25\pi$

Q48. Equation of a circle which passes through (3, 6) and touches the axes is
(a) $x^2+y^2+6x+6y+3=0$
(b) $x^2+y^2-6x-6y-9=0$
(c) $x^2+y^2-6x-6y+9=0$
(d) None of these

Solution: (c)
Let the centre of the circle be (a, a); then equation of the circle is $(x-a)^2+(y-a)^2=a^2$

The point (3, 6) lies on this circle; therefore $(3-a)^2+(6-a)^2=a^2$

$$\Rightarrow a^2+9-6a+36-12a+a^2=a^2$$
$$\Rightarrow a^2-18a+45=0\Rightarrow (a-3)(a-15)=0\Rightarrow a=3,15$$

Thus, the equation of circle is $(x-3)^2+(y-3)^2=9$

$$\Rightarrow x^2-6x+9+y^2-6y+9=9$$
$$\Rightarrow x^2+y^2-6x-6y+9=0$$

Q49. Equation of the circle with the centre on the Y-axis and passing through the origin and the point (2, 3) is
(a) $x^2+y^2+13y=0$
(b) $3x^2+3y^2+13x+3=0$
(c) $6x^2+6y^2-13y=0$
(d) $x^2+y^2+13x+3=0$

Solution: (c)
Let the general equation of the circle be
$x^2+y^2+2gx+2fy+c=0$
The circle passes through the origin (0, 0) and the point (2, 3); therefore $c=0$ and
$4+9+4g+6f=0\Rightarrow 2g+3f=-13/2$
The centre lies on the Y-axis; therefore $g=0$
$\Rightarrow 3f=-13/2\Rightarrow f=-13/6$

Thus, the equation of the circle is $x^2+y^2-\frac{13y}{6}=0$

$\Rightarrow 6x^2+6y^2-13y=0$

Q50. The equation of a circle with the origin as the centre and passing through the vertices of an equilateral triangle whose median is of length $3a$ is
(a) $x^2+y^2=9a^2$ (b) $x^2+y^2=16a^2$
(c) $x^2+y^2=4a^2$ (d) $x^2+y^2=a^2$

Solution: (c)
The length of the median $AD=3a$

Therefore, radius of the circle $=\frac{3}{2}\times$ length of median

$=\frac{2}{3}\times 3a=2a$

Thus, the equation of the circle is $x^2+y^2=4a^2$

Q51. If the focus of a parabola is (0, –3) and its directrix is $y=3$, then its equation is
(a) $x^2=-12y$ (b) $x^2=12y$
(c) $y^2=-12x$ (d) $y^2=12x$

Solution: (a)

Q6. The perimeter of the circle with centre at (1, 2) and passing through the point (4, 6) is

(a) 5π (b) 10π
(c) 25π (d) None of these

Solution: (b)
Radius of the circle = distance between points (1, 2) and (4, 6) $r = \sqrt{9+16} = 5$
$\therefore$ Required perimeter $= 2\pi r = 10\pi$

Q7. The equations of the circle which passes through (1, 2) and touches the axes are … … … and … …

Solution:
Let the centre of the circle be (a, a); then equation of the circle is $(x-a)^2+(y-a)^2 = a^2$
The point (1, 2) lies on this circle; therefore $(1-a)^2+(2-a)^2 = a^2$
$\Rightarrow a^2+1-2a+4-4a+a^2 = a^2$
$\Rightarrow a^2-6a+5 = 0 \Rightarrow (a-1)(a-5) = 0 \Rightarrow a = 1,5$
Thus, the equations of the circle are $(x-1)^2+(y-1)^2 = 1$ and $(x-5)^2+(y-5)^2 = 25$.

Q8. The equations of the circle with centre on the X-axis and passing through the origin and the point (2, 3) are

(a) $2x^2+2y^2-13x = 0$
(b) $3x^2+3y^2+13x+3 = 0$
(c) $6x^2+6y^2-13y = 0$
(d) $x^2+y^2+13x+3 = 0$

Solution: (a)
Let the general equation of the circle be
$x^2+y^2+2gx+2fy+c = 0$.
The circle passes through the origin (0, 0) and the point (2, 3); therefore $c = 0$ and
$4+9+4g+6f = 0 \Rightarrow 2g+3f = -13/2$
The centre lies on the X-axis; therefore $f = 0$
$\Rightarrow 2g = -13/2 \Rightarrow g = -13/4$
Thus, the equations of the circle are $x^2+y^2-\frac{13x}{2} = 0$
$\Rightarrow 2x^2+2y^2-13x = 0$

Q9. The equation of a circle with origin as centre and passing through the vertices of an equilateral whose median is of length 3 unit is

(a) $x^2+y^2 = 9$ (b) $x^2+y^2 = 16$
(c) $x^2+y^2 = 4$ (d) $x^2+y^2 = 1$

Solution: (c)
The length of the median $AD = 3$
Therefore, radius of the circle $= \frac{2}{3} \times$ length of median
$= \frac{2}{3} \times 3 = 2$
Thus, the equation of the circle is $x^2+y^2 = 4$

Q10. Find the equation of the circle with centre $(-1,-1)$ and radius $\sqrt{2}$.

Solution:
The equation of a circle with centre (h, k) and radius r is $(x-h)^2+(y-k)^2 = r^2$
Therefore, the equation of the circle with centre $(-1,-1)$ and radius $\sqrt{2}$ is $(x+1)^2+(y+1)^2 = (\sqrt{2})^2$
$\Rightarrow x^2+2x+1+y^2+2y+1 = 2$
$\Rightarrow x^2+y^2+2x+2y = 0$

Q11. Find the equation of the circle with centre $(a,-b)$ and radius $\sqrt{a^2+b^2}$.

Solution:
The equation of a circle with centre (h, k) and radius r is $(x-h)^2+(y-k)^2 = r^2$.
Therefore, the equation of the circle with centre $(a,-b)$ and radius $\sqrt{a^2+b^2}$ is
$(x-a)^2+(y+b)^2 = \left(\sqrt{a^2+b^2}\right)^2$
$\Rightarrow x^2-2ax+a^2+y^2+2by+b^2 = a^2+b^2$
$\Rightarrow x^2+y^2-2ax+2by = 0$

Q12. Find the centre and radius of the circle $x^2+y^2-8x+10y-8 = 0$.

Solution:
The given equation $x^2+y^2-8x+10y-8 = 0$ of the circle can be written as $(x^2-8x)+(y^2+10y) = 8$
$\Rightarrow \{x^2-2(x)(4)+4^2\}+\{y^2+2(y)(5)+5^2\}-16-25 = 8$
$\Rightarrow (x-4)^2+(y+5)^2 = 49$
$\Rightarrow (x-4)^2+\{y-(-5)\}^2 = (7)^2$
Comparing this with the equation of the circle $(x-h)^2+(y-k)^2 = r^2$, we get
$h = 4, k = -5, r = 7$
Therefore, the centre of the given circle is (4, –5) and radius is 7.

Q13. Find the centre and radius of the circle $2x^2+2y^2-x+y = 0$

Solution:

The given equation $2x^2+2y^2-x+y=0$ of the circle can be written as $2x^2+2y^2-x+y=0$

$\Rightarrow (2x^2-x)+(2y^2+y)=0$

$\Rightarrow 2\left[\left(x^2-\frac{x}{2}\right)+\left(y^2+\frac{y}{2}\right)\right]=0$

$\Rightarrow \left\{x^2-2x\left(\frac{1}{4}\right)+\left(\frac{1}{4}\right)^2\right\}$

$+\left\{y^2+2y\left(\frac{1}{4}\right)+\left(\frac{1}{4}\right)^2\right\}-\left(\frac{1}{4}\right)^2-\left(\frac{1}{4}\right)^2=0$

$\Rightarrow \left(x-\frac{1}{4}\right)^2+\left(y+\frac{1}{4}\right)^2=\frac{1}{8}=\left(\frac{1}{2\sqrt{2}}\right)^2$

Comparing this with the equation of the circle $(x-h)^2+(y-k)^2=r^2$, we get

$h=\frac{1}{4},\ k=-\frac{1}{4},\ r=\frac{1}{2\sqrt{2}}.$

Therefore, the centre of the given circle is $\left(\frac{1}{4},-\frac{1}{4}\right)$ and radius is $\frac{1}{2\sqrt{2}}$.

Q14. Find the equation of the circle passing through the points (2, 3) and (–1, 1) and whose centre is on the line $2x+4y-5=0$.

Solution:

Let the equation of the required circle be $(x-h)^2+(y-k)^2=r^2$

The circle passes through points (2, 3) and (–1, 1); therefore

$(2-h)^2+(3-k)^2=r^2$ (i)

$(-1-h)^2+(1-k)^2=r^2$ (ii)

The centre (h, k) of the circle lies on the line $2x+4y-5=0$; therefore $2h+4k-5=0$ (iii)

Solving equations (i) and (ii), we get

$(2-h)^2+(3-k)^2=(-1-h)^2+(1-k)^2$

$\Rightarrow 4-4h+h^2+9-6k+k^2=1+2h+h^2+1-2k+k^2$

$\Rightarrow 6h+4k=11$ (iv)

Now, solving equations (iii) and (iv), we get $h=\frac{3}{2}, k=\frac{1}{2}$.

Substituting the values of h and k in Equation (i), we get

$\left(2-\frac{3}{2}\right)^2+\left(3-\frac{1}{2}\right)^2=r^2$

$\Rightarrow \left(\frac{1}{2}\right)^2+\left(\frac{5}{2}\right)^2=r^2 \Rightarrow r^2=\frac{26}{4}=\frac{13}{2}$

Thus, the equation of the required circle is

$\left(x-\frac{3}{2}\right)^2+\left(y-\frac{1}{2}\right)^2=\frac{13}{2}$

$\Rightarrow \left(\frac{2x-3}{2}\right)^2+\left(\frac{2y-1}{2}\right)^2=\frac{13}{2}$

$\Rightarrow 4x^2-12x+9+4y^2-4y+1=26$

$\Rightarrow 4(x^2+y^2-3x-y-4)=0$

$\Rightarrow x^2+y^2-3x-y-4=0$

Q15. Find the equation of a circle which touches both the axes and the line $3x-4y+12=0$ and lies in the third quadrant.

Solution:

Let a be the radius of the circle. The circle touches both the axes and lies in the third quadrant therefore the coordinates of the circle will be $(-a,-a)$.

Now, Radius of the circle = Perpendicular distance of centre $(-a,-a)$ from the tangent $3x-4y+12=0$. Thus,

$a=\left|\frac{-3a+4a+12}{\sqrt{9+16}}\right|$

$\Rightarrow a=\left|\frac{a+12}{5}\right| \Rightarrow a=\pm\left(\frac{a+12}{5}\right)$

Taking the positive sign, we get

$a=\frac{a+12}{5} \Rightarrow 5a=a+12 \Rightarrow 4a=12 \Rightarrow a=3$

Taking the negative sign, we get

$a=\frac{-a-12}{5} \Rightarrow 5a=-a-12 \Rightarrow 6a=-12 \Rightarrow a=-2$

But since $a\neq -2$, $\therefore a=3$

Hence, the equation of the circle is $(x+3)^2+(y+3)^2=3^2$ $[\because a=3]$

$\Rightarrow x^2+6x+9+y^2+6y+9=9$

$\Rightarrow x^2+y^2+6x+6y+9=0$

Q16. Show that the point (x, y) given by $x=\frac{4t}{1+t^2}, y=\frac{2-2t^2}{1+t^2}$ lies on a circle.

Solution:

Since $x=\frac{4t}{1+t^2}, y=\frac{2-2t^2}{1+t^2}$

Therefore, squaring and adding, we get

$x^2+y^2=\frac{16t^2}{(1+t^2)^2}+\frac{(2-2t^2)^2}{(1+t^2)^2}$

$=\frac{16t^2+4+4t^4-8t^2}{(1+t^2)^2}$

$\Rightarrow x^2+y^2=\frac{4+4t^4+8t^2}{(1+t^2)^2}=4\frac{(1+t^2)^2}{(1+t^2)^2}=2^2$

$\Rightarrow x^2 + y^2 = a^2$ which is the equation of a circle.

Q17. If a circle passes through the points (0, 0), (2, 0) and (0, 4), then find the coordinates of its centre.

Solution:
Let the equation of the circle be
$x^2 + y^2 + 2gx + 2fy + c = 0$
Since this circle passes through the points
$(0,0), (2,0), (0,4)$, these points must satisfy the equation of a circle. Thus,
$c = 0, \quad 4 + 4g = 0 \quad$ and $\quad 16 + 8f = 0$
Solving the above equations, we get
$g = -4/4 = -1$ and $f = -16/8 = -2$

Hence, the coordinates of the circle are $(-g, -f) = (1, 2)$

Q18. Find the equation of the circle which touches the X-axis and whose centre is (1, –1).

Solution:
The centre of the circle is (1, –1) and the circle touches the X-axis; therefore its radius = 1.
Thus, the equation of the circle is
$(x-1)^2 + (y+1)^2 = 1^2$
$\Rightarrow x^2 - 2x + 1 + y^2 + 2y + 1 = 1$
$\Rightarrow x^2 + y^2 - 2x + 2y + 1 = 0$

Day 159

Long Answer-I (LA-I): 4-Mark and Long Answer-II (LA-II): 6-Mark Questions

Q1. Find the coordinates of the focus, axis of the parabola, the equation of directrix and the length of the latus rectum for $y^2 = 100x$.

Solution:
In the given equation $y^2 = 100x$ of the parabola, the coefficient of x is positive; therefore the parabola opens towards the right.
Comparing this equation with $y^2 = 4ax$ we get
$4a = 100 \Rightarrow a = 25$
Thus, coordinates of the focus $= (a, 0) = (25, 0)$
The given equation involves y^2; therefore the axis of the parabola is the x-axis.
Equation of the directrix, $x = -a \Rightarrow x = -25 \Rightarrow x + 25 = 0$
Length of the latus rectum $= 4a = 4 \times 25 = 100$

Q2. Find the equation of the parabola that satisfies the following conditions: focus (0, –6); directrix $y = 6$.

Solution:
The focus (0, –6) lies on the y-axis; therefore the axis of the parabola is the y-axis.
Thus, the equation of the parabola is of the form $x^2 = 4ay$ or $x^2 = -4ay$.
Now, the directrix is $y = 6$ and focus is (0, –6); therefore the parabola is of the form $x^2 = -4ay$ with $a = 6$. Thus, the equation of the parabola is $x^2 = -24y$.

Q3. Find the equation of the parabola that satisfies the following conditions: vertex (0, 0), focus (4, 0).

Solution:
The focus (4, 0) lies on the x-axis; therefore the axis of the parabola is the x-axis.
Thus, the equation of the parabola is of the form $y^2 = 4ax$ or $y^2 = -4ax$.
Now, the focus is (4, 0); therefore the parabola is of the form $y^2 = 4ax$ with $a = 4$. Thus, the equation of the parabola is $y^2 = 16x$.

Q4. Find the equation of the parabola that satisfies the following conditions: vertex (0, 0), focus (–1, 0).

Solution:
Since the focus (–1, 0) lies on the x-axis, the axis of the parabola is the x-axis.
Thus, the equation of the parabola is of the form $y^2 = 4ax$ or $y^2 = -4ax$.
Now, the focus is (–1, 0); therefore the parabola is of the form $y^2 = -4ax$ with $a = 1$. Thus, the equation of the parabola is $y^2 = -4x$.

Q5. Find the equation of the parabola that satisfies the following conditions: vertex (0, 0) passing through (2, 4) and axis is along the x-axis.

Solution:
Since the vertex of the parabola is (0, 0) and the axis of the parabola is the x-axis, the equation of the parabola is of the form $y^2 = 4ax$ or $y^2 = -4ax$.
Now, the parabola passes through the point (2, 4) which lies in the first quadrant; therefore the equation of the parabola is of the form $y^2 = 4ax$. Now since point (2, 4) must satisfy the equation $y^2 = 4ax, 4^2 = 4a(2) \Rightarrow a = 2$.
Thus, the equation of the parabola is
$y^2 = 4 \times 2 \times x \Rightarrow y^2 = 8x$

Q6. Find the coordinates of the foci, the vertices, the length of major axis, the minor axis, the eccentricity and the latus rectum of the ellipse $\frac{x^2}{81}+\frac{y^2}{9}=1.$

Solution:

In the given equation $\frac{x^2}{81}+\frac{y^2}{9}=1$ of the ellipse, the denominator of x^2 is greater than the denominator of y^2; therefore the major axis is along the x-axis and the minor axis is along the y-axis.

Comparing the given equation with $\frac{x^2}{a^2}+\frac{y^2}{b^2}=1$, we get $a=9$, $b=3$ and $c=\sqrt{a^2-b^2}=\sqrt{81-9}=6\sqrt{2}$

Thus, the coordinates of the foci are $(6\sqrt{2},0),(-6\sqrt{2},0)$.

The coordinates of the vertices are (9, 0), (–9, 0).

Length of major axis = $2a = 19$

Length of minor axis = $2b = 6$

Eccentricity $e=\frac{c}{a}=\frac{6\sqrt{2}}{9}=\frac{2\sqrt{2}}{3}$

Length of latus rectum $=\frac{2b^2}{a}=\frac{2\times 9}{9}=2$

Q7. Find the equation of the ellipse that satisfies the given conditions: vertices $(0,\pm 15)$, foci $(0,\pm 5)$.

Solution:

Since the vertices $(0,\pm 15)$ are on the y-axis, the equation of the ellipse will be of the form $\frac{x^2}{b^2}+\frac{y^2}{a^2}=1$, and the semi-major axis $a=15$.

The foci are $(0,\pm 5)$; therefore $c=5$.

Now, from equation $a^2=b^2+c^2$

$\Rightarrow 15^2=b^2+5^2 \Rightarrow b^2=225-25 \Rightarrow b=10\sqrt{2}$

Thus, the equation of the ellipse is $\frac{x^2}{(10\sqrt{2})^2}+\frac{y^2}{15^2}=1$ or $\frac{x^2}{200}+\frac{y^2}{225}=1.$

Q8. Find the equation of the ellipse that satisfies the given conditions: the ends of the major axis are $(\pm 4,0)$ and the ends of the minor axis are $(0,\pm 2)$.

Solution:

The ends of the major axis are $(\pm 4,0)$; therefore the major axis is along the x-axis and $a=4$.

The ends of the minor axis are $(0,\pm 2)$; therefore the minor axis is along the y-axis and $b=2$.

Thus, the equation of the ellipse will be of the form

$$\frac{x^2}{a^2}+\frac{y^2}{b^2}=1 \Rightarrow \frac{x^2}{4^2}+\frac{y^2}{2^2}=1 \Rightarrow \frac{x^2}{16}+\frac{y^2}{4}=1$$

Q9. Find the equation of the ellipse that satisfies the given conditions: the ends of the major axis are $(0,\pm\sqrt{5})$ and the ends of the minor axis are $(\pm\sqrt{2},0)$.

Solution:

The ends of the major axis are $(0,\pm\sqrt{5})$; therefore the major axis is along the y-axis and $a=\sqrt{5}$.

The ends of the minor axis are $(\pm\sqrt{2},0)$; therefore the minor axis is along the y-axis and $b=\sqrt{2}$.

Thus, the equation of the ellipse will be of the form

$$\frac{x^2}{b^2}+\frac{y^2}{a^2}=1 \Rightarrow \frac{x^2}{(\sqrt{2})^2}+\frac{y^2}{(\sqrt{5})^2}=1 \Rightarrow \frac{x^2}{2}+\frac{y^2}{5}=1$$

Q10. Find the equation of the ellipse that satisfies the given conditions: the length of the major axis 20, foci $(\pm 5,0)$.

Solution:

Length of major axis, $2a=20 \Rightarrow a=10$

The foci $(\pm 5,0)$ are on the x-axis; therefore the major axis is along the x-axis and $c=5$.

Now, from the equation $a^2=b^2+c^2$

$\Rightarrow 10^2=b^2+5^2 \Rightarrow b^2=100-25 \Rightarrow b=\sqrt{75}=5\sqrt{3}$

Thus, the equation of the ellipse will be of the form

$$\frac{x^2}{a^2}+\frac{y^2}{b^2}=1 \Rightarrow \frac{x^2}{10^2}+\frac{y^2}{(5\sqrt{3})^2}=1 \Rightarrow \frac{x^2}{100}+\frac{y^2}{75}=1$$

Q11. Find the equation of the ellipse that satisfies the given conditions: the length of minor axis 36, foci $(0,\pm 6)$.

Solution:

Length of minor axis, $2b=36 \Rightarrow b=18$

The foci $(0,\pm 6)$ are on the y-axis; therefore the major axis is along the y-axis and $c=6$.

Now, from the equation $a^2=b^2+c^2$

$\Rightarrow a^2=8^2+6^2 \Rightarrow a^2=360 \Rightarrow a=6\sqrt{10}$

Thus, the equation of the ellipse will be of the form

$$\frac{x^2}{b^2}+\frac{y^2}{a^2}=1 \Rightarrow \frac{x^2}{18^2}+\frac{y^2}{(6\sqrt{10})^2}=1 \Rightarrow \frac{x^2}{324}+\frac{y^2}{360}=1$$

Q12. Find the equation of the ellipse that satisfies the given conditions: foci $(\pm 2,0)$, $a=4$.

Solution:

The foci $(\pm 2,0)$ are on the x-axis; therefore the major axis is along the x-axis and $c=2$.

Now, from the equation $a^2=b^2+c^2$

$\therefore 4^2=b^2+2^2 \Rightarrow b^2=16-4 \Rightarrow b=2\sqrt{3}$

Therefore, the equation of the ellipse will be of the form

$$\frac{x^2}{a^2}+\frac{y^2}{b^2}=1 \Rightarrow \frac{x^2}{4^2}+\frac{y^2}{(2\sqrt{3})^2}=1 \Rightarrow \frac{x^2}{16}+\frac{y^2}{12}=1$$

Q13. Find the coordinates of the foci and the vertices, the eccentricity, and the length of the latus rectum of the hyperbola $x^2 - 4y^2 = 100$.

Solution:
Comparing the given equation $x^2 - 4y^2 = 100$ $\Rightarrow \frac{x^2}{100} - \frac{y^2}{25} = 1$ with the standard equation of the hyperbola $\frac{x^2}{a^2} - \frac{y^2}{b^2} = 1$, we get $a = 10$, $b = 5$.

Now, from the equation $a^2 + b^2 = c^2$
$\Rightarrow c = \sqrt{a^2 + b^2} = \sqrt{100 + 25} = 5\sqrt{5}$
Thus, the coordinates of the foci are $(\pm 5\sqrt{5}, 0)$
The coordinates of the vertices are $(\pm 10, 0)$

Eccentricity $e = \frac{c}{a} = \frac{5\sqrt{5}}{10} = \frac{\sqrt{5}}{2}$

Length of the latus rectum $= \frac{2b^2}{a} = \frac{2 \times 25}{10} = 5$

Q14. Find the equation of the hyperbola satisfying the given conditions: vertices $(0, \pm 4)$, foci $(0, \pm 8)$.

Solution:
The vertices are on the y-axis; therefore the equation of the hyperbola is of the form $\frac{y^2}{a^2} - \frac{x^2}{b^2} = 1$.

The vertices are $(0, \pm 4) \Rightarrow a = 4$ and the foci are $(0, \pm 8) \Rightarrow c = 8$, therefore from the equation $a^2 + b^2 = c^2 \Rightarrow b^2 = 8^2 - 4^2 \Rightarrow b^2 = 64 - 16 = 48$

Thus, the equation of the hyperbola is $\frac{y^2}{16} - \frac{x^2}{48} = 1$

Q15. Find the equation of the hyperbola satisfying the give conditions: foci $(\pm 5, 0)$ and the transverse axis is of length 6.

Solution:
The foci $(\pm 5, 0)$ are on the x-axis; therefore the equation of the hyperbola is of the form $\frac{x^2}{a^2} - \frac{y^2}{b^2} = 1$ and $c = 5$.

Now, the length of the transverse axis is 8; therefore $2a = 6 \Rightarrow a = 3 \Rightarrow a^2 = 9$.
From the equation $a^2 + b^2 = c^2$
$\Rightarrow 3^2 + b^2 = 5^2 \Rightarrow b^2 = 25 - 9 = 16$

Thus, the equation of the hyperbola $\frac{x^2}{9} - \frac{y^2}{16} = 1$

Q16. Find the equation of the hyperbola satisfying the given conditions: foci $(0, \pm 16)$ and the conjugate axis is of length 30.

Solution:
The foci $(0, \pm 16)$ are on the y-axis; therefore the equation of the hyperbola is of the form $\frac{y^2}{a^2} - \frac{x^2}{b^2} = 1$ and $c = 16$.

Now, the conjugate axis is of length 24; therefore $2b = 30 \Rightarrow b = 15 \Rightarrow b^2 = 225$.
From the equation $a^2 + b^2 = c^2$
$\Rightarrow a^2 + 15^2 = 16^2 \Rightarrow a^2 = 31$

Thus, the equation of the hyperbola is $\frac{y^2}{31} - \frac{x^2}{225} = 1$.

Q17. Find the equation of the hyperbola satisfying the given conditions: foci $(\pm 2\sqrt{6}, 0)$ and the latus rectum is of length 10.

Solution:
The foci $(\pm 2\sqrt{6}, 0)$ are on the x-axis; therefore the equation of the hyperbola is of the form $\frac{x^2}{a^2} - \frac{y^2}{b^2} = 1$ and $c = 2\sqrt{6}$.

Now, the length of the latus rectum is 8; therefore $\frac{2b^2}{a} = 10 \Rightarrow b^2 = 5a$
From the equation $a^2 + b^2 = c^2$
$\Rightarrow a^2 + 5a = 24 \Rightarrow a^2 + 5a - 24 = 0$
$\Rightarrow (a + 8)(a - 3) = 0 \Rightarrow a = -8, 3$
Since a is non-negative, $a = 3$.
$\Rightarrow b^2 = 6a = 6 \times 3 = 18$

Thus, the equation of the hyperbola is $\frac{x^2}{9} - \frac{y^2}{18} = 1$.

Q18. Find the equation of the hyperbola satisfying the given conditions: vertices $(\pm 6, 0)$ and eccentricity $e = \frac{4}{3}$.

Solution:
The vertices $(\pm 6, 0)$ are on the x-axis; therefore the equation of the hyperbola is of the form $\frac{x^2}{a^2} - \frac{y^2}{b^2} = 1$ and $a = 6$.

Now, since eccentricity $e = \frac{4}{3}$

$\Rightarrow \frac{c}{a} = \frac{4}{3} \Rightarrow \Rightarrow \frac{c}{6} = \frac{4}{3} \Rightarrow c = 8$
From the equation $a^2 + b^2 = c^2$
$\Rightarrow 6^2 + b^2 = 8^2 \Rightarrow b^2 = 64 - 36 = 28$

Thus, the equation of the hyperbola is $\frac{x^2}{36} - \frac{y^2}{28} = 1$

CHAPTER 12

Introduction to Three-Dimensional Geometry

Day 160

Coordinate Axes, Coordinate Planes and Coordinates of a Point in Three-Dimensional Space

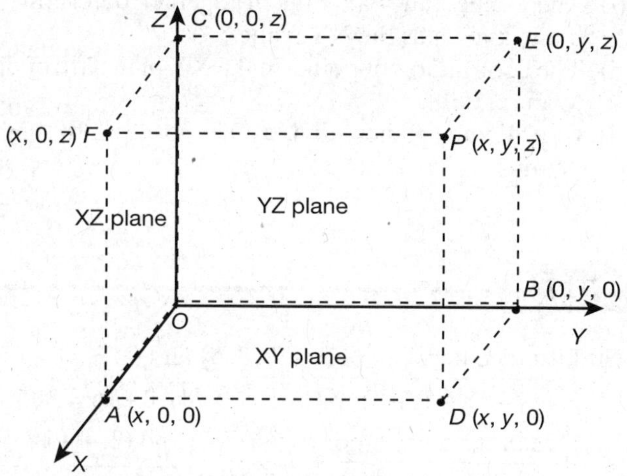

Fig. 12.1 Three-dimensional space

- Let the three mutually perpendicular planes (XY plane, YZ plane and ZX plane) along the coordinate axes OX, OY and OZ (x, y and z axes, respectively) intersect at a point O (Fig. 12.1).
- The point O is called the origin of the coordinate system and its coordinates are given by (0, 0, 0).
- The XY plane, YZ plane and ZX plane known as the three coordinate planes.
- The three coordinate planes divide the space into eight parts known as octants.
- Any point $P(x, y, z)$ in the coordinate system (three dimensional space) is given by a triplet of numbers (x, y, z) along the three coordinate axes. x, y and z are perpendicular distances from the YZ, ZX and XY planes, respectively.
- The value of the coordinate (x, y or z) is taken as positive if it is along the positive axes and negative if it is along the negative axes.
- The coordinates of any point on the x-axis will be given by (x, 0, 0), on the y-axis will be given by (0, y, 0) and on the z-axis will be given by (0, 0, z).
- The z-coordinate is zero in the XY plane, the x-coordinate is zero in the YZ plane and the y-coordinate is zero in the ZX plane.
- The coordinates of any point on the XY plane will be given by (x, y, 0), on the YZ plane will be given by (0, y, z), and on the ZX plane will be given by (x, 0, z).
- The sign of the coordinates of a point determine the octant in which the point lies as given in the table below.

Coordinates / Octants	x	y	z
I	+	+	+
II	-	+	+
III	-	-	+
IV	+	-	+
V	+	+	-
VI	-	+	-
VII	-	-	-
VIII	+	-	-

NCERT Exercise 12.1 (With Solutions)

Q1. A point is on the x-axis. What are its y-coordinate and z-coordinate?

Solution:
If a point is on the x-axis, then its y and z coordinates are zero.

Q2. A point is in the XZ-plane. What can you say about its y-coordinate?

Solution:
If a point is in the XZ plane, then its y-coordinate is zero.

Q3. Name the octants in which the following points lie: $(1,2,3),(4,-2,3),(4,-2,-5),(4,2,-5),$ $(-4,2,-5),(-4,2,5),(-1,-3,6),(2,-4,-7)$

Solution:
Since the x-coordinate, y-coordinate and z-coordinate of point (1, 2, 3) are all positive, the point lies in octant I.
The x-coordinate, y-coordinate and z-coordinate of point (4, –2, 3) are positive, negative and positive, respectively. Therefore, the point lies in octant IV.
The x-coordinate, y-coordinate and z-coordinate of point (4, –2, –5) are positive, negative and negative, respectively. Therefore, the point lies in octant VIII.
The x-coordinate, y-coordinate and z-coordinate of point (4, 2, –5) are positive, positive and negative, respectively. Therefore, the point lies in octant V.
The x-coordinate, y-coordinate and z-coordinate of point (–4, 2, –5) are negative, positive and negative, respectively. Therefore, the point lies in octant VI.
The x-coordinate, y-coordinate and z-coordinate of point (–4, 2, 5) are negative, positive and positive, respectively. Therefore, the point lies in octant II.
The x-coordinate, y-coordinate and z-coordinate of point (–3, –1, 6) are negative, negative and positive, respectively. Therefore, the point lies in octant III.
The x-coordinate, y-coordinate and z-coordinate of point (2, –4, –7) are positive negative and negative, respectively. Therefore, the point lies in octant VIII.

Q4. Fill in the blanks
(i) The x-axis and y-axis taken together determine a plane known as_______________.
(ii) The coordinates of points in the XY plane are of the form___________.
(iii) Coordinate planes divide the space into___________ octants.

Solution:
(i) The x-axis and y-axis taken together determine a plane known as the XY plane.
(ii) The coordinates of points in the XY plane are of the form $(x, y, 0)$.
(iii) Coordinate planes divide the space into eight octants.

Day 161

Distance between Two Points

- Let $P(x_1, y_1, z_1)$, $Q(x_2, y_2, z_2)$ be the two points in the coordinate system of rectangular axes OX, OY and OZ.
- Then the distance between two points $P(x_1, y_1, z_1)$, $Q(x_2, y_2, z_2)$ is given by $|PQ| = \sqrt{(x_2-x_1)^2+(y_2-y_1)^2+(z_2-z_1)^2}$.
- The distance of point $P(x, y, z)$ from the origin $O(0,0,0)$ is $|OP| = \sqrt{x^2+y^2+z^2}$.

NCERT Exercise 12.2 (With Solutions)

Q1. Find the distance between the following pairs of points:
(i) (2, 3, 5) and (4, 3, 1)
(ii) (–3, 7, 2) and (2, 4, –1)
(iii) (–1, 3, –4) and (1, –3, 4)
(iv) (2, –1, 3) and (–2, 1, 3)

Solution:
Distance between two points $A(x_1, y_1, z_1)$ and $B(x_2, y_2, z_2)$ is given by
$$AB = \sqrt{(x_2-x_1)^2+(y_2-y_1)^2+(z_2-z_1)^2}$$

(i) Distance between points (2, 3, 5) and (4, 3, 1)
$$= \sqrt{(4-2)^2+(3-3)^2+(1-5)^2} = \sqrt{4+0+16} = 2\sqrt{5}$$
(ii) Distance between points (–3, 7, 2) and (2, 4, –1)
$$= \sqrt{(2+3)^2+(4-7)^2+(-1-2)^2} = \sqrt{25+9+9} = \sqrt{43}$$
(iii) Distance between points (–1, 3, –4) and (1, –3, 4)
$$= \sqrt{(1+1)^2+(-3-3)^2+(4+4)^2} = \sqrt{4+36+64} = 2\sqrt{26}$$
(iv) Distance between points (2, –1, 3) and (–2, 1, 3)
$$= \sqrt{(-2-2)^2+(1+1)^2+(3-3)^2} = \sqrt{16+4+0} = 2\sqrt{5}$$

Q2. Show that the points $(-2,3,5),(1,2,3),(7,0,-1)$ are collinear.

Solution:
Let $A(-2,3,5)$, $B(1,2,3)$, $C(7,0,-1)$ be the given points; then points A, B and C are collinear if they lie on a line.
$$AB = \sqrt{(1+2)^2+(2-3)^2+(3-5)^2}$$
$$= \sqrt{(3)^2+(-1)^2+(-2)^2} = \sqrt{9+1+4} = \sqrt{14}$$
$$BC = \sqrt{(7-1)^2+(0-2)^2+(-1-3)^2}$$
$$= \sqrt{(6)^2+(-2)^2+(-4)^2} = \sqrt{36+4+16} = 2\sqrt{14}$$
$$AC = \sqrt{(7+2)^2+(0-3)^2+(-1-5)^2}$$
$$\sqrt{(9)^2+(-3)^2+(-6)^2} = \sqrt{81+9+36} = 3\sqrt{14}$$

As $AB+BC=\sqrt{14}+2\sqrt{14}=3\sqrt{14}=AC$, points $A(-2,3,5)$, $B(1,2,3)$, $C(7,0,-1)$ are collinear.

Q3. Verify the following:

(i) $(0,7,-10),(1,6,-6),(4,9,-6)$ are the vertices of an isosceles triangle.

(ii) $(0,7,10),(-1,6,6),(-4,9,6)$ are the vertices of a right-angled triangle.

(iii) $(-1,2,1),(1,-2,5),(4,-7,8),(2,-3,4)$ are the vertices of a parallelogram.

Solution:

(i) Given points $A(0,7,-10)$, $B(1,6,-6)$, $C(4,9,-6)$ are the vertices of an isosceles triangle if two sides of triangle are equal.

$AB=\sqrt{(1-0)^2+(6-7)^2+(-6+10)^2}$

$=\sqrt{(1)^2+(-1)^2+(4)^2}=\sqrt{1+1+6}=\sqrt{18}=3\sqrt{2}$

$BC=\sqrt{(4-1)^2+(9-6)^2+(-6+6)^2}$

$=\sqrt{(3)^2+(3)^2}=\sqrt{9+9}=\sqrt{18}=3\sqrt{2}$

$CA=\sqrt{(0-4)^2+(7-9)^2+(-10+6)^2}$

$=\sqrt{(-4)^2+(-2)^2+(-4)^2}=\sqrt{16+4+16}$

$=\sqrt{36}=6$

As $AB=BC\neq CA$, the given points $A(0,7,-10)$, $B(1,6,-6)$, $C(4,9,-6)$ are the vertices of an isosceles triangle.

(ii) Given points $A(0,7,10)$, $B(-1,6,6)$, $C(-4,9,6)$ are the vertices of a right-angled triangle if the square of the sum of two sides is equal to square of the third side.

$AB=\sqrt{(-1-0)^2+(6-7)^2+(6-10)^2}$

$=\sqrt{(-1)^2+(-1)^2+(-4)^2}=\sqrt{1+1+16}=\sqrt{18}=3\sqrt{2}$

$BC=\sqrt{(-4+1)^2+(9-6)^2+(6-6)^2}$

$=\sqrt{(-3)^2+(3)^2+(0)^2}=\sqrt{9+9}=\sqrt{18}=2\sqrt{3}$

$CA=\sqrt{(0+4)^2+(7-9)^2+(10-6)^2}$

$=\sqrt{(4)^2+(-2)^2+(4)^2}=\sqrt{16+4+16}=\sqrt{36}=6$

Since, $AB^2+BC^2=(3\sqrt{2})^2+(3\sqrt{2})^2=36=AC^2$, the given points $A(0,7,10)$, $B(-1,6,6)$, $C(-4,9,6)$ are the vertices of a right-angled triangle.

(iii) Given points $A(-1,2,1)$, $B(1,-2,5)$, $C(4,-7,8)$, $D(2,-3,4)$ are the vertices of a parallelogram if both opposite sides are equal.

$AB=\sqrt{(1+1)^2+(-2-2)^2+(5-1)^2}$

$=\sqrt{4+16+16}=\sqrt{36}=6$

$BC=\sqrt{(4-1)^2+(-7+2)^2+(8-5)^2}$

$=\sqrt{9+25+9}=\sqrt{43}$

$CD=\sqrt{(2-4)^2+(-3+7)^2+(4-8)^2}$

$=\sqrt{4+16+16}=\sqrt{36}=6$

$DA=\sqrt{(-1-2)^2+(2+3)^2+(1-4)^2}$

$=\sqrt{9+25+9}=\sqrt{43}$

Since opposite sides $AB=CD=6$ and opposite sides $BC=AD=\sqrt{43}$, the given points are the vertices of a parallelogram.

Q4. Find the equation of the set of points which are equidistance from the points (1, 2, 3) and (3, 2, –1).

Solution:

Let $P(x, y, z)$ be the point that is equidistant from points $A(1, 2, 3)$ and $B(3, 2, -1)$, then $PA = PB$

$\Rightarrow PA^2 = PB^2$

$\Rightarrow (x-1)^2+(y-2)^2+(z-3)^2$

$=(x-3)^2+(y-2)^2+(z+1)^2$

$\Rightarrow x^2-2x+1+y^2-4y+4+z^2-6z+9$

$=x^2-6x+9+y^2-4y+4+z^2+2z+1$

$\Rightarrow -2x-4y-6z+14=-6x-4y+2z+14$

$\Rightarrow -2x-6z+6x-2z=0$

$\Rightarrow 4x-8z=0 \Rightarrow x-2z=0$

Thus, the required equation is $x-2z=0$.

Q5. Find the equation of the set of points P, the sum of the whose distances from $A(4,0,0)$ and $B(-4,0,0)$ is equal to 10.

Solution:

Let the coordinates of point P be (x, y, z), and the sum of distances of point $P(x,y,z)$ from $A(4,0,0)$ and $B(-4,0,0)$ be equal to 10; therefore $PA+PB=10$

$\Rightarrow \sqrt{(x-4)^2+y^2+z^2}+\sqrt{(x+4)^2+y^2+z^2}=10$

$\Rightarrow \sqrt{(x-4)^2+y^2+z^2}=10-\sqrt{(x+4)^2+y^2+z^2}$

Squaring both sides, we get

$\Rightarrow (x-4)^2+y^2+z^2=100-20$

$\sqrt{(x+4)^2+y^2+z^2}+(x+4)^2+y^2+z^2$

$\Rightarrow x^2-8x+16+y^2+z^2=100-20$

$\sqrt{x^2+8x+16+y^2+z^2}+x^2+8x+16+y^2+z^2$

$\Rightarrow 5\sqrt{x^2+8x+16+y^2+z^2}=(25+4x)$

Again, squaring both sides, we get

$25(x^2+8x+16+y^2+z^2)=625+16x^2+200x$

$\Rightarrow 25x^2+200x+400+25y^2+25z^2$

$=625+16x^2+200x$

$\Rightarrow 9x^2+25y^2+25z^2-225=0$

Thus, the required equation is $9x^2+25y^2+25z^2-225=0$

Day 162

Section Formula

- Let $P(x_1, y_1, z_1)$, $Q(x_2, y_2, z_2)$ be the two points in the coordinate system of rectangular axes OX, OY and OZ.
- The coordinates of a point which divides the line segment joining two points $P(x_1, y_1, z_1)$, $Q(x_2, y_2, z_2)$ internally in the ratio $m:n$ is given by $\left(\frac{mx_2+nx_1}{m+n}, \frac{my_2+ny_1}{m+n}, \frac{mz_2+nz_1}{m+n}\right)$.
- The coordinates of a point which divides the line segment joining two points $P(x_1, y_1, z_1)$, $Q(x_2, y_2, z_2)$ externally in the ratio $m:n$ is given by $\left(\frac{mx_2-nx_1}{m-n}, \frac{my_2-ny_1}{m-n}, \frac{mz_2-nz_1}{m-n}\right)$.
- The coordinates of the mid-point of the line segment joining two points $P(x_1, y_1, z_1), Q(x_2, y_2, z_2)$ is given by $\left(\frac{x_1+x_2}{2}, \frac{y_1+y_2}{2}, \frac{z_1+z_2}{2}\right)$.
- The coordinates of a point which divides the line segment joining two points $P(x_1, y_1, z_1), Q(x_2, y_2, z_2)$ in the ratio $k:1$ is given by $\left(\frac{kx_2+x_1}{1+k}, \frac{ky_2+y_1}{1+k}, \frac{kz_2+z_1}{1+k}\right)$.
- If, $k > 0$ the point divides the line segment PQ internally and if $k < 0$ the point divides the line segment PQ externally.
- The point $R\left(\frac{kx_2+x_1}{1+k}, \frac{ky_2+y_1}{1+k}, \frac{kz_2+z_1}{1+k}\right)$ gives the general point on the line passing through two given points $P(x_1, y_1, z_1), Q(x_2, y_2, z_2)$. That is, the points P, Q and R, are collinear points.
- The coordinates of the centroid of the triangle PQR, whose vertices are $P(x_1, y_1, z_1)$, $Q(x_2, y_2, z_2)$ and $R(x_3, y_3, z_3)$ is given by $\left(\frac{x_1+x_2+x_2}{3}, \frac{y_1+y_2+y_3}{3}, \frac{z_1+z_2+z_3}{3}\right)$.

NCERT Exercise 12.3 (With Solutions)

Q1. Find the coordinates of the point which divides the line segment joining the points $(-2,3,5),(1,-4,6)$ in the ratio
(i) 2:3 internally
(ii) 2:3 externally

Solution:

(i) The coordinates of a point $R(x, y, z)$ that divides the line segment joining points $P(x_1, y_1, z_1), Q(x_2, y_2, z_2)$ internally in the ratio $m:n$ are $R(x, y, z) = \left(\frac{mx_2+nx_1}{m+n}, \frac{my_2+ny_1}{m+n}, \frac{mz_2+nz_1}{m+n}\right)$.

Thus, the point that divides the line segment joining the points $(-2,3,5),(1,-4,6)$ internally in the ratio 2:3 is $x = \frac{2(1)+3(-2)}{2+3}, y = \frac{2(-4)+3(3)}{2+3}$, $z = \frac{2(6)+3(5)}{2+3} \Rightarrow x = \frac{-4}{5}, y = \frac{1}{5}, z = \frac{27}{5}$

Thus, the coordinates of the required point are $\left(-\frac{4}{5}, \frac{1}{5}, \frac{27}{5}\right)$.

(ii) The coordinates of a point $R(x, y, z)$ that divides the line segment joining points $P(x_1, y_1, z_1), Q(x_2, y_2, z_2)$ internally in the ratio $m:n$ are $R(x, y, z) = \left(\frac{mx_2-nx_1}{m-n}, \frac{my_2-ny_1}{m-n}, \frac{mz_2-nz_1}{m-n}\right)$

Thus, the point that divides the line segment joining the points $(-2,3,5),(1,-4,6)$ externally in the ratio 2:3 is $x = \frac{2(1)-3(-2)}{2-3}, y = \frac{2(-4)-3(3)}{2-3}$,

$z = \frac{2(6)-3(5)}{2+3} \Rightarrow x = -8, y = 17, z = 3$

Thus, the coordinates of the required point are (–8, 17, 3).

Q2. Given that $P(3,2,-4), Q(5,4,-6), R(9,8,-10)$ are collinear, find the ratio in which Q divides PR.

Solution:

Let point $Q(5,4,-6)$ divide the line segment joining the points $P(3,2,-4), R(9,8,-10)$ in the ratio $k:1$; then, by the section formula,

$(5,4,-6) = \left(\frac{k(9)+3}{k+1}, \frac{k(8)+2}{k+1}, \frac{k(-10)-4}{k+1}\right)$

$\Rightarrow \frac{9k+3}{k+1} = 5 \Rightarrow 9k+3 = 5k+5 \Rightarrow 4k = 2$

$\Rightarrow k = \frac{2}{4} = \frac{1}{2}$

Thus, the point Q divides PR in the ratio 1:2.

Q3. Find the ratio in which the YZ-plane divides the line segment formally by joining the points $(-2,4,7),(3,-5,8)$.

Solution:

Let the YZ-plane divide the line segment joining points $(-2,4,7),(3,-5,8)$ in the ratio $k:1$; then by the section formula, the coordinates of the point of intersection are $\left(\frac{k(3)-2}{k+1}, \frac{k(-5)+4}{k+1}, \frac{k(8)+7}{k+1}\right)$.

On the YZ-plane, the x-coordinate of any point is zero; therefore $\frac{3k-2}{k+1}=0 \Rightarrow 3k-2=0 \Rightarrow k=2/3.$

Thus, the YZ-plane divides the line segment formed by joining the given points in the ratio 2:3.

Q4. Using the section formula, show that the points $A(2,-3,4), B(-1,2,1), C(0,1/3,2)$ are collinear.

Solution:

Let P be a point that divides line segment AB in the ratio $k:1$; then by the section formula, the coordinates of the point P are

$$\left(\frac{k(-1)+2}{k+1}, \frac{k(2)-3}{k+1}, \frac{k(1)+4}{k+1}\right)$$

Now, equating the x-coordinate of P and C, we get

$$\frac{-k+2}{k+1}=0 \Rightarrow k=2$$

For $k=2$, the coordinates of point P are $(0,1/3,2)$ which is the same as of point $C(0,1/3,2)$; therefore, point C is a point that divides AB externally in the ratio 2:1, which is the same point P.

Thus, the points A, B, C are collinear.

Q5. Find the coordinates of the points which trisect the line segment joining the points $P(4,2,-6), Q(10,-16,6)$.

Solution:

Let A and B be the points that trisect the line segment joining points $P(4,2,-6), Q(10,-16,6)$.

Point A divides line segment PQ in the ratio 1:2; therefore by the section formula, the coordinates of point A are given by

$$\left(\frac{1(10)+2(4)}{1+2}, \frac{1(-16)+2(2)}{1+2}, \frac{1(6)+2(-6)}{1+2}\right)$$

$= A(6,-4,-2)$

Similarly, point B divides the line segment PQ in the ratio 2:1; therefore by the section formula, the coordinates of point B are given by

$$\left(\frac{2(10)+1(4)}{2+1}, \frac{2(-16)+1(2)}{2+1}, \frac{2(6)-1(6)}{2+1}\right)$$

$= B(8,-10,2)$

Thus, $A(6,-4,-2)$, $B(8,-10,2)$ are the points that trisect the line segment joining the points $P(4,2,-6), Q(10,-16,6)$.

Day 163

NCERT Miscellaneous Exercise (With Solutions)

Q1. Three vertices of a parallelogram $ABCD$ are $A(3,-1,2), B(1,2,-4), C(-1,1,2)$. Find the coordinates of the fourth vertex.

Solution:

Let the coordinates of the fourth vertex of parallelogram $ABCD$ be $D(x,y,z)$.

The diagonals of a parallelogram bisect each other; therefore, in parallelogram $ABCD$, diagonals AC and BD bisect each other.

Thus, mid-point of AC = mid-point of BD

$$\Rightarrow \left(\frac{3-1}{2}, \frac{-1+1}{2}, \frac{2+2}{2}\right)=\left(\frac{x+1}{2}, \frac{y+2}{2}, \frac{z-4}{2}\right)$$

$$\Rightarrow (1,0,2)=\left(\frac{x+1}{2}, \frac{y+2}{2}, \frac{z-4}{2}\right)$$

$$\Rightarrow \frac{x+1}{2}=1, \frac{y+2}{2}=0, \frac{z-4}{2}=2$$

$$\Rightarrow x=1, y=-2, z=8$$

Hence, the coordinates of the fourth vertex are D (1, –2, 8).

Q2. Find the length of the median of the triangle with vertices $A(0,0,6), B(0,4,0), C(6,0,0)$.

Solution:

Let AD, BE and CF be the medians of the given triangle ABC. For median AD, D is the mid-point of BC; therefore the coordinates of point D are

$$=\left(\frac{0+6}{2}, \frac{4+0}{2}, \frac{0+0}{2}\right)=(3,2,0)$$

Thus, length of the median AD is

$$=\sqrt{(0-3)^2+(0-2)^2+(6-0)^2}=\sqrt{9+4+36}=7$$

For median BE, E is the mid-point of AC; therefore coordinates of point E are

$$=\left(\frac{0+6}{2}, \frac{0+0}{2}, \frac{6+0}{2}\right)=(3,0,3)$$

Thus, length of the median BE is

$$=\sqrt{(3-0)^2+(0-4)^2+(3-0)^2}=\sqrt{9+16+9}=\sqrt{34}$$

For median CF, F is the mid-point of AB; therefore coordinates of point F are

$$=\left(\frac{0+0}{2}, \frac{0+4}{2}, \frac{6+0}{2}\right)=(0,2,3)$$

Thus, length of the median CF is

$CF = \sqrt{(6-0)^2 + (0-2)^2 + (0-3)^2} = \sqrt{36+4+9} = 7$

Hence, the lengths of the medians of ΔABC are $7, \sqrt{34}, 7$.

Q3. If the origin is the centroid of the triangle PQR with vertices $P(2a, 2, 6), Q(-4, 3b, -10), R(8, 12, 2c)$, then find the values of a, b, c.

Solution:

The coordinates of the centroid of the triangle, whose vertices are $(x_1, y_1, z_1), (x_2, y_2, z_2), (x_3, y_3, z_3)$ are

$$\left(\frac{x_1+x_2+x_3}{3}, \frac{y_1+y_2+y_3}{3}, \frac{z_1+z_2+z_3}{3}\right)$$

Therefore, coordinates of the centroid of ΔPQR

$$= \left(\frac{2a-4+8}{3}, \frac{2+3b+14}{3}, \frac{6-10+2c}{3}\right)$$

$$= \left(\frac{2a+4}{3}, \frac{3b+16}{3}, \frac{2c-4}{3}\right)$$

It is given that the origin is the centroid of ΔPQR; therefore,

$$\left(\frac{2a+4}{3}, \frac{3b+16}{3}, \frac{2c-4}{3}\right) = (0,0,0)$$

$$\Rightarrow \frac{2a+4}{3} = 0, \ \frac{3b+16}{3} = 0, \ \frac{2c-4}{3} = 0$$

Solving we get, $a = -2, \ b = -16/3, \ c = 2$.

Q4. Find the coordinates of a point on the y-axis which is at a distance of $5\sqrt{2}$ from the point P (3, –5, 5).

Solution:

If a point is on the y-axis, then the x-coordinate and the z-coordinate of the point are zero.

Thus, let A (0, b, 0) be the point on the y-axis at a distance of $5\sqrt{2}$ from point (3, –2, 5); then

$AP = 5\sqrt{2} \Rightarrow AP^2 = 50$

$\Rightarrow (3-0)^2 + (-2-b)^2 + (5-0)^2 = 50$

$\Rightarrow 9 + 4 + b^2 + 4b + 25 = 50 \Rightarrow b^2 + 4b - 12 = 0$

$\Rightarrow (b+6)(b-2) = 0 \Rightarrow b = -6, 2$

Thus, the coordinates of the required points are (0, –6, 0) and (0, 2, 0).

Q5. A point R with x-coordinate 4 lies on the segment joining the points $P(2, -3, 4), Q(8, 0, 10)$. Find the coordinates of the point R.

Solution:

Let point R divide the line segment PQ in the ratio $k:1$; then by the section formula, the coordinates of the point R are given by,

$$\left(\frac{k(8)+2}{k+1}, \frac{k(0)-3}{k+1}, \frac{k(10)+4}{k+1}\right)$$

$$= \left(\frac{8k+2}{k+1}, \frac{-3}{k+1}, \frac{10k+4}{k+1}\right)$$

It is given that the x-coordinate of point R is 4; therefore

$$\frac{8k+2}{k+1} = 4 \Rightarrow 8k + 2 = 4k + 4 \Rightarrow k = \frac{1}{2}$$

Therefore, the coordinates of point R are

$$\left(4, \frac{-3}{\frac{1}{2}+1}, \frac{10(1/2)+4}{\frac{1}{2}+1}\right) = (4, -2, 6)$$

Q6. If A and B are the points $(3, 4, 5), (-1, 3, -7)$, respectively, find the equation of the set of points P such that $PA^2 + PB^2 = k^2$, where k is a constant.

Solution:

Let the coordinates of point P be (x, y, z); then using the distance formula,

$PA^2 = (x-3)^2 + (y-4)^2 + (z-5)^2$

$\Rightarrow PA^2 = x^2 - 6x + y^2 - 8y + z^2 - 10z + 50$

Similarly, $PB^2 = (x+1)^2 + (y-3)^2 + (z+7)^2$

$\Rightarrow PB^2 = x^2 + 2x + y^2 - 6y + z^2 + 14z + 59$

Now, it is given that $PA^2 + PB^2 = k^2$, then

$(x^2 - 6x + y^2 - 8y + z^2 - 10z + 50) +$

$(x^2 + 2x + y^2 - 6y + z^2 + 14z + 59) = k^2$

$\Rightarrow 2x^2 + 2y^2 + 2z^2 - 4x + 14y + 4z + 109 = k^2$

$\Rightarrow x^2 + y^2 - 2x - 7y + 2z = \dfrac{k^2 - 109}{2}$ is the required equation of the set of points P.

Day 164

NCERT Exemplar Problems: Solutions (Q1–Q18)

Q1. Locate the following points

(i) $(1,-1,3)$ (ii) $(-1,2,4)$
(iii) $(-2,-4,-7)$ (iv) $(-4,2,-5)$

Solution:

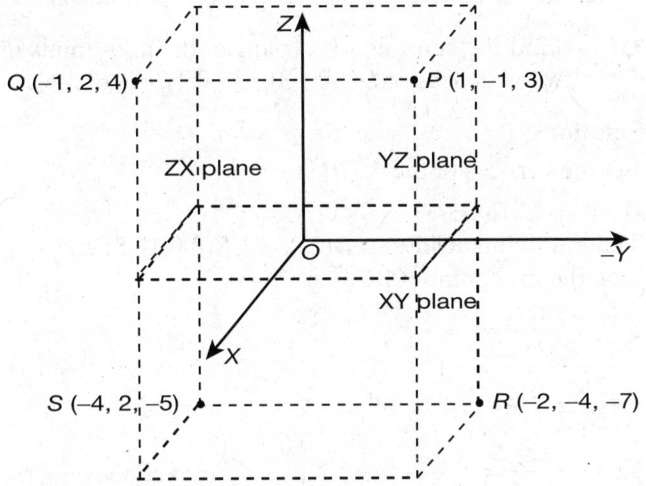

The given coordinates are

(i) $P(1,-1,3)$ (ii) $Q(-1,2,4)$
(iii) $R(-2,-4,-7)$ (iv) $S(-4,2,-5)$

Q2. Name the octant in which each of the following points lies.

(i) $(1, 2, 3)$ (ii) $(4,-2,3)$ (iii) $(4,-2,-5)$
(iv) $(4,2,-5)$ (v) $(-4,2,5)$ (vi) $(-3,-1,6)$
(vii) $(2,-4-7)$ (viii) $(-4,2,-5)$

Solution:

(i) $(1, 2, 3)$ lies in the first quadrant
(ii) $(4,-2,3)$ lies in the fourth octant
(iii) $(4,-2,-5)$ lies in the eighth octant
(iv) $(4,2,-5)$ lies in the fifth octant
(v) $(-4,2,5)$ lies in the second octant
(vi) $(-3,-1,6)$ lies in the third octant
(vii) $(2,-4-7)$ lies in the eighth octant
(viii) $(-4,2,-5)$ lies in the sixth octant

Q3. If A, B, C are the feet of the perpendicular from a point P on the X, Y and Z-axes, respectively, then find the coordinates of A, B, C in each of the following where the point P is

(i) $A(3, 4, 2)$ (ii) $B(-5, 3, 7)$
(iii) $C(4, -3, -5)$

Solution:
The coordinates of the feet of the perpendiculars from a point P on the X, Y and Z-axes, respectively, of points A, B and C are as follows:

(i) $A(3,0,0), B(0,4,0), C(0,0,2)$
(ii) $A(-5,0,0), B(0,3,0), C(0,0,7)$
(iii) $A(4,0,0), B(0,-3,0), C(0,0,-5)$

Q4. If A, B, C are the feet of the perpendiculars from a point P on the XY, YZ and XY-planes, respectively, then find the coordinates of A, B and C in each of the following where the point P is

(i) $(3, 4, 5)$ (ii) $(-5, 3, 7)$ (iii) $(4, -3, -5)$

Solution:
On the XY plane, $z = 0$, on the YZ plane, $x = 0$ and on the ZX plane, $y = 0$. Therefore, the coordinates of the feet of the perpendiculars from a point P on the XY, YZ and XY-planes, respectively, of points A, B and C are following

(i) $A(3,4,5), B(0,4,5), C(3,0,5)$
(ii) $A(-5,3,0), B(0,3,7), C(-5,0,7)$
(iii) $A(4,-3,0), B(0,-3,-5), C(4,0,-5)$

Q5. How far apart are the points $(2, 0, 0)$ and $(-3, 0, 0)$?

Solution:
The distance between the points $A(2,0,0), B(-3,0,0)$ is

$AB = \sqrt{(2+3)^2 + 0^2 + 0^2} = 5$

Q6. Find the distance from the origin to $(6, 6, 7)$.

Solution:
Distance from the origin to the point $(6, 6, 7)$ is

$d = \sqrt{(x_1 - x_2)^2 + (y_1 - y_2)^2 + (z_1 - z_2)^2}$

$= \sqrt{(0-6)^2 + (0-6)^2 + (0-7)^2}$

$= \sqrt{36+36+49} = \sqrt{121} = 11$

Q7. Show that, if $x^2 + y^2 = 1$, then the point $(x, y, \sqrt{1-x^2-y^2})$ is at a distance of 1 unit from the origin.

Solution:

Distance of the point $(x, y, \sqrt{1-x^2-y^2})$ from the origin is

$d = \left| \sqrt{x^2 + y^2 + (\sqrt{1-x^2-y^2})^2} \right|$

$= \left| \sqrt{x^2 + y^2 + 1 - x^2 - y^2} \right| = 1 \quad (\because x^2 + y^2 = 1)$

Q8. Show that the points $A(1,-1,3), B(2,-4,5), C(5,-13,11)$ are collinear.

Solution:

$AB=\sqrt{(1-2)^2+(-1+4)^2+(3-5)^2}$

$=\sqrt{1+9+4}=\sqrt{14}$

$BC=\sqrt{(2-5)^2+(-4+13)^2+(5-11)^2}$

$=\sqrt{9+81+36}=\sqrt{126}=3\sqrt{14}$

$AC=\sqrt{(1-5)^2+(-1+13)^2+(3-11)^2}$

$=\sqrt{16+144+64}=\sqrt{224}=4\sqrt{14}$

$\Rightarrow AB+BC=AC$; therefore, the points A, B and C are collinear.

Q9. Three consecutive vertices of a parallelogram $ABCD$ are $A(6,-2,4), B(2,4,-8), C(-2,2,4)$. Find the coordinates of the fourth vertex.

Solution:

Let the coordinates of the fourth vertex be $D(x,y,z)$.

The mid-point of diagonal AC

$x=\frac{x_1+x_2}{2}=\frac{6-2}{2}=2$

$y=\frac{y_1+y_2}{2}=\frac{-2+2}{2}=0$

$z=\frac{z_1+z_2}{2}=\frac{4+4}{2}=4$

And the mid-point of diagonal BD is

$\left(\frac{x+2}{2},\frac{y+4}{2},\frac{z-8}{2}\right)$

The mid-points of the diagonals of a parallelogram coincide at a point; therefore,

$\left(\frac{x+2}{2},\frac{y+4}{2},\frac{z-8}{2}\right)=(2,0,4)$

$\Rightarrow x=2,\ y=-4$ and $z=16$

Thus, the coordinates of fourth vertex D are (2, –4, 16).

Q10. Show that the ΔABC with vertices $A(0,4,1), B(2,2,-1), C(4,5,0)$ is right-angled.

Solution:

$AB=\sqrt{(0-2)^2+(4-3)^3+(1+1)^2}=3$

$BC=\sqrt{(2-4)^2+(3-5)^2+(-1-0)^2}=3$

$AC=\sqrt{(0-4)^2+(4-5)^2+(1-0)^2}=\sqrt{18}$

$\therefore AC^2=AB^2+BC^2 \Rightarrow 18=9+9$

Thus, ΔABC is a right-angled triangle at vertex B.

Q11. Find the third vertex of a triangle whose centroid is the origin and two vertices are (2, 4, 6) and (0, –2, 5).

Solution:

Let the two vertices of ΔABC be A (2, 4, 6) and B (0, –2, 5) and the third vertex C be (x, y, z).

The coordinates of centroid G are (0,0,0).

$\therefore 0=\frac{x+2+0}{3}\Rightarrow x=-2$

$0=\frac{y+4-2}{3}\Rightarrow y=-2$

$0=\frac{z+6-5}{2}\Rightarrow z=-1$

Thus, the third vertex of the triangle is $C\,(-2,-2,-1)$.

Q12. Find the centroid of a triangle, the mid-points of whose sides are $D(1,2,-3), E(3,0,1), F(-1,1,-4)$.

Solution:

Let the vertices of the ΔABC be $A(x_1,y_1,z_1), B(x_2,y_2,z_2), C(x_3,y_3,z_3)$.

The mid-points of sides are $D(1,2,-3), E(3,0,1), F(-1,1,-4)$; then the mid-point of BC is (1, 2, –3).

$\therefore 1=\frac{x_2+x_3}{2}\Rightarrow x_2+x_3=2$ (i)

$2=\frac{y_2+y_3}{2}\Rightarrow y_2+y_3=4$ (ii)

$-3=\frac{z_2+z_3}{2}\Rightarrow z_2+z_3=-6$ (iii)

Similarly, for the sides AB and AC,

$\Rightarrow -1=\frac{x_1+x_2}{2}\Rightarrow x_1+x_2=-2$ (iv)

$\Rightarrow 1=\frac{y_1+y_2}{2}\Rightarrow y_1+y_2=2$ (v)

$\Rightarrow -4=\frac{z_1+z_2}{2}\Rightarrow z_1+z_2=-8$ (vi)

$\Rightarrow 3=\frac{x_1+x_3}{2}\Rightarrow x_1+x_3=6$ (vii)

$\Rightarrow 0=\frac{y_1+y_3}{2}\Rightarrow y_1+y_3=0$ (viii)

$\Rightarrow 1=\frac{z_1+z_3}{2}\Rightarrow z_1+z_3=2$ (ix)

On adding equations (i) and (iv), we get

$x_1+2x_2+x_3=0$ (x)

On adding equations (ii) and (v), we get

$y_1+2y_2+y_3=0$ (xi)

On adding equations (iii) and (vi), we get

$z_1+2z_2+z_3=0$ (xii)

From equations (vii) and (x), $2x_2=-6\Rightarrow x_2=-3$

If $x_2=-3$, then $x_3=5$

If $x_3=5$, then $x_1=1, x_2=-3, x_3=5$

From equations (xi) and (viii), $2y_2=6\Rightarrow y_2=3$

If $y_2 = 3$, then $y_1 = -1$

If $y_1 = -1$, then $y_3 = 1, y_2 = 3, y_3 = 1$

From equations (xii) and (ix), $2z_2 = -16 \Rightarrow z_2 = -8$

$z_2 = -8$, then $z_1 = 0$

$z_1 = 0$, then $z_3 = 2$

$z_1 = 0, z_2 = -8, z_3 = 2.$

Thus, the points are $A(1-1,0), B(-3,3,-8), C(5,1,2)$.

$\therefore$ Centroid of the triangle

$= G\left(\frac{1-3+5}{3}, \frac{-1+3+1}{3}, \frac{0-8+2}{3}\right) = G(1,1,-2)$

Q13. The mid-points of the sides of a triangle are $(5,7,11), (0,8,5), (2,3,-5)$. Find its vertices.

Solution:

Let the vertices of the ΔABC be $A(x_1, y_1, z_1), B(x_2, y_2, z_2), C(x_3, y_3, z_3)$; then the mid-point of BC is (5, 7, 11).

$\therefore 5 = \frac{x_2 + x_3}{2} \Rightarrow x_2 + x_3 = 10$ (i)

$7 = \frac{y_2 + y_3}{2} \Rightarrow y_2 + y_3 = 14$ (ii)

$11 = \frac{z_2 + z_3}{2} \Rightarrow z_2 + z_3 = 22$ (iii)

Similarly for the sides AB and AC,

$2 = \frac{x_1 + x_2}{2} \Rightarrow x_1 + x_2 = 4$ (iv)

$3 = \frac{y_1 + y_2}{2} \Rightarrow y_1 + y_2 = 6$ (v)

$-1 = \frac{z_1 + z_2}{2} \Rightarrow z_1 + z_2 = -2$ (vi)

$0 = \frac{x_1 + x_3}{2} \Rightarrow x_1 + x_3 = 0$ (vii)

$8 = \frac{y_1 + y_3}{2} \Rightarrow y_1 + y_3 = 16$ (viii)

$5 = \frac{z_1 + z_3}{2} \Rightarrow z_1 + z_3 = 10$ (ix)

From equations (i) and (iv), $x_1 + 2x_2 + x_3 = 14$ (x)

From equations (ii) and (v), $y_1 + 2y_2 + y_3 = 20$ (xi)

From equations (iii) and (vi), $z_1 + 2z_2 + z_3 = 20$ (xii)

From equations (vii) and (x), $2x_2 = 14 \Rightarrow x_2 = 7$

$x_2 = 7$, then $x_3 = 10 - 7 = 3$

$x_3 = 3$, then $x_1 = -3$

$x_1 = -3, x_2 = 7, x_3 = 3$

From equations (viii) and (xi), $2y_2 = 4 \Rightarrow y_2 = 2$

$y_2 = 2$, then $y_1 = 4$

$y_1 = 4$, then $y_3 = 12$

$y_1 = 4, y_2 = 2, y_3 = 12$

From equations (ix) and (xii), $2z_2 = 10 \Rightarrow z_2 = 5$

$z_2 = 5$, then $z_1 = -7 \Rightarrow z_1 = -7$, then $z_3 = 17$

$z_1 = -7, z_2 = 5, z_3 = 17$

Thus, the vertices are $A(-3,4,-7), B(7,2,5), C(3,12,17)$.

Q14. If the vertices of a parallelogram $ABCD$ are $A(1,2,3), B(-1,-2,-1), C(2,3,2)$, then find the fourth vertex D.

Solution:

Let the fourth vertex of the parallelogram $ABCD$ be D (x, y, z). Then, the mid-point of AC is

$P\left(\frac{1+2}{2}, \frac{2+3}{2}, \frac{3+2}{2}\right) = P\left(\frac{3}{2}, \frac{5}{2}, \frac{5}{2}\right)$.

The mid-points of the diagonals of a parallelogram coincide at a point; therefore the mid-point of BD is

$\frac{3}{2} = \frac{-1+x}{2} \Rightarrow x = 4$

$\frac{5}{3} = \frac{-2+y}{2} \Rightarrow y = 7$

$\frac{5}{2} = \frac{-1+z}{2} \Rightarrow z = 6$

Thus, the coordinates of the fourth vertex are (4, 7, 6).

Q15. Find the coordinates of the points which trisect the line segment joining the points $A(2,1,-3), B(5,-8,3)$.

Solution:

Let the points $P(x_1, y_1, z_1), Q(x_2, y_2, z_2)$ trisect the line segment AB.

If the point P divides the line AB in 1:2 internally, then

$x_1 = \frac{2\times 2 + 1\times 5}{1+2} = \frac{9}{3} = 3$

$y_1 = \frac{2\times 1 + 1\times(-8)}{3} = \frac{-6}{3} = -2$

$z_1 = \frac{2\times(-3) + 1\times 3}{3} = \frac{-6+3}{3} = \frac{-3}{3} = -1$

The point Q divides the line segment AB in 2:1 then

$x_2 = \frac{1\times 2 + 2\times 5}{3} = 4$

$y_2 = \frac{1\times 1 + (-8\times 2)}{3} = -5$

$z_2 = \frac{1\times(-3) + 2\times 3}{3} = -1$

Thus, the coordinates of P are $(3,-2,-1)$ and the coordinates of Q are $(4,-5,1)$.

Q16. If the origin is the centroid of a ΔABC having vertices $A(a,1,3), B(-2,b,-5), C(4,7,c)$, then find the values of a, b, c.

Solution:
The origin is the centroid of the ΔABC, that is, $G(0, 0, 0)$.

$\therefore 0 = \frac{a-2+4}{3} \Rightarrow a = -2$

$0 = \frac{1+b+7}{3} \Rightarrow b = -8$

$0 = \frac{3-5+c}{3} \Rightarrow c = +2$

$\Rightarrow a = -2, b = -8, c = 2$

Q17. Let $A(2,2,-3), B(5,6,9), C(2,7,9)$ be the vertices of a triangle; if the internal bisector of the angle A meets BC at the point D, then find the coordinates of D.

Solution:
Let the coordinates of D be (x, y, z).

$AB = \sqrt{9+16+144} = \sqrt{169} = 13$

$AC = \sqrt{0+25+144} = \sqrt{169} = 13$

$\Rightarrow \frac{AB}{AC} = \frac{13}{13} \Rightarrow AB = AC, \frac{BD}{DC} = \frac{1}{1} \Rightarrow BD = DC$

Let D divide the line BC in two equal parts. Therefore, D is the mid-point of BC.

Thus, $x = \frac{5+2}{2} = \frac{7}{2}$

$y = \frac{6+7}{2} = \frac{13}{2}, \quad z = \frac{9+9}{2} = 9$

Thus, the coordinates of D are $\left(\frac{7}{2}, \frac{13}{2}, 9\right)$.

Q18. Show that the three points $A(2,3,4), B(1,2,-3), C(-4,1,-10)$ are collinear and find the ratio in which D divides AB.

Solution:

$AB = \sqrt{(2+1)^2 + (3-2)^2 + (4+3)^2}$

$= \sqrt{9+1+49} = \sqrt{59}$

$BC = \sqrt{(-1+4)^2 + (2-1)^2 + (-3+10)^2}$

$= \sqrt{9+1+49} = \sqrt{59}$

$AC = \sqrt{(2+4)^2 + (3-1)^2 + (4+10)^2}$

$= \sqrt{36+4+196} = \sqrt{236} = 2\sqrt{59}$

Since $AB + BC = \sqrt{59} + \sqrt{59} = 2\sqrt{59} = AC$, the points A, B and C are collinear.

Now, $AC : BC = 2\sqrt{59} : \sqrt{59} = 2:1$

Therefore, C divides AB in the ratio 2:1 externally.

Day 165

NCERT Exemplar Problems: Solutions (Q19–Q50)

Q19. The mid-points of the sides of a triangle are $(1,5,-1), (0,4,-2), (2,3,4)$. Find its vertices and also find the centroid of the triangle.

Solution:
Let the vertices of ΔABC be $A(x_1, y_1, z_1), B(x_2, y_2, z_2), C(x_3, y_3, z_3)$.

The mid-point of side BC is $D(1,5,-1)$; therefore

$\frac{x_2+x_3}{2} = 1 \Rightarrow x_2 + x_3 = 2$ (i)

$\frac{y_2+y_3}{2} = 5 \Rightarrow y_2 + y_3 = 10$ (ii)

$\frac{z_2+z_3}{2} = -1 \Rightarrow z_2 + z_3 = -2$ (iii)

Similarly, the mid-points of AB and AC are $F(2,3,4), E(0,4,-2)$,

$\frac{x_1+x_2}{2} = 2 \Rightarrow x_1 + x_2 = 4$ (iv)

$\frac{y_1+y_2}{2} = 3 \Rightarrow y_1 + y_2 = 6$ (v)

$\frac{z_1+z_2}{2} = 4 \Rightarrow z_1 + z_2 = 8$ (vi)

Now,

$\frac{x_1+x_3}{2} = 0 \Rightarrow x_1 + x_3 = 0$ (vii)

$\frac{y_1+y_3}{2} = 4 \Rightarrow y_1 + y_3 = 8$ (viii)

$\frac{z_1+z_3}{2} = -2 \Rightarrow z_1 + z_3 = -4$ (ix)

From equations (i) and (iv), $x_1 + 2x_2 + x_3 = 6$ (x)

From equations (ii) and (v), $y_1 + 2y_2 + y_3 = 16$ (xi)

From equations (iii) and (vi), $z_1 + 2z_2 + z_3 = 6$ (xii)

From equations (vii) and (x), $2x_2 = 6 \Rightarrow x_2 = 3$

$x_2 = 3$, then $x_3 = -1$

$x_3 = -1$

Then, $x_1 = 1 \Rightarrow x_1 = 1, x_2 = 3, x_2 = -1$

From equations (vii) and (xi), $2y_2 = 8 \Rightarrow y_2 = 4, y_2 = 4$

Then, $y_1 = 3, y_1 = 2,$

Then, $y_3 = 6 \Rightarrow y_1 = 2, y_2 = 4, y_3 = 6$

From equations (ix) and (xii), $2z_2 = 10 \Rightarrow z_2 = 5, z_2 = 5,$

Then, $z_1 = 3, z_1 = 3,$

Then, $z_3 = -7 \Rightarrow z_1 = 3, z_2 = 5, z_3 = -7$
Thus, the vertices of the triangle
$A(1,2,3), B(3,4,5), C(-1,6,-7)$.
Hence, the centroid of the triangle
$G\left(\frac{1+3-1}{3}, \frac{2+4+6}{3}, \frac{3+5-7}{3}\right) = G(1,4,1/3)$.

Q20. Prove that the points $(0,-1,-7),(2,1,-9),(6,5,-13)$ are collinear. Find the ratio in which the first point divides the meeting point of the other two.

Solution:
$AB = \sqrt{(0-2)^2 + (-1-1)^2 + (-7+9)^2}$
$= \sqrt{4+4+4} = 2\sqrt{3}$
$BC = \sqrt{(2-6)^2 + (1-5)^2 + (-9+13)^2}$
$= \sqrt{16+16+16} = 4\sqrt{3}$
$AC = \sqrt{(0-6)^2 + (-1-5)^2 + (-7+13)^2}$
$= \sqrt{36+36+36} = 6\sqrt{3}$
$\because AB + BC = 2\sqrt{3} + 4\sqrt{3} = 6\sqrt{2} = AC$
Thus, the points A, B and C are collinear, and the point A divides BC in the ratio of 1:3 externally.

Q21. What are the coordinates of the vertices of a cube whose edge is 2 units, one of whose vertices coincides with the origin and the three edges passing through the origin coincide with the positive direction of the axes through the origin?

Solution:
The coordinates of the cube with edge 2 units, one of the vertex on the origin and sides are on the positive direction of the axes through the origin are $(0,0,0),(2,0,0),(0,2,0),(0,0,2)$,
$(2,2,0),(2,0,2),(0,2,2),(2,2,2)$.

Q22. The distance of point $P(3, 4, 5)$ from the YZ-plane is
(a) 3 units (b) 4 units (c) 5 units (d) 10 units

Solution: (a)
The given point is $P(3, 4, 5)$.
The distance of P from the YZ-plane
= coordinate of x-axis = 3

Q23. What is the length of the foot of the perpendicular drawn from the point $P(3, 4, 5)$ on the Y-axis?
(a) $\sqrt{41}$ (b) $\sqrt{34}$
(c) 5 (d) none of these

Solution: (b)
The foot of the perpendicular from the point $P(3, 4, 5)$ on the Y-axis is $A(0, 4, 0)$.
$\therefore PA = \sqrt{(0-3)^2 + (4-4)^2 + (0-5)^2} = \sqrt{34}$

Q24. Distance of the point (3, 4, 5) from the origin (0, 0, 0) is ______.
(a) $\sqrt{50}$ (b) 3 (c) 4 (d) 5

Solution: (a)
Given points $A(3, 4, 5)$ and O(0, 0, 0),
$\therefore AO = \sqrt{(0-3)^2 + (0-4)^2 + (0-5)^2} = \sqrt{50}$

Q25. If the distance between the points $(a,0,1),(0,1,2)$ is $\sqrt{27}$, then the value of a is
(a) 5 (b) ±5
(c) –5 (d) none of these

Solution: (b)
Given the points are $A(a,0,1), B(0,1,2)$.
$\therefore AB = \sqrt{(a-0)^2 + (0-1)^2 + (1-2)^2} = \sqrt{27}$
$\Rightarrow \sqrt{a^2+1+1} = \sqrt{27} \Rightarrow a^2 + 2 = 27$
$\Rightarrow a^2 = 25 \Rightarrow a = \pm 5$

Q26. X-axis is the intersection of the two planes
(a) XY and XZ (b) YZ and ZX
(c) XY and YZ (d) none of these

Solution: (a)
The X-axis is the line of intersection of the XY and XZ-planes.

Q27. Equation of the Y-axis is considered as
(a) $x = 0, y = 0$ (b) $y = 0, z = 0$
(c) $z = 0, x = 0$ (d) none of these

Solution: (c)
On the Y-axis, $z = 0, x = 0$

Q28. The point (–2, –3, –4) lies in the
(a) first octant (b) seventh
(c) second octant (d) eight octant

Solution: (b)
The point (–2, –3, –4) lies in seventh octant.

Q29. A plane is parallel to the YZ-plane, so it is perpendicular to the
(a) X-axis (b) Y-axis
(c) Z-axis (d) none of these

Solution: (a)
A plane parallel to the YZ-plane will be perpendicular to the X-axis.

Q30. The locus of a point for which $y = 0, z = 0$ is
(a) equation of X-axis (b) equation of Y-axis
(c) equation at Z-axis (d) none of these

Solution: (a)
The locus of a point for which $y = 0, z = 0$, is the equation of the X-axis.

Q31. The locus of a point for which $x = 0$, is the
(a) XY-plane (b) YZ-plane
(c) ZX-plane (d) none of these

Solution: (b)
The locus of a point for which $x = 0$, is the YZ-plane.

Q32. If a parallelepiped is formed by planes drawn through the points (5, 8, 15) and (3, 6, 8) parallel to the coordinate planes, then the length of the diagonal of the parallelepiped is
(a) $2\sqrt{3}$ (b) $3\sqrt{2}$ (c) $\sqrt{2}$ (d) $\sqrt{3}$

Solution: (a)
The parallelepiped is formed by planes drawn through the points (5, 8, 15) and (3, 6, 8) parallel to the coordinate planes; therefore the length of the diagonal of the parallelepiped is
$= \sqrt{(5-3)^2 + (6-8)^2 + (10-8)^2}$
$= \sqrt{4+4+4} = 2\sqrt{3}$

Q33. L is the foot of the perpendicular drawn from a point $P(3, 4, 5)$ on the XY-plane. The coordinates of the point L are
(a) (3, 0, 0) (b) (0, 4, 5)
(c) (3, 0, 5) (d) None of these

Solution: (d)
On the XY-plane, $z = 0$
Therefore, the coordinates of the points L are (3, 4, 0).

Q34. L is the foot of the perpendicular drawn from a point (3, 4, 5) on the X-axis. The coordinates of L are
(a) (3, 0, 0) (b) (0, 4, 0)
(c) (0,0,5) (d) none of these

Solution: (a)
On the X-axis, $y = 0, z = 0$.
Therefore, the coordinates of L are (3, 0, 0).

Q35. The three axes OX, OY and OZ determine................

Solution:
The three axes OX, OY and OZ determine the three coordinate planes.

Q36. The three planes determine a rectangular parallelepiped which has .... of rectangular faces.

Solution:
Six rectangular faces

Q37. The coordinates of a point are the perpendicular distance from the .... on the respective axes.

Solution:
Given point

Q38. The three coordinate planes divide the space into .... parts.

Solution:
Eight parts

Q39. If a point P lies in the YZ-plane, then the coordinates of a point on the YZ-plane is of the form................

Solution:
On the YZ-plane $x = 0$.
Therefore, the coordinates of the required point are (0, y, z).

Q40. The equation of the YZ-plane is..........

Solution:
The equation of the YZ-plane is $x = 0$.

Q41. If the point P lies on the Z-axis, then the coordinates of P are of the form.................

Solution:
On the Z-axis, $x = 0, y = 0$.
Therefore, the required coordinates are (0, 0, z).

Q42. The equation of the Z-axis, is ....

Solution:
The equation of the Z-axis is $x = 0, y = 0$.

Q43. A line is parallel to the XY-plane if all the points on the line have equal..............

Solution:
Z-coordinates.

Q44. A line is parallel to the X-axis if all the points on the line have equal...............

Solution:
Y- and Z-coordinates.

Q45. $x = a$ represents a plane parallel to.........

Solution:
$x = a$ represents a plane parallel to the YZ-plane.

Q46. A plane parallel to the YZ-plane is perpendicular to.......

Solution:
The plane parallel to the YZ-plane is perpendicular to the X-axis.

Q47. The length of the longest piece of a string that can be stretched straight in a rectangular room whose dimensions are 10, 13 and 8 units are............

Solution:

Length of the longest piece of a string is

$= \sqrt{a^2+b^2+c^2} = \sqrt{100+169+64} = \sqrt{333}$

Q48. If the distance between the points $(a,2,1),(1,-1,1)$ is 5, then a is equal to ....

Solution:

$\because \sqrt{(a-1)^2+(2+1)^2+(1-1)^2} = 5$

$\Rightarrow (a-1)^2+9+0=25$

$\Rightarrow a^2-2a+1+9=25$

$\Rightarrow a^2-2a-15=0$

$\Rightarrow (a-5)(a+3)=0$

$\Rightarrow a=+5,-3$

Q49. If the mid-points of the sides of a triangle AB, BC and CA are $D(1,2,-3), E(3,0,1), F(-1,1,-4)$, then the centroid of the ΔABC is............

Solution:

Let the vertices of ΔABC be $A(x_1,y_1,z_1), B(x_2,y_2,z_2), C(x_3,y_3,z_3)$.

Since D is the mid-point of AB, then

$\frac{x_1+x_2}{2}=1 \Rightarrow x_1+x_2=2$ (i)

$\frac{y_1+y_2}{2}=2 \Rightarrow y_1+y_2=4$ (ii)

$\frac{z_1+z_2}{2}=-3 \Rightarrow z_1+z_2=-6$ (iii)

Similarly, E and F are the mid-points of the sides BC and AC, respectively.

$\frac{x_2+x_3}{2}=3 \Rightarrow x_2+x_3=6$ (iv)

$\frac{y_2+y_3}{2}=0 \Rightarrow y_2+y_3=0$ (v)

$\frac{z_2+z_3}{2}=1 \Rightarrow z_2+z_3=2$ (vi)

$\frac{x_1+x_3}{2}=-1 \Rightarrow x_1+x_3=-2$ (vii)

$\frac{y_1+y_3}{2}=1 \Rightarrow y_1+y_3=2$ (viii)

$\frac{z_1+z_3}{2}=-4 \Rightarrow z_1+z_3=-8$ (ix)

From equations (i) and (iv), $x_1+2x_2+x_3=8$ (x)

From equations (ii) and (v), $y_1+2y_2+y_3=4$ (xi)

From equations (iii) and (vi), $z_1+2z_2+z_3=-4$ (xii)

From equations (vii) and (x), $2x_2=10 \Rightarrow x_2=5 \Rightarrow x_2=5$

Then $x_3=1$

If $x_3=1$, then $x_1=-3 \Rightarrow \because x_1=-3, x_2=5, x_3=1$

From equations (viii) and (xi), $2y_2=2 \Rightarrow y_2=1$

If $y_2=1$, then $y_3=-1$

If $y_3=-1$, then $y_1=3 \Rightarrow \because y_1=3, y_2=1, y_3=-1$

From equations (ix) and (xii), $2z_2=4 \Rightarrow z_2=2$

If $z_2=2$, then $z_3=0$

If $z_3=0$, then $z_1=-8 \Rightarrow \because z_1=-8, z_2=2, z_3=0$

Therefore, the vertices of ΔABC are $A(-3,3,-8), B(5,1,2), C(1,-1,0)$.

Hence, coordinates of the centroid of ΔABC

$G\left(\frac{-3+5+1}{3},\frac{3+1-1}{3},\frac{-8+2+0}{3}\right)=G(1,1,-2)$.

Q50. Match each item given under Column I to its correct answer given under Column II.

Column I

(i) In the XY-plane
(ii) Point (2, 3, 4) lies in the
(iii) Locus of the points having X-coordinate 0 is
(iv) A line is parallel to the X-axis if and only
(v) If $x = 0$, $y = 0$ taken together will represent the
(vi) $z = c$ represents the plane
(vii) Planes $x = a$, $y = b$ represent the line
(viii) Coordinates of a point are the distance from the origin to the feet of the perpendiculars
(ix) A ball is the solid region in the space enclosed by a
(x) Region in the plane enclosed by a circle is known as a

Column II

(a) 1st octant
(b) YZ-plane
(c) z-coordinate is zero
(d) Z-axis
(e) Plane parallel to the XY-plane
(f) If all the points on the line have equal y- and z-coordinates
(g) From the point on the respective
(h) Parallel to the Z-axis
(i) Disc
(j) Sphere

Solution:

(i) —(c): In the XY-plane, the z-coordinate is zero.

(ii) –(a): The point (2, 3, 4) lies in the 1st octant.
(iii) –(b): Locus of the points having the x-coordinate as zero is in the YZ-plane.
(iv) –(f): A line is parallel to the X-axis if and only if all the points on the line have equal y- and z-coordinates.
(v) –(d): $x = 0, y = 0$ represents the Z-axis.
(vi) –(e): $z = c$ represents the plane parallel to the XY-plane.
(vii) –(h): The planes $x = a, y = b$ represent the line parallel to the Z-axis.
(viii) –(g): Coordinates of a point are the distance from the origin to the feet of perpendicular from the point on the respective.
(ix) –(j): A ball is the solid region in the space enclosed by a sphere.
(x) –(i): The region in the plane enclosed by a circle is known as a disc.

Day 166

Very Short Answer (VSA: 1-Mark) and Short Answer (SA: 2-Mark) Questions

Q1. A plane is parallel to the XY-plane; so it is perpendicular to the
(a) X-axis (b) Y-axis
(c) Z-axis (d) none of these

Solution: (c)
A plane parallel to the XY-plane will be perpendicular to the Z-axis.

Q2. The locus of a point for which $x = 0, y = 0$, is
(a) equation of X-axis (b) equation of Y-axis
(c) equation of Z-axis (d) none of these

Solution: (c)
The locus of a point for which $x = 0, y = 0$, is the equation of the Z-axis.

Q3. The locus of a point for which $z = 0$, is
(a) XY-plane (b) YZ-plane
(c) ZX-plane (d) none of these

Solution: (a)
The locus of a point for which $z = 0$, is the XY-plane.

Q4. If a parallelepiped is formed by planes drawn through the points (–2, 3, 0) and (4, 0, –1) parallel to the coordinate planes, then the length of the diagonal of the parallelepiped is
(a) $\sqrt{32}$ (b) $\sqrt{23}$ (c) $\sqrt{46}$ (d) $2\sqrt{33}$

Solution: (c)
The parallelepiped is formed by planes drawn through the points (–2, 3, 0) and (4, 0, –1) parallel to the coordinate planes; therefore, the length of the diagonal of the parallelepiped is

$$= \sqrt{(-2-4)^2 + (3-0)^2 + (0+1)^2}$$
$$= \sqrt{36+9+1} = \sqrt{46}$$

Q5. Fill in the blanks.
(a) The y-axis and z-axis taken together determine a plane known as______________
(b) The coordinates of points in the YZ-plane are of the form__________

Solution:
(a) The y-axis and z-axis taken together determine a plane known as YZ-pane.
(ii) The coordinates of the points in the YZ-plane are of the form $(0, y, z)$.

Q6. A point is on the y-axis. What are its x-coordinate and z-coordinate?

Solution:
If a point is on the y-axis, then its x and z coordinates are zero.

Q7. A point is in the YZ-plane. What can you say about its x-coordinates?

Solution:
If a point is in the YZ-plane, then its x-coordinate is zero.

Q8. Name the octants in which the following points lie:$(1,-2,3),(3,-2,-1),(3,2,-1),(-3,2,-1)$.

Solution:
The x-coordinate, y-coordinate and z-coordinate of point (1, –2, 3) are positive, negative and positive, respectively. Therefore, the point lies in octant IV.
The x-coordinate, y-coordinate and z-coordinate of point (3, –2, –1) are positive, negative and negative, respectively. Therefore, the point lies in octant VIII.
The x-coordinate, y-coordinate and z-coordinate of point (3, 2, –1) are positive, positive and negative, respectively. Therefore, the point lies in octant V.
The x-coordinate, y-coordinate and z-coordinate of point (–3, 2, –1) are negative, positive and negative, respectively. Therefore, the point lies in octant VI.

Q9. Find the distance between the following pairs of points:

(a) (–3, 1, 2) and (2, 4, –1)
(b) (–1, 3, –4) and (1, 0, 4)

Solution:
Distance between two points $A(x_1, y_1, z_1)$ and $B(x_2, y_2, z_2)$ is given by

$$AB = \sqrt{(x_2 - x_1)^2 + (y_2 - y_1)^2 + (z_2 - z_1)^2}$$

(i) Distance between the points (–3, 1, 2) and (2, 4, –1)

$$= \sqrt{(2+3)^2 + (4-1)^2 + (-1-2)^2} = \sqrt{25+9+9} = \sqrt{43}$$

(ii) Distance between the points (–1, 3, –4) and (1, 0, 4)

$$= \sqrt{(1+1)^2 + (0-3)^2 + (4+4)^2} = \sqrt{4+9+64} = \sqrt{77}$$

Q10. Find the equation of the set of points which are equidistant from the points (1, –2, –3) and (3, 2, –1).

Solution:
Let $P(x, y, z)$ be the point that is equidistant from points $A(1, -2, -3)$ and $B(3, 2, -1)$; then $PA = PB$.

$\Rightarrow PA^2 = PB^2$

$\Rightarrow (x-1)^2 + (y+2)^2 + (z+3)^2$

$= (x-3)^2 + (y-2)^2 + (z+1)^2$

$\Rightarrow x^2 - 2x + 1 + y^2 + 4y + 4 + z^2 + 6z + 9$

$= x^2 - 6x + 9 + y^2 - 4y + 4 + z^2 + 2z + 1$

$\Rightarrow -2x + 4y + 6z + 10 = -6x - 4y + 2z + 14$

$\Rightarrow 4x + 8y + 4z = 4 \Rightarrow x + 2y + z = 1$ is the required equation.

Q11. Find the equation of the set of points P, the sum of the whose distances from $A(4,0,0)$ and $B(-4,0,0)$ is equal to 8.

Solution:
Let the coordinates of point P be (x, y, z). The sum of distances of point $P(x, y, z)$ from $A(4,0,0)$ and $B(-4,0,0)$ is equal to 8; therefore $PA + PB = 10$.

$\Rightarrow \sqrt{(x-4)^2 + y^2 + z^2} + \sqrt{(x+4)^2 + y^2 + z^2} = 8$

$\Rightarrow \sqrt{(x-4)^2 + y^2 + z^2} = 8 - \sqrt{(x+4)^2 + y^2 + z^2}$

Squaring both sides, we get

$\Rightarrow (x-4)^2 + y^2 + z^2 = 64 - 16$

$\sqrt{(x+4)^2 + y^2 + z^2} + (x+4)^2 + y^2 + z^2$

$\Rightarrow x^2 - 8x + 16 + y^2 + z^2 = 64 - 16$

$\sqrt{x^2 + 8x + 16 + y^2 + z^2} + x^2 + 8x + 16 + y^2 + z^2$

$\Rightarrow 16\sqrt{x^2 + 8x + 16 + y^2 + z^2} = (64 + 16x)$

$\Rightarrow \sqrt{x^2 + 8x + 16 + y^2 + z^2} = (4 + x)$

Again squaring both sides, we get

$(x^2 + 8x + 16 + y^2 + z^2) = 16 + x^2 + 8x$

$\Rightarrow y^2 + z^2 = 0$ is the required equation.

Q12. Find the coordinates of the point which divides the line segment joining the points $(-2,3,5), (1,-4,0)$ in the ratio
(a) 2:3 internally (b) 2:3 externally

Solution:
(a) The coordinates of a point $R(x, y, z)$ that divides the line segment joining points $P(x_1, y_1, z_1), Q(x_2, y_2, z_2)$ internally in the ratio $m:n$ are $R(x, y, z) = \left(\frac{mx_2 + nx_1}{m+n}, \frac{my_2 + ny_1}{m+n}, \frac{mz_2 + nz_1}{m+n}\right)$.

Thus, the point that divides the line segment joining the points $(-2,3,5), (1,-4,0)$ internally in the ratio 2:3 is $x = \frac{2(1) + 3(-2)}{2+3}, y = \frac{2(-4) + 3(3)}{2+3}$,

$z = \frac{2(0) + 3(5)}{2+3} \Rightarrow x = \frac{-4}{5}, y = \frac{1}{5}, z = \frac{15}{5}$

Thus, the coordinates of the required point are $\left(-\frac{4}{5}, \frac{1}{5}, 3\right)$.

(b) The coordinates of a point $R(x, y, z)$ that divides the line segment joining points $P(x_1, y_1, z_1), Q(x_2, y_2, z_2)$ internally in the ratio $m:n$ are $R(x, y, z) = \left(\frac{mx_2 - nx_1}{m-n}, \frac{my_2 - ny_1}{m-n}, \frac{mz_2 - nz_1}{m-n}\right)$.

Thus, the point that divides the line segment joining the points $(-2,3,5), (1,-4,6)$ externally in the ratio 2:3 is $x = \frac{2(1) - 3(-2)}{2-3}, y = \frac{2(-4) - 3(3)}{2-3}$,

$z = \frac{2(0) - 3(5)}{2+3} \Rightarrow x = -8, y = 17, z = -3$

Thus, the coordinates of the required point are (–8, 17, –3).

Q13. Given that $P(3,2,-4), Q(5,4,-5), R(9,8,-7)$ are collinear, find the ratio in which Q divides PR.

Solution:
Let point $Q(5,4,-5)$ divide the line segment joining the points $P(3,2,-4), R(9,8,-7)$ in the ratio $k:1$; then, by the section formula,

$$(5,4,-5) = \left(\frac{k(9)+3}{k+1}, \frac{k(8)+2}{k+1}, \frac{k(-7)-4}{k+1}\right)$$

$\Rightarrow \frac{9k+3}{k+1} = 5 \Rightarrow 9k + 3 = 5k + 5 \Rightarrow 4k = 2$

$\Rightarrow k = \frac{2}{4} = \frac{1}{2}$

Thus, the point Q divides PR in the ratio 1:2.

Q14. Find the ratio in which the XZ-plane divides the line segment by joining the points $(-2,10,7),(3,-5,8)$.

Solution:
Let the YZ-plane divide the line segment joining points $(-2,10,7),(3,-5,8)$ in the ratio $k:1$; then by the section formula, the coordinates of the point of intersection are $\left(\frac{k(3)-2}{k+1},\frac{k(-5)+10}{k+1},\frac{k(8)+7}{k+1}\right)$.
On the YZ-plane, the y-coordinate of any point is zero; therefore $\frac{-5k+10}{k+1}=0 \Rightarrow 5k=10 \Rightarrow k=2$.
Thus, the XZ-plane divides the line segment formed by joining the given points in the ratio 2:1.

Q15. Using the section formula, show that the points $A(2,-1,4), B(-1,2,1), C(0,1,2)$ are collinear.

Solution:
Let P be a point that divides the line segment AB in the ratio $k:1$; then by the section formula, the coordinates of point P are $\left(\frac{k(-1)+2}{k+1},\frac{k(2)-1}{k+1},\frac{k(1)+4}{k+1}\right)$
Now, equating the x-coordinate of P and C, we get
$\frac{-k+2}{k+1}=0 \Rightarrow k=2$
For $k=2$, the coordinates of point P are $(0,1,2)$ which is same as of point $C(0,1,2)$; therefore, point C is a point that divides AB externally in the ratio 2:1, which is the same point P.
Thus, the points A, B, C are collinear.

Q16. Three vertices of a parallelogram $ABCD$ are $A(3,-1,2), B(1,2,0), C(-1,1,2)$. Find the coordinates of the fourth vertex.

Solution:
Let the coordinates of the fourth vertex of parallelogram $ABCD$ be $D(x,y,z)$.
The diagonals of a parallelogram bisect each other; therefore in parallelogram $ABCD$, diagonals AC and BD bisect each other.
Thus, mid-point of AC = mid-point of BD.

$$\Rightarrow \left(\frac{3-1}{2},\frac{-1+1}{2},\frac{2+2}{2}\right)=\left(\frac{x+1}{2},\frac{y+2}{2},\frac{z+0}{2}\right)$$

$$\Rightarrow (1,0,2)=\left(\frac{x+1}{2},\frac{y+2}{2},\frac{z}{2}\right)$$

$$\Rightarrow \frac{x+1}{2}=1, \frac{y+2}{2}=0, \frac{z}{2}=2$$

$$\Rightarrow x=1, y=-2, z=4$$

Hence, the coordinates of the fourth vertex are $D(1,-2,4)$.

Q17. If A, B, C are the feet of the perpendiculars from a point P on the XY, YZ and ZX-planes, respectively, then find the coordinates of A, B and C in each of the following where the point P is
(a) $(-5, 3, 4)$ (b) $(4, -3, -2)$

Solution:
On the XY-plane $z = 0$, on the YZ-plane $x = 0$ and on the ZX-plane $y = 0$; therefore the coordinates of the feet of the perpendiculars from a point P on the XY, YZ and ZX-planes, respectively, of the points A, B and C are the following.
(a) $A(-5,3,0), B(0,3,4), C(-5,0,4)$
(b) $A(4,-3,0), B(0,-3,-2), C(4,0,-2)$

Q18. Find the distance from the origin to (1, 2, 3).

Solution:
Distance from origin to the point (1, 2, 3) is

$$d=\sqrt{(x_1-x_2)^2+(y_1-y_2)^2+(z_1-z_2)^2}$$

$$=\sqrt{(0-1)^2+(0-2)^2+(0-3)^2}$$

$$=\sqrt{1+4+9}=\sqrt{14}$$

Q19. Find the third vertex of the triangle whose centroid is the origin and two vertices are (2, 3, 6) and (0, –2, 1).

Solution:
Let the two vertices of ΔABC be A (2, 3, 6) and B (0, –2, 1) and the third vertex C be (x, y, z).
The coordinates of the centroid G are (0,0,0).

$$\therefore 0=\frac{x+2+0}{3} \Rightarrow x=-2$$

$$0=\frac{y+3-2}{3} \Rightarrow y=-1$$

$$0=\frac{z+6-1}{2} \Rightarrow z=-5$$

Thus, the third vertex of triangle is $C(-2,-1,-5)$.

Q20. If $A(2,2,-3), B(5,6,9), C(2,7,9)$ are the vertices of a triangle, the internal bisector of the angle A meets BC at the point D, then find the cordinates of D.

Solution:
Let the coordinates of D be (x, y, z).

$$AB=\sqrt{9+16+144}=\sqrt{169}=13$$

$$AC=\sqrt{0+25+144}=\sqrt{169}=13$$

$$\Rightarrow \frac{AB}{AC}=\frac{13}{13} \Rightarrow AB=AC, \frac{BD}{DC}=\frac{1}{1} \Rightarrow BD=DC$$

D divides the line BC into two equal parts. Therefore D is the mid-point of BC.

Thus, $x = \frac{5+2}{2} = \frac{7}{2}$

$y = \frac{6+7}{2} = \frac{13}{2}$, $z = \frac{9+9}{2} = 9$

Thus, the coordinates of D are $\left(\frac{7}{2}, \frac{13}{2}, 9\right)$.

Day 167

Long Answer-I ((LA-I: 4-Mark) and Long Answer-II (LA-II: 6-Mark) Questions

Q1. Prove the following:

(a) $(-2,3,5),(-1,2,3),(1,0,-1)$ are collinear points.

(b) $(0,7,-10),(1,6,-6),(4,9,-6)$ are the vertices of an isosceles triangle.

(c) $(0,7,10),(-1,6,6),(-4,9,6)$ are the vertices of a right-angled triangle.

(d) $(-1,2,1),(1,-2,5),(4,-7,8),(2,-3,4)$ are the vertices of a parallelogram.

Solution:

(a) Let $A(-2,3,5)$, $B(-1,2,3)$, $C(1,0,-1)$ be the given points; then points A, B and C are collinear if they lie on a line.

$AB = \sqrt{(-1+2)^2 + (2-3)^2 + (3-5)^2}$

$= \sqrt{(1)^2 + (-1)^2 + (-2)^2} = \sqrt{6}$

$BC = \sqrt{(1+1)^2 + (0-2)^2 + (-1-3)^2}$

$= \sqrt{(2)^2 + (-2)^2 + (-4)^2} = \sqrt{24} = 2\sqrt{6}$

$AC = \sqrt{(1+2)^2 + (0-3)^2 + (-1-5)^2}$

$\sqrt{(3)^2 + (-3)^2 + (-6)^2} = \sqrt{54} = 3\sqrt{6}$

Since $AB + BC = \sqrt{6} + 2\sqrt{6} = 3\sqrt{6} = AC$, the points $(-2,3,5),(-1,2,3),(1,0,-1)$ are collinear.

(b) Given points $A(0,7,-10)$, $B(1,6,-6)$, $C(4,9,-6)$ are the vertices of an isosceles triangle if two sides of the triangle are equal.

$AB = \sqrt{(1-0)^2 + (6-7)^2 + (-6+10)^2}$

$= \sqrt{(1)^2 + (-1)^2 + (4)^2} = \sqrt{1+1+6} = \sqrt{18} = 3\sqrt{2}$

$BC = \sqrt{(4-1)^2 + (9-6)^2 + (-6+6)^2}$

$= \sqrt{(3)^2 + (3)^2} = \sqrt{9+9} = \sqrt{18} = 3\sqrt{2}$

$CA = \sqrt{(0-4)^2 + (7-9)^2 + (-10+6)^2}$

$= \sqrt{(-4)^2 + (-2)^2 + (-4)^2} = \sqrt{16+4+16}$

$= \sqrt{36} = 6$

Since $AB = BC \neq CA$, the given points $A(0,7,-10)$, $B(1,6,-6)$, $C(4,9,-6)$ are the vertices of an isosceles triangle.

(c) Given points $A(0,7,10)$, $B(-1,6,6)$, $C(-4,9,6)$ are the vertices of a right-angled triangle if the square of the sum of two sides is equal to the square of the third side.

$AB = \sqrt{(-1-0)^2 + (6-7)^2 + (6-10)^2}$

$= \sqrt{(-1)^2 + (-1)^2 + (-4)^2} = \sqrt{1+1+16} = \sqrt{18} = 3\sqrt{2}$

$BC = \sqrt{(-4+1)^2 + (9-6)^2 + (6-6)^2}$

$= \sqrt{(-3)^2 + (3)^2 + (0)^2} = \sqrt{9+9} = \sqrt{18} = 2\sqrt{3}$

$CA = \sqrt{(0+4)^2 + (7-9)^2 + (10-6)^2}$

$= \sqrt{(4)^2 + (-2)^2 + (4)^2} = \sqrt{16+4+16} = \sqrt{36} = 6$

Since, $AB^2 + BC^2 = (3\sqrt{2})^2 + (3\sqrt{2})^2 = 36 = AC^2$, the given points $A(0,7,10)$, $B(-1,6,6)$, $C(-4,9,6)$ are the vertices of a right-angled triangle.

(d) The given points $A(-1,2,1)$, $B(1,-2,5)$, $C(4,-7,8)$, $D(2,-3,4)$ are the vertices of a parallelogram if both sets of opposite sides are equal.

$AB = \sqrt{(1+1)^2 + (-2-2)^2 + (5-1)^2}$

$= \sqrt{4+16+16} = \sqrt{36} = 6$

$BC = \sqrt{(4-1)^2 + (-7+2)^2 + (8-5)^2}$

$= \sqrt{9+25+9} = \sqrt{43}$

$CD = \sqrt{(2-4)^2 + (-3+7)^2 + (4-8)^2}$

$= \sqrt{4+16+16} = \sqrt{36} = 6$

$DA = \sqrt{(-1-2)^2 + (2+3)^2 + (1-4)^2}$

$= \sqrt{9+25+9} = \sqrt{43}$

Since the opposite sides $AB = CD = 6$ and opposite sides $BC = AD = \sqrt{43}$, the given points are the vertices of a parallelogram.

Q2. If the origin is the centroid of the triangle PQR with vertices $P(2a,2,6), Q(-4,2b,-10), R(8,12,2c)$ then find the values of a, b, c.

Solution:

The coordinates of the centroid of the triangle, whose vertices are $(x_1, y_1, z_1), (x_2, y_2, z_2), (x_3, y_3, z_3)$ are $\left(\frac{x_1+x_2+x_3}{3}, \frac{y_1+y_2+y_3}{3}, \frac{z_1+z_2+z_3}{3}\right)$.

Therefore, coordinates of the centroid of ΔPQR

$$=\left(\frac{2a-4+8}{3}, \frac{2+2b+14}{3}, \frac{6-10+2c}{3}\right)$$

$$=\left(\frac{2a+4}{3}, \frac{2b+16}{3}, \frac{2c-4}{3}\right)$$

It is given that the origin is the centroid of ΔPQR, therefore,

$$\left(\frac{2a+4}{3}, \frac{2b+16}{3}, \frac{2c-4}{3}\right) = (0,0,0)$$

$$\Rightarrow \frac{2a+4}{3} = 0, \frac{2b+16}{3} = 0, \frac{2c-4}{3} = 0$$

Solving, we get $a = -2,\ b = -8,\ c = 2$

Q3. Find the coordinates of a point on the y-axis which are at a distance of $\sqrt{22}$ from the point $P(3, -5, 5)$.

Solution:

If a point is on the y-axis, then the x-coordinate and the z-coordinate of the point are zero.

Thus, let $A(0,\ b,\ 0)$ be the point on the y-axis at a distance of $\sqrt{22}$ from point $(3,\ -1,\ 2)$; then

$AP = \sqrt{22} \Rightarrow AP^2 = 22$

$\Rightarrow (3-0)^2 + (-1-b)^2 + (2-0)^2 = 22$

$\Rightarrow 9+1+b^2+2b+4 = 22 \Rightarrow b^2+2b-8 = 0$

$\Rightarrow (b+4)(b-2) = 0 \Rightarrow b = -4, 2$

Thus, the coordinates of the required points are $(0, -4, 0)$ and $(0, 2, 0)$.

Q4. A point R with x-coordinate 4 lies on the segment joining the points $P(2,-3,4), Q(8,0,6)$. Find the coordinates of the point R.

Solution:

Let point R divide the line segment PQ in the ratio $k:1$; then by the section formula, the coordinates of the point R are given by $\left(\frac{k(8)+2}{k+1}, \frac{k(0)-3}{k+1}, \frac{k(6)+4}{k+1}\right)$

$$=\left(\frac{8k+2}{k+1}, \frac{-3}{k+1}, \frac{6k+4}{k+1}\right)$$

It is given that the x-coordinate of the point R is 4; therefore $\frac{8k+2}{k+1} = 4 \Rightarrow 8k+2 = 4k+4 \Rightarrow k = \frac{1}{2}$.

Therefore, the coordinates of point R are

$$\left(4, \frac{-3}{\frac{1}{2}+1}, \frac{6(1/2)+4}{\frac{1}{2}+1}\right) = (4, -2, 14/3)$$

Q5. If A and B are the points $(3,4,5), (-1,3,-7)$, respectively, find the equation of the set of points P such that $PA^2 + PB^2 = 1$.

Solution:

Let the coordinates of point P be (x, y, z); then using the distance formula,

$PA^2 = (x-3)^2 + (y-4)^2 + (z-5)^2$

$\Rightarrow PA^2 = x^2 - 6x + y^2 - 8y + z^2 - 10z + 50$

Similarly, $PB^2 = (x+1)^2 + (y-3)^2 + (z+7)^2$

$\Rightarrow PB^2 = x^2 + 2x + y^2 - 6y + z^2 + 14z + 59$

Now, it is given that $PA^2 + PB^2 = 1$; then

$(x^2 - 6x + y^2 - 8y + z^2 - 10z + 50) +$

$(x^2 + 2x + y^2 - 6y + z^2 + 14z + 59) = 1$

$\Rightarrow 2x^2 + 2y^2 + 2z^2 - 4x + 14y + 4z + 109 = 1$

$\Rightarrow x^2 + y^2 - 2x - 7y + 2z + 54 = 0$ is the required equation.

Q6. Three consecutive vertices of a parallelogram $ABCD$ are $A(6,-2,4), B(2,4,0), C(-2,2,4)$. Find the coordinates of the fourth vertex.

Solution:

Let the coordinates of the fourth vertex be $D(x, y, z)$.

Mid-point of diagonal AC is,

$$x = \frac{x_1+x_2}{2} = \frac{6-2}{2} = 2$$

$$y = \frac{y_1+y_2}{2} = \frac{-2+2}{2} = 0$$

$$z = \frac{z_1+z_2}{2} = \frac{4+4}{2} = 4$$

And mid-point of diagonal BD is,

$$\left(\frac{x+2}{2}, \frac{y+4}{2}, \frac{z}{2}\right)$$

The mid-points of the diagonal of a parallelogram coincide at a point; therefore

$$\left(\frac{x+2}{2}, \frac{y+4}{2}, \frac{z}{2}\right) = (2,0,4)$$

$\Rightarrow x = 2,\ y = -4$ and $z = 8$

Thus, the coordinates of fourth vertex D is $(2, -4, 8)$

Q7. Find the coordinates of the points which trisect the line segment joining the points $A(2,1,-3), B(5,0,3)$.

Solution:

Let the $P(x_1, y_1, z_1), Q(x_2, y_2, z_2)$ trisect the line segment AB. The point P divides the line AB in a 1:2 ratio internally; then

$$x_1 = \frac{2\times 2+1\times 5}{1+2} = \frac{9}{3} = 3$$

$$y_1 = \frac{2\times 1+1\times (0)}{3} = \frac{2}{3}$$

$$z_1 = \frac{2\times(-3)+1\times 3}{3} = \frac{-6+3}{3} = \frac{-3}{3} = -1$$

The point Q divides the line segement AB in a 2:1 ratio, then

$$x_2 = \frac{1\times 2+2\times 5}{3} = 4$$

$$y_2 = \frac{1\times 1+(0\times 2)}{3} = \frac{1}{3}$$

$$z_2 = \frac{1\times(-3)+2\times 3}{3} = -1$$

Thus, the coordinates of P are $(3, 2/3, -1)$ and the coordinates of Q are $(4, 1/3, 1)$.

Q8. Find the length of the median of the triangle with vertices $A(0,0,2), B(0,4,0), C(2,0,0)$.

Solution:

Let AD, BE and CF be the medians of the given triangle ABC.

For median AD, D is the mid-point of BC; therefore the coordinates of the point D are,

$$= \left(\frac{0+2}{2}, \frac{4+0}{2}, \frac{0+0}{2}\right) = (1,2,0)$$

Thus, length of the median AD is $= \sqrt{(0-1)^2+(0-2)^2+(2-0)^2} = \sqrt{1+4+4} = 3$

For median BE, E is the mid-point of AC; therefore the coordinates of the point E are,

$$= \left(\frac{0+2}{2}, \frac{0+0}{2}, \frac{2+0}{2}\right) = (1,0,1)$$

Thus, length of the median BE is

$$= \sqrt{(1-0)^2+(0-4)^2+(1-0)^2} = \sqrt{1+16+1} = 3\sqrt{2}$$

For median CF, F is the mid-point of AB; therefore the coordinates of the point F are,

$$= \left(\frac{0+0}{2}, \frac{0+4}{2}, \frac{2+0}{2}\right) = (0,2,1)$$

Thus, length of the median CF is

$$CF = \sqrt{(2-0)^2+(0-2)^2+(0-1)^2} = \sqrt{4+4+1} = 3$$

Hence, the lengths of the medians of ΔABC are $3, 3\sqrt{2}, 3$.

Q9. If the origin is the centroid of a ΔABC having vertices $A(a,1,3), B(-2,b,-5), C(4,3,c)$; then find the values of a, b, c.

Solution:

The origin is the centroid of the ΔABC, that is, G(0, 0, 0).

$$\therefore 0 = \frac{a-2+4}{3} \Rightarrow a = -2$$

$$0 = \frac{1+b+3}{3} \Rightarrow b = -4$$

$$0 = \frac{3-5+c}{3} \Rightarrow c = +2$$

$\Rightarrow a = -2,\ b = -4,\ c = 2$

Q10. Find the centroid of a triangle, the mid-point of whose sides are $D(1,2,-3), E(3,0,1), F(-1,1,0)$.

Solution:

Let the vertices of the ΔABC be $A(x_1, y_1, z_1), B(x_2, y_2, z_2), C(x_3, y_3, z_3)$.

The mid-points of the sides are $D(1,2,-3), E(3,0,1), F(-1,1,-4)$; then the mid-point of BC is (1, 2, –3).

$$\therefore 1 = \frac{x_2+x_3}{2} \Rightarrow x_2+x_3 = 2 \qquad \text{(i)}$$

$$2 = \frac{y_2+y_3}{2} \Rightarrow y_2+y_3 = 4 \qquad \text{(ii)}$$

$$-3 = \frac{z_2+z_3}{2} \Rightarrow z_2+z_3 = -6 \qquad \text{(iii)}$$

Similarly, for the sides AB and AC,

$$\Rightarrow -1 = \frac{x_1+x_2}{2} \Rightarrow x_1+x_2 = -2 \qquad \text{(iv)}$$

$$\Rightarrow 1 = \frac{y_1+y_2}{2} \Rightarrow y_1+y_2 = 2 \qquad \text{(v)}$$

$$\Rightarrow 0 = \frac{z_1+z_2}{2} \Rightarrow z_1+z_2 = 0 \qquad \text{(vi)}$$

$$\Rightarrow 3 = \frac{x_1+x_3}{2} \Rightarrow x_1+x_3 = 6 \qquad \text{(vii)}$$

$$\Rightarrow 0 = \frac{y_1+y_3}{2} \Rightarrow y_1+y_3 = 0 \qquad \text{(viii)}$$

$$\Rightarrow 1 = \frac{z_1+z_3}{2} \Rightarrow z_1+z_3 = 2 \qquad \text{(ix)}$$

On adding equations (i) and (iv), we get

$$x_1+2x_2+x_3 = 0 \qquad \text{(x)}$$

On adding equations (ii) and (v), we get

$$y_1+2y_2+y_3 = 0 \qquad \text{(xi)}$$

On adding equations (iii) and (vi), we get

$z_1 + 2z_2 + z_3 = -6$ (xii)

From equations (vii) and (x), $2x_2 = -6 \Rightarrow x_2 = -3$

If $x_2 = -3$, then $x_3 = 5$

If $x_3 = 5$, then $x_1 = 1, x_2 = -3, x_3 = 5$

From equations (xi) and (viii), $2y_2 = 6 \Rightarrow y_2 = 3$

If $y_2 = 3$, then $y_1 = -1$

If $y_1 = -1$ then $y_1 = -1, y_2 = 3, y_3 = 1$

From equations (xii) and (ix), $2z_2 = -4 \Rightarrow z_2 = -2$

$z_2 = -2$ then $z_1 = 2$

$z_1 = 2$ then $z_3 = -8$

$z_1 = 2, z_2 = -2, z_3 = -8.$

Thus, the points are

$A(1 -1, 2), B(-3, 3, -2), C(5, 1, -8).$

$\therefore$ The centroid of the triangle

$= G\left(\frac{1-3+5}{3}, \frac{-1+3+1}{3}, \frac{2-2-8}{3}\right) = G(1, 1, -8/3)$

CHAPTER

13 Limits and Derivatives

Day 168

Limit of a Function

A function $y = f(x)$ tends to limit l as $x \to a$ (read as 'x tends to a', meaning x is very very close to a), written as $\lim_{x\to a} f(x) = l$.

Left-hand and Right-hand Limits

$\lim_{x\to a^-} f(x) = \lim_{h\to 0} f(a-h)$ is the left-hand limit (LHL), meaning x is very very close to a but less than a.

$\lim_{x\to a^+} f(x) = \lim_{h\to 0} f(a+h)$ is the right-hand limit (RHL), meaning x is very very close to a but greater than a.

Existence of Limit

Limit of $f(x)$ exists if and only if

$$\lim_{x\to a^-} f(x) = \lim_{x\to a^+} f(x)$$

Example 1: Consider the function $f(x) = 2$. This is a constant function, and therefore, it takes the same value $f(x) = 2$ everywhere. That is, $\lim_{x\to a} f(x) = 2$ for any real number a.

Example 2: Consider the function $f(x) = |x|$, $x \neq 0$. Now as x takes values very close to 0, the value of $f(x) \to 0$, that is, $\lim_{x\to 0} f(x) = 0$.

Example 3: Consider the function $f(x) = x^2$.
Now as x takes values very close to 0, the value of $f(x) \to 0$, that is, $\lim_{x\to 0} f(x) = 0$.

And as x takes values very close to 1, the value of $f(x) \to 1$, that is, $\lim_{x\to 1} f(x) = 1$.

And as x takes values very close to –1, the value of $f(x) \to 1$, that is, $\lim_{x\to -1} f(x) = 1$.

Example 4: Consider the function $f(x) = x^3$.
Now as x takes values very close to 0, the value of $f(x) \to 0$, that is, $\lim_{x\to 0} f(x) = 0$.

And as x takes values very close to 1, the value of $f(x) \to 1$, that is, $\lim_{x\to 1} f(x) = 1$.

And as x takes values very close to –1, the value of $f(x) \to 1$, that is, $\lim_{x\to -1} f(x) = -1$.

Example 5: Consider the function $f(x) = x^2 + x + 1$. Now as x takes values very close to 0, the value of $f(x) \to 1$, that is, $\lim_{x\to 0} f(x) = 1$.

And as x takes values very close to 1, the value of $f(x) \to 3$, that is, $\lim_{x\to 1} f(x) = 3$.

And as x takes values very close to –1, the value of $f(x) \to 1$, that is, $\lim_{x\to -1} f(x) = 1$.

Example 6: Consider the function $f(x) = \dfrac{x^2 - 1}{x - 1}$, $x \neq 1$.
Now as x takes values very close to 1, the value of $f(x) \to 2$, that is, $\lim_{x\to 1} f(x) = 2$, is as shown in the table below:

x	0.9	0.99	0.999	1.001	1.01	1.1
$f(x)$	1.9	1.99	1.999	2.001	2.01	2.1

Example 7: Consider the function $f(x) = \dfrac{x^2 - x - 2}{x - 2}$, $x \neq 2$

$\Rightarrow f(x) = \dfrac{(x-2)(x+1)}{x-2}$, $x \neq 2$

$\Rightarrow f(x) = x + 1$, $x \neq 2$

Now as x takes values very close to 2, the value of $f(x) \to 3$, that is, $\lim_{x\to 2} f(x) = 3$, is as shown in the table below:

x	1.9	1.99	1.999	2.001	2.01	2.1
$f(x)$	2.9	2.99	2.999	3.001	3.01	3.1

Example 8: Consider the function

$$f(x) = \begin{cases} \dfrac{|x|}{x}, & x < 0 \\ \dfrac{|x|}{x}, & x \geq 0 \end{cases} = \begin{cases} -1, & x < 0 \\ 1, & x \geq 0 \end{cases}$$

Now as x takes values very close to 0 but less than 0, the value of $f(x) \to -1$, that is, LHL $= \lim_{x\to 0^-} f(x) = -1$.

And as x takes values very close to 0 but greater than 0, the value of $f(x) \to 1$, that is, RHL $= \lim_{x \to 0^+} f(x) = 1$.

Example 9: Consider the function $f(x) = \begin{cases} x-1, & x<0 \\ 0, & x=0 \\ 1-x, & x>0 \end{cases}$

Now as x takes values very close to 0 but less than 0, the value of $f(x) \to -1$, that is, LHL $= \lim_{x \to 0^-} f(x) = -1$.

And as x takes values very close to 0 but greater than 0, the value of $f(x) \to 1$, that is, RHL $= \lim_{x \to 0^+} f(x) = 1$.

And since $\lim_{x \to 0} f(x) = 0 \neq$ LHL $\neq$ RHL, the limit at $x = 0$ does not exist.

Example 10: Consider the function $f(x) = \begin{cases} x-1, & x<1 \\ 0, & x=1. \\ 1-x, & x>1 \end{cases}$

Now as x takes values very close to 1 but less than 1, the value of $f(x) \to 0$, that is, LHL $= \lim_{x \to 1^-} f(x) = 1-1 = 0$.

And as x takes values very close to 1 but greater than 1, the value of $f(x) \to 0$, that is, RHL $= \lim_{x \to 1^+} f(x) = 1-1 = 0$.

And since $\lim_{x \to 1} f(x) = 0 =$ LHL $=$ RHL, the limit at $x = 0$ exists and is equal to 0.

Example 11: Consider the function $f(x) = \sin x$.

Now as x (measured in radians) takes values very close to 0, the value of $f(x) \to 0$, that is, $\lim_{x \to 0} f(x) = 0$, is as shown in the table below:

x	−0.1	−0.01	−0.001	0.001	0.01	0.1
$f(x)$	−0.0998	−0.0099	−0.0009	0.0998	0.0099	0.0998

Example 12: Consider the function $f(x) = \cos x$.

Now as x (measured in radian) takes values very close to 0, the value of $f(x) \to 1$, that is, $\lim_{x \to 0} f(x) = 1$, is as shown in the table below:

x	−0.1	−0.01	0.01	0.1
$f(x)$	0.9950	0.99995	0.99995	0.9950

Example 13: Consider the function $f(x) = \sin x$.

Now as x (measured in radians) takes values very close to $\frac{\pi}{2}$, the value of $f(x) \to 1$, that is, $\lim_{x \to \frac{\pi}{2}} f(x) = 1$, is as shown in the table below:

x	$\frac{\pi}{2} - 0.1$	$\frac{\pi}{2} - 0.01$	$\frac{\pi}{2} + 0.01$	$\frac{\pi}{2} + 0.1$
$f(x)$	0.9950	0.9999	0.9999	0.9950

Example 14: Consider the function $f(x) = \cos x$.

Now as x (measured in radians) takes values very close to $\frac{\pi}{2}$, the value of $f(x) \to 0$, that is, $\lim_{x \to \frac{\pi}{2}} f(x) = 0$, is as shown in the table below:

x	$\frac{\pi}{2} - 0.1$	$\frac{\pi}{2} - 0.01$	$\frac{\pi}{2} + 0.01$	$\frac{\pi}{2} + 0.1$
$f(x)$	0.0017	0.00017	−0.00017	−0.0017

Example-15: Consider the function $f(x) = \frac{1}{x^2}$.

Now as x takes values very close to 0, the value of $f(x) \to \infty$, that is, $\lim_{x \to 0} f(x) = \infty$, is as shown in the table below:

x	−0.1	−0.01	−0.001	0.001	0.01	0.1
$f(x)$	10^2	10^4	10^6	10^6	10^4	10^2

Thus, as x tends to 0, $f(x)$ becomes larger and larger, that is, $\lim_{x \to 0} f(x) = +\infty$. (This type of limit is beyond the scope of the syllabus.)

Day 169

Algebra of Limits

Let f and g be two functions such that both $\lim_{x \to a} f(x)$ and $\lim_{x \to a} g(x)$ exist. Then

- Limit of the sum of two functions is the sum of the limits of the functions, that is, $\lim_{x \to a}[f(x) + g(x)] = \lim_{x \to a} f(x) + \lim_{x \to a} g(x)$.
- Limit of the difference of two functions is the difference of the limits of the functions, that is, $\lim_{x \to a}[f(x) - g(x)] = \lim_{x \to a} f(x) - \lim_{x \to a} g(x)$.
- Limit of the product of two functions is the product of the limits of the functions, that is, $\lim_{x \to a}[f(x) \cdot g(x)] = \lim_{x \to a} f(x) \cdot \lim_{x \to a} g(x)$.

- Limit of the product of a constant and a function is the product of the constant and limit of the function, that is, $\lim_{x\to a}[\lambda\, g(x)] = \lambda \cdot \lim_{x\to a} g(x)$.
- Limit of the quotient of two functions is the quotient of the limits of the functions, that is, $\lim_{x\to a}\frac{f(x)}{g(x)} = \frac{\lim_{x\to a} f(x)}{\lim_{x\to a} g(x)},\ \left(\lim_{x\to a} g(x) \neq 0\right)$

Limits of a Polynomial Function

Let a polynomial function be $f(x) = a_0 + a_1x + a_2x^2 + ... + a_nx^n$, where all a_i are real numbers such that $a_n \neq 0$ for some natural number n. Then,

$$\lim_{x\to a} f(x) = \lim_{x\to a}[a_0 + a_1x + a_2x^2 + ... + a_nx^n]$$

$$= \lim_{x\to a} a_0 + \lim_{x\to a} a_1x + \lim_{x\to a} a_2x^2 + ... + \lim_{x\to a} a_nx^n$$

$$= a_0 + a_1 \lim_{x\to a} x + a_2 \lim_{x\to a} x^2 + ... + a_n \lim_{x\to a} x^n$$

$$= a_0 + a_1a + a_2a^2 + ... + a_na^n = f(a)$$

Limits of a Rational Function

Let a rational function be $f(x) = \frac{g(x)}{h(x)}$, where $g(x), h(x)$ are polynomial functions such that $h(x) \neq 0$. Then

$$\lim_{x\to a} f(x) = \lim_{x\to a}\frac{g(x)}{h(x)} = \frac{\lim_{x\to a} g(x)}{\lim_{x\to a} h(x)} = \frac{g(a)}{h(a)}$$

Case 1: If $h(a) = 0$ and $g(a) \neq 0$, then $\lim_{x\to a} f(x)$ does not exist.

Case 2: If $h(a) = 0$ and $g(a) = 0$, then $\lim_{x\to a} f(x) = \frac{0}{0}$ is of indeterminate form and can be solved by rationalising and simplifying.

Some Important Limits

- $\lim_{x\to 0}\frac{x^n - a^n}{x - a} = na^{n-1}$
- $\lim_{x\to 0}\frac{(1+x)^n - 1}{x} = n$

Example 1: Evaluate the given limit: $\lim_{x\to\pi}(x-\pi)$.

Solution:

$$\lim_{x\to\pi}(x-\pi) = \pi - \pi = 0$$

Example 2: Evaluate the given limit: $\lim_{x\to 1}\frac{x+2}{x-2}$.

Solution:

$$\lim_{x\to 1}\frac{x+2}{x-2} = \frac{1+2}{1-2} = -3$$

Example 3: Evaluate the given limit: $\lim_{x\to -1}\frac{x^{10}+x^5+x}{x}$.

Solution:

$$\lim_{x\to -1}\frac{x^{10}+x^5+x}{x} = \frac{(-1)^{10}+(-1)^5-1}{-1}$$

$$= \frac{1-1-1}{-1} = 1$$

Example 4: Evaluate the given limit: $\lim_{x\to 0}\frac{(x+1)^6-1}{x}$.

Solution:

Substituting $x+1 = y$, we get $y \to 1$ as $x \to 0$; therefore

$$\lim_{x\to 0}\frac{(x+1)^6-1}{x} = \lim_{y\to 1}\frac{y^6-1}{y-1} = 6\cdot 1^{6-1} = 6$$

$$\left[\because \lim_{x\to 0}\frac{x^n - a^n}{x-a} = na^{n-1}\right]$$

Example 5: Evaluate the given limit: $\lim_{x\to 2}\frac{3x^2-4x-4}{x^2-4}$.

Solution:

$$\lim_{x\to 2}\frac{3x^2-4x-4}{x^2-4} = \lim_{x\to 2}\frac{(x-2)(3x+2)}{(x-2)(x+2)}$$

$$= \lim_{x\to 2}\frac{3x+2}{x+2} = \frac{3(2)+2}{2+2} = 2$$

Example 6: Evaluate the given limit: $\lim_{x\to 3}\frac{x^4-81}{x^2-2x-3}$.

Solution:

$$\lim_{x\to 3}\frac{x^4-81}{x^2-2x-3}$$

$$= \lim_{x\to 3}\frac{(x-3)(x+3)(x^2+9)}{(x-3)(x+1)}$$

$$= \lim_{x\to 3}\frac{(x+3)(x^2+9)}{x+1}$$

$$= \frac{(3+3)(3^2+9)}{3+1} = \frac{6\times 18}{4} = 27$$

Example 7: Evaluate the given limit: $\lim_{x\to 1}\frac{ax+b}{cx-d}$.

Solution:

$$\lim_{x\to 1}\frac{ax+b}{cx-d} = \frac{a(1)+b}{c(1)-d} = \frac{a+b}{c-d}$$

Example 8: Evaluate the given limit: $\lim_{z\to 1}\frac{z^{\frac{1}{3}}-1}{z^{\frac{1}{9}}-1}$.

Solution:

Substituting $z^{1/9} = x$, we get $z \to 1$ as $x \to 1$; therefore

$$\lim_{z\to1}\frac{z^{\frac{1}{3}}-1}{z^{\frac{1}{9}}-1} = \lim_{x\to1}\frac{x^3-1}{x-1} = 3\cdot1^{3-1} = 3$$

$$\left[\because \lim_{x\to a}\frac{x^n-a^n}{x-a} = na^{n-1}\right]$$

Example 9: Evaluate the given limit: $\lim_{x\to-1}\dfrac{1+\frac{1}{x}}{1+x}$.

Solution:

$$\lim_{x\to-1}\frac{1+\frac{1}{x}}{1+x} = \lim_{x\to-1}\frac{\left(\frac{x+1}{x}\right)}{1+x} = \lim_{x\to-1}\frac{1}{x} = \frac{1}{(-1)} = -1$$

Example 10: Find $\lim_{x\to0} f(x)$, $\lim_{x\to1} f(x)$, where

$$f(x) = \begin{cases} 2x+3, & x \le 0 \\ 2(x+1), & x > 0 \end{cases}.$$

Solution:

The given function is $f(x) = \begin{cases} x+2, & x \le 0 \\ 2(x+1), & x > 0 \end{cases}$

$\lim_{x\to0^-} f(x) = \lim_{x\to0}[x+2] = 0+2 = 2$

$\lim_{x\to0^+} f(x) = \lim_{x\to0} 2(x+1) = 2(0+1) = 2$

$\therefore \lim_{x\to0^-} f(x) = \lim_{x\to0^+} f(x) = \lim_{x\to0} f(x) = 2$

$\lim_{x\to1^-} f(x) = \lim_{x\to0} 2(x+1) = 2(1+1) = 4$

$\lim_{x\to1^+} f(x) = \lim_{x\to1} 2(x+1) = 2(1+1) = 4$

$\therefore \lim_{x\to1^-} f(x) = \lim_{x\to1^+} f(x) = \lim_{x\to1} f(x) = 4$

Example 11: Find $\lim_{x\to-1} f(x)$, where $f(x) = \begin{cases} x^2-1, & x \le -1 \\ -x^2+1; & x > -1 \end{cases}$.

Solution:

The given function is $f(x) = \begin{cases} x^2-1, & x \le -1 \\ -x^2+1, & x > -1 \end{cases}$

$\lim_{x\to-1^-} f(x) = \lim_{x\to-1}[x^2-1] = (-1)^2-1 = 1-1 = 0$

$\lim_{x\to-1^+} f(x) = \lim_{x\to-1}[-x^2-1] = -(-1)^2-1 = -1-1 = -2$

Since $\lim_{x\to1^-} f(x) \ne \lim_{x\to1^+} f(x)$, $\lim_{x\to1} f(x)$ does not exist.

Example 12: Find $\lim_{x\to1} f(x)$, where $f(x) = \dfrac{|x|-1}{|x|+1}$.

Solution:

The given function is $f(x) = \dfrac{|x|-1}{|x|+1}$

$\lim_{x\to1^-} f(x) = \lim_{x\to1^-}\dfrac{|x|-1}{|x|+1} = \dfrac{|1|-1}{|1|+1} = \dfrac{0}{2} = 0$

$[\because |x| = x, \text{ when } x > 0]$

$\lim_{x\to1^+} f(x) = \lim_{x\to1^+}\dfrac{|x|-1}{|x|+1} = \dfrac{|1|-1}{|1|+1} = \dfrac{0}{2} = 0$

$\because \lim_{x\to1^-} f(x) = \lim_{x\to1^+} f(x) = 0$

Therefore, $\lim_{x\to1} f(x) = 0$

Example 13: Suppose $f(x) = \begin{cases} a-bx, & \text{if } x<1 \\ 2, & \text{if } x=1 \\ b+ax, & \text{if } x>1 \end{cases}$ and if $\lim_{x\to1} f(x) = f(1)$, what are the possible values of a and b?

Solution:

The given function is $f(x) = \begin{cases} a-bx, & \text{if } x<1 \\ 2, & \text{if } x=1 \\ b+ax, & \text{if } x>1 \end{cases}$

$\lim_{x\to1^-} f(x) = \lim_{x\to1}(a-bx) = a-b$

$\lim_{x\to1^+} f(x) = \lim_{x\to1}(b+ax) = b+a$

and $f(1) = 2$

Since $\lim_{x\to1} f(x) = f(1) = 2$

$\therefore \lim_{x\to1^-} f(x) = \lim_{x\to1^+} f(x) = \lim_{x\to1} f(x) = f(1)$

$\Rightarrow a-b = 2,\ \ b+a = 2$

Solving the above two equations, we get $a = 2,\ b = 0$.

Example 14: The function $f(x)$ satisfies $\lim_{x\to1}\dfrac{f(x)-1}{x^2-1} = \pi$. Evaluate $\lim_{x\to1} f(x)$.

Solution:

$$\lim_{x\to1}\frac{f(x)-1}{x^2-1} = \pi$$

$$\Rightarrow \frac{\lim_{x\to1}(f(x)-1)}{\lim_{x\to1}(x^2-1)} = \pi$$

$$\Rightarrow \lim_{x\to1}(f(x)-1) = \pi\lim_{x\to1}(x^2-1)$$

$$\Rightarrow \lim_{x\to1}(f(x)-1) = \pi(1^2-1)$$

$$\Rightarrow \lim_{x\to1}(f(x)-1) = 0 \Rightarrow \lim_{x\to1} f(x) - \lim_{x\to1} 1 = 0$$

$$\Rightarrow \lim_{x\to1} f(x) - 1 = 0 \Rightarrow \lim_{x\to1} f(x) = 1$$

Example 15: $f(x)=\begin{cases} mx+n, & x<0 \\ nx^2+m, & 0\le x\le 1 \\ mx^3+m, & x>1 \end{cases}$

For what integers m and n does $\lim_{x\to 0} f(x), \lim_{x\to 1} f(x)$ exist?

Solution:

The given function is $f(x)=\begin{cases} mx+n, & x<0 \\ nx^2+m, & 0\le x\le 1 \\ mx^3+m, & x>1 \end{cases}$

$\lim_{x\to 0^-} f(x)=\lim_{x\to 0}(mx+n)=m(0)+n=n$

$\lim_{x\to 0^+} f(x)=\lim_{x\to 0}(nx^2+m)=n(0)^2+m=m$

Thus, $\lim_{x\to 0} f(x)$ exist if $m=n$

$\lim_{x\to 1^-} f(x)=\lim_{x\to 1}(nx^2+m)=n(1)^2+m=m+n$

$\lim_{x\to 1^+} f(x)=\lim_{x\to 1}(nx^3+m)=n(1)^3+m=m+n$

$\therefore \lim_{x\to 1^-} f(x)=\lim_{x\to 1^+} f(x)=\lim_{x\to 1} f(x)=m+n$

Thus, $\lim_{x\to 1} f(x)$ exists for any integral value of m and n.

Example 16: Evaluate the given limits at $x = 5$:

(a) $f(x)=x-5$

(b) $f(x)=\frac{1}{x-5}, x\ne 5$

(c) $f(x)=\frac{x^2-25}{x+5}, x\ne -5$

(d) $f(x)=|x-5|$

Solution:

(a) The given function is $f(x)=x-5$

LHL $=\lim_{x\to 5^-} f(x)=\lim_{x\to 5^-}(x-5)=5-5=0$

RHL $=\lim_{\to 5^+} f(x)=\lim_{x\to 5^+}(x-5)=5-5=0$

At $x=5$, LHL = RHL = 0; therefore, $\lim_{x\to 5}(x-5)=0$

(b) The given function $f(x)=\frac{1}{x-5}, x\ne 5$

LHL $=\lim_{x\to 5^-} f(x)=\lim_{x\to 5^-}\left(\frac{1}{x-5}\right)=\frac{1}{5-5}=\frac{1}{0}$

RHL $=\lim_{x\to 5^+} f(x)=\lim_{x\to 5^+}\left(\frac{1}{x-5}\right)=\frac{1}{5-5}=\frac{1}{0}$

Since $\lim_{x\to 5}\left(\frac{1}{x-5}\right)=\frac{1}{0}$, the limit does not exist.

(c) The given function is $f(x)=\frac{x^2-25}{x-5}, x\ne 5$

LHL $=\lim_{x\to 5^-} f(x)=\lim_{x\to 5^-}\left(\frac{x^2-25}{x-5}\right)$

$=\lim_{x\to 5^-}\left(\frac{(x-5)(x+5)}{x-5}\right)$

$=\lim_{x\to 5^-}(x+5)=5+5=10$

RHL $=\lim_{x\to 5^+} f(x)=\lim_{x\to 5^+}\left(\frac{x^2-25}{x-5}\right)$

$=\lim_{x\to 5^+}\left(\frac{(x-5)(x+5)}{x-5}\right)$

$=\lim_{x\to 5^+}(x+5)=5+5=10$

Since at $x=5$, LHL = RHL = 10, $\lim_{x\to 5}\frac{x^2-25}{x-5}=10$ (except at $x = 5$).

(d) $f(x)=|x-5|=\begin{cases} -(x-5)=5-x, & x<5 \\ x-5, & x\ge 5 \end{cases}$

LHL $=\lim_{x\to 5^-} f(x)=\lim_{x\to 5^-}(5-x)=5-5=0$

RHL $=\lim_{x\to 5^+} f(x)=\lim_{x\to 5^+}(x-5)=5-5=0$

Since at $x = 5$, LHL = RHL = 0, $\lim_{x\to 5}|x-5|=0$

Day 170

Limits of Trigonometric Functions

Theorem: Let f and g be two real-valued functions with the same domain such that $f(x)\le g(x)$ for all x in the domain of definition. For some a, if both $\lim_{x\to a} f(x), \lim_{x\to a} g(x)$ exist, then $\lim_{x\to a} f(x)\le \lim_{x\to a} g(x)$.

Theorem (Sandwich Theorem): Let f, g, h be real functions such that $f(x)\le g(x)\le h(x)$ for all x in the common domain of definition. For some real number a, if $\lim_{x\to a} f(x)=l=\lim_{x\to a} h(x)$, then $\lim_{x\to a} g(x)=l$.

Limits of Trigonometric Functions

Some important trigonometric limits are:

(i) $\lim_{x\to 0}\frac{\sin x}{x}=1$

(ii) $\lim_{x\to 0}\frac{\tan x}{x}=1$

(The above two limits can be proved by the sandwich theorem).

(iii) $\lim_{x\to 0}\frac{1-\cos x}{x}=0$

Proof:

$$\lim_{x\to 0}\frac{1-\cos x}{x}=\lim_{x\to 0}\frac{2\sin^2(x/2)}{x}$$

$$=\lim_{x\to 0}\frac{\sin(x/2)}{(x/2)}\sin(x/2)$$

$$=1\times\sin(x/2)\left(\because \lim_{h\to 0}\frac{\sin h}{h}=1\right)$$

$$=\sin(0/2)=0$$

Some Important Series

- $\sin x = x-\frac{x^3}{3!}+\frac{x^5}{5!}-\frac{x^7}{7!}+\ldots$
- $\cos x = 1-\frac{x^2}{2!}+\frac{x^4}{4!}-\frac{x^4}{6!}+\ldots$
- $\tan x = x+\frac{1}{3}x^3+\frac{2}{15}x^5+\ldots$

Example 1: Evaluate the given limit: $\lim_{x\to 0}\frac{\sin x}{\sin 2x}, a, b\neq 0.$

Solution:

$$\lim_{x\to 0}\frac{\sin x}{\sin 2x}=\lim_{x\to 0}\frac{\left(\frac{\sin x}{x}\right)\times x}{\left(\frac{\sin 2x}{2x}\right)\times 2x}$$

$$=\left(\frac{1}{2}\right)\times\frac{\lim_{x\to 0}\left(\frac{\sin x}{x}\right)}{\lim_{2x\to 0}\left(\frac{\sin 2x}{2x}\right)}\quad [\because x\to 0\Rightarrow 2x\to 0]$$

$$=\left(\frac{1}{2}\right)\times\frac{1}{1}=\frac{1}{2}\quad\left[\because \lim_{x\to 0}\frac{\sin x}{x}=1\right]$$

Example 2: Evaluate the given limit: $\lim_{x\to \pi}\frac{\sin(\pi-x)}{(\pi-x)}$.

Solution:
Substituting $\pi - x = y$, we get $x\to\pi\Rightarrow y\to 0$; therefore

$$\lim_{x\to\pi}\frac{\sin(\pi-x)}{(\pi-x)}$$

$$=\lim_{y\to 0}\frac{\sin y}{y}=1\left[\because \lim_{y\to 0}\frac{\sin y}{y}=1\right]$$

Example 3: Evaluate the given limit: $\lim_{x\to 0}\frac{\cos 2x-1}{4(\cos x-1)}$.

Solution:

$$\lim_{x\to 0}\frac{\cos 2x-1}{4\cos x-4}=\lim_{x\to 0}\frac{1-2\sin^2 x-1}{4-8\sin^2\frac{x}{2}-4}$$

$$\left[\because \cos 2\theta = 1-2\sin^2\theta\right]$$

$$=\lim_{x\to 0}\frac{\sin^2 x}{4\sin^2 x/2}=\lim_{x\to 0}\frac{\left(\frac{\sin^2 x}{x^2}\right)}{\left(\frac{\sin^2 x/2}{(x/2)^2}\right)}$$

$$=\frac{1}{1}=1\quad\left[\because \lim_{y\to 0}\frac{\sin y}{y}=1\right]$$

Example 4: Evaluate the given limit: $\lim_{x\to 0}\frac{2x+x\cos x}{3\sin x}$.

Solution:

$$\lim_{x\to 0}\left(\frac{2x+x\cos x}{3\sin x}\right)=\frac{1}{3}\lim_{x\to 0}\left\{\frac{x(2+\cos x)}{\sin x}\right\}$$

$$=\frac{1}{3}\lim_{x\to 0}\left(\frac{x}{\sin x}\right)\times\lim_{x\to 0}(2+\cos x)$$

$$=\frac{1}{3}\times\frac{1}{\left(\lim_{x\to 0}\frac{\sin x}{x}\right)}\times\lim_{x\to 0}(2+\cos x)$$

$$=\frac{1}{3}\times\frac{1}{1}\times(2+\cos 0)=\frac{2+1}{3}=1\quad\left[\because \lim_{x\to 0}\frac{\sin x}{x}=1\right]$$

Example 5: Evaluate the given limit: $\lim_{x\to 0}\frac{\operatorname{cosec} x-\cot x}{x}$.

Solution:

$$\lim_{x\to 0}\left(\frac{\operatorname{cosec} x-\cot x}{x}\right)=\lim_{x\to 0}\frac{1}{x}\left(\frac{1}{\sin x}-\frac{\cos x}{\sin x}\right)$$

$$=\lim_{x\to 0}\frac{1}{x}\left(\frac{1-\cos x}{\sin x}\right)$$

$$=\lim_{x\to 0}\frac{1}{x}\left(\frac{2\sin^2 x/2}{2\sin x/2\cos x/2}\right)\quad\left[\begin{array}{l}\because \cos 2x = 1-2\sin^2 x\\ \sin 2x = 2\sin x\cos x\end{array}\right]$$

$$=\lim_{x\to 0}\frac{1}{x}\left(\frac{\sin x/2}{\cos x/2}\right)$$

$$=\lim_{x\to 0}\frac{1}{2}\left(\frac{\sin(x/2)}{x/2}\right)\lim_{x\to 0}\left(\frac{1}{\cos x/2}\right)=\frac{1}{2}\times 1\times 1=\frac{1}{2}$$

Example 6: $\lim_{x\to\pi/2}\frac{\tan 2x}{2x-\pi}$

Solution:

$$\lim_{x\to\pi/2}\frac{\tan 2x}{2x-\pi}=\frac{1}{2}\lim_{x\to\pi/2}\frac{\tan 2x}{x-\pi/2}$$

Substituting $x-\frac{\pi}{2}=y$, we get $x\to\frac{\pi}{2}\Rightarrow y\to 0$; therefore

$$\frac{1}{2}\lim_{x\to\pi/2}\frac{\tan 2x}{x-\frac{\pi}{2}}=\frac{1}{2}\lim_{y\to 0}\frac{\tan 2\left(y+\frac{\pi}{2}\right)}{y}$$

$$= \frac{1}{2}\lim_{y\to 0}\frac{\tan(\pi+2y)}{y}$$

$$= \frac{1}{2}\lim_{y\to 0}\frac{\tan 2y}{y} \quad [\because \tan(\pi+2y)=\tan 2y]$$

$$= \lim_{y\to 0}\frac{\tan 2y}{2y} = 1 = 1 \quad \left[\because \lim_{x\to 0}\frac{\tan x}{x} = 1\right]$$

Example 7: Evaluate the given limit: $\lim_{x\to\pi}\left(x^2 - \sin x + 5\right)$.

Solution:

$$\lim_{x\to\pi}(x^2 - \sin x + 5) = \pi^2 - \sin\pi + 5$$

$$= \pi^2 - 0 + 5 = \pi^2 + 5$$

Example 8: Evaluate the given limit:

$$\lim_{x\to 0} f(x) = \begin{cases} \dfrac{\sin x}{x}, & \text{if } x < 0 \\ x+1, & \text{if } x \ge 0 \end{cases}.$$

Solution:

$$\text{LHL} = \lim_{x\to 0^-} f(x) = \lim_{x\to 0^-}\left(\frac{\sin x}{x}\right) = 1$$

$$\text{RHL} = \lim_{x\to 0^+} f(x) = \lim_{x\to 0^+}(x+1) = 0+1 = 1$$

At $x = 0, \text{LHL} = \text{RHL}$

Therefore, $\lim_{x\to 0} f(x) = 1$.

Example 9: Evaluate the given limit:

$$f(x) = \begin{cases} x^2 \sin\dfrac{1}{x}, & \text{if } x \ne 0 \\ 0, & \text{if } x = 0 \end{cases}.$$

Solution:

$$\text{LHL} = \lim_{x\to 0^-} f(x) = \lim_{x\to 0^-}\left(x^2 \sin\frac{1}{x}\right)$$

$$= \lim_{x\to 0^-}\left((-0)^2 \sin\frac{1}{0}\right)$$

$$= 0\times(-1,1) = 0 \ (\because -1 \le \sin\theta \le 1)$$

$$\text{RHL} = \lim_{x\to 0^+} f(x) = \lim_{x\to 0^+}\left(x^2 \sin\frac{1}{x}\right)$$

$$= \lim_{x\to 0^+}\left((0)^2 \sin\frac{1}{0}\right)$$

$$= 0\times(-1,1) = 0 \ (\because -1 \le \sin\theta \le 1)$$

At $x = 0, \text{LHL} = \text{RHL} = f(0)$

Therefore, $\lim_{x\to 0} f(x) = 0$

Example 10: Evaluate the given limit

$$f(x) = \begin{cases} \sin x - \cos x, & \text{if } x \ne 0 \\ -1, & \text{if } x = 0 \end{cases}.$$

Solution:

$$\text{LHL} = \lim_{x\to 0^-} f(x) = \lim_{x\to 0^-}(\sin x - \cos x) = 0 - 1 = -1$$

$$\text{RHL} = \lim_{x\to 0^+} f(x) = \lim_{x\to 0^+}(\sin x - \cos x) = 0 - 1 = -1$$

For $x \ne 0, \text{LHL} = \text{RHL} = f(x)$

Therefore, $\lim_{x\to 0} f(x) = -1$.

Day 171

NCERT Exercise 13.1 (With Solutions Q1–Q22)

Q1. Evaluate the given limit: $\lim_{x\to 3}(x+3)$

Solution: $\lim_{x\to 3}(x+3) = 3+3 = 6$

Q2. Evaluate the given limit: $\lim_{x\to\pi}\left(x - \frac{22}{7}\right)$

Solution:

$$\lim_{x\to\pi}\left(x - \frac{22}{7}\right) = \pi - \frac{22}{7} \approx \frac{22}{7} - \frac{22}{7} = 0$$

Q3. Evaluate the given limit: $\lim_{x\to t}\pi r^2$

Solution: $\lim_{r\to 1}\pi r^2 = \pi(1)^2 = \pi$

Q4. Evaluate the given limit: $\lim_{x\to 4}\frac{4x+3}{x-2}$

Solution:

$$\lim_{x\to 4}\frac{4x+3}{x-2} = \frac{4(4)+3}{4-2} = \frac{16+3}{2} = 19/2$$

Q5. Evaluate the given limit: $\lim_{x\to -1}\frac{x^{10}+x^5+1}{x-1}$

Solution:

$$\lim_{x\to -1}\frac{x^{10}+x^5+1}{x-1} = \frac{(-1)^{10}+(-1)^5+1}{-1-1}$$

$$= \frac{1-1+1}{-2} = -\frac{1}{2}$$

Q6. Evaluate the given limit: $\lim_{x\to 0}\frac{(x+1)^5-1}{x}$

Solution:
Substituting $x+1=y$, we get $y\to 1$ as $x\to 0$. Therefore,

$$\lim_{x\to 0}\frac{(x+1)^5-1}{x}=\lim_{y\to 1}\frac{y^5-1}{y-1}=5\cdot 1^{5-1}=5$$

$$\left[\because \lim_{x\to 0}\frac{x^n-a^n}{x-a}=na^{n-1}\right]$$

Q7. Evaluate the given limit: $\lim_{x\to 2}\frac{3x^2-x-10}{x^2-4}$

Solution:

$$\lim_{x\to 2}\frac{3x^2-x-10}{x^2-4}=\lim_{x\to 2}\frac{(x-2)(3x+5)}{(x-2)(x+2)}$$

$$=\lim_{x\to 2}\frac{3x+5}{x+2}=\frac{3(2)+5}{2+2}=11/4$$

Q8. Evaluate the given limit: $\lim_{x\to 3}\frac{x^4-81}{2x^2-5x-3}$

Solution:

$$\lim_{x\to 3}\frac{x^4-81}{2x^2-5x-3}$$

$$=\lim_{x\to 3}\frac{(x-3)(x+3)(x^2+9)}{(x-3)(2x+1)}$$

$$=\lim_{x\to 3}\frac{(x+3)(x^2+9)}{2x+1}$$

$$=\frac{(3+3)(3^2+9)}{2(3)+1}=\frac{6\times 18}{7}=\frac{108}{7}$$

Q9. Evaluate the given limit: $\lim_{x\to 0}\frac{ax+b}{cx+1}$

Solution: $\lim_{x\to 0}\frac{ax+b}{cx+1}=\frac{a(0)+b}{c(0)+1}=b$

Q10. Evaluate the given limit: $\lim_{z\to 1}\frac{z^{\frac{1}{3}}-1}{z^{\frac{1}{6}}-1}$

Solution:
Substituting $z^{1/6}=x$, we get $z\to 1$ as $x\to 1$. Therefore,

$$\lim_{z\to 1}\frac{z^{\frac{1}{3}}-1}{z^{\frac{1}{6}}-1}=\lim_{x\to 1}\frac{x^2-1}{x-1}=2\cdot 1^{2-1}=2$$

$$\left[\because \lim_{x\to a}\frac{x^n-a^n}{x-a}=na^{n-1}\right]$$

Q11. Evaluate the given limit:

$$\lim_{x\to 1}\frac{ax^2+bx+c}{cx^2+bx+a}, a+b+c\neq 0$$

Solution:

$$\lim_{x\to 1}\frac{ax^2+bx+c}{cx^2+bx+a}=\frac{a(1)^2+b(1)+c}{c(1)^2+b(1)+a}=\frac{a+b+c}{a+b+c}=1$$

Q12. Evaluate the given limit: $\lim_{x\to -2}\frac{\frac{1}{x}+\frac{1}{2}}{x+2}$

Solution:

$$\lim_{x\to -2}\frac{\frac{1}{x}+\frac{1}{2}}{x+2}=\lim_{x\to -2}\frac{\left(\frac{2+x}{2x}\right)}{x+2}=\lim_{x\to -2}\frac{1}{2x}=\frac{1}{2(-2)}=-\frac{1}{4}$$

Q13. Evaluate the given limit: $\lim_{x\to 0}\frac{\sin ax}{bx}$

Solution:

$$\lim_{x\to 0}\frac{\sin ax}{bx}=\lim_{x\to 0}\frac{\sin ax}{bx}\times\frac{a}{a}=\lim_{x\to 0}\left(\frac{\sin ax}{ax}\right)\times\left(\frac{a}{b}\right)$$

$$=\frac{a}{b}\lim_{ax\to 0}\left(\frac{\sin ax}{ax}\right)\quad[\because x\to 0\Rightarrow ax\to 0]$$

$$=\frac{a}{b}\times 1=\frac{a}{b}\quad\left[\because \lim_{x\to 0}\frac{\sin x}{x}=1\right]$$

Q14. Evaluate the given limit: $\lim_{x\to 0}\frac{\sin ax}{\sin bx}, a,b\neq 0$

Solution:

$$\lim_{x\to 0}\frac{\sin ax}{\sin bx}=\lim_{x\to 0}\frac{\left(\frac{\sin ax}{ax}\right)\times ax}{\left(\frac{\sin bx}{bx}\right)\times bx}$$

$$=\left(\frac{a}{b}\right)\times\frac{\lim_{ax\to 0}\left(\frac{\sin ax}{ax}\right)}{\lim_{bx\to 0}\left(\frac{\sin bx}{bx}\right)}$$

$[\because x\to 0\Rightarrow ax\to 0,\ x\to 0\Rightarrow bx\to 0]$

$$=\left(\frac{a}{b}\right)\times\frac{1}{1}=\frac{a}{b}\quad\left[\because \lim_{x\to 0}\frac{\sin x}{x}=1\right]$$

Q15. Evaluate the given limit: $\lim_{x\to\pi}\frac{\sin(\pi-x)}{\pi(\pi-x)}$

Solution:
Substituting $\pi-x=y$, we get $x\to\pi\Rightarrow y\to 0$. Therefore,

$$\lim_{x\to\pi}\frac{\sin(\pi-x)}{\pi(\pi-x)}=\frac{1}{\pi}\lim_{y\to 0}\frac{\sin y}{y}$$

$$= \frac{1}{\pi} \times 1 = \frac{1}{\pi} \quad \left[\because \lim_{y \to 0} \frac{\sin y}{y} = 1 \right]$$

Q16. Evaluate the given limit: $\lim_{x \to 0} \frac{\cos x}{\pi - x}$

Solution: $\lim_{x \to 0} \left(\frac{\cos x}{\pi - x} \right) = \frac{\cos 0}{\pi - 0} = \frac{1}{\pi}$

Q17. Evaluate the given limit: $\lim_{x \to 0} \frac{\cos 2x - 1}{\cos x - 1}$

Solution:

$$\lim_{x \to 0} \frac{\cos 2x - 1}{\cos x - 1} = \lim_{x \to 0} \frac{1 - 2\sin^2 x - 1}{1 - 2\sin^2 \frac{x}{2} - 1}$$

$$\left[\because \cos 2\theta = 1 - 2\sin^2 \theta \right]$$

$$= \lim_{x \to 0} \frac{\sin^2 x}{\sin^2 x/2} = \lim_{x \to 0} \frac{\left(\frac{\sin^2 x}{x^2} \right)}{\left(\frac{\sin^2 x/2}{(x/2)^2} \right) \times \frac{1}{4}}$$

$$= 4 \frac{\lim_{x \to 0} \left(\frac{\sin^2 x}{x^2} \right)}{\lim_{x \to 0} \left(\frac{\sin^2 x/2}{(x/2)^2} \right)} = 4 \times \frac{1}{1} = 4 \quad \left[\because \lim_{y \to 0} \frac{\sin y}{y} = 1 \right]$$

Q18. Evaluate the given limit: $\lim_{x \to 0} \frac{ax + x\cos x}{b \sin x}$

Solution:

$$\lim_{x \to 0} \left(\frac{ax + x\cos x}{b \sin x} \right) = \frac{1}{b} \lim_{x \to 0} \left\{ \frac{x(a + \cos x)}{\sin x} \right\}$$

$$= \frac{1}{b} \lim_{x \to 0} \left(\frac{x}{\sin x} \right) \times \lim_{x \to 0} (a + \cos x)$$

$$= \frac{1}{b} \times \frac{1}{\left(\lim_{x \to 0} \frac{\sin x}{x} \right)} \times \lim_{x \to 0} (a + \cos x)$$

$$= \frac{1}{b} \times \frac{1}{1} \times (a + \cos 0) = \frac{a+1}{b} \quad \left[\because \lim_{x \to 0} \frac{\sin x}{x} = 1 \right]$$

Q19. Evaluate the given limit: $\lim_{x \to 0} x \sec x$

Solution:

$$\lim_{x \to 0} (x \sec x) = \lim_{x \to 0} \frac{x}{\cos x} = \frac{0}{\cos 0} = \frac{0}{1} = 0$$

Q20. Evaluate the given limit:

$$\lim_{x \to 0} \frac{\sin ax + bx}{ax + \sin bx}, \quad a, b, a + b \neq 0$$

Solution:

$$\lim_{x \to 0} \frac{\sin ax + bx}{ax + \sin bx} = \lim_{x \to 0} \frac{\left(\frac{\sin ax}{ax} \right) ax + bx}{ax + bx(\sin bx / bx)}$$

$$= \frac{\left(\lim_{ax \to 0} \frac{\sin ax}{ax} \right) \times \lim_{x \to 0} (ax) + \lim_{x \to 0} bx}{\lim_{x \to 0} ax + \lim_{x \to 0} bx \left(\lim_{x \to 0} \frac{\sin ax}{bx} \right)}$$

$$[\because x \to 0 \Rightarrow ax \to 0, bx \to 0]$$

$$= \frac{\lim_{x \to 0} (ax) + \lim_{x \to 0} bx}{\lim_{x \to 0} ax + \lim_{x \to 0} bx} \quad \left[\because \lim_{x \to 0} \frac{\sin x}{x} = 1 \right]$$

$$= \frac{\lim_{x \to 0} (ax + bx)}{\lim_{x \to 0} (ax + bx)}$$

$$= \lim_{x \to 0} \frac{ax + bx}{ax + bx} = \lim_{x \to 0} (1) = 1$$

Q21. Evaluate the given limit: $\lim_{x \to 0} (\operatorname{cosec} x - \cot x)$

Solution:

$$\lim_{x \to 0} (\operatorname{cosec} x - \cot x) = \lim_{x \to 0} \left(\frac{1}{\sin x} - \frac{\cos x}{\sin x} \right)$$

$$= \lim_{x \to 0} \left(\frac{1 - \cos x}{\sin x} \right)$$

$$= \lim_{x \to 0} \left(\frac{2\sin^2 x/2}{2\sin x/2 \cos x/2} \right) \quad \left[\begin{array}{l} \because \cos 2x = 1 - 2\sin^2 x \\ \sin 2x = 2\sin x \cos x \end{array} \right]$$

$$= \lim_{x \to 0} \left(\frac{\sin x/2}{\cos x/2} \right) = \frac{0}{1} = 0$$

Q22. $\lim_{x \to \pi/2} \frac{\tan 2x}{x - \frac{\pi}{2}}$

Solution:

Substituting $x - \frac{\pi}{2} = y$, we get $x \to \frac{\pi}{2} \Rightarrow y \to 0$. Therefore,

$$\lim_{x \to \pi/2} \frac{\tan 2x}{x - \frac{\pi}{2}} = \lim_{y \to 0} \frac{\tan 2\left(y + \frac{\pi}{2} \right)}{y}$$

$$= \lim_{y \to 0} \frac{\tan(\pi + 2y)}{y}$$

$$= \lim_{y \to 0} \frac{\tan 2y}{y} \quad [\because \tan(\pi + 2y) = \tan 2y]$$

$$= \lim_{y \to 0} \frac{\tan 2y}{2y} \times 2$$

$$= 1 \times 2 = 2 \quad \left[\because \lim_{x \to 0} \frac{\tan x}{x} = 1 \right]$$

Day 172

NCERT Exercise 13.1 (With Solutions Q23–Q32)

Q23. Find $\lim_{x\to 0} f(x)$, $\lim_{x\to 1} f(x)$, where

$$f(x)=\begin{cases}2x+3, & x\le 0\\ 3(x+1), & x>0\end{cases}$$

Solution:

The given function is $f(x)=\begin{cases}2x+3, & x\le 0\\ 3(x+1), & x>0\end{cases}$

$\lim_{x\to 0^-} f(x)=\lim_{x\to 0}[2x+3]=2(0)+3=3$

$\lim_{x\to 0^+} f(x)=\lim_{x\to 0} 3(x+1)=3(0+1)=3$

$\therefore \lim_{x\to 0^-} f(x)=\lim_{x\to 0^+} f(x)=\lim_{x\to 0} f(x)=3$

$\lim_{x\to 1^-} f(x)=\lim_{x\to 0} 3(x+1)=3(1+1)=6$

$\lim_{x\to 1^+} f(x)=\lim_{x\to 1} 3(x+1)=3(1+1)=6$

$\therefore \lim_{x\to 1^-} f(x)=\lim_{x\to 1^+} f(x)=\lim_{x\to 1} f(x)=6$

Q24. Find $\lim_{x\to 1} f(x)$, where $f(x)=\begin{cases}x^2-1, & x\le 1\\ -x^2+1, & x>1\end{cases}$

Solution:

The given function is $f(x)=\begin{cases}x^2-1, & x\le 1\\ -x^2+1, & x>1\end{cases}$

$\lim_{x\to 1^-} f(x)=\lim_{x\to 1}[x^2-1]=1^2-1=1-1=0$

$\lim_{x\to 1^+} f(x)=\lim_{x\to 1}[-x^2-1]=-1^2-1=-1-1=-2$

Since $\lim_{x\to 1^-} f(x)\ne \lim_{x\to 1^+} f(x)$, $\lim_{x\to 1} f(x)$ does not exist.

Q25. Evaluate $\lim_{x\to 0} f(x)$, where $f(x)=\begin{cases}\dfrac{|x|}{x}, & x\ne 0\\ 0, & x=0\end{cases}$

Solution:

The given function is $f(x)=\begin{cases}\dfrac{|x|}{x}, & x\ne 0\\ 0, & x=0\end{cases}$

$\lim_{x\to 0^-} f(x)=\lim_{x\to 0^-}\dfrac{|x|}{x}=\lim_{x\to 0}\left(\dfrac{-x}{x}\right)$

$[\because |x|=-x, \text{ when } x<0]$

$=\lim_{x\to 0}(-1)=-1$

$\lim_{x\to 0^+} f(x)=\lim_{x\to 0^+}\left[\dfrac{|x|}{x}\right]=\lim_{x\to 0^+}\left(\dfrac{x}{x}\right)$

$[\because |x|=x, \text{ when } x>0]$

$=\lim_{x\to 0}(1)=1$

Since $\lim_{x\to 0^-} f(x)\ne \lim_{x\to 0^+} f(x)$, $\lim_{x\to 0} f(x)$ does not exist.

Q26. Find $\lim_{x\to 0} f(x)$, where $f(x)=\begin{cases}\dfrac{x}{|x|}, & x\ne 0\\ 0, & x=0\end{cases}$

Solution:

The given function is $f(x)=\begin{cases}\dfrac{x}{|x|}, & x\ne 0\\ 0, & x=0\end{cases}$

$\lim_{x\to 0^-} f(x)=\lim_{x\to 0^-}\dfrac{x}{|x|}=\lim_{x\to 0}\dfrac{x}{-x}=-1$

$[\because |x|=-x, \text{ when } x<0]$

$\lim_{x\to 0^+} f(x)=\lim_{x\to 0^+}\dfrac{x}{|x|}=\lim_{x\to 0}\dfrac{x}{x}=1$

$[\because |x|=x, \text{ when } x>0]$

Since $\lim_{x\to 0^-} f(x)\ne \lim_{x\to 0^+} f(x)$, $\lim_{x\to 0} f(x)$ does not exist.

Q27. Find $\lim_{x\to 5} f(x)$, where $f(x)=|x|-5$

Solution:

The given function is $f(x)=|x|-5$

$\lim_{x\to 5^-} f(x)=\lim_{x\to 5^-}[|x|-5]=\lim_{x\to 5}(x-5)=5-5=0$

$[\because |x|=x, \text{ when } x>0]$

$\lim_{x\to 5^+} f(x)=\lim_{x\to 5^+}(|x|-5)=\lim_{x\to 5}(x-5)=5-5=0$

$\because \lim_{x\to 5^-} f(x)=\lim_{x\to 5^+} f(x)=0$

Therefore, $\lim_{x\to 5} f(x)=0$

Q28. Suppose $f(x)=\begin{cases}a+bx, & \text{if } x<1\\ 4, & \text{if } x=1\\ b-ax, & \text{if } x>1\end{cases}$ and if $\lim_{x\to 1} f(x)=f(1)$, what are possible values of a and b?

Solution:

The given function is $f(x)=\begin{cases}a+bx, & \text{if } x<1\\ 4, & \text{if } x=1\\ b-ax, & \text{if } x>1\end{cases}$

$\lim_{x\to1^-} f(x) = \lim_{x\to1}(a+bx) = a+b$

$\lim_{x\to1^+} f(x) = \lim_{x\to1}(b-ax) = b-a$

and $f(1) = 4$

$\lim_{x\to1} f(x) = f(1) = 4$

$\therefore \lim_{x\to1^-} f(x) = \lim_{x\to1^+} f(x) = \lim_{x\to1} f(x) = f(1)$

$\Rightarrow a+b=4,\quad b-a=4$

Solving the above two equations, we get $a = 0,\ b = 4$.

Q29. Let $a_1, a_2, ..., a_n$ be fixed real numbers and let us define a function $f(x) = (x-a_1)(x-a_2)...(x-a_n)$. What is $\lim_{x\to a_1} f(x)$? For some $a \neq a_1, a_2 ..., a_n$, compute $\lim_{x\to a} f(x)$.

Solution:

The given function $f(x) = (x-a_1)(x-a_2)...(x-a_n)$

$\lim_{x\to a_1} f(x) = [(x-a_1)(x-a_2)...(x-a_n)]$

$= [\lim_{x\to a_1}(x-a_1)][\lim_{x\to a_1}(x-a_2)]...[\lim_{x\to a_1}(x-a_n)]$

$= (a_1-a_1)(a_1-a_2)...(a_1-a_n) = 0$

$\therefore \lim_{x\to a_1} f(x) = 0$

Now, $\lim_{x\to a} f(x) = [(x-a_1)(x-a_2)...(x-a_n)]$

$= [\lim_{x\to a}(x-a_1)][\lim_{x\to a}(x-a_2)]...[\lim_{x\to a}(x-a_n)]$

$= (a-a_1)(a-a_2)...(a-a_n)$

$\therefore \lim_{x\to a} f(x) = (a-a_1)(a-a_2)...(a-a_n)$

Q30. If $f(x) = \begin{cases} |x|+1, & x<0 \\ 0, & x=0 \\ |x|-1, & x>0 \end{cases}$

for what value of a does $\lim_{x\to a} f(x)$ exist?

Solution:

The given function is $f(x) = \begin{cases} |x|+1, & x<0 \\ 0, & x=0 \\ |x|-1, & x>0 \end{cases}$

Case 1: When $a = 0$,

$\lim_{x\to0^-} f(x) = \lim_{x\to0^-}(|x|+1) = \lim_{x\to0}(-x+1) = -0+1 = 1$

$[\because |x| = -x, \text{ when } x<0]$

$\lim_{x\to0^+} f(x) = \lim_{x\to0^+}(|x|-1) = \lim_{x\to0}(x-1) = 0-1 = -1$

Since $\lim_{x\to0^-} f(x) \neq \lim_{x\to0^+} f(x)$

$\lim_{x\to0} f(x)$ does not exist for $a = 0$

Case 2: When $a < 0$

$\lim_{x\to a^-} f(x) = \lim_{x\to a^-}(|x|+1) = \lim_{x\to a}(-x+1) = -a+1$

$\lim_{x\to a^+} f(x) = \lim_{x\to a^+}(|x|+1) = \lim_{x\to a}(-x+1) = -a+1$

$[\because |x| = -x, \text{ when } x<0]$

$\because \lim_{x\to a^-} f(x) = \lim_{x\to a^+} f(x) = -a+1$

Therefore, limit of $f(x)$ exists at $x = a$, where $a < 0$.

Case 3: When $a > 0$

$\lim_{x\to a^-} f(x) = \lim_{x\to a^-}(|x|+1) = \lim_{x\to a}(x-1) = a-1$

$\lim_{x\to a^+} f(x) = \lim_{x\to a^+}(|x|+1) = \lim_{x\to a}(x-1) = a-1$

$[\because |x| = x, \text{ when } x>0]$

$\because \lim_{x\to a^-} f(x) = \lim_{x\to a^+} f(x) = a-1$

Therefore, limit of $f(x)$ exists at $x = a$, where $a > 0$.

Thus, limit of $f(x)$ exists for all $a \neq 0$.

Q31. If the function $f(x)$ satisfies $\lim_{x\to1} \frac{f(x)-2}{x^2-1} = \pi$, evaluate $\lim_{x\to1} f(x)$.

Solution: $\lim_{x\to1} \frac{f(x)-2}{x^2-1} = \pi$

$\Rightarrow \frac{\lim_{x\to1}(f(x)-2)}{\lim_{x\to1}(x^2-1)} = \pi$

$\Rightarrow \lim_{x\to1}(f(x)-2) = \pi \lim_{x\to1}(x^2-1)$

$\Rightarrow \lim_{x\to1}(f(x)-2) = \pi(1^2-1)$

$\Rightarrow \lim_{x\to1}(f(x)-2) = 0$

$\Rightarrow \lim_{x\to1} f(x) - \lim_{x\to1} 2 = 0$

$\Rightarrow \lim_{x\to1} f(x) - 2 = 0 \Rightarrow \lim_{x\to1} f(x) = 2$

Q32. If $f(x) = \begin{cases} mx^2+n, & x<0 \\ nx+m, & 0\le x\le 1 \\ mx^3+m, & x>1 \end{cases}$

for what integers m and n does $\lim_{x\to0} f(x), \lim_{x\to1} f(x)$ exist?

Solution:

The given function is $f(x) = \begin{cases} mx^2+n, & x<0 \\ nx+m, & 0\le x\le 1 \\ mx^3+m, & x>1 \end{cases}$

$\lim_{x\to0^-} f(x) = \lim_{x\to0}(mx^2+n) = m(0)^2+n = n$

$\lim_{x\to0^+} f(x) = \lim_{x\to0}(nx+m) = n(0)+m = m.$

Thus, $\lim_{x\to0} f(x)$ exists if $m = n$.

$\lim_{x\to1^-} f(x) = \lim_{x\to1}(nx+m) = n(1)+m = m+n$

$\lim_{x\to1^+} f(x) = \lim_{x\to1}(nx^3+m) = n(1)^3+m = m+n$

$\therefore \lim_{x\to1^-} f(x) = \lim_{x\to1^+} f(x) = \lim_{x\to1} f(x) = m+n$

Thus, $\lim_{x\to1} f(x)$ exists for any integral value of m and n.

Day 173

Derivative: First Principle of Derivative

Let there exist a real-valued function $f(x)$. Then, the derivative of function $f(x)$ is given by $f'(x) = \lim_{h\to0}\frac{f(x+h)-f(x)}{h}$ (provided this limit exists). This is known as the first principle of derivative.

The derivative $f'(x_0) = \lim_{h\to0}\frac{f(x_0+h)-f(x_0)}{h}$ denotes the change in $f(x)$ at x_0 with respect to x.

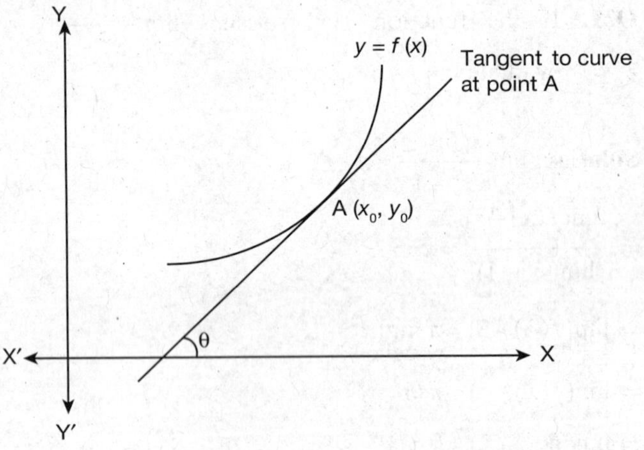

Fig. 13.1 Derivative of a function

Geometrically (Fig. 13.1), the derivative of function $y = f(x)$ at $x = x_0$, that is, $f'(x_0) = \lim_{h\to0}\frac{f(x_0+h)-f(x_0)}{h}$ represents the slope $\left.\frac{dy}{dx}\right|_{x_0} = f'(x_0) = \tan\theta$ of the tangent to the curve $y = f(x)$ at $x = x_0$.

Example 1: Find the derivative of x^2+x at $x = 1$.

Solution: Let $f(x) = x^2+x$.

Then, according to the first principle

$$f'(x) = \lim_{h\to0}\frac{f(x+h)-f(x)}{h}$$

$$= \lim_{h\to0}\frac{[(x+h)^2+(x+h)]-(x^2+x)}{h}$$

$$= \lim_{h\to0}\frac{x^2+2\cdot x\cdot h+h^2+x+h-x^2-x}{h}$$

$$= \lim_{h\to0}\frac{2\cdot x\cdot h+h^2+h}{h} = \lim_{h\to0}(2x+h+1) = 2x+1$$

$$\therefore f'(1) = 2\cdot1+1 = 3$$

Example 2: Find the derivative of x^3+1 at $x = 1$.

Solution: Let $f(x) = x^3+1$.

Then, according to the first principle

$$f'(x) = \lim_{h\to0}\frac{f(x+h)-f(x)}{h}$$

$$= \lim_{h\to0}\frac{[(x+h)^3+1]-(x^3+1)}{h}$$

$$= \lim_{h\to0}\frac{x^3+h^3+3x^2h+3xh^2+1-x^3-1}{h}$$

$$= \lim_{h\to0}\frac{h^3+3x^2h+3xh^2}{h}$$

$$= \lim_{h\to0}(h^2+3x^2+3xh) = 0+3x^2+0 = 3x^2$$

$$\therefore f'(1) = 3\cdot1^2 = 3$$

Example 3: Find the derivative of $(x-a)(x+b)$ at $x = 1$.

Solution: Let $f(x) = (x-a)(x+b)$.

Then, according to the first principle

$$f'(x) = \lim_{h\to0}\frac{f(x+h)-f(x)}{h}$$

$$= \lim_{h\to0}\frac{(x+h-a)(x+h+b)-(x-a)(x+b)}{h}$$

$$= \lim_{h\to0}\frac{(x^2+hx+bx+hx+h^2+bh-ax-ah-ab) - (x^2-ax+bx-ab)}{h}$$

$$= \lim_{h\to0}\frac{(hx+hx+h^2+bh-ah)}{h}$$

$$= \lim_{h\to 0}\frac{2hx+h^2-ah+bh}{h}$$
$$= \lim_{h\to 0}(2x+h-a+b) = 2x-a+b$$
$$\therefore f'(1) = 2\cdot 1-a+b = 2-a+b$$

Example 4: Find the derivative of $\frac{1}{x^2}+\frac{1}{x}$ at $x = 1$.

Solution: Let $f(x) = \frac{1}{x^2}+\frac{1}{x}$.

Then, according to the first principle

$$f'(x) = \lim_{h\to 0}\frac{f(x+h)-f(x)}{h}$$
$$= \lim_{h\to 0}\frac{\left\{\frac{1}{(x+h)^2}+\frac{1}{(x+h)}\right\}-\left\{\frac{1}{x^2}+\frac{1}{x}\right\}}{h}$$
$$= \lim_{h\to 0}\frac{1}{h}\left[\frac{x^2+x^2(x+h)-(x+h)^2-x(x+h)^2}{x^2(x+h)^2}\right]$$
$$= \lim_{h\to 0}\frac{1}{h}\left[\frac{x^2+x^3+x^2h-x^2-h^2-2hx-x^3-xh^2-2hx^2}{x^2(x+h)^2}\right]$$
$$= \lim_{h\to 0}\frac{1}{h}\left[\frac{-h^2-2hx-xh^2-hx^2}{x^2(x+h)^2}\right]$$
$$= \lim_{h\to 0}\left[\frac{-h-2x-xh-x^2}{x^2(x+h)^2}\right]$$
$$= \left[\frac{-2x-x^2}{x^4}\right] = -\frac{2}{x^3}-\frac{1}{x^2}$$
$$\therefore f'(1) = -\frac{2}{1^3}-\frac{1}{1^2} = -3$$

Example 5: Find the derivative of $\frac{x+a}{x-a}$ at $x = 1$.

Solution: According to the first principle

$$f'(x) = \lim_{h\to 0}\frac{f(x+h)-f(x)}{h}$$
$$= \lim_{h\to 0}\frac{\left(\frac{x+h+a}{x+h-a}-\frac{x+a}{x-a}\right)}{h}$$
$$= \lim_{h\to 0}\frac{1}{h}\left[\frac{(x-a)(x+h+a)-(x+a)(x+h-a)}{(x-a)(x+h-a)}\right]$$
$$= \lim_{h\to 0}\frac{1}{h}\left[\frac{(x^2+hx+ax-ax-ah-a^2)-(x^2+hx-ax+ax+ah-a^2}{(x-a)(x+h-a)}\right]$$
$$= \lim_{h\to 0}\frac{1}{h}\left[\frac{-2ah}{(x-a)(x+h-a)}\right]$$
$$= \lim_{h\to 0}\left[\frac{-2a}{(x-a)(x+h-a)}\right]$$
$$= \frac{-2a}{(x-a)(x-a)} = \frac{-2a}{(x-a)^2}$$
$$\therefore f'(1) = \frac{-2a}{(1-a)^2}$$

Example 6: Find the derivative of $\sin x$ from the first principle.

Solution: Let $f(x) = \sin x$.

Then, according to the first principle

$$f'(x) = \lim_{h\to 0}\frac{f(x+h)-f(x)}{h}$$
$$= \lim_{h\to 0}\frac{\sin(x+h)-\sin x}{h}$$
$$= \lim_{h\to 0}\left[\frac{\sin x\cosh+\cos x\sinh-\sin x}{h}\right]$$
$$= \sin x\left(\lim_{h\to 0}\frac{\cosh-1}{h}\right)+\cos x\lim_{h\to 0}\left(\frac{\sinh}{h}\right)$$
$$= \sin x(0)+\cos x(1) = \cos x$$
$$\left[\because \lim_{h\to 0}\frac{1-\cosh}{h} = 0, \lim_{h\to 0}\frac{\sinh}{h} = 1\right]$$
$$\therefore f'(x) = \cos x$$

Example 7: Find the derivative of $\cot x$ from the first principle.

Solution: Let $f(x) = \cot x$.

Then, according to the first principle

$$f'(x) = \lim_{h\to 0}\frac{f(x+h)-f(x)}{h}$$
$$= \lim_{h\to 0}\frac{\cot(x+h)-\cot x}{h}$$
$$= \lim_{h\to 0}\frac{1}{h}[\cot(x+h)-\cot x]$$
$$= \lim_{h\to 0}\frac{1}{h}\left[\frac{\cos(x+h)}{\sin(x+h)}-\frac{\cos x}{\sin x}\right]$$
$$= \lim_{h\to 0}\frac{1}{h}\left[\frac{\cos(x+h)\sin x-\cos x\sin(x+h)}{\sin x\sin(x+h)}\right]$$
$$= \lim_{h\to 0}\frac{1}{h}\left[\frac{\sin(x-x-h)}{\sin x\sin(x+h)}\right]$$
$$= \lim_{h\to 0}\frac{1}{h}\left[\frac{\sin(-h)}{\sin x\sin(x+h)}\right]$$
$$= -\left(\lim_{h\to 0}\frac{\sinh}{h}\right)\cdot\left(\lim_{h\to 0}\frac{1}{\sin x\cdot\sin(x+h)}\right)$$

$$= -1 \cdot \frac{1}{\sin x \cdot \sin(x+0)} = \frac{-1}{\sin^2 x} = -\operatorname{cosec}^2 x$$

$\therefore f'(x) = -\operatorname{cosec}^2 x$

Example 8: Find the derivative of $\tan x$ from the first principle.

Solution: Let $f(x) = \tan x$.

Then, according to the first principle

$$f'(x) = \lim_{h\to 0} \frac{f(x+h) - f(x)}{h}$$

$$= \lim_{h\to 0} \frac{1}{h}[\tan(x+h) - \tan x]$$

$$= \lim_{h\to 0} \frac{1}{h}\left[\frac{\sin(x+h)}{\cos(x+h)} - \frac{\sin x}{\cos x}\right]$$

$$= \lim_{h\to 0} \frac{1}{h}\left[\frac{\sin(x+h)\cos x - \sin x \cos(x+h)}{\cos x \cos(x+h)}\right]$$

$$= \lim_{h\to 0} \frac{1}{h}\left[\frac{\sin(x+h-x)}{\cos x \cos(x+h)}\right]$$

$$= \lim_{h\to 0}\left[\left(\frac{\sin h}{h}\right)\frac{1}{\cos x \cos(x+h)}\right]$$

$$= \left(\lim_{h\to 0} \frac{\sin h}{h}\right)\left(\lim_{h\to 0} \frac{1}{\cos x \cos(x+h)}\right)$$

$$= \frac{1}{\cos x \cos x} = \sec^2 x$$

Day 174

Algebra of Derivatives of Functions

Let $f(x)$ and $g(x)$ be two functions such that their derivatives are defined in a common domain. Then

- Derivative of the sum of two functions is equal to the sum of the derivatives of the functions $\frac{d}{dx}[f(x)+g(x)] = \frac{d}{dx}f(x) + \frac{d}{dx}g(x)$.
- Derivative of the difference of two functions is equal to the difference of the derivatives of the function $\frac{d}{dx}[f(x)-g(x)] = \frac{d}{dx}f(x) - \frac{d}{dx}g(x)$.
- Derivative of the product of two functions is given by the product rule $\frac{d}{dx}[f(x)\cdot g(x)] = f(x)\cdot\frac{d}{dx}g(x) + g(x)\cdot\frac{d}{dx}f(x)$
- Derivative of the quotient of two functions is given by the quotient rule $(g(x) \neq 0)$

$$\frac{d}{dx}\left(\frac{f(x)}{g(x)}\right) = \frac{g(x)\cdot\frac{d}{dx}f(x) - f(x)\frac{d}{dx}g(x)}{(g(x))^2}$$

Derivatives of Some Standard Functions:

(i) Derivative of the function $f(x) = x$

By the first principle of derivative

$$f'(x) = \lim_{h\to 0}\frac{f(x+h)-f(x)}{h} = \lim_{h\to 0}\frac{x+h-x}{h}$$

$$\Rightarrow \frac{d(x)}{dx} = f'(x) = \lim_{h\to 0} 1 = 1$$

(ii) Derivative of the function $f(x) = nx$

By the first principle of derivative

$$f'(x) = \lim_{h\to 0}\frac{f(x+h)-f(x)}{h} = \lim_{h\to 0}\frac{n(x+h)-nx}{h}$$

$$\Rightarrow \frac{d(nx)}{dx} = f'(x) = \lim_{h\to 0} n = n$$

(iii) Derivative of the function $f(x) = x^n$, where n is positive integer

$$f'(x) = \lim_{h\to 0}\frac{f(x+h)-f(x)}{h} = \lim_{h\to 0}\frac{(x+h)^n - x^n}{h}$$

Using Binomial Theorem

$$(x+h)^n = ({}^nC_0)x^n + ({}^nC_1)x^{n-1}h + \dots + ({}^nC_n)h^n$$

$$\therefore (x+h)^n - x^n = \left(x^n + nx^{n-1}h + \dots + h^n\right) - x^n$$

$$= h(nx^{n-1} + \dots + h^{n-1})$$

Thus, $\frac{d(x^n)}{dx} = \lim_{h\to 0}\frac{h(nx^{n-1} + \dots + h^{n-1})}{h}$

$$\Rightarrow \frac{d(x^n)}{dx} = \lim_{h\to 0}(nx^{n-1} + \dots + h^{n-1}) = nx^{n-1}.$$

(iv) Derivative of the polynomial function $f(x) = a_n x^n + a_{n-1}x^{n-1} + \dots + a_1 x + a_0$

Using the result $\frac{d(x^n)}{dx} = nx^{n-1}$, the derivative of the function is given by

$$\frac{df(x)}{dx} = na_n x^{n-1} + (n-1)a_{n-1}x^{x-2} + \dots + 2a_2 x + a_1.$$

Some Standard Results

- $\frac{d}{dx}x^n = nx^{n-1}$
- $\frac{d}{dx}\frac{1}{x^n} = \frac{-n}{x^{n+1}}$
- $\frac{d}{dx}(ax+b)^n = na(ax+b)^{n-1}$

- $\frac{d}{dx}\sin x = \cos x$
- $\frac{d}{dx}\cos x = -\sin x$
- $\frac{d}{dx}\tan x = \sec^2 x$
- $\frac{d}{dx}\cot x = -\text{cosec}^2 x$
- $\frac{d}{dx}\sec x = \sec x \tan x$
- $\frac{d}{dx}\text{cosec}\, x = -\text{cosec}\, x \cot x$

Day 175

NCERT Exercise 13.2 (With Solutions Q1–Q9)

Q1. Find the derivative of $x^2 - 2$ at $x = 10$.

Solution: Let $f(x) = x^2 - x$.

Then, according to the first principle

$$f'(10) = \lim_{h\to 0}\frac{f(10+h)-f(10)}{h}$$

$$= \lim_{h\to 0}\frac{[(10+h)^2-2]-(10^2-2)}{h}$$

$$= \lim_{h\to 0}\frac{10^2+2\cdot 10\cdot h+h^2-2-10^2+2}{h}$$

$$= \lim_{h\to 0}\frac{20h+h^2}{h} = \lim_{h\to 0}(20+h)$$

$$= (20+0) = 20$$

Thus, the derivative of $x^2 - 2$ at $x = 10$ is 20.

Q2. Find the derivative of 99 x at $x = 100$.

Solution: Let $f(x) = 99x$.

Then, according to the first principle

$$f'(100) = \lim_{h\to 0}\frac{f(100+h)-f(100)}{h}$$

$$= \lim_{h\to 0}\frac{99(100+h)-99(100)}{h}$$

$$= \lim_{h\to 0}\frac{99\times 100+99h-99\times 100}{h}$$

$$= \lim_{h\to 0}\frac{99h}{h} = \lim_{h\to 0}(99) = 99$$

Thus, the derivative of $99x$ at $x = 100$ is 99.

Q3. Find the derivative of x at $x = 1$.

Solution: Let $f(x) = x$.

Then, according to the first principle

$$f'(1) = \lim_{h\to 0}\frac{f(1+h)-f(1)}{h}$$

$$= \lim_{h\to 0}\frac{(1+h)-1}{h} = \lim_{h\to 0}\frac{h}{h} = \lim_{h\to 0}(1) = 1$$

Thus, the derivative of x at $x = 1$ is 1.

Q4. Find the derivative of the following functions from the first principle:

(i) $x^3 - 27$ (ii) $(x-1)(x-2)$

(iii) $1/x^2$ (iv) $\frac{x+1}{x-1}$

Solution: (i) Let $f(x) = x^3 - 27$.

Then, according to the first principle

$$f'(x) = \lim_{h\to 0}\frac{f(x+h)-f(x)}{h}$$

$$= \lim_{h\to 0}\frac{[(x+h)^3-27]-(x^3-27)}{h}$$

$$= \lim_{h\to 0}\frac{x^3+h^3+3x^2h+3xh^2-x^3}{h}$$

$$= \lim_{h\to 0}\frac{h^3+3x^2h+3xh^2}{h}$$

$$= \lim_{h\to 0}(h^2+3x^2+3xh) = 0+3x^2+0 = 3x^2$$

(ii) Let $f(x) = (x-1)(x-2)$.

Then, according to the first principle

$$f'(x) = \lim_{h\to 0}\frac{f(x+h)-f(x)}{h}$$

$$= \lim_{h\to 0}\frac{(x+h-1)(x+h-2)-(x-1)(x-2)}{h}$$

$$= \lim_{h\to 0}\frac{(x^2+hx-2x+hx+h^2-2h-x-h+2) -(x^2-2x-x+2)}{h}$$

$$= \lim_{h\to 0}\frac{(hx+hx+h^2-2h-h)}{h}$$

$$= \lim_{h\to 0}\frac{2hx+h^2-3h}{h}$$

$$= \lim_{h\to 0}(2x+h-3)$$

$$= (2x+0-3) = 2x-3$$

(iii) Let $f(x) = \frac{1}{x^2}$.

Then, according to the first principle

$$f'(x)=\lim_{h\to 0}\frac{f(x+h)-f(x)}{h}$$

$$=\lim_{h\to 0}\frac{\frac{1}{(x+h)^2}-\frac{1}{x^2}}{h}$$

$$=\lim_{h\to 0}\frac{1}{h}\left[\frac{x^2-(x+h)^2}{x^2(x+h)^2}\right]$$

$$=\lim_{h\to 0}\frac{1}{h}\left[\frac{x^2-x^2-h^2-2hx}{x^2(x+h)^2}\right]$$

$$=\lim_{h\to 0}\left[\frac{-h^2-2hx}{x^2(x+h)^2}\right]=\lim_{h\to 0}\left[\frac{-h-2x}{x^2(x+h)^2}\right]$$

$$=\frac{0-2x}{x^2(x+0)^2}=\frac{-2}{x^3}$$

(iv) Let $f(x)=\frac{x+1}{x-1}$.

Then, according to the first principle

$$f'(x)=\lim_{h\to 0}\frac{f(x+h)-f(x)}{h}$$

$$=\lim_{h\to 0}\frac{\left(\frac{x+h+1}{x+h-1}-\frac{x+1}{x-1}\right)}{h}$$

$$=\lim_{h\to 0}\frac{1}{h}\left[\frac{(x-1)(x+h+1)-(x+1)(x+h-1)}{(x-1)(x+h-1)}\right]$$

$$=\lim_{h\to 0}\frac{1}{h}\left[\frac{(x^2+hx+x-x-h-1)\\-(x^2+hx-x+x+h-1}{(x-1)(x+h-1)}\right]$$

$$=\lim_{h\to 0}\frac{1}{h}\left[\frac{-2h}{(x-1)(x+h-1)}\right]$$

$$=\lim_{h\to 0}\left[\frac{-2}{(x-1)(x+h-1)}\right]$$

$$=\frac{-2}{(x-1)(x-1)}=\frac{-2}{(x-1)^2}$$

Q5. For the function $f(x)=\frac{x^{100}}{100}+\frac{x^{99}}{99}+\ldots+\frac{x^2}{2}+x+1$, prove that $f'(1)=100f'(0)$.

Solution: The given function is

$$f(x)=\frac{x^{100}}{100}+\frac{x^{99}}{99}+\ldots+\frac{x^2}{2}+x+1$$

$$\frac{d}{dx}f(x)=\frac{d}{dx}\left[\frac{x^{100}}{100}+\frac{x^{99}}{99}+\ldots+\frac{x^2}{2}+x+1\right]$$

$$\frac{d}{dx}f(x)=\frac{d}{dx}\left(\frac{x^{100}}{100}\right)+\frac{d}{dx}\left(\frac{x^{99}}{99}\right)+\ldots$$

$$+\frac{d}{dx}\left(\frac{x^2}{2}\right)+\frac{d}{dx}(x)+\frac{d}{dx}(1)$$

Using theorem $\frac{d}{dx}(x^n)=nx^{n-1}$, we get

$$\frac{d}{dx}f(x)=\frac{100x^{99}}{100}+\frac{99x^{98}}{99}+\ldots+\frac{2x}{2}+1+0$$

$$=x^{99}+x^{98}+\ldots+x+1$$

$$\therefore f'(x)=x^{99}+x^{98}+\ldots+x+1$$

At $x=0, f'(0)=1$

At $x=1,\ f'(1)=1^{99}+1^{98}+\ldots+1^1+1^0$

$=[1+1+\ldots+1+1]_{100\text{ terms}}=1\times 100=100$

Thus, $f'(1)=100\times f^{-1}(0)$

Q6. Find the derivative of $x^n+ax^{n-1}+a^2x^{n-2}+\ldots+a^{n-1}x+a^n$ for some fixed real number a.

Solution:

Let $f(x)=x^n+ax^{n-1}+a^2x^{n-2}+\ldots+a^{n-1}x+a^n$

$$\therefore f'(x)=\frac{d}{dx}(x^n+ax^{n-1}+a^2x^{n-2}+\ldots+a^{n-1}x+a^n)$$

$$=\frac{d}{dx}(x^n)+a\frac{d}{dx}(x^{n-1})+a^2\frac{d}{dx}(x^{n-2})+\ldots$$

$$+a^{n-1}\frac{d}{dx}(x)+a^n\frac{d}{dx}(1)$$

Using theorem $\frac{d}{dx}x^n=nx^{n-1}$, we get

$$f'(x)=nx^{n-1}+a(n-1)x^{n-2}+a^2(n-2)x^{n-3}+$$
$$\ldots+a^{n-1}+a^n(0)$$

$$=nx^{n-1}+a(n-1)x^{n-2}+a^2(n-2)x^{n-3}+\ldots+a^{n-1}$$

Q7. For some constants a and b, find the derivative of:

(i) $(x-a)(x-b)$ (ii) $(ax^2+b)^2$ (iii) $\frac{x-a}{x-b}$

Solution: (i) Let $f(x)=(x-a)(x-b)$

$$\Rightarrow f(x)=x^2-(a+b)x+ab$$

$$\Rightarrow f'(x)=\frac{d}{dx}(x^2-(a+b)x+ab)$$

$$=\frac{d}{dx}(x^2)-(a+b)\frac{d}{dx}(x)+\frac{d}{dx}(ab)$$

Using theorem $\frac{d}{dx}(x^n)=nx^{n-1}$, we get

$$f'(x)=2x-(a+b)+0=2x-a-b$$

(ii) Let $f(x)=(ax^2+b)^2=a^2x^4+2abx^2+b^2$

$$\Rightarrow f'(x)=\frac{d}{dx}(a^2x^4+2abx^2+b^2)$$

$$=a^2\frac{d}{dx}(x^4)+2ab\frac{d}{dx}(x^2)+\frac{d}{dx}(b^2)$$

Using theorem $\frac{d}{dx}x^n=nx^{n-1}$, we get

$$f'(x)=a^2(4x^3)+2ab(2x)+b^2(0)$$

$$=4a^2x^3+4abx=4ax(ax^2+b).$$

(iii) Let $f(x)=\frac{(x-a)}{(x-b)}\Rightarrow f'(x)=\frac{d}{dx}\left(\frac{x-a}{x-b}\right).$

Using the quotient rule, we get

$$f'(x)=\frac{(x-b)\frac{d}{dx}(x-a)-(x-a)\frac{d}{dx}(x-b)}{(x-b)^2}$$

$$=\frac{(x-b)(1)-(x-a)(1)}{(x-b)^2}$$

$$=\frac{x-b-x+a}{(x-b)^2}=\frac{a-b}{(x-b)^2}$$

Q8. Find the derivative of $\frac{x^n-a^n}{x-a}$ for some constant a.

Solution:

Let $f(x)=\frac{x^n-a^n}{x-a}\Rightarrow f'(x)=\frac{d}{dx}\left(\frac{x^n-a^n}{x-a}\right).$

Using the quotient rule, we get

$$f'(x)=\frac{(x-a)\frac{d}{dx}(x^n-a^n)-(x^n-a^n)\frac{d}{dx}(x-a)}{(x-a)^2}$$

$$=\frac{(x-a)(nx^{n-1}-0)-(x^n-a^n)}{(x-a)^2}$$

$$=\frac{nx^n-anx^{n-1}-x^n+a^n}{(x-a)^2}$$

Q9. Find the derivative of:

(i) $2x-\frac{3}{4}$ (ii) $(5x^3+3x-1)(x-1)$

(iii) $x^{-3}(5+3x)$ (iv) $x^5(3-6x^{-9})$

(v) $x^{-4}(3-4x^{-5})$ (vi) $\frac{2}{x+1}-\frac{x^2}{3x-1}$

Solution:

(i) Let $f(x)=2x-\frac{3}{4}\Rightarrow f'(x)=\frac{d}{dx}\left(2x-\frac{3}{4}\right)$

$$=2\frac{d}{dx}(x)-\frac{d}{dx}\left(\frac{3}{4}\right)=2-0=2$$

(ii) Let $f(x)=(5x^3+3x-1)(x-1).$

Using the Leibniz product rule

$$f'(x)=(5x^3+3x-1)\frac{d}{dx}(x-1)$$

$$+(x-1)\frac{d}{dx}(5x^3+3x-1)$$

$$=(5x^3+3x-1)(1)+(x-1)(5\cdot 3x^2+3-0)$$

$$=(5x^3+3x-1)+(x-1)(15x^2+3)$$

$$=5x^3+3x-1+15x^3+3x-15x^2-3$$

$$=20x^3-15x^2+6x-4$$

(iii) Let $f(x)=x^{-3}(5+3x).$

Using the Leibniz product rule

$$f'(x)=x^{-3}\frac{d}{dx}(5+3x)+(5+3x)\frac{d}{dx}(x^{-3})$$

$$=x^{-3}(0+3)+(5+3x)(-3x^{-3-1})$$

$$=x^{-3}(3)+(5+3x)(-3x^{-4})$$

$$=3x^{-3}-15x^{-4}-9x^{-3}$$

$$=-6x^{-1}-15x^{-4}=-3x^{-3}\left(2+\frac{5}{x}\right)$$

$$=\frac{-3x^{-3}}{x}(2x+5)=\frac{-3}{x^4}(5+2x)$$

(iv) Let $f(x)=x^5(3-6x^{-9}).$

Using the Leibniz product rule

$$f'(x)=x^5\frac{d}{dx}(3-6x^{-9})+(3-6x^{-9})\frac{d}{dx}(x^5)$$

$$=x^5\{0-6(-9)x^{-9-1}\}+(3-6x^{-9})(5x^4)$$

$$=x^5(54x^{-10})+15x^4-30x^{-5}$$

$$=54x^{-5}+15x^4-30x^{-5}=24x^{-5}+15x^4$$

$$=15x^4+24/x^5$$

(v) Let $f(x)=x^{-4}(3-4x^{-5}).$

Using the Leibniz product rule

$$f'(x)=x^{-4}\frac{d}{dx}(3-4x^{-5})+(3-4x^{-5})\frac{d}{dx}(x^{-4})$$

$$=x^{-4}\{0-4(-5)x^{-5-1}\}+(3-4x^{-5})(-4)x^{-4-1}$$

$$=x^{-4}(20x^{-6})+(3-4x^{-5})(-4x^{-5})$$

$$=20x^{-10}-12x^{-5}+16x^{-10}$$

$$=36x^{-10}-12x^{-5}=-\frac{12}{x^5}+\frac{36}{x^{10}}$$

(vi) Let $f(x)=\frac{2}{x+1}-\frac{x^2}{3x-1}$

$$f'(x)=\frac{d}{dx}\left(\frac{2}{x+1}\right)-\frac{d}{dx}\left(\frac{x^2}{3x-1}\right)$$

Using the quotient rule

$$f'(x)=\left[\frac{(x+1)\frac{d}{dx}(2)-2\frac{d}{dx}(x+1)}{(x+1)^2}\right]$$

$$-\left[\frac{(3x-1)\frac{d}{dx}(x^2)-x^2\frac{d}{dx}(3x-1)}{(3x-1)^2}\right]$$

$$=\left[\frac{(x+1)(0)-2(1)}{(x+1)^2}\right]-\left[\frac{(3x-1)(2x)-(x^2)(3)}{(3x-1)^2}\right]$$

$$=\frac{-2}{(x+1)^2}-\left[\frac{6x^2-2x-3x^2}{(3x-1)^2}\right]$$

$$=\frac{-2}{(x+1)^2}-\left[\frac{3x^2-2x^2}{(3x-1)}\right]=\frac{-2}{(x+1)^2}-\frac{x(3x-2)}{(3x-1)}$$

Example 1: For some constants a, b and c, find the derivative of:

(i) $x(x-a)(x-b)$ (ii) $(ax^2+bx+c)^2$

(iii) $\frac{x^2-a}{x-b}$ (iv) $\frac{x^{-2}(x-a)}{x-b}$

(v) $\frac{a}{x-b}+\frac{x}{x-a}$

Solution: (i) Let $f(x)=x(x-a)(x-b)$

$\Rightarrow f(x)=x^3-(a+b)x^2+abx$

$\Rightarrow f'(x)=\frac{d}{dx}(x^3-(a+b)x^2+abx)$

$=\frac{d}{dx}(x^3)-(a+b)\frac{d}{dx}(x^2)+ab\frac{d}{dx}(x)$

Using theorem $\frac{d}{dx}(x^n)=nx^{n-1}$, we get

$f'(x)=3x^2-2(a+b)x+ab$

(ii) Let $f(x)=(ax^2+bx+c)^2$

$=a^2x^4+b^2x^2+c^2+2abx^3+2acx^2+2bcx$

$=a^2x^4+2abx^3+(b^2+2ac)x^2+2bcx+c^2$

$$\Rightarrow f'(x)=\frac{d}{dx}\begin{pmatrix}a^2x^4+2abx^3+(b^2+2ac)x^2\\+2bcx+c^2\end{pmatrix}$$

$=\frac{d}{dx}(a^2x^4)+\frac{d}{dx}(2abx^3)+\frac{d}{dx}\{(b^2+2ac)x^2\}$

$+\frac{d}{dx}(2bcx)+\frac{d}{dx}(c^2)$

Using theorem $\frac{d}{dx}x^n=nx^{n-1}$ and $\frac{d}{dx}(\text{constant})=0$,

we get

$f'(x)=a^2\frac{d}{dx}(x^4)+2ab\frac{d}{dx}(x^3)+(b^2+2ac)\frac{d}{dx}(x^2)$

$+2bc\frac{d}{dx}(x)+\frac{d}{dx}(c^2)$

$=a^2 4x^3+2ab\cdot 3x^2+(b^2+2ac)2x+2bc\cdot 1+0$

$=4a^2x^3+6abx^2+2(b^2+2ac)x+2bc$

(iii) Let $f(x)=\frac{x^2-a}{x-b}\Rightarrow f'(x)=\frac{d}{dx}\left(\frac{x^2-a}{x-b}\right)$.

Using the quotient rule, we get

$$f'(x)=\frac{(x-b)\frac{d}{dx}(x^2-a)-(x^2-a)\frac{d}{dx}(x-b)}{(x-b)^2}$$

$$=\frac{(x-b)(2x)-(x^2-a)(1)}{(x-b)^2}$$

$$=\frac{2x^2-2bx-x^2+a}{(x-b)^2}=\frac{x^2-2bx+a}{(x-b)^2}$$

(iv) Let $f(x)=\frac{x^{-2}(x-a)}{x-b}$

$\Rightarrow f(x)=\frac{x^{-1}-ax^{-2}}{x-b}$

$\Rightarrow f'(x)=\frac{d}{dx}\left(\frac{x^{-1}-ax^{-2}}{x-b}\right)$

Using the quotient rule, we get

$$f'(x)=\frac{(x-b)\frac{d}{dx}(x^{-1}-ax^{-2})-(x^{-1}-ax^{-2})\frac{d}{dx}(x-b)}{(x-b)^2}$$

Using theorem $\frac{d}{dx}x^n=nx^{n-1}$ and $\frac{d}{dx}(\text{constant})=0$,

we get

$$f'(x)=\frac{(x-b)(-1x^{-2}+2ax^{-3})-(x^{-1}-ax^{-2})(1-0)}{(x-b)^2}$$

$$=\frac{(x-b)(-x^{-2}+2ax^{-3})-(x^{-1}-ax^{-2})}{(x-b)^2}$$

$$=\frac{-x^{-1}+2ax^{-2}+bx^{-2}-2abx^{-3}-x^{-1}+ax^{-2}}{(x-b)^2}$$

$$=\frac{-2x^{-1}+3ax^{-2}+bx^{-2}-2abx^{-3}}{(x-b)^2}$$

(v) Let $f(x)=\frac{a}{x-b}+\frac{x}{x-a}$

$f'(x)=\frac{d}{dx}\left(\frac{a}{x-b}\right)+\frac{d}{dx}\left(\frac{x}{x-a}\right)$

Using the quotient rule, we get

$$f'(x)=\left[\frac{(x-b)\frac{d}{dx}(a)-a\frac{d}{dx}(x-b)}{(x-b)^2}\right]+\left[\frac{(x-a)\frac{d}{dx}(x)-x\frac{d}{dx}(x-a)}{(x-a)^2}\right]$$

$$=\left[\frac{(x-b)(0)-a(1-0)}{(x-b)^2}\right]+\left[\frac{(x-a)(1)-x(1-0)}{(x-a)^2}\right]$$

$$=\left[\frac{-a}{(x-b)^2}\right]+\left[\frac{-a}{(x-a)^2}\right]$$

$$=-a\left[\frac{1}{(x-b)^2}+\frac{1}{(x-a)^2}\right]$$

Example 2: Find the derivative of the following functions:

(i) $\sec x+\cos x$ (ii) $\sin^2 x$ (iii) $x^2\tan x$

(iv) $\frac{1+x}{\cot x}$ (iv) $\frac{\sin x+\cos x}{x}$

Solution: (i) Let $f(x)=\sec x+\cos x$

$$f'(x)=\frac{d}{dx}(\sec x+\cos x)$$

$$=\frac{d}{dx}(\sec x)+\frac{d}{dx}(\cos x)$$

$$=\sec x\tan x-\sin x$$

(ii) Let $f(x)=\sin^2 x$

$$f'(x)=\frac{d}{dx}(\sin x\sin x)$$

Using the product rule, we get

$$f'(x)=\sin x\frac{d}{dx}(\sin x)+\sin x\frac{d}{dx}(\sin x)$$

$$=\sin x\cos x+\sin x\cos x=2\sin x\cos x=\sin 2x$$

(iii) Let $f(x)=x^2\tan x$

$$f'(x)=\frac{d}{dx}(x^2\tan x)$$

Using the product rule, we get

$$f'(x)=x^2\frac{d}{dx}(\tan x)+\tan x\frac{d}{dx}(x^2)$$

$$=x^2\sec^2 x+\tan x\cdot 2x=x^2\sec^2 x+2x\tan x$$

(iv) Let $f(x)=\frac{1+x}{\cot x}$

$$f'(x)=\frac{d}{dx}\left(\frac{1+x}{\cot x}\right)$$

Using the quotient rule, we get

$$f'(x)=\frac{\cot x\frac{d}{dx}(1+x)-(1+x)\frac{d}{dx}(\cot x)}{\cot^2 x}$$

$$=\frac{\cot x(0+1)-(1+x)(-\operatorname{cosec}^2 x)}{\cot^2 x}$$

$$=\frac{\cot x+(1+x)\operatorname{cosec}^2 x}{\cot^2 x}$$

(v) Let $f(x)=\frac{\sin x+\cos x}{x}$

$$f'(x)=\frac{d}{dx}\left(\frac{\sin x+\cos x}{x}\right)$$

Using the quotient rule, we get

$$f'(x)=\frac{x\frac{d}{dx}(\sin x+\cos x)-(\sin x+\cos x)\frac{d}{dx}(x)}{x^2}$$

$$=\frac{x(\cos x-\sin x)-(\sin x+\cos x)(1)}{x^2}$$

$$=\frac{x(\cos x-\sin x)-(\sin x+\cos x)}{x^2}$$

Day 176

NCERT Exercise 13.2 (With Solutions Q10–Q11)

Q10. Find the derivative of $\cos x$ from the first principle.

Solution: Let $f(x)=\cos x$.

Then, according to the first principle

$$f'(x)=\lim_{h\to 0}\frac{f(x+h)-f(x)}{h}$$

$$=\lim_{h\to 0}\frac{\cos(x+h)-\cos x}{h}$$

$$=\lim_{h\to 0}\left[\frac{\cos x\cos h-\sin x\sin h-\cos x}{h}\right]$$

$$=\lim_{h\to 0}\left[\frac{-\cos x(1-\cos h)-\sin x\sin h}{h}\right]$$

$$=\lim_{h\to 0}\left[\frac{-\cos x(1-\cos h)}{h}-\frac{\sin x\sin h}{h}\right]$$

$$=-\cos x\left(\lim_{h\to 0}\frac{1-\cos h}{h}\right)-\sin x\lim_{h\to 0}\left(\frac{\sin h}{h}\right)$$

$$=-\cos x(0)-\sin x(1)=-\sin x$$

$$\left[\because \lim_{h\to 0}\frac{1-\cos h}{h}=0,\ \lim_{h\to 0}\frac{\sin h}{h}=1\right]$$

$$\therefore f'(x)=-\sin x$$

Q11. Find the derivative of the following functions:

(i) $\sin x\cos x$ (ii) $\sec x$

(iii) $5\sec x+4\cos x$ (iv) $\operatorname{cosec} x$

(v) $3\cot x+5\operatorname{cosec} x$

(vi) $5\sin x-6\cos x+7$ (vii) $2\tan x-7\sec x$

Solution: (i) Let $f(x)=\sin x\cos x$.

Then, according to the first principle

$$f'(x)=\lim_{h\to 0}\frac{f(x+h)-f(x)}{h}$$

$$=\lim_{h\to 0}\frac{\sin(x+h)\cos(x+h)-\sin x\cos x}{h}$$

$$=\lim_{h\to 0}\frac{1}{2h}[2\sin(x+h)\cos(x+h)-2\sin x\cos x]$$

$$=\lim_{h\to 0}\frac{1}{2h}[\sin 2(x+h)-\sin 2x]$$

$$=\lim_{h\to 0}\frac{1}{2h}\left[2\cos\frac{2x+2h+2x}{2}\cdot\sin\frac{2x+2h-2x}{2}\right]$$

$$=\lim_{h\to 0}\frac{1}{h}\left[\cos\frac{4x+2h}{2}\sin\frac{2h}{2}\right]$$

$$=\lim_{h\to 0}\frac{1}{h}[\cos(2x+h)\sinh]$$

$$=\lim_{h\to 0}\cos(2x+h)\cdot\lim_{h\to 0}\frac{\sinh}{h}=\cos(2x+0)=\cos 2x$$

(ii) Let $f(x)=\sec x$.

Then, according to the first principle

$$f'(x)=\lim_{h\to 0}\frac{f(x+h)-f(x)}{h}$$

$$=\lim_{h\to 0}\frac{\sec(x+h)-\sec x}{h}$$

$$=\lim_{h\to 0}\frac{1}{h}\left[\frac{1}{\cos(x+h)}-\frac{1}{\cos x}\right]$$

$$=\lim_{h\to 0}\frac{1}{h}\left[\frac{\cos x-\cos(x+h)}{\cos x\cos(x+h)}\right]$$

$$=\frac{1}{\cos x}\cdot\lim_{h\to 0}\frac{1}{h}\left[\frac{-2\sin\left(\frac{x+x+h}{2}\right)\sin\left(\frac{x-x-h}{2}\right)}{\cos(x+h)}\right]$$

$$=\frac{1}{\cos x}\lim_{h\to 0}\frac{1}{h}\left[\frac{-2\sin\left(\frac{2x+h}{2}\right)\sin\left(-\frac{h}{2}\right)}{\cos(x+h)}\right]$$

$$=\frac{1}{\cos x}\lim_{h\to 0}\frac{\left[\sin\left(\frac{2x+h}{2}\right)\frac{\sin(h/2)}{(h/2)}\right]}{\cos(x+h)}$$

$$=\frac{1}{\cos x}\lim_{h\to 0}\frac{\sin(h/2)}{(h/2)}\cdot\lim_{h\to 0}\frac{\sin\left(\frac{2x+h}{2}\right)}{\cos(x+h)}$$

$$=\frac{1}{\cos x}\cdot 1\cdot\frac{\sin x}{\cos x}=\sec x\tan x$$

(iii) Let $f(x)=5\sec x+4\cos x$.

Then, according to the first principle

$$f'(x)=\lim_{h\to 0}\frac{f(x+h)-f(x)}{h}$$

$$=\lim_{h\to 0}\frac{5\sec(x+h)+4\cos(x+h)-[5\sec x+4\cos x]}{h}$$

$$=\lim_{h\to 0}\frac{[\sec(x+h)-\sec x]}{h}+4\lim_{h\to 0}\frac{[\cos(x+h)-\cos x]}{h}$$

$$=5\lim_{h\to 0}\frac{1}{h}\left[\frac{1}{\cos(x+h)}-\frac{1}{\cos x}\right]$$

$$+4\lim_{h\to 0}\frac{1}{h}[\cos(x+h)-\cos x]$$

$$=5\lim_{h\to 0}\frac{1}{h}\left[\frac{\cos x-\cos(x+h)}{\cos x\cos(x+h)}\right]$$

$$+4\lim_{h\to 0}\frac{1}{h}[\cos x\cosh-\sin x-\cos x]$$

$$=\frac{5}{\cos x}\lim_{h\to 0}\frac{1}{h}\left[\frac{-2\sin\left(\frac{x+x+h}{2}\right)\sin\left(\frac{x-x-h}{2}\right)}{\cos(x+h)}\right]$$

$$+4\lim_{h\to 0}\frac{1}{h}[-\cos x(1-\cosh)-\sin x\sinh]$$

$$=\frac{5}{\cos x}\lim_{h\to 0}\frac{1}{h}\left[\frac{-2\sin\left(\frac{2x+h}{2}\right)}{\cos(x+h)}\right]$$

$$+4[-\cos x\lim_{h\to 0}\frac{(1-\cosh)}{h}-\sin x\lim_{h\to 0}\frac{\sinh}{h}]$$

$$=\frac{5}{\cos x}\lim_{h\to 0}\left[\frac{\sin\left(\frac{2x+h}{2}\right)\cdot\frac{\sin(h/2)}{h/2}}{\cos(x+h)}\right]$$

$$+4[(-\cos x)\cdot(0)-(\sin x)\cdot 1]$$

$$=\frac{5}{\cos x}\lim\cdot\left[\lim_{h\to 0}\frac{\sin\left(\frac{2x+h}{2}\right)}{\cos(x+h)}\cdot\lim_{h\to 0}\frac{\sin(h/2)}{h/2}\right]-4\sin x$$

$$=\frac{5}{\cos x}\cdot\frac{\sin x}{\cos x}\cdot 1-4\sin x=5\sec x\tan x-4\sin x$$

(iv) Let $f(x)=\operatorname{cosec} x$.

Then, according to the first principle

$$f'(x)=\lim_{h\to 0}\frac{f(x+h)-f(x)}{h}$$

$$f'(x)=\lim_{h\to 0}\frac{1}{h}[\operatorname{cosec}(x+h)-\operatorname{cosec} x]$$

$$= \lim_{h\to 0}\frac{1}{h}\left[\frac{1}{\sin(x+h)}-\frac{1}{\sin x}\right]$$

$$= \lim_{h\to 0}\frac{1}{h}\left[\frac{\sin x-\sin(x+h)}{\sin(x+h)\sin x}\right]$$

$$= \lim_{h\to 0}\frac{1}{h}\left[\frac{2\cos\left(\frac{x+x+h}{2}\right)\cdot\sin\left(\frac{x-x-h}{2}\right)}{\sin(x+h)\sin x}\right]$$

$$= \lim_{h\to 0}\frac{1}{h}\left[\frac{2\cos\left(\frac{2x+h}{2}\right)\sin\left(-\frac{h}{2}\right)}{\sin(x+h)\sin x}\right]$$

$$= \lim_{h\to 0}\frac{-\cos\left(\frac{2x+h}{2}\right)\cdot\frac{\sin(h/2)}{(h/2)}}{\sin(x+h)\sin x}$$

$$= \lim_{h\to 0}\left(\frac{-\cos\left(\frac{2x+h}{2}\right)}{\sin(x+h)\sin x}\right)\cdot\lim_{h\to 0}\frac{\sin(h/2)}{(h/2)}$$

$$= \left(\frac{-\cos x}{\sin x\sin x}\right)\cdot 1 = -\operatorname{cosec} x\cot x$$

(v) Let $f(x) = 3\cot x + 5\operatorname{cosec} x$.

Then, according to the first principle

$$f'(x) = \lim_{h\to 0}\frac{f(x+h)-f(x)}{h}$$

$$= \lim_{h\to 0}\frac{3\cot(x+h)+5\operatorname{cosec}(x+h)-3\cot x-5\operatorname{cosec} x}{h}$$

$$= 3\lim_{h\to 0}\frac{1}{h}[\cot(x+h)-\cot x]$$

$$+5\lim_{h\to 0}\frac{1}{h}[\operatorname{cosec}(x+h)-\operatorname{cosec} x] \quad ...(1)$$

Now, $\lim_{h\to 0}\frac{1}{h}[\cot(x+h)-\cot x]$

$$= \lim_{h\to 0}\frac{1}{h}\left[\frac{\cos(x+h)}{\sin(x+h)}-\frac{\cos x}{\sin x}\right]$$

$$= \lim_{h\to 0}\frac{1}{h}\left[\frac{\cos(x+h)\sin x-\cos x\sin(x+h)}{\sin x\sin(x+h)}\right]$$

$$= \lim_{h\to 0}\frac{1}{h}\left[\frac{\sin(x-x-h)}{\sin x\sin(x+h)}\right]$$

$$= \lim_{h\to 0}\frac{1}{h}\left[\frac{\sin(-h)}{\sin x\sin(x+h)}\right]$$

$$= -\left(\lim_{h\to 0}\frac{\sin h}{h}\right)\cdot\left(\lim_{h\to 0}\frac{1}{\sin x\cdot\sin(x+h)}\right)$$

$$= -1\cdot\frac{1}{\sin x\cdot\sin(x+0)} = \frac{-1}{\sin^2 x} = -\operatorname{cosec}^2 x \quad ...(2)$$

Now, $\lim_{h\to 0}\frac{1}{h}[\operatorname{cosec}(x+h)-\operatorname{cosec} x]$

$$= \lim_{h\to 0}\frac{1}{h}\left[\frac{1}{\sin(x+h)}-\frac{1}{\sin x}\right]$$

$$= \lim_{h\to 0}\frac{1}{h}\left[\frac{\sin x-\sin(x+h)}{\sin(x+h)\sin x}\right]$$

$$= \lim_{h\to 0}\frac{1}{h}\left[\frac{2\cos\left(\frac{x+x+h}{2}\right)\cdot\sin\left(\frac{x-x-h}{2}\right)}{\sin(x+h)\sin x}\right]$$

$$= \lim_{h\to 0}\frac{1}{h}\left[\frac{2\cos\left(\frac{2x+h}{2}\right)\sin\left(-\frac{h}{2}\right)}{\sin(x+h)\sin x}\right]$$

$$= \lim_{h\to 0}\frac{-\cos\left(\frac{2x+h}{2}\right)\cdot\frac{\sin(h/2)}{(h/2)}}{\sin(x+h)\sin x}$$

$$= \lim_{h\to 0}\left(\frac{-\cos\left(\frac{2x+h}{2}\right)}{\sin(x+h)\sin x}\right)\lim_{h/2\to 0}\frac{\sin(h/2)}{(h/2)}$$

$$= \left(\frac{-\cos x}{\sin x\sin x}\right)\cdot 1 = -\operatorname{cosec} x\cot x \quad ...(3)$$

From equations (1), (2) and (3), we get

$f'(x) = \operatorname{cosec}^2 x - 5\operatorname{cosec} x\cot x$

(vi) Let $f(x) = 5\sin x - 6\cos x + 7$.

Accordingly, from the first principle

$$f'(x) = \lim_{h\to 0}\frac{f(x+h)-f(x)}{h}$$

$$= \lim_{h\to 0}\frac{1}{h}\left[\begin{matrix}5\sin(x+h)-6\cos(x+h)+7\\ -5\sin x+6\cos x-7\end{matrix}\right]$$

$$= \lim_{h\to 0}\frac{1}{h}[5\{\sin(x+h)-\sin x\}-6\{\cos(x+h)-\cos x\}$$

$$= 5\lim_{h\to 0}\frac{1}{h}[\sin(x+h)-\sin x]-6\lim_{h\to 0}\frac{1}{h}[\cos(x+h)-\cos x]$$

$$= 5\lim_{h\to 0}\frac{1}{h}\left[2\cos\left(\frac{x+x+h}{2}\right)\sin\left(\frac{x+h-x}{2}\right)\right]$$

$$-6\lim_{h\to 0}\frac{\cos x\cos h-\sin x\sin h-\cos x}{h}$$

$$= 5\lim_{h\to 0}\frac{1}{h}\left[2\cos\left(\frac{2x+h}{2}\right)\sin\frac{h}{2}\right]$$

$$-6\lim_{h\to 0}\left[\frac{-\cos x(1-\cos h)}{h}-\frac{\sin x\sin h}{h}\right]$$

$$= 5\lim_{h\to 0}\left(\cos\left(\frac{2x+h}{2}\right)\frac{\sin\frac{h}{2}}{h/2}\right)$$

$$-6\lim_{h\to 0}\left[\frac{-\cos x(1-\cos h)}{h}-\frac{\sin x\sin h}{h}\right]$$

$$= 5\left[\lim_{h\to 0}\cos\left(\frac{2x+h}{2}\right)\right]\left[\lim_{h/2\to 0}\frac{\sin\frac{h}{2}}{h/2}\right]$$

$$-6\left[(-\cos x)\left(\lim_{h\to 0}\frac{1-\cos h}{h}\right)-\sin x\lim_{h\to 0}\left(\frac{\sin h}{h}\right)\right]$$

$$= 5\cos x\cdot 1-6[(-\cos x)\cdot(0)-\sin x]$$

$$= 5\cos x+6\sin x$$

(vii) Let $f(x)=2\tan x-7\sec x$.

Then, according to the first principle

$$f'(x)=\lim_{h\to 0}\frac{f(x+h)-f(x)}{h}$$

$$=\lim_{h\to 0}\frac{1}{h}[2\tan(x+h)-7\sec(x+h)-2\tan x+7\sec x]$$

$$=\lim_{h\to 0}\frac{1}{h}[2\{\tan(x+h)-\tan x\}-7\{\sec(x+h)-\sec x\}]$$

$$=2\lim_{h\to 0}\frac{1}{h}[\tan(x+h)-\tan x]$$

$$-7\lim_{h\to 0}\frac{1}{h}[\sec(x+h)-\sec x]$$

$$=2\lim_{h\to 0}\frac{1}{h}\left[\frac{\sin(x+h)}{\cos(x+h)}-\frac{\sin x}{\cos x}\right]$$

$$-7\lim_{h\to 0}\frac{1}{h}\left[\frac{1}{\cos(x+h)}-\frac{1}{\cos x}\right]$$

$$=2\lim_{h\to 0}\frac{1}{h}\left[\frac{\sin(x+h)\cos x-\sin x\cos(x+h)}{\cos x\cos(x+h)}\right]$$

$$-7\lim_{h\to 0}\frac{1}{h}\left[\frac{\cos x-\cos(x+h)}{\cos x\cos(x+h)}\right]$$

$$=2\lim_{h\to 0}\frac{1}{h}\left[\frac{\sin(x+h-x)}{\cos x\cos(x+h)}\right]$$

$$-7\lim_{h\to 0}\frac{1}{h}\left[\frac{-2\sin\left(\frac{x+x+h}{2}\right)\sin\left(\frac{x-x-h}{2}\right)}{\cos x\cos(x+h)}\right]$$

$$=2\lim_{h\to 0}\left[\left(\frac{\sin h}{h}\right)\frac{1}{\cos x\cos(x+h)}\right]$$

$$-7\lim_{h\to 0}\frac{1}{h}\left[\frac{-2\sin\left(\frac{2x+h}{2}\right)\sin\left(-\frac{h}{2}\right)}{\cos x\cos(x+h)}\right]$$

$$=2\left(\lim_{h\to 0}\frac{\sin h}{h}\right)\left(\lim_{h\to 0}\frac{1}{\cos x\cos(x+h)}\right)$$

$$-7\left(\lim_{h/2\to 0}\frac{\sin\frac{h}{2}}{h/2}\right)\left(\lim_{h\to 0}\frac{\sin\left(\frac{2x+h}{2}\right)}{\cos x\cos(x+h)}\right)$$

$$=2\cdot 1\frac{1}{\cos x\cos x}-7\cdot 1\left(\frac{\sin x}{\cos x\cos x}\right)$$

$$=2\sec^2 x-7\sec x\tan x$$

Day 177

Limits of Exponential and Logarithmic Functions

Limit of an Exponential Function

$$\lim_{x\to 0}\frac{e^x-1}{x}=1$$

Proof: $\lim_{x\to 0}\frac{e^x-1}{x}$

$$=\lim_{x\to 0}\frac{1+x+\frac{x^2}{2!}+\frac{x^3}{3!}+\ldots-1}{x}=\lim_{x\to 0}\frac{x+\frac{x^2}{2!}+\frac{x^3}{3!}+\ldots}{x}$$

$$=\lim_{x\to 0}\left(1+\frac{x}{2!}+\frac{x^2}{3!}+\ldots\right)=1$$

Limit of a Logarithmic Function

$$\lim_{x\to 0}\frac{\log_e(1+|x|)}{|x|}=1$$

Proof: $\lim_{x\to 0}\frac{\log_e(1+|x|)}{|x|}$

$$=\lim_{x\to 0}\frac{|x|-\frac{|x|^2}{2}+\frac{|x|^3}{3}-\frac{|x|^4}{4}+\ldots}{|x|}$$

$$=\lim_{x\to 0}\left(1-\frac{|x|}{2}+\frac{|x|^2}{3}-\frac{|x|^3}{4}+\ldots\right)=1$$

Some Important Series

- $e^x=1+x+\frac{x^2}{2!}+\frac{x^3}{3!}+\ldots$

- $e^{-x} = 1 - x + \frac{x^2}{2!} - \frac{x^3}{3!} + \ldots$
- $\log(1+x) = x - \frac{x^2}{2} + \frac{x^3}{3} - \frac{x^4}{4} + \ldots, (|x| < 1)$
- $\log(1-x) = -\left(x + \frac{x^2}{2} + \frac{x^3}{3} + \frac{x^4}{4} + \ldots\right); (|x| < 1)$

Example 1: $\lim\limits_{x \to 0} \frac{e^{nx} - 1}{x}$

Solution:

$l = \lim\limits_{x \to 0} \frac{e^{nx} - 1}{x} = \lim\limits_{x \to 0} \frac{e^{nx} - 1}{x} \times n$

Substituting $nx = y \Rightarrow x \to 0 \Rightarrow nx \to 0 \Rightarrow y \to 0$

$\therefore l = \lim\limits_{x \to 0} \frac{e^y - 1}{y} \times n = 1 \times n = n \left(\because \lim\limits_{y \to 0} \frac{e^y - 1}{y} = 1\right)$

Example 2: $\lim\limits_{x \to 0} \frac{e^{n+x} - e^n}{x}$

Solution:

$l = \lim\limits_{x \to 0} \frac{e^{n+x} - e^n}{x} = \lim\limits_{x \to 0} \frac{e^n (e^x - 1)}{x} = e^n \times 1 = e^n$

$\left(\because \lim\limits_{x \to 0} \frac{e^x - 1}{x} = 1\right)$

Example 3: $\lim\limits_{x \to n} \frac{e^x - e^n}{x - n}$

Solution:

$l = \lim\limits_{x \to n} \frac{e^x - e^n}{x - n}$

Put $x = n + h$; then as $x \to n \Rightarrow h \to 0$. Therefore

$l = \lim\limits_{h \to 0} \frac{e^{n+h} - e^n}{h} = \lim\limits_{h \to 0} \frac{e^n (e^h - 1)}{h}$

$= e^n \times 1 = e^n \quad \left(\because \lim\limits_{x \to 0} \frac{e^x - 1}{x} = 1\right)$

Example 4: $\lim\limits_{x \to 0} \frac{e^{\tan x} - 1}{x}$

Solution:

$l = \lim\limits_{x \to 0} \frac{e^{\tan x} - 1}{x} = \lim\limits_{x \to 0} \frac{e^{\tan x} - 1}{x} \times \frac{\tan x}{\tan x}$

$= \lim\limits_{x \to 0} \frac{e^{\tan x} - 1}{\tan x} \times \frac{\tan x}{x} = \lim\limits_{y \to 0} \frac{e^y - 1}{y} \times \lim\limits_{x \to 0} \frac{\tan x}{x}$

Substituting $\sin x = y \Rightarrow x \to 0 \Rightarrow \sin x \to 0 \Rightarrow y \to 0$

$\therefore l = 1 \times 1 = 1 \left(\because \lim\limits_{y \to 0} \frac{e^y - 1}{y} = 1, \lim\limits_{x \to 0} \frac{\tan x}{x} = 1\right)$

Example 5: $\lim\limits_{x \to n} \frac{e^x - e^n}{x - n}$

Solution:

$l = \lim\limits_{x \to n} \frac{e^x - e^n}{x - n}$

Substituting $x = n + h$; then as $x \to n \Rightarrow h \to 0$.

$\therefore l = \lim\limits_{x \to n} \frac{e^{n+h} - e^n}{h} = \lim\limits_{x \to n} \frac{e^n (e^h - 1)}{h} = e^n \times 1 = e^n$

$\left(\because \lim\limits_{h \to 0} \frac{e^h - 1}{h} = 1\right)$

Example 6: $\lim\limits_{x \to 0} \frac{\log_e (1 + nx)}{x}$

Solution:

$l = \lim\limits_{x \to 0} \frac{\log_e (1 + nx)}{x}$

$= \lim\limits_{x \to 0} \frac{\log_e (1 + nx)}{nx} \times n$

Substituting $nx = y \Rightarrow x \to 0 \Rightarrow nx \to 0 \Rightarrow y \to 0$

$\therefore l = \lim\limits_{y \to 0} \frac{\log_e (1 + y)}{y} \times n = 1 \times n = n$

$\left(\because \lim\limits_{y \to 0} \frac{\log_e (1 + y)}{y} = 1\right)$

Example 7: $\lim\limits_{x \to 0} \frac{\log(1 + x^3)}{\tan^3 x}$

Solution:

$l = \lim\limits_{x \to 0} \frac{\log(1 + x^3)}{\tan^3 x}$

$= \lim\limits_{x \to 0} \frac{\log(1 + x^3)}{\tan^3 x} \times \frac{x^3}{x^3}$

$= \lim\limits_{x \to 0} \frac{\log(1 + x^3)}{x^3} \times \frac{x^3}{\tan^3 x}$

$= \lim\limits_{y \to 0} \frac{\log(1 + y)}{y} \times \lim\limits_{x \to 0} \frac{1}{\left(\frac{\tan x}{x}\right)^3}$

Substituting $x^3 = y \Rightarrow x \to 0 \Rightarrow x^3 \to 0 \Rightarrow y \to 0$

$= 1 \times \frac{1}{1^3} = 1 \left(\because \lim\limits_{y \to 0} \frac{\log(1 + y)}{y} = 1, \lim\limits_{x \to 0} \frac{\sin x}{x} = 1\right)$

Example 8: $\lim\limits_{x \to 0} \frac{\log_e (1 + \sin x)}{\sin x}$

Solution: $l = \lim\limits_{x \to 0} \frac{\log_e (1 + \sin x)}{\sin x}$

Substituting $\sin x = y \Rightarrow x \to 0 \Rightarrow \sin x \to 0 \Rightarrow y \to 0$

$\therefore l = \lim_{y\to 0}\frac{\log_e(1+y)}{y} = 1 \left(\because \lim_{y\to 0}\frac{\log_e(1+y)}{y} = 1\right)$

Example 9: $\lim_{x\to 0}\frac{\log_e(1+x^2)}{x\sin x}$

Solution: $l = \lim_{x\to 0}\frac{\log_e(1+x^2)}{x\sin x}$

$= \lim_{x\to 0}\frac{\log_e(1+x^2)}{x^2 \sin x/x} = \lim_{x\to 0}\frac{\log_e(1+x^2)}{x^2}\lim_{x\to 0}\frac{1}{\sin x/x}$

Substituting $x^2 = y \Rightarrow x\to 0 \Rightarrow x^2 \to 0 \Rightarrow y \to 0$

$\therefore l = 1\times 1 = 1\left(\because \lim_{y\to 0}\frac{\log_e(1+y)}{y} = 1,\ \lim_{x\to 0}\frac{1}{\sin x/x}\right)$

NCERT Exercise 13.3 (With Solutions)

Q1. $\lim_{x\to 0}\frac{e^{4x}-1}{x}$

Solution: $l = \lim_{x\to 0}\frac{e^{4x}-1}{x} = \lim_{x\to 0}\frac{e^{4x}-1}{4x}\times 4$

Substituting $4x = y \Rightarrow x\to 0 \Rightarrow 4x\to 0 \Rightarrow y\to 0$

$\therefore l = \lim_{x\to 0}\frac{e^y-1}{y}\times 4 = 1\times 4 = 4\left(\because \lim_{y\to 0}\frac{e^y-1}{y} = 1\right)$

Q2. $\lim_{x\to 0}\frac{e^{2+x}-e^2}{x}$

Solution:

$l = \lim_{x\to 0}\frac{e^{2+x}-e^2}{x} = \lim_{x\to 0}\frac{e^2(e^x-1)}{x} = e^2\times 1 = e^2$

$\left(\because \lim_{x\to 0}\frac{e^x-1}{x} = 1\right)$

Q3. $\lim_{x\to 5}\frac{e^x-e^5}{x-5}$

Solution: $l = \lim_{x\to 5}\frac{e^x-e^5}{x-5}$

Put $x = 5+h$; then as $x\to 5 \Rightarrow h\to 0$. Therefore

$\lim_{x\to 5}\frac{e^x-e^5}{x-5} = \lim_{h\to 0}\frac{e^{5+h}-e^5}{h} = \lim_{h\to 0}\frac{e^5(e^h-1)}{h}$

$= e^5\times 1 = e^5 \quad \left(\because \lim_{x\to 0}\frac{e^x-1}{x} = 1\right)$

Q4. $\lim_{x\to 0}\frac{e^{\sin x}-1}{x}$

Solution:

$l = \lim_{x\to 0}\frac{e^{\sin x}-1}{x} = \lim_{x\to 0}\frac{e^{\sin x}-1}{x}\times\frac{\sin x}{\sin x}$

$= \lim_{x\to 0}\frac{e^{\sin x}-1}{\sin x}\times\frac{\sin x}{x} = \lim_{y\to 0}\frac{e^y-1}{y}\times\lim_{x\to 0}\frac{\sin x}{x}$

Substituting $\sin x = y \Rightarrow x\to 0 \Rightarrow \sin x\to 0 \Rightarrow y\to 0$

$\therefore l = 1\times 1 = 1\left(\because \lim_{y\to 0}\frac{e^y-1}{y} = 1,\ \lim_{x\to 0}\frac{\sin x}{x} = 1\right)$

Q5. $\lim_{x\to 3}\frac{e^x-e^3}{x-3}$

Solution: $l = \lim_{x\to 3}\frac{e^x-e^3}{x-3}$

Substituting $x = 3+h$; then as $x\to 3 \Rightarrow h\to 0$.

$\therefore l = \lim_{x\to 3}\frac{e^{3+h}-e^3}{h} = \lim_{x\to 3}\frac{e^3(e^h-1)}{h} = e^3\times 1 = e^3$

$\left(\because \lim_{h\to 0}\frac{e^h-1}{h} = 1\right)$

Q6. $\lim_{x\to 0}\frac{x(e^x-1)}{1-\cos x}$

Solution:

$l = \lim_{x\to 0}\frac{x(e^x-1)}{1-\cos x}$

$= \lim_{x\to 0}\frac{x(e^x-1)}{1-\cos x}\times\frac{1+\cos x}{1+\cos x}\times\frac{x}{x}$

$= \lim_{x\to 0}\frac{(e^x-1)}{x}\times\frac{1+\cos x}{1}\times\frac{x^2}{1-\cos^2 x}$

$= \lim_{x\to 0}\frac{(e^x-1)}{x}\times\lim_{x\to 0}\frac{1+\cos x}{1}\times\lim_{x\to 0}\frac{x^2}{\sin^2 x}$

$= \lim_{x\to 0}\frac{(e^x-1)}{x}\times\lim_{x\to 0}\frac{1+\cos x}{1}\times\lim_{x\to 0}\frac{1}{(\sin x/x)^2}$

$= 1\times(1+1)\times\frac{1}{1^2} = 2$

$\left(\because \lim_{x\to 0}\frac{e^x-1}{x} = 1,\ \lim_{x\to 0}\frac{\sin x}{x} = 1\right)$

Q7. $\lim_{x\to 0}\frac{\log_e(1+2x)}{x}$

Solution:

$l = \lim_{x\to 0}\frac{\log_e(1+2x)}{x}$

$= \lim_{x\to 0}\frac{\log_e(1+2x)}{2x}\times 2$

Substituting $2x = y \Rightarrow x\to 0 \Rightarrow 2x\to 0 \Rightarrow y\to 0$

$\therefore l = \lim_{y\to 0}\frac{\log_e(1+y)}{y}\times 2 = 1\times 2 = 2$

$\left(\because \lim_{y\to 0}\frac{\log_e(1+y)}{y}=1\right)$

Q8. $\lim_{x\to 0}\frac{\log(1+x^3)}{\sin^3 x}$

Solution:

$$l=\lim_{x\to 0}\frac{\log(1+x^3)}{\sin^3 x}$$

$$=\lim_{x\to 0}\frac{\log(1+x^3)}{\sin^3 x}\times\frac{x^3}{x^3}$$

$$=\lim_{x\to 0}\frac{\log(1+x^3)}{x^3}\times\frac{x^3}{\sin^3 x}$$

$$=\lim_{y\to 0}\frac{\log(1+y)}{y}\times\lim_{x\to 0}\frac{1}{\left(\frac{\sin x}{x}\right)^3}$$

Substituting $x^3=y\Rightarrow x\to 0\Rightarrow x^3\to 0\Rightarrow y\to 0$

$$=1\times\frac{1}{1^3}=1\left(\because \lim_{y\to 0}\frac{\log(1+y)}{y}=1,\ \lim_{x\to 0}\frac{\sin x}{x}=1\right)$$

Day 178

NCERT Miscellaneous Exercise (With Solutions Q1–Q12)

Q1. Find the derivative of the following functions from the first principle:

(i) –x (ii) $(-x)^{-1}$

(iii) $\sin(x+1)$ (iv) $\cos\left(x-\frac{\pi}{8}\right)$

Solution: (i) Let $f(x)=-x$.

Then, according to the first principle

$$f'(x)=\lim_{h\to 0}\frac{f(x+h)-f(x)}{h}$$

$$=\lim_{h\to 0}\frac{-(x+h)-(-x)}{h}$$

$$=\lim_{h\to 0}\frac{-x-h-x}{h}$$

$$=\lim_{h\to 0}\frac{-h}{h}=\lim_{h\to 0}(-1)=-1$$

(ii) Let $f(x)=(-x)^{-1}=\frac{1}{-x}=\frac{-1}{x}$.

Then, according to the first principle

$$f'(x)=\lim_{h\to 0}\frac{f(x+h)-f(x)}{h}$$

$$=\lim_{h\to 0}\frac{1}{h}\left[\frac{-1}{x+h}-\left(\frac{-1}{x}\right)\right]$$

$$=\lim_{h\to 0}\frac{1}{h}\left[\frac{-1}{x+h}+\frac{1}{x}\right]$$

$$=\lim_{h\to 0}\frac{1}{h}\left[\frac{-x+(x+h)}{x(x+h)}\right]$$

$$=\lim_{h\to 0}\frac{1}{h}\left[\frac{-x+x+h}{x(x+h)}\right]$$

$$=\lim_{h\to 0}\frac{1}{h}\left[\frac{h}{x(x+h)}\right]$$

$$=\lim_{h\to 0}\frac{1}{x(x+h)}=\frac{1}{x\cdot x}=\frac{1}{x^2}$$

(iii) Let $f(x)=\sin(x+1)$.

Then, according to the first principle

$$f'(x)=\lim_{h\to 0}\frac{f(x+h)-f(x)}{h}$$

$$=\lim_{h\to 0}\frac{1}{h}[\sin(x+h+1)-\sin(x+1)]$$

$$=\lim_{h\to 0}\frac{1}{h}\left[2\cos\left(\frac{x+h+1+x+1}{2}\right)\sin\left(\frac{x+h+1-x-1}{2}\right)\right]$$

$$=\lim_{h\to 0}\frac{1}{h}\left[2\cos\left(\frac{2x+h+2}{2}\right)\sin(h/2)\right]$$

$$=\lim_{h\to 0}\frac{1}{h}\left[\cos\left(\frac{2x+h+2}{2}\right)\cdot\frac{\sin(h/2)}{(h/2)}\right]$$

$$=\lim_{h\to 0}\cos\left(\frac{2x+h+2}{2}\right)\cdot\lim_{h/2\to 0}\frac{\sin(h/2)}{(h/2)}$$

$\left[\text{As } h\to 0\Rightarrow\frac{h}{2}\to 0\right]$

$$=\cos\left(\frac{2x+0+2}{2}\right)\cdot 1\left[\lim_{x\to 0}\frac{\sin x}{x}=1\right]$$

$$=\cos(x+1)$$

(iv) Let $f(x)=\cos\left(x-\frac{\pi}{8}\right)$.

Then, according to the first principle

$$f'(x)=\lim_{h\to 0}\frac{f(x+h)-f(x)}{h}$$

$$=\lim_{h\to 0}\frac{1}{h}\left[\cos\left(x+h-\frac{\pi}{8}\right)-\cos\left(x-\frac{\pi}{8}\right)\right]$$

$$= \lim_{h\to 0}\frac{1}{h}\left[-2\sin\left(\frac{x+h-\frac{\pi}{8}+x-\frac{\pi}{8}}{2}\right)\times\sin\left(\frac{x+h-\frac{\pi}{8}-x+\frac{\pi}{8}}{2}\right)\right]$$

$$= \lim_{h\to 0}\frac{1}{h}\left[-2\sin\left(\frac{2x+h-\frac{\pi}{8}}{2}\right)\sin\frac{h}{2}\right]$$

$$= \lim_{h\to 0}\left[-\sin\left(\frac{2x+h-\frac{\pi}{8}}{2}\right)\frac{\sin(h/2)}{(h/2)}\right]$$

$$= \lim_{h\to 0}\left[-\sin\left(\frac{2x+h-\frac{\pi}{8}}{2}\right)\right]\lim_{h/2\to 0}\frac{\sin(h/2)}{(h/2)}$$

$\left[\text{As } h\to 0 \Rightarrow \frac{h}{2}\to 0\right]$

$$= -\sin\left(\frac{2x+0-\frac{\pi}{4}}{2}\right)\cdot 1 = -\sin\left(x-\frac{\pi}{8}\right)$$

Find the derivative of the following functions (it is to be understood that *a, b, c, d, p, q, r, s* are fixed non-zero constants and *m* and *n* are integers).

Q2. $(x+a)$

Solution: Let $f(x)=x+a$.

Then, $f'(x)=\frac{d}{dx}(x+a)=\frac{d}{dx}(x)+\frac{d}{dx}(a)=1+0=1$

Q3. $(px+q)\left(\frac{r}{x}+s\right)$

Solution: Let $f(x)=(px+q)\left(\frac{r}{x}+s\right)$.

By the product rule

$$f'(x)=(px+q)\frac{d}{dx}\left(\frac{r}{x}+s\right)+\left(\frac{r}{s}+s\right)\frac{d}{dx}(px+q)$$

$$=(px+q)(-rs^{-2}+0)+\left(\frac{r}{s}+s\right)(p)$$

$$=(px+q)(-rx^{-2})+\left(\frac{r}{x}+s\right)p$$

$$=(px+q)\left(\frac{-r}{x^2}\right)+\left(\frac{r}{x}+s\right)p$$

$$=\frac{-pr}{x}-\frac{qr}{x^2}+\frac{pr}{x}+ps=ps-\frac{qr}{x^2}$$

Q4. $f(x)=(ax+b)(cx+d)^2$

Solution: Let $f(x)=(ax+b)(cx+d)^2$.

By the product rule

$$f'(x)=(ax+b)\frac{d}{dx}(cx+d)^2+(cx+d)^2\frac{d}{dx}(ax+d)$$

$$=(ax+b)\frac{d}{dx}(c^2x^2+2cdx+d^2)+(cx+d)^2(a+0)$$

$$=(ax+b)\left[\frac{d}{dx}(c^2x^2)+\frac{d}{dx}(2cdx)+\frac{d}{dx}d^2\right]+a(cx+d)^2$$

$$=(ax+b)(2c^2x+2cd+0)+a(cx+d^2)$$

$$=2c(ax+b)(cx+d)+a(cx+d)^2$$

$$=(cx+d)\{2c(ax+b)+a(cx+d)\}$$

Q5. $\frac{ax+b}{cx+d}$

Solution: Let $f(x)=\frac{ax+b}{cx+d}$.

By the quotient rule

$$f'(x)=\frac{(cx+d)\frac{d}{dx}(ax+b)-(ax+b)\frac{d}{dx}(cx+d)}{(cx+d)^2}$$

$$=\frac{(cx+d)(a)-(ax+b)(c)}{(cx+d)^2}$$

$$=\frac{acx+ad-acx-bc}{(cx+d)^2}=\frac{ad-bc}{(cx+d)^2}$$

Q6. $\frac{1+\frac{1}{x}}{1-\frac{1}{x}}$

Solution: Let $f(x)=\frac{1+\frac{1}{x}}{1-\frac{1}{x}}=\frac{\frac{x+1}{x}}{\frac{x-1}{x}}=\frac{x+1}{x-1}$, where $x\neq 0,1$.

By the quotient rule

$$f'(x)=\frac{(x-1)\frac{d}{dx}(x+1)-(x+1)\frac{d}{dx}(x-1)}{(x-1)^2}$$

$$=\frac{(x-1)(1)-(x+1)(1)}{(x-1)^2}$$

$$=\frac{x-1-x-1}{(x-1)^2}$$

$$\Rightarrow f'(x)=\frac{-2}{(x-1)^2},\quad x\neq 0,1$$

Q7. $\frac{1}{ax^2+bx+c}$

Solution: Let $f(x)=\dfrac{1}{ax^2+bx+c}$.

By the quotient rule

$$f'(x)=\frac{(ax^2+bx+c)\dfrac{d}{dx}(1)-(1)\dfrac{d}{dx}(ax^2+bx+c)}{(ax^2+bx+c)^2}$$

$$=\frac{(ax^2+bx+c)(0)-(2ax+b)}{(ax^2+bx+c)^2}=\frac{-(2ax+b)}{(ax^2+bx+c)^2}$$

Q8. $\dfrac{ax+b}{px^2+qx+r}$

Solution: Let $f(x)=\dfrac{ax+b}{px^2+qx+r}$.

By the quotient rule

$$f'(x)=\frac{(px^2+qx+r)\dfrac{d}{dx}(ax+b)-(ax+b)\dfrac{d}{dx}(px^2+qx+c)}{(px^2+qx+r)^2}$$

$$=\frac{(px^2+qx+r)(a)-(ax+b)(2px+q)}{(px^2+qx+r)^2}$$

$$=\frac{apx^2+aqx+ar-2apx^2-apx-2bpx-bq}{(px^2+qx+r)^2}$$

$$=\frac{-apx^2-2bpx+ar-bq}{(px^2+qx+r)^2}$$

Q9. $\dfrac{px^2+qx+r}{ax+b}$

Solution: Let $f(x)=\dfrac{px^2+qx+r}{ax+b}$.

By the quotient rule

$$f'(x)=\frac{(ax+b)\dfrac{d}{dx}(px^2+qx+r)-(px^2+qx+r)\dfrac{d}{dx}(ax+b)}{(ax+b)^2}$$

$$=\frac{(ax+b)(2px+q)-(px^2+qx+r)(a)}{(ax+b)^2}$$

$$=\frac{2apx^2+aqx+2bpx+bq-apx^2-apx^2-aqx-ar}{(ax+b)^2}$$

$$=\frac{apx^2+2bpx+bq-ar}{(ax+b)^2}$$

Q10. $\dfrac{a}{x^4}-\dfrac{b}{x^2}+\cos x$

Solution: Let $f(x)=\dfrac{a}{x^4}-\dfrac{b}{x^2}+\cos x$

$$f'(x)=\frac{d}{dx}\left(\frac{a}{x^4}\right)-\frac{d}{dx}\left(\frac{b}{x^2}\right)+\frac{d}{dx}(\cos x)$$

$$=a\frac{d}{dx}(x^{-4})-b\frac{d}{dx}(x^{-2})+\frac{d}{dx}(\cos x)$$

$$=a(-4x^{-5})-b(-2x^{-3})+(-\sin x)$$

$$\left[\because \frac{d}{dx}(x^n)=nx^{n-1},\frac{d}{dx}(\cos x)=-\sin x\right]$$

$$=\frac{-4a}{x^5}+\frac{2b}{x^3}-\sin x$$

Q11. $4\sqrt{x}-2$

Solution: Let $f(x)=4\sqrt{x}-2$

$$f'(x)=\frac{d}{dx}(4\sqrt{x}-2)$$

$$=\frac{d}{dx}(4\sqrt{x})-\frac{d}{dx}(2)$$

$$=4\frac{d}{dx}(x^{1/2})-0=4\left(\frac{1}{2}x^{\frac{1}{2}-1}\right)$$

$$=\left(2x^{-\frac{1}{2}}\right)=\frac{2}{\sqrt{x}}$$

Q12. $(ax+b)^n$

Solution: Let $f(x)=(ax+b)^n$.

Then, according to the first principle

$$f'(x)=\lim_{h\to 0}\frac{f(x+h)-f(x)}{h}$$

$$=\lim_{h\to 0}\frac{(ax+ah+b)^n-(ax+b)^n}{h}$$

$$=\lim_{h\to 0}\frac{(ax+b)^n\left(1+\dfrac{ah}{ax+b}\right)^n-(ax+b)^n}{h}$$

$$=(ax+b)^n\lim_{h\to 0}\frac{\left(1+\dfrac{ah}{ax+b}\right)^n-1}{h}$$

Using Binomial Theorem

$=(ax+b)^n\times$

$$\lim_{h\to 0}\frac{1}{h}\left[\left\{1+n\left(\frac{ah}{ax+b}\right)+\frac{n(n-1)}{\dots}\left(\frac{ah}{ax+b}\right)^2+\dots\right\}-1\right]$$

$$=(ax+b)^n\lim_{h\to 0}\frac{1}{h}\left[n\left(\frac{ah}{ax+b}\right)+\frac{n(n-1)a^2h^2}{\dots(ax+b)^2}+\dots\right.$$

(Other terms contain higher degrees of h, and therefore, we ignore them)

$$= (ax+b)^n \lim_{h\to 0}\left[\frac{na}{(ax+b)} + \frac{n(n-1)a^2h}{...(ax+b)^2} + ...\right]$$

$$= (ax+b)^n \left[\frac{na}{(ax+b)} + 0\right]$$

$$= na\frac{(ax+b)^n}{(ax+b)} = na(ax+b)^{n-1}$$

Day 179

NCERT Miscellaneous Exercise (With Solutions Q13–Q30)

Q13. $(ax+b)^n(cx+d)^m$

Solution: Let $f(x) = (ax+b)^n(cx+d)^m$.

By the product rule

$$f'(x) = (ax+b)^n \frac{d}{dx}(cx+d)^m + (cx+d)^m \frac{d}{dx}(ax+b)^n$$

Now, using the above proof, we get

$$f'(x) = (ax+b)^n mc(cx+d)^{m-1} + (cx+d)^m na(ax+b)^{n-1}$$

$$\Rightarrow f'(x) = (ax+b)^{n-1}(cx+d)^{m-1} \times \{(ax+b)mc + (cx+d)na\}$$

Q14. $\sin(x+a)$

Solution: Let $f(x) = \sin(x+a)$,

Then, according to the first principle

$$f'(x) = \lim_{h\to 0}\frac{f(x+h)-f(x)}{h}$$

$$= \lim_{h\to 0}\frac{1}{h}\left[2\cos\left(\frac{x+h+a+x+a}{2}\right)\sin\left(\frac{x+h+a-x-a}{2}\right)\right]$$

$$= \lim_{h\to 0}\frac{1}{h}\left[2\cos\left(\frac{2x+2a+h}{2}\right)\sin\left(\frac{h}{2}\right)\right]$$

$$= \lim_{h\to 0}\left[\cos\left(\frac{2x+2a+h}{2}\right)\left\{\frac{\sin(h/2)}{(h/2)}\right\}\right]$$

$$= \lim_{h\to 0}\cos\left(\frac{2x+2a+h}{2}\right) = \lim_{h/2\to 0}\left\{\frac{\sin(h/2)}{(h/2)}\right\}$$

$$\left[\text{As } h\to 0 \Rightarrow \frac{h}{2}\to 0\right]$$

$$= \cos\left(\frac{2x+2a}{2}\right)\times 1\left[\lim_{x\to 0}\frac{\sin x}{x} = 1\right]$$

$$= \cos(x+a)$$

Q15. $\operatorname{cosec} x \cot x$

Solution: Let $f(x) = \operatorname{cosec} x \cot x$.

By the product rule

$$f'(x) = \operatorname{cosec} x\frac{d}{dx}(\cot x) + \cot x\frac{d}{dx}(\operatorname{cosec} x) ...(1)$$

Now, $f_1(x) = \cot x$.

Then, according to the first principle

$$f_1'(x) = \lim_{h\to 0}\frac{f_1(x+h)-f_1(x)}{h}$$

$$= \lim_{h\to 0}\frac{\cot(x+h)-\cot x}{h}$$

$$= \lim_{h\to 0}\frac{1}{h}\left(\frac{\cos(x+h)}{\sin(x+h)} - \frac{\cos x}{\sin x}\right)$$

$$= \lim_{h\to 0}\frac{1}{h}\left[\frac{\sin x\cos(x+h)-\cos x\sin(x+h)}{\sin x\sin(x+h)}\right]$$

$$= \lim_{h\to 0}\frac{1}{h}\left[\frac{\sin(x-x-h)}{\sin x\sin(x+h)}\right]$$

$$= \frac{1}{\sin x}\cdot\lim_{h\to 0}\frac{1}{h}\left[\frac{\sin(-h)}{\sin(x+h)}\right]$$

$$= \frac{-1}{\sin x}\cdot\left(\lim_{h\to 0}\frac{\sin h}{h}\right)\left(\lim_{h\to 0}\frac{1}{\sin(x+h)}\right)$$

$$= \frac{-1}{\sin x}\cdot 1\cdot\left(\frac{1}{\sin(x+0)}\right) = \frac{-1}{\sin^2 x} = -\operatorname{cosec}^2 x$$

$$\therefore \frac{d}{dx}(\cot x) = -\operatorname{cosec}^2 x ...(2)$$

Now, let $f_2(x) = \operatorname{cosec} x$.

Then, according to the first principle

$$f_2'(x) = \lim_{h\to 0}\frac{f_2(x+h)-f_2(x)}{h}$$

$$= \lim_{h\to 0}\frac{1}{h}[\operatorname{cosec}(x+h) - \operatorname{cosec} x]$$

$$= \lim_{h\to 0}\frac{1}{h}\left[\frac{1}{\sin(x+h)} - \frac{1}{\sin x}\right]$$

$$= \lim_{h\to 0}\frac{1}{h}\left[\frac{\sin x - \sin(x+h)}{\sin x\sin(x+h)}\right]$$

$$= \frac{1}{\sin x}\cdot\lim_{h\to 0}\frac{1}{h}\left[\frac{2\cos\left(\frac{x+x+h}{2}\right)\sin\left(\frac{x-x-h}{2}\right)}{\sin(x+h)}\right]$$

$$= \frac{1}{\sin x} \cdot \lim_{h \to 0} \frac{1}{h} \left[\frac{2\cos\left(\frac{2x+h}{2}\right)\sin(-h/2)}{\sin(x+h)} \right]$$

$$= \frac{1}{\sin x} \cdot \lim_{h \to 0} \left[\frac{-\sin(h/2)}{(h/2)} \cdot \frac{\cos\left(\frac{2x+h}{2}\right)}{\sin(x+h)} \right]$$

$$= \frac{-1}{\sin x} \cdot \lim_{h \to 0} \frac{\sin(h/2)}{(h/2)} \cdot \lim_{h \to 0} \frac{\cos\left(\frac{2x+h}{2}\right)}{\sin(x+h)}$$

$$= \frac{-1}{\sin x} \cdot 1 \cdot \frac{\cos\left(\frac{2x+0}{2}\right)}{\sin(x+0)} = \frac{-1}{\sin x} \cdot \frac{\cos x}{\sin x}$$

$$= -\operatorname{cosec} x.\cot x$$

$$\frac{d}{dx}(\operatorname{cosec} x) = -\operatorname{cosec} x \cot x ...(3)$$

From equations (1), (2) and (3), we get

$$f'(x) = \operatorname{cosec} x(-\operatorname{cosec}^2 x) + \cot x(-\operatorname{cosec} x \cot x)$$

$$= -\operatorname{cosec}^3 x - \cot^2 x \operatorname{cosec} x$$

Q16. $\frac{\cos x}{1+\sin x}$

Solution: Let $f(x) = \frac{\cos x}{1+\sin x}$.

By the quotient rule

$$f'(x) = \frac{(1+\sin x)\frac{d}{dx}(\cos x) - (\cos x)\frac{d}{dx}(1+\sin x)}{(1+\sin x)^2}$$

$$= \frac{(1+\sin x)(-\sin x) - (\cos x)(\cos x)}{(1+\sin x)^2}$$

$$= \frac{-\sin x - \sin^2 x - \cos^2 x}{(1+\sin x)^2}$$

$$= \frac{-\sin x - (\sin^2 x + \cos^2 x)}{(1+\sin x)^2}$$

$$= \frac{-\sin x - 1}{(1+\sin x)^2} = \frac{-(1+\sin x)}{(1+\sin x)^2} = \frac{-1}{(1+\sin x)}$$

Q17. $\frac{\sin x + \cos x}{\sin x - \cos x}$

Solution: Let $f(x) = \frac{\sin x + \cos x}{\sin x - \cos x}$.

By the quotient rule

$$f'(x) = \frac{(\sin x - \cos x)(\sin x + \cos x)' - (\sin x + \cos x)(\sin x - \cos x)'}{\sin x - \cos x}$$

$$= \frac{(\sin x - \cos x)(\cos x - \sin x) - (\sin x + \cos x)(\cos x + \sin x)}{\sin x - \cos x}$$

$$= \frac{-(\sin x - \cos x)^2 - (\sin x + \cos x)^2}{(\sin x - \cos x)^2}$$

$$= \frac{-\sin^2 x - \cos^2 x + 2\sin x \cos x - \sin^2 x - \cos^2 x - 2\sin x \cos x}{(\sin x - \cos x)^2}$$

$$= \frac{-[1+1]}{(\sin x - \cos x)^2} = \frac{-2}{(\sin x - \cos x)^2}$$

Q18. $\frac{\sec x - 1}{\sec x + 1}$

Solution:

Let $f(x) = \frac{\sec x - 1}{\sec x + 1}, f(x) = \frac{\frac{1}{\cos x} - 1}{\frac{1}{\cos x} + 1} = \frac{1-\cos x}{1+\cos x}$.

By the quotient rule

$$f'(x) = \frac{(1+\cos x)\frac{d}{dx}(1-\cos x) - (1-\cos x)\frac{d}{dx}(1+\cos x)}{(1+\cos x)^2}$$

$$= \frac{(1+\cos x)(\sin x) - (1-\cos x)(-\sin x)}{(1+\cos x)^2}$$

$$= \frac{\sin x + \cos x \sin x + \sin x - \sin x \cos x}{(1+\cos x)^2}$$

$$= \frac{2\sin x}{(1+\cos x)^2} = \frac{2\sin x}{\left(1+\frac{1}{\sec x}\right)^2}$$

$$= \frac{2\sin x}{\frac{(\sec x + 1)}{\sec^2 x}} = \frac{2\sin x \sec^2 x}{(\sec x + 1)^2}$$

$$= \frac{\frac{2\sin x}{\cos x}\sec x}{(\sec x + 1)^2} = \frac{2\sec x \tan x}{(\sec x + 1)^2}$$

Q19. $\sin^n x$

Solution: Let $y = \sin^n x$.

For $n = 1, y = \sin x$.

$$\therefore \frac{d}{dx}\sin x = \cos x$$

For $n = 2, y = \sin^2 x$.

$$\therefore \frac{dy}{dx} = \frac{d}{dx}(\sin x \sin x) = (\sin x)' \sin x + \sin x(\sin x)'$$

[By the Leibniz product rule]

$$= \cos x \sin x + \sin x \cos x = 2\sin x \cos x ...(1)$$

For $n = 3, y = \sin^3 x$.

$\therefore \frac{dy}{dx} = \frac{d}{dx}(\sin x \sin^2 x) = (\sin x)' \sin^2 x + \sin(\sin^2 x)'$

[By the Leibniz product rule]

$= \cos x \sin^2 x + \sin x(2\sin x \cos x)$ [Using (1)]

$= \cos x \sin^2 x + 2\sin^2 x \cos x = 3\sin^2 x \cos x$

Similarly, we can prove that

$\frac{d}{dx}(\sin^n x) = n\sin^{(n-1)} x \cos x$

Q20. $\frac{a + b\sin x}{c + d\cos x}$

Solution: Let $f(x) = \frac{a + b\sin x}{c + d\cos x}$.

By the quotient rule

$$f'(x) = \frac{(c + d\cos x)(a + b\sin x)' - (a + b\sin x)(c + d\cos x)}{(c + d\cos x)^2}$$

$$= \frac{(c + d\cos x)(b\cos x) - (a + b\sin x)(-d\sin x)}{(c + d\cos x)^2}$$

$$= \frac{cd\cos x + bd\cos^2 x + ad\sin x + bd\sin^2 x}{(c + d\cos x)^2}$$

$$= \frac{bc\cos x + ad\sin x + bd(\cos^2 x + \sin^2 x)}{(c + d\cos x)^2}$$

$$= \frac{bc\cos x + ad\sin x + bd}{(c + d\cos x)^2}$$

Q21. $\frac{\sin(x + a)}{\cos x}$

Solution: Let $f(x) = \frac{\sin(x + a)}{\cos x}$.

By the quotient rule

$$f'(x) = \frac{\cos x \frac{d}{dx}[\sin(x + a)] - \sin(x + a)\frac{d}{dx}\cos x}{\cos^2 x}$$

$$= \frac{\cos x \frac{d}{dx}[\sin(x + a)] - \sin(x + a)(-\sin x)}{\cos^2 x} \quad \text{...(i)}$$

$$f'(x) = \frac{\cos x.\cos(x + a) + \sin x \sin(x + a)}{\cos^2 x}$$

$$= \frac{\cos(x + a - x)}{\cos^2 x} = \frac{\cos a}{\cos^2 x}$$

Q22. $x^4(5\sin x - 3\cos x)$

Solution: Let $f(x) = x^4(5\sin x - 3\cos x)$.

By the product rule

$$f'(x) = x^4 \frac{d}{dx}(5\sin x - 3\cos x) + (5\sin x - 3\cos x)\frac{d}{dx}(x^4)$$

$$= x^4\left[5\frac{d}{dx}(\sin x) - 3\frac{d}{dx}(\cos x)\right] + (5\sin x - 3\cos x)\frac{d}{dx}(x^4)$$

$$= x^4[5\cos x - 3(-\sin x)] + (5\sin x - 3\cos x)(4x^3)$$

$$= x^3[5x\cos x + 3x\sin x + 20\sin x - 12\cos x]$$

Q23. $(x^2 + 1)\cos x$

Solution: Let $f(x) = (x^2 + 1)\cos x$.

By the product rule

$$f'(x) = (x^2 + 1)\frac{d}{dx}(\cos x) + \cos x \frac{d}{dx}(x^2 + 1)$$

$$= (x^2 + 1)(-\sin x) + \cos x(2x)$$

$$= -x^2 \sin x - \sin x + 2x\cos x$$

Q24. $(ax^2 + \sin x)(p + q\cos x)$

Solution: Let $f(x) = (ax^2 + \sin x)(p + q\cos x)$.

By the product rule

$$f'(x) = (ax^2 + \sin x)\frac{d}{dx}(p + q\cos x) + (p + q\cos x)\frac{d}{dx}(ax^2 + \sin x)$$

$$= (ax^2 + \sin x)(-q\sin x) + (p + q\cos x)(2ax + \cos x)$$

$$= -q\sin x(ax^2 + \sin x) + (p + q\cos x)(2ax + \cos x)$$

Q25. $(x + \cos x)(x - \tan x)$

Solution: Let $f(x) = (x + \cos x)(x - \tan x)$.

By the product rule

$$f'(x) = (x + \cos x)\frac{d}{dx}(x - \tan x) + (x - \tan x)\frac{d}{dx}(x + \cos x)$$

$$= (x + \cos x)(1 - \sec^2 x) + (x - \tan x)(1 - \sin x)$$

$$= (x + \cos x)(-\tan^2 x) + (x - \tan x)(1 - \sin x)$$

$$= -\tan^2 x(x + \cos x) + (x - \tan x)(1 - \sin x)$$

Q26. $\frac{4x + 5\sin x}{3x + 7\cos x}$

Solution: Let $f(x) = \frac{4x + 5\sin x}{3x + 7\cos x}$.

By the quotient rule

$$f'(x) = \frac{(3x+7\cos x)(4x+5\sin x)' - (4x+5\sin x)(3x+7\cos x)'}{(3x+7\cos x)^2}$$

$$= \frac{(3x+7\cos x)(4+5\cos x)-(4x+5\sin x)(3-7\sin x)}{(3x+7\cos x)^2}$$

$$= \frac{35+15x\cos x+28\cos x+28x\sin x-15\sin x}{(3x+7\cos x)^2}$$

Q27. $\dfrac{x^2\cos(\pi/4)}{\sin x}$.

Solution: Let $f(x) = \dfrac{x^2\cos(\pi/4)}{\sin x}$.

By the quotient rule

$$f'(x) = \cos\frac{\pi}{4}\cdot\left[\frac{\sin x\frac{d}{dx}(x^2) - x^2\frac{d}{dx}(\sin x)}{\sin^2 x}\right]$$

$$= \cos\frac{\pi}{4}\cdot\left[\frac{\sin x\cdot 2x - x^2\cos x}{\sin^2 x}\right]$$

$$= \frac{x\cos\frac{\pi}{4}[2\sin x - x\cos x]}{\sin^2 x}$$

Q28. $\dfrac{x}{1+\tan x}$

Solution: Let $f(x) = \dfrac{x}{1+\tan x}$

$$f'(x) = \frac{(1+\tan x)\frac{d}{dx}(x) - x\frac{d}{dx}(1+\tan x)}{(1+\tan x)^2}$$

$$= \frac{(1+\tan x) - x\cdot(0+\sec^2 x)}{(1+\tan x)^2}$$

$$= \frac{1+\tan x - x\sec^2 x}{(1+\tan x)^2}$$

Q29. $(x+\sec x)(x-\tan x)$

Solution: Let $f(x) = (x+\sec x)(x-\tan x)$.

By the product rule

$$f'(x) = (x+\sec x)(x-\tan x)' + (x-\tan x)(x+\sec x)'$$

$$= (x+\sec x)(1-\sec^2 x) + (x-\tan x)(1+\sec x\tan x)$$

Q30. $\dfrac{x}{\sin^n x}$

Solution: Let $f(x) = \dfrac{x}{\sin^n x}$.

By the quotient rule

$$f'(x) = \frac{\sin^n x\frac{d}{dx}x - x\frac{d}{dx}\sin^n x}{\sin^{2n} x}$$

$$\frac{d}{dx}\sin^n x = n\sin^{n-1} x\cos x$$

Therefore, $f'(x) = \dfrac{\sin^n x\times 1 - x n\sin^{n-1} x\cos x}{\sin^{2n} x}$

$$= \frac{\sin^{n-1} x(\sin x - nx\cos x)}{\sin^{2n} x}$$

$$= \frac{\sin x - nx\cos x}{\sin^{n+1} x}$$

Day 180

NCERT Exemplar Problems: Solutions (Q1–Q15)

Q1. Evaluate $\lim\limits_{x\to 3}\dfrac{x^2-9}{x-3}$.

Solution:

$$\lim_{x\to 3}\frac{x^2-9}{x-3} = \lim_{x\to 3}\frac{x^2-(3)^2}{x-3}$$

$$= \lim_{x\to 3}\frac{(x+3)(x-3)}{x-3} = \lim_{x\to 3}(x+3) = 3+3 = 6$$

Q2. Evaluate $\lim\limits_{x\to 1/2}\dfrac{4x^2-1}{2x-1}$.

Solution:

$$\lim_{x\to 1/2}\frac{4x^2-1}{2x-1} = \lim_{x\to 1/2}\frac{(2x)^2-(1)^2}{2x-1}$$

$$= \lim_{x\to 1/2}\frac{(2x+1)(2x-1)}{2x-1} = \lim_{x\to 1/2}(2x+1)$$

$$= 2\times\frac{1}{2}+1 = 1+1 = 2$$

Q3. Evaluate $\lim\limits_{h\to 0}\dfrac{\sqrt{x+h}-\sqrt{x}}{h}$

Solution:

$$\lim_{h\to 0}\frac{\sqrt{x+h}-\sqrt{x}}{h} = \lim_{h\to 0}\frac{(x+h)^{1/2}-(x)^{1/2}}{x+h-x}$$

$$= \lim_{h\to 0}\frac{(x+h)^{1/2}-(x)^{1/2}}{(x+h)-x}\left[\because \lim_{x\to a}\frac{x^n-a^n}{x-a}=na^{n-1}\right]$$

$$=\frac{1}{2}x^{\frac{1}{2}-1}=\frac{1}{2}x^{-1/2}[\because h\to 0\Rightarrow x+h\to x]$$

$$=\frac{1}{2\sqrt{x}}$$

Q4. Evaluate $\lim_{x\to 0}\frac{(x+2)^{1/3}-2^{1/3}}{x}$

Solution:

$$\lim_{x\to 0}\frac{(x+2)^{1/3}-2^{1/3}}{x}=\lim_{x\to 0}\frac{(x+2)^{1/3}-2^{1/3}}{(x+2)-2}$$

$$=\frac{1}{3}\times 2^{\frac{1}{3}-1}=\frac{1}{3}(2)^{-2/3}\left[\because \lim_{x\to a}\frac{x^n-a^n}{x-a}=na^{n-1}\right]$$

$$=\frac{1}{3(2)^{2/3}}[\because x\to 0\Rightarrow x+2\to 2]$$

Q5. Evaluate $\lim_{x\to 0}\frac{(1+x)^6-1}{(1+x)^2-1}$

Solution:

$$\lim_{x\to 0}\frac{(1+x)^6-1}{(1+x)^2-1}=\lim_{x\to 0}\frac{\frac{(1+x)^6-1}{x}}{\frac{(1+x)^2-1}{x}}$$

$$=\lim_{x\to 0}\frac{\frac{(1+x)^6-1}{(1+x)-1}}{\frac{(1+x)^2-1}{(1+x)-1}}[\because x\to 0\Rightarrow 1+x\to 1]$$

$$=\lim_{x\to 0}\frac{\frac{(1+x)^6-(1)^6}{(1+x)-1}}{\frac{(1+x)^2-(1)^2}{(1+x)-1}}\left[\because \lim_{x\to a}\frac{f(x)}{g(x)}=\frac{\lim_{x\to a}f(x)}{\lim_{x\to a}g(x)}\right]$$

$$=\frac{6(1)^{6-1}}{2(1)^{2-1}}\left[\therefore \lim_{x\to a}\frac{x^n-a^n}{x-a}=na^{n-1}\right]$$

$$=\frac{6\times 1}{2\times 1}=\frac{6}{2}=3.$$

Q6. Evaluate $\lim_{x\to a}\frac{(2+x)^{5/2}-(a+2)^{5/2}}{x-a}$

Solution:

$$\lim_{x\to a}\frac{(2+x)^{5/2}-(a+2)^{5/2}}{x-a}$$

$$=\lim_{x\to a}\frac{(2+x)^{5/2}-(a+2)^{5/2}}{(2+x)-(a+2)}=\frac{5}{2}(a+2)^{\frac{5}{2}-1}$$

$$\left[\because \lim_{x\to a}\frac{x^n-a^n}{x-a}=na^{n-1}\right]$$

$$=5/2(a+2)^{3/2}[\because x\to a\Rightarrow x+2\to a+2]$$

Q7. Evaluate $\lim_{x\to 1}\frac{x^4-\sqrt{x}}{\sqrt{x}-1}$

Solution:

$$\lim_{x\to 1}\frac{x^4-\sqrt{x}}{\sqrt{x}-1}=\lim_{x\to 1}\frac{\sqrt{x}[(x)^{7/2}-1]}{\sqrt{x}-1}$$

$$=\lim_{x\to 1}\frac{(x)^{7/2}-1}{\sqrt{x}-1}\cdot\lim_{x\to 1}\sqrt{x}$$

$$\left[\because \lim_{x\to a}f(x)\cdot g(x)=\lim_{x\to a}f(x)\cdot\lim_{x\to a}g(x)\right]$$

$$=\lim_{x\to 1}\frac{\frac{x^{7/2}-1}{x-1}}{\frac{(x)^{1/2}-1}{x-1}}=\frac{\lim_{x\to 1}\frac{x^{7/2}-1}{x-1}}{\lim_{x\to 1}\frac{(x)^{1/2}-1}{x-1}}$$

$$\left[\because \lim_{x\to a}\frac{f(x)}{g(x)}=\frac{\lim_{x\to a}f(x)}{\lim_{x\to a}g(x)}\right]$$

$$=\frac{\frac{7}{2}(1)^{\frac{7}{2}-1}}{\frac{1}{2}(1)^{\frac{1}{2}-1}}=\frac{7/2}{1/2}=7$$

Q8. Evaluate $\lim_{x\to 2}\frac{x^2-4}{\sqrt{3x-2}-\sqrt{x+2}}$

Solution:

$$\lim_{x\to 2}\frac{x^2-4}{\sqrt{3x-2}-\sqrt{x+2}}$$

$$=\lim_{x\to 2}\frac{(x^2-4)(\sqrt{3x-2}+\sqrt{x+2})}{(\sqrt{3x-2}-\sqrt{x+2})(\sqrt{3x-2}-\sqrt{x+2})}$$

$$=\lim_{x\to 2}\frac{(x^2-4)(\sqrt{3x-2}+\sqrt{x+2})}{(\sqrt{3x-2})^2-(\sqrt{x+2})^2}$$

$$[\because (a+b)(a-b)=a^2-b^2]$$

$$=\lim_{x\to 2}\frac{(x^2-4)(\sqrt{3x-2}+\sqrt{x+2})}{(3x-2)-(x+2)}$$

$$=\lim_{x\to 2}\frac{(x^2-4)(\sqrt{3x-2}+\sqrt{x+2})}{3x-2-x-2}$$

$$=\lim_{x\to 2}\frac{(x^2-4)(\sqrt{3x-2}+\sqrt{x+2})}{2x-4}$$

$$=\lim_{x\to 2}\frac{(x+2)(x-2)(\sqrt{3x-2}+\sqrt{x+2})}{2(x-2)}$$

$$= \lim_{x\to 2} \frac{(x+2)(\sqrt{3x-2}+\sqrt{x+2})}{2}$$

$$= \frac{(2+2)(\sqrt{6-2}+\sqrt{2+2})}{2} = \frac{4(2+2)}{2} = 8$$

Q9. Evaluate $\lim_{x\to\sqrt{2}} \frac{x^2-4}{x^2+3\sqrt{2}x-8}$

Solution:

$$\lim_{x\to\sqrt{2}} \frac{x^2-4}{x^2+3\sqrt{2}x-8} = \lim_{x\to\sqrt{2}} \frac{(x^2)^2-(2)^2}{x^2+3\sqrt{2}x-8}$$

$$= \lim_{x\to\sqrt{2}} \frac{(x^2-2)(x^2+2)}{x^2+4\sqrt{2}x-\sqrt{2}x-8}$$

$$= \lim_{x\to\sqrt{2}} \frac{(x-\sqrt{2})(x+\sqrt{2})(x^2+2)}{x(x+4\sqrt{2})-\sqrt{2}(x+4\sqrt{2})}$$

$$= \lim_{x\to\sqrt{2}} \frac{(x-\sqrt{2})(x+\sqrt{2})(x^2+2)}{(x-\sqrt{2})(x+4\sqrt{2})}$$

$$= \lim_{x\to\sqrt{2}} \frac{(x+\sqrt{2})(x^2+2)}{(x+4\sqrt{2})} = \frac{(\sqrt{2}+\sqrt{2})[(\sqrt{2})^2+2]}{(\sqrt{2}+4\sqrt{2})}$$

$$= \frac{2\sqrt{2}(2+2)}{5\sqrt{2}} = 8/5$$

Q10. Evaluate $\lim_{x\to 1} \frac{x^7-2x^5+1}{x^3+3x^2+2}$

Solution:

$$\lim_{x\to 1} \frac{x^7-2x^5+1}{x^3+3x^2+2} = \lim_{x\to 1} \frac{x^7-x^5-x^5+1}{x^3+x^2-2x^2+2}$$

$$= \lim_{x\to 1} \frac{x^5(x^2-1)-1(x^2-1)}{x^2(x-1)-2(x^2-1)}$$

$$= \lim_{x\to 1} \frac{\frac{x^5(x^2-1)}{(x-1)} - \frac{1(x^2-1)}{(x-1)}}{\frac{x^2(x-1)}{(x-1)} - \frac{2(x^2-1)}{(x-1)}}$$

$$= \frac{\lim_{x\to 1} x^5(x+1) - \lim_{x\to 1}\left(\frac{x^5-1}{x-1}\right)}{\lim_{x\to 1} x^2 - \lim_{x\to 1}(x+1)}$$

$$= \frac{1\times 2-5(1)^4}{1-2\times 2} = \frac{2-5}{1-4} = \frac{-3}{-3} = 1$$

Q11. Evaluate $\lim_{x\to 0} \frac{\sqrt{1+x^3}-\sqrt{1-x^3}}{x^2}$

Solution:

$$\lim_{x\to 0} \frac{\sqrt{1+x^3}-\sqrt{1-x^3}}{x^2}$$

$$= \lim_{x\to 0} \frac{\sqrt{1+x^3}-\sqrt{1-x^3}}{x^2} \cdot \frac{\sqrt{1+x^3}+\sqrt{1-x^3}}{\sqrt{1+x^3}+\sqrt{1-x^3}}$$

$$= \lim_{x\to 0} \frac{(1+x^3)-(1-x^3)}{x^2(\sqrt{1+x^3}+\sqrt{1-x^3})}$$

$$= \lim_{x\to 0} \frac{1+x^3-1+x^3}{x^2(\sqrt{1+x^3}+\sqrt{1-x^3})}$$

$$= \lim_{x\to 0} \frac{2x^3}{x^2(\sqrt{1+x^3}+\sqrt{1-x^3})}$$

$$= \lim_{x\to 0} \frac{2x}{(\sqrt{1+x^3}+\sqrt{1-x^3})} = 0$$

Q12. Evaluate $\lim_{x\to -3} \frac{x^3+27}{x^5+243}$

Solution:

$$\lim_{x\to -3} \frac{x^3+27}{x^5+243} = \lim_{x\to -3} \frac{\frac{x^3+27}{x+3}}{\frac{x^5+243}{x+3}}$$

$$= \lim_{x\to -3} \frac{\frac{x^3-(-3)^3}{x-(-3)}}{\frac{x^5-(-3)^5}{x-(-3)}} = \frac{\lim_{x\to -3} \frac{x^3-(-3)^3}{x-(-3)}}{\lim_{x\to -3} \frac{x^5-(-3)^5}{x-(-3)}}$$

$$\left[\because \lim_{x\to a} \frac{f(x)}{g(x)} = \frac{\lim_{x\to a} f(x)}{\lim_{x\to a} g(x)}\right]$$

$$= \frac{3(-3)^{3-1}}{5(-3)^{5-1}} = \frac{3(-3)^2}{5(-3)^4} \left[\because \lim_{x\to a} \frac{x^n-a^n}{x-a} = na^{n-1}\right]$$

$$= \frac{3}{5(-3)^2} = \frac{3}{45} = 1/15$$

Q13. Evaluate $\lim_{x\to 1/2} \left(\frac{8x-3}{2x-1} - \frac{4x^2+1}{4x^2-1}\right)$

Solution:

$$\lim_{x\to 1/2} \left(\frac{8x-3}{2x-1} - \frac{4x^2+1}{4x^2-1}\right)$$

$$= \lim_{x\to 1/2} \left[\frac{(8x-3)(2x+1)-(4x^2+1)}{(4x^2-1)}\right]$$

$$= \lim_{x\to 1/2} \left[\frac{16x^2+8x-6x-3-4x^2-1}{4x^2-1}\right]$$

$$= \lim_{x\to 1/2} \left[\frac{12x^2+2x-4}{4x^2-1}\right] = \lim_{x\to 1/2} \frac{2(6x^2+x-2)}{4x^2-1}$$

$$= \lim_{x\to 1/2} \frac{2(6x^2+4x-3x-2)}{4x^2-1}$$

$$= \lim_{x\to 1/2} \frac{2[2x(3x+2)-1(3x+2)]}{4x^2-1}$$

$$= \lim_{x\to 1/2} \frac{2[(3x+2)(2x-1)]}{(2x)^2-(1)^2} = \lim_{x\to 1/2} \frac{2(3x+2)(2x-1)}{(2x-1)(2x+1)}$$

$$\lim_{x\to 1/2} \frac{2(3x+2)}{2x-1} = \frac{2\left(3\times\frac{1}{2}+2\right)}{2\times\frac{1}{2}+1} = \frac{3}{2}+2 = 7/2$$

Q14. Find the value of n if $\lim_{x\to 2} \frac{x^n-2^n}{x-2} = 80, n \in N$.

Solution:

$$\lim_{x\to 2} \frac{x^n-2^n}{x-2} = 90 \Rightarrow n(2)^{n-1} = 80$$

$$\left[\because \lim_{x\to a} \frac{x^n-a^n}{x-a} = na^{n-1}\right]$$

$$\Rightarrow n(2)^{n-1} = 5\times 16 \Rightarrow n\times 2^{n-1} = 5\times(2)^4$$

$$\Rightarrow n\times 2^{n-1} = 5\times(2)^{5-1} \Rightarrow n = 5$$

Q15. Evaluate $\lim_{x\to 0} \frac{\sin 3x}{\sin 7x}$

Solution:

$$\lim_{x\to 0} \frac{\frac{\sin 3x}{3x}\cdot 3x}{\frac{\sin 7x}{7x}\cdot 7x} = \frac{\lim_{x\to 0}\frac{\sin 3x}{3x}}{\frac{\sin 7x}{7x}}\cdot\frac{3x}{7x}$$

$$= \frac{3}{7}\cdot\frac{\lim_{x\to 0}\frac{\sin 3x}{3x}}{\lim_{x\to 0}\frac{\sin 7x}{7x}}\left[\because \lim_{x\to 0}\frac{\sin x}{x} = 1\right]$$

$$= \frac{3}{7}[\because x\to 0 \Rightarrow (kx\to 0);\text{ here } k \text{ is a real number}]$$

Day 181

NCERT Exemplar Problems: Solutions (Q16–Q28)

Q16. Evaluate $\lim_{x\to 0} \frac{\sin^2 2x}{\sin^2 4x}$

Solution:

$$\lim_{x\to 0} \frac{\sin^2 2x}{\sin^2 4x} = \lim_{x\to 0} \frac{\sin^2 2x}{[\sin 2(2x)]^2}$$

$$= \lim_{x\to 0} \frac{\sin^2 2x}{(2\sin 2x\cos 2x)^2} = \lim_{x\to 0} \frac{\sin^2 2x}{4\sin^2 2x\cos^2 2x}$$

$[\because \sin 2\theta = 2\sin\theta\cos\theta]$

$$= \lim_{x\to 0} \frac{1}{4\cos^2 2x} = \frac{1}{4}[\because \cos 0 = 1]$$

Q17. Evaluate $\lim_{x\to 0} \frac{1-\cos 2x}{x^2}$

Solution:

$$\lim_{x\to 0} \frac{1-\cos 2x}{x^2} = \lim_{x\to 0} \frac{1-1+2\sin^2 x}{x^2}$$

$[\because \cos 2x = 1-2\sin^2 x]$

$$= \lim_{x\to 0} \frac{2\sin^2 x}{x^2} = 2\lim_{x\to 0} \frac{\sin^2 x}{x^2} = 2\lim_{x\to 0}\left(\frac{\sin x}{x}\right)^2$$

$$\left[\lim_{x\to 0} \frac{\sin x}{x} = 1\right]$$

$$= 2\times 1 = 2$$

Q18. Evaluate $\lim_{x\to 0} \frac{2\sin x - \sin 2x}{x^3}$

Solution:

$$\lim_{x\to 0} \frac{2\sin x-\sin 2x}{x^3} = \lim_{x\to 0} \frac{2\sin x - 2\sin x\cos x}{x^3}$$

$[\because \sin 2x = 2\sin x\cos x]$

$$= \lim_{x\to 0} \frac{2\sin x(1-\cos x)}{x^3} = 2\lim_{x\to 0}\frac{\sin x}{x}\cdot\lim_{x\to 0}\left(\frac{1-\cos x}{x^2}\right)$$

$$= 2\cdot 1\lim_{x\to 0}\frac{1-\cos x}{x^2}\left[\because \lim_{x\to 0}\frac{\sin x}{x} = 1\right]$$

$$= 2\lim_{x\to 0}\frac{1-1+2\sin^2\frac{x}{2}}{x^2} = 2\lim_{x\to 0}\frac{2\sin^2\frac{x}{2}}{4\times\frac{x^2}{4}}$$

$$= \frac{2\cdot 2}{4}\lim_{x\to 0}\left(\frac{\sin\frac{x}{2}}{\frac{x}{2}}\right) = \lim_{x\to 0}\left(\frac{\sin\frac{x}{2}}{x/2}\right)^2 = 1$$

Q19. Evaluate $\lim_{x\to 0} \frac{1-\cos mx}{1-\cos nx}$

Solution: $$\lim_{x\to 0} \frac{1-\cos mx}{1-\cos nx} = \lim_{x\to 0} \frac{1-1+2\sin^2 mx/2}{1-1+2\sin nx/2}$$

$$\left[\because \cos mx = 1-2\sin^2\frac{mx}{2}, \sin nx = 1-2\sin^2\frac{nx}{2}\right]$$

$$= \lim_{x\to 0}\frac{\sin^2 mx/2}{\sin^2 nx/2} = \lim_{x\to 0}\frac{\dfrac{\sin^2 \frac{mx}{2}}{(mx/2)^2}\cdot(mx/2)^2}{\dfrac{\sin^2\frac{nx}{2}}{(nx/2)^2}\cdot\left(\frac{nx}{2}\right)^2}$$

$$= \frac{\lim_{x\to 0}\left(\dfrac{\sin\frac{mx}{2}}{mx/2}\right)^2}{\lim_{x\to 0}\left(\dfrac{\sin\frac{nx}{2}}{nx/2}\right)^2}\cdot\frac{m^2\frac{x^2}{4}}{n^2\frac{x^2}{4}}$$

$$= \frac{m^2}{n^2}\cdot\frac{\lim_{x\to 0}\left(\dfrac{\sin\frac{mx}{2}}{mx/2}\right)^2}{\lim_{x\to 0}\left(\dfrac{\sin\frac{nx}{2}}{nx/2}\right)^2} = m^2/n^2\left[\because \lim_{x\to 0}\frac{\sin x}{x}=1\right]$$

$[\because x\to 0 \Rightarrow kx\to 0]$

Q20. Evaluate $\lim_{x\to\pi/3}\dfrac{\sqrt{1-\cos 6x}}{\sqrt{2}\left(\frac{\pi}{3}-x\right)}$

Solution:

$$\lim_{x\to\pi/3}\frac{\sqrt{1-\cos 6x}}{\sqrt{2}\left(\frac{\pi}{3}-x\right)}$$

$$= \lim_{x\to\pi/3}\frac{\sqrt{1-1+2\sin^2 3x}}{\sqrt{2}\left(\frac{\pi}{3}-x\right)}[\because \cos 2x = 1-2\sin^2 x]$$

$$= \lim_{x\to\pi/3}\frac{\sqrt{2}\sin 3x}{\sqrt{2}\left(\frac{\pi}{3}-x\right)} = \lim_{x\to\pi/3}\frac{\sin 3x}{\frac{\pi}{3}-x}$$

$$= \lim_{x\to\pi/3}\frac{\sin(\pi-3x)}{\frac{\pi-3x}{3}}[\because \sin(\pi-\theta)=\sin\theta]$$

$$= 3\lim_{x\to\pi/3}\frac{\sin(\pi-3x)}{(\pi-3x)} = 3\times 1\left[\because \lim_{x\to 0}\frac{\sin x}{x}=1\right]$$

$$= 3\left[\because x\to\frac{\pi}{3}\Rightarrow\left(x-\frac{\pi}{3}\right)\to 0\right]$$

Q21. Evaluate $\lim_{x\to\pi/4}\dfrac{\sin x-\cos x}{x-\frac{\pi}{4}}$

Solution:

$$\lim_{x\to\pi/4}\frac{\sqrt{2}(\sin x\cdot\frac{1}{\sqrt{2}}-\cos x\cdot\frac{1}{\sqrt{2}})}{\left(x-\frac{\pi}{4}\right)}$$

$$= \lim_{x\to\pi/4}\frac{\sqrt{2}\left(\sin x\cos\frac{\pi}{4}-\cos x\cdot\sin\frac{\pi}{4}\right)}{\left(x-\frac{\pi}{4}\right)}$$

$$= \lim_{x\to\pi/4}\frac{\sqrt{2}\left\{\sin\left(x-\frac{\pi}{4}\right)\right\}}{\left(x-\frac{\pi}{4}\right)} = \sqrt{2}\lim_{x\to\pi/4}\frac{\sin\left(x-\frac{\pi}{4}\right)}{\left(x-\frac{\pi}{4}\right)} = \sqrt{2}$$

$$\left[\because x\to\frac{\pi}{4}\Rightarrow\left(x-\frac{\pi}{4}\right)\to 0,\ \lim_{x\to 0}\frac{\sin x}{x}=1\right]$$

Q22. Evaluate $\lim_{x\to\pi/6}\dfrac{\sqrt{3}\sin x-\cos x}{x-\frac{\pi}{6}}$

Solution:

$$\lim_{x\to\pi/6}\frac{\sqrt{3}\sin x-\cos x}{x-\frac{\pi}{6}}$$

$$= \lim_{x\to\pi/6}\frac{2\left(\frac{\sqrt{3}}{2}\sin x-\frac{1}{2}\cos x\right)}{\left(x-\frac{\pi}{6}\right)}$$

$$= \lim_{x\to\pi/6}\frac{2\left(\sin x\cos\frac{\pi}{6}-\cos x\sin\frac{\pi}{6}\right)}{x-\frac{\pi}{6}}$$

$[\because \sin A\cos B-\cos A\sin B=\sin(A-B)]$

$$= 2\lim_{x\to\pi/6}\frac{\sin\left(x-\frac{\pi}{6}\right)}{\left(x-\frac{\pi}{6}\right)} = 2$$

$$\left[\because x\to\frac{\pi}{6}\Rightarrow\left(x-\frac{\pi}{6}\right)\to 0,\ \lim_{x\to 0}\frac{\sin x}{x}=1\right]$$

Q23. Evaluate $\lim_{x\to 0}\dfrac{\sin 2x-3x}{2x+\tan 3x}$

Solution:

$$\lim_{x\to 0}\frac{\sin 2x-3x}{2x+\tan 3x} = \lim_{x\to 0}\frac{\dfrac{\sin 2x-3x}{2x}\cdot 2x}{\dfrac{2x+\tan 3x}{3x}\cdot 3x}$$

$$= \lim_{x\to 0}\frac{\left(\frac{\sin 2x}{2x}-\frac{3x}{2x}\right)2x}{\left(\frac{2x}{3x}+\frac{\tan 3x}{3x}\right)3x} = \lim_{x\to 0}\frac{\frac{\sin 2x}{2x}+3/2}{2/3+\frac{\tan 3x}{3x}}\cdot\frac{2}{3}$$

$$= \frac{2}{3}\lim_{x\to 0}\frac{\frac{\sin 2x}{2x}+\frac{3}{2}}{\frac{2}{3}+\lim_{x\to 0}\frac{\tan 3x}{3x}} = \frac{2}{3}\left(\frac{1+\frac{3}{2}}{\frac{2}{3}+1}\right)$$

$$\left[\because \lim_{x\to 0}\frac{\sin x}{x}=1, \lim_{x\to 0}\frac{\tan x}{x}=1\right]$$

$$= \frac{2}{3}\times\frac{5/2}{5/3} = \frac{2}{3}\times\frac{5}{2}\times\frac{3}{5} = 1$$

Q24. Evaluate $\lim_{x\to a}\frac{\sin x-\sin a}{\sqrt{x}-\sqrt{a}}$

Solution:

$$\lim_{x\to a}\frac{\sin x-\sin a}{\sqrt{x}-\sqrt{a}} = \lim_{x\to a}\frac{2\cos\left(\frac{x+a}{2}\right)\sin\left(\frac{x-a}{2}\right)}{\sqrt{x}-\sqrt{a}}$$

$$\left[\because \sin C-\sin D = 2\cos\frac{C+D}{2}\cdot\sin\frac{C-D}{2}\right]$$

$$= \lim_{x\to a}\frac{2\cos\left(\frac{x+a}{2}\right)\sin\left(\frac{x-a}{2}\right)(\sqrt{x}+\sqrt{a})}{(\sqrt{x}-\sqrt{a})(\sqrt{x}+\sqrt{a})}$$

$$= \lim_{x\to a}\frac{2\cos\left(\frac{x+a}{2}\right)\sin\left(\frac{x-a}{2}\right)(\sqrt{x}+\sqrt{a})}{x-a}$$

$$= 2\lim_{x\to a}\cos\left(\frac{x+a}{2}\right)(\sqrt{x}+\sqrt{a})\lim_{x\to 0}\frac{\sin\left(\frac{x-a}{2}\right)}{2\left(\frac{x-a}{2}\right)}$$

$$= 2\lim_{x\to 0}\cos\left(\frac{x+a}{2}\right)(\sqrt{x}+\sqrt{a})\cdot\frac{1}{2}\lim_{x\to 0}\frac{\sin\left(\frac{x-a}{2}\right)}{\left(\frac{x-a}{2}\right)}$$

$$= 2.\cos\frac{a}{2}\cdot\sqrt{a}\cdot\frac{1}{2}\left[\because \lim_{x\to 0}\frac{\sin x}{x}=1\right]$$

$$= \sqrt{a}\cos\frac{a}{2}$$

Q25. Evaluate $\lim_{x\to\pi/6}\frac{\cot^2 x-3}{\operatorname{cosec} x-2}$

Solution:

$$\lim_{x\to\pi/6}\frac{\cot^2 x-3}{\operatorname{cosec} x-2} = \lim_{x\to\pi/6}\frac{\operatorname{cosec}^2 x-1-3}{\operatorname{cosec} x-2}$$

$$[\because \operatorname{cosec}^2 x = 1+\cot^2 x]$$

$$= \lim_{x\to\pi/6}\frac{\operatorname{cosec}^2 x-4}{\operatorname{cosec} x-2} = \lim_{x\to\pi/6}\frac{(\operatorname{cosec} x)^2-(2)^2}{\operatorname{cosec} x-2}$$

$$= \lim_{x\to\pi/6}\frac{(\operatorname{cosec} x+2)(\operatorname{cosec} x-2)}{(\operatorname{cosec} x-2)}$$

$$= \lim_{x\to\pi/6}(\operatorname{cosec} x+2) = \operatorname{cosec}\frac{\pi}{6}+2 = 2+2 = 4$$

Q26. Evaluate $\lim_{x\to 0}\frac{\sqrt{2}-\sqrt{1+\cos x}}{\sin^2 x}$

Solution:

$$\lim_{x\to 0}\frac{\sqrt{2}-\sqrt{1+\cos x}}{\sin^2 x}$$

$$= \lim_{x\to 0}\frac{\sqrt{2}-\sqrt{1+2\cos^2\frac{x}{2}-1}}{\sin^2 x}\left[\because \cos x = 2\cos^2\frac{x}{2}-1\right]$$

$$= \lim_{x\to 0}\frac{\sqrt{2}-\sqrt{2\cos^2\frac{x}{2}}}{\sin^2 x}\left[\because \sin x = 2\sin\frac{x}{2}\cos\frac{x}{2}\right]$$

$$= \lim_{x\to 0}\frac{\sqrt{2}\left(1-\cos\frac{x}{2}\right)}{\sin^2 x} = \lim_{x\to 0}\frac{\sqrt{2}\left(1-1+2\sin^2\frac{x}{2}\right)}{\sin^2 x}$$

$$= \lim_{x\to 0}\frac{\sqrt{2}\left(2\sin^2\frac{x}{2}\right)}{\sin^2 x} = \lim_{x\to 0}2\sqrt{2}\frac{\sin^2\frac{x}{2}}{(x/2)^2}\cdot\frac{\left(\frac{x}{4}\right)^2}{\sin^2 x}$$

$$= 2\sqrt{2}\lim_{x\to 0}\left(\frac{\sin\frac{x}{4}}{x/4}\right)^2\cdot\lim_{x\to 0}\left(\frac{x}{\sin x}\right)^2\cdot\frac{1}{16}$$

$$= 2\sqrt{2}\cdot 1\cdot 1\cdot\frac{1}{16} = \frac{1}{4\sqrt{2}}$$

Q27. Evaluate $\lim_{x\to 0}\frac{\sin x-2\sin 3x+\sin 5x}{x}$

Solution:

$$\lim_{x\to 0}\frac{\sin x-2\sin 3x+\sin 5x}{x}$$

$$= \lim_{x\to 0}\frac{\sin 5x+\sin x-2\sin 3x}{x}$$

$$= \lim_{x\to 0}\frac{2\sin 3x\cos 2x-2\sin 3x}{x}$$

$$\lim_{x\to 0}\frac{2\sin 3x(\cos 2x-1)}{x} = \lim_{x\to 0}\frac{2\sin 3x}{\frac{1}{3}\times 3x}(\cos 2x-1)$$

$$\lim_{x\to 0}\frac{\sin 3x}{3x}(\cos 2x-1)=6\lim_{x\to 0}\frac{\sin 3x}{3x}\cdot\lim_{x\to 0}(\cos 2x-1)$$
$$=6\times 1\times 0=0$$

Q28. If $\lim_{x\to 1}\frac{x^4-1}{x-1}=\lim_{x\to k}\frac{x^3-k^3}{x^2-k^2}$, find the value of k.

Solution:

$$\lim_{x\to 1}\frac{x^4-1}{x-1}=\lim_{x\to k}\frac{x^3-k^3}{x^2-k^2}$$

$$\Rightarrow 4(1)^{4-1}=\lim_{x\to}\frac{\frac{x^3-k^3}{x-k}}{\frac{x^2-k^2}{x-k}}\left[\because \lim_{x\to 1}\frac{x^n-a^n}{x-a}=na^{n+1}\right]$$

$$\Rightarrow 4=\frac{\lim_{x\to k}\frac{x^3-k^3}{x-k}}{\frac{x^2-k^2}{x-k}}1\left[\because \lim_{x\to a}\frac{f(x)}{g(x)}=\frac{\lim_{x\to a}f(x)}{\lim_{x\to a}g(x)}\right]$$

$$\Rightarrow 4=\frac{3k^2}{2k}\Rightarrow 4=\frac{3}{2}k\Rightarrow\therefore k=\frac{4\times 2}{3}=8/3$$

Day 182

NCERT Exemplar Problems: Solutions (Q29–Q46)

Differentiate with respect to x in the following questions (Q29–Q42).

Q29. $\frac{x^4+x^3+x^2+1}{x}$

Solution: $\frac{d}{dx}\left(\frac{x^4+x^3+x^2+1}{x}\right)=\frac{d}{dx}\left(x^3+x^2+x+\frac{1}{x}\right)$

$$=\frac{d}{dx}x^3+\frac{d}{dx}x^2+\frac{d}{dx}x+\frac{d}{dx}(1/x)$$

$$=3x^2+2x+1+\left(-\frac{1}{x^2}\right)$$

$$=3x^2+2x+1-\frac{1}{x^2}=\frac{3x^2+2x^3+x^2-1}{x^2}$$

Q30. $\left(x+\frac{1}{x}\right)^3$

Solution: Let $y=\left(x+\frac{1}{x}\right)^3$

$$\therefore\frac{dy}{dx}=\frac{d}{dx}\left(x+\frac{1}{x}\right)^3=3\left(x+\frac{1}{x}\right)^{3-1}\frac{d}{dx}\left(x+\frac{1}{x}\right)$$

$$\left(\because\frac{d}{dx}y^n=ny^{n-1}\frac{dy}{dx}\right)$$

$$=3\left(x+\frac{1}{x}\right)^2\left(1-\frac{1}{x^2}\right)=3x^2-\frac{3}{x^2}-\frac{3}{x^4}+3$$

Q31. $(3x+5)(1+\tan x)$

Solution: Let $y=(3x+5)(1+\tan x)$

$$\therefore\frac{dy}{dx}=\frac{d}{dx}[(3x+5)(1+\tan x)]$$

$$=(3x+5)\frac{d}{dx}(1+\tan x)+(1+\tan x)\frac{d}{dx}(3x+5)$$

[By the product rule]

$$=(3x+5)(\sec^2 x)+(1+\tan x).3$$

$$=(3x+5)\sec^2 x+3(1+\tan x)$$

$$=3x\sec^2 x+5\sec^2 x+\tan x+3$$

Q32. $(\sec x-1)(\sec x+1)$

Solution: Let $y=(\sec x-1)(\sec x+1)$

$$y=(\sec^2-1)=\tan^2 x$$

$$\Rightarrow\frac{dy}{dx}=2\tan x\cdot\frac{d}{dx}\tan x$$

$$\left(\because\frac{d}{dx}y^n=ny^{n-1}\frac{dy}{dx}\right)$$

$$=2\tan x\cdot\sec^2 x$$

Q33. $\frac{3x+4}{5x^2-7x+9}$

Solution: Let $y=\frac{3x+4}{5x^2-7x+9}$

$$\therefore\frac{dy}{dx}=\frac{(5x^2-7x+9)\frac{d}{dx}(3x+4)-(3x+4)\frac{d}{dx}(5x^2-7x+9)}{(5x^2-7x+9)^2}$$

[By the quotient rule]

$$=\frac{(5x^2-7x+9)\cdot 3-(3x+4)(10x-7)}{(5x^2-7x+9)^2}$$

$$=\frac{15x^2-21x+27-30x^2+21x-40x+28}{(5x^2-7x+9)^2}$$

$$=\frac{-15x^2-40x+55}{(5x^2-7x+9)^2}=\frac{55-15x^2-40x}{(5x^2-7x+9)^2}$$

Q34. $\dfrac{x^5-\cos x}{\sin x}$

Solution: Let $y=\dfrac{x^5-\cos x}{\sin x}$

$$\therefore \frac{dy}{dx}=\frac{\sin x\dfrac{d}{dx}(x^5-\cos x)-(x^5-\cos x)\dfrac{d}{dx}\sin x}{(\sin x)^2}$$

[By the quotient rule]

$$=\frac{\sin x(5x^4+\sin x)-(x^5-\cos x)\cos x}{\sin^2 x}$$

$$=\frac{5x^4\sin x+\sin^2 x-x^5\cos x+\cos^2 x}{\sin^2 x}$$

$$=\frac{5x^4\sin x-x^5\cos x+\sin^2 x+\cos^2 x}{\sin^2 x}$$

$$=\frac{5x^4\sin x-x^5\cos x+1}{\sin^2 x}$$

Q35. $\dfrac{x^2-\cos\dfrac{\pi}{4}}{\sin x}$

Solution:

$$\text{Let } y=\frac{x^2-\cos\dfrac{\pi}{4}}{\sin x}=\frac{\dfrac{x^2}{\sqrt{2}}}{\sin x}\Rightarrow y=\frac{1}{\sqrt{2}}\cdot\frac{x^2}{\sin x}$$

$$\therefore \frac{dy}{dx}=\frac{1}{\sqrt{2}}\left[\frac{\sin x\cdot\dfrac{d}{dx}x^2-x^2\dfrac{d}{dx}\sin x}{\sin^2 x}\right]$$

[By the quotient rule]

$$=\frac{1}{\sqrt{2}}\left[\frac{\sin x\cdot 2x-x^2\cdot\cos x}{\sin^2 x}\right]=\frac{1}{\sqrt{2}}\frac{2x\sin x-x^2\cos x}{\sin^2 x}$$

$$=\frac{x}{\sqrt{2}}[2\operatorname{cosec} x-x\cot x\operatorname{cosec} x]$$

$$=\frac{x}{\sqrt{2}}\operatorname{cosec}[2-x\cot x]$$

Q36. $(ax^2+\cot x)(p+q\cos x)$

Solution: Let $y=(ax^2+\cot x)(p+q\cos x)$

$$\therefore \frac{y}{dx}=(ax^2+\cot x)\frac{d}{dx}(p+q\cos x)+(p+q\cos x)$$

$$\frac{d}{dx}(ax^2+\cot x)$$

[By the product rule]

$$=(ax^2+\cot x)(-q\sin x)+(p+q\cos x)(2ax-\operatorname{cosec}^2 x)$$

$$=-q\sin x(ax^2+\cot x)+(p+q\cos x)(2ax-\operatorname{cosec}^2 x)$$

Q37. $\dfrac{a+b\sin x}{c+d\cos x}$

Solution: Let $y=\dfrac{a+b\sin x}{c+d\cos x}$

$$=\frac{(c+d\cos x)\dfrac{d}{dx}(a+b\sin x)-(a+b\sin x)\dfrac{d}{dx}(c+d\cos x)}{(c+d\cos x)^2}$$

[By the quotient rule]

$$=\frac{(c+d\cos x)(b\cos x)-(a+b\sin x)(-d\sin x)}{(c+d\cos x)^2}$$

$$=\frac{bc\cos x+bd\cos^2 xad\sin s+bd\sin^2 x}{(c+d\cos x)^2}$$

$$=\frac{bc\cos x+ad\sin x+bd(\cos^2 x+\sin^2 x)}{(c+d\cos x)^2}$$

$$=\frac{bc\cos x+ad\sin x+bd}{(c+d\cos x)^2}$$

Q38. $(\sin x+\cos x)^2$

Solution: Let $y=(\sin x+\cos x)^2$

$$\therefore \frac{dy}{dx}=2(\sin x+\cos x)(\cos x-\sin x)$$

$$\left(\because \frac{d}{dx}y^n=ny^{n-1}\frac{dy}{dx}\right)$$

$$=2(\cos^2 x-\sin^2 x)$$

$$=2\cos 2x[\because \cos 2x=\cos^2 x-\sin^2 x]$$

Q39. $(2x-7)^2(3x+5)^3$

Solution: Let $y=(2x-7)^2(3x+5)^3$

$$\frac{dy}{dx}=(2x-7)^2\frac{d}{dx}(3x+5)^3+(3x+5)^3\frac{d}{dx}(2x-7)^2$$

[By the product rule]

$$=(2x-7)^2(3)(3x+5)^2(3)+(3x+5)^3 2(2x-7)(2)$$

$$\left(\because \frac{d}{dx}y^n=ny^{n-1}\frac{dy}{dx}\right)$$

$$=9(2x-7)^2(3x+5)^2+4(3x+5)^3(2x-7)$$

$$=(2x-7)(3x+5)^2[9(2x-7)+4(3x+5)]$$

$$=(2x-7)(3x+5)^2(18x-63+12x+20)$$

$$=(2x-7)(3x+5)^2(30x-43)$$

Q40. $x^2\sin x+\cos 2x$

Solution: Let $y=x^2\sin x+\cos 2x$

$\frac{dy}{dx}=\frac{d}{dx}(x^2\sin x)+\frac{d}{dx}\cos 2x$

$=x^2\cdot\cos x+\sin x 2x+(-\sin 2x)\cdot 2$

[By the product rule]

$=x^2\cos x+2x\sin x-2\sin 2x$

$\left(\because \frac{d}{dx}y^n=ny^{n-1}\frac{dy}{dx}\right)$

Q41. $\sin^3 x\cos^3 x$

Solution: Let $y=\sin^3 x\cos^3 x$

$\therefore \frac{dy}{dx}=\sin^3 x\cdot\frac{d}{dx}\cos^3 x+\cos^3 x\frac{d}{dx}\sin^3 x$

[By the product rule]

$=\sin^3 x\cdot 3\cos^3 x(-\sin x)+\cos^3 x\cdot 3\sin^2 s\cos x$

$\left(\because \frac{d}{dx}y^n=ny^{n-1}\frac{dy}{dx}\right)$

$=-3\cos^2 x\sin^4 x+3\sin^2 x\cos^4 x$

$=3\sin^2 x\cos^2 x(\cos^2 x-\sin^2 x)$

$=3\sin^2 x\cos^2 x\cos 2x=\frac{3}{4}(2\sin x\cos x)^2\cos 2x$

$=\frac{3}{4}\sin^2 2x\cos 2x$

Q42. $\frac{1}{ax^2+bx+c}$

Solution:

Let $y=\frac{1}{ax^2+bx+c}=(ax^2+bx+c)^{-1}$

$\therefore \frac{dy}{dx}=-(ax^2+bx+c)^{-2}(2ax+b)$

$\left(\because \frac{d}{dx}y^n=ny^{n-1}\frac{dy}{dx}\right)$

$=\frac{-(2ax+b)}{(ax^2+bx+c)^2}$

Differentiate with respect to *x* in the following questions (Q43–Q46) using first principle.

Q43. $\cos(x^2+1)$

Solution: Let

$f(x)=\cos(x^2+1), f(x+h)=\cos\{(x+h)^2+1\}$

$\therefore \frac{d}{dx}f(x)=\lim_{h\to 0}\frac{f(x+h)-f(x)}{h}$

$=\lim_{h\to 0}\frac{\cos\{(x+h)^2+1\}-\cos(x^2+1)}{h}$

$=\lim_{h\to 0}\frac{-2\sin\left\{\frac{(x+h)^2+1+x^2+1}{2}\right\}\sin\left\{\frac{(x+h)^2+1-x^2-1}{2}\right\}}{h}$

$\left[\because \cos C-\cos D=-2\sin\frac{C+D}{2}\cdot\sin\frac{C-D}{2}\right]$

$=\lim_{h\to 0}\frac{1}{h}\left[-2\sin\left\{\frac{(x+h)^2+x^2+2}{2}\right\}\sin\left\{\frac{(x+h)^2-x^2}{2}\right\}\right]$

$=\lim_{h\to 0}\frac{1}{h}\left[-2\sin\left\{\frac{(x+h)^2+x^2+2}{2}\right\}\sin\left\{\frac{h^2+2hx}{2}\right\}\right]$

$=-2\lim_{h\to 0}\sin\left\{\frac{(x+h)^2+x^2+2}{2}\right\}$

$\lim_{h\to 0}\left\{\frac{\sinh\left(\frac{h+2x}{2}\right)}{h\left(\frac{h+2x}{2}\right)}\right\}\times\left(\frac{h+2x}{2}\right)$

$=-2\lim_{h\to 0}\sin\left\{\frac{(x+h)^2+x^2+2}{2}\right\}\lim_{h\to 0}\left(\frac{h+2x}{2}\right)$

$\left[\because \lim_{x\to 0}\frac{\sin x}{x}=1\right]$

$=-2x\sin(x^2+1)[\because x\to 0\Rightarrow kx\to 0]$

Q44. $\frac{ax+b}{cx+d}$

Solution:

Let $f(x)=\frac{ax+b}{cx+d}\Rightarrow f(x+h)=\frac{a(x+h)+b}{c(x+h)+d}$

$\therefore \frac{d}{dx}f(x)=\lim_{h\to 0}\frac{1}{h}[f(x+h)-f(x)]$

$=\lim_{h\to 0}\frac{1}{h}\left[\frac{a(x+h)+b}{c(x+h)+d}-\frac{ax+b}{cx+d}\right]$

$=\lim_{h\to 0}\frac{1}{h}\left[\frac{ax+b+ah}{c(x+h)+d}-\frac{ax+b}{cx+d}\right]$

$=\lim_{h\to 0}\frac{1}{h}\left[\frac{(ax+ah+b)(cx+d)-(ax+b\{c(x+h)-d\}}{\{c(x+h)+d\}(cx+d)}\right]$

$=\lim_{h\to 0}\frac{1}{h}\left[\frac{(ax+ah+b)(cx+d)-(ax+b)(cx+ch+d)}{\{c(x+h)+d\}(cx+d)}\right]$

$=\lim_{h\to 0}\left[\frac{acx^2+achx+bcx+adx+adh+bd \\ -\{acx^2+achx+adx+bcx+bch+bd}{\{c(x+h)+d\}\{cx+d)}\right]$

$=\lim_{h\to 0}\frac{1}{h}\left[\frac{acx^2+achx+bcx+adx+adhbd \\ -acx^2-achx+adx-bcx-bch-bd}{\{c(x+h)+d\}(cx+d)}\right]$

$$= \lim_{h\to 0}\frac{1}{h}\left[\frac{adh-bch}{\{c(x+h)+d\}(cx+d)}\right]$$

$$= \lim_{h\to 0}\frac{ac-bd}{\{c(x+h)+d\}(cx+d)} = \frac{ac-bd}{(cx+d)^2}$$

Q45. $x^{2/3}$

Solution: Let $f(x)=x^{2/3} \Rightarrow f(x+h)=(x+h)^{2/3}$

Now, $\frac{d}{dx}f(x)=\lim_{h\to 0}\frac{f(x+h)-f(x)}{h}$

$$= \lim_{h\to 0}\frac{1}{h}[(x+h)^{2/3}-x^{2/3}] = \lim_{h\to 0}\frac{1}{h}\left[x^{2/3}\left(1+\frac{h}{x}\right)^{2/3}-x^{2/3}\right]$$

$$= \lim_{h\to 0}\frac{1}{h}\left[x^{2/3}\left(1+\frac{h}{x}\cdot\frac{2}{3}+\frac{2}{3}\left(\frac{2}{3}-1\right)\frac{h^2}{x^2}+\ldots\right)-1\right]$$

$$\left[\because (1+x)^n = 1+nx+\frac{n(n-1)}{2!}x^2+\ldots\ldots\right]$$

$$= \lim_{h\to 0}\frac{1}{h}\left[x^{2/3}\left(\frac{2}{x}\cdot\frac{h}{x}-\frac{2}{9}\cdot\frac{h^2}{x^2}+\ldots\right)\right]$$

$$= \lim_{h\to 0}\frac{x^{2/3}}{h}\cdot\frac{2}{3}\frac{h}{x}\left(1-\frac{1}{3}\cdot\frac{h}{x}+\ldots\ldots\right) = \frac{2}{3}x^{2/3-1} = \frac{2}{3}x^{-1/3}$$

Q46. $x\cos x$

Solution: Let $f(x)=x\cos x$

$\therefore f(x+h)=(x+h)\cos(x+h)$

$$\therefore \frac{d}{dx}f(x)=\lim_{h\to 0}\frac{f(x+h)-f(x)}{h}$$

$$= \lim_{h\to}\frac{1}{h}[(x+h)\cos(x+h)-x\cos x]$$

$$= \lim_{h\to 0}\frac{1}{h}[x\cos(x+h)+h\cos(x+h)-x\cos x]$$

$$= \lim_{h\to 0}\frac{1}{h}[x\{\cos(x+h)-\cos x\}+h\cos(x+h)]$$

$$= \lim_{h\to 0}\frac{1}{h}\left[x\left\{-2\sin\left(\frac{2x+h}{2}\right)\sin\frac{h}{2}\right\}+h\cos(x+h)\right]$$

$$= \lim_{h\to 0}\left[-2x\sin\left(x+\frac{h}{2}\right)\frac{\sin h/2}{h}+\cos(x+h)\right]$$

$$\left[\because \cos C-\cos D=-2\sin\frac{C+D}{2}\cdot\sin\frac{C-D}{2}\right]$$

$$= -2\lim_{h\to 0}x\sin\left(x+\frac{h}{2}\right)\lim_{h\to 0}\frac{\sin h/2}{h/2}\cdot\frac{1}{2}+\lim_{h\to 0}\cos(x+h)$$

$$= -2\cdot\frac{1}{2}x\sin x+\cos x = \cos x - x\sin x$$

Day 183

NCERT Exemplar Problems: Solutions (Q47–Q62)

Q47. Evaluate $\lim_{y\to 0}\frac{(x+y)\sec(x+y)-x\sec x}{y}$

Solution:

$$\lim_{y\to 0}\frac{(x+y)\sec(x+y)-x\sec x}{y}$$

$$= \lim_{y\to 0}\frac{\frac{x+y}{\cos(x+y)}-\frac{x}{\cos x}}{y}$$

$$= \lim_{y\to 0}\frac{(x+y)\cos x - x\cos(x+y)}{y\cos x\cos(x+y)}$$

$$= \lim_{y\to 0}\left[\frac{x\cos x+y\cos x-x\cos(x+y)}{y\cos x\cos(x+y)}\right]$$

$$= \lim_{y\to 0}\left[\frac{x\cos x-x\cos(x+y)+y\cos x}{y\cos x\cos(x+y)}\right]$$

$$= \lim_{y\to 0}\frac{x\{\cos x-x\cos(x+y)\}+y\cos x}{y\cos x\cos(x+y)}$$

$$= \lim_{y\to 0}\frac{x\left[-2\sin\left(x+\frac{y}{2}\right)\sin\left(\frac{-y}{2}\right)\right]+y\cos x}{y\cos x\cos(x+y)}$$

$$\left[\because \cos C-\cos D=-2\sin\frac{C+D}{2}\cdot\sin\frac{C-D}{2}\right]$$

$$= \lim_{y\to 0}\left[\frac{x\left\{2\sin\left(x+\frac{y}{2}\right)\sin\frac{y}{2}\right\}+y\cos x}{y\cos x\cos(x+y)}\right]$$

$$= \lim_{y\to 0}\frac{2x\sin\left(x+\frac{y}{2}\right)}{\cos x\cos(x+y)}\cdot\lim_{y\to 0}\frac{\sin\frac{y}{2}}{y/2}\cdot\frac{1}{2}+\lim_{y\to 0}\sec(x+y)$$

$$\left[\because \lim_{y\to 0}\frac{\sin x}{x}=1, x\to 0 \Rightarrow kx\to 0\right]$$

$$= \lim_{y\to 0}\frac{2x\sin\left(x+\frac{y}{2}\right)}{\cos x\cos(x+y)}\cdot\frac{1}{2}+\lim_{y\to 0}\sec(x+y)$$

$$= \frac{2x\sin x}{\cos x\cos x}\cdot\frac{1}{2}+\sec x = x\tan x\sec x+\sec x$$

$$= \sec x(x\tan x+1)$$

Q48. $\lim_{x\to 0}\dfrac{\sin(\alpha+\beta)+\sin(\alpha-\beta)x+\sin 2\alpha x}{\cos 2\beta x-\cos 2\alpha x}\cdot x$

Solution:

$\lim_{x\to 0}\dfrac{\sin(\alpha+\beta)+\sin(\alpha-\beta)x+\sin 2\alpha x}{\cos 2\beta x-\cos 2\alpha x}\cdot x$

$= \lim_{x\to 0}\dfrac{[2\sin\alpha x.\cos\beta x+\sin 2\alpha x]}{\cos 2\beta x-\cos 2\alpha x}\cdot x$

$\left[\because \sin C+\sin D = 2\sin\dfrac{C+D}{2}\cos\dfrac{C-D}{2}\right]$

$= \lim_{x\to 0}\dfrac{[2\sin\alpha x\cos\beta x+\sin 2\alpha x]x}{2\sin(\alpha+\beta)x\sin(\alpha-\beta)x}$

$\left[\because \cos C-\cos D = 2\sin\dfrac{C+D}{2}\cdot\sin\dfrac{D-C}{2}\right]$

$= \lim_{x\to 0}\dfrac{[2\sin\alpha x\cos\beta x+2\sin\alpha x\cos\alpha x]x}{2\sin(\alpha+\beta)x\sin(\alpha-\beta)x}$

$= \lim_{x\to 0}\dfrac{2\sin\alpha x[\cos\beta x+\cos\alpha x]x}{2\sin(\alpha+\beta)x\sin(\alpha-\beta)x}$

$= \lim_{x\to 0}\dfrac{\sin\alpha x\cdot x}{2\sin\left(\dfrac{\alpha+\beta}{2}\right)x\sin\left(\dfrac{\alpha-\beta}{2}\right)x}$

$= \dfrac{\dfrac{1}{2}\lim_{x\to 0}\dfrac{\sin\alpha x}{\alpha x}\cdot\alpha x^2}{\lim_{x\to 0}\dfrac{\sin\left(\dfrac{\alpha+\beta}{2}\right)x}{\left(\dfrac{\alpha+\beta}{2}\right)x}\lim_{x\to 0}\dfrac{\left(\dfrac{\alpha-\beta}{2}\right)x}{\left(\dfrac{\alpha-\beta}{2}\right)x}\cdot\left(\dfrac{\alpha^2-\beta^2}{4}\right)x^2}$

$\left[\because \lim_{x\to 0}\dfrac{\sin x}{x}=1, x\to\Rightarrow kx\to 0\right]$

$= \dfrac{1}{2}\cdot\dfrac{\alpha.4}{\alpha^2-\beta^2}\left[\dfrac{\lim_{x\to 0}\dfrac{\sin\alpha x}{\alpha x}}{\lim_{x\to 0}\sin\dfrac{\left(\dfrac{\alpha+\beta}{2}\right)x}{\left(\dfrac{\alpha+\beta}{2}\right)x}\lim_{x\to 0}\dfrac{\left(\dfrac{\alpha-\beta}{2}\right)x}{\left(\dfrac{\alpha-\beta}{2}\right)x}}\right]$

$= \dfrac{1}{2}\cdot\dfrac{4\alpha}{\alpha^2-\beta^2} = \dfrac{2\alpha}{\alpha^2-\beta^2}$

Q49. $\lim_{x\to\pi/4}\dfrac{\tan^3 x-\tan x}{\cos\left(x+\dfrac{\pi}{4}\right)}$

Solution:

Given $\lim_{x\to\pi/4}\dfrac{\tan^3 x-\tan x}{\cos\left(x+\dfrac{\pi}{4}\right)}\left[\dfrac{0}{0}\text{form}\right]$

$= \lim_{x\to\pi/4}\dfrac{\tan x(\tan^2 x-1)}{\cos\left(x+\dfrac{\pi}{4}\right)}$

$= \lim_{x\to\pi/4}\tan x\cdot\lim_{x\to\pi/4}\left(\dfrac{1-\tan^2 x}{\cos\left(x+\dfrac{gx}{4}\right)}\right)$

$= -1\times\lim_{x\to\pi/4}\dfrac{(1+\tan x)(1-\tan x)}{\cos\left(x+\dfrac{\pi}{4}\right)}$

$[\because a^2-b^2=(a+b)(a-b)]$

$= -\lim_{x\to\pi/4}(1+\tan x)\left[\dfrac{\cos x-\sin x}{\cos x\cdot\cos\left(x+\dfrac{\pi}{4}\right)}\right]$

$= -(1+1)\times\lim_{x\to\pi/4}\dfrac{\sqrt{2}\left[\dfrac{1}{\sqrt{2}}\cdot\cos x-\dfrac{1}{\sqrt{2}}\cdot\sin x\right]}{\cos x\cdot\cos\left(x+\dfrac{\pi}{4}\right)}$

$= -2\sqrt{2}\lim_{x\to\pi/4}\left[\dfrac{\cos\dfrac{\pi}{4}\cdot\cos x-\sin\dfrac{\pi}{4}\cdot\sin x}{\cos x\cos\left(x+\dfrac{\pi}{4}\right)}\right]$

$[\because \cos A\cdot\cos B-\sin A\sin B=\cos(A+B)]$

$= -2\sqrt{2}\lim_{x\to\pi/4}\dfrac{\cos\left(x+\dfrac{\pi}{4}\right)}{\cos x\cdot\cos\left(x+\dfrac{\pi}{4}\right)}$

$= -2\sqrt{2}\times\dfrac{1}{1/\sqrt{2}} = -2\sqrt{2}\times\sqrt{2} = -4$

Q50. $\lim_{x\to\pi}\dfrac{1-\sin\dfrac{x}{2}}{\cos\dfrac{x}{2}\left(\cos\dfrac{x}{4}-\sin\dfrac{x}{4}\right)}$

Solution:

$\lim_{x\to\pi}\dfrac{1-\sin\dfrac{x}{2}}{\cos\dfrac{x}{2}\left(\cos\dfrac{x}{4}-\sin\dfrac{x}{4}\right)}$

$= \lim_{x\to\pi}\dfrac{\cos^2\dfrac{x}{2}+\sin^2\dfrac{x}{2}-2\cdot\sin\dfrac{x}{4}\cdot\cos\dfrac{x}{4}}{\cos\dfrac{x}{2}\cdot(\cos\dfrac{x}{4}-\sin\dfrac{x}{4})}$

$[\because \sin^2\theta+\cos^2\theta=1\sin 2\theta=2\sin\theta\cos\theta]$

$$= \lim_{x\to\pi} \frac{\left(\cos\frac{x}{2} - \sin\frac{x}{4}\right)^2}{\left(\cos^2\frac{x}{4} - \sin^2\frac{x}{4}\right)\left(\cos\frac{x}{4} - \sin\frac{x}{4}\right)}$$

$[\because \cos^2 2\theta = \cos^2\theta - \sin^2\theta]$

$$\lim_{x\to\pi} \frac{\left(\cos\frac{x}{4} - \sin\frac{x}{4}\right)}{\left(\cos\frac{x}{4} + \sin\frac{x}{4}\right)\left(\cos\frac{x}{4} - \sin\frac{x}{4}\right)}$$

$[\because a^2 - b^2 = (a+b)(a-b)]$

$$\lim_{x\to\pi} \frac{1}{\cos\frac{x}{4} + \sin\frac{x}{4}} = \frac{1}{\frac{1}{\sqrt{2}} + \frac{1}{\sqrt{2}}} = \frac{\sqrt{2}}{2} = \frac{1}{\sqrt{2}}$$

Q51. Show that $\lim_{x\to\pi/4} \frac{|x-4|}{x-4}$ does not exist.

Solution: Given that $\lim_{x\to\pi/4} \frac{|x-4|}{x-4}$

$$\text{LHL} = \lim_{x\to\pi^-/4} \frac{-(x-4)}{x-4} = 1\ [\because |x-4| = -(x-4), x < 4]$$

$$\text{RHL} = \lim_{x\to\pi^+/4} \frac{(x-4)}{x-4} = 1[\because |x-4| = (x-4), x > 4]$$

$\because$ LHL $\neq$ RHL

Therefore, limit does not exist.

Q52. If

$$f(x) = \begin{cases} \frac{k\cos x}{\pi - 2x}, & \text{when } x \neq \frac{\pi}{2} \\ 3, & \text{when } x = \pi/2 \end{cases},$$

$\lim_{x\to\pi/2} f(x) = f(\pi/2)$, find the value of k.

Solution:

$$\text{Given } f(x) = \begin{cases} \frac{k\cos x}{\pi - 2x}, \text{when } x \neq \frac{\pi}{2} \\ 3, \text{when } x = \pi/2 \end{cases}$$

$$\therefore \text{LHL} = \lim_{x\to\frac{\pi^-}{2}} \frac{k\cos x}{\pi - 2x} = \lim_{h\to 0} \frac{k\cos\left(\frac{\pi}{2} - h\right)}{\pi - 2\left(\frac{\pi}{2} - h\right)}$$

$$= \lim_{h\to 0} \frac{k\sin h}{\pi - \pi + 2h} = \lim_{h\to 0} \frac{k\sin h}{2h} = \frac{k}{2}\lim_{h\to 0}\frac{\sin h}{h} = \frac{k}{2}\cdot 1 = \frac{k}{2}$$

$$\left[\because \lim_{h\to 0} \frac{\sin x}{x} = 1\right]$$

$$\text{RHL} = \lim_{x\to\frac{\pi^+}{2}} \frac{k\cos x}{\pi - 2x} = \lim_{x\to\frac{\pi}{2}} + \frac{k\cos\left(\frac{\pi}{2} + h\right)}{\pi - 2\left(\frac{\pi}{2} + h\right)}$$

$$= \lim_{h\to 0} \frac{-k\sin h}{\pi - \pi - 2h} = \lim_{h\to 0} \frac{k\sin h}{2h} = \frac{k}{2}\lim_{h\to 0}\frac{\sin h}{2h} = \frac{k}{2}, f\left(\frac{\pi}{2}\right) = 3$$

Since $\lim_{x\to\pi/2} f(x) = f(\pi/2) \Rightarrow \frac{k}{2} = 3 \Rightarrow k = 6$

Q53. If $f(x) = \begin{cases} x+2, x \le -1 \\ cx^2, x > -1 \end{cases}$, find c when $\lim_{x\to -1} f(x)$ exists.

Solution: Given that $f(x) = \begin{cases} x+2, x \le -1 \\ cx^2, x > -1 \end{cases}$,

$$\text{LHL} = \lim_{x\to -1^-} f(x) = \lim_{x\to -1^-} (x+2)$$

$$= \lim_{h\to 0}(-1-h+2) = \lim_{h\to 0}(1-h) = 1$$

$$\text{RHL} = \lim_{x\to -1^+} f(x) = \lim_{x\to -1^+} cx^2 = \lim_{h\to 0} c(-1+h)^2 = c$$

Since $\lim_{x\to -1} f(x)$ exists, LHL = RHL $\Rightarrow c = 1$

Q54. $\lim_{x\to\pi} \frac{\sin x}{x-\pi}$ is equal to:

(a) 1 (b) 2 (c) -1 (d) -2

Solution: (c)

$$\lim_{x\to\pi} \frac{\sin x}{x-\pi} = \lim_{x\to\pi} \frac{\sin(\pi - x)}{x-\pi}$$

$[\because \sin\theta = \sin(\pi - \theta)]$

$$= -\lim_{x\to\pi} \frac{\sin(\pi - x)}{x - \pi} = -1$$

$$\left[\because \lim_{x\to 0} \frac{\sin x}{x} = 1, \pi - x \to 0 \Rightarrow x \to \pi\right]$$

Q55. $\lim_{x\to 0} \frac{x^2\cos x}{1-\cos x}$ is equal to:

(a) 2 (b) 3/2 (c) -3/2 (d) 1

Solution: (a)

$$\lim_{x\to 0} \frac{x^2\cos x}{1-\cos x} = \lim_{x\to 0} \frac{x^2\cos x}{2\sin^2 x/2}$$

$$\left[\because 1 - \cos x = 2\sin^2 x/2\right]$$

$$= 2\lim_{x\to 0} \frac{(x/2)^2}{\sin^2 x/2} \cdot \lim_{x\to 0} \cos x = 2\cdot 1 = 2$$

Q56. $\lim_{x\to 0} \frac{(1+x)^n - 1}{x}$ is equal to:

(a) n (b) 1 (c) –n (d) 0

Solution: (a)

$$\lim_{x\to 0} \frac{(1+x)^n - 1}{x} = \lim_{x\to 0} \frac{(1+x)^n - 1}{(1+x) - 1}$$

$$= \lim_{x\to 0}\frac{(1+x)^n-1}{(1+x)-1} = \lim_{x\to 0}\frac{(1+x)^n-1^n}{(1+x)-1}$$

$$= \lim_{x\to 0}\frac{(1+x)^n-1^n}{(1+x)-x} = n\cdot(1)^{n-1} = n$$

$$\left[\because \lim_{x\to a}\frac{x^n-a^n}{x-a} = na^{n-1}\right]$$

Q57. $\lim_{x\to 1}\frac{x^m-1}{x^n-1}$ is equal to:

(a) 1 (b) m/n (c) –m/n (d) m^2/n^2

Solution: (b)

$$\lim_{x\to 1}\frac{x^m-1}{x^n-1} = \lim_{x\to 1}\frac{\frac{x^m-1}{x-1}}{\frac{x^n-1}{x-1}}$$

$$\frac{\lim_{x\to 1}\frac{x^m-1^m}{x-1}}{\lim_{x\to 1}\frac{x^n-1^n}{x-1}} = \frac{m(1)^{m-1}}{n(1)^{n-1}} = m/n$$

$$\left[\because \lim_{x\to a}\frac{x^n-a^n}{x-a} = na^{n-1}\right]$$

Q58. $\lim_{\theta\to 0}\frac{1-\cos 4\theta}{1-\cos 6\theta}$ is equal to:

(a) 4/9 (b) 1/2 (c) -1/2 (d) -1

Solution: (a)

$$\lim_{\theta\to 0}\frac{1-\cos 4\theta}{1-\cos 6\theta} = \lim_{\theta\to 0}\frac{2\sin^2 2\theta}{2\sin^2 3\theta}$$

$[\because 1-\cos 2\theta = 2\sin^2\theta]$

$$= \frac{\lim_{\theta\to 0}\frac{\sin^2 2\theta}{(2\theta)^2}\cdot(2\theta)^2}{\lim_{\theta\to 0}\frac{\sin^2 3\theta}{(3\theta)^2}} = \frac{4}{9}\frac{\lim_{\theta\to 0}\left(\frac{\sin 2\theta}{2\theta}\right)^2}{\lim_{\theta\to 0}\left(\frac{\sin 3\theta}{3\theta}\right)^2}$$

$$\left[\because \lim_{x\to 0}\frac{\sin x}{x} = 1, x\to 0 \Rightarrow kx\to 0\right]$$

$= 4/9$

Q59. $\lim_{x\to 0}\frac{\operatorname{cosec} x - \cot x}{x}$ is equal to:

(a) -1/2 (b) 1 (c) 1/2 (d) 1

Solution: (c)

$$\lim_{x\to 0}\frac{\operatorname{cosec} x - \cot x}{x}$$

$$= \lim_{x\to 0}\frac{\frac{1}{\sin x}-\frac{\cos x}{\sin x}}{x} = \lim_{x\to 0}\frac{1-\cos x}{x\cdot\sin x}$$

$$= \lim_{x\to 0}\frac{2\sin^2 x/2}{x\cdot 2\sin x/2\cos x/2} = \lim_{x\to 0}\frac{\tan x/2}{x}$$

$$= \lim_{x\to 0}\frac{\tan x/2}{x/2}\cdot\frac{1}{2} = \frac{1}{2}\left[\because \lim_{\theta\to 0}\frac{\tan\theta}{\theta} = 1\right]$$

Q60. $\lim_{x\to 0}\frac{\sin x}{\sqrt{x+1}-\sqrt{1-x}}$ is equal to:

(a) 2 (b) 0 (c) 1 (d) -1

Solution: (c)

$$\lim_{x\to 0}\frac{\sin x}{\sqrt{x+1}-\sqrt{1-x}}$$

$$= \lim_{x\to 0}\frac{\sin x}{\sqrt{x+1}-\sqrt{1-x}}\cdot\frac{\sqrt{x+1}+\sqrt{1-x}}{\sqrt{x+1}+\sqrt{1-x}}$$

$$= \lim_{x\to 0}\frac{\sin x(\sqrt{x+1}+\sqrt{1-x})}{(x+1)-(1-x)}$$

$$= \lim_{x\to 0}\frac{\sin x(\sqrt{x+1}+\sqrt{1-x})}{x+1-1+x}$$

$$= \frac{1}{2}\lim_{x\to 0}\frac{\sin x}{x}\lim_{x\to 0}(\sqrt{x+1}+\sqrt{1-x}) = \frac{1}{2}\cdot 1\cdot 2 = 1$$

$$\left[\because \lim_{x\to 0}\frac{\sin x}{x} = 1\right]$$

Q61. $\lim_{x\to\pi/4}\frac{\sec^2 x - 2}{\tan x - 1}$ is:

(a) 3 (b) 1 (c) 0 (d) 2

Solution: (d)

$$\lim_{x\to\pi/4}\frac{\sec^2 x-2}{\tan x-1}$$

$$= \lim_{x\to\pi/4}\frac{1+\tan^2 x-2}{\tan x-1} = \lim_{x\to\pi/4}\frac{\tan^2 x-1}{\tan x-1}$$

$$= \lim_{x\to\pi/4}\frac{(\tan x+1)(\tan x-1)}{\tan x-1} = \lim_{x\to\pi/4}(\tan x-1) = 2$$

Q62. $\lim_{x\to 1}\frac{(\sqrt{x}-1)(2x-3)}{2x^2+x-3}$ is equal to:

(a) 1/10 (b) –1/10 (c) 1

(d) none of these

Solution: (b)

$$\lim_{x\to 1}\frac{(\sqrt{x}-1)(2x-3)}{2x^2+x-3}$$

$$= \lim_{x\to 1}\frac{(\sqrt{x}-1)(2x-3)}{(2x+3)(x-1)} = \lim_{x\to 1}\frac{(\sqrt{x}-1)(2x-3)}{(2x+3)(\sqrt{x}-1)(\sqrt{x}+1)}$$

$$= \lim_{x\to 1}\frac{2x-3}{(2x+3)(\sqrt{x}+1)} = \frac{-1}{5\times 2} = \frac{-1}{10}$$

Day 184

NCERT Exemplar Problems: Solutions (Q63–Q80)

Q63. If $f(x)=\begin{cases}\dfrac{\sin[x]}{[x]}, & [x]\neq 0\\ 0, & [x]=0\end{cases}$, where [.] denotes the greatest integer function, then $\lim\limits_{x\to 0} f(x)$ is equal to:

(a) 1 (b) 0 (c) –1
(d) Does not exist

Solution: (d)

$$f(x)=\begin{cases}\dfrac{\sin[x]}{[x]}, & [x]\neq 0\\ 0, & [x]=0\end{cases},$$

$$\therefore \text{LHL}=\lim_{x\to 0^-} f(x)=\lim_{x\to 0^-}\frac{\sin[x]}{[x]}=\lim_{h\to 0}\frac{\sin[0-h]}{[0-h]}$$

$$=\lim_{h\to 0}\frac{-\sin[-h]}{[-h]}=-1$$

$$\text{RHL}=\lim_{x\to 0^+} f(x)=\lim_{x\to 0^+}\frac{\sin[x]}{[x]}=\lim_{x\to 0^+}\frac{\sin[0+h]}{[0+h]}$$

$$=\lim_{h\to 0}\frac{\sin[h]}{[h]}=1 \because \text{LHL}\neq\text{RHL}$$

Therefore, limit does not exist.

Q64. $\lim\limits_{x\to 0}\dfrac{|\sin x|}{x}$ is equal to:

(a) 1 (b) –1
(c) Does not exist
(d) None of these

Solution: (c)

$$\lim_{x\to 0}\frac{|\sin x|}{x}$$

$$\therefore \text{LHL}=\lim_{x\to 0}\left(\frac{-\sin x}{x}\right)=-\lim_{x\to 0^-}\frac{\sin x}{x}=-1$$

$$\text{RHL}=\lim_{x\to 0^+}\frac{\sin x}{x}=1 \because \text{LHL}\neq\text{RHL}$$

Therefore, limit does not exist.

Q65. If $f(x)=\begin{cases}x^2-1, & 0<x<2\\ 2x+3, & 2\le x<3\end{cases}$, then the quadratic equation whose roots are $\lim\limits_{x\to 2^-} f(x), \lim\limits_{x\to 2^+} f(x)$ is:

(a) $x^2-6x+9=0$ (b) $x^2-7x+8=0$
(c) $x^2-14x+49=0$ (d) $x^2-10x+21=0$

Solution: (d)

$$f(x)=\begin{cases}x^2-1, & 0<x<2\\ 2x+3, & 2\le x<3\end{cases},$$

$$\therefore \lim_{x\to 2^-} f(x)=\lim_{x\to 2^-}(x^2-1)$$

$$=\lim_{h\to 0}[(2-h)^2-1]=\lim_{h\to 0}(4+h^2-4h-1)$$

$$=\lim_{h\to 0}(h^2-4h+3)=3$$

$$\lim_{x\to 2^+} f(x)=\lim_{x\to 2^+}(2x+3)$$

$$=\lim_{h\to 0}[2(2+h)+3]=\lim_{h\to 0}(4+2h+3)=7$$

Thus, the quadratic equation whose roots are 3 and 7 is $x^2-(3+7)x+3\times 7=0 \Rightarrow x^2-10x+21=0$

Q66. $\lim\limits_{x\to 0}\dfrac{\tan 2x-x}{3x-\sin x}$ is equal to:

(a) 2 (b) 1/2 (c) –1/2 (d) 1/4

Solution: (b)

$$\lim_{x\to 0}\frac{\tan 2x-x}{3x-\sin x}=\lim_{x\to 0}\frac{x\left[\dfrac{\tan 2x}{x}-1\right]}{x\left[3-\dfrac{\sin x}{x}\right]}$$

$$\frac{\lim\limits_{x\to 0} 2\times\dfrac{\tan 2x}{2x}-1}{3-\lim\limits_{x\to 0}\dfrac{\sin x}{x}}=\frac{2-1}{3-1}=1/2$$

Q67. If $f(x)=x-[x], \in R$, then $f'(1/2)$ is equal to:

(a) 3/2 (b) 1 (c) 0 (d) –1

Solution: (b)

$f(x)=x-[x]$

$$\therefore \text{Lf}'(1/2)=\text{LHD}=\lim_{h\to 0}\frac{f\left(\frac{1}{2}-h\right)-f(1/2)}{-h}$$

$$=\lim_{n\to 0}\frac{\left(\frac{1}{2}-h\right)-\left(\frac{1}{2}-h\right)-\frac{1}{2}+\left(\frac{1}{2}\right)}{h}$$

$$=\lim_{h\to 0}\frac{\frac{1}{21}-h-0-\frac{1}{2}+0=1}{h}$$

$$\text{Rf}'\left(\frac{1}{2}\right)=\text{RHD}\lim_{h\to 0}\frac{f\left(\frac{1}{2}+h\right)-f\left(\frac{1}{2}\right)}{h}$$

$$= \lim_{h \to 0} \frac{\left(\frac{1}{2}+h\right) - \left(\frac{1}{2}+h\right) - \frac{1}{2} + \left(\frac{1}{2}\right)}{h}$$

$$= \lim_{h \to 0} \frac{\frac{1}{2}+h-0-\frac{1}{2}+0}{h} = 1 \because \text{LHD} = \text{RHD}$$

$\therefore f'(1/2) = 1$

Q68. If $y = \sqrt{x} + \frac{1}{\sqrt{x}}$, then $\frac{dy}{dx}$ at $x = 1$ equal to:

(a) 1 (b) 1/2 (c) $1/\sqrt{2}$ (d) 0

Solution: (d)

$$y = \sqrt{x} + \frac{1}{\sqrt{x}},$$

$$\frac{dy}{dx} = \frac{1}{2\sqrt{x}} - \frac{1}{2x^{3/2}}$$

$$\therefore \left(\frac{dy}{dx}\right)_{x=1} = \frac{1}{2} - \frac{1}{2} = 0$$

Q69. If $f(x) = \frac{x-4}{2\sqrt{x}}$, then $f'(1)$ is equal to:

(a) 5/4 (b) 4/5 (c) 1 (d) 0

Solution: (a)

$$f(x) = \frac{x-4}{2\sqrt{x}},$$

$$f'(x) = \frac{2\sqrt{x} - (x-4)\cdot 2 \cdot \frac{1}{2\sqrt{x}}}{4x}$$

$$= \frac{2x-(x-4)}{4x^{3/2}} = \frac{2x-x+4}{4x^{3/2}} = \frac{x+4}{4x^{3/2}}$$

$$\therefore f'(1) = \frac{1+4}{4 \times (1)^{3/2}} = 5/4$$

Q70. If $y = \dfrac{1+\frac{1}{x^2}}{1-\frac{1}{x^2}}$, then $\frac{dy}{dx}$ is equal to:

(a) $\frac{-4x}{(x^2-1)^2}$ (b) $\frac{-4x}{x^2-1}$

(c) $\frac{1-x^2}{4x}$ (d) $\frac{4x}{x^2-1}$

Solution: (a)

$$y = \frac{1+\frac{1}{x^2}}{1-\frac{1}{x^2}} \Rightarrow y = \frac{x^2+1}{x^2-1}$$

$$\therefore \frac{dy}{dx} = \frac{(x^2-1)2x-(x^2+1)(2x)}{(x^2-1)^2}$$

[By the quotient rule]

$$\frac{dy}{dx} = \frac{2x(x^2-1-x^2-1)}{(x^2-1)^2} = \frac{2x(-2)}{(x^2-1)^2} = \frac{-4x}{(x^2-1)^2}$$

Q71. If $y = \frac{\sin x + \cos x}{\sin x - \cos x}$, then $\frac{dy}{dx}$ at $x = 0$ is equal to:

(a) –2 (b) 0 (c) 1/2

(d) Does not exist

Solution: (a)

$$y = \frac{\sin x + \cos x}{\sin x - \cos x},$$

$$= \frac{-(\sin x - \cos x)^2 - (\sin x + \cos x)^2}{(\sin x - \cos x)^2}$$

$$= \frac{-[(\sin x - \cos x)^2 + (\sin x + \cos x)^2]}{(\sin x - \cos x)^2}$$

$$= \frac{-2}{(\sin x - \cos x)^2} \Rightarrow \left(\frac{dy}{dx}\right)_{x=0} = -2$$

Q72. If $y = \frac{\sin(x+9)}{\cos x}$, then $\frac{dy}{dx}$ at $x = 0$ is equal to:

(a) cos 9 (b) sin 9 (c) 0 (d) 1

Solution: (a)

$$y = \frac{\sin(x+9)}{\cos x}$$

$$\therefore \frac{dy}{dx} = \frac{\cos x \cos(x+9) - \sin(x+9)(-\sin x)}{(\cos x)^2}$$

[By the quotient rule]

$$= \frac{\cos x \cos(x+9) + \sin x \sin(x+9)}{\cos^2 x}$$

$$\therefore \left(\frac{dy}{dx}\right)_{x=0} = \frac{\cos 9}{1} = \cos 9$$

Q73. If $f(x) = 1 + x + \frac{x^2}{2} + \ldots + \frac{x^{100}}{100}$, then $f'(1)$ is equal to:

(a) 1/100 (b) 100 (c) 0

(d) Does not exist

Solution: (b)

$$f(x) = 1 + x + \frac{x^2}{2} + \ldots + \frac{x^{100}}{100}$$

$$\therefore f'(x) = 0 + 1 + 2 + \frac{x}{2} + \ldots + 100\frac{x^{99}}{100}$$

$$f'(x) = 1 + x + x^2 + \ldots + x^{99}$$

$$f'(1) = 1 + 1 + 1 + \ldots + 1(100 \text{ times}) = 100$$

Q74. If $f(x)=\dfrac{x^n-a^n}{x-a}$ for some constant a, then $f'(a)$ is equal to:
(a) 1 (b) 0 (c) 1/2
(d) Does not exist

Solution: (d)

$$f(x)=\frac{x^n-a^n}{x-a}$$

$$\therefore f'(x)=\frac{(x-a)nx^{n-1}-(x^n-a^n)}{(x-a)^2}$$ [By the quotient rule]

$$\Rightarrow f(x)=\frac{nx^{n-1}(x-a)-x^n+a^n}{(x-a)^2}$$

$$f'(a)=\frac{na^{n-1}(0)-a^n+a^n}{(x-a)^2}\Rightarrow f'(a)=0/0$$

Thus, $f'(a)$ does not exist.

Q75. If $f(x)=x^{100}+x^{99}+.....+x+1$, then $f'(1)$ is equal to:
(a) 5050 (b) 5049 (c) 5051 (d) 50051

Solution: (a)

$f(x)=x^{100}+x^{99}+.....+x+1,$

$\therefore f(x)=100x^{99}+99x^{98}+.....+1+0$

$=100x^{99}+99x^{98}+...+1$

$f'(1)=100+99+....+1$

$=\dfrac{100}{2}[2\times100+(100-1)(-1)]=50[200-99]$

$\left[\because S_n=\dfrac{n}{2}\{2a+(n-1)d\}\right]=50\times101=5050$

Q76. If $f(x)=1-x+x^2-x^3+...-x^{99}+x^{100}$, then $f'(1)$ is equal to:
(a) 150 (b) -50 (c) -150 (d) 50

Solution: (d)

Given $f(x)=1-x+x^2-x^3+...-x^{99}+x^{100}$,

$f'(x)=0-1+2x-3x^2+....-99x^{98}+100x^{99}$

$=-1+2x-3x^2+...-99x^{98}+100x^{99}$

$\therefore f'(1)=-1+2-3+....-99+100$

$=(-1-3-5-.....-99)(2+4+...+100)$

$[\because S_n=n/2(2a+(n-1)d]$

$=-\dfrac{50}{2}[2\times1+(50-1)2]+\dfrac{50}{2}[2\times2+(50-1)2]$

$=-25[2+49\times2]+25[4+49\times2]$

$=-25(2+98)+25(4+98)=-25\times100+25\times102$

$=-2500+2550=50$

Q77. If $f(x)=\dfrac{\tan x}{x-\pi}$, then $\lim\limits_{x\to\pi}f(x)=..........$

Solution: $f(x)=\dfrac{\tan x}{x-\pi}=\lim\limits_{x\to\pi}\dfrac{\tan x}{x-\pi}$

$=\lim\limits_{\pi-x\to0}\dfrac{-\tan(\pi-x)}{-(\pi-x)}$ $\left[\because=\lim\limits_{x\to0}\dfrac{\tan x}{x}=1\right]$

Q78. $\lim\limits_{x\to0}\left(\sin mx\cot\dfrac{x}{\sqrt3}\right)=2$, then $m=$............

Solution:

$\lim\limits_{x\to0}\left(\sin mx\cot\dfrac{x}{\sqrt3}\right)=2,$

$\Rightarrow\lim\limits_{x\to0}\dfrac{\sin mx}{mx}\cdot mx\cdot\dfrac{1}{\tan x/\sqrt3}=2$

$\Rightarrow\lim\limits_{x\to0}\dfrac{\sin mx}{mx}\cdot mx\cdot\dfrac{x/\sqrt3}{\tan x/\sqrt3}\cdot\dfrac{1}{x/\sqrt3}=2$

$\Rightarrow\lim\limits_{x\to0}\dfrac{\sin mx}{mx}\cdot\lim\limits_{x\to0}\dfrac{x/\sqrt3}{\tan\dfrac{x}{\sqrt3}}\cdot\lim\limits_{x\to0}\dfrac{mx}{x/\sqrt3}=2$

$\Rightarrow\sqrt3x=2\therefore m=2\sqrt3/3$

Q79. If $y=1+\dfrac{x}{1!}+\dfrac{x^2}{2!}+\dfrac{x^3}{3!}+....,$ then $\dfrac{dy}{dx}=.......$

Solution:

$y=1+\dfrac{x}{1!}+\dfrac{x^2}{2!}+\dfrac{x^3}{3!}+....,$

$\therefore\dfrac{dy}{dx}=0+1+\dfrac{2x}{2}+\dfrac{3x^2}{6}+\dfrac{4x^3}{4!}$

$=1+x+\dfrac{x^2}{2}+\dfrac{x^3}{6}+......=1+\dfrac{x}{1!}+\dfrac{x^2}{2!}+\dfrac{x^3}{3!}+.....=y$

Q80. $\lim\limits_{x\to3^+}\dfrac{x}{[x]}=........$

Solution:

$\lim\limits_{x\to3^+}\dfrac{x}{[x]}=\lim\limits_{h\to0}\dfrac{(3+h)}{(3+h)}=\lim\limits_{h\to0}\dfrac{(3+h)}{3}=1$

Day 185

Very Short Answer (VSA): 1-Mark and Short Answer (SA): 2-Mark Questions

Q1. Find the derivative of $\dfrac{1-\frac{1}{x}}{1+\frac{1}{x}}$

Solution:

Let $f(x)=\dfrac{1-\frac{1}{x}}{1+\frac{1}{x}}=\dfrac{\frac{x-1}{x}}{\frac{x+1}{x}}=\dfrac{x-1}{x+1}$, where $x\neq 0,1$.

By the quotient rule

$$f'(x)=\frac{(x+1)\frac{d}{dx}(x-1)-(x-1)\frac{d}{dx}(x+1)}{(x+1)^2}$$

$$=\frac{(x+1)(1)-(x-1)(1)}{(x+1)^2}$$

$$=\frac{x+1-x+1}{(x-1)^2}$$

$$\Rightarrow f'(x)=\frac{2}{(x+1)^2},\quad x\neq 0,-1$$

Q2. Find the derivative of $\dfrac{1}{x^2-2x+1}$

Solution: Let $f(x)=\dfrac{1}{x^2-2x+1}$.

By the quotient rule

$$f'(x)=\frac{(x^2-2x+1)\frac{d}{dx}(1)-(1)\frac{d}{dx}(x^2-2x+1)}{(x^2-2x+1)^2}$$

$$=\frac{(x^2-2x+1)(0)-(2x-2)}{(x^2-2x+1)^2}=\frac{2(1-x)}{(x^2-2x+1)^2}$$

Q3. Find the derivative of $-\dfrac{1}{4x^4}-\dfrac{1}{2x^2}$

Solution: Let $f(x)=-\dfrac{1}{4x^4}-\dfrac{1}{2x^2}$

$$f'(x)=\frac{d}{dx}\left(-\frac{1}{4x^4}\right)+\frac{d}{dx}\left(-\frac{1}{2x^2}\right)$$

$$=\frac{-1}{4}\frac{d}{dx}(x^{-4})+\frac{-1}{2}\frac{d}{dx}(x^{-2})$$

$$=\frac{-1}{4}(-4x^{-5})+\frac{-1}{2}(-2x^{-3})$$

$$\left[\because \frac{d}{dx}(x^n)=nx^{n-1}\right]$$

$$=\frac{1}{x^5}+\frac{1}{x^3}$$

Q4. Find the derivative of $(x+1)^2(x+2)^3$

Solution: Let $f(x)=(x+1)^2(x+2)^3$.

By the product rule

$$f'(x)=(x+1)^2\frac{d}{dx}(x+2)^3+(x+2)^3\frac{d}{dx}(x+1)^2$$

Now, using the above proof, we get

$$f'(x)=(x+1)^2\cdot 3(x+2)^2+(x+2)^3\cdot 2(x+1)$$

$$\Rightarrow f'(x)=3(x+1)^2(x+2)^2+2(x+2)^3(x+1)$$

$$=(x+1)(x+2)^2\{3(x+1)+2(x+2)\}$$

$$=(x+1)(x+2)^2(5x+7)$$

Q5. Find the derivative of $\sin^3 x$.

Solution: Let $y=\sin^2 x$

$$\therefore \frac{dy}{dx}=\frac{d}{dx}(\sin x\sin x)=(\sin x)'\sin x+\sin x(\sin x)'$$

[By the Leibniz product rule]

$$=\cos x\sin x+\sin x\cos x=2\sin x\cos x\ldots(1)$$

Now, for $y=\sin^3 x$.

$$\therefore \frac{dy}{dx}=\frac{d}{dx}(\sin x\sin^2 x)=(\sin x)'\sin^2 x+\sin(\sin^2 x)'$$

[By the Leibniz product rule]

$$=\cos x\sin^2 x+\sin x(2\sin x\cos x)\ \text{[Using (1)]}$$

$$=\cos x\sin^2 x+2\sin^2 x\cos x=3\sin^2 x\cos x$$

Q6. Find the derivative of $(x^2+1)\tan x$

Solution: Let $f(x)=(x^2+1)\tan x$.

By the product rule

$$f'(x)=(x^2+1)\frac{d}{dx}(\tan x)+\tan x\frac{d}{dx}(x^2+1)$$

$$=(x^2+1)(\sec^2 x)+\tan x(2x)$$

$$=(x^2+1)\sec^2 x+2x\tan x$$

Q7. Find the derivative of $(x+\sec x)(x-\cot x)$

Solution: Let $f(x)=(x+\sec x)(x-\cot x)$.

By the product rule

$$f'(x)=(x+\sec x)\frac{d}{dx}(x-\cot x)+(x-\cot x)\frac{d}{dx}(x+\sec x)$$

$= (x+\sec x)(1+\text{cosec}^2 x)+(x-\cot x)(1+\sec x\tan x)$

Q8. Find the derivative of $\frac{x}{\sin^4 x}$

Solution: Let $f(x)=\frac{x}{\sin^4 x}$.

By the quotient rule

$$f'(x)=\frac{\sin^4 x\frac{d}{dx}x - x\frac{d}{dx}\sin^4 x}{(\sin^4 x)^2}$$

$$\frac{d}{dx}\sin^n x = n\sin^{n-1} x\cos x$$

Therefore, $f'(x)=\frac{\sin^4 x\times 1 - x4\sin^{4-1}x\cos x}{\sin^8 x}$

$$=\frac{\sin^3 x(\sin x - 4x\cos x)}{\sin^8 x}$$

Q9. Find the derivative of $\frac{x^4+x^3+x^2+1}{x^2}$

Solution: $\frac{d}{dx}\left(\frac{x^4+x^3+x^2+1}{x^2}\right)=\frac{d}{dx}\left(x^2+x+1+\frac{1}{x^2}\right)$

$$=\frac{d}{dx}x^2+\frac{d}{dx}x+\frac{d}{dx}(1)+\frac{d}{dx}\left(\frac{1}{x^2}\right)$$

$$=2x+1+0+\left(\frac{-2}{x^3}\right)=1+2x-\frac{2}{x^3}$$

Q10. Find the derivative of $\left(x+\frac{1}{x}\right)^4$

Solution: Let $y=\left(x+\frac{1}{x}\right)^4$

$$\therefore \frac{dy}{dx}=\frac{d}{dx}\left(x+\frac{1}{x}\right)^4=4\left(x+\frac{1}{x}\right)^{4-1}\frac{d}{dx}\left(x+\frac{1}{x}\right)$$

$$\left(\because \frac{d}{dx}y^n=ny^{n-1}\frac{dy}{dx}\right)$$

$$=4\left(x+\frac{1}{x}\right)^3\left(1-\frac{1}{x^2}\right)=4\left(x+\frac{1}{x}\right)^4\left(1-\frac{1}{x}\right)$$

Q11. Find the derivative of $\tan x(\sec x-1)(\sec x+1)$

Solution: Let $y=\tan x(\sec x-1)(\sec x+1)$

$y=\tan x(\sec^2-1)=\tan^3 x$

$$\Rightarrow \frac{dy}{dx}=3\tan^2 x\cdot\frac{d}{dx}\tan x$$

$$\left(\because \frac{d}{dx}y^n=ny^{n-1}\frac{dy}{dx}\right)$$

$$=3\tan^2 x\sec^2 x$$

Q12. Find the derivative of $(\sin x+\cos x)^3$

Solution: Let $y=(\sin x+\cos x)^3$

$$\therefore \frac{dy}{dx}=3(\sin x+\cos x)^2(\cos x-\sin x)$$

$$\left(\because \frac{d}{dx}y^n=ny^{n-1}\frac{dy}{dx}\right)$$

$$=3(\cos^2 x-\sin^2 x)(\sin x+\cos x)$$

$$=3\cos 2x(\sin x+\cos x)$$

$[\because \cos 2x=\cos^2 x-\sin^2 x]$

Q13. Find the derivative of $(2x-3)^2(3x-2)^3$

Solution: Let $y=(2x-3)^2(3x-2)^3$

$$\frac{dy}{dx}=(2x-3)^2\frac{d}{dx}(3x-2)^3+(3x-2)^3\frac{d}{dx}(2x-3)^2$$

[By the product rule]

$$=(2x-3)^2(3)(3x-2)^2(3)+(3x-2)^3 2(2x-3)(2)$$

$$\left(\because \frac{d}{dx}y^n=ny^{n-1}\frac{dy}{dx}\right)$$

$$=9(2x-3)^2(3x-2)^2+4(3x-2)^3(2x-3)$$

$$=(2x-3)(3x-2)^2[9(2x-3)+4(3x-2)]$$

$$=(2x-3)(3x-2)^2(30x-35)$$

Q14. Evaluate $\lim_{x\to 0}\frac{(x+3)^{1/3}-3^{1/3}}{x}$

Solution:

$$\lim_{x\to 0}\frac{(x+3)^{1/3}-3^{1/3}}{x}=\lim_{x\to 0}\frac{(x+3)^{1/3}-3^{1/3}}{(x+3)-3}$$

$$=\frac{1}{3}\times 3^{\frac{1}{3}-1}=\frac{1}{3}(3)^{-2/3}\left[\because \lim_{x\to a}\frac{x^n-a^n}{x-a}=na^{n-1}\right]$$

$$=(3)^{-5/3}=\frac{1}{(3)^{5/3}}[\because x\to 0\Rightarrow x+3\to 3]$$

Q15. Evaluate $\lim_{x\to 0}\frac{(1+x)^4-1}{x\{(1+x)^2+1\}}$

Solution: $\lim_{x\to 0}\frac{(1+x)^4-1}{x\{(1+x)^2+1\}}$

$$=\lim_{x\to 0}\frac{\{(1+x)^2+1\}\{(1+x)^2-1\}}{x\{(1+x)^2+1\}}$$

$$=\lim_{x\to 0}\frac{(1+x)^2-1}{x}$$

$$=\lim_{x\to 0}\frac{(1+x)^2-1}{(1+x)-1}[\because x\to 0\Rightarrow 1+x\to 1]$$

$$=2(1)^{2-1}=2\left[\therefore \lim_{x\to a}\frac{x^n-a^n}{x-a}=na^{n-1}\right]$$

Q16. Evaluate $\lim_{x\to 1}\frac{(2+x)^{5/2}-(3)^{5/2}}{x-1}$

Solution: $\lim_{x\to 1}\frac{(2+x)^{5/2}-(3)^{5/2}}{x-1}$

$=\lim_{x\to 1}\frac{(2+x)^{5/2}-(3)^{5/2}}{(2+x)-(3)}=\frac{5}{2}(3)^{\frac{5}{2}-1}$

$\left[\because \lim_{x\to a}\frac{x^n-a^n}{x-a}=na^{n-1}\right]$

$=\frac{5}{2}(3)^{3/2}$ $[\because x\to 1\Rightarrow x+2\to 3]$

Q17. Find the value of n, if $\lim_{x\to 2}\frac{x^n-2^n}{x-2}=80, n\in N$

Solution:

$\lim_{x\to 3}\frac{x^n-3^n}{x-3}=108\Rightarrow n(3)^{n-1}=108$

$\left[\because \lim_{x\to a}\frac{x^n-a^n}{x-a}=na^{n-1}\right]$

$\Rightarrow n(3)^{n-1}=4\times 27\Rightarrow n\times 3^{n-1}=4\times(3)^3$

$\Rightarrow n\times 3^{n-1}=4\times(3)^{4-1}\Rightarrow n=4$

Q18. Evaluate $\lim_{x\to 0}\frac{\sin 3x}{\sin 5x}$

Solution:.

$\lim_{x\to 0}\frac{\frac{\sin 3x}{3x}\cdot 3x}{\frac{\sin 5x}{5x}\cdot 5x}=\frac{\lim_{x\to 0}\frac{\sin 3x}{3x}}{\frac{\sin 5x}{5x}}\cdot\frac{3x}{5x}$

$=\frac{3}{5}\cdot\frac{\lim_{x\to 0}\frac{\sin 3x}{3x}}{\lim_{x\to 0}\frac{\sin 5x}{5x}}\left[\because \lim_{x\to 0}\frac{\sin x}{x}=1\right]$

$=\frac{3}{5}[\because x\to 0\Rightarrow (kx\to 0)$, where k is a real number$]$

Q19. Evaluate $\lim_{x\to 0}\frac{\sin^2 x}{\sin^2 2x}$

Solution:

$\lim_{x\to 0}\frac{\sin^2 x}{\sin^2 2x}=\lim_{x\to 0}\frac{\sin^2 x}{[\sin(2x)]^2}$

$=\lim_{x\to 0}\frac{\sin^2 x}{(2\sin x\cos x)^2}=\lim_{x\to 0}\frac{\sin^2 x}{4\sin^2 x\cos^2 x}$

$[\because \sin 2\theta=2\sin\theta\cos\theta]$

$=\lim_{x\to 0}\frac{1}{4\cos^2 x}=\frac{1}{4}[\because \cos 0=1]$

Q20. Evaluate $\lim_{x\to 0}\frac{1-\cos 2x}{2x^2}$

Solution:

$\lim_{x\to 0}\frac{1-\cos 2x}{2x^2}=\lim_{x\to 0}\frac{1-1+2\sin^2 x}{2x^2}$

$[\because \cos 2x=1-2\sin^2 x]$

$=\lim_{x\to 0}\frac{2\sin^2 x}{2x^2}=\lim_{x\to 0}\frac{\sin^2 x}{x^2}=\lim_{x\to 0}\left(\frac{\sin x}{x}\right)^2=1$

$\left[\because \lim_{x\to 0}\frac{\sin x}{x}=1\right]$

Q21. If $f(x)=\begin{cases}x-2, & x\le 1\\ cx^2, & x>1\end{cases}$, find c when $\lim_{x\to 1}f(x)$ exists.

Solution: Given that $f(x)=\begin{cases}x-2, & x\le 1\\ cx^2, & x>1\end{cases}$,

$\text{LHL}=\lim_{x\to 1^-}f(x)=\lim_{x\to 1^-}(x-1)$

$=\lim_{h\to 0}(1-h-2)=\lim_{h\to 0}(-1-h)=-1$

$\text{RHL}=\lim_{x\to 1^+}f(x)=\lim_{x\to 1^+}cx^2=\lim_{h\to 0}c(1+h)^2=c$

Since, $\lim_{x\to 1}f(x)$ exists, LHL = RHL $\Rightarrow c=-1$

Q22. Evaluate $\lim_{x\to 0}\frac{(1+x)^5-1}{x}$

Solution: $\lim_{x\to 0}\frac{(1+x)^5-1}{x}=\lim_{x\to 0}\frac{(1+x)^5-1}{(1+x)-1}$

$=\lim_{x\to 0}\frac{(1+x)^5-1}{(1+x)-1}=\lim_{x\to 0}\frac{(1+x)^5-1^5}{(1+x)-1}$

$=\lim_{x\to 0}\frac{(1+x)^5-1^5}{(1+x)-1}=5\cdot(1)^{5-1}=5$ $\left[\because \lim_{x\to a}\frac{x^n-a^n}{x-a}=na^{n-1}\right]$

Day 186

Long Answer-I (LA-I): 4-Mark and Long Answer-II (LA-II): 6-Mark Questions

Q1. Find the derivative of the following functions with respect to x (θ is a fixed non-zero constant) from the first principle:

(i) $\sin(x+\theta)$ (ii) $\cos(x-\theta)$ (iii) $\dfrac{\sin(x+\theta)}{\cos x}$

(i) $\sin(x+\theta)$

Solution: Let $f(x)=\sin(x+\theta)$.

Then, according to the first principle

$$f'(x)=\lim_{h\to0}\frac{f(x+h)-f(x)}{h}$$

$$=\lim_{h\to0}\frac{1}{h}[\sin(x+\theta+h)-\sin(x+\theta)]$$

$$=\lim_{h\to0}\frac{1}{h}\left[2\cos\left(\frac{x+h+\theta+x+\theta}{2}\right)\sin\left(\frac{x+h+\theta-x-\theta}{2}\right)\right]$$

$$=\lim_{h\to0}\frac{1}{h}\left[2\cos\left(\frac{2x+h+2\theta}{2}\right)\sin(h/2)\right]$$

$$=\lim_{h\to0}\frac{1}{h}\left[\cos\left(\frac{2x+h+2\theta}{2}\right)\cdot\frac{\sin(h/2)}{(h/2)}\right]$$

$$=\lim_{h\to0}\cos\left(\frac{2x+h+2\theta}{2}\right)\cdot\lim_{h/2\to0}\frac{\sin(h/2)}{(h/2)}$$

$$\left[\text{As } h\to0\Rightarrow\frac{h}{2}\to0\right]$$

$$=\cos\left(\frac{2x+0+2\theta}{2}\right)\cdot1\left[\lim_{x\to0}\frac{\sin x}{x}=1\right]$$

$$=\cos(x+\theta)$$

(ii) $\cos(x-\theta)$

Solution: Let $f(x)=\cos(x-\theta)$.

Then, according to the first principle

$$f'(x)=\lim_{h\to0}\frac{f(x+h)-f(x)}{h}$$

$$=\lim_{h\to0}\frac{1}{h}[\cos(x+h-\theta)-\cos(x-\theta)]$$

$$=\lim_{h\to0}\frac{1}{h}\left[\begin{array}{l}-2\sin\dfrac{(x+h-\theta+x-\theta)}{2}\\ \times\sin\left(\dfrac{x+h-\theta-x+\theta}{2}\right)\end{array}\right]$$

$$=\lim_{h\to0}\frac{1}{h}\left[-2\sin\left(\frac{2x+h-2\theta}{2}\right)\sin\frac{h}{2}\right]$$

$$=\lim_{h\to0}\left[-\sin\left(\frac{2x+h-2\theta}{2}\right)\frac{\sin(h/2)}{(h/2)}\right]$$

$$=\lim_{h\to0}\left[-\sin\left(\frac{2x+h-2\theta}{2}\right)\right]\lim_{h/2\to0}\frac{\sin(h/2)}{(h/2)}$$

$$\left[\text{As } h\to0\Rightarrow\frac{h}{2}\to0\right]$$

$$=-\sin\left(\frac{2x+0-2\theta}{2}\right)\cdot1=-\sin(x-\theta)$$

(iii) $\dfrac{\sin(x+\theta)}{\cos x}$

Solution: Let $f(x)=\dfrac{\sin(x+\theta)}{\cos x}$.

By the quotient rule

$$f'(x)=\frac{\cos x\dfrac{d}{dx}[\sin(x+\theta)]-\sin(x+\theta)\dfrac{d}{dx}\cos x}{\cos^2 x}$$

$$=\frac{\cos x\dfrac{d}{dx}[\sin(x+\theta)]-\sin(x+\theta)(-\sin x)}{\cos^2 x}$$

$$f'(x)=\frac{\cos x.\cos(x+\theta)+\sin x\sin(x+\theta)}{\cos^2 x}$$

$$=\frac{\cos(x+\theta-x)}{\cos^2 x}=\frac{\cos\theta}{\cos^2 x}$$

Q2. Differentiatiate $\cos(x^2)$ with respect to x using the first principle.

Solution: Let

$f(x)=\cos(x^2), f(x+h)=\cos(x+h)^2$

$$\therefore\frac{d}{dx}f(x)=\lim_{h\to0}\frac{f(x+h)-f(x)}{h}$$

$$=\lim_{h\to0}\frac{\cos\{(x+h)\}-\cos(x^2)}{h}$$

$$=\lim_{h\to0}\frac{-2\sin\left\{\dfrac{(x+h)^2+x^2}{2}\right\}\sin\left\{\dfrac{(x+h)^2-x^2}{2}\right\}}{h}$$

$$\left[\because\cos C-\cos D=-2\sin\frac{C+D}{2}\cdot\sin\frac{C-D}{2}\right]$$

$$=\lim_{h\to0}\frac{1}{h}\left[-2\sin\left\{\frac{(x+h)^2+x^2}{2}\right\}\sin\left\{\frac{(x+h)^2-x^2}{2}\right\}\right]$$

$$=\lim_{h\to0}\frac{1}{h}\left[-2\sin\left\{\frac{(x+h)^2+x^2}{2}\right\}\sin\left\{\frac{h^2+2hx}{2}\right\}\right]$$

$$= -2\lim_{h\to 0}\sin\left\{\frac{(x+h)^2+x^2}{2}\right\}$$

$$\lim_{h\to 0}\left\{\frac{\sinh\left(\frac{h+2x}{2}\right)}{h\left(\frac{h+2x}{2}\right)}\right\}\times\left(\frac{h+2x}{2}\right)$$

$$= -2\lim_{h\to 0}\sin\left\{\frac{(x+h)^2+x^2}{2}\right\}\lim_{h\to 0}\left(\frac{h+2x}{2}\right)$$

$$\left[\because \lim_{x\to 0}\frac{\sin x}{x}=1\right]$$

$$= -2x\sin(x^2)\ \ [\because x\to 0 \Rightarrow hx\to 0]$$

Q3. Differentiatiate $x^{1/3}$ with respect to x using the first principle.

Solution: Let $f(x)=x^{1/3} \Rightarrow f(x+h)=(x+h)^{1/3}$.

Now, $\frac{d}{dx}f(x)=\lim_{h\to 0}\frac{f(x+h)-f(x)}{h}$

$$= \lim_{h\to 0}\frac{1}{h}[(x+h)^{1/3}-x^{1/3}] = \lim_{h\to 0}\frac{1}{h}\left[x^{1/3}\left(1+\frac{h}{x}\right)^{1/3}-x^{1/3}\right]$$

$$= \lim_{h\to 0}\frac{1}{h}\left[x^{1/3}\left(1+\frac{h}{x}\cdot\frac{1}{3}+\frac{1}{3}\left(\frac{1}{3}-1\right)\frac{h^2}{x^2}+\ldots\right)-1\right]$$

$$\left[\because (1+x)^n = 1+nx+\frac{n(n-1)}{2!}x^2+\ldots\ldots\right]$$

$$= \lim_{h\to 0}\frac{1}{h}\left[x^{1/3}\left(\frac{1}{3}\cdot\frac{h}{x}-\frac{2}{9}\cdot\frac{h^2}{x^2}+\ldots\right)\right]$$

$$= \lim_{h\to 0}\frac{x^{1/3}}{h}\cdot\frac{1}{3}\frac{h}{x}\left(1-\frac{2}{3}\cdot\frac{h}{x}+\ldots\ldots\right) = \frac{1}{3}x^{1/3-1} = \frac{1}{3}x^{-2/3}$$

Q4. Find the derivative of $\frac{x^4-\cos x}{\sin x}$

Solution: Let $y=\frac{x^4-\cos x}{\sin x}$

$$\therefore \frac{dy}{dx}=\frac{\sin x\frac{d}{dx}(x^4-\cos x)-(x^4-\cos x)\frac{d}{dx}\sin x}{(\sin x)^2}$$

[By the quotient rule]

$$= \frac{\sin x(4x^3+\sin x)-(x^4-\cos x)\cos x}{\sin^2 x}$$

$$= \frac{4x^3\sin x+\sin^2 x-x^4\cos x+\cos^2 x}{\sin^2 x}$$

$$= \frac{4x^3\sin x-x^4\cos x+\sin^2 x+\cos^2 x}{\sin^2 x}$$

$$= \frac{4x^3\sin x-x^4\cos x+1}{\sin^2 x}$$

Q5. Find the derivative of $\frac{x^2\cos(\pi/3)}{\sin x}$

Solution: Let $f(x)=\frac{x^2\cos(\pi/3)}{\sin x}$.

By the quotient rule

$$f'(x)=\cos\frac{\pi}{3}\cdot\left[\frac{\sin x\frac{d}{dx}(x^2)-x^2\frac{d}{dx}(\sin x)}{\sin^2 x}\right]$$

$$= \cos\frac{\pi}{3}\cdot\left[\frac{\sin x\cdot 2x-x^2\cos x}{\sin^2 x}\right]$$

$$= \frac{x\cos\frac{\pi}{3}[2\sin x-x\cos x]}{\sin^2 x}$$

$$= \frac{x\frac{1}{2}[2\sin x-x\cos x]}{\sin^2 x}$$

$$= \frac{x[\sin x-\frac{1}{2}x\cos x]}{\sin^2 x}$$

Q6. Find the derivative of $x^4(2\sin x-3\cos x)$

Solution: Let $f(x)=x^4(2\sin x-3\cos x)$.

By the product rule

$$f'(x)=x^4\frac{d}{dx}(2\sin x-3\cos x)$$
$$+(2\sin x-3\cos x)\frac{d}{dx}(x^4)$$

$$= x^4\left[2\frac{d}{dx}(\sin x)-3\frac{d}{dx}(\cos x)\right]$$
$$+(2\sin x-3\cos x)\frac{d}{dx}(x^4)$$

$$= x^4[2\cos x-3(-\sin x)]+(2\sin x-3\cos x)(4x^3)$$

$$= x^3[2x\cos x+3x\sin x+8\sin x-12\cos x]$$

Q7. Find the derivative of $\frac{1+\cos x}{1+\sin x}$

Solution: Let $f(x)=\frac{1+\cos x}{1+\sin x}$.

By the quotient rule

$$f'(x)=\frac{(1+\sin x)\frac{d}{dx}(1+\cos x)-(1+\cos x)\frac{d}{dx}(1+\sin x)}{(1+\sin x)^2}$$

$$= \frac{(1+\sin x)(-\sin x)-(1+\cos x)(\cos x)}{(1+\sin x)^2}$$

$$= \frac{-\sin x-\sin^2 x-\cos x-\cos^2 x}{(1+\sin x)^2}$$

$$= \frac{-\sin x-\cos x-(\sin^2 x+\cos^2 x)}{(1+\sin x)^2}$$

$$= \frac{-\sin x-\cos x-1}{(1+\sin x)^2} = \frac{-(1+\sin x+\cos x)}{(1+\sin x)^2}$$

Q8. Evaluate $\lim\limits_{x\to 2}\frac{x-2}{\sqrt{3x-2}-\sqrt{x+2}}$

Solution: $\lim\limits_{x\to 2}\frac{x-2}{\sqrt{3x-2}-\sqrt{x+2}}$

$$= \lim_{x\to 2}\frac{(x-2)(\sqrt{3x-2}+\sqrt{x+2})}{(\sqrt{3x-2}-\sqrt{x+2})(\sqrt{3x-2}-\sqrt{x+2})}$$

$$= \lim_{x\to 2}\frac{(x-2)(\sqrt{3x-2}+\sqrt{x+2})}{(\sqrt{3x-2})^2-(\sqrt{x+2})^2}$$

$[\because (a+b)(a-b)=a^2-b^2]$

$$= \lim_{x\to 2}\frac{(x-2)\sqrt{3x-2}+\sqrt{x+2}}{(3x-2)-(x+2)}$$

$$= \lim_{x\to 2}\frac{(x-2)(\sqrt{3x-2}+\sqrt{x+2})}{3x-2-x-2}$$

$$= \lim_{x\to 2}\frac{(x-2)(\sqrt{3x-2}+\sqrt{x+2})}{2x-4}$$

$$= \lim_{x\to 2}\frac{(x-2)(\sqrt{3x-2}+\sqrt{x+2})}{2(x-2)}$$

$$= \lim_{x\to 2}\frac{(\sqrt{3x-2}+\sqrt{x+2})}{2}$$

$$= \frac{(\sqrt{6-2}+\sqrt{2+2})}{2} = \frac{(2+2)}{2} = 4$$

Q9. Evaluate $\lim\limits_{x\to 1}\frac{x^3-\sqrt{x}}{\sqrt{x}-1}$

Solution: $\lim\limits_{x\to 1}\frac{x^3-\sqrt{x}}{\sqrt{x}-1} = \lim\limits_{x\to 1}\frac{\sqrt{x}[(x)^{5/2}-1]}{\sqrt{x}-1}$

$$= \lim_{x\to 1}\frac{(x)^{5/2}-1}{\sqrt{x}-1}\cdot\lim_{x\to 1}\sqrt{x}$$

$$\left[\because \lim_{x\to a} f(x)\cdot g(x) = \lim_{x\to a} f(x)\cdot \lim_{x\to a} g(x)\right]$$

$$= \lim_{x\to 1}\frac{\frac{x^{5/2}-1}{x-1}}{\frac{(x)^{1/2}-1}{x-1}} = \frac{\lim\limits_{x\to 1}\frac{x^{5/2}-1}{x-1}}{\lim\limits_{x\to 1}\frac{(x)^{1/2}-1}{x-1}}$$

$$\left[\because \lim_{x\to a}\frac{f(x)}{g(x)} = \frac{\lim\limits_{x\to a} f(x)}{\lim\limits_{x\to a} g(x)}\right]$$

$$= \frac{\frac{5}{2}(1)^{\frac{5}{2}-1}}{\frac{1}{2}(1)^{\frac{1}{2}-1}} = \frac{5/2}{1/2} = 5$$

Q10. Evalaute $\lim\limits_{x\to 0}\frac{1-\cos 3x}{1-\cos 5x}$

Solution:

$$\lim_{x\to 0}\frac{1-\cos 3x}{1-\cos 5x} = \lim_{x\to 0}\frac{1-1+2\sin^2 3x/2}{1-1+2\sin 5x/2}$$

$[\because \cos 3x = 1-2\sin^2\frac{3x}{2}, \sin 5x = 1-2\sin^2\frac{5x}{2}]$

$$= \lim_{x\to 0}\frac{\sin^2 3x/2}{\sin^2 5x/2} = \lim_{x\to 0}\frac{\frac{\sin^2\frac{3x}{2}}{(3x/2)^2}\cdot(3x/2)^2}{\frac{\sin^2\frac{5x}{2}}{(5x/2)^2}\cdot\left(\frac{5x}{2}\right)^2}$$

$$= \frac{\left(\lim\limits_{x\to 0}\frac{\sin\frac{3x}{2}}{3x/2}\right)^2}{\left(\lim\limits_{x\to 0}\frac{\sin\frac{5x}{2}}{5x/2}\right)^2}\cdot\frac{3^2\frac{x^2}{4}}{5^2\frac{x^2}{4}}$$

$$= \frac{m^2}{n^2}\cdot\frac{\left(\lim\limits_{x\to 0}\frac{\sin\frac{3x}{2}}{3x/2}\right)^2}{\left(\lim\limits_{x\to 0}\frac{\sin\frac{5x}{2}}{5x/2}\right)^2} = \frac{3^2}{5^2} = \frac{9}{25}\left[\because \lim_{x\to 0}\frac{\sin x}{x} = 1\right]$$

$[\because x\to 0 \Rightarrow kx\to 0]$

Revision and Unit Test 3

Day 187

Revision (Chapters X–XI)

Cartesian System

The Cartesian system, also known as coordinate system, specifies each point uniquely using numerical coordinates.

The two-dimensional coordinate system, also known as rectangular coordinate system, represents a point in a plane while a three-dimensional coordinate represent a point in a space.

Rectangular Coordinate System

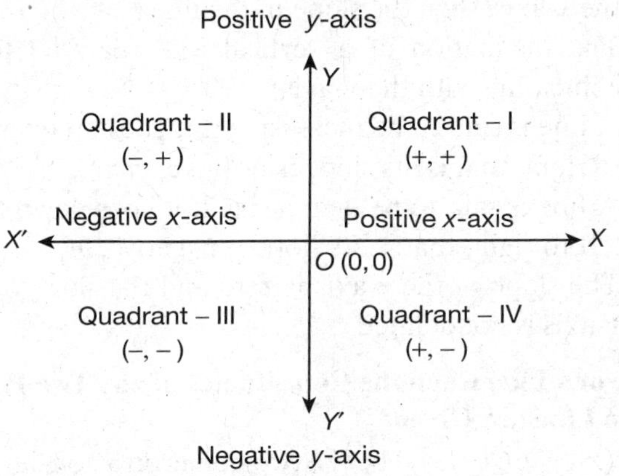

Fig. 10.1 Coordinate system

This system (Fig. 10.1) is shown as two perpendicular lines, known as the coordinate axes. The horizontal line represents the x-axis and the vertical represents the y-axis. The origin is the intersection of the x- and y-axes, given by $O(0,0)$.

In stating the coordinates of a point in the coordinate plane, the x-coordinate comes first, and then the y-coordinate, inside a small bracket, as (x, y)

The position of a point in a rectangular coordinate system is given by a pair of numbers written as (x, y), with a specific sign. This gives the distances of the point from the x- and y-axes, respectively.

The values of the x- and y-coordinates of a point are known as abscissa and ordinate, respectively.

The horizontal x-axis is represented by $X'X$, where OX' gives the negative direction of the x-axis and OX gives the positive direction of the x-axis.

The vertical y-axis is represented by $Y'Y$, where OY' gives the negative direction of the y-axis and OY gives the positive direction of the y-axis.

The x-coordinate of a point is its perpendicular distance from the y-axis measured along the x-axis (positive along the positive direction of the x-axis and negative along the negative direction of the x-axis).

The y-coordinate of a point is its perpendicular distance from the x-axis measured along the y-axis (positive along the positive direction of the y-axis and negative along the negative direction of the y-axis).

These two coordinate axes divide the plane into four parts. These four parts are called the quadrants (one fourth part), numbered I, II, III and IV, anti-clockwise from OX.

Quadrant Plane I: This is the quadrant plane XOY, where both coordinate axes are positive, and thus, for a point (x, y) in this plane, the signs of both coordinates x and y are positive, that is, $(+,+)$.

Quadrant Plane II: This is the quadrant plane YOX', where the x-axis is negative, while the y-axis is positive, and thus, for a point (x, y) in this plane, the sign of coordinate x is negative and that of y is positive, that is, $(-,+)$.

Quadrant Plane III: This is the quadrant plane $X'OY'$, where both coordinate axes are negative, and thus, for a point (x, y) in this plane, the signs of both coordinates x and y are negative, that is, $(-,-)$.

Quadrant Plane IV: This is the quadrant plane $Y'OX$, where the x-axis is positive, while the y-axis is negative, and thus, for a point (x, y) in this plane, the sign of coordinate x is positive and that of y is negative, that is, $(+,-)$.

Every point on the x-axis has no distance (zero distance) from the x-axis; therefore, the y-coordinate of every point

lying on the x-axis is always zero. Thus, the coordinates of any point on the x-axis are of the form $(x, 0)$, where x is the distance of the point from the y-axis.
Every point on the y-axis has no distance (zero distance) from the y-axis; therefore, the x-coordinate of every point lying on the y-axis is always zero. Thus, the coordinates of any point on the y-axis are of the form $(0, y)$, where y is the distance of the point from the x-axis.
Coordinates of the origin are (0, 0), that is, it has zero distance from both the axes so that its abscissa and ordinate are both zero.

Note: If $x \neq y$ then $(x,y) \neq (y,x)$, and $(x,y) = (y,x)$ if and only if $x = y$.

Distance Formula

The distance between any two points $P(x_1, y_1)$ and $Q(x_2, y_2)$ in a plane is given by

$PQ = \sqrt{(x_2 - x_1)^2 + (y_2 - y_1)^2}$

Distance of a point $P(x, y)$ from the origin $O(0,0)$ is $PO = \sqrt{x^2 + y^2}$

Note:

Square: All sides are equal and both diagonals are also equal.
Parallelogram: All sides are equal but both diagonals are not equal.
Rectangle: Opposite sides are equal and diagonals are also equal.
Now, since $AB = BC = CD = DA, AC = BD$, all the four sides of the quadrilateral $ABCD$ are equal and its diagonals AC and BD are also equal. Therefore, $ABCD$ is a square.

Section Formula

The coordinates of the point P (x, y) that divides the line segment joining the points $A(x_1, y_1)$ and $B(x_2, y_2)$ internally in the ratio $m : n$ are $\left(\frac{mx_2 + nx_1}{m+n}, \frac{my_2 + ny_1}{m+n}\right)$

Note: If the ratio in which P divides AB is 1:k, then the coordinates of the point P will be $\left(\frac{x_2 + kx_1}{1+k}, \frac{y_2 + ky_1}{1+k}\right)$

Note: If P is the mid-point of the line segment AB, then P divides AB in the ratio 1:1; thus the coordinates of the point P will be $\left(\frac{x_2 + x_1}{1+1}, \frac{y_2 + y_1}{1+1}\right) = \left(\frac{x_2 + x_1}{2}, \frac{y_2 + y_1}{2}\right)$

Area of a Triangle

The area of a triangle whose vertices are $A(x_1, y_1), B(x_2, y_2), C(x_3, y_3)$ is given by

$= \frac{1}{2}|x_1(y_2 - y_3) + x_2(y_3 - y_1) + x_3(y_1 - y_2)|$

Note: If we change the order in which the vertices are taken, it will only alter the sign, but the numerical value will remain the same. Therefore, we take absolute numerical values since the area is a positive quantity.

Note: To find the area of a polygon, we divide it into triangular regions that have no common area and add the areas of these regions.

Slope of a Line

In two dimensions, the angle (θ) made by the line l with the positive direction of the x-axis and measured anti-clockwise is called the inclination of the line. The value $\tan\theta$ is called the gradient of the line or the slope of the line.
The slope of a line is denoted by m; thus, $m = \tan\theta$, $\theta \neq 90°$, that is, the slope of a line whose inclination is 90° is not defined.

Note:

(i) $0° \leq \theta \leq 180°$
(ii) If the line is parallel to the x-axis, or coinciding with the x-axis, then the slope of the line $\theta = 0°$.
(iii) The inclination of a vertical line (parallel to or coinciding with the y-axis) is 90°.
(iv) A line is said to be increasing if it goes up from left to right, that is, its slope is positive, $m > 0$.
(v) A line is said to be decreasing if it goes down from left to right, that is, its slope is negative, $m < 0$.
(vi) The slope of the x-axis is zero and the slope of the y-axis is not defined.

Slope of a Line when the Coordinates of any Two Points on the Line are Given

Let $P(x_1, y_1), Q(x_2, y_2)$ be the two points on a non-vertical line l whose inclination is θ. Then the slope of the line l is given by $m = \tan\theta = \frac{y_2 - y_1}{x_2 - x_1}$.

Conditions for Parallelism and Perpendicularity of Lines in Terms of their Slopes

In a coordinates plane, suppose the non-vertical lines l_1, l_2 have slopes m_1, m_2, respectively, and inclination α, β, respectively.

Parallel Lines: Two lines are parallel if their inclinations are equal, and thus, their slopes are equal. That is, $\tan\alpha = \tan\beta \Rightarrow m_1 = m_2$.

Perpendicular Lines: Two lines are perpendicular if the difference of inclination of the two lines is equal to 90°. That is, $\beta = \alpha + 90°$

$\Rightarrow \tan\beta = \tan(\alpha + 90°) \Rightarrow \tan\beta = -\cot\alpha$

$\Rightarrow \tan\beta = -\dfrac{1}{\tan\alpha} \Rightarrow m_2 = -\dfrac{1}{m_1} \Rightarrow m_1 m_2 = -1$

Thus, two non-vertical lines are perpendicular to each other if and only if the product of their slopes is equal to –1.

Angle between Two Lines

Let l_1, l_2 be two non-vertical lines with slopes m_1, m_2, respectively. Then the angle between two lines is given by $\left|\dfrac{m_2 - m_1}{1 + m_1 m_2}\right|$, $(1 + m_1 m_2 \neq 0)$.

Thus, the acute angle between the lines is $\tan\theta = \dfrac{m_2 - m_1}{1 + m_1 m_2}$

And the obtuse angle between the lines is $\tan\phi = -\left(\dfrac{m_2 - m_1}{1 + m_1 m_2}\right)$

Note: $(\theta + \phi = 180° \Rightarrow \tan\phi = \tan(180° - \theta) = -\tan\theta)$

In general, if $\dfrac{m_2 - m_1}{1 + m_1 m_2}$ is positive, then $\tan\theta$ will be positive and $\tan\phi$ will be negative, which means θ will be acute and ϕ will be obtuse.

And if $\dfrac{m_2 - m_1}{1 + m_1 m_2}$ is negative, then $\tan\theta$ will be negative and $\tan\phi$ will be positive, which means that θ will be obtuse and ϕ will be acute.

Collinearity of Three Points (Using Slope)

Three points are collinear in a plane if the slope of the line segment AB = slope of the line segment BC (since two lines having the same slope passing through a common point are coincident lines). Thus, the three given points A, B and C are collinear.

Horizontal and Vertical Lines

If a horizontal line L is at a distance a from the x-axis, then the ordinate of every point lying on the line is either a or $-a$. Therefore, the equation of the line L is either $y = a$ or $y = -a$.

Similarly, if a vertical line L is at a distance b from the y-axis, then the abscissa of every point lying on the line is either b or $-b$. Therefore, the equation of the line L is either $x = b$ or $x = -b$.

Point-Slope Form

Let there be a non-vertical line L whose slope is m and passing through a fixed point $P_0(x_0, y_0)$ and an arbitrary point $Q(x, y)$ on the line. Then by the definition of the slope of a line passing through two points, we get

$\tan\theta = \dfrac{y - y_0}{x - x_0} = m \Rightarrow y - y_0 = m(x - x_0)$

Thus, the equation of a line whose slope is m and passing through a fixed point is $y - y_0 = m(x - x_0)$.

Two-Point Form

The equation of a line passing through two given points (x_1, y_1) and (x_2, y_2) is given by $y - y_1 = \dfrac{y_2 - y_1}{x_2 - x_1}(x - x_1)$

Slope-Intercept Form

y-Intercept: The equation of the line having slope m and passing through a fixed point (0, c) is $y - c = m(x - 0) \Rightarrow y = mx + c$

Note: The value of c will be positive or negative according to whether the intercept is made on the positive or negative side of the y-axis.

x-Intercept: The equation of the line having slope m and passing through a fixed point (d, 0) is $y - 0 = m(x - d) \Rightarrow y = m(x - d)$

Note: The value of d will be positive or negative according to whether the intercept is made on the positive or negative side of the x-axis.

Intercept-Form

Let a line L makes x-intercept equal to a and y-intercept equal to b on the axes. Therefore, line L meets the x-axis at the point (a, 0) and the y-axis at the point (0, b).

Thus, the equation of the line making intercepts a and b on the x- and y-axes, respectively, is $\dfrac{x}{a} + \dfrac{y}{b} = 1$

Normal Form

The equation of the line having normal distance p from the origin and angle ω, which the normal makes with the positive direction of x-axis, is given by $x\cos\omega + y\sin\omega = p$

General Equation of a Line

Any equation of the form $Ax + By + C = 0$, where A, B and C are real constants such that A and B are not zero simultaneously, is called the general equation of a line. Various forms of the general equation of a line are as follows:

(a) Slope-intercept form

The equation $Ax + By + C = 0$ is the slope-intercept form of the equation of a line whose slope is $-\dfrac{A}{B}$ and y-intercept is $-\dfrac{C}{B}$.

Note: If $B = 0$, then $x = -\frac{C}{A}$, which is a vertical line whose slope is undefined and x-intercept is $-\frac{C}{A}$.

(b) Intercept form

The equation $Ax + By + C = 0$ is the intercept form of the equation of a line whose x-intercept is $-\frac{C}{A}$ and y-intercept is $-\frac{C}{B}$.

Note: If $C = 0$, then $Ax + By + C = 0 \Rightarrow Ax + By = 0$, which is a line passing through the origin. Therefore, it has zero intercepts on the axes.

(c) Normal form

The normal form of the general equation $Ax + By + C = 0$ is $x\cos\omega + y\sin\omega = p$, where

$$\cos\omega = \pm\frac{A}{\sqrt{A^2+B^2}}, \quad \sin\omega = \pm\frac{B}{\sqrt{A^2+B^2}},$$

$$p = \pm\frac{C}{\sqrt{A^2+B^2}}.$$

Note: Manage the signs in such a way that p is positive.

Distance of a Point from a Line

The perpendicular distance (d) of a line $Ax + By + C = 0$ from a point (x_0, y_0) is given by

$$d = \frac{|Ax_0 + By_0 + C|}{\sqrt{A^2+B^2}}$$

Distance between Two Parallel Lines

The distance d between two parallel lines $y = mx + c_1$ and $y = mx + c_2$ is $d = \frac{|c_1 - c_2|}{\sqrt{1+m^2}}$

Note: If the parallel lines are given in general form, that is, $Ax + By + C_1 = 0, Ax + By + C_2 = 0$, then the distance between these parallel is given by $d = \frac{|C_1 + C_2|}{\sqrt{A^2+B^2}}$.

Note:

(I) Two lines $A_1x + B_1y + C_1 = 0, A_2x + B_2y + C_2 = 0$, in general form, where $B_1, B_2 \neq 0$ are

(i) Parallel if $\frac{A_1}{B_1} = \frac{A_2}{B_2}$

(ii) Perpendicular if $A_1A_2 + B_1B_2 = 0$

(II) A line through the point (x_1, y_1) and parallel to the line $Ax + By + C = 0$ is $A(x - x_1) + B(y - y_1) = 0$.

Equation of Family of Lines Passing through the Point of Intersection of Two Lines

Let the general equation of two intersecting lines l_1, l_2 be given by

$l_1 : A_1x + B_1y + C_1 = 0$

$l_2 : A_2x + B_2y + C_2 = 0$

Then with the above two equations, we can form an equation $l_1 + k\,l_2 = 0$

$\Rightarrow A_1x + B_1y + C_1 + k(A_2x + B_2y + C_2) = 0$ represents a family of lines, where k is an arbitrary constant called the parameter. For a particular value of k, we obtain a family of equations and the value of k may be obtained from the other given condition.

Translation of Axes

This is a special case of transfer of axes, in which a new system of coordinate axes is obtained by transferring the coordinate axes parallel to the original axes (Fig. 10.13). Thus, in the new system of coordinate axes, each point (including origin) shifts to a new point.

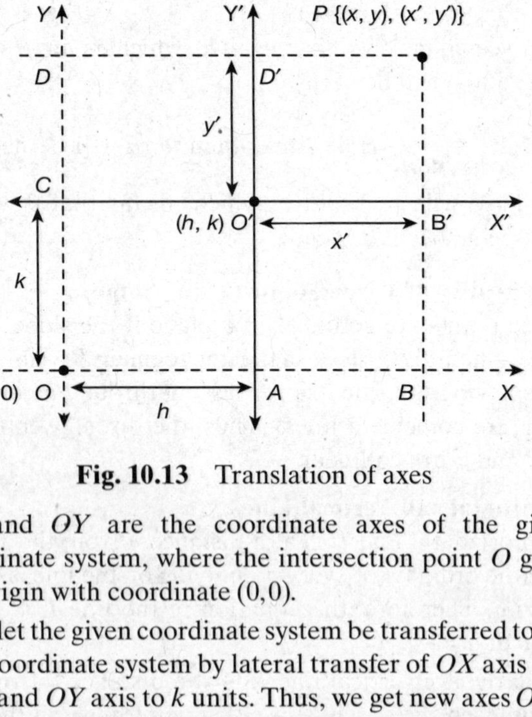

Fig. 10.13 Translation of axes

OX and OY are the coordinate axes of the given coordinate system, where the intersection point O gives the origin with coordinate $(0,0)$.

Now, let the given coordinate system be transferred to the new coordinate system by lateral transfer of OX axis to h units and OY axis to k units. Thus, we get new axes $O'X'$ and $O'Y'$, and intersection point O' will be the origin in the new coordinate system with coordinate (h, k).

Therefore, any point $P(x, y)$ in the given coordinate system will be changed to $P(x', y')$ in the new coordinate system.

The relation between the given and new coordinate systems will be obtained as

$\because OB = OA + AB \Rightarrow x = h + x' \Rightarrow x' = x - h$

Thus, Abscissa in new coordinate system = Abscissa in given coordinate system – Lateral transfer of OX axis (h units).

Similarly, $OD = OC + CD \Rightarrow y = k + y' \Rightarrow y' = y - k$

Thus, Ordinate in new coordinate system = Ordinate in given coordinate system – Lateral transfer of OY axis (k units).

Hence, any point $P(x, y)$ in the given coordinate system will be given by $P(x', y') = P(x - h, y - k)$ in the new coordinate system.

Circle: A circle is a set of all points in a plane that are equidistant from a fixed point in the plane.

- The fixed point is called the centre of the circle and the distance from the centre to a point on the circle is called the radius of the circle.
- Let P (x, y) be any point on the circle. Then the equation of the circle with a given centre C (h, k) and radius r is given by $|CP| = r$.
- Using the distance formula, we get $(x - h)^2 + (y - k)^2 = r^2$ as the required equation of the circle with centre at (h, k) and radius r.
- The equation of the circle with a given centre at origin O (0, 0) and radius r is given by $|OP| = r$.
- Using the distance formula, we get $x^2 + y^2 = r^2$ as the required equation of the circle with centre at origin (0, 0) and radius r.

Parabola: A parabola is a set of all points in a plane that are equidistant from a fixed line and a fixed point in the plane.

Note:

- The fixed line is called the directrix of the parabola.
- The fixed point is called the focus (F) of the parabola.
- A line through the focus and perpendicular to the directrix is called the axis of the parabola.
- The point of intersection of the parabola with the axis is called the vertex (O) of the parabola.

Standard Equations of Parabola

- The standard equation of a parabola is simplest if the vertex is at the origin and the axis of symmetry is along the x-axis or y-axis.
- The standard equations of parabola have focus on one of the coordinates axes, the vertex at the origin and the directrix parallel to the other coordinate axis.
- The four equations $y^2 = 4ax$, $y^2 = -4ax$, $x^2 = 4ay$ and $x^2 = -4ay$ are known as standard equations of parabolas.

  (i) Parabola $y^2 = 4ax$: Focus at $(a, 0)$ and Directrix $x = -a$

  (ii) Parabola $y^2 = -4ax$: Focus at $(-a, 0)$ and Directrix $x = a$

  (iii) Parabola $x^2 = 4ay$: Focus at $(0, a)$ and Directrix $y = -a$

  (iv) Parabola $x^2 = -4ay$: Focus at $(0, -a)$ and Directrix $y = a$

Note: From the standard equations of the parabolas, we may observe that:

- Parabola is symmetric with respect to the axis of the parabola.
- If the equation has an y^2 term, then the axis of symmetry is along the x-axis and if the equation has an x^2 term, then the axis of symmetry is along the y-axis.
- When the axis of symmetry is along the x-axis, the parabola opens to the right if the coefficient of x is positive and to the left if the coefficient of x is negative.
- When the axis of symmetry is along the y-axis, the parabola opens upwards if the coefficient of y is positive and downwards if the coefficient of y is negative.

Latus Rectum: The latus rectum of a parabola is a line segment perpendicular to the axis of the parabola, passing through the focus and whose end points lie on the parabola.

Length of the latus rectum = $4a$.

Ellipse: An ellipse is a set of all points (P) in a plane, the sum of whose distances from two fixed points (F_1 and F_2) in the plane is a constant (which is always greater than the distance between the two fixed points).

Note:

- The two fixed points (F_1 and F_2) are called the foci (plural of 'focus') of the ellipse.
- The mid-point of the line segment joining the foci is called the centre (O) of the ellipse.
- The line segment through the foci of the ellipse is called the major axis and the line segment through the centre and perpendicular to the major axis is called the minor axis.

- The end points of the major axis are called the vertices (A and B) of the ellipse.

Relationship between Semi-major Axis (a), Semi-minor Axis (b) and the Distance of the Focus from the Centre of the Ellipse (c)

$c = \sqrt{a^2 - b^2}$

Eccentricity (e): The eccentricity of an ellipse is the ratio of the distances from the centre of the ellipse to one focus and to one of the vertices of the ellipse, that is,

eccentricity, $e = \frac{c}{a}$.

Thus, we can say that the focus of an ellipse is at a distance of $c = ae$. from the centre.

And, we can say that the length of minor axis is $b = \sqrt{a^2 - c^2} = \sqrt{a^2 - (ae)^2} = a\sqrt{1-e^2}$

Special Cases of an Ellipse

From the equation $b = a\sqrt{1-e^2}$, if we keep a fixed and eccentricity (e) from 0 to 1, the resulting ellipses will vary from circle to line segment.

Case (i): When $e = 0 \Rightarrow b = a\sqrt{1-0^2} = a$, both foci merge together with the centre of the ellipse, and thus, the ellipse becomes circle with radius $r = a = b$.

Case (ii): When $e = 1 \Rightarrow b = a\sqrt{1-1^2} = 0$, the minor axis becomes zero and major axis equal to the distance between the foci = $2a$. Therefore, the ellipse reduces to the line segment F_1F_2 joining the two foci.

Standard Equations of an Ellipse

The equation of an ellipse is simplest if the centre of the ellipse is at the origin and the foci are on the x-axis or on the y-axis.

$\frac{x^2}{a^2} + \frac{y^2}{b^2} = 1$ is the equation of ellipse with foci on the x-axis.

$\frac{x^2}{b^2} + \frac{y^2}{a^2} = 1$ is the equation of the ellipse with foci on the y-axis.

These two equations are known as standard equations of the ellipse.

Note:

- The ellipse is symmetric with respect to both the coordinates axes, since if point (x, y) is on the ellipse, then $(-x, y), (x, -y), (-x, -y)$ are also points on the ellipse.
- The foci always lie on the major axis.
- The major axis can be determined by finding the intercepts on the axes of symmetry.
- In standard equations of the ellipse, the major axis is along the x-axis if the coefficient of x^2 has the larger denominator and it is along the y-axis if the coefficient of y^2 has the larger denominator.

Latus Rectum: The latus rectum of an ellipse is a line segment perpendicular to the major axis through any of the foci and whose end points lie on the ellipse.

$y = \pm\frac{b^2}{a}$, which is the intersection point of the latus rectum and the ellipse, that is, coordinates of A and B are $\left(c, \frac{b^2}{a}\right)$ and $\left(c, -\frac{b^2}{a}\right)$, respectively.

Since the ellipse is symmetric with respect to the y-axis, $\frac{b^2}{a}$ and thus the length of the latus rectum is $\frac{2b^2}{a}$.

Hyperbola: A hyperbola is a set of all points in a plane, the magnitude of difference of whose distances from two fixed points (F_1 and F_2) in the plane is a constant (which is always less than the distance between the two fixed points).

Note:

- The two fixed points (F_1 and F_2) are called the foci (plural of 'focus') of the hyperbola.
- The mid-point of the line segment joining the foci is called the centre (O) of the hyperbola.
- The line through the foci of the hyperbola is called the transverse axis and the line through the centre and perpendicular to the transverse axis is called the conjugate axis.
- The points at which the hyperbola intersect the transverse axis are called the vertices of the hyperbola.

Relationship between Length of the Transverse Axis ($2a$), Length of the Conjugate Axis ($2b$) and the Distance between the Two Foci of the Hyperbola ($2c$)

Let the length of the transverse axis be $2a$, the length of the conjugate axis be $2b$ and the distance between the foci (F_1 and F_2) be $2c$. Then, $b = \sqrt{c^2 - a^2}$.

Eccentricity (e): The eccentricity of a hyperbola is the ratio of the distances from the centre of the hyperbola to one foci and to one of the vertices of the hyperbola, that is, eccentricity, $e = \frac{c}{a}$.

Thus, we can say that the focus of an hyperbola is at a distance of $c = ae$. from the centre.

And, $b=\sqrt{c^2-a^2}=\sqrt{(ae)^2-a^2}-=a\sqrt{e^2-1}$

Standard Equations of a Hyperbola

The equation of a hyperbola is simplest if the centre of the hyperbola is at the origin and the foci are on the x-axis or y-axis.

$\frac{x^2}{a^2}-\frac{y^2}{b^2}=1$ is the equation of a hyperbola with foci on the x-axis.

$\frac{y^2}{a^2}-\frac{x^2}{b^2}=1$ is the equation of the hyperbola with foci on the y-axis.

Note:

- A hyperbola in which $a = b$ is called an equilateral hyperbola.
- A hyperbola is symmetric with respect to both the coordinates axes, since if point (x, y) is on the ellipse, then $(-x, y), (x, -y), (-x, -y)$ are also points on the hyperbola.
- The foci are always on the transverse axis.
- In standard equations of the hyperbola, the transverse axis is along the x-axis if the denominator of coefficient of x^2 is positive and it is along the y-axis if the denominator of coefficient of y^2 is positive.
- From the equation of the hyperbola $\frac{x^2}{a^2}=1+\frac{y^2}{b^2}\geq 1$ $\Rightarrow \left|\frac{x}{a}\right|\geq 1 \Rightarrow x\leq -a$ or $x\geq a$.
- Therefore, no portion of the curve lies between the lines $x=-a$, $x=a$, (that is, no intercept on the conjugate axis).

Latus Rectum: The latus rectum of a hyperbola is a line segment perpendicular to the transverse axis through any of the foci and whose end points lie on the hyperbola.

- $y=\pm\frac{b^2}{a}$, which is the intersection point of the latus rectum and the hyperbola, that is, coordinates of A and B are $\left(c, \frac{b^2}{a}\right)$ and $\left(c, -\frac{b^2}{a}\right)$, respectively.
- Since the hyperbola is symmetric with respect to the y-axis, the length of the latus rectum is $\frac{2b^2}{a}$.

Day 188

Revision (Chapters XII–XIII)

Coordinate Axes, Coordinate Planes and Coordinates of a Point in Three-Dimensional Space

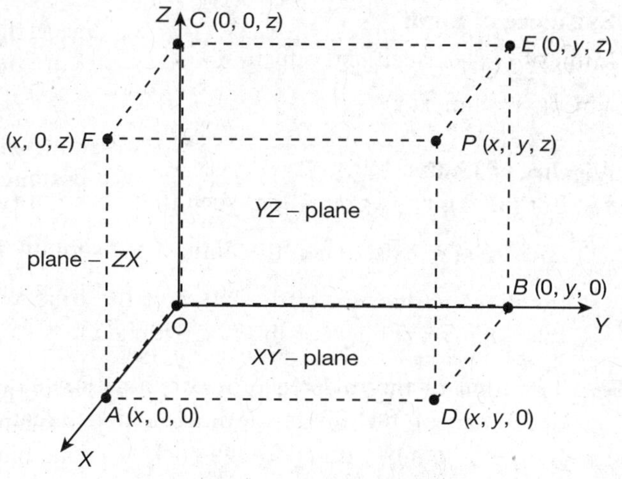

Fig. 12.1 Three-dimensional space

- Let the three mutually perpendicular planes (XY-plane, YZ-plane and ZX-plane) along the coordinate axes OX, OY and OZ (x-, y- and z-axes, respectively) intersect at a point O (Fig. 12.1).
- The point O is called the origin of the coordinate system and its coordinates are given by (0, 0, 0).
- The XY-plane, YZ-plane and ZX-plane are known as the three coordinate planes. They divide the space into eight parts, known as octants.
- Any point $P(x, y, z)$ in the coordinate system (three-dimensional space) is given by a triplet of three numbers (x, y, z) along the three coordinate axes. And x, y and z are perpendicular distances from the YZ-, ZX- and XY-planes, respectively.
- The value of a coordinate (x, y or z) is taken as positive if it is along positive axes and negative if it is along negative axes.
- The coordinates of any point on the x-axis will be given by $(x, 0, 0)$, on the y-axis will be given by $(0, y, 0)$ and on the z-axis will be given by $(0, 0, z)$.

- The z-coordinate is zero in the XY-plane, in the YZ-plane the x-coordinate is zero and in the ZX-plane the y-coordinate is zero.
- The coordinates of any point on the XY-plane will be given by $(x, y, 0)$, on the YZ-plane will be given by $(0, y, z)$, and on the ZX-plane will be given by $(x, 0, z)$,
- The signs of the coordinates of a point determine the octant in which the point lies, as given in the table below:

Coordinates / Octants	x	y	z
I	+	+	+
II	-	+	+
III	-	-	+
IV	+	-	+
V	+	+	-
VI	-	+	-
VII	-	-	-
VIII	+	-	-

Section Formula

- Let $P(x_1, y_1, z_1)$, $Q(x_2, y_2, z_2)$ be the two points in the coordinate system of rectangular axes OX, OY and OZ.
- The coordinates of a point that divides the line segment joining two points $P(x_1, y_1, z_1)$, $Q(x_2, y_2, z_2)$ internally in the ratio $m{:}n$ is given by $\left(\frac{mx_2+nx_1}{m+n}, \frac{my_2+ny_1}{m+n}, \frac{mz_2+nz_1}{m+n}\right)$
- The coordinates of a point that divides the line segment joining two points $P(x_1, y_1, z_1)$, $Q(x_2, y_2, z_2)$ externally in the ratio $m{:}n$ is given by $\left(\frac{mx_2-nx_1}{m-n}, \frac{my_2-ny_1}{m-n}, \frac{mz_2-nz_1}{m-n}\right)$
- The coordinates of the mid-point of the line segment joining two points $P(x_1, y_1, z_1)$,$Q(x_2, y_2, z_2)$ is given by $\left(\frac{x_1+x_2}{2}, \frac{y_1+y_2}{2}, \frac{z_1+z_2}{2}\right)$
- The coordinates of a point that divides the line segment joining two points $P(x_1, y_1, z_1)$,$Q(x_2, y_2, z_2)$ in the ratio $k:1$ is given by $\left(\frac{kx_2+x_1}{1+k}, \frac{ky_2+y_1}{1+k}, \frac{kz_2+z_1}{1+k}\right)$
- If $k>0$, the point divides the line segment PQ internally and if $k<0$, the point divides the line segment PQ externally.
- The point $R\left(\frac{kx_2+x_1}{1+k}, \frac{ky_2+y_1}{1+k}, \frac{kz_2+z_1}{1+k}\right)$ gives the general point on the line passing through two given points $P(x_1, y_1, z_1)$,$Q(x_2, y_2, z_2)$. That is, the points P, Q and R are collinear points.
- The coordinates of the centroid of the triangle PQR, whose vertices are $P(x_1, y_1, z_1)$, $Q(x_2, y_2, z_2)$ and $R(x_3, y_3, z_3)$, is given by $\left(\frac{x_1+x_2+x_2}{3}, \frac{y_1+y_2+y_3}{3}, \frac{z_1+z_2+z_3}{3}\right)$

Distance between Two Points

- Let $P(x_1, y_1, z_1)$, $Q(x_2, y_2, z_2)$ be the two points in the coordinate system of rectangular axes OX, OY and OZ.
- Then the distance between two points $P(x_1, y_1, z_1)$, $Q(x_2, y_2, z_2)$ is given by $|PQ| = \sqrt{(x_2-x_1)^2+(y_2-y_1)^2+(z_2-z_1)^2}$
- The distance of point $P(x,y,z)$ from the origin $O(0,0,0)$ is $|OP| = \sqrt{x^2+y^2+z^2}$.

Limit of a Function

A function $y = f(x)$ tends to limit l as $x \to a$(read as 'x tends to a' means x is very close to a), written as $\lim_{x\to a} f(x) = l$

Left-hand and Right-hand Limits

$\lim_{x\to a^-} f(x) = \lim_{h\to 0} f(a-h)$ is the left-hand limit (LHL), which means x is very close to a but less than a.

$\lim_{x\to a^+} f(x) = \lim_{h\to 0} f(a+h)$ is the right-hand limit (RHL), which means x is very close to a but greater than a.

Existence of Limit

Limit of $f(x)$ exists if and only if

$\lim_{x\to a^-} f(x) = \lim_{x\to a^+} f(x)$

Algebra of Limits

- Let f and g be two functions such that both $\lim_{x\to a} f(x)$ and $\lim_{x\to a} g(x)$ exist. Then the limit of the sum of two functions is the sum of the limits of the functions, that is,$\lim_{x\to a}[f(x)+g(x)] = \lim_{x\to a} f(x) + \lim_{x\to a} g(x)$.
- The limit of the difference of two functions is the difference of the limits of the functions, that is, $\lim_{x\to a}[f(x)-g(x)] = \lim_{x\to a} f(x) - \lim_{x\to a} g(x)$.
- The limit of the product of two functions is the product of the limits of the functions, that is, $\lim_{x\to a}[f(x)\cdot g(x)] = \lim_{x\to a} f(x)\cdot \lim_{x\to a} g(x)$.

- The limit of the product of a constant and a function is the product of the constant and the limit of the function, that is, $\lim_{x\to a}[\lambda\, g(x)] = \lambda \cdot \lim_{x\to a} g(x)$.
- The limit of the quotient of two functions is the quotient of the limits of the functions, that is, $\lim_{x\to a}\frac{f(x)}{g(x)} = \frac{\lim_{x\to a} f(x)}{\lim_{x\to a} g(x)}, \quad \left(\lim_{x\to a} g(x) \neq 0\right)$

Limits of Polynomial Function

Let a polynomial function $f(x) = a_0 + a_1x + a_2x^2 + \ldots + a_nx^n$, where all a_i are real numbers such that $a_n \neq 0$ for some natural number n. Then

$$\lim_{x\to a} f(x) = \lim_{x\to a}[a_0 + a_1x + a_2x^2 + \ldots + a_nx^n]$$
$$= \lim_{x\to a} a_0 + \lim_{x\to a} a_1x + \lim_{x\to a} a_2x^2 + \ldots + \lim_{x\to a} a_nx^n$$
$$= a_0 + a_1 \lim_{x\to a} x + a_2 \lim_{x\to a} x^2 + \ldots + a_n \lim_{x\to a} x^n$$
$$= a_0 + a_1a + a_2a^2 + \ldots + a_na^n = f(a)$$

Limits of Rational Function

Let a rational function $f(x) = \frac{g(x)}{h(x)}$, where $g(x), h(x)$ are polynomial functions such that $h(x) \neq 0$. Then

$$\lim_{x\to a} f(x) = \lim_{x\to a}\frac{g(x)}{h(x)} = \frac{\lim_{x\to a} g(x)}{\lim_{x\to a} h(x)} = \frac{g(a)}{h(a)}$$

Case 1: If $h(a) = 0$ and $g(a) \neq 0$, then $\lim_{x\to a} f(x)$ does not exist.

Case 2: If $h(a) = 0$ and $g(a) = 0$, then $\lim_{x\to a} f(x) = \frac{0}{0}$ is of indeterminate form and can be solved by rationalising and simplifying.

Some Important Limits

- $\lim_{x\to 0}\frac{x^n - a^n}{x - a} = na^{n-1}$
- $\lim_{x\to 0}\frac{(1+x)^n - 1}{x} = n$

Theorem: Let f and g be two real-valued functions with the same domain such that $f(x) \leq g(x)$ for all x in the domain of definition. For some a, if both $\lim_{x\to a} f(x), \lim_{x\to a} g(x)$ exist, then $\lim_{x\to a} f(x) \leq \lim_{x\to a} g(x)$.

Sandwich Theorem: Let f, g, h be real functions such that $f(x) \leq g(x) \leq h(x)$ for all x in the common domain of definition. For some real number a, if $\lim_{x\to a} f(x) = l = \lim_{x\to a} h(x)$, then $\lim_{x\to a} g(x) = l$.

Limits of Trigonometric Functions

Some important trigonometric limits are:

(i) $\lim_{x\to 0}\frac{\sin x}{x} = 1$ (ii) $\lim_{x\to 0}\frac{\tan x}{x} = 1$

(The above two limits can be proved by the Sandwich Theorem)

(iii) $\lim_{x\to 0}\frac{1-\cos x}{x} = 0$.

Proof:

$$\lim_{x\to 0}\frac{1-\cos x}{x} = \lim_{x\to 0}\frac{2\sin^2(x/2)}{x}$$
$$= \lim_{x\to 0}\frac{\sin(x/2)}{(x/2)}\sin(x/2)$$
$$= 1 \times \sin(x/2)\left(\because \lim_{h\to 0}\frac{\sin h}{h} = 1\right)$$
$$= \sin(0/2) = 0$$

Some Important Series

- $\sin x = x - \frac{x^3}{3!} + \frac{x^5}{5!} - \frac{x^7}{7!} + \ldots$
- $\cos x = 1 - \frac{x^2}{2!} + \frac{x^4}{4!} - \frac{x^4}{6!} + \ldots$
- $\tan x = x + \frac{1}{3}x^3 + \frac{2}{15}x^5 + \ldots$

Derivative: First Principle of Derivative

Let there exist a real-valued function $f(x)$. Then the derivative of function $f(x)$ is given by $f'(x) = \lim_{h\to 0}\frac{f(x+h) - f(x)}{h}$ (provided this limit exists) and this is known as the first principle of derivative.

The derivative $f'(x_0) = \lim_{h\to 0}\frac{f(x_0+h) - f(x_0)}{h}$ denotes the change in $f(x)$ at x_0 with respect to x.

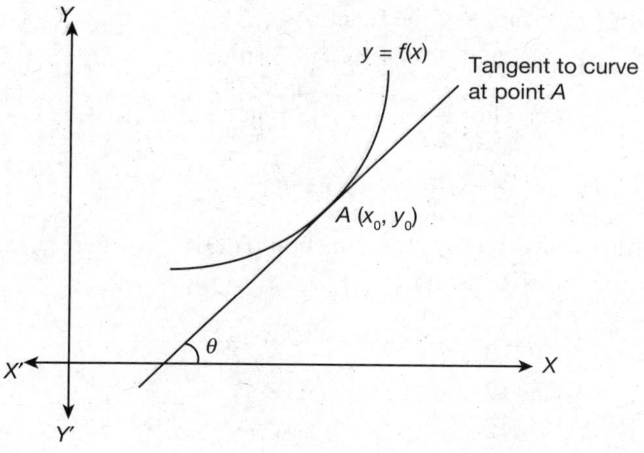

Fig. 13.1 Derivative of a function

Geometrically (Fig. 13.1), the derivative of function $y = f(x)$ at $x = x_0$, that is, $f'(x_0) = \lim_{h\to 0}\frac{f(x_0+h)-f(x_0)}{h}$ represents the slope $\left.\frac{dy}{dx}\right|_{x_0} = f'(x_0) = \tan\theta$ of the tangent to the curve $y = f(x)$ at $x = x_0$.

Algebra of Derivative of Functions

Let $f(x)$ and $g(x)$ be two functions such that their derivatives are defined in a common domain. Then

- Derivative of the sum of two functions is equal to the sum of the derivatives of the functions $\frac{d}{dx}[f(x)+g(x)] = \frac{d}{dx}f(x)+\frac{d}{dx}g(x)$.
- Derivative of the difference of two functions is equal to difference of the derivatives of the function $\frac{d}{dx}[f(x)-g(x)] = \frac{d}{dx}f(x)-\frac{d}{dx}g(x)$.
- Derivative of the product of two functions is given by the product rule $\frac{d}{dx}[f(x)\cdot g(x)] = f(x)\cdot\frac{d}{dx}g(x)+g(x)\cdot\frac{d}{dx}f(x)$
- Derivative of the quotient of two functions is given by the quotient rule $(g(x) \neq 0)$

$$\frac{d}{dx}\left(\frac{f(x)}{g(x)}\right) = \frac{g(x)\cdot\frac{d}{dx}f(x)-f(x)\frac{d}{dx}g(x)}{(g(x))^2}$$

Derivatives of Some Standard Functions

(i) Derivative of the function $f(x) = x$

By the first principle of derivative

$$f'(x) = \lim_{h\to 0}\frac{f(x+h)-f(x)}{h} = \lim_{h\to 0}\frac{x+h-x}{h}$$

$$\Rightarrow \frac{d(x)}{dx} = f'(x) = \lim_{h\to 0} 1 = 1$$

(ii) Derivative of the function $f(x) = nx$

By the first principle of derivative

$$f'(x) = \lim_{h\to 0}\frac{f(x+h)-f(x)}{h} = \lim_{h\to 0}\frac{n(x+h)-nx}{h}$$

$$\Rightarrow \frac{d(nx)}{dx} = f'(x) = \lim_{h\to 0} n = n$$

(iii) Derivative of the function $f(x) = x^n$, where n is positive integer.

$$f'(x) = \lim_{h\to 0}\frac{f(x+h)-f(x)}{h} = \lim_{h\to 0}\frac{(x+h)^n - x^n}{h}$$

Using Binomial Theorem

$$(x+h)^n = ({}^nC_0)x^n + ({}^nC_1)x^{n-1}h + \ldots + ({}^nC_n)h^n$$

$$\therefore (x+h)^n - x^n = \left(x^n + nx^{n-1}h + \ldots + h^n\right) - x^n$$

$$= h(nx^{n-1} + \ldots + h^{n-1})$$

Thus, $\frac{d(x^n)}{dx} = \lim_{h\to 0}\frac{h(nx^{n-1}+\ldots+h^{n-1})}{h}$

$$\Rightarrow \frac{d(x^n)}{dx} = \lim_{h\to 0}(nx^{n-1}+\ldots+h^{n-1}) = nx^{n-1}.$$

(iv) Derivative of the polynomial function $f(x) = a_n x^n + a_{n-1}x^{n-1} + \ldots + a_1 x + a_0$

Using the above result $\frac{d(x^n)}{dx} = nx^{n-1}$, the derivative of the function is given by

$$\frac{df(x)}{dx} = na_n x^{n-1} + (n-1)a_{n-1}x^{x-2} + \ldots + 2a_2 x + a_1.$$

Some Standard Results

- $\frac{d}{dx}x^n = nx^{n-1}$
- $\frac{d}{dx}\frac{1}{x^n} = \frac{-n}{x^{n+1}}$
- $\frac{d}{dx}(ax+b)^n = na(ax+b)^{n-1}$
- $\frac{d}{dx}\sin x = \cos x$
- $\frac{d}{dx}\cos x = -\sin x$
- $\frac{d}{dx}\tan x = \sec^2 x$
- $\frac{d}{dx}\cot x = -\text{cosec}^2 x$
- $\frac{d}{dx}\sec x = \sec x \tan x$
- $\frac{d}{dx}\text{cosec}\, x = -\text{cosec}\, x \cot x$

Limit of Exponential Function

$$\lim_{x\to 0}\frac{e^x-1}{x} = 1$$

Proof: $\lim_{x\to 0}\frac{e^x-1}{x}$

$$= \lim_{x\to 0}\frac{1+x+\frac{x^2}{2!}+\frac{x^3}{3!}+\ldots-1}{x}$$

$$= \lim_{x\to 0}\frac{x+\frac{x^2}{2!}+\frac{x^3}{3!}+\ldots}{x}$$

$$= \lim_{x\to 0}\left(1+\frac{x}{2!}+\frac{x^2}{3!}+\ldots\right) = 1$$

Limit of Logarithmic Function

$$\lim_{x\to 0}\frac{\log_e(1+|x|)}{|x|}=1$$

Proof: $\lim_{x\to 0}\frac{\log_e(1+|x|)}{|x|}$

$$=\lim_{x\to 0}\frac{|x|-\frac{|x|^2}{2}+\frac{|x|^3}{3}-\frac{|x|^4}{4}+\ldots}{|x|}$$

$$=\lim_{x\to 0}\left(1-\frac{|x|}{2}+\frac{|x|^2}{3}-\frac{|x|^3}{4}+\ldots\right)=1$$

Some Important Series

- $e^x=1+x+\frac{x^2}{2!}+\frac{x^3}{3!}+\ldots$
- $e^{-x}=1-x+\frac{x^2}{2!}-\frac{x^3}{3!}+\ldots$
- $\log(1+x)=x-\frac{x^2}{2}+\frac{x^3}{3}-\frac{x^4}{4}+\ldots,(|x|<1)$
- $\log(1-x)=-\left(x+\frac{x^2}{2}+\frac{x^3}{3}+\frac{x^4}{4}+\ldots\right);(|x|<1)$

Day 189

Unit Test 3 (Chapters X–XIII)

Time: 3 Hrs Maximum **Marks:** 80

General Instructions:

(i) All the questions are compulsory.
(ii) The question paper consists of 36 questions divided into 4 sections A, B, C and D.
(iii) Section A comprises 20 questions of 1 mark each. Section B comprises 6 questions of 2 marks each. Section C comprises 6 questions of 4 marks each. Section D comprises 4 questions of 6 marks each.
(iv) There is no overall choice. However, an internal choice has been provided in four questions of 1 mark each, two questions of 2 marks each, two questions of 4 marks each, and two questions of 6 marks each. You have to attempt only one of the alternatives in all such questions.
(v) Use of calculators is not permitted.

SECTION A

Q1–Q10 are multiple choice type questions. Select the correct option. (1 mark each)

Q1. If $y=\sqrt{x}-\frac{1}{\sqrt{x}}$, then $\frac{dy}{dx}$ at $x=1$ is equal to

(a) 1 (b) 1/2 (c) $1/\sqrt{2}$ (d) 0

Q2. If $f(x)=\frac{x-2}{2\sqrt{x}}$, then $f'(1)$ is equal to

(a) 5/4 (b) 3/4 (c) 1 (d) 0

Q3. If $y=\frac{x+1}{1-x}$ then $\frac{dy}{dx}$ is equal to

(a) $\frac{1}{(1-x)^2}$ (b) $\frac{x}{(1-x)^2}$

(c) $\frac{2}{(1-x)^2}$ (d) $\frac{2x}{(1-x)^2}$

Q4. If $f(x)=1+x+x^2+x^3\ldots+x^{100}$, then $f'(1)$ is equal to

(a) 100 (b) 1010 (c) 4900 (d) 5050

Q5. Evaluate the given limit: $\lim_{x\to 0}\frac{(x+1)^8-1}{x}$

(a) 5 (b) 4 (c) 8 (d) 10

Q6. Evaluate the given limit: $\lim_{x\to 1}\frac{1-\frac{1}{x}}{1-x}$

(a) 1 (b) 0 (c) 1/2 (d) –1

Q7. Evaluate the given limit: $\lim_{x\to 0}\frac{\sin 2x}{x}$

(a) 5 (b) 4 (c) 2 (d) 1

Q8. Evaluate the given limit: $\lim_{x\to 0} x\operatorname{cosec} x$

(a) 1 (b) 1/2 (c) 0 (d) –1

Q9. Find the distance between parallel lines $3x+4y-2=0, 3x+4y+5=0$

(a) 1/2 (b) 5/7 (c) 7/5 (d) 2

Q10. Evaluate the given limit:

$\lim_{x\to 0}\frac{ax^2+bx+c}{cx^2+bx+a}$, $a\neq 0$

(a) b/a (b) c/a (c) a/b (d) a/c

Q11–Q15 are fill in the blanks. (1 mark each)

Q11. $\lim_{z\to 1}\frac{z^{\frac{1}{2}}-1}{z^{\frac{1}{16}}-1}$ is equal to ... ...

Q12. $\lim_{x \to \pi} \frac{\cos x}{\pi}$ is equal to ... ...

Q13. The equation of the line through the intersection of lines $x-2y+3=0$, $2x+y-3=0$ and which passes through the origin is ... ...

Q14. If three points $(h,1),(0,0)(-1,k)$ lie on a line, then the value of kh is equal to ... ...

OR

Q14. If the line $y=\sqrt{3}x+k$ touches the circle $x^2+y^2=9$, then the value of k is equal to ... ...

Q15. The equation of the hyperbola satisfying the given conditions: vertices: $(\pm 2,0)$, foci: $(\pm 4,0)$ is ... ...

OR

Q15. The equation of the circle which touches both the axes in the second quadrant and whose radius is a is ... ...

(Q16–Q20) Answer the following questions

(1 mark each)

Q16. Find the coordinates of the point that divides the line segment joining the points (4, –3) and (8, 5) in the ratio 3:2 internally.

Q17. Find the equation of the ellipse that satisfies the given conditions: $b=2$, $c=3$, centre at the origin; foci on the x-axis.

Q18. Find a relation between x and y such that the point (x, y) is equidistance from the points $(-2,6)$ and $(3,-4)$

Q19. Find the slope of a line that passes through the origin and the mid-point of the line segment joining the points $P(2,-4)$, $B(6,0)$.

OR

Q19. Find the equation of the line which passes through the points $(1,0),(2,-4)$.

Q20. Find the equation of the parabola that satisfies the following conditions: focus (12, 0); directrix $x=-12$

OR

Q20. Find the equation of the ellipse that satisfies the given conditions: vertices $(\pm 6,0)$, foci $(\pm 4,0)$

SECTION B

(Q21–Q26) Solve the following questions.

(2 marks each)

Q21. If A, B, C be the feet of the perpendicular from a point P on the X-, Y- and Z-axes, respectively, then find the coordinates of A, B, C in each of the following where the point P is

(i) $A(3,-4,2)$ (ii) $C(4,-3,-2)$

Q22. Find the coordinates of the focus, axis of the parabola, the equation of directrix and the length of the latus rectum for $x^2=-2y$.

Q23. Find the equations of the parabola that satisfy the following conditions: vertex (0, 0), passing through (6, 2) and symmetric with respect to the y-axis.

Q24. Find the equation of a line that cuts off equal intercepts on the coordinates axes and passes through the point (1, –2).

Q25. Reduce the equation $6x+3y-5=0$ to slope-intercept form and find its slope and the y-intercept.

OR

Q25. Find equation of the line perpendicular to the line $x-2y+4=0$ and given the x-intercept equal to 3.

Q26. If one end of a diameter of the circle $x^2+y^2-4x-6y+11=0$ is (1, –1), then find the coordinates of the other end of the diameter.

OR

Q26. Find the equation of the circle having (0, –1) as its centre and passing through the point of intersection of the lines $3x+y=14, 2x+5y=18$.

SECTION C

(Q27–Q32) Solve the following questions.

(4 marks each)

Q27. Differentiate cosec x with respect to x using the first principle.

Q28. Evaluate $\lim_{x \to 0} f(x)$, where $f(x)=\begin{cases} \frac{|x|}{x}, & x \neq 0 \\ 0, & x=0 \end{cases}$

Q29. Find $\lim_{x\to -3} f(x),\ \lim_{x\to 3} f(x)$, where

$$f(x)=\begin{cases}-x+3, & \text{if } x\le -3\\ -2x, & \text{if } -3<x<3\\ 6x+2, & \text{if } x\ge 3\end{cases}$$

Q30. Evaluate the given limit:

$$\lim_{x\to 0}\frac{\sin ax+bx}{ax+\sin bx},\quad a,b,a+b\ne 0$$

Q31. Find the equation of a circle concentric with the circle $x^2+y^2-6x+12y+42=0$ and that has three times of its area.

OR

Q31. Find the equation of the circle which passes through the points (1, 0) and (0, –1) and the centre lies on the straight line $x-y+2=0$.

Q32. Find the equation of the ellipse that satisfies the given conditions: centre at (0, 0), major axis on the x-axis and passing through the points (4, 3) and (5, 2).

OR

Q32. Find the equation of the ellipse that satisfies the given conditions: centre at (0, 0), major axis on the y-axis and passing through the points (3, 2) and (1, 4).

SECTION D

(Q33–Q36) Solve the following questions. (6 marks each)

Q33. Differentiate $\cos(x^2)$ with respect to $x\dfrac{\sec x-1}{\sec x+1}$

Q34. Differentiate $\cos(x^2)$ with respect to x using the first principle.

Q35. Find the coordinates of the foci, the vertices the length of major axis, the minor axis, the eccentricity and the length of the latus rectum of the ellipse $25x^2+y^2=25$.

OR

Q35. Find the coordinates of the foci and the vertices, the eccentricity, and the length of the latus rectum of the hyperbola $\dfrac{y^2}{9}-\dfrac{x^2}{16}=1$.

Q36. Find the equation of the hyperbola with the given conditions: foci $(0,\pm\sqrt{5})$ and passing through (2, 3).

OR

Q36. The mid-points of the sides of a triangle are $(5,7,0),(0,8,5),(2,3,-5)$. Find its vertices.

Day 190

Unit Test 3 (Solutions)

SECTION A

Q1–Q10 are multiple choice type questions. Select the correct option. (1 mark each)

Q1. Solution: (a)

$$y=\sqrt{x}+\frac{1}{\sqrt{x}},$$

$$\frac{dy}{dx}=\frac{1}{2\sqrt{x}}+\frac{1}{2x^{3/2}}$$

$$\therefore\left(\frac{dy}{dx}\right)_{x=1}=\frac{1}{2}+\frac{1}{2}=1$$

Q2. Solution: (b)

$$f(x)=\frac{x-2}{2\sqrt{x}},$$

$$f'(x)=\frac{2\sqrt{x}(1)-(x-2)\cdot 2\cdot\frac{1}{2\sqrt{x}}}{4x}$$

$$=\frac{2x-(x-2)}{4x^{3/2}}=\frac{2x-x+2}{4x^{3/2}}=\frac{x+2}{4x^{3/2}}$$

$$\therefore f'(1)=\frac{1+2}{4\times(1)^{3/2}}=3/4$$

Q3. Solution: (c)

$$y=\frac{x+1}{1-x}$$

$$\therefore\frac{dy}{dx}=\frac{(1-x)(1)-(x+1)(-1)}{(1-x)^2}$$

$$=\frac{1-x+x+1}{(1-x)^2}=\frac{2}{(1-x)^2}$$

[By the quotient rule]

Q4. Solution: (d)

$f(x) = 1 + x + x^2 + x^3 ... + x^{100}$,

$\therefore f'(x) = 0 + 1 + 2x + 3x^2 ... + 100x^{99}$

$f'(1) = 1 + 2 + 3 + ... + 100 = \dfrac{100(100+1)}{2} = 5050$

Q5. Solution: (c)

Substituting, $x + 1 = y$, we get $y \to 1$ as $x \to 0$

Therefore, $\lim_{x\to 0} \dfrac{(x+1)^8 - 1}{x} = \lim_{y\to 1} \dfrac{y^8 - 1}{y - 1} = 8 \cdot 1^{8-1} = 8$

$\left[\because \lim_{x\to 0} \dfrac{x^n - a^n}{x - a} = na^{n-1}\right]$

Q6. Solution: (d)

$\lim_{x\to 1} \dfrac{1 - \frac{1}{x}}{1 - x} = \lim_{x\to 1} \dfrac{\left(\frac{x-1}{x}\right)}{1 - x} = \lim_{x\to 1} \dfrac{-1}{x} = \dfrac{-1}{1} = -1$

Q7. Solution: (c)

$\lim_{x\to 0} \dfrac{\sin 2x}{x} = \lim_{x\to 0} \dfrac{\sin 2x}{2x} \times 2$

$= 2 \lim_{2x\to 0} \left(\dfrac{\sin 2x}{2x}\right) \quad [\because x \to 0 \Rightarrow 2x \to 0]$

$= 2 \times 1 = 2 \left[\because \lim_{x\to 0} \dfrac{\sin x}{x} = 1\right]$

Q8. Solution: (a)

$\lim_{x\to 0} (x \operatorname{cosec} x) = \lim_{x\to 0} \dfrac{x}{\sin x} = \dfrac{1}{\lim_{x\to 0} (\sin x / x)} = \dfrac{1}{1} = 1a$

Q9. Solution: (c)

The distance (d) between parallel lines $Ax + By + C_1 = 0$, $Ax + By + C_2 = 0$ is given by

$d = \dfrac{|C_1 - C_2|}{\sqrt{A^2 + B^2}}$

Therefore, the distance between the given parallel lines $3x + 4y - 2 = 0, 3x + 4y + 5 = 0$ is (where $A = 3, B = 4, C_1 = -2, C_2 = 5$)

$d = \dfrac{|C_1 - C_2|}{\sqrt{A^2 + B^2}} = \dfrac{|-2-5|}{\sqrt{(3)^2 + (4)^2}} = \dfrac{|-7|}{5} = \dfrac{7}{5}$ units

Q10. Solution: (b)

$\lim_{x\to 0} \dfrac{ax^2 + bx + c}{cx^2 + bx + a} = \dfrac{a(0)^2 + b(0) + c}{c(0)^2 + b(0) + a} = \dfrac{c}{a}$

Q11–Q15 are fill in the blanks. (1 mark each)

Q11. Solution: Substituting $z^{1/16} = x$, we get $z \to 1$ as $x \to 1$

Therefore, $\lim_{z\to 1} \dfrac{z^{\frac{1}{2}} - 1}{z^{\frac{1}{16}} - 1} = \lim_{x\to 1} \dfrac{x^8 - 1}{x - 1} = 8 \cdot 1^{8-1} = 8$

$\left[\because \lim_{x\to a} \dfrac{x^n - a^n}{x - a} = na^{n-1}\right]$

Q12. Solution: $\lim_{x\to \pi} \dfrac{\cos x}{\pi} = \dfrac{\cos \pi}{\pi} = \dfrac{-1}{\pi}$

Q13. Solution: The equation of any line through the point of intersection of the given lines is of the form

$x - 2y + 3 + k(2x + y - 3) = 0$

$\Rightarrow (1 + 2k)x + (-2 + k)y + 3 - 3k = 0$

Since this line passes through origin, (0,0) must satisfy the equation, that is, $3 - 3k = 0 \Rightarrow k = 1$

Substituting the value of $k = 1$, we get $(1 + 2\times 1)x + (-2 + 1)y + 3 - 3\times 1 = 0 \Rightarrow 3x - y = 0$ as the required equation of the line.

Q14. Solution: The points $A(h,1), B(0,0), C(-1,k)$ lie on a line. Then the slope of the line segment AB = slope of the line segment BC. Therefore

$\dfrac{0-1}{0-h} = \dfrac{k-0}{-1-0} \Rightarrow 1 = -kh \Rightarrow kh = -1$

OR

Q14. Solution: For the given equation of a circle, $x^2 + y^2 = 9$, radius = 3 and centre = (0, 0)

Now, since the line $y = \sqrt{3}x + k$ touches the circle, the perpendicular from (0, 0) to line $y = \sqrt{3}x + k$ = radius of the circle $\Rightarrow \left|\dfrac{0 - 0 + k}{\sqrt{3+1}}\right| = 3$

$\Rightarrow \pm\dfrac{k}{2} = 3 \Rightarrow k = \pm 6$

(Note: Distance from the point (m, n) to the line $Ax + By + k = 0$ is $d = \left|\dfrac{Am + Bn + C}{A^2 + B^2}\right|$)

Q15. Solution: The vertices $(\pm 2, 0)$ are on the x-axis. Therefore, the equation of the hyperbola is of the form $\dfrac{x^2}{a^2} - \dfrac{y^2}{b^2} = 1$

Now, since the vertices are $(\pm 2,0) \Rightarrow a=2$ and the foci are $(\pm 4,0) \Rightarrow c=4$, from the equation $a^2+b^2=c^2 \Rightarrow b^2=4^2-2^2 \Rightarrow b^2=12$

Thus, the equation of the hyperbola is $\frac{x^2}{4}-\frac{y^2}{12}=1$

OR

Q15. Solution: Given that, the radius of the circle is a and it touches both the axes in the second quadrant

that is $(h,k)=(-a,a)$

So, the equation of required circle is

$(x+a)^2+(y-a)^2=a^2$

$\Rightarrow x^2+2ax+a^2+y^2+2ay+a^2=a^2$

$\Rightarrow x^2+y^2+2ax-2ay+a^2=0$

(Q16–Q20) Answer the following questions. (1 mark each)

Q16. Solution: Let $P(x, y)$ be the point which divides the line segment joining the points (4, –3), and (8, 5) in the ratio 3:2 internally. Then by section formula,

we get $x=\frac{3(8)+2(4)}{3+2}=\frac{32}{5}$, $y=\frac{3(5)+2(-3)}{3+2}=\frac{9}{5}$

Therefore, P (32/5, 9/5) is the required point.

Q17. Solution: It is given that $b=2$, $c=3$. Therefore, from the equation $a^2=b^2+c^2$

$\Rightarrow a^2=1^2+2^2=5 \Rightarrow a=\sqrt{5}$

The foci are on the x-axis. Therefore, the major axis is along the x-axis.

Thus, the equation of the ellipse will be of the form

$\frac{x^2}{a^2}+\frac{y^2}{b^2}=1 \Rightarrow \frac{x^2}{\sqrt{5}^2}+\frac{y^2}{2^2}=1 \Rightarrow \frac{x^2}{5}+\frac{y^2}{4}=1$

Q18. Solution: Let $P(x,y)$ be equidistant from the points $A(-2,6)$ and $B(3,-4)$. Then $AP=BP \Rightarrow AP^2=BP^2$

By the distance formula, we get

$(x+2)^2+(y-6)^2=(x-3)^2+(y+4)^2$

$\Rightarrow x^2+4+4x+y^2+36-12y$

$=x^2+9-6x+y^2+16+8y$

$\Rightarrow 10x-20y+15=0$

$\Rightarrow 2x-4y+3=0$ is the required relation.

Q19. Solution: The coordinates of the mid-point of the segment joining the points $P(2,-4), B(6,0)$ are $\left(\frac{2+6}{2},\frac{-4+0}{2}\right)=(4,-2)$

Then, the slope line passing through the points (0, 0) and (4, –2) is $\frac{-2-0}{4-0}=\frac{-2}{4}=-\frac{1}{2}$

(since slope of the line passing through the points $(x_1,y_1),(x_2,y_2)$ is given by $m=\frac{y_2-y_1}{x_2-x_1}, x_2 \neq x_1$)

OR

Q19. Solution: The equation of the line passing through points $(x_1,y_1),(x_2,y_2)$ is $y-y_1=\frac{y_2-y_1}{x_2-x_1}(x-x_1)$

Therefore, the equation of the line passing through the points $(1,0),(2,-4)$ is $(y-0)=\frac{-4-0}{2-1}(x-1)$

$\Rightarrow y=-4(x-1) \Rightarrow y=-4x+4 \Rightarrow 4x+y-4=0$

Q20. Solution: The focus (12, 0) lies on the x-axis. Therefore, the axis of the parabola is the x-axis. Thus, the equation of the parabola is either of the form $y^2=4ax$ or $y^2=-4ax$.

Now, since the directrix is $x=-12$ and focus is (12, 0), the parabola is of the form $y^2=4ax$ with $a=12$.

Thus, the equation of the parabola is $y^2=48x$.

OR

Q20. Solution: The vertices $(\pm 6,0)$ are on the x-axis. Therefore, the equation of the ellipse will be of the form $\frac{x^2}{a^2}+\frac{y^2}{b^2}=1$ and semi-major axis $a=6$

The foci are $(\pm 4,0)$. Therefore, $c=4$

Now, from equation $a^2=b^2+c^2$

$\Rightarrow 6^2=b^2+4^2 \Rightarrow b^2=20 \Rightarrow b=2\sqrt{5}$

Thus, the equation of the ellipse is $\frac{x^2}{6^2}+\frac{y^2}{(2\sqrt{5})^2}=1$

or $\frac{x^2}{36}+\frac{y^2}{20}=1$

SECTION B

(Q21–Q26) Solve the following questions.

(2 marks each)

Q21. Solution: The coordinates of the feet of the perpendicular from a point P on the X-, Y- and Z-axes of points A, B and C are as follows:

(i) $A(3,0,0), B(0,-4,0), C(0,0,2)$

(ii) $A(4,0,0), B(0,-3,0), C(0,0,-2)$

Q22. Solution: In the given equation $x^2 = -2y$ of the parabola, the coefficient of y is negative. Therefore the parabola opens downwards.

Comparing this equation with $x^2 = -2y$, we get

$-4a = -2 \Rightarrow a = 1/2$

Thus, coordinates of the focus $= (0,-a) = (0,-1/2)$

The given equation involves x^2. Therefore, the axis of the parabola is the y-axis.

Equation of directrix

$y = -1/2 \Rightarrow y = 1/2 \Rightarrow 2y - 1 = 0$

Length of latus rectum $= 4a = 2$

Q23. Solution: The vertex of the parabola is (0, 0) and the axis of the parabola is the y-axis. Therefore, the equation of the parabola is either of the form $x^2 = 4ay$ or $x^2 = -4ay$.

Now, since the parabola passes through point (6, 2) which lies in the first quadrant, the equation of the parabola is of the form $x^2 = 4ay$. Now since point (6, 2) must satisfy the equation $x^2 = 4ay$

$(6)^2 = 4 \times a \times 2 \Rightarrow a = 9/2$

Thus, the equation of the parabola is

$x^2 = 4(9/2)y \Rightarrow x^2 = 18y$

Q24. Solution: The equation of a line in the intercept form is $\frac{x}{a}+\frac{y}{b} = 1$, where, a and b are the intercepts on the x- and y-axis, respectively.

The line cuts off equal intercepts on both the axes. Therefore, $a=b$, and thus, the equation of line must be written as $\frac{x}{a}+\frac{y}{b} = 1 \Rightarrow x + y = a$

Since the given line passes through point (1, –2), it satisfies the equation of line. Hence,

$1 - 2 = a \Rightarrow a = -1$

Thus, $x + y + 1 = 0$ is the required equation of the line.

Q25. Solution: The given equation $2x + 3y - 4 = 0$ in slope-intercept form can be written as

$y = \frac{1}{3}(-2x+4) \Rightarrow y = -\frac{2}{3}x + \frac{4}{3}$

Comparing above equation with $y = mx + c$, we get

$m = -\frac{2}{3}, c = \frac{4}{3}$

Therefore, in the slope-intercept form, the slope and the y-intercept are $-\frac{2}{3}, \frac{4}{3}$.

OR

Q25. Solution: The given equation of the line is $x - 2y + 3 = 0 \Rightarrow y = \frac{1}{2}x + \frac{3}{2}$

Comparing with equation $y = mx + c$, we get the slope of the given line =1/2

Thus, the slope of the line perpendicular to the line having a slope of 1/2 is $m = -\frac{1}{(1/2)} = -2$

Therefore, the equation of the line with slope –2 and x-intercept equal to 3 is given by $y = m(x - d)$

$\Rightarrow y = -2(x-3) \Rightarrow 2x + y - 6 = 0$

Q26. Solution: Comparing the given equation $x^2 + y^2 - 4x - 6y + 11 = 0$ of the circle with $x^2 + y^2 + 2gx + 2fy + c = 0$, we get

$2g = -4,\ 2f = -6$

Thus, the centre of the circle is $(-g,-f) = (2,3)$.

Let the other end of the diameter be (x_1, y_1), and since the mid-points of two diametrical points lie on the centre of the circle

$2 = \frac{1+x_1}{2} \Rightarrow x_1 = 3$ and $3 = \frac{-1+y_1}{2} \Rightarrow y_1 = 7$

Hence, the coordinates of the other end of the diameter will be (3, 7).

OR

Q26. Solution: The circle passes through the point of intersection of the lines $3x + y = 14, 2x + 5y = 18$.

Therefore, solving the above equations of lines, we get

$2x + 5y = 18 \Rightarrow 2x + 5(14 - 3x) = 18$

$\Rightarrow -13x = -52 \Rightarrow x = 4$

Putting the value of $x = 4$, we get $12 + y = 14 \Rightarrow y = 2$

Now, since the centre of the circle is (0, –1) and the point (4, 2) lies on the circle

Radius of the circle = $\sqrt{(4-0)^2+(2+1)^2} = 5$

Thus, equation of the circle is $(x-0)^2+(y+1)^2 = 5^2$

$\Rightarrow x^2+y^2+2y+1 = 25 \Rightarrow x^2+y^2+2y-24 = 0$

SECTION C

(Q27–Q32) Solve the following questions. (4 marks each)

Q27. Solution: Let $f(x) = \cot x$

Then, according to the first principle

$$f'(x) = \lim_{h\to 0}\frac{f(x+h)-f(x)}{h}$$

$$= \lim_{h\to 0}\frac{\cot(x+h)-\cot x}{h}$$

$$= \lim_{h\to 0}\frac{1}{h}\left(\frac{\cos(x+h)}{\sin(x+h)}-\frac{\cos x}{\sin x}\right)$$

$$= \lim_{h\to 0}\frac{1}{h}\left[\frac{\sin x\cos(x+h)-\cos x\sin(x+h)}{\sin x\sin(x+h)}\right]$$

$$= \lim_{h\to 0}\frac{1}{h}\left[\frac{\sin(x-x-h)}{\sin x\sin(x+h)}\right]$$

$$= \frac{1}{\sin x}\cdot\lim_{h\to 0}\frac{1}{h}\left[\frac{\sin(-h)}{\sin(x+h)}\right]$$

$$= \frac{-1}{\sin x}\cdot\left(\lim_{h\to 0}\frac{\sinh}{h}\right)\left(\lim_{h\to 0}\frac{1}{\sin(x+h)}\right)$$

$$= \frac{-1}{\sin x}\cdot 1\cdot\left(\frac{1}{\sin(x+0)}\right) = \frac{-1}{\sin^2 x} = -\operatorname{cosec}^2 x$$

$$\therefore \frac{d}{dx}(\cot x) = -\operatorname{cosec}^2 x$$

Q28. Solution: The given function is $f(x) = \begin{cases}\frac{|x|}{x}, & x\neq 0\\ 0, & x = 0\end{cases}$

$$\lim_{x\to 0^-} f(x) = \lim_{x\to 0^-}\frac{|x|}{x} = \lim_{x\to 0}\left(\frac{-x}{x}\right)$$

$[\because |x| = -x, \text{ when } x<0]$

$$= \lim_{x\to 0}(-1) = -1$$

$$\lim_{x\to 0^+} f(x) = \lim_{x\to 0^+}\left[\frac{|x|}{x}\right] = \lim_{x\to 0^+}\left(\frac{x}{x}\right)$$

$[\because |x| = x, \text{ when } x>0]$

$$= \lim_{x\to 0}(1) = 1$$

Since, $\lim_{x\to 0^-} f(x) \neq \lim_{x\to 0^+} f(x)$, $\lim_{x\to 0} f(x)$ does not exist.

Q29. Solution:

Given $f(x) = \begin{cases}-x+3, & \text{if } x\le -3\\ -2x, & \text{if } -3<x<3\\ 6x+2, & \text{if } x\ge 3\end{cases}$

LHL = $\lim_{x\to -3^-} f(x) = \lim_{x\to -3^-}(-x+3) = -(-3)+3 = 6$

RHL = $\lim_{x\to -3^+} f(x) = \lim_{x\to -3^+}(-2x) = -2(-3) = 6.$

Thus, $\lim_{x\to -3} f(x) = 6$ Since, $\lim_{x\to k} f(x) = f(k)$

LHL = $\lim_{x\to 3^-} f(x) = \lim_{x\to 3^-}(-2x) = -2(3) = -6$

RHL = $\lim_{x\to 3^+} f(x) = \lim_{x\to 3^+}(6x+2) = 6(3)+2 = 20,$

LHL $\neq$ RHL. Therefore at $x = 3$, limit does not exist.

Q30. Solution:

$$\lim_{x\to 0}\frac{\sin ax+bx}{ax+\sin bx} = \lim_{x\to 0}\frac{\left(\frac{\sin ax}{ax}\right)ax+bx}{ax+bx(\sin bx/bx)}$$

$$= \frac{\left(\lim_{ax\to 0}\frac{\sin ax}{ax}\right)\times\lim_{x\to 0}(ax)+\lim_{x\to 0}bx}{\lim_{x\to 0}ax+\lim_{x\to 0}bx\left(\lim_{x\to 0}\frac{\sin ax}{bx}\right)}$$

$[\because x\to 0 \Rightarrow ax\to 0, bx\to 0]$

$$= \frac{\lim_{x\to 0}(ax)+\lim_{x\to 0}bx}{\lim_{x\to 0}ax+\lim_{x\to 0}bx}\quad\left[\because \lim_{x\to 0}\frac{\sin x}{x} = 1\right]$$

$$= \frac{\lim_{x\to 0}(ax+bx)}{\lim_{x\to 0}(ax+bx)}$$

$$= \lim_{x\to 0}\frac{ax+bx}{ax+bx} = \lim_{x\to 0}(1) = 1$$

Q31. Solution: For the given equation of the circle $x^2+y^2-6x+12y+36 = 0$, we get

$2g = -6 \Rightarrow g = -3,\ 2f = 12 \Rightarrow f = 6,\ c = 42$

$\therefore$ Centre $(-g,-f) = (3,-6)$

Radius of the given circle

$= \sqrt{g^2+f^2-c} = \sqrt{9+36-42} = \sqrt{3}$

Thus, the centre of the required circle will be (3, −6).

(since the circles are concentric)

And let the radius of the required circle be = R

$\therefore$ Area of the required circle = 3 x Area of the given circle $\Rightarrow \pi R^2 = 3[\pi(\sqrt{3})^2] \Rightarrow R^2 = 9 \Rightarrow R = 3$

$\therefore \sqrt{g^2+f^2-C} = 3 \Rightarrow 9+36-C = 9$

$\Rightarrow C = 36$

Hence, the required equation of the circle is
$x^2+y^2-6x+12y+36=0$

OR

Q31. Solution: Let the general equation of the circle be
$x^2+y^2+2gx+2fy+c=0$ ...(i)
This circle passes through the points (1, 0) and (0, –1). Therefore
$1+0+2g+0+c=0 \Rightarrow 2g+c=-1$ ...(ii)
and
$0+1+0-2f+c=0 \Rightarrow -2f+c=-1$ ...(iii)
The centre of the circle $(-g,-f)$ lies on the straight line $x-y+2=0$. Therefore
$-g+f+2=0$ ...(iv)
Solving above three equations, we get
$g=1,\ f=-1$ and $c=-3$
Thus, the equation of the circle is
$x^2+y^2+2x-2y-3=0$

Q32. Solution: Since the centre is at (0, 0) and the major axis is on the x-axis, the equation of the ellipse will be of the form $\frac{x^2}{a^2}+\frac{y^2}{b^2}=1$, where a is the semi-major axis.
Since the ellipse passes through points (4, 3) and (5, 2), these points must satisfy the equation of ellipse $\frac{x^2}{a^2}+\frac{y^2}{b^2}=1$.

Thus, $\frac{16}{a^2}+\frac{9}{b^2}=1$ and $\frac{25}{a^2}+\frac{4}{b^2}=1$
Solving the above equations, we get $a^2=161/5,\ b^2=161/9$.
Thus, the equation of the ellipse is
$\frac{x^2}{161/5}+\frac{y^2}{161/9}=1 \Rightarrow 5x^2+9y^2=161$

OR

Q32. Solution: Since the centre is at (0, 0) and the major axis is on the y-axis, the equation of the ellipse will be of the form $\frac{x^2}{b^2}+\frac{y^2}{a^2}=1$, where a is the semi-major axis.
Since the ellipse passes through points (3, 2) and (1, 4), these points must satisfy the equation of ellipse $\frac{x^2}{b^2}+\frac{y^2}{a^2}=1$.

Thus, $\frac{9}{b^2}+\frac{4}{a^2}=1$ and $\frac{1}{b^2}+\frac{16}{a^2}=1$

Solving the above equations, we get $b^2=\frac{35}{3},\ a^2=\frac{35}{2}$.

Thus, the equation of the ellipse is $\frac{x^2}{35/3}+\frac{y^2}{35/2}=1$
$\Rightarrow 3x^2+2y^2=35$

SECTION D

(Q33–Q36) Solve the following questions. (6 marks each)

Q33. Solution:

Let $f(x)=\frac{\sec x-1}{\sec x+1}, f(x)=\frac{\frac{1}{\cos x}-1}{\frac{1}{\cos x}+1}=\frac{1-\cos x}{1+\cos x}$

By the quotient rule

$$f'(x)=\frac{(1+\cos x)\frac{d}{dx}(1-\cos x)-(1-\cos x)\frac{d}{dx}(1+\cos x)}{(1+\cos x)^2}$$

$$=\frac{(1+\cos x)(\sin x)-(1-\cos x)(-\sin x)}{(1+\cos x)^2}$$

$$=\frac{\sin x+\cos x\sin x+\sin x-\sin x\cos x}{(1+\cos x)^2}$$

$$=\frac{2\sin x}{(1+\cos x)^2}=\frac{2\sin x}{\left(1+\frac{1}{\sec x}\right)^2}$$

$$=\frac{2\sin x}{\frac{(\sec x+1)}{\sec^2 x}}=\frac{2\sin x\sec^2 x}{(\sec x+1)^2}$$

$$=\frac{\frac{2\sin x}{\cos x}\sec x}{(\sec x+1)^2}=\frac{2\sec x\tan x}{(\sec x+1)^2}$$

Q34. Solution: Let
$f(x)=\cos(x^2), f(x+h)=\cos(x+h)^2$

$$\therefore \frac{d}{dx}f(x)=\lim_{h\to 0}\frac{f(x+h)-f(x)}{h}$$

$$=\lim_{h\to 0}\frac{\cos\{(x+h)\}-\cos(x^2)}{h}$$

$$=\lim_{h\to 0}\frac{-2\sin\left\{\left[\frac{(x+h)^2+x^2}{2}\right]\right\}\sin\left\{\left[\frac{(x+h)^2-x^2}{2}\right]\right\}}{h}$$

$$\left[\because \cos C-\cos D=-2\sin\frac{C+D}{2}\cdot\sin\frac{C-D}{2}\right]$$

$$= \lim_{h\to 0}\frac{1}{h}\left[-2\sin\left\{\frac{(x+h)^2+x^2}{2}\right\}\sin\left\{\frac{(x+h)^2-x^2}{2}\right\}\right]$$

$$= \lim_{h\to 0}\frac{1}{h}\left[-2\sin\left\{\frac{(x+h)^2+x^2}{2}\right\}\sin\left\{\frac{h^2+2hx}{2}\right\}\right]$$

$$= -2\lim_{h\to 0}\sin\left\{\frac{(x+h)^2+x^2}{2}\right\}$$

$$\lim_{h\to 0}\left\{\frac{\sinh\left(\frac{h+2x}{2}\right)}{h\left(\frac{h+2x}{2}\right)}\right\}\times\left(\frac{h+2x}{2}\right)$$

$$= -2\lim_{h\to 0}\sin\left\{\frac{(x+h)^2+x^2}{2}\right\}\lim_{h\to 0}\left(\frac{h+2x}{2}\right)$$

$$\left[\because \lim_{x\to 0}\frac{\sin x}{x}=1\right]$$

$$= -2x\sin(x^2)\quad[\because x\to 0 \Rightarrow hx\to 0]$$

Q35. Solution: Since in the given equation $25x^2+y^2=25 \Rightarrow \frac{x^2}{1}+\frac{y^2}{25}=1$ of ellipse, the denominator of y^2 is greater than the denominator of x^2, the major axis is along the y-axis and minor axis is along the x-axis.

Comparing the given equation with

$\frac{x^2}{b^2}+\frac{y^2}{a^2}=1$, we get $a=5$, $b=1$ and

$c=\sqrt{a^2-b^2}=\sqrt{25-1}=\sqrt{24}=2\sqrt{6}$

Thus, the coordinates of the foci are $(0,2\sqrt{6}),(0,-2\sqrt{6})$

The coordinates of the vertices are $(0,5),(0,-5)$

Length of major axis $= 2a = 10$

Length of minor axis $= 2b = 2$

Eccentricity $e=\frac{c}{a}=\frac{2\sqrt{6}}{5}$

Length of latus rectum $=\frac{2b^2}{a}=\frac{2\times 1}{5}=2/5$

OR

Q35. Solution: Comparing the given equation $\frac{y^2}{9}-\frac{x^2}{16}=1 \Rightarrow \frac{y^2}{3^2}-\frac{x^2}{4^2}=1$ with the standard equation of hyperbola $\frac{y^2}{a^2}-\frac{x^2}{b^2}=1$, we get $a=3$, $b=4$

Now, from the equation $a^2+b^2=c^2$

$\Rightarrow c=\sqrt{3^2+4^2}=\sqrt{9+16}=5$

Thus, the coordinates of the foci are $(0,\pm 5)$

The coordinates of the vertices are $(0,\pm 3)$

Eccentricity $e=\frac{c}{a}=\frac{5}{3}$

Length of latus rectum $=\frac{2b^2}{a}=\frac{2\times 16}{3}=\frac{32}{3}$

Q36. Solution: The foci $(0,\pm\sqrt{5})$ are on the y-axis. Therefore, the equation of the hyperbola is of the form $\frac{y^2}{a^2}-\frac{x^2}{b^2}=1$ and $c=\sqrt{5}$.

From the equation $a^2+b^2=c^2$.

$\Rightarrow a^2+b^2=5 \Rightarrow b^2=5-a^2$

Now, since the hyperbola passes through point (2, 3), $\frac{9}{a^2}-\frac{4}{b^2}=1 \Rightarrow \frac{9}{a^2}-\frac{4}{(5-a^2)}=1$

$\Rightarrow 9(5-a^2)-4a^2=a^2(5-a^2)$

$\Rightarrow 45-9a^2-4a^2=5a^2-a^4$

$\Rightarrow a^4-18a^2+45=0$

$\Rightarrow (a^2-3)(a^2-15)=0 \Rightarrow a^2=3,15$

Since in the hyperbola $a<c \Rightarrow a^2<c^2 \Rightarrow a^2<5$, $a^2=3 \Rightarrow b^2=5-a^2=5-3=2$

Thus, the equation of the hyperbola is $\frac{y^2}{3}-\frac{x^2}{2}=1$

OR

Q36. Solution: Let the vertices of the ΔABC be $A(x_1,y_1,z_1), B(x_2,y_2,z_2), C(x_3,y_3,z_3)$. Then the mid-point of BC is (5, 7, 11).

$\therefore 5=\frac{x_2+x_3}{2} \Rightarrow x_2+x_3=10$ ...(i)

$7=\frac{y_2+y_3}{2} \Rightarrow y_2+y_3=14$ ...(ii)

$0=\frac{z_2+z_3}{2} \Rightarrow z_2+z_3=0$ ...(iii)

Similarly, for the sides AB and AC

$2=\frac{x_1+x_2}{2} \Rightarrow x_1+x_2=4$ ...(iv)

$3=\frac{y_1+y_2}{2} \Rightarrow y_1+y_2=6$ ...(v)

$-1=\frac{z_1+z_2}{2} \Rightarrow z_1+z_2=-2$ ...(vi)

$0=\frac{x_1+x_3}{2} \Rightarrow x_1+x_3=0$ ...(vii)

$8 = \frac{y_1 + y_3}{2} \Rightarrow y_1 + y_3 = 16$ ...(viii)

$5 = \frac{z_1 + z_3}{2} \Rightarrow z_1 + z_3 = 10$ ...(ix)

From equations (i) and (iv),

$x_1 + 2x_2 + x_3 = 14$ ...(x)

From equations (ii) and (v),

$y_1 + 2y_2 + y_3 = 20$ ...(xi)

From equations (iii) and (vi)

$z_1 + 2z_2 + z_3 = -2$ ...(xii)

From equations (vii) and (x), $2x_2 = 14 \Rightarrow x_2 = 7$

$x_2 = 7, \text{then } x_3 = 10 - 7 = 3$

$x_3 = 3, \text{then } x_1 = -3$

$x_1 = -3, x_2 = 7, x_3 = 3$

From equations (viii) and (xi) $2y_2 = 4 \Rightarrow y_2 = 2$

$y_2 = 2, \text{then } y_1 = 4$

$y_1 = 4, \text{then } y_3 = 12$

$y_1 = 4, y_2 = 2, y_3 = 12$

From equations (ix) and (xii), $2z_2 = -12 \Rightarrow z_2 = -6$

$z_2 = -6, \text{then } z_1 = -7 \Rightarrow z_1 = -7, \text{then } z_3 = 17$

$z_1 = -7, z_2 = 5, z_3 = 17$

Thus, the vertices are

$A(-3, 4, -7), B(7, 2, 5), C(3, 12, 17).$

CHAPTER

14 Mathematical Reasoning

Day 191

Statements

Mathematical Statements: The basic unit of mathematical reasoning is a mathematical statement. That is, a sentence is called a mathematically acceptable statement if it is either true or false, but not both.

Example 1: Consider the following:
7 is greater than 4 is correct and is a statement.
7 is less than 4 is incorrect but it is also a statement.
$7x$ is less than $4x$ is not a statement because we do not know about x; that is, for some values of x, the statement will be correct while for others, the statement will be incorrect. Therefore, it is mathematically not acceptable as a statement.

Example 2: Consider the following sentences:
The sum of two positive numbers is positive.
The sum of two negative numbers is positive.
The sum of one negative number and one positive number is positive.
Of these sentences, the first one is true and the second one is false. There is no ambiguity regarding these sentences. Therefore, they are statements.
The third sentence may be true or false, depending on the numbers. Therefore, this sentence is not a statement.

Example 3: Consider the following sentences:
How bad!
Open the book.
Where are you doing?
These are not statements because the first one is an exclamation, the second an order and the third a question. None of these is considered as a statement in mathematical language.
Note: Sentences involving variable time such as "today", "tomorrow" or "yesterday" are not statements. This is because it is not known what time they refer to.

Example 4: Consider the following sentences:
Tomorrow is Monday.
The day after Sunday is Monday.
The first sentence is not a statement because the sentence is correct (true) on a Sunday but not on other days. The second sentence is a statement; it is true that Monday comes after Sunday.
Note: Sentences involving variable places such as "here", "there" or "where" are not statements. This is because it is not known what place they refer to.

Example 5: Consider the following sentences:
He is a mathematics graduate.
Mughalsarai is very near Varanasi.
These are not statements.

Example 5: Consider the following sentences:
There are 31 days in January.
There are 40 days in a month.
There are 30 days in a month.
Of these sentences, the first one is true and the second one is false (since, irrespective of the month, the maximum number of days in a month can never exceed 31). There is no ambiguity regarding these sentences. Therefore, they are statements.
The third sentence may be true or false, depending on which month is referred to here. Therefore, this sentence is not a statement.
Note: While dealing with statements, we usually denote them with small letters $p, q, r, \ldots$
For example, we denote the statement "3 is a prime number" by p. This is also written as
p: 3 is a prime number.

NCERT Exercise 14.1 (With Solutions)

Q1. Which of the following sentences are statements? Give reasons for your answer.
(i) There are 35 days in a month.
(ii) Mathematics is difficult.
(iii) The sum of 5 and 7 is greater than 10.
(iv) The square of a number is an even number.
(v) The sides of a quadrilateral have equal length.

(vi) Answer this question.
(vii) The product of (–1) and 8 is 8.
(viii) The sum of all interior angles of a triangle is 180°.
(ix) Today is a windy day.
(x) All real numbers are complex numbers.

Solution:
(i) This sentence is incorrect because the maximum number of days in a month is 31. Thus, it is a statement.
(ii) This sentence is biased in the sense that for some people, mathematics can be easy and for others, it can be difficult. Thus, it is not a statement.
(iii) The sum of 5 and 7 is 12, which is greater than 10. Therefore, this sentence is always correct. Thus, it is a statement.
(iv) This sentence is sometimes correct and sometimes incorrect depending on the number. For example, the square of 3 is an odd number. However, the square of 4 is an even number. Thus, it is not a statement.
(v) This sentence is sometimes correct and sometimes incorrect. For example, square and rhombus have sides of equal length. However, trapezium and rectangle have sides of unequal length. Thus, it is not a statement.
(vi) It is an order. Therefore, it is not a statement.
(vii) The product of (–1) and 8 is (–8). Therefore, this sentence is always incorrect. Thus, it is a statement.
(viii) This sentence is always correct. Thus, it is a statement.
(ix) The day that is being referred to here is not a fixed date. Thus, it is not a statement.
(x) All real numbers can be written as $x+0i$. Therefore, the given sentence is always correct. Thus, it is a statement.

Q2. Give three examples of sentences that are not statements. Give reasons for your answers.

Solution: Three examples of sentences that are not statements are as follows:
(i) He is a mathematician.
It is not evident from the sentence who 'he' refers to. Therefore, it is not a statement.
(ii) Algebra is difficult.
This is not a statement because for some people, algebra can be easy and for others, it can be difficult.
(iii) What she is doing?
This is question, which also contains 'she' and it is not evident who 'she' is. Hence, it is not a statement.

Day 192

Negation of a Statement and Compounds Statements

Negation of a Statement: The denial of a statement is called the negation of the statement. If p is a statement, then the negation of p is also a statement and it is denoted by $\sim p$ and read as 'not p'.

Example 1: Consider the statement:
p: Varanasi is a city.
The negation of this statement is
$\sim p$: It is not the case that Varanasi is a city.
This can also be written as
$\sim p$: It is false that Varanasi is a city.
This can simply be expressed as
$\sim p$: Varanasi is not a city.

Note: While forming the negation of a statement, phrases such as "It is not the case" or "It is false that" are also used. The word 'not' can also be used in the statement.

Example 2: Consider the statement:
p: Everyone in America speaks English.
The denial of this sentence tells us that not everyone in America speaks English. This does not mean that no person in America speaks English. It merely states that at least one person in America does not speak English.

Compound Statements: A compound statement is one that is made up of two or more statements. In this case, each statement is called a component statement.
Many mathematical statements are obtained by combining one or more statements using connecting words such as "and", "or", etc.

Example 3: Consider the following statements:
p: There is something wrong with the computer or with the connections.
This statement tells us that there is something wrong with the computer or there is something wrong with the connections.
That means the given statement is actually made up of two component statements:

q: There is something wrong with the computer.
r: There is something wrong with the connections.

Example 4: Consider the following statements:
p: 2 is an even number.
q: 2 is a prime number.
These two statements can be combined with "and"
r: 2 is both even and a prime number.
This is a compound statement.

NCERT Exercise 14.2 (With Solutions)

Q1. Write the negation of the following statements:
(i) Chennai is the capital of Tamil Nadu.
(ii) $\sqrt{2}$ is not a complex number.
(iii) All triangles are not equilateral.
(iv) The number 2 is greater than 7.
(iv) Every natural number is an integer.

Solution: (i) Chennai is not the capital of Tamil Nadu.
(ii) $\sqrt{2}$ is a complex number.
(iii) All triangles are equilateral.
(iv) The number 2 is not greater than 7.
(v) Every natural number is not an integer.

Q2. Are the following pairs of statements negations of each other?
(i) The number x is not a rational number.
The number x is not an irrational number.
(ii) The number x is a rational number.
The number x is an irrational number.

Solution:
(i) The negation of the first statement is "the number x is a rational number". This is the same as the second statement. This is because if a number is not an irrational number, then it is a rational number. Thus, the given statements are negations of each other.
(ii) The negation of the first statement is "the number x is not a rational number". This means that the number x is an irrational number, which is the same as the second statement. Thus, the given statements are negations of each other.

Q3. Find the component statements of the following compound statements and check whether they are true or false:
(i) Number 3 is prime or it is odd.
(ii) All integers are positive or negative.
(iii) 100 is divisible by 3, 11 and 5.

Solution:
(i) The component statements are as follows:
p: Number 3 is prime
q: Number 3 is odd
Both the statements are true.
(ii) The component statements are as follows:
P: All integers are positive.
q: All integers are negative.
Both the statements are false.
(iii) The component statements are as follows:
P: 100 is divisible by 3.
q: 100 is divisible by 11.
r: 100 is divisible by 5.
The statements *p* and *q* are false and statement *r* is true.

Day 193

Special Words/Phases

Rules Regarding the Connective Word "and ($\wedge$)"

1. The compound statement with 'and' is true if all its component statements are true.
2. The compound statement with 'and' is false if any one of its component statements is false or all its component statements are false.

Example 1: Consider the compound statement with the connective word "and".
p: A line can be drawn with two given points and its slope can be determined.
q: A line can be drawn with two given points.
r: Its slope can be determined.
In this compound statement, both component statements are true. Therefore, the compound statement is true.

Example 2: Consider the compound statement with the connective word "and".
p: 10 is divisible by 2 and 3.
This statement has the following component statements:
q: 10 is divisible by 2.
r: 10 is divisible by 3.
In this compound statement, the first component statement is true while the second component statement is false. Therefore, the compound statement is false (since a minimum of one statement is false).

Example 3: Consider the compound statement with the connective word "and".
p: 10 is divisible by 3 and 7.

This statement has following the component statements:
q: 10 is divisible by 3.
r: 10 is divisible by 7.
In this compound statement, both component statements are false. Therefore, the compound statement is false.

Rules Regarding the Connective Word "or (∨)"

1. The compound statement with 'or' is true if any one of its component statements is true or all its component statements are true.
2. The compound statement with 'or' is false if all its component statements are false.

Example 4: Consider the compound statement with the connective word "or".
p: A line can be drawn either with two given points or with one given point and its slope.
q: A line can be drawn with two given points.
r: A line can be drawn with one given point and its slope.
In this compound statement, both component statements are true. Therefore, the compound statement is true.

Example 5: Consider the compound statement with the connective word "or".
p: A natural number is either even or odd.
This statement has the following component statements:
q: A natural number is even.
r: A natural number is odd.
In this compound statement, both component statements are true. Therefore, the compound statement is true.

Example 6: Consider the compound statement with the connective word "or".
p: 10 is divisible by 2 or 3.
This statement has the following component statements:
q: 10 is divisible by 2.
r: 10 is divisible by 3.
In this compound statement, the first component statement is true while the second component statement is false. Therefore, the compound statement is true (since a minimum of one statement is true).

Example 7: Consider the compound statement with the connective word "or".
p: 10 is divisible by 3 or 7.
This statement has the following component statements:
q: 10 is divisible by 3.
r: 10 is divisible by 7.
In this compound statement, both component statements are false. Therefore, the compound statement is false.

Exclusive and Inclusive Connective Word "or"
'Exclusive or' means the person has only one choice from all of the choices given as 'or' while 'Inclusive or' means the person can choose any one of all the options given as 'or'.

Example 8: Consider the compound statement with the connective word "or".
p: In the science stream, after Class X, a student can take either Maths or Biology along with Physics and Chemistry for admission to Class XI.
In this statement, 'or' is used as an exclusive or. This means the student can take either Maths or Biology for admission, but not both the subjects.
q: A line can be drawn either with two given points or with one given point and its slope.
In this statement, 'or' is used as inclusive or. This means a line can be drawn either with two given points or with one given point and its slope.

Quantifiers: Quantifiers are phrases such as "There exists" and "For all". Consider the examples given below.

Example 9: *p*: There exists a triangle, all the sides of which are equal. This means that there is at least one triangle that has all its sides equal.

Example 10: *q*: For every prime number p (except 2), $p \pm 1$ is an even number.
This means that if S denotes the set of all prime numbers (except 2), then for all the members p of the set S, $p \pm 1$ is an even number.
Note: Consider the following statements:

1. For every positive number x, there exists a positive number y such that $y < x$.
2. There exists a positive number y such that for every positive number x, we have $y < x$.

These statements do not say the same thing. Here, statement (1) is true and statement (2) is false. This is because in statement (1), we have a minimum of one y that is less than every positive number x, while in statement (2), it says that there exists a given positive number y that is less than every positive number x, which may not be possible for many numbers.

NCERT Exercise 14.3 (With Solutions)

Q1. For each of the following compound statements, first identify the connecting words and then break it into component statements:

(i) All rational numbers are real and all real numbers are not complex.
(ii) The square of an integer is positive or negative.
(iii) The sand heats up quickly in the sun and does not cool down fast at night.
(iv) $x = 2, x = 3$ are the roots of the equation $3x^2 - x - 10 = 0$

Solution:
(i) The connecting word is 'and'. The component statements are as follows:
p: All rational numbers are real.
q: All real number are not complex.
(ii) The connecting word is 'or'. The component statements are as follows:
p: The square of an integer is positive.
q: The square of an integer is negative.
(iii) The connecting word is 'and'. The component statements are as follows:
p: The sand heats up quickly in the sun.
q: The sand does not cool down fast at night.
(iv) The connecting word is 'and'. The component statements are as follows:
$p: x = 2$ is a root of the equation $3x^2 - x - 10 = 0$
$q: x = 3$ is a root of the equation $3x^2 - x - 10 = 0$

Q2. Identify the quantifier in the following statements and write the negation of the statements:
(i) There exists a number that is equal to its square.
(ii) For every real number x, x is less than $x+1$.
(iii) There exists a capital for every state in India.

Solution:
(i) The quantifier is "There exists".
The negation of this statement is as follows:
There does not exist a number that is equal to its square.
(ii) The quantifier is "For every".
The negation of this statement is as follows:
There exists a real number x such that x is not less than $x+1$.
(iii) The quantifier is "There exists".
The negation of this statement is as follows:
There exists a state in India that does not have a capital.

Q3. Check whether the following pairs of statements are negations of each other. Give reasons for the answer:
(i) $x + y = y + x$ is true for real numbers x and y.
(ii) There exist real numbers x and y for which $x + y = y + x$.

Solution: The negation of statement (i) is as follows:
There exist real numbers x and y for which $x + y \neq y + x$. This is not the same as statement (ii). Thus, the given statements are not negations of each other.

Q4. State whether the "or" used in the following statement is "exclusive or" or "inclusive or". Give reasons for your answer:
(i) The sun rises or the moon sets.
(ii) To apply for a driving licence, you should have a ration card or a passport.
(iii) All integers are positive or negative.

Solution:
(i) Here, "or" is exclusive because it is not possible for the sun to rise and the moon to set together.
(ii) Here, "or" is inclusive since a person can have both a ration card and a passport to apply for a driving licence.
(iii) Here, "or" is exclusive because all integers cannot be both positive and negative.

Day 194

Implication, Contrapositive and Converse

Implication: Implications such as "if-then", "only if" and "if and only if" are very common in mathematics.

Example 1: Consider the statement:
r: If a number is a multiple of 4, then it is a multiple of 2.
Let *p* and *q* denote the statements:
p: A number is a multiple of 4.
q: A number is a multiple of 2.

Note: The sentence "if *p* then *q*" means that in the event that *p* is true, then *q* must be true. It is the same as the following:

- *p* implies *q* is denoted by $p \Rightarrow q$. The symbol $\Rightarrow$ stands for implies. Here, a number is a multiple of 4 implies that it is a multiple of 2.
- *p* is a sufficient condition for *q*.
This means that knowing that a number is a multiple of 4 is sufficient for us to conclude that it is a multiple of 2.
- *p* only if *q*.

This means that a number is a multiple of 4 only if it is a multiple of 2.

4. q is a necessary condition for p.
This means that when a number is a multiple of 4, it is necessarily a multiple of 2.
5. $\sim q$ implies $\sim p$.
This means that if a number is not a multiple of 2, then it is not a multiple of 4.
6. Non-occurrence of p has no effect on the occurrence of q.

<u>**Example 2:**</u> Consider the statement:
r: If two line are not parallel, then they must intersect at some point.
The component statements are given by:
p: Two lines are not parallel
q: They must intersect at some point.
Then the sentence "if p then q" says that in the event that p is true, then q must be true.

Contrapositive: Contrapositives are certain statements that can be formed from a given statement with "if then". The contrapositive of statement "if p then q" is "if $\sim q$ then $\sim p$".

<u>**Example 3:**</u> Consider the statement:
p: If a number is divisible by 4, then it is a divisible by 2.
The contrapositive of this statement is
q: If a number is not divisible by 2, then it is not divisible by 4.

Converse: Converses are certain statements that can be formed from a given statement "if p then q" as "if q then p".

<u>**Example 4:**</u> Consider the following "if p then q" as "if q then p" statement:
p: If a number is divisible by 2, then it is an even number.
q: If a number is even, then it is divisible by 2.
Statements p and q are the converse of each other.

If and only if (⇔): Converses also comprise statements that can be formed from a given statement "p if and only if q" and "q if and only if p", symbolically written as '⇔'. p is a necessary and sufficient condition for q and vice-versa.

<u>**Example 5:**</u> Consider the following statements:
p: If a number is divisible by 3, then the sum of its digits is divisible by 3.
q: If the sum of the digits of a number is divisible by 3, then the number is divisible by 3.
Statements p and q, using "if and only if" can be written as:

A number is divisible by 3 if and only if the sum of its digits is divisible by 3.

NCERT Exercise 14.4 (With Solutions)

Q1. Rewrite the following statement with "if-then" in different ways conveying the same meaning:
If a natural number is odd, then its square is also odd.

Solution: The given statement can be written in five different ways as follows:

(i) A natural number is odd implies that its square is odd.
(ii) A natural number is odd only if its square is odd.
(iii) For a natural number to be odd, it is necessary that its square is odd.
(iv) For the square of a natural number to be odd, it is sufficient that the number is odd.
(v) If the square of a natural number is not odd, then the natural number is not odd.

Q2. Write the contrapositive and converse of the following statements:

(i) If x is a prime number, then x is odd.
(ii) If two lines are parallel, they do not intersect in the same plane.
(iii) Something is cold implies that it has low temperature.
(iv) You cannot comprehend geometry if you do not know how to reason deductively.
(v) x is an even number implies that x is divisible by 4

Solution:

(i) The contrapositive is as follows:
If a number x is not odd, then x is not a prime number.
The converse is as follows:
If a number x is odd, then it is a prime number.
(ii) The contrapositive is as follows:
If two lines intersect in the same plane, then they are not parallel.
The converse is as follows:
If two lines do not intersect in the same plane, then they are parallel.
(iii) The contrapositive is as follows:
If something does not have low temperature, then it is not cold.
The converse is as follows:
If something is at low temperature, then it is cold.

(iv) The contrapositive is as follows:
If you know how to reason deductively, then you can comprehend geometry.
The converse is as follows:
If you do not know how to reason deductively, then you cannot comprehend geometry.

(v) The given statement can be written as follows:
If x is an even number, then x is divisible by 4.
The contrapositive is as follows:
If x is not divisible by 4, then x is not an even number.
The converse is as follows:
If x is divisible by 4, then x is an even number.

Q3. Write each of the following statements in the form "if-then":
(i) You get a job implies that your credentials are good.
(ii) The banana trees will bloom if they stay warm for a month.
(iii) A quadrilateral is a parallelogram if its diagonals bisect each other.
(iv) To obtain A+ in class, it is necessary that you work out the exercises in the book.

Solution:
(i) If you get a job, then your credentials are good.
(ii) If the banana trees stay warm for a month, then they will bloom.
(iii) If the diagonals of a quadrilateral bisect each other, then it is a parallelogram.
(iv) If you want to obtain an A+ in class, then you must work out all the exercises in the book.

Q4. Identify the statements given below as the contrapositive or converse of each other:
(a) If you live in Delhi, then you have winter clothes.
  (i) If you do not have winter clothes, then you do not live in Delhi.
  (ii) If you have winter clothes, then you live in Delhi.
(b) If a quadrilateral is a parallelogram, then its diagonals bisect each other.
  (i) If the diagonals of a quadrilateral do not bisect each other, then the quadrilateral is not a parallelogram.
  (ii) If the diagonals of a quadrilateral bisect each other, then it is a parallelogram.

Solution:
(a) (i) This is the contrapositive of the given statement (a).
  (ii) This is the converse of the given statement (a).
(b) (i) This is the contrapositive of the given statement (b).
  (ii) This is the converse of the given statement (b).

Day 195

Validating Statements

To validate a statement as true, one must answer all of the following questions:
What does the statement mean?
What would it mean to say that this statement is true?
When is this statement not true?
The answers to these questions depend upon which of the special words "and", "or" and which of the implications "if and only", "if-then", and which of the quantifiers "for every", "there exists", appear in the given statement.

Rule 1: Statement with "and"
If p and q are mathematical statements, then in order to show that "p and q" is true, we have to show that:
Step 1: The statement p is true.
Step 2: The statement q is true.

Rule 2: Statement with "or"
If p and q are mathematical statements, then in order to show that "p or q" is true, we must take two cases:

Case 1: By assuming that p is false, show that q must be true.

Case 2: By assuming that q is false, show that p must be true.

Rule 3: Statement with "if-then"
In order to prove "if p then q", we must show that any one of the following cases is true:

Case 1: By assuming that p is true, show that q must be true. (Direct method)

Case 2: By assuming that q is false, show that p must be false. (Contrapositive method)

Rule 4: Statement with "if and only if"
In order to prove "p if and only if q", we must show that:

(i) If p is true, then q is true and
(ii) If q is true, then p is true

By Contradiction: To check whether a statement p is true, we assume that p is not true, that is, $\sim p$ is true. Then we prove the result by contradicting our assumption. Thus, we conclude that p is true.

NCERT Exercise 14.5 (With Solutions)

Q1. Show that the statement p: "If x is a real number such that $x^3 + 4x = 0$, then x is 0" is true by:
(i) direct method
(ii) method of contradiction
(iii) method of contrapositive

Solution: p: "If x is a real number such that $x^3 + 4x = 0$, then x is 0".

Let q: x is a real number such that $x^3 + 4x = 0$ or x is 0.

(i) To show that statement p is true, assume that q is true and then show that p is true.
Therefore, let statement q be true.
$\therefore x^3 + 4x = 0 \Rightarrow x(x^2 + 4) = 0$
$\Rightarrow x = 0,\ x^2 + 4 = 0$
However, since x is real, it is 0.
Thus, statement p is true.
Therefore, the given statement is true.

(ii) To show that statement p is true by contradiction, we assume that p is not true.
Let x be a real number such that $x^3 + 4x = 0$, and let x not be 0. Therefore, $x^3 + 4x = 0 \Rightarrow x(x^2 + 4) = 0$ $\Rightarrow x = 0, x^2 + 4 = 0 \Rightarrow x = 0, x^2 = -4$. However, x is real. Therefore, $x = 0$, which is a contradiction since we have assumed that x is not 0.
Thus, the given statement p is true.

(iii) To prove statement p to be true by contrapositive method, we assume that p is false and prove that q must be false.
p is false implies that we need to consider the negation of statement p. This gives the following statement: $\sim p : x$ is not 0.
It can be seen that $(x^3 + 4)$ will always be positive.
$x \neq 0$ implies that the product of any positive real number with x is not zero.
Let us consider the product of x with $(x^3 + 4)$
$\therefore x(x^3 + 4) \neq 0 \Rightarrow x^3 + 4x \neq 0$
This shows that statement q is not true.
Thus, it has been proved that $\sim p \Rightarrow \sim q$
Therefore, the given statement p is true.

Q2. Show that the statement "For any real numbers a and b, $a^2 = b^2$ implies that $a = b$ is not true" by giving a counter-example.

Solution: The given statement can be written in the form of "if-then" as follows:
If a and b are real numbers such that $a^2 = b^2$, then $a = b$
Let p: a and b be real numbers such that $a^2 = b^2$
$q : a = b$
The given statement must be proved false.
For this, it needs to be proved that if p, then $\sim q$.
To show this, two real numbers a and b with $a^2 = b^2$ are required such that $a^2 \neq b^2$. Let
$a = 1, b = -1;\ \ a^2 = (1)^2 = 1,\ \ b^2 = (-1)^2 = 1$
$\therefore a^2 = b^2$ However, $a \neq b$
Thus, it can be concluded that the given statement is false.

Q3. Show that the following statement is true by the method of contrapositive.
p: If x is an integer and x^2 is even, then x is also even

Solution: p: If x is an integer and x^2 is even, then x is also even.
Let q: x be an integer and x^2 be even
r: x is even.
To prove that p is true by the contrapositive method, we assume that r is false and that q is also false.
Let x not be even.
To prove that q is false, it must be proved that x is not an integer or x^2 is not even.
x is not even implies that x^2 is also not even. Therefore, statement q is false. Thus, the given statement p is true.

Q4. By giving a counter-example, show that the following statements are not true:
(i) p: If all the angles of a triangle are equal, then the triangle is an obtuse-angled triangle.
(ii) q: The equation $x^2 - 1 = 0$ does not have a root lying between 0 and 2.

Solution:
(i) The given statement is of the form "if q then r".
q: All the angles of a triangle are equal.
r. The triangle is an obtuse-angled triangle.
The given statement p must be proved false.
For this purpose, it needs to be proved that if q, then $\sim r$.

To show this, angles of a triangle are required such that none of them is an obtuse angle.
It is known that the sum of all the angles of a triangle is 180°. Therefore, if all the three angles of the triangle are of measure 60°, then they are not obtuse angles.
In an equilateral triangle, the measure of all the angles is equal. However, the triangle is not an obtuse-angled triangle.
Thus, it can be concluded that the given statement p is false.

(ii) The given statement is as follows:
q: The equation $x^2 - 1 = 0$ does not have a root lying between 0 and 2.
This statement must be proved false. To show this, a counter-example is required.
Consider $x^2 - 1 = 0, x^2 = 1, x = \pm 1$
One root of the equation $x^2 - 1 = 0$, that is, the root $x = 1$ lies between 0 and 2.
Thus, the given statement is false.

Q5. Which of the following statements are true and which are false? In each case, give a valid reason for saying so:

(i) p: Each radius of a circle is a chord of the circle.
(ii) q: The centre of a circle bisects each chord of the circle.
(iii) r: The circle is a particular case of an ellipse.
(iv) s: If x and y are integers such that $x > y$, then $-x < -y$.
(v) t: $\sqrt{11}$ is a rational number.

Solution:

(i) The given statement p is false.
According to the definition of chord, it should intersect the circle at two distinct points.

(ii) The given statement q is false.
If the chord is not the diameter of the circle, then the centre will not bisect that chord. In other words, the centre of a circle only bisects the diameter, which is the chord of the circle.

(iii) The equation of an ellipse is $\frac{x^2}{a^2} + \frac{y^2}{b^2} = 1$
If we put $a = b = 1$, then we obtain $x^2 + y^2 = 1$, which is an equation of a circle.
Therefore, the circle is a particular case of an ellipse.
Thus, statement r is true.

(iv) $x > y \Rightarrow -x < -y$ (by a rule of identity)
Thus, the given statement s is true.

(v) 11 is a prime number and we know that the square root of any prime number is an irrational number.
Therefore, $\sqrt{11}$ is an irrational number.
Thus, the given statement t is false.

Day 196

NCERT Miscellaneous Exercise (With Solutions)

Q1. Write the negation of the following statements:

(i) p: For every positive real number x, the number $x - 1$ is also positive.
(ii) q: All cats scratch.
(iii) r: For every real number x, either $x > 1$ or $x < 1$.
(iv) s: There exists a number x such that $0 < x < 1$.

Solution: (i) $\sim p$: There exists a positive real number x, such that $x - 1$ is not positive.

(ii) $\sim q$: There exists a minimum of one cat that does not scratch.

(iii) $\sim r$: There exists a real number $x = 1$.

(iv) $\sim s$: There exists a number x, such that $x \geq 1$ or $x \leq 0$.

Q2. State the converse and contrapositive of each of the following statements:

(i) p: A positive integer is prime only if it has no divisors other than 1 and itself.
(ii) q: I go to a beach whenever it is a sunny day.
(iii) r: If it is hot outside, then you feel thirsty.

Solution:

(i) Statement p can be written as follows:
If a positive integer is prime, then it has no divisors other than 1 and itself.
The converse of the statement is as follows:
If a positive integer has no divisors other than 1 and itself, then it is prime.
The contrapositive of the statement is as follows:
If a positive integer has divisors other than 1 and itself, then it is not prime.

(ii) The given statement can be written as follows:
If it is a sunny day, then I go to a beach.
The converse of the statement is as follows:
If I go to a beach, then it is a sunny day
The contrapositive of the statement is as follows:

If I do not go to a beach, then it is not a sunny day.

(iii) The converse of the statement is as follows:
If you feel thirsty, then it is hot outside.
The contrapositive of the statement is as follows:
If you do not feel thirsty, then it is not hot outside.

Q3. Write each of the statements in the form "if p, then q":

(i) p: It is necessary to have a password to log on to the server.

(ii) q: There is a traffic jam whenever it rains.

(iii) r: You can access the website only if you pay a subscription fee.

Solution:

(i) A statement in the form "if p, then q" can be written as follows:
If you log on to the server, then you have a password.

(ii) The statement in the form "if p, then q" can be written as follows:
If it rains, then there is a traffic jam.

(iii) The statement in the form "if p, then q" can be written as follows:
If you access the website, then you pay a subscription fee.

Q4. Re-write each of the following statements in the form "p if and only if q":

(i) p: If you watch television, then your mind is free and if your mind is free, then you watch television.

(ii) q: For you to get an A grade, it is necessary and sufficient that you do all the homework regularly.

(iii) r: If a quadrilateral is equiangular, then it is a rectangle and if a quadrilateral is a rectangle, then it is equiangular.

Solution: (i) You watch television if and only if your mind is free.

(ii) You get an A grade if and only if you do all the homework regularly.

(iii) A quadrilateral is equiangular if and only it is a rectangle.

Q5. Given below are two statements:
p: 25 is a multiple of 5.
q: 25 is a multiple of 8.
Write the compound statements connecting these two statement with "and" and "or".
In both cases, check the validity of the compound statement.

Solution: The compound statement with 'and' is "25 is a multiple of 5 and 8". This is a false statement, since 25 is not a multiple of 8.

The compound statement with 'or' is "25 is a multiple of 5 or 8". This is a true statement, since 25 is not a multiple of 8 but it is a multiple of 5.

Q6. Check the validity of the statements given below using the method given against it:

(i) p: The sum of an irrational number and a rational number is irrational (by contradiction method).

(ii) q: If n is a real number with $n > 3$, then $n^2 > 9$ (by contradiction method).

Solution:

(i) The given statement is as follows:
p: The sum of an irrational number and a rational number is irrational.
Let us assume that the given statement p is false. That is, we assume that the sum of an irrational number and a rational number is rational.
Therefore, $\sqrt{a} + \frac{b}{c} = \frac{d}{e}$, where $\sqrt{a}$ is irrational and b, c, d, e are integers.
$\frac{d}{e} - \frac{b}{c}$ is a rational number and $\sqrt{a}$ is an irrational number.
This is a contradiction. Therefore, our assumption is wrong. Therefore, the sum of an irrational number and a rational number is rational.
Thus, the given statement is true.

(ii) The given statement q is as follows:
If n is a real number with $n > 3$, then $n^2 > 9$.
Let us assume that n is a real number with $n > 3$, but $n^2 > 9$ is not true. That is, $n^2 < 9$
Then $n > 3$ and n is a real number.
Squaring both the sides, we obtain $n^2 > (3)^2$
$\Rightarrow n^2 > 9$, which is a contradiction, since we have assumed that $n^2 < 9$
Thus, the given statement is true, that is, if n is a real number with $n>3$, then $n^2 > 9$

Q7. Write the following statement in five different ways, conveying the same meaning:
p: If a triangle is equiangular, then it is an obtuse-angled triangle.

Solution: The given statement can be written in five different ways as follows:

(i) For a triangle to be equiangular, it is necessary that the triangle is an obtuse-angled triangle.

(ii) A triangle is equiangular implies that it is an obtuse-angled triangle.

(iii) For a triangle to be an obtuse-angled triangle, it is sufficient that the triangle is equiangular.

(iv) A triangle is equiangular only if it is an obtuse-angled triangle.

(v) If a triangle not an obtuse-angled triangle, then the triangle is not equiangular.

Day 197

NCERT Exemplar Problems: Solutions (Q1–Q11)

Q1. Which of the following sentences are statements? Justify:

(i) A triangle has three sides.
(ii) 0 is a complex number.
(iii) The sky is red.
(iv) Every set is an infinite set.
(v) $15+8>23$
(vi) $y+9=7$
(vii) Where is your bag?
(viii) Every square is a rectangle.
(ix) The sum of the opposite angles of a cyclic quadrilateral is 180°.
(x) $\sin^2 x+\cos^2 x=0$

Solution: A statement is a sentence that is either true or false, but not both simultaneously.

(i) It is a true statement.
(ii) $0+0i$. Thus, the given statement is true.
(iii) The sky is blue. Thus, the given statement is false.
(iv) Not every set is an infinite set. Thus, the given statement is false.
(v) $15+8=23$. Thus, the given statement is false.
(vi) $y+9=7$ will be true for some value and it will be false for some value. Thus, it is not a statement.
(vii) It is a question. Thus, it is not a statement.
(viii) Square is a subset of rectangle. Thus, the given statement is true.
(ix) It is true.
(x) $\sin^2 x+\cos^2 x=1$. Thus, the given statement is false.

Q2. Find the component statements of the following compound statements:

(i) The number 7 is prime and odd.
(ii) Chennai is in India and is the capital of Tamil Nadu.
(iii) The number 100 is divisible by 3, 11 and 5.
(iv) Chandigarh is the capital of Haryana and UP.
(v) $\sqrt{7}$ is a rational number or an irrational number.
(vi) 0 is less than every positive integer and every negative integer.
(vii) Plants use sunlight, water and carbon dioxide for photosynthesis.
(viii) Two lines in a plane either intersect at one point or they are parallel.
(ix) A rectangle is a quadrilateral or four-sided polygon.

Solution: (i) p: The number 7 is prime.
q: The number 7 is odd.

(ii) p: Chennai is in India.
q: Chennai is the capital of Tamil Nadu.

(iii) p: 100 is divisible by 3.
q: 100 is divisible by 11.
r: 100 is diviisble by 5.

(iv) p: Chandigarh is the capital of Haryana.
q: Chandigarh is the capital of UP.

(v) $p:\sqrt{7}$ is a rational number.
$q:\sqrt{7}$ is an irrational number.

(vi) p: 0 is less than every positive integer.
q: 0 is less than every negative integer.

(vii) p: Plants use sunlight for photosynthesis.
q: Plants use water for photosynthesis.
r: Plants use carbon dioxide for photosynthesis.

(viii) p: Two lines in a plane intersect at one point.
q: Two lines in a plane are parallel.

(ix) p: A rectangle is a quadrilateral.
q: A rectangle is a four-sided polygon.

Q3. Write the component statements of the following compound statements and check whether the compound statement is true or false:

(i) 57 is divisible by 2 or 3.
(ii) 24 is a multiple of 4 and 6.
(iii) All living things have two eyes and two legs.
(iv) 2 is an even number and a prime number.

Solution:

(i) The component statements are:
$p:57$ is divisible by 2. [false]
$q:57$ is divisible by 3. [true]
Since the given compound statement is of the form '$p\vee q$', and one of the statements is true, the compound statement is true.

(ii) The component statements are:
$p:24$ is a multiple of 4. [true]

$q: 24$ is a multiple of 6. [true]
Since the given compound statement is of the form '$p \wedge q$', and both the component statements are true, the compound statement is true.

(iii) The component statements are:
p: All living things have two eyes. [false]
q: All living things have two legs. [false]
Since the given compound statement is of the form '$p \wedge q$', and both the component statements are false, the compound statement is false.

(iv) The component statements are:
p: 2 is an even number. [true]
q: 2 is a prime number. [true]
Since the given compound statement is of the form '$p \wedge q$', and both the component statements are true, the compound statement is true.

Q4. Write the negation of the following simple statements:
(i) The number 17 is prime.
(ii) $2+7=6$.
(iii) Violets are blue.
(iv) $\sqrt{5}$ is a rational number.
(v) 2 is not a prime number.
(vi) Every real number is an irrational number.
(vii) A cow has four legs.
(viii) A leap year has 366 days.
(ix) All similar triangles are congruent.
(x) The area of a circle is the same as the perimeter of the circle.

Solution: (i) The number 17 is not prime.
(ii) $2+7 \neq 6$.
(iii) Violets are not blue.
(iv) $\sqrt{5}$ is not a rational number.
(v) 2 is a prime number.
(vi) Every real number is not an irrational number.
(vii) A cow does not have four legs.
(viii) A leap year does not have 366 days.
(ix) There exist similar triangles that are not congruent.
(x) The area of a circle is not the same as the perimeter of the circle.

Q5. Translate the following statements into symbolic form:
(i) Rahul passed in Hindi and English.
(ii) x and y are even integers.
(iii) 2, 3 and 6 are factors of 12.
(iv) Either x or x+1 is an odd integer.
(v) A number is either divisible by 2 or 3.
(vi) Either $x = 2$ or $x = 3$ is a root of $3x^2 - x - 10 = 0$.
(vii) Students can take Hindi or English as an optional paper.

Solution: (i) p: Rahul passed in Hindi.
q: Rahul passed in English.
$p \wedge q$: Rahul passed in Hindi and English.

(ii) $p: x$ is even integer.
$q: y$ is even integer.
$p \wedge q: x$ and y are even integers.

(iii) p: 2 is a factor of 12.
q: 3 is a factor of 12.
r: 6 is a factor of 12.
$p \wedge q \wedge r$: 2,3,6 are factors of 12.

(iv) $p: x$ is an odd integer.
$q: (x+1)$ is an odd integer.
$p \vee q$: Either x or $(x+1)$ is an odd integer.

(v) p: A number is divisible by 2.
q: A number is divisible by 3.
$p \vee q$: A number is either divisible by 2 or 3.

(vi) $p: x = 2$ is a root of $3x^2 - x - 10 = 0$
$q: x = 3$ is a root of $3x^2 - x - 10 = 0$
$p \vee q$: Either $x = 2$ or $x = 3$ is a root of $3x^2 - x - 10 = 0$

(vii) p: Students can take Hindi as an optional paper.
q: Students can take English as an optional subject.
$p \vee q$: Students can take Hindi or English as an optional paper.

Q6. Write down the negation of the following compound statements:
(i) All rational numbers are real and complex.
(ii) All real numbers are rational or irrational.
(iii) $x = 2, x = 3$ are roots of the quadratic equation $x^2 - 5x + 6 = 0$.
(iv) A triangle has either 3 sides or 4 sides.
(v) 35 is a prime number or a composite number.
(vi) All prime integers are either even or odd.
(vii) $|x|$ is equal to either x or $-x$.
(viii) 6 is divisible by 2 and 3.

Solution:
(i) p: All rational numbers are real.
q: All rational numbers are complex.
$\sim p$: All rational number are not real.
$\sim q$: All rational number are not complex.
$\sim (p \wedge q)$: All rational numbers are not real or not complex. $[\because \sim (p \wedge q) = \sim p \vee \sim q]$

(ii) p: All real numbers are rational.
q: All real numbers are irrational.
Then, the negation of the above statements is given by:
$\sim(p \vee q)$: All real numbers are not rational and all real numbers are not irrational. [$\because \sim(p \vee q) = \sim p \wedge \sim q$]

(iii) $p: x = 2$ is the root of quadratic equation $x^2 - 5x + 6 = 0$
$q: x = 3$ is the root of quadratic equation $x^2 - 5x + 6 = 0$
Then, the negation of the above statements is given by:
$\sim(p \wedge q): x = 2$ is not a root of the quadratic equation $x^2 - 5x + 6 = 0$ or $x = 3$ is not a root of the quadratic equation $x^2 - 5x + 6 = 0$

(iv) p: A triangle has 3 sides.
q: A triangle has 4 sides.
Then, negation of the disjunction of the above statements is given by:
$\sim(p \vee q)$: A triangle has neither 3 sides nor 4 sides.

(v) p: 35 is a prime number.
q: 35 is a composite number.
Then, negation of the disjunction of the above statements is given by:
$\sim(p \vee q)$: 35 is not a prime number and it is not a composite number.

(vi) p: All prime integers are even.
q: All prime integers are odd.
Then, negation of the disjunction of the above statements is given by:
$\sim(p \vee q)$: All prime integers are not even and all prime integers are not odd.

(vii) $p: |x|$ is equal to x.
$q: |x|$ is equal to $-x$.
Then, negation of the disjunction of the above statements is given by:
$\sim(p \vee q): |x|$ is not equal to x and it is not equal to $-x$.

(viii) p: 6 is divisible by 2.
q: 6 is divisible by 3
Then, negation of the conjunction of above statements is given by:
$\sim(p \wedge q)$: 6 is not divisible by 2 or it is not divisible by 3

Q7. Rewrite each of the following statements in the form of conditional statements:
(i) The square of an odd number is odd.
(ii) You will get a sweet dish after dinner.
(iii) You will fail, if you will not study.
(iv) The unit digit of an integer is 0 or 5, if it is divisible by 5.
(v) The square of a prime number is not prime.
(vi) $2b = a + c$, if a, b, c are in AP.

Solution: (i) If the number is an odd number, then its square is an odd number.
(ii) If you take dinner, then you will get a sweet dish.
(iii) If you will not study, then you will fail.
(iv) If the integer is divisible by 5, then its units digits are 0 or 5.
(v) If the number is prime, then its square is not prime.
(vi) If a, b, c are in AP, then $2b = a + c$.

Q8. Form the bi-conditional statement $p \leftrightarrow q$ where:
(i) p: The units digit of an integer is zero.
q: It is divisible by 5.
(ii) p: A natural number n is odd.
q: Natural number n is not divisible by 2
(iii) p: A triangle is an equilateral triangle.
q: All three sides of a triangle are equal.

Solution: (i) $p \leftrightarrow q$: The units digit of an integer is zero, if and only if it is divisible by 5.
(ii) $p \leftrightarrow q$: A natural number odd if and only if it is not divisible by 2.
(iii) $p \leftrightarrow q$: A triangle is an equilateral triangle if and only if all three sides of the triangle are equal.

Q9. Write down the contrapositive of the following statements:
(i) If $x = y, y = 3$, then $x = 3$.
(ii) If n is a natural number, then n is an integer.
(iii) If all three sides of a triangle are equal, then the triangle is equilateral.
(iv) If x and y are negative integers, then xy is positive.
(v) If natural number n is divisible by 6, then n is divisible by 2 and 3.
(vi) If it snows, then the weather will be cold.
(vii) If x is a real number such that $0 < x < 1$, then $x^2 < 1$.

Solution: (i) If $x \neq 3$, then $x \neq y, y \neq 3$.
(ii) If n is not an integer, then it is not a natural number.
(iii) If the triangle is not equilateral, then all three sides of the triangle are not equal.
(iv) If xy is not a positive integer, then either x or y are not positive.
(vi) The weather will not be cold, if it does not snow.
(vii) If $x^2 .. 1$, then x is not a real number such that $0 < x < 1$.

Q10. Write down the converse of following statements:

(i) If a rectangle 'R' is a square, then R is a rhombus.

(ii) If today is Monday, then tomorrow is Tuesday.

(iii) If you go to Agra, then you must visit the Taj Mahal.

(iv) If the sum of the squares of two sides of a triangle is equal to the square of the third side of a triangle, then the triangle is right angled.

(v) If all three angles of a triangle are equal, then the triangle is equilateral.

(vi) If $x : y = 3 : 2$, then $2x = 3y$.

(vii) If S is a cyclic quadrilateral, then the opposite angles of S are supplementary.

(viii) If x is zero, then x is neither positive nor negative.

(ix) If two triangles are similar, then the ratio of their corresponding sides are equal.

Solution: (i) If the rectangle 'R' is rhombus, then it is a square.

(ii) If tomorrow is Tuesday, then today is Monday.

(iii) If you must visit the Taj Mahal, you must go to Agra.

(iv) If the triangle is right angled, then the sum of the square of two sides of the triangle must be equal to the square of the third side.

(v) If the triangle is equilateral, then all three angles of the triangle are equal.

(vi) If $2x = 3y$, then $x : y = 3 : 2$

(vii) If the opposite angles of a quadrilateral are supplementary, then S is cyclic.

(viii) If x is neither positive nor negative, then x is 0.

(ix) If the ratio of the corresponding sides of two triangles are equal, then the triangles are similar.

Q11. Identity the quantifiers in the following statements:

(i) There exists a triangle which is not equilateral.

(ii) For all real number x and y, $xy = yx$.

(iii) There exists a real number which is not a rational number.

(iv) For every natural number $x, x+1$ is also a natural number.

(vi) There exists a triangle which is not an isosceles triangle.

(vii) For all negative integers x, x^3 is also a negative integer.

(viii) There exists a statement in the above statements which is not true.

(ix) There exists an even prime number other than 2.

(x) There exists a real number x such that $x^2 + 1 = 0$.

Solution: The quantifiers in the above statements are:

(i) There exists	(ii) For all
(iii) There exists	(iv) For every
(v) For all	(vi) There exists
(vii) For all	(viii) There exists
(ix) There exists	(x) There exists

Day 198

NCERT Exemplar Problems: Solutions (Q12–Q37)

Q12. Prove by direct method that for any integer 'n', $n^3 - n$ is always even.

Solution: Two cases arise:

Case I: When n is even

Let $n = 2K, K \in N$

$\Rightarrow n^3 - n = (2K)^3 - (2K) = 2K(4K^2 - 1)$

$= 2m$, where $m = K(4K^2 - 1)$

Thus, $(n^3 - n)$ is even when n is even.

Case II: When n is odd

Let $n = 2K + 1,\ K \in N$

$\Rightarrow n^3 - n = (2K+1)^3 - (2K+1)$

$= (2K+1)[(2K+1)^2 - 1]$

$= (2K+1)[4k^2 + 1 + 4K - 1] = (2K+1)(4K^2 + 4K)$

$= 4K(2K+1)(K+1)$

$= 2p$, when $p = 2K(K+1)(2K+1)$

Then $n^3 - n$ is even when n is odd.

So, $n^3 - n$ is always even.

Q13. Check the validity of the following statements:

(i) $p : 125$ is divisible by 5 and 7.

(ii) $q : 131$ is a multiple of 3 or 11.

Solution: (i) $p : 125$ is divisible by 5 and 7.

$q : 131$ is divisible by 5.

$r: 125$ is divisible by 7.
q is true, r is false
$\Rightarrow q \wedge r$ is false.
[Since $p \wedge q$ is false whenever either p or q or both are false]
Hence, p is not valid.

(ii) $q: 131$ is a multiple of 3 or 11.

$p: 131$ is multiple of 3.

$r: 131$ is a multiple of 11.
p is true, r is false
$\Rightarrow p \vee r$ is true.

[Since, $p \vee r$ is true whenever either p or r or both are true]

Hence, q is valid.

Q14. Prove the following statement by contradiction method:
p: The sum of an irrational number and a rational number is irrational.

Solution: Let p be false, that is, the sum of an irrational and a rational number is rational.
Let $\sqrt{m}$ be an irrational and n be a rational number.
$\Rightarrow \sqrt{m} + n = r \Rightarrow \sqrt{m} = r - n$ [rational]
$\sqrt{m}$ is irrational, whereas $(r-n)$ is rational. This is a contradiction.
Then, our supposition is wrong.
Thus, p is true.

Q15. Prove by direct method that for any real number x, y if $x = y$, then $x^2 = y^2$

Solution: Let $p: x = y, x, y \in R$
On squaring both sides, $x^2 = y^2 : q, p \Rightarrow q$ [say]

Q16. Using contrapositive method, prove that if n^2 is an even integer, then n is also an even integer.

Solution: Let $p: n^2$ be an even integer.
$q: n$ is also an even integer.
Let $\sim p$ be true, that is, n is not an even integer.
$\Rightarrow n^2$ is not an even integer [since the square of an odd integer is odd]
$\Rightarrow \sim p$ is true.
Therefore, $\sim q$ is true $\Rightarrow \sim p$ is true.

Q17. Which of the following is a statement?
(a) x is a real number.
(b) Switch off the fan.
(c) 6 is a natural number.
(d) Let me go.

Solution: (c)
6 is a natural number, which is true.
Thus, it is a statement.

Q18. Which of the following is not a statement?
(a) Smoking is injurious to health.
(b) $2+2=4$
(c) 2 is the only even prime number.
(d) Come here.

Solution: (d)
"Come here" is not a statement. It is an order.

Q19. The connective in the statement '$2+7>9$ or $2+7<9$' is:
(a) and (b) or (c) > (d) <

Solution: (b)
In '$2+7>9$ or $2+7<9$', or is the connective.

Q20. The connective in the statement "Earth revolves round the sun and moon is a satellite of earth" is:
(a) or (b) Earth (c) Sun (d) and

Solution: (d)
The connective word is 'and'.

Q21. The negation of the statement "A circle is an ellipse" is:
(a) An ellipse is a circle
(b) An ellipse is not a circle
(c) A circle is not an ellipse
(d) A circle is an ellipse

Solution: (c)
p: A circle is an ellipse
$\sim p$: A circle is not an ellipse

Q22. The negation of the statement "7 is greater than 8" is:
(a) 7 is equal to 8
(b) 7 is not greater than 8
(c) 8 is less than 7
(d) None of these

Solution: (b)
$p: 7$ is greater than 8.
$\sim p: 7$ is not greater than 8.

Q23. The negation of the statement "72 divisible by 2 and 3" is:

(a) 72 is not divisible by 2 or 72 is not divisible by 3
(b) 72 is not divisible by 2 and 72 is not divisible by 3
(c) 72 is divisible by 2 and 72 is not divisible by
(d) 72 is not divisible by 2 and 72 is divisible by 3

Solution: (b)
p: 72 is divisible by 2 and 3.
q: 72 is divisible by 2
r: 72 is divisible by 3.
$\sim q$: 72 is not divisible by 2.
$\sim r$: 72 is not divisible by 3.
$\sim(q \wedge r) :\sim q \vee \sim r$
$\Rightarrow$ 72 is not divisible by 2 or 72 is not divisible by 3

Q24. The negation of the statement "Plants take in CO_2 and give out O_2" is:
(a) Plants do not take in CO_2 and do not given out O_2
(b) Plants do not take in CO_2 or do not given out O_2
(c) Plants take is CO_2 and do not given out O_2
(d) Plants take in CO_2 or do not give out O_2

Solution: (b)
p: Plants take in CO_2 and given out O_2
q: Plants take in CO_2
r: Plants give out O_2
$\sim q$: Plants do not take in CO_2
$\sim r$: Plants do not give out O_2
$\sim(q \wedge r)$: Plants do not take in CO_2 or do not give out O_2.

Q25. The negation of the statement "Rajesh or Rajni lived in Bengaluru" is:
(a) Rajesh did not live Bengaluru or Rajni lived in Bengaluru
(b) Rajesh lived in Bangaluru and Rajni did not live in Bangaluru
(c) Rajesh did not live in Bengaluru and Rajni did not live in Bengaluru
(d) Rajesh did not live in Bengaluru or Rajni did not live in Bengaluru

Solution: p: Rajesh or Rajni lived in Bengaluru.
q: Rajesh lived in Bengaluru.
r: Rajni lived in Bengaluru
$\sim q$: Rajesh did not live in Bengaluru
$\sim r$: Rajni did not live in Bengaluru
$\sim(q \vee r)$: Rajesh did not live in Bengaluru and Rajni did not live in Bengaluru.

Q26. The negation of the statement "101 is not a multiple of 3" is:
(a) 101 is a multiple of 3
(b) 101 is multiple of 2
(c) 101 is an odd number
(d) 101 is an even number

Solution: (a)
p: 101 is not a multiple of 3.
$\sim p$: 101 is a multiple of 3.

Q27. The contrapositive of the statement "If 7 is greater than 5, then 8 is greater than 6" is:
(a) If 8 is greater than 6, then 7 is greater than 5
(b) If 8 is not greater than 6, then 7 is greater than 5
(c) If 8 is not greater than 6, then 7 is not greater than 5
(d) If 8 us greater than 6, then 7 is not greater than 5

Solution: (c)
p: 7 is greater than 5 and q: 8 is greater than 6.
$\sim p \rightarrow q$
$\sim p$: 7 is not greater than 5.
$\sim q$: 8 is not greater than 6.
$(\sim q) \rightarrow (\sim p)$; that is, if 8 is not greater than 6, then 7 is not greater than 5.

Q28. The converse of the statement "if $x>y$, then $x+a>y+a$" is:
(a) If $x<y$, then $x+a<y+a$
(b) If $x+a>y+a$, then $x>y$
(c) If $x<y$, then $x+a<y+a$
(d) If $x>y$ then $x+a<y+a$

Solution: (b)
Let $p: x>y, q: x+a>y+a \Rightarrow p \rightarrow q$
The converse of the above statement is $q \rightarrow p$
That is, if $x+a>y+a$, then $x>y$

Q29. The converse of the statement "If the sun is not shining, then the sky is filled with clouds" is:
(a) If the sky is filled with clouds, then the sun is not shining
(b) If the sun is shining, then the sky is filled with clouds
(c) If the sky is clear, then the sun is shining
(d) If the sun is not shining, then the sky is not filled with clouds.

Solution: (a)
Let p: Sun is not shining and q: Sky is filled with clouds.
The converse of the above statement $p \rightarrow q$ is $q \rightarrow p$
If the sky is filled with clouds, then the sun is not shining.

Q30. The contrapositive of the statement "if p, then q" is:
(a) if q, then p (b) if p, then $\sim q$
(c) if $\sim q$ then $\sim p$ (d) if $\sim p$ then $\sim q$

Solution: (c)
$p \rightarrow q$
If p, then q
The contrapositive of the statement $p \rightarrow q$ is
$(\sim q) \rightarrow (\sim p)$.
If $\sim q$ then $\sim p$

Q31. The statement "If x^2 is not even, then x is not even" is the converse of the statement:
(a) If x^2 is odd, then x is even
(b) If x is not even, then x^2 is not even
(c) If x is even, then x^2 is even
(d) If x is odd, then x^2 is even

Solution: (b)
Let $p : x^2$ is not even and $q : x$ is not even.
The converse of the statement $p \rightarrow q$ is $q \rightarrow p$
That is, if x is not even, then x^2 is not even.

Q32. The contrapositive of the statement 'If Chandigarh is the capital of Punjab, then Chandigarh is in India' is:
(a) If Chandigarh is not in India, then Chandigarh is not the capital of Punjab
(b) If Chandigarh is not the capital of Punjab, then Chandigarh is not the capital of India
(d) If Chandigarh is the capital of Punjab, then Chandigarh is not in India
(d) If Chandigarh is the capital of Punjab, then Chandigarh is in India

Solution: (a)
Let p: Chandigarh is the capital of Punjab.
And q: Chandigarh is in India.
$\sim p$: Chandigarh is not capital of Punjab
$\sim q$: Chandigarh is not in India.
The contrapositive of the statement $p \rightarrow q$ is if $(\sim q)$, then $(\sim p)$.
If Chandigarh is not in India, then Chandigarh is not the capital of Punjab.

Q33. Which of the following is the conditional $p \rightarrow q$?
(a) q is sufficient for p (b) p is necessary for q
(c) p only if q (d) if q then p

Solution: (c)
$p \rightarrow q$ is the same as 'p only if q'.

Q34. The negation of the statement "The product of 3 and 4 is 9" is:
(a) It is false that the product of 3 and 4 is 9
(b) The product of 3 and 4 is 12
(c) The product of 3 and 4 is not 12
(d) It is false that the product of 3 and 4 is not 9

Solution: (a)
The negation of the above statement is "It is false that the product of 3 and 4 is 9".

Q35. Which of the following is not a negation of "A natural number is greater than zero"?
(a) A natural number is not greater than zero
(b) It is false that a natural number is greater than zero
(c) It is false that a natural number is not greater than zero
(d) None of the above

Solution: (c)
The false negation of the given statement is "It is false that a natural number is not greater than zero".

Q36. Which of the following statements is a conjunction?
(a) Ram and Shyam are friends
(b) Both Ram and Shyam are tall
(c) Both Ram and Shyam are enemies
(d) None of the above

Solution: (d)
If two simple statements p and q are connected by the word 'and', then the resulting compound statement is called a conjunction of p and q.
Here, none of the given statements is a conjunction.

Q37. State whether the following sentences are statements or not:
(i) The angles opposite to equal sides of a triangle are equal.
(ii) The moon is a satellite of the earth.
(iii) May God bless you!
(iv) Asia is a continent.

(v) How are you?

Solution: (i) It is a statement.
(ii) It is a statement.
(iii) It is not a statement, since it is an exclamation.
(iv) It is a statement.
(v) It is not a statement, since it is a question.

Day 199

Very Short Answer (VSA): 1-Mark and Short Answer (SA): 2-Mark Questions

Q1. Which of the following is a statement?
(a) x is a rational number.
(b) What are you doing?
(c) 1 is a natural number.
(d) Please, may I come?

Solution: (c)
6 is a natural number, which is true.
Thus, it is a statement.

Q2. Which of the following is not a statement?
(a) Drinking is injurious to health.
(b) $2 \times 2 = 4$
(c) 3 is an odd prime number.
(d) Go there!

Solution: (d)
"Go there!" is not a statement. It is an order.

Q3. The connective in the statement '$2+6<9$ or $2+6>9$' is:
(a) and (b) or (c) > (d) <

Solution: (b)
'or' is the connective.

Q4. The connective in the statement "Earth revolves round the sun and the moon revolves round the earth" is:
(a) or (b) Earth (c) Sun (d) and

Solution: (d)
The connective word is 'and'.

Q5. The negation of the statement "A circle is a rectangle" is:
(a) A rectangle is a square
(b) A rectangle is not a square
(c) A square is not a rectangle
(d) A square is a rectangle

Solution: (c)
Let p: A square is a rectangle
$\sim p$: A square is not a rectangle

Q6. The negation of the statement "6 is greater than 8" is:
(a) 6 is equal to 8
(b) 6 is not greater than 8
(c) 8 is less than 6
(d) None of these

Solution: (b)
Let p: 6 is greater than 8.
$\sim p$: 6 is not greater than 8.

Q7. The negation of the statement "72 divisible by 3 and 6" is:
(a) 72 is not divisible by 3 or 72 is not divisible by 6
(b) 72 is not divisible by 3 and 72 is not divisible by 6
(c) 72 is diviisble by 3 and 72 is not divisible by 6
(d) 72 is not divisible by 3 and 72 is divisible by 6

Solution: (b)
Let p: 72 is divisible by 3 and 6.
Let q: 72 is divisible by 3.
r: 72 is divisible by 6.
$\sim q$: 72 is not divisible by 3.
$\sim r$: 72 is not divisible by 6.
$\sim (q \wedge r) :\sim q \vee \sim r$
$\Rightarrow$ 72 is not divisible by 3 or 72 is not divisible by 6

Q8. The negation of the statement "11 is not a multiple of 3" is:
(a) 11 is a multiple of 3
(b) 11 is a prime number
(c) 11 is an odd number
(d) 11 is an even number

Solution: (a)
Let p: 11 is not a multiple of 3.
$\sim p$: 11 is a multiple of 3.

Q9. The contrapositive of the statement "If 7 is greater than 5, then 8 is greater than 6" is:

(a) If 8 is greater than 6, then 7 is greater than 5
(b) If 8 is not greater than 6, then 7 is greater than 5
(c) If 8 is not greater than 6, then 7 is not greater than 5
(d) If 8 is greater than 6, then 7 is not greater than 5

Solution: (c)
Let p: 7 is greater than 5.
q: 8 is greater than 6.
$\sim p \rightarrow q$
$\sim p: 7$ is not greater than 5.
$\sim q: 8$ is not greater than 6.
$(\sim q) \rightarrow (\sim p)$; that is, if 8 is not greater than 6, then 7 is not greater than 5.

Q10. The converse of the statement "if $x < y$, then $x + a < y + a$" is:

(a) If $x < y$, then $x + a < y + a$
(b) If $x + a > y + a$, then $x > y$
(c) If $x + a < y + a$ then $x < y$,
(d) If $x > y$ then $x + a < y + a$

Solution: (c)
Let $p : x < y, q : x + a < y + a \Rightarrow p \rightarrow q$
The converse of the above statement is $q \rightarrow p$
That is, if $x + a < y + a$, then $x < y$

Q11. The contrapositive of the statement "if p, then q" is:

(a) if q, then p
(b) if p, then $\sim q$
(c) if $\sim q$ then $\sim p$
(d) if $\sim p$ then $\sim q$

Solution: (c)
$p \rightarrow q$
If p, then q
The contrapositive of the statement $p \rightarrow q$ is
$(\sim q) \rightarrow (\sim p)$. If $\sim q$ then $\sim p$

Q12. The statement "If x^2 is not odd, then x is not odd" is the converse of the statement:

(a) If x^2 is odd, then x is odd
(b) If x is not odd, then x^2 is not odd
(c) If x is odd, then x^2 is odd
(d) If x is even, then x^2 is even

Solution: (b)

Let $p : x^2$ is not odd and $q : x$ is not odd.
The converse of the statement $p \rightarrow q$ is $q \rightarrow p$
That is, if x is not odd, then x^2 is not odd.

Q13. The negation of the statement "The product of 3 and 4 is 12" is:

(a) It is false that the product of 3 and 4 is 12
(b) The product of 3 and 4 is 9
(c) The product of 3 and 4 is not 9
(d) It is false that the product of 3 and 4 is not 12

Solution: (a)
The negation of the above statement is "It is false that the product of 3 and 4 is 12".

Q14. Which of the following is not a negation of "A natural number is greater than zero":

(a) A natural number is not greater than zero
(b) It is false that a natural number is greater than zero
(c) It is false that a natural number is not greater than zero
(d) None of the above

Solution: (c)
The false negation of the given statement is "It is false that a natural number is not greater than zero".

Q15. Write the negation of the following statements:

(i) p: For every positive real number x, the number $2x + 1$ is also positive.
(ii) r: For every real number x, either $x > 1$ or $x < -1$

Solution: (i) $\sim p$: There exists a positive real number x, such that $2x + 1$ is not positive.
(ii) $\sim r$: There exists a real number x, such that $-1 \leq x \leq 1$.

Q16. State the converse and contrapositive of the following statement:
q: I go for a walk whenever it is a sunny day.

Solution: The given statement can be written as follows:
If it is a sunny day, then I go for a walk.
The converse of the statement is as follows:
If I go for a walk, then it is a sunny day
The contrapositive of the statement is as follows:
If I do not go for a walk, then it is not a sunny day.

Q17. Write each of the statements in the form "if p, then q":

(i) p: It is necessary to have a password to log on to the computer.

(ii) r: You can access the online library only if you pay a subscription fee.

Solution:

(i) Statement in the form "if p, then q" can be written as follows:
If you log on to the computer, then you have a password.

(ii) Statement in the form "if p, then q" can be written as follows:
If you access the online library, then you pay a subscription fee.

Q18. Identity the quantifiers in the following statements:

(i) There exists a triangle which is not equilateral.

(ii) For all real numbers a and b, $ab = ba$.

(iii) There exists a real number which is not a rational number.

(iv) For every natural number $x, x+2$ is also a natural number.

(vi) There exists a triangle which is an isosceles triangle.

(vii) For all negative integers x, x^2 is a positive integer.

(ix) There exists no even prime number other than 2.

(x) There exists no real number x such that $x^2+1=0$.

Solution: The quantifiers in the above statements are:

(i) There exists	(ii) For all
(iii) There exists	(iv) For every
(v) For all	(vi) There exists
(vii) For all	(viii) There exists
(ix) There exists	(x) There exists

Q19. Rewrite each of the following statements in the form "p if and only if q":

(i) p: If you study Math, then your mind is free and if your mind is free, then you study Math.

(ii) q: For you to get an A grade, it is necessary and sufficient that you do all the work regularly.

(iii) r: If a triangle is equiangular, then it is an equilateral triangle and if a triangle is an equilateral triangle, then it is equiangular.

Solution: (i) You study Math if and only if your mind is free.

(ii) You get an A grade if and only if you do all the work regularly.

(iii) A triangle is equiangular if and only it is an equilateral triangle.

Q20. Write the negation of the following simple statements:

(i) The number 7 is prime.

(ii) $\sqrt{3}$ is an irrational number.

(iii) 6 is not a prime number.

(iv) Every real number is a rational number.

(v) All similar triangles are congruent.

(vi) The area of a rectangle is the same as the perimeter of the circle.

Solution: (i) The number 7 is not prime.

(ii) $\sqrt{5}$ is not an irrational number.

(iii) 6 is a prime number.

(iv) Every real number is not a rational number.

(v) There exist similar triangles which are not congruent.

(vi) The area of a rectangle is not the same as the perimeter of the circle.

Q21. Which of the following sentences are statements? Justify:

(i) A quadrilateral has four sides.

(ii) i is a complex number.

(iii) Every set is a finite set.

(iv) $15+7>23$.

(v) Open your book!

(vi) Every rectangle is a quadrilateral.

(vii) The sum of the internal angles of a quadrilateral is 360°.

(viii) $\sin^2 x+\cos^2 x=2$

Solution: A statement is a sentence which is either true or false but not both simultaneously.

(i) It is a true statement.

(ii) $0+i$. Thus, the given statement is true.

(iii) Not every set is a finite set. Thus, the given statement is false.

(iv) $15+7<23$. Thus, the given statement is false.

(v) It is an order. Thus, the given sentence is not a statement.

(vi) Rectangle is a subset of quadrilateral. Thus, the given statement is true.

(vii) It is a true statement.

(viii) $\sin^2 x+\cos^2 x=1$. Thus, the given statement is false.

Q22. Find the component statements of the following compound statements:

(i) The number 2 is prime and even.

(ii) The number 100 is divisible by 2, 10 and 5.

(iii) $\sqrt{2}$ is a rational number or an irrational number.

(iv) 0 is less than every positive integer and greater than every negative integer.

(v) Two lines in a plane either intersect at one point or they are parallel.
(vi) A rectangle is a quadrilateral or 4-sided polygon.

Solution: (i) p: The number 2 is prime.
q: The number 2 is even.
(ii) p: 100 is divisible by 2.
q: 100 is divisible by 10.
r: 100 is diviisble by 5.
(iii) $p: \sqrt{2}$ is a rational number.
$q: \sqrt{2}$ is an irrational number.
(iv) p: 0 is less than every positive integer.
q: 0 is greater than every negative integer.
(v) p: Two lines in a plane intersect at one point.
q: Two lines in a plane are parallel.
(vi) p: A rectangle is a quadrilateral.
q: A rectangle is a 4-sided polygon.

Day 200

Long Answer-I (LA-I): 4-Mark and Long Answer-II (LA-II): 6-Mark Questions

Q1. Write the component statements of the following compound statements and check whether the compound statement is true or false:
(i) 10 is divisible by 2 or 3.
(ii) 12 is a multiple of 2 and 3.
(iii) 11 is divisible by 2 and 3.
(iv) 3 is an odd number and a prime number.

Solution:
(i) Its component statements are:
p: 10 is divisible by 2. [true]
q: 10 is divisible by 3. [false]
Since the given compound statement is of the form '$p \vee q$', and one of the statements is true, the compound statement is true.
(ii) Its component statements are:
p: 12 is multiple of 2 [true]
q: 12 is multiple of 3. [true]
Since the given compound statement is of the form '$p \wedge q$', and both the component statements are true, the compound statement is true.
(iii) Its component statements are:
p: 11 is divisible by 2. [false]
q: 11 is divisible by 3. [false]
Since the given compound statement is of the form '$p \wedge q$', and both the component statements are false, the compound statement is false.
(iv) Its component statements are:
p: 3 is an odd number. [true]
q: 3 is a prime number. [true]
Since the given compound statement is of the form '$p \wedge q$', and both the component statements are true, the compound statement is true.

Q2. Translate the following statements into symbolic form:
(i) Rahul passed in Math and English.
(ii) 4 and 8 are even integers.
(iii) 3 and 5 are factors of 15.
(iv) Either x or x+1 is an even integer.
(v) Either $x = 2$ or $x = 3$ is a root of $x^2 - x - 2 = 0$.
(vi) Students can take Math or Hindi as an optional paper.

Solution:
(i) p: Rahul passed in Math.
q: Rahul passed in English.
$p \wedge q$: Rahul passed in Math and English.
(ii) p: 4 is even integer.
q: 8 is even integer.
$p \wedge q$: 4 and 8 are even integers.
(iii) p: 3 is factor of 15.
q: 5 is factor of 15.
$p \wedge q$: 3,5 are factors of 15.
(iv) p: x is an even integer.
q: $(x+1)$ is an even integer.
$p \vee q$: Either x or $(x+1)$ is an even integer.
(v) $p: x = 2$ is a root of $x^2 - x - 2 = 0$.
$q: x = 3$ is a root of $x^2 - x - 2 = 0$.
$p \vee q$: Either $x = 2$ or $x = 3$ is a root of $x^2 - x - 2 = 0$.
(vi) p: Students can take Math as an optional paper.
q: Students can take Hindi as an optional subject.
$p \vee q$: Students can take Math or Hindi as an optional paper.

Q3. Write down the negation of following compound statements:

(i) All rational numbers are real and complex.
(ii) All real numbers are rational or irrational.
(iii) $x = 1, x = 2$ are roots of the quadratic equation $x^2 - 3x + 2 = 0$.
(iv) 25 is a prime number or a composite number.
(v) All prime integers are either even or odd.
(vi) $|x|$ is equal to either x or $-x$.
(vii) 10 is divisible by 2 and 5.

Solution:

(i) Let p: All rational numbers are real.
q: All rational numbers are complex.
$\sim p$: All rational number are not real.
$\sim q$: All rational number are not complex.
$\sim (p \wedge q)$: All rational numbers are not real or not complex. $[\because \sim (p \wedge q) = \sim p \vee \sim q]$

(ii) Let p: All real numbers are rational.
q: All real numbers are irrational.
Then, the negation of the above statement is given by:
$\sim (p \vee q)$:Allrealnumbersarenotrationalandallreal numbers are not irrational. $[\because \sim (p \vee q) = \sim p \wedge \sim q]$

(iii) Let $p : x = 1$ is root of quadratic equation $x^2 - 3x + 2 = 0$.
$q : x = 2$ is root of quadratic equation $x^2 - 3x + 2 = 0$.
Then, the negation of the above statement is given by:
$\sim (p \wedge q) : x = 1$ is not a root of the quadratic equation $x^2 - 3x + 2 = 0$. or $x = 2$ is not a root of the quadratic equation $x^2 - 3x + 2 = 0$.

(iv) Let p: 25 is a prime number.
q: 25 is a composite number.
Then, negation of the disjunction of the above statements is given by:
$\sim (p \vee q)$: 25 is not a prime number and it is not a composite number.

(v) Let p: All prime integers are even.
q: All prime integers are odd.
Then, negation of the disjunction of the above statements is given by:
$\sim (p \vee q)$: All prime integers are not even and all prime integers are not odd.

(vi) Let $p : |x|$ is equal to x.
$q : |x|$ is equal to $-x$.
Then, negation of disjunction of the above statements is given by:
$\sim (p \vee q) : |x|$ is not equal to x and it is not equal to $-$x.

(vii) Let p: 10 is divisible by 2.
q: 10 is divisible by 5
Then, negation of the conjunction of above statements is given by:
10~ $(p \wedge q)$: 10 is not divisible by 2 or it is not divisible by 5

Q4. Rewrite each of the following statements in the form of conditional statements:

(i) The cube of an odd number is odd.
(ii) The units digit of an integer is 0 or 5, if it is divisible by 5.
(iii) The square of a prime number is not prime.
(iv) $b^2 = a \cdot c$, if a, b, c are in GP.

Solution:

(i) If the number is an odd number, then its cube is an odd number.
(ii) If the integer is divisible by 5, then its units digits are 0 or 5.
(iii) If the number is prime, then its square is not prime.
(iv) If a, b, c are in GP, then $b^2 = a \cdot c$

Q5. Form the bi-conditional statement $p \leftrightarrow q$ where:

(i) p: The units digit of an integer is zero or 5.
q: It is divisible by 5.
(ii) p: A natural number n is even.
q: Natural number n is divisible by 2
(iii) p: A triangle is an equilateral triangle.
q: All three sides of a triangle are equal.

Solution: (i) $p \leftrightarrow q$: The units digit of an integer is zero or 5, if and only if it is divisible by 5.

(ii) $p \leftrightarrow q$: A natural number number is even if and only if it is divisible by 2.

(iii) $p \leftrightarrow q$: A triangle is an equilateral triangle if and only if all three sides of the triangle are equal.

Q6. Write down the contrapositive of the following statements:

(i) If $x = y, y = 2$, then $x = 2$.
(ii) If n is a natural number, then n is an integer.
(iii) If all three sides of a triangle are equal, then the triangle is equilateral.
(iv) If x is a real number such that $0 < x < 1$, then $x^2 < 1$.

Solution: (i) If $x \neq 2$, then $x \neq y, y \neq 2$.

(ii) If n is not an integer, then it is not a natural number.

(iii) If the triangle is not equilateral, then all three sides of the triangle are not equal.

(iv) If $x^2 > 1$, then x is not a real number such that $0 < x < 1$.

Q7. Given below are two statements:
p: 25 is a multiple of 5.
q: 25 is a multiple of 7.
Write the compound statements connecting these two statements with "and" and "or".
In both cases, check the validity of the compound statement.

Solution: The compound statement with 'and' is "25 is a multiple of 5 and 7". This is a false statement, since 25 is not a multiple of 7.
The compound statement with 'or' is "25 is a multiple of 5 or 7". This is a true statement, since 25 is not a multiple of 7 but it is a multiple of 5.

Q8. Check the validity of the statement given below using the method given against it:
q: If n is a real number with $n > 2$, then $n^2 > 4$
(by contradiction method).

Solution: The given statement, *q*, is as follows.
If n is a real number with $n > 2$, then $n^2 > 4$
Let us assume that n is a real number with $n > 2$, but $n^2 > 4$ is not true.
That is $n^2 < 4$
Then $n > 2$ and n is a real number.
Squaring both the sides, we obtain $n^2 > (2)^2$
$\Rightarrow n^2 > 4$ which is a contradiction, since we have assumed that $n^2 < 4$
Thus, the given statement is true; that is, if n is a real number with $n > 2$, then $n^2 > 4$

Q9. Prove by direct method that for any integer 'n' $n^3 + n$ is always even.

Solution: Two cases arise:

Case I: When n is even

Let $n = 2K, K \in N$

$\Rightarrow n^3 + n = (2K)^3 + (2K) = 2K(4K^2 + 1)$

$= 2m$, where $m = K(4K^2 + 1)$

Thus, $(n^3 + n)$ is even when n is even.

Case II: When n is odd,

Let $n = 2K + 1, \quad K \in N$

$\Rightarrow n^3 + n = (2K+1)^3 + (2K+1)$

$= (2K+1)[(2K+1)^2 + 1]$

$= (2K+1)[4k^2 + 1 + 4K + 1]$

$= (2K+1)(4K^2 + 4K + 2)$

$= 2(2K+1)(2K^2 + 2K + 1)$

$= 2p$, when $p = (2K+1)(2K^2 + 2K + 1)$

Then $n^3 - n$ is even when n is odd.

So, $n^3 - n$ is always even.

Q10. Check the validity of the following statements:
(i) $p: 125$ is divisible by 5 and 11.
(ii) $q: 131$ is a multiple of 3 or 13.

Solution: (i) $p: 125$ is divisible by 5 and 11.
Let $q: 125$ is divisible by 5.
$r: 125$ is divisible by 11.
q is true, r is false
$\Rightarrow q \wedge r$ is false.
[since $p \wedge q$ is false whenever either p or q or both are false]
Hence, p is not valid.

(ii) $q: 131$ is a multiple of 3 or 13.
Let $p: 131$ is multiple of 3.
$r: 131$ is a multiple of 13.
p is true, r is false
$\Rightarrow p \vee r$ is true.
[since $p \vee r$ is true whenever either p or r or both are true]
Hence, q is valid.

Q11. Write down the converse of following statements:
(i) If a rectangle 'R' is a square, then R is a rhombus.
(ii) If today is Sunday, then tomorrow is Monday.
(iii) If you go to Varanasi, then you must visit BHU.
(iv) If the sum of the squares of two sides of a triangle is equal to the square of the third side of a triangle, then the triangle is right angled.
(v) If all three angles of a triangle are equal, then the triangle is equilateral.
(vi) If $x : y = 3 : 1$, then $x = 3y$.
(vii) If x is zero, then x is neither positive nor negative.

Solution: (i) If the rectangle 'R' is a rhombus, then it is a square.

(ii) If tomorrow is Monday, then today is Sunday.

(iii) If you must visit BHU, you must go to Varanasi.

(iv) If the triangle is right angled, then the sum of the square of two sides of the triangle is equal to the square of the third side.

(v) If the triangle is equilateral, then all three angles of the triangle are equal.

(vi) If $x = 3y$, then $x : y = 3 : 1$

(vii) If x is neither positive nor negative, then x is 0.

CHAPTER

15 Statistics

Day 201

Mean Deviation

Range: The range of a series is the difference of the maximum and minimum values of the series.

Deviation: The deviation of an observation x from a fixed value 'a' is the difference $x - a$.

Mean Deviation: Mean deviation about a central value 'a' is the mean of the absolute values of the deviations of the observations from 'a'. The mean deviation from 'a' is denoted as MD (a). Thus,

$$\text{MD}(a) = \frac{\text{Sum of absolute values of deviations from '}a\text{'}}{\text{Number of observations}}$$

Mean Deviation for Ungrouped Data: Let n observations be $x_1, x_2, x_3, \ldots, x_n$. The following steps are taken in the calculation of mean deviation about mean or median:

Step 1: Calculate the measure of central tendency 'a' about which we need to find mean deviation.

Step 2: Find the deviation of each x_i from a, that is, $x_1 - a, x_2 - a, x_3 - a, \ldots, x_n - a$

Step 3: Find the absolute values of the deviations, that is, $|x_1 - a|, |x_2 - a|, |x_3 - a|, \ldots, |x_n - a|$

Step 4: Find the mean of the absolute values of the deviations, known as mean deviation about a, that is

$$\text{MD}(a) = \frac{\sum_{i=1}^{n} |x_i - a|}{n}$$

Thus $\text{MD}(\bar{x}) = \frac{1}{n}\sum_{i=1}^{n} |x_i - \bar{x}|$, where $\bar{x}$ = mean

And $\text{MD}(M) = \frac{1}{n}\sum_{i=1}^{n} |x_i - M|$, where M = median

Mean Deviation for Grouped Data

(a) Discrete frequency distribution: Let $x_1, x_2, \ldots, x_n$ be the n distinct values occurring with frequencies $f_1, f_2, \ldots, f_n$, respectively. This discrete frequency distribution data can be represented in tabulate form as given below:

$x : x_1 \ x_2 \ x_3 \ \ldots \ x_n$

$f : f_1 \ f_2 \ f_3 \ \ldots \ f_n$

(i) Mean deviation about mean

Step 1: Find the mean $\bar{x}$ of the given data using the formula $\bar{x} = \frac{\sum_{i=1}^{n} x_i f_i}{\sum_{i=1}^{n} f_i} = \frac{1}{N}\sum_{i=1}^{n} x_i f_i$, where $\sum_{i=1}^{n} x_i f_i$ denotes the sum of the products of observations x_i with their respectively frequencies f_i and $N = \sum_{i=1}^{n} f_i$ is the sum of the frequencies.

Step 2: Find the deviation of observations x_i from the mean $\bar{x}$ and their absolute values, that is, $|x_i - \bar{x}|$ for all $i = 1, 2, \ldots, n$.

Step 3: Find the mean of the absolute values of the deviations, which is the required mean deviation about the mean. Thus

$$\text{MD}(\bar{x}) = \frac{\sum_{i=1}^{n} f_i |x_i - \bar{x}|}{\sum_{i=1}^{n} f_i} = \frac{1}{N}\sum_{i=1}^{n} f_i |x_i - \bar{x}|$$

(ii) Mean deviation about median

Step 1: Arrange the observations in ascending order.

Step 2: Identify the observation whose cumulative frequency is equal to or just greater than N/2, where N is the sum of frequencies. This value of the observation lies in the middle of the data. Therefore, it is the required median.

Step 3: Obtain the mean of the absolute values of the deviation from median. That is

$$\text{MD}(M) = \frac{1}{N}\sum_{i=1}^{n} f_i |x_i - M|$$

(b) Continuous frequency distribution: A continuous frequency distribution is a series in which the data is classified into different class intervals without gaps along with their respective frequencies.

(i) Mean deviation about the mean

For calculating the mean of a continuous frequency distribution, we assume that the frequency in each class is centred at its mid-point, that is, the mid-point of each given class, and proceed as for a discrete frequency distribution to find mean deviation.

Shortcut method for calculating mean deviation about mean

In the shortcut method, we avoid the tedious calculations of computing $\bar{x}$ by following the step-deviation method.

In this method, we take an assumed mean which is in the middle or just close to it in the data. Then, deviations of the observations (or mid-points of classes) are taken from the assumed mean.

If there is a common factor to all the deviations, we divide them by using the common factor to further simplify the deviations. These steps are known as step-deviations.

Let the new variable be denoted by $d_i = \frac{x_i - a}{h}$, where 'a' is the assumed mean and h is the common factor.

Then, the mean $\bar{x}$ by step-deviation method is given by $\bar{x} = a + \frac{\sum_{i=1}^{n} f_i d_i}{N} \times h$

Thus, mean deviation about the mean is

$$\text{MD}(\bar{x}) = \frac{\sum_{i=1}^{n} f_i |x_i - \bar{x}|}{\sum_{i=1}^{n} f_i} = \frac{1}{N}\sum_{i=1}^{n} f_i |x_i - \bar{x}|$$

(ii) Mean deviation about the median

The process of finding the mean deviation about the median for a continuous frequency distribution is similar to mean deviation about mean.

The only difference lies in the replacement of mean with median while taking deviations.

The data is first arranged in ascending order.

The median of a continuous frequency distribution is obtained by first identifying the class in which median lies (median class) and then applying the formula where median class is the class interval whose cumulative frequency is just greater than or equal to N/2, N is the sum of frequencies l, f, h, C, respectively, the lower limit, the frequency, the width of the median class and the cumulative frequency of the class just preceding the median class.

After finding the median, the absolute values of the deviations of mid-point x_i of each class from the median, that is, $|x_i - M|$ are obtained.

Then $\text{MD}(M) = \frac{1}{N}\sum_{i=1}^{n} f_i |x_i - M|$

Example 1: Find the mean deviation about the mean for the data 2, 4, 7, 8, 9,

Solution: Mean of the data

$$\bar{x} = \frac{2+4+7+8+9}{5} = \frac{30}{5} = 6$$

The deviations of the respective observations from the mean $\bar{x} = x_i - \bar{x}$ are $-4, -2, 1, 2, 3$

The absolute values of the deviation $= |x_i - \bar{x}|$ are $4, 2, 1, 2, 3$

Thus, the mean deviation about the mean is

$$\text{MD}(\bar{x}) = \frac{\sum_{i=1}^{5} |x_i - \bar{x}|}{5}$$

$$= \frac{4+2+1+2+3}{5} = \frac{12}{5} = 2.4$$

Example 2: Find the mean deviation about the median for the data 6, 2, 9, 4, 5, 8.

Solution: The number of observations is 6, which is even. The given data 6, 2, 9, 4, 5, 8 can be arranged in ascending order as 2, 4, 5, 6, 8, 9.

The number of observations is 6, which is even. Therefore

$$\text{Median } M = \frac{\left(\frac{6}{2}\right)^{\text{th}} \text{observation} + \left(\frac{6}{2}+1\right)^{\text{th}} \text{observation}}{2}$$

$$= \frac{3^{\text{rd}} \text{ observation} + 4^{\text{th}} \text{ observation}}{2}$$

$$= \frac{5+6}{2} = \frac{11}{2} = 5.5$$

The deviations of the respective observations from the median $= x_i - M$ are

$-3.5, -1.5, -0.5, 0.5, 2.5, 3.5$

The absolute values of the deviations $= |x_i - M|$ are

$3.5, 1.5, 0.5, 0.5, 2.5, 3.5$

Thus, the mean deviation about the median is

$$\text{MD}(M) = \frac{\sum_{i=1}^{6} |x_i - M|}{6}$$

$$= \frac{3.5+1.5+0.5+0.5+2.5+3.5}{6} = \frac{12}{6} = 2$$

Example 3: Find the mean deviation about the mean for the data given below:

x_i	10	20	30	40	50
f_i	4	2	8	6	5

Solution:

x_i	f_i	$x_i f_i$	$\lvert x_i - \bar{x}\rvert = \lvert x_i - 32\rvert$	$f_i \lvert x_i - \bar{x}\rvert$
10	4	40	22	88
20	3	60	12	36
30	7	210	2	14
40	6	240	8	48
50	5	250	18	90
	25	800		276

$$N = \sum_{i=1}^{5} f_i = 25,$$

$$\sum_{i=1}^{5} f_i x_i = 800$$

$$\therefore \bar{x} = \frac{1}{N}\sum_{i=1}^{5} f_i x_i = \frac{1}{25} \times 800 = 32$$

$$\text{MD}(\bar{x}) \frac{1}{N}\sum_{i=1}^{5} f_i \lvert x_i - \bar{x}\rvert = \frac{1}{25} \times 276 = 11.04$$

Example 4: Find the mean deviation about the median for the data given below:

x_i	5	7	9	10	12
f_i	8	6	2	2	2

Solution: The given observations in ascending order and cumulative frequencies of the given data are shown in the table below:

x_i	f_i	$c.f.$
5	8	8
7	6	14
9	2	16
10	2	18
12	2	20

Since $n = 20$, which is even, the median is the mean of $\left(\frac{20}{2}\right)^{\text{th}} = 10^{\text{th}}$ and $\left(\frac{20}{2}+1\right)^{\text{th}} = 11^{\text{th}}$ observations.

Both of these observations lie in the cumulative frequency 14, for which the corresponding observation is 7.

$$\therefore \text{Median} = \frac{10^{\text{th}}\,\text{observation} + 11^{\text{th}}\,\text{observation}}{2}$$

$$= \frac{7+7}{2} = 7$$

The absolute values of the deviation from median $= \lvert x_i - M\rvert$ are

$\lvert x_i - M\rvert$	2	0	2	3	5
f_i	8	6	2	2	2
$f_i \lvert x_i - M\rvert$	16	0	4	6	10

$$\sum_{i=1}^{5} f_i = 20, \quad \sum_{i=1}^{5} f_i \lvert x_i - M\rvert = 36$$

$$\text{MD }(M) = \frac{1}{N}\sum_{i=1}^{6} f_i \lvert x_i - M\rvert = \frac{1}{20} \times 36 = 1.8$$

Example 5: Find the mean deviation about the mean for the data given below:

Height in cm	**Number of boys**
105–115	12
115–125	20
125–135	10
135–145	15
145–155	8

Solution: The table is formed as:

Height in cm	**Number of boys f_i**	**Mid-point x_i**	$f_i x_i$	$\lvert x_i - \bar{x}\rvert$	$f_i \lvert x_i - \bar{x}\rvert$
105–115	12	110	1320	18	216
115–125	20	120	2400	8	160
125–135	10	130	1300	2	20
135–145	15	140	2100	12	180
145–155	8	150	1200	22	176

$$N = \sum_{i=1}^{5} f_i = 65,$$

$$\sum_{i=1}^{5} f_i x_i = 8320$$

$$\therefore \bar{x} = \frac{1}{N}\sum_{i=1}^{5} f_i x_i = \frac{1}{65} \times 8320 = 128$$

$$\text{MD }(\bar{x}) = \frac{1}{N}\sum_{i=1}^{5} f_i \lvert x_i - \bar{x}\rvert = \frac{1}{65} \times 752 = 11.57$$

Example 6: Find the mean deviation about median for the following data:

Marks	Number of girls
0–10	9
10–20	7
20–30	6
30–40	5
40–50	3

Solution: The table is formed as:

Marks	Number of f_i	Cumulative frequency (c.f.)	Mid-point x_i	$\lvert x_i - \text{Med.}\rvert$	$f_i\lvert x_i - \text{Med.}\rvert$
0–10	9	9	5	13.57	122.13
10–20	7	16	15	3.57	24.99
20–30	6	22	25	7.57	45.42
30–40	5	27	35	17.57	87.85
40–50	3	30	45	27.57	82.71
	30				363.1

The class interval containing the $\left(\frac{N}{2}\right)^{th}$ or 15th item is 10–20.

Therefore, 10–20 is the median class.

Thus, Median $= l + \dfrac{\frac{N}{2} - c}{f} \times h$

Here, $l = 10, C = 9, f = 7, h = 10, N = 30$

$\therefore$ Median

$= 10 + \dfrac{15-9}{7} \times 10 = 10 + \dfrac{60}{7} = 10 + 8.57 = 18.57$

Thus, mean deviation about the median is given by

$$\text{MD}(M) = \frac{1}{N}\sum_{i=1}^{5} f_i\lvert x_i - M\rvert = \frac{1}{30} \times 363.1 = 12.10$$

Day 202

NCERT Exercise 15.1 (With Solutions)

Q1. Find the mean deviation about the mean for the data 4, 7, 8, 9, 10, 12, 13, 17.

Solution: Mean of the data

$$\bar{x} = \frac{4+7+8+9+10+12+13+17}{8} = \frac{80}{8} = 10$$

The deviations of the respective observations from the mean $\bar{x} = x_i - \bar{x}$ are $-6, -3, -2, -1, 0, 2, 3, 7$

The absolute values of the deviation $= \lvert x_i - \bar{x}\rvert$ are

6, 3, 2, 1, 0, 2, 3, 7

Thus, the mean deviation about the mean is

$$\text{MD}(\bar{x}) = \frac{\sum_{i=1}^{8}\lvert x_i - \bar{x}\rvert}{8}$$

$$= \frac{6+3+2+1+0+2+3+7}{8} = \frac{24}{8} = 3$$

Q2. Find the mean deviation about the mean for the data 38, 70, 48, 42, 55, 63, 46, 54, 44.

Solution: Mean of the given data

$$\bar{x} = \frac{38+70+48+40+42+55+63+46+54+44}{10}$$

$$= \frac{500}{10} = 50$$

The deviations of the respective observations from the mean $\bar{x} = x_i - \bar{x}$ are

$-12, 20, -2, -10, -8, 5, 13, -4, 4, -6$

The absolute values of the derivations $= \lvert x_i - \bar{x}\rvert$ are

12, 20, 2, 10, 8, 5, 13, 4, 4, 6

Thus, the mean deviation about the mean is

$$\text{MD}(\bar{x}) = \frac{\sum_{i=1}^{10}\lvert x_i - \bar{x}\rvert}{10}$$

$$= \frac{12+20+2+10+8+5+13+4+4+6}{10} = \frac{84}{10} = 8.4$$

Q3. Find the mean deviation about the median for the data 13, 17, 16, 14, 11, 13, 10, 16, 11, 18, 12, 17.

Solution: The given data 13, 17, 16, 14, 11, 13, 10, 16, 11, 18, 12, 17 can be arrange in ascending order as

10, 11, 11, 12, 13, 14, 16, 16, 17, 17, 18

The number of observations is 12, which is even.

$$\text{Median } M = \frac{\left(\frac{12}{2}\right)^{th}\text{observation} + \left(\frac{12}{2}+1\right)^{th}\text{observation}}{2}$$

$$= \frac{6^{th}\text{observation} + 7^{th}\text{observation}}{2}$$

$$= \frac{13+14}{2} = \frac{27}{2} = 13.5$$

The deviations of the respective observations from the median, $= x_i - M$ are

$-3.5, -2.5, -2.5, -1.5, -0.5, -0.5, 0.5, 2.5, 2.5, 3.5, 3.5, 4.5$ The absolute values of the deviations $= |x_i - M|$ are

3.5, 2.5, 2.5, 1.5, 0.5, 0.5, 0.5, 2.5, 2.5, 3.5, 3.5, 3.5, 4.5

Thus, the mean deviation about the median is

$$\text{MD}(M) = \frac{\sum_{i=1}^{12} |x_i - M|}{12}$$

$$= \frac{3.5+2.5+2.5+1.5+0.5+0.5+0.5+2.5+2.5+3.5+3.5+4.5}{12}$$

$$= \frac{28}{12} = 2.33$$

Q4. Find the mean deviation about the median for the data 36, 72, 46, 42, 60, 45, 53, 46, 51, 49.

Solution: The number of observations is 10, which is even.

The given data 36, 72, 46, 42, 60, 45, 53, 46, 51, 49 can be arrange in ascending order as

36, 42, 45, 46, 46, 49, 51, 53, 60, 72

The number of observations is 10, which is even.

$$\text{Median } M = \frac{\left(\frac{10}{2}\right)^{\text{th}} \text{observation} + \left(\frac{10}{2}+1\right)^{\text{th}} \text{observation}}{2}$$

$$= \frac{5^{\text{th}} \text{observation} + 6^{\text{th}} \text{observation}}{2}$$

$$= \frac{46+49}{2} = \frac{95}{2} = 47.5$$

The deviations of the respective observations from the median $= x_i - M$ are

$-11.5, -5.5, -2.5, -1.5, -1.5, 1.5, 3.5, 5.5, 12.5, 24.5$

The absolute values of the deviations $= |x_i - M|$ are

11.5, 5.5, 2.5, 1.5, 1.5, 1.5, 3.5, 5.5, 12.5, 24.5

Thus, the mean deviation about the median is

$$\text{MD}(M) = \frac{\sum_{i=1}^{10} |x_i - M|}{10}$$

$$= \frac{11.5+5.5+2.5+1.5+1.5+1.5+3.5+5.5+12.5+24.5}{10} = \frac{70}{10} = 7$$

Q5. Find the mean deviation about the mean for the data.

x_i	5	10	15	20	25
f_i	7	4	6	3	5

Solution:

x_i	f_i	$f_i x_i$	$\lvert x_i - \bar{x}\rvert$	$f_i\lvert x_i - \bar{x}\rvert$
5	7	35	9	63
10	4	40	4	16
15	6	90	1	6
20	3	60	6	18
25	5	125	11	55
	25	350		158

$$N = \sum_{i=1}^{5} f_i = 25$$

$$\sum_{i=1}^{5} f_i x_i = 350$$

$$\therefore \bar{x} = \frac{1}{N}\sum_{i=1}^{5} f_i x_i = \frac{1}{25} \times 350 = 14$$

$$\therefore \text{MD}(\bar{x}) = \frac{1}{N}\sum_{i=1}^{5} f_i |x_i - \bar{x}| = \frac{1}{25} \times 158 = 6.32$$

Q6. Find the mean deviation about the mean for the data.

x_i	10	30	50	70	90
f_i	4	24	28	16	8

Solution:

x_i	f_i	$x_i f_i$	$\lvert x_i - \bar{x}\rvert = \lvert x_i - 10\rvert$	$f_i\lvert x_i - \bar{x}\rvert$
10	4	40	40	160
30	24	720	20	480
50	28	1400	0	0
70	16	1120	20	320
90	8	720	40	320
	80	4000		1280

$N = \sum_{i=1}^{5} f_i = 80,$

$\sum_{i=1}^{5} f_i x_i = 4000$

$\therefore \bar{x} = \frac{1}{N}\sum_{i=1}^{5} f_i x_i = \frac{1}{80} \times 4000 = 50$

$\text{MD}(\bar{x}) \frac{1}{N}\sum_{i=1}^{5} f_i |x_i - \bar{x}| = \frac{1}{80} \times 1280 = 16$

Q7. Find the mean deviation about the median for the data.

x_i	5	7	9	10	12	15
f_i	8	6	2	2	2	6

Solution: The given observations in ascending order and cumulative frequencies of the given data are shown in the table below:

x_i	f_i	$c.f.$
5	8	8
7	6	14
9	2	16
10	2	18
12	2	20
15	6	26

$n = 26$, which is even. Therefore, median is the mean of $\left(\frac{26}{2}\right)^{\text{th}} = 13^{\text{th}}$ and $\left(\frac{26}{2}+1\right)^{\text{th}} = 14^{\text{th}}$ observations.

Both of these observations lie in the cumulative frequency 14, for which the corresponding observation is 7.

$\therefore \text{Median} = \frac{13^{\text{th}}\text{observation} + 14^{\text{th}}\text{observation}}{2}$

$= \frac{7+7}{2} = 7$

The absolute values of the deviation from median, $= |x_i - M|$ are

$\lvert x_i - M\rvert$	2	0	2	3	5	8
f_i	8	6	2	2	2	6
$f_i\lvert x_i - M\rvert$	16	0	4	6	10	48

$\sum_{i=1}^{6} f_i = 26, \sum_{i=1}^{6} f_i |x_i - M| = 84$

$\text{MD}(M) = \frac{1}{N}\sum_{i=1}^{6} f_i |x_i - M| = \frac{1}{26} \times 84 = 3.23$

Q8. Find the mean deviation about the median for the data.

x_i	15	21	27	30	35
f_i	3	5	6	7	8

Solution: The given observations in ascending order and cumulative frequencies of the given data are shown in the table below:

x_i	f_i	$c.f.$
15	3	3
21	5	8
27	6	14
30	7	21
35	8	29

$n = 29$ which is odd.

$\therefore \text{Median} = \left(\frac{29+1}{2}\right)^{\text{th}}$ observation $= 15^{\text{th}}$ observation

This observation lies in the cumulative frequency 21, for which the corresponding observation is 30.

$\therefore$ Median =30

The absolute values of the deviations from median, that is, $|x_i - M|$, are

$\lvert x_i - M\rvert$	15	9	3	0	5
f_i	3	5	6	7	8
$f_i\lvert x_i - M\rvert$	45	45	18	0	40

$\sum_{i=1}^{5} f_i = 29,$

$\sum_{i=1}^{5} f_i |x_i - M| = 148$

$\therefore \text{MD}(M) = \frac{1}{N}\sum_{i=1}^{5} f_i |x_i - M| = \frac{1}{29} \times 148 = 5.1$

Q9. Find the mean deviation about the mean for the data.

Income per day	Number of persons
0–100	4
100–200	8
200–300	9
300–400	10
400–500	7
500–600	5
600–700	4
700–800	3

Solution: The table is formed as:

Income per day	Number of persons f_i	Mid-point x_i	$f_i x_i$	$\lvert x_i - \bar{x}\rvert$	$f_i\lvert x_i - \bar{x}\rvert$
0–100	4	50	200	308	1232
100–200	8	150	1200	208	1664
200–300	9	250	2250	108	972
300–400	10	350	3500	8	80
400–500	7	450	3150	92	644
500–600	5	550	2750	192	960
600–700	4	650	2600	292	1168
700–800	3	750	2250	392	1176
	50		17900		7896

$$N = \sum_{i=1}^{8} f_i = 50,$$

$$\sum_{i=1}^{8} f_i x_i = 17900$$

$$\therefore \bar{x} = \frac{1}{N}\sum_{i=1}^{8} f_i x_i = \frac{1}{50}\times 17900 = 358$$

$$\text{MD}(\bar{x}) = \frac{1}{N}\sum_{i=1}^{8} f_i\lvert x_i - \bar{x}\rvert = \frac{1}{50}\times 7896 = 157.92$$

Q10. Find the mean deviation about the mean for the data.

Height in cm	Number of boys
95–100	9
105–115	13
115–125	26
125–135	30
135–145	12
145–155	10

Solution: The table is formed as:

Height in cm	Number of boys f_i	Mid-point x_i	$f_i x_i$	$\lvert x_i - \bar{x}\rvert$	$f_i\lvert x_i - \bar{x}\rvert$
95–105	9	100	900	25.3	227.7
105–115	13	110	1430	15.3	198.9
115–125	26	120	3120	5.3	137.8
125–135	30	130	3900	4.7	141
135–145	12	140	1680	14.7	176.4
145–155	10	150	1500	24.7	247

$$N = \sum_{i=1}^{6} f_i = 100,$$

$$\sum_{i=1}^{6} f_i x_i = 12530$$

$$\therefore \bar{x} = \frac{1}{N}\sum_{i=1}^{6} f_i x_i = \frac{1}{100}\times 12530 = 125.3$$

$$\text{MD}(\bar{x}) = \frac{1}{N}\sum_{i=1}^{6} f_i\lvert x_i - \bar{x}\rvert = \frac{1}{100}\times 1128.8 = 11.28$$

Q11. Find the mean deviation about median for the following data.

Marks	Number of girls
0–10	6
10–20	8
20–30	14
30–40	16
40–50	4
50–60	2

Solution: The table is formed as:

Marks	Number of f_i	Cumulative frequency (c.f.)	Mid-point x_i	$\lvert x_i - \text{Med.}\rvert$	$f_i\lvert x_i - \text{Med.}\rvert$
0–10	6	6	5	22.85	137.1
10–20	8	14	15	12.85	102.8
20–30	14	28	25	2.85	39.5
30–40	16	44	35	7.15	114.4
40–50	4	48	45	17.15	68.6
50–60	2	50	55	27.15	54.3
	50				517.1

The class interval containing the $\left(\frac{N}{2}\right)^{\text{th}}$ or 25th item is 20–30.

Therefore, 20–30 is the median class.

Thus, Median $= l + \dfrac{\frac{N}{2} - c}{f} \times h$

Here, $l = 20, C = 14, f = 14, h = 10, N = 50$

$\therefore$ Median

$= 20 + \dfrac{25-14}{14} \times 10 = 20 + \dfrac{110}{14} = 20 + 7.85 = 27.85$

Thus, mean deviation about the median is given by

$$\text{MD}(M) = \frac{1}{N}\sum_{i=1}^{6} f_i\lvert x_i - M\rvert = \frac{1}{50} \times 517.1 = 10.34$$

Q12. Calculate the mean deviation about median age for the age distribution of 100 persons given below:

Age	Number
16–20	5
21–25	6
26–30	12
31–35	14
3-–40	26
41–45	12
46–50	16
51–55	9

Solution: The given data is not continuous. Therefore, it must be converted into continuous frequency distribution by subtracting 0.5 from the lower limit and adding 0.5 to the upper limit of each class interval.
The table is formed as follows:

Age	Number f_i	Cumulative frequency (c.f.)	Mid-point x_i	$\lvert x_i - \text{Med.}\rvert$	$f_i\lvert x_i - \text{Med.}\rvert$
15.5–20.5	5	5	18	20	100
20.5–25.5	6	11	23	15	90
25.5–30.5	12	23	28	10	120
30.5–35.5	14	37	33	5	70
35.5–40.5	26	63	38	0	0
40.5–45.5	12	75	43	5	60
45.5–50.0	16	91	48	10	160
50.0–55.5	9	100	53	15	135
	100				735

The class interval containing the $\frac{N^{\text{th}}}{2}$ or 50th item is 35.5–40.5.

Therefore, 35.5–40.5 is the median class.

Median $= l + \dfrac{\frac{N}{2} - C}{f} \times h$

And, $l = 35.5, C = 37, f = 26, h = 5, N = 100$

$\therefore$ Median

$= 35.5 + \dfrac{50-37}{26} \times 5 = 35.5 + \dfrac{13\times 5}{26} = 35.5 + 2.5 = 3.8$

Thus, mean deviation about the median is given by

$$\text{MD}(M) = \frac{1}{N}\sum_{i=1}^{8} f_i\lvert x_i - M\rvert = \frac{1}{100} \times 735 = 7.35$$

Day 203

Variance and Standard Deviation

Degree of Dispersion: Let $x_1, x_2, ..., x_n$ be n observations and $\bar{x}$ be their mean. Then

$$(x_1-\bar{x})^2+(x_2-\bar{x})^2+...+(x_n-\bar{x})^2=\sum_{i=1}^{n}(x_i-\bar{x})^2$$

Note:

If $\sum_{i=1}^{n}(x_i-\bar{x})^2$ is zero, then each $(x_i-\bar{x})$ must be zero. This implies that there is no dispersion at all as all observations are equal to the mean x.

If $\sum_{i=1}^{n}(x_i-\bar{x})^2$ is small, this indicates that the observations $x_1, x_2, x_3, ..., x_n$ are close to the mean $\bar{x}$, and therefore, there is a lower degree of dispersion.

If $\sum_{i=1}^{n}(x_i-\bar{x})^2$ is large, there is a higher degree of dispersion of the observations from the mean $\bar{x}$, and we say that the sum $\sum_{i=1}^{n}(x_i-\bar{x})^2$ is a reasonable indicator of the degree of dispersion or scatter.

Standard Deviation and Variance: The proper measure of dispersion about the mean of a set of observations is expressed as the positive square root of the variance and is called standard deviation. Therefore, the standard deviation, usually denoted by σ, is given by

$$\sigma=\sqrt{\frac{1}{n}\sum_{i=1}^{n}(x_i-\bar{x})^2} \quad ...(1)$$

And the mean of the squares of the deviations from mean is called the variance and is denoted by σ^2 (read as sigma square). Therefore, the variance of n observations $x_1, x_2, ..., x_n$ is given by $\sigma^2=\frac{1}{n}\sum_{i=1}^{n}(x_i-\bar{x})^2$

Standard Deviation of a Discrete Frequency Distribution: Let the given discrete frequency distribution be

$x: x_1, x_2, x_3, ..., x_n$

$f: f_1, f_2, f_3, ..., f_n$

$$(\sigma)=\sqrt{\frac{1}{N}\sum_{i=1}^{n}f_i(x_i-\bar{x})^2}, \text{ where } N=\sum_{i=1}^{n}f_i.$$

Standard Deviation of a Continuous Frequency Distribution: The given continuous frequency distribution can be represented as a discrete frequency distribution by replacing each class with its mid-point. Then, the standard deviation is calculated by the technique adopted in the case of a discrete frequency distribution.

If there is a frequency distribution of n classes, each class defined by its mid-point x_i with frequency f_i, the standard deviation will be obtained by the formula

$$\sigma=\sqrt{\frac{1}{N}\sum_{i=1}^{n}f_i(x_i-\bar{x})^2},$$ where $\bar{x}$ is the mean of the distribution and $N=\sum_{i=1}^{n}f_i$.

Direct Formula for Standard Deviation

$$(\sigma^2)=\frac{1}{N}\sum_{i=1}^{n}f_i(x_i-\bar{x})^2$$

$$=\frac{1}{N}\sum_{i=1}^{n}f_i(x_i^2+\bar{x}^2-2\bar{x}x_i)$$

$$=\frac{1}{N}\left[\sum_{i=1}^{n}f_ix_i^2+\sum_{i=1}^{n}\bar{x}^2f_i-\sum_{i=1}^{n}2\bar{x}f_ix_i\right]$$

$$=\frac{1}{N}\left[\sum_{i=1}^{n}f_ix_i^2+\bar{x}^2\sum_{i=1}^{n}f_i-2\bar{x}\sum_{i=1}^{n}x_if_i\right]$$

$$=\frac{1}{N}\left[\sum_{i=1}^{n}f_ix_i+\bar{x}^2N-2\bar{x}.N\bar{x}\right]$$

$$\left[\text{Here }\frac{1}{N}\sum_{i=1}^{n}x_if_i=\bar{x}\text{ or }\sum_{i=1}^{n}x_if_i=N\bar{x}\right]$$

$$=\frac{1}{N}\sum_{i=1}^{n}f_ix_i^2+\bar{x}^2-2\bar{x}^2=\frac{1}{N}\sum_{i=1}^{n}f_ix_i^2-\bar{x}^2$$

$$\text{or }\sigma^2=\frac{1}{N}\sum_{i=1}^{n}f_ix_i^2-\left(\frac{\sum_{i=1}^{n}f_ix_i}{N}\right)^2$$

$$=\frac{1}{N^2}\left[N\sum_{i=1}^{n}f_ix_i^2-\left(\sum_{i=1}^{n}f_ix_i\right)^2\right]$$

Thus, standard deviation

$$(\sigma) = \frac{1}{N}\sqrt{N\sum_{i=1}^{n} f_i x_i^2 - \left(\sum_{i=1}^{n} f_i x_i\right)^2}$$

Shortcut Method to find Variance and Standard Deviation: Let the assumed mean be 'A' and the scale be reduced to $1/h$ times (where h is the width of the class interval).

Let the step-deviations or new values be y_i.

$$y_i = \frac{x_i - A}{h} \text{ or } x_i = A + hy_i \quad ...(1)$$

$$\bar{x} = \frac{\sum_{i=1}^{n} f_i x_i}{N} \quad ...(2)$$

Replacing x_i in (1) and (2), we get

$$\bar{x} = \frac{\sum_{i=1}^{n} f_i(A + hy_i)}{N} = \frac{1}{N}\left(\sum_{i=1}^{n} f_i A + \sum_{i=1}^{n} hf_i y_i\right)$$

$$= \frac{1}{N}\left(A\sum_{i=1}^{n} f + h\sum_{i=1}^{n} f_i y_i\right) = A \cdot \frac{N}{N} + h\frac{\sum_{i=1}^{n} f_i y_i}{N}$$

$$\left(\text{since} \sum_{i=1}^{n} f_i = N\right)$$

Thus, $\bar{x} = A + h\bar{y}$ ...(3)

Now, variance of the variable x

$$\sigma_x^2 = \frac{1}{N}\sum_{i=1}^{n} f_i(x_i - \bar{x})^2$$

$$= \frac{1}{N}\sum_{i=1}^{n} f_i(A + hy_i - A - h\bar{y})^2$$

(Using (1) and (3)) $= \frac{1}{N}\sum_{i=1}^{n} f_i h^2 (y_i - \bar{y})^2$

$= \frac{h^2}{N}\sum_{i=1}^{n} f_i(y_i - \bar{y})^2 = h^2 \times$ Variance of the variable y_i, that is, $\sigma_x^2 = h^2\sigma_y^2 \Rightarrow \sigma_x = h\sigma_y$ ...(4)

From (3) and (4), we have

$$\sigma_x = \frac{1}{N}\sqrt{N\sum_{i=1}^{n} f_i y_i^2 - \left(\sum_{i=1}^{n} f_i y_i\right)^2}$$

Example 1: Find the standard deviation and variance for the data 5, 8, 2, 9, 7.

Solution: Mean, $\bar{x} = \frac{\sum_{i=1}^{5} x_i}{n}$

$$= \frac{5+8+2+9+6}{5} = \frac{30}{5} = 6$$

The following table is obtained as:

x_i	$(x_i - \bar{x})$	$(x_i - \bar{x})^2$
5	–1	1
8	2	4
2	–4	16
9	3	9
6	0	0
		30

$$\text{Variance } (\sigma^2) = \frac{1}{n}\sum_{i=1}^{5}(x_i - \bar{x})^2 = \frac{1}{5} \times 30 = 6$$

Standard deviation $(\sigma) = \sqrt{6} = 2.45$

Example 2: Find the standard deviation and variance for the data.

x_i	9	10	14	18	20
f_i	3	4	7	10	6

Solution: The data is obtained in tabular form as follows:

x_i	f_i	$f_i x_i$	$(x_i - \bar{x})$	$(x_i - \bar{x})^2$	$f_i(x_i - \bar{x})^2$
9	3	27	–6.5	42.25	126.75
10	4	40	–5.5	30.25	121
14	7	98	–1.5	2.25	15.75
18	10	180	2.5	6.25	62.5
20	6	120	4.5	20.25	121.5
	30	465			447.5

$N = 30$, 65

$$\therefore \bar{x} = \frac{\sum_{i=1}^{5} f_i x_i}{N} = \frac{465}{30} = 15.5$$

$$\text{Variance} = (\sigma^2) = \frac{1}{N}\sum_{i=1}^{5} f_i(x_i - \bar{x})^2 = \frac{1}{30} \times 447.5 = 14.92$$

Standard deviation $(\sigma) = \sqrt{14.92} = 3.86$

Example 3: Find the mean and standard deviation using the shortcut method.

x_i	11	12	13	14	15
f_i	4	5	1	7	3

Solution: The data is obtained in tabular form as follows:

x_i	f_i	$y_i = \frac{x_i - 13}{1}$	y_i^2	$f_i y_i$	$f_i y_i^2$
11	4	–2	4	–8	16
12	5	–1	1	–5	5
13	1	0	0	0	0
14	7	1	1	7	7
15	3	2	4	6	12
	20			0	40

$$\text{Mean } \bar{x} = A + \frac{\sum_{i=1}^{5} f_i y_i}{N} \times h = 13 + \frac{0}{20} \times 1 = 13$$

$$\text{Variance, } \sigma^2 = \frac{h^2}{N^2}\left[N\sum_{i=1}^{9} f_i y_i^2 - (\sum_{i=1}^{9} f_i y_i)^2\right]$$

$$= \frac{1}{20^2}[20 \times 40 - 0^2] = 2.0$$

$\therefore$ Standard deviation $(\sigma) = \sqrt{2.0} = 1.41$

Example 4: Find the mean and variance for the following frequency distribution:

Classes	0–10	10–20	20–30	30–40	40–50
Frequencies	5	8	2	4	6

Solution:

Class	Frequency f_i	Mid-point x_i	$y_i = \frac{x_i - 25}{10}$	y_i^2	$f_i y_i$	$f_i y_i^2$
0–10	1	5	–2	4	–2	4
10–20	7	15	–1	1	–7	7
20–30	2	25	0	0	0	0
30–40	4	35	1	1	4	4
40–50	6	45	2	4	10	24
	25				5	39

$$\text{Mean } \bar{x} = A + \frac{\sum_{i=1}^{5} f_i y_i}{N} \times h = 25 + \frac{5}{25} \times 10 = 25 + 2 = 27$$

$$\text{Variance } (\sigma^2) = \frac{h^2}{N^2}\left[N\sum_{i=1}^{5} f_i y_i^2 - \left(\sum_{i=1}^{5} f_i y_i\right)^2\right]$$

$$= \frac{(10)^2}{(25)^2}[25 \times 39 - (5)^2] = \frac{1}{25}[975 - 25] = 38$$

Day 204

NCERT Exercise 15.2 (With Solutions)

Q1. Find the mean and deviation for the data 6, 7, 10, 12, 13, 4, 8, 12.

Solution:

$$\text{Mean, } \bar{x} = \frac{\sum_{i=1}^{8} x_i}{n}$$

$$= \frac{6+7+10+12+13+4+8+12}{8} = \frac{72}{8} = 9$$

The following table is obtained as:

x_i	$(x_i - \bar{x})$	$(x_i - \bar{x})^2$
6	–3	9
7	–2	4
10	–1	1
12	3	9
13	4	16
4	–5	25
8	–1	1
12	3	9
		74

$$\text{Variance } (\sigma^2) = \frac{1}{n}\sum_{i=1}^{8}(x_i - \bar{x})^2 = \frac{1}{8} \times 74 = 9.25$$

Q2. Find the mean and variance for the first n natural numbers.

Solution: The mean of the first n natural is calculated as follows:

$$\text{Mean} = \frac{\text{Sum of all observations}}{\text{Number of observations}}$$

$$\therefore \text{Mean} = \frac{\frac{n(n+1)}{2}}{n} = \frac{n+1}{2}$$

$$\text{Variance } (\sigma^2) = \frac{1}{n}\sum_{i=1}^{n}(x_i - \bar{x})^2 = \frac{1}{n}\sum_{i=1}^{n}\left[x_i - \left(\frac{n+1}{2}\right)\right]^2$$

$$= \frac{1}{n}\sum_{i=1}^{n} x_i^2 - \frac{1}{n}\sum_{i=1}^{n} 2\left(\frac{n+1}{2}\right)x_i + \frac{1}{n}\sum_{i=1}^{n}\left(\frac{n+1}{2}\right)^2$$

$$= \frac{1}{n}\frac{n(n+1)(2n+1)}{6} - \left(\frac{n+1}{n}\right)\left[\frac{n(n+1)}{2}\right] + \frac{(n+1)^2}{4n} \times n$$

$$= \frac{(n+1)(2n+1)}{6} - \frac{(n+1)^2}{2} + \frac{(n+1)^2}{4}$$

$$= \frac{(n+1)(2n+1)}{6} - \frac{(n+1)^2}{4}$$

$$= (n+1)\left[\frac{4n+2-3n-3}{12}\right]$$

$$= \frac{(n+1)(n-1)}{12} = \frac{n^2-1}{12}$$

Q3. Find the mean and variance for the first 10 multiples of 3.

Solution: The first 10 multiples of 3 are
3, 6, 9, 12, 15, 18, 21, 24, 27, 30
Since the number of observations, $n = 10$

$$\text{Mean } \bar{x} = \frac{\sum_{i=1}^{10} x_i}{10} = \frac{165}{10} = 16.5$$

The following table is obtained as:

x_i	$(x_i - \bar{x})$	$(x_i - \bar{x})^2$
3	–13.5	182.25
6	–10.5	110.25
9	–7.5	56.25
12	–4.5	20.25
15	–1.5	2.25
18	1.5	2.25
21	4.5	20.25
24	7.5	56.25
27	10.5	110.25
30	13.5	182.25
		742.5

$$\text{Variance } (\sigma^2) = \frac{1}{n}\sum_{i=1}^{10}(x_i - \bar{x})^2 = \frac{1}{10} \times 742.5 = 74.25$$

Q4. Find the mean and variance for the data.

x_i	6	10	14	18	24	28	30
f_i	2	4	7	12	8	4	3

Solution: The data is obtained in tabular form as follows:

x_i	fi	$f_i x_i$	$(x_i - \bar{x})$	$(x_i - \bar{x})^2$	$f_i(x_i - \bar{x})^2$
6	2	12	–13	169	338
10	4	40	–9	81	324
14	7	98	–5	25	175
18	12	216	–1	1	12
24	8	192	5	25	200
28	4	112	9	81	324
30	3	90	11	121	363
	40	760			1736

$$N = 40, \quad \sum_{i=1}^{7} f_i x_i = 760$$

$$\therefore \bar{x} = \frac{\sum_{i=1}^{7} f_i x_i}{N} = \frac{760}{40} = 19$$

$$\text{Variance} = (\sigma^2) = \frac{1}{N}\sum_{i=1}^{7} f_i(x_i - \bar{x})^2 = \frac{1}{40} \times 1736 = 43.4$$

Q5. Find the mean and variance for the data.

x_i	92	93	97	98	102	104	109
f_i	3	2	3	2	6	3	3

Solution: The data is obtained in tabular form as follows:

x_i	f_i	$f_i x_i$	$x_i - \bar{x}$	$(x_i - \bar{x})^2$	$f_i(x_i - \bar{x})^2$
9	3	276	–8	64	192
93	2	186	–7	49	98
97	3	291	–3	9	27
98	2	196	–2	4	8
102	6	612	2	4	24
104	3	312	4	16	48
109	3	327	9	81	243
	22	2200			640

$N = 22, \quad \sum_{i=1}^{7} f_i x_i = 2200$

$\therefore \bar{x} = \frac{1}{N}\sum_{i=1}^{7} f_i x_i = \frac{1}{22} \times 2200 = 100$

$\text{Variance } (\sigma^2) = \frac{1}{N}\sum_{i=1}^{7} f_i (x_i - \bar{x})^2 = \frac{1}{22} \times 640 = 29.09$

Q6. Find the mean and standard deviation using shortcut method.

x_i	60	61	62	63	64	65	66	67	68
f_i	2	1	12	29	25	12	10	4	5

Solution: The data is obtained in tabular form as follows:

x_i	f_i	$y_i = \frac{x_i - 64}{1}$	y_i^2	$f_i y_i$	$f_i y_i^2$
60	2	–4	16	–8	32
61	1	–3	9	–3	9
62	12	–2	4	–24	48
63	29	–1	1	–29	29
64	25	0	0	0	0
65	12	1	1	12	12
66	10	2	4	20	40
67	4	3	9	12	36
68	5	4	16	20	80
	100			0	286

$\text{Mean } \bar{x} = A + \frac{\sum_{i=1}^{9} f_i y_i}{N} \times h = 64 + \frac{0}{100} \times 1 = 64 + 0 = 64$

$\text{Variance, } \sigma^2 = \frac{h^2}{N^2}\left[N\sum_{i=1}^{9} f_i y_i^2 - (\sum_{i=1}^{9} f_i y_i)^2\right]$

$= \frac{1}{100^2}[100 \times 286 - 0] = 2.86$

$\therefore \text{Standard deviation } (\sigma) = \sqrt{2.86} = 1.69$

Q7. Find the mean and variance for the following frequency distribution.

Classes	0–30	30–60	60–90	90–120	120–150	150–180	180–210
Frequency	2	3	5	10	3	5	2

Solution:

Class	**Frequency f_i**	**Mid-point x_i**	$y_i = \frac{x_i - 105}{30}$	y_i^2	$f_i y_i$	$f_i y_i^2$
0–30	2	15	–3	9	–6	18
30–60	3	45	–2	4	–6	12
60–90	5	75	–1	1	–5	5
90–120	10	105	0	0	0	0
120–150	3	135	1	1	3	3
150–180	5	165	2	4	10	20
180–210	2	195	3	9	6	18
	30				2	76

$\text{Mean, } \bar{x} = A + \frac{\sum_{i=1}^{7} f_i y_i}{N} \times h = 105 + \frac{2}{30} \times 30 = 105 + 2 = 107$

$\text{Variance } (\sigma^2) = \frac{h^2}{N^2}\left[N\sum_{i=1}^{7} f_i y_i^2 - \left(\sum_{i=1}^{7} f_i y_i\right)^2\right]$

$= \frac{(30)^2}{(30)^2}[30 \times 76 - (2)^2] = 2280 - 4 = 2276$

Q8. Find the mean and variance for the following frequency distribution.

Classes	0–10	10–20	20–30	30–40	40–50
Frequencies	5	8	15	16	6

Solution:

Class	**Frequency f_i**	**Mid-point x_i**	$y_i = \frac{x_i - 25}{10}$	y_i^2	$f_i y_i$	$f_i y_i^2$
0–10	5	5	–2	4	–10	20
10–20	8	15	–1	1	–8	8
20–30	15	25	0	0	0	0
30–40	16	35	1	1	16	16
40–50	6	45	2	4	12	24
	50				10	68

$$\text{Mean } \bar{x} = A + \frac{\sum_{i=1}^{5} f_i y_i}{N} \times h = 25 + \frac{10}{50} \times 10 = 25 + 2 = 27$$

$$\text{Variance } (\sigma^2) = \frac{h^2}{N^2}\left[N\sum_{i=1}^{5} f_i y_i^2 - \left(\sum_{i=1}^{5} f_i y_i\right)^2\right]$$

$$= \frac{(10)^2}{(50)^2}[50 \times 68 - (10)^2] = \frac{1}{25}[3400 - 100] = 132$$

Q9. Find the mean, variance and standard deviation using the shortcut method.

Height in cm	No. of children
70–75	3
75–80	4
80–85	7
85–90	7
90–95	15
95–100	9
100–105	6
105–110	6
110–115	3

Solution:

Class interval	Frequency f_i	Mid-point x_i	$y_i = \frac{x_i - 92.5}{5}$	y_i^2	$f_i y_i$	$f_i y_i^2$
70–75	3	72.5	−4	16	−12	48
75–80	4	77.5	−2	9	−12	36
80–85	7	82.5	−2	4	−14	28
85–90	7	87.5	−1	1	−7	7
90–95	15	92.5	0	0	0	0
95–100	9	97.5	1	1	9	9
100–105	6	102.5	2	4	12	24
105–110	6	107.5	3	9	18	54
110–115	3	112.5	4	16	12	48
	60				6	254

Mean

$$\bar{x} = A + \frac{\sum_{i=1}^{9} f_i y_i}{N} \times h = 92.5 + \frac{6}{60} \times 5 = 92.5 + 0.5 = 93$$

$$\text{Variance } (\sigma^2) = \frac{h^2}{N^2}\left[N\sum_{i=1}^{9} f_i y_i^2 - (\sum_{i=1}^{9} f_i y_i)^2\right]$$

$$= \frac{(5)^2}{(60)^2}[60 \times 254 - (6)^2] = \frac{25}{3600}(15204) = 105.58$$

$\therefore$ Standard deviation $(\sigma) = \sqrt{105.58} = 10.27$

Q10. The diameters of circles (in mm) drawn in a design are given below:

Diameters	No. of children
33–36	15
37–40	17
41–44	21
45–48	22
49–52	25

Solution:

Class interval	Frequency f_i	Mid-point x_i	$y_i = \frac{x_i - 42.5}{4}$	f_i^2	$f_i y_i$	$f_i y_i^2$
32.5–36.5	15	34.5	−2	4	−30	60
36.5–44.5	17	38.5	−1	1	−17	17
40.5–44.5	21	42.5	0	0	0	0
44.5–48.5	22	46.6	1	1	1	22
48.5–52.5	25	50.5	2	4	4	100
	100					199

$N = 100, h = 4$

Let assumed mean A be 42.5

$$\text{Mean } \bar{x} = A + \frac{\sum_{i=1}^{5} f_i y_i}{N} \times h = 42.5 + \frac{25}{100} \times 4 = 43.5$$

$$\text{Variance } (\sigma^2) = \frac{h^2}{N^2}\left[N\sum_{i=1}^{5} f_i y_i^2 - (\sum_{i=1}^{5} f_i y_i)^2\right]$$

$$= \frac{16}{10000}[100 \times 199 - (25)^2] = \frac{16}{10000}[19900 - 625]$$

$$= \frac{16}{10000} \times 19275 = 30.84$$

$\therefore$ Standard deviation $(\sigma) = 5.55$

Day 205

Analysis of Frequency Distributions

Coefficient of Variation: The measure of variability which is independent of units is called the coefficient of variation (denoted as CV).

The coefficient of variation is defined as

$CV = \frac{\sigma}{\bar{x}} \times 100, \bar{x} \neq 0$, where $\sigma, \bar{x}$ are the standard deviation and mean of the data.

Note: The series with greater CV is said to be more variable than the other. The series with lesser CV is said to be more consistent than the other.

Comparison of Two Frequency Distributions with the Same Mean: Let $\bar{x}_1, \sigma_1$ be the mean and standard deviation of the first distribution, and $\bar{x}_2, \sigma_2$ be the mean and standard deviation of the second distribution.

Then CV (1st distribution) $= \frac{\sigma_1}{\bar{x}_1} \times 100$

And CV (2nd distribution) $= \frac{\sigma_2}{\bar{x}_2} \times 100$

Let assume $\bar{x}_1 = \bar{x}_2 = \bar{x}$

Therefore, CV (1st distribution) $= \frac{\sigma_1}{\bar{x}} \times 100$ ...(1)

And CV (2nd distribution) $= \frac{\sigma_2}{\bar{x}} \times 100$ ...(2)

It is clear from (1) and (2) that the two CVs can be compared on the basis of values of σ_1, σ_2 only.

Thus, we say that for two series with equal means, the series with greater standard deviation (or variance) is called more variable or dispersed than the other. Also, the series with lesser value of standard deviation (or variance) is said to be more consistent than the other.

NCERT Exercise 15.3 (With Solutions)

Q1. For the data given below, state which group is more variable, A or B.

Marks	10–20	20–30	30–40	40–50	50–60	60–70	70–80
Group A	9	17	32	33	40	10	9
Group B	10	20	30	25	43	15	7

Solution: The standard deviation of group A is calculated as follows:

Marks	Group A, f_i	Mid-point x_i	$y_i = \frac{x_i - 45}{10}$	y_i^2	$f_i y_i$	$f_i y_i^2$
10–20	9	15	–3	9	–27	81
20–30	17	25	–2	4	–34	68
30–40	32	35	–1	1	–32	32
40–50	33	45	0	0	0	0
50–60	40	55	1	1	40	40
60–70	10	65	2	4	20	40
70–80	9	75	3	9	27	81
	150				–6	342

$h = 10,\ N = 150,\ A = 45$

$$\text{Mean}\ \bar{x} = A + \frac{\sum_{i=1}^{5} f_i y_i}{N} \times h = 45 + \frac{(-6) \times 10}{150} = 45 - 0.4 = 44.6$$

$$\sigma_1^2 = \frac{h^2}{N^2}\left(N\sum_{i=1}^{7} f_i y_i^2 - \left(\sum_{i=1}^{7} f_i y_i\right)^2\right)$$

$$= \frac{100}{22500}(150 \times 342 - (-6)^2) = \frac{1}{225}(51264) = 227.84 \quad \therefore$$

Standard deviation $(\sigma_1) = \sqrt{227.84} = 15.09$

The standard deviation of group B is calculated as follows:

Marks	Group B, f_i	Mid-point x_i	$y_i = \frac{x_i - 45}{10}$	y_i^2	$f_i y_i$	$f_i y_i^2$
10–20	10	15	–3	9	–30	90
20–30	20	25	–2	4	–40	80
30–40	30	35	–1	1	–30	30
40–50	25	45	0	0	0	0
50–60	43	55	1	1	43	43
60–70	15	65	2	4	30	60
70–80	7	75	3	9	21	63
	150				–6	366

$$\text{Mean}\ \bar{x} = A + \frac{\sum_{i=1}^{5} f_i y_i}{N} \times h = 45 + \frac{(-6) \times 10}{150} = 45 - 0.4 = 44.6$$

$$\sigma_2^2 = \frac{h^2}{N^2}\left(N\sum_{i=1}^{7} f_i y_i^2 - \left(\sum_{i=1}^{7} f_i y_i\right)^2\right)$$

$$= \frac{100}{22500}[150 \times 366 - (-6)^2] = \frac{1}{225}[54864] = 243.84$$

$\therefore$ Standard deviation $(\sigma_2) = \sqrt{243.84} = 15.61$

Since the mean of both the groups is the same, the group with greater standard deviation will be more variable. Thus, group B has more variability in marks.

Q2. From the prices of shares X and Y below, find out which is more stable in value:

X	35	54	52	53	56	58	52	50	51	49
Y	108	107	105	105	106	107	104	103	104	101

Solution: The prices of the share X are 35, 54, 52, 53, 56, 58, 52, 50, 51, 49

And the number of observations, N = 10

$$\therefore \text{Mean, } \bar{x} = \frac{1}{N}\sum_{i=1}^{10} x_i = \frac{1}{10} \times 510 = 51$$

The following table is obtained corresponding to share X:

x_i	$(x_i - \bar{x})$	$(x_i - \bar{x})^2$
35	−16	256
54	3	9
52	1	1
53	2	4
56	5	25
58	7	49
52	1	1
50	−1	1
51	0	0
49	−2	4
		350

$$\text{Variance } (\sigma_1^2) = \frac{1}{N}\sum_{i=1}^{10}(x_i - \bar{x})^2 = \frac{1}{10} \times 350 = 35$$

$\therefore$ Standard deviation $(\sigma_1) = \sqrt{35} = 5.91$

$$\text{CV (Share X)} = \frac{\sigma_1}{x} \times 100 = \frac{5.91}{51} \times 100 = 11.58$$

Now, the prices of share Y are
108, 107, 105, 105, 106, 107, 104, 103, 104, 101

$$\therefore \text{Mean, } \bar{y} = \frac{1}{N}\sum_{i=1}^{10} y_i = \frac{1}{10} \times 1050 = 105$$

The following table is obtained corresponding to share Y:

y_i	$(y_i - \bar{y})$	$(y_i - \bar{y})^2$
108	3	9
107	2	4
105	0	0
105	0	0
106	1	1
107	2	4
104	−1	1
103	−2	4
104	−1	1
101	−4	16
		40

$$\text{Variance } (\sigma_2^2) = \frac{1}{N}\sum_{i=1}^{10}(y_i - \bar{y})^2 = \frac{1}{10} \times 40 = 4$$

$\therefore$ Standard deviation $(\sigma_2) = \sqrt{4} = 2$

$$\therefore \text{CV(Shares } Y) = \frac{\sigma_2}{y} \times 100 = \frac{2}{105} 100 = 1.9 = 11.58$$

CV of the prices of share X is greater than the CV of the prices of share Y.

Thus, the prices of share Y are more stable than the prices of share X.

Q3. An analysis of monthly wages paid to workers in two firms *A* and *B*, belonging to the same industry, gives the following results:

	Firms A	**Firm B**
No. of wage earners	586	648
Mean of monthly wages	Rs 5253	Rs 5253
Variance of the distribution of wages	100	121

(i) Which firm A or B plays the larger amount as monthly wages?

(ii) Which firm A or B shows greater variability in individual wages?

Solution:

(i) Monthly wages of firm A = 5253
Number of wage earners in firm A=586
$\therefore$ Total amount paid = Rs. 5253 × 586
Monthly wages of firm B = 5253
Number of wage earners in firm B =648

$\therefore$ Total amount paid Rs 5253×648

Thus, firm B pays the larger amount as monthly wages as the number of wage earners in firm B is more than the number of wage earners in firm A.

(ii) Variance of the distribution of wages in firm $A(\sigma_1^2) = 100$

$\therefore$ Standard deviation of the distribution of wages in firm $A((\sigma_1) = \sqrt{100} = 10$

Variance of the distribution of wages in firm $B(\sigma_2^2) = 121$

$\therefore$ Standard deviation of the distribution of wages in firm $B(\sigma_2^2) = \sqrt{121} = 11$

The mean of monthly wages of both the firms is the same, that is, 5253. Therefore, the firm with greater standard deviation will have more variability.

Thus, firm B has greater variability in the individual wages.

Q4. The following is the record of goals scored by team A in a football session:

No. of goals scored	0	1	2	3	4
No. of matches	1	9	7	5	3

For team B, mean number of goals scored per match was 2 with a standard deviation 1.25 goals. Find which team will be considered more consistent.

Solution: The mean and the standard deviation of goals scored by team A are calculated as follows.

No. of goals scored	No. of matched	$f_i x_i$	x_i^2	$f_i x_i^2$
0	1	0	0	0
1	9	9	1	9
2	7	14	4	28
3	5	15	9	45
4	3	12	16	48
	25	50		130

$$\text{Mean} = \frac{\sum_{i=1}^{5} f_i x_i}{\sum_{i=1}^{5} f_i} = 50/25 = 2$$

Thus, the mean of both the teams is the same.

$$\sigma = \frac{1}{N}\sqrt{N\sum f_i x_i^2 - (\sum f_i x_i)^2}$$

$$= \frac{1}{25}\sqrt{25 \times 130 - (50)^2} = \frac{1}{25}\sqrt{750} = \frac{1}{25} \times 27.38 = 1.09$$

The standard deviation of team B is 1.25 goals.

The average number of goals scored by both the teams is the same, that is, 2. Therefore, the team with lower standards will be more consistent.

Thus, team A is more consistent than team B.

Q5. The sum and sum of squares corresponding to length x (in cm) and weight y (in g) of 50 plant products are given below:

$$\sum_{i=1}^{50} x_i = 212, \quad \sum_{i=1}^{50} x_i^2 = 902.8,$$

$$\sum_{i=1}^{50} y_1 = 261, \quad \sum_{i=1}^{50} y_i^2 = 1457.6$$

Which varies more, the length or weight?

Solution:

Since $\sum_{i=1}^{50} x_i = 212, \quad \sum_{i=1}^{50} x_i^2 = 902.8$

And, N = 50

$$\therefore \text{Mean } \bar{x} = \frac{\sum_{i=1}^{50} y_i}{N} = 212/50 = 4.25$$

$$\text{Variance } (\sigma_1^2) = \frac{1}{N}\sum_{i=1}^{50}(x_i - \bar{x})^2$$

$$= \frac{1}{50}\sum_{i=1}^{50}(x_i - 4.24)^2 = \frac{1}{50}\sum_{i=1}^{50}[x_i^2 - 8.48x_i + 17.97]$$

$$= \frac{1}{50}\left[\sum_{i=1}^{50} x_i^2 - 8.48\sum_{i=1}^{50} x_i + 17.97 \times 50\right]$$

$$= \frac{1}{50}[902.8 - 8.48 \times (212) + 898.5]$$

$$= \frac{1}{50}[1801.3 - 1797.76] = \frac{1}{50} \times 3.54 = 0.07$$

Standard deviation, σ_1 (Length) $= \sqrt{0.07} = 0.26$

$$\text{CV(Length)} = \frac{\text{Standard deviation}}{\text{Mean}} \times 100$$

$$= \frac{0.26}{4.26} \times 100 = 6.13$$

Now, since $\sum_{i=1}^{50} y_i = 261, \quad \sum_{i=1}^{50} y_i^2 = 1457.6$

Mean $\bar{y} = \frac{1}{N}\sum_{i=1}^{50} y_i = \frac{1}{50}261 = 5.22$

Variance $(\sigma_2^2) = \frac{1}{N}\sum_{i=1}^{50}(y_i - \bar{y})^2$

$= \frac{1}{50}\sum_{i=1}^{50}(y_i - 5.22)^2 = \frac{1}{50}\sum_{i=1}^{50}[y_i^2 - 10.44y_i + 27.24]$

$= \frac{1}{50}\left[\sum_{i=1}^{50} y_i^2 - 10.44\sum_{i=1}^{50} y_i + 27.24\times 50\right]$

$= \frac{1}{50}[1457.6 - 10.44\times(261) + 1326]$

$= \frac{1}{50}[2819.6 - 2724.84] = \frac{1}{50}\times 94.76 = 1.89$

$\therefore$ Standard deviation, σ_2 (Weight) $= \sqrt{1.89} = 1.37$

$\therefore$ CV(Weight) $= \frac{\text{Standard deviation}}{\text{Mean}}\times 100$

$= \frac{1.37}{5.22}\times 100 = 26.24$

Thus, the CV of weights is greater than the CV of lengths. Therefore, weights vary more than the lengths.

Day 206

NCERT Miscellaneous Exercise (With Solutions)

Q1. The mean and variance of eight observations are 9 and 9.25, respectively. If six of the observations are 6, 7, 10, 12, 13, find the remaining two observations.

Solution: Let the remaining two observations be x and y. Therefore, the observation will be 6, 7, 10, 12, 13, x, y.

Mean $\bar{x} = \frac{6+7+10+12+13+x+y}{8} = 9$

$\Rightarrow 60 + x + y = 72 \Rightarrow x + y = 12$ ...(1)

Variance $= 9.25 = \frac{1}{n}\sum_{i=1}^{8}(x_i - \bar{x})^2$

$9.25 = \frac{1}{8}\left[\begin{matrix}(-3)^2+(-2)^2+(1)^2+(3)^2+(4)^2\\ +x^2+y^2-2\times 9(x+y)+2\times(9)^2\end{matrix}\right]$

$\Rightarrow 9.25 = \frac{1}{8}\left[\begin{matrix}9+4+1+9+9+16\\ +x^2+y^2-18(12)+162\end{matrix}\right]$ [Using (1)]

$\Rightarrow 9.25 = \frac{1}{8}[48 + x^2 + y^2 - 216 + 162]$

$\Rightarrow 9.25 = \frac{1}{8}[x^2 + y^2 - 6] \Rightarrow x^2 + y^2 = 80$ ...(2)

Squaring (1), we get $x^2 + y^2 + 2xy = 144$ ...(3)

From (2) and (3), we get $2xy = 64$ ...(4)

Subtracting (4) from (2), we get

$x^2 + y^2 - 2xy = 80 - 64 = 16 \Rightarrow x - y = \pm 4$ ...(5)

Therefore, solving (1) and (5), we get

$x = 8, y = 4$, when $x - y = 4$

$x = 4, y = 8$, when $x - y = -4$

Thus, the remaining observations are 4 and 8.

Q2. The mean and variance of 7 observations are 8 and 16, respectively. If five of the observations are 2, 4, 10, 12 and 14, find the remaining two observations.

Solutions: Let the remaining two observations be x and y.

The observations are 2, 4, 10, 12, 14, x, y

Mean, $\bar{x} = \frac{2+4+10+12+14+x+y}{7} = 8$

$\Rightarrow 56 = 42 + x + y \Rightarrow x + y = 14$ ...(1)

Variance $= 16 = \frac{1}{n}\sum_{i=1}^{7}(x_i - \bar{x})^2$

$16 = \frac{1}{7}\left[\begin{matrix}(-6)^2+(-4)^2+(2)^2+(4)^2+(6)^2\\ +x^2+y^2-2\times 8(x+y)+2\times(8)^2\end{matrix}\right]$

$\Rightarrow 16 = \frac{1}{7}\left[\begin{matrix}36+16+4+16+36\\ +x^2+y^2-16(14)+2(64)\end{matrix}\right]$ ...[Using (1)]

$\Rightarrow 16 = \frac{1}{7}[108 + x^2 + y^2 - 224 + 128]$

$\Rightarrow 16 = \frac{1}{7}[12 + x^2 + y^2]$

$\Rightarrow x^2 + y^2 = 112 - 12 = 100 \Rightarrow x^2 + y^2 = 100$ ...(2)

Squaring (1), we get $x^2 + y^2 + 2xy = 196$ ...(3)

From (2) and (3), we get

$2xy = 196 - 100 \Rightarrow 2xy = 196$ ...(4)

Subtracting (4) from (2), we get

$x^2 + y^2 - 2xy = 100 - 96 \Rightarrow (x - y)^2 = 4$

$\Rightarrow x - y = \pm 2$ ...(5)

Therefore, solving (1) and (5). we get

$x = 8, y = 6$ when $x - y = 2$

$x = 6, y = 8$ when $x - y = -2$

Thus, the remaining observations are 6 and 8.

Q3. The mean and standard deviation of six observations are 8 and 4, respectively. If each observation is multiplied by 3, find the new mean and new standard deviation of the resulting observations.

Solution: Let the observations be $x_1, x_2, x_3, x_4, x_5, x_6$.

Mean $\bar{x} = \dfrac{x_1 + x_2 + x_3 + x_4 + x_5 + x_6}{6} = 8$ ...(1)

If each observation is multiplied by 3 and the resulting observations are y_i, then $y_i = 3x \Rightarrow x_i = \dfrac{1}{3}y_i$, for $i = 1$ to 6

$\therefore$ New mean, $\bar{y} = \dfrac{y_1 + y_2 + y_3 + y_4 + y_5 + y_6}{6}$

$= \dfrac{3(x_1 + x_2 + x_3 + x_4 + x_5 + x_6)}{6} = 3 \times 8 = 24$ [Using (1)]

Standard deviation, $\sigma = \sqrt{\dfrac{1}{n}\sum_{i=1}^{6}(x_i - \bar{x})^2}$

$\therefore (4)^2 = \dfrac{1}{6}\sum_{i=1}^{6}(x_i - \bar{x})^2 \Rightarrow \sum_{i=1}^{6}(x_i - \bar{x})^2 = 96$ ...(2)

From (1) and (2), get,

$\bar{y} = 3\bar{x} \Rightarrow \bar{x} = \dfrac{1}{3}\bar{y}$

Substituting the values of $x_i, \bar{x}$ in (2), we get

$\sum_{i=1}^{6}\left(\dfrac{1}{3}y_i - \dfrac{1}{3}\bar{y}\right)^2 = 96 \Rightarrow \sum_{i=1}^{6}(y_i - \bar{y})^2 = 864$

Therefore, variance of new observations

$= \left(\dfrac{1}{6} \times 864\right) = 144$

Thus, the standard deviation of new observations is $\sqrt{144} = 12$

Q4. Given that $\bar{x}$ is the mean and σ^2 is the variance of n observations $x_1, x_2 ... x_n$. Prove that the mean and variance of the observations $ax_1, ax_2, ax_3 ... ax_n$ are $a\bar{x}, a^2\sigma^2$, respectively $(a \neq 0)$.

Solution: The given n observation are $x_1, x_2 ... x_n$.

Mean $= \bar{x}$

Variance $= \sigma^2$

$\therefore \sigma^2 = \dfrac{1}{n}\sum_{i=1}^{n} y_i(x_i - \bar{x})^2$ ...(1)

If each observation is multiplied by a and the new observations are y_i, then $y_i = ax_i \Rightarrow x_1 = \dfrac{1}{a}y_i$

$\therefore \bar{y} = \dfrac{1}{n}\sum_{i=1}^{n} y_i = \dfrac{1}{n}\sum_{i=1}^{n} ax_i = \dfrac{a}{n}\sum_{i=1}^{n} x_i = a\bar{x}\left(\bar{x} = \dfrac{1}{n}\sum_{i=1}^{n} x_i\right)$

Therefore, mean of the observations $ax_1, ax_2, ax_3 ... ax_n$ is $a\bar{x}$.

Substituting the values of $x_i, \bar{x}$ in (1), we get

$\sigma^2 = \dfrac{1}{n}\sum_{i=1}^{n}\left(\dfrac{1}{a}y_i - \dfrac{1}{a}\bar{y}\right)^2 \Rightarrow a^2\sigma^2 = \dfrac{1}{n}\sum_{i=1}^{n}(y_i - \bar{y})^2$

Thus, the variance of the observations $ax_1, ax_2, ax_3 ... ax_n$ is $a^2\sigma^2$.

Q5. The mean and standard deviation of 20 observations are found to be 10 and 2, respectively. On rechecking, it was found that an observation 8 was incorrect. Calculate the correct mean and standard deviation in each of the following cases:

(i) If the wrong item is omitted.

(ii) If it is replaced by 12.

Solution: (i) Number of observations $(n) = 20$

Incorrect mean =10

Incorrect standard deviation =2

$\bar{x} = \dfrac{1}{n}\sum_{i=1}^{20} x_i \Rightarrow 10 = \dfrac{1}{20}\sum_{i=1}^{20} x_i \Rightarrow \sum_{i=1}^{20} x_i = 200$

That is, incorrect sum of observations =200

Correct sum of observations $= 200 - 8 = 192$

Correct mean $= \dfrac{\text{correct sum}}{19} = \dfrac{192}{19} = 10.1$

Standard deviation

$\sigma = \sqrt{\dfrac{1}{n}\sum_{i=1}^{n} x_i^2 - \dfrac{1}{n^2}\left(\sum_{i=1}^{n} x_i\right)^2} = \sqrt{\dfrac{1}{n}\sum_{i=1}^{n} x_i^2 - (\bar{x})^2}$

$\Rightarrow 2 = \sqrt{\dfrac{1}{20}\text{incorrect}\sum_{i=1}^{n} x_i^2 - (10)^2}$

$\Rightarrow 4 = \dfrac{1}{20}\text{incorrect}\sum_{i=1}^{n} x_i^2 - 100$

$\Rightarrow \text{Incorrect}\sum_{i=1}^{n} x_i^2 = 2080$

$\therefore \text{Correct}\sum_{i=1}^{n} x_i^2 = \text{Incorrect}\sum_{i=1}^{n} x_i^2 - (8)^2$

$= 2080 - 64 = 2016$

$\therefore$ Correct standard deviation

$= \sqrt{\dfrac{\text{correct}\sum x_i^2}{n} - (\text{correct mean})^2}$

$= \sqrt{\dfrac{2016}{19} - (10.1)^2} = \sqrt{106.1 - 102.02} = \sqrt{4.09} = 2.02$

(ii) When 8 is replaced by 12.

Incorrect sum of observations =200

$\therefore$ Correct sum of observations $= 200 - 8 + 12 = 204$

$\therefore$ Correct mean $= \dfrac{\text{correct sum}}{20} = \dfrac{204}{20} = 10.2$

Standard deviation

$$\sigma = \sqrt{\frac{1}{n}\sum_{i=1}^{n} x_i^2 - \frac{1}{n^2}\left(\sum_{i=1}^{n} x_i\right)^2} = \sqrt{\frac{1}{n}\sum_{i=1}^{n} x_i^2 - (\bar{x})^2}$$

$$\Rightarrow 2 = \sqrt{\frac{1}{20}\text{incorrect}\sum_{i=1}^{n} x_i^2 - (10)^2}$$

$$\Rightarrow 4 = \frac{1}{20}\text{incorrect}\sum_{i=1}^{n} x_i^2 - 100$$

$$\Rightarrow \text{Incorrect}\sum_{i=1}^{n} x_i^2 = 2080$$

$$\therefore \text{Correct}\sum_{i=1}^{n} x_i^2 = \text{Incorrect}\sum_{i=1}^{n} x_i^2 - (8)^2 + (12)^2$$

$= 2080 - 64 + 144 = 2160$

$\therefore$ Correct standard deviation

$$= \sqrt{\frac{\text{correct}\sum x_i^2}{n} - (\text{correct mean})^2}$$

$$= \sqrt{\frac{2016}{20} - (10.2)^2} = \sqrt{108 - 104.04} = \sqrt{3.96} = 1.98$$

Q6. The mean and standard deviation of marks obtained by 50 students of a class in three subjects, Mathematics, Physics and Chemistry, are given below:

Subject	Mathematics	Physics	Chemistry
Mean	42	32	40.9
Standard deviation	12	15	20

Which of the three subjects shows the highest variability in marks and which shows the lowest?

Solution: Standard deviation of Mathematics =12
Standard deviation of Physics =15
Standard deviation of Chemistry =20
The coefficient of variation (CV) is given by
Standard deviation/Mean x 100.

$$\text{CV (in Mathematics)} = \frac{12}{42} \times 100 = 28.57$$

$$\text{CV (in physics)} = \frac{15}{32} \times 100 = 46.87$$

$$\text{CV (in Chemistry)} = \frac{20}{40.9} \times 100 = 48.89$$

The subject with greater CV is more variable than others. Therefore, the highest variability in marks is in Chemistry and the lowest variability in marks is in Mathematics.

Q7. The mean and standard deviation of a group of 100 observations were found to be 20 and 3, respectively. Later on, it was found that three observations were incorrect, which were recorded as 21, 21, 18. Find the mean and standard deviation if the incorrect observations are omitted.

Solution: Number of observations (n) =100
Incorrect mean ($\bar{x}$) = 20
Incorrect standard deviation (σ) = 3

$$\Rightarrow 20 = \frac{1}{100}\sum_{i=1}^{100} x_i \Rightarrow \sum_{i=1}^{100} x_i = 20 \times 100 = 2000$$

$\therefore$ Incorrect sum of observations =2000
$\Rightarrow$ Correct sum of observations
$= 2000 - 21 - 21 - 18 = 2000 - 60 = 1940$

$$\therefore \text{Correct mean} = \frac{\text{Correct sum}}{100 - 3} = \frac{1940}{97} = 20$$

Standard deviation

$$\sigma = \sqrt{\frac{1}{n}\sum_{i=1}^{n} x_i^2 - \frac{1}{n^2}\left(\sum_{i=1}^{n} x_i\right)^2} = \sqrt{\frac{1}{n}\sum_{i=1}^{n} x_i^2 - (\bar{x})^2}$$

$$\Rightarrow 3 = \sqrt{\frac{1}{100} \times \text{incorrect}\sum_{i=1}^{n} x_i^2 - (20)^2}$$

$$\Rightarrow \text{incorrect}\sum_{i=1}^{n} x_i^2 = 100(9 + 400) = 409900$$

$$\text{Correct}\sum_{i=1}^{n} x_i^2$$

$$= \text{Incorrect}\sum_{i=1}^{n} x_i^2 - (21)^2 - (21)^2 - (18)^2$$

$= 40900 - 441 - 441 - 324 = 39694$

$\therefore$ Correct standard deviation

$$= \sqrt{\frac{\text{correct}\sum x_i^2}{n} - (\text{correct mean})^2}$$

$$= \sqrt{\frac{39694}{97} - (20)^2} = \sqrt{409.216 - 400}$$

$= \sqrt{9.216} = 3.036$

Day 207

NCERT Exemplar Problems: Solutions (Q1–Q19)

Q1. Find the mean deviation about the mean of the distribution.

Size	20	21	22	23	24
Frequency	6	4	5	1	4

Solution:

Size	Frequency	$f_i x_i$	$d_i = x_i = x$	$f_i d_i$
20	6	120	1.65	9.90
21	4	84	0.65	2.60
22	5	110	0.35	1.75
23	1	23	1.35	1.35
24	4	96	2.35	9.40
Total	20	433		25

Now, $\bar{x} = \dfrac{\sum f_i x_i}{\sum f_i} = \dfrac{433}{20} = 21.65$

$\therefore \text{MD} = \dfrac{\sum f_i |x_i - \bar{x}|}{\sum f_i} = 25/20 = 1.25$

Q2. Find the mean deviation about the median of the following distribution.

Marks obtained	10	11	12	14	15
Number of students	2	3	8	3	4

Solution:

Marks obtained	f_i	Cf	$d_i = \lvert x_i - Me \rvert$	$f_i d_i$
10	2	2	2	4
11	3	5	1	3
12	8	13	0	0
14	3	16	2	6
15	4	20	3	12
Total	$\sum f_i$ = 20			$\sum f_i d_i = 25$

Now, $M_e = \left(\dfrac{20+1}{2}\right)$th item $= \left(\dfrac{21}{2}\right) = 10.5$ item

$\therefore M_e = 12$

$\therefore \text{MD} = \dfrac{\sum f_i d_i}{\sum f_i} = 25/20 = 1.25$

Q3. Calculate the mean deviation about the mean of the set of first n natural numbers when n is an odd number.

Solution: Let the first n natural numbers when n is odd be 1, 2, 3, 4,.........n [odd].

Mean $\bar{x} = \dfrac{1+2+3+...+n}{n} = \dfrac{n(n+1)}{2n} = \dfrac{n+1}{2}$

$$\therefore \text{MD} = \frac{\left|1-\frac{n+1}{2}\right| + \left|2-\frac{n+1}{2}\right| + \left|3-\frac{n+1}{2}\right| + ... + \left|n-\frac{n+1}{2}\right|}{n}$$

$$= \frac{2}{n}\left[1+2+.....+\frac{n-3}{2}+\frac{n-1}{2}\right]\left(\frac{n-1}{2}\right) \text{ terms}$$

$$= \frac{2}{n}\left[\frac{\left(\frac{n-1}{2}\right)\left(\frac{n-1}{2}+1\right)}{2}\right]$$

$\left[\because \text{Sum of first } n \text{ natural numbers} = \dfrac{n(n+1)}{2}\right]$

$$= \frac{2}{n}\cdot\frac{1}{2}\left[\left(\frac{n-1}{2}\right)\left(\frac{n+1}{2}\right)\right] = \frac{1}{n}\left(\frac{n^2-1}{4}\right) = \frac{n^2-1}{4n}$$

Q4. Calculate the mean deviation about the mean of the set of first n natural numbers when n is an even number.

Solution: Let the first n natural numbers when n is even be 1, 2, 3, 4,.................n [even]

$\therefore$ Mean $\bar{x} = \dfrac{1+2+3+...+n}{n} = \dfrac{n(n+1)}{2n} = \dfrac{n+1}{2}$

$$\text{MD} = \frac{1}{n}\left[\left|1-\frac{n+1}{2}\right| + \left|2-\frac{n+1}{2}\right| + \left|3-\frac{n+1}{2}\right|\right]$$

$$= \frac{2}{n}\left[\frac{1}{2}+\frac{3}{2}+...+\frac{n-1}{2}\right]\left(\frac{n}{2}\right) \text{ terms}$$

$$= \frac{1}{n}\cdot\left(\frac{n}{2}\right)^2 \quad [\because \text{Sum of first } n \text{ natural numbers} = n^2]$$

$$= \frac{1}{n}\cdot\frac{n^2}{4} = n/4$$

Q5. Find the standard deviation of the first n natural numbers.

Solution:

x_i	1	2	3	4	5	...	n
x_i^2	1	4	9	16	25	...	n^2

Now, $\sum x_i = 1+2+3+4+....+n = \dfrac{n(n+1)}{2}$

$\sum x_i^2 = 1^2+2^2+3^2+....+n^2 = \dfrac{n(n+1)(2n+1)}{6}$

$$\because \sigma = \sqrt{\frac{\sum x_i^2}{N} - \left(\frac{\sum x_i}{N}\right)^2}$$

$$= \sqrt{\frac{n(n+1)(2n+1)}{6n} - \frac{n^2(n+1)^2}{4n^2}}$$

$$= \sqrt{\frac{(n+1)(2n+1)}{6} - \frac{(n+1)^2}{4}}$$

$$= \sqrt{\frac{2(2n^2+3n+1)-3(n^2+2n+1)}{12}}$$

$$= \sqrt{\frac{4n^2+6n+2-3n^2-6n-3}{12}} = \sqrt{\frac{n^2-1}{12}}$$

Q6. The mean and standard deviation of some data for the time taken to complete a test are calculated with the following results:
Number of observations = 25, mean = 18.2 s, standard deviation = 3.25 s
Further, another set of 15 observations $x_1 x_2 ..... x_{15}$, also in seconds, is now available and we have $\sum_{i=1}^{15} x_i = 279, \sum_{i=1}^{15} x_i^2 = 5524$. Calculate the standard deviation based on all 40 observations.

Solution: $n_i = 25, \overline{x_i} = 18.2, \sigma_1 = 3.25, n_2 = 15,$

$\sum_{i=1}^{15} x_i = 279, \sum_{i=1}^{15} x_1^2 = 5524$

For the first set, $\sum x_i = 25 \times 18.2 = 455$

$$\therefore \sigma_1^2 = \frac{\sum x_i^2}{25} - (18.2)^2 \Rightarrow (3.25)^2 = \frac{\sum x_i^2}{25} - 331.24$$

$$\Rightarrow 10.5625 + 331.24 = \frac{\sum x_i^2}{25}$$

$$\Rightarrow \sum x_i^2 = 25 \times (10.5625 + 331.24)$$

$= 25 \times 341.8025 = 8545.0625$

For the combined SD of the 40 observations $n = 40$.

Now $\sum x_i^2 = 5524 + 8545.0625 = 14069.0625$

$\sum x_i = 455 + 279 = 734$

$$\therefore \text{SD} = \sqrt{\frac{14069.0625}{40} - \left(\frac{734}{40}\right)^2}$$

$= \sqrt{351.726 - (18.35)^2}$

$= \sqrt{351.726 - 336.7225} = \sqrt{15.0035} = 3.87$

Q7. The mean and standard deviation of a set of n_1 observations are $\overline{x_1}, s_1$, respectively, while the mean and standard deviation of another set of n_2 observation are $\overline{x_2}, s_2$, respectively. Show that the standard deviation of the combined set of $(n_1 + n_2)$ observations is given by

$$\text{SD} = \sqrt{\frac{n_1(s_1)^2 + n_2(s_2)^2}{n_1+n_2} + \frac{n_1 n_2(\overline{x_1} - \overline{x_2})^2}{(n_1 - n_2)^2}}$$

Solution: Let the observations be x_i and y_j, where $i = 1,2,3,....n_1,\ j = 1,2,3,...,n_2$

$$\therefore \overline{x_1} = \frac{1}{n_1}\sum_{i=1}^{n_1} x_i,\quad \overline{x_2} = \frac{1}{n_2}\sum_{i=1}^{n_2} y_j$$

$$\Rightarrow \sigma_1^2 = \frac{1}{n_1}\sum_{i=1}^{n_1}(x_i - \overline{x_1})^2, \sigma_2^2 = \frac{1}{n_2} = \frac{1}{n_2}\sum_{j=1}^{n}(y_j - \overline{x_2})^2$$

Now, mean $\overline{x}$ of the given series is given by

$$\overline{x} = \frac{1}{n_1+n_2}\left[\sum_{i=1}^{n_1} x_i + \sum_{j=1}^{n_2} y_j\right] = \frac{n_1\overline{x_1} + n_2\overline{x_2}}{n_1+n_2}$$

The variance σ^2 of the combined series is given by

$$\sigma^2 = \frac{1}{n_1+n_2}\left[\sum_{i=1}^{n_1}(x_i - \overline{x})^2 + \sum_{j=1}^{n_2}(y_j - \overline{x})^2\right]$$

Now, $\sum_{i=1}^{n_1}(x_i - \overline{x})^2 = \sum_{i=1}^{n_1}(x_i - \overline{x}_j + \overline{x}_j - \overline{x})^2$

$$= \sum_{i=1}^{n_1}(x_i - \overline{x}_j)^2 + n_1(\overline{x}_j - \overline{x})^2 + 2(\overline{x}_j - \overline{x})\sum_{i=1}^{n_1}(x_i - \overline{x}_j)^2$$

But $\sum_{i=1}^{n_1}(x_i - \overline{x}_j) = 0$

[Since algebraic sum of the deviation of values of the first series from their mean is zero]

Also, $\sum_{i=1}^{n_1}(x_i - \overline{x})^2 = n_1 s_1^2 + n_1(\overline{x_1} - \overline{x})^2 = n_1 d_1^2$ where,

$d_1 = (\overline{x_1} - \overline{x})$

S i m i l a r l y ,

$\sum_{j=1}^{n_2}(y_j - \overline{x})^2 = \sum_{j=1}^{n_2}(y_j - \overline{x}_i + \overline{x}_i - \overline{x})^2 = n_2 s_2^2 + n_2 d_2^2$

where $d_2 = \overline{x_2} - \overline{x}$

Combined SD, $\sigma = \sqrt{\dfrac{[n_1(s_1^2+d_1^2)+n_2(s_2^2+d_2^2)]}{n_1+n_2}}$

where $d_1 = \bar{x}_1 - \bar{x} = \bar{x}_1 - \left(\dfrac{n_1\bar{x}_1 + n_2\bar{x}_2}{n_1+n_2}\right) = \dfrac{n_2(\bar{x}_1 - \bar{x}_2)}{n_1+n_2}$

$d_2 = \bar{x}_2 - \bar{x} = \bar{x}_2 - \left(\dfrac{n_1\bar{x}_1 + n_2\bar{x}_2}{n_1+n_2}\right) = \dfrac{n_1(\bar{x}_2 - \bar{x}_1)}{n_1+n_2}$

$\therefore \sigma^2 = \dfrac{1}{n_1+n_2}$

$\left[n_1 s_1^2 + n_2 s_2^2 + \dfrac{n_1 n_2(\bar{x}_1 - \bar{x}_2)^2}{(n_1+n_2)^2} + \dfrac{n_2 n_1(\bar{x}_2 - \bar{x}_1)^2}{(n_1+n_2)}\right]$

Thus, $\sigma = \sqrt{\dfrac{n_1 s_1^2 + n_2 s_2^2}{n_1+n_2} + \dfrac{n_1 n_2(\bar{x}_1 - \bar{x}_2)^2}{(n_1+n_2)^2}}$

Q8. Two sets each of 20 observations have the same standard deviation, 5. The first set has mean 17 and the second mean 22.
Determine the standard deviation of the x sets obtained by combining the given two sets.

Solution:

$n_1 = 20, \sigma_1 = 5, \bar{x}_1 = 17, n_2 = 20, \sigma_2 = 5, \bar{x}_2 = 22$

Therefore, $\sigma = \sqrt{\dfrac{n_1 s_1^2 + n_2 s_2^2}{n_1+n_2} + \dfrac{n_1 n_2(\bar{x}_1 - \bar{x}_2)^2}{(n_1+n_2)^2}}$

$= \sqrt{\dfrac{20\times(5)^2 + 20\times(5)^2}{20+20} + \dfrac{20\times 20(17-22)^2}{(20+20)^2}}$

$= \sqrt{\dfrac{1000}{40} + \dfrac{400\times 25}{1600}} = \sqrt{25 + \dfrac{25}{4}} = \sqrt{\dfrac{125}{4}}$

$= \sqrt{31.25} = 5.59$

Q9. The frequency distribution

X	A	2A	3A	4A	5A	6A
F	2	1	1	1	1	1

where A is a positive integer, has a variance of 160. Determine the value of A.

Solution:

x	f_i	$f_i x_i$	$f_i x_i^2$
A	2	2A	$2A^2$
2A	1	2A	$4A^2$
3A	1	3A	$9A^2$
4A	1	4A	$16A^2$
5A	1	5A	$25A^2$
6A	1	6A	$36A^2$
Total	7	22A	$94A^2$
	$n = 7$	$\sum f_i n_i = 22A$	$\sum f_i n_i^2 = 92A^2$

$\therefore \sigma^2 = \dfrac{\sum f_i x_i^2}{n} - \left(\dfrac{\sum f_i x_i}{n}\right)^2$

$\Rightarrow 160 = \dfrac{92A^2}{7} - \left(\dfrac{22A}{7}\right)^2 \Rightarrow 160 = \dfrac{92A^2}{7} - \dfrac{484A^2}{49}$

$\Rightarrow 160 = (644-484)\dfrac{A^2}{49} \Rightarrow 160 = \dfrac{160A^2}{49}$

$\Rightarrow A^2 = 49 \Rightarrow \therefore A = 7$

Q10. For the frequency distribution

x	2	3	4	5	6	7
f	4	9	16	14	11	6

find the standard distribution.

Solution:

x_i	f_i	$d_i = x_i - 4$	$f_i d_i$	$f_i d_i^2$
2	4	−2	−8	16
3	9	−1	−9	9
4	16	0	0	0
5	14	1	14	14
6	11	2	22	44
7	6	3	18	54
Total	60		$\sum f_i d_i = 37$	$\sum f_i d_i^2 = 137$

$\therefore \text{SD} = \sqrt{\dfrac{f_i d_i^2}{N} - \left(\dfrac{f_i d_i}{N}\right)^2} = \sqrt{\dfrac{137}{60} - \left(\dfrac{37}{60}\right)^2}$

$= \sqrt{2.2833 - (0.616)^2} = \sqrt{2.283 - 0.3794} = 1.38$

Q11. There are 60 students in a class. The following is the frequency distribution of the marks obtained by the students in a test.

Marks	0	1	2	3	4	5
Frequency	$x-2$	x	x^2	$(x+1)^2$	$2x$	$x+1$

where x is a positive integer. Determine the mean and standard deviation of the marks

Solution: Sum of frequencies is

$x-2+x+x^2+(x+1)^2+2x+x+1=60$

$\Rightarrow 2x-2+x^2+x^2+1+2x+2x+x+1=60$

$\Rightarrow 2x^2+7x=60 \Rightarrow 2x^2+7x-60=0$

$\Rightarrow 2x^2+15x-8x-60=0$

$\Rightarrow x(2x+15)-4(2x+15)=0 \Rightarrow (2x+15(x-4)=0$

$\Rightarrow x=-\frac{15}{2},4 \Rightarrow x=-\frac{15}{2}$

[Inadmissible] [$\because x \in I^+$] Therefore

x_i	f_i	$d_i = x_i - 3$	$f_i d_i$	$f_i d_i^2$
0	2	–3	–6	18
1	4	–2	–8	16
2	16	–1	–16	16
A = 3	25	0	0	0
4	8	1	8	8
5	5	2	10	20
Total	$\sum f_i = 60$		$\sum f_i d_i = -12$	$\sum f_i d_i^2 = 78$

$$\text{Mean} = A + \frac{\sum f_i d_i}{\sum f_i} = 3 + \left(\frac{-12}{60}\right) = 2.8$$

$$\sigma = \sqrt{\frac{f_i d_i^2}{\sum f_i} - \left(\frac{f_i d_i}{\sum f_i}\right)^2} = \sqrt{\frac{78}{60} - \left(\frac{-12}{60}\right)^2}$$

$$= \sqrt{1.3 - 0.04} = \sqrt{1.26} = 1.12$$

Q12. The mean life of a sample of 60 bulbs was 650 h and the standard deviation was 8 h. If a second sample of 80 bulbs has a mean life of 650 h and standard deviation 7 h, then find the overall standard deviation.

Solution: $n_1 = 60, \bar{x}_1 = 650, s_1 = 8$

$n_2 = 80, \bar{x}_2 = 660, s_2 = 7$

$$\therefore \sigma = \sqrt{\frac{n_1 s_1^2 + n_2 s_2^2}{n_1 + n_2} + \frac{n_1 n_2 (\bar{x}_1 - \bar{x}_2)^2}{(n_1 + n_2)^2}}$$

$$= \sqrt{\frac{60 \times (8)^2 + 80 \times (7)^2}{60 + 80} + \frac{60 \times 80 (650 - 660)^2}{(60 + 80)^2}}$$

$$= \sqrt{\frac{6 \times 64 + 8 \times 49}{14} + \frac{60 \times 80 \times 100}{140 \times 140}}$$

$$= \sqrt{\frac{192 + 196}{7} + \frac{1200}{49}} = \sqrt{\frac{383}{7} + \frac{1200}{49}}$$

$$= \sqrt{\frac{2716 + 1200}{49}} = \sqrt{\frac{3916}{49}} = \frac{62.58}{7} = 8.9$$

Q13. If the mean and standard deviation of 100 items are 50 and 4, respectively, then find the sum of all the items and the sum of the squares of the items.

Solution: $\bar{x} = 50, n = 100, \sigma = 4$

$$\therefore \frac{\sum x_i}{100} = 50 \Rightarrow \sum x_i = 5000,$$

$$\sigma^2 = \frac{\sum f_i x_i^2}{\sum f_i} - \left(\frac{\sum f_i x_i}{\sum f_i}\right)^2$$

$$\Rightarrow (4)^2 = \frac{\sum f_i x_i^2}{100} - (50)^2 \Rightarrow 16 = \frac{\sum f_i x_i^2}{100} - 2500$$

$$\Rightarrow \frac{\sum f_i x_i^2}{100} = 16 + 2500 = 2516$$

$$\Rightarrow \sum f_i x_i^2 = 251600$$

Q14. If for distribution $\sum (x-5) = 3, \sum (x-5)^2 = 43$ and total number of items is 18, find the mean and standard deviation.

Solution: $n = 18, \sum (x-5) = 3, \sum (x-5)^2 = 43$

$$\therefore \text{Mean} = A + \frac{\sum (x-5)}{18} = 5 + \frac{3}{18} = 5 + 0.166 = 5.17$$

$$\text{SD} = \sqrt{\frac{\sum (x-5)^2}{n} - \left(\frac{\sum (x-5)}{n}\right)^2} = \sqrt{\frac{43}{18} - \left(\frac{3}{18}\right)^2}$$

$$= \sqrt{.3944 - (0.166)^2} = \sqrt{2.3944 - 0.2755} = 1.59$$

Q15. Find the mean deviation of the frequency distribution given below.

x	$1 \le x \le 3$	$3 \le x \le 5$	$5 \le x \le 7$	$7 \le x \le 10$
f	6	4	5	1

Solution:

x	f_i	x_i	$f_i x_i$	$f_i x_i^2$
1–3	6	2	12	24
3–5	4	4	16	64
5–7	5	6	30	180
7–10	1	8.5	8.5	72.25
Total	$n = 16$		$\sum f_i x_i = 66.5$	$\sum f_i x_i^2 = 340.25$

$\therefore$ Mean $= \frac{\sum f_i x_i}{\sum f_i} = \frac{66.5}{16} = 4.15$

Variance $= \sigma^2 = \frac{\sum f_i x_i^2}{\sum f_i} - \left(\frac{\sum f_i x_i}{\sum f_i}\right)^2$

$= \frac{340.25}{16} - (4.15)^2 = 21.2656 - 17.2225 = 4.043$

Q16. Calculate the mean deviation about the mean for the following frequency distribution.

Class interval	0–4	4–8	8–12	12–16	16–20
Frequency	4	6	8	5	2

Solution:

Class interval	f_i	x_i	$f_i x_i$	$d_i = \lvert x_i - \bar{x}\rvert$	$f_i d_i$
0–4	4	2	8	7.2	28.8
4–8	6	6	36	3.2	19.2
8–12	8	10	80	0.3	6.4
12–16	5	14	70	4.8	24.0
16–20	2	18	36	8.8	17.6
Total	$\sum f_i = 25$		$\sum f_i x_i = 230$		$\sum f_i d_i = 96$

$\therefore$ Mean $= \frac{\sum f_i x_i}{\sum f_i} = \frac{230}{25} = 9.2$

Mean deviation $= \frac{\sum f_i d_i}{\sum f_i} = \frac{96}{25} = 3.84$

Q17. Calculate the mean deviation from the median of the following data.

Class interval	0–6	6–12	12–18	18–24	24–30
Frequency	4	5	3	6	2

Solution:

Class interval	f_i	x_i	cf	$d_i = \lvert x_i - m_d\rvert$	$f_i d_i$
0–6	4	3	4	11	44
6–12	5	9	9	5	25
12–18	3	15	12	1	3
18–24	6	21	18	7	42
24–30	2	27	20	13	26
Total	N=20				$\sum f_i d_i = 140$

$\because \frac{N}{2} = \frac{20}{2} = 10$

Therefore, the median class is 12–18.

Median $= l + \frac{\frac{N}{2} - cf}{f} \times i = 12 + \frac{6}{3}(10 - 9) = 14$

$\Rightarrow$ MD $= \frac{\sum f_i d_i}{\sum f_i} = 140/20 = 7$

Q18. Determine the mean and standard deviation for the following distribution.

Marks	2	3	4	5	6	7	8	9	10	11	12	13	14	15	16
Frequency	1	6	6	8	8	2	2	3	0	2	1	0	0	0	1

Solution:

Marks	f_i	$f_i x_i$	$d_i = x_i - \bar{x}$	$f_i d_i$	$f_i d_i^2$
2	1	2	$2 - 6 = -4$	–4	16
3	6	18	$3 - 6 = -3$	–18	54
4	6	24	$4 - 6 = -2$	–12	24
5	8	40	$5 - 6 = -1$	–8	8
6	8	48	$6 - 6 = 0$	0	0
7	2	14	$7 - 6 = 1$	2	2
8	2	16	$8 - 6 = 2$	4	8
9	3	27	$9 - 6 = 3$	9	27
10	0	0	$10 - 6 = 4$	0	0
11	2	22	$11 - 6 = 5$	10	50
12	1	12	$12 - 6 = 6$	6	36
13	0	0	$13 - 6 = 7$	0	0

Marks	f_i	$f_i x_i$	$d_i = x_i - \bar{x}$	$f_i d_i$	$f_i d_i^2$
14	0	0	$14-6=8$	0	0
15	0	0	$15-6=9$	0	0
16	1	16	$16-6=10$	10	100
Total	$\sum f_i = 40$	$\sum f_i x_i = 239$		$\sum f_i d_i = -1$	$\sum f_i x_i^2 = 325$

$\therefore$ Mean $\bar{x} = \dfrac{\sum f_i x_i}{\sum f_i} = \dfrac{239}{40} = 5.975 \approx 6$ and

$$\sigma = \sqrt{\frac{\sum f_i d_i^2}{\sum f_i} - \left(\frac{\sum f_i d_i}{\sum f_i}\right)^2} = \sqrt{\frac{325}{40} - \left(\frac{-1}{40}\right)^2}$$

$= \sqrt{8.125 - 0.000625} = \sqrt{8.124375} = 2.85$

Q19. The weight of coffee in 70 jars is shown in the following table:

Weight (in g)	Frequency
200–201	13
201–202	27
202–203	18
203–204	10
204–205	1
205–206	1

Determine variance and standard deviation of the above distribution.

Solution:

CI	f_i	x_i	$d_i = x_i - \bar{x}$	$f_i d_i$	$f_i d_i^2$
200–201	13	200.5	–2	–26	52
201–202	27	201.5	–1	–27	27
202–203	18	202.5	0	0	0
203–204	10	203.5	1	10	10
204–205	1	204.5	2	2	4
205–206	1	205.5	3	3	9
	$\sum f_i = 70$			$\sum f_i d_i = -38$	$\sum f_i d_i^2 = 102$

$$\therefore \sigma^2 = \frac{\sum f_i d_i^2}{\sum f_i} - \left(\frac{\sum f_i d_i}{\sum f_i}\right)^2 = \frac{102}{70} - \left(\frac{-38}{70}\right)^2$$

$= 1.4571 - 0.2916 = 1.1655 \Rightarrow \sigma = \sqrt{1.1655} = 1.08g$

Day 208

NCERT Exemplar Problems: Solutions (Q20–Q46)

Q20. Determine the mean and standard deviation of the first n items of an AP whose first term is a and common difference is d.

Solution:

x_i	$x_i = a$	$(x_i - a)^2$
a	0	0
$a+d$	d	d^2
$a+2d$	$2d$	$4d^2$
...	...	$9d^2$

x_i	$x_i = a$	$(x_i - a)^2$
...	...	...
...	...	...
$= a+(n-1)d$	$(n-1)d$	$(n-1)^2 d^2$
$\sum x_i = \frac{n}{2}[2a+(n-1]$		

$\because$ Mean $= \dfrac{\sum x_i}{n} = \dfrac{1}{n}\left[\dfrac{n}{2}(2a+(n-1)d\right]$

$= a + \dfrac{(n-1)}{2}d$

$\therefore \sum (x_i - a) = d[1+2+3+\ldots+(n-1)d]$

$= d\dfrac{(n-1)n}{2}$

$$\sum(x_i-a)^2 = d^2[1^2+2^2+3^2+\ldots+(n-1)^2]$$

$$= \frac{d^2(n-1)n(2n-1)}{6}$$

$$\sigma = \sqrt{\frac{(x_i-a)^2}{n} - \left(\frac{x_i-a)}{n}\right)^2}$$

$$= \sqrt{\frac{d^2(n-1)(n)(2n-1)}{6n} - \left[\frac{d(n-1)n}{4}\right]^2}$$

$$= \sqrt{\frac{d^2(n-1)(2n-1)}{6} - \frac{d^2(n-1)^2}{4}}$$

$$= d\sqrt{\frac{(n-1)(2n-1)}{6} - \frac{(n-1)^2}{4}}$$

$$= d\sqrt{\frac{(n-1)}{2}\left(\frac{2n-1}{3} - \frac{n-1}{4}\right)}$$

$$= d\sqrt{\frac{(n-1)}{2}\left[\frac{2n-2-3n+3}{3}\right]}$$

$$= d\sqrt{\frac{(n-1)(n+1)}{12}} = d\sqrt{\frac{(n^2-1)}{12}}$$

Q21. The following are the marks obtained, out of 100, by two students Ravi and Hashina in 10 tests:

Ravi	25	50	45	30	70	42	36	48	35	60
Hashina	10	70	50	20	95	55	42	60	48	80

Who is more intelligent and who is more consistent?

Solution: For Ravi

x_i	$d_i = x_i - 45$	d_i^2
25	−20	400
50	5	25
45	0	0
30	−15	225
70	25	625
42	−3	9
36	−9	81
48	3	9
35	−10	100
60	15	225
Total	$\sum d_i = -14$	$\sum d^2 i = 1699$

$$\sigma = \sqrt{\frac{\sum d^2 i}{n} - \left(\frac{\sum d_i}{n}\right)^2} = \sqrt{\frac{1699}{10} - \left(\frac{-14}{10}\right)^2}$$

$$= \sqrt{16.9 - 0.0196} = \sqrt{169.88} = 13.03$$

Now, $\bar{x} = A + \frac{\sum d_i}{\sum f_i} = 45 - \frac{14}{10} = 43.6$

For Hashina

x_i	$d_i = x_i - 55$	d_i^2
10	−45	2025
70	25	625
50	−5	25
20	−35	1225
95	40	1600
55	0	0
42	−13	169
60	5	25
48	−7	49
80	25	625
Total	$\sum d_i = 0$	$\sum d_i^2 = 6368$

$\because$ Mean = 55

$$\therefore \sigma = \sqrt{\frac{6368}{10}} = \sqrt{636.8} = 25.2$$

For Ravi, CV $= \frac{\sigma}{x} 100 = \frac{13.03}{43.6} 100 = 29.88$

For Hashina, CV $= \frac{\sigma}{x} 100 = \frac{25.2}{55} 100 = 45.89$

Hence, Hashina is more consistent and intelligent.

Q22. The mean and standard deviation of 100 observations were found to be 40 and 10, respectively. If at the time of calculation, two observations were wrongly taken as 30 and 70 in place of 3 and 27, respectively, find the correct standard deviation.

Solution: $n = 100, \bar{x} = 40, \sigma = 10, \bar{x} = 40$

$$\therefore \frac{\sum x_i}{n} = 40 \Rightarrow \frac{\sum x_i}{100} = 40 \Rightarrow \sum x_i = 4000$$

Now, corrected $\sum x_i = 4000 - 30 - 70 + 3 + 27$

$= 3930$

Corrected mean $= \dfrac{2930}{100} = 39.3$

Now, $\sigma^2 = \dfrac{\sum x_i^2}{n} - (40)^2 \Rightarrow 100 = \dfrac{\sum x_i^2}{100} - 1600$

$\Rightarrow \sum x_i^2 = 170000$

Now, corrected

$\sum x_i^2 = 170000 - (30)^2 - (70)^2 + 3^2 + (27)^2$

$\therefore$ Corrected $\sigma = \sqrt{\dfrac{164939}{100} - (39.3)^2}$

$= \sqrt{1649.39 - 39.3 \times 39.3} = \sqrt{1649.39 - 1544.49}$

$= \sqrt{104.9} = 10.24$

Q23. While calculating the mean and deviation of 10 readings, a student wrongly used the reading 52 for the correct reading 25. He obtained the mean and variance as 45 and 16, respectively. Find the correct mean and the variance.

Solution: $n = 10, \bar{x} = 45, \sigma^2 = 16$

$\therefore \bar{x} = 45 \Rightarrow \dfrac{\sum x_i}{n} = 45 \Rightarrow \dfrac{\sum x_i}{10} = 45 \Rightarrow \sum x_i = 450$

Corrected $\sum x_i = 450 - 52 + 25 = 423$

$\therefore \bar{x} = \dfrac{423}{10} = 42.3 \Rightarrow \sigma^2 = \dfrac{\sum x_i^2}{n} - \left(\dfrac{\sum x_i}{n}\right)^2$

$\Rightarrow 16 = \dfrac{\sum x_i^2}{10} - (45)^2 \Rightarrow \sum x_i^2 = 10(2025 + 16)$

$\Rightarrow \sum x_i^2 = 20410$

$\therefore$ Corrected $\sum x_i^2 = 20410 - (52)^2 + (25)^2 = 18331$

and corrected $\sigma^2 = \dfrac{18331}{10} = 43.81$

Q24. The mean deviation of the data 3, 10, 10, 4, 7, 10, 5 from the mean is:
(a) 2 (b) 2.57 (c) 3 (d) 3.75

Solution: (b)
Observations are 3, 10, 10, 4, 7, 10 and 5.

$\therefore \bar{x} = \dfrac{3+10+10+4+7+10+5}{7} = \dfrac{49}{7} = 7$

x_i	$d_i = \lvert x_i - \bar{x} \rvert$
3	4
10	3
10	3
4	3
7	0
10	3
5	2
Total	$\sum d_i = 18$

Now, MD $= \dfrac{\sum d_i}{N} = \dfrac{18}{7} = 2.57$

Q25. Mean deviation for n observations $x_1, x_2, \ldots x_n$ from their mean $\bar{x}$ is given by:

(a) $\sum_{i=1}^{n} (x_i - \bar{x})$ (b) $\dfrac{1}{n}\sum_{i=1}^{n} \lvert x_i - \bar{x} \rvert$

(c) $\sum_{i=1}^{n} (x_i - \bar{x})^2$ (d) $\dfrac{1}{n}\sum_{i=1}^{n} (x_i - \bar{x})^2$

Solution: (b) MD $= \dfrac{1}{n}\sum_{i=1}^{n} \lvert x_i - \bar{x} \rvert$

Q26. When tested, the lives (in hours) of bulbs were noted as follows: 1357, 1090, 1666, 1494, 1623
The mean deviations (in hours) from their mean is:
(a) 178 (b) 179 (c) 220 (d) 356

Solution: (a)
The lives of 5 bulbs are 1357, 1090, 1666, 1494 and 1623.

$\therefore$ Mean $= \dfrac{1357 + 1090 + 1666 + 1494 + 1623}{5} = 1446$

x_i	$d_i = \lvert x_i - \bar{x} \rvert$
1357	89
1090	356
1666	220
1494	48
1623	177
Total	$\sum d_i = 890$

$$\text{MD} = \frac{\sum d_i}{N} = 890/5 = 178$$

Q27. The following are the marks obtained by 9 students in a mathematics test: 50, 69, 20, 33, 53, 39, 40, 65, 59. The mean deviation from the median is:
(a) 9 (b) 10.5 (c) 12.67 (d) 14.76

Solution: (c)
The marks obtained by 9 students in mathematics are 50, 69, 20, 33, 53, 39, 40, 65 and 59.
Rewrite the given data in ascending order.
20, 33, 39, 40, 50, 53, 59, 65, 69,
$n = 9$ [odd]

$$\therefore \text{Median} = \left(\frac{9+1}{2}\right)\text{term} = 5\text{th term}$$

Median = 50

x_i	$d_i = \lvert x_i - Me \rvert$
20	30
33	17
39	11
40	10
50	0
53	3
59	9
65	15
69	19
Total	$\sum d_i = 114$

$\therefore \text{MD} = 114/9 = 12.67$

Q28. The standard deviation of data 6, 5, 9, 13, 12, 8 and 10 is:
(a) $\sqrt{\frac{52}{7}}$ (b) 52/7 (c) $\sqrt{6}$ (d) 6

Solution: (a)
The given data is 6, 5, 9, 13, 12, 8 and 10.

x_i	x_i^2
6	36
5	25
9	81
13	169
12	144
8	64
10	100
Total	$\sum x_i^2 = 619$

$$\therefore \text{SD} = \sigma = \sqrt{\frac{\sum x_i^2}{N} - \left(\frac{\sum x_i}{N}\right)^2} = \sqrt{\frac{619}{7} - \left(\frac{63}{7}\right)^2}$$

$$= \sqrt{\frac{7 \times 619 - 3969}{49}} = \sqrt{\frac{4333 - 3969}{49}} = \sqrt{\frac{364}{49}} = \sqrt{\frac{52}{7}}$$

Q29. If $x_1, x_2, \ldots x_n$ be n observations and $\bar{x}$ be their arithmetic mean, the formula for standard deviation is given by:
(a) $\sum (x_i - \bar{x})^2$ (b) $\frac{\sum (x_i - \bar{x})^2}{n}$
(c) $\sqrt{\frac{\sum (x_i - \bar{x})^2}{n}}$ (d) $\sqrt{\frac{\sum x_i^2}{n} + \bar{x}^{-2}}$

Solution: (c) SD is given by $\sigma = \sqrt{\frac{\sum (x_i - \bar{x})^2}{n}}$

Q30. If the mean of 100 observations is 50 and their standard deviation is 5, then the sum of all the squares of all the observations is:
(a) 50000 (b) 250000
(c) 252500 (d) 255000

Solution: (c)
Given $\bar{x} = 50, n = 100, \sigma = 5$

$$\because \bar{x} = \frac{\sum x_i}{n} \Rightarrow 50 = \frac{\sum x_i}{100} \Rightarrow \therefore \sum x_i = 50 \times 100 = 5000$$

$$\text{Now, } \sigma = \sqrt{\frac{\sum x_i^2}{n} - \left(\frac{\sum x_i}{n}\right)^2} \Rightarrow \sigma^2 = \frac{\sum x_i^2}{n} - (\bar{x})^2$$

$$\Rightarrow 25 = \frac{\sum x_i^2}{100} - (50)^2 \Rightarrow 25 = \frac{\sum x_i^2}{100} - 2500$$

$$\Rightarrow 2525 = \frac{\sum x_i^2}{100} \Rightarrow \therefore \sum x_i^2 = 252500$$

Q31. If a, b, c, d and e be the observations with mean m and standard deviation s, then find the standard deviation of the observation $a+k, b+k, c+k, d+k, e+k$.

Solution: The given observations are *a, b, c, d, e*.

$$\text{Mean} = m = \frac{a+b+c+d+e}{5}$$

$$\sum x_i = a+b+c+d+e = 5m$$

$$\text{Now, mean} = \frac{a+k, b+k, c+k, d+k, e+k}{5}$$

$$= \frac{(a+b+c+d+e)+5k}{5} = m+k$$

$$\therefore \text{SD} = \sqrt{\frac{\sum (x_i+k)^2}{n} - (m+k)^2}$$

$$= \sqrt{\frac{\sum (x_i^2 - k^2 + 2kx_i)}{n} - (m^2+k^2+2mk)}$$

$$= \sqrt{\frac{\sum x_i^2}{n} - m^2 + \frac{2k\sum x_i}{n} - 2mk}$$

$$= \sqrt{\frac{\sum x_i^2}{n} - m^2 + 2km - 2km} \left[\because \frac{\sum x_i}{n} = m\right]$$

$$= \sqrt{\frac{\sum x_i^2}{n} - m^2} = s$$

Q32. If x_1, x_2, x_3, x_4, x_5 be the observations with mean *m* and standard deviation *s*, then the standard deviation of the observations $kx_1, kx_2, kx_3, kx_4, kx_5$ is:

(a) $k+s$ (b) s/k (c) ks (d) s

Solution: (c)

$$\text{Here, } m = \frac{\sum x_i}{5}, s = \sqrt{\frac{\sum x_i^2}{5} - \left(\frac{\sum x_i}{5}\right)^2}$$

$$\therefore \text{SD} = \sqrt{\frac{k^2\sum x_i^2}{5} - \left(\frac{k\sum x_i}{5}\right)^2}$$

$$= \sqrt{\frac{k^2\sum x_i^2}{5} - k^2\left(\frac{\sum x_i}{5}\right)^2} = \sqrt{\left(\frac{\sum x_i^2}{5}\right) - \left(\frac{\sum x_i}{5}\right)^2} = ks$$

Q33. Let $x_1, x_2, \ldots x_n$ be *n* observations. Let $w_i = lx_i + k$ for $i = 1, 2, \ldots, n$, where *l* and *k* are constants. If the mean of x_i is 48 and their standard deviation is 12, the mean of w_i is 55 and standard deviation of w_i is 15, then the value of l, k should be:

(a) $l = 1.25, k = -5$ (b) $l = 1.25, k = 5$

(c) $l = 2.5, k = -5$ (d) $l = 2.5, k = 5$

Solution: (a)

$w_i = x_i + k_1, \bar{x1} = 48, sx_1 = 12, w_i = 55, sw_1 = 15$

Then, $\bar{w_i} = \bar{x_i} + k \Rightarrow 55 = 48 + k$

[where $\bar{w_i}$ is the mean w_i and x_i is the mean of x_i]

Now, SD of w_i = 50 of $x_i \Rightarrow 15 = 12 \Rightarrow l = \frac{15}{12} = 1.25$

Solving, we get $k = 55 - 1.25 \times 48 = -5$

Q34. The standard deviation for the first natural number is:

(a) 5.5 (b) 3.87 (c) 2.97 (d) 2.87

Solution: (d)

$$\text{SD of first } n \text{ natural numbers} = \sqrt{\frac{n^2-1}{12}}$$

$$\therefore \text{SD of first 10 natural numbers} = \sqrt{\frac{(10)^2-1}{12}}$$

$$= \sqrt{\frac{100-1}{12}} = \sqrt{\frac{99}{12}} = \sqrt{8.25} = 2.87$$

Q35. Consider the numbers 1, 2, 3, 4, 5, 6, 7, 8, 9 and 10. If 1 is added to each number, the variance of the numbers so obtained is:

(a) 6.5 (b) 2.87 (c) 3.87 (d) 8.25

Solution: (d)

The given numbers are 1, 2, 3, 4, 5, 6, 7, 8, 9, 10.

If 1 is added to each number, then the observations will be 2, 3, 4, 5, 6, 7, 8, 9, 10, 11.

$$\therefore \sum x_1 = 2+3+4+\ldots+11$$

$$= \frac{10}{2}[2\times2+9\times1] = 5[4+9] = 65$$

$$\sum x_i^2 = 2^2+3^2+4^2+5^2+\ldots+11^2$$

$$= (1^2+2^2+3^2+\ldots+11^2) - (1)^2$$

$$= \frac{11\times12\times23}{6} - 1 = \frac{11\times12\times23-6}{6} = 505$$

$$\therefore s^2 = \frac{\sum x_i^2}{n} - \left(\frac{\sum x_i}{n}\right)^2 = \frac{505}{10} - \left(\frac{65}{10}\right)^2$$

$$= 505 - (65)^2 = 505 - 42.25 = 8.25$$

Q36. Consider the first 10 positive integers. If we multiply each number by –1 and then add 1 to each number, the variance of the number so obtained will be:

(a) 8.25 (b) 6.5 (c) 3.87 (d) 2.87

Solution: (a)

The first 10 positive integers are 2, 3, 4, 5, 6, 7, 8, 9 and 10.
If we multiply each number by –1, we get

$-1, -2, -3, -4, -5, -6, -7, -8, -9, -10$

On adding 1 to each number, we get

$0, -1, -2, -3, -4, -5, -6, -7, -8, -9$

$\therefore \sum x_i = 0-1-2-3--4-5-6-7-8-9$

$= -\dfrac{9 \times 10}{2} = -45$

$\sum x_i^2 = 0^2 + (-1)^2 + (-2)^2 + ... + (-9)^2$

$= \dfrac{9 \times 10 \times 19}{6} = 285$

$\therefore \text{SD} = \sqrt{\dfrac{285}{10} - \left(\dfrac{-45}{10}\right)^2} = \sqrt{\dfrac{285}{10} - \dfrac{2025}{100}}$

$= \sqrt{\dfrac{2850-2025}{100}} = \sqrt{8.25}$

Now, variance $= (\text{SD})^2 = (\sqrt{8.25})^2 = 8.25$

Q37. The following information relates to a sample of size 60, $\sum x^2 = 18000, \sum x = 960$. Then, the variance is:

(a) 6.63 (b) 16 (c) 22 (d) 44

Solution: (d)

$\text{Variance} \dfrac{\sum x_i^2}{n} - \left(\dfrac{\sum x_i}{n}\right)^2$

$= \dfrac{18000}{60} - \left(\dfrac{960}{60}\right)^2 = 300 - 256 = 44$

Q38. If the coefficient of variance of two distributions are 50, 60 and their arithmetic means are 30 and 25, respectively, then the difference of their standard deviation is:

(a) 0 (b) 1 (c) 1.5 (d) 2.5

Solution: (a)

$\text{CV}_1 = 50, \text{CV}_2 = 60, \bar{x}_1 = 30, \bar{x}_2 = 25$

$\therefore \text{CV}_1 = \dfrac{\sigma_1}{\bar{x}_1} \times 100 \Rightarrow 50 = \dfrac{\sigma_1}{30} \times 100$

$\therefore \sigma_1 = \dfrac{30 \times 50}{100} = 15, \text{CV}_2 = \dfrac{\sigma_2}{\bar{x}_2} \times 100$

$\Rightarrow 60 = \dfrac{\sigma_2}{25} \times 100 \Rightarrow \therefore \sigma_2 = \dfrac{60 \times 25}{100} = 15$

Now, $\sigma_1 - \sigma_2 = 15 - 15 = 0$

Q39. The standard deviation of some temperature data in °C is 5. If the data were converted into °F, then the variance would be:

(a) 81 (b) 57 (c) 36 (d) 25

Solution: (a)

$\sigma_C = 5 \Rightarrow \dfrac{5}{9}(F-32) = C$

$\Rightarrow F = \dfrac{9C}{5} + 32$

$\sigma_F = \dfrac{9}{5}\sigma_C = \dfrac{9}{5} \times 5 = 9$

$\Rightarrow \sigma_F^2 = (9)^2 = 81$

Q40. Coefficient of variation $= \dfrac{...}{\text{Mean}} \times 100$

Solution: $\text{CV} = \dfrac{\text{SD}}{\text{Mean}} \times 100$

Q41. If $\bar{x}$ is the mean of n values of x, then $\sum_{i=1}^{n}(x_1 - \bar{x})$ is always equal to … if a has any value other than $\sum_{i=1}^{n}(x_1 - \bar{x})^2$ is......than $\sum_{i=1}^{n}(x_1 - a)^2$

Solution: If $\bar{x}$ is the mean of n values of x, then $\sum_{i=1}^{n}(x_1 - \bar{x}) = 0$ and if a has any value other $\bar{x}$, then

$\sum_{i=1}^{n}(x_1 - \bar{x})^2$ is less than $\sum_{i=1}^{n}(x_1 - a)^2$

Q42. If the variance of a data is 121, then the standard deviation of the data is......

Solution: If the variance of data is 121.
Then, $\text{SD} = \sqrt{\text{Variance}} = \sqrt{121} = 11$

Q43. The standard deviation of data is... of any change in origin but is .... of change of scale.

Solution: The standard deviation of data is independent of any change in origin but is dependent of change of scale.

Q44. The sum of the squares of the deviations of the value of the variable is … … … when taken about their arithmetic mean.

Solution: The sum of the squares of the deviations of the values of the variable is minimum when taken about their arithmetic mean.

Q45. The mean deviation of the data is......when measured from the median.

Solution: The mean deviation of the data is least when measured from the median.

Q46. The standard deviation is......to the mean deviation taken from the arithmetic mean.

Solution: The SD is greater than or equal to the mean deviation taken from the arithmetic mean.

CHAPTER

16 Probability

Day 209

Sample Space

Random Experiment: An experiment is called a random experiment if it satisfies the following two conditions:

(i) It has more than one possible outcome.
(ii) It is not possible to predict the outcome in advance.

Outcomes and Sample Space

- A possible result of a random experiment is called its outcome.
- The set of all possible outcomes of a random experiment is called the sample space of the experiment, denoted by the symbol S.
- Each element or outcome of the sample space of a random experiment is called a sample point.
- The total number of possible outcomes of the sample space of a random experiment is denoted by $n(S)$.

Example 1: Describe the sample space for the indicated experiment: Tossing of three coins.

Solution: A coin has one head H and one tail T.
When three coins are tossed, the total number of possible outcomes is $2^3 = 8$
Thus, the sample space is given by $S = \{HHH, HHT, HTH, THH, HTT, THT, TTH, TTT\}$

Example 2: Describe the sample space for the indicated experiment: Two dice are thrown.

Solution: When a die is thrown, the possible outcomes are 1, 2, 3, 4, 5, 6
When two dice are thrown, the sample space is given by $S = \{(x, y) : x, y = 1, 2, 3, 4, 5, 6\}$
The number of elements in this sample space is 6 x 6 = 36, given by

$$S = \{(1,1),(1,2),(1,3),(1,4),(1,5),(1,6),(2,1),\\ (2,2)(2,3),(2,4),(2,5),(2,5),(2,6),(3,1),(3,2),\\ (3,3),(3,4),(3,5),(3,6),(4,1),(4,2),(4,3),(4,4),\\ (4,5),(4,6),(5,1),(5,2),(5,3),(5,4),(5,5),(5,6),\\ (6,1),(6,2),(6,3),(6,4),(6,5),(6,6)\}$$

Example 3: Describe the sample space for the indicated experiment: A coin is tossed and then a die is rolled only if a head is shown on the coin, or a coin is tossed again if a tail is shown on the coin.

Solution: A coin has two faces: head (H) and tail (T).
A die has six faces that are numbered from 1 to 6, with one number on each face.
Thus, when a coin is tossed and then a die is rolled only if a head is shown on the coin, the sample space is given by $S_1 = \{H1, H2, H3, H4, H5, H6\}$
When a coin is tossed and then a coin is tossed again if a tail is shown on the coin, the sample space is given by $S_2 = \{TH, TT\}$
Thus, the total sample space is given by $S = S_1 \cup S_2 = \{H1, H2, H3, H4, H5, H6, TH, TT\}$

Example 4: One die of red and one of white are placed in a bag. One die is selected at random and rolled, its colour and the number on its uppermost face is noted.
Describe the sample space.

Solution: A die has six faces that are numbered from 1 to 6, with one number on each face.
Let us denote the red and white dices as R and W, respectively.
Thus, when a die is selected and then rolled, the sample space is given by

$$S = \{R1, R2, R3, R4, R5, R6,\\ W1, W2, W3, W4, W5, W6,\}$$

Example 5: An experiment consists of rolling a die and then tossing a coin once if the number on the die is even. If the number on the die is odd, a die is thrown again. Write the sample space for this experiment.

Solution: A die has six faces numbered from 1 to 6, with one number on each face. Among these numbers, 2, 4 and 6 are even numbers, while 1, 3, 5 are odd numbers.
A coin has two faces: head (H) and tail (T).
Thus, the sample space of this experiment is given by

$$S = \{2H, 2T, 4H, 4T, 6H, 6T, 11, 12, 13, 14, 15, 16,\\ 31, 32, 33, 34, 35, 36, 51, 52, 53, 54, 55, 56\}$$

Example 6: A coin is tossed. If the outcome is a head, a die is thrown. If the die shows up an even number, the coin is thrown again. What is the sample space for the experiment?

Solution: When a coin is tossed, the possible outcomes are head (H) and tail (T).
When a die is thrown, the possible outcomes are 1, 2, 3, 4, 5, 6
Thus, the sample space of this experiment is given by

$$S = \{T, H1, H3, H5, H2H, H2T, H4H, H4T, H6H, H6T\}$$

Example 7: A coin is thrown repeatedly until a head comes up. What is the sample space for this experiment?

Solution: In this experiment, head may come up on the first throw, the second throw, the third throw, and so on till a head is obtained.
Thus, the sample space of this experiment is given by $S = \{H, (T, H), (T, T, H) ... (T, T, T, ..., H)\}$

Example 8: A coin is tossed. If it shows a tail, we draw a ball from a box that contains 1 red and 2 black balls. If it shows a head, we draw a ball from a box that contains 2 white and 1 green balls. Find the sample space for this experiment.

Solution: The box contains 1 red ball and 2 black balls. Let denote 1 red ball as R_1 and the 2 black balls as B_1, B_2
Let us denote 2 white balls as W_1, W_2 and the 1 green ball as G_1
The sample space of this experiment is given by
$S = \{TR_1, TB_1, TB_2, HW_1, HW_2, HG_1\}$

Example 9: One boy and 2 girls are in Class I, and 2 boys and 1 girl in Class II. A coin is tossed. If it shows a head, a student from Class I is selected or if it shows a tail, a student from Class II is selected. Find the sample space for this experiment.

Solution: Let us denote 1 boy and 2 girls in Class I as B_1 and G_1, G_2, respectively. Let us denote 2 boys and 1 girl in Class II as B_2, B_3 and G_3, respectively.
$S = \{HB_1, HG_1, HG_2, TB_2, TB_3, TG_3\}$

Example 10: An experiment consists of recording the boy-girl composition of families with 3 children.
(i) What is the sample space if we are interested in knowing whether it is a boy or girl in the order of their births?
(ii) What is the sample space if we are interested in the number of girls in the family?

Solution: (i) When the order of the birth of a girl or a boy is considered, the sample space is given by
$S = \{GGG, GGB, GBG, BGG, BBG, BGB, GBB, BBB\}$
(ii) Since the maximum number of children in each family is 3, a family can have 3 girls or 2 girls or 1 girl or no girl. Hence, the required sample space is $S = \{0, 1, 2, 3\}$

Example 11: A box contains 2 red balls and 1 white ball. Two balls are drawn at random in succession, (i) with replacement and (ii) without replacement. Write the sample space for this experiment.

Solution: The box contains 2 red balls and 1 white ball. Let us denote the red ball as R and the white ball as W.
(i) When two balls are drawn at random in succession with replacement, the sample space is given by $S = \{RW, WR, WW, RR\}$
(ii) When two balls are drawn at random in succession without replacement, the sample space is given by $S = \{RW, WR, RR\}$

Example 12: Suppose two bulbs are selected at random from a lot. Each bulb is tested and classified as defective (D) or non-defective (N). Write the sample space of this experiment.

Solution: Two bulbs are to selected at random from the lot. Each bulb in the lot is tested and classified as defective (D) or non-defective (N).
The sample space of this experiment is given by
$S = \{DD, DN, ND, NN\}$

Example 13: The numbers 1, 2 and 3 are written separately on four slips of paper. The slips are put in a box and mixed thoroughly. A person draws two slips from the box, one after the other, (i) without replacement and (ii) with replacement. Describe the sample space for the experiment.

Solution: (i) The sample space of this experiment without replacement is given by
$S = \{(1,2), (1,3), (2,1), (2,3), (3,1), (3,2)\}$
(ii) The sample space of this experiment with replacement is given by

$$S = \{(1,1), (1,2), (1,3), (2,1), (2,2), (2,3), (3,1), (3,2), (3,3)\}$$

Day 210

NCERT Exercise 16.1 (With Solutions)

Q1. Describe the sample space for the indicated experiment: A coin is tossed three times.

Solution: A coin has one head H and one tail T.
When a coin is tossed three times, the total number of possible outcomes is $2^3 = 8$
Thus, when a coin is tossed three times, the sample space is given by $S = \{HHH, HHT, HTH, THH, HTT, THT, TTH, TTT\}$

Q2. Describe the sample space for the indicated experiment: A die is thrown two times.

Solution: When a die is thrown, the possible outcomes are 1, 2, 3, 4, 5, 6
When a die is thrown two times, the sample space is given by $S = \{(x, y) : x, y = 1,2,3,4,5,6\}$
The number of elements in this sample space is 6 x 6 = 36, given by

$$S = \{(1,1),(1,2),(1,3),(1,4),(1,5),(1,6),(2,1),\\(2,2)(2,3),(2,4),(2,5),(2,5),(2,6),(3,1),(3,2),\\(3,3),(3,4),(3,5),(3,6),(4,1),(4,2),(4,3),(4,4),\\(4,5),(4,6),(5,1),(5,2),(5,3),(5,4),(5,5),(5,6),\\(6,1),(6,2),(6,3),(6,4),(6,5),(6,6)\}$$

Q3. Describe the sample space for the indicated experiment: A coin is tossed four times.

Solution: When a coin is tossed once, there are two possible outcomes: head (H) and tail (T).
When a coin is tossed four times, the total number of possible outcomes is $2^4 = 16$
Thus, when a coin is tossed four times, the sample space is given by

$$S = \{HHHH, HHHT, HHTH, HHTT, HTHH,\\HTHT, HTTH, HTTT, THHH, THHT, THTH,\\THTT, TTHH, TTHT, TTTH, TTTT\}$$

Q4. Describe the sample space for the indicated experiment: A coin is tossed and a die is thrown.

Solution: A coin has two faces: head (H) and tail (T).
A die has six faces numbered from 1 to 6, with one number on each face.
Thus, when a coin is tossed and a die is thrown, the sample space is given by

$$S = \{H1, H2, H3, H4, H5, H6,\\T1, T2, T3, T4, T5, T6\}$$

Q5. Describe the sample space for the indicated experiment: A coin is tossed and then a die is rolled only if a head is shown on the coin.

Solution: A coin has two faces: head (H) and tail (T). A die has six faces numbered from 1 to 6, with one number on each face.
Thus, when a coin is tossed and then a die is rolled only in case a head is shown on the coin, the sample space is given by $S = \{H1, H2, H3, H4, H5, H6, T\}$

Q6. Two boys and two girls are in Room X, and one boy and three girls in Room Y. Specify the sample space for the experiment in which a room is selected and then a person.

Solution: Let us denote 2 boys and 2 girls in Room X as B_1, B_2 and G_1, G_2, respectively. Let us denote 1 boy and 3 girls in Room Y as B_3 and G_3, G_4, G_5, respectively.
$S = \{XB_1 XB_2, XG_1, XG_2, YB_3, YG_3, YG_4, YG_5\}$

Q7. One die of red, one of white and one of blue are placed in a bag. One die is selected at random and rolled, its colour and the number on its uppermost face is noted.
Describe the sample space.

Solution: A die has six faces numbered from 1 to 6, with one number on each face.
Let us denote the red, white and blue dices as R, W and B, respectively.
Thus, when a die is selected and then rolled, the sample space is given by

$$S = \{R1, R2, R3, R4, R5, R6, W1, W2, W3,\\W4, W5, W6, B1, B2, B3, B4, B5, B6\}$$

Q8. An experiment consists of recording the boy-girl composition of families with 2 children.
(i) What is the sample space if we are interested in knowing whether it is a boy or girl in the order of their births?
(ii) What is the sample space if we are interested in the number of girls in the family?

Solution: (i) When the order of the birth of a girl or a boy is considered, the sample space is given by
$S = \{GG, GB, BG, BB\}$
(ii) Since the maximum number of children in each family is 2, a family can either have 2 girls or 1 girl or no girl. Hence, the required sample space is $S = \{0, 1, 2\}$

Q9. A box contains 1 red and 3 identical white balls. Two balls are drawn at random in succession without replacement. Write the sample space for this experiment.

Solution: The box contains 1 red ball and 3 identical white balls. Let us denote the red ball as R and a white ball as W.
When two balls are drawn at random in succession without replacement, the sample space is given by
$S = \{RW, WR, WW\}$

Q10. An experiment consists of tossing a coin and them throwing it a second time if a head occurs. If a tail occurs on the first toss, then a die is rolled once. Find the sample space.

Solution: A coin has two faces: head (H) and tail (T). A die has six faces numbered from 1 to 6, with one number on each face.
Thus, in the given experiment, the sample space is given by $S = \{HH, HT, T1, T2, T3, T4, T5, T6\}$

Q11. Suppose 3 bulbs are selected at random from a lot. Each bulb is tested and classified as defective (D) or non-defective (N). Write the sample space of this experiment.

Solution: Three bulbs are to be selected at random from the lot. Each bulb in the lot is tested and classified as defective (D) or non-defective (N).
The sample space of this experiment is given by
$S = \{DDD, DDN, DND, DNN, NDD,$
$NDN, NND, NNN\}$

Q12. A coin is tossed. If the outcome is a head, a die is thrown. If the die shows up an even number, the die is thrown again. What is the sample space for the experiment?

Solution: When a coin is tossed, the possible outcomes are head (H) and tail (T).
When a die is thrown, the possible outcomes are 1, 2, 3, 4, 5, 6
Thus, the sample space of this experiment is given by
$S = \{T, H1, H3, H5, H21, H22, H23, H24,$
$H26, H41, H42, H43, H44, H45,$
$H46, H61, H62, H63, H64, H65, H66\}$

Q13. The numbers 1, 2, 3 and 4 are written separately on four slips of paper. The slips are put in a box and mixed thoroughly. A person draws two slips from the box, one after the other without replacement. Describe the sample space for the experiment.

Solution: If 1 appears on the first drawn slip, then the possibilities for the number appearing on the second drawn slip are 2, 3, 4. Similarly, if 2 appears on the first drawn slip, then the possibilities for the number appearing on the second drawn slip are 1, 3, 4. The same holds true for the remaining numbers.
Thus, the sample space of this experiment is given by
$S = \{(1,2), (1,3), (1,4), (2,1), (2,3), (2,4),$
$(3,1), (3,2), (3,4), (4,1), (4,2), (4,3)\}$

Q14. An experiment consists of rolling a die and then tossing a coin once if the number on the die is even. If the number on the die is odd, the coin is tossed twice. Write the sample space for this experiment.

Solution: A die has six faces numbered from 1 to 6, with one number on each face. Among these numbers, 2, 4 and 6 are even numbers, while 1, 3, 5 are odd numbers.
A coin has two faces: head (H) and tail (T).
Hence, the sample space of this experiment is given by
$S = \{2H, 2T, 4H, 4T, 6H, 6T, 1HH, 1HT, 1TH,$
$1TT, 3HH, 3HT, 3TH, 3TT, 5HH, 5HT, 5TH, 5TT\}$

Q15. A coin is tossed. If it shows a tail, we draw a ball from a box that contains 2 red and 3 black balls. If it shows head, we throw a die. Find the sample space for this experiment.

Solution: The box contains 2 red balls and 3 black balls. Let denote 2 red balls as R_1, R_2 and the 3 black balls as B_1, B_2, B_3.
The sample space of this experiment is given by
$S = \{TR_1, TR_2, TB_1, TB_2, TB_3, H1, H2, H3, H4, H5, H6\}$

Q16. A die is thrown repeatedly until a six comes up. What is the sample space for this experiment?

Solution: In this experiment, six may come up on the first throw, the second throw, the third throw, and so on, till six is obtained.
Thus, the sample space of this experiment is given by
$S = \{6, (1,6), (2,6), (3,6), (4,6), (5,6),$
$(1,1,6), (1,2,6)\ldots\ldots, (1,5,6), (2,1,6), (2,2,6), \ldots,$
$(2,5,6), \ldots\ldots (5,1,6), (5,2,6)\ldots\ldots\}$

Day 211

Event

Any subset E of a sample space S associated with a random experiment is called an event.

Occurrence of an Event: Let there be an event E in a sample space S associated with a random experiment. Event E of the sample space S is said to have occurred if an element (or outcome) ω of the experiment is such that $\omega \in E$. And if the outcome ω is such that $\omega \notin E$, then we say that the event E has not occurred.

Types of Events

(i) Impossible and sure events: The empty set ϕ is called an impossible event of the sample space S and the whole sample space S is called the sure event.

(ii) Simple event: If an event E has only one sample point (or outcome) of a sample space, it is called a simple (or elementary) event.
Note: In a sample space containing n distinct elements, there are exactly n simple events.

(iii) Compound event: If an event has more than one sample point (or outcome), it is called a compound event.

Algebra of Events: Let A, B, C be events associated with a random experiment whose sample space is S.

(i) Complementary event: For every event A, there exists another event A' called the complementary event to A. It is also called the event 'not A' and is given by $A' = S - A$; that is, if an element (or outcome) ω of the experiment is such that $\omega \in A$, then $\omega \notin A'$.
Note: $S' = \phi$ and $\phi' = S$

(ii) The event 'A or B': Let there be events A and B associated with a sample space S. Then the union of two sets A and B is denoted by $A \cup B$, which contains all the elements that are either in A or in B or in both.
Event 'A or B' $= A \cup B = \{\omega : \omega \in A$ or $\omega \in B\}$

(iii) The event 'A and B': Let the events A and B be associated with a sample space S. Then the intersection of two sets A and B is denoted by $A \cap B$, which contains all the elements that are common to both A and B; that is, which belong to both 'A and B'.
Event 'A and B' $= A \cap B = \{\omega : \omega \in A$ and $\omega \in B\}$

(iv) The event 'A but not B': Let the events A and B be associated with a sample space S. Then the set 'A but not B' denoted by $A - B$ is the set of all the elements that are in A but not in B.
Event 'A but not B' $= A - B = \{\omega : \omega \in A$ but $\omega \notin B\}$

Mutually Exclusive Events: Two events A and B of a sample space are called mutually exclusive events if the occurrence of any one of them excludes the occurrence of the other event; that is, they cannot occur simultaneously.
Thus, mutually exclusive events A and B are disjoint sets; that is, $A \cap B = \phi$.

Note: Simple events of a sample space are always mutually exclusive.

Exhaustive Events: Let $E_1, E_2, ..., E_n$ be n events of a sample space S such that $E_1 \cup E_2 \cup E_3 \cup ... \cup E_n = \bigcup_{i=1}^{n} E_i = S$.
Then $E_1, E_2, ..., E_n$ are called exhaustive events.
That is, events $E_1, E_2, ..., E_n$ are said to be exhaustive if at least one of them necessarily occurs whenever the experiment is performed.

Also, if $E_i \cap E_j = \phi$ for $i \neq j$, that is, events E_i, E_j are pair-wise disjoint sets and $\bigcup_{i=1}^{n} E_i = S$, then events $E_1, E_2, ..., E_n$ are called mutually exhaustive events.

Example 1: Consider the experiment of tossing a coin two times. The sample space of this random experiment is $S = \{HH, HT, TH, TT\}$.
Now, consider the events

E_1 = occurrence of exactly one head

E_2 = occurrence of exactly two tails

E_3 = occurrence of at least one head

E_4 = occurrence of three heads

E_5 = occurrence of two tails or at least one head

Which events are

(i) impossible?
(ii) sure?
(iii) simple?
(iv) compound?
(v) complementary events?
(vi) event $E_1 \cup E_2$?
(vii) event $E_1 \cap E_3$?
(vii) event $E_3 - E_1$?
(ix) mutually exclusive events?
(x) mutually exhaustive events?

Solution: $E_1 = \{HT, TH\}$

$E_2 = \{TT\}$

$E_3 = \{HT, TH, HH\}$

$E_4 = \phi$

$E_5 = \{TT, HT, TH, HH\}$

(i) Event $E_4 = \phi$ is impossible because occurrence of three heads is not possible since only two coins are tossed.

(ii) Event $E_5 = \{TT, HT, TH, HH\} = S$ is a sure event

(iii) Event $E_2 = \{TT\}$ is a simple event

(iv) Events $E_1 = \{HT, TH\}$, $E_3 = \{HT, TH, HH\}$ and $E_5 = \{TT, HT, TH, HH\}$ are compound events

(v) Events E_2 and E_3 are complementary to each other. Events E_4 and E_5 are complementary to each other.

(vi) $E_1 \cup E_2 = \{HT, TH\} \cup \{TT\} = \{HT, TH, TT\}$

(vii) $E_1 \cap E_3 = \{HT, TH\} \cap \{HT, TH, HH\} = \{HT, TH\}$

(vii) $E_3 - E_1 = \{HT, TH, HH\} - \{HT, TH\} = \{HH\}$

(ix) Events E_1 and E_2 are mutually exclusive events. Events E_2 and E_3 are mutually exclusive events.

(x) Events E_2 and E_3 are mutually exhaustive events since $E_2 \cup E_3 = \{TT, HT, TH, HH\} = S$

Events E_4 and E_5 are mutually exhaustive events since $E_4 \cup E_5 = \{TT, HT, TH, HH\} = S$

Example 2: Consider the experiment of rolling a die. The sample space of this random experiment is $S = \{1, 2, 3, 4, 5, 6\}$.

Now, consider the events

E_1 = an odd number appears

E_2 = an even number appears

E_3 = a number less than 4 appears

E_4 = a number that is a multiple of 3 appears

E_5 = a number greater than 6 appears

E_6 = a prime number appears

E_7 = a prime number or multiple of 2 appears

E_8 = an even prime number appears

Which events are

(i) impossible?

(ii) sure?

(iii) simple?

(iv) compound?

(v) complementary events?

(vi) event $E_1 \cup E_3$?

(vii) event $E_1 \cap E_6$?

(vii) event $E_6 - E_4$?

(ix) mutually exclusive events?

(x) mutually exhaustive events?

Solution:

$E_1 = \{1, 3, 5\}$ $\quad E_2 = \{2, 4, 6\}$

$E_3 = \{1, 2, 3\}$ $\quad E_4 = \{3, 6\}$

$E_5 = \phi$ $\quad E_6 = \{1, 2, 3, 5\}$

$E_7 = \{1, 2, 3, 4, 5, 6\}$ $\quad E_8 = \{2\}$

(i) Event $E_5 = \phi$ is impossible because occurrence of a number greater than 6 is not possible

(ii) Event $E_7 = \{1, 2, 3, 4, 5, 6\}$ is a sure event

(iii) Event $E_8 = \{2\}$ is a simple event

(iv) $E_1 = \{1, 3, 5\}$, $E_2 = \{2, 4, 6\}$, $E_3 = \{1, 2, 3\}$, $E_4 = \{3, 6\}$, $E_6 = \{1, 2, 3, 5\}$, $E_7 = \{1, 2, 3, 4, 5, 6\}$ are compound events.

(v) Events E_1 and E_2 are complementary to each other. Events E_5 and E_7 are complementary to each other.

(vi) $E_1 \cup E_3 = \{1, 3, 5\} \cup \{1, 2, 3\} = \{1, 2, 3, 5\}$

(vii) $E_1 \cap E_6 = \{1, 3, 5\} \cap \{1, 2, 3, 5\} = \{1, 2, 3\}$

(vii) $E_6 - E_4 = \{1, 2, 3, 5\} - \{3, 6\} = \{1, 2, 5\}$

(ix) Events E_1 and E_2 are mutually exclusive events. Events E_1 and E_8 are mutually exclusive events. Events E_5 and E_7 are mutually exclusive events.

(x) Events E_1 and E_2 are mutually exhaustive events since $E_1 \cup E_2 = \{1, 2, 3, 4, 5, 6\} = S$

Events E_2 and E_6 are mutually exhaustive events since $E_2 \cup E_6 = \{1, 2, 3, 4, 5, 6\} = S$

Events E_5 and E_7 are mutually exhaustive events since $E_5 \cup E_7 = \{1, 2, 3, 4, 5, 6\} = S$

Day 212

NCERT Exercise 16.2 (With Solutions)

Q1. A die is rolled. Let E be the event "die shows 4" and F be the event "die shows even number". Are E and F mutually exclusive?

Solution: When a die is rolled, the sample space is given by $S = \{1, 2, 3, 4, 5, 6\}$

Then, $E = \{4\}$ and $F = \{2, 4, 6\}$

$E \cap F = \{4\} \neq \phi$. Therefore, E and F are not mutually exclusive events.

Q2. A die is thrown. Describe the following events:

(i) A: a number less than 7

(ii) B: a number greater than 7

(iii) C: a multiple of 3

(iv) D: a number less than 4
(v) E: an even number greater than 4
(vi) F: a number not less than 3

Also find $A \cup B, A \cap B, B \cup C, E \cap F, D \cap E,$
$A - C, D - E, E \cap F', F'$

Solution: When a die is thrown, the sample space is given by $S = \{1,2,3,4,5,6\}$

Now

(i) $A = \{1,2,3,4,5,6\}$
(ii) $B = \phi$
(iii) $C = \{3,6\}$
(iv) $D = \{1,2,3\}$
(v) $E = \{6\}$
(vi) $F = \{3,4,5,6\}$

$A \bigcup B = \{1,2,3,4,5,6\},$
$A \bigcap B = \phi$
$B \bigcup C = \{3,6\},$
$E \bigcap F = \{6\}$
$D \bigcap E = \phi,$
$A - C = \{1,2,4,5\}$
$D - E = \{1,2,3\}$
$F' = \{1,2\}$
$E \bigcap F' = \phi$

Q3. An experiment involves rolling a pair of dice and recording the numbers that come up. Describe the following events:
A: the sum is greater than 8
B: 2 occurs on either die
C: The sum is at least 7 and a multiple of 3
Which pairs of these events are mutually exclusive?

Solution: When a pair of dice is rolled, the sample space is given by $S = \{(x, y) : x, y = 1,2,3,4,5,6\}$

$$= \begin{Bmatrix} (1,1),(1,2),(1,3),(1,4),(1,5),(1,6) \\ (2,1),(2,2),(2,3),(2,4),(2,5),(2,6) \\ (3,1),(3,2),(3,3),(3,4),(3,5),(3,6) \\ (4,1),(4,2),(4,3),(4,4),(4,5),(4,6) \\ (5,1),(5,2),)5,3),(5,4),(5,5),(5,6) \\ (6,1),(6,2),(6,3),(6,4),(6,5),(6,6) \end{Bmatrix}$$

Now

$A = \{(3,6),(4,5),(4,6),(5,4),(5,5),(5,6),$
$(6,3),(6,4),(6,5),(6,6)\}$

$B = \{(2,1),(2,2),(2,3),(2,4),(2,5),(2,6),$
$(1,2),(3,2),(4,2),(5,2),(6,2)\}$

$C = \{(3,6),(4,5),(5,4),(6,3),(6,6)\}$

It is observed that

$A \bigcap B = \phi$
$B \bigcap C = \varphi$
$C \cap A = \{(3,6),(4,5),(5,4),(6,3),(6,6)\} \neq \phi$

Hence, events A and B and events B and C are mutually exclusive.

Q4. Three coins are tossed once. Let A denote the event 'three heads show', B denote the event "two heads and one tail show", C denote the event "three tails show" and D denote the event 'a head shows on the first coin". Which events are
(i) mutually exclusive?
(ii) simple?
(iii) compound?

Solution: When three coins are tossed, the samples space is given by
$S = \{HHH, HHT, HTH, HTT, THH, THT, TTH, TTT\}$
Then
$A = \{HHH\}$
$B = \{HHT, HTH, THH\}$
$C = \{TTT\}$
$D = \{HHH, HHT, HTH, HTT\}$
Then
$A \bigcap B = \phi, A \bigcap C = \phi, A \bigcap D = \{HHH\} \neq \phi$
$B \bigcap C = \phi, B \bigcap D = \{HHT, \{HTH\} \neq \phi$
$C \bigcap D = \phi$

(i) Events A and B, events A and C, events B and C and events C and D are all mutually exclusive.
(ii) If an event has only one sample point of a sample space, it is called a simple event. Thus, A and C are simple events.
(iii) If an event has more than one sample point of a sample space, it is called a compound event. Thus, B and D are compound events.

5. Three coins are tossed. Describe
(i) Two events that are mutually exclusive
(ii) Three events that are mutually exclusive and exhaustive
(iii) Two events that are not mutually exclusive
(iv) Two events that are mutually exclusive but not exhaustive
(v) Three events that are exclusive but not exhaustive

Solution: When three coins are tossed, the sample space is given by
$S = \{HHH, HHT, HTH, HTT, THH, THT, TTH, TTT\}$ (i)
Two events that are mutually exclusive can be
A: getting all tails
B: getting all heads
This is because sets $A = \{TTT\}, B = \{HHH\}$ are disjoint.

(ii) Three events that are mutually exclusive and exhaustive can be
A: getting all tails
B: getting exactly one head
C: getting at least two heads
That is
$A = \{TTT\}$
$B = \{HTT, THT, TTH\}$
$C = \{HHH, HHT, HTH, THH\}$
This is because
$A \cap B = B \cap C = C \cap A = \phi, A \cup B \cup C = S$

(iii) Two events that are not mutually exclusive can be
A: getting no tails
B: getting at least 2 heads
That is
$A = \{HHH\}$
$B = \{HHH, HHT, HTH, THH\}$
This is because $A \cap B = \{HHH\} \neq \phi$

(iv) Two events that are mutually exclusive but not exhaustive can be
A: getting exactly one head
B: getting exactly one tail
That is
$A = \{HTT, THT, THH\}$
$B = \{HHT, HTH, THH\}$
It is because $A \cap B \neq \phi$ but $A \cup B \neq S$

(v) Three events that are mutually exclusive but not exhaustive can be
A: getting exactly three heads
B: getting one head and two tails
C: getting one tail and two heads
That is
$A = \{HHH\}$
$B = \{HTT, THT, TTH\}$
$C = \{HHT, HTH, THH\}$
This is because $A \cap B = B \cap C = C \cap A = \phi$, but $A \cup B \cup C \neq S$

Q6. Two dice are thrown. The events *A, B, C* are as follows:
A: getting an even number on the first die
B: getting an odd number on the first die.
C: getting the sum of the numbers on the dice ≤ 5
Describe the events.
(i) A′
(ii) not B
(iii) A or B
(iv) A and B
(v) A but not C
(vi) B or C
(vii) B and C
(viii) $A \cap B' \cap C'$

Solution: When two dice are thrown, the sample space is given by $S = \{(x, y) : x, y = 1, 2, 3, 4, 5, 6\}$

$$= \begin{Bmatrix} (1,1),(1,2),(1,3),(1,4),(1,5),(1,6) \\ (2,1),(2,2),(2,3),(2,4),(2,5),(2,6) \\ (3,1),(3,2),(3,3),(3,4),(3,5),(3,6) \\ (4,1),(4,2),(4,3),(4,4),(4,5),(4,6) \\ (5,1),(5,2),(5,3),(5,4),(5,5),(5,6) \\ (6,1),(6,2),(6,3),(6,4),(6,5),(6,6) \end{Bmatrix}$$

Now

$$A = \begin{Bmatrix} (2,1),(2,2),(2,3),(2,4),(2,5),(2,6), \\ (4,1),(4,2),(4,3)(4,4),(4,5),(4,6), \\ (6,1),(6,2),(6,3),(6,4),(6,5),(6,6) \end{Bmatrix}$$

$$B = \begin{Bmatrix} (1,1),(1,2),(1,3),(1,4),(1,5),(1,6) \\ (3,1),(3,2),(3,3),(3,4),(3,5),(3,6), \\ (5,1),(5,2),(5,3),(5,4),(5,5),(5,6) \end{Bmatrix}$$

$C = \{(1,1),(1,2),(1,3),(1,4),(2,1),$
$(2,2),(2,3)(3,1),(3,2),(3,3)\}$

(i) $$A' = \begin{Bmatrix} (1,1),(1,2),(1,3),(1,4),(1,5),(1,6), \\ (3,1),(3,2),(3,3),(3,4),(3,5),(3,6), \\ (5,1),(5,2),(5,3),(5,4),(5,5),(5,6) \end{Bmatrix} = B$$

(ii) Not $B =$
$$B' = \begin{Bmatrix} (2,1),(2,2),(2,3),(2,4),(2,5),(2,6) \\ (4,1),(4,2),(4,3),(4,4),(4,5),(4,6), \\ (6,1),(6,2),(6,3),(6,4),(6,5),(6,6) \end{Bmatrix} = A$$

(iii) A or $B = A \cup B =$
$$\begin{Bmatrix} (1,1),(1,2),(1,3),(1,4),(1,5),(1,6) \\ (2,1),(2,2),(2,3),(2,4),(2,5),(2,6) \\ (3,1),(3,2),(3,3),(3,4),(3,5),(3,6) \\ (4,1),(4,2),(4,3),(4,4),(4,5),(4,6) \\ (5,1),(5,2),(5,3),(5,4),(5,5),(5,6) \\ (6,1),(6,2),(6,3),(6,4),(6,5),(6,6) \end{Bmatrix} = S$$

(iv) A and $B = A \cap B = \phi$

(v) A but not $C = A - C$
$$= \begin{Bmatrix} (2,4),(2,5),(2,6),(4,2),(4,3),(4,4),(4,5) \\ (4,6),(6,1),(6,2),(6,3),(6,4),(6,5),(6,6) \end{Bmatrix}$$

(vi) B or $C = B \cup C$
$$= \begin{Bmatrix} (1,1),(1,2),(1,3),(1,4),(1,5),(1,6),(2,1),(2,2), \\ (2,3),(3,1),(3,2),(3,3),(3,4),(3,5),(3,6), \\ (4,1),(5,1),(5,2),(5,3),(5,4),(5,5),(5,6) \end{Bmatrix}$$

(vii) B and $C = B \cap C$
$= \{(1,1),(1,2),(1,3),(1,4),(3,1),(3,2)\}$

(viii) C'

$$= \begin{Bmatrix} (1,5),(1,6),(2,4)(2,5),(2,6),(3,3),(3,4),(3,5), \\ (3,6),(4,2)(4,3),(4,4),(4,5),(4,6), \\ (5,1),(5,2),(5,3),(5,4),(5,5),(5,6) \\ (6,1),(6,2),(6,3),(6,4),(6,5),(6,6) \end{Bmatrix}$$

$\therefore A \cap B' \cap C' = A \cap A \cap C' = A \cap C'$

$$= \begin{Bmatrix} (2,4),(2,5),(2,6),(4,2),(4,3),(4,4),(4,5), \\ (4,6),(6,1),(6,3),(6,4),(6,5),(6,6) \end{Bmatrix}$$

Q7. Two dice are thrown. The events *A*, *B* and *C* are as follows:

A: getting an even number on the first die

B: getting an odd number on the first die

C: getting the sum of the numbers on the dice ≤ 5

State true or false: (give reasons for your answer)

(i) *A* and *B* are mutually exclusive

(ii) *A* and *B* are mutually exclusive and exhaustive

(iii) $A = B'$

(iv) *A* and *C* are mutually exclusive

(v) *A* and B' are mutually exclusive

(vi) A' B' and C' are mutually exclusive and exhaustive

Solution:

$$A = \begin{Bmatrix} (2,1),(2,2),(2,3),(2,4),(2,5),(2,6),(4,1), \\ (4,2),(4,3),(4,4),(4,5),(4,6), \\ (6,1),(6,2),(6,3),(6,4),(6,5),(6,6) \end{Bmatrix}$$

$$B = \begin{Bmatrix} (1,1),(1,2),(1,3),(1,4),(1,5),(1,6), \\ (3,1),(3,2),(3,3),(3,4),(3,5),(3,6), \\ (5,1),(5,2),(5,3),(5,4),(5,5),(5,5) \end{Bmatrix}$$

$$C = \begin{Bmatrix} (1,1),(1,2),(1,3),(1,4),(2,1), \\ (2,2),(2,3),(3,1),(3,2),(4,1) \end{Bmatrix}$$

(i) It is observed that $A \cap B = \phi$

$\therefore A, B$ are mutually exclusive. Thus, the given statement is true.

(ii) It is observed that $A \cap B = \phi$ and $A \cup B = S \therefore A, B$ are mutually exclusive and exhaustive. Thus, the given statement is true.

(iii) It is observed that

$$B' = \begin{Bmatrix} (2,1),(2,2),(2,3),(2,4),(2,5),(2,6) \\ (4,1),(4,2),(4,3),(4,4),(4,5),(4,6) \\ (6,1),(6,2),(6,3),(6,4),(6,5),(6,6) \end{Bmatrix} = A$$

Thus, the given statement is true.

(iv) It is observed that

$A \cap C = \{(2,1),(2,2),(2,3),(4,1)\} \ne \phi$

$\therefore A, C$ are not mutually exclusive. Thus, the given statement is false.

(v) $A \cap B' = A \cap A = A$

$\therefore A \cap B' \ne \phi$

$\therefore A, B'$ are not mutually exclusive. Thus, the given statement is false.

(vi) It can be observed that: $A' \cup B' \cup C = S$

However, $B' \cap C = \{(2,1),(2,2),(2,3),(4,1)\} \ne \phi$

Therefore, events A', B' and C are not mutually exclusive and exhaustive. Thus, the given statement is false.

Day 213

Probability of an Event

Axiomatic Approach to Probability: Let *S* be the sample space of a random experiment. Then the probability *P* is a real-valued function whose domain is the power set of *S* and whose range is the interval [0, 1] satisfying the following axioms:

(i) For any event *E*, $P(E) \ge 0$

(ii) $P(S) = P(E_1) + P(E_2) + P(E_3) ... + P(E_n) = 1$

(iii) If *E* and *F* are mutually exclusive events, then $P(E \cup F) = P(E) + P(F)$, that is, $P(E \cap F) = \phi$

In general, let *S* be a sample space containing outcomes $\omega_1, \omega_2, ..., \omega_n$, that is, $S = \{\omega_1, \omega_2, ..., \omega_n\}$. Then

(i) $0 \le P(\omega_i) \le 1$ for each $\omega_i \in S$

(ii) $P(\omega_1) + P(\omega_2) + ... + P(\omega_n) = 1$

(iii) For any event *A*, $P(A) = \sum P(\omega_i),\ \omega_i \in A$.

Note: The singleton $\{\omega_i\}$ is called the elementary event and is written as $P(\omega_i)$ for $P(\{\omega_i\})$.

Probability of an Event: Let *S* be a sample space associated with a random experiment, and *n*(*S*) be the number of elements (or outcomes) of the sample space *S*. Let *E* be the event of a sample space *S* associated with the random experiment, and *n*(*E*) be the number of elements (or outcomes) of the event *E*.

Then, the probability of event *E* is given by

$$P(E) = \frac{\text{Number of outcomes of event } E}{\text{Number of outcomes of sample space } S}$$

$$\Rightarrow P(E) = \frac{n(E)}{n(S)}$$

Probabilities of Equally Likely Outcomes: Let a sample space of an experiment be $S = \{\omega_1, \omega_2, ..., \omega_n\}$.

All the outcomes are equally likely to occur. This means the chance of occurrence of each simple event must be the same, that is, $P(\omega_i) = p$, for all $\omega_i \in S$, where $0 \le p \le 1$

Since $\sum_{i=1}^{n} P(\omega_i) = 1 \Rightarrow p + p + ... + p(n\text{ times}) = 1$

$\Rightarrow np = 1 \Rightarrow p = \frac{1}{n}$

Probability of the Event 'A or B'

$P(A \cup B) = P(A) + P(B) - P(A \cap B)$.

Note: If A and B are disjoint sets, that is, they are mutually exclusive events, then $A \cap B = \phi$

Therefore, $P(A \cap B) = P(\phi) = 0$

Thus, for mutually exclusive events A and B, we get

$P(A \cup B) = P(A) + P(B)$

Probability of the Event 'not A' (A')

$P(A) + P(A') = 1 \Rightarrow P(A') = P(\text{not } A) = 1 - P(A)$

Note: A' and A are mutually exclusive and exhaustive events, that is

$A \cap A' = \phi,\ A \cup A' = S$

$\Rightarrow P(A \cap A') = 0,\ P(A \cup A') = P(S)$

Example 1: Which of the following cannot be a valid assignment of probabilities for outcomes of sample space $S = \{\omega_1, \omega_2, \omega_3, \omega_4, \omega_5\}$?

Assignment	ω_1	ω_2	ω_3	ω_4	ω_5
(a)	0.1	0.15	0.5	0.2	0.05
(b)	0.2	0.2	0.2	0.2	0.2
(c)	0.1	0.2	0.3	0.4	0.5
(d)	–0.1	0.6	0.3	0.4	–0.2
(e)	0.1	0.2	1.3	0.4	0.5

Solution:

(a)

ω_1	ω_2	ω_3	ω_4	ω_5
0.1	0.15	0.5	0.2	0.05

Each of the numbers $p(\omega_1)$ is positive and less than 1.
Also, the sum of the probabilities
$= p(\omega_1) + p(\omega_2) + p(\omega_3) + p(\omega_4) + p(\omega_5)$
$= 0.1 + 0.15 + 0.5 + 0.2 + 0.05 = 1$
Therefore, the assignment is valid.

(b)

ω_1	ω_2	ω_3	ω_4	ω_5
0.2	0.2	0.2	0.2	0.2

Each of the numbers $p(\omega_1)$ is positive and less than 1. Each of the numbers $p(\omega_1) = 0.2$. Therefore, this represents equally likely events.
Also, the sum of the probabilities
$= p(\omega_1) + p(\omega_2) + p(\omega_3) + p(\omega_4) + p(\omega_5)$
$= 0.2 + 0.2 + 0.2 + 0.2 + 0.2 = 5 \times 0.2 = 1$
Therefore, the assignment is valid.

(c)

ω_1	ω_2	ω_3	ω_4	ω_5
0.1	0.2	0.3	0.4	0.5

Each of the numbers $p(\omega_1)$ is positive and less than 1.
However, the sum of the probabilities
$= p(\omega_1) + p(\omega_2) + p(\omega_3) + p(\omega_4) + p(\omega_5)$
$= 0.1 + 0.2 + 0.3 + 0.4 + 0.5 = 1.5 \neq 1$
Therefore, the assignment is not valid.

(d)

ω_1	ω_2	ω_3	ω_4	ω_5
–0.1	0.6	0.3	0.4	–0.2

$p(\omega_1)$ and $p(\omega_5)$ are negative. Therefore, the assignment is not valid.

(e)

ω1	ω2	ω3	ω4	ω5
0.1	0.2	1.3	0.4	0.5

$p(\omega_3) = 1.3 > 1$. Therefore, the assignment is not valid.

Example 2: A coin is tossed three times. Find the probability of following events:

(i) Exactly one tail occurs
(ii) At least two tails occur
(iii) At most one tail occurs

Solution: When a coin is tossed three times, the sample space is given by $S = \{HHH, HHT, HTH, THH, HTT, THT, TTH, TTT\}$

Thus, the total number of possible outcomes is $n(S) = 8$

(i) Let A be the event of the occurrence of exactly one tail. Then, $A = \{HHT, HTH, THH\} \Rightarrow n(A) = 3$

$\therefore P(A) = \frac{\text{Number of outcomes favourable to A}}{\text{Total number of possible outcomes}}$

$= \frac{n(A)}{n(S)} = \frac{3}{8}$

(ii) Let B be the event of the occurrence of at least two tails. Then

$B = \{TTH, THT, HTT, TTT\} \Rightarrow n(B) = 4$

$\therefore P(B) = \dfrac{\text{Number of outcomes favourable to } B}{\text{Total number of possible outcomes}}$

$= \dfrac{n(B)}{n(S)} = \dfrac{4}{8} = \dfrac{1}{2}$

(iii) Let C be the event of the occurrence of at most one tail. Then

$C = \{HHH, HHT, HTH, THH\} \Rightarrow n(C) = 4$

$\therefore P(C) = \dfrac{\text{Number of outcomes favourable to } C}{\text{Total number of possible outcomes}}$

$= \dfrac{n(C)}{n(S)} = \dfrac{4}{8} = \dfrac{1}{2}$

Example 3: Two dice are thrown and the sum of the numbers that appears on two dice are noted. Find the probability of the following events:

(i) Sum of numbers is prime
(ii) Sum is greater than or equal to 6
(iii) Sum is less than 6
(iv) Sum is an odd number
(v) Sum is divisible by 3

Solution: When two dice are thrown, the sample space for the sum of numbers that appears on the two dice is given by $S = \{2,3,4,5,6,7,8,9,10,11,12\} \Rightarrow n(S) = 11$

(i) Let A be the event of the occurrence of the sum as a prime number. Then $A = \{2,3,5,7,11\} \Rightarrow n(A) = 5$

$\therefore P(A) = \dfrac{\text{Number of outcomes favourable to A}}{\text{Total number of possible outcomes}}$

$= \dfrac{n(A)}{n(S)} = \dfrac{5}{11}$

(ii) Let B be the event of the occurrence of a sum greater than or equal to 6. Then $B = \{6,7,8,9,10,11,12\}$

$\therefore P(B) = \dfrac{\text{Number of outcomes favourable to } B}{\text{Total number of possible outcomes}}$

$= \dfrac{n(B)}{n(S)} = \dfrac{7}{11}$

(iii) Let C be the event of the occurrence of a sum less than 6. Then $C = \{1,2,3,4,5\}$

$\therefore P(C) = \dfrac{\text{Number of outcomes favourable to } C}{\text{Total number of possible outcomes}}$

$= \dfrac{n(C)}{n(S)} = \dfrac{5}{11}$

(iv) Let D be the event of the occurrence of the sum as an odd number. Then $D = \{1,3,5,7,9,11\}$

$\therefore P(D) = \dfrac{\text{Number of outcomes favourable to } D}{\text{Total number of possible outcomes}}$

$= \dfrac{n(D)}{n(S)} = \dfrac{6}{11} = 0$

(v) Let E be the event of the occurrence of a sum divisible by 3. $E = \{3,6,9,12\}$

$\therefore P(E) = \dfrac{\text{Number of outcomes favourable to } E}{\text{Total number of possible outcomes}}$

$= \dfrac{n(E)}{n(S)} = \dfrac{4}{11}$

Example 4: A card is selected from a pack of 52 cards.

(i) Calculate the probability that the card drawn is a black.
(ii) Calculate the probability that the card drawn is a diamond.
(iii) Calculate the probability that the card drawn is a jack.
(iv) Calculate the probability that the card drawn is an ace or a king of hearts.
(v) Calculate the probability that the card drawn is a queen of clubs or spades.

Solution: When a card is selected from a pack of 52 cards, the number of possible outcomes is $52 \Rightarrow n(S) = 52$

(i) Let A be the event in which the card drawn is a black. Then $n(A) = 13 + 13 = 26$ (since in a pack, there are 13 black clubs and 13 black spades)

$\therefore P(A) = \dfrac{n(A)}{n(S)} = \dfrac{26}{52} = \dfrac{1}{2}$

(ii) Let B be the event in which the card drawn is a diamond. Then $n(B) = 13$ (since in a pack, there are 13 diamonds)

$\therefore P(B) = \dfrac{n(B)}{n(S)} = \dfrac{13}{52} = \dfrac{1}{4}$

(iii) Let C be the event in which the card drawn is a jack. Then $n(C) = 4$ (since in a pack, there are 4 jacks)

$\therefore P(C) = \dfrac{n(C)}{n(S)} = \dfrac{4}{52} = \dfrac{1}{13}$

(iv) Let D be the event in which the card drawn is an ace or a king of hearts. Then $n(A) = 1 + 1 = 2$ (since in a pack, there is 1 ace of hearts and 1 king of hearts)

$\therefore P(D) = \dfrac{n(D)}{n(S)} = \dfrac{2}{52} = \dfrac{1}{26}$

(v) Let A be the event in which the card drawn is a queen of clubs or spades. Then $n(A) = 1 + 1 = 2$(since in a pack, there is one queen of clubs and one queen of spades)

$\therefore P(E) = \dfrac{n(E)}{n(S)} = \dfrac{2}{52} = \dfrac{1}{26}$

Example 5: If $P(A) = 0.5$, $P(B) = 0.6$, $P(A \cap B) = 0.3$ Find (i) $P(A \cup B)$, (ii) $P(\text{not } A)$, (iii) $P(\text{not } B)$,

(iv) $P(\text{not } A \text{ or not } B)$

(v) $P(\text{not } A \text{ and not } B)$

Solution: (i) $P(A \cup B) = P(A) + P(B) - P(A \cap B)$

$= 0.5 + 0.6 - 0.3 = 0.8$

(ii) $P(\text{not } A) = P(A') = 1 - P(A) = 1 - 0.5 = 0.5$

(iii) $P(\text{not } B) = P(B') = 1 - P(B) = 1 - 0.6 = 0.4$

(iv) $P(\text{not } A \text{ or not } B) = P(A' \cup B')$

$= P(A \cap B)' = 1 - P(A \cap B) = 1 - 0.3 = 0.7$

(v) $P(\text{not } A \text{ and not } B) = P(A' \cap B')$

$= P(A \cup B)' = 1 - P(A \cup B) = 1 - 0.8 = 0.2$

Day 214

NCERT Exercise 16.3 (With Solutions)

Q1. Which of the following cannot be valid assignment of probabilities for outcomes of sample space $S = \{\omega_1, \omega_2, \omega_3, \omega_4, \omega_5, \omega_6, \omega_7\}$?

Assignment	ω_1	ω_2	ω_3	ω_4	ω_5	ω_6	ω_7
(a)	0.1	0.01	0.05	0.03	0.01	0.2	0.6
(b)	$\frac{1}{7}$	$\frac{1}{7}$	$\frac{1}{7}$	$\frac{1}{7}$	$\frac{1}{7}$	$\frac{1}{7}$	$\frac{1}{7}$
(c)	0.1	0.2	0.3	0.4	0.5	0.6.	0.7
(d)	−0.1	0.2	0.3	0.4	−0.2	0.1	0.3
(e)	$\frac{1}{14}$	$\frac{2}{14}$	$\frac{3}{14}$	$\frac{4}{14}$	$\frac{5}{14}$	$\frac{6}{14}$	$\frac{15}{14}$

Solution:

(a)

ω_1	ω_2	ω_3	ω_4	ω_5	ω_6	ω_7
0.1	0.01	0.05	0.03	0.01	0.2	0.6

Each of the numbers $p(\omega_1)$ is positive and less than 1. Also, the sum of the probabilities

$= p(\omega_1) + p(\omega_2) + p(\omega_3) + p(\omega_4) + p(\omega_5)$

$+ p(\omega_6) + p(\omega_7)$

$= 0.1 + 0.01 + 0.05 + 0.03 + 0.01 + 0.2 + 0.6 = 1$

Therefore, the assignment is valid.

(b)

ω_1	ω_2	ω_3	ω_4	ω_5	ω_6	ω_7
$\frac{1}{7}$	$\frac{1}{7}$	$\frac{1}{7}$	$\frac{1}{7}$	$\frac{1}{7}$	$\frac{1}{7}$	$\frac{1}{7}$

Each of the numbers $p(\omega_1)$ is positive and less than 1. Also, the sum of the probabilities

$= p(\omega_1) + p(\omega_2) + p(\omega_3) + p(\omega_4) + p(\omega_5)$

$+ p(\omega_6) + p(\omega_7)$

$= \frac{1}{7} + \frac{1}{7} + \frac{1}{7} + \frac{1}{7} + \frac{1}{7} + \frac{1}{7} + \frac{1}{7} = 7 \times \frac{1}{7} = 1$

Therefore, the assignment is valid.

(c)

ω_1	ω_2	ω_3	ω_4	ω_5	ω_6	ω_7
0.1	0.2	0.3	0.4	0.5	0.6	0.7

Each of the numbers $p(\omega_1)$ is positive and less than 1. However, the sum of the probabilities

$= p(\omega_1) + p(\omega_2) + p(\omega_3) + p(\omega_4) + p(\omega_5)$

$+ p(\omega_6) + p(\omega_7)$

$= 0.1 + 0.2 + 0.3 + 0.4 + 0.5 + 0.6 + 0.7$

$= 2.8 \neq 1$

Therefore, the assignment is not valid.

(d)

ω_1	ω_2	ω_3	ω_4	ω_5	ω_6	ω_7
−0.1	0.2	0.3	0.4	−0.2	0.1	0.3

$p(\omega_1)$ and $p(\omega_5)$ are negative. Therefore, the assignment is not valid.

(e)

ω1	ω2	ω3	ω4	ω5	ω6	ω7
$\frac{1}{14}$	$\frac{2}{14}$	$\frac{3}{14}$	$\frac{4}{14}$	$\frac{5}{14}$	$\frac{6}{14}$	$\frac{15}{14}$

$p(\omega_7) = \frac{15}{14} > 1$. Therefore, the assignment is not valid.

Q2. A coin is tossed twice. What is the probability that at least one tail occurs?

Solution: When a coin is tossed twice, the sample space is given by $S = \{HH, HT, TH, TT\}$

Let A be the event of the occurrence of at least one tail. Then $A = \{HT, TH, TT\}$

$\therefore P(A) = \frac{\text{Number of outcomes favourable to A}}{\text{Total number of possible outcomes}}$

$= \frac{n(A)}{n(S)} = \frac{3}{4}$

Q3. A die is thrown. Find the probability of the following events:

(i) A prime number will appear
(ii) A number greater than or equal to 3 will appear
(iii) A number less than or equal to 1 will appear
(iv) A number greater than 6 will appear
(v) A number less than 6 will appear

Solution: The sample space of the given experiment is given by $S = \{1,2,3,4,5,6\}$

(i) Let A be the event of the occurrence of a prime number. Then $A = \{2,3,5\}$

$$\therefore P(A) = \frac{\text{Number of outcomes favourable to A}}{\text{Total number of possible outcomes}}$$

$$= \frac{n(A)}{n(S)} = \frac{3}{6} = \frac{1}{2}$$

(ii) Let B be the event of the occurrence of a number greater than or equal to 3. Then $B = \{3,4,5,6\}$

$$\therefore P(B) = \frac{\text{Number of outcomes favourable to } B}{\text{Total number of possible outcomes}}$$

$$= \frac{n(B)}{n(S)} = \frac{4}{6} = \frac{2}{3}$$

(iii) Let C be the event of the occurrence of a number less than or equal to 1. Then $C = \{1\}$

$$\therefore P(C) = \frac{\text{Number of outcomes favourable to } C}{\text{Total number of possible outcomes}}$$

$$= \frac{n(C)}{n(S)} = \frac{1}{6}$$

(iv) Let D be the event of the occurrence of a number greater than 6. Then $D = \phi$

$$\therefore P(D) = \frac{\text{Number of outcomes favourable to } D}{\text{Total number of possible outcomes}}$$

$$= \frac{n(D)}{n(S)} = \frac{0}{6} = 0$$

(v) Let E be the event of the occurrence of a number less than 6. Accordingly, $E = \{1,2,3,4,5\}$

$$\therefore P(E) = \frac{\text{Number of outcomes favourable to } E}{\text{Total number of possible outcomes}}$$

$$= \frac{n(E)}{n(S)} = \frac{5}{6}$$

Q4. A card is selected from a pack of 52 cards.
(a) How many points are there in the sample space?
(b) Calculate the probability that the card is an ace of spades.
(c) Calculate the probability that the card is (i) an ace (ii) a black card.

Solution: (a) When a card is selected from a pack of 52 cards, the number of possible outcomes is 52, that is, the sample space contains 52 elements.
Therefore, there are 52 points in the sample space.

(b) Let A be the event in which the card drawn is an ace of spades. Accordingly, $n(A) = 1$

$$\therefore P(A) = \frac{\text{Number of outcomes favourable to } A}{\text{Total number of possible outcomes}}$$

$$= \frac{n(A)}{n(S)} = \frac{1}{52}$$

(c) (i) Let E be the event in which the card drawn is an ace. There are 4 aces in a pack of 52 cards. Therefore $n(E) = 4$

$$\therefore P(E) = \frac{\text{Number of outcomes favourable to } E}{\text{Total number of possible outcomes}}$$

$$= \frac{n(E)}{n(S)} = \frac{4}{52} = \frac{1}{13}$$

(ii) Let F be the event in which the card drawn is black. There are 26 black cards in a pack of 52 cards. Therefore $n(F) = 26$

$$\therefore P(F) = \frac{\text{Number of outcomes favourable to } F}{\text{Total number of possible outcomes}}$$

$$= \frac{n(F)}{n(S)} = \frac{26}{52} = \frac{1}{2}$$

Q5. A fair coin with 1 marked on one face and 6 on the other and a fair die are both tossed. Find the probability that the sum of numbers that turn up is (1) 3 (ii) 12

Solution: Since the fair coin has 1 marked on one face and 6 on the other, and the die has six faces that are numbered 1, 2, 3, 4, 5 and 6, the sample space is given by
$S = \{(1,1),(1,2),(1,3),(1,4),(1,5),(1,6),$
$(6,1),(6,2),(6,3),(6,4),(6,5),(6,6)\}$
Then $n(S) = 12$

(i) Let A be the event in which the sum of numbers that turn up is 3. Then $A = \{(1,2)\}$

$$\therefore P(A) = \frac{\text{Number of outcomes favourable to } A}{\text{Total number of possible outcomes}}$$

$$= \frac{n(A)}{n(S)} = \frac{1}{12}$$

(ii) Let B be the event in which the sum of numbers that turn up is 12. Then $B = \{(6,6)\}$

$$\therefore P(B) = \frac{\text{Number of outcomes favourable to } B}{\text{Total number of possible outcomes}}$$

$$= \frac{n(B)}{n(S)} = \frac{1}{12}$$

Q6. There are four men and six women on the city council. If one council member is selected for a committee at random, how likely is it that it is a woman?

Solution: There are four men and six women on the city council. As one council member is to be selected for a committee at random, the sample space contains 10 (4+6) elements.

Let A be the event in which the selected council member is a woman. Then $n(A) = 6$

$$\therefore P(A) = \frac{\text{Number of outcomes favourable to } A}{\text{Total number of possible outcomes}}$$

$$= \frac{n(A)}{n(S)} = \frac{6}{10} = \frac{3}{5}$$

Q7. A fair coin is tossed four times, and a person wins Re 1 for each head and loses Rs 1.50 for each tail that turns up. From the sample space, calculate how many different amounts of money you can have after four tosses and the probability of having each of these amounts.

Solution: Since the coin is tossed four times, there can be a maximum of 4 heads or tails.

When 4 heads turn up, $\text{Re}1 + \text{Re}1 + \text{Re}1 + \text{Re}1 = \text{Rs } 4$ is the gain.

When 3 heads and 1 tail turn up.

$\text{Re}1 + \text{Re}1 + \text{Re}1 - \text{Rs } 1.50 = \text{Rs } 3 - \text{Rs } 1.50 = \text{Rs } 1.50$ is the gain.

When 1 head and 2 tails turn up,

$\text{Re}1 + \text{Re}1 - \text{Rs } 1.50 - \text{Rs } 1.50 = -\text{Re } 1$; that is, Re 1 is the loss.

When 1 head and 3 tails turn up,

$\text{Re}1 - \text{Rs } 1.50 - \text{Rs } 1.50 - \text{Rs } 1.50 = -\text{Rs } 3.50$,

That is, Rs 3.50 is the loss.

When 4 tails turn up,

$\text{Rs } 1.50 - \text{Rs } 1.50 - \text{Rs } 1.50 - \text{Rs } 1.50 = -\text{Rs } 6.00$

That is, Rs 6.00 is the loss.

There are $2^4 = 16$ elements in the sample space S, which is given by

$S = \{HHHH, HHHT, HHTH, HTHH, THHH, HHTT, HTTH, TTHH, HTHT, THTH, THHT, HTTT, THTT, TTHT, TTTH, TTTT\}$

$\therefore n(S) = 16$

The person wins Rs 4.00 when 4 heads and one tail turn up, that is, when the event {HHHH} occurs.

∴ Probability (of winning Rs 4.00) = 1/16

The person wins Rs 1.50 when 3 heads and one tail turn up, that is, when the event {HHHT, HHTH, HTHH, THHH} occurs.

∴ Probability (of winning Rs 1.50) =4/16=1/4

The person loses Re 1.00 when 2 heads and 2 tails turn up, that is, when the event {HHTT, HTTH, TTHH, HTHT, THTH, THHT} occurs.

∴ Probability (of losing Re 1.00) =6/16=3/8

The person loses Rs 3.50 when 1 head and 3 tails turn up, that is, when the event {HTTT, THTT, TTHT, TTTH} occurs.

Probability (of losing Rs 3.50) =4/16=1/4

The person loses Rs 6.00 when 4 tails turn up, that is, when the event {TTTT} occurs.

Probability (of losing Rs 6.00) =1/16

Q8. Three coins are tossed once. Find the probability of getting

(i) 3 heads
(ii) 2 heads
(iii) at least 2 heads
(iv) at most 2 heads
(v) no head
(vi) 3 tails
(vii) exactly two tails
(viii) no tail
(ix) at most two tails

Solution: When three coins are tossed once, the sample space is given by

$S = \{HHH, HHT, HTH, THH, HTT, THT, TTH, TTT\}$ ∴ $n(S) = 8$

The probability of an event A is given by

$$P(A) = \frac{\text{Number of outcomes favourable to } A}{\text{Total number of possible outcomes}}$$

$$= \frac{n(A)}{n(S)}$$

(i) Let B be the event of the occurrence of 3 heads. Then B = {HHH}

$$\therefore P(B) = \frac{n(B)}{n(S)} = \frac{1}{8}$$

(ii) Let C be the event of the occurrence of 12 heads. Then C = {HHT, HTH, THH}

$$\therefore P(C) = \frac{n(C)}{n(S)} = \frac{3}{8}$$

(iii) Let D be the event of the occurrence of at least 2 heads. Then D = {HHH, HHT, HTH, THH}

$$\therefore P(D) = \frac{n(D)}{n(S)} = \frac{4}{8} = \frac{1}{2}$$

(iv) Let E be the event of the occurrence of at most 2 heads. Then E = {HHT, HTH, THH, HTT, THT, TTH, TTT}

$$\therefore P(E) = \frac{n(E)}{n(S)} = \frac{7}{8}$$

(v) Let F be the event of the occurrence of no head. Then F = {TTT}

$$\therefore P(F) = \frac{n(F)}{n(S)} = \frac{1}{8}$$

(vi) Let G be the event of the occurrence of 3 tails. Then G = {TTT}

$\therefore P(G) = \dfrac{n(G)}{n(S)} = \dfrac{1}{8}$

(vii) Let H be the event of the occurrence of exactly 2 tails. Then H = {HTT, THT, TTH}

$\therefore P(H) = \dfrac{n(H)}{n(S)} = \dfrac{3}{8}$

(viii) Let I be the event of the occurrence of no tail. Then I = {HHH}

$\therefore P(I) = \dfrac{n(I)}{n(S)} = \dfrac{1}{8}$

(ix) Let J be the event of the occurrence of at most 2 tails. Then J = {HHH, HHT, HTH, THH, HTT, THT, TTH}

$\therefore P(J) = \dfrac{n(J)}{n(S)} = \dfrac{7}{8}$

Q9. If 2/11 is the probability of an event A, what is the probability of the event 'not A'.

Solution: P(A) = 2/11.
Then $P(\text{not } A) = 1 - P(A) = 1 - 2/11 = 9/11$

Q10. A letter is chosen at random from the word 'ASSASSINATION'. Find the probability that the letter is (i) a vowel (ii) a consonant.

Solution: There are 13 letters in the word ASSASSINATION.
∴ Hence, $n(S) = 13$

(i) There are 6 vowels in the given word.
∴ Probability (vowel) = 6/13

(ii) There are 7 consonants in the given word.
∴ Probability (consonant) = 7/13

Q11. In a lottery, a person chose six different natural numbers at random from 1 to 20, and if these six numbers match with the six numbers already fixed by the lottery committee, he wins the prize. What is the probability of winning the prize in the game? [*Hint*: Order of the numbers is not important.]

Solution: The total number of ways in which one can choose six different numbers from 1 to 20

$= {}^{20}C_6 = \dfrac{\lfloor 20}{\lfloor 6 \lfloor 20-6} = \dfrac{\lfloor 20}{\lfloor 6 \lfloor 14}$

$= \dfrac{20 \times 19 \times 18 \times 17 \times 16 \times 15}{1 \cdot 2 \cdot 3 \cdot 4 \cdot 5 \cdot 6} = 38760$

Hence, there are 38760 combinations of 6 numbers.
Of these combinations, one combination is already fixed by the lottery committee.
∴ Required probability of winning the prize in the game

$= \dfrac{1}{38760}$

Q12. Check whether the following probabilities $P(A)$ and $P(B)$ are consistently defined:

(i) $P(A) = 0.5, P(B) = 0.7, P(A \cap B) = 0.6$

(ii) $P(A) = 0.5, P(B) = 0.4, P(A \cup B) = 0.8$

Solution:

(i) $P(A) = 0.5, P(B) = 0.7, P(A \cap B) = 0.6$
It is known that if E and F are two events such that $E \subset F$, then $P(E) \leq P(F)$
However, here, $P(A \cap B) > P(A)$.
Thus, $P(A)$ and $P(B)$ are not consistently defined.

(ii) $P(A) = 0.5, P(B) = 0.4, P(A \cup B) = 0.8$
It is known that if E and F are two events such that $E \subset F$,. Then $P(E) \leq P(F)$
Here, it is seen that
$P(A \cup B) > P(A), P(A \cup B) > P(B)$.
Hence, $P(A)$ and $P(B)$ are consistently defined.

Q13. Fill in the blanks in following table:

	$P(A)$	$P(B)$	$P(A \cap B)$	$P(A \cup B)$
(i)	$\frac{1}{3}$	$\frac{1}{5}$	$\frac{1}{15}$	....
(ii)	0.35	....	0.25	0.6
(iii)	0.5	0.35	......	0.7

Solution: $P(A) = \dfrac{1}{3}, P(B) = \dfrac{1}{5}, P(A \cap B) = \dfrac{1}{15}$

Also $P(A \cup B) = P(A) + P(B) - P(A \cap B)$

$\therefore P(A \cup B) = \dfrac{1}{3} + \dfrac{1}{5} - \dfrac{1}{15} = \dfrac{5+3-1}{15} = \dfrac{7}{15}$

(ii) $P(A) = 0.35, P(A \cap B) = 0.25, P(A \cup B) = 0.6$

$P(A \cup B) = P(A) + P(B) - P(A \cap B)$
$\therefore 0.7 = 0.5 + 0.35 - P(A \cap B)$
$\Rightarrow P(A \cap B) = 0.5 + 0.35 - 0.7$
$\Rightarrow P(A \cap B) = 0.15$

Q14. Given $P(A) = 3/5, P(B) = 1/5$. Find $P(A)$ or $P(B)$, if A and B are mutually exclusive events.

Solution: $P(A) = 3/5, P(B) = 1/5$
For mutually exclusive A and B
$P(A \text{ or } B) = P(A) + P(B)$
$\therefore P(A \text{ or } B) = 3/5 + 1/5 = 4/5$

Q15. If E and F are events such that
$P(E) = 1/4, P(F) = 1/2, P(E \cap F) = 1/8$, find
(i) P (E or F) (ii) P (not E and not F)

Solution:
$P(E) = 1/4, P(F) = 1/2, P(E \cap F) = 1/8$

(i) $P(E \text{ or } F) = P(E) + P(F) - P(E \cap F)$

$\therefore P(E \text{ or } F) = \frac{1}{4} + \frac{1}{2} - \frac{1}{8} = \frac{2+4-1}{8} = \frac{5}{8}$

(ii) From (i), $P(E \text{ or } F) = P(E \cup F) = \frac{5}{8}$

We have $(E \cup F)' = (E' \cap F')$
[By De Morgan's law]
$\therefore P(E' \cap F') = P(E \cup F)'$

Now $P(E \cup F)' = 1 - P(E \cup F) = 1 - \frac{5}{8} = \frac{3}{8}$

$\therefore P(E' \cap F') = \frac{3}{8}$

Thus P *(not E and not F)* $= \frac{3}{8}$

Q16. Events E and F are such that P *(not E or not F)* = 0.25. State whether E and F are mutually exclusive.

Solution: P *(not E or not F)* = 0.25
That is, $P(E' \cup F') = 0.25$
$\Rightarrow P(E \cap F)' = 0.25 [E' \cup F') = (E \cap F)']$
Now $P(E \cap F) = 1 - P(E \cap F)'$
$\Rightarrow P(E \cap F) = 1 - 0.25$
$\Rightarrow P(E \cap F) = 0.75 \neq 0$
$\Rightarrow E \cap F \neq \phi$
Thus, E and F are not mutually exclusive.

Q17. A and B are events such that $P(A) = 0.42$, $P(B) = 0.48$ and $P(A \text{ and } B) = 0.16$. Determine
(i) P *(not A)*
(ii) P *(not B)*
(iii) P *(A or B)*

Solution: P(A) = 0.42, P(B) = 0.48, P(A and B) = 0.16
(i) $P(\text{not } A) = 1 - P(A) = 1 - 0.42 = 0.58$
(ii) $P(\text{not } B) = 1 - P(B) = 1 - 0.48 = 0.52$
(iii) $P(A \text{ or } B) = P(A) + P(B) - P(A, B)$
$\therefore P(A \text{ or } B) = 0.42 + 0.48 - 0.16 = 0.74$

Q18. In Class XI of a school, 40% of the students study Mathematics, 30% study Biology and 10% study both Mathematics and Biology. If a student is selected at random from the class, find the probability that he will be studying Mathematics or Biology.

Solution: Let A be the event in which the selected student studies Mathematics and B be the event in which the selected student studies Biology.
Now

$P(A) = 40\% = 40/100 = 2/5$
$P(B) = 30\% = 30/100 = 3/10$
$P(A \cap B) = 10\% = 10/100 = 1/10$
We know that $P(A \text{ or } B) = P(A) + P(B) - P(A \cap B)$

$\therefore P(A \text{ or } B) = \frac{2}{5} + \frac{3}{10} - \frac{1}{10} = \frac{6}{10} = 0.6$

Thus, the probability that the selected student will be studying Mathematics or Biology is 0.6.

Q19. In an entrance test that is graded on the basis of two examinations, the probability of a randomly chosen student passing the first examination is 0.8 and the probability of the student passing the second examination is 0.7. The probability of the student passing at least one of them is 0.95. What is the probability of the student passing both?

Solution: Let A and B be the events of the student passing the first and second examinations, respectively. Accordingly
$P(A) = 0.8, P(B) = 0.7, P(A \text{ or } B) = 0.95$
We know that $P(A \text{ or } B) = P(A) + P(B) - P(A \text{ and } B)$
$\therefore 0.95 = 0.8 + 0.7 - P(A \text{ and } B)$
$\Rightarrow P(A \text{ and } B) = 0.8 + 0.7 - 0.95 = 0.55$
Thus, the probability of the student passing both the examinations is 0.55.

Q20. The probability that a student will pass the final examination in both English and Hindi is 0.5 and the probability of the student passing neither is 0.1. If the probability of the student passing the English examination is 0.75, what is the probability of the student passing the Hindi examination?

Solution: Let A and B be the events of the student passing the English and Hindi examinations, respectively. Accordingly
$P(A, B) = 0.5 P(\text{not } A, \text{not } B) = 0.1$, that is, $P(A' \cap B') = 0.1$
$P(A) = 0.75$
Now, $(A \cup B)' = (A' \cap B')$
[De Morgan's law]
$\therefore P(A \cup B)' = P(A' \cap B') = 0.1$
$A(A \cup B) = 1 - P(A \cup B)' = 1 - 0.9 = 0.9$
$P(A \text{ or } B) = P(A) + P(B) - P(A \cap B)$
$\therefore 0.9 = 0.75 + P(B) - 0.5$
$\Rightarrow P(B) = 0.9 - 0.75 + 0.5$
$\Rightarrow P(B) = 0.65$
Thus, the probability of the student passing the Hindi examination is 0.65.

Q21. In a class of 60 students, 30 opted for NCC, 32 for NSS and 24 for both NCC and NSS. If one of these students is selected at random, find the probability that:

(i) The student opted for NCC or NSS
(ii) The student opted for neither NCC nor NSS
(iii) The student opted for NSS but not NCC

Solution: Let A be the event in which the selected student opted for NCC and B be the event in which the selected student opted for NSS.

Total number of students = 60

Number of students who opted for NCC = 30

$\therefore P(A) = 30.60 = 1/2$

Number of students who opted for NSS = 32

$\therefore P(B) = \frac{32}{60} = \frac{8}{15}$

Number of students who opted for NCC and NSS = 24

$\therefore P(A \cap B) = \frac{24}{60} = \frac{2}{5}$

(i) We know that $P(A \text{ or } B) = P(A) + P(B) - P(A \cap B)$

$\therefore P(A \text{ or } B) = \frac{1}{2} + \frac{8}{15} - \frac{2}{5} = \frac{15+16-12}{30} = \frac{19}{30}$

Thus, the probability that the selected student opted for NCC or NSS is 19/30.

(ii) $P(\text{not } A \text{ and not } B)$

$= P(A', B')$

$= P(A' \cap B')$

$= P(A \cup B)$

$[(A' \cap B') = (A \cup B)'$ (by De Morgan's law)]

$= 1 - P(A \cup B)$

$= 1 - P(A \text{ or } B)$

$= 1 - \frac{19}{30} = \frac{11}{30}$

Thus, the probability that the selected students opted neither for NCC nor NSS is 11/30.

(iii) This can be represented by a Venn diagram. It is clear that

Number of students who opted for NSS but not NCC

$= n(B - A) = n(B) - n(A \cap B) = 32 - 24 = 8$

Thus, the probability that the selected student opted for NSS but not for NCC $= \frac{8}{60} = \frac{2}{15}$

Day 215

NCERT Miscellaneous Exercise (With Solutions)

Q1. A box contains 10 red marbles, 20 blue marbles and 30 green marbles. Five marbles are drawn from the box. What is the probability that:

(i) All will be blue?
(ii) At least one will be green?

Solution: Total number of marbles =10+20+30=60

Number of ways of drawing 5 marbles from 60 marbles $= {}^{60}C_5$

(i) All the drawn marbles will be blue if we draw 5 marbles out of 20 blue marbles. Five blue marbles can be drawn from 20 blue marbles in ${}^{20}C_5$ ways.

$\therefore$ Probability that all marbles will be blue $= \frac{{}^{20}C_5}{{}^{60}C_5}$

(ii) Number of ways in which the drawn marble is not green $= {}^{(20+10)}C_5$

$\therefore$ Probability that no drawn marble is green $= \frac{{}^{30}C_5}{{}^{60}C_5}$

Probability that at least one marble is green

$= 1 - \frac{{}^{30}C_5}{{}^{60}C_5}$

Q2. Four cards are drawn from a well-shuffled deck of 52 cards. What is the probability of obtaining 3 diamonds and 1 spade?

Solution: Number of ways of drawing 4 cards from 52 cards $= {}^{52}C_4$

In a deck of 52 cards, there are 13 diamonds and 13 spades.

$\therefore$ Number of ways of drawing 3 diamonds and one spade

$= \frac{{}^{13}C_3 \times {}^{13}C_1}{{}^{52}C_4}$

Q3. A die has two faces, each with number '1', three faces each with number '2' and one face with number '3'. If the die is rolled once, determine:

(i) $P(2)$ (ii) $P(1 \text{ or } 3)$ (iii) $P(\text{not } 3)$

Solution: Total number of faces = 6

(i) Number faces with number '2' = 3

$\therefore P(2) = \frac{3}{6} = \frac{1}{2}$

(ii) $P(1 \text{ or } 3) = P(\text{not } 2) = 1 - P(2) = 1 - \frac{1}{2} = \frac{1}{2}$

(iii) Number of faces with number '3' = 1

$\therefore P(3) = \frac{1}{6}$

Thus, $P(\text{not } 3) = 1 - P(3) = 1 - \frac{1}{6} = \frac{5}{6}$

Q4. In a certain lottery, 10,000 tickets are sold and 10 equal prizes are awarded. What is the probability of not getting a prize if you buy:
(a) 1 ticket?
(b) 2 tickets?
(c) 10 tickets?

Solution: Total number of tickets sold = 10000
Number of prizes awarded = 10

(i) If we buy one ticket, then

$$P\ (getting\ a\ prize) = \frac{10}{10000} = \frac{1}{1000}$$

$$\therefore P\ (not\ getting\ a\ prize) = 1 - \frac{1}{1000} = \frac{999}{1000}$$

(ii) If we buy two tickets, then
Number of tickets not awarded =10000 – 10 = 9990

$$P\ (not\ getting\ a\ prize) = \frac{^{9990}C_2}{^{10000}C_2}$$

(iii) If we buy 10 tickets, then

$$P\ (not\ getting\ a\ prize) = \frac{^{9990}C_{10}}{^{10000}C_{10}}$$

Q5. Out of 100 students, two sections of 40 and 60 are formed. If you and your friend are among the 100 students, what is the probability that:
(i) You both enter the same section?
(ii) You both enter different sections?

Solution: My friend and I are among the 100 students.
Total number of ways of selecting 2 students out of 100 students = $^{100}C_2$

(i) The two of us will enter same section if both of us are among 40 students or among 60 students.
$\therefore$ Number of ways in which both of us enter the same section = $^{40}C_2 + {}^{60}C_2$
$\therefore$ Probability that both of us enter the same section

$$= \frac{^{40}C_2 + {}^{60}C_2}{^{100}C_2} = \frac{\frac{\lfloor 40}{\lfloor 2 \lfloor 38} + \frac{\lfloor 60}{\lfloor 2 \lfloor 58}}{\frac{\lfloor 100}{\lfloor 2 \lfloor 98}}$$

$$= \frac{(39\times 40)+(59\times 60)}{99\times 100} = \frac{17}{33}$$

(ii) P *(we enter different sections)*

$= 1 - P$ *(we enter the same section)*

$$= 1 - \frac{17}{33} = \frac{16}{33}$$

Q6. Three letters are dictated to three persons and an envelope is addressed to each of them. The letters are inserted into the envelopes at random so that it contains exactly one letter. Find the probability that at least one letter is in its proper envelope.

Solution: Let L_1, L_2, L_3 be the three letters and E_1, E_2, E_3 be their corresponding envelopes.
There are 6 ways of inserting 3 letters in 3 envelopes. These are as follows:

$$\left.\begin{matrix} L_1E_1, L_2E_3, L_3E_2 \\ L_2E_2, L_1E_3, L_3E_1 \\ L_3E_3, L_1E_2, L_2E_1 \\ L_1E_1, L_2E_2, L_3E_3 \end{matrix}\right]$$

$$L_1E_2, L_2E_2, L_3E_1$$

$$L_1E_3, L_2E_1, L_3E_3$$

There are 4 ways in which at least one letter is inserted in a proper envelope.

Thus, the required probability is $\frac{4}{6} = \frac{2}{3}$

Q7. A and B are two events such that $P(A) = 0.54, P(B) = 0.69, P(A \cap B) = 0.35$. Find
(i) $P(A \cap B)$ (ii) $P(A' \cap B')$
(iii) $P(A \cap B')$ (iv) $P(B \cap A')$

Solution:
$P(A) = 0.54, P(B) = 0.69, P(A \cap B) = 0.35$

(i) $P(A \cup B) = P(A) + P(B) - P(A \cap B)$
$\therefore P(A \cup B) = 0.54 + 0.69 - 0.35 = 0.88$

(ii) $A' \cap B' = (A \cup B)'$ [By De Morgan's law]
$\therefore P(A' \cap B') = P(A \cup B)' = 1 - P(A \cup B)$
$= 1 - 0.88 = 0.12$

(iii) $P(A \cap B') = P(A) - P(A \cap B) = 0.54 - 0.35 = 0.19$

(iv) $n(B \cap A') = n(B) - n(A \cap B)$

$$\Rightarrow \frac{n(B \cap A')}{n(S)} = \frac{n(B)}{n(S)} - \frac{n(A \cap B)}{n(S)}$$

$\therefore P(B \cap A') = P(B) - P(A \cap B)$
$\therefore P(B \cap A') = 0.69 - 0.35 = 0.34$

Q8. From the employees of a company, 5 persons are selected to represent them in the managing committee of the company. The particulars of the five persons are as follows:

S. No.	Name	Sex	Age in years
1.	Harish	M	30
2.	Roshan	M	33
3.	Sheetal	F	46
4.	Alis	F	28
5.	Salim	M	41

A person is selected at random from this group to act as a spokesperson. What is the probability that the spokesperson will be either male or over 35 years?

Solution: Let E be the event in which the spokesperson will be a male and F be the event in which the spokesperson will be over 35 years of age.

Then, $P(E) = 3/5, P(F) = 2/5$

Since there is only one male who is over 35 years of age,

$P(E \cap F) = \frac{1}{5}$

$P(E \cup F) = P(E) + P(F) - P(E \cap F)$

$\therefore P(E \cup F) = \frac{3}{5} + \frac{2}{5} - \frac{1}{5} = \frac{4}{5}$

Thus, the probability that the spokesperson will either be a male or over 35 years of age is 4/5.

Q9. If four-digit numbers greater than 5000 are randomly formed from the digits 0, 1, 3, 5, and 7, what is the probability of forming a number divisible by 5 when:

(i) The digits are repeated?

(ii) The repetition of digits is not allowed?

Solution: (i) When the digits are repeated

Since four-digit numbers greater than 5000 are formed, the leftmost digit is either 7 or 5. The remaining 3 places can be filled by any of the digits 0, 1, 3, 5 or 7 as repetition of digits is allowed.

$\therefore$ Total number of four-digit numbers greater than $5000 = 2 \times 5 \times 5 \times 5 - 1 = 250 - 1 = 249$

[In this case, 5000 cannot be counted, so 1 is subtracted]

A number is divisible by 5 if the digit at the unit's place is either 0 or 5.

$\therefore$ Total number of four-digit numbers greater than 5000 that are divisible by 5

$= 2 \times 5 \times 5 \times 2 - 1 = 100 - 1 = 99$

Thus, the probability of forming a number divisible by 5 when the digits are repeated is $= \frac{99}{249} = \frac{33}{83}$

(ii) When repetition of digits is not allowed

The thousands place can be filled with 5 or 7. The remaining 3 places can filled with any of the remaining 4 digits.

$\therefore$ Total number of four-digit numbers greater than $5000 = 2 \times 4 \times 3 \times 2 = 48$

When the digits at the thousands place is 5, the units place can be filled only with 0 and the tens and hundreds places can be filled with any two of the remaining 3 digits.

Here, the number of four-digit numbers starting with 5 and divisible by 5= $3 \times 2 = 6$

When the digit at the thousands place is 7, the units place can be filled in two ways (0 or 5) and the tens and hundreds places can be filled with any two of the remaining 3 digits.

$\therefore$ The number of four-digit numbers starting with 7 and divisible by $5 = 1 \times 2 \times 3 \times 2 = 12$

$\therefore$ Total number of four-digit numbers greater than 5000 that are divisible by 5= $6 + 12 = 18$

Thus, the probability of getting a number divisible by 5 when the repetition of digits is not allowed is $\frac{18}{48} = \frac{3}{8}$

Q10. The number lock of a suitcase has four wheels, each labelled with 10 digits, that is, from 0 to 9. The lock opens with a sequence of four digits with no repeats. What is the probability of a person getting the right sequence to open the suitcase?

Solution: The number lock has four wheels, each labelled from 0 to 9.

Number of ways of selecting 4 different digits out of the 10 digits = ${}^{10}C_4$

Now, each combination of 4 different digits can be arranged in 4! ways.

$\therefore$ Number of four digits with no repetitions $= 4!\,{}^{10}C_4 = 5040$

There is only one number that can open the suitcase.

Thus, the required probability is $\frac{1}{5040}$

Day 216

NCERT Exemplar Problems: Solutions (Q1–Q15)

Q1. If the letters of the word "ALGORITHM" are arranged at random in a row, what is the probability that the letters 'GOR' remain together as a unit?

Solution: Number of letters in the word 'ALGORITHM' = 9

If 'GOR' remain together, then consider it as one group.

$\therefore$ Number of letters = 6+1 = 7

Number of words if 'GOR' remain together = 7!

Total number of words from the letters of the word 'ALGORITHM' = 9!

$\therefore$ Required probability $= \dfrac{7!}{9!} = \dfrac{1}{72}$

Q2. Six new employees, two of whom are married to each other, are to be assigned six desks that are lined up in a row. If the assignment of employees to desks is made randomly, what is the probability that the married couple will have non-adjacent desks?

Solution: Let the couple occupying adjacent desks be considered as one group.
There are (4 + 1), that is, 5 persons to be assigned.
$\therefore$ Number of ways of assigning these five persons = 5! x 2!
Total number of ways of assigning 6 persons = 6!
$\therefore$ Probability that the couple has adjacent desks
$= \dfrac{5!\times 2!}{6!} = \dfrac{2}{6} = \dfrac{1}{3}$
Probability that the married couple will have non-adjacent desks $= 1 - \dfrac{1}{3} = \dfrac{2}{3}$

Q3. If an integer from 1 through 1000 is chosen at random, find the probability that the integer is a multiple of 2 or a multiple of 9.

Solution: Multiples of 2 from 1 to 1000 are 2, 4, 6, 8, ..., 1000
Let n be the number of terms of the above series.
$\therefore n^{\text{th}}$ term = 1000
$\Rightarrow 2+(n-1)2 = 1000$
$\Rightarrow 2n = 1000 \Rightarrow n = 500$
The number of multiples of 2 is 500.
The multiples of 9 are 9, 18, 27,.....,999
Let m be the number of terms in the above series.
$\therefore m^{\text{th}}$ term = 999
$\Rightarrow 9+(m-1)9 = 999$
$\Rightarrow 9m = 999 \Rightarrow m = 111$
The number of multiples of 9 is 111.
Now, the multiples of 2 and 9 are 18, 36,.......990
Let p be the number of terms in the above series.
$\therefore p^{\text{th}}$ term = 990
$\Rightarrow 18+(p-1)18 = 990$
$\Rightarrow 18p = 990 \Rightarrow p = \dfrac{990}{18} = 55$
The number of multiples of 2 and 9 is 55.
$\therefore$ Number of multiples of 2 or 9
$= 500+111-55 = 556$
$\therefore$ Required probability $= \dfrac{n(E)}{n(S)} = \dfrac{556}{1000} = 0.556$

Q4. An experiment consists of rolling a die until a 2 appears.
(i) How many elements of the sample space correspond to the event that the 2 appears on the kth roll of the die?
(ii) How many elements of the sample space correspond to the event that 2 appears not later than the kth roll of the die?

Solution: In a throw of a die, there is 6 sample points.
(i) If 2 appears on the kth roll of the die.
The first $(k-1)$ roll can have 5 outcomes each, and the kth roll results in 2, that is, 1 outcome.
$\therefore$ Number of elements of the sample space corresponding to the event that 2 appears on the kth roll of the die $= 5^{k-1}$
(ii) If we consider that 2 appears not later than the kth roll of the die, then it is possible that 2 occurs in the first throw, that is, 1 outcome.
If 2 does not appear in the first throw, then the outcomes will be 5 and 2 will occur in the second throw, that is, 1 outcome. Possible outcomes $= 5\times 1 = 5$
Similarly, if 2 does not appear in the second throw but appears in the third throw
$\therefore$ Possible outcomes $5\times 5\times 1$
Given the series $1+5+5\times 5+5\times 5\times 5+....+5^{k-1}$
$= 1+5+5^2+5^3+...+5^{k-1} = \dfrac{1(5^k-1)}{5-1} = \dfrac{5^k-1}{4}$

Q5. A die is loaded in such a way that each odd number is twice as likely to occur as each even number. Find $P(G)$, where G is the event that a number greater than 3 occurs on a single roll of the die.

Solution: Since 2 × probability of even number = Probability of odd number
$\Rightarrow P(O) = 2P(E)$
$\Rightarrow P(O):P(E) = 2:1$
$\therefore$ Probability of an odd number occurring
$P(O) = \dfrac{2}{2+1} = \dfrac{2}{3}$

Probability of each number being a 5 $P(E) = \dfrac{1}{2+1} = \dfrac{1}{3}$
Now, let G be the even that a number greater than 3 occurs in a single roll of the die.
So, the possible outcomes are 4, 5 and 6, out of which two are even and one odd.
$\therefore$ Required probability $= P(G) = 2\times P(E)\times P(O)$
$= 2\times\dfrac{1}{3}\times\dfrac{2}{3} = \dfrac{4}{9}$

Q6. In a large metropolitan area, the probabilities are 0.87, 0.36, 0.30 that a family (randomly chosen for a sample survey) owns a colour television set, a black and white television set or both kinds of sets. What is the probability that a family owns any one or both kinds of sets?

Solution: Let E_1 be the event that the family owns a colour television set and E_2 be the event that the family owns a black and white television set.

Therefore

$P(E_1) = 0.87, P(E_2) = P(E_1 \cap E_2) = 0.30$

Probability that a family owns either one or both kind of sets, that is, $P(E_1 \cup E_2)$.

Now, $P(E_1 \cup E_2) = P(E_1) + P(E_2) - P(E_1 \cap E_2)$

$= 0.87 + 0.36 - 0.30 = 0.93$

Q7. If A and B are mutually exclusive events $P(A) = 0.35, P(B) = 0.45$, then find

(i) $P(A')$

(ii) $P(B')$

(iii) $P(A \cup B)$

(iv) $P(A \cap B)$

(v) $P(A \cap B')$

(vi) $P(A' \cap B')$

Solution: It is given that A and B are mutually exclusive events.

$\therefore P(A \cap B) = 0\ [\because A \cap B = \phi]$

$P(A) = 0.35, P(B) = 0.45$

(i) $P(A') = 1 - P(A) = 1 - 0.35 = 0.65$

(ii) $P(B') = 1 - P(B) = 1 - 0.45 = 0.55$

(iii) $P(A \cup B) = P(A) + P(B) - P(A \cap B)$

$= 0.35 + 0.45 = 0.80$

(iv) $P(A \cap B) = 0$

(v) $P(A \cap B') = P(A) - P(A \cap B) = .35 - 0 = 0.35$

(vi) $P(A' \cap B') = P(A \cup B)' = 1 - P(A \cup B) = 1 - 0.8 = 0.2$

Q8. A team of medical students doing their internship have to assist during surgeries at a city hospital. The probabilities of surgeries rated as very complex, complex, routine, simple or very simple are, respectively, 0.15, 0.20, 0.31, 0.26 and 0.08. Find the probabilities that a particular surgery will be rated:

(i) complex or very comple.

(ii) neither very complex nor very simple

(iii) routine or complex

(iv) routine or simple

Solution: Let E_1, E_2, E_3, E_4, E_5 will be the events that surgeries are rated very complex, complex routine, simple or very simple.

$\therefore P(E_1) = 0.15, P(E_2) = 0.20, P(E_3) = 0.31,$

$P(E_4) = 0.26, P(E_5) = 0.08$

(i) P *(complex or very complex)*

$= P(E_1 \text{ or } E_2)$

$= P(E_1 \cup E_2) = P(E_1) + P(E_2) - P(E_1 \cap E_2)$

$= 0.15 + 0.20 - 0\ [\because P(E_1 \cap E_2) = 0]$

$= 0.35$

Because all events are independent

(ii) P *(neither very complex or very simple)*

$(P(E_1' \cap E_5') = P(E_1 \cup E_5)'$

$= 1 - P(E_1 \cup E_5)$

$= 1 - [P(E_1) + P(E_5)]$

$= 1 - (0.15 + 0.08)$

$= 1 - 0.23 = 0.77$

(iii) P *(routine or complex)*

$= P(E_3 \cup E_2) = P(E_3) + P(E_2)$

$= 0.31 + 0.20 = 0.51$

(iv) P *(routine or simple)*

$= P(E_3 \cup E_4) = P(E_3) + P(E_4)$

$= 0.31 + 0.28 = 0.57$

Q9. Four candidates A, B, C and D have applied for the assignment to coach a school cricket team. If A is twice as likely to be selected as B and B and C are given about the same chance of being selected, while C is twice as likely to be selected as D, then what are the probabilities that:

(i) C will be selected?

(ii) A will not be selected?

Solution: A is twice as likely to be selected as D

$\therefore P(A) = 2P(B) \Rightarrow \frac{P(A)}{2} = P(B)$

C is twice as likely to be selected as D

$P(C) = 2P(D) \Rightarrow P(B) = 2P(D)$

$\Rightarrow \frac{2(A)}{2} = 2P(D) \Rightarrow P(D) = \frac{P(A)}{4}$

B and C are given about the same chance of being selected.

$P(B) = P(C)$

Now, sum of probability =1

$P(A) + P(B) + P(C) + P(D) = 1$

$P(A) + \frac{P(A)}{2} + P\frac{(A)}{2} + \frac{P(A)}{2} = 1$

$\Rightarrow \frac{4P(A) + 2P(A) + 2P(A) + P(A)}{4} = 1$

$\Rightarrow 9P(A) = 4 \Rightarrow P(A) = 4/9$

(i) *P (C will be selected)*

$= P(C) = P(B) = \frac{P(A)}{2}$

$= \frac{4}{9\times 2} = \frac{2}{9}\left[\because P(A) = \frac{9}{4}\right]$

(ii) *P (A will not be selected)*

$= P(A') = 1 - P(A) = 1 - \frac{4}{9} = \frac{5}{9}$

Q10. One of the four persons John, Rita, Aslam or Gurpreet will be promoted next month. Consequently, the sample space consists of four elementary outcomes:
S = {John promoted, Rita promoted, Aslam promoted, Gurpreet promoted}.
You are told that the chance of John's promotion is same as that of Gurpreet, Rita's chances of promotion are twice as likely as John's. Aslam's chances are four times that of John.

(i) Determine:
P (John promoted), P (Rita promoted), P (Aslam promoted), P (Gurpreet promoted)

(ii) If A = {John promoted or Gurpreet promoted}, find *P(A)*.

Solution: Let E_1 = John promoted

E_2 = Rita promoted

E_3 = Aslam promoted

E_4 = Gurpreet promoted

Given the sample space S = {John promoted, Rita promoted, Aslam promoted, Gurpreet promoted}.

That is, $S = \{E_1, E_2, E_3, E_4\}$

The chances of John's promotion are the same as that of Gurpreet.

$P(E_1) = P(E_4)$

Rita's chances of promotion are as likely as John.

$P(E_2) = 2P(E_1)$

And Aslam's chances of promotion are four times that of John.

$P(E_3) = 4P(E_1)$

Now, $P(E_1) + P(E_2) + P(E_3) + P(E_4) = 1$

$\Rightarrow P(E_1) + 2P(E_1) + 4P(E_1) + P(E_1) = 1$

$\Rightarrow 8P(E_1) = 1$

$\therefore P(E_1) = 1/8$

(i) *P (John promoted)* $= P(E_1) = 1/8$

P (Rita promoted) $= P(E_2) = 2P(E_1) = 2\times\frac{1}{8} = \frac{2}{8} = \frac{1}{4}$

P (Aslam promoted) $= P(E_3) = 4P(E_1) = 4\times\frac{1}{8} = \frac{1}{2}$

P (Gurpreet promoted) $= P(E_4) = P(E_1) = \frac{1}{8}$

(ii) A = John promoted or Gurpreet promoted

$P(A) = P(E_1 \cup E_4) = P(E_1) + P(E_4) - P(E_1 \cap E_4)$

$= \frac{1}{8} + \frac{1}{8} - 0 = \frac{2}{8} = \frac{1}{4}$ $[\because P(E_1 \cap E_4) = 0]$

Q11. The accompanying Venn diagram shows three events *A, B, C* and the probabilities of the various intersections [for instance, $P(A \cap B) = 0.07$].

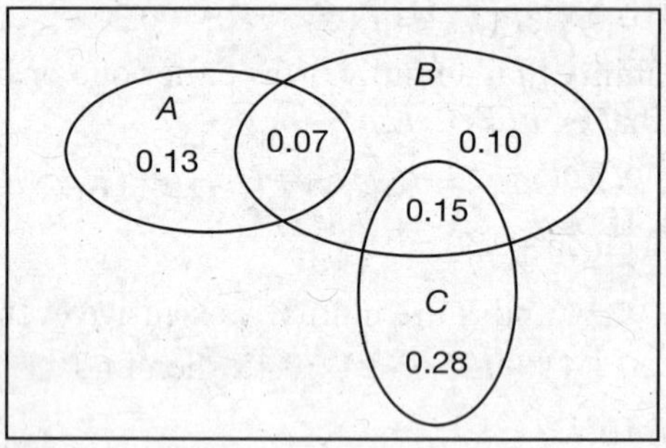

Determine:

(i) $P(A)$

(ii) $P(B \cap \overline{C})$

(iii) $P(A \cup B)$

(iv) $P(A \cap \overline{B})$

(v) $P(B \cap C)$

(vi) Probability of exactly one of the three occuring

Solution: From the above Venn diagram

(i) $P(A) = 0.13 + 0.07 = 0.20$

(ii) $P(B \cap \overline{C}) = P(B) - P(B \cap C) = 0.07$
$+ 0.10 + 0.15 - 0.15 = 0.07 + 0.10 = 0.17$

(iii) $P(A \cup B) = P(A) + P(B) - P(A \cap B)$
$= 0.13 + 0.07 + 0.07 + 0.10 + 0.15 - 0.07$
$= 0.13 + 0.07 + 0.10 + 0.15 = 0.45$

(iv) $P(A \cap \overline{B}) = P(A) - P(A \cap B)$
$= 0.13 + 0.07 - 0.07 = 0.13$

(v) $P(B \cap C) = 0.15$

(vi) *P (exactly one of the three occurs)*
$= 0.13 + 0.10 + 0.28 = 0.51$

Q12. One urn contains two black balls (labelled B_1, B_2) and one white ball. A second urn contains one black ball and two white balls (labelled W_1, W_2). Suppose the following experiment is performed. One of the two urns is chosen at random. Next a ball is randomly chosen from the urn. Then, a second ball is chosen at random from the same urn without replacing the first ball.

(i) Write the sample space showing all possible outcomes.
(ii) What is the probability that two black balls are chosen?
(iii) What is the probability that two balls of opposite colour are chosen?

Solution: One of the two urns is chosen, and a ball is randomly chosen from the urn. Then a second ball is chosen at random from the same urn without replacing the first ball.

(i) ∴ Sample space
$S = \{B_1B_2, B_1W, B_2B_1, B_2W, WB_1, WB_2, BW_1,$
$BW_2, W_1B, W_1W_2, W_2B, W_2W_1\}$
∴ Total sample points =12

(ii) If two black balls are chosen.
So, the favourable events are B_1B_2, B_2B_1, that is, 2
$\therefore$ Required probability $= \frac{2}{12} = \frac{1}{6}$

(iii) If two balls of opposite colour are chosen.
So, the favourable events are
$\{B_1W_1, B_2W_1, WB_1, WB_2, BW_1BW_2, W_1B_1, W_2B\}$
That is, 8.
$\therefore$ Required probability $= \frac{8}{12} = \frac{2}{3}$

Q13. A bag contains 8 red and 5 white balls. Three balls are drawn at random. Find the probability that:
(i) All the three balls are white
(ii) All the three balls are red
(iii) One ball is red and two balls are white

Solution: ∵ Number of red balls =8
And number of white balls =2

(i) *P (all the three balls are white)*
$$= \frac{{}^5C_3}{{}^{13}C_3} = \frac{\frac{5!}{3!2!}}{\frac{13!}{3!10!}} = \frac{5!}{3!2!} \times \frac{3!10!}{13!}$$
$$= \frac{5\times4\times3\times2!}{2!} \times \frac{10!}{13\times12\times11\times10!} = \frac{5\times4\times3}{13\times12\times11} = \frac{5}{143}$$

(ii) *P (all the three balls are red)*
$$= \frac{{}^8C_3}{{}^{13}C_3} = \frac{\frac{8!}{3!5!}}{\frac{13!}{3!10!}} = \frac{8!}{3!\times5!} \times \frac{3!10!}{13!}$$
$$= \frac{8\times7\times6\times5!}{5!} \times \frac{10!}{13\times12\times11\times10!} = \frac{8\times7\times6}{13\times12\times11} = \frac{28}{143}$$

(ii) *P (one ball is red and two balls are white)*
$$= \frac{{}^8C_1 \times {}^5C_2}{{}^{13}C_3} = \frac{8\times10}{13\times6\times11} = \frac{40}{143}$$

Q14. The letters of the word 'ASSASSINATION' are arranged at random. Find the probability that:
(i) Four Ss occur consecutively in the word
(ii) Two Is and two Ns come together
(iii) All As do not occur together
(iv) No two As come together

Solution: Total number of letters in the word 'ASSASSINATION' is 13.
Of which there are, 3 As, 4 Ss, 2 Is, 2 Ns, 1 T and 1 O.

(i) If four Ss occur consecutively in the word, then we considers these 4 Ss as 1 group.
Now, the number of letters is 10.

S S S S	A A A I I N N T O
1	9

Number of words when all Ss are together $= \frac{10!}{3!2!2!}$
Total number of words obtained using letters of the word
'ASSASSINATION'
$$= \frac{13!}{3!4!2!2!}$$
$$\therefore \text{Required probability} = \frac{\frac{10!}{3!2!2!\times13!}}{3!4!2!2!}$$
$$= \frac{10!\times4!}{13!} = \frac{4!}{13\times12\times11} = \frac{24}{1716} = \frac{2}{143}$$

(ii) If 2 Is and 2 Ns come together, then there as 10 alphabets.
Number of words when 2 Is and 2 Ns come together
$$= \frac{10!}{3!4!} \times \frac{4!}{2!2!}$$
∴ Required probability
$$= \frac{\frac{10!4!}{3!4!2!2!}}{\frac{13!}{3!4!2!2!}} = \frac{4!01!}{2!2!3!4!} \times \frac{3!4!2!21}{13!}$$
$$= \frac{4!10!}{13!} = \frac{4!}{13\times12\times11} = \frac{24}{13\times12\times11} = \frac{2}{143}$$

(iii) If all *A*s come together, then there are 11 alphabets.

Number of words when all *A*s come together
$$= \frac{11!}{4!2!2!}$$
Probability that all As come together
$$= \frac{\frac{11!}{4!2!2!}}{\frac{13!}{4!3!2!2!}} = \frac{11!}{4!2!2!} \times \frac{4!3!2!2!}{13!} = \frac{11!\times3!}{13!} = \frac{1}{26}$$
Required probability when all As does not come together $= 1 - \frac{1}{26} = \frac{25}{26}$

(iv) If no two As are together, then first we arrange the alphabets except the As.

S	S	S	S	I	N	T	I	O	N

All the alphabets except As are arranged in $\frac{10!}{4!2!2!}$

There are 11 vacant places between these alphabets.

So, 3 As can be place in 11 places in ${}^{11}C_3$ ways $= \frac{11!}{3!8!}$

∴ Total number of words when no two As come together

$= \frac{11!}{3!8!} \times \frac{10!}{4!2!2!}$

Required probability

$= \frac{11! \times 10!}{3!8!4!2!2!} \times \frac{4!3!2!2!}{13!} = \frac{10!}{8! \times 13! \times 12}$

$= \frac{10 \times 9}{13 \times 12} = \frac{90}{156} = \frac{15}{26}$

Q15. If a card is drawn from a deck of 52 cards, then find the probability of getting a king or a heart or a red card.

Solution: ∵ Number of possible events = 52

Favourable events = 4 kings + 13 hearts + 26 red cards – 13 – 2 = 28

∴ Required probability $= \frac{28}{52} = \frac{7}{13}$

Day 217

NCERT Exemplar Problems: Solutions (Q16–Q43)

Q16. A sample space consists of 9 elementary outcomes $E_1, E_2, ...E_9$ whose probabilities are

$P(E_1) = P(E_2) = 0.08, P(E_3) = P(E_5) = 0.1$

$P(E_6) = P(E_7) = 0.2, P(E_8) = P(E_9) = 0.07$

Suppose $A = \{E_1, E_5, E_8\}, B = \{E_2, E_5, E_8, E_9\}$

(i) Calculate $P(A)$, $P(B)$ and $P(A \cap B)$.

(ii) Using the addition law of probability, calculate $P(A \cup B)$.

(iii) List the composition of the event $A \cup B$ and calculate $P(A \cup B)$ by adding the probabilities of the elementary outcomes.

(iv) Calculate $P(\overline{B})$ from $P(B)$, also calculate $P(\overline{B})$ directly from the elementary outcomes of $\overline{B}$.

Solution:

Given $S = \{E_1, E_2, E_3, E_4, E_5, E_6, E_7, E_8, E_9\}$

$A = \{E_1, E_5, E_8\}, B = \{E_2, E_5, E_8, E_9\}$

$P(E_1) = P(E_2) = 0.08$

$P(E_3) = P(E_4) = P(E_5) = 0.1$

$P(E_6) = P(E_7) = 2, P(E_8) = P(E_9) = 0.07$

(i) $P(A) = P(E_1) + P(E_5) + P(E_8)$

$= 0.08 + 0.1 + 0.07 = 0.25$

(ii) $P(A \cup B) = P(A) + P(B) - P(A \cap B)$ ...(i)

Now, $P(B) = P(E_2) + (E_5) + P(E_8) + P(E_9)$

$= 0.08 + 0.1 + 0.07 + 0.07 = 0.32$

$A \cap B = \{E_5, E_8\}$

$P(A \cap B) = P(E_5) + P(E_8) = 0.1 + 0.7 = 0.17$

On substituting these values in Equation (i), we get

$P(A \cup B) = 0.25 + 0.32 - 0.17 = 0.40$

(iii) $A \cup B = \{E_1, E_2, E_5, E_8, E_9\}$

$P(A \cup B) = P(E_1) + P(E_2) + P(E_5) + P(E_8) + P(E_9)$

$= 0.08 + 0.08 + 0.1 + 0.07 + 0.07 = 0.40$

(iv) ∵ $P(\overline{B}) = 1 - P(B) = 1 - 0.32 = 0.68$

$\overline{B} = \{E_1, E_3, E_4, E_6, E_7\}$

∴ $P(\overline{B}) = P(E_1) + P(E_3) + P(E_4) + P(E_6) + P(E_7)$

$= 0.08 + 0.1 + 0.1 + 0.2 + 0.2 = 0.68$

Q17. Determine the probability p, for each of the following events:

(i) An odd number appears in a single toss of a fair die.

(ii) At least one head appears in two tosses of a fair coin.

(iii) A king, 9 of hearts or 3 of spades appears in drawing a single card from a well-shuffled ordinary deck of 52 cards.

(iv) The sum of 6 appears in a single toss of a pair of fair dice.

Solution: (i) When a die is thrown, the possible outcomes are

$S = \{1, 2, 3, 4, 5, 6\}$, of which 1, 3, 5 are odd.

∴ Required probability $= \frac{3}{6} = \frac{1}{2}$

(ii) When a fair coin is tossed two times, the sample space is S = {HH, HT, TH, TT}

In at least one head, the favourable events are HH, HT, TH

$\therefore$ Required probability =3/4

(iii) Total cards =52

Favourable = 4 king + 2 of hearts + 3 of spades = 4+1+1 = 6

$\therefore$ Required probability $= \frac{6}{52} = \frac{3}{26}$

(iv) When a pair of dice is rolled, the total sample points are 36. Out of which, (1, 5), (5, 1), (2, 4), (4, 2) and (3, 3).

$\therefore$ Required probability =5/36

Q18. In a non-leap year, the probability of having 53 Tuesdays or 53 Wednesdays is:

(a) 1/7 (b) 2/7
(c) 3/7 (d) None of these

Solution: (b)

In a non-leap year, there are 365 days which have 52 weeks and 1 day. If this day is a Tuesday or Wednesday, then the year will have 53 Tuesdays or 53 Wednesdays.

$\therefore$ Required probability = 2/7

Q19. Three numbers are chosen from 1 to 20. Find the probability that they are not consecutive:

(a) $\frac{186}{190}$ (b) $\frac{187}{190}$ (c) $\frac{188}{190}$ (d) $\frac{18}{^{20}C_3}$

Solution: (b)

The set of three consecutive numbers from 1 to 20 is 1,2,3; 2,3,4; 3,4,5;.......;18, 19, 20, that is, 18.

$P \text{ (numbers are consecutive)} = \frac{18}{^{20}C_3} = \frac{18}{1140} = \frac{3}{190}$

P (three numbers are not consecutive)

$= 1 - \frac{3}{190} = \frac{187}{190}$

Q20. While shuffling a pack of 52 playing cards, 2 are accidentally dropped. Find the probability that the missing cards are of different colours:

(a) 29/52 (b) 1/2 (c) 26/51 (d) 27/51

Solution: (c)

In a pack of 52 cards, 26 are red and 26 are black.

$\therefore$ *P (both cards of possible colour)*

$= \frac{26}{52} \times \frac{26}{51} + \frac{26}{52} \times \frac{26}{51} = 2 \times \frac{26}{52} \times \frac{26}{51} = \frac{26}{51}$

Q21. Seven persons are to be seated in a row. Then, the probability that two particular persons sit next to each other is:

(a) 1/3 (b) 1/6 (c) 2/7 (d) 1/2

Solution: (c)

If two persons sit next to each other, consider these two persons as 1 group.

Now, we have to arrange 6 persons.

$\therefore$ Number of arrangements $= 2! \times 6!$

Total number of arrangements of 7 persons = 7!

Required probability $= \frac{2!6!}{7!} = \frac{2}{7}$

Q22. If without repetition, four-digit numbers are formed with the number 0, 2, 3, 5, then the probability of such a number being divisible by 5 is:

(a) 1/5 (b) 4/5 (c) 1/30 (d) 5/9

Solution: (d)

Four-digit numbers using the digit 0, 2, 3, 5 which are divisible by 5 are

If 0 is fixed at the units place $= 3 \times 2 \times 1 = 6$

If 5 is fixed at the units place $= 2 \times 2 \times 1 = 4$

Total four-digit numbers divisible by $5 = 6 + 4 = 10$

$\therefore$ Required probability $= \frac{10}{18} = \frac{5}{9}$

Q23. If A and B are mutually exclusive events, then:

(a) $P(A) \le P(\overline{B})$ (b) $P(A) \ge P(\overline{B})$
(c) $P(A) < P(\overline{B})$ (d) None of these

Solution: (a)

For mutually exclusive events

$P(A \cap B) = 0$

$P(A \cup B) = P(A) + P(B) - P(A \cap B)$

$\Rightarrow P(A \cup B) = P(A) + P(B)$

$\Rightarrow P(A) + P(B) \le 1$

$\Rightarrow P(A) + 1 - P(\overline{B}) \le 1 \quad [\because P(B) = 1 - P(\overline{B})]$

$\therefore P(A) \le P(\overline{B})$

Q24. If $P(A \cup B) = P(A \cap B)$ for any two events A and B, then:

(a) $P(A) = P(B)$ (b) $P(A) > P(B)$
(c) $P(A) < P(B)$ (d) None of these

Solution: (a)

$P(A \cup B) = P(A \cap B)$

$P(A) + P(B) - P(A \cap B) = P(A \cap B)$

$\Rightarrow [P(A) - P(A \cap B)] + [P(B) - P(A \cap B)] = 0$

However $P(A) - P(A \cap B) \ge 0, P(B) - P(A \cap B) \ge 0$

$[\because P(A \cap B) \le P(A) \text{or } P(B)]$

$\Rightarrow P(A) - P(A \cap B) = 0$

$\Rightarrow P(B) - P(A \cap B) = 0$

[Since the sum of two non-negative numbers can be zero only when these numbers are zero]

$\Rightarrow P(A) = P(A \cap B)$ and $P(B) = P(A \cap B)$

$\therefore P(A) = P(B)$

Q25. If 6 boys and 6 girls sit in a row at random, then the probability that all the girls sit together is:

(a) $\frac{1}{432}$ (b) $\frac{12}{431}$

(c) $\frac{1}{132}$ (d) None of these

Solution: (c)

If all the girls sit together, then consider it as 1 group.

$\therefore$ Arrangement of $6 + 1 = 7$ persons in a row is 7! And the girls interchange their seats in 6! ways

$\therefore$ Required probability $= \frac{6!7!}{12!} = \frac{1}{132}$

Q26. If a single letter is selected at random from the word 'PROBABILITY', then the probability that it is a vowel is:

(a) 1/3 (b) 4/11 (c) 2/11 (d) 3/11

Solution: (b)

Total number of alphabets in the word PROBABILITY = 11

Number of vowels = 4

P (letter is vowel) = 4/11

Q27. If the probabilities for *A* to fail in an examination is 0.2 and that for *B* is 0.3, then the probability that either *A* or *B* fails is:

(a) >0.5 (b) 0.5 (c) ≤ 0.5 (d) 0

Solution: (c)

P (A fails) = 0.2

P (B fails) = 0.3

$\therefore$ *P (either A or B fail)* ≤ *P (A fails)* + *P (B fails)*

$\leq 0.2 + 0.3 \leq 0.5$

Q28. The probability that at least one of the events *A* and *B* occurs is 0.6. If *A* and *B* occur simultaneously with probability 0.2, then $P(\overline{A}) + P(\overline{B})$ is equal to:

(a) 0.4 (b) 0.8 (c) 1.2 (d) 1.6

Solution: (c)

$P(A \cup B) = 0.6, P(A \cap B) = 0.2$

$\therefore P(A \cup B) = P(A) + P(B) - P(A \cap B)$

$\Rightarrow 0.6 = P(A) + P(B) - 0.2$

$\Rightarrow P(A) + P(B) = 0.8$

$\therefore P(\overline{A}) + P(\overline{B}) = 1 - P(A) + 1 - P(B)$

$= 2 - [P(A) + P(B)] = 2 - 0.8 = 1.2$

Q29. If *M* and *N* are any two events, the probability that at least one of them occurs is:

(a) $P(M) + P(N) - 2P(M \cap N)$

(b) $P(M) + P(N) - P(M \cap N)$

(c) $P(M) + P(N) + P(M \cap N)$

(d) $P(M) + P(N) + 2P(M \cap N)$

Solution: (b)

If *M* and *N* are any two events

$\therefore P(M \cup N) = P(M) + P(N) - P(M \cap N)$

Q30. The probability that a person visiting a zoo will see a giraffe is 0.72, the probability that he will see a bear is 0.84 and the probability that he will see both is 0.52.

Solution: False

P (to see giraffe) = 0.72

P (to see bear) = 0.84

P (to see giraffe and bear) = 0.52

P (to see giraffe or bear)

= *P (giraffe)* + *P (giraffe and bear)*

= 0.72+0.84−0.52 = 1.04, which is not possible. Thus, the statement is false.

Q31. The probability that a student will pass his examination is 0.73, the probability of the student getting a compartment is 0.13 and the probability that the student will either pass or get a compartment is 0.96.

Solution: False

Let *A* = Student will pass the examination

B = Student will get a compartment

P (A) = 0.73 and *P(A or B)* = 0.96 and *P(B)* = 0.13

$\therefore P(A \text{ or } B) = P(A) + P(B) = 0.73 + 0.13 = 0.86$

But *P (A or B)* = 0.96

Thus, the statement is false.

Q32. The probabilities that a typist will make 0, 1, 2, 3, 4, 5 or more mistakes in typing a report are, respectively, 0.12, 0.25, 0.36, 0.14, 0.08 and 0.11.

Solution: False

Sum of these probabilities must be equal to 1.

$P(0) + P(1) + P(2) + P(3) + P(4) + P(5)$

$= 0.12 + 0.25 + 0.36 + 0.14 + 0.08 + 0.11$

= 1.06, which is greater than 1.

Thus, the statement is false.

Q33. *A* and *B* are two candidates seeking admission in an engineering collage. The probability that *A* is selected is 0.5 and the probability that both *A* and *B* are selected is at most 0.3. Is it possible that the probability of *B* getting selected is 0.7?

Solution: False

$P(A) = 0.5, P(A \cap B) \leq 0.3$

Now $P(A) \times P(B) \leq 0.33$

$\Rightarrow 0.5 \times P(B) \leq 0.3 \Rightarrow P(B) \leq 0.6$

Thus, the statement is false.

Q34. The probability of the intersection of two events A and B is always less than or equal to those favourable to the event A.

Solution: True

$P(A \cap B) \leq P(A)$

Thus, the statement is true.

Q35. The probability of the occurrence of event A is 0.7 and that of the occurrence of event B is 0.3 and the probability occurrence of both is 0.4.

Solution: False

$P(A) = 0.7, P(B) = 0.3$

$\therefore P(A \cap B) = P(A) \times P(B)$

$= 0.7 \times 0.3 = 0.21$

Thus, the statement is false.

Q36. The sum of the probabilities of two students getting distinction in their final examinations is 1.2.

Solution: True

These two events not related to the same sample space. So, the sum of the probabilities of two students getting distinction in their final examination may be 1.2.

Thus, the statement is true.

Q37. The probability that the home team will win an upcoming football game is 0.77, the probability that it will tie the game is 0.08 and the probability that it will lose the game is...................

Solution: P *(losing)* $= 1-(0.77+0.08) = 0.15$

Q38. If e_1, e_2, e_3, e_4 are the four elementary outcomes in a sample space and $P(e_1) = 0.1, P(e_2) = 0.5, P(e_3) = 0.1,$ then the probability of e_4 is.....

Solution:

$\therefore P(e_1) + P(e_2) + P(e_3) + P(e_4) = 1$

$\Rightarrow 0.1 + 0.5 + 0.1 + P(e_4) = 1$

$\Rightarrow 0.7 + P(e_4) = 1 \Rightarrow P(e_4) = 0.3$

Q39. If $S = \{1,2,3,4,5,6\}, E = \{1,3,5\}$, then $\overline{E}$ is..........

Solution: $S = \{1,2,3,4,5,6\}, E = \{1,3,5\}$

$\therefore \overline{E} = S - E = \{2,4,6\}$

Q40. If A and B are two events associated with a random experiment such that $P(A) = 0.3, P(B) = 0.2, P(A \cap B) = 0.1,$ then the value of $P(A \cap \overline{B})$ is.........

Solution:

$P(A \cap \overline{B}) = P(A) - P(A \cap B) = 0.3 - 0.1 = 0.2$

Q41. The probability of an event A occurring is 0.5 and that of B is 0.3. If A and B are mutually exclusive events, then the probability of neither A nor B occurring is........

Solution: $P(\overline{A} \cap \overline{B}) = P(\overline{A \cup B}) = 1 - P(A \cup B)$

$= 1 - [P(A) + P(B)]$

[Since A and B are mutually exclusive]

$= 1 - [0.5 + 0.3] = 1 - 0.8 = 0.2$

Q42. Match the following.

Column I	Column II
(i) 0.95	(a) An incorrect assignment
(ii) 0.02	(b) No chance of happening
(iii) −0.3	(c) As much chance of happening as not
(iv) 0.5	(d) Very little chance of happening
(v) 0	(e) Is very likely happen, so it is close to 1

Solution: (i) 0.95 is very likely happen, so it is close to 1.

(ii) 0.02 means very little chance of happening because probability is very low.

(iii) −0.3 is an incorrect assignment because the probability of any event lies between 0 and 1.

(iv) 0.5 means as much chance of happening as not because the sum of chances of happening and not happening is zero.

(v) 0, so no chance of happening.

Q43. Match the following.

Column I

(i) If E_1, E_2 are the two mutually exclusive events

(ii) If E_1, E_2 are mutually exclusive and exhaustive events

(iii) If E_1, E_2 have common outcomes

(iv) If E_1, E_2 are two events such that $E_1 \subset E_2$

Column II

(a) $E_1 \cap E_2 = E_1$

(b) $(E_1 - E_2) \cup (E_1 \cap E_2) = E_1$

(c) $E_1 \cap E_2 = \phi, E_1 \cup E_2 = 5$

(d) $E_1 \cap E_2 = \phi$

Solution: (i) If E_1, E_2 are two mutually exclusive event, then $E_1 \cap E_2 = \phi$.

(ii) If E_1, E_2 are mutually exclusive and exhaustive events, then $E_1 \cap E_2 = \phi, E_1 \cup E_2 = S$.

(iii) If E_1, E_2 have common outcomes, then

$$(E_1 - E_2) \cup (E_1 \cap E_2) = E_1$$

(iv) If E_1, E_2 are two events such that

$$E_1 \subset E_2 \Rightarrow E_1 \cap E_2 = E_1$$

Day 218

Very Short Answer (VSA): 1-Mark and Short Answer (SA): 2-Mark Questions

Q1. Describe the sample space for the indicated experiment: A coin is tossed three times.

Solution: Since a coin has either head H or tail T, when a coin is tossed three times, the total number of possible outcomes or sample space is $2^3 = 8$, given by:
$S = \{HHH, THH, HTH, HHT, TTH, THT, HTT, TTT\}$

Q2. Describe the sample space for the indicated experiment: A die is thrown two times.

Solution: A die thrown has six possible outcomes: 1, 2, 3, 4, 5, 6.
Therefore, when a die is thrown two times, the total number of possible outcomes or sample space is $6^2 = 36$, given by $S = \{(x, y) : x, y = 1, 2, 3, 4, 5, 6\}$

$$\begin{aligned} S = \{&(1,1),(1,2),(1,3),(1,4),(1,5),(1,6),\\ &(2,1),(2,2),(2,3),(2,4),(2,5),(2,6),\\ &(3,1),(3,2),(3,3),(3,4),(3,5),(3,6),\\ &(4,1),(4,2),(4,3),(4,4),(4,5),(4,6),\\ &(5,1),(5,2),(5,3),(5,4),(5,5),(5,6),\\ &(6,1),(6,2),(6,3),(6,4),(6,5),(6,6)\} \end{aligned}$$

Q3. Describe the sample space for the indicated experiment: A coin is tossed and a die is thrown.

Solution: Since a coin has two faces: head (H) and tail (T), and a die has six faces numbered from 1 to 6, when a coin is tossed and a die is thrown, the sample space is given by

$$\begin{aligned} S = \{&H1, H2, H3, H4, H5, H6,\\ &T1, T2, T3, T4, T5, T6\} \end{aligned}$$

Q4. Describe the sample space for the indicated experiment: A die is thrown and then a coin is tossed only in case an even number is shown on the die.

Solution: Since a coin has two faces: head (H) and tail (T), and a die has six faces numbered from 1 to 6, when a die is thrown and then a coin is tossed only in case an even number is shown on the die, it is given by:

$S = \{1, 2H, 2T, 3, 4H, 4T, 5, 6H, 6T\}$

Q5. Two boys and 1 girl are in Room *A*, and 1 boy and 2 girls are in Room *B*. Specify the sample space for the experiment in which a room is selected and then a person.

Solution: Let us denote 2 boys and 1 girl in Room X as B_1, B_2 and G_1, respectively, and denote 1 boy and 2 girls in room Y as B_3 and G_2, G_3, respectively. The sample space is given by:
$S = \{XB_1, XB_2, XG_1, YB_3, YG_2, YG_3\}$

Q6. One die of white colour and one of black colour are placed in a bag. One die is selected at random and rolled, its colour and the number on its uppermost face is noted. Describe the sample space.

Solution: Let us denote the white and black dices as *W* and *B*, respectively.
When a die is selected and then rolled, the sample space is given by:

$$\begin{aligned} S = \{&W1, W2, W3, W4, W5, W6,\\ &B1, B2, B3, B4, B5, B6\} \end{aligned}$$

Q7. A box contains 1 red and 2 green balls. Two balls are drawn at random in succession, (i) with replacement and (ii) without replacement. Write the sample space for this experiment.

Solution: Let us denote the red ball as *R* and a green ball as *G*.

(i) When two balls are drawn at random in succession with replacement, the sample space is given by:
$S = \{RR, RG, GR, GG\}$

(ii) When two balls are drawn at random in succession without replacement, the sample space is given by:
$S = \{RG, GR, GG\}$

Q8. Suppose three machines are selected at random from a lot. Each machine is tested and classified as defective (*D*) or non-defective (*N*). Write the sample space of this experiment.

Solution: The sample space of this experiment is given by:

$S = \{DDD, NDD, DND, DDN,$
$NND, NDN, DNN, NNN\}$

Q9. A coin is tossed. If it shows a tail, we draw a ball from a box which contains 1 red and 2 blue balls. If it shows a head, we throw a die. Find the sample space for this experiment.

Solution: Let us denote 1 red and 2 blue balls as R_1 and B_1, B_2, respectively. The sample space is given by:
$S = \{TR_1, TB_1, TB_2, H1, H2, H3, H4, H5, H6\}$

Q10. A coin is thrown repeatedly until a head comes up. What is the sample space for this experiment?

Solution: In this experiment, head may come up on the first throw, or on the second throw, or on the third throw, and so on.
The sample space is given by:
$S = \{H, TH, TTH, TTTH, \ldots\ TTT\ldots TH\}$

Q11. Three coins are tossed once. Let
A denote the event 'three heads show',
B denote the event "two heads one tail show",
C denote the event "at least two tails show",
D denote the event 'a head shows on the first coin"
Which events are
(i) simple?
(ii) compound?
(iii) mutually exclusive?
(iv) mutually exclusive and exhaustive?
(v) mutually exclusive and but not exhaustive?

Solution: When three coins are tossed, the sample space is given by:
$S = \{HHH, HHT, HTH, HTT, THH, THT, TTH, TTT\}$
The different events are given by:
$A = \{HHH\}$ $B = \{HHT, HTH, THH\}$
$C = \{TTH, THT, HTT, TTT\}$
$D = \{HHH, HHT, HTH, HTT\}$

(i) If an event has only one sample element of a sample space, it is called a simple event. Thus, event $A = \{HHH\}$ is a simple event.
(ii) If an event has more than one sample element of a sample space, it is called a compound event. Thus, events B, C and D are compound events.
(iii) $A \cap B = \phi$, $A \cap C = \phi$ and $B \cap C = \phi$ $\therefore$ Events A and B, A and C, B and C all are mutually exclusive.
(iv) $\because A \cap B = B \cap C = C \cap A = \phi$ and $A \cup B \cup C = S$
Events A, B and C are mutually exclusive and exhaustive events.
(v) $\because A \cap B = \phi$ but $A \cup B \neq S$
Events A and B are mutually exclusive but not exhaustive.

Q12. A die is thrown. Describe the following events:
(i) A: a number less than 7
(ii) B: a number greater than 7
(iii) C: a multiple of 3
(iv) D: an even number
(v) E: an odd number
Also find
$A \cup B, A \cap B, C \cup D, C \cap E, A - C, C \cap D', C \cup E'$

Solution: When a die is thrown, the sample space is given by:
$S = \{1, 2, 3, 4, 5, 6\}$
The different events are given by
$A = \{1, 2, 3, 4, 5, 6\}$ $B = \phi$
$C = \{3, 6\}$ $D = \{2, 4, 6\}$
$E = \{1, 3, 5\}$
$A \cup B = \{1, 2, 3, 4, 5, 6\}$
$A \cap B = \{\phi\}$
$C \cup D = \{2, 3, 4, 6\}$
$C \cap E = \{3\}$
$A - C = \{1, 2, 4, 5\}$
$C \cap D' = \{3\}$
$C \cup E' = \{2, 3, 4, 6\}$

Q13. An experiment involves rolling a pair of dice and recording the numbers that come up. Describe the following events:
A: 1 or 6 occurs on either die
B: The sum is greater than 8
C: The sum is at most 10 and a multiple of 3
D: Getting an even number on the first dice
Also find $A \cup B, C \cap D, A - D, B \cap C'$,

Solution: When a pair of dice is rolled, the sample space is given by $S = \{(x, y) : x, y = 1, 2, 3, 4, 5, 6\}$

$$S = \begin{Bmatrix} (1,1),(1,2),(1,3),(1,4),(1,5),(1,6), \\ (2,1),(2,2),(2,3),(2,4),(2,5),(2,6), \\ (3,1),(3,2),(3,3),(3,4),(3,5),(3,6), \\ (4,1),(4,2),(4,3),(4,4),(4,5),(4,6), \\ (5,1),(5,2),(5,3),(5,4),(5,5),(5,6), \\ (6,1),(6,2),(6,3),(6,4),(6,5),(6,6) \end{Bmatrix}$$

The different events are given by

$$A = \begin{Bmatrix} (1,1),(1,2),(1,3),(1,4),(1,5),(1,6), \\ (2,1),(2,6),(3,1),(3,6),(4,1),(4,6),(5,1),(5,6), \\ (6,1),(6,2),(6,3),(6,4),(6,5),(6,6) \end{Bmatrix}$$

$$B = \begin{Bmatrix} (3,6),(4,5),(4,6),(5,4),(5,5),(5,6) \\ (6,3),(6,4),(6,5),(6,6) \end{Bmatrix}$$

$$C = \begin{Bmatrix} (1,2),(1,5),(2,1),(2,4),(3,3),(3,6), \\ (4,2),(4,5),(5,1),(5,4),(6,3), \end{Bmatrix}$$

$$D = \begin{Bmatrix} (2,1),(2,2),(2,3),(2,4),(2,5),(2,6), \\ (4,1),(4,2),(4,3),(4,4),(4,5),(4,6), \\ (6,1),(6,2),(6,3),(6,4),(6,5),(6,6) \end{Bmatrix}$$

Now

$$A \cup B = \begin{Bmatrix} (1,1),(1,2),(1,3),(1,4),(1,5),(1,6), \\ (2,1),(2,6),(3,1),(3,6),(4,1),(4,5), \\ (4,6),(5,1),(5,4),(5,5),(5,6), \\ (6,1),(6,2),(6,3),(6,4),(6,5),(6,6) \end{Bmatrix}$$

$$C \cap D = \begin{Bmatrix} (2,1),(2,4), \\ (4,2),(4,5),(6,3), \end{Bmatrix}$$

$$A - D = \begin{Bmatrix} (1,1),(1,2),(1,3),(1,4),(1,5),(1,6), \\ (3,1),(3,6),(5,1),(5,6), \end{Bmatrix}$$

$$B \cap C' = \begin{Bmatrix} (4,6),(5,5),(5,6) \\ (6,4),(6,5),(6,6) \end{Bmatrix}$$

Q14. A coin is tossed three times. Find the probability of the following events:
(i) At least two tails occur
(ii) At most 2 heads occur

Solution: Sample space
$S = \{HHH, HHT, HTH, THH, HTT, THT, TTH, TTT\}$

(i) Let E be the event of the occurrence of at least two tails, $E = \{HTT, THT, TTH, TTT\}$

$$P(E) = \frac{\text{Number of outcomes favourable to } E}{\text{Total number of possible outcomes}}$$

$$= \frac{n(E)}{n(S)} = \frac{4}{8} = \frac{1}{2}$$

(ii) Let F be the event of the occurrence of at most 1 head, $F = \{HHT, HTH, THH, HTT, THT, TTH, TTT\}$

$$P(F) = \frac{\text{Number of outcomes favourable to } F}{\text{Total number of possible outcomes}}$$

$$= \frac{n(F)}{n(S)} = \frac{7}{8}$$

Q15. A die is thrown. Find the probability of the following events:
(i) A number greater than or equal to 4 will appear
(ii) A number less than 3 will appear
(iii) A multiple of 2 or 3 appear

Solution: Sample space $S = \{1,2,3,4,5,6\}$

(i) Let A be the event of the occurrence of a number greater than or equal to 4, $A = \{4,5,6\}$

$$P(A) = \frac{\text{Number of outcomes favourable to } A}{\text{Total number of possible outcomes}}$$

$$= \frac{n(A)}{n(S)} = \frac{3}{6} = \frac{1}{2}$$

(ii) Let B be the event of the occurrence of a number less than 3, $B = \{1,2\}$

$$P(B) = \frac{\text{Number of outcomes favourable to } B}{\text{Total number of possible outcomes}}$$

$$= \frac{n(B)}{n(S)} = \frac{2}{6} = \frac{1}{3}$$

(iii) Let C be the event of the occurrence of a number multiple of 2 or 3, $C = \{2,3,4,6\}$

$$P(C) = \frac{\text{Number of outcomes favourable to } C}{\text{Total number of possible outcomes}}$$

$$= \frac{n(C)}{n(S)} = \frac{4}{6} = \frac{2}{3}$$

Q16. A card is selected from a pack of 52 cards. Find the probability of the following events:
(i) Card is an ace
(ii) Card is a spade
(iii) Card is an ace or spade

Solution: When a card is selected from a pack of 52 cards, the number of possible outcomes is 52, that is, the sample space contains 52 elements, $n(S) = 52$.

(i) Let A be the event in which the card drawn is an ace, $n(A) = 4$

$$P(A) = \frac{n(A)}{n(S)} = \frac{4}{52} = \frac{1}{13}$$

(ii) Let B be the event in which the card drawn is a spade, $n(B) = 13$

$$P(B) = \frac{n(B)}{n(S)} = \frac{13}{52} = \frac{1}{4}$$

(iii) Let C be the event in which the card drawn is an ace or spade $n(C) = 4 + 13 - 1 = 16$ ($\because$ 1 ace of spade)

$$P(C) = \frac{n(C)}{n(S)} = \frac{16}{52} = \frac{4}{13}$$

Q17. A fair coin and a fair die are tossed together. Find the probability of following events:
(i) Head on the coin and multiple of 3 on the die
(ii) Tail on the coin and an even number on the die

Solution: Sample space
$S = \{(H,1),(H,2),(H,3),(H,4),(H,5),(H,6),$
$(T,1),(T,2),(T,3),(T,4),(T,5),(T,6)\}$

$\Rightarrow n(S) = 12$

(i) Let A be the event of getting head on the coin and a multiple of 3 on the die, $A = \{(H,3),(H,6)\}$
$\Rightarrow n(A) = 2$

$$P(A) = \frac{n(A)}{n(S)} = \frac{2}{12} = \frac{1}{6}$$

(ii) Let B be the event of getting tail on the coin and an even number on the die, $B = \{(T,2),(T,4),(T,6)\}$
$\Rightarrow n(B) = 3$

$$P(B) = \frac{n(B)}{n(S)} = \frac{3}{12} = \frac{1}{4}$$

Q18. Three coins are tossed once. Find the probability of getting:
(i) No head (ii) At most 1 head
(iii) Exactly 2 heads (iv) At least 2 heads

Solution: Sample space
$S = \{HHH, HHT, HTH, THH, HTT, THT, TTH, TTT\}$
$\Rightarrow n(S) = 8$

(i) Let A be the event of getting no head, $A = \{TTT\}$
$\Rightarrow n(A) = 1$

$$P(A) = \frac{n(A)}{n(S)} = \frac{1}{8}$$

(ii) Let B be the event of getting at most 1 head, $B = \{HTT, THT, TTH, TTT\} \Rightarrow n(B) = 4$

$$P(B) = \frac{n(B)}{n(S)} = \frac{4}{8} = \frac{1}{2}$$

(iii) Let C be the event of getting exactly 2 heads, $C = \{HHT, HTH, THH\} \Rightarrow n(C) = 3$

$$P(C) = \frac{n(C)}{n(S)} = \frac{3}{8}$$

(iv) Let D be the event of getting at least 2 heads, $D = \{HHH, HHT, HTH, THH\} \Rightarrow n(D) = 4$

$$P(B) = \frac{n(B)}{n(S)} = \frac{4}{8} = \frac{1}{2}$$

Day 219

Long Answer-I (LA-I): 4-Mark and Long Answer-II (LA-II): 6-Mark Questions

Q1. There are three men and four women on the city council. If one council member is selected for a committee at random, how likely is that it is a woman?

Solution: Since one council member is to be selected for a committee at random, the sample space $n(S) = 3 + 4 = 7$ (three men and four women)
Let A be the event in which the selected council member is a woman. Therefore $n(A) = 4$

$$P(A) = \frac{n(A)}{n(S)} = \frac{4}{7}$$

Q2. A fair coin is tossed three times, and a person will win Rs 2 for each head and lose Re 1 for each tail that turns up. Find the probability of a person winning Rs. 3.

Solution: Since the coin is tossed three times, there can be a maximum of 3 heads or tails.
A person wins Rs. 5 if he gets 2 heads and 1 tail, that is, Rs 2 + Rs 2 – Re 1 = Rs. 5
Sample space
$S = \{HHH, HHT, HTH, THH, HTT, THT, TTH, TTT\}$
$\Rightarrow n(S) = 8$
Let A be the event of getting exactly 2 heads, $A = \{HHT, HTH, THH\} \Rightarrow n(A) = 3$

$$P(A) = \frac{n(A)}{n(S)} = \frac{3}{8}$$

∴ Probability (of winning Rs 3) = 3/8

Q3. If 2/5 is the probability of an event A, what is the probability of the event 'not 'A.

Solution: $P(A) = 2/5$
$\therefore P(\text{not } A) = 1 - P(A) = 1 - 2/5 = 3/5$

Q4. A letter is chosen at random from the word 'MATHEMATICS'. Find the probability that the letter is (i) a vowel (ii) an consonant.

Solution: There are 11 letters in the word 'MATHEMATICS'. $\therefore P(S) = 11$
(i) There are 4 vowels in the given word.
∴ Probability (vowel) = 4/11
(ii) There are 7 consonants in the given word.
∴ Probability (consonant) = 7/11

Q5. In a lottery, a person choses four different natural numbers at random from 1 to 10, and if these four numbers match with the four numbers already fixed by the lottery committee, he wins the prize. What is the probability of him winning the prize in the game? (Given: Order of the numbers is not considered)

Solution: Total number of ways in which one can choose four different numbers from 1 to 10

$$n(S) = {}^{10}C_4 = \frac{10!}{4!6!} = 210$$

Of these 210 combinations, only one combination is already fixed by the lottery committee.
∴ Required probability of winning the prize in the game is $= \frac{1}{210}$

Q6. If $P(A)=\frac{1}{3}$, $P(B)=\frac{1}{5}$, $P(A\cup B)=\frac{1}{2}$, then find $P(\text{not } A)$ and $P(A\cap B)$ and state whether A and B are mutually exclusive events or not.

Solution:

$P(\text{not } A)=1-P(A)=1-\frac{1}{3}=\frac{2}{3}$

$P(A\cup B)=P(A)+P(B)-P(A\cap B)$

$\Rightarrow P(A\cap B)=P(A)+P(B)-P(A\cup B)$

$=\frac{1}{3}+\frac{1}{5}-\frac{1}{2}=\frac{10+6-15}{30}=\frac{1}{30}$

$\because P(A\cap B)\neq 0$, $\therefore$ A and B are not mutually exclusive events.

Q7. In an entrance test that is graded on the basis of two examinations, the probability of a randomly chosen student passing the first examination is 0.6 and the probability of the student passing the second examination is 0.8. The probability of the student passing at least one of them is 0.9. What is the probability of the student passing both exams?

Solution: Let A and B be the events of the student passing the first and second examinations, respectively. Then

$P(A)=0.6$, $P(B)=0.8$, $P(A \text{ or } B)=P(A\cup B)=0.9$

Probability of passing both exams $=P(A \text{ and } B)=P(A\cap B)$

$\because P(A\cap B)=P(A)+P(B)-P(A\cup B)$

$\therefore P(A\cap B)=0.6+0.8-0.9$

$\Rightarrow P(A\cap B)=0.5$

Thus, the probability of the student passing both the examinations is 0.5.

Q8. In a class of 50 students, 20 opted for Math, 30 opted for Biology and 10 opted for both Math and Biology. If one of these students is selected at random, find the probability that:

(i) The student opted for Math or Biology

(ii) The student opted neither for Math nor for Biology

(iii) The student opted for Math but not Biology

Solution: Let A be the event in which the selected student opted for Math and B be the event in which the selected student opted for Biology.

Total number of students = 60

Number of students who opted for Math = 20

$\therefore P(A)=20/50=2/5$

Number of students who opted for Biology = 30

$\therefore P(B)=\frac{30}{50}=\frac{3}{5}$

Number of students who opted for both Math and Biology = 10

$\therefore P(A\cap B)=\frac{10}{50}=\frac{1}{5}$

(i) $P(A \text{ or } B)=P(A\cup B)=P(A)+P(B)-P(A\cap B)$

$\Rightarrow P(A\cup B)=\frac{2}{5}+\frac{3}{5}-\frac{1}{5}=\frac{4}{5}$

Thus, the probability that the selected student opted for Math or Biology is 4/5.

(ii) $P(\text{not } A \text{ and not } B)=P(A'\cap B')=P(A\cup B)'$

$=1-P(A\cup B)=1-\frac{4}{5}=\frac{1}{5}$

Thus, the probability that the selected students neither opted for Math nor for Biology is 1/5.

(iii) $n(A-B')=n(A)-n(A\cap B)=20-10=10$

Thus, the probability that the selected student opted for Math but not for Biology $=\frac{10}{50}=\frac{1}{5}$

Q9. A box contains 8 red marbles, 12 blue marbles and 10 green marbles. Four marbles are drawn from the box. What is the probability that:

(i) All will be red?

(ii) Exactly two will be red?

(iii) At least one will be red?

(iv) At most one will be red?

(v) Exactly one red and 1 blue?

Solution: Total number of marbles = 8 + 12 + 10 = 30

Number of ways of drawing 4 marbles from 30 marbles $={}^{30}C_4$

(i) All the drawn marbles will be red if we draw 4 marbles out of 8 red marbles, that is, ${}^{8}C_4$ ways.

$\therefore$ Probability that all marbles will be red $=\frac{{}^{8}C_4}{{}^{30}C_4}$

(ii) Exactly two drawn marbles will be red if we draw 2 marbles out of 8 red marbles and 2 from remaining 22 (blue + green) marbles, that is, ${}^{8}C_2\,{}^{22}C_2$ ways.

$\therefore$ Probability that exactly two will be red marbles $=\frac{{}^{8}C_2\,{}^{22}C_2}{{}^{30}C_4}$

(iii) At least one drawn marble will be red means total minus no marbles from red.

Thus, drawing no red marbles is possible when we select all 4 marbles from the remaining 22 (blue + green) marbles, that is, ${}^{22}C_4$ ways.

$\therefore$ Probability of at least one red marble $=1-\frac{{}^{22}C_4}{{}^{30}C_4}$

(iv) At most one two drawn marbles will be red means no red marble + one red marble.
Thus, drawing no red marbles is possible when we select all 4 marbles from the remaining 22 (blue + green) marbles, that is, ${}^{22}C_4$ ways.
Exactly one red drawn marble is possible if we draw 1 marble out of 8 red marbles and 3 from remaining 22 (blue + green) marbles, that is, ${}^{8}C_1\,{}^{22}C_3$ ways.
∴ Probability of at most one red marble
$= \dfrac{{}^{22}C_4}{{}^{30}C_4} + \dfrac{{}^{8}C_1\,{}^{22}C_3}{{}^{30}C_4}$

(v) Exactly one red and one blue is possible if we draw 1 marble out of 8 red marbles and 1 marble out of 12 blue marbles and the remaining 2 marbles out of 10 green marbles, that is, ${}^{8}C_1\,{}^{12}C_1\,{}^{10}C_2$ ways.
∴ Probability that exactly one red marble and one blue marble $= \dfrac{{}^{8}C_1\,{}^{12}C_1\,{}^{10}C_2}{{}^{30}C_4}$

Q10. In a certain lottery, 1000 tickets are sold and 10 equal prizes are awarded. What is the probability of not getting a prize if you buy:
(i) 1 ticket (ii) 10 tickets

Solution: Total number of tickets sold =1000
Number prizes awarded =10

(i) If we buy one ticket, then

$$P \text{ (not getting a prize)} = \frac{{}^{990}C_1}{{}^{1000}C_1} = \frac{990}{1000} = \frac{99}{100}$$

(ii) If we buy ten tickets, then

$$P \text{ (not getting a prize)} = \frac{{}^{990}C_{10}}{{}^{1000}C_{10}}$$

Q11. Out of 100 students, two sections of 30 and 70 are formed. If you and your friend are among the 100 students, what is the probability that:
(i) You both enter the same sections?
(ii) You both enter different sections?

Solution: Total number of ways of selecting 2 students out of 100 students $= {}^{100}C_2$

(i) The two of us will enter the same section if both of us are among 30 students or among 70 students.
∴ Number of ways in which both of us enter the same section $= {}^{30}C_2 + {}^{70}C_2$
∴ Probability that both of us enter the same section

$$P = \frac{{}^{30}C_2 + {}^{70}C_2}{{}^{100}C_2}$$

(ii) Probability that both of us enter the different sections

$$= 1 - P = 1 - \frac{{}^{30}C_2 + {}^{70}C_2}{{}^{100}C_2}$$

Q12. Three letters are dictated to three persons and an envelope is addressed to each of them. The letters are inserted into the envelopes at random so that each contains exactly one letter. Find the probability that no letter is in its proper envelope.

Solution: Let a, b, c be the three letters and A, B, C be their corresponding envelopes.
There are 6 ways of inserting 3 letters in 3 envelopes as follows:
aA, bB, cC; aA, bC, cB; aC, bB, cC
aB, bA, cC; aC, bA, cB; aB, bC, cA
There are 2 ways (aC, bA, cB; aB, bC, cA) in which no letter is inserted in a proper envelope.
Thus, the probability that no letter is in its proper envelope is $= 2/6 = 1/3$

Q13. The number lock of a suitcase has 3 wheels, each labelled with 10 digits, from 0 to 9. The lock opens with a sequence of three digits with no repeats. What is the probability of a person getting the right sequence to open the suitcase?

Solution: The number lock has 3 wheels, each labelled with 10 digits, from 0 to 9.
Number of ways of selecting 3 different digits out of the 10 digits $= {}^{10}C_3$
Now, each combination of 3 different digits can be arranged in 3! ways.
∴ Number of four digits with no repetitions
$= 3!\,{}^{10}C_3 = 720$
Since there is only one number that can open the suitcase, the probability of a person getting the right sequence to open the suitcase is $\dfrac{1}{720}$

Q14. If the letters of the word 'VIKASH' are arranged at random in a row, what is the probability that the letters A and K remain together?

Solution: Number of letters in the word 'VIKASH' = 6
If A and K remain together, then consider it as 1 group.
∴ Number of words formed, if A and K remain together
= 5!2!
Total number of words formed from the letters of the word 'VIKASH'= 6!
∴ Probability that the letters A and K remain together
$= \dfrac{5!2!}{6!} = \dfrac{1}{3}$

Q15. If an integer from 1 through 100 is chosen at random, then find the probability that the integer is a multiple of 2 or a multiple of 3.

Solution: nth term of a series is given by

$t_n = a+(n-1)d \Rightarrow n = 1+\frac{t_n - a}{d}$

Multiples of 2 from 1 to 100 are 2, 4, 6, ..., 100

$\therefore n = 1+\frac{t_n - a}{d} = 1+\frac{100-2}{2} = 50$

Multiples of 3 from 1 to 100 are 3, 6, 9, ..., 99

$\therefore n = 1+\frac{t_n - a}{d} = 1+\frac{99-3}{3} = 33$

Multiples of 6 from 1 to 100 are 6, 12, 18, ..., 96

$\therefore n = 1+\frac{t_n - a}{d} = 1+\frac{96-6}{6} = 16$

Thus, the total number of multiples of 2 or 3 must be equal to the number of multiples of 2 + number of multiples of 3 - number of multiples of 6 (since 6 is a multiple of 2 and 3)

$= 55+33-16 = 72$

Therefore, the probability that the integer is a multiple of 2 or a multiple of 3 $= \frac{72}{100} = 0.72$

Q16. Urn A contains two black balls (labelled B_1, B_2) and one white ball and urn B contains one black ball and two white balls (labelled W_1, W_2). Suppose the following experiment is performed. One of the two urns is chosen at random. Next a ball is randomly chosen from the urn. If the ball choosen is black then a second ball is chosen at random from urn A or if it is white then a second ball is chosen at random from urn B without replacing the first ball.

(i) What is the probability that two black balls are chosen?

(ii) What is the probability that two balls of opposite colour are chosen?

Solution: Sample space

If urn A is selected

$S_A = \{B_1B_2, B_1W, B_2B_1, B_2W, WB, WW_1, WW_2\}$

If urn B is selected

$S_B = \{BW, BB_1, BB_2, W_1B, W_1W_2, W_2B, W_2W_1\}$

$\therefore$ Total sample space

$S = \{B_1B_2, B_1W, B_2B_1, B_2W, WB, WW_1, WW_2, BW, BB_1, BB_2, W_1B, W_1W_2, W_2B, W_2W_1\} = 14$

(i) Favourable events that two black balls are chosen are $\{B_1B_2, B_2B_1, BB_1, BB_2\} = 4$

$\therefore$ Probability that two black balls are chosen $= \frac{4}{14} = \frac{2}{7}$

(ii) Favourable events that two balls of opposite colour are chosen are $= \{B_1W, B_2W, WB, BW, W_1B, W_2B\} = 6$

$\therefore$ Probability that two balls of opposite colour are chosen $= \frac{6}{14} = \frac{3}{7}$

Q17. A bag contains 6 black and 4 white balls. Three balls are drawn at random. Find the probability that:

(i) All the three balls are black

(ii) One ball is white and two balls are black

(iii) At least 2 black balls are chosen

Solution:

(i) P *(all the three balls are black)*$= \frac{^6C_3}{^{10}C_3}$

(ii) P *(one ball is white and two balls are black)*$= \frac{^4C_1\, ^6C_2}{^{10}C_3}$

(iii) P *(at least two black balls are chosen)*$= \frac{^6C_2\, ^4C_1 + ^6C_3}{^{10}C_3}$

Q18. If the letters of the word 'MATHEMATICA' are arranged at random, find the probability that:

(i) Three As are together

(ii) Two Ms are never together

Solution: Total number of letters in the word 'MATHEMATICA' is 11.

Number of words with 'MATHEMATICA' $= \frac{11!}{3!2!2!}$

(i) Number of words when all As are together $= \frac{9!}{3!2!2!}$

$\therefore$ Probability that all three As are together

$= \frac{\frac{9!}{3!2!2!}}{\frac{11!}{3!2!2!}} = \frac{1}{110}$

(ii) Number of words when two Ms are together $= \frac{10!}{3!2!2!}$

Number of words when two Ms are never together

$= \frac{11!}{3!2!2!} - \frac{10!}{3!2!2!} = \frac{10\cdot 10!}{3!2!2!}$

$\therefore$ Probability that two Ms are never together

$= \frac{\frac{10\cdot 10!}{3!2!2!}}{\frac{11!}{3!2!2!}} = \frac{10}{11}$

Day 220

Revision (Chapters XIV–XVI)

Mathematical Statement: The basic unit involved in mathematical reasoning is a mathematical statement. That is, a sentence is called a mathematically acceptable statement if it is either true or false, but not both.

Negation of a Statement: The denial of a statement is called the negation of the statement. If p is a statement, then the negation of p is also a statement and it is denoted by ~ p and read as 'not p'.

Note: While forming the negation of a statement, phrases such as "It is not the case" or "It is false that" are also used, or a simple 'not' occurs in the statement.

Compound Statement: A compound statement is one that is made up of two or more statements. In this case, each statement is called a component statement.
Many mathematical statements are obtained by combining one or more statements using connecting words such as "and", "or", etc.

Rules Regarding the Connective Word "and "

1. The compound statement with 'and' is true if all its component statements are true.
2. The compound statement with 'and' is false if any one of its component statements is false or all its component statements are false.

Rules Regarding the Connective Word "or "

1. The compound statement with 'or' is true if any one of its component statements is true or all its component statements are true.
2. The compound statement with 'or' is false if all its component statements are false.

Exclusive and Inclusive Connective Word "or"
'Exclusive or' means the person has only one choice from all of the choices given as 'or' while 'Inclusive or' means the person can choose any one of all the options given as 'or'.
Quantifiers: Quantifiers are phrases such as "There exists" and "For all". Another phrase that appears in mathematical statements is "There exists".

Implications: Implications such as "if-then", "only if" and "if and only if" are very common in mathematics.

Note: The sentence "if p then q" means that in the event that p is true, then q must be true. It is the same as the following:

1. p implies q is denoted by $p \Rightarrow q$. The symbol $\Rightarrow$ stands for implies. Here, a number is a multiple of 4 implies that it is a multiple of 2.
2. p is a sufficient condition for q.
   This means that knowing that a number is a multiple of 4 is sufficient for us to conclude that it is a multiple of 2.
3. p only if q.
   This means that a number is a multiple of 4 only if it is a multiple of 2.
4. q is a necessary condition for p.
   This means that when a number is a multiple of 4, it is necessarily a multiple of 2.
5. :q implies :p
   This means that if a number is not a multiple of 2, then it is not a multiple of 4.
6. Non-occurrence of p has no effect on the occurrence of q.

Contrapositive: Contrapositives are certain statements that can be formed from a given statement with "if then". The contrapositive of statement "if p then q" is "if ~q then ~p".
Converse: Converses are certain statements that can be formed from a given statement "if p then q" as if q then p".

If and only if (⇔): Converses also comprise statements that can be formed from a given statement "p if and only if q" and "q if and only if p", symbolically written as ''. p is a necessary and sufficient condition for q and vice-versa.

Validating Statements: To validate a statement as true, one must answer all of the following questions:
What does the statement mean?
What would it mean to say that this statement is true?
When is this statement not true?
The answers to these questions depend upon which of the special words "and", "or" and which of the implications "if and only", "if-then", and which of the quantifiers "for every", "there exists", appear in the given statement.

Rule 1: Statement with "and"
If p and q are mathematical statements, then in order to show that the statement "p and q" is true, we have to show that:
Step 1: The statement p is true.
Step 2: The statement q is true.

Rule 2: Statement with "or"
If p and q are mathematical statements, then in order to show that the statement "p ot q" is true, we must take two cases:
Case 1: By assuming that p is false, show that q must be true.
Case 2: By assuming that q is false, show that p must be true.

Rule 3: Statement with "if-then"
In order to prove the statement "if p then q", we must show that any one of the following cases is true:
Case 1: By assuming that p is true, show that q must be true. (Direct method)
Case 2: By assuming that q is false, show that p must be false. (Contrapositive method)
Rule 4: Statement with "if and only if"
In order to prove the statement "p if and only if q", we must show that:
(i) If p is true, then q is true and
(ii) If q is true, then p is true

By Contradiction: To check whether a statement p is true, we assume that p is not true, that is, ~ p is true. Then we prove the result by contradicting our assumption. Thus, we conclude that p is true.

Range: The range of a series is the difference of the maximum and minimum values of the series. This difference is called the 'range' of the data.

Deviation: The deviation of an observation x from a fixed value 'a' is the difference $x - a$.

Mean Deviation: Mean deviation about a central value 'a' is the mean of the absolute values of the deviations of the observations from 'a'. The mean deviation from 'a' is denoted as MD (a). Thus,

$$\text{MD }(a) = \frac{\text{Sum of absolute values of deviations from 'a'}}{\text{Number of observations}}$$

Mean Deviation for Ungrouped Data: Let n observations be $x_1, x_2, x_3, ..., x_n$. The following steps are taken in the calculation of mean deviation about mean or median:

Step 1: Calculate the measure of central tendency 'a' about which we need to find mean deviation.

Step 2: Find the deviation of each x_i from a, that is, $x_1 - a, x_2 - a, x_3 - a, ..., x_n - a$

Step 3: Find the absolute values of the deviations, That is, $|x_1 - a|, |x_2 - a|, |x_3 - a|, ..., |x_n - a|$

Step 4: Find the mean of the absolute values of the deviations, known as mean deviations about a, that is

$$\text{MD }(a) = \frac{\sum_{i=1}^{n} |x_i - a|}{n}$$

Thus $\text{MD }(\bar{x}) = \frac{1}{n}\sum_{i=1}^{n} |x_i - \bar{x}|$, where $\bar{x}$ = Mean

And $\text{MD }(M) = \frac{1}{n}\sum_{i=1}^{n} |x_i - M|$, where M = median

Mean Deviation for Grouped Data
(a) Discrete frequency distribution: Let $x_1, x_2, ..., x_n$ be the n distinct values occurring with frequencies $f_1, f_2, ..., f_n$, respectively. This discrete frequency distribution data can be represented in tabulate form as given below:
$x: x_1 \; x_2 \; x_3 \; ... \; x_n$
$f: f_1 \; f_2 \; f_3 \; ... \; f_n$

(i) Mean deviation about the mean
Step 1: Mean $\bar{x}$ of the given data using the formula

$$\bar{x} = \frac{\sum_{i=1}^{n} x_i f_i}{\sum_{i=1}^{n} f_i} = \frac{1}{N}\sum_{i=1}^{n} x_i f_i,$$ where $\sum_{i=1}^{n} x_i f_i$ denotes the sum of the products of observations x_i with their respectively frequencies f_i and $N = \sum_{i=1}^{n} f_i$ is the sum of the frequencies.

Step 2: Deviation of observations x_i from the mean $\bar{x}$ and their absolute values, that is, $|x_i - \bar{x}|$ for all $i = 1, 2, .., n$.

Step 3: Find the mean of the absolute values of the deviations, which is the required mean deviation about the mean. Thus

$$\text{MD }(\bar{x}) = \frac{\sum_{i=1}^{n} f_i |x_i - \bar{x}|}{\sum_{i=1}^{n} f_i} = \frac{1}{N}\sum_{i=1}^{n} f_i |x_i - \bar{x}|$$

(ii) Mean deviation about the median

Step 1: Arrange the observations in ascending order.

Step 2: Identify the observation whose cumulative frequency is equal to or just greater than N/2, where N is the sum of frequencies. This value of the observation lies in the middle of the data. Therefore, it is the required median.

Step 3: Obtain the mean of the absolute values of the deviations from median. That is

$$\text{MD}(M) = \frac{1}{N}\sum_{i=1}^{n} f_i |x_i - M|$$

(b) Continuous frequency distribution: A continuous frequency distribution is a series in which the data is classified into different class intervals without gaps along with their respective frequencies.

(i) Mean deviation about the mean

For calculating the mean of a continuous frequency distribution, we assume that the frequency in each class is centred at its mid-point, that is, the mid-point of each given class, and proceed as for a discrete frequency distribution to find mean deviation.

Shortcut method for calculating mean deviation about mean

In the shortcut method, we avoid the tedious calculations of computing $\bar{x}$ by following the step-deviation method.

In this method, we take an assumed mean which is in the middle or just close to it in the data. Then, deviations of the observations (or mid-points of classes) are taken from the assumed mean.

If there is a common factor to all the deviations, we divide them by using the common factor to further simplify the deviations. These steps are known as step-deviations.

Let the new variable be denoted by $d_i = \frac{x_i - a}{h}$, where '*a*' is the assumed mean and *h* is the common factor.

Then, the mean $\bar{x}$ by step-deviation method is given

by $\bar{x} = a + \frac{\sum_{i=1}^{n} f_i d_i}{N} \times h$

Thus, mean deviation about the mean is

$$\text{MD}(\bar{x}) = \frac{\sum_{i=1}^{n} f_i |x_i - \bar{x}|}{\sum_{i=1}^{n} f_i} = \frac{1}{N}\sum_{i=1}^{n} f_i |x_i - \bar{x}|$$

(ii) Mean deviation about the median

The process of finding the mean deviation about median for a continuous frequency distribution is similar to mean deviation about mean.

The only difference lies in the replacement of mean with median while taking deviations.

The data is first arranged in ascending order.

The median of continuous frequency distribution is obtained by first identifying the class in which median lies (median class) and then applying the formula where median class is the class interval whose cumulative frequency is just greater than or equal to N/2, N is the sum of frequencies *l, f, h, C,* respectively, the lower limit, the frequency, the width of the median class and the cumulative frequency of the class just preceding the median class.

After finding the median, the absolute values of the deviations of mid-point x_i of each class from the median, that is, $|x_i - M|$ are obtained.

$$\text{Then MD}(M) = \frac{1}{N}\sum_{i=1}^{n} f_i |x_i - M|$$

Degree of Dispersion: Let $x_1, x_2, \ldots, x_n$ be *n* observations and $\bar{x}$ be their mean. Then

$$(x_1 - \bar{x})^2 + (x_2 - \bar{x})^2 + \ldots + (x_n - \bar{x})^2 = \sum_{i=1}^{n} (x_i - \bar{x})^2$$

Note:

If $\sum_{i=1}^{n} (x_i - \bar{x})^2$ is zero, then each $(x_i - \bar{x})$ must be zero.

This implies that there is no dispersion at all as all observations are equal to the mean *x*.

If $\sum_{i=1}^{n} (x_i - \bar{x})^2$ is small, this indicates that the observations $x_1, x_2, x_3, \ldots, x_n$ are close to the mean $\bar{x}$, and therefore, there is a lower degree of dispersion.

If $\sum_{i=1}^{n} (x_i - \bar{x})^2$ is large, there is a higher degree of dispersion of the observations from the mean $\bar{x}$, and we say that the sum $\sum_{i=1}^{n} (x_i - \bar{x})^2$ is a reasonable indicator of the degree of dispersion or scatter.

Standard Deviation and Variance: The proper measure of dispersion about the mean of a set of observations is expressed as the positive square root of the variance and is called standard deviation. Therefore, the standard deviation, usually denoted by σ, is given by

$$\sigma = \sqrt{\frac{1}{n}\sum_{i=1}^{n} (x_i - \bar{x})^2} \quad \ldots(1)$$

And the mean of the squares of the deviations from mean is called the variance and is denoted by σ^2 (read as sigma square). Therefore, the variance of n observations $x_1, x_2, ..., x_n$ is given by $\sigma^2 = \frac{1}{n}\sum_{i=1}^{n}(x_i - \bar{x})^2$

Standard Deviation of a Discrete Frequency Distribution: Let the given discrete frequency distribution be

$x: x_1, x_2, x_3, ..., x_n$

$f: f_1, f_2, f_3, ..., f_n$

$$(\sigma) = \sqrt{\frac{1}{N}\sum_{i=1}^{n} f_i(x_i - \bar{x})^2} \text{ where } N = \sum_{i=1}^{n} f_i.$$

Standard Deviation of a Continuous Frequency Distribution: The given continuous frequency distribution can be represented as a discrete frequency distribution by replacing each class by its mid-point. Then, the standard deviation is calculated by the technique adopted in the case of a discrete frequency distribution.

If there is a frequency distribution of n classes, each class defined by its mid-point x_i with frequency f_i, the standard deviation will be obtained by the formula

$\sigma = \sqrt{\frac{1}{N}\sum_{i=1}^{n} f_i(x_i - \bar{x})^2}$, where $\bar{x}$ is the mean of the distribution and $N = \sum_{i=1}^{n} f_i$.

Direct Formula for Standard Deviation

$$(\sigma) = \frac{1}{N}\sqrt{N\sum_{i=1}^{n} f_i x_i^2 - \left(\sum_{i=1}^{n} f_i x_i\right)^2}$$

Shortcut Method to find Variance and Standard Deviation: Let the assumed mean be 'A' and the scale be reduced to $1/h$ times (where h is the width of class intervals).

Let the step-deviations or new values be y_i.

That is, $y_i = \frac{x_i - A}{h}$ or $x_i = A + hy_i$ ...(1)

$$\bar{x} = \frac{\sum_{i=1}^{n} f_i x_i}{N} \quad ...(2)$$

Replacing x_i from (1) and (2), we get

$$\bar{x} = \frac{\sum_{i=1}^{n} f_i(A + hy_i)}{N} = \frac{1}{N}\left(\sum_{i=1}^{n} f_i A + \sum_{i=1}^{n} hf_i y_i\right)$$

$$= \frac{1}{N}\left(A\sum_{i=1}^{n} f + h\sum_{i=1}^{n} f_i y_i\right) = A\cdot\frac{N}{N} + h\frac{\sum_{i=1}^{n} f_i y_i}{N}$$

$$\left(\text{since} \sum_{i=1}^{n} f_i = N\right)$$

Thus, $\bar{x} = A + h\bar{y}$ ...(3)

Now, variance of the variable x

$$\sigma_x^2 = \frac{1}{N}\sum_{i=1}^{n} f_i(x_i - \bar{x})^2$$

$$= \frac{1}{N}\sum_{i=1}^{n} f_i(A + hy_i - A - h\bar{y})^2$$

(Using (1) and (3)) $= \frac{1}{N}\sum_{i=1}^{n} f_i h^2 (y_i - \bar{y})^2$

$= \frac{h^2}{N}\sum_{i=1}^{n} f_i(y_i - \bar{y})^2 = h^2 \times$ Variance of the variable y_i, that is, $\sigma_x^2 = h^2\sigma_y^2 \Rightarrow \sigma_x = h\sigma_y$ ...(4)

From (3) and (4), we have

$$\sigma_x = \frac{1}{N}\sqrt{N\sum_{i=1}^{n} f_i y_i^2 - \left(\sum_{i=1}^{n} f_i y_i\right)^2}$$

Analysis of Frequency Distributions: The measure of variability which is independent of units is called the coefficient of variation (denoted as CV)

The coefficient of variation is defined as

CV $= \frac{\sigma}{\bar{x}} \times 100, \bar{x} \neq 0$, where $\sigma, \bar{x}$ are the standard deviation and mean of the data.

Note: The series with greater CV is said to be more variable than the other. The series with lesser CV is said to be more consistent than the other.

Comparison of Two Frequency Distributions with the Same Mean: Let $\bar{x}_1, \sigma_1$ be the mean and standard deviation of the first distribution, and $\bar{x}_2, \sigma_2$ be the mean and standard deviation of the second distribution.

Then CV (1st distribution) $= \frac{\sigma_1}{\bar{x}_1} \times 100$

And CV (2nd distribution) $= \frac{\sigma_2}{\bar{x}_2} \times 100$

Let assume $\bar{x}_1 = \bar{x}_2 = \bar{x}$

Therefore, CV (1st distribution) $= \frac{\sigma_1}{\bar{x}} \times 100$ ...(1)

And CV (2nd distribution) $= \frac{\sigma_2}{\bar{x}} \times 100$ ...(2)

It is clear from (1) and (2) that the two CVs can be compared on the basis of values of σ_1, σ_2 only.

Thus, we say that for two series with equal means, the series with greater standard deviation (or variance) is called more variable or dispersed than the other. Also, the series with lesser value of standard deviation (or variance) is said to be more consistent than the other.

Random Experiment: An experiment is called a random experiment if it satisfies the following two conditions:
(i) It has more than one possible outcome.
(ii) It is not possible to predict the outcome in advance.

Outcomes and Sample Space

- A possible result of a random experiment is called its outcome.
- The set of all possible outcomes of a random experiment is called the sample space of the experiment, denoted by the symbol S.
- Each element or outcome of the sample space of a random experiment is called a sample point.
- The total number of possible outcomes of the sample space of a random experiment is denoted by $n(S)$.

Event: Any subset E of a sample space S associated with a random experiment is called an event.

Occurrence of an Event: Let there be an event E in a sample space S associated with a random experiment. Event E of the sample space S is said to have occurred if an element (or outcome) ω of the experiment is such that $\omega \in E$. And if the outcome ω is such that $\omega \notin E$, then we say that the event E has not occurred.

Types of Events

(i) Impossible and sure events: The empty set ϕ is called an impossible event of the sample space S and the whole sample space S is called the sure event.

(ii) Simple event: If an event E has only one sample point (or outcome) of a sample space, it is called a simple (or elementary) event.
Note: In a sample space containing n distinct elements, there are exactly n simple events.

(iii) Compound event: If an event has more than one sample point (or outcome), it is called a compound event.

Algebra of Events: Let A, B, C be events associated with a random experiment whose sample space is S.

(i) Complementary event: For every event A, there exists another event A' called the complementary event to A. It is also called the event 'not A' and is given by $A' = S - A$; that is, if an element (or outcome) ω of the experiment is such that $\omega \in A$, then $\omega \notin A'$.
Note: $S' = \phi$ and $\phi' = S$

(ii) The event 'A or B': Let there be events A and B associated with a sample space S. Then the union of two sets A and B is denoted by $A \cup B$, which contains all the elements that are either in A or in B or in both.
Event 'A or B' $= A \cup B = \{\omega : \omega \in A \text{ or } \omega \in B\}$

(iii) The event 'A and B': Let the events A and B be associated with a sample space S. Then the intersection of two sets A and B is denoted by $A \cap B$, which contains all the elements that are common to both A and B; that is, which belong to both 'A and B'.
Event 'A and B' $= A \cap B = \{\omega : \omega \in A \text{ and } \omega \in B\}$

(iv) The event 'A but not B': Let the events A and B be associated with a sample space S. Then the set 'A but not B' denoted by $A - B$ is the set of all the elements that are in A but not in B.
Event 'A but not B' $= A - B = \{\omega : \omega \in A \text{ but } \omega \notin B\}$

Mutually Exclusive Events: Two events A and B of a sample space are called mutually exclusive events if the occurrence of any one of them excludes the occurrence of the other event; that is, they cannot occur simultaneously. Thus, mutually exclusive events A and B are disjoint sets; that is, $A \cap B = \phi$.

Note: Simple events of a sample space are always mutually exclusive.

Exhaustive Events: Let $E_1, E_2, ..., E_n$ be n events of a sample space S such that $E_1 \cup E_2 \cup E_3 \cup ... \cup E_n = \bigcup_{i=1}^{n} E_i = S$.
Then $E_1, E_2, ..., E_n$ are called exhaustive events.
That is, events $E_1, E_2, ..., E_n$ are said to be exhaustive if at least one of them necessarily occurs whenever the experiment is performed.
Also, if $E_i \cap E_j = \phi$ for $i \neq j$, that is, events E_i, E_j are pair-wise disjoint sets and $\bigcup_{i=1}^{n} E_i = S$, then events $E_1, E_2, ..., E_n$ are called mutually exhaustive events.

Axiomatic Approach to Probability: Let S be the sample space of a random experiment. Then the probability P is a real-valued function whose domain is the power set of S and whose range is the interval [0, 1] satisfying the following axioms:

(i) For any event E, $P(E) \geq 0$

(ii) $P(S) = P(E_1) + P(E_2) + P(E_3) ... + P(E_n) = 1$

(iii) If E and F are mutually exclusive events, then
$P(E \cup F) = P(E) + P(F)$, that is, $P(E \cap F) = \phi$

In general, let S be a sample space containing outcomes $\omega_1, \omega_2, ..., \omega_n$, that is, $S = \{\omega_1, \omega_2, ..., \omega_n\}$. Then

(i) $0 \leq P(\omega_i) \leq 1$ for each $\omega_i \in S$

(ii) $P(\omega_1) + P(\omega_2) + ... + P(\omega_n) = 1$

(iii) For any event A, $P(A) = \sum P(\omega_i)$, $\omega_i \in A$.

Note: The singleton $\{\omega_i\}$ is called the elementary event and is written as $P(\omega_i)$ for $P(\{\omega_i\})$.

Probability of an Event: Let S be a sample space associated with a random experiment, and $n(S)$ be the number of elements (or outcomes) of the sample space S. Let E be the event of a sample space S associated with the random experiment, and $n(E)$ be the number of elements (or outcomes) of the event E.

Then, the probability of event E is given by

$$P(E) = \frac{\text{Number of outcomes of event } E}{\text{Number of outcomes of sample space } S}$$

$$\Rightarrow P(E) = \frac{n(E)}{n(S)}$$

Probabilities of Equally Likely Outcomes: Let a sample space of an experiment be $S = \{\omega_1, \omega_2, ..., \omega_n\}$.

All the outcomes are equally likely to occur. This means the chance of occurrence of each simple event must be the same, that is, $P(\omega_i) = p$, for all $\omega_i \in S$, where $0 \le p \le 1$

Since $\sum_{i=1}^{n} P(\omega_i) = 1 \Rightarrow p + p + ... + p\,(n \text{ times}) = 1$

$$\Rightarrow np = 1 \Rightarrow p = \frac{1}{n}$$

Probability of the Event 'A or B'

$P(A \cup B) = P(A) + P(B) - P(A \cap B)$.

Note: If A and B are disjoint sets, that is, they are mutually exclusive events, then $A \cap B = \phi$

Therefore, $P(A \cap B) = P(\phi) = 0$

Thus, for mutually exclusive events A and B, we get

$P(A \cup B) = P(A) + P(B)$

Probability of the Event 'not A' (A′)

$P(A) + P(A') = 1 \Rightarrow P(A') = P(\text{not } A) = 1 - P(A)$

Note: A' and A are mutually exclusive and exhaustive events, that is

$A \cap A' = \phi,\ \ A \cup A' = S$

$\Rightarrow P(A \cap A') = 0,\ P(A \cup A') = P(S)$

Sample Paper - I

Day 221

Time: 3 Hrs Maximum **Marks:** 80

General Instructions:

(i) All the questions are compulsory.
(ii) The question paper consists of 36 questions divided into 4 sections A, B, C and D.
(iii) Section A comprises 20 questions of 1 mark each. Section B comprises 6 questions of 2 marks each. Section C comprises 6 questions of 4 marks each. Section D comprises 4 questions of 6 marks each.
(iv) There is no overall choice. However, an internal choice has been provided in four questions of 1 mark each, two questions of 2 marks each, two questions of 4 marks each, and two questions of 6 marks each. You have to attempt only one of the alternatives in all such questions.
(iv) Use of calculators is not permitted.

SECTION A

Q1–Q10 are multiple choice type questions. Select the correct option. (1 mark each)

Q1. Which of the following is a statement?
(a) x is a rational number
(b) What are you doing?
(c) 0 is a whole number
(d) Please, may I come?

Q2. The value of $\sqrt{-4} \times \sqrt{-16}$ is
(a) 8 (b) −8
(c) $8i$ (d) Not defined

Q3. If $\left(\frac{1+i}{1-i}\right)^{101} = a+ib$, then (a, b) is equal to
(a) (1,0) (b) (1,1)
(c) (0,1) (d) (−1,−1)

Q4. If $\sin\theta + \operatorname{cosec}\theta = 2$, then $\sin^2\theta + \operatorname{cosec}^2\theta$ is equal to
(a) 1 (b) 4 (c) 2 (d) 0

Q5. The value of $\sin\frac{\pi}{12}\sin\frac{19\pi}{12}$ is
(a) 1/2 (b) −1/2 (c) −1/4 (d) 1

Q6. The domain for which the functions defined by $f(x) = x(x-1)$, $g(x) = x-1$ are equal to
(a) $\{0\}$ (b) $\{-1,1\}$
(c) $\{1\}$ (d) $[0,1]$

Q7. Let $T = \left\{x \middle| \frac{x+6}{x-2} = \frac{2-x}{6-x}\right\}$. Then find set is
(a) $\{4\}$ (b) $\{8\}$ (c) $\{10\}$ (d) $\{12\}$

Q8. Find the radian measure corresponding to the following degree measure: 18°
(a) $\frac{\pi}{10}$ (b) $\frac{\pi}{20}$ (c) $\frac{\pi}{5}$ (d) $\frac{\pi}{15}$

Q9. Out of 10 points in a plane, 6 are in the same line and are collinear. Find the number of lines that can be formed by joining two points.
(a) 210 (b) 130 (c) 31 (d) 30

Q10. The middle term (terms) in the expansion of $\left(\frac{x}{a}-\frac{a}{x}\right)^{10}$ is
(a) ${}^{10}C_5$ (b) $-{}^{10}C_5$ (c) ${}^{10}C_6$ (d) $-{}^{10}C_6$

Q11–Q15 are fill in the blanks. (1 mark each)

Q11. The set $\{x \in N : 0 \le x < 3\}$ can be written as ....

Q12. The coefficient of $a^{-5}b^5$ in the expansion of $\left(\frac{4}{a}-\frac{b}{2}\right)^{10}$ is...............

Q13. The line $2x-3y=0$ is a ... ... of the circle $x^2+y^2+6x+4y=0$.

Q14. The distance between the following pairs of points: (3, 1, 2) and (2, 4, –1) is ... ...

Q15. If $z_1 = 1-i$, $z_2 = 1+i$, then the value of $\left|\frac{z_1+z_2-i}{z_1-z_2+i}\right|$ is equal to ... ...

(Q16–Q20) Answer the following questions.
(1 mark each)

Q16. Two boys are in Room A, and 2 girls are in Room B. Specify the sample space for the experiment in which a room is selected and then a person.

Q17. Find the equation of the circle with centre $(-a, b)$ and radius $\sqrt{a^2 + b^2}$

Q18. Find the equation of the parabola that satisfies the following conditions: vertex (0, 0) passing through (1, 4) and axis is along the x-axis.

Q19. Find the equation of the hyperbola satisfying the give conditions: foci $(\pm 5, 0)$ and the transverse axis is of length 8.

Q20. Find the number terms of GP $2, 2^2, 2^3 \ldots$ needed to give the sum 254.

SECTION B

(Q21–Q26) Solve the following questions.
(2 marks each)

Q21. Find the mean deviation about the mean for the data 5, 8, 10, 12, 15

Q22. If $A = \{1, 0\}, B = \{0, 1\}$, then determine
(i) $A \times B$ (ii) $B \times A$ (iii) $B \times B$
(iv) $A \times A$

Q23. Prove that
$$\cos\left(\frac{\pi}{3} + x\right) - \cos\left(\frac{\pi}{3} - x\right) = -\sin x$$

Q24. Find the real numbers x and y if $(x - iy)(3 + 5i)$ is the conjugate of $11 - 7i$.

Q25. Expand $e^{(x^2+1)}$ up to the power of x^2.

OR

Q25. Find the derivative of $\tan^2 x(\sec x - 1)(\sec x + 1)$

Q26. Find the derivative of $\frac{x-1}{x}$

OR

Q26. Find the value of k if the points $A(2, 3), B(4, k), C(6, -3)$ are collinear.

SECTION C

(Q27–Q32) Solve the following questions.
(4 marks each)

Q27. Solve the system of equations $\text{Re}(z^2) = 0$, $|z| = 4\sqrt{2}$.

Q28. Find the seventh term of the expansion $(y^{1/3} + x^{1/2})^n$, if the binomial coefficient of the fourth term from the end is 220.

Q29. Find the equation of the line passing through the point (3/2, 3/2) and cutting off intercepts on the axes whose sum is 6.

Q30. Solve the following system of inequalities graphically: $x + y \geq 4, x - y > 0$

Q31. If $A(2, 2, -3), B(5, 6, 9), C(2, 7, 9)$ be the vertices of a triangle and the internal bisector of the angle A meets BC at the point D, then find the coordinates of D.

OR

Q31. The perpendicular from the origin to the line $y = mx + c$ meets it at the point (2, 3). Find the values of m and c.

Q32. Find the equation of the line which passes through $(2, \sqrt{3})$ and is inclined with the x-axis at an angle of 75°.

OR

Q32. In an AP, if the first term is 5 and the sum of the first five terms is one-third of the next five terms, then find the 15th term.

SECTION D

(Q33–Q36) Solve the following questions.
(6 marks each)

Q33. $|x - 1| \leq 2$, $|x| > 2$

Q34. A bag contains 4 black and 6 red balls. Determine the number of ways in which 2 black and 3 red balls or 3 black and 2 red balls or 1 black and 4 red balls can be selected from the lot.

Q35. A committee of 8 is to be chosen from 10 men and 6 women, so as to contain at least 4 men and 3 women. In how many different ways can this be done, if one particular woman and man refuse to serve on the same committee?

OR

Q35. Differentiate $\cos(x^2)$ with respect to x using the first principle.

Q36. Find the coefficient of x^6 in the product $x(1 - x)^4(1 + 2x)^6$ using binomial theorem.

OR

Q36. In a group of 200 students of the Indian Institute of Science, the number of students who like Coffee, Milk, Tea were found to be as follows: Coffee = 50%, Milk = 75%, Tea = 30%, Coffee and Milk = 35%, Milk and Tea = 25%, Coffee and Tea = 15%, Coffee, Milk and Tea = 10%. Find the number of students who like:

(i) Only Coffee
(ii) Only Milk
(iii) Only Tea
(iv) Coffee and Milk but not Tea
(v) Milk and Tea but not Coffee
(vi) Coffee and Tea but not Milk
(vii) At least one of the drinks
(viii) None of the three drinks

Day 222

Sample Paper – I (Solutions)

SECTION A

Q1–Q10 are multiple choice type questions. Select the correct option. (1 mark each)

Q1. Solution: (c)
0 is a whole number, which is true.
Thus, it is a statement.

Q2. Solution: (b)
$\sqrt{-4}\times\sqrt{-16} = i\sqrt{4}\times i\sqrt{16} = i^2(2\times 4) = -8$

Q3. Solution: (c)

$$\left(\frac{1+i}{1-i}\right)^{101} \Rightarrow \left[\frac{(1+i)}{(1-i)}\times\frac{(1+i)}{(1+i)}\right]^{101}$$

$$= \left(\frac{1+i^2+2i}{1-i^2}\right)^{101} = \left(\frac{2i}{2}\right)^{101} \quad (\because i^2 = -1)$$

$$= (i)^{101} = (i^4)^{25}\cdot i = i \quad (\because i^4 = 1)$$

$$= 0 + 1i = a + ib \Rightarrow a = 0,\ b = 1$$

Q4. Solution: (d)

$\sin\theta + \operatorname{cosec}\theta = 2$

$\Rightarrow \sin^2\theta + \operatorname{cosec}^2\theta + 2\sin\theta\cdot\operatorname{cosec}\theta = 2$

$\Rightarrow \sin^2\theta + \operatorname{cosec}^2\theta + 2\cdot 1 = 2$

$\Rightarrow \sin^2\theta + \operatorname{cosec}^2\theta = 2 - 2$

$\Rightarrow \sin^2\theta + \operatorname{cosec}^2\theta = 0$

Q5. Solution: (c)

$$\sin\frac{\pi}{12}\sin\frac{19\pi}{12} = \sin\frac{\pi}{12}\sin\left(2\pi - \frac{5\pi}{12}\right)$$

$$= -\sin\frac{\pi}{12}\sin\frac{5\pi}{12} = -\sin 15°\cdot\sin 75°$$

$$= -\sin 15°\cdot\cos 15° = -\frac{1}{2}\sin 30° = -\frac{1}{4}$$

Q6. Solution: (c)
The domain for which the functions defined by $f(x) = x(x-1)$, $g(x) = x-1$ are equal means all values of x for which functions $f(x) = g(x)$. Thus

$x(x-1) = x-1 \Rightarrow (x-1)^2 = 0 \Rightarrow x = 1$

Thus, the domain for which $f(x) = g(x)$ is equal is $\{1\}$

Q7. Solution: (c)

$$T = \left\{x \,\middle|\, \frac{x+8}{x-4} = \frac{4-x}{8-x}\right\}$$

$$\therefore \frac{x+6}{x-2} = \frac{2-x}{6-x} \Rightarrow (x^2 - 6^2) = (x-2)^2$$

$$\Rightarrow x^2 - 36 = x^2 - 4x + 4 \Rightarrow 4x = 40 \Rightarrow x = 10$$

$$\therefore T = \{10\}$$

Q8. Solution: (a)
$180° = \pi$ radians

Therefore, $18° = \frac{\pi}{180}\times 18$ radian $= \frac{\pi}{10}$ radians

Q9. Solution: (c)
The number of lines that can be formed by joining two points from the given 10 points is = ${}^{10}C_2$
Six points are collinear, and these 6 collinear points, when joined pair-wise will give only one line.
Therefore, the total number of straight lines formed
$= {}^{10}C_2 - {}^{6}C_2 + 1 = 31$

Q10. Solution: (b)
The power of binomial expansion $\left(\frac{x}{a} - \frac{a}{x}\right)^{10}$ is $n =$ 10 is even.

Therefore, it has one middle term, that is, $\left(\frac{10}{2} + 1\right)$

term = 6th term.

Thus, $T_6 = T_{5+1} = {}^{10}C_5\left(\frac{x}{a}\right)^{10-5}\left(\frac{-a}{x}\right)^5 = -{}^{10}C_5$

Q11–Q15 are fill in the blanks. **(1 mark each)**

Q11. Solution: The set $\{x \in N : 0 \le x < 3\}$ can be written as $x = \{1, 2\}$.

Q12. Solution: Let the rth term have coefficient $a^{-5}b^5$. Then $T_{r+1} = {}^{10}C_r\left(\frac{4}{a}\right)^{10-r}\left(-\frac{b}{2}\right)^r$. Thus, for the coefficient of $a^{-5}b^5$, $10 - r = 5 \Rightarrow r = 5$

$\therefore$ Coefficient of $a^{-5}b^5 = {}^{10}C_5(4)^5\left(-\frac{1}{2}\right)^5 = -2^5\,{}^{10}C_5$

Q13. Solution: Equation of the circle is $x^2 + y^2 + 6x + 4y = 0$.

$\therefore$ Centre $(-3, -2)$

Since the diameter passes through the centre, the centre $(-3, -2)$ must satisfy the equation of diameter.

Now, since for given line $2x - 3y = 0 \Rightarrow 2\times(-3) - 3\times(-2) = 0$, the line is the diameter of the circle.

Q14. Solution: Distance between two points $A(x_1, y_1, z_1)$ and $B(x_2, y_2, z_2)$ is given by

$AB = \sqrt{(x_2 - x_1)^2 + (y_2 - y_1)^2 + (z_2 - z_1)^2}$

Distance between points (3, 1, 2) and (2, 4, –1)

$= \sqrt{(2-3)^2 + (4-1)^2 + (-1-2)^2} = \sqrt{1+9+9} = \sqrt{19}$

Q15. Solution:

$\left|\frac{z_1 + z_2 - i}{z_1 - z_2 + i}\right| = \left|\frac{(1-i)+(1+i)-i}{(1-i)-(1+i)+i}\right|$

$= \left|\frac{2-i}{-i}\right| = \left|\frac{2-i}{-i} \times \frac{i}{i}\right|$

$= \left|\frac{2i+1}{1}\right| \quad (\because i^2 = -1)$

$\sqrt{2^2 + 1^2} = \sqrt{5}$

(Q16–Q20) Answer the following questions. **(1 mark each)**

Q16. Solution: Letus denote 2 boys in Room A as B_1, B_2 and 2 girls in room B as G_1, G_2, respectively. Sample space is given by $S = \{AB_1, AB_2, BG_1, BG_2\}$

Q17. Solution: The equation of a circle with centre (h, k) and radius r is $(x-h)^2 + (y-k)^2 = r^2$

Therefore, equation of the circle with centre $(-a, b)$ and radius $\sqrt{a^2 + b^2}$ is

$(x+a)^2 + (y-b)^2 = \left(\sqrt{a^2+b^2}\right)^2$

$\Rightarrow x^2 + 2ax + a^2 + y^2 - 2by + b^2 = a^2 + b^2$

$\Rightarrow x^2 + y^2 + 2ax - 2by = 0$

Q18. Solution: The vertex of the parabola is (0, 0) and the axis of the parabola is the x-axis. Therefore, the equation of the parabola is of the form $y^2 = 4ax$ or $y^2 = -4ax$.

Now, since the parabola passes through point (1, 4) which lies in the first quadrant, the equation of the parabola is of the form $y^2 = 4ax$. Now, since point (1, 4) must satisfy the equation $y^2 = 4ax$, $4^2 = 4a(1) \Rightarrow a = 4$

Thus, the equation of the parabola is

$y^2 = 4\times4\times x \Rightarrow y^2 = 16x$

Q19. Solution: The foci $(\pm5, 0)$ are on the x-axis. Therefore, the equation of the hyperbola is of the form $\frac{x^2}{a^2} - \frac{y^2}{b^2} = 1$ and $c = 5$.

Now, since the length of the transverse axis is 8, $2a = 8 \Rightarrow a = 4 \Rightarrow a^2 = 16$.

From the equation $a^2 + b^2 = c^2$

$\Rightarrow b^2 = 25 - 16 = 9$

Thus, the equation of the hyperbola $\frac{x^2}{16} - \frac{y^2}{9} = 1$

Q20. Solution: The given GP is $2, 2^2, 2^3 \ldots$ then, $a = 2$, $r = 2$.

Let the sum of n terms of this GP equal 252. Then

$S_n = \frac{a(1-r^n)}{1-r} \Rightarrow 254 = \frac{2(2^n - 1)}{2-1}$

$\Rightarrow 127 = 2^n - 1$

$\Rightarrow 2^n = 128 = 2^7 \Rightarrow n = 7$

Thus, seven terms of the given GP are required to obtain the sum as 254.

SECTION B

(Q21–Q26) Solve the following questions. **(2 marks each)**

Q21. Solution: Mean of the data

$\bar{x} = \frac{5+8+10+12+15}{5} = \frac{35}{5} = 7$

The deviations of the respective observations from the mean $\bar{x} = x_i - \bar{x}$, are $-2, 1, 3, 5, 8$

The absolute values of the deviation $= |x_i - \bar{x}|$ are 2,1,3,5,8

Thus, the mean deviation about the mean is

$$\text{M.D.}(\bar{x}) = \frac{\sum_{i=1}^{5} |x_i - \bar{x}|}{5}$$

$$= \frac{2+1+3+5+8}{5} = \frac{19}{5} = 3.8$$

Q22. Solution:

(i) $A \times B = \{(1,0),(1,1),(0,0),(0,1)\}$

(ii) $B \times A = \{(0,1),(0,0),(1,1),(1,0)\}$

(iii) $B \times B = \{(0,0),(1,0),(0,1),(1,1)\}$

(iv) $A \times A = \{(1,1),(1,0),(0,0),(0,1)\}$

Q23. Solution:

$$\text{LHS} = \cos\left(\frac{\pi}{3}+x\right) - \cos\left(\frac{\pi}{3}-x\right)$$

$$= -2\sin\left\{\frac{\left[\left(\frac{\pi}{3}+x\right)+\left(\frac{\pi}{3}-x\right)\right]}{2}\right\}\sin\left\{\frac{\left[\left(\frac{\pi}{3}+x\right)-\left(\frac{\pi}{3}-x\right)\right]}{2}\right\}$$

$$\left[\because \cos A - \cos B = -2\sin\left(\frac{A+B}{2}\right)\cdot\sin\left(\frac{A-B}{2}\right)\right]$$

$$= -2\sin(\pi/6)\sin x = -2\times\frac{1}{2}\sin x = -\sin x$$

Q24. Solution:

Let $z = (x-iy)(3+5i)$

$z = 3x + 5xi - 3yi - 5yi^2 = 3x + 5xi - 3yi + 5y$

$= (3x-5y) + i(5x-3y)$

$\Rightarrow \bar{z} = (3x+5y) - i(5x-3y)$

Since conjugate of $(x-iy)(3+5i)$ is $8-2i$,

$(3x+5y) - i(5x-3y) = 8-2i$

Equating real imaginary parts, we get

$3x+5y = 8$ and $5x-3y = 2$

Solving the above two equations, we get

$x = 2$ and $y = 1$

Q25. Solution:

$$e^x = 1 + \frac{x}{1!} + \frac{x^2}{2!} + \frac{x^3}{3!} + \frac{x^4}{4!} + \ldots$$

Therefore, $e^{(x^2+1)}$

$$= 1 + \frac{x^2+1}{1!} + \frac{(x^2+1)^2}{2!} + \frac{(x^2+1)^3}{3!}\ldots$$

$$= 1 + x^2 + 1 + \frac{x^4+2x^2+1}{2} + \frac{x^6+3x^4+3x^2+1}{6}\ldots$$

$$= 1 + x^2 + 1 + x^2 + \frac{1}{2} + \frac{1}{2}x^2 + \frac{1}{6}\ldots$$

$$= \frac{8}{3} + \frac{5}{2}x^2 + \ldots$$

OR

Q25. Solution:

Let $y = \tan^2 x(\sec x - 1)(\sec x + 1)$

$y = \tan^2 x(\sec^2 - 1) = \tan^4 x$

$$\Rightarrow \frac{dy}{dx} = 4\tan^3 x \cdot \frac{d}{dx}\tan x$$

$$\left(\because \frac{d}{dx}y^n = ny^{n-1}\frac{dy}{dx}\right)$$

$$= 4\tan^3 x \sec^2 x$$

Q26. Solution: Let $f(x) = \frac{x-1}{x}$, where $x \neq 0$

By the quotient rule

$$f'(x) = \frac{(x)\frac{d}{dx}(x-1) - (x-1)\frac{d}{dx}(x)}{(x)^2}$$

$$= \frac{(x)(1)-(x-1)(1)}{(x)^2} = \frac{1}{x^2}$$

$$\Rightarrow f'(x) = \frac{1}{x^2}, \quad x \neq 0$$

OR

Q26. Solution: The given points are collinear. The area of the triangle formed by them must be 0, that is

$$\frac{1}{2}[2(k+3) + 4(-3-3) + 6(3-k)] = 0$$

$$\Rightarrow \frac{1}{2}(2k+6-24+18-6k) = 0$$

$\Rightarrow -2k = 0 \Rightarrow k = 0$

Thus, for $k = 0$, the points $A(2,3), B(4,k), C(0,-3)$ are collinear.

SECTION C

(Q27–Q32) Solve the following questions.

(4 marks each)

Q27. Solution: Let $z = x + iy$

$\text{Re}(z^2) = 0$

$\Rightarrow \text{Re}(x^2 + 2ixy - y^2) = 0$

$\Rightarrow \text{Re}\{(x^2-y^2) + 2ixy\} = 0$

$\Rightarrow x^2 - y^2 = 0 \Rightarrow x^2 = y^2$

Also, $|z| = \sqrt{x^2+y^2} = 4\sqrt{2}$

$\Rightarrow x^2 + y^2 = 32$

Solving the above two equations, we get
$x^2 + y^2 = 32 \Rightarrow 2x^2 = 32 \Rightarrow x^2 = 16$

$\Rightarrow x = \pm 4 \Rightarrow y = \pm 4 \quad (\because x^2 = y^2)$

$\Rightarrow z = x + iy = \pm 4 \pm 4i$

Q28. Solution: The seventh term of the given expansion $(y^{1/3} + x^{1/2})^n$, is $T_7 = T_{6+1} = {}^nC_6 (y^{1/3})^{n-6} (x^{1/2})^6$

The binomial coefficient of the fourth term from the end = binomial coefficient of the fourth term from the beginning = ${}^nC_3$

And it is given that the binomial coefficient of the fourth term from the end is 220

$\therefore {}^nC_3 = 220 \Rightarrow \dfrac{n(n-1)(n-2)}{3.2} = 220$

$\Rightarrow n(n-1)(n-2) = 1320 = 12 \times 11 \times 10$

$\Rightarrow n = 12$

Thus, $T_7 = {}^{12}C_6 (y^{1/3})^{12-6} (x^{1/2})^6 = {}^{12}C_6 y^2 x^3$

Q29. Solution: The equation of a line in the intercept form is $\dfrac{x}{a} + \dfrac{y}{b} = 1$, where, a and b are the intercepts on the x- and y-axes, respectively.

It is given that $a + b = 6 \Rightarrow b = 6 - a$. Therefore, the equation of line can be written as $\dfrac{x}{a} + \dfrac{y}{6-a} = 1$

The line passes through point (3/2, 3/2). Therefore, it satisfies the equation of the line, and hence

$\dfrac{3/2}{a} + \dfrac{3/2}{6-a} = 1 \Rightarrow \dfrac{3}{2}\left(\dfrac{1}{a} + \dfrac{1}{6-a}\right) = 1$

$\Rightarrow 1\left(\dfrac{9-a+a}{a(6-a)}\right) = 1 \Rightarrow \dfrac{9}{6a - a^2} = 1$

$\Rightarrow a^2 - 6a + 9 = 0 \Rightarrow (a-3)^2 \Rightarrow a = 3$

Thus, the equation of the line is

$\dfrac{x}{3} + \dfrac{y}{6-3} = 1 \Rightarrow x + y - 3 = 0$

Q30. Solution: The graph of the lines, $x + y = 4$ and $x - y = 0$ are shown in the figure given below.

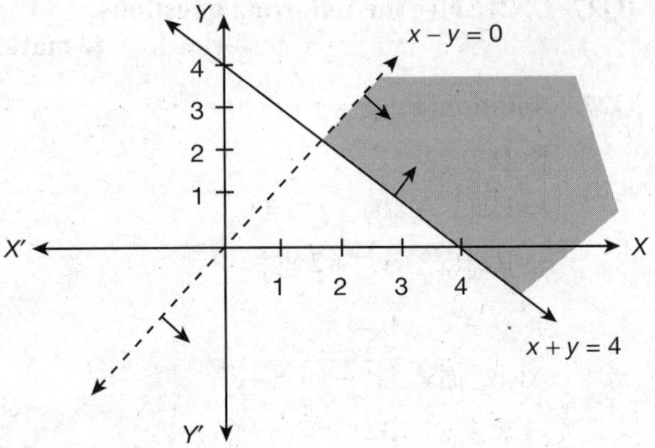

Inequality $x + y \geq 4$ represents the shaded region above the line $x + y = 4$ (including all the points on the line).

Inequality $x - y > 0$ represents the shaded region below the line $x - y = 0$ (excluding all the points on the line). This can be checked by taking any point like (1,0), other than (0,0) because the line passes through (0,0).

Thus, the solution of the given system of linear inequalities is represented by the common shaded region excluding the points on the line $x - y = 0$.

Q31. Solution: Let the coordinates of D be (x, y, z).

$AB = \sqrt{9 + 16 + 144} = \sqrt{169} = 13$

$AC = \sqrt{0 + 25 + 144} = \sqrt{169} = 13$

$\Rightarrow \dfrac{AB}{AC} = \dfrac{13}{13} \Rightarrow AB = AC, \dfrac{BD}{DC} = \dfrac{1}{1} \Rightarrow BD = DC$

D divides the line BC in two equal parts. Therefore, D is the mid-point of BC.

Thus, $x = \dfrac{5+2}{2} = \dfrac{7}{2}$

$y = \dfrac{6+7}{2} = \dfrac{13}{2}, \quad z = \dfrac{9+9}{2} = 9$

Thus, the coordinates of D are $\left(\dfrac{7}{2}, \dfrac{13}{2}, 9\right)$.

OR

Q31. Solution: The perpendicular from the origin to the line $y = mx + c$ meets it at the point (2, 3). Therefore, the slope of the line joining $(0,0),(2,3) = \dfrac{3-0}{2-0} = 3/2$

It is perpendicular to the given equation of line $y = mx + c$. Therefore $m = -2/3$

Since the point (2, 3) lies on the given line, it will satisfy the equation $y = mx + c$. Therefore

$\therefore 3 = (-2/3) \times 2 + c \Rightarrow c = 13/3$

Thus, the respective values of m and c are $-\dfrac{2}{3}$ and $\dfrac{13}{3}$.

Q32. Solution: Slope of the line that inclines with the x-axis at an angle of 75° is $m = \tan 75°$

$\Rightarrow m = \tan(45° + 30°) = \dfrac{\tan 45° + \tan 30°}{1 - \tan 45° \cdot \tan 30°}$

$= \dfrac{1 + \dfrac{1}{\sqrt{3}}}{1 - 1 \cdot \dfrac{1}{\sqrt{3}}} = \dfrac{\dfrac{\sqrt{3}+1}{\sqrt{3}}}{\dfrac{\sqrt{3}-1}{\sqrt{3}}} = \dfrac{\sqrt{3}+1}{\sqrt{3}-1}$

Since the equation of the line passing through point (x_0, y_0) whose slope is m, is $(y - y_0) = m(x - x_0)$, the equation of the line passes through $(2, \sqrt{3})$ and inclines with the x-axis at an angle of 75°, that is, whose slope is $\frac{\sqrt{3}+1}{\sqrt{3}-1}$ is given by

$$\Rightarrow (y - \sqrt{3}) = \frac{\sqrt{3}+1}{\sqrt{3}-1}(x-2)$$

$$\Rightarrow (y - \sqrt{3})(\sqrt{3}-1) = (\sqrt{3}+1)(x-2)$$

$$\Rightarrow y(\sqrt{3}-1) - \sqrt{3}(\sqrt{3}-1) = x(\sqrt{3}+1) - 2(\sqrt{3}+1)$$

$$\Rightarrow (\sqrt{3}+1)x - (\sqrt{3}-1)y = -2\sqrt{3} - 2 + 3 - 2\sqrt{3}$$

$$\Rightarrow (\sqrt{3}+1)x - (\sqrt{3}-1)y = 1 - 4\sqrt{3}$$

OR

Q32. Solution: First term, $a = 5$

Let d be the common difference of the AP

Therefore, AP is $5, 5+d, 5+2d, 5+3d...$

Sum of first five terms is

$$S_5 = \frac{5}{2}[2 \cdot 5 + (5-1)d] = 25 + 10d$$

Sum of next five terms is

$$S_{10-5} = \frac{10}{2}[2 \cdot 5 + (10-1)d] - [25 + 10d]$$

$$[50 + 45d] - [25 + 10d] = 25 + 35d$$

The sum of the first five terms is one-third of the next five terms. Therefore $S_5 = \frac{1}{3} S_{10-5}$

$$\Rightarrow 25 + 10d = \frac{1}{3}(25 + 35d)$$

$$\Rightarrow 75 + 30d = 25 + 35d \Rightarrow 5d = 50 \Rightarrow d = 10$$

Thus, the 15th term of the AP is

$$a_{15} = a + (n-1)d = 5 + (15-1)(10) = 145$$

SECTION D

(Q33–Q36) Solve the following questions. (6 marks each)

Q33. Solution: Consider case I: $x \geq 1$

Then, $|x-1| = x - 1$ and $|x| = x$

$\therefore |x-1| \leq 2$ and $|x| > 2$

$\Rightarrow x - 1 \leq 2$ and $x > 2$

$\Rightarrow x \leq 3$ and $x > 2$

$\Rightarrow 2 < x \leq 3$ is the solution, since $x \geq 1$

Consider case II: $0 \leq x < 1$

Then, $|x-1| = -(x-1) = 1 - x$ and $|x| = x$

$\therefore |x-1| \leq 2$ and $|x| > 2$

$\Rightarrow 1 - x \leq 2$ and $x > 2$

$\Rightarrow -1 \leq x$ and $x > 2$

Does not give any solution, since intersection of $-1 \leq x$ and $x > 2$ is empty

Consider case III: $x < 0$

Then, $|x-1| = -(x-1) = 1 - x$ and $|x| = -x$

$\therefore |x-1| \leq 2$ and $|x| > 2$

$\Rightarrow 1 - x \leq 2$ and $-x > 2$

$\Rightarrow -1 \leq x$ and $-2 > x$

Does not give any solution, since intersection of $-1 \leq x$ and $-2 > x$ is empty

Thus, the solution set of the given inequality is $x \in (2, 3]$

Q34. Solution: The bag contains 4 black and 6 red balls. Two black balls are selected from 4 blacks balls in $^4C_2$ ways.

Three red balls are selected from 6 red balls in $^6C_3$ ways.

Three black balls are selected from 4 black balls in $^4C_3$ ways.

Two red balls are selected from 6 red balls in $^6C_2$ ways.

One black ball is selected from 4 black balls in $^4C_1$ ways.

Four red balls are selected from 6 red balls in $^6C_4$ ways.

$\therefore$ Total number of ways in which balls are selected is

$$= {}^4C_2 \times {}^6C_3 + {}^4C_3 \times {}^6C_2 + {}^4C_1 \times {}^6C_4$$

$$= 6 \times 20 + 4 \times 15 + 4 \times 15 = 360$$

Q35. Solution: Total number of men = 10

Total number of women = 6

Selection of committee containing at least 4 men and 3 women can be done in the following ways: (4 men and 4 women or 5 men and 3 women)

Total number of ways = ${}^{10}C_4 \times {}^6C_4 + {}^{10}C_5 \times {}^6C_3$

$= (210 \times 15 + 252 \times 20 = 8190$

If one particular woman and man must be present in the committee, then total number of ways $= {}^9C_3 \times {}^5C_3 + {}^9C_4 \times {}^5C_2 = 840 + 1260 = 2100$

Thus, total number of committee when one particular woman and man are never present together

= Total– Together = 8190 – 2100 = 6090

OR

Q35. Solution: Let

$f(x) = \cos(x^2), f(x+h) = \cos(x+h)^2$

$\therefore \frac{d}{dx}f(x) = \lim_{h\to 0}\frac{f(x+h)-f(x)}{h}$

$= \lim_{h\to 0}\frac{\cos\{(x+h)\}-\cos(x^2)}{h}$

$= \lim_{h\to 0}\frac{-2\sin\left\{\frac{(x+h)^2+x^2}{2}\right\}\sin\left\{\frac{(x+h)^2-x^2}{2}\right\}}{h}$

$\left[\because \cos C - \cos D = -2\sin\frac{C+D}{2}\cdot\sin\frac{C-D}{2}\right]$

$= \lim_{h\to 0}\frac{1}{h}\left[-2\sin\left\{\frac{(x+h)^2+x^2}{2}\right\}\sin\left\{\frac{(x+h)^2-x^2}{2}\right\}\right]$

$= \lim_{h\to 0}\frac{1}{h}\left[-2\sin\left\{\frac{(x+h)^2+x^2}{2}\right\}\sin\left\{\frac{h^2+2hx}{2}\right\}\right]$

$= -2\lim_{h\to 0}\sin\left\{\frac{(x+h)^2+x^2}{2}\right\}$

$\lim_{h\to 0}\left\{\frac{\sinh\left(\frac{h+2x}{2}\right)}{h\left(\frac{h+2x}{2}\right)}\right\}\times\left(\frac{h+2x}{2}\right)$

$= -2\lim_{h\to 0}\sin\left\{\frac{(x+h)^2+x^2}{2}\right\}\lim_{h\to 0}\left(\frac{h+2x}{2}\right)$

$\left[\because \lim_{x\to 0}\frac{\sin x}{x} = 1\right]$

$= -2x\sin(x^2)\quad [\because x\to 0 \Rightarrow hx \to 0]$

Q36. Solution: Using binomial theorem, the expansions of $(1+2x)^6$ is

$(1+2x)^6 = {}^6C_0 + {}^6C_1(2x) + {}^6C_2(2x)^2 + {}^6C_3(2x)^3 + {}^6C_4(2x)^4 + {}^6C_5(2x)^5 + {}^6C_6(2x)^6$

$= 1+12x+60x^2+160x^3+240x^4+192x^5+64x^6$

Using binomial theorem, the expansions of $(1-x)^4$ is

$(1-x)^4 = {}^4C_0 - {}^4C_1(x) + {}^4C_2(x)^2 - {}^4C_3(x)^3 + {}^4C_4(x)^4$

$= 1-4x+6x^2-4x^3+x^4$

$\therefore x(1-x)^4(1+2x)^6 = x(1-4x+6x^2-4x^3+x^4)\times (1+12x+60x^2+160x^3+240x^4+192x^5+64x^6)$

Since we need the terms which involve x^6, complete multiplication of the two brackets is not required.

Thus, the terms containing x^6 are

$= x(1-4x+6x^2-4x^3+x^4)\times (1+12x+60x^2+160x^3+240x^4+192x^5+64x^6)$

$= x(1)(192x^5) + x(-4x)(240x^4) + x(6x^2)(160x^3) + x(-4x^3)(60x^2) + x(x^4)(12x)$

$= (192-960+960-240+12)x^6 = -36x^6$

Thus, the coefficient of x^6 in the given product is -36

OR

Q36. Solution: Let C be the set of students who like Coffee, M be the set of Students who like Milk and T be the set of students who like Tea.

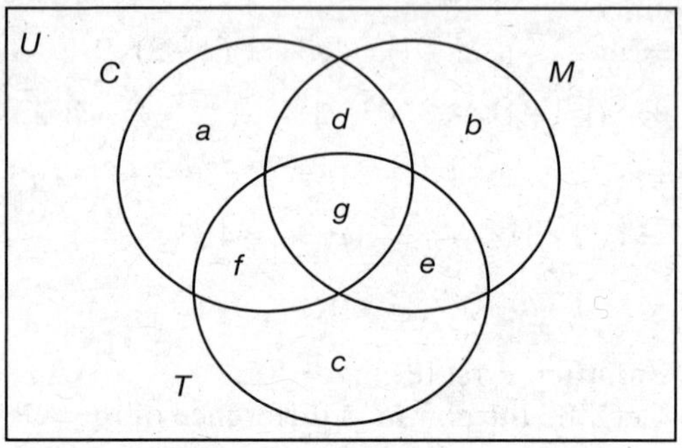

Then, $n(U) = 200$

$n(C) = a+d+f+g = 50\% = 50\times\frac{200}{100} = 100$

$n(M) = b+d+e+g = 75\% = 75\times\frac{200}{100} = 150$

$n(T) = c+e+f+g = 30\% = 30\times\frac{200}{100} = 60$

$n(C\cap M) = d+g = 35\% = 35\times\frac{200}{100} = 70$

$n(M\cap T) = e+g = 25\% = 25\times\frac{200}{100} = 50$

$n(C\cap T) = f+g = 15\% = 15\times\frac{200}{100} = 30$

$n(C\cap M\cap T) = g = 10\% = 10\times\frac{200}{100} = 20$

(i) Number of students who like Coffee only

$a = n(C\cap M'\cap T')$

$= (a+d+f+g)-(d+g)-(f+g)+g$

$= n(C)-n(C\cap M)-n(C\cap T)+n(C\cap M\cap T)$

$= 100-70-30+20 = 20$

(ii) Number of students who like Milk only

$b = n(M\cap C'\cap T')$

$= (b+d+e+g)-(d+g)-(e+g)+g$

$= n(M)-n(M\cap C)-n(M\cap T)+n(C\cap M\cap T)$

$= 150-70-50+20 = 50$

(iii) Number of students who like Tea only

$c = n(T\cap C'\cap M')$

$= (c+e+f+g)-(e+g)-(f+g)+g$

$= n(T)-n(T\cap C)-n(T\cap M)+n(C\cap M\cap T)$

$= 60-30-50+20 = 0$

(iv) Number of students who like Coffee and Milk but not Tea

$d = n(C \cap M \cap T') = (d+g) - g$

$= n(C \cap M) - n(C \cap M \cap T) = 70 - 20 = 50$

(v) Number of students who like Milk and Tea but not Coffee

$e = n(M \cap T \cap C') = (e+g) - g$

$= n(M \cap T) - n(C \cap M \cap T) = 50 - 20 = 30$

(vi) Number of students who like Coffee and Tea but not Milk

$f = n(C \cap T \cap M') = (f+g) - g$

$= n(C \cap T) - n(C \cap M \cap T) = 30 - 20 = 10$

(vii) Number of students who like at least one of the three drinks $= a+b+c+d+e+f+g$

$= 20 + 50 + 0 + 50 + 30 + 10 + 20 = 180$

(viii) Number of students who like none of three drinks = Total students – Students who like at least one of the three drinks $= 200 - 180 = 20$

Sample Paper – II

Day 223

Time: 3 Hrs Maximum **Marks:** 80

General Instructions:

(i) All the questions are compulsory.
(ii) The question paper consists of 36 questions divided into 4 sections A, B, C and D.
(iii) Section A comprises 20 questions of 1 mark each. Section B comprises 6 questions of 2 marks each. Section C comprises 6 questions of 4 marks each. Section D comprises 4 questions of 6 marks each.
(iv) There is no overall choice. However, an internal choice has been provided in four questions of 1 mark each, two questions of 2 marks each, two questions of 4 marks each, and two questions of 6 marks each. You have to attempt only one of the alternatives in all such questions.
(iv) Use of calculators is not permitted.

SECTION A

Q1–Q10 are multiple choice type questions. Select the correct option. (1 mark each)

Q1. Which of the following is not a statement?
(a) $2+6>9$
(b) $2\times 2=4$
(c) 3 is odd prime number
(d) Go there

Q2. When $A=\{-1,0,1\}$, then number of elements in P (A) is
(a) 3 (b) 8 (c) 9 (d) 6

Q3. The value of $(z+i)(\bar{z}-i)$ is equivalent to
(a) $|z+2|^2$ (b) $|z-2|^2$
(c) $|z|^2+2x+1$ (d) $|z|^2+2y+1$

Q4. The total number of terms in the expansion of $(x+a)^{99}+(x-a)^{99}$ after simplification is
(a) 50 (b) 198
(c) 51 (d) None of these

Q5. If the sum of n terms of an AP is given by $S_n=2n-n^2$, then the common difference of the AP is
(a) 2 (b) –2 (c) 4 (d) –4

Q6. If $[x]^2-[x]-6=0$, where [.] denotes the greatest integer function, then
(a) $x\in[2,3)$ (b) $x\in[3,4)\cup[-2,-1)$
(c) $x\in[-3,-2)$ (d) $x\in(2,3)\cup(-3,-2)$

Q7. If $\tan\theta=\frac{1}{2},\tan\phi=\frac{1}{4}$, then the value of $\tan(\theta-\phi)$ is
(a) $\frac{1}{7}$ (b) $\frac{1}{9}$ (c) $\frac{2}{9}$ (d) $\frac{6}{7}$

Q8. If $\sin\theta=\frac{-4}{5},\theta$ lies in fourth quadrant, then the value of $\cos\frac{\theta}{2}$ is
(a) $\frac{1}{\sqrt{5}}$ (b) $-\frac{2}{\sqrt{5}}$
(c) $-\frac{1}{\sqrt{5}}$ (d) $\frac{1}{\sqrt{10}}$

Q9. The sum of the digits in the units place of all the numbers formed with the help of 1, 2, 3, 4, 5 taken all at a time is
(a) 40 (b) 60 (c) 80 (d) 90

Q10. The sum of the series $i+i^2+i^3+\ldots$ up to 101 terms is
(a) 0 (b) –1 (c) i (d) $-i$

Q11–Q15 are fill in the blanks. (1 mark each)

Q11. If $P=\{1,-1\}$, then $P\times P$ is … … …

Q12. The degree measures corresponding to the $\frac{3\pi}{4}$ radian measures (use $\pi=\frac{22}{7}$) is … … …

Q13. If $z=x+iy$, then $z\bar{z}+(z+\bar{z})=0$, where $b\in R$, represents a... … …

Q14. The middle term in the expansion of $(a^2b+b^2a)^{20}$ is……..

Q15. The number of terms of the AP $-6,-\frac{9}{2},-3,-\frac{3}{2}....$ needed to give the sum 165 is ... ...

(Q16–Q20) Answer the following questions.
(1 mark each)

Q16. If 10 lines are drawn in a plane such that no two of them are parallel and no three are concurrent, in how many points will they intersect each other?

Q17. If the points $A(6,1), B(q,2), C(9,4), D(7,3)$ are the vertices of a parallelogram, taken in order, find the value of q.

Q18. Find the derivative of $\frac{1}{x^2-x-1}$

Q19. Find the derivative of $(\sin x+\cos x)^2$

Q20. If $Y=\{1,2,3,4,5,6\}, a$ represents any element of Y, write the following sets, containing all the elements satisfying the given conditions $a \in Y$ and $a^2-a \in Y$

SECTION B

(Q21–Q26) Solve the following questions.
(2 marks each)

Q21. Find the equation of the set of points which are equidistance from the points (1, 0, 4) and (3, 2, –1)

Q22. A box contains 1 red and 2 identical green balls. Two balls are drawn at random in succession (i) with replacement and (ii) without replacement. Write the sample space for this experiment.

Q23. Let $A=\{6,8,9,10\}$ and let $f: A \to N$ be defined by $f(n)=$ the highest prime factor of n. Find the range of f.

Q24. Prove that $\sin^2 5x-\sin^2 3x=\sin 2x \sin 4x$

Q25. Let $z_1=2-i,\ z_2=-2+i$. Find $\left(\frac{z_1 z_2}{\bar{z}_1}\right)$

OR

Q25. Find the centre and radius of the circle $x^2+y^2-6x+8y-11=0$.

Q26. Find the equation of the line through the intersection of lines $x-y+3=0, 2x+y-4=0$ and parallel to the y-axis.

OR

Q26. Find the shortest distance from the point (4, 1) to the circle $x^2+y^2-14x-10y-151=0$

SECTION C

(Q27–Q32) Solve the following questions.
(4 marks each)

Q27. A bag contains 6 black and 4 white balls. Two balls are drawn at random. Find the probability that
(i) all the two balls are black.
(ii) one ball is white and one ball is black.
(iii) at least one ball is black.

Q28. Find the equation of the hyperbola satisfying the give conditions: foci $(0,\pm 16)$ and the conjugate axis of length 28.

Q29. Differentiate $x^{1/3}$ with respect to x using the first principle.

Q30. Find the number of positive integers greater than 1000 and less than 4000 which are divisible by 5, provided that no digit is to be repeated.

Q31. Evaluate $(\sqrt{6}+\sqrt{2})^5+(\sqrt{6}-\sqrt{2})^5$

OR

Q31. Find the sum of the series $0.2+0.22+0.222+.........$ up to n terms.

Q32. The perpendicular from the origin to a line meets it at the point (–2, 6).Find the equation of the line.

OR

Q32. Find the value of r, if the coefficient of $(2r+5)^{\text{th}}$, $(2r-3)^{\text{th}}$ terms in the expansion of $(1+x)^{12}$ are equal.

SECTION D

(Q33–Q36) Solve the following questions.
(6 marks each)

Q33. Find the mean deviation about the median for the data 6, 2, 10, 4, 5, 8

Q34. Convert the given complex number to polar form: $-i$

Q35. Solve the following system of inequalities
$\frac{3}{x-2}>-1, \frac{2x-1}{x+1} \leq -1.$

OR

Q35. The vertices of triangle ΔPQR are $P(2,1), Q(-2,3), R(3,1)$. Find the equation of the median through the vertex R.

Q36. Find the coordinates of the foci, the vertices, the length of major axis, the minor axis, the eccentricity and the latus rectum of the ellipse $\frac{x^2}{49}+\frac{y^2}{9}=1$

OR

Q36. Solve the following system of inequalities graphically: $3x+4y \geq 12, x+2y \geq 4$

Day 224

Sample Paper – II (Solutions)

SECTION A

Q1–Q10 are multiple choice type questions. Select the correct option. (1 mark each)

Q1. Solution: (d)

Go there is not a statement. It is an order.

Q2. Solution: (b)

$\because A=\{-1,0,1\} \Rightarrow n(A)=3$

Therefore, the number of elements in $P(A)$ is $n\{P(A)\}=2^{n(A)}=2^3=8$

Q3. Solution: (d)

Let $z=x+iy$ then $(z+i)(\bar{z}-i)$

$=(x+iy+i)(x-iy-i)$

$=\{x+i(y+1)\}\{(x-i(y+1)\}$

$=(x)^2+(y+1)^2=x^2+y^2+2y+1$

$=|z|^2+2y+1$

Q4. Solution: (a)

The 50 terms of $(x+a)^{99}$ cancel out the 50 terms of $(x-a)^{99}$ and the 50 terms of $(x+a)^{99}$ are added to the 50 terms of $(x-a)^{99}$.

$(x+a)^{99}+(x-a)^{99}$

$={}^{99}C_0x^{99}+{}^{99}C_1x^{98}a+{}^{99}C_2x^{97}a^2+\ldots+{}^{99}C_{98}a^{98}x$

$+{}^{99}C_{99}a^{99}+{}^{99}C_0x^{99}-{}^{99}C_1x^{98}a+{}^{99}C_2x^{97}a^2-\ldots+$

${}^{99}C_{98}a^{98}x-{}^{99}C_{99}a^{99}$

$=\underbrace{2[{}^{99}C_0x^{99}+{}^{99}C_2x^{97}a^2\ldots+{}^{99}C_{98}a^{98}x]}_{50\,terms}$

Thus, in the expansion of $(x+a)^n+(x-a)^n$, the total number of terms will be $\frac{n+1}{2}$ when n is odd.

Q5. Solution: (b)

$\because S_n=2n-n^2$

$\therefore S_1=2\times1-(1)^2=1=T_1$

$S_2=2\times2-(2)^2=0$

$T_2=S_2-S_1=0-1=-1$

Thus, common difference $(d)=T_2-T_1=-1-1=-2$

Q6. Solution: (b)

$[x]^2-[x]-6=0,$

$\Rightarrow [x]^2-3[x]+2[x]-6=0$

$\Rightarrow [x]([x]-3)+2([x]-3)=0$

$\Rightarrow ([x]-3)([x]+2)=0$

$\Rightarrow [x]=3,-2$

$\because$ For $[x]=3 \Rightarrow x\in[3,4)$

and for $[x]=-2 \Rightarrow x\in[-2,-1)$

Q7. Solution: (c)

$\tan(\theta-\phi)=\frac{\tan\theta-\tan\phi}{1+\tan\theta\cdot\tan\phi}$

$\Rightarrow \tan(\theta-\phi)=\frac{\frac{1}{2}-\frac{1}{4}}{1+\frac{1}{2}\cdot\frac{1}{4}}=\frac{\frac{2-1}{4}}{\frac{8+1}{8}}=\frac{1/4}{9/8}=\frac{2}{9}$

Q8. Solution: (b)

$\cos\theta=\sqrt{1-\frac{16}{25}}=\sqrt{\frac{25-16}{25}}=\pm\frac{3}{5}\Rightarrow\cos\theta=\frac{3}{5}$

[Since θ lies in the fourth quadrant]

$\therefore \cos\theta=2\cos^2\frac{\theta}{2}-1\Rightarrow\frac{3}{5}\Rightarrow 2\cos^2\frac{\theta}{2}-1$

$\Rightarrow 2\cos^2\frac{\theta}{2}=\frac{8}{5}\Rightarrow\cos\frac{\theta}{2}=\pm\frac{2}{\sqrt{5}}$

$\Rightarrow \cos\frac{\theta}{2}=-\frac{2}{\sqrt{5}}$ [Since θ lies in the fourth quadrant, $\theta/2$ lies in the second quadrant]

Q9. Solution: (b)

If we fixed digit 1 at the units place, then the total possible number formed with the remaining 4 digits is 4!

Similarly, if we fixed 2, 3, 4 and 5 at the units place, in each case, the total possible numbers are 4!.

Thus, the sum of the digits in the units place of all these numbers $=1\times4!+2\times4!+3\times4!+4\times4!+5\times4!$ $=360$

Q10. Solution: (c)

$101=4\times25+1$ and $i^n+i^{n+1}+i^{n+2}+i^{n+3}=0$

$\therefore i+i^2+i^3+i^4+\ldots i^{101}=i\left(\because i^2+i^3+i^4+\ldots i^{101}=0\right)$

Q11–Q15 are fill in the blanks. **(1 mark each)**

Q11. Solution: $P = \{1, -1\} \Rightarrow n(P) = 2$

$\therefore n(P \times P) = n(P) \times n(P) = 2 \times 2 = 4$ elements are given by $P \times P = \{(1,1), (1,-1), (-1,1), (-1,-1)\}$

Q12. Solution: π radians = 180°

Therefore, $\frac{3\pi}{4}$ radians $= \frac{180}{\pi} \times \frac{3\pi}{4}$ degree

= 135 degree = 135^0

Thus, $\frac{3\pi}{4}$ radians $= 135^0$

Q13. Solution: Let $z = x + iy$,

$z\bar{z} + (z + \bar{z}) = 0$,

$\Rightarrow (x+iy)(x-iy) + \{(x+iy)+(x-iy)\} = 0$

$\Rightarrow x^2 + y^2 + 2x = 0$, represents the equation of a circle.

Q14. Solution: Middle term of the expansion $(a^2b + b^2a)^{20}$ is

$= \left(\frac{20}{2}+1\right)$th term $= 11^{th}$ term

$\therefore T_{11} = T_{10+1} = {}^{20}C_{10}(a^2b)^{20-10}(b^2a)^{10} = {}^{20}C_{10}a^{30}b^{30}$

Q15. Solution: Common difference of the given AP is

$d = -\frac{9}{2} - (-6) = \frac{-9+12}{2} = \frac{3}{2}$

Let the sum of n terms of the given AP be 165. Then

$$S_n = \frac{n}{2}[2a + (n-1)d]$$

$$\Rightarrow 165 = \frac{n}{2}\left[2 \times (-6) + (n-1)\left(\frac{3}{2}\right)\right]$$

$$\Rightarrow 330 = n\left[-12 + \frac{3n}{2} - \frac{3}{2}\right]$$

$$\Rightarrow 660 = n(-27 + 3n) \Rightarrow 3n^2 - 27n - 660 = 0$$

$$\Rightarrow n^2 - 9n - 220 = 0$$

$$\Rightarrow (n+11)(n-20) = 0 \Rightarrow n = -11 \text{ or } 20$$

$\therefore n = 20 \ (\because n \neq -11)$

(Q16–Q20) Answer the following questions. **(1 mark each)**

Q16. Solution: No two lines are parallel. Therefore, all lines intersect pair-wise, and since no three lines are concurrent, no three lines will intersect at a point.

Thus, for one point of intersection, we require two lines. Therefore, the number of points of intersection = ${}^{10}C_2 = 45$

Q17. Solution: The diagonals of a parallelogram bisect each other. Therefore, the coordinates of the mid-point of diagonal AC = coordinates of the mid-point of diagonal BD

$$\Rightarrow \left(\frac{6+9}{2}, \frac{1+4}{2}\right) = \left(\frac{q+7}{2}, \frac{2+3}{2}\right)$$

$$\Rightarrow \left(\frac{15}{2}, \frac{5}{2}\right) = \left(\frac{q+7}{2}, \frac{2+3}{2}\right)$$

$$\Rightarrow \frac{15}{2} = \frac{7+q}{2} \Rightarrow q = 8$$

Q18. Solution: Let $f(x) = \frac{1}{x^2 - x - 1}$

By the quotient rule

$$f'(x) = \frac{(x^2-x-1)\frac{d}{dx}(1) - (1)\frac{d}{dx}(x^2-x-1)}{(x^2-x-1)^2}$$

$$= \frac{(x^2-x-1)(0) - 1(2x-1)}{(x^2-x-1)^2} = \frac{1-2x}{(x^2-x-1)^2}$$

Q19. Solution: Let $y = (\sin x + \cos x)^2$

$$\therefore \frac{dy}{dx} = 2(\sin x + \cos x)(\cos x - \sin x)$$

$$\left(\because \frac{d}{dx}y^n = ny^{n-1}\frac{dy}{dx}\right)$$

$$= 3(\cos^2 x - \sin^2 x)$$

$$= 3\cos 2x \quad [\because \cos 2x = \cos^2 x - \sin^2 x]$$

Q20. Solution: $Y = \{1, 2, 3, 4, 5, 6\}$,

$\therefore \{a : a \in Y, a^2 - a \in Y\} = \{2, 3\}$

SECTION B

(Q21–Q26) Solve the following questions. **(2 marks each)**

Q21. Solution: Let P (x, y, z) be the point that is equidistance from points A (1, 0, 4) and B (3, 2, –1). Then $PA = PB$

$$\Rightarrow PA^2 = PB^2$$

$$\Rightarrow (x-1)^2 + (y-0)^2 + (z-4)^2$$

$$= (x-3)^2 + (y-2)^2 + (z+1)^2$$

$$\Rightarrow x^2 - 2x + 1 + y^2 + z^2 - 8z + 16$$

$$= x^2 - 6x + 9 + y^2 - 4y + 4 + z^2 + 2z + 1$$

$$\Rightarrow -2x - 8z + 17 = -6x - 4y + 2z + 14$$

$\Rightarrow 4x + 4y - 10z + 3 = 0$ is the required equation.

Q22. Solution: Let us denote the red ball as R and a green ball as G.

(i) When two balls are drawn at random in succession with replacement, the sample space is given by

$S = \{RR, RG, GR, GG\}$

(ii) When two balls are drawn at random in succession without replacement, the sample space is given by

$S = \{RG, GR, GG\}$

Q23. Solution: $A = \{6, 8, 9\}$ and $f : A \to N$ is defined as

$f(n)$ = The highest prime factor of n

Now, since prime factor of 6 = 2, 3

Prime factor of 8 = 2

Prime factors of 9 = 3

$\therefore f(6)$ = The highest prime factor of 6 = 3

$f(8)$ = The highest prime factor of 8 = 2

$f(9)$ = The highest prime factor of 9 = 3

$f(10)$ = The highest prime factor of 10 = 5

Thus, the range of f is the set of all $f(n)$, where $n \in A$., that is, range of $f = \{2, 3, 5\}$

Q24. Solution:

$$\text{LHS} = \sin^2 5x - \sin^2 3x$$

$$= (\sin 5x + \sin 3x)(\sin 5x - \sin 3x)$$

$$= \left[2\sin\left(\frac{5x+3x}{2}\right)\cos\left(\frac{5x-3x}{2}\right)\right] \times \left[2\cos\left(\frac{5x+3x}{2}\right)\sin\left(\frac{5x-3x}{2}\right)\right]$$

$$\left[\begin{array}{l}\because \sin A + \sin B = 2\sin\left(\frac{A+B}{2}\right)\cos\left(\frac{A-B}{2}\right) \\ \sin A - \sin B = 2\cos\left(\frac{A+B}{2}\right)\sin\left(\frac{A-B}{2}\right)\end{array}\right]$$

$$= (2\sin 4x \cos x)(2\cos 4x \sin x)$$

$$= (2\sin x \cos x)(2\sin 4x \cos 4x)$$

$$= \sin 2x \sin 8x = \text{RHS}.$$

Q25. Solution: $z_1 = 2 - i,\ z_2 = -2 + i.$

$$\because z_1 z_2 = (2-i)(-2+i) = -4 + 4i - i^2$$

$$= -3 + 4i \qquad (\because i^2 = -1)$$

and $z_1 = 2 - i \Rightarrow \bar{z}_1 = 2 + i$

$$\therefore \frac{z_1 z_2}{\bar{z}_1} = \frac{-3+4i}{2+i} = \frac{(-3+4i)(2-i)}{(2+i)(2-i)}$$

$$= \frac{-6+3i+8i-4i^2}{2^2+1^2} = \frac{-2+11i}{5}$$

$$\Rightarrow \frac{z_1 z_2}{\bar{z}_1} = \frac{-2}{5} + \frac{11}{5}i$$

OR

Q25. Solution: The given equation $x^2 + y^2 - 6x + 8y - 11 = 0$ of the circle can be written as $(x^2 - 6x) + (y^2 + 8y) = 11$

$$\Rightarrow \{x^2 - 2(x)(3) + 3^2\} + \{y^2 + 2(y)(4) + 4^2\} - 9 - 16 = 11$$

$$\Rightarrow (x-3)^2 + (y+4)^2 = 36$$

$$\Rightarrow (x-3)^2 + \{y-(-4)\}^2 = (6)^2$$

Comparing this with the equation of the circle $(x-h)^2 + (y-k)^2 = r^2$, we get

$h = 3, k = -4, r = 6$

Therefore, the centre of the given circle is (3, –4) and radius is 6.

Q26. Solution: The equation of any line through the point of intersection of the given lines is of the form

$$x - y + 3 + k(2x + y - 4) = 0$$

$$\Rightarrow (1+2k)x + (-1+k)y + 3 - 4k = 0$$

Since this line is parallel to the y-axis, the coefficient of y should be zero, that is, $-1 + k = 0 \Rightarrow k = 1$

Substituting the value of $k = 1$, we get $(1 + 2\times1)x + (-1+1)y + 3 - 4\times1 = 0 \Rightarrow 3x - 1 = 0$ as the required equation of the line.

OR

Q26. Solution: For the given circle

$x^2 + y^2 - 14x - 10y - 151 = 0,$

centre = (7, 5) and

radius $= \sqrt{49 + 25 + 151} = \sqrt{225} = 15$

Thus, the distance between the point (4, 1) and centre of the circle is given by

$$= \sqrt{(7-4)^2 + (5-1)^2} = \sqrt{9+16} = 5$$

$\therefore$ Shortest distance, $d = |15 - 5| = 10$

SECTION C

(Q27–Q32) Solve the following questions. (4 marks each)

Q27. Solution: (i) P (*the two balls are black*)$= \dfrac{{}^6C_2}{{}^{10}C_2}$

(ii) P (*one ball is white and one ball is black*)

$$= \frac{{}^4C_1\,{}^6C_1}{{}^{10}C_2}$$

(iii) P (*at least one ball is black*)$= \dfrac{{}^6C_1\,{}^4C_1 + {}^6C_2}{{}^{10}C_2}$

Q28. Solution: The foci (0,±16) are on the y-axis. Therefore, the equation of the hyperbola is of the form $\frac{y^2}{a^2}-\frac{x^2}{b^2}=1$ and $c = 16$.

Now, since the conjugate axis is of length 24,

$2b = 28 \Rightarrow b = 14 \Rightarrow b^2 = 196$

From the equation $a^2+b^2=c^2$

$\Rightarrow a^2+14^2=16^2 \Rightarrow a^2=60$

Thus, the equation of the hyperbola is $\frac{y^2}{60}-\frac{x^2}{196}=1$

Q29. Solution:

Let $f(x)=x^{1/3} \Rightarrow f(x+h)=(x+h)^{1/3}$

Now, $\frac{d}{dx}f(x)=\lim_{h\to 0}\frac{f(x+h)-f(x)}{h}$

$$=\lim_{h\to 0}\frac{1}{h}[(x+h)^{1/3}-x^{1/3}]=\lim_{h\to 0}\frac{1}{h}\left[x^{1/3}\left(1+\frac{h}{x}\right)^{1/3}-x^{1/3}\right]$$

$$=\lim_{h\to 0}\frac{1}{h}\left[x^{1/3}\left(1+\frac{h}{x}\cdot\frac{1}{3}+\frac{1}{3}\left(\frac{1}{3}-1\right)\frac{h^2}{x^2}+\ldots\right)-1\right]$$

$$\left[\because (1+x)^n=1+nx+\frac{n(n-1)}{2!}x^2+\ldots\ldots\right]$$

$$=\lim_{h\to 0}\frac{1}{h}\left[x^{1/3}\left(\frac{1}{3}\cdot\frac{h}{x}-\frac{2}{9}\cdot\frac{h^2}{x^2}+\ldots\right)\right]$$

$$=\lim_{h\to 0}\frac{x^{1/3}}{h}\cdot\frac{1}{3}\frac{h}{x}\left(1-\frac{2}{3}\cdot\frac{h}{x}+\ldots\ldots\right)=\frac{1}{3}x^{1/3-1}=\frac{1}{3}x^{-2/3}$$

Q30. Solution: The number is divisible by 5. Therefore, at the units place, the number will be 0 or 5. Therefore, the units digit can be filled in two ways. Now, we need a four-digit number greater than 1000 and less than 4000. Therefore, the thousands digit can be filled in three ways (digit 1 or 2 or 3).

Since repetition is not allowed, the tens place can be filled in 8 ways, and the hundreds place can be filled in 7 ways.

Thus, the total number of integers formed is

$= 2\times 3\times 8\times 7 = 336$

Q31. Solution: Using Binomial Theorem

$$(a+b)^5={}^5C_0a^5+{}^5C_1a^4b+{}^5C_2a^3b^2+{}^5C_3a^2b^3+{}^5C_4ab^4+{}^5C_5b^5$$

$$=a^5+5a^4b+10a^3b^2+10a^2b^3+5ab^4+b^5$$

And,

$$(a-b)^5={}^5C_0a^5-{}^5C_1a^4b+{}^5C_2a^3b^2-{}^5C_3a^2b^3+{}^5C_4ab^4-{}^5C_5b^5$$

$$=a^5-5a^4b+10a^3b^2-10a^2b^3+5ab^4-b^5$$

$$\therefore (a+b)^5+(a-b)^5$$

$$=2[a^5+10a^3b^2+5ab^4]$$

Putting $a=\sqrt{6}, b=\sqrt{2}$, we get

$$(\sqrt{6}+\sqrt{2})^5+(\sqrt{6}-\sqrt{2})^5$$

$$=2[5(\sqrt{6})^5+10(\sqrt{6})^3(\sqrt{2})^2+5(\sqrt{6})(\sqrt{2})^4]$$

$$=2[5\times 36\sqrt{6}+120\sqrt{6}+20\sqrt{6}]=640\sqrt{6}$$

OR

Q31. Solution: Let $S_n = 0.2+0.22+0.222+\ldots\ldots$ to n terms

$$=2[0.1+0.11+0.111+\ldots.\text{to } n \text{ terms}]$$

$$=\frac{2}{9}[0.9+0.99+0.999+\ldots\text{to } n \text{ terms}]$$

$$=\frac{2}{9}\left[\left(1-\frac{1}{10}\right)+\left(1-\frac{1}{10^2}\right)+\left(1-\frac{1}{10^3}\right)+\ldots\text{to } n \text{ terms}\right]$$

$$=\frac{2}{9}\left[(1+1+\ldots n\text{ terms})-\frac{1}{10}\left(1+\frac{1}{10}+\frac{1}{10^2}+\ldots n\text{ terms}\right)\right]$$

$$=\frac{2}{9}\left[n-\frac{1}{10}\left(\frac{1-\left(\frac{1}{10}\right)^n}{1-\frac{1}{10}}\right)\right]$$

$$=\frac{2}{9}n-\frac{1}{45}\times\frac{10}{9}(1-10^{-n})=\frac{2}{9}n-\frac{2}{81}(1-10^{-n})$$

Q32. Solution: Let the slope of the line joining the origin (0, 0) and point (–2, 6) be $m=\frac{6-0}{-2-0}=-3$

Then, the slope of the line perpendicular to the line joining the origin and point (–2, 6) is

$$=-\frac{1}{m}=-\left(\frac{1}{-3}\right)=\frac{1}{3}$$

Thus, the equation of the line passing through (–2, 6) and having a slope $\frac{1}{3}$ is $(y-6)=\frac{1}{3}(x+2)$

$3y-18=x+2 \Rightarrow x-3y+20=0$

OR

Q32. Solution: The $(2r+5)^{th}$ term of the expansion $(1+x)^{12}$ is

$$T_{2r+4+1}={}^{12}C_{2r+4}(x)^{2r+4}={}^{12}C_{2r+4}x^{2r+4}$$

The $(2r-3)^{th}$ term of the expansion $(1+x)^{10}$ is

$$T_{2r-4+1}={}^{12}C_{2r-4}(x)^{2r-4}={}^{12}C_{2r-4}x^{2r-4}$$

The coefficients of $(2r+5)^{th}$, $(2r-3)^{th}$ terms in the expansion of $(1+x)^{12}$ are equal. Therefore

$^{12}C_{2r+4} = {}^{12}C_{2r-4}$

$\Rightarrow {}^{12}C_{2r+4} = {}^{12}C_{12-2r+4} \quad [\because {}^nC_r = {}^nC_{n-r}]$

$\Rightarrow {}^{12}C_{2r+4} = {}^{12}C_{16-2r}$

$\Rightarrow 2r+4 = 16-2r \Rightarrow 4r = 12 \Rightarrow r = 3$

SECTION D

(Q33–Q36) Solve the following questions.

(6 marks each)

Q33. Solution: The number of observations is 6, which is even.

The given data 6, 2, 10, 4, 5, 8 can be arranged in ascending order as 2, 4, 5, 6, 8, 10

The number of observations is 6, which is even. Therefore, median

$$M = \frac{\left(\frac{6}{2}\right)^{\text{th}} \text{observation} + \left(\frac{6}{2}+1\right)^{\text{th}} \text{observation}}{2}$$

$$= \frac{3^{\text{rd}} \text{observation} + 4^{\text{th}} \text{observation}}{2}$$

$$= \frac{5+6}{2} = \frac{11}{2} = 5.5$$

The deviations of the respective observations from the median $= x_i - M$ are

$-3.5, -1.5, -0.5, 0.5, 2.5, 4.5$

The absolute values of the deviations $= |x_i - M|$ are

$3.5, 1.5, 0.5, 0.5, 2.5, 4.5$

Thus, the mean deviation about the median is

$$\text{M.D.}(M) = \frac{\sum_{i=1}^{6} |x_i - M|}{6}$$

$$= \frac{3.5+1.5+0.5+0.5+2.5+4.5}{6} = \frac{13}{6} = 2.17$$

Q34. Solution: Complex number $z = i = 0 - i$

Therefore, $r\cos\theta = 0, \quad r\sin\theta = -1$

On squaring and adding, we get

$r^2\cos^2\theta + r^2\sin^2\theta = (0)^2 + (-1)^2$

$\Rightarrow r^2(\cos^2\theta + \sin^2\theta) = 1 \quad (\because \sin^2\theta + \cos^2\theta = 1)$

$\Rightarrow r^2 = 1 \Rightarrow r = 1 \Rightarrow |z| = 1 \qquad (\because r > 0)$

$\because \cos\theta = 0, \quad \sin\theta = -1$

$\therefore |\tan\theta| = \infty \Rightarrow \theta = \frac{\pi}{2}$

$\sin\theta$ is negative and $\cos\theta$ is positive (or zero). Therefore, θ lies in the fourth quadrant.

Thus, $\text{Arg}(z) = -\theta = -\frac{\pi}{2}$

Hence, the polar form of $-i$ is

$r(\cos\theta + i\sin\theta) = \cos\left(-\frac{\pi}{2}\right) + i\sin\left(-\frac{\pi}{2}\right)$

Q35. Solution: Inequality $\frac{3}{x-2} > -1 \Rightarrow \frac{3}{x-2} + 1 > 0$

$\Rightarrow \frac{3+1(x-2)}{x-2} > 0$

$\Rightarrow \frac{x+1}{x-2} > 0$

$\Rightarrow x < -1$ or $x > 2$

$\Rightarrow x \in (-\infty, -1) \cup (2, \infty)$

Inequality $\frac{2x-1}{x+1} \le -1 \Rightarrow \frac{2x-1}{x+1} + 1 \le 0$

$\Rightarrow \frac{2x-1+1(x+1)}{x+1} \le 0$

$\Rightarrow \frac{3x}{x+1} \le 0$

$\Rightarrow -1 \le x \le 0$

$\Rightarrow x \in [-1, 0]$

Since the intersection of the solutions, $x \in (-\infty, -1) \cup (2, \infty)$ and $x \in [-1, 0]$ is a null set, the given system of equations has no solution.

OR

Q35. Solution: It is given that the vertices of triangle ΔPQR are $P(2,1), Q(-2,3), R(3,1)$.

Let RL be the median through vertex R, where L is the mid-point of PQ.

Then, the coordinates of point L are given by $\left(\frac{2-2}{2}, \frac{1+3}{2}\right) = (0,2)$

Since the equation of the line passing through points $(x_1, y_1), (x_2, y_2)$ is $y - y_1 = \frac{y_2 - y_1}{x_2 - x_1}(x - x_1)$

The equation of median RL, where $R(3,1)$ and $L(0,2)$ is $y - 1 = \frac{2-1}{0-3}(x-3)$

$\Rightarrow y - 1 = \frac{-1}{3}(x-3) \Rightarrow 3(y-1) = -x + 3$

$\Rightarrow x + 3y - 6 = 0$

Thus, the required equation of the median through the vertex R is $x + 3y - 6 = 0$

Q36. Solution: In the given equation $\frac{x^2}{49}+\frac{y^2}{9}=1$ of ellipse, the denominator of x^2 is greater than the denominator of y^2. Therefore, the major axis is along the x-axis and the minor axis is along the y-axis.

Comparing the given equation with $\frac{x^2}{a^2}+\frac{y^2}{b^2}=1$, we get $a=7$, $b=3$ and $c=\sqrt{a^2-b^2}=\sqrt{49-9}=2\sqrt{10}$

Thus, the coordinates of the foci are $(2\sqrt{10},0),(-2\sqrt{10},0)$

The coordinates of the vertices are (7, 0), (–7, 0).

Length of major axis = $2a = 14$

Length of minor axis = $2b = 6$

Eccentricity $e=\frac{c}{a}=\frac{2\sqrt{10}}{7}$

Length of latus rectum $=\frac{2b^2}{a}=\frac{2\times 9}{7}=\frac{18}{7}$

OR

Q36. Solution: The graphs of the lines $3x+4y=12$ and $x+2y=4$ are shown in the figure.

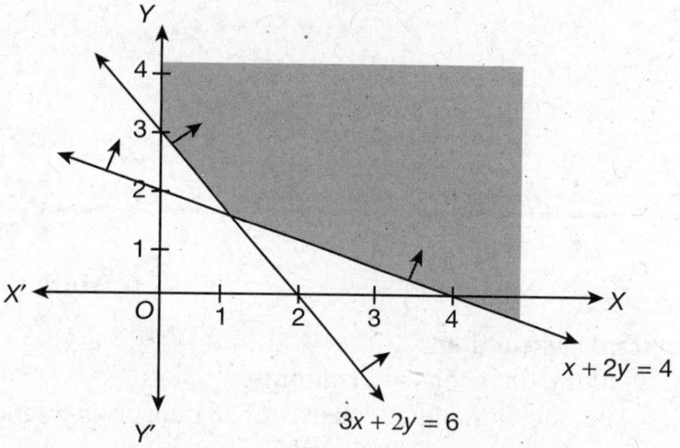

Inequality $3x+4y\geq 12$ represents the shaded region above the line $3x+4y=12$ (including all the points on the line).

Inequality $x+2y\geq 4$ represents the shaded region above the line $x+2y=4$ (including all the points on the line).

Thus, the solution of the given system of linear inequalities is represented by the common shaded region, including the points on the respective lines.

Sample Paper – III

Day 225

Time: 3 Hrs Maximum **Marks:** 80

General Instructions:

(i) All the questions are compulsory.

(ii) The question paper consists of 36 questions divided into 4 sections A, B, C and D.

(iii) Section A comprises 20 questions of 1 mark each. Section B comprises 6 questions of 2 marks each. Section C comprises 6 questions of 4 marks each. Section D comprises 4 questions of 6 marks each.

(iv) There is no overall choice. However, an internal choice has been provided in four questions of 1 mark each, two questions of 2 marks each, two questions of 4 marks each, and two questions of 6 marks each. You have to attempt only one of the alternatives in all such questions.

(iv) Use of calculators is not permitted.

SECTION A

Q1–Q10 are multiple choice type questions. Select the correct option. (1 mark each)

Q1. The negation of the statement "6 is greater than 9" is

(a) 6 is equal to 8

(b) 6 is not greater than 8

(c) 8 is less than 6

(d) None of these

Q2. If $\left(\frac{1+i}{1-i}\right)^x = -1$, then

(a) $x = 2n+1$ (b) $x = 4n+2$

(c) $x = 2n$ (d) $x = 4n+1$

Q3. The number of solutions of the equation $\Rightarrow 1+\sin x = \operatorname{cosec} x(1+\operatorname{cosec} x)$, x lying in the interval $[0, 2\pi]$, is

(a) 0 (b) 1 (c) 2 (d) 3

Q4. If the integer $r > 1, y > 2$ and coefficient of $(r–2)^{\text{th}}$ and $(r+2)^{\text{th}}$ terms in the binomial expansion of $(1+x)^{2n}$ are equal, then

(a) $n = 2r-1$ (b) $n = r-1$

(c) $n = 2r+1$ (d) None of these

Q5. Range of $f(x) = \frac{1}{1-\sin x}$ is

(a) $[-1,1]$ (b) $[-1,2]$

(c) $[1/2, \infty)$ (d) $(-\infty, -1)$

Q6. Which of the following is correct?

(a) $\sin\theta = \frac{1}{2}$ (b) $\tan\theta = \frac{1}{2}$

(c) $\sec\theta = \frac{1}{2}$ (d) $\cos\theta = \frac{1}{2}$

Q7. The sum of all the numbers formed with the help of digits 1, 2, 3 taken all at a time is

(a) 980 (b) 1228 (c) 1332 (d) 1456

Q8. If the third term of a GP is 3, then the product of its first 5 terms is

(a) 9 (b) 27 (c) 81 (d) 243

Q9. If the letters of the word 'VIKASH' are arranged at random in a row, what is the probability that the letters A and I remain together?

(a) 1/2 (b) 1/3 (c) 1/4 (d) 1/5

Q10. Out of 10 points in a plane, 6 are in the same line, which are collinear. Find the number of triangles that can be formed by joining the points.

(a) 25 (b) 50 (c) 100 (d) 200

Q11–Q15 are Fill in the blanks. (1 mark each)

Q11. A coin is tossed. If it shows a tail, we draw a ball from a box, which contains 1 red and 2 blue balls. If it shows head, we throw a coin again. The sample space for this experiment is … … …

Q12. If $A \times B = \{(0,1),(0,-1),(2,1),(2,-1)\}$, then sets A and B are … … …

Q13. The coefficient of the term independent of x in the expansion in the expansion $\left(\frac{\sqrt{x}}{3} + \frac{3}{x}\right)^{15}$ is................

Q14. The power set of the set $A = \{-1, 1\}$ is.........

Q15. If the line $lx + my = 1$ is a tangent to the circle $x^2 + y^2 = 9$, then the value of $l^2 + m^2$ = ...

(Q16–Q20) Answer the following questions.
(1 mark each)

Q16. If $\arg(z+1)=\arg(z+i)$, then find the value of $x+y$, where $z=x+iy$.

Q17. Find the angle in radian and degree through which a pendulum swings if its length is 70 cm and the tip describes an arc of length 22 cm.

Q18. If $|z+2|\le 4$, then find the greatest value of $|z+1|$

Q19. Find the derivative of $4x^3-\frac{1}{2x^2}$

Q20. Evaluate $\lim_{x\to 0}\frac{(x+2)^{1/3}-2^{1/3}}{x}$

SECTION B

(Q21–Q26) Solve the following questions.
(2 marks each)

Q21. Write the following sets in the roster form.
(i) $A=\{x: x\in I, x^2+x-6=0\}$
(ii) $B=\{x | x^2+x=0, x\in I\}$

Q22. Find the equation of the ellipse that satisfies the given conditions: vertices $(0,\pm 10)$, foci $(0,\pm 5)$

Q23. Find the equation of the hyperbola satisfying the given conditions: vertices $(\pm 6,0)$ and eccentricity $e=\frac{4}{3}$

Q24. $A=\{0,1,2,3\}, S=\{(x,y): x\in A, y\in A\}$,. Then find the ordered set which satisfies the conditions given below:
(i) $x+y=3$ (ii) $x\cdot y=3$ (iii) $x+y\le 3$

OR

Q24. The length L (in centimetre) of a copper rod is a linear function of its Celsius temperature C. In an experiment, if $L=103.5$ when $C=20$, and $L=105.0$ when $C=120$, express L in terms of C.

Q25. Find the slope of the line, which makes an angle of 60^0 with the positive direction of the y-axis measured clockwise.

OR

Q25. Show that the locus represented by $|z+1|=|z+i|$ is a line.

Q26. Find derivative of $\frac{x^2-\cos x}{x-\sin x}$

OR

Q26. Ravi obtained 60, 70 and 80 marks in the first three unit tests. Find the minimum marks he should obtain in the fourth test to have an average of at least 60 marks and at most 75 marks.

SECTION C

(Q27–Q32) Solve the following questions.
(4 marks each)

Q27. Find the general solution of the equation $\tan^2 2x=\tan 2x$

Q28. Find the modulus and argument of the complex number $\frac{1+i}{1-i}$

Q29. There are 12 persons named $P_1, P_2, P_3, ..., P_{12}$. Out of 12 persons, 6 persons are to be arranged in a line such that in each arrangement P_1 must occur whereas P_{12} must not occur. Find the number of such possible arrangements.

Q30. Find the value of $(\sqrt{a+1}+\sqrt{a-1})^4+(\sqrt{a+1}-\sqrt{a-1})^4$

Q31. Find the sum of the following series up to 100 terms $\frac{1^2}{1}+\frac{1^2+2^2}{1+2}+\frac{1^2+2^2+3^2}{1+2+3}+....$

OR

Q31. If the origin is the centroid of the triangle PQR with vertices $P(a,2,6), Q(-4,b,-10), R(8,12,c)$, then find the values of a, b, c.

Q32. A line perpendicular to the line segment joining the points (1, 0) and (2, 4) divides it in the ratio 1:2. Find the equation of the line.

OR

Q32. Find the equation of the line through the intersection of lines $x-2y+3=0$, $2x+y-3=0$ and parallel to the line $y=3x+1$.

SECTION D

(Q33–Q36) Solve the following questions.
(6 marks each)

Q33. Find the general solution of the equation $\sin 3x+\sin 5x=0$

Q34. Solve the following system of inequalities graphically: $x-2y\le 3$, $3x+4y\ge 12$, $y\ge 1$

OR

Q34. Find the equation of the circle passing through the points (2, 3) and (–1, 1) and whose centre is on the line $2x+4y-1=0$.

Q35. Find the coordinates of the point that divides the line segment joining the points $(-2,3,6),(1,-4,0)$ in the ratio
(i) 2:3 internally (ii) 2:3 externally

OR

Q35. If the coefficient of the first, second and third terms in the expansion of $(1+x)^{2n}$ are in HP, then show that $2n^2-3n+2=0$

Q36. Let S_n denote the sum of the first n terms of an AP with first term 2. If $S_{3n}=2S_n$, then find S_{2n}.

OR

Q36. Find the mean deviation about the mean for the data

x_i	10	20	30	40	50
f_i	3	2	6	4	5

Day 226

Sample Paper – III (Solutions)

SECTION A

Q1–Q10 are multiple choice type questions. Select the correct option. (1 mark each)

Q1. Solution: (b)
Let p: 6 be greater than 9.
$\sim p$: 6 is not greater than 9.

Q2. Solution: (b)

$$\left(\frac{1+i}{1-i}\right)^x=-1\Rightarrow\left(\frac{(1+i)(1+i)}{(1-i)(1-i)}\right)^x=-1$$

$$\Rightarrow\left(\frac{1+2i+i^2}{1-i^2}\right)^x=-1\Rightarrow\left(\frac{2i}{2}\right)^x=-1$$

$$\Rightarrow i^x=-1\Rightarrow i^x=(i^{4n+2})$$

$$\Rightarrow x=4n+2\quad(\therefore i^{4n}=1,\ n\in N)$$

Q3. Solution: (c)

$$\Rightarrow \sin x-\operatorname{cosec}x=\operatorname{cosec}^2x-1$$

$$\Rightarrow \sin x-\frac{1}{\sin x}=\frac{1}{\sin^2 x}-1$$

$$\Rightarrow \sin^2 x-1=1-\sin^2 x$$

$$\Rightarrow \sin^2 x=1\Rightarrow \sin x=\pm1$$

$$\Rightarrow x=\frac{\pi}{2},\frac{3\pi}{2}$$

Thus, number of possible solutions is 2.

Q4. Solution: (b)
The coefficients of the $(r–2)^{th}$ and $(r+2)^{th}$ terms are equal in the expansion of $(1+x)^{2n}$. Therefore

$$T_{r-2}=T_{r+2}\Rightarrow T_{r-3+1}=T_{r+1+1}$$

$$\Rightarrow {}^{2n}C_{r-3}x^{r-3}={}^{2n}C_{r+1}x^{r+1}\Rightarrow {}^{2n}C_{r-3}={}^{2n}C_{r+1}$$

$$\Rightarrow r-3+r+1=2n\ (\because {}^nC_x={}^nC_y\Rightarrow x+y=n)$$

$$\Rightarrow 2r=2n+2\Rightarrow n=r-1$$

Q5. Solution: (c)

Let $f(x)=\dfrac{1}{1-\sin x}=y$

$$\Rightarrow 1-\sin x=\frac{1}{y}\Rightarrow \sin x=1-\frac{1}{y}$$

$$\because -1\le \sin x\le 1$$

$$\therefore -1\le 1-\frac{1}{y}\le 1$$

$$\Rightarrow -1-1\le -\frac{1}{y}\le 1-1$$

$$\Rightarrow -2\le -\frac{1}{y}\le 0\Rightarrow -2\le -\frac{1}{y}\le 0$$

$$\Rightarrow 2\ge \frac{1}{y}\ge 0\Rightarrow y\ge 1/2\ \text{ and }\ y<\infty$$

$$\Rightarrow 1/2\le y<\infty$$

Thus, range of $f=[1/2,\infty)$

Q6. Solution: (c)
The range of $\sec\theta$ is $R-(-1,1)$.
Therefore, $\sec\theta$ cannot be equal to 1/2.

Q7. Solution: (b)
The total number of three-digit numbers formed by using all the digits at a time is $3!=6$
In these 6 numbers, each digit occurs $6/3=2$ times at each (units, tens and hundreds) place.

Thus, the sum of digits at each place is $= 3\times2+2\times2+1\times2 = 12$
Hence, the sum of all the numbers formed is
$= 12\times100+12\times10+12\times1 = 1332$

Q8. Solution: (d)
Let a and r be the first term and common ratio, respectively, then $T_3 = 3 \Rightarrow ar^2 = 3$
Thus, product of first 5 terms is
$= a\cdot ar\cdot ar^2\cdot ar^3\cdot ar^4 = a^5r^{10} = (ar^2)^5 = (3)^5 = 243$

Q9. Solution: (b)
Number of letters in the word 'VIKASH' = 6
If A and I remain together, then consider it as one group.
∴ Number of words formed, if A and I remain together = 5!2!
Total number of words formed from the letters of the word 'VIKASH'= 6!
∴ Probability that the letters A and I remain together $= \frac{5!2!}{6!} = \frac{1}{3}$

Q10. Solution: (c)
The number of triangles that can be formed by joining three points from the given 10 points $= {}^{10}C_3$
Six points are collinear, and out of these 6 collinear points, no three points will give any triangle.
Therefore, total number of triangles formed
$= {}^{10}C_3 - {}^6C_3 = 120 - 20 = 100$

Q11–Q15 are fill in the blanks. (1 mark each)

Q11. Solution: Let denote 1 red and 2 blue balls as R_1 and B_1, B_2, respectively. The sample space is given by
$S = \{TR_1, TB_1, TB_2, HH, HT\}$

Q12. Solution: $A\times B = \{(0,1),(0,-1),(2,1),(2,-1)\}$,
Therefore
A = set of first element of ordered pairs in $A\times B$
$= \{0,2\}$
B = set of second element of ordered pairs in $A\times B$
$= \{1,-1\}$

Q13. Solution: Let the constant term in expansion $\left(\frac{\sqrt{x}}{3}+\frac{3}{x}\right)^{15}$ be

$$T_{r+1} = {}^{15}C_r\left(\frac{\sqrt{x}}{3}\right)^{15-r}\left(\frac{3}{x}\right)^r$$

$$= {}^{15}C_r\cdot x^{\frac{15-r}{2}}\cdot 3^{r-15}\cdot 3^r\cdot x^{-r} = {}^{13}C_r 3^{2r-15} x^{\frac{15-3r}{2}}$$

For constant term, $15 - 3r = 0 \Rightarrow r = 5$
Coefficient of the term independent of x is
$${}^{13}C_5 3^{2\cdot5-15} = \frac{{}^{13}C_5}{3^5}$$

Q14. Solution: ∵ $A = \{-1,1\}$
∴ Power set of the set $A = \{-1,1\}$ is
$P(A) = \{\phi,\{-1\},\{1\},\{-1,1\}\}$

Q15. Solution: For the given circle $x^2 + y^2 = a^2$
Radius of circle = 3 and centre = (0, 0)
The distance from point (l,m) to the centre is
$\sqrt{(l-0)^2+(m-0)^2} = 3 \Rightarrow l^2 + m^2 = 9$

(Q16–Q20) Answer the following questions. (1 mark each)

Q16. Solution:
$\arg(z+1) = \arg(z+i)$
$\Rightarrow \arg(x+1+iy) = \arg(x+iy+i)$
$\Rightarrow \arg[x+1+iy] = \arg[x+i(y+1)]$
$\Rightarrow \tan^{-1}\frac{y}{x+1} = \tan^{-1}\frac{y+1}{x}$
$\Rightarrow \frac{y}{x+1} = \frac{y+1}{x} \Rightarrow xy = xy + x + y + 1$
$\Rightarrow x + y = -1$

Q17. Solution: Given the length of arc $l = 22$ cm
Length of the pendulum = radius of circle, $r = 70$ cm

Thus, using the relations $\theta = \frac{l}{r}$, we get
$\theta = \frac{22}{70}$ radians
π radians = 180°
Therefore, 22/70 radians $= \frac{180}{\pi}\times\frac{22}{70}$ degree
$= \frac{180\times7}{22}\times\frac{22}{70}$ degree = 18 degree.
Thus, the angle formed by thependulum is 22/70 radians or 18°

Q18. Solution: For the greatest value of $|z+1|$
$|z+1| = |z+2-1| \le |z+2| + |-1|$
$\Rightarrow |z+1| \le 4+1 \Rightarrow |z+1| \le 5 \quad (\because |z+2| \le 3)$
Thus, the greatest value of $|z+1|$ is 5.

Q19. Solution: Let $f(x) = 4x^3 - \frac{1}{2x^2}$

$f'(x) = \frac{d}{dx}\left(4x^3\right) - \frac{d}{dx}\left(\frac{1}{2x^2}\right)$

$= 4\frac{d}{dx}(x^3) - \frac{1}{2}\frac{d}{dx}(x^{-2})$

$= 4(3x^2) - \frac{1}{2}(-2x^{-3})$

$\left[\because \frac{d}{dx}(x^n) = nx^{n-1}\right]$

$= 12x^2 + \frac{1}{x^3}$

Q20. Solution:

$\lim_{x\to 0}\frac{(x+2)^{1/3} - 2^{1/3}}{x} = \lim_{x\to 0}\frac{(x+2)^{1/3} - 2^{1/3}}{(x+2)-2}$

$= \frac{1}{3}\times 2^{\frac{1}{3}-1} = \frac{1}{3}(2)^{-2/3}\left[\because \lim_{x\to a}\frac{x^n - a^n}{x-a} = na^{n-1}\right]$

$= \frac{1}{3(2)^{2/3}}[\because x\to 0 \Rightarrow x+2\to 2]$

SECTION B

(Q21–Q26) Solve the following questions.

(2 marks each)

Q21. Solution: (i) $A = \{x : x \in I, x^2 + x - 6 = 0\}$

$\because x^2 + x - 6 = 0 \Rightarrow (x+3)(x-2) = 0 \Rightarrow x = -3, 2$

Therefore, $A = \{-3, 2\}$

(ii) $B = \{x | x^2 + x = 0, x \in I\}$

$\because x^2 + x = 0 \Rightarrow x(x-1) = 0 \Rightarrow x = 0, 1$

Therefore, $B = \{0, 1\}$

Q22. Solution: The vertices $(0, \pm 10)$ are on the y-axis. Therefore, the equation of the ellipse will be of the form $\frac{x^2}{b^2} + \frac{y^2}{a^2} = 1$, and semi-major axis $a = 10$

Foci are $(0, \pm 5)$. Therefore $c = 5$

Now, from equation $a^2 = b^2 + c^2$

$\Rightarrow 10^2 = b^2 + 5^2 \Rightarrow b^2 = 100 - 25 \Rightarrow b = 5\sqrt{3}$

Thus, the equation of the ellipse is $\frac{x^2}{(5\sqrt{3})^2} + \frac{y^2}{10^2} = 1$

or $\frac{x^2}{75} + \frac{y^2}{100} = 1$

Q23. Solution: The vertices $(\pm 6, 0)$ are on the x-axis. Therefore, the equation of the hyperbola is of the form $\frac{x^2}{a^2} - \frac{y^2}{b^2} = 1$ and $a = 6$.

Now, since eccentricity $e = \frac{5}{3}$

$\Rightarrow \frac{c}{a} = \frac{5}{3} \Rightarrow\Rightarrow \frac{c}{6} = \frac{5}{3} \Rightarrow c = 10$

From the equation $a^2 + b^2 = c^2$

$\Rightarrow 6^2 + b^2 = 10^2 \Rightarrow b^2 = 100 - 36 = 64$

Thus, the equation of the hyperbola is $\frac{x^2}{36} - \frac{y^2}{64} = 1$

Q24. Solution:

$A = \{0, 1, 2, 3\},\ S = \{(x, y) : x \in A, y \in A\}$

(i) The set of ordered pairs satisfying $x + y = 3$ is $\{(0, 3), (1, 2), (2, 1), (3, 0)\}$

(ii) The set of ordered pairs satisfying $x \cdot y = 3$ is $\{(1, 3), (3, 1)\}$

(iii) The set of ordered pairs satisfying $x + y \le 3$ is $\{(0, 0), (0, 1), (0, 2), (0, 3), (1, 0), (1, 1), (1, 2), (2, 0), (2, 1), (3, 0)\}$

OR

Q24. Solution: It is given that when C=20, the value of L is 103.5, whereas when C= 120, the value of L is 105.0.

The points (C, L), assuming C along the x-axis and L along the y-axis, satisfy the linear relation between C and L. Therefore, the equation of the line passing through the given points (20, 103.5) and (120, 105.0) in the XY-plane is given by

$(L - 103.5) = \frac{105.0 - 103.5}{120 - 20}(C - 20)$

$L - 103.5 = \frac{1.5}{100}(C - 20)$

$\Rightarrow L = 0.015(C - 20) + 103.5$, is the required linear relation between C and L.

Q25. Solution: The line makes an angle of 60^0 with the positive direction of the y-axis measured clockwise. Therefore, the angle made by the line with the positive direction of the x-axis measured anticlockwise is $90° - 60° = 30°$.

Thus, the slope of the given line is $m = \tan 30° = \frac{1}{\sqrt{3}}$

OR

Q25. Solution: Let $z = x + iy$. Then

$|z + 1| = |z + i|$

$\Rightarrow |x + iy + 1| = |x + iy + i|$

$\Rightarrow |x + 1 + iy| = |x + i(y + 1)|$

$\Rightarrow (x+1)^2 + y^2 = x^2 + (y+1)^2$

$\Rightarrow x^2 + 2x + 1 + y^2 = x^2 + 1 + y^2 + 2y$

$\Rightarrow y = x$ represents the equation of a line.

Q26. Solution: Let $y = \dfrac{x^2 - \cos x}{x - \sin x}$

$$\therefore \frac{dy}{dx} = \frac{(x-\sin x)\frac{d}{dx}(x^2-\cos x) - (x^2-\cos x)\frac{d}{dx}(x-\sin x)}{(x-\sin x)^2}$$

[By the quotient rule]

$$= \frac{(x-\sin x)(2x+\sin x)-(x^2-\cos x)(1-\cos x)}{(x-\sin x)^2}$$

$$= \frac{x^2 - x\sin x - \sin^2 x + \cos x + x^2\cos x - \cos^2 x}{(x-\sin x)^2}$$

$$= \frac{x^2 - x\sin x - 1 + \cos x + x^2 \cos x}{(x-\sin x)^2}$$

OR

Q26. Solution: Let x be the marks obtained by Ravi in the fourth unit test, so that the student should have an average of at least 60 marks and at most 75 marks. Then

$$60 \le \frac{60+70+80+x}{4} \le 75$$

$\Rightarrow 240 \le 210 + x \le 300$

$\Rightarrow 30 \le x \le 90$

Thus, the student should obtain a minimum of 30 marks and a maximum of 90 marks to have an average of at least 60 marks and at most 75 marks.

SECTION C

(Q27–Q32) Solve the following questions. (4 marks each)

Q27. Solution:

$\tan^2 2x - \tan 2x = 0$

$\Rightarrow \tan 2x(\tan 2x - 1) = 0$

$\Rightarrow \tan 2x = 0$ or $\tan 2x - 1 = 0$

Now, $\tan 2x = 0 \Rightarrow \tan 2x = \tan(n\pi)$

$\Rightarrow 2x = n\pi \Rightarrow x = \dfrac{n\pi}{2}$, where $n \in Z$

$\tan 2x - 1 = 0$

$\Rightarrow \tan 2x = 1 = \tan\dfrac{\pi}{4} \Rightarrow 2x = n\pi + \dfrac{\pi}{4} \Rightarrow x = \dfrac{n\pi}{2} + \dfrac{\pi}{8}$,

where $n \in Z$

Therefore, the general solution is

$\dfrac{n\pi}{2}$ or $\dfrac{n\pi}{2} + \dfrac{\pi}{8}$, $n \in Z$

Q28. Solution:

$$z = \frac{1+i}{1-i} = \frac{1+i}{1-i} \times \frac{1+i}{1+i}$$

$$= \frac{1+i+i+i^2}{1^2+1^2} = \frac{0+2i}{2} = 0+i$$

Let $r\cos\theta = 0$, $r\sin\theta = 1$

On squaring and adding, we get

$r^2(\cos^2\theta + \sin^2\theta) = (0)^2 + (1)^2$

$\Rightarrow r^2 = 1 \quad (\because \sin^2\theta + \cos^2\theta = 1)$

$\Rightarrow r = 1 \quad (\because r > 0)$

Since, $|\tan q| = Y \;\; q = \dfrac{p}{2}$

Since $\sin\theta$ is positive and $\cos\theta$ is zero, θ lies in the first quadrant.

Thus, $\text{Arg}(z) = (\theta) = \dfrac{\pi}{2}$

Therefore, the modulus and argument of the given complex number are $1, \dfrac{\pi}{2}$, respectively.

Q29. Solution: Out of 12 persons $P_1, P_2, P_3, \ldots\ldots. P_{12}$, 6 persons are to be arranged in which P_1 must occur whereas P_{12} never occur. Therefore, we need to select only 5 persons out of remaining 10 persons $(P_2, P_3, P_6, \ldots, P_{11})$

Thus, the total number of selections is

$= {}^{10}C_5$

The number of arrangements of 6 persons is

$$= {}^{10}C_5 \times 6! = \frac{10!}{5!(10-5)!} \times 6! = 181440$$

Q30. Solution: Using Binomial Theorem

$(x+y)^4 = {}^4C_0x^4 + {}^4C_1x^3y + {}^4C_2x^2y^2 + {}^4C_3xy^3 + {}^4C_4y^4$

$= x^4 + 4x^3y + 6x^2y^2 + 4xy^3 + y^4$

$(x-y)^4 = {}^4C_0x^4 - {}^4C_1x^3y + {}^4C_2x^2y^2 - {}^4C_3xy^3 + {}^4C_4y^4$

$= x^4 - 4x^3y + 6x^2y^2 - 4xy^3 + y^4$

And,

$\therefore (x+y)^4 + (x-y)^4 = 2(x^4 + 6x^2y^2 + y^4)$

Putting $x = \sqrt{a+1}, y = \sqrt{a-1}$, we get

$(\sqrt{a+1} + \sqrt{a-1})^4 + (\sqrt{a+1} - \sqrt{a-1})^4$

$= 2[(\sqrt{a+1})^4 + 6(\sqrt{a+1})^2(\sqrt{a-1})^2 + (\sqrt{a-1})^4]$

$= 2[(a+1)^2 + 6(a^2-1) + (a-1)^2]$

$= 2[a^2 + 2a + 1 + 6a^2 - 6 + a^2 - 2a + 1]$

$= 16a^2 - 8$

Q31. Solution:

The nth of the given series

$$\frac{1^2}{1}+\frac{1^2+2^2}{1+2}+\frac{1^2+2^2+3^2}{1+2+3}+\dots \text{ is}$$

$$a_n=\frac{1^2+2^2+3^2+\dots+n^2}{1+2+3+\dots+n}=\frac{\frac{n(n+1)(2n+1)}{6}}{\frac{n(n+1)}{2}}=\frac{2n+1}{3}$$

$$\therefore S_n=\sum_{k=1}^{n}a_k=\sum_{k=1}^{n}\left(\frac{2}{3}k+\frac{1}{3}\right)$$

$$=\frac{2}{3}\frac{n(n+1)}{2}+\frac{n}{3}=\frac{n(n+1)+n}{3}=\frac{n(n+2)}{3}$$

$$\Rightarrow S_{100}=\frac{n(n+2)}{3}=\frac{100(100+2)}{3}=3400$$

OR

Q31. Solution: The coordinates of the centroid of the triangle, whose vertices are $(x_1,y_1,z_1),(x_2,y_2,z_2),(x_3,y_3,z_3)$, are $\left(\frac{x_1+x_2+x_3}{3},\frac{y_1+y_2+y_3}{3},\frac{z_1+z_2+z_3}{3}\right)$

Therefore, coordinates of the centroid of ΔPQR

$$=\left(\frac{a-4+8}{3},\frac{2+b+14}{3},\frac{6-10+2}{3}\right)$$

$$=\left(\frac{a+4}{3},\frac{b+16}{3},\frac{c-4}{3}\right)$$

It is given that the origin is the centroid of ΔPQR. Therefore

$$\left(\frac{a+4}{3},\frac{b+16}{3},\frac{c-4}{3}\right)=(0,0,0)$$

$$\Rightarrow \frac{a+4}{3}=0,\ \frac{b+16}{3}=0,\ \frac{c-4}{3}=0$$

Solving, we get $a=-4,\ b=-16,\ c=4$

Q32. Solution: Using the section formula, the coordinates of the point that divides the line segment joining the points (1, 0) and (2, 4) in the ratio $1:2$ is given by

$$\left(\frac{1\times2+2\times1}{1+2},\frac{0\times2+4\times1}{1+2}\right)=\left(\frac{4}{3},\frac{4}{3}\right)$$

The slope of the line joining the points (1, 0) and (2, 4) is $m=\frac{4-0}{2-1}=4$. Therefore, the slope of the line that is perpendicular to the line joining the points (1, 0) and (2, 4) $=-\frac{1}{m}=-\frac{1}{4}$(since two non-vertical lines are perpendicular to each other if and only if their slopes are negative reciprocals of each other).

Thus, the equation of the line passing through $\left(\frac{4}{3},\frac{4}{3}\right)$ and whose slope is $-\frac{1}{4}$ is

$$\left(y-\frac{4}{3}\right)=\frac{-1}{4}\left(x-\frac{4}{3}\right)\Rightarrow\frac{3y-4}{3}=\frac{-(3x-4)}{12}$$

$$\Rightarrow 12y-16=-3x+4\Rightarrow 3x+12y-20=0$$

OR

Q32. Solution: The equation of any line through the point of intersection of the given lines is of the form

$$x-2y+3+k(2x+y-3)=0$$

$$\Rightarrow(1+2k)x+(-2+k)y+3-3k=0$$

$$\Rightarrow y=\frac{(1+2k)}{(2-k)}x+\frac{3-3k}{(2-k)}=0$$

Since this line is parallel to line $y=x+1$, the slope of both lines must be equal, that is,

$$3=\frac{(1+2k)}{(2-k)}\Rightarrow 6-3k=1+2k\Rightarrow k=1$$

Substituting the value of $k=1$, we get

$$(1+2\times1)x+(-2+1)y+3-3\times1=0$$

$\Rightarrow 3x-y=0$ as required equation of the line.

SECTION D

(Q33–Q36) Solve the following questions. **(6 marks each)**

Q33. Solution:

$$\sin 3x+\sin 5x=0$$

$$\Rightarrow 2\sin\left(\frac{3x+5x}{2}\right)\cos\left(\frac{3x-5x}{2}\right)=0$$

$$\left[\sin A+\sin B=2\sin\left(\frac{A+B}{2}\right)\cos\left(\frac{A-B}{2}\right)\right]$$

$$\Rightarrow 2\sin 4x\cos(-x)=0$$

$$\Rightarrow 2\sin 4x\cos x=0$$

$$\Rightarrow \sin 4x=0 \text{ or } \cos x=0$$

Now, $\sin 4x=0\Rightarrow 4x=n\pi\Rightarrow x=\frac{n\pi}{4}$ where $n\in Z$

And, $\cos x=0\Rightarrow \cos x=\cos\frac{\pi}{2}$

$\Rightarrow x=2n\pi\pm\frac{\pi}{2}$ where $n\in Z$

Therefore, the general solution is

$\frac{n\pi}{4}$ or $2n\pi\pm\frac{\pi}{2}, n\in Z$

Q34. Solution: The graphs of the lines, $x-2y=3$, $3x+4y=12, x=0$ and $y=1$ are shown in the figure.

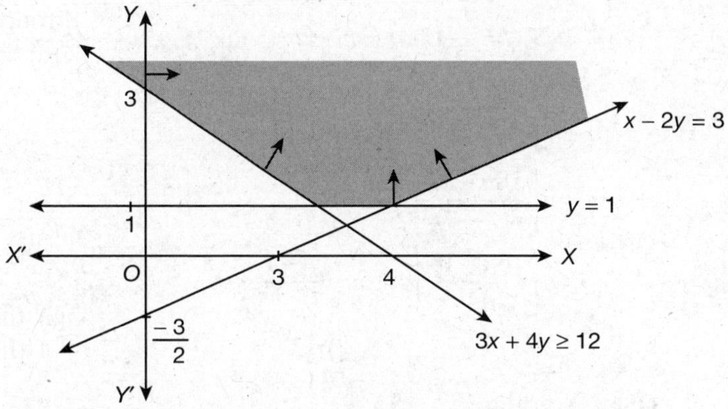

Inequality $x-2y \le 3$ represents the shaded region below the line $x-2y=3$ (including all the points on the line).

Inequality $3x+4y \ge 12$ represents the shaded region below the line $3x+4y=12$ (including all the points on the line).

Inequality $y \ge 1$ represents the shaded region above the line $y=1$ (including all the points on the line).

Thus, the solution of the given system of linear inequalities is represented by the common shaded region, including the points on the respective lines.

OR

Q34. Solution: Let the equation of the required circle be $(x-h)^2+(y-k)^2=r^2$

Since the circle passes through points (2, 3) and (–1, 1)

$(2-h)^2+(3-k)^2=r^2$ (1)

$(-1-h)^2+(1-k)^2=r^2$ (2)

Since the centre (*h*, *k*) of the circle lies on line on the line $2x+4y-1=0$,

$2h+4k-1=0$ (3)

Solving equations (1) and (2), we get

$(2-h)^2+(3-k)^2=(-1-h)^2+(1-k)^2$

$\Rightarrow 4-4h+h^2+9-6k+k^2=1+2h+h^2+1-2k+k^2$

$\Rightarrow 6h+4k=11$ (4)

Now, solving equations (3) and (4), we get $h=\frac{5}{2}, k=-1$

Substituting the values of *h* and *k* in equation (1),

we get $\left(2-\frac{5}{2}\right)^2+(3-(-1))^2=r^2$

$\Rightarrow \left(-\frac{1}{2}\right)^2+(4)^2=r^2 \Rightarrow r^2=\frac{65}{4}$

Thus, the equation of the required circle is

$\left(x-\frac{5}{2}\right)^2+(y+1)^2=\frac{65}{4}$

$\Rightarrow \left(\frac{2x-5}{2}\right)^2+(y+1)^2=\frac{65}{4}$

$\Rightarrow 4x^2-20x+25+4y^2+8y+4=65$

$\Rightarrow 4x^2+4y^2-20x+8y-36=0$

$\Rightarrow x^2+y^2-5x+2y-9=0$

Q35. Solution: (i) The coordinates of a point $R(x,y,z)$ that divides the line segment joining points $P(x_1,y_1,z_1), Q(x_2,y_2,z_2)$ internally in the ratio $m:n$ are $R(x,y,z)=\left(\frac{mx_2+nx_1}{m+n}, \frac{my_2+ny_1}{m+n}, \frac{mz_2+nz_1}{m+n}\right)$

Thus, the point that divides the line segment joining the points $(-2,3,6),(1,-4,0)$ internally in the ratio 2:3 is $x=\frac{2(1)+3(-2)}{2+3}, y=\frac{2(-4)+3(3)}{2+3}$,

$z=\frac{2(0)+3(6)}{2+3} \Rightarrow x=\frac{-4}{5}, y=\frac{1}{5}, z=\frac{18}{5}$

Thus, the coordinates of the required point are $\left(-\frac{4}{5},\frac{1}{5},\frac{18}{5}\right)$

(ii) The coordinates of a point $R(x,y,z)$ that divides the line segment joining points $P(x_1,y_1,z_1), Q(x_2,y_2,z_2)$ internally in the ratio $m:n$ are $R(x,y,z)=\left(\frac{mx_2-nx_1}{m-n}, \frac{my_2-ny_1}{m-n}, \frac{mz_2-nz_1}{m-n}\right)$

Thus, the point that divides the line segment joining the points $(-2,3,6),(1,-4,0)$ externally in the ratio 2:3 is

$x=\frac{2(1)-3(-2)}{2-3}, y=\frac{2(-4)-3(3)}{2-3}$,

$z=\frac{2(0)-3(6)}{2+3} \Rightarrow x=-8, y=17, z=\frac{-18}{5}$

Thus, the coordinates of the required point are $\left(-8,17,\frac{-18}{5}\right)$

OR

Q35. Solution: In the given expansion $(1+x)^{2n}$

Coefficient of 1st term = ${}^{2n}C_0$

Coefficient of 2nd term = ${}^{2n}C_1$

Coefficient of 3rd term = ${}^{2n}C_2$

Since the coefficient of first, second and third terms are in AP, ${}^{2n}C_0, {}^{2n}C_1, {}^{2n}C_2$ are in AP.

Thus, $\frac{2}{{}^{2n}C_1} = \frac{1}{{}^{2n}C_0} + \frac{1}{{}^{2n}C_2}$

$\Rightarrow \frac{2}{2n} = 1 + \frac{2}{(2n)(2n-1)}$

$\Rightarrow 2(2n-1) = 2n(2n-1)+2$

$\Rightarrow 4n-2 = 4n^2 - 2n + 2$

$\Rightarrow 4n^2 - 6n + 4 = 0 \Rightarrow 2n^2 - 3n + 2 = 0$

Q36. Solution: Let a, d be the first term and common difference of AP, respectively. Then

$S_n = \frac{n}{2}[2a+(n-1)d]$

$\Rightarrow S_{2n} = \frac{2n}{2}[2a+(2n-1)d] = n[2a+(2n-1)d]$

$\Rightarrow S_{3n} = \frac{3n}{2}[2a+(3n-1)d]$

Since $S_{3n} = 2S_n$

$\Rightarrow \frac{3n}{2}[2a+(3n-1)d] = 2\frac{n}{2}[2a+(n-1)d]$

$\Rightarrow 6a + 3(3n-1)d = 4a + 2(n-1)d$

$\Rightarrow 2a + (7n-5)d = 0$

$\Rightarrow d = \frac{-2a}{7n-5} = \frac{-2\times 2}{7n-5} = \frac{-4}{7n-5}$

Now, $S_{2n} = \frac{2n}{2}[2a+(3n-1)d]$

$\Rightarrow S_{2n} = \frac{2n}{2}[2a+(3n-1)d]$

$= n\left[2\cdot 2 + (3n-1)\left(\frac{-4}{7n-5}\right)\right]$

$= n\left[\frac{28n-20-12n+4}{7n-5}\right]$

$= n\left[\frac{16n-16}{7n-5}\right] = 16n\left(\frac{n-1}{7n-5}\right)$

OR

Q36. Solution:

x_i	f_i	$x_i f_i$	$\lvert x_i - \bar{x}\rvert = \lvert x_i - 33\rvert$	$f_i \lvert x_i - \bar{x}\rvert$
10	3	30	23	69
20	2	40	13	26
30	6	180	3	18
40	4	160	7	28
50	5	250	17	85
	20	660		226

$N = \sum_{i=1}^{5} f_i = 20,$

$\sum_{i=1}^{5} f_i x_i = 660$

$\therefore \bar{x} = \frac{1}{N}\sum_{i=1}^{5} f_i x_i = \frac{1}{20}\times 660 = 33$

$\text{MD}(\bar{x}) \frac{1}{N}\sum_{i=1}^{5} f_i \lvert x_i - \bar{x}\rvert = \frac{1}{20}\times 226 = 11.3$

Sample Paper – IV

Day 227

Time: 3 Hrs Maximum **Marks:** 80

General Instructions:

(i) All the questions are compulsory.
(ii) The question paper consists of 36 questions divided into 4 sections A, B, C and D.
(iii) Section A comprises 20 questions of 1 mark each. Section B comprises 6 questions of 2 marks each. Section C comprises 6 questions of 4 marks each. Section D comprises 4 questions of 6 marks each.
(iv) There is no overall choice. However, an internal choice has been provided in four questions of 1 mark each, two questions of 2 marks each, two questions of 4 marks each, and two questions of 6 marks each. You have to attempt only one of the alternatives in all such questions.
(iv) Use of calculators is not permitted.

SECTION A

Q1–Q10 are multiple choice type questions. Select the correct option. (1 mark each)

Q1. A plane is parallel to the XY-plane. So it is perpendicular to
(a) X-axis (b) Y-axis
(c) Z-axis (d) None of these

Q2. The negation of the statement "The product of 2 and 3 is 6" is
(a) It is false that the product of 2 and 3 is 6
(b) The product of 2 and 3 is 6
(c) The product of 2 and 3 is not 6
(d) It is false that the product of 2 and 3 is not 6

Q3. The domain of $\frac{1}{\sqrt{x^2-a^2}}$ $(a>0)$ is
(a) $(-a,a)$ (b) $[-a,a]$
(c) $[0, a]$ (d) $(-\infty,-a)\cup(a,\infty)$

Q4. If $f(x)=ax-b$, where a and b are integers, $f(2)=1,\ f(1)=2$, then a and b are equal to
(a) $a=-1, b=3$ (b) $a=2, b=-3$
(c) $a=0, b=2$ (d) $a=2, b=3$

Q5. The value of $\frac{1-\tan^2 15°}{1+\tan^2 15°}$ is
(a) 1 (b) $\sqrt{3}$ (c) $\frac{\sqrt{3}}{2}$ (d) 2

Q6. The point represented by the complex number $(1-i)$ is rotated about the origin through an angle π in the anti-clockwise direction. The new position of the points is
(a) $1-i$ (b) $-1-i$
(c) $1+i$ (d) $-1+i$

Q7. The two successive terms in the expansion of $(1+x)^{15}$ whose coefficient are equal is
(a) 3rd and 4th (b) 4th and 5th
(c) 5th and 6th (d) 9th and 10th

Q8. The coefficients of x^n in the expansion of $(1+x)^{2n}, (1+x)^{2n-1}$ are in the ratio
(a) $\frac{2n}{n-1}$ (b) $\frac{(2n+1)}{(n+1)}$
(c) $\frac{2(2n+1)}{(2n-1)}$ (d) $\frac{(2n+1)}{(2n-1)}$

Q9. If t_n denotes the nth term of the series $2+4+8+14+22+...$, then t_{11} is
(a) 112 (b) 2^{11} (c) 504 (d) 11^2

Q10. Find the number of parallelograms that can be formed from a set of 10 parallel lines intersecting another set of 6 parallel lines.
(a) 15 (b) 45 (c) 600 (d) 675

Q11–Q15 are fill in the blankss (1 mark each)

Q11. The number of terms in the expansion of $[\{(x+y)^1\}^2]^3$ is ... ...

Q12. If $\left|\frac{z+1}{z-1}\right| = \tan\frac{\pi}{4}$, then the locus of z is.........

Q13. The point (1, 1) lies ... ... the circle $x^2+y^2-2x-6y+6=0$

Q14. If the sets $A=\{0,1\},\ B=\{1,2\}$, the universal set of all the three sets A, B and C can be ......

Q15. The value of $\frac{\sin 120°}{\sin 240°}$ is............

Q16–Q20) Answer the following questions.

(1 mark each)

Q16. In any triangle ABC, if $a = 5$, $b = 3$, $c = 4$, find: $\cos B$

Q17. Find the value of x for which the points $(1,-1),(2,1),(x,5)$ are collinear.

Q18. Find the derivative of $(x+1)^2(x+2)$

Q19. A, B and C are subsets of universal set U. If $U = \{x : x \in N, x < 10\}$
$A = \{1,4,7,8\}$, $B = \{2,4,6,8\}$, $C = \{1,3,6,9\}$,
draw the diagram showing the relation of U, A, B and C.

Q20. Given that $P(4,3,-2), Q(8,7,-6), R(9,8,-7)$ are collinear, find the ratio in which Q divides PR.

OR

Q20. Does the ordered pair (3, 2) and (2, 1) belong to the relation $R = \{(x, y) : y = x-1, x, y \in Z\}$

SECTION B

(Q21–Q26) Solve the following questions.

(2 marks each)

Q21. Show that the inequality $|z-1| < |z+2|$ represents the region given by $x > -3/2$.

Q22. If $\left(\frac{1+i}{1-i}\right)+\left(\frac{1-i}{1+i}\right)^2 = x+iy$, then find (x, y).

Q23. The longest side of a triangle is 3 times the shortest side and the third side is 3 cm shorter than the longest side. If the perimeter of the triangle is at most 53 cm, find the maximum length of the shortest side.

Q24. A box contains two white, three black and four red balls. In how many ways can four balls be drawn from the box, if at least one ball of each colour must be included in the draw?

OR

Q24. Find the coordinates of a point on y-axis that is at a distance of 3 from the point P (2, –1, 1).

Q25. Evaluate $\lim_{x\to 0}\frac{(1+x)^4-1}{x\{(1+x)^2+1\}}$

OR

Q25. Twelve guests are to be seated, half on each side of a long table. Three particular guests desire to sit on one particular side and four others on other side of the table. Find the number of ways in the which the seating arrangement can be made.

Q26. Find the derivative of $\frac{x\cos(\pi/3)}{\sin x}$

OR

Q26. Find the equation of the ellipse that satisfies the given conditions: wnds of major axis are $(\pm 5,0)$ and ends of minor axis are $(0,\pm 2)$

SECTION C

(Q27–Q32) Solve the following questions.

(4 marks each)

Q27. The number lock of a suitcase has 3 wheels, each labelled with 9 digits, from 1 to 9. The lock opens with a sequence of three digits with no repeats. What is the probability of a person getting the right sequence to open the suitcase?

Q28. Let $f = \{(0,1),(1,2),(2,3)\}$ and $g = \{(1,1),(2,0),(3,2)\}$ be two real functions. Then find

(i) $f-g$ (ii) $f+g$

(iii) $f\cdot g$ (iv) $\frac{f}{g}$

Q29. Find the values of the other five trigonometric functions if $\cos x = -\frac{1}{2}$, x lies in the third quadrant.

Q30. If $\frac{z+1}{z-1}$ is a purely imaginary number $(z \neq -1)$, then find the value of $|z|$.

OR

Q30. Find the equation of the line through the point (0, 2) making an angle $\frac{2\pi}{3}$ with the positive x-axis. Also find the equation of the line parallel to it and crossing the y-axis at a distance of 2 units above the origin.

Q31. Find the equation of a circle that touches both the axes and the line $3x-4y+12=0$ and lies in the fourth quadrant.

OR

Q31. Solve the following system of inequalities graphically:
$x+2y \le 4, x+y \ge 1, x-y \le 0, x \ge 0, y \ge 0$

Q32. Find the sum of the series $(3^2-2^2)+(5^2-4^2)+(7^2-6^2)+\ldots$ up to 20 terms.

OR

Q32. Find the equation of the line through the intersection of the lines $3x+4y=7$, $x-y+2=0$ and whose slope is –1.

SECTION D

(Q33–Q36) Solve the following questions. **(6 marks each)**

Q33. Find n, if the fourth term from the beginning and the fourth term from the end in the expansion of $\left(\sqrt[3]{2}+\sqrt[2]{3}\right)^n$ are equal.

Q34. Find the coefficient of x^5 in the expansion of $(1+x+x^2+x^3)^6$.

OR

Q34. Find the mean deviation about the median for the data.

x_i	5	7	9	10	12
f_i	8	6	2	2	2

Q35. What is the number of ways of choosing 4 cards from a pack of 52 playing cards, if:
(i) 4 cards are of the same suit
(ii) 4 cards are of the same colour
(iii) 4 cards belong to four different suits
(iv) 2 are red and 2 are black
(v) 2 are belong to diamonds and 2 belong to clubs
(vi) they contain at least 2 kings
(vii) they contain at most 1 king and 1 black joker

OR

Q35. If the sum of n terms of an AP is $2n^2+3n$ and its mth term is 205, find the value of m.

Q36. By using the concept of equation of a line, prove that the three points $(3,0),(-2,-2),(8,2)$ are collinear.

OR

Q36. A die is thrown. Describe the following events:
(i) A: a number less than 5
(ii) B: a number greater than 3
(iii) C: a multiple of 3
(iv) D: an even number
(v) E: an odd number
Also find
$A\cup B, A\cap B, C\cup D, C\cap E, A-C, C\cap D', C\cup E'$

Day 228

Sample Paper – IV (Solutions)

SECTION A

Q1–Q10 are multiple choice type questions. Select the correct option. **(1 mark each)**

Q1. Solution: (c)
A plane parallel to the XY-plane will be perpendicular to the Z-axis.

Q2. Solution: (a)
The negation of the statement is "It is false that the product of 2 and 3 is 6".

Q3. Solution: (d)
$f(x)=\dfrac{1}{\sqrt{x^2-a^2}}$ is defined, if
$x^2-a^2>0 \Rightarrow (x-a)(x+a)>0$
$\Rightarrow x<-a$ or $x>a$ $(\because a>0)$
$\therefore$ Domain of $f=(-\infty,-a)\cup(a,\infty)$

Q4. Solution: (a)
$\because f(x)=ax-b$
$\therefore f(2)=2a+b=1$
and $f(1)=a+b=2$

Solving the above two equations, we get $a=-1,\ b=3$

Q5. Solution: (c)
Let $\theta=15°$
$$\cos 2\theta=\frac{1-\tan^2\theta}{1+\tan^2\theta}$$
$$\therefore \cos 30°=\frac{1-\tan^2 15°}{1+\tan^2 15°}=\frac{\sqrt{3}}{2}$$

Q6. Solution: (d)
If a complex number is rotated about the origin through an angle θ in the anti-clockwise direction, the new position of z is $z'=ze^{i\theta}$
Thus, the new position of $(1-i)$ when rotated about the origin through an angle π in the anti-clockwise direction is $z'=ze^{i\pi}=(1-i)e^{i\pi}$
$=(1-i)\left[\cos(\pi)+i\sin(\pi)\right]$
$=(1-i)(-1+0i)=-1+i$

Q7. Solution: (d)
Let two successive terms in the expansion of $(1+x)^{15}$ be the $(r+1)^{\text{th}}$ and $(r+2)^{\text{th}}$ terms. Then

$\therefore T_{r+1} = {}^{15}C_r x^r, \quad T_{r+2} = {}^{15}C_{r+1} x^{r+1}$

It is given that the coefficients of two successive terms are equal. Therefore

$\therefore {}^{15}C_r = {}^{15}C_{r+1} \Rightarrow \dfrac{(15)!}{r!(15-r)!} = \dfrac{(15)!}{(r+1)!(15-r-1)!}$

$\Rightarrow 15 - r = r + 1 \Rightarrow r = 8$

$\therefore T_{8+1} = T_9, \quad T_{8+2} = T_{10}$

Thus, the 9th and 10th terms are equal.

Q8. Solution: (a)

Coefficient of x^n in the expansion of $(1+x)^{2n} = {}^{2n}C_n$ and coefficient of x^n in the expansion of $(1+x)^{2n-1} = {}^{2n-1}C_n$

$$\therefore \frac{{}^{2n}C_n}{{}^{2n-1}C_n} = \frac{\dfrac{(2n)!}{n!n!}}{\dfrac{(2n-1)!}{n!(n-1)!}} = \frac{2n}{n-1}$$

Q9. Solution: (a)

Let S_n be the sum of the series $2+4+8+14+\ldots+t_{11}$, then

$S_n = 2+4+8+14+22+\ldots+t_{11}\ldots$

$\Rightarrow S_n = 2+4+8+14+\ldots+t_{10}+t_{11}\ldots$

Subtracting, we get $0 = 2+2+4+6+8+\ldots-t_{11}$

$\Rightarrow t_{11} = 2+2+4+6+8+\ldots$ up to 10 terms

$\therefore t_{11} = 2+[2+4+6+8+\ldots$ up to 10 terms$]$

$= 2+\dfrac{10}{2}[2\times 2+9\times 2] = 2+5\times 22 = 112$

Q10. Solution: (d)

To form a parallelogram, we require a pair of lines from a set of 10 parallel lines and another pair of lines from another set of 6 parallel lines.

Thus, total number of parallelograms formed is $= {}^{10}C_2 \times {}^6C_2 = 675$

Q11–Q15 are fill in the blanks. (1 mark each)

Q11. Solution: Since the given expansion is $[\{(x+y)^1\}^2]^3 = (x+y)^{1\cdot 2\cdot 3} = (x+y)^6$, the number of terms is 7.

Q12. Solution: Let $z = x+iy$ then $\left|\dfrac{z+1}{z-1}\right| = \tan\dfrac{\pi}{4}$,

$\Rightarrow \left|\dfrac{x+iy+1}{x+iy-1}\right| = 1 \Rightarrow \left|\dfrac{x+1+iy}{x-1+iy}\right| = 1$

$\Rightarrow |x+1+iy| = |x-1+iy|$

$\Rightarrow (x+1)^2 + y^2 = (x-1)^2 + y^2$

$\Rightarrow x^2+1+2x+y^2 = x^2-2x+1+y^2$

$\Rightarrow 4x = 0 \Rightarrow x = 0$ represents the imaginary axis.

Q13. Solution: For the given circle $S: x^2+y^2-2x-6y+6=0$, the point (1, 1) gives $S: 1+1-2-6+6=0$. Thus, the point (1, 1) lies on the circle.

Q14. Solution: Universal set for A, B and C is given by $U = \{0,1,2\}$

Q15. Solution:

$$\frac{\sin 120^\circ}{\sin 240^\circ} = \frac{\sin(180^\circ-60^\circ)}{\sin(180^\circ+60^\circ)} = \frac{\sin 60^\circ}{-\sin 60^\circ} = -1$$

(Q16–Q20) Answer the following questions. (1 mark each)

Q16. Solution: Using the cosine formula:

$$\cos B = \frac{c^2+a^2-b^2}{2ca} = \frac{5^2+4^2-3^2}{2\times 5\times 4} = \frac{32}{40} = \frac{4}{5}$$

Q17. Solution: If the points $A(1,-1), B(2,1), C(x,5)$ are collinear, the slope of line segment AB = slope of line segment BC

$\Rightarrow \dfrac{1-(-1)}{2-1} = \dfrac{5-1}{x-2} \Rightarrow 2 = \dfrac{4}{x-2} \Rightarrow x = 4$

Q18. Solution: Let $f(x) = (x+1)^2(x+2)$

By the product rule

$f'(x) = (x+1)^2\dfrac{d}{dx}(x+2)$

$+(x+2)\dfrac{d}{dx}(x+1)^2$

$\Rightarrow f'(x) = (x+1)^2\cdot(1)+(x+2)\cdot 2(x+1)$

$= (x+1)\{(x+1)+2(x+2)\}$

$= (x+1)(3x+5)$

Q19. Solution:

$\because U = \{x : x \in N, x < 10\} = \{1,2,3,4,5,6,7,8,9\}$

Therefore, the Venn diagram showing the relation of U, A, B and C is given below:

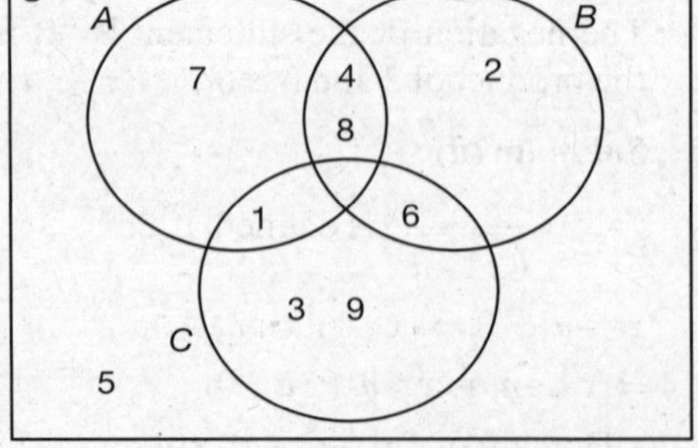

Q20. Solution: Let point $Q(8,7,-6)$ divide the line segment joining the points $P(4,3,-2), R(9,8,-7)$ in the ratio $k:1$. Then, by the section formula

$$(8,7,-6)=\left(\frac{k(9)+4}{k+1},\frac{k(8)+3}{k+1},\frac{k(-7)-2}{k+1}\right)$$

$$\Rightarrow \frac{9k+4}{k+1}=8\Rightarrow 9k+4=8k+8\Rightarrow k=4$$

$$\Rightarrow k=\frac{4}{1}=4$$

Thus, the point Q divides PR in the ratio 4:1.

OR

Q20. Solution: $R=\{(x,y): y=x-1,\ x,y\in Z\}$

If $x=3$, then $y=3-1=2$

Thus, (3, 2) belongs to R.

If $x=2$, then $y=2-1=1$

Thus, (2, 1) belongs to R.

SECTION B

(Q21–Q26) Solve the following questions. (2 marks each)

Q21. Solution:

$$|z-1|<|z+2|\Rightarrow |x-1+iy|<|x+2+iy|$$

$$\Rightarrow \sqrt{(x+1)^2+y^2}<\sqrt{(x+2)^2+y^2}$$

$$\Rightarrow (x+1)^2+y^2<(x+2)^2+y^2$$

$$\Rightarrow x^2+2x+1+y^2<x^2+4x+4+y^2$$

$$\Rightarrow 2x>-3\Rightarrow x>-3/2$$

Q22. Solution:

Given that $\left(\frac{1+i}{1-i}\right)+\left(\frac{1-i}{1+i}\right)^2=x+iy$..........(i)

$$\left(\frac{1+i}{1-i}\right)+\left(\frac{1-i}{1+i}\right)^2$$

$$=\left(\frac{1+i}{1-i}\times\frac{1+i}{1+i}\right)+\left(\frac{1-i}{1+i}\times\frac{1-i}{1-i}\right)^2$$

$$=\left(\frac{1+i^2+2i}{1-i^2}\right)+\left(\frac{1+i^2-2i}{1-i^2}\right)^2$$

$$=\left(\frac{1+(-1)+2i}{1-(-1)}\right)+\left(\frac{1+(-1)-2i}{1-(-1)}\right)^2 \quad (\because i^2=-1)$$

$$=\left(\frac{2i}{2}\right)+\left(\frac{-2i}{2}\right)^2$$

$$=i+(-i)^2=i+1 \quad (\because i^2=-1)$$

$$=1+i=x+iy\Rightarrow x=1,\quad y=1$$

Thus, $(x,y)=(1,1)$

Q23. Solution: Let the length of the shortest side of the triangle be x cm.

Then, length of the longest side = $3x$ cm

And length of the third side = $(3x-3)$ cm

Since the perimeter of the triangle is at most 53 cm

$$x+3x+(3x-3)\le 53$$

$$\Rightarrow 7x-3\le 53$$

$$\Rightarrow 7x\le 56\Rightarrow x\le 8$$

Thus, the maximum length of the shortest side is 8 cm.

Q24. Solution: Selection is to draw 4 balls out of these 9 balls, in which at least one ball of each colour is included. This can be done in the following ways:

2W1B1R + 1W2B1R + 1W1B2R

Therefore, total number of selections is

$$={}^2C_2\times{}^3C_1\times{}^4C_1+{}^2C_1\times{}^3C_2\times{}^4C_1+{}^2C_1\times{}^3C_1\times{}^4C_2$$

$$=1\times3\times4+2\times3\times4+2\times3\times6=12+24+36=72$$

OR

Q24. Solution: If a point is on the y-axis, then the x-coordinates and z-coordinates of the point are zero.

Thus, let $A(0, b, 0)$ be the point on the y-axis at a distance of 3 from point (2, –1, 1). Then

$$AP=3\Rightarrow AP^2=9$$

$$\Rightarrow (2-0)^2+(-1-b)^2+(1-0)^2=9$$

$$\Rightarrow 4+1+b^2+2b+1=9\Rightarrow b^2+2b-3=0$$

$$\Rightarrow (b+3)(b-1)=0\Rightarrow b=-3,1$$

Thus, the coordinates of the required points are (0, –3, 0) and (0, 1, 0).

Q25. Solution: $\lim\limits_{x\to 0}\frac{(1+x)^4-1}{x\{(1+x)^2+1\}}$

$$=\lim_{x\to0}\frac{\{(1+x)^2+1\}\{(1+x)^2-1\}}{x\{(1+x)^2+1\}}$$

$$=\lim_{x\to0}\frac{(1+x)^2-1}{x}$$

$$=\lim_{x\to0}\frac{(1+x)^2-1}{(1+x)-1}[\because x\to0\Rightarrow 1+x\to1]$$

$$=2(1)^{2-1}=2\left[\therefore \lim_{x\to a}\frac{x^n-a^n}{x-a}=na^{n-1}\right]$$

OR

Q25. Solution: After seating 3 on one side and 4 on the other side, we have to select out of 5, 3 on one side and 2 on the other side.

Thus, total arrangement is

$$={}^5C_3\times6!+{}^2C_2\times6!=\frac{5!}{3!2!}\times6!+\frac{2!}{2!0!}\times6!=7920$$

Q26. **Solution:** Let $f(x) = \frac{x\cos(\pi/3)}{\sin x}$

By the quotient rule

$$f'(x) = \cos\frac{\pi}{3}\cdot\left[\frac{\sin x\frac{d}{dx}(x) - x\frac{d}{dx}(\sin x)}{\sin^2 x}\right]$$

$$= \cos\frac{\pi}{3}\cdot\left[\frac{\sin x.1 - x\cos x}{\sin^2 x}\right]$$

$$= \frac{\cos\frac{\pi}{3}[\sin x - x\cos x]}{\sin^2 x}$$

$$= \frac{\sin x - x\cos x}{2\sin^2 x}$$

OR

Q26. **Solution:** The ends of the major axis are $(\pm 5, 0)$. Therefore, the major axis is along the x-axis and $a = 5$.

Since the ends of the minor axis are $(0, \pm 2)$, the minor axis is along the y-axis and $b = 2$.

Thus, the equation of the ellipse will be of the form

$$\frac{x^2}{a^2} + \frac{y^2}{b^2} = 1 \Rightarrow \frac{x^2}{5^2} + \frac{y^2}{2^2} = 1 \Rightarrow \frac{x^2}{25} + \frac{y^2}{4} = 1$$

SECTION C

(Q27–Q32) Solve the following questions. **(4 marks each)**

Q27. **Solution:** The number lock has 3 wheels, each labelled with 9 digits, from 1 to 9.

Number of ways of selecting 3 different digits out of the 9 digits = ${}^9C_3$

Now, each combination of 3 different digits can be arranged in 3! ways.

∴ Total three-digit numbers with no repetitions $= 3!\,{}^9C_3 = 504$

Since there is only one number that can open the suitcase, the probability of a person getting the right sequence to open the suitcase is $\frac{1}{504}$

Q28. **Solution:** $f = \{(0,1),(1,2),(2,3)\}$

Domain of $f = \{0,1,2\}$

and $g = \{(1,1),(2,0),(3,2)\}$

Domain of $g = \{1,2,3\}$

The domain of $f-g,\ f+g,\ f\cdot g,\ \frac{f}{g}$ is the domain of $f \cap$ domain of g.

∴ Domain of $f-g,\ f+g,\ f\cdot g,\ \frac{f}{g}$ is

= Domain of $f \cap$ Domain of g

$= \{0,1,2\} \cap \{1,2,3\} = \{1,2\}$

Thus

(i) $(f-g)(1) = f(1) - g(1) = 2 - 1 = 1$

$(f-g)(2) = f(2) - g(2) = 3 - 0 = 3$

$\therefore f - g = \{(1,1),(2,3)\}$

(ii) $(f+g)(1) = f(1) + g(1) = 2 + 1 = 3$

$(f+g)(2) = f(2) + g(2) = 3 + 0 = 3$

$\therefore f - g = \{(1,3),(2,3)\}$

(iii) $(f\cdot g)(1) = f(1)\cdot g(1) = 2\cdot 1 = 2$

$(f\cdot g)(2) = f(2)\cdot g(2) = 3\cdot 0 = 0$

$\therefore f\cdot g = \{(1,2),(2,0)\}$

Q29. **Solution:** For x lies in the third quadrant:

$\sin x = \sqrt{1-\cos^2 x}$

$$= \sqrt{1-\left(-\frac{1}{2}\right)^2} = \sqrt{\frac{3}{4}} = \pm\frac{\sqrt{3}}{2}$$

Therefore, $\sin x = -\frac{\sqrt{3}}{2}$

($\because \sin x$ is negative in the third quadrant)

$$\operatorname{cosec} x = \frac{1}{\sin x} = -\frac{2}{\sqrt{3}}$$

($\because \operatorname{cosec} x$ is negative in the third quadrant)

$$\sec x = \frac{1}{\cos x} = -\frac{2}{1} = -2$$

($\because \cos x$ is negative in the third quadrant)

$$\tan x = \frac{\sin x}{\cos x} = \frac{-\sqrt{3}/2}{-1/2} = \sqrt{3}$$

($\because \tan x$ is positive in the third quadrant)

$$\cot x = \frac{1}{\tan x} = \frac{1}{\sqrt{3}}$$

($\because \cot x$ is positive in the third quadrant)

Q30. **Solution:** $\frac{z+1}{z-1}$ is a purely imaginary number.

Therefore

$$\operatorname{Re}\left(\frac{z+1}{z-1}\right) = 0$$

Let $z = x + iy$, then

$$\operatorname{Re}\left(\frac{z+1}{z-1}\right) = 0$$

$$\Rightarrow \operatorname{Re}\left(\frac{x+iy+1}{x+iy-1}\right) = 0 \Rightarrow \operatorname{Re}\left(\frac{x+1+iy}{x-1+iy}\right) = 0$$

$$\Rightarrow \text{Re}\left(\frac{(x+1+iy)(x-1-iy)}{(x-1+iy)(x-1-iy)}\right)=0$$

$$\Rightarrow \text{Re}\left(\frac{(x^2-1)+y^2-2iy}{(x-1)^2+y^2}\right)=0$$

$$\Rightarrow (x^2-1)+y^2=0 \Rightarrow x^2+y^2=1$$

$$\Rightarrow |z|=1$$

OR

Q30. Solution: The slope of the line making angle $\frac{2\pi}{3}$ with the positive x-axis is $m=\tan\left(\frac{2\pi}{3}\right)=-\sqrt{3}$.

Therefore, the equation of the line passing through point (0, 2) and having a slope $-\sqrt{3}$ is $(y-2)=-\sqrt{3}(x-0) \Rightarrow \sqrt{3}x+y-2=0$

Now, the slope of the line parallel to line $\sqrt{3}x+y-2=0$ is $-\sqrt{3}$.

It is given that the line crosses the y-axis 2 units above the origin, that is, it passes through point (0, 2).

Therefore, the equation of the line passing through point (0, 2) and having a slope $-\sqrt{3}$ is $y-2=-\sqrt{3}(x-0) \Rightarrow \sqrt{3}x+y-2=0$

Q31. Solution: Let a be the radius of the circle. Since the circle touches both the axes and lies in the third quadrant, the coordinates of the circle will be $(a,-a)$.

Now, radius of the circle = perpendicular distance of centre $(a,-a)$ from the tangent $3x-4y+12=0$.

Thus, $a=\left|\frac{3a-4a+12}{\sqrt{9+16}}\right|$

$$\Rightarrow a=\left|\frac{-a+12}{5}\right| \Rightarrow a=\pm\left(\frac{-a+12}{5}\right)$$

Taking the positive sign, we get

$$a=\frac{-a+12}{5} \Rightarrow 5a=-a+12 \Rightarrow 6a=12 \Rightarrow a=2$$

Taking the negative sign, we get

$$a=\frac{a-12}{5} \Rightarrow 5a=a-12 \Rightarrow 4a=-12 \Rightarrow a=-3$$

$a \neq -3, \therefore a=2$

Hence, the equation of circle is $(x-2)^2+(y+2)^2=2^2 \quad [\because a=2]$

$$\Rightarrow x^2-4x+4+y^2+4y+4=4$$

$$\Rightarrow x^2+y^2-4x+4y+4=0$$

OR

Q31. Solution: The graphs of the lines $x+2y \le 10$, $x+y \ge 1$ and $x-y \le 0$ are shown in the figure.

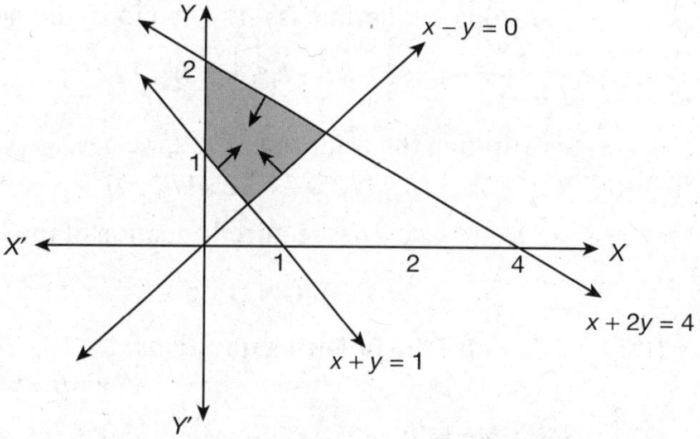

Inequality $x+2y \le 4$ represents the shaded region below the line $x+2y=4$ (including all the points on the line).

Inequality $x+y \ge 1$ represents the shaded region above the line $x+y=1$ (including all the points on the line).

Inequality $x-y \le 0$ represents the shaded region above the line $x-y=0$ (including all the points on the line).

Inequality $x \ge 0$, $y \ge 0$ represents every point in the common shaded region in the first quadrant including the points on the axes.

Thus, the solution of the given system of linear inequalities is represented by the common shaded region, including the points on the respective lines in the first quadrant.

Q32. Solution: Let T_n be the nth term of the series. Then

$$T_n=(2n+1)^2-(2n)^2$$

$$=(2n+1-2n)[(2n+1)+(2n)]=4n+1$$

$$[\because a^2-b^2=(a-b)(a+b)]$$

Thus, the sum of the n term of series is

$$S_n=\sum T_n=\sum(4n+1)$$

$$=4\sum n+\sum 1=4\cdot\frac{n(n+1)}{2}+n$$

$$=2n(n+1)+n=n(2n+3)$$

$$\Rightarrow S_{20}=20(2\cdot 20+3)=860$$

OR

Q32. Solution: The equation of any line through the point of intersection of the given lines is of the form

$$3x+4y-7+k(x-y+2)=0$$

$$\Rightarrow (3+k)x+(4-k)y-7+2k=0$$

$$\Rightarrow (k-4)y=(3+k)x-7+2k=0$$

$$\Rightarrow y=\frac{(3+k)}{(k-4)}x+\frac{-7+2k}{(k-4)}=0$$

The slope of the line is –1. Therefore, the slope $\frac{(3+k)}{(k-4)} = -1 \Rightarrow 3+k = -k+4 \Rightarrow k = 1/2$

Substituting the value of $k = 1/2$, we get

$(3+1/2)x+(4-1/2)y-7+2\times 1/2 = 0$

$\Rightarrow 7x+7y-12 = 0$ is required equation of the line.

SECTION D

(Q33–Q36) Solve the following questions.

(6 marks each)

Q33. Solution: In the expansion

$(a+b)^n = {}^nC_0a^n + {}^nC_1a^{n-1}b + ... + {}^nC_{n-1}ab^{n-1} + {}^nC_nb^n$

Fourth term from the beginning $= {}^nC_4a^{n-4}b^4$

Fourth term from the end $= {}^nC_{n-4}a^4b^{n-4}$

Therefore, in the expansion of $\left(\sqrt[3]{2}+\sqrt[2]{3}\right)^n$

Fourth term from the beginning is ${}^nC_3(\sqrt[3]{2})^{n-3}\left(\sqrt[2]{3}\right)^3$

Fourth term from the end is ${}^nC_{n-3}(\sqrt[3]{2})^3\left(\sqrt[2]{3}\right)^{n-3}$

The fourth term from the beginning and the fourth term from the end are equal. Therefore

${}^nC_3(\sqrt[3]{2})^{n-3}\left(\sqrt[2]{3}\right)^3 = {}^nC_{n-3}(\sqrt[3]{2})^3\left(\sqrt[2]{3}\right)^{n-3}$

$\Rightarrow (\sqrt[3]{2})^{n-3}\left(\sqrt[2]{3}\right)^3 = (\sqrt[3]{2})^3\left(\sqrt[2]{3}\right)^{n-3} \quad (\because {}^nC_3 = {}^nC_{n-3})$

$\Rightarrow (2)^{n/3-2} = (3)^{n/2-3} \Rightarrow (2)^0 = (3)^0 \Rightarrow n = 6$

Q34. Solution: The given expansion can be written as

$= (1+x+x^2+x^3)^6 = [(1+x)+x^2(1+x)]^6$

$= [(1+x)(1+x^2)]^6 = (1+x)^6 \cdot (1+x^2)^6$

Using Binomial Theorem

$= ({}^6C_0 + {}^6C_1x + {}^6C_2x^2 + {}^6C_3x^3 + {}^6C_4x^4 + {}^6C_5x^5 + ...)$

$\times({}^6C_0 + {}^6C_1x^2 + {}^6C_2x^4 + ...)$

$= {}^6C_0\,{}^6C_5x^5 + {}^6C_1\,{}^6C_3x^5 + {}^6C_2\,{}^6C_1x^5$

Thus, coefficient of x^5 is

$= {}^6C_0\,{}^6C_5 + {}^6C_1\,{}^6C_3 + {}^6C_2\,{}^6C_1 = 6+6\cdot 20+15\cdot 6 = 216$

OR

Q34. Solution: The given observations in ascending order and cumulative frequencies of the given data are shown in the table below:

x_i	f_i	$c.f.$
5	8	8
7	6	14
9	2	16
10	3	19
12	1	20

$n = 20$, which is even. Therefore, the median is the mean of $\left(\frac{20}{2}\right)^{th} = 10^{th}$ and $\left(\frac{20}{2}+1\right)^{th} = 11^{th}$ observations.

Both of these observations lie in the cumulative frequency 14, for which the corresponding observation is 7.

$\therefore \text{Median} = \frac{10^{th}\text{ observation} + 11^{th}\text{ observation}}{2}$

$= \frac{7+7}{2} = 7$

The absolute values of the deviation from median, $= |x_i - M|$ are

$\lvert x_i - M\rvert$	2	0	2	3	5
f_i	8	6	2	3	1
$f_i\lvert x_i - M\rvert$	16	0	4	9	5

$\sum_{i=1}^{5} f_i = 20, \quad \sum_{i=1}^{5} f_i|x_i - M| = 34$

$\text{M.D.}(M) = \frac{1}{N}\sum_{i=1}^{6} f_i|x_i - M| = \frac{1}{20}\times 34 = 1.7$

Q35. Solution: In a pack of 52 playing cards:

There are 4 suits (clubs, diamonds, hearts and spades) containing 13 cards each.

There are 2 colours, black (clubs and spades) and red (diamonds and hearts) containing 26 cards each.

There are 4 cards of king, queen and joker.

(i) Number of ways of choosing 4 cards that are of the same suit is = ${}^4C_1 \times {}^{13}C_4$(${}^4C_1$ is for choosing 1 suit out of four suits and ${}^{13}C_4$ is for choosing 4 cards from the chosen suit).

(ii) Number of ways of choosing 4 cards of the same colour is = ${}^2C_1 \times {}^{26}C_4$(${}^2C_1$ is for choosing 1 colour out of two colours and ${}^{26}C_4$ is for choosing 4 cards from the chosen colour)

(iii) Number of ways of choosing 4 cards from four different suits is = ${}^{13}C_1 \times {}^{13}C_1 \times {}^{13}C_1 \times {}^{13}C_1$(${}^{13}C_1$ is for choosing 1 card from a suit and it is four times because of choosing 1 card from each of the 4 suits).

(iv) Number of ways of choosing 2 red cards and 2 black cards is = ${}^{26}C_2 \times {}^{26}C_2$(${}^{26}C_2$ is for choosing 2 cards from a colour and it is two times because of choosing 2 cards from each of the 2 colours).

(v) Number of ways of choosing 2 diamond cards and 2 club cards is $= {}^{13}C_2 \times {}^{13}C_2$ (${}^{13}C_2$ is for choosing 2 cards from diamond and club each).

(vi) Number of ways of choosing 4 cards with at-least 2 kings is $= {}^4C_2 \times {}^{48}C_2 + {}^4C_3 \times {}^{48}C_1 + {}^4C_4 \times {}^{48}C_0$ (2 kings and 2 other or 3 kings and 1 other or 4 kings).

(vii) Number of ways of choosing 4 cards with at most 1 king and 1 black joker is $= {}^4C_0 \times {}^2C_1 \times {}^{46}C_3 + {}^4C_1 \times {}^2C_1 \times {}^{46}C_2$ (no king, 1 black joker and 3 other or 1 king, 1 black joker and 2 other).

OR

Q35. Solution: Let a and d be the first term and common difference of the AP, respectively. Then, sum of n terms of an AP is

$$S_n = \frac{n}{2}[2a + (n-1)d]$$

$$\Rightarrow \frac{n}{2}[2a - d + nd] = 2n^2 + 3n$$

$$\Rightarrow \frac{2a-d}{2}n + \frac{d}{2}n^2 = 3n + 2n^2$$

Comparing the coefficient of n and n^2, we get

$$\Rightarrow \frac{2a-d}{2} = 3 \text{ and } \frac{d}{2} = 2 \Rightarrow d = 4$$

Also, $a - \frac{d}{2} = 3 \Rightarrow a - 2 = 3 \Rightarrow a = 5$

The mth term is equal to 205. Therefore $a_m = a + (m-1)d = 205$

$$\Rightarrow 5 + (m-1)4 = 205 \Rightarrow (m-1)4 = 200$$

$$\Rightarrow m - 1 = 50 \Rightarrow m = 51$$

Q36. Solution: If the points $(3,0), (-2,-2), (8,2)$ are collinear, then the line passing through points $(3,0), (-2,-2)$ must also be pass through point $(8,2)$.

Now, the equation of the line passing through points

$(3,0), (-2,-2)$ is $(y - 0) = \frac{(-2-0)}{(-2-3)}(x-3)$

$$y = \frac{-2}{-5}(x-3) \Rightarrow 2x - 5y = 6$$

Since the third point $x = 8, y = 2$ also satisfies the equation of line passing through points (3, 0) and (–2,–2), the three points $(3,0), (-2,-2), (8,2)$ lie on a line, and hence, they are collinear.

OR

Q36. Solution: When a die is thrown, the sample space is given by

$S = \{1,2,3,4,5,6\}$

The different events are given by

$A = \{1,2,3,4\}$ $B = \{4,5,6\}$

$C = \{3,6\}$ $D = \{2,4,6\}$

$E = \{1,3,5\}$

$A \cup B = \{1,2,3,4,5,6\}$

$A \cap B = \{4\}$

$C \cup D = \{2,3,4,6\}$

$C \cap E = \{6\}$

$A - C = \{1,2,4\}$

$D' = \{1,3,5\}$

Sample Paper – V

Day 229

Time: 3 Hrs Maximum **Marks:** 80

General Instructions:

(i) All the questions are compulsory.
(ii) The question paper consists of 36 questions divided into 4 sections A, B, C and D.
(iii) Section A comprises of 20 questions 1 mark each. Section B comprises 6 questions of 2 marks each. Section C comprises 6 questions of 4 marks each. Section D comprises 4 questions of 6 marks each.
(iv) There is no overall choice. However, an internal choice has been provided in four questions of 1 mark each, two questions of 2 marks each, two questions of 4 marks each, and two questions of 6 marks each. You have to attempt only one of the alternatives in all such questions.
(iv) Use of calculators is not permitted.

SECTION A

Q1–Q10 are multiple choice type questions. Select the correct option. (1 mark each)

Q1. The locus of a point for which $x = 0, y = 0$, is
(a) Equation of X-axis (b) Equation of Y-axis
(c) Equation at Z-axis (d) None of these

Q2. The domain of the function f defined by
$f(x) = \sqrt{1-x} + \dfrac{1}{\sqrt{x^2-1}}$ is equal to
(a) $(-\infty,-1)\cup(1,2]$ (b) $(-\infty,-1)$
(c) $(-\infty,-1)\cup[1,2]$ (d) $(-\infty,-1)\cup[1,2)$

Q3. Let S_n denote the sum of the squares of the first n natural number and s_n denote the sum of the first n natural numbers. Then $\sum_{r=1}^{6} \dfrac{S_r}{s_r}$ is equal to
(a) 26 (b) 44 (c) 88 (d) 104

Q4. If the sum of the coefficient of the 1st and 3rd terms is equal to the coefficient of the 2nd term in the expansion of $(1+x)^n$, then value of n is
(a) 2 (b) 6 (c) 11 (d) 14

Q5. If $\tan\theta = -2$, θ lies in the second quadrant, then the value of $\cos\theta$ is
(a) $\dfrac{1}{\sqrt{10}}$ (b) $-\dfrac{1}{\sqrt{10}}$ (c) $\dfrac{-2}{\sqrt{5}}$ (d) $\dfrac{-1}{\sqrt{5}}$

Q6. If $f(z) = \dfrac{1-z}{1-z^2}$, where $z = 1+i$, then $|f(z)|$ is equal to
(a) $\dfrac{1}{\sqrt{5}}$ (b) $\dfrac{2}{\sqrt{5}}$ (c) $2|z|$ (d) $|z|$

Q7. If 23^{25} is divided by 11, then the remainder is
(a) 1 (b) 2 (c) 4 (d) $-\sqrt{2}$

Q8. Coefficient of x^2 in the expansion of $e^{(2x)}$ is
(a) 1 (b) 2 (c) 4 (d) $-\sqrt{2}$

Q9. The equation of the circle having its centre at (4, 3) and touching the line $4x+5y-5=0$ is
(a) 1 (b) 2 (c) 4 (d) $-\sqrt{2}$

Q10. The value of the trigonometric function cosec(−1125°) is equal to
(a) 1 (b) 2 (c) 4 (d) $-\sqrt{2}$

Q11–Q15 are fill in the blanks. (1 mark each)

Q11. Let $A = \{x \in N | 0 \le x \le 2\}$, $B = \{x \in R | 0 \le x \le 2\}$. Then A..........B

Q12. $8\cos\left(\dfrac{4\pi}{9}\right)\cos\left(\dfrac{2\pi}{9}\right)\cos\left(\dfrac{\pi}{9}\right)$ is equal to...........

Q13. If $|z| = \sqrt{2}$, $\arg(z) = \dfrac{3\pi}{4}$, then $z =$ .....

Q14. The conjugate of $\dfrac{-4+3i}{(1-2i)^2}$ is … … …

Q15. If $|x-1| \le -1$, then $x \in$ ....

(Q16–Q20) Answer the following questions. (1 mark each)

Q16. Write the negation of the following statement:
p: For every positive real number x, the number $2x-1$ is also positive.

Q17. How many chords can be drawn through 10 points on a circle?

Q18. Find the angle between the x-axis and the line joining the points $(2,1),(4,3)$.

Q19. If $P=\{x:x\le 1,\ x\in W\}$, $Q=\{x:-1\le x\le 1,\ x\in I\}$, then find $(P\cup Q)\times(P\cap Q)$, where N is the set of natural numbers and W is the set of whole numbers.

Q20. Evaluate $\lim\limits_{x\to 1}\dfrac{(2+x)^{3/2}-(3)^{3/2}}{x-1}$

SECTION B

(Q21–Q26) Solve the following questions.

(2 marks each)

Q21. Solve the equation $|z|=\bar{z}-1+i$.

Q22. Reduce the equation $2x-5y-10=0$ to intercept form and find its intercepts on the axes.

Q23. Find the new coordinates of the point $(0,-2)$ if the origin is shifted to the point $(2,-1)$ by a translation of axes.

Q24. If a circle passes through the points (0, 0), (2, 0) and (0, 3), then find the coordinates of its centre.

Q25. Find the equation of the ellipse that satisfies the given conditions: ends of major axis $(0,\pm\sqrt{6})$ and ends of minor axis $(\pm\sqrt{2},0)$

OR

Q25. Point $R(3,2)$ divides a line segment between the axes in the ratio 2:3. Find the equation of the line.

Q26. Out of 100 students, two sections of 40 and 60 are formed. If you and your friend are among the 100 students, what is the probability that

(i) You both enter the same section?
(ii) You both enter different sections?

OR

Q26. Find the derivative of $\sin^4 x$

SECTION C

(Q27–Q32) Solve the following questions.

(4 marks each)

Q27. If $f(x)=x,\ g(x)=x^2$ be two functions defined in the domain R, then find

(i) $(f+g)(x)$ (ii) $(f-g)(x)$

(iii) $(fg)(x)$ (iv) $\left(\dfrac{f}{g}\right)(x)$

(v) $\left(\dfrac{g}{f}\right)(x)$

Q28. Find the derivative of $x^2(\sin x-2\cos x)$

Q29. Find the ratio in which the XZ-plane divides the line segment by joining the points $(-2,15,7),(3,-5,8)$

Q30. The sum of the first three terms of a GP is $\dfrac{39}{10}$ and their product is 1. Find the common ratio and the terms.

Q31. Find the number of integers greater than 6000 that can be formed with the digits 3, 4, 5, 6, 7, where no digits are repeated.

OR

Q31. Expand using binomial theorem $\left(2+x+\dfrac{1}{x}\right)^3, x\neq 0$

Q32. A die is thrown. Find the probability of the following events:

(i) A number greater than 3 will appear
(ii) A number less than 3 will appear
(iii) A number multiple of 3 appear

OR

Q32. A point R with x-coordinate 5 lies on the segment joining the points $P(2,-3,4),Q(6,0,6)$. Find the coordinates of the point R.

SECTION D

(Q33–Q36) Solve the following questions.

(6 marks each)

Q33. Find $\sin\dfrac{x}{2}$, $\cos\dfrac{x}{2}$, $\tan\dfrac{x}{2}$, if $\tan x=\dfrac{3}{4}$, x is in quadrant II.

Q34. A solution of 8% boric acid is to be diluted by adding a 2% boric acid solution to it. The resulting mixture is to be more than 4% but less than 6% boric acid. If we have 100 litres of the 8% solution, how many litres of the 2% solution will need to be added?

OR

Q34. A farmer buys a used tractor for Rs 5000, in an annual installment of Rs 500 plus 5% interest on the unpaid amount. How much will the tractor cost him?

Q35. Let $A=\{\phi,1,\{2\}\}$. Which of the following statements is correct and why?

(i) $\phi\in A$ (ii) $\{\phi\}\subset A$
(iii) $\phi\subset A$ (iv) $1\in A$
(v) $\{1\}\subset A$ (vi) $\{1\}\in A$
(vii) $2\in A$ (viii) $\{2\}\subset A$
(ix) $\{2\}\in A$ (x) $\{\{2\}\}\subset A$

OR

Q35. Find the linear inequalities for which the shaded region in the given figure is the solution set.

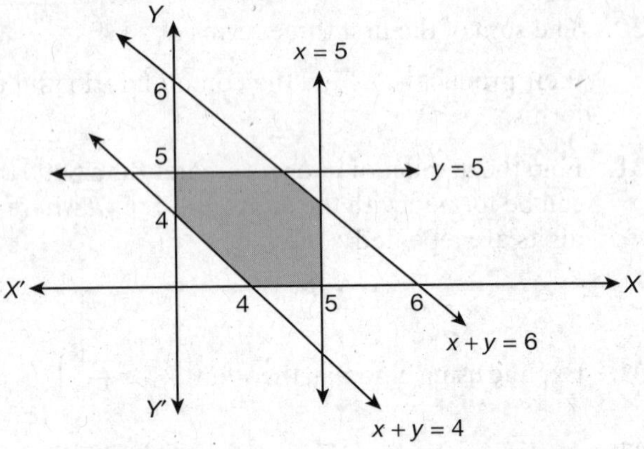

Q36. There are 4 women and 5 men, of whom a committee of 3 persons is to be formed. Find the number of ways in which the committee can be formed

(i) If it consists of at least 1 woman

(ii) If a particular woman is included and a particular man is excluded

(iii) If a particular woman is included, then a particular man is excluded, and vice versa.

OR

Q36. Find the mean deviation about the mean for the data.

Height in cms	Number of boys
105–115	12
115–125	20
125–135	10
135–145	15
145–155	8

Day 230

Sample Paper – V (Solutions)

Time: 3 Hrs Maximum **Marks:** 80

General Instructions:

(i) All the questions are compulsory.

(ii) The question paper consists of 36 questions divided into 4 sections A, B, C and D.

(iii) Section A comprises 20 questions of 1 mark each. Section B comprises 6 questions of 2 marks each. Section C comprises of 6 questions 4 marks each. Section D comprises 4 questions of 6 marks each.

(iv) There is no overall choice. However, an internal choice has been provided in four questions of 1 mark each, two questions of 2 marks each, two questions of 4 marks each, and two questions of 6 marks each. You have to attempt only one of the alternatives in all such questions.

(iv) Use of calculators is not permitted.

SECTION A

Q1–Q10 are multiple choice type questions. Select the correct option. (1 mark each)

Q1. Solution: (c)

The locus of a point for which $x = 0, y = 0$, is the equation of the Z-axis.

Q2. Solution: (b)

$f(x) = \sqrt{1-x} + \dfrac{1}{\sqrt{x^2-1}}$, $f(x)$ is defined, if

$1 - x \geq 0$ and $x^2 - 1 > 0$

$\Rightarrow x - 1 \leq 0$ and $(x+1)(x-1) > 0$

$\Rightarrow x \leq 1$ and $x < -1$ or $x > 1$

$\therefore$ Domain of $(-\infty, -1)$

Q3. Solution: (b)

$$\sum_{r=1}^{n} \frac{S_r}{s_r} = \frac{S_1}{s_1} + \frac{S_2}{s_2} + \frac{S_3}{s_3} + ... + \frac{S_n}{s_n}$$

Let T_n be the nth term of the above series. Then

$$T_n = \frac{S_n}{s_n} = \frac{n(n+1)(2n+1)/6}{n(n+1)/2} = \frac{(2n+1)}{3} = \frac{2}{3}n + \frac{1}{3}$$

Thus, summation $\displaystyle\sum_{r=1}^{n} \frac{S_r}{s_r} = \sum T_n = \frac{2}{3}\sum n + \sum \frac{1}{3}$

$$= \frac{2}{3}\left[\frac{n(n+1)}{2}\right]^2 + \frac{1}{3}n = \frac{1}{3}n\left[\frac{n(n+1)}{2} + 1\right]$$

$$= \frac{1}{3}n\left[\frac{n^2+n+2}{2}\right] = \frac{n(n^2+n+2)}{6}$$

$$\Rightarrow \sum_{r=1}^{6} \frac{S_r}{s_r} = \frac{6(6^2+6+2)}{6} = 44$$

Q4. Solution: (a)

The expansion of $(1+x)^n$ is

${}^nC_0 + {}^nC_1x + {}^nC_2x^2 + {}^nC_3x^3 + ... + {}^nC_nx^n$

$\therefore$ Coefficient of the 1st term = ${}^nC_0$,

Coefficient of the 2nd term = ${}^nC_1$,

Coefficient of the 3rd term = ${}^nC_2$,

The sum of the coefficients of the 1st and 3rd terms is equal to the coefficient of the 2nd term. Therefore

$^nC_0 + {^nC_2} = {^nC_1}$

$\Rightarrow 1 + \frac{(n)!}{2!(n-2)!} = \frac{(n)!}{(n-1)!}$

$\Rightarrow 1 + \frac{n(n-1)}{2} = n$

$\Rightarrow n(n-1) - 2(n-1) = 0$

$\Rightarrow (n-1)(n-2) = 0$

$n = 1$ is not possible. Therefore $n = 2$

Q5. Solution: (d)

$\because \sec\theta = \sqrt{1+\tan^2\theta} = \sqrt{1+(-2)^2} = \pm\sqrt{5}$

$\Rightarrow \sec\theta = -\sqrt{5} \Rightarrow \cos\theta = -\frac{1}{\sqrt{5}}$

[Since sec is negative in the second quadrant]

Q6. Solution: (a)

$f(z) = \frac{1-z}{1-z^2} = \frac{1-(1+i)}{1-(1+i)^2}$

$= \frac{-i}{1-1-i^2-2i} = \frac{-i}{1-2i}$

$\Rightarrow f(z) = \frac{-i}{1-2i} \times \frac{1+2i}{1+2i} = \frac{2-i}{5}$

$\Rightarrow |f(z)| = \frac{\sqrt{5}}{5} = \frac{1}{\sqrt{5}} \quad (\because |z| = \sqrt{2^2+1^2} = \sqrt{5})$

Q7. Solution: (a)

$23^{25} = (1+22)^{25}$

$= {^{25}C_0}22^0 + {^{25}C_1}22^1 + {^{25}C_2}22^2 \ldots {^{25}C_{25}}22^{25}$

$= 1 + 11.2.({^{25}C_1} + {^{25}C_2}22^1 \ldots {^{25}C_{25}}22^{25})$

$= 1 + 11k$ {where, $k = 2\cdot({^{25}C_1} + {^{25}C_2}22^1 \ldots {^{25}C_{25}}22^{25})$}

Therefore, when 23^{25} is divided by 11, the remainder will be 1.

Q8. Solution: (b)

$e^x = 1 + \frac{x}{1!} + \frac{x^2}{2!} + \frac{x^3}{3!} + \frac{x^4}{4!} + \ldots$

Therefore, $e^{(2x)}$

$= 1 + \frac{2x}{1!} + \frac{(2x)^2}{2!} + \frac{(2x)^3}{3!} \ldots$

$= 1 + 2x + 2x^2 + \frac{4}{3}x^3 \ldots$

Thus, coefficient of x^2 in the expansion of $e^{(2x)}$ is 2

Q9. Solution: (c)

Radius of the circle = the perpendicular distance from the centre (4, 3) to the line =

$r = \left|\frac{16+15-5}{\sqrt{25+144}}\right| = 26/13 = 2$

Thus, the required equation of the circle is $(x-4)^2 + (y-3)^2 = 2^2 = 4.$

Q10. Solution: (d)

$\text{cosec}(-1125°) = -\text{cosec}(1125°)$

$(\because \text{cosec}(-\theta) = -\text{cosec}\,\theta)$

$= -\text{cosec}(3 \times 360° + 45°) = -\text{cosec}(+45°)$

$= -\sqrt{2}$

Q11–Q15 are fill in the blanks. **(1 mark each)**

Q11. Solution:

$A = \{1, 2\},$

$B = \{x \in R | 0 \le x \le 2\} \Rightarrow x \in [0, 2]$

A is finite with only two elements 1 and 2, while in set B, there are infinite real numbers from 0 to 2. Therefore $A \subset B$

Q12. Solution:

$8\cos\left(\frac{4\pi}{9}\right)\cos\left(\frac{2\pi}{9}\right)\cos\left(\frac{\pi}{9}\right)$

$= 8\cos 80° \cos 40° \cdot \cos 20°$

$= 8\sin 10° \cos 40° \cdot \cos 20°$

$= 4\sin 10°[2\cos 40° \cdot \cos 20°]$

$= 4\sin 10°[\cos 60° + \cos 20°]$

$[\because 2\cos x \cdot \cos y = \cos(x+y) + \cos(x-y)]$

$= 4\sin 10° \cdot \frac{1}{2} + 4\sin 10° \cos 20°$

$= 2\sin 10° + 2 \cdot 2\sin 10° \cos 20°$

$= 2\sin 10° + 2[\sin 30° - \sin 10°] = 1$

Q13. Solution: Let $z = x + iy = r(\cos\theta + i\sin\theta)$ Then

$|z| = r = \sqrt{2}$ and $\arg(z) = \theta = \frac{3\pi}{4} \Rightarrow \tan\theta = \frac{3\pi}{4}$

$\therefore z = \sqrt{2}\left(\cos\frac{3\pi}{4} + i\sin\frac{3\pi}{4}\right)$

$= \sqrt{2}[\cos(\pi - \pi/4) + i\sin(\pi - \pi/4)]$

$= \sqrt{2}\left[-\cos\frac{\pi}{4} + i\sin\frac{\pi}{4}\right] = \sqrt{2}\left[\frac{-1}{\sqrt{2}} + i\frac{1}{\sqrt{2}}\right] = -1 + i$

Q14. Solution:

$z = \frac{-4+3i}{(1-2i)^2} = \frac{-4+3i}{1+4i^2-4i}$

$= \frac{-4+3i}{1-4-4i} = \frac{-4+3i}{-3-4i} = \frac{i(3+4i)}{-(3+4i)} = -i$

$\Rightarrow z = 0 - i \Rightarrow \bar{z} = 0 + i$

Q15. Solution: The modulus function is always greater then or equal to zero. Therefore, the solution set will be empty, $x \in \phi$.

(Q16–Q20) Answer the following questions.

(1 mark each)

Q16. Solution: $\sim p$: There exists a positive real number x, such that $2x-1$ is not positive.

Q17. Solution: For drawing one chord on a circle, only 2 points are required.
Therefore, the number of chords that can be draw through the given 10 points on a circle is $= {}^{10}C_2 = 45$

Q18. Solution: The slope of the line joining the points $(2,1),(4,3)$ is $m = \dfrac{3-1}{4-2} = 1$. Then the inclination (θ) given by $\tan\theta = m = 1 \Rightarrow \theta = 45°$
Thus, the angle between the x-axis and the line joining the points $(2,1),(4,3)$ is 45^0.

Q19. Solution: $P = \{x : x \le 1,\ x \in W\} = \{0,1\}$

$Q = \{x : -1 \le x \le 1,\ x \in I\} = \{-1,0,1\}$

$\therefore P \cup Q = \{-1,0,1),\ P \cap Q = \{0,1\}$

Thus, $(P \cup Q) \times (P \cap Q)$

$= \{-1,0,1\} \times \{0,1\}$

$= \{(-1,0),(0,0),(1,0),(-1,1),(0,1),(1,1)\}$

Q20. Solution: $\lim\limits_{x\to 1} \dfrac{(2+x)^{3/2}-(3)^{3/2}}{x-1}$

$= \lim\limits_{x\to 1} \dfrac{(2+x)^{3/2}-(3)^{3/2}}{(2+x)-(3)} = \dfrac{3}{2}(3)^{\frac{3}{2}-1}$

$\left[\because \lim\limits_{x\to a} \dfrac{x^n-a^n}{x-a} = na^{n-1}\right]$

$= \dfrac{3}{2}(3)^{1/2} = \dfrac{3\sqrt{3}}{2} \quad [\because x \to 1 \Rightarrow x+2 \to 3]$

SECTION B

(Q21–Q26) Solve the following questions.

(2 marks each)

Q21. Solution: Let $z = x+iy$, then $\bar{z} = x-iy$

Thus

$|z| = \bar{z} - 1 + i \Rightarrow |x+iy| = x - iy - 1 + i$

$\Rightarrow \sqrt{x^2+y^2} = x-1+i(1-y)$

$\Rightarrow \sqrt{x^2+y^2} - (x-1) + i(y-1) = 0 + 0i$

On comparing real and imaginary parts, we get

$y-1=0 \Rightarrow y=1$

Also, $\sqrt{x^2+y^2} - (x-1) = 0$

$\Rightarrow x^2+1-2x = x^2+y^2$

$\Rightarrow x = \dfrac{1-y^2}{2} = \dfrac{1-(1)^2}{2} = 0$

Thus, $z = x+iy = 0+i$

Q22. Solution: The given equation $2x-5y-10=0$ in intercepts form can be written as $2x-5y-10=0 \Rightarrow \dfrac{2x}{10} + \dfrac{-5y}{10} = 1 \Rightarrow \dfrac{x}{5} + \dfrac{y}{(-2)} = 1$

Comparing the above equation with $\dfrac{x}{a}+\dfrac{y}{b}=1$, we get $a=5, b=-2$.
Therefore, in the intercept form, the x- and y-intercepts are 5 and –2, respectively.

Q23. Solution: The coordinates of the new origin are $h=2,\ k=-1$ and the original coordinates are $x=0,\ y=-2$.
Since any point $P(x,y)$ in given coordinate system will be changed to $P(x',y')$ in the new coordinate system, by the relation $x' = x-h$ and $y' = y-k$
Therefore, the point $P(0,-2)$ in the given coordinate system will be transferred to $x' = x-h = 0-2 = -2$ and $y' = y-k = -2-(-1) = -1$ in the new coordinate system.
Thus, the coordinates of the point $(0,-2)$ in the new coordinate system are $(1,-1)$.

Q24. Solution: Let the equation of circle be
$x^2+y^2+2gx+2fy+c=0$
Since this circle passes through the points $(0,0)$, $(2,0)$, $(0,3)$, these points must satisfy the equation of circle. Thus
$c=0, \quad 4+4g=0 \quad$ and $\quad 9+6f=0$
Solving the above equations, we get
$g = -4/4 = -1$ and $f = -3/2 =$
Hence, the coordinates of the circle are $= (-g,-f) = (1,3/2)$

Q25. Solution: The ends of the major axis are $(0,\pm\sqrt{6})$. Therefore, the major axis is along the y-axis and $a=\sqrt{6}$.
The ends of the minor axis are $(\pm\sqrt{2},0)$. Therefore, the minor axis is along the y-axis and $b=\sqrt{2}$.
Thus, the equation of the ellipse will be of the form

$$\frac{x^2}{b^2}+\frac{y^2}{a^2}=1\Rightarrow\frac{x^2}{(\sqrt{2})^2}+\frac{y^2}{(\sqrt{6})^2}=1\Rightarrow\frac{x^2}{2}+\frac{y^2}{6}=1$$

OR

Q25. Solution: Let AB be the line segment between the axes such that point $R(3,2)$ divides AB in the ratio 2:3.

Let AB be the line segment with coordinates $A(x,0)$ on the x-axis and $B(0,y)$ on the y-axis

Since point $R(3,2)$ divides AB in the ratio 2:3, according to the section formula

$$(3,2)=\left(\frac{2\times0+3\times x}{2+3},\frac{2\times y+3\times0}{2+3}\right)\Rightarrow(3,2)=\left(\frac{3x}{5},\frac{2y}{5}\right)$$

$$\Rightarrow 3=\frac{3x}{5},\ 2=\frac{2y}{5}\Rightarrow x=5,\ y=5$$

Therefore, the respective coordinates of A and B are

$(5,0),(0,5)$.

Thus, the equation of line AB passing through points

$(5,0),(0,5)$. is $(y-0)=\dfrac{5-0}{0-5}(x-5)$

$y=-1(x-5)\Rightarrow x+y-5=0$ is the required equation of the line.

Q26. Solution: Total number of ways of selecting 2 students out of 100 students = ${}^{100}C_2$

(i) The two of us will enter the same section if both of us are among 40 students or among 60 students.

∴ Number of ways in which both of us enter the same section = ${}^{40}C_2+{}^{60}C_2$

∴ Probability that both of us enter the same section

$$P=\frac{{}^{40}C_2+{}^{60}C_2}{{}^{100}C_2}$$

(ii) Probability that both of us enter different sections

$$=1-P=1-\frac{{}^{40}C_2+{}^{60}C_2}{{}^{100}C_2}$$

OR

Q26. Solution: Let $y=\sin^2 x$

$$\therefore\frac{dy}{dx}=\frac{d}{dx}(\sin x\sin x)=(\sin x)'\sin x+\sin x(\sin x)'$$

[By the Leibniz product rule]

$=\cos x\sin x+\sin x\cos x=2\sin x\cos x$ ...(1)

Now, for $y=\sin^4 x$

$$\therefore\frac{dy}{dx}=\frac{d}{dx}(\sin^2 x\sin^2 x)$$

$=(\sin^2 x)'\sin^2 x+\sin^2 x(\sin^2 x)'$

[By the Leibniz product rule]

$=2\sin^2 x(2\sin x\cos x)$ [Using (1)]

$=4\sin^3 x\cos x$

SECTION C

(Q27–Q32) Solve the following questions.

(4 marks each)

Q27. Solution: (i) $(f+g)(x)=f(x)+g(x)=x+x^2$

(ii) $(f-g)(x)=f(x)-g(x)=x-x^2$

(iii) $(fg)(x)=f(x)\cdot g(x)=x\cdot x^2=x^3$

(iv) $\left(\dfrac{f}{g}\right)(x)=\dfrac{f(x)}{g(x)}=\dfrac{x}{x^2}=\dfrac{1}{x}$ defined in the domain R^+

(v) $\left(\dfrac{g}{f}\right)(x)=\dfrac{g(x)}{f(x)}=\dfrac{x^2}{x}=x$

Q28. Solution: Let $f(x)=x^2(\sin x-2\cos x)$

By the product rule

$$f'(x)=x^2\frac{d}{dx}(\sin x-2\cos x)+(\sin x-2\cos x)\frac{d}{dx}(x^2)$$

$$=x^2\left[\frac{d}{dx}(\sin x)-2\frac{d}{dx}(\cos x)\right]+(\sin x-2\cos x)\frac{d}{dx}(x^2)$$

$=x^2[\cos x-2(-\sin x)]+(\sin x-2\cos x)(2x)$

$=x[x\cos x+2x\sin x+2\sin x-4\cos x]$

Q29. Solution: Let the YZ-plane divide the line segment joining points $(-2,15,7),(3,-5,8)$ in the ratio $k:1$. Then by the section formula, the coordinates of point of intersection are

$$\left(\frac{k(3)-2}{k+1},\frac{k(-5)+15}{k+1},\frac{k(8)+7}{k+1}\right)$$

Since on the YZ-plane, the y-coordinate of any point is zero, $\dfrac{-5k+15}{k+1}=0\Rightarrow 5k=15\Rightarrow k=3$

Thus, the XZ-plane divides the line segment formed by joining the given points in the ratio 3:1.

Q30. Solution: Let $\dfrac{a}{r},a,ar$ be the first three terms of the GP. Then

$\frac{a}{r}+a+ar=\frac{39}{10}\Rightarrow a\left(\frac{1}{r}+1+r\right)=\frac{39}{10}$

and $\left(\frac{a}{r}\right)(a)(ar)=1\Rightarrow a^3=1\Rightarrow a=1$

Thus, we get

$\frac{1}{r}+1+r=\frac{39}{10}\Rightarrow 1+r+r^2=\frac{39}{10}r$

$\Rightarrow 10r^2+10r+10=39r\Rightarrow 10r^2-29r+10=0$

$\Rightarrow (5r-2)(2r-5)=0\Rightarrow r=\frac{2}{5}$ or $r=\frac{5}{2}$

Hence, the first three terms of the GP are $\frac{2}{5},1,\frac{5}{2}$ or $\frac{5}{2},1,\frac{2}{5}$

Q31. Solution: The number greater than 6000 will be of 4 digits and of 5 digits, with the digits 3, 4, 5, 6, 7 because repetition is not allowed.

Since all the five-digit numbers are greater than 6000, the number of ways of forming 5-digit numbers is $5!=5\times4\times3\times2\times1=120$

Now, four-digit numbers greater than 6000 can be formed, when the thousands place is filled in 2 ways (6, 7), the hundreds place is filled in 4 ways and the tens place is filled in 3 ways. The units place can be filled in 2 ways.

Thus, total number of 4-digit numbers greater than 5000 is $=2\times4\times3\times2=48$

Thus, total number of integers greater than 6000 $=120+48=168$

OR

Q31. Solution: The given expression $\left(2+x+\frac{1}{x}\right)^3$ can be written as $\left[2+\left(x+\frac{1}{x}\right)\right]^3$. Using Binomial Theorem, $\left[2+\left(x+\frac{1}{x}\right)\right]^3$

$={}^3C_0 2^3+{}^3C_1 2^2\left(x+\frac{1}{x}\right)+{}^3C_2 2\left(x+\frac{1}{x}\right)^2+{}^3C_3\left(x+\frac{1}{x}\right)^3$

$=8+12\left(x+\frac{1}{x}\right)+6\left(x+\frac{1}{x}\right)^2+\left(x+\frac{1}{x}\right)^3$

Again using Binomial Theorem, $\left(x+\frac{1}{x}\right)^3$

$={}^3C_0 x^3+{}^3C_1 x^2\left(\frac{1}{x}\right)+{}^3C_2 x\left(\frac{1}{x}\right)^2+{}^3C_3\left(\frac{1}{x}\right)^3$

$=x^3+3x+\frac{3}{x}+\frac{1}{x^3}$

Thus, $\left(2+x-\frac{1}{x}\right)^3$

$=8+12x+\frac{12}{x}+6x^2+\frac{1}{x^2}+12+x^3+3x+\frac{3}{x}+\frac{1}{x^3}$

$=x^3+6x^2+15x+20+\frac{15}{x}+\frac{1}{x^2}+\frac{1}{x^3}$

Q32. Solution: Sample space, $S=\{1,2,3,4,5,6\}$

(i) Let A be the event of the occurrence of a number greater than 3, $A=\{4,5,6\}$

$P(A)=\frac{\text{Number of outcomes favorable to } A}{\text{Total number of possible outcomes}}$

$=\frac{n(A)}{n(S)}=\frac{3}{6}=\frac{1}{2}$

(ii) Let B be the event of the occurrence of a number less than 3, $B=\{1,2\}$

$P(B)=\frac{\text{Number of outcomes favorable to } B}{\text{Total number of possible outcomes}}$

$=\frac{n(B)}{n(S)}=\frac{2}{6}=\frac{1}{3}$

(iii) Let C be the event of the occurrence of a number multiple of 3, $C=\{3,6\}$

$P(C)=\frac{\text{Number of outcomes favorable to } C}{\text{Total number of possible outcomes}}$

$=\frac{n(C)}{n(S)}=\frac{2}{6}=\frac{1}{2}$

OR

Q32. Solution: Let point R divide the line segment PQ in the ratio $k:1$. Then by the section formula, the coordinates of point R are given by

$\left(\frac{k(6)+2}{k+1},\frac{k(0)-3}{k+1},\frac{k(6)+4}{k+1}\right)$

$=\left(\frac{6k+2}{k+1},\frac{-3}{k+1},\frac{6k+4}{k+1}\right)$

Since it is given that the x-coordinate of point R is 5, $\frac{6k+2}{k+1}=5\Rightarrow 6k+2=5k+5\Rightarrow k=3$

The coordinates of point R are $\left(5,\frac{-3}{4},\frac{11}{2}\right)$

SECTION D

(Q33–Q36) Solve the following questions.

(6 marks each)

Q33. Solution: x is in quadrant II. Therefore

$\frac{\pi}{2}<x<\pi\Rightarrow\frac{\pi}{4}<\frac{x}{2}<\frac{\pi}{2}$

Thus, $\sin\frac{x}{2}$, $\cos\frac{x}{2}$, $\tan\frac{x}{2}$, lie in the first quadrant.

And in first quadrant sin, cos and tan all are positive.

Now, since $\tan x = \frac{3}{4}$

$\Rightarrow \sec x = \sqrt{1+\tan^2 x} = \sqrt{1+\left(\frac{3}{4}\right)^2} = \sqrt{1+\frac{9}{16}} = \pm\frac{5}{4}$

$\Rightarrow \cos x = -\frac{4}{5}$ ($\because$ cos is negative in the second quadrant)

Now, $\because \cos x = 2\cos^2\frac{x}{2} - 1 \Rightarrow \frac{-4}{5} = 2\cos^2\frac{x}{2} - 1$

$\Rightarrow 2\cos^2\frac{x}{2} = 1 - \frac{4}{5} \Rightarrow 2\cos^2\frac{x}{2} = \frac{1}{5}$

$\Rightarrow \cos^2\frac{x}{2} = \frac{1}{10} \Rightarrow \cos\frac{x}{2} = \frac{1}{\sqrt{10}}$

($\because$ cos is positive in the first quadrant)

$\sin\frac{x}{2} = \sqrt{1-\cos^2\frac{x}{2}} = \sqrt{1-\left(\frac{1}{\sqrt{10}}\right)^2} = \sqrt{\frac{9}{10}} = \frac{3}{\sqrt{10}}$

($\because$ sin is positive in the first quadrant)

And, $\tan\frac{x}{2} = \frac{\sin\frac{x}{2}}{\cos\frac{x}{2}} = \frac{(3/\sqrt{10})}{(1/\sqrt{10})} = 3$

($\because$ tan is positive in the first quadrant)

Thus, the values of $\sin\frac{x}{2}, \cos\frac{x}{2}, \tan\frac{x}{2}$ are

$\frac{3}{\sqrt{10}}, \frac{1}{\sqrt{10}}, 3$

Q34. Solution: Let x litres of 2% boric acid solution be required to be added.

Then, total mixture = $(x+100)$ litres

This resulting mixture is to be more than 3% but less than 5% boric acid.

Therefore

$3\% \text{ of } (x+100) < 2\% \text{ of } x + 8\% \text{ of } 100$

$\Rightarrow \frac{3}{100}(x+100) < \frac{2}{100}x + \frac{8}{100}(100)$

$\Rightarrow \frac{1}{100}x < 8-3$

$\Rightarrow x < 500$

Also

$2\% \text{ of } x + 8\% \text{ of } 100 < 5\% \text{ of } (x+100)$

$\frac{2}{100}x + \frac{8}{100}(100) < \frac{5}{100}(x+100)$

$\Rightarrow 8-5 < \frac{3}{100}x \Rightarrow 100 < x$

Thus, $100 < x < 500$

Hence, the number of litres of 2% of boric solution that is to be added must be more than 100 litres but less than 500 litres.

OR

Q34. Solution: The farmer pays Rs 500 per month. Therefore, the interest paid annually is 10% of 5000, 10% of 4500, 10% of 4000, …, 10% of 500

Therefore, total interest to be paid is

= 10% of (5000 + 4500 + 4000 + … +500)

= 10% of (500 + 1000 + 1500 + … + 5000)

Since 500, 1000, 1500, …, 5000 forms an AP series with first term equal to 500 and common difference equal to 500, and the number of terms is n

$5000 = 500 + (n-1)(500) \Rightarrow n = 10$

Therefore,

$\therefore 500 + 1000 + ... + 1000$

$= \frac{10}{2}[2(500) + (10-1)(500)] = 27500$

Thus, total interest to be paid is

= 10% of 27500 = 2750

Hence, cost of the tractor is (5000 + 2750) = Rs. 7750

Q35. Solution: (i) and (ii): The statements $\phi \in A$ and $\{\phi\} \subset A$ are correct because ϕ is an element of A implies that $\{\phi\} \subset A$.

(iii) The statement $\phi \subset A$ is correct because ϕ represents an empty set and every empty set is a subset of a set.

(iv) and (v): The statements $1 \in A$ and $\{1\} \subset A$ are correct because 1 is an element of A implies that $\{1\} \subset A$

(vi) The statement $\{1\} \in A$ is incorrect because $\{1\}$ is not an element of A.

(vii) and (viii): The statement $2 \in A$ and $\{2\} \subset A$ is incorrect because 2 is not an element of A.

(ix) and (x): The statements $\{2\} \in A$ and $\{\{2\}\} \subset A$ are correct because {2} is an element of A implies that $\{\{2\}\} \subset A$

OR

Q35. Solution: Consider the line $x + y = 4$.

Since the shaded region and the origin lie on the opposite side of this line, (0, 0) does not satisfy the linear constraint $x + y \geq 4$.

Therefore, we get one linear inequality as $x + y \geq 4$.

Now, consider the line $x + y = 6$,. Since the shaded region and origin lie on the same side of this line, (0, 0) satisfies the linear constraints $x + y \leq 6$.

Therefore, we get the second linear inequality as $x + y \le 6$

Now, consider the line $x = 5$. Since the shaded region and origin are on the left of this line, (0, 0) satisfies the constraints $x \le 5$.

Therefore, we get the third linear inequality as $x \le 5$.

Now, consider the line $y = 5$. Since the shaded region and origin are below this line, (0, 0) satisfies the constraints $y \le 5$.

Therefore, we get the fourth linear inequality as $y \le 5$.

Also the shaded region is above the X-axis and on the right of the Y-axis, that is, theshaded region is in the first quadrant. So, we must have $x \ge 0, y \ge 0$.

Thus, the linear inequalities corresponding to the given solution set are $x + y \ge 4$; $x + y \le 6$; $x \le 5$; $y \le 5$;$x \ge 0$, $y \ge 0$

Q36. Solution: (i) Selection of 3 persons with at least 1 woman consist of the following cases:

1 woman and 2 men + 2 women and 1 man + 3 women and no man

Thus, by the multiplication principle, the required number of ways is

$= {}^4C_1 \times {}^5C_2 + {}^4C_2 \times {}^5C_1 + {}^4C_3 \times {}^5C_0$

$= \frac{4!}{1!3!} \times \frac{5!}{2!3!} + \frac{4!}{2!2!} \times \frac{5!}{1!4!} + \frac{4!}{3!1!} \times \frac{5!}{0!5!}$

$= 4 \times 10 + 6 \times 5 + 4 \times 1 = 40 + 30 + 4 = 74$

(ii) Let a woman W_1 be included in the committee and a man M_1 be excluded from the selection (or committee). The remaining 2 persons (since woman W_1 is already selected) can be selected from the remaining 7 persons (since man M_1 is excluded from the selection) in ${}^7C_2 = 21$ ways.

(iii) If woman W_1 is included in the committee, then man M_1 is excluded from the committee, and if man M_1 is included in the committee, then woman W_1 is excluded from the committee.

This can be solved in two different ways:

(a) W_1 is included but M_1 is not + M_1 is included but W_1 is not + neither W_1 nor M_1 is included in the committee

Thus, by the multiplication principle, the required number of ways is $= {}^7C_2 + {}^7C_2 + {}^7C_3 = 21 + 21 + 35 = 77$

(b) Total selection with no condition – both W_1 and M_1 are included in the committee

Thus, by the multiplication principle, the required number of ways is $= {}^9C_3 - {}^7C_1 = 84 - 7 = 77$

OR

Q36. Solution: The following table is formed as:

Height in cms	**Number of boys** f_i	**Mid-point** x_i	$f_i x_i$	$\lvert x_i - \bar{x} \rvert$	$f_i \lvert x_i - \bar{x} \rvert$
105–115	12	110	1320	17	204
115–125	20	120	2400	7	140
125–135	10	130	1300	3	30
135–145	10	140	1400	13	130
145–155	8	150	1200	23	184

$N = \sum_{i=1}^{5} f_i = 60,$

$\sum_{i=1}^{5} f_i x_i = 7620$

$\therefore \bar{x} = \frac{1}{N} \sum_{i=1}^{5} f_i x_i = \frac{1}{60} \times 7620 = 127$

$\text{M.D.}(\bar{x}) = \frac{1}{N} \sum_{i=1}^{5} f_i \lvert x_i - \bar{x} \rvert = \frac{1}{60} \times 688 = 11.47$

Sample Paper – VI

Day 231

Time: 3 Hrs Maximum **Marks:** 80

General Instructions:

(i) All the questions are compulsory.

(ii) The question paper consists of 36 questions divided into 4 sections A, B, C and D.

(iii) Section A comprises 20 questions of 1 mark each. Section B comprises 6 questions of 2 marks each. Section C comprises 6 questions of 4 marks each. Section D comprises 4 questions of 6 marks each.

(iv) There is no overall choice. However, an internal choice has been provided in four questions of 1 mark each, two questions of 2 marks each, two questions of 4 marks each, and two questions of 6 marks each. You have to attempt only one of the alternatives in all such questions.

(iv) Use of calculators is not permitted.

SECTION A

Q1–Q10 are multiple choice type questions. Select the correct option. (1 mark each)

Q1. If 4*x*, 3*y* and 2*z* are in AP, where the distinct number *x, y, z* are in GP, then the common ratio of the GP is

(a) 3 (b) 1/3 (c) 2 (d) 1/2

Q2. The perimeter of the circle with centre at (0, 1) and passing through the point (3, 5) is

(a) 5π (b) 10π

(c) 25π (d) None of these

Q3. The locus of a point for which $z = 0$, is

(a) *XY*-plane (b) *YZ*-plane

(c) *ZX*-plane (d) None of these

Q4. If $|x-1| > 2$, then

(a) $x \in (0,2)$ (b) $x \in [0,2]$

(c) $x \in (-\infty,-1) \cup (3,\infty)$ (d) $x \in [-\infty,0) \cup [2,\infty)$

Q5. If the middle term of $\left(\frac{1}{x} - x\cos x\right)^8$ is equal to $4\frac{3}{8}$, then the value of *x* is

(a) $2n\pi$ (b) $n\pi + \frac{\pi}{6}$

(c) $n\pi$ (d) $2n\pi \pm \frac{\pi}{3}$

Q6. The domain and range of the real function *f* defined by $f(x) = \frac{2-x}{x-1}$ is given by

(a) Domain = R, Range = $\{-1, 1\}$

(b) Domain = $R - \{1\}$, Range = $R - \{2\}$,

(c) Domain = $R - \{2\}$, Range = $\{-1\}$

(d) Domain = $R - \{-2\}$, Range = $\{-1, 1\}$

Q7. The value of $\tan 75° + \cot 75°$ is

(a) 4 (b) 3 (c) 2 (d) 1

Q8. The value of *n*, if $\lim_{x\to 3} \frac{x^n - 3^n}{x-3} = 27, n \in N$

(a) 4 (b) 3 (c) 2 (d) 1

Q9. If $\sin x + \cos x = 1$, then $\sin^3 x + \cos^3 x$ is equal to

(a) 4 (b) 3 (c) 2 (d) 1

Q10. The value of $\left|\frac{(1+i)(2+i)}{(2-i)}\right|$ is equal to

(a) $-\sqrt{2}$ (b) 0 (c) $\sqrt{2}$ (d) 1

Q11–Q15 are fill in the blanks. (1 mark each)

Q11. Let sets R and T be defined as

$R = \{x \in Z | x \text{ is divisible by } 2\}$

$T = \{x \in Z | x \text{ is divisible by } 6\}$.

Then *T*.........*R*

Q12. The sum of the series $\sum_{r=0}^{20} {}^{41}C_r$ is … … …

Q13. The derivative of $(x^2 + x)\tan x$ is … … …

Q14. The multiple inverse of the complex number $6 + 8i$ is … … …

Q15. The given complex number $i^{21} + i^{12}$ in the form $a + ib$ is equal to ... ...

(Q16–Q20) Answer the following questions.

(1 mark each)

Q16. Identity the quantifiers in the following statement: There exists a triangle which is not obtuse.

Q17. Find a and b if $(a, 1+b) = (2b, a-1)$

Q18. Do the points $(-4,0), (4,0), (0,4)$ form a triangle? If yes, name the type of triangle formed.

Q19. Find the equation of the circle which touches the X-axis and whose centre is $(2, -2)$.

Q20. Find the equation of the line which intersects the x-axis at a distance of 4 units to the left of the origin with slope -2.

SECTION B

(Q21–Q26) Solve the following questions.

(2 marks each)

Q21. Given below are two statements:
p: 25 is a multiple of 5.
q: 25 is a multiple of 3.
Write the compound statements connecting these two statements with "and" and "or".
In both cases, check the validity of the compound statement.

Q22. In a certain city, all telephone numbers have six digits, the first two digits always being 91. How many telephone numbers have all six digits distinct?

Q23. Find the coefficient of x^2 in the expansion of $\log(1+2x^2)$.

Q24. The IQ of a person is given by the formula $IQ = \frac{MA}{CA} \times 100$, where MA is mental age and CA is chronological age. If $60 \leq IQ \leq 100$ for a group of 12-year-old children, find the range of their mental age.

OR

Q24. A candidate is required to answer 6 questions out of 10 questions, which are divided into two groups, each containing 5 questions. Find the number of different ways of answering questions if at least 1 question is compulsory from either group?

Q25. How many diagonals can be drawn through the vertices of a 10-sided polygon?

OR

Q25. Find the equation of the ellipse that satisfies the given conditions: length of major axis 30, foci $(\pm 5, 0)$

Q26. Using the section formula, show that the points $A(2,-1,4), B(-1,2,1), C(0,1,2)$ are collinear.

OR

Q26. Find the term independent of x, where $x \neq 0$, in the expansion of $\left(2x^2 - \frac{1}{2x}\right)^9$.

SECTION C

(Q27–Q32) Solve the following questions.

(4 marks each)

Q27. Given a GP with $a = 64$ and 7th term 729, determine summation S_6.

Q28. A card is selected from a pack of 52 cards. Find the probability of the following events:
(i) Card is an ace (ii) Card is a spade
(iii) Card is an ace or spade

Q29. Prove that
$$(\sin x - \operatorname{cosec} x)^2 + (\cos x - \sec x)^2 - (\tan x - \cot x)^2 = 1$$

Q30. Let $f = \left\{\left(x, \frac{x^2-1}{x^2+1}\right) : x \in R\right\}$ be a function from R into R. Determine the range of f.

OR

Q30. Find the distance of the point $(2, -1)$ from the line $12(x+6) = 5(y-2)$

Q31. In what ratio does the point $(-4, y)$ divide the line segment joining the points $A(-6,1), B(3,-8)$? Also find the value of y.

OR

Q31. If A and B are the points $(3,4,5), (-1,3,-2)$, respectively, find the equation of the set of points P such that $PA^2 + PB^2 = 1$.

Q32. Find the derivative of $\frac{1-\cos x}{x+\sin x}$

OR

Q32. Show that the solution set of the following system of linear inequalities is an unbounded region:
$2x + y \geq 8, x + 2y \geq 8, x \geq 0, y \geq 0$.

SECTION D

(Q33–Q36) Solve the following questions.
(6 marks each)

Q33. A box contains 8 red marbles, 12 blue marbles and 10 green marbles. Six marbles are drawn from the box. What is the probability that
(i) All will be red?
(ii) Exactly two will be red?
(iii) At least one will be red?
(iv) At most one will be red?
(v) Exactly one red and 1 blue?

Q34. If $A = \{1,2,3\}, B = \{2,3,4\}$ and $U = \{1,2,3,4,5,6\}$, then prove that
(i) $(A \cup B)' = A' \cap B'$
(ii) $(A \cap B)' = A' \cup B'$
(iii) $(A - B)' \neq A' - B'$

OR

Q34. Reduce the equation $x - \sqrt{3}y - 4 = 0$ to normal form. Find its perpendicular distance from the origin and the angle between the perpendicular and the positive x-axis.

Q35. Find the mean deviation about median for the following data.

Marks	Number of girls
0–10	9
10–20	7
20–30	6
30–40	5
40–50	3

OR

Q35. Prove the following by using the principle of mathematical induction for all $n \in N$

$$\frac{1}{\sqrt{1}+\sqrt{2}} + \frac{1}{\sqrt{2}+\sqrt{3}} + \ldots + \frac{1}{\sqrt{n}+\sqrt{n+1}} = \frac{n}{\sqrt{1}+\sqrt{n+1}}$$

Q36. Find the modulus and the argument of the complex number $z = 1 - i$

OR

Q36. If z_1, z_2 are two complex numbers such that $|z_1| = |z_2|$ and $\arg(z_2) - \arg(z_1) = \frac{\pi}{2}$, then show that $z_2 = iz_1$

Day 232

Sample Paper – VI (Solutions)

SECTION A

Q1–Q10 are multiple choice type questions. Select the correct option. (1 mark each)

Q1. Solution: (c)
Let x, y, z be in GP, with common ratio r. Then $y = xr$ and $z = xr^2$
x, $3/2y$, $2z$ are in AP. Therefore

$$3y = \frac{4x+2z}{2} \Rightarrow 3y = 2x + z$$

$$3(xr) = 2x + (xr^2) \Rightarrow 3r = 2 + r^2$$

$$\Rightarrow r^2 - 3r + 2 = 0 \Rightarrow (r-2)(r-1) = 0$$

$$\Rightarrow r = 2 \text{ or } r = 1$$

Q2. Solution: (b)
Radius of the circle = distance between points (0, 1) and (3, 5), $r = \sqrt{9+16} = 5$
$\therefore$ Required perimeter $= 2\pi r = 10\pi$

Q3. Solution: (a)
The locus of a point for which $z = 0$, is the XY-plane.

Q4. Solution: (c)

$$|x-1| > 2 \Rightarrow (x-1) < -2 \text{ or } (x-1) > 2$$

$$(\because |x| > a \Rightarrow x < -a \text{ or } x > a)$$

$$\Rightarrow x < -2+1 \text{ or } x > 2+1$$

$$\Rightarrow x < -1 \text{ or } x > 3 \Rightarrow x \in (-\infty, -1) \cup (3, \infty)$$

Q5. Solution: (c)
In the expansion $\left(\frac{1}{x} - x\cos x\right)^8$, $n = 8$ (even).
Therefore, it has only one middle term, that is, the 5th term.

$$\therefore T_5 = T_{4+1} = {}^8C_4 \left(\frac{1}{x}\right)^{8-4} (-x\cos x)^4$$

$$\Rightarrow (-1)^4\, {}^8C_4 x^{-4} x^4 \cos^4 x = 70$$

$$\Rightarrow \frac{8\cdot7\cdot6\cdot5}{4\cdot3\cdot2\cdot1}\cos^4 x = 70$$

$$\Rightarrow 70\cos^4 x = 70$$

$$\Rightarrow \cos^4 x = 1$$

$$\Rightarrow \cos x = \pm 1 \Rightarrow x = n\pi$$

Q6. Solution: (b)

$f(x) = \dfrac{2-x}{x-1}$ is defined if $x - 1 \neq 0 \Rightarrow x \neq 1$

$\therefore$ Domain of $f = R - \{1\}$

Let $f(x) = \dfrac{2-x}{x-1} = y$

$\Rightarrow xy - y = 2 - x$

$\Rightarrow xy + x = 2 + y$

$\Rightarrow x(y+1) = 2 + y$

$\Rightarrow x = \dfrac{1+y}{y+2}$

x will be real for all values of y, except when $y + 2 = 0 \Rightarrow y = 2$

Thus, range of $f = R - \{2\}$

Q7. Solution: (a)

$$\tan 75° + \cot 75° = \frac{\sin 75°}{\cos 75°} + \frac{\cos 75°}{\sin 75°}$$

$$= \frac{\sin^2 75° + \cos^2 75°}{\sin 75° \cdot \cos 75°} = \frac{2}{2\sin 75° \cdot \cos 75°}$$

$$= \frac{2}{\sin 150°} = \frac{2}{\sin(180-30)°} = \frac{2}{\sin 30°} = 4$$

Q8. Solution: (b)

$$\lim_{x\to 3}\frac{x^n - 3^n}{x-3} = 27 \Rightarrow n(3)^{n-1} = 27$$

$$\left[\because \lim_{x\to a}\frac{x^n - a^n}{x-a} = na^{n-1}\right]$$

$$\Rightarrow n(3)^{n-1} = 3\times 9 \Rightarrow n\times 3^{n-1} = 3\times(3)^2$$

$$\Rightarrow n\times 3^{n-1} = 3\times(3)^{3-1} \Rightarrow n = 3$$

Q9. Solution: (d)

$\because \sin x + \cos x = 1,$

Squaring both sides, we get

$(\sin x + \cos x)^2 = 1^2$

$\Rightarrow \sin^2 x + \cos^2 x + 2\sin x\cos x = 1^2$

$\Rightarrow \sin x \cdot \cos x = \dfrac{1}{2}(1^2 - 1) = 0$

Therefore, $\sin^3 x + \cos^3 x$

$= (\sin x + \cos x)^3 - 3\sin x\cos x(\sin x + \cos x)$

$= 1^3 - 3\cdot 0\cdot 1 = 1$

Q10. Solution: (c)

$$\left|\frac{(1+i)(2+i)}{(2-i)}\right|.$$

$$= \left|\frac{(1+3i)(2+i)}{(2+i)(2+i)}\right|$$

$$= \left|\frac{2+6i+i+3i^2}{4-i^2}\right| = \left|\frac{-1+7i}{5}\right| = \frac{\sqrt{50}}{5} = \sqrt{2}$$

Q11–Q15 are fill in the blanks. **(1 mark each)**

Q11. Solution: $R = \{x \in Z | x \text{ is divisible by } 2\}$

$= \{..., -6, -4, -2, 0, 2, 4, 6...\}$

$T = \{x \in Z | x \text{ is divisible by } 6\}$

$\{...., -12, -6, 0, 6, 12....\}$

Every element of T is also in R. Therefore $T \subset R$

Q12. Solution:

$$\sum_{r=0}^{20} {}^{41}C_r = {}^{41}C_0 + {}^{41}C_1 + {}^{41}C_2 + ... + {}^{41}C_{20}$$

$$= \frac{1}{2}\begin{pmatrix} {}^{41}C_0 + {}^{41}C_1 + {}^{41}C_2 + ... + {}^{41}C_{20} \\ + {}^{41}C_0 + {}^{41}C_1 + {}^{41}C_2 + ... + {}^{41}C_{20} \end{pmatrix}$$

$$= \frac{1}{2}\begin{pmatrix} {}^{41}C_0 + {}^{41}C_1 + {}^{41}C_2 + ... + {}^{41}C_{20} \\ + {}^{41}C_{41} + {}^{41}C_{40} + {}^{41}C_{39} + ... + {}^{41}C_{21} \end{pmatrix}$$

$(\because {}^nC_r = {}^nC_{n-r})$

$$= \frac{1}{2}(2^{41}) = 2^{40} \quad (\because {}^nC_0 + {}^nC_1 + ... + {}^nC_n = 2^n)$$

Q13. Solution: Let $f(x) = (x^2 + x)\tan x$

By the product rule

$$f'(x) = (x^2 + x)\frac{d}{dx}(\tan x) + \tan x\frac{d}{dx}(x^2 + 1)$$

$$= (x^2 + x)(\sec^2 x) + \tan x(2x + 1)$$

$$= (x^2 + 1)\sec^2 x + (2x + 1)\tan x$$

Q14. Solution:

$z = 6 + 8i \Rightarrow \bar{z} = 6 - 8i,$

$|z|^2 = 6^2 + (-8)^2 = 36 + 64 = 100$

Therefore, the multiple inverse of $6 + 8i$ is

$$z^{-1} = \frac{\bar{z}}{|z|^2} = \frac{6-8i}{100} = \frac{6}{100} - \frac{8}{100}i = \frac{3}{50} - \frac{2}{25}i$$

Q15. Solution:

$i^{21} + i^{12} = i^{4\times5+1} + i^{4\times3} = (i^4)^5 \cdot i + (i^4)^3$

$= i + 1 \quad (\because i^4 = 1, i^3 = -i, i^2 = -1)$

$\Rightarrow a + ib = 1 + i$

(Q16–Q20) Answer the following questions. **(1 mark each)**

Q16. Solution: The quantifier in the statement is 'There exists'.

Q17. Solution: Two ordered pairs are equal if their corresponding elements are equal. Therefore

$(a,1+b)=(2b,a-1)$

$\Rightarrow a=2b$ and $1+b=a-1$

$\Rightarrow a=2b$ and $a-b=2$

$\Rightarrow 2b-b=2 \Rightarrow b=2 \Rightarrow a=4$

Q18. Solution: The given points are $P(-4,0)$, $Q(4,0)$, $R(0,4)$. Then by the distance formula

$PQ=\sqrt{(4+4)^2+(0-0)^2}=\sqrt{8^2}=8$

$QR=\sqrt{(0-4)^2+(4-0)^2}=\sqrt{16+16}=4\sqrt{2}$

$PR=\sqrt{(0+4)^2+(4-0)^2}=\sqrt{16+16}=4\sqrt{2}$

$PR=QR$. Therefore, triangle PQR is an isosceles triangle.

Q19. Solution: Since the centre of the circle is (2, –2) and the circle touches the X-axis, its radius = 2

Thus, the equation of circle is

$(x-2)^2+(y+2)^2=2^2$

$\Rightarrow x^2-4x+4+y^2+4y+4=4$

$\Rightarrow x^2+y^2-4x+4y+4=0$

Q20. Solution: The equation of a line with slope m and x-intercept is a is given by $y=m(x-a)$

Thus, equation of the line intersecting the x-axis at a distance of 4 units to the left of the origin, that is,$a=-4$ and the slope of the line is $m=-2$ is given by $y=-2[x-(-4)] \Rightarrow y=-2x-8 \Rightarrow 2x+y+8=0$

SECTION B

(Q21–Q26) Solve the following questions.

(2 marks each)

Q21. Solution: The compound statement with 'and' is "25 is a multiple of 5 and 3".

This is a false statement, since 25 is not a multiple of 3.

The compound statement with 'or' is "25 is a multiple of 5 or 3".

This is a true statement, since 25 is not a multiple of 3 but it is a multiple of 5.

Q22. Solution: The first two digits are always 91.

And if the first two digits are 91, the remaining 4 digits can be arranged in

$={}^8P_4=\frac{8!}{8-4!}=\frac{8!}{4!}=\frac{8\times7\times6\times5\times4!}{4!}=1680$

Thus, total number of telephone numbers having all six digits distinct = 1680

Q23. Solution:

$\log_e(1+x)=x-\frac{x^2}{2}+\frac{x^3}{3}-\frac{x^4}{4}+\ldots$

Therefore, $\log(1+2x^2)$

$=(2x^2)-\frac{(2x^2)^2}{2}+\frac{(2x^2)^3}{3}+\ldots$

$=2x^2-2x^4+\frac{8}{3}x^6+\ldots$

Thus, coefficient of x^2 in the expansion of $\log(1+2x^2)$ is 2.

Q24. Solution: For a group of 12-year-old children

$60 \le IQ \le 100$

And for a group of 12-year-old children

CA = 12 years

$\therefore IQ=\frac{MA}{CA}\times100=\frac{MA}{12}\times100=\frac{25}{3}MA$

Thus,

$60 \le IQ \le 100$

$\Rightarrow 60 \le \frac{25}{3}MA \le 100$

$\Rightarrow \frac{36}{5} \le MA \le 12$

Thus, the range of mental age of the group of 12-year-old children is $\frac{36}{5} \le MA \le 12$

OR

Q24. Solution: At least 2 questions are compulsory from either group. Therefore, the total number of questions attempted from each group is given as:

Group I	1	2	3	4	5
Group II	5	4	3	2	1

Thus, total number of possible ways

$={}^5C_1\times{}^5C_5+{}^5C_2\times{}^5C_4+{}^5C_3\times{}^5C_3+{}^5C_4\times{}^5C_2+{}^5C_5\times{}^5C_1=5\times1+10\times5+10\times10+5\times10+5\times1=210$

Q25. Solution: For drawing one diagonal of a polygon, only two points are required.

Therefore, the number of diagonals that can be drawn through 10 vertices of a polygon is $={}^{10}C_2-10=45-10=35$ (10 is subtracted because two adjacent vertices do not give the diagonal of a polygon while it gives the side of a polygon)

OR

Q25. Solution: Length of major axis, $2a=30 \Rightarrow a=15$

Since the foci $(\pm5,0)$ are on the x-axis, the major axis is along the x-axis and $c=5$.

Now, from the equation $a^2=b^2+c^2$

$\Rightarrow 15^2=b^2+5^2 \Rightarrow b^2=225-25 \Rightarrow b=10\sqrt{2}$

Thus, the equation of the ellipse will be of the form

$$\frac{x^2}{a^2}+\frac{y^2}{b^2}=1 \Rightarrow \frac{x^2}{15^2}+\frac{y^2}{(10\sqrt{2})^2}=1 \Rightarrow \frac{x^2}{225}+\frac{y^2}{200}=1$$

Q26. Solution: Let P be a point that divides line segment AB in the ratio $k:1$. Then by the section formula, the coordinates of point P are

$$\left(\frac{k(-1)+2}{k+1}, \frac{k(2)-1}{k+1}, \frac{k(1)+4}{k+1}\right)$$

Now, equating the x-coordinate of P and C, we get

$$\frac{-k+2}{k+1}=0 \Rightarrow k=2$$

For $k=2$, the coordinates of point P are $(0,1,2)$ which is the same as of point $C(0,1,2)$. Therefore, point C is a point that divides AB externally in the ratio 2:1, which is the same point P.

Thus, the points A, B, C are collinear.

OR

Q26. Solution:

The general term of the given expansion $\left(2x^2-\frac{1}{2x}\right)^9$ is $T_{r+1}={}^9C_r\left(2x^2\right)^{9-r}\left(-\frac{1}{2x}\right)^r$

$$={}^9C_r 2^{9-r}x^{18-2r}(-1)^r 2^{-r}\cdot x^{-r}$$

$$=(-1)^r\,{}^9C_r 2^{9-2r}x^{18-3r}$$

For the term independent of x, $18-3r=0 \Rightarrow r=6$

Therefore, $T_{r+1}=T_{6+1}=7^{\text{th}}$ term is independent of x Thus $T_7=T_{6+1}=(-1)^6\,{}^9C_6 2^{9-2\cdot 6}$

$$=\frac{{}^9C_6}{2^3}=\frac{9\cdot 8\cdot 7}{2^3\cdot 3\cdot 2}=\frac{21}{2}$$

SECTION C

(Q27–Q32) Solve the following questions.

(4 marks each)

Q27. Solution: $a=64$, $a_7=729$ and let r be the common ratio of the GP. Then $a_n=ar^{n-1}$

$$\Rightarrow a_7=ar^{7-1} \Rightarrow 729=64r^6$$

$$\Rightarrow r^6=\frac{729}{64} \Rightarrow r^6=\left(\frac{3}{2}\right)^6 \Rightarrow r=\frac{3}{2}$$

Therefore, summation $S_n=\dfrac{a(r^n-1)}{r-1}$

$$\Rightarrow S_6=\frac{64\left[\left(\frac{3}{2}\right)^6-1\right]}{\frac{3}{2}-1}=2\times 64\left[\left(\frac{3}{2}\right)^6-1\right]$$

$$\Rightarrow S_6=(2)^7\left[\frac{(3)^6-(2)^6}{(2)^6}\right]=1330$$

Q28. Solution: When a card is selected from a pack of 52 cards, the number of possible outcomes is 52, that is, the sample space contains 52 elements, $n(S)=52$.

(i) Let A be the event in which the card drawn is an ace, $n(A)=4$

$$P(A)=\frac{n(A)}{n(S)}=\frac{4}{52}=\frac{1}{13}$$

(ii) Let B be the event in which the card drawn is a spade, $n(B)=13$

$$P(B)=\frac{n(B)}{n(S)}=\frac{13}{52}=\frac{1}{4}$$

(iii) Let C be the event in which the card drawn is an ace or spade $n(C)=4+13-1=16$ ($\because$ 1 ace of spade)

$$P(C)=\frac{n(C)}{n(S)}=\frac{16}{52}=\frac{4}{13}$$

Q29. Solution:

$$\text{LHS}=(\sin x-\operatorname{cosec} x)^2+(\cos x-\sec x)^2-(\tan x-\cot x)^2$$

$$=\sin^2 x+\operatorname{cosec}^2 x-2\sin x\operatorname{cosec} x+\cos^2 x+\sec^2 x-2\cos x\sec x-\tan^2 x-\cot^2 x+2\tan x\cot x$$

$$=(\sin^2 x+\cos^2 x)+(\operatorname{cosec}^2 x-\cot^2 x)+(\sec^2 x-\tan^2 x)-2\sin x\operatorname{cosec} x-2\cos x\sec x+2\tan x\cot x$$

$$=1+1+1-2-2+2=1=\text{RHS}$$

($\because \sin^2 x+\cos^2 x=1$, $\operatorname{cosec}^2 x-\cot^2 x=1$, $\sec^2 x-\tan^2 x=1$, $\sin x\operatorname{cosec} x=1$, $\cos x\sec x=1$, $\tan x\cot x=1$)

Q30. Solution: For $x\in R$, $y=\dfrac{x^2-1}{x^2+1}$, where y is the range of f

Now, rearranging the equation in terms of x, we get

$$y=\frac{x^2-1}{x^2+1} \Rightarrow x^2y+y=x^2-1$$

$$\Rightarrow x^2-x^2y=1+y$$

$$\Rightarrow x^2(1-y)=1+y$$

$$\Rightarrow x^2=\frac{1+y}{1-y} \Rightarrow x=\sqrt{\frac{1+y}{1-y}}$$

Now, the function $g(y)=\sqrt{\frac{1+y}{1-y}}$ is defined when

$\frac{1+y}{1-y}\ge 0$ and $1-y\ne 0$

$\Rightarrow -1\le y\le 1$ and $y\ne 1$

$\Rightarrow -1\le y<1$

Thus, the range of $f=[-1,1)$

OR

Q30. Solution: The given equation of the line is $12(x+6)=5(y-2)$

$\Rightarrow 12x+72=5y-10 \Rightarrow 12x-5y+82=0$

Comparing the above equation with the general equation of the line $Ax+By+C=0$, we get $A=12$, $B=-5$, $C=82$.

The perpendicular distance (d) of a line $Ax+By+C=0$, from a point (x_1,y_1) is given by

$$d=\frac{|Ax_1+By_1+C|}{\sqrt{A^2+B^2}}$$

The given point is $(x_1,y_1)=(2,-1)$

Therefore, the distance of point $(2,-1)$ from the given line $12x-5y+82=0$ is

$$=\frac{|12(2)+(-5)(-1)+82|}{\sqrt{(12)^2+(-5)^2}}=\frac{|24+5+82|}{13}=\frac{111}{13}\text{ units}$$

Q31. Solution: Let $(-4, y)$ divide AB internally in the ratio $m:n$. Then by the section formula, we get

$$(-4,y)=\left(\frac{3m-6n}{m+n},\frac{-8m+1n}{m+n}\right)$$

$$\Rightarrow -4=\frac{3m-6n}{m+n},\quad y=\frac{-8m+1n}{m+n}$$

From the 1st equation, $-4m-4n=3m-6n$

$\Rightarrow 7m=2n \Rightarrow m/n=2/7 \Rightarrow m:n=2:7$

Now from the 2nd equation, $y=\frac{-8m+1n}{m+n}$

$$\Rightarrow y=\frac{-8(m/n)+1}{(m/n)+1}=\frac{-8(2/7)+1}{(2/7)+1}$$

$$\Rightarrow y=\frac{-16+7}{2+7}=-1$$

Therefore, the point $(-4, -1)$ divides the line segment joining the points $A(-6,10), B(3,-8)$ in the ratio 2:7.

OR

Q31. Solution: Let the coordinates of point P be (x, y, z). Then using the distance formula $PA^2=(x-3)^2+(y-4)^2+(z-5)^2$

$\Rightarrow PA^2=x^2-6x+y^2-8y+z^2-10z+50$

Similarly, $PB^2=(x+1)^2+(y-3)^2+(z+2)^2$

$\Rightarrow PB^2=x^2+2x+y^2-6y+z^2+4z+14$

Now, it is given that $PA^2+PB^2=1$. Then

$(x^2-6x+y^2-8y+z^2-10z+50)+$

$(x^2+2x+y^2-6y+z^2+4z+14)=1$

$\Rightarrow 2x^2+2y^2+2z^2-4x-14y-6z+63=0$ is the required equation.

Q32. Solution: Let $f(x)=\frac{1-\cos x}{x+\sin x}$

By the quotient rule,

$$f'(x)=\frac{(x+\sin x)\frac{d}{dx}(1-\cos x)-(1-\cos x)\frac{d}{dx}(x+\sin x)}{(x+\sin x)^2}$$

$$=\frac{(x+\sin x)(\sin x)-(1-\cos x)(1-\cos x)}{(x+\sin x)^2}$$

$$=\frac{x\sin x+\sin^2 x-1+2\cos x-\cos^2 x}{(x+\sin x)^2}$$

$$=\frac{x\sin x+2\cos x-1-\cos 2x}{(x+\sin x)^2}$$

OR

Q32. Solution: The graphs of the lines, $2x+y=8$, $x+2y=8$, $x=0$ and $y=0$ are shown in the figure.

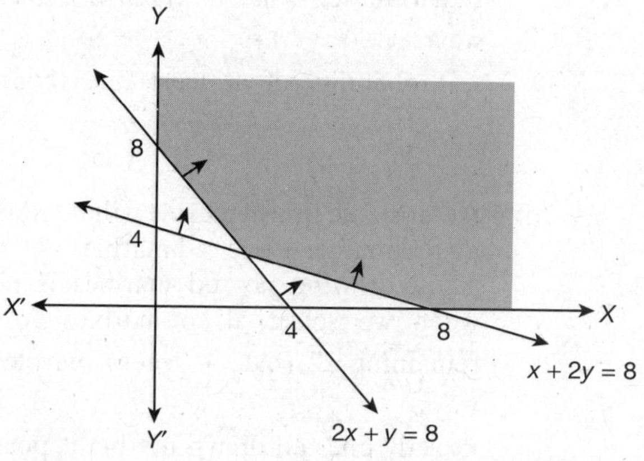

Inequality $2x+y\ge 8$ represents the shaded region above the line $2x+y=8$ (including all the points on the line).

Inequality $x+2y\ge 8$ represents the shaded region above the line $x+2y=8$ (including all the points on the line).

Inequalities $x \geq 0,\ y \geq 0$ represent the region in the first quadrant.
Thus, the graph of the common shaded portion is unbounded.

SECTION D

(Q33–Q36) Solve the following questions.

(6 marks each)

Q33. Solution: Total number of marbles = 8 + 12 + 10 = 30

Number of ways of drawing 6 marbles from 30 marbles = $^{30}C_6$

(i) All the drawn marbles will be red if we draw 6 marbles out of 8 red marbles, $^8C_6$ ways.

∴ Probability that all marbles will be red

$$= \frac{^8C_6}{^{30}C_6}$$

(ii) Exactly two drawn marbles will be red if we draw 2 marbles out of 8 red marbles and 4 from remaining 22 (blue + green) marbles, $^8C_2\,^{22}C_4$ ways.

∴ Probability that exactly two will be red marbles $= \dfrac{^8C_2\,^{22}C_4}{^{30}C_6}$

(iii) At least one drawn marble will be red means total minus no marbles from red.
Thus, drawing no red marbles is possible when we select all 6 marbles from the remaining 22 (blue + green) marbles, $^{22}C_6$ ways.

∴ Probability of at least one red marble

$$= 1 - \frac{^{22}C_6}{^{30}C_6}$$

(iv) At most one drawn marble will be red means no red marble + one red marble.
Thus, drawing no red marbles is possible when we select all 6 marbles from the remaining 22 (blue + green) marbles, $^{22}C_6$ ways.
Exactly one red drawn marble is possible if we draw 1 marble out of 8 red marbles and 5 from remaining 22 (blue + green) marbles, $^8C_1\,^{22}C_5$ ways.

∴ Probability of at most one red marble

$$= \frac{^{22}C_6}{^{30}C_6} + \frac{^8C_1\,^{22}C_5}{^{30}C_6}$$

(v) Exactly one red and one blue is possible if we draw 1 marble out of 8 red marbles and 1 marble out of 12 blue marbles and the remaining 4 marbles out of 10 green marbles, $^8C_1\,^{12}C_1\,^{10}C_4$ ways.

∴ Probability that exactly one red marble and one blue marble $= \dfrac{^8C_1\,^{12}C_1\,^{10}C_4}{^{30}C_6}$

Q34. Solution: $A = \{1,2,3\} \Rightarrow A' = U - A$

$\Rightarrow A' = \{1,2,3,4,5,6\} - \{1,2,3\} = \{4,5,6\}$

$B = \{2,3,4\} \Rightarrow B' = U - B$

$\Rightarrow B' = \{1,2,3,4,5,6\} - \{2,3,4\} = \{1,5,6\}$

(i) LHS $= (A \cup B)' = (\{1,2,3\} \cup \{2,3,4\})'$

$= (\{1,2,3,4\})' = \{1,2,3,4,5,6\} - \{1,2,3,4\} = \{5,6\}$

RHS $= A' \cap B' = \{4,5,6\} \cap \{1,5,6\} = \{5,6\}$

(ii) LHS $= (A \cap B)' = (\{1,2,3\} \cap \{2,3,4\})'$

$= (\{2,3\})' = \{1,2,3,4,5,6\} - \{2,3\} = \{1,4,5,6\}$

RHS $= A' \cup B' = \{4,5,6\} \cap \{1,5,6\} = \{1,4,5,6\}$

Thus,LHS = RHS, verify above two results.

(iii) LHS $= (A - B)' = (\{1,2,3\} - \{2,3,4\})'$

$= (\{1\})' = \{1,2,3,4,5,6\} - \{1\} = \{2,3,4,5,6\}$

RHS $= A' - B' = \{4,5,6\} - \{1,5,6\} = \{4\}$

LHS ≠ RHS

Therefore $(A - B)' \neq A' - B'$.

OR

Q34. Solution: The given equation $x - \sqrt{3}y - 4 = 0$ can be written as $x - \sqrt{3}y - 4 = 0 \Rightarrow x - \sqrt{3}y = 4$

Dividing both sides by $\sqrt{(-1)^2 + (\sqrt{3})^2} = \sqrt{4} = 2$,

we get $\dfrac{x}{2} - \dfrac{\sqrt{3}}{2}y = \dfrac{4}{2} \Rightarrow \left(\dfrac{1}{2}\right)x + \left(-\dfrac{\sqrt{3}}{2}\right)y = 2$

$\Rightarrow x\cos 300° + y\sin 300° = 2$ which is the normal form of the given equation.

Comparing the above equation with the normal form of equation of line by $x\cos\alpha + y\sin\alpha = p$, we get $\alpha = 300°$, $p = 2$

Thus, the perpendicular distance of the line from the origin is 2, while the angle between the perpendicular and the positive x-axis is 300°.

Q35. Solution: The following table is formed as:

Marks	Number of f_i	Cumulative frequency (c.f.)	Mid-point x_i	$\|x_i - \text{Med.}\|$	$f_i\|x_i - \text{Med.}\|$
0–10	9	9	5	13.57	122.13
10–20	7	16	15	3.57	24.99
20–30	6	22	25	7.57	45.42
30–40	4	26	35	17.57	70.28
40–50	4	30	45	27.57	110.28
	30				373.1

The class interval containing the $\left(\frac{N}{2}\right)^{\text{th}}$ or 15th item is 10–20.

Therefore, 10–20 is the median class.

Thus, median $= l + \dfrac{\frac{N}{2} - c}{f} \times h$

Here, $l = 10, C = 9, f = 7, h = 10, N = 30$

$\therefore$ Median

$= 10 + \dfrac{15-9}{7} \times 10 = 10 + \dfrac{60}{7} = 10 + 8.57 = 18.57$

Thus, mean deviation about the median is given by

$$\text{M.D.}(M) = \frac{1}{N}\sum_{i=1}^{5} f_i |x_i - M| = \frac{1}{30} \times 373.1 = 12.44$$

OR

Q35. Solution: The given statement is $P(n)$:

$$\frac{1}{\sqrt{1}+\sqrt{2}} + \frac{1}{\sqrt{2}+\sqrt{3}} + \ldots + \frac{1}{\sqrt{n}+\sqrt{n+1}} = \frac{n}{\sqrt{1}+\sqrt{n+1}}$$

For $n = 1$

$$P(1) = \frac{1}{\sqrt{1}+\sqrt{1+1}} = \frac{1}{\sqrt{1}+\sqrt{2}} \text{ is true.}$$

Let $P(k)$ be true for some positive integer k. Then

$$P(k) = \frac{1}{\sqrt{1}+\sqrt{2}} + \frac{1}{\sqrt{2}+\sqrt{3}} + \ldots + \frac{1}{\sqrt{k}+\sqrt{k+1}}$$

$$= \frac{k}{\sqrt{1}+\sqrt{k+1}}$$

Now we shall prove that $P(k+1)$ is true

$$P(k+1) = \frac{1}{\sqrt{1}+\sqrt{2}} + \frac{1}{\sqrt{2}+\sqrt{3}} + \ldots + \frac{1}{\sqrt{k}+\sqrt{k+1}} + \frac{1}{\sqrt{k+1}+\sqrt{k+2}} = \frac{k+1}{\sqrt{1}+\sqrt{k+2}}$$

LHS $= P(k+1)$

$$= \frac{1}{\sqrt{1}+\sqrt{2}} + \ldots + \frac{1}{\sqrt{k}+\sqrt{k+1}} + \frac{1}{\sqrt{k+1}+\sqrt{k+2}}$$

$$= \frac{k}{\sqrt{1}+\sqrt{k+1}} + \frac{1}{\sqrt{k+1}+\sqrt{k+2}} \quad (\because P(k) \text{ is true})$$

$$= \left(\frac{k}{\sqrt{1}+\sqrt{k+1}}\right)\left(\frac{\sqrt{1}-\sqrt{k+1}}{\sqrt{1}-\sqrt{k+1}}\right) + \left(\frac{1}{\sqrt{k+1}+\sqrt{k+2}}\right)\left(\frac{\sqrt{k+1}-\sqrt{k+2}}{\sqrt{k+1}-\sqrt{k+2}}\right)$$

$$= \frac{k(\sqrt{1}-\sqrt{k+1})}{1-(k+1)} + \frac{\sqrt{k+1}-\sqrt{k+2})}{(k+1)-(k+2)}$$

$$= \frac{k - k\sqrt{k+1}}{-k} + \frac{\sqrt{k+1}-\sqrt{k+2})}{-1}$$

$$= \sqrt{k+1} - 1 + \sqrt{k+2} - \sqrt{k+1}$$

$$= \sqrt{k+2} - 1 = \left(\sqrt{k+2}-1\right)\left(\frac{\sqrt{k+2}+1}{\sqrt{k+2}+1}\right)$$

$$= \frac{k+2-1}{\sqrt{k+2}+1} = \frac{k+1}{\sqrt{1}+\sqrt{k+2}} = \text{RHS}$$

$P(k+1)$ is true whenever $P(k)$ is true.

Therefore by the principle of mathematical induction, statement $P(n)$ is true for all $n \in N$

Q36. Solution: $r\cos\theta = 1, \quad r\sin\theta = -1$

On squaring and adding, we obtain

$r^2\cos^2\theta + r^2\sin^2\theta = 1^2 + (-1)^2$

$\Rightarrow r^2(\cos^2\theta + \sin^2\theta) = 1 + 1 \quad (\because \sin^2\theta + \cos^2\theta = 1)$

$\Rightarrow r^2 = 2 \Rightarrow r = \sqrt{2} \quad (\because r > 0)$

Modulus $|z| = \sqrt{2}$

$\because \sqrt{2}\cos\theta = 1, \quad \sqrt{2}\sin\theta = -1$

$\therefore \cos\theta = \dfrac{1}{\sqrt{2}}, \quad \sin\theta = -\dfrac{1}{\sqrt{2}}$

$|\tan\theta| = 1 \Rightarrow \theta = \dfrac{\pi}{4}$

Since $\sin\theta$ is negative and $\cos\theta$ is positive, θ lies in IV quadrant.

Thus, $\text{Arg}(z) = -(\theta) = -\dfrac{\pi}{4}$

Hence, the modulus and argument of the complex number $1 - i$ are $\sqrt{2}$ and $-\dfrac{\pi}{4}$, respectively.

OR

Q36. Solution:

Let $z_1 = r_1(\cos\theta_1 + i\sin\theta_1)$, $z_2 = r_2(\cos\theta_2 + i\sin\theta_2)$ be two complex numbers.

Since $|z_1| = |z_2|$ and $\arg(z_2) - \arg(z_1) = \frac{\pi}{2}$,

, $r_1 = r_2$ and $\theta_2 - \theta_1 = \frac{\pi}{2}$

Now,

RHS $= z_2 = r_2(\cos\theta_2 + i\sin\theta_2)$

$= r_1[\cos(\frac{\pi}{2} + \theta_1) + i\sin(\frac{\pi}{2} + \theta_1)]$ $(\because r_1 = r_2, \theta_2 = \frac{\pi}{2} + \theta_1)$

$= r_1(-\sin\theta_1 + i\cos\theta_1)$

$= r_1(i^2 \sin\theta_1 + i\cos\theta_1)$ $(\because i^2 = -1)$

$= i r_1(i\sin\theta_1 + \cos\theta_1)$

$= i r_1(\cos\theta_1 + i\sin\theta_1) = i z_1 =$ LHS

Sample Paper – VII

Day 233

Time: 3 Hrs Maximum **Marks:** 80

General Instructions:

(i) All the questions are compulsory.

(ii) The question paper consists of 36 questions divided into 4 sections A, B, C and D.

(iii) Section A comprises 20 questions of 1 mark each. Section B comprises 6 questions of 2 marks each. Section C comprises 6 questions of 4 marks each. Section D comprises 4 questions of 6 marks each.

(iv) There is no overall choice. However, an internal choice has been provided in four questions of 1 mark each, two questions of 2 marks each, two questions of 4 marks each, and two questions of 6 marks each. You have to attempt only one of the alternatives in all such questions.

(iv) Use of calculators is not permitted.

SECTION A

Q1–Q10 are multiple choice type questions. Select the correct option. (1 mark each)

Q1. Two finites sets have m and n elements. The number of subsets of the first set is 30 more than of the second set. The values of m and n are, respectively

(a) 2, 5 (b) 5, 1 (c) 3, 4 (d) 4, 4

Q2. If $\tan\alpha = \frac{m+1}{m}$, $\tan\beta = \frac{m-1}{m}$, $\tan(\alpha+\beta) = 2$, then m is equal to

(a) $\pm\sqrt{2}$ (b) ± 1 (c) $\pm\frac{1}{\sqrt{2}}$ (d) 0

Q3. The domain and range of real function f defined by $f(x) = \sqrt{x-1}$ is given by

(a) Domain $= (1, \infty)$, Range $= (0, \infty)$

(b) Domain $= [1, \infty)$, Range $= (0, \infty)$

(c) Domain $= (1, \infty)$, Range $= [0, \infty)$

(d) Domain $= [1, \infty)$, Range $= [0, \infty)$

Q4. If a parallelepiped is formed by planes drawn through the points (–2, 3, 0) and (4, 0, 1) parallel to the coordinates planes, then the length of diagonal of the parallelepiped is

(a) $\sqrt{32}$ (b) $\sqrt{23}$ (c) $\sqrt{46}$ (d) $2\sqrt{33}$

Q5. The approximation of $(1.01)^4$ using the first three terms of its expansion is

(a) 1.0512 (b) 1.0406 (c) 1.5122 (d) 1.4124

Q6. Find the value of $\frac{\sin^2 30^\circ + \cos^2 45^\circ}{\text{cosec}^2 30^\circ - \sec^2 45^\circ}$

(a) $\frac{3}{8}$ (b) $2\sqrt{2}$ (c) $\frac{1}{2}$ (d) 2

Q7. The value of the expansion $2^1 \times 2^{\frac{1}{3}} \times 2^{\frac{1}{9}} \times 2^{\frac{1}{27}} \times \ldots$ is

(a) $\frac{3}{8}$ (b) $2\sqrt{2}$ (c) $\frac{1}{2}$ (d) 2

Q8. Evaluate $\lim_{x\to 0} \frac{\sin 3x}{\sin 6x}$

(a) $\frac{3}{8}$ (b) $2\sqrt{2}$ (c) $\frac{1}{2}$ (d) 2

Q9. Evaluate $\lim_{x\to 0} \frac{2\sin^2 x}{\sin^2 2x}$

(a) $\frac{3}{8}$ (b) $2\sqrt{2}$ (c) $\frac{1}{2}$ (d) 2

Q10. The modulus of $\frac{1+i}{1-i} - \frac{1-i}{1+i}$ is … … …

(a) $\frac{3}{8}$ (b) $2\sqrt{2}$ (c) $\frac{1}{2}$ (d) 2

Q11–Q15 are fill in the blanks. (1 mark each)

Q11. The number of triangles that can be drawn through the vertices of a 10-sided polygon is … … …

Q12. The sum of the coefficient of the two middle terms in the expansion of $(1+x)^{2n+1}$ is equal to … … …

Q13. If $(\sin x - \cos x)^4 + 2(\sin x + \cos x)^2 = 3$, then $x =$ .... where $x \in [0, 2\pi]$

Q14. The equation of a circle which passes through (0, 2) and touches the axes is … … …

Q15. The value of $\lim_{x\to 0} \frac{1-\cos 2x}{x^2}$ is equal to … … …

(Q16–Q20) Answer the following questions. **(1 mark each)**

Q16. Rewrite the following statement in the form "p if and only if q".
p: If you read books, then your mind is intelligent and if your mind is intelligent, then you study books.

Q17. Write down the contrapositive of the following statement: If n is a whole number, then n is an integer.

Q18. Ten chairs are numbered 1 to 10. Three women and 4 men wish to occupy one chair each. First the women choose the chairs from amongst the chairs 1 to 6 and then the men select from the remaining chairs. Find the total number of possible arrangements.

OR

Q18. There are five men and four women on the city council. If one council member is selected for a committee at random, how likely is that it is a woman?

Q19. Find the derivative of $(\operatorname{cosec} x)(x - \cot x)$

OR

Q19. Find the region represented by $|z+i| \geq 1$

Q20. A bag contains five black marbles and six red marbles. Find the number of ways in which four marbles can be drawn from the bag, if they must all be of the same colour.

OR

Q20. Show that the points $A(3, 1)$, $B(4, 2)$ and $C(2, 0)$ are collinear (form a straight line).

SECTION B

(Q21–Q26) Solve the following questions. **(2 marks each)**

Q21. Find the equation of the line which passes through the point $(-4, 3)$ with slope $3/2$

Q22. A letter is chosen at random from the word 'VIKASH'. Find the probability that letter is (i) a vowel (ii) an consonant

Q23. If $R = \{(x, y) : x + y \leq 3,\ x \in W \text{ and } y \in N\}$ then find the domain and range of R.

Q24. Find the domain and the range of the real function f defined by $f(x) = \dfrac{1}{|x(x-1)|}$.

Q25. Find the equation of the ellipse that satisfies the given conditions: length of the minor axis 16, foci $(0, \pm 6)$

OR

Q25. Solve the given inequality for real x:
$$\frac{2(x-1)}{5} \leq \frac{5(1-x)}{2}$$

Q26. If the letters of the word 'EIGHT' are arranged in all possible ways as listed in the dictionary, what is the rank of the word 'EIGHT'?

OR

Q26. If $f(x) = \begin{cases} x-1, & x \leq 1 \\ cx^2, & x > 1 \end{cases}$, then find c when $\lim_{x \to 1} f(x)$ exists.

SECTION C

(Q27–Q32) Solve the following questions. **(4 marks each)**

Q27. $A = \{1, 2, 3\}$, $B = \{2, 3, 4\}$ and $C = \{0, 1, 4\}$. Then prove that

(i) $A - (B \cup C) = (A - B) \cap (A - C)$

(ii) $A - (B \cap C) = (A - B) \cup (A - C)$

Q28. Find the square root of $3 - 4i$

Q29. Three vertices of a parallelogram $ABCD$ are $A(3, -1, 2)$, $B(2, 1, 0)$, $C(-1, 1, 2)$. Find the coordinates of the fourth vertex.

OR

Q29. Write the complex number $z = \dfrac{1-i}{\cos\dfrac{\pi}{4} + i\sin\dfrac{\pi}{4}}$ in polar form.

Q30. If $A(-5, 0)$, $B(-4, -5)$, $C(-1, -3)$, $D(4, 5)$ are the vertices of a quadrilateral, find the area of the quadrilateral $ABCD$.

OR

Q30. The longest side of a rectangle is 10 cm longer than the shortest side. If the perimeter of the rectangle is more than 100 cm and less than 120 cm, then find the minimum and maximum length of the shortest side.

Q31. Find the coordinates of the focus, axis of the parabola, the equation of directrix and the length of the latus rectum for $y^2 = 20x$

OR

Q31. Find a GP for which sum of the first two terms is -4 and the fifth term is 9 times the third term.

Q32. If the term free from x in the expansion of $\left(\sqrt{x}+\frac{k}{x}\right)^9$ is 84, then find the value of k.

SECTION D

(Q33–Q36) Solve the following questions. **(6 marks each)**

Q33. Find the standard deviation and variance for the data 5, 8, 6, 9, 7

Q34. Find the points on the x-axis whose distances from the line $\frac{x}{3}+\frac{y}{4}=1$ are 4 units.

OR

Q34. n^3-n is divisible by 3, for all natural numbers n.

Q35. Three consecutive vertices of a parallelogram $ABCD$ are $A(6,-2,4), B(2,4,1), C(-2,2,4)$. Find the coordinates of the fourth vertex.

OR

Q35. Insert two numbers between 27 and 64 so that the resulting sequence is a GP.

Q36. Evaluate $\lim_{x\to 2}\frac{x-2}{\sqrt{3x-2}-\sqrt{x+2}}$

OR

Q36. Find the equation of the line parallel to the line $3x-4y+2=0$ and passing through the point (2, –3).

Day 234

Sample Paper – VII (Solutions)

SECTION A

Q1–Q10 are multiple choice type questions. Select the correct option. **(1 mark each)**

Q1. Solution: (b)

The number of subsets of a set containing m elements is 120 more than the subsets of the set containing n elements. Then

$\because 2^m-2^n=30 \Rightarrow 2^n\cdot(2^{m-n}-1)=2\cdot 15$

$\Rightarrow 2^n=2^1$ and $2^{m-n}-1=15$

$\Rightarrow n=1$ and $2^{m-n}=16=2^4$

$\Rightarrow m-n=4 \Rightarrow m-1=4 \Rightarrow m=5$.

Q2. Solution: (b)

$$\tan(\alpha+\beta)=\frac{\tan\alpha+\tan\beta}{1-\tan\alpha\cdot\tan\beta}$$

$$\Rightarrow 2=\frac{\frac{m+1}{m}+\frac{m-1}{m}}{1-\left(\frac{m+1}{m}\right)\left(\frac{m-1}{m}\right)}$$

$$\Rightarrow 2-2\left(\frac{m+1}{m}\right)\left(\frac{m-1}{m}\right)=\frac{m+1}{m}+\frac{m-1}{m}$$

$$\Rightarrow 2-2\left(\frac{m^2-1}{m^2}\right)=\frac{2m}{m}\Rightarrow 1-\left(\frac{m^2-1}{m^2}\right)=1$$

$$\Rightarrow m^2-1=0 \Rightarrow m=\pm 1$$

Q3. Solution: (b)

$f(x)=\sqrt{x-1}$ is defined, if $x-1\geq 0 \Rightarrow x\geq 1$

$\therefore$ Domain of $f=[1,\infty)$

Let $f(x)=\sqrt{x-1}=y$

x is defined for all real values of y

But $y=\sqrt{x-1}$ gives only non-negative values. Therefore, range of $f=[0,\infty)$

Q4. Solution: (c)

The parallelepiped is formed by planes drawn through the points (–2, 3, 0) and (4, 0, 1) parallel to the coordinates planes. Therefore, the length of the diagonal of the parallelepiped is

$=\sqrt{(-2-4)^2+(3-0)^2+(0-1)^2}$

$=\sqrt{36+9+1}=\sqrt{46}$

Q5. Solution: (b)

We can write $1.01=1+0.01$

$\therefore (1.01)^4=(1+0.01)^4$

$={}^4C_0(1)^4+{}^4C_1(1)^3(0.01)+{}^4C_2(1)^2(0.01)^2+...$

(Approximately neglecting smaller value terms)

$=1+4(0.01)+6(0.01)^2=1+0.04+0.0006=1.0406$

Thus, the value of $(1.01)^4$ is approximately 1.0406.

Q6. Solution: (a)

$$\frac{\sin^2 30^\circ+\cos^2 45^\circ}{\text{cosec}^2 30^\circ-\sec^2 45^\circ}=\frac{(1/2)^2+(1/\sqrt{2})^2}{(2)^2-(\sqrt{2})^2}$$

$$=\frac{1/4+1/2}{4-2}=\frac{3/4}{2}=\frac{3}{8}$$

Q7. Solution: (b)

$\text{LHS} = 2^1 \times 2^{\frac{1}{3}} \times 2^{\frac{1}{9}} \times 2^{\frac{1}{27}} \times .... = 2^{1+\frac{1}{3}+\frac{1}{9}+\frac{1}{27}+...}$

The power of 3 forms a GP with $a = 1, r = \frac{1}{3}$

$\therefore \text{LHS} = 2^{1+\frac{1}{3}+\frac{1}{9}+\frac{1}{27}+...} = 2^{\left(\frac{1}{1-1/3}\right)} = 2^{\left(\frac{3}{2}\right)} = 2\sqrt{2}$

$(\because S_\infty = 1 + \frac{1}{3} + \frac{1}{9} + \frac{1}{27} + ... = \frac{a}{1-r} = \frac{1}{1-1/3} = 3/2)$

Q8. Solution: (c)

$$\lim_{x\to 0} \frac{\frac{\sin 3x}{3x}\cdot 3x}{\frac{\sin 6x}{6x}\cdot 6x} = \frac{\lim_{x\to 0}\frac{\sin 3x}{3x}}{\frac{\sin 6x}{6x}}\cdot\frac{3x}{6x}$$

$$= \frac{3}{6}\cdot\frac{\lim_{x\to 0}\frac{\sin 3x}{3x}}{\lim_{x\to 0}\frac{\sin 6x}{6x}}\left[\because \lim_{x\to 0}\frac{\sin x}{x} = 1\right]$$

$= \frac{1}{2}$ $[\because x \to 0 \Rightarrow (kx \to 0)$, where k is a real number]

Q9. Solution: (c)

$$\lim_{x\to 0}\frac{2\sin^2 x}{\sin^2 2x} = \lim_{x\to 0}\frac{2\sin^2 x}{[\sin(2x)]^2}$$

$$= \lim_{x\to 0}\frac{2\sin^2 x}{(2\sin x\cos x)^2} = \lim_{x\to 0}\frac{2\sin^2 x}{4\sin^2 x\cos^2 x}$$

$[\because \sin 2\theta = 2\sin\theta\cos\theta]$

$$= \lim_{x\to 0}\frac{1}{2\cos^2 x} = \frac{1}{2}[\because \cos 0 = 1]$$

Q10. Solution: (d)

$$\left|\frac{1+i}{1-i} - \frac{1-i}{1+i}\right| = \left|\frac{(1+i)^2-(1-i)^2}{(1-i)(1+i)}\right|$$

$$= \left|\frac{1+i^2+2i-1-i^2+2i}{1^2+1^2}\right| = \left|\frac{4i}{2}\right| = |2i| = 2$$

Q11–Q15 are fill in the blanks. (1 mark each)

Q11. Solution: For drawing one triangle of a polygon, 3 points are required.
Therefore, number of diagonals that can be drawn through 10 vertices of a polygon = ${}^{10}C_3 = 120$

Q12. Solution: The expansion $(1+x)^{2n+1}$ has two middle terms. Therefore $\left(\frac{2n+1+1}{2}\right)$ and $\left(\frac{2n+1+1}{2}+1\right)$, that is, $(n+1)^{\text{th}}$ term and $(n+2)^{\text{th}}$ terms are the middle terms.

Thus, coefficient of the $(n+1)^{\text{th}}$ term = ${}^{2n+1}C_{n+1}$

Coefficient of the $(n+2)^{\text{th}}$ term = ${}^{2n+1}C_{n+2}$

Hence, sum of the coefficients = ${}^{2n+1}C_{n+1} + {}^{2n+1}C_{n+2}$

$= {}^{2n+2}C_{n+2} = {}^{2n+2}C_n$

$(\because {}^nC_r + {}^nC_{r-1} = {}^{n+1}C_r$ and ${}^nC_r = {}^nC_{n-r})$

Q13. Solution:

$(\sin x - \cos x)^4 + 2(\sin x + \cos x)^2 = 3$

$\Rightarrow [\sin^2 x + \cos^2 x - 2\sin x\cos x]^2$

$+2[\sin^2 x + \cos^2 x + 2\cdot\sin x\cos x] = 3$

$\Rightarrow (1-\sin 2x)^2 + 2(1+\sin 2x) = 3$

$\Rightarrow 1 + \sin^2 2x - 2\sin 2x + 2 + 2\sin 2x = 3$

$\Rightarrow \sin^2 2x = 0 \Rightarrow \sin 2x = 0$

$\Rightarrow 2x = 0, \pi, 2\pi \Rightarrow x = 0, \frac{\pi}{2}, \pi$

Q14. Solution: Let the centre of the circle be (a, a). Then equation of the circle is $(x-a)^2 + (y-a)^2 = a^2$

Since, the point (0, 2) lies on this circle $(0-a)^2 + (2-a)^2 = a^2$

$\Rightarrow a^2 + 4 - 4a + a^2 = a^2$

$\Rightarrow a^2 - 4a + 4 = 0 \Rightarrow (a-2)^2 = 0 \Rightarrow a = 2$

Thus, the equation of circle is $(x-1)^2 + (y-1)^2 = 1$ and $(x-5)^2 + (y-5)^2 = 25$

Q15. Solution: $\lim_{x\to 0}\frac{1-\cos 2x}{x^2} = \lim_{x\to 0}\frac{1-1+2\sin^2 x}{x^2}$

$[\because \cos 2x = 1 - 2\sin^2 x]$

$$= \lim_{x\to 0}\frac{2\sin^2 x}{x^2} = 2\lim_{x\to 0}\frac{\sin^2 x}{x^2} = 2\lim_{x\to 0}\left(\frac{\sin x}{x}\right)^2 = 2$$

$\left[\because \lim_{x\to 0}\frac{\sin x}{x} = 1\right]$

(Q16–Q20) Answer the following questions. (1 mark each)

Q16. Solution: You study books if and only if your mind is intelligent.

Q17. Solution: If n is not a whole number, then it is not a natural number.

Q18. Solution: Women choose the chairs first from among 1 to 6 chairs, and there are 3 women. Therefore, the number of arrangements = ${}^6P_3$ ways.

Now, arrangement of 4 men out of remaining 7 chairs is ${}^7P_4$ ways.

Thus, total number of possible arrangements is

$$= {}^6P_3 \times {}^7P_4 = \frac{6!}{(6-3)!}\times\frac{7!}{(7-4)!} = \frac{6!}{3!}\times\frac{7!}{3!} = 100800$$

OR

Q18. Solution: One council member is to be selected for a committee at random. Therefore, the sample space $n(S) = 5+4 = 9$(three men and four women)

Let A be the event in which the selected council member is a women. Therefore $n(A) = 4$

$$P(A) = \frac{n(A)}{n(S)} = \frac{4}{9}$$

Q19. Solution: Let $f(x) = (\text{cosec}\, x)(x - \cot x)$

By the product rule

$$f'(x) = (\text{cosec}\, x)\frac{d}{dx}(x - \cot x) + (x - \cot x)\frac{d}{dx}(\text{cosec}\, x)$$

$$= (\text{cosec}\, x)(1 + \text{cosec}^2 x) + (x - \cot x)(-\text{cosec}\, x \cot x)$$

$$= (\text{cosec}\, x)(1 + \text{cosec}^2 x) - (x - \cot x)(\text{cosec}\, x \cot x)$$

OR

Q19. Solution:

$|z+i| \geq 1 = |x+iy+i| \geq 1$

$\Rightarrow |x+i(y+1)| \geq 1 \Rightarrow \sqrt{x^2+(y+1)^2} \geq 1$

$\Rightarrow x^2+(y+1)^2 \geq 1 \Rightarrow x^2+y^2+2y \geq 0$

This represents a circle, on or outside, having its centre at (0, –1) and radius 1.

Q20. Solution: The marbles must all be of the same colour. Therefore, selection of 4 marbles out of 5 black marbles is ${}^5C_4$ and selection of 4 marbles out of 6 red marbles is ${}^6C_4$

$\therefore$ Total number of ways = ${}^5C_4 + {}^6C_4 = 20$

OR

Q20. Solution: Using the distance formula, we get

$AB = \sqrt{(4-3)^2+(2-1)^2} = \sqrt{1+1} = \sqrt{2}$

$BC = \sqrt{(2-4)^2+(0-2)^2} = \sqrt{4+4} = 2\sqrt{2}$

$AC = \sqrt{(2-3)^2+(0-1)^2} = \sqrt{1+1} = \sqrt{2}$

Now, since, $AB + AC = \sqrt{2}+\sqrt{2} = 2\sqrt{2} = BC$, we can say that the points A, B and C are on a straight line and thus collinear.

SECTION B

(Q21–Q26) Solve the following questions.

(2 marks each)

Q21. Solution: The equation of the line passes through point (x_0, y_0) whose slope is m, is $(y - y_0) = m(x - x_0)$

Therefore, the equation of the line passing through point (–4, 3) whose slope is 3/2 is

$(y-3) = \frac{3}{2}(x+4) \Rightarrow 2(y-3) = 3x+12$

$\Rightarrow 3x - 2y + 18 = 0$

Q22. Solution: There are 6 letters in the word 'VIKASH', $\therefore P(S) = 6$

(i) There are 2 vowels in the given word.

$\therefore$ Probability (vowel) = 2/6=1/3

(ii) There are 4 consonant in the given word.

$\therefore$ Probability (consonant) = 4/6=2/3

Q23. Solution: The set of ordered pairs satisfying $R = \{(x, y) : x + y \leq 3,\ x \in W \text{ and } y \in N\}$ is

$= \{(0,1), (0,2), (0,3), (1,1), (1,2), (2,1)\}$

Domain of R = Set of first elements of ordered pairs in $R = \{0,1,2,3\}$

Range of R = Set of second elements of ordered pairs in $R = \{1,2,3\}$

Q24. Solution: Function $f(x) = \frac{1}{|x(x-1)|}$ is defined for all real numbers except when $x(x-1) = 0 \Rightarrow x = 0, 1$

$\therefore$ Domain of $f = R - \{0,1\}$

Also since $|x(x-1)|$ will take all non-negative real numbers, the range of f is the set of all positive real numbers, that is, range of $f = (0, \infty)$.

Q25. Solution: Length of minor axis, $2b = 16 \Rightarrow b = 8$

The foci $(0, \pm 6)$ are on the y-axis. Therefore, the major axis is along the y-axis and $c = 6$.

Now, from the equation $a^2 = b^2 + c^2$

$\Rightarrow a^2 = 8^2 + 6^2 \Rightarrow a^2 = 100 \Rightarrow a = 10$

Thus, the equation of the ellipse will be of the form $\frac{x^2}{b^2} + \frac{y^2}{a^2} = 1 \Rightarrow \frac{x^2}{8^2} + \frac{y^2}{(10)^2} = 1 \Rightarrow \frac{x^2}{64} + \frac{y^2}{100} = 1$

OR

Q25. Solution:

$\frac{2(x-1)}{5} > \frac{5(1-x)}{2} \Rightarrow 4(x-1) > 25(1-x)$

$\Rightarrow 4x - 4 > 25 - 25x \Rightarrow 4x + 25x > 25 + 4$

$\Rightarrow 29x > 29 \Rightarrow x > 1$

Therefore, all real numbers x, which are greater than 1, are the solutions of the given inequality. Thus, the solution set of the given inequality is $x \in (1, \infty)$.

Q26. Solution: The letters of the word 'EIGHT' in alphabetical order are E, G, H, I, T.

Words starting with EG = 3!

Word starting with EH = 3!
Next word will be EIGHT = 1
∴ Rank of the word 'EIGHT' in dictionary is
$= 3! + 3! + 1 = 6 + 6 + 1 = 13$

OR

Q26. Solution: Given that $f(x) = \begin{cases} x-1, & x \le 1 \\ cx^2, & x > 1 \end{cases}$,

$\text{LHL} = \lim_{x\to 1^-} f(x) = \lim_{x\to 1^-} (x-1)$

$= \lim_{h\to 0}(1-h-1) = \lim_{h\to 0}(-h) = 0$

$\text{RHL} = \lim_{x\to 1^+} f(x) = \lim_{x\to 1^+} cx^2 = \lim_{h\to 0} c(1+h)^2 = c$

Since, $\lim_{x\to 1} f(x)$ exist, then LHL = RHL $\Rightarrow c = 0$

SECTION C

(Q27–Q32) Solve the following questions. (4 marks each)

Q27. Solution: $A - B = \{1,2,3\} - \{2,3,4\} = \{1\}$

$A - C = \{1,2,3\} - \{0,1,4\} = \{2,3\}$

$B \cap C = \{2,3,4\} \cap \{0,1,4\} = \{4\}$

$B \cup C = \{2,3,4\} \cup \{0,1,4\} = \{0,1,2,3,4\}$

Thus,

(i) For $A-(B\cup C) = (A-B)\cap(A-C)$

LHS: $A-(B\cup C) = \{1,2,3\} - \{0,1,2,3,4\} = \{\phi\}$

RHS: $(A-B)\cap(A-C) = \{1\}\cap\{2,3\} = \{\phi\}$

(ii) For $A-(B\cap C) = (A-B)\cup(A-C)$

LHS: $A-(B\cap C) = \{1,2,3\} - \{4\} = \{1,2,3\}$

RHS: $(A-B)\cup(A-C) = \{1\}\cup\{2,3\} = \{1,2,3\}$

Thus,LHS = RHS, verify above two results.

Q28. Solution: Let $x+iy = \sqrt{a+ib}$, where $a = 3$ and $b = -4$.

Then, $\text{Re}(z) = a = 3$ and

$|z| = \sqrt{a^2+b^2} = \sqrt{(3)^2+(-4)^2} = 5$

The product xy (or b) is negative. Therefore we take the value of x as positive and y as negative, or vice-versa.

Thus, square root of $3-4i$ is given by $\sqrt{3-4i}$

$= \pm\sqrt{\frac{|z|+\text{Re}(z)}{2}} \mp i\sqrt{\frac{|z|-\text{Re}(z)}{2}}$

$= \pm\sqrt{\frac{5+3}{2}} \mp i\sqrt{\frac{5-3}{2}}$

$= \pm\sqrt{4} \mp i\sqrt{1} = \pm 2 \mp i$

$= (2-i, -2+i)$

Q29. Solution: Let the coordinates of the fourth vertex of parallelogram $ABCD$ be $D(x,y,z)$..
The diagonals of a parallelogram bisect each other. Therefore, in parallelogram $ABCD$, diagonals AC and BD bisect each other.
Thus, mid-point of AC = Mid-point of BD

$\Rightarrow \left(\frac{3-1}{2}, \frac{-1+1}{2}, \frac{2+2}{2}\right) = \left(\frac{x+2}{2}, \frac{y+1}{2}, \frac{z+0}{2}\right)$

$\Rightarrow (1,0,2) = \left(\frac{x+2}{2}, \frac{y+1}{2}, \frac{z}{2}\right)$

$\Rightarrow \frac{x+2}{2} = 1, \frac{y+1}{2} = 0, \frac{z}{2} = 2$

$\Rightarrow x = 0, y = -1, z = 4$

Hence, the coordinates of the fourth vertex are $D(0, -1, 4)$.

OR

Q29. Solution:

$z = \dfrac{1-i}{\cos\frac{\pi}{4} + i\sin\frac{\pi}{4}}$

$= \dfrac{\sqrt{2}\left[\frac{1}{\sqrt{2}} - i\frac{1}{\sqrt{2}}\right]}{\cos\frac{\pi}{4} + i\sin\frac{\pi}{4}}$

$= \dfrac{\sqrt{2}[\cos(-\pi/4) + i\sin(-\pi/4)]}{\cos\pi/4 + i\sin\pi/4}$

$= \sqrt{2}\left[\cos\left(\frac{-\pi}{4} - \frac{\pi}{4}\right) + i\sin\left(\frac{-\pi}{4} - \frac{\pi}{4}\right)\right]$

$= \sqrt{2}\left[\cos\left(\frac{-2\pi}{4}\right) + i\sin\left(\frac{-2\pi}{4}\right)\right]$

$= \sqrt{2}\left[\cos\left(\frac{\pi}{2}\right) - i\sin\left(\frac{\pi}{2}\right)\right]$

$= \sqrt{2}\left[-\cos\left(\pi - \frac{\pi}{2}\right) - i\sin\left(\pi - \frac{\pi}{2}\right)\right]$

$= -\sqrt{2}\left[\cos\frac{\pi}{2} + i\sin\frac{\pi}{2}\right]$

Q30. Solution: By joining $B$and D, we get two triangles ABD and BCD.

Now, area of ΔABD

$= \frac{1}{2}[-5(-5-5) + (-4)(5-0) + 4(0+5)]$

$= \frac{1}{2}(50 - 20 + 20) = 25$ square units

And area of ΔBCD

$= \frac{1}{2}[-4(-3-5) - 1(5+5) + 4(-5+3)]$

$= \frac{1}{2}(32 - 10 - 8) = 7$ square units

Thus, the area of quadrilateral $ABCD$ = area of ΔABD+ area of ΔBCD= 25 + 7 = 32 square units.

OR

Q30. Solution: Let the length of the shortest side be x cm.

Longest side = $(x + 10)$ cm

Perimeter of rectangle $= 2 \times (x + (x + 10)) = 4x + 20$

According to the question,100 < Perimeter < 120

$\Rightarrow 100 < 4x + 20 < 120$

$\Rightarrow 80 < 4x < 120$

$\Rightarrow 20 < x < 30$

Thus, the minimum length of the shortest side is 20 cm and the maximum side is 30 cm.

Q31. Solution: In the given equation, $y^2 = 20x$ of the parabola, the coefficient of x is positive. Therefore, the parabola opens towards the right.

Comparing this equation with $y^2 = 4ax$, we get $4a = 20 \Rightarrow a = 5$

Thus, coordinates of the focus = $(a, 0) = (5, 0)$

The given equation involves y^2. Therefore, the axis of the parabola is the x-axis.

Equation of the directrix, $x = -a \Rightarrow x = -5 \Rightarrow x + 5 = 0$

Length of the latus rectum $= 4a = 4 \times 5 = 20$

OR

Q31. Solution: Let a be the first term and r be the common ratio of the GP. Then $S_n = \frac{a(1-r^n)}{1-r}$

The fifth term of the GP is 4 times the third term.

$\therefore a_5 = 9 \times a_3 \Rightarrow ar^4 = 9ar^2 \Rightarrow r^2 = 9 \Rightarrow r = \pm 3$

Since sum of the first two terms of a GP is – 4,

$S_2 = -4 = \frac{a(1-r^2)}{1-r}$

For $r = 3$

$\Rightarrow -4 = \frac{a[1-(3)^2]}{1-3} \Rightarrow -4 = \frac{a(1-9)}{-2} \Rightarrow a = \frac{8}{-8} = -1$

For $r = -3$

$-4 = \frac{a[1-(-3)^2]}{1-(-3)} \Rightarrow -4 = \frac{a(1-9)}{1+3} \Rightarrow a = 2$

Thus, the required GP is

$-1, -3, -9, -27, -81...$ or $2, -6, 18, -54, 162, ...$

Q32. Solution: The general term of the given expansion $\left(\sqrt{x} + \frac{k}{x}\right)^9$ is

$T_{r+1} = {}^9C_r(\sqrt{x})^{9-r}\left(\frac{k}{x}\right)^r = {}^9C_r(k)^r x^{\frac{9-3r}{2}}$

For the term free from x, $\frac{9-3r}{2} = 0 \Rightarrow r = 3$

Thus, $T_{3+1} = T_4$ term is free from x.

The term free from x is equal to 84. Therefore

$T_{3+1} = {}^9C_3(k)^3 = 84 \Rightarrow \frac{9 \cdot 8 \cdot 7}{3 \cdot 2}(k)^3 = 84$

$\Rightarrow 84k^3 = 84 \Rightarrow k^3 = 1 \Rightarrow k = 1$

SECTION D

(Q33–Q36) Solve the following questions.

(6 marks each)

Q33. Solution: Mean, $\bar{x} = \frac{\sum_{i=1}^{5} x_i}{n}$

$= \frac{5+8+6+9+7}{5} = \frac{35}{5} = 7$

The following table is obtained as:

x_i	$(x_i - \bar{x})$	$(x_i - \bar{x})^2$
5	–2	4
8	3	9
6	–1	1
9	2	4
7	0	0
		18

Variance $(\sigma^2) = \frac{1}{n}\sum_{i=1}^{5}(x_i - \bar{x})^2 = \frac{1}{5} \times 18 = 3.6$

Standard deviation $(\sigma) = \sqrt{3.6} = 1.9$

Q34. Solution: The given equation of the line is $\frac{x}{3} + \frac{y}{4} = 1$

$\Rightarrow 4x + 3y - 12 = 0$

Comparing the above equation with the general equation of line $Ax + By + C = 0$, we get $A = 4$, $B = 3$, $C = -12$.

Let $(a, 0)$ be the point on the x-axis whose distance from the given line is 4 units.

The perpendicular distance (p) of a line $Ax + By + C = 0$ from a point (x_1, y_1) is given by

$\frac{|Ax_1 + By_1 + C|}{\sqrt{A^2 + B^2}}$

Therefore $4 = \frac{|4a + 3 \times 0 - 12|}{\sqrt{4^2 + 3^2}}$

$\Rightarrow 4 = \frac{|4a-12|}{5} \Rightarrow |4a-12| = 20 \Rightarrow \pm(4a-12) = 20$

$\Rightarrow (4a-12) = 20 \text{ or } -(4a-12) = 20$

$\Rightarrow 4a = 20+12 \text{ or } -4a = 20-12 \Rightarrow a = 8 \text{ or } -2$

Thus, the required points on the x-axis are $(8,0),(-2,0)$.

OR

Q34. Solution: The statement $P(n): n^3 - n$ is divisible by 3 for all natural number n.

For n = 1

$P(1) = (1)^3 - 1 = 0$ is divisible by 3, thus P(1) is true.

Let $P(k)$ be true for some positive integer k. Then $k^3 - k$ is divisible by 3. Thus

$k^3 - k = 3m$, where $m \in N$

Now we shall prove that $P(k+1)$ is true.

$(k+1)^3 - (k+1)$

$= k^3 + 1 + 3k^2 + 3k - k - 1$

$= k^3 + 3k^2 + 2k$

$= 3m + k + 3k^2 + 2k$ $(\because P(k) \text{ is true})$

$= 3m + 3k^2 + 3k$

$= 3(m + k^2 + 1)$ is divisible by 3

$P(k+1)$ is true whenever $P(k)$ is true.

Therefore, by the principle of mathematical induction, statement $P(n)$ is true for all $n \in N$

Q35. Solution: Let the coordinates of the fourth vertices be $D(x, y, z)$.

Mid-point of diagonal AC

$x = \frac{x_1 + x_2}{2} = \frac{6-2}{2} = 2$

$y = \frac{y_1 + y_2}{2} = \frac{-2+2}{2} = 0$

$z = \frac{z_1 + z_2}{2} = \frac{4+4}{2} = 4$

And the mid-point of diagonal i is

$\left(\frac{x+2}{2}, \frac{y+4}{2}, \frac{z+1}{2}\right)$

Since the mid-points of the diagonals of a parallelogram coincide at a point

$\left(\frac{x+2}{2}, \frac{y+4}{2}, \frac{z+1}{2}\right) = (2,0,4)$

$\Rightarrow x = 2,\ y = -4 \text{ and } z = 7$

Thus, the coordinates of the fourth vertex D is (2, –4, 7)

OR

Q35. Solution: Let G_1 and G_2 be two numbers between 27 and 64 such that the series 27, G_1, G_2, 64 forms a GP.

Let a be the first term and r be the common ratio of the GP. Then $a = 27$ and $64 = (27)(r)^3 \Rightarrow r^3 = \frac{64}{27} \Rightarrow r = \frac{4}{3}$

Thus, $G_1 = ar = 27 \times \frac{4}{3} = 36$

$G_2 = ar^2 = 27 \times \left(\frac{4}{3}\right)^2 = 48$

Thus, the two numbers are 36 and 48.

Q36. Solution:

$\lim_{x\to 2} \frac{x-2}{\sqrt{3x-2} - \sqrt{x+2}}$

$= \lim_{x\to 2} \frac{(x-2)(\sqrt{3x-2} + \sqrt{x+2})}{(\sqrt{3x-2} - \sqrt{x+2})(\sqrt{3x-2} - \sqrt{x+2})}$

$= \lim_{x\to 2} \frac{(x-2)(\sqrt{3x-2} + \sqrt{x+2})}{(\sqrt{3x-2})^2 - (\sqrt{x+2})^2}$

$[\because (a+b)(a-b) = a^2 - b^2]$

$= \lim_{x\to 2} \frac{(x-2)\sqrt{3x-2} + \sqrt{x+2}}{(3x-2) - (x+2)}$

$= \lim_{x\to 2} \frac{(x-2)(\sqrt{3x-2} + \sqrt{x+2})}{3x-2-x-2}$

$= \lim_{x\to 2} \frac{(x-2)(\sqrt{3x-2} + \sqrt{x+2})}{2x-4}$

$= \lim_{x\to 2} \frac{(x-2)(\sqrt{3x-2} + \sqrt{x+2})}{2(x-2)}$

$= \lim_{x\to 2} \frac{(\sqrt{3x-2} + \sqrt{x+2})}{2}$

$= \frac{(\sqrt{6-2} + \sqrt{2+2})}{2} = \frac{(2+2)}{2} = 4$

OR

Q36. Solution: The equation of the given line is $3x - 4y + 2 = 0$

$\Rightarrow y = \frac{3}{4}x + \frac{1}{2}$

Comparing with the slope intercept form $y = mx + c$, we get slope of the given line = 3/4

Thus, slope of the other parallel lines is $= m = \frac{3}{4}$

Now, the equation of the line that has a slope of 3/4 and passes through the point (2, –3) is

$(y - (-3)) = \frac{3}{4}\{x - 2\} \Rightarrow 4y + 12 = 3x - 6$

$\Rightarrow 3x - 4y - 18 = 0$

Sample Paper – VIII

Day 235

Time: 3 Hrs Maximum **Marks:** 80

General Instructions:

(i) All the questions are compulsory.

(ii) The question paper consists of 36 questions divided into 4 sections A, B, C and D.

(iii) Section A comprises 20 questions of 1 mark each. Section B comprises 6 questions of 2 marks each. Section C comprises 6 questions of 4 marks each. Section D comprises 4 questions of 6 marks each.

(iv) There is no overall choice. However, an internal choice has been provided in four questions of 1 mark each, two questions of 2 marks each, two questions of 4 marks each, and two questions of 6 marks each. You have to attempt only one of the alternatives in all such questions.

(iv) Use of calculators is not permitted.

SECTION A

Q1–Q10 are multiple choice type questions. Select the correct option. (1 mark each)

Q1. If the expression $4^{16} - 6^{14}$ is divided by 25, then the remainder is

(a) 0 (b) 1 (c) 2 (d) 5

Q2. The equation of the circle with centre on the Y-axis and passing through the origin and the point (2, 3) is

(a) $2x^2 + 2y^2 - 13x = 0$

(b) $3x^2 + 3y^2 - 13y = 0$

(c) $6x^2 + 6y^2 - 13y = 0$

(d) $x^2 + y^2 + 13x + 3 = 0$

Q3. If $X = \{6n - 1 \mid n \in N\}$, and $Y = \{6n + 1 \mid n \in N\}$, then

(a) $X \subset Y$ (b) $Y \subset X$

(c) $X = Y$ (d) $X \cap Y \neq \phi$

Q4. Domain of the function f given by $f(x) = \dfrac{x-1}{x(x+1)}$

(a) $f = R - \{-1, 1\}$ (b) $R - (-1, 1)$

(c) $R - [-1, 1]$ (d) $R - \{-1, 0\}$

Q5. If $\tan\theta = 2$, θ lies in the first quadrant, then $\cos 2\theta + \sin 2\theta$ is equal to

(a) $\dfrac{1}{15}$ (b) $\dfrac{2}{15}$ (c) $\dfrac{1}{5}$ (d) $\dfrac{2}{5}$

Q6. The value of $\sin 390° - \cos 1330°$ is

(a) 0 (b) 1 (c) 2 (d) 5

Q7. The value of $(1-i)^5 \left(1 - \dfrac{1}{i}\right)^5$ is

(a) 0 (b) 1 (c) 32 (d) 49

Q8. The maximum distance of a point on the graph of the function $y = \dfrac{\sin x + \cos x}{\sqrt{2}}$ from the x-axis is

(a) 0 (b) 1 (c) 2 (d) 5

Q9. Express the given complex number in the form $a + ib : (1-i)^{12}$

(a) $-32 + 0i$ (b) $32 + 0i$

(c) $0 + 32i$ (d) $0 - 32i$

Q10. The last two digits of the numbers 49^{35} are

(a) 0 (b) 1 (c) 32 (d) 49

Q11–Q15 are fill in the blanks. (1 mark each)

Q11. The y-axis and z-axis taken together determine a plane known as______________

Q12. The coordinates of points in the YZ-plane are of the form__________

Q13. Summation of series $-3, 1, \dfrac{-1}{3}, \ldots\ldots$ up to infinite terms is … … …

Q14. The largest coefficient in the expansion of $(1+x)^{22}$ is………

Q15. The coefficient of x^2 in the expansion of $(1 - 2x + 3x^2)(1-x)^{10}$ is … … …

(Q16–Q20) Answer the following questions. (1 mark each)

Q16. Form the bi-conditional statement $p \leftrightarrow q$ where

p: The units digit of an integer is zero or 5.

q: It is divisible by 5.

Q17. Find the component statements of the following compound statement: Number 2 is prime and natural.

Q18. A fair coin is tossed three times, and a person will win Rs 2 for each head and lose Re 1 for each tail that turns up. Find the probability of a person winning Rs. 3.

Q19. If *A*, *B*, *C* be the feet of the perpendicular from a point *P* on the *XY*-, *YZ*- and *ZX*-planes, respectively, then find the coordinates of *A*, *B* and *C* for the point *P* (–1, 3, 4)

Q20. Find the equation of the parabola that satisfies the following conditions: focus (0, –3); directrix $y = 3$

SECTION B

(Q21–Q26) Solve the following questions.

(2 marks each)

Q21. In an entrance test that is graded on the basis of two examinations, the probability of a randomly chosen student passing the first examination is 0.6 and the probability of thr student passing the second examination is 0.6. The probability of passing at least one of them is 0.9. What is the probability of the student passing both exams?

Q22. Find the domain and the range of the real function *f* defined by $f(x) = \sqrt{x(x-1)}$

Q23. We wish to select 4 persons from 6 persons, but if person A is chosen, then B must not be chosen. In how many ways can the selection be made?

Q24. In how many ways can a cricket team of 11 players be selected from 15 players? How many of them will

(i) Include 2 particular players?

(ii) Exclude 2 particular players?

OR

Q24. Find the equation of the ellipse that satisfies the given conditions: foci $(\pm 2, 0)$, $a = 6$

OR

Q25. Find the complex number satisfying the equation $|(z+i)| + z - 1 = 0$.

OR

Q25. Find the sum of all natural numbers from 40 to 1000, which are multiples of 4.

Q26. Find a point on the *y*-axis which is equidistant from the points *A*(–2, 1) and *B*(4, –3).

OR

Q26. If $R_1 = \{(x, y) | y = 2x - 1, \text{ where } x \in R, -1 \le x \le 1\}$ is a relation, then find the range of R_1.

SECTION C

(Q27–Q32) Solve the following questions.

(4 marks each)

Q27. Solve the given inequality for real x:

$$\frac{2x}{15} \le \frac{(5x-2)}{3} - \frac{(7x-3)}{5}$$

Q28. Find all pairs of consecutive even positive integers, both of which are smaller than or equal to 12, such that their sum is greater than or equal to 16.

Q29. Find the equation of the line which intersects the *y*-axis at a distance of 3 units above the origin and makes an angle 30^0 with the positive direction of the *x*-axis.

Q30. Find the coordinates of the foot of perpendicular from the point (2, 1) to the line $3x - 4y - 5 = 0$

OR

Q30. Find the angles between the lines $x + y = 2$, $y = 2$

Q31. Find the coordinates of the points that trisect the line segment joining the points $A(2,1,-3), B(5,4,3)$.

OR

Q31. Evaluate $\lim_{x \to 1} \frac{x^4 - \sqrt{x}}{\sqrt{x} - 1}$

Q32. $A = \{1\}$, $B = \{2, 3\}$ and $C = \{3\}$, then prove that

(i) $A \times (B \cap C) = (A \times B) \cap (A \times C)$

(ii) $A \times (B \cup C) = (A \times B) \cup (A \times C)$

OR

Q32. Find the derivative of $\frac{x^2}{\sin^4 x}$

SECTION D

(Q33–Q36) Solve the following questions.

(6 marks each)

Q33. $\frac{1}{\sqrt{1}} + \frac{1}{\sqrt{2}} + ... + \frac{1}{\sqrt{n}} > 2(\sqrt{n+1} - 1)$ for all natural numbers $n \in N$.

Q34. Find the standard deviation and variance for the data.

x_i	9	10	14	18	20
f_i	3	4	7	10	6

OR

Q34. If the real part of $\frac{\bar{z}+1}{\bar{z}-1}$ is 2, then show that the locus of the point representing z in the complex plane is a circle.

Q35. Find the derivative of function $\sin(x+\theta)$ with respect to x (θ is fixed non-zero constant) from the first principle.

OR

Q35. If the letters of the word 'DURABLE' are arranged in all possible ways as listed in the dictionary, what is the rank of the word 'DURABLE'?

Q36. The lengths of three unequal edges of a rectangular solid block are in GP. If the volume of the block is 27 cm^3 and the total surface area is 63 cm^2, then find the dimensions of the rectangular solid block.

OR

Q36. Without using Pythagoras' Theorem, find the value of a, such that the points $A(4,1), B(3,a), C(-1,2)$ are the vertices of a right-angled triangle at angle B.

Day 236

Sample Paper – VIII (Solutions)

SECTION A

Q1–Q10 are multiple choice type questions. Select the correct option. (1 mark each)

Q1. Solution: (a)

$4^{16}+6^{14}=(5-1)^{16}-(1+5)^{14}$

$=({}^{16}C_0 5^{16}-{}^{16}C_1 5^{15}+\ldots {}^{16}C_{14}5^2-{}^{16}C_{15}5+{}^{16}C_{16})$

$-({}^{14}C_0+{}^{14}C_1 5+{}^{14}C_2 5^2+{}^{14}C_{13}5^{13}+{}^{14}C_{14}5^{14})$

$=({}^{16}C_0-{}^{16}C_1 5+{}^{16}C_2 5^2-\ldots-{}^{16}C_{15}5^{15}+{}^{16}C_{16}5^{16})$

$-({}^{14}C_0+{}^{14}C_1 5+{}^{14}C_2 5^2+\ldots+{}^{14}C_{13}5^{13}+{}^{14}C_{14}5^{14})$

$=({}^{16}C_0-{}^{16}C_1 5-{}^{14}C_0-{}^{14}C_1 5)+25k$, where,

$k=({}^{16}C_2-\ldots+{}^{16}C_{16}5^{14}-{}^{14}C_2-\ldots-{}^{14}C_{14}5^{12})$

$\Rightarrow 4^{16}-6^{14}=({}^{16}C_0-{}^{16}C_1 5-{}^{14}C_0-{}^{14}C_1 5)+25k$

$=1-16.5-1-14.5)+25k$

$=-30.5+25k=25(-6+k)=25k'$

which is divisible by 25.

Q2. Solution: (c)

Let the general equation of the circle be

$x^2+y^2+2gx+2fy+c=0$.

Since the circle passes through the origin (0, 0) and the point (2, 3), $c=0$ and

$4+9+4g+6f=0 \Rightarrow 2g+3f=-13/2$

And since the centre lie on the Y-axis, $g=0$

$\Rightarrow 3f=-13/2 \Rightarrow f=-13/6$

Thus, the equation of circle is $x^2+y^2-\frac{13y}{3}=0$

$\Rightarrow 3x^2+3y^2-13y=0$

Q3. Solution: (a)

$X=\{6n-1 | n\in N\}=\{5,11,17,23,\ldots\}$

$Y=\{6n+1 | n\in N\}=\{7,13,19,25,\ldots\}$

Since every element of X is not in Y, and every element of Y is not in X

$X \not\subset Y$ and $Y \not\subset X$ and, $X\cap Y \neq \phi$

Q4. Solution: (d)

$\because f(x)=\frac{x-1}{x(x+1)}$,

$\therefore f(x)$ is defined, if $x(x+1)\neq 0 \Rightarrow x\neq 0,-1$

Thus, domain is $R-\{-1,0\}$

Q5. Solution: (c)

$\cos 2\theta+\sin 2\theta$

$=\left(\frac{1-\tan^2\theta}{1+\tan^2\theta}\right)+\left(\frac{2\tan\theta}{1+\tan^2\theta}\right)$

$=\left(\frac{1-4}{1+4}\right)+\left(\frac{4}{1+4}\right)$

$=\left(\frac{-3}{5}\right)+\left(\frac{4}{5}\right)=\frac{1}{5}$

Q6. Solution: (b)

$\sin 390°-\cos 1330°$

$=\sin(360°+30°)-\cos(1080°+250°)$

$=\sin(360°+30°)-\cos(3\times 360°+250°)$

$=\sin 30°+\cos(180°+60°)$

$=\sin 30°+\cos 60°=\frac{1}{2}+\frac{1}{2}=1$

Q7. Solution: (c)

$$(1-i)^5\left(1-\frac{1}{i}\right)^5=\left(\frac{(1-i)(i-1)}{i}\right)^5$$

$$=\left(\frac{i-1-i^2+i}{i}\right)^5=\left(\frac{2i}{i}\right)^5=2^5=32 \quad (\because i^2=-1)$$

Q8. Solution: (b)

$$y=\left[\frac{1}{\sqrt{2}}\sin x+\frac{1}{\sqrt{2}}\cos x\right]$$

$$=[\sin x\cdot\cos\frac{\pi}{4}+\cos x\sin\frac{\pi}{4}]$$

$$=\sin(x+\pi/4)$$

Thus, the maximum distance of $y=\sin x+\cos x$ from x-axis is 1 unit.

Q9. Solution: (a)

$$(1-i)^{12}=[(1-i)^2]^6=[1^2+i^2-2i]^6$$

$$=[1-1-2i]^6=32i^6=32i^4\cdot i^2=-32$$

$(\because i^2=-1,\ i^4=1)$

$\Rightarrow a+ib=-32+0i$

Q10. Solution: (d)

$$49^{35}=(50-1)^{35}$$

$$={}^{35}C_0 50^{35}-{}^{35}C_1 50^{34}+....-{}^{35}C_{33}50^2+{}^{35}C_{34}50^1-{}^{35}C_{35}$$

$$=\underbrace{({}^{35}C_0 50^{35}-{}^{35}C_1 50^{34}+....-{}^{35}C_{33}50^2)}_{\text{last two digits be 00}}+35.50-1$$

= number with last two digits 00 + 1750 − 1

= number with last two digits 00 + 1749

Thus, the last two digits are 49.

Q11–Q15 are fill in the blanks. **(1 mark each)**

Q11. Solution: The y-axis and z-axis taken together determine a plane known as the yz-pane.

Q12. Solution: The coordinates of the points in the YZ-plane are of the form (0, y, z).

Q13. Solution: $a=-3,\ r=\frac{-1}{3}\Rightarrow|r|=\frac{1}{3}<1,$

Therefore, summation, $S=-3+1-\frac{1}{3}+\frac{1}{9}-...$

$$\Rightarrow S_\infty=\frac{a}{1-r}=\frac{-3}{1-(-1/3)}=\frac{-3}{5/3}=\frac{-9}{5}$$

Q14. Solution: The middle terms will be the largest coefficient in the expansion. Therefore, in the expansion of $(1+x)^{22}$, the largest coefficient is $={}^{22}C_{22/2}={}^{22}C_{11}$

Q15. Solution: $(1-2x+3x^2)(1-x)^{10}$

$$=(1-2x+3x^2)\begin{pmatrix}{}^{10}C_0-{}^{10}C_1x^1-{}^{10}C_2x^2\\+\text{terms of higher power of }x\end{pmatrix}$$

$$=(1-2x+3x^2)\begin{pmatrix}1-10x-45x^2\\+\text{terms of higher power of }x\end{pmatrix}$$

$$=-45x^2+20x^2+3x^2+\text{terms of higher power of }x$$

$$=-22x^2+\text{terms of higher power of }x$$

Therefore, coefficient of x^2 is –22.

(Q16–Q20) Answer the following questions. **(1 mark each)**

Q16. Solution: $p\leftrightarrow q$: The units digit of an integer is zero or 5, if and only if it is divisible by 5.

Q17. Solution: p: Number 2 is prime.
q: Number 2 is even.

Q18. Solution: The coin is tossed three times. There can be a maximum of 3 heads or tails.
Person win Rs. 3 if he gets 2 heads and 1 tail, Rs 2 + Rs 2 – Re 1 = Rs. 3
Sample space
$S=\{HHH,HHT,HTH,THH,HTT,THT,TTH,TTT\}$
$\Rightarrow n(S)=8$
Let A be the event of getting exactly 2 heads, $A=\{HHT,HTH,THH\}\Rightarrow n(A)=3$

$$P(A)=\frac{n(A)}{n(S)}=\frac{3}{8}$$

∴ Probability (of winning Rs 3) = 3/8

Q19. Solution: Since on the XY-plane $z=0$, on the YZ-plane, $x=0$ and on ZX-plane, $y=0$, the coordinates of the feet of the perpendicular from a point P on the XY-, YZ- and ZX-planes of points A, B and C are as follows:
$A(-1,3,0), B(0,3,4), C(-1,0,4)$

Q20. Solution: The focus (0, –3) lies on the y-axis. Therefore the axis of the parabola is the y-axis. Thus, the equation of the parabola is of theform $x^2=4ay$ or $x^2=-4ay$.
Now, since the directrix is $y=3$ and focus is $(0,-3)$, the parabola is of the form $x^2=-4ay$ with $a=3$. Thus, the equation of the parabola is $x^2=-12y$.

SECTION B

(Q21–Q26) Solve the following questions. **(2 marks each)**

Q21. Solution: Let A and B be the events of passing the first and second examination, respectively. Then $P(A)=0.6$, $P(B)=0.7$, $P(A\text{ or }B)=P(A\cup B)=0.9$

Probability of passing both exams $= P(A \text{ and } B) = P(A \cap B)$
$\because P(A \cap B) = P(A) + P(B) - P(A \cup B)$
$\therefore P(A \cap B) = 0.6 + 0.7 - 0.9$
$\Rightarrow P(A \cap B) = 0.4$
Thus, the probability of passing both the examinations is 0.4

Q22. Solution: The given function $f(x) = \sqrt{x(x-1)}$ is defined when $x(x-1) \geq 0 \Rightarrow x \leq 0$ or $x \geq 1$
Therefore, the domain of f is the set of all real numbers less than or equal to 0 or greater than or equal to 1, that is, the domain of $f = (-\infty, 0] \cup [1, \infty)$.
Now, since in the domain of x, $x(x-1) \geq 0 \Rightarrow \sqrt{x(x-1)} \geq 0 \Rightarrow f(x) \geq 0$, the range of f is the set of all real numbers greater than or equal to 0, that is, range of $f = [0, \infty)$.

Q23. Solution: Total number of persons = 6
Number of persons to be selected = 4
If A is chosen, then B must not be chosen.
Therefore, the following cases will arise:
Case I: When A is chosen, but B must not be chosen
Number of ways $= {}^{6-2}C_3 = {}^4C_3 = 4$
Case II: When A is not chosen, but B may or may not be chosen
Number of ways $= {}^{6-1}C_4 = {}^5C_4 = 5$
Thus, total number of ways is $= 4 + 5 = 9$

Q24. Solution: (i) If 3 particular players are included $= {}^{15-2}C_{11-2} = {}^{13}C_9$
[Selection of n objects taken r at a time in which p objects are always included is ${}^{n-p}C_{r-p}$]

(ii) If 2 particular players are excluded $= {}^{15-2}C_{11} = {}^{13}C_{11}$
[Selection of n objects taken r at a time in which p objects are never included is ${}^{n-p}C_r$]

OR

Q24. Solution: Since the foci $(\pm 2, 0)$ are on the x-axis, the major axis is along the x-axis and $c = 2$
Now, from the equation $a^2 = b^2 + c^2$
$\therefore 6^2 = b^2 + 2^2 \Rightarrow b^2 = 36 - 4 \Rightarrow b = 4\sqrt{2}$
Therefore, the equation of the ellipse will be of the form $\frac{x^2}{a^2} + \frac{y^2}{b^2} = 1 \Rightarrow \frac{x^2}{6^2} + \frac{y^2}{(4\sqrt{2})^2} = 1 \Rightarrow \frac{x^2}{36} + \frac{y^2}{32} = 1$

OR

Q25. Solution: Let $z = x + iy$. Then $|(z+i)| + z - 1 = 0$.
$\Rightarrow |x + iy + i| + x + iy - 1 = 0$
$\Rightarrow \sqrt{x^2 + (1+y)^2} + x - 1 + iy = 0 = 0 + i0$
Equating real and imaginary parts, we get
$\Rightarrow \sqrt{x^2 + (1+y)^2} + x - 1 = 0$ and $y = 0$
Also,
$\Rightarrow \sqrt{x^2 + (1+y)^2} + x - 1 = 0$
$\Rightarrow \sqrt{x^2 + 1} + x - 1 = 0 \quad (\because y = 0)$
$\Rightarrow x^2 + 1 = x^2 + 1 - 2x \Rightarrow x = 0$
$\therefore z = x + iy = 0 + 0i$

OR

Q25. Solution: The natural numbers from 40 to 1000, which are multiples of 4, are 40, 44, 48,..., 1000.
This sequence forms an AP, with first term $a = 40$, last term $l = 1000$ and common difference $d = 4$.
Let n be the total number of terms. Then
$a + (n-1)d = l \Rightarrow 40 + (n-1)4 = 1000$
$\Rightarrow (n-1)4 = 960 \Rightarrow n - 1 = 240 \Rightarrow n = 241$
Thus, the sum of all natural numbers from 40 to 1000, which are multiples of 4, is
$S_n = \frac{n}{2}[a + l] = \frac{241}{2}[40 + 1000] = 125320$

Q26. Solution: Let a point on the y-axis be $P(0, y)$, which is equidistant from the points $A(-2, 1)$ and $B(4, -3)$. Then $AP = BP \Rightarrow AP^2 = BP^2$
By distance formula,
$(-2-0)^2 + (1-y)^2 = (4-0)^2 + (-3-y)^2$
$\Rightarrow 4 + 1 + y^2 - 2y = 16 + 9 + y^2 + 6y$
$\Rightarrow 8y = -20 \Rightarrow y = -5/2$
Thus, the point (0, –5/2) is equidistance from the points $A(-2, 1)$ and $B(4, -3)$.

OR

Q26. Solution: $R_1 = \{(x, y) | y = 2x - 1$, where $x \in R, -1 \leq x \leq 1\}$
Since, domain of R is $-1 \leq x \leq 1$
Therefore, $-1 \leq x \leq 1$
$\Rightarrow -2 \leq 2x \leq 2$
$\Rightarrow -2 - 1 \leq 2x - 1 \leq 2 - 1$
$\Rightarrow -2 \leq 2x - 1 \leq 1$
$\Rightarrow -2 \leq y \leq 1$
Thus, the range of $R_1 = \{-2 \leq y \leq 1, x \in R) = [-2, 1]$

SECTION C

(Q27–Q32) Solve the following questions.

(4 marks each)

Q27. Solution:

$\frac{2x}{15} \le \frac{(5x-2)}{3} - \frac{(7x-3)}{5}$

$\Rightarrow \frac{2x}{15} \le \frac{5(5x-2)-3(7x-3)}{15}$

$\Rightarrow \frac{2x}{15} \le \frac{25x-10-21x+9}{15}$

$\Rightarrow \frac{2x}{15} \le \frac{4x-1}{15} \Rightarrow 2x \le 4x-1$

$\Rightarrow 2x \ge 1 \Rightarrow x \ge \frac{1}{2}$

Therefore, all real numbers x, which are greater than or equal to $\frac{1}{2}$, are the solutions of the given inequality.

Thus, the solution set of the given inequality is $x \in \left[\frac{1}{2}, \infty\right)$.

Q28. Solution: Let x be the smaller of the two consecutive even positive integers. Then, the other larger integer must be $x+2$.

Since both the integers are smaller than 16

$x+2 \le 12 \Rightarrow x \le 12-2 \Rightarrow x \le 10$

Also, since the sum of the two integers is more than 16

$x+(x+2) \ge 16$

$\Rightarrow 2x+2 \ge 16$

$\Rightarrow 2x \ge 16$

$\Rightarrow x \ge 8$

Therefore, from inequalities, $x \le 10$ and $x \ge 8$, we get $x = 8, 9, 10$

But since x is an even number, x can take only values, 8 and 10.

Thus, all pairs of consecutive even positive integers are (8, 10) and (10, 12).

Q29. Solution: The equation of the line with slope m and which makes y-intercept equal to b is given by $y - b = mx$

Thus, equation of the line intersecting the y-axis at a distance of 2 units above the origin, that is, $b = 3$ and the slope of the line is $m = \tan 30° = \frac{1}{\sqrt{3}}$ is given by

$y = \frac{1}{\sqrt{3}}x + 3 \Rightarrow y = \frac{x+3\sqrt{3}}{\sqrt{3}}$

$\Rightarrow x - \sqrt{3}y + 3\sqrt{3} = 0$

Q30. Solution: Let (a, b) be the coordinates of the foot of the perpendicular from the point (2, 1) to the line $3x-4y-5=0$. Then

Slope of the line joining (2, 1) and (a, b), is $m_1 = \frac{b-1}{a-2}$

Slope of the line $3x-4y-5=0 \Rightarrow y = \frac{3}{4}x - \frac{5}{4}$ is $m_2 = 3/4$

Since above two lines are perpendicular to each other, $m_1 m_2 = -1$

$\Rightarrow \left(\frac{b-1}{a-2}\right) \times \left(\frac{3}{4}\right) = -1 \Rightarrow \frac{3b-3}{4a-8} = -1$

$\Rightarrow 3b-3 = -4a+8 \Rightarrow 4a+3b = 11$

Also, since the point (a, b) lies on line $3x-4y-5=0$, $3a-4b=5$

Solving the above two equations, we get $a = 59/25$, $b = 13/25$ which are the required coordinates of the foot of the perpendicular to the line $3x-4y-5=0$

OR

Q30. Solution: The given equation of lines can be written as

$x + y = 2 \Rightarrow y = -x+2$

and $y = 0 \cdot x + 2$

Thus, we get slope first line, $m_1 = -1$ and slope of other line, $m_2 = 0$

Therefore, the acute angle (θ) between the two lines is given by

$\tan\theta = \frac{|m_1 - m_2|}{|1+m_1 m_2|} \Rightarrow \tan\theta = \left|\frac{-1-0}{1+(-1)\times 0}\right| = 1$

$\Rightarrow \tan\theta = 1 \Rightarrow \theta = 45°$

Thus, the angle between the given lines is either 45° or $180° - 45° = 135°$.

Q31. Solution: Let the $P(x_1, y_1, z_1), Q(x_2, y_2, z_2)$ trisect line segment AB.

Since the point P divided line AB in 1:2 internally

$x_1 = \frac{2\times 2+1\times 5}{1+2} = \frac{9}{3} = 3$

$y_1 = \frac{2\times 1+1\times(4)}{3} = 2$

$$z_1 = \frac{2\times(-3)+1\times 3}{3} = \frac{-6+3}{3} = \frac{-3}{3} = -1$$

Since the point Q divides the line segment AB in 2:1

$$x_2 = \frac{1\times 2+2\times 5}{3} = 4,$$

$$y_2 = \frac{1\times 1+(4\times 2)}{3} = 3$$

$$z_2 = \frac{1\times(-3)+2\times 3}{3} = -1$$

Thus, the coordinates of P are $(3,2,-1)$ and the coordinates of Q are $(4,3,-1)$.

OR

Q31. Solution:

$$\lim_{x\to 1}\frac{x^4-\sqrt{x}}{\sqrt{x}-1} = \lim_{x\to 1}\frac{\sqrt{x}[(x)^{7/2}-1]}{\sqrt{x}-1}$$

$$= \lim_{x\to 1}\frac{(x)^{7/2}-1}{\sqrt{x}-1}\cdot\lim_{x\to 1}\sqrt{x}$$

$$\left[\because \lim_{x\to a} f(x)\cdot g(x) = \lim_{x\to a} f(x)\cdot\lim_{x\to a} g(x)\right]$$

$$= \lim_{x\to 1}\frac{\frac{x^{7/2}-1}{x-1}}{\frac{(x)^{1/2}-1}{x-1}} = \frac{\lim_{x\to 1}\frac{x^{7/2}-1}{x-1}}{\lim_{x\to 1}\frac{(x)^{1/2}-1}{x-1}}$$

$$\left[\because \lim_{x\to a}\frac{f(x)}{g(x)} = \frac{\lim_{x\to a} f(x)}{\lim_{x\to a} g(x)}\right]$$

$$= \frac{\frac{7}{2}(1)^{\frac{7}{2}-1}}{\frac{1}{2}(1)^{\frac{1}{2}-1}} = \frac{7/2}{1/2} = 7$$

Q32. Solution: $B\cap C = \{2,3\}\cap\{3\} = \{3\}$

$B\cup C = \{2,3\}\cup\{3\} = \{2,3\}$

Thus

(i) For $A\times(B\cap C) = (A\times B)\cap(A\times C)$

LHS: $A\times(B\cap C) = \{1\}\times\{3\} = \{(1,3)\}$

RHS: $(A\times B)\cap(A\times C)$

$= \{(1,2),(1,3)\}\cap\{(1,3)\} = \{(1,3)\}$

(ii) For $A\times(B\cup C) = (A\times B)\cup(A\times C)$

LHS: $A\times(B\cup C) = \{1\}\times\{2,3\} = \{(1,2),(1,3)\}$

RHS: $(A\times B)\cup(A\times C)$

$= \{(1,2),(1,3)\}\cup\{(1,3)\} = \{(1,2),(1,3)\}$

Thus, LHS = RHS, verify above two results.

$f = R-\{-1,0\}$

OR

Q32. Solution: Let $f(x) = \dfrac{x^2}{\sin^4 x}$

By the quotient rule

$$f'(x) = \frac{\sin^4 x\frac{d}{dx}x^2 - x^2\frac{d}{dx}\sin^4 x}{(\sin^4 x)^2}$$

$$\frac{d}{dx}\sin^n x = n\sin^{n-1} x\cos x$$

Therefore, $f'(x) = \dfrac{\sin^4 x\times 2x - x^2\, 4\sin^{4-1} x\cos x}{\sin^8 x}$

$$= \frac{2x\sin^3 x(\sin x - 2x\cos x)}{\sin^8 x}$$

SECTION D

(Q33–Q36) Solve the following questions.

(6 marks each)

Q33. Solution: The statement is

$P(n): \dfrac{1}{\sqrt{1}}+\dfrac{1}{\sqrt{2}}+....+\dfrac{1}{\sqrt{n}} > 2(\sqrt{n+1}-1)$ for all natural numbers $n\in N$

For $n = 1$, $P(1): 1 > 2(\sqrt{2}-1)$ is true.

Let $P(k)$ is true for some positive integer k.

$$\frac{1}{\sqrt{1}}+\frac{1}{\sqrt{2}}+....+\frac{1}{\sqrt{k}} > 2(\sqrt{k+1}-1)$$

Now we shall prove that $P(k+1)$ is true

$$\frac{1}{\sqrt{1}}+\frac{1}{\sqrt{2}}+....+\frac{1}{\sqrt{k}}+\frac{1}{\sqrt{k+1}} > 2(\sqrt{k+2}-1)$$

$$\because \frac{1}{\sqrt{1}}+\frac{1}{\sqrt{2}}+....+\frac{1}{\sqrt{k}} > 2(\sqrt{k+1}-1)$$

($\because P(k)$ is true)

$$\Rightarrow \frac{1}{\sqrt{1}}+\frac{1}{\sqrt{2}}+....+\frac{1}{\sqrt{k}}+\frac{1}{\sqrt{k+1}} > 2(\sqrt{k+1}-1)+\frac{1}{\sqrt{k+1}}$$

Now we have to prove that

$$\Rightarrow 2(\sqrt{k+1}-1)+\frac{1}{\sqrt{k+1}} > 2(\sqrt{k+2}-1)$$

$$\Rightarrow \frac{2(\sqrt{k+1}-1)\sqrt{k+1}+1}{\sqrt{k+1}} > 2(\sqrt{k+2}-1)$$

$$\Rightarrow \frac{2(k+1)-2\sqrt{k+1}+1}{\sqrt{k+1}} > 2(\sqrt{k+2}-1)$$

$$\Rightarrow 2k-2\sqrt{k+1}+3 > 2(\sqrt{k+2}-1)\sqrt{k+1}$$

$$\Rightarrow 2k-2\sqrt{k+1}+3 > 2\sqrt{k+2}\sqrt{k+1}-2\sqrt{k+1}$$

$$\Rightarrow 2k+3 > 2\sqrt{k+2}\sqrt{k+1} > 2\sqrt{k+1}\sqrt{k+1}$$

$$\Rightarrow 2k+3 > 2(k+1) \Rightarrow 3 > 2 \text{ is true}$$

$P(k+1)$ is true whenever $P(k)$ is true.

Therefore by the principle of mathematical induction, statement $P(n)$ is true for all $n \in N$

Q34. Solution:
The data is obtained in tabular form as follows:

x_i	f_i	$f_i x_i$	$(x_i - \bar{x})$	$(x_i - \bar{x})^2$	$f_i(x_i - \bar{x})^2$
9	3	27	−6.5	42.25	126.75
10	4	40	−5.5	30.25	121
14	7	98	−1.5	2.25	15.75
18	10	180	2.5	6.25	62.5
21	6	120	5.5	30.25	181.5
	30	465			507.5

$N = 30,$

$$\therefore \bar{x} = \frac{\sum_{i=1}^{5} f_i x_i}{N} = \frac{465}{30} = 15.5$$

$$\text{Variance} = (\sigma^2) = \frac{1}{N}\sum_{i=1}^{5} f_i (x_i - \bar{x})^2 = \frac{1}{30} \times 507.5 = 16.92$$

Standard deviation $(\sigma) = \sqrt{16.92} = 4.11$

OR

Q34. Solution: Let $z = x + iy$,

$$\text{Re}\left(\frac{\bar{z}+1}{\bar{z}-1}\right) = 2$$

$$\Rightarrow \text{Re}\left(\frac{x - iy + 1}{x + iy - 1}\right) = 2$$

$$\Rightarrow \text{Re}\left(\frac{[(x+1) - iy][(x-1) + iy]}{[(x-1) - iy][(x-1) + iy]}\right) = 2$$

$$\Rightarrow \text{Re}\left(\frac{(x+1)(x-1) + y^2 + i[y(x+1) - y(x-1)]}{(x-1)^2 + y^2}\right) = 2$$

$$\Rightarrow \text{Re}\left(\frac{x^2 - 1 + y^2 + i(2y)}{(x-1)^2 + y^2}\right) = 2$$

$$\Rightarrow \frac{x^2 - 1 + y^2}{(x-1)^2 + y^2} = 2$$

$$\Rightarrow x^2 - 1 + y^2 = 2x^2 + 2 - 4x + 2y^2$$

$$\Rightarrow x^2 + y^2 - 4x + 3 = 0$$

$\Rightarrow x^2 + y^2 - 4x + 3 = 0$ represents the equation of a circle.

Q35. Solution: Let $f(x) = \sin(x + \theta)$

Then, according to the first principle

$$f'(x) = \lim_{h \to 0} \frac{f(x+h) - f(x)}{h}$$

$$= \lim_{h \to 0} \frac{1}{h}[\sin(x + \theta + h) - \sin(x + \theta)]$$

$$= \lim_{h \to 0} \frac{1}{h}\left[2\cos\left(\frac{x + h + \theta + x + \theta}{2}\right)\sin\left(\frac{x + h + \theta - x - \theta}{2}\right)\right]$$

$$= \lim_{h \to 0} \frac{1}{h}\left[2\cos\left(\frac{2x + h + 2\theta}{2}\right)\sin(h/2)\right]$$

$$= \lim_{h \to 0} \frac{1}{h}\left[\cos\left(\frac{2x + h + 2\theta}{2}\right) \cdot \frac{\sin(h/2)}{(h/2)}\right]$$

$$= \lim_{h \to 0} \cos\left(\frac{2x + h + 2\theta}{2}\right) \cdot \lim_{h/2 \to 0} \frac{\sin(h/2)}{(h/2)}$$

$$\left[\text{As } h \to 0 \Rightarrow \frac{h}{2} \to 0\right]$$

$$= \cos\left(\frac{2x + 0 + 2\theta}{2}\right) \cdot 1 \left[\lim_{x \to 0} \frac{\sin x}{x} = 1\right]$$

$$= \cos(x + \theta)$$

OR

Q35. Solution: The letters of the word 'DURABLE' in alphabetical order are A, B, D, E, L, R, U.
Words starting with A = 6!
Words starting with B = 6!
Word starting with DA = 5!
Word starting with DB = 5!
Word starting with DE = 5!
Word starting with DL = 5!
Word starting with DR = 5!
Word starting with DUA = 4!
Word starting with DUB = 4!
Word starting with DUE = 4!
Word starting with DUL = 4!
Next word will be DURABEL = 1
Next word will be DURABLE = 1
$\therefore$ Rank of the word 'DURABLE' in dictionary is
$= 2 \times 6! + 5 \times 5! + 4 \times 4! + 1 + 1 = 2138$

Q36. Solution: Let the length, breadth and height of rectangular solid block be $\frac{a}{r}, a, ar$. Then

$$\text{Volume of solid block} = \frac{a}{r} \times a \times ar = 27 \text{ cm}^3$$

$$\Rightarrow a^3 = 27 \Rightarrow a^3 = 3^3 \Rightarrow a = 3$$

$$\text{Surface area} = 2\left(\frac{a^2}{r} + a^2 + a^2 r\right) = 63$$

$$\Rightarrow 2a^2\left(\frac{1}{r} + r + 1\right) = 63 \Rightarrow 2 \times 3^2\left(\frac{1 + r + r^2}{r}\right) = 63$$

$$\Rightarrow 1 + r + r^2 = \frac{7}{2}r \Rightarrow 2r^2 - 5r + 2 = 0$$

$$\Rightarrow (2r - 1)(r - 2) = 0 \Rightarrow r = \frac{1}{2}, 2$$

For $r = \frac{1}{2}$:

Length $= \frac{a}{r} = \frac{3}{1/2} = 6$

Breadth $= a = 3$

Height $= ar = 3 \times \frac{1}{2} = \frac{3}{2}$

For $r = 2$:

Length $= \frac{a}{r} = \frac{3}{2}$

Breadth $= a = 3$

Height $= ar = 3 \times 2 = 6$

OR

Q36. Solution: Let the vertices of the given triangle be $A(4,1), B(3,a), C(1,2)$

The slope of a non-vertical passing through the points $(x_1, y_1), (x_2, y_2)$ is given by $m = \frac{y_2 - y_1}{x_2 - x_1}, x_2 \neq x_1$

$\therefore$ Slope of line segment $AB\ (m_1) = \frac{a-1}{3-4} = 1-a$

Slope of segment $BC\ (m_2) = \frac{2-a}{1-3} = \frac{a-2}{2}$

Slope of segment $CA\ (m_3) = \frac{2-1}{1-4} = -\frac{1}{3}$

Since the points $A(4,1), B(3,a), C(1,2)$ are the vertices of a right-angled triangle at angle B, the product of slopes of the line segments AB and BCmust be equal to –1.

Thus, $m_1 m_2 = -1 \Rightarrow \left(\frac{a-2}{2}\right) \times (1-a) = -1$

$\Rightarrow a^2 - 3a + 2 = 2 \Rightarrow a^2 - 3a = 0 \Rightarrow a = 0, 3$

Therefore, line segments AB and BC are perpendicular to each other, that is, the given triangle is right-angled at vertex B (3, 0) or B (3, 3).

Sample Paper – IX

Day 237

Time: 3 Hrs Maximum **Marks:** 80

General Instructions:

(i) All the questions are compulsory.
(ii) The question paper consists of 36 questions divided into 4 sections A, B, C and D.
(iii) Section A comprises 20 questions of 1 mark each. Section B comprises 6 questions of 2 marks each. Section C comprises 6 questions of 4 marks each. Section D comprises 4 questions of 6 marks each.
(iv) There is no overall choice. However, an internal choice has been provided in four questions of 1 mark each, two questions of 2 marks each, two questions of 4 marks each, and two questions of 6 marks each. You have to attempt only one of the alternatives in all such questions.
(iv) Use of calculators is not permitted.

SECTION A

Q1–Q10 are multiple choice type questions. Select the correct option. (1 mark each)

Q1. If $S = \{x | x$ is a positive multiple of 5 less than or equal to 100$\}$ and $P = \{x | x$ is a prime number less than 25$\}$, then, $n(S) - n(P)$ is equal to

(a) 14 (b) 11 (c) 13. (d) 15

Q2. If X and Y are two sets that $X \cap Y$ has 5 elements, X has 10 elements and Y has 20 elements, how many elements does $X \cup Y$ have?

(a) 14 (b) 11 (c) 10 (d) 25

Q3. The domain and range of the function f is given by $f(x) = 1 - |x - 1|$ is

(a) Domain = R, Range = $(-\infty, 1]$
(b) Domain = R, Range = $[-\infty, 2]$
(c) Domain = R, Range = $(-\infty, 2]$
(d) Domain = $R+$, Range = $(-\infty, 2]$

Q4. If a, b and c are in GP with common ratio 2, then the value of $\frac{b-a}{c-b}$ is equal to

(a) 1 (b) 0 (c) 1/2 (d) 2

Q5. The equation of a circle with origin as centre and passing through the vertices of an equilateral whose median is of length 6 unit is

(a) $x^2 + y^2 = 9$ (b) $x^2 + y^2 = 16$
(c) $x^2 + y^2 = 4$ (d) $x^2 + y^2 = 1$

Q6. The number of terms in the expansion of $(x + y + z)^6$ is

(a) 18 (b) 28 (c) 36 (d) 196

Q7. If the expansion of $\left(\sqrt{x} + \frac{1}{x^2}\right)^{2n}$ contains a term independent of x, then n is a multiple of

(a) 4 (b) 2 (c) 3 (d) 5

Q8. The coefficient of x^{12} in the product $(x^2 - x)^4 (1 + x)^6$ using Binomial Theorem is

(a) 2 (b) –2 (c) 3 (d) –3

Q9. The value of $\sin(45° + \theta) - \cos(45° - \theta)$ is

(a) $2\cos\theta$ (b) $2\sin\theta$
(c) $\sqrt{2}(\cos\theta + \sin\theta)$ (d) $\frac{1}{\sqrt{2}}(\cos\theta - \sin\theta)$

Q10. Express the following functions as set of ordered pairs and determine their range. $f : x \to R$, $f(x) = x(x - 1)(x + 1)$, where $x = \{-1, 0, 1\}$

(a) $\{0\}$ (b) $\{-1, 1\}$ (c) $\{-1, 0, 1\}$ (d) $\{0, 1\}$

Q11–Q15 are fill in the blanks. (1 mark each)

Q11. If $\tan 2\theta + \tan 3\theta + \tan 2\theta \tan 3\theta = 1$, then $\theta = \ldots\ldots\ldots$

Q12. The value of $\sum_{n=1}^{15} (i^n + i^{n+1})$, is ... ...

Q13. The derivative of $\frac{x^4 + x^3 + x^2 + x + 1}{x^2}$ is ... ...

Q14. The value of $\left[i^{25} - \left(\frac{1}{i}\right)^{25}\right]^3$ is ... ...

Q15. If 3/5 is the probability of an event, then the probability of the event not A is ... ...

(Q16–Q20) Answer the following questions.
(1 mark each)

Q16. Write the component statements of the following compound statement and check whether the compound statement is true or false: 10 is divisible by 2 or 7.

Q17. Rewrite the following statement in the form of conditional statement: The square of an odd number is odd.

Q18. Solve the equation $x^2 + 2 = 0$

Q19. A point is on the y-axis. What are its x-coordinate and z-coordinate?

Q20. Find the distance from the origin to (–1, 3, 2).

SECTION B

(Q21–Q26) Solve the following questions.
(2 marks each)

Q21. If $|z_1| = |z_2| = 1$, then show that

$$|z_1 + z_2| = \left|\frac{1}{z_1} + \frac{1}{z_2}\right|.$$

Q22. Solve the given inequality for real x:
$8-(3x+5) \geq 9x - 8(x-3)$

Q23. $\sin\frac{13\pi}{6} + \cos\frac{11\pi}{3} - \tan\frac{15\pi}{4}$

Q24. Find the middle term (terms) in the expansion of

$$\left(3x - \frac{x^3}{3}\right)^6$$

Q25. The sum of some terms of GP is 315, whose first term and the common ratio are 5 and 2, respectively. Find last term and the number of terms.

OR

Q25. Find the equation of the parabola that satisfies the following conditions: vertex (0, 0); focus (2, 0)

Q26. Find the area of a triangle whose vertices are $(1,-1),(4,2),(-3,0)$.

OR

Q26. If the origin is the centroid of a ΔABC having vertices $A(a,1,3), B(-2,b,-1), C(2,3,c)$, then find the values of a, b, c.

SECTION C

(Q27–Q32) Solve the following questions.
(4 marks each)

Q27. How many committees of seven persons with a chairperson and a chair leader can be selected from 10 persons?

Q28. A cricket team of 11 students is to be constituted, choosing at least 4 from Class XI and at least 5 from Class XII. If there are 10 students in each of these classes, in how many ways can the team be constituted?

Q29. If the letters of the word 'MAXIMA' are arranged in all possible ways as listed in dictionary, what is the rank of the word 'MAXIMA'?

Q30. Find the coordinates of the foci and the vertices, the eccentricity, and the length of the latus rectum of the hyperbola $x^2 - 4y^2 = 200$

OR

Q30. Find the sum to n terms of the series $\frac{1}{2\times4} + \frac{1}{4\times6} + \frac{1}{6\times8} + \ldots$

Q31. Find the equation of the line which is at a perpendicular distance of $1/\sqrt{2}$ units from the origin and where the angle made by the perpendicular with the positive x-axis is 45^0

OR

Q31. Solve the given inequality graphically in two-dimensional plane: $2x - y \leq 8$

Q32. Find the equation of the line through the intersection of lines $2x+3y-4=0$, $x-5y=7$ that has its x-intercept equal to 1.

OR

Q32. Find the equation of the right bisector of the line segment joining the points (2, 3) and (3, 4).

SECTION D

(Q33–Q36) Solve the following questions.
(6 marks each)

Q33. Find the value of x such that the points $(-2,-1),(4,0),(3,3),(x,y)$ are vertices of a parallelogram.

Q34. Find the derivative of function $\cos(x-\theta)$ with respect to x (θ is fixed non-zero constant) from the first principle.

OR

Q34. In a class of 50 students, 20 opted for Math, 20 opted for Biology and 10 opted for both Math and Biology. If one of these students is selected at random, find the probability that

(i) The student opted for Math or Biology.

(ii) The student has opted neither Math nor Biology.

(iii) The student has opted Math but not Biology.

Q35. Find domain of the following functions:

(i) $f(x)=\frac{1}{\sqrt{1-\sin^2 x}}$ (ii) $f(x)=\frac{x-1}{x^2-x-6}$

(iii) $f(x)=\frac{1}{x+|x|}$

OR

Q35. Evaluate $\lim_{x\to 0}\frac{1-\cos 4x}{1-\cos 5x}$

Q36. $\frac{1}{2}\cdot\frac{3}{4}\cdot\frac{5}{6}\ldots\frac{2n-1}{2n}\le\frac{1}{\sqrt{3n+1}}$ for natural numbers $n\in N$.

OR

Q36. Find the mean and standard deviation using shortcut method.

x_i	1	2	3	4	5
f_i	4	5	1	7	3

Day 238

Sample Paper – IX (Solutions)

SECTION A

Q1–Q10 are multiple choice type questions. Select the correct option. (1 mark each)

Q1. Solution: (b)

$S=\{x|x$ is a positive multiple of 5 less than or equal to 100$\}$

$\Rightarrow S=\{5,10,15,...,95,100\}$

$\because t_n=a+(n-1)d\Rightarrow 100=5+(n-1)5\Rightarrow n=20$

Thus, $n(S)=20$

$P=\{x|x$ is a prime number less than 25$\}$

$\Rightarrow P=\{2,3,5,7,11,13,17,19,23\}\Rightarrow n(P)=9$

Thus, $n(S)-n(P)=20-9=11$

Q2. Solution: (d)

$\because n(X\cap Y)=5,\ n(X)=10,\ n(Y)=20$

$\therefore n(X\cup Y)=n(X)+n(Y)-n(X\cap Y)$

$\Rightarrow n(X\cup Y)=10+20-5=25$

Q3. Solution: (a)

$f(x)=1-|x-1|$ is defined for all $x\in R$

$\therefore$ Domain of $f=R$

$|x-1|\ge 0\Rightarrow -|x-1|\le 0$

$\Rightarrow 1-|x-1|\le 1\Rightarrow f(x)\le 1$

$\therefore$ Range of $f=(-\infty,1]$

Q4. Solution: (c)

$a,\ b,\ c$ are in GP, with common ratio 2. $b=2a,\ c=2^2a=4a$

$\therefore\ \frac{b-a}{c-b}=\frac{2a-1}{4a-2a}=\frac{2a-1}{2(2a-1)}=\frac{1}{2}$

Q5. Solution: (b)

The length of the median $AD=6$

Therefore, radius of the circle $=\frac{2}{3}\times$ length of median $=\frac{2}{3}\times 6=4$

Thus, the equation of the circle is $x^2+y^2=16$

Q6. Solution: (b)

The number of terms in the expansion of $(x+y+z)^n$ is $=\frac{(n+1)(n+2)}{2}$. Therefore, the number of terms in the expansion of $(x+y+z)^6$ is $=\frac{(6+1)(6+2)}{2}=28$

Q7. Solution: (d)

Let the independent term of x in the binomial expansion $\left(\sqrt{x}+\frac{1}{x^2}\right)^{2n}$ be

$T_{r+1}={}^{2n}C_r(\sqrt{x})^{2n-r}\left(\frac{1}{x^2}\right)^r$

$={}^{2n}C_r x^{\frac{2n-r}{2}}x^{-2r}={}^{2n}C_r x^{\frac{2n-5r}{2}}$

For independent of x, $2n-5r=0\Rightarrow r=\frac{2n}{5}$ which will be an integer, only when n is a multiple of 5

Q8. Solution: (d)

Since the product $(x^2-x)^4(x^2+x)^6$

$= x^{10}(x-1)^4(1+x)^6$
$= x^{10}(-1)^4(1-x)^4(1+x)^6$
$= x^{10}(1-x^2)^4(1+x)^2$
$= x^{10}(1-x^2)^4(1+2x+x^2)$
$= (x^{10}+2x^{11}+x^{12})(1-x^2)^4$

Using Binomial Theorem, the expansions of $(1-x^2)^4$ is

$= {}^4C_0 - {}^4C_1(x^2) + {}^4C_2(x^2)^2 - {}^4C_3(x^2)^3 - {}^4C_4(x^2)^4$
$= 1-4x^2+6x^4-4x^6-x^8$
$\therefore (x^2-x)^4(x^2+x)^6$
$= (x^{10}+2x^{11}+x^{12})(1-x^2)^4$
$= (x^{10}+2x^{11}+x^{12})(1-4x^2+6x^4-4x^6-x^8)$

Since we need the terms which involve x^{12}, complete multiplication of the two brackets is not required.

Thus, the terms containing x^{12} are
$= (x^{10})(-4x^2)+(x^{12})(1) = -3x^{12}$

Thus, the coefficient of x^{12} in the given product is –3.

Q9. Solution: (c)

$\sin(45°+\theta)+\cos(45°-\theta)$
$= \sin 45°\cdot\cos\theta + \cos 45°\cdot\sin\theta$
$+\cos 45°\cdot\cos\theta + \sin 45°\cdot\sin\theta$
$= \sqrt{2}(\cos\theta+\sin\theta)$

Q10. Solution: (a)

$f : x \to R, f(x) = x(x-1)(x+1),$
$f(-1) = -1(-1-1)(-1+1) = 0$
$f(0) = 0(0-1)(0+1) = 0$
$f(1) = 1(1-1)(1+1) = 0$
$\therefore f = \{(-1,0),(0,0),(1,0)\}$

Thus, range of $f = \{0\}$

Q11–Q15 are fill in the blanks. **(1 mark each)**

Q11. Solution:

$\tan 2\theta + \tan 3\theta + \tan 2\theta \tan 3\theta = 1$
$\Rightarrow \tan 2\theta + \tan 3\theta = 1 - \tan 2\theta \tan 3\theta$

$$\Rightarrow \frac{\tan 2\theta + \tan 3\theta}{1-\tan 2\theta \tan 3\theta} = 1$$

$$\Rightarrow \tan(2\theta+3\theta) = \tan\frac{\pi}{4} \Rightarrow \tan 5\theta = \tan\frac{\pi}{4}$$

$$\therefore 5\theta = n\pi + \frac{\pi}{4} \Rightarrow \theta = \frac{n\pi}{5} + \frac{\pi}{20}$$

Q12. Solution: The summation of four consecutive power of i is equal to zero, and there are 15 terms in the summation. Therefore, the summation of 12 (4x3) terms will be equal to zero, and we will have with only three terms (either from beginning or from end).

$$\text{Thus, } \sum_{n=1}^{14}(i^n + i^{n+1})$$

$$= i^1 + i^{1+1} + i^2 + i^{2+1} + i^3 + i^{3+1} + \sum_{n=4}^{15}(i^n + i^{n+1})$$

$= i^1 + i^2 + i^2 + i^3 + i^3 + i^4 + 0 \quad (\because i^n + i^{n+1} + i^{n+2} + i^{n+3} = 0)$

$= i - 1 - 1 - i - i + 1 = -1 - i \quad (\because i^2 = -1)$

Q13. Solution:

$$\frac{d}{dx}\left(\frac{x^4+x^3+x^2+x+1}{x^2}\right) = \frac{d}{dx}\left(x^2+x+1+\frac{1}{x}+\frac{1}{x^2}\right)$$

$$= \frac{d}{dx}x^2 + \frac{d}{dx}x + \frac{d}{dx}(1) + \frac{d}{dx}\left(\frac{1}{x}\right) + \frac{d}{dx}\left(\frac{1}{x^2}\right)$$

$$= 2x+1+0+\left(\frac{-1}{x^2}\right)+\left(\frac{-2}{x^3}\right) = 1+2x-\frac{1}{x^2}-\frac{2}{x^3}$$

Q14. Solution:

$$\left[i^{25} - \left(\frac{1}{i}\right)^{25}\right]^3 = \left[i^{4\times4+2} - \frac{1}{i^{4\times6+1}}\right]^3$$

$$= \left[(i^4)^6 \cdot i - \frac{1}{(i^4)^6 \cdot i}\right]^3$$

$$= \left[i - \frac{1}{i}\right]^3 \quad (\because i^4 = 1)$$

$$= \left[i - \frac{1}{i}\times\frac{i}{i}\right]^3 \quad (\because i^2 = -1)$$

$$= [i+i]^3 = [2i]^3 = -8i \quad (\because i^3 = -i)$$

Q15. Solution: $P(A) = 3/5$

$\therefore P(\text{not } A) = 1 - P(A) = 1 - 3/5 = 2/5$

(Q16–Q20) Answer the following questions. **(1 mark each)**

Q16. Solution: Its component statements are

p : 10 is divisible by 2. [true]

q : 10 is divisible by 7. [false]

Since the given compound statement is of the form '$p \vee q$', and one of the statement is true, the compound statement is true.

Q17. Solution: If the number is an odd number, then its square will be an odd number.

Q18. Solution: Since discriminant,
$D = b^2 - 4ac = 0^2 - 4\times1\times2 = -8 < 0$

The solution of the given equation is

$$x = \frac{-b \pm \sqrt{D}}{2a} = \frac{\pm\sqrt{-8}}{2\times 1} = \frac{\pm 2\sqrt{2}\,i}{2} = \pm\sqrt{2}\,i \ (\because \sqrt{-1} = i)$$

Q19. Solution: If a point is on the y-axis, then its x and z coordinates are zero.

Q20. Solution: Distance from origin to the point (–1, 3, 2) is $d = \sqrt{(x_1 - x_2)^2 + (y_1 - y_2)^2 + (z_1 - z_2)^2}$

$= \sqrt{(0+1)^2 + (0-3)^2 + (0-2)^2}$

$= \sqrt{1+9+4} = \sqrt{14}$

SECTION B

(Q21–Q26) Solve the following questions.

(2 marks each)

Q21. Solution:

LHS $= |z_1 + z_2|$

$= |\overline{z_1 + z_2}| = |\bar{z}_1 + \bar{z}_2| \quad (\because |z| = |\bar{z}|)$

$= \left|\frac{z_1\bar{z}_1}{z_1} + \frac{z_2\bar{z}_2}{z_2}\right| = \left|\frac{|z_1|}{z_1} + \frac{|z_2|}{z_2}\right|$

$= \left|\frac{1}{z_1} + \frac{1}{z_2}\right| =$ RHS

$(\because |z_1| = |z_2| = 1 \Rightarrow z_1\bar{z}_1 = z_2\bar{z}_2 = 1)$

Q22. Solution:

$5 - (3x+5) \geq 9x - 8(x-3)$

$\Rightarrow 5 - 3x - 5 \geq 9x - 8x + 24$

$\Rightarrow -3x \geq x + 24$

$\Rightarrow -24 \geq 4x \Rightarrow x \leq -6$

Therefore, all real numbers x, which are less than or equal to –6, are the solutions of the given inequality.

Thus, the solution set of the given inequality is $x \in (-\infty, -6]$.

Q23. Solution:

$\sin\frac{13\pi}{6} + \cos\frac{11\pi}{3} - \tan\frac{15\pi}{4}$

$= \sin\left(2\pi + \frac{\pi}{6}\right) + \cos\left(4\pi - \frac{\pi}{3}\right) - \tan\left(3\pi + \frac{3\pi}{4}\right)$

$= \sin\left(2\pi + \frac{\pi}{6}\right) + \cos\left(4\pi - \frac{\pi}{3}\right) - \tan\left(3\pi + \frac{\pi}{2} + \frac{\pi}{4}\right)$

$\left(\begin{array}{l}\because \text{sin and cosine function repeats after } 2\pi \\ \text{and tan function repeats after } \pi\end{array}\right)$

$= \sin\left(\frac{\pi}{6}\right) + \cos\left(\frac{\pi}{3}\right) + \cot\left(\frac{\pi}{4}\right)$

$= \frac{1}{2} + \frac{1}{2} + 1 = 2$

Q24. Solution: The power of binomial expansion $\left(3x - \frac{x^3}{3}\right)^6$ is $n = 6$ is even.

Therefore, it has only one middle term $\left(\frac{6}{2} + 1\right) = 4^{\text{th}}$ term.

$T_4 = T_{(3+1)} = {}^6C_3 (3x)^{6-3}\left(-\frac{x^3}{3}\right)^3$

$= 20 \cdot 3^3 x^3 x^9 3^{-3} (-1)^3 = -20x^{12}$

Q25. Solution: Let the first term a be 5 and common ratio r be 2, and sum of n terms of the GP be 315.

Therefore, $S_n = \frac{a(r^n - 1)}{r - 1}$

$\Rightarrow 315 = \frac{5(2^n - 1)}{2 - 1} \Rightarrow 2^n - 1 = 63$

$\Rightarrow 2^n = 64 = (2)^6 \Rightarrow n = 6$

Thus, last term of the GP is the 6th term, given by

$a_5 = ar^{6-1} = (5)(2)^5 = (5)(32) = 160$

OR

Q25. Solution: The focus (2, 0) lies on the x-axis, and therefore, the axis of the parabola is the x-axis. Thus, the equation of the parabola is either of the form $y^2 = 4ax$ or $y^2 = -4ax$.

Now, since the focus is (2, 0), the parabola is of the form $y^2 = 4ax$ with $a = 2$. Thus, the equation of the parabola is $y^2 = 8x$.

Q26. Solution: The area of the triangle formed by the vertices $A(1,-1), B(4,2), C(-3,0)$, is given by

$\frac{1}{2}[1(2+0) + (4)(-0+1) + (-3)(-1-2)]$

$= \frac{1}{2}(2+4+9) = \frac{15}{2}$

Thus, the area of the triangle is 15/2 square units.

OR

Q26. Solution: The origin is the centroid of the ΔABC i.e., G(0, 0, 0).

$\therefore 0 = \frac{a - 2 + 2}{3} \Rightarrow a = 0,$

$0 = \frac{1 + b + 3}{3} \Rightarrow b = -4$

$0 = \frac{3 - 1 + c}{3} \Rightarrow c = -2$

$\Rightarrow a = 0, b = -4, c = -2$

SECTION C

(Q27–Q32) Solve the following questions. (4 marks each)

Q27. Solution: Total number of persons =10

And number of persons to be selected = 7

Out of 10 persons, a chairperson is selected $= {}^{10}C_1 = 10$ ways

Out of remaining 9 persons, a chair leader is selected $= {}^{9}C_1 = 9$ ways

And remaining 5 persons are selected out of 8 persons as $= {}^{8}C_5 = 56$

Thus, total number of ways to form a committee of 7 persons $= 10 \times 9 \times 56 = 5040$

Q28. Solution: Total students in each class = 10

Choosing at least 4 from Class XI and at least 6 from Class XII can be done in the following ways:

Class XI	4	5	6
Class XII	7	6	5

Therefore, total number of ways of selecting a team of 11 players is

$= {}^{10}C_4 \times {}^{10}C_7 + {}^{10}C_5 \times {}^{10}C_6 + {}^{10}C_6 \times {}^{10}C_5$

$= 210 \times 120 + 252 \times 210 + 210 \times 252 = 131040$

Q29. Solution: The letters of the word 'MAXIMA' in alphabetical order are 2A, 2M, 1I, 1X.

Words starting with AA $= \frac{4!}{2!} = 12$

Words starting with AM $= \frac{4!}{2!} = 12$

Words starting with AI $= \frac{4!}{2!} = 12$

Words starting with AX $= \frac{4!}{2!} = 12$

Words starting with MAA $= 3! = 6$

Words starting with MAM $= 3! = 6$

Words starting with MAI $= 3! = 6$

Words starting with MAXA $= 2! = 2$

Words starting with MAXM $= 2! = 2$

Next word will be MAXIAM = 1

Next word will be MAXIMA = 1

∴ Rank of the word 'MAXIMA' in dictionary is $= 4 \times 12 + 3 \times 6 + 2 \times 2 + 2 = 72$

Q30. Solution: Comparing the given equation $x^2 - 4y^2 = 200 \Rightarrow \frac{x^2}{200} - \frac{y^2}{50} = 1$ with the standard equation of hyperbola $\frac{x^2}{a^2} - \frac{y^2}{b^2} = 1$, we get

$a = 10\sqrt{2},\ b = 5\sqrt{2}$

Now, from the equation $a^2 + b^2 = c^2$

$\Rightarrow c = \sqrt{a^2 + b^2} = \sqrt{200 + 50} = 5\sqrt{10}$

Thus, the coordinates of the foci are $(\pm 5\sqrt{10}, 0)$

The coordinates of the vertices are $(\pm 10\sqrt{2}, 0)$

Eccentricity $e = \frac{c}{a} = \frac{5\sqrt{10}}{10\sqrt{2}} = \frac{\sqrt{5}}{2}$

Length of the latus rectum $= \frac{2b^2}{a} = \frac{2 \times 50}{10\sqrt{2}} = 5\sqrt{2}$

OR

Q30. Solution: The nth term of the series $\frac{1}{2\times 4} + \frac{1}{4\times 6} + \frac{1}{6\times 8} + \ldots$ is

$a_n = \frac{1}{(2n)(2n+2)} = \left(\frac{1}{2n} - \frac{1}{2n+2}\right)$

Therefore, $a_1 = \frac{1}{2} - \frac{1}{4}$, $\quad a_2 = \frac{1}{4} - \frac{1}{6}$

$a_3 = \frac{1}{6} - \frac{1}{8}, \ldots \quad a_n = \frac{1}{2n} - \frac{1}{2n+2}$

Thus, $S_n = \sum_{k=1}^{n} a_k = a_1 + a_2 + \ldots + a_n$

$= \left(\frac{1}{2} - \frac{1}{4}\right) + \left(\frac{1}{4} - \frac{1}{6}\right) + \ldots + \left(\frac{1}{2n} - \frac{1}{2n+2}\right)$

$\Rightarrow S_n = \frac{1}{2} - \frac{1}{2n+2} = \frac{1}{2}\left(\frac{n+1-1}{n+1}\right) = \frac{n}{2(n+1)}$

Q31. Solution: If p is the length of the normal from the origin to a line and α is the angle made by the normal with the positive direction of the x-axis, then the equation of the line is given by $x\cos\alpha + y\sin\alpha = p$.

Thus, the required equation of the given line is $x\cos 45° + y\sin 45° = 1/\sqrt{2}$

$\left(\because p = 1/\sqrt{2} \text{ and } \alpha = 45°\right)$

$\Rightarrow x\cdot\frac{1}{\sqrt{2}} + y\cdot\frac{1}{\sqrt{2}} = \frac{1}{\sqrt{2}} \Rightarrow x + y = 1$

OR

Q31. Solution: The line $2x - y = 8$ divides the xy-plane in two half planes, I and II.

Let select point (0, 0) (not on the line), which lies in one of the half planes.

Substituting this point in the given inequality, we get $2(0) - 0 \le 8 \Rightarrow 0 \le 8$, which is true.

Therefore, the solution region of the given inequality is the half plane I (containing the point (0, 0)).
Also, all the points on the line satisfy the given inequality. Therefore, the solution region of the given inequality is the shaded half plane I (including the points on the line), as shown in the figure.

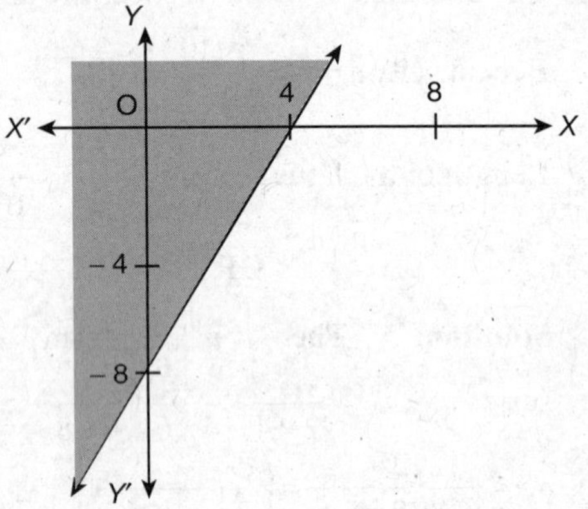

Q32. Solution: The equation of any line through the point of intersection of the given lines is of the form

$2x+3y-4+k(x-5y-7)=0$

$\Rightarrow (2+k)x+(3-5k)y-4-7k=0$

This line has its x-intercept equal to 1. Therefore

$\Rightarrow (2+k)\times(1)+(3-5k)\times 0-4-7k=0$

$\Rightarrow 2+k-4-7k=0 \Rightarrow k=-2/6=-1/3$

Substituting the value of $k=-1/3$, we get

$(2+-1/3)x+(3-5\times(-1/3))y-4-7\times(-1/3)=0$

$\Rightarrow 5x+14y-5=0$ is required equation of the line.

OR

Q32. Solution: The slope of the line segment passing through the points (2, 3) and (3, 4) is $\dfrac{4-3}{3-2}=1$

Therefore, slope of the line perpendicular to this line is $-\dfrac{1}{(1)}=-1$

The right bisector of a line segment bisects the line segment at 90^0 and passes through the mid-point.

Now, the mid-point of the line is $=\left(\dfrac{2+3}{2},\dfrac{3+4}{2}\right)=(5/2,7/2)$.

Therefore, the equation of the line passing through (5/2, 7/2) and having a slope of –1 is $(y-7/2)=-1(x-5/2)\Rightarrow x+y-6=0$

SECTION D

(Q33–Q36) Solve the following questions.

(6 marks each)

Q33. Solution: Let the vertices of the given parallelogram be $A(-2,-1)$, $B(4,0)$, $C(3,3)$, $D(x,y)$ then

Slope of line segment $AB=\dfrac{0+1}{4+2}=\dfrac{1}{6}$

Slope of line segment $CD=\dfrac{y-3}{x-3}$

Slope of line segment $BC=\dfrac{3-0}{3-4}=-3$

Slope of line segment $AD=\dfrac{y+1}{x+2}$

Since ABCD represents a parallelogram, slope of line segment AB = slope of line segment CD

$\Rightarrow \dfrac{y-3}{x-3}=\dfrac{1}{6}\Rightarrow 6y-x-15=0$

Also, the slope of line segment BC = slope of line segment $AD \Rightarrow \dfrac{y+1}{x+2}=-3\Rightarrow y+3x+7=0$

Solving, above two equations we get, $x=-3$ and $y=2$.

Thus, the points $(-2,-1),(4,0),(3,3),(-3,2)$ are the vertices of a parallelogram.

Q34. Solution: Let $f(x)=\cos(x-\theta)$.

Then, according to the first principle

$$f'(x)=\lim_{h\to 0}\frac{f(x+h)-f(x)}{h}$$

$$=\lim_{h\to 0}\frac{1}{h}\left[\cos(x+h-\theta)-\cos(x-\theta)\right]$$

$$=\lim_{h\to 0}\frac{1}{h}\left[\begin{array}{l}-2\sin\dfrac{(x+h-\theta+x-\theta)}{2}\\ \times\sin\left(\dfrac{x+h-\theta-x+\theta}{2}\right)\end{array}\right]$$

$$=\lim_{h\to 0}\frac{1}{h}\left[-2\sin\left(\frac{2x+h-2\theta}{2}\right)\sin\frac{h}{2}\right]$$

$$=\lim_{h\to 0}\left[-\sin\left(\frac{2x+h-2\theta}{2}\right)\frac{\sin(h/2)}{(h/2)}\right]$$

$$=\lim_{h\to 0}\left[-\sin\left(\frac{2x+h-2\theta}{2}\right)\right]\lim_{h/2\to 0}\frac{\sin(h/2)}{(h/2)}$$

$\left[\text{As } h\to 0\Rightarrow \dfrac{h}{2}\to 0\right]$

$$=-\sin\left(\frac{2x+0-2\theta}{2}\right)\cdot 1=-\sin(x-\theta)$$

OR

Q34. Solution: Let A be the event in which the selected student has opted for Math and B be the event in which the selected student has opted for Biology.

Total number of students = 50

Number of students who has opted for Math = 20

$\therefore P(A) = 20/50 = 2/5$

Number of students who has opted for Biology = 20

$\therefore P(B) = \frac{20}{50} = \frac{2}{5}$

Number of students who have opted for both Math and Biology = 10

$\therefore P(A \cap B) = \frac{10}{50} = \frac{1}{5}$

(i)

$P(A \text{ or } B) = P(A \cup B) = P(A) + P(B) - P(A \cap B)$

$\Rightarrow P(A \cup B) = \frac{2}{5} + \frac{2}{5} - \frac{1}{5} = \frac{3}{5}$

Thus, the probability that the selected student has opted for Math or Biology is 3/5.

(ii)

$P(\text{not } A \text{ and not } B) = P(A' \cap B') = P(A \cup B)'$

$= 1 - P(A \cup B) = 1 - \frac{3}{5} = \frac{2}{5}$

Thus, the probability that the selected students has neither opted for Math nor Biology is 1/5.

(iii) $n(A - B') = n(A) - n(A \cap B) = 20 - 10 = 10$

Thus, the probability that the selected student has opted for Math but not for Biology $= \frac{10}{50} = \frac{1}{5}$

Q35. Solution:

(i) $f(x) = \frac{1}{\sqrt{1-\sin^2 x}}$ is defined when

$1 - \sin^2 x > 0 \Rightarrow \sin^2 x < 1$

Also, $\sin^2 x \geq 0$ and since $-1 \leq \sin x < 1$

Thus, $f(x)$ is defined, for all x, except

when $1 - \sin^2 x \neq 0 \Rightarrow \sin x \neq \pm 1$

$\Rightarrow x \neq (2n+1)\frac{\pi}{2} \ \forall n \in Z$

Hence, domain of $f = R - \{(2n+1)\frac{\pi}{2} : n \in Z\}$

(ii) The given function $f(x) = \frac{x-1}{x^2 - x - 6}$ can be simplified as $f(x) = \frac{(x-1)}{(x-3)(x+2)}$

Function f is defined for all real numbers except when $x - 3 = 0$ or $x + 2 = 0$

Thus, at $x = 3, -2$ function f is not defined

Hence, the domain of f is $R - \{-2, 3\}$.

(iii) $f(x) = \frac{1}{x + |x|}$

$$\Rightarrow x + |x| = \begin{cases} x + x = 2x, & x \geq 0 \\ x - x = 0, & x < 0 \end{cases}$$

$f(x)$ is not defined when $x + |x| = 0$. Therefore Domain of $f = R^+$

OR

Q35. Solution:

$$\lim_{x\to 0} \frac{1-\cos 4x}{1-\cos 5x} = \lim_{x\to 0} \frac{1-1+2\sin^2 4x/2}{1-1+2\sin 5x/2}$$

$[\because \cos 4x = 1 - 2\sin^2 \frac{4x}{2}, \sin 5x = 1 - 2\sin^2 \frac{5x}{2}]$

$$= \lim_{x\to 0} \frac{\sin^2 4x/2}{\sin^2 5x/2} = \lim_{x\to 0} \frac{\frac{\sin^2 \frac{4x}{2}}{(4x/2)^2} \cdot (4x/2)^2}{\frac{\sin^2 \frac{5x}{2}}{(5x/2)^2} \cdot \left(\frac{5x}{2}\right)^2}$$

$$= \frac{\lim_{x\to 0} \left(\frac{\sin \frac{4x}{2}}{4x/2}\right)^2}{\lim_{x\to 0} \left(\frac{\sin \frac{5x}{2}}{5x/2}\right)^2} \cdot \frac{4^2 \frac{x^2}{4}}{5^2 \frac{x^2}{4}}$$

$$= \frac{4^2}{5^2} \cdot \frac{\lim_{x\to 0} \left(\frac{\sin \frac{4x}{2}}{4x/2}\right)^2}{\lim_{x\to 0} \left(\frac{\sin \frac{5x}{2}}{5x/2}\right)^2} = \frac{4^2}{5^2} = \frac{16}{25} \left[\because \lim_{x\to 0} \frac{\sin x}{x} = 1\right]$$

$[\because x \to 0 \Rightarrow kx \to 0]$

Q36. Solution: The statement is $P(n) : \frac{1}{2} \cdot \frac{3}{4} \cdot \frac{5}{6} \ldots \frac{2n-1}{2n} \leq \frac{1}{\sqrt{3n+1}}$ for natural numbers $n \in N$.

For n = 1, $P(1) : \frac{1}{2} = \frac{1}{\sqrt{3 \cdot 1 + 1}} = \frac{1}{2}$, thus P(1) is true.

Let $P(k)$ is true for some positive integer k, then

$$\frac{1}{2}\cdot\frac{3}{4}\cdot\frac{5}{6}\cdots\frac{2k-1}{2k}\le\frac{1}{\sqrt{3k+1}} \text{ is true.}$$

Now we shall prove that $P(k+1)$ is true

$$\frac{1}{2}\cdot\frac{3}{4}\cdot\frac{5}{6}\cdots\frac{2k-1}{2k}\cdot\frac{2k+1}{2k+2}\le\frac{1}{\sqrt{3(k+1)+1}}$$

$$\because \frac{1}{2}\cdot\frac{3}{4}\cdot\frac{5}{6}\cdots\frac{2k-1}{2k}\le\frac{1}{\sqrt{3k+1}}$$

($\because P(k)$ is true)

$$\therefore \frac{1}{2}\cdot\frac{3}{4}\cdot\frac{5}{6}\cdots\frac{2k-1}{2k}\cdot\frac{2k+1}{2k+2}\le\frac{1}{\sqrt{3k+1}}\cdot\frac{2k+1}{2k+2}$$

Now, we have to prove that

$$\Rightarrow \frac{1}{\sqrt{3k+1}}\cdot\frac{2k+1}{2k+2}\le\frac{1}{\sqrt{3(k+1)+1}}$$

$$\Rightarrow (2k+1)\sqrt{3(k+1)+1}\le 2(k+1)\sqrt{3k+1}$$

$$\Rightarrow (2k+1)\sqrt{3k+4}\le 2(k+1)\sqrt{3k+1}$$

$$\Rightarrow (2k+1)^2(3k+4)\le 4(k+1)^2(3k+1)$$

$$\Rightarrow (4k^2+4k+1)(3k+4)\le(4k^2+8k+4)(3k+1)$$

$$\Rightarrow 12k^3+12k^2+3k+16k^2+16k+4$$

$$\le 12k^3+24k^2+12k+4k^2+8k+4$$

$\Rightarrow 19k \le 20k$ is true.

$P(k+1)$ is true whenever $P(k)$ is true.

Therefore, by the principle of mathematical induction, statement $P(n)$ is true for natural numbers $n \in N$.

OR

Q36. Solution: The data is obtained in tabular form as follows:

x_i	f_i	$y_i = \frac{x_i - 3}{1}$	y_i^2	$f_i y_i$	$f_i y_i^2$
1	4	–2	4	–8	16
2	5	–1	1	–5	5
3	1	0	0	0	0
4	7	1	1	7	7
5	3	2	4	6	12
	20			0	40

$$\text{Mean } \bar{x} = A + \frac{\sum_{i=1}^{5} f_i y_i}{N}\times h = 3 + \frac{0}{20}\times 1 = 3$$

$$\text{Variance, } \sigma^2 = \frac{h^2}{N^2}\left[N\sum_{i=1}^{5} f_i y_i^2 - \left(\sum_{i=1}^{5} f_i y_i\right)^2\right]$$

$$= \frac{1}{20^2}[20\times 40 - 0^2] = 2.0$$

$\therefore$ Standard deviation $(\sigma) = \sqrt{2.0} = 1.41$

Sample Paper – X

Day 239

Time: 3 Hrs Maximum **Marks:** 80

General Instructions:

(i) All the questions are compulsory.

(ii) The question paper consists of 36 questions divided into 4 sections A, B, C and D.

(iii) Section A comprises 20 questions of 1 mark each. Section B comprises 6 questions of 2 marks each. Section C comprises 6 questions of 4 marks each. Section D comprises 4 questions of 6 marks each.

(iv) There is no overall choice. However, an internal choice has been provided in four questions of 1 mark each, two questions of 2 marks each, two questions of 4 marks each, and two questions of 6 marks each. You have to attempt only one of the alternatives in all such questions.

(iv) Use of calculators is not permitted.

SECTION A

Q1–Q10 are multiple choice type questions. Select the correct option (1 mark each)

Q1. If $\tan A = \tan B = \frac{1}{2}$, then $\tan(A+2B)$ is equal to

(a) $\frac{11}{2}$ (b) 11 (c) $\frac{9}{2}$ (d) 9

Q2. The range of the function f is given by $f(x) = |1-x| - |1+x|$ is

(a) $(-\infty,-1)\cup(1,\infty)$ (b) $(2,\infty)$

(c) $(-\infty,-1)\cup(1,\infty)$ (d) $[2,\infty)$

Q3. A survey shows that 50% of the people watch A news channel whereas 80% watch B news channel. If x% of the people watch both channel, then

(a) $x = 30$ (b) $x = 15$

(c) $30 \le x \le 50$ (d) $50 \le x \le 80$

Q4. The maximum value of $3\cos x + 4\sin x + 8$ is

(a) $\frac{11}{2}$ (b) 11 (c) $\frac{9}{2}$ (d) 13

Q5. The number $\frac{(1-i)^3}{1-i^3}$ is equal to

(a) 2 (b) –2 (c) 1 (d) –1

Q6. The coefficient of x^{20} in the expansion of $(x^3 - x^2)^{10}$ is

(a) 1 (b) 2 (c) 4 (d) 9

Q7. A man started saving Rs. 100. If he increases the installment by Rs 5 every month, the amount he will save in 10 years is … … …

(a) 17576 (b) 45600

(c) 47700 (d) 73200

Q8. How many secret codes can be made if each code has three small letters?

(a) 17576 (b) 45600 (c) 47700 (d) 73200

Q9. A group consists of 4 girls and 6 boys. In how many ways can a team of 5 members be selected, if the team has at least two girls and at least two boys?

(a) 60 (b) 120 (c) 180 (d) 240

Q10. Find the coefficient of $\frac{1}{x^5}$ in the expansion of $\left(x^2 - \frac{1}{x^3}\right)^{10}$.

(a) –100 (b) –252 (c) 252 (d) 300

Q11–Q15 are fill in the blanks. (1 mark each)

Q11. In the expansion of $\left(x^3 - \frac{1}{x^2}\right)^{15}$, the value of constant term is..............

Q12. The value of $\frac{\sin(\pi + x)\cos\left(\frac{\pi}{2} - x\right)}{\cos(\pi - x)\sin\left(\frac{\pi}{2} + x\right)}$ is … … ….

Q13. In a lottery, a person chose three different natural numbers at random from 1 to 9, and if these three numbers match with a number already fixed by the lottery committee, he wins the prize. The probability of winning the prize in the game is … … … (Given: Order of the numbers is not considered)

Q14. If the sum of three numbers in AP is 15 and their product is 45, then the numbers are... ... ...

Q15. The line through the points $(h,3),(2,1)$ intersects the line $2x-3y-6=0$ at right angles. The value of h is ... ...

(Q16–Q20) Answer the following questions.

(1 mark each)

Q16. Translate the following statements into symbolic form: 7 and 5 are factors of 35.

Q17. Write down the negation of the following compound statement: 20 is divisible by 2 and 5.

Q18. A point is in the YZ-plane. What can you say about its x-coordinates?

Q19. If $P(A)=\frac{1}{3}$, $P(B)=\frac{1}{4}$, $P(A\cup B)=\frac{1}{2}$, then find $P(\text{not } A)$ and $P(A\cap B)$ and state whether A and B are mutually exclusive events or not.

Q20. Reduce $\frac{1+2i}{(2-i)^2}$ to the standard form

SECTION B

(Q21–Q26) Solve the following questions.

(2 marks each)

Q21. Let $f,g:R\to R$ be defined, respectively, by $f(x)=x,\ g(x)=x-1$.

Find $f+g, f-g, fg, \frac{f}{g}$

Q22. If the complex numbers z_1,z_2 such that $\arg(z_1)-\arg(z_2)=0$, then show that $|z_1+z_2|=|z_1|+|z_2|$

Q23. Find the equation of the parabola that satisfies the following conditions: vertex (0, 0) focus (–2, 0).

Q24. Find the equation of the hyperbola satisfying the given conditions: vertices: $(0,\pm4)$, foci: $(0,\pm6)$

Q25. Find the third vertex of triangle whose centroid is origin and two vertices are (2, 3, 4) and (3, –2, 1).

Q26. Find the sum to n terms of the series $2\times3+3\times5+4\times7+...$

OR

Q26. Find the derivative of $\left(x-\frac{1}{x}\right)^3$

SECTION C

(Q27–Q32) Solve the following questions.

(4 marks each)

Q27. Reduce $\left(\frac{1}{1-4i}-\frac{1}{1+i}\right)$ to the standard form.

Q28. Solve the given inequality for real x:

$$\frac{(2x+1)}{20}\geq\frac{(3x-2)}{4}-\frac{(2-x)}{5}+\frac{1}{10}$$

Q29. Find the equation of the circle with centre $(-1,1)$ and radius 2

Q30. Find what the equation $2x-3y+xy-1=0$ becomes when the origin is shifted to the point $(1,1)$ by a translation of axes.

OR

Q30. Find the coordinates of the points of trisection (points dividing in three equal parts) of the line segment joining the points $A(2,-2)$ and $B(-3,3)$.

Q31. Find the equation of the line passing through (2, 1) and perpendicular to the line through the points $(2,5),(-3,6)$.

OR

Q31. Find the derivative of function $\frac{\sin(x+\theta)}{\cos x}$ with respect to x (θ is fixed non-zero constant) from the first principle.

Q32. Solve the following system of inequalities graphically: $x\geq2, y\geq1$.

OR

Q32. The owner of a milk store finds that he can sell 980 litres of milk each week at Rs 14/litre and 1220 litres of milk each week at Rs 16/litre. Assuming a linear relationship between selling price and demand, how many litres could be sell weekly at Rs 20/litre?

SECTION D

(Q33–Q36) Solve the following questions.

(6 marks each)

Q33. In a group of 20 people, the number playing Cricket and Football were found to be as follows: Cricket = 10, Football = 7, Cricket and Football both = 5. Find the number of peoples who play

(i) Only Cricket

(ii) Only Football
(iii) At least one game
(iv) No game

Q34. Find the range of the following functions:

(i) $f(x)=\frac{1}{4-x^2}$ (ii) $f(x)=2-|x-2|$

(iii) $f(x)=2+\cos 2x$

Q35. $\left(1-\frac{1}{\sqrt{2}}\right)\left(1-\frac{1}{\sqrt{3}}\right)\ldots\left(1-\frac{1}{\sqrt{n}}\right)<\frac{2}{n^2}$ for natural numbers $n \geq 2$.

OR

Q35. Find the mean and variance for the following frequency distribution:

Classes	0–20	20–40	40–60	60–80	80–100
Frequencies	5	8	2	4	6

Q36. Find a if the coefficient of x^2, x^3 in the expansion of $(2a+x)^9$ are equal.

OR

Q36. If the letters of the word 'MINIMIZES' are arranged in all possible ways as listed in dictionary, what is the rank of the word 'MINIMIZES'?

Day 240

Sample Paper – X (Solutions)

SECTION A

Q1–Q10 are multiple choice type questions. Select the correct option. (1 mark each)

Q1. Solution: (a)

$$\tan 2B=\frac{2\tan B}{1-\tan^2 B}=\frac{2\cdot\frac{1}{2}}{1-\frac{1}{4}}=\frac{4}{3}$$

Therefore, $\tan(A+2B)=\frac{\tan A+\tan 2B}{1-\tan A\cdot\tan 2B}$

$$=\frac{\frac{1}{2}+\frac{4}{3}}{1-\frac{1}{2}\cdot\frac{4}{3}}=\frac{\frac{11}{6}}{\frac{2}{6}}=\frac{11}{2}$$

Q2. Solution: (d)

$$f(x)=\begin{cases}(1-x)+(1+x), & x\leq -1\\ (1-x)-(1+x), & -1\leq x\leq 1\\ -(1-x)-(1+x), & x\geq 1\end{cases}$$

$$\Rightarrow f(x)=\begin{cases}2, & x\leq -1\\ -2x, & -1\leq x\leq 1\\ -2, & x\geq 1\end{cases}$$

$\therefore$ Range of $f=[-2,2]$

Q3. Solution: (c)

Let A be the set of percentages of those people who watch A news channel and B be the set of percentages of those people who watch B news channel. Then

$n(A)=50,\ n(B)=80,\ n(A\cap B)=x$

$\because n(A\cup B)\leq 100$

$\therefore n(A)+n(B)-n(A\cap B)\leq 100$

$\Rightarrow 50+80-x\leq 100\Rightarrow 130-x\leq 100$

$\Rightarrow 130-100\leq x\Rightarrow 30\leq x$

$\because n(A\cap B)\leq n(A)$

$\therefore x\leq 50$

Thus, $30\leq x\leq 50$

Q4. Solution: (d)

Let $y=3\cos x+4\sin x+8\Rightarrow y-8=3\cos x+4\sin x$

Maximum and minimum value of $a\cos x+b\sin x$ is $\pm\sqrt{a^2+b^2}$

$\therefore$ Maximum value of $y-8=\sqrt{9+16}$

$\Rightarrow y-8=5\Rightarrow y_{\max}=5+8=13$

Q5. Solution: (b)

$$\frac{(1-i)^3}{1-i^3}=\frac{1-i^3+3i^2-3i}{1-i^3}$$

$$=\frac{1+i-3-3i}{1+i}=\frac{-2-2i}{1+i}=\frac{-2(1+i)}{1+i}=-2$$

Q6. Solution: (a)

The general term of the given expansion is $T_{r+1}={}^{10}C_r(x^3)^{10-r}(-x^2)^r=(-1)^r\cdot{}^{10}C_r\cdot x^{30-3r+2r}$

$=(-1)^r\,{}^{10}C_r x^{30-r}$

For the coefficient of x^{20}, $30-r=20\Rightarrow r=10$

Thus, $T_{10+1}=(-1)^{10}\,{}^{10}C_{10}x^{20}$

Hence, coefficient of x^{20} is ${}^{10}C_{10}=1$

Q7. Solution: (c)

In the first month, he saves Rs 100.
Therefore, in the second month, he will save Rs 105, and so on.
Thus, the amount that the man saves every month forms an AP, given by 100, 105, 110,..., where first term $a = 100$, common difference, $d = 5$.
Hence, the amount he will save in 10 years (or 120 months) is

$$S_{120} = \frac{120}{2}[2\times100+(120-1)5]$$
$$= 60[200+595] = 47700$$

Q8. Solution: (a)
There are 26 English alphabets.
Therefore, arrangement of three small letters out of 26 letters is = all three letters are different + two letters are same + all letters are same

$$= {}^{26}C_3\times3!+{}^{26}C_1\times{}^{25}C_1\times\frac{3!}{2!}+{}^{26}C_1\times1!$$
$$= 15600+1950+26 = 17576$$

Thus, total number of secret codes contains three small letters is 17576

Q9. Solution: (c)
Number of girls = 4 and number of boys = 6
Selection of a team of 5 members with at least two girl and at least two boys is

$$= {}^4C_2\times{}^6C_3+{}^4C_3\times{}^6C_2$$
$$= 6\times20+4\times15 = 180$$

Q10. Solution: (b)
The general term of the given expansion $\left(x^2-\frac{1}{x^3}\right)^{10}$ is $T_{r+1} = {}^{10}C_r(x^2)^{10-r}\left(-\frac{1}{x^3}\right)^r$

$$= {}^{10}C_rx^{20-2r}(-1)^rx^{-3r} = (-1)^r\,{}^{10}C_rx^{20-5r}$$

For the coefficient of $\frac{1}{x^5}$ or x^{-5}, $20-5r=-5 \Rightarrow r=5$

Thus, $T_{5+1} = (-1)^5\,{}^{10}C_5x^{20-25} = -{}^{10}C_5x^{-5}$

Hence, coefficient of $\frac{1}{x^5}$ or x^{-5} is

$$= -{}^{10}C_5 = \frac{-10\times9\times8\times7\times6}{5\times4\times3\times2\times1} = -252$$

Q11–Q15 are fill in the blanks. (1 mark each)

Q11. Solution: Let r^{th} term be the constant term. Then

$$T_{r+1} = {}^{15}C_r(x^3)^{15-r}\left(-\frac{1}{x^2}\right)^r = (-1)^r\,{}^{15}C_rx^{45-3r-2r}$$
$$= (-1)^r\,{}^{15}C_r\,x^{45-5r}$$

For the constant term, $45-5r=0 \Rightarrow r=9$
.Therefore, the value of constant term is

$$T_{9+1} = (-1)^9\,{}^{15}C_9 = -{}^{15}C_9$$

Q12. Solution:

$$\text{LHS} = \frac{\sin(\pi+x)\cos\left(\frac{\pi}{2}-x\right)}{\cos(\pi-x)\sin\left(\frac{\pi}{2}+x\right)}$$
$$= \frac{[-\sin x][\sin x]}{(-\cos x)(\cos x)} = \tan^2 x$$

Q13. Solution: Total number of ways in which one can choose three different numbers from 1 to 9

$$n(S) = {}^9C_3 = \frac{9!}{3!6!} = 84$$

Out of these 84 combinations, only one combination is already fixed by the lottery committee.
∴ Required probability of winning the prize in the game is $= \frac{1}{84}$

Q14. Solution: Let the three numbers in AP be $a-d, a, a+d$, then
$(a-d)+(a)+(a+d) = 15 \Rightarrow 3a = 15 \Rightarrow a = 5$
Also,
$(a-d)a(a+d) = 45 \Rightarrow (5-d)(5)(5+d) = 45$

$$\Rightarrow 25-d^2 = 9$$
$$\Rightarrow d^2 = 16 \Rightarrow d = \pm4$$

Therefore, for $d = 4$, the numbers are 1, 5, and 9 and for $d = -4$ the numbers are 9, 5 and 1.
Thus, the three numbers are 1, 5, 9

Q15. Solution: The slope of the line passing through points $(h,3),(2,1)$ is $m_1 = \frac{1-3}{2-h} = \frac{-2}{2-h}$
And the slope of the line $2x-3y-6=0$

$$\Rightarrow y = \frac{2}{3}x-2 \text{ is } m_2 = 2/3$$

Therefore, the two lines are perpendicular if

$$m_1\times m_2 = -1 \Rightarrow \left(\frac{-2}{2-h}\right)\times\left(\frac{2}{3}\right) = -1 \Rightarrow \frac{-4}{6-3h} = -1$$
$$\Rightarrow -8 = -6+3h \Rightarrow 3h = -2 \Rightarrow h = -2/3$$

(Q16–Q20) Answer the following questions. (1 mark each)

Q16. Solution: p: 7 is factor of 35.
q: 5 is factor of 35.
$p\wedge q$: 7,5 are factors of 35.

Q17. Solution: Let p: 20 is divisible by 2.
q: 20 is divisible by 5
Then, negation of the conjunction of the above statement is given by 20~ $(p\wedge q)$: 20 is not divisible by 2 or it is not divisible by 5

Q18. Solution: If a point is in the YZ-plane, then its x-coordinate is zero.

Q19. Solution:

$$P(\text{not } A) = 1 - P(A) = 1 - \frac{1}{3} = \frac{2}{3}$$

$$P(A \cup B) = P(A) + P(B) - P(A \cap B)$$

$$\Rightarrow P(A \cap B) = P(A) + P(B) - P(A \cup B)$$

$$= \frac{1}{3} + \frac{1}{4} - \frac{1}{2} = \frac{4+3-6}{12} = \frac{1}{12}$$

$\because P(A \cap B) \neq 0$, $\therefore$ A and B are not mutually exclusive events.

Q20. Solution:

$$\frac{1+2i}{(2-i)^2} = \frac{1+2i}{4+i^2-4i} = \frac{1+2i}{3-4i} \qquad (\because i^2 = -1)$$

$$= \frac{1+2i}{3-4i} \times \frac{3+4i}{3+4i} = \frac{3+6i+4i+8i^2}{3^2+4^2}$$

$$= \frac{3+10i-8}{25} = \frac{-5+10i}{25} = \frac{-1}{5} + \frac{2}{5}i$$

SECTION B

(Q21–Q26) Solve the following questions. (2 marks each)

Q21. Solution: (i) $(f+g)(x) = f(x) + g(x)$

$$\Rightarrow (f+g)(x) = (x) + (x-1) = 2x - 1$$

(ii) $(f-g)(x) = f(x) - g(x)$

$$\Rightarrow (f-g)(x) = (x) - (x-1) = 1$$

(iii) $(fg)(x) = f(x) \cdot g(x) = (x)(x-1) = x^2 - x$

(iv) $\left(\frac{f}{g}\right)(x) = \frac{f(x)}{g(x)},\ g(x) \neq 0,\ x \in R$

$$\Rightarrow \left(\frac{f}{g}\right)(x) = \frac{x}{x-1},\ x - 1 \neq 0 \Rightarrow x \neq 1$$

Q22. Solution:

Let $z_1 = r_1(\cos\theta + i\sin\theta), z_2 = r_2(\cos\theta + i\sin\theta)$

$\because \arg(z_1) - \arg(z_2) = 0 \Rightarrow \theta_1 = \theta_2 = \theta$

$$\text{LHS} = |z_1 + z_2|$$

$$= |r_1(\cos\theta + i\sin\theta) + r_2(\cos\theta + i\sin\theta)|$$

$$= |(r_1 + r_2)\cos\theta + i(r_1 + r_2)\sin\theta|$$

$$= (r_1 + r_2)|\cos\theta + i\sin\theta|$$

$$= (r_1 + r_2)\sqrt{\cos^2\theta + \sin^2\theta}$$

$$= (r_1 + r_2) = |z_1| + |z_2| \qquad (\because r = |z|)$$

Q23. Solution: The focus (–2, 0) lies on the x-axis, and therefore, the axis of the parabola is the x-axis. Thus, the equation of the parabola is either of the form $y^2 = 4ax$ or $y^2 = -4ax$.

Now, since the focus is (–2, 0). Therefore, the parabola is of the form $y^2 = -4ax$ with $a = 2$. Thus, the equation of the parabola is $y^2 = -8x$.

Q24. Solution: The vertices are on the y-axis, and therefore, the equation of the hyperbola is of the form $\frac{y^2}{a^2} - \frac{x^2}{b^2} = 1$

Now, since the vertices are $(0, \pm 4) \Rightarrow a = 4$ and the foci are $(0, \pm 6) \Rightarrow c = 6$, from the equation $a^2 + b^2 = c^2 \Rightarrow b^2 = 6^2 - 4^2 \Rightarrow b^2 = 36 - 16 = 20$

Thus, the equation of the hyperbola is $\frac{y^2}{16} - \frac{x^2}{20} = 1$

Q25. Solution: Let the two vertices of ΔABC be A (2, 3, 4) and B (3, –2, 1) and the third vertex C be (x, y, z).

The coordinates of centroid G are (0,0,0)

$$\therefore 0 = \frac{x+2+3}{3} \Rightarrow x = -5$$

$$0 = \frac{y+3-2}{3} \Rightarrow y = -1$$

$$0 = \frac{z+4+1}{2} \Rightarrow z = -5$$

Thus, the third vertex of triangle is $C\,(-5, -1, -5)$.

Q26. Solution: The nth term of the series $2 \times 3 + 3 \times 5 + 4 \times 7 + \ldots$ is

$$a_n = (n+1)(2n+1) = 2n^2 + 3n + 1$$

$$\text{Therefore, } S_n = \sum_{k=1}^{n} a_k$$

$$= \sum_{k=1}^{n}(2k^2 + 3k + 1) = 2\sum_{k=1}^{n} k^2 + 3\sum_{k=1}^{n} k + n$$

$$= 2\left[\frac{n(n+1)(2n+1)}{6}\right] + 3\left[\frac{n(n+1)}{2}\right] + n$$

$$= \left[\frac{2n(n+1)(2n+1) + 9n(n+1) + 6n}{6}\right]$$

$$= \frac{n}{6}\left[2(2n^2 + 3n + 1) + 9n + 9 + 6\right]$$

$$= \frac{n}{6}\left[4n^2 + 15n + 17\right]$$

OR

Q26. Solution: Let $y = \left(x - \frac{1}{x}\right)^3$

$$\therefore \frac{dy}{dx} = \frac{d}{dx}\left(x - \frac{1}{x}\right)^3 = 3\left(x - \frac{1}{x}\right)^{3-1} \frac{d}{dx}\left(x - \frac{1}{x}\right)$$

$$\left(\because \frac{d}{dx} y^n = ny^{n-1} \frac{dy}{dx}\right)$$

$$= 3\left(x - \frac{1}{x}\right)^2 \left(1 + \frac{1}{x^2}\right)$$

SECTION C

(Q27–Q32) Solve the following questions.

(4 marks each)

Q27. Solution:

$$\left(\frac{1}{1-4i} - \frac{1}{1+i}\right)$$

$$= \left[\frac{(1+i)-(1-4i)}{(1-4i)(1+i)}\right]$$

$$= \left[\frac{1+i-1+4i}{1+i-4i-4i^2}\right]$$

$$= \left[\frac{5i}{5-3i}\right] \qquad (\because i^2 = -1)$$

$$= \frac{5i}{5-3i} \times \frac{5+3i}{5+3i}$$

$$= \frac{25i + 15i^2}{\left[(5)^2 - (3i)^2\right]}$$

$$= \frac{-15+25i}{34} \qquad (\because i^2 = -1)$$

$$= \frac{-15}{34} + \frac{25}{34} i$$

Q28. Solution:

$$\frac{(2x+1)}{20} \geq \frac{(3x-2)}{4} - \frac{(2-x)}{5} + \frac{1}{10}$$

$$\Rightarrow \frac{(2x+1)}{20} \geq \frac{5(3x-2)-4(2-x)+2}{20}$$

$$\Rightarrow \frac{(2x+1)}{20} \geq \frac{15x-10-8+4x+2}{20}$$

$$\Rightarrow \frac{(2x+1)}{20} \geq \frac{19x-16}{20}$$

$$\Rightarrow 2x+1 \geq 19x-16$$

$$\Rightarrow 17 \geq 17x \Rightarrow x \leq 1$$

Therefore, all real numbers x, which are less than or equal to 1, are the solutions of the given inequality.

Thus, the solution set of the given inequality is $x \in (-\infty, 1]$.

Q29. Solution: The equation of a circle with centre (h, k) and radius r is $(x-h)^2 + (y-k)^2 = r^2$

Therefore, equation of the circle with centre $(-1,1)$ and radius 2 is $(x+1)^2 + (y-1)^2 = (2)^2$

$$\Rightarrow x^2 + 2x + 1 + y^2 - 2y + 1 = 4$$

$$\Rightarrow x^2 + y^2 + 2x - 2y - 2 = 0$$

Q30. Solution: The coordinates of the new origin are $h = -1$, $k = 1$ and any point $P(x, y)$ in given coordinate system will be changed to $P(x', y')$ in new coordinate system, by the relation $x = x' + h \Rightarrow x = x' - 1$ and $y = y' + k \Rightarrow y = y' - 1$

Therefore, substituting the value of x and y in the given equation $2x - 3y + xy - 1 = 0$, we get

$$2(x'-1) - 3(y'-1) + (x'-1)(y'-1) - 1 = 0$$

$$\Rightarrow 2x' - 2 - 3y' + 3 + x'y' - x' - y' + 1 - 1 = 0$$

$$\Rightarrow x' - 4y' + x'y' + 1 = 0$$

Thus, the required equation of line in new coordinate system is $x - 4y + xy + 1 = 0$.

OR

Q30. Solution: Let P and Q be the points of trisection of AB, that is $AP = PQ = QB$

Therefore, P divides AB internally in the ratio 1:2. Thus, the coordinates of P, using the section formula, are

$$\left(\frac{1(-3)+2(2)}{1+2}, \frac{1(3)+2(-2)}{1+2}\right) = (1/3, -1/3)$$

Similarly, Q divides AB internally in the ratio 2:1. Thus the coordinates of Q, using the section formula, are $\left(\frac{2(-3)+1(2)}{2+1}, \frac{2(3)+1(-2)}{2+1}\right) = (-4/3, 4/3)$

Therefore, the coordinates of the points of trisection of the line segment joining A and B are (1/3, –1/3) and (–4/3, 4/3).

Q31. Solution: The slope of the line joining the points $(2,5), (-3,6)$ is $m = \frac{6-5}{-3-2} = \frac{1}{-5}$

And we know that two non-vertical lines are perpendicular to each other if and only if their slopes are negative reciprocal of each other.

Therefore, slope of the line perpendicular to the line through the points $(2,5), (-3,6)$ is

$$= -\frac{1}{m} = -\frac{1}{(-1/5)} = 5$$

Thus, the equation of the line passing through point

(2, 1), whose slope is 5, is $(y-1) = 5(x-2)$

$$\Rightarrow y - 1 = 5x - 10 \Rightarrow 5x - y - 9 = 0$$

OR

Q31. Solution: Let $f(x) = \frac{\sin(x+\theta)}{\cos x}$

By the quotient rule

$$f'(x) = \frac{\cos x \frac{d}{dx}[\sin(x+\theta)] - \sin(x+\theta)\frac{d}{dx}\cos x}{\cos^2 x}$$

$$= \frac{\cos x \frac{d}{dx}[\sin(x+\theta)] - \sin(x+\theta)(-\sin x)}{\cos^2 x}$$

$$f'(x) = \frac{\cos x.\cos(x+\theta) + \sin x \sin(x+\theta)}{\cos^2 x}$$

$$= \frac{\cos(x+\theta-x)}{\cos^2 x} = \frac{\cos\theta}{\cos^2 x}$$

Q32. Solution: The graphs of the lines, $x = 2$ and $y = 1$, are shown in the figure.

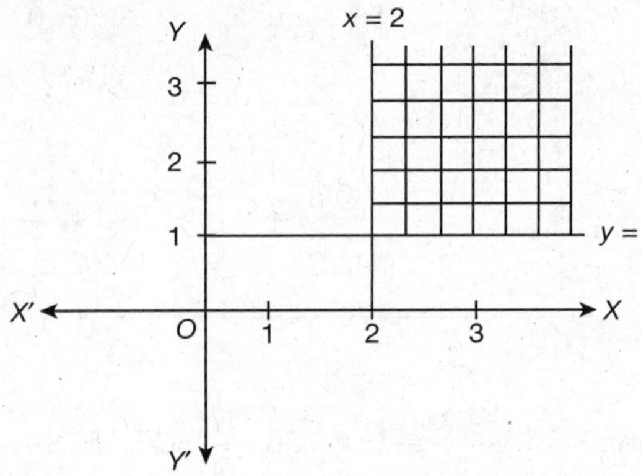

Inequality $x \geq 2$ represents the shaded region on the right of line $x = 2$ (including all the points on the line).
Inequality $y \geq 1$ represents the shaded region above the line $y = 1$(including all the points on the line).
Thus, the solution of the given system of linear inequalities is represented by the common double shaded region including the points on the respective lines.

OR

Q32. Solution: The relationship between selling price and demand is linear. Therefore, we assume selling price per litre along the x-axis and demand along the y-axis. Thus, the two given points that is (14,980) and (16, 1220) in the XY plane will satisfy the linear relationship between selling price and demand.
Thus, the linear relationship between selling price per litre and demand is the equation of the line passing through points (14, 980) and (16, 1220), given by

$$y - 980 = \frac{1220-980}{16-14}(x-14) \Rightarrow y - 980 = \frac{240}{2}(x-14)$$

$$\Rightarrow y - 980 = 120(x-14) \Rightarrow y = 120(x-14) + 980$$

When x = Rs 20/litre, $y = 120(20-14) + 980 = 1700$
Thus, the owner of the milk store could sell 1700 litres of milk weekly at Rs 20/litre.

SECTION D

(Q33–Q36) Solve the following questions.
(6 marks each)

Q33. Solution: Let C be the set of people who play Cricket and F be the set of people who play Football.

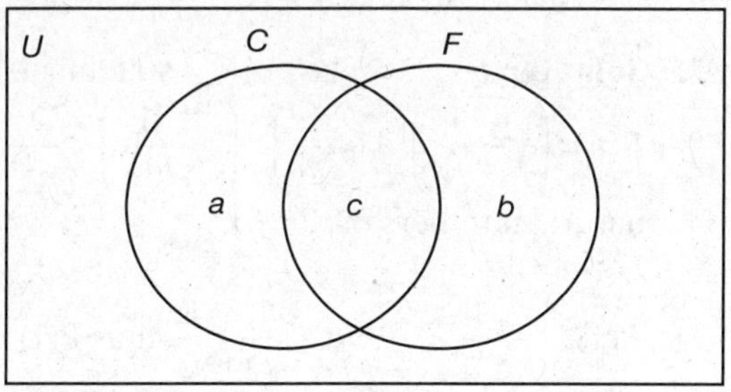

Then, $n(U) = 20$
$n(C) = a + c = 10$
$n(F) = b + c = 7$
$n(C \cap F) = c = 5$

(i) Number of peoples who play Cricket only
$a = n(C \cap F') = n(C) - n(C \cap F)$
$= (a+c) - (c) = 10 - 5 = 5$

(ii) Number of peoples who play Football only
$b = n(F \cap C') = n(F) - n(C \cap F)$
$= (b+c) - (c) = 7 - 5 = 2$

(iii) Number of peoples who play at least one game $= a + b + c$
$= 5 + 2 + 5 = 12$

(iv) Number of eoples who play no game = Total People – People who play at least one game $= 20 - 12 = 8$

Q34. Solution: (i) $f(x) = \dfrac{1}{4-x^2}$

Let $y = f(x)$

$$\Rightarrow y = \frac{1}{4-x^2} \Rightarrow 4 - x^2 = \frac{1}{y}$$

$$\Rightarrow x^2 = 4 - \frac{1}{y} \Rightarrow x = \sqrt{\frac{4y-1}{y}}$$

Since x is real, $\dfrac{4y-1}{y} \geq 0$ and $y \neq 0$

$$\Rightarrow y \geq \frac{1}{4} \text{ or } y \leq 0 \text{ and } y \neq 0$$

$$\Rightarrow y \geq \frac{1}{4} \text{ or } y < 0$$

Thus, range of $f = (-\infty, 0) \cup \left[\frac{1}{4}, \infty\right)$

(ii) $\because |x-2| \geq 0 \Rightarrow -|x-2| \leq 0$

$\Rightarrow 2-|x-2| \leq 2 \Rightarrow f(x) \leq 2$

$\therefore$ Range of $f = (-\infty, 2]$

(iii) $-1 \leq \cos 2x \leq 1$

$\Rightarrow 2-1 \leq 2+\cos 2x \leq 1+2$

$\Rightarrow 1 \leq 2+\cos 2x \leq 3 \Rightarrow 1 \leq f(x) \leq 3$

Thus, range of $f = [1, 3]$

Q35. Solution: The statement is $P(n): \left(1-\frac{1}{\sqrt{2}}\right)\left(1-\frac{1}{\sqrt{3}}\right)\ldots\left(1-\frac{1}{\sqrt{n}}\right) < \frac{2}{n^2}$ for natural numbers $n \geq 2$.

For $n = 1$

$P(1): 1-\frac{1}{\sqrt{2}} < \frac{2}{4} \Rightarrow 1-\frac{1}{\sqrt{2}} < \frac{1}{2}$, thus P(1) is true.

Let $P(k)$ be true for some positive integer k

$\left(1-\frac{1}{\sqrt{2}}\right)\left(1-\frac{1}{\sqrt{3}}\right)\ldots\left(1-\frac{1}{\sqrt{k}}\right) < \frac{2}{k^2}$ is true.

Now we shall prove that $P(k+1)$ is true

$$\left(1-\frac{1}{\sqrt{2}}\right)\left(1-\frac{1}{\sqrt{3}}\right)\ldots\left(1-\frac{1}{\sqrt{k}}\right)\left(1-\frac{1}{\sqrt{k+1}}\right) < \frac{2}{(k+1)^2}$$

$$\because \left(1-\frac{1}{\sqrt{2}}\right)\left(1-\frac{1}{\sqrt{3}}\right)\ldots\left(1-\frac{1}{\sqrt{k}}\right) < \frac{2}{k^2}$$

($\because P(k)$ is true)

$$\therefore \left(1-\frac{1}{\sqrt{2}}\right)\ldots\left(1-\frac{1}{\sqrt{k}}\right)\left(1-\frac{1}{\sqrt{k+1}}\right) < \frac{2}{k^2}\left(1-\frac{1}{\sqrt{k+1}}\right)$$

Now, we have to prove that

$$\Rightarrow \frac{2}{k^2}\left(1-\frac{1}{\sqrt{k+1}}\right) < \frac{2}{(k+1)^2}$$

$$\Rightarrow 1-\frac{1}{\sqrt{k+1}} < \frac{k^2}{(k+1)^2}$$

$$\Rightarrow 1-\frac{k^2}{(k+1)^2} < \frac{1}{\sqrt{k+1}}$$

$$\Rightarrow \frac{k^2+1+2k-k^2}{(k+1)^2} < \frac{1}{\sqrt{k+1}}$$

$$\Rightarrow 1+2k < (k+1)^{3/2}$$

$$\Rightarrow (1+2k)^2 < (k+1)^3$$

$$\Rightarrow 1+4k^2+4k < k^3+1+3k^2+3k$$

$\Rightarrow k^2+k < k^3$ is true ($\because k \geq 2$)

$P(k+1)$ is true whenever $P(k)$ is true.

Therefore, by the principle of mathematical induction, statement $P(n)$ is true for natural numbers $n \in N$.

OR

Q35. Solution:

Class	Frequency f_i	Mid-point x_i	$y_i = \frac{x_i - 50}{20}$	y_i^2	$f_i y_i$	$f_i y_i^2$
0–20	1	10	–2	4	–2	4
20–40	7	30	–1	1	–7	7
40–60	2	50	0	0	0	0
60–80	4	70	1	1	4	4
80–100	6	90	2	4	12	24
	25				7	39

$$\text{Mean } \bar{x} = A + \frac{\sum_{i=1}^{5} f_i y_i}{N} \times h = 50 + \frac{7}{50} \times 20 = 52.8$$

$$\text{Variance } (\sigma^2) = \frac{h^2}{N^2}\left[N\sum_{i=1}^{5} f_i y_i^2 - \left(\sum_{i=1}^{5} f_i y_i\right)^2\right]$$

$$= \frac{(20)^2}{(50)^2}[50 \times 39 - (7)^2] = \frac{4}{25}[1950 - 49] = 304.16$$

Q36. Solution: The $(r+1)^{\text{th}}$ term in the binomial expansion of $(a+b)^n$ is $T_{r+1} = {}^nC_r a^{n-r} b^r$

Let us assume that x^2 occurs in the $(r+1)^{\text{th}}$ term in the expansion of $(2a+x)^9$. We get

$T_{r+1} = {}^9C_r (2a)^{9-r} (x)^r$

Comparing the indices of x in x^2 and in (T_{r+1}), we get $r = 2$

Thus, the coefficient of x^2 is ${}^9C_2 (2a)^{9-2} = {}^9C_2 2^7 a^7$

Similarly, assume that x^3 occurs in the $(k+1)^{\text{th}}$ term in the expansion of $(2a+x)^9$. We get

$T_{k+1} = {}^9C_k (2a)^{9-k} (x)^k = {}^9C_k (2a)^{9-k} x^k$

Comparing the indices of x in x^3 and in (T_{r+1}), we get $k = 3$

Thus, the coefficient of x^3 is

${}^9C_3 (2a)^{9-3} = {}^9C_3 2^6 a^6$

Since the coefficient of x^2 and x^3 are equal

$${}^9C_2 2^7 a^7 = {}^9C_3 2^6 a^6$$

$$\Rightarrow a = \frac{9\cdot 8\cdot 7}{3\cdot 2} \times \frac{2}{9\cdot 8} \times \frac{1}{2} \Rightarrow a = \frac{7}{6}$$

OR

Q36. Solution: The letters of the word 'MINIMIZES' in alphabetical order are 1E, 3I, 2M, 1N, 1S, 1Z.

Words starting with E $= \frac{8!}{3!2!} = 3360$

Words starting with I $= \frac{8!}{2!2!} = 5040$

Words starting with ME $= \frac{7!}{3!} = 840$

Words starting with MIE $= \frac{6!}{2!} = 360$

Words starting with MII $= 6! = 720$

Words starting with MIM $= \frac{6!}{2!} = 360$

Words starting with MINE $= \frac{5!}{2!} = 60$

Words starting with MINIE $= 4! = 24$

Words starting with MINII $= 4! = 24$

Words starting with MINIME $= 3! = 6$

Words starting with MINIMIE $= 2! = 2$

Words starting with MINIMIS $= 2! = 2$

Next word will be MINIMIZES $= 1$

∴ Rank of the word 'MINIMIZES' in dictionary is

$$= 3360 + 5040 + 840 + 360 + 720 + 360 + 60 + 24 + 24 + 6 + 2 + 2 + 1 = 10799$$